Handbook of Reagents for Organic Synthesis

Reagents for Glycoside, Nucleotide, and Peptide Synthesis

OTHER TITLES IN THIS COLLECTION

Reagents for High-Throughput Solid-Phase and Solution-Phase Organic Synthesis
Edited by Peter Wipf
ISBN 0 470 86298 X

Chiral Reagents for Asymmetric Synthesis
Edited by Leo A. Paquette
ISBN 0 470 85625 4

Oxidizing and Reducing Agents
Edited by Steven D. Burke and Rick L. Danheiser
ISBN 0 471 97926 0

Acidic and Basic Reagents
Edited by Hans J. Reich and James H. Rigby
ISBN 0 471 97925 2

Activating Agents and Protecting Groups
Edited by Anthony J. Pearson and William R. Roush
ISBN 0 471 97927 9

Reagents, Auxiliaries, and Catalysts for C-C Bond Formation
Edited by Robert M. Coates and Scott E. Denmark
ISBN 0 471 97924 4

LDA	lithium diisopropylamide
LDMAN	lithium 1-(dimethylamino)naphthalenide
LHMDS	= LiHMDS
LICA	lithium isopropylcyclohexylamide
LiHMDS	lithium hexamethyldisilazide
LiTMP	lithium 2,2,6,6-tetramethylpiperidide
LTMP	= LiTMP
LTA	lead tetraacetate
lut	lutidine
m-CPBA	*m*-chloroperbenzoic acid
MA	maleic anhydride
MAD	methylaluminum bis(2,6-di-*t*-butyl-4-methylphenoxide)
MAT	methylaluminum bis(2,4,6-tri-*t*-butylphenoxide)
Me	methyl
MEK	methyl ethyl ketone
MEM	(2-methoxyethoxy)methyl
MIC	methyl isocyanate
MMPP	magnesium monoperoxyphthalate
MOM	methoxymethyl
MoOPH	oxodiperoxomolybdenum(pyridine)-(hexamethylphosphoric triamide)
mp	melting point
MPM	= PMB
Ms	mesyl (methanesulfonyl)
MS	mass spectrometry; molecular sieves
MTBE	methyl *t*-butyl ether
MTM	methylthiomethyl
MVK	methyl vinyl ketone
n	refractive index
NaHDMS	sodium hexamethyldisilazide
Naph	naphthyl
NBA	*N*-bromoacetamide
nbd	norbornadiene (bicyclo[2.2.1]hepta-2,5-diene)
NBS	*N*-bromosuccinimide
NCS	*N*-chlorosuccinimide
NIS	*N*-iodosuccinimide
NMO	*N*-methylmorpholine *N*-oxide
NMP	*N*-methyl-2-pyrrolidinone
NMR	nuclear magnetic resonance
NORPHOS	bis(diphenylphosphino)bicyclo[2.2.1]-hept-5-ene
Np	= Naph
PCC	pyridinium chlorochromate
PDC	pyridinium dichromate
Pent	*n*-pentyl
Ph	phenyl
phen	1,10-phenanthroline
Phth	phthaloyl
Piv	pivaloyl
PMB	*p*-methoxybenzyl
PMDTA	*N,N,N′,N″,N″*-pentamethyldiethylenetriamine
PPA	polyphosphoric acid
PPE	polyphosphate ester
PPTS	pyridinium *p*-toluenesulfonate
Pr	*n*-propyl
PTC	phase transfer catalyst/catalysis
PTSA	*p*-toluenesulfonic acid
py	pyridine
RAMP	(R)-1-amino-2-(methoxymethyl)pyrrolidine
rt	room temperature
salen	bis(salicylidene)ethylenediamine
SAMP	(*S*)-1-amino-2-(methoxymethyl)pyrrolidine
SET	single electron transfer
Sia	siamyl (3-methyl-2-butyl)
TASF	tris(diethylamino)sulfonium difluorotrimethylsilicate
TBAB	tetrabutylammonium bromide
TBAF	tetrabutylammonium fluoride
TBAD	= DBAD
TBAI	tetrabutylammonium iodide
TBAP	tetrabutylammonium perruthenate
TBDMS	*t*-butyldimethylsilyl
TBDPS	*t*-butyldiphenylsilyl
TBHP	*t*-butyl hydroperoxide
TBS	= TBDMS
TCNE	tetracyanoethylene
TCNQ	7,7,8,8-tetracyanoquinodimethane
TEA	triethylamine
TEBA	triethylbenzylammonium chloride
TEBAC	= TEBA
TEMPO	2,2,6,6-tetramethylpiperidinoxyl
TES	triethylsilyl
Tf	triflyl (trifluoromethanesulfonyl)
TFA	trifluoroacetic acid
TFAA	trifluoroacetic anhydride
THF	tetrahydrofuran
THP	tetrahydropyran; tetrahydropyranyl
Thx	thexyl (2,3-dimethyl-2-butyl)
TIPS	triisopropylsilyl
TMANO	trimethylamine *N*-oxide
TMEDA	*N,N,N′,N′*-tetramethylethylenediamine
TMG	1,1,3,3-tetramethylguanidine
TMS	trimethylsilyl
Tol	*p*-tolyl
TPAP	tetrapropylammonium perruthenate
TBHP	*t*-butyl hydroperoxide
TPP	tetraphenylporphyrin
Tr	trityl (triphenylmethyl)
Ts	tosyl (*p*-toluenesulfonyl)
TTN	thallium(III) nitrate
UHP	urea–hydrogen peroxide complex
Z	= Cbz

Handbook of Reagents for Organic Synthesis

Reagents for Glycoside, Nucleotide, and Peptide Synthesis

Edited by

David Crich

University of Illinois at Chicago, Chicago, IL, USA

WILEY

Telephone (+44) 1243 779777

E-mail (for orders and customer service enquiries): cs-books@wiley.co.uk
Visit our Home Page on www.wileyeurope.com or www.wiley.com

Other Wiley Editorial Offices

John Wiley & Sons Inc., 111 River Street,
Hoboken, NJ 07030, USA

Jossey-Bass, 989 Market Street,
San Francisco, CA 94103-1741, USA

Wiley-VCH Verlag GmbH, Boschstr. 12,
D-69469 Weinheim, Germany

John Wiley & Sons Australia Ltd, 42 McDougall Street,
Milton, Queensland 4064, Australia

John Wiley & Sons (Asia) Pte Ltd, 2 Clementi Loop #02-01,
Jin Xing Distripark, Singapore 129809

John Wiley & Sons Canada Ltd, 5353 Dundas Street West, Suite 400
Etobicoke, Ontario, Canada M9B 6HB

Wiley also publishes its books in a variety of electronic formats. Some content that appears in print may not be available in electronic books.

Library of Congress Cataloging-in-Publication Data

Handbook of reagents for organic synthesis.
p. cm.
Includes bibliographical references.
Contents: [1] Reagents, auxiliaries, and catalysts for C–C bond formation/edited by Robert M. Coates and Scott E. Denmark [2] Oxidising and reducing agents/edited by Steven D. Burke and Riek L. Danheiser [3] Acidic and basic reagents/edited by Hans J. Reich and James H. Rigby [4] Activating agents and protecting groups/edited by Anthony J. Pearson and William R. Roush [5] Chiral Reagents for Asymmetric Synthesis/edited by Leo A. Paquette
ISBN 0-471-97924-4 (v. 1) ISBN 0-471-97926-0 (v. 2)
ISBN 0-471-97925-2 (v. 3) ISBN 0-471-97927-9 (v. 4)
ISBN 0-470-85625-4 (v. 5) ISBN 0-470-86298-X (v. 6)
ISBN 0-470-02304-X (v. 7)
1. Chemical tests and reagents. 2. Organic compounds–Synthesis.
QD77.H37 1999
547′.2 dc 21 98-53088
CIP

British Library Cataloguing in Publication Data
A catalogue record for this book is available from the British Library

ISBN 10: 0-470-02304-X (Hb)
ISBN 13: 978-0470-02304-4
Typeset in 9½/11½ pt Times Roman by Thomson Press (India) Ltd., New Delhi.
Printed and bound in Great Britain by Antony Rowe, Chippenham, Wiltshire
This book is printed on acid-free paper responsibly manufactured from sustainable forestry in which at least two trees are planted for each one used for paper production.

e-EROS Editorial Board

Contents

Preface

As stated in its Preface, the major motivation for our undertaking publication of the *Encyclopedia of Reagents for Organic Synthesis* was "to incorporate into a single work a genuinely authoritative and systematic description of the utility of all reagents used in organic chemistry." By all accounts, this reference compendium succeeded admirably in approaching this objective. Experts from around the globe contributed many relevant facts that define the various uses characteristic of each reagent. The choice of a masthead format for providing relevant information about each entry, the highlighting of key transformations with illustrative equations, and the incorporation of detailed indexes serve in tandem to facilitate the retrieval of desired information.

Notwithstanding these accomplishments, the editors came to recognize that the large size of this eight-volume work and its cost of purchase often deterred the placement of copies of the *Encyclopedia* in or near laboratories where the need for this type of information is most critical. In an effort to meet this demand in a cost-effective manner, the decision was made to cull from the major work that information having the highest probability for repeated consultation and to incorporate the same into a set of handbooks. The latter would also be purchasable on a single unit basis.

The ultimate result of these deliberations was the publication of the *Handbook of Reagents for Organic Synthesis*, the first four volumes of which appeared in 1999:

Oxidizing and Reducing Agents
Edited by Steven D. Burke and Rick L. Danheiser

Acidic and Basic Reagents
Edited by Hans J. Reich and James H. Rigby

Activating Agents and Protecting Groups
Edited by Anthony J. Pearson and William R. Roush

Reagents, Auxiliaries, and Catalysts for C-C Bond Formation
Edited by Robert M. Coates and Scott E. Denmark

Since then, the fifth and sixth members of this series have made their appearance in 2003 and 2005, respectively:

Chiral Reagents for Asymmetric Synthesis
Edited by Leo A. Paquette

Reagents for High-Throughput Solid-Phase and Solution-Phase Organic Synthesis
Edited by Peter Wipf

Each of the volumes contains a selected compilation of those entries from the original *Encyclopedia* that bear on the specific topic. The coverage of the last two handbooks also extends to the electronic sequel *e-EROS*. Ample listings can be found to functionally related reagents contained in the original work. For the sake of current awareness, references to recent reviews and monographs have been included, as have relevant new procedures from *Organic Syntheses*.

The present volume, entitled *Reagents for Glycoside, Nucleotide, and Peptide Synthesis*, constitutes the seventh entry in this continuing series of utilitarian reference works. As with its predecessors, this handbook is intended to be an affordable, enlightening compilation that will hopefully find its way into the laboratories of all practicing synthetic chemists. Every attempt has been made to be of the broadest possible relevance and the expectation is that our colleagues will share in this opinion.

Leo A. Paquette
Department of Chemistry
The Ohio State University
Columbus, OH, USA

Introduction

With the advent of Chemical Biology, mainstream organic and medicinal chemists are increasingly aware of the need for, and are turning their attention to, the synthesis of bioconjugates. These might take the form of more classical peptides, nucleotides, or oligosaccharides, or some hybrid thereof, and may be required in straightforward soluble forms or appended to soluble or insoluble polymeric supports, nanoparticles, dendrimers, gold or silicon surfaces, or microtiter plates. Whatever the system or eventual use, the need for the efficient synthesis of glycoside, nucleotide, or peptide bonds has never been greater. The synthesis of each of these three main classes of biopolymers has evolved into a fine and sophisticated art in recent years with many reagents available to the experienced practitioner. The aim of this handbook is to collect the more important of these reagents together into a convenient, single volume and so aid the non-specialist in locating information and in making an informed choice when the need for the synthesis of a glycoside, nucleotide, or peptide bond arises. This handbook focuses on the chemical reagents required for glycoside, nucleotide, and peptide bond: information on polymeric supports and a broader selection of supported reagents, obviously very important considerations in this area, can be found in the recent *Handbook of Reagents for High-Throughput Solid-Phase and Solution-Phase Organic Synthesis* edited by Professor Peter Wipf.

Of the reagents featured in this volume, approximately one third are taken from the *Encyclopedia of Reagents for Organic Synthesis* (*EROS*), published in 1995. Many of these are classical reagents in the field whose principal use has not changed in the intervening period. The remainder, and indeed the bulk, of the entries are comprised either of completely new articles, or of updated versions of original *EROS* articles taking into account recent developments, written by experts in the field for the continually-expanding online encyclopedia (*e-EROS*). The volume opens with a selection of recent monographs and review articles intended to help the reader in searching the literature. This list is divided into three sections: Glycoside, Nucleotide, and Peptide Bond Formation. The main sequence of reagents in this volume is alphabetical in keeping with the *EROS* and *e-EROS* format.

Finally, it is hoped that this volume will be a useful and convenient handbook for any researcher involved in the synthesis of bioconjugates, perhaps even helping to identify present shortcomings in the field and, so, spur the design and development of even more and better reagents for the same.

David Crich
Department of Chemistry
University of Illinois at Chicago
Chicago, IL, USA

Recent Review Articles and Monographs

Glycoside Bond Formation

Lockhoff, O. Acetals at the Anomeric Cemter of Carbohydrates, In *Houben-Weyl. Methods of Organic Chemistry*, 4th ed.; Hagemann, H.; Klamann, D., Eds.; Georg Thieme Verlag: Stuttgart, 1992; Vol. E14a/3, pp 621–1077.

Ogura, H.; Hasegawa, A.; Suami, T., Eds., *Carbohydrates*; Kondansha: Tokyo, 1992

van Boeckel, C. A. A. Protective Group Strategies in the Synthesis of Functionalized Carbohydrates, In *Modern Synthetic Methods*; Scheffold, R., Ed.; Verlag Helvetica Chimica Acta: Basel, 1992; Vol. 6, pp 439–481.

Toshima, K.; Tatsuta, K. Recent Progress in O-Glycosylation Methods and Its Application to Natural Products Synthesis, *Chem. Rev.* **1993**, *93*, 1503–1531.

Barresi, F.; Hindsgaul, O. Chemically Synthesized Oligosaccharides, 1994. A Searchable Table of Glycosidic Linkages, *J. Carbohydr. Chem.* **1995**, *14*, 1043–1087.

Bols, M. *Carbohydrate Building Blocks*; Wiley: New York, 1995.

Collins, P. M.; Ferrier, R. J. *Monosaccharides: Their Chemistry and Their Roles in Natural Products*; Wiley: Chichester, 1995.

Levy, D. E.; Tang, C. *The Chemistry of C-Glycosides*; Pergamon: Oxford, 1995.

Montreuil, J.; Vliegenthart, J. F. G.; Schachter, H., Eds., *Glycoproteins*; Elsevier: Amsterdam, 1995

Paulsen, H. Progress in Oligosaccharide Synthesis through a New Orthogonal Glycosylation Strategy, *Angew. Chem., Int. Ed. Engl.* **1995**, *34*, 1432–1434.

Postema, M. H. D., Ed. *C-Glycoside Synthesis*; CRC: Boca Raton, 1995

Wilson, L. J.; Hager, M. W.; Elkattan, Y. A.; Liotta, D. C. Nitrogen Glycosylation Reactions Involving Pyrimidine and Purine Nucleoside Bases with Furanoside Sugars, *Synthesis* **1995**, 1465–1479.

Boons, G. J. Strategies in Oligosaccharide Synthesis, *Tetrahedron* **1996**, *52*, 1095–1121.

Danishefsky, S. J.; Bilodeau, M. T. Glycals in Organic Synthesis: The Evolution of Comprehensive Strategies for the Assembly of Oligosaccharides and Glycoconjugates of Biological Consequence, *Angew. Chem., Int. Ed. Engl.* **1996**, *35*, 1380–1419.

Gijsen, H. J. M.; Qiao, L.; Fitz, W.; Wong, C. H. Recent Advances in the Chemoenzymatic Synthesis of Carbohydrates and Carbohydrate Mimetics, *Chem. Rev.* **1996**, *96*, 443–473.

Khan, S. H.; O'Neill, R. A., Eds., *Modern Methods in Carbohydrate Synthesis*; Harwood: Amsterdam, 1996

Schelhaas, M.; Waldmann, H. Protecting Group Strategies in Organic Synthesis, *Angew. Chem., Int. Ed. Engl.* **1996**, *35*, 2056–2083.

Arya, P.; Ben, R. N. Combinatorial Chemistry for the Synthesis of Carbohydrate Libraries, *Angew. Chem., Int. Ed. Engl.* **1997**, *36*, 1280–1282.

Beau, J.-M.; Gallagher, T. Nucleophilic C-Glycosyl Donors for C-Glycoside Synthesis, *Top. Curr. Chem.* **1997**, *187*, 1–54.

David, S. *Molecular and Supramolecular Chemistry of Carbohydrates: Chemical Introduction to the Glyco-Sciences*; Oxford University Press: Oxford, 1997.

Fernandez Mayoralas, A. Synthesis and Modification of Carbohydrates Using Glycosidases and Lipases, *Top. Curr. Chem.* **1997**, *186*, 1–20.

Gambert, U.; Thiem, J. Chemical Transformations Employing Glycosyl Transferases, *Top. Curr. Chem.* **1997**, *186*, 21–43.

Garegg, P. J. Thioglycosides as Glycosyl Donors in Oligosaccharide Synthesis, *Adv. Carbohydr. Chem. Biochem.* **1997**, *52*, 179–205.

Guibé, F. Allylic Protecting Groups and Their Use in a Complex Environment.1. Allylic Protection of Alcohols, *Tetrahedron* **1997**, *53*, 13509–13556.

Hanessian, S., Ed. *Preparative Carbohydrate Chemistry*; Dekker: New York, 1997

Imperiali, B. Protein Glycosylation: The Clash of the Titans, *Acc. Chem. Res.* **1997**, *30*, 452–459.

Kirschning, A.; Bechthold, A. F. W.; Rohr, J. Chemical and Biochemical Aspects of Deoxysugars and Deoxysugar Oligosaccharides, *Top. Curr. Chem.* **1997**, *188*, 1–84.

Kren, V. Enzymatic and Chemical Glycosylations of Ergot Alkaloids and Biological Aspects of New Compounds, *Top. Curr. Chem.* **1997**, *186*, 45–64.

Kren, V.; Thiem, J. Glycosylation Employing Bio-Systems: From Enzymes to Whole Cells, *Chem. Soc. Rev.* **1997**, *26*, 463–473.

McGarvey, G. J.; Wong, C. H. Chemical, Enzymatic and Structural Studies in Molecular Glycobiology, *Liebigs Ann. Chem.* **1997**, 1059–1074.

Montreuil, J.; Vliegenthart, J. F. G.; Schachter, H., Eds., *Glycoproteins II*; Elsevier: Amsterdam, 1997

Nicotra, F. Synthesis of C-Glycosides of Biological Interest, *Top. Curr. Chem.* **1997**, *1987*, 55–83.

Ziegler, T. Pyruvated Saccharides - Novel Strategies for Oligosaccharide Synthesis, *Top. Curr. Chem.* **1997**, *186*, 203–229.

Boons, G. J., Ed. *Carbohydrate Chemistry*; Blackie Academic & Professional: Glasgow, 1998

Chapleur, Y., Ed. *Carbohydrate Mimics*; Wiley-VCH: Weinheim, 1998

Grindley, T. B. Applications of Tin-Containing Intermediates to Carbohydrate Chemistry, *Adv. Carbohydr. Chem. Biochem.* **1998**, *53*, 17–142.

Guibé, F. Allylic Protecting Groups and Their Use in a Complex Environment. Part 2: Allylic Protecting Groups and Their Removal through Catalytic π-Allyl Methodology, *Tetrahedron* **1998**, *54*, 2967–3042.

Kirschning, A. Hypervalent Iodine and Carbohydrates - a New Liaison, *Eur. J. Org. Chem.* **1998**, 2267–2274.

Lehmann, J. *Carbohydrates: Structure and Biology*; Thieme: Stuttgart, 1998.

Seeberger, P. H.; Danishefsky, S. J. Solid-Phase Synthesis of Oligosaccharides and Glycoconjugates by the Glycal Assembly Method: A Five Year Retrospective, *Acc. Chem. Res.* **1998**, *31*, 685–695.

Seitz, O. New Protecting Group Strategies for the Solid-Phase Synthesis and Modification of Peptides, Oligonucleotides, and Oligosaccharides, *Angew. Chem., Int. Ed. Engl.* **1998**, *37*, 3109–3111.

Taylor, C. M. Glycopeptides and Glycoproteins: Focus on the Glycosidic Linkage, *Tetrahedron* **1998**, *54*, 11317–11362.

Arsequell, G.; Valencia, G. Recent Advances in the Synthesis of Complex N-Glycopeptides, *Tetrahedron: Asymmetry* **1999**, *10*, 3045–3094.

Davis, B. G. Recent Developments in Glycoconjugates, *J. Chem. Soc., Perkin Trans. 1* **1999**, 3215–3237.

Finch, P., Ed. *Carbohydrates, Structures, Syntheses and Dynamics*; Kluwer: Dordrecht, Netherlands, 1999

Hecht, S. M., Ed. *Bioorganic Chemistry: Carbohydrates*; Oxford University Press: New York, 1999

Imperiali, B.; O'Connor, S. E.; Hendrickson, T.; Kellenberger, C. Chemistry and Biology of Asparagine-Linked Glycosylation, *Pure Appl. Chem.* **1999**, *71*, 777–787.

Ohrlein, R. Glycosyl Transferase-Catalyzed Synthesis of Non-Natural Oligosaccharides, *Top. Curr. Chem.* **1999**, *200*, 227–254.

Boons, G.-J.; Demchenko, A. V. Recent Advances in O-Sialylation, *Chem. Rev.* **2000**, *100*, 4539–4565.

Danishefsky, S. J.; Allen, J. R. From the Laboratory to the Clinic: A Retrospective on Fully Synthetic Carbohydrate-Based Anticancer Vaccines, *Angew. Chem., Int. Ed. Engl.* **2000**, *39*, 837–863.

Davis, B. G. Recent Developments in Oligosaccharide Synthesis, *J. Chem. Soc., Perkin Trans. 1* **2000**, 2137–2160.

Dondoni, A.; Marra, A. Methods for Anomeric Carbon-Linked and Fused Sugar Amino Acid Synthesis: The Gateway to Artificial Glycopeptides, *Chem. Rev.* **2000**, *100*, 4395–4422.

Ernst, B.; Hart, G. W.; Sinaÿ, P., Eds., *Carbohydrates in Chemistry and Biology*; Wiley-VCH: Weinheim, 2000

Fernandez, J. M. G.; Mellet, C. O. Chemistry and Developments of N-Thiocarbonyl Carbohydrate Derivatives: Sugar Isothiocyanates, Thioamides, Thioureas, Thiocarbamates, and Their Conjugates, *Adv. Carbohydr. Chem. Biochem.* **2000**, *55*, 35–135.

Gridley, J. J.; Osborn, H. M. I. Recent Advances in the Construction of β-D-Mannose and β-D-Mannosamine Linkages, *J. Chem. Soc., Perkin Trans. 1* **2000**, 1471–1491.

Haase, W. C.; Seeberger, P. H. Recent Progress in Polymer-Supported Synthesis of Oligosaccharides and Carbohydrate Libraries, *Curr. Org. Chem.* **2000**, *4*, 481–511.

Herzner, H.; Reipen, T.; Schultz, M.; Kunz, H. Synthesis of Glycopeptides Containing Carbohydrate and Peptide Recognition Motifs, *Chem. Rev.* **2000**, *100*, 4495–4537.

Jung, K.-H.; Müller, M.; Schmidt, R. R. Intramolecular O-Glycoside Bond Formation, *Chem. Rev.* **2000**, *100*, 4423–4442.

Kartha, K. P. R.; Aloui, M.; Cura, P.; Marsh, S. J.; Field, R. A. Iodine, A Versatile Reagent in Carbohydrate Chemistry: Activation of Thioglycosides and Glycosyl Sulfoxides, In *Advances in Sulfur Chemistry*; Rayner, C. M., Ed.; Jai Press: Stamford, 2000; Vol. 2, pp 37–56.

Kochetkov, N. K. Solid-Phase Synthesis of Oligosaccharides and Glycoconjugates, *Russ. Chem. Rev. (Engl. Transl.)* **2000**, *69*, 795–820.

Koeller, K. M.; Wong, C. H. Synthesis of Complex Carbohydrates and Glycoconjugates: Enzyme-Based and Programmable One-Pot Strategies, *Chem. Rev.* **2000**, *100*, 4465–4493.

Kuberan, B.; Lindhardt, R. J. Carbohydrate Based Vaccines, *Curr. Org. Chem.* **2000**, *4*, 653–677.

Marzabadi, C. H.; Franck, R. W. The Synthesis of 2-Deoxyglycosides: 1988–1999, *Tetrahedron* **2000**, *56*, 8385–8417.

Seeberger, P. H.; Haase, W. C. Solid-Phase Oligosaccharide Synthesis and Combinatorial Carbohydrate Libraries, *Chem. Rev.* **2000**, *100*, 4349–4393.

St Hilaire, P. M.; Meldal, M. Glycopeptide and Oligosaccharide Libraries, *Angew. Chem., Int. Ed. Engl.* **2000**, *39*, 1163–1179.

Toshima, K. Glycosyl Fluorides in Glycosidations, *Carbohydr. Res.* **2000**, *327*, 15–26.

Vankar, Y. D.; Schmidt, R. R. Chemistry of Glycosphingolipids-Carbohydrate Molecules of Biological Significance, *Chem. Soc. Rev.* **2000**, *29*, 201–216.

Williams, S. J.; Withers, S. G. Glycosyl Fluorides in Enzymatic Reactions, *Carbohydr. Res.* **2000**, *327*, 27–46.

Barkley, A.; Arya, P. Combinatorial Chemistry toward Understanding the Function(s) of Carbohydrates and Carbohydrate Conjugates, *Chem. Eur. J.* **2001**, *7*, 555–563.

Bertozzi, C. R.; Kiessling, L. L. Chemical Glycobiology, *Science* **2001**, *291*, 2357–2364.

Fraser-Reid, F.; Kuniaki, T.; Thiem, J., Eds., *Glycoscience: Chemistry and Chemical Biology*; Springer-Verlag: Berlin, 2001

Hang, H. C.; Bertozzi, C. R. Chemoselective Approaches to Glycoprotein Assembly, *Acc. Chem. Res.* **2001**, *34*, 727–736.

Nicolaou, K. C.; Mitchell, H. J. Adventures in Carbohydrate Chemistry: New Synthetic Technologies, Chemical Synthesis, Molecular Design, and Chemical Biology, *Angew. Chem., Int. Ed. Engl.* **2001**, *40*, 1576–1624.

Ritter, T. K.; Wong, C. H. Carbohydrate-Based Antibiotics: A New Approach to Tackling the Problem of Resistance, *Angew. Chem., Int. Ed. Engl.* **2001**, *40*, 3509–3533.

Seeberger, P. H., Ed. *Solid Support Oligosaccharide Synthesis and Combinatorial Carbohydrate Libraries*; Wiley: New York, 2001

Stick, R. V. *Carbohydrates: The Sweet Molecules of Life*; Academic Press: San Diego, 2001.

Thorson, J. S.; Hosted, T. J.; Jiang, J. Q.; Biggins, J. B.; Ahlert, J. Nature's Carbohydrate Chemists: The Enzymatic Glycosylation of Bioactive Bacterial Metabolites, *Curr. Org. Chem.* **2001**, *5*, 139–167.

Vaino, A. R.; Szarek, W. A. Iodine in Carbohydrate Chemistry, *Adv. Carbohydr. Chem. Biochem.* **2001**, *56*, 9–63.

Wang, P.; Bertozzi, C. R., Eds., *Glycochemistry: Principles, Synthesis, and Applications*; Dekker: New York, 2001

Zhang, J. B.; Wu, B. Y.; Liu, Z. Y.; Kowal, P.; Chen, X.; Shao, J.; Wang, P. G. Large-Scale Synthesis of Carbohydrates for Pharmaceutical Development, *Curr. Org. Chem.* **2001**, *5*, 1169–1176.

Angata, T.; Varki, A. Chemical Diversity in the Sialic Acids and Related α-Keto Acids: An Evolutionary Perspective, *Chem. Rev.* **2002**, *102*, 439–469.

Bogusiak, J. Synthesis of Glycofuranosides, *Pol. J. Chem.* **2002**, *76*, 1–12.

Crich, D. Chemistry of Glycosyl Triflates: Synthesis of β-Mannopyranosides, *J. Carbohydr. Chem.* **2002**, *21*, 667–690.

Davis, B. G. Synthesis of Glycoproteins, *Chem. Rev.* **2002**, *102*, 579–601.

Gin, D. Dehydrative Glycosylation with 1-Hydroxy Donors, *J. Carbohydr. Chem.* **2002**, *21*, 645–665.

Halcomb, R. L.; Chappell, M. D. Recent Developments in Technology for Glycosylation with Sialic Acid, *J. Carbohydr. Chem.* **2002**, *21*, 723–768.

Jensen, K. J. O-Glycosylations under Neutral or Basic Conditions, *J. Chem. Soc., Perkin Trans. 1* **2002**, 2219–2233.

Klefel, M. J.; von Itzstein, M. Recent Advances in the Synthesis of Sialic Acid Derivatives and Sialyl Mimetics as Biological Probes, *Chem. Rev.* **2002**, *102*, 471–490.

Lowary, T. L. D-Arabinofuranosides from Mycobacteria: Synthesis and Conformation, *J. Carbohydrate Chem.* **2002**, *21*, 691–722.

Palmacci, E. R.; Plante, O. J.; Seeberger, P. H. Oligosaccharide Synthesis in Solution and on Solid Support with Glycosyl Phosphates, *Eur. J. Org. Chem.* **2002**, 595–606.

Ritchie, G. E.; Moffatt, B. E.; Sim, R. B.; Morgan, B. P.; Dwek, R. A.; Rudd, P. M. Glycosylation and the Complement System, *Chem. Rev.* **2002**, *102*, 305–319.

Vankayalapati, H.; Jiang, S. D.; Singh, G. Glycosylation Based on Glycosyl Phosphates as Glycosyl Donors, *Synlett* **2002**, 16–25.

Wong, C.-H., Ed. *Carbohydrate-Based Drug Discovery*; Wiley-VCH: Weinheim, 2002

Demchenko, A. V. 1,2-Cis-O-Glycosylation: Methods, Strategies, Principles, *Curr. Org. Chem.* **2003**, *7*, 35–79.

Demchenko, A. V. Stereoselective Chemical 1,2-Cis-O-Glycosylation: From "Sugar Ray" to Modern Techniques of the 21st Century, *Synlett* **2003**, 1225–1240.

Ferrier, R. J.; Hoberg, J. O. Synthesis and Reactions of Unsaturated Sugars, *Adv. Carbohydr. Chem. Biochem.* **2003**, *58*, 55–119.

Ferrier, R. J.; Zubkov, O. A. Transformation of Glycals into 2,3-Unsaturated Glycosyl Derivatives, *Org. React.* **2003**, *62*, 569–736.

Lindhorst, T. K. *Essentials of Carbohydrate Chemistry and Biochemistry*; Wiley-VCH: Weinheim, 2003.

Osborn, H. M. I.; Harwood, L. M. *Best Synthetic Methods: Carbohydrates*; Academic Press: San Diego, CA, 2003.

Plante, O. J.; Palmacci, E. R.; Seeberger, P. H. Development of an Automated Oligosaccharide Synthesizer, *Adv. Carbohydr. Chem. Biochem.* **2003**, *58*, 35–54.

Seeberger, P. H. Automated Carbohydrate Synthesis to Drive Chemical Glycomics, *J. Chem. Soc., Chem. Commun.* **2003**, 1115–1121.

Shoda, S.-I.; Izumi, R.; Fujita, M. Green Process in Glycotechnology, *Bull. Chem. Soc. Jpn.* **2003**, *76*, 1–13.

Virta, P.; Katajisto, J.; Niittymaki, T.; Lonnberg, H. Solid-Supported Synthesis of Oligomeric Bioconjugates, *Tetrahedron* **2003**, *59*, 5137–5174.

Corsaro, A.; Chiacchio, U.; Pistara, V.; Romeo, G. Microwave-Assisted Chemistry of Carbohydrates, *Curr. Org. Chem.* **2004**, *8*, 511–538.

Crich, D.; Lim, L. B. L. Glycosylation with Sulfoxides and Sulfinates as Donors or Promoters, *Org. React.* **2004**, *64*, 115–251.

Das, S. K. Application of Microwave Irradition in the Synthesis of Carbohydrates, *Synlett* **2004**, 915–932.

Fraser-Reid, B.; López, J. C.; Gómez, A. M.; Uriel, C. Reciprocal Donor Acceptor Selectivity (RDAS) and Paulsen's Concept of "Match" in Saccharide Coupling, *Eur. J. Org. Chem.* **2004**, 1387–1395.

Pellissier, H. The Glycosylation of Steroids, *Tetrahedron* **2004**, *60*, 5123–5162.

Peri, F.; Nicotra, F. Chemoselective Ligation in Glycochemistry, *Chem Comm* **2004**, 623–627.

Petitou, M.; van Boeckel, C. A. A. A Synthetic Antithrombin III Binding Pentasaccharide Is Now a Drug! What Comes Next?, *Angew. Chem. Int. Ed. Engl.* **2004**, *43*, 3118–3133.

Taillefumier, C.; Chapleur, Y. Synthesis and Uses of Exo-Glycals, *Chem. Rev.* **2004**, *104*, 263–292.

Yu, B.; Yang, Z.; Cao, H. One-Pot Glycosylation (OPG) for the Chemical Synthesis of Oligosaccharides, *Curr. Org. Chem.* **2005**, *9*, 179–194.

Nucleotide Bond Formation

Beaucage, S. L.; Iyer, R. P. Advances in the Synthesis of Oligonucleotides by the Phosphoramidite Approach, *Tetrahedron* **1992**, *48*, 2223–2311.

Dueholm, K. L.; Pedersen, E. B. 2,3-Dideoxyfuranoses in Convergent Syntheses of 2',3'-Dideoxy Nucleosides, *Synthesis* **1992**, 1–22.

Engel, R., Ed. *Handbook of Organophosphorus Chemistry*; Dekker: New York, 1992

Huryn, D. M.; Okabe, M. Aids-Driven Nucleoside Chemistry, *Chem. Rev.* **1992**, *92*, 1745–1768.

Westheimer, F. H. Phosphorus Chemistry, In *Phosphorus Chemistry Developments in American Sciences*; Walsh, E. N.; Griffith, E. J.; Parry, R. W.; Quin, L. D., Eds.; American Chemical Society: Washington, 1992; Vol. 486, pp 1–18.

Baraniak, J.; Stec, W. J. Stereocontrolled Syntheses of P-Chiral Analogues of Nucleoside Cyclic 3',5'-Phosphates, *Rev. Heteroat. Chem.* **1993**, *8*, 143–164.

Beaucage, S. L.; Iyer, R. P. The Functionalization of Oligonucleotides via Phosphoramidite Derivatives, *Tetrahedron* **1993**, *49*, 1925–1963.

Beaucage, S. L.; Iyer, R. P. The Synthesis of Modified Oligonucleotides by the Phosphoramidite Approach and Their Applications, *Tetrahedron* **1993**, *49*, 6123–6194.

Beaucage, S. L.; Iyer, R. P. The Synthesis of Specific Ribonucleotides and Unrelated Phosphorylated Biomolecules by the Phosphoramidite Method, *Tetrahedron* **1993**, *49*, 10441–10488.

Billington, D. C. *The Inositol Phosphates: Chemical Synthesis and Biological Significance*; VCH: Weinheim, 1993.

Dorfman, Y. A.; Aleshkova, M. M.; Polimbetova, G. S.; Levina, L. V.; Petrova, T. V.; Abdreimova, R. R.; Doroshkevich, D. M.

New Reactions Involving the Oxidative O-, N-, and C-Phosphorylation of Organic Compounds by Phosphorus and Phosphides in the Presence of Metal Complexes, *Russ. Chem. Rev. (Engl. Transl.)* **1993**, *62*, 877–896.

Frauendorf, A.; Engels, J. W. Automated Synthesis, Structure and Biological Activity of Backbone-Modified Oligonucleotides - the Antisense Approach, In *Studies in Natural Products Chemistry*; Atta-ur-Rahman; Basha, F. Z., Eds.; Elsevier: Amsterdam, 1993; Vol. 13, pp 257–294.

Givens, R. S.; Kueper, L. W., III Photochemistry of Phosphate Esters, *Chem. Rev.* **1993**, *93*, 55–66.

Suska, A.; Grajkowski, A.; Wilk, A.; Uznanski, B.; Blaszczyk, J.; Wieczorek, M.; Stec, W. J. Antisense Oligonucleotides: Stereocontrolled Synthesis of Phosphorothioate Oligonucleotides, *Pure Appl. Chem.* **1993**, *65*, 707–714.

Troev, K. Dialkyl Hydrogen Phosphonates. 1. Structures of Dialkyl Hydrogen Phosphonates, *Rev. Heteroat. Chem.* **1993**, *8*, 165–181.

Varma, R. S. Synthesis of Oligonucleotide Analogs with Modified Backbones, *Synlett* **1993**, 621–637.

Wnuk, S. F. Sulfur- and Seleno-Sugar Modified Nucleosides. Synthesis, Chemical Transformations and Biological Properties, *Tetrahedron* **1993**, *49*, 9877–9936.

Stec, W. J.; Wilk, A. Stereocontrolled Synthesis of Oligo(Nucleoside Phosphorothioate)s, *Angew. Chem., Int. Ed. Engl.* **1994**, *33*, 709–722.

Timoshchuk, V. A. The Synthesis and Transformations of Uronic Acid Nucleosides, *Russ. Chem. Rev. (Engl. Transl.)* **1994**, *63*, 695–710.

Townsend, L. B., Ed. *Chemistry of Nucleosides and Nucleotides*; Plenum: New York, 1994; Vol. 3.

Troev, K. Dialkyl Hydrogen Phosphonates. 2. Substitution Reactions at Phosphorus Atom, *Rev. Heteroat. Chem.* **1994**, *10*, 89–119.

Knapp, S. Synthesis of Complex Nucleoside Antibiotics, *Chem. Rev.* **1995**, *95*, 1859–1876.

Vorbrüggen, H. Adventures in Silicon - Organic Chemistry, *Acc. Chem. Res.* **1995**, *28*, 509–520.

Wilson, L. J.; Hager, M. W.; El-Kattan, Y. A.; Liotta, D. C. Nitrogen Glycosylation Reactions Involving Pyrimidine and Purine Nucleoside Bases with Furanoside Sugars, *Synthesis* **1995**, 1465–1479.

De Clercq, E., Ed. *Advances in Antiviral Drug Design*; Jai Press Inc: Greenwich, CT, 1996; Vol. 2

Egli, M. Structural Aspects of Nucleic Acid Analogs and Antisense Nucleotides, *Angew. Chem. Int. Ed. Engl.* **1996**, *35*, 1894–1909.

Hecht, S. M., Ed. *Bioorganic Chemistry: Nucleic Acids*; Oxford University Press: New York, 1996

Watanabe, Y. Selective Reactions and Total Synthesis of Inositol Phosphates, In *Studies in Natural Products Chemistry*; Atta-Ur-Rahman, Ed.; Elsevier: Amsterdam, 1996; Vol. 18, pp 391–456.

Nielsen, P. E.; Haaima, G. Peptide Nucleic Acid (PNA). A DNA Mimic with a Pseudopeptide Backbone, *Chem. Soc. Rev.* **1997**, *26*, 73–78.

Osborne, S. E.; Ellington, A. D. Nucleic Acid Selection and the Challenge of Combinatorial Chemistry, *Chem. Rev.* **1997**, *97*, 349–370.

Perreault, D. M.; Anslyn, E. V. Unifying the Current Data on the Mechanism of Cleavage - Transesterification of RNA, *Angew. Chem., Int. Ed. Engl.* **1997**, *36*, 432–450.

Sugimura, H. Methodology for the Stereoselective Synthesis of Nucleosides Utilizing 1-Thioglycoside as Glycosyl Donor, In *Recent Research Developments in Organic Chemistry*; Pandalai, S. G., Ed.; Transworld Research Network: Trivandrum, 1997; Vol. 1, pp 25–34.

Frauendorf, C.; Jaschke, A. Catalysis of Organic Reactions by RNA, *Angew. Chem., Int. Ed. Engl.* **1998**, *37*, 1378–1381.

Givens, R. S.; Weber, J. F. W.; Jung, A. H.; Park, C.-H. New Photoprotecting Groups: Desyl and *p*-Hydroxyphenacyl Phosphate and Carboxylate Esters, *Meth. Enzym.* **1998**, *291*, 1–29.

Izawa, K.; Shiragami, H. Practical Syntheses of Antiviral Nucleosides, *Pure Appl. Chem.* **1998**, *70*, 313–318.

Meier, C. Pro-Nucleotides - Recent Advances in the Design of Efficient Tools for the Delivery of Biologically Active Nucleoside Monophosphates, *Synlett* **1998**, 233–242.

Seitz, O. New Protecting Group Strategies for the Solid-Phase Synthesis and Modification of Peptides, Oligonucleotides, and Oligosaccharides, *Angew. Chem., Int. Ed. Engl.* **1998**, *37*, 3109–3111.

Uhlmann, E.; Peyman, A.; Breipohl, G.; Will, D. W. PNA: Synthetic Polyamide Nucleic Acids with Unusual Binding Properties, *Angew. Chem., Int. Ed. Engl.* **1998**, *37*, 2797–2823.

Yu, H. W.; Zhang, H. Y.; Yang, Z. J.; Min, J. M.; Ma, L. T.; Zhang, L. H. Studies on the Syntheses and Biological Activities of Isonucleosides, *Pure Appl. Chem.* **1998**, *70*, 435–438.

Butenandt, J.; Burgdorf, L. T.; Carell, T. Synthesis of DNA Lesions and DNA-Lesion-Containing Oligonucleotides for DNA-Repair Studies, *Synthesis* **1999**, 1085–1105.

Elhalabi, J. M.; Rice, K. G. Synthesis and Applications for Unnatural Sugar Nucleotides, *Curr. Med. Chem.* **1999**, *6*, 93–116.

Lamberth, C. The Chemistry of 4',5'-Unsaturated Nucleosides. A Review, *Org. Prep. Proced. Int.* **1999**, *31*, 379–397.

Vorbrüggen, H.; Ruh-Pohlenz, C. Synthesis of Nucleosides, *Org. React.* **1999**, *55*, 1–630.

Wengel, J. Synthesis of 3 'C- and 4 'C-Branched Oligodeoxynucleotides and the Development of Locked Nucleic Acid (LNA), *Acc. Chem. Res.* **1999**, *32*, 301–310.

Wozniak, L. A. Stereodefined Methanephosphonate Analogues of Nucleic Acids, *Rev. Heteroat. Chem.* **1999**, *19*, 173–202.

Yang, Z. J.; Zhang, H. Y.; Min, J. M.; Ma, L. T.; Zhang, L. H. Synthesis and Enzymatic Stability of Oligonucleotides Consisting of Isonucleosides, *Pure Appl. Chem.* **1999**, *71*, 1143–1146.

Yokoyama, M.; Momotake, A. Synthesis and Biological Activity of Azanucleosides, *Synthesis* **1999**, 1541–1554.

Burgess, K.; Cook, D. Syntheses of Nucleoside Triphosphates, *Chem. Rev.* **2000**, *100*, 2047–2059.

Couvreur, P.; Malvy, C., Eds., *Pharmaceutical Aspects of Oligonucleotides*; Taylor & Francis: London, 2000

Davies, M. J.; Shah, A.; Bruce, I. J. Synthesis of Fluorescently Labelled Oligonucleotides and Nucleic Acids, *Chem. Soc. Rev.* **2000**, *29*, 97–107.

Ferrero, M.; Gotor, V. Biocatalytic Selective Modifications of Conventional Nucleosides, Carbocyclic Nucleosides, and C-Nucleosides, *Chem. Rev.* **2000**, *100*, 4319–4347.

Ganesh, K. N.; Nielsen, P. E. Peptide Nucleic Acids: Analogs and Derivatives, *Curr. Org. Chem.* **2000**, *4*, 931–943.

Kates, S. A.; Albericio, F., Eds., *Solid Phase Synthesis*; Dekker: New York, 2000

Kisaurek, M. V.; Rosemeyer, H., Eds., *Perspectives in Nucleoside and Nucleic Acid Chemistry*; Wiley-VCH: Weinheim, 2000

Meldgaard, M.; Wengel, J. Bicyclic Nucleosides and Conformational Restriction of Oligonucleotides, *J. Chem. Soc., Perkin Trans. 1* **2000**, 3539–3554.

Nifantiev, E. E.; Grachev, M. K.; Burmistrov, S. Y. Amides of Trivalent Phosphorus Acids as Phosphorylating Reagents for Proton-Donating Nucleophiles, *Chem. Rev.* **2000**, *100*, 3755–3799.

Pankiewicz, K. W. Fluorinated Nucleosides, *Carbohydr. Res.* **2000**, *327*, 87–105.

Quin, L. D. *A Guide to Organophosphorus Chemistry*; Wiley: New York, 2000.

Yokoyama, M. Synthesis and Biological Activity of Thionucleosides, *Synthesis* **2000**, 1637–1655.

Hayakawa, Y. Toward an Ideal Synthesis of Oligonucleotides: Development of a Novel Phosphoramidite Method with High Capability, *Bull. Chem. Soc. Jpn.* **2001**, *74*, 1547–1565.

Hoffman, R. V. Catalysts, Anticatalysts, and Receptors for Unactivated Phosphate Diesters in Water, *Chemtracts: Org. Chem.* **2001**, *14*, 500–504.

Inamoto, N. Studies on Low-Coordinated Nitrogen, Phosphorus, Sulfur, and Selenium Compounds, *Heteroat. Chem.* **2001**, *12*, 183–194.

Jackson, M. D.; Denu, J. M. Molecular Reactions of Protein Phosphatases - Insights from Structure and Chemistry, *Chem. Rev.* **2001**, *101*, 2313–2340.

Johnson, L. N.; Lewis, R. J. Structural Basis for Control by Phosphorylation, *Chem. Rev.* **2001**, *101*, 2209–2242.

Leumann, C. J. Design and Evaluation of Oligonucleotide Analogues, *Chimia* **2001**, *55*, 295–301.

Leumann, C. J. Conformationally Restricted Oligonucleotide Analogues 'Made in Bern': A Mini Review, *Chimia* **2001**, *55*, 1041–1044.

Micklefield, J. Backbone Modification of Nucleic Acids: Synthesis, Structure and Therapeutic Applications, *Curr. Med. Chem.* **2001**, *8*, 1157–1179.

Pirrung, M. C. Discovery of a Catalytic Asymmetric Phosphorylation through Selection of a Minimal Kinase Mimic: A Concise Total Synthesis of D-Myo-Inositol-1-Phosphate, *Chemtracts: Org. Chem.* **2001**, *14*, 802–804.

Rowlands, G. J. Ambifunctional Cooperative Catalysts, *Tetrahedron* **2001**, *57*, 1865–1882.

Vorbrüggen, H.; Ruh-Pohlenz, C. *Handbook of Nucleoside Synthesis*; Wiley: Chichester, 2001.

Bochet, C. G. Photolabile Protecting Groups and Linkers, *J. Chem. Soc., Perkin Trans. 1* **2002**, 125–142.

Chambert, S.; Decout, J.-L. Recent Developments in the Synthesis, Chemical Modifications and Biological Applications of Sulfur Modified Nucleosides, Nucleotides and Oligonucleotides, *Org. Prep. Proced. Int.* **2002**, *34*, 27–85.

Chu, C. K. *Recent Advances in Nucleosides: Chemistry and Chemotherapy*; Elsevier: Amsterdam, 2002.

Guga, P.; Okruszek, A.; Stec, W. J. Recent Advances in Stereocontrolled Synthesis of P-Chiral Analogues of Biophosphates, *Top. Curr. Chem.* **2002**, *220*, 169–200.

Imanishi, T.; Obika, S. BNAS: Novel Nucleic Acid Analogs with a Bridged Sugar Moiety, *J. Chem. Soc., Chem. Commun.* **2002**, 1653–1659.

Kachalova, A. V.; Zubin, E. M.; Oretskaya, T. S. Methods for the Synthesis of Oligonucleotides Containing Reactive Electrophilic Groups, *Russ. Chem. Rev. (Engl. Transl.)* **2002**, *71*, 1041–1059.

Reese, C. B. The Chemical Synthesis of Oligo- and Poly-Nucleotides: A Personal Commentary, *Tetrahedron* **2002**, 58, 8893–8920.

Stawinski, J.; Kraszewski, A. How to Get the Most out of Two Phosphorus Chemistries. Studies on H-Phosphonates, *Acc. Chem. Res.* **2002**, *35*, 952–960.

Zatsepin, T. S.; Romanova, E. A.; Oretskaya, T. S. Synthesis of 2'-O-Alkyl-Nucleosides, *Russ. Chem. Rev. (Engl. Transl.)* **2002**, *71*, 513–534.

Zubin, E. M.; Romanova, E. A.; Oretskaya, T. S. Modern Methods for the Synthesis of Peptide - Oligonucleotide Conjugates, *Russ. Chem. Rev. (Engl. Transl.)* **2002**, *71*, 239–264.

Agrofoglio, L. A.; Gillaizeau, I.; Saito, Y. Palladium-Assisted Routes to Nucleosides, *Chem. Rev.* **2003**, *103*, 1875–1916.

Virta, P.; Katajisto, J.; Niittymaki, T.; Lonnberg, H. Solid-Supported Synthesis of Oligomeric Bioconjugates, *Tetrahedron* **2003**, *59*, 5137–5174.

Benner, S. A. Understanding Nucleic Acids Using Synthetic Chemistry, *Acc. Chem. Res.* **2004**, *37*, 784–797.

Marx, A.; Detmer, I.; Gaster, J.; Summerer, D. Probing DNA Polymerase Function with Synthetic Nucleotides, *Synthesis* **2004**, 1–14.

Michalski, J.; Dabkowski, W. State of the Art Chemical Synthesis of Biophosphates and Their Analogues Via P(III) Derivatives, *Top. Curr. Chem.* **2004**, *232*, 93–144.

Summerer, D.; Marx, A. 4'C-Modified Nucleotides as Chemical Tools for Investigation and Modulation of DNA Polymerase Function, *Synlett* **2004**, 217–224.

Wu, Q.; Simons, C. Synthetic Methodologies for C-Nucleosides, *Synthesis* **2004**, 1533–1553.

Hengge, A. C.; Onyido, I. Physical Organic Perspectives on Phospho Group Transfer from Phosphates and Phosphinates, *Curr. Org. Chem.* **2005**, *9*, 61–74.

Herdewijn, P., Ed. *Oligonucleotide Synthesis: Methods and Applications*; Humana Press: Totowa, 2005; Vol. 288

Peptide Bond Formation

Grant, G. A. *Synthetic Peptides: A User's Guide*; Freeman: New York, 1992.

Jones, J. *Amino Acid and Peptide Synthesis*; Oxford University Press: Oxford, 1992.

Jung, G.; Beck-Sickinger, A. G. Multiple Peptide Synthesis Methods and Their Applications, *Angew. Chem., Int. Ed. Engl.* **1992**, *31*, 367–383.

Wong, C. H.; Liu, K. K. C.; Kajimoto, T.; Chen, L.; Zhong, Z.; Dumas, D. P.; Liu, J. L. C.; Ichikawa, Y.; Shen, G. J. Enzymes for Carbohydrate and Peptide Syntheses, *Pure Appl. Chem.* **1992**, *64*, 1197–1202.

Bodanszky, M. *Peptide Chemistry. A Practical Textbook*, 2nd ed.; Springer: Berlin, 1993.

Chorev, M.; Goodman, M. A Dozen Years of Peptidoimetics, *Acc. Chem. Res.* **1993**, *26*, 266–273.

Lloyd-Williams, P.; Albericio, F.; Giralt, E. Convergent Solid-Phase Peptide Synthesis, *Tetrahedron* **1993**, *49*, 11065–11133.

Merrifield, R. B. *Life During a Golden Age of Peptide Chemistry. The Concept and Development of Solid-Phase Peptide Synthesis*; American Chemical Society: Washington, 1993.

Basava, C.; Anantharamaiah, G. M., Eds., *Peptides: Design, Synthesis and Biological Activity*; Birkhaeuser: Boston, 1994

Benz, H. The Role of Solid-Phase Fragment Condensation (SPFC) in Peptide Synthesis, *Synthesis* **1994**, 337–358.

Bodanszky, M.; Bodanszky, A. *The Practice of Peptide Synthesis*, 2nd ed.; Springer-Verlag: Heidelberg, 1994.

Gutte, B., Ed. *Peptides: Synthesis, Structures, and Applications*; Academic Press: San Diego, 1995

Jakubke, H. D. Peptide Ligases -Tools for Peptide Synthesis, *Angew. Chem., Int. Ed. Engl.* **1995**, *34*, 175–177.

Montreuil, J.; Vliegenthart, J. F. G.; Schachter, H., Eds., *Glycoproteins*; Elsevier: Amsterdam, 1995

Bernard, J. M.; Bouzid, K.; Casati, J. P.; Galvez, M.; Gervais, C.; Meilland, P.; Pevere, V.; Vandewalle, M. F.; Badey, J. P.; Enderlin, J. M. Peptide Synthesis by Sappho Technology, In *Roots of Organic Development*; Desmurs, J. R.; Ratton, S., Eds.; Elsevier: Amsterdam, 1996, pp 405–415.

Carpino, L. A.; Beyermann, M.; Wenschuh, H.; Bienert, M. Peptide Synthesis Via Amino Acid Halides, *Acc. Chem. Res.* **1996**, *29*, 268–274.

Jung, G., Ed. *Combinatorial Peptide and Non-Peptide Libraries*; VCH: Weinheim, 1996

Stepanov, V. M. Proteinases as Catalysts in Peptide Synthesis, *Pure Appl. Chem.* **1996**, *68*, 1335–1339.

Arsequell, G.; Valencia, G. O-Glycosyl α-Amino Acids as Building Blocks for Glycopeptide Synthesis, *Tetrahedron: Asymmetry* **1997**, *8*, 2839–2876.

Humphrey, J. M.; Chamberlin, A. R. Chemical Synthesis of Natural Product Peptides: Coupling Methods for the Incorporation of Noncoded Amino Acids into Peptides, *Chem. Rev.* **1997**, *97*, 2243–2266.

Kunz, H. O- and N-Glycopeptides: Synthesis of Selectively Deprotected Building Blocks, In *Preparative Carbohydrate Chemistry*; Hanessian, S., Ed.; Dekker: New York, 1997, pp 265–281.

Lloyd-Williams, P.; Albericio, F.; Giralt, E. *Chemical Approaches to the Synthesis of Peptides and Proteins*; CRC: Boca Raton, 1997.

Montreuil, J.; Vliegenthart, J. F. G.; Schachter, H., Eds., *Glycoproteins II*; Elsevier: Amsterdam, 1997

Nielsen, P. E.; Haaima, G. Peptide Nucleic Acid (PNA). A DNA Mimic with a Pseudopeptide Backbone, *Chem. Soc. Rev.* **1997**, *26*, 73–78.

Seebach, D.; Matthews, J. L. β-Peptides: A Surprise at Every Turn, *J. Chem. Soc., Chem. Commun.* **1997**, 2015–2022.

von Dohren, H.; Keller, U.; Vater, J.; Zocher, R. Multifunctional Peptide Synthetases, *Chem. Rev.* **1997**, *97*, 2675–2705.

Walker, M. A. Protein Synthesis by Chemical Ligation of Unprotected Peptides in Aqueous Solution, *Angew. Chem., Int. Ed. Engl.* **1997**, *36*, 1069–1071.

Fletcher, M. D.; Campbell, M. M. Partially Modified Retro-Inverso Peptides: Development, Synthesis, and Conformational Behavior, *Chem. Rev.* **1998**, *98*, 763–795.

Gellman, S. H. Foldamers: A Manifesto, *Acc. Chem. Res.* **1998**, *31*, 173–180.

Hecht, S. M., Ed. *Bioorganic Chemistry: Peptides and Proteins*; OUP: Oxford, 1998

Jarowicki, K.; Kocienski, P. Protecting Groups, *J. Chem. Soc., Perkin Trans. 1* **1998**, 4005–4037.

Seitz, O. New Protecting Group Strategies for the Solid-Phase Synthesis and Modification of Peptides, Oligonucleotides, and Oligosaccharides, *Angew. Chem., Int. Ed. Engl.* **1998**, *37*, 3109–3111.

Uhlmann, E.; Peyman, A.; Breipohl, G.; Will, D. W. PNA: Synthetic Polyamide Nucleic Acids with Unusual Binding Properties, *Angew. Chem., Int. Ed. Engl.* **1998**, *37*, 2797–2823.

Vital'evna, K. E.; Vladimirovna, S. M.; Porfir'evna, E. R. Peculiarities of Synthesis of the Cysteine-Containing Peptides, *Russ. Chem. Rev. (Engl. Transl.)* **1998**, *67*, 611–630.

Abell, A., Ed. *Advances in Amino Acid Mimetics and Peptidomimetics*; Jai Press Inc: Stamford, 1999; Vol. 2.

Alsina, J.; Jensen, K. J.; Albericio, F.; Barany, G. Solid-Phase Synthesis with Tris(Alkoxy)Benzyl Backbone Amide Linkage (Bal), *Chem. Eur. J.* **1999**, *5*, 2787–2795.

Kochendoerfer, G. G.; Kent, S. B. H. Chemical Protein Synthesis, *Curr. Opinion. Chem. Biol.* **1999**, *3*, 665–671.

Kulathila, R.; Merkler, K. A.; Merkler, D. J. Enzymatic Formation of C-Terminal Amides, *Nat. Prod. Rep.* **1999**, *16*, 145–154.

North, M. Amines and Amides, *J. Chem. Soc., Perkin Trans. 1* **1999**, 2209–2229.

Abele, S.; Seebach, D. Preparation of Achiral and of Enantiopure Geminally Disubstituted β-Amino Acids for β-Peptide Synthesis, *Eur. J. Org. Chem.* **2000**, 1–15.

Chan, W. C.; White, P. D., Eds., *Fmoc Solid Phase Peptide Synthesis: A Practical Approach*; Oxford University Press: Oxford, 2000.

Coltart, D. M. Peptide Segment Coupling by Prior Ligation and Proximity-Induced Intramolecular Acyl Transfer, *Tetrahedron* **2000**, *56*, 3449–3491.

Dawson, P. E.; Kent, S. B. H. Synthesis of Native Proteins by Chemical Ligation, *Ann. Rev. Biochem.* **2000**, *69*, 923–960.

Domling, A.; Ugi, I. Multicomponent Reactions with Isocyanides, *Angew. Chem., Int. Ed. Engl.* **2000**, *39*, 3169–3210.

Fischer, G. Chemical Aspects of Peptide Bond Isomerisation, *Chem. Soc. Rev.* **2000**, *29*, 119–127.

Kates, S. A.; Albericio, F., Eds., *Solid Phase Synthesis*; Dekker: New York, 2000.

Kunz, H.; Schultz, M. Glycopeptide Synthesis in Solution and on the Solid Phase, In *Carbohydrate Chemistry and Biology*; Ernst, B.; Hart, G. W.; Sinaÿ, P., Eds.; Wiley-VCH: Weinheim, 2000; Vol. 2.

Li, P.; Xu, J.-C. Highly Efficient Synthesis of Peptides by Rational Utilization of Novel Coupling Reagents, *Chin. J. Chem.* **2000**, *18*, 456–466.

St Hilaire, P. M.; Meldal, M. Glycopeptide and Oligosaccharide Libraries, *Angew. Chem., Int. Ed. Engl.* **2000**, *39*, 1163–1179.

Theodoridis, G. Nitrogen Protecting Groups: Recent Developments and New Applications, *Tetrahedron* **2000**, *56*, 2339–2358.

Aimoto, S. Contemporary Methods for Peptide and Protein Synthesis, *Curr. Org. Chem.* **2001**, *5*, 45–87.

Albericio, F.; Chinchilla, R.; Dodsworth, D. J.; Najera, C. New Trends in Peptide Coupling Reagents, *Org. Prep. Proced. Int.* **2001**, *33*, 205–303.

Bianco, A.; Da Ros, T.; Prato, M.; Toniolo, C. Fullerene-Based Amino Acids and Peptides, *J. Pept. Sci.* **2001**, *7*, 208–219.

Cheng, R. P.; Gellman, S. H.; DeGrado, W. F. β-Peptides: From Structure to Function, *Chem. Rev.* **2001**, *101*, 3219–3232.

Goodman, M.; Zapf, C.; Yosup, R. New Reagents, Reactions, and Peptidomimetics for Drug Design, *Biopolymers* **2001**, *60*, 229–245.

Lambert, J. N.; Mitchell, J. P.; Roberts, K. D. The Synthesis of Cyclic Peptides, *J. Chem. Soc., Perkin Trans. 1* **2001**, 471–484.

Muir, T. W. Development and Application of Expressed Protein Ligation, *Synlett* **2001**, 733–740.

Okada, Y. Synthesis of Peptides by Solution Methods, *Curr. Org. Chem.* **2001**, *5*, 1–43.

Seitz, O.; Heinemann, I.; Mattes, A.; Waldmann, H. Synthetic Peptide Conjugates - Tailor-Made Probes for the Biology of Protein Modification and Protein Processing, *Tetrahedron* **2001**, *57*, 2247–2277.

Agami, C.; Couty, F. The Reactivity of the N-Boc Protecting Group: An Underrated Feature, *Tetrahedron* **2002**, *58*, 2701–2724.

Antsypovitch, S. I. Peptide Nucleic Acids: Structure, Properties, and Synthesis, *Russ. Chem. Rev. (Engl. Transl.)* **2002**, *71*, 71–83.

Davis, B. G. Synthesis of Glycoproteins, *Chem. Rev.* **2002**, *102*, 579–601.

Dinsmore, C. J.; Beshore, D. C. Recent Advances in the Synthesis of Diketopiperazines, *Tetrahedron* **2002**, *58*, 3297–3312.

Dinsmore, C. J.; Beshore, D. C. Syntheses and Transformations of Piperazinone Rings. A Review, *Org. Prep. Proced. Int.* **2002**, *34*, 367, 369–404.

Najera, C. From α-Amino Acids to Peptides: All You Need for the Journey, *Synlett* **2002**, 1388–1403.

Sewald, N. Efficient, Racemization-Free Peptide Coupling of N-Alkyl Amino Acids by Using Amino Acid Chlorides Generated in Situ - Total Syntheses of the Cyclopeptides Cyclosporin O and Omphalotin A, *Angew. Chem., Int. Ed. Engl.* **2002**, *41*, 4661–4663.

Zubin, E. M.; Romanova, E. A.; Oretskaya, T. S. Modern Methods for the Synthesis of Peptide - Oligonucleotide Conjugates, *Russ. Chem. Rev. (Engl. Transl.)* **2002**, *71*, 239–264.

Goodman, M.; Mahwah, A. F.; Moroder, L.; Toniolo, C., Eds., *Houben-Weyl. Methods of Organic Chemistry. Synthesis of Peptides and Peptidomimetics*; Thieme: Stuttgart, 2002-2003; Vol. E22a-e

Kotha, S. The Building Block Approach to Unusual α-Amino Acid Derivatives and Peptides, *Acc. Chem. Res.* **2003**, *36*, 342–351.

Marshall, G. R. Solid-Phase Synthesis: A Paradigm Shift, *J. Pept. Sci.* **2003**, *9*, 534–544.

Virta, P.; Katajisto, J.; Niittymaki, T.; Lonnberg, H. Solid-Supported Synthesis of Oligomeric Bioconjugates, *Tetrahedron* **2003**, *59*, 5137–5174.

Albericio, F. Developments in Peptide and Amide Synthesis, *Curr. Op. Chem. Biol* **2004**, *8*, 211–221.

Han, S.-Y.; Kim, Y.-A. Recent Development of Peptide Coupling Reagents in Organic Synthesis, *Tetrahedron* **2004**, *60*, 2447–2467.

Hodgson, D. R. W.; Sanderson, J. M. The Synthesis of Peptides and Proteins Containing Non-Natural Amino Acids, *Chem. Soc. Rev.* **2004**, *33*, 422–430.

Köhn, M.; Breinbauer, R. The Staudinger Ligation - a Gift to Chemical Biology, *Angew. Chem. Int. Ed., Engl.* **2004**, *43*, 3106–3116.

Lelais, G.; Seebach, D. β2-Amino Acids - Syntheses, Occurrence in Natural Products, and Components of β-Peptides, *Biopolymers* **2004**, *76*, 206–243.

Rademann, J. Organic Protein Chemistry: Drug Discovery through the Chemical Modifcation of Proteins, *Angew. Chem. Int. Ed., Engl.* **2004**, *43*, 4554–4556.

Yeo, D. S. Y.; Srinivasan, R.; Chen, G. Y. J.; Yao, S. Q. Expanded Utility of the Native Chemical Ligation Reaction, *Chem. Eur. J.* **2004**, *10*, 4664–4672.

Yiotakis, A.; Georgiadis, D.; Matziari, M.; Makaritis, A.; Dive, V. Phosphinic Peptides: Synthetic Approaches and Biochemical Evaluation as Zn-Metalloprotease Inhibitors, *Curr. Org. Chem.* **2004**, *8*, 1135–1158.

Zhang, W. Fluorous Tagging Strategy for Solution-Phase Synthesis of Small Molecules, Peptides and Oligosaccharides, *Curr. Op. Drug Dis. Dev.* **2004**, *7*, 784–797.

Paul, M.; van der Donk, W. A. Chemical and Enzymatic Synthesis of Lanthionines, *Mini-Rev. Org. Chem.* **2005**, *2*, 23–37.

van Swieten, P. F.; Leeuwenburgh, M. F.; Kessler, B. M.; Overkleeft, H. S. Bioorthogonal Organic Chemistry in Living Cells: Novel Strategies for Labeling Biomolecules, *Org. Biomol. Chem.* **2005**, *3*, 20–27.

A

Acetonitrile[1]

MeCN

[75-05-8] C_2H_3N (MW 41.05)

(solvent for inorganic and organometallic reagents; undergoes addition reactions to nitrile group; anion is nucleophilic reagent; Ritter reaction reagent; oxidizing agent in combination with H_2O_2)

Alternate Name: methyl cyanide; ethanenitrile; cyanomethane.
Physical Data: mp −45 °C; bp 81.6 °C; *d* 0.787 g cm^{-3} (20 °C).
Solubility: miscible with water and most organic solvents (but not petroleum); forms constant boiling mixture containing 16% water, bp 76 °C.
Form Supplied in: colorless liquid; widely available.
Drying: fractional distillation from CaH_2 or P_2O_5.
Analysis of Reagent Purity: GLC; HPLC grade (99.9+% purity) commercially available.
Handling, Storage, and Precautions: flammable volatile liquid; keep dry; toxic vapor; mild lachrymator; possible skin irritant; evolves toxic fumes if heated to decomposition; incompatible with fuming HNO_3, N_2O_4/In, 2-cyano-2-propylnitrate, ClF/F_2, Cl_3HSi/diphenyl sulfoxide, metal perchlorates, $HClO_4$, fuming H_2SO_4, heat/conc. H_2SO_4, and *N*-fluoro compounds.[2]

Original Commentary

Roger Bishop
The University of New South Wales, Kensington, NSW, Australia

Solvent Properties. Acetonitrile is neutral, aprotic, polar (3.9 D) with a dielectric constant of 38 at 20 °C; it promotes ionic processes in solution. It is suitable for a wide range of physical organic techniques including HPLC, UV ($\lambda > 190$ nm), and electrochemistry.[3] However, its outstanding ability is to facilitate reaction between organic substrates and inorganic materials (eq 1).[4] These reagents include: Group 1 and 2 salts; Ag^{I}, Ce^{IV}, Cu^{I}, Cu^{II}, Fe^{III}, Hg^{II}, Mn^{IV}, Mn^{VII}, Pd^{II}, Rh^{III}, Ru^{IV}, Sn^{II}, Te^{IV}, and Tl^{III} salts; R_4N^+, ArN_2^+, $MeCO^+$, NO_2^+, and Ph_3RP^+ salts. In some cases, such as $CuCl_2\cdot(MeCN)_2$ and $PdCl_2\cdot(MeCN)_2$, the complexes involved are isolable.[5]

Reactions at Methyl.[6] Deprotonation of acetonitrile using strong base affords the nitrile-stabilized anion,[7a] which can add to other nitrile molecules producing diacetonitrile[7b] or the trimer 4-amino-2,6-dimethylpyrimidine (eq 2).[7c] The dimerization process is the forerunner of the Thorpe–Ziegler synthesis of large ring systems.[8] Mixed β-enaminonitriles[9a] or diacetonitrile itself[9b] undergo condensation with carbonyl compounds, producing pyridine derivatives.

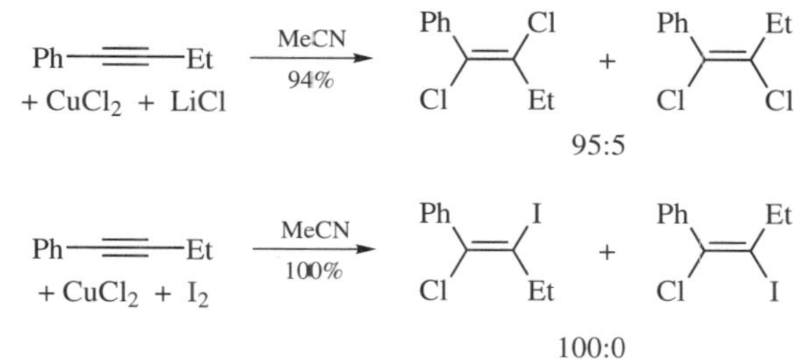

(1)

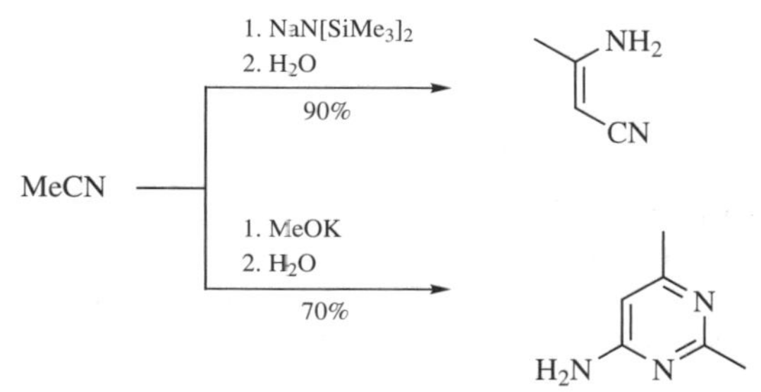

(2)

The anion performs conventional substitution reactions with alkyl halides and epoxides,[6] but Cyanomethylcopper is especially effective for cyanoalkylation of allylic halide reagents (eq 3).[10] Esters react with the monoanion yielding β-ketonitriles, which are readily converted by Boron Trifluoride or Polyphosphoric Acid into the corresponding β-ketoamides (eq 4).[11]

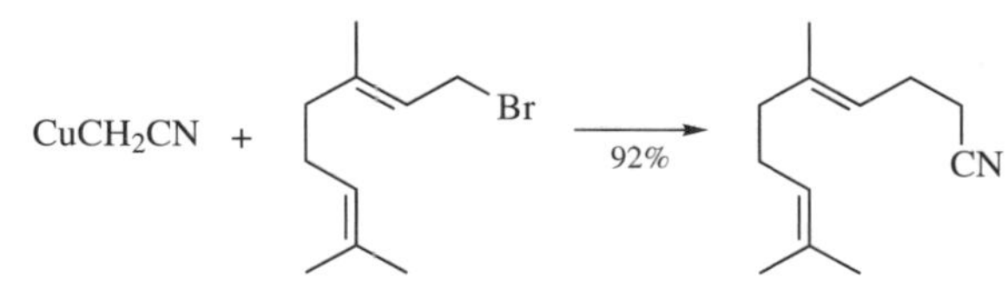

(3)

$$NaCH_2CN + PhCO_2Et \xrightarrow{94\%} PhCOCH_2CN \xrightarrow[87\%]{PPA} PhCOCH_2CONH_2 \quad (4)$$

Aldol-type condensations with aldehydes and ketones produce the cyanomethyl alcohols[12a] or unsaturated nitriles (eq 5),[12b] depending on the chosen conditions. Acetonitrile anion results in exclusively 1,2-addition to conjugated enones in THF at −78 °C, in contrast to the behavior of other nitriles.[12c]

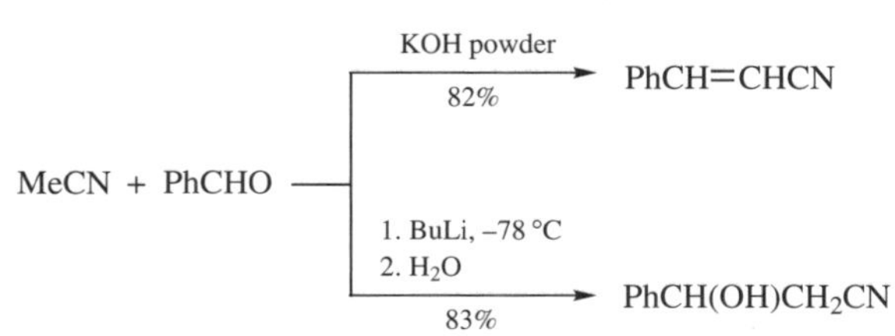

(5)

Addition Reactions.[13] Acetonitrile yields acetophenone derivatives (eq 6) when reacted with phenols (or phenolic ethers) and Lewis acids (Hoesch reaction).[14] It reacts with alcohols in the presence of dry HCl to yield imidates (Pinner reaction)[15a] which can react with further alcohol to produce orthoacetate esters,[15b] or with amines to yield acetamidines (eq 7).[15c,d]

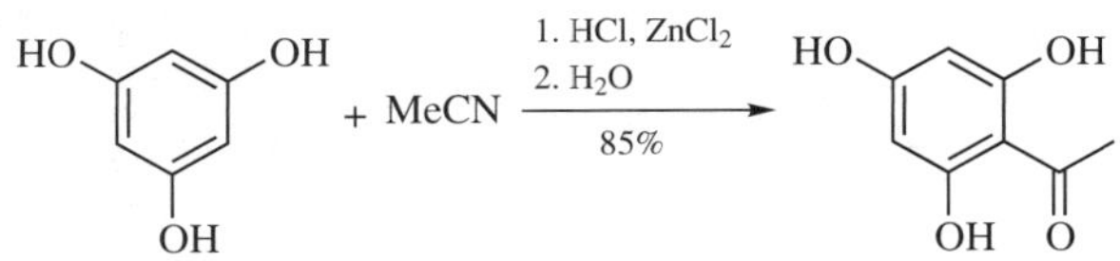

(6)

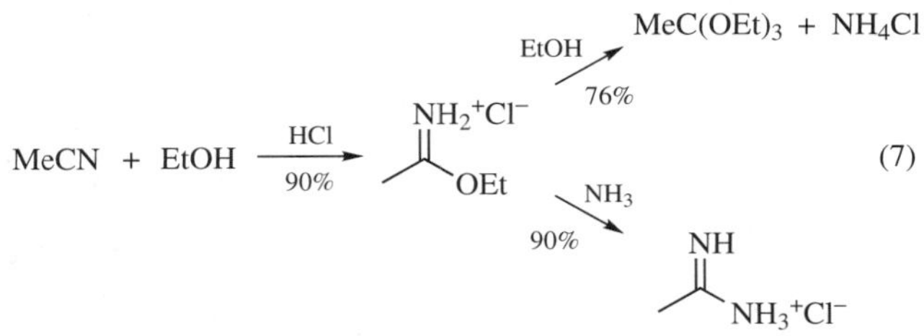

(7)

The Blaise reaction[16] of bromo esters with acetonitrile produces β-keto esters (eq 8) in a process analogous to the Reformatsky reaction. Acetonitrile reacts with ammonia and hydrogen sulfide to yield thioacetamide. Higher thioamides can be produced by reacting this with nitriles and distilling the lower boiling MeCN from the resulting equilibrium mixture.[17]

$$\text{MeCHBrCO}_2\text{Et} + \text{MeCN} \xrightarrow[\text{2. H}_3\text{O}^+ \quad 70\%]{\text{1. Zn}} \text{MeCOCH(Me)CO}_2\text{Et} \quad (8)$$

Acetonitrile also reacts with thiols to yield thioimidates,[15a] with amines to give acetamidines,[18] and with 1,2-amino alcohols to produce 2-oxazolines (eq 9).[19]

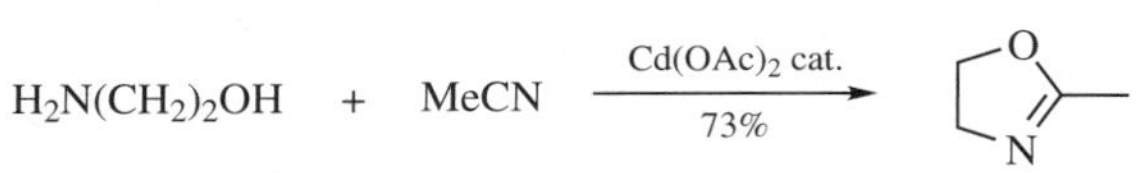

(9)

Ritter-Type Reactions.[20] The acetonitrile nitrogen atom is an extremely weak nucleophile, but attacks carbenium ions to yield a nitrilium ion, which proceeds to an imidate and then undergoes hydrolysis to the amide (eq 10).[21] Since carbenium ions may be generated from a very wide range of functional groups under strongly acidic conditions, the Ritter reaction provides a facile one-flask route to *N*-alkylacetamides. As expected, the carbenium ions may participate in rearrangement or transannular reactions before undergoing substitution (eq 11).[22] Alternatively, the nitrilium ion may be generated via other electrophilic reagents[23] (eq 12),[24] and it can also undergo intramolecular closure to yield heterocyclic products (eq 13).[25]

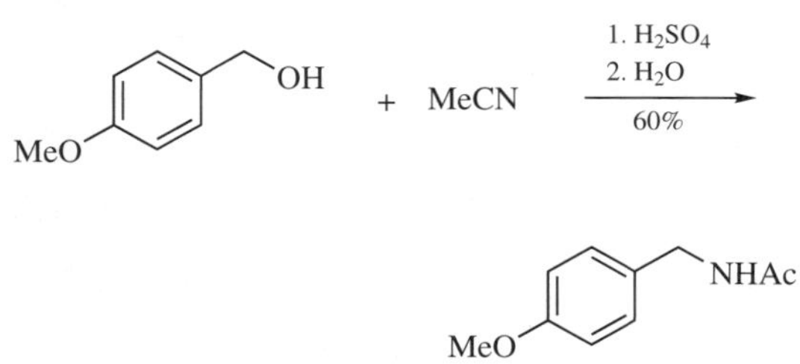

(10)

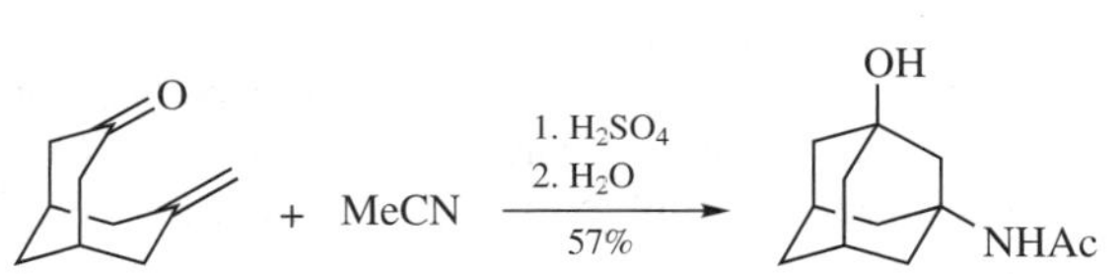

(11)

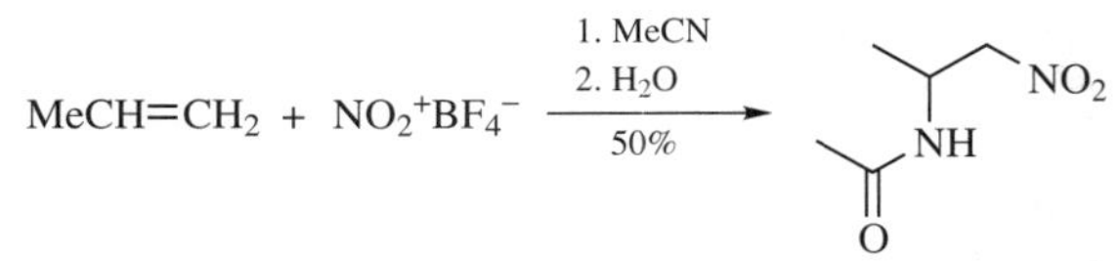

(12)

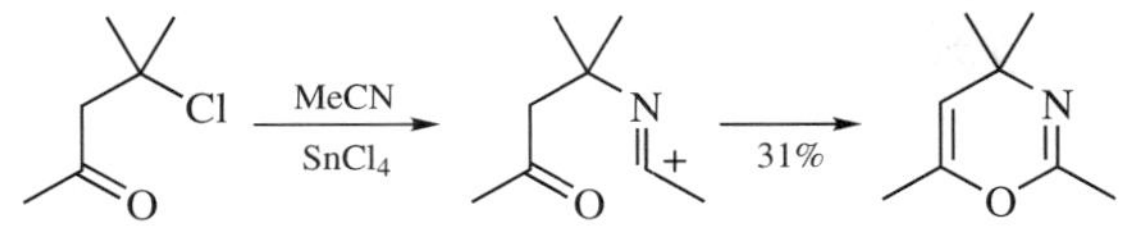

(13)

Reactions with Hydrogen Peroxide and Base. The Radziszewski reaction[26] describes the formation of amide products and dioxygen when nitriles are treated with Hydrogen Peroxide solution under basic conditions. Yields of aliphatic amides[27a] are 50–70%, while those of aromatic amides[27b] are typically 80–95%. This reaction proceeds via the intermediacy of the peroxyimidic acid (eq 14).[28] If an alkene functionality is present, then it is epoxidized with concomitant reduction of the peroxyimidic acid to the amide (the Payne oxidation)[29] in a simple one-flask process. A convenient procedure is to react the alkene and nitrile with H_2O_2 in methanol using potassium hydrogen carbonate as the base (pH ca. 8).[30a]

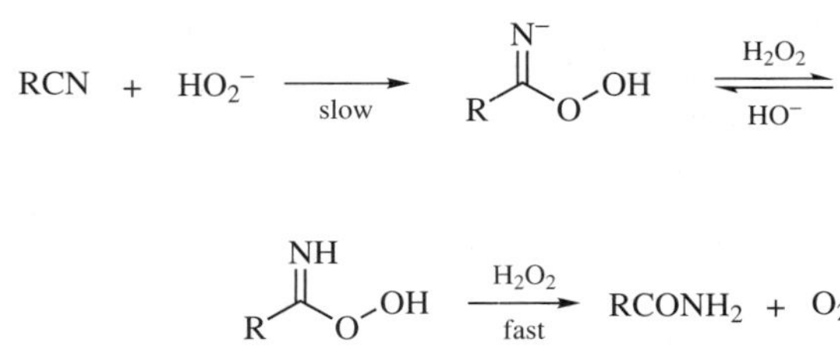

(14)

The Payne oxidation has the advantage that the oxidant is generated in situ, while the mildly basic conditions ensure that labile epoxides do not undergo acid-catalyzed degradation (eq 15)[30b] and competing Baeyer–Villiger oxidations may be suppressed (eq 16).[29c] In some cases the peroxycarboxylic acid and Payne processes may give differing outcomes,[31a] stereochemistry,[31b] or regioselectivity.[31c] While both benzonitrile and acetonitrile are commonly employed, the latter reagent has operational advantages. Its oxidation products are water soluble and easily removed during reaction workup, whereas the benzamide coproduct requires a more complex separation.

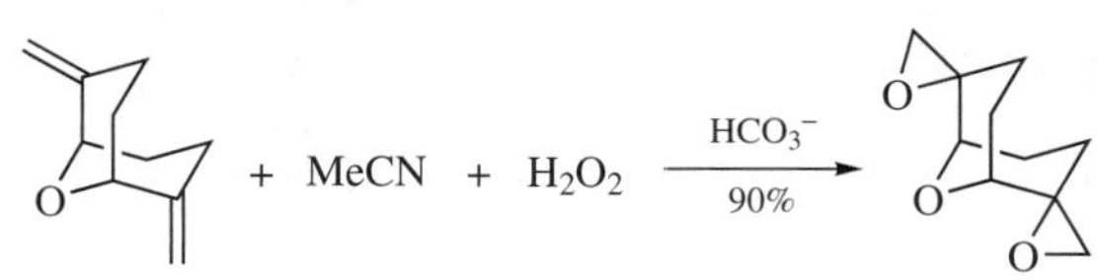

(15)

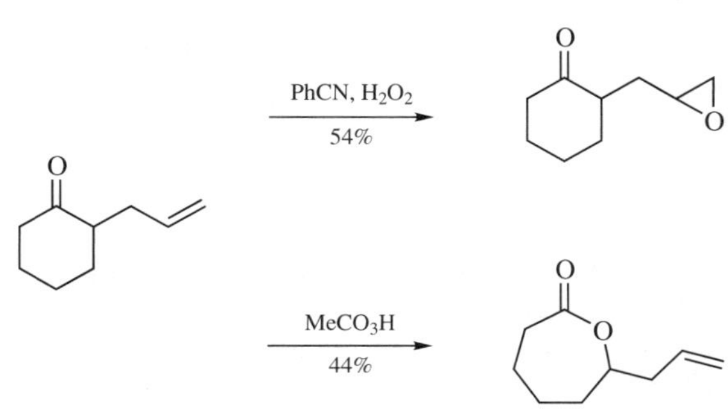

(16)

First Update

Debatosh Majumdar & Geert-Jan Boons
Complex Carbohydrate Research Center, University of Georgia, Athens, GA, USA

Use as a Stereodirecting Solvent in Glycosylations. The influence that solvents exert on the anomeric outcome of glycosylations is well known.[32] The most remarkable example of this influence is the effect of acetonitrile, which often leads to the formation of mainly equatorial glycosides (e.g., β-glucosides and β-galactosides). Several studies have indicated that a cyclic oxocarbenium ion (**2**), formed after the departure of a glycosyl donor's leaving group (**1**), can react with acetonitrile to generate an axial α-acetonitrilium ion (**3**). Displacement of the acetonitrilium ion with an alcohol or other nucleophile in an S_N2 fashion will then give an equatorially substituted product (**4**) (eq 17).

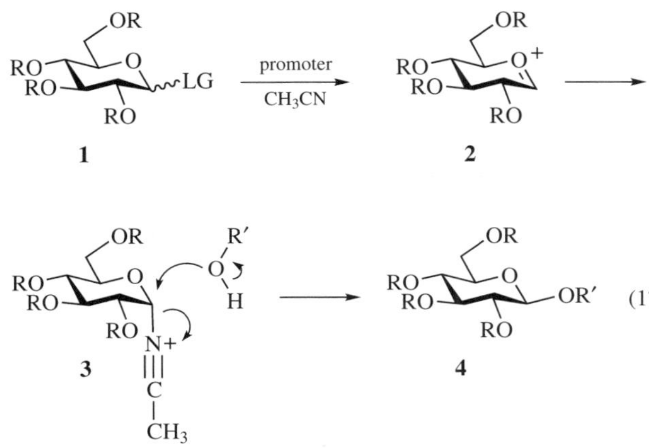

1, 2, 3, 4 (17)

Lemieux and co-workers were the first to observe that azido nitration of 3,4,6-tri-*O*-acetyl-galactal (**5**) with CAN and NaN_3 in acetonitrile gave mainly the β-D-galactopyranosyl nitrate (**7**) and *N*-acetyl 2-azido-2-deoxy-3,4,6-tri-*O*-acetyl-α-D-galactopyranosyl amine (**8**) (eq 18).[33] They rationalized these results by assuming that an initially formed oxocarbenium ion would react with acetonitrile to predominantly give an α-acetonitrilium ion. This intermediate can react either with water to give the *N*-acetyl α-D-galactopyranosyl amine (**8**) or be displaced by nitrate to give the β-D-galactopyranosyl nitrate (**7**).

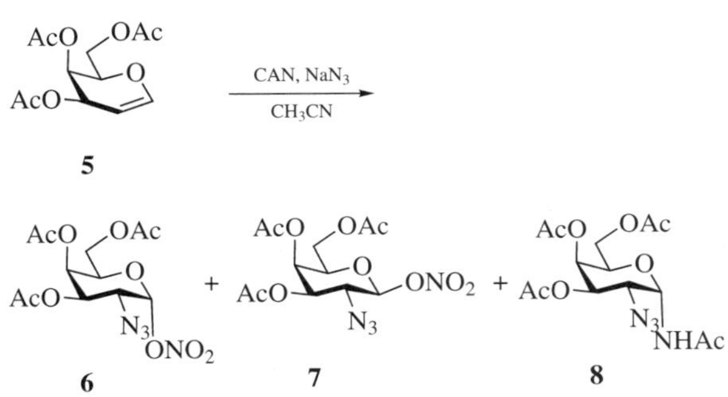

5, 6, 7, 8 (18)

In an attempt to trap the intermediate α-acetonitrilium ion **10**, Sinaÿ and co-workers treated (*N*-phenyl)-benzimidoyl glucopyranose (**9**) with 2-chlorobenzoic acid to give chlorobenzimide (**12**) as the major compound (eq 19).[34] Compound **12** displayed a large $J_{1,2}$ of 7.3 Hz, and therefore was initially assigned as a β-anomer. Fraser Reid[35] reinterpreted these results and showed convincingly that **12** had an α-anomeric configuration. In this study, treatment of the α- or β-anomers of pent-4-enyl 2,3,4,6-tetra-*O*-benzyl-D-glucopyranoside with NBS and 2-chlorobenzoic acid in acetronitrile lead to formation of the α-imide **12**. The anomeric configuration of imide was confirmed by an alternative preparation of this compound. The unusually large $J_{1,2}$ of 7.3 Hz coupling constant of the imide was rationalized with the aid of a computational study showing a substantial flattening of the pyranose ring at C-1 and C-2. This flattening resulted primarily from the steric bulk of the imide moiety. Removal of the *N*-acetyl group to give the 2-chlorobenzamide (**13**) showed $J_{1,2}$ of 5.1 Hz; a value more akin to a standard $^4C_1$ conformation. NMR measurements and molecular mechanics calculations yielded further support for the α-configuration of the glucopyranosyl acetonitrilium ion.[36] Thus, the nitrilium intermediate **10** was generated at −30 °C and ^1H, ^{13}C, and ^{15}N chemical shifts and coupling constants were measured. The conformational properties of α- and β-chair, boat, and skewed forms of **10** were computed using the CHARM force field. In the case of the α-glucopyranosyl acetonitrilium ion, minimizations with constraints gave low energy conformations for a flattened $^4C_1$ and $^0S_2$ form. On the contrary, the constrained β-starting structures evolved into high energy products upon minimization.

Schmidt and co-workers have extensively explored the influence of the solvent acetonitrile on the anomeric outcome of glycosylations.[37] A neccessary requirement of this type of anomeric control is the absence of a neighboring participating functionality at C-2. It was shown that a trimethylsilyl triflate (TMSOTf) mediated coupling of 2,3,4,6-tetra-*O*-benzyl glucosopyranosyl trichloroacetimidate with several glycosyl acceptors in acetonitrile at −35 °C lead mainly to the formation of β-linked disaccharides (Table 1). The α/β ratio of the glycosylation was independent of the anomeric configuration of the starting material supporting a reaction mechanism with a common intermediate. The use of other nitriles as solvent had similar effects but lead to lower β-selectivities. The use of propionitrile as a solvent gave, however, similar anomeric outcomes to the use of acetonitrile. This solvent has the additional advantage in that it can be cooled to −78 °C without freezing. Unfortunately, the acetonitrile effect could not be extended to the stereoselective introduction of β-mannosides. Thus, coupling of 2,3,4,6-tetra-*O*-benzyl-D-mannosyl trichloroacetimidate with several glycosyl acceptors in acetonitrile at low temperature led, in each case, to the formation of anomeric mixtures. In general, stereoselective formation of β-mannosides is difficult because the axial C-2 substituent sterically blocks the β-face of the glycosyl donor. Furthermore, an axial alkoxy group at C(2) of a sugar derivative increases the strength of the anomeric effect. This effect has been rationalized by the destabilization of β-anomer of D-mannose by the close proximity of the endocyclic oxygen and the C(1) and C(2) oxygen atoms, resulting in unfavorable dipole-dipole interactions. This effect, named the Δ2 effect,[38,39] has also been explained in stereoelectronic terms. It has been proposed[40] that the anomeric effect of α-D-mannose is significantly larger because of lowering of the antibonding orbital of the C(1)-O(1) bond as a result of secondary orbital overlap between the antibonding orbitals of C(1)-O(1) and C(2)-O(2).

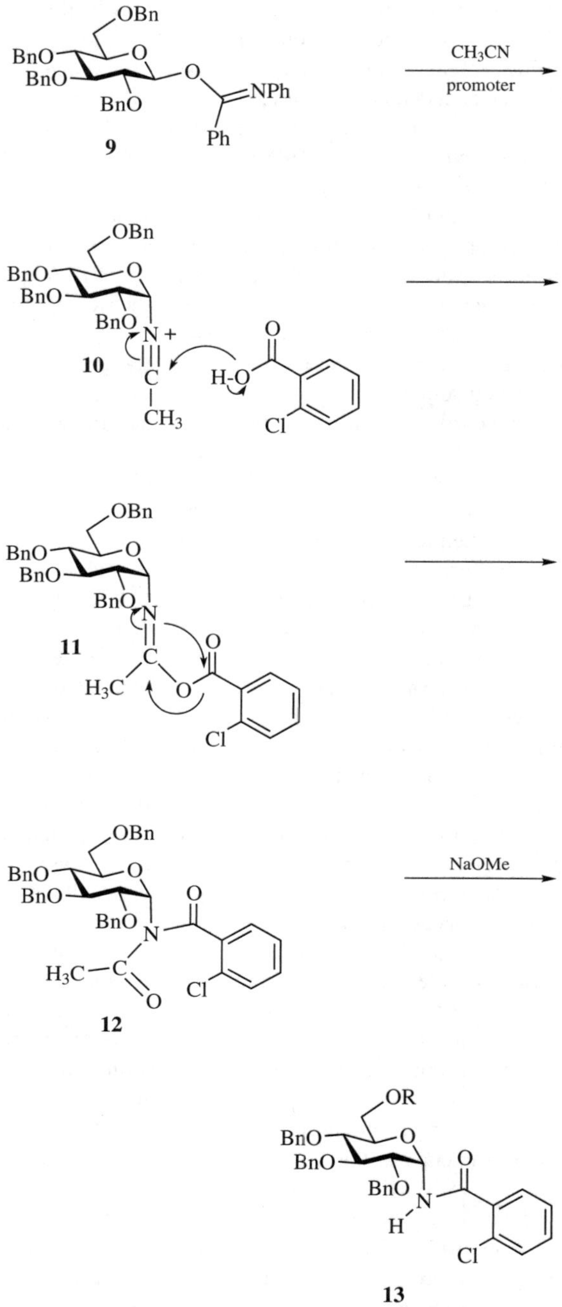

(19)

Glycosylations with bromo 2,3,4-tri-*O*-benzyl glucopyranoic esters in the presence of silver perchlorate in acetonitrile gave mainly disaccharides as α-glycosides.[41]

It has been shown that many different types of glycosyl donors such as trichloroacetimidates,[37] halides,[42,43] phosphates,[42,44] and pentenyl-,[45] vinyl-[46,47] and thioglycosides[48,49] feature the ability to form highly reactive nitrilium intermediates with acetonitrile. Representative examples are depicted in Table 1. The methodology is most commonly applied for the preparation of β-glucosides and β-galactosides. In general, the best β-selectivities are obtained with reactive alcohols at low reaction temperatures.

Acetonitrile has also been used for the preparation of α-glycosides of *N*-acetyl-neuraminic acid (Neu5Ac) (eq 20).[50,51]

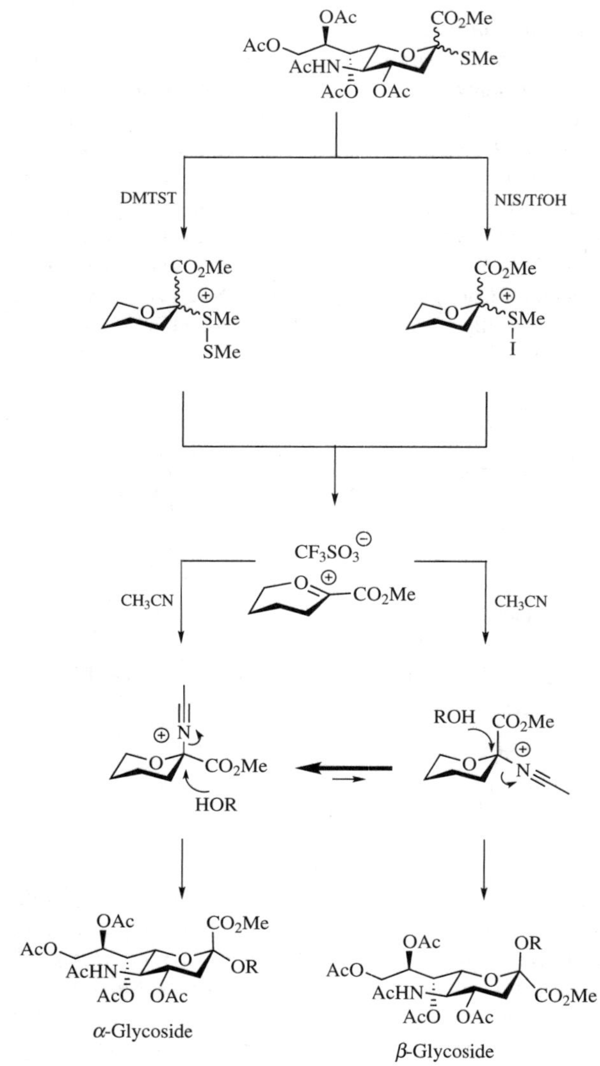

(20)

N-Acetyl-α-neuraminic acid frequently terminates oligosaccharide chains of glycoproteins and glycolipids of cell membranes. The sugar ring of Neu5Ac adopts a $^4C_1$ conformation, placing the bulky side chain and the C-5 acetamido moiety in equatorial orientations. In this particular case, the equatorial glycoside is the α-anomer and the axial one the β-glycoside. In natural oligosaccharides, the glycosides of Neu5Ac are in α-configurations. The use of Neu5Ac derivatives as glycosyl donors is complicated by the fact that no C-3 functionality is present to direct the stereochemical outcome of the glycosylations. In addition, the electron withdrawing carboxylic acid at the anomeric center makes these derivatives prone to elimination. Finally, the glycosylation of Neu5Ac is performed at a tertiary oxocarbenium ion. α-Glycosides of Neu5Ac are most commonly synthesized by a glycosylation of a thioglycoside of Neu5Ac in acetonitrile at a

Donor	Acceptor	Condition	Product/selectivity
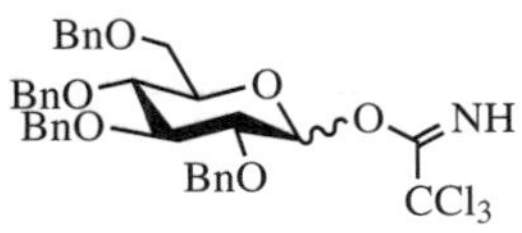	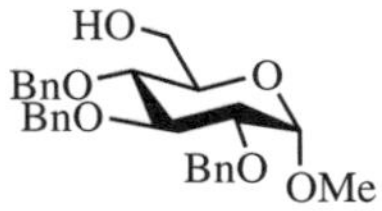	TMSOTf CH_3CN, –20 °C	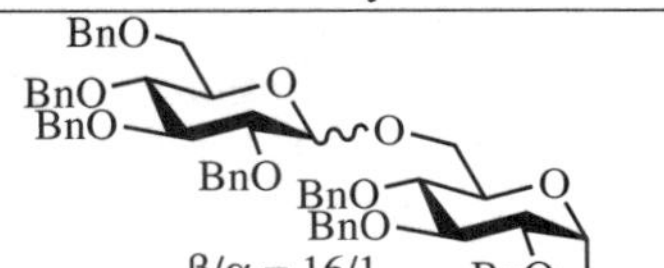 β/α = 16/1
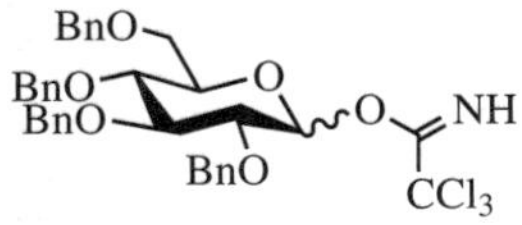	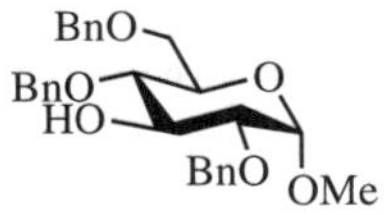	TMSOTf CH_3CN, –20 °C	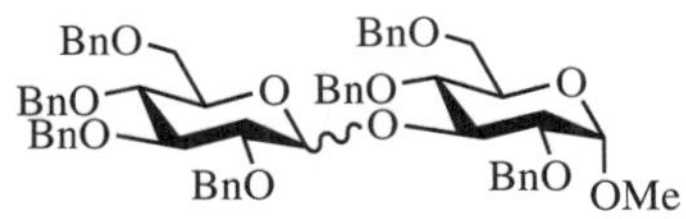 β/α = 24/1
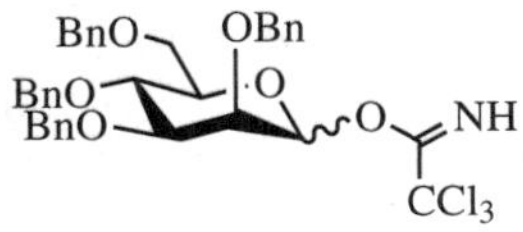	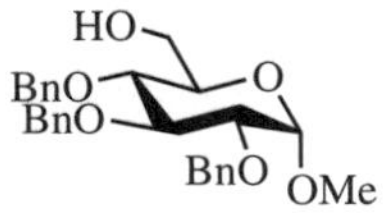	TMSOTf CH_3CN, –20 °C	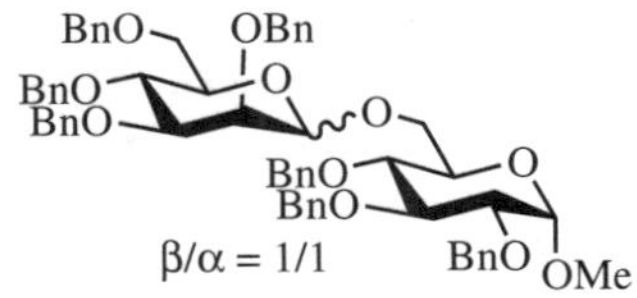 β/α = 1/1
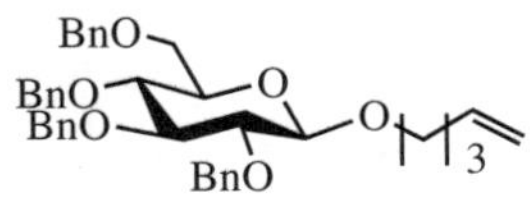	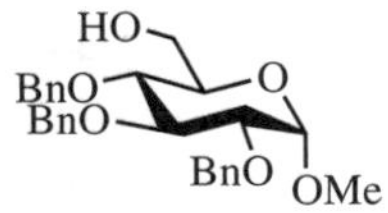	NBS CH_3CN, rt	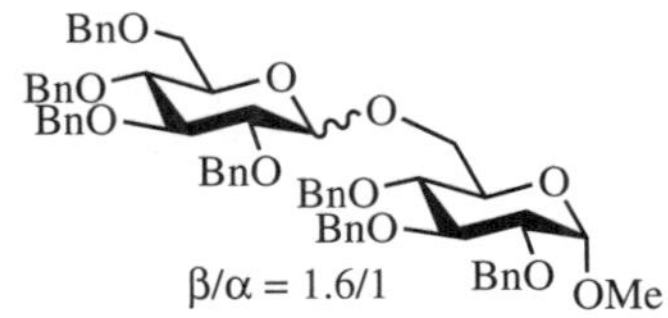 β/α = 1.6/1
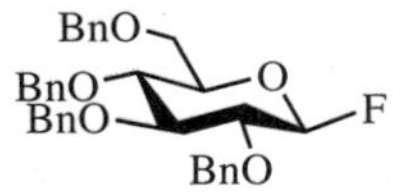	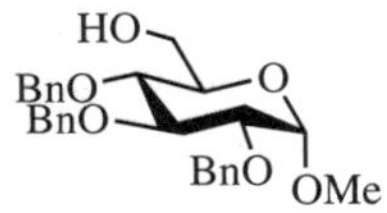	SiF_4 CH_3CN, 0 °C	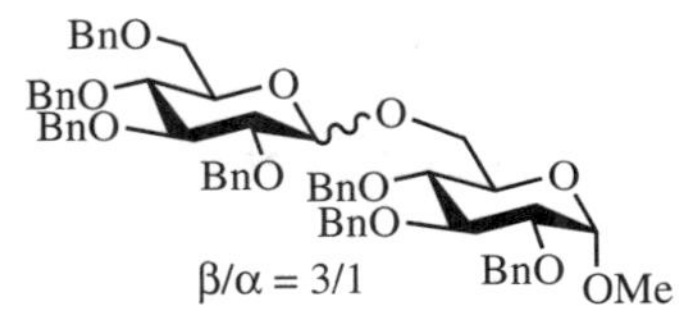 β/α = 3/1
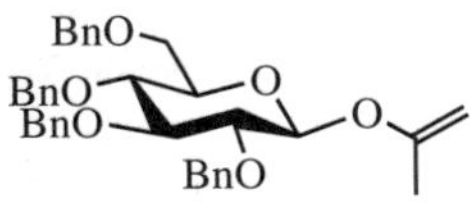	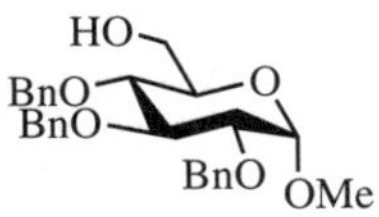	TMSOTf CH_3CN, –20 °C	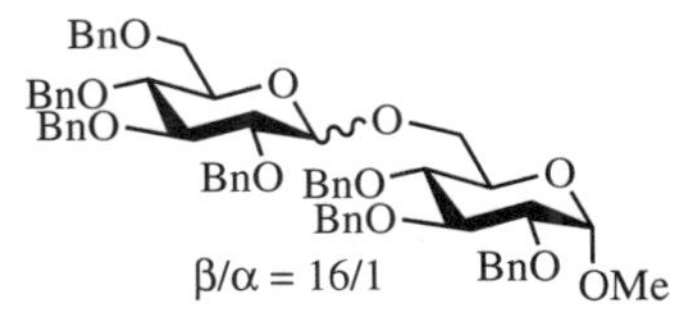 β/α = 16/1
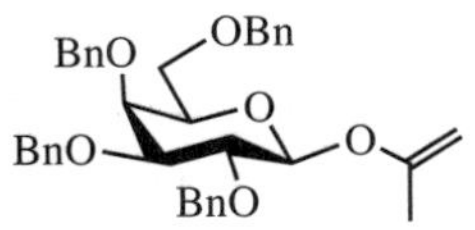	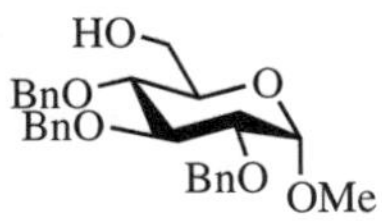	TMSOTf CH_3CN, –20 °C	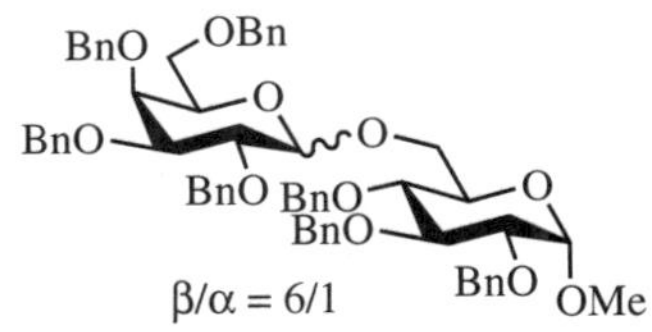 β/α = 6/1
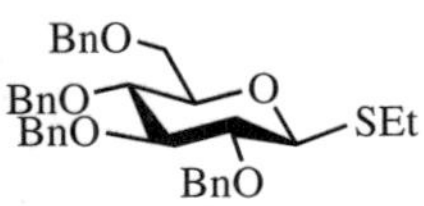	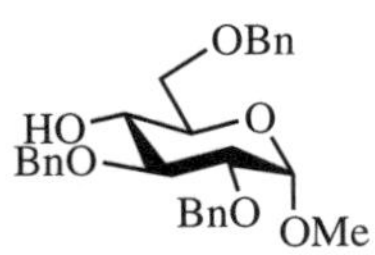	$TBPA^+$, CH_3CN, –20 °C	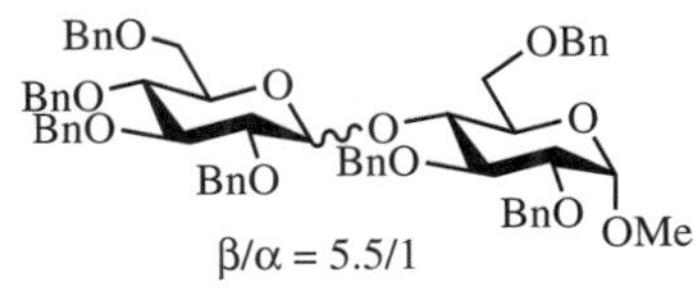 β/α = 5.5/1
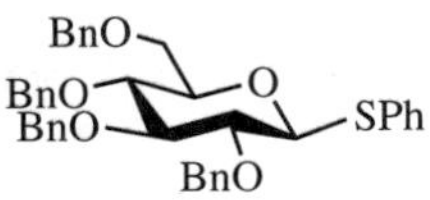	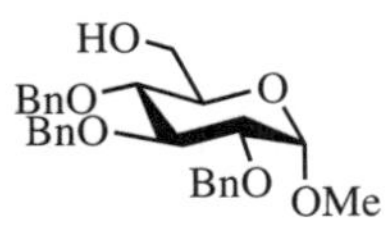	PhSeOTf CH_3CN, –20 °C	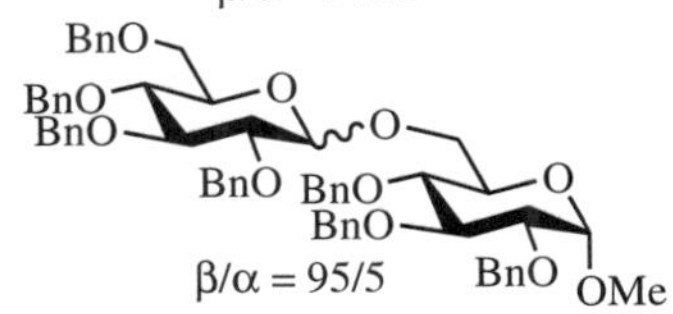 β/α = 95/5
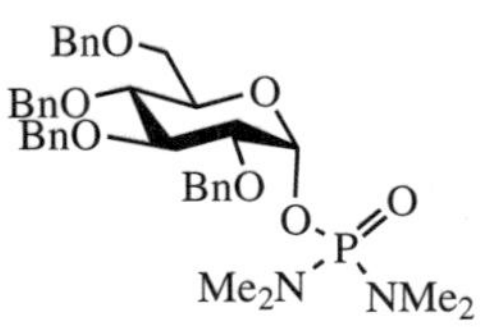	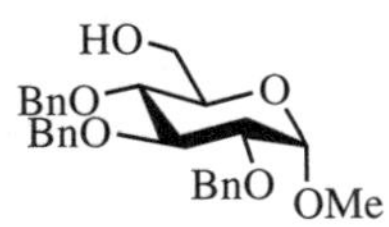	TMSOTf CH_3CH_2CN, –78 °C	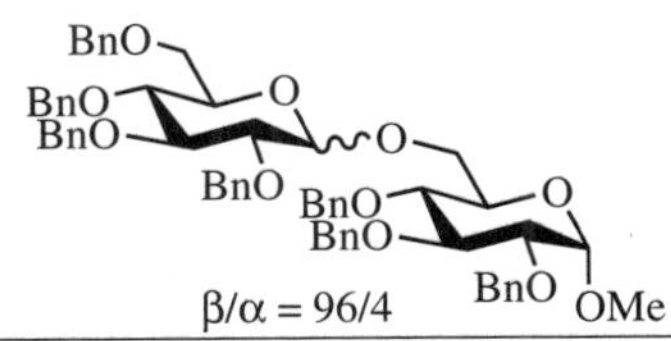 β/α = 96/4

low temperature in the presence of activators such as *N*-iodosuccinimide–triflic acid (NIS-TfOH), dimethyl(methylthio) sulfonium trifluoromethanesulfonate (DMTST), and phenylselenenyl triflate (PhSeOTf). It has been proposed that these glycosylations proceed through a reactive β-nitrile intermediate. Sialylations in acetonitrile have been successfully performed using phosphites[52,53] or xanthates[54] as the anomeric leaving group. Furthermore, the reactivity of these donors can be improved by a conversion of the *N*-acetyl group into an *N*-acetyl acetamido functionality ((NAc_2))[55] or a *N*-trifluoracetamido group (NHTFA).[56,57]

Glucosamines, which are precursors for the preparation of *N*-linked glycopeptides, can be synthesized by a reaction of 2,3,4,6-tetra-*O*-benzyl-glucopyranose (**14**) with acetonitrile in the presence of trifluormethanesulfonic anhydride (eq 21).[58] This reaction is thought to proceed through an α-acetonitrilium ion **15**, which reacted to oxazoline **16**. An unusual hydrolysis of the oxazoline intermediate gave glucosamine **17**, which anomerized to β-amine **18**. In a similar fashion, treatment of a 1,2-anhydropyranose with $ZnCl_2$ in acetonitrile gave an oxazoline, which could be hydrolyzed with 1N HCl to an anomeric amine.[59] This derivative could be acylated with properly protected aspartic acid derivatives to give a *N*-glycosylated amino acid.

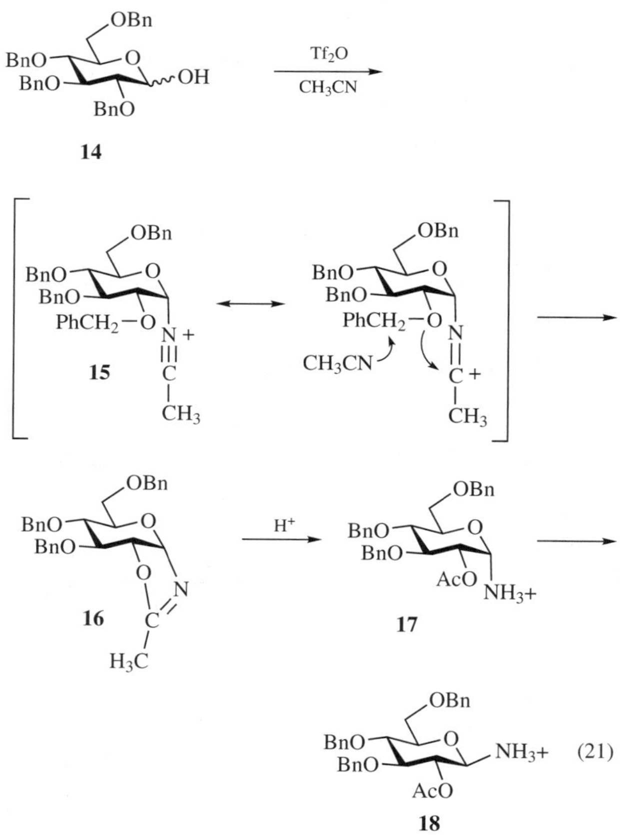

C-glycosylations have also been performed in acetonitrile as a solvent.[60–65] In general, these procedures gave mainly α-glycosides (axial glycosides), and therefore probably do not proceed through an α-acetonitriliumion.

1. (a) *The Chemistry of the Cyano Group*; Rappoport, Z., Ed.; Interscience: New York, 1970. (b) Smith, P. A. S., *The Chemistry of Open-Chain Nitrogen Compounds*; Benjamin: New York, 1965; Vol. 1.
2. Bretherick, L., *Handbook of Reactive Chemical Hazards*, 3rd ed.; Butterworths: London, 1985.
3. Coetzee, J. F., *Progr. Phys. Org. Chem.* **1967**, *4*, 45.
4. Uemura, S.; Okazaki, H.; Onoe, A.; Okano, M., *J. Chem. Soc., Perkin Trans. 1* **1977**, 676.
5. Endres, H., *Comp. Coord. Chem.* **1987**, *2*, 261.
6. Arseniyadis, S.; Kyler, K. S.; Watt, D. S., *Org. React.* **1984**, *31*, 1.
7. (a) Gornowicz, G. A.; West, R., *J. Am. Chem. Soc.* **1971**, *93*, 1714. (b) Krüger, C., *J. Organomet. Chem.* **1967**, *9*, 125. (c) Ronzio, A. R.; Cook, W. B., *Org. Synth., Coll. Vol.* **1955**, *3*, 71.
8. (a) Schaefer, J. P.; Bloomfield, J. J., *Org. React.* **1967**, *15*, 1. (b) Davis, B. R.; Garratt, P. J., *Comprehensive Organic Synthesis* **1991**, *2*, 848.
9. (a) Meyer, E. von, *J. Prakt. Chem.* **1908**, [ii] *78*, 497. (b) Shibata, K.; Urano, K.; Matsui, M., *Bull. Chem. Soc. Jpn.* **1988**, *61*, 2199.
10. Corey, E. J.; Kuwajima, I., *Tetrahedron Lett.* **1972**, 487.
11. (a) Eby, C. J.; Hauser, C. R., *J. Am. Chem. Soc.* **1957**, *79*, 723. (b) Hauser, C. R.; Eby, C. J., *J. Am. Chem. Soc.* **1957**, *79*, 725.
12. (a) Crouse, D. N.; Seebach, D., *Ber. Dtsch. Chem. Ges./Chem. Ber.* **1968**, *101*, 3113. (b) DiBiase, S. A.; Lipisko, B. A.; Haag, A.; Wolak, R. A.; Gokel, G. W., *J. Org. Chem.* **1979**, *44*, 4640. (c) Sauvetre, R.; Roux-Schmitt, M.-C.; Seyden-Penne, J., *Tetrahedron* **1978**, *34*, 2135.
13. Zil'berman, E. N., *Russ. Chem. Rev. (Engl. Transl.)* **1962**, *31*, 615.
14. (a) Spoerri, P. E.; DuBois, A. S., *Org. React.* **1949**, *5*, 387. (b) Gulati, K. C.; Seth, S. R.; Venkataraman, K., *Org. Synth., Coll. Vol.* **1943**, *2*, 522.
15. (a) Roger, R.; Neilson, D. G., *Chem. Rev.* **1961**, *61*, 179. (b) McElvain, S. M.; Nelson, J. W., *J. Am. Chem. Soc.* **1942**, *64*, 1825. (c) Dox, A. W., *Org. Synth., Coll. Vol.* **1941**, *1*, 5. (d) Shriner, R. L.; Neumann, F. W., *Chem. Rev.* **1944**, *35*, 351.
16. (a) Kagan, H. B.; Suen, Y.-H., *Bull. Soc. Chem. Fr.* **1966**, 1819. (b) Hannick, S. M.; Kishi, Y., *J. Org. Chem.* **1983**, *48*, 3833.
17. (a) Hurd, R. N.; DeLaMater, G., *Chem. Rev.* **1961**, *61*, 45. (b) Taylor, E. C.; Zoltewicz, J. A., *J. Am. Chem. Soc.* **1960**, *82*, 2656.
18. Oxley, P.; Partridge, M. W.; Short, W. F., *J. Chem. Soc.* **1947**, 1110.
19. Witte, H.; Seeliger, W., *Angew Chem., Int. Ed. Engl.* **1972**, *11*, 287.
20. (a) Krimen, L. I.; Cota, D. J., *Org. React.* **1969**, *17*, 213. (b) Bishop, R., *Comprehensive Organic Synthesis* **1991**, *6*, 261. (c) Johnson, F.; Madroñero, R., *Adv. Heterocycl. Chem.* **1966**, *6*, 95.
21. Parris, C. L.; Christenson, R. M., *J. Org. Chem.* **1960**, *25*, 331.
22. Stetter, H.; Gärtner, J.; Tacke, P., *Ber. Dtsch. Chem. Ges./Chem. Ber.* **1965**, *98*, 3888.
23. Meerwein, H.; Laasch, P.; Mersch, R.; Spille, J., *Ber. Dtsch. Chem. Ges./Chem. Ber.* **1956**, *89*, 209.
24. Scheinbaum, M. L.; Dines, M., *J. Org. Chem.* **1971**, *36*, 3641.
25. Lora-Tamayo, M.; Madroñero, R.; Muñoz, G. G.; Leipprand, H., *Ber. Dtsch. Chem. Ges./Chem. Ber.* **1964**, *97*, 2234.
26. (a) Radziszewski, Br., *Ber. Dtsch. Chem. Ges./Chem. Ber.* **1855**, *18*, 355. (b) McMaster, L.; Langreck, F. B., *J. Am. Chem. Soc.* **1917**, *39*, 103.
27. (a) McMaster, L.; Noller, C. R., *J. Indian Chem. Soc.* **1935**, *12*, 652. (b) Noller, C. R., *Org. Synth., Coll. Vol.* **1943**, *2*, 586.
28. (a) Wiberg, K. B., *J. Am. Chem. Soc.* **1955**, *77*, 2519. (b) McIsaac, J. E.; Ball, R. E.; Behrman, E. J., *J. Org. Chem.* **1971**, *36*, 3048.
29. (a) Payne, G. B.; Williams, P. H., *J. Org. Chem.* **1961**, *26*, 651. (b) Payne, G. B.; Deming, P. H.; Williams, P. H., *J. Org. Chem.* **1961**, *26*, 659. (c) Payne, G. B., *Tetrahedron* **1962**, *18*, 763.

30. (a) Bach, R. D.; Knight, J. W., *Org. Synth., Coll. Vol.* **1990**, *7*, 126. (b) Pich, K. C.; Bishop, R.; Craig, D. C.; Dance, I. G.; Rae, A. D.; Scudder, M. L., *Struct. Chem.* **1993**, *4*, 41.
31. (a) Chaudhuri, N. K.; Ball, T. J., *J. Org. Chem.* **1982**, *47*, 5196. (b) Carlson, R. G.; Behn, N. S., *J. Org. Chem.* **1967**, *32*, 1363. (c) Carlson, R. G.; Behn, N. S.; Cowles, C., *J. Org. Chem.* **1971**, *36*, 3832.
32. Boons, G.-J., *Contemp. Org. Synth.* **1996**, *3*, 173–200.
33. Lemieux, R. U.; Ratcliffe, R. M., *Can. J. Chem.* **1979**, *57*, 1244–1251.
34. Pougny, J.-R.; Sinaÿ, P., *Tetrahedron Lett.* **1976**, *17*, 4076.
35. Ratcliffe, A. J.; Fraser Reid, B., *J. Chem. Soc., Perkin Trans. 1* **1990**, 747–750.
36. Braccini, I.; Derouet, C.; Esnault, J.; de Penhoat, C. H.; Mallet, J.-M.; Michon, V.; Sinaÿ, P., *Carbohydr. Res.* **1993**, *246*, 23–41.
37. Schmidt, R. R.; Behrendt, M.; Toepfer, A., *Synlett* **1990**, 694–696.
38. Reeves, R. E., *J. Am. Chem. Soc.* **1950**, *72*, 1499–1506.
39. Reeves, R. E., *Ann. Rev. Biochem.* **1958**, *27*, 15–34.
40. Steinlin, H.; Camarda, L.; Vasella, A., *Helv. Chim. Acta* **1979**, *62*, 378–390.
41. Schmidt, R. R.; Rücker, E., *Tetrahedron Lett.* **1980**, *21*, 1421–1424.
42. Hashimoto, S.; Yanagiya, Y.; Honda, T.; Harada, H.; Ikegami, S., *Tetrahedron Lett.* **1992**, *33*, 3523–3526.
43. Hashimoto, S.; Hayashi, M.; Noyori, R., *Tetrahedron Lett.* **1984**, *25*, 1379–1382.
44. Hashimoto, S.; Honda, T.; Ikegami, S., *J. Chem. Soc., Chem. Comm.* **1989**, 685–687.
45. Fraser-Reid, B.; Udodong, U. E.; Wu, Z.; Ottoson, H.; Merritt, J. R.; Rao, C. S.; Roberts, C.; Madsen, R., *Synlett* **1992**, 927–942.
46. Boons, G. J.; Isles, S., *J. Org. Chem.* **1996**, *61*, 4262–4271.
47. Marra, A.; Esnault, J.; Veyrieres, A.; Sinaÿ, P., *J. Am. Chem. Soc.* **1992**, *114*, 6354–6360.
48. Ito, Y.; Ogawa, T., *Tetrahedron Lett.* **1987**, *28*, 4701–4704.
49. Marra, A.; Mallet, J.-M.; Amatore, C.; Sinaÿ, P., *Synlett* **1990**, 572–574.
50. Boons, G. J.; Demchenko, A. V., *Chem. Rev.* **2000**, *100*, 4539–4566.
51. Murase, T.; Kameyama, A.; Kartha, K. P. R.; Ishida, H.; Kiso, M.; Hasegawa, A., *J. Carbohydr. Chem.* **1989**, *8*, 265–283.
52. Martin, T. J.; Schmidt, R. R., *Tetrahedron Lett.* **1992**, *33*, 6123–6126.
53. Kondo, H.; Ichikawa, Y.; Wong, C.-H., *J. Am. Chem. Soc.* **1992**, *114*, 8748–8750.
54. Marra, A.; Sinaÿ, P., *Carbohydr. Res.* **1990**, *195*, 303–308.
55. Demchenko, A. V.; Boons, G. J., *J. Org. Chem.* **2001**, *66*, 2547–2554.
56. De Meo, C.; Demchenko, A. V.; Boons, G. J., *J. Org. Chem.* **2001**, *66*, 5490–5497.
57. De Meo, C.; Demchenko, A. V.; Boons, G. J., *Aust. J. Chem.* **2002**, *55*, 131–134.
58. Pavia, A. A.; Ung-Chhun, S. N.; Durand, J. L., *J. Org. Chem.* **1981**, *46*, 3158–3160.
59. Gordon, D. M.; Danishefsky, S. J., *J. Org. Chem.* **1991**, *56*, 3713–3715.
60. Kudelska, W., *Z. Naturforsch., B* **2002**, *57*, 243–247.
61. Girard, C.; Miramon, M.-L.; de Solminihac, T.; Herscovici, J., *Carbohydr. Res.* **2002**, *337*, 1769–1774.
62. Dondoni, A.; Marra, A.; Massi, A., *J. Org. Chem.* **1999**, *64*, 933–944.
63. Bertozzi, C.; Bednarski, M., *Carbohydr. Res.* **1992**, *223*, 243–253.
64. Brenna, E.; Fuganti, C.; Grasselli, P.; Serra, S.; Zambotti, S., *Chem. Eur. J.* **2002**, *8*, 1872–1878.
65. Hosomi, A.; Sakata, Y.; Sakurai, H., *Tetrahedron Lett.* **1984**, *25*, 2383–2386.

p-Acetoxybenzyl Bromide

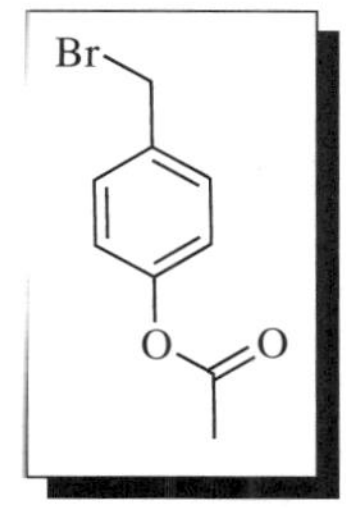

[52727-95-4] $C_9H_9BrO_2$ (MW 229.07)

(reagent used as a hydroxyl-protecting group and a benzylating agent)

Alternate Name: 4-acetoxybenzyl bromide, 4-(acetyloxy)benzyl bromide, 1-acetoxy-4-bromomethyl benzene, *p*-(bromomethyl)-phenyl acetate, PAB-Br.

Physical Data: mp 48–50 °C (from *n*-hexane); bp 100–105 °C (0.4 mm Hg).

Solubility: limited specific data.

Form Supplied in: low melting solid; not commercially available.

Analysis of Reagent Purity: ^{1}H NMR ($CDCl_3$) δ 7.1–7.6 (q), 4.53 (s), 2.33 (s).

Preparative Methods: bromination of *p*-cresyl acetate [140-39-6] in CCl_4 using NBS,[1] or selective tritylation of 4-hydroxybenzyl alcohol followed by phenol acylation, deprotection, and bromination via CBr_4/PPh_3.[2]

Purification: recrystallization from hexane or distillation at 100–105 °C (0.4 mm Hg).[1]

Protection of Alcohols. This relatively new entry in the arsenal of hydroxyl-protecting groups showcases a complimentary cleavage protocol to the existing benzyl ether protection strategies. Although the scope and versatility of this reagent has not yet been fully defined, one method for the introduction of *p*-acetoxybenzyl (PAB) ether via the AgOTf-promoted reaction of a primary alcohol and *p*-acetoxybenzyl bromide has been reported (eq 1).[2]

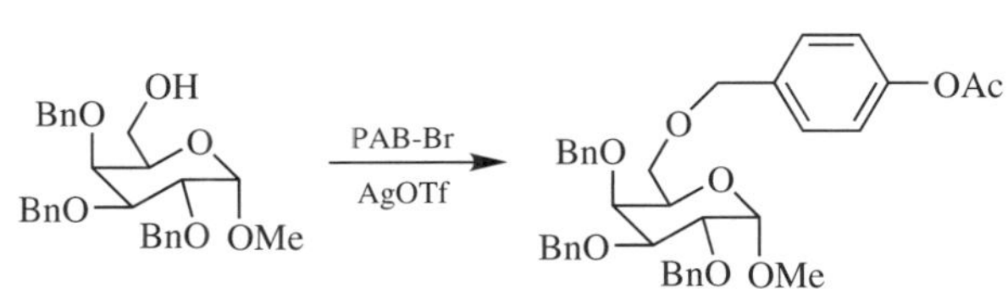

(1)

Deprotection occurs via an initiation step in which the substrate is treated with sodium methoxide and the resulting phenoxide liberates the alcohol by either thermal decomposition or mild oxidation (eq 2). Presumably, the mechanism of decomposition occurs via methylene quinone extrusion at a moderate temperature (65 °C).[2]

Oxidizing agents known to effect this deprotection include DDQ (1.0 equiv), $FeCl_3$, Ag_2CO_3/celite, and iodobenzene diacetate in methylene chloride or diethyl ether with reaction times ranging from 5.0 min to 18 h at room temperature, and proceed in yields from 80–95%. However, without initiation via phenoxide generation, PAB ether is stable towards oxidative conditions (DDQ) which cleave the *p*-methoxybenzyl (PMB) ether as shown in eq 3.[3]

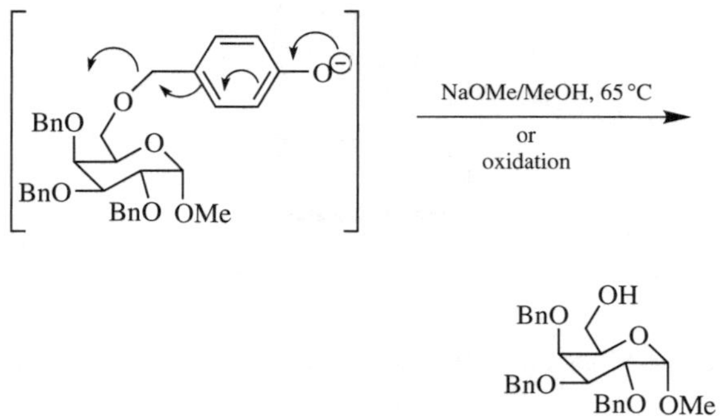

(2)

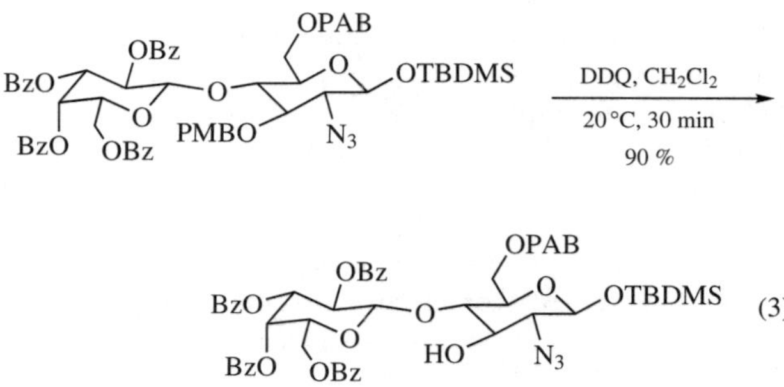

(3)

The removal of the PAB group occurs chemoselectively in the presence of benzoate-, benzyl-, and TBDMS-protected alcohols by treatment with NaOMe (5.0 min) and subsequent oxidation with $FeCl_3$, as shown in eq 4. Alternatively, the PAB group can also be removed under typical hydrogenation conditions.[2]

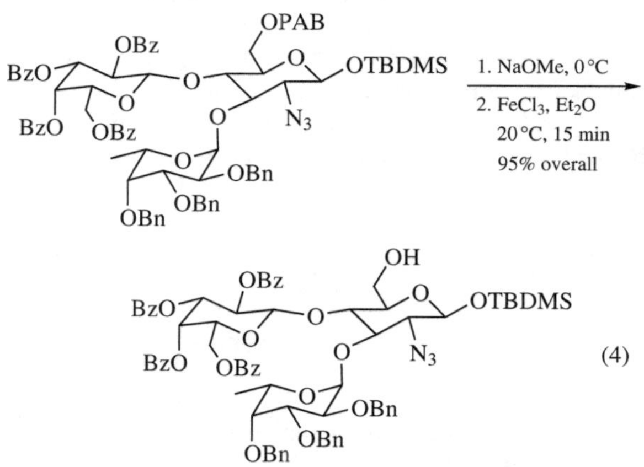

(4)

Benzylation of Heteroatomic Functional Groups. Several references to the use of *p*-acetoxybenzyl bromide have appeared in the literature including quaternization of amines,[1,4] reaction with pyridine *N*-oxides generating the corresponding aldehyde under microwave irradiation,[5] treatment with phenoxide giving diaryl benzyl ethers,[6] and reaction with benzothiazoles[7] and substituted imidazoles.[8]

Related Reagents. Benzyl bromide; *p*-methoxybenzyl bromide; *p*-acetoxybenzyl trichloroacetimidate; 4-(2-trimethylsilyl) ethoxymethoxybenzyl bromide.

1. Moss, R. A.; Dix, F. M., *J. Org. Chem.* **1981**, *46*, 3029.
2. Jobron, L.; Hindsgaul, O., *J. Am. Chem. Soc.* **1999**, *121*, 5835.
3. Green, T. W.; Wuts, P. G. M., *Protective Groups in Organic Synthesis*, 2nd ed; Wiley: New York, 1991.
4. Nakano, M.; Sato, Y., *J. Org. Chem.* **1987**, *52*, 1844.
5. Barbry, D.; Champagne, P., *Tetrahedron Lett.* **1996**, *37*, 7725.
6. Barnes, J. H.; Elks, J.; Stephens, F. F.; Waller, G. J., *J. Chem. Soc.* **1953**, 764.
7. Cossey, H. D.; Judd, J.; Stephens, F. F., *J. Chem. Soc.* **1965**, 954.
8. Yashioka, H.; Choshi, T.; Sugino, E.; Hibino, S., *Heterocycles* **1995**, *41*, 161.

Thomas E. Mabry
Eli Lilly and Company, Indianapolis, IN, USA

2-[(1-Acetyl-2-oxopropyl)thio]-1*H*-isoindole-1,3(2*H*)-dione

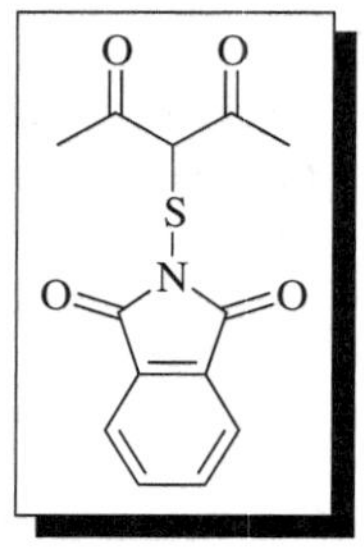

[107552-89-6] $C_{13}H_{11}NO_4S$ (MW 277.30)

(reagent used as a stable, diacylthione precursor)

Physical Data: mp 180 °C (dec).
Solubility: soluble in $CHCl_3$, CH_2Cl_2, THF, and ethyl acetate.
Form Supplied in: white solid; not commercially available.
Analysis of Reagent Purity: ^{1}H-NMR.
Preparative Methods: can be prepared from phthalimidosulfenyl chloride and 2,4-pentanedione,[1] or from the corresponding silylenol ether of 2,4-pentanedione and phthalimidosulfenyl chloride.[1b]
Purification: recrystallized from acetone or from chloroform/*n*-hexanes.
Handling, Storage, and Precautions: may be kept at room temperature for more than one year without any appreciable decomposition. Care should be exercised when preparing the reagent. The 2,4-pentanedione precursor is flammable, toxic [(oral) rat LD_{50}: 55 mg kg^{-1}], a known neurologic and teratogenic hazard, and is readily absorbed by the skin.

Reaction as a Dienophile in Diels–Alder Reactions. Treatment of the title compound with 2 equiv of base (pyridine or 2,6-lutidine) generates a reactive diacylthione that undergoes

Diels–Alder reactions with a variety of dienes to afford diacylthiine products.[1b,2] Electron-rich butadienes react at room temperature (3 h) to give modest to good yields of cycloadducts (64–88%). 1,3-Cyclohexadiene forms the bicyclic adduct in 69% yield (eq 1).[1b] When the Diels–Alder reaction was carried out with 2-methoxyfuran, the expected dihydrothiopyran regioisomers were not obtained.[2] Another possible product (the oxathiine), from the reaction of the diacylthione as a diene, also was not present. Instead, the five-membered oxathiolene was obtained. Its formation was rationalized by the nucleophilic attack of the thiocarbonyl sulfur on the C_5 position of the furan and the subsequent attack of the carbonyl oxygen at the same carbon.

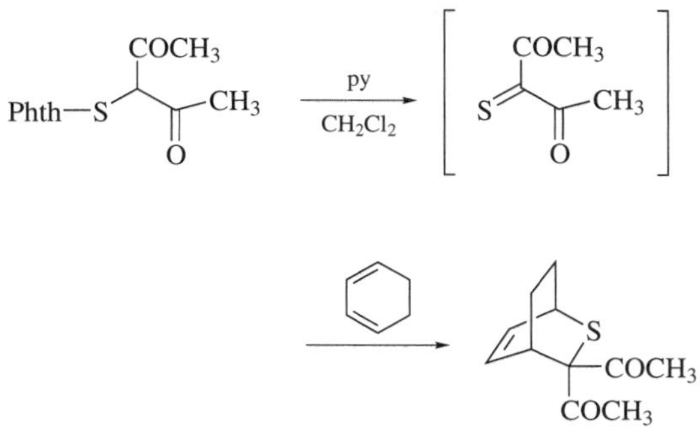

(1)

Reaction as a Diene in Inverse Electron Demand Diels–Alder Reactions. The title compound, when activated with a base, also serves as a heterodiene undergoing [4 + 2] cycloaddition reactions with a variety of electron-rich alkenes.[2,3] Ethyl vinyl ether affords an 80% yield of the 1,4-oxathiine after a 2 h reaction time (rt) in $CHCl_3$ (eq 2). Dihydropyran yields only the *cis*-fused bicycle after 20 h at room temperature in 90% yield (80:20 mixture of regioisomers; O at C2 position versus S at C2 position). The trimethylsilylenol ether of cyclohexanone also undergoes cycloaddition reactions with the dioxothione.[2]

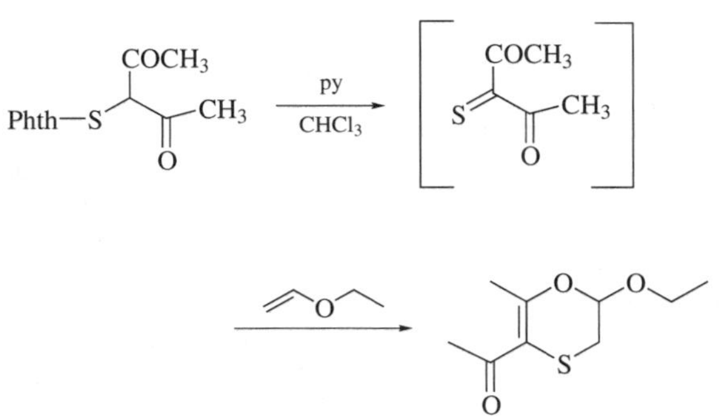

(2)

α,α′-Dioxothione also reacts with unsaturated sugars to form bicylic adducts.[3–5] When the vinyl ether double bond is endocyclic, the reactions are highly regioselective. The facial selectivity of the cycloaddition depends on the location and orientation of the neighboring sugar substituents. For example, when 3,4,6-tri-*O*-benzyl-D-galactal is reacted with α,α′-dioxothione, only the adduct with an axial C1 oxygen substituent is isolated (eq 3). For 3,4,6-tri-*O*-benzyl-D-allal, with an axial substituent at the allylic position, only the opposite adduct with an equatorial C1 oxygen substituent was isolated (68%).[4a,b] The ratio of diastereomeric adducts is affected by changes in solvent, by hydrogen bonding effects with allylic free hydroxyls, and by substitution on the double bond. The addition of powdered, activated, 3 Å molecular sieves has been demonstrated to improve the yields in these reactions. Methanol trapping experiments suggest that a polar mechanism may be involved in the cycloaddition reaction.[4b] Further transformations of the adducts produce glycosyl donors that have been used for the preparation of 2-deoxy-glycosides.[6]

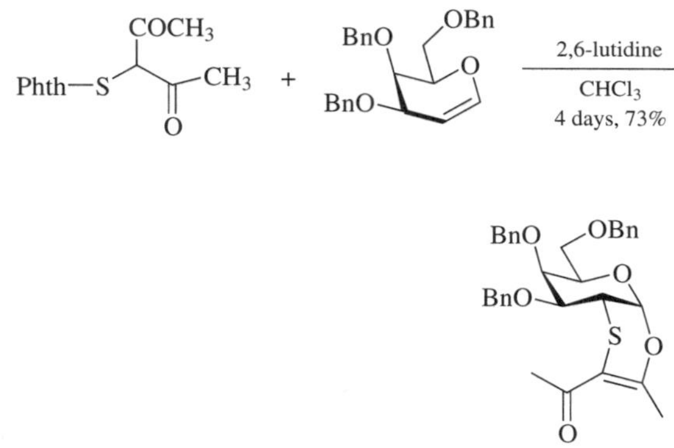

(3)

Thiaspiroacetal derivatives have been prepared from the reaction of exo-methylene sugars with the title compound and excess pyridine.[5] Lower diastereoselectivities were obtained compared to the cycloaddition reactions of the endo-glycals. For example, 2,6-anhydro-3,4,5,7-tetra-*O*-benzyl-1-deoxy-D-gluco-hept-1-enitol, when reacted with α,α′-dioxothione, afforded a 2.5:1 mixture of diastereomers (eq 4).

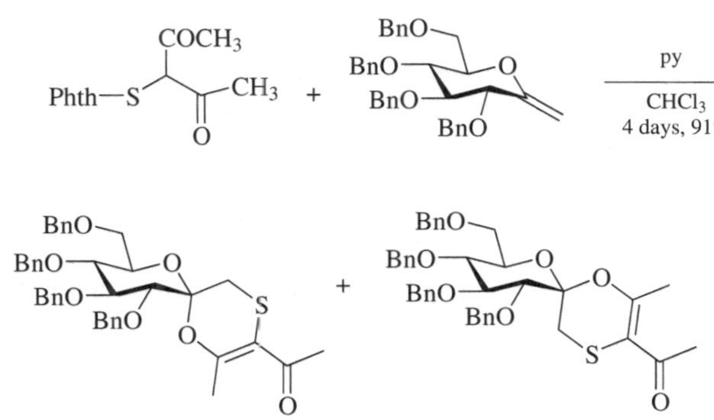

(4)

2.5:1

The cycloaddition reaction is general and somewhat predictable by quantum mechanical calculations and by examination of the energy gap between the HOMO of the dienophile and the LUMO of the diene.[2] Thus, vinyl sulfides, 2-vinyl-pyrrolidone, and electron-rich styrenes are all reactive partners in the cycloaddition reaction. Allenes undergo reaction with α,α′-dioxothione to give vinyl oxathiines, which can decompose to produce α,β-unsaturated aldehydes. Alkynes react with 2 equiv of dioxothione to give the bicyclic adducts (eq 5). Presumably the electron-rich monoadduct reacts faster than the starting alkyne with the electron-poor diene.[2]

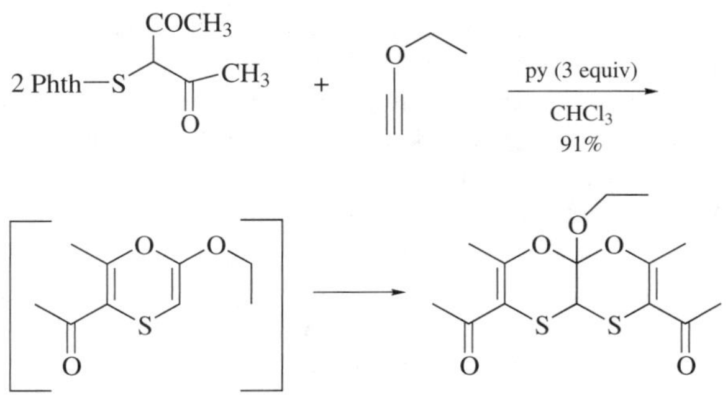

(5)

Ene Reactions. Vinyl sulfides are formed exclusively in the reaction of sterically hindered alkenes with α,α′-dioxothione generated in situ from the title compound.[2,3,7] Single products are obtained in the room temperature reactions of 2,3-dimethylbutene (eq 6) and β-pinene with the title compound in the presence of excess pyridine.[3] When the allylic position of the double bond is substituted with an alkyl group, high (*E*) stereoselectivity is observed in the reaction product.[7] With α-methyl styrene, a mixture of oxathiine cycloadduct and vinyl sulfide is realized, with the thiophilic ene product predominating (58% ene; 10% [4 + 2]).[3]

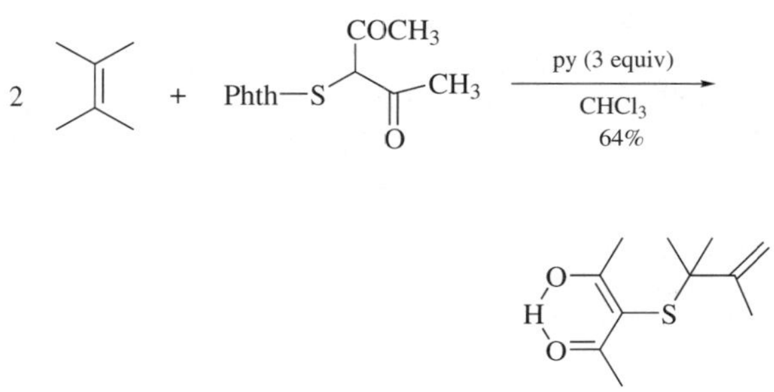

(6)

Tandem 'Ene-Cycloaddition' Reactions. When allyl ethyl ether (10 equiv) was allowed to react with the title compound (1 equiv) and pyridine (2 equiv) in chloroform for 5 days (rt), the substituted oxathiine was isolated (eq 7). The formation of an oxathiine structure was rationalized by an initial ene reaction followed by a fast $(4\pi + 2\pi)$ cycloaddition reaction affording the final product with retention of stereochemistry of the intermediate enol ether. A similar tandem reaction sequence was observed for allyl *tert*-butyldimethylsilyl ether and diallyl ether from which the bis-adducts also were isolated.[7]

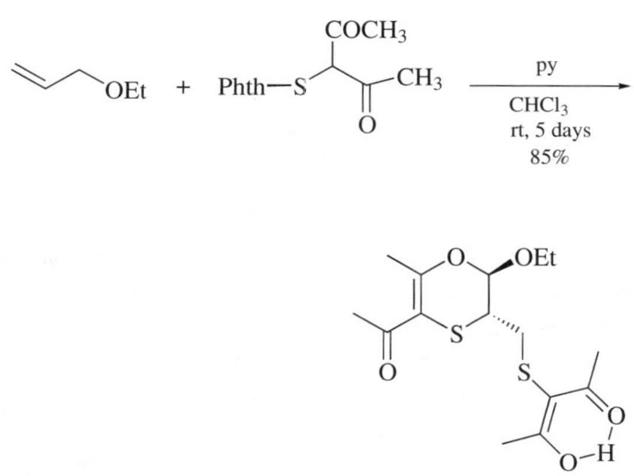

(7)

When dioxothione was treated with 1,1-dimethylallene, a single cycloadduct was isolated in 73% yield. The formation of this product was attributed to the generation of a reactive, thio-substituted butadiene intermediate from the initial ene reaction, followed by cycloaddition with acyl thione acting as a dienophile.[7]

Electrophilic Aromatic Substitution Reactions. α,α′-Dioxothione generated from the title compound reacts with activated arenes affording alkyl aryl or diaryl sulfides.[8] These reactions proceed at room temperature in chloroform over a period of 1–5 days. *p*-Substituted products are obtained exclusively for *o*-cresol, 2,6-dimethylphenol, and *N,N*-dimethylaniline. β-Naphthol affords a 64% yield of a single product (eq 8). 2,5-Dimethylphenol, 2-methoxyphenol, and 3-methoxyphenol afford mixtures of regioisomers.

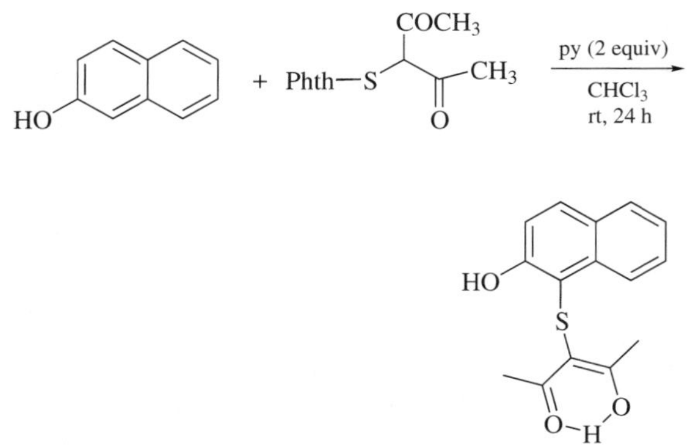

(8)

1. (a) Mullin, V. A.; Zolotov, A. N., *Zh. Obshch. Khim.* **1986**, *56*, 898 (CAN 106:156209). (b) Capozzi, G.; Menichetti, S.; Nativi, C.; Rosi, A.; Vale, G., *Tetrahedron* **1992**, *48*, 9023.
2. Capozzi, G.; Franck, R. W.; Mattoli, M.; Menichetti, S.; Nativi, C.; Valle, G., *J. Org. Chem.* **1995**, *60*, 6416.
3. Capozzi, G.; Menichetti, S.; Nativi, C.; Rosi, A.; Franck, R. W., *Tetrahedron Lett.* **1993**, *34*, 4253.
4. (a) Capozzi, G.; Dios, A.; Franck, R. W.; Geer, A.; Marzabadi, C.; Menichetti, S.; Nativi, C.; Tamarez, M., *Angew. Chem., Int. Ed. Engl.* **1996**, *35*, 777. (b) Dios, A.; Geer, A.; Marzabadi, C. H.; Franck, R. W., *J. Org. Chem.* **1998**, *63*, 6673. (c) Capozzi, G.; Catelani, G.; D'Andrea, F.; Menichetti, S.; Nativi, C., *Carbohydr. Res.* **2003**, *338*, 123.
5. (a) Capozzi, G.; Bartolozzi, A.; Falciani, C.; Menichetti, S.; Nativi, C., *Phosphorus Sulfur Silicon* **1999**, *153–154*, 309. (b) Bartolozzi, A.; Capozzi, G.; Falciani, C.; Menichetti, S.; Nativi, C.; Bacialli, A. P., *J. Org. Chem.* **1999**, *64*, 6490.
6. (a) Marzabadi, C. H.; Franck, R. W., *Chem. Commun.* **1996**, 2651. (b) Capozzi, G.; Mannocci, F.; Menichetti, S.; Nativi, C.; Paoletti, S., *Chem. Commun.* **1997**, 2291. (c) Franck, R. W.; Marzabadi, C. H., *J. Org. Chem.* **1998**, *63*, 2197. (d) Dios, A.; Nativi, C.; Capozzi, G.; Franck, R. W., *Eur. J. Org. Chem.* **1999**, 1869. (e) Bartolozzi, A.; Capozzi, G.; Menichetti, S.; Nativi, C., *Eur. J. Org. Chem.* **2001**, 2083.
7. Capozzi, G.; Fragai, M.; Menichetti, S.; Nativi, C., *Eur. J. Org. Chem.* **1999**, 3375.
8. Capozzi, G.; Menichetti, S.; Nativi, C., *Eur. J. Org. Chem.* **2000**, 3653.

Cecilia H. Marzabadi
Seton Hall University, South Orange, NJ, USA

2-Adamantyl Chloroformate

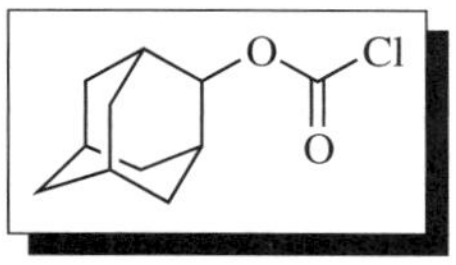

[53120-53-9] $C_{11}H_{15}ClO_2$ (MW 214.69)

(reagent for introducing the 2-adamantyloxycarbonyl protecting group)

Alternate Name: 2-Adoc-Cl.
Physical Data: mp < 40 °C.
Solubility: soluble in 1,4-dioxane, DMF.
Form Supplied in: colorless crystals. Commercially available from Watanabe Chemical Industries (Hiroshima, Japan).
Preparative Methods: 2-adamantyl chloroformate (2-Adoc-Cl) is synthesized from 2-adamantanol and phosgene in a manner similar to 1-adamantyl chloroformate.[1]
Handling, Storage, and Precautions: irritant; avoid contact with skin, eyes, and clothes; avoid breathing dust, vapor, moisture, or gas; corrosive; combustible; moisture sensitive. Store under nitrogen. Keep tightly closed and refrigerated. A bottle stored for a long-term may develop pressure.

Protection of the N^ϵ-amino Group of Lys. 2-Adamantyloxycarbonyl (2-Adoc) was first developed as a protecting group for the ϵ-amino function of Lys.[2,3] H-Lys(2-Adoc)-OH is prepared by the Schotten–Baumann reaction of the Lys-copper complex and 2-adamantyl chloroformate followed by removal of copper with EDTA (eq 1). Recrystallization of the crude product from boiling water containing a small amount of EDTA affords the chromatographically pure material. N^α-Boc- and N^α-Fmoc-protected derivatives are synthesized in the usual manners using di-*tert*-butyl bicarbonate and 9-fluorenylmethyl succinimidyl carbonate, respectively. The products are crystallized as dicyclohexylammonium salts from diethyl ether. The salts are recommended for long-term storage. The free acids are amorphous and can be stored at 4 °C for several months. The *O*-Bn protected derivative can be obtained by esterification of the Cs salt of Boc-Lys(2-Adoc)-OH with benzyl bromide followed by removal of Boc with HCl in 1,4-dioxane. The stability characteristics of N^ϵ-2-Adoc group are summarized in Table 1. The N^ϵ-2-Adoc is fairly stable towards the reagents used to remove *tert*-butyl-based and 9-fluorenylmethyl-based protecting groups. It is also stable towards some reagents for the removal of Bn-based protecting groups, such as 25% HBr in AcOH, 1 M TMSBr–thioanisole in TFA and H_2/Pd. The N^ϵ-2Adoc is removed with HF or 1 M trifluoromethanesulfonic acid (TFMSA)-thioanisole in TFA. The solubilities of peptides protected with 2-adamantane-based protecting groups including 2-Adoc are often better than those of Bn-protected peptides.[3,4]

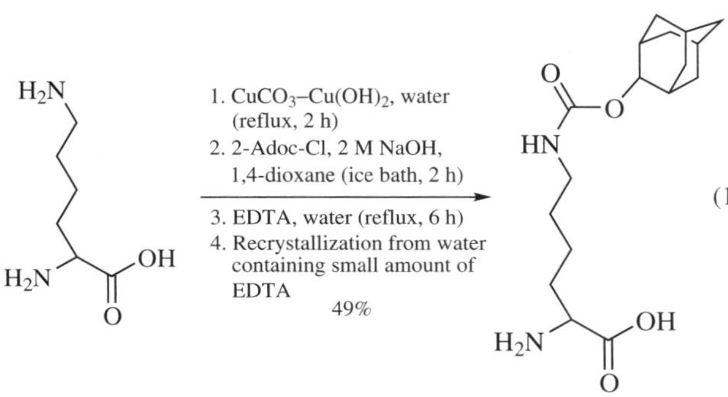

(1)

Table 1 Stability characteristics of N^ϵ-2-Adoc group

Conditions	Stability/Removability
7.6 M HCl in 1,4-dioxane	No removal (24 h, rt)
TFA	No removal (24 h, rt)
25% HBr in AcOH	No removal (2 h, rt)
1 M TMSBr–thioanisole in TFA	No removal (2 h, rt)
Methanesulfonic acid	Slow removal ($t_{1/2}$ 2.1 h, rt)
1 M TFMSA–thioanisole in TFA	Complete removal (<20 min, rt)
HF	Complete removal (<20 min, rt)
2 M NaOH	No removal (24 h, rt)
20% Piperdine in DMF	No removal (24 h, rt)

Protection of the Imidazole Group of His. 2-Adoc is the first $N^{im}(\tau)$-protecting group compatible with Boc-chemistry, HOBt-mediated coupling, and one-step final deprotection.[5,6] Boc-His(2-Adoc)-OH can be synthesized readily from 2-adamantyl chloroformate and Boc-His-OH (eq 2). The reaction mixture should be neutralized before work-up, since N^{im}-2-Adoc is unstable in basic aqueous solutions. Z-His(2-Adoc)-OH is synthesized in a similar manner. The stability characteristics of N^{im}-2-Adoc group are summarized in Table 2. The N^{im}-2-Adoc group is fairly stable towards reagents used to remove *tert*-butyl-based protecting groups. However, His(2-Adoc) derivatives, unlike Lys(2-Adoc), are not suitable for Fmoc-chemistry, since the N^{im}-2-Adoc group is susceptible to an Fmoc-deprotecting reagent, i.e., 20% piperidine in DMF ($t_{1/2}$ 13.5 min).[6] N^{im}-2-Adoc is deprotected with 25% HBr in AcOH, 1 M TFMSA–thioanisole in TFA or HF within 10 min. The N^{im}-2-Adoc group is fully stable towards Et_3N in anhydrous DMF, but not in an aqueous solution. The N^{im}-2-Adoc group is stable towards a common coupling additive, i.e. HOBt, using which the popular N^{im}-protecting group Ts is partially removed. N^{im}-2-Adoc suppresses the stereomutation of His to a similar extent to Ts and dinitrophenyl. An analogous protecting group 2,4-dimethylpent-3-yloxycarbonyl has somewhat improved stability towards nulceophiles.[7]

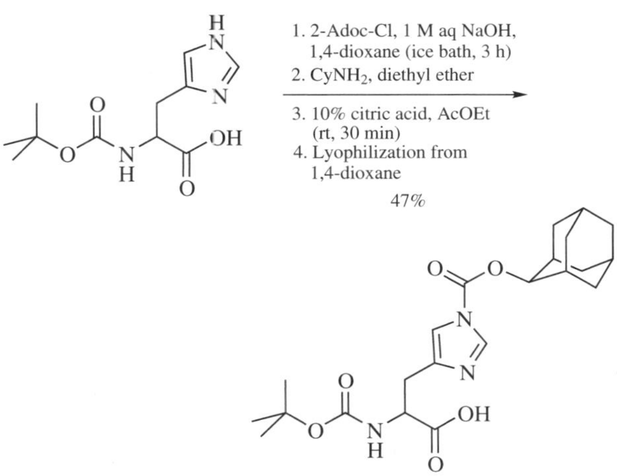

(2)

Table 2 Stability characteristics of N^{im}-2-Adoc protecting group

Conditions	Stability/Removability
7.6 M HCl in 1,4-dioxane	No removal (24 h, rt)
TFA	No removal (24 h, rt)
25% HBr in AcOH	Complete removal (<10 min, rt)
1 M TMSBr–thioanisole in TFA	Complete removal (<10 min, rt)
HF	Complete removal (<10 min, rt)
2 M NaOH	Complete removal (<10 min, rt)
20%Piperidine in DMF	Complete removal (<2 h, rt)
10% Et_3N in DMF	No removal (24 h, rt)
10% Et_3N in water	Complete removal ($t_{1/2}$<10 min, rt)
20%HOBt in DMF	No removal (24 h, rt)

Related Reagents. 2-Adamantyl fluoroformate [163798-91-2]; 1-adamantyl chloroformate [5854-52-4]; 1-adamantyl fluoroformate [62087-82-5].

1. Haas, W. L.; Krumkalns, E. V.; Gerzon, K., *J. Am. Chem. Soc.* **1966**, *88*, 1988.
2. Nishiyama, Y.; Okada, Y., *J. Chem. Soc., Chem. Commun.* **1993**, 1083.
3. Nishiyama, Y.; Shintomi, N.; Kondo, Y.; Okada, Y., *J. Chem. Soc., Perkin Trans. 1* **1994**, 3201.
4. Nishiyama, Y.; Okada, Y., *Tetrahedron Lett.* **1994**, *35*, 7409.
5. Nishiyama, Y.; Shintomi, N.; Kondo, Y.; Okada, Y., *J. Chem. Soc., Chem. Commun.* **1994**, 2515.
6. Nishiyama, Y.; Shintomi, N.; Kondo, Y.; Izumi, T.; Okada, Y., *J. Chem. Soc., Perkin Trans. 1* **1995**, 2309.
7. Karlstrom, A.; Unden, A., *Chem. Commun.* **1996**, 959.

Yasuhiro Nishiyama
University of Texas-Houston Medical School, Houston, TX, USA

N-[(Allyloxycarbonyl)oxy]maleimide-styrene Copolymer (Alloc-resin)

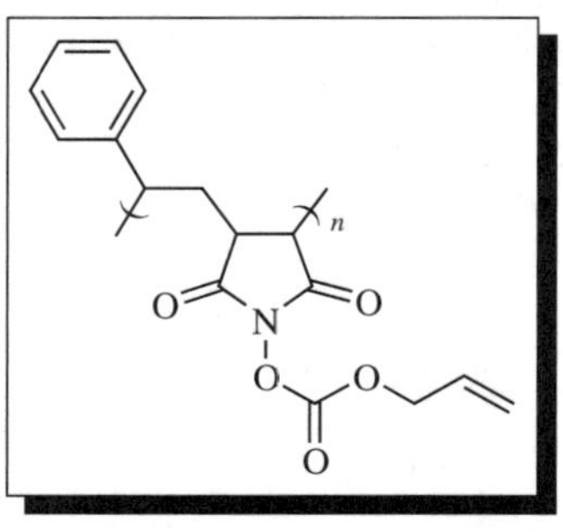

[1344-28-1] $C_{16}H_{15}NO_5$ (MW 301.29)

(reagent used for the *N*-protection of amines and amino acids with the allyloxycarbonyl (Alloc or Aloc) group)

Alternate Name: Alloc-P-OSu.
Solubility: soluble in DMSO and DMF; partially soluble in acetone, CH_2Cl_2, THF, CH_3CN, and dioxane; insoluble in toluene, water, $CHCl_3$, AcOEt, and hexanes.
Form Supplied in: white powder.
Preparative Methods: prepared by reaction of alternating *N*-hydroxymaleimide-styrene copolymer (P-HOSu)[1] with allyl chloroformate in the presence of triethylamine, in dichloromethane as solvent and at room temperature, followed by precipitation after addition of toluene.[2]
Handling, Storage, and Precautions: very stable, not hygroscopic, remains active after extended periods of time stored under ambient conditions.

Introduction. The allyloxycarbonyl (Alloc or Aloc) function is employed as a protecting group for primary and secondary amines, and especially for amino acids in the synthesis of peptides due to its stability to acidic and some basic conditions, thus allowing selective deprotection of other groups such as Boc or Fmoc.[3] The removal of the Alloc group is usually achieved under mild reaction conditions using palladium(0)-catalyzed allyl transfer,[1] which represents an especially useful method for the synthesis of highly sensitive structures such as glycopeptides.[4]

The Alloc protection of the amino group has been frequently achieved by reaction with allyl chloroformate,[5] a moisture-sensitive reagent which can contain variable amounts of very toxic phosgene;[6] this reagent can also induce the formation of protected dipeptide by-products from amino acid substrates.[3d] Other more stable reagents for the introduction of the Alloc group have also been developed, such as diallyl dicarbonate,[7] allyl 1-benzotriazolyl carbonate (Alloc-OBt),[7,8] allyl 1-[6-(trifuoromethyl)] benzotriazolyl carbonate[9] or allyl succinimidyl carbonate (Alloc-OSu);[10] the *N*-[(allyloxycarbonyl)oxy]maleimide-styrene copolymer (Alloc-P-OSu)[2] can be considered as an *N*-allyloxycarbonylated copolymeric succinimide, a polymeric analogue of Alloc-OSu.

***N*-Protection of amines and amino acids.**[2] The reaction of Alloc-P-OSu (**1**) (1 equiv) with primary amines or amino acid esters in dicholoromethane as a solvent at room temperature during 1 day afforded the corresponding *N*-Alloc-protected derivatives (eq 1). When the *N*-protection was carried out on free α-amino acids, a solution of the corresponding amino acid and K_2CO_3in water was added to the suspension of (**1**) in dichloromethane. The solid present when the reaction was finished was filtered off from the reaction mixture consisting of polymeric *N*-hydroxysuccinimide (P-HOSu, **2**),[1] allowing thus an easy separation of this residue from the reaction mixture, and its recycling for the regeneration of Alloc-P-OSu.[2]

Although the yields of the Alloc-protected amino acids are generally lower than with non-polymeric Alloc-protecting reagents, they are considerably improved when working with amino esters instead of the corresponding free amino acids, as in the case of the Alloc-protection of phenylalanine and phenylalanine benzyl ester (eq 2).[2] The main advantage using this polymeric reagent is the easy removal of the polymeric *N*-hydroxysuccinimide by-product; this is especially valuable when working on small scale.

Related Reagents. The polymeric *N*-hydroxybenzotriazole-derived Alloc-protecting reagent Alloc-OBt-6-carboxamidomethyl polystyrene (**3**) has recently been commercialized.

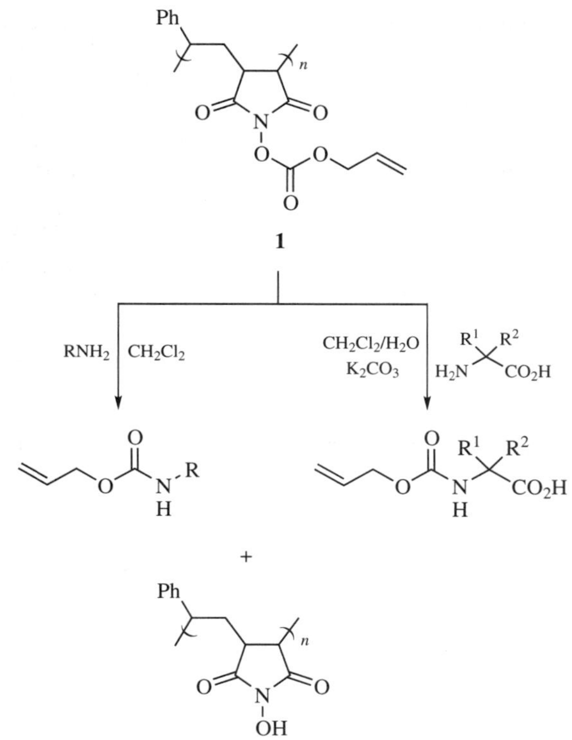

(1)

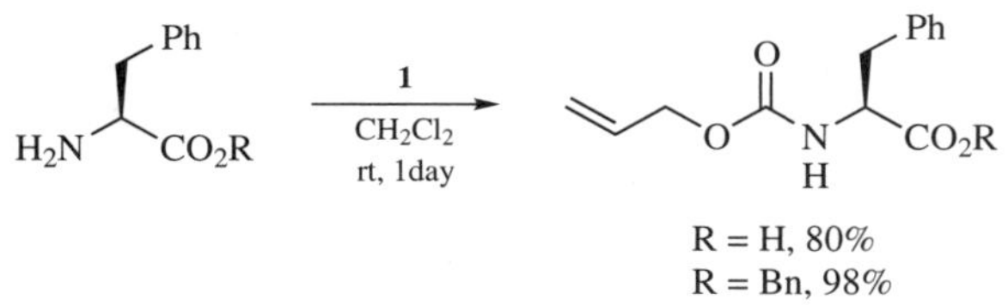

(2)

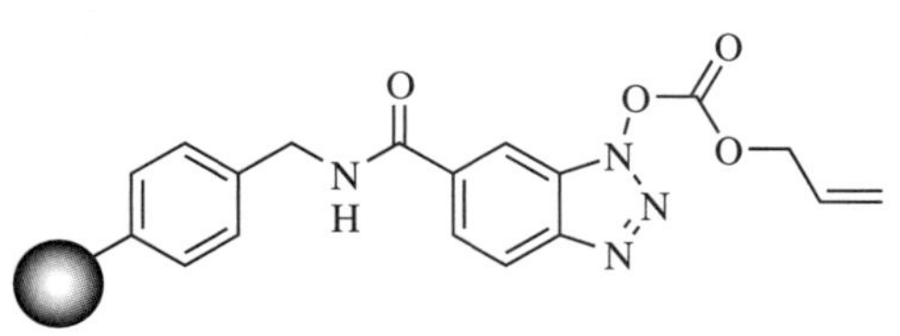

3

1. Chinchilla, R.; Dodsworth, D. J.; Nájera, C.; Soriano, J. M., *Tetrahedron Lett.* **2001**, *42*, 4487.

2. Chinchilla, R.; Dodsworth, D. J.; Nájera, C.; Soriano, J. M., *Synlett* **2003**, 809.

3. (a) Green, T. W.; Wuts, P. G. M. *Protective Groups in Organic Synthesis*; Wiley: New York 1999, 526. (b) Kocienski, P. J. *Protecting Groups*; Thieme: Stuttgart 1994, 199. (c) Bodanszky, M.; Bodanszky, A. *The Practice of Peptide Synthesis*, Springer: Berlin, 1984. (d) Albericio, F., *Biopolymers (Peptide Science)* **2000**, *55*, 123.

4. Some examples: (a) Paulsen, H.; Merz, G.; Brockhausen, I., *Liebigs Ann. Chem.* **1990**, 719. (b) Attal, S.; Bay, S.; Cantacuzene, D., *Tetrahedron* **1992**, *48*, 9251. (c) Unverzagt, C.; Kunz, H., *Bioorg. Med. Chem.* **1994**, *2*, 1189. (d) Meinjohanns, E.; Vargas-Berenguel, A.; Meldal, M.; Paulsen, H.; Bock, K., *J. Chem. Soc., Perkin Trans. 1* **1995**, 2165.

5. (a) Sinha, S.; Ilankumaran, P.; Chandrasekaran, S., *Tetrahedron Lett.* **1999**, *40*, 771. (b) Fukase, Y.; Fukase, K.; Kusumoto, S., *Tetrahedron Lett.* **1999**, *40*, 1169. (c) Fukase, Y.; Fukase, K.; Kusumoto, S., *Pept. Sci.* **1999**, *36*, 93. (d) Bhat, R. G.; Sinha, S.; Chandrasekaran, S., *Chem. Commun.* **2002**, 812.

6. Herrinton, P. M. in *Encyclopedia of Reagents for Organic Synthesis*, Paquette, L. A., Ed., Wiley: Chichester, 1995, 95.

7. Sennyey, G.; Barcelo, G.; Senet, J.-P., *Tetrahedron Lett.* **1987**, *28*, 5809.

8. Hayakawa, Y.; Kato, H.; Uchiyama, M.; Kajino, H.; Noyori, R., *J. Org. Chem.* **1986**, *51*, 2400.

9. Takeda, K.; Tsuboyama, K.; Hoshino, M.; Kishino, M.; Ogura, H., *Synthesis* **1987**, 557.

10. Blaakmeer, J.; Tijsse-Klasen, T.; Tesser, G. I., *Int. J. Pept. Protein Res.* **1991**, *37*, 556.

Carmen Nájera & Rafael Chinchilla
Universidad de Alicante, Spain

Allyl Tetraisopropylphosphoramidite

$$\mathrm{OCH_2CH{=}CH_2}$$
$$(i\text{-}C_3H_7)_2N{-}P{-}N(i\text{-}C_3H_7)_2$$

[108554-72-9] $C_{15}H_{33}N_2OP$ (MW 288.42)

(a useful reagent for the synthesis of various organophosphoric acid derivatives such as nucleotides, nucleotide–peptide conjugates, nucleotide–carbohydrate conjugates, and sugar phosphates)

Physical Data: bp 114–117 °C/0.4 mm Hg, ca. 120 °C/0.6 mm Hg; *d* 0.903.

Solubility: highly soluble in various aprotic organic solvents including acetonitrile, dichloromethane, tetrahydrofuran, pyridine, etc.

Form Supplied in: colorless liquid.

Analysis of Reagent Purity: IR ($CHCl_3$): 2970, 1360, 1185, 1025, 955 cm^{-1}; ^1H NMR ($CDCl_3$) 1.17 (dd, 24H, J = 1.5 and 6.6 Hz), 3.55 (sept d, 4H, J = 6.6 and 10.5 Hz), 4.11 (ddt, 2H, J = 5.1, 7.5, and 1.8 Hz), 4.99–5.43 (m, 2H), 5.98 (ddt, 1H, J = 10.2, 17.4, and 5.1 Hz); ^{31}P NMR ($CDCl_3$) 125.7.

Preparative Methods: the title reagent can be prepared by the following two methods. (a)[1] In the first method, allyl alcohol (1.0 mol) is slowly added to PCl_3 (1.0 mol) in the presence of pyridine (1.0 mol) in ether (200 mL) at −78 °C over 90 min. The mixture is stirred at 25 °C overnight. The precipitates are removed by filtration under argon atmosphere. Diisopropylamine (7.1 mol) is added to the filtrate at −10 °C over 60 min, followed by stirring at 25 °C overnight. The resulting precipitates are removed by filtration. The filtrate is evaporated and the residual liquid is distilled over CaH_2 under 0.6 mm Hg. Fractions appearing around 120 °C are collected. (b)[2] Second method involves addition of allyl alcohol (1.0 equiv) slowly to PCl_3 (1.0 equiv) at −20 °C (over 2 h on the 0.94-mol scale), followed by stirring at 25 °C overnight under lightly reduced pressure in order to remove the HCl liberated. The resulting product (allyl phosphorodichloridite) is dissolved in ether and the same solution (1.0 equiv) is added to diisopropylamine (4.1 equiv) in ether at −15 °C. The mixture is stirred at 25 °C for 20 h. The resulting precipitates are removed by filtration. The filtrate is concentrated. The residual liquid is distilled under 0.4 mm Hg. The fractions are collected at bp 114–117 °C.

Purification: distilled under reduced pressure.

Handling, Storage, and Precautions: the phosphoramidite is sensitive to moisture, particularly in the presence of an acidic substance. Further, this reagent is not quite stable in the presence of oxygen. Should be stored under argon below −20 °C. Toxicity unknown.

General Aspects. Allyl tetraisopropylphosphorodiamidite reacts successively with two kinds of alcoholic substances with the assistance of an activator, such as 1*H*-tetrazole,[1] 5-(*p*-nitrophenyl)-1*H*-tetrazole,[2] benzimidazolium triflate,[3] or *N*-phenylimidazolium triflate,[3] to give the corresponding phosphite triester **1**. This phosphite can be converted to a variety of organophosphoric acid derivatives (**2**) including phosphate, phosphorothioate, and phosphoroselenoate through a suitable oxidative treatment (eq 1). Further, the allyl group can be removed by various methods, e.g., treatment with $Pd(PPh_3)_4$ or $Pd_2[(PhCH{=}CH)_2CO]_3{\cdot}CHCl_3$ in the presence of PPh_3 and a nucleophile including butylammonium formate,[4] diethylammonium carbonate,[4] diethylammonium hydrogencarbonate,[4] dimedone, formic acid,[4] or triphenyltin hydride;[5] exposure to $RhCl(PPh_3)_3$ (catalyst) in the presence diisopropylethylamine;[6] treatment with aqueous pyridine,[7] concentrated ammonium hydroxide containing 2% mercaptoethanol,[8] hot concentrated ammonium hydroxide;[9] or heating with sodium iodide in acetonitrile[10] (eq 1). Accordingly, allyl tetraisopropylphosphorodiamidite serves as a building block for the synthesis of biologically important and attractive organophosphoric acid derivatives including oligonucleotides, nucleotide–peptide conjugates, nucleotide–carbohydrate conjugates, sugar phosphates, etc. As an alternative reagent to allyl tetraisopropylphosphorodiamidite, allyl diisopropyl (chlorophosphoramidite) $\{(CH_2{=}\ CHCH_2O)PCl[N(i\text{-}C_3H_7)_2]\}$ is occasionally used for construction of the allyl phosphate linkage.[11] However, the chlorophosphoramidite seems to be less useful as it is highly sensitive to moisture and thus requires great care in its treatment.

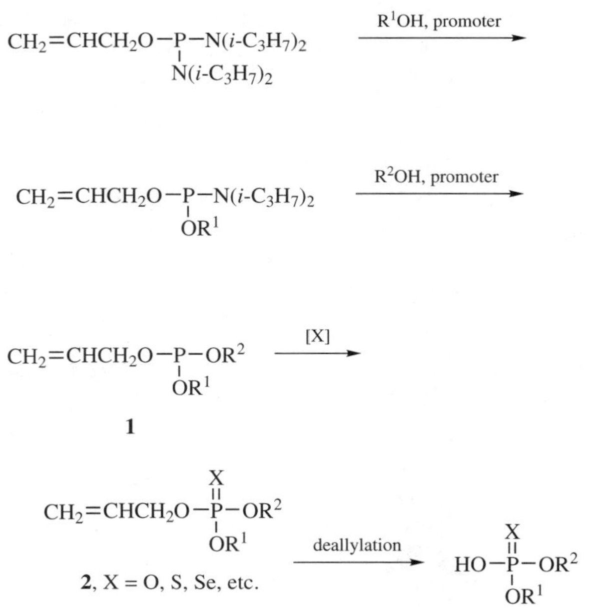

(1)

Synthesis of Oligonucleotides. In the presence of the activator, allyl tetraisopropylphosphorodiamidite reacts with 3′-*O*- or 5′-*O*-free 2′-deoxyribonucleosides to afford 2′-deoxyribonucleoside 3′- or 5′-(allyl diisopropylphosphoramidite)s, respectively[2,3] (eqs 2 and 3). In a similar manner, ribonucleoside 2′-, 3′-, or 5′-(allyl diisopropylphosphoramidite)s can be prepared via the condensation of allyl tetraisopropylphosphorodiamidite and 2′-*O*-, 3′-*O*-, or 5′-*O*-free ribonucleosides, respectively.[3] These nucleoside phosphoramidites also react with nucleosides by the assistance of a promoter to form, after oxidation, dinucleoside phosphates (eq 4).

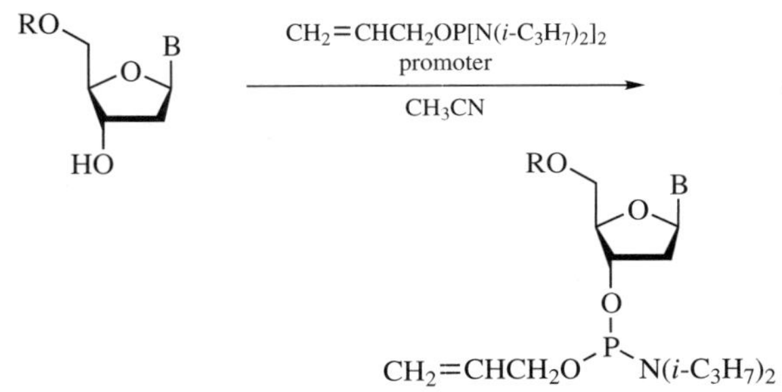

(2)

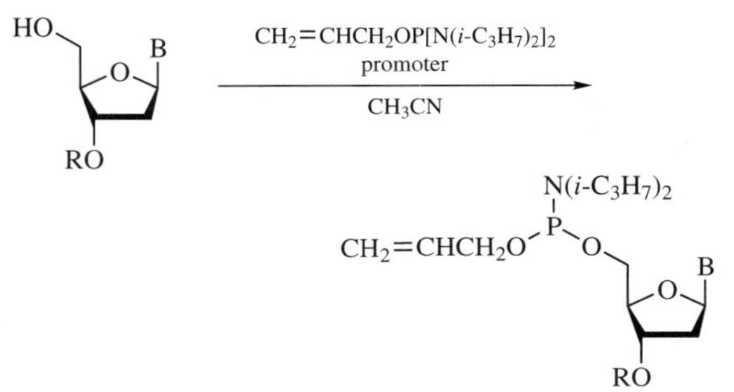

(3)

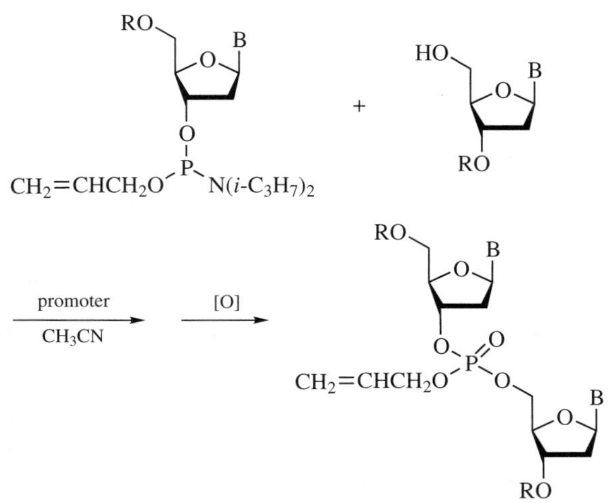

(4)

B = protected nucleoside base

When the allyloxycarbonyl group is used for the protection of nucleoside bases and sugar hydroxyls in oligonucleotide synthesis, the allylic protecting groups are cleanly removed all at once by means of the Pd(0)-catalyzed reaction to give the target product in excellent yields.[4,12] The strategy using *N*-allyloxycarbonyl and *P-O*-allyl protectors allows clean preparation of biologically important nucleotides such as adenylyl(2′–5′)adenylyl(2′–5′)adenylic acid (2–5A core),[3] adenyldiyl(2′–5′)(3′–5′) dinucleosides,[13] cyclic bis(3′–5′)diguanylic acid,[14] and a covalently linked dimensional analog of A·dT (**3**)[15] (eq 5) in a solution phase, and oligodeoxyribonucleotides and oligoribonucleotides on solid supports.[2,3]

It is noteworthy that this allyl-protected method can be used effectively to prepare chemically labile compounds, because removal of the allylic groups can be achieved under almost neutral conditions according to the Pd(0)-catalyzed reaction using a suitable nucleophile such as diethylammonium carbonate or dimedone. For example, high-yield and clean preparation of base-labile artificial oligonucleotides having biologically attractive properties—such as oligoribonucleotides containing *N*-acetylcytidine;[16]

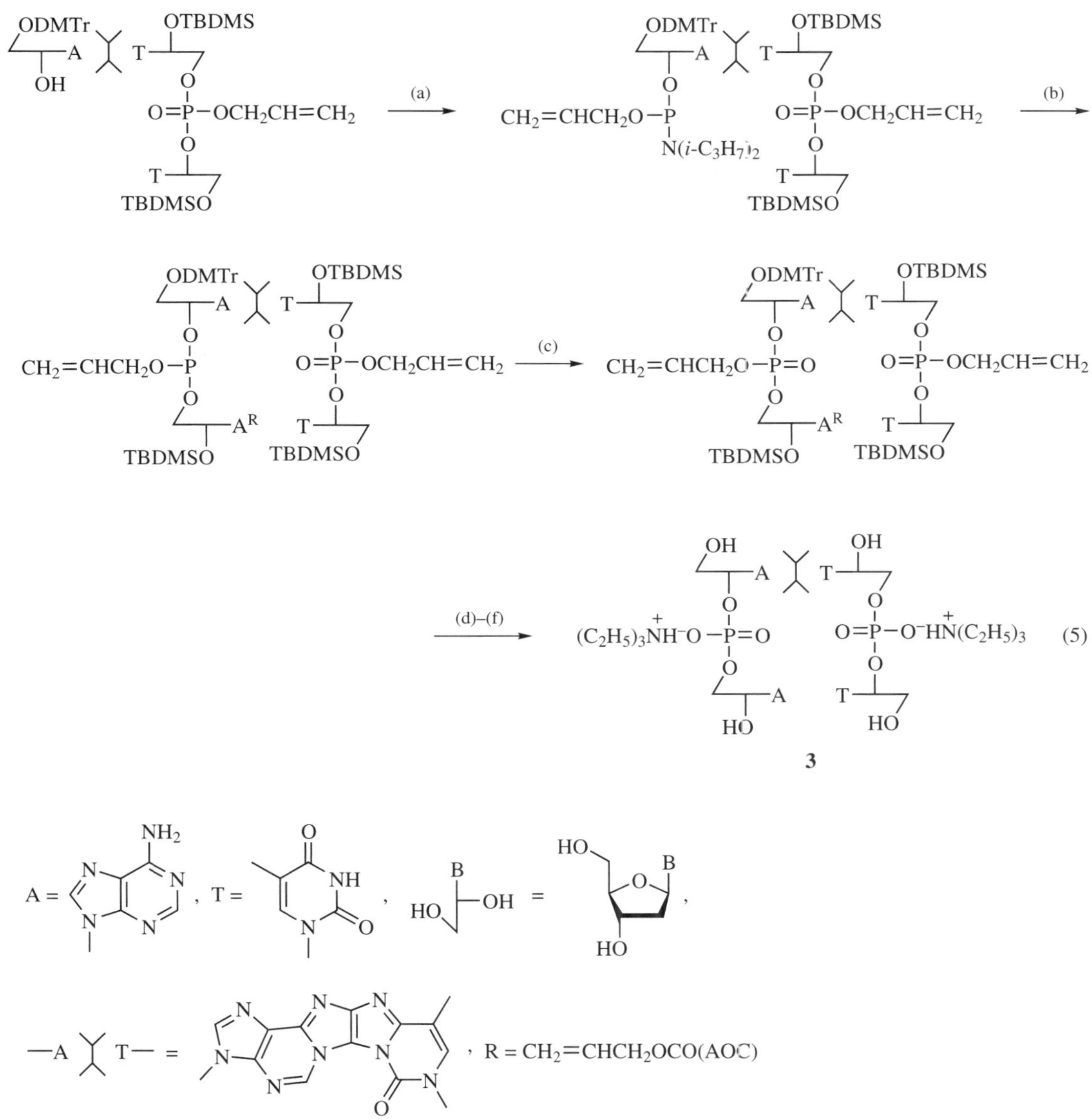

(5)

3

Conditions:

(a) $CH_2{=}CHCH_2OP[N(i\text{-}C_3H_7)_2]_2$, $(i\text{-}C_3H_7)_2NH$, 1*H*-tetrazole, CH_3CN, 25 °C, 82%

(b) 3′-*O*-TBDMS-A^{AOC}, 1*H*-tetrazole, CH_3CN, 25 °C, 83%

(c) TBHP, CH_2Cl_2, 0–25 °C, 94%

(d) TBAF on silica gel, THF, 14 h, 25 °C

(e) PPh_3, $BuNH_2$, HCOOH, $Pd_2[(PhCH{=}CH)_2CO]_3{\cdot}CHCl_3$, 5 h, 25 °C

(f) $Cl_2CHCOOH$, CH_2Cl_2, 3 min, 25 °C, 3 steps, 50%.

oligodeoxyribonucleotides site-specifically containing (5*R*)-5,6-dihydro-5-hydroxylthymidine;[17] and oligodeoxyribonucleotides with partially modified backbones, such as methyl[18] or *S*-acetyl-2-thioethyl[19] phosphotriester linkage and methyl-[18] or phenylphosphonodiester[18] linkage—have been achieved via this method using *N*-allyloxycarbonyl-protected nucleoside 3′-(allyl diisopropylphosphoramidite)s as key building blocks. An extremely base-labile artificial nucleotide, 2′-*C*-cyano-2′-deoxy-1-β-D-*arabino*-pentafuranosylcytosinylyl(3′–5′)thymidine (**4**),[20] which is completely decomposed in aqueous solutions at pH 7.3 and 8.4 for 8 and 2 h, respectively, is also prepared via the approach using allyl tetraisopropylphosphorodiamidite for construction of the phosphate linkage (eq 6). Efficient synthesis

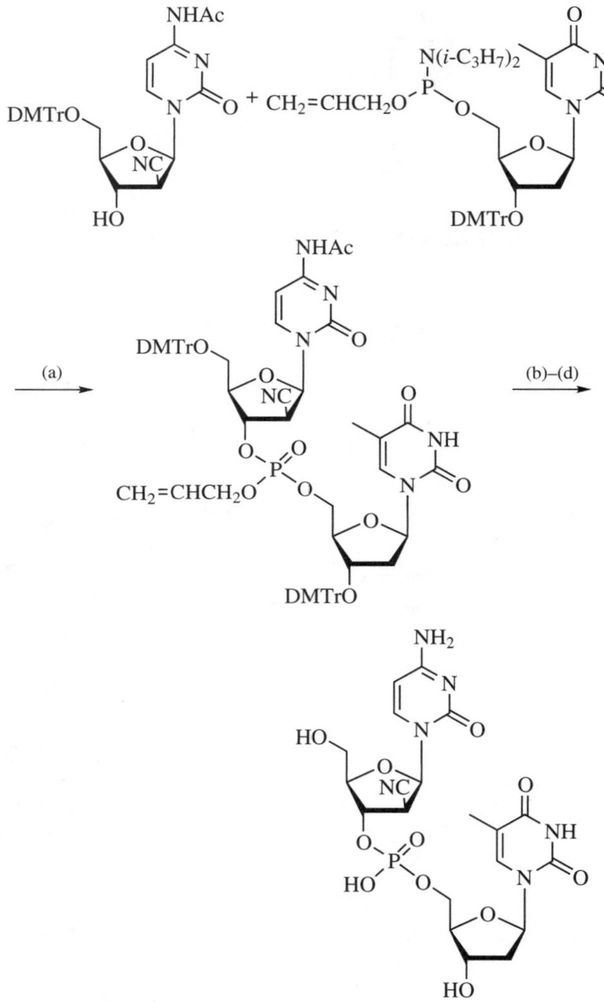

4

(6)

Conditions:
(a) (i) 1*H*-tetrazole, CH_3CN, molecular seives 3 Å, 25 °C;
(ii) bis(trimethylsilyl)peroxide, TMSOTf, CH_3CN, 25 °C
(b) 90% acetic acid, 25 °C, 76%
(c) 1% hydrochloric acid, 25 °C
(d) $Pd_2[(PhCH{=}CH)_2CO]_3$ $CHCl_3$, THF–CH_3OH, 25 °C, quant.

of these artificial nucleotides sensitive to bases is difficult or impossible using the representative existing methods which employ acyl protectors for nucleoside bases and 2-cyanoethyl protector of internucleotide bonds.[21]

Synthesis of Nucleoside/Nucleotide–Amino Acid/Peptide Conjugates. The adenosine 5′-(allyl diisopropylphosphoramidite) obtained as described above reacts with *S*-methyl-L-cysteine sulfoximine with the aid of 1*H*-tetrazole; the subsequent oxidation affords a nucleoside–amino acid conjugate, which serves as a precursor for the synthesis of biologically important *N*-adenylated *S*-methyl-L-cysteine (**5**) (eq 7).[22]

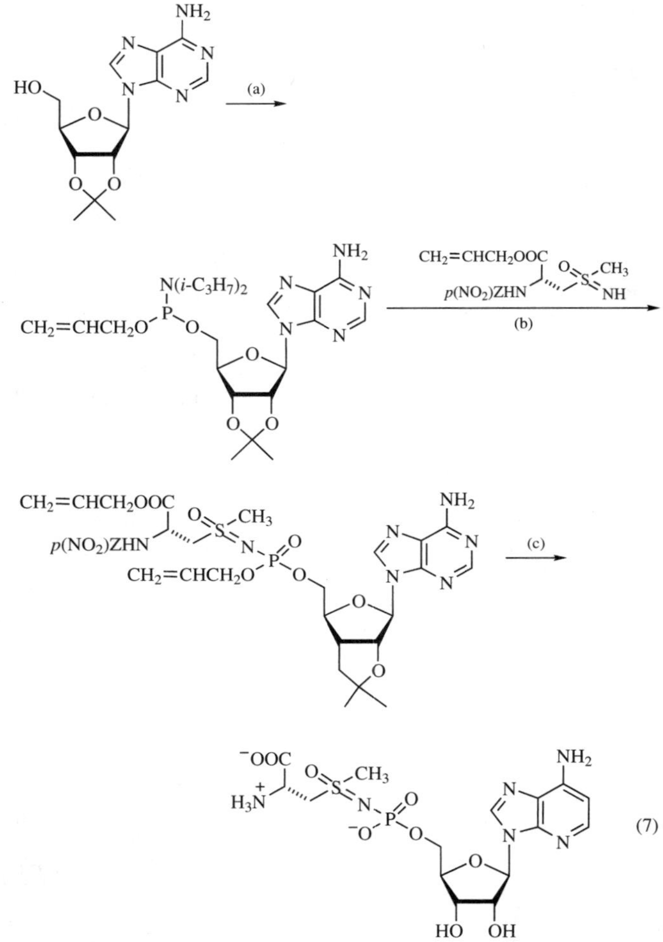

5

(7)

Conditions:
(a) $CH_2{=}CHCH_2OP[N(i\text{-}C_3H_7)_2]_2$, diisopropylammonium tetrazolide, pyridine, 25 °C, 76%
(b) (i) 1*H*-tetrazole, CH_3CN, 25 °C; (ii) TBHP, 56%
(c) (i) TFA–H_2O (8/3), 25 °C; (ii) $Pd(PPh_3)_4$, PPh_3, HCOOH, $(C_2H_5)_3N$, THF, 25 °C;
(iii) H_2, 10% Pd–C, CH_3CN–H_2O, 93%

Allyl tetraisopropylphosphorodiamidite also serves as a building unit for the preparation of a variety of nucleoside/nucleotide–amino acid/peptide conjugates having a base-labile serine- or threonine-phosphate linker between the two components; these are fundamental compounds in a class of naturally occurring biopolymers that play decisive roles in important biological process such as viral replication. Thus, a hydroxyl group of serine or threonine undergoes phosphitylation by treatment with allyl tetraisopropylphosphorodiamidite and 1*H*-tetrazole to give allyl diisopropylphosphoramidite of the corresponding amino acid or peptide. This amino acid phosphoramidite or peptide phosphoramidite subsequently reacts with a nucleoside[5,23,24] or nucleotide[23] to give the target conjugate compound (eq 8).

are important substances acting as a source of sialic acid in the sialyltransferase-mediated biosynthesis of sialyl oligosuccarides.

Synthesis of Other Biologically Attractive Compounds. The reaction of allyl tetraisopropylphosphorodiamidite and *o*-nitrobenzyl alcohol in the presence of diisopropylethylammonium 1*H*-tetrazolide[27] forms allyl *o*-nitrobenzyl diisopropylphos

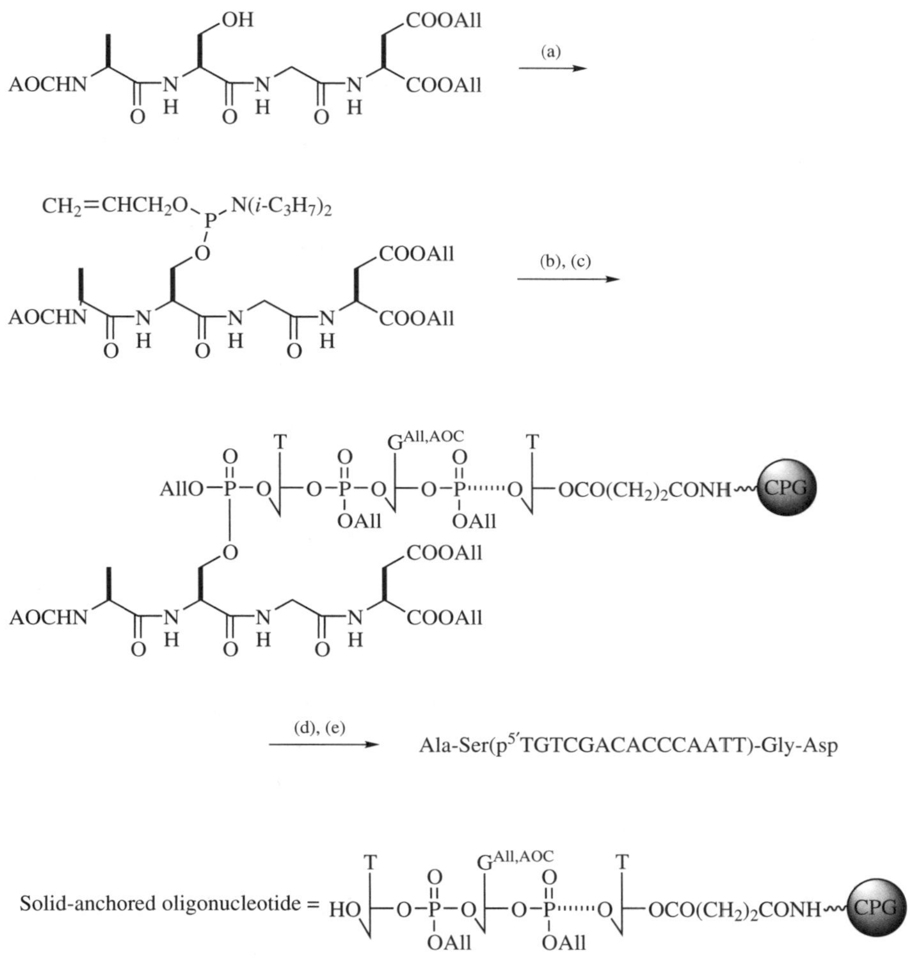

(8)

All = $CH_2{=}CHCH_2$; AOC = $CH_2{=}CHCH_2OCO$

Conditions:
(a) $CH_2{=}CHCH_2OP[N(i\text{-}C_3H_7)_2]_2$, 1*H*-tetrazole, $(i\text{-}C_3H_7)_2NH$, CH_3CN, 0 °C, 53%
(b) Solid-anchored oligonucleotide, benzimidazolium triflate, CH_3CN, 1 min, 25 °C
(c) TBHP, toluene, 0.5 min, 25 °C
(d) PPh_3, $(C_2H_5)_2NH$, HCOOH, $Pd_2[(PhCH{=}CH_2)_2CO]_3{\cdot}CHCl_3$, THF, 60 min, 50 °C
(e) NH_4OH, 60 min, 25 °C.

Synthesis of Nucleoside–Carbohydrate Conjugates. Cytidine 5′-(allyl diisopropylphosphoramidite)s are employed as key building blocks for the synthesis of cytidine-5′-monophosphono-*N*-acetylneuraminic acid (**6**)[25] (eq 9) and its analog,[26] which

phoramidite, which is used as a building unit for construction of *o*-nitrobenzyl phosphate moieties of photolabile P(1,2)- and P(4,5)-(*o*-nitrobenzyl) esters of *myo*-inositol 1,2,3,4,5,6-hexaphosphoric acids (**7**)[6] (eq 10).

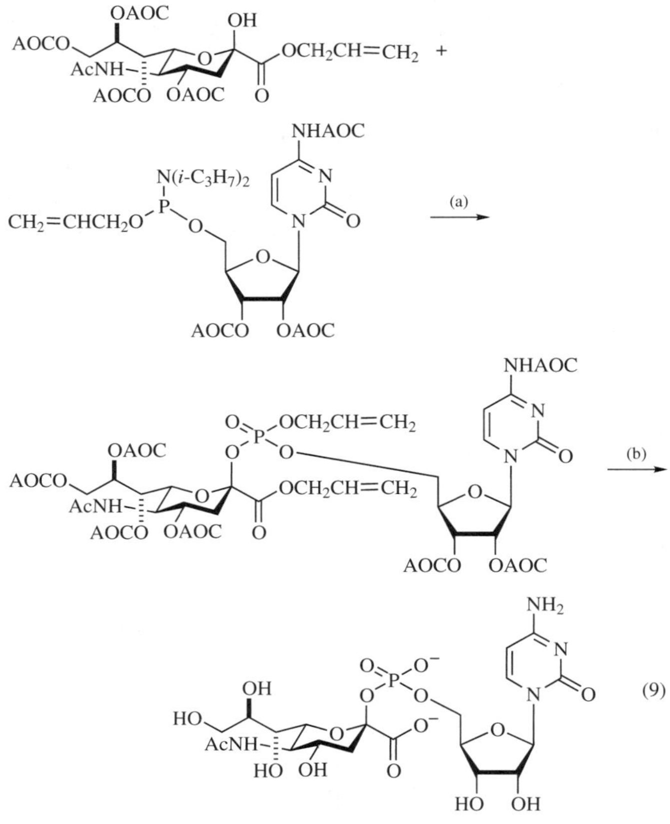

(9)

6

Conditions:
(a) (i) 1*H*-tetrazole, CH_3CN, 0 °C; (ii) $(C_2H_5)_3N$, 80% TBHP, 0–25 °C
(b) (i) $BuNH_2$–HCOOH; (ii) PPh_3, $Pd(PPh_3)_4$, THF, 25%

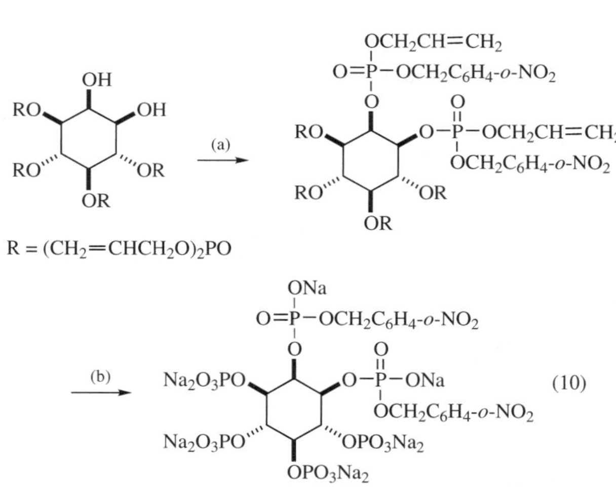

R = $(CH_2{=}CHCH_2O)_2PO$

(10)

7

Conditions:
(a) (i) $(CH_2{=}CHCH_2O)P[N(i\text{-}C_3H_7)_2](OCH_2C_6H_4\text{-}o\text{-}NO_2)$, 1*H*-tetrazole, CH_2Cl_2, 25 °C;
(ii) *m*-CPBA, –40 to 25 °C, 88%
(b) (i) $RhCl(PPh_3)_3$, $(i\text{-}C_3H_7)_2NC_2H_5$, 95% C_2H_5OH, 90 °C;
(ii) ion exchange (Na^+ form) column, 83%

1. Bannwarth, W.; Küng, E., *Tetrahedron Lett.* **1989**, *30*, 4219–4222.
2. Hayakawa, Y.; Wakabayashi, S.; Kato, H.; Noyori, R., *J. Am. Chem. Soc.* **1990**, *112*, 1691–1696.
3. Hayakawa, Y.; Kawai, R.; Hirata, A.; Sugimoto, J.; Kataoka, M.; Sakakura, A.; Hirose, M.; Noyori, R., *J. Am. Chem. Soc.* **2001**, *123*, 8165–8176.
4. (a) Hayakawa, Y.; Uchiyama, M.; Kato, H.; Noyori, R., *Tetrahedron Lett.* **1985**, *26*, 6505–6508. (b) Hayakawa, Y.; Kato, H.; Uchiyama, M.; Kajino, H.; Noyori, R., *J. Org. Chem.* **1986**, *51*, 2400–2402.
5. Flohr, S.; Jungmann, V.; Waldmann, H., *Chem. Eur. J.* **1999**, *5*, 669–681.
6. Chen, J.; Prestwich, G. D., *Tetrahedron Lett.* **1997**, *38*, 969–972.
7. Kuyl-Yeheskiely, E.; Tromp, C. M.; Lefeber, A. W. M.; van der Marel, G. A.; van Boom, J. H., *Tetrahedron* **1988**, *44*, 6515–6523.
8. Manoharan, M.; Lu, Y.; Casper, M. D.; Just, G., *Org. Lett.* **2000**, *2*, 243–246.
9. Bergmann, F.; Kueng, E.; Iaiza, P.; Bannwarth, W., *Tetrahedron* **1995**, *51*, 6971–6976.
10. Hayakawa, Y.; Hirose, M.; Noyori, R., *Nucleosides Nucleotides* **1989**, *8*, 867–870.
11. (a) Bolli, M.; Lubini, P.; Leumann, C., *Helv. Chim. Acta* **1995**, *78*, 2077–2096. (b) Pitsch, S.; Wendeborn, S.; Jaun, B.; Eschenmoser, A., *Helv. Chim. Acta* **1993**, *76*, 2161–2183.
12. Leonard, N. J.; Neelima, *Nucleosides Nucleotides* **1996**, *15*, 1369–1381.
13. Hayakawa, Y.; Hirose, M.; Noyori, R., *Tetrahedron* **1995**, *51*, 9899–9916.
14. Hayakawa, Y.; Nagata, R.; Hirata, A.; Hyodo, M.; Kawai, R., *Tetrahedron* **2003**, *59*, 6465–6471.
15. Bhat, B.; Leonard, N. J., Robinson, H.; Wang, A. H.-J., *J. Am. Chem. Soc.* **1996**, *118*, 10744–10751.
16. Bogdan, F. M.; Chow, C. S., *Tetrahedron Lett.* **1998**, *39*, 1897–1900.
17. (a) Greenberg, M. M.; Barvian, M. R.; Cook, G. P.; Goodman, B. K.; Matray, T. J.; Tronche, C.; Venkatesan, H., *J. Am. Chem. Soc.* **1997**, *119*, 1828–1839. (b) Greenberg, M. M.; Matray, T. J., *Biochemistry* **1997**, *36*, 14071–14079. (c) Matray, T. J.; Greenberg, M. M., *J. Am. Chem. Soc.* **1994**, *116*, 6931–6932.
18. Hayakawa, Y.; Hirose, M.; Hayakawa, M.; Noyori, R., *J. Org. Chem.* **1995**, *60*, 925–930.
19. Spinelli, N.; Meyer, A.; Hayakawa, Y.; Imbach, J.-L.; Vasseur, J.-J., *Eur. J. Org. Chem.* **2002**, 49–56.
20. Hayakawa, Y.; Kawai, R.; Otsuki, K.; Kataoka, M.; Matsuda, A., *Bioorg. Med. Chem. Lett.* **1998**, *8*, 2559–2562.
21. (a) Beaucage, S. L.; Caruthers, M. H., *Tetrahedron Lett.* **1981**, *22*, 1859–1862. (b) Matteucci, M. D.; Caruthers, M. H., *J. Am. Chem. Soc.* **1981**, *103*, 3185–3191. (c) McBride, L. J.; Caruthers, M. H., *Tetrahedron Lett.* **1983**, *24*, 245–248. (d) Sinha, N. D.; Biernat, J. P.; Köster, H., *Tetrahedron Lett.* **1983**, *24*, 5843–5846.
22. Koizumi, M.; Hiratake, J.; Nakatsu, T.; Kato, H.; Oda, J., *J. Am. Chem. Soc.* **1999**, *121*, 5799–5800.
23. (a) Sakakura, A.; Hayakawa, Y.; Harada, H.; Hirose, M.; Noyori, R., *Tetrahedron Lett.* **1999**, *40*, 4359–4362. (b) Sakakura, A.; Hayakawa, Y., *Tetrahedron* **2000**, *56*, 4427–4435.
24. (a) Jeyaraj, D. A.; Waldmann, H., *Tetrahedron Lett.* **2001**, *42*, 835–837. (b) Jeyaraj, D. A.; Prinz, H.; Waldmann, H., *Chem. Eur. J.* **2002**, *8*, 1879–1887.

25. Makino, S.; Ueno, Y.; Ishikawa, M.; Hayakawa, Y.; Hata, T., *Tetrahedron Lett.* **1993**, *34*, 2775–2778.

26. Cohen, S. B.; Halcomb, R. L., *J. Org. Chem.* **2000**, *65*, 6145–6152.

27. (a) Barone, A. D.; Tang, J.-Y.; Caruthers, M. H., *Nucleic Acids Res.* **1984**, *12*, 4051–4061. (b) Hamamoto, S.; Takaku, H., *Chem. Lett.* **1986**, 1401–1404.

Yoshihiro Hayakawa & Rie Kawai
Nagoya University, Nagoya, Japan

Allyl Trichloroacetimidate[1]

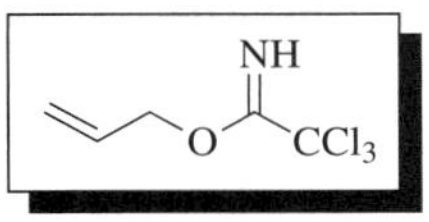

[51479-73-3] $C_5H_6Cl_3NO$ (MW 202.47)

(delivers allyl group under mild conditions;[2] used in the preparation of N-containing intermediates[1])

Physical Data: bp 74 °C/11 mmHg.
Solubility: sol hexane, ether, dichloromethane.
Preparative Methods: typical procedure:[2] Sodium Hydride (0.5 g, 21 mmol) is slurried in anhydrous ether (20 mL) under a nitrogen blanket. Allyl alcohol (210 mmol) is introduced in ether (30 mL) dropwise with stirring to the slurry. After 20 min, the homogeneous solution is cooled to 0 °C and Trichloroacetonitrile (20 mL, 200 mmol) is added over 15 min. The mixture is allowed to warm to 20 °C over 60 min and then concentrated to a syrup. Pentane (20 mL) containing methanol (0.8 mL, 21 mmol) is added followed by vigorous shaking, filtration, and concentration of the filtrate and pentane washings (2 × 20 mL). The product imidate is obtained as a clear liquid (90–97%). No further purification is required. Alternatively, the imidate can be prepared by using sodium methoxide as a catalyst instead of sodium hydride. Other procedures for the preparation of allyl trichloroacetimidate have been reported.[3,4]
Handling, Storage, and Precautions: allyl trichloroacetimidate can be stored in solutions of hexane for up to 2 months at 0 °C. Preparation of allyl trichloroacetimidate should be performed in a well ventilated hood to avoid exposure to trichloroacetonitrile vapors.

O-Allylation and Protecting Group Delivery. Protecting groups have become an important part of natural and unnatural product chemistry. In oligosaccharide synthesis, they are vital in obtaining the desired compounds because of the various sites at which reaction can occur if hydroxyl groups, for example, are left unprotected. Allyl groups are one of the many substituents used as a protecting group. This group can be delivered to a free hydroxyl group via a trichloroacetimidate derivative. More importantly, the use of allyl trichloroacetimidate allows the protection of an alcohol to take place in the presence of an ester. This is due, in part, to the fact that this reagent is applied under mildly acidic conditions. For example, a rhamnoside is dissolved in dichloromethane and to it is added allyl trichloroacetimidate in cyclohexane followed by a catalytic amount of triflic acid. After 18 h at 20 °C the allylated product is isolated after neutralization with $NaHCO_3$ and chromatography (eq 1).[2,5]

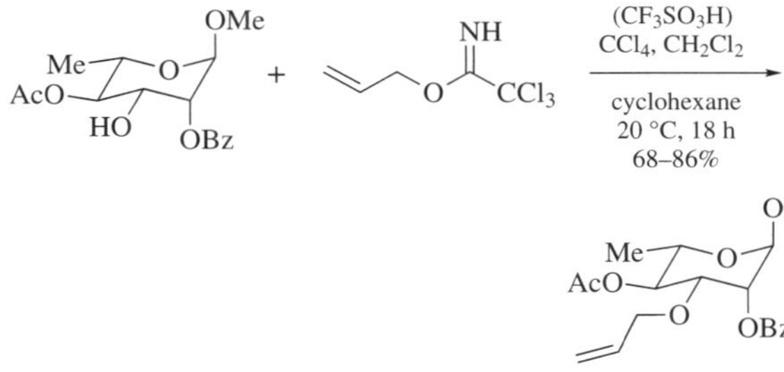

(1)

Substituted tetrahydrofurans have been prepared by using allyl trichloroacetimidate to generate allyl ether intermediates. For example, the hydroxy ester prepared from Methyl Bromoacetate and isobutyraldehyde reacts with allyl trichloroacetimidate and a catalytic amount of Trifluoromethanesulfonic Acid to give the allylic ether ester in 80% yield (eq 2).[6] The resulting tetrahydrofuran-3-one is prepared in four steps with high stereoselection, which is attributed to the use of Copper(II) Acetylacetonate as the cyclization catalyst.

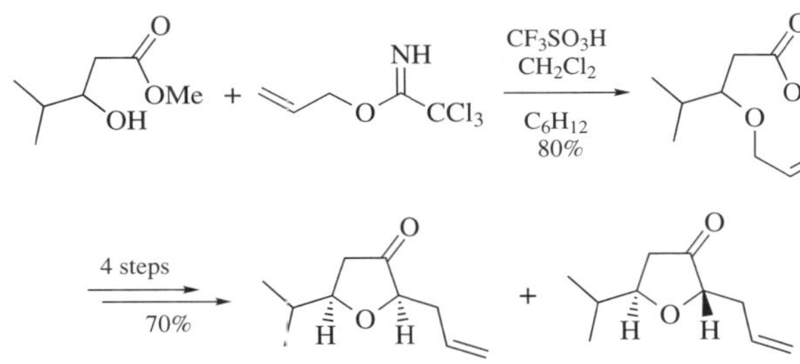

(2)

7:93

α-Amino acids can be prepared from allylic alcohols using allylic trichloroacetimidates. Thermal 1,3-transposition of the imidate group of allyl trichloroacetimidate is accomplished in refluxing xylene. Oxidation and acidic hydrolysis afford the corresponding α-amino acid (eq 3).[7] Other acids prepared by this methodology include imidates derived from benzyl alcohol, 4,4-dimethyl-2-pentenol, and 2-butenol.

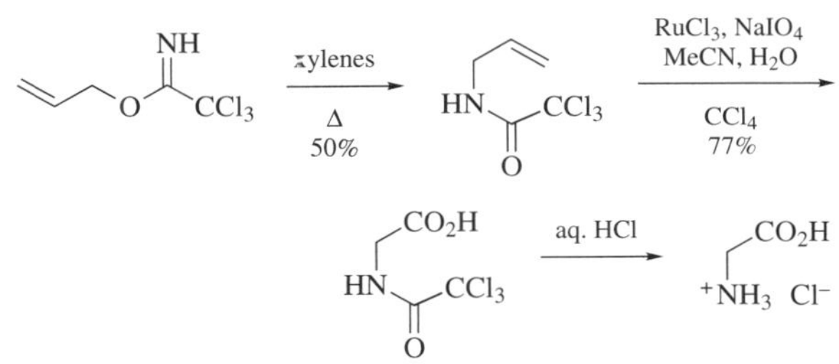

(3)

Allyl trichloroacetimidate has also been used to deliver an acetaldehyde equivalent in the preparation of β-hydroxy-α-amino acid derivatives. An *anti*-Phe-Leu mimetic was prepared in four steps from the *O*-allyl ether of *N*-phthaloyl-L-phenylalanyl-L-serine. This allyl ether was prepared from allyl trichloroacetimidate and the β-hydroxy-α-amino acid in 60–74% yield (eq 4).[8]

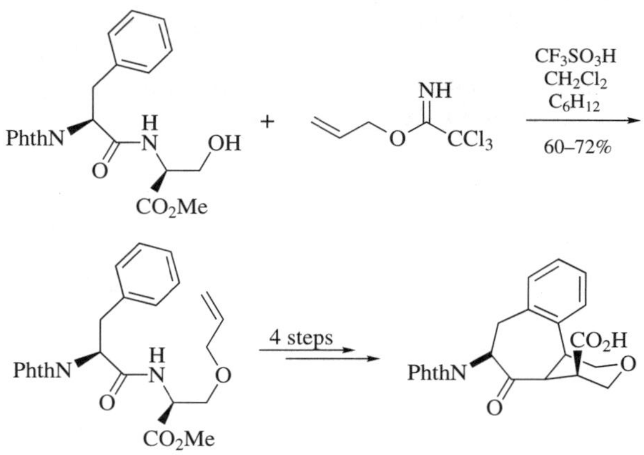

(4)

Benzyl groups can be used to protect alcohols in place of the allyl group with comparable yields. For example, under similar reaction conditions, an isopropylidene protected pyranoside was treated with benzyl trichloroacetimidate at room temperature to give a benzyl ether in 82% yield (eq 5).[1,9]

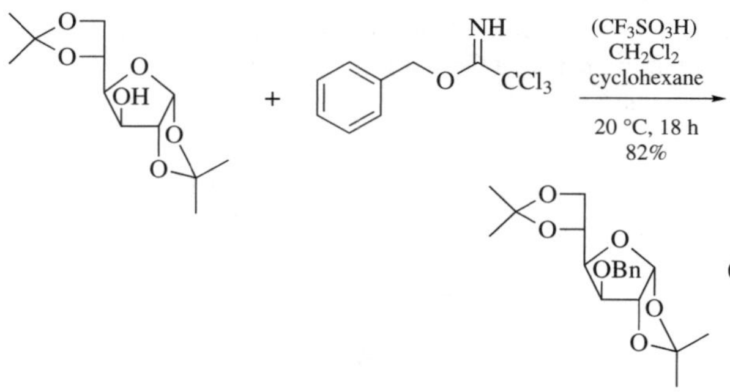

(5)

N-Containing Compounds from Imidates. The synthetic procedure to make allyl trichloroacetimidate has been useful for other allylic imidate systems.[1] For example, geraniol trichloroimidate is prepared by reacting geraniol with sodium hydride in ether. The cooled (−10 °C) solution is treated with Trichloroacetonitrile followed by work-up with a methanolic–pentane solution to give the corresponding imidate. Amines and amides are prepared from these intermediates. Thermal re arrangement of the imidate, followed by hydrolysis, gives rise to 1,3-transposition of hydroxyl and amino groups. The amine is produced in 61% yield (eq 6).[4,10]

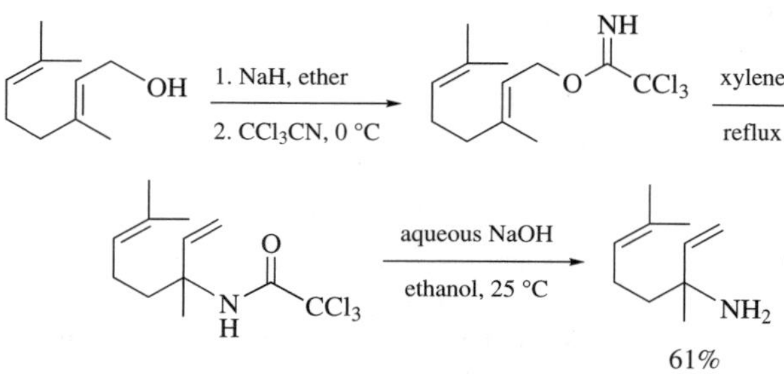

(6)

Related imidates can undergo [3,3]-sigmatropic rearrangements by using a palladium(II) catalyst instead of heat. For example, an allyl trichloroacetimidate rearranges to the corresponding amide when catalyzed by Bis(benzonitrile)dichloropalladium(II) in THF (eq 7).[11] The reaction goes in 8% yield with complete chirality transfer.

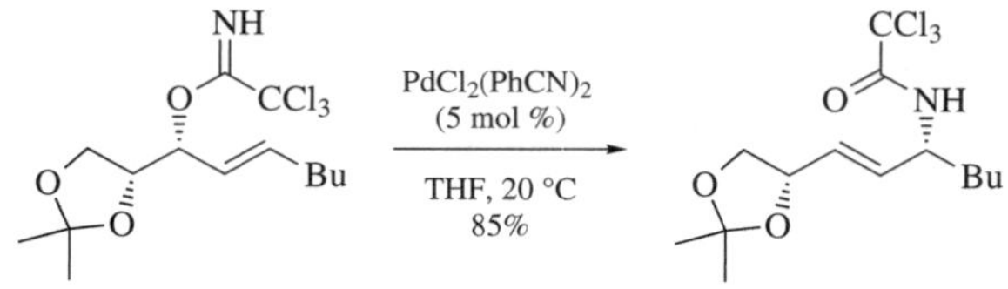

(7)

Allylic trichloroacetimidates undergo iodocyclization to give the corresponding 4,5-dihydro-1,3-oxazines. The regioselection of the reaction depends on the alkene geometry of the starting allyl imidate (eq 8).[12]

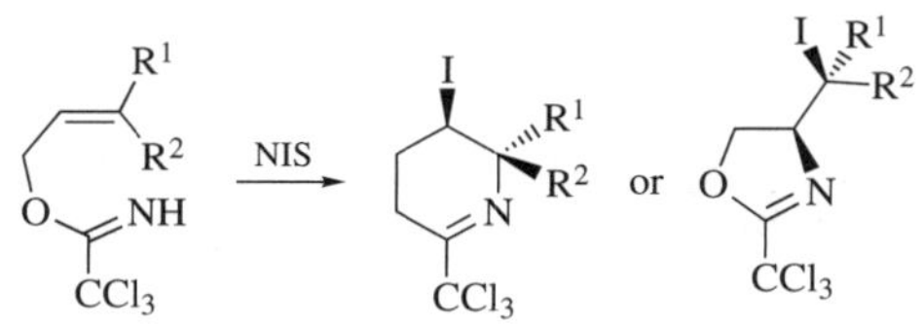

(8)

$R^1 = Et, R^2 = H$	90%	–
$R^1 = H, R^2 = Et$	–	88%
$R^1 = Pr, R^2 = H$	87%	–
$R^1 = H, R^2 = C_{15}H_{31}$	–	90%
$R^1 = H, R^2 = Ph$	–	92%

Imidates have been used in the synthesis of other organic systems, such as coupling reactions in carbohydrate synthesis[13] and Claisen-type rearrangements (see also Trichloroacetonitrile).[14]

1. For a general review of allylic imidic esters in organic synthesis see Overman, L. E. *Acc. Chem. Res.* **1980**, *13*, 218.
2. Wessel, H.-P.; Iversen, T.; Bundle, D. R. *J. Chem. Soc., Perkin Trans. 1* **1985**, 2247.
3. (a) Cramer, F.; Pawelzik, K.; Baldauf, H. J. *Ber. Dtsch. Chem. Ges./Chem. Ber.* **1985**, *91*, 1049. (b) Cramer, F.; Hennrich, N. *Ber. Dtsch. Chem. Ges./Chem. Ber.* **1961**, *94*, 976.
4. (a) Overman, L. E. *J. Am. Chem. Soc.* **1974**, *96*, 597. (b) Overman, L. E. *J. Am. Chem. Soc.* **1976**, *98*, 2901.
5. Wessel, H.-P.; Bundle, D. R. *J. Chem. Soc., Perkin Trans. 1* **1985**, 2251.
6. Clark, J. S. *Tetrahedron* **1992**, *33*, 6193.
7. Takano, S.; Akiyama, M.; Ogasawara, K. *Chem. Commun./J. Chem. Soc., Chem. Commun.* **1984**, 770.
8. Burkholder, T. P.; Le, T.-B.; Giroux, E. L.; Flynn, G. A. *Bioorg. Med. Chem. Lett.* **1992**, *2*, 579.
9. Iversen, T.; Bundle, D. R. *Chem. Commun./J. Chem. Soc., Chem. Commun.* **1981**, 1240.
10. Clizbe, L. A.; Overman, L. E. *Org. Synth.* **1978**, *58*, 4.
11. Metz, P.; Mues, C.; Schoop, A. *Tetrahedron* **1992**, *48*, 1071.
12. Bongini, A.; Cardillo, G.; Orena, M.; Sandri, S.; Tomasini, C. *J. Org. Chem.* **1986**, *51*, 4905.
13. (a) Schmidt, R. R. *Pure Appl. Chem.* **1989**, *61*, 1257. (b) Schmidt, R. R.; Michel, J. *Angew. Chem. Int. Ed. Engl.* **1982**, *21*, 72. (c) Urban, F. J.; Moore, B. S.; Breitenbach, R. *Tetrahedron* **1990**, *31*, 4421. (d) Paulsen, H. *Angew. Chem. Int. Ed. Engl.* **1982**, *21*, 155.

14. (a) Cramer, F.; Baldauf, H.-J. *Ber. Dtsch. Chem. Ges./Chem. Ber.* **1959**, *92*, 370. (b) Mumm, O.; Möller, F. *Ber. Dtsch. Chem. Ges./Chem. Ber.* **1937**, *70*, 2214. (c) Lauer, W. M.; Lockwood, R. G. *J. Am. Chem. Soc.* **1954**, *76*, 3974. (d) Lauer, W. M.; Benton, C. S. *J. Org. Chem.* **1959**, *24*, 804. (e) Roberts, R. M.; Hussein, F. A. *J. Am. Chem. Soc.* **1960**, *82*, 1950. (f) Black, D. St.C.; Eastwood, F. W.; Okraglik, R.; Poynton, A. J.; Wade, A. M.; Welker, C. H. *Aust. J. Chem.* **1972**, *25*, 1483.

Mark R. Sivik
Procter & Gamble, Cincinnati, OH, USA

Anthracenesulfonamide

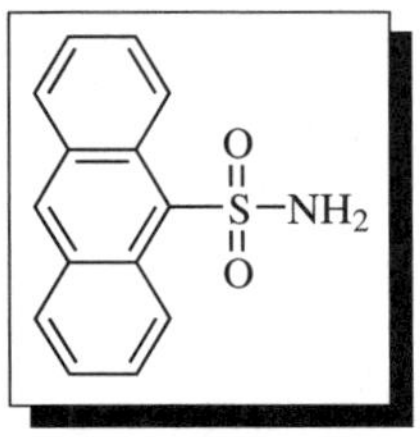

[155495-93-5] $C_{14}H_{11}NO_2S$ (MW 257.31)

(readily cleavable reagent for the synthesis of β-amino acids and for the iodosulfonamidation of glycals)

Physical Data: mp 200.5–202.5 °C.
Form Supplied in: yellow solid; synthetically available.
Analysis of Reagent Purity: IR (KBr): $\nu = 3409, 3300, 1312, 1144$ cm^{-1}; ^1H NMR [250 MHz, $(CD_3)_2CO$]: $\delta = 7.0$ (br s, 2H), 7.56–7.72 (m, 4H) 8.19 (‘d’, $J = 9$ Hz, 2H), 8.91 (s, 1H), 9.37 (‘d’, $J = 9$ Hz, 2H).
Preparative Methods: synthesized from anthracene by sulfonation with chlorosulfonic acid followed by chlorination with $POCl_3$, then treatment of the crude anthracenesulfonyl chloride with aq NH_3 (eq 1).[1]

(1)

Synthesis of Protected β-Amino Acids. The synthesis of protected β-amino acids can be accomplished by treatment of an imine, generated by treatment of an aldehyde with anthracenesulfonamide in the presence of either $TiCl_4$ or Amberlyst 15 (eq 2), with $BrZnCH_2CO_2{}^tBu$ in a Reformatsky-like reaction (eq 3).[1]

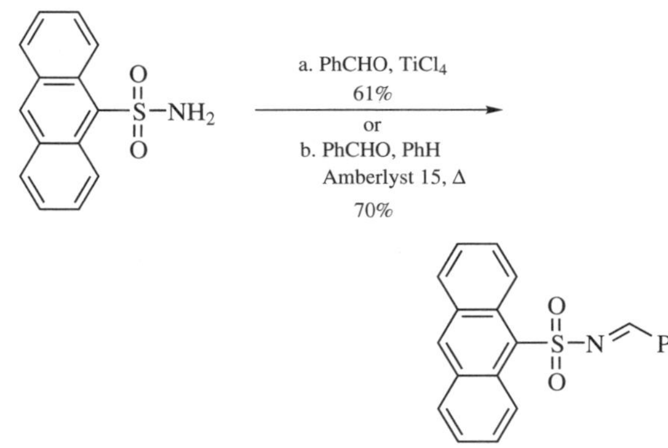
(2)

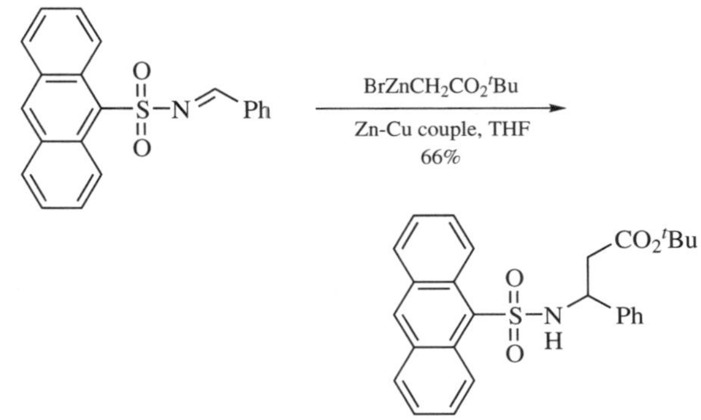
(3)

The advantage of using anthracenesulfonamide lies in the more facile reductive cleavage of this group compared to other more traditional *N*-sulfonyl protecting groups, such as toluenesulfonamide or 4-methoxycarbonylbenzenesulfonamide (4-MCBS) (eq 4).[1,2]

Optically active β-amino acids have been produced by reaction of the imine generated from anthracenesulfonamide and benzaldehyde using lithium and titanium enolates of *N*-acyloxazolidinones with limited success (eq 5).[3] In this case, however, better results were obtained with the less desirable *N*-tosyl imine (eq 6).

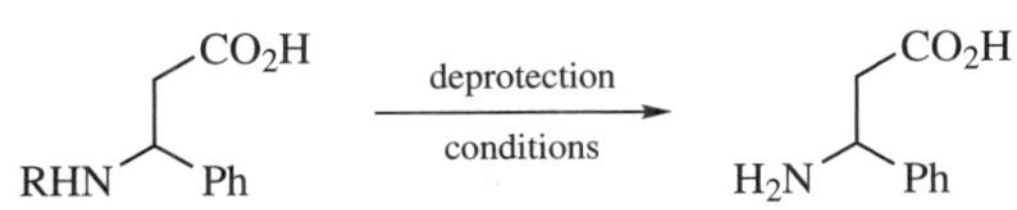
(4)

R	Conditions	Yield (%)
Tosyl	TFA then Na/naphth	11
4-MCBS	TFA then Na/naphth	14
Anthryl	Al/Hg then 1M HCl	63

Iodosulfonamidation of Glycals. Anthracenesulfonamide has been used extensively in the iodosulfonamidation of glycals.[4–6] This method involves the *trans*-diaxial addition of *N*-iodoanthracenesulfonamide to a glycal (eq 7). The so-formed iodosulfonamide can then be converted to a 2-α-anthracenglycosamide derivative by the addition of a nucleophile under the appropriate conditions (eq 8).[7,8]

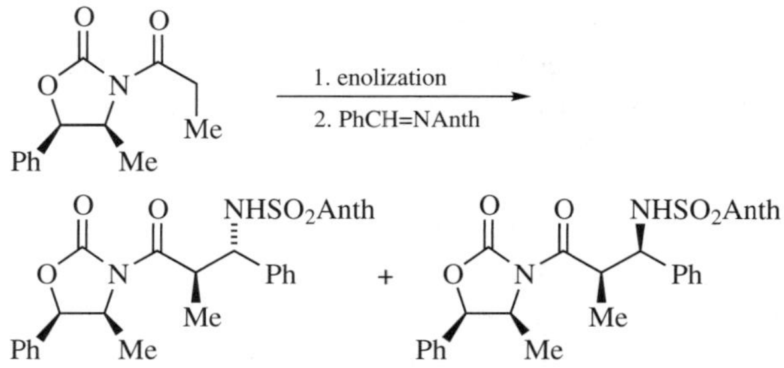

(5)

Enolization conditions	Result
LDA, THF	39% Combined yield (unseparable)
$TiCl_4$, EtN^iPr_2	No reaction

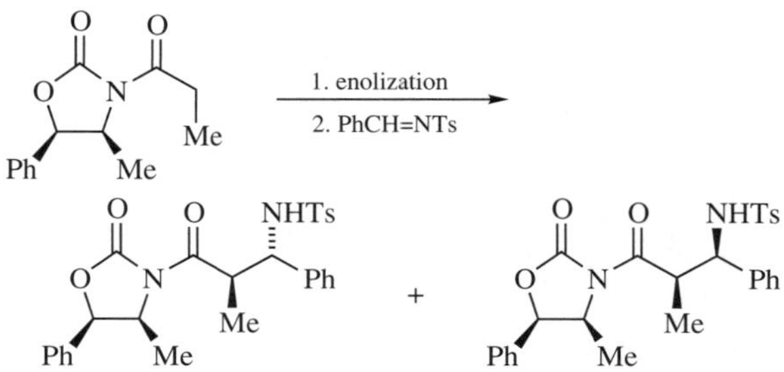

(6)

Enolization conditions	Result
$TiCl_4$, EtN^iPr_2	66% *anti*, 12.5% *syn*

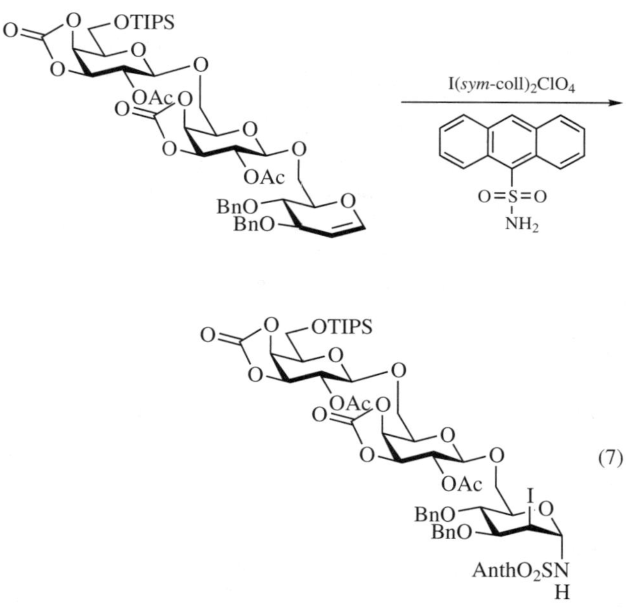

(7)

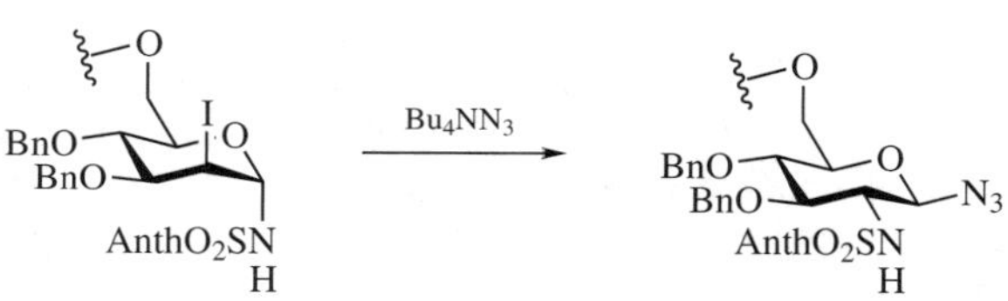

(8)

The principle advantage to using anthracenesulfonamide, rather than the more widely employed benzenesulfonamide, is that the nitrogen-sulfur linkage can be cleaved under mild conditions, e.g., by thiophenol or 1,3-propanedithiol and Hünig's base. Such mild conditions render this reagent more amenable to solid phase synthesis (eq 9).[9–11] In addition, anthracenesulfonamide is more soluble than benzenesulfonamide in THF, a good swelling solvent for the polymer, resulting in a more efficient and complete reaction.

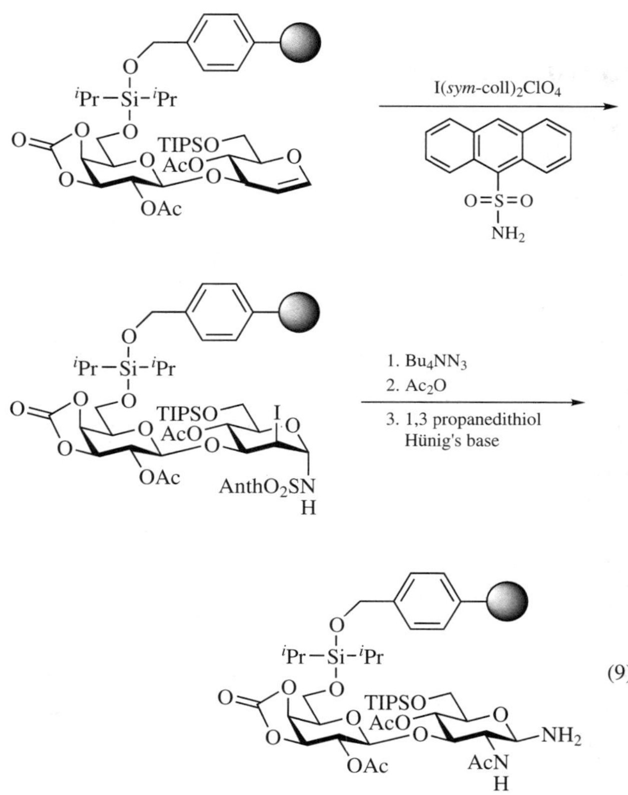

(9)

Related Reagents. Benzenesulfonamide; toluenesulfonamide; mesitylsulfonamide.

1. Robinson, A. J.; Wyatt, P. B., *Tetrahedron* **1993**, *49*, 11329–11340.
2. Arzeno, H. B.; Kemp, D. S., *Synthesis* **1988**, 32–36.
3. Abrahams, I.; Motevalli, M.; Robinson, A. J.; Wyatt, P. B., *Tetrahedron* **1994**, *50*, 12755–12772.
4. Danishefsky, S. J.; Bilodeau, M. T., *Angew. Chem. Int. Ed. Engl.* **1996**, *35*, 1381–1419.
5. Arsequell, G.; Valencia, G., *Tetrahedron: Asymmetry* **1999**, *10*, 3045–3094.
6. Herzner, H.; Reipen, T.; Schultz, M.; Kunz, H., *Chem. Rev.* **2000**, *100*, 4495–4537.
7. Danishefsky, S. J.; Roberge, J. Y., *Pure & Appl. Chem.* **1995**, *67*, 1647–1662.
8. Danishefsky, S. J.; Hu, S.; Cirillo, P. F.; Eckhardt, M.; Seeberger, P. H., *Chem. Eur. J.* **1997**, *3*, 1617–1628.
9. Roberge, J. Y.; Beebe, X.; Danishefsky, S. J., *Science* **1995**, *269*, 202–204.

10. Roberge, J. Y.; Beebe, X.; Danishefsky, S. J., *J. Am. Chem. Soc.* **1998**, *120*, 3915–3927.

11. Savin, K. A.; Woo, J. C. G.; Danishefsky, S. J., *J. Org. Chem.* **1999**, *64*, 4183–4186.

J. David Warren
Sloan-Kettering Institute for Cancer Research, New York NY, USA

O-(7-Azabenzotriazol-1-yl)-1,3-dimethyl-1,3-dimethyleneuronium Hexafluorophosphate

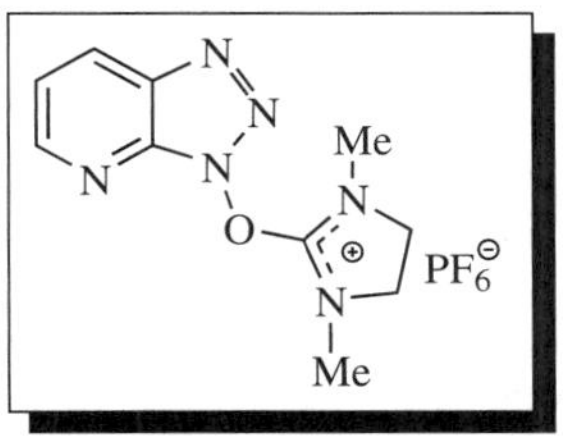

[154561-27-0] $C_{10}H_{13}F_6N_6OP$ (MW 378.214)

(coupling reagent for peptide synthesis)

Alternate Name: HAMDU.
Physical Data: mp: sample decomposed above 185 °C.
Solubility: soluble in DMF, MeCN, and NMP.
Form Supplied in: white solid, commercially available.
Purification: can be purified by crystallization from a mixture of MeCN and CH_2Cl_2 or MeCN and ether.
Handling, Storage, and Precautions: HAMDU is not hygroscopic and can be stored indefinitely at 0 °C. A solution of HAMDU in DMF is very sensitive to moisture and cannot be stored in DMF,[1] it is therefore less stable than its analog HAMTU. The synthesis of peptides should be carried out with a freshly prepared solution. CAUTION: Violent decomposition can occur when dried at elevated temperature.

General Consideration. HAMDU and its analog **O-(7-Azabenzotriazol-1-yl)-1,3-dimethyl-1,3-trimethyleneuronium hexafluorophosphate** (HAMTU), have been used as coupling reagents for the preparation of peptides by both solution- and solid-phase techniques.[1–6] In solution, the reaction of α-amino protected amino acids or dipeptides with equimolar amounts of α-carboxyl protected amino acid hydrochlorides or amino amides, and a base (2 equiv) such as **Diisopropylethylamine** (DIEA), *N*-methylmorpholine (NMM), **2,4,6-Collidine** (TMP) or **1,8-Bis(dimethylamino)naphthalene** (proton sponge, PS) in DMF proceeds cleanly, providing the corresponding peptides in good yields. The level of epimerization in HAMDU-mediated couplings was reduced compared to O-**(Benzotriazol-1-yl)-1,3-dimethyl-1,3-dimethyleneuronium hexafluorophosphate** (HBMDU)[2] - promoted acylation (eq 1) (Table 1).

$$RCOOH + H_2NR' \xrightarrow[\text{Base}]{\text{HAMDU}} RCONHR' \quad (1)$$

The results shown in Table 1 clearly indicated that (i) HAMDU is superior to HBMDU (entries 1 vs. 2, 3 vs. 4, 5 vs. 6, 7 vs. 8, 9 vs. 10, 11 vs. 12, and 13 vs. 14); (ii) the use of collidine considerably reduces the level of epimerization (entries 1 vs. 5, 7 vs. 9 and 11 vs. 13).

HAMDU activation has been adapted for automated stepwise solid-phase peptide synthesis for both fluorenylmethyloxycarbonyl (Fmoc) and *t*-butyloxycarbonyl (Boc) protective group strategies. For the former, the fragment 65–74 (H-Val-Gln-Ala-Ala-Ile-Asp-Tyr-Ile-Asn-Gly-NH_2) of the acyl carrier protein was chosen as a model to compare the performance of HAMDU, HAMTU, and HBMDU. Results from these studies indicated that uronium salt coupling is superior to carbodiimide and pentafluorophenyl ester-mediated methods, and that the aza derivatives are more efficient than HBMDU.[3,5]

To compare the relative stability of Xt- derivatives, Table 2 shows that aza derivatives are less stable than the benzotriazole analogous (entry 1 vs. 2, and 3 vs. 4). Furthermore, the nature of the carbon skeleton is of marked importance in connection with the stability of a compound. Both HAMDU and HBMDU are very unstable, while the salts from dimethylamino derivatives (HATU and HBTU) are the most stable, and the pyrrolidino derivative

Table 1 Racemization during HAMDU/HBMDU coupling in solution and solid phase

Entry	RCO_2H	$R'NH_2$ or $R'NH_2\cdot HCl$	Conditions	DL-isomer (%)
1	Z-Phe-Val-OH	H-Ala-OMe·HCl	HAMDU/DIEA (3 equiv), 6 h	2.2
2	Z-Phe-Val-OH	H-Ala-OMe·HCl	HBMDU/DIEA (3 equiv), 6 h	14.9
3	Z-Phe-Val-OH	H-Ala-OMe·HCl	HAMDU/PS (3 equiv), 6 h	0.4
4	Z-Phe-Val-OH	H-Ala-OMe·HCl	HBMDU/PS (3 equiv), 6 h	1.3
5	Z-Phe-Val-OH	H-Ala-OMe·HCl	HAMDU/TMP (3 equiv), 6 h	0.1
6	Z-Phe-Val-OH	H-Ala-OMe·HCl	HBMDU/TMP (3 equiv), 6 h	1.0
7	Z-Phe-Val-OH	H-Pro-O*t*Bu·HCl	HAMDU/DIEA (3 equiv), 8 h	7.8
8	Z-Phe-Val-OH	H-Pro-O*t*Bu·HCl	HBMDU/DIEA (3 equiv), 8 h	25.1
9	Z-Phe-Val-OH	H-Pro-O*t*Bu·HCl	HAMDU/TMP (3 equiv), 8 h	2.3
10	Z-Phe-Val-OH	H-Pro-O*t*Bu·HCl	HBMDU/TMP (3 equiv), 8 h	11.9
11	Z-Gly-Gly-Val-OH	H-Ala-Gly-Gly-OMe·HCl	HAMDU/DIEA (3 equiv), 8 h	8.7
12	Z-Gly-Gly-Val-OH	H-Ala-Gly-Gly-OMe·HCl	HBMDU/DIEA (3 equiv), 8 h	16.8
13	Z-Gly-Gly-Val-OH	H-Ala-Gly-Gly-OMe·HCl	HAMDU/TMP (3 equiv), 8 h	2.3
14	Z-Gly-Gly-Val-OH	H-Ala-Gly-Gly-OMe·HCl	HBMDU/TMP (3 equiv), 8 h	7.9

(HAPyU) is of intermediate stability, while HAMTU is considered slightly more stable than HAMDU.[1]

Table 2 Hydrolytic stability of Xt derivatives[a] in DMF

		Stability (%) open vial			Stability (%) closed vial
Entry	Reagent	5 h	20 h	42 h	48 h
1	HATU	>99	95	78	>99
2	HBTU	100	98	88	100
3	HAMDU	0	–	–	0
4	HBMDU	0	–	–	0
5	HAMTU	71	31	0	52
6	HBMTU	96	79	22	93

[a]X = OA or OB.

These results should be considered mainly for those syntheses carried out in automated synthesizers where the coupling reagents are stored in open vessels.

For comparisons of the activation step of carboxyl groups, Fmoc-diethylglycine (Fmoc-Deg-OH) was chosen due to its inherent steric hindrance. The results in Table 3 are in agreement with the previous observations.

Table 3 Activation of Fmoc-Deg-OH

Entry	Activation method	Fmoc-Deg-OXt[a]		
		2 min	1 h	2 h
1	DCC-HOAt	60	83	89
2	HBTU-DIEA	70	95	96
3	HATU-DIEA	88	90	90
4	HAPyU-DIEA	89	92	92
5	HAMDU-DIEA	60	62	–

[a]X = A or B.

Examination of the activation of Fmoc-Deg-OH after 2 min indicates that the aza derivatives are more reactive than their corresponding benzotriazole analogues (entry 2 vs. 3). Furthermore, formation of the active species competes with hydrolysis of the coupling reagent. Thus, the most reactive uronium salt, HAMDU, gives poorer activation when compared with the less reactive compounds HATU and HAPyU, due to the fact that after a few seconds no activating reagent is present in the medium.[7]

In the absence of the carboxylic component, uronium salts such as HAMDU as well as HBMDU react with amino groups leading to the formation of guanidinium derivatives (eq 2).[8] In syntheses conducted in solution, an excess of either uronium salt or amino component should be avoided. In both solution- and solid-phase strategies, the order of addition of reagents is critical; HAMDU should be delivered to the carboxylic component for the preactivation step prior to addition of the amine.[8c]

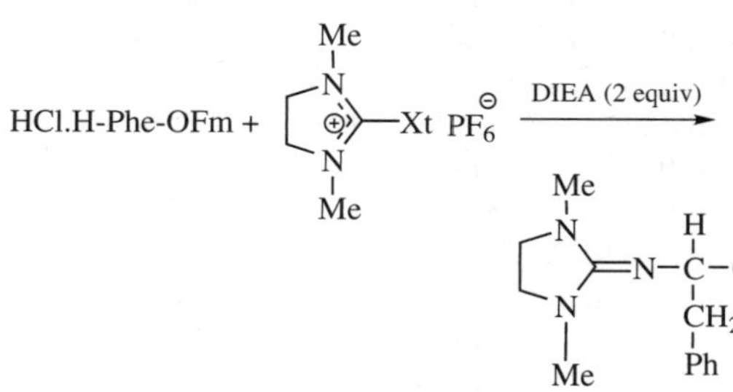

(2)

X = OAt or OBt

Related Regents. Recently, other uronium salts, such as bis(tetramethylene)pentafluorophenoxyformamidinium hexafluorophosphate[9] (HPyOPfp); 5-(3′,4′-dihydro-4′-oxo-1′,2′,3′-benzotriazin-3′-yloxy)-3,4-dihydro-1-methyl-2*H*-pyrrolium hexachloroantimonate (DOMP); 5-pentafluorophenyl-3,4-dihydro-1-methyl-2*H*-pyrrolium hexafluoroantimonate (FOMP); 5-(1*H*-benzotriazol-1-yloxy)-3,4-dihydro-1-methyl-2*H*-pyrrolium hexachloroantimonate (BDMP); *N*-(1*H*-benzotriazol-1-ylmethylene)-*N*-methylmethanaminium hexachloroantimonate (BOMI);[10] and 2-(benzotriazol-1-yloxy)-1,3-dimethyl-2-pyrrolidin-1-yl-1,2,3-diazaphospholidinium hexafluorophosphate (BOMP)[10] have been described. The uronium salts derived from 7-aza-1-hydroxybenzotriazole have been shown to be superior to their benzotriazole analogs, with regard to coupling efficiency[9] and racemization,[8] in both solution and solid-phase strategies.[11]

1. Albericio, F.; Bofill, J. M.; El-Faham, A.; Kates, S. A., *J. Org. Chem.* **1998**, *63*, 9678.
2. Carpino, L. A.; El-Faham, A., *J. Org. Chem.* **1994**, *59*, 695.
3. Carpino, L. A.; El-Faham, A.; Minor, C. A.; Albericio, F., *J. Chem. Soc., Chem. Commun.* **1994**, 201.
4. Carpino, L. A.; El-Faham, A.; Albericio, F., *Tetrahedron Lett.* **1994**, *35*, 2279.
5. Carpino, L. A.; El-Faham, A.; Truran, G.; Minor, C. A.; Kates, S. A.; Griffin, W.; Shroff, H.; Triolo, S. A.; Albericio, F., In *Innovation and Perspective in Solid Phase Synthesis: Peptides, Proteins, and Nucleic Acids*; Epton, R., Ed.; Mayflower Worldwide Limited: Brmingham, 1994, pp 95–104.
6. Kates, S. A.; Diekmann, E.; El-Faham, A.; Herman, L. W.; Ionescu, D.; McGuinnes, B. F.; Triolo, S. A.; Albericio, F.; Carpino, L. A., *Techniques in Protein Chemistry VII*, Marchak, D. R., Ed.; Academic Press: New York, 1996, pp 515–523.
7. Humphrey, J. M.; Chamberlin, A. R., *Chem. Rev.* **1997**, *97*, 2243.
8. (a) Kates, S. A.; Minor, C. A.; Shroff, H.; Haaseth, R. C.; Triolo, S. A.; El-Faham, A.; Carpino, L. A.; Albericio, F., In *Peptides 1994: Proceedings of the Twenty-Third European Peptide Symposium*; Maia, H. L. S., Ed.; ESCOM: Leiden, 1995; pp 248–249. (b) Gausepohl, H.; Pieles, U.; Frank, R. W., In *Proceedings of the 12th American Peptide Symposium*, Smith, J. A.; Rivier, J. E., Eds; ESCOM: Leiden, 1992, pp 523–524. (c) del Frenso, M.; El-Faham, A.; Carpino, L. A.; Royo, M.; Albericio, F., *Org. Lett.* **2000**, *2*, 3539.
9. (a) Habermann, J.; Kunz, H., *Tetrahedron Lett.* **1998**, *39*, 265. (b) El-Faham, A., *Lett. Pept. Sci.* **2000**, *7*, 113.
10. Li, P.; Xu J. C., *Tetrahedron* **2000**, *56*, 4437.
11. Wada, T.; Sato, Y.; Honda, F.; Kawahara, S. I.; Sekine, M., *J. Am. Chem. Soc.* **1997**, *119*, 12710.

Ayman El-Faham
Alexandria University, Egypt

(7-Azabenzotriazol-1-yloxy)tris (pyrrolidino)phosphonium Hexafluorophosphate

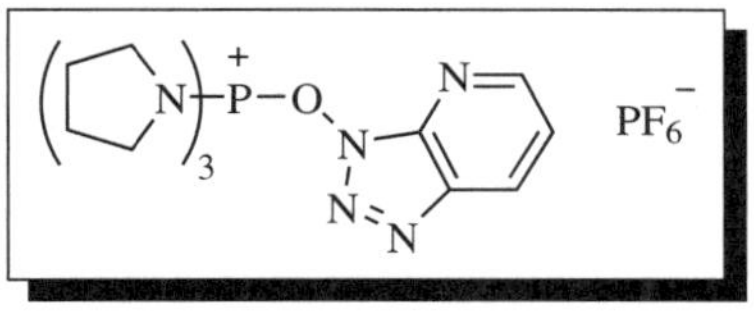

[156311-83-0] $C_{17}H_{27}F_6N_7OP_2$ (MW 521)

(reagent for carboxylic acid activation, particularly used for peptide synthesis)

Alternate Name: PyAOP; tri-Pyrrolidinyl[3*H*-1,2,3-triazolo (4,5-b)pyridin-3-yloxy]phosphonium hexafluorophosphate.
Physical Data: mp 172–174 °C.[1]
Solubility: soluble in CH_2Cl_2, $CHCl_3$, DMF, DMSO, NMP, THF, CH_3CN, and acetone.
Form Supplied in: colorless crystals.
Analysis of Reagent Purity: ^{13}C NMR; ^1H NMR; FAB-MS.
Preparative Methods: commercially available. Prepared by reaction of tripyrrolidinophosphine with CCl_4, or trispyrrolidinophosphine oxide with $POCl_3$, then with KPF_6/H_2O,[2] and reaction of $HOAt/NEt_3$ (1-hydroxy-7-azabenzotriazole/triethyl amine) on PyBroP (bromotripyrrolidinophosphonium hexafluorophosphate).[1]
Purification: recrystallization is done using acetone/Et_2O.
Handling, Storage, and Precautions: the toxicological properties of this reagent have not yet been investigated, and no special instructions for storage and handling are mentioned in the literature.

Introduction. PyAOP is one of the more effective reagents for activation of carboxylic acids. HOAt has been reported to be more effective than HOBt as an additive in peptide coupling reactions.[3,4] Accordingly, uronium/aminium (HATU, HAPyU) or phosphonium (AOP, PyAOP) reagents containing an –OAt residue are considered to be more effective than their OBt-based analogues (HBTU, HBPyU, BOP, PyBOP) in peptide coupling reactions. PyAOP gives comparable or slightly better results than the tris(dimethylamino)phosphonium counterpart AOP [(7-azabenzotriazol-1-yloxy)tris(dimethylamino)phosphonium hexafluorophosphate] with the advantage that the carcinogen HMPA is not produced as a side product of the reaction.[5,6]

Formation of pyrrolidide derivatives has been observed due to the presence of small amounts of pyrrolidine (0.45%) in the commercial phosphonium salt and can be easily avoided by recrystallization of the reagent.[7]

Peptide synthesis. PyAOP is widely used in solution-and solid-phase peptide synthesis. The phosphonium reagents do not react with primary amines, in contrast to the corresponding uronium/aminium salts which lead to guanidino derivatives, a process that terminates peptide chain elongation.[6] The phosphonium salts are slightly less stable than the corresponding uronium/aminium derivatives in stock solutions used for automated solid-phase peptide synthesis.[6] As with other reagents, HOAt is often used as an additive with PyAOP but is not always required.[8] PyAOP is among the most reactive coupling reagents, as shown (Table 1) by the synthesis of ACP(65-74) under carefully controlled conditions (coupling time decreased to 1.5 min and use of small excesses of reagent) chosen to accentuate the differences between OAt- and OBt-based reagents.[5,8] This high reactivity has been used to suppress diketopiperazine formation while coupling the third amino acid to dipeptides.[9]

PyAOP is effective in difficult coupling reactions, for example involving the activation and coupling of diethylglycine[6] or during sequential introduction of Aib residues on solid phase.[8] PyAOP is useful to minimize epimerization when coupling peptide fragments.[10] In fragment or fragment-like coupling, preactivation is detrimental,[11] and the presence of excess HOAt enhances racemization (eq 1).[11]

Macrolactams. As shown in Table 2, PyAOP is very effective for macrolactamization.[8,12–18] Although slightly more epimerization was observed, PyAOP is preferred to analogous uronium/aminium coupling agents because reaction of the reagent with the amino terminus is avoided.[8]

Synthesis of phosphoesters. PyAOP activation of *H*-phosphonate monomers without amino protection was found to be highly efficient for the synthesis of oligodeoxynucleotides (eq 2), although 2-(benzotriazol-1-yloxy)-1,1-dimethyl-2-pyrrolidin-1-yl-1,3,2-diazaphospholidinium hexafluorophosphate (BOMP reagent) is slightly more efficient for rapid condensation.[19]

Synthesis of *N,N′*-substituted Acylguanidines. PyAOP was used in the solid-phase synthesis of *N,N′*-substituted acylguanidines (eq 3).[20]

Table 1 Assembly of ACP(65–74): distribution of products including deletion peptides[a]

Entry	Method	ACP	-2Ile	-Ile^{72}	-Ile^{69}	-Val	-Asn	Reference
1	PyBOP	10	22	13	13	3	<1	8
2	PyBOP–HOBt	11	21	14	14	3	<1	8
3	PyAOP	46	3	5	5	2	18	8
4	PyAOP–HOAt	60	3	6	7	4	2	8
5	HATU	53	3	6	12	3	16	5
6	HATU–HOAt	50	5	9	12	2	8	5

[a]Solid-phase synthesis of ACP(65–74): Val-Gln-Ala-Ala-Ile^{69}-Asp-Tyr-Ile^{72}-Asn-Gly-NH_2. 1.5 equiv of reagent; reaction time 1 min. Yields of ACP and deletion peptides (-Xxx) based on HPLC analysis.

Table 2 Cyclization of H-Arg(NO_2)-Lys(Z)-Asp(OBn)Val-Tyr-OH[12]

Reagent	Cyclomonomer (%)[a]	D-Tyr-isomer (%)[a]	Cyclodimer (%)[a]
PyBOP	52	13	17
PyAOP	56	11	20
HAPyU	55	9	25
DPPA	23[b]	20[b]	7[b]

[a]Reaction time 1 h.
[b]Reaction time 24 h.

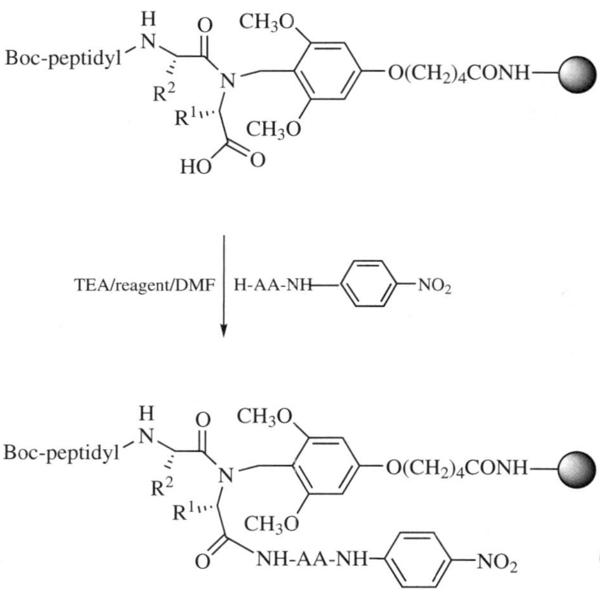

(1)

Boc-peptidyl = Boc-Val-Tyr, $R^1 = CH_3$, $R^2 = CH_2Ph$,
AA = Val-H-Val-Tyr-Phe-Ala-Val-pNA

Reagent	Product purity (%)	LLLDL-Isomer (%)
PyAOP	90	2.7
PyAOP/HOAt	89	7.0

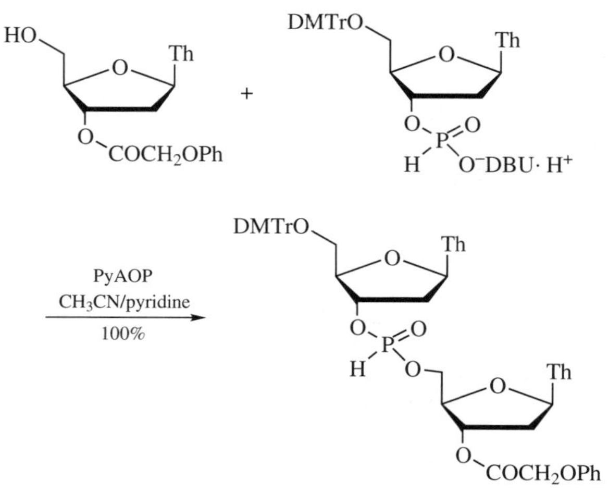

(2)

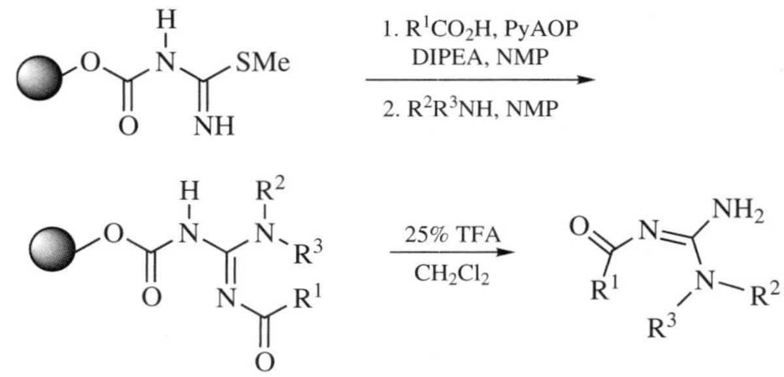

(3)

R^1 = *s*-Bu, Ph, 2-MePh, 2,6-diCl-Ph, 3,4,5-triOMe-Ph;
R^2 = H; R^3 = CH_2Ph: 75–85%

R^1= 2,6-diMe-Ph; R^2 = H;
R^3 = CH_2Ph: fail

Related Reagents. AOP [7-azabenzotriazol-1-yloxytris (dimethylamino)phosphonium hexafluorophosphate]; BOI [*O*-(benzotriazol-1-yl)-1, 3-dimethyl-1,3-dimethyleneuronium hexafluorophosphate]; BOP [1*H*-benzotriazol-1-yloxytris(dimethylamino)phosphonium hexafluorophosphate]; BroP [bromotris (dimethylamino)phosphonium hexafluorophosphate]; BTFFH [bis(tetramethylene) fluoroformamidium hexafluorophosphate]; CF_3-PyBOP[[6-(trifluoromethyl)benzotriazol-1-yl]-*N*-oxy-tris (pyrrolidino)phosphonium hexafluorophosphate]; CF_3-NO_2-PyBOP[[4-nitro-6-(trifluoromethyl)benzotriazol-1-yl]-*N*-oxy-tris(pyrrolidino)phosphonium hexafluorophosphate]; HAMDU [*O*-(7-azabenzotriazol-1-yl)-1, 3-dimethyl-1,3-dimethyleneuronium hexafluorophosphate]; HAPyU [*O*-(7-azabenzotriazol-1-yl)-1,1,3,3-bis(tetramethylene)uronium hexafluorophosphate]; HATU [*O*-(7-azabenzotriazol-1-yl)-1,1,3,3-tetramethyluronium hexafluorophosphate]; HBPyU [*O*-(benzotriazol-1-yl)-1,1,3,3-bis(tetramethylene)uronium hexafluorophosphate]; HBTU [*O*-(benzotriazol-1-yl)-1,1,3,3-tetramethyluronium hexafluorophosphate]; PyBOP (1*H*-benzotriazol-1-yloxytripyrrolidinophosphonium hexafluorophosphate]; PyCloP [chlorotripyrrolidinophosphonium hexafluorophosphate]; PyClU [1,1,3,3-bis(tetramethylene)chlorouronium hexafluorophosphate];PyNOP [[6-nitrobenzotriazol-1-yl]-*N*-oxy-tris(pyrrolidino)phosphonium hexafluorophosphate].

1. Hoeg-Jensen, T.; Holm, A., Sorensen, H., *Synthesis* **1996**, 383.
2. U. S. Pat. 5,580,981 (to Carpino, L. A.) (1996).
3. Carpino, L. A., *J. Am. Chem. Soc.* **1993**, *115*, 4397.
4. Albericio, F.; Chinchilla, R.; Dodsworth, D. J.; Najera, C., *Organic, Preparations and Procedures Int.* **2001**, *33*, 205.
5. Carpino, L. A.; El-Faham, A.; Minor, C. A.; Albericio, F., *J. Chem. Soc., Chem. Commun.* **1994**, 201.
6. Albericio, F.; Bofill, J. M.; El-Faham, A.; Kates, S. A., *J. Org. Chem.* **1998**, *63*, 9678.
7. Alsina, J.; Barany, G.; Albericio, F.; Kates, S. A., *Lett. Pept. Sci.* **1999**, *6*, 243.
8. Albericio, F.; Cases, M.; Alsina, J.; Triolo, S. A.; Carpino, L. A.; Kates, S. A., *Tetrahedron, Lett.* **1997**, *38*, 4853.
9. Alsina, J.; Giralt, E.; Albericio, F., *Tetrahedron, Lett.* **1996**, *37*, 4195.
10. Carpino, L. A.; El-Faham, A.; Albericio, F., *Tetrahedron, Lett.* **1994**, *35*, 2279.
11. Alsina, J.; Yokum, T. S.; Albericio, F.; Barany, G., *J. Org. Chem.* **1999**, *64*, 8761.

12. Ehrlich, A.; Heyne, H.-U.; Winter, R.; Beyermann, M.; Haber, H.; Carpino, L. A.; Bienert, M., *J. Org. Chem.* **1996**, *61*, 8831.
13. Jou, G.; González, I.; Albericio, F.; Lloyd-Williams, P.; Giralt, E., *J. Org. Chem.* **1997**, *62*, 354.
14. Belshaw, P. J.; Schreiber, S., *J. Am. Chem. Soc.* **1997**, *119*, 1805.
15. Jensen, K. J.; Alsina, J.; Songster, M. F.; Vagner, J.; Albericio, F.; Barany, G., *J. Am. Chem. Soc.* **1998**, *120*, 5441.
16. Hubler, F.; Rückle, T.; Patiny, L.; Muamba, T.; Guichou, J.-F.; Mutter, M.; Wenger, R., *Tetrahedron, Lett.* **2000**, *41*, 7193.
17. Pettit, G. R.; Toki, B. E.; Xu, J. P.; Brune, D., *J. Nat. Prod.* **2000**, *63*, 22.
18. Planas, M.; Bardají E.; Barany, G., *Tetrahedron, Lett.* **2000**, *41*, 4097.
19. Wada, T.; Sato, Y.; Honda, F.; Kawahara, S.-I.; Sekine, M., *J. Am. Chem. Soc.* **1997**, *119*, 12710.
20. Dodd, D. S.; Zhao, Y., *Tetrahedron, Lett.* **2001**, *42*, 1259.

Jacques Coste, Patrick Jouin
CNRS-UPR9023, Montpellier, France

7-Azabenzotriazol-1-yloxytris (dimethylamino)phosphonium Hexafluorophosphate

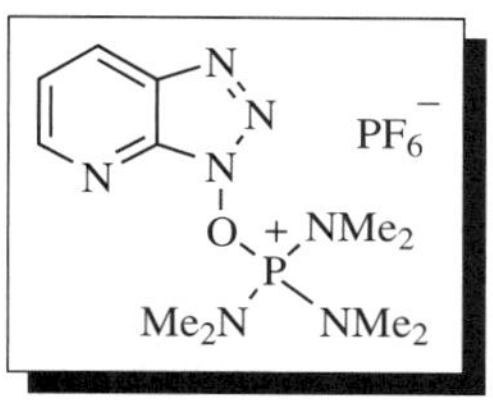

[156311-85-2] $C_{11}H_{21}F_6N_7OP_2$ (MW 443.27)

(peptide coupling agent)

Alternate Name: AOP.
Solubility: soluble in *N,N*-dimethylformamide.
Form Supplied in: not commercially available.
Preparative Methods: AOP is prepared by reaction of 2-chlorotri(dimethylamino)phosphonium chloride with 1-hydroxy-7-azabenzotriazole (**HOAt**) to give 7-azabenzotriazol-1-yloxytris (dimethylamino)phosphonium chloride. The chloride is then exchanged with KPF_6 to give AOP in a way similar to that used for the preparation of **benzotriazol-1-yloxytris(dimethylamino)phosphonium hexafluorophosphate (BOP)**.[1]
Handling, Storage, and Precautions: follow general recommendations for uronium/aminium type of reagents such as BOP or **N-[(dimethylamino)-1H-1,2,3-triazolo-[4,5-b]pyridin-1-yl-methylene]-N-methylmethanaminium hexafluorophosphate N-oxide (HATU)**. A main by-product of AOP reaction is carcinogenic **HMPA**, and therefore the reagent must be handled with caution in a fume hood.[2]

The preparation and use of 7-azabenzotriazol-1-yloxytris (dimethylamino)phosphonium hexafluorophosphate (AOP) as a coupling reagent for peptide synthesis was first reported in 1994.[3] It is a HOAt-derived phosphonium salt[4] and hence it is a nitrogenated version of BOP.[1] When mixed with a carboxylic acid in the presence of a tertiary amine such as **DIEA**, the azabenzotriazolyl ester is formed, which is presumably the active intermediate that yields the amide bond by reaction with the amino component.[3,5,6]

AOP has been used in the solid-phase synthesis of difficult peptide sequences such as the 65-74 fragment of ACP (acyl carrier protein) using fast coupling protocols (90 s) and a small excess of acylating species (1.5 equiv).[3] The yield of ACP 65-74 under these demanding conditions was 49%, quite similar to other HOAt-containing phosphonium and aminium salt derivatives such as 7-azabenzotriazol-1-yloxytris(pyrrolidino)phosphonium hexafluorophosphate (PyAOP) (45% yield) or **HATU** (53% yield). The same study shows that HOAt-derived reagents are far superior than HOBt-based ones, which gave 12–18% yields for ACP synthesis under those conditions. Side reactions affecting the chiral integrity of the α-carbon during the activation of amino acids are also minimized by AOP. Thus, in the model coupling of Fmoc-Ser(tBu)-Phe-OH onto Pro-PAL-PEG-PS resin, activation with AOP afforded only a 5% of tripeptide epimer (PyAOP and HATU gave 7% and 5%, respectively), whereas the classical DCC/HOBt method gave 18% of epimerization (eq 1).[7]

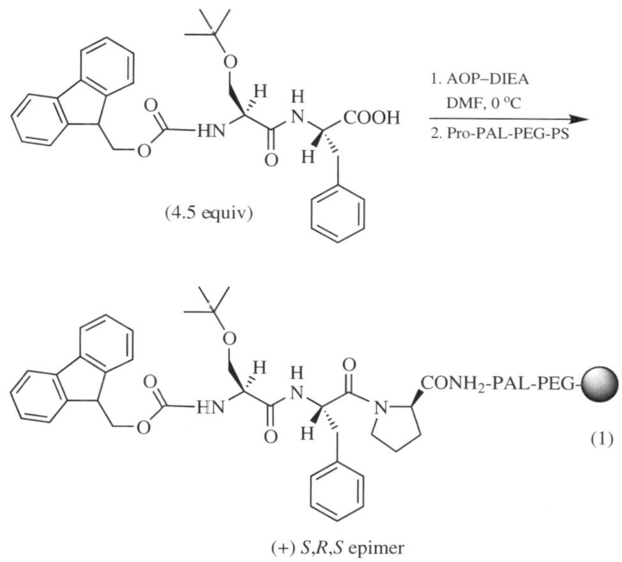

From a mechanistic point of view, AOP, like other phosphonium salts, is thought to function in a similar way as aminium and uronium salts.[6] However, aminium/uronium salts, unlike phosphonium derivatives, may react with the *N*-terminal amino function leading to a guanidino by-product, which terminates the peptide chain growth.[5,8,9] This side reaction may have some importance in reactions involving hindered amino acids or in peptide cyclization since coupling in these cases tends to be rather slow. In a thorough study carried out with several activating reagents, AOP gave practically undetectable amounts of any by-product (1%).[5]

In comparison with its pyrrolidino derivative (PyAOP), AOP shows a much better hydrolytic stability in DMF (78% vs 13% after 2 d at 25 °C).[5] This makes AOP specially suitable to be used in automatic synthesizers where the coupling reagent is kept in solution.

Related Reagents. Although AOP is an effective amino acid activating reagent devoid of common unwanted side reactions in peptide synthesis, it has found little application due to the generation of the carcinogenic by-product HMPA. Other phosphonium and aminium HOAt-containing reagents (PyAOP and HATU) offer similar activating and racemization-suppressing properties and thus are widely used and commercially available.[10,11]

1. Rabanal, F.; Giralt, E., This Handbook pp. 86-88.
2. Dykstra, R. R., In *Encyclopedia of Reagents for Organic Synthesis*; Paquette, L. A., Ed.; Wiley: Chichester, UK, 1995; Vol. 4, pp. 2668-2675.
3. Carpino, L. A.; El-Faham, A.; Minor, C. A.; Albericio, F., *J. Chem Soc. Chem. Comm* **1994**, 201–203.
4. Carpino, L. A., *J. Am. Chem. Soc.* **1993**, *115*, 4397–4398.
5. Albericio, F.; Bofill, J. M.; El-Faham, A.; Kates, S. A., *J. Org. Chem.* **1998**, *63*, 9678–9683.
6. Lloyd-Williams, P.; Albericio, F.; Giralt, E., *Chemical Approaches to the Synthesis of Peptides and Proteins*; CRC: Boca Raton, FL, 1997.
7. Carpino, L. A.; El-Faham, A.; Albericio, F., *Tetrahedron Lett.* **1994**, *35*, 2279–2282.
8. Gausepohl, H.; Pieles, U.; Frank, R. W., In *Peptides, Chemistry and Biology: Proceedings of the 12th American Peptide Symposium*; Smith, J. A.; Rivier, J. E., Eds.; ESCOM, Science: Leiden, The Netherlands, 1992, pp. 523–524.
9. Story, S. C.; Aldrich, J. V., *Int. J. Pept. Prot. Res.* **1994**, *43*, 292–296.
10. Albericio, F., Carpino, L. A., *In Methods in Enzymology, Solid-Phase Peptide Synthesis*; Fields, G. B., Ed.; Academic Press: Orlando, 1997, Vol. 289, pp. 104–126.
11. Albericio, F.; Chinchilla, R.; Dodsworth, D.; Nájera, C., *Org. Prep. & Proc. Int.* **2001**, *33*, 203–303.

Francesc Rabanal & Ernest Giralt
Universitat de Barcelona, 08028-Barcelona, Spain

Azidotris(dimethylamino)phosphonium Hexafluorophosphate

$$(Me_2N)_3\overset{+}{P}{-}N_3 \quad PF_6^-$$

[50281-51-1] $C_6H_{18}F_6N_6P_2$ (MW 350.19)

(coupling reagent for amides and peptide synthesis via an acyl azide;[1] in the absence of a nucleophile undergoes a Curtius rearrangement to give the corresponding isocyanate;[1,2] by irradiation gives the iminophosphonium salt;[3,4] diazo transfer reagent[5])

Physical Data: mp >250 °C; IR, $\nu(N_3)$ 2176 cm^{-1}.
Solubility: sol acetone, acetonitrile, DMF.
Form Supplied in: crystalline colorless solid; not available from commercial sources.
Preparative Methods: Bromine (1 equiv) is added dropwise to Hexamethylphosphorous Triamide (1 equiv) in dry diethyl ether. After the addition is complete, Potassium Hexafluorophosphate (1 equiv), dissolved in H_2O is added, and the solid formed is filtered and washed with H_2O. After drying under vacuum, the solid is dissolved in acetone and Sodium Azide (1.5 equiv) is added. The mixture is stirred overnight at 25 °C, then the sodium bromide formed and the excess of sodium azide is filtered off and the solvent is removed under vacuum.
Purification: by crystallization in a mixture of acetone and diethyl ether.
Handling, Storage, and Precautions: is very stable, not hygroscopic, and can be stored indefinitely under vacuum. The related bromine derivative is an exceptionally safe azide, that does not detonate by shock, friction, rapid heating, or even flame.[5] When used in combination with carboxylic acids, the potential carcinogenic Hexamethylphosphoric Triamide is a side-product and the reagent must be handled with caution.[6]

Coupling Reagent. Azidotris(dimethylamino)phosphonium hexafluorophosphate has been used as a condensing reagent for the preparation of peptides.[1] Reaction of α-amino protected amino acids or dipeptides with equimolar amounts of α-carboxyl protected amino acids or dipeptides and Triethylamine (2 equiv) in DMF proceed cleanly overnight at −10 to 0 °C, giving the corresponding peptides in good yields (81–97%) (eq 1). Neither acylation of the hydroxy group of serine nor dehydration of the side chain of asparagine is reported during the coupling.[1]

$$RCO_2H + R'NH_2 \xrightarrow[\substack{Et_3N \\ 81-97\%}]{(Me_2N)_3\overset{+}{P}{-}N_3\ PF_6^-} RCONHR' \qquad (1)$$

Racemization from azidotris(dimethylamino)phosphonium hexafluorophosphate-mediated couplings is analyzed by different methods. While no racemization is detected by the Anderson[7] and Weinstein[8] methods, the Young[9] method gave some degree (3%) of racemization.

Although azidotris(dimethylamino)phosphonium hexafluorophosphate has not been used in solid-phase peptide synthesis,[10] other commercial phosphonium salt derivatives such as Benzotriazol-1-yloxytris(dimethylamino)phosphonium Hexafluorophosphate (BOP),[11] benzotriazol-1-yloxytris(pyrrolidino) phosphonium hexafluorophosphate (PyBOP),[12] or bromotris (pyrrolidino)phosphonium hexafluorophosphate (PyBroP)[13] are commonly used in that strategy.

Isocyanate Formation. The coupling reaction described above occurs via an acyl azide intermediate, which at 50 °C suffers a Curtius rearrangement to give an isocyanate (eq 2),[1] which can further react with amines to give the corresponding ureas.

Iminophosphonium Salt Formation. Irradiation of azidotris (dimethylamino)phosphonium hexafluorophosphate in acetonitrile at 254 nm using a Rayonnet photochemical reactor for 15 h at 25 °C gives the iminophosphonium salt with release of nitrogen (eq 3). This is the first example reported of a Curtius-type rearrangement involving a charged atom.[3,4]

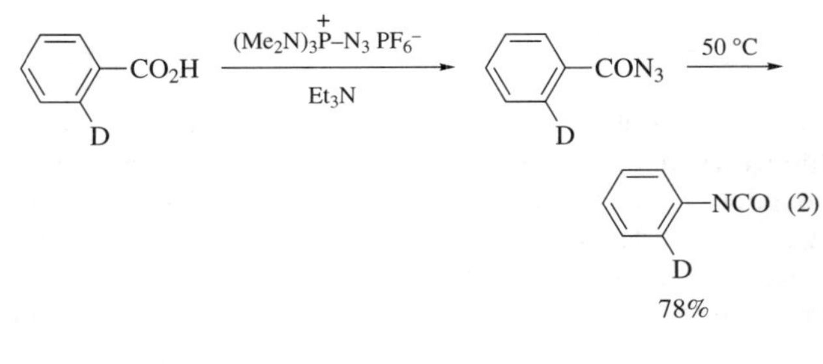

Diazo Transfer Reagent. The related compound azidotris (diethylamino)phosphonium bromide has been used for the preparation of diazo products.[5] Active methylene compounds react with azidotris(diethylamino)phosphonium bromide in dry diethyl ether in the presence of only catalytic amounts of base, such as Potassium t-Butoxide or other alkoxide, yielding the corresponding diazo products in good yields (70–78%) (eq 4).

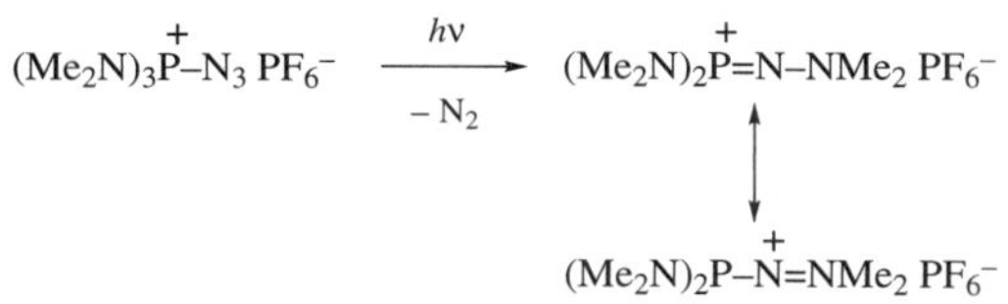

(3)

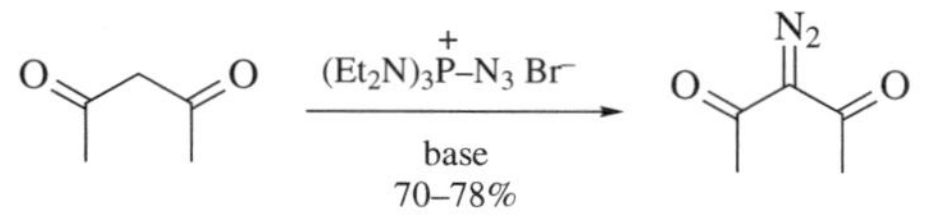

(4)

1. Castro, B.; Dormoy, J. R. *Bull. Soc. Claim. Fr., Part 2* **1973**, *12*, 3359.
2. Masson, M. A.; Dormoy, J. R. *J. Labelled Compd. Radiopharm.* **1979**, *16*, 785.
3. Majoral, J. P.; Bertrand, G.; Baceiredo, A.; Mulliez, M.; Schmutzler, R. *Phosphorus Sulfur/Phosphorus Sulfur Silicon* **1983**, *18*, 221.
4. Mulliez, M.; Majoral, J. P.; Bertrand, G. *J. Chem. Soc. (C)* **1984**, 284.
5. McGuiness, M.; Shechter, H. *Tetrahedron Lett.* **1990**, *31*, 4987.
6. Zapp, J. A., Jr. *Science* **1975**, *190*, 422.
7. (a) Anderson, G. W.; Young, R. W. *J. Am. Chem. Soc.* **1952**, *74*, 5307. (b) Anderson, G. W.; Zimmermann, J. E.; Callanan, F. M. *J. Am. Chem. Soc.* **1967**, *89*, 5012.
8. Weinstein, B.; Pritchard, A. E. *J. Chem. Soc. (C)* **1972**, 1015.
9. Williams, M. W.; Young, G. T. *J. Chem. Soc. (C)* **1963**, 881.
10. Fields, G. B.; Tian, Z.; Barany, G. In *Synthetic Peptides: A User's Guide*; Grant, G. A., Ed.; Freeman:New York, **1992**; p 77.
11. Castro, B.; Dormoy, J. R.; Evin, G. *Tetrahedron Lett.* **1975**, 1219.
12. Coste, J.; Le-Nguyen, D.; Castro, B. *Tetrahedron Lett.* **1990**, *31*, 205.
13. Frérot, E.; Coste, J.; Pantaloni, A.; Dufour, M. N.; Jouin, P. *Tetrahedron* **1991**, *47*, 259.

Fernando Albericio & Steven A. Kates
Millipore Corporation, Bedford, MA, USA

B

Benzeneselenenyl Bromide[1]

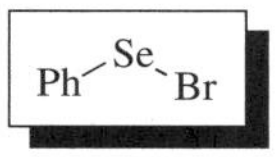

[34837-55-3] C_6H_5BrSe (MW 235.97)

(electrophilic selenium reagent for alkene synthesis[2–6])

Physical Data: mp 60–62 °C; bp 134 °C/35 mmHg, 107–108 °C/15 mmHg; sublimes at 25 °C/0.001 mmHg.
Solubility: soluble in organic solvents such as hexane, CH_2Cl_2, ether, THF.
Form Supplied in: dark-red crystals; commercially available.
Preparative Method: can be quantitatively prepared in situ by treatment of Diphenyl Diselenide with an equimolar amount of Bromine.[2,3,7,66]
Purification: recrystallization from petroleum ether or sublimation.
Handling, Storage, and Precautions: highly toxic, unpleasant smell.

Original Commentary

M. Iwaoka
The University of Tokyo, Japan

Reaction with Ketones and Enols. Benzeneselenenyl bromide (PhSeBr) is a useful electrophilic organoselenium reagent which exhibits similar reactivity to Benzeneselenenyl Chloride (PhSeCl). Although the reagent can be substituted by PhSeCl in many reactions, there is a significant difference in reactivity between these reagents.[2] Ketones can be efficiently converted into α-phenylseleno ketones through their lithium or copper enolates.[3,8] This reaction is useful for the transformation of saturated ketones to α,β-unsaturated ketones and the β-alkylation of α,β-unsaturated ketones (eqs 1 and 2). Similarly, enol acetates are

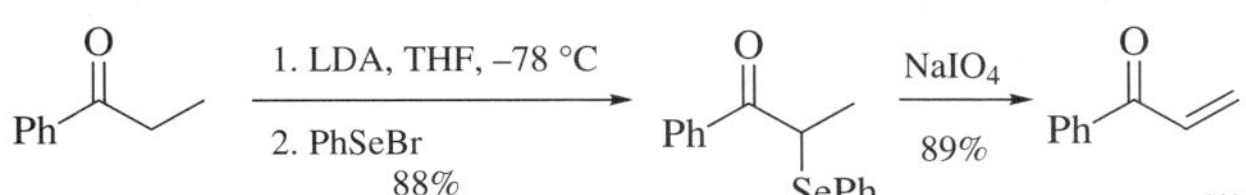

(1)

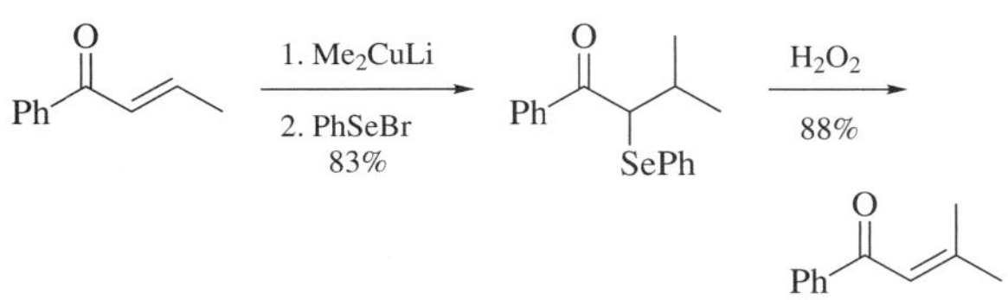

(2)

converted into α,β-unsaturated ketones by using benzeneselenenyl trifluoroacetate ($PhSeOCOCF_3$),[9] which is prepared in situ by treatment of PhSeCl or PhSeBr with Silver(I) Trifluoroacetate (eq 3).[3,4] Silyl enol ethers react with PhSeBr to provide α-phenylseleno ketones and aldehydes in good yields (eq 4).[10] On the other hand, PhSeBr adds to enol ethers forming bromoselenenylation adducts regioselectively (eq 5).[11] Unlike PhSeCl, PhSeBr does not usually react well with ketones and aldehydes to form α-phenylseleno carbonyl compounds.[3]

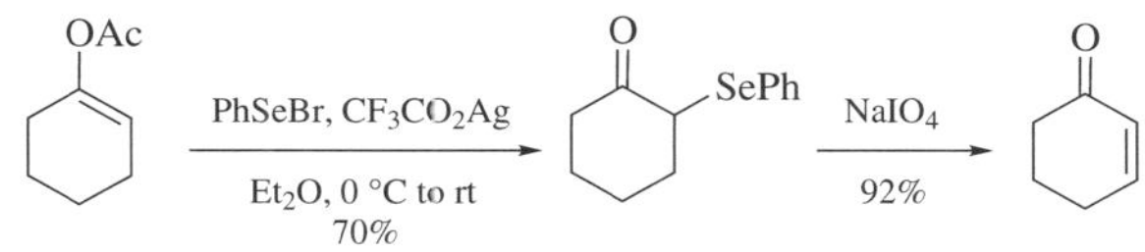

(3)

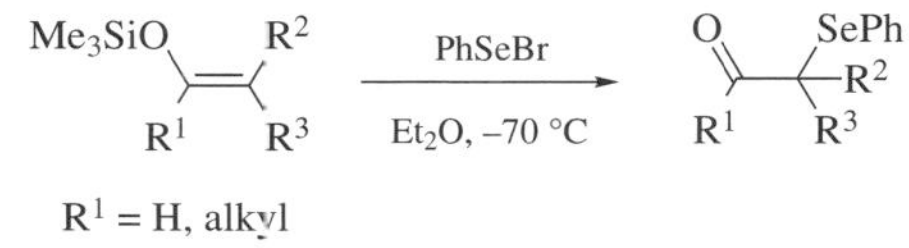

(4)

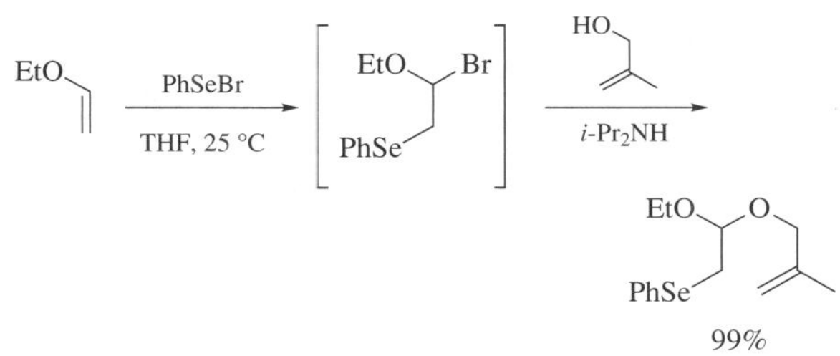

(5)

Reaction with Carbanions. α-Lithiated nitriles[12] and nitroalkanes[13] rapidly react with PhSeBr to afford the corresponding α-phenylseleno compounds, which are oxidized to α,β-unsaturated nitriles and nitroalkanes, respectively (eqs 6 and 7).

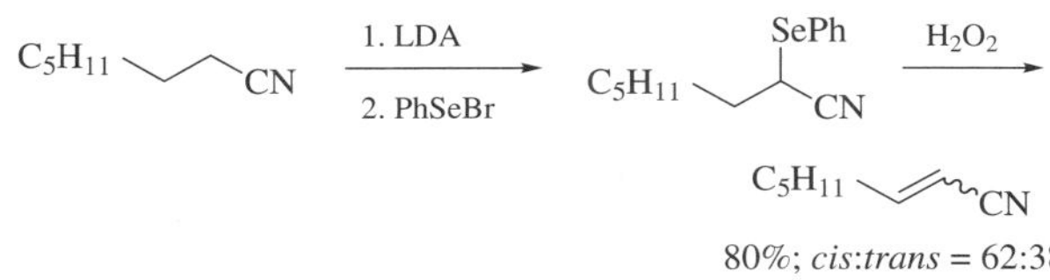

(6)

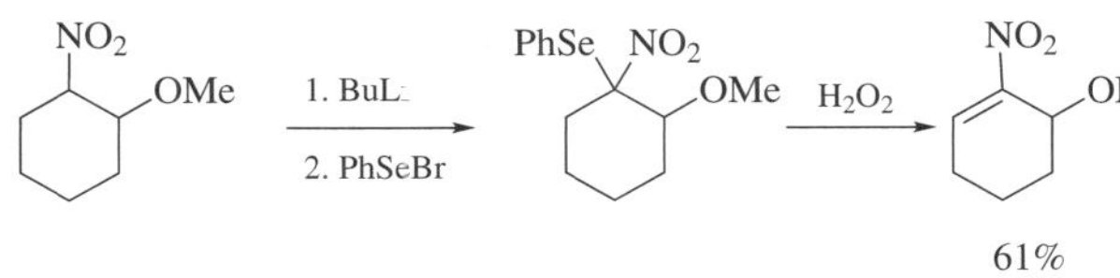

(7)

Addition to Alkenes. PhSeBr reacts with alkenes in aprotic organic solvents to provide bromoselenenylation adducts efficiently.[5] This addition is considered to proceed via a seleniranium ion, a selenium-containing 3-membered cyclic cation, and *trans* adducts are obtained exclusively. For terminal alkenes the regioselectivity of the addition can be controlled by changing the reaction conditions (eq 8).[6] The products can be easily transformed to vinylic selenides,[6] vinylic and allylic bromides,[14a] or α-phenylseleno ketones.[14b] When the reaction is carried out in alcohol or carboxylic acid, oxyselenenylated products of alkenes are obtained in high yields (eq 9).[5,15]

Alkynyl Selenide Synthesis. Terminal alkynes undergo electrophilic substitution by treatment with the reagent prepared from PhSeBr and Silver(I) Nitrite, providing alkynyl selenides in good yields (eq 10).[16]

Reaction with Ketonic Hydrazones. In the presence of t-Butyltetramethylguanidine, a strong base, ketonic hydrazones smoothly react with PhSeBr to give phenyl vinyl selenides (eq 11).[17] This reaction requires an excess amount of PhSeBr.

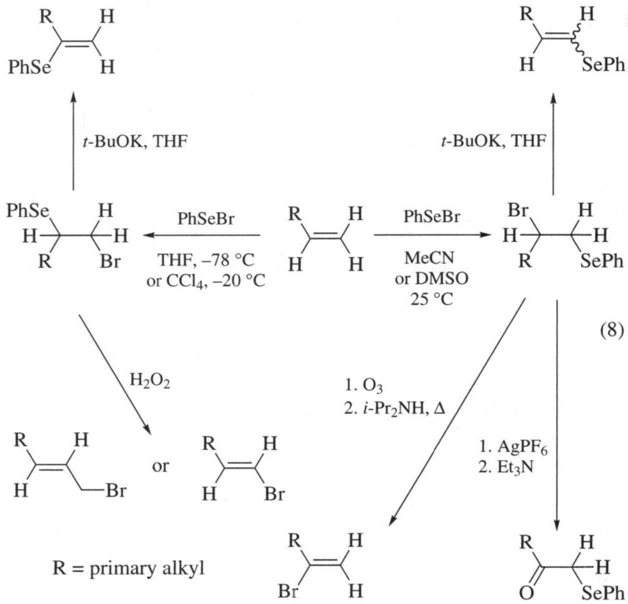

(8)

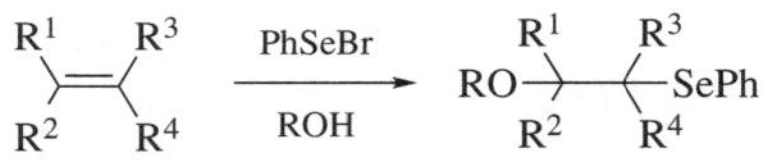

(9)

R = Me, Et, Ac, etc.

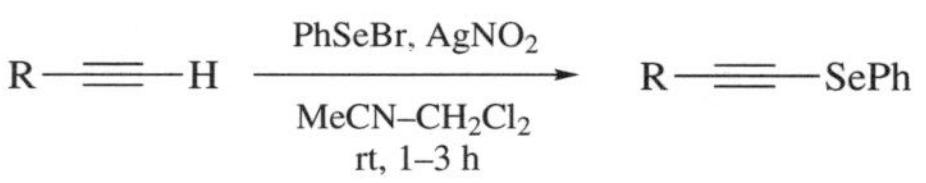

(10)

R = Bu, Ph, OCOEt etc.

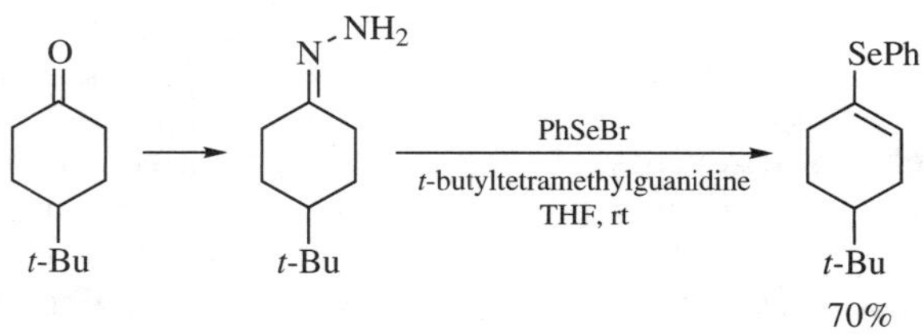

(11)

First Update

Jesús Sandoval Ramírez
Universidad Autónoma de Puebla, Puebla, México

Reaction with Ketones and Aldehydes. In contrast to the usual approach involving enols or enolates, a benzeneselenenyl derivative bearing an electron withdrawing group (eq 12)[18] has been shown to react directly with acetone (76%). The use of more vigorous conditions led to subsequent acylation. With cyclohexanone, the selenenylation reaction required a longer reaction time (48 h) and similar internal acylation was observed (59%). With cyclic and acyclic 1,3-diones, the arylselenenyl reagent is added at the 2-position and treatment with water resulted in fragmentation products.

Unlike PhSeCl and *o*-(chlorocarbonyl)phenylselanyl chloride, PhSeBr does not react well with ketones and aldehydes;[3] an exception being the α-selenenylation of isovaleraldehyde, in the presence of minute amounts of (*S*)-prolinamide (eq 13), which turned out to be very slow (1 day, 24%), it was faster with PhSeCl (15 h, 75%), and it occurs immediately (5 min, 96%) with *N*-phenylselenophthalimide (NPSP). No enantioselective induction in the α-selenenylation reaction was observed.[19]

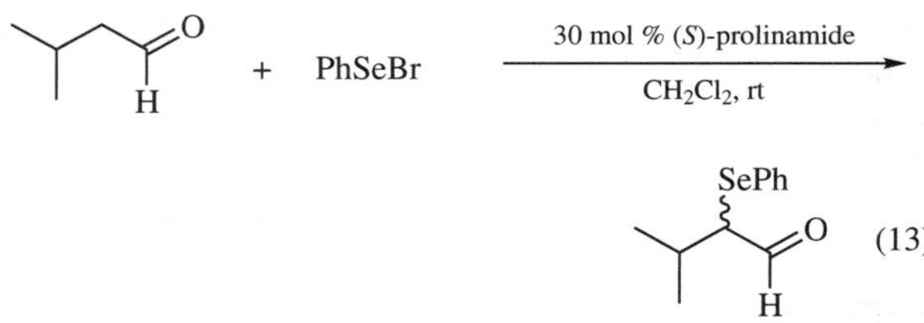

(13)

Reaction with Enol Ethers, Derived from α,β-Unsaturated Ketones. The reaction of PhSeBr with a silyl enol ether derived from an α,β-unsaturated ketone, gave the α′-phenylselenated-α,β-unsaturated ketone (eq 14).[20] The phenylseleno group permitted a stereoselective reduction with deuterated sodium borohydride in the presence of cerium^{III} chloride.

In reactions forming phenylselanyl compounds, a large excess of PhSeBr leads to a very reactive selanylselenonium intermediate (eq 15).

Activation of the phenyl seleno group has become a useful driving force to achieve nucleophilic displacement.[21] The phenylselanyl group can also be activated by methyl iodide into methylselenonium salts or by bromine, chlorine, sulfuryl chloride,

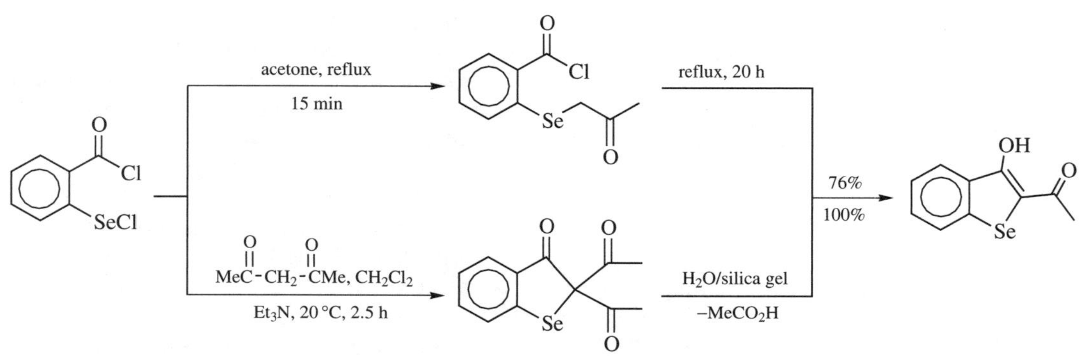

(12)

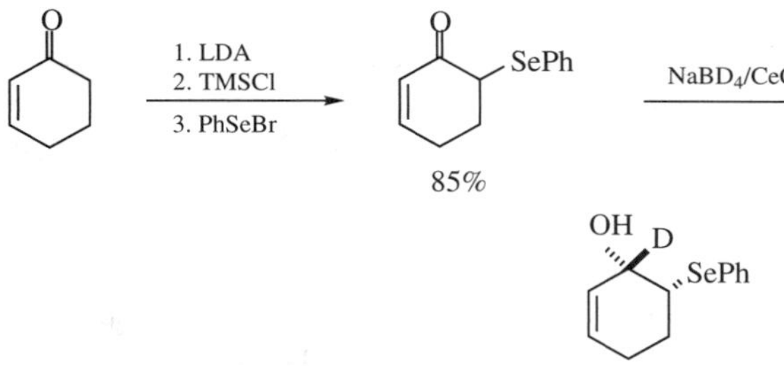
(14)

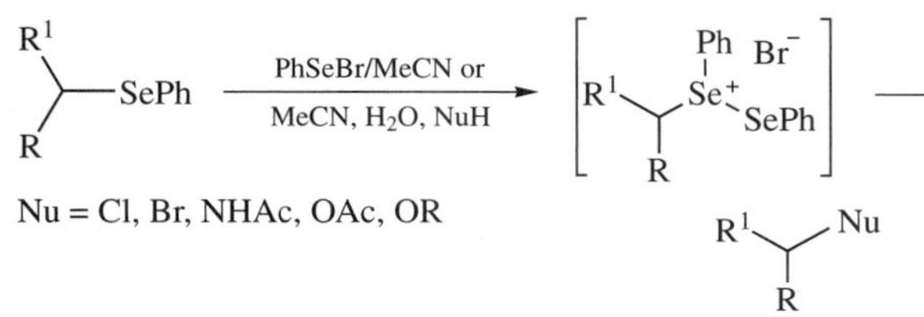
(15)

N-bromosuccinimide , and *N*-iodosuccinimide to produce the corresponding haloselenonium species. The selenonium derivatives can be transformed into selenium free compounds in two steps processes (eqs 16 and 17),[22,23] or through a one step process using thallium^{I} ethoxide.

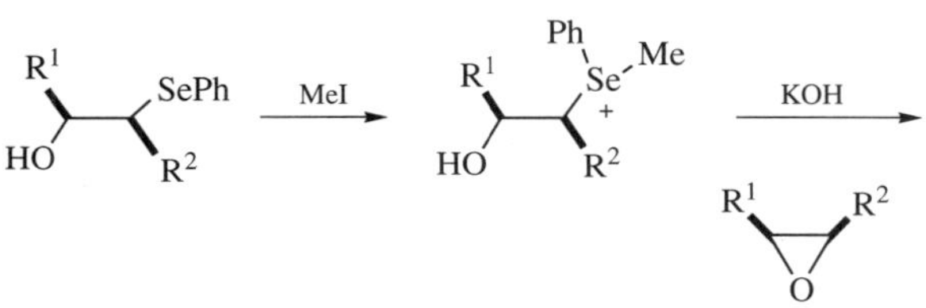
(16)

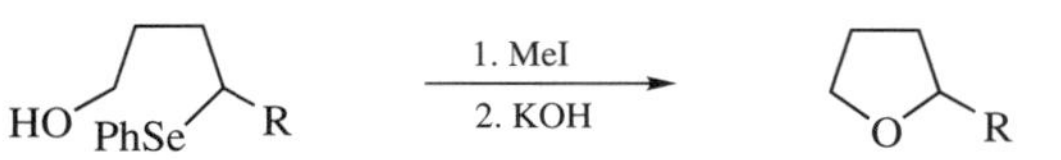
(17)

When the selanylselenonium ion is located in the allylic position its elimination occurs with rearrangement (eq 18).[24]

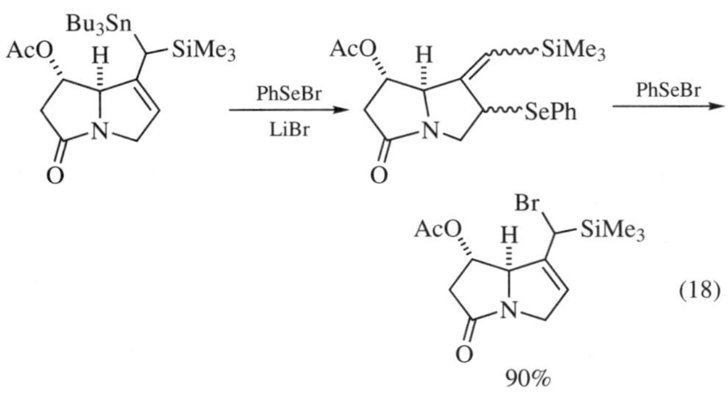
(18)

Chiral bromides can be obtained from chiral phenylselanyl substrates under treatment with bromine (eq 19).[25]

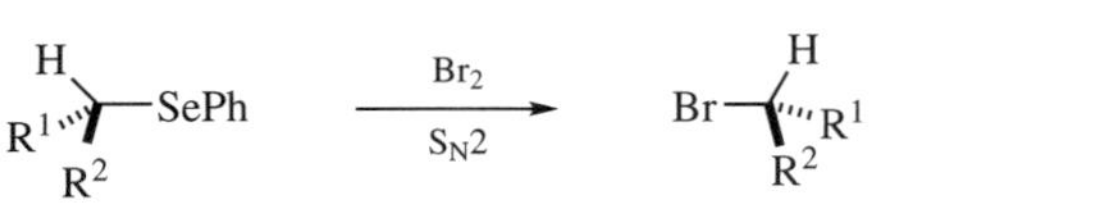
(19)

Similar deselenenylation procedures were observed when PhSeBr, PhSeCl, diphenyl diselenide, *N*-phenylselenophthalimide, and other selenenyl compounds were used in catalytic amounts, in the solid-phase, with $(NH_4)_2S_2O_8$, $Na_2S_2O_8$, or $K_2S_2O_8$ as oxidizing agents.[26]

The regioselectivity of elimination, following oxidation of alkyl phenyl selenides depends on groups at the β-position (eq 20) of the selenoxides, has been supported by theoretical studies.[27] In the case of allyl selenoxides, the elimination directs to allylic alcohols, via a selenenic ester, by a [2,3] sigmatropic rearrangement and subsequent hydrolysis.

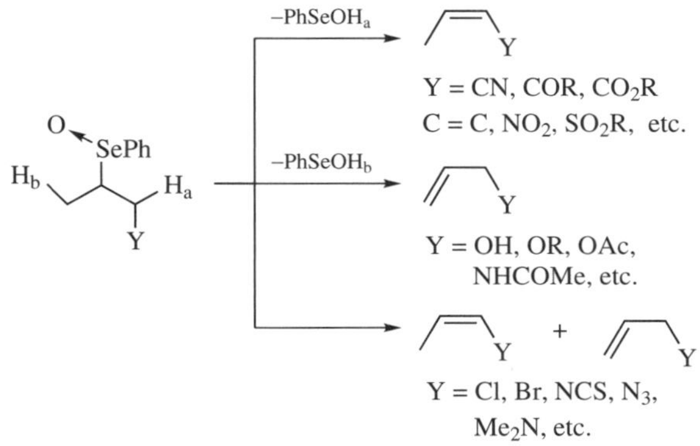
(20)

The phenylseleno group can be also activated as a phenylselenone group using an excess of *m*-chloroperbenzoic acid. The phenylselenone can be nucleophilically displaced;[28] in this way β-hydroxy epoxides (eq 21) can be obtained with good yields (65–78%).

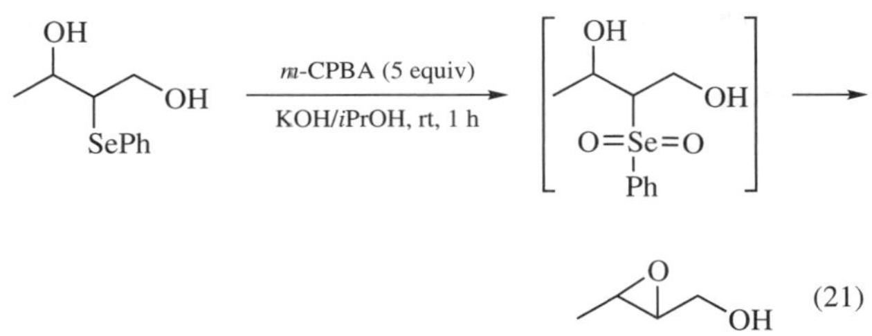
(21)

Reaction with Enolates. It has been found that α-bromoketones are transformed into α-phenylselenoketones in yields between 60% and 70% (eq 22) by PhSeBr and samarium^{II} or samarium^{III} iodide.[29] Samarium enolates are formed in situ.

$$ArCOCH_2Br + PhSeBr \xrightarrow[CH_3CN]{SmI_2 \text{ or } SmI_3} ArCOCH_2SePh \quad (22)$$

Phenylselanation of ester enolates allows the preparation of α-phenylselanyl esters which on treatment with a base afford enolates stabilized by the phenylseleno group (eq 23). These enolates were alkylated quantitatively; the use of allyl bromides as electrophile ultimately gave α,β,γ,δ-unsaturated esters owing to a final elimination of the phenylselanyl group.[30]

A variety of metallated derivatives have been treated with PhSeBr leading to alkyl aryl selenides in a wide range of yields. The anion of a *N,N*-dialkylhydrazone (eq 24),[31] derived from an aliphatic aldehyde, led to α-phenylselanyl hydrazones or to α-phenylselanyl nitriles depending on the amounts of base and PhSeBr employed.

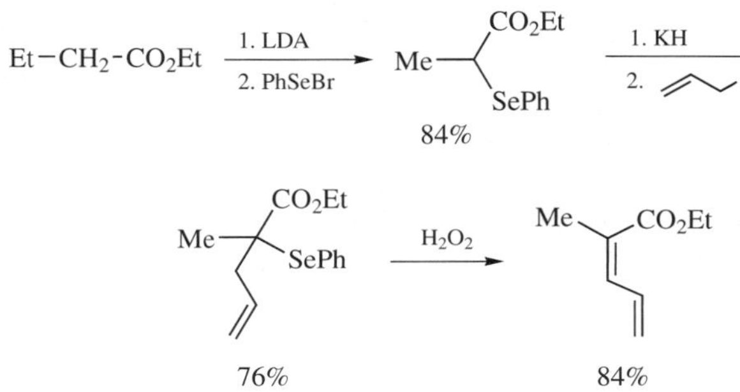

(23)

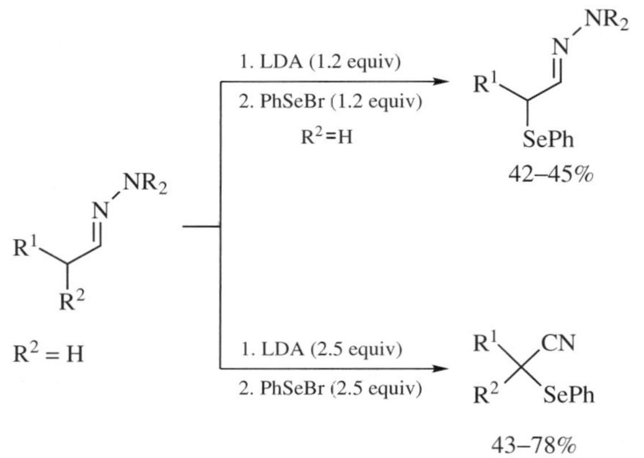

(24)

The anion of an α-sulfone ester (eq 25)[32] gave access to the α-phenylseleno derivative which after oxidation generated 2-sulfonylacrylates. The latter reacted stereoselectively with 1-seleno-2-silylethene to produce sulfonyl cyclopropanes by means of a [2 + 1] cycloaddition.

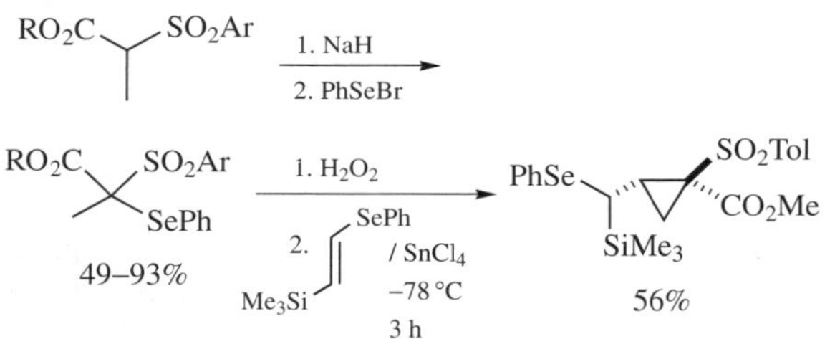

(25)

The preparation of α,α-bis(phenylseleno)ketones has been achieved easily from α-phenylselenoketones upon a second treatment with PhSeBr (eq 26).[33] This same reaction with NPSP and triethylamine took 2 days, whereas 3–8 days were required with *N*-phenylselenomorpholine, both at rt. The formation of α,α-bis(phenylseleno)ketones was not observed when PhSeCl was added to the silyl enol ethers of similar starting materials.

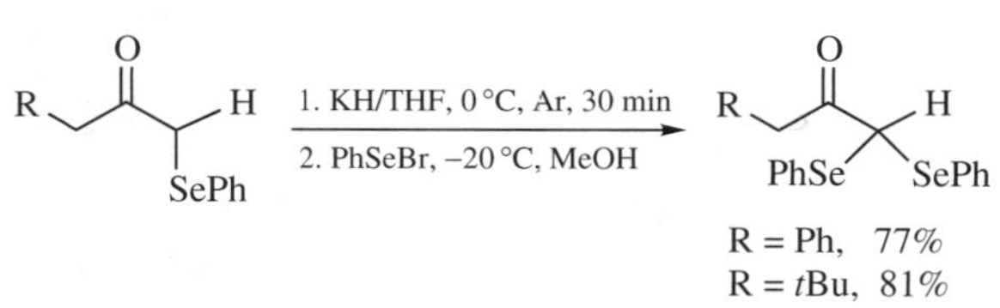

(26)

Ethyl α,α-bis(phenylseleno)acetate has been obtained by adding PhSeBr to a solution of ethyl diazoacetate. The best yield of the bis(phenylseleno) compound was produced at −78 °C, in THF, with addition of PhSeBr to the diazoester (eq 27).[34] The yield of the alternative monobromophenylseleno derivative could be improved to 89% by choosing the appropriate solvent, temperature, and order of addition of the reactants. Starting with diazomethane, bromo(phenylseleno)methane was obtained. Insertion reactions of diazo compounds with other selenenyl electrophiles also were observed.[35]

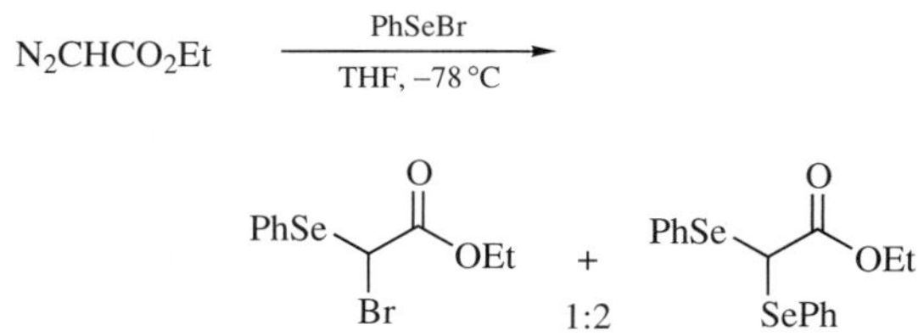

(27)

The sunlamp photolysis of methyl or ethyl 2-(phenylseleno) malonates in the presence of a variety of alkenes and alkynes yielded addition products (eq 28).[36] These transformations involve free radical mechanisms with a transfer of the phenylseleno group.

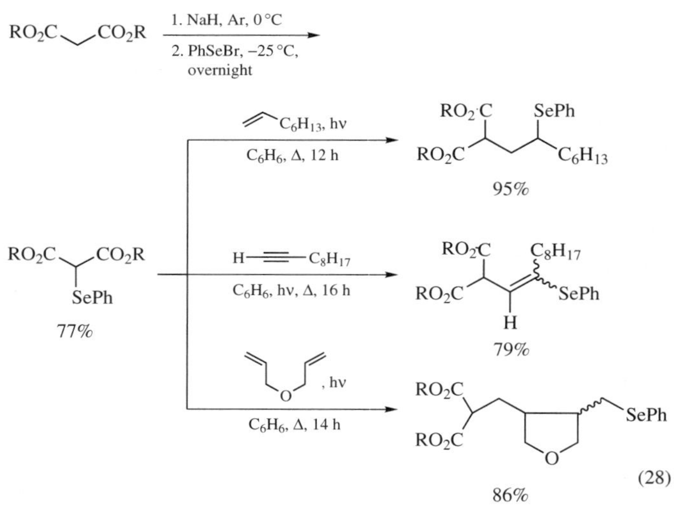

(28)

Reaction with Carbanions Present in Aromatic Frameworks. Aromatic systems have been less commonly used as substrates for selenide formation because stronger bases are needed and the phenylseleno group is introduced in low yields. Nevertheless, the treatment of the OMOM ether of α-naphtol by *sec*-butyllithium, followed by addition of PhSeBr, resulted in the addition of the phenylseleno group at the β position of the naphtol skeleton in moderate yield (eq 29).[37]

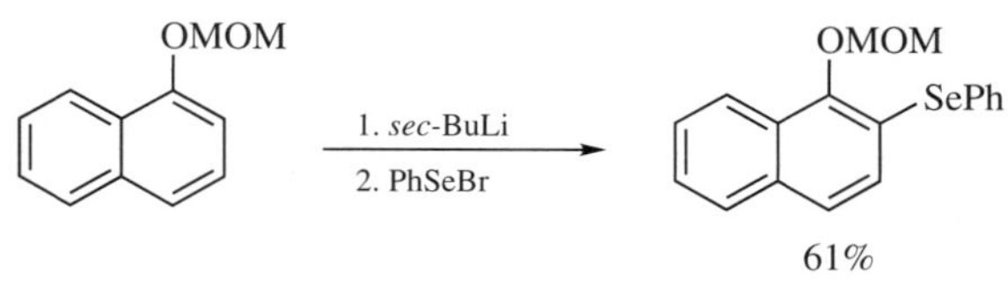

(29)

A lithiated pyridazine nucleus (eq 30) was quenched with PhSeBr affording the corresponding phenyl pyridazinyl selenide.[38]

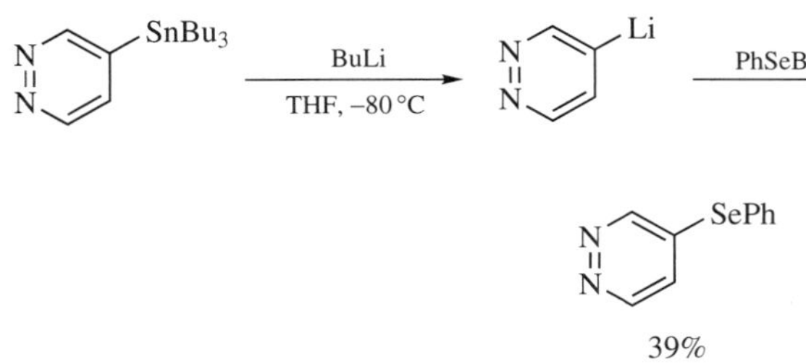

(30)

Nitroselenation of Conjugated Dienes. Regioselective nitroselenation of (1-alkenyl)trialkylsilanes was achieved by the use of benzeneselenenyl bromide or benzeneselenenyl chloride in the presence of silver[I] nitrite and mercury[II] chloride.[39] The nitroalkylphenylselenides, were oxidized with hydrogen peroxide inducing their *syn*-elimination and producing 2-nitroalkyl phenylselenides (eq 31).

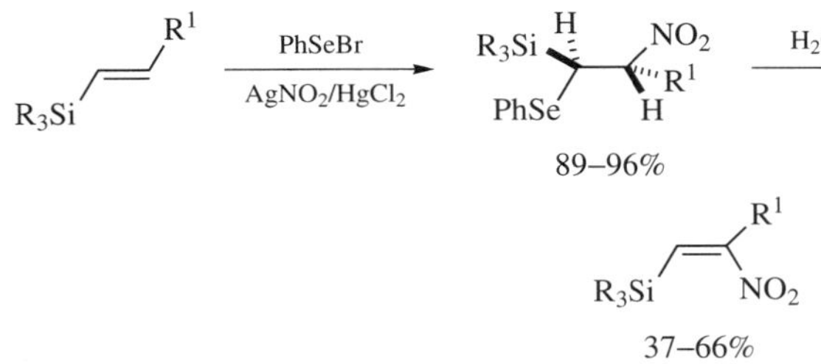

(31)

Using a similar procedure the nitro and seleno groups were introduced stereospecifically at the 1,2-positions of cyclic and acyclic conjugated dienes[40] (eq 32).

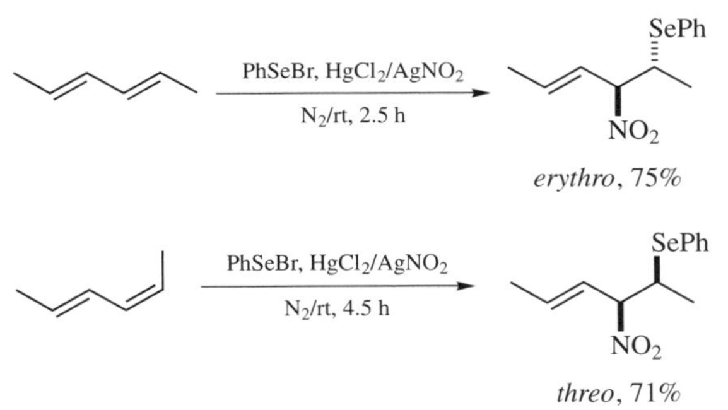

(32)

Reactions with Alkynes. The electrophilic addition of benzeneselenenyl bromide, benzeneselenenyl chloride, and other selenenylating reagents to acetylene in different solvents as AcOEt, AcOH, $CHCl_3$, and DMF, showed that the stereochemistry of the adducts was always *E* (eq 33).[41] The *anti* addition suggested the formation of an intermediate selenirenium ion.

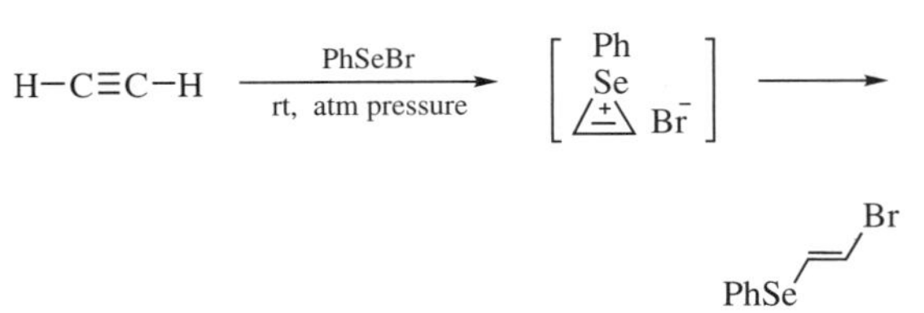

(33)

PhSeBr has been added to alkynes producing diversely functionalized alkenes (eqs 34–38),[42–46] useful for further transformations. For example, the addition of PhSeBr to chalcogen substituted alkynes is highly stereoselective affording β-bromovinyl ketene chalcogenoacetals (eq 38).[46] These in turn, can be coupled with other alkynes to generate chalcogenoenyne compounds.

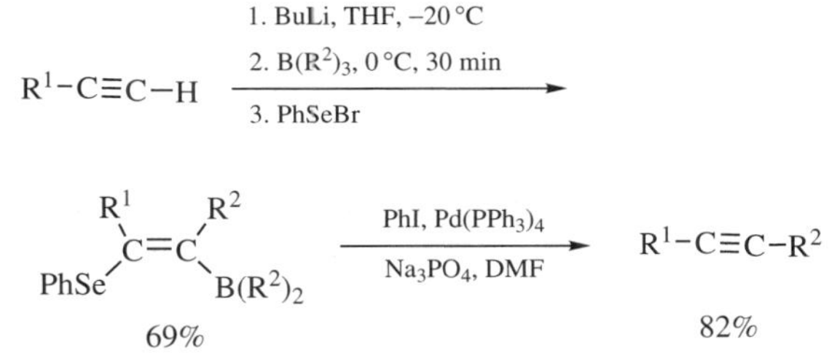

(34)

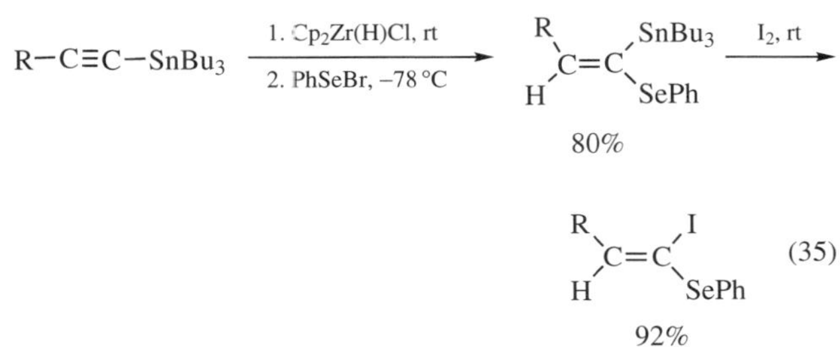

(35)

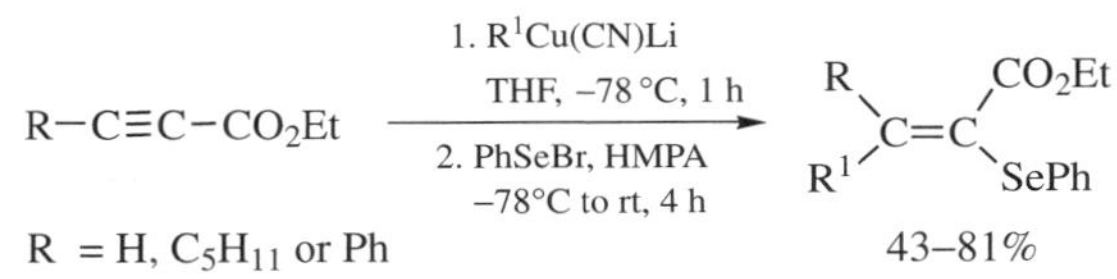

(36)

R = H, C_5H_{11} or Ph

R^1 = Me, Bu, sec-Bu

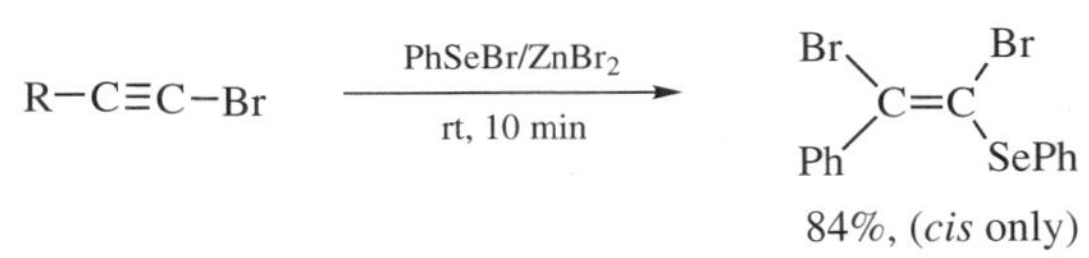

(37)

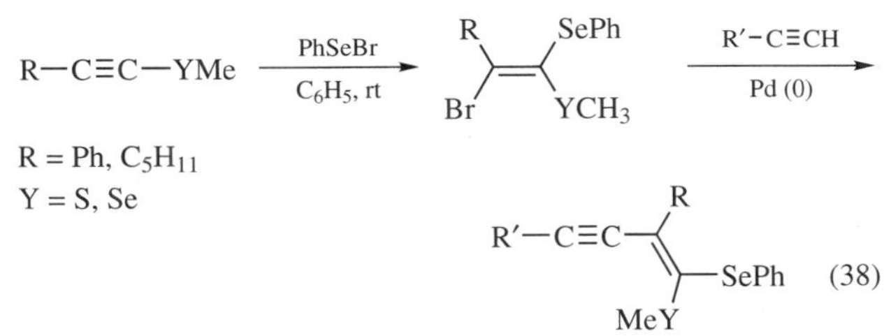

R = Ph, C_5H_{11}

Y = S, Se

(38)

R′ = CH_2OH , $C(Me_2)OH$

C(Me)(Et)OH

Mild conditions for the formation of alkynylselenides from alkynes with the assistance of copper[1] iodide in place of $AgNO_3$ (eq 10)[16] have been described and are compatible with the presence of numerous groups including halogens, double bonds, hydroxyls, amino, or ester groups.[47] Lower yields were obtained with PhSeCl.

A selenocyclization reaction of a deactivated di-substituted alkyne was performed using PhSeBr (eq 39).[48] Better cyclization yields were obtained using $p\text{-}NO_2C_6H_4SCl$ as electrophile.

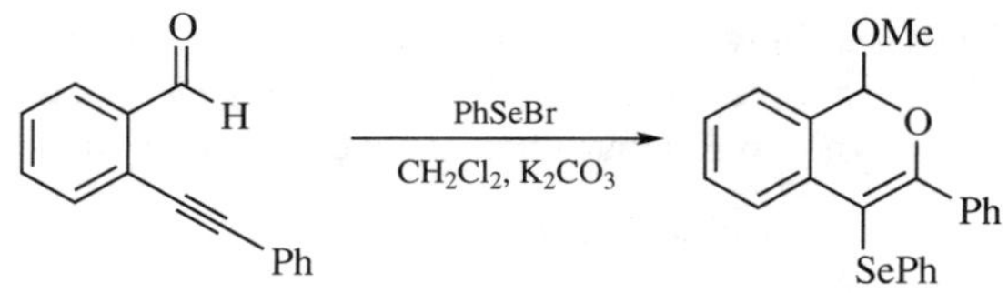

(39)

Synthesis of Selenoesters. Selenoesters have been prepared with the aid of PhSeCl[49] or PhSeBr[50] and serve as precursors to acyl radicals, intermediates that can be added to cyclic alkenes (eqs 40 and 41).[51–53]

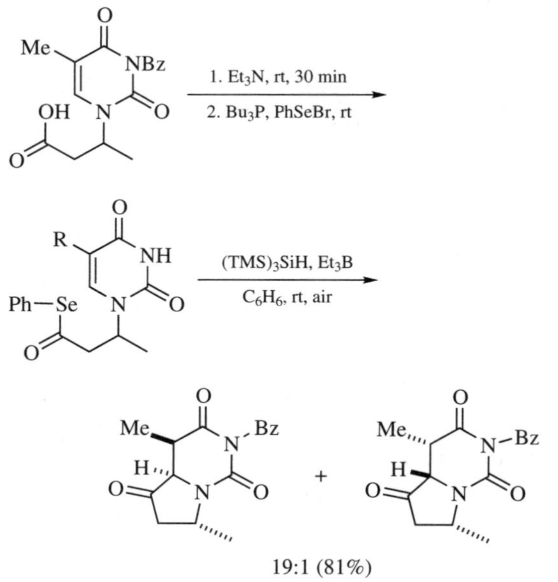

(40)

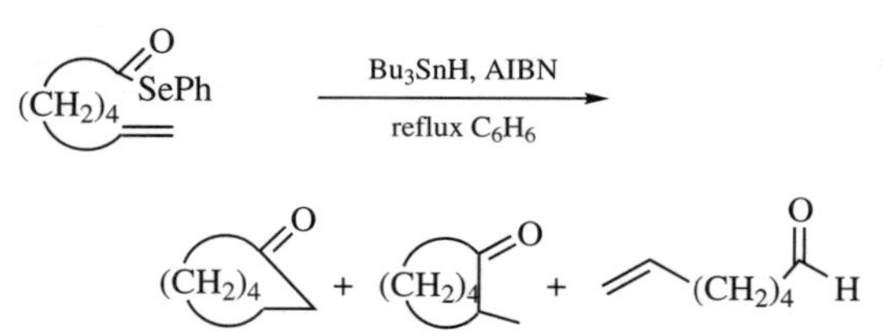

(41)

Transmetalation Reactions. Benzeneselenenyl bromide reacts with organozirconium metallacycles and with silyl-substituted alkenyl zirconium complexes producing (alkanedienyl) phenylselenides (eq 42).[54] Hydrozirconation of alkylsilyl acetylenes followed by PhSeBr treatment yielded phenylseleno alkenes (eq 43).[55] The best results were obtained when a stabilizing group, phenyl or trimethylsilyl groups, were placed at the α position of the phenylseleno group. The 1-silyl-1-phenylselanyl alkene produced is regarded as the equivalent of a geminal cation-anion synthon.

Addition to Coordination Compounds. PhSeBr has been added to planar platinum^{II} coordination compounds with good yields (eq 44).[56] Oxidative addition directs to octahedral products bearing bromo and phenylseleno ligands in *trans*-positions. Depending on the nature of the hydrocarbyl ligands, the complexes may undergo reductive elimination of a Se–C bond with the formation of new Pt^{II} complexes and organoselenium compounds.

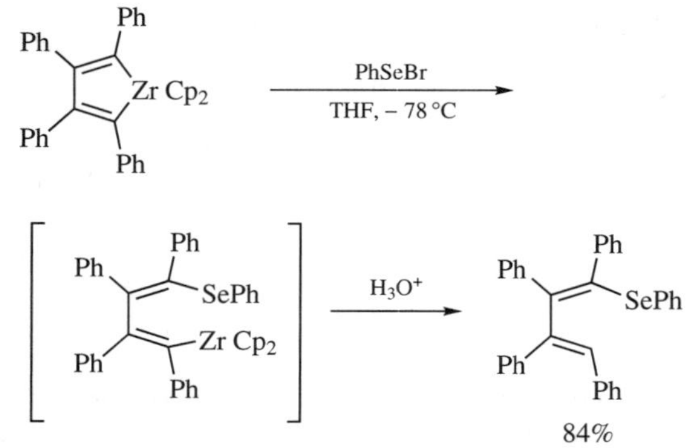

(42)

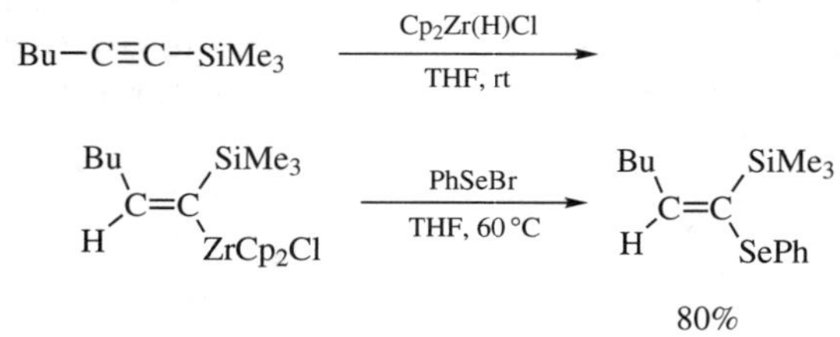

(43)

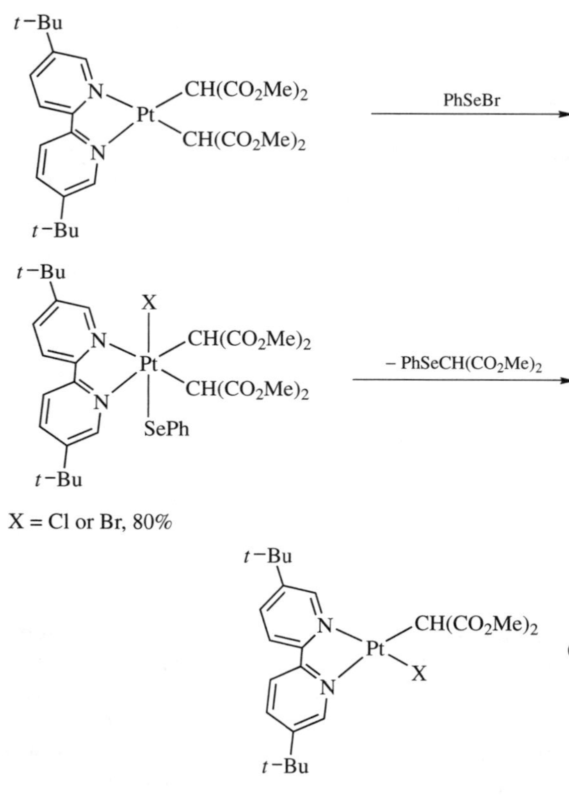

(44)

Reaction with Phosphoranes and Phosphaalkynes. Phosphorus ylides react with PhSeBr to generate alkyl(phenylselanyl) phosphonium salts (eq 45).[57] In particular, the synthesis and

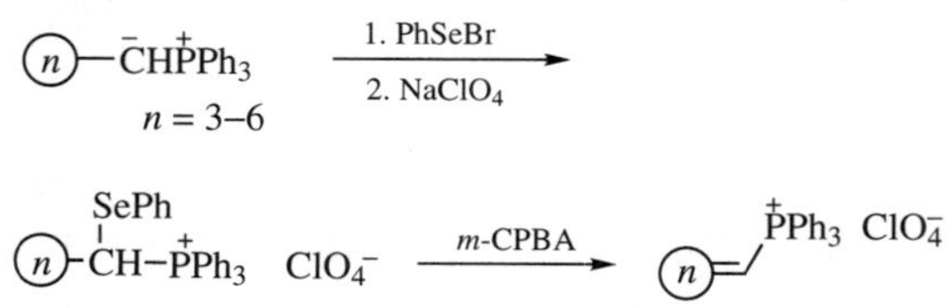

(45)

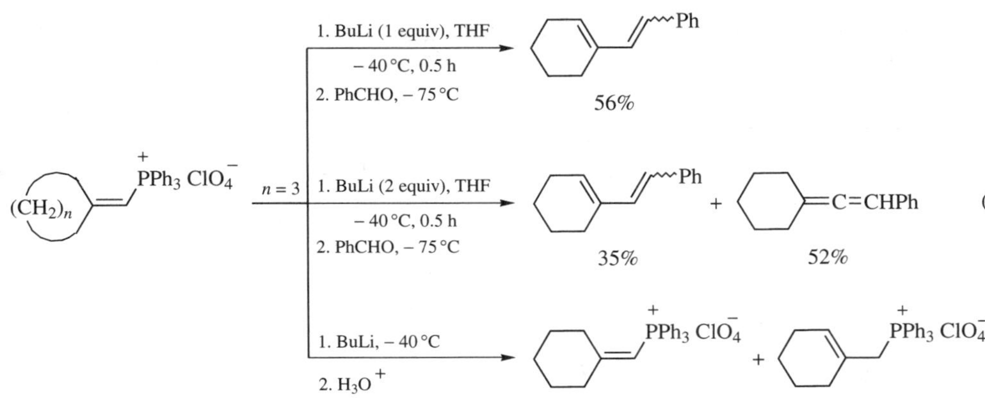

(46)

reactions of [cycloalkyl(phenylselanyl)methyl]triphenylphosphonium salts is of synthetic value. The oxidative elimination of the phenylseleno group provides the corresponding (cycloalkylydenemethyl) triphenylphosphonium salts which in turn can be employed in the formation of conjugated dienes, allenes, and other interesting alkenes (eq 46). The transylidation reactions of PhSeBr with two equivalents of an alkylidenetriphenylphosphorane gave selenophosphoranes, $Ph_3P{=}CRSePh$,[58,59] used in Wittig reactions (eq 48).[59]

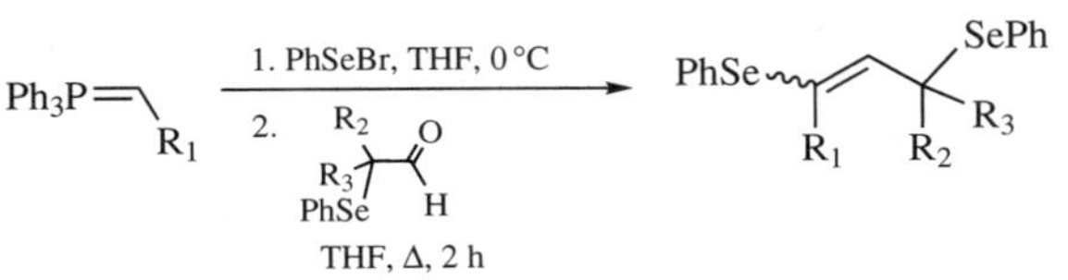

(47)

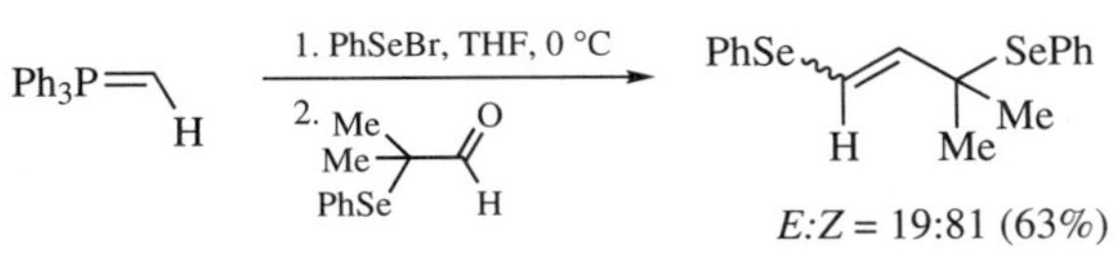

(48)

The anions of α-phosphono esters reacted with PhSeBr to give α-phosphono(phenylseleno) esters (eq 49)[60] which enable the synthesis of alkenes bearing three electron withdrawing groups.

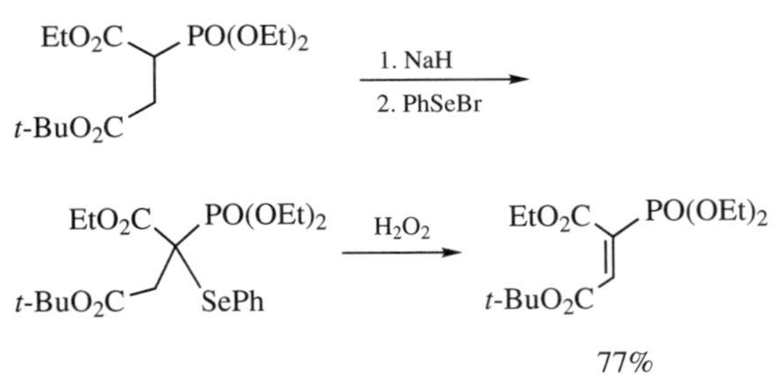

(49)

The anions of α-phenylseleno-α-phosphono esters underwent Wittig-Horner reactions. Yields in the range of 65–89% were obtained with aliphatic and aromatic aldehydes (eq 50). The best results were attained using bases with sodium as counterion, such as sodium hexamethyldisilazide and sodium hydride.[61]

In these reactions, the E isomer was marginally favored. When the Z isomer was required, an alternative protocol was developed using the coupling of the (phenylseleno)methylidene phosphorane with aldehydes under microwave irradiation (eq 51).[61]

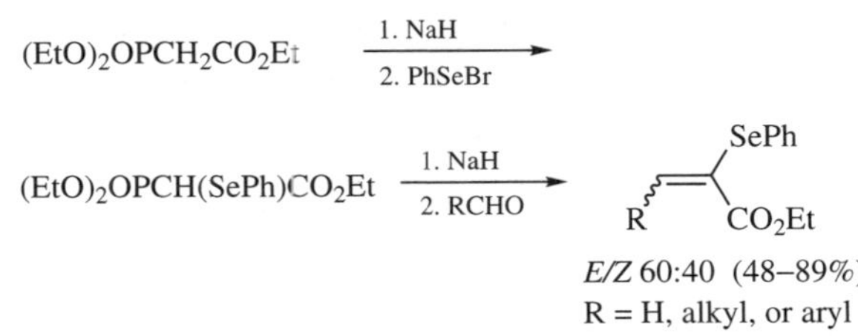

(50)

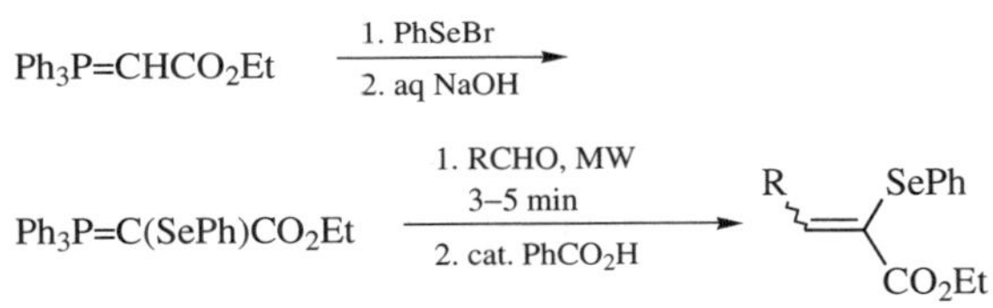

(51)

Treatment of alkyl phosphonates with three equivalents of lithium diisopropylamide and 2 equivalents of benzeneselenenyl bromide provided the diethyl 1,1-bis(phenylseleno)methyl phosponate anion, which in the presence of an aldehyde (aromatic or aliphatic) produced ketene selenoacetals, in a one pot procedure (eq 52).[62] When the phosphonate anion was quenched with a ketone (aliphatic or aromatic), a transfer of a phenylseleno group to the α-position of the ketone was observed.

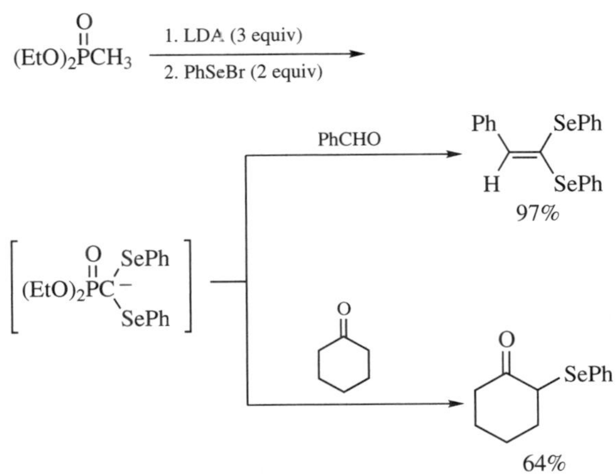

(52)

Addition of PhSeBr or PhSeCl to phosphaalkynes (eq 53)[63] led to selenophosphaalkenes stereo- and regioselectively.

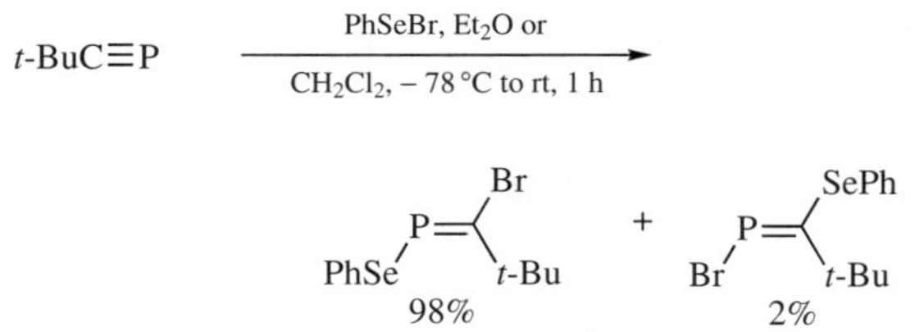

(53)

Addition to Alkenes: Selenocyclofunctionalization Reactions. Olefins containing an oxygen substituent (acyloxy or aryloxy group) in allylic or homoallylic position showed a marked preference for anti-Markovnikov addition (eqs 54 and 55).[28,64] The regioisomeric relationship can be reversed by the use of catalytic amounts of boron trifluoride etherate to produce mixtures containing more than 90% of the Markovnikov product.[64]

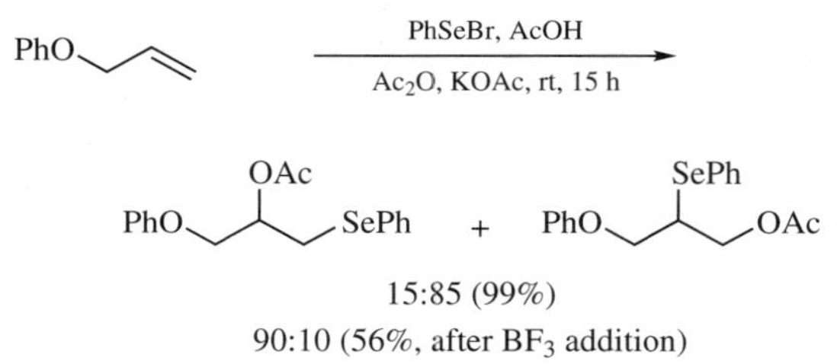

(54)

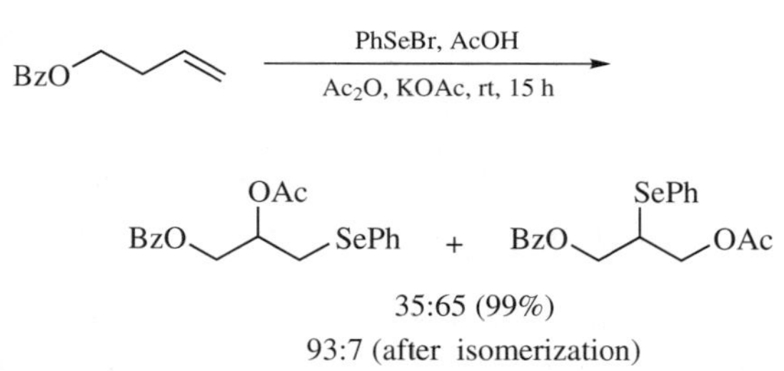

(55)

The addition of PhSeBr to acyclic allylic alcohols, ethers, and esters, in the presence of 2,6-di-*t*-butylpyridine (DTBP) and methanol proceeded in a highly regio- and stereoselective manner (eq 56)[65] with good yields (70–90%). In the absence of DTBP, the reaction was less clean and yields were diminished.

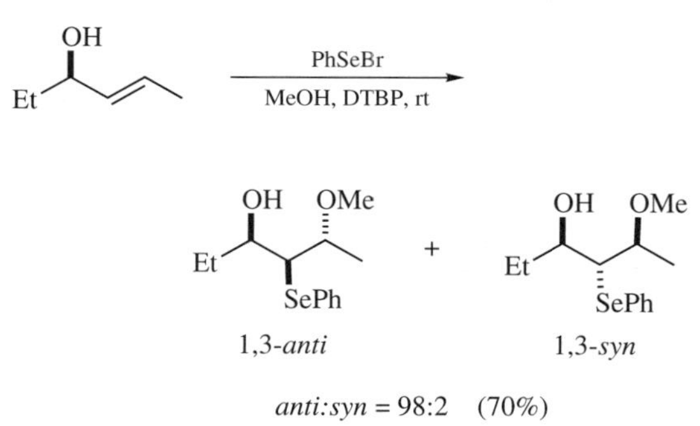

(56)

Mixed acetals of benzeneselenenylacetaldehyde were prepared from allylic alcohols, ethyl vinyl ether, and PhSeBr, in the presence of diisopropylamine (eq 57).[66] The oxidative elimination of the phenylseleno group produced vinyl allyl ethers which afforded γ,δ-unsaturated esters via an ester Claisen rearrangement.

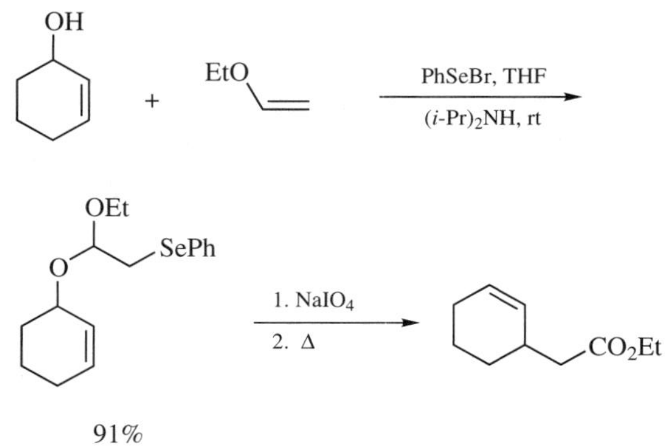

(57)

Azanucleosides have been rapidly prepared with excellent regio- and stereoselectivity with the aid of PhSeBr (eq 58).[67]

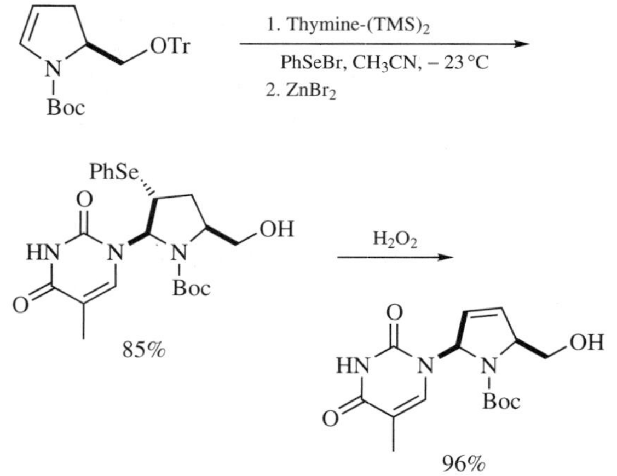

(58)

PhSeBr adds readily to one of the double bonds of allenes. The adduct often rearranges on heating or in acidic media (eq 59).[68,69]

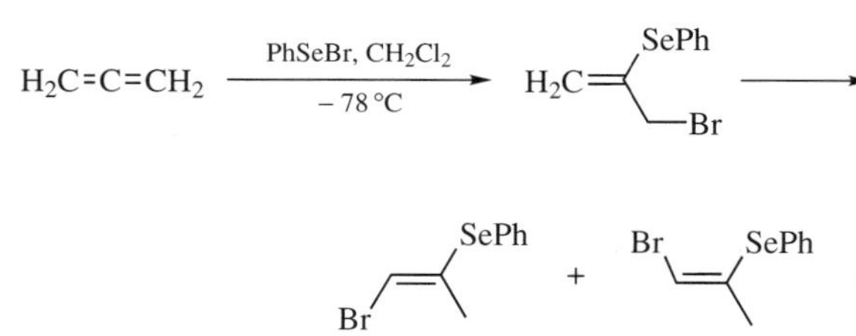

(59)

Phenylselenenyl halides induce cyclization of alkenes or alkynes bearing an internal nucleophile (eq 60); this general

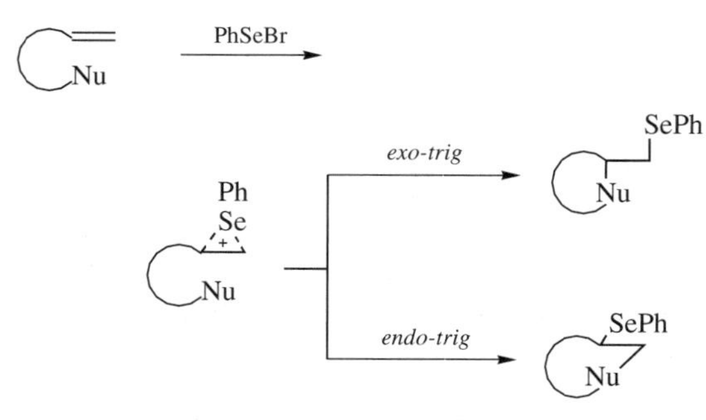

(60)

reaction has been widely applied in the last decade and most of the examples exhibit the formation of five- and six-membered heterocycles, but four- and seven-membered rings are also often produced.

A review concerning the broad general area of selenocyclofunctionalization reactions is available;[70] others cover examples using PhSeBr and other pioneering phenylselenenic reagents for selenolactonization[71,72] and for cyclization of amides.[73]

Phenylselenoetherification of Δ^4-alkenols affords a mixture of five- and six-membered rings. The regioselectivity and the yield of the addition reactions depend on the structural features and on the reaction conditions. Normally, a high number of substituents on the double bond and next to the carbinol group decreases the yield. Addition of pyridine (py) to the reaction mixture improves the yield of the cyclization and even allows reaction of tertiary alcohols with tri-substituted alkenes instantaneously and quantitatively, at room temperature (eqs 61 and 62).[74] There was no improvement in the reaction when $NaHCO_3$ or NEt_3 were used instead of pyridine. In some cases, in the absence of pyridine, there is no cyclization reaction when substrates are highly substituted. PhSeCl and PhSeBr worked equally well in the presence of equimolecular quantities of pyridine.

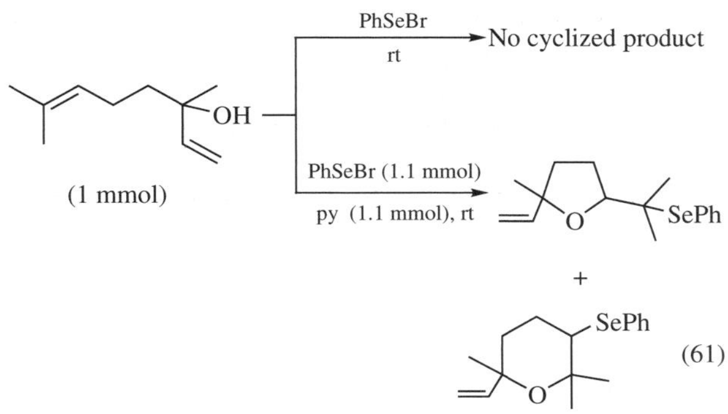

(61)

88:12 (100%)

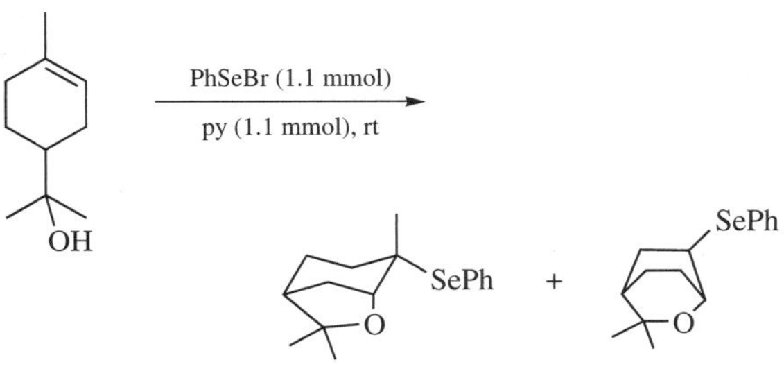

(62)

5:95 (100%)

Δ^5-Alkenols are cyclized to give the 6-*exo*-trig adduct (eq 63).[75] The corresponding benzyl ethers of the unsaturated alcohols work equally well in these phenylselenoetherification reactions.

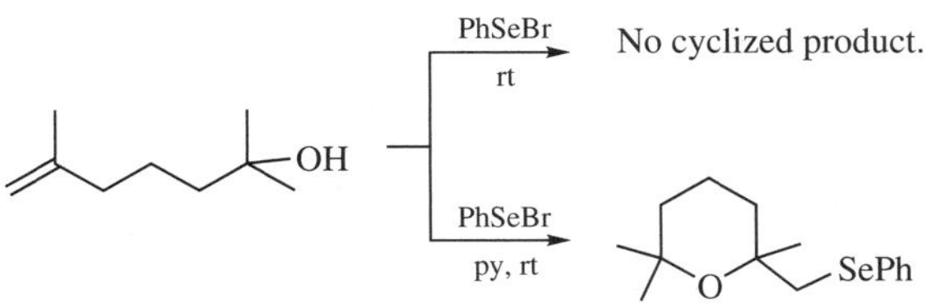

(63)

An interesting study on the regioselectivity of the selenoetherification of a pair of diastereoisomeric acyclic Δ^4-alkenols using PhSeCl or PhSeBr showed a high dependence on alkene geometry of the reaction (eqs 64 and 65).[76] Changes in product composition are obtained when reactions are carried out at different temperatures. A significant influence is exerted by the selenenyl agent. For example, when the (*Z*)-alkenol was treated with PhSeCl, at $-78\,^\circ$C, the *threo*/*erythro* ratio was 2:98 whereas the ratio was 67:33 when PhSeBr was used.

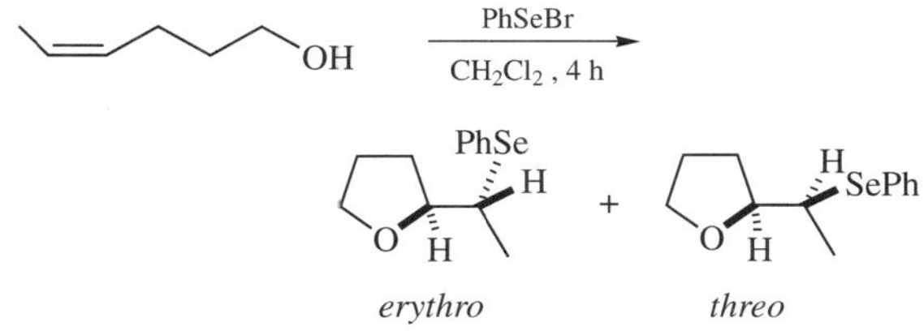

(64)

−78 °C	67:33 (83%)
rt	70:30 (75%)

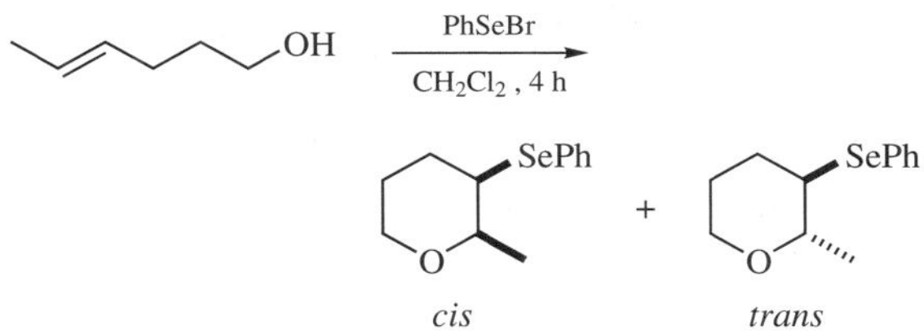

(65)

−78 °C	20:80 (77%)
rt	35:65 (65%)

Easily enolizable unsaturated β-keto esters or β-diketones can also provide an unsaturated ether ring (eqs 66–68).[77,78]

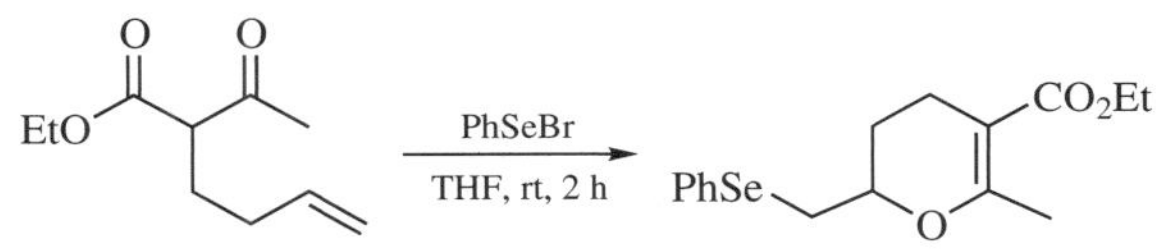

(66)

65%

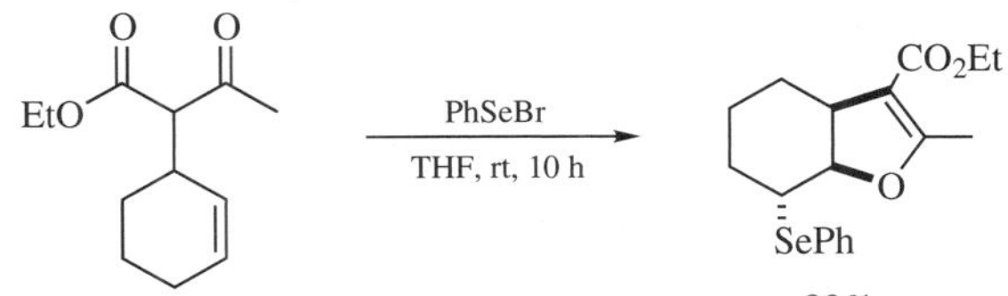

(67)

80%

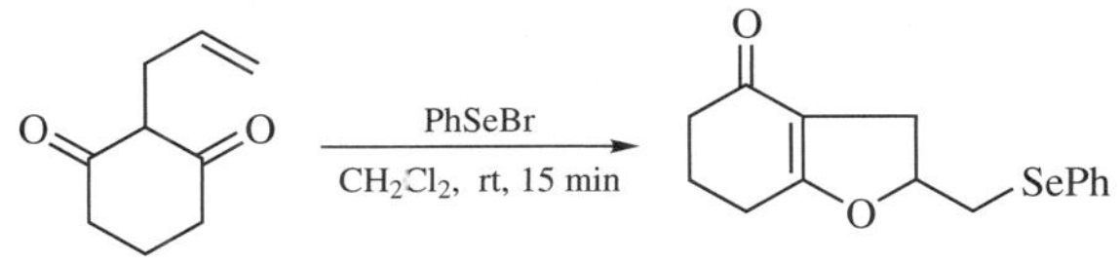

(68)

80%

γ-Unsaturated acids, on exposure to PhSeBr afforded γ-lactones (eqs 69 and 70)[79,80] through a 5-*exo* cyclization step. A different pathway was followed by β-hydroxy-γ-unsaturated acids that gave the product of a 6-*endo* cyclization (eq 71).[81] A high yield (94%) and an excellent diastereoselectivity could be obtained using 2 equivalents of PhSeBr. When one equivalent was introduced, the yield attained was only 48%. Surprisingly, poor yields and low diastereoselectivity were observed when PhSeCl replaced PhSeBr.

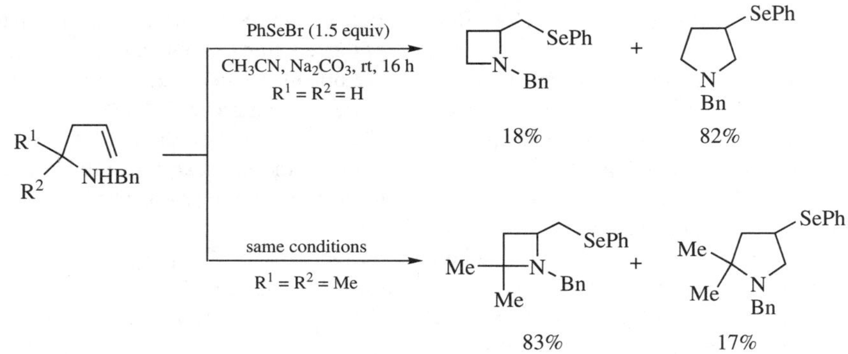

(73)

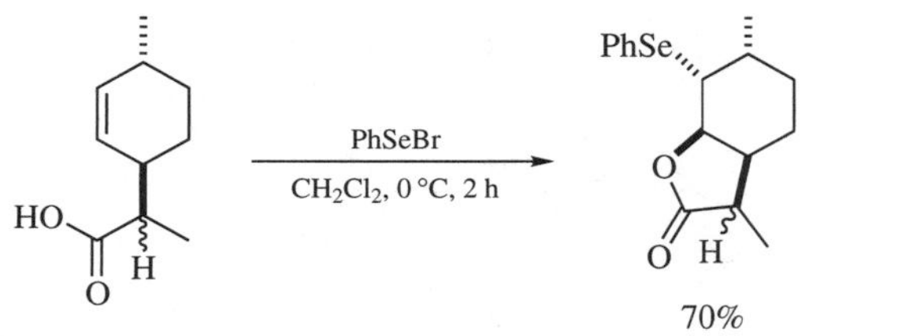

(69)

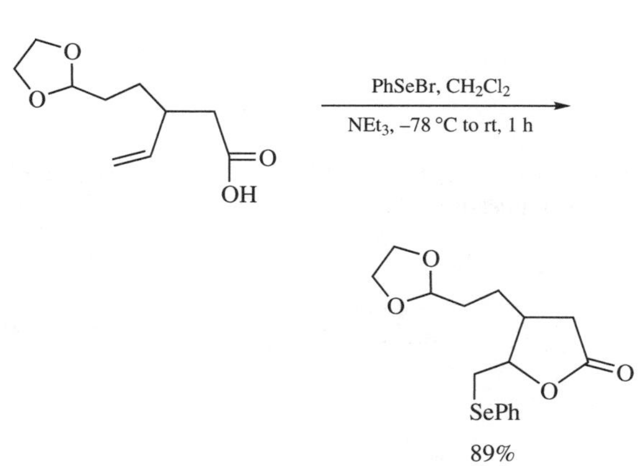

(70)

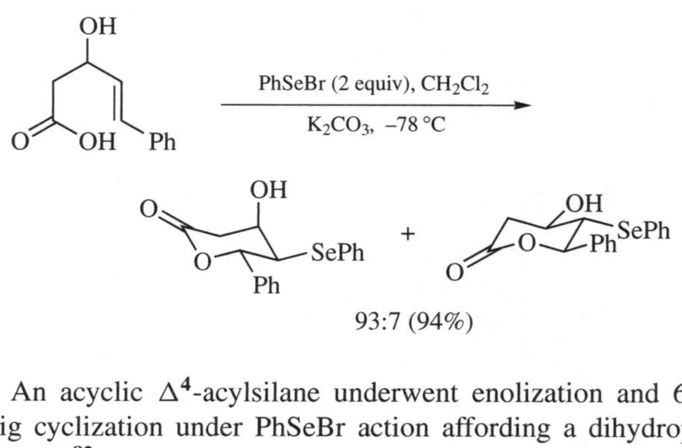

(71)

An acyclic Δ^4-acylsilane underwent enolization and 6-*exo*-trig cyclization under PhSeBr action affording a dihydropyran (eq 72).[82]

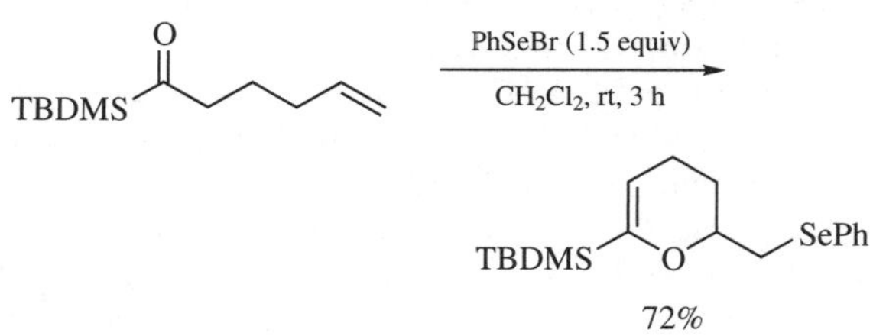

(72)

Electrophilic cyclization of unsaturated nitrogen compounds has been widely applied to yield four-, five-, six-, and seven-membered heterocycles. Homoallyl benzylamines were treated with PhSeCl, PhSeBr, and PhSeI leading to mixtures of azetidines and pyrrolidines (eq 73).[83,84] The azetidine derivatives were found to be the major products when the solvent was acetonitrile and when two substituents were present at the homoallylic carbon. When no substituents were placed at the allylic or homoallylic carbons, the pyrrolidine ring compounds were formed preferentially. The synthesized α-alkyl- and α,α-dialkyl pyrrolidines could be transformed into alkylated-4-halopyrrolidines using an excess of the selenenylating agent.[85]

Amides work as ambident nucleophiles in selenocyclofunctionalization giving *N*- and *N,O*-heterocycles, via *N*- or *O*-cyclization (eqs 74 and 75).

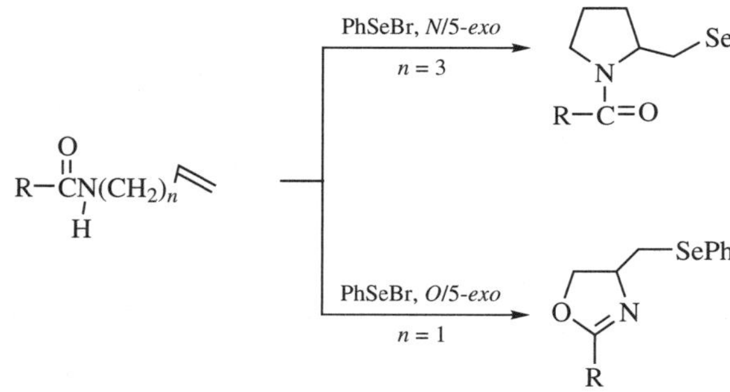

(74)

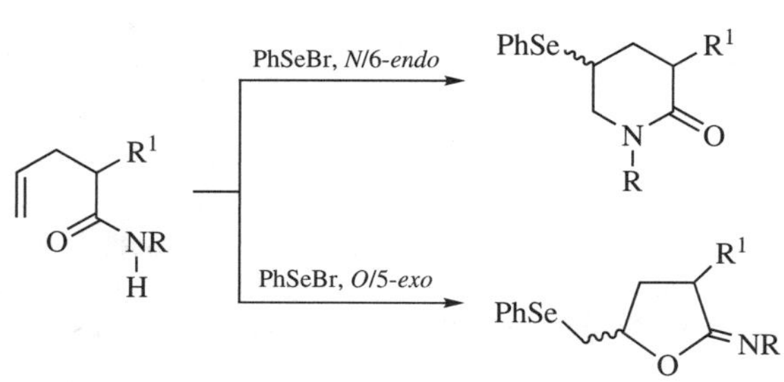

(75)

The regioselectivity of the cyclization depends on several factors, mainly structural ones. For example, alkyl groups on the alkenyl residue modify the relative rate of 5-*exo* and 6-*endo* cyclization (eq 76);[86] PhSeBr afforded better yields of the selenocyclization than PhSeCl.

N-Allyl amides underwent *O*-cyclization via a 5-*exo* procedure producing a nitrogenated salt which was opened on hydrolysis (eq 77).[87] The *threo* compound can be converted into the *erythro*

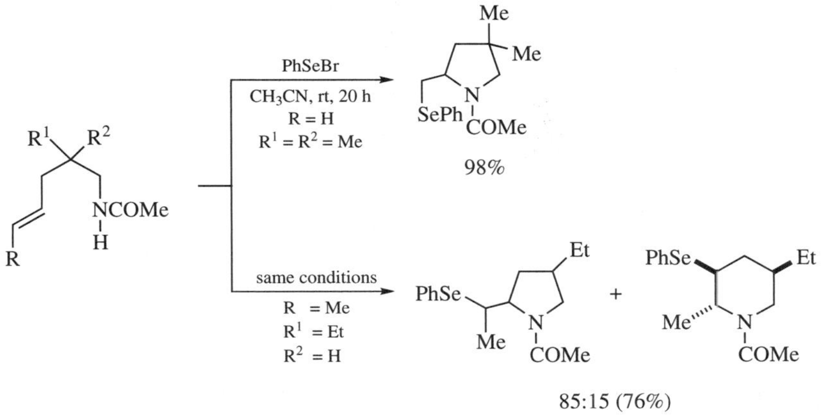

(76)

derivative, in a thermodynamically controlled reaction. Electron deficient unsaturated amides react very slowly with PhSeBr, at room temperature, and poor yields are obtained. With PhSeCl or N-PSP more forcing conditions were required and yields were diminished (eq 78).[88]

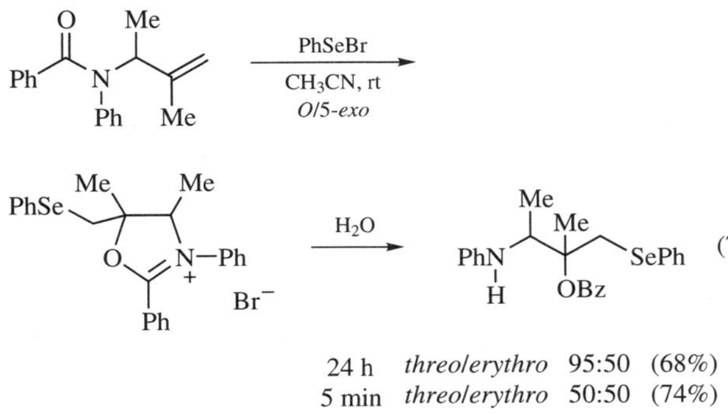

(77)

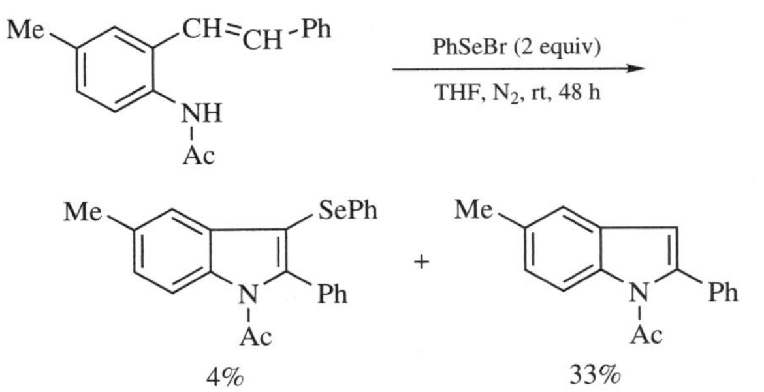

(78)

Besides amines and amides, a variety of unsaturated nitrogenated substrates have been used in selenocyclization reactions; for example: lactams (eq 79),[86] 1,3-oxazolines (eq 80),[89]

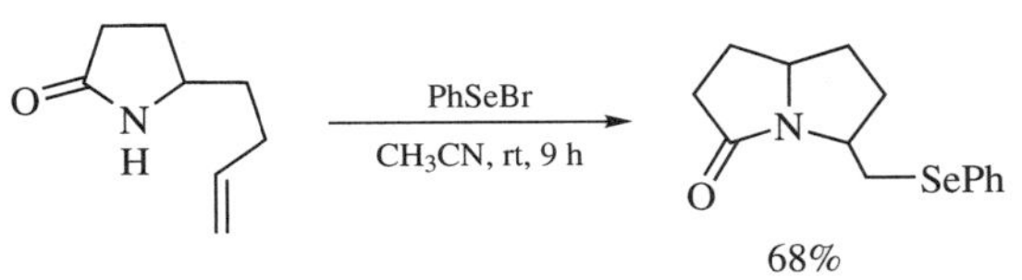

(79)

imidates (eqs 81 and 82),[86,89] carbamates and sulfonamides (eqs 83 and 84),[90–92] hydrazides (eq 85),[93] hydrazones (eqs 86–88),[94] nitrones (eq 89),[95] oximes (eqs 90–93),[96–99] enamines (eq 94),[100] thiazolines (eq 95),[101] and thioamides (eq 96).[102] Valuable heterocycles are produced in good yields; among them,

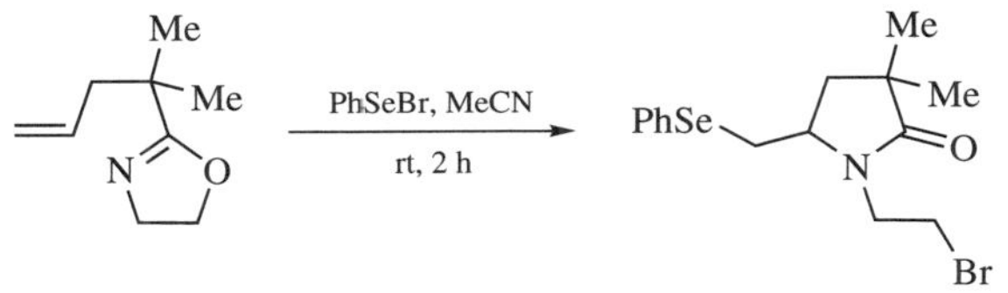

(80)

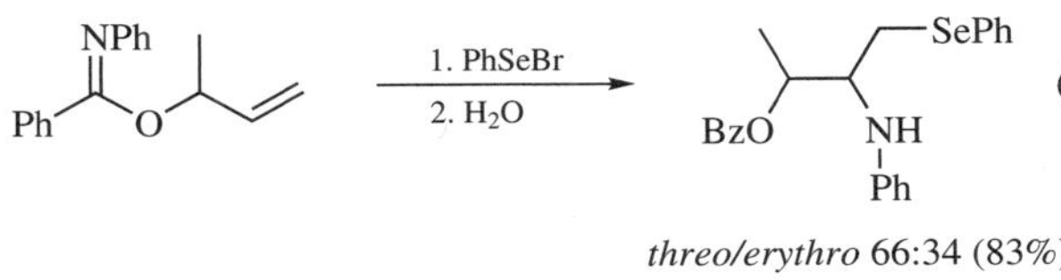

(81)

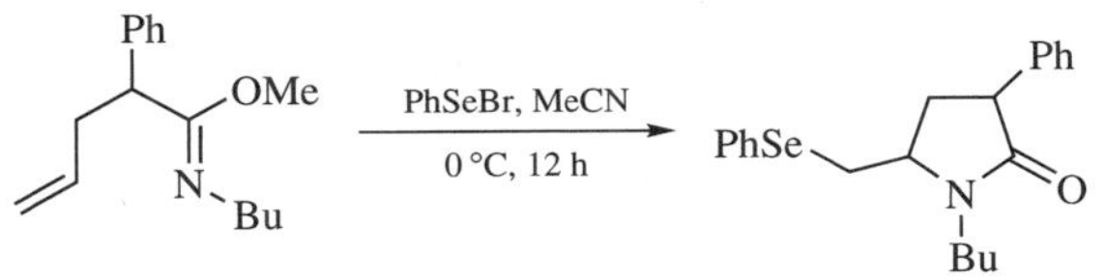

(82)

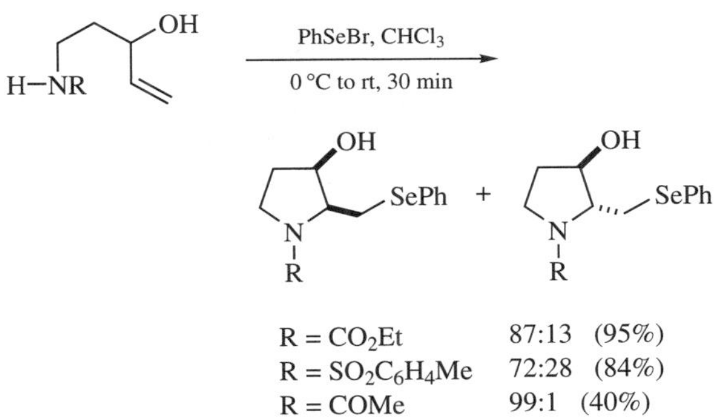

(83)

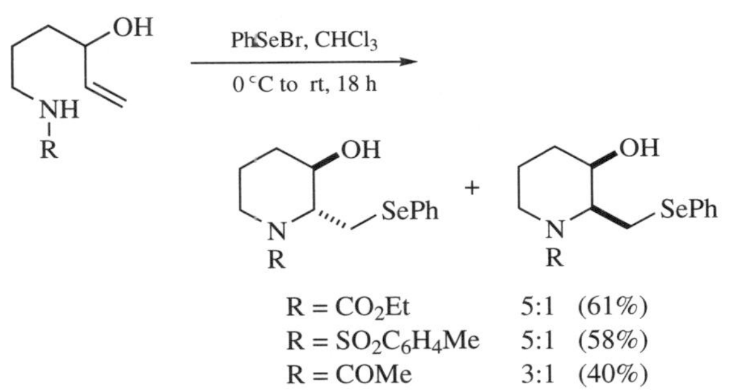

(84)

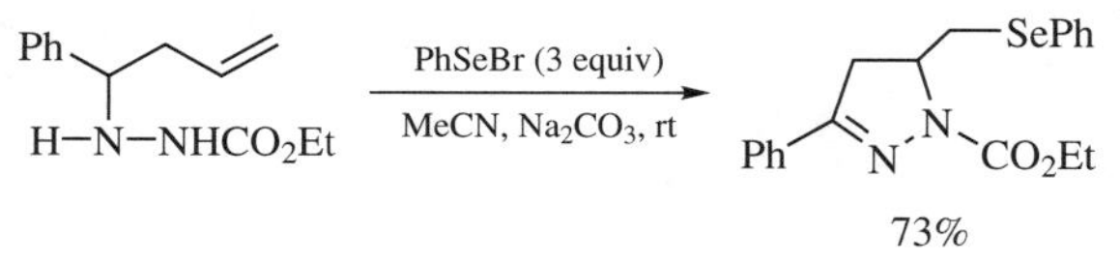

(85)

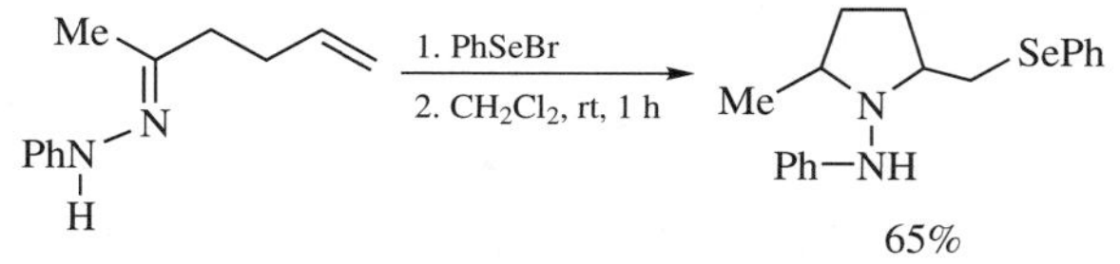

(86)

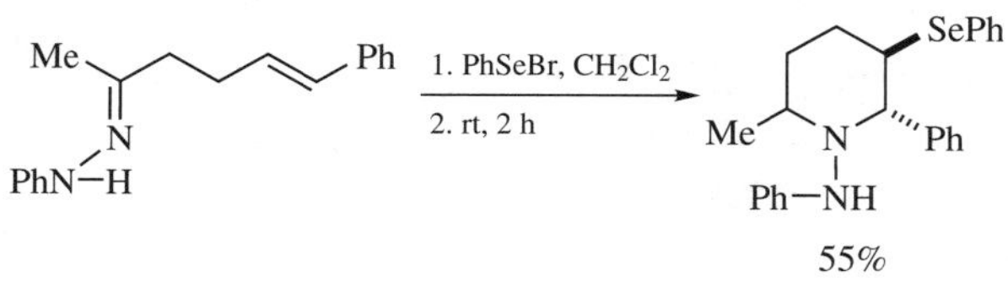

(87)

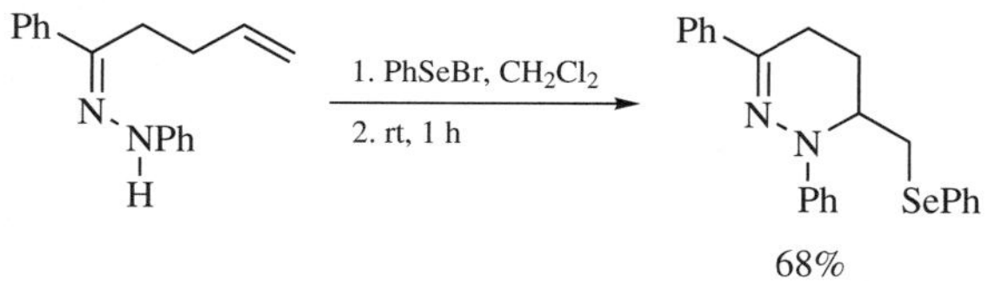

(88)

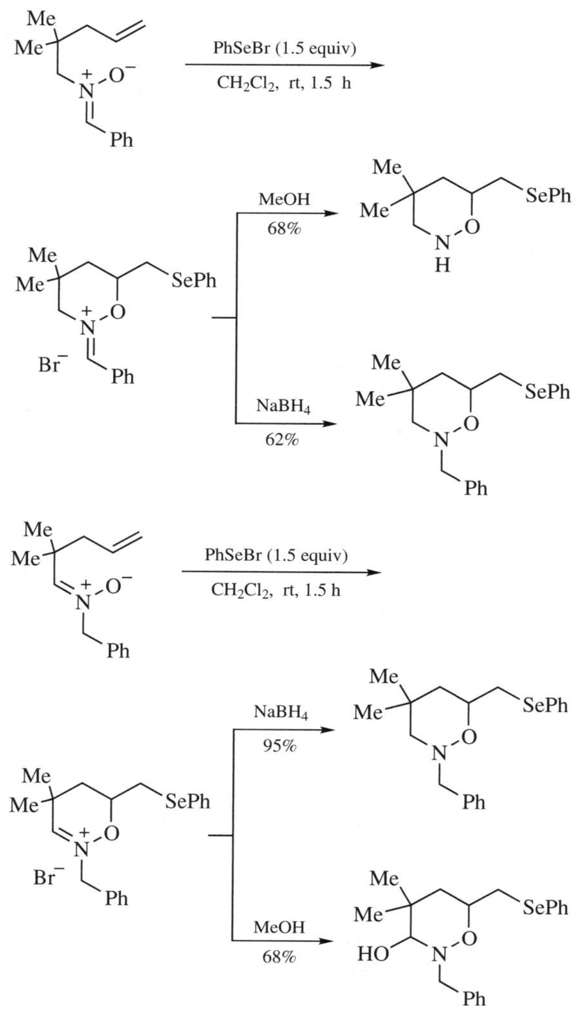

(89)

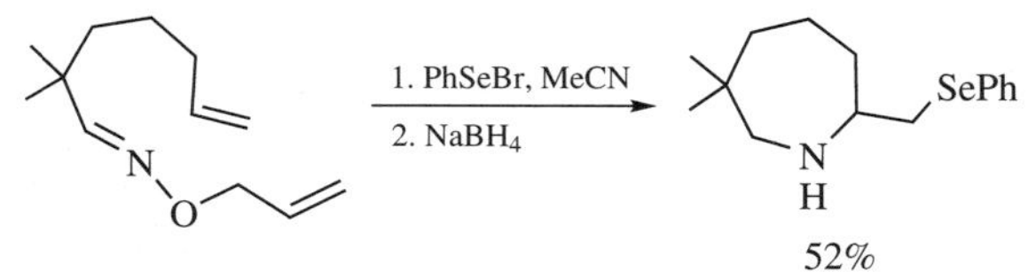

(90)

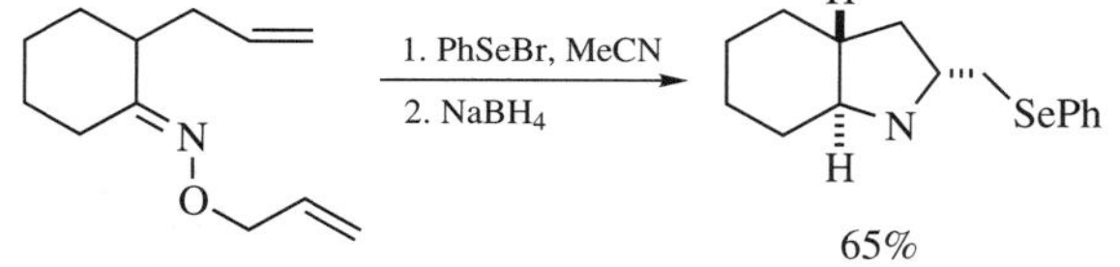

(91)

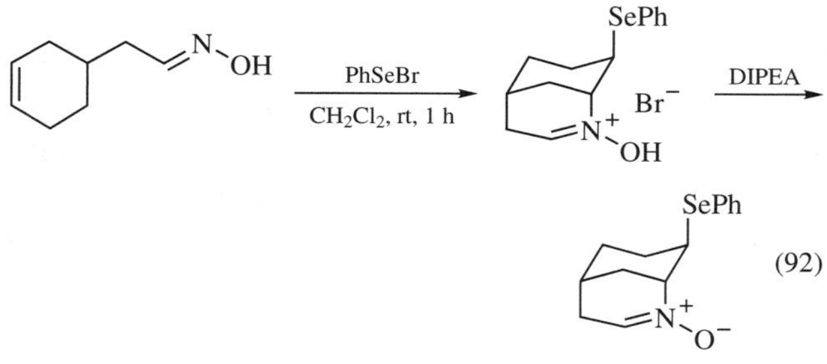

(92)

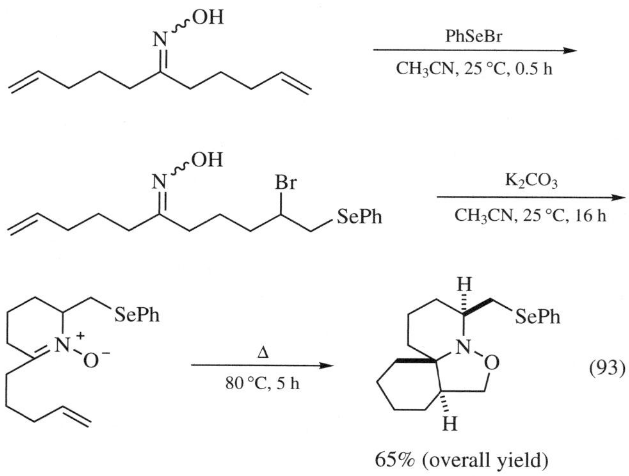

(93)

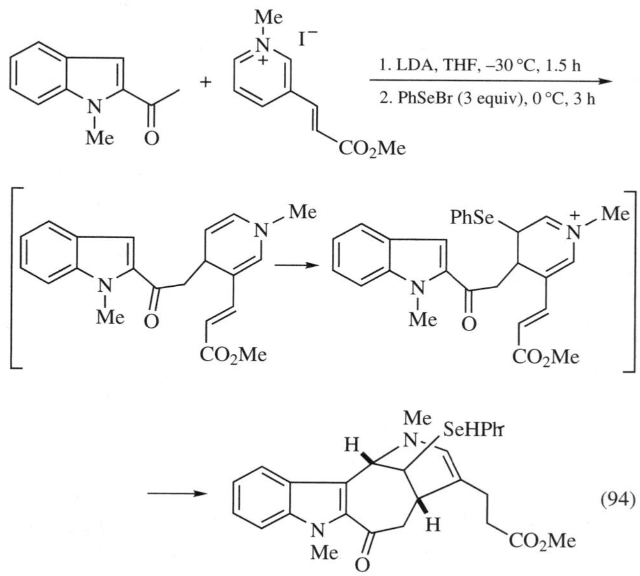

(94)

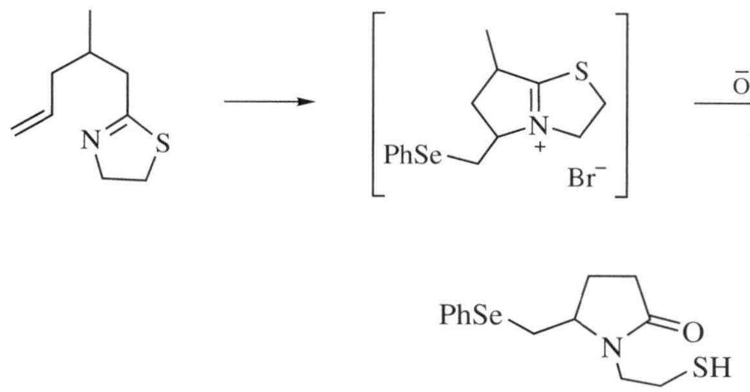

(95)

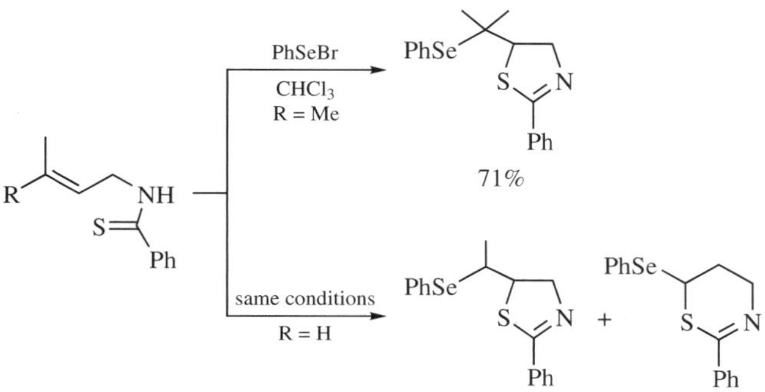

87:13 (85)% (96)

azetidines, pyrrolidines, piperidines, lactams, pyrrolizidines, pyrazolidines, pyrazolines, pyridazines, 1,3-oxazines, azepines, hydroquinolines, perhydroindoles, diazepines, and many others.

Formation of Highly Selective Glycosylating Agents. Glycosyl selenenylsufides (glyco-SeS) have been revealed to be excellent glycosylating agents, allowing glycoconjugation of peptides, proteins, and alkane thiols. Glyco-SeS were prepared by the action of PhSeBr on the thiol group present in proteine cysteine and saccharides (eqs 97 and 98).[103]

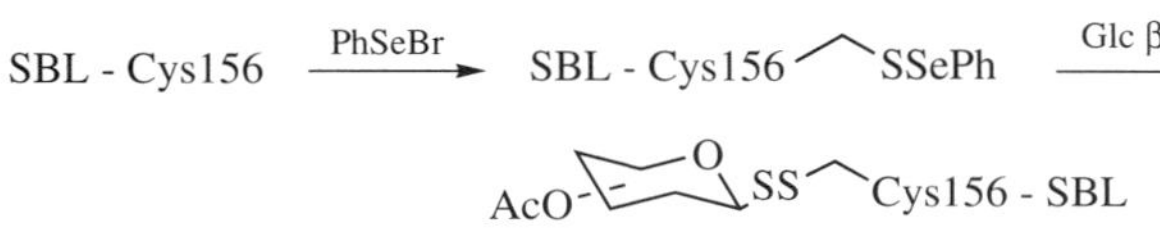

(97)

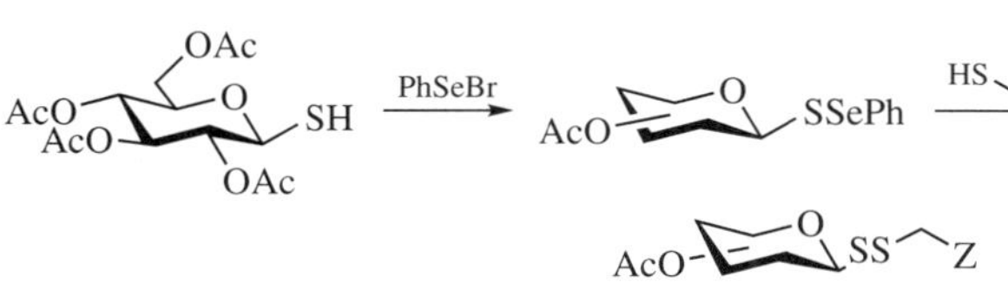

(98)

Z = Me (82%)
Z = Boc-Cys-Thr-OMe (75%)
Z = SBL-Cys156 (95%)
Z = BSA-Cys58 (95%)

Related Reagents. A number of chiral electrophilic selenium reagents have been tested for their possible use in asymmetric synthesis (addition to alkenes, eq 9). Diastereomeric excesses as high as 79% have been achieved for (**1**) (R = *N*-(2,4-dinitrophenyl)prolinoyl),[104] and 93% for the triflate derived from (**2**).[105] Little asymmetric induction was seen in the addition of (**3**) to alkenes.[106]

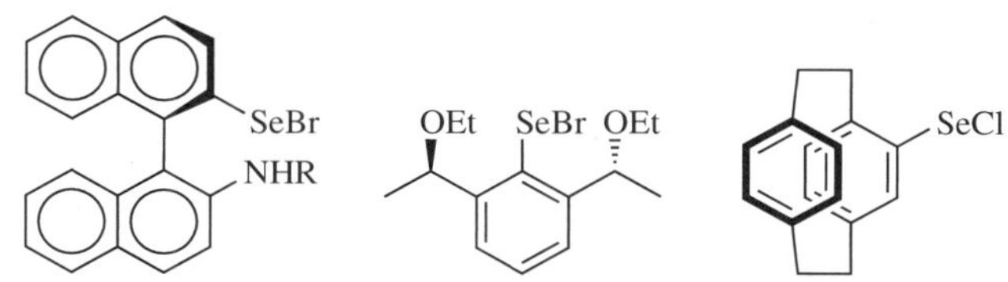

1 **2** **3**

In the last decade many new arylselenenyl compounds have been synthesized mainly with a view to avoiding competition reactions from the halide nucleophiles. Some of the more important ones are: benzeneselenenyl trifluoromethanesulfonate; *N*-phenylselenosuccinimide (NPSS),[107,108] *N*-phenylseleno phthalimide (NPSP); phenylselenenyl sulfate,[109,110] benzeneselenenyl trialkylsilanes,[111,112] optically active areneselenenyl reagents[113–119] as (**4**)–(**8**), and polystyrene-supported phenylselanyl bromide derivatives.[117–123] Some selenocyclofunctionalizations have been performed using a polymer-supported arenylselenenyl bromide in aqueous medium, although yields are lower than those obtained in an organic solvent.[124] Catalytic selenolactonization-deselenenylation reactions were carried out with several seleno reagents.[29] The family of chiral phenyl- and arylselenyl halides increases continuously.[38,125–130] See also Benzeneselenenyl Chloride, Benzenesulfenyl Chloride, Phenyl Selenocyanate, 2-Pyridineselenenyl Bromide.

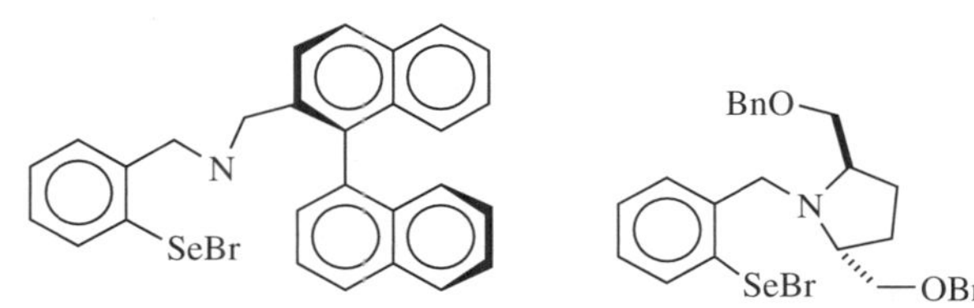

4 **5**

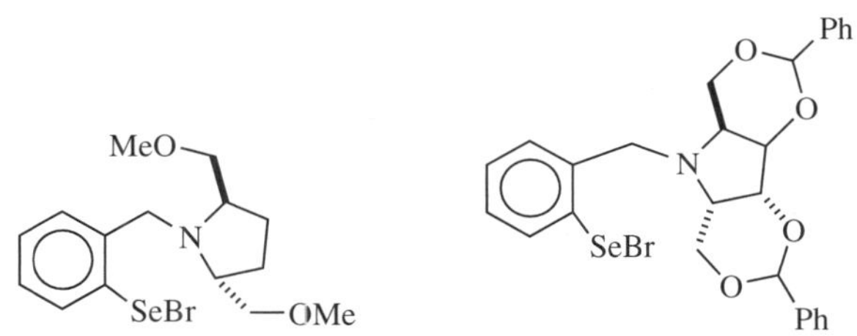

6 **7**

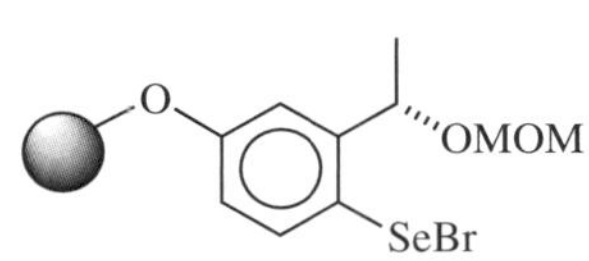

8

1. (a) Clive, D. L. J., *Tetrahedron* **1978**, *34*, 1049. (b) Reich, H. J., *Ann. Chim. (Rome)* **1979**, *12*, 22. (c) Nicolaou, K. C.; Petasis, N. A. Selenium in Natural Products Synthesis; CIS:Philadelphia, **1984**. (d) The Chemistry of Organic Selenium and Tellurium Compounds; Patai, S.; Rappoport, Z., Eds.; Wiley:New York, **1986**; Vol. *1*. (e) Paulmier, C. Selenium Reagents and Intermediates in Organic Synthesis; Pergamon:Oxford, **1986**. (f) Organoselenium Chemistry; Liotta, D., Ed.; Wiley:New York, **1987**.

2. Sharpless, K. B.; Lauer, R. F.; Teranishi, A. Y., *J. Am. Chem. Soc.* **1973**, *95*, 6137.
3. Reich, H. J.; Renga, J. M.; Reich, I. L., *J. Am. Chem. Soc.* **1975**, *97*, 5434.
4. Clive, D. L. J., *Chem. Commun./J. Chem. Soc.*, **1973**, 695.
5. Sharpless, K. B.; Lauer, R. F., *J. Org. Chem.* **1974**, *39*, 429.
6. Raucher, S., *J. Org. Chem.* **1977**, *42*, 2950.
7. Behaghel, O.; Seibert, H., *Ber. Dtsch. Chem. Ges./Chem. Ber.* **1932**, *65*, 812.
8. Reich, H. J.; Renga, J. M.; Reich, I. L., *J. Org. Chem.* **1974**, *39*, 2133.
9. Reich, H. J., *J. Org. Chem.* **1974**, *39*, 428.
10. Ryu, I.; Murai, S.; Niwa, I.; Sonoda, N., *Synthesis* **1977**, 874.
11. Petrzilka, M., *Helv. Chim. Acta* **1978**, *61*, 2286.
12. Brattesani, D. N.; Heathcock, C. H., *Tetrahedron Lett.* **1974**, 2279.
13. Sakakibara, T.; Takai, I.; Ohara, E.; Sudoh, R., *Chem. Commun./J. Chem. Soc.*, **1981**, 261.
14. (a) Raucher, S., *Tetrahedron Lett.* **1977**, 3909. (b) Raucher, S., *Tetrahedron Lett.* **1978**, 2261.
15. Takahashi, T.; Nagashima, H.; Tsuji, J., *Tetrahedron Lett.* **1978**, 799.
16. Hayama, T.; Tomoda, S.; Takeuchi, Y.; Nomura, Y., *Chem. Lett.* **1982**, 1249.
17. Barton, D. H. R.; Bashiardes, G.; Fourrey, J.-L., *Tetrahedron Lett.* **1984**, *25*, 1287.
18. Kloc, K.; Mlochowski, J., *Tetrahedron Lett.* **2001**, *42*, 4899.
19. Wang, W.; Wang, J.; Li, H., *Org. Lett.* **2004**, *6*, 2817.
20. Crich, D.; Yao, Q.; Filzen, G. F., *J. Am. Chem. Soc.* **1995**, *117*, 11455.
21. Paulmier, C., *Phosphorus Sulfur Silicon* **2001**, *171–172*, 279.
22. Labar, D.; Krief, A.; Norberg, B.; Evrard, G.; Durant, F., *Bull. Soc. Chim. Belg.* **1985**, *94*, 1083.
23. Sevrin, M.; Krief, A., *Tetrahedron Lett.* **1980**, *21*, 585.
24. Huang, J. M.; Hong, S. C.; Wu, K. L.; Tsai, Y. M., *Tetrahedron Lett.* **2004**, *45*, 3047.
25. Sevrin, M.; Krief, A., *J. Chem. Soc., Chem. Commun.* **1980**, 656.
26. Fujita, K.; Taka, H.; Oishi, A.; Ikeda, Y.; Taguchi, Y.; Fujie, K.; Saeki, T.; Sakuma, M., *Synlett* **2000**, 1509.
27. Kondo, N.; Fueno, H.; Fujimoto, H.; Makino, M.; Nakaoka, H.; Aoki, I.; Uemura, S., *J. Org. Chem.* **1994**, *59*, 5254.
28. Cooper, M. A.; Ward, A. D., *Tetrahedron* **2004**, *60*, 7963.
29. Ying, T.; Bao, W.; Zhang, Y., *Synth. Commun.* **1996**, *26*, 1517.
30. Lebarillier, L.; Outurquin, F.; Paulmier, C., *Tetrahedron* **2000**, *56*, 7483.
31. Ternon, M.; Pannecoucke, X.; Outurquin, F.; Paulmier, C., *Tetrahedron* **2002**, *58*, 3275.
32. Yamazaki, S.; Yanase, Y.; Tanigawa, E.; Yamabe, S.; Tamura, H., *J. Org. Chem.* **1999**, *64*, 9521.
33. Ponthieux, S.; Outurquin, F.; Paulmier, C., *Tetrahedron Lett.* **1995**, *36*, 6453.
34. Dabdoub, M. J.; Guerrero, Jr., P. G.; Silveira, C. C., *J. Organomet. Chem.* **1993**, *460*, 31.
35. Back, T. G.; Kerr, R. G., *J. Organomet. Chem.* **1985**, *286*, 171.
36. Byers, J. H.; Lane, G. C., *J. Org. Chem.* **1993**, *58*, 3355.
37. Engman, L.; Stern, D.; Frisell, H.; Vessman, K.; Berglund, M.; Ek, B.; Andersson, C. M., *Bioorg. Med. Chem.* **1995**, *3*, 1255.
38. Heldmann, D. K.; Sauer, J., *Tetrahedron Lett.* **1997**, *38*, 5791.
39. Hayama, T.; Tomoda, S.; Takeuchi, Y.; Nomura, Y., *J. Org. Chem.* **1984**, *49*, 3235.
40. Nájera, C.; Yus, M.; Karlsson, U.; Gogoll, A.Bäckvall, J. E., *Tetrahedron Lett.* **1990**, *31*, 4199.
41. Potapov, V. A.; Amosova, S. V.; Petrov, B. V.; Starkova, A. A.; Malyushenko, R. N., *Sulfur Lett.* **1998**, *21*, 109.
42. Gerard, J.; Bietlot, E.; Hevesi, L., *Tetrahedron Lett.* **1998**, *39*, 8735.
43. Dabdoub, M. J.; Baroni, A. C. M., *J. Org. Chem.* **2000**, *65*, 54.
44. Silveira, C. C.; Braga, A. L.; Guerra, R. B., *Tetrahedron Lett.* **2002**, *43*, 3395.
45. Braga, A. L.; Marchi, M. I.; de Andrade, L. H.; Silveira, C. C., *Synth. Commun.* **2000**, *30*, 407.
46. Zeni, G.; Stracke, M. P.; Lissner, E.; Braga, A. L., *Synlett* **2003**, 1880.
47. Braga, A. L.; Silveira, C. C.; Reckziegel, A.; Menezes, P. H., *Tetrahedron Lett.* **1993**, *34*, 8041.
48. Yue, D.; della Cà, N.; Larock, R. C., *Org. Lett.* **2004**, *6*, 1581.
49. Batty, D.; Crich, D., *Synthesis* **1990**, 273.
50. Huang, X.; Liang, C. G., *J. Chem. Res. (S)* **1999**, 634.
51. Evans, P. A.; Manangan, T., *Tetrahedron Lett.* **2001**, *42*, 6637.
52. Crich, D.; Fortt, S. M., *Tetrahedron Lett.* **1988**, *29*, 2585.
53. Batty, D.; Crich, D.; Fortt, S. M., *J. Chem. Soc., Perkin Trans. 1* **1990**, 2875.
54. Chung, H. S.; Oh, D. Y., *Tetrahedron Lett.* **1992**, *33*, 5097.
55. Xu, X. H.; Zheng, W. X.; Huang, X., *Synth. Commun.* **1998**, *28*, 4165.
56. Panunzi, A.; Roviello, G.; Ruffo, F., *Inorg. Chem. Commun.* **2003**, *6*, 1282.
57. Minami, T.; Shikita, S.; So, S.; Nakayama, M.; Yamamoto, I., *J. Org. Chem.* **1988**, *53*, 2937.
58. Petragnani, N.; Rodrigues, R.; Comasseto, J. V., *J. Organomet. Chem.* **1976**, *114*, 281.
59. Renard, M.; Hevesi, L., *Tetrahedron* **1985**, *41*, 5939.
60. Yamazaki, S.; Yamada, K.; Yamamoto, K., *Org. Biomol. Chem.* **2004**, *2*, 257.
61. Silveira, C. C.; Nunes, M. R. S.; Wendling, E.; Braga, A. L., *J. Organomet. Chem.* **2001**, *623*, 131.
62. Shin, W. S.; Lee, K.; Oh, D. Y., *Tetrahedron Lett.* **1992**, *37*, 5375.
63. Francis, M. D.; Jones, C.; Junk, P. C.; Roberts, J. L., *Phosphorus Sulfur Silicon* **1997**, *130*, 23.
64. Engman, L., *J. Org. Chem.* **1989**, *54*, 884.
65. Kim, K. S.; Park, H. B.; Kim, J. Y.; Ahn, Y. H.; Jeong, I. H., *Tetrahedron Lett.* **1996**, *37*, 1249.
66. Pitteloud, R.; Petrzilka, M., *Helv. Chim. Acta.* **1979**, *62*, 1319.
67. Costenero, E. R.; Fontoura, L. A. M.; Oliveira, D. F.; Correia, C. R. D., *Tetrahedron Lett.* **2001**, *42*, 1599.
68. Halazy, S.; Hevesi, L., *Tetrahedron Lett.* **1983**, *24*, 2689.
69. Garratt, D. G.; Beaulieu, P. L.; Morisset, V. M.; Ujjainwalla, M., *Can. J. Chem.* **1980**, *58*, 2737.
70. Petragnani, N.; Stefani, H. A.; Valduga, C. J., *Tetrahedron* **2001**, *57*, 1411.
71. Ranganathan, S.; Muraleedharan, K. M.; Vaish, N. K.; Jayaraman, N., *Tetrahedron* **2004**, *60*, 5273.
72. Frederickson, M.; Grigg, R., *Org. Prep. Proced. Int.* **1997**, *29*, 33.
73. Robin, S.; Rousseau, G., *Tetrahedron* **1998**, *54*, 13681.
74. Mojsilovic, B. M.; Bugarcic, Z. M., *Heteroatom Chem.* **2001**, *12*, 475.
75. Bugarcic, Z.; Konstantinovic, S.; Mojsilovic, B., *Indian J. Chem.* **1999**, *38B*, 728.
76. Konstantinovic, S.; Bugarcic, Z.; Milosavljevic, S.; Schroth, G.; Mihailovic, M. L., *Liebigs Ann. Chem.* **1992**, 261.
77. Stefani, H. A.; Petragnani, N.; Brandt, C. A.; Rando, D. G.; Valduga, C. J., *Synth. Commun.* **1999**, *29*, 3517.
78. Ferraz, H. M. C.; Sano, M. K.; Nunes, M. R. S.; Bianco, G. G., *J. Org. Chem.* **2002**, *67*, 4122.
79. Yadav, J. S.; Babu, R. S.; Sabitha, G., *Arkivoc* **2003**, 125.
80. Husinec, S.; Savic, V., *J. Serb. Chem. Soc.* **2000**, *65*, 87.
81. Aprile, C.; Gruttadauria, M.; Amato, M. E.; D'Anna, F.; Lo Meo, P.; Riela, S.; Noto, R., *Tetrahedron* **2003**, *59*, 2241.
82. Tsai, Y. M.; Nieh, H. C.; Cherng, C. D., *J. Org. Chem.* **1992**, *57*, 7010.
83. Berthe, B.; Outurquin, F.; Paulmier, C., *Tetrahedron Lett.* **1997**, *38*, 1393.

84. Pannecoucke, X.; Outurquin, F.; Paulmier, C., *Eur. J. Org. Chem.* **2002**, 995.
85. Outurquin, F.; Pannecoucke, X.; Berthe, B.; Paulmier, C., *Eur. J. Org. Chem.* **2002**, 1007.
86. Toshimitsu, A.; Terao, K.; Uemura, S., *J. Org. Chem.* **1986**, *51*, 1724.
87. Engman, L., *J. Org. Chem.* **1993**, *58*, 2394.
88. Izumi, T.; Sugano, M.; Konno, T., *J. Heterocyclic Chem.* **1992**, *29*, 899.
89. Toshimitsu, A.; Terao, K.; Uemura, S., *J. Chem. Soc., Chem. Commun.* **1986**, 530.
90. Cooper, M. A.; Ward, A. D., *Tetrahedron Lett.* **1992**, *33*, 5999.
91. Cooper, M. A.; Ward, A. D., *Tetrahedron Lett.* **1994**, *35*, 5065.
92. Cooper, M. A.; Francis, C. L.; Holman, J. W.; Kasum, B.; Taverner, T.; Tiekink, E. R. T.; Ward, A. D., *Aust. J. Chem.* **2000**, *53*, 123.
93. Ternon, M.; Outurquin, F.; Paulmier, C., *Tetrahedron* **2001**, *57*, 10259.
94. Tiecco, M.; Testaferri, L.; Marini, F.; Santi, C.; Bagnoli, L.; Temperini, A., *Tetrahedron* **1997**, *53*, 7311.
95. Tiecco, M.; Testaferri, L.; Bagnoli, L., *Tetrahedron* **1996**, *52*, 6811.
96. Dondas, H. A.; Grigg, R.; Markandu, J.; Perrior, T.; Suzuki, T.; Thibault, S.; Thomas, W. A.; Thornton-Pett, M., *Tetrahedron* **2002**, *58*, 161.
97. Dondas, H. A.; Grigg, R.; Thibault, S.; Thomas, W. A.; Thornton-Pett, M., *Tetrahedron* **2002**, *58*, 5827.
98. Dondas, H. A.; Yaktubay, N., *Heterocyclic Commun.* **2003**, *9*, 337.
99. Grigg, R.; Hadjisoteriou, M.; Kennewell, P.; Markandu, J., *J. Chem. Soc., Chem. Commun.* **1992**, 1537.
100. Bennasar, M. L.; Vidal, B.; Bosch, J., *J. Org. Chem.* **1997**, *62*, 3597.
101. Terao, K.; Kunishima, M.; Tani, S., *Synlett* **1999**, 733.
102. Engman, L., *J. Org. Chem.* **1991**, *56*, 3425.
103. Gamblin, D. P.; Garnier, P.; van Kasteren, S.; Oldham, N. J.; Fairbanks, A. J.; Davis, B. G., *Angew. Chem. Int. Ed.* **2004**, *43*, 828.
104. Tomoda, S.; Fujita, K.; Iwaoka, M., *Chem. Commun./J. Chem. Soc., Chem. Commun.* **1990**, 129.
105. Deziél, R.; Goulet, S.; Grenier, L.; Bordleau, J.; Bernier, J., *J. Org. Chem.* **1993**, *58*, 3619.
106. Reich, H. J.; Yelm, K. E., *J. Org. Chem.* **1991**, *56*, 5672.
107. Frejd, T.; Sharpless, K. B., *Tetrahedron Lett.* **1978**, 2239.
108. Nicolaou, K. C.; Claremon, D. A.; Barnette, W. E.; Seitz, S. P., *J. Am. Chem. Soc.* **1979**, *101*, 3704.
109. Tiecco, M.; Testaferri, L.; Marini, F.; Bagnoli, L.; Santi, C.; Temperini, A., *Tetrahedron* **1997**, *53*, 4441.
110. Tiecco, M.; Testaferri, L.; Bagnoli, L.; Marini, F.; Santi, C.; Temperini, A., *Heterocycles* **1996**, *43*, 2679.
111. Pandey, G.; Rao, K. S. S. P.; Palit, D. K.; Mittal, J. P., *J. Org. Chem.* **1998**, *63*, 3968.
112. Pandey, G.; Rao, K. S. S. P.; Rao, K. V. N., *J. Org. Chem.* **2000**, *65*, 4309.
113. Fujita, K.; Murata, K.; Iwaoka, M.; Tomoda, S., *Tetrahedron* **1997**, *53*, 2029.
114. Back, T. G.; Dyck, B. P.; Nan, S., *Tetrahedron* **1999**, *55*, 3191.
115. Back, T. G.; Moussa, Z., *Org. Lett.* **2000**, *2*, 3007.
116. Back, T. G.; Moussa, Z.; Parvez, M., *J. Org. Chem.* **2002**, *67*, 499.
117. Tiecco, M.; Testaferri, L.; Santi, C.; Tomassini, C.; Marini, F.; Bagnoli, L.; Temperini, A., *Chem. Eur. J.* **2002**, *8*, 1118.
118. Uehlin, L.; Wirth, T., *Org. Lett.* **2001**, *3*, 2931.
119. Wirth, T., *Tetrahedron* **1999**, *55*, 1.
120. Nicolaou, K. C.; Pfefferkorn, J. A.; Cao, G. Q., *Angew. Chem. Int. Ed.* **2000**, *39*, 734.
121. Nicolaou, K. C.; Pfefferkorn, J. A.; Barluenga, S.; Mitchell, H. J.; Roecker, A. J.; Cao, G. Q., *J. Am. Chem. Soc.* **2000**, *122*, 9968.
122. Nicolaou, K. C.; Pfefferkorn, J. A.; Mitchell, H. J.; Roecker, A. J.; Barluenga, S.; Cao, G. Q.; Affleck, R. L.; Lillig, J. E., *J. Am. Chem. Soc.* **2000**, *122*, 9954.
123. Nicolaou, K. C.; Pfefferkorn, J. A.; Roecker, A. J.; Cao, G. Q.; Barluenga, S.; Mitchell, H. J., *J. Am. Chem. Soc.* **2000**, *122*, 9939.
124. Fujita, K.; Hashimoto, S.; Oishi, A.; Taguchi, Y., *Tetrahedron Lett.* **2003**, *44*, 3793.
125. Usuki, Y.; Iwaoka, M.; Tomoda, S., *Chem. Lett.* **1992**, 1507.
126. Uneyama, K., *J. Organomet. Chem.* **2000**, *611*, 158.
127. Zhao, H.; Cai, M., *Synthesis* **2002**, 1347.
128. Klapötke, T. M.; Krumm, B.; Mayer, P.; Piotrowski, H.; Vogt, M., *Z. Anorg. Allg. Chem.* **2003**, *629*, 1117.
129. Austad, T., *Acta Chem. Scand.* **1976**, *30A*, 579.
130. Iwaoka, M.; Komatsu, H.; Tomoda, S., *J. Organomet. Chem.* **2000**, *611*, 164.

Benzeneselenenyl Trifluoromethanesulfonate

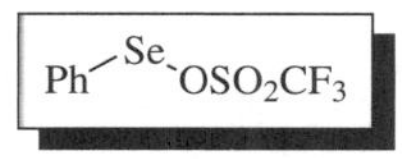

[112238-64-9] $C_7H_5F_3O_3SSe$ (MW 305.13)

(a highly electrophilic selenium reagent[2–8])

Alternate Name: benzeneselenenyl triflate.

Preparative Methods: usually prepared in situ by the reaction of Benzeneselenenyl Chloride with Silver(I) Trifluoromethanesulfonate in dichloromethane or toluene.[2] Two other economical methods are available. The first is the removal of hydrogen chloride from a mixture of benzeneselenenyl chloride and Trifluoromethanesulfonic Acid.[3] The second is the comproportionation reaction of benzeneseleninic anhydride (see Benzeneseleninic Acid) and Diphenyl Diselenide in the presence of trifluoromethanesulfonic anhydride.[4]

Handling, Storage, and Precautions: moisture sensitive. The dichloromethane solution free of silver chloride and water can be stored at rt for several weeks in the dark.[3]

Original Commentary

M. Iwaoka & S. Tomoda
The University of Tokyo, Japan

Addition Reactions. Benzeneselenenyl triflate is a highly electrophilic selenium reagent due to the poor nucleophilicity of triflate anion. It adds to alkenes having an intramolecular carboxy or hydroxy group to give cyclic products. For example, γ,δ-unsaturated carboxylic acids react with benzeneselenenyl triflate in dichloromethane to give γ-lactones in high yields (eq 1).[2]

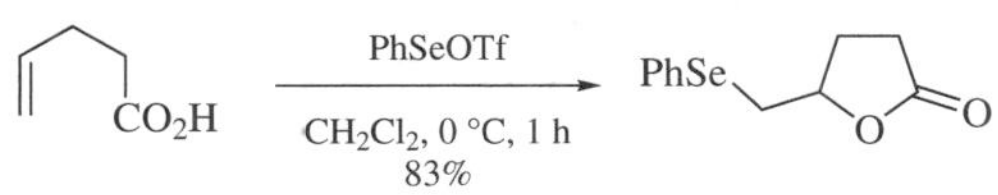

(1)

Under similar reaction conditions, 5- and 6-hydroxyalkenes are transformed to *exo*-cyclized tetrahydrofuran or -pyran derivatives (eq 2).[5,6] This reaction is useful for one-step conversion of

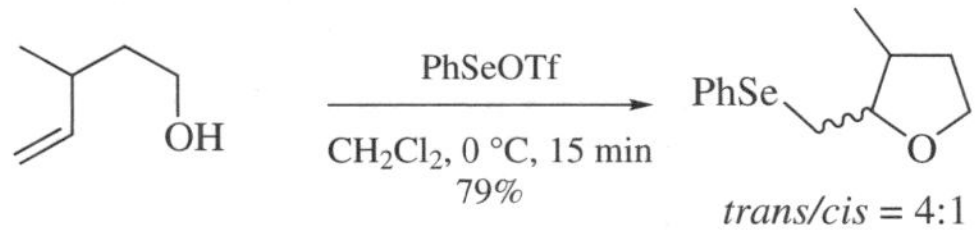

(2)

trans/cis = 4:1

acyclic terpene alcohols and derivatives into mono- or polycyclic compounds.[7] In the reaction with phenylacetylene, benzeneselenenyl triflate gives a *trans* adduct, a vinyl triflate, in high yield, which can be easily hydrolyzed to the ketone (eq 3).[4]

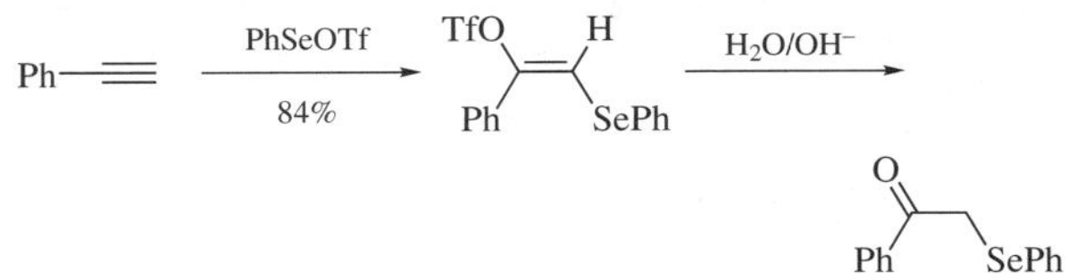

(3)

Selenium Polonovski Reaction. Amine *N*-oxides react with benzeneselenenyl triflate to form selenoxyammonium salts, which rearrange to α-selenoxyamines in the presence of organic bases.[3] The products can be easily converted into secondary or tertiary amine derivatives (eq 4).

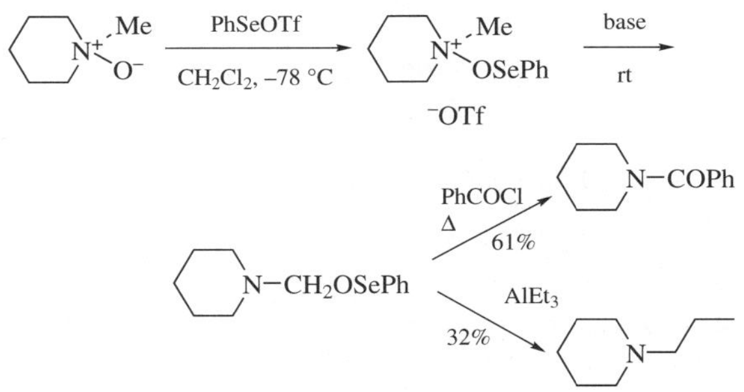

(4)

***O*-Glycosylation.** Alcohols efficiently undergo *O*-glycosylation under mild conditions by treatment with 1-thioglycosides and benzeneselenenyl triflate (eq 5).[8] This reaction can be applied to stereo- and regioselective synthesis of α-cyclodextrin derivatives.[9]

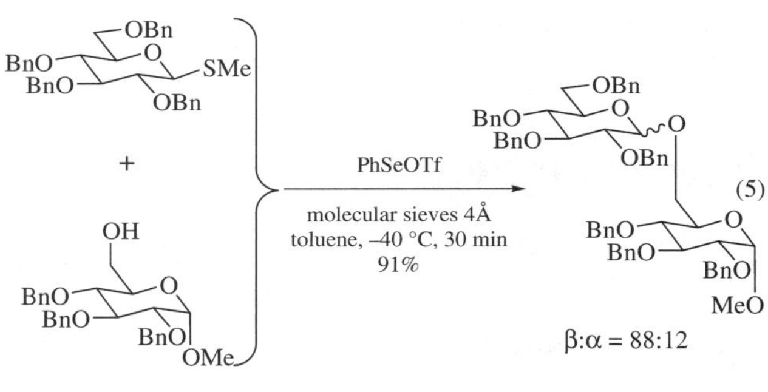

(5)

First Update

Robert J. Kerns & Peng Wei
University of Iowa, Iowa City, IA, USA
Jesús Sandoval Ramírez
Universidad Autónoma de Puebla, Puebla, México

Introduction. Electrophilic selenium reagents including benzeneselenenyl triflate have been increasingly studied and employed over the last decade in an expanding number of intermolecular and intramolecular addition reactions. A number of detailed discussions and reviews have been published regarding addition reactions of electrophilic selenium reagents, including benzeneselenenyl triflate.[10]

Addition Reactions. The benzeneselenenyl triflate-mediated formation of lactones and lactams (eq 1) has been expanded to include a number of regio- and stereoselective cyclization reactions. For example, the diastereoselective synthesis of hydroxy substituted γ- and δ-lactones was achieved through 5-*endo* or 6-*endo* cyclization of α- or β-hydroxy acids or 5-*exo* or 6-*exo* cyclization of hydroxy esters obtained from the diastereoselective hydroxy selenenylation of α- or β-hydroxy esters.[11] The formation of γ-lactones from 5-*exo* ring closure was shown to compete effectively against 6-*exo* and 6-*endo* cyclization pathways.[12] A potential route to α- and β- amino acids from *N*-(α,β-unsaturated-acyl)prolinamides has also been proposed based on the diastereoselective benzeneselenenyl triflate-induced lactamization of olefinic amides (eq 6).[13,14]

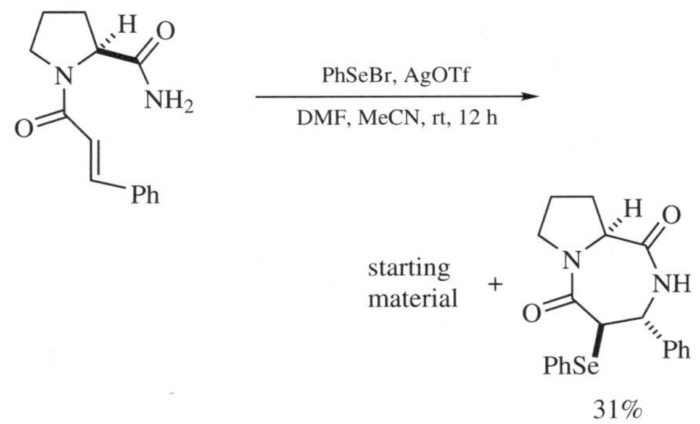

(6)

Benzeneselenenyl triflate-mediated cyclization of hydroxy alkenes to furans and pyrans (eq 2) has also been applied towards the stereoselective synthesis of more highly substituted cyclic ethers.[15] Four-membered heteroaromatic ring systems including β-lactones and oxetanes have been observed in the benzeneselenenyl triflate-promoted cyclization of 3-butenoic acid and 3-butenol, respectively.[10c]

A number of newer intramolecular addition reactions promoted by benzeneselenenyl triflate have been reported, affording access to a variety of bicyclic and heterocyclic ring systems. *cis*-Fused bicyclic dioxanes and oxazines have been prepared through intramolecular oxyselenenylation (eq 7).[16,17]

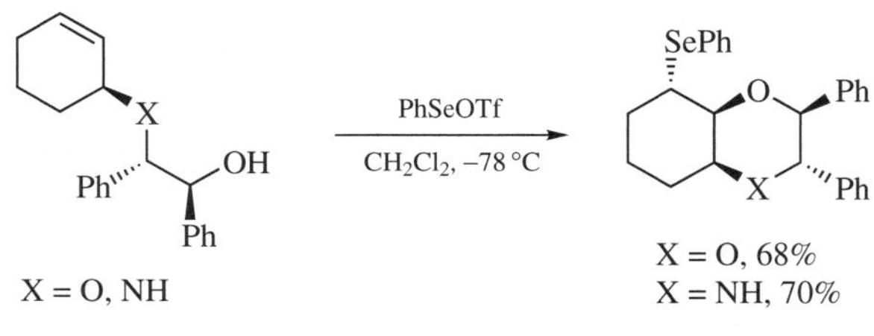

(7)

Cyclization of allylic *O*-methyl isoureas promoted by benzeneselenenyl triflate in the presence of silica gel and trifluoromethanesulfonic acid has been shown to afford access to a variety of 5,6-dihydro-1,3-oxazines (eq 8).[18] Electrophilic cyclization of allylic 1,1-dimethylisoureas with benzeneselenenyl triflate affords substituted oxazolines (eq 9).[19]

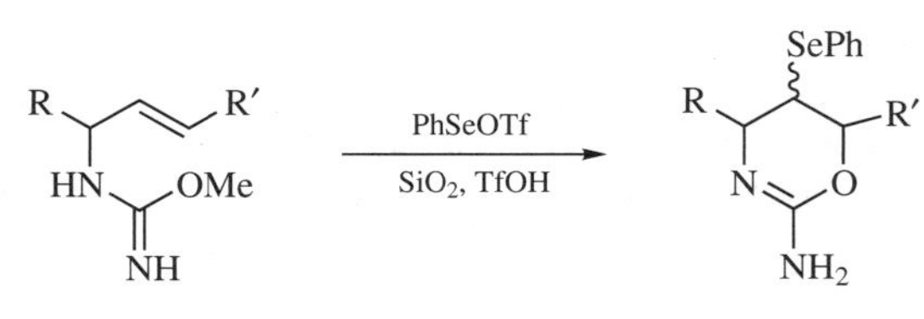

(8)

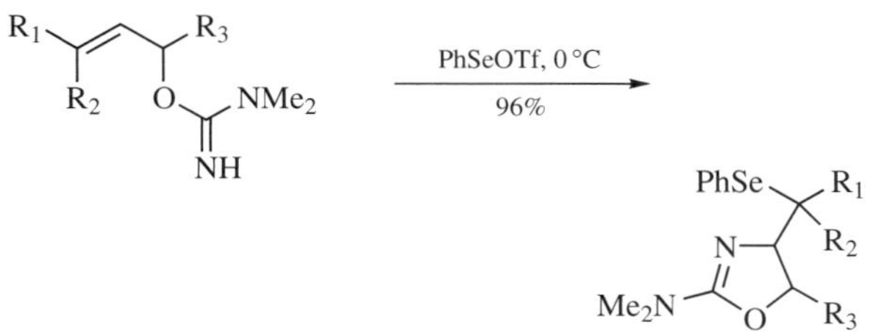
(9)

An alternate route to enantiopure oxazolines employs the amidoselenenylation of alkenes with camphorselenenyl sulfate followed by benzeneselenenyl triflate-mediated intramolecular nucleophilic substitution and deselenenylation of the optically pure acetamido selenides (eq 10).[20] Benzeneselenenyl triflate-induced intramolecular nucleophilic deselenenylation has also been employed to obtain a variety of substituted heterocycles (eq 11).[21]

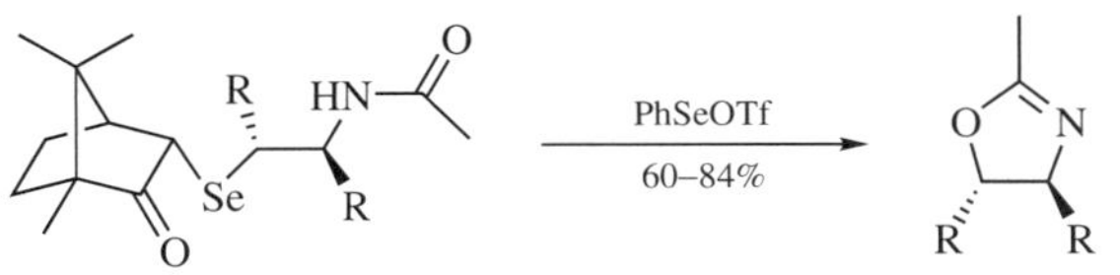
(10)

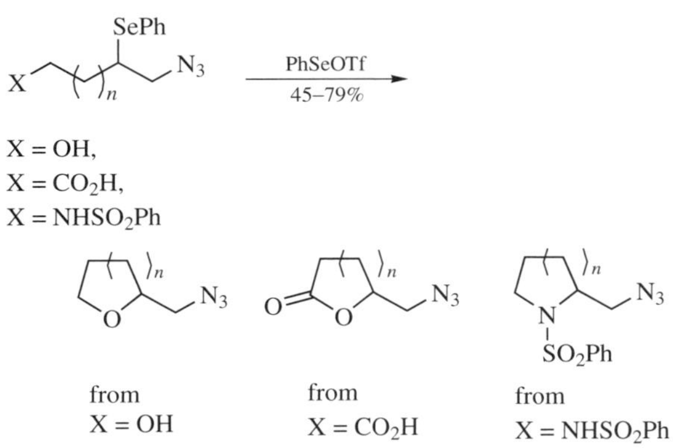

(11)

This chemistry proceeds via the formation of a very reactive selanylselenonium salt,[22] which can be trapped by internal or external nucleophiles (eq 12).[21,23] The diphenyl diselenide generated is easily recovered and can be reused.

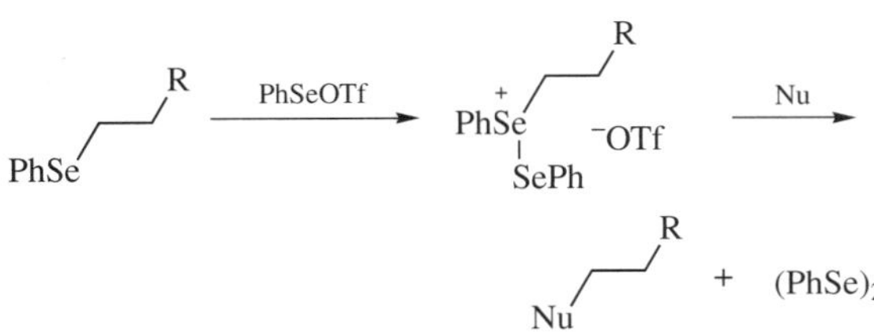
(12)

Whereas the adducts of PhSeOTf to most alkenes are unstable, an isolable 2-phenylselenoalkyl triflate adduct was obtained from the addition of PhSeOTf (formed in situ) to a γ-unsaturated oxime (eq 13). The Markovnikov adduct, on treatment with potassium carbonate, produced nitrone and oxazine heterocyles through 5-*exo*-trig and 6-*endo*-trig processes, respectively. When the same γ-unsaturated oxime reacted with PhSeBr the Markovnikov adduct was formed quantitatively (CH_2Cl_2, 25 °C, 16 h), but the treatment with K_2CO_3 afforded mainly the nitrone.[24]

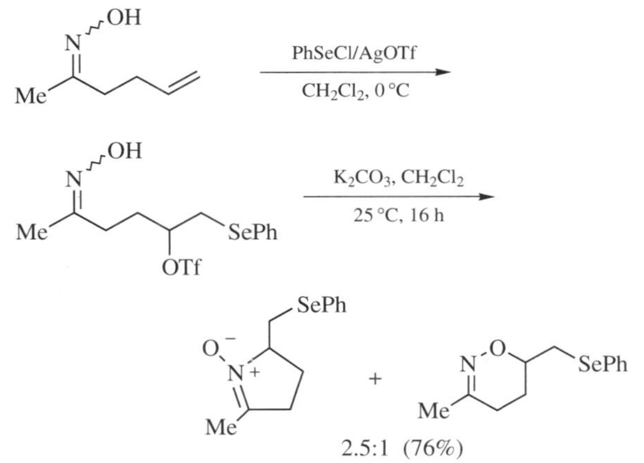
(13)

Reaction between PhSeOTf and a δ-unsaturated oxime has been shown to produce the Markovnikov adduct and a six-membered nitrone, after basic treatment (eq 14).[24]

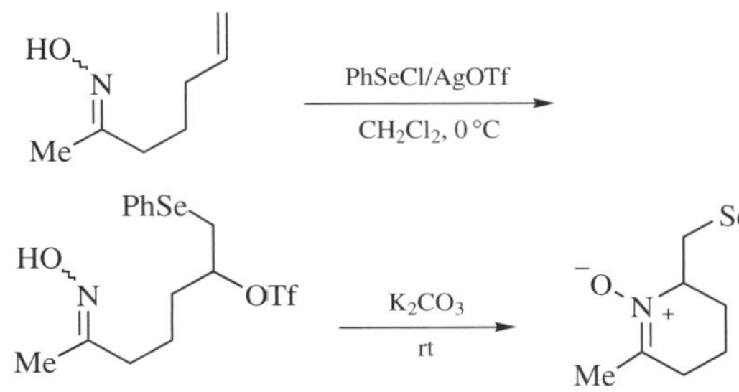
(14)

Reaction of PhSeOTf with 1,3-dialkoxypent-4-ene-2-ol derivatives afforded the 5-*endo* cyclization product; the 4-*exo* cyclization has been rarely observed.[25] In the case of selenocyclization of 3-alkoxypent-4-ene-1,2-diols both alcohols may compete in the cyclization process. A study on the diastereoselectivity of the addition of PhSeOTf over pent-4-ene-1,2,3-triol in the presence of diisopropylethylamine (DIPEA), revealed remarkable features (eqs 15 and 16).[15a] The PhSeOTf activity was compared with that of *N*-phenylselenophthalimide and iodine showing; first, iodine always provided 5-*exo* cyclization products, independently of the protecting groups. Second, selenium reagents could direct the cyclization reaction towards either 5-*endo* or 5-*exo*, depending on whether the primary hydroxy group was protected or not. Third, when the primary alcohol was protected, cyclization could be directed in either 5-*exo* or the 5-*endo* direction, by using iodine or selenium reagents, respectively.

β,γ-Unsaturated hydroxamic acids react with PhSeOTf to produce *N*-hydroxy imidates or *N*-hydroxy γ–lactams depending on the substituents of the double bond and the reaction temperature (eq 17).[25] At −20 °C the reaction is under kinetic control with faster trapping of the seleniranium ion by the more nucleophilic oxygen leading to the five-membered imidate; at 50 °C the reaction is thermodynamically controlled and the *N*-hydroxy γ–lactam is the favored product.

In addition to several studies discussing the influence of a variety of reaction conditions, such as solvents, additives, temperature, olefin substituents, and counterions on phenylselenofunctionalization reactions with PhSeOTf and other selenenylating

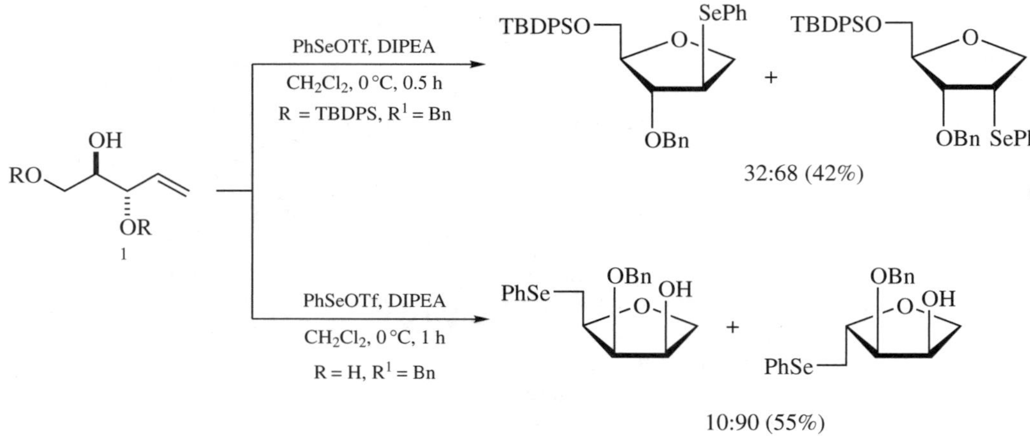

(15)

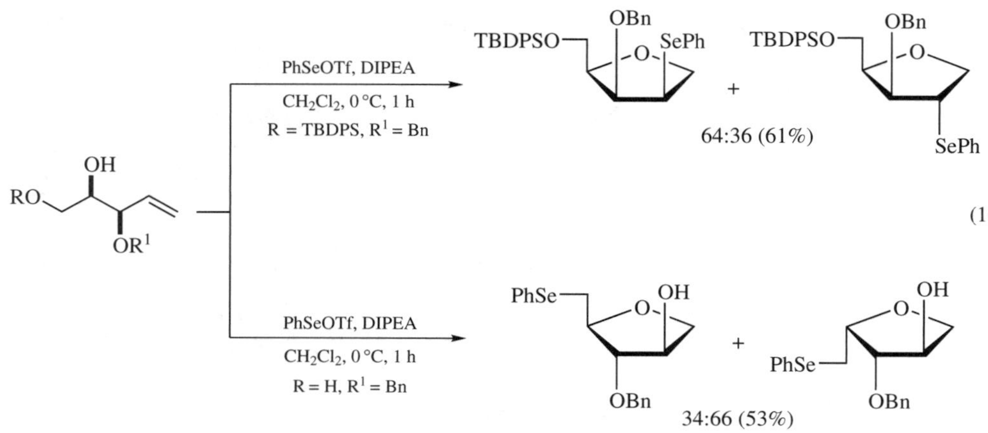

(16)

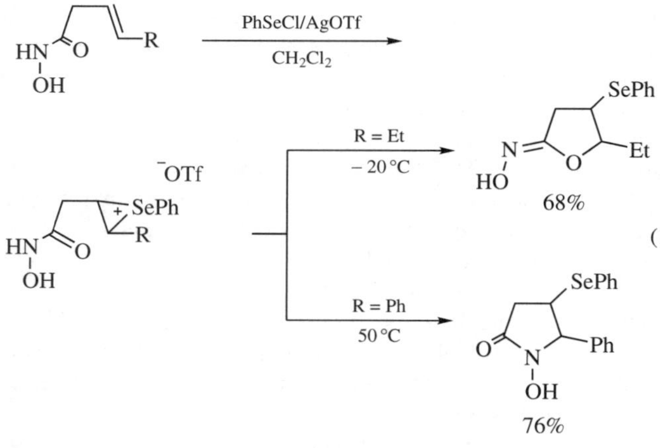

(17)

Intermolecular addition reactions promoted by benzeneselenenyl triflate have also found utility. The stereospecific *trans* addition of NaN_3 to phenyl substituted alkenes affords the regioselective formation of β-phenylseleno azides (eq 18).[31] However,the regioselectivity of this reaction varies greatly when there is not a dominant substituent on the alkene to direct electrophilic addition.

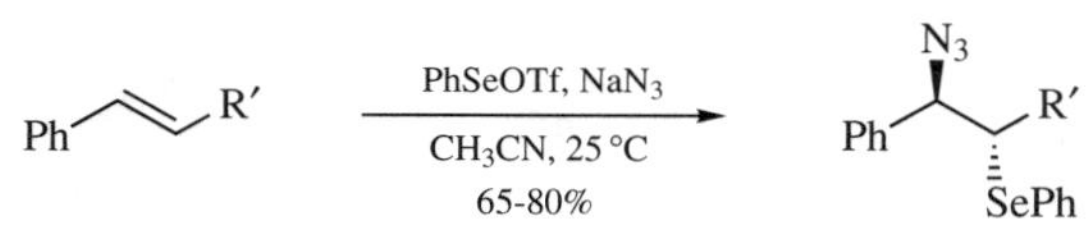

(18)

PhSeOTf adds to alkynes generating an alkenyl selenirenium ion, which has been characterized by ^1H NMR in the reaction of $PhSe^+$ SbF_6^- with but-2-yne.[32] The attack of an external nucleophile yields the product from *trans* addition, while unsymmetrical alkynes give mixtures of regioisomers. Since PhSeOTf prepared from PhSeBr and AgOTf contains some bromide ion that may act as a competitive nucleophile, pure PhSeOTf for these studies is obtained by the comproportionation reaction of benzeneseleninic anhydride and diphenyl diselenide in the presence of trifluoromethanesulfonic anhydride.[4] The fluoroselenenylation of acetylenes has been achieved employing benzeneselenenyl tri-

agents,[10b,c,22,26–28] a systematic study about the influence of nucleophiles, solvents, additives, and counterions on the reaction of PhSeOTf with 2-hydroxymethyl-4-phenylpent-4-enoic acid or 2-hydroxymethyl-4-phenylpent-4-enol has been published.[29] The mechanism of phenylselenoetherification of unsaturated alcohols involving seleniranium cationic intermediates was studied by semiempirical molecular-orbital MNDO-PM3 method. Results of calculations on several Δ^4-alkenols attacked by $PhSe^+$, revealed that the reaction should be highly regio- and stereospecific.[30]

flate in combination with $Et_3N{\cdot}3HF$ (eq 19).[33] The addition of $PhSe^+$ and F^- to alkynes has also been carried out successfully using diphenyl diselenide and xenon^{II} fluoride.[34,35] The PhSeOTf-$Et_3N{\cdot}HF$ reagent showed somewhat higher yields and shorter reaction times in the fluoroselenenylation of alkynes than other procedures.

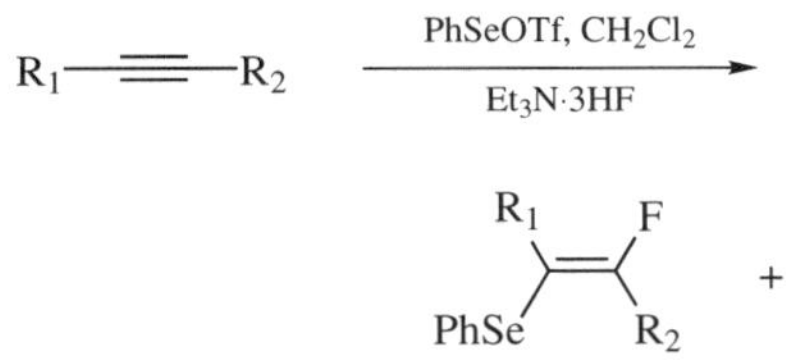

(19)

Reaction with Diazo Compounds. Diazo compounds react with PhSeOTf displacing a nitrogen molecule with the formation of phenylseleno-, and α,α-bis-phenylseleno derivatives (eq 20).[36] The selectivity was dependent on the substitutents in the diazo compound (See benzeneselenenyl bromide for an alternative procedure).

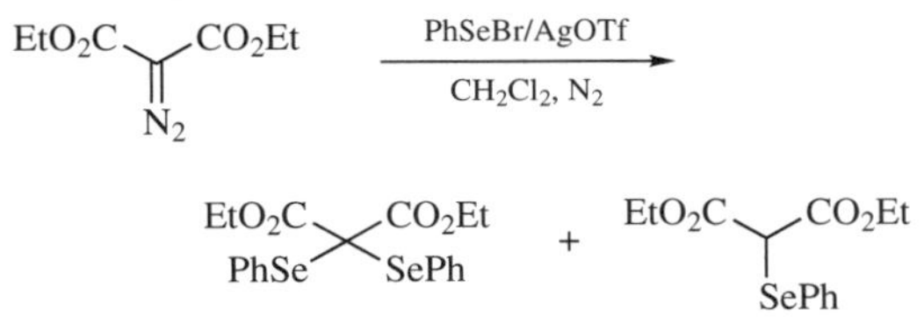

(20)

Temp (°C)	Yield (%)
0	28:58
25	15:62
− 78	45

***O*-Glycoside Formation.** Since the introduction of benzeneselenenyl triflate as a thioglycoside activating agent in *O*-glycoside formation, dozens of reports over the last decade have described the use of benzeneselenenyl triflate to activate a variety of thioglycosides (eq 5).[8,9,37] However, the yields and stereoselectivity reported for benzeneselenenyl triflate-promoted glycosylation reactions have varied greatly, with benzeneselenenyl triflate often being outperformed by other thiophilic thioglycoside activating agents. In addition, the toxicity and complexity of preparing benzeneselenenyl triflate, as compared to many other thiophilic thioglycoside activating agents, are also factors to be considered when employing this reagent in thioglycoside activation. Despite these factors, benzeneselenenyl triflate-mediated activation of thioglycosides has shown excellent success and utility in the synthesis of a number of important glycans including *O*-sialylation,[38] synthesis of the β-mannosyl-chitobiosyl component of *N*-glycoproteins,[39] synthesis of inositol glycan cyclic phosphates,[40] and the formation of cyclo-oligosaccharide structures, which is based on early success of this reagent in preparing cyclodextrin systems.[9]

In addition to thioglycoside activation, benzeneselenenyl triflate-promoted *O*-glycosylation has been reported for a number of other glycosyl donor systems. Deselenylation of 2-azido-2-deoxy-1-selenoglycosides employing benzeneselenenyl triflate has been reported to afford *O*-glycoside formation (eq 21).[23] Direct activation and glycosidation of a glycal using benzeneselenenyl triflate is exemplified by the stereospecific formation of 3-deoxy-D-manno-2-octulosonic acid (Kdo)-containing disaccharides (eq 22).[41]

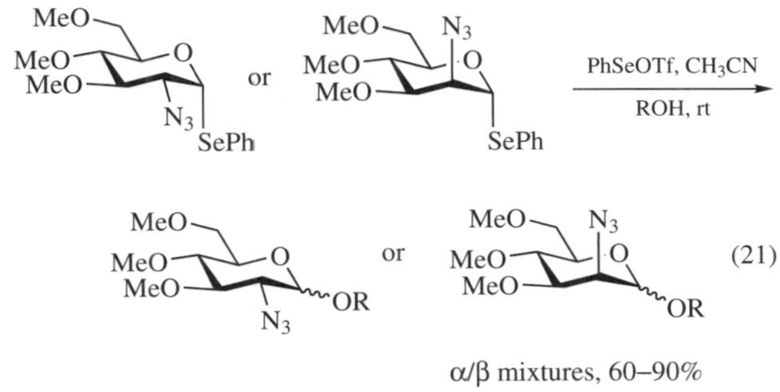

(21)

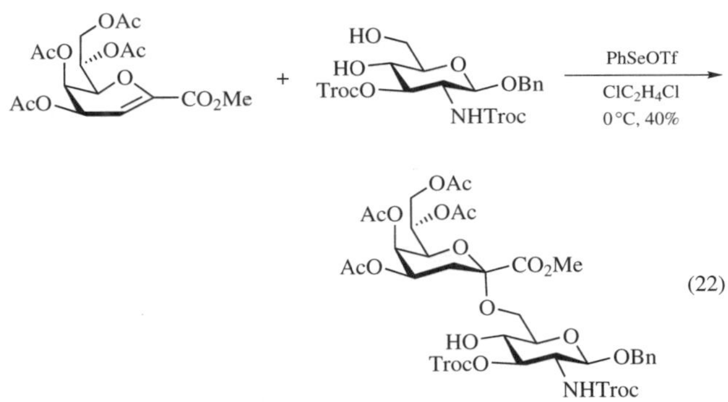

(22)

Carbohydrate Addition Reactions: Ring-closing Glycoside Formation. Based on early work showing that the benzeneselenenyl triflate-mediated intramolecular cycloaddition of hydroxy alkenes afforded pyran and furan ring systems, a number of complex polyhydroxylated alkenes have been employed to obtain carbohydrate-based structures including *O*- and *C*-glycosides. The stereoselective synthesis of β-glycosides of 3-deoxy-2-hexulosonates has been achieved by benzeneselenenyl triflate-promoted cyclization of appropriately protected *Z*-enol ether precursors (eq 23).[42] In the same system, an unprotected *Z*-enol ether precursor afforded the 3-deoxy-D-*erythro*-2-hexulofuranosylonate product due to kinetically preferred formation of the five-membered ring (eq 23).[42] The formation of *C*-linked ketosides of *N*-acetylneuraminic acid has similarly been achieved through stereoselective 6-*exo-trig* electrophilic cyclization reactions (eq 24).[43,44]

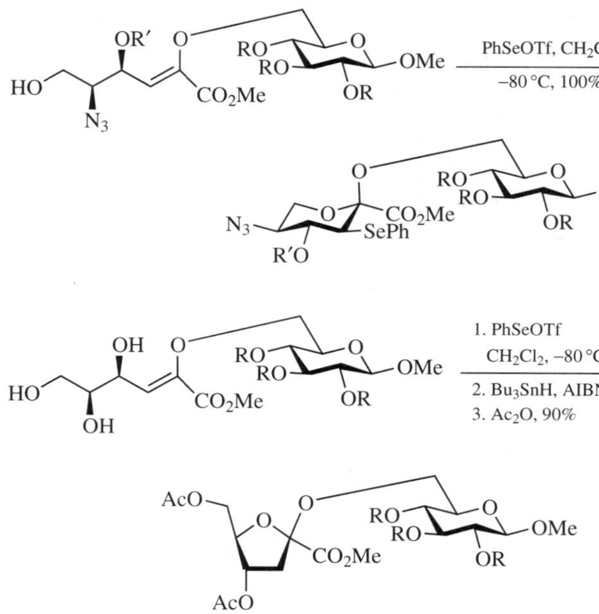

(23)

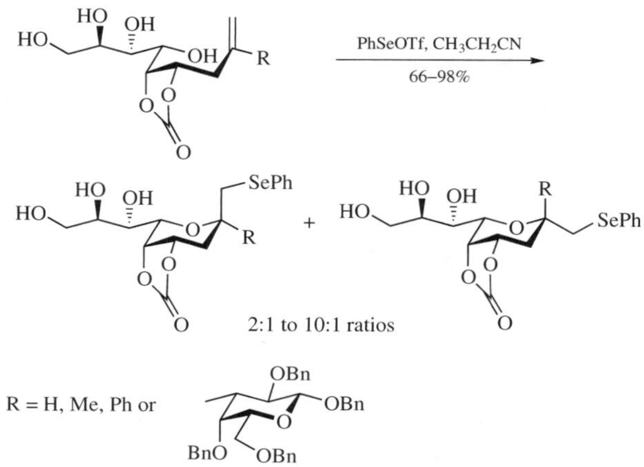

(24)

Related Reagents. 1-Benzenesulfinyl piperidine; diphenyl sulfoxide; benzenesulfenyl triflate; dimethyl(methylthio) sulfonium triflate. A number of other sources of electrophilic benzeneselenium (phenylselenium, $PhSe^+$) have been reported in the literature and shown to promote a variety of intramolecular and intermolecular addition and elimination reactions. These reagents include benzeneselenenyl chloride; benzeneselenenyl bromide; *N*-(phenylseleno)phthalimide; *N*-(phenylseleno)succinimide; phenyl selenocyanate; diphenyl diselenide and various other diselenides, diphenyl diselenide-ammonium peroxydisulfate; phenylselenylazide. Attempts to prepare PhSeOTf from phenylselenocyanate and copper(II) trifluoromethanesulfonate failed, even in the presence of trifluoromethanesulfonic acid.[45] However, the new organoselenium reagent generated from PhSePh and copper(I) trifluoromethanesulfonate is highly electrophilic and presents some advantages over PhSeOTf.[46] Thus, the regio- and stereoselectivity were almost the same as for reactions carried out with PhSeOTf, but the yields are higher. Other combinations of ArSeCN-(metal)triflates have also been tested.[47] In order to develop asymmetric selenofuncionalizations, chiral selenenyl triflate agents have been synthesized and tested.[48–61]

Benzeneselenenyl *m*-nitrobenzenesulfonate,[62a] benzeneselenenyl *p*-toluenesulfonate,[62b] (phenylseleno)dimethylsulfonium ($PhSeS^+Me_2$) fluoroborate,[63] and (phenylseleno)methylphenylselenonium ($PhSeS^+eMePh$) fluoroborate[64] can be considered as analogs of benzeneselenenyl triflate. (Phenylseleno)dimethylsulfonium ($PhSeS^+Me_2$) fluoroborate reacts with electron-rich aromatic compounds to give the products of electrophilic substitution (eq 25).[63] On the other hand, (phenylseleno)methylphenylselenonium ($PhSeS^+eMePh$) fluoroborate reacts selectively with allylic silanes to give the corresponding allylic selenides, which easily undergo 2,3-sigmatropic rearrangement to allylic alcohols by oxidation (eq 26).[64]

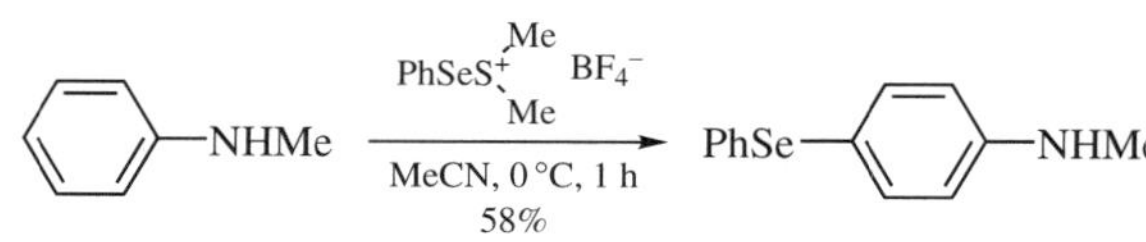

(25)

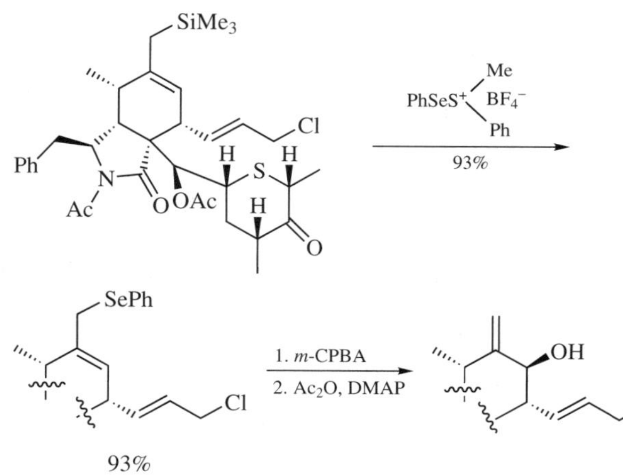

(26)

See also Benzeneselenenyl Bromide, Benzeneselenenyl Chloride, Benzenesulfenyl Chloride, and 2-Pyridinesulfenyl Bromide.

Attempts to prepare PhSeOTf from phenylselenocyanate and copper[II] trifluoromethanesulfonate failed, even in the presence of trifluoromethanesulfonic acid.[65] However, the new organoselenium reagent was highly electrophilic. Thus, the regio- and stereoselectivity were almost the same as for reactions carried out with PhSeOTf, but the yields are higher. Other combinations of ArSeCN-(metal)triflates have also been tested.[66] The binary reagent PhSeSePh-CuOTf is also a powerful phenylselenenylating agent and presents some advantages over PhSeOTf.[67] In order to develop asymmetric selenofuncionalizations, chiral selenenyl triflate agents have been synthesized and tested.[68–81]

---

1. Reviews and books on organoselenium reagents. (a) Clive, D. L. J., *Tetrahedron* **1978**, *34*, 1049. (b) Reich, H. J., *Acc. Chem. Res.* **1979**, *12*, 22. (c) Nicolaou, K. C.; Petasis, N. A. *Selenium in Natural Products Synthesis*; CIS:Philadelphia, 1984. (d) *The Chemistry of Organic Selenium and Tellurium Compounds*; Patai, S.; Rappoport, Z.,

Eds.; Wiley:New York, 1986; Vol. 1. (e) Paulmier, C. *Selenium Reagents and Intermediates in Organic Synthesis*; Pergamon:Oxford, 1986. (f) *Organoselenium Chemistry*; Liotta, D. Ed.; Wiley:New York, 1987.

2. Murata, S.; Suzuki, T., *Chem. Lett.* **1987**, 849.
3. Okazaki, R.; Itoh, Y., *Chem. Lett.* **1987**, 1575.
4. Kutateladze, A. G.; Kice, J. L.; Kutateladze, T. G.; Zefirov, N. S.; Zyk, N. V., *Tetrahedron Lett.* **1992**, *33*, 1949.
5. Murata, S.; Suzuki, T., *Tetrahedron Lett.* **1987**, *28*, 4297.
6. Murata, S.; Suzuki, T., *Tetrahedron Lett.* **1987**, *28*, 4415.
7. Murata, S.; Suzuki, T., *Tetrahedron Lett.* **1990**, *31*, 6535.
8. Ito, Y.; Ogawa, T., *Tetrahedron Lett.* **1988**, *29*, 1061.
9. Mori, M.; Ito, Y.; Ogawa, T., *Tetrahedron Lett.* **1990**, *31*, 3029.
10. (a) *Organoselenium Chemistry: A Practical Approach*; Back, T. G., Ed.; Oxford University Press:Oxford, 1999. (b) Petragnani, N.; Stefani, H. A.; Valduga, C. J., *Tetrahedron* **2001**, *57*, 1411. (c) Robin, S.; Rousseau, G., *Eur. J. Org. Chem.* **2002**, 3099. (d) Wessjohann, L. A.; Sinks, U., *J. Prakt. Chem.* **1998**, *340*, 189. (e) Tiecco, M.; Testaferri, L.; Tingoli, M.; Bagnoli, L.; Marini, F.; Santi, C.; Temperini, A., *Gazz. Chim. Ital.* **1996**, *126*, 635.
11. Aprile, C.; Gruttadauria, M.; Amato, M. E.; D'Anna, F.; Lo Meo, P.; Riela, S.; Noto, R., *Tetrahedron* **2003**, *59*, 2241.
12. Ramana, C. V.; Reddy, K. R.; Nagarajan, M., *Indian J. Chem., Sect. B* **1996**, *35*, 534.
13. Chung, S.-K.; Jeong, T.-H.; Kang, D.-H., *Tetrahedron: Asymmetry* **1997**, *8*, 5.
14. Chung, S. K.; Jeong, T. H.; Kang, D. H., *J. Chem. Soc., Perkin Trans. 1* **1998**, 969.
15. (a) Bravo, F.; Castillón, S., *Eur. J. Org. Chem.* **2001**, 507. (b) Lipshutz, B. H.; Barton, J. C., *J. Am. Chem. Soc.* **1992**, *114*, 1084. (c) Inoue, H.; Murata, S.; Suzuki, T., *Liebigs Ann. Chem.* **1994**, 901.
16. Kim, K. S.; Park, J. I.; Moon, H. K.; Yi, H., *Chem. Commun.* **1998**, 1945.
17. Kim, K. S.; Choi, S. O.; Park, J. M.; Lee, Y. J.; Kim, J. H., *Tetrahedron: Asymmetry* **2001**, *12*, 2649.
18. Freire, R.; León, E. I.; Salazar, J. A.; Suárez, E., *J. Chem. Soc., Chem. Commun.* **1989**, 452.
19. Giuliano, R. M.; Bigos, C., *Synth. Commun.* **1995**, *25*, 1145.
20. Tiecco, M.; Testaferri, L.; Santi, C.; Tomassini, C.; Marini, F.; Bagnoli, L.; Temperini, A., *Eur. J. Org. Chem.* **2000**, 3451.
21. Tingoli, M.; Testaferri, L.; Temperini, A.; Tiecco, M., *J. Org. Chem.* **1996**, *61*, 7085.
22. Paulmier, C., *Phosphorus Sulfur Silicon* **2001**, *172*, 25.
23. Tingoli, M.; Tiecco, M.; Testaferri, L.; Temperini, A., *J. Chem. Soc., Chem. Commun.* **1994**, 1883.
24. Grigg, R.; Hadjisoteriou, M.; Kennewell, P.; Markandu, J., *J. Chem. Soc., Chem. Commun.* **1992**, 1537.
25. Tiecco, M.; Testaferri, L.; Tingoli, M.; Marini, F., *J. Chem. Soc., Chem. Commun.* **1994**, 221.
26. Ranganathan, S.; Muraleedharan, K. M.; Vaish, N. K.; Jayaraman, N., *Tetrahedron* **2004**, *60*, 5273.
27. Robin, S.; Rousseau, G., *Tetrahedron* **1998**, *54*, 13681.
28. Frederickson, M.; Grigg, R., *Org. Prep. Proced. Int.* **1997**, *29*, 33.
29. Khokhar, S.; Wirth, T., *Angew. Chem. Int. Ed.* **2004**, *43*, 631.
30. Markovic, Z.; Konstantinovic, S.; Juranic, I.; Dosen-Micovic, L., *Gazz. Chim Ital.* **1997**, *127*, 429.
31. Tiecco, M.; Testaferri, L.; Temperini, A.; Bagnoli, L.; Marini, F.; Santi, C., *Synth. Commun.* **1998**, *28*, 2167.
32. Schmid, G. H.; Garratt, D. G., *Tetrahedron Lett.* **1975**, *16*, 3991.
33. Poleschner, H.; Seppelt, K., *J. Chem. Soc., Perkin Trans. 1* **2002**, 2668.
34. Poleschner, H.; Heydenreich, M.; Spindler, K.; Haufe, G., *Synthesis* **1994**, 1043.
35. Usuki, Y.; Iwaoka, M.; Tomoda, S., *Chem. Lett.* **1992**, 1507.
36. Curini, M.; Epifano, F.; Marcotullio, M. C.; Rosati, O., *Synlett* **2000**, 1813.
37. Ito, Y.; Ogawa, T.; Numata, M.; Sugimoto, M., *Carbohydr. Res.* **1990**, *202*, 165.
38. Boons, G.-J.; Demchenko, A. V., *Chem. Rev.* **2000**, *100*, 4539.
39. Günther, W.; Kunz, H., *Angew. Chem. Int. Ed. Engl.* **1990**, *29*, 1050.
40. Jaworek, C. H.; Iacobucci, S.; Calias, P.; d'Alarcao, M., *Carbohydr. Res.* **2001**, *331*, 375.
41. Ikeda, K.; Akamatsu, S.; Achiwa, K., *Carbohydr. Res.* **1989**, *189*, c1.
42. Vlahov, I. R.; Vlahova, P. I.; Schmidt, R. R., *Tetrahedron:Asymmetry* **1993**, *4*, 293.
43. Notz, W.; Hartel, C.; Waldscheck, B.; Schmidt, R. R., *J. Org. Chem.* **2001**, *66*, 4250.
44. Khodair, A. I.; Schmidt, R. R., *Carbohydr. Res.* **2002**, *337*, 1967.
45. Inoue, H.; Murata, S., *Heterocycles* **1997**, *45*, 847.
46. Miyachi, N.; Satoh, H.; Shibasaki, M., *J. Chem. Soc., Perkin Trans. 1* **1991**, 2049.
47. Murata, S.; Suzuki, C.; Inoue, H.; Andoh, Y.; Hayashi, Y.; Suzuki, T., *Heterocycles* **2000**, *52*, 621.
48. Tiecco, M.; Testaferri, L.; Santi, C.; Tomassini, C.; Marini, F.; Bagnoli, L.; Temperini, A., *Angew. Chem. Int. Ed.* **2003**, *42*, 3131.
49. Tiecco, M.; Testaferri, L.; Santi, C.; Tomassini, C.; Marini, F.; Bagnoli, L.; Temperini, A., *Chem. Eur. J.* **2002**, *8*, 1118.
50. Back, T. G.; Moussa, Z.; Parvez, M., *J. Org. Chem.* **2002**, *67*, 499.
51. Uehlin, L.; Fragale, G.; Wirth, T., *Chem. Eur. J.* **2002**, *8*, 1125.
52. Wirth, T., *Angew. Chem. Int. Ed.* **2000**, *39*, 3740.
53. Back, T. G.; Moussa, Z., *Org. Lett.* **2000**, *2*, 3007.
54. Tiecco, M.; Testaferri, L.; Santi, C.; Tomassini, C.; Marini, F.; Bagnoli, L.; Temperini, A., *Tetrahedron: Asymmetry* **2000**, *11*, 4645.
55. Back, T. G.; Dyck, B. P.; Nan, S., *Tetrahedron* **1999**, *55*, 3191.
56. Back, T. G.; Nan, S., *J. Chem. Soc., Perkin Trans. 1* **1998**, 3123.
57. Fragale, G.; Wirth, T., *Eur. J. Org. Chem.* **1998**, 1361.
58. Déziel, R.; Malenfant, E.; Thibault, C., *Tetrahedron Lett.* **1998**, *39*, 5493.
59. Wirth, T., *Phosphorus Sulfur Silicon* **1998**, *136–138*, 235.
60. Wirth, T.; Fragale, G., *Chem. Eur. J.* **1997**, 1894.
61. Fujita, K.; Murata, K.; Iwaoka, M.; Tomoda, S., *Tetrahedron* **1997**, *53*, 2029.
62. (a) Yoshida, M.; Satoh, N.; Kamigata, N., *Chem. Lett.* **1989**, 1433. (b) Back, T. G.; Muralidharan, K. R. , *J. Org. Chem.* **1991**, *56*, 2781.
63. Gassman, P. G.; Miura, A.; Miura, T., *J. Org. Chem.* **1982**, *47*, 951.
64. (a) Vedejs, E.; Rodgers, J. D.; Wittenberger, S. J. , *Tetrahedron Lett.* **1988**, *29*, 2287. (b) Vedejs, E.; Ahmad, S., *Tetrahedron Lett.* **1988**, *29*, 2291.
65. Inoue, H.; Murata, S., *Heterocycles* **1997**, *45*, 847.
66. Murata, S.; Suzuki, C.; Inoue, H.; Andoh, Y.; Hayashi, Y.; Suzuki, T., *Heterocycles* **2000**, *52*, 621.
67. Miyachi, N.; Satoh, H.; Shibasaki, M., *J. Chem. Soc., Perkin Trans. 1* **1991**, 2049.
68. Tiecco, M.; Testaferri, L.; Santi, C.; Tomassini, C.; Marini, F.; Bagnoli, L.; Temperini, A., *Angew. Chem. Int. Ed.* **2003**, *42*, 3131.
69. Tiecco, M.; Testaferri, L.; Santi, C.; Tomassini, C.; Marini, F.; Bagnoli, L.; Temperini, A., *Chem. Eur. J.* **2002**, *8*, 1118.
70. Back, T. G.; Moussa, Z.; Parvez, M., *J. Org. Chem.* **2002**, *67*, 499.
71. Uehlin, L.; Fragale, G.; Wirth, T., *Chem. Eur. J.* **2002**, *8*, 1125.
72. Wirth, T., *Angew. Chem. Int. Ed.* **2000**, *39*, 3740.
73. Back, T. G.; Moussa, Z., *Org. Lett.* **2000**, *2*, 3007.
74. Tiecco, M.; Testaferri, L.; Santi, C.; Tomassini, C.; Marini, F.; Bagnoli, L.; Temperini, A., *Tetrahedron: Asymmetry* **2000**, *11*, 4645.
75. Back, T. G.; Dyck, B. P.; Nan, S., *Tetrahedron* **1999**, *55*, 3191.
76. Back, T. G.; Nan, S., *J. Chem. Soc., Perkin Trans. 1* **1998**, 3123.
77. Fragale, G.; Wirth, T., *Eur. J. Org. Chem.* **1998**, 1361.

78. Déziel, R.; Malenfant, E.; Thibault, C., *Tetrahedron Lett.* **1998**, *39*, 5493.
79. Wirth, T., *Phosphorus Sulfur Silicon* **1998**, *136–138*, 235.
80. Wirth, T.; Fragale, G., *Chem. Eur. J.* **1997**, 1894.
81. Fujita, K.; Murata, K.; Iwaoka, M.; Tomoda, S., *Tetrahedron* **1997**, *53*, 2029.

Benzenesulfenyl Chloride[1]

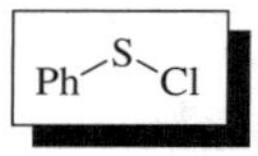

[931-59-9] C_6H_5ClS (MW 144.62)

(useful reagent in regio- and stereoselective syntheses,[2] directed alkylation,[3] carbonyl transposition,[4] functionalization of unsaturated groups,[5] sulfenylation of silyl enolates,[6] cycloaddition[7])

Alternate Names: phenylsulfenyl chloride.
Physical Data: bp 88 °C/17 mmHg; 55 °C/3 mmHg; *d* 1.252 g cm^{-3}; n_D^{20} 1.6120.
Solubility: soluble in pentane, benzene, acetone, $CHCl_3$, CH_2Cl_2, CCl_4.
Form Supplied in: oily, dark red liquid; fumes in air.
Preparative Methods: easily obtained in high yields (90–100%) from Thiophenol or Diphenyl Diselenide by chlorination at −5 to 0 °C with chlorine gas[8] or Sulfuryl Chloride.[9] Arenesulfenyl chlorides, including PhSCl, are conveniently prepared from arenethiols and N-Chlorosuccinimide in dichloromethane solution in quantitative yields and used without isolation in addition reactions to alkenes.[10]
Handling, Storage, and Precautions: extremely hygroscopic and must be carefully protected from moisture. It may be stored for months under dry inert gas in a refrigerator. However, there are two reports on the spontaneous explosion of PhSCl on storage, even in a refrigerator. It is recommended that PhSCl be prepared fresh as needed. Use in a fume hood.

Original Commentary

Gunars Zelčans
Institute of Organic Synthesis, Riga, Latvia

Sulfenylation of Allylic Alcohols. Sulfenyl (as well as sulfoxide and sulfonyl) groups favor formation of a stabilized carbanion at the neighboring carbon atom by metalation with strong bases, and thus direct sequential alkylation with alkyl halides to the α-position. However, directed γ-alkylation of allylic alcohols can be realized by application of a reversible sulfenate–sulfoxide [2,3]-sigmatropic rearrangement of their benzenesulfenate esters (eq 1).[3,11]

Sequential treatment of an allylic alcohol with n-Butyllithium and benzenesulfenyl chloride forms the sulfenate ester which spontaneously rearranges to the sulfoxide. The latter is metalated with Lithium Diisopropylamide (LDA) (*n*-BuLi affords 5–10% of byproducts) and treated with an alkyl iodide to give predominantly an α-alkylated (relative to the sulfoxide group) product. Finally, the alkylated sulfoxide on treatment with Trimethyl Phosphite and methanol rearranges back to the sulfenate ester and liberates the γ-alkylated allylic alcohol.

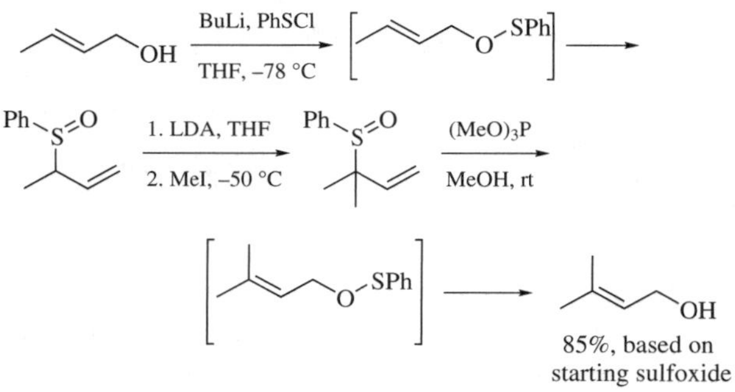

(1)

Formation of the *trans* isomer is favored, the *trans*:*cis* ratio being 2.7–30.0. Along with the α-alkylation of the sulfoxide, some of the γ-isomer usually forms, the α:γ ratio being 1.1 to >>10 depending on the structure of the sulfoxide. Cyclic allylic alcohols can be alkylated in the same way. The stereochemistry of [2,3]-sigmatropic rearrangements is discussed in Hoffmann's review.[12] The alkylative sulfenate–sulfoxide rearrangement is widely used in stereoselective syntheses of natural products, e.g. to form the side chains of coenzyme ubiquinone-10 and the growth-promoting steroid brassinolide,[13,14] for epimerization of 17-hydroxy-17-vinyl steroids.[15] A transformation of a prostaglandin with 13-*cis*-15β-hydroxy configuration to the inverted natural 13-*trans*-15α-hydroxy configuration (PGE_1) has been achieved employing this rearrangement and *p*-toluenesulfenyl chloride.[16]

The sulfenate–sulfoxide rearrangement can be used for alkylative carbonyl 1,3-transposition in vinyl carbonyl compounds, both acyclic and cyclic. In the case of saturated aldehydes or ketones they are treated at first with Vinylmagnesium Bromide (or Vinyllithium) to form the allylic alcohol and then sulfenylated as shown in eq 2.[4]

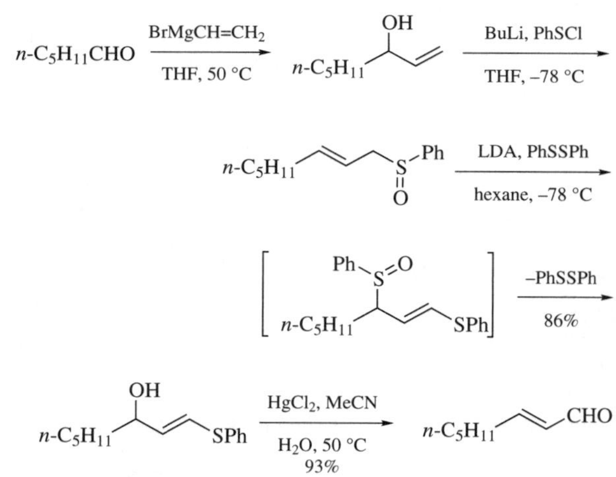

(2)

The sulfoxide formed is lithiated with LDA and sulfenylated with diphenyl disulfide to undergo simultaneous rearrangement and elimination of PhSSPh to yield the enol sulfide. The latter, after hydrolysis with Mercury(II) Chloride, gives the α,β-unsaturated aldehyde. Another example, for a cyclic alkenone, is shown in eq 3.

Phenyl sulfoxides, formed by reaction of 3-trimethylsilylallylic alcohols with PhSCl, have been used for the synthesis of unsaturated aldehydes (eq 4).[17] The α-silyl sulfoxide undergoes a sila-Pummerer rearrangement.

4-Trimethylsilyl-1-buten-3-ol by alkylation of the phenyl sulfoxide produced by reaction with PhSCl, and subsequent desulfonylation and desilylation by Tetra-n-butylammonium Fluoride, affords (*E*)-1,3-alkadienes in yields >90%.[18]

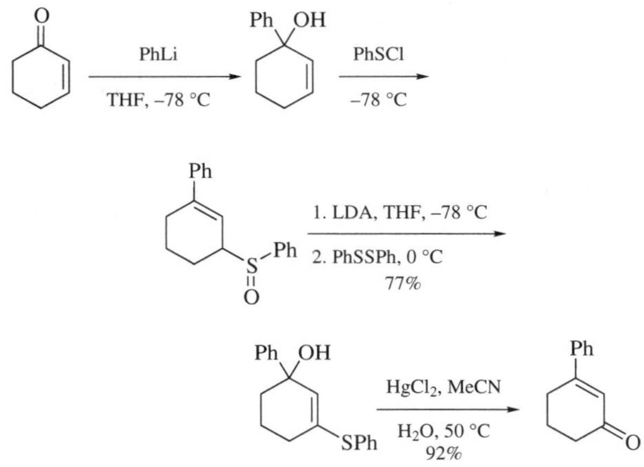

(3)

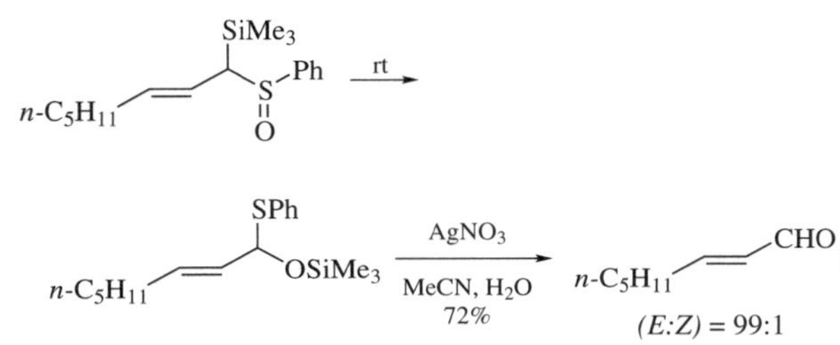

(4)

Sulfenylation of Propargylic Alcohols. Benzenesulfenates of propargylic alcohols form allene sulfoxides by [2,3]-sigmatropic rearrangement. These can be easily transformed to useful derivatives[2d,e] or cyclized.[2f,h] The starting propargylic alcohols are readily accessible from the reaction of aldehydes or ketones with metal acetylides. Thus β-cyclocitral treated with propynyllithium, followed by PhSCl, after rearrangement affords the triene sulfoxide. Reduction with t-Butyllithium and methanol (as an internal proton source) yields 72% of the desulfurized hydrocarbon with retention of the double bond configuration (eq 5).[19]

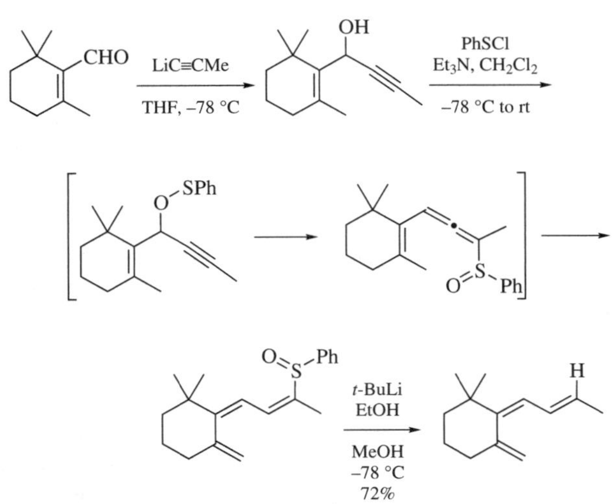

(5)

A steroid 17-methylallene sulfoxide has been desulfurized under similar conditions with Methyllithium to give 72% of the β-methylallene derivative.[20] Trienyl and allenyl phenyl sulfoxides can therefore be desulfurized stereospecifically with retention of configuration using *t*-BuLi or MeLi.

The allene sulfoxide obtained when androst-4-ene-3,17-dione is treated with Dipotassium Acetylide and PhSCl, is functionalized as shown in eq 6 (only the D-ring of the steroid is shown) to form the corticosteroid structure.[21]

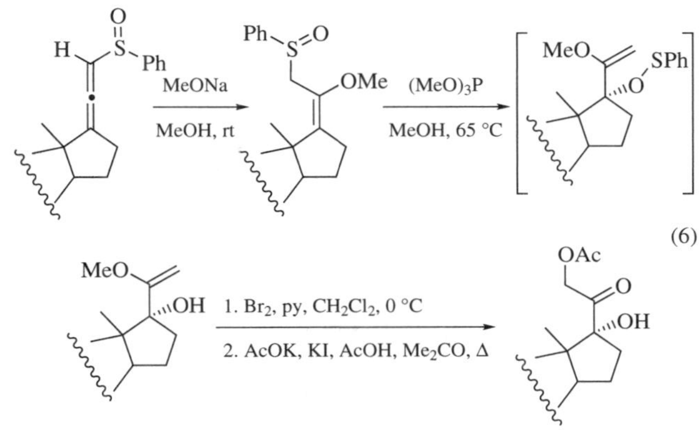

(6)

Phenyl sulfoxides with the *cis*-allenyldiene structure, obtained from corresponding propargylic alcohols, undergo spontaneous electrocyclic ring closure (eq 7).[22] As shown by detailed studies,[22b] the chiral center in the starting propynyl alcohol is stereospecifically transferred to the bridgehead carbon in the cyclized product drimatriene sulfoxide. This type of rearrangement has been used for an enantioselective synthesis of the sesquiterpene (+)-sterpurene.[2h,23]

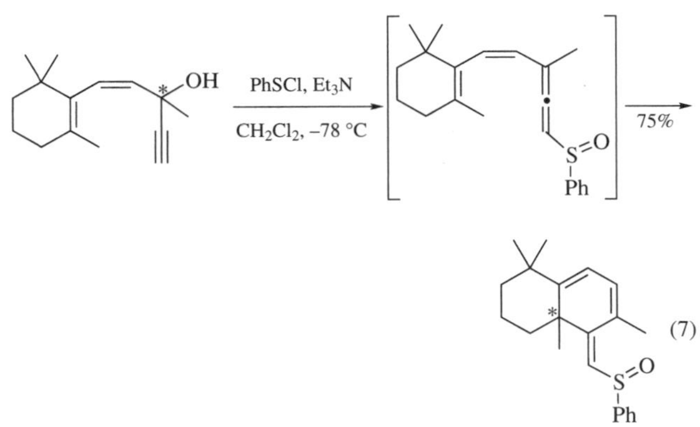

(7)

diastereomeric mixture 3:2

A tandem [2,3]-sigmatropic rearrangement/Michael addition sequence occurs when the hydroxyphenylpropynol is sulfenylated with PhSCl (eq 8).[24] Other substituted benzofurans are obtained in this way.

Ketones from Arylacetonitriles.[2a] Secondary aryl, alkyl, or diaryl acetonitriles, on sulfenylation with PhSCl and subsequent oxidative decyanation by N-Bromosuccinimide and NaOH, afford aryl, alkyl, or diaryl ketones in good yields (eq 9).[25] Four other oxidative decyanation methods were tested;[25b] the one shown is the best.

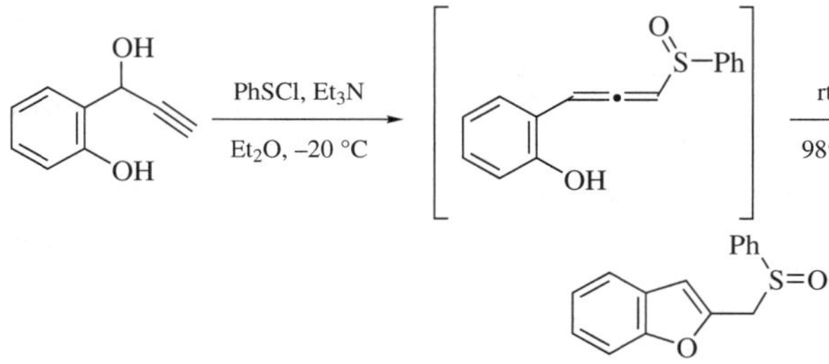

(8)

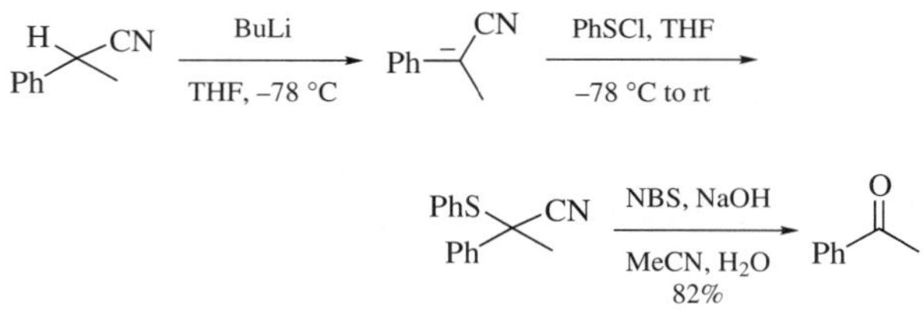

(9)

Reaction with α-Diazo Ketones.[2d] The mild displacement of diazo nitrogen in α-diazo ketones by PhSCl represents a convenient and regiospecific method for functionalization of both cyclic and acyclic ketones at the α-position with halogen and phenylthio substituents (eqs 10–12).[26] These functionalized ketones are useful for a variety of syntheses, particularly phenylthiocyclopentenone for the synthesis of natural products based on 2,3-disubstituted cyclopentanone.[27]

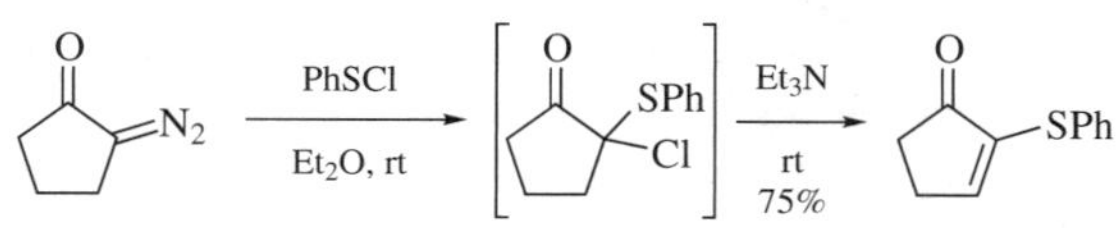

(10)

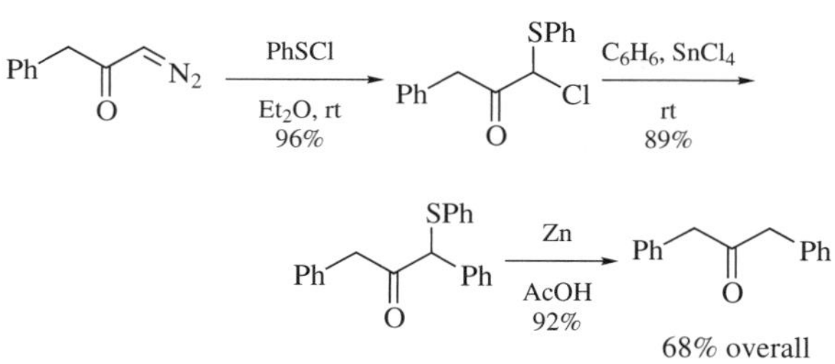

(11)

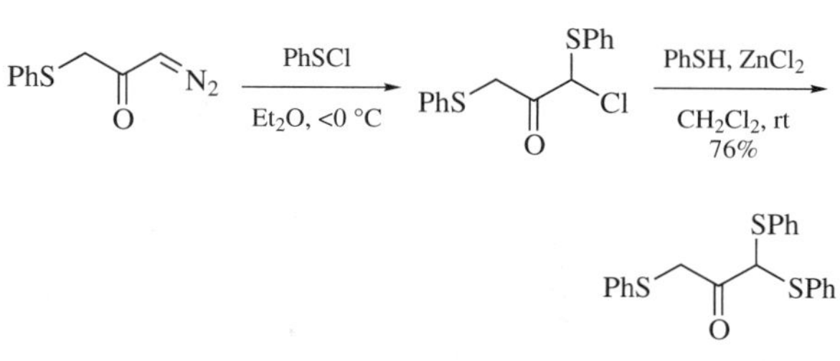

(12)

Addition to Alkenic and Alkynic Compounds. The most commonly used reaction of PhSCl is the addition to double and triple bonds. The regioselectivity of this reaction is not high and is determined mainly by the steric bulk of neighboring groups. Mixtures of regioisomers are usually formed with anti-Markovnikov adducts predominating, which on standing or refluxing in MeCN undergo conversion to Markovnikov products. Addition to cycloalkenes and alkynes is stereoselective and *trans* isomers form almost exclusively. This fact is consistent with an episulfonium ion intermediate.

Combined chlorosulfenylation–dehydrochlorination reactions have great synthetic utility for the conversion of alkenes to cyclic and acyclic vinyl, allyl, or dienyl phenyl sulfides (eqs 13 and 14).[10] The addition reaction is achieved by simply adding the alkene to a cooled (−50 to −78 °C) solution of PhSCl in CH_2Cl_2 and allowing the mixture to warm to rt. The dechlorination is carried out by briefly heating the β-chloroalkyl sulfides with the amidine base 1,8-Diazabicyclo[5.4.0]undec-7-ene. Higher yields on dechlorination are achieved with sulfones quantitatively formed by oxidation of the β-chlorosulfides with peroxy acids.

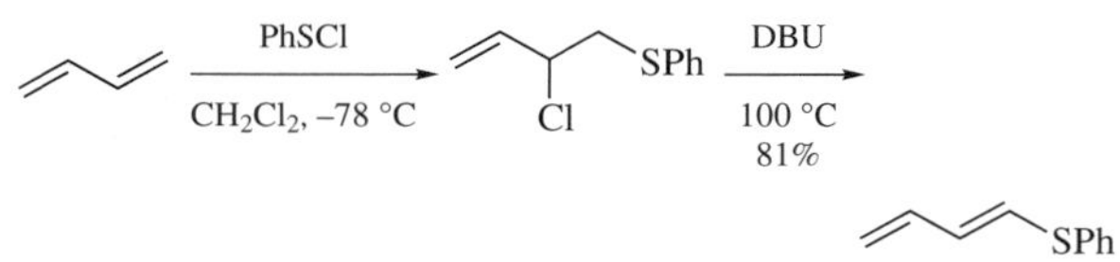

(13)

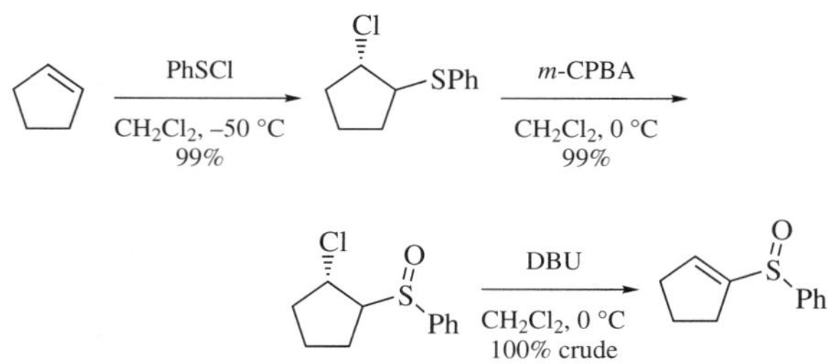

(14)

The addition of PhSCl can be used for functionalization of unsaturated groups,[2d] especially the isopropylidene terminus of natural isoprenoids. The synthetic sequence, as shown for the model compound 2-methylbut-2-ene (eq 15), involves addition of PhSCl, dehydrochlorination of isomeric adducts with Et_3N in DMF to methallyl sulfide, oxidation to the sulfoxide, and its rearrangement to the terminal *trans*-allylic alcohol.[5,28a,c] This procedure is used for the synthesis of the biologically important oxygenated terpenes nuciferal, *ar*-turmerone, ipsdienol, neotorreyol, 6-hydroxydendrolasin, 6-oxodendrolasin, and solanesol.[28a,b]

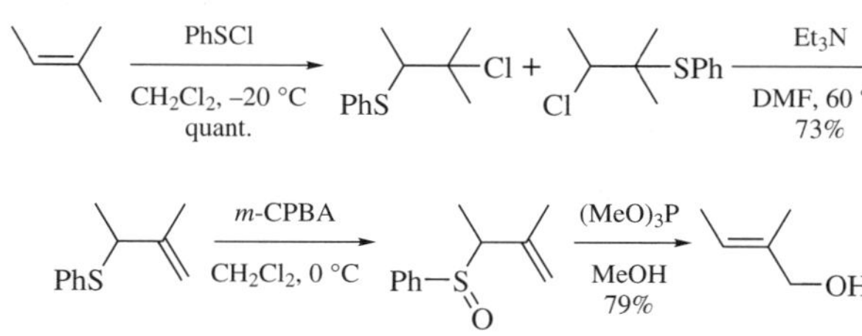

(15)

Complete *cis–trans* alkene isomerization can be achieved by PhSCl addition and dehydrochlorination with *t*-BuOK in DMSO, which causes *trans* elimination of HCl and formation of vinyl sulfide with inverted configuration.[29]

Selective functionalization of 1,3-dienes by PhSCl involves formation of 1,2-adducts (which convert on heating or storage to 1,4-adducts) and hydrolysis, alcoholysis, or acetolysis of the chlorine in these adducts. The allyl alcohol obtained from the 1,2-adduct may be converted to an epoxide and aldehyde as shown in eq 16.[30]

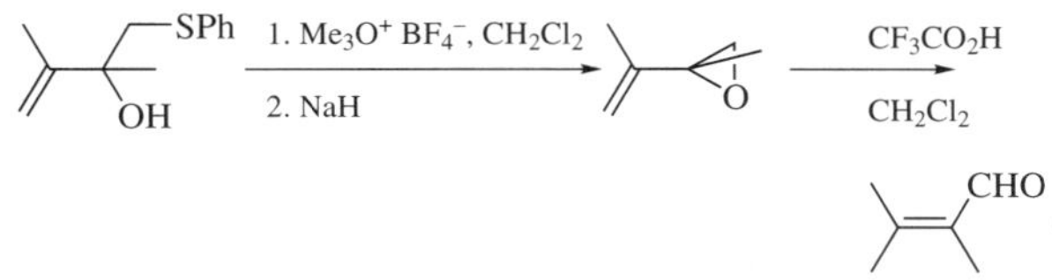

(16)

Adducts of PhSCl to alkenes and cycloalkenes undergo methylation with $Me_2Zn/TiCl_4$ at the chlorine position with complete retention of configuration (controlled by the episulfonium ion intermediate).[2h,31] This one-pot process is termed carbosulfenylation.

Addition to O-Silylated Enols and Alkenols.[2d,i] Silyl enol ethers react smoothly with PhSCl to give α-keto sulfides.[32] The reaction proceeds by addition of PhSCl to the enol double bond and elimination of chlorosilane.[33] *O*-Silylated dienolates (from unsaturated esters, ketones, and aldehydes) undergo γ-sulfenylation with this reagent,[34] with the exception of γ-chloro-substituted dienolates, which give α-products.[35] This type of reaction may be used for stereoselective synthesis of α-sulfenyl-β-hydroxy esters with the *anti* configuration (eq 17).[6] The latter can be converted into either *syn*- or *anti*-epoxides.

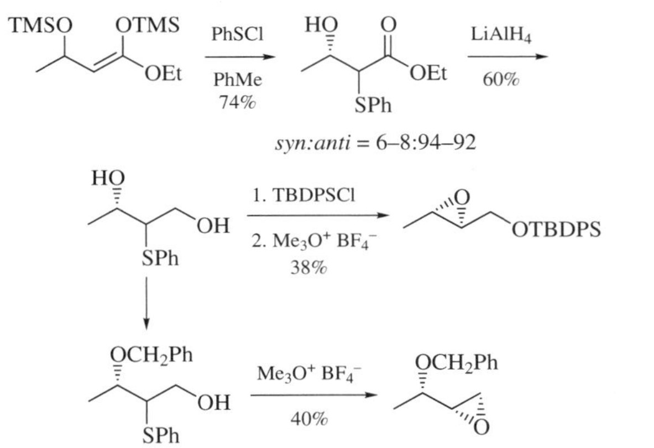

(17)

1-Phenylthiocyclohexanecarbaldehyde, obtained via PhSCl addition to trimethylsiloxymethylenecyclohexane, has been used in stereocontrolled synthesis of allylic alcohols.[36] Benzenesulfenyl chloride adducts of mono- and disubstituted alkenes and cycloalkenes, under Zinc Bromide catalysis, give stereospecific (*anti*) alkylation of *O*-silylated ester enolates.[37]

t-Butyldimethylsilyl ethers of 1-alkenylcycloalkanols, when treated with PhSCl, readily rearrange to the ring-expanded α-1-phenylthioalkylcycloalkanones via the episulfonium ion (eq 18).[38]

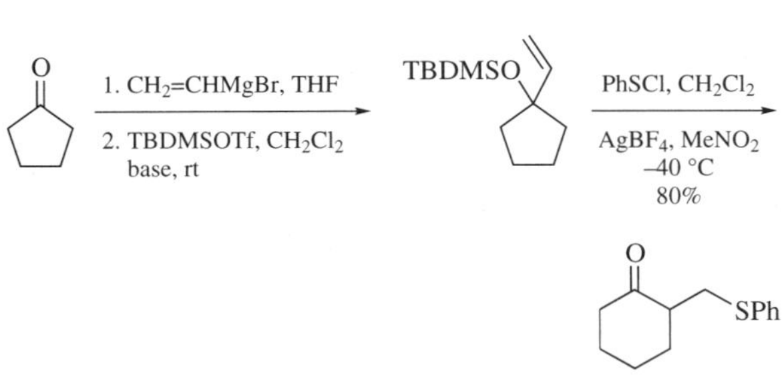

(18)

Cycloaddition. A range of benzenesulfenyl chloride addition reactions to unsaturated bonds proceed with simultaneous cyclization to form important compounds, examples of which are depicted below. These reactions include cyclization of isoprenoid β-keto esters (eq 19),[2f,39] homoconjugative cycloaddition to form tricyclenes (precursors of natural sesquiterpenes) (eq 20),[2d,e,7] cyclization of aryloxyphenylalkynes (eq 21),[40] lactonization (eq 22),[2c,41,42] and sulfenocycloamination of unsaturated amides and amines to form lactams, and pyrrolidine and piperidine rings (eqs 23 and 24).[2f,g,43]

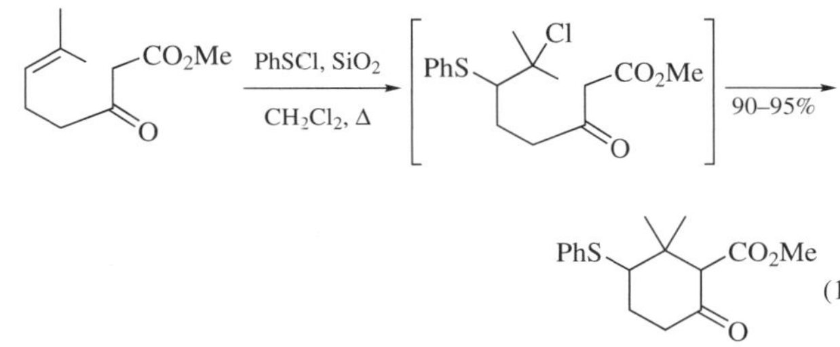

(19)

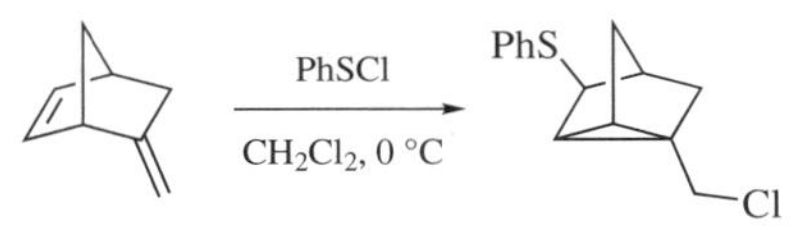

(20)

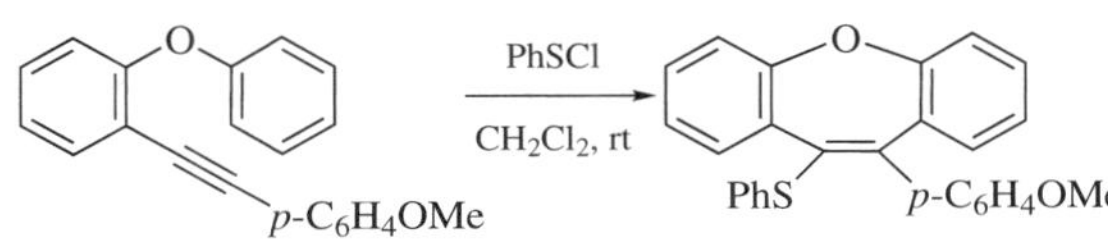

(21)

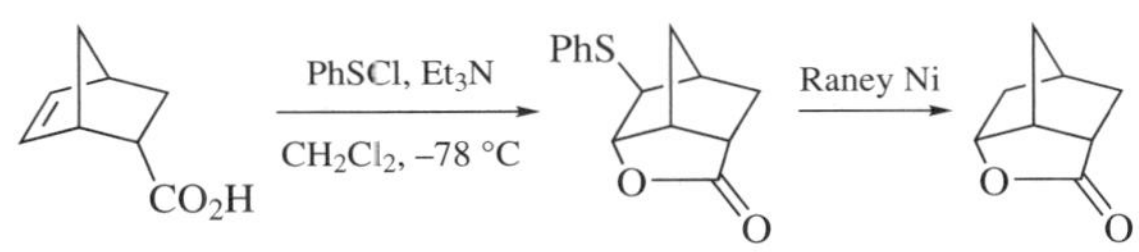

(22)

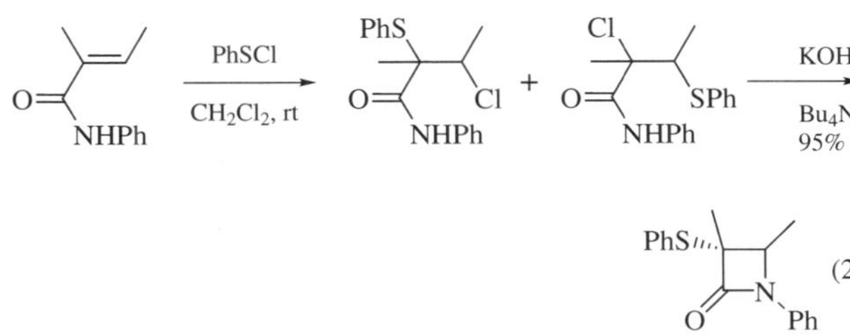

(23)

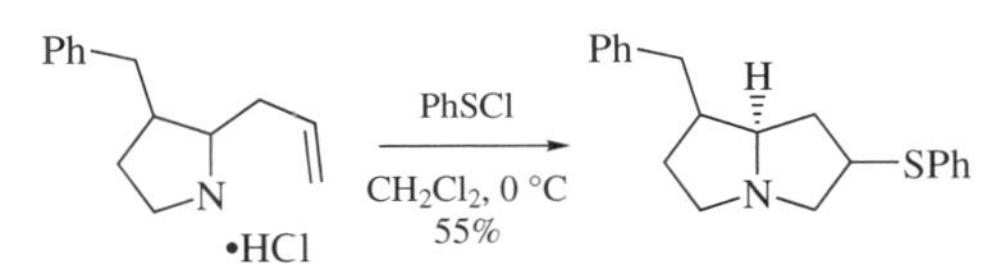

(24)

First Update

Thomas K. Hutton
University of Illinois at Chicago, Chicago, IL, USA

Addition to Alkenes and Alkynes. Reaction of PhSCl with imines derived from 7-oxanorbornene gave substituted furans, proposed to result from sulfonium ion formation followed by Grob-type fragmentation (eq 25).[44] Hydrolysis and aromatization then gave the observed furan product.

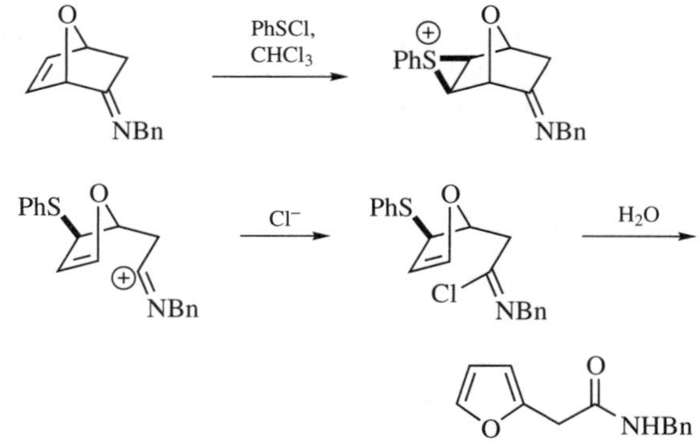

(25)

The addition of PhSCl to allene furnishes the known compound 2-(phenylthio)-3-chloropropene.[45] This was condensed with the Schöllkopf bis-lactim ether[46,47] then hydrolyzed to (*S*)-4-(phenylthio)allylglycine, which was protected as the *N*-Cbz derivative. A Lewis-acid mediated addition of methyl glyoxylate gave the coupled product, from which the olefin and thiophenyl group were removed using nickel boride (eq 26). In this case, use of the vinylic thioether activates the terminal olefin, otherwise inert to the coupling conditions, while still allowing simple removal to give a traceless coupling.[48]

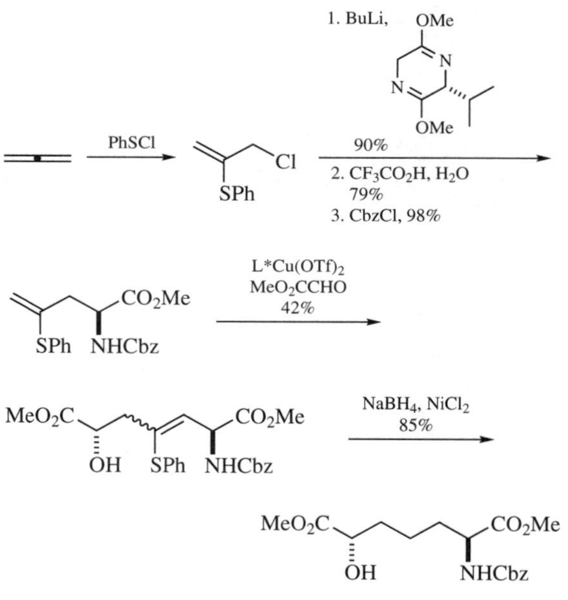

(26)

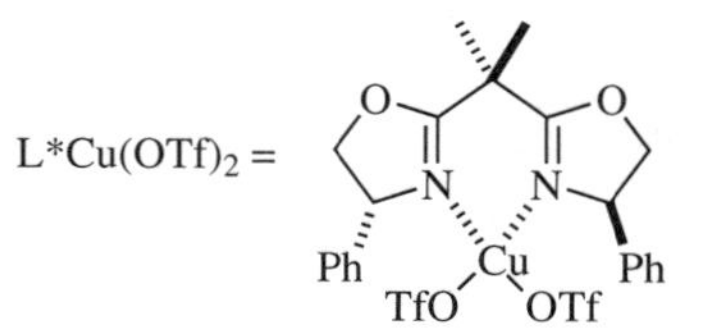

Reaction of PhSCl with vinyl stannanes has been shown to effect an electrophilic destannylation, with retention of configuration, to form the vinyl sulfides.[49] During an assessment of binding affinities of estradiol derivatives, *E*- and *Z*-vinyl stannanes were prepared and destannylated to give (*E*/*Z*)-vinyl thioethers in good yield.

Acetylenic stannanes react with $Cp_2Zr(H)Cl$ to give the corresponding vinylic Zr^{IV} complex, which can react with PhSCl, to give 1-phenylthio vinyl stannanes.[50] These intermediates may be conveniently transformed into *Z*-vinyl sulfides upon reaction with ^{n}BuLi, or can be coupled in a Stille-type reaction leading to more complex vinylic sulfides, which are then hydrolyzed to ketones.

PhSCl addition to alkynyl boronate[51–53] or aluminate[54] complexes resulted in electrophilic addition to the triple bond followed by rearrangement to form vinyl boranes or vinyl aluminium compounds. The vinyl borane products were subsequently coupled in Suzuki-Miyaura type reactions or underwent a boron-copper transmetallation followed by addition of a suitable electrophile. The vinyl sulfide then underwent a nickel-catalyzed Grignard addition, resulting in a tetrasubstituted olefin (eq 27).

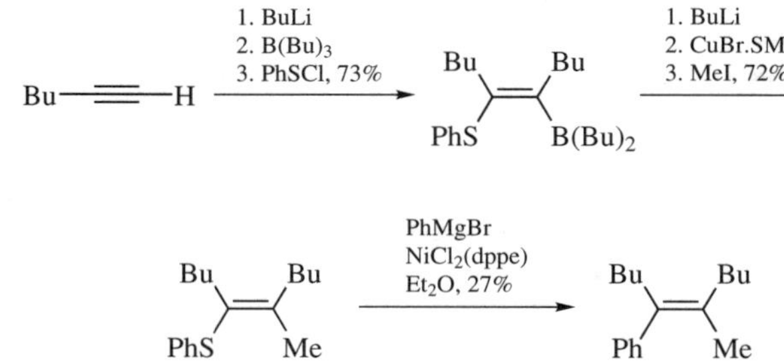

(27)

Addition to Silyl Enol Ethers and Ketenimines. Addition to a silyl enol ether was used as a key step in a 1,2-carbonyl transposition of optically pure bicyclo[2.2.2]octanes.[55] Reaction of the silyl enol ether with PhSCl gave the α-sulfenylated compound, which was transformed into the thio enol ether and then hydrolyzed to give the 1,2-transposed carbonyl group (eq 28).

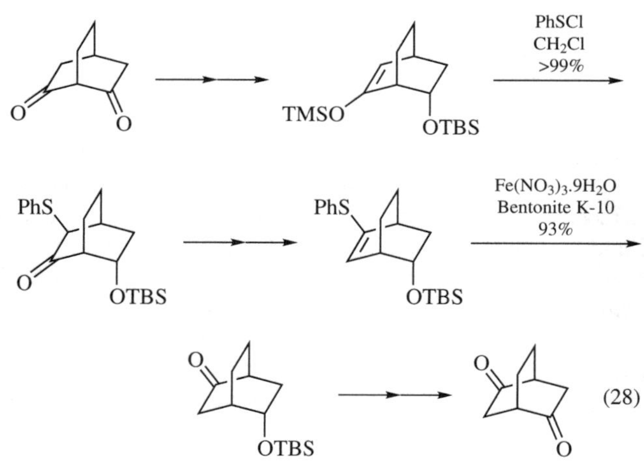

(28)

Addition of PhSCl to a cyclohexylidene silyl enol ether led to an aldehyde, which underwent *anti*-stereoselective aldol reaction followed by $LiAlH_4$ reduction to give a diol (eq 29). Upon reaction with TsOH in pyridine, the diol underwent a [1,4] PhS shift to give an allylic alcohol. Alternatively, treatment of the diol with catalytic TsOH in refluxing CH_2Cl_2 gave a stereospecific [1,2] PhS shift to yield a spirocyclic tetrahydrofuran.[56]

Addition of PhSCl to another silyl enol ether was used in a synthesis of the Taxol A-ring. Reaction of the silyl enol ether with PhSCl gave the corresponding cyclohexanone, which was alkylidenated to give a protected A-ring subunit (eq 30).[57]

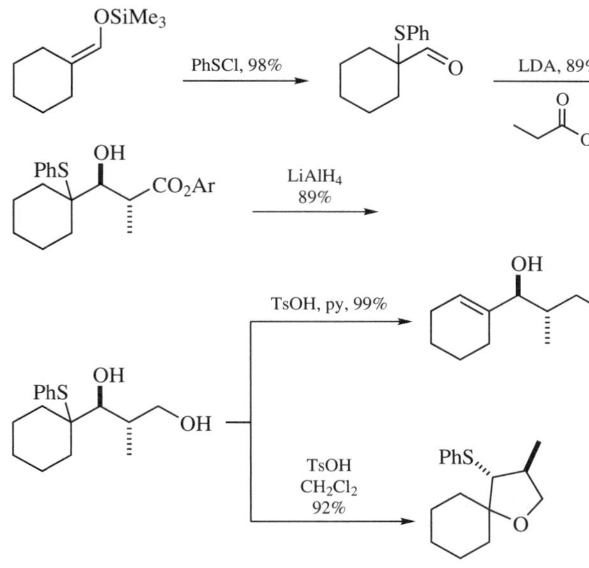

(29)

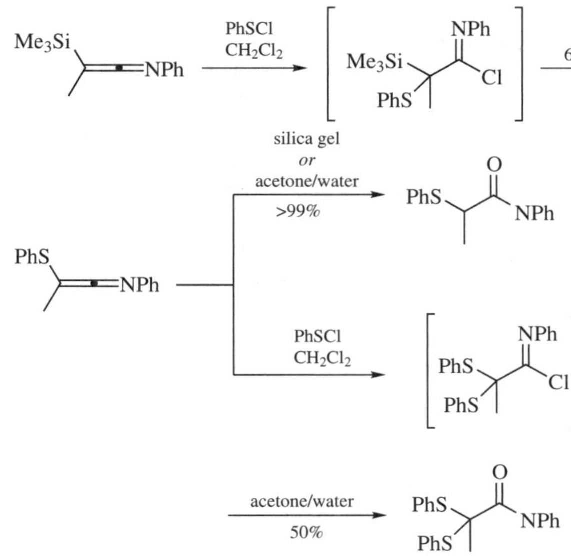

(31)

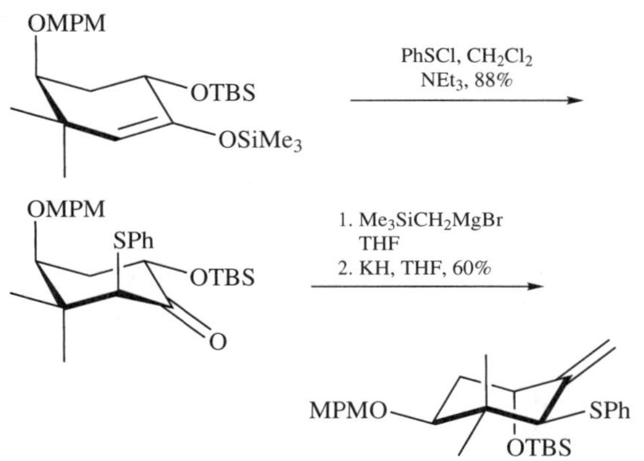

(30)

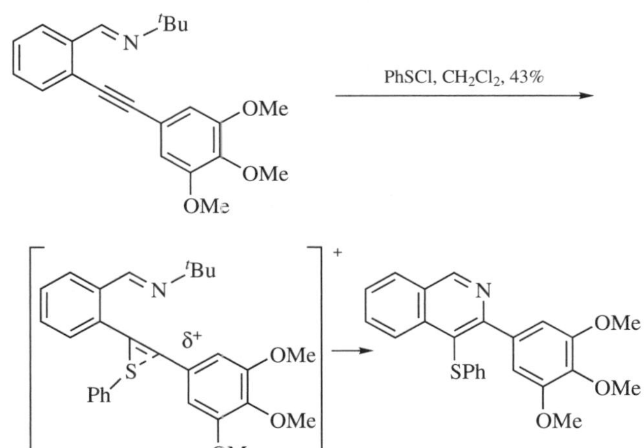

(32)

Tetrahydrofurans have also been formed using PhSCl addition to silyl alkenols via a 5-*endo*-trig cyclization.[58] Pyramidilization of the electrophilic centre and benzylic stabilization of the nascent carbocation formed after addition of PhSCl to the olefin allows the otherwise disfavored cyclization to take place.

Dependant on the reaction conditions, PhSCl can display varied behavior towards silylated ketenimines.[59] Addition to the olefin takes place and chlorosilane elimination leads to an *ipso*-substitution of SPh for $SiMe_3$. Due to the loss of silyl stabilization, the resulting ketenimine slowly hydrolyzes to the amide; this decomposition can be expedited using silica gel or an acetone/water mixture. Alternatively, addition of 2 equiv of PhSCl results in a second addition to the intermediate ketenimine to give an imidoyl chloride which is again hydrolyzed to the corresponding amide (eq 31).

Cycloadditions. Upon reaction with a variety of electrophiles, including PhSCl, aryl iminoalkynes undergo a 6-*endo* type cyclization to yield isoquinolines and naphthyridines.[60,61] Addition of PhSCl across the triple bond gives a sulfonium ion intermediate, and reaction of the internal imine nucleophile with the developing carbocation yields the corresponding substituted heterocycle (eq 32). As a consequence of the need for benzylic stabilization, the reaction performs best when the alkyne bears an electron-rich aromatic group and does not proceed at all with alkyl-substituted alkynes with PhSCl as the electrophile, although other reagents can promote reaction of such substrates.

Reaction of PhSCl with the Diels-Alder adduct of quinone with cyclopentadiene gives the sulfonium ion from addition to the olefin, which, in the presence of $AgBF_4$, undergoes an cyclization to give the tetracyclic enone illustrated (eq 33).[62] This reaction is thought to proceed via fluoride addition to the carbonyl group, with the resulting oxyanion attacking the sulfonium ion, and hydrolytic workup generating the hemiacetal center.

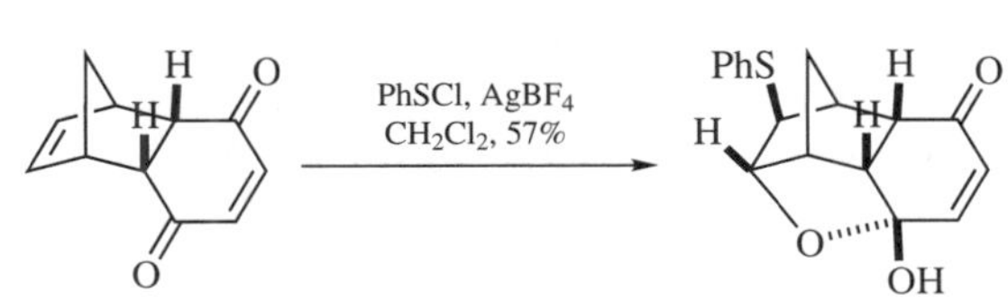

(33)

Addition to Silylated Alkenes and Alkynes. The synthesis of (−)-isocitric acid from D-malic acid dimethyl ester via an alkynyl silane is straightforward, however synthesis of (−)-homoisocitric acid is more complex due to the need to transform the alkynyl silane to a carboxylic acid while maintaining the carbon chain length. Reaction of the alkynyl silane with PhSCl and chlorosilane elimination gives the corresponding alkynyl sulfide, which in turn may be hydrolyzed to a carboxylic acid (eq 34).[64] This method allows transformation of a silylated alkyne to an acid of equal chain length.

Reaction of *o*-alkenylbenzyltributyl stannanes with PhSCl led to in situ generation of *o*-quinodimethane intermediates which could undergo intermolecular cycloaddition with a variety of dieneophiles to yield the corresponding tetrahydronaphthalene derivative.[63] This approach permits generation of *o*-quinodimethane cycloaddition products with the promoter incorporated in the final product.

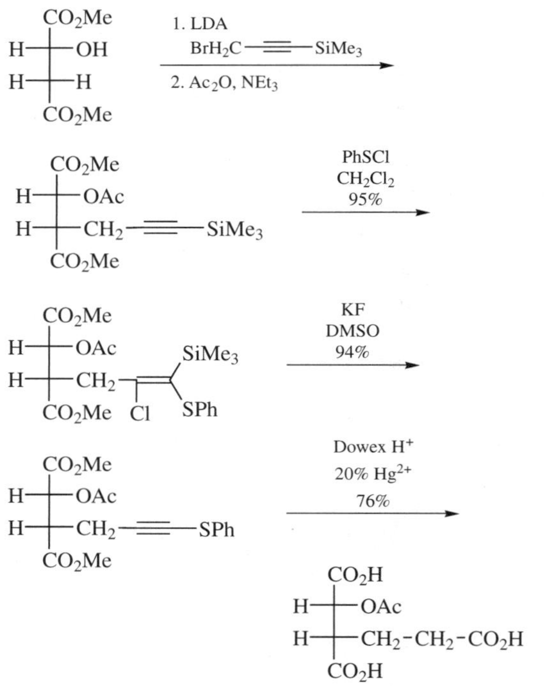

(34)

Reaction of disilylated dienes with PhSCl was found to give monosubstitution products with inversion of configuration.[65] Thus, (1*E*,3*E*)-1,4-bis(trimethylsilyl)buta-1,3-diene, upon reaction with PhSCl gave the silylated sulfide with inversion (*Z,E*/*E,E* = 95:5) in 70% yield. On extension to (1*E*,3*E*,5*E*)-1,6-bis(trimethylsilyl)hexa-1,3,5-triene, the same reaction was observed, but the yield and selectivity dropped (*Z,E,E*/*E,E,E* = 90:10, 51% yield).

Addition to Enol Ethers, Enones, and Enamines. PhSCl addition adducts of alkenes gave *anti*-alkylation of *O*-silylated ester enolates under Zn^{2+} catalysis.[37] This concept has been extended to include cyclic enol ethers, again giving exclusively *trans*-addition of silyl enol ethers (eq 35).[66]

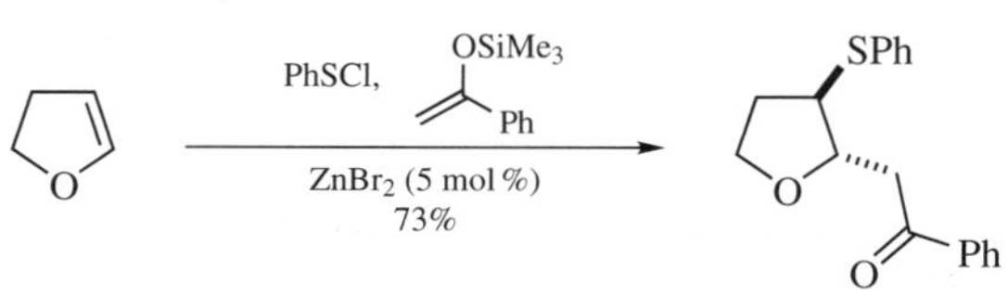

(35)

Cyclization of E-α′-trialkylsilyloxy-α,β-unsaturated ketones was achieved by addition of PhSCl to give the expected regioisomeric mixture of addition products, whereupon treatment with dry silica gel promoted cyclization to give the corresponding furanone in good yield, again with predominantly *trans*-stereochemistry (eq 36).[67]

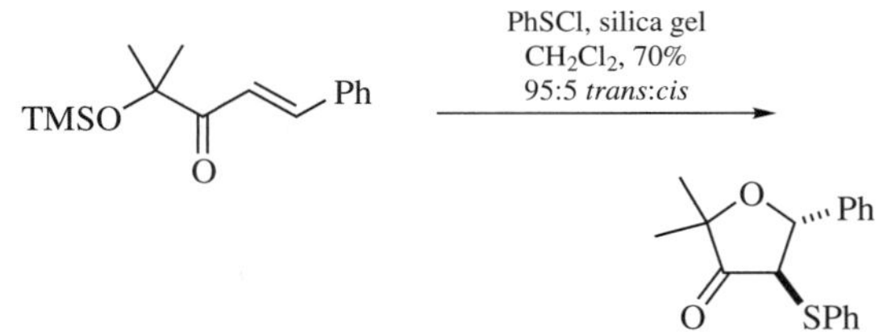

(36)

A number of studies have been made into the sulfenylation of indole and its derivatives in recent years.[68–70] The initial sulfenylation of indole by PhSCl occurs regioselectively at the 3-position, and particular attention has been focused on the mechanism of further sulfenylation, which is observed, with this reagent, to give addition to the 2-position. Isolation of a 3,3-disubstituted indolinium intermediate would seem to point to an addition/migration mechanism (eq 37).

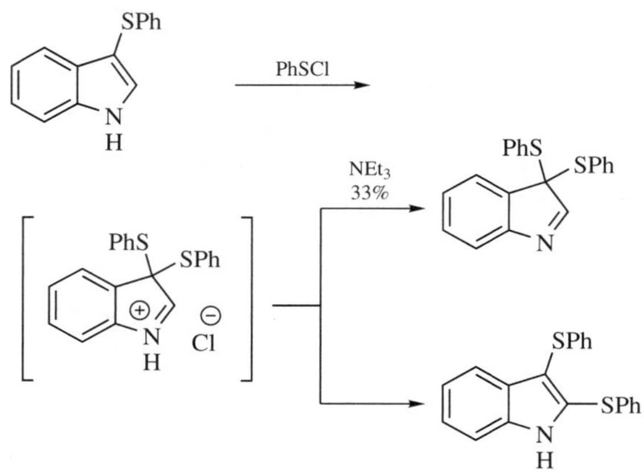

(37)

This work has found use in synthesis, in the formation of benzothiopyrano[2,3-*b*]indol-11-one from indole.[71] Attempted addition of thiophenol to the 2-position of the ester-substituted indole failed, thus the addition/migration of PhSCl was used to install the desired moiety, with polyphosphoric acid cyclization giving the desired indole derivative (eq 38).

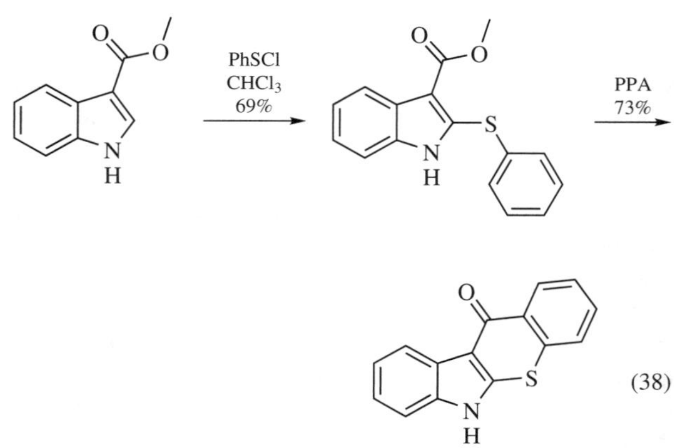

(38)

Addition to Glycals. Furanoid glycals were employed as starting materials in the formation of nucleoside derivatives through the use of PhSCl.[72] Substitution of appropriate furanoid glycals gave the corresponding addition products which had the dual benefit of the chloride moiety as a coupling point and the thiophenyl group as a stereocontrol element. Without isolation, these intermediates were coupled to protected nucleic bases by means of an $SnCl_4$ promoted reaction, giving the desired products in good yield and high stereoselectivities (eq 39). The thiophenyl group was removed postcoupling by reaction with tributyltin hydride, generated in situ from bis(tributyltin)oxide and polymethylhydrosiloxane.

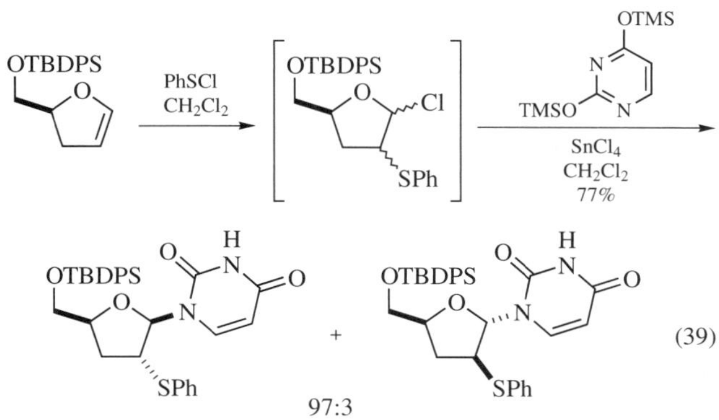

(39)

Addition of PhSCl to glycals has been employed in order to form new types of glycosyl donors.[73] Thus, dehydration of a neuraminate afforded the corresponding glycal, which gave an anomeric chloride on addition of PhSCl. Activation of this species with AgOTf allowed coupling to a number of carbohydrate based acceptors (eq 40).

Addition of PhSCl to D-glucal derivatives gave a mixture of "manno" and "gluco" anomeric chlorides; this was followed by reaction with AgOAc to form the corresponding anomeric acetates (eq 41).[74] Some interesting observations were noted, such as the improvement in facial selectivity of PhSCl addition in the peracetylated vs. the unprotected glucal. This led to the hypothesis that the glycal conformation largely determines facial selectivity, with the conformation being in turn influenced by the protecting groups.

Formation of Heteroatom Radical Precursors. Due to the susceptibility of the thiophenyl group towards radical attack, PhSCl has found use in forming precursors to heteroatom-centred radicals. Relatively few techniques exist for the generation of nitrogen based radicals, one of the most popular one being the use of PTOC (*N*-hydroxypyridine-2-thione) carbamates, popularized by Newcomb.[75,76] In a manner analogous to the formation of iminyl radicals from sulphenylimines,[77–79] *N*-centered aminyl radicals are formed upon reaction of Bu_3SnH with sulfenamides, themselves formed from secondary amines and PhSCl. Initially, formation of aminyl radicals was investigated by ring opening of an aminocyclopropane, leading, after imine reduction, to a linear amine.[80] This method was subsequently applied to form a variety of polycyclic pyrrolizidine products, for example the synthesis of an aphanorphine analog,[81] by a tandem 5-*exo*-trig/6-*endo*-trig cyclization (eq 42),[82] and formation of more complex polycycles (eq 43).[82]

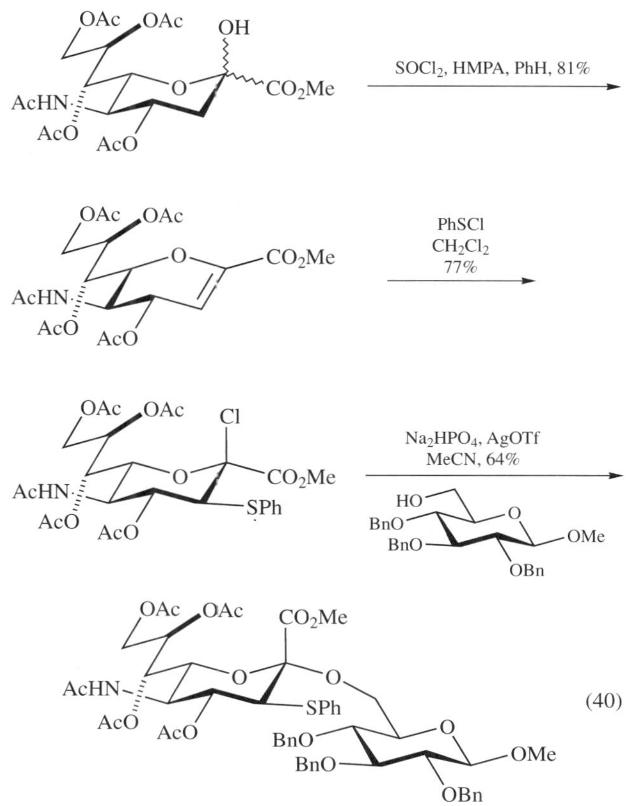

(40)

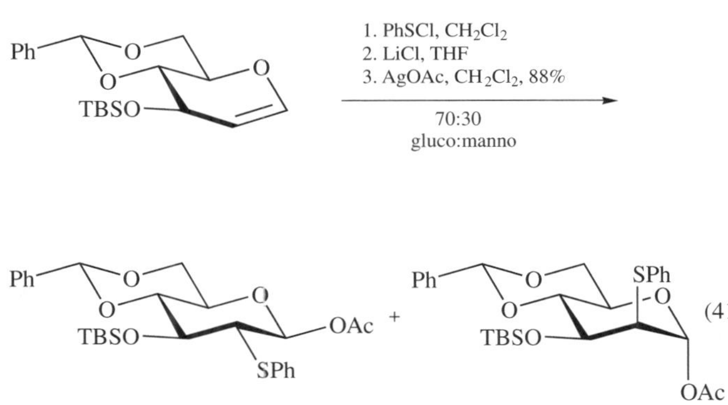

(41)

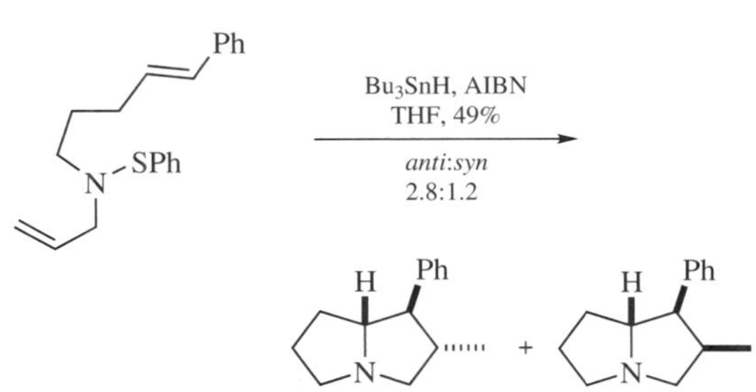

(42)

Disulfenamides were investigated as a means of generating two *N*-centered radicals, however no definite conclusions could be

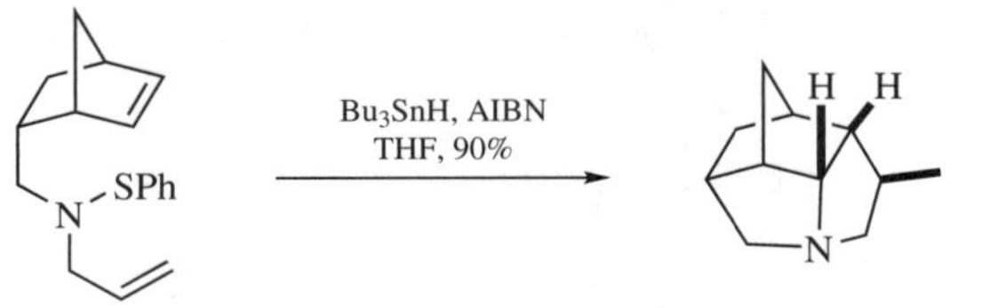

(43)

drawn about this hypothesis.[83] Attempts were also made to employ amino acid-derived aminyl radicals to impart diastereoselectivity on the reaction, with some success, although diastereomeric excesses were moderate at best.[84]

This concept was extended to the synthesis and cyclization of amidyl radicals by Newcomb. These species enjoy the combined benefits of high reactivity and neutral reaction conditions. *N*-(Phenylthio)amides were prepared from a range of secondary amides, and were used in Bu_3SnH reactions as precursors of amidyl radicals.[85] These were utilized in a variety of 5-*exo*-trig/6-*endo*-trig radical cyclizations and tandem cyclizations to form pyrrolidinones (eq 44) and bicyclic pyrrolidinones (eq 45).

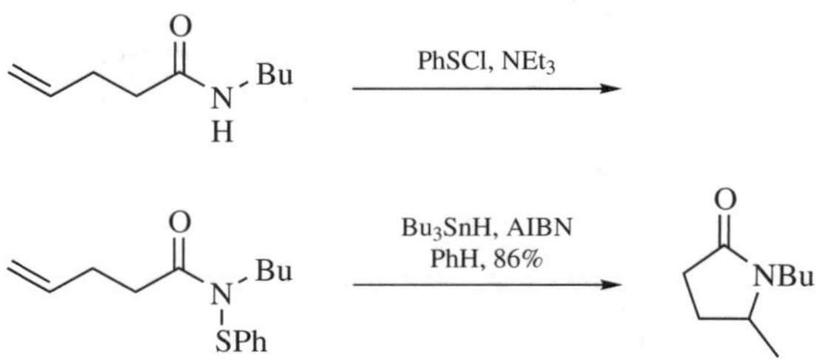

(44)

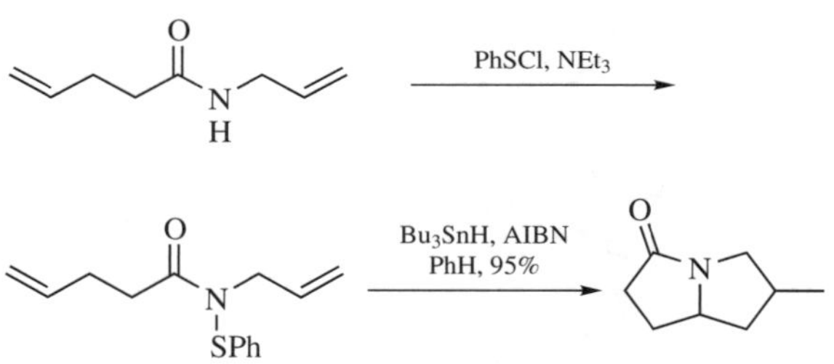

(45)

Reaction of PhSCl with substituted anilines gives benzenesulfenanilides, which on reaction with Lewis acids such as $BF_3{\cdot}Et_2O$ form *N*-centered radical cation intermediates.[86] This is evidenced by the isolation of addition products when the reaction was carried out in the presence of cyclohexene, and by the ESR spectrum of the stable bicyclic radical cation.

Substituted derivatives of PhSCl have been used to from stable thioaminyl radicals in order to assess their magnetic properties.[87] *p*-Nitrobenzenesulfenyl chloride was coupled to a substituted aniline, then oxidized with PbO_2 to form the stable radical. This crystalline substance was analyzed by ESR; its magnetic susceptibility also was measured, and it was found to be ferromagnetic.

This method of heteroatom radical generation has also found application in the generation of *O*-centered radicals from the corresponding SPh derivative. Tetrahydrofurans were synthesized by cyclization of 4-pentenyl-1-oxyl radicals, formed from homolytic decomposition of the corresponding phenylsulfenic acid *O*-ester.[88] Thus, reaction of PhSCl with substituted 4-penten-1-ols gave the desired precursors, with photolysis in the presence of Bu_3SnH affording tetrahydrofurans (eq 46).

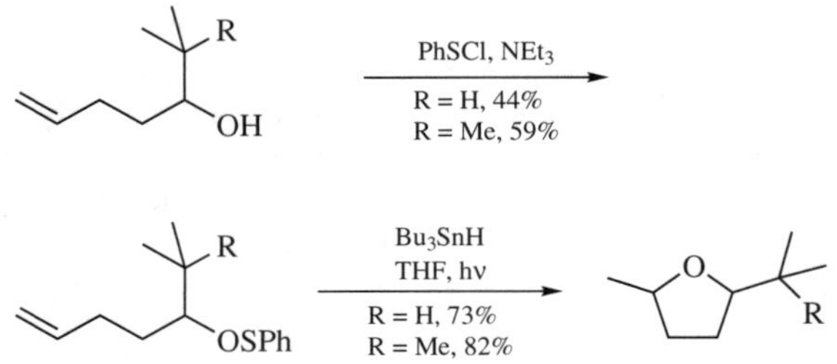

(46)

A similar process employed more reactive α,β-unsaturated esters as the radical acceptors. Phenylsulfenic acid *O*-esters were thus cyclized to form tetrahydrofurans under Bu_3SnH promotion in a tandem radical addition/hydrogen atom transfer reaction (eq 47). It is postulated that a combination of steric and electronic effects lead to the stereochemical outcome illustrated.[89]

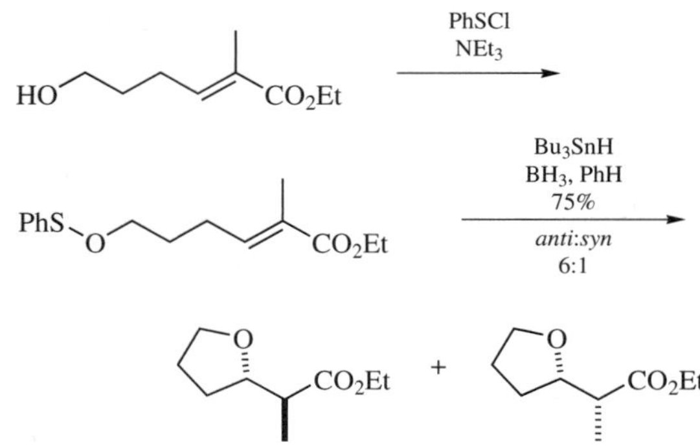

(47)

It has been observed that arylsulfenic acid *O*-esters can form radicals which undergo 1,5-*H* atom abstraction and subsequent reincorporation of the thiophenyl group. Initially observed in reactions of the tropane skeleton (eq 48),[90] this process was found to occur in a variety of substrates.[91] In general yields are moderate, however in systems where the migration occurs between points of fixed geometry, i.e., across rings, yields are improved.

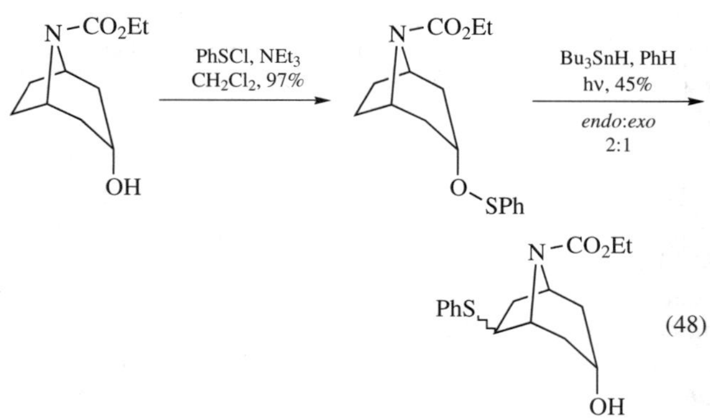

(48)

Formation of α-Phenylsulfenyl Acetonitriles. Linear aliphatic aldehydes were used to form *N,N*-dialkylhydrazones which, upon treatment with LDA, form azaenolates. Reaction of these intermediates with PhSCl led to a range of α-phenylsulfenyl acetonitriles (eq 49).[92]

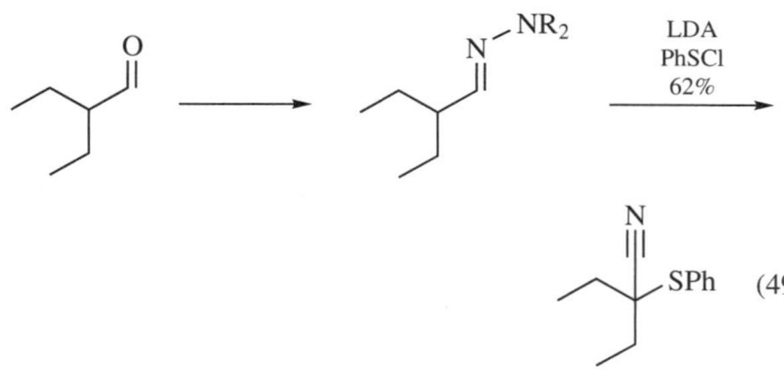

(49)

Formation of Sulfenamide Oxidants. Oxidation of alcohols to carbonyl compounds has been achieved using a catalytic oxidant formed from the reaction of PhSCl with *tert*-butylamine. Previously it was noted that a stoichiometric quantity of *N-tert*-butylbenzenesulfinimidoyl chloride was effective in oxidizing a range of primary and secondary alcohols.[93] This process was made catalytic through the use of *N-tert*-butylbenzenesulfenamide with *N*-chlorosuccinimide as a stoichiometric chlorinating agent.[94,95] The catalytic oxidation was found to be effective for a wide range of alcohols bearing a variety of functionality, both primary and secondary, with yields for the oxidation of primary alcohols to aldehydes ranging from 88–>99% and 90–>99% for the formation of ketones (eq 50). In addition, it was noted that the reaction effectively promoted the oxidation of primary alcohols in the presence of secondary alcohols, albeit in slightly lower yields of 69–72%. Nonetheless, this is a powerful oxidation protocol with wide applicability and good selectivity.

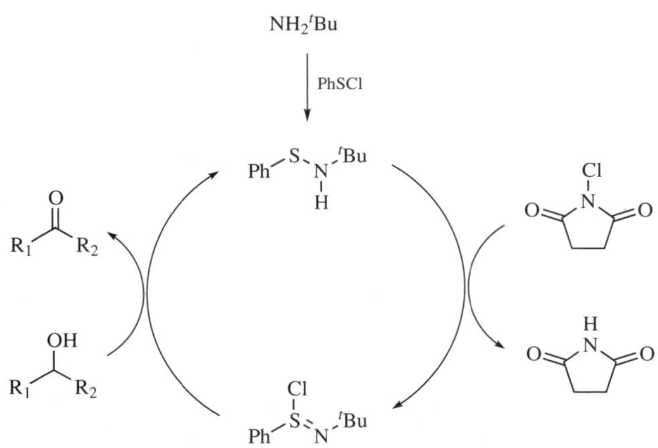

R_1, R_2 = Alkyl, Allylic, Benzylic
Yields = 90-99%

(50)

Related Reagents. Benzeneselenenyl bromide; benzeneselenenyl chloride; 2,4-dinitrobenzenesulfenyl chloride; diphenyl disulfide; o-nitrobenzenesulfenyl chloride; n-phenylthiophthalimide; 2-pyridinesulfenyl bromide; benzenesulfenyl triflate.

1. (a) Kuehle, E., *Synthesis* **1970**, 561. (b) Kuehle, E., *Synthesis* **1971**, 536, 617. (c) Kuehle, E. *The Chemistry of Sulfenic Acids*; Thieme: Stuttgart, **1973**. (d) Rasteikiene, L.; Greiciute, D.; Knunyants, I. L., *Usp. Khim.* **1977**, *46*, 1041. (e) *The Chemistry of Sulfenic Acids and their Derivatives*; Patai, S., Ed.; Wiley: Chichester, **1990**.
2. (a) Fieser, L. F.; Fieser, M., *Fieser & Fieser* **1975**, *5*, 523. (b) **1977**, *6*, 30. (c) **1980**, *8*, 32. (d) **1981**, *9*, 35. (e) **1982**, *10*, 3, 24. (f) **1984**, *11*, 34, 39. (g) **1986**, *12*, 42. (h) **1989**, *14*, 27, 152. (i) **1990**, *15*, 19.
3. Evans, D. A.; Andrews, G. C., *Acc. Chem. Res.* **1974**, *7*, 147.
4. Trost, B. M.; Stanton, J. L., *J. Am. Chem. Soc.* **1975**, *97*, 4018.
5. Masaki, Y.; Hashimoto, K.; Sakuma, K.; Kaji, K., *Chem. Commun./J. Chem. Soc., Chem. Commun.* **1979**, 855.
6. Guanti, G.; Banfi, L.; Narisano, E.; Thea, S., *Chem. Lett.* **1988**, 1683.
7. (a) Heissler, D.; Riehl, J. J., *Tetrahedron Lett.* **1979**, 3957. (b) Heissler, D.; Riehl, J. J., *Tetrahedron Lett.* **1980**, *21*, 4707, 4711.
8. (a) Kuliev, A. M.; Kyazim-zade, A. K.; Guseinov, K. Z., *J. Org. Chem. USSR (Engl. Transl.)* **1970**, *6*, 2117. (b) Stepanov, B. I.; Rodionov, V. Ya.; Chibisova, T. A.; Yagodina, L. A.; Stankevich, A. D., *J. Org. Chem. USSR (Engl. Transl.)* **1977**, *13*, 334.
9. (a) Barrett, A. G. M.; Dhanak, D.; Graboski, G. G.; Taylor, S. J., *Org. Synth.* **1990**, *68*, 8. (b) Harpp, D. N.; Friedlander, B. T.; Larsen, C.; Steliou, K.; Stockton, A., *J. Org. Chem.* **1978**, *43*, 3481.
10. Hopkins, P. B.; Fuchs, P. L., *J. Org. Chem.* **1978**, *43*, 1208.
11. (a) Evans, D. A.; Andrews, G. C.; Sims, C. L., *J. Am. Chem. Soc.* **1971**, *93*, 4956. (b) Evans, D. A.; Andrews, G. C.; Fujimoto, T. T.; Wells, D., *Tetrahedron Lett.* **1973**, 1385, 1389.
12. Hoffmann, R. W., *Angew. Chem., Int. Ed. Engl.* **1979**, *18*, 563.
13. Terao, S.; Kato, K.; Shiraishi, M.; Morimoto, H., *J. Org. Chem.* **1979**, *44*, 868.
14. Zhou, W.-S.; Shen, Z.-W., *J. Chem. Soc., Perkin Trans. 1* **1991**, 2827.
15. Morera, E.; Ortar, G., *J. Org. Chem.* **1983**, *48*, 119.
16. Miller, J. G.; Kurz, W.; Untch, K. B.; Stork, G., *J. Am. Chem. Soc.* **1974**, *96*, 6774.
17. Cutting, I.; Parsons, P. J., *Tetrahedron Lett.* **1981**, *22*, 2021.
18. Hsiao, C. N.; Shechter, H., *Tetrahedron Lett.* **1984**, *25*, 1219.
19. Theobald, P. G.; Okamura, W. H., *J. Org. Chem.* **1990**, *55*, 741.
20. Neef, G.; Eder, U.; Seeger, A., *Tetrahedron Lett.* **1980**, *21*, 903.
21. VanRheenan, V.; Shephard, K. P., *J. Org. Chem.* **1979**, *44*, 1582.
22. (a) Reischl, W.; Okamura, W. H., *J. Am. Chem. Soc.* **1982**, *104*, 6115. (b) Okamura, W. H.; Peter, R.; Reischl, W., *J. Am. Chem. Soc.* **1985**, *107*, 1034.
23. Gibbs, R. A.; Okamura, W. H., *J. Am. Chem. Soc.* **1988**, *110*, 4062.
24. Gray, M.; Parsons, P. J.; Neary, A. P., *Synlett* **1992**, 597.
25. (a) Selikson, S. J.; Watt, D. S., *Tetrahedron Lett.* **1974**, 3029. (b) Freerksen, R. W.; Selikson, S. J.; Wroble, R. R., *J. Org. Chem.* **1983**, *48*, 4087.
26. (a) McKervey, M. A.; Ratananukul, P., *Tetrahedron Lett.* **1983**, *24*, 117. (b) Cronin, J. P.; Dilworth, B. M.; McKervey, M. A., *Tetrahedron Lett.* **1986**, *27*, 757.
27. Monteiro, H. J., *J. Org. Chem.* **1977**, *42*, 2324.
28. (a) Masaki, Y.; Hashimoto, K.; Kaji, K., *Tetrahedron Lett.* **1978**, 4539, 5123. (b) Masaki, Y.; Hashimoto, K.; Sakuma, K.; Kaji, K., *J. Chem. Soc., Perkin Trans. 1* **1984**, 1289. (c) Masaki, Y.; Hashimoto, K.; Kaji, K., *Tetrahedron* **1984**, *40*, 3481.
29. Masaki, Y.; Sakuma, K.; Kaji, K., *Chem. Lett.* **1979**, 1235.
30. Golding, B. T.; Pombo-Villar, E.; Samuel, C. J., *Chem. Commun./J. Chem. Soc., Chem. Commun.* **1985**, 1444.
31. Reetz, M. T.; Seitz, T., *Angew. Chem., Int. Ed. Engl.* **1987**, *26*, 1028.
32. Murai, S.; Kuroki, Y.; Hasegawa, K.; Tsutsumi, S., *Chem. Commun./J. Chem. Soc., Chem. Commun.* **1972**, 946.
33. Shchepin, V. V.; Petukhova, I. Yu.; Nedugov, A. H.; Vakhrin, M. I., *J. Gen. Chem. USSR (Engl. Transl.)* **1992**, *62*, 1510.
34. Fleming, I.; Goldhill, J.; Paterson, I., *Tetrahedron Lett.* **1979**, 3205.
35. Shchepin, V. V.; Petukhova, I. Yu.; Russkikh, N. Yu., *J. Gen. Chem. USSR (Engl. Transl.)* **1992**, *62*, 1072.
36. Aggarwal, V. K.; Warren, S., *Tetrahedron Lett.* **1986**, *27*, 101.
37. Patel, S. K.; Paterson, I., *Tetrahedron Lett.* **1983**, *24*, 1315.
38. Kim, S.; Park, J. H., *Tetrahedron Lett.* **1989**, *30*, 6181.
39. Alderdice, M.; Weiler, L., *Can. J. Chem.* **1981**, *59*, 2239.

40. Kitamura, T.; Takachi, T.; Kawasato, H.; Taniguchi, H., *J. Chem. Soc., Perkin Trans. 1* **1992**, 1969.
41. (a) Nicolaou, K. C.; Lysenko, Z., *Chem. Commun./J. Chem. Soc.* **1977**, 293. (b) Nicolaou, K. C.; Seitz, S. P.; Sipio, W. J.; Blount, J. F., *J. Am. Chem. Soc.* **1979**, *101*, 3884.
42. Toru, T.; Fujita, S.; Saito, M.; Maekawa, E., *J. Chem. Soc., Perkin Trans. 1*, **1986**, 1999.
43. (a) Ihara, M.; Fukumoto, K.; Kametani, T., *Heterocycles* **1982**, *19*, 1435. (b) Ihara, M.; Haga, Y.; Yonekura, M.; Ohsawa, T.; Fukumoto, K.; Kametani, T., *J. Am. Chem. Soc.* **1983**, *105*, 7345. (c) Ohsawa, T.; Ihara, M.; Fukumoto, K.; Kametani, T., *J. Org. Chem.* **1983**, *48*, 3644.
44. Arjona, O.; Csákÿ, A. G.; Murcia, M. C.; Plumet, J., *Helv. Chim. Acta.* **2001** *84*, 3667.
45. Mueller, W. H.; Butler, P. E., *J. Org. Chem.* **1968**, *33*, 1533.
46. Schöllkopf, U.; Hartwig, W.; Groth, U., *Angew. Chem. Int. Ed. Engl.* **1979**, *18*, 863.
47. Schöllkopf, U.; Groth, U.; Deng, C., *Angew. Chem. Int. Ed. Engl.* **1981**, *20*, 798.
48. Gao, Y.; Lane-Bell, P.; Vederas, J. C., *J. Org. Chem.* **1998**, *63*, 2133.
49. Napolitano, E.; Fiaschi, R.; Herman, L. W.; Hanson, R. N., *Steroids* **1996**, *61*, 384.
50. Huang, X.; Zhong, P., *J. Chem. Res. (S)* **1999**, 290.
51. Gérard, J.; Bietlot, E.; Hevesi, L., *Tetrahedron Lett.* **1998**, *39*, 8735.
52. Gérard, J.; Hevesi, L., *Tetrahedron* **2001**, *57*, 9109.
53. Gérard, J.; Hevesi, L., *Tetrahedron* **2004**, *60*, 367.
54. Debuigne, A.; Gérard, J.; Hevesi, L., *Tetrahedron Lett.* **1999**, *40*, 5943.
55. Almqvist, F.; Ekman, N.; Frejd, T., *J. Org. Chem.* **1996**, *61*, 3794.
56. Eames, J.; Jones, R. V. H.; Warren, S., *Tetrahedron Lett.* **1996**, *37*, 4823.
57. Ermolenko, M. S.; Shekharam, T.; Lukacs, G.; Potier, P., *Tetrahedron Lett.* **1995**, *36*, 2461.
58. Andrey, O.; Ducry, L.; Landais, Y.; Planchenault, D.; Weber, V., *Tetrahedron* **1997**, *53*, 4339.
59. Fromont, C.; Masson, S., *Tetrahedron* **1999**, *55*, 5405.
60. Huang, Q.; Hunter, J. A.; Larock, R. C., *Org. Lett.* **2001**, *3*, 2973.
61. Huang, Q.; Hunter, J. A.; Larock, R. C., *J. Org. Chem.* **2002**, *67*, 3437.
62. Marchand, A. P.; Sorokin, V. D.; Rajagopal, D.; Bott, S. G., *Tetrahedron* **1994**, *50*, 9933.
63. Sano, H.; Kawata, K.; Kosugi, M., *Synlett* **1993**, 831.
64. Schmitz, C.; Rouanet-Dreyfuss, A.-C.; Tueni, M.; Biellmann, J.-F., *J. Org. Chem.* **1996**, *61*, 1817.
65. Fiandanese, V.; Mazzone, L., *Tetrahedron Lett.* **1992**, *33*, 7067.
66. Studley, J. R.; Wills, M., *J. Organomet. Chem.* **1993**, *455*, C3.
67. Colins, C. C.; Cronin, M. F.; Moynihan, H. A.; McCarthy, D. G., *J. Chem. Soc. Perkin Trans. 1* **1997**, 1267.
68. Hamel, P.; Préville, P., *J. Org. Chem.* **1996**, *61*, 1573.
69. Hamel, P., *Tetrahedron Lett.* **1997**, *38*, 8473.
70. Hamel, P., *J. Org. Chem.* **2002**, *67*, 2854.
71. Engqvist, R.; Bergman, J., *Tetrahedron* **2003**, *59*, 9649.
72. Kawakami, H.; Ebata, T.; Koseki, K.; Okano, K.; Matsumoto, K.; Matsushita, H., *Heterocycles* **1993**, *36*, 2765.
73. Tomoo, T.; Kondo, T.; Abe, H.; Tsukamoto, S.; Isobe, M.; Goto, T., *Carbohydr. Res.* **1996**, *284*, 207.
74. Roush, W. R.; Sebesta, D. P.; Bennett, C. E., *Tetrahedron* **1997**, *53*, 8825.
75. Newcomb, M.; Marquardt, D. J.; Deeb, T. M., *Tetrahedron* **1990**, *46*, 2329.
76. Newcomb, M.; Esker, J. L., *Tetrahedron Lett.* **1991**, *32*, 1035.
77. Boivin, J.; Fouquet, E.; Zard, S. Z., *Tetrahedron Lett.* **1990**, *31*, 85.
78. Boivin, J.; Fouquet, E.; Zard, S. Z., *Tetrahedron Lett.* **1990**, *31*, 3545.
79. Boivin, J.; Fouquet, E.; Zard, S. Z., *J. Am. Chem. Soc.* **1991**, *113*, 1055.
80. Bowman, W. R.; Clark, D. N.; Marmon, R. J., *Tetrahedron Lett.* **1991**, *32*, 6441.
81. Bowman, W. R.; Clark, D. N.; Marmon, R. J., *Tetrahedron* **1994**, *50*, 1275.
82. Bowman, W. R.; Clark, D. N.; Marmon, R. J., *Tetrahedron* **1994**, *50*, 1295.
83. Bowman, W. R.; Clark, D. N.; Marmon, R. J., *J. Chem. Res. (S)* **1995**, 514.
84. Bowman, W. R.; Broadhurst, M. J.; Coghlan, D. R.; Lewis, K. A., *Tetrahedron Lett.* **1997**, *38*, 6301.
85. Esker, J. L.; Newcomb, M., *Tetrahedron Lett.* **1993**, *34*, 6877.
86. Grossi, L.; Montevecchi, P. C., *Tetrahedron* **1993**, *49*, 9095.
87. Nakatsuji, M.; Miura, Y.; Teki, Y., *Polyhedron* **2001**, *20*, 1355.
88. Hartung, J.; Gallou, F., *J. Org. Chem.* **1995**, *60*, 6706.
89. Guindon, Y.; Denis, R. C., *Tetrahedron Lett.* **1998**, *39*, 339.
90. Petrović, G.; Saičić, R. N.; Čeković, Ž, *Synlett* **1999**, 635.
91. Petrović, G.; Saičić, R. N.; Čeković, Ž., *Tetrahedron* **2003**, *59*, 187.
92. Ternon, M.; Pannecoucke, X.; Outurquin, F.; Paulmier, C., *Tetrahedron* **2002**, *58*, 3275.
93. Matsuo, J.; Iida, D.; Tatani, K.; Mukaiyama, T., *Bull. Chem. Soc. Japan* **2002**, *75*, 223.
94. Kitagawa, H.; Mukaiyama, T., *Chem. Pharm. Bull.* **2002**, *50*, 1276.
95. Matsuo, J.; Iida, D.; Yamanaka, H.; Mukaiyama, T., *Tetrahedron* **2003**, *59*, 6739.

Benzenesulfenyl Trifluoromethanesulfonate

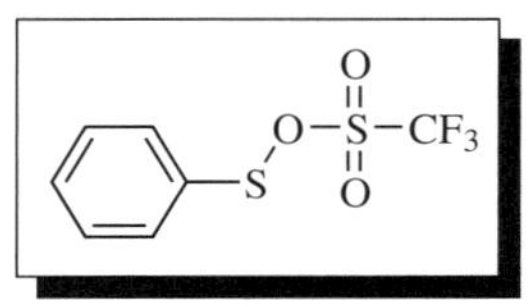

[84224-65-7] $C_7H_5F_3O_3S_2$ (MW 258.23)

(the title compound is a strong electrophilic reagent in organic synthesis)

Alternate Name: BST, PST.

Preparative Methods: BST is prepared in situ by reacting benzenesulfenyl chloride with silver triflate.[1,2] A convenient protocol for this is as follows: To silver triflate (83.2 mg, 0.32 mmol) in dichloromethane (1.5 mL) at −78 °C is added a cold (0 °C) solution of benzenesulfenyl chloride (39.1 mg, 0.27 mmol) in dichloromethane (1.5 mL) (eq 1). After vigorous stirring at −78 °C for 5 min, the BST generated is ready for use in further reactions.[3–6] The starting material, benzenesulfenyl chloride, can be synthesized from phenyl thioacetate, thiophenol, or diphenyl disulfide. Benzenesulfenyl chloride can be prepared by the following typical protocol: To a solution of phenyl thioacetate (25 g, 164 mmol) in CCl_4 (10 mL) at 0 °C is added SO_2Cl_2 (13.3 mL, 164 mmol) within 10 min. The mixture is stirred at room temperature for 30 min, concentrated in vacuo, and distilled to afford benzenesulfenyl chloride as a red liquid.[2,7] From thiophenol or diphenyl disulfide, benzenesulfenyl chloride can be prepared following the same protocol.[8,9] Benzenesulfenyl chloride is stable at 4 °C under nitrogen for several months.[2]

(1)

Use as a Glycosylation Promoter with Thioglycosides. BST acts as a powerful electrophilic activating reagent in the activation of thioglycosides in glycosylation reactions (eq 2).

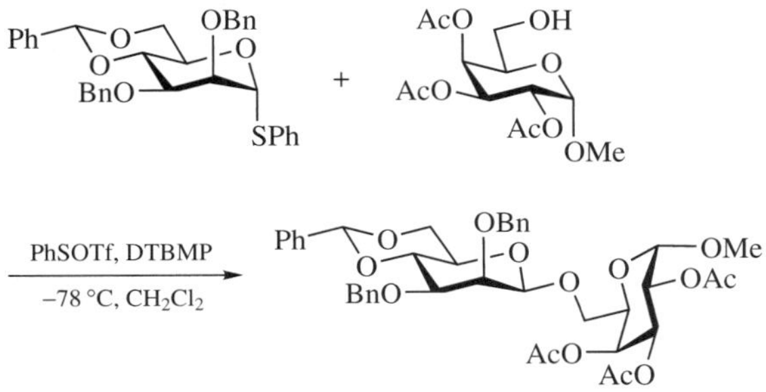

(2)

82%, $\beta:\alpha > 25:1$

This protocol, which enables the β-mannosylation of even hindered and unreactive alcohols in high yields (eq 3), has been determined to proceed through the formation of glycosyl triflates as intermediates after the activation of thioglycosides by BST.[4]

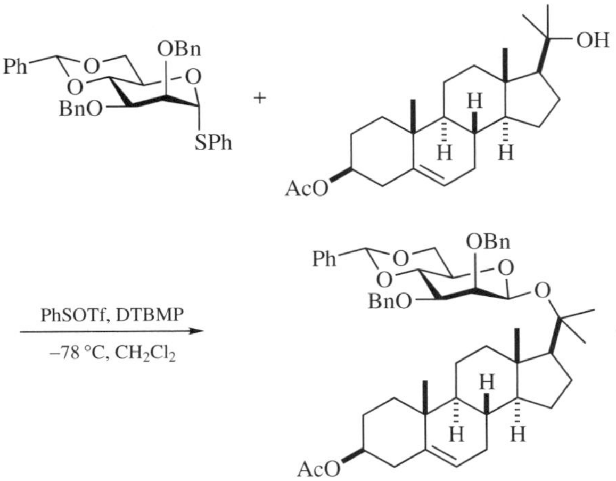

(3)

90%, $\beta:\alpha > 25:1$

A typical protocol is as follows: To a stirred mixture of AgOTf (83.2 mg, 0.32 mmol) in a chosen solvent (1.5 mL) at −78 °C is added a solution of PhSCl (39.1 mg, 0.27 mmol) in the solvent (1.5 mL) and a solution of the thioglycoside (0.11 mmol) and 2,6-di-*tert*-butyl-4-methylpyridine (66.8 mg, 0.33 mmol) in the solvent (1.5 mL) dropwise 5 min later. After another 5 min, a solution of the glycosyl acceptor (0.22 mmol) in the solvent (1.5 mL) is added slowly. The reaction mixture is allowed to warm to −20 °C in 40 min and is then quenched with saturated aqueous sodium bicarbonate solution. Filtration through Celite, washing with brine, and purification by chromatography on silica gel gives the glycoside in over 80% yield in most cases.[4–6]

In addition to mannosyl thioglycosides, BST has also been used to activate sialyl xanthates for use in glycosylation (eq 4),[2] and a number of disarmed thioglycosides under an essentially standard set of conditions. Particularly interesting in this respect was the activation of a strongly disarmed, oxazolidinone protected glucosamine donor which afforded exclusively α-coupled disaccharides (eq 5).[10]

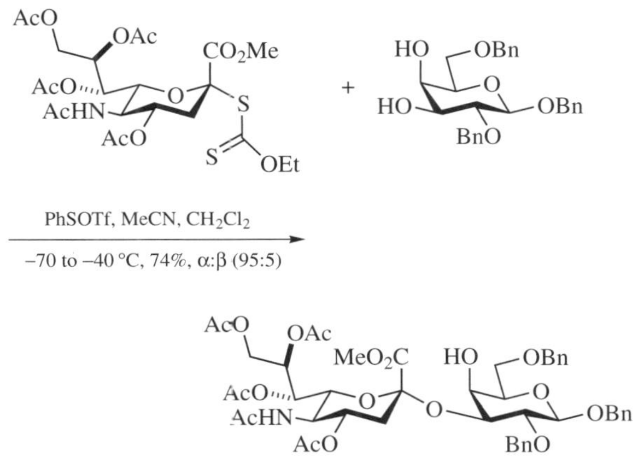

(4)

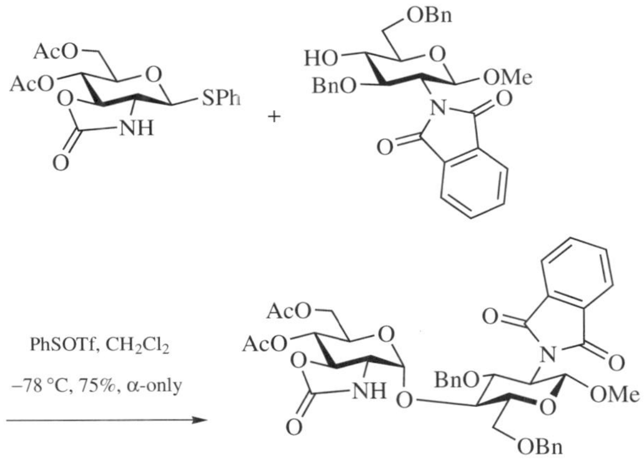

(5)

Related Reagents. MeSOTf;[11,12] PhSeOTf;[13,14] 1-benzenesulfinyl piperidine;[15] and diphenyl sulfoxide.[16]

1. Effenberger, F.; Russ, W., *Chem. Ber.* **1982**, *115*, 3719.
2. Martichonok, V.; Whitesides, G. M., *J. Org. Chem.* **1996**, *61*, 1702.
3. Crich, D.; Sun, S., *J. Am. Chem. Soc.* **1997**, *119*, 11217.
4. Crich, D.; Sun, S., *J. Am. Chem. Soc.* **1998**, *120*, 435.
5. Crich, D.; Cai, W.; Dai, Z., *J. Org. Chem.* **2000**, *65*, 1291.
6. Crich, D.; Sun, S., *Tetrahedron* **1998**, *54*, 8321.
7. Thea, S.; Cevasco, G., *Tetrahedron Lett.* **1988**, *29*, 2865.
8. Barrett, A. G. M.; Dhanak, D.; Graboski, G. G.; Taylor, S. J., *Org. Synth.* **1989**, *68*, 8.
9. Hursthouse, M. B.; Malik, K. M. A.; Hibbs, D. E.; Roberts, S. M.; Seago, A. J. H.; Sik, V.; Storer, R., *J. Chem. Soc., Perkin Trans. 1* **1995**, 2419.
10. Benakli, K.; Zha, C.; Kerns, R. J., *J. Am. Chem. Soc.* **2001**, *123*, 9461.
11. Dasgupta, F.; Garegg, P. J., *Carbohydr. Res.* **1988**, *177*, C13.
12. Dasgupta, F.; Garegg, P. J., *Carbohydr. Res.* **1990**, *202*, 225.
13. Ito, Y.; Ogawa, T., *Tetrahedron Lett.* **1988**, *29*, 1061.
14. Ito, Y.; Ogawa, T.; Numata, M.; Sugimoto, M., *Carbohydr. Res.* **1990**, *202*, 165.

15. Crich, D.; Smith, M., *J. Am. Chem. Soc.* **2001**, *123*, 9015.
16. Codée, J. D.; Litjens, R. E. J. N.; den Heeten, R.; Overkleeft, H. S.; van Boom, J. H.; van der Marel, G. A., *Org. Lett.* **2003**, *5*, 1519.

Sanxing Sun
Boehringer Ingelheim, Ridgefield, CT, USA

1-Benzenesulfinyl Piperidine

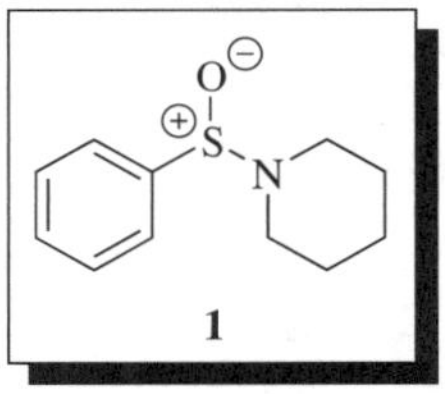

1

[4972-31-0] $C_{11}H_{15}NOS$ (MW 209.31)

(reagent used in combination with trifluoromethanesulfonic anhydride to form a powerful electrophilic salt capable of activating thioglycosides and selenoglycosides for the construction of glycosidic linkages; can also be used for the *C*-alkylation of β-ketoester enolates)

Alternate Name: BSP.
Physical Data: mp 84–85 °C.
Solubility: soluble in dichloromethane, diethyl ether, toluene, and most organic solvents.
Form Supplied in: off-white solid.
Analysis of Reagent Purity: melting point, NMR spectrum.
Preparative Methods: a solution of PhSOCl (58.0 g, 0.365 mol) in anhydrous diethyl ether (200 mL) is slowly added to a cooled solution (5 °C) of piperidine (72 mL, 0.73 mmol) in anhydrous diethyl ether (200 mL). The reaction mixture is stirred at room temperature for 1 h, filtered, and then concentrated under reduced pressure. The solid residue is triturated with hexanes to give 1-benzenesulfinyl piperidine (53.4 g, 70%).
Handling, Storage, and Precautions: unknown toxicity. Should be used within a fumehood.

Activation of Thioglycosides in Glycosylation Reactions.[1] The combination of benzenesulfinyl piperidine (BSP, **1**) and trifluoromethanesulfonic anhydride (Tf_2O) is a powerful tool for the activation of both armed and disarmed thioglycosides in a matter of minutes at low temperature, allowing for the clean conversion to glycosides, upon treatment with alcohols. This reagent combination compares favorably with other methods for the activation of thioglycosides as it allows direct access to otherwise difficult glycosidic linkages such as β-mannosides and α-glucosides in high yield and stereoselectivity. Thus, reaction of thiomannoside (**2**), bearing 4,6-*O*-benzylidene and 2,3-di-*O*-benzyl protecting groups, with **1**/Tf_2O, and the non-nucleophilic base, tri-*tert*-butylpyrimidine (TTBP)[3] reagent combination followed by addition of the glucoside (**3**) provided β-mannoside (**4**) within 5 min in dichloromethane at –60 °C in 77% yield (eq 1). On the other hand, reaction of the thioglucoside (**5**), also bearing the 4,6-*O*-benzylidene and 2,3-di-*O*-benzyl protecting groups, under standard conditions with adamantanol (**6**) provided α-glucoside (**7**) in 74% yield as a single isomer (eq 2).

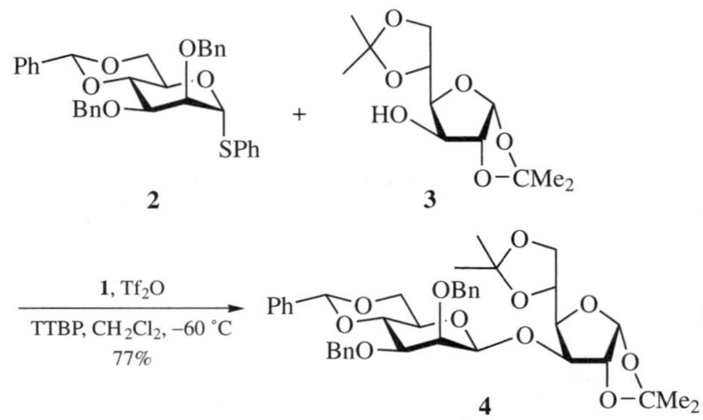

(1)

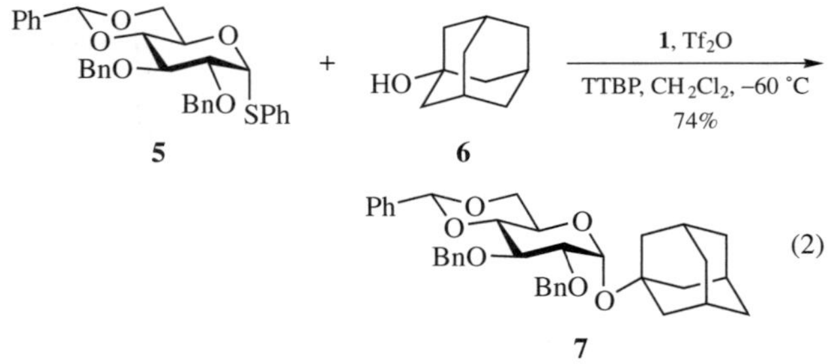

(2)

Low temperature studies have determined that the reaction of **1** and Tf_2O is an equilibrium that favors the starting materials over salt **8**. However, salt **8** is a very potent thiophile, capable of converting thioglycosides to glycosyl triflates in a matter of minutes at low temperature. In the presence of thioglycosides, salt **8** is therefore constantly removed from the equilibrium and the reaction is driven to completion. Finally, the thioglycoside reacts with the mannosyl triflate in an S_N2-like manner to give the β-mannoside (eq 3).[1]

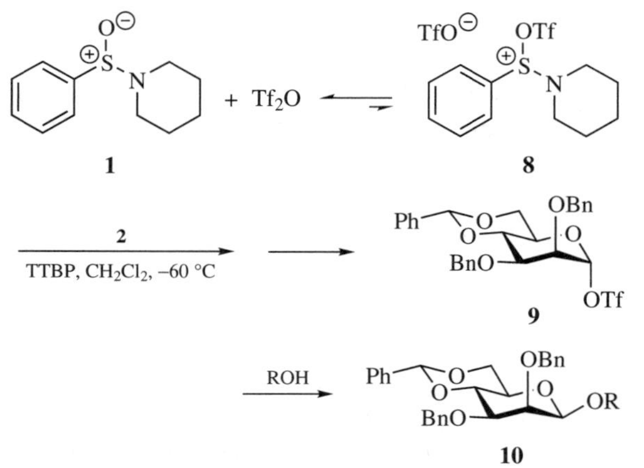

(3)

The BSP/Tf_2O reagent combination is not limited to the activation of ether protected thioglycosides. It also activates a wide range of thioglycosides to give standard glycosidic linkages. For example, reaction of tetrabenzoyl thiogalactoside (**11**) under standard conditions, with the glucoside acceptor **3** furnished β-galactoside (**12**) in 78% yield (eq 4).[1] In another example, reaction of tetrabenzyl thiogalactoside (**13**) with 3β-cholestanol (**14**), under standard conditions, provided a 1:1 mixture of α:β-galactoside (**15**) (eq 5). Repeating the reaction in propionitrile as a solvent yielded a 1:4 mixture of the galactoside **15** favoring the β-anomer (eq 5).[1]

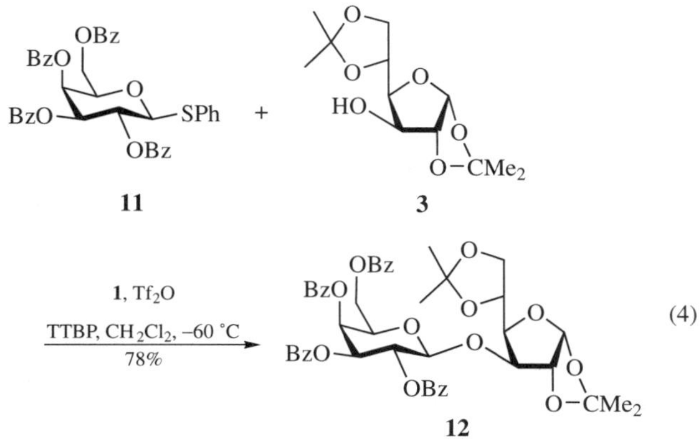

(4)

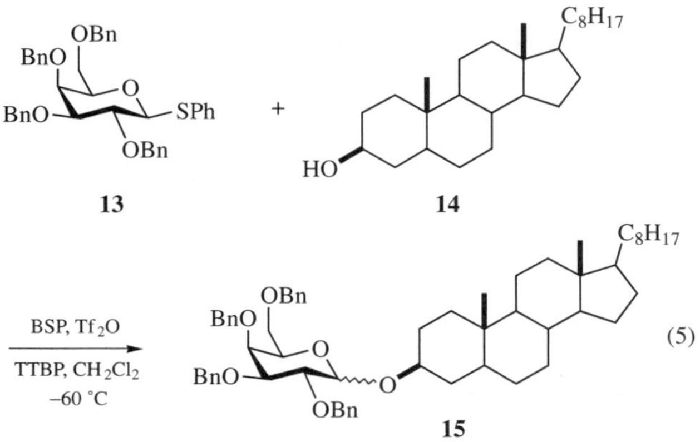

(5)

CH_2Cl_2, 78%, α:β = 1:1
CH_3CH_2CN, 73%, α:β = 1:4

The direct synthesis of trisaccharides has also been realized using the standard reagent combination. Thus, treatment of the glucosamine donor **16** with BSP/Tf_2O/TTBP at −60 °C in dichloromethane, followed by addition of 4,6-glucopyranosyl diol (**17**) furnished the trisaccharide **18** in 85% yield as a single isomer (eq 6).[1]

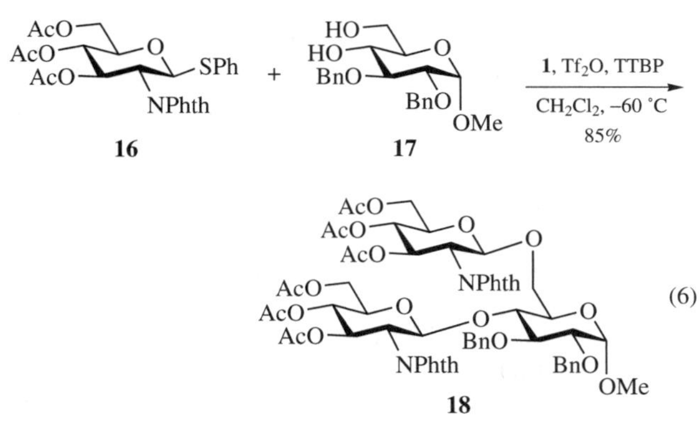

(6)

The generality of the BSP/Tf_2O glycosylation method has been displayed in Crich's synthesis of the salmonella type E_1-core trisaccharide (**25**), wherein the three glycosidic bonds were all prepared using this methodology.[4] Thus, reaction of the thiogalactoside (**19**) with BSP/Tf_2O, in the absence of TTBP, with the alcohol **20** provided the thiorhamnoside (**22**) as a single isomer in 80% yield. The acidic conditions of this coupling reaction suppressed any orthoester formation and affected the removal of the PMB protecting group revealing the 3-OH. Treatment of thiorhamnoside (**22**) with the acceptor **21** under standard coupling conditions, followed by deprotection of the silyl group, furnished the disaccharide **23** as a single isomer. Coupling of the acceptor **23** under standard conditions with the thiomannoside donor **24** preceded smoothly in 92% yield as a 9.6:1 mixture favoring the β-anomer. Deprotection with sodium methoxide followed by 5% trifluoroacetic acid provided the desired trisaccharide (**25**) (eq 7).

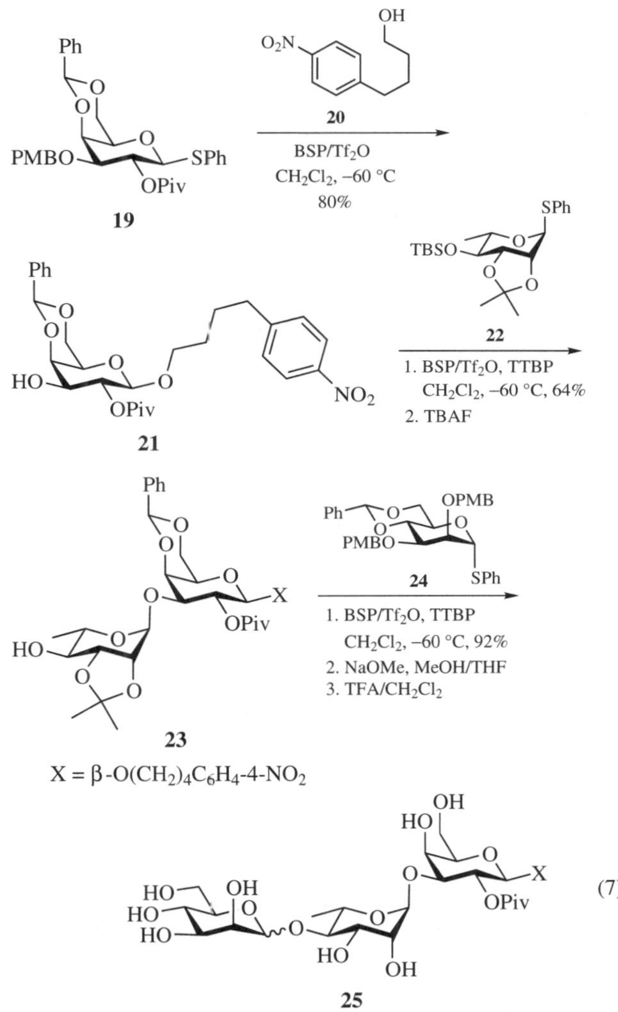

(7)

The BSP/Tf_2O methodology has been used in the first example of the direct solid phase synthesis of β-mannosides.[5] Treatment of thiomannoside (**26**), bearing a 4,6-*O*-polystyrylphenylboronate group, with BSP/Tf_2O/TTBP at −60 °C in dichloromethane, followed by the addition of the rhamnosyl acceptor **27** provided β-mannoside (**28**). The β-mannoside (**29**) was obtained from the resin by heating in acetone and water in 77% overall yield (eq 8). Isolated yields of the β-mannosides from the polymer supported protocol were comparable with those obtained using the analogous solution phase methodology.

Activation of Selenoglycosides in Glycosylation Reactions.[1] The standard BSP/Tf_2O/TTBP reagent combination activates selenoglycosides in a manner analogous to thioglycosides. Thus, reaction of selenoglycoside (**30**) with the glucosyl acceptor **3** under standard conditions resulted in the isolation of α-glucoside (**31**) in 79% yield as a single isomer (eq 9).[1]

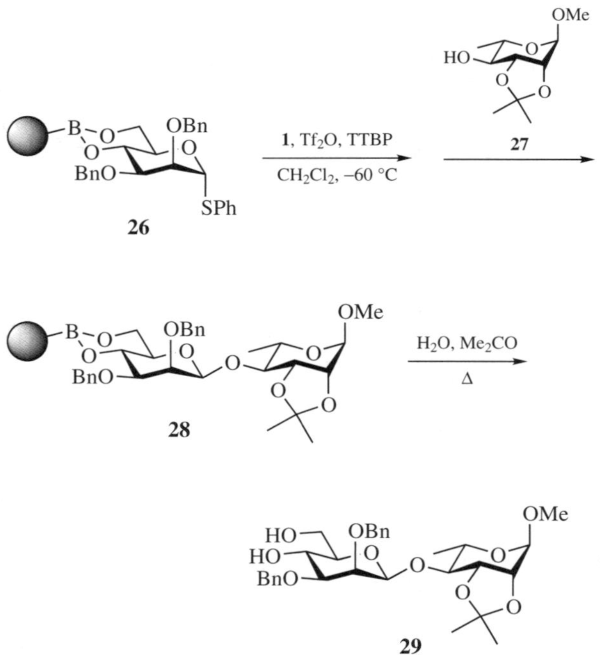

(8)

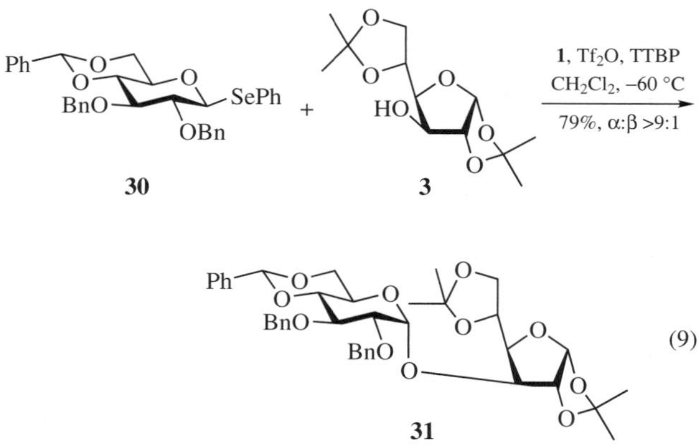

(9)

C-Alkylation of β-Ketoester Enolates.[6] Treatment of **1** with MeOTf in $MeNO_2$, followed by anion exchange with sodium tetraphenylborate gave the sulfoxonium salt **32** (eq 10). Reaction of this sulfoxonium salt with the β-ketoester **33** provided a 3:1 mixture favoring the *C*-alkylated product **34** over the *O*-alkylated material **35** (eq 11).[6]

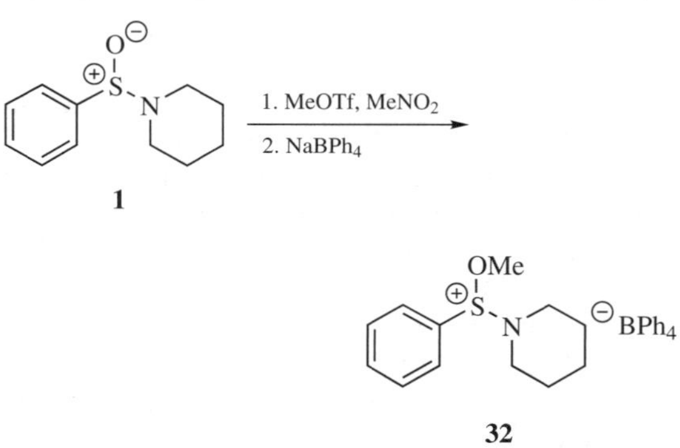

(10)

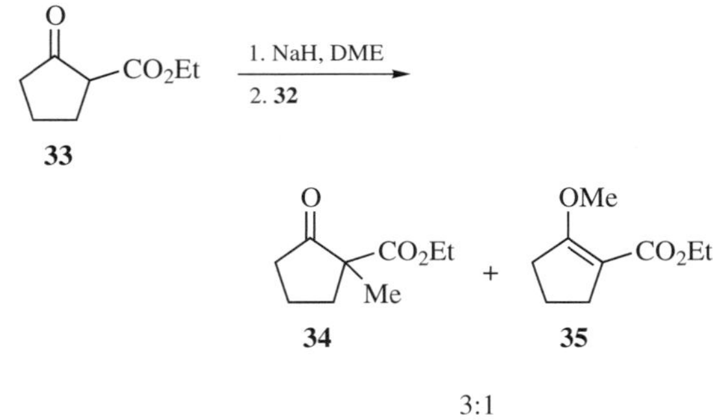

(11)

Related Reagents. Benzenesulfenyl triflate (PhSOTf); *S*-(4-methoxyphenyl) benzenethiosulfinate (MPBT).[2]

---

1. Crich, D.; Smith, M., *J. Am. Chem. Soc.* **2001**, *123*, 9015.
2. Crich, D.; Smith, M., *Org. Lett.* **2000**, *2*, 4067.
3. Crich, D.; Smith, M.; Yao, Q.; Picione, J., *Synthesis* **2001**, 323.
4. Crich, D.; Li, H., *J. Org. Chem.* **2002**, *67*, 4640.
5. Crich, D.; Smith, M., *J. Am. Chem. Soc.* **2002**, *124*, 8867.
6. Pickersgill, I. F.; Marchington, A. P.; Rayner, C. M., *J. Chem. Soc., Chem. Commun.* **1994**, 2597.

Mark Smith
Roche Bioscience, Palo Alto, CA, USA

Benzimidazolium Trifluoromethanesulfonate

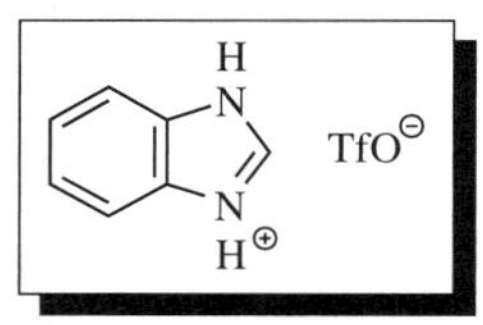

[99257-95-1] $C_8H_7F_3N_2O_3S$ (MW 268.21)

(example of a series of closely related reagents serving as promoters for condensation of a nucleoside phosphoramidite and a nucleoside/nucleotide, constructing internucleotide linkage[1–5])

Physical Data: mp 188–190 °C; density unknown.

Solubility: soluble in H_2O (>0.1 mol/L), methanol (>1 mol/L), and acetonitrile (ca. 0.4 mol/L); insoluble in dichloromethane, ether, and toluene.

Form Supplied in: colorless crystals; not commercially available.

Analysis of Reagent Purity: UV (CH_3OH) λ_{max} 242, 272 nm; ^1H NMR (CD_3OD) 7.53–7.56 (m, 2H), 7.74–7.76 (m, 2H), 9.24 (s, 1H); ^{13}C NMR (CD_3OD) 114.4, 122.4, 127.0, 130.8, 139.8.

Preparative Methods: TfOH (1.0 equiv) is added to benzimidazole (1.0 equiv) in dichloromethane at 0 °C, followed by stirring at 0 °C for 30 min. The precipitated product is collected by filtration.

Purification: crystallized from dichloromethane.

Handling, Storage, and Precautions: this reagent is non-hygroscopic and is stable in air. Special care is not necessary during usual handling, but, to keep high quality, store under argon atmosphere. Toxicity unknown.

Synthesis of Nucleotides via the Phosphoramidite Method. Benzimidazolium triflate[1–5] serves as a highly reactive promoter for the condensation of a nucleoside phosphoramidite and a nucleoside (or a nucleotide), which is a key step in the synthesis of oligonucleotides by the phosphoramidite strategy.[6,7] Nucleoside phosphoramidites include 2′-deoxyribonucleoside 3′-phosphoramidites, 2′-deoxyribonucleoside 5′-phosphoramidites, ribonucleoside 2′-phosphoramidites, and ribonucleoside 3′-phosphoramidites. Further, the following benzimidazolium triflate–related compounds also could be used as promoters: imidazolium triflate [mp 197–198 °C; soluble in acetonitrile (>1 mol/L); available from Aldrich],[3,5,8] imidazolium perchlorate [mp >300 °C; soluble in acetonitrile (>1 mol/L); not commercially available],[3] imidazolium tetrafluoroborate [mp 175–177 °C; soluble in acetonitrile (>1 mol/L); not commercially available],[3] *N*-methylimidazolium triflate [mp 127–129 °C; soluble in acetonitrile (>1 mol/L); not commercially available],[3] *N*-phenylimidazolium triflate [mp 126–127 °C; soluble in acetonitrile (>1 mol/L); not commercially available],[3–5,9] *N*-phenylimidazolium perchlorate [mp 142–144 °C; soluble in acetonitrile (>1 mol/L); not commercially available],[3] *N*-phenylimidazolium tetrafluoroborate [mp 96–98 °C; soluble in acetonitrile (>1 mol/L); not commercially available],[3] *N*-(*p*-acetylphenyl)imidazolium triflate [mp 129–131 °C; soluble in acetonitrile (>1 mol/L); not commercially available],[3] 2-phenylimidazolium triflate [mp 106–107 °C; soluble in acetonitrile (>1 mol/L); not commercially available],[3] 4-methylimidazolium triflate [mp 92–93 °C; soluble in acetonitrile (>1 mol/L); not commercially available],[3] 4-phenylimidazolium triflate [mp 168–169 °C; soluble in acetonitrile (>1 mol/L); not commercially available],[3] benzimidazolium tetrafluoroborate [mp 168–170 °C; soluble in acetonitrile (ca. 0.5 mol/L); not commercially available],[3] and *N*-methylbenzimidazolium triflate [mp 129–130 °C; soluble in acetonitrile (ca. 0.5 mol/L); not commercially available].[3]

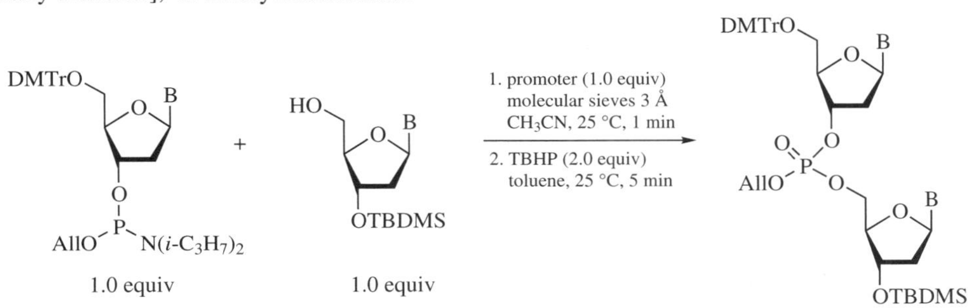

All = $CH_2{=}CHCH_2$;
B = protected nucleobase;
DMTr = $C_6H_5(p\text{-}CH_3OC_6H_4)_2C$

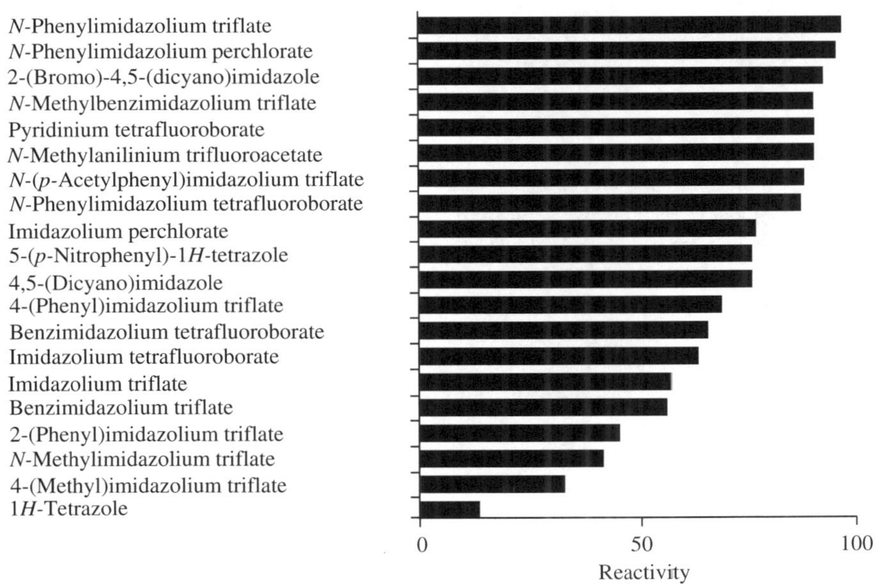

Figure 1 Reactivity of promoters estimated by product yield (^{31}P-NMR assay) obtained by the condensation of a nucleoside phosphoramidite (B = Thy) and a nucleoside (B = Thy) using equimolar amounts (0.02 mmol) of these reactants and the promoter in a 0.02 M acetonitrile solution (25 °C, 1 min), followed by TBHP oxidation (25 °C, 5 min).

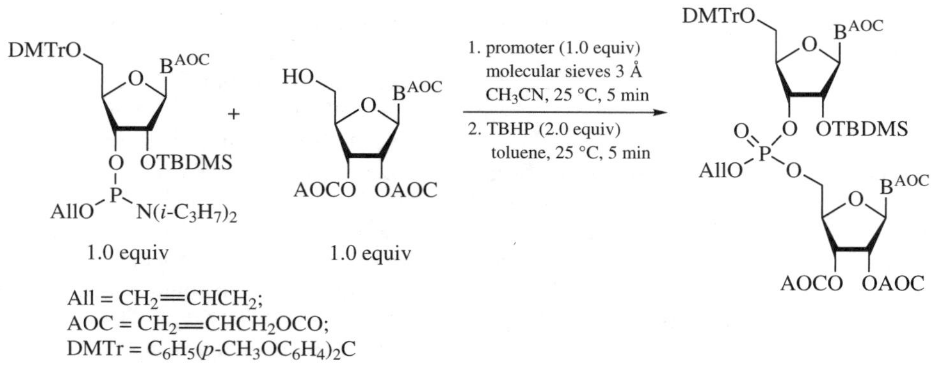

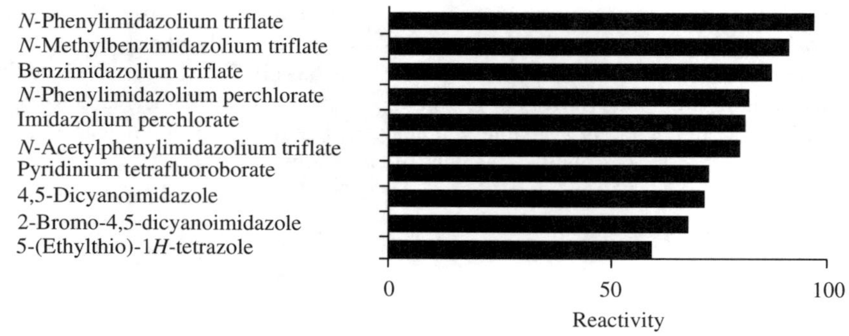

Figure 2 Reactivity of promoters estimated by product yield (^{31}P-NMR assay) obtained by the condensation of a nucleoside phosphoramidite (B = Cyt^{AOC}) and a nucleoside (B = Ade^{AOC}) by the stoichiometric (0.02 mmol each) use of these reactants and the promoter in a 0.1 M acetonitrile solution (25 °C, 5 min), followed by TBHP oxidation (25 °C, 5 min).

In general, condensation is smoothly achieved and the subsequent oxidation provides a nucleotide in a high yield. For example, the condensation of 2′-deoxyribonucleoside 3′-phosphoramidite and 5′-*O*-free 2′-deoxyribonucleoside promoted by *N*-phenylimidazolium triflate, which is the most reactive among benzimidazolium triflate-related compounds, in a 0.02 M acetonitrile solution is completed in 1–10 min (1–5 min for allyl phosphoramidites; 3–10 min for 2-cyanoethyl phosphoramidites).[3] The reaction using other imidazolium promoters with less reactivity than *N*-phenylimidazolium triflate requires a longer period, usually 10-20 min, for completion. The subsequent oxidation with TBHP[10] gives a dinucleoside phosphate in an almost quantitative yield; bis(trimethylsilyl) peroxide (TMSOOTMS) in the presence of a catalytic amount of TMSOTf[10,11] or 2-butanone peroxide[12] can also be used for this oxidation. Similar results are observed with ribonucleoside 2′- or 3′-phosphoramidites, leading to the construction of 2′–5′- or 3′–5′-inter-ribonucleotidic linkages, respectively. The reaction using these ribonucleoside phosphoramidites is slightly slower than that of 2′-deoxyribonucleoside phosphoramidites. For example, the reaction using ribonucleoside 3′-(allyl phosphoramidite) in a 0.1 M acetonitrile solution requires 5–15 min for completion, even when *N*-phenylimidazolium triflate is used as the promoter.[3] Figures 1 and 2 indicate the relative reactivity of benzimidazolium triflate–related promoters and representative alternative promoters[7,13–19] in the solution-phase preparation of a dideoxyribonucleoside phosphate and a diribonucleoside phosphate, respectively. The imidazolium reagents with high reactivity, such as phenylimidazolium triflate and benzimidazolium triflate, are particularly useful for the synthesis of nucleic acid–related compounds using a phosphoramidite with low reactivity as the building block.[2,3] The approach using imidazolium promoters can be applied to the solid-phase synthesis of oligodeoxyribonucleotides and oligoribonucleotides.

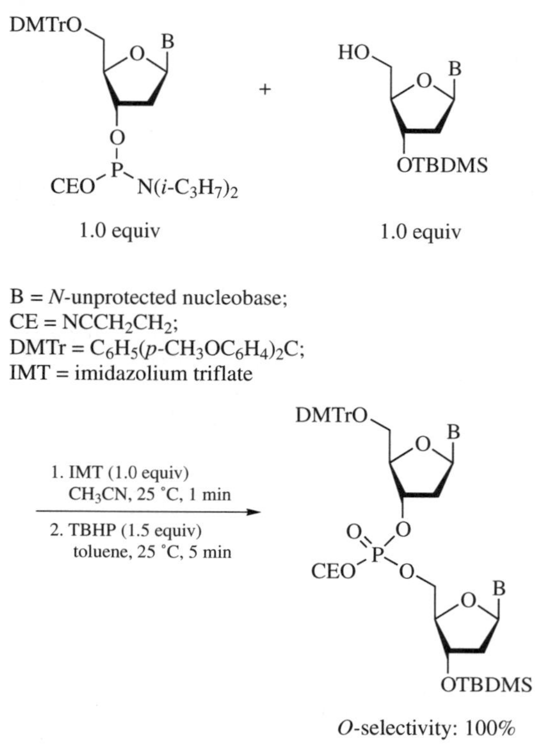

(1)

O-selectivity: 100%

Further, it is noteworthy that imidazolium triflate allows *O*-selective condensation of a nucleoside and a nucleoside phosphoramidite without protection of the nucleobase (eq 1).[5,7]

1. (a) Hayakawa, Y.; Kataoka, M.; Noyori, R., *J. Org. Chem.* **1996**, *61*, 7996–7997. (b) Sakakura, A.; Hayakawa, Y., *Tetrahedron* **2000**, *56*, 4427–4435.
2. (a) Iwai, S.; Mizukoshi, T.; Fujisawa, Y.; Matsutani, C.; Hanaoka, F.; Hayakawa, Y., *Nucleic Acids Res.* **1999**, *27*, 2299–2303. (b) Matsuura, K.; Hibino, M.; Kataoka, M.; Hayakawa, Y.; Kobayashi, K., *Tetrahedron Lett.* **2000**, *41*, 7529–7533.
3. Hayakawa, Y.; Kawai, R.; Hirata, A.; Sugimoto, J.; Kataoka, M.; Sakakura, A.; Hirose, M.; Noyori, R., *J. Am. Chem. Soc.* **2001**, *123*, 8165–8176.
4. Hayakawa, Y.; Hirata, A.; Sugimoto, J.; Kawai, R.; Sakakura, A.; Kataoka, M., *Tetrahedron* **2001**, *57*, 8823–8826.
5. Hayakawa, Y., *Bull. Chem. Soc. Jpn.* **2001**, *74*, 1547–1565.
6. A comprehensive review: Beaucage, S. L.; Iyer, R. P., *Tetrahedron* **1992**, *48*, 2223–2311.
7. Matteucci, M. D.; Caruthers, M. H., *J. Am. Chem. Soc.* **1981**, *103*, 3185–3191.
8. Hayakawa, Y.; Kataoka, M., *J. Am. Chem. Soc.* **1998**, *120*, 12395–12401.
9. Kawai, R.; Sugimoto, J.; Kataoka, M.; Hayakawa, Y., *Nucleosides, Nucleotides, and Nucleic Acids* **2001**, *20*, 1047–1048.
10. Hayakawa, Y.; Uchiyama, M.; Noyori, R., *Tetrahedron Lett.* **1986**, *27*, 4191–4194.
11. Hayakawa, Y.; Uchiyama, M.; Noyori, R., *Tetrahedron Lett.* **1986**, *27*, 4195–4196.
12. Kataoka, M.; Hattori, A.; Okino, S.; Hyodo, M.; Asano, M.; Kawai, R.; Hayakawa, Y., *Org. Lett.* **2001**, *3*, 815–818.
13. Beaucage, S. L.; Caruthers, M. H., *Tetrahedron Lett.* **1981**, *22*, 1859–1862.
14. Froehler, B. C.; Matteucci, M. D., *Tetrahedron Lett.* **1983**, *24*, 3171–3174.
15. Vinayak, R.; Colonna, F.; Tsou, D.; Mullah, B.; Andrus, A.; Sproat, B., *Nucleic Acids Symp. Ser.* **1994**, *31*, 165–166.
16. Fourrey, J.-L., Varrene, J., *Tetrahedron Lett.* **1984**, *25*, 4511–4514.
17. Brill, W. K.-D., Nielsen, J.; Caruthers, M. H., *J. Am. Chem. Soc.* **1991**, *113*, 3972–3980.
18. Vergeese, C.; Carter, J.; Yegge, J.; Krivjansky, S.; Settle, A.; Kropp, E.; Peterson, K.; Pieken, W., *Nucleic Acids Res.* **1998**, *26*, 1046–1050.
19. Jin, Y.; Just, G., *J. Org. Chem.* **1998**, *63*, 3647–3654.

Yoshihiro Hayakawa
Nagoya University, Nagoya, Japan

3*H*-1,2-Benzodithiol-3-one 1,1-Dioxide

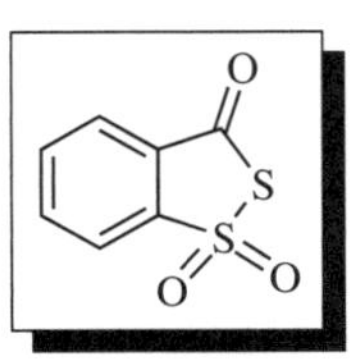

[66304-01-6] $C_7H_4O_3S_2$ (MW 200.24)

(reagent extensively used in the efficient conversion of phosphite triesters to phosphorothioate triesters in solid-phase oligonucleotide synthesis)

Alternate Name: Beaucage reagent.
Physical Data: mp 102.5–103 °C.
Solubility: soluble in most organic solvents except nonpolar solvent such as alkanes.
Form Supplied in: white solid.
Analysis of Reagent Purity: ^{13}C NMR, ^{1}H NMR, EI-MS.
Preparative Methods: the reagent is prepared from the oxidation of precursor 3*H*-1,2-benzodithiol-3-one with trifluoroperoxyacetic acid.
Purification: recrystallization from methylene chloride-hexanes.
Handling, Storage, and Precautions: the reagent can be stored indefinitely as a crystalline material in amber glass bottles at ambient temperature. The reagent decomposes rapidly when in contact with inorganic and organic bases. The reagent will also decompose in solution when exposed to glass and/or metallic surfaces.
Disposal: the reagent can be safely discarded in a chemical waste container.

Preparation and Handling of 3*H*-1,2-benzodithiol-3-one 1,1-dioxide. The sulfur-transfer reagent 3*H*-1,2-benzodithiol-3-one 1,1-dioxide (**2**, eq 1)[1] is widely used in the preparation of oligonucleoside phosphorothioates, which are especially valuable in the development of therapeutic oligonucleotides against various types of cancer and infectious diseases in humans.[2,3]

The scope of this report is not to provide an exhaustive list of all oligonucleoside phosphorothioates that have been prepared using **2**, but to review those critical issues stemming from the preparation and handling of the reagent along with specific applications of 3*H*-1,2-benzodithiol-3-one 1,1-dioxide as a sulfur-transfer reagent.

In regard to the preparation of **2**, oxidation of 3*H*-1,2-benzodithiol-3-one (**1**) to its 1,1-dioxide (eq 1) has initially been performed by treatment with 3-chloroperbenzoic acid.[4] While such an oxidation reaction appears satisfactory for the preparation of **2** on a small scale, the generation of hydrophobic 3-chlorobenzoic acid during the course of the reaction complicates the purification of the desired product on large scale.

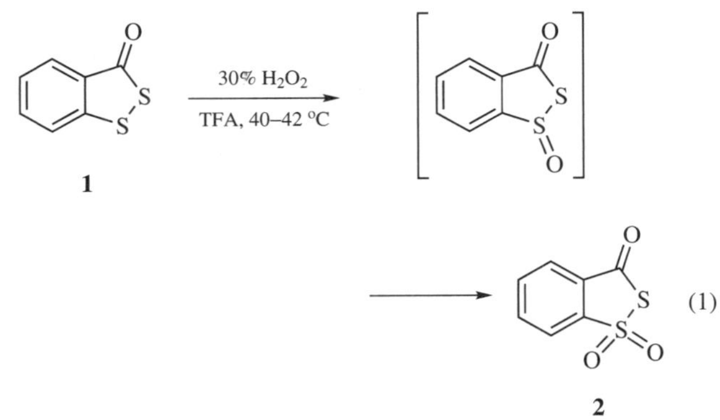

The search for an oxidizing reagent that would produce hydrophilic side-products has identified trifluoroperoxyacetic acid as a promising oxidizer. Oxidation of **1** with trifluoroperoxyacetic acid is efficient only when the internal temperature of the reaction mixture is maintained at 40–42 °C.[1a] As the reaction temperature exceeds 42 °C, the yield of **2** decreases presumably as a result of concurrent hydrolysis of the reagent. At a reaction temperature below 37 °C, the oxidation of **1** becomes sluggish and also results in poorer isolated yields of the 1,1-dioxide (**2**). Thus, when using trifluoroperoxyacetic acid as an oxidizing reagent at 40–42 °C, the preparation of 3*H*-1,2-benzodithiol-3-one 1,1-dioxide can be scaled up to permit facile isolation of the sulfur-transfer reagent in quantities exceeding 125 g, free from peroxide contaminants, upon precipitation on ice.[1c]

In addition to 3-chloroperoxybenzoic acid and trifluoroperoxyacetic acid, dimethyldioxirane is also adequate for the conversion of **1** to its 1,1-dioxide.[5] The reaction of 3*H*-1,2-benzodithiol-3-one with 4 equiv of 70–90 mM dimethyldioxirane in acetone at ambient temperature affords a quantitative yield of the corresponding 1,1-dioxide.[5] While handling dimethyldioxirane on a large scale may be inconvenient, the production of 3*H*-1,2-benzodithiol-3-one 1,1-dioxide in high yields under anhydrous conditions further suggests that hydrolysis is likely responsible for the lower yields of **2** obtained under aqueous conditions.

Sulfurization of tricoordinated phosphorus compounds by treatment with 3*H*-1,2-benzodithiol-3-one 1,1-dioxide results from the nucleophilic attack of, for example, a phosphite triester on the electrophilic thiosulfonate function of **2** to produce the phosphonium sulfinate intermediate **3** (eq 2).

Intramolecular condensation of the sulfinate anion with the carbonyl group of the activated thiol ester function releases the phosphorothioate triester (**4**) with concomitant formation of **5**.[1a] Because of the intramolecular nature of the sulfur-transfer reaction, sulfurization of phosphite triesters in oligodeoxyribonucleoside phosphorothioate syntheses is rapid and complete within 30 s at 25 °C. Formation of the intermediate **3** is rate-limiting, as the nucleophilicity of phosphite triesters depends on steric and electronic factors. A striking example of such subtle effects relates to the sulfurization of N^4-benzoyl-5′-*O*-(4,4′-dimethoxytrityl)-2′-*O*-*tert*-butyldimethylsilyl cytidylyl-(3′ → 5′)-N^4-benzoyl-2′-*O*-*tert*-butyldimethylsilylcytidine methylphosphite triester being complete within 30 s upon reaction with **2**,[6] whereas sulfurization of 5′-*O*-(4,4′-dimethoxytrityl)-2′-*O*-methyluridylyl-(3′ → 5′)-N^4-benzoyl-2′-*O*-methylcytidine(2-cyanoethyl) phosphite triester or its 2′-*O*-*tert*-butyldimethylsilyl congener under the same conditions reportedly takes 300 s for completion of the reaction.[7]

The production of 3*H*-1,2-benzoxathiolan-3-one 1-oxide (**5**) during the course of the sulfur-transfer reaction effected by **2** appears counterproductive considering that **5** is a potent oxidizing reagent.[1a] It should, however, be noted that the concentration of **5** generated during sulfurization is at least 50-fold lower than that of **2** at all times. Thus, oxidation of phosphite to phosphate triesters caused by **5** during sulfurization is negligible.

Stability of the sulfur-transfer reagent **2** in solution has been questioned by us[1a] and others.[8–12] The reagent decomposes slowly when in contact with glass and/or metallic surfaces. Consequently, any amber glass containers suitable for storing acetonitrile solutions of **2** must first be completely immersed in concentrated sulfuric acid for at least 15 h before being rinsed with distilled water and dried. The container is then filled to ~20% of its volume with a 10% solution of dichlorodimethylsilane in dichloromethane and vigorously shaken for ~5 min to ensure uniform distribution of the solution all over its internal surface area. The solution is discarded and the siliconized glass container is thoroughly rinsed with 50% aqueous methanol before being dried in an oven at 110 °C for ~1 h. Prior to oligonucleotide synthesis, the siliconized glass container is carefully rinsed with acetonitrile to remove any foreign particulates. A 0.05 M solution of **2** in HPLC grade acetonitrile is then filtered through a 45 μm teflon membrane into the siliconized container and stored under a positive pressure of argon on the DNA/RNA synthesizer. Given the incompatibility of **2** with DNA/RNA synthesis reagents, namely, the aqueous iodine solution, the capping solutions, and ammonium hydroxide, it is critically important that the teflon reagent line be carefully cleaned prior to coupling the siliconized container filled with reagent **2** to the instrument. By fulfilling these precautionary measures, acetonitrile solutions of **2** can be used for months on a DNA/RNA synthesizer without noticeable precipitation.

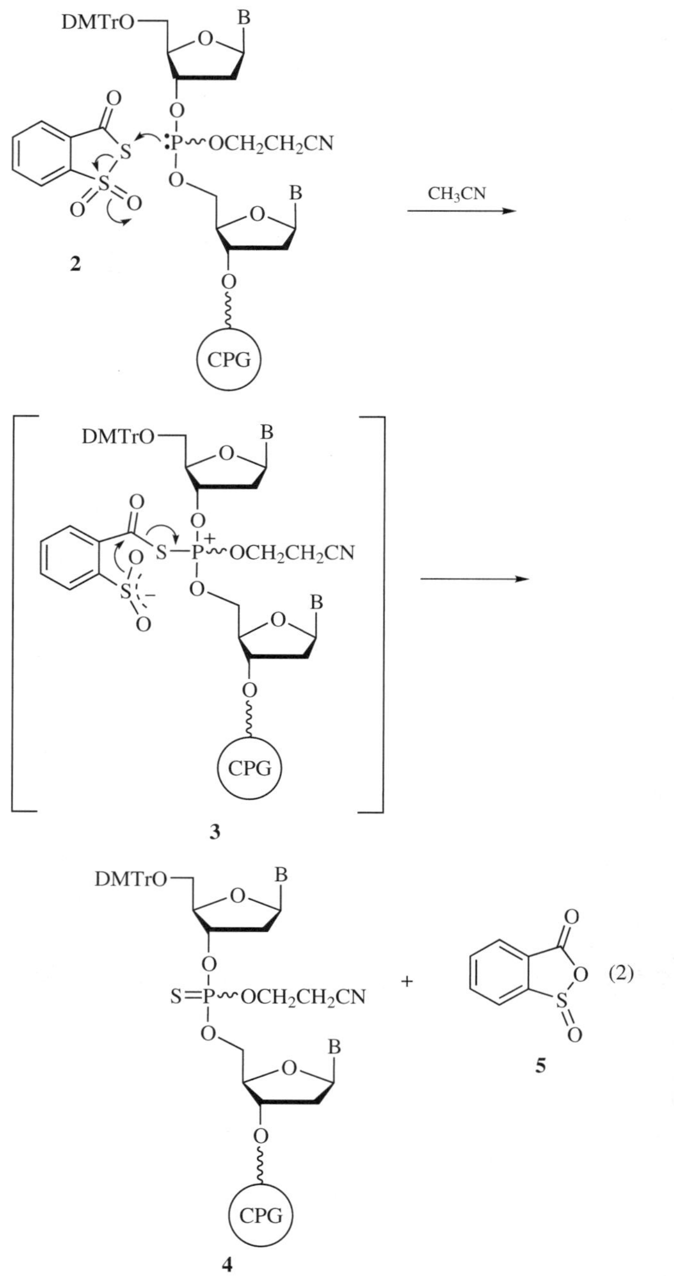

DMTr:4,4′-dimethoxytrityl; B: thymine or any N-protected nucleobase; CPG: controlled pore glass

[^{35}S]-3*H*-1,2-benzodithiol-3-one 1,1-dioxide in the Synthesis of ^{35}S-Labeled Oligonucleoside Phosphorothioates. To facilitate in vivo pharmacokinetic studies of potential therapeutic oligonucleotides such as those DNA, RNA or chimeric RNA–DNA sequences modified with internucleotidic phosphorothioate linkages, mixed phosphodiester-phosphorothioate linkages or mixed phosphorothioate-methyl phosphonate linkages, ^{35}S-labeling of these oligonucleotide analogues is a viable approach. In an attempt to prepare the sulfur-transfer reagent [^{35}**S**]-**2** for this purpose, the synthesis of [^{35}**S**]-**1** is accomplished according to eq 3.[13]

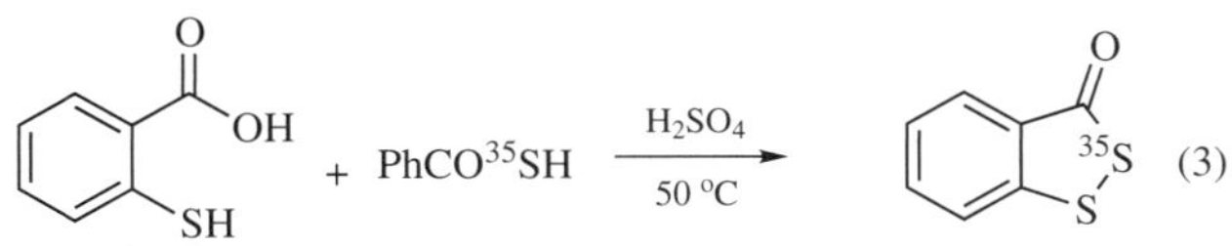

6 **7** [^{35}S]–1

Typically, thiosalicylic acid (**6**) is heated with ^{35}S-thiobenzoic acid (**7**) and concentrated sulfuric acid at 50 °C to afford [^{35}S]-3*H*-1,2-benzodithiol-3-one in 60% yield. Conversion of [^{35}**S**]-**1** to its 1,1-dioxide [^{35}**S**]-**2** is performed according to eq 1. The radiolabeled sulfur-transfer reagent exhibits a specific activity of 91 μCi/μmol and is isolated in ~30% yield from **6**.[13]

Sulfurization of *H*-phosphonate and *H*-phosphonothioate Diesters using 3*H*-1,2-benzodithiol-3-one 1,1-dioxide. Considering the efficiency with which phosphite triesters are converted to the corresponding phosphorothioate triesters upon reaction with **2**, investigations on the use of 3*H*-1,2-benzodithiol-3-one 1,1-dioxide in the sulfurization of *H*-phosphonate and *H*-phosphonothioate diesters were undertaken.[14] Conversion of the dinucleoside *H*-phosphonothioate diester (**8**) to its dithioated diester (**9**) is accomplished quantitatively within 30 s when a 0.02 M solution of **2** or **1** in 2% aqueous pyridine is used as a sulfurization reagent.

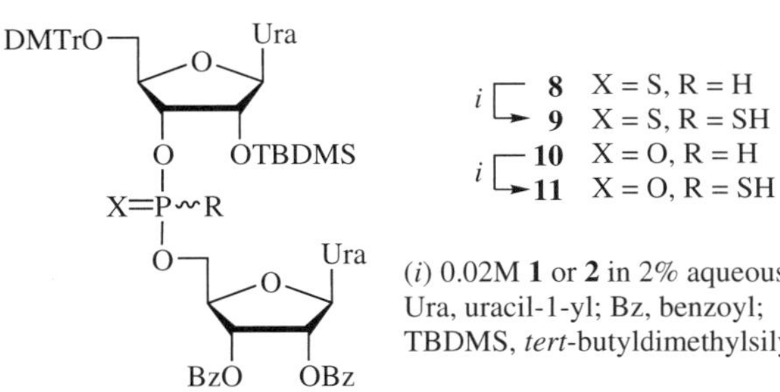

(*i*) 0.02M **1** or **2** in 2% aqueous pyridine
Ura, uracil-1-yl; Bz, benzoyl;
TBDMS, *tert*-butyldimethylsilyl

However, conversion of the dinucleoside *H*-phosphonate diester **10** to the phosphorothioate analogue **11** effected by a 0.02 M solution of **2** in 2% aqueous pyridine is relatively slow; it reportedly takes ~3 h for a complete reaction.[14a] While a solution of **2** in 2% aqueous acetonitrile containing triethylamine can completely transform **10** into **11** within 30 s, the use of this sulfurization mixture is incompatible with automated solid-phase oligonucleotide synthesis given the rapid formation of a yellow precipitate caused by triethylamine. In the absence of triethylamine, **10** is not sulfurized under these conditions. Nonetheless, **10** is completely converted to **11** within 20 min when a 0.02 M solution of **1** in 2% aqueous pyridine is used for the sulfurization reaction. Thus, 3*H*-1,2-benzodithiol-3-one in aqueous pyridine is compatible with automated solid-phase synthesis of both DNA and RNA oligonucleoside phosphorodithioates or phosphorothioates from appropriate *H*-phosphonate derivatives.[14a]

3*H*-1,2-benzodithiol-3-one 1,1-dioxide and other Sulfur-Transfer Reagents. Because of its efficiency and rapid sulfurization kinetics, 3*H*-1,2-benzodithiol-3-one 1,1-dioxide has served over the years as a reference standard in the evaluation of a number of sulfurizing reagents for the synthesis of oligonucleoside phosphorothioates.[9,11,15,16] These include phenylacetyl disulfide,[17] tetraethylthiuram disulfide,[18] dibenzoyl tetrasulfide,[19] bis(*O,O*-diisopropoxyphosphinothioyl)disulfide,[20] benzyltriethylammonium tetrathiomolybdate,[12] sulfur-triethylamine,[15b] bis(ethoxythiocarbonyl)tetrasulfide,[10] bis(arylsulfonyl)disulfide,[11] pyridinium tetrathionate,[11] bis(isopropylsulfonyl) disulfide,[11] thiiranes,[22] 1,2,4-dithiazoline-3,5-dione,[8,21] 3-ethoxy-1,2,4-dithiazoline-5-one,[8,21] and 3-methyl-1,2,4-dithiazoline-5-one.[16]

Additional Applications of 3*H*-1,2-benzodithiol-3-one 1,1-dioxide. In addition to being a potent sulfurizing reagent for phosphite triesters and being particularly useful in the synthesis of phosphorothioated oligonucleotides, 3*H*-1,2-benzodithiol-3-one 1,1-dioxide has also been used in the preparation of phosphorothioated phospholipid derivatives.[23] Also noteworthy is the reaction of 3*H*-1,2-benzodithiol-3-one 1,1-dioxide with excess thiols to generate unstable hydrosulfides, which have been demonstrated to efficiently cause single strand breaks in duplex DNA.[24] Incidentally, the reaction of thiols with **2** is directly analogous to that of phosphite triester with 3*H*-1,2-benzodithiol-3-one 1,1-dioxide.

1. (a) Iyer, R. P.; Phillips, L. R.; Egan, W.; Regan, J. B.; Beaucage, S. L., *J. Org. Chem.* **1990**, *55*, 4693. (b) Iyer, R. P.; Egan, W.; Regan, J. B.; Beaucage, S. L., *J. Am. Chem. Soc.* **1990**, *112*, 1253. (c) Regan, J. B.; Phillips, L. R.; Beaucage S. L., *Org. Prep. Proc. Int.* **1992**, *24*, 488.
2. Crooke, S. T.; Bennett, C. F., *Annu. Rev. Pharmacol. Toxicol.* **1996**, *36*, 107.
3. Beaucage, S. L.; Iyer R. P., *Tetrahedron* **1993**, *49*, 6123.
4. Hortmann, A. G.; Aron, A. J.; Bhattacharya A. K., *J. Org. Chem.* **1978**, *43*, 3374.
5. Marchán, V.; Gibert, M.; Messeguer, A.; Pedroso, E.; Grandas, A., *Synthesis* **1999**, 43.
6. Morvan, F.; Rayner, B.; Imbach, J.-L., *Tetrahedron Lett.* **1990**, *31*, 7149.
7. DiRenzo, A.; Grimm, S.; Levy, K.; Haeberli, P.; Maloney, L.; Usman, N.; Wincott, F., *Nucleosides Nucleotides* **1996**, *15*, 1687.
8. Xu, Q.; Musier-Forsyth, K.; Hammer, R. P.; Barany, G., *Nucleic Acids Res.* **1996**, *24*, 3643.
9. Zhang, Z.; Nichols, A ; Tang, J. X.; Alsbeti, M.; Tang, J. Y., *Nucleosides Nucleotides* **1997**, *16*, 1585.
10. Zhang, Z.; Nichols, A ; Alsbeti, M.; Tang, J. X.; Tang, J. Y., *Tetrahedron Lett.* **1998**, *39*, 2467.
11. Efimov, V. A.; Kalinkina, A. L.; Chakhmakhcheva, O. G.; Hill, T. S.; Jayaraman, K., *Nucleic Acids Res.* **1995**, *23*, 4029.
12. Rao, M. V.; Macfarlane, K., *Tetrahedron Lett.* **1994**, *35*, 6741.
13. Iyer, R. P.; Tan, W.; Yu, D.; Agrawal, S., *Tetrahedron Lett.* **1994**, *35*, 9521.
14. (a) Stawiński, J.; Thelin, M., *J. Org. Chem.* **1991**, *56*, 5169. (b) Stawinski, J.; Thelin, M.; von Stedingk, E., *Nucleosides Nucleotides* **1991**, *10*, 517.
15. (a) Wyrzykiewicz, T. K.; Ravikumar, V. T., *Bioorg. Med. Chem. Lett.* **1994**, *4*, 1519. (b) Cheruvallath, Z. S.; Cole, D. L.; Ravikumar, V. T., *Nucleosides Nucleotides* **1996**, *15*, 1441.

16. Zhang, Z. D.; Nichols, A.; Tang, J. X.; Han, Y. X.; Tang, J. Y., *Tetrahedron Lett.* **1999**, *40*, 2095.
17. Kamer, P. C. J.; Roelen, H. C. P. F.; van den Elst, H.; van der Marel, G. A.; van Boom, J. H., *Tetrahedron Lett.* **1989**, *30*, 6757.
18. Vu, H.; Hirschbein, B. L., *Tetrahedron Lett.* **1991**, *32*, 3005.
19. Rao, M. V.; Reese, C. B.; Zhengyun, Z., *Tetrahedron Lett.* **1992**, *33*, 4839.
20. Stec, W. J.; Uznanski, B.; Wilk, A.; Hirschbein, B. L.; Fearon, K. L.; Bergot, B. J., *Tetrahedron Lett.* **1993**, *34*, 5317.
21. Xu, Q.; Barany, G.; Hammer, R. P.; Musier-Forsyth, K., *Nucleic Acids Res.* **1996**, *24*, 3643.
22. Arterburn, J. B.; Perry M. C., *Tetrahedron Lett.* **1997**, *38*, 7701.
23. Martin, S. F.; Josey, J. A.; Wong, Y.-L.; Dean, D. W., *J. Org. Chem.* **1994**, *59*, 4805.
24. Breydo, L.; Gates, K. S., *Bioorg. Med. Chem. Lett.* **2000**, *10*, 885.

Serge L. Beaucage
Food and Drug Administration, Bethesda, MD, USA

1,3-Benzodithiol-1-ium Tetrafluoroborate[1]

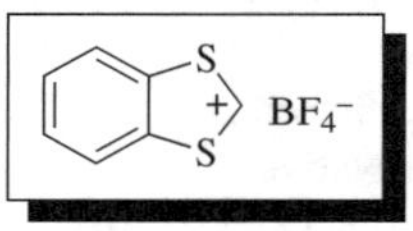

[57842-27-0] $C_7H_5BF_4S_2$ (MW 240.07)

(formyl[2] and methyl cation[3] equivalent; nucleoside hydroxyl pro tection group;[4] 1,1-dipole equivalent[11])

Physical Data: mp 150–150.5 °C (dec).
Solubility: sol MeCN and DMF (slowly reacts); insol ether.
Form Supplied in: light brown powder.
Preparative Method: treatment of 2-(3-methylbutoxy)-1,3-benzodithiole[2] with Tetrafluoroboric Acid in Acetic Anhydride at 0 °C.
Purification: recrystallized from acetic acid or acetonitrile/ether.
Handling, Storage, and Precautions: can be stored in the absence of moisture and light; no toxicity data is currently available; 1,3-benzodithiol-1-ium perchlorate *[32283-21-9]* is also known,[2] but poses a risk of explosion.

Acyl Cation Equivalent. Benzodithiolium ions react with a wide range of carbon nucleophiles to afford masked aldehyde and ketone products.[2] Electron-rich arenes (eq 1)[5] and active methylene compounds[6] are useful nucleophiles and 2-substituted benzodithiolium ions lead to good yields of 1-deuteroaldehydes upon reaction with $LiAlD_4$ and subsequent hydrolysis.[7]

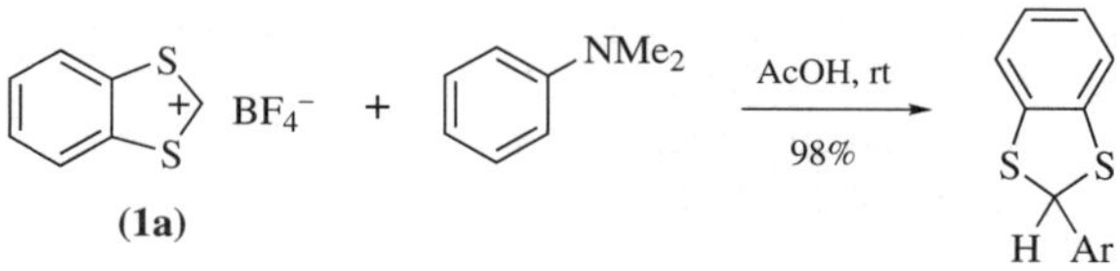

(1a) (1)

Ketone products are available by a reaction sequence involving Grignard reagent addition to the parent cation followed by regeneration of the 2-substituted 1,3-benzodithiolium ion via Triphenylmethyl Perchlorate mediated hydride abstraction. A second Grignard reagent addition and dithiole hydrolysis completes the sequence (eq 2).[8] Chloramine-T in conjunction with Mercury(II) Chloride is a superior method for benzodithiole hydrolysis.[7,8a] Reductive cleavage of the benzodithiole unit can be achieved by treatment with Sodium–Ammonia.[8]

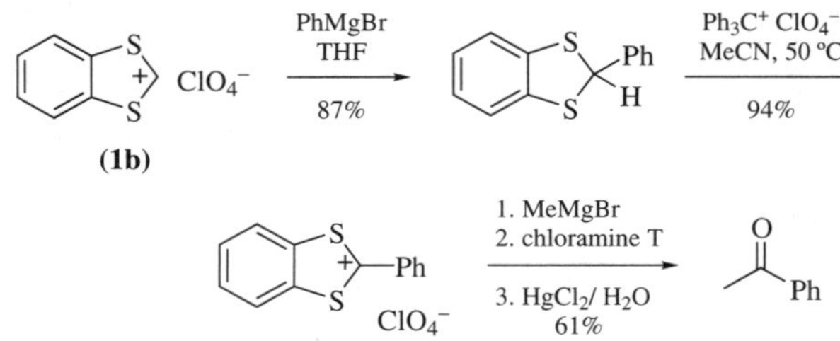

(1b) (2)

Methyl Cation Equivalent. Mild *N*-methylation of guanosine has been achieved employing 1,3-benzodithiol-1-ium tetrafluoroborate as a masked methyl group. Treatment of the *O*-silylated guanosine derivative (**2**) with (**1a**) provided high yields of (**3**), which on exposure to Tri-n-butylstannane provided the *N*-methyl compound (**4**) (eq 3).[3] This process can be repeated to afford the corresponding *N,N*-dimethyl derivative. A related *O*-methylation sequence is not as effective.[3,9]

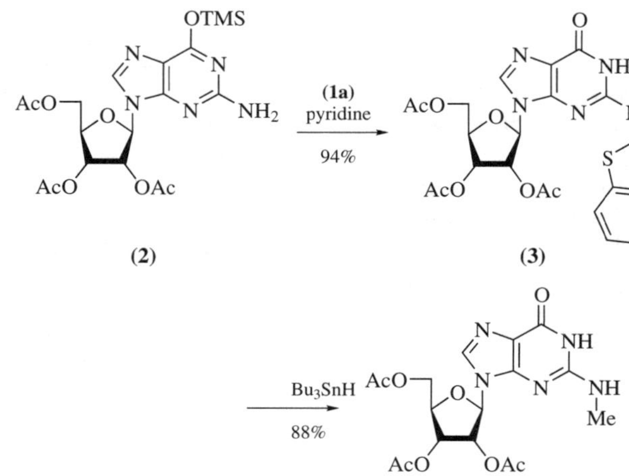

(2) (3) (4) (3)

Nucleoside Hydroxyl Group Protection. The benzodithiole group has been used as a selective protecting group for the 5′-hydroxyl group of thymidine (eq 4).[4] No competitive reactions with base residues were noted with any of the nucleosides studied. The benzodithiole protection can be cleaved to reveal the free hydroxyl group with 2% aqueous TFA.

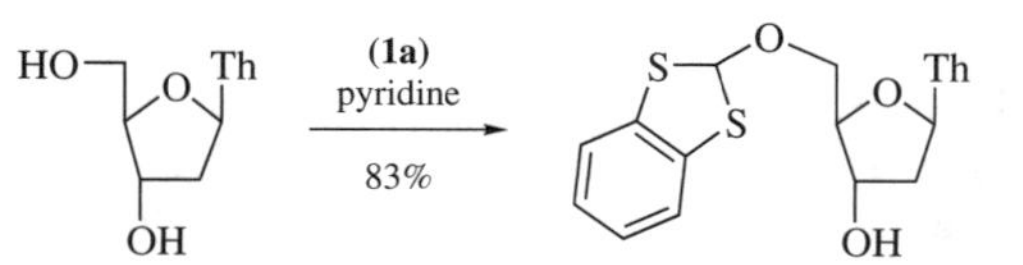

(4)

1,1-Dipole Equivalents. The parent 1,3-benzodithiole species can be rendered nucleophilic by deprotonation at the 2-position with strong base. A useful reagent for expressing this nucleophilic character is readily available 2-diethoxyphosphinyl-1,3-benzodithiole.[10] Reaction of the 2-lithiated species (**5**) with aldehydes affords excellent yields of ketene dithioacetals suitable for accessing the corresponding substituted benzodithiolium cations for subsequent carbon–carbon bond formation (eq 5).[11]

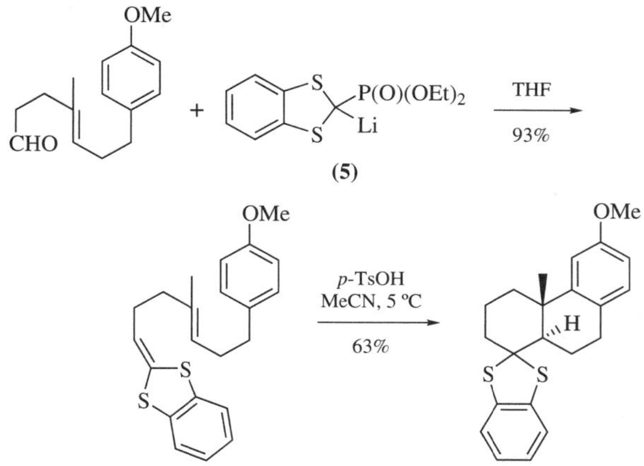

(5)

See also 2-(3-Methylbutoxy)-1,3-benzodithiolane.

1. (a) Nakayama, J. *Sulfur Rep.* **1985**, *4*, 159. (b) Lozac'h, N.; Stavaux, M. In *Advances in Heterocyclic Chemistry*; Katritzky, A. R.; Boulton, A. J.; Eds.; Academic Press:New York, **1980**; Vol. *27*, p 151.
2. (a) Nakayama, J.; Fujiwara, K.; Hoshino, M. *Bull. Chem. Soc. Jpn.* **1976**, *49*, 3567. (b) Nakayama, J. *Synthesis* **1975**, 38.
3. Sekine, M.; Satoh, T. *J. Org. Chem.* **1991**, *56*, 1224.
4. Sekine, M.; Hata, T. *J. Am. Chem. Soc.* **1983**, *105*, 2044.
5. Nakayama, J. *Synthesis* **1975**, 170.
6. (a) Nakayama, J. *J. Chem. Soc., Perkin Trans. 1* **1976**, 540. (b) Scherowsky, G.; Weiland, J. *Justus Liebigs Ann. Chem./Liebigs Ann. Chem.* **1974**, 403.
7. Degani, I.; Fochi, R. *Synthesis* **1976**, 759.
8. (a) Degani, I.; Fochi, R. *J. Chem. Soc., Perkin Trans. 1* **1976**, 1886. (b) Degani, I.; Fochi, R. *Synthesis* **1976**, 471.
9. Sekine, M.; Nakanishi, T. *J. Org. Chem.* **1990**, *55*, 924.
10. Akiba, K.; Ishikawa, K.; Inamoto, N. *Bull. Chem. Soc. Jpn.* **1978**, *51*, 2674.
11. (a) Rigby, J. H.; Kotnis, A.; Kramer, J. *J. Org. Chem.* **1990**, *55*, 5078. (b) Rigby, J. H.; Kotnis, A. S. *Tetrahedron Lett.* **1987**, *28*, 4943.

James H. Rigby
Wayne State University, Detroit, MI, USA

Benzothianthrene Oxide

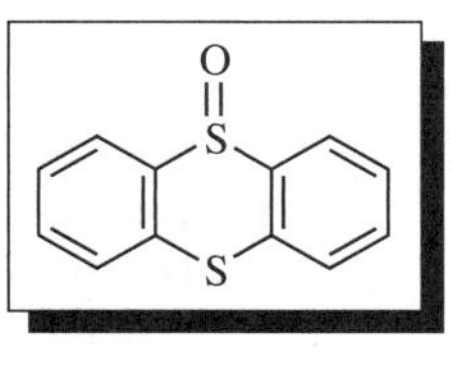

[2362-50-7] $C_{12}H_8OS_2$ (MW 232.32)

(reagent used to activate glycals for acylamidoglycosylation and as a probe of the electronic character of oxygen transfer reagents)

Alternate Name: thianthrene monosulfoxide, thianthrene-5-oxide, thianthrene monoxide, Adam probe.
Physical Data: mp 142–144 °C.
Solubility: very soluble in chloroform and acetone, slightly less soluble in glacial acetic acid and benzene, slightly soluble in methanol and ethanol, very slightly soluble in petroleum ether and diethyl ether. Very soluble in boiling water.
Form Supplied in: off-white crystals.
Preparative Methods: the title reagent has been prepared by the oxidation of thianthrene via numerous methods. The most often cited method is by Gilman and Swayampati,[1] which is actually a modified version of the first reported synthesis by Fries and Vogt.[2] In 86.4 g (0.4 mol) of thianthrene was added 1.4 L of glacial acetic acid and refluxed. To this mixture 150 mL of dilute nitric acid (40%) was added dropwise over 1.5 h. The mixture was further allowed to reflux for 15 min. The yellow solution was diluted with 5 L of ice water and the product was filtered, washed with water, and dried to yield 90.5 g (98%) of off-white crystals of benzothianthrene oxide. Another method is the oxidation of the commercially available thianthrene by *m*-chloroperbenzoic acid (*m*-CPBA).[3,4] To a solution of thianthrene (37 mmol) in CH_2Cl_2 at room temperature, a solution of 88% m-CPBA (50 mmol) in CH_2Cl_2 was slowly added. Once all of the thianthrene got dissolved, the mixture was washed with 2 L of aqueous $NaHCO_3$, then with 1 L of H_2O, dried over $MgSO_4$, and finally concentrated under reduced pressure. The sample was then chromatographed on silica gel utilizing cyclohexane/ethyl acetate (9:1) as eluent to give thianthrene oxide in a good yield.
Purification: title compound forms long off-white needles when recrystallized from methanol, ethanol, or benzene.

Preparation of C2-Acetamidoglycosides. Numerous natural oligosaccharide structures contain a C2-acetamidoglycosidic linkage. Due to the biological importance of these carbohydrates, their preparation has been the focus of much research. Glycals have been used to introduce the aza functionality at C2 as an oxime, azide, or sulfonamide. The most notable aspect of these approaches is that they require additional functional group manipulation after introduction of the nitrogen at C2 to form the acetamide group. A direct one-pot preparation of C2-acetamidoglycosides using thianthrene oxide has been reported by Gin.[4] The procedure is one of the most efficient ways to introduce the C2-acetamide linkage in carbohydrates.

Trifluoromethane sulfonic anhydride and thianthrene oxide form the sulfonium reagent (eq 1) which can activate glycals.

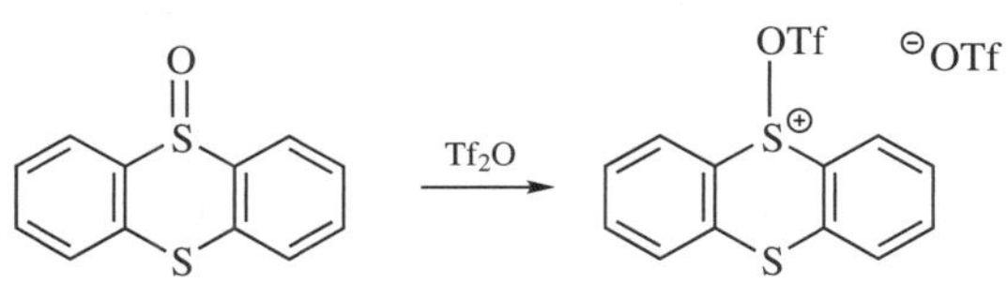

(1)

Triflic anhydride (2 equiv) is added to a solution of tri-*O*-benzyl glucal (1 equiv) and thianthrene oxide (2 equiv) at −78 °C (eq 2). After activation of the glycal, *N,N'*-diethylaniline (4 equiv) and *N*-trimethylsilyl-acetamide (3 equiv) are added and the reaction is stirred at room temperature for 2 h. This sequence converts the glycal to the bicyclic acetimidate intermediate which serves as a glycosyl donor in the subsequent glycosylation step. Introduction of a nucleophile (3 equiv) and Amberlyst-15 acidic resin in the same pot effects glycosylation to form β-C2-acetamidoglycoside in good yields (Table 1). The reaction selectively gave the β-glycosides using both glucal and galactal donors. Activation using thianthrene oxide was found to give enhanced yields relative to other sulfoxide activating agents. The ability of the

Table 1 Preparation of C2-acetamido glycosides

Glycal	Acceptor	Product	Yield (%)
Tri-*O*-benzyl glucal	Benzyl alcohol	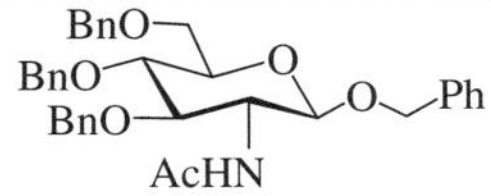	69
Tri-*O*-benzyl glucal	Cholestanol	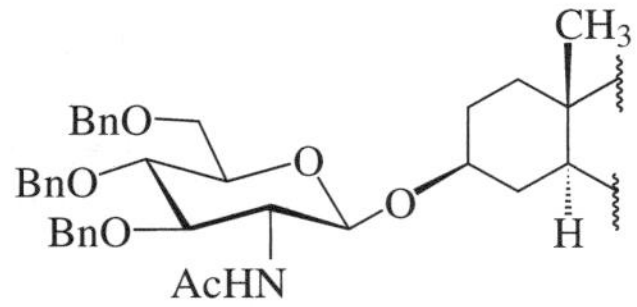	70
Tri-*O*-benzyl glucal	Methyl 2,3,4-tri-*O*-benzyl glucoside	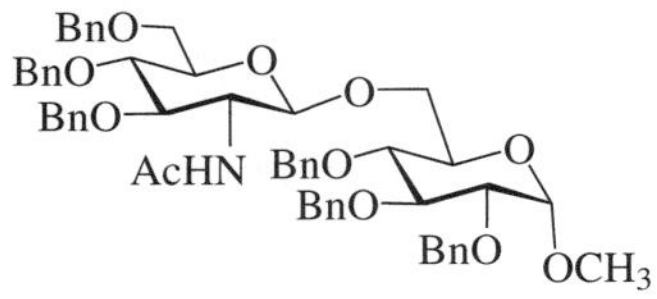	69
Tri-*O*-benzyl glucal	*N*-Cbz D-threonine methyl ester	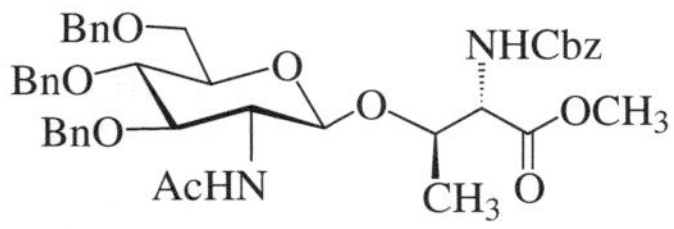	65
Tri-*O*-benzyl glucal	Sodium azide	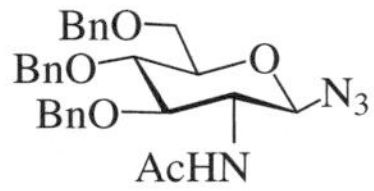	70
6-*O*-allyl-2,3-*O*-benzyl glucal	Methyl 2,3,4-tri-*O*-benzyl glucoside	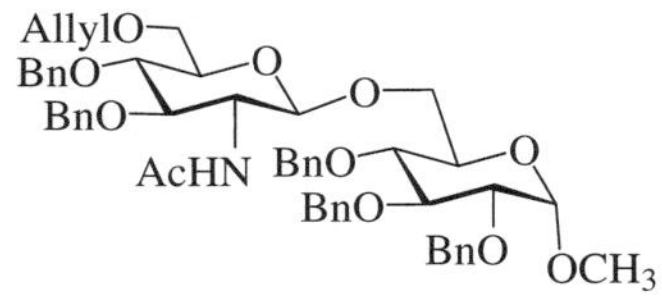	55
Tri-*O*-benzyl galactal	Cholestanol	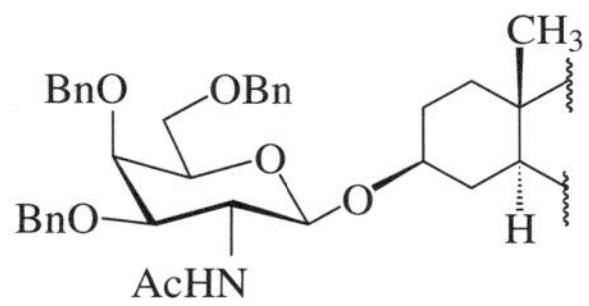	51
Tri-*O*-benzyl galactal	Sodium azide	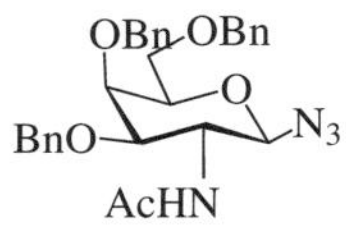	62

remote sulfur atom of thianthrene to stabilize the C2-sulfonium and C1-oxocarbenium ions by resonance and direct participation, respectively, was advanced in rationalization of the increased yields.

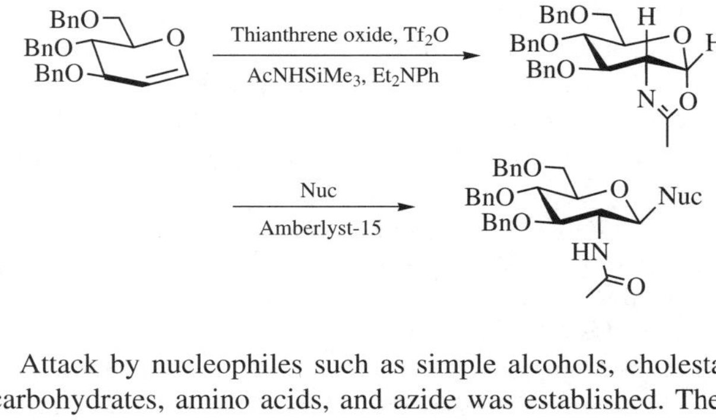

(2)

Attack by nucleophiles such as simple alcohols, cholestarol, carbohydrates, amino acids, and azide was established. The reaction was amenable to various protecting groups on the donor and the acceptor. Camphorsulfonic acid was used in some examples in place of Amberlyst-15 as acidic promoter. Also, with azide as the nucleophile, $Cu(OTf)_2$ was used as the acid promoter. Table 1 collects a number of examples of this procedure showing the glycal and the nucleophile acceptor along with the yield.

The procedure described was extended to the synthesis of other C2-amido glycosides.[4b] Primary amides such as benzamide, 2-benzyloxyacetamide, and pent-4-enamide were used in place of *N*-trimethylacetamide to form the corresponding imidates which were subsequently glycosylated under the same conditions as the acetimidate. This allowed for the introduction of additional acyl groups at C2 of glucose. The use of trimethylsilyl protected amides was not necessary in the synthesis of these modified glucosamine derivatives as the primary amides themselves were sufficiently reactive as nucleophiles. Seemingly, the trimethylsilyl group in *N*-trimethylsilyl acetamide served only to solubilize this species in the reaction mixture.

Mechanistic Probe in Oxygen Transfer Reactions. It has been extensively demonstrated that benzothianthrene oxide can be chemoselectively oxidized. This has been exploited to determine the electronic character of oxygen transfer agents by quantifying the nucleophilic oxidation at the sulfoxide site of the compound versus total oxidation given by X_{SO} (eq 3).[5] Benzothianthrene oxide is therefore an excellent and widely used probe to determine the nature of various oxidizing agents. The first study employed benzothianthrene oxide (later known as the Adam probe) to distinguish carbonyl oxides from dioxiranes.[6] Subsequently many diverse oxygen transfer systems have been studied using the Adam probe. Some recent examples include dioxygen and peroxides catalyzed by phthalocyanine,[7] peroxy acids,[8] peroxometal complexes,[3] metalloporphyrin catalysts,[9] hemoprotein oxidizing species,[10] heteropolyoxometalate oxidants,[11] persulfoxides,[9] titanium catalysts in heterogenous and homogenous media,[12,13] and surface mediated oxidants.[14]

$$X_{SO} = \frac{\text{nucleophilic oxidation}}{\text{total oxidation}} = \frac{SSO_2 + SOSO_2}{SSO_2 + SOSO + 2SOSO_2} \quad (3)$$

The rationale for the determination of the electronic character of the oxidant is based on the assumption that electrophilic oxidants attack the sulfide group, producing 5,10-dioxides (SOSO), while nucleophilic oxidants prefer the sulfoxide moiety to yield the sulfone (SSO_2) (eq 4). Over-oxidation of either species gives the

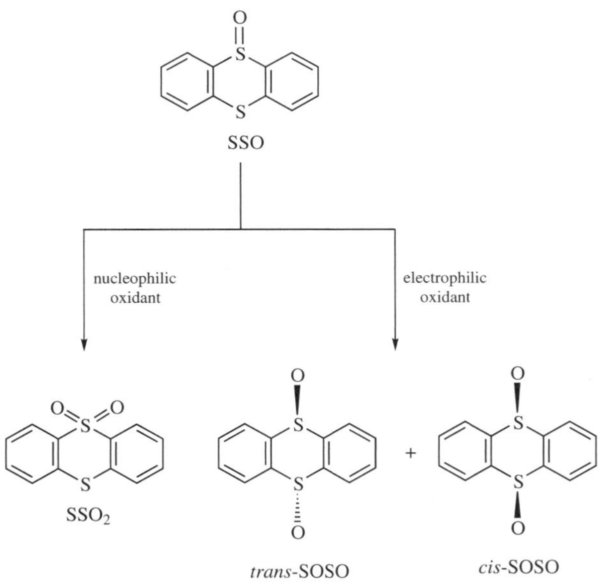

(4)

sulfone/sulfoxide ($SOSO_2$). Oxidants are then assigned a numerical value, termed the X_{SO} value, to quantify their chemical nature in terms of nucleophilic oxidation. Originally, a value greater than 0.85 constituted a nucleophilic oxidant, while one with a value less than 0.15 was considered electrophilic. Recent work suggested that this basic analysis is slightly more complex than first thought. The later analyses have found that one of the major oxidation products, *trans*-SOSO, was not being seen due to unusually long HPLC retention times. Once this was taken into account, the X_{SO} values for electrophilic oxidants were shifted slightly higher to ≤ 0.3, while the nucleophilic values were shifted slightly lower to ≥ 0.7.[5] The predominant product of electrophilic oxidants is the *trans*-SOSO product. This is explained by noting that the axial lone pair present at the sulfide site is preferentially attacked over the equatorial lone pair.

Interpretation of the stereo- and chemoselectivity of oxidation reactions using thianthrene oxide must consider its low barrier to ring-inversion, steric effects, and the role that solvents may play in the product distribution.[5,15] All of these possibilities could influence the X_{SO} value of oxidants. Specifically, mechanistic consideration of electrophilic oxidation of thianthrene oxide in the presence of protic solvents or acid leads to diminished yields of SSO_2. Despite all of these concerns, benzothianthrene oxide is still considered a reliable probe for the electronic character of oxidants.

1. Gilman, H.; Swayampati, D. R., *J. Am. Chem. Soc.* **1955**, *77*, 3387.
2. Fries, K.; Vogt, W., *Liebigs Ann. Chem.* **1911**, *381*, 312.
3. Ballistreri, F. P.; Tomaselli, G. A.; Toscano, R. M.; Conte, V.; Di Furia, F., *J. Am. Chem. Soc.* **1991**, *113*, 6209.
4. (a) Di Bussolo, V.; Liu, J.; Huffman, Jr, L. G.; Gin, D. Y., *Angew. Chem. Int. Ed.* **2000**, *39*, 204. (b) Liu, J.; Gin, D. Y., *J. Am. Chem. Soc.* **2002**, *124*, 9789.
5. Adam, W.; Golsch, D.; Görth, F. C., *Chem. Eur. J.* **1996**, *2*, 255.
6. Adam, W.; Haas, W.; Sieker, G., *J. Am. Chem. Soc.* **1984**, *106*, 5020.
7. Sorokin, A. B.; Mangematin, S.; Pergrale, C., *J. Mol. Catal. A.* **2002**, *182*, 267.
8. Eržen, E.; Koller, J.; Plesničar, B., *J. Org. Chem.* **2001**, *66*, 5155.
9. Akasaka, T.; Haranaka, M.; Ando, W., *J. Am. Chem. Soc.* **1991**, *113*, 9898.
10. Alvarez, J. C.; Ortiz de Montellano, P. R., *Biochemistry* **1992**, *31*, 8315.
11. Hamamoto, M.; Nakayama, K.; Nishiyama, Y.; Ishii, Y., *J. Org. Chem.* **1993**, *58*, 6421.
12. Adam, W.; Mitchell, C. M.; Saha-Möller, C. R.; Selvam, T.; Weichold, O., *J. Mol. Catal. A.* **2000**, *154*, 251.
13. Adam, W.; Corma, A., García, H.; Weichold, O., *J. Catal.* **2000**, *196*, 339.
14. Kropp, P. J.; Breton, G. W.; Fields, J. D.; Tung, J. C.; Loomis, B. R., *J. Am. Chem. Soc.* **2000**, *122*, 4280.
15. Deubel, D. V., *J. Org. Chem.* **2001**, *66*, 2686.

Mark W. Peczuh & Chrysa F. Malosh
University of Connecticut, Storrs, CT, USA

3*H*-1,2-Benzothiaselenol-3-one

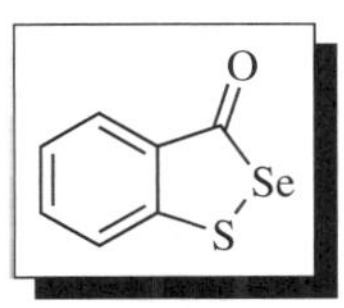

[28774-69-8] C_7H_4OSSe (MW 215.91)

(selenium transferring reagent; selenization of *H*-phosphonate diesters, selenization of *H*-phosphonothioate diesters, selenization of phosphite triesters, giving phosphoroselenoates and phosphoroselenothioates)

Physical Data: yellow crystals; mp 79 °C.[1]
Preparative Methods: the title reagent can be prepared by reacting 2,2′-dithiosalicylic acid with (phenylseleno)phosphonic dichloride in the presence of pyridine.[1]
Purification: recrystallization from ethanol.

Selenization of *H*-Phosphonate and *H*-Phosphonothioate Diesters. *H*-Phosphonate (**1**, X = O) and *H*-phosphonothioate (**1**, X = S) diesters readily react with 3*H*-1,2-benzothiaselenol-3-one (BTSe) to produce the corresponding phosphoroselenoate and phosphorothioselenoate (eq 1).[1]

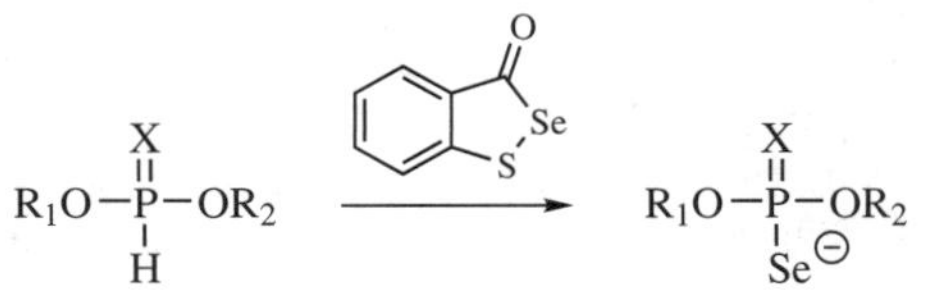

1 2 (1)

X = O or S;
R_1 and R_2 = nucleoside moieties

The reagent is soluble in common organic solvents and shows an enhanced rate of selenium transfer compared to elemental selenium or potassium selenocyanate. The reactions in acetonitrile in the presence of triethylamine are rapid and furnish a clean and quantitative conversion of *H*-phosphonate diester into the corresponding phosphoroselenoate within 60 s. A yellow precipitate is usually formed during the reaction, but it can be avoided by performing the reaction in a 95:5 (v/v) mixture of pyridine/triethylamine as a solvent.

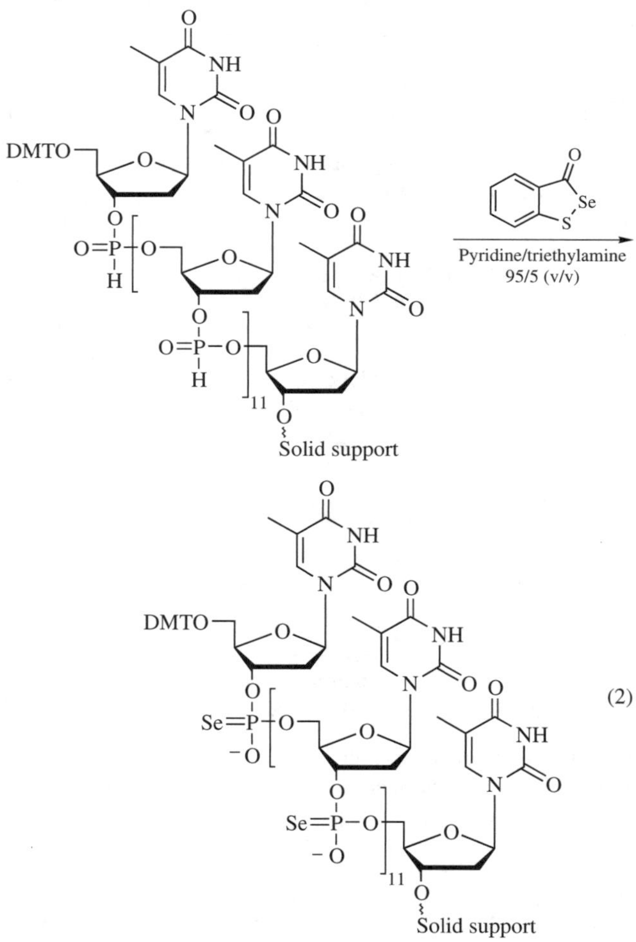

(2)

BTSe is compatible with the solid phase synthesis of oligonucleotide analogs as demonstrated by efficient transformation of a resin-attached dodecathymidine undeca-*H*-phosphonate into the corresponding dodecathymidine undecaphosphoroselenoate (eq 2).[2] The transfer of selenium to *H*-phosphonate and *H*-phosphonothioate diesters mediated by BTSe is stereospecific and occurs with retention of configuration at the phosphorus center.[2]

Selenization of Phosphites. BTSe also transfers selenium to phosphite triesters, producing the corresponding phosphoroselenoates,[2] but the reaction is usually slower than that of the corresponding *H*-phosphonate diesters. Selenization of triethyl phosphite with 3*H*-1,2-benzothiaselenol-3-one in acetonitrile with 5% triethylamine yields triethyl phosphoroselenoate within 5 min (eq 3), while for the reaction in pyridine/triethylamine, it takes ca. 1.5 h for completion.

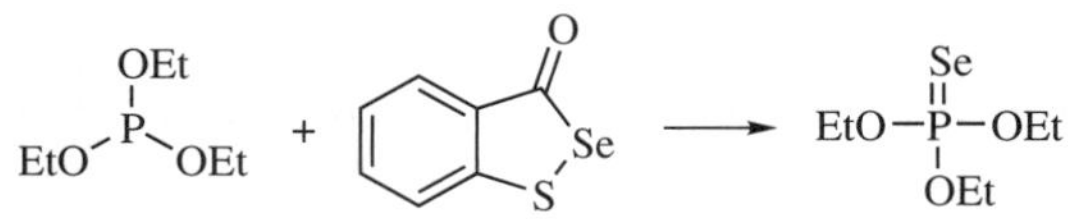

(3)

Related Reagents. Elemental selenium; potassium selenocyanate; triphenyl phosphoroselenoate.

1. Stawinski, J.; Thelin, M., *Tetrahedron Lett.* **1992**, *33*, 7255–58.
2. Stawinski, J.; Thelin, M., *J. Org. Chem.* **1994**, *59*, 130–36.

Jacek Stawinski
Stockholm University, Stockholm, Sweden

Mats Thelin
AstraZeneca, Södertälje, Sweden

Benzothiazole-2-sulfonyl Chloride (BtsCl)

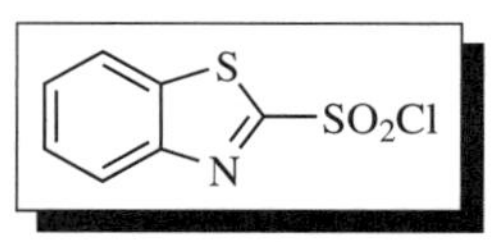

[2824-46-6] $C_7H_4ClNO_2S_2$ (MW 233.7)

(reagent used as a nitrogen protecting group, in ligand preparation, and in combinatorial sulfonamide libraries)

Physical Data: mp 108–110 °C.[1]
Solubility: soluble in most common organic solvents.
Form Supplied in: not commercially available.
Handling, Storage, and Precautions: this material should not be stored at room temperature for any length of time. The author has had room temperature samples spontaneously decompose with liberation of SO_2. The rate at which this occurs is quality dependent.[2] Dilute solutions in $CHCl_3$ (0.1 M) are relatively stable but a 1 M solution decomposes within 3 days. The stability in solution is improved by the addition of 1% BHT.

Reagent Preparation. The reagent is easily prepared by a chlorine oxidation of commercially available 2-mercaptobenzothiazole.[2]

Use as a Protecting Group for Nitrogen. Vedejs, who showed that benzothiazole-2-sulfonyl–aminoacids (Bts-aminoacids) could readily be converted to their acid chlorides without racemization, pioneered the use of the Bts group for the protection of amino acids.[3] The acid chloride of Bts-protected amino acids were found to be effective for racemization free peptide couplings.[2] The sulfonamide is readily prepared using an aq NaOH maintained at pH 10–10.5 by slow addition of 1.3 M NaOH. For simple amines an aq THF mixture of the amine and $NaHCO_3$ is treated with BtsCl to form the sulfonamide. A two-phase system using CH_2Cl_2, H_2O, and $Na_2CO_3$is also effective.[4] Primary Bts derivatives of amines can be alkylated readily (eq 1).[4]

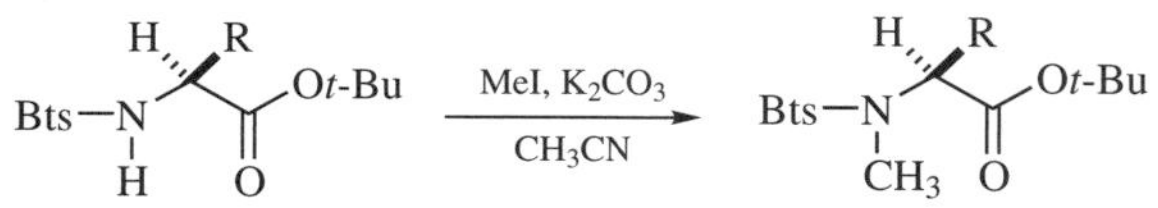

(1)

R = Me, *i*-Bu, *i*-Pr

A number of protocols have been developed for the deprotection of Bts derivatives, and these are listed below. The preferred and probably the mildest methods are those based on the use of a thiol and base.

1. 6 N HCl reflux 2 h.[1]
2. H_3PO_2, DMF, or THF. Best conversions are achieved using lower concentrations (0.05 M, 90%; 0.3 M, 77%; 0.8 M, 43%) for the deprotection of peptides.[3]
3. Zn, AcOH-EtOH.[3]
4. Al(Hg), ether, H_2O.[3]
5. $Na_2S_2O_4$ or $NaHSO_3$, EtOH, and water reflux. With peptides these conditions cause racemization.[3]
6. TFA and PhSH, 25% conversion after 2 days.[3]
7. H_2, Pd/C, and EtOH. Some cleavage occurs before the catalyst is poisoned.[3]
8. NaOH, rt, 12 h can be used for cleavage of secondary amines such as the Bts-proline derivative, but primary derivatives require 90–100 °C for 24 h and result in racemization of the amino acid.[3]
9. $NaBH_4$ and EtOH. This method works for Bts derivatives of 2° amines. With 1° amines the reaction fails to go to completion.[2]
10. PhSH, DIPEA, DMF or PhSH, *t*-BuOK, DMF.[5] K_2CO_3 has also been used as a base.[2] These conditions were effective for both primary and secondary Bts derivatives.
11. Glutathione S-transferase has also been shown to cleave the Bts group.[6] This has considerable significance when using this group as part of a drug candidate.

Other applications. The Bts group has also been used in the development of ligands[7] and the preparation of sulfonamide libraries for drug development.[8]

1. Roblin, R. O., Jr.; Clapp, J. W., *J. Am. Chem. Soc.* **1950**, *72*, 4890.
2. Vedejs, E.; Kongkittingam, C., *J. Org. Chem.* **2000**, *65*, 2309.
3. Vedejs, E.; Lin, S.; Klapars, A.; Wang, J., *J. Am. Chem. Soc.* **1996**, *118*, 9796.
4. Vedejs, E.; Kongkittingam, C., *J. Org. Chem.* **2001**, *66*, 7355.
5. Wuts, P. G. M.; Gu, R. L.; Northuis, J. M.; Thomas, C. L., *Tetrahedron. Lett.* **1998**, *39*, 9155.
6. Zhao, Z.; Koeplinger, K. A.; Peterson, T.; Conradi, R. A.; Burton, P. S.; Suarato, A.; Heinrikson, R. L.; Tomasselli, A. G., *Drug Met. Disp.* **1999**, *27*, 992.
7. Diltz, S.; Aguirre, G.; Ortega, F.; Walsh, P. J., *Tetrahedron Asym.* **1997**, *8*, 3559.
8. (a) Raghavan, S.; Yang, Z.; Mosley, R. T.; Schleif, W. A.; Gabryelski, L.; Olsen, D. B.; Stahlhut, M.; Kuo, L. C.; Emini, E. A.; Chapman, K. T.; Tat, J. R., *Bioorg. Med. Chem. Lett.* **2002**, *12*, 2855. Hafez, A. A. A.; Geies, A. A.; Hozien, A.; Khalil, Z. H., *Collect. Czech. Chem. Commun.* **1994**, *59*, 957.

Peter G. M. Wuts
Pfizer, Kalamazoo, Michigan, USA

1-Benzotriazolyl Diethyl Phosphate (BDP)

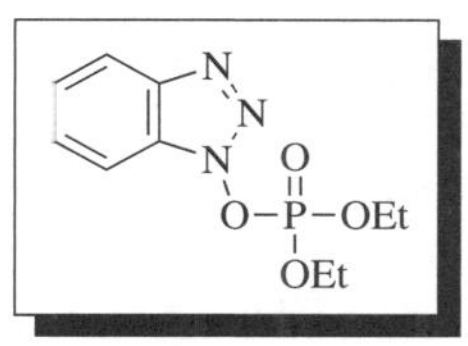

[96338-11-3] $C_{10}H_{14}N_3O_4P$ (MW 271.21)

(coupling reagent for amides and peptide synthesis[1,2])

Alternate Name: BDP.
Solubility: sol MeCN, DMF, EtOAc, CH_2Cl_2, THF.
Form Supplied in: oil; not available from commercial sources.
Preparative Method: by reaction of diethyl chlorophosphate (1 equiv) and 1-hydroxybenzotriazole (1 equiv) in the presence of triethylamine (1 equiv) in dry THF at 0 °C for 30 min, with 90–95% yields.
Purification: by filtration through a short column of silica gel or cellulose.
Handling, Storage, and Precautions: the oil can be stored in a desiccator with drierite at 4 °C for several weeks without decomposition. Distillation in vacuo leads to decomposition.

1-Benzotriazolyl diethyl phosphate (BDP) has been used as a condensing reagent for the preparation of amides. Typically, equimolar amounts of carboxylic acids, amines, and 1-benzotriazolyl diethyl phosphate, in the presence of triethylamine (2.1 equiv) in a broad range of solvents (the relative coupling efficiency is DMF > MeCN = THF > CH_2Cl_2) react rapidly and cleanly with coupling times in the range of 10–60 min at 25 °C, yielding the corresponding amides in > 90% yields (eq 1).

$$R^1CO_2H + R^2NH_2 \xrightarrow[Et_3N \; >90\%]{BDP} R^1CONHR^2 \quad (1)$$

1-Benzotriazolyl diethyl phosphate is also suitable for hindered carboxylic acids and amines, as well as poorly nucleophilic amines. Thus *N*-cyclohexylbenzamide is obtained from benzoic acid and cyclohexylamine with the addition of 1-benzotriazolyl diethyl phosphate, whereas the use of 1,3-Dicyclohexylcarbodiimide did not yield any trace of the target product. Benzoic acid is readily converted to the corresponding amides by the reaction with 1-benzotriazolyl diethyl phosphate and benzylamine or aniline. Similarly, reaction of acetic acid with sterically hindered *t*-butylamine yielded the corresponding acetamide. Finally, crotonic acid is cleanly converted to the corresponding amides without 1,4-addition side-products.

Additionally, 1-benzotriazolyl diethyl phosphate has been used for the preparation, in solution, of dipeptides and tripeptides. The reaction is carried out as described above with *N*-protected amino acids or dipeptides, and amino acid ester hydrochlorides, in DMF for 4 h, at 25 °C, with a yield superior to 75%. Racemization from 1-benzotriazolyl diethyl phosphate mediated couplings is checked by different methods. Young[3] and Anderson[4] tests gave low degree (<4%) of racemization when DMF is used as a solvent. Other solvents such as THF, EtOAc, and CH_2Cl_2 gave more racemization. Furthermore, the Shiori[5] test gave a higher degree of racemization even in DMF (18%). A related analog, 1-benzotriazolyl diphenyl phosphate, led to 39% of racemization, indicating that the degree of racemization is also dependent on the nature of organophosphate.

1-Benzotriazolyl diethyl phosphate has not been used in solid-phase peptide synthesis,[6] although other commercialtriazole derivatives such as Benzotriazol-1-yloxytris (dimethylamino)phosphonium Hexafluorophosphate (BOP)[7] or O-Benzotriazol-1-yl-N,N,N′, N′-tetramethyluronium Hexafluorophosphate (HBTU)[8] are commonly used in that strategy.

1. Kim, S.; Chang, H.; Ko, Y. K. *Tetrahedron Lett.* **1985**, *26*, 1341.
2. Kim, S.; Chang, H.; Ko, Y. K. *Bull. Korean Chem. Soc.* **1987**, *8*, 471.
3. Williams, M. W.; Young, G. T. *J. Chem. Soc. (C)* **1963**, 881.
4. (a) Anderson, G. W.; Young, R. W. *J. Am. Chem. Soc.* **1952**, *74*, 5307. (b) Anderson, G. W.; Zimmermann, J. E.; Callanan, F. M. *J. Am. Chem. Soc.* **1967**, *89*, 5012.
5. Takuma, S.; Hamada, Y.; Shiori, T. *Chem. Pharm. Bull.* **1982**, *30*, 3147.
6. Fields, G. B.; Tian, Z.; Barany, G. In *Synthetic Peptides: A User's Guide*; Grant, G. A., Ed.; Freeman:New York, **1992**; pp77–183.
7. Castro, B.; Dormoy, J. R.; Evin, G. *Tetrahedron Lett.* **1975**, 1219.
8. (a) Dourtoglou, V.; Ziegler, J. C.; Gross, B. *Tetrahedron Lett.* **1978**, 1269. (b) Knorr, R.; Trzeciak, A.; Bannwarth, W.; Gillessen, D. *Tetrahedron Lett.* **1989**, *30*, 1927.

Fernando Albericio & Steven A. Kates
Millipore Corporation, Bedford, MA, USA

Benzotriazol-1-yloxytris(dimethylamino) phosphonium Hexafluorophosphate (BOP)

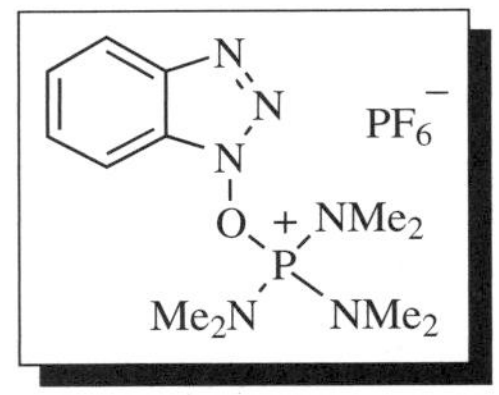

[566002-33-6] $C_{12}H_{22}F_6N_6OP_2$ (MW 442.29)

(peptide coupling agent, general condensating reagent)

Alternate Name: BOP; Castro's reagent.[1]

Physical Data: mp 138 °C (dec); soluble in most common solvents (DMF, CH_2Cl_2, THF, MeCN, acetone, DMSO, NMP) and insoluble in water.

Form Supplied in: white solid, widely available.

Preparative Methods: BOP was initially prepared by reaction of tris(dimethylamino)phosphine with CCl_4 in the presence of **1-Hydroxybenzotriazole** (HOBt) and precipitation of the phosphonium salt by anion exchange with potassium hexafluorophosphate. A more economical synthesis was reported later from $[(CH_3)_2N]_3PO$ and **Phosgene** followed by treatment of the phosphonium salt with HOBt and triethylamine.[2] Other preparations similarly employed **Triphosgene**[3] or $POCl_3$.[4] A resin-bound version of BOP has been recently reported.[5]

Purification: by crystallization in a mixture of acetone–ether or CH_2Cl_2–ether.

Handling, Storage, and Precautions: avoid contact with eyes, skin, and breathing dust as it is irritant and harmful. Store in a refrigerator. Light sensitive, protect from light. A main by-product of BOP reaction is carcinogenic **HMPA**. Handle it with caution in a fume hood.

Coupling Reagent for Peptide Synthesis. The synthesis and use of benzotriazol-1-yloxytris(dimethylamino)phosphonium hexafluorophosphate (BOP) as a coupling reagent for peptide synthesis was first reported by Castro in 1975 with the aim of developing a reagent that comprise both activating and racemization suppressing properties.[1]

BOP has been widely used as a condensating reagent for the preparation of peptides. It promotes peptide bond formation in high yields with barely any side reaction and short reaction times. Generally speaking, peptide bond formation is performed by adding BOP (1.1 equiv) to a concentrated solution of the protected amino acid (Boc or Fmoc protected) in an organic solvent such as DMF, CH_2Cl_2, THF, or MeCN. Then, the tertiary base TEA or DIEA is added (eq 1).

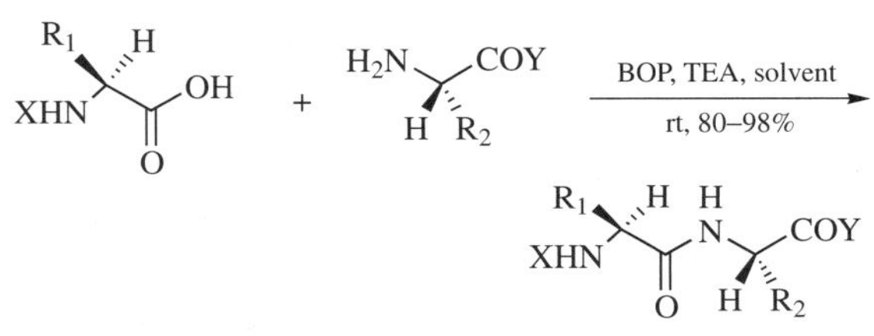

(1)

BOP is also effective in the coupling of bulky amino acids and has shown selective peptide bond formation with trifunctional amino acids (Tyr, Ser, Thr) bearing their side chain functional groups unprotected.[1,2,4,6]

According to mechanistic studies,[7,8] BOP reacts with the carboxylate form of acids (in fact, dicyclohexylammonium salts of amino acids may be used directly).[9] Hence, the reaction mixture needs to be kept rather basic for the reaction to proceed at fast rates. Typically, 3 equiv of a tertiary base (although 1 equiv will suffice) are used to neutralize the carboxylic acid component, the ammonium salt (conjugated acid of the amine component), and the generated HOBt.

Racemization levels in *N*-α-urethane protected amino acids are usually low.[10] Special protocols involving BOP activation have been developed for protected amino acids that are prone to racemize, such as *N*-, *S*-protected cysteine derivatives.[11] However, the chiral integrity of non-urethane protected amino acids, such as acetylvaline,[12] is still greatly compromised on reacting with BOP, as it is with other activating reagents, due to the higher tendency of these derivatives to yield proton-labile oxazolone intermediates. BOP has also been used for the coupling of protected peptide segments in convergent synthetic strategies. In this context, low levels of epimerization of the *C*-terminal residue are usually reported, although care should be taken when bulky or *N*-methyl amino acids are involved in segment coupling.[9a,13,14]

BOP-mediated acylation protocols have also been developed for Boc/Bn protecting group strategies to prevent dipeptide intermediates from giving internal cyclization side reactions such as diketopiperazine formation. Thus, it is possible to promote peptide bond formation directly onto the α-ammonium salt by carrying out the neutralization step in situ. [9a,b,13b,c,15]

BOP has also found wide application in solid-phase peptide synthesis,[16] manually or automatically performed, using either Boc/Bn[17,18] or Fmoc/*t*-Bu[8,19] protecting group strategies. As a coupling agent, it offers faster acylation kinetics and fewer side reactions than standard carbodiimide methods. However, the higher cost of BOP and particularly the generation of the carcinogenic by-product HMPA has progressively reduced its use in favor of other reagents. Thus, carbodiimide/HOBt is usually the method of choice for regular non-problematic peptide syntheses. In addition, other phosphonium analogs such as the pyrrolidine derivative **PyBOP**, HOBt-derived uronium salts (i.e., *O*-benzotriazol-1-yl-*N,N,N′,N′*-tetramethyluronium hexafluorophosphate, **HBTU**), or **HOAt** (1-hydroxy-7-azabenzotriazole)-derived aminium salts (i.e., *N*-[(dimethylamino)-1*H*-1,2,3-triazolo-[4,5-*b*]pyridin-1-ylmethylene]-*N*-methylmethanaminium hexafluorophosphate *N*-oxide, **HATU**) have emerged as valuable alternatives to the use of BOP both in manual and automated peptide synthesis.[19b]

Peptide Cyclization and Lactamization. BOP has also been used satisfactorily to promote ring formation in peptides (head-to-tail, side-chain–side-chain, etc).[20,21] BOP compares equally or even favorably to other condensating reagents such as DPPA (**diphenyl phosphoroazidate**), NDPP (norborn-5-ene-2,3-dicarboxamidodephenyl phosphate), and DCC/HOBt. In addition, BOP, unlike uronium salts (i.e. HBTU), is devoid of guanidyl-type by-product formation.[19b,22] BOP has proven to be useful in difficult cyclization reactions such as didemnins B depsipeptide ring formation[23] and in cyclosporine ring closure (where it was also used to acylate hindered *N*-methyl amino acids)[24] as well as in β-lactam formation.[25] Macrocyclic polyether tetralactam rings have also been synthesized using BOP.[26]

Oligonucleotide Synthesis and Related Phosphorous Chemistry. BOP has also found application in the formation of nucleotide phosphate bonds, for instance, in the preparation of phosphotriester DNA biopolymers.[27] Recently, BOP and other phosphonium and uronium salts have been applied to the rapid (4 s) automated attachment of nucleosides to CPG (controlled pore glass) support via succinyl or hydroquinone-*O,O′*-diacetyl linkers.[28]

Phosphonamide derivatized dipeptides[29] and mixed phosphonate diester analogs of dipeptides (from phosphonate monoesters) have been succesfully synthesized using BOP (eq 2).[30a,b] However, BOP-mediated coupling between the phosphinic acid group and amino esters has been reported to give poor results.[31] In fact, this lack of reactivity has been used to prepare phosphinopeptides using unprotected phosphinic acid moieties.[30c] Phosphinate esters though have been prepared by phosphinic acid activation with BOP.[30c]

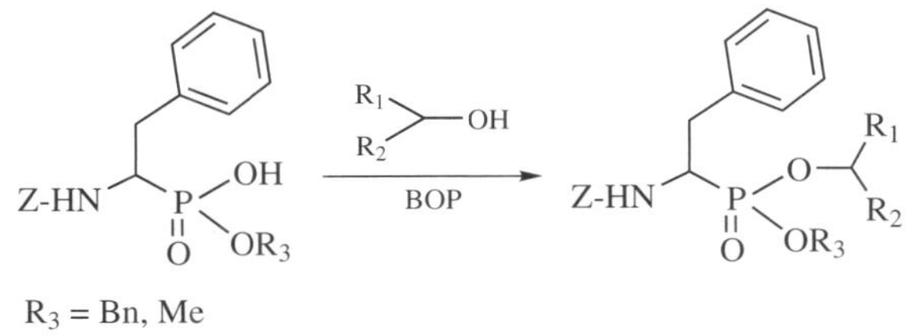

(2)

General Amidation, Esterification, and other Applications. The capability of BOP to activate the carboxylic group has also been extended to other type of organic compounds. Thus, aromatic amides,[12,32] *N*, *O*-dimethylhydroxamates of amino acids or peptides,[33] and esters[34] (phenol esters,[34b] esters between fatty acids and saccharides[34c]) have been prepared using BOP. Furthermore, reaction of trityl amino acids with methoxymethylamine in the presence of BOP afforded the stable hydroxybenzotriazolyl amide, which showed similar reaction properties to the Weinreb amide.[35] Chemoselective reduction of carboxylic acids to alcohols is also possible using BOP and sodium borohydride.[36] Finally, BOP has also found application in the preparation of isocyanates in high yields and purities by reacting aliphatic amines with carbon disulfide and BOP.[37]

1. Castro, B.; Dormoy, J. R.; Evin, G.; Selve, C., *Tetrahedron Lett.* **1975**, 1219–1222.
2. Castro, B.; Dormoy, J. R.; Dortouglou, B.; Evin, G.; Selve, C.; Ziegler, J. C., *Synthesis* **1976**, 751–752.
3. Rivero, I. A.; Somanathan, R.; Hellberg, L. H., *Synth. Comm.* **1995**, 2185–2188.
4. (a) Dormoy, J. R.; Castro, B., *Tetrahedron Lett.* **1979**, 3321–3322. (b) Dormoy, J. R.; Castro, B., *Tetrahedron* **1981**, *37*, 3699–3706.
5. Rivero, I. A.; Aceves, R., *Rev Soc. Quim. Mex.* **2000**, *44*, 97–100.
6. Fournier, A.; Danho, W.; Felix, A. M., *Int. J. Pept. Prot. Res.* **1989**, *33*, 133–139.
7. (a) Castro, B.; Dormoy, J. R.; Dortouglou, B.; Evin, G.; Selve, C., *J. Chem. Res.(S)* **1977**, *7*, 182. (b) Castro, B.; Dormoy, J. R.; Dortouglou, B.; Evin, G.; Selve, C., *J. Chem. Res. (M)* **1977**, *8*, 2118–2125.
8. Hudson, D., *J. Org. Chem.* **1988**, *53*, 617–624.

9. (a) Le Nguyen, D.; Seyer, R.; Heitz, A.; Castro, B., *J. Chem. Soc. Perkin Trans. 1* **1985**, 1025–1031. (b) Le Nguyen, D.; Heitz, A.; Castro, B., *J. Chem. Soc. Perkin Trans.1* **1987**, 1915–1919. (c) Seyer, R.; Aumelas, A.; Caraty, A.; Rivaille, P.; Castro, B., *Int. J. Pept. Prot. Res.* **1990**, *35*, 465–472.
10. (a) Castro, B.; Dormoy, J. R.; Le Nguyen, D., *Tetrahedron Lett.* **1978**, 4419–4422. (b) Le Nguyen, D.; Castro, B.; Dormoy, J. R.; Prevot, D., *Tetrahedron* **1981**, *37*, 4229–4238.
11. Han, Y.; Albericio, F.; Barany, G., *J. Org. Chem.* **1997**, *62*, 4307–4312.
12. Beaulieu, P. L.; Lavallé, P.; Abraham, A.; Anderson, P. C.; Boucher, C.; Bousquet, Y.; Duceppe, J. S.; Gillard, J.; Gorys, V.; Grand-Maître, C.; Grenier, L.; Guindon, Y.; Guse, I.; Plamondon, L.; Soucy, F.; Valois, S.; Wernic, D.; Yoakim, C., *J. Org. Chem.* **1997**, *62*, 3440–3448.
13. (a) Fehrentz, J. A.; Heitz, A.; Fulcrand, P.; Castro, B.; Corvol, P., *Int. J. Pept. Prot. Res.* **1986**, *28*, 620–627. (b) Seyer, R.; Aumelas, A.; Tence, M.; Marie, J.; Bonnafous, J. C.; Jard, S.; Castro, B., *Int. J. Pept. Prot. Res.* **1989**, *34*, 235–245. (c) Seyer, R.; Aumelas, A.; Marie, J.; Bonnafous, J. C.; Jard, S.; Castro, B., *Helvetica Chim. Acta* **1989**, *72*, 678–689.
14. (a) Jarrett, J. T.; Lansbury, P. T., *Tetrahedron Lett.* **1990**, *31*, 4561–4564. (b) Nishiyama, Y.; Tanaka, M.; Saito, S.; Ishizuka, S.; Mori, T.; Kurita, K., *Chem. Pharm. Bull.* **1999**, *47*, 576–578.
15. Gairí, M.; Lloyd-Williams, P.; Albericio, F.; Giralt, E., *Tetrahedron* **1990**, *31*, 7363–7366.
16. Lloyd-Williams, P.; Albericio, F.; Giralt, E., *Chemical Approaches to the Synthesis of Peptides and Proteins*; CRC: Boca Raton (FL), 1997.
17. (a) Fournier, A.; Wang, C. T.; Felix, A. M., *Int. J. Pept. Prot. Res.* **1988**, *31*, 86–97. (b) Forest, M.; Felix, A. M., *Int. J. Pept. Prot. Res.* **1990**, *35*, 89–94.
18. Jezek, J.; Houghten, R. A., *Collect. Czech. Chem. Comm.* **1994**, *59*, 691–706.
19. (a) Gausepohl, H.; Kraft, M.; Frank, R. W., *Int. J. Pept. Prot. Res.* **1989**, *34*, 287–291. (b) Albericio, F.; Bofill, J. M.; El-Faham, A.; Kates, S. A., *J. Org. Chem.* **1998**, *63*, 9678–9683.
20. (a) Schmidt, R.; Klaus, N., *Int. J. Pept. Prot. Res.* **1991**, *37*, 502–507. (b) El Haddadi, M.; Cavelier, F.; Vives, E.; Azmani, A.; Verducci, J.; Martinez, J., *J. Pept. Sci.* **2000**, *6*, 560–570.
21. Crusi, E.; Huerta, J. M.; Andreu, D.; Giralt, E., *Tetrahedron Lett.* **1990**, *31*, 4191–4194.
22. Gausepohl, H.; Pieles, U.; Frank, R. W., In *Peptides, Chemistry and Biology: Proceedings of the 12th American Peptide Symposium*; Smith, J. A.; Rivier, J. E., Eds.; ESCOM, Science: Leiden, The Netherlands, 1992, 523–524.
23. Jou, G.; González, I.; Albericio, F.; Lloyd-Williams, P.; Giralt, E., *J. Org. Chem.* **1997**, *62*, 354–366.
24. Wenger, R. M., *Helvetica Chim. Acta* **1984**, *67*, 502–525.
25. Kim, S.; Lee, T. A., *Bull. Korean Chem. Soc.* **1988**, *9*, 189–190.
26. Duriez, M. C.; Pigot, T.; Picard, C.; Cazaux, L.; Tisnes, P. T., *Tetrahedron* **1992**, *48*, 4347–4358.
27. (a) Molko, D.; Guy, A.; Teoule, R.; Castro, B.; Dormoy, J. R., *Nouv. J. Chim.* **1982**, *6*, 277–281. (b) Molko, D.; Guy, A.; Teoule, R., *Nucleosides and Nucleotides* **1982**, *1*, 65–80. (c) Wada, T.; Sato, Y.; Honda, F.; Kawahara, S.; Sekine, M. G., *J. Am. Chem. Soc.* **1997**, *119*, 12710–12721.
28. Pon, R. T.; Yu, S., *Tetrahedron Lett.* **1997**, *38*, 3331–3334.
29. Dumy, P.; Escale, R.; Vidal, J.; Girard, J. P.; Parello, J., *Comptes Rendues de l'Academie de Sciences (C)* **1991**, *312*, 235–240.
30. (a) Campagne, J.-M.; Coste, J.; Jouin, P., *Tetrahedron Lett.* **1993**, *34*, 6743–6744. (b) Campagne, J.-M.; Coste, J.; Jouin, P., *J. Org. Chem.* **1995**, *60*, 5214–5223. (c) Campagne, J.-M.; Coste, J.; Guillou, L.; Heitz, A.; Jouin, P., *Tetrahedron Lett.* **1993**, *34*, 4181–4184.
31. El Haddadi, M.; Jacquier, R.; Petrus, C.; Petrus, F., *Phosphorous, Sulfur, Silicon Relat. Elem.* **1991**, *63*, 255–259.
32. Somers, P. K.; Wandless, T. J.; Schreiber, S. L., *J. Am. Chem. Soc.* **1991**, *113*, 8045–8056.
33. (a) Fehrentz, J. A.; Castro, B., *Synthesis* **1983**, 676–678. (b) Fehrentz, J. A.; Heitz, A.; Castro, B., *Int. J. Pept. Prot. Res.* **1985**, *26*, 236–241.
34. (a) Kim, M. H.; Patel, D. V., *Tetrahedron Lett.* **1994**, *35*, 5603–5606. (b) Castro, B.; Evin, G.; Selve, C.; Seyer, R., *Synthesis* **1977**, 413. (c) Chapleur, Y.; Castro, B.; Toubiana, R., *J. Chem. Soc. Perkin Trans.1* **1980**, 1940–1943.
35. Sliedregt, K. M.; Schouten, A; Kroon, J.; Liskamp, R. M., *Tetrahedron Lett.* **1996**, *37*, 4237–4240.
36. Mc Geary, R. P., *Tetrahedron Lett.* **1998**, *39*, 3319–3322.
37. Boas, U.; Jakobsen, M. H., *J. Chem. Soc. Chem. Comm.* **1995**, 1995–1996.

Francesc Rabanal & Ernest Giralt
Universitat de Barcelona, 08028-Barcelona, Spain

(1*H*-Benzotriazol-1-yloxy) tris(pyrrolidino)phosphonium Hexafluorophosphate

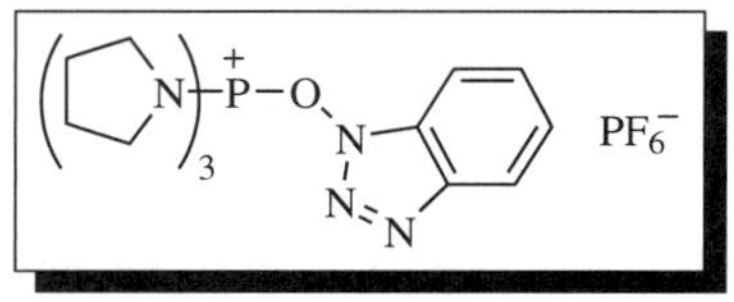

[128625-52-5] $C_{18}H_{28}F_6N_6OP_2$ (MW 520)

(reagent for carboxylic acid activation, particularly used for peptide synthesis)

Alternate Name: benzotriazol-1-yl-*N*-oxy-tris(pyrrolidino)phosphonium hexafluorophosphate; PyBOP.
Physical Data: mp 156–157 °C.[1,2]
Solubility: soluble in CH_2Cl_2, $CHCl_3$, DMF, DMSO, NMP, THF, CH_3CN, and acetone.
Form Supplied in: colorless crystals.
Analysis of Reagent Purity: ^{31}P NMR; ^{1}H NMR.
Preparative Methods: commercially available. Prepared by addition of Cl_2 or Br_2 to a dichloromethane solution of tripyrrolidino phosphine,[1] or addition of tripyrrolidino phosphine oxide to a solution of $POCl_3$,[1] or triphosgene,[3] then addition of HOBt/TEA and pour into a KPF_6 water solution. Alternatively, by reaction of HOBt/NEt_3 with PyCloP (chlorotripyrrolydinophosphonium hexafluorophosphate) or PyBroP (bromotripyrrolydinophosphonium hexafluorophosphate).[1]
Purification: recrystallization is done using CH_2Cl_2/Et_2O.
Handling, Storage, and Precautions: prolonged storage: ≤ + 8 °C; keep cool and dry. Risk phrases: R 36/37/38; safety phrase: S 24/25. The toxicological properties of this reagent have not yet been determined.[4]

Introduction. Treatment of a carboxylic acid with PyBOP in the presence of base produces the unstable acylphosphonium intermediate. This species reacts immediately with a nucleophile or decomposes into the corresponding benzotriazolyl ester reactive intermediates (eq 1).[5–7] This reagent was designed for coupling in peptide synthesis [1,2] and is also used for the synthesis of amide and amide-like linkages, as well as for macrocyclization and for the synthesis of esters and nitriles. PyBOP gives results comparable, or even better than those obtained with BOP [(1*H*-benzotriazol-1-yloxy)tris(dimethylamino)phosphonium hexafluorophosphate],[2,8–10] with the advantage that the carcinogen HMPA is not produced as a side product of the coupling reaction. In numerous applications PyBOP is preferred to the corresponding uronium or aminium salts to avoid undesired guanidine formation between the coupling reagent and a primary amine.[9] PyBOP is preferably used without preactivation in a one-pot reaction. The formation of pyrrolidide derivatives has been observed due to the presence of small amounts of pyrrolidine (0.14%) in the commercial phosphonium salt and can be easily avoided by recrystallization of the reagent before use.[11]

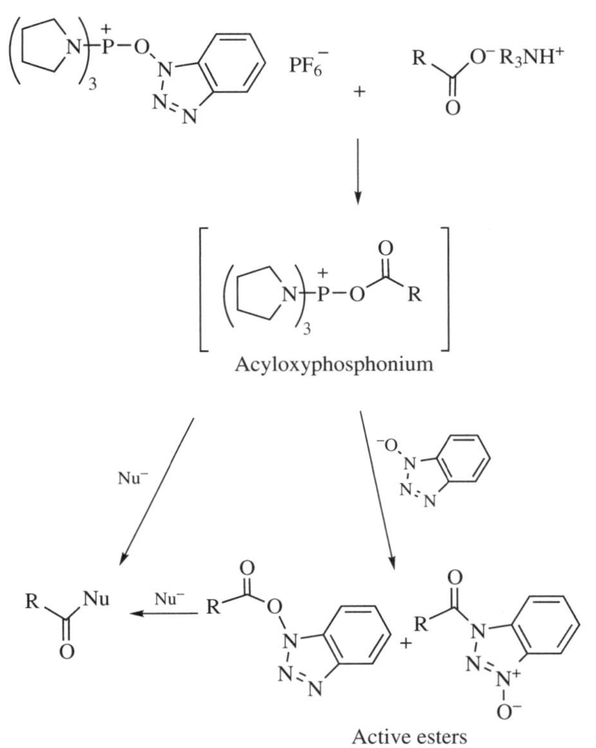

(1)

Reagent for Peptide Synthesis. PyBOP has been widely used as a classic peptide coupling reagent giving epimerization-free peptides in excellent yields and short reaction times, in solution- (eq 2)[2] or solid-phase peptide synthesis employing manual,[2,12]

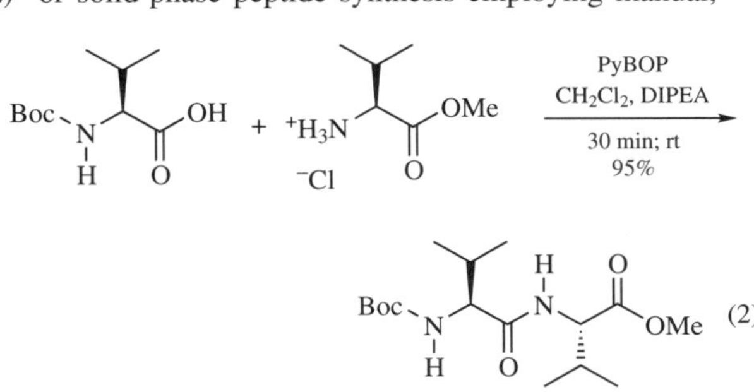

(2)

automated[13,14] or pin[15] methods. It has been noticed that repetitive couplings without purification is possible in solution synthesis[16] and that the stock solutions used for automated solid-phase peptide synthesis have good stability.[9] With PyBOP, extra HOBt is sometimes used.[13,14,17,18] Like most coupling reagents, PyBOP can lead to epimerization when used for fragment or fragment-like condensation,[2,10] or in difficult coupling reactions.[5,18] Therefore, in fragment condensation, addition of HOBt,[10] $CuCl_2$[19] or working at low temperature[18] generally reduces epimerization; the opposite effect has also been observed by addition of HOBt for the difficult acylation of a sulfonamide support. In that case, the limited reactivity of the OBt-ester is probably the explanation.[18] *N*-methyl[5] or α,α-disubstituted[20] amino acids can be successfully coupled and microwave irradiation can improve Aib coupling.[21] However, in the case of such difficult coupling reactions, halogenophosphonium (see BroP, PyBroP, PyCloP) or uronium (see PyClU) reagents, acyl fluorides; and OAt- or substituted OBt-based (see HATU, PyAOP) coupling reagents can be more effective.[17,22]

PyBOP is also efficient in the synthesis of oligonucleotide-peptide conjugates,[23] PNAs (polyamides or peptide nucleic acids)[24] or peptide porphyrins.[25]

Synthesis of Amides. PyBOP has been used extensively in the preparation of non-substituted amides (eq 3),[26] and *N*-monosubstituted amides[27,28] as, for example, in the determination of the absolute configuration of α,α or β,β-disubstituted carboxylic acids (eq 4).[29] It has also been used to synthesize *N*-monosubstituted amides of conjugated acids and/or aromatic amines (eq 5)[30,31] and di-substituted amides (eq 6),[32] as well as Weinreb amides (eq 7).[33,34] PyBOP has also been used for attachment of carboxylic acids to immobilized amino groups[35–37] for synthesis of libraries,[28,35] or oligonucleotide conjugates[38] on resins, and to prepare dendrimers in solution.[39]

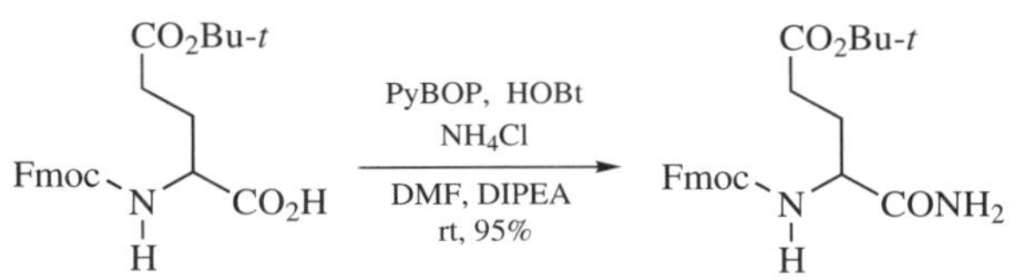

(3)

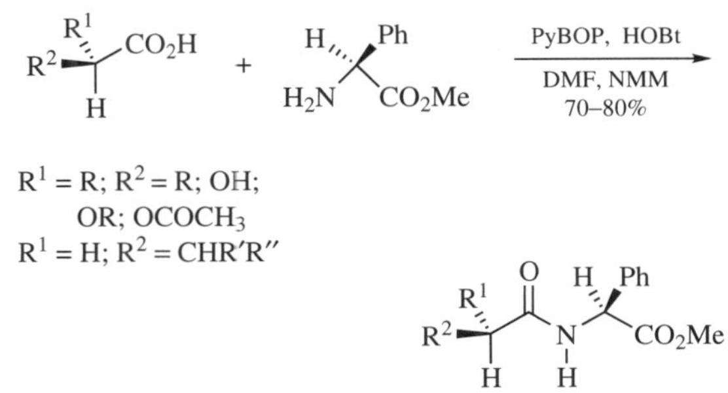

(4)

Macrolactamization. PyBOP also gives good results when used for the synthesis of macrolactams (eq 8),[40] including cyclic peptides in solution or on solid phase.[41,42]

Formation of Amide-like linkages. PyBOP has been used (1) to activate monothiocarboxylic acids to prepare thioamides,[43] (2) to prepare the sulfonamide 'safety-catch' linker in high yield and without racemization (eq 9),[18] and (3) to prepare *N*-(methoxyphenylacetyl)sulfoximines to determine the absolute configuration of sulfoxides (eq 10).[44]

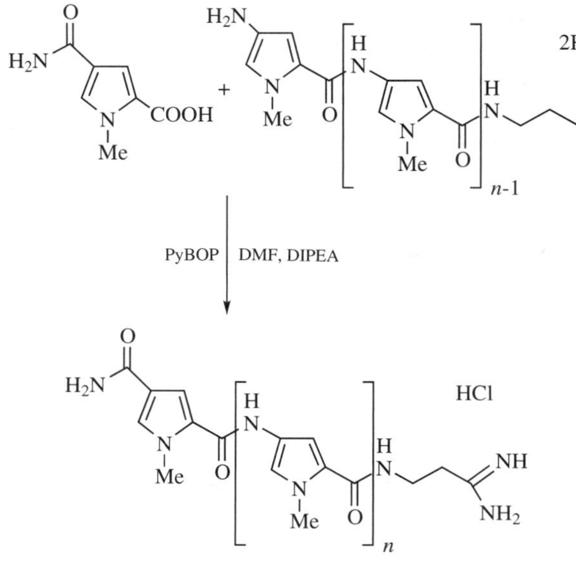

(5)

$n = 2$, 63%
$n = 3$, 59%

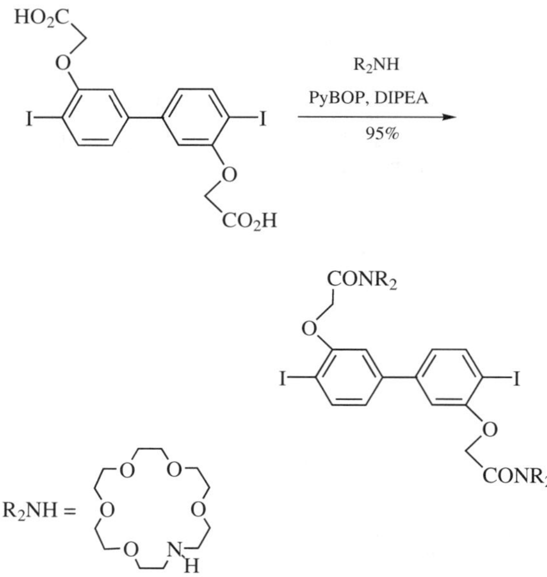

(6)

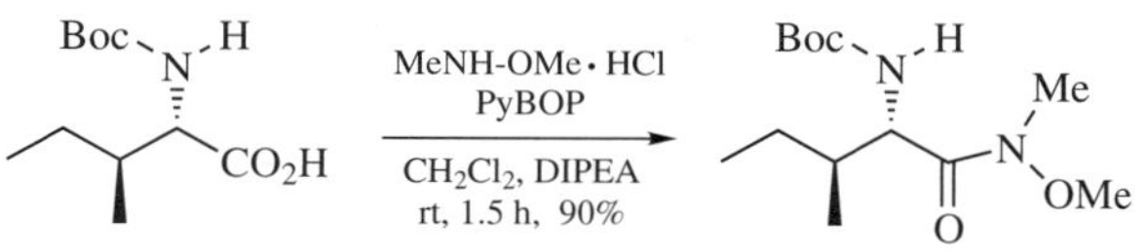

(7)

PyBOP has also been used as a mild, selective reagent for the conversion of primary amines to phthalimides (eq 11),[45] and to obtain excellent yields of 3,5-pyrazolidinediones (eq 12).[46]

Synthesis of carboxylic Esters. Esters from PyBOP-mediated carboxylic acid activation are reported for primary [6,47] and secondary alcohols,[6,47,48] with addition of DMAP[49] or 1-methylimidazole (eq 13).[48] The reagent can also be used to attach carboxylic acids to resins through an ester linkage.[50]

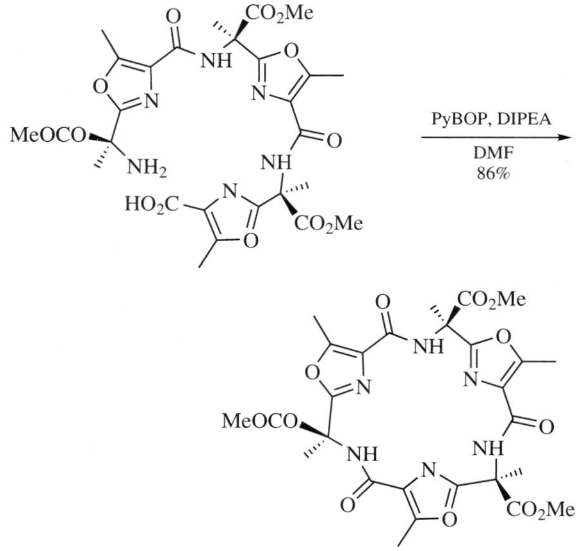

(8)

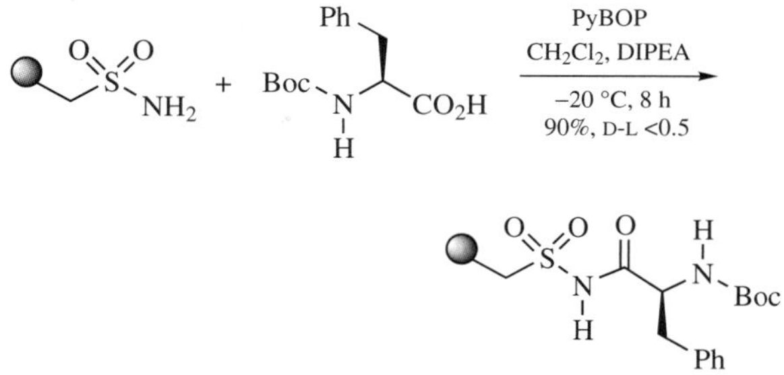

(9)

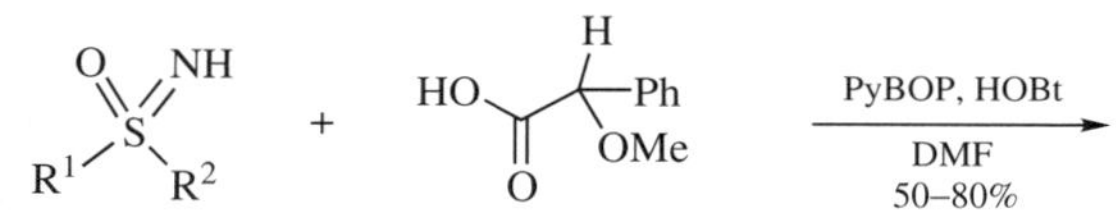

$R^1 = CH_2Ph$
R^2 = Pr, Bu, *i*-Pr, octyl

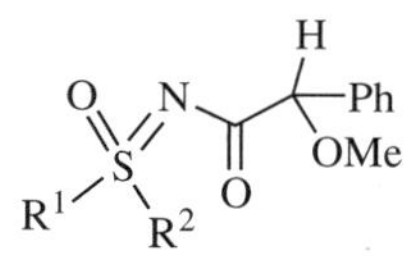

(10)

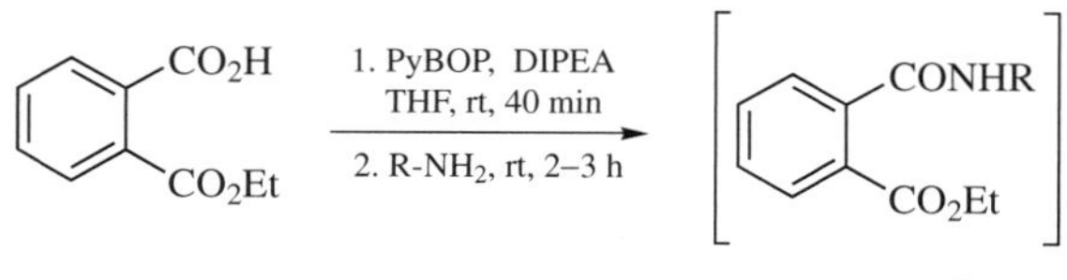

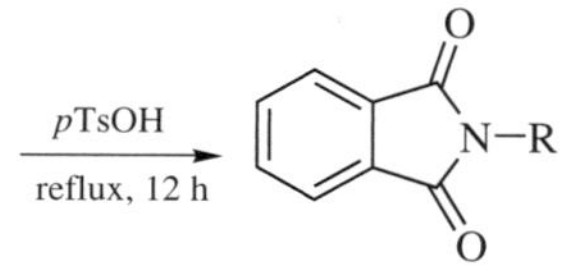

(11)

R = *n*-butyl: 77%
R = 1-phenylethyl: 87%
R = *t*-butyl: 67%

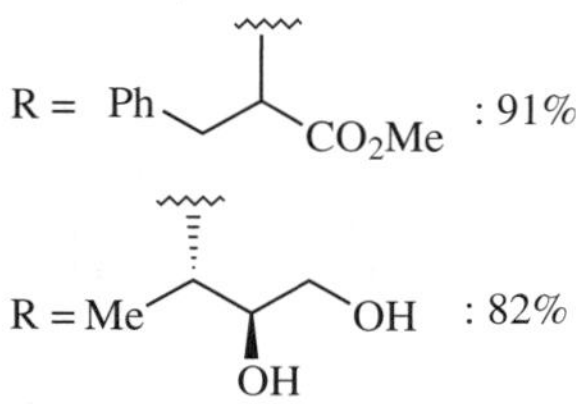

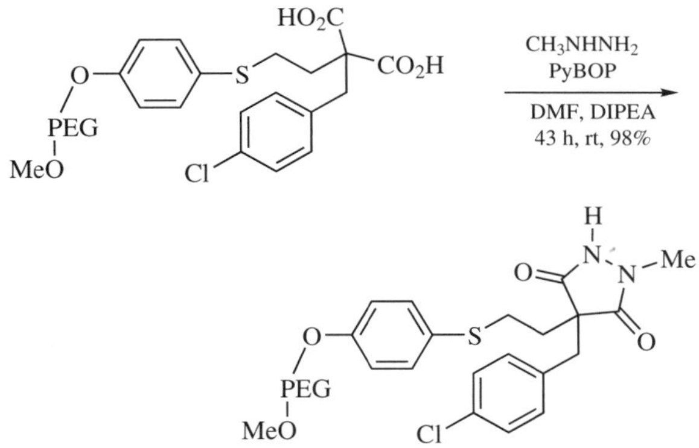

(12)

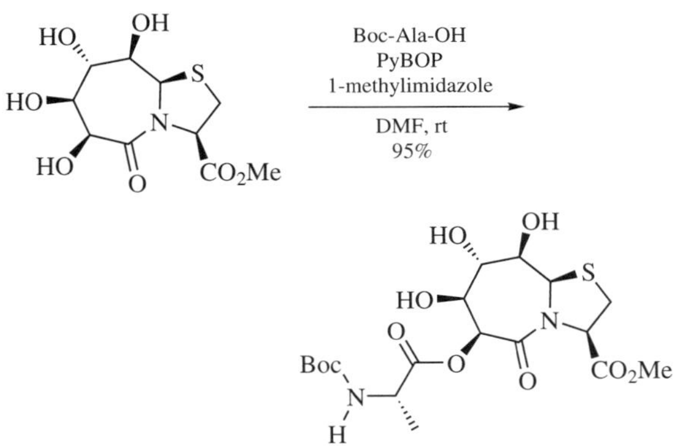

(13)

Synthesis of phosphoesters. PyBOP has been used for esterification of *H*-phosphonic acid (eq 14).[51] The addition of *N*-methylimidazole accelerates the reaction in the synthesis of oligodeoxyribonucleotides.[8] Esters of phosphinic acid,[52] mixed phosphonates from monophosphonates and primary or secondary alcohols (eq 15),[47] as well as cyclic phosphonates[53] have also been obtained.

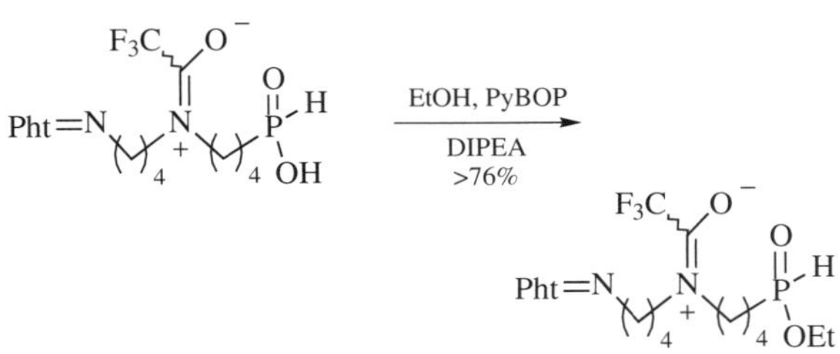

(14)

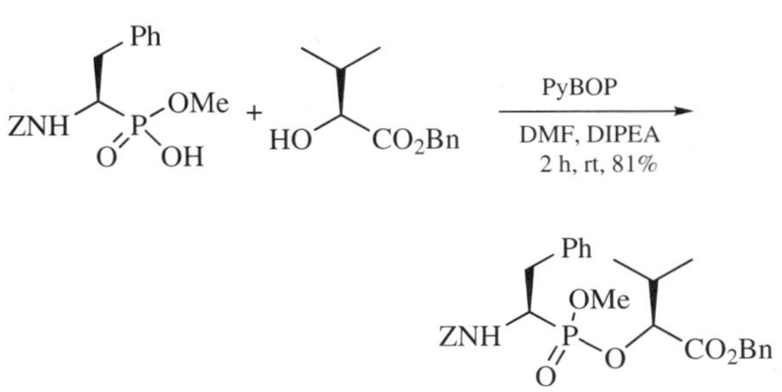

(15)

Synthesis of nitriles. PyBOP is also an excellent activator for converting unsubstituted amides or thioamides to nitriles (eq 16).[54]

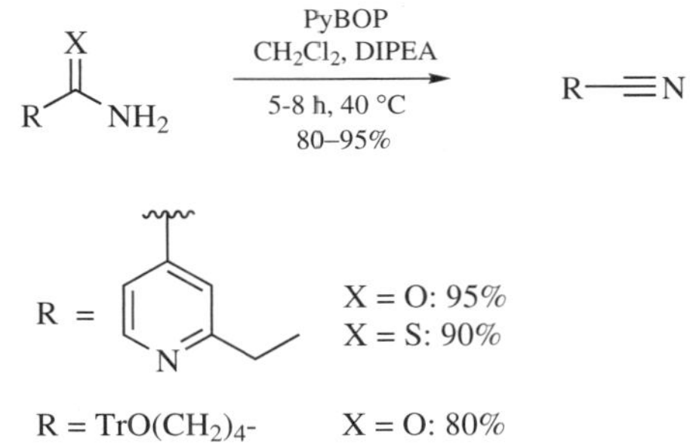

(16)

Related Reagents. AOP [7-azabenzotriazol-1-yloxytris (dimethylamino)phosphonium hexafluorophosphate]; BOI [*O*-(benzotriazol-1-yl)-1, 3-dimethyl-1,3-dimethyleneuronium hexafluorophosphate]; BOP [1*H*-benzotriazol-1-yloxytris(dimethylamino)phosphonium hexafluorophosphate]; BroP [bromotris (dimethylamino)phosphonium hexafluorophosphate]; BTFFH [bis(tetramethylene)fluoroformamidium hexafluorophosphate]; CF_3-PyBOP[[6-(trifluoromethyl)benzotriazol-1-yl]-*N*-oxy-tris (pyrrolidino)phosphonium hexafluorophosphate]; CF_3-NO_2-PyBOP[[4-nitro-6-(trifluoromethyl)benzotriazol-1-yl]-*N*-oxy-tris (pyrrolidino)phosphonium hexafluorophosphate]; HAMDU [*O*-(7-azabenzotriazol-1-yl)-1, 3-dimethyl-1,3-dimethyleneuronium hexafluorophosphate]; HAPyU [*O*-(7-azabenzotriazol-1-yl)-1,1,3,3-bis(tetramethylene)uronium hexafluorophosphate]; HATU [*O*-(7-azabenzotriazol-1-yl)-1,1,3,3-tetramethyluronium hexafluorophosphate]; HBPyU [*O*-(benzotriazol-1-yl)-1,1,3, 3-bis(tetramethylene)uronium hexafluorophosphate]; HBTU [*O*-(benzotriazol-1-yl)-1,1,3,3-tetramethyluronium hexafluorophosphate]; PyAOP [7-azabenzotriazol-1-yloxytripyrrolidinophosphonium hexafluorophosphate]; PyCloP [chlorotripyrrolidinophosphonium hexafluorophosphate]; PyClU [1,1,3,3-bis(tetramethylene)chlorouronium hexafluorophosphate]; PyNOP[[6-nitrobenzotriazol-1-yl]-*N*-oxy-tris(pyrrolidino) phosphonium hexafluorophosphate].

1. International Patent WO 90/10009, (Castro, B.; Coste, J.) (1990).
2. Coste, J.; Le-Nguyen, D.; Castro, B., *Tetrahedron Lett.* **1990**, *31*, 205.
3. Rivero, I. A.; Somanathan, R.; Hellberg, L. H., *Synth. Commun.* **1995**, *25*, 2185.
4. Sheet, M. S. D.; ChemWeb.com; http//www.fishersci.ca/msds.nsf
5. Coste, J.; Frérot, E.; Jouin, P., *J. Org. Chem.* **1994**, *59*, 2437.
6. Coste, J.; Campagne, J.-M., *Tetrahedron Lett.* **1995**, *36*, 4253.
7. Pivonka, D. E.; Palmer, D. L.; Gero, T., *Appl. Spectrosc.* **1999**, *53*, 1027.
8. Wada, T.; Sato, Y.; Honda, F.; Kawahara, S.-I., Sekine, M., *J. Am. Chem. Soc.* **1997**, *119*, 12710.
9. Albericio, F.; Bofill, J. M.; El-Faham, A.; Kates, S. A., *J. Org. Chem.* **1998**, *63*, 9678.
10. Benoiton, N. L.; Lee, Y. C.; Steinaur, R.; Chen, F. M. F., *Int. J. Peptide Protein Res.* **1992**, *40*, 559.
11. Alsina, J.; Barany, G.; Albericio, F.; Kates, S. A., *Lett. Pept. Sci.* **1999**, *6*, 243.
12. Sueiras-Diaz, J., *Protein and Peptide Lett.* **1994**, *1*, 33.
13. Neimark, J.; Briand, J. P., *Pept. Res.* **1993**, *6*, 219.
14. Luu, T.; Pham, S.; Deshpande, S., *Int. J. Peptide Protein Res.* **1996**, *47*, 91.
15. Wang, J.-X.; Bray, A. M.; Dipasquale, A. J.; Maeji, N. J.; Geysen, H. M., *Int. J. Peptide Protein Res.* **1993**, *42*, 384
16. Hoeg-Jensen, T.; Jakobsen, M. H.; Holm, A., *Tetrahedron Lett.* **1991**, *32*, 6387.

17. Albericio, F.; Chinchilla, R.; Dodsworth, D. J.; Najera, C., *Organic Preparations and Procedures Int.* **2001**, *33*, 205.
18. Backes, B. J.; Ellman, J. A., *J. Org. Chem.* **1999**, *64*, 2322.
19. Fischer, P. M.; Solbakken, M.; Undheim, K., *Tetrahedron* **1994**, *50*, 2277.
20. Frérot, E.; Coste, J.; Pantaloni, A.; Dufour, M.-N.; Jouin, P., *Tetrahedron* **1991**, *47*, 259.
21. Santagada, V.; Fiorino, F.; Perissutti, E.; Severino, B.; De Filippis, V.; Vivenzio, B.; Caliendo, G., *Tetrahedron Lett.* **2001**, *42*, 5171.
22. Humphrey, J. M.; Chamberlin, A. R., *Chem. Rev.* **1997**, *97*, 2243.
23. Kahl, J. D.; Greenberg, M. M., *J. Org. Chem.* **1999**, *64*, 507.
24. Uhlmann, E.; Peyman, A.; Breipohl, G.; Will, D. W., *Angew. Chem. Int. Ed. Engl.* **1998**, *37*, 2796.
25. Ushiyama, M.; Yoshino, A.; Yamamura, T.; Shida, Y.; Arisaka, F., *Bull. Chem. Soc. Jpn.* **1999**, *72*, 1351.
26. Wang, W.; McMurray, J. S., *Tetrahedron Lett.* **1999**, *40*, 2501.
27. Tan, D. S.; Foley, M. A.; Shair, M. D.; Schreiber, S. L., *J. Am. Chem. Soc.* **1998**, *120*, 8565.
28. Boger, D. L.; Ducray, P.; Chai, W.; Jiang, W.; Goldberg, J., *Bioorg. Med. Chem. Lett.* **1998**, *8*, 2339.
29. Yabuuchi, T.; Kusumi, T., *J. Org. Chem.* **2000**, *65*, 397.
30. Animati, F.; Arcamone, F. M.; Conte, M. R.; Felicetti, P.; Galeone, A.; Lombardi, P.; Mayol, L.; Paloma, L. G.; Rossi, C., *J. Med. Chem.* **1995**, *38*, 1140.
31. Dudouit, F.; Goossens, J.-F., Houssin, R.; Hénichart, J.-P.; Colson, P.; Houssier, C.; Gelus, N.; Bailly, C., *Bioorg. Med. Chem. Lett.* **2000**, *10*, 553.
32. Robert, F.; Winum, J.-Y.; Sakai, N.; Gerard, D.; Matile, S., *Org. Lett.* **2000**, *2*, 37.
33. Roux, F.; Maugras, I.; Poncet, J.; Niel, G.; Jouin, P., *Tetrahedron* **1994**, *50*, 5345.
34. Bluet, G.; Campagne, J.-M., *Synlett* **2000**, 221.
35. Nielsen, J.; Brenner, S.; Janda, K. D., *J. Am. Chem. Soc.* **1993**; *115*, 9812.
36. Brown, B. B.; Wagner, D. S.; Geysen, H. M., *Mol. Div.* **1995**, *1*, 4.
37. Souers, A. J.; Virgilio, A. A.; Rosenquist, A.; Fenuik, W.; Ellman, J. A., *J. Am. Chem. Soc.* **1999**, *121*, 1817.
38. Kahl, J. D.; Greenberg, M. M., *J. Am. Chem. Soc.* **1999**, *121*, 597.
39. Kim, R. M.; Manna, M.; Hutchins, S. M.; Griffin, P. R.; Yates, N. A.; Bernick, A. M.; Champman, K. T., *Proc. Natl. Acad. Sci. USA* **1996**, *93*, 10012.
40. Haberhauer, G.; Somogyi, L.; Rebek, J., Jr, *Tetrahedron Lett.* **2000**, *41*, 5013.
41. Brugghe, H. F.; Timmermans, H. A. M.; Van Unen, L. M. A.; Ten Hove, G. J.; Van de Werken, G.; Poolman, J. T.; Hoogerhout, P., *Int. J. Pept. Protein Res.* **1994**, *43*, 166.
42. Boger, D. L.; Miyazaki, S.; Kim, S. H.; Wu, J. H.; Castle, S. L.; Loiseleur, O.; Jin Q., *J. Am. Chem. Soc.* **1999**, *121*, 10004.
43. Le, H.-T.; Gallard, J.-F.; Mayer, M.; Guittet, E.; Michelot, R., *Bioorg. Med. Chem.* **1996**, *4*, 2201.
44. Yabuuchi, T.; Kusumi, T., *J. Am. Chem. Soc.* **1999**, *121*, 10646.
45. Aguilar, N.; Moyano, A.; Pericas, M. A.; Riera, A., *Synthesis* **1998**, 313.
46. Zhao, X.-Y.; Metz, W. A.; Sieber, F.; Janda, K. D., *Tetrahedron Lett.* **1998**, *39*, 8433.
47. Campagne, J.-M.; Coste, J.; Jouin P., *J. Org. Chem.* **1995**, *60*, 5214.
48. Geyer, A.; Moser, F., *Eur. J. Org. Chem.* **2000**, 1113.
49. Yoshimura, S.; Otsuka, T.; Takase, S.; Okamoto, M.; Okada, S.; Hemmi, K., *J. Antibiot.* **1998**, *51*, 655.
50. Hoemann, M. Z.; Melikian-Badalian, A.; Kumaravel, G.; Hauske, J. R., *Tetrahedron Lett.* **1998**, *39*, 4749.
51. Chen, S. J.; Lin, C.-H.; Walsh, C. T.; Coward, J. K., *Bioorg. Med. Chem. Lett.* **1997**, *7*, 505.
52. Campagne, J.-M.; Coste, J.; Guillou, L.; Heitz, A.; Jouin, P., *Tetrahedron Lett.* **1993**, *34*, 4181.
53. Hillman, J. M. L.; Roberts, S. M., *J. Chem. Soc., Perkin Trans. 1* **1997**, 3601.
54. Bose, D. S.; Narsaiah, A. V., *Synthesis* **2001**, 373.

Jacques Coste & Patrick Jouin
CNRS-UPR9023, Montpellier, France

O-Benzotriazol-1-yl-*N,N,N′,N′*-tetramethyluronium Hexafluorophosphate

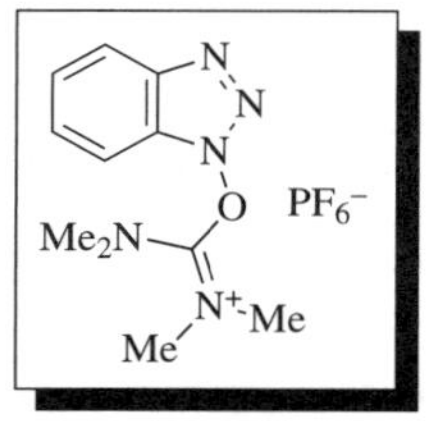

[94790-37-1] $C_{11}H_{16}F_6N_5OP$ (MW 379.29)

(coupling reagent for peptide synthesis[1])

Alternate Name: HBTU
Physical Data: mp 206–207 °C (dec).
Solubility: sol DMF (0.5 mol L^{-1}). With the tetrafluoroborate counterion the corresponding salt (TBTU) is slightly more soluble (0.6 mol L^{-1}).
Form Supplied in: white solid; widely available.
Purification: by crystallization in a mixture of MeCN and CH_2Cl_2.
Handling, Storage, and Precautions: very stable, not hygroscopic, and can be stored indefinitely. Solutions in DMF (0.45 M) can be stored in an inert atmosphere for weeks. Syntheses of peptides carried out with freshly prepared, 6 and 13 week stored solutions show similar quality of the crude product.[3] Violent decomposition can occur when dried at elevated temperature.[4]

Both *O*-benzotriazol-1-yl-*N,N,N′,N′*-tetramethyluronium hexafluorophosphate (HBTU) and tetrafluoroborate (TBTU) salts have been used as condensing reagents for the preparation of peptides in both solution and solid-phase strategies.[1,4] In solution, reaction of α-amino protected amino acids or dipeptides with equimolar amounts of α-carboxyl protected amino acids or dipeptides and Diisopropylethylamine (DIEA) (2 equiv) in DMF proceeds cleanly for 15 min at 0 °C, yielding the corresponding peptides in good yields (80–96%) (eq 1).[1,4]

$$R^1CO_2H + R^2NH_2 \xrightarrow[\text{DIEA}\ 80\text{–}96\%]{\text{HBTU}} R^1CONHR^2 \quad (1)$$

HBTU activation has been adapted for automated stepwise solid-phase peptide synthesis for both *t*-butoxycarbonyl (Boc) and fluorenylmethoxycarbonyl (Fmoc) strategies.[5] For the former, a simple, effective protocol has been developed, which involves simultaneous in situ neutralization with coupling (extra equivalents of base are required). This protocol is particularly suitable for assembling complex peptides, arising from sequence-dependent peptide chain aggregation. Since aggregation occurs when protonated α-ammonium peptide-resin intermediates are neutralized, simultaneous neutralization and acylation can help to overcome this phenomenon.[6] For Fmoc chemistry, efficient protocols are also available for both batch and continuous-flow systems.[3,7]

Various protected peptide segments have been coupled on solid-phase using a combination of TBTU, HOBt, and DIEA for the syntheses of several small proteins. Coupling yields, using three-fold excess of peptide segment and coupling times of ~12 h, are greater than 95%.[8]

Racemization from uronium salt mediated couplings has been determined by analysis of the epimeric products by HPLC using different models. In all cases, racemization is similar or lower than that obtained with carbodiimide and benzotriazole based phosphonium salt methods.[1,2,4,9]

In the absence of the carboxylic component, HBTU reacts with amino groups leading to the formation of a Schiff base (eq 2).[10] Thus in syntheses conducted in solution, the excess of both HBTU and amino component should be avoided. In both solution and solid-phase strategies the sequence of reagent addition is critical. HBTU should be delivered to the carboxylic component for preactivation, prior to the addition of the amine. The Schiff base formation can also occur during slow reactions, such as the preparation of cyclic peptides, where both amino and carboxylic components are in equimolar amounts and an excess of the uronium salt can block the amino group.[11] For the synthesis of cyclic peptides the phosphonium derivatives Benzotriazol-1-yloxytris(dimethylamino)phosphonium Hexafluorophosphate (BOP)[12] and benzotriazol-1-yloxytris(pyrrolidino)phosphonium hexafluorophosphate (PyBOP),[13] can be more useful.

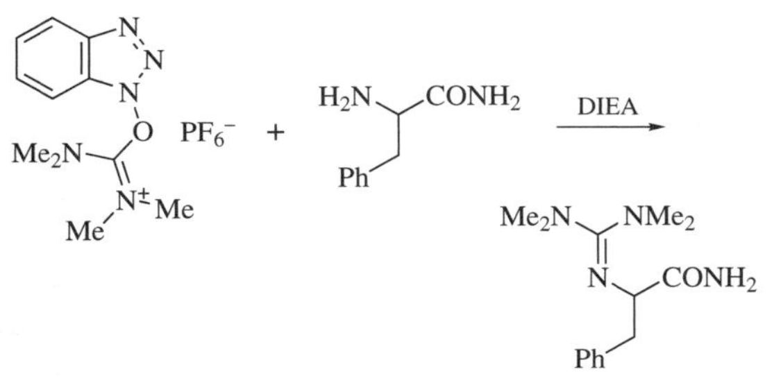

(2)

Uronium salts have been used for cyclization of linear peptides in both solution and solid-phase modes. Since this reaction is sequence dependent, there are no general conditions. Thus cyclization in solution with uronium salt methods give, for some peptides, better results than the classical reagent Diphenyl Phosphorazidate (DPPA),[14] but for another case the converse is true.[15] Furthermore, dehydration of *C*-terminal aspartylamide peptides during cyclization with HBTU has been described. This side-reaction can be prevented by the addition of one equivalent of HOBt.[16] Although good results have been obtained in the solid-phase,[17] guanidino formation side-reactions have also been reported.[11]

Recently, other uronium salts, such as *O*-(*N*-succinimidyl)-*N,N,N′,N′*-tetramethyluronium tetrafluoroborate (TSTU),[4,18] *O*-(*N*-5-norbornene-*endo*-2,3-dicarboximidyl)-*N,N,N′,N′*-tetramethyluronium tetrafluoroborate (TNTU),[4,18] *O*-(2-oxo-1(2*H*)-pyridyl)-*N,N,N′,N′*-bis(pentamethylene)uronium tetrafluoroborate (TOPPipU),[19] N-[(Dimethylamino)-1 H-1,2,3-triazolo[4,5-b] pyridin-1-ylmethylene]-N-methylmethanaminium Hexafluorophosphate N-Oxide (HATU),[20] and *O*-(7-azabenzotriazol-1-yl)-*N,N,N′,N′*-bis(tetramethylene)uronium hexafluorophosphate (HAPyU)[21] have been described and are commerically available. The uronium salts derived from 7-aza-1-hydroxybenzotriazole (HATU and HAPyU) have been shown to be superior to their benzotriazole analogs in terms of coupling efficiency,[20,21] racemization,[22] and cyclization,[23] in both solution and solid-phase strategies.

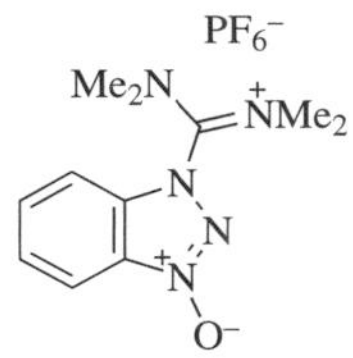

(3)

Finally, an X-ray structure determination of HBTU revealed that the solid-state structure differs considerably from the formulation commonly presented in the literature. The solid-state structure is not the *N,N,N′,N′*-tetramethyluronium salt but rather the guanidinium *N*-oxide (**3**).[24]

1. Dourtoglou, V.; Ziegler, J. C.; Gross, B. *Tetrahedron Lett.* **1978**, 1269.
2. Dourtoglou, V.; Gross, B.; Lambropoulou, V.; Zioudrou, C. *Synthesis* **1984**, 572.
3. Fields, C. G.; Lloyd, D. H.; Macdonald, R. L.; Otteson, K. M.; Noble, R. L. *Peptide Res.* **1991**, *4*, 95.
4. Knorr, R.; Trzeciak, A.; Bannwarth, W.; Gillessen, D. *Tetrahedron Lett.* **1989**, *30*, 1927.
5. Fields, G. B.; Tian, Z.; Barany, G. In *Synthetic Peptides: A User's Guide*, Grant, G. A. Ed.; Freeman:New York, **1992**; pp 77–183.
6. (a) Schnoelzer, M.; Alewood, P; Jones, A.; Alewood, D.; Kent, S. B. H. *Int. J. Peptide Protein Res.* **1992**, *40*, 180. (b) Reid, G. E.; Simpson, R. J. *Anal. Biochem.* **1992**, *200*, 301. (c) Sueiras-Diaz, J.; Horton, J. *Tetrahedron Lett.* **1992**, *33*, 2721.
7. Poulat, F.; Guichard, G.; Gozé, C.; Heitz, F.; Calas, B.; Berta, P. *FEBS Lett.* **1992**, *309*, 385.
8. Surovoy, A.; Metzger, J.; Jung, G. In *Innovation and Perspectives in Solid-Phase Synthesis: Peptides, Polypeptides and Oligonucleotides*, Epton, R. Ed.; Intercept:Andover, **1992**, pp 467–473 (*Chem. Abstr.* **1993**, *118*, 234 476s).
9. Benoiton, N. L.; Lee, Y. C.; Steinauer, R.; Chen, F. M. F. *Int. J. Peptide Protein Res.* **1992**, *40*, 559.
10. Gausepohl, H.; Pieles, U.; Frank, R. W. In *Proceedings of the Twelth American Peptide Symposium*; Smith, J. A.; Rivier, J. E. Eds.; ESCOM:Leiden, **1992**; pp 523–524. (*Chem. Abstr.* **1992**, *117*, 171 981j).
11. Story, S. G.; Aldrich, J. V. *Int. J. Peptide Protein Res.* **1994**, *43*, 292.
12. Castro, B.; Dormoy, J. R.; Evin, G. *Tetrahedron Lett.* **1975**, 1219.
13. Coste, J.; Le-Nguyen, D.; Castro, B. *Tetrahedron Lett.* **1990**, *31*, 205.
14. (a) Hoffmann, E.; Beck-Sickinger, A. G.; Jung, G. *Justus Liebigs Ann. Chem./Liebigs Ann. Chem.* **1991**, 585. (b) Zimmer, S.; Hoffmann, E.; Jung, G.; Kessler, H *Justus Liebigs Ann. Chem./Liebigs Ann. Chem.* **1993**, 497.
15. Schmidt, R.; Neubert, K. *Int. J. Peptide Protein Res.* **1991**, *37*, 502.
16. Rovero, P.; Pegoraro, S.; Bonelli, F.; Triolo, A. *Tetrahedron Lett.* **1993**, *34*, 2199.
17. (a) Trzeciak, A.; Bannwarth, W. *Tetrahedron Lett.* **1992**, *33*, 4557. (b) Neugebauer, W.; Willick, G. In *Proceedings of the Twenty-Second European Peptide Symposium*; Schneider, C. H.; Eberle, A. N. Eds.; ESCOM:Leiden, **1993**; pp 393–394.

18. Bannwarth, W.; Schmidt, D.; Stallard, R. L.; Hornung, C.; Knorr, R.; Müller, F. *Helv. Chim. Acta* **1988**, *71*, 2085.
19. Henklein, P.; Beyermann, M.; Bienert, M.; Knorr, R. In *Proceedings of the Twenty-First European Peptide Symposium*; Giralt, E.; Andreu, D. Eds.; ESCOM:Leiden, **1991**; pp 67–68 (*Chem. Abstr.* **1991** *115*, 232 811m) .
20. Carpino, L. A. *J. Am. Chem. Soc.* **1993**, *115*, 4397.
21. Carpino, L. A.; El-Faham, A.; Minor, C. A.; Albericio, F. *J. Chem. Soc., Perkin Trans. 1* **1994**, 201.
22. Carpino, L. A.; E.-Faham, A. *J. Org. Chem.* **1994**, *59*, 695.
23. Ehrlich, A.; Rothemund, S.; Brudel, M.; Beyermann, M.; Carpino, L. A.; Bienert, M. *Tetrahedron Lett.* **1993**, *34*, 4781.
24. Abdelmoty, I.; Albericio, F.; Carpino, L. A.; Foxman, B. M.; Kates, S. A. *Lett. Peptide Sci.* **1994**, *1*, 57.

Fernando Albericio & Steven A. Kates
Millipore Corporation, Bedford, MA, USA

O-Benzotriazol-1-yl-*N,N,N',N'*-tetramethyluronium Tetrafluoroborate

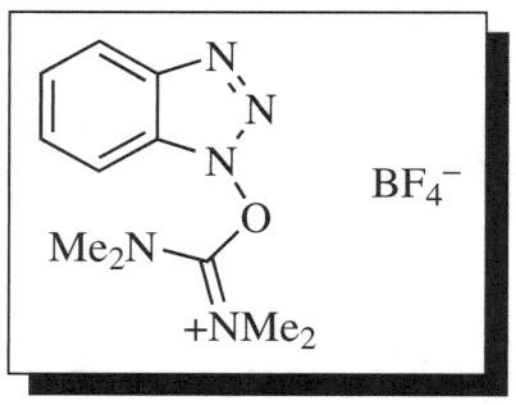

[125700-67-6] $C_{11}H_{16}BF_4N_5O$ (MW 321.09)

(reagent used for peptide coupling)

Alternate Name: TBTU.
Physical Data: mp 205 °C.
Solubility: soluble in DMF (0.6 mol L^{-1}).
Form Supplied in: white solid; widely available.
Purification: can be purified by crystallization from a mixture of MeOH and *i*-PrOH.
Handling, Storage, and Precautions: very stable, not hygroscopic and can be stored more than 2 years. Violent decomposition can occur when dried at elevated temperature.[1]

Peptide Coupling Reagent. *O*-Benzotriazol-1-yl-*N,N,N',N'*-tetramethyluronium tetrafluoroborate (TBTU) and hexafluorophosphate (HBTU) have been found to behave similarly in solution and in solid-phase peptide synthesis.[1,2,3] This reagent gives lower racemization in solid-phase than in solution-phase[1] and has been implemented in automated stepwise peptide synthesis on different resins by means of Fmoc/*t*-Bu[4] and Boc/Bn[5] strategies employing a combination of TBTU:HOBt:DIEA (1:1:1.5, threefold excess) in DMF. Thus, coupling yields are higher than 95% by using coupling times of ca. 12 h.[6] The protected segments 74-99 of HIV 1-protease have been prepared using TBTU:NMM and Fmoc/*t*-Bu strategy on TentaGel resin.[7] However, in conventional Fmoc solid-phase peptide synthesis (SPPS) with TBTU:HOBt: DIEA, considerable racemization has been detected for Fmoc-Cys (Trt) residue, which requires a base-free activation step.[8]

The preparation of cyclic peptides is focused on the total synthesis of natural products with important biological activities, being carried out both in solid- or solution-phase by a lactamization process with condensing reagents.[9] Thus, in a comparison study of different reagents during the cyclization of hexapeptides, TBTU was 5 to 70 times faster than diphenylphosphoryl azide (DPPA) or HBTU.[4a,4b] The solid-phase head-to-tail cyclization using TBTU:DIEA activation gave quantitative yields for cyclic peptides containing Asp, Asn or Glx residues.[10] The solution-phase macrolactamization of a cyclopeptide analog of neuropeptide Y has been carried out with TBTU:HOBt:DIEA under high dilution conditions.[4a] TBTU has shown to be faster than diisopropylcarbodiimide (DIC):HOBt in the synthesis of tyrosine-containing cyclic 10 to 12-membered cyclopeptides.[11] For the solid-phase cyclization of the sponge origin cyclopeptide stylopeptide 1, TBTU: HOBt:DIEA in dichloromethane gave higher yields than HBTU, *N,N*-bis(2-oxo-3-oxazolidinyl)phosphinic chloride (BOP-Cl) or diethyl phosphorocyanidate (DEPC).[12] Moreover, 1-hydroxy-7-azabenzotriazole (HOAt) containing uronium or phosphonium salts and TBTU were used as coupling reagents in the cyclization of thymopentin-derived pentapeptides, the later giving lower degrees of epimerization on the *C*-terminal Tyr residue.[13] The combination TBTU:HOAt afforded better yields in the cyclization of the (40*R*,45*R*)-isomer of the marine metabolite trunkamide A than the usually used reagent *O*-(7-azabenzotriazol-1-yl)-*N,N,N',N'*-tetramethyluronium hexafluorophosphate (HATU).[14] Another selected application of TBTU was in the total synthesis of the sponge origin cyclotheonamide B.[15] This coupling reagent was used in the convergent synthesis of the segments Pro-hArg-D-Phe and vinylogous Tyr-Dpr to give the linear pentapeptide, as well as in its lactamization step under dilution conditions (eq 1). In addition, for the synthesis of the antineoplastic constituent of marine sponge phakellistatin 2 (cyclo-Tyr-Pro-Phe-Pro-Ile-Ile-Pro), the final cyclization has been performed using several coupling reagents, such as TBTU, BOP-Cl, PyBroP [bromotris(pyrrolidino) phosphonium hexafluorophosphate], and TBTU:HOAt, the former giving the highest yield (55%).[16] However, in the case of phakellistatin 5 (cyclo-Phe-Asn-Ala-Met-Ala-Ile-Pro), the cyclization step gave better results with HATU, whereas only a 15% yield was obtained employing TBTU:DIEA.[17] Cyclic hexapeptides composed by alternating L-proline and substituted 3-aminobenzoic acid subunits, able to bind monosaccharides, have been prepared by means of TBTU:DIEA.[18]

In the solution-phase synthesis of sterically hindered aminoisobutyric acid (Aib) containing thiopeptides, TBTU has been employed as coupling reagent in the presence of HOBt and DIEA at room temperature (eq 2).[19]

Amidation. In the preparation of a library of the tumor-promoting teleocidin, the synthesis of the N^{13}-des(methyl)indolactam V core structure has been carried out by means of TBTU: HOBt in *N*-methylpyrrolidone (NMP) in solution-phase and in solid-phase (eq 3).[20] In the synthesis of seco forms of the peptide metabolite diazonamide A, a potent tumor-growth inhibitor, the amide bond of the ring and the valine appendage have been incorporated by means of the combination TBTU:DIEA.[21]

Peptide nucleic acid (PNA) oligomers replace the oligonucleotide backbone of DNA with a poly[*N*-(2-aminoethyl)glycine] backbone. For solid-phase synthesis of dithiasuccinoyl

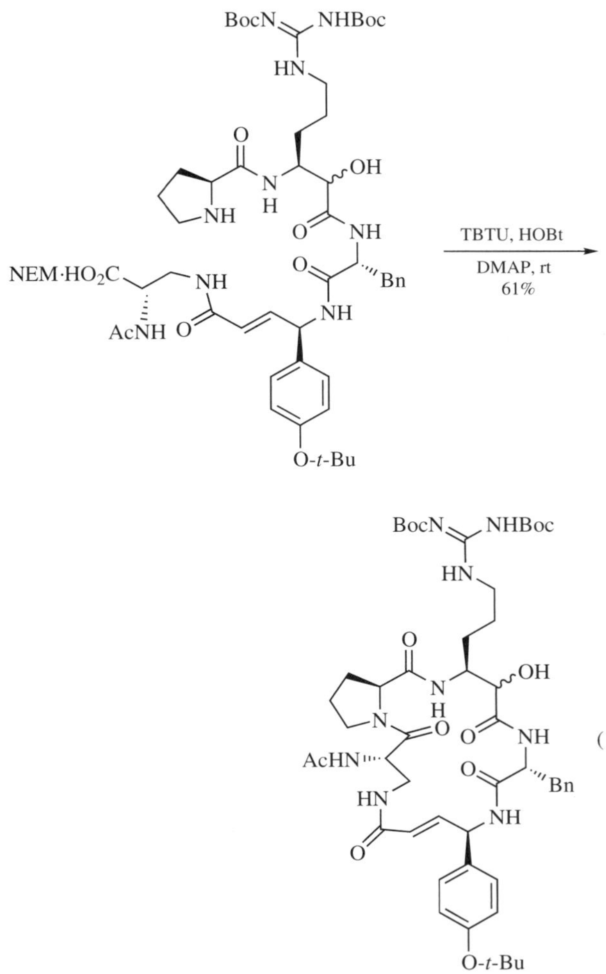

(1)

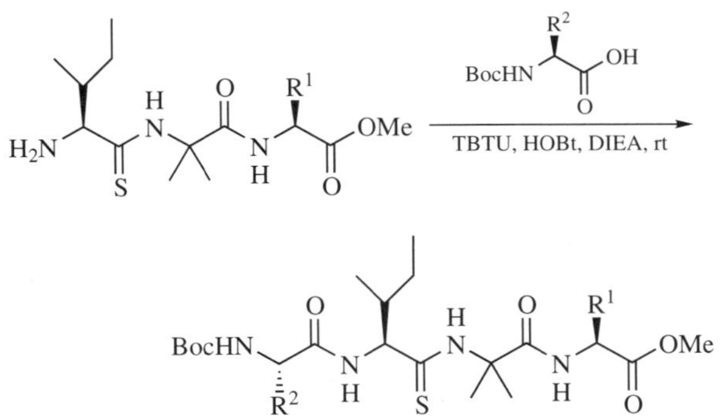

(2)

$R^1 = R^2 = H$; 50 min, 76%
$R^1 = R^2 = Bn$; 2 h, 73%

(Dts)-protected PNA oligomers, the corresponding monomer has been incorporated by means of HBTU in NMP, whereas TBTU gave similar results by reducing DIEA:coupling reagent ratio to 1:3.[22] The TBTU:TEA combination is a very fast reagent (10 min) for the anchoring of nucleoside-3′-*O*-succinates to a widely used alkylamine-controlled pore glass (LCAA-CPG) for solid-phase synthesis of oligonucleotides.[23] Another resin for solid-phase *C*-glycopeptide synthesis, polyethylene glycol polyacrylamide (PEGA), has been derivatized with *p*-(hydroxymethyl)benzoic acid linker by means of TBTU:NEM (*N*-ethylmorpholine), which has also been employed in the incorporation step of the unprotected *C*-glycosyl azido acid building block.[24]

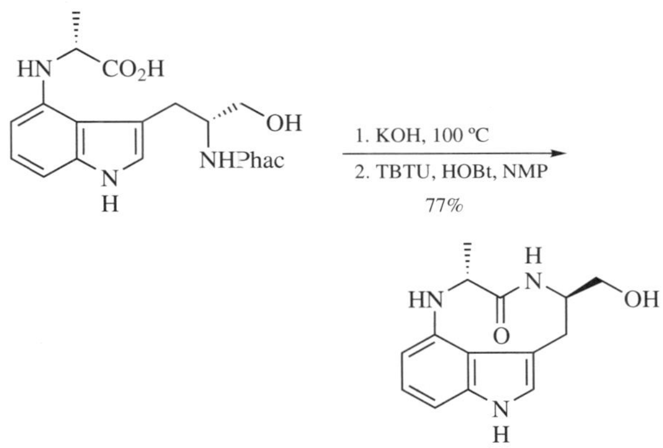

(3)

Iron(III) hexadentate 3-hydroxy-2(1*H*)pyridinone ligands have been prepared under solution-phase conditions by coupling of its benzyl ether with tris(2-aminoethyl)amine (TREN), 1,3,5-tris(aminomethyl)benzene (TRAM) or 1,3,5-tris(methylaminomethyl)benzene (TRMAM) in the presence of TBTU and NMM as base.[25] Moreover, (*E*)-*N*-isopropyl-5-tosyl-4-pentenamide, a precursor of a δ-acyldienyl anion equivalent, has been prepared in 80% yield from the corresponding acid by means of TBTU, whereas it was obtained in 70% overall yield through the acyl chloride generated with oxalyl chloride.[26] The tosylated crotonamide (*E*)-*N*-isobutyl-4-tosyl-2-butenamide (**1**) has been obtained in 80% yield from the acid by means of TBTU, but only in 50% by using the acyl chloride.[27] In the preparation of the chiral auxiliaries (*R*)- and (*S*)-α-aminoisovalerophenone hydrochlorides (**2**), which have been used for the construction of pyrazinone alanine[28a] and glycine[28b] templates, their *N*,*N*-dimethylamides of *N*-Boc-D- and L-valine precursors have been synthesized by using TBTU.

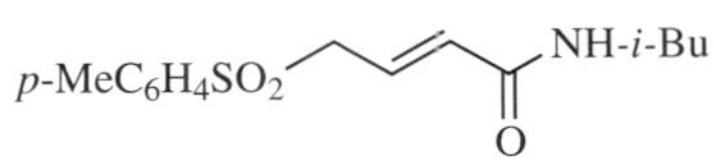

1

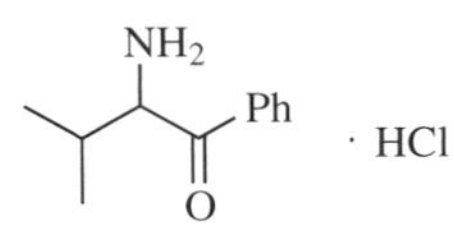

2

Weinreb amides can be prepared in high yield and without purification by treatment of carboxylic acids with TBTU:DIEA and MeONHMe-HCl in acetonitrile at room temperature.[29] However, this coupling with 1-[(3-dimethylamino)propyl]-3-ethylcarbodiimide hydrochloride (WSC or EDC) gave lower yields of the corresponding hydroxamates, chromatographic purification being necessary.[29a]

A new application of TBTU is the use as cleavage reagent for tetrahydropyranyl, silyl, and 4,4′-dimethoxytrityl ethers in a mixture of acetonitrile and water. THP and DMT ethers can be removed in the presence of the TBDMS group.[30]

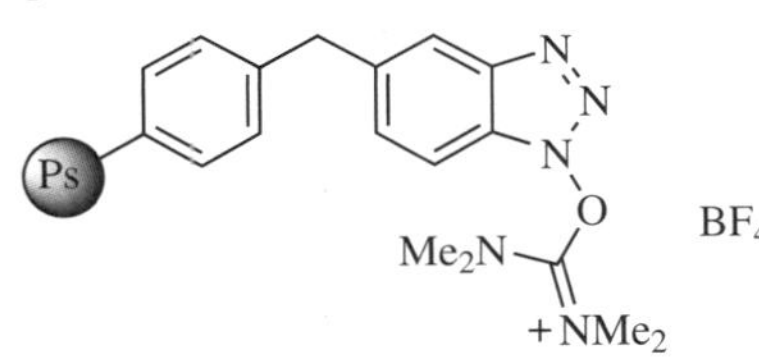

3

The polymer-supported TBTU (PS-TBTU, **3**) has been prepared[31] from polystyrene-supported 1-hydroxybenzotriazole (PS-HOBt)[32] and has been used for di and tripeptide synthesis with pyridine as base, even in wet acetonitrile. Yields are similar or slightly lower than those using TBTU, lower racemization ratios than with HATU being obtained.

Related Reagents. HBTU.

1. Knorr, R.; Trzeciak, A.; Bannwarth, W.; Gillessen, D., *Tetrahedron Lett.* **1989**, *30*, 1927.
2. Albericio, F.; Bates, S. A., *Encyclopedia of Reagents for Organic Synthesis*; Paquette, L. A., Ed., J. Wiley & Sons: Chichester, 1995, Vol. 1, p 304.
3. (a) Albericio, F.; Carpino, L. A., *Methods Enzymol.* **1997**, *289*, 104. (b) Humphrey, J. M.; Chamberlin, A. R., *Chem. Rev.* **1997**, *97*, 2243. (c) Albericio, F.; Kates, S. A.; In *Solid-Phase Synthesis, A Practical Guide*; Kates, S. A.; Albericio, F., Eds; Marcel Dekker: New York, 2000, p 275. (d) Albericio, F.; Chinchilla, R.; Dodsworth, D. J.; Nájera, C., *Org. Prep. Proced. Int.* **2001**, *33*, 203.
4. (a) Hoffmann, E.; Beck-Sickinger, A.; Jung, G., *Liebigs Ann. Chem.* **1991**, 585. (b) Zimmer, S.; Hoffmann E.; Jung, G.; Kessler, H., *Liebigs Ann. Chem.* **1993**, 497. (c) Frank, R.; Jakob, M.; Thunecke, F.; Fischer, G.; Schutkowski, M., *Angew. Chem. Int. Ed.* **2000**, *39*, 1120.
5. Sueiras-Diaz, J.; Horton J., *Tetrahedron Lett.* **1992**, *33*, 2721.
6. Surovoy, A.; Metzger, J.; Jung, G., In *Innovation and Perspectives in Solid-Phase Synthesis: Peptides, Polypeptides, and Oligonucleotides*; Epton, R., Ed.; Intercept: Andover, 1992; pp 467–473.
7. Henkel, B.; Zhang, L.; Goldammer, C.; Bayer, E., *Z. Naturforsch.* **1996**, *51b*, 1339.
8. Kaiser, T.; Nicholson, G. J.; Kohlbau, H. J.; Voelter, W., *Tetrahedron Lett.* **1996**, *37*, 1187.
9. Lambert, J. N.; Mitchell, J. P.; Roberts, K. D., *J. Chem. Soc., Perkin Trans. 1* **2001**, 471.
10. Trzeciak A.; Bannwarth, W., *Tetrahedron Lett.* **1992**, *33*, 4557.
11. Cabrele, C.; Langer, M.; Beck-Sickinger, A. G., *J. Org. Chem.* **1999**, *64*, 4353.
12. Pettit, G. R.; Taylor, S. R., *J. Org. Chem.* **1996**, *61*, 2322.
13. Ehrlich, A.; Heyne, H.-U.; Winter, R.; Beyermann, M.; Haber, H.; Carpino, L. A.; Bienert, M., *J. Org. Chem.* **1996**, *61*, 8831.
14. Wipf, P.; Uto, Y., *J. Org. Chem.* **2000**, *65*, 1037.
15. Bastiaans, H. M. M.; Van der Baan, J. L.; Ottenheijm, H. C. J., *J. Org. Chem.* **1997**, *62*, 3880.
16. Pettit, G. R.; Rhodes, M. R.; Tan, R., *J. Nat. Prod.* **1999**, *62*, 409.
17. Pettit, G. R.; Toki, B. E.; Xu, J. P.; Brune, D. C., *J. Nat. Prod.* **2000**, *63*, 22.
18. Bitta, J.; Kubik S., *Org. Lett.* **2001**, *3*, 2637.
19. (a) Lehmann, J.; Linden, A.; Heimgartner, H., *Tetrahedron* **1999**, *55*, 5359. (b) Lehmann, J.; Linden, A.; Heimgartner, H., *Helv. Chim. Acta* **1999**, *82*, 888.
20. Meseguer, B.; Alonso-Díaz, D.; Griebenow, N.; Herget, T.; Waldmann, H., *Chem. Eur. J.* **2000**, *6*, 3943.
21. (a) Jeong, S.; Chen, X.; Harran, P. G., *J. Org. Chem.* **1998**, *63*, 8640. (b) Chen, X.; Esser, L.; Harran, P. G., *Angew. Chem. Int. Ed.* **2000**, *39*, 937. (c) Li, J.; Chen, X.; Burgett, A. W. G.; Harran, P. G., *Angew. Chem. Int. Ed.* **2001**, *40*, 2682.
22. Planas, M.; Bardají E.; Jensen, K. J.; Barany, G., *J. Org. Chem.* **1999**, *64*, 7281.
23. Gupta, K. C.; Kumar, P.; Bhatia, D.; Sharma, A. K., *Nucleosides & Nucleotides* **1995**, *14*, 829.
24. Tedebark, U.; Meldal, M.; Panza, L.; Bock, K., *Tetrahedron Lett.* **1998**, *39*, 1815.
25. Rai, B. L.; Khodr, H.; Hider, R. C., *Tetrahedron* **1999**, *55*, 1129.
26. (a) Caturla, F.; Nájera, C., *Tetrahedron Lett.* **1996**, *37*, 2833. (b) Caturla, F.; Nájera C., *Tetrahedron* **1998**, *54*, 11255.
27. Caturla, F.; Nájera, C., *Tetrahedron Lett.* **1996**, *37*, 4787.
28. (a) Nájera, C.; Abellán, T.; Sansano, J. M., *Eur. J. Org. Chem.* **2000**, 2809. (b) Abellán, T.; Mancheño, B.; Nájera, C.; Sansano, J. M., *Tetrahedron* **2001**, *57*, 6627.
29. (a) Mannekens, E.; Tourwé D.; Lubell, W. D., *Synthesis* **2000**, 1214. (b) Yus, M.; Radivoy, G.; Alonso, F., *Synthesis* **2001**, 914.
30. Ramasamy, K. S.; Averett, D., *Synlett* **1999**, 709.
31. Chinchilla, R.; Dodsworth, D. J.; Nájera, C.; Soriano, J. M., *Tetrahedron Lett.* **2000**, *41*, 2463.
32. Kalir, R.; Warshawsky, A.; Fridkin, M.; Patchornik, A., *Eur. J. Biochem.* **1975**, *59*, 55.

Carmen Nájera
Universidad de Alicante, Spain

Benzyl Bromide

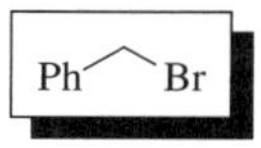

[100-39-0] C_7H_7Br (MW 171.04)

(benzylating agent for a variety of heteroatomic functional groups as well as carbon nucleophiles)

Physical Data: mp −3 to −1 °C; bp 198–199 °C; *d* 1.438 g cm^{-3}.
Solubility: sol ethereal, chlorinated, and dipolar aprotic solvents.
Form Supplied in: 98–99% pure liquid.
Handling, Storage, and Precautions: the reagent is a potent lachrymator and should be handled in a fume hood.

Benzylation of Heteroatomic Functional Groups. Benzylation of various heteroatomic functional groups is readily achieved with this reagent under a variety of conditions and finds widespread application in organic synthesis, primarily as a protecting group.[1]

Alcohols and phenols are benzylated upon treatment with benzyl bromide under basic conditions. For example, treatment of alcohols with Sodium Hydride or Potassium Hydride in ethereal solvent[2] or DMF[3] generates alkoxides, which subsequently undergo Williamson reactions with benzyl bromide. Selective benzylation of a primary alcohol in the presence of a secondary alcohol has been accomplished in DMF at low temperature.[4]

Benzylation of alcohols using Potassium Fluoride–Alumina and benzyl bromide in acetonitrile at room temperature is effective.[5] Silver oxide in DMF is yet another base system.[6] Of particular interest in carbohydrate applications is the reaction of benzyl bromide with carbohydrate derivatives which have been pretreated with tin reagents. Thus it is possible to benzylate an equatorial alcohol in the presence of an axial alcohol (eq 1)[7] and also to selectively benzylate an anomeric hydroxy through Di-n-butyltin Oxide.[8]

In some instances the sluggish reactivity of sterically hindered alcohols toward benzyl bromide may be overcome through addition of a catalytic iodide source such as Tetra-n-butylammonium

Iodide, which generates the more reactive benzyl iodide in situ (see Benzyl Iodide). Benzylation of phenols proceeds well under the conditions described for aliphatic alcohols. Owing to the greater acidity of phenols it is possible to use weaker bases such as Potassium Carbonate for these reactions.[9]

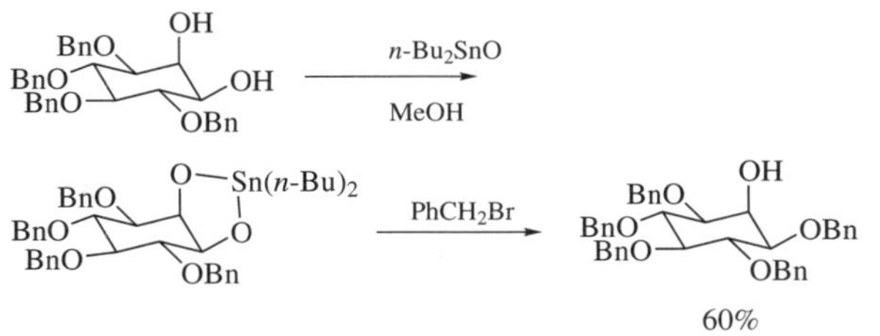

(1)

Benzyl bromide will readily alkylate amino groups. Reactions are normally carried out in the presence of additional base and dibenzylation of primary amines is usually predominant.[10] Selective quaternization of a less hindered tertiary amine in the presence of a more hindered tertiary amine has been described.[11] Amide and lactam nitrogens can be benzylated under basic conditions,[12] as can those of sulfonamides[13] and nitrogen heterocycles.[14]

Thiols,[15] silyl thioethers,[16] and thiosaccharins[17] may be benzylated with benzyl bromide under basic conditions. Thus L-cysteine is *S*-benzylated under basic conditions (eq 2).[18] Benzylation of selenols is likewise possible.[19] A synthesis of benzylic sulfones is possible using Benzenesulfonyl Chloride and Sodium O,O-Diethyl Phosphorotelluroate with benzyl bromide.[20]

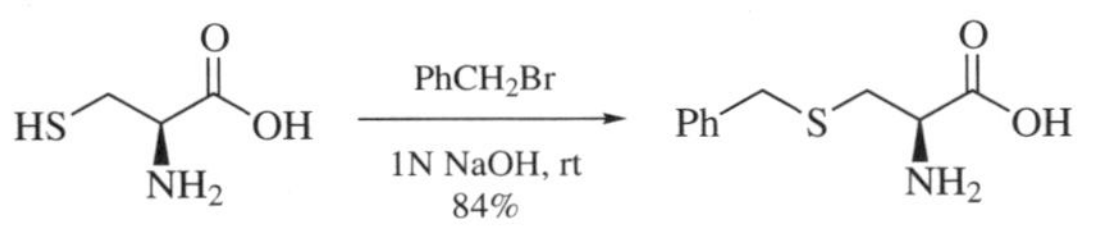

(2)

Although the preparation of benzyl carboxylate esters from benzyl bromide and carboxylate anions is not the most common route to these compounds, the reaction is possible when carried out in DMF[20] or using zinc carboxylates.[21]

Nucleophilic attack on benzyl bromide by cyanide and azide anions is feasible with ion-exchange resins or with the corresponding salts.[22]

Reactions with Active Methylene Compounds. Enolates of ketones,[23] esters,[24] enediolates,[25] 1,3-dicarbonyl compounds,[26] amides and lactams,[27] as well as nitrile-stabilized carbanions,[28] can be alkylated with benzyl bromide. Cyclohexanone may be benzylated in 92% ee using a chiral amide base.[29] Amide bases as well as alkoxides have been employed in the case of nitrile alkylations.[28b] Benzylation of metalloenamines may be achieved[30a] and enantioselective reactions are possible using a chiral imine (eq 3).[30b] However, reactions between benzyl bromide and enamines proceed in low yield.[31] The benzylation of a ketone via its enol silyl ether, promoted by fluoride, has been observed.[32]

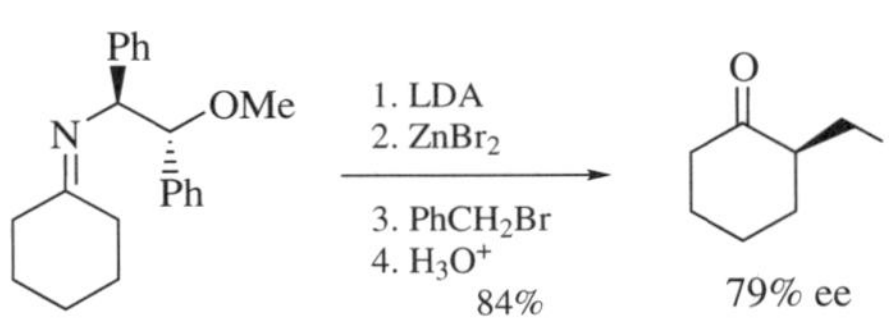

(3)

Reactions with Metals and Organometallics. Difficulties encountered in the preparation of benzylic metal compounds with active metals are due primarily to the tendency of these compounds to undergo Wurtz coupling (self condensation).[33] Benzylmagnesium bromide may nevertheless be prepared from benzyl bromide and used under standard[34] or Barbier conditions.[35] Benzyllithium cannot be obtained practically from benzyl bromide. Benzylzinc bromide and the cyanocuprate BnCu(CN)ZnBr have both been prepared. The cuprate undergoes 1,2-additions with aldehydes and ketones.[36]

The propensity of benzyl bromide to undergo coupling with organometallic reagents may be used to advantage, as organolithiums,[37] Grignard reagents,[38] organocuprates,[39] organocadmiums,[40] organochromiums,[41] and organoiron reagents[42] are all known to give coupling products. An interesting coupling of benzyl bromide with *N*-methylphthalimide under dissolving metal conditions has been reported (eq 4).[43]

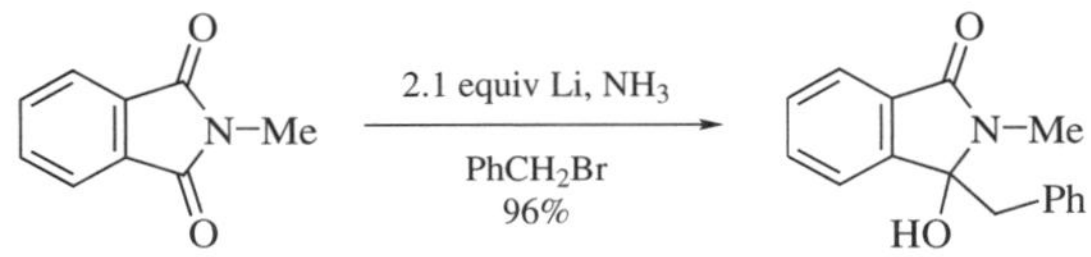

(4)

1. (a) Greene, T. W.; Wuts, P. G. M. *Protective Groups in Organic Synthesis*, 2nd ed.; Wiley:New York, **1991**. (b) *Protective Groups in Organic Chemistry*; McOmie, J. F. W., Ed.; Plenum:New York, **1973**.
2. Nicolaou, K. C.; Pavia, M. R.; Seitz, S. P. *J. Am. Chem. Soc.* **1981**, *103*, 1224.
3. Hanessian, S.; Liak, T. J.; Dixit, D. M. *Carbohydr. Res.* **1981**, *88*, C14.
4. Fukuzawa, A.; Sato, H.; Masamune, T. *Tetrahedron Lett.* **1987**, *28*, 4303.
5. Ando, T.; Yamawaki, J.; Kawate, T.; Sumi, S.; Hanafusa, T. *Bull. Chem. Soc. Jpn.* **1982**, *55*, 2504.
6. Kuhn, R.; Low, I.; Trischmann, H. *Ber. Dtsch. Chem. Ges./Chem. Ber.* **1957**, *90*, 203.
7. (a) Nashed, M. A.; Anderson, L. *Tetrahedron Lett.* **1976**, 3503. (b) Cruzado, C.; Bernabe, M.; Martin-Lomas, M. *J. Org. Chem.* **1989**, *54*, 465.
8. Bliard, C.; Herczegh, P.; Olesker, A.; Lukacs, G. *Carbohydr. Res.* **1989**, *8*, 103.
9. Schmidhammer, H.; Brossi, A. *J. Org. Chem.* **1971**, *93*, 746.
10. (a) Yamazaki, N.; Kibayashi, C. *J. Am. Chem. Soc.* **1989**, *111*, 1396. (b) Gray, B. D.; Jeffs, P. W. *Chem. Commun./J. Chem. Soc., Chem. Commun.* **1987**, 1329.
11. (a) Chung, B.-H.; Zymalkowski, F. *Arch. Pharm. (Weinheim, Ger.)* **1984**, *317*, 307. (b) Chung, B.-H.; Zymalkowski, F. *Arch. Pharm. (Weinheim, Ger.)* **1984**, *317*, 323.
12. (a) Landini, D.; Penso, M. *Synth. Commun.* **1988**, *18*, 791. (b) Staskun, B. *J. Org. Chem.* **1979**, *44*, 875. (c) Sato, R.; Senzaki, T.; Goto, T.; Saito, M. *Bull. Chem. Soc. Jpn.* **1986**, *59*, 2950.
13. Bergeron, R. J.; Hoffman, P. G. *J. Org. Chem.* **1979**, *44*, 1835.
14. Chivikas, C. J.; Hodges, J. C. *J. Org. Chem.* **1987**, *52*, 3591.
15. Harpp, D. N.; Kobayashi, M. *Tetrahedron Lett.* **1986**, *27*, 3975.
16. Ando, W.; Furuhata, T.; Tsumaki, H.; Sekiguchi, A. *Synth. Commun.* **1982**, *12*, 627.
17. Yamada, H.; Kinoshita, H.; Inomata, K.; Kotake, H. *Bull. Chem. Soc. Jpn.* **1983**, *56*, 949.
18. Dymicky, M.; Byler, D. M. *Org. Prep. Proced. Int.* **1991**, *23*, 93.
19. Mitchell, R. H. *Chem. Commun./J. Chem. Soc., Chem. Commun.* **1974**, 990.

20. Huang, X.; Pi, J.-H. *Synth. Commun.* **1990**, *20*, 2291.
21. (a) Comber, M. F.; Sargent, M. V.; Skelton, B. W.; White, A. H. *J. Chem. Soc., Perkin Trans. 1* **1989**, 441. (b) Shono, T.; Ishige, O.; Uyama, H.; Kashimura, S. *J. Org. Chem.* **1986**, *51*, 546.
22. (a) Gordon, M.; Griffin, C. E. *Chem. Ind. (London)* **1962**, 1019. (b) Hassner, A.; Stern, M. *Angew. Cheml* **1986**, *98*, 479. (c) Bram, G.; Loupy, A.; Pedoussaut, M. *Bull. Soc. Claim. Fr., Part 2* **1986**, 124. (d) Ravindranath, B.; Srinivas, P. *Tetrahedron* **1984**, *40*, 1623.
23. (a) Gall, M.; House, H. O. *Org. Synth., Coll. Vol.* **1988**, *6*, 121. (b) Sato, T.; Watanabe, T.; Hayata, T.; Tsukui, T. *Chem. Commun./J. Chem. Soc., Chem. Commun.* **1989**, 153.
24. (a) Seebach, D.; Estermann, H. *Tetrahedron Lett.* **1987**, *28*, 3103. (b) Lerner, L. M. *J. Org. Chem.* **1976**, *41*, 2228.
25. Duhamel, L.; Poirier, J.-M. *Bull. Soc. Claim. Fr., Part 2* **1982**, 297.
26. (a) Berry, N. M.; Darey, M. C. P.; Harwood, L. M. *Synthesis* **1986**, 476. (b) Bassetti, M.; Cerichelli, G.; Floris, B. *Gazz. Chim. Ital.* **1986**, *116*, 583. (c) Asaoka, M.; Miyake, K.; Takei, H. *Chem. Lett.* **1975**, 1149. (d) Ogura, K.; Yahata, N.; Minoguchi, M.; Ohtsuki, K.; Takahashi, K.; lida, H. *J. Org. Chem.* **1986**, *51*, 508.
27. (a) Woodbury, R. P.; Rathke, M. W. *J. Org. Chem.* **1977**, *42*, 1688. (b) Klein, U.; Sucrow, W. *Ber. Dtsch. Chem. Ges./Chem. Ber.* **1977**, *110*, 1611. (c) Meyers, A. I.; Harre, M.; Garland, R. *J. Am. Chem. Soc.* **1984**, *106*, 1146.
28. (a) Arseniyadis, S.; Kyler, K. S.; Watt, D. S. *Org. React.* **1984**, *31*, 1. (b) Cope, A. C.; Holmes, H. L.; House, H. O. *Org. React.* **1957**, *9*, 107.
29. Murakata, M.; Nakajima, M.; Koga, K. *Chem. Commun./J. Chem. Soc., Chem. Commun.* **1990**, 1657.
30. (a) Stork, G.; Dowd, S. R. *Org. Synth., Coll. Vol.* **1988**, *6*, 526. (b) Saigo, K.; Kashahara, A.; Ogawa, S.; Nohira, H. *Tetrahedron Lett.* **1983**, *24*, 511.
31. (a) *Enamines: Synthesis, Structure, and Reactions*, 2nd ed.; Cook, A. G., Ed.; Dekker:New York, **1988**. (b) Brannock, K. C.; Burpitt, R. D. *J. Org. Chem.* **1961**, *26*, 3576. (c) Opitz, G.; Hellmann, H.; Mildenberger, M.; Suhr, H. *Justus Liebigs Ann. Chem./Liebigs Ann. Chem.* **1961**, *649*, 36.
32. (a) Kuwajima, I.; Nakamura, E. *J. Am. Chem. Soc.* **1975**, *97*, 3257. (b) Binkley, E. S.; Heathcock, C. H. *J. Org. Chem.* **1975**, *40*, 2156.
33. Wakefield, B. J. *Organolithium Methods*; Academic:New York, **1988**.
34. (a) Kharasch, M. S.; Reinmuth, O. *Grignard Reactions of Nonmetallic Substances*; Constable:London, **1954**. (b) Reuvers, A. J. M.; van Bekkum, H.; Wepster, B. M. *Tetrahedron* **1970**, *26*, 2683.
35. Blomberg, C.; Hartog, F. A. *Synthesis* **1977**, 18.
36. Berk, S. C.; Knochel, P.; Yeh, M. C. P. *J. Org. Chem.* **1988**, *53*, 5789.
37. (a) Hirai, K.; Matsuda, H.; Kishida, Y. *Tetrahedron Lett.* **1971**, 4359. (b) Hirai, K.; Kishida, Y. *Tetrahedron Lett.* **1972**, 2743. (c) Villieras, J.; Rambaud, M.; Kirschleger, B.; Tarhouni, R. *Bull. Soc. Claim. Fr., Part 2* **1985**, 837.
38. Rahman, M. T.; Nahar, S. K. *J. Organomet. Chem.* **1987**, *329*, 133.
39. (a) Kobayashi, Y.; Yamamoto, K.; Kumadaki, I. *Tetrahedron Lett.* **1979**, 4071. (b) Furber, M.; Taylor, R. J. K.; Burford, S. C. *Tetrahedron Lett.* **1985**, *26*, 3285.
40. (a) Emptoz, G.; Huet, F. *Bull. Soc. Claim. Fr., Part 2* **1974**, 1695.
41. Wellmann, J.; Steckhan, E. *Synthesis* **1978**, 901.
42. (a) Sawa, Y.; Ryang, M.; Tsutsumi, S. *J. Org. Chem.* **1970**, *35*, 4183. (b) Cookson, R. C.; Farquharson, G. *Tetrahedron Lett.* **1979**, 1255. (c) Sawa, Y.; Ryang, M.; Tsutsumi, S. *Tetrahedron Lett.* **1969**, 5189.
43. Flynn, G. A. *Chem. Commun./J. Chem. Soc., Chem. Commun.* **1980**, 862.

William E. Bauta
Sandoz Research Institute, East Hanover, NJ, USA

Benzyl Chloroformate[1]

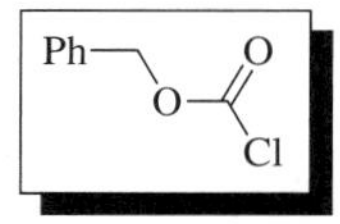

[501-53-1] $C_8H_7ClO_2$ (MW 170.60)

(protecting agent for many functional groups, especially for the N-protection of amino acids;[7] activates pyridine rings toward nucleophilic attack;[53] reagent for the preparation of other benzyloxycarbonylating agents[3])

Alternate Name: benzyloxycarbonyl chloride (CbzCl).
Physical Data: colorless oily liquid, bp 103 °C/20 mmHg (with slow decarboxylation to benzyl chloride at this temperature); bp 85–87 °C/7 mmHg; mp 0 °C; *d* (20 °C) 1.195 g cm^{-3}; n_D^{20} 1.5190.
Form Supplied in: widely available.
Preparative Methods: CbzCl can be readily prepared by the reaction of benzyl alcohol and phosgene, either in toluene solution[4] or neat.[1b,5,6] Purification of such freshly prepared CbzCl is often not necessary; indeed, ca. 25% residual toluene solvent does not interfere with derivatization reactions.[4]
Purification: commercial CbzCl is typically better than 95% pure and is often contaminated by benzyl chloride, benzyl alcohol, toluene, and HCl.[2,3] On storage it can undergo a slow HCl-catalyzed decomposition, even at low temperature.[1a,3] It has been recommended[1b,2] that CbzCl which has been stored for long periods be purified by flushing with a stream of dry air to remove dissolved CO_2 and HCl impurities. Filtration and storage over Na_2SO_4 may be followed by distillation; it is important that low temperatures (oil bath ≤85 °C)[1b] and high vacuum be used to minimize thermal decomposition.[3]
Handling, Storage, and Precautions: CbzCl is highly toxic and is a cancer suspect agent; it is a lachrymator with an acrid odor and should be handled with caution in a fume hood.

Protection of Amines. The most widespread application of CbzCl is in the protection of primary and secondary amines as the corresponding benzyl carbamates. These reactions are usually performed either under Schotten–Baumann conditions,[7] by using aq $NaHCO_3$ or aq Na_2CO_3 as base in an organic solvent (eq 1),[8] or in the presence of an organic base (typically Et_3N) in CH_2Cl_2.[9] The isolation and/or crystallization of amines is often facilitated by Cbz derivatization of the crude amine generated in another reaction.[10] A wide variety of primary and secondary aliphatic amines, including aziridines,[11] azetidines,[12] and other sensitive[7,9] and electron-deficient[8c] amines, can be selectively protected in the presence of alcohol,[8b,13] phenol,[14] and thiol[15] functionality; however, the selective protection of hindered secondary amines over secondary alcohols has proved difficult.[16]

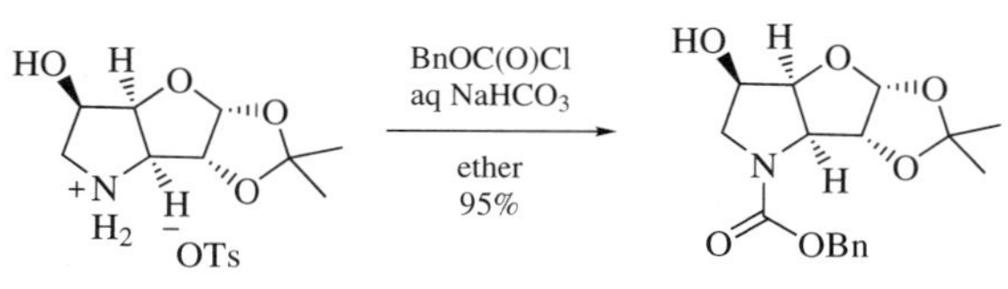

(1)

The N-protection of amino acids during peptide synthesis constitutes a major application of this methodology.[1,8a] References to the N-protection of individual amino acids using CbzCl have been tabulated.[1b,e] Use of the *N*-Cbz protecting group generally affords crystalline amino acid derivatives and suppresses racemization at the α-stereocenter during peptide bond formation. The *N*-Cbz group is stable to conditions used for the formation of peptide bonds. Benzyl carbamates are highly complementary to *t*-butyl (i.e. Boc) carbamates in peptide synthesis (see t-Butyl Azidoformate). Schotten–Baumann conditions are most commonly employed for the preparation of *N*-Cbz amino acids (eq 2);[1b,4] racemization at the α stereocenter is suppressed under these basic conditions by the adjacent negatively charged carboxylate group. The use of a lower reaction temperature (−20 °C) has been shown in one case to afford a higher yield of pure *N*-Cbz derivative.[17] One disadvantage of the standard Schotten–Baumann protocol is the need for either the simultaneous addition of CbzCl and aq NaOH, or the sequential addition of small aliquots of these reagents, throughout the course of the reaction. A modified procedure[1b] using $NaHCO_3$ as the base[8d] avoids this inconvenience. A recent report indicates that *N*-Cbz derivatization of a range of α-amino acids can be conveniently achieved, albeit in only moderate yield, by refluxing with CbzCl in ethyl acetate without added base; interestingly, no racemization was observed.[18]

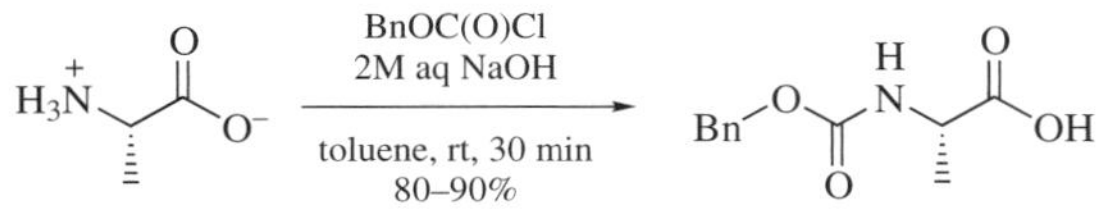

(2)

The selective protection of the ω-amine in α,ω-diamino acids can be achieved by chelation of the α-amino and α-carboxylate functionality by Cu^{II} ion (eq 3).[1b,f,18,19] The Cu^{II} chelate can be cleaved by H_2S,[1f] by thioacetamide,[19] or by use of a chelating ion-exchange resin.[17] Selective protection of the ε-amino group in lysine, the imidazole ring nitrogen in histidine, and the α-amino function in serine, can all be accomplished in homogeneous aqueous medium in the absence of added base using Cbz-imidazolium chloride, a water-soluble benzyloxycarbonylating agent which is prepared in situ from CbzCl and *N*-methylimidazole.[20] Benzyl succinimidyl carbonate, best prepared by reaction of the dicyclohexylammonium salt of *N*-hydroxysuccinimide with CbzCl,[21] is also useful for the *N*-Cbz protection of α-amino acids without the need for added base.

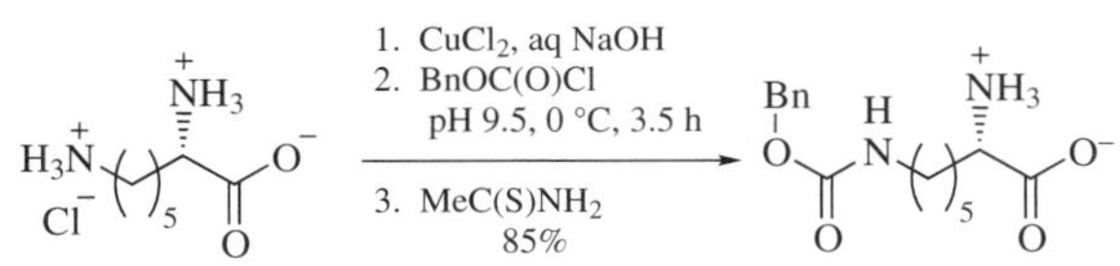

(3)

The *N*-Cbz group is stable to a variety of weakly acidic and basic conditions.[22] Benzyl carbamates are most commonly cleaved to the free amine by catalytic hydrogenolysis (H_2/Pd–C)[23] although a plethora of alternate reductive (e.g. transfer hydrogenolysis,[24] Li/liq NH_3[25]), strongly acidic (e.g. HBr/HOAc[26]), and neutral conditions have been employed.[1] The use of transfer hydrogenolysis (cyclohexadiene, 10% Pd–C) allows for the clean deprotection of *N*-Cbz pyrimidines without concomitant reduction of the 5,6-double bond that is observed under standard catalytic hydrogenolysis conditions.[27] The acid stability of the *N*-Cbz group can be manipulated by the introduction of electron-withdrawing or electron-donating substituents onto the phenyl ring. Selective deprotection of an *N*-Cbz group in the presence of an *S*-Cbz moiety can be achieved using HBr/HOAc at rt.[28]

Primary and secondary arylamines can also be *N*-carbamoylated using CbzCl, typically in the presence of aq Na_2CO_3 (eq 4)[29] or pyridine.[30] The protection of electron-deficient anilines can be accomplished using MgO in acetone.[31] Protection of a pyrrole nitrogen can be achieved by initial deprotonation to the potassium salt.[32]

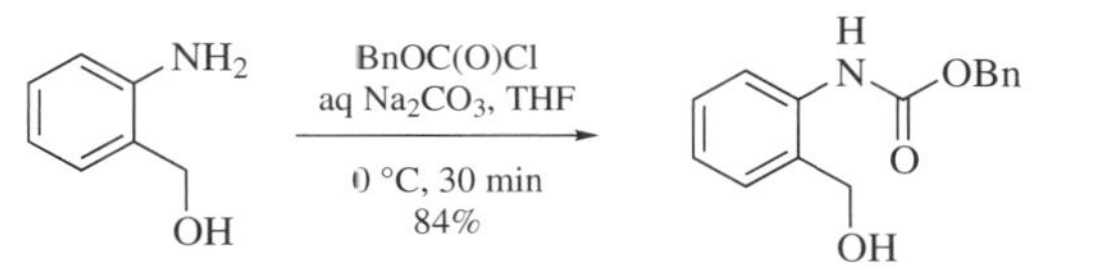

(4)

N-Benzyl carbamates prepared as outlined above can also serve as useful intermediates in the overall *N*-methylation of secondary amines, since $LiAlH_4$ reduction of a benzyl carbamate affords the corresponding tertiary *N*-methylamine (eq 5).[33]

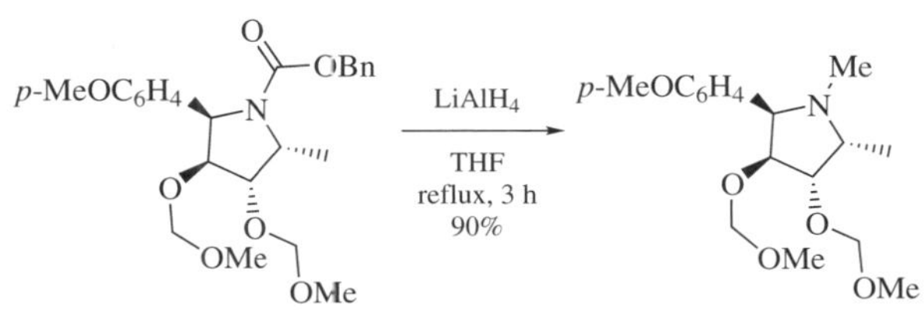

(5)

Other N-Protection Reactions. The protection of an amide as its *N*-Cbz derivative can be achieved using CbzCl with Et_3N/DMAP,[34] or by initial deprotonation of the amide using BuLi or $(Me_3Si)_2NLi$.[35] The N-protection of nucleosides is usually ineffective with CbzCl but can be selectively and efficiently accomplished using *N*-Cbz imidazolium salts generated in situ from the reaction of CbzCl with *N*-alkylimidazoles (eq 6).[27,36]

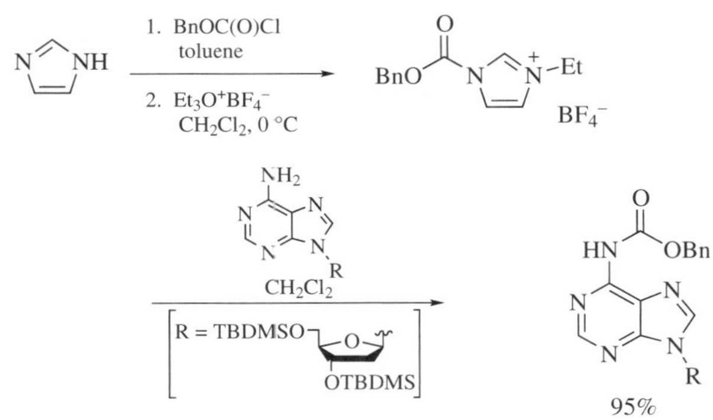

(6)

Protection of Alcohols and Phenols. Protection of alcohols and phenols as the corresponding benzyl carbonates is typically achieved using CbzCl in the presence of an organic base (e.g. Et_3N,[37] pyridine[38]) in CH_2Cl_2 or ether at low temperature (−20 °C to rt). The selective protection of one secondary alcohol in the presence of other secondary and tertiary alcohols has been accomplished using DMAP at very low temperatures (−40 to −20 °C) for 3 days (eq 7).[39]

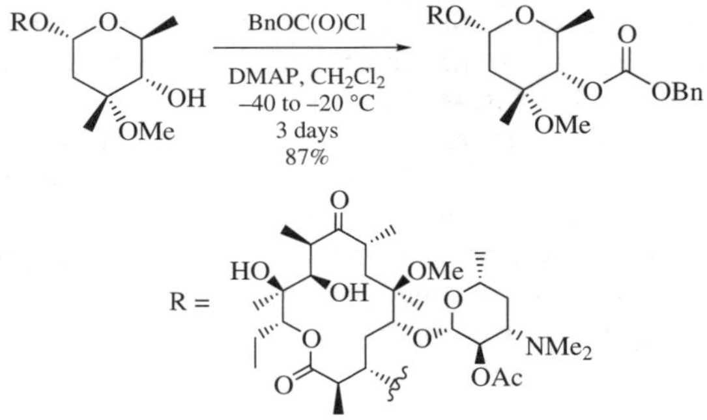

(7)

The use of NaH as base allows for the efficient protection of even hindered electron-deficient alcohols (eq 8).[40] A secondary alcohol can be selectively protected over an indole nitrogen.[41] 1-*O*-Methylated 2-deoxyribose sugars can be protected at the 3- and 5-hydroxyl groups[38] while the hemiacetal function is selectively protected over secondary alcohols in 1-unprotected sugars.[42] Both alcohol-[38,43] and phenol-derived[44] benzyl carbonates are efficiently deprotected by catalytic hydrogenolysis; benzyl carbonates are more readily cleaved under these conditions than benzyl ethers.[38] Removal of the *O*-Cbz group can also be achieved by electrolytic reduction.[45]

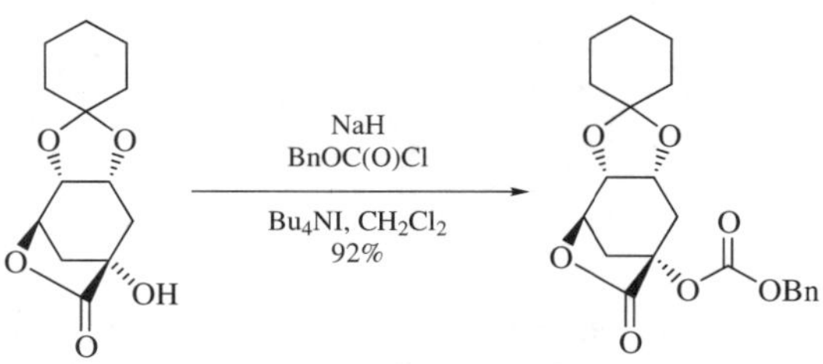

(8)

Protection of Thiols. The protection of thiols using CbzCl can be achieved using Et_3N as base.[46] The use of aq $NaHCO_3$ as base allows for the selective protection of the SH group in cysteine (eq 9), while *N*,*S*-bisprotection is observed under Schotten–Baumann conditions (eq 10).[47] Selective protection of the cysteine SH moiety can also be achieved using Cbz-imidazolium chloride, a water-soluble benzyloxycarbonylating agent prepared in situ from CbzCl and *N*-methylimidazole.[20] Both *N*- and *S*-Cbz groups are deprotected using refluxing CF_3CO_2H[48] or by electrolytic reduction.[45] The use of NaOMe/MeOH at rt for 5–10 min allows for the selective cleavage of *S*-Cbz over *N*-Cbz groups.[28]

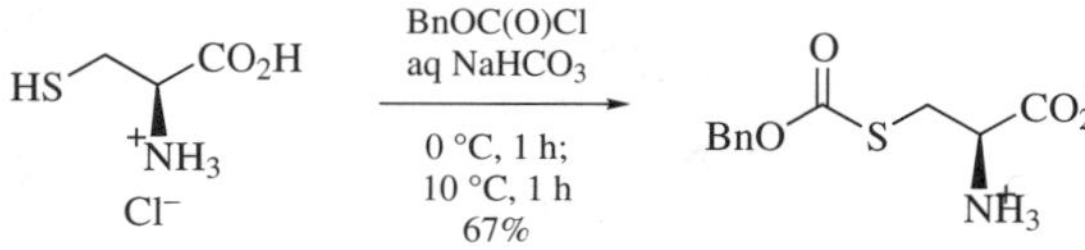

(9)

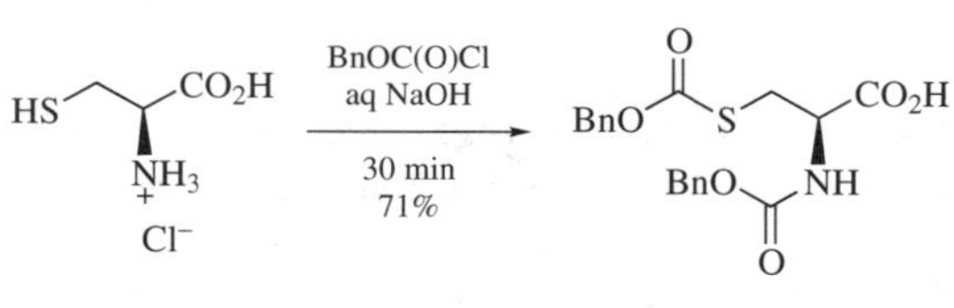

(10)

Decarboxylative Esterification of Carboxylic Acids. The preparation of benzyl esters from sterically uncongested carboxylic acids,[49] including α-keto acids[37,50] and malonic acid half esters,[51] can be achieved using CbzCl in the presence of an organic base (Et_3N,[37,50,51] pyridine,[52] or Et_3N/DMAP[49] in CH_2Cl_2 or THF). The lack of reactivity of sterically crowded carboxylic acids[49] allows for the selective esterification of the less crowded of two acid moieties (eq 11).[52] A major attraction of the benzyl ester group is its susceptibility to cleavage to the free carboxylic acid by catalytic hydrogenolysis[1d,23] or transfer hydrogenolysis.[1d]

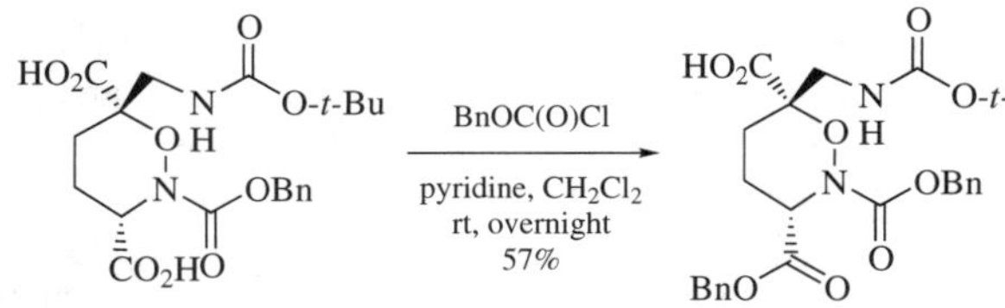

(11)

Activation of Pyridine Ring Towards Nucleophilic Attack. Pyridine and various 4-substituted pyridines can be activated by CbzCl toward regioselective nucleophilic attack by $NaBH_4$[53] and by alkyl,[54] aryl,[55] and alkynyl[56] Grignard reagents, affording the corresponding 2-substituted 1,2-dihydropyridines (eq 12). Treatment of a 3-substituted pyridine with CbzCl and $NaBH_4$ can afford a mixture of 1,2- and 1,4-reduction products; however, reaction with CbzCl and Red-Al proceeds via regioselective 1,2-reduction at the less crowded α-carbon (eq 13).[57]

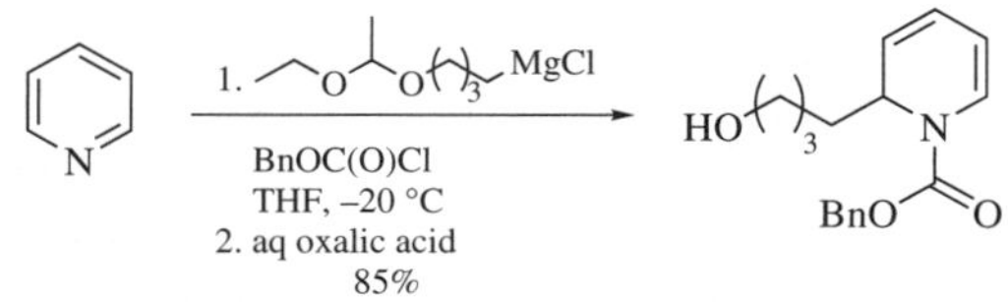

(12)

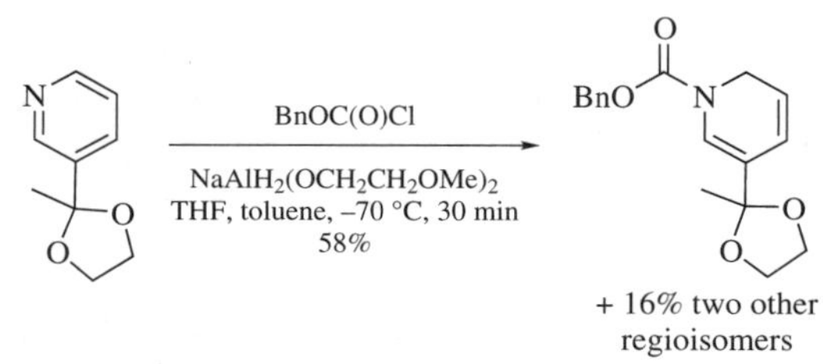

(13)

Generation of Other Benzyloxycarbonylating Agents. A variety of other benzyloxycarbonylating agents can be prepared from CbzCl.[3] For example, reaction of CbzCl with sodium benzyl carbonate affords dibenzyl carbonate, a promising stable crystalline alternative to CbzCl for achieving clean and efficient *N*-benzyloxycarbonylation.[3] *N*-Cbz-imidazole, prepared from CbzCl and imidazole, has proved a suitable reagent for the selective protection of primary over secondary amines.[58]

Related Reagents. 4-Bromobenzyl Chloroformate; Methyl Chloroformate; and 2,2,2-Trichloroethyl Chloroformate.

---

1. (a) Bodanszky, M. *Peptide Chemistry A Practical Textbook*; Springer:Berlin, **1988**. (b) Greenstein, J. P.; Winitz, M. *Chemistry of the Amino Acids*; Wiley:New York, **1961**; Vol. 2, pp 887–895. (c) Wunsch, E. *Methoden Org. Chem. (Houben-Weyl)* **1974**, *15 (I)* , 47–69. (d) Meienhofer, J. In *Chemistry and Biochemistry of the Amino Acids*,

Barrett, G. C., Ed.; Chapman & Hall:London, **1985**; pp 299–302. (e) Pettit, G. R. *Synthetic Peptides*; Van Nostrand Reinhold:New York, **1970–71**; Vols. 1, 2, Academic:New York, **1975**; Vol. *3*. (f) Bailey, P. D. *An Introduction to Peptide Chemistry*; Wiley:Chichester, **1990**.

2. Perrin, D. D.; Armerago, W. L. F. *Purification of Laboratory Chemicals*, 3rd ed.; Pergamon:Oxford, **1988**; p 97.
3. Wunsch, E.; Graf, W.; Keller, O.; Keller, W.; Wersin, G. *Synthesis* **1986**, 958.
4. Carter, H. E.; Frank, R. L.; Johnston, H. W. *Org. Synth., Coll. Vol.* **1955**, *3*, 167.
5. Katchalski, E. *Methods Enzymol.* **1957**, *3*, 541.
6. Farthing, A. C. *J. Chem. Soc* **1950**, 3213.
7. Krow, G. R.; Cannon, K. C.; Carey, J. T.; Ma, H.; Raghavachari, R.; Szczepanski, S. W. *J. Org. Chem.* **1988**, *53*, 2665.
8. (a) Bergmann, M.; Zervas, L. *Ber. Dtsch. Chem. Ges./Chem. Ber.* **1932**, *65*, 1192 (*Chem. Abstr.* **1932**, *26*, 5072) . (b) Austin, G. N.; Baird, P. D.; Fleet, G. W. J.; Peach, J. M.; Smith, P. W.; Watkin, D. J. *Tetrahedron* **1987**, *43*, 3095. (c) Takahashi, Y.; Yamashita, H.; Kobayashi, S.; Ohno, M. *Chem. Pharm. Bull.* **1986**, *34*, 2732. (d) Akhtar, M.; Gani, D. *Tetrahedron* **1987**, *43*, 5341.
9. Meyers, A. I.; Dupre, B. *Heterocycles* **1987**, *25*, 113.
10. Bashyal, B. P.; Fleet, G. W. J.; Gough, M. J.; Smith, P. W. *Tetrahedron* **1987**, *43*, 3083.
11. Senter, P. D.; Pearce, W. E.; Greenfield, R. S. *J. Org. Chem.* **1990**, *55*, 2975.
12. Baldwin, J. E.; Adlington, R. M.; Jones, R. H.; Schofield, C. J.; Zaracostas, C.; Greengrass, C. W. *Tetrahedron* **1986**, *42*, 4879.
13. Patel, A.; Poller, R. C. *Recl. Trav. Chim. Pays-Bas* **1988**, *107*, 182.
14. Van der Eycken, J.; et al. *Tetrahedron Lett.* **1989**, *30*, 3873.
15. Atkinson, J. G.; Girard, Y.; Rokach, J.; Rooney, C. S. *J. Med. Chem.* **1979**, *22*, 99.
16. Iida, H.; Watanabe, Y.; Kibayashi, C. *J. Am. Chem. Soc.* **1985**, *107*, 5534.
17. Scott, J. W.; Parker, D.; Parrish, D. R. *Synth. Commun.* **1981**, *11*, 303.
18. Kruse, C. H.; Holden, K. G. *J. Org. Chem.* **1985**, *50*, 2792.
19. Payne, L. S.; Boger, J. *Synth. Commun.* **1985**, *15*, 1277.
20. Guibe-Jampel, E.; Bram, G.; Vilkas, M. *Bull. Soc. Claim. Fr., Part 2* **1973**, 1021 (*Chem. Abstr.* **1973**, *79*, 31 982g) .
21. Paquet, A. *Can. J. Chem.* **1982**, *60*, 976.
22. (a) Boisonnas, R. A. *Adv. Org. Chem.* **1963**, *3*, 159. (b) Greene, T. W.; Wuts, P. G. M. *Protective Groups in Organic Synthesis*, 2nd ed.; Wiley:New York, **1991**; pp 441–444.
23. Hartung, W. H.; Simonoff, R. *Org. React.* **1953**, *7*, 263.
24. Jackson, A. E.; Johnstone, R. A. W. *Synthesis* **1976**, 685.
25. Boutin, R. H.; Rapoport, R. H. *J. Org. Chem.* **1986**, *51*, 5320.
26. Ben-Ishai, D.; Berger, A. *J. Org. Chem.* **1952**, *17*, 1564.
27. Watkins, B. E.; Kiely, J. S.; Rapoport, H. *J. Am. Chem. Soc.* **1982**, *104*, 5702.
28. Sokolovsky, M.; Wilchek, M.; Patchornik, A. *J. Am. Chem. Soc.* **1964**, *86*, 1202.
29. Jones, G. B.; Moody, C. J. *J. Chem. Soc., Perkin Trans. 1* **1989**, 2455.
30. Torrini, I.; Zecchini, G. P.; Agrosi, F.; Paradisi, M. P. *J. Heterocycl. Chem.* **1986**, *23*, 1459.
31. Rewcastle, G. W.; Baguley, B. C.; Cain, B. F. *J. Med. Chem.* **1982**, *25*, 1231.
32. Drew, M. G. B.; George, A. V.; Isaacs, N. S.; Rzepa, H. S. *J. Chem. Soc., Perkin Trans. 1* **1985**, 1277.
33. Iida, H.; Yamazaki, N.; Kibayashi, C. *J. Org. Chem.* **1987**, *52*, 1956.
34. Fukuyama, T.; Nunes, J. J. *J. Am. Chem. Soc.* **1988**, *110*, 5196.
35. Nagasaka, T.; Koseki, Y.; Hamaguchi, F. *Tetrahedron Lett.* **1989**, *30*, 1871.
36. Watkins, B. E.; Rapoport, H. *J. Org. Chem.* **1982**, *47*, 4471.
37. Arnoldi, A.; Merlini, L.; Scaglioni, L. *J. Heterocycl. Chem.* **1987**, *24*, 75.
38. Bobek, M.; Bloch, A.; Kuhar, S. *Tetrahedron Lett.* **1973**, 3493.
39. Baker, W. R.; Clark, J. D.; Stephens, R. L.; Kim, K. H. *J. Org. Chem.* **1988**, *53*, 2340.
40. Sutherland, J. K.; Watkins, W. J.; Bailey, J. P.; Chapman, A. K.; Davies, G. M. *Chem. Commun./J. Chem. Soc., Chem. Commun.* **1989**, 1386.
41. Kuehne, M. E.; Podhorez, D. E.; Mulamba, T.; Bornmann, W. G. *J. Org. Chem.* **1987**, *52*, 347.
42. Saito, H.; Nishimura, Y.; Kondo, S.; Umezawa, H. *Chem. Lett.* **1987**, 799.
43. Daubert, B. F.; King, C. G. *J. Am. Chem. Soc.* **1939**, *61*, 3328.
44. Kuhn, M.; von Wartburg, A. *Helv. Chim. Acta* **1969**, *52*, 948.
45. Mairanovsky, V. G. *Angew. Chem., Int. Ed. Engl.* **1976**, *15*, 281.
46. Barton, D. H. R.; Manly, D. P.; Widdowson, D. A. *J. Chem. Soc., Perkin Trans. 1* **1975**, 1568.
47. Berger, A.; Noguchi, J ; Katchalski, E. *J. Am. Chem. Soc.* **1956**, *78*, 4483.
48. Zervas, L.; Photaki, I.; Ghelis, N. *J. Am. Chem. Soc.* **1963**, *85*, 1337.
49. (a) Kim, S.; Kim, Y. C.; Lee, J. I. *Tetrahedron Lett.* **1983**, *24*, 3365. (b) Kim, S.; Lee, J. I.; Kim, Y. C. *J. Org. Chem.* **1985**, *50*, 560.
50. Domagala, J. M. *Tetrahedron Lett.* **1980**, *21*, 4997.
51. Gutman, A. L.; Boltanski, A. *Tetrahedron Lett.* **1985**, *26*, 1573.
52. Baldwin, J. E.; Otsuka, M.; Wallace, P. M. *Tetrahedron* **1986**, *42*, 3097.
53. Sundberg, R. J.; Amat, M.; Fernando, A. M. *J. Org. Chem.* **1987**, *52*, 3151.
54. (a) Comins, D. L.; Abdullah, A. H.; Smith, R. K. *Tetrahedron Lett.* **1983**, *24*, 2711. (b) Comins, D. L.; Brown, J. D. *Tetrahedron Lett.* **1986**, *27*, 4549.
55. Kurita, J.; Iwata, K.; Tsuchiya, T. *Chem. Pharm. Bull.* **1987**, *35*, 3166.
56. Natsume, M.; Ogawa, M. *Heterocycles* **1983**, *20*, 601.
57. Natsume, M.; Utsonomiya, I.; Yamaguchi, K.; Sakai, S.-I. *Tetrahedron* **1985**, *41*, 2115.
58. Sharma, S. K.; Miller, M. J.; Payne, S. M. *J. Med. Chem.* **1989**, *32*, 357.

Paul Sampson
Kent State University, Kent, OH, USA

Benzyl 2-Hydroxymethylbenzoate

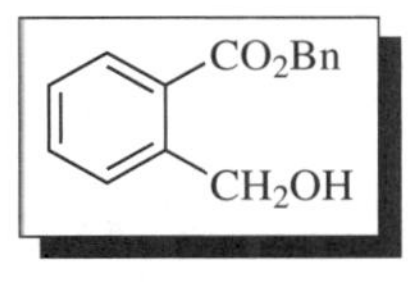

[365534-57-2] $C_{15}H_{14}O_3$ (MW 242.27)

(precursor to 2′-carboxybenzyl glycosides which are used as glycosyl donors for efficient and stereoselective glycosylations)

Physical Data: mp 62–64 °C.
Solubility: soluble in CH_2Cl_2, EtOAc.
Form Supplied in: white solid.
Analysis of Reagent Purity: [1] IR, ^{1}H NMR, ^{13}C NMR.
Preparative Methods: readily prepared in large quantities by the reaction of benzyl bromide and sodium 2-hydroxymethylbenzoate, which was obtained from inexpensive phthalide (eq 1).

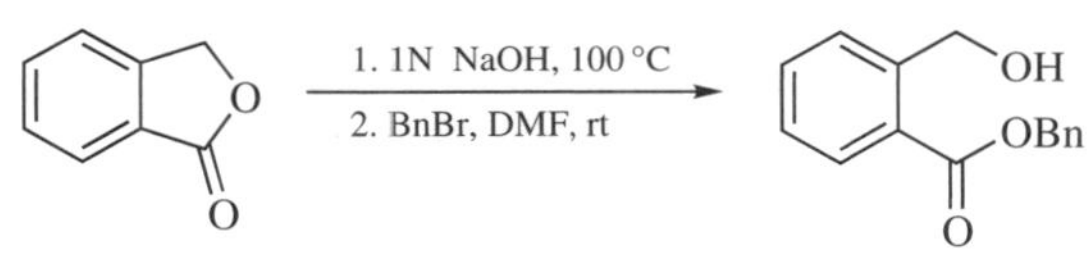

(1)

Purification: recrystallization from DMF/ H_2O.
Handling, Storage, and Precautions: stable to air and moisture. The solid may be stored in a refrigerator for extended periods.

Introduction. The synthesis of biologically important, complex oligosaccharides requires the efficient and stereoselective glycosylation reaction between the glycosyl donor and the glycosyl acceptor.[2,3] The choice of the glycosyl donor is crucial for the stereoselective glycosylation. Benzyl 2-hydroxymethylbenzoate has been employed for the preparation of 2′-carboxybenzyl (CB) glycosides that serve as a novel type of glycosyl donors for stereoselective glycosylation.

Preparation of 2′-Carboxybenzyl Glycosides as Glycosyl Donors. Reaction of the peracetylmannosyl bromide and benzyl 2-hydroxymethylbenzoate gave 2′-(benzyloxycarbonyl)benzyl (BCB) D-mannopyranoside (eq 2).[1] The BCB 2-deoxyglucoside, on the other hand, can be prepared by the reaction of tribenzylglucal and benzyl 2-hydroxymethylbenzoate (eq 3).[4]

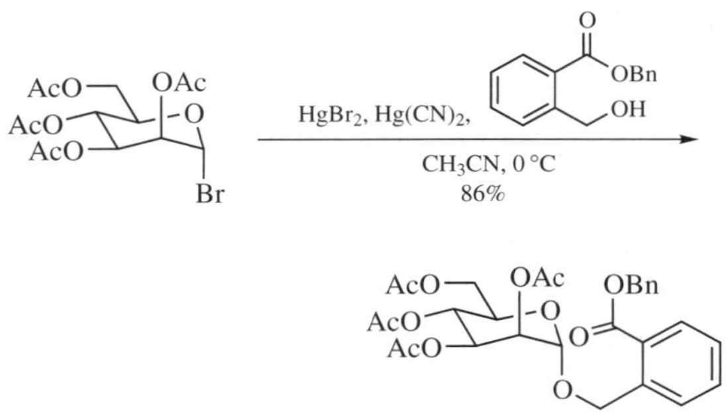

(2)

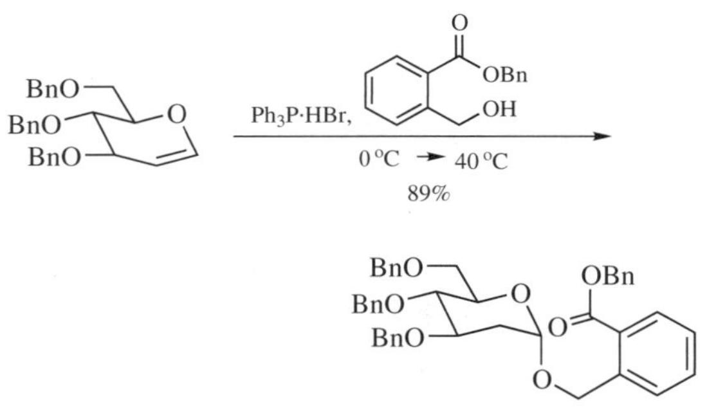

(3)

The BCB glycosides are converted almost quantitatively into the corresponding CB glycosides by selective hydrogenolysis of the benzyl ester functionality without affecting the benzylidene acetal and the benzyl ether groups (eq 4).[1]

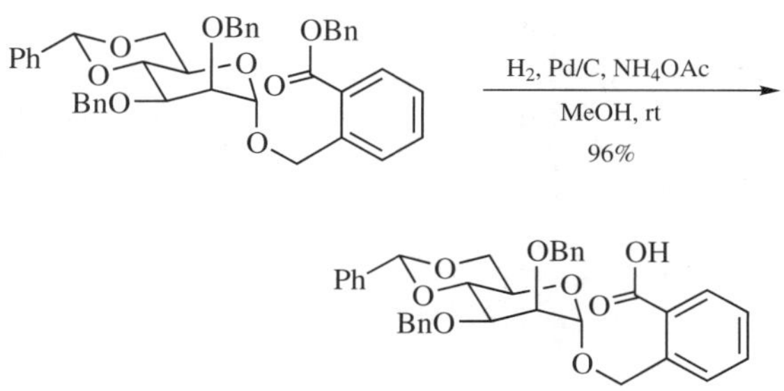

(4)

Stereoselective Glycosylation Using 2′-Carboxybenzyl Glycosides as Glycosyl Donors. Glycosylation of the CB 4,6-*O*-benzylidenemannopyranoside with various glycosyl acceptors (ROH) affords stereoselectively the β-D-mannopyranoside (eq 5).[1] In fact, the construction of the 1,2-*cis*-β-mannopyranosyl linkages is one of the challenges in the glycoside synthesis.[5,6] On the other hand, glycosylation of CB 4,6-*O*-benzylideneglucopyranoside with glycosyl acceptors provides the α-D-glucopyranoside (eq 6).[1,7] The 4,6-*O*-benzylidene group has been found to be essential for this type of stereocontrol in both β-mannopyranosylation and α-glucopyranosylation employing glycosyl triflates as glycosyl donors.[8]

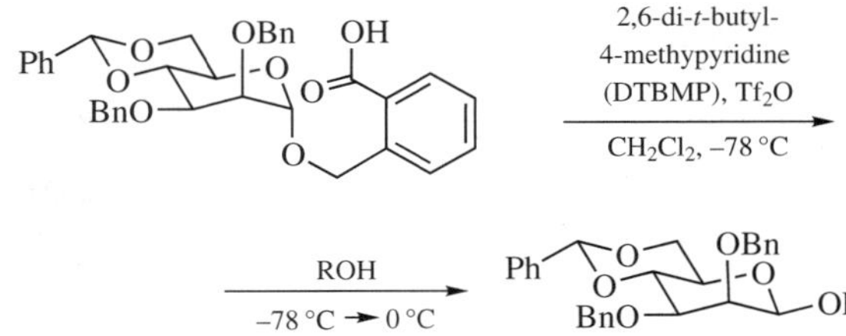

(5)

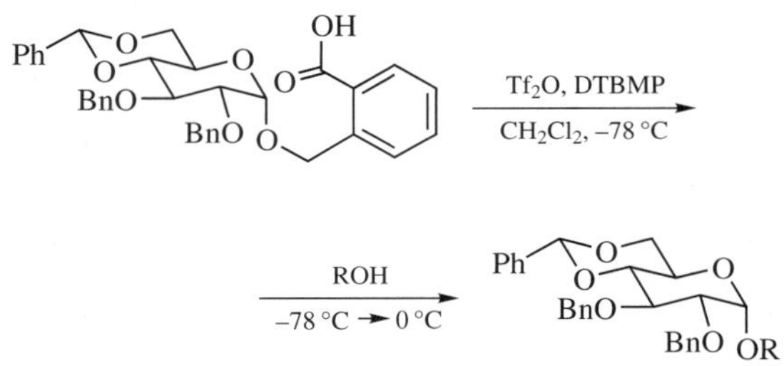

(6)

Use of CB 2-deoxyglycosides as glycosyl donors with secondary alcohol glycosyl acceptors can be either highly α- or β-stereoselective depending upon the protective groups in the glycosyl donor. The direct construction of the β-2-deoxyglycosyl linkage is also one of the difficult problems in the glycoside synthesis.[9] Thus, glycosylation of the benzyl-protected CB 2-deoxyglycoside with a secondary alcohol glycosyl acceptor (R′OH) gives the α-disaccharide (eq 7). On the other hand, glycosylation of the 4,6-*O*-benzylidene-protected CB 2-deoxyglycoside with a secondary alcohol glycosyl acceptor (R′OH) affords the β-disaccharide (eq 8).[4]

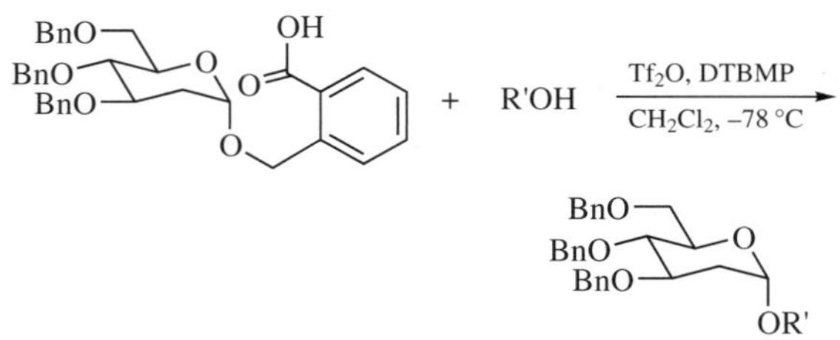

(7)

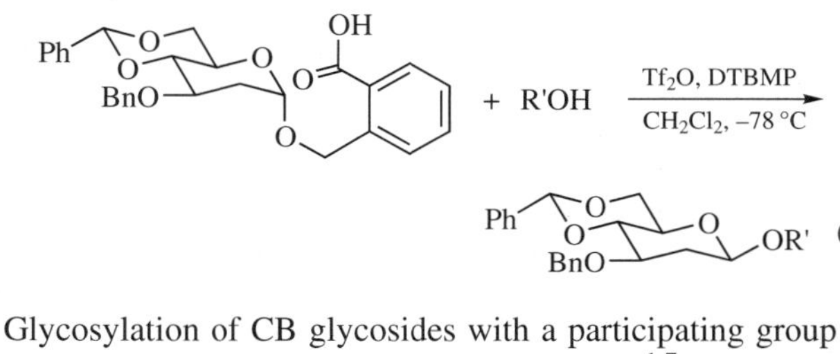

(8)

Glycosylation of CB glycosides with a participating group at C-2 works well to give 1,2-*trans* glycosides,[1,7] and the CB glycosylation methodology has been applied to synthesis of trisaccharides[1] and a tetrasaccharide.[7]

1. Kim, K. S.; Kim, J. H.; Lee, Y. J.; Lee, Y. J.; Park, J., *J. Am. Chem. Soc.* **2001**, *123*, 8477.
2. Toshima, K.; Tatsuta, K., *Chem. Rev.* **1993**, *93*, 1503–1531.
3. Boons, G. J., *Tetrahedron* **1996**, *52*, 1095–1121.
4. Kim, K. S.; Park, J.; Lee, Y. J.; Seo, Y. S., *Angew. Chem., Int. Ed. Engl.* **2003**, *42*, 459.
5. Barresi, F.; Hindsgaul, O., In *Modern Methods in Carbohydrate Synthesis*; Khan, S. H.; O'Neil, R. A., Eds.; Harwood Academic Publishers: Amsterdam, 1996, pp 251–276.
6. Gridley, J. J.; Osborn, H. M. I., *J. Chem. Soc., Perkin Trans. 1* **2000**, 1471–1491.
7. Kim, K. S.; Kang, S. S.; Seo, Y. S.; Kim, H. J.; Lee, Y. J.; Jeong, K. S., *Synlett.* **2003**, 1311–1314.
8. Crich, D., In *Glycochemistry: Principles, Synthesis, and Applications*; Wang, P. G.; Bertozzi, C. R., Eds.; Marcel Dekker: New York, 2001, pp 53–75.
9. Marzabadi, C. H.; Franck, R. W., *Tetrahedron* **2000**, *56*, 8385–8417.

Kwan Soo Kim & Yong Joo Lee
Yonsei University, Seoul, Korea

Benzyl Isopropenyl Ether

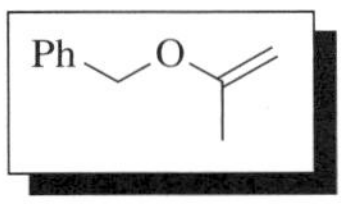

[32783-20-3] $C_{10}H_{12}O$ (MW 148.20)

(protecting group for compounds with multiple hydroxy groups;[1] acetone enolate equivalent[2])

Physical Data: bp 191–192 °C.[3]

Solubility: sol acetone, benzene.

Preparative Methods: the first reported synthesis of the reagent was accomplished by the pyrolytic decarboxylation of β-benzyloxycrotonic acid.[3] An alternate method for the preparation proceeds via dehydrohalogenation of 2-benzyloxy-1-chloropropane with potassium *t*-butoxide in DMSO (82% yield).[1]

Alcohol Protection. Benzyl isopropenyl ether (**1**) has been shown to react with free hydroxy groups at rt in benzene in the presence of 10 mol % Dichloro(1,5-cyclooctadiene)palladium(II) to produce the corresponding mixed benzyl acetals in 85–98% isolated yield (eq 1).[1] The protecting group can also be introduced in comparable yields with catalytic Phosphorus Oxychloride or p-Toluenesulfonic Acid. Cleavage of the mixed acetal took place at rt in the presence of catalytic 5% Palladium on Carbon under 1 atm of hydrogen in 92–99% yield. The protecting group was stable to hydride reducing agents (LAH, DIBAL), alkyllithium, and Grignard reagents, and was stable to alkaline hydrolysis conditions (1 M NaOH/THF, 1/1). Exposure to 1 M Acetic Acid in THF (1/1) effects cleavage of the protecting group.

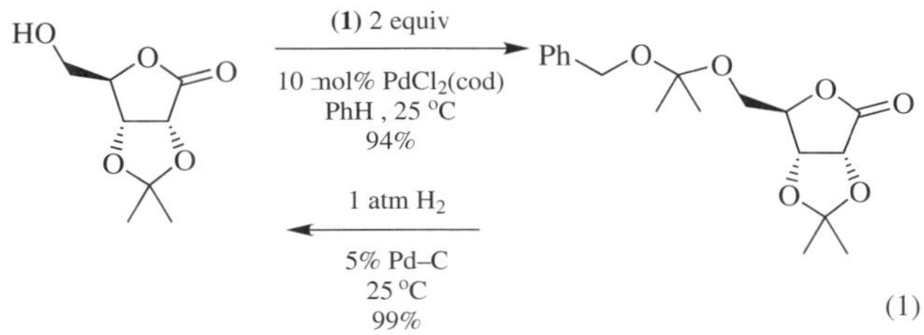

(1)

The primary advantage of this alcohol-protecting group over many similar reagents[4] is the neutral conditions used for its introduction and removal. An alternative method for protecting alcohols as mixed acetals makes use of catalytic Cobalt(II) Chloride in MeCN at rt.[5] The reported range of yields for this method was slightly lower (68–91%); however, the ready availability and low cost of the cobalt catalyst recommends its use.

A closely related reagent for alcohol protection is 2-phenoxypropene, which was easily prepared from phenyl acetate and Tebbe's reagent in 57% yield.[6] Conversion of alcohols to the corresponding phenyl mixed acetals took place with catalytic phosphorus trichloride. Hydrolytic cleavage with 0.1 M HCl and ether was complete in 30 s.

Acetone Enolate Equivalent. An early application in which (**1**) was used as an acetone enolate equivalent in a Lewis acid-catalyzed condensation with ketones has been reported.[2] For example (eq 2), condensation of (**1**) with acetone in the presence of Boron Trifluoride Etherate produced 4-benzyloxy-4-methyl-2-pentanone in 52% yield. This early methodology has largely been eclipsed by advances involving the use of silyl enol ethers.

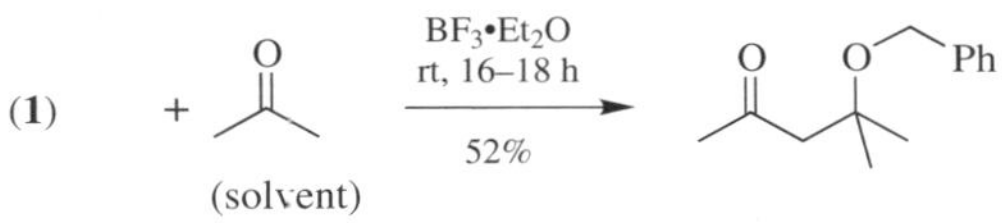

(2)

1. Mukaiyama, T.; Ohshima, M.; Murakami, M. *Chem. Lett.* **1984**, 265.
2. Albaiges, J.; Camps, F; Castells, J.; Fernandez, J.; Guerrero, A. *Synthesis* **1972**, 378.
3. Autenrieth, W. *Ber. Dtsch. Chem. Ges./Chem. Ber.* **1896**, *29*, 1639.
4. Greene, T. W.; Wuts, P. G. M. *Protective Groups in Organic Synthesis*, 2nd ed.; Wiley:New York, **1991**; pp 10–68.
5. Iqbal, J.; Srivastava, R. R.; Gupta, K. B.; Khan, M. A. *Synth. Commun.* **1989**, *19*, 901.
6. Zandbergen, P.; Willems, H. M. G.; van der Marel, G. A.; Brussee, J.; van der Gen, A. *Synth. Commun.* **1992**, *22*, 2781.

Marcus A. Tius
University of Hawaii, Honolulu, HI, USA

Benzyl 2,2,2-Trichloroacetimidate

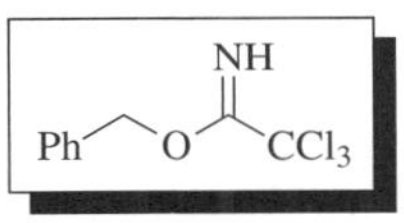

[81927-55-1] $C_9H_8Cl_3NO$ (MW 252.53)

(reagent for protection of hydroxy groups as benzyl ethers under mildly acidic conditions[1])

Alternate Name: BTCA.
Physical Data: bp 106–114 °C/0.5 mmHg; *d* 1.359 g cm^{-3}.
Solubility: reactions are most often carried out in a mixture of cyclohexane/dichloromethane;[1] decomposition by rearrangement to *N*-benzyl trichloroacetamide is accelerated in more polar solvents.
Form Supplied in: colorless oil.
Purification: distillation under reduced pressure.
Handling, Storage, and Precautions: moisture sensitive.

Hydroxy Group Protection. The benzyl ether is one of the most versatile and frequently used protecting groups in organic synthesis.[2] The most common way of introducing this group is with a strong base and Benzyl Bromide using the classical Williamson ether synthesis methodology. Recently, benzyl trichloroacetimidate[3] has found widespread use to protect hydroxy functionalities as their benzyl ethers under mildly acidic conditions,[1,4,5a] when the more conventional strong base-mediated methods either failed or gave lower yields due to side reactions and/or starting material decomposition.

The imidate can be prepared from the sodium alkoxide ion of benzyl alcohol and trichloroacetonitrile according to procedures based on those developed by Cramer[3] in the late 1950s. Substituted benzyl ethers have also been prepared this way, for example using 4-methoxybenzyl trichloroacetimidate,[5] 3,4-dimethoxybenzyl trichloroacetimidate,[6] and 2,6-dichlorobenzyl trichloroacetimidate,[7] and the method has also been applied for the synthesis of allyl[1b] and *t*-butyl ethers,[8] as well as for 2-phenylisopropyl esters[9] for peptide synthesis. The benzyl and 4-methoxybenzyl 2,2,2-trichloroacetimidates are available commercially.

The reaction is performed in nonpolar media to prevent rearrangement of the acetimidate to *N*-benzyl trichloroacetamide, and as an added advantage the solid trichloroacetamide byproduct can simply be removed by filtration. These constraints limit the substrate range to those molecules soluble in this type of solvent; for example, a sugar triol which was insoluble in dichloromethane, when reacted in DMSO yielded very little of the benzylated product desired.[10] It was also noted that differential (primary versus secondary) hydroxy protection in carbohydrates (eq 1) was poor,[10] and the ratio of the 6-OBn to 4-OBn products was at best 2:1.

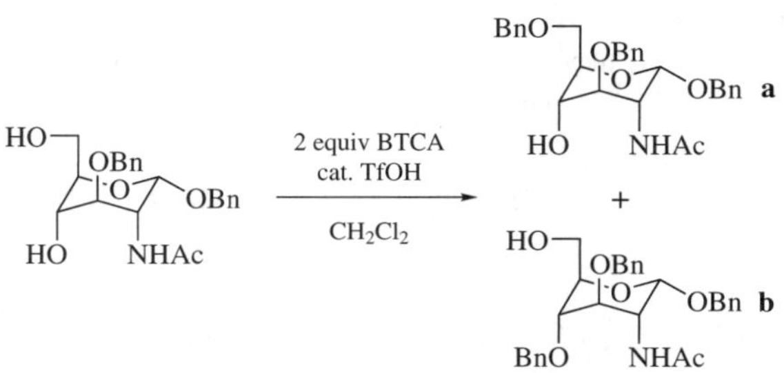

(1)

a:**b** = 2:1, 45–60% total

The reagent has been used widely in carbohydrate[1,10–14] chemistry. The method was found to be useful in cases where *O*-benzoyl group migration had been observed under basic conditions (eq 2),[11] and has also been applied to inositol derivatives,[15] when the same problem[15a] was encountered.

When *N*-acetyl-protected sugar amines were under investigation[1,12] the *N*-acetyl group was observed to undergo transformation to the *O*-benzyl imidate in competition with OH benzylation (eq 3). It appears for best yields at least 1 equiv of reagent per OH and NAc group is needed, and subsequent hydrolysis and reacetylation of the amine are necessary to obtain are the originally desired material. Similar problems were encountered with the morpholine derived lactams (eq 4),[16a] where mixed products were obtained, but with β-lactams[16b] with *N*-amide protection (eq 5), *O*-benzylation with benzyl trichloroacetimidate proceeded smoothly. BTCA has recently been used to convert diketopiperazines into their bis-benzyl imino ethers.[17]

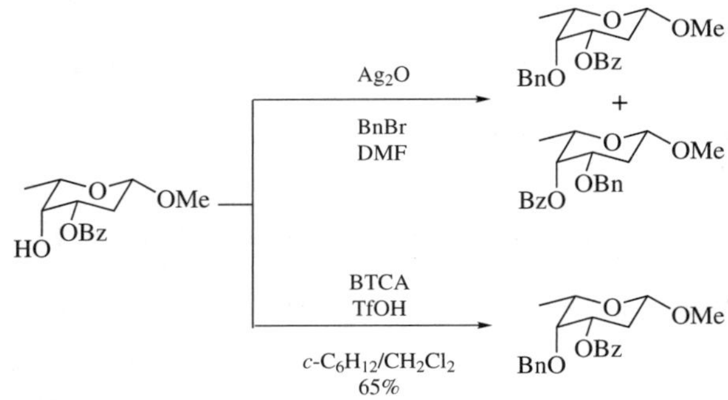

(2)

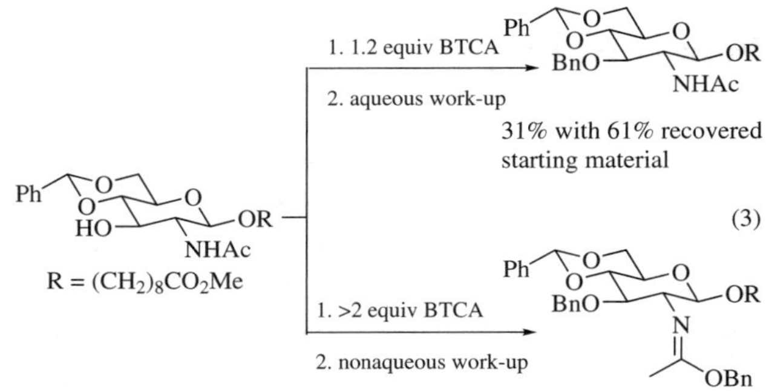

(3)

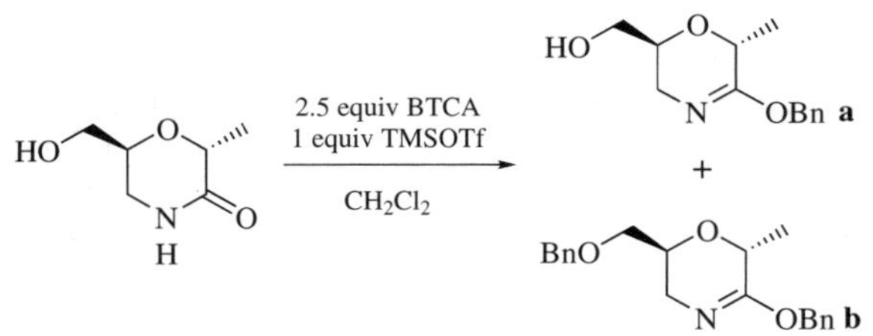

(4)

a:**b** = 6:4, 80% total

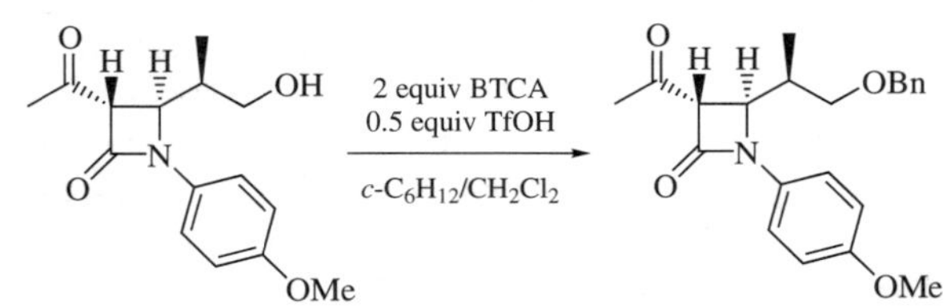

(5)

44% after N-deprotection

Other classes of molecules which have been benzylated using this reagent include hydroxy ketones,[4] hydroxy halides[4,7,18] and esters[19–23] such as tartrates,[19] hydroxy propionates or butanoates,[20] long alkyl chain hydroxy esters,[21] and hydroxy lactones.[4] Under basic conditions problems such as enolization, aldol reactions, elimination and epoxide formation can cause problems and when 2-substituted esters, such as ethyl (*S*)-lactate,[22] are required as homochiral starting materials, then protection using this method circumvents the considerable risk of racemization that accompanies the use of bases (eq 6).

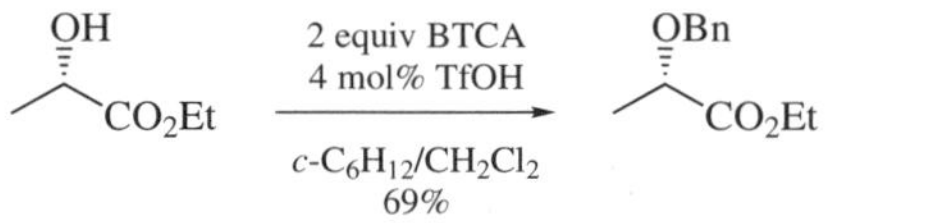

(6)

$[\alpha]_D^{20}$ −74.5° (c 2.94, $CHCl_3$)

Under these benzylation reaction conditions, it has been shown that the *N-t*-butoxycarbonyl (N-Boc) (eq 7) and *N*-(2-trimethylsilyl)ethoxycarbonyl (*N*-TEOC) (eq 8) protected amino acid derivatives are converted into their *N*-benzyloxycarbonyl (N-Z) analogs.[24] In most cases yields were moderate, but surprisingly the *t*-butyldimethylsilyloxy and *t*-butyl ester functions are relatively unaffected by the acidic conditions.

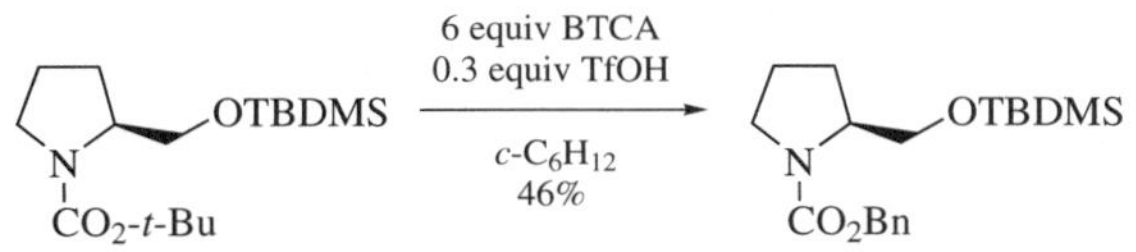

(7)

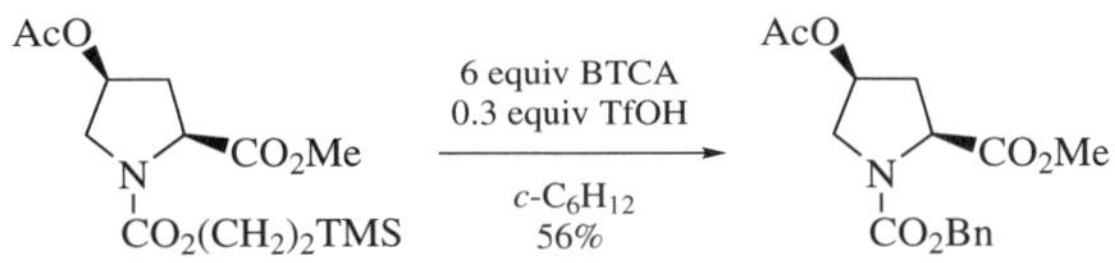

(8)

The most frequently employed acid catalyst is TfOH (Trifluoromethanesulfonic Acid, or triflic acid), but others that have been used include $BF_3{\cdot}OEt_2$,[4,8] TMSOTf,[4,16] *p*-TsOH,[6] and TFA.[23,16] Camphorsulfonic acid, pyridinium *p*-toluenesulfonate, and trityl perchlorate have also been investigated in conjunction with the more reactive 4-methoxybenzyl trichloroacetimidate.[5a]

Functional groups sometimes sensitive to acidic conditions such as simple esters,[1,19–23] lactones,[4,5d] acetals,[1,5a,15] epoxides,[5a,13] 2-methoxyethyloxymethyl (MEM) groups,[6] *t*-butyldimethylsilyl ethers,[14,24–27] and orthoformates[15b] are generally unaffected by the catalytic amounts of acid used, although some tertiary alcohols[4] and 1-trimethylsilylalkynes[4] have been noted to give poor yields.

Related Reagents. Benzyl Bromide; Benzyl Chloride; Benzyl Iodide; Benzyl Trifluoromethanesulfonate; 3,4-Dimethoxybenzyl Bromide; 4-Methoxybenzyl 2,2,2-Trichloroacetimidate.

---

1. (a) Wessel, H.-P.; Iversen, T.; Bundle, D. R. *J. Chem. Soc., Perkin Trans. 1* **1985**, 2247. (b) Iversen, T.; Bundle, D. R. *Chem. Commun./J. Chem. Soc., Chem. Commun.* **1981**, 1240.
2. Greene, T. W.; Wuts, P. G. M. *Protective Groups in Organic Synthesis*, 2nd ed.; Wiley:New York, **1991**.
3. (a) Cramer, F.; Pawelzik, K.; Baldauf, H. J. *Ber. Dtsch. Chem. Ges./Chem. Ber.* **1958**, *91*, 1049. (b) Cramer, F.; Hennrich, N. *Ber. Dtsch. Chem. Ges./Chem. Ber.* **1961**, *94*, 976.
4. Eckenberg, P.; Groth, U.; Huhn, T.; Richter, N.; Schmeck, C. *Tetrahedron* **1993**, *49*, 1619.
5. (a) Nakajima, N.; Horita, K.; Abe, R.; Yonemitsu, O. *Tetrahedron Lett.* **1988**, *29*, 4139. (b) Takaku, H.; Ueda, S.; Ito, T. *Tetrahedron Lett.* **1983**, *24*, 5363. (c) Romo, D.; Johnson, D. D.; Plamondon, L.; Miwa, T.; Schreiber, S. L. *J. Org Chem.* **1992**, *57*, 5060. (d) Adams, E.; Hiegemann, M.; Duddeck, H.; Welzel, P. *Tetrahedron* **1990**, *46*, 5975.
6. Takaku, H.; Ito, T.; Imai, K. *Chem. Lett.* **1986**, 1005.
7. Amouroux, R.; Gerin, B.; Chastrette, M. *Tetrahedron* **1985**, *41*, 5321.
8. Armstrong, A.; Brackenridge, I.; Jackson, R. F. W.; Kirk, J. M. *Tetrahedron Lett.* **1988**, *29*, 2483.
9. Yue, C.; Thierry, J.; Potier, P. *Tetrahedron Lett.* **1993**, *34*, 323.
10. Bovin, N. V.; Musina, L. Y.; Khorlin, A. Y. *Bull. Acad. Sci. USSR, Div. Chem. Sci.* **1986**, *35*, 614.
11. Yasumori, T.; Sato, K.; Hashimoto, H.; Yoshimura, J. *Bull. Chem. Soc. Jpn.* **1984**, *57*, 2538.
12. (a) Kusumoto, S.; Yamamoto, M.; Shiba, T. *Tetrahedron Lett.* **1984**, *25*, 3727. (b) Imoto, M.; Yoshimura, H.; Yamamoto, M.; Shimamoto, T.; Kusumoto, S.; Shiba, T. *Tetrahedron Lett.* **1984**, *25*, 2667.
13. Schubert, J.; Schwesinger, R.; Knothe, L.; Prinzbach, H. *Justus Liebigs Ann. Chem./Liebigs Ann. Chem.* **1986**, 2009.
14. Takahashi, S.; Terayama, H.; Kuzuhara, H. *Tetrahedron Lett.* **1992**, *33*, 7565.
15. (a) Ozaki, S.; Kondo, Y.; Nakahira, H.; Yamaoka, S.; Watanabe, Y. *Tetrahedron Lett.* **1987**, *28*, 4691. (b) Baudin, G.; Glanzer, B. I.; Swaminathan, K. S.; Vasella, A.; *Helv. Chim. Acta* **1988**, *71*, 1367. (c) Estevez, V. A.; Prestwich, G. D. *Tetrahedron Lett.* **1991**, *32*, 1623.
16. (a) Danklmaier, J.; Honig, H. *Justus Liebigs Ann. Chem./Liebigs Ann. Chem.* **1989**, 665. (b) Kawabata, T.; Kimura, Y.; Ito, Y.; Terashima, S.; Sasaki, A.; Sunagawa, M. *Tetrahedron* **1988**, *44*, 2149.
17. Groth, U.; Schmeck, C.; Schöllkopf, U. *Justus Liebigs Ann. Chem./Liebigs Ann. Chem.* **1993**, 321.
18. Voss, G.; Gerlach, H. *Helv. Chim. Acta* **1983**, *66*, 2294.
19. Hansson, T. G.; Kihlberg, J. O. *J. Org. Chem.* **1986**, *51*, 4490.
20. (a) White, J. D.; Kawasaki, M. *J. Org. Chem.* **1992**, *57*, 5292. (b) Keck, G. E.; Murry, J. A. *J. Org. Chem.* **1991**, *56*, 6606.
21. (a) Imoto, M.; Yoshimura, H.; Yamamoto, M.; Shimamoto, T.; Kusumoto, S.; Shiba, T. *Bull. Chem. Soc. Jpn.* **1987**, *60*, 2197. (b) Widmer, U. *Synthesis* **1987**, 568.
22. Ito, Y.; Kobayashi, Y.; Kawabata, T.; Takase, M.; Terashima, S. *Tetrahedron* **1989**, *45*, 5767.
23. Keck, G. E.; Andrus, M. B.; Romer, D. R.; *J. Org. Chem.* **1991**, *56*, 417.
24. Barrett, A. G. M.; Pilipauskas, D. *J. Org. Chem.* **1990**, *55*, 5170.
25. Hong, C. Y.; Kishi, Y.; *J. Am. Chem. Soc.* **1992**, *114*, 7001.
26. Lawrence, N. J.; Fleming, I. *Tetrahedron Lett.* **1990**, *31*, 3645.
27. Danishefsky, S. J.; Deninno, S.; Lartey, P. *J. Am. Chem. Soc.* **1987**, *109*, 2082.

Andrew N. Boa & Paul R. Jenkins
Leicester University, UK

N,N′-Bis(*t*-butoxycarbonyl)-*N′′*-Trifluoromethanesulfonyl Guanidine

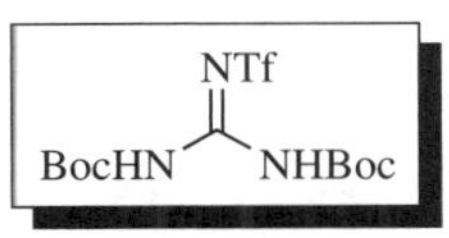

[145013-06-5] $C_{12}H_{20}F_3N_3O_6S$ (MW 391.36)

(electrophilic reagent that specifically reacts with amines to yield substituted guanidines[1–4])

Alternate Name: (*tert*-butoxycarbonylamino-trifluoromethanesulfonylimino-methyl)-carbamic acid *tert*-butyl ester.

Physical Data: mp 114–115 °C.

Solubility: soluble in $CHCl_3$, CH_2Cl_2, MeOH, DMF, and most organic solvents; insoluble in H_2O.

Form Supplied in: nonhygroscopic, white crystalline solid.

Analysis of Reagent Purity: NMR, HPLC, and elemental analysis.

Preparative Methods: A solution of *N,N′*-diBoc-guanidine (29 mmol) and NEt_3 (36 mmol) in anhydrous CH_2Cl_2 (100 mL) was cooled to −78 °C under N_2. Triflic anhydride (35 mmol) was added dropwise over a period of 20 min, and the resulting mixture was allowed to warm to −20 °C for 4 h. A 2 M aq $NaHSO_4$ solution was added at −20 °C such that the reaction temperature does not rise above −10 °C, and the resulting layers were stirred vigorously for 5 min. The organic layer was washed with 2 M $NaHSO_4$, H_2O, and brine and then dried ($MgSO_4$).[1,2]

Purification: flash column chromatography (20% hexanes in CH_2Cl_2).

Handling, Storage, and Precautions: indefinitely stable at rt.

Synthesis of Substituted Guanidines. *N,N′*-Bis (*tert*-butoxycarbonyl)-*N″*-trifluromethanesulfonyl guanidine (**1**) serves as an efficient electrophilic species for the guanidinylation of a variety of amines under mild conditions to give high yields (75–100%) of substituted, protected guanidines (eq 1).[1,2] Its use is limited, however, to the guanidinylation of primary and secondary amines. Aromatic amines react somewhat slower and hindered secondary amines such as diisopropylamine do not react at all. The reaction proceeds most efficiently in nonpolar solvents such as CH_2Cl_2 and $CHCl_3$ although reactions have been successfully carried out in polar solvents such as DMF and MeOH. In a typical reaction, reagent **1** is added as a solid to a slight excess of amine followed by 1 equiv of NEt_3.[3,4] This transformation is not moisture sensitive. After completion of the reaction determined by TLC, the excess starting amines, NEt_3 and triflic amide are removed by aqueous work up. Kinetic studies of several guanidinylating reagents showed compound **1** and *N,N′*-diBoc-thiourea with the Mukaiyama reagent (2-chloro-1-methylpyridinium iodide)[5] gave very rapid product formation.[1] Between these two reagents, the thiourea seems to be superior for sterically hindered amines, but experimental setup and product isolation are much less demanding for reagent **1**. The reaction using *N,N′*-diBoc-1*H*-pyrazole-1-carboxamidine[6,7] proceeded much slower and *N,N′*-diBoc-isothiourea[8] did not react at all. (All reactions were carried out using the recommended solvent.) The *N,N′*-diCbz[1,2] and *N,N′*-diAlloc derivatives of reagent **1** have also been prepared starting from guanidine hydrochloride. These compounds work similar to reagent **1** and their use is advantageous when different protecting groups are required.

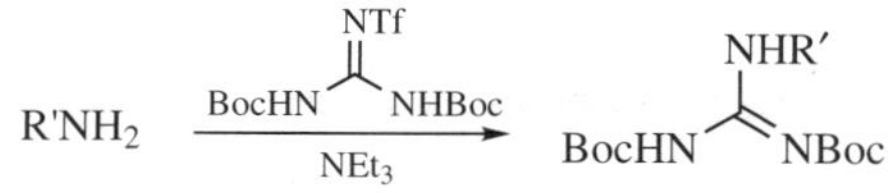

(1)

Modification of Amino Acids and Peptides in Solution. Orthogonally protected amino acids and peptides react in the same manner as described above; however, a different protocol is used for the guanidinylation of *N*-α-ornithine and lysine derivatives that are insoluble in CH_2Cl_2 (eq 2).[1] It is necessary to first convert

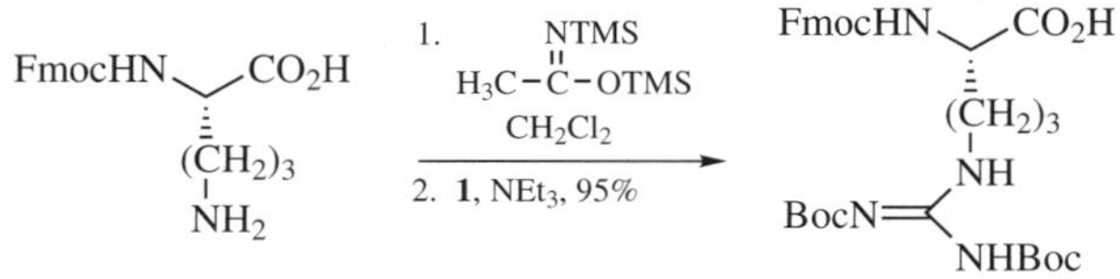

(2)

these compounds into soluble derivatives by silylation with methyl(trimethylsilyl)trifluoroacetamide in refluxing CH_2Cl_2 under N_2. A slight excess of reagent **1** is then added at room temperature followed by NEt_3. Yields of greater than 90% are typical.

Modification of Peptides on Solid Support. As with 4-nitro-1*H*-pyrazole-*N,N′*-diBoc-1-carboxamidine,[9] reagent **1** can guanidinylate ornithine and lysine residues of a peptide on solid support using a similar protocol as described in the first section (eq 3).[2] Once the desired peptide is constructed on the resin of choice and the amino-protecting group is removed, the free amino function is guanidinylated by treatment with a solution of reagent **1** and NEt_3 overnight. This reaction can be accomplished for a peptide with multiple amine-containing residues and even when the amine-containing amino acid of the peptide is proximal to a sterically demanding resin.

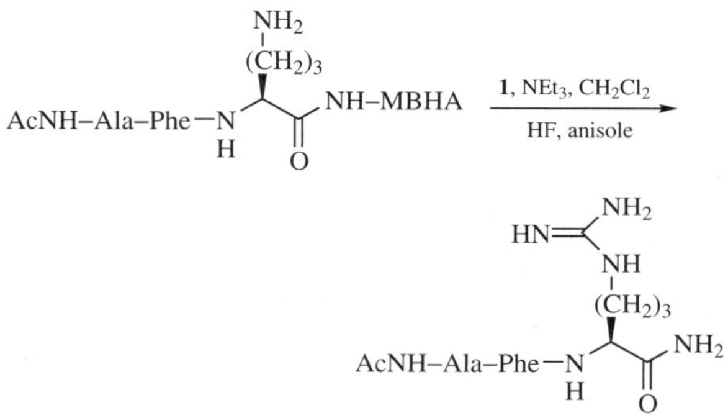

(3)

Preparation of Guanidinoglycosides. Reagent **1** is also used to convert aminoglycosides such as tobramycin to the corresponding guanidinoglycosides (eq 4).[10] This transformation is carried

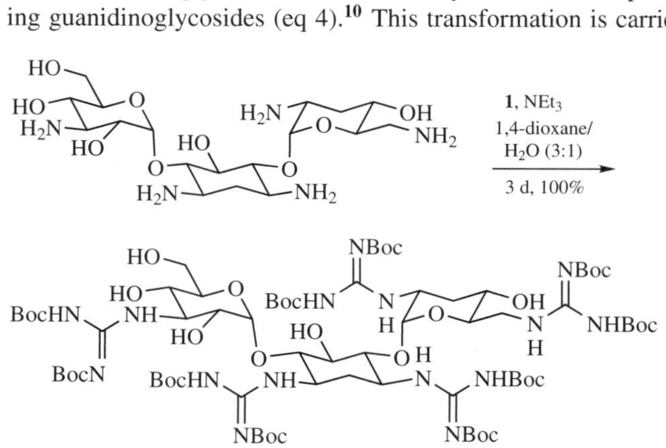

(4)

out in a mixture of 1,4-dioxane and H_2O with a three fold excess of reagent **1** and NEt_3 to each amino group. The reaction is carried out at room temperature and takes approximately 3 d for complete conversion. The reaction can be terminated before complete conversion and the partially guanidinylated derivatives can be isolated using flash column chromatography. Other syntheses employ reagents such as 3,5-dimethylpyrazoylformamidinium nitrate[11] and 1*H*-pyrazole-1-carboxamidine hydrochloride[12,13] at elevated temperatures in DMF, but all show limited success presumably because of limited solubility of the aminoglycosides in DMF.

Related Reagents. 1*H*-Pyrazole-*N,N′*-diBoc-1-carboxamidine [*152120-54-2*]; *N*-[bis(methylthio)methylene]-*p*-toluenesulfonamide [*2651-15-2*]; 2-methyl-2-thiopseudourea sulfate [*867-44-7*]; *O*-methylisourea sulfate [*52328-05-9*]; 1*H*-pyrazole-1-carboxamidine hydrochloride [*4023-02-3*].

---

1. Feichtinger, K.; Zapf, C.; Sings, H. L.; Goodman, M., *J. Org. Chem.* **1998**, *63*, 3804.
2. Feichtinger, K.; Sings, H. L.; Baker, T. J.; Matthews, K.; Goodman, M., *J. Org. Chem.* **1998**, *63*, 8432.
3. Baker, T. J.; Goodman, M., *Synthesis Special Issue*, **1999**, 1423.
4. Baker, T. J.; Tomioka, M.; Goodman, M., *Organic Syntheses* **2001**, *78*, 91–98.
5. Yong, Y. F.; Kowalski, J. A.; Lipton, M. A. J., *Org. Chem.* **1997**, *62*, 1540.
6. Bernatowicz, M. S.; Wu, Y.; Matsueda, G. R., *Tetrahedron Lett.* **1993**, *34*, 3389.
7. Drake, B.; Patek, M.; Lebl, M., *Synthesis* **1994**, 579.
8. Bergeron, R. J.; McManis, J. S., *J. Org. Chem.* **1987**, *52*, 1700.
9. Yong, Y. F.; Kowalski, J. A.; Thoen, J. C.; Lipton, M. A., *Tetrahedron Lett.* **1999**, *40*, 53.
10. Baker, T. J.; Luedtke, N. W.; Tor, Y.; Goodman, M., *J. Org. Chem.* **2000**, *65*, 9054.
11. Wessel, H. P.; Banner, D.; Gubernator, K.; Hilpert, K.; Müller, K.; Tschopp, T., *Angew. Chem. Int. Ed. Engl.* **1997**, *36*, 751.
12. Cotner, E. S.; Smith, P. J., *J. Org. Chem.* **1998**, *63*, 1737.
13. Hauser, S. L.; Cotner, E. S.; Smith, P. J., *Tetrahedron Lett.* **1999**, *40*, 2865.

Murray Goodman
University of California, San Diego, La Jolla, CA, USA

1,3-Bis(2,2-dimethyl-1,3-dioxolan-4-ylmethyl)carbodiimide (BDDC)

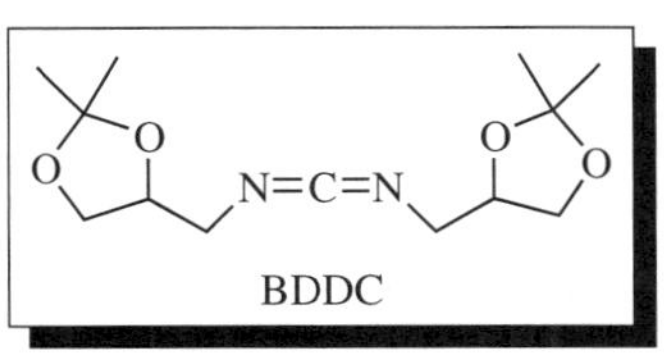

[159390-26-8] $C_{13}H_{22}N_2O_4$ (MW 270.32)

(reagent for clean, efficient peptide coupling, esterification and dehydration)

Physical Data: bp 115–125 °C (<1 mmHg); density 1.062 g mL^{-1}.
Solubility: most organic solvents.
Form Supplied in: pale yellow oil; commercially available.
Purification: distill at <1 mmHg.
Handling, Storage, and Precautions: material should be stored under a dry atmosphere of Ar or N_2 at 0 °C. Many carbodiimides are strong sensitizers, therefore 1,3-bis(2,2-dimethyl-1,3-dioxolan-4-ylmethyl)carbodiimide (BDDC) should be considered as such. Avoid all skin contact, and use proper personal protection equipment (gloves, goggles, lab coat, etc.). BDDC should only be used in a well-ventilated fume hood. BDDC is incompatible with strong acids, including aqueous acids. BDDC is flammable; avoid fire or sparks. Decomposes upon prolonged storage at room temperature and exposure to air and moisture.
Preparative Methods: BDDC can be purchased commercially or it can be easily made in the laboratory from 3-aminomethyl-2,2-dimethyl-1,3-dioxolane.[1]

Residue-free Reactions. BDDC performs peptide couplings, esterifications, and dehydrations similar to the well known carbodiimides dicyclohexylcarbodiimide (DCC), diisopropylcarbodiimide (DIC), and 1-[3-(dimethylamino)propyl]-3-ethylcarbodiimide hydrochloride (EDC),[2] but is designed to be easily removed from reaction mixtures with dilute acidic washes (eq 1).[1] BDDC and its by-products are highly hydrophilic allowing for easy removal from organic systems with dilute aqueous acid washes, unlike DCC or DIC. In addition, the neutral structure of BDDC gives it enhanced solubility in organic solvents, and makes it less sensitive to moisture as compared to EDC.

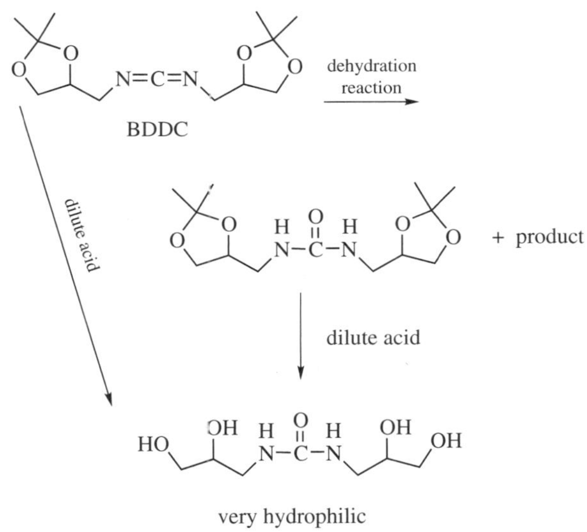

(1)

Peptide Couplings. BDDC is an excellent reagent for the coupling of amino acids to form peptide bonds (eq 2). The reagent

$$\text{CBz-Ala-Phe} + \text{Gly-OBn-TsOH} \xrightarrow[\text{DMF, rt}]{\textbf{BDDC}\text{, HOBT, NMM, CuCl}_2} \text{CBz-Ala-Phe-Gly-OBn} \quad (2)$$

78%, <0.1% LD epimer

affords good yields of dipeptides and tripeptides under typical coupling conditions with minimal racemization. When combined with $CuCl_2$, even amide protected *C*-terminus couplings give no racemization. All carbodiimide residues are removed from the reaction mixture with a simple aqueous acid wash, eliminating the need for chromatography.

Esterifications. Many methods are known for esterifying carboxylic acids.[3] Strongly acidic or basic methods dominate these methods. One essentially neutral method is to employ an isourea (eq 3).[4] BDDC easily forms isoureas in nearly quantitative yield under mild conditions, unlike DCC, DIC, or EDC.[1]

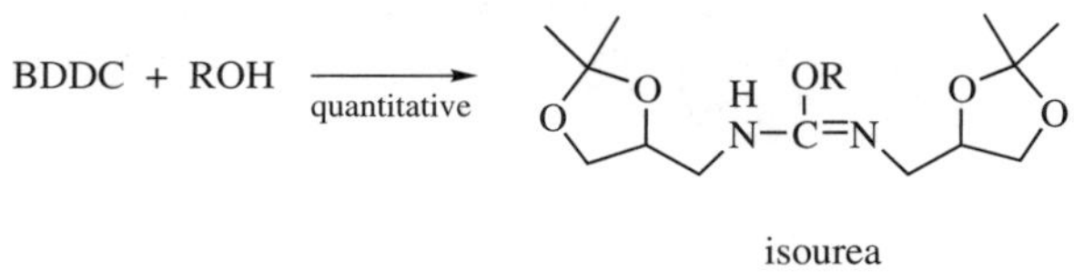

(3)

ROH = primary, secondary, tertiary alcohol

The resulting isoureas convert carboxylic acids cleanly to the corresponding esters without the use of harshly acidic or basic reagents (eq 4).

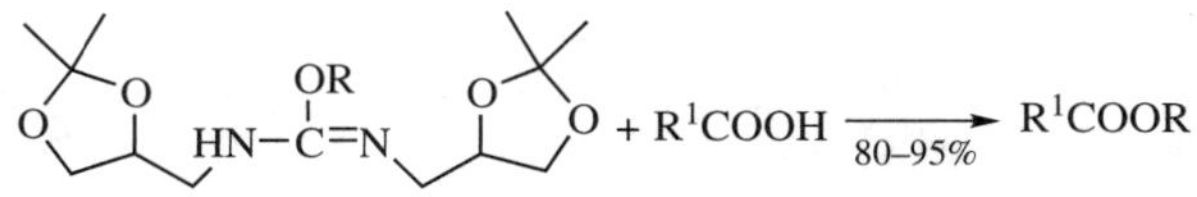

(4)

This is desirable when sensitive functional groups are present in the molecule and racemization must be avoided. The isourea reacts at the oxygen center, and not the carbonyl, so α-racemization cannot occur.

Dehydrations. Carbodiimides are also employed as general dehydrating agents.[2] BDDC was used to dehydrate benzyl 3-hydroxybutanoate to benzyl crotonate in 83% yield.

Related Reagents. dicyclohexylcarbodiimide (DCC); diisopropylcarbodiimide (DIC); 1-[3-(dimethylamino)propyl]-3-ethylcarbodiimide hydrochloride (EDC).

1. Gibson, F.; Park, M.; Rapoport, H., *J. Org. Chem.* **1994**, *59*, 7503.
2. (a) Williams, A.; Ibrahim, I. T., *Chem. Rev.* **1981**, *81*, 589. (b) Mikolajczyk, M.; Kielbasinski, P., *Tetrahedron* **1981**, *37*, 233. (c) Kurzer, F.; Douraghi-Zahdeh, D., *Chem. Rev.* **1967**, *67*, 107.
3. Greene, T.; Wuts, P., *Protective Groups in Organic Synthesis*, 3rd edn; Wiley Interscience: Chichester, 1999, pp 372-428.
4. Mathias, L., *Synthesis* **1979**, 561.

Frank Gibson
Bristol-Myers Squibb, New Brunswick, New Jersey, USA

Bis(2,4-dinitrophenyl) Carbonate[1]

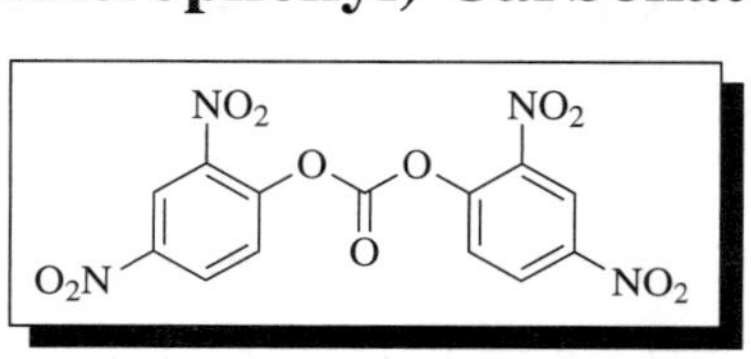

[7497-12-3] $C_{13}H_6N_4O_{11}$ (MW 394.21)

(peptide and azapeptide coupling reagent that functions by activating amines, forming derivatives suitable for condensation with N-protected amino acids;[1–4] converts carboxylic acids to activated esters[2])

Alternate Names: carbonic acid bis(2,4-dinitrophenyl) ester; 2,4-dinitrophenyl 2,4-dinitrophenoxyformate.

Physical Data: yellow solid, mp 130–131 °C,[5] mp 148 °C;[2] hygroscopic; X-ray structure.[6]

Solubility: sol EtOAc, pyridine, DMF, $CHCl_3$; moderately sol benzene; insol hexane; dec hot H_2O, EtOH.

Preparative Method: diphenyl carbonate (20 g) is added to a stirred mixture of conc HNO_3 (100 mL) and conc H_2SO_4 (120 mL) at 50–55 °C over 1 h. The resulting suspension is stirred at 50 °C for 3 h, left at rt for 90 h and poured onto ice. The precipitate is washed with iced water and dried in vacuo over P_2O_5. This affords bis(2,4-dinitrophenyl) carbonate (31.2 g, 85%) which recrystallizes from hexane/benzene giving crystals, mp 144–146 °C.[3]

Handling, Storage, and Precautions: store in a cool dry place to preclude contact with moisture; the hazardous and toxicological properties have not been investigated; may explode if heated rapidly.

Peptide and Azapeptide Coupling. The reagent is used to activate amino groups (eq 1) for subsequent coupling with N-protected amino acids and hydrazides, affording peptide (eq 2) and azapeptide (eq 3) derivatives, respectively.[1–4] Cbz-protected amino acids are coupled without racemization but *N*-acylated amino acids lose all optical activity.[2]

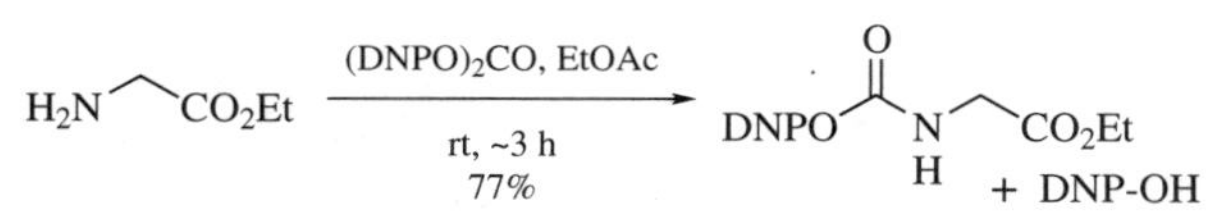

(1)

DNP = 2,4-dinitrophenyl

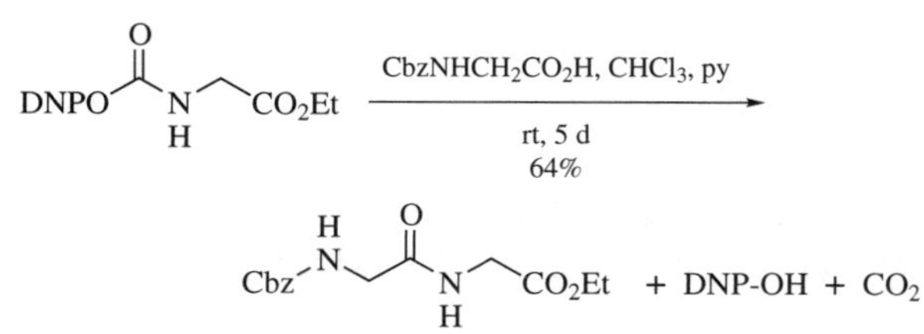

(2)

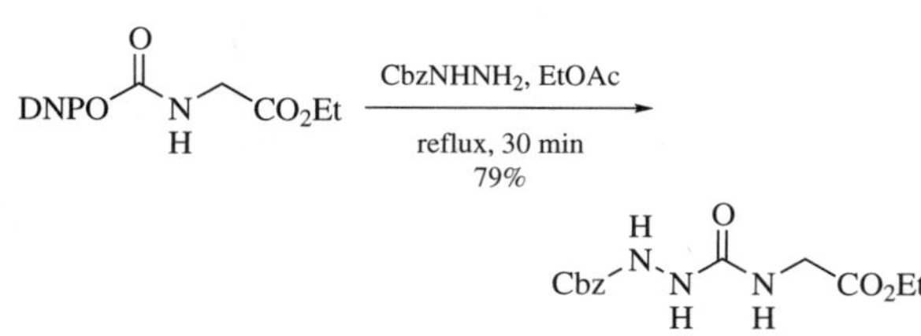

(3)

Attempts to prepare azapeptides by activating hydrazides have proved unsuccessful because of the propensity of the intermediate to undergo base-catalyzed cyclization and elimination to stable oxadiazolones (eq 4).[3]

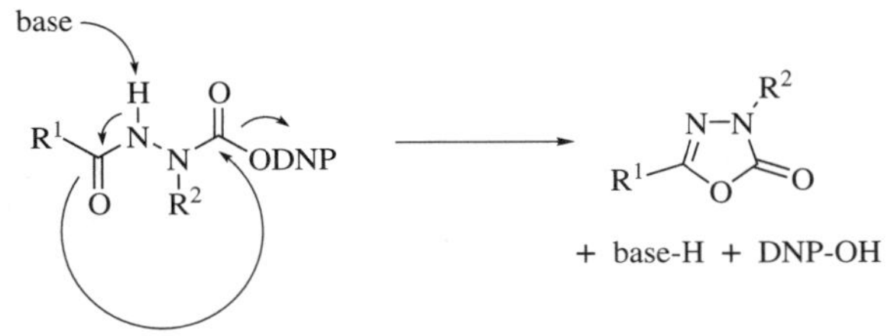

(4)

Activated Esters from Carboxylic Acids. The reagent is attacked by carboxylate ions, forming unstable intermediates which spontaneously decarboxylate to dinitrophenyl esters (eq 5).[2] These activated esters are good acylating agents and undergo transesterification on reaction with alkoxide or aryloxide ions (eq 6).[2] The mild conditions under which these activated dinitrophenyl esters can be formed makes them preferable intermediates to the analogous 4-mononitrophenyl esters which are less active towards nucleophiles; also 2,4-dinitrophenol is more acidic than 4-nitrophenol and consequently may be easier to separate from the desired reaction product during workup.

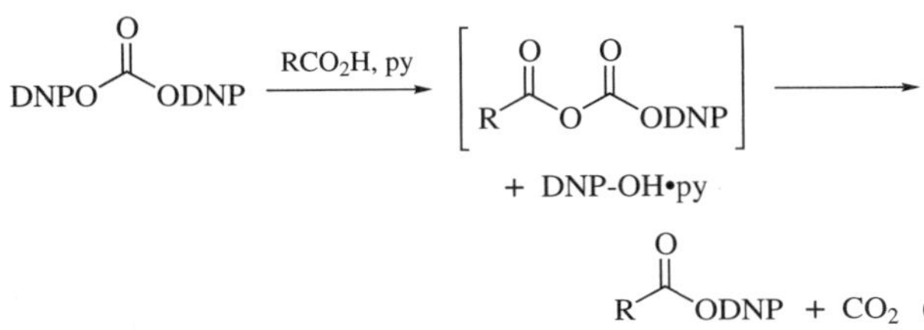

(5)

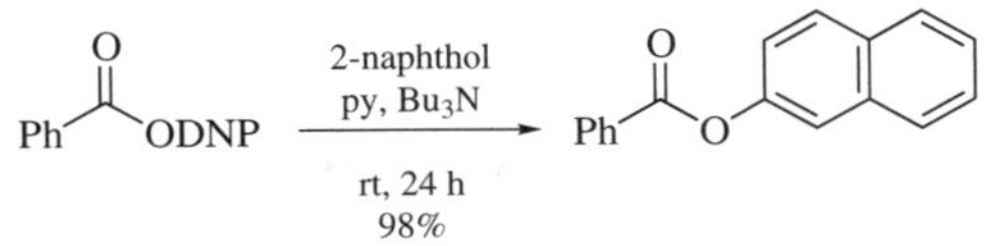

(6)

1. (a) Gante, J. *Ber. Dtsch. Chem. Ges./Chem. Ber.* **1966**, *99*, 1576 (*Chem. Abstr.* **1966**, *65*, 2346b) . (b) *Fieser & Fieser* **1967**, *1*, 56; **1969**, *2*, 29.
2. Glattard, R.; Matter, M. *Helv. Chim. Acta* **1963**, *46*, 795 (*Chem. Abstr.* **1963**, *59*, 2939g) .
3. Gray, C. J.; Ireson, J. C.; Parker, R. C. *Tetrahedron* **1977**, *33*, 739.
4. Gray, C. J.; Quibell, M; Jiang, K.-L.; Baggett, N. *Synthesis* **1991**, 141.
5. Gresser, M. J.; Jencks, W. P. *J. Am. Chem. Soc.* **1977**, *99*, 6970.
6. King, J. A., Jr.; Bryant, G. L., Jr.; *Acta Crystallogr. Sect. C* **1990**, *C46*, 1330.

Niall M. Hamilton

Organon Laboratories Ltd, Newhouse, Lanarkshire, UK

Bis(1-methoxy-2-methyl-1-propenyloxy)dimethylsilane[1]

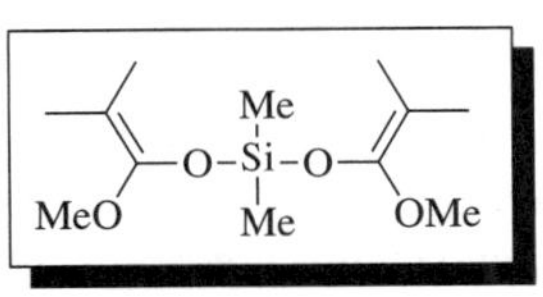

[86934-32-9] $C_{12}H_{24}O_4Si$ (MW 260.40)

(very useful bifunctional protecting agent for various types of H-acidic materials under mild conditions)

Physical Data: 89–90 °C/6.5 mmHg.

Form Supplied in: colorless liquid; not commercially available.

Preparative Method: prepared by treating 2 equiv of methyl lithioisobutyrate with 1 equiv of dichlorodimethylsilane at low temperature in dry THF, followed by distillation of the colorless liquid.

Handling, Storage, and Precautions: moisture sensitive; should be stored under nitrogen in a refrigerator.

The Dialkylsilylene Protecting Group. Bifunctional compounds containing alcohol, thiol, acid, and amine units may be protected as cyclic dimethylsilylene derivatives, which are analogous to isopropylidene derivatives, by treatment with the title reagent (**1**).[2] Cleavage of the silylene protective group is easily achieved by hydrolysis or solvolysis. It should be noted that the dimethylsilylene group is highly sensitive to hydrolysis and is unstable to column chromatography.

Silylene derivatives have become very useful in synthetic organic chemistry. For example, the cyclic dimethylsilylene derivative (**2**). of *o*-hydroxyacetophenone is useful for [4 + 2] cycloadditions (eq 1).[3] A glycosyl donor and acceptor have been linked as the dimethylsilylene derivative (**3**) and used for stereoselective glycosylation (eq 2).[4] Also, an efficient and stereocontrolled aldol reaction of silylene enolate (**4**) with aldehydes has been reported (eq 3).[5]

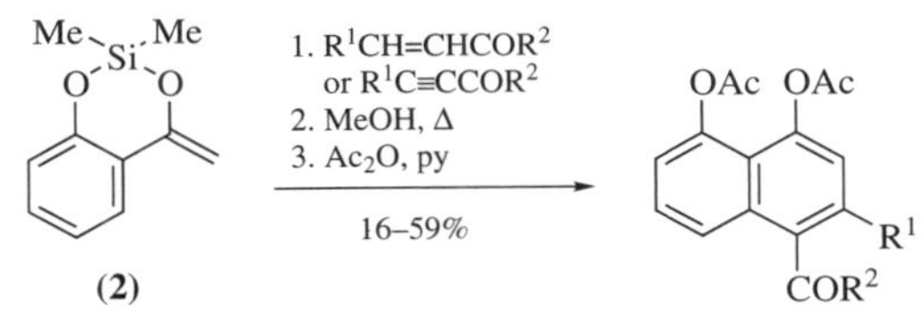

(1)

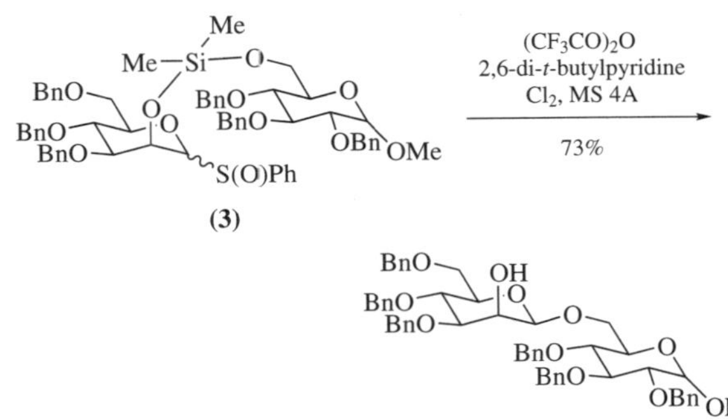

(2)

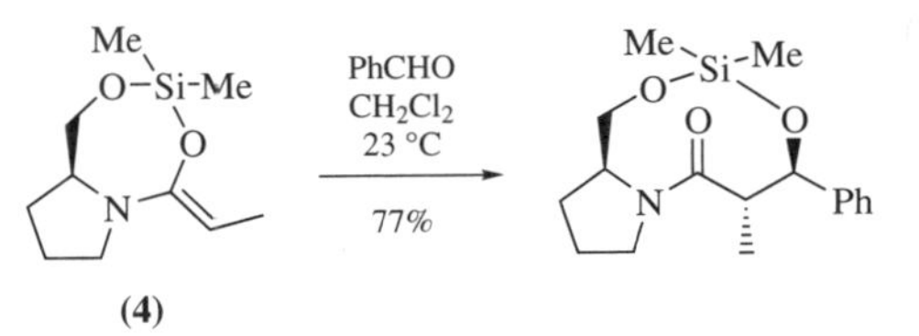

(3)

Advantages of Bis(1-methoxy-2-methyl-1-propenyloxy)dimethylsilane. Dichlorosilanes,[6] bis(trifluoromethanesulfonyl) silanes,[7] and dimethoxysilanes[8] have been used for the synthesis of dialkylsilylene derivatives. These methods, however, have some disadvantages: (i) a strong base or acid catalyst is required; (ii) forcing reaction conditions (high temperature for a long period) are sometimes required; (iii) large amounts of side-products such as inorganic salts, or amine or acid salts, are produced that must be separated from the moisture-sensitive silylene derivatives; and (iv) the yields of the products are not always high. Reagent (**1**), which is much more reactive than other difunctional silyl reagents, allows the ready silylenation of H-acidic materials in the absence of base, and allows isolation of pure products in almost quantitative yields without aqueous workup. The only reaction side-product is methyl isobutyrate, which is volatile.

Bifunctional Protection of H-Acidic Materials[1,9]. The reaction of (**1**) with H-acidic materials is generally carried out by employing a slight excess of the reagent in an inert solvent such as CH_2Cl_2 or MeCN and usually gives a high yield of the corresponding silylene derivatives. Furthermore, addition of catalytic amounts of dichlorodimethylsilane induces a slightly exothermal reaction, leading to the rapid formation of the silylenes. A wide range of 1,2-, 1,3-, and 1,4-difunctional compounds give the corresponding silylenes in high yields (eqs 4–8). Monofunctional alcohols and carboxylic acids can also be linked via the dimethylsilylene group forming dialkoxy- and diacyloxydimethylsilane, respectively (eqs 9 and 10).

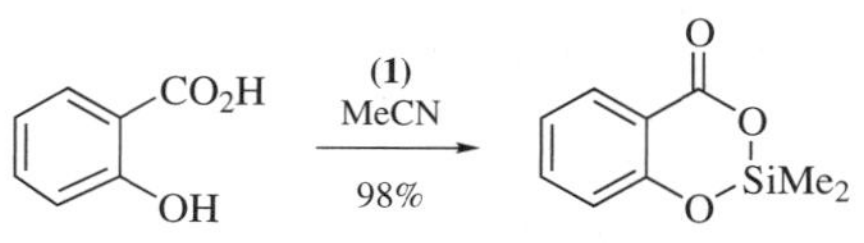

(4)

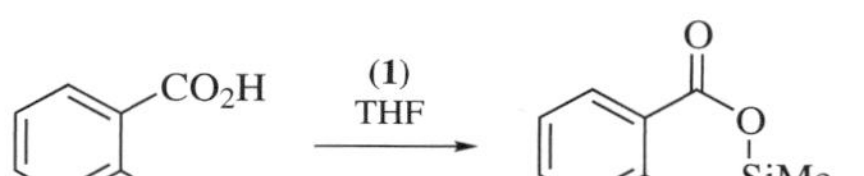

(5)

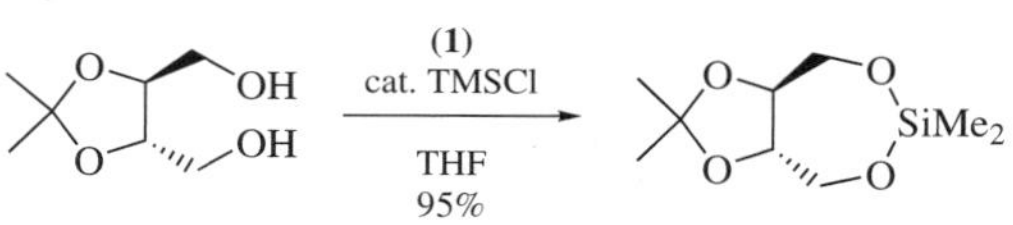

(6)

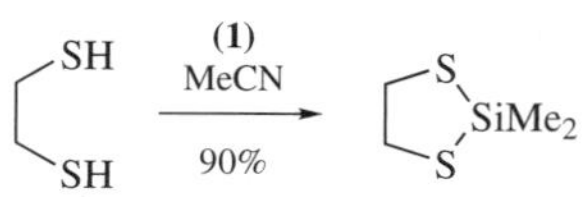

(7)

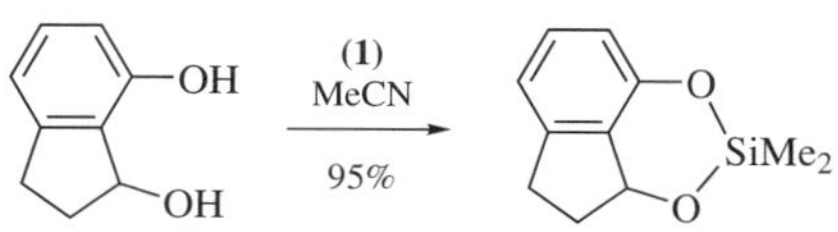

(8)

$$\text{PhOH} \xrightarrow[\text{98\%}]{(\mathbf{1}),\ CH_2Cl_2} (\text{PhO})_2\text{SiMe}_2 \qquad (9)$$

$$\text{PhCO}_2\text{H} \xrightarrow[\text{96\%}]{(\mathbf{1}),\ CH_2Cl_2} (\text{PhCO}_2)_2\text{SiMe}_2 \qquad (10)$$

1. Kita, Y.; Yasuda, H.; Sugiyama, Y.; Fukata, F.; Haruta, J.; Tamura, Y. *Tetrahedron Lett.* **1983**, *24*, 1273.
2. Lalonde, M.; Chan, T. H. *Synthesis* **1985**, 817.
3. (a) Kita, Y.; Yasuda, H.; Tamura, O.; Tamura, Y. *Tetrahedron Lett.* **1984**, *25*, 1813. (b) Kita, Y.; Ueno, H.; Kitagaki, S.; Kobayashi, K.; Iio, K.; Akai, S. *Chem. Commun./J. Chem. Soc., Chem. Commun.* **1994**, 701.
4. Stork, G.; Kim, G. *J. Am. Chem. Soc.* **1992**, *114*, 1087.
5. (a) Myers, A. G.; Widdowson, K. L. *J. Am. Chem. Soc.* **1990**, *112*, 9672. (b) Myers, A. G.; Widdowson, K. L.; Kukkola, P. J. *J. Am. Chem. Soc.* **1992**, *114*, 2765.
6. (a) Spassky, N. *C.R. Hebd. Seances Acad. Sci.* **1960**, *251*, 2371. (b) Wieber, M.; Schmidt, M. *Z. Naturforsch., Tell B* **1963**, *18b*, 846. (c) Wieber, M.; Schmidt, M. *J. Organomet. Chem.* **1963**, *1*, 93. (d) Abel, E. W.; Bush, R. P. *J. Organomet. Chem.* **1965**, *3*, 245. (e) Kober, F.; Ruhl, W. J. *J. Organomet. Chem.* **1975**, *101*, 57. (f) Meyer, H.; Nagorsen, G.; Weiss, A. *Z. Naturforsch., Tell B* **1975**, *30b*, 488. (g) Markiewicz, W. T.; Wiewiorowski, M. *Nucl. Acid Res. Special Publ.* **1978**, *4*, 185. (h) Markiewicz, W. T. *J. Chem. Res. (S)* **1979**, 24; Markiewicz, W. T.; Padyukova, N. Sh.; Samek, Z.; Smrt, J. *Collect. Czech. Chem. Commun.* **1980**, *45*, 1860. (i) van Boeckel, S. A. A.; van Boom, J. H. *Chem. Lett.* **1981**, 581. (j) Djuric, S.; Venit, J.; Magnus, P. *Tetrahedron Lett.* **1981**, *22*, 1787. (k) Cragg, R. H.; Lane, R. D. *J. Organomet. Chem.* **1981**, *212*, 301. (l) Trost, B. M.; Caldwell, C. G. *Tetrahedron Lett.* **1981**, *22*, 4999.
7. Corey, E. J.; Hopkins, P. B. *Tetrahedron Lett.* **1982**, *23*, 4871.
8. Jenner, M. R.; Khan, R. *Chem. Commun./J. Chem. Soc., Chem. Commun.* **1980**, 50.
9. Brown, R. F. C.; Coulston, K. J.; Eastwood, F. W.; Hill, M. P. *Aust. J. Chem.* **1988**, *41*, 215.

Yasuyuki Kita
Osaka University, Japan

N-Bis(methylthio)methylene-*p*-toluenesulfonamide

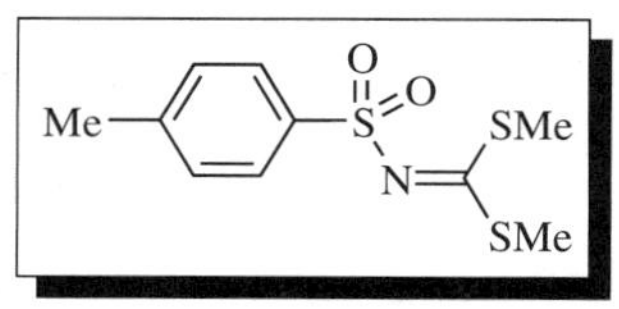

[2651-15-2] $C_{10}H_{13}NO_2S_3$ (MW 275.41)

(the title reagent reacts with a range of different nucleophilic reagents to give medicinally useful guanidinyl derivates and various heterocyclic compounds. In addition, a unique application of this reagent in 1,3-dipolar cycloaddition reactions has been developed)

Physical Data: mp 110–112 °C.

Preparation. *N*-Bis(methylthio)methylene-*para*-methylbenzenesulfonamide (**1**) can be prepared by condensing *para*-toluenesulfonamide with carbon disulfide (CS_2) in an alkaline medium using either DMSO as solvent and dimethylsulfate (Me_2SO_4) as the methylating agent (eq 1),[1] or DMF as solvent and MeI as the methylating agent (eq 2).[2]

A typical procedure involves dissolving 1 mol of *para*-toluenesulfonamide in sufficient DMSO or DMF (~150 mL per

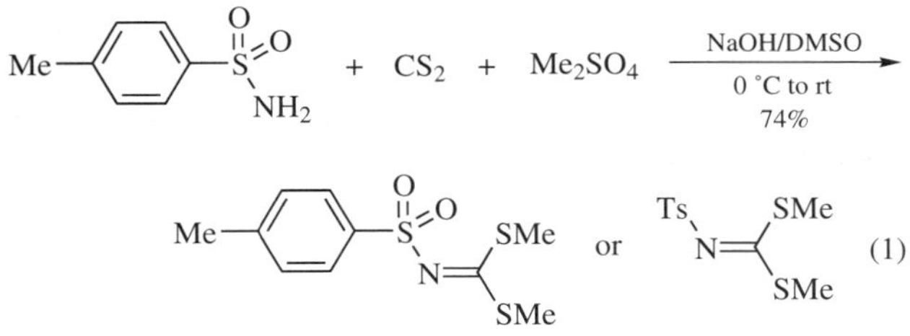

1

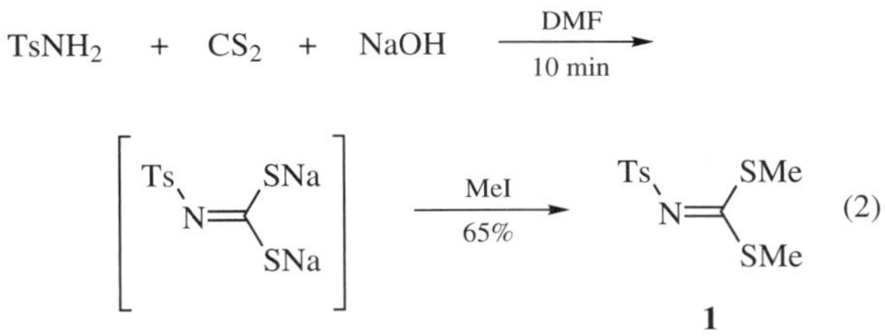

1

0.2 mol[2]) at room temperature. After adding and dissolving 2 mol of NaOH, 1 mol of carbon disulfide was added carefully at 0 °C over a few minutes. After stirring for another 30 min at 0 °C, 2 mol of Me_2SO_4 was added dropwise over 1 h. After warming up to room temperature, ample water was added and the resulting crystalline precipitate was collected via filtration. The filter cake was washed with water, dried, and recrystallized using MeOH. The temperature needed to be raised to 50–60 °C when using MeI as the alkylating agent.

Applications In Organic Synthesis. *N*-Bis(methylthio) methylene-*para*-toluenesulfonamide has been extensively employed as an electrophilic agent for condensation with different nucleophiles such as amines, alcohols, thiols, and anionic species to synthesize medicinally useful guanidinyl derivates and heterocycles.

Synthesis of Guanidine Derivatives

Using Two Different Amines

The sodium salt of *N*-Boc ornithine could be reacted with **1** to give a mono-addition product (eq 3).[3] Subsequent addition of a second amine (HNRR[1]) allowed the formation of *N*-Boc arginine derivatives in good yields. These derivatives are useful for solid-phase peptide synthesis.

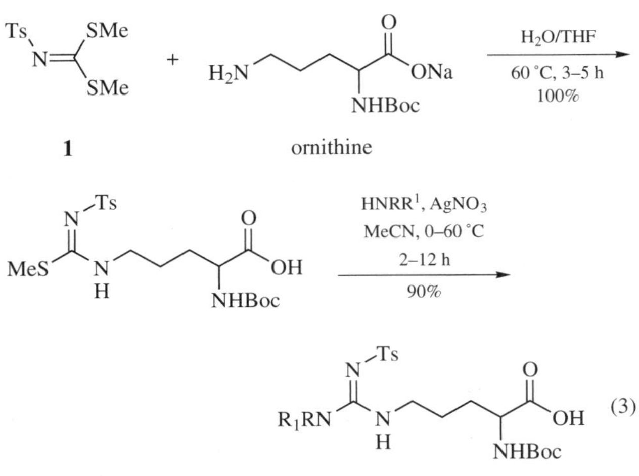

1 ornithine

arginines

Burgess and co-workers have illustrated the use of reagent **1** in the total syntheses of all four stereoisomers of carnosadine, a cyclopropane amino acid based natural product isolated from red marine alga (eqs 4 and 5).[4,5] These synthetic efforts ultimately led to preparations of a series of novel 2,3-methanologs of natural amino acids for constructing peptidomimetics with unique conformational motifs and potential medicinal value.

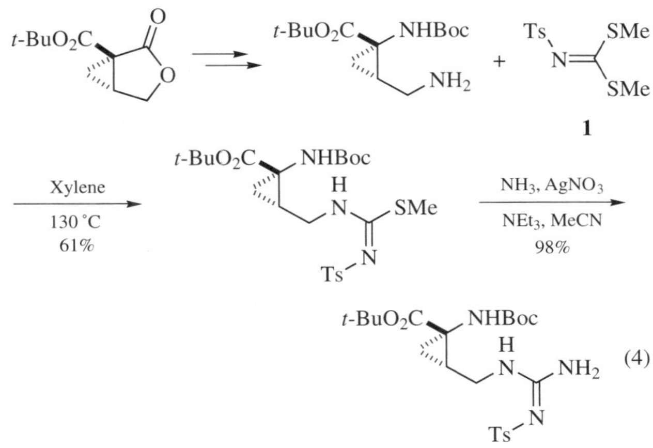

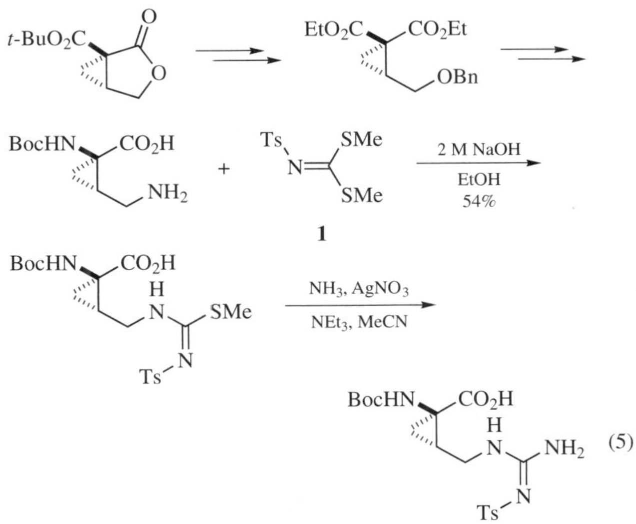

In a related manner, reagent **1** has been used to prepare thymidine dimers that are linked by a guanidine unit (eq 6).[6,7] These dimers were subsequently incorporated into oligothymidylates to study base pairing properties.[6,7] Another example of guanidination of amines is shown in eq 7 for the synthesis of conformationally constrained arginine derivatives starting from 4-hydroxy-prolines.[8] Although only one stereoisomer is shown, all four possible stereoisomers of 4-hydroxy-prolines were employed in this manner In general, by comparing all the examples shown thus far, unless $AgNO_3$ is used, reactions of amine with **1** require refluxing conditions in solvents such as CH_3CN, H_2O/THF, $CHCl_3$, and alcoholic solvents to displace one of the two methyl sulfide groups.

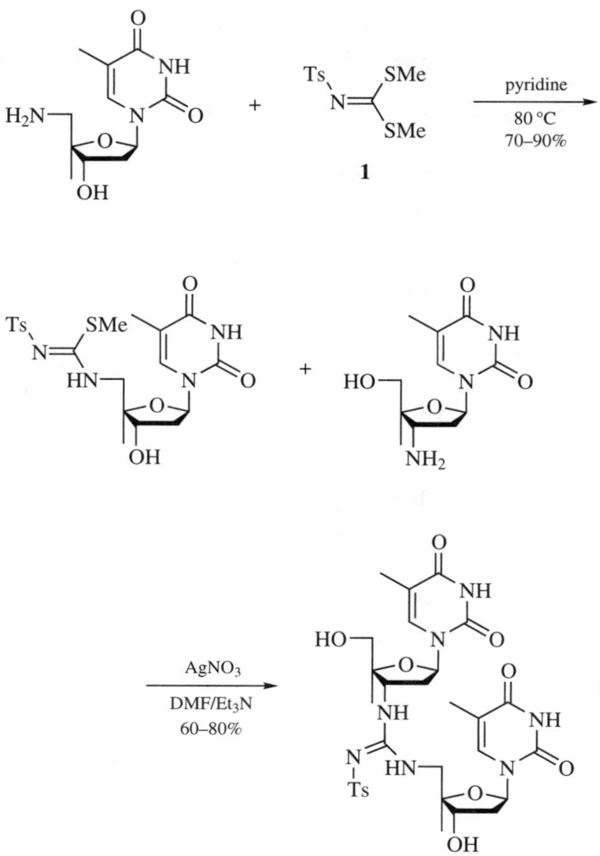

(6)

Some guanidinyl diesters could be prepared using reagent **1**.[9a] As shown in eq 8, symmetric guanidinyl diesters can be derived directly via chlorination of **1** in glacial acetic acid[9b,9c] at rt followed by addition of 2.0 equiv of an amino ester NH_2E^1 (or NH_2E^2) to dichloro imine intermediate **2** at rt. In addition to preparing guanidines via addition of amines, it is noteworthy that dichloroimine **2** has also been employed in various reactions with thiols, thiolates, and alkoxides to give the corresponding imido-dithiocarbamates and imido-carbamates.[9b,9c] Unsymmetric ones can be prepared by first refluxing the sodium salt of the first amino ester NH_2E^1 with reagent **1** to give mono-ester intermediate **3**. Subsequent elimination of HSMe in **3** using $AgNO_3/Et_3N$ followed by addition of a second amino ester NH_2E^2 to the resulting carbodiimide leads to unsymmetric guanidinyl diesters. Alternatively, mono-ester intermediate **3** could be chlorinated carefully using Cl_2 and HOAc at low temperature to give chloro amidate **4** followed by addition of the second amino ester NH_2E^2. Chloro amidate **4** could also be prepared directly from **1** via refluxing with SO_2Cl_2 followed by addition of the first amino ester NH_2E^1.

Using Diamines

Diamines could also be used to construct guanidines employing reagent **1**. As shown in eq 9, depending upon the molar ratio, reactions of **1** with the diamine derived from triethylene glycol led to various different macrocycles potentially useful in host-guest chemistry.[10]

Reactions of reagent **1** with other simple diamines such as 1,2- (eqs 10 and 11),[11,12] 1,3- (eqs 12 and 13),[13,14] and 1,4-diamines (eqs 14 and 15)[15,16] have all been reported to give interesting cyclic guanidines. Particularly, they have been featured in the synthesis of natural products such as DL-capreomycidine (eq 13),[14] and in Rappoport's bicyclic guanidines with a bridgehead nitrogen atom (eq 14).[15] All these studies were carried out using racemic diamines.

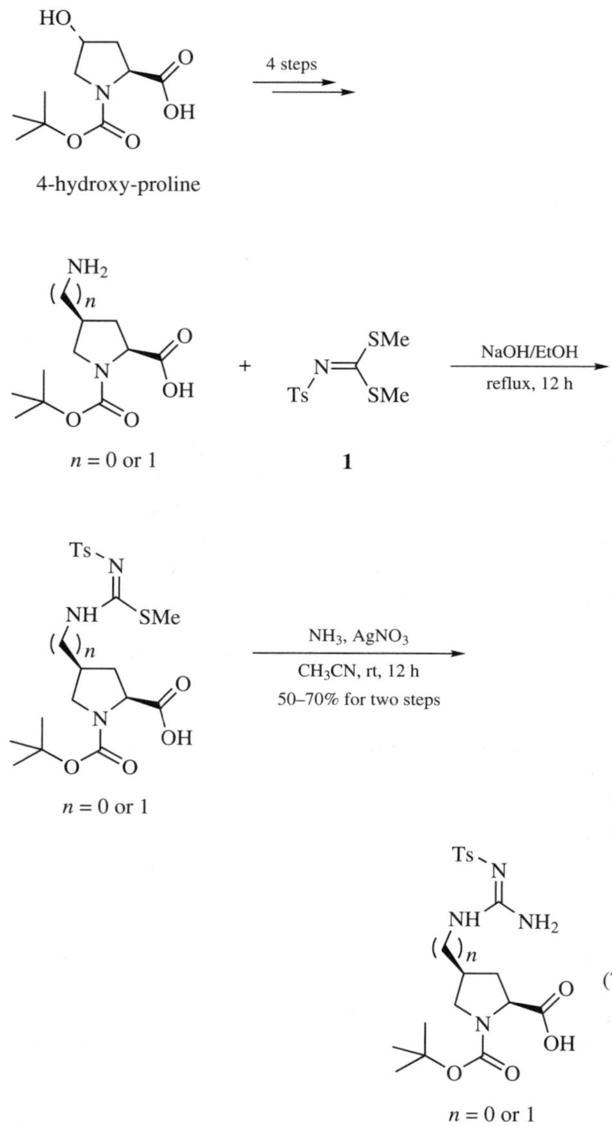

(7)

Synthesis of Triazoles

N-Bis(methylthio)methylene-*para*-toluenesulfonamide has been utilized in the synthesis of substituted triazoles that possess antimicrobial activities (eqs 16–18).[17–19] The substituted triazoles in eq 17 were specifically used as precursors for constructing several fused heterocycles that exhibit leishmanicidal activity against *L. donovani*. The level of regioselectivity in these annulation reactions is also noteworthy.

Other Nucleophilic Addition Reactions

Addition of Neutral Nucleophiles The previous two sections have already showcased many of the additions of neutral nucelophiles such as amines and alcohols to **1**. The remaining examples, mostly mono-addition of amines, are summarized below.

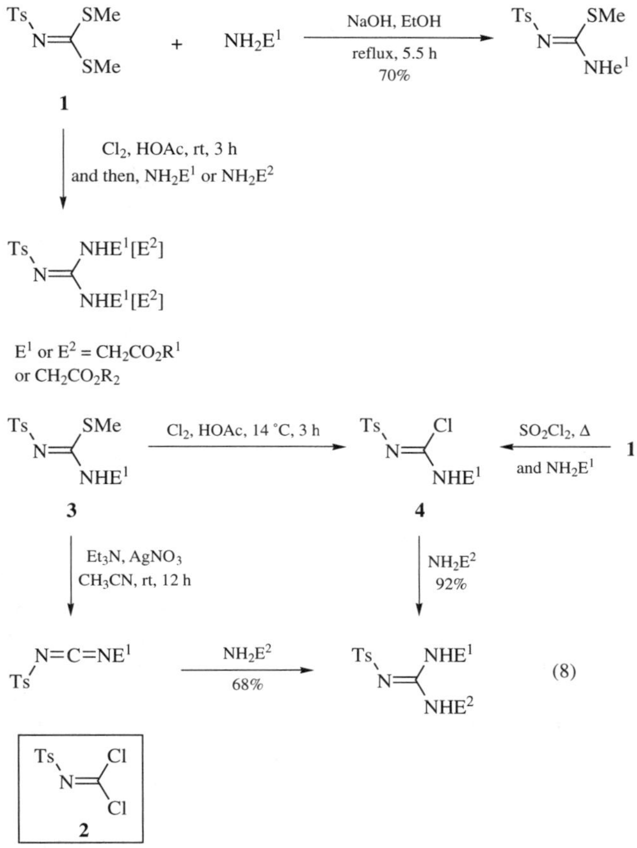

(8)

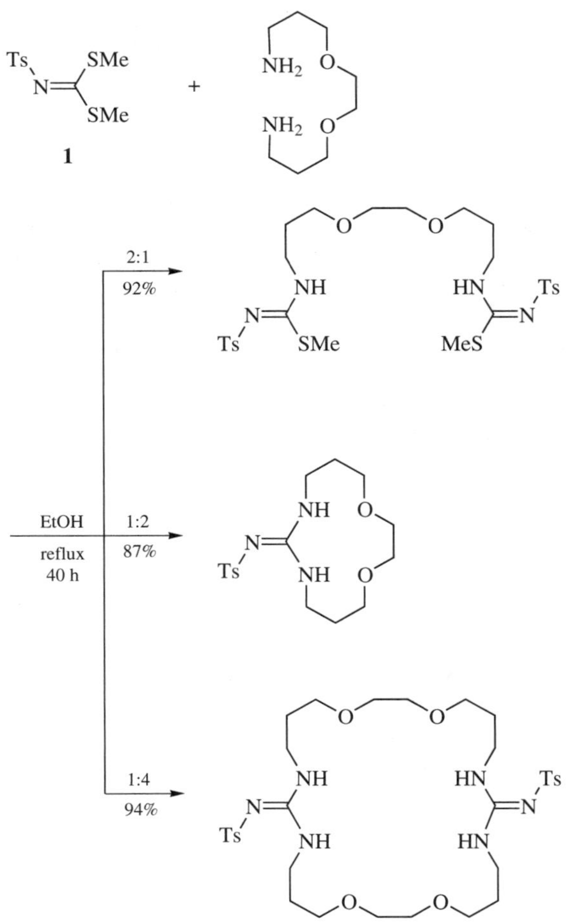

(9)

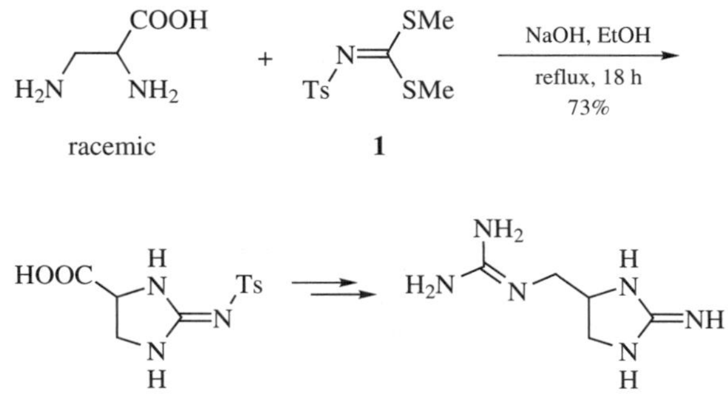

(10)

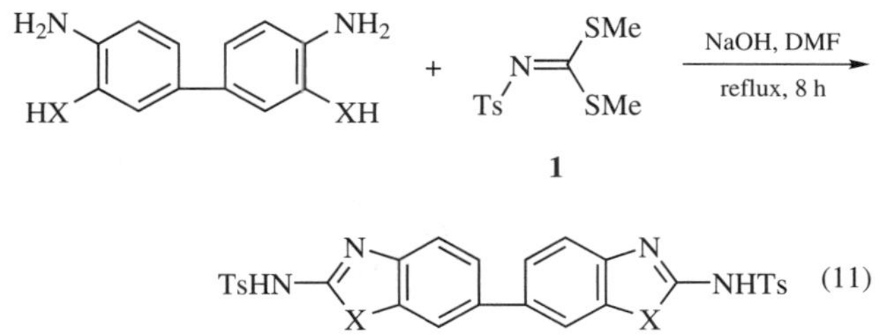

(11)

X	Yield (%)
O	73
S	73
NH	—

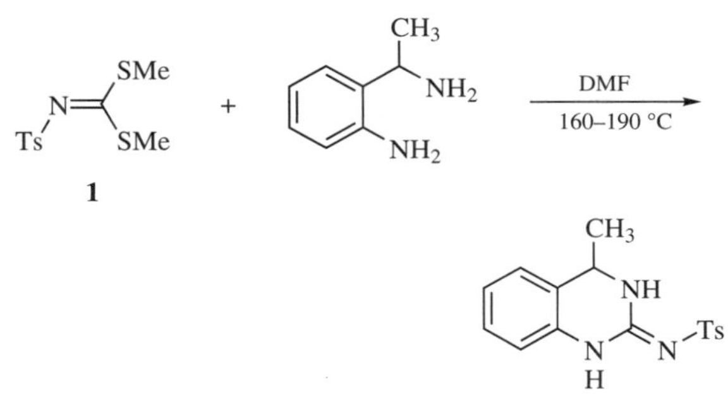

(12)

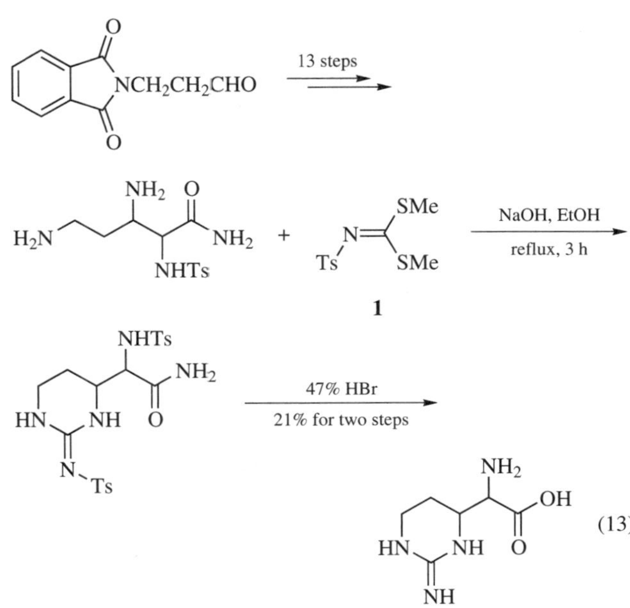

(13)

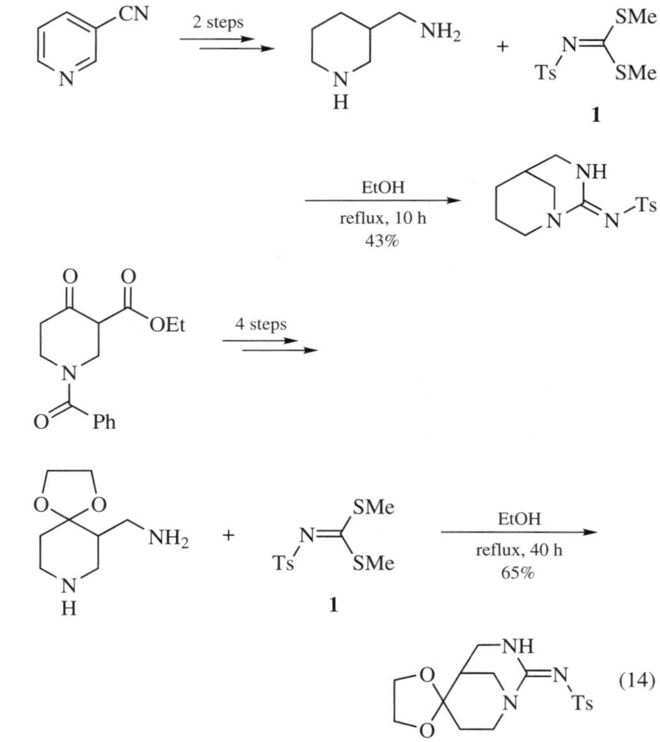

(14)

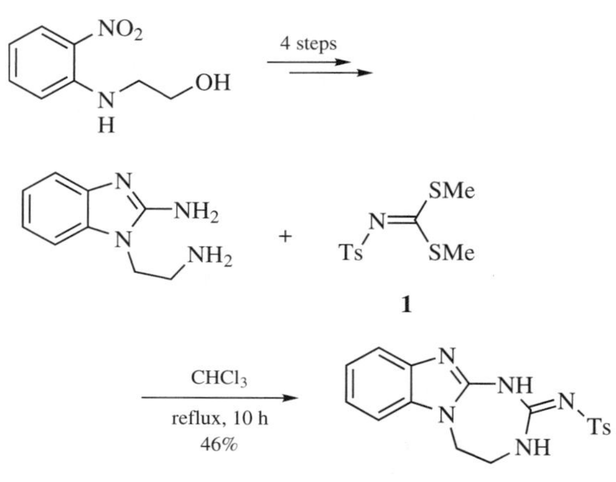

(15)

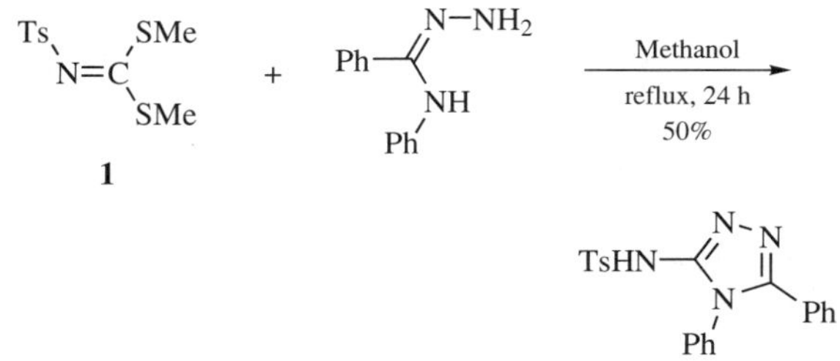

(16)

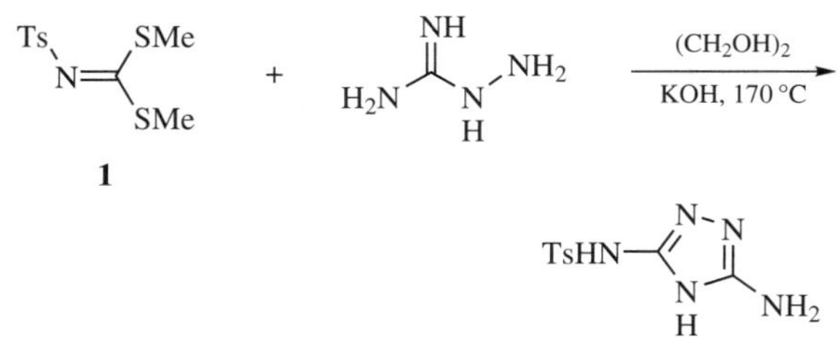

(17)

In Kuwayama and Kataoka's preparations of **1**, the addition of ammonia or ethylene diamine under refluxing conditions in MeOH was also reported.[1] In addition, Gompper also illustrated addition of ammonia to **1** in heated phenol (eq 19).[2]

In addition, a series of acyclic and cyclic amines have been added to reagent **1** in refluxing MeOH to afford compounds that possess promising activities against Gram +ve and Gram −ve bacteria (eq 20).[20]

Rapoport reported preparations of amino guanidines for studies of UV-*p*H profiles. It was found that the addition of a substituted amine must precede the addition of ammonia, otherwise no desired amino guanidines will be found (eq 21).[21] Finally, another example involved addition of the dianion of α-amino acetic acid to reagent **1** (eq 22).[22]

Addition of Ylides

N-Bis(methylthio)methylene-*para*-toluenesulfonamide has been added to a series of ylides with some of these reactions leading to heterocyclic compounds in a formal cycloaddition manner. As shown in eq 23, reactions of reagent **1** with the ylide generated by deprotonating the bromide salt of α-pyridinium ethyl acetate led to imidazo[1,2-a]pyridines in modest yields presumably via another ylide intermediate **5**.[23a] The product resulting from transfer of the Ts group dominated when the reaction was carried out in refluxing xylene. This process can be considered as a formal

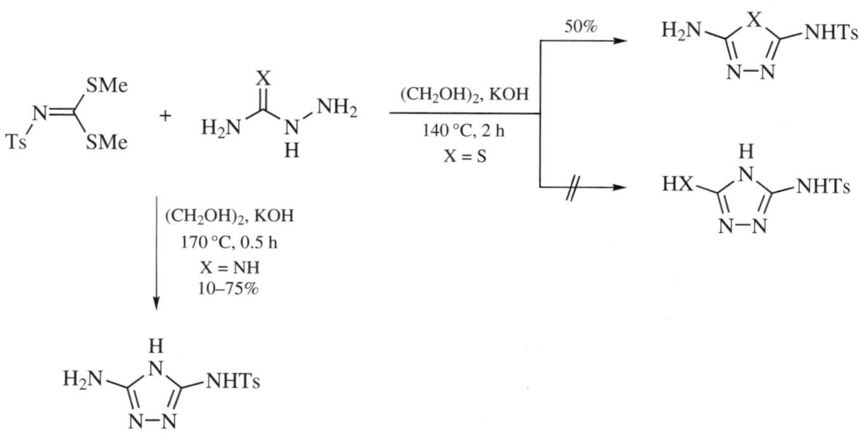

(18)

[3+2] cycloaddition. The same reaction can also provide imidazo [1,2-a]isoquinolines when the bromide salt of α-isoquinolinium ethyl acetate was used.[23b]

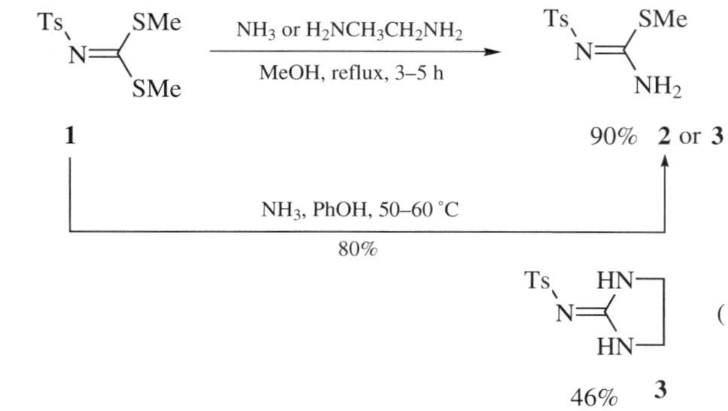

(19)

Tominaga and Kobayashi also reported that addition of reagent **1** to the ylide derived from the bromide salt of α-thiazolium methyl acetate led to imidazo[2,-b]thiazoles (eq 24).[24] This constitutes a formal [4+1] cycloaddition. The loss of the methoxycarbonyl group occurred presumably through a decarboxylative pathway when using NaOMe. When NaH was used, formally 2 equiv of reagent **1** were incorporated.

Finally, triphenylphosphonium methylide could be added to reagent **1** to give a thioimidate substituted triphenylphosphonium methylide (eq 25).[25]

Addition of Other Activated Methylenes

Tominaga also embarked on a series of studies involving addition of activated methylenes to **1**. These activated methylenes are substituted mostly α to the nitriles (eqs 26 and 27),[26,27] esters (eq 28),[28] or ketones (eqs 29 and 30).[29,30] Almost all of these reactions involved mono-addition of the carbanion intermediate to reagent **1** followed by elimination of one of the two SMe groups, leading to highly substituted push-pull types of enamides (substituted with electron donating and withdrawing groups at the two terminal olefinic carbons) which could be further transformed to various heterocycles such as pyrazoles (eq 26)[26] and isoquinolones and quinolones (eqs 27 and 28, respectively).[27,28]

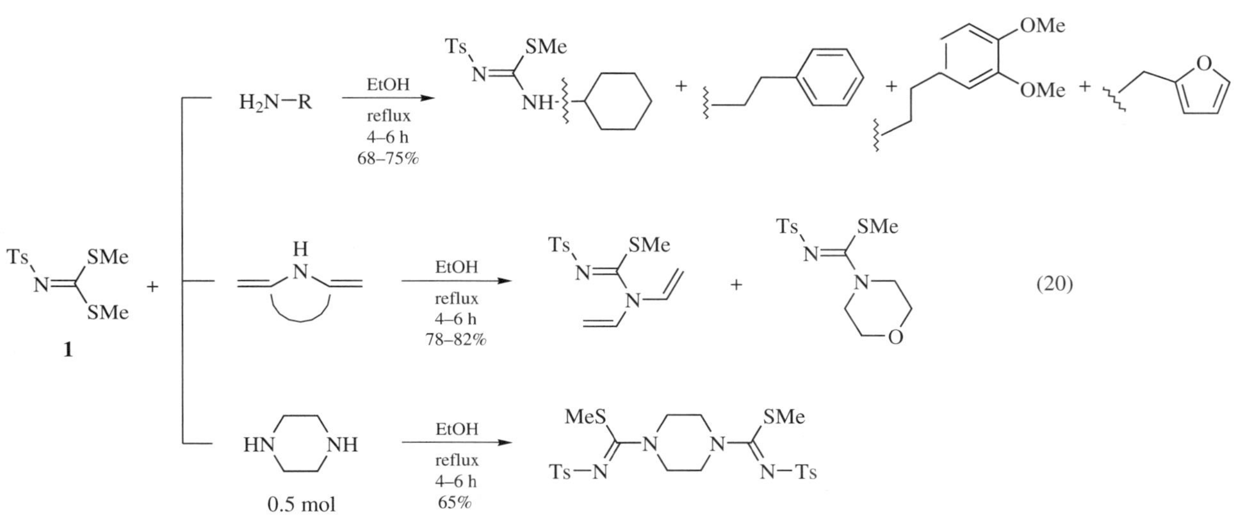

(20)

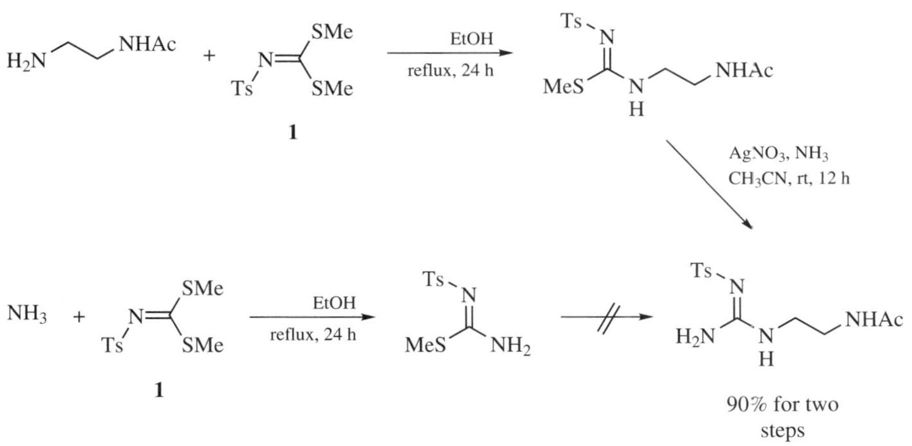

(21)

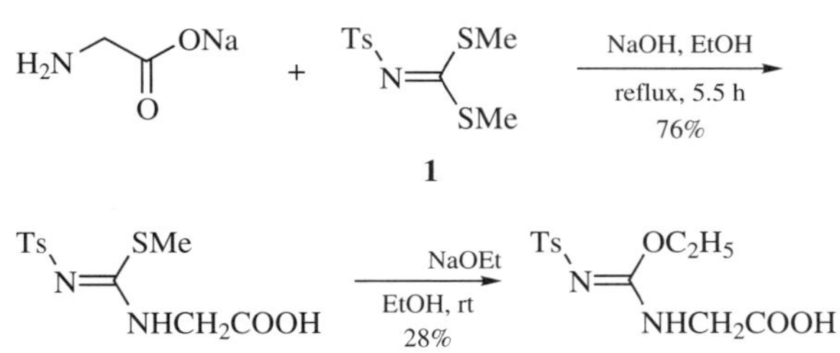

(22)

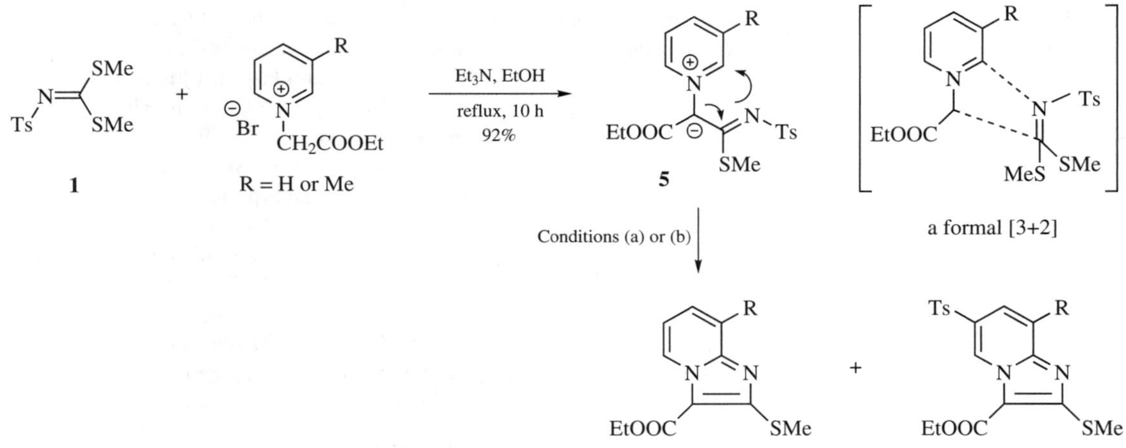

Conditions	Yield (%)	
(a) xylene reflux, 30 h	7	38
(b) DMSO reflux, 20 h	35	0

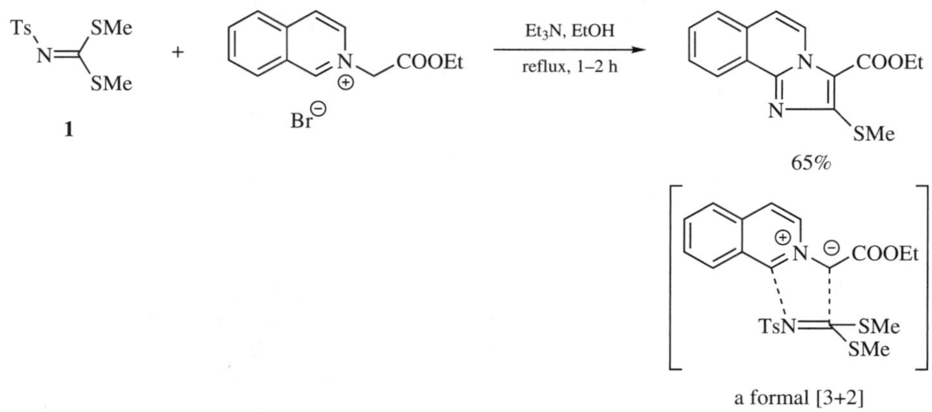

(23)

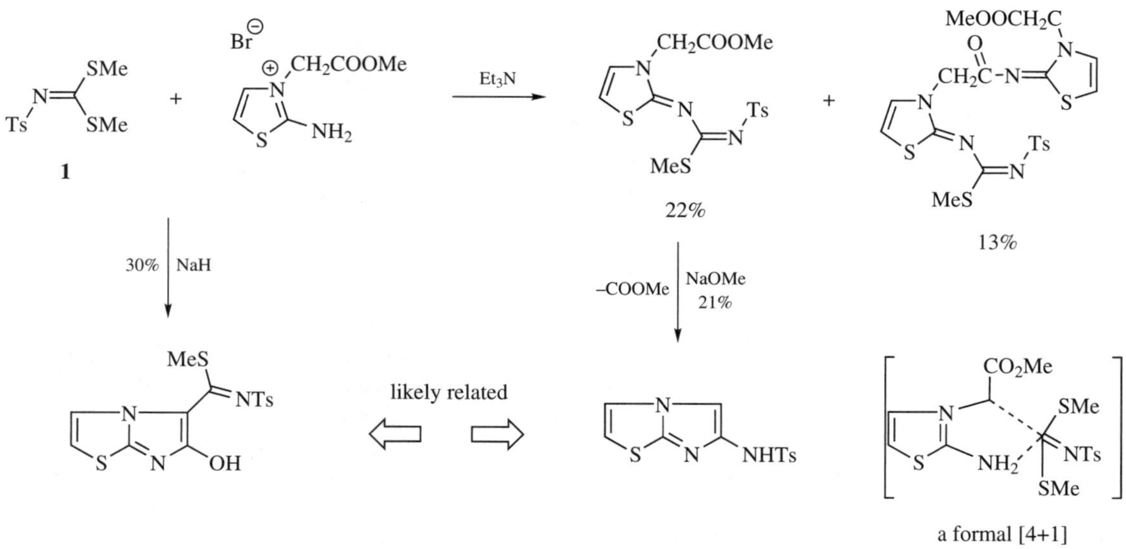

(24)

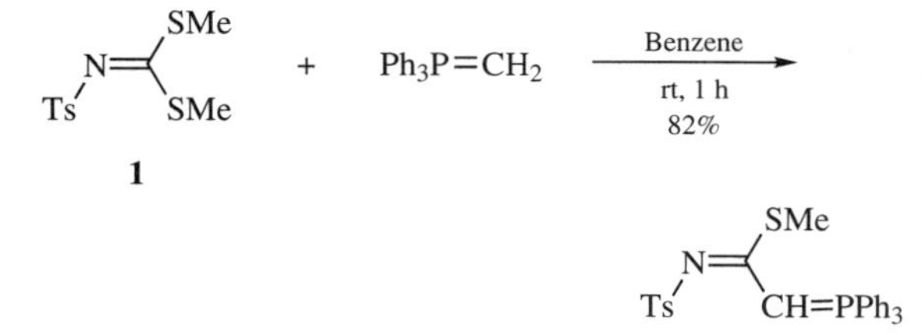

(25)

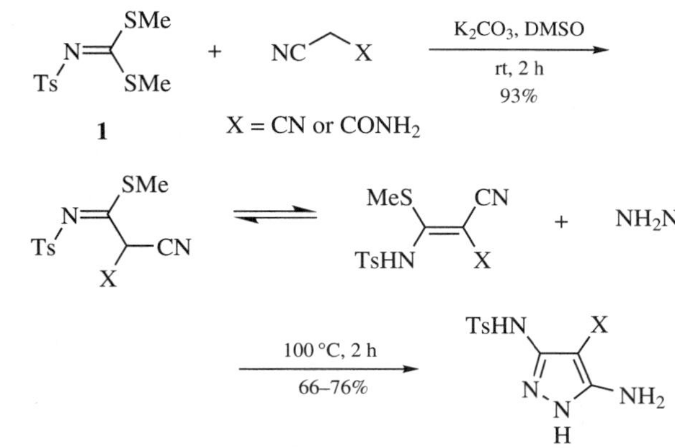

(26)

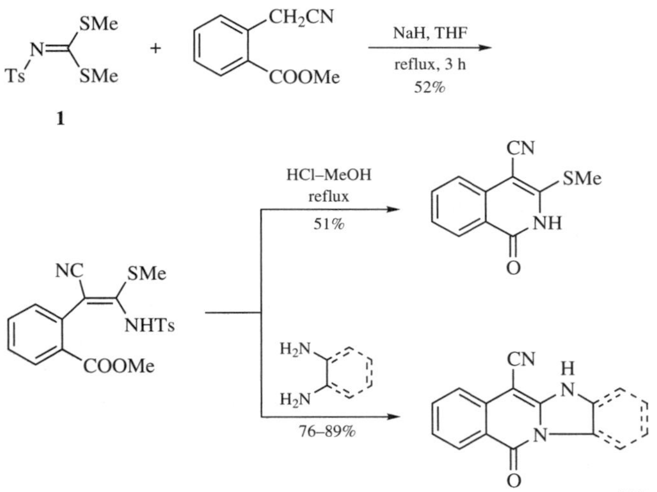

(27)

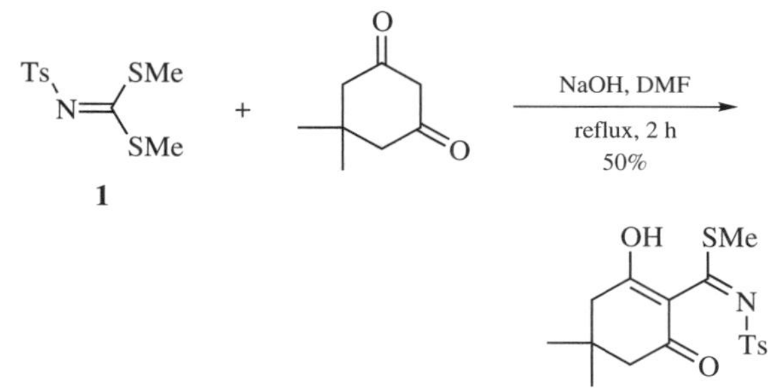

(29)

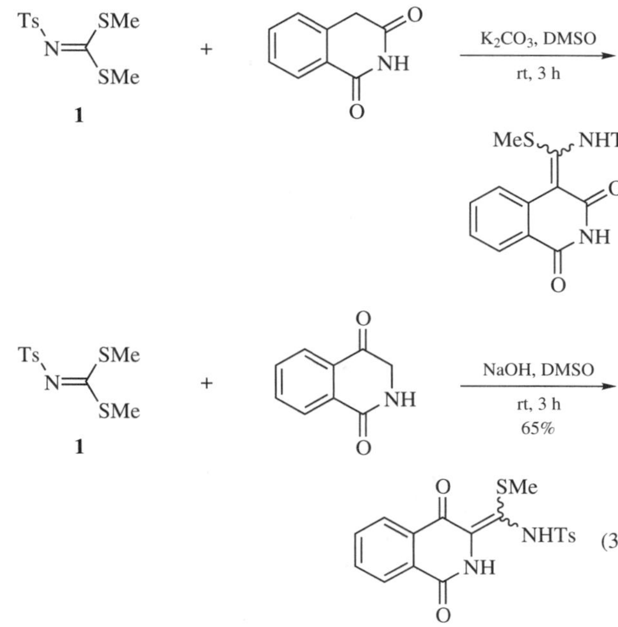

(30)

Other examples are summarized in eqs 31–33. In particular, addition of the α-anion of *N*-nitrile imines to reagent **1** led to 2-amino pyrimidines (eq 31). The α-anion of α-phenyl nitromethane could also be added to reagent **1**, and subsequent hydrolysis of thioimidate intermediate gave amide-substituted nitro alkanes (eq 32).[31–33]

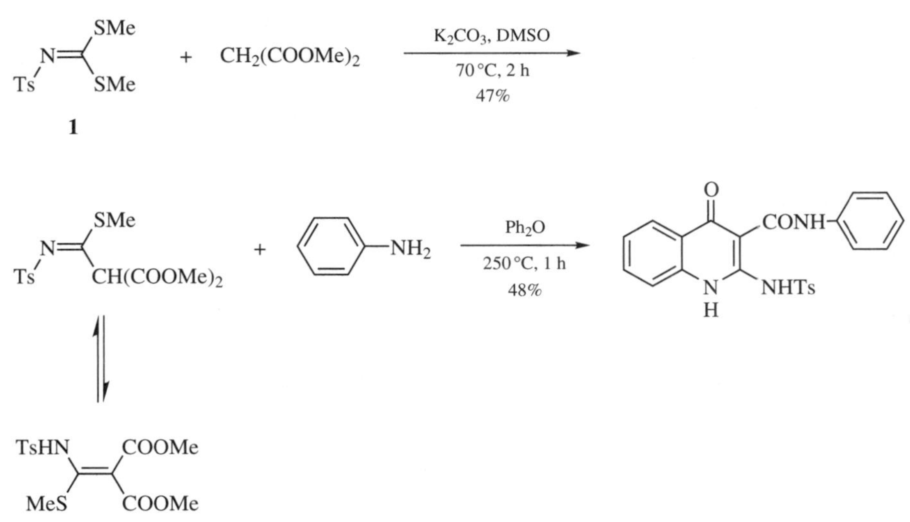

(28)

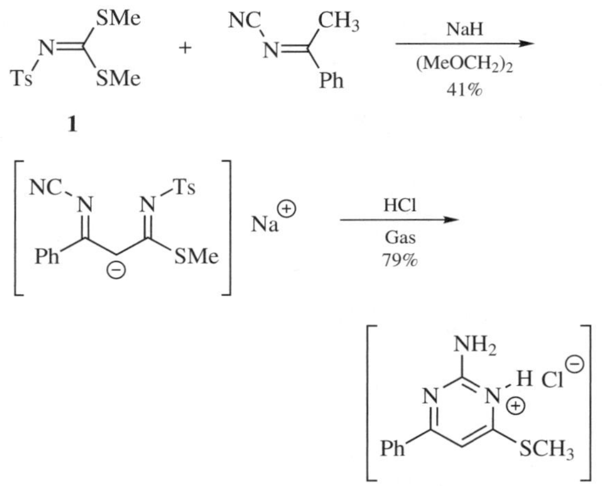

(31)

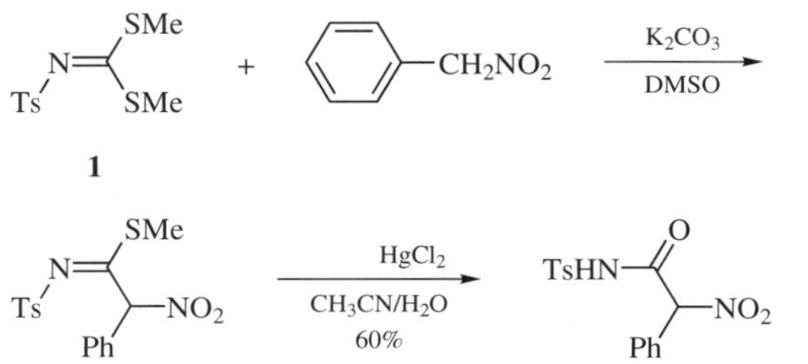

(32)

[3+2] Cycloadditions

Perhaps the most innovative use of **1** is the application in [3+2] cycloadditions that also was developed by Tominaga and Hosomi (eq 34).[34] Condensing reagent **1** with trimethylsilylmethyl amine or trimethylsilylmethanethiol in refluxing MeOH led to mono-addition products **6**. Treatment of **6a** (or **6b**) with CsF in MeCN at rt in the presence of ketones or aldehydes led to either 2-imino-oxazolidines or thiiranes.

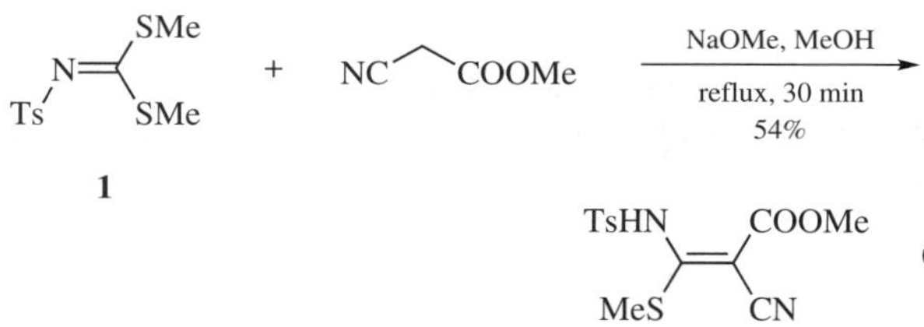

(33)

Mechanistically, desilylation of **6a** using CsF clearly led to anion **6c**. A loss of methylsulfide would lead to 1,3-dipolar intermediate **7a** that could be subsequently trapped with a ketone or an aldehyde in a [3+2] cycloaddition manner. This would then provide 2-imino-oxazolidines in a highly regioselective manner with a new stereocenter at the C5 position (eq 35).

On the other hand, CsF treatment of **6b** would lead to 1,3-dipolar intermediate **7b**. Subsequently, the initial [3+2] cycloadduct **8** undergoes fragmentation to give the observed thiiranes if pathway [a] occurs. When pathway [b] occurs with a simple loss of methylsulfide, 2-imino-thiazolidines can be isolated.

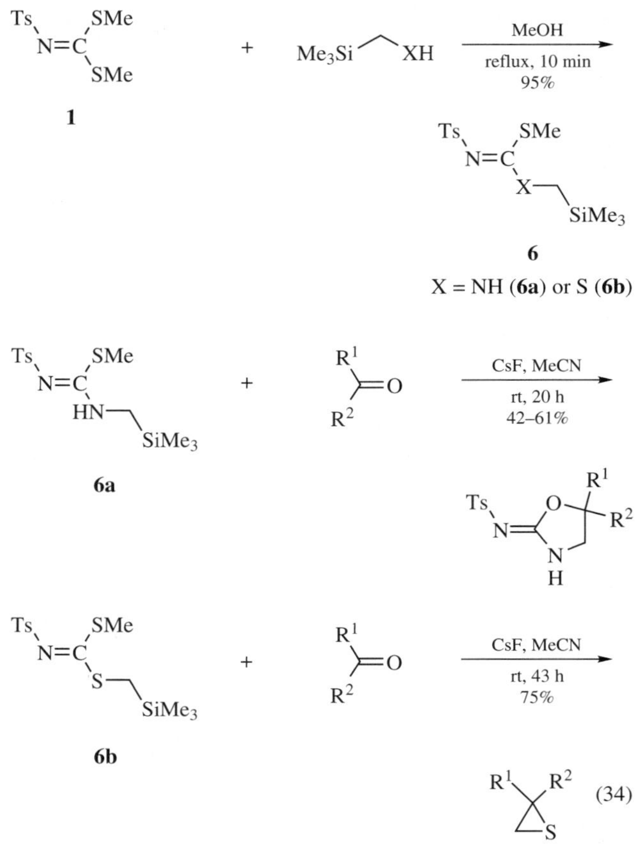

(34)

1. Kuwayama, Y.; Kataoka, S., *Yakugaku Zasshi* **1965**, *85*, 387.
2. (a) Gompper, R.; Hägele, W., *Chem. Ber.* **1966**, *99*, 2885. (b) Field, L.; Buckman, J. D., *J. Org. Chem.* **1968**, *33*, 3865.
3. Székely, Z.; Zakhariev, S.; Guarnaccia, C.; Antcheva, N.; Pongor, S., *Tetrahedron Lett.* **1999**, *40*, 4439.
4. Burgess, K.; Lim, D.; Ho, K. K.; Ke, C. Y., *J. Org. Chem.* **1994**, *59*, 2179.
5. Burgess, K.; Ho, K. K., *Tetrahedron Lett.* **1992**, *33*, 5677.
6. Vandendriessche, F.; Van Aerschot, A.; Voortmans, M.; Janssen, G.; Busson, R.; Van Overbeke, A.; Van den Bossche, W.; Hoogmartens, J.; Herdewijn, P., *J. Chem. Soc., Perkin Trans. 1* **1993**, *14*, 1567.
7. Vandendriessche, F.; Voortmans, M.; Hoogmartens, J.; Van Aerschot, A.; Herdewijn, P., *Bioorg. Med. Chem. Lett.* **1993**, *3*, 193.
8. Webb, T. R.; Eigenbrot, C., *J. Org. Chem.* **1991**, *56*, 3009.
9. (a) Bosin, T. R.; Hanson, R. N.; Rodricks, J. V.; Simpson, R. A.; Rapoport, H., *J. Org. Chem.* **1973**, *38*, 1591. (b) Neidlein, R.; Haussmann, W.; Heukelbach, E., *Chem. Ber.* **1966**, *99*, 1252. (c) Gompper, R.; Kunz, R., *Chem. Ber.* **1966**, *99*, 2900.
10. Fukada, N.; Takano, M.; Nakai, K.; Kuboike, S.; Takeda, Y., *Bull. Chem. Soc. Jpn.* **1993**, *66*, 148.
11. Wegner, M. M.; Rapoport, H., *J. Org. Chem.* **1977**, *42*, 3065.
12. Garín, J.; Meléndez, E.; Merchán, F. L.; Merino, P.; Orduna, J.; Tejero, T., *J. Heterocycl. Chem.* **1990**, *27*, 321.
13. Spindler, J., *Chem.-Zeitung* **1991**, *115*, 224.
14. Wakamiya, T.; Mizuno, K.; Ukita, T.; Teshima, T.; Shiba, T., *Bull. Chem. Soc. Jpn.* **1978**, *51*, 850.
15. Houghten, R. A.; Simpson, R. A.; Hanson, R. N.; Rapoport, H., *J. Org. Chem.* **1979**, *44*, 4536.
16. Ágai, B.; Doleschall, G.; Hornyák, G.; Lempert, K.; Simig, G., *Tetrahedron* **1976**, *32*, 839.
17. Modzelewska-Banachiewics, B.; Matosiuk, D., *Pharmazie* **1999**, *54*, 588.

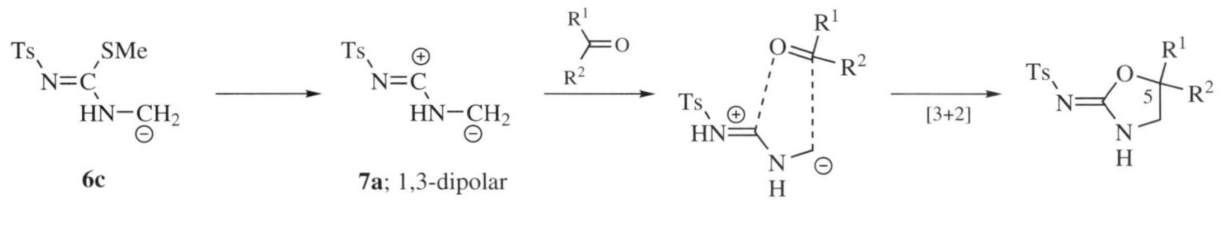

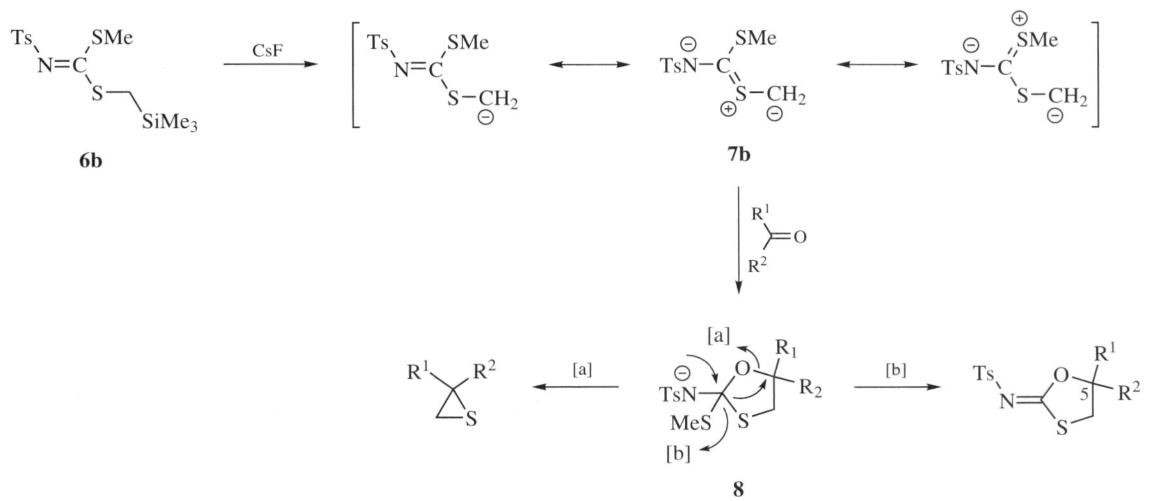

(35)

18. Ram, V. J.; Srivastava, P.; Singh, S. K.; Kandpal, M.; Tekwani, B. L., *Bioorg. Med. Chem. Lett.* **1997**, *7*, 1087.
19. Maybhate, S. P.; Rajamohanan, P. P.; Rajappa, S., *Synthesis* **1991**, *3*, 220.
20. Seshadri, S.; Sanghavi, N. M.; Naik, R. V.; Tawate, S. R.; Trivedi, M. N.; Fruitwala, M. A., *India. J. Chem.* **1993**, *32B*, 688.
21. Barker, P. L.; Gendler, P. L.; Rapoport, H., *J. Org. Chem.* **1981**, *46*, 2455.
22. Gante, J., *Chem. Ber.* **1968**, *101*, 1195.
23. (a) Tominaga, Y.; Motokawa, S.; Shiroshita, Y.; Hosomi, A., *J. Heterocycl. Chem.* **1987**, *24*, 1365. (b) Fujito, H.; Tominaga, Y.; Matsuda, Y.; Kobayashi, G., *Heterocycles* **1976**, *4*, 939.
24. Kurata, K.; Awaya, H.; Tominaga, Y.; Matsuda, Y.; Kobayashi, G., *Yakugaku Zasshi* **1981**, *101*, 991.
25. Yoshida, H.; Ogata, T.; Inokawa, S., *Bull. Chem. Soc. Jpn.* **1977**, *50*, 3315.
26. Tominaga, Y.; Hara, M.; Honkawa, H.; Hosomi, A., *J. Heterocycl. Chem.* **1990**, *27*, 1245.
27. Tominaga, Y.; Hidaki, S.; Matsuda, Y.; Kobayashi, G.; Sakemi, K., *Yakugaku Zasshi* **1980**, *100*, 456.
28. Tominaga, Y.; Michioka, T.; Moriyama, K.; Hosomi, A., *J. Heterocycl. Chem.* **1990**, *27*, 1217.
29. Hatada, T.; Sone, M.; Tominaga, Y.; Natsuki, R.; Matsuda, Y.; Kobayashi, G., *Yakugaku Zasshi* **1975**, *95*, 623.
30. Ueno, S.; Tominaga, Y.; Matsuda, Y.; Kobayashi, G., *Chem. Pharm. Bull.* **1974**, *22*, 2624.
31. Hartke, K.; Sauerbier, M.; Richter, W. F., *Arch. Pharm.* **1992**, *325*, 279.
32. Maybhate, S. P.; Deshmukh A. R. A. S.; Rajappa, S., *Tetrahedron* **1991**, *47*, 3887.
33. Peseke, K., *Pharmazie* **1976**, *31*, 607.
34. (a) Tominaga, Y.; Ogata, K.; Ueda, H.; Kohra, S.; Hosomi, A., *Chem. Pharm. Bull.* **1995**, *43*, 1425. (b) Tominaga, Y.; Ueda, H.; Ogata, K.; Kohra, S.; Hojo, M.; Ohkuma, M.; Tomita, K.; Hosomi, A., *Tetrahedron Lett.* **1992**, *33*, 85. (c) Tominaga, Y.; Ogata, K.; Kohra, S.; Hojo, M.; Hosomi, A., *Tetrahedron Lett.* **1991**, *32*, 5987.

Jia Liu & Richard P. Hsung
University of Minnesota, Minneapolis, MN, USA

Bis(4-nitrophenyl) Carbamate

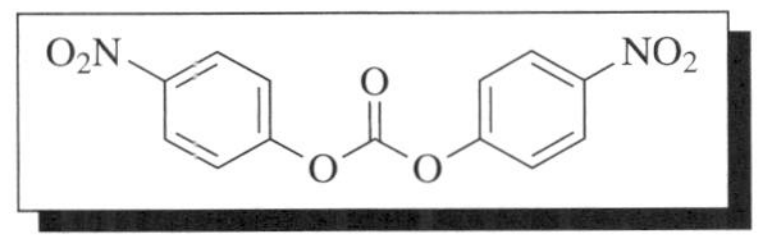

[5070-13-3] $C_{13}H_8N_2O_7$ (MW 304.22)

(peptide coupling reagent;[1] reagent for formation of carbamate linked cytosines[2])

Physical Data: mp 144–145 °C.
Solubility: sol chloroform and THF.
Form Supplied in: off-white powder.

Peptide Coupling Reagent. N-Protected amino acids can be converted into activated 4-nitrophenyl esters suitable for peptide coupling reactions. Reaction of the N-protected amino acid in acetic acid and pyridine with 1.5 equiv of the reagent under reflux for 3 h affords the activated ester following loss of CO_2 from the intermediate mixed anhydride (eq 1).[1]

Formation of Carbamate-Linked Cytosine Oligomers.[2] Reaction of the 5′-protected aminocytosine derivative (**1a**) with the reagent in DMF and Et_3N afforded an activated 4-nitrophenyl carbonate derivative. Following purification, reaction of this compound with another monomer unit, (**1b**), provided the linked dimer (**2**) (eq 2). This compound was elaborated further into a carbamate-linked hexameric cytosine oligomer which was found to conjugate strongly to native DNA strands.

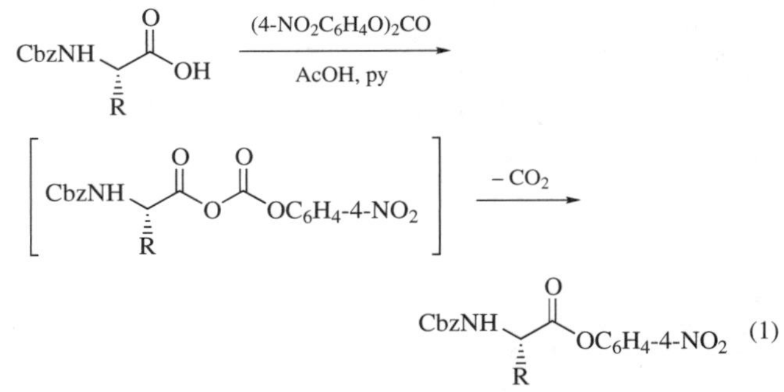

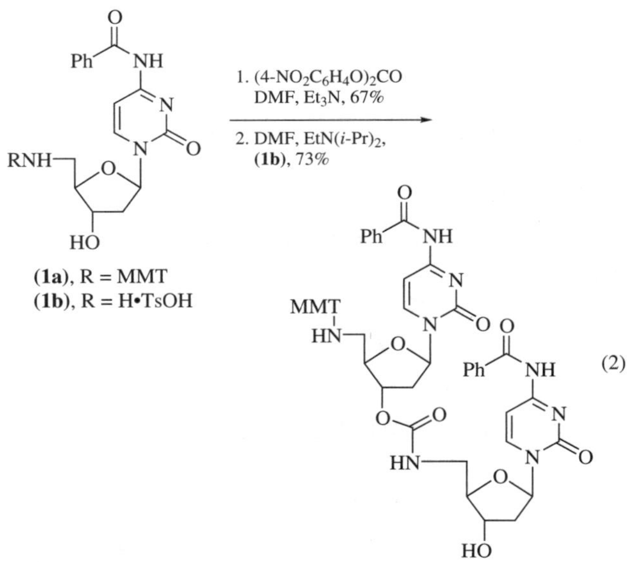

(1a), R = MMT
(1b), R = H•TsOH

(2)

Formation of Ureido Sugars.[3] A variety of 2-amino sugars can be converted into ureido derivatives in reasonable yields (50–83% for two steps). For example, methyl 3,4,6-tri-*O*-acetyl-2-amino-2-deoxy-β-D-glucopyranoside is first reacted with the reagent to afford a 4-nitrophenyl carbamate intermediate which, upon further reaction with a variety of secondary amines in CH_2Cl_2 at room temperature for 5–24 h, provides the ureido sugar. In a similar manner, symmetrical and unsymmetrical *N,N′*-disubstituted ureas can be prepared by the sequential addition of amines to the reagent.[4]

---

1. (a) Wieland, T.; Heinke, B.; Vogeler, K.; Morimoto, H. *Justus Liebigs Ann. Chem./Liebigs Ann. Chem.* **1962**, *655*, 189. (b) Glatthard, R.; Matter, M. *Helv. Chim. Acta* **1963**, *46*, 795.
2. Stirchak, E. P.; Summerton, J. E.; Weller, D. D. *J. Org. Chem.* **1987**, *52*, 4202.
3. Piekarska-Bartoszewicz, B.; Temeriusz, A. *Carbohydr. Res.* **1990**, *203*, 302.
4. Izdebski, J.; Pawlak, D. *Synthesis* **1989**, 423.

Eric D. Edstrom
Utah State University, Logan, UT, USA

Bis(*p*-nitrophenyl) Hydrogen Phosphate

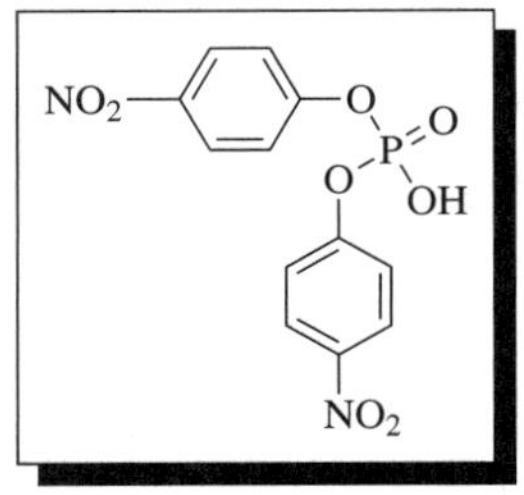

[645-15-8] $C_{12}H_9N_2O_8P$ (MW 340.19)

(acid catalyst[1,2] for preparation of nucleosides[3,4] and nucleoside acetals[1,5])

Physical Data: mp 175–175.5 °C;[6] pK_a 2.79 (H_2O),[7] pK_a 1.7 (50% EtOH).[8]
Form Supplied in: yellow crystals.
Preparative Methods: treat freshly distilled $POCl_3$ with finely powdered sodium *p*-nitrophenoxide, followed by partial alkaline hydrolysis of the corresponding tris(*p*-nitrophenyl) phosphate;[6] *p*-nitrophenol treated with $POCl_3$ in acetonitrile.[9]
Purification: recrystallization from H_2O to yield white needles.[6]

Nucleoside Synthesis. Effective catalytic reagent for the preparation of nucleosides by fusion of purines with fully *O*-acetylated sugars.[3] This is superior to p-Toluenesulfonic Acid (except for synthesis of phenylborate nucleosides[10]), ethyl polyphosphate, or Zinc Chloride. An example of the method[4] is the fusion of purine (**1**) with tetra-*O*-acetyl-D-ribofuranose (**2**), at 170–180 °C for 15 min under reduced pressure (water pump), to yield two nucleosides (**3**) and (**4**) (eq 1). The major product (**3**) is the triacetyl derivative of a naturally occurring nucleoside antibiotic nebularine.

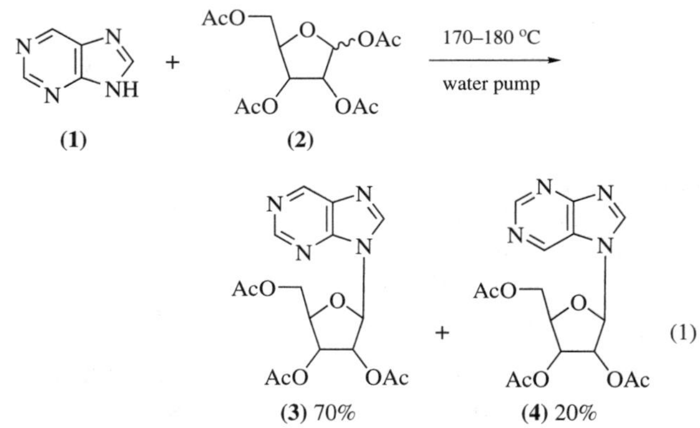

Acetal Synthesis. The reagent is used as an acid catalyst for the conversion of purines and pyrimidine ribonucleosides to their corresponding 2′,3′-*O*-isopropylidene (acetonide) derivatives, in the presence of 2,2-dimethoxypropane (effective for maintaining the anhydrous conditions necessary of such conversions). An example of the method[1] is the preparation of 2′,3′-*O*-isopropylidene uridine (eq 2).

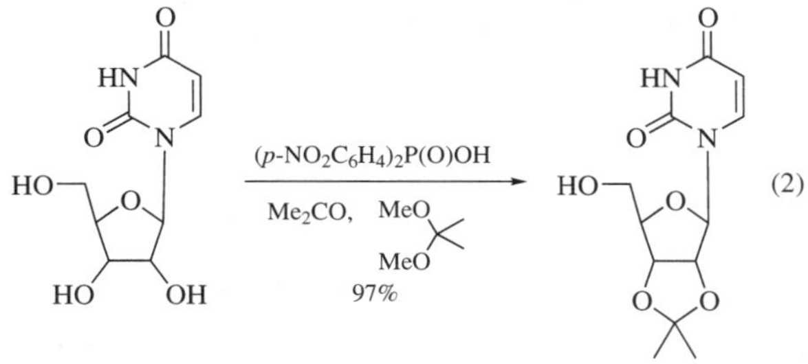

Bis(*p*-nitrophenyl) hydrogen phosphate is superior to p-Toluenesulfonic Acid as a catalyst for the reaction because it has twice the lipophilic area, making its salts with nucleosides correspondingly more soluble in acetone. This procedure has been extended to a range of acetals other than isopropylidene, with greater or lesser stability to acid.[5]

1. Hampton, A. *J. Am. Chem. Soc.* **1961**, *83*, 3640.
2. Hampton, A. *Biochem. Prep.* **1963**, *10*, 91.
3. Hashizume, T.; Iwamura, H. *Tetrahedron* **1965**, *35*, 3095.
4. Iwamura, H.; Hashizume, T. *J. Org. Chem.* **1968**, *33*, 1796.
5. Hampton, A.; Frakantoni, J. C.; Caroll, P. M.; Wang, S. *J. Am. Chem. Soc.* **1965**, *87*, 5481.
6. Moffat, J. G.; Khorana, H. G. *J. Am. Chem. Soc.* **1957**, *79*, 3741.
7. *Dictionary of Organophosphorus Compounds*; Chapman and Hall: New York, **1988**; p 86.
8. Corby, N. S.; Kenner, G. W.; Todd, A. R. *J. Chem. Soc* **1952**, 1234.
9. Hashizume, T.Jpn. Patent 18 132, **1964** (*Chem. Abstr.* **1965**, *62*, 6432) .
10. McCormick, J. E.; McElhinney, R. S. *J. Chem. Soc., Perkin Trans. 1* **1975**, 500.

Harjinder S. Bansal
Zeneca Agrochemicals, Bracknell, UK

Bis(*o*-nitrophenyl) Phenylphosphonate

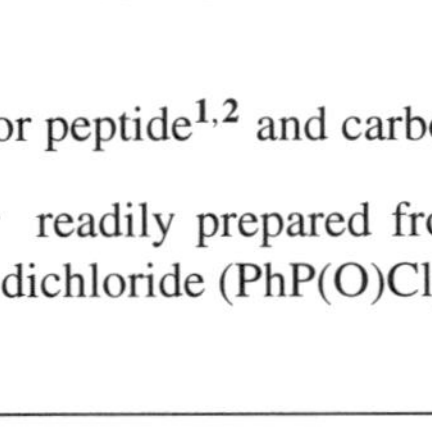

[73404-18-9] $C_{18}H_{13}N_2O_7P$ (MW 400.28)

(coupling reagent for peptide[1,2] and carboxamide synthesis[3])

Preparative Method: readily prepared from *o*-nitrophenol and phenylphosphonic dichloride ($PhP(O)Cl_2$).[1]

Peptide and Carboxamide Synthesis. Condensation of the tetrabutylammonium salt of *N*-benzoyl-L-leucine with glycine ethyl ester, the components in the Young's test,[4] in DMF in the presence of bis(*o*-nitrophenyl) phenylphosphonate proceeds to give the corresponding dipeptide in 82–93% yield without racemization.[1] Employment of quarternary ammonium salts[5] enables the reaction to proceed smoothly because of increased solubility[6] and nucleophilicity of the carboxyl component. The coproducts are *o*-nitrophenol and the phosphonate ammonium salt, which means that no free base such as tertiary amines need to be utilized, which often cause serious problems such as racemization during condensation (eq 1).[4] This method has been efficiently applied to more complex hexaglycine and leucine-enkephalin (L-Tyr-Gly-Gly-L-Phe-L-Leu) peptides.[2]

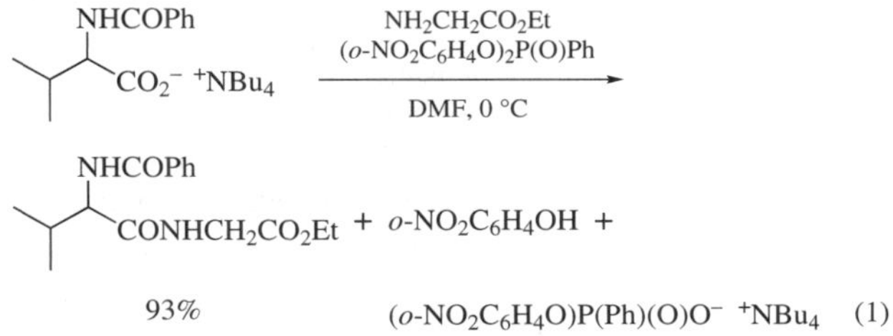

Preparation of both peptides and carboxamides have been simplified by the use of phase-transfer conditions in which the tetrabutylammonium carboxylic acid salts are formed in situ.[3] Thus treatment of free acids and amines in aqueous CH_2Cl_2 (or aqueous THF) with bis(*o*-nitrophenyl) phenylphosphonate in the presence of $n\text{-}Bn_4N^+HSO_4^-$ (or $n\text{-}Bn_4N^+Br^-$) gave corresponding carboxamides and peptides. An example of the method[3] is the preparation of an arylamide in quantative yield (eq 2).

Related Reagents. Bis(p-nitrophenyl) Phenylphosphonate.

1. Mukaiyama, T.; Morito, N.; Watanabe, Y. *Chem. Lett.* **1979**, *11*, 1305.
2. Watanabe, Y.; Morito, N.; Kamekawa, K.; Mukaiyama, T. *Chem. Lett.* **1981**, *1*, 65.
3. Watanabe, Y.; Mukaiyama, T. *Chem. Lett.* **1981**, *3*, 285.
4. Williams, M. W.; Young, G. T. *J. Chem. Soc* **1963**, 881.
5. Wakselman, M.; Acher, F. *Tetrahedron Lett.* **1980**, *21*, 2705.
6. Luisi, P. L.; Bonner, F. J.; Pellegrini, A.; Wiget, P.; Wolf, R. *Helv. Chim. Acta* **1979**, *62*, 740.

Harjinder S. Bansal
Zeneca Agrochemicals, Bracknell, UK

Bis(*p*-nitrophenyl) Phenylphosphonate

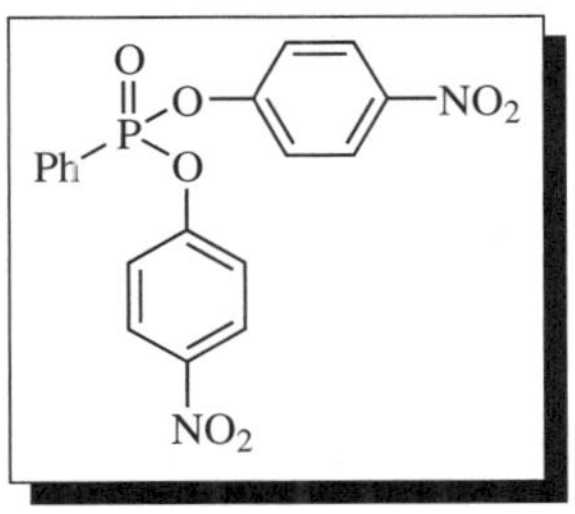

[38873-91-5] $C_{18}H_{13}N_2O_7P$ (MW 400.28)

(efficient peptide coupling agent causing little racemization[1,2])

Physical Data: mp 99–101 °C.[3]

Form Supplied in: off-white to tan granules.[3]
Preparative Method: heat phenylphosphonic dichloride ($PhP(O)Cl_2$) with 2 equiv of 4-nitrophenol at 160 °C for 4 h.[4]
Purification: recrystallize from CCl_4 (mp 94–96 °C).[4]
Handling, Storage, and Precautions: hygroscopic; store in a cool dry place. Incompatible with strong oxidizing agents, acids, and bases.[3]

Peptide Synthesis. Bis(*p*-nitrophenyl) phenylphosphonate (see also Bis(o-nitrophenyl) Phenylphosphonate) has been used as a coupling reagent in the condensation of the tetrabutylammonium salt of *N*-benzoyl-L-leucine with glycine ethyl ester. These are the reactants used in the Young's test,[5] which is known as the most severe racemization test. The most favorable solvent used in the coupling was DMF and the reaction proceeded to the corresponding dipeptide in 93% yield, at below 0 °C with little racemization (95–97% L-isomer) (eq 1).

The advantage of using the tetrabutylammonium salt in the coupling reaction was that it increased both solubility[6] and nucleophilicity of the carboxylic component, enabling the reaction to take place without using tertiary amines, which play a major role in causing racemization. The quarternary ammonium salts were prepared by the reaction of alkaline n-$Bn_4N^+HSO_4^-$ or Bu_4N^+ OH^- on the carboxylic acid.[1]

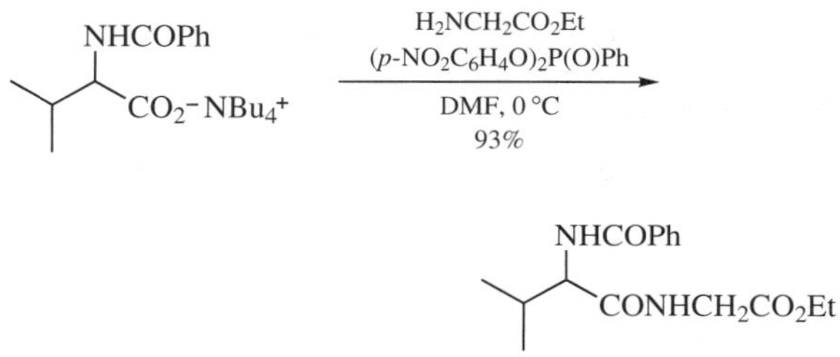

(1)

This method has also been efficiently applied to the synthesis of opioid agonist leucine-enkephalin[2] (L-Tyr-Gly-Gly-L-Ppe-L-Leu) without protection of the phenolic hydroxy group in the tyrosine residue.[2]

1. Mukaiyama, T.; Morito, N.; Watanabe, Y. *Chem. Lett.* **1979**, *11*, 1305.
2. Watanabe, T.; Morito, N.; Kamekawa, K.; Mukaiyama, T. *Chem. Lett.* **1981**, *1*, 65.
3. *The Sigma-Aldrich Library of Chemical Safety Data*, 2nd ed.; Vol 1, p 443.
4. Loran, J. S.; Naylor, R. A.; Williams, A. *J. Chem. Soc., Perkin Trans. 2* **1977**, 418.
5. Williams, M. W.; Young, G. T. *J. Chem. Soc* **1963**, 881.
6. Luisi, P. L.; Bonner, F. J.; Pellegrini, A.; Wiget, P.; Wolf, R. *Helv. Chim. Acta* **1979**, *62*, 740.

Harjinder S. Bansal
Zeneca Agrochemicals, Bracknell, UK

Bis(2-oxo-3-oxazolidinyl)phosphinic Chloride[1]

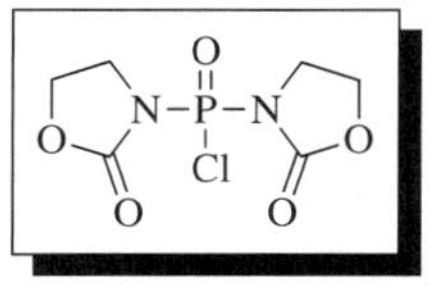

[68641-49-6] $C_6H_8ClN_2O_5P$ (MW 254.57)

(reagent for activating carboxyl groups,[1] converting acids to esters[2] (including thioesters[3] and phosphoesters[4]), amides[5] (including peptides[6–8] and β-lactams[9,10]), and anhydrides;[11] reagent for kinetic resolution of racemic carboxylic acids and alcohols[12])

Alternate Names: *N,N*-bis(2-oxo-3-oxazolidinyl)phosphordiamidic chloride; *N,N*-bis(2-oxo-3-oxazolidinyl)phosphorodiamidic chloride; phosphoric acid bis(2-oxooxazolidide)chloride; BOP-Cl; BOPDCl.
Physical Data: white powder, mp 195–198 °C; hygroscopic.
Solubility: slightly sol CH_2Cl_2, MeCN, THF, DMF; dec H_2O.
Form Supplied in: white to off-white powder, typically 97% pure.
Analysis of Reagent Purity: IR (KBr): 1775 (C=O), 765 (P–Cl) cm^{-1}.
Purification: the outcome of reactions can be greatly affected by the purity of the reagent; commercial samples may require to be washed with cold water and recrystallized from MeCN prior to use.[13]
Handling, Storage, and Precautions: store in a cool dry place under nitrogen to preclude contact with moisture and oxygen. Use only in a chemical fume hood. Wear suitable protective clothing and eye/face protection. Causes burns and is harmful if swallowed, inhaled, or absorbed through skin.

Carboxyl Group Activator. BOP-Cl[1] is used for a direct, one-pot conversion of carboxylic acids into esters (eq 1)[2] (including thioesters[3] and phosphoesters[4]), amides (eq 2),[5] and anhydrides (eq 3).[11] Conditions for minimizing side reactions and

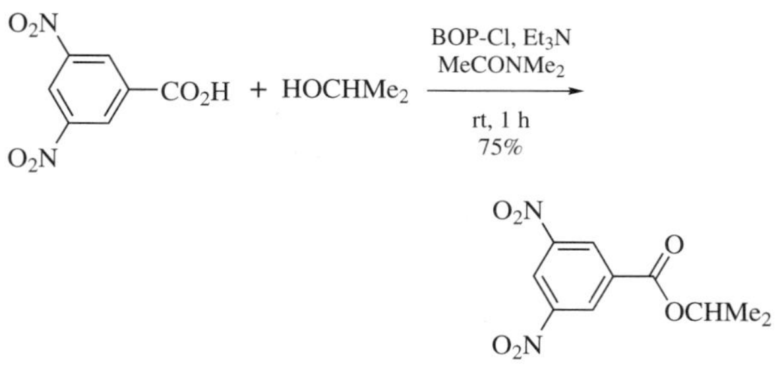

(1)

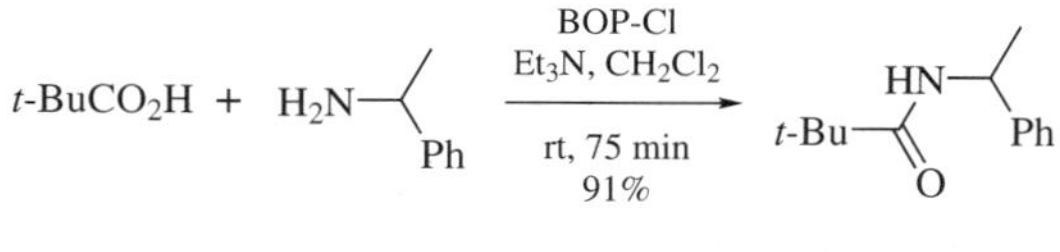

(2)

optimizing yields of both esters and amides are reported.[2,5] Advantages offered over previously used reagents, such as N,N′-Carbonyldiimidazole, carbodiimides, and unsymmetrical anhydrides, include good yields (typically >80%), mild reaction conditions, and tolerance of both steric bulk and a wide range of functional groups.

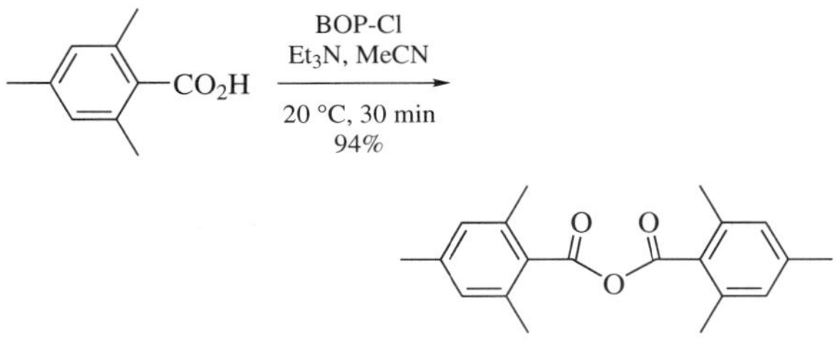

(3)

The mechanism of these reactions is assumed to involve displacement of chloride by the carboxylate anion to give a phosphorodiamidate (eq 4). Subsequent nucleophilic attack of an alcohol, amine, or a carboxylate anion affords phosphorodiamidic acid and esters, amides, and anhydrides, respectively (eq 5).

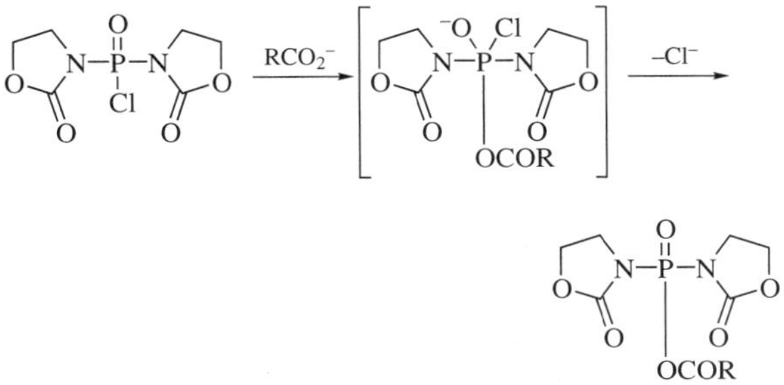

(4)

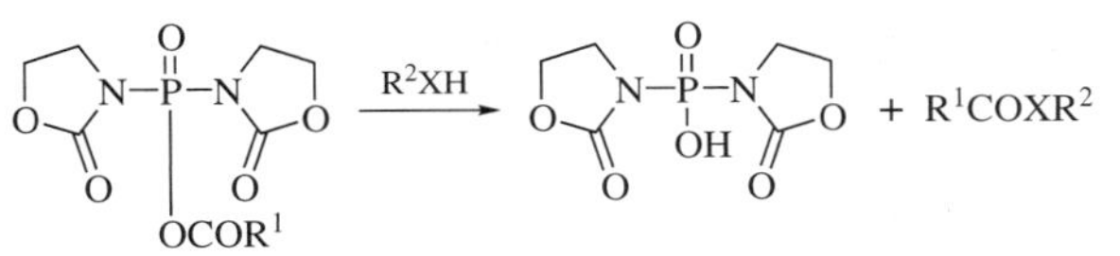

(5)

X = O, NH, OCO

For lactonization (eq 6),[14] including the synthesis of macrolactones,[15] BOP-Cl is a useful reagent.

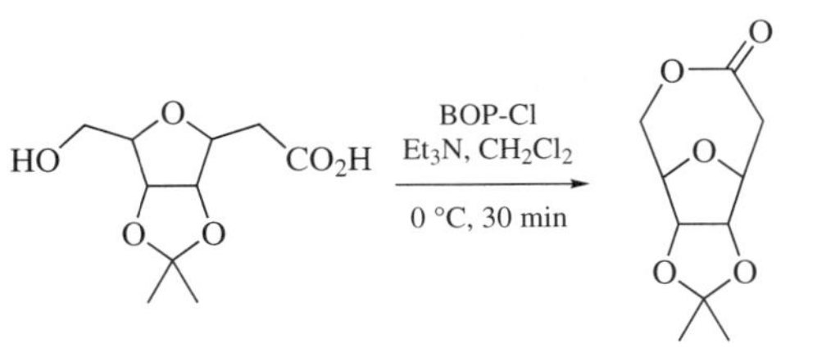

(6)

Chemo- and regioselective acylation with BOP-Cl is illustrated by the formation of 5′-benzoyldeoxyadenosine in good yield (eq 7).[16]

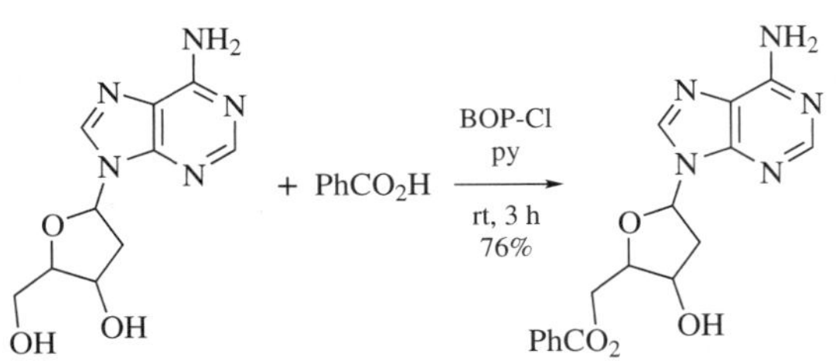

(7)

For the synthesis of oligonucleotides, BOP-Cl is an effective coupling reagent, allowing phosphotriesters to be prepared in high yields without detectable side reactions (eq 8).[17]

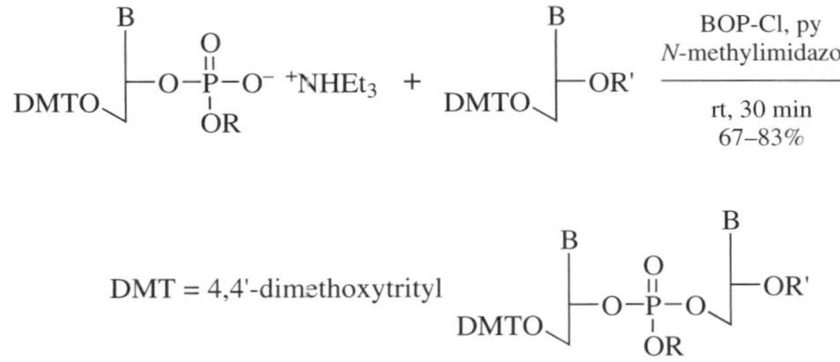

(8)

β-Lactams may be prepared either by condensation of imines with carboxylic acids (eq 9)[9] or by cyclization of β-amino acids (eq 10).[10]

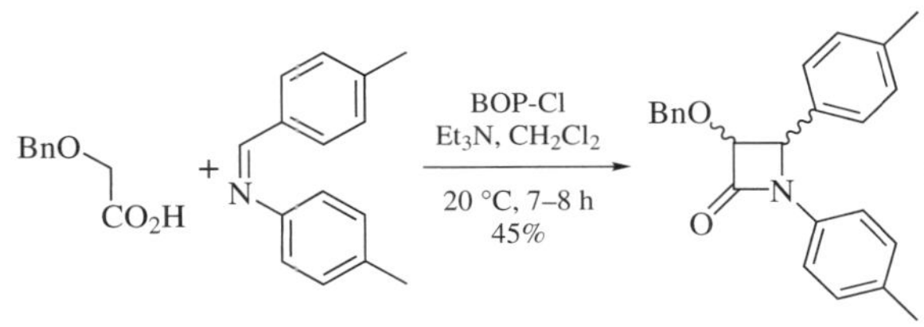

(9)

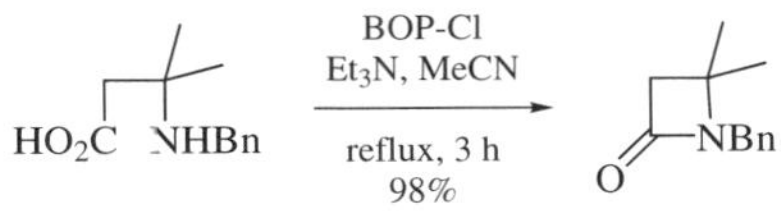

(10)

BOP-Cl has been used quite extensively for the synthesis of peptides.[6–8] It is particularly useful for coupling *N*-alkyl-α-amino acids with minimal racemization (eq 11),[18,19] although the quality of BOP-Cl employed in these reactions can significantly affect yields,[13] and the analogous azide, BOP-N_3, has been proposed as an alternative reagent.[20] Racemization can sometimes be suppressed by inclusion of additives such as 1-Hydroxybenzotriazole (HOBt) or Imidazole (eq 12),[21] although under certain conditions HOBt can induce epimerization.[22]

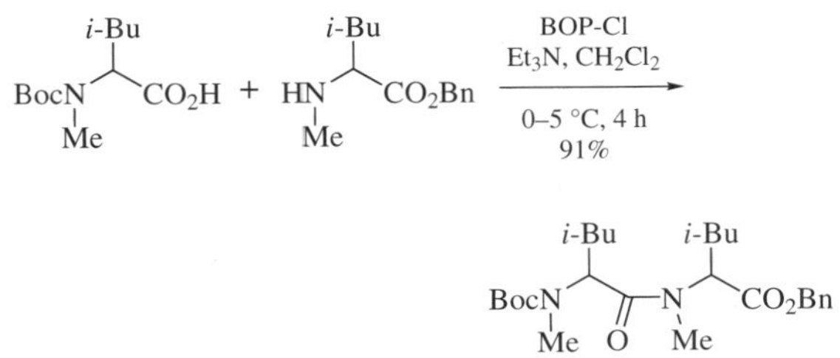

(11)

Problems, notably racemization and poor yields due to oxazolone formation, are encountered when coupling *N*-acyl amino acids with BOP-Cl;[6] 1,3-Dicyclohexylcarbodiimide (DCC) is a better reagent to use in such cases. Some limitations of BOP-Cl for peptide couplings have been discussed;[7] it is not recommended for β-branched α-amino acids (e.g. Boc-Val) or for couplings in which the nucleophile is a primary amine.[8]

Kinetic Resolution. BOP-Cl is a condensation agent for the kinetic resolution of racemic carboxylic acids and alcohols with chiral alcohols (eq 13) and carboxylic acids, respectively.[12]

Cbz-Gly-L-Phe + L-Val-OMe $\xrightarrow[\text{0 °C to rt, 20 h}]{\text{BOP-Cl, N-methylmorpholine, THF}}$

Cbz-Gly-D/L-Phe-L-Val-OMe (12)

	Yield (%)	Epimerization (%)
with HO-*t*-Bu	88	0.0
with imidazole	96	0.8
without HO-*t*-Bu or imidazole	78	5.6

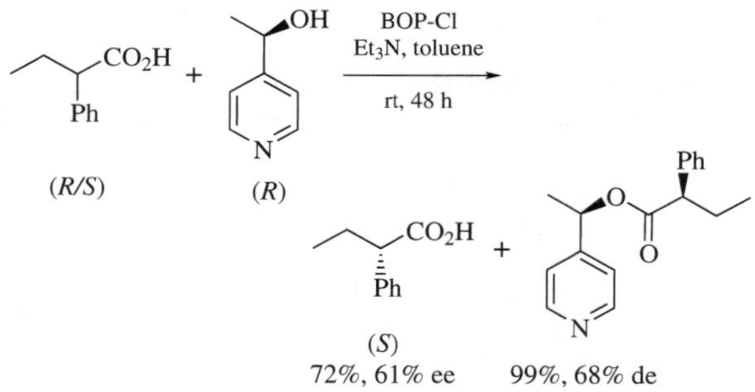
(13)

1. (a) Diago-Meseguer, J.; Palomo-Coll, A. L. *Synthesis* **1980**, 547. (b) *Fieser & Fieser* **1982**, *10*, 41; **1984**, *11*, 57; **1988**, *13*, 39.
2. Ballester-Rodes, M.; Palomo-Coll, A. L. *Synth. Commun.* **1984**, *14*, 515.
3. Arrieta, A.; Garcia, T.; Lago, J. M.; Palomo, C. *Synth. Commun.* **1983**, *13*, 471.
4. Katti, S. B.; Agarwal, K. *Tetrahedron Lett.* **1986**, *27*, 5327.
5. Cabre, J.; Palomo, A. L. *Synthesis* **1984**, 413.
6. Kolodziejczyk, A. M.; Wodecki, Z. J. *Peptides, Proceedings of the 19th European Peptide Symposium, 1986*; Theodoropoulos, D., Ed.; de Gruyter:Berlin, **1987**; p 115.
7. Van der Auwera, C.; Anteunis, M. J. O. *Int. J. Pept. Protein Res.* **1987**, *29*, 574.
8. Colucci, W. J.; Tung, R. D.; Petri, J. A.; Rich, D. H. *J. Org. Chem.* **1990**, *55*, 2895.
9. Shridhar, D. R.; Ram, B.; Narayana, V. L. *Synthesis* **1982**, 63.
10. Chung, B. Y.; Goh, W.; Nah, C. S. *Bull. Korean Chem. Soc.* **1991**, *12*, 457.
11. Cabre-Castellvi, J.; Palomo-Coll, A.; Palomo-Coll, A. L. *Synthesis* **1981**, 616.
12. Mazon, A.; Najera, C.; Yus, M.; Heumann, A. *Tetrahedron: Asymmetry* **1992**, *3*, 1455.
13. Van der Auwera, C.; Anteunis, M. J. O. *Bull. Soc. Chim. Belg.* **1986**, *95*, 203.
14. Cruickshank, K. A.; Reese, C. B. *Synthesis* **1983**, 199.
15. Corey, E. J.; Hua, D. H.; Pan, B.-C.; Seitz, S. P. *J. Am. Chem. Soc.* **1982**, *104*, 6818.
16. Liguori, A.; Perri, E.; Sindona, G.; Uccella, N. *Tetrahedron* **1988**, *44*, 229.
17. Katti, S. B.; Agarwal, K. L. *Tetrahedron Lett.* **1985**, *26*, 2547.
18. Tung, R. D.; Dhaon, M. K.; Rich, D. H. *J. Org. Chem.* **1986**, *51*, 3350.
19. Tung, R. D.; Rich, D. H. *J. Am. Chem. Soc.* **1985**, *107*, 4342.
20. Katti, S. B.; Misra, P. K.; Haq, W.; Mathur, K. B. *Indian J. Chem., Sect. B* **1988**, *27*, 3.
21. Van der Auwera, C.; Van Damme, S.; Anteunis, M. J. O. *Int. J. Pept. Protein Res.* **1987**, *29*, 464.
22. Anteunis, M. J. O.; Sharma, N. K. *Bull. Soc. Chim. Belg.* **1988**, *97*, 281.

David C. Rees & Niall M. Hamilton
Organon Laboratories Ltd, Newhouse, Lanarkshire, UK

Bis(pentafluorophenyl) Carbonate

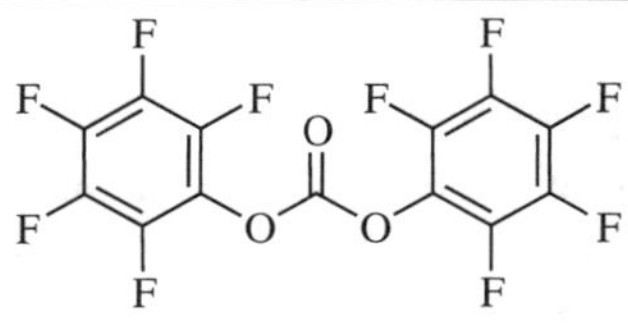

1

[59483-84-0] $C_{13}F_{10}O_3$ (MW 394.1)

(reagent used as a carbonyl equivalent and for various coupling reactions, e.g., for azapeptides)

Alternate Name: pentafluorophenyl carbonate, dipentafluorophenyl carbonate.
Physical Data: mp 48–51 °C.
Solubility: readily soluble in most inert organic solvents, e.g., chloroform, ethyl acetate, or methylene chloride, but may be crystallized/purified by addition of heptane to a solution in methylene chloride. Hydrolyses slowly in the presence of water.
Form Supplied in: available as almost colorless non-hygroscopic crystalline compound from several suppliers.
Preparative Methods: the title reagent may be prepared[1] by slow addition (30 min) of a solution of phosgene (*Caution!*) in toluene (1 equiv) to pentafluorophenol (2 equiv) in methylene chloride/triethylamine (2 equiv) (rt, 3 h), followed by aqueous work-up and purification as above or by sublimation.[2]
Handling, Storage, and Precautions: stored dry at room temperature, the compound is stable for several months, even years. Slight coloration occurs after years of storage. Being a reactive equivalent of phosgene, any contact between the reagent and the skin should be avoided. Use only in a well-ventilated hood.

Bis(pentafluorophenyl)carbonate (**1**) is related in its reactions to compounds such as *N,N′*-carbonyldiimidazole and phosgene. It is, however, less reactive toward nucleophiles. On the other hand, the reagent is considered more reactive/less hindered than reagents such as bis-(2,4-dinitrophenyl)carbonate or 2,4-dinitrophenyl chloroformate. It is particularly well suited for reactions where advantage can be taken from reagent's balanced reactivity and convenient handling as well as its tendency to form crystalline products.

Azapeptides[3] are analogs of peptides where the α-carbon of one or more amino acids have been substituted with a nitrogen. If all amino acids are substituted, the name azatide has been coined for the compound.

These peptide analogs may have beneficial pharmacological properties relative to the native peptide. Bis(pentafluorophenyl) carbonate has turned out to be a particularly suitable reagent for the coupling of protected hydrazine derivatives to form azapeptides, since other more reactive reagents suffer from side reactions whereas less reactive reagents often are too slow (eqs 1–3).

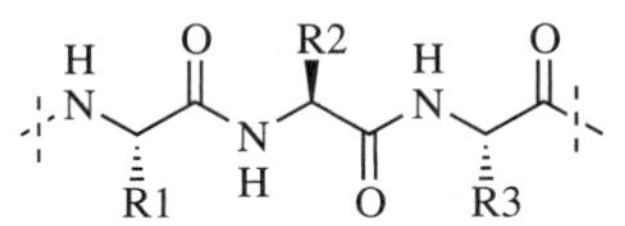

Peptide

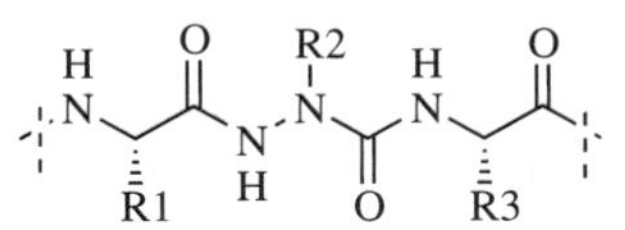

Azapeptide

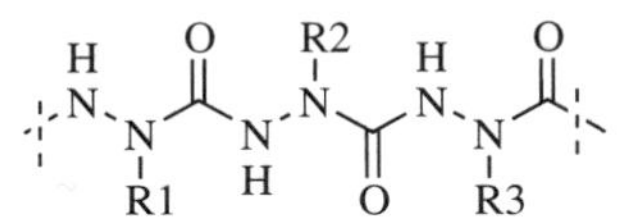

Azatide

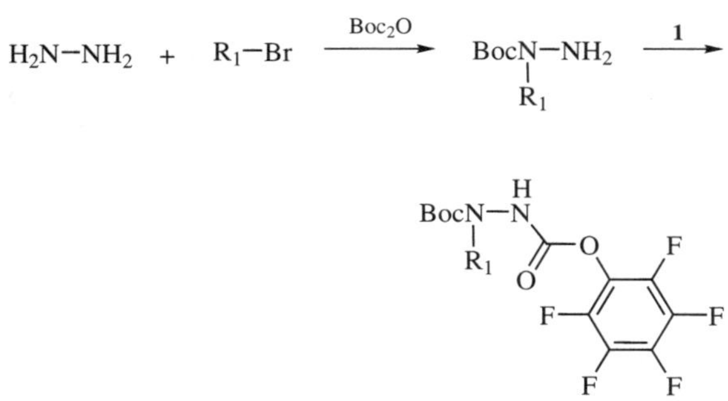

(1)

2

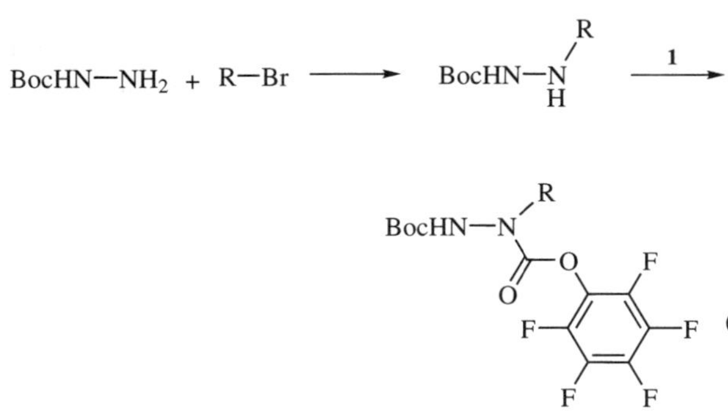

(2)

3

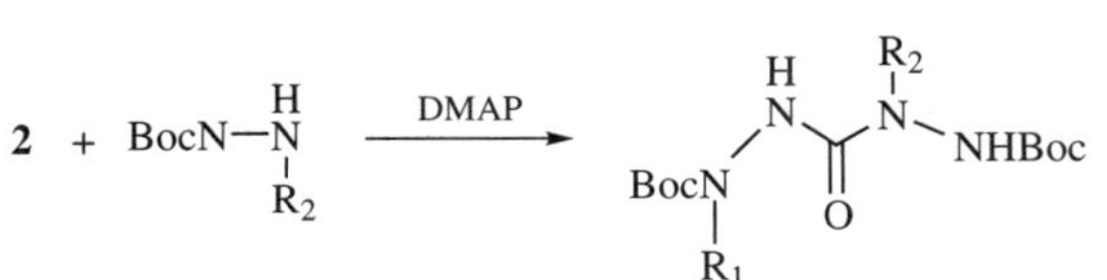

(3)

4

In a pioneering work, Han and Janda[4] used **1** to couple azaamino acids and reported the ready coupling of reactive intermediate **2** with the poor nucleophile 2-(R_2)-Boc-hydrazine to produce azatide **4** (eqs 1–3). Surprisingly, the opposite reaction using **3** and 1-(R_2)-Boc-hydrazine did not result in coupling. This fact led Han and Janda to surmise isocyanate **5** as a reactive intermediate generated in situ from **2**. Such an intermediate cannot be obtained from **3**.

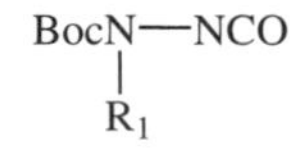

5

However, Hansen[5] has used analogs of **3** successfully with primary as well as secondary amines leading to the growth hormone secretagogues such as **6**, which are highly substituted around the central urea functionality. Although **3** is more crowded than **2**, it is still possible to substitute the second pentafluorophenol.

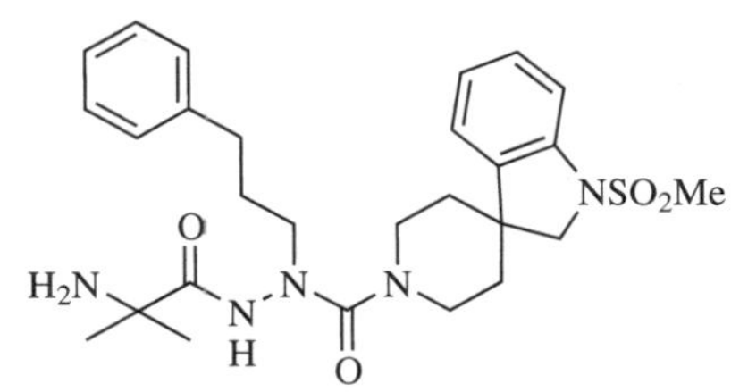

6

Efimov and Kalinkina[6] have advocated the use of the reagent in oligonucleotide synthesis, but Zain and Stawinski[7] later found that it was somewhat slow for this particular purpose and that better choices with regard to side reactions were available. Finally, bis(pentafluorophenyl)carbonate has been used for preparing active esters useful for anchoring polymeric carriers to immunoglobulins.[8]

1. Unpublished.
2. Falardeau, E. R.; Desmarteau, D. D., *J. Fluorine Chem.* **1976**, *7*, 409.
3. (a) Zega, A.; Urleb, U., *Acta Chim. Slov.* **2002**, *49*, 649. (b) Gante, J., *Synthesis* **1989**, 405.
4. Han, H.; Janda, K. D., *J. Am. Chem. Soc.* **1996**, *118*, 2539.
5. Hansen, T. K., *Tetrahedron Lett.* **1999**, *40*, 9119.
6. Efimov, V. A.; Kalinkina, A. L.; Chakhmakhcheva, O. G., *Nucleic Acids Res.* **1993**, *21*, 5337.
7. Zain, R.; Stawinski, J., *J. Org. Chem.* **1996**, *61*, 6617.
8. Taliene, V. R.; Vidziunaite, R.; Mikulskis, P.; Kulys, J., *Biologija* **1994**, *1*, 30.

Thomas Kruse Hansen
Novo Nordisk, Novo Nordisk Park, Maaloev, Denmark

Bis(*o*-phenylene) Pyrophosphite

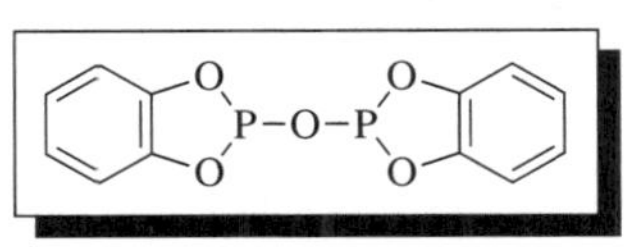

[16421-86-6] $C_{12}H_8O_5P_2$ (MW 294.14)

(reagent for peptide synthesis[1])

Physical Data: mp 72 °C; bp 156 °C/0.25 mmHg; $n^{20}{}_D$ 1.5090 (supercooled liquid).

Preparative Method: reaction of Catechol with Phosphorus(III) Chloride gives *o*-phenylene phosphorochloridite, which upon hydrolysis affords bis(*o*-phenylene) pyrophosphite (**1**).[2]
Purification: purified by distillation at reduced pressure.
Handling, Storage, and Precautions: may be stored indefinitely in a stoppered bottle.

Peptide Synthesis. The reagent (**1**) was introduced as a substitute in peptide synthesis for the widely used, but less easily prepared, Tetraethyl Pyrophosphate. Heating a mixture of an *N*-protected (Cbz or phthaloyl) amino acid and an amino acid or peptide ester with 1.1 equiv of (**1**) in pyridine effects amide bond formation (eq 1).[1] The amino acid or peptide ester may be used as its hydrohalide or tosylate salt. The method is experimentally simple and has the additional advantage that the crude product is generally of sufficient purity for preparative purposes. Prereaction of (**1**) with either coupling partner to give the active ester or the phosphoramidite intermediate is possible but offers no advantage.

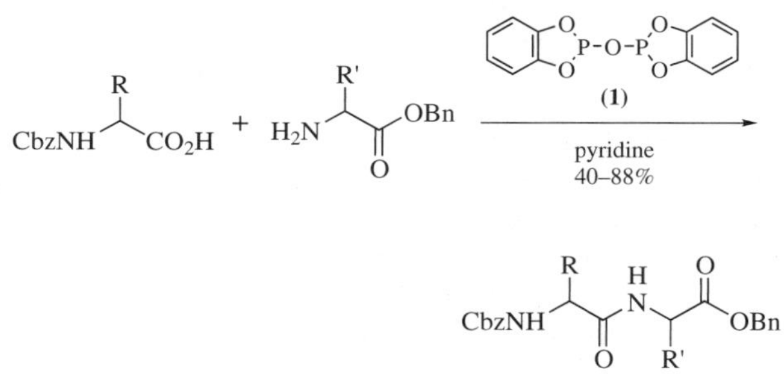

(1)

While no racemization was detected in couplings involving *N*-Cbz-protected amino acids, use of *N*-protected dipeptides as carboxyl components resulted in extensive racemization, presumably by the oxazolone mechanism. The method is also unsuitable for the synthesis of tyrosine peptides without prior protection of the phenolic hydroxy group, since *O*-phosphorylation may occur.

Synthesis of cyclic peptides using (**1**) was studied by Miller and Smith.[3] Use of diethyl phosphite as solvent rather than pyridine was found to be necessary; the method gave better results than 1,3-Dicyclohexylcarbodiimide (DCC) for cyclization of penta- and hexapeptides, but was not suitable for cyclization of tripeptides.

---

1. Crofts, P. C.; Markes, J. H. H.; Rydon, H. M. *J. Chem. Soc* **1959**, 3610.
2. Crofts, P. C.; Markes, J. H. H.; Rydon, H. M. *J. Chem. Soc* **1958**, 4250.
3. Miller, A. W.; Smith, P. W. G. *J. Chem. Soc. (C)* **1967**, 2140.

Alan Armstrong
University of Bath, UK

1,2-Bis(phenylsulfonyl)ethylene[1]

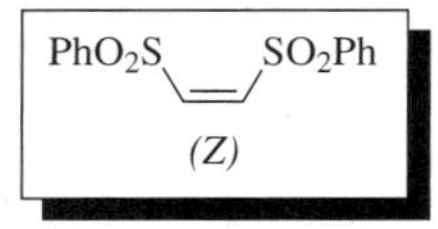

(*Z*)
[*963-15-5*] $C_{14}H_{12}O_4S_2$ (MW 308.37)
(*E*)
[*963-16-6*]

(powerful dienophiles and synthetic equivalents of acetylene and ethylene in Diels–Alder cycloadditions;[2] react as Michael acceptors with substitution of a phenylsulfonyl group[3])

Alternate Name: BPSE.
Physical Data: (*Z*) mp 100–101 °C; (*E*) mp 219.5 °C.
Solubility: the (*Z*)-isomer is more soluble than the (*E*)-isomer in most solvents. In Diels–Alder cycloadditions conducted in dichloromethane, the (*Z*)-isomer affords quite insoluble adducts; conversely, the (*E*)-compound leads to more soluble products.
Form Supplied in: colorless solids; commercially available.
Analysis of Reagent Purity: purity with respect to one another can be checked by ^1H NMR. The vinyl protons of the (*Z*)-isomer appear at 6.82 ppm and those of the (*E*)-isomer at 7.37 ppm in $CDCl_3$.
Preparative Methods: (*Z*)-1,2-bis(phenylsulfonyl)ethylene is prepared by oxidation of the corresponding thioether,[2] which is prepared from (*Z*)-1,2-dichloroethylene[4] or, for large quantities, from 1,1-dichloroethylene.[4b] A mixture of the (*Z*)- and (*E*)-1,2-dichloroethylenes can also be used, but the (*E*)-isomer is unreactive under standard conditions.[4b] Under more drastic reaction conditions (DMF, 70–90 °C), (*E*)-1,2-dichloroethylene also affords the substitution product [57% (*E*), 14% (*Z*)].[4c] The preparation of the (*E*)-isomer is accomplished by sunlight–iodine catalyzed isomerization of the (*Z*)-isomer.[2]
Handling, Storage, and Precautions: the (*Z*)-isomer should be stored protected from direct light because it isomerizes to the (*E*)-form.

Original Commentary

Ottorino De Lucchi
Università di Venezia, Italy

Lucia Pasquato
CNR-CMRO, Padova, Italy

Cycloadditions. 1,2-Bis(phenylsulfonyl)ethylenes [(*Z*)- and (*E*)-BPSE] are most commonly used as synthetic equivalents of acetylene in cycloaddition reactions (eq 1).[2,5] A number of dienes have been prepared via this route which are otherwise not readily accessible.[6]

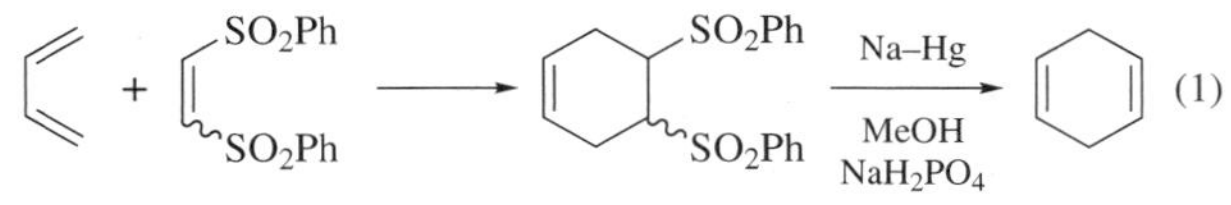

The preparation of both the *syn*- and the *anti*-sesquinorbornatrienes was possible because of the different stereochemistry by which the (*Z*)- and (*E*)-isomers add to 8,9-dehydroisodicyclopentadiene (eq 2).[7] Further studies on these molecules have been reported, including the preparation of deutero derivatives via the deuterated (*Z*)-BPSE.[8] Use of (*E*)-BPSE in the preparation of a natural product is described in eq 3.[9]

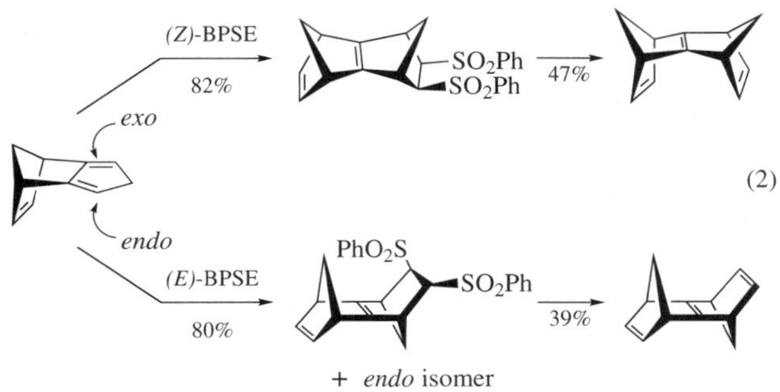

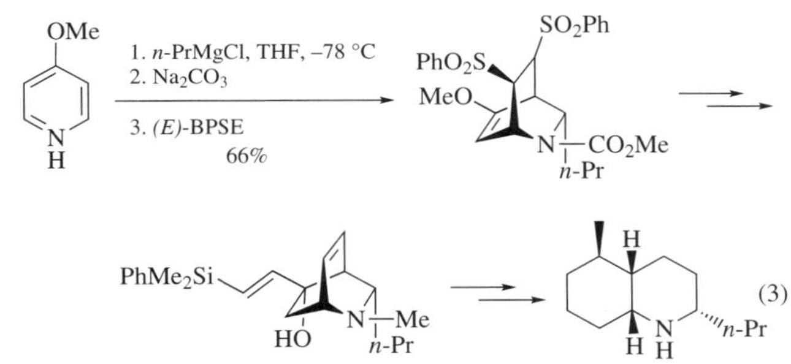

The choice of the solvent for cycloadditions is often crucial. While both reagents are not reactive towards furan in refluxing benzene,[10] they react smoothly at rt in quantitative yields in dichloromethane.[2,11]

Generally, the (*E*)-isomer is more reactive in cycloadditions than the (*Z*)-isomer. For example, the reaction of (*E*)-BPSE with cyclopentadiene in dioxane is 100 times faster than that of the (*Z*)-isomer.[12] A few dienes (1,3-cycloheptadiene and cyclooctatetraene are examples) react only with the former. With cyclopentadiene in dioxane, (*E*)-BPSE is twice as reactive as Maleic Anhydride,[12] but cases have been reported where (*E*)-BPSE failed to react while maleic anhydride did react. Styrene, homobarrelene, and other polycyclic dienes are examples.[13] High pressure (10–12 kbar) can be used to increase reactivity.[14] Examples are the preparations of derivatives of *syn*-sesquinorbornatriene[14a] and of the C_{4v} [4]peristylane tetraketone (eq 4).[14c]

The desulfonylation of cycloadducts can be performed with Sodium Amalgam (1.5–6%) in methanol buffered with suspended sodium dihydrogen phosphate.[2] Because the reaction is heterogeneous, sonication has been proved beneficial.[13a] Samarium(II) Iodide in HMPA has also been used.[15] Other reducing agents and reaction conditions, such as Magnesium–Methanol,[16] afford mixtures of saturated and unsaturated hydrocarbons. Lithium or Sodium metal in liquid ammonia effectively gives solely the saturated adduct, making the title reagent equivalent to ethylene in cycloaddition reactions.[17]

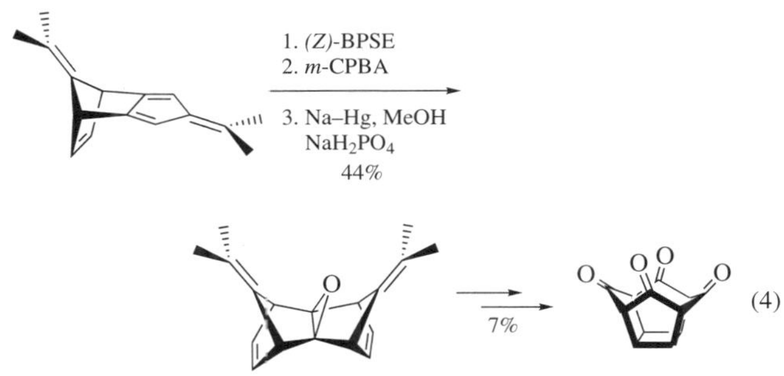

When treated with base, the Diels–Alder adducts give vinyl sulfones. In one example, the vinyl sulfone was further reacted with nucleophiles in a synthesis of ibogamine (eq 5).[18]

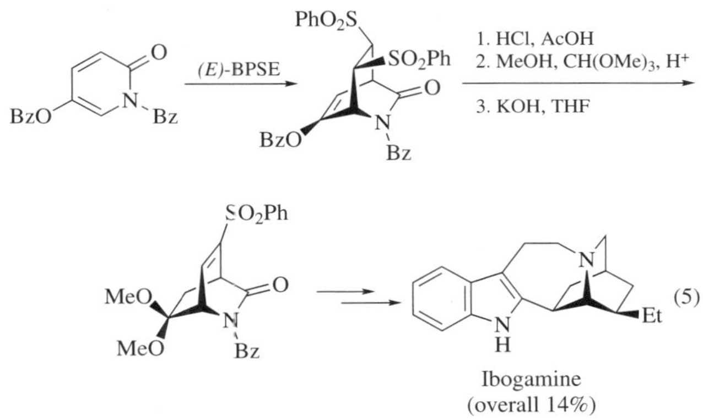

Substitution of one vinyl hydrogen with chlorine gives reagents that can be considered equivalents of the unstable bis(arylsulfonyl) alkynes (eq 6).[19]

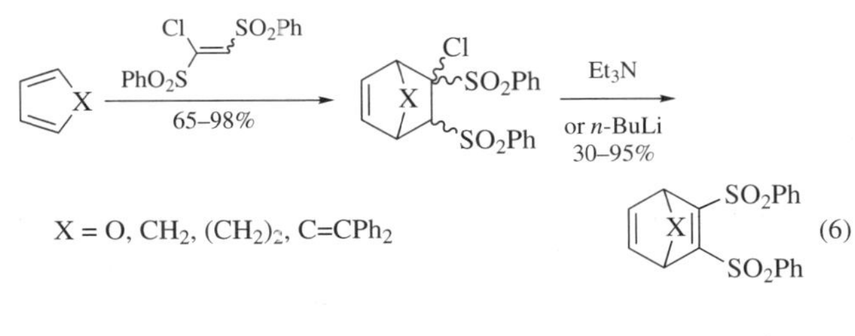

The parent and substituted bis(arylsulfonyl)alkenes, such as the one prepared in eq 6, are good Michael acceptors[20] and react with heteroatomic nucleophiles and Grignard reagents.[3,21] The products correspond to the formal substitution of one of the phenylsulfonyl groups. In eq 7, the product of the Michael addition cyclizes to a dihydrofuran upon deprotonation with Lithium Diisopropylamide.[21a]

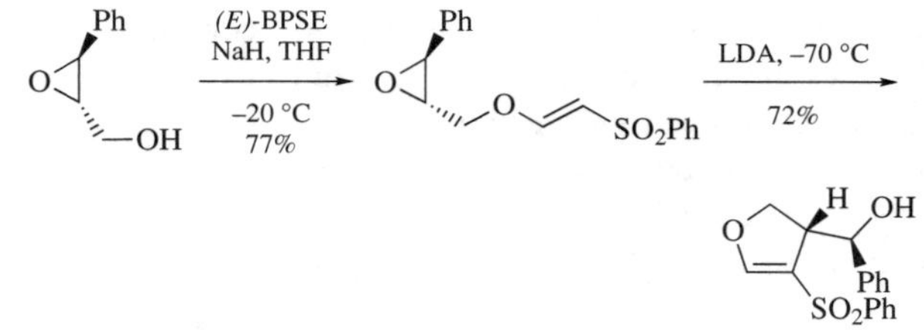

(7)

Both isomeric reagents are also reactive dipolarophiles.[22] For example, they react with azomethine ylides,[22a,b] nitrones,[22c] and diazoalkanes (eq 8).[22d]

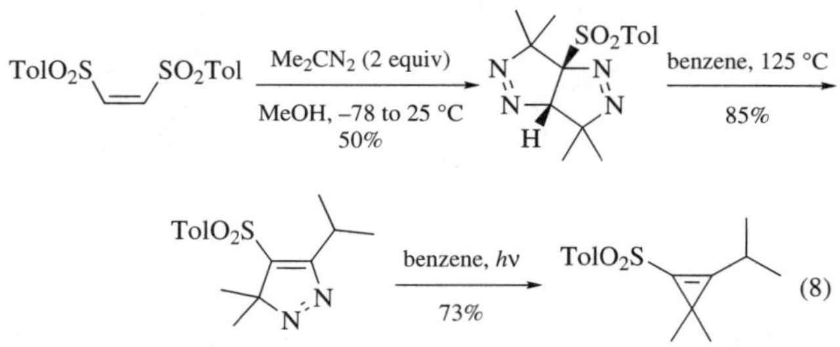

(8)

C_2-Symmetrical chiral versions of these reagents (**1**) and (**2**), which maintain the chemical properties associated with the achiral reagents, are available (see 1,1′-Binaphthalene-2,2′-dithiol).[23]

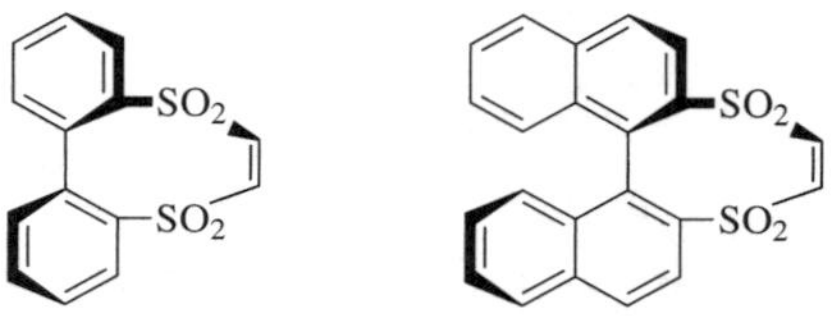

(**1**) (**2**)

These reagents are stable towards racemization and are readily available from the respective dibenzo- or dinaphtho-2,2′-dithiols. In cycloadditions with nonsymmetrical dienes they afford a single diastereoisomer in most cases.

First Update

Patrick Rollin & Arnaud Tatibouët
Université d'Orléans, France

Cycloadditions. The use of 1,2-bis(phenylsulfonyl)ethylenes [(*Z*)- and (*E*)-BPSE] as synthetic equivalents of acetylene in cycloaddition reactions was further developed, with applications to either simple[24a] or more complex[24b–e] dienic templates providing sophisticated polycyclic systems. Cycloaddition of BPSE—or closely related reagents—to heterocyclic dienes was also performed, using furanic[25] or pyrrolic systems[26] as well as vinyl pyridones,[27] vinyl indoles,[28] and indolo-pyrones.[29] Finally, some dipole cycloadditions were explored by Bhaskar Reddy et al. using diazomethane,[30] nitrile oxides, and nitrilimines.[31]

Michael Additions. The strong electrophilic character of BPSE has been exploited in Michael-type additions, generally followed by an elimination step (eq 9).

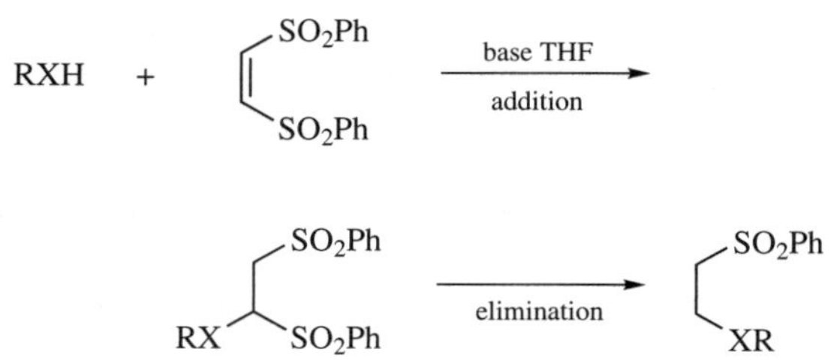

(9)

With a view to constructing new acceptors for radical cyclizations, Evans et al. have developed a controlled process to β-alkoxyvinyl sulfones—also called "vinylogous sulfonates"—involving addition to BPSE of an alkoxide anion generated by Lithium or potassium Hexamethyldisilyl Azide in THF at low temperature and subsequent expulsion of a phenylsulfonate anion (eq 10).[32] Application of such a radical cyclization was considered in the course of a synthetic route to the antitumor agent mucocin[33] and in a synthesis of garsubellin A, a potent inducer of choline acetyltransferase.[34]

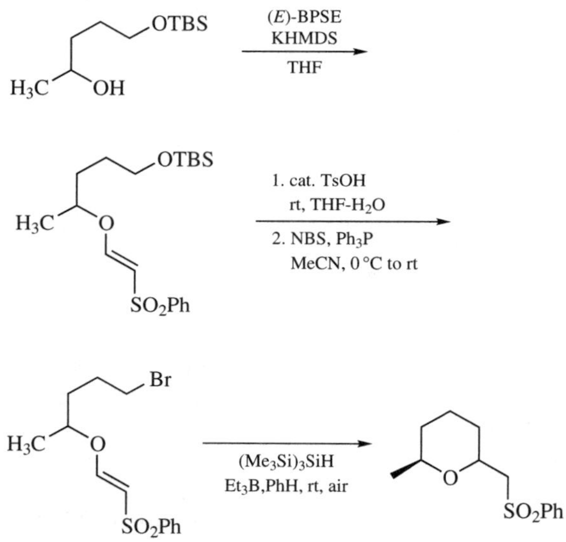

(10)

The above vinylogous sulfonates also function as leaving groups in palladium-catalyzed allylic alkylations and rearrangements (eq 11).[35]

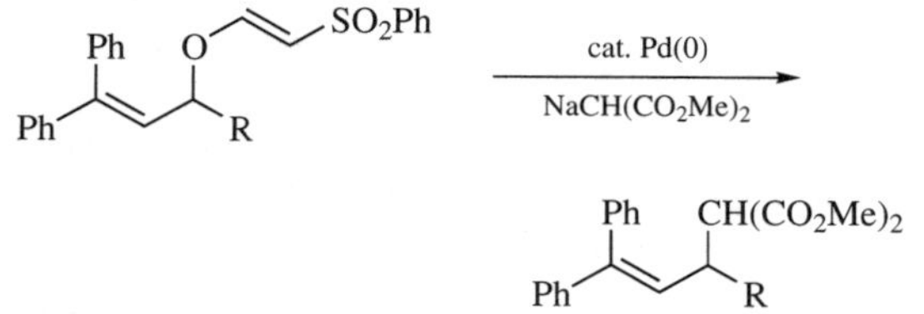

(11)

Selective Tributyltin Hydride reduction of BPSE was demonstrated by De Lucchi as an alternative preparation of Phenyl Vinylsulfone.[36] Further reduction of the previously obtained β-alkoxyvinyl sulfones using single-electron transfer reagents—Sodium Amalgam in particular—smoothly afforded the corresponding vinyl ethers: this process was applied by Rollin and co-workers to both simple alcohols[37] and many diverse carbohydrate derivatives.[38] Sugar-based vinyl ethers were investigated by the same group as templates for stereoselective dipolar cycloadditions (eq 12).[39]

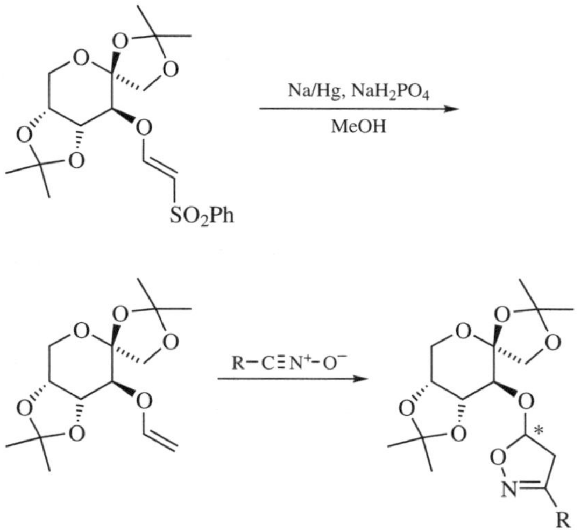

(12)

Some nitrogen nucleophiles were also shown to afford *N*-phenylsulfonylvinyl) derivatives, precursors to the *N*-vinylated counterparts: such a reaction was first observed serendipitously on a γ-lactam by Knapp[40] and later applied, with triethylamine catalysis, to simple or sugar-derived 1,3-oxazolidine-2-thiones (eq 13).[41]

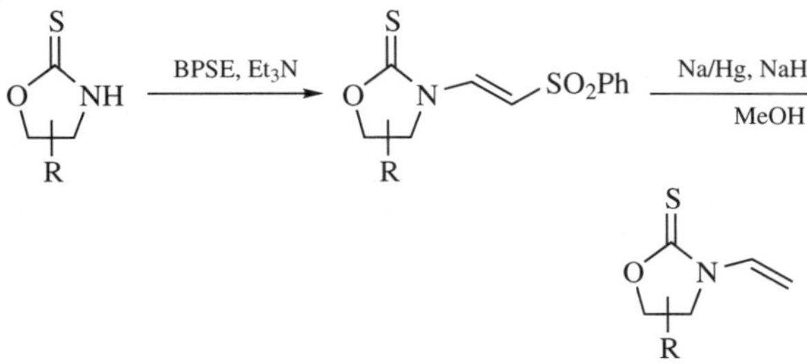

(13)

The case of thiols was mentioned[25b] then thoroughly examined: under triethylamine catalysis, it was shown that more complex reactions occurred (eq 14).[42]

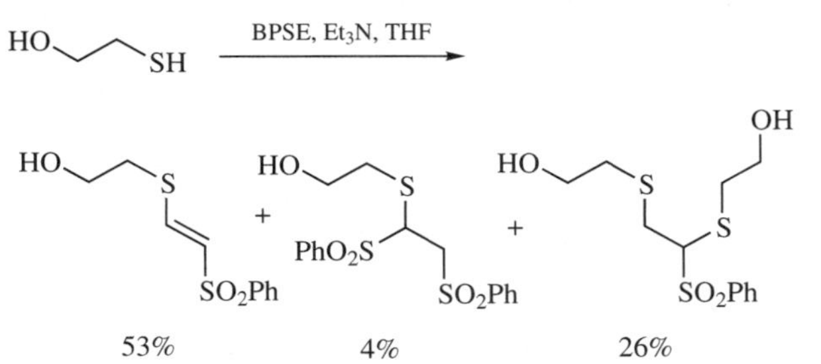

(14)

Addition of a sulfone-stabilized carbanion to BPSE constituted the initial step in the synthesis of highly substituted naphthalenes (eq 15).[43]

Radical monoaddition resulting in a vinylsulfonyl group transfer to alkyl radicals generated from alkyl allyl sulfones was demonstrated by Kim (eq 16).[44]

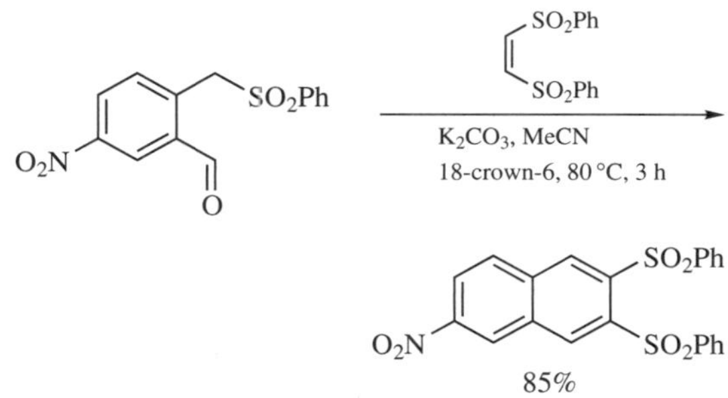

(15)

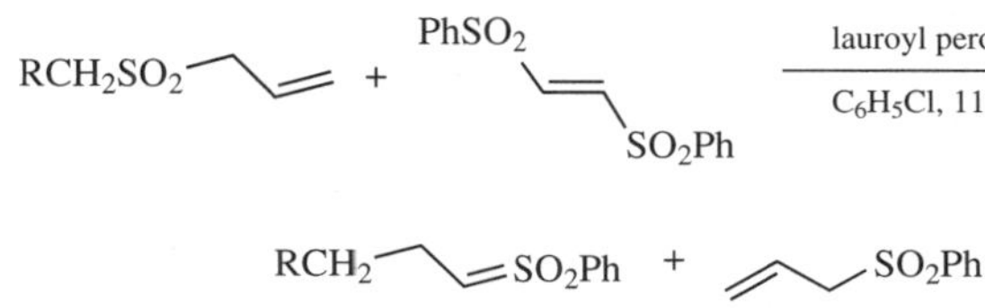

(16)

Michael Double Additions. Following a double Michael addition process on BPSE, diols were readily converted into a novel family of cyclic acetals: 2-phenylsulfonylethylidene acetals (PSE acetals) (eq 17). Pioneering results from De Lucchi's group[45] were extended to the carbohydrate field, in which both 1,3-dioxolanes and 1,3-dioxanes could be prepared in good yield and in most cases with high regio- and stereoselectivity.[46] Syntheses of thio-analog (PSE thioacetals) were similarly investigated from simple substrates, then adapted to the case of carbohydrate-derived 1,3-oxathianes.[42]

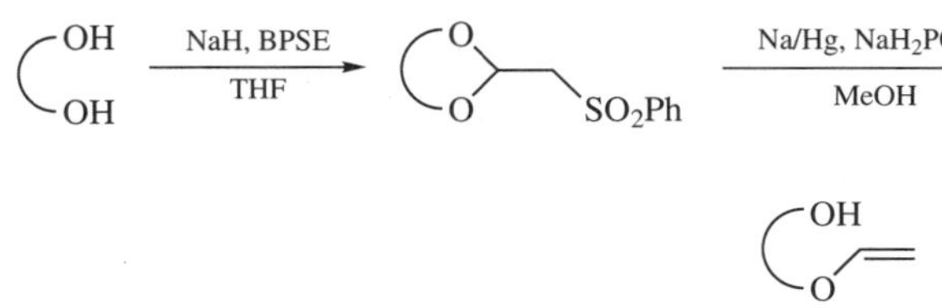

(17)

PSE acetals—which are prepared in basic media—show very unusual properties in that they are strongly resistant to acidic conditions but can in contrast undergo ring-opening reactions under basic or reductive conditions (eq 18). A general selective synthesis of ω-hydroxyethenyl ethers was devised through sodium amalgam reductive opening of these novel protecting groups (eq 17).[47] PSE acetals can be applied as a protecting group particularly in carbohydrate chemistry: for example, regioselective de-*O*-benzylation using Triisobutylaluminum (TIBAL) was successfully performed on monosaccharidic glycosides (eq 18).[48]

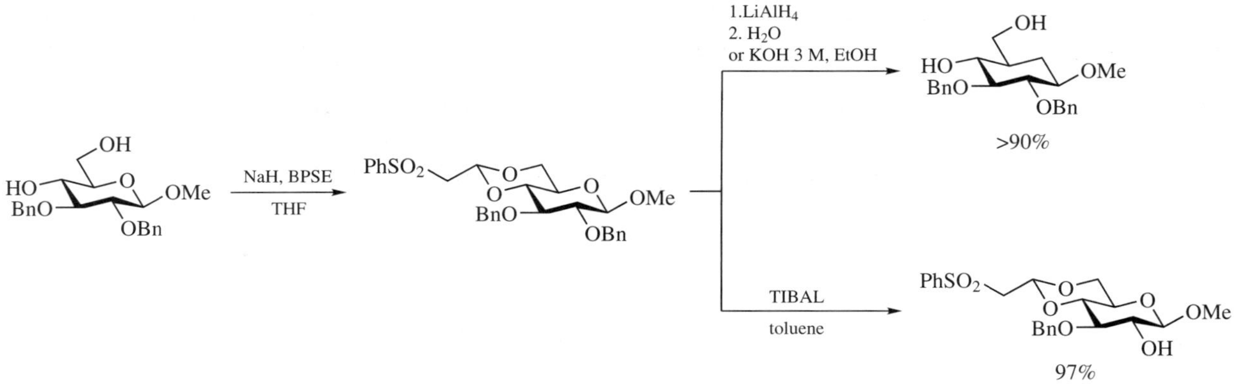

(18)

1. (a) De Lucchi, O.; Pasquato, L., *Tetrahedron* **1988**, *44*, 6755. (b) Cossu, S.; De Lucchi, O.; Fabbri, D.; Licini, G.; Pasquato, L., *Org. Prep. Proced. Int.* **1991**, *23*, 571. (c) Tanaka, K.; Kaji, A. In *The Chemistry of Sulphones and Sulphoxides*, Patai, S.; Rappoport, Z.; Stirling, C. J. M., Eds.; Wiley: Chichester, 1988; Chapter 15, pp 791–799. (d) Simpkins, N. S. *Sulphones in Organic Synthesis*; Pergamon: Oxford, 1993.
2. De Lucchi, O.; Lucchini, V.; Pasquato, L.; Modena, G., *J. Org. Chem.* **1984**, *49*, 596.
3. Meek, J. S.; Fowler, J. S., *J. Org. Chem.* **1968**, *33*, 985.
4. (a) Cusa, N. W.; McCombie, H., *J. Chem. Soc.* **1937**, 767. (b) Truce, W. E.; Boudakian, M. M.; Heine, R. F.; McManimie, R. J., *J. Am. Chem. Soc.* **1956**, *78*, 2743. (c) Tanimoto, S.; Taniyasu, R.; Takahashi, T.; Miyake, T.; Okano, M., *Bull. Chem. Soc. Jpn.* **1976**, *49*, 1931. (d) Tiecco, M.; Testaferri, L.; Tingoli, M.; Chianelli, D.; Montanucci, M., *J. Org. Chem.* **1983**, *48*, 4795.
5. De Lucchi, O.; Modena, G., *Tetrahedron* **1984**, *40*, 2585.
6. (a) Mirsadeghi, S.; Rickborn, B., *J. Org. Chem.* **1985**, *50*, 4340. (b) Kazimirchik, I. V.; Lukin, K. A.; Taranyuk, I. L.; Bebikh, G. F.; Zefirov, N. S., *Zh. Org. Khim.* **1985**, *21*, 313. (c) Paquette, L. A.; Racherla, U. S., *J. Org. Chem.* **1987**, *52*, 3250. (d) Kobayashi, T.; Miki, S.; Yoshida, Z.; Asako, Y.; Kajimoto, C., *J. Am. Chem. Soc.* **1988**, *110*, 5622. (e) Krebs, J.; Guggisberg, D.; Stämpfli, U.; Neuenschwander, M., *Helv. Chim. Acta* **1986**, *69*, 835. (f) Azzena, U.; Cossu, S.; De Lucchi, O.; Melloni, G., *Synth. Commun.* **1988**, *18*, 351. (g) Patney, H. K., *J. Org. Chem.* **1988**, *53*, 6106. (h) Paquette, L. A. Gugelchuk, M.; McLaughlin, M. L., *J. Org. Chem.* **1987**, *52*, 4732.
7. (a) De Lucchi, O.; Licini, G.; Pasquato, L., *Chem. Commun./J. Chem. Soc., Chem. Commun.* **1985**, 418. (b) Paquette, L. A.; Künzer, H.; Green, K. E., *J. Am. Chem. Soc.* **1985**, *107*, 4788. (c) Paquette, L. A.; Künzer, H.; Green, K. E.; De Lucchi, O.; Licini, G.; Pasquato, L.; Valle, G., *J. Am. Chem. Soc.* **1986**, *108*, 3453.
8. (a) Künzer, H.; Cottrell, C. E.; Paquette, L. A., *J. Am. Chem. Soc.* **1986**, *108*, 8089. (b) Künzer, H.; Litterst, E.; Gleiter, R.; Paquette, L. A., *J. Org. Chem.* **1987**, *52*, 4740. (c) Paquette, L. A.; Künzer, H.; Kesselmayer, M. A., *J. Am. Chem. Soc.* **1988**, *110*, 6521.
9. Polniaszek, R. P.; Dillard, L. W., *J. Org. Chem.* **1992**, *57*, 4103.
10. Snyder, H. R.; Hallada, D. P., *J. Am. Chem. Soc.* **1952**, *74*, 5595.
11. Cossu, S.; De Lucchi, O.; Dilillo, F., *Gazz. Chim. Ital.* **1989**, *119*, 519.
12. (a) Sauer, J.; Wiest, H.; Mielert, A., *Ber. Dtsch. Chem. Ges./Chem. Ber.* **1964**, *97*, 3183. (b) Sauer, J.; Lang, D.; Wiest, H., *Ber. Dtsch. Chem. Ges./Chem. Ber.* **1984**, *97*, 3208.
13. (a) De Lucchi, O.; Pasquato, L. unpublished. (b) Fessner, W.-D.; Sedelmeier, G.; Spurr, P. R.; Rihs, G.; Prinzbach, H., *J. Am. Chem. Soc.* **1987**, *109*, 4626. (c) Fessner, W.-D.; Grund, C.; Prinzbach, H., *Tetrahedron Lett.* **1989**, *30*, 3133.
14. (a) Paquette, L. A.; Waykole, L.; Shen, C.-C.; Racherla, U. S.; Gleiter, R.; Litterst, E., *Tetrahedron Lett.* **1988**, *29*, 4213. (b) Paquette, L. A.; Shen, C.-C., *J. Am. Chem. Soc.* **1990**, *112*, 1159. (c) Paquette, L. A.; Shen, C.-C.; Engel, P., *J. Org. Chem.* **1989**, *54*, 3329.
15. Künzer, H.; Stahnke, M.; Sauer, G.; Wiechert, R., *Tetrahedron Lett.* **1991**, *32*, 1949.
16. Brown, A. C.; Carpino, L. A., *J. Org. Chem.* **1985**, *50*, 1749.
17. De Lucchi, O.; Pasquato, L., *Gazz. Chim. Ital.* **1984**, *114*, 349.
18. (a) Herdeis, C.; Hartke, C., *Heterocycles* **1989**, *29*, 287. (b) Herdeis, C.; Hartke-Karger, C., *Justus Liebigs Ann. Chem./Liebigs Ann. Chem.* **1991**, 99.
19. Pasquato, L.; De Lucchi, O.; Krotz, L., *Tetrahedron Lett.* **1991**, *32*, 2177.
20. (a) Azzena, U.; Cossu, S.; De Lucchi, O.; Melloni, G., *Tetrahedron Lett.* **1989**, *30*, 1845. (b) Cossu, S.; De Lucchi, O., *Gazz. Chim. Ital.* **1990**, *120*, 569. (c) Azzena, U.; Cossu, S.; De Lucchi, O.; Melloni, G., *Gazz. Chim. Ital.* **1989**, *119*, 357.
21. (a) McCombie, S. W.; Shankar, B. B.; Ganguly, A. K., *Tetrahedron Lett.* **1985**, *26*, 6301. (b) McCombie, S. W.; Shankar, B. B.; Ganguly, A. K., *Tetrahedron Lett.* **1987**, *28*, 4123. (c) Padwa, A.; Bullock, W. H.; Dyszlewski, A. D.; McCombie, S. W.; Shankar, B. B.; Ganguly, A. K., *J. Org. Chem.* **1991**, *56*, 3556. (d) Flitsch, W.; Lubisch, W., *Ber. Dtsch. Chem. Ges./Chem. Ber.* **1982**, *115*, 1547.
22. (a) Padwa, A.; Fryxell, G. E.; Gasdaska, J. R.; Venkatramanan, M. K.; Wong, G. S. K., *J. Org. Chem.* **1989**, *54*, 644. (b) Pandey, G.; Lakshmaiah, G.; Kumaraswamy, G., *Chem. Commun./J. Chem. Soc., Chem. Commun.* **1992**, 1313. (c) Burdisso, M.; Gandolfi, R.; Grünanger, P.; Rastelli, A., *J. Org. Chem.* **1990**, *55*, 3427. (d) Padwa, A.; Wannamaker, M. W., *Tetrahedron* **1991**, *47*, 6139.
23. (a) De Lucchi, O., *Phosphorus Sulfur/Phosphorus Sulfur Silicon* **1993**, *74*, 195. (b) Cossu, S.; Delogu, G.; De Lucchi, O.; Fabbri, D.; Licini, G., *Angew. Chem., Int. Ed. Engl.* **1989**, *28*, 766. (c) De Lucchi, O.; Fabbri, D.; Cossu, S.; Valle, G., *J. Org. Chem.* **1991**, *56*, 1888 and unpublished results.
24. (a) Cossu, S.; Battaggia, S.; De Lucchi, O., *J. Org. Chem.* **1997**, *62*, 4162. (b) Griesbeck, A. G.; Deufel, T.; Peters, K.; Peters, E.-M.; von Schnering, H. G., *Angew. Chem. Int. Ed. Engl.* **1993**, *32*, 97. (c) Shen, C. C.; Paquette, L. A., *Tetrahedron* **1994**, *50*, 4949. (d) Lin, C.-T.; Wang, N.-J.; Yeh, Y.-L.; Chou, T.-C., *Tetrahedron* **1995**, *51*, 2907. (e) Hirai, Y.; Ito, K.; Nagaoka, H., *Heterocycles* **1998**, *48*, 235.
25. (a) Arjona, O.; Iradier, F.; Mañas, R. M.; Plumet, J.; Grabuleda, X.; Jaime, C., *Tetrahedron* **1998**, *54*, 9095. (b) Arjona, O.; Iradier, F.; Medel, R.; Plumet, J., *J. Org. Chem.* **1999**, *64*, 6090.
26. Ando, K.; Kankake, M.; Suzuki, T.; Takayama, H., *Synlett* **1994**, 741.
27. Kuethe, J. T.; Brooks, C. A.; Comins, D. L., *Org. Lett.* **2003**, *5*, 321.
28. Pindur, U.; Lutz, G.; Fischer, G.; Schollmeyer, D.; Masser, W.; Schroeder, L., *Tetrahedron* **1993**, *49*, 2863.
29. Pindur, U.; Erfanian-Abdoust, H., *Heterocycles* **1990**, *31*, 1751.

30. Bashkar Reddy, D.; Rajagopala Sarma, M.; Padmaja, A.; Padmavathi, V., *Phosphorous Sulfur Silicon* **2000**, *164*, 23.
31. Padmavathi, V.; Sumathi, R. P.; Venugopal Reddy, K.; Somasekar Reddy, A.; Bashkar Reddy, D., *Synth. Commun.* **2000**, *30*, 4007.
32. (a) Evans, P. A.; Manangan, T., *Tetrahedron Lett.* **1997**, *38*, 8165. (b) Evans, P. A.; Manangan, T., *J. Org. Chem.* **2000**, *65*, 4523.
33. Evans, P. A.; Roseman, J. D., *Tetrahedron Lett.* **1997**, *38*, 5249.
34. Nicolaou, K. C.; Pfefferkorn, J. A.; Kim, S.; Wei, H. X., *J. Am. Chem. Soc.* **1999**, *121*, 4724.
35. (a) Evans, P. A.; Brandt, T. A., *Org. Lett.* **1999**, *1*, 1563. (b) Evans, P. A.; Brandt, T. A.; Robinson, J. E., *Tetrahedron Lett.* **1999**, *40*, 3105.
36. Cossu, S.; De Lucchi, O.; Durr, R.; Fabris, F., *Synth. Commun.* **1996**, *26*, 211.
37. Cabianca, E.; Chéry, F.; Rollin, P.; Cossu, S.; De Lucchi, O., *Synlett* **2001**, 1962.
38. Chéry, F.; Desroses, M.; Tatibouët, A.; De Lucchi, O.; Rollin, P., *Tetrahedron* **2003**, *59*, 4563.
39. Desroses, M.; Chéry, F.; Tatibouët, A.; De Lucchi, O., Rollin, P., *Tetrahedron: Asymmetry* **2002**, *13*, 2535.
40. Knapp, S.; Levorse, A. T., *J. Org. Chem.* **1988**, *53*, 4006.
41. Girniene, J.; Tardy, S.; Tatibouët, A.; Sackus, A. ; Rollin, P., *Tetrahedron Lett.* **2004**, *45*, 6443.
42. Cabianca, E.; Tatibouët, A.; Chéry, F.; Pillard, C.; De Lucchi, O.; Rollin, P., *Tetrahedron Lett.* **2003**, *44*, 5723.
43. Tyrala, A.; Makosza, M., *Synthesis* **1994**, 264.
44. Kim, S.; Lim, C. J., *Angew. Chem. Int. Ed. Engl.* **2002**, *41*, 3265.
45. Cossu, S.; De Lucchi, O.; Fabris, F. ; Ballini, R.; Bosica, G., *Synthesis* **1996**, 1481.
46. Chéry, F.; Rollin, P.; De Lucchi, O.; Cossu, S., *Synthesis* **2001**, 286.
47. Cabianca, E.; Chéry, F.; Rollin, P.; Tatibouët, A.; De Lucchi, O., *Tetrahedron Lett.* **2002**, *43*, 585.
48. Chevalier-Du Roizel, B.; Cabianca, E.; Rollin, P.; Sinaÿ, P., *Tetrahedron* **2002**, *58*, 9579.

1,1,3,3-Bis(tetramethylene)chlorouronium Hexafluorophosphate

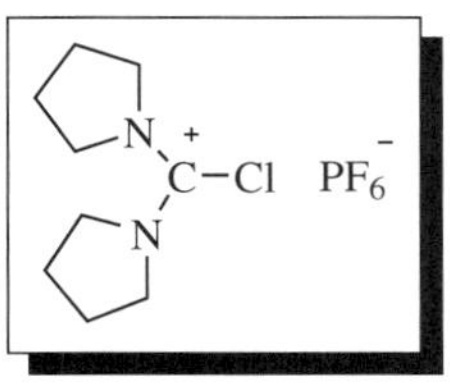

[207915-96-6; 135540-11-3] $C_9H_{16}ClF_6N_2P$ (MW 332)

(peptide synthesis coupling reagent)

Alternate Name: PyClU,[1] BTCFH.[2]
Physical Data: mp 151–153 °C,[1] 146–148 °C.[3]
Solubility: soluble in CH_2Cl_2, $CHCl_3$, DMF, DMSO, NMP, acetone.
Form Supplied in: colorless crystals.
Analysis of Reagent Purity: ^{1}H-NMR; elemental analysis.
Preparative Methods: addition of $POCl_3$[1] or a 20% solution of phosgene in toluene[3,4] to the corresponding urea and then addition of KPF_6/H_2O.
Purification: recrystallization (acetone/Et_2O).
Handling, Storage, and Precautions: the toxicological properties of this reagent have not been investigated and no special instructions for storage and handling are mentioned in the literature.

Reagent for Peptide Synthesis. This reagent, which was designed for coupling *N*-methylated amino acids (eq 1), gives results comparable to those obtained with **PyBroP**.[1] Compared with the hydroxybenzotriazol-based activator **HBTU**, which produces the moderately activated benzotriazolyl esters from carboxylic acids, PyClU gives high yields and low racemization.[1,5] However, significant epimerization is observed during segment condensation.[6]

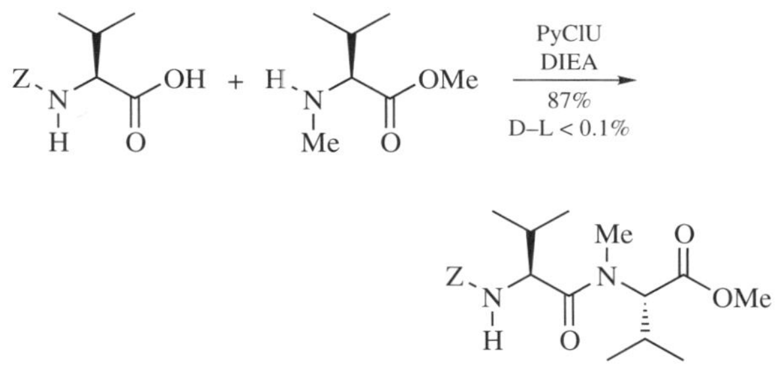

(1)

In addition, the homologous tetrafluoroborate salt TPyClU proved to be efficient for monoester formation from *N*-protected α-amino phosphonic acid (eq 2).[7]

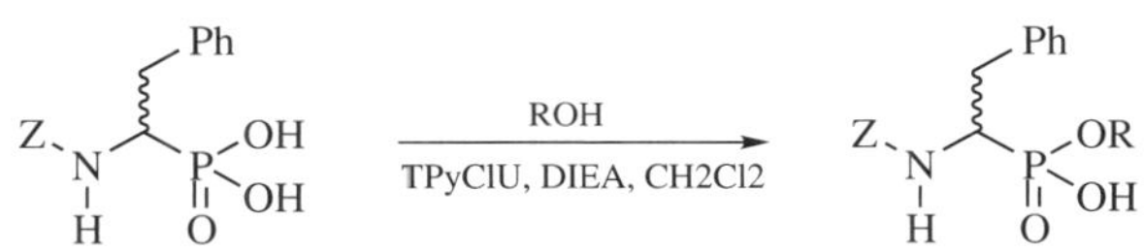

(2)

R = *i*-PrOH (83%), PhCH2OH (86%), *t*-BuOH (65%)

Preparation of Coupling Reagents for Peptide Synthesis. Several related uronium coupling reagents have been prepared from PyClU:

1. Bis(tetramethylene)phenoxyformamidinium hexafluorophosphate derivatives PfPyU,[4] HPyOPFp, HpyOTcp, and HpyONp[8] (eq 3).

2. Benzotriazolyl-1,1,3,3-bis(tetramethylene)chlorouronium hexafluorophosphate (HBPyU[1] or BBC[3]) and its 7-azabenzotriazole[9] analog, reagents for peptide synthesis (eq 4).

3. Bis(tetramethylene)fluoroformamidinium hexafluorophosphate (BTFFH), reagent for in situ conversion of amino acids into amino acid fluorides in peptide synthesis (eq 5).[2]

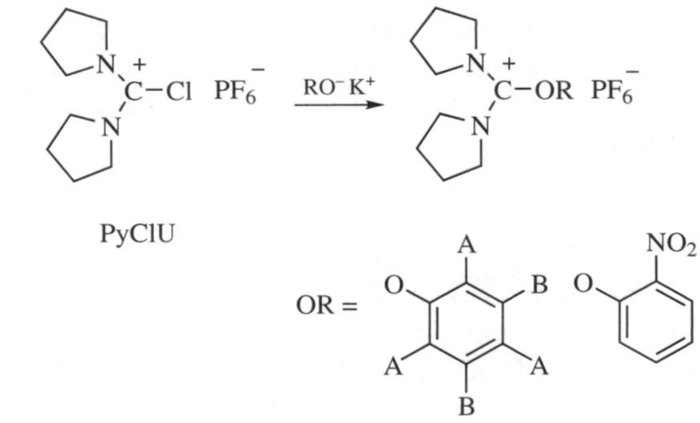

(3)

A = B = F HPyONp (88%)
PfPyU(90%)
(HPyOPFp)
A = Cl, B = H
HPyOTcp (86%)

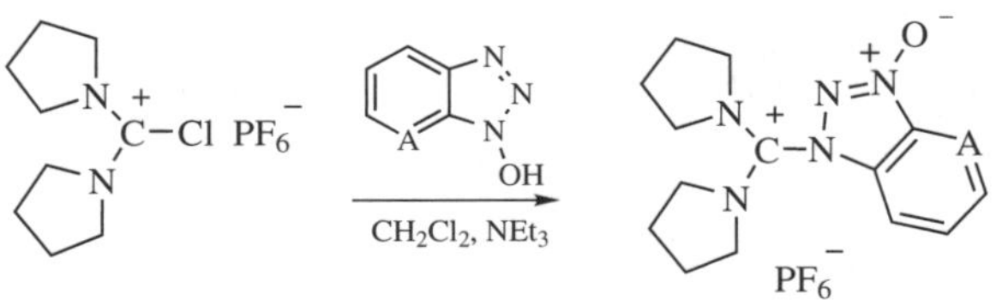

(4)

PyClU

A = C, HBPyU (BBC) (93%)
A = N, HAPyU (88%)

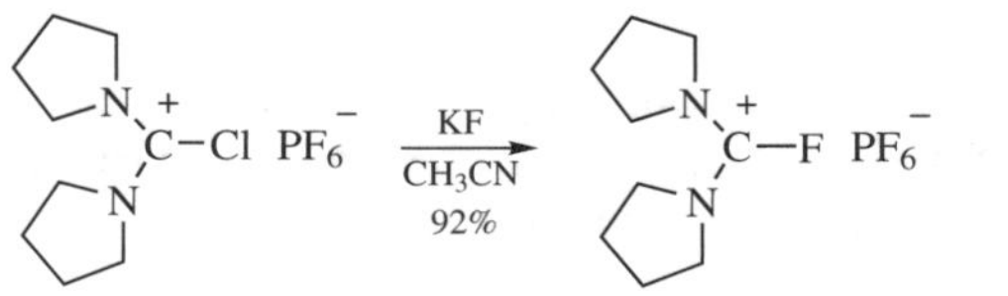

(5)

PyClU BTFFH

Related Reagents. PyCloP (chlorotripyrrolidinophosphonium hexafluorophosphate); PyBroP (bromotripyrrolidinophosphonium hexafluorophosphate); HBPyU [*O*-(benzotriazol-1-yl)-1,1,3,3-bis(tetramethylene)uronium hexafluorophosphate]; HAPyU [*O*-(7-azabenzotriazol-1-yl)-1,1,3,3-bis(tetramethylene)uronium hexafluorophosphate]; TpyClU [1,1,3,3-bis(tetramethylene)chlorouronium tetrafluoroborate]; CIP (2-chloro-1,3-dimethylimidazolidium hexafluorophosphate); HPyOPFp [(pentafluorophenyl)oxy-1,1,3,3-bis(tetramethylene)uronium hexafluorophosphate]; HpyOTcp [(2,4,5-trichlorophenyl)oxy-1,1,3,3-bis(tetramethylene)uronium hexafluorophosphate]; HPyONp [(2-nitrophenyl)oxy-1,1,3,3-bis(tetramethylene)uronium hexafluorophosphate]; BTFFH [bis(tetramethylene)fluoroformamidium hexafluorophosphate]; TFFH (tetramethylfluoroformamidium hexafluorophosphate).

---

1. Coste, J.; Frérot, E.; Jouin, P.; Castro, B., *Tetrahedron Lett.* **1991**, *32*, 1967.
2. Carpino, L. A.; El-Faham, A., *J. Am. Chem. Soc.* **1995**, *117*, 5401.
3. Chen, S.; Xu, J., *Tetrahedron Lett.* **1992**, *33*, 647.
4. Habermann, J.; Kunz, H., *J. Prakt. Chem.* **1998**, *340*, 233.
5. Humphrey, J. M.; Chamberlin, A. R., *Chem. Rev.* **1997**, *97*, 2243.
6. Li, P.; Xu, J.-C.; *Tetrahedron* **2000**, *56*, 9949.
7. Galéotti, N.; Coste, J.; Bedos, P.; Jouin, P., *Tetrahedron Lett.* **1996**, *37*, 3997.
8. El-Faham, A., *Lett. Peptide Sci.* **2000**, *7*, 113.
9. Carpino, L. A.; El-Faham, A., *J. Org. Chem.* **1994**, *59*, 695.

Jacques Coste & Patrick Jouin
CNRS-UPR9023, Montpellier, France

Bis(tributyltin) Oxide

$(n\text{-}Bu_3Sn)_2O$

[56-35-9] $C_{24}H_{54}OSn_2$ (MW 596.20)

promotes the oxidation of secondary alcohols and sulfides with Br_2; O- and N-activations; dehydrosulfurizations; hydrolysis catalyst)

Physical Data: bp 180 °C/2 mmHg; *d* 1.170 g cm^{-3}.
Solubility: sol ether and hexane.
Form Supplied in: colorless oil.
Handling, Storage, and Precautions: $(Bu_3Sn)_2O$ should be stored in the absence of moisture. Owing to the toxicity of organostannanes, this reagent should be handled in a well-ventilated fume hood. Contact with the eyes and skin should be avoided.

Oxidations. Benzylic, allylic, and secondary alcohols are oxidized to the corresponding carbonyl compounds by using $(Bu_3Sn)_2O$– Bromine.[1] This procedure is quite useful for selective oxidation of secondary alcohols in the presence of primary alcohols, which are inert under these conditions (eqs 1–3).[2] $(Bu_3Sn)_2O$– N-Bromosuccinimide can also be applied to the selective oxidation of secondary alcohols (eq 4).[3]

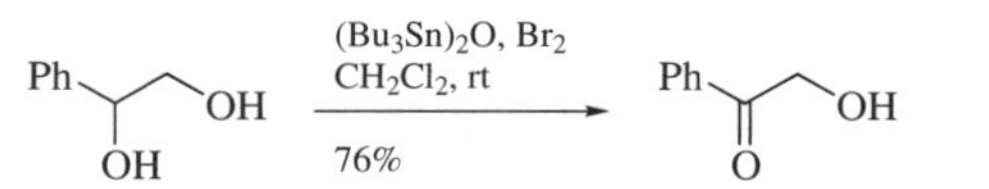

(1)

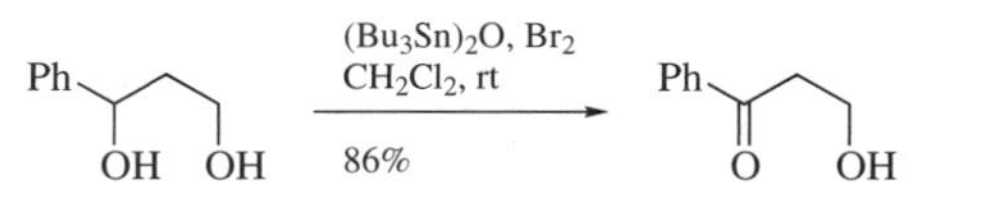

(2)

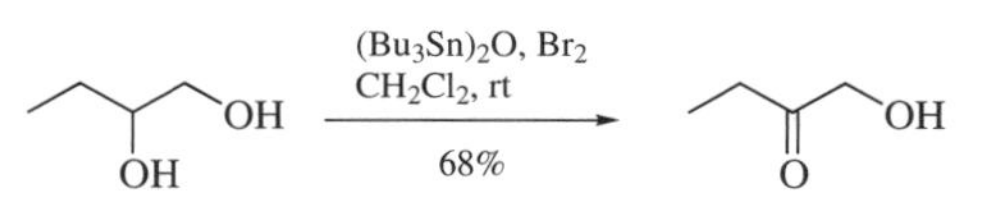

(3)

$(Bu_3Sn)_2O$–Br_2 oxidizes sulfides to sulfoxides in CH_2Cl_2 without further oxidation to sulfones, even in the presence of excess reagent (eq 5).[4] This procedure is especially useful for sulfides having long, hydrophobic alkyl chains, for which solubility problems are often encountered in the Sodium Periodate oxidation in aqueous organic solvents. Oxidation of sulfenamides to sulfinamides can be achieved without formation of sulfonamides using the reagent (eq 6).[4]

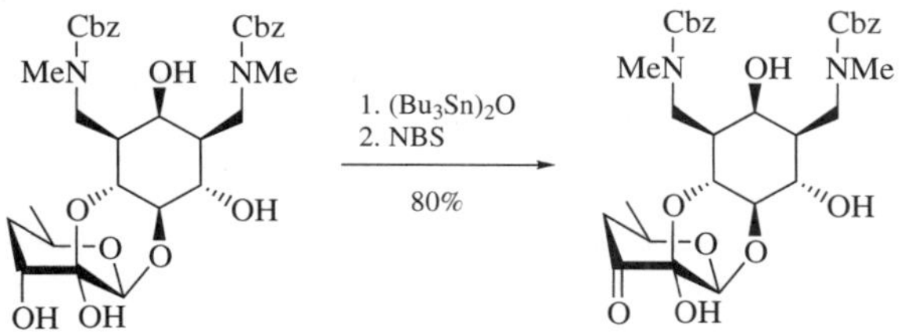

(4)

$$PhCH_2SCH_2Ph \xrightarrow[92\%]{(Bu_3Sn)_2O,\ Br_2,\ CH_2Cl_2,\ rt} PhCH_2SOCH_2Ph \quad (5)$$

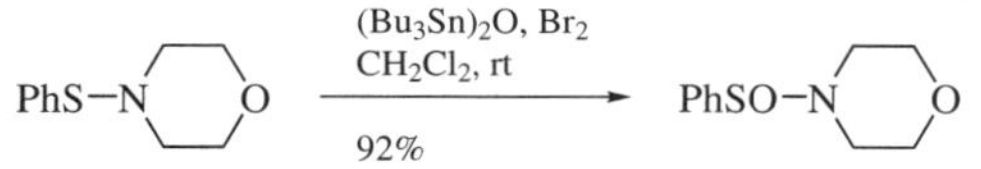

(6)

$(Bu_3Sn)_2O$–Br_2– Diphenyl Diselenide in refluxing $CHCl_3$ transforms alkenes into α-seleno ketones (eq 7).[5]

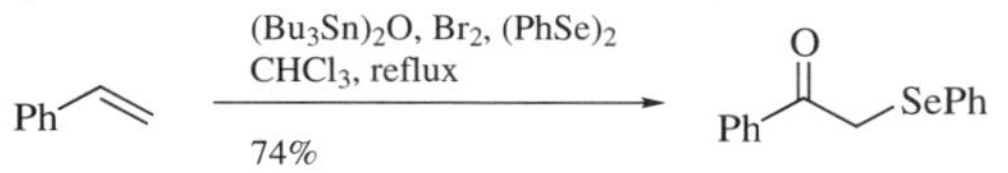

(7)

O- and N-activations. $(Bu_3Sn)_2O$ has been used in the activation of hydroxy groups toward sulfamoylations, acylations, carbamoylations, and alkylations because conversion of alcohols to stannyl ethers enhances the oxygen nucleophilicity. Tributylstannyl ethers are easily prepared by heating the alcohol and $(Bu_3Sn)_2O$, with azeotropic removal of water. Sulfamoylation of alcohols can be achieved via tributyltin derivatives in high yields, whereas direct sulfamoylation gives low yields (eq 8).[6] This activation can be used for selective

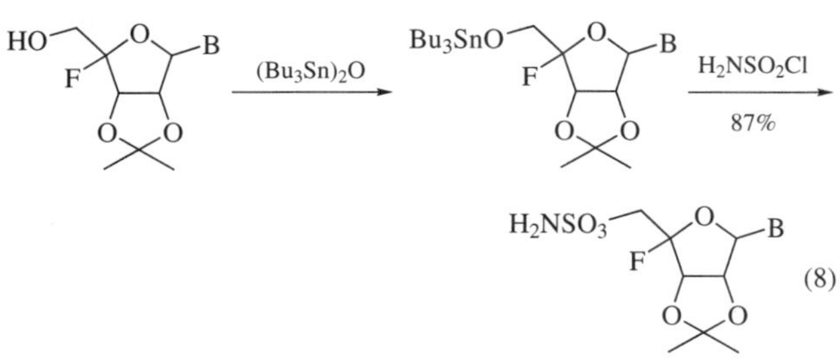

(8)

acylation of vicinal diols (eq 9).[7] In carbohydrate chemistry this approach is extremely useful for the regioselective acylation without the use of a blocking–deblocking technique (eq 10).[8] The order of the activation of hydroxy

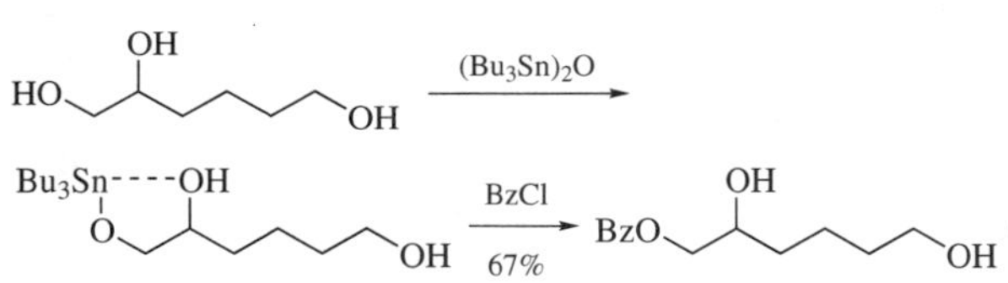

(9)

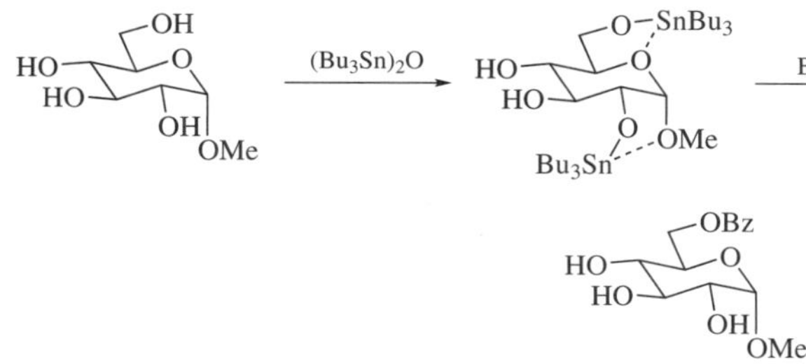

(10)

groups on carbohydrates has been investigated, and is shown in partial structures (**1**), (**2**), and (**3**).[9] Regioselective carbamoylation can also be accomplished by changing experimental conditions (eq 11).[10] On the other hand, alkylations of the tin derivatives are sluggish and less selective than acylations under similar conditions. Regioselective alkylation of sugar compounds, however, can be carried out in high yield by conversion to a tributyltin ether followed by addition of alkylating agent and quaternary ammonium halide catalysts (eq 12).[11]

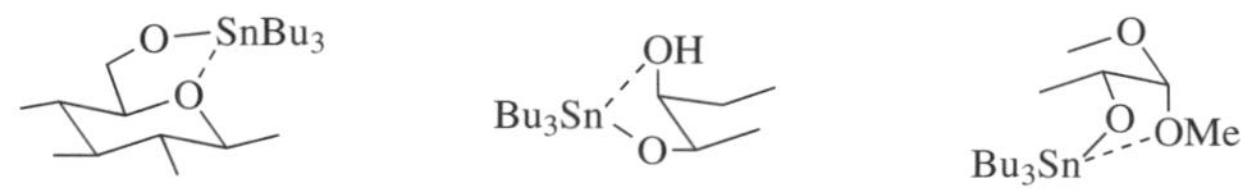

(**1**) most reactive | (**2**) next most reactive | (**3**) least reactive

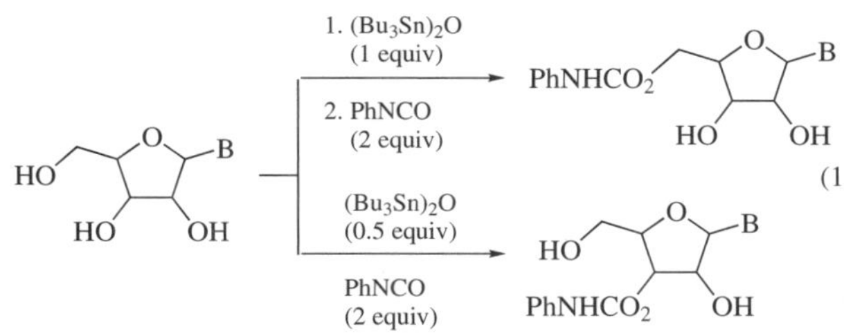

(11)

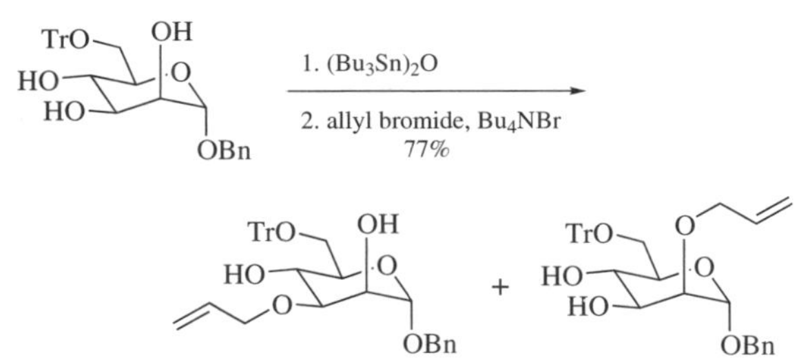

80:20 (12)

This O-activation is also effective for intramolecular alkylations such as oxetane synthesis (eq 13).[12] Similar N-activation has been used in the synthesis of pyrimidine nucleosides (eq 14).[13]

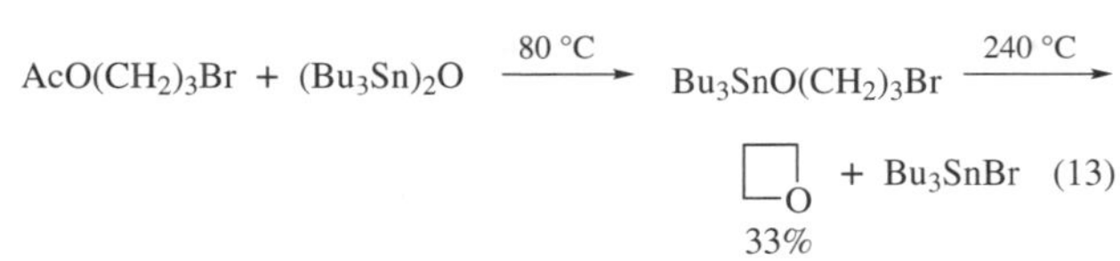

(13)

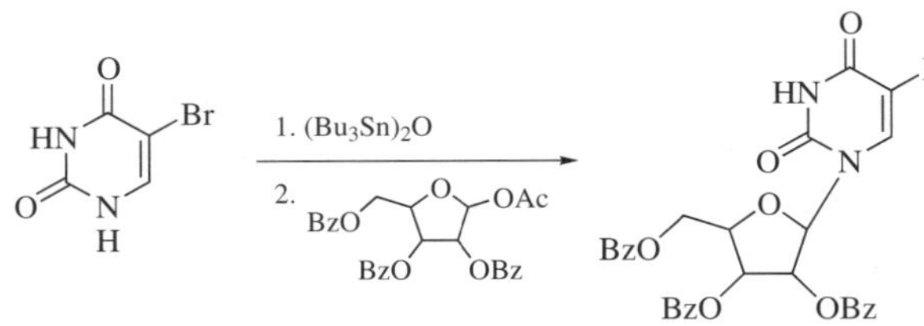

(14)

Dehydrosulfurizations. The thiophilicity of tin compounds is often utilized in functional group transformations. Thus conversion of aromatic and aliphatic thioamides to the corresponding nitriles can be accomplished by using $(Bu_3Sn)_2O$ in boiling benzene under azeotropic conditions (eq 15).[14]

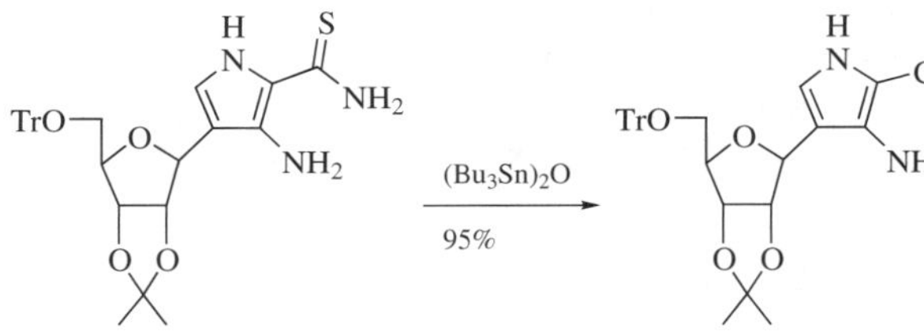

(15)

Hydrolysis. Esters are efficiently hydrolyzed with $(Bu_3Sn)_2O$ under mild conditions (eq 16).[15]

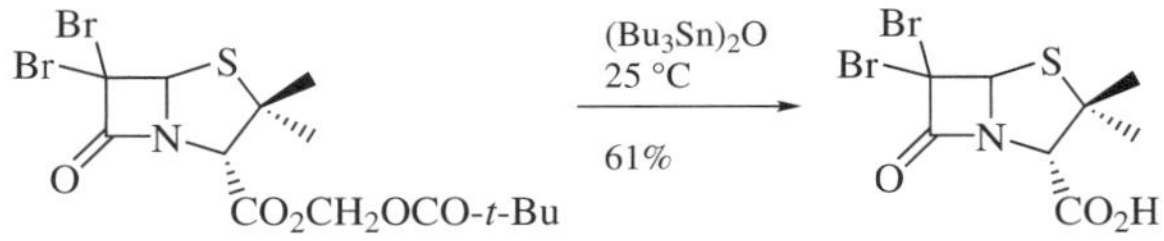

(16)

Transformation of primary alkyl bromides or iodides to the corresponding primary alcohols is achieved in good yield by using $(Bu_3Sn)_2O$– Silver(I) Nitrate (or Silver(I) p-Toluenesulfonate) (eq 17),[16] whereas this method is not applicable to secondary halides due to elimination.

$$MeCO_2(CH_2)_4I \xrightarrow[96\%]{(Bu_3Sn)_2O,\ AgTos,\ DMF,\ 20\ ^\circ C} MeCO_2(CH_2)_4OH \quad (17)$$

$(Bu_3Sn)_2O$ is a useful starting material for the preparation of tributyltin hydride, which is a convenient radical reducing reagent in organic synthesis. Thus Tri-n-butylstannane is easily prepared by using exchange reactions of $(Bu_3Sn)_2O$ with polysiloxanes (eq 18).[17]

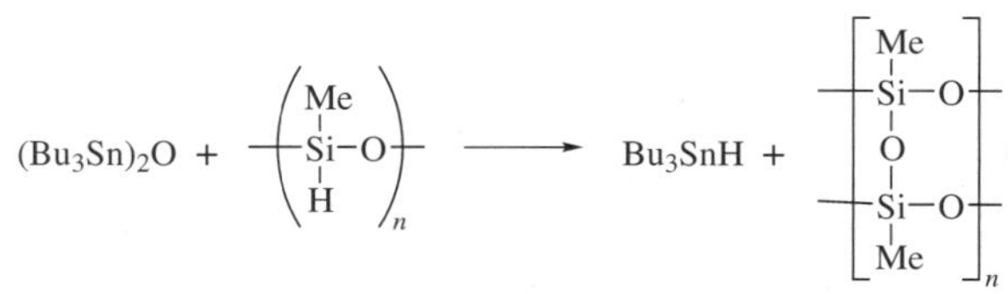

(18)

Related Reagents. Tri-n-butyl(methoxy)stannane.

1. Saigo, K.; Morikawa, A.; Mukaiyama, T. *Chem. Lett.* **1975**, 145.
2. Ueno, Y.; Okawara, M. *Tetrahedron Lett.* **1976**, 4597.
3. Hanessian, S.; Roy, R. *Synlett* **1985**, *63*, 163.
4. Ueno, Y.; Inoue, T.; Okawara, M. *Tetrahedron Lett.* **1977**, 2413.
5. Kuwajima, I.; Shimizu, M. *Tetrahedron Lett.* **1978**, 1277.
6. Jenkins, I. D.; Verheyden, J. P. H.; Moffatt, J. G. *J. Am. Chem. Soc.* **1971**, *93*, 4323.
7. (a) Ogawa, T.; Matsui, M. *Tetrahedron* **1981**, *37*, 2363. (b) David, S.; Hanessian, S. *Tetrahedron* **1985**, *41*, 643.
8. (a) Crowe, A. J.; Smith, P. J. *J. Organomet. Chem.* **1976**, *110*, C57. (b) Blunden, S. J.; Smith, P. J.; Beynon, P. J.; Gillies, D. G. *Carbohydr. Res.* **1981**, *88*, 9. (c) Ogawa, T.; Matsui, M. *Carbohydr. Res.* **1977**, *56*, C1. (d) Hanessian, S.; Roy, R. *J. Am. Chem. Soc.* **1979**, *101*, 5839. (e) Arnarp, J.; Loenngren, J. *Chem. Commun./J. Chem. Soc., Chem. Commun.* **1980**, 1000. (f) Ogawa, T.; Nakabayashi, S.; Sasajima, K. *Carbohydr. Res.* **1981**, *96*, 29.
9. Tsuda, Y.; Haque, M. E.; Yoshimoto, K. *Chem. Pharm. Bull.* **1983**, *31*, 1612.
10. (a) Ishido, Y.; Hirao, I.; Sakairi, N.; Araki, Y. *Heterocycles* **1979**, *13*, 181. (b) Hirao, I.; Itoh, K.; Sakairi, N.; Araki, Y.; Ishido, Y. *Carbohydr. Res.* **1982**, *109*, 181.
11. (a) Alais, J.; Veyrières, A. *J. Chem. Soc., Perkin Trans. 1* **1981**, 377. (b) Veyrières, A. *J. Chem. Soc., Perkin Trans. 1* **1981**, 1626.
12. Biggs, J. *Tetrahedron Lett.* **1975**, 4285.
13. Ogawa, T.; Matsui, M. *J. Organomet. Chem.* **1978**, *145*, C37.
14. Lim, M.-I.; Ren, W.-Y.; Klein, R. S. *J. Org. Chem.* **1982**, *47*, 4594.
15. Mata, E. G.; Mascaretti, O. A. *Tetrahedron Lett.* **1988**, *29*, 6893.
16. Gingras, M.; Chan, T. H. *Tetrahedron Lett.* **1989**, *30*, 279.
17. Hayashi, K.; Iyoda, J.; Shiihara, I. *J. Organomet. Chem.* **1967**, *10*, 81.

Hiroshi Sano
Gunma University, Kiryu, Japan

Bis(2,2,2-trichloroethyl) Azodicarboxylate

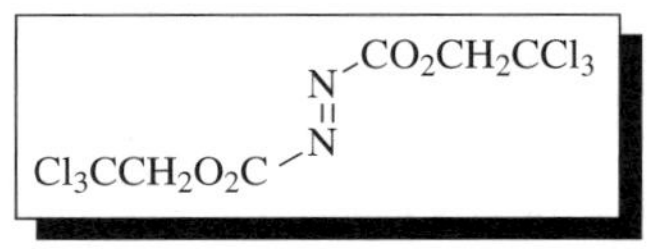

[38857-88-4] $C_6H_4Cl_6N_2O_4$ (MW 380.83)

(a reactive dienophile,[2] diene,[7] and enophile[8])

Physical Data: mp 110–111 °C.
Solubility: sol in a wide range of organic solvents.
Form Supplied in: slightly yellow plates.
Analysis of Reagent Purity: ^{1}H NMR ($CDCl_3$) displays a singlet at δ 5.05.
Preparative Method: commercially available.[1]
Purification: recrystallize from pentane.
Handling, Storage, and Precautions: as with all azo compounds, caution should be exercised to avoid exposure of the material to heat. Store in a brown bottle.

Reaction as a Diels–Alder Dienophile. Bis(2,2,2-trichloroethyl) azodicarboxylate (**1**) is most often used as a Diels–Alder dienophile.[2] Its relatively high reactivity occasionally makes it preferable to the less reactive diethyl or dimethyl analogs, since the times and temperatures needed to achieve reaction are often reduced. For example, the

Diels–Alder reaction illustrated (eq 1) required 5 days to reach completion when dimethyl azodicarboxylate was used, in contrast to just 1 h with (**1**).[2]

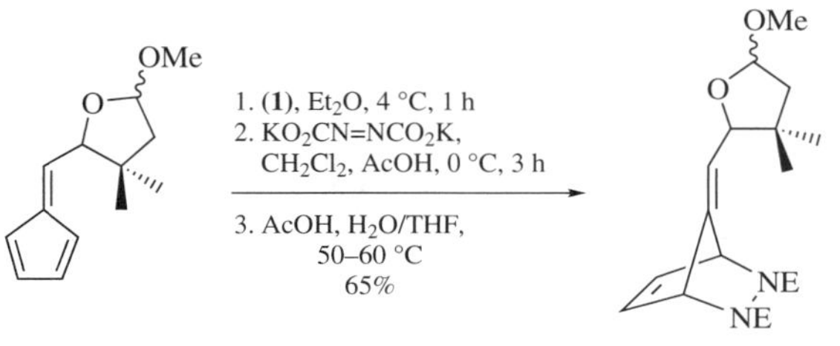

(1)

$E = CO_2CH_2CCl_3$

The bis(carbamate) cycloadducts are often highly viscous materials. Attempts to characterize them by ^1H NMR can be difficult by virtue of the several dynamic processes which occur on the NMR timescale.[3] To obtain a well-resolved spectrum requires that it be recorded at an elevated temperature.[3]

Most often, bis(carbamates) serve as precursors to azo compounds and the latter, in turn, to a host of theoretically interesting materials, reactive intermediates, and precursors to natural products.[2,4] While hydrolytic methods can also be used, the trichloroethoxy group is most often cleaved reductively, usually with Zinc–Acetic Acid,[5] or with a Zinc/Copper Couple in the presence of a proton donating solvent,[2b,4] or via an electroreductive process.[6] The presence of a proton donor leads to the formation of an intermediate hydrazo compound, which requires further oxidation to generate the azo linkage. Occasionally, the azo compound is initially isolated as a copper complex and then freed by making the solution basic, often by using aqueous ammonia.[5] In the absence of a proton donor, the hydrazo compound isbypassed. In this case, oxidation of the intermediate dicarboxylate salt leads directly to the azo functionality.[2,6] Regardless of the method, yields are sometimes variable, and experimentation is often required to achieve optimal conversion.

Reaction as a Diene. Interestingly, (**1**) can also serve as a diene. Thus when irradiated (eq 2) (350 nm), a novel [4 + 2]

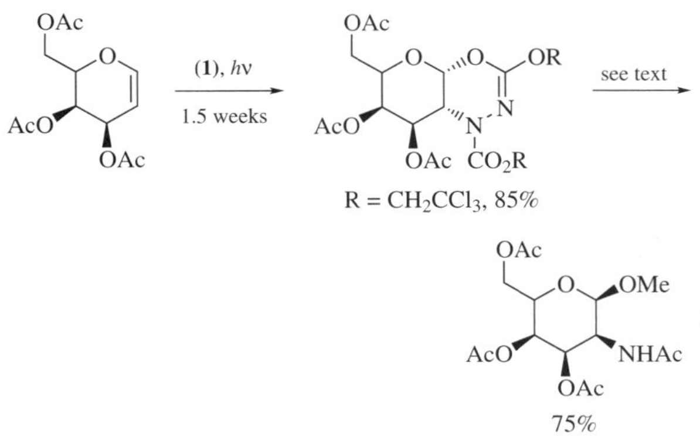

(2)

cycloaddition occurs between (**1**) and a variety of glycals.[7] Reaction times are quite long (e.g. 18 h to 2 weeks), but the yields are uniformly good (73–85%). Once again, the increased reactivity of (**1**) relative to other commonly used azodicarboxylates is apparent. For example, while the photoreaction (eq 2) requires 1.5 weeks, it fails to occur using dibenzyl azodicarboxylate. Treatment of the adducts with a catalytic amount of *p*-TsOH in methanol affords the corresponding methyl glycosides which, when exposed to zinc dust in acetic acid and acetone, followed by acetic anhydride/pyridine, generate 2-amino glycosides in good to excellent yields (60–97%; 75% for the case shown).

Reaction as an Enophile. When used as an enophile, (**1**) provides a novel route to allylic amines.[8] Ene reactions with diethyl and dimethyl azodicarboxylate have also been reported, but the yields are low. In contrast, use of the trichloroethyl derivative affords ene adducts in 60–88% yield and allows generation of both cyclic and acyclic *N*-acylamines under mild conditions (eq 3).

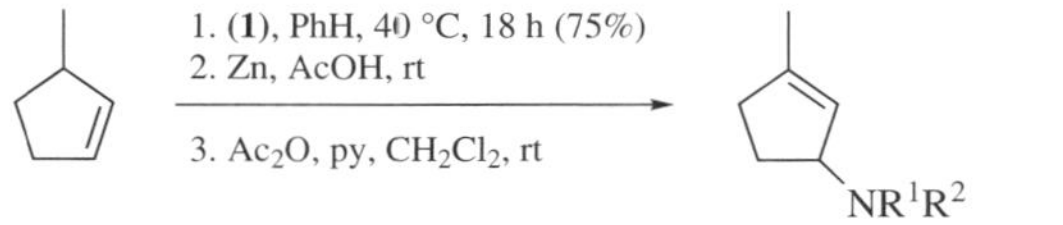

(3)

a, $R^1 = CO_2CH_2CCl_3$, $R^2 = NHCO_2CH_2CCl_3$
b, $R^1 = H$, $R^2 = Ac$, 88%

Related Reagents. Di-*t*-butyl Azodicarboxylate; Diethyl Azodicarboxylate.

---

1. (a) Little, R. D.; Venegas, M. G. *Q. Rev., Chem. Soc.* **1983**, *61*, 17. (b) Mackay, D.; Pilger, C. W.; Wong, L. L. *J. Org. Chem.* **1973**, *38*, 2043.
2. (a) van Hijfte, L.; Little, R. D.; Petersen, J. L.; Moeller, K. D. *J. Org. Chem.* **1987**, *52*, 4647. (b) Stone, K. J.; Little, R. D. *J. Am. Chem. Soc.* **1985**, *107*, 2495.
3. (a) Gibbard, A. C.; Moody, C. J.; Rees, C. W. *J. Chem. Soc., Perkin Trans. 1* **1986**, 145. (b) Warrener, R. N.; Russell, R. A.; Tan, R. Y. S. *Aust. J. Chem.* **1981**, *34*, 855.
4. Semmelhack, M. F.; Foos, J. S.; Katz, S. *J. Am. Chem. Soc.* **1973**, *95*, 7325.
5. Rastetter, W. H. *J. Am. Chem. Soc.* **1976**, *98*, 6350.
6. Little, R. D.; Carroll, G. L. *J. Org. Chem.* **1979**, *44*, 4720.
7. Leblanc, Y.; Fitzsimmons, B. J. *Tetrahedron Lett.* **1989**, *30*, 2889.
8. Leblanc, Y.; Zamboni, R.; Bernstein, M. A. *J. Org. Chem.* **1991**, *56*, 1971.

R. Daniel Little & Therese M. Bregant
University of California, Santa Barbara, CA, USA

Bis(2,2,2-trichloroethyl) Phosphorochloridate

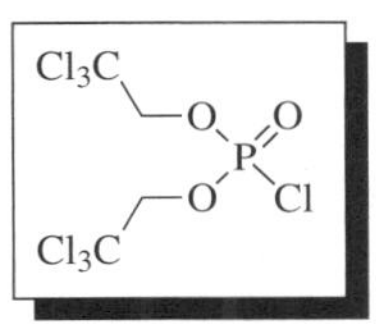

[17172-53-6] $C_4H_4Cl_7O_3P$ (MW 379.22)

(phosphorylating reagent)

Physical Data: mp 45–47 °C.
Form Supplied in: colorless solid.
Handling, Storage, and Precautions: may decompose on exposure to air or water. Store in a dry, cool environment. Corrosive material and it should be handled in a well-ventilated fume hood.

Phosphorylating Reagent. Phosphorochlorides have been widely used for the direct introduction of a phosphate group in hydroxy amino acids.[1] The use of the 2,2,2-trichloroethyl group for the phosphate protection[2] in phosphoserine derivatives[3] constitutes an efficient method for effective phosphopeptide synthesis.[4] Nucleotides are selectively protected as the 5′-OH phosphate ester derivatives[5,6] with bis(2,2,2-trichloroethyl) phosphorochloridate (eq 1).[7]

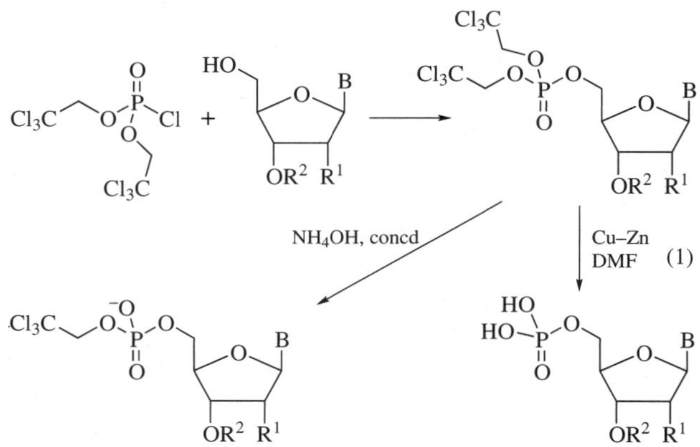

B = thymine, 4-thiothymine, and uracyl; R^1 = H, OH; R^2 = H, COMe, COPh

The isolated esters are stable in alkaline medium,[8] and selective cleavage of the 2,2,2-trichloroethyl group to afford the phosphate nucleotides occurs upon treatment with Zinc metal[9,10] (in DMF or acetic acid media) in presence of other protecting groups.[11] Bis(2,2,2-trichloroethyl) phosphorochloridate has also been successfully used as a phosphorylating agent for carbohydrates (eq 2),[12,13] and for the preparation of forskolin derivatives.[14]

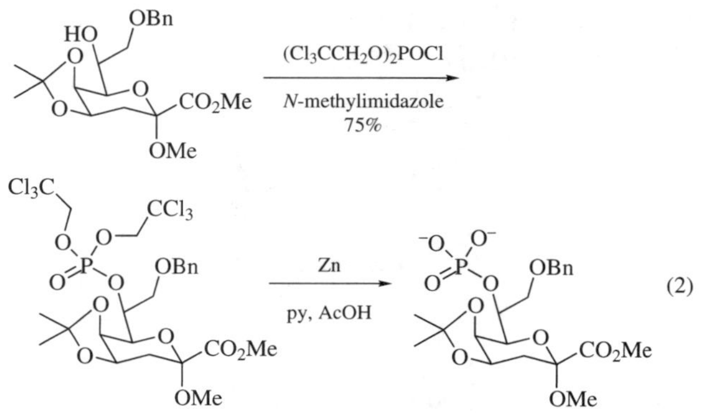

Preparation of Mixed Trialkyl Phosphates. In a general transesterification method designed to prepare mixed trialkyl phosphates,[15] reactions of bis(2,2,2-trichloroethyl) phosphorochloridate with different alcohols, in pyridine solvent, and in presence of fluoride ion (Cesium Fluoride or Tetra-n-butylammonium Fluoride), afford the mixed trialkyl phosphates (eq 3).[16] Reactions with diols give cyclic phosphates (eq 4).[16] The 2,2,2-trichloroethyl group has been also conveniently used for the preparation of mixed triesters of nucleotides.[17]

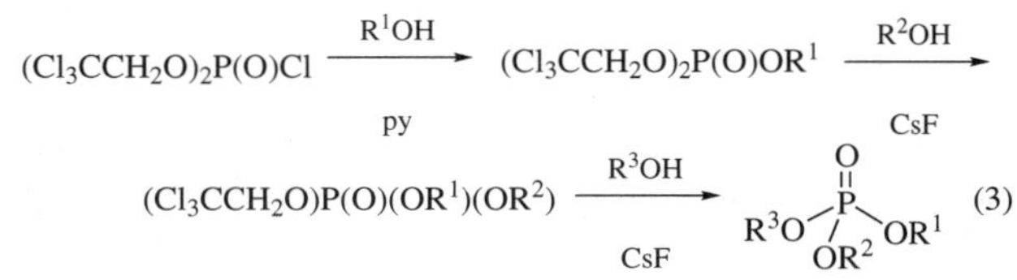

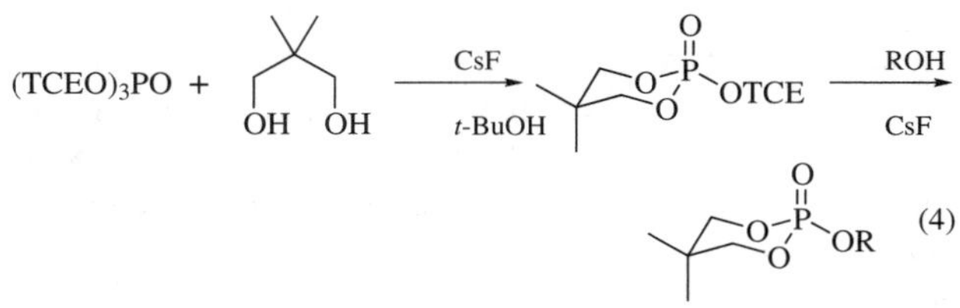

TCE = Cl_3CCH_2-; R = Me, Et, *i*-Pr

---

1. Paquet, A.; Johns, M. **1990**, *36*, 97.
2. Paquet, A. *Tetrahedron Lett.* **1990**, *31*, 5269.
3. Paquet, A.; *Int. J. Peptide Protein Res.* **1992**, *39*, 82.
4. (a) Perich, J. W.; Johns, R. B. *Aust. J. Chem.* **1990**, *43*, 1633. (b) Perich, J. W. *Peptides and Protein Phosphorylation*; Kemp, B. E., Ed.; CRC Press:Boca Raton, FL, **1990**; pp 289–314.
5. (a) Woodward, R. B. *Science* **1966**, *153*, 487. (b) Franke, A.; Scheit, K.-H.; Eckstein, F. *Ber. Dtsch. Chem. Ges./Chem. Ber.* **1968**, *101*, 2998.
6. Eckstein, F.; Scheit, K.-H. *Angew. Chem., Int. Ed. Engl.* **1967**, *6*, 362.
7. Preparation by analogy with bis (2, 2, 2-trifluoroethyl) phosphorochloridate:*Methoden Org. Chem. (Houben-Weyl)* **1964**, Part 2, 276.
8. (a) Cherbuliez, E.; Gabbai, A.; Probst, H.; Yazgi, A.; Rabinowitz, J. *Helv. Chim. Acta* **1962**, *45*, 2282. (b) Lichtenthaler, F. W. *Chem. Rev.* **1961**, *61*, 607.
9. (a) LeGoff, E. *J. Org. Chem.* **1964**, *29*, 2048. (b) Eckstein, F.; *Angew. Chem., Int. Ed. Engl.* **1966**, *5*, 671.
10. van Boom, J. H.; Burgers, P. M. J.; van Deursen, P. H. *Tetrahedron Lett.* **1976**, 869.
11. Eckstein, F. *Ber. Dtsch. Chem. Ges./Chem. Ber.* **1967**, *100*, 2228.
12. Auzanneau, F.-I.; Charon, D.; Szabó, L. *J. Chem. Soc., Perkin Trans. 1* **1990**, 2831.
13. Lammers, J. G.; van Boom, J. H. *Recl. Trav. Chim. Pays-Bas* **1979**, *98*, 243.
14. Lal, B.; Gangopadhyay, A. K. *J. Chem. Soc., Perkin Trans. 1* **1992**, 1993.
15. Ramirez, F.; Marecek, J. F.; Ugi, I. *J. Am. Chem. Soc.* **1975**, *97*, 3809.
16. Ogilvie, K. K.; Beaucage, S. L.; Theriault, N.; Entwistle, D. W. *J. Am. Chem. Soc.* **1977**, *99*, 1277.
17. Letsinger, R. L.; Lunsford, W. B. *J. Am. Chem. Soc.* **1976**, *98*, 3655.

Isabel Mathieu-Pelta & Slayton A. Evans
University of North Carolina, Chapel Hill, NC, USA

Bis(trichloromethyl) Carbonate[1]

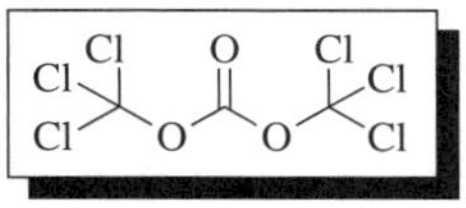

[32315-10-9] $C_3Cl_6O_3$ (MW 296.73)

(a phosgene surrogate)

Alternate Name: triphosgene.
Physical Data: mp 81–83 °C; bp 203–206 °C.
Solubility: sol methanol, ethanol, benzene, diethyl ether, hexane, THF, ethyl acetate; dec slowly in cold water.
Form Supplied in: white crystalline solid
Handling, Storage, and Precautions: the reagent is somewhat moisture sensitive, but scrupulously anhydrous conditions are not necessary. Rapid handling of the reagent in open air, in the absence of a glove bag or dry box, is usually satisfactory. This reagent should only be handled in a fume hood.

Introduction. Phosgene gas is a versatile reagent for organic synthesis, which has been used in carbonylation, chloroformylation, chlorination, and dehydration reactions, to name a few.[1] However, because of the high toxicity of Phosgene, it has not been used widely. Surrogates for phosgene have long been sought. Trichloromethyl Chloroformate (diphosgene) was introduced as one such phosgene substitute.[2] To the extent that diphosgene is a highly toxic, hygroscopic, and noxious liquid, it is not a fully satisfactory phosgene substitute. On the other hand, bis(trichloromethyl) carbonate (triphosgene; **1**) is a crystalline solid compound, is not very hygroscopic, and can substitute for phosgene in a plethora of chemical reactions. Triphosgene has been known since 1880,[3] but its synthetic utility had not been exploited until recently.

Carbonylation. The formation of oxazoles,[4] quinazolinediones (eq 1),[5] carbonates (eq 2),[6] urea analogs,[7] and isocyanates (eq 3)[1] have been reported with triphosgene. The reactions give good to excellent yields and often require only 1/3 equiv of triphosgene.

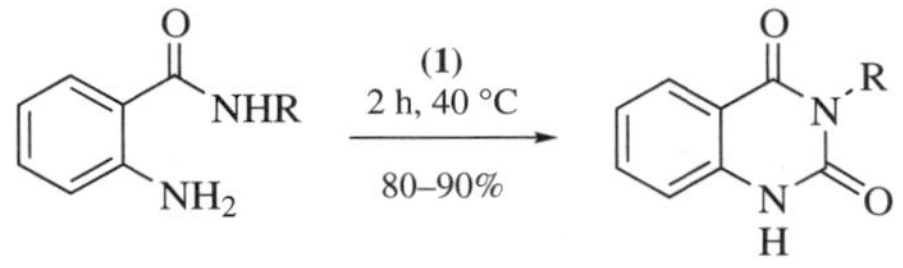

(1)

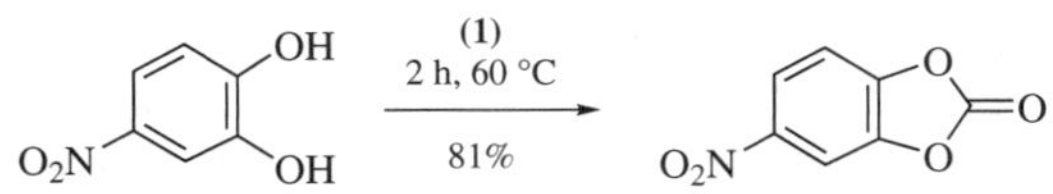

(2)

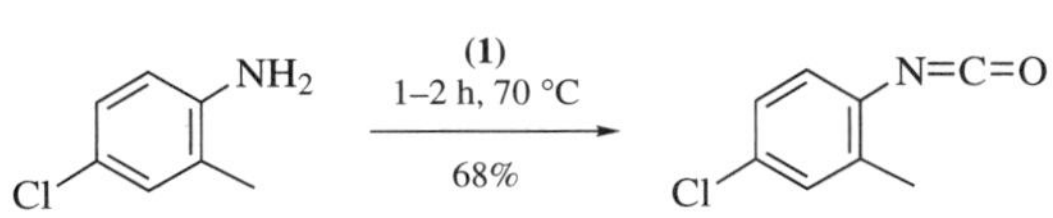

(3)

Chloroformylation. The use of triphosgene in the formation of chloroformates with both hydroxy compounds (eq 4) and substituted amines (eq 5) has been reported.[1]

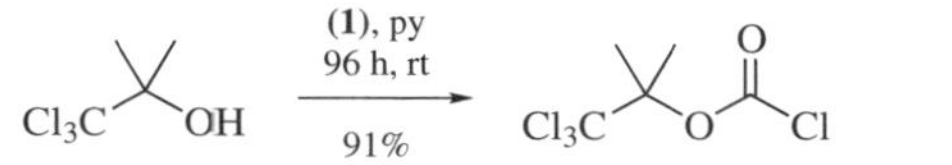

(4)

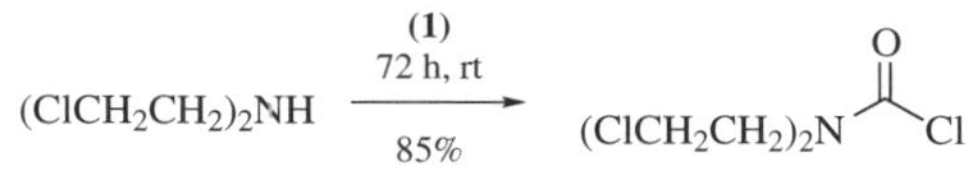

(5)

Dehydration. Synthesis of isocyanides from formamides (eq 6) is the only example of a dehydration reaction reported for triphosgene to date.[1]

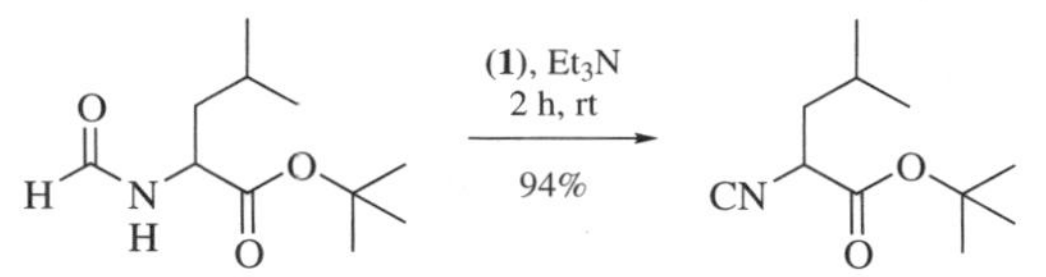

(6)

N-Carboxyamino Acid Anhydrides. Synthesis of *N*-carboxyamino acid anhydrides (NCAs) is accomplished by the reaction of the zwitterionic unprotected amino acid with triphosgene at elevated temperatures (eq 7).[8] Wilder and Mobashery reported a milder version of this reaction, utilizing *N-t*-butoxycarbonyl-α-amino acids; in the presence of a stoichiometric quantity of Triethylamine and 1/3 equiv of triphosgene (eq 8) the reaction is believed to proceed via a chloroformic *N-t*-butoxycarbonyl-α-amino acid anhydride intermediate to give the corresponding NCA.[9] There are many synthetic routes to NCAs;[10] however, the use of triphosgene in the preparation of NCAs provides a facile, safe, mild, and practical entry to these important molecules.

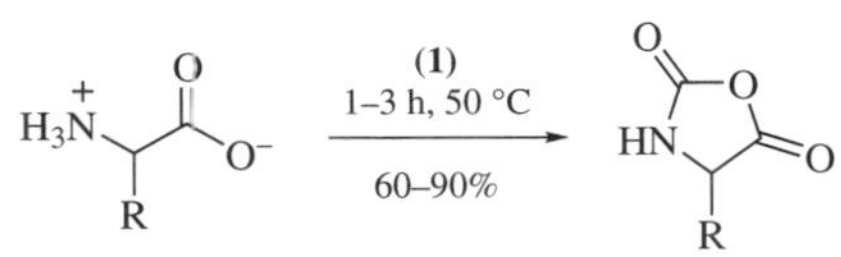

(7)

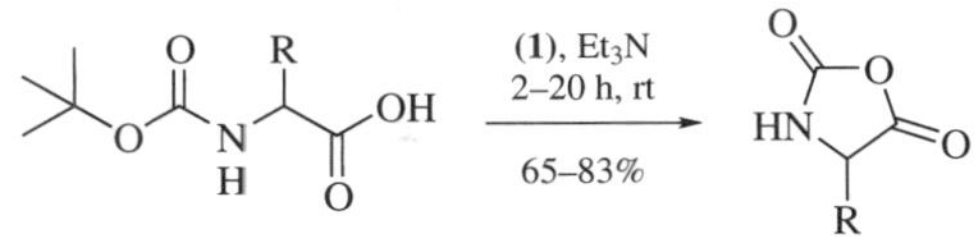

(8)

Chlorination. Triphosgene has been used in the synthesis of acyl chlorides (eq 9).[1] Also, it has been shown to react with a variety of aldehydes to give α-chloro chloroformates (eq 10).[11] Goren et al. have shown that triphosgene mediates chloride substitution reactions on activated alcohols (eq 11);[12] this reaction involves the formation of an intermediary chloroformate species en route to the chlorinated product. The generality of the reaction was shown for benzylic (eq 12), allylic, and propargylic systems. This reaction proceeds primarily via S_N2, with some contribution by S_N1 and/or S_Ni mechanisms.[12] Chlorination with triphosgene is considerably milder than the typical chlorination reactions with Thionyl Chloride, Phosphorus(III) Chloride, Phosphorus(V) Chloride, Oxalyl Chloride, and the like; for example, acid-labile functionalities, such as *t*-butyl carbamate (BOC), benzhydryl, and *p*-methoxybenzyl groups, can be tolerated in the triphosgene-mediated reactions, whereas they do not normally survive reactions with other chlorination reagents.

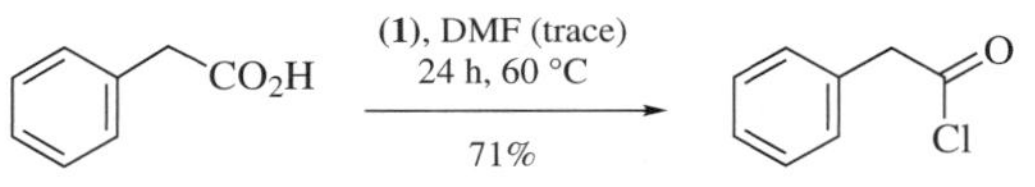

(9)

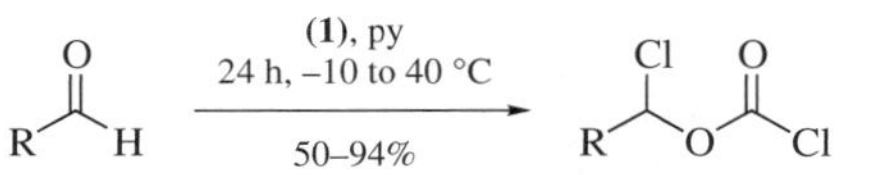

(10)

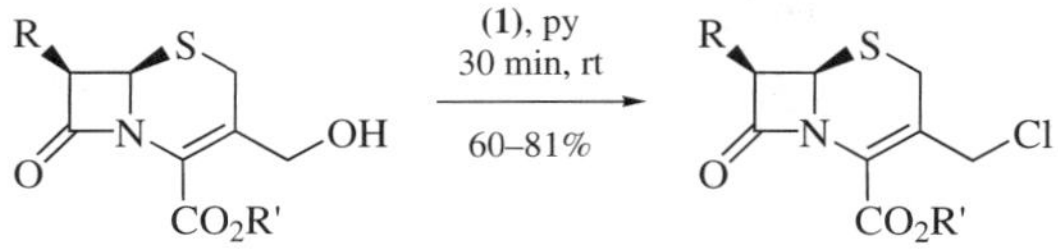

(11)

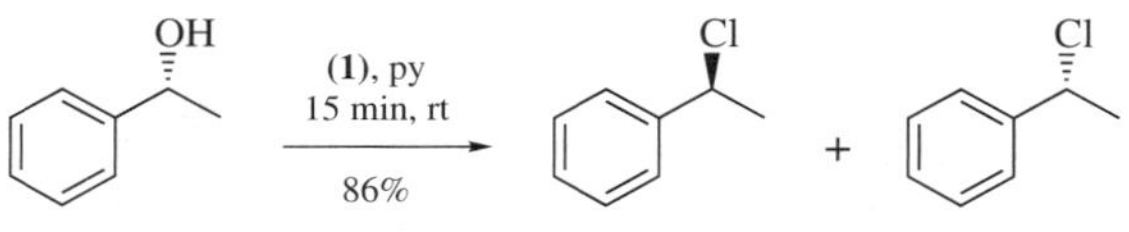

(12)

99% ee 7:3

Oxidation. Palomo et al. have demonstrated that primary and secondary alcohols are readily oxidized by triphosgene in the presence of Dimethyl Sulfoxide (DMSO) in good to excellent yields;[13a] the reaction is applicable to both activated (eq 13) and

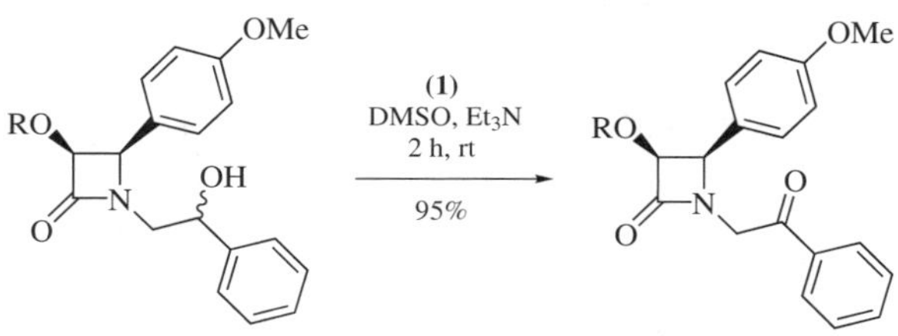

(13)

unactivated alcohols (eq 14). This reaction is a variation of the one reported by Barton with phosgene gas,[14] and it is an excellent alternative to Swern-type oxidations. Indeed, in some respects, oxidation of alcohols by triphosgene is superior to the widely used Swern oxidation. The triphosgene reaction tolerates acid-labile functionalities and is not as sensitive to residual moisture as oxalyl chloride used in the Swern oxidation. An additional advantage of the method of Palomo et al. is the ease of handling of triphosgene, and the fact that the reaction is amenable to large-scale synthesis. The breadth and scope of this reaction have been investigated with a large variety of alcohol substrates.[13b] See also Dimethyl Sulfoxide–Triphosgene.

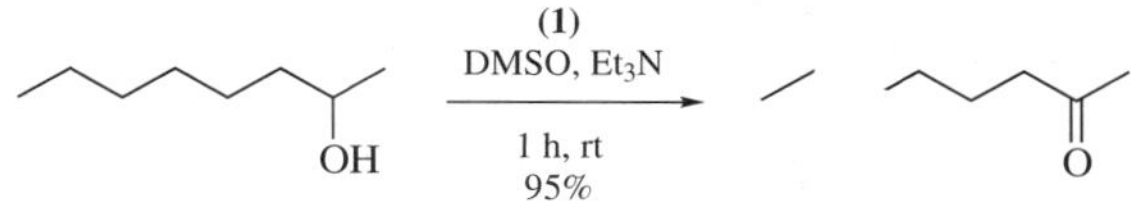

(14)

1. Eckert, H.; Foster, B. *Angew. Chem., Int. Ed. Engl.* **1987**, *26*, 894.
2. Kurita, K.; Iwakura, Y. *Org. Synth., Coll. Vol.* **1988**, *6*, 715.
3. Councler, C. *Ber. Dtsch. Chem. Ges./Chem. Ber.* **1880**, *13*, 1697.
4. Sicker, D. *Synth. Commun.* **1989**, 875.Flouzat, C.; Blanchet, M.; Guillaumet, G. *Tetrahedron Lett.* **1992**, *33*, 4571.
5. Cortez, R.; Rivero, J. A.; Samanathan, R.; Aguire, G.; Ramirez, F. *Synth. Commun.* **1991**, *21*, 285.
6. Laufer, D. A.; Doyle, K.; Zhang, X. *Org. Prep. Proced. Int.* **1989**, *21*, 771.
7. Zhao, X.; Chang, Y.-L.; Fowler, F. W. *J. Am. Chem. Soc.* **1990**, *112*, 6627.Cotarca, L.; Bacaloglu, R.; Csunderlik, C.; Marcu, N.; Tarnaveanu, A. *Methoden Org. Chem. (Houben-Weyl)* **1987**, *329*, 1052.
8. Daly, W. H.; Poché, D. *Tetrahedron Lett.* **1988**, *29*, 5859.
9. Wilder, R.; Mobashery, S. *J. Org. Chem.* **1992**, *57*, 2755.
10. Kricheldorf, H. R. *α-Aminoacid-N-Carboxy-Anhydride and Related Heterocycles*; Springer:Berlin, **1987**; pp 3–58.Blacklock, T. J.; Hirschmann, R.; Veber, D. F. *The Peptides*; Academic:New York, **1987**; Vol. *9*, pp 39–102.
11. Coghlan, M. J.; Caley, B. A. *Tetrahedron Lett.* **1989**, *30*, 2033.
12. Goren, Z.; Heeg, M. J.; Mobashery, S. *J. Org. Chem.* **1991**, *56*, 7188.
13. (a) Palomo, C.; Cossio, F. P.; Ontoria, J. M.; Odrizola, J. M. *J. Org. Chem.* **1991**, *56*, 5948. (b) Rivero, I. A.; Samanathan, R.; Hellberg, L. H. *OPPI Briefs* **1992**, *24*, 363.
14. Barton, D. H. R.; Garner, B. J.; Wightman, R. H. *J. Chem. Soc* **1964**, 1855.

Juliatiek Roestamadji & Shahriar Mobashery
Wayne State University, Detroit, MI, USA

2,7-Bis(trimethylsilyl)fluorenylmethyl Chloroformate[1]

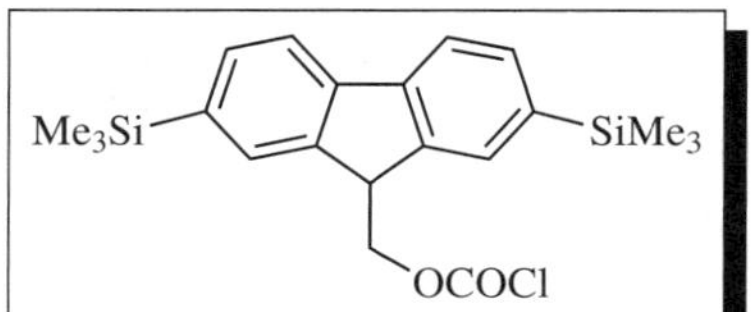

1

[136453-62-8] $C_{21}H_{27}ClO_2Si_2$ (MW 403.06)

(reagent for the introduction of a more soluble Fmoc protecting group for amines and alcohols)

Alternate Name: 2,7-bis(trimethylsilyl)fluorenylmethoxycarbonyl chloride; Bts-Fmoc-Cl.
Physical Data: mp 64–66 °C, colorless crystals.[1]
Solubility: soluble in CH_2Cl_2, $CHCl_3$, and THF.
Analysis of Reagent Purity: ^{1}H NMR, IR, elemental analysis.[1]
Preparative Methods: the title compound **1** is obtained by the action of phosgene on 2,7-bis(trimethylsilyl)-9-fluorenylmethanol (**6**). This alcohol can be obtained from 9,9-bis(trimethylsilyl)fluorene (**2**)[2] which, on treatment with NBS/HBr, is converted to the corresponding 2,7-dibromo compound (**3**). Silylation with Me_3SiCl followed by selective protodesilylation at the 9-position effected by KOH/EtOH, leads to the 2,7-bis(trimethylsilyl)fluorene (**5**). Reaction of **5** with *n*-BuLi and formaldehyde then generates the required alcohol (eq 1).[1]

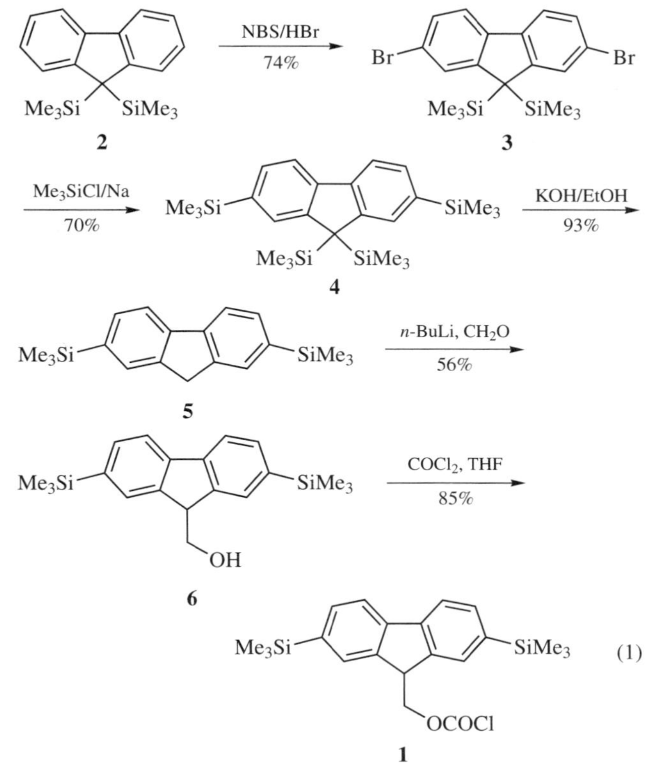

Purification: recrystallization from Skelly F (saturated hydrocarbon solvent, bp range 30–60 °C).[1]

Handling, Storage, and Precautions: moisture sensitive; protect from heat. Prolonged storage of the reagent can lead to extrusion of CO_2, causing pressure to develop inside the bottle. Converted to the Fmoc function itself upon treatment with trifluoroacetic acid. Probably toxic. Use in a well-ventilated fumehood.

Amino Group Protection. The Fmoc function has been widely adopted for the protection/blocking of amino functionality in peptide synthesis over the past two decades.[3] This is due to the exceptional acid stability of this group combined with its facile deprotection/deblocking by exposure to simple amines.[4] In addition, the intense fluorescence of the fluorenyl chromophore facilitates HPLC analysis and fluorescence detection.[5] Unfortunately, however, Fmoc derivatives are often relatively insoluble. The solubilizing influence of 2,7-bis(trimethylsilyl) substitution on the Fmoc residue circumvents this problem. Indeed it has been shown that Bts-Fmoc derivatives are some 2–3 times more soluble than the corresponding 2,7-di-*tert*-butyl-9-fluorenylmethyl (Dtb-Fmoc or Fmoc*) protected substances.[6] Through quantification, the solubilities of *p*-chlorophenyl carbanilates (**7a–7c**) derived from Fmoc, Dtb-Fmoc, and Bts-Fmoc per 100 mL of DCM were known to be 3.34, 6.74, and 22.3 g, respectively.

Another advantage of the Bts-Fmoc system is the similarity of the deblocking time to that of the corresponding Fmoc derivatives. This may be due to the steric factors being balanced by anionic stabilization provided by the silicon residue.[7] In an investigation of the deblocking time with various bases, interestingly, it was found that the Bts-Fmoc required 2–3 times less time than the Dtb-Fmoc analog, and was comparable to Fmoc itself (Table 1). The Bts-Fmoc function may also be removed under conditions of catalytic hydrogenolysis as previously reported for Fmoc.[8]

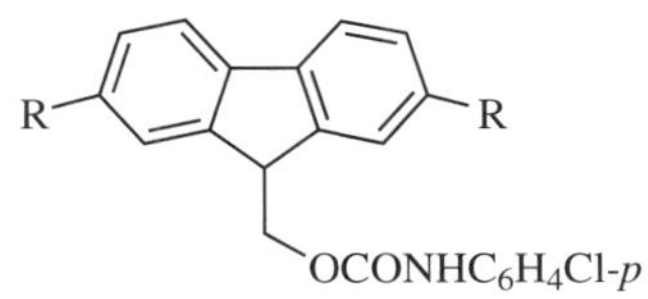

7a, R=H
7b, R=*t*-Bu
7c, R=Me_3Si

Table 1 Deblocking time for urethanes (**7a–7c**) by various amines

Base	Fmoc	Bts-Fmoc	Dtb-Fmoc
Piperidine	<3 min	<3 min	12 min
Ethanolamine	45 min	90 min	4 h
Morpholine	75 min	190 min	10 h
tert-Butylamine	5 h	4.5 h	20 h

Related Reagents. 9-Fluorenylmethyl chloroformate (Fmoc-Cl);[4] 2,7-di-*tert*-butyl-9-fluorenylmethyl chloroformate (Dtb-Fmoc-Cl);[6] 9-trimethylsilyl-9-fluorenylmethyl chloroformate;[9] 2,7-dibromo-9-fluorenylmethyl chloroformate.[10]

---

1. Carpino, L. A.; Wu, A.-C., *J. Org. Chem.* **2000**, *65*, 9238–9240.
2. Eaborn, C.; Shaw, R. A., *J. Chem. Soc.* **1955**, 1420–1425.
3. *The Practice of Peptide Synthesis*; Bodanszky, M.; Bodanszky, A., Eds.; Springer-Verlag: Berlin, 1994.
4. Carpino, L. A., *Acc. Chem. Res.* **1987**, *20*, 401–407.
5. (a) Einarsson, S.; Folestad, S.; Josefsson, B.; Lagerkvist, S., *Anal. Chem.* **1986**, *58*, 1638–1643. (b) Einarsson, S., *J. Chromatogr.* **1985**, *348*, 213–220.
6. Stigers, K. D.; Koutroulis, M. R.; Chung, D. M.; Nowick, J. S., *J. Org. Chem.* **2000**, *65*, 3858–3860.
7. (a) Gornowicz, G. A., West, R., *J. Am. Chem. Soc.* **1968**, *90*, 4478–4479. (b) Eaborn, C.; Eidenschink, R.; Jackson, P. M.; Walton, D. R. M., *J Organomet. Chem.* **1975**, *101*, C40–C42. (c) Zhang, S.; Zhang, X.-M.; Bordwell, F. G., *J. Am. Chem. Soc.* **1995**, *117*, 602–606.
8. Carpino, L. A.; Tunga, A., *J. Org. Chem.* **1986**, *51*, 1930–1932.
9. U.S. Pat. 5,101,059 (to Carpino, L. A.; Wu, A.-C.) (1992).
10. Carpino, L. A., *J. Org. Chem.* **1980**, *45*, 4250–4252.

A. U. Vinod
University of Illinois, Chicago, Illinois, USA

p-Bromobenzyl Bromide

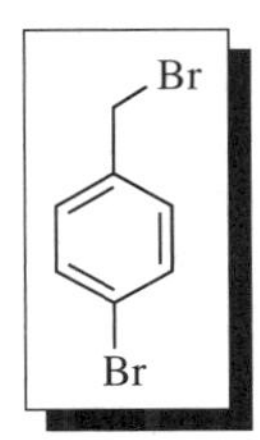

[589-15-1] $C_7H_6Br_2$ (MW 249.94)

(reagent used as a stable protecting group for hydroxyl groups removable by Pd-catalyzed amination followed by acid treatment, reagent used for the *p*-bromobenzylation of a variety of heterofunctional groups and carbon nucleophiles, substrate for Suzuki type coupling reactions)

Alternate Name: α-bromo-4-bromotoluene, 1-bromomethyl-4-bromobenzene, 1-bromo-4-bromomethyl-benzene.
Physical Data: mp 60–62 °C, bp 116–124 °C (at 12 Torr).
Solubility: soluble in ether, benzene, glacial acetic acid, and most other organic solvents; decomposes in hot water.
Form Supplied in: light orange crystalline solid. Commercially available.
Analysis of Reagent Purity: ^{1}H NMR, IR, HPLC, and elemental analysis.
Preparative Methods: can be prepared by photobromination of 4-bromotoluene;[1] bromination of 4-bromotoluene with NBS in CCl_4 and catalytic dibenzoyl peroxide.[2]
Purification: crystallize from ethanol.
Handling, Storage, and Precautions: corrosive; lachrymator; moisture sensitive; stored under nitrogen; causes burns; and irritant to eyes, respiratory system, and skin.
Disposal: contact a licensed professional waste disposal service to dispose off this material. Dissolve or mix it with a combustible solvent and burn in a chemical incinerator equipped with an afterburner and scrubber.

---

p-Bromobenzyl Bromide (PBB) as a Hydroxyl Protecting Group. Alcohols can be *p*-bromobenzylated by alkylation with *p*-bromobenzyl bromide under basic conditions. The base, usually sodium hydride in ether, THF, or DMF, deprotonates the alcohol and the resulting alkoxide reacts with *p*-bromobenzyl bromide in a Williamson reaction to give the corresponding *p*-bromobenzyl ether. *p*-Bromobenzyl ethers are stable under basic, acidic, and oxidative conditions. Thus they can be used as orthogonal protecting groups in the presence of numerous other protecting groups such as silyl ethers, acetates, benzoates, or PMB-acetals.

The *p*-bromobenzyl group can be efficiently cleaved to reveal the free hydroxyl group in a two-step reaction:[3] in the first step, palladium-catalyzed coupling of the aryl bromide with an amine such as benzylamine or methyl-phenylamine yields the corresponding benzylamine derivative.[4] As a catalyst $Pd_2(dba)_3$ in the presence of 2-(di-*tert*-butylphosphino)biphenyl and NaO*t*-Bu as base has proven effective. For base sensitive substrates, K_3PO_4 can be used instead. In the second step, the resulting aminated benzyl ether is cleaved by brief exposure to a Lewis acid, protic acid, or an oxidant, releasing the unprotected hydroxyl compound (eq 1).[3]

Because of these mild deprotection conditions, halobenzyl ethers can also be used for molecules where other protection groups such as silyl ethers or esters are present. In contrast, other *para*-substituted benzyl ethers, e.g., *p*-nitrobenzyl, *p*-acetoxybenzyl, or 2-(trimethylsilyl)ethoxymethoxybenzyl groups require Zn/AcOH, base or fluoride treatment for cleavage rendering them uncompatible with azido, ester, or silyl ether groups.[5]

Apart from the ease of cleavage and their excellent stability, which is comparable to benzyl ethers, *p*-halobenzyl ethers can be used to introduce three more degrees of orthogonality as they can be sequentially removed.[3] In a molecule that contains *p*-iodobenzyl ethers, *p*-bromobenzyl ethers, and *p*-chlorobenzyl ethers, the *p*-iodobenzyl ether can be selectively cleaved because it reacts much faster in the Pd-catalyzed amination step than *p*-chlorobenzyl or *p*-bromobenzyl ethers (eq 2).[6]

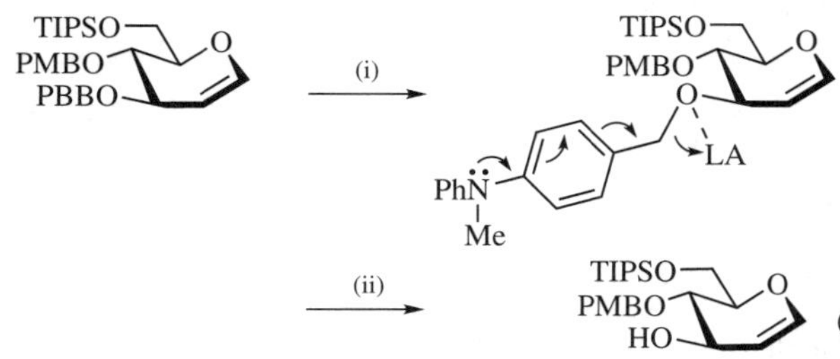

(1)

Conditions:
(i) cat. Pd_2dba_3, (*o*-biphenyl)P(*t*-Bu)$_2$, *N*-methylaniline, NaO*t*-Bu, rt, 12 h.
(ii) $ZnCl_2$, CH_3CN:H_2O (9:1), rt, 30 min.

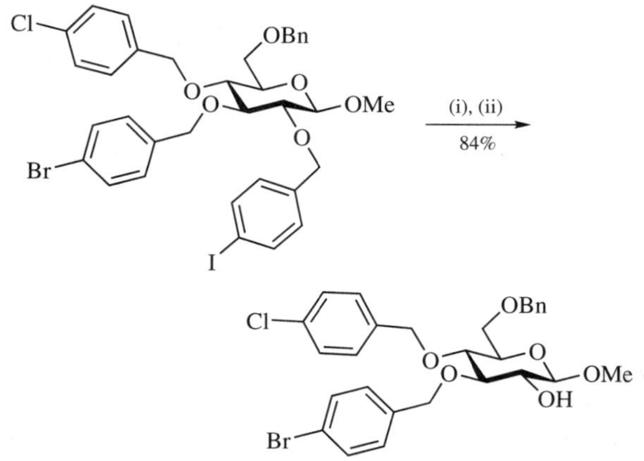

(2)

Conditions:
(i) Pd_2dba_3, NaO*t*-Bu, PhN(H)Me, 1-*N*,*N*-dimethylamino-1′-(dicyclohexylphosphino)biphenyl, rt, 4 h.
(ii) $Cl_2HCCOOH$.

Similarly, the *p*-bromobenzyl ether is selectively removed in the presence of the *p*-chlorobenzyl group (eq 3).[7]

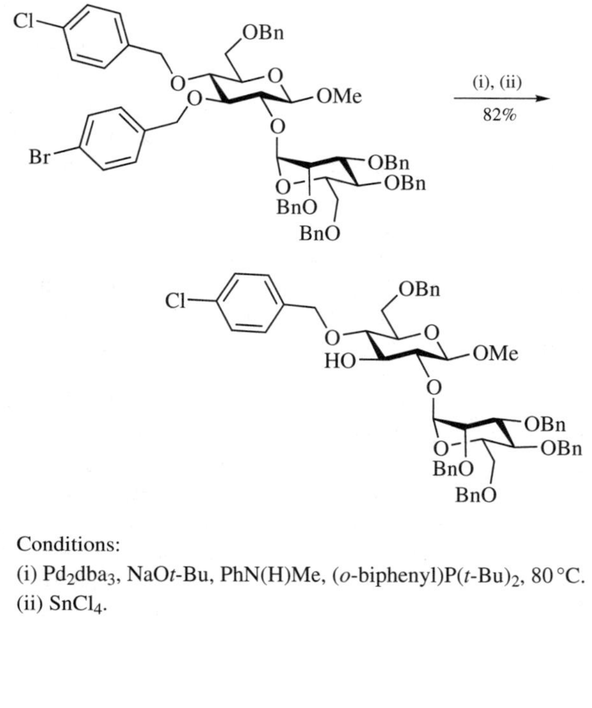

(3)

Conditions:
(i) Pd_2dba_3, NaO*t*-Bu, PhN(H)Me, (*o*-biphenyl)P(*t*-Bu)$_2$, 80 °C.
(ii) $SnCl_4$.

The rate limiting step for the amination reaction is the oxidative addition of the halobenzyl group to the palladium catalyst (eq 4), which decreases in the order I>Br>Cl>F.[8] By choosing appropriate ligands for the Pd(0) catalyst, the selectivity of the amination reaction in the presence of different halobenzyl ether groups can be enhanced (compare reaction conditions in eq 2 to the conditions in eq 3).

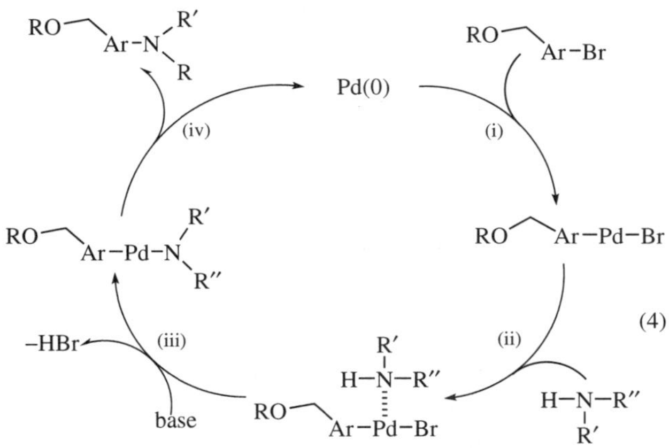

(i) Oxidative addition
(ii) Coordination of the amine
(iii) Dehydrohalogenation
(iv) Reductive elimination

Related Reagents. *p*-Iodobenzyl bromide; *p*-chlorobenzyl bromide; benzyl bromide; *p*-methoxybenzyl bromide.

1. Weizmann, M.; Patai, S., *J. Am. Chem. Soc.* **1946**, *68*, 150.
2. Gassman, P. G.; Macomber, D. W.; Willging, S. M., *J. Am. Chem. Soc.* **1985**, *107*, 2380.
3. Plante, O. J.; Buchwald, S. L.; Seeberger, P. H., *J. Am. Chem. Soc.* **2000**, *122*, 7148.
4. Hartwig, J. F., *Angew. Chem. Int. Ed.* **1998**, *37*, 2046.
5. Jobron, L.; Hindsgaul, O., *J. Am. Chem. Soc.* **1999**, *121*, 5835.
6. Wolfe, J. P.; Tomori, H.; Sadighi, J. P.; Yin, J.; Buchwald, S. L., *J. Org. Chem.* **2000**, *65*, 1158.
7. Wolfe, J. P.; Buchwald, S. L., *J. Org. Chem.* **2000**, *65*, 1144.
8. Collman, J. P.; Hegedus, L. S.; Norton, J. R.; Finke, R. G., *Principles and Applications of Organotransition Metal Chemistry*; University Science Books: Mill Valley, 1987, pp. 306–320.

Holger Herzner & Peter H. Seeberger
Massachusetts Institute of Technology, Cambridge, MA, USA

3-Bromocyclohexene

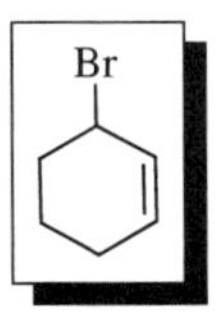

[1521-51-3] C_6H_9Br (MW 161.04)

(reagent for the introduction of the cyclohex-2-enyl moiety, which can be converted to cyclohexyl by hydrogenation)

Physical Data: bp 89–91 °C/13 mm Hg; d 1.380 g cm^{-3}.
Solubility: soluble in THF, DMF, and diethyl ether.
Form Supplied in: colorless to light yellow liquid.
Analysis of Reagent Purity: GS/MS.
Handling, Storage, and Precautions: irritant; avoid contact with eyes, skin, and clothes. Use adequate ventilation to avoid breathing vapor, mist, or gas. Chemical safety goggles are recommended. Combustible. Incompatible with strong oxidizing agents. Moisture sensitive. Store in a cool, dry place. Decomposition may produce hydrogen bromide.

***C*-Alkylation.** 3-Bromocyclohexene can be regarded as an equivalent of allyl bromide, a superior electrophile. For example, 3-bromocyclohexene reacts with Grignard reagents to afford *C*-cyclohex-2-enyl derivatives (eq 1).[1] The substrate may be a primary, secondary, or tertiary alkyl or aryl derivative. The product yield was reported to vary considerably according to the mode of addition; an appreciable increase in the product was observed when 3-bromocyclohexene was added to the Grignard reagent as compared to the reversed addition.[1] Additional examples of *C*-alkylation with 3-bromocyclohexene are shown in eqs 2 and 3.[2,3]

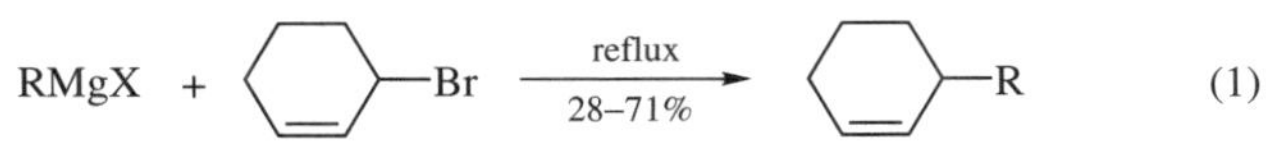

R= *i*-Pr, *n*-Bu, *t*-Bu, *p*-methoxyphenyl

The introduction of the cyclohexyl moiety is often difficult because of the poor reactivity of cyclohexyl halides. Cyclohexen-2-yl derivatives are useful precursors for cyclohexyl derivatives. For instance, *trans*-4-cyclohexyl-L-proline has been synthesized via 4-(cyclohex-2-enyl)-*N*-benzyl-L-pyroglutamic acid.[4] Relevant steps are shown in eq 4. The lithium complex of *N*-benzyl-L-pyroglutamic acid was allowed to react with 3-bromocyclohexene to afford the 4-cyclohex-2-enyl derivative, which was hydrogenated over Pd/C to yield the 4-cyclohexyl product. The stereoselectivity is held to be attributable to the formation of the dilithium complex shown in eq 4.

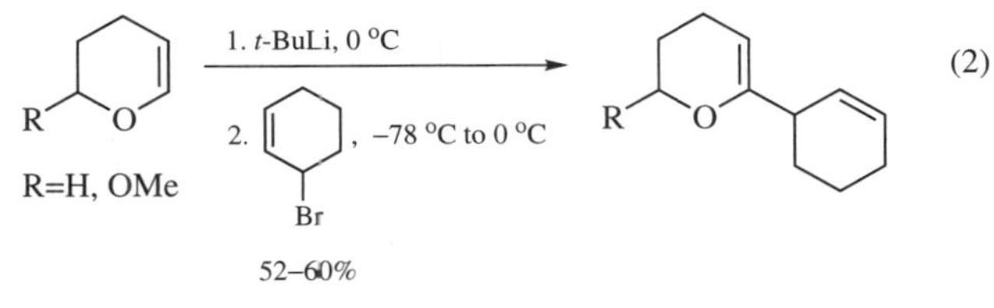

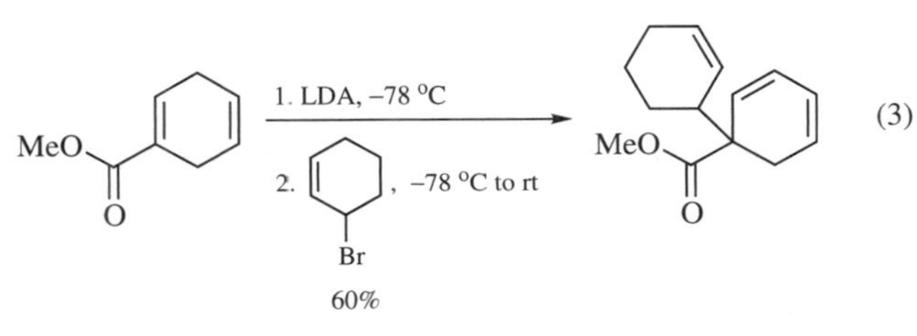

***O*-Alkylation.** 3-Bromocyclohexene is used to introduce the cyclohexyl (Chx) protecting group to the side-chain hydroxyl group of Ser and Thr.[5,6] The *N*-Boc amino acids are treated with NaH, and the resulting alkoxides are allowed to react with 3-bromocyclohexene to give *O*-cyclohex-2-enyl derivatives which are hydrogenated to *N*-Boc-*O*-Chx derivatives with the aid of PtO_2 (eq 5). Note that Pd-catalysts do not work for conversion of *O*-cyclohex-2-enyl to *O*-Chx derivatives. The *O*-Chx group is fairly stable towards reagents used to remove *tert*-butyl-based (TFA and 4 M HCl in AcOEt) and 9-fluorenyl-based protecting groups (20% piperidine in DMF). It is also stable towards 1 M TMSBr–thioanisole in TFA, which is used to remove Bn-based protecting groups. The *O*-Chx protecting group can be removed with 1 M trifluoromethanesulfonic acid-thioanisole in TFA within a short period. Boc-Tyr(Chx)-OH can also be synthesized by essentially the same manner in a better yield than the original method.[7]

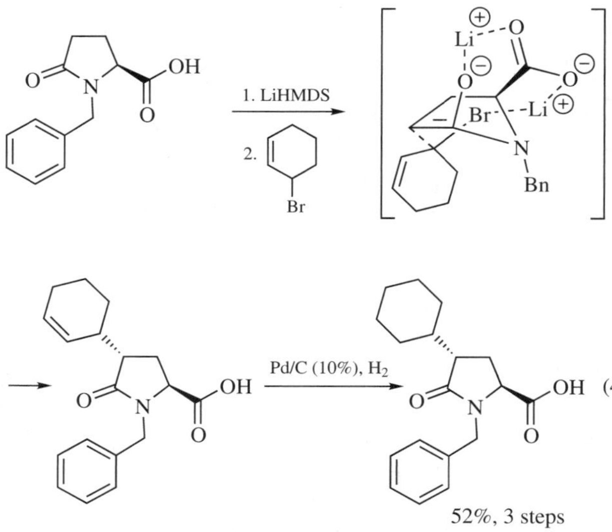

(4)

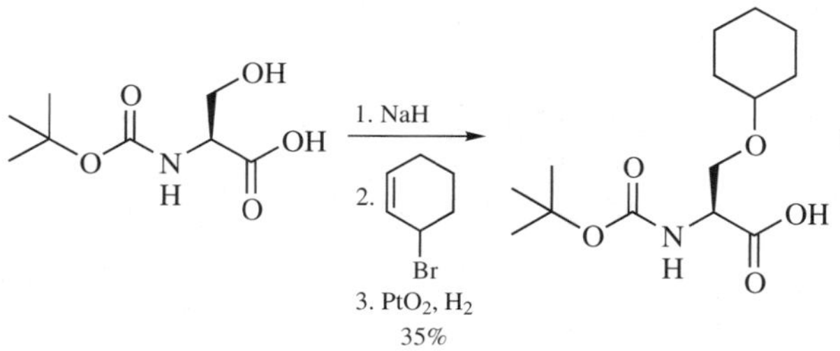

(5)

1. Biggerstaff, W. R.; Menditto, A. P.; Yokoyama, I., *J. Org. Chem.* **1954**, *19*, 934–939.
2. Boeckman, R. K.; Bruza, K. J., *Tetrahedron Lett.* **1977**, 4187–4190.
3. Boeckman, R. K.; Ramaiah, M.; Medwid, J. B., *Tetrahedron Lett.* **1977**, 4485–4488.
4. Chen, X.; Du, D. M.; Hua, W. T., *Tetrahedron Asymmety* **2002**, *13*, 43–46.
5. Nishiyama, Y.; Kurita, K., *Tetrahedron Lett.* **1999**, *40*, 927–930.
6. Nishiyama, Y.; Shikama, S.; Morita, K.; Kurita, K., *J. Chem. Soc., Perkin Trans. 1* **2000**, 1949–1954.
7. Nishiyama, Y.; Ishizuka, S.; Shikama, S.; Kurita, K., *Chem. Pharm. Bull.* **2001**, *49*, 233–235.

Hiroaki Taguchi & Yasuhiro Nishiyama
University of Texas-Houston Medical School, Houston, TX, USA

2-Bromo-1-ethylpyridinium Tetrafluoroborate (BEP)

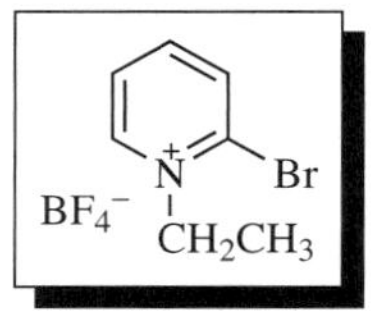

[878-23-9] $C_7H_9BBrF_4N$ (MW 273.86)

(coupling reagent used for the synthesis of various amides and esters)

Alternate Name: α-bromo-*N*-ethylpyridinium tetrafluoroborate; BEP.
Physical Data: mp 103–104 °C;[1] 102–104 °C.[2]
Solubility: soluble in most organic solvents.
Form Supplied in: white crystalline solid or colorless crystals.
Analysis of Reagent Purity: ^{1}H NMR; FAB–MS.
Preparative Methods: the reagent can be prepared by reaction of 2-bromo pyridine with triethyloxonium tetrafluoroborate in $ClCH_2CH_2Cl$ solution (eq 1).[1]

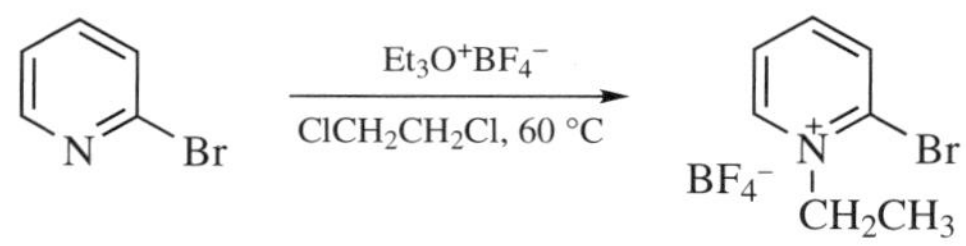

(1)

Purification: recrystallization from acetone–ether.
Handling, Storage, and Precautions: the reagent is both base and strong acid sensitive and should be stored at 0 °C in the dark.
Disposal: as irritant chemical.

Introduction. 2-Bromo-1-ethylpyridinium tetrafluoroborate (BEP) has been utilized as an efficient coupling reagent for the synthesis of amides and esters, especially for the synthesis of hindered peptides containing *N*-methylated or C_α, C_α-dialkylated amino acid residues.

Amidation. 2-Bromo-1-ethylpyridinium tetrafluoroborate (BEP) is a very powerful coupling reagent for the synthesis of sterically hindered amides, especially peptides.[1,3] Generally, the coupling reaction of a carboxylic acid with an amine can be carried out at −10 °C to 25 °C in the presence of BEP and a base such as diisopropylethyl amine.[1] The yields of products are usually above 90%. For example, the extremely hindered dipeptide Fmoc-MeLeu-MeVal-OCH_3 was easily synthesized in 95% yield using BEP (eq 2).[1] In terms of reactivity and racemization-suppressing capability, this pyridinium-type reagent, BEP, proved to be more efficient than commonly used uronium- and phosphonium-type reagents, such as HBTU,[4] BOP,[5] and PyBroP,[6] for the synthesis of hindered peptides containing *N*-methylated or C_α,C_α-dialkylated amino acid residues.[1,3] The efficiency and usefulness of BEP was also demonstrated by the total synthesis of the immunosuppressive cyclic undecapeptide Cyclosporin O.[7]

BEP can also be used to synthesize peptides on solid-phase. For instance, the highly hindered 8–11 tetrapeptide segment of Cyclosporin A, Fmoc-D-Ala-MeLeu-Meleu-MeVal-OH, can be synthesized on Wang resin in 85% overall yield using BEP, and the purity of the crude product exceeded 95% (eq 3).[1]

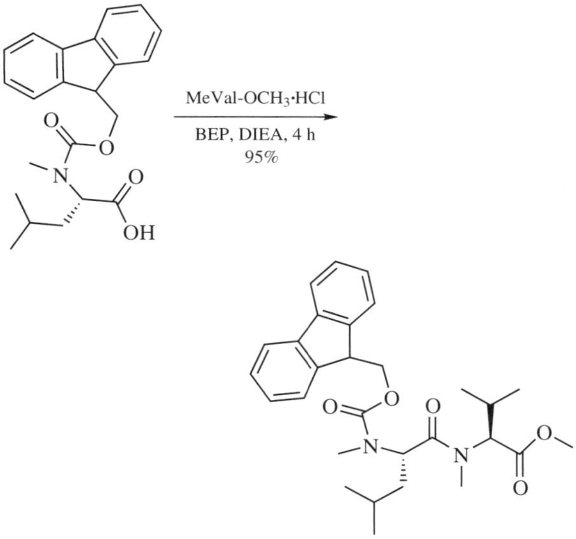

(2)

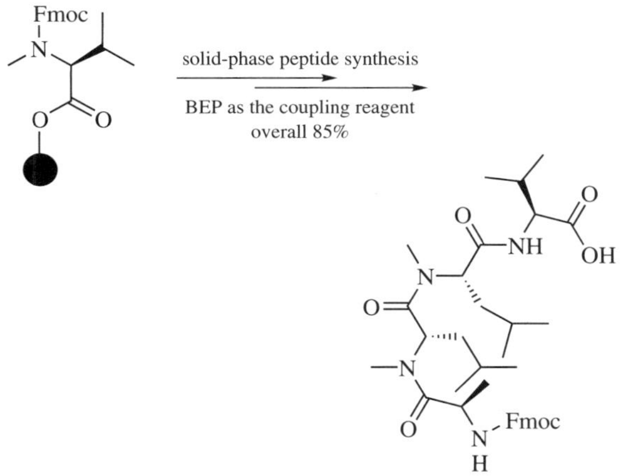

(3)

Esterification. BEP was also proven to be an efficient reagent for the synthesis of esters, especially active esters and hindered esters, such as benzotriazolyl esters, pentafluorophenyl esters, succinimidyl esters, 2,3-dihydro-4-oxobenzotriazinyl esters (eq 4), and *tert*-butyl esters.[1] These active esters are commonly used in the synthesis of macrolactones and macrolactams.

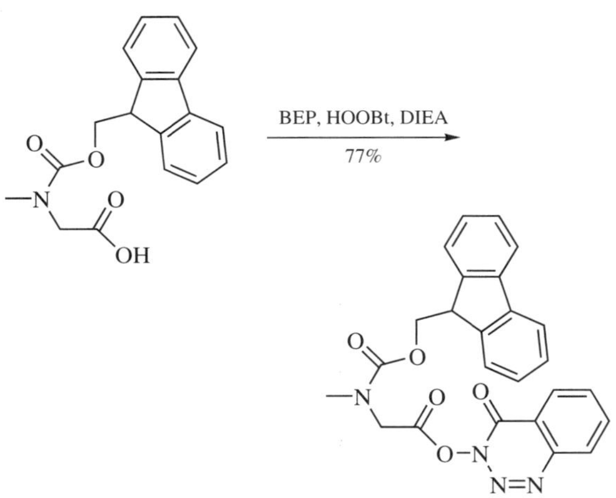

(4)

Related Reagents. 2-Fluoro-1-ethylpyridinium tetrafluoroborate (FEP).

1. Li, P.; Xu, J.-C., *Tetrahedron* **2000**, *56*, 8119.
2. Balli, H.; Kersting, F., *Liebigs Ann. Chem.* **1961**, *647*, 1.
3. Li, P.; Xu, J.-C., *Chem. Lett.* **2000**, 204.
4. Dourtoglou, V.; Ziegler, J. C.; Gross, B., *Tetrahedron Lett.* **1978**, *19*, 1269.
5. Castro, B.; Dormoy, J. R.; Evin, G.; Selve, C., *Tetrahedron Lett.* **1975**, *16*, 1219.
6. Coste, J.; Frerot, E.; Jouin, P., *J. Org. Chem.* **1994**, *59*, 2437.
7. Li, P.; Xu, J.-C., *J. Org. Chem.* **2000**, *65*, 2951.

Peng Li
National Institute of Health, Frederick, Maryland, USA

Bromonium Di-*sym*-collidine Perchlorate[1]

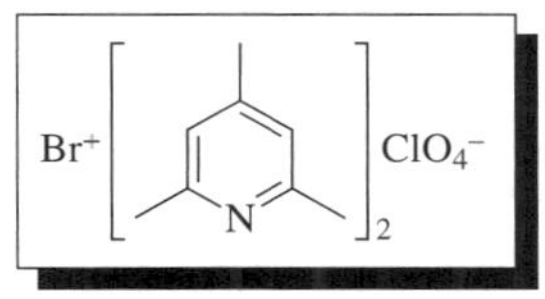

[4836-12-8] $C_{16}H_{22}BrClN_2O_4$ (MW 421.72)

(reactive electrophilic brominating agent; activates alkenes towards nucleophilic reagents;[2] reagent for bromocyclization,[3,4] useful in the preparation of *cis*-α-alkylamino alcohols)

Alternate Names: bromonium dicollidine perchlorate; bis(collidine)bromonium perchlorate.
Physical Data: mp 230 °C.
Solubility: sol chloroform; insol ether.
Form Supplied in: fine colorless crystalline powder.
Preparative Method: it is important in the preparation of this reagent to be sure that conditions are anhydrous, otherwise bromine tends to be reliberated. Silver di-*sym*-collidine perchlorate is prepared by adding *sym*-collidine (20 mL) to a solution of silver nitrate (9 g) and sodium perchlorate (11 g) in 100 mL of water, while stirring vigorously, to give a curdy white precipitate. After washing repeatedly with water, the product is washed with ethanol and ether and finally dried under vacuum over P_2O_5. The yield is quantitative. Silver di-*sym*-collidine perchlorate (13.47 g, 30 mmol) is suspended in 150 mL of chloroform and 1 mL of *sym*-collidine. Bromine (1.65 mL, 30 mmol) is added. Silver bromide rapidly precipitates and is filtered through Celite. The clear pale-yellow filtrate gives a fine, white crystalline precipitate on the addition of ether. The crystals are dried under high vacuum.
Handling, Storage, and Precautions: perchlorates are potentially powerful explosives. This compound detonates on strong heating and so due precautions[1] must be taken. The compound is stored under anhydrous conditions. This reagent should be handled in a fume hood.

Preparation of Glycosides and Nonreducing Disaccharides.[1] Reaction of D-glucal triacetate (**2**) with equimolar amounts of methanol and bromonium di-*sym*-collidine complex (**1**) dissolved in chloroform gives virtually quantitative yields of acetylated methyl 2-deoxy-2-bromoglycosides. Unlike pyridine or 2-methylpyridine, the steric hindrance to *N*-glycoside formation provided by the methyl groups at the 2- and 6-positions of *sym*-collidine is required for successful *O*-glycoside synthesis. If 2,3,4,6-tetra-*O*-acetyl-β-D-glucose (**3**) is used as the alcohol, the 2-deoxy-2-bromo disaccharide heptaacetate (**4**) is formed (eq 1). Deacetylation with Triethylamine in aqueous methanol followed by hydrogenolysis with Palladium on Carbon as the catalyst affords β-D-glucopyranosyl-2-deoxy-α-D-*arabino*-hexopyranoside (2-deoxyneotrehalose), isolated as the crystalline heptaacetate (**5**).

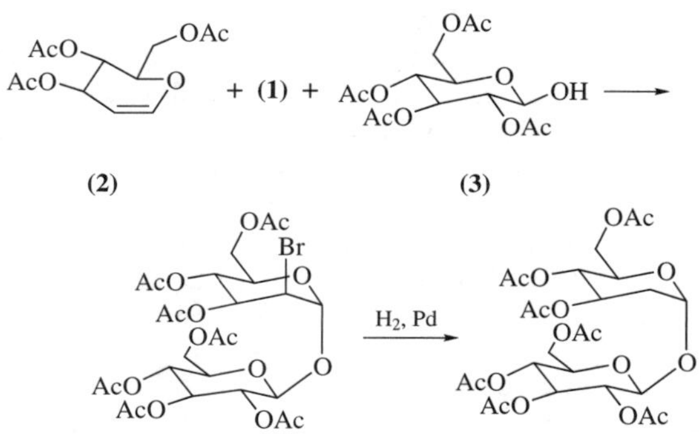
(2) (3) (4) (5) (1)

Bromocyclization[3] of Unsaturated Thiocarbamidates. This reagent can effect the stereo- and regiospecific functionalization of alkenes to furnish *cis*-1,2-methylamino alcohols in protected form. For example, unsaturated thiocarbamates, prepared as shown in eq 2, on treatment with bromonium dicollidine perchlorate cyclize to bromocarbamates. This cyclization has been successfully applied to several cyclic and acyclic unsaturated alcohols using methyl, benzyl, and *t*-butyl isothiocyanates. In all cases the condensation step (**6** to **7**) is essentially quantitative. The product (**8**) can be converted into *cis*-α-alkylamino alcohols by reduction of the bromine with Tri-n-butylstannane followed by alkaline hydrolysis. When R is a *t*-butyl group it can be removed in high yield by treatment with Trifluoroacetic Acid at 25 °C. The value of these transformations for the synthesis of aminocyclitols is illustrated by the conversion of bromocarbamate (**9**) to (±)-sporamine (**10**) in three steps (TFA; *n*-Bu_3SnH; aqueous NaOH, 25 °C) in better than 90% overall yield (eq 3). Sporamine is the aminocyclitol portion of the broad-spectrum antibiotic sporacin A.

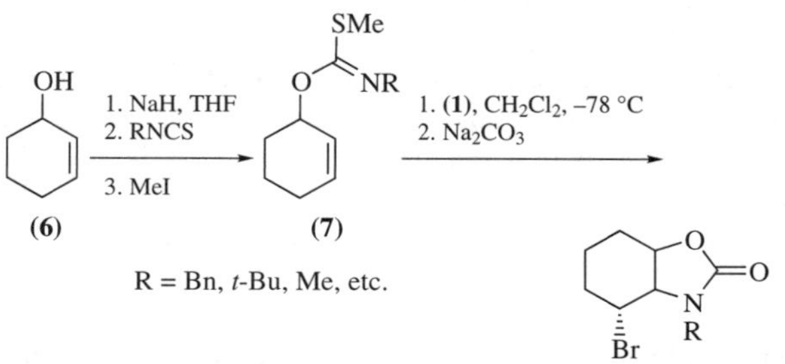

(6) (7)
R = Bn, *t*-Bu, Me, etc.
(**8**) 65% overall (2)

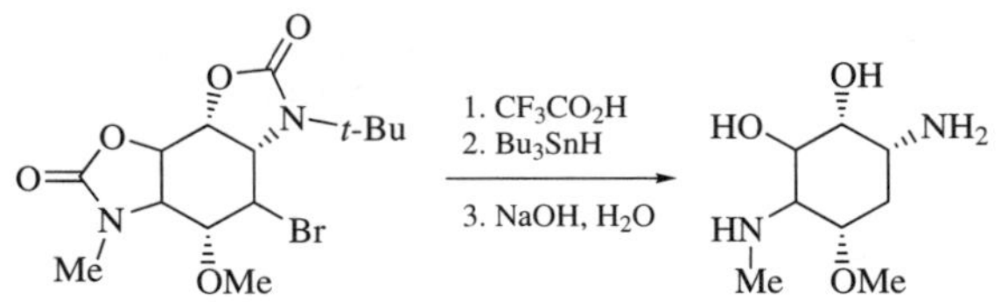

(**9**) (**10**) (3)

Bromolactamization. Esoaminuroic acid (**11**), an amino sugar, is present in esomycin A, which is an antifungal antibiotic isolated from *Streptomyces*. The key step in its synthesis is the transannular bromolactamization of an unsaturated amide (eq 4).[4]

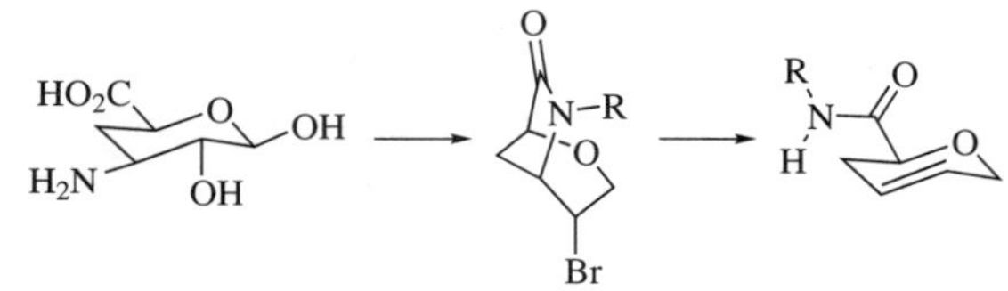
(4)

N-(*R*)-*p*-Methoxyphenethylamides were subjected to the bromolactamization sequence involving the use of Trimethylsilyl Trifluoromethanesulfonate as the amide silylating agent, and bromonium dicollidine perchlorate as electrophile. The desired [3.2.1] lactams were the only cyclization products, with yields exceeding 75% (eq 5). No [2.2.2] lactams were formed in detectable quantities. It may be noted that the *N,O*-bis(*t*-butyldimethylsilyl) derivative gave more [2.2.2] product, particularly at higher temperature.[5]

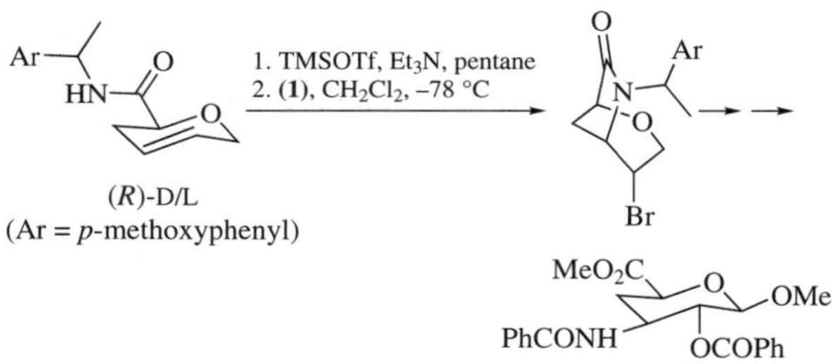

(*R*)-D/L
(Ar = *p*-methoxyphenyl)
(5)

1. Lemieux, R. U.; Levine, S. *Synlett* **1964**, *42*, 1473.
2. *Perchlorates, their Properties, Manufacture and Uses*; Schumacher, J. C., Reinhold: Ed.; New York, **1960**; p 187.
3. Knapp, S.; Patel, D. V. *J. Am. Chem. Soc.* **1983**, *105*, 6985.
4. Knapp, S.; Levorse, A. T.; Potenza, J. A. *J. Org. Chem.* **1988**, *53*, 4773.
5. (a) Takaoka, K.; Kuwayama, T.; Aoki, A. *Jpn. Pat.* 615 332, 1971. (b) Sakata, K.; Sakurai, A.; Tamura, S. *Agric. Biol. Chem.* **1973**, *37*, 697.

Tapan Ray
Sigma Radiochemical Co., St. Louis, MO, USA

N-Bromosuccinimide[1]

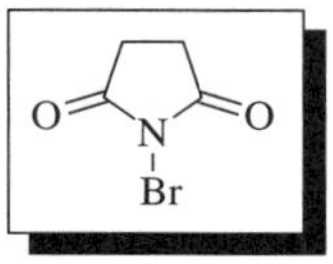

[128-08-5] $C_4H_4BrNO_2$ (MW 177.99)

(radical bromination of allylic and benzylic positions; electrophilic bromination of ketones, aromatic and heterocyclic compounds; bromohydration, bromoetherification, and bromolactonization of alkenes)

Alternate Names: NBS; 1-bromo-2,5-pyrrolidinedione.
Physical Data: mp 173–175 °C (dec); *d* 2.098 g cm^{-3}.
Solubility: sol acetone, THF, DMF, DMSO, MeCN; slightly sol H_2O, AcOH; insol ether, hexane, CCl_4 (at 25 °C).
Form Supplied in: white powder or crystals having a faint odor of bromine when pure; widely available.
Purification: in many applications the use of unrecrystallized material has led to erratic results. Material stored for extended periods often contains significant amounts of molecular bromine and is easily purified by recrystallization from H_2O (AcOH has also been used). In an efficient fume hood (caution: bromine evolution), an impure sample of NBS (200 g) is dissolved as quickly as possible in 2.5 L of preheated water at 90–95 °C. As filtration is usually unnecessary, the solution is then chilled well in an ice bath to effect crystallization. After most of the aqueous portion has been decanted, the white crystals are collected by filtration through a bed of ice and washed well with water. The crystals are dried on the filter and then in vacuo. The purity of NBS may be determined by the standard iodide–thiosulfate titration method.
Handling, Storage, and Precautions: should be stored in a refrigerator and protected from moisture to avoid decomposition. One of the advantages of using NBS is that it is easier and safer to handle than bromine; however, the solid is an irritant and bromine may be released during some operations. Therefore, precautions should be taken to avoid inhalation of the powder and contact with skin. All operations with this reagent are best conducted in an efficient fume hood. In addition, since reactions involving NBS are generally quite exothermic, large-scale operations (>0.1 mol) should be approached with particular caution.

Introduction. *N*-Bromosuccinimide is a convenient source of bromine for both radical substitution and electrophilic addition reactions. For radical substitution reactions, NBS has several advantages over the use of molecular Bromine, while 1,3-Dibromo-5,5-dimethylhydantoin is another reagent of use. N-Chlorosuccinimide and N-Iodosuccinimide generally do not facilitate analogous substitution reactions. For electrophilic substitutions, Bromine, N-Bromoacetamide, Bromonium Di-sym-collidine Perchlorate, 1,3-Dibromoisocyanuric Acid, and 2,4,4,6-Tetrabromo-2,5-cyclohexadienone also have applicability and the analogous halogenation reactions are generally possible using NCS, NIS, and I_2. Possible impurities generated during NBS brominations include conjugates of succinimide and, if basic conditions are employed, β-alanine (formed by the Hofmann reaction) and its derivatives may be isolated.

Allylic Bromination of Alkenes.[2] Standard conditions for allylic bromination involve refluxing of a solution of the alkene and recrystallized NBS in anhydrous CCl_4 using Dibenzoyl Peroxide, irradiation with visible light (ordinary 100 W light bulb or sunlamp[3]), or both to effect initiation. Both NBS and the co-product succinimide are insoluble in CCl_4 and succinimide collects at the surface of the reaction mixture as the reaction proceeds.[4] High levels of regioselectivity operate during the hydrogen-abstraction step of the chain mechanism, such that allylic methylene groups are attacked much more rapidly than allylic methyl groups.[5] However, a thermodynamic mixture of allylic bromides is generally isolated since both the allylic radical and the allylic bromide are subject to isomerization under the reaction conditions.[6] High levels of functional group selectivity are characteristic of this reaction, for example alkenic esters may be converted to allylic bromides prior to intramolecular cyclization (eq 1).[7] Brominations of α,β-unsaturated esters (eq 2)[8] and lactones (eq 3) are also successful.[9]

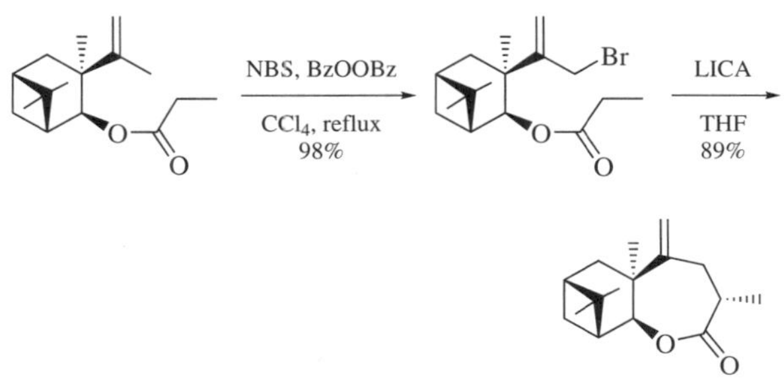

(1)

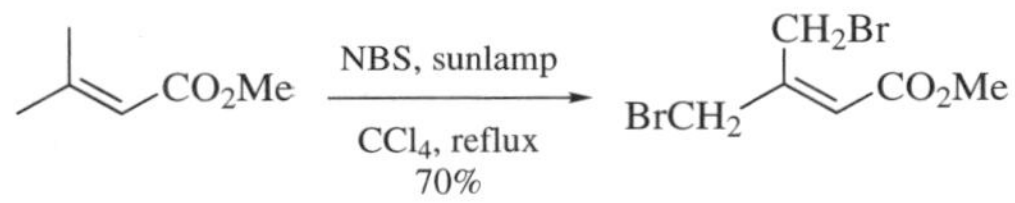

(2)

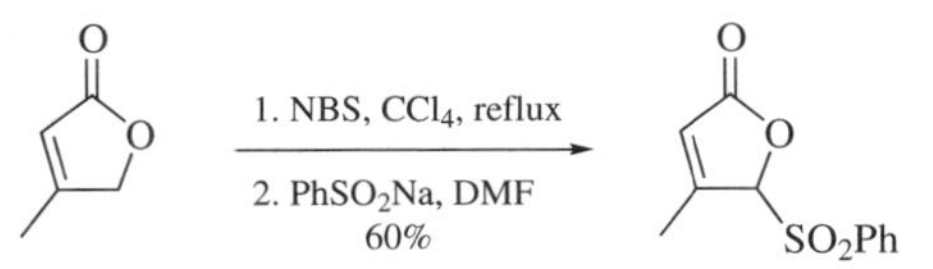

(3)

Benzylic Bromination of Aromatic Compounds. Using the conditions described above, NBS also effects the bromination of benzylic positions.[10] Bromine is also regularly used for benzylic bromination (eq 4);[11] however, many functional groups are sensitive to the generation of HBr during the reaction, including carbonyl groups which suffer competing acid-catalyzed bromination. These considerations render NBS as the reagent of choice for bromination of polyfunctional aromatic compounds. Selectivity can be anticipated with polyfunctional molecules based on the predicted stabilities of the radical intermediates (eq 5).[12] Accordingly, the use of NBS allows the bromination of alkyl groups attached to sensitive heterocyclic compounds (eq 6).[13] Complications which may arise from this method include *gem*-dibromination (eq 7)[14] of methyl substituents as well as in situ elimination of the product benzylic bromide (see also 1,3-Dibromo-5,5-dimethylhydantoin).

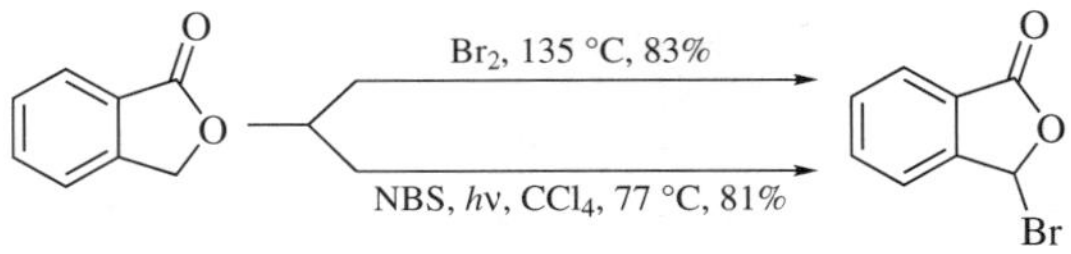

(4)

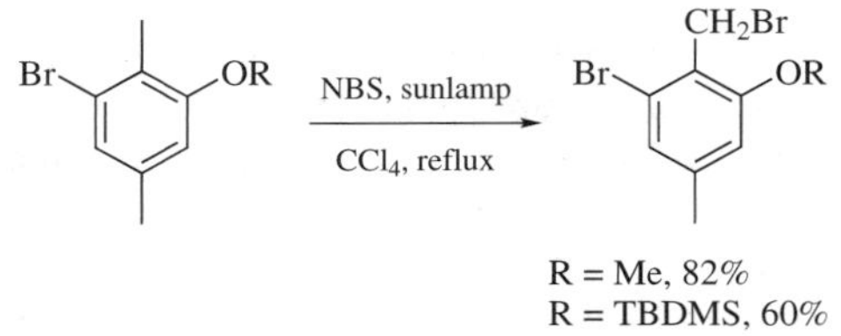

(5)

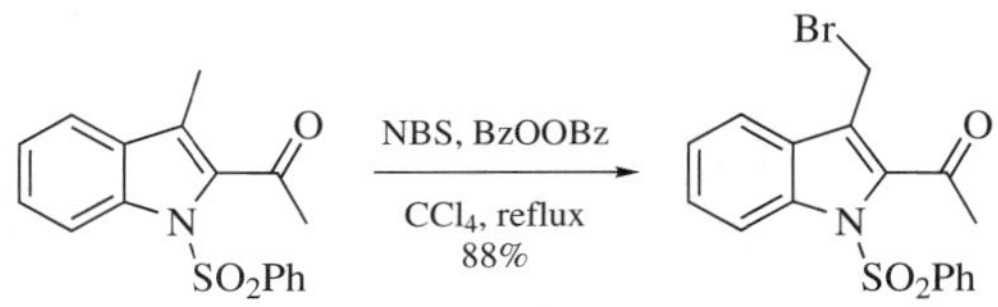

(6)

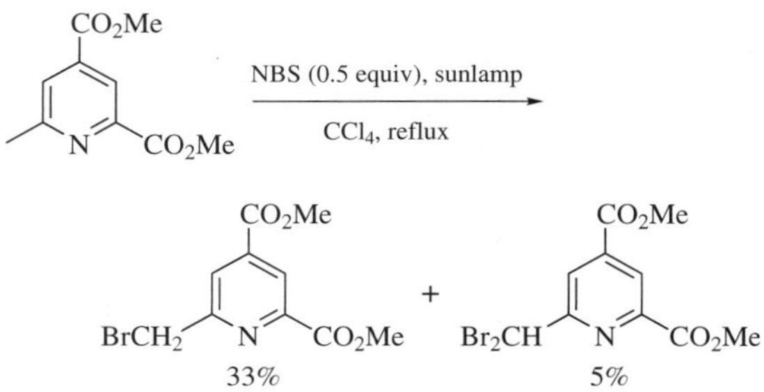

(7)

The regioselective cleavage of benzylidene acetals using NBS has been used widely in the synthesis of natural products from carbohydrates (eq 8)[15] and other chiral materials (eq 9).[16] It is rather important that the reaction be conducted in anhydrous CCl_4 (passage through activated alumina is sufficient), since in the presence of water the hydroxy benzoate is formed.[17] Barium carbonate is generally added to maintain anhydrous and acid-free conditions, and the addition of $Cl_2CHCHCl_2$ often improves solubility of the substrate. Selectivity is usually very high in cases in which a primary bromide can be produced, but may also be obtained in systems such as shown in eq 10.[18] As alkoxy substituents serve to further stabilize the adjacent radicals, these reactions proceed with high selectivity in the presence of other functional groups. Other applications in the carbohydrate field include the cleavage of benzyl ethers and benzyl glycosides (to the corresponding glycosyl bromides) and the bromination of pyranoses in the 5-position.[19]

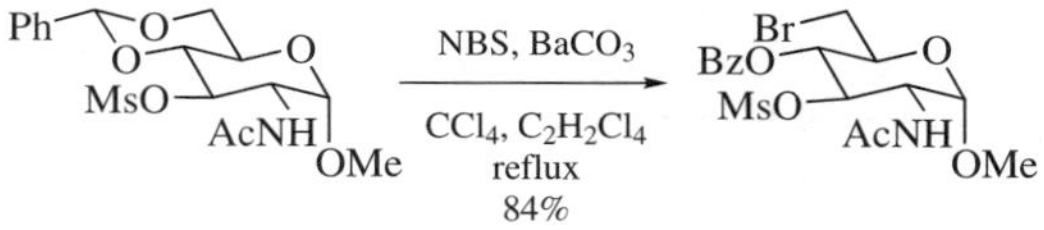

(8)

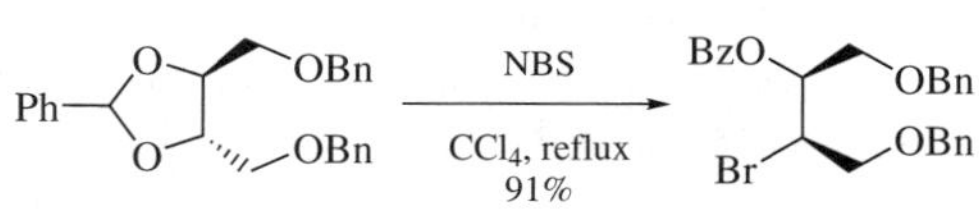

(9)

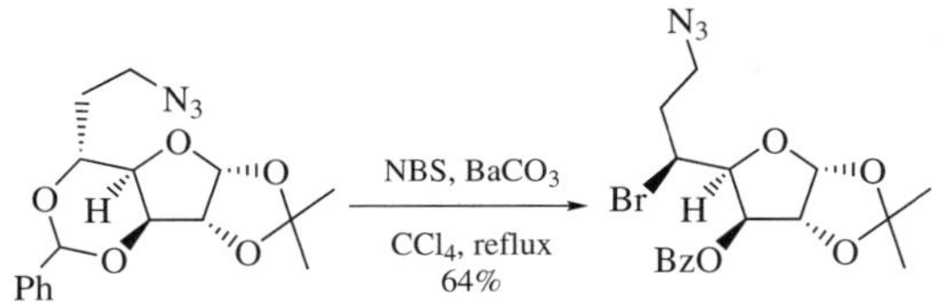

(10)

Unsaturation and Aromatization Reactions.[20] Unsaturated aldehydes, esters, and lactones can be accessed via strategies involving radical bromination and subsequent elimination. The allylic bromination of unsaturated lactones may be followed by elimination with base to obtain dienoic and trienoic lactones (eqs 11 and 12).[21] Conversion of an aldehyde to the enol acetate allows the radical bromination at the C_β position to proceed smoothly and, upon ester hydrolysis, the α,β-unsaturated aldehyde is obtained (eq 13).[22]

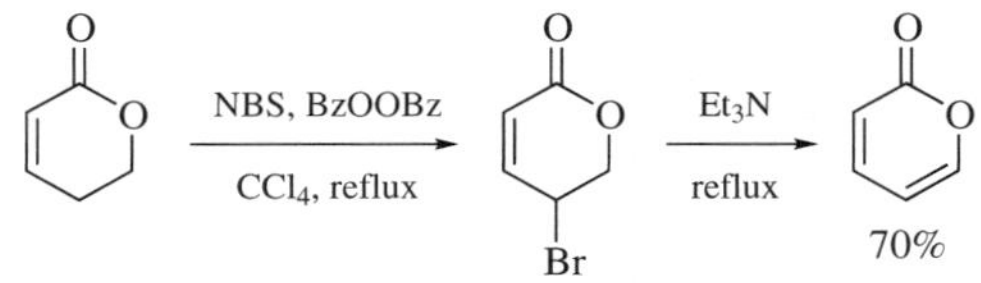

(11)

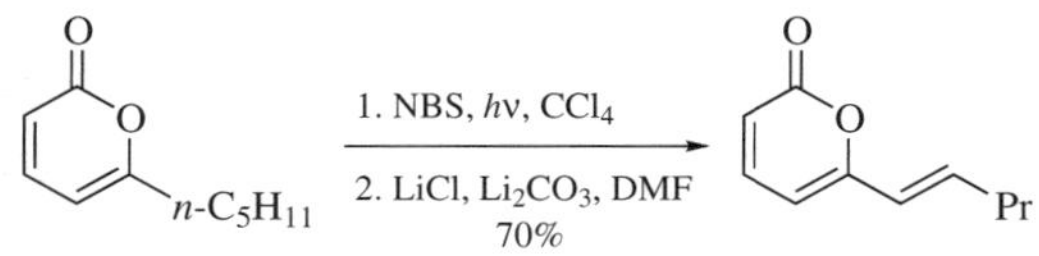

(12)

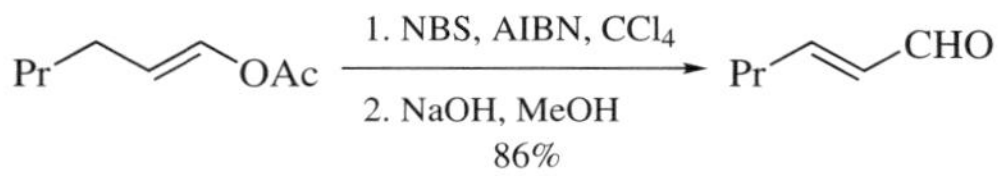

(13)

The direct bromination of β-alkoxylactones at the β position initially generates the α,β-unsaturated lactones (eq 14); however, the required radical abstraction is not so facile and further bromination of the α,β-unsaturated lactone proceeds competitively to afford the mono- and dibrominated products.[23] NBS is also used for the oxidative aromatization of polycyclic compounds, including steroids and anthraquinone precursors (eq 15).[24]

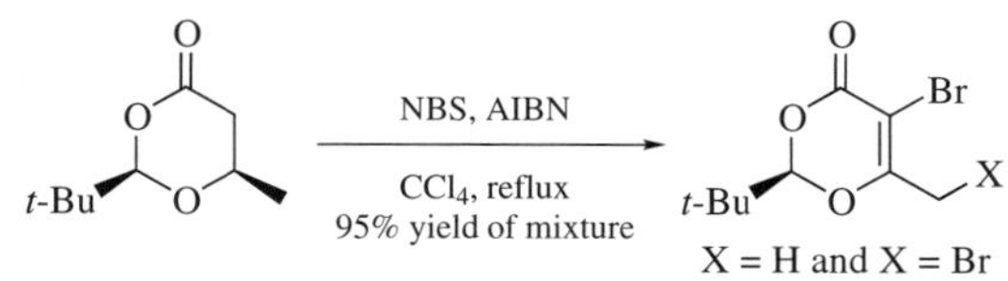

(14)

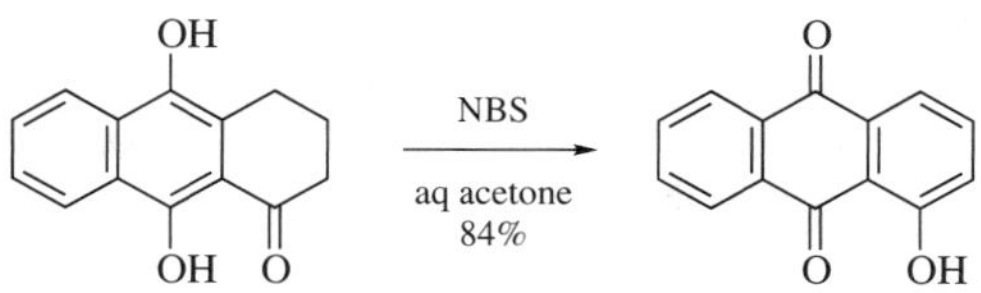

(15)

α-Bromination of Carbonyl Derivatives. Although simple carbonyl derivatives are not attacked in the α-position under radical bromination conditions, substitution by electron-donating groups stabilizes the radical intermediates by the capto-dative effect[25] and thus facilitates the substitution reaction which has been applied to a number of useful synthetic strategies. Protected glycine derivatives are easily brominated by NBS and benzoyl peroxide in $CHCl_3$ or CCl_4 at reflux to afford the corresponding α-bromoglycine derivatives.[26] These compounds are stable precursors of *N*-acyliminoacetates, which may be alkylated by silyl enol ethers in the presence of Lewis acids, organometallic reagents, and other nucleophiles to afford novel α-amino acids (eq 16).[27] Diketopiperazines and related heterocycles are also substituted in good yields (eq 17).[28] Furthermore, in contrast to aldehydes which undergo abstraction of the aldehydic hydrogen (see below), *O*-trimethylsilylaldoximes are readily brominated at the α-position under radical bromination conditions and can be converted to substituted nitrile oxides (*O*-trimethylsilylketoximes react similarly).[29]

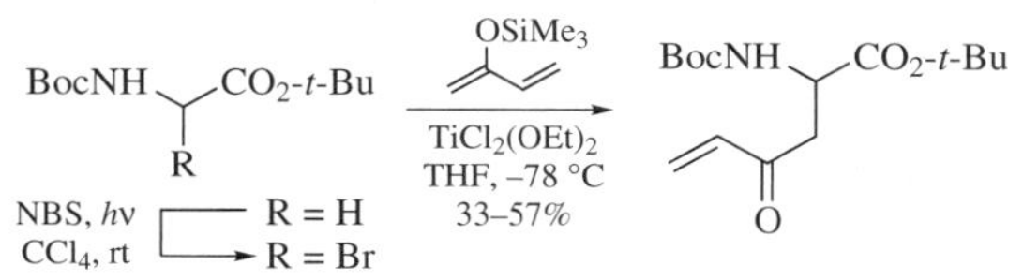

(16)

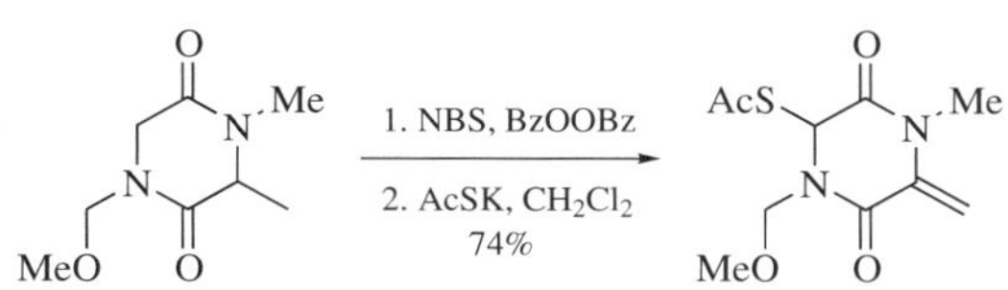

(17)

The use of NBS in the presence of catalytic Hydrogen Bromide has proven to be more convenient than Br_2 for the conversion of acid chlorides to α-bromo acid chlorides.[30] The reaction of the corresponding enolates, enol ethers, or enol acetates with NBS (and other halogenating agents) offers considerable advantages over direct acid-catalyzed halogenation of ketones and esters.[31] Although both reagents may afford the α-brominated products in high yields, NBS is more compatible than is bromine with sensitive functional groups and has been used in the asymmetric synthesis of α-amino acids.[32] The bromination of cyanoacetic acid proceeds rapidly with NBS to afford dibromoacetonitrile[33] and, similarly, β-keto esters, β-diketones, and β-sulfonyl ketones may be reacted with NBS in the presence of base to afford the products of bromination and in situ deacylation (see N-Chlorosuccinimide).[34] (5*E*)-Bromovinyluridine derivatives are readily prepared by bromodecarboxylation of the corresponding α,β-unsaturated acids with NBS (eq 18).[35]

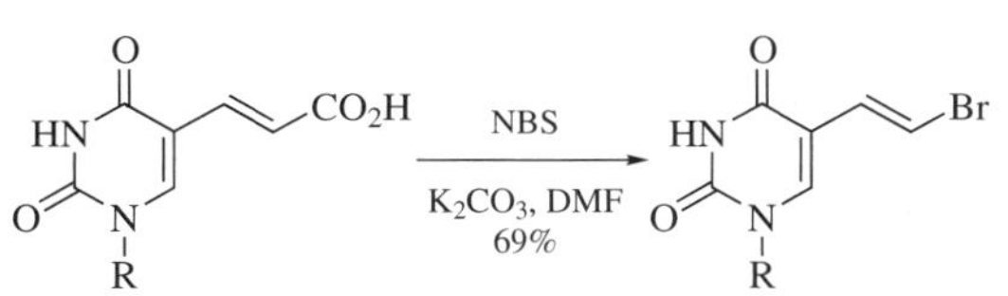

(18)

R = 2-deoxyribosyl

Reaction with Vinylic and Alkynic Derivatives. NBS is a suitable source of bromine for the conversion of vinylcopper and other organometallic derivatives to the corresponding vinyl bromides.[36] Vinylsilanes, prepared from the corresponding 1-trimethylsilylalkyne by reduction with Diisobutylaluminum Hydride, can be isomerized from the (*Z*) to the (*E*) geometry by irradiation with NBS and Pyridine, thus making (*E*)-vinylsilanes readily available stereoselectively in three steps from the corresponding alkyne (eq 19).[37] Allylsilane can be brominated by NBS under radical conditions, whereas more reactive allylsilanes are bromodesilated by NBS in CH_2Cl_2 at −78 °C.[38] 1-Bromoalkynes can be prepared under mild conditions by reaction with NBS in acetone in the presence of catalytic Silver(I) Nitrate.[39]

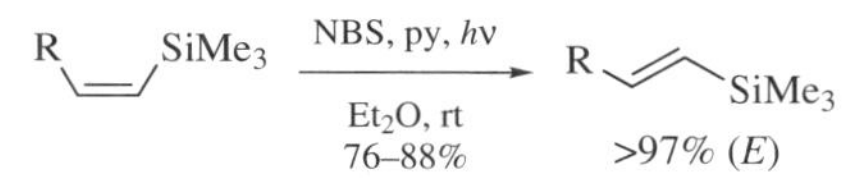

(19)

Bromination of Aromatic Compounds. Phenols, anilines, and other electron-rich aromatic compounds can be monobrominated using NBS in DMF with higher yields and higher levels of *para* selectivity than with Br_2.[40] *N*-Trimethylsilylanilines and aromatic ethers are also selectively brominated by NBS in $CHCl_3$ or CCl_4.[41] *N*-Substituted pyrroles are brominated with NBS in THF to afford 2-bromopyrroles (1 equiv) or 2,5-dibromopyrroles (2 equiv) with high selectivity, whereas bromination with Br_2 affords the thermodynamically more stable 3-bromopyrroles.[42] The use of NBS in DMF also achieves the controlled bromination of imidazole and nitroimidazole.[43] Thiophenes are also selectively brominated in the 2-position using NBS in acetic acid–chloroform.[44]

Bromohydration, Bromolactonization, and Other Additions to C=C.[45] The preferred conditions for the bromohydration of alkenes involves the portionwise addition of solid or predissolved NBS (recrystallized) to a solution of the alkene in 50–75% aqueous DME, THF, or *t*-butanol at 0 °C. The formation of dibromide and α-bromo ketone byproducts can be minimized by using recrystallized NBS. High selectivity for Markovnikov addition and *anti* stereochemistry results from attack of the bromonium ion intermediate by water. Aqueous DMSO can also be used as the solvent; however, since DMSO is readily oxidized under the reaction conditions, significant amounts of the dibromide byproduct may be produced.[46,47] In the bromohydration of polyalkenic compounds, high selectivity is regularly achieved for attack of the most electron-rich double bond (eq 20).[48] With farnesol acetate, squalene, and other polyisoprenes, choice of the optimum proportion of water is used to effect the selective bromohydration at the terminal double bond (eq 21),[49] and the two-step sequence shown is often the method of choice for the preparation of the corresponding epoxides.[50]

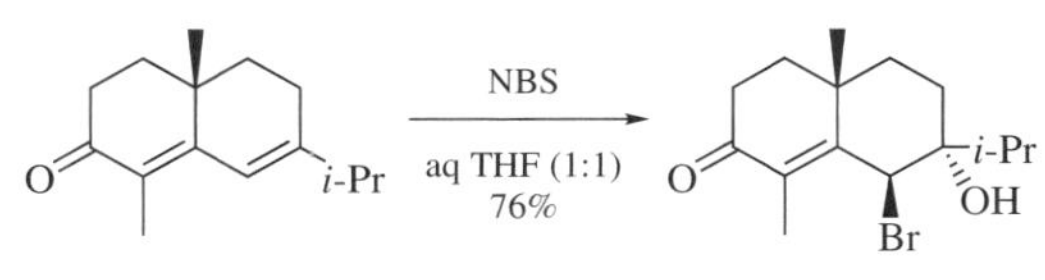

(20)

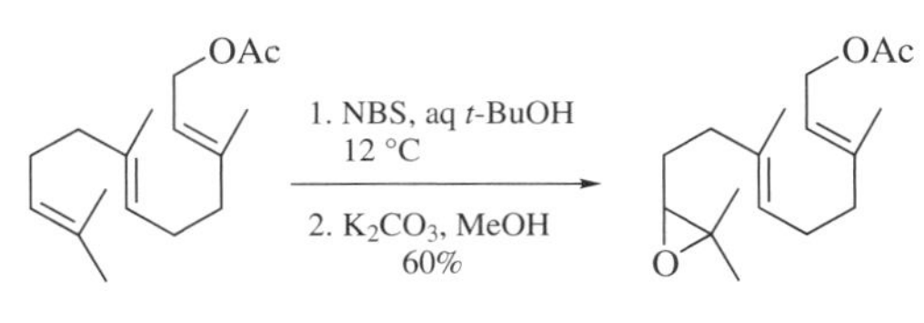

(21)

Bromoetherification of alkenes can be achieved using NBS in the desired alcohol as the solvent. The reaction of 1,3-dichloropropene with NBS in methanol yields an α-bromo dimethyl acetal in the first step in a convenient synthesis of cyclopropenone.[51] Using propargyl alcohol the reaction depicted in eq 22 has been extended to an annulation method for the synthesis of α-methylene-γ-butyrolactones.[52] Intramolecular bromoetherification and bromoamination reactions are generally very facile (eq 23).[53] In natural products synthesis, bromoetherification has been used for the synthesis of cyclic ethers (by subsequent debromination, see Tri-n-butylstannane) and for the protection of alkene appendages as cyclic bromoethers (regenerated by reaction with zinc).[54]

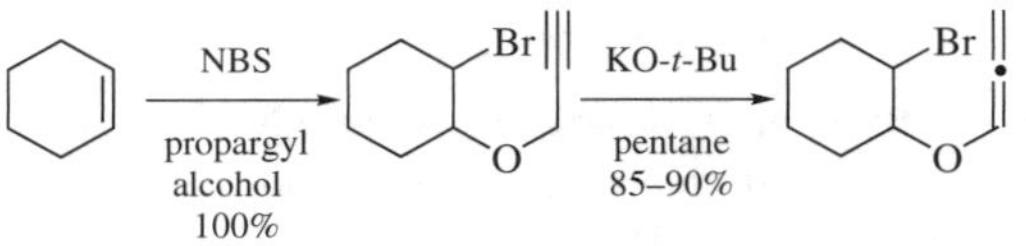

(22)

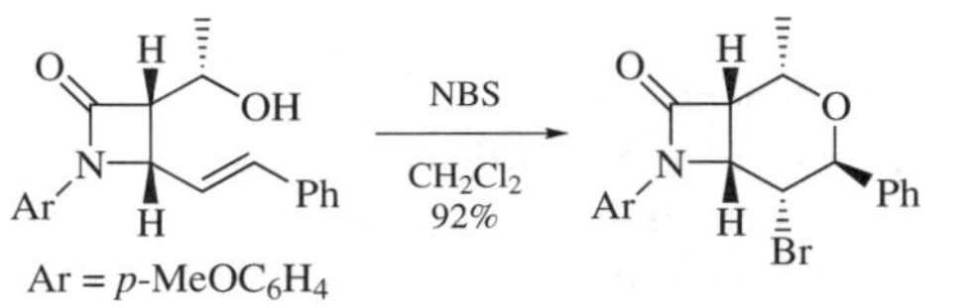

(23)

NBS is also an effective reagent for bromolactonization of unsaturated acids and acid derivatives with the same high stereo- and Markovnikov selectivity (see also Iodine). Dienes, such as the cycloheptadiene derivative shown, may react exclusively via *syn*-1,4-addition (eq 24).[55] Alkynic acids are converted to the (*E*)-bromo enol lactones by NBS in a biphasic medium, whereas the combination of bromine and silver nitrate afford the (*Z*)-bromo enol lactones (eq 25).[56] α,β-Unsaturated acylprolines react with NBS in anhydrous DMF to afford the corresponding bromolactones having diastereomeric excesses up to 93%, which can be converted to chiral α-hydroxy acids by debromination followed by acidic hydrolysis (eq 26).[57] In contrast to alkenic amides, which generally react with NBS to afford bromolactones (via the cyclic iminoether derivatives), alkenic sulfonamides readily undergo cyclization on nitrogen when reacted with NBS to afford the bromosulfonamides in high yields.[58] *N*-Methoxyamides have also proven effective for bromolactamization, leading to diketopiperazines (eq 27)[59] (see also Bromonium Di-sym-collidine Perchlorate).

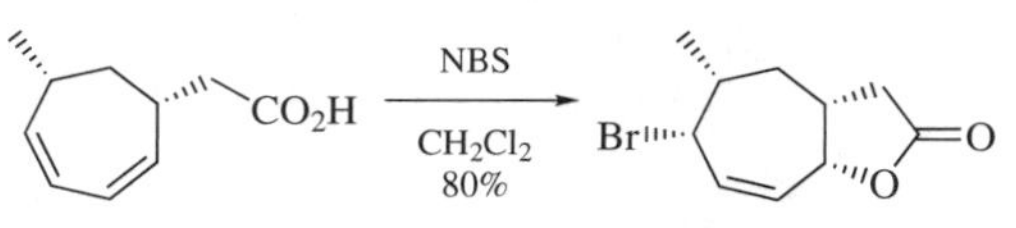

(24)

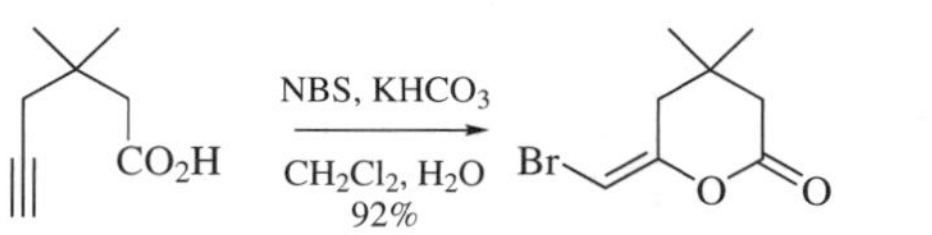

(25)

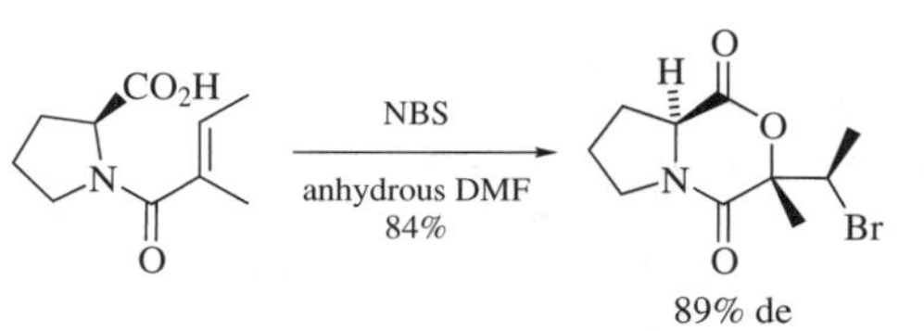

(26)

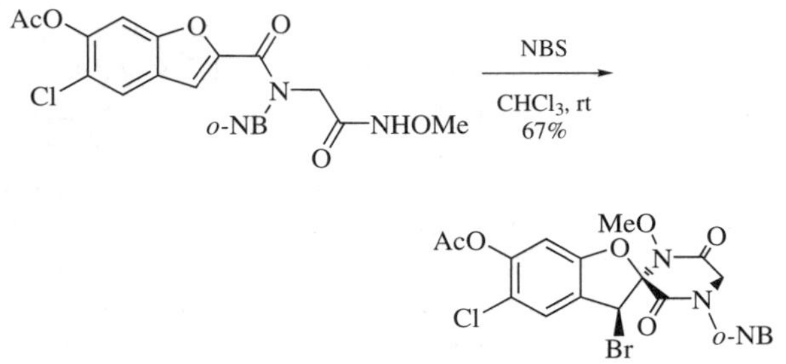

(27)

Addition of NBS to an alkene in the presence of aqueous Sodium Azide affords fair yields of the corresponding β-bromoazides, which can be converted by Lithium Aluminum Hydride reduction to aziridines.[60] Intermolecular reactions of alkenes with NBS and weaker nucleophiles can be achieved if conducted under anhydrous conditions to avoid the facile bromohydration reaction. In this manner, bromofluorination of alkenes has been extensively studied using Pyridinium Poly(hydrogen fluoride), triethylammonium dihydrogentrifluoride or tetrabutylammonium hydrogendifluoride as the fluoride ion source.[61]

Oxidation and Bromination of Other Functional Groups. Conjugate bases of other functional groups can be α-brominated with NBS. Nitronate anions of aliphatic nitro compounds react with NBS to afford the *gem*-bromonitro compounds in high yield.[62] The α-bromination of sulfoxides can be performed in the presence of pyridine and proceeds more satisfactorily using NBS in the presence of catalytic Br_2 than with either reagent alone.[63] NBS also reacts with sulfides to afford sulfoxides when methanol is used as a solvent, or to form α-bromo sulfides in anhydrous solvents.[64] NBS is a favored reagent for the deprotection of dithianes and dithioacetals to regenerate carbonyl groups (eq 28)[65] (see also N-Chlorosuccinimide and 1,3-Diiodo-5,5-dimethylhydantoin).

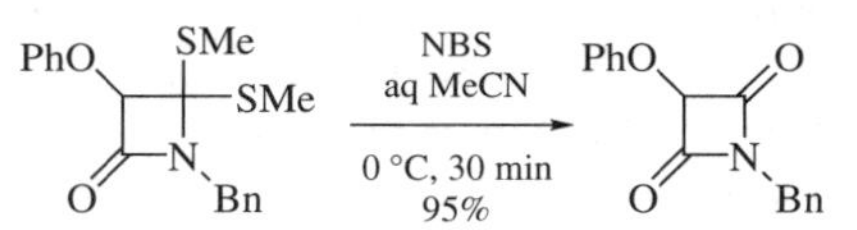

(28)

In polar media, NBS effectively oxidizes primary and secondary alcohols to carbonyl compounds via hypobromite or alkoxysuccinimide intermediates. Although this transformation is more commonly effected by the use of chromium reagents or activated Dimethyl Sulfoxide, the most notable application of NBS and related reagents lies in its selectivity for the oxidation of axial vs. equatorial hydroxy groups in steroid systems (see N-Bromoacetamide).[66] Often, a single secondary alcohol may be converted to the ketone in the presence of many other alcohol groups.

Under radical conditions, aldehydes are readily oxidized by NBS to acid bromides.[67] The oxidation of aldoximes to nitrile oxides using NBS and Triethylamine in DMF is superior to the use of aqueous hypochlorite.[68] Tosylhydrazones are cleaved by reaction with NBS in methanol,[69] and hydrazines and hydrazides are oxidized to azo compounds.[70]

Related Reagents. *N*-Bromosuccinimide–Dimethylformamide; *N*-Bromosuccinimide–Dimethyl Sulfide; *N*-Bromosuccinimide–Hydrogen Fluoride; *N*-Bromosuccinimide–Sodium

Azide; Triphenylphosphine– *N*-Bromosuccinimide.

---

1. Pizey, J. S. *Synthetic Reagents*; Wiley:New York, **1974**; Vol. *2*, p 1.
2. (a) Djerassi, C. *Chem. Rev.* **1948**, *43*, 271. (b) Horner, L.; Winkelmann, E. H. *Angew. Cheml* **1959**, *71*, 349.
3. UV irradiation through Pyrex (λ > 313 nm) can lead to Cl- and Cl_3C-substituted products from the solvent CCl_4. Futamura, S.; Zong, Z.-M. *Bull. Chem. Soc. Jpn.* **1992**, *65*, 345.
4. Greenwood, F. L.; Kellert, M. D.; Sedlak, J. *Org. Synth., Coll. Vol.* **1963**, *4*, 108.
5. (a) Ziegler, K.; Spaeth, A.; Schaaf, E.; Schumann, W.; Winkelmann, E. *Justus Liebigs Ann. Chem./Liebigs Ann. Chem.* **1942**, *551*, 80. (b) Using the solvents $CHCl_3$ and MeCN, different selectivities are observed. Day, J. C.; Lindstrom, M. J.; Skell, P. S. *J. Am. Chem. Soc.* **1974**, *96*, 5616.
6. Accordingly, the product obtained in Ref. [4] is almost certainly a mixture of isomers.
7. Inokuchi, T.; Asanuma, G.; Torii, S. *J. Org. Chem.* **1982**, *47*, 4622.
8. (a) Franck-Neumann, M.; Martina, D.; Heitz, M.-P. *Tetrahedron Lett.* **1989**, *30*, 6679. (b) Martin, R.; Chapleo, C. B.; Svanholt, K. L.; Dreiding, A. S. *Helv. Chim. Acta* **1976**, *59*, 2724.
9. Yoda, H.; Shirakawa, K.; Takabe, K. *Chem. Lett.* **1989**, 1391.
10. (a) Corbin, T. F.; Hahn, R. C.; Shechter, H. *Org. Synth., Coll. Vol.* **1973**, *5*, 328. (b) Kalir, A. *Org. Synth., Coll. Vol.* **1973**, *5*, 825.
11. (a) Koten, I. A.; Sauer, R. J. *Org. Synth., Coll. Vol.* **1973**, *5*, 145. (b) Shriner, R. L.; Wolf, F. J. *Org. Synth., Coll. Vol.* **1955**, *3*, 737.
12. (a) Leed, A. R.; Boettger, S. D.; Ganem, B. *J. Org. Chem.* **1980**, *45*, 1098. (b) Goldberg, Y.; Bensimon, C.; Alper, H. *J. Org. Chem.* **1992**, *57*, 6374.
13. (a) Gribble, G. W.; Keavy, D. J.; Davis, D. A.; Saulnier, M. G.; Pelcman, B.; Barden, T. C.; Sibi, M. P.; Olson, E. R.; BelBruno, J. J. *J. Org. Chem.* **1992**, *57*, 5878. (b) Campaigne, E.; Tullar, B. F. *Org. Synth., Coll. Vol.* **1963**, *4*, 921.
14. Hendrickson, J. B.; de Vries, J. G. *J. Org. Chem.* **1985**, *50*, 1688.
15. (a) Hanessian, S. *Q. Rev., Chem. Soc.* **1987**, *65*, 243; *Org. Synth., Coll. Vol.* **1993**, *8*, 363. (b) Hanessian, S. *Methods Carbohydr. Chem.* **1972**, *6*, 183. (c) Hanessian, S.; Plessas, N. R. *J. Org. Chem.* **1969**, *34*, 1035. (d) Hanessian, S.; Plessas, N. R. *J. Org. Chem.* **1969**, *34*, 1045.
16. (a) Wenger, R. M. *Helv. Chim. Acta* **1983**, *66*, 2308. (b) Machinaga, N.; Kibayashi, C. *J. Org. Chem.* **1992**, *57*, 5178.
17. Binkley, R. W.; Goewey, G. S.; Johnston, J. C. *J. Org. Chem.* **1984**, *49*, 992.
18. Hendry, D.; Hough, L.; Richardson, A. C. *Tetrahedron Lett.* **1987**, *28*, 4597.
19. (a) Binkley, R. W.; Hehemann, D. G. *J. Org. Chem.* **1990**, *55*, 378. (b) Hashimoto, H.; Kawa, M.; Saito, Y.; Date, T.; Horito, S.; Yoshimura, J. *Tetrahedron Lett.* **1987**, *28*, 3505. (c) Giese, B.; Linker, T. *Synthesis* **1992**, 46. (d) Ferrier, R. J.; Tyler, P. C. *J. Chem. Soc., Perkin Trans. 1* **1980**, 2767.
20. Filler, R. *Chem. Rev.* **1963**, *63*, 21.
21. (a) Nakagawa, M.; Saegusa, J.; Tonozuka, M.; Obi, M.; Kiuchi, M.; Hino, T.; Ban, Y. *Org. Synth., Coll. Vol.* **1988**, *6*, 462. (b) Jones, T. H.; Fales, H. M. *Tetrahedron Lett.* **1983**, *24*, 5439.
22. Jung, F.; Ladjama, D.; Riehl, J. J. *Synthesis* **1979**, 507.
23. (a) Zimmermann, J.; Seebach, D. *Helv. Chim. Acta* **1987**, *70*, 1104. (b) Lange, G. L.; Organ, M. G.; Roche, M. R. *J. Org. Chem.* **1992**, *57*, 6000. (c) Seebach, D.; Gysel, U.; Job, K.; Beck, A. K. *Synthesis* **1992**, 39.
24. Hauser, F. M.; Prasanna, S. *J. Org. Chem.* **1982**, *47*, 383.
25. Viehe, H. G.; Merényi, R.; Stella, L.; Janousek, Z. *Angew. Chem., Int. Ed. Engl.* **1979**, *18*, 917.
26. (a) Yamaura, M.; Suzuki, T.; Hashimoto, H.; Yoshimura, J.; Shin, C. *Bull. Chem. Soc. Jpn.* **1985**, *58*, 2812. (b) Lidert, Z.; Gronowitz, S. *Synthesis* **1980**, 322.
27. (a) Bretschneider, T.; Miltz, W.; Münster, P.; Steglich, W. *Tetrahedron* **1988**, *44*, 5403. (b) Mühlemann, C.; Hartmann, P.; Odrecht, J.-P. *Q. Rev., Chem. Soc.* **1992**, *71*, 200. (c) Allmendinger, T.; Rihs, G.; Wetter, H. *Helv. Chim. Acta* **1988**, *71*, 395. (d) Ermert, P.; Meyer, J.; Stucki, C.; Schneebeli, J.; Obrecht, J.-P. *Tetrahedron Lett.* **1988**, *29*, 1265.
28. (a) Kishi, Y.; Fukuyama, T.; Nakatsuka, S.; Havel, M. *J. Am. Chem. Soc.* **1973**, *95*, 6493. (b) Zimmermann, J.; Seebach, D. *Helv. Chim. Acta* **1987**, *70*, 1104.
29. Hassner, A.; Murthy, K. *Tetrahedron Lett.* **1987**, *28*, 683.
30. (a) Harpp, D. N.; Bao. L. Q.; Coyle, C.; Gleason, J. G.; Horovitch, S. *Org. Synth., Coll. Vol.* **1988**, *6*, 190. (b) Harpp, D. N.; Bao, L. Q.; Black, C. J.; Gleason, J. G.; Smith, R. A. *J. Org. Chem.* **1975**, *40*, 3420.
31. (a) Stotter, P. L.; Hill, K. A. *J. Org. Chem.* **1973**, *38*, 2576. (b) Blanco, L.; Amice, P.; Conia, J. M. *Synthesis* **1976**, 194. (c) Hooz, J.; Bridson, J. N. *Synlett* **1972**, *50*, 2387. (d) Lichtenthaler, F. W.; Kläres, U.; Lergenmüller, M.; Schwidetzky, S. *Synthesis* **1992**, 179.
32. (a) Evans, D. A.; Ellman, J. A.; Dorow, R. L. *Tetrahedron Lett.* **1987**, *28*, 1123. (b) Oppolzer, W.; Dudfield, P. *Tetrahedron Lett.* **1985**, *26*, 5037.
33. Wilt, J. W.; Diebold, J L. *Org. Synth., Coll. Vol.* **1963**, *4*, 254.
34. Mignani, G.; Morel, D.; Grass, F. *Tetrahedron Lett.* **1987**, *28*, 5505.
35. (a) Izawa, T.; Nishiyama, S.; Yamamura, S.; Kato, K.; Takita, T. *J. Chem. Soc., Perkin Trans. 1* **1992**, 2519. (b) Jones, A. S.; Verhelst, G.; Walker, R. T. *Tetrahedron Lett* **1979**, 4415.
36. Levy, A. B.; Talley, P.; Dunford, J. A. *Tetrahedron Lett.* **1977**, 3545.
37. (a) Zweifel, G.; On, H. P. *Synthesis* **1980**, 803. (b) Camps, F.; Chamorro, E.; Gasol, V.; Guerrero, A. *Synth. Commun.* **1989**, *19*, 3211.
38. (a) Fleming, I.; Dunogues, J.; Smithers, R. *Org. React.* **1989**, *37*, 57. (b) Angell, R.; Parsons, P. J.; Naylor, A. *Synlett* **1993**, 189. (c) Weng, W.-W.; Luh, T.-Y. *J. Org. Chem.* **1992**, *57*, 2760.
39. Hofmeister, H.; Annen, K.; Laurent, H.; Wiechert, R. *Angew. Chem., Int. Ed. Engl.* **1984**, *23*, 727.
40. Mitchell, R. H.; Lai, Y-H.; Williams, R. V. *J. Org. Chem.* **1979**, *44*, 4733.
41. (a) Ando, W.; Tsumaki, H. *Synthesis* **1982**, 263. (b) Townsend, C. A.; Davis, S. G.; Christensen, S. B.; Link, J. C.; Lewis, C. P. *J. Am. Chem. Soc.* **1981**, *103*, 6885.
42. (a) Gilow, H. M.; Burton, D. E. *J. Org. Chem.* **1981**, *46*, 2221. (b) Martina, S.; Enkelmann, V.; Wegner, G.; Schlüter, A.-D. *Synthesis* **1991**, 613.
43. Palmer, B. D.; Denny. W. A. *J. Chem. Soc., Perkin Trans. 1* **1989**, 95.
44. (a) Kellogg, R. M.; Schaap, A. P.; Harper, E. T.; Wynberg, H. *J. Org. Chem.* **1968**, *33*, 2902. (b) Goldberg, Y.; Alper, H. *J. Org. Chem.* **1993**, *58*, 3072.
45. (a) Bartlett, P. A. *Asymmetric Synthesis*; Morrison, J. D., Ed.; Academic:New York **1984**; Vol. *3*, Chapter 6. (b) Beger, J. *Methoden Org. Chem. (Houben-Weyl)* **1991**, *333*, 677.
46. (a) Dalton, D. R.; Dutta, V. P.; Jones, D. C. *J. Am. Chem. Soc.* **1968**, *90*, 5498. (b) Langman, A. W.; Dalton, D. R. *Org. Synth., Coll. Vol.* **1988**, *6*, 184.
47. NBS in anhydrous DMSO converts dihydropyrans to α-bromolactones. Berkowitz, W. F.; Sasson, I.; Sampathkumar, P. S.; Hrabie, J.; Choudhry, S.; Pierce, D. *Tetrahedron Lett.* **1979**, 1641.
48. Kutney, J. P.; Singh, A. K. *Synlett* **1982**, *60*, 1842.
49. (a) van Tamelen, E. E.; Curphey, T. J. *Tetrahedron Lett.* **1962**, 121. (b) van Tamalen, E. E.; Sharpless, K. B. *Tetrahedron Lett.* **1967**, 2655. (c) Hanzlik, R. P. *Org. Synth., Coll. Vol.* **1988**, *6*, 560. (d) Nadeau, R.; Hanzlik, R. *Methods Enzymol.* **1969**, *15*, 346.
50. (a) Jennings, R. C.; Ottridge, A. P. *Chem. Commun./J. Chem. Soc., Chem. Commun.* **1979**, 920. (b) Gold, A.; Brewster, J.; Eisenstadt, E. *Chem. Commun./J. Chem. Soc., Chem. Commun.* **1979**, 903.
51. Breslow, R.; Pecoraro, J.; Sugimoto, T. *Org. Synth., Coll. Vol.* **1988**, *6*, 361.
52. Dulcere, J. P.; Mihoubi, M. N.; Rodriguez, J. *Chem. Commun./J. Chem. Soc., Chem. Commun.* **1988**, 237.

53. (a) Demole, E.; Enggist, P. *Helv. Chim. Acta* **1971**, *54*, 456. (b) Hart, D. J.; Leroy, V.; Merriman, G. H.; Young, D. G. J. *J. Org. Chem.* **1992**, *57*, 5670. (c) Michael, J. P.; Ting, P. C.; Bartlett, P. A. *J. Org. Chem.* **1985**, *50*, 2416. (d) Baskaran, S.; Islam, I.; Chandrasekaran, S. *J. Org. Chem.* **1990**, *55*, 891.
54. (a) Corey, E. J.; Pearce, H. L. *J. Am. Chem. Soc.* **1979**, *101*, 5841. (b) Schlessinger, R. H.; Nugent, R. A. *J. Am. Chem. Soc.* **1982**, *104*, 1116.
55. Pearson, A. J.; Ray, T. *Tetrahedron Lett.* **1986**, *27*, 3111.
56. Dai, W.; Katzenellenbogen, J. A. *J. Org. Chem.* **1991**, *56*, 6893.
57. (a) Jew, S-s.; Terashima, S.; Koga, K. *Tetrahedron* **1979**, *35*, 2337. (b) Hayashi, M.; Terashima, S.; Koga, K. *Tetrahedron* **1981**, *37*, 2797.
58. (a) Tamaru, Y.; Kawamura, S.; Tanaka, K.; Yoshida, Z. *Tetrahedron Lett.* **1984**, *25*, 1063. (b) Balko, T. W.; Brinkmeyer, R. S.; Terando, N. H. *Tetrahedron Lett.* **1989**, *30*, 2045.
59. Miknis, G. F.; Williams, R. M. *J. Am. Chem. Soc.* **1993**, *115*, 536.
60. (a) Van Ende, D.; Krief, A. *Angew. Chem., Int. Ed. Engl.* **1974**, *13*, 279. (b) Nagorski, R. W.; Brown, R. S. *J. Am. Chem. Soc.* **1992**, *114*, 7773.
61. (a) Olah, G. A.; Welch, J. T.; Vankar, Y. D.; Nojima, M.; Kerekes, I.; Olah, J. A. *J. Org. Chem.* **1979**, *44*, 3872. (b) Alvernhe, G.; Laurent, A.; Haufe, G. *Synthesis* **1987**, 562. (c) Camps, F.; Chamorro, E.; Gasol, V.; Guerrero, A. *J. Org. Chem.* **1989**, *54*, 4294. (d) Kuroboshi, M.; Hiyama, T. *Tetrahedron Lett.* **1991**, *32*, 1215.
62. Amrollah-Madjdabadi, A.; Beugelmans, R.; Lechevallier, A. *Synthesis* **1986**, 828.
63. (a) Iriuchijima, S.; Tsuchihashi, G. *Synthesis* **1970**, 588. (b) Drabowicz, J. *Synthesis* **1986**, 831.
64. Harville, R.; Reed, Jr., S. F. *J. Org. Chem.* **1968**, *33*, 3976.
65. (a) Corey, E. J.; Erickson, B. W. *J. Org. Chem.* **1971**, *36*, 3553. (b) Bari, S. S.; Trehan, I. R.; Sharma, A. K.; Manhas, M. S. *Synthesis* **1992**, 439.
66. Filler, R. *Chem. Rev.* **1963**, *63*, 21.
67. Cheung, Y.-F. *Tetrahedron Lett.* **1979**, 3809.
68. Grundmann, C.; Richter, R. *J. Org. Chem.* **1968**, *33*, 476.
69. Rosini, G. *J. Org. Chem.* **1974**, *39*, 3504.
70. (a) Carpino, L. A.; Crowley, P. J. *Org. Synth., Coll. Vol.* **1973**, *5*, 160. (b) Bock, H.; Rudolph, G.; Baltin, E. *Ber. Dtsch. Chem. Ges./Chem. Ber.* **1965**, *98*, 2054.

Scott C. Virgil
Massachusetts Institute of Technology, Cambridge, MA, USA

Bromotripyrrolidinophosphonium Hexafluorophosphate[1]

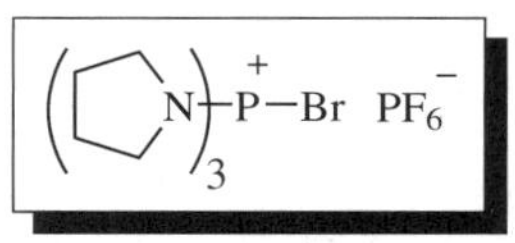

[132705-51-2] $C_{12}H_{24}BrF_6N_3P_2$ (MW 466)

(reagent for carboxylic acid activation, particularly used for peptide synthesis; reagent for formamidine synthesis)

Alternate Name: PyBroP.
Physical Data: mp 138–139 °C.[1]
Solubility: soluble in CH_2Cl_2, $CHCl_3$, DMF, DMSO, NMP, acetone.
Form Supplied in: colorless crystals.
Analysis of Reagent Purity: ^{31}P NMR, HPLC.
Preparative Methods: addition of Br_2 to tripyrrolidinophosphine[2] or addition of tripyrrolidinophosphine oxide to an acetonitrile solution of phosphoryl tribromide,[1,2] then addition to a KPF_6 solution in water.
Purification: recrystallization from CH_2Cl_2/Et_2O; CH_2Cl_2/hexane; acetone/Et_2O; EtOAc.
Handling, Storage, and Precautions: the toxicological properties of this reagent have not been investigated and no special instructions for storage and handling are mentioned in the literature.

Introduction. Treatment of a carboxylic acid with bromotripyrrolidinophosphonium hexafluorophosphate (PyBroP) in the presence of base produces the unstable acylphosphonium intermediate. Such species directly react with a nucleophile or decompose into a different activated form (e.g. carboxylic anhydride).[1] The reagent was first designed for coupling in peptide synthesis but is also widely used for the synthesis of amides or esters, and for macrocyclization. This reagent gives results that are comparable to those obtained with BroP [bromotris(dimethylamino)phosphonium hexafluorophosphate][3] with the advantage of not producing the carcinogen **HMPA** as a side product of the coupling reaction.[4] By-products have been observed from slowly reacting activated carboxylates: (i) the formation of pyrrolidide derivatives due to the presence of small amounts of pyrrolidine in the commercial phosphonium salt (avoided by recrystallization of the reagent),[5] and (ii) a side reaction in DMF but favorable for the preparation of formamidines.[6]

Reagent for Peptide Synthesis. PyBroP was efficiently used in both liquid and solid phase peptide synthesis, but it can lead to epimerization when used for segment condensation.[7] Compared with hydroxybenzotriazol-based activators like BOP, PyBOP, HBTU or DCC/HOBT, which produce the moderately activated benzotriazolyl esters from carboxylic acids, PyBroP is highly recommended for coupling *N*-methylated amino acids,[1,8–14] as illustrated by the difficult coupling of NMeVal (eq 1),[1,8] and other *N*-alkylated amino acids[15–17] as well as α-disubstituted amino acids (e.g. α-MeSer,[18] Aib[19] or α-MeCys (eq 2).[20]

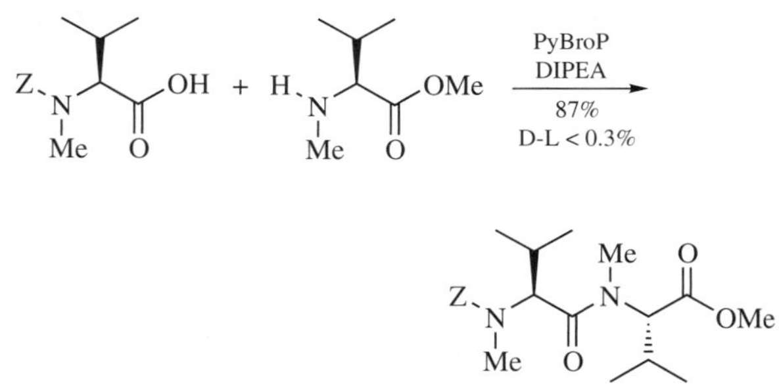

(1)

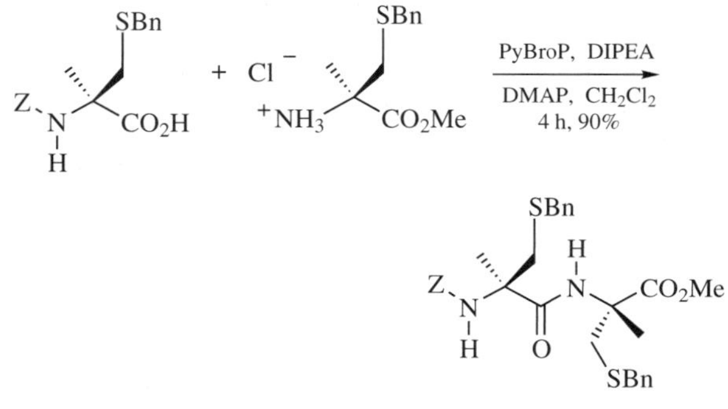

(2)

Coupling of *N*-alkylated amino acids can be improved by addition of HOAt.[21] The addition of a catalytic amount of **DMAP** or **4-pyrrolidinopyridine** is recommended to improve both the rate and the yield of the coupling reaction for non-epimerizable substrates.[10,13,18–20] PyBroP activation of Boc-amino acids can, however, give moderate yields due to the formation of the *N*-carboxyanhydride (NCA).[1,22]

Synthesis of Amides. PyBroP has been employed in the synthesis of unsubstituted amides,[23] *N*-monosubstituted amides (eq 3),[24] and disubstituted amides, even from conformationally constrained derivatives (eq 4)[25] and anilines.[19,26,27] It has also been used for activation of conjugated carboxylic acids (eq 5).[28]

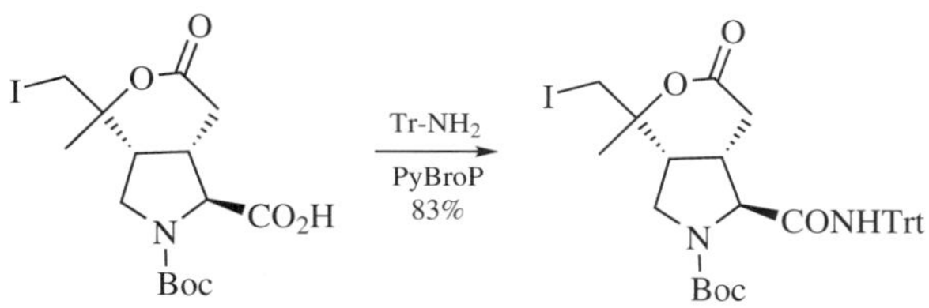

(3)

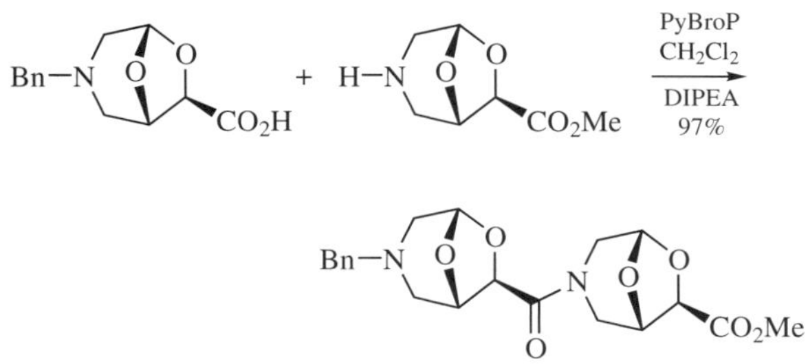

(4)

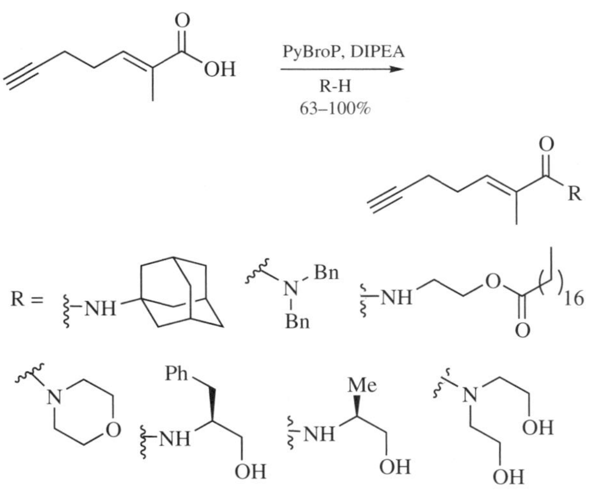

(5)

PyBroP also gives good results for the synthesis of macrocyclic lactams (eq 6),[29] including cyclic peptides at high dilution[13] with addition of HOAt[30] or on solid phase.[31]

Synthesis of Carboxylic Esters. Ester linkage from carboxylic acid activation with PyBroP was reported for primary and secondary alcohols. This reagent gives impressive results for depsipeptide synthesis (eq 7).[32] PyBroP has been used to attach nucleosides to a hydroxylated solid support.[33]

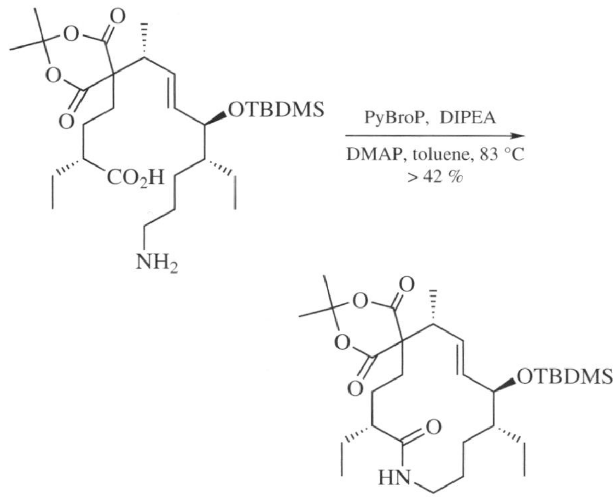

(6)

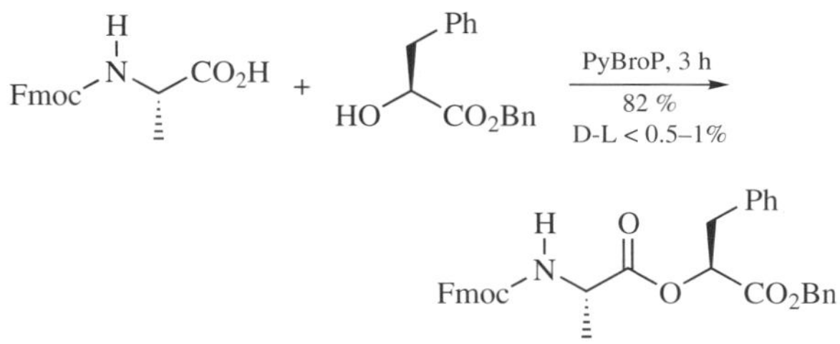

(7)

In addition, this activation was selected over carbodiimide-derived reagents for the preparation of polymer-bound activated esters derived from **1-hydroxybenzotriazole** (eq 8)[34,35] and for the preparation of the storable *p*-nitrophenyl esters in solution.[36]

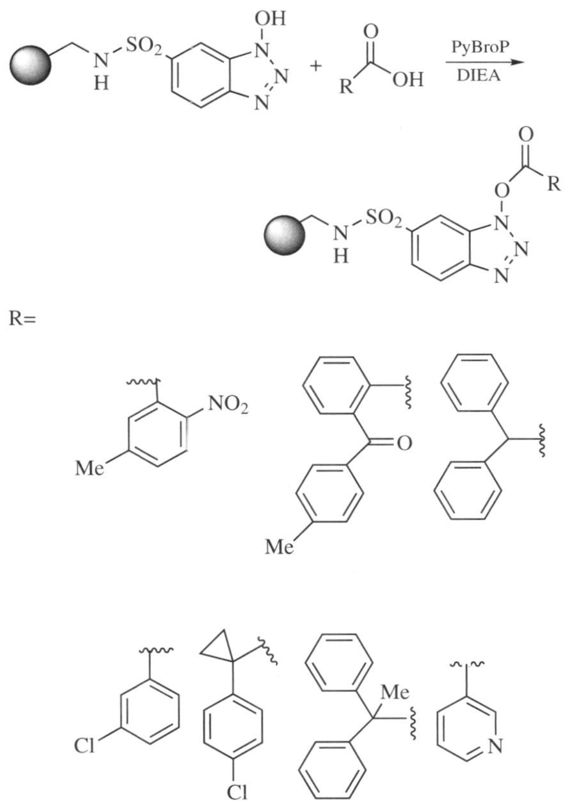

(8)

PyBroP gives improved yields for the cyclization step in the synthesis of the macrocyclic dynemicin A, when compared with the Yamaguchi macrolactonization protocol (eq 9).[37]

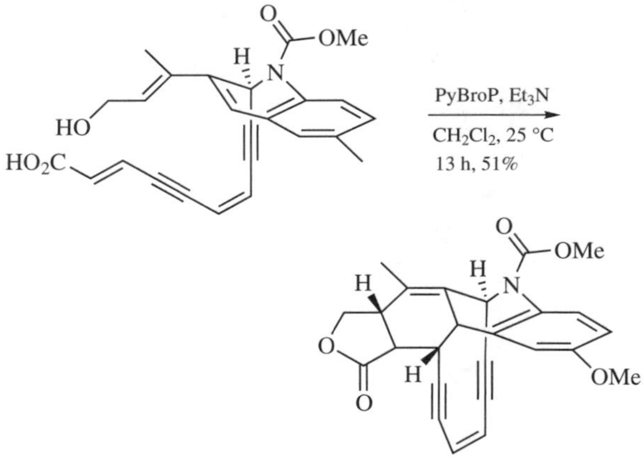

(9)

Synthesis of Amidines. Formamidinium formation has been reported as a side reaction of PyBroP with DMF in solid-phase peptide synthesis. This reaction proved to be suitable for the synthesis of formamidines from aliphatic or aromatic amines and DMF (eq 10).[6]

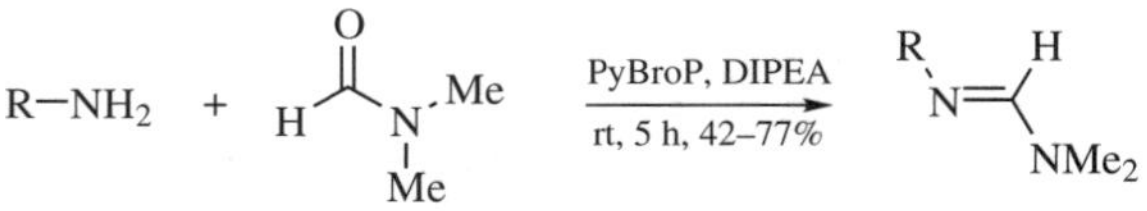

(10)

$R\text{-}NH_2$ = aniline (56%), benzyl amine (77%), benzene-1,3-diamine (53%), pyridine-2,6-diamine (48%)

Preparation of Coupling Reagents for Peptide Synthesis. Several phosphonium-related coupling reagents for peptide synthesis have been prepared from PyBroP (eq 11): PyBOP,[2] PyAOP,[38] and CF_3PyBOP.[39]

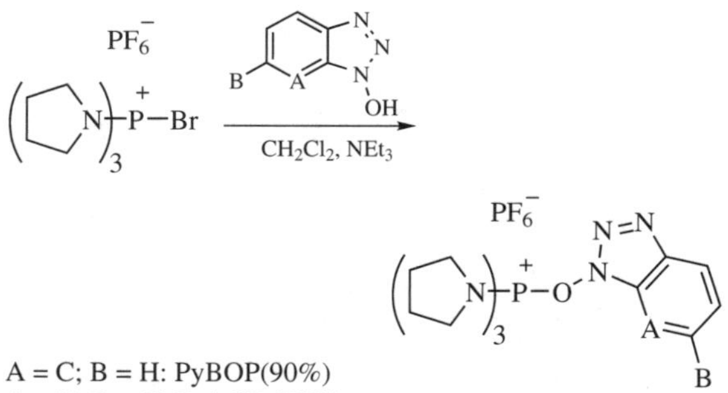

(11)

A = C; B = H: PyBOP(90%)
A = N; B = H: PyAOP (76%)
A = C; B = CF_3: CF_3PyBOP (69%)

Related Reagents. BroP, bromotris(dimethylamino)phosphonium hexafluorophosphate; PyCloP, chlorotripyrrolidinophosphonium hexafluorophosphate; BOP, 1*H*-benzotriazol-1-yloxytris(dimethylamino)phosphonium hexafluorophosphate; PyBOP, 1*H*-benzotriazol-1-yloxytripyrrolidinophosphonium hexafluorophosphate; AOP, 7-azabenzotriazol-1-yloxytris (dimethylamino)phosphonium hexafluorophosphate; PyAOP, 7-azabenzotriazol-1-yloxytripyrrolidinophosphonium hexafluorophosphate; HBTU, *O*-(benzotriazol-1-yl)-1,1,3,3-tetramethyluronium hexafluorophosphate; HATU, *O*-(7-azabenzotriazol-1-yl)-1,1,3,3-tetramethyluronium hexafluorophosphate; HBPyU, *O*-(benzotriazol-1-yl)-1,1,3,3-bis(tetramethylene) uronium hexafluorophosphate; HAPyU, *O*-(7-azabenzotriazol-1-yl)-1,1,3,3-bis(tetramethylene)uronium hexafluorophosphate; BOP-Cl, bis(2-oxo-3-oxazolidinyl)phosphinic chloride; PyClU, 1,1,3,3-bis(tetramethylene)chlorouronium hexafluorophosphate; TpyClU, 1,1,3,3-bis(tetramethylene)chlorouronium tetrafluoroborate; CIP, 2-chloro-1,3-dimethylimidazolidium hexafluorophosphate; TFFH, tetramethylfluoroformamidium hexafluorophosphate; BTFFH, bis(tetramethylene)fluoroformamidium hexafluorophosphate.

1. Coste, J.; Frérot, E.; Jouin, P., *J. Org. Chem.* **1994**, *59*, 2437.
2. Castro, B.; Coste, J., I Patent WO 90/10009, 1990.
3. Castro, B.; Dormoy, J. R., *Tetrahedron Lett.* **1973**, *35*, 3243.
4. Humphrey, J. M.; Chamberlin, A. R., *Chem. Rev.* **1997**, *97*, 2243.
5. Alsina, J.; Barany, G.; Albericio, F.; Kates, S. A., *Lett. Peptide Sci.* **1999**, *6*, 243.
6. Delarue, S.; Sergheraert, C., *Tetrahedron Lett.* **1999**, *40*, 5487.
7. Li, P.; Xu, J. C., *Tetrahedron* **2000**, *56*, 9949.
8. Coste, J.; Frérot, E.; Jouin, P.; Castro, B., *Tetrahedron Lett.* **1991**,*32*, 1967.
9. Ward, D. E.; Lazny, R.; Pedras, M. S. C., *Tetrahedron Lett.* **1997**, *38*, 339.
10. Andersen, R. J.; Coleman, J. E.; Piers, E.; Wallace, D. J., *Tetrahedron Lett.* **1997**, *38*, 317.
11. Mattern, R.-H.; Gunasekera, S. P.; McConnell, O., *Tetrahedron* **1996**, *52*, 425.
12. Mattern, R.-H.; Gunasekera, S. P.; McConnell, O. J., *Tetrahedron Lett.* **1997**, *38*, 2197.
13. Kurome, T.; Inami, K.; Inoue, T.; Ikai, K.; Takesako, K.; Kato, I.; Shiba, T., *Tetrahedron* **1996**, *52*, 4327.
14. Belshaw, P. J.; Schoepfer, J. G.; Liu, K.-Q.; Morrison, K. L.; Schreiber, S. L., *Angew. Chem., Int. Ed. Engl.* **1995**, *34*, 2129.
15. Kruijtzer, J. A. W.; Hofmeyer, L. J. F.; Heerma, W.; Versluis, C.; Liskamp, R. M. J., *Chem. Eur. J.* **1998**, *4*, 1570.
16. Cudic, M.; Horvat, J.; Elofsson, M.; Bergquist, K.-E.; Kihlberg, J.; Horvat, S., *J. Chem. Soc., Perkin Trans. 1* **1998**, 1789.
17. Schmitt, J.; Bernd, M.; Kutscher, B.; Kessler, H., *Bioorg. Med. Chem. Lett.* **1998**, *8*, 385.
18. Wipf, P.; Venkatraman, S., *Synlett* **1997**, 1.
19. Frérot, E.; Coste, J.; Dufour, M.-N.; Pantaloni, A.; Jouin, P., *Tetrahedron* **1991**, *47*, 259.
20. Walker, M. A.; Heathcock, C. H., *J. Org. Chem.* **1992**, *57*, 5566.
21. Reichwein, J. F.; Liskamp, R. M. J., *Tetrahedron Lett.* **1998**, *39*, 1243.
22. Frérot, E.; Coste, J.; Poncet, J.; Jouin, P.; Castro, B., *Tetrahedron Lett.* **1992**, *33*, 2815.
23. Parsons Jr, R. L.; Heathcock, C. H., *Tetrahedron Lett.* **1994**, *35*, 1383.
24. Artuso, F.; Sindona, G.; Athanassopoulos, C.; Stavropoulos, G.; Papaioannou, D., *Tetrahedron Lett.* **1995**, *36*, 9309.
25. Machetti, F.; Ferrali, A.; Menchi, G.; Occhiato, E. G.; Guarna, A., *Org. Lett.* **2000**, *2*, 3987.
26. Chodorowski-Kimmes, S.; Meyer, T. J., *Tetrahedron Lett.* **1997**, *38*, 3659.
27. Gibson, S. E.; Guillo, N.; Kalindjian, S. B.; Tozer, M. J., *Bioorg. Med. Chem. Lett.* **1997**, *7*, 1289.
28. Andrus, M. B.; Turner, T. M.; Asgari, D.; Li, W., *J. Org. Chem.* **1999**, *64*, 2978.

29. Trost, B. M.; Ceschi, M. A.; König, B., *Angew. Chem., Int. Ed. Engl.* **1997**, *36*, 1486.
30. Sutton, P. W.; Bradley, A.; Farras, J.; Romea, P.; Urpi, F.; Vilarrasa, J., *Tetrahedron* **2000**, *56*, 7947.
31. Davies, M.; Bradley, M., *Angew. Chem., Int. Ed. Engl.* **1997**, *36*, 1097.
32. Davies, J. S.; Howe, J.; Le Breton, M., *J. Chem. Soc., Perkin Trans. 2* **1995**, 2335.
33. Pon, R. T.; Yu, S.; Sanghvi, Y. S., *Bioconjugate Chem.* **1999**, *10*, 1051.
34. Pop, I. E.; Déprez, B. P.; Tartar, A. L., *J. Org. Chem.* **1997**, *62*, 2594.
35. Baxendale, I. R.; Ley, S. V., *Bioorg. Med. Chem. Lett.* **2000**, *10*, 1983.
36. Norman, T. C.; Gray, N. S.; Koh, J. T.; Schultz, P. G., *J. Am. Chem. Soc.* **1996**, *118*, 7430.
37. Wood, J. L.; Porco Jr, J. A.; Taunton, J.; Lee, A. Y.; Clardy, J.; Schreiber, S. L., *J. Am. Chem. Soc.* **1992**, *114*, 5898.
38. Hoeg-Jensen, T.; Holm, A.; Sorensen, H., *Synthesis* **1996**, 383.
39. Wijkmans, J. C. H. M.; Kruijtzer, J. A. W.; van der Marel, G. A.; van Boom, J. H.; Bloemhoff, W., *Recl. Trav. Chim. Pays Bas* **1994**, *113*, 394

Jacques Coste & Patrick Jouin
CNRS-UPR9023, Montpellier, France

1-*N*-(*t*-Butoxycarbonyl)-1*H*-benzotriazole-3*N*-oxide

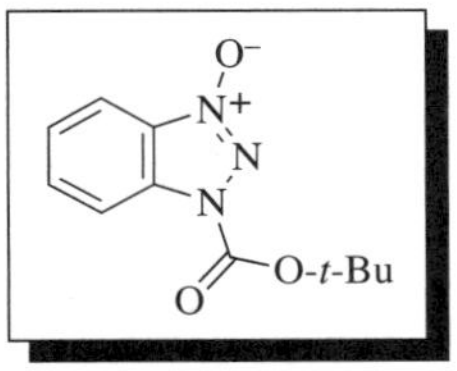

[111615-82-8] $C_{11}H_{13}N_3O_3$ (MW 235.27)

(a reagent for protection of amines with the *t*-butoxycarbonyl group[1])

Physical Data: mp 105 °C.

Solubility: sol DMF, CH_2Cl_2, and dioxane, among others.

Form Supplied in: the reagent is a white solid; however, it is not commercially available.

Handling, Storage, and Precautions: anhydrous conditions are not necessary for reactions, and the reagent can be stored at rt.

Introduction. The subject of this review was believed initially to be 1-*N*-(*t*-butoxycarbonyloxy)benzotriazole. However, a single-crystal X-ray analysis established the structure unequivocally as 1-*N*-(*t*-butoxycarbonyl)-1*H*-benzotriazole 3-*N*-oxide (**1**). It is likely that initially the carbonate, i.e. 1-*N*-(*t*-butoxycarbonyloxy)benzotriazole, is formed, and it then rearranges to the more stable carbamate (**1**). Two syntheses for (**1**), which actually had been originally targeted for 1-*N*-(*t*-butoxycarbonyloxy)benzotriazole, are reported. In the first synthesis, 1-Hydroxybenzotriazole is converted to its chloroformate, which in turn is allowed to react with *t*-butanol to give 1-*N*-(*t*-butoxycarbonyloxy)benzotriazole (eq 1).[1] In the second synthesis, 1-hydroxybenzotriazole is deprotonated by Triethylamine, and subsequent mixing with Di-t-butyl Dicarbonate affords the carbonate which rearranges to (**1**) (eq 2).[2]

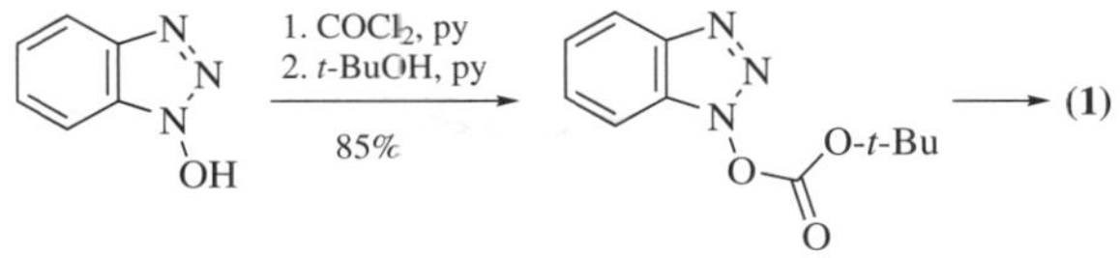

(1)

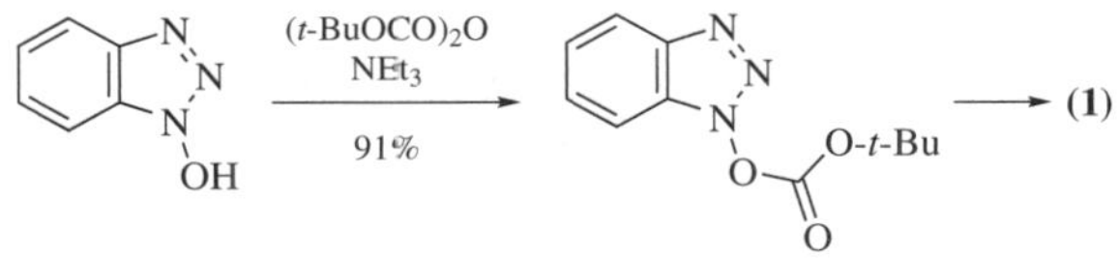

(2)

***t*-Butoxycarbonylation.** Reagent (**1**) is exceptionally effective in N-protection of amino acids. For example, when 1 equiv of the reagent was added to a solution of L-proline and triethylamine in aqueous dioxane, an essentially quantitative conversion to the corresponding N-Boc amino acid took place in under 1 min.[1] The reaction in aqueous DMF proceeds somewhat more sluggishly; however, the reaction reaches completion within 30 min. The speed of these reactions is in sharp contrast to the more commonly used reagents for *N*-*t*-butoxycarbonylation, such as Di-t-butyl Dicarbonate, 2-(t-Butoxycarbonyloxyimino)-2-phenylacetonitrile ('Boc-ON'), 1,2,2,2-tetrachloroethyl *t*-butyl carbonate, or t-Butyl Azidoformate, which usually require from a few hours to overnight reaction duration. The isolated yields for the products of the reactions of several typical amino acids with (**1**) have been reported in the range of 80–99% (eq 3).

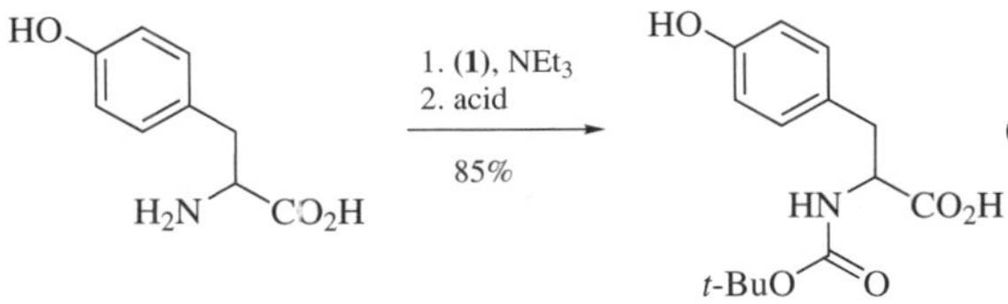

(3)

The reaction of (**1**) with simple amines, such as Piperidine or benzylamine, or even less reactive amines, such as aniline (eq 4) or Imidazole, have been reported.[1] In contrast to the reactions with amino acids, simpler amines appear to require somewhat longer reaction times.

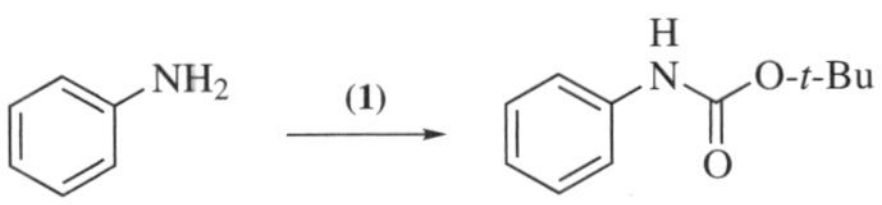

(4)

---

1. Kim, S.; Chang, H. *Chem. Commun./J. Chem. Soc., Chem. Commun.* **1983**, 1357.Kim, S.; Chang, H. *Bull. Korean Chem. Soc.* **1986**, *7*, 70.
2. Singh, J.; Fox, R.; Wong, M.; Kissick, T.; Moniot, J.; Gougoutas, J.; Malley, M.; Kocy, O. *J. Org. Chem.* **1988**, *53*, 205.

Ioannis Grapsas & Shahriar Mobashery
Wayne State University, Detroit, MI, USA

1-(*t*-Butoxycarbonyl)imidazole

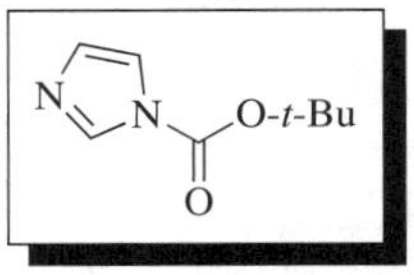

[49761-82-2] $C_8H_{12}N_2O_2$ (MW 168.20)

(a reagent for the introduction of the *t*-butoxycarbonyl group at nucleophilic sites[1])

Physical Data: mp 44.5–47 °C; bp 64 °C/1 mmHg.
Solubility: sol CH_2Cl_2, $CHCl_3$, THF, hexane, acetone, and DMF, among others.
Form Supplied in: not commercially available.
Preparative Method: 1-(*t*-butoxycarbonyl)imidazole (**1**) can be prepared from the reaction of imidazole and 1,2,2,2-tetrachloroethyl-*t*-butyl carbonate[1a] or the reaction of *t*-butanol and *N,N'*-carbonyldiimidazole[1b] in high yield. The reagent has been reported to be stable at rt.[2]
Handling, Storage, and Precautions: anhydrous conditions are recommended.

***t*-Butoxycarbonylation.** The amine functions of various amino acids have been protected with the *t*-butoxycarbonyl ('Boc') group by (**1**) (eq 1).[1b,3] Polar anhydrous organic solvents have been used in general, although traces of moisture can be tolerated.

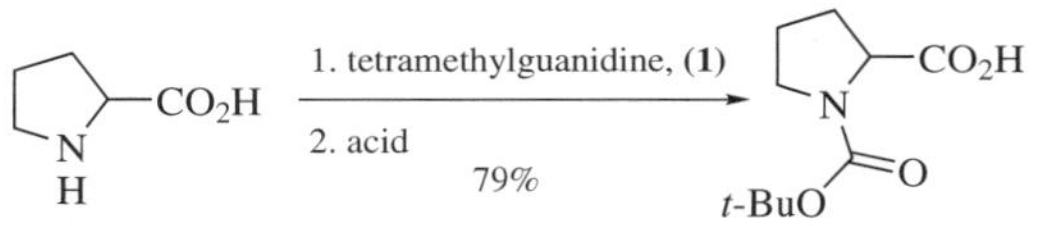
(1)

Alkylation of (**1**), or its silver perchlorate complex, produces water-soluble imidazolium salts with enhanced electrophilic character (eq 2).[4] The perchlorate and tetrafluoroborate salts of this type are remarkably stable and they have been used to protect amino acids in aqueous media (eq 3).[4]

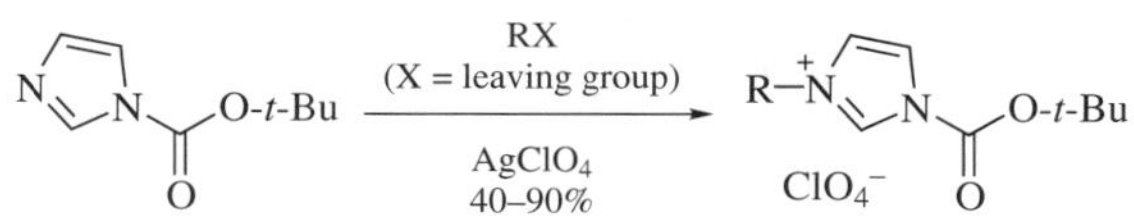
(2)

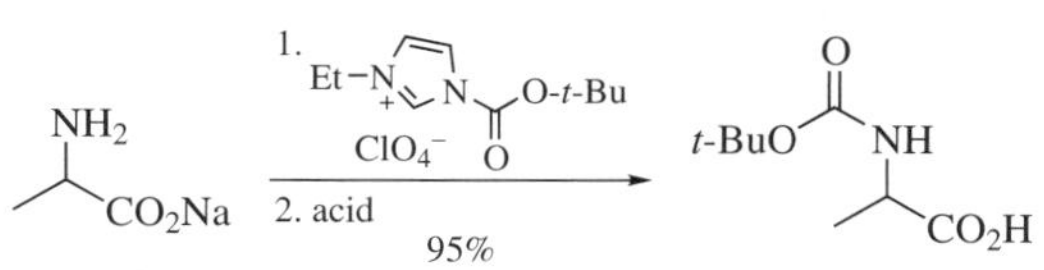
(3)

Selective protection of a single amine in alkyl diamines by (**1**) is reported (eq 4).[5] Furthermore, the reaction of Hydrazine with (**1**) affords *t*-butoxycarbonylhydrazide, which is a useful reagent in peptide chemistry.[1b]

Incorporation of the Boc group at the aromatic nitrogen of the pyridine ring system (eq 5)[6] and at a carbon nucleophile (eq 6)[7] by (**1**) has also been reported.

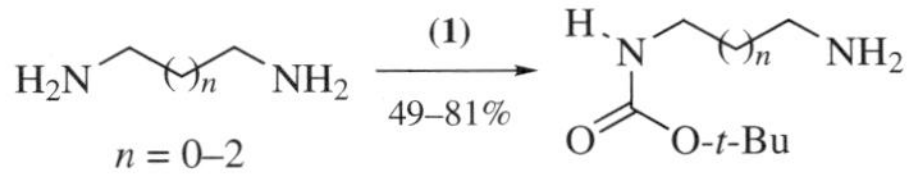
(4)

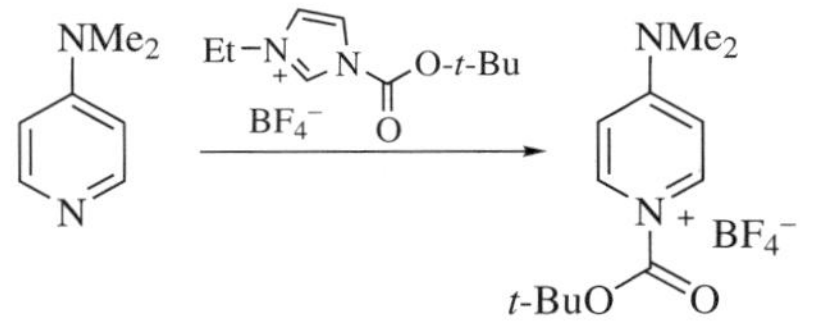
(5)

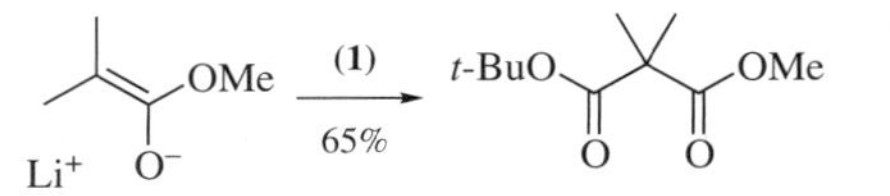
(6)

Formation of O-Alkylperoxycarbonic Acids. Epoxidation of alkenes in a biphasic system could be performed by in situ generation of *O*-alkylperoxycarbonic acids (eq 7).[2] The reaction is typically carried out by the addition of 35% Hydrogen Peroxide to a solution of (**1**) and the alkene substrate in an organic solvent. The epoxide of cholesteryl acetate was prepared quantitatively in a 3:1 ratio of α- to β-epoxide by this method.

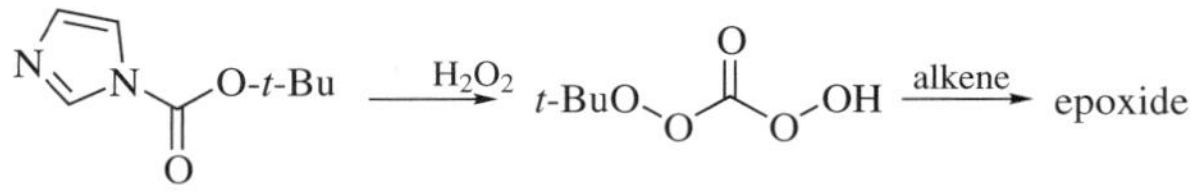
(7)

Related Reagents. 1-*t*-Butoxycarbonyl-1,2,4-triazole.

---

1. Barcelo, G.,; Senet, J. P.; Sennyey, G. *J. Org. Chem.* 1985,*50*, 3951. (b) Klee, W.; Brenner, M. *HCA* **1961**, 2151.
2. Tsunokawa, Y.; Iwasaki, S.; Okuba, S. *Tetrahedron Lett.* **1982**, *23*, 2113.
3. Chou, C. Y. U. S. Patent 4837332, **1989** (*Chem. Abstr.* **1988**, *109*, 6966k).
4. Guibé-Jampel, E.; Bram, G.; Wakselman, M.; Vilkas, M. *Synth. Commun.* **1973**, 111.
5. Devan, P. B.; Hertzberg, R. P.U.S. Patent 4 942 227, **1990** (*Chem. Abstr.* **1987**, *107*, 232 672b).
6. Wakselman, M.; Guibé-Jampel, E. *Pept., Proc. Eur. Pept. Symp., 14th* **1976**, 131.
7. Bourgeois, M. J.; Montaudon, E.; Campagnole, M.; Maillard, B. *Bull. Soc. Chim. Belg.* **1982**, *91*, 871.

Ioannis Grapsas & Shahriar Mobashery
Wayne State University, Detroit, MI, USA

2-(*t*-Butoxycarbonyloxyimino)-2-phenylacetonitrile

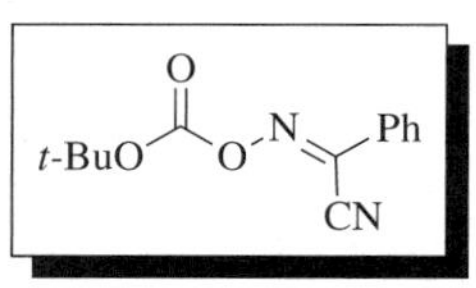

[58632-95-4] $C_{13}H_{14}N_2O_3$ (MW 246.27)

(reagent for protection of amino acids as *t*-butyl carbamates[1–3] and for protection of alcohols as *t*-butyl carbonates[4])

Alternate Name: Boc-ON.
Physical Data: mp 87–89 °C.
Solubility: very sol ethyl acetate, ether, benzene, chloroform, dioxane and acetone; sol methanol, 2-propanol, and *t*-butanol; insol petroleum ether and water.
Form Supplied in: solid.
Handling, Storage, and Precautions: contact with the reagent may cause irritation. The toxicological properties of the reagent have not been thoroughly investigated. The reagent should be stored in a brown bottle at −20 °C to prolong shelf life. After several weeks at rt the reagent undergoes gradual decomposition with evolution of CO_2.

Amine Protection. The reagent, informally referred to as Boc-ON, has been widely employed for the protection of amines as their *t*-butoxycarbonyl (Boc) derivatives. The original description of Boc-ON reports its utility for the N-protection of α-amino acids (eq 1),[1] and this use is still the major application.[5]

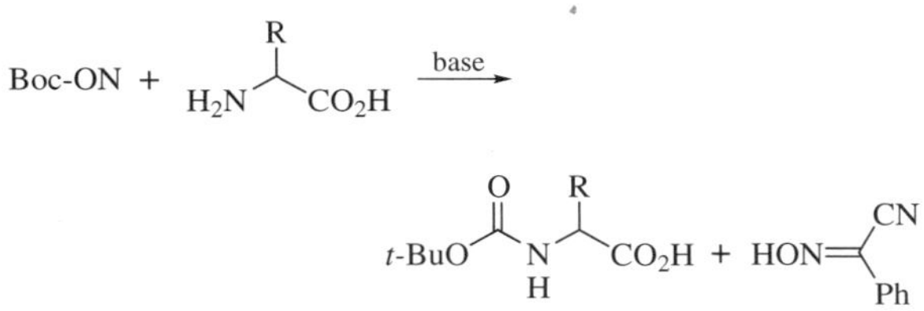

(1)

The reaction is generally carried out using 1.1 equiv of Boc-ON and 1.5 equiv of Triethylamine in either 50% aqueous dioxane or 50% aqueous acetone. Boc derivatives are obtained in high yields after 4–5 h at 20 °C or 1 h at 45 °C. The oxime byproduct (eq 1) can be easily and completely removed from the reaction mixture by extraction with ether, ethyl acetate, or benzene. To secure high purity and high yield, distillation of the dioxane or acetone prior to extraction of the oxime byproduct is recommended, though direct extraction may be feasible in most cases. Yields for various amino acids range from good to virtually quantitative (Table 1).

Table 1 Examples of Boc-Amino Acids Prepared with Boc-ON[2]

Boc-amino acid	Reaction time (h)	Oxime extraction solvent	Yield (%)
Ala-OH	4	Ether	80
Asn-OH	20	EtOAc	86
$Glu(OH)_2$	3	EtOAc	78
Gly-OH	2	EtOAc	87
Leu-OH·$0.5H_2O$	3	Ether	99
Met-OH·DCHA[a]	3	EtOAc	82
Phe-OH·DCHA[a]	5	Ether	98
Pro-OH	1.5	Benzene	88
Trp-OH	3	EtOAc	99

[a]DCHA = dicyclohexylamine.

Boc-ON offers a distinct advantage over *t*-Boc azide which can require reaction temperatures of 50–60 °C.[6] In addition, *t*-Boc azide is thermally unstable and decomposes with apparent detonation at temperatures above 80 °C. Boc-ON is comparable to Di-t-butyl Dicarbonate in terms of reaction times, yields, convenience, availability, and work-up procedure.[7]

Boc-ON has also been employed for the protection of secondary amines (eqs 2 and 3).[8,9]

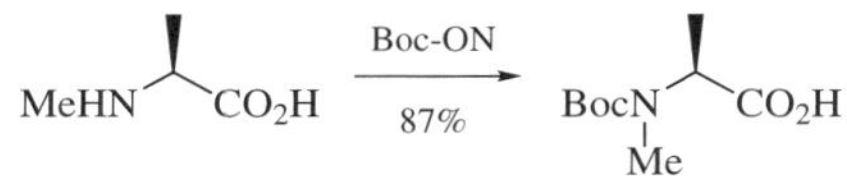

(2)

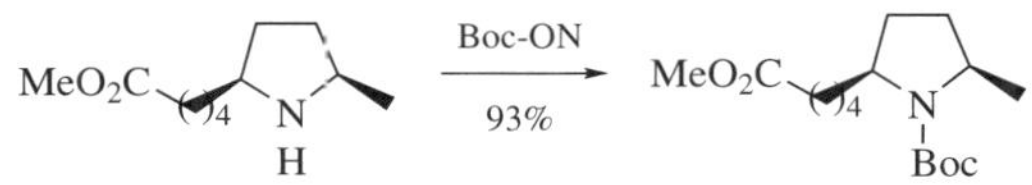

(3)

The reagent will selectively protect the amino group in the presence of alcohols and phenols as well as carboxylic acids. Examples include the Boc protection of a kanamycin A derivative (eq 4)[10] and of the glycopeptide antibiotic A41030A (eq 5).[11]

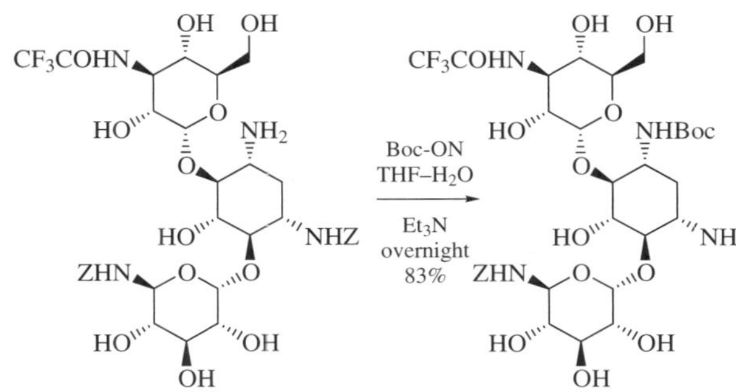

(4)

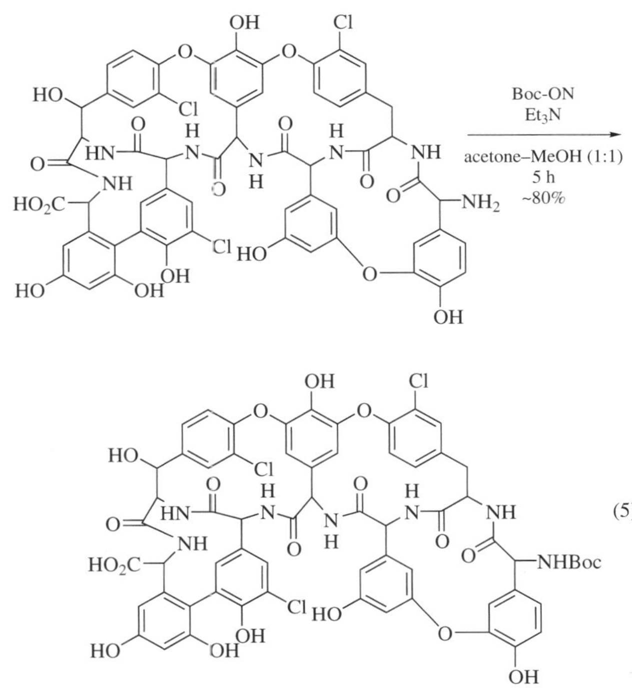

(5)

In the case of diamino acids, such as ornithine or lysine, Boc-ON has been used to effect selective protection of N^W at pH 11 (eqs 6 and 7).[12,13]

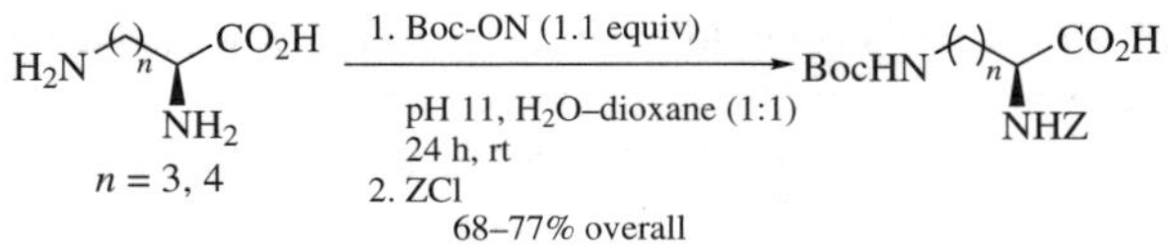

(6)

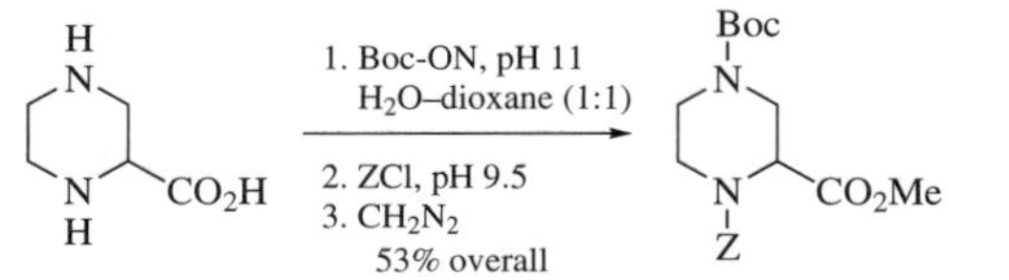

(7)

This selective protection has also been accomplished with copper-chelated diamino acids (eq 8).[14]

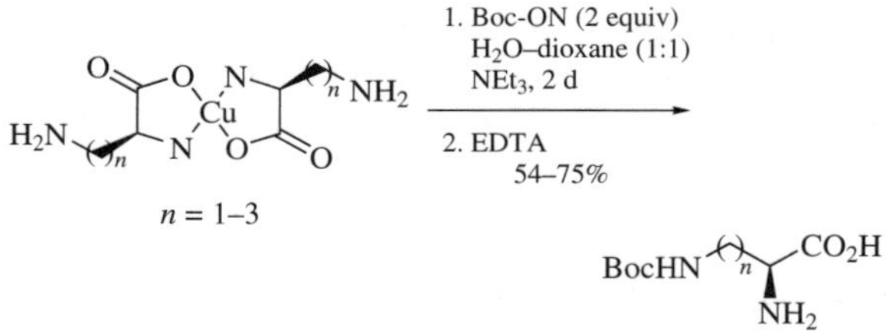

(8)

Alcohol Protection. Boc-ON has also been employed to protect alcohols as their *t*-butyl carbonates.[4,15,16] For this application the alcohol must first be deprotonated to its alkoxide conjugate base (eqs 9 and 10).

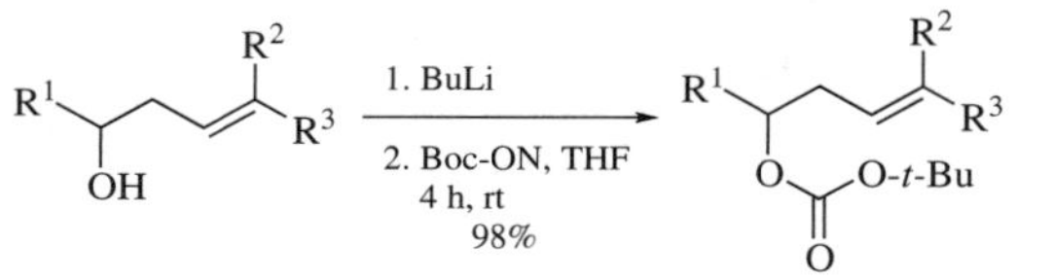

(9)

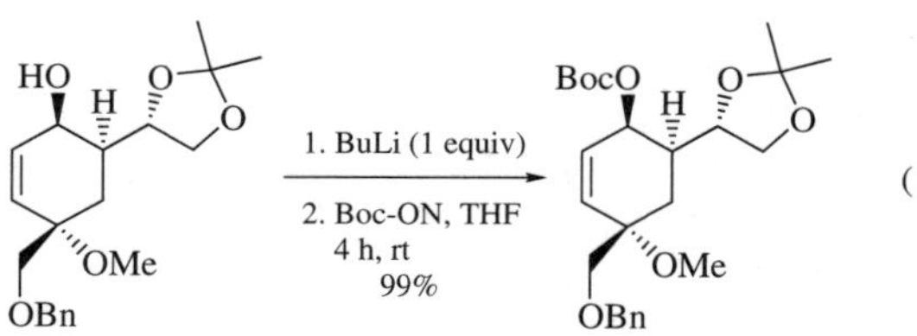

(10)

Related Reagents. *N*-(*t*-Butoxycarbonyloxy)phthalimide; *N*-(*t*-Butoxycarbonyloxy)succinimide; *t*-Butyl Azidoformate; Di-*t*-butyl Dicarbonate.

1. Itoh, M.; Hagiwara, D.; Kamiya, T. *Tetrahedron Lett.* **1975**, *49*, 4393.
2. Itoh, M.; Hagiwara, D.; Kamiya, T. *Bull. Chem. Soc. Jpn.* **1977**, *50*, 718.
3. Itoh, M.; Hagiwara, D.; Kamiya, T. *Q. Rev., Chem. Soc.* **1980**, *59*, 95.
4. Barlett, P. A.; Meadows, J. D.; Ottow, E. *J. Am. Chem. Soc.* **1984**, *106*, 5304.
5. *Fieser & Fieser* **1977**, *6*, 91.
6. *Q. Rev., Chem. Soc.* **1977**, *57*, 122.
7. Persio, G.; Piani, S.; De Castiglione, R. *Int. J. Peptide Protein Res.* **1983**, *21*, 227.
8. Galardy, R. E. *Biochemistry* **1987**, *21*, 5777.
9. Vaurecka, M.; Hesse, M. *Helv. Chim. Acta* **1989**, *72*, 847.
10. Matsuda, K.; Yasuda, N.; Tsutsumi, H.; Takaya, T. *J. Antibiot.* **1985**, *38*, 1719.
11. Hunt, A. H.; Dorman, D. E.; Debono, M.; Molloy, R. M. *J. Org. Chem.* **1985**, *50*, 2031.
12. Heimer, E. P.; Wang, C.-T.; Lambros, T. J.; Felix, A. M. *Org. Prep. Proced. Int.* **1983**, *15*, 379.
13. Bigge, C. F.; Hays, S. J.; Novak, P. M.; Drummond, J. T.; Johnson, G.; Bobovski, T. P. *Tetrahedron Lett.* **1989**, *30*, 5193.
14. Rosowsky, A.; Freisham, J. H.; Moran, R. G.; Solan, V. C.; Bader, H.; Wright, J. E.; Radike-Smith, M. *J. Med. Chem.* **1986**, *29*, 655.
15. Bartlett, P. A.; Meadows, J. D.; Brown, E. G.; Morimoto, A.; Jernstedt, K. K. *J. Org. Chem.* **1982**, *47*, 4013.
16. Martin, S. F.; Zinke, P. W. *J. Org. Chem.* **1991**, *56*, 6600.

Michael S. Wolfe
National Institutes of Health, Bethesda, MD, USA

Jeffrey Aubé
University of Kansas, Lawrence, KS, USA

t-Butyl Chloroformate

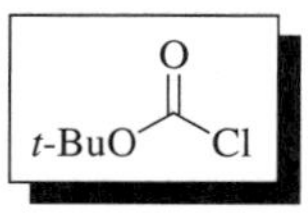

[24608-52-4] $C_5H_9ClO_2$ (MW 136.58)

(reagent for the protection of amino groups, including amino acids[2])

Alternate Name: Boc-Cl.
Physical Data: bp 3–4 °C/0.9–1.7 mmHg.[1]
Solubility: sol ether, toluene, and most organic aprotic solvents.
Form Supplied in: not commercially available; should be prepared from *t*-butanol[2] or potassium *t*-butoxide[1,3] and phosgene shortly before use.
Handling, Storage, and Precautions: unstable at rt. *t*-Butyl chloroformate should be kept as a solution in anhydrous solvent at low temperature (deep freezer) and for a period of time not exceeding a few days. Decomposition leads to high pressure in the storage vessel. May contain some residual phosgene. Reacts with DMSO or DMF.

Protection of Amino Groups. *t*-Butyl chloroformate (Boc-Cl), first prepared by Choppin and Rogers in 1948,[1] has been used, despite the reagent's instability, for the *t*-butyloxycarbonyl (Boc) protection of amino compounds and especially of amino acids. Boc protection has gained considerable importance because of the resistance of the protecting group to strong base hydrolysis and catalytic hydrogenation, as well as the ease of deprotection under mildly acidic conditions. Boc protection has become a fundamental tool of modern peptide synthesis and particularly of the Merrifield strategy for solid-phase peptide synthesis. Most amino acids have been protected with Boc-Cl using slightly modified Schotten–Baumann conditions (eq 1).[2]

Reaction of pyridine compounds with Boc-Cl and Sodium Borohydride in methanol at −65 °C afforded the corresponding *N*-*t*-butyloxycarbonyldihydropyridine derivatives in good yield (eq 2).[4]

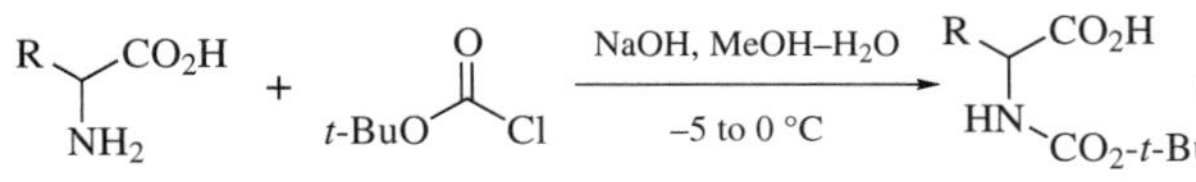

(1)

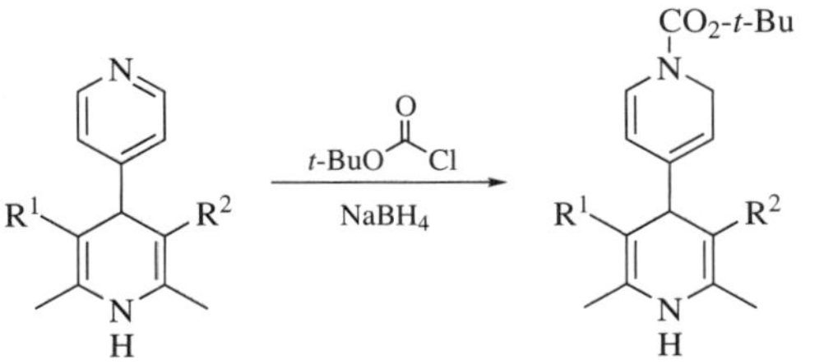

(2)

The main drawback of the reagent remains its very low stability and the difficulties encountered in its preparation. Various preparative procedures are currently available in the literature.[5] It is also necessary to determine the reagent concentration in solution accurately, which can be done by reacting an aliquot with a selected amino compound before use. These liabilities notwithstanding, *t*-butyl chloroformate can be a cheap and valuable reagent for Boc protection and in some cases it is the best reagent available. This is, for instance, the case in the synthesis of Boc-protected *N*-carboxy anhydrides, which are used as building blocks in peptide synthesis (eq 3).[6] Di-t-butyl Dicarbonate required much longer reaction times in this case. These Boc-UNCAs (urethane-protected *N*-carboxy anhydrides) are, in fact, best prepared from the reaction of Boc-Cl with Leuchs anhydride in the presence of *N*-methylmorpholine or any nonnucleophilic organic base.

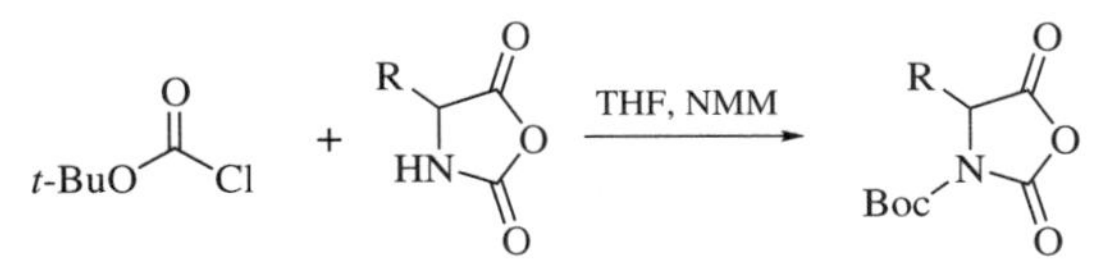

(3)

Preparation of Other Reagents for Boc Protection. Because of its low stability, *t*-butyl chloroformate has mainly been used as a source of various other reagents that are easier to handle under ordinary conditions. t-Butyl Azidoformate, the most widely employed reagent before the discovery of Di-t-butyl Dicarbonate, is conveniently obtained from Boc-Cl through the formation of *t*-butyl carbazate (eq 4).[7]

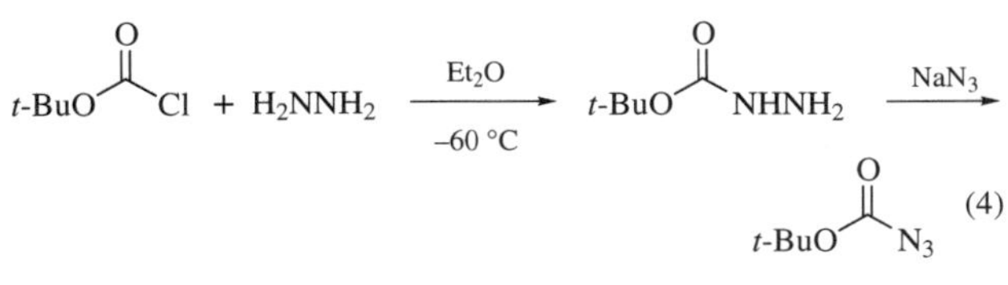

(4)

t-Butyl azidoformate is also obtained more directly from Boc-Cl and hydrazoic acid (eq 5).[8]

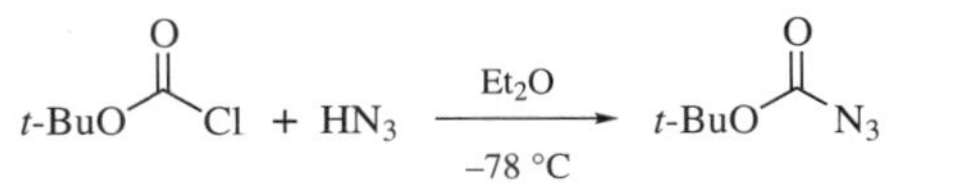

(5)

Various active carbonates, which are primarily used for the preparation of stable reagents for the introduction of the Boc protection, have been prepared from Boc-Cl.[9]

Exchange reactions utilizing hydrogen fluoride[10] and various salts such as sodium[11] and silver fluoride do not lead to the more stable *t*-butyl fluoroformate, which has to be prepared instead from carbonyl chlorofluoride or fluorophosgene.[12]

***t*-Butyl Carbonates and Carbamates.** Because Boc-Cl is potentially the cheapest reagent for the synthesis of *t*-butyl carbonates, e.g. (**1**) and (**2**), or carbamates, it has been claimed in numerous patents and may have growing utility in industrial processes, such as for the production of carbonates as components in thermal recording materials.[13]

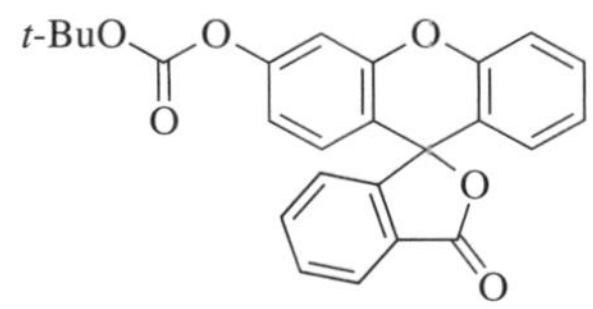

(1)

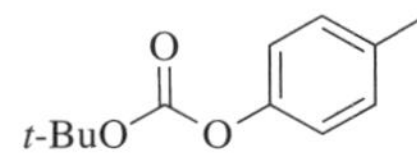

(2)

1. Choppin, A. R.; Rogers, J. W. *J. Am. Chem. Soc.* **1948**, *70*, 2967.
2. Sakakibara, S.; Shin, M.; Fujino, Y.; Shimonishi, Y.; Inouye, S.; Inukai, N. *BCS* **1969**, *42*, 809.
3. Howard, J. C. *J. Org. Chem.* **1981**, *46*, 1720.
4. Dagnino, L.; Li-Kwong-Ken, M. C.; Wynn, H.; Wolowyk, M. W.; Triggle, C. R.; Knaus, E. E. *J. Med. Chem.* **1987**, *30*, 640.
5. Ger. Patent 2 108 782 (*Chem. Abstr.* **1967**, *76*, 24 715).
6. Fuller, W. D.; Cohen, M. P.; Shabankareh, M.; Blair, R. B.; Goodman, M.; Naider, F. *J. Am. Chem. Soc.* **1990**, *112*, 7414.
7. Ovchinnikov, Y. A.; Kiryushkin, A. A.; Miroshnikov, A. I. *E* **1965**, *21*, 418.
8. Yajima, H.; Kawatani, H. *Chem. Pharm. Bull.* **1968**, *16*, 183.
9. Rzeszotarska, B.; Wiejak, S.; Pawelczak, K. *Org. Prep. Proced. Int.* **1973**, *5*, 71.
10. Olah, G. A.; Kuhn, S. J. *J. Org. Chem.* **1961**, *26*, 237.
11. Olah, G. A.; Kuhn, S. J. *Ber. Dtsch. Chem. Ges./Chem. Ber.* **1956**, *89*, 862.
12. Schnabel, E.; Herzog, H.; Hoffmann, P.; Klauke, E.; Ugi, I. *Angew. Chem., Int. Ed. Engl.* **1968**, *7*, 380.
13. Jpn. Patent 04 213 368 (*Chem. Abstr.* **1993**, *118*, 14 048).

G. Sennyey
SNPE, Vert-le-Petit, France

C

Cadmium Carbonate

[513-78-0] $CCdO_3$ (MW 172.42)

(used in the preparation of aryl glycosides[1–8] and also for cleavage of thioacetals[9])

Physical Data: mp 357 °C; *d* 4.258 g cm^{-3}.
Solubility: sol concentrated ammonia and dilute acids.
Form Supplied in: white powder and chunks.
Handling, Storage, and Precautions: should be stored in a cool dry place. It is a carcinogen and hence should be handled carefully.

Synthesis of Glycosides. $CdCO_3$ has been used as a catalyst in the Koenigs–Knorr synthesis of aryl glycosides.[1–3] Better yields are reported with $CdCO_3$ than with the usual catalyst, silver carbonate. For example, the 3-β-D-glucuronide triacetate methyl ester of estrone was prepared in 71% yield (eq 1), compared to 7% yield using silver carbonate.[1] Similar preparations of the β-D-glucuronides of 17β-estradiol,[1] estriol,[1,4] equilin,[1] equilenin,[1] phenolphthalein,[5] cholesterol,[6] and anthracyclines[7] are documented in the literature.

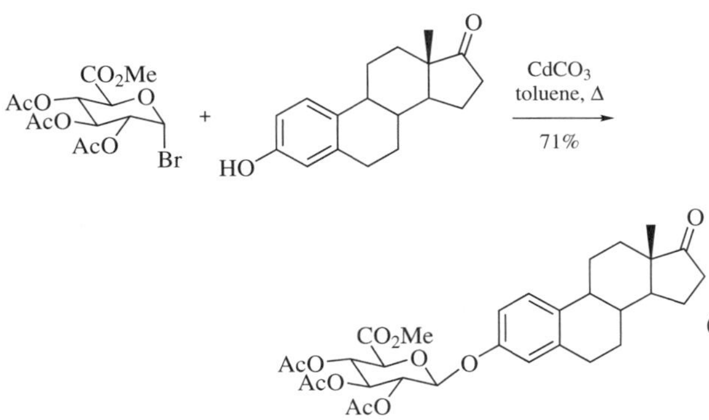

(1)

This methodology has also been used to prepare 2-aminoethyl glycosides (eq 2) for the preparation of carbohydrate–protein conjugates.[8]

Cleavage of Thioacetals. Thioacetals of aldosugars are cleaved by $CdCO_3$ in the presence of $HgCl_2$ (see Mercury(II) Chloride–Cadmium Carbonate). This method has been used to prepare acyclic derivatives of monosaccharides (eq 3).[9]

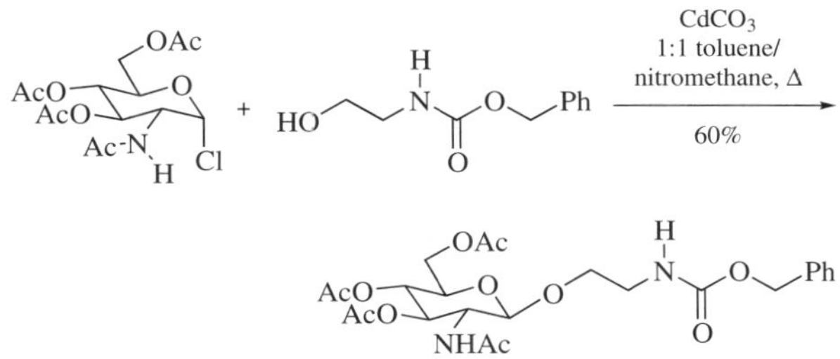

(2)

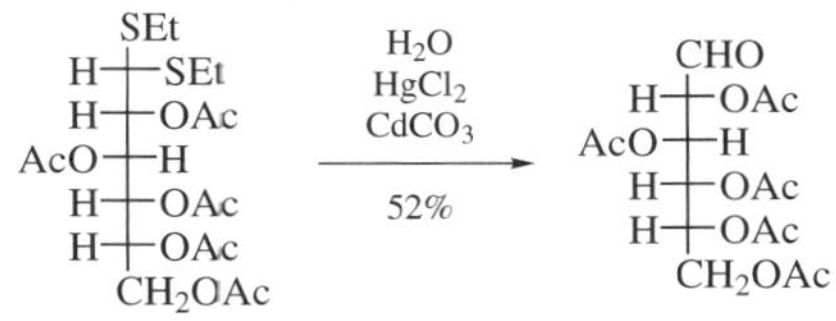

(3)

---

1. (a) Conrow, R. B.; Bernstein, S. *J. Org. Chem.* **1971**, *36*, 863. (b) Fieser, M.; Fieser, L. F. *Fieser & Fieser* **1974**, *4*, 67.
2. Dick, W. E. *Carbohydr. Res.* **1979**, *70*, 313.
3. Wallace, J. E.; Schroeder, L. R. *J. Chem. Soc., Perkin Trans. 1* **1976**, 1938.
4. Shimada, K.; Xie, F.; Nambara, T. *Chem. Pharm. Bull.* **1986**, *34*, 179.
5. Nambara, T.; Takizawa, N.; Goto, J.; Shimada, K. *Chem. Pharm. Bull.* **1976**, *24*, 2869.
6. Goto, J.; Suzaki, K.; Nambara, T. *Chem. Pharm. Bull.* **1979**, *27*, 1926.
7. Lown, J. W.; Sondhi, S. M.; Mandal, S. B.; Murphy, J. *J. Org. Chem.* **1982**, *47*, 4304.
8. King, R. R.; Cooper, F. P.; Bishop, C. *Carbohydr. Res.* **1977**, *55*, 83.
9. (a) Wolfrom, M. L. *J. Am. Chem. Soc.* **1929**, *51*, 2188. (b) Wolfrom, M. L. *J. Am. Chem. Soc.* **1930**, *52*, 2464. (c) Wolfrom, M. L.; Newlin, M. R.; Stahly, E. E. *J. Am. Chem. Soc.* **1931**, *53*, 4379. (d) Wolfrom, M. L.; Weisblat, D. I.; Zophy, W. H.; Waisbrot, S. W. *J. Am. Chem. Soc.* **1941**, *63*, 201.

Prabu Prabhakaran
Bristol-Myers Squibb Pharmaceutical Research Institute, New Brunswick, NJ, USA

Carbomethoxysulfenyl Chloride

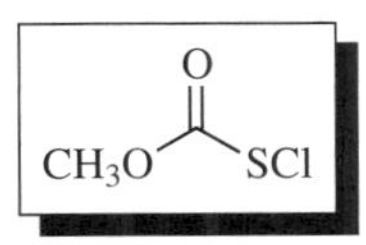

[26555-40-8] $C_2H_3ClO_2S$ (MW 126.56)

(peptide synthesis by prior thiol capture, solid phase peptide synthesis, thiol protective groups, formation of *S*-aryl thiocarbonates via Friedel-Crafts, preparation of unsymmetrical disulfides, formation of trisulfides)

Alternate Name: methoxycarbonylsulfenyl chloride, *S*-chloro-*O*-methylthiocarbonate.

Introduction. Methoxycarbonylsulfenyl chloride (Scm-Cl),[1,2] easily prepared from equimolar amounts of methanol and chlorocarbonylsulfenyl chloride (eq 1),[2,3] finds many applications, based on the reactivity of alkoxycarbonylsulfenyl chlorides against thiols.[1] Brois and co-workers[4] realized the value of methoxycarbonylsulfenyl chloride in the synthesis of asymmetric disulfides, which led to important applications in the chemistry of cysteine and its derivatives,[5,6] particularly cysteinyl peptides.[7,8] Other applications include preparation of aromatic thiols, symmetric trisulfides, etc. The applicability of Scm-Cl surpasses its drawbacks which include relatively short storage life (1–2 months at −20 °C in sealed ampules)[9] and relatively rapid hydrolysis in the protic media that some of its most important uses require.

$$\text{ClS-C(O)-Cl} \xrightarrow{\text{ROH}} \text{ClS-C(O)-OR} + \text{RSH} \xrightarrow{0\,^{\circ}\text{C}}$$

$$\text{RSS-C(O)-OR} + \text{R'SH} \longrightarrow$$

$$\text{RSSR'} + \text{COS} + \text{ROH} \quad (1)$$

Formation of Unsymmetrical Disulfides. This facile protocol (eq 1)[4,10] proceeds through sulfenyl thiocarbonate intermediates obtained in very good yields from equimolar amounts of thiols and alkoxycarbonylsulfenyl chlorides (of which methoxycarbonylsulfenyl chloride is widely used) at low temperature (generally 0–10 °C), in solvents such as methanol, chloroform, dichloromethane, and dioxane. An oxygen-free atmosphere of nitrogen is recommended.[3,11]

The thiol mediated heterolytic fragmentation of the sulfenyl thiocarbonates occurs at ambient temperature or lower, and most of the common functional groups are compatible with the procedure. The fragmentation is catalyzed by small amounts of tertiary amines (e.g., TEA).[4] Mechanistically, both cyclic and noncyclic transition states could be proposed for the fragmentation.[4]

A particular case is the use of a hydrogensulfide salt instead of the thiol, to effect the fragmentation. Treatment of the thiocarbonate derived from cysteine with 2 equiv of potassium hydrogensulfide in methanol results in the formation of the potassium salt of thiocysteine in 85% yield (eq 2).[12]

$$\text{HSCH}_2\text{CH(CO}_2^-\text{)NH}_3^+ \xrightarrow{\text{Scm-Cl}}$$

$$\text{CH}_3\text{OC(O)SSCH}_2\text{CH(CO}_2\text{H)NH}_3^+\,\text{Cl}^- \xrightarrow[\text{2 equiv KSH}]{-\text{COS},\ -\text{MeOH},\ -\text{HCl}}$$

$$\text{H}_3\text{N}^+\text{(CO}_2^-\text{)CHCH}_2\text{SS}^-\,\text{K}^+ \quad (2)$$

When the substrate thiol is suitably substituted in such a way that a nucleophile is present in the γ position with respect to the thiol, the sulfenyl thiocarbonate intermediate derived from the thiol and Scm-Cl may cyclize in the absence of acid via intramolecular nucleophilic attack at the pre-existing thiol S (with elimination of COS and methanol). The final product forms through the intermolecular nucleophilic attack of a second thiol at the same sulfur atom, with displacement of the "internal" nucleophile.[13]

Applications of the Asymmetric Disulfides in the Chemistry of Cysteine and Cysteinyl Peptides.[14] The Brois protocol can be applied[7a] to the formation of the disulfide bridges in unsymmetrical cysteine peptides, e.g., human insulin,[7b] vasopressin, and oxytocin.[15] In the Scm method (eq 3), a substrate in which the thiol of the cysteine is protected as a trityl (Tr) or acetamidomethyl (Acm) derivative reacts with Scm-Cl to generate a sulfenyl thionocarbonate. The second thiol group, after removal of its Tr protecting group, reacts with the sulfenyl thionocarbonate to induce its fragmentation, with the corresponding formation of the asymmetric disulfide bridge.[7b]

$$\text{R}_1\text{-S-Tr or R}_1\text{-S-Acm} \xrightarrow{\text{ClS-C(O)-OCH}_3} \text{R}_1\text{-S-S-C(O)-OCH}_3$$

$$\text{R}_2\text{-S-Tr} \xrightarrow{\text{1. Hg(OAc)}_2\text{; 2. H}_2\text{S}} \text{R}_2\text{-SH}$$

$$\longrightarrow \text{R}_1\text{-S-S-R}_2 + \text{COS} + \text{ROH} \quad (3)$$

This is exemplified by the construction of the disulfide bridge between A20 and B19 in human insulin (eq 4).

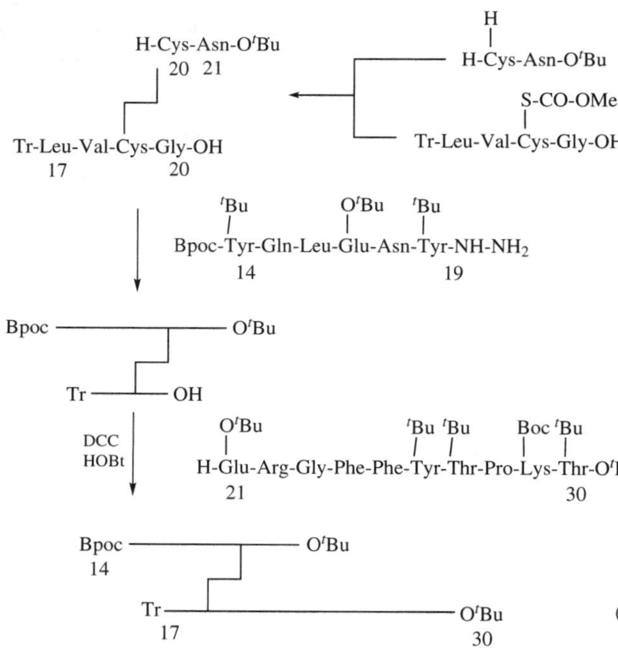

(4)

The reaction of cysteine hydrochloride H-Cys(H)-OH · HCl with 2 equiv Scm-Cl in methanol (1 h) provides the HCl salt of Scm-protected cysteine, H-Cys(Scm)-OH · HCl. The presence of excess HCl effectively prevents the nucleophilic attack by the amine function of cysteine to Scm-Cl; cystine is a by-product if less Scm-Cl is used.[16] The intermediate reacts with excess alkyl thiols and 1 equiv triethylamine in methanol or methanol-chloroform to afford *S*-alkylsulfenylcysteines, H-Cys(SR)-OH.[16]

The *S*-Carbomethoxysulfenyl Group as a Protective Group for Thiols, Particularly Cysteine.[8,14] The Scm group may be used as a thiol protective group, as well as a labile intermediate in the conversion of a cysteine residue to cystine in the late stages of a peptide synthesis.[5] The crystallinity of the sulfenylthiocarbonate intermediates is an added advantage of the method, facilitating handling and purification. The S-Scm intermediates can be prepared by the reaction of Scm-Cl in methanol with thiols or with their trityl (Tr), *S*-benzhydryl (Bzh), *S*-acetamidomethyl (Acm),[17] or (in lower yield) *S*-benzyl (Bn) derivatives. When working with cysteine in its free carboxylic acid form, esterification of the carboxy group or transesterification may occur. This problem can be minimized by working at low temperature (0 °C) or by addition of calcium carbonate[5] or diethylamine[7] to the reaction mixture. Acid labile protecting groups, such as *tert*-butyl (^{t}Bu) or 2,4,6-trimethylbenzyl, are stable under the conditions of the Scm derivative formation.[5] The Scm group itself is stable to strong acids (e.g., anhydrous hydrogen fluoride and trifluoromethanesulfonic acid). The S–S bond is stable under certain peptide coupling conditions, such as DCC/NMM.[5]

Unlike sulfenyl iodides[18] or sulfenyl thiocyanates[19] derived from cysteine compounds, which react readily with cysteine, as well as with its thioethers, hemithioacetals, and *S*-acetamidomethyl derivatives, the *S*-carbomethoxysulfenyl derivatives generate unsymmetrical disulfides only when treated with free thiols, thus are more selective in their reactivity towards sulfur nuclophiles.[5] With respect to solvents, chloroform is a frequent choice (especially in the presence of free carboxyl groups, see above). When esterification is not a concern, methanol is frequently used. Good results (79% yield) are obtained with a mixture of chloroform/methanol (1:1) for the reaction described in eq 5.[5] For the same reaction, using *N,N*-dimethylacetamide as a solvent dramatically slows down the reaction, whereas the use of dimethylformamide not only drastically reduces the product yield but also induces the formation of large amounts of symmetric disulfide.[5]

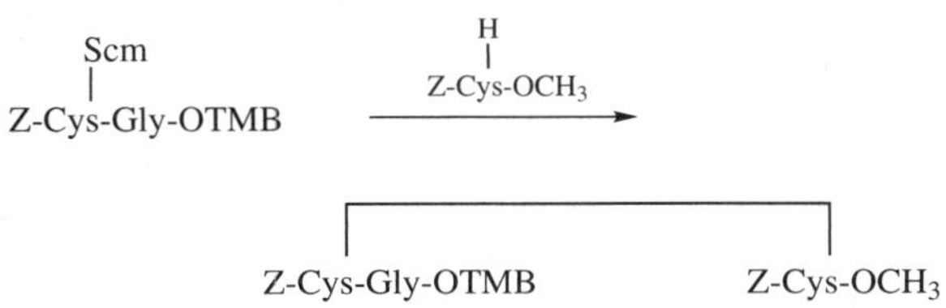

(5)

For the process described in eq 6, use of pyridine or of TFA:AcOH (1:1) as a solvent slows down the reaction and lowers the yield.[5]

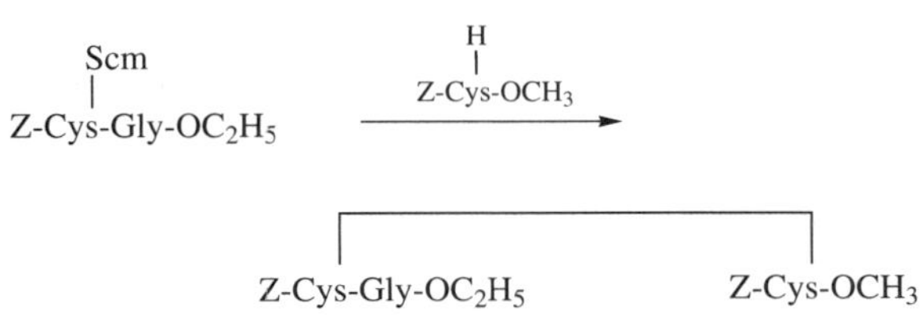

(6)

The sulfur atom in an Scm-protected thiophenol is stable to oxidation by *m*-chloroperbenzoic acid.[20]

As a transient protective group, the Scm group easily replaces the Acm group[17] or the diphenylmethyl group (eq 7).[21,22] After further manipulations of the substrate, the Scm group itself can be cleaved with dithiothreitol.[8]

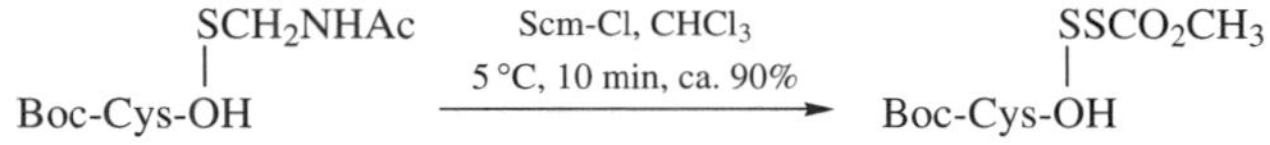

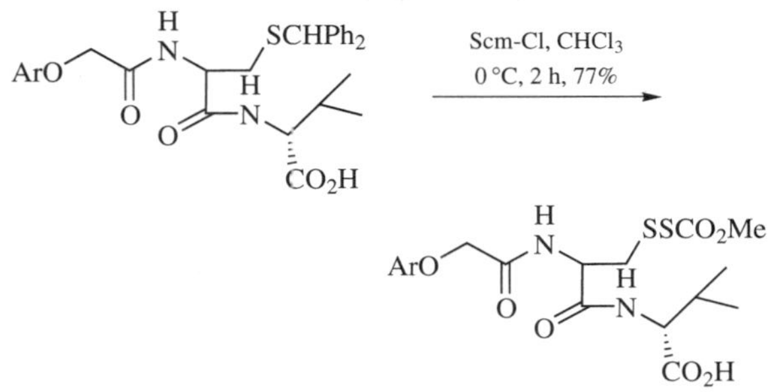

(7)

Formation of Symmetric Trisulfides via Sulfenylthiocarbonates. Treatment of a thiol with Scm-Cl followed by the decomposition of the resulting sulfenyl thiocarbonate (eq 8) with equimolar amount of alkoxide generates a symmetric trisulfide in yields that vary with the nature of the base.[23]

$$\text{RSH} \xrightarrow{\text{Scm-Cl}} \text{RSSC(O)CH}_3 \xrightarrow{\text{R}'\text{O}^-} \text{RSSSR} \quad (8)$$

In the case of R = Et, *i*-Pr, ^{t}Bu, sodium methoxide produces ca. 1:1 trisulfide:disulfide mixture. The yield of trisulfide increases as the base is changed: ethoxide <*iso*-propoxide <*tert*-butoxide. The yield is also influenced by the nature/bulkiness of the alkyl R in the alkylsulfenyl group (Me < Et < Pr < *i*-Pr < ^{t}Bu < Ph). Attempts to prepare monomeric cyclic trisulfides starting from bis(sulfenyl) thiocarbonates were unsuccessful, resulting instead in the formation of various bis(trisulfide) macrocycles, accompanied by variable but significant amounts of polymers.[24] Work in acidic media (pH 2) facilitates trisulfide formation.[25]

Preparation of Aromatic Thiols. Reaction of methoxycarbonylsulfenyl chloride with benzene and substituted benzenes in the presence of Lewis acids ($AlCl_3$, BF_3) in dichloromethane, carbon disulfide, or excess substrate as a solvent generates aryl thiocarbonates in low to good yields. Hydrolysis of the aryl thiocarbonates with aqueous methanolic potassium hydroxide generates the corresponding thiophenols in high yield (eq 9).[26]

$$\text{Ar-H} \xrightarrow{\text{Scm-Cl}} \text{Ar-S-C(O)-OCH}_3 \longrightarrow \text{Ar-SH} \quad (9)$$

Preparation of Diarylthiophenes. Reaction of acetophenones, under acidic conditions, with Scm-Cl generates 2-oxoalkylthiocarbonates in 60–80% yield. These, in turn, when heated to ca. 100 °C in the presence of sulfuric acid, produce 2,5-diarylthiophenes in 40–74% yield (eq 10).[26]

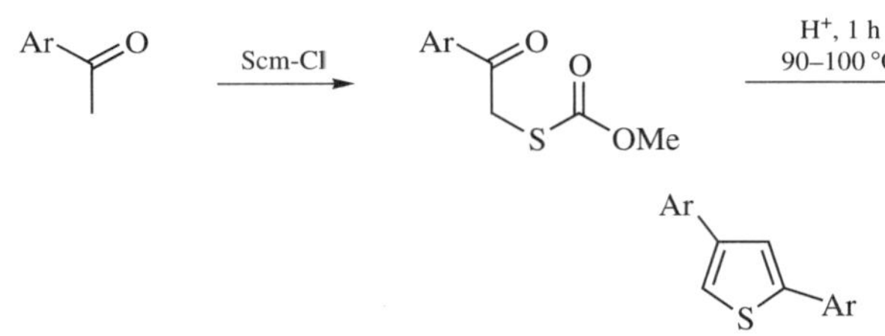

(10)

The yield of the last step appears to be favorably influenced by the presence of electron-donating groups (*p*-Me >> *p*-Br > H).[26]

Reaction with Diazo Compounds. Treatment of a diazolactam derived from penicillin with Scm-Cl in dichloromethane containing an excess of methanol gives a 6α-methoxy-6-methoxycarbothio derivative in 78% yield, accompanied by 3% of the corresponding 6α-chloro-6-methoxycarbothio derivative.[27] When the reaction is run in the absence of methanol, the 6α-chloro derivative is obtained in 88% yield. The reaction proceeds via a common diazonium intermediate (eq 11).[27]

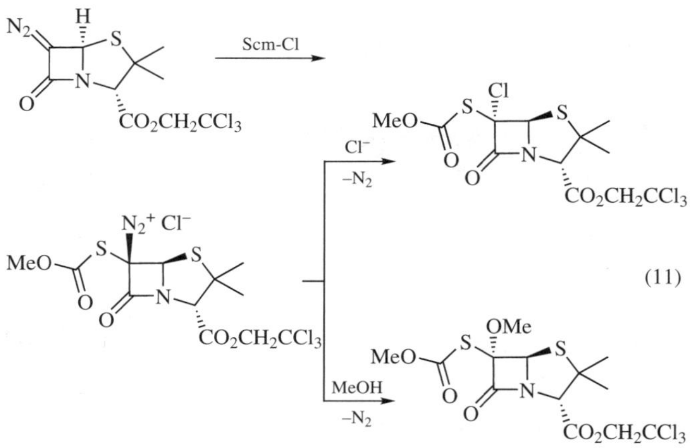

(11)

Preparation of Sulfinyl and Sulfonyl Thiocarbonates. Treatment of the sulfenyl thiocarbonates formed from thiols and Scm-Cl with an equimolar amount of *m*-chloroperbenzoic acid in chloroform generates the corresponding sulfinyl thiocarbonate which, unlike its precursor, is reactive towards ethanol in the presence of catalytic amount of imidazole.[28] The latter reaction also establishes the regiospecificity of the oxidation reaction. The oxidation of the same sulfenyl thiocarbonate substrate with excess *m*CPBA (min. 2.4 mol *m*CPBA/mol substrate) produces the corresponding sulfonyl thiocarbonate (eq 12).[28]

Desulfurization of Alkylsulfenyl Methyl Thiocarbonates. Treatment of alkylsulfenyl methyl thiocarbonates derived from alkyl thiols and Scm-Cl with triphenylphosphine at 0 °C in benzene generates *S*-alkyl methyl thiocarbonates in good yield (78% for alkyl = benzyl) (eq 13).[29] Arylsulfenyl methyl thiocarbonates do not parallel this behavior vs. triphenylphosphine[29] but they react cleanly with tributylphosphine (e.g., in dioxane-water) to generate the free arylthiol (see below).[30] The reductive cleavage of alkylsulfenyl methyl thiocarbonates with trialkylphosphines (Me_3P, Et_3P, Bu_3P) can be achieved with practically quantitative yields in protic solvents (THF-water) in the presence of acids (TFA).[30–32]

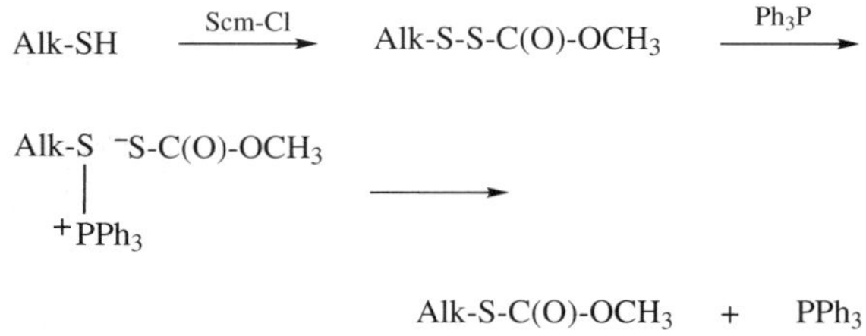

(13)

In eq 14, a Boc-protected dimethylketal derivative of cysteine is first cleaved with Scm-Cl, in acetate buffer. The resulting intermediate has the amino function Boc-protected and the thiol function Scm-protected. Acidic cleavage of the Boc group followed by reductive cleavage of the sulfenyl thiocarbonate with trimethylphosphine in aq THF in the presence of trifluoroacetic acid frees both functionalities.[31–33]

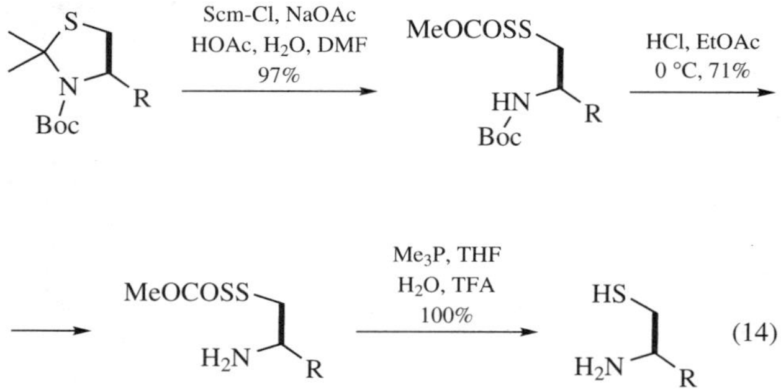

(14)

Peptide Synthesis by Prior Thiol Capture. A coupling method has been designed to be used at the late stages of fragment condensation syntheses of large peptides that contain cysteine residues spaced 5–20 amino acid residues apart.[6,30] The method takes advantage of an N-terminal cysteine residue and centers around a 6-hydroxy-4-mercaptodibenzofuran template that puts the participating fragments in the right proximity, with the right orientation. Other templates examined for the same process were not as successful.[34,35] The 4-mercapto functionality is used to form an unsymmetric disulfide with the cysteine residue via the Brois protocol involving Scm-Cl as described above (eq 1). The fragments containing the template are first prepared by solid-phase synthesis and purified. They are then linked together in a three-step sequence known as the "thiol capture strategy" (eq 15).[6,30]

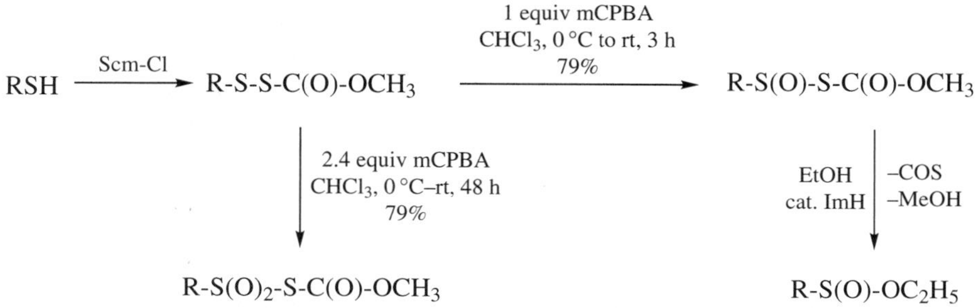

(12)

Step A involves disulfide bond formation between a peptide fragment derivatized at the C-terminus as 4-acyloxy-6-mercaptodibenzofuran and a peptide fragment bearing an S-Scm activated cysteine residue at its N-terminus. This step was designed to occur cleanly and very rapidly at submillimolar concentrations in protic solvents or solvent mixtures. The intramolecular acyl transfer (step B) is highly efficient (effective molarity > 1 M),[35,36] to the effect that, for most cases, a very weakly activated phenolic ester can be employed and many quite reactive side-chain functions in R and R′ can be left unprotected.[30] These high reactivity criteria are met best in the protein-solubilizing solvent hexafluoro-2-propanol (HFIP) and its mixtures with water and acetonitrile.[6]

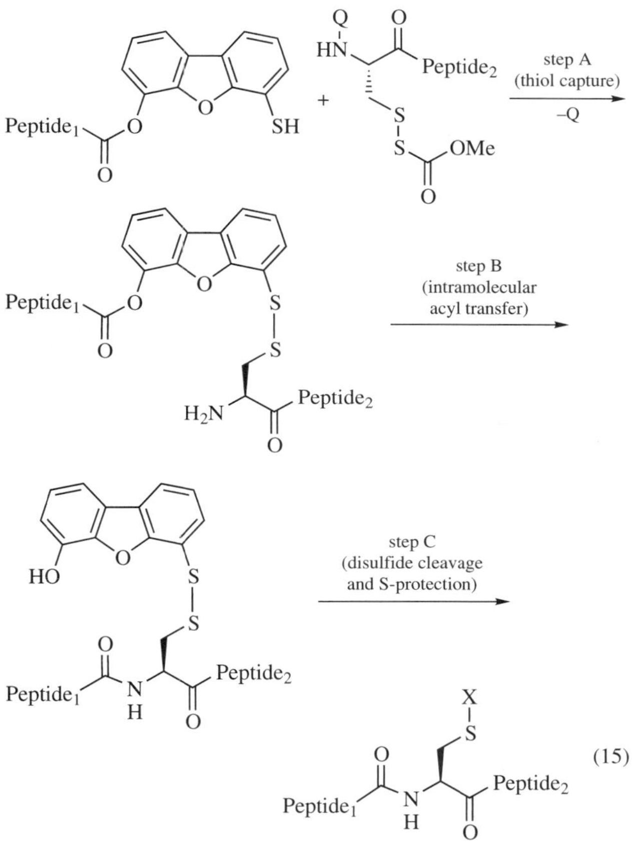

(15)

X = acetyl, 2,4-dinitrophenyl

The use of a phenolic ester as anchoring group limits the choice of the protective groups. An appropriate choice is *tert*-butyloxycarbonyl (Boc, trifluoroacetic acid cleavable) for N^{α}-blocking of the amino acid esterified with the phenol and functionalized benzyl groups for side-chain functions.[30] Alternatives include *tert*-butyl-derived side-chain blocking groups and 2-*p*-biphenylyl-2-propyloxycarbonyl (Bpoc) for N^{α}-protection.[6]

The resin-bound fragment preparation starts with the chlorine displacement from the chloromethylene function on a chloromethylated polystyrene by the cesium salt of Z-L-Cys(Tr)-OH. The resulting benzyl ester is treated with Scm-Cl to form the active intermediate sulfenyl thiocarbonate which is allowed to react with 4-acyloxy-6-mercaptodibenzofuran, preferably in dichloromethane– hexafluoro-2-propanol, with or without catalytic triethylamine (eq 16).[30]

An alternative to Acm and Tr for the temporary protection of cysteine is 2,2-dimethylthiazolidine-4-carboxylic acid (H-Dmt-OH, used as the anhydride derived from Boc-Dmt-OH or as Boc-Dmt-OPfp).[9]

It is often simpler to modify the sequence in eq 16 to a sequence in which the resin-bound sulfenyl thiocarbonate intermediate is first reacted with 4-hydroxy-6-mercaptodibenzofuran. The free phenol is then acylated by 3–4 equiv of a symmetrical anhydride derived from an N^{α}-carbamate-blocked amino acid, in the presence of diisopropylethylamine in dichloromethane (ca. 10–30 min reaction time at 0 °C).

The reductive cleavage of the disulfide bond (eq 15, step C) can be performed cleanly and with high reaction rates (almost instantaneously) by treatment with 1 equiv of tributylphosphine in dioxane-water at room temperature.[30]

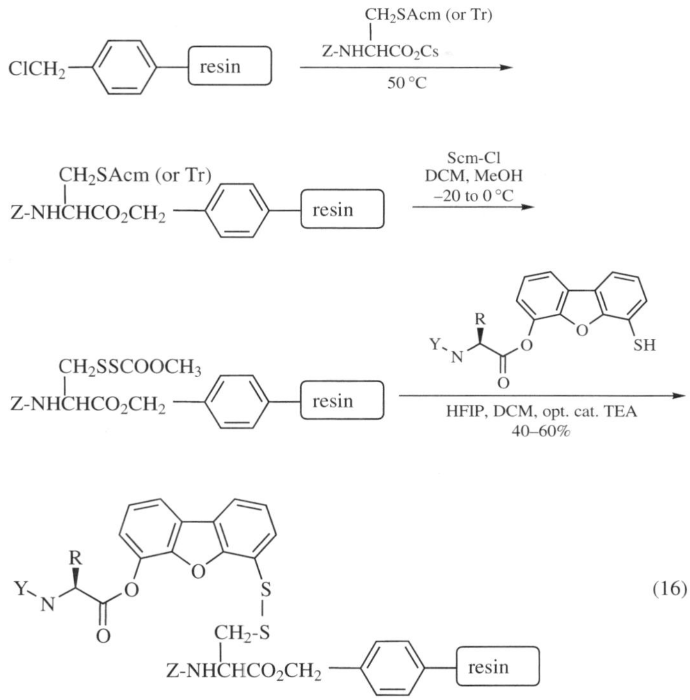

(16)

Reaction with Nitrogen Nucleophiles. The reaction of Scm-Cl with arylsulfonylhydrazines (or aralkylsulfonylhydrazines) occurs in minutes at 0 °C in anhydrous ether to provide stable, recrystallizable products (eq 17).[37]

The relatively high reactivity of Scm-Cl, however, lowers the selectivity towards substrates with more than one nitrogen nucleophile.[38] Nevertheless, with less reactive nucleophiles, such as the two nitrogens in unsymmetrically substituted ureas, selectivity is possible. In the reaction shown in eq 18, only the nitrogen adjacent to the oxazole ring reacts with the electrophile Scm-Cl (albeit in low yield, 34%).[39]

α-Methoxycarbonylsulfenylation of Carbonyl Compounds as a Route to Heterocycles. Ketones and aldehydes react with equimolar amounts of Scm-Cl in chloroform to form, in moderate to good yields, α-methoxycarbonylsulfenyl ketones or aldehydes that, upon condensation with amines, alkylhydrazines, or arylhydrazines afford, respectively, thiazolones, thiadiazinones, and 3-indolethiols (eq 19).[40]

In more complex substrates, the regio- and stereoselectivity of the reaction of Scm-Cl with carbon nucleophiles may be hard to predict.[41]

$$ArSO_2N(CH_3)NH_2 \xrightarrow{ClSC(O)OCH_3} CH_3OC(O)S\text{-}NH\text{-}N(CH_3)SO_2Ar \quad (17)$$

Other Heterocycles. Reaction of 2,2′-biindolyl in dichloromethane with 2 equiv Scm-Cl generates a 3,3′-bis(methoxycarbonylsulfenyl) derivative that, upon heating with base and exposure to air, produces a highly insoluble bis-disulfide "dimer" (eq 20).[42]

Other Reactions. The ring opening of the functionalized 2-alkyl-3-ethoxycarbonyl-1,1-dimethoxypropanes in eq 21 with 1 equiv Scm-Cl in carbon tetrachloride generates the corresponding *syn*- and *anti*-2-sulfenylbutanedioates in low to fair yields, with 70–80% de in favor of the *syn* isomer. When the alkyl substituent in the substrate is replaced by hydrogen, the reaction takes a different pathway.[43]

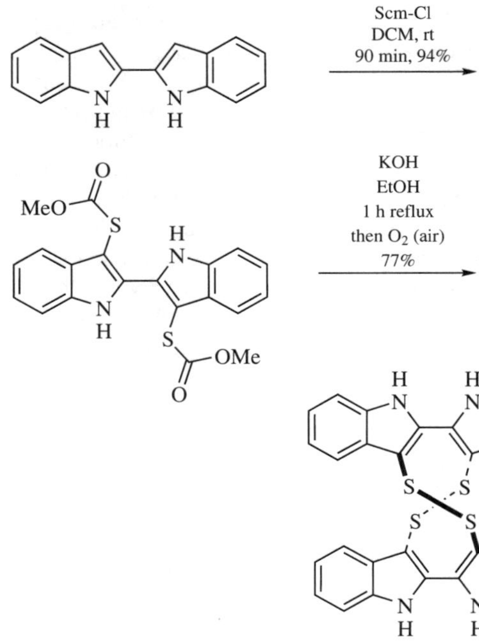

(20)

Related Reagents. benzenesulfenyl chloride; 2-(methylthio)-1*H*-isoindole-1,3(2*H*)-dione; *N-tert*-butylthiophthalimide; *N*-phenylthiophthalimide; dimethyl(methylthio)sulfonium tetrafluoroborate; ethyl thiosulfinate; disulfur dichloride.

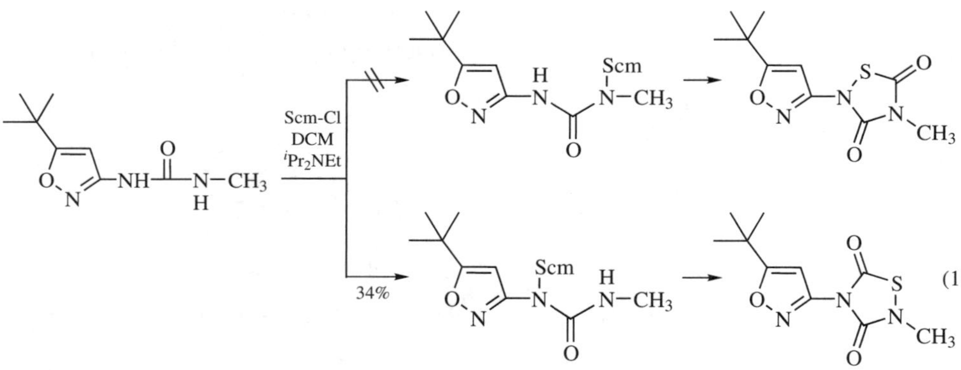

(18)

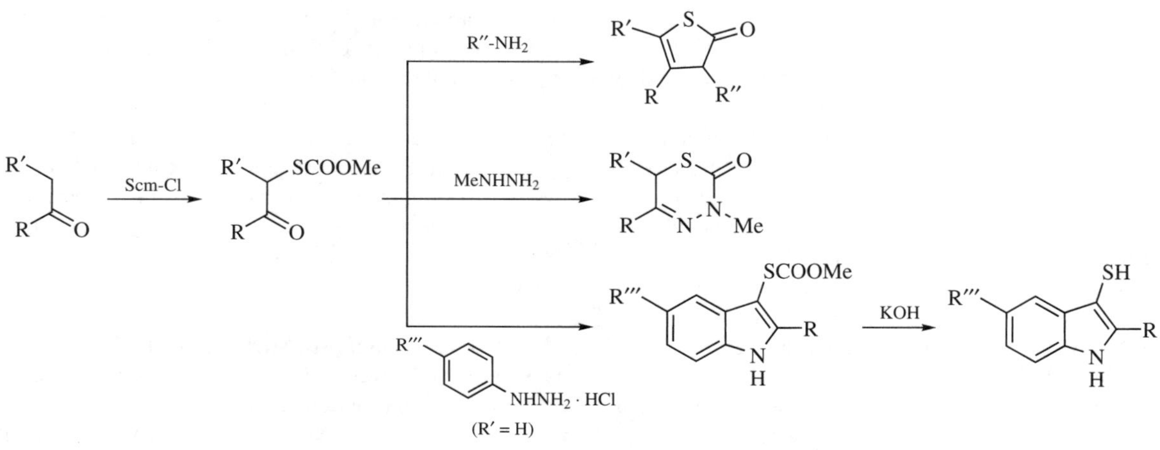

(19)

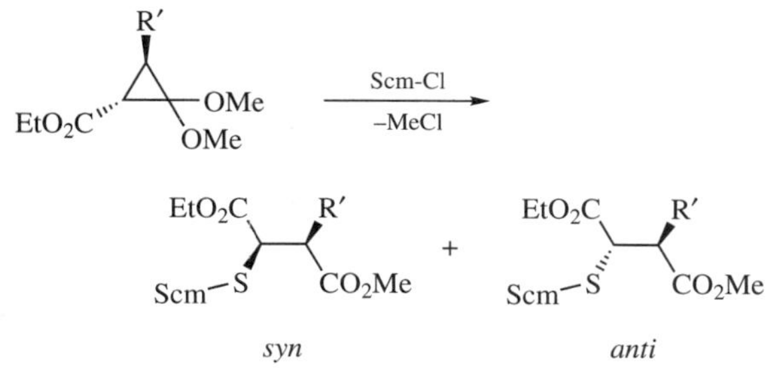

(21)

1. Bohme, H.; Steudel, H.-P., *Liebigs Ann. Chem.* **1969**, *730*, 121.
2. Zumach, G.; Kühle, E., *Angew. Chem. Int. Ed. Engl.* **1970**, *9*, 54 and references therein.
3. Barany, G.; Schroll, A. L.; Mott, A. W.; Halsrud, D. A., *J. Org. Chem.* **1983**, *48*, 4750.
4. Brois, S. J.; Pilot, J. F.; Barnum, H. W., *J. Am. Chem. Soc.* **1970**, *92*, 7629.
5. Hiskey, R. G.; Muthukumaraswamy, N.; Vunnam, R. R., *J. Org. Chem.* **1975**, *40*, 950.
6. Fotouhi, N.; Galakatos, N. G.; Kemp, D. S., *J. Org. Chem.* **1989**, *54*, 2803.
7. (a) Kamber, B., *Helv. Chim. Acta* **1973**, *56*, 1370. (b) Kamber, B., *Z. Naturforsch.* **1981**, *36b*, 508.
8. Greene, T. W.; Wuts, P. G., *Protective Groups in Organic Chemistry*, 2nd ed.; Wiley: Chichester, 1991, pp. 303–304.
9. Kemp, D. S.; Carey, R. I., *J. Org. Chem.* **1989**, *54*, 3640.
10. Field, L.; Ravichandran, R., *J. Org. Chem.* **1979**, *44*, 2624.
11. Brasili, L.; Giannella, M.; Melchiorre, C.; Belleau, B.; Benfey, B. G., *Eur. J. Med. Chem.* **1981**, *16*, 115.
12. Smith, D. J.; Venkatraghavan, V., *Synth. Commun.* **1985**, *15*, 945.
13. Sanchez, J. P., *J. Heterocylic Chem.* **1997**, *34*, 1463.
14. Barany, G.; Kneib-Cordonnier, N.; Mullen, D. G., *Int. J. Peptide Protein Res.* **1987**, *30*, 705.
15. van Nispen, J. W.; Hannink, J. A. J.; Schoffelmeer, M. S.; Janssen, W. P. A.; Polderdijk, J. P.; Greven, H. M., *Recl. Trav. Chim. Pays-Bas* **1984**, *103*, 68.
16. Rietman, B. H.; Peters, R. F. R.; Tesser, G. I., *Synth. Commun.* **1994**, *24*, 1323.
17. Schroll, A. L.; Barany, G., *J. Org. Chem.* **1989**, *54*, 244.
18. Kamber, B., *Helv. Chim. Acta* **1971**, *54*, 927.
19. Hiskey, R. G.; Ward, B. F., *J. Org. Chem.* **1970**, *35*, 1118.
20. Kemp, D. S.; Buckler, D. R., *J. Org. Chem.* **1989**, *54*, 3647.
21. Baldwin, J. E.; Pratt, A. J.; Moloney, M. G., *Tetrahedron* **1987**, *43*, 2565.
22. Baldwin, J. E.; Jesudason, C. D.; Moloney, M. G.; Morgan, D. R.; Pratt, A. J., *Tetrahedron* **1991**, *47*, 5603.
23. Harpp, D. N.; Granata, A., *Tetrahedron Lett.* **1976**, 3001.
24. Harpp, D. N.; Granata, A., *J. Org. Chem.* **1979**, *44*, 4144.
25. Chen, L.; Zoulikova, I.; Slaninova, J.; Barany, G., *J. Med. Chem.* **1997**, *40*, 864.
26. Schroth, W.; Hassfeld, M.; Schiedewitz, W.; Pfotenhauer, C., *Z. Chem.* **1977**, *17*, 411.
27. Sheehan, J. C.; Commons, T. J., *J. Org. Chem.* **1978**, *43*, 2203.
28. Harpp, D. N.; Granata, A., *Synthesis* **1978**, 782.
29. Harpp, D. N.; Granata, A., *J. Org. Chem.* **1980**, *45*, 271.
30. Kemp, D. S.; Galakatos, N. G., *J. Org. Chem.* **1986**, *51*, 1821, and references therein.
31. Yang, H.; Sheng, X. C.; Harrington, E. M.; Ackermann, K.; Garcia, A. M.; Lewis, M. D., *J. Org. Chem.* **1999**, *64*, 242.
32. O'Connell, C. E.; Ackermann, K.; Rowell, C. A.; Garcia, A. M.; Lewis, M. D.; Schwartz, C. E., *Bioorg. Med. Chem. Lett.* **1999**, *9*, 2095.
33. O'Connell, C. E.; Rowell, C. A.; Ackermann, K.; Garcia, A. M.; Lewis, M. D.; Kowalczyk, J., *Chem. Pharm. Bull.* **2000**, *48*, 740.
34. Kemp, D. S.; Galakatos, N. G., Bowen, B.; Tan, K. *J. Org. Chem.* **1986**, *51*, 1829.
35. Kemp, D. S.; Carey, R. I.; Dewan, J. C.; Galakatos, N. G.; Kerkman, D.; Leung, S.-L., *J. Org. Chem.* **1989**, *54*, 1589.
36. Kemp, D. S.; Carey, R. I., *J. Org. Chem.* **1993**, *58*, 2216.
37. Hrubiec, R. T.; Shyam, K.; Cosby, L. A.; Sartorelli, A. C., *J. Med. Chem.* **1986**, *29*, 1777.
38. Pilgram, K. H.; Gale, L. H., *J. Heterocyclic Chem.* **1988**, *25*, 193.
39. Pilgram, K. H.; Gale, L. H., *J. Heterocyclic Chem.* **1988**, *25*, 607.
40. Sanemitsu, Y.; Kawamura, S.; Tanabe, Y., *J. Org. Chem.* **1992**, *57*, 1053.
41. Kozikowski, A. P.; Campiani, G.; Tückmantel, W., *Heterocycles* **1994**, *39*, 101.
42. Janosik, T.; Bergman, J.; Romero, I.; Stensland, B.; Stalhandske, C.; Marques, M. M. B.; Santos, M. M. M.; Lobo, A. M.; Prabhakar, S.; Duarte, F. M.; Florencio, M. H., *Eur. J. Org. Chem.* **2002**, 1392.
43. Graziano, M. L.; Iesce, M. R.; Cermola, F., *Synthesis* **1999**, 1944.

Livia Enache
deCode Genetics, Woodridge, Illinois, USA

1,1′-Carbonylbis(3-methylimidazolium) Bis(trifluoromethanesulfonate)[1]

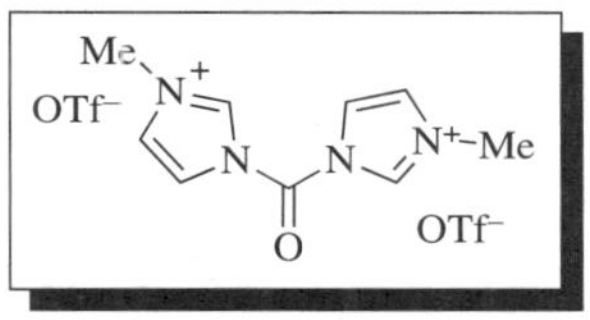

[120418-31-7] $C_{11}H_{12}F_6N_4O_7S_2$ (MW 490.36)

(versatile coupling reagent for aminoacylations[2])

Alternate Name: 1,1-carbonylbis(3-methylimidazolium) bistriflate.

Physical Data: pale white solid mp 78–80 °C; ^{1}H δ ($CDCl_3$) 8.86 (s, 2H), 7.44 (m, 2H), 7.16 (m, 2H), 4.00 (s, 6H).

Solubility: sol nitromethane, chloroform; insol ether; reacts with water.

Analysis of Reagent Purity: freshly prepared sample is treated with water: 200 mol % of *N*-methylimidazolium triflate is produced along with CO_2.

Preparative Methods: *fresh* methyl triflate is added to 1,1-carbonylbisimidazole in nitromethane at 10 °C. Solvent is then removed in vacuo, or the reagent used in situ.[2]

Handling, Storage, and Precautions: usually prepared under anhydrous conditions and used immediately. Can be stored in dry nitromethane solution or in solid state in absence of moisture. Preparation using aged methyl triflate should be avoided, owing to possible contamination of the coupling reagent with triflic acid.[2]

Coupling Reactions. When *N*-Cbz derived amino acids are exposed to a freshly prepared solution of carbonylbis(3-methylimidazolium) triflate in nitromethane, coupling ensues with loss of CO_2 to give the derived amide and one equivalent of methylimidazolium triflate (eq 1).[2]

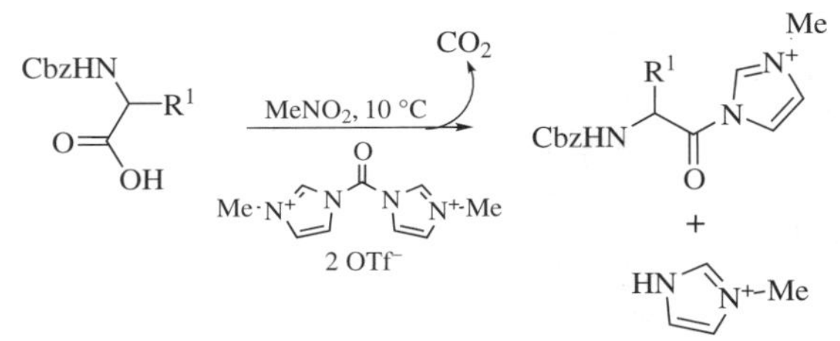

(1)

The acylated species so prepared can be coupled with alcohols to yield esters or with amines to form amides. The yields of coupled product are typically high and formation of the coupled products takes place within 15 min in the case of amides (eq 2) and within 3 h for esters.[2]

Of particular note is the fact that sterically hindered alcohols (e.g. menthol) couple in nearly quantitative yield (eq 3).

Racemization is not prevalent under these neutral conditions: routine analysis of the diastereomeric excess of coupled products revealed a >99.75:0.25 ratio, confirming both the mildness of the system and the near neutral conditions that persist during the coupling. The system compares most favorably against conventional coupling agents such as *N,N'*-carbonyldiimidazole,[3,4] where sluggish *O*-acylations are common[5] and racemization of the acyl component in the couplings has been observed.[6]

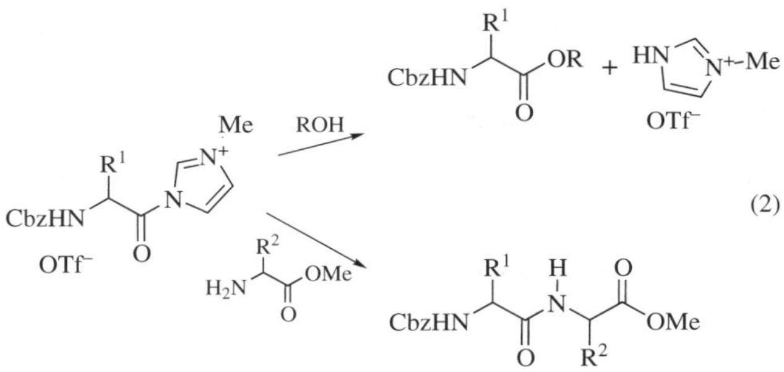

(2)

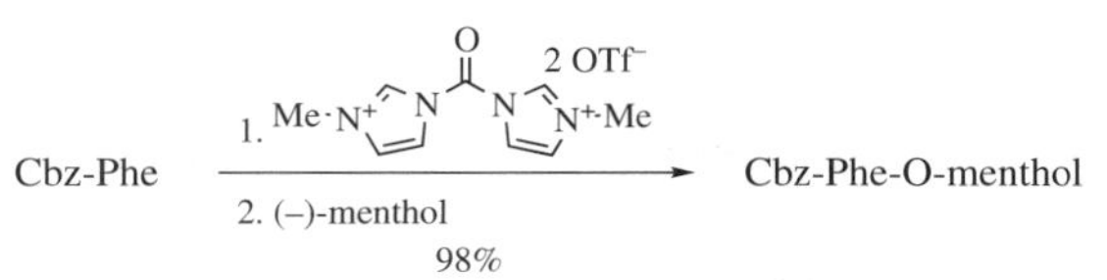

(3)

1. Bodanszky, M.; Bodanszky, A. *The Practice of Peptide Synthesis*, Springer: New York, 1984.
2. Saha, A. K.; Schultz, P.; Rapoport, H. *J. Am. Chem. Soc.*, **1989**, *111*, 4856.
3. Staab, H. A. *Justus Liebigs Ann. Chem./Liebigs Ann. Chem.*, **1957**, *609*, 75.
4. For related carbonyl bis-imidazole system see: Kamijo, T.; Harada, H.; Iizuka, K. *Chem. Pharm. Bull.*, **1984**, *32*, 5044.
5. Staab, H. A. *Angew. Chem., Int. Ed. Engl.*, **1962**, *1*, 351.
6. Weygand, F.; Prox, A.; Konig, W. *Ber. Dtsch. Chem. Ges./Chem. Ber.*, **1966**, *99*, 1451.

Graham B. Jones & Robert S. Huber
Clemson University, SC, USA

N,N'-carbonyldiimidazole[1]

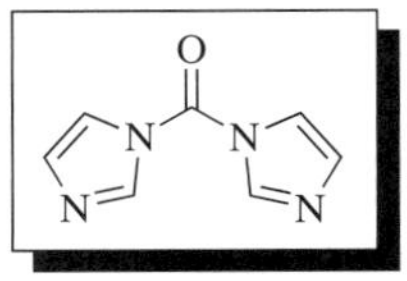

[530-62-1] $C_7H_6N_4O$ (MW 162.15)

(reagent for activation of carboxylic acids;[1] synthesis of esters,[2] amides,[3] peptides,[4] aldehydes,[5] ketones,[6] β-keto thioesters,[7] tetronic acids;[8] ureas,[9] isocyanates,[10] carbonates;[9] halides from alcohols;[11] glycosidation;[12] dehydration[13–16])

Physical Data: mp 116–118 °C.
Solubility: no quantitative data available. Inert solvents such as THF, benzene, $CHCl_3$, DMF are commonly used for reactions.
Form Supplied in: commercially available white solid.
Analysis of Reagent Purity: purity can be determined by measuring the amount of CO_2 evolved on hydrolysis.
Preparative Method: prepared by mixing phosgene with four equivalents of imidazole in benzene/THF.[17]
Purification: may be purified by recrystallization from hot, anhydrous THF with careful exclusion of moisture.[17]
Handling, Storage, and Precautions: moisture sensitive; reacts readily with water with evolution of carbon dioxide. May be kept for long periods either in a sealed tube or in a desiccator over P_2O_5.

Activation of Carboxylic Acids: Synthesis of Acyl Imidazoles. *N,N'*-Carbonyldiimidazole (**1**) converts carboxylic acids into the corresponding acylimidazoles (**2**) (eq 1).[1] The method can be applied to a wide range of aliphatic, aromatic, and heterocyclic carboxylic acids, including some examples (such as formic acid and vitamin A acid) where acid chloride formation is difficult. The reactivity of (**2**) is similar to that of acid chlorides, but the former have the advantage that they are generally crystalline and easily handled. Isolation of (**2**) is simple, but often unnecessary; further reaction with nucleophiles is usually performed in the same reaction vessel. Conversion of (**2**) into acid chlorides (via reaction with HCl),[18] hydrazides,[3] hydroxamic acids,[3] and peroxy esters[19] have all been described. Preparation of the more important carboxylic acid derivatives is described below.

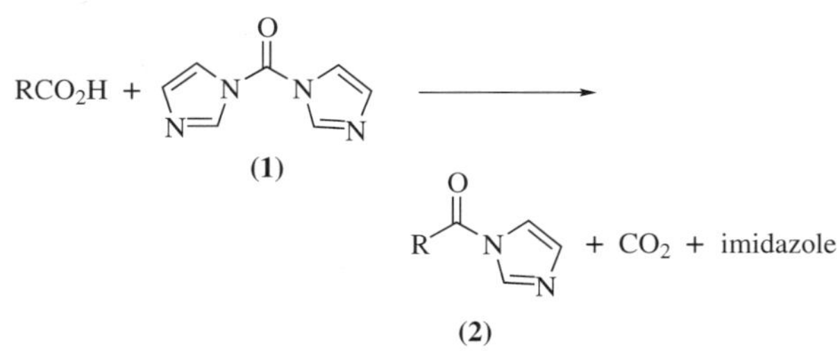

(1)

Esters from Carboxylic Acids. Reaction of equimolar amounts of carboxylic acid, alcohol, and (**1**) in an inert solvent (e.g. THF, benzene, or chloroform) results in ester formation (eq 2). Since alcoholysis of the intermediate acylimidazole is relatively slow, the reaction mixture must be heated at 60–70 °C for some time. However, addition of a catalytic amount of a base such as Sodium Amide to convert the alcohol to the alkoxide, or a catalytic amount of the alkoxide itself, allows rapid and complete formation of the ester at room temperature.[2] The base catalyst must of course be added after formation of (**2**) from the acid is complete, as indicated by cessation of evolution of carbon dioxide.

$$\mathrm{RCO_2H} + \mathrm{R^1OH} \xrightarrow[\text{2. cat. } \mathrm{R^1ONa} \text{ or } \mathrm{NaNH_2}]{\text{1. } (\mathbf{1})} \mathrm{RCO_2R^1} \quad (2)$$

Esters of tertiary alcohols may not be prepared from carboxylic acids containing acidic α-protons using this modified procedure, since deprotonation and subsequent condensation, competes. However, the use of stoichiometric 1,8-Diazabicyclo[5.4.0]undec-7-ene as base has been shown to provide good yields of *t*-butyl esters even for acids with acidic α-protons (eq 3).[20] This procedure was unsuccessful for pivalic acid or for *N*-acyl-α-amino acids.

$$\mathrm{Ph(CH_2)_3CO_2H} \xrightarrow[\text{2. 1 equiv } t\text{-BuOH, 1 equiv DBU, 40 °C, 24 h}]{\text{1. 1 equiv } (\mathbf{1})\text{, DMF, 40 °C, 1 h}} \mathrm{Ph(CH_2)_3CO_2}\text{-}t\text{-Bu} \quad (3)$$

An alternative approach to increasing the rate of esterification is to activate further the intermediate (**2**). N-Bromosuccinimide has been used for this purpose,[21] but unsaturation in the carboxylic acid or alcohol is not tolerated. More generally useful is the addition of an activated halide, usually Allyl Bromide, to a chloroform solution of (**1**) and a carboxylic acid, resulting in formation of the acylimidazolium salt (**3**) (eq 4).[22] Addition of the alcohol and stirring for 1–10 h at room temperature or at reflux affords good yields of ester in a one-pot procedure. These conditions work well for the formation of methyl, ethyl, and *t*-butyl esters of aliphatic, aromatic, and α,β-unsaturated acids. Hindered esters such as *t*-butyl pivalate can be prepared cleanly (90% yield). The only limitation is that substrates must not contain functionality that can be alkylated by the excess of the reactive halide.

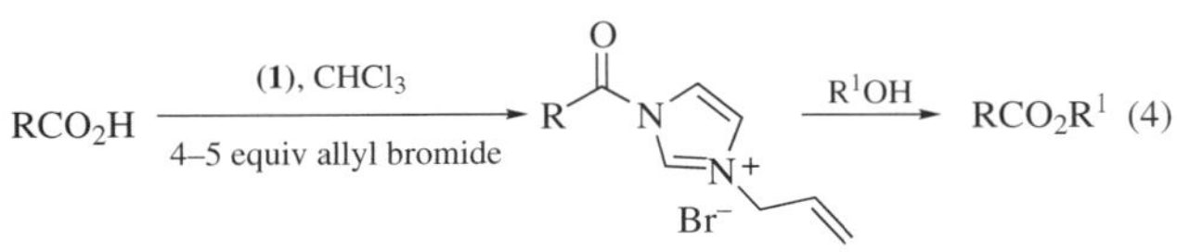

(**3**)

Since the purity of commercial (**1**) may be variable due to its water sensitivity, it is common to employ an excess in order to ensure complete conversion of the carboxylic acid to the acylimidazole. It has been suggested that alcohols react faster with residual (**1**) than with the acylimidazole (**2**), thus reducing the yield of ester. A procedure has been developed for removal of excess (**1**) before addition of the alcohol.[23]

Macrolactonization has been accomplished using (**1**).[24] Thiol and selenol esters can also be prepared in one pot from carboxylic acids using (**1**);[25] reaction of the intermediate (**2**) with aromatic thiols or selenols is complete within a few minutes in DMF, while aliphatic thiols require a few hours. Formation of the phenylthiol ester of *N*-Cbz-L-phenylalanine was accompanied by only slight racemization. N,N′-Carbonyldi-sym-triazine can be used in place of (**1**), with similar results.

Amides from Carboxylic Acids: Peptide Synthesis. Analogous to ester formation, reaction of equimolar amounts of a carboxylic acid and (**1**) in THF, DMF, or chloroform, followed by addition of an amine, allows amide bond formation.[3] The method has been applied to peptide synthesis (eq 5).[4] One equivalent of (**1**) is added to a 1 M solution of an acylamino acid in THF, followed after 1 h by the desired amino acid or peptide ester. The amino acid ester hydrochloride may be used directly instead of the free amino acid ester. An aqueous solution of the amino acid salt can even be used, but yields are lower.

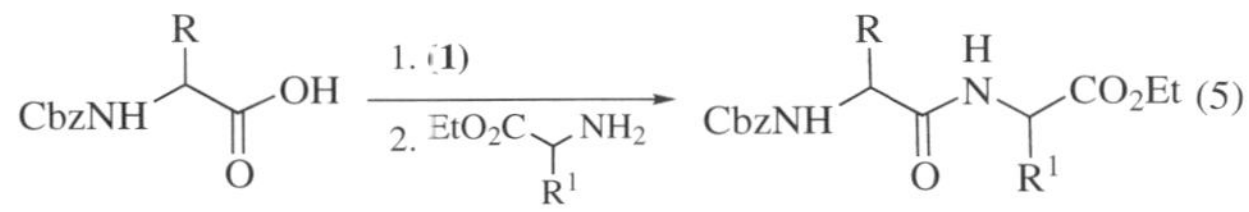

As is the case in esterification reactions, the presence of unreacted (**1**) can cause problems since the amine reacts with this as quickly as it does with the acylimidazole, forming urea byproducts that can be difficult to separate. Use of exactly one equivalent of (**1**) is difficult due to its moisture sensitivity, and also because of the tendency of some peptides or amino acids to form hydrates. Paul and Anderson solved this problem by use of an excess of (**1**) to form the acylimidazole, then cooling to −5 °C and adding a small amount of water to destroy the unreacted (**1**) before addition of the amine.[26]

For the sensitive coupling of Cbz-glycyl-L-phenylalanine and ethyl glycinate (the Anderson test), Paul and Anderson reported the level of racemization as 5% using THF as solvent at −10 °C, but as <0.5% in DMF.[4] Performing the same coupling reaction at room temperature, Beyerman and van den Brink later claimed that the degree of racemization in DMF was in fact as high as 17%, and reported better results (no detectable racemization) using the related reagent *N,N′*-carbonyl di-*sym*-triazine in place of (**1**).[27] In a comparative study of several reagents, Weygand and co-workers also observed extensive racemization using (**1**).[28] In the formation of tyrosine esters, Paul reported that the mixed anhydride method is to be preferred to use of (**1**), since *O*-acylation is a major side reaction with the latter.[29]

Aldehydes and Ketones from Carboxylic Acids. Reduction of the derived acylimidazole (**2**) with Lithium Aluminum Hydride achieves conversion of an aliphatic or aromatic carboxylic acid to an aldehyde (eq 6).[5] Diisobutylaluminum Hydride has also been used, allowing preparation of α-acylamino aldehydes from *N*-protected amino acids.[30] Similarly, reaction of (**2**) with Grignard reagents affords ketones,[6] with little evidence for formation of tertiary alcohol.

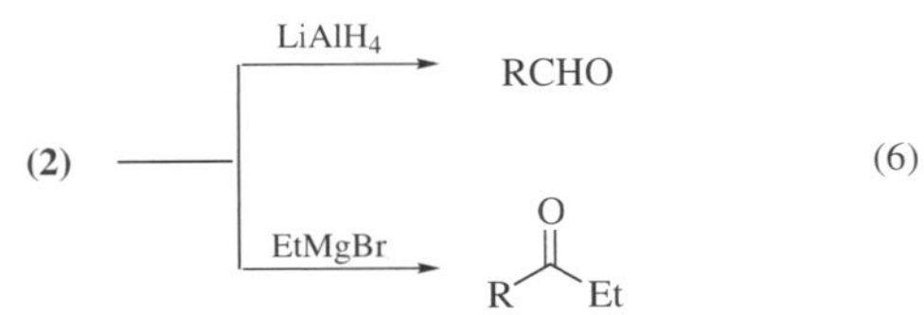

Reaction of acylimidazoles with the appropriate carbon nucleophile has also been used for the preparation of α-nitro ketones[31] and β-keto sulfoxides.[32]

***C*-Acylation of Active Methylene Compounds.** Treatment of an acylimidazole, derived from a carboxylic acid and (**1**), with the magnesium salt of a malonic or methylmalonic half thiol ester results in *C*-acylation under neutral conditions (eq 7).[7] The presence of secondary hydroxyl functionality in the carboxylic acid is tolerated, but primary alcohols require protection. Magnesium salts of malonic esters may be used equally effectively. Intramolecular *C*-acylation of ketones has also been reported.[33]

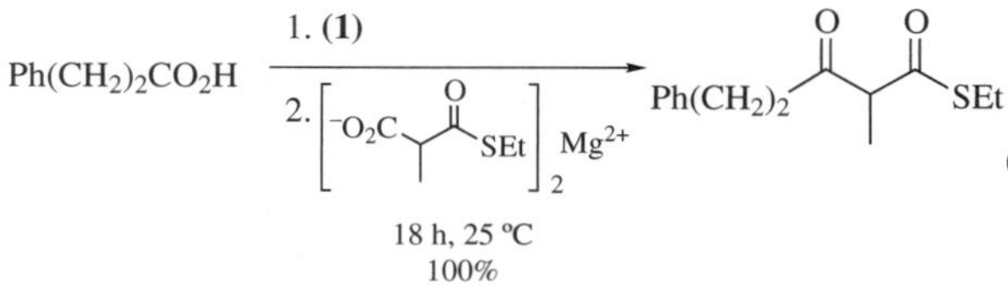

(7)

Tetronic Acids. A synthesis of tetronic acids reported by Smith and co-workers relies on the reaction between (**1**) and the dianion derived from an α-hydroxy ketone (eq 8).[8] The reaction proceeds in moderate yield (31–57%).

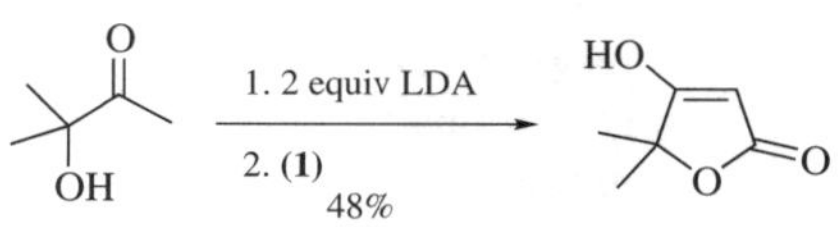

(8)

Ureas and Carbonates. Reagent (**1**) may be used as a direct replacement for the highly toxic Phosgene in reactions with alcohols and amines. Reaction of (**1**) with two equivalents of a primary aliphatic or aromatic amine at room temperature rapidly yields a symmetrical urea (eq 9).[9] If only one equivalent of a primary amine is added to (**1**), then the imidazole-*N*-carboxamide (**4**) is formed (eq 10). These compounds can dissociate into isocyanates and imidazole, even at room temperature, and distillation from the reaction mixture provides a useful synthesis of isocyanates (eq 10).[10] Secondary amines react only at one side of (**1**) at room temperature, again giving the imidazole-*N*-carboxamide of type (**4**).

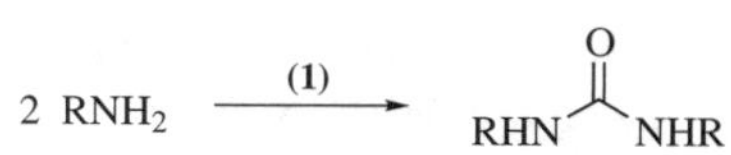

(9)

R = alkyl, aryl

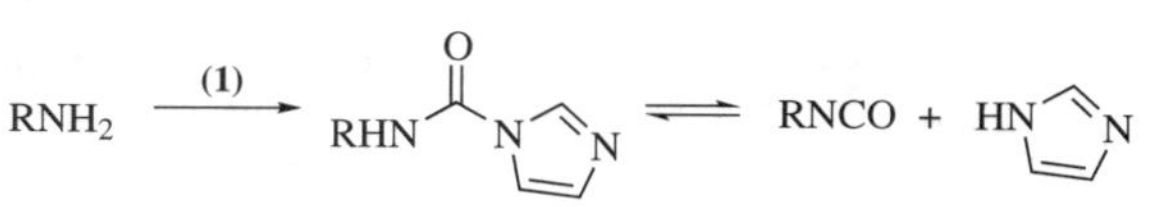

(10)

R = alkyl, aryl (**4**)

Reaction of (**1**) with one equivalent of an alcohol provides the imidazole-*N*-carboxylic ester (**5**) (eq 11).[34] Further treatment with another alcohol or phenol yields an unsymmetrical carbonate; alternatively, reaction with an amine affords a carbamate (eq 12).[34]

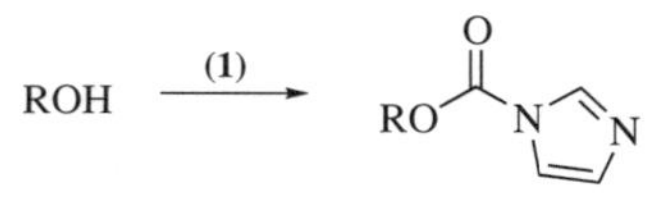

(11)

(**5**)

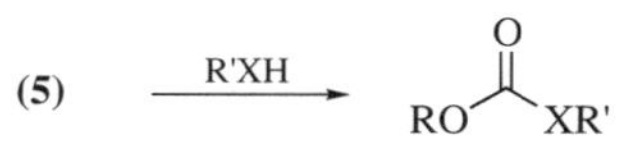

(12)

X = O, NH

Heating (**1**) with an excess of an alcohol or phenol gives the symmetrical carbonate.[9] This reaction can be accelerated dramatically by the presence of catalytic base (e.g. Sodium Ethoxide). Reaction under these conditions is so exothermic even at room temperature that only *t*-butanol stops at the imidazole-*N*-carboxylic ester stage.

1,2-Diamines, 1,2-diols, or 1,2-amino alcohols react with (**1**) to form cyclic ureas,[35] carbonates,[36] or oxazolidinones,[35] respectively. In the case of cyclohexane-1,2-diols, the *cis*-diol reacts much more rapidly than the *trans*, as would be expected.[36] Thiazolidinones can also be prepared using (**1**).[37]

Halides from Alcohols. Treatment of an alcohol with (**1**) and an excess (at least three equivalents) of an activated halide results in its conversion to the corresponding halide (eq 13).[11] Any halide more reactive than the product halide may be used, but in practice Allyl Bromide or Iodomethane give best results as they are effective and readily removed after the reaction. Acetonitrile is the best solvent and yields are generally high (>80%). Bromide or iodide formation work well, but not chlorination. Optically active alcohols are racemized.

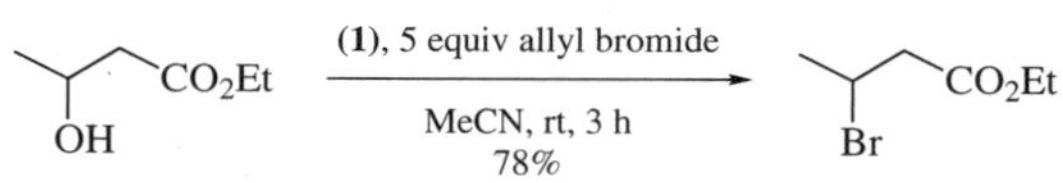

(13)

Glycosidation. A mild glycosidation procedure involving (**1**) has been reported by Ford and Ley.[12] A carbohydrate derivative in ether or dichloromethane reacts with (**1**) through the anomeric C-1-hydroxyl to give the (1-imidazolylcarbonyl) glycoside (IMG) (**6**) (eq 14). Isolation of (**6**) is not usually necessary; treatment with one equivalent of an alcohol and two equivalents of Zinc Bromide in ether at reflux gives the glycoside. Generally, higher α:β ratios are obtained for more hindered alcohols and when ether is used as solvent rather than the less polar dichloromethane. Along with the fact that the α:β ratio is independent of the configuration of (**6**), this suggests an S_N1-type mechanism. In contrast, treatment of (**6**) with Acetyl Chloride provides the anomeric chloride with essentially exclusive inversion.

Dehydration. Reagent (**1**) has been used for the dehydration of various substrates, including aldoximes (to give nitriles),[13] β-hydroxy amino acids,[14] and β-hydroxy sulfones.[15] 3-Aryl-2-hydroxyiminopropionic acids undergo dehydration and decarboxylation, to give 2-aryl acetonitriles, upon reaction with (**1**).[16]

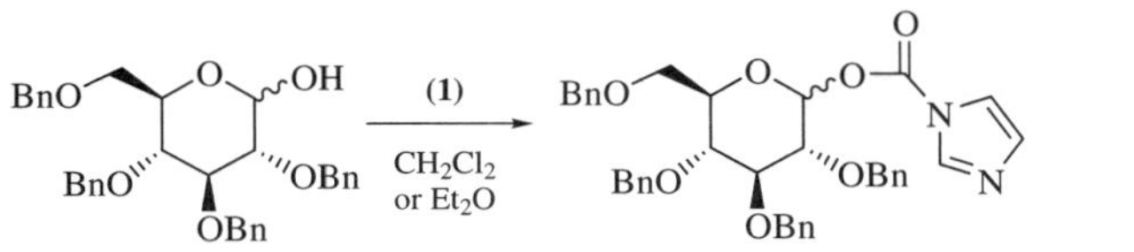

(6)

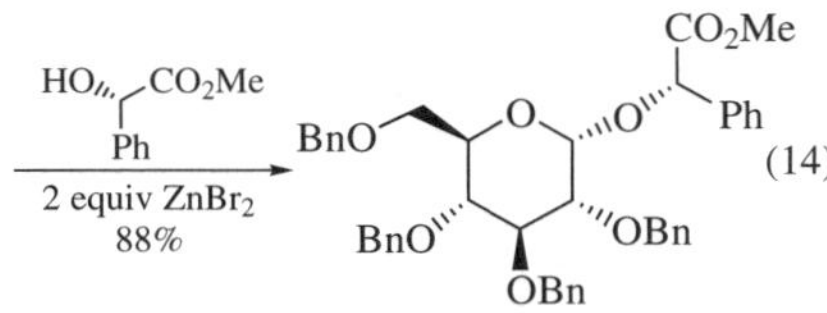

(14)

$\alpha:\beta = 10:1$

1. Staab, H. A. *Angew. Chem., Int. Ed. Engl.* **1962**, *1*, 351.
2. Staab, H. A.; Mannschreck, A. *Ber. Dtsch. Chem. Ges./Chem. Ber.* **1962**, *95*, 1284 (*Chem. Abstr.* **1962**, *57*, 5846d) .
3. Staab, H. A.; Lüking, M.; Dürr, F. H. *Ber. Dtsch. Chem. Ges./Chem. Ber.* **1962**, *95*, 1275 (*Chem. Abstr.* **1962**, *57*, 5908a) .
4. Paul, R.; Anderson, G. W. *J. Am. Chem. Soc.* **1960**, *82*, 4596.
5. Staab, H. A.; Bräunling, H. *Justus Liebigs Ann. Chem./Liebigs Ann. Chem.* **1962**, *654*, 119 (*Chem. Abstr.* **1962**, *57*, 5906c) .
6. Staab, H. A.; Jost, E. *Justus Liebigs Ann. Chem./Liebigs Ann. Chem.* **1962**, *655*, 90 (*Chem. Abstr.* **1962**, *57*, 15 090g) .
7. Brooks, D. W.; Lu, L. D. -L.; Masamune, S. *Angew. Chem., Int. Ed. Engl.* **1979**, *18*, 72.
8. Jerris, P. J.; Wovkulich, P. M.; Smith, A. B., III. *Tetrahedron Lett.* **1979**, 4517.
9. Staab, H. A. *Justus Liebigs Ann. Chem./Liebigs Ann. Chem.* **1957**, *609*, 75 (*Chem. Abstr.* **1958**, *52*, 7332e) .
10. Staab, H. A.; Benz, W. *Justus Liebigs Ann. Chem./Liebigs Ann. Chem.* **1961**, *648*, 72 (*Chem. Abstr.* **1962**, *57*, 4649g) .
11. Kamijo, T.; Harada, H.; Iizuka, K. *Chem. Pharm. Bull.* **1983**, *31*, 4189.
12. Ford, M. J.; Ley, S. V. *Synlett* **1990**, 255.
13. Foley, H. G.; Dalton, D. R. *Chem. Commun./J. Chem. Soc., Chem. Commun.* **1973**, 628.
14. Andruszkiewicz, R.; Czerwinski, A. *Synthesis* **1982**, 968.
15. Kang, S. -K.; Park, Y. -W.; Park, Y. -W.; Kim, S.-G.; Jeon, J. -H. *J. Chem. Soc., Perkin Trans. 1* **1992**, 405.
16. Kitagawa, T.; Kawaguchi, M.; Inoue, S.; Katayama, S. *Chem. Pharm. Bull.* **1991**, *39*, 3030.
17. Staab, H. A.; Wendel, K. *Q. Rev., Chem. Soc.* **1968**, *48*, 44.
18. Staab, H. A.; Datta, A. P. *Angew. Chem., Int. Ed. Engl.* **1964**, *3*, 132.
19. Hecht, R.; Rüchardt, C. *Ber. Dtsch. Chem. Ges./Chem. Ber.* **1963**, *96*, 1281 (*Chem. Abstr.* **1963**, *59*, 1523h) .
20. Ohta, S.; Shimabayashi, A.; Aono, M.; Okamoto, M. *Synthesis* **1982**, 833.
21. Katsuki, T. *Bull. Chem. Soc. Jpn.* **1976**, *49*, 2019.
22. Kamijo, T.; Harada, H.; Iizuka, K. *Chem. Pharm. Bull.* **1984**, *32*, 5044.
23. Morton, R. C.; Mangroo, D.; Gerber, G. E. *Synlett* **1988**, *66*, 1701.
24. (a) White, J. D.; Lodwig, S. N.; Trammell, G. L.; Fleming, M. P. *Tetrahedron Lett.* **1974**, 3263. (b) Colvin, E. W.; Purcell, T. A.; Raphael, R. A. *Chem. Commun./J. Chem. Soc., Chem. Commun.* **1972**, 1031.
25. Gais, H. -J. *Angew. Chem., Int. Ed. Engl.* **1977**, *16*, 244.
26. Paul, R.; Anderson, G. W. *J. Org. Chem.* **1962**, *27*, 2094.
27. Beyerman, H. C.; Van Den Brink, W. M. *Recl. Trav. Chim. Pays-Bas* **1961**, *80*, 1372.
28. Weygand, F.; Prox, A.; Schmidhammer, L.; König, W. *Angew. Chem., Int. Ed. Engl.* **1963**, *2*, 183.
29. Paul, R. *J. Org. Chem.* **1963**, *28*, 236.
30. Khatri, H.; Stammer, C. H. *Chem. Commun./J. Chem. Soc., Chem. Commun.* **1979**, 79.
31. Baker, D. C.; Putt, S. R. *Synthesis* **1978**, 478.
32. Ibarra, C. A.; Rodríguez, R. C.; Monreal, M. C. F.; Navarro, F. J. G.; Tesorero, J. M. *J. Org. Chem.* **1989**, *54*, 5620.
33. Garigipati, R. S.; Tschaen, D. M.; Weinreb, S. M. *J. Am. Chem. Soc.* **1985**, *107*, 7790.
34. Staab, H. A. *Justus Liebigs Ann. Chem./Liebigs Ann. Chem.* **1957**, *609*, 83 (*Chem. Abstr.* **1957**, *52*, 16341e) .
35. Wright, W. B., Jr. *J. Heterocycl. Chem.* **1965**, *2*, 41.
36. Kutney, J. P.; Ratcliffe, A. H. *Synth. Commun.* **1975**, *5*, 47.
37. D'Ischia, M.; Prota, G.; Rotteveel, R. C.; Westerhof, W. *Synth. Commun.* **1987**, *17*, 1577.

Alan Armstrong
University of Bath, UK

Chloroacetyl Chloride

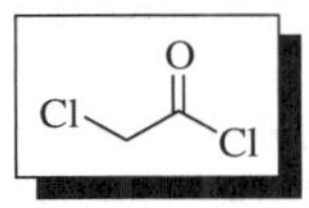

[79-04-9] $C_2H_2Cl_2O$ (MW 112.94)

(acylating agent;[1–34] chlorinating agent;[39] a two-carbon building block for cyclization;[6,13,25–27,33] a precursor to monochloroketene;[35–38])

Physical Data: mp −22 °C; bp 105–106 °C; *d* 1.418 g cm^{-3}.
Solubility: soluble in ether, acetone, and methylene chloride.
Form Supplied in: liquid; commercially available.
Handling, Storage, and Precautions: is corrosive and moisture sensitive. It should be used in an anhydrous atmosphere. Pure $ClCH_2COCl$ can be obtained by simple distillation of the commercial product. This reagent should be used in a fume hood.

Original Commentary

George A. Olah, G. K. Surya Parkash, Qi Wang & Xing-ya Li
University of Southern California, Los Angeles, CA, USA

Introduction. Chloroacetyl chloride is a versatile reagent and has been extensively used in organic synthesis. It is primarily used as an acylating agent. The multi-functional nature of $ClCH_2COCl$ has made it a suitable two-carbon building block for cyclization. Numerous cyclic compounds, especially heterocycles, have been prepared through the use of the reagent. Chloroacetyl chloride is also a valuable precursor for monochloroketene and occasionally serves as a chlorinating agent.

***C*-Acylation.** Acylation of carbon functionalities using $ClCH_2COCl$ as an acylating reagent generally falls into two categories: acylation under Friedel–Crafts conditions and under anionic conditions. Friedel–Crafts acylation of arenes using $ClCH_2COCl$ as an acylating agent is usually carried out in CS_2, nitromethane, sulfolane, etc. solution with Aluminum Chloride as a promoter, although other Lewis acids have also been employed. Benzene, toluene, naphthalene, *s*-hydrindacene, fluorene, and fluorenone react readily with $ClCH_2COCl$ under the reaction conditions.[1] Heterocyclic arenes, phenols, and derivatives of phenols are also similarly acylated.[2,3] In the case of phenols, the major acylation products are the *para* isomers (eq 1). However, excellent *ortho* regioselectivity can be achieved by combining $ClCH_2COCl$ with phenoxides of aluminum and magnesium (eq 2),[4] a condition similar to that used in the Kolbe–Schmitt reaction.[5]

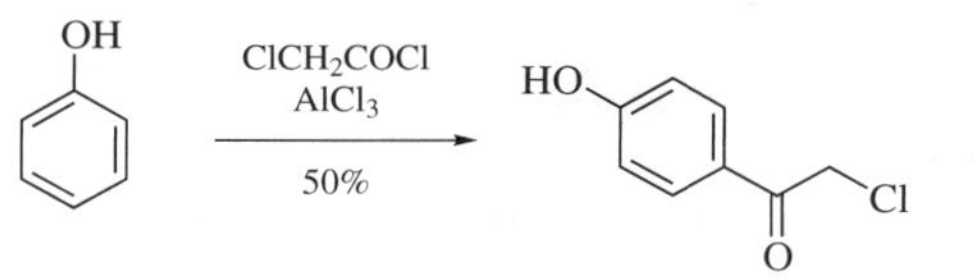
(1)

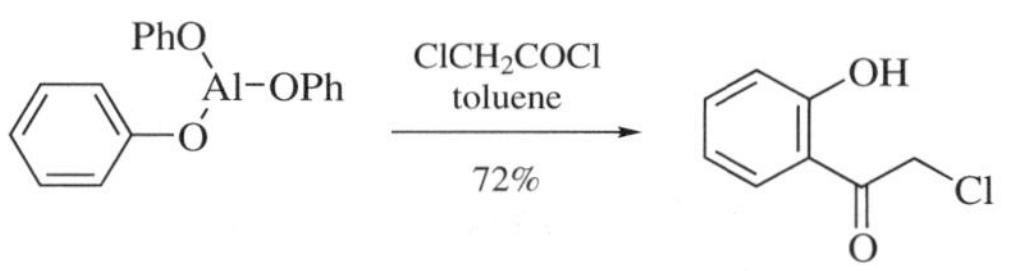
(2)

In the Friedel–Crafts reaction, both the acyl and the alkyl chloride functionalities of $ClCH_2COCl$ can react with an arene, resulting in the formation of a cyclic compound. For example, treatment of 1,2,3,4-tetrahydronaphthacene with $ClCH_2COCl/AlCl_3$ led to formation of 6-oxo-1,2,3,4,5,6-hexahydrobenz[*d*]aceanthrylene in 22% yield (eq 3).[6a] When 3,5-dimethoxytoluene is used as the substrate, a 2,3-dihydrobenzofuran-3-one is obtained almost quantitatively (eq 4).[6b,c] The carbonyl group of $ClCH_2COCl$ may also participate in cyclization under appropriate conditions.[6d,e] This behavior is illustrated by the formation of 2,7-difluoro-9-chloromethylenexanthene in the reaction of 4,4′-difluorophenyl ether with $ClCH_2COCl$ under $AlCl_3$ activation (eq 5).

(3)

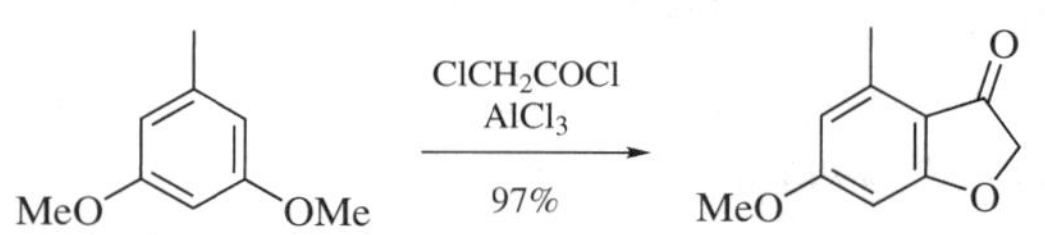
(4)

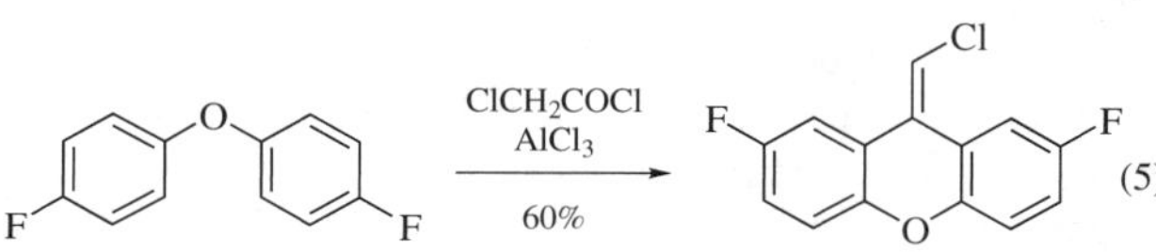
(5)

The Friedel–Crafts acylation reaction of $ClCH_2COCl$ has been extended to alkenes and alkynes.[7,8] When alkenes serve as substrates, addition or addition–elimination products can be obtained selectively depending on the reaction conditions. The addition–elimination reactions led to formation of α,β-unsaturated ketones (eq 6). In the case of 1,8-bis(trimethylsilyl)-2,6-octadiene acylation, using the $ClCH_2COCl$/ Titanium(IV) Chloride system at low temperature yielded chloromethylcyclopentanols (eq 7).[7e] When acetylene is treated with $ClCH_2COCl$ in the presence of $AlCl_3$, 1,4-dichlorobut-3-en-2-one was obtained in 75% yield (eq 8).[8a] In the presence of Sodium Tetracarbonylcobaltate, alkynes were cyclized by $ClCH_2COCl$ (eq 9).[8b]

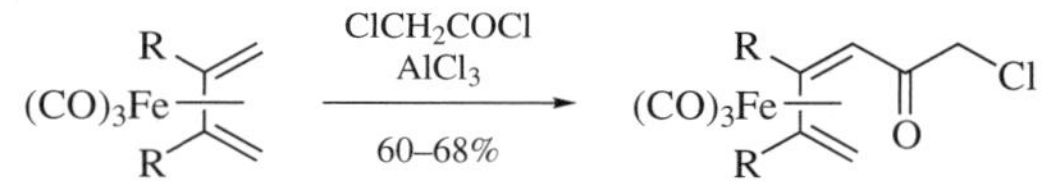
(6)

R = H, Me

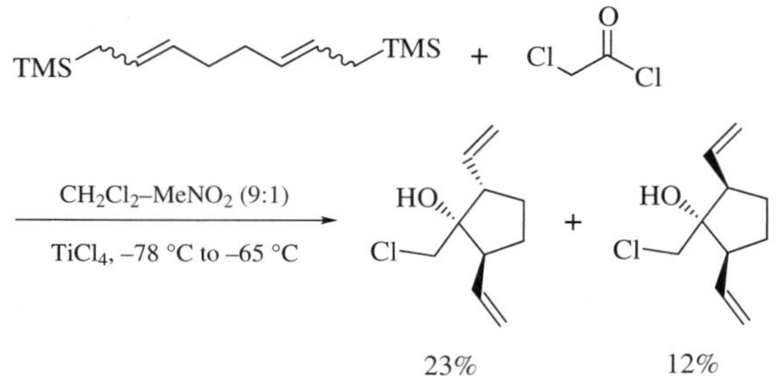
(7)

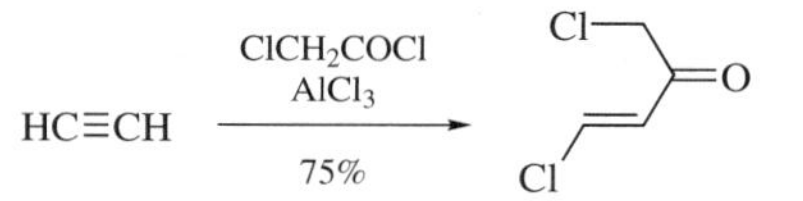
(8)

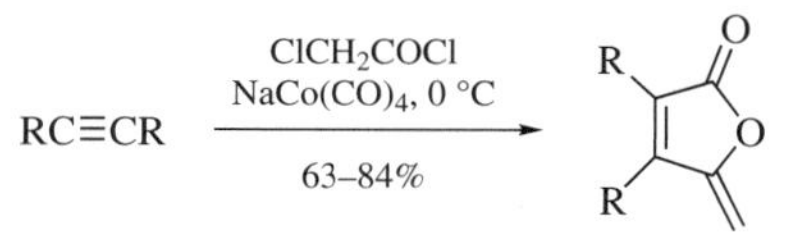
(9)

R = Et, *t*-Bu, CH_2SCN, TMS, CH_2OCH_2OMe

Organocadmium and organozinc compounds are generally the reagents of choice in the preparation of chloromethyl ketones from $ClCH_2COCl$,[9] even though Grignard reagents are sometimes also used for the preparation.[10] Strong bases such as Lithium Diisopropylamide (LDA) and Lithium Hexamethyldisilazide (LHMDS) have been introduced as deprotonation agents for the *C*-acylation reaction of a number of enolizable carbonyl compounds.[11] For the strongly acidic C–H bond of 1,3-cyclohexanedione, a weak base such as pyridine is sufficient to achieve this purpose.[12] An interesting reaction involving enolate *C*-acylation is the synthesis of 3-furanones by organocopper conjugate addition to α,β-unsaturated ketones. Trapping of the enolate with $ClCH_2COCl$ and subsequent base-promoted ring closure offers a very simple route to the construction of a furan ring (eq 10).[13]

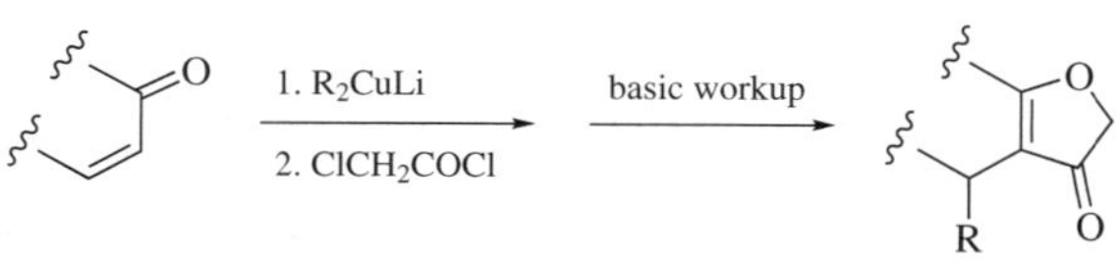
(10)

***N*-Acylation.** Acylation of nitrogen groups in organic compounds is the most frequently used reaction of $ClCH_2COCl$ in organic synthesis.[14–28] Primary and secondary amines react readily with $ClCH_2COCl$ to form amides.[14–17] In the case of diamines, both nitrogen atoms can be acylated. In fact, the diacylation has been utilized in the synthesis of cyclams[17a,b] and enantiomerically pure *trans*-2,3-diphenyl-DABCO.[17] Other nitrogen containing compounds such as *O,N*-dimethylhydroxylamine hydrochloride,[18a] benzyloxyguanidine,[18b] hydrazines,[18c] and amides[19] can also be acylated by $ClCH_2COCl$, under appropriate conditions. A one-pot procedure has been developed for reductive acylation of nitro compounds to chloroacetamides.[18d] If both amino and hydroxyl functionalities exist in the same molecular, *N*-acylation generally proceeds preferentially.[20]

Chloroacetyl chloride is able to add across the C=N double bonds of imines to form *N*-chloroalkylchloroacetamide.[21] These addition products generally are unstable and are used for further reaction without isolation. The C=N double bonds of imines can be reductively acylated in the presence of Trichlorosilane.[21f,g] The addition reaction is also applicable to carbodiimides[22] and other multiply bonded heteroatom functions such as P=N and S=N.[23]

1,3,5-Triisopropylhexahydrotriazine can undergo ring opening with $ClCH_2COCl$ in the presence of KSCN to form *N*-isopropyl-*N*-isothiocyanatomethylchloroacetamide in 67% yield (eq 11).[24]

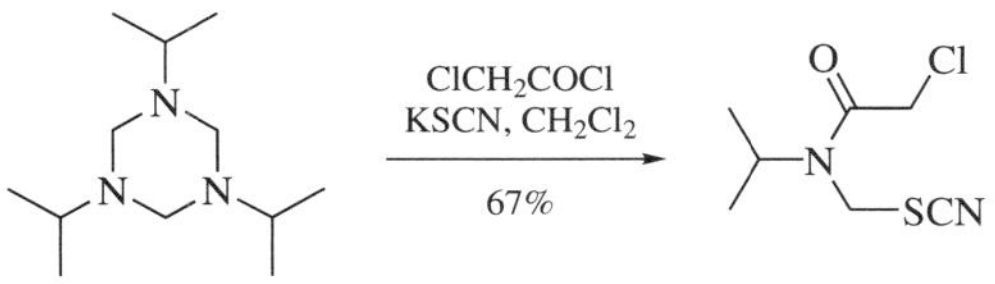

(11)

Similar to *C*-acylation, cyclocondensation can also proceed in *N*-acylation to form cyclic compounds.[25] The α-chlorine atom of $ClCH_2COCl$ is also reactive in this case.

The cyclization is not limited to the formation of only C–N and C–C bonds, but can also occur through forming C–N and C–N bonds.[26] This transformation was exemplified by the preparation of 3-hydroxy-1-phenylpyrido[1,2-α]pyrazinium chloride in the reaction of $ClCH_2COCl$ with phenyl(2-pyridyl)ketimine (eq 12).[26b] Besides the α-chlorine, the carbonyl group of $ClCH_2COCl$ can also be utilized for the cyclization.

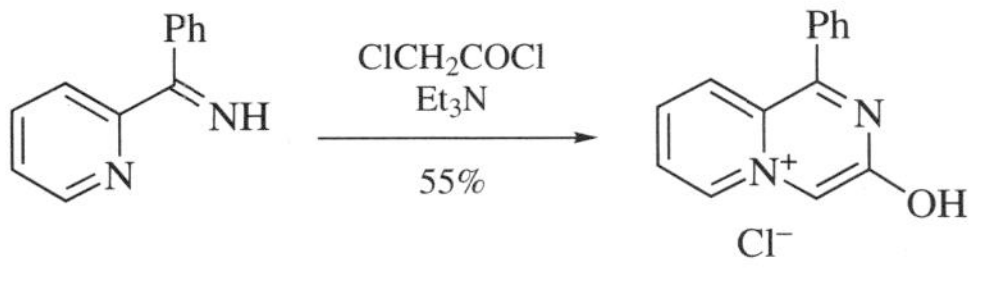

(12)

The chlorine atom at the α position of $ClCH_2COCl$ may also be displaced by an oxygen to form a ring containing both O and N atoms.[27] Cyclization of 4-amino-1-phenylisoquinolin-3-ol hydrochloride with $ClCH_2COCl$ generated 6-phenyl-2,3-dihydro-1*H*-[1,4]oxazino[2,3-*c*]isoquinolin-2-one in 40% yield (eq 13).[27b]

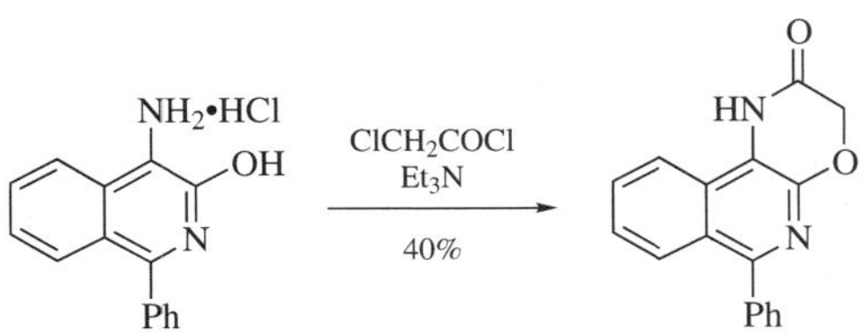

(13)

If a substrate has both N and S atoms, a heterocycle containing both atoms can be obtained upon treatment with $ClCH_2COCl$.[28] Preparation of thiazoles from thiobenzamide and $ClCH_2COCl$ has been accomplished through this process.[28c]

***O*-Acylation.** Like *N*-acylation, *O*-acylation using $ClCH_2COCl$ can be achieved either with or without a base.[29–33] Acylation has been successfully carried out with various hydroxy group-containing compounds such as alcohols,[29] phenols,[31] and saccharides[30] as substrates. Acylation of an enolate by $ClCH_2COCl$ has also been documented.[32] It is reported that treatment of salicylic hydrazide with $ClCH_2COCl$ in the presence of $MeSO_3H$ yielded 4,5-dihydro-5*H*-1,3,4-benzoxadiazepin-5-one in 59% yield (eq 14).[33]

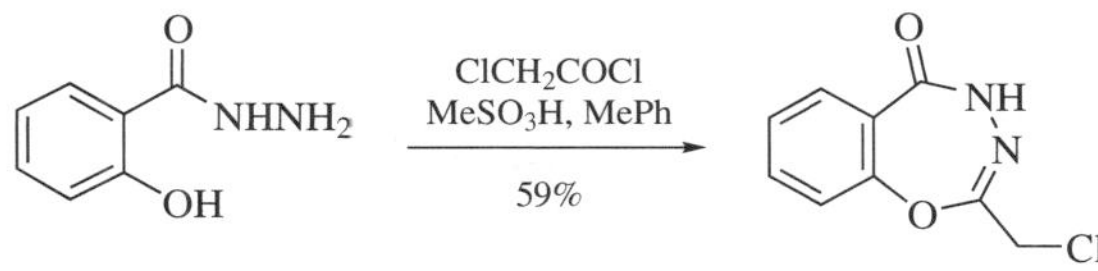

(14)

***S*-Acylation.** *S*-Acylation is a less well-studied reaction of $ClCH_2COCl$.[34] However, it has found use in the preparation of 2-oxo-1,4-dithiane and its derivatives, which can be further transformed to the analogs of tetrathiafulvalene for the study of conducting molecular solids.[34b,c]

Precursor to Monochloroketene. By treatment with Et_3N, $ClCH_2COCl$ loses HCl to yield monochloroketene.[35–38] The ketene generated by this method has been extensively used to perform [2 + 2] cycloadditions. The most widely studied reaction is the addition of the ketene to imines for the construction of the valuable β-lactam ring (eq 15).[35] This reaction proceeds readily with a wide range of imines with varying structures. Even diimines can be doubly cyclized. The in situ generated ketene was also trapped by allenes.[36a] An interesting method regarding the preparation of squaric acid is based the use of the monochloroketene.[36b]

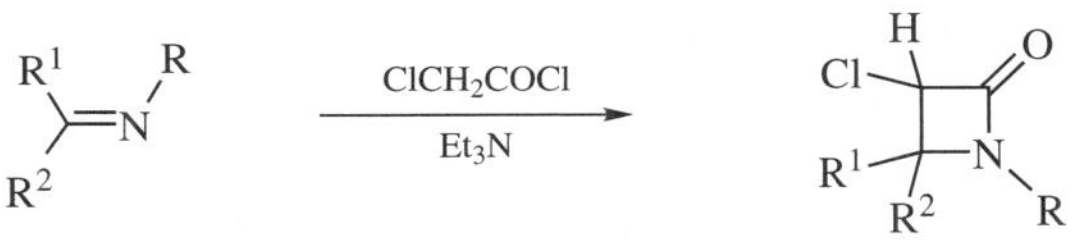

(15)

Beside [2 + 2] cycloadditions, [4 + 2] and [8 + 2] cycloadditions involving the in situ generated chloroketene have also been studied.[35a,37,38]

Chlorination. Chloroacetyl chloride occasionally serves as a chlorinating agent.[39] 2,4-Pentanedione was chlorinated at the 1- and 5-positions upon treatment with $ClCH_2COCl/AlCl_3$ in the presence of $Cu(OAc)_2$.[39a,b] Chloroacetyl chloride has also been used to transform 2-picoline *N*-oxide and 2-oxo-2-dimethylamino-5-methyl-1,2-oxophospholane to 2-chloromethylpyridine and 2-oxo-2-chloro-5-methyl-1,2-oxophospholane, respectively.[39c,d]

First Update

Thomas Ziegler
University of Tübingen, Germany

***C*-Acylation.** Friedel–Crafts acylations with chloroacetyl chloride catalyzed by Lewis acids can be enhanced in ionic solvents such as 1-butyl-3-methylimidazolium heptachlorodialuminate.[40] The regioselective acylation of indole derivatives under various Friedel–Crafts conditions has been extensively studied.[41] Indole itself may be acylated efficiently with chloroacetyl chloride at the 3-position when nitromethane is added as cosolvent.[41b] A rather uncommon Friedel–Crafts acylation with chloroacetyl chloride is observed under Montmorillonite-$FeCl_3$ catalysis which gives dichlorostyrene beside the regular acylation product.[42] Fries rearrangements of *p*-tolyl chloroacetates catalyzed by $FeCl_3$ have also been reported.[43] *C*-Acylations under anionic conditions have been performed with Meldrum's acid,[44] β-zinc iodides of amino acids under palladium catalysis (eq 16),[45] and with titanium(IV)ene-diolates.[46]

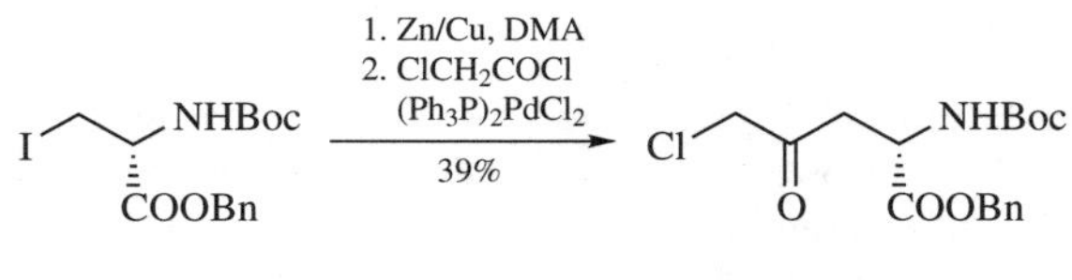

(16)

A chemoselective acylation of alkoxy functionalized organolithium compounds was achieved when lithiated isochromane was reacted successively with zinc dibromide-copper cyanide, benzoic acid anhydride and chloroacetyl chloride (eq 17).[47]

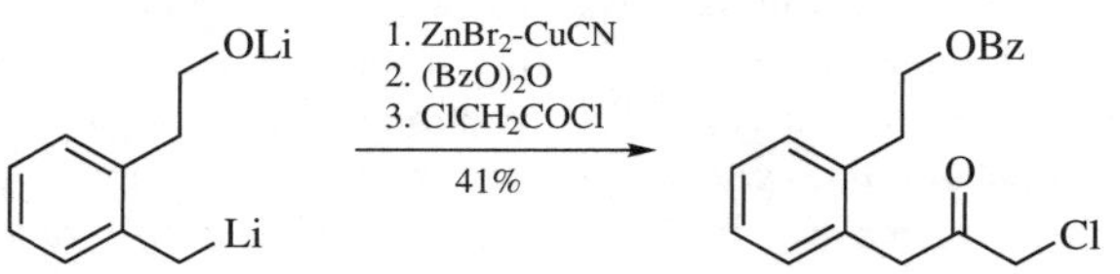

(17)

***N*-Acylation.** *N*-Acylations with chloroacetyl chloride can be achieved either by simply stirring an amine with the acid chloride in a suitable solvent[48] or by adding a base.[49] In the case of *N*-acylation of 1,2-thiazetidine 1,1-dioxides, deprotonation with butyl lithium is required.[49c] Zeolites have also been applied to the promotion of *N*-acylations with chloroacetyl chloride.[50]

Most *N*-acyaltions with chloroacetyl chloride involve ring closure reaction in a succeeding step to form various heterocycles: pyranazines,[51] pyrimidines,[52] and pyridines[53] can be prepared easily in this manner. For example, in a straightforward synthesis of Cialis® *N*-chloroacetylation of a piperonal imine derived from tryptophan methyl ester provokes a Pictet-Sprengler reaction, which is followed by methyl amine induced closure of the annelated pyrazine ring (eq 18).[51a]

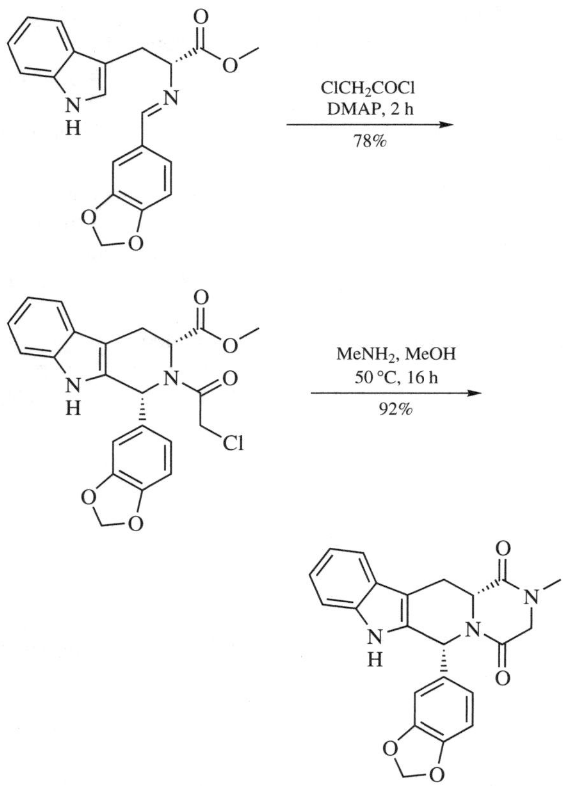

(18)

Oxindoles are accessible by palladium-catalyzed ring closure of α-chloroacetanilides which, in turn, are easily prepared by condensation of the corresponding anilines with chloroacetyl chloride (eq 19).[54]

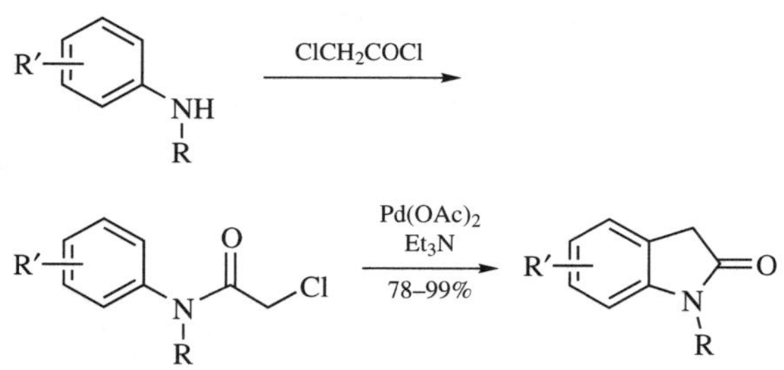

(19)

***O*-Acylation.** Regioselective *O*-acylation with chloroacetyl chloride has gained significant importance for temporary protection of carbohydrates since the resulting chloroacetyl esters can be selectively cleaved in the presence of other esters.[55a] For example, (eq 20) preliminary to the construction of an oligosaccharide library, it was demonstrated that a chloroacetate ester could be cleanly removed in the presence of both a standard carboxylate ester and a levulinate ester. At the same time, it was possible to cleave the levulinate with hydrazine hydrate, while leaving the chloroacetate in tact (eq 20).[55b]

Likewise it has been demonstrated that a chloroacetate ester may be selectively hydrolyzed in the presence of a pair of thiol esters, albeit hindered ones (eq 21).[55g]

O-Chloroacetylation can also serve for the introduction of a glycolic acid ester residue by substitution of chloride with formate followed by acidic methanolysis of the formate moiety.[56] Treatment of 2-methyltetrahydrofuran with chloroacetyl chloride and zinc chloride gives rise to 4-chloropentyl chloroacetate in 97% yield.[57] α,β-Dehydroamino acid esters form when hydroxylamino esters are treated with chloroacetyl chloride and triethylamine or DBU (eq 22).[58]

Similar to *N*-acylations, *O*-acylations with chloroacetyl chloride with subsequent ring closure reactions can be applied to the synthesis of heterocyclic compounds.[53] Yet another application of chloroacetyl chloride is its reaction with polyethyleneglycol or other polymeric supports in combinatorial syntheses.[59]

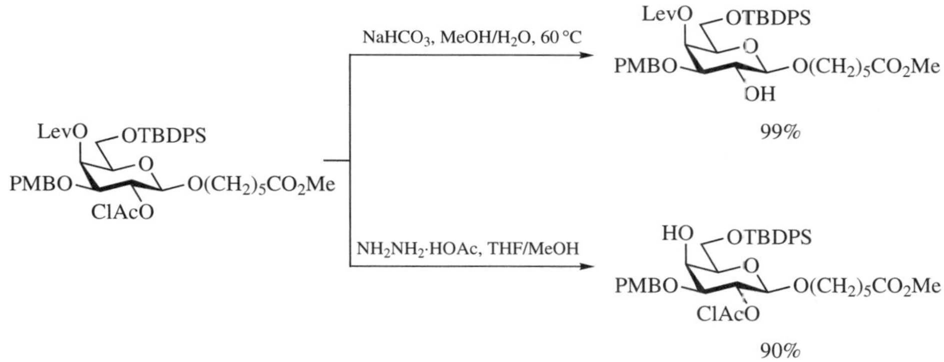

(20)

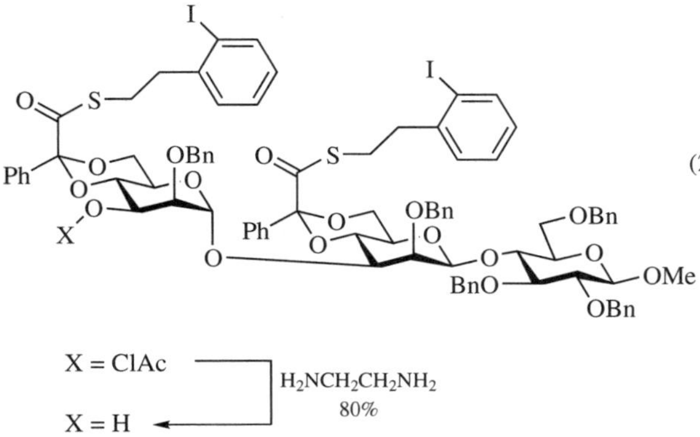

(21)

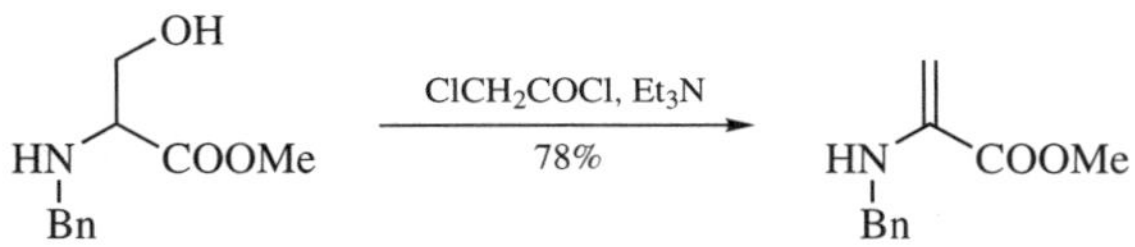

(22)

S-Acylation. *S*-Acylation with chloroacetyl chloride has not been well-studied. However, it was shown that chloroacetyl chloride reacts with sodium disulfide and sulphur to form rubber-like thiocol polysulfide polymers.[60] In a rather uncommon reaction heating of selenazadienes with chloroacetyl chloride results in alkylation of the selenium atom followed by ring closure condensation to form 1,3-selenazole-5-carboxylic acids in moderate yield (eq 23).[61]

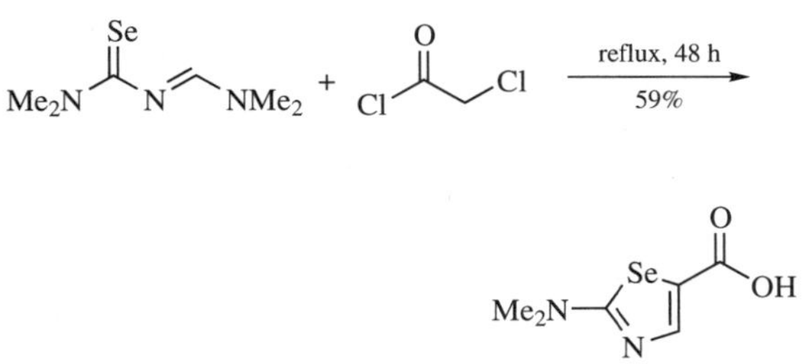

(23)

Precursor to Monochloroketene. [2 + 2] Cycloadditions of imines and monochloroketene generated in situ from chloroacetyl chloride with a suitable base to give the corresponding β-lactams were studied extensively.[62] An alternative method of preparation of β-lactam from chloroacetyl chloride involves an efficient two step procedure in which imines are first acylated with choroacetyl chloride, followed by ring closure with trialkyl phosphates.[63] In situ generated monochloroketene can also add to the amino groups of adenine, cytosine, and guanine,[64] and to cyclic *N*-benzoyl hydrazones forming oxadiazoles.[65] A remarkable reaction is the [2 + 2] cycloaddition of in situ generated monochloroketene to olefins catalyzed by soft Lewis-acids like $SnBr_4$ (eq 24).[66]

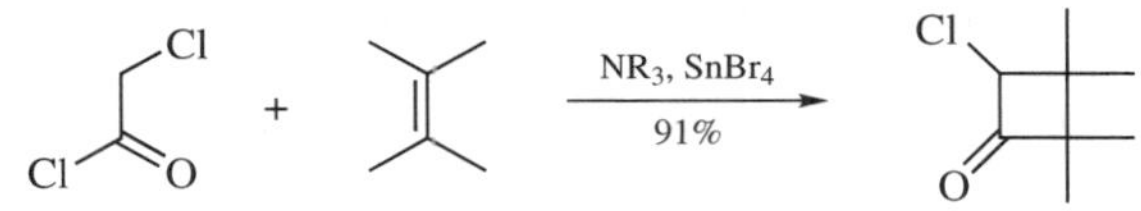

(24)

1-Azaazulenes-2(1*H*)-ones are formed in high yield from 2,4,6-cycloheptatrien-1-imines upon reaction with monochloroketene generated in situ from chloroacetyl chloride (eq 25).[67]

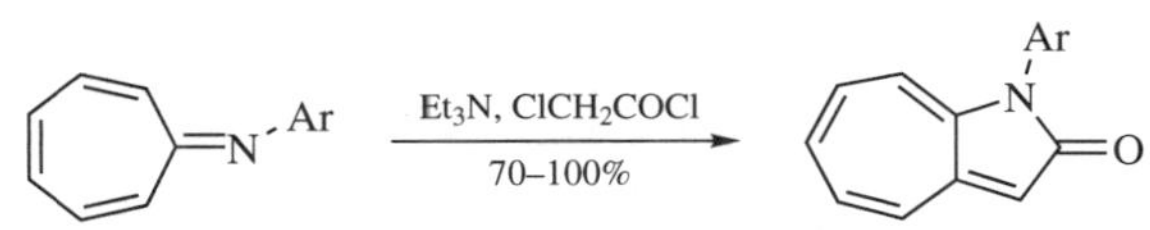

(25)

Related Reagents. Chloroacetyl fluoride; dichloroketene.

1. (a) Olah, G. A.; Kobayashi, S., *J. Am. Chem. Soc.* **1971**, *93*, 6964. (b) Dumpis, M. A.; Kudryashova, N. I.; Khromov-Borisov, N. V., *J. Org. Chem. USSR (Engl. Transl.)* **1980**, *16*, 1799. (c) Vejdelek, Z. J.; Bartosova, M.; Capek, A.; Protiva, M., *Collect. Czech. Chem. commun.* **1978**, *43*, 970. (d) Albrecht, W. L.; Fleming, R. W.; Horgan, S. W.; Kihm, J. C.; Mayer, G. D., *J. Med. Chem.* **1974**, *17*, 886.
2. (a) Belen'kii, L. I.; Yakubov, A. P.; Gol'dfarb, Ya. L., *J. Org. Chem. USSR (Engl. Transl.)* **1970**, *6*, 2531. (b) Belen'kii, L. I.; Yakubov, A. P.; Gol'dfarb, Ya. L., *J. Org. Chem. USSR (Engl. Transl.)* **1970**, *6*, 2537. (c) Belen'kii, L. I.; Yakubov, A. P.; Gol'dfarb, Ya. L., *J. Org. Chem. USSR (Engl. Transl.)* **1975**, *11*, 412. (d) del Agua, M. J.; Alvarez-Insua, A. S.; Conde, S., *J. Heterocycle Chem.* **1981**, *18*, 1345. (e) Murakami, Y.; Tani, M.; Tanaka, K.; Yokoyama, Y., *Heterocycles* **1984**, *22*, 241. (f) Nakayama., J.; Murabayashi, S.; Hoshino, M., *Heterocycles* **1987**, *26*, 2599. (g) Nakatsuka, S.; Asano, O.; Goto, T., *Chem. Lett.* **1987**, 1225. (h) Murakami, Y.; Tani, M.; Ariyasu, T.; Nishiyama, C.; Watanabe, T.; Yokoyama, Y., *Heterocycles* **1988**, *27*, 1855.
3. (a) Giordano, C.; Castaldi, G.; Casagrande, F.; Belli, A., *J. Chem. Soc., Perkin Trans. 1* **1982**, 2575. (b) Babad, E.; Carruthers, N. I.; Jaret, R. S.; Steinman, M., *Synthesis* **1988**, 966. (c) Gurjar, M. K.; Joshi, S. V.; Sastry, B. S.; Rao, A. V. R., *Synth. Commun.* **1990**, *20*, 3489. (d) Jolly, C. A.; Wang, F.; Krichene, S.; Reynolds, J. R.; Cassoux, P.; Faulmann, C., *Synth. Met.* **1989**, *29*, F189.
4. Sartori, G.; Casnati, G.; Bigi, F., *J. Org. Chem.* **1990**, *55*, 4371.
5. Lindsey, A. S.; Jeskey, H., *Chem. Rev.* **1957**, *57*, 583.
6. (a) Sangaiah, R.; Gold, A., *J. Org. Chem.* **1987**, *52*, 3205. (b) Murai, A.; Sato, S.; Masamune, T., *Chem. Commun./J. Chem. Soc.* **1982**, 511. (c) Murai, A.; Sato, S.; Masamune, T., *Bull. Chem. Soc. Jpn.* **1984**, *57*,

2286. (d) Granoth, I.; Pownall, H. J., *J. Org. Chem.* **1975**, *40*, 2088. (e) Granoth, I.; Kalir, A., *J. Org. Chem.* **1973**, *38*, 841.

7. (a) Baddeley, G.; Taylor, H. T.; Pickles, W., *J. Chem. Soc.* **1953**, 124. (b) Sieglitz, A.; Horn, O., *Ber. Dtsch. Chem. Ges./Chem. Ber.* **1951**, *84*, 607. (c) Franck-Neumann, M.; Abdali, A.; Colson, P. J.; Sedrati, M., *Synlett* **1991**, 331. (d) Quijano, M. L.; Nogueras, M.; Sanchez, A.; de Cienfuegos, G. A.; Melgarejo, M., *J. Heterocycle Chem.* **1990**, *27*, 1079. (e) Tubul, A.; Ouvrard, P.; Santelli, M., *Synthesis* **1991**, 173.

8. (a) Clayton, J. P.; Guest, A. W.; Taylor, A. W., *Chem. Commun./J. Chem. Soc.* **1979**, 500. (b) Krafft, M. E.; Pankowski, J., *Synlett* **1991**, 865.

9. (a) Cason, J., *Chem. Rev.* **1947**, *40*, 15. (b) Bergman, J.; Venemalm, L., *Tetrahedron* **1990**, *46*, 6061. (c) Jackson, R. F. W.; Wood, A.; Wythes, M., *Synlett* **1990**, 735. (d) Jackson, R. F. W.; Wishart, N.; Wood, A.; Wythes, M.; James, K., *J. Org. Chem.* **1992**, *57*, 3397.

10. (a) Karaev, S. F.; Guseinov, S. O.; Akhundov, E. A., *J. Gen. Chem. USSR (Engl. Transl.)* **1981**, *51*, 1163. (b) Furstner, A.; Kollegger, G.; Weidmann, H., *J. Organomet. Chem.* **1991**, *414*, 295.

11. (a) Mastalerz, H.; Vinet, V., *Chem. Commun./J. Chem. Soc.* **1987**, 1283. (b) Vedejs, E.; Reid, J. G.; Rodgers, J. D.; Wittenberger, S. J., *J. Am. Chem. Soc.* **1990**, *112*, 4351. (c) Kumar, V.; Daum, S. J.; Bell, M. R.; Alexander, M. A.; Christiansen, R. G.; Ackerman, J. H.; Krolski, M. J.; Pilling, G. M.; Herrmann, J. L.; Winneker, R. C.; Wagner, M. M., *Tetrahedron* **1991**, *47*, 5099. (d) Sakaki, J.; Sugita, Y.; Sato, M.; Kaneko, C., *Chem. Commun./J. Chem. Soc.* **1991**, 434.

12. Pshenichnyi, V. N.; Gulyakevich, O. V.; Khripach, V. A., *J. Org. Chem. USSR (Engl. Transl.)* **1989**, *25*, 1699.

13. (a) Bernasconi, S.; Jommi, G.; Montanari, S.; Sisti, M., *Synthesis* **1987**, 1126. (b) Bernasconi, S.; Gariboldi, P.; Jommi, G.; Sisti, M.; Tavecchia, P., *J. Org. Chem.* **1981**, *46*, 3719.

14. (a) Chenault, H. K.; Dahmer, J.; Whitesides, G. M., *J. Am. Chem. Soc.* **1989**, *111*, 6354. (b) Hermecz, I., *J. Heterocycle Chem.* **1987**, *24*, 1473. (c) Lukanov, L. K.; Venkov, A. P., *Synthesis* **1992**, 263. (d) Heckendorn, R., *Helv. Chim. Acta* **1990**, *73*, 1700. (e) Tsuge, O.; Kanemasa, S.; Yoshioka, M., *J. Org. Chem.* **1988**, *53*, 5649. (f) Lemus, R. H.; Skibo, E. B., *J. Org. Chem.* **1992**, *57*, 5649. (g) Massa, S.; Mai, A.; Corelli, F., *Tetrahedron Lett.* **1988**, *29*, 6471. (h) Kruizinga, W.; Kingma, R.; Leegstra, L.; Rokaszewski, E.; Gruppen, G.; Kellogg, R. M., *Chem. Commun./J. Chem. Soc.* **1990**, 1647. (i) Orena, M.; Porzi, G.; Sandri, S., *J. Org. Chem.* **1992**, *57*, 6532. (j) Compernolle, F.; Saleh, M. A.; Toppet, S.; Hoornaert, G., *J. Org. Chem.* **1991**, *56*, 5192. (k) White, B. D.; Mallen, J. M.; Arnold, K. A.; Fronczek, F. R.; Gandour, R. D.; Gehrig, L. M. B.; Gokel, G. W., *J. Org. Chem.* **1989**, *54*, 937. (l) El-Shafei, A. K.; El-Sayed, A. M.; Khalaf, A. A., *Bull. Chem. Soc. Jpn.* **1983**, *56*, 918. (m) Korshin, E. E.; Soboleva, G. I.; Levin, Y. A.; Zakharova, L. G.; Litvinov, I. A.; Naumov, V. A.; Podval'nyi, E. A.; Efremov, Y. Y., *J. Org. Chem. USSR (Engl. Transl.)* **1992**, *28*, 973.

15. (a) Charles, E. S.; Sharma, S., *Indian J. Chem., Sect. B* **1987**, *26B*, 752. (b) Katz, H. E.; Schilling, M. L., *J. Am. Chem. Soc.* **1989**, *111*, 7554.

16. (a) Wanner, M. J.; Koomen, G. J., *Tetrahedron Lett.* **1989**, *30*, 2301. (b) Carelli, I.; Inesi, A., *Synthesis* **1986**, 591. (c) Krafft, G. A.; Sutton, W. R.; Cummings, R. T., *J. Am. Chem. Soc.* **1988**, *110*, 301. (d) Lee, M.; Rhodes, A. L.; Wyatt, M. D.; D'Incalci, M.; Forrow, S.; Hartley, J. A., *J. Med. Chem.* **1993**, *36*, 863. (e) Smith, B. D.; Paugam, M-F.; Haller, K. J., *J. Chem. Soc., Perkin Trans. 2* **1993**, 165. (f) Fetter, J.; Czuppon, T.; Hornyak, G.; Feller, A., *Tetrahedron* **1991**, *47*, 9393. (g) Ananthanarayanan, C.; Ramakrishnan, V. T., *Indian J. Chem., Sect. B* **1988**, *27B*, 156. (h) El-Ezbawy, S. R.; Alshaikh, M. A., *Phosphorus Sulfur/Phosphorus Sulfur Silicon* **1990**, *48*, 111.

17. (a) Bradshaw, J. S.; Krakowiak, K. E.; Wu, G.; Izatt, R. M., *Tetrahedron Lett.* **1988**, *29*, 5589. (b) Bradshaw, J. S.; Krakowiak, K. E.; Izatt, R. M.; Zamecka-Krakowiak, D. J., *Tetrahedron Lett.* **1990**, *31*, 1077. (c) Oi, R.; Sharpless, K. B., *Tetrahedron Lett.* **1991**, *32*, 4853. (d) Tsizin, Y. S.; Sergovskaya, N. L.; Chernyak, S. A., *Chem. Heterocycl. Compd.* **1986**, 421.

18. (a) Nuzillard, J-M.; Boumendjel, A.; Massiot, G., *Tetrahedron Lett.* **1989**, *29*, 3779. (b) Campbell, M. M.; Campbell, A. C.; Peace, A.; Pick, J.; Woods, G. F., *Chem. Commun./J. Chem. Soc.* **1985**, 1164. (c) Taylor, E. C.; Hinkle, J. S., *J. Org. Chem.* **1987**, *52*, 4107. (d) Krowicki, K.; Lown, J. W., *J. Org. Chem.* **1987**, *52*, 3493.

19. (a) Orena, M.; Porzi, G.; Sandri, S., *Tetrahedron Lett.* **1992**, *33*, 3797. (b) Takeuchi, H.; Hagiwara, S.; Eguchi, S., *Tetrahedron* **1989**, *45*, 6375. (c) Rigo, B.; Lespagnol, C.; Pauly, M., *J. Heterocycle Chem.* **1988**, *25*, 59.

20. (a) Krakowiak, K. E.; Bradshaw, J. S.; Forsnes, E. V.; Izatt, R. M., *J. Heterocycle Chem.* **1989**, *26*, 661. (b) Gennari, C.; Poli, G.; Scolastico, C.; Vassallo, M., *Tetrahedron: Asymmetry* **1991**, *2*, 793. (c) Kikelj, D.; Kidric, J.; Pristovsek, P.; Pecar, S.; Urleb, U.; Krbavcic, A.; Honig, H., *Tetrahedron* **1992**, *48*, 5915.

21. (a) Hassner, A.; Burke, S. S.; , *J. Am. Chem. Soc.* **1975**, *97*, 4692. (b) Venkov, A. P.; Nikolova, M. S.; Mollov, N. M., *Chem. Ind. (London)* **1982**, 808. (c) Venkov, A. P.; Statkova, S. M.; Ivanov, I. I., *Synth. Commun.* **1992**, *22*, 125. (d) Venkov, A.; Lukanov, L. K., *Synthesis* **1989**, 59. (e) Bayard, P.; Sainte, F.; Beaudeginies, R.; Ghosez, L., *Tetrahedron Lett.* **1988**, *31*, 3799. (f) Okamoto, H.; Kato, S., *Bull. Chem. Soc. Jpn.* **1991**, *64*, 2128. (g) Okamoto, H.; Kobutani, T.; Kato, S., *Bull. Chem. Soc. Jpn.* **1992**, *65*, 674.

22. (a) Brady, W. T.; Owens, R. A., *J. Heterocycle Chem.* **1977**, *14*, 179. (b) Brady, W. T.; Owens, R. A., *J. Org. Chem.* **1977**, *42*, 3220.

23. (a) Yonezawa, Y.; Kisuno, A.; Shin, C., *Bull. Chem. Soc. Jpn.* **1986**, *59*, 3696. (b) Bruche, L.; Garanti, L.; Zecchi, G., *Synthesis* **1986**, 772. (c) Jones, S. D.; Kennewell, P. D.; Tulley, W. R.; Westwood, R.; Sammes, P. G., *J. Chem. Soc., Perkin Trans. 1* **1990**, 447.

24. Vass, A.; Szalontai, G., *Synthesis* **1986**, 817.

25. (a) Hogale, M. B.; Salunkhe, V. K.; Kachare, D. S., *J. Indian Chem. Soc.* **1989**, *66*, 484. (b) Sagitdinova, K. F.; Ibatullin, U. G.; Gallyamova, Z. R.; Rakhmatullina, G. A.; Akmanova, N. A.; Safarov, M. G., *Chem. Heterocycl. Compd.* **1990**, 932.

26. (a) Walser, A.; Flynn, T.; Mason, C., *J. Heterocycle Chem.* **1991**, *28*, 1121. (b) Batori, S.; Messmer, A., *J. Heterocycle Chem.* **1990**, *27*, 1673. (c) Okawara, T.; Kato, R.; Yamasaki, T.; Yasuda, N.; Furukawa, M., *Heterocycles* **1986**, *24*, 885. (d) Liu, K-C.; Shih, B-J.; Chern, J-W., *J. Heterocycle Chem.* **1989**, *26*, 457. (e) Szilagyi, G.; Matyus, P.; Sohar, P., *Tetrahedron* **1989**, *45*, 7921. (f) Misztal, S.; Dukat, M.; Mokrosz, J. L., *J. Chem. Soc., Perkin Trans. 1* **1990**, 2311. (g) Ichinari, M.; Nakayama, K.; Hayase, Y., *Heterocycles* **1988**, *27*, 2635.

27. (a) Brown, G. R.; Foubister, A. J., *J. Chem. Soc., Perkin Trans. 1* **1989**, 1401. (b) Hussein, A. Q., *Heterocycles* **1987**, *26*, 163. (c) Tikk, I.; Deak, G.; Sohar, P.; Tamas, J., *J. Heterocycle Chem.* **1988**, *25*, 273. (d) Agirbas, H.; Sumengen, D.; Durust, Y.; Durust, N.; , *Synth. Commun.* **1992**, *22*, 209. (e) Rao, M. N.; Holkar, A. G.; Ayyangar, N. R., *Tetrahedron Lett.* **1990**, *31*, 3343. (f) Melillo, D. G.; Larsen, R. D.; Mathre, D. J.; Shukis, W. F.; Wood, A. W.; Colleluori, J. R., *J. Org. Chem.* **1987**, *52*, 5143. (g) Brown, G. R.; Foubister, A. J.; Wright, B., *J. Chem. Soc., Perkin Trans. 1* **1987**, 557. (h) Brown, G. R.; Foubister, A. J.; Stribling, D., *J. Chem. Soc., Perkin Trans. 1* **1987**, 547.

28. (a) Hammouda, H. A.; Abd-Allah, S. O.; Hussain, S. M.; Yousef., N. M., *Gazz. Chim. Ital.* **1984**, *114*, 201. (b) El-Bahaie, S.; Bayoumy, E. E.; Youssif, S., *J. Indian Chem. Soc.* **1988**, *65*, 695. (c) Kopka, I. E., *Tetrahedron Lett.* **1988**, *29*, 3765. (d) Molina, P.; Arques, A.; Alias, A., *Tetrahedron* **1992**, *48*, 1285. (e) Okawara, T.; Kato, R.; Yasuda, N.; Yamasaki, T.; Furukawa, M., *J. Chem. Res. (S)* **1987**, 254.

29. (a) Tanikaga, R.; Jun, T. X.; Kaji, A., *J. Chem. Soc., Perkin Trans. 1* **1990**, 1185. (b) Ngooi, T. K.; Scilimati, A.; Guo, Z.; Sih, C., *J. Org. Chem.* **1989**, *54*, 911. (c) Schmidt, U.; Kroner, M.; Griesser, H., *Synthesis* **1991**, 294. (d) Fasth, K-J.; Antoni, G.; Langstrom, B., *J. Chem. Soc., Perkin Trans. 1* **1988**, 3081. (e) Genet, J. P.; Kopola, N.; Juge, S.; Ruiz-Montes, J.; Antunes, O. A. C.; Tanier, S., *Tetrahedron Lett.* **1990**, *31*, 3133. (f) Gokel, G. W.; Hernandez, J. C.; Viscariello, M. J.; Arnold, K. A.; Campana, C. F.; Echegoyen, L.; Fronczek, F. R.; Gandour, R. D.; Morgan, C. R.; Trafton, J. E.; Miller, S. R.; Minganti, C.; Eiband, D.; Schultz, R. A.; Tamminen, M., *J. Org. Chem.* **1987**, *52*, 2663. (g) Schwartz, A.; Madan, P. B.; Mohacsi, E.; O'Brien, J. P.; Todaro, L. J.;

Coffen, D. L., *J. Org. Chem.* **1992**, *57*, 851. (h) Franck-Neumann, M.; Briswalter, C.; Chemla, P.; Martina, D., *Synlett* **1990**, 637.

30. (a) Pozsgay, V.; Glaudemans, C. P. J.; Robbins, J. B.; Schneerson, R., *Tetrahedron* **1992**, *48*, 10249. (b) Nilsson, S.; Lonn, H.; Norberg, T., *J. Carbohydr. Chem.* **1991**, *10*, 1023. (c) Lubineau, A.; Meter, E.; Place, P., *Carbohydr. Res.* **1992**, *228*, 191.
31. (a) Himbert, G.; Fink, D.; Diehl, K., *Ber. Dtsch. Chem. Ges./Chem. Ber.* **1988**, *121*, 431. (b) Briody, J. M.; Satchell, D. P. N., *J. Chem. Soc.* **1965**, 168. (c) Strazzolini, P.; Verardo, G.; Giumanini, A. G., *J. Org. Chem.* **1988**, *53*, 3321.
32. (a) Velezheva, V. S.; Ryabova, S. Y., *Chem. Heterocycl. Compd.* **1990**, 333. (b) Cunico, R.; Kuan, C. P., *J. Org. Chem.* **1992**, *57*, 3331.
33. Lee, K-J.; Kim, S. H.; Kim, S.; Cho, Y. R., *Synthesis* **1992**, 929.
34. (a) Kang, J.; Lim, D. S., *Synlett* **1990**, 611. (b) Larsen, J.; Lenoir, C., *Synthesis* **1989**, 134. (c) Muller, H.; Fritz, H. P.; Nemetschek, R.; Hackl, R.; Biberacher, W.; Heidmann, C-P., *Z. Naturforsch., Teil B* **1992**, *47B*, 718.
35. (a) Duran, F.; Ghosez, L., *Tetrahedron Lett.* **1970**, 245. (b) Bentley, R. L.; Suschitzky, H., *J. Chem. Soc., Perkin Trans. 1* **1976**, 1725. (c) Cardellini, M.; Claudi, F.; Moracci, M. F., *Synthesis* **1984**, 1070. (d) Yamamoto, I.; Abe, I.; Nozawa, M.; Kotani, M.; Motoyoshiya, J.; Gotoh, H.; Matsuzaki, K., *J. Chem. Soc., Perkin Trans. 1* **1983**, 2297. (e) Wharton, C. J.; Wrigglesworth, R.; Rowe, M., *J. Chem. Soc., Perkin Trans. 1* **1984**, 29. (f) Alcaide, B.; Dominguez, G.; Escobar, G.; Parreno Plumet, J., *Heterocycles* **1986**, *24*, 1579. (g) Fodor, L.; Szabo, J.; Szucs, E.; Bernath, G.; Sohar, P.; Tamas, J., *Tetrahedron* **1984**, *40*, 4089. (h) Sohar, P.; Stajer, G.; Pelczer, I.; Szabo, A. E.; Szunyog, J.; Bernath, G., *Tetrahedron* **1985**, *41*, 1721. (i) Szollosy, A.; Kotovych, G.; Toth, G.; Levai, A., *Can. J. Chem.* **1988**, *66*, 279. (j) Alcaide, B.; Martin-Cantalejo, Y.; Perez-Castells, J.; Rodriguez-Lopez, J.; Sierra, M.; Monge, A.; Perez-Garcia, V., *J. Org. Chem.* **1992**, *57*, 5921. (k) Hogale, M. B.; Uthale, A. C., *Indian J. Chem., Sect. B* **1990**, *29B*, 590. (l) Srivastava, V. K.; Palit, G.; Singh, S.; Dhawan, R.; Shanker, K., *J. Indian Chem. Soc.* **1990**, *67*, 335. (m) El-Ezbawy, S. R.; Abdel-Wahab, A. A., *Phosphorus Sulfur/Phosphorus Sulfur Silicon* **1991**, *57*, 127. (n) Vansdadia, R. N.; Roda, K. P.; Parekh, H., *J. Indian Chem. Soc.* **1989**, *66*, 56.
36. (a) Brady, W. T.; Stockton, J. D.; Patel, A. D., *J. Org. Chem.* **1974**, *39*, 236. (b) Bellus, D., *J. Org. Chem.* **1979**, *44*, 1208.
37. (a) Brady, W. T.; Agho, M. O., *Synthesis* **1982**, 500. (b) Manrao, M. R.; Kumar, A.; Kalsi, P. S., *J. Indian Chem. Soc.* **1986**, *63*, 410.
38. (a) Machiguchi, T.; Yamabe, S., *Chem. Lett.* **1990**, 1511. (b) Ito, K.; Saito, K.; Takahashi, K., *Bull. Chem. Soc. Jpn.* **1992**, *65*, 812.
39. (a) Rychnovsky, S. D.; Griesgraber, G.; Zeller, S.; Skalitzky, D. J., *J. Org. Chem.* **1991**, *56*, 5161. (b) Matsui, K.; Motoi, M.; Nojiri, T., *Bull. Chem. Soc. Jpn.* **1973**, *46*, 562. (c) Ash, M. L; Pews, R. G., *J. Heterocycle Chem.* **1981**, *18*, 939. (d) Neretdinova, O. N.; Guseva, F. F., *Izv. Akad. Nauk SSSR. Ser. Khim.* **1990**, 1433.
40. Krayushkin, M. M.; Yarovenko, V. N.; Semenov, S. L.; Zavarzin, I. V.; Ignatenko, A. V.; Martynkin, A. Y.; Uzhinov, B. M., *Org. Lett.* **2002**, *4*, 3879.
41. (a) Nakatsuka, S-i.; Teranishi, K.; Goto, T, *Tetrahedron Lett.* **1994**, *35*, 2699. (b) Ottoni, O.; Neder, A. de V. F.; Dias, A. K. B.; Cruz, R. P. A.; Aquino, L. B., *Org. Lett.* **2001**, *3*, 1005. (c) Jiang, Y.; Ma, D., *Tetrahedron Lett* **2002**, *43*, 7013 (d) Li, J.; Li, B.; Chen, X.; Zhang, G., *Synlett* **2003**, 1447.
42. Bastock, T. W.; Clark, J. H.; Landon, P.; Martin, K., *J. Chem. Res. (S)* **1994**, 104.
43. Mamatkulov, N. N.; Abdushukurov, A. K.; Khidirov, Sh.; Rakhmonova, S., *Russ. J. Org. Chem.* **2001**, *37*, 1668.
44. Toth, G.; Tamas, T.; Borbely, I., *Synth. Commun.* **2002**, *32*, 3659.
45. Jackson, R. F. W.; Wishart, N.; Wood, A.; James, K.; Wythes, M. J., *J. Org. Chem.* **1992**, *57*, 3397.
46. Clerici, A.; Clerici, L.; Porta, O., *Tetrahedron* **1996**, *52*, 11037.
47. Yus, M.; Gomis, J., *Tetrahedron* **2003**, *59*, 4967.
48. (a) Abbas, A. A., *Synthesis* **2004**, 419. (b) Soyer, Z.; Kilic, F. S.; Erol, K.; Pabuccuoglu, V., *Arch. Pharm.* **2004**, *337*, 105.
49. (a) Fitt, J.; Prasad, K.; Repic, O.; Blacklock, T. J., *Tetrahedron Lett.* **1998**, *39*, 6991. (b) Middel, O.; Verboom, W.; Reinhoudt, D. N., *J. Org. Chem.* **2001**, *66*, 3998. (c) Röhrich, T.; Thaher, B. A.; Otto, H-H., *Monatsh. Chem.* **2004**, *135*, 55.
50. Hell, Z.; Cwik, A.; Finta, Z.; Horvath, Z., *J. Mol. Catal. A.* **2002**, *184*, 191.
51. (a) Revell, J. D.; Srinivasan, N.; Ganesan, A., *Synlett* **2004**, *8*, 1428. (b) Pawlowska, J.; Czarnocki, Z.; Wojtasiewicz, K.; Maurin, J. K., *Tetrahedron: Asymmetry* **2003**, *14*, 3335. (c) Chang, W.-J.; Yeh, W.-B.; Sun, C.-M., *Synlett* **2003**, 1688. (d) Viso, A.; Fernandez de la Pradilla, R.; López-Rodríguez, M. L.; García, A.; Tortosa, M., *Synlett* **2002**, *5*, 755. (e) Soukara, S.; Wünsch, B., *Synthesis* **1999**, 1739.
52. Gangwal, N. A.; Narasimhan, B.; Mourya, V. K.; Dhake, A. S., *Ind. J. Heterocycl. Chem.* **2002**, *12*, 201
53. Desai, V. G.; Shet, J. B.; Tilve, S. G.; Mali, R. S., *J. Chem. Res. (S)* **2003**, 628.
54. Hennessy, E. J.; Buchwald, S. L., *J. Am. Chem. Soc.* **2003**, *125*, 12084.
55. (a) Ziegler, T.; Pantkowski, G., *J. Carbohydr. Chem.* **1993**, *12*, 357. (b) Wong, C.-H.; Ye, X. S.; Zhang, Z., *J. Am. Chem. Soc.* **1998**, *120*, 7137. (c) Zhang, S-Q.; Li, Z-J.; Wang, A-B.; Cai, M-S.; Feng, R., *Carbohydr. Res.* **1997**, *299*, 281. (d) Carpov, A.; Mocanu, G.; Vizitiu, D., *Angew. Makromol. Chem.* **1998**, *256*, 75. (e) Jenkins, D. W.; Hudson, S. M., *J. Polymer Sci. A* **2001**, *39*, 4174. (f) Takeo, K.; Ohguchi, Y.; Hasegawa, R.; Kitamura, S., *Carbohydr. Res.* **1995**, *278*, 301. (g) Crich, D.; Yao, Q., *J. Am. Chem. Soc.* **2004**, *126*, 8232.
56. Georg, G. I.; Harriman, G. C. B.; Hepperle, M.; Clowers, J. S.; Vander Velde, D. G.Himes, R. H., *J. Org. Chem.* **1996**, *61*, 2664.
57. Mimero, P.; Saluzzo, C.; Amouroux, R., *Tetrahedron Lett.* **1994**, *35*, 1553.
58. Goodall, K.; Parsons, A. F., *J. Chem. Res. (S)* **2000**, 54.
59. Lee, M.-J.; Sun, C.-M., *Tetrahedron Lett.* **2004**, *45*, 437.
60. Deryagina, E. N.; Russavaskaya, N. V.; Grabel'nykh, V. A., *Russ. J. Gen. Chem.* **2004**, *74*, 346.
61. Koketsu, M.; Mio, T.; Ishihara, H., *Synthesis* **2004**, 233.
62. (a) Lecea, B.; Arrastina, I.; Arrieta, A.; Roa, G.; Lopez, X.; Arriortua, M. I.; Ugalde, J. M.; Cossio, F. P., *J. Org. Chem.* **1996**, *61*, 3070. (b) Bhusare, S. R.; Pawar, V. G.; Shinde, S. B.; Pawar, R. P.; Vibhute, Y. B., *Int. J. Chem. Sci.* **2003**, *1*, 31. (c) Patel, H.; Patel, V. K., *Ind. J. Heterocycl. Chem.* **2003**, *12*, 253. (d) Choudhari, B. P.; Mulwad, V. V., *Ind. J. Heterocycl. Chem.* **2003**, *12*, 197. (e) Abe, N.; Otani, A.; Nagai, T.; Fujii, H., *Heterocycles* **2003**, *59*, 199. (f) Mulwad, V. V.; Shirodkar, J. M., *Indian J. Heterocycl. Chem.* **2002**, *11*, 291. (g) Bharmal, F.; Kaneriya, D.; Parekh, H., *Oriental J. Chem.* **2002**, *18*, 89. (h) Abbiati, G.; Rossi, E., *Tetrahedron* **2001**, *57*, 7205. (i) Jayakumar, S.; Kumar, V.; Mahajan, M. P., *Tetrahedron Lett.* **2001**, *42*, 2235. (j) Dey, P. D.; Sharma, A. K.; Rai, S. N.; Mahajan, M. P., *Tetrahedron* **1995**, *51*, 7459.
63. Stevens, C. V.; Vekemans, W.; Moonen, K.; Rammeloo, T., *Tetrahedron Lett.* **2003**, *44*, 1619.
64. Müller, M.; Birner, G.; Dekant, W., *Chem. Res. Toxicol.* **1998**, *11*, 454.
65. Tsoleridis, C. A.; Stephanidou-Stephanatou, J.; Gounaridis, P.; Zika, H.; Pozarentzi, M., *Tetrahedron* **2003**, *59*, 4591.
66. Nakyama, Y.; Nagase, M., *J. Org. Chem.* **1995**, *60*, 1878.
67. Ito, K.; Saito, K.; Takahashi, K., *Bull. Chem. Soc. Jpn.* **1992**, *65*, 812.

2-Chloro-4*H*-1,3,2-benzodioxaphosphorin-4-one

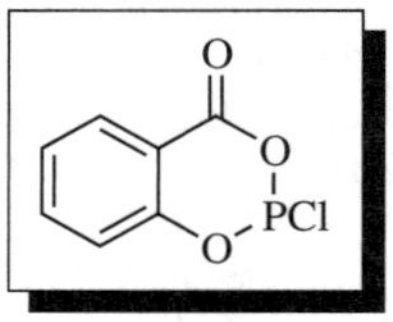

[5381-99-7] $C_7H_4ClO_3P$ (MW 202.54)

(reagent used in phosphorylation and phosphitylation of alcohols and in the formation of *H*-phosphonates)

Alternate Name: salicylchlorophosphite, SalPCl.
Physical Data: mp 36–40 °C, bp 127–128 °C/11 mm Hg.
Solubility: soluble in CH_2Cl_2, THF, CH_3CN, DMF, and most other organic solvents.
Preparative Methods: the reagent[1] is prepared by heating a mixture of salicylic acid and redistilled reagent grade phosphorus trichloride (1.1 equiv) at 85 °C in a Claisen flask under an inert atmosphere with a provision to absorb hydrogen chloride by sodium hydroxide solution. The reaction mixture is heated overnight and the product purified by vacuum distillation.
Handling, Storage, and Precautions: reagent is hygroscopic, corrosive, and lachrymatory; it should be protected from moisture and handled with care in a fume hood. The toxicity of the reagent is unknown.

Introduction. The primary application of salicylchlorophosphite is in the phosphitylation of alcohols. The alkyl salicylphosphites so-obtained serve as precursors to a variety of mono-, di-, and triphosphates, as well as their thia and bora analogs depending on the mode of work-up. Hydrolysis of the alkyl salicylphosphites provides access to the *H*-phosphonates[2] which are widely employed in nucleotide synthesis.

Synthesis of Phosphates and Thiophosphates. Perhaps the simplest application of salicylchlorophosphite is in the preparation of phosphate and thiophosphate esters. Such conversions are readily achieved by oxidative work-up of the first formed alkyl salicylphosphates with reagents such as *tert*-butyl hydroperoxide (eq 1)[3] or elemental sulfur (eq 2).[4] Additionally, eq 1 illustrates the typical ease with which primary alcohols are selectively phosphitylated in the presence of secondary alcohols with this reagent. In the synthesis of an allylic thiophosphate (eq 2), mention was made of the superiority of salicylchlorophosphite over a number of other phosphitylation reagents, each of which resulted in lower yields and/or complex reaction mixtures.[4]

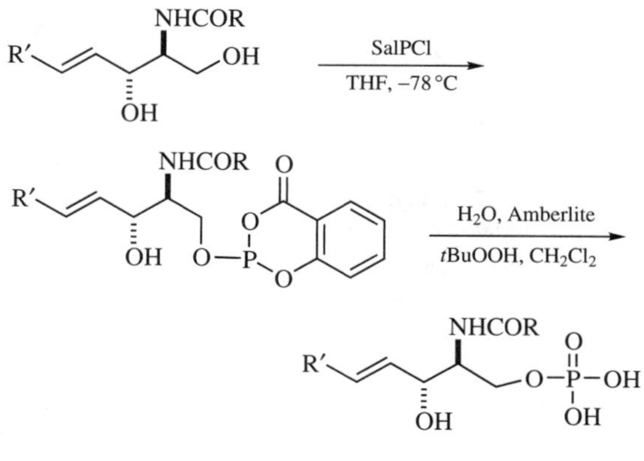

(1)

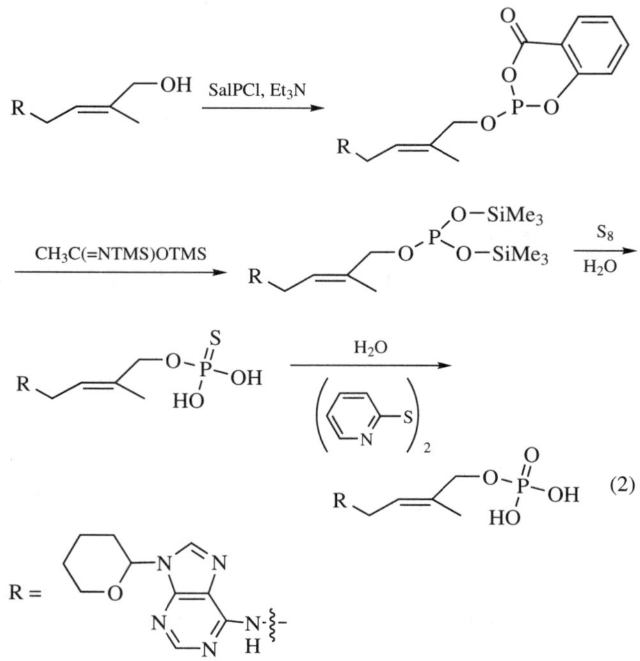

(2)

Synthesis of Di- and Triphosphates and their Thia and Bora Analogs. Treatment of an alkyl salicylphosphite, derived from salicylchlorophosphite in the normal manner, with pyrophosphate yields a cyclotriphosphite. As illustrated in eq 3, oxidation of

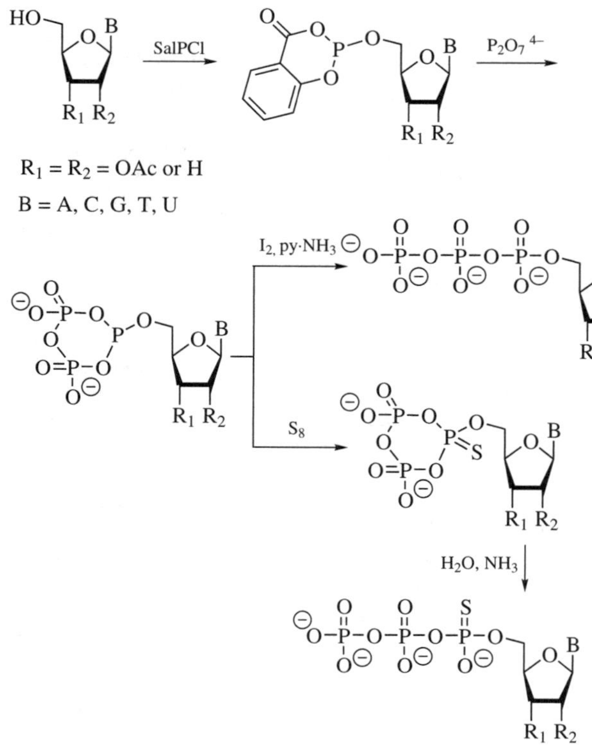

(3)

this derivative with iodine and water leads to the triphosphate, whereas treatment with sulfur affords a cyclic monothiophosphate which can be hydrolyzed to a monothiotriphosphate of established regiochemistry.[5] As is obvious from eq 3, there is no need for protection of standard pyrimidine or purine bases in the application of this chemistry to ribonucleosides.

In an alternative work-up protocol, the cyclic triphosphite is exposed to an amine borane resulting, after hydrolysis of the cyclic derivative, in a bora-triphosphate again with complete control of regiochemistry (eq 4).[6]

Nucleoside dithiodiphosphates may be obtained from exposure of the initial alkyl salicylphosphite to tri-*n*-butylammonium *S*-cyanoethylthiophosphate followed by oxidative work-up with sulfur (eq 5),[7] while dithiotriphosphates are obtained by reaction of the alkyl salicylphosphite with monothiopyrophosphate and then sulfur (eq 6).[8] In this latter case, however, a mixture of 1,2- and 1,3-dithiotriphosphates is obtained.

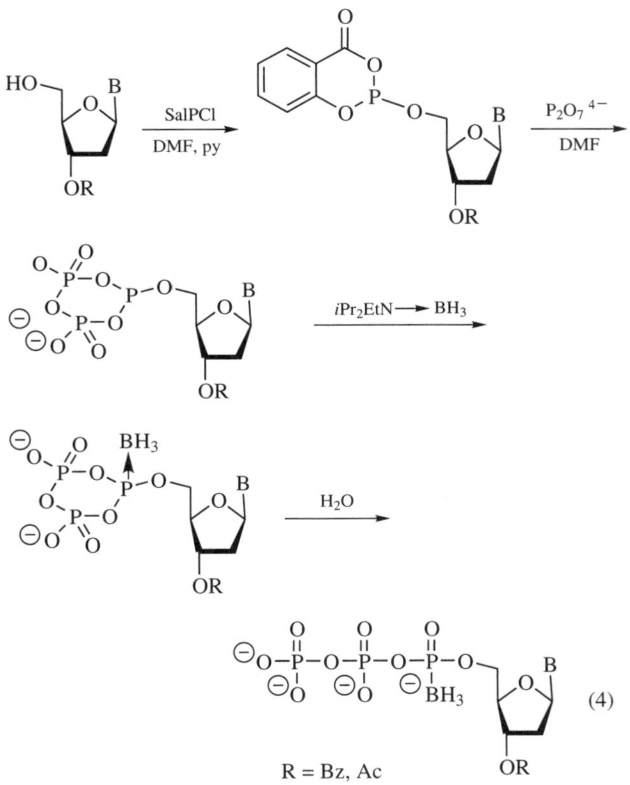

(4)

R = Bz, Ac
B = A, T, G, C, U

Synthesis and Applications of *H*-Phosphonates. The synthesis of *H*-phosphonates is achieved simply by hydrolysis of salicylphosphites. This is illustrated by the preparation of the novel 3′-*O*-*H*-phosphonate of a tetracyclic 2-deoxyadenosine analog (eq 7).[9]

The power of the *H*-phosphonate methodology is clearly evident in eq 8 in which a conjugate of 7β-hydroxycholesterol with mannose is obtained with no requirement for protection of anomeric and ring oxygens of the carbohydrate.[10] Conjugates of the steroid with galactose[10] and azidothymidine[11] were similarly obtained.

With appropriately protected sugars, the *H*-phosphonate methodology may be applied to the synthesis of glycosyl

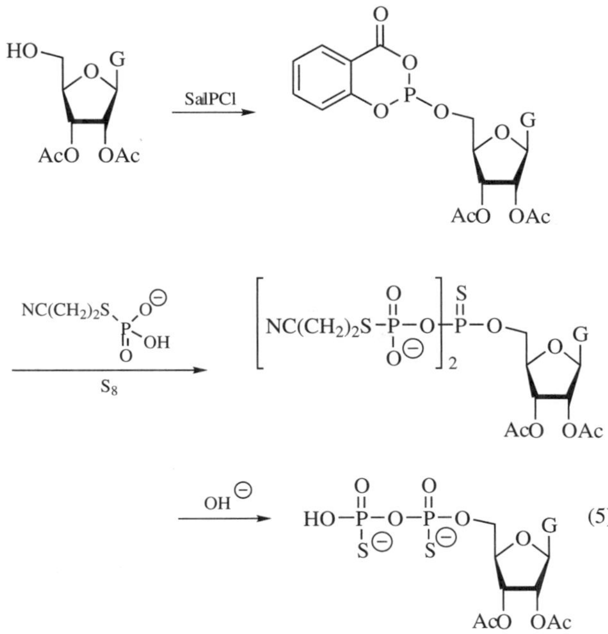

(5)

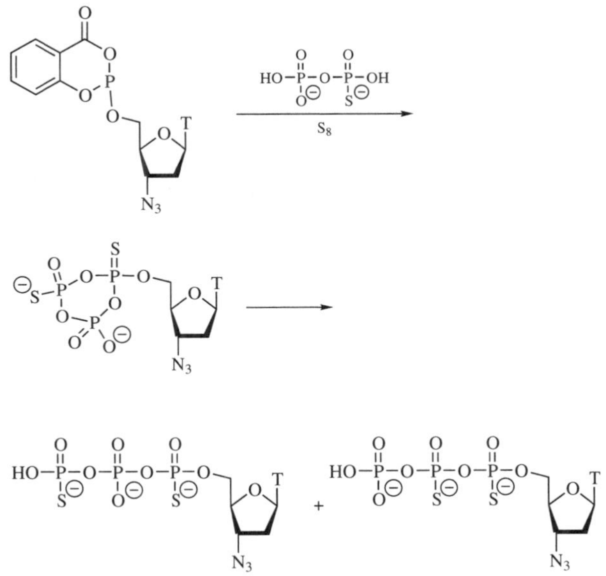

(6)

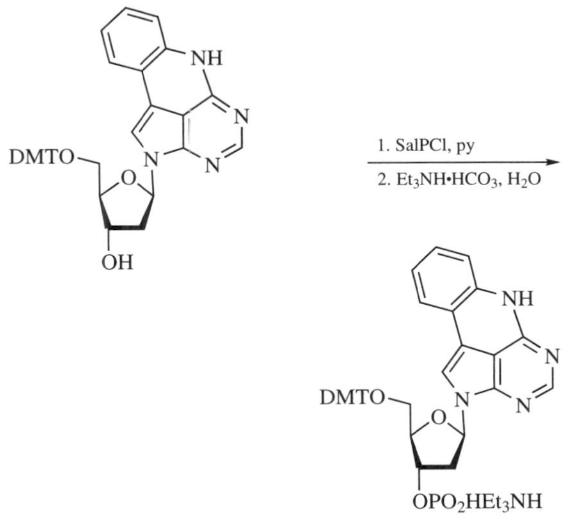

(7)

phosphodiesters such as the repeating unit (eq 9) of the oligomeric *N*-acetylglucosamine phosphate found in the cell wall of *Staphylococcus lactis*.[12]

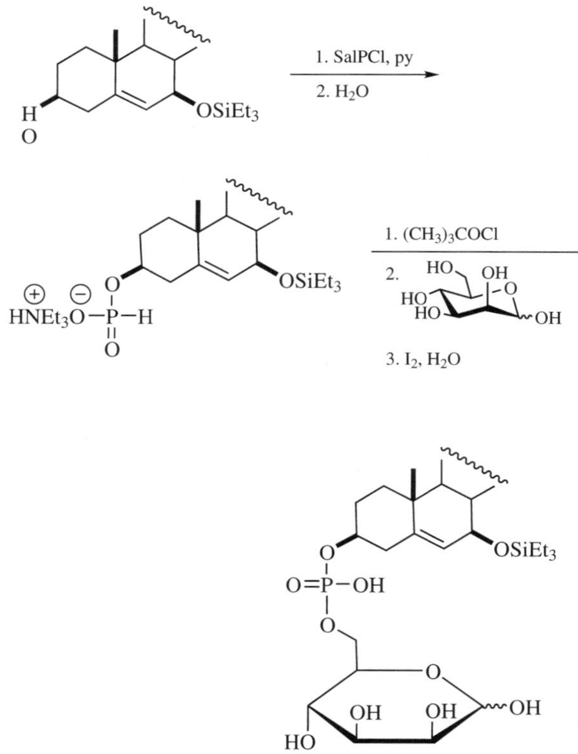

(8)

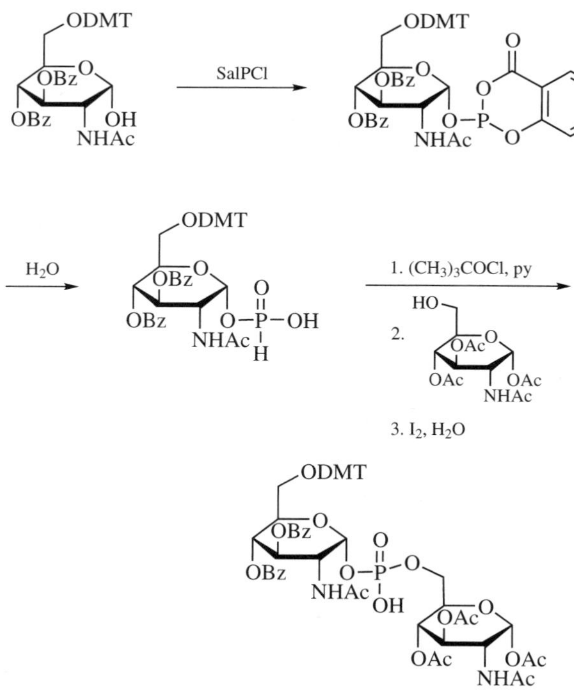

(9)

6-*O*-(2-Aminoethylphosphate) esters of galactose, constituents of the galactan from the albumen glands of the snail megalobulimus paranagnensis, have also been obtained via the aegis of salicylchlorophosphite and *H*-phosphonates.[13] The key steps here (eq 10) being the conversion of the *H*-phosphonate to a bis(trimethylsilyl)phosphite with *N*,*O*-bis(trimethylsilyl)acetamide and subsequent Arbuvoz type reaction.[13]

Finally, salicylchlorophosphite has also found application in the generation of *H*-phosphonates of polymer-supported alcohols. Coupling to a range of alcohols followed by DBU-mediated condensation with a series of aldehydes and cleavage from the resin generated a small library of α-hydroxyphosphonates for screening as enzyme inhibitors (eq 11).[14]

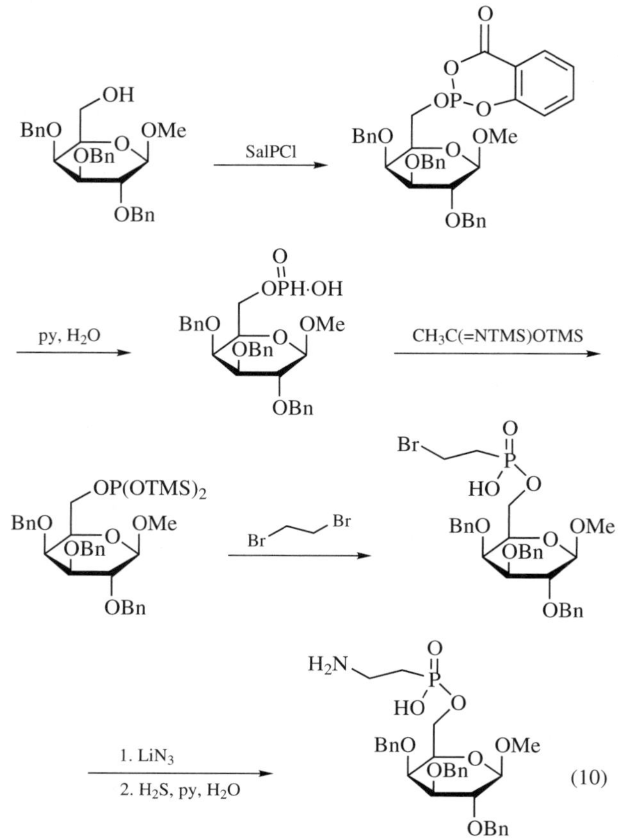

(10)

Related Reagents. Hypophosphorus acid; hypodiphosphoric acid; phosphorous acid; hypophosphoric acid; *H*-pyrophosphonic acid; diphenyl *H*-phosphonate.

1. Pinkus, A. G.; Waldrep, P. G.; Collier, W. J., *J. Org. Chem.* **1961**, *26*, 682–686.
2. Stawinski, J.; Kraszewski, A., *Acc. Chem. Res.* **2002**, *35*, 952–960.
3. Byun, H.-S.; Erukulla, R. K.; Bittman, R., *J. Org. Chem.* **1994**, *59*, 6495–6498.
4. Shadid, B.; van der Plas, H. C.; Vonk, R. C.; Davelaar, E.; Ribot, S. A., *Tetrahedron* **1989**, *45*, 3889–3896.
5. Ludwig, J.; Eckstein, F., *J. Org. Chem.* **1989**, *54*, 631–635.
6. Krzyzanowska, B. K.; He, K.; Hasan, A.; Ramsay Shaw, B., *Tetrahedron* **1998**, *54*, 5119–5128.
7. Ludwig, J.; Eckstein, F., *J. Org. Chem.* **1991**, *56*, 5860–5865.
8. Ludwig, J.; Eckstein, F., *J. Org. Chem.* **1991**, *56*, 1777–1783.
9. Buhr, C. A.; Matteucci, M. D.; Froehler, B. C., *Tetrahedron Lett.* **1999**, *40*, 8969–8970.
10. Pannecoucke, X.; Schmitt, G.; Luu, B., *Tetrahedron* **1994**, *50*, 6569–6578.

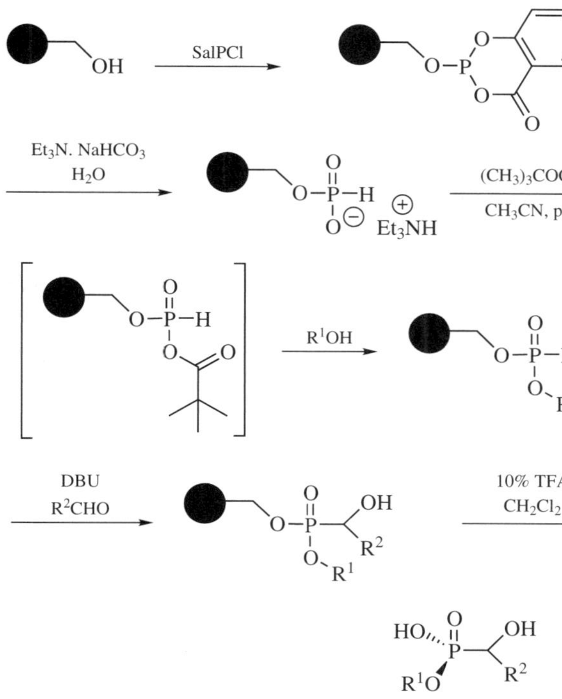

(11)

11. Pannecoucke, X.; Parmentier, G.; Schmitt, G.; Dolle, F.; Luu, B., *Tetrahedron* **1994**, *50*, 1173–1178.
12. Westerduin, P.; Veeneman, G. H.; van der Marel, G. A.; van Boom, J. H., *Tetrahedron Lett.* **1986**, *27*, 6271–6274.
13. Dreef, C. E.; Valentijn, A. R. P. M.; de Vroom, E.; van der Marel, G. A.; van Boom, J. H., *Tetrahedron Lett.* **1988**, *29*, 1199–1202.
14. Cao, X.; Mjalli, A. M. M., *Tetrahedron Lett.* **1996**, *37*, 6073–6076.

Krishnakumar Ranganathan
University of Illinois at Chicago, IL, USA

p-Chlorobenzyl Bromide

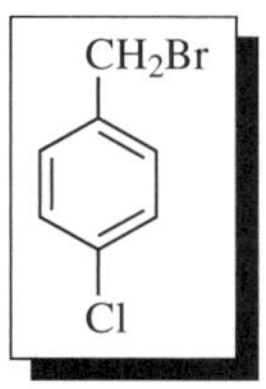

[622-95-7] C_7H_6BrCl (MW 205.48)

(reagent used as a stable protecting group for hydroxyl functions, removable by Pd-catalyzed amination followed by acid treatment, as well as for *p*-chlorobenzylation of heterofunctional groups and nucleophilic carbons)

Alternate Name: α-bromo-4-chlorotoluene, 1-bromomethyl-4-chloro-benzene, 1-chloro-4-bromomethyl-benzene.

Physical Data: mp 48–52 °C, bp 108–110 °C (at 12 Torr).

Solubility: soluble in ether and most organic solvents, insoluble in water.

Form Supplied in: light yellow, crystalline solid.

Analysis of Reagent Purity: ^{1}H NMR, IR, and elemental analysis.

Preparative Methods: bromination of 4-chlorotoluene with NBS in CCl_4, using benzoyl peroxide as initiator [1] or with $NaBrO_3$/$NaHSO_3$ reagent in ethyl acetate.[2] Commercially available.

Purification: crystallization from ethanol, methanol, and petroleum ether.

Handling, Storage, and Precautions: corrosive, lachrymator, moisture sensitive, stored under nitrogen, causes burns, irritant to eyes, respiratory system, and skin.

Disposal: contact a licensed professional waste disposal service for disposal of this material. Dissolve or mix with a combustible solvent and burn in a chemical incinerator equipped with an afterburner and a scrubber.

***p*-Chlorobenzyl Bromide as a Protecting Group.** *p*-Chlorobenzyl bromide is most commonly used as a protective group for hydroxyl and amino functions. *p*-Chlorobenzylation of alcohols can be readily achieved by treatment with *p*-Chlorobenzyl bromide in the presence of a base, in general sodium hydride, and in solvents such as THF, diethyl ether, or DMF (eq 1).[3] The

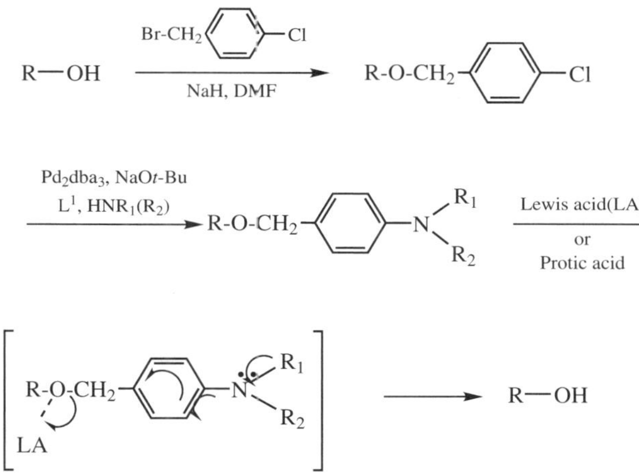

(1)

L^1 = 1-(*N*,*N*-dimethylamino)-1′-(dicyclohexylphosphino)biphenyl

p-chlorobenzyl (PCB) ethers thus obtained are particularly stable under a range of reaction conditions, including acidic and basic media, normally used for the removal of other common protective groups such as esters, silyl ethers, and *p*-methoxybenzyl (PMB) ethers. Cleavage of PCB ethers can be accomplished smoothly by conversion into labile aryl amines by Pd-catalyzed amination. The reaction mechanism involves the coupling of the *p*-chlorobenzyl group with an amine, in general benzylamine or methylphenylamine, using $Pd_2(dba)_3$ (1 mol % Pd) and 2-(*N*,*N*-dimethylamino)-2′-(dicyclohexylphosphino)biphenyl in presence of sodium *tert*-butoxide as a catalyst (eq 1). The substituted *p*-aminobenzyl ethers thus formed can be easily cleaved by treatment with a protic or a Lewis acid or by oxidation.[4]

p-Chlorobenzyl ethers are equally stable, but more easily removable than most protecting groups. Other substituted benzylethers, such as *p*-nitrobenzyl, *p*-acetobenzyl, or 2-(trimethylsilyl)ethoxymethoxybenzyl ethers require forcing conditions (e.g., treatment with Zn/AcOH, base, or fluoride) for cleavage. This renders substituted benzyl ethers incompatible with many other functional groups.[5] *p*-Methoxybenzyl ethers, on the other hand, are often too labile under acidic conditions and thus are not the ideal protective groups. Halo-substituted benzyl ethers provide, without substantial changes in the stability, three more degrees of orthogonality. *p*-Iodobenzyl ethers, *p*-bromobenzyl ethers, and *p*-chlorobenzyl ethers can be selectively and sequentially removed in presence of one another (eq 2).[3] This characteristic can be attributed to the higher affinity for oxidative addition to Pd(0) of iodide compared to bromide and of bromide compared to chloride. Halobenzyl ethers are, therefore, the preferred protective groups in situations where many hydroxyl group of similar reactivity need to be

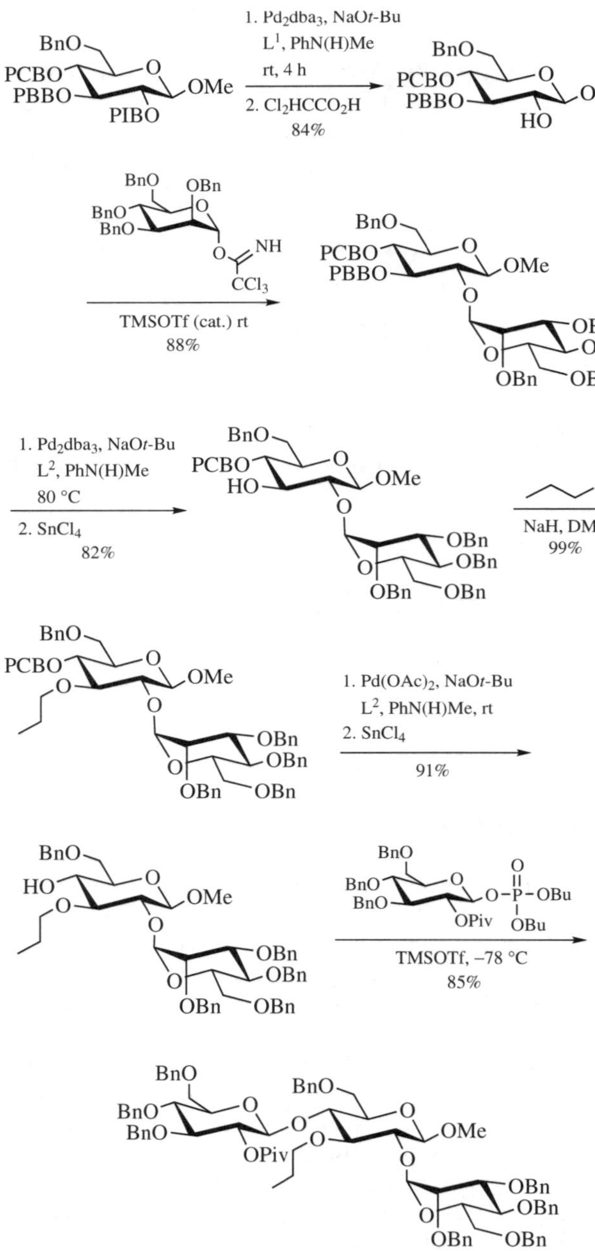

(2)

PCB = *p*-chlorobenzyl, PBB = *p*-bromobenzyl, PIB = *p*-iodobenzyl;

L^1 = 1-(*N*,*N*-dimethylamino)-1′-(dicyclohexylphosphino)biphenyl,

L^2 = (*o*-biphenyl)P(*t*-Bu)$_2$

differentially functionalized. *p*-Chlorobenzyl ethers, however, cannot be removed in presence of the brominated or iodinated analogs.

p-Chlorobenzylation of amino groups can also be accomplished in alkaline conditions (K_2CO_3 or NaH) at reflux.[6] Cleavage is effected with H_2/Pd-C, although it is often very slow.[7]

p-Chlorobenzyl bromide can also serve as an electrophile in the formation of new carbon-carbon bonds.[7] Strongly nucleophilic carbons, such as organometallic reagents, are required.[8]

Related Reagents. *p*-Bromobenzyl bromide; *p*-iodobenzyl bromide; benzyl bromide; *p*-methoxybenzyl bromide.

1. Ashcroft, M. R.; Bougeard, P.; Bury, A.; Cooksey, C. J.; Johnson, M. D., *J. Organomet. Chem.* **1985**, *289*, 403.
2. Kikuchi, D.; Sakaguchi, S.; Ishii, Y., *J. Org. Chem.* **1998**, *63*, 6023.
3. Greene, T. W.; Wuts, P. G. M., *Protective Groups in Organic Synthesis*; Wiley: New York, 1999, pp. 86–95.
4. Plante, O. J.; Buchwald, S. L.; Seeberger, P. H., *J. Am. Chem. Soc.* **2000**, *122*, 7148.
5. Jobron, L.; Hindsgaul, O., *J. Am. Chem. Soc.* **1999**, *121*, 5835.
6. Wu, S.; Lee, S.; Beak, P., *J. Am. Chem. Soc.* **1996**, *118*, 715.
7. Lygo, B.; Crosby, J.; Peterson, J. A., *Tetrahedron Lett.* **1999**, *40*, 8671.
8. Sarges, R.; Hank, R. F.; Blake, J. F.; Bordner, J.; Bussolotti, D. L.; Hargrove, D. M.; Treadway, J. L.; Gibbs, E. M., *J. Med. Chem.* **1996**, *39*, 4783.

Eva Castagnetti & Peter H. Seeberger
Massachusetts Institute of Technology,
Cambridge, Massachusetts

6-Chloro-2,4-dimethoxy-s-triazine

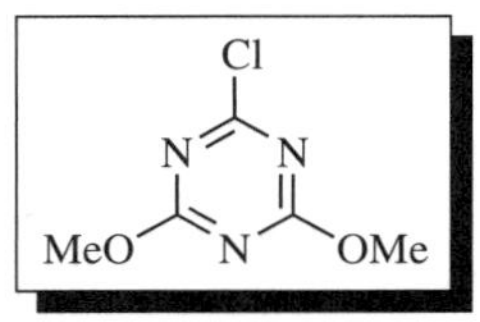

[3140-73-6] $C_5H_6ClN_3O_2$ (MW 175.58)

(peptide coupling reagent; used to prepare active esters which can be used for the preparation of amides and esters;[1] used in the reduction of carboxylic acids to aldehydes or alcohols[2])

Alternate Name: CDMT, 2-chloro-4,6-dimethoxy-1,3,5-triazine.
Physical Data: mp 71–74°C; bp 125–128°C/66.661 Pa.
Solubility: soluble in most organic solvents.
Form Supplied in: white solid; commercially available.
Preparative Methods: commercially available; prepared from cyanuric chloride and MeOH.[3] An improved procedure for a large-scale preparation has been reported.[4]
Purification: commercial reagent is generally used as received.
Handling, Storage, and Precautions: irritant; sternutator. Avoid inhalation. Should be handled with gloves in a fume hood. Relatively stable but should be stored under anhydrous conditions.

Amide Formation. 6-Chloro-2,4-dimethoxy-s-triazine (CDMT) is utilized primarily as a coupling reagent in the condensation of amines and carboxylic acids for the preparation of peptides and other amides. Treatment of CDMT with a tertiary amine, preferably *N*-methylmorpholine (NMM), and a carboxylic acid yields an intermediate 6-acyloxy-2,4-dimethoxy-s-triazine (**1**) which serves as the acylating agent.[1,5] A typical example is shown in eq 1. The procedure involves adding *N*-methylmorpholine (1.02 equiv) to a solution of the corresponding carboxylic acid (1.02 equiv) and CDMT (1.0 equiv) at −5°C to 0°C in CH_2Cl_2, CH_3CN, THF, or DMF. The mixture is stirred for 1–4 h until formation of **1** is complete. The corresponding amine (1.0 equiv) and NMM (1.0 equiv) are then added, the mixture is allowed to warm to room temperature and stirred until the reaction is complete. Yields typically range from 75–98%. One major advantage offered by CDMT over other peptide coupling reagents, such as the carbodiimide reagents (see **1,3-dicyclohexylcarbodiimide**),[6] is the ease of purification. Since the triazine ring is weakly basic, a dilute acid wash will remove any by-products as well as any excess reagent from the reaction mixture. Furthermore, protected functional groups, as well as the unprotected hydroxyl group in serine, do not interfere with the coupling reactions, and racemization is not commonly observed. It is also worth noting that CDMT is inexpensive and has been used on kilogram scale.[7]

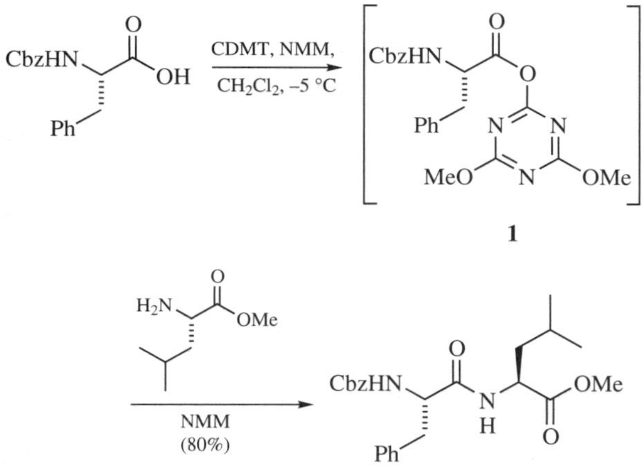

(1)

Studies directed toward a better understanding of the mechanism of activation of carboxylic acids by CDMT revealed that the reaction proceeds through quarternary triazinylammonium salts such as 4-(4,6-dimethoxy-1,3,5-triazin-2-yl)-4-methylmorpholinium chloride (DMTMM) (**2**)[8,9] prior to formation of acyloxytriazines shown in eq 1. In fact, DMTMM, synthesized quantitatively by treatment of CDMT with NMM in THF (eq 2), is a white solid and is stable for a month at ambient temperature. It can be used directly in condensation reactions of amines and carboxylic acids,[10,11] thus providing a useful, "one-step" variant of CDMT. The example in eq 2 illustrates the complexity of amines and carboxylic acids that can be coupled using DMTMM or CDMT.[11,12] It should be noted that the stability of DMTMM varies in different solvents. For instance, DMTMM completely demethylates in CH_2Cl_2 in 3 h. Therefore, lower yields of amides are observed when CH_2Cl_2 is used as solvent for the condensation reactions. DMTMM is, however, very stable in THF and alcoholic solvents. Amides can be obtained from the condensation of carboxylic acids and amines with DMTMM in THF with yields similar to those observed in the CDMT-NMM system. A typical procedure involves adding DMTMM (1.1 equiv) to a mixture of the corresponding acid (1.0 equiv) and amine (1.1 equiv) in THF and stirring at room temperature for 3–24 h. Furthermore, it is not necessary to dry the THF or perform the reaction under a N_2 atmosphere.

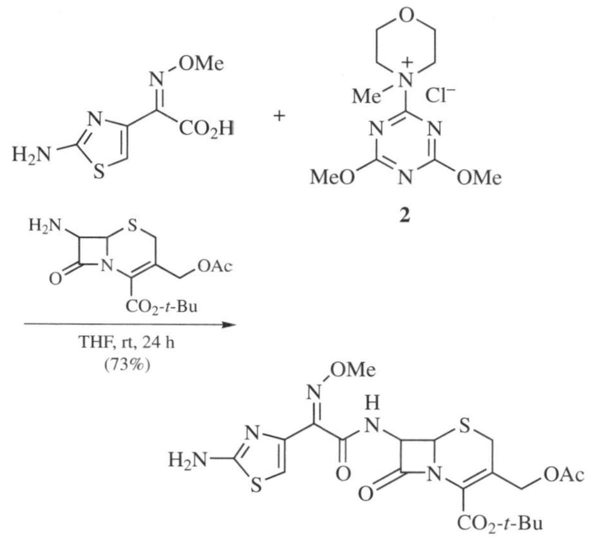

(2)

Ester Formation. Alcohols react much more slowly than amines with the 6-acyloxy-2,4-dimethoxy-s-triazine intermediates similar to **1**. Using typical CDMT-NMM conditions, ethyl benzoate was prepared by condensation of benzoic acid and ethanol in 65% yield, but the reaction time was 10 days.[1] It was later determined that ester formation proceeds much faster and in higher yields when the intermediate acyloxytriazine is isolated.[13] It was also determined that the rate of esterification can be increased by catalysis with 5 mol% $MgBr_2$ (eq 3). A typical procedure involves preparing the acyloxytriazine with CDMT, NMM and the corresponding carboxylic acid as described in the previous section. After work-up, the corresponding alcohol (10 equiv) and $MgBr_2$ (0.05 equiv) are then added to the crude, intermediate acyloxytriazine (1.0 equiv), and the mixture is stirred until the reaction is complete. Primary, secondary, and tertiary alcohols can be acylated under these conditions with yields between 44–92%. The reaction conditions are mild, but reaction times are somewhat long (typically 2–4 days), and the alcohol must be used in excess.

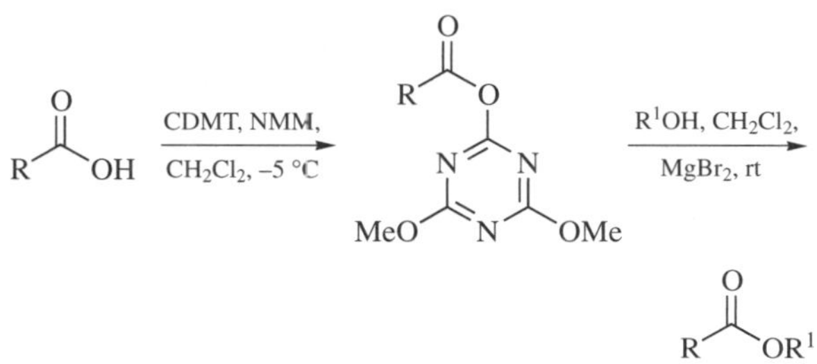

(3)

Carboxylic acids can also be esterified using DMTMM and the corresponding alcohols.[10,11] A typical procedure involves adding NMM (0.1–1.2 equiv) to a mixture of the carboxylic acid (1.0 equiv) and DMTMM (1.1–2.0 equiv) in the corresponding alcohol (solvent). Reaction times range from 4–21 h. The yields are quite good, but in most cases, the alcohol must be used in large excess.

Reduction of Carboxylic Acids. Active esters prepared from CDMT (e.g. 6-acyloxy-2,4-dimethoxy-s-triazines) can be reduced directly to aldehydes by treatment of hydrogen (1 atm) at room temperature in the presence of 10 mol% Pd/C (eq 4).[2] A typical procedure involves adding a carboxylic acid (1.0 equiv) to a mixture of DMTMM, prepared from CDMT (1.0 equiv) and NMM (1.02 equiv), in DME. The active ester is filtered, added to a mixture of Pd/C (10 mol%) in ethanol and stirred under a hydrogen atmosphere until the reaction is complete. Yields range from 75–84% when aliphatic carboxylic acids are used, but yields of aldehydes drop dramatically with aromatic carboxylic acids as substrates. The main product in these cases is the corresponding alcohol. Even with aliphatic carboxylic acids, the reactions have to be monitored carefully to avoid further reduction to the corresponding alcohols. In fact, if the hydrogen pressure is increased to 3-5 atm, alcohols can obtained in high yields.

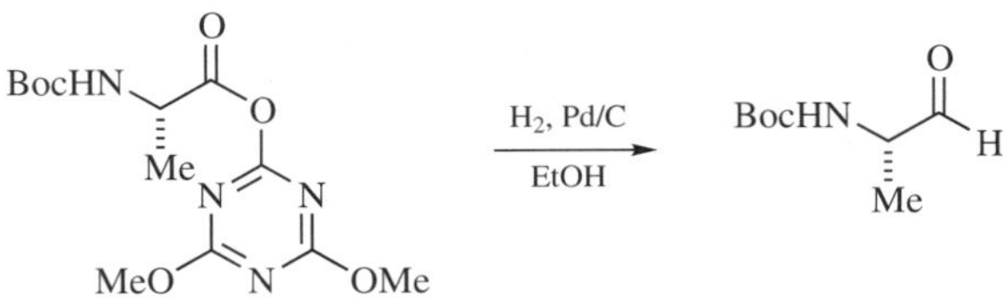

(4)

Other Uses. CDMT can be used as a source of 2,4-dimethoxy-s-triazine for the preparation of 6-substituted-2,4-dimethoxy-s-triazine derivatives. In these cases, substitution or coupling reactions occur at the carbon bearing the chlorine. Some examples include modified Sonogashira Pd-mediated couplings to form 6-alkynyl-2,4-dimethoxy-s-triazine derivatives,[14] glycosolations of thioglycosides,[15] nickel(0)-catalyzed cross-coupling reactions with arylboronic acids[16] and addition of amines to form 6-amino-2,4-dimethoxy-s-triazine derivatives.[17]

Related Reagents. cyanuric chloride, benzotriazol-1-yloxy-tris(dimethylamino)phosphinium hexafluoro-phosphate (BOP), *O*-benzotriazol-1-yl-*N,N,N′,N′*-tetramethyluronium hexafluorophosphate (HBTU), bis(2-oxo-3-oxazolidinyl)phosphinic chloride (BOP), 1,3-dicyclohexylcarbodiimide (DCC), and 1-ethyl-3-(3′-dimethylaminopropyl)carbodiimide hydrochloride (EDCI).

1. Kamiński, Z. J., *Tetrahedron Lett.* **1985**, *26*, 2901.
2. Falorni, M.; Giacomelli, G.; Porcheddu, A.; Taddei, M., *J. Org. Chem.* **1999**, *64*, 8962.
3. Dudley, J. R.; Thurston, J. T.; Schaefer, F. C.; Holm-Hansen, D.; Hull, C. J.; Adams, P., *J. Am. Chem. Soc.* **1951**, *73*, 2986.
4. Cronin, J. S.; Ginah, F. O.; Murray, A. R.; Copp, J. D., *Synth. Commun.* **1996**, *26*, 3491.
5. Kamiński, Z. J., *Synthesis* **1987**, 917.
6. Sheehan, J. C.; Hess, G. P., *J. Am. Chem. Soc.* **1955**, *77*, 1067.
7. Barnett, C. J.; Wilson, T. M.; Kobierski, M. E., *Org. Process Res. Dev.* **1999**, *3*, 184.
8. Kamiński, Z. J., *J. Prakt. Chem.* **1990**, *332*, 579.
9. Kamiński, Z. J.; Paneth, P.; Rudzinski, J., *J. Org. Chem.* **1998**, *63*, 4248.
10. Kunishima, M.; Kawachi, C.; Iwasaki, F.; Terao, K.; Tani, S., *Tetrahedron Lett.* **1999**, *40*, 5327.
11. Kunishima, M.; Kawachi, C.; Morita, J.; Terao, K.; Iwasaki, F.; Tani, S., *Tetrahedron* **1999**, *55*, 13159.
12. Lee, H.-W.; Kang, T. W.; Cha, K. H.; Kim, E.-N.; Choi, N.-H.; Kim, J.-W.; Hong, C. I., *Synth. Commun.* **1998**, *28*, 1339.
13. Kaminska, J. E.; Kaminski, Z. J.; Gora, J., *Synthesis* **1999**, 593.
14. Menicagli, R.; Samaritani, S.; Gori, S., *Tetrahedron Lett.* **1999**, *40*, 8419.
15. Sugimura, H.; Motegi, M.; Sujino, K., *Nucleosides Nucleotides* **1995**, *14*, 413.
16. Saito, S.; Oh-tani, S.; Miyaura, N., *J. Org. Chem.* **1997**, *62*, 8024.
17. Tucker, JA.; Allwine, DA.; Grega, KC.; Barbachyn, MR.; Klock, JL.; Adamski, JL.; Brickner, SJ.; Hutchinson, DK.; Ford, CW.; Zurenko, GE.; Conradi, RA.; Burton, PS.; Jensen, RM., *J. Med. Chem.* **1998**, *41*, 3727.

Dr D. Scott Coffey
Eli Lilly and Company, Indianapolis, IN, USA

2-Chloro-1,3-dimethylimidazolidium Hexafluorophosphate

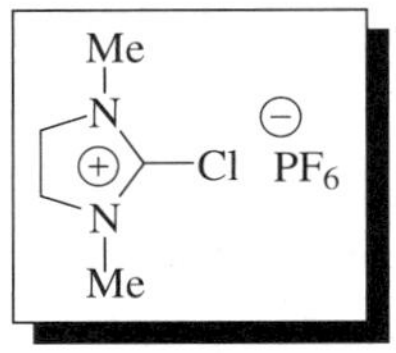

[101385-69-7] $C_5H_{10}N_2Cl{\cdot}PF_6$ (MW 278.57)

(reagent used for the coupling and esterification of sterically hindered amino acids[1])

Physical Data: mp 227–229 °C.
Solubility: soluble in DMF; partially soluble in H_2O, MeOH, EtOAc, CH_2Cl_2, and $CHCl_3$.
Form Supplied in: colorless solid; available from Aldrich, Fluka, and Watanabe Chemical Industries.
Analysis of Reagent Purity: IR (KBr), 1650, 1555, 1420, 1345, 1310, 840 cm^{-1}. ^{1}H NMR (270 MHz, DMSO-d_6) δ 3.07 (s, 6H), 3.64 (s, 4H).
Handling, Storage, and Precautions: can be stored refrigerated for at least 5 years without loss of reactivity. 2-Chloro-1,3-dimethylimidazolidium hexafluorophosphate (CIP) is an irritant and should be used in a fumehood, when possible.

Scope. 2-Chloro-1,3-dimethylimidazolidium hexafluorophosphate (CIP), in the presence of an additive, is best suited for the coupling of sterically hindered α,α-dialkylamino acids, *N*-methylamino acids, and peptide fragments. The coupling reaction proceeds under slightly basic conditions, with diisopropylethylamine generally being used as an organic base. A catalytic amount of the additive is sufficient for the coupling reaction to proceed, and a suitable additive can be selected for each reaction type. 1-Hydroxy-7-azabenzotriazole (HOAt, **1**)[2] or *N*-dimethylaminopyridine (DMAP, **2**)[3] is generally best suited for use as an additive. An optically active α-amino acid can be activated by CIP to yield the coupling product with negligible racemization.

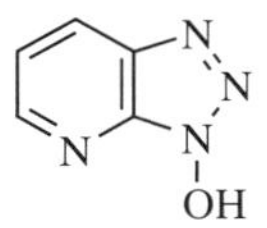

HOAt **1**

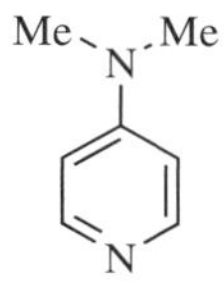

DMAP **2**

Coupling of α, α-Dialkylamino Acid. The efficient coupling of an α,α-dialkylamino acid is particularly difficult when the carboxyl and amino components are both α,α-dialkylamino acids.[4] In the presence of 0.25 equiv of HOAt as an additive, CIP mediates the solution-phase coupling of an α,α-dimethylamino acid (Aib, α-aminoisobutyric acid) to yield Z-Aib-Aib-OMe, in 82% yield (eq 1).[5] The reaction proceeds at 25 °C in less than 60 min. The same dipeptide ester was prepared using PyBrop,[6] a typical coupling reagent in peptide synthesis, under the same conditions in 17% yield. The amount of D-Val in the Z-Val-Aib-OMe, which was prepared by the CIP/HOAt procedure, was less than 0.5%. In the CIP-mediated Aib coupling, a Z or Fmoc group as an amino protecting group gave much better results than a Boc protecting group, probably due to the facile formation of the *N*-carboxy anhydride during the activation of the Boc amino acid.[7]

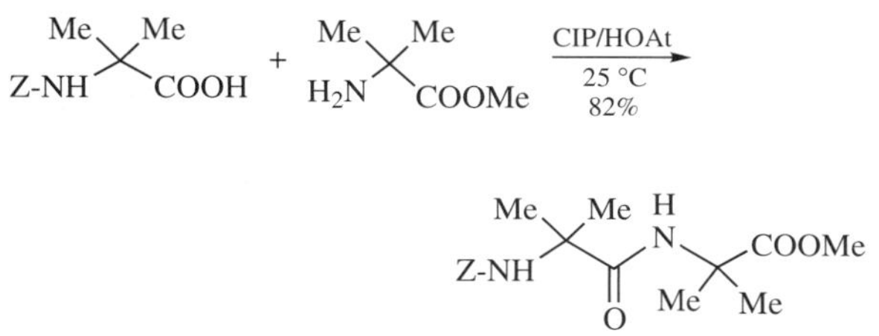

(1)

The CIP-mediated coupling of Aib proceeds through the oxazolone,[8] followed by its transformation to a highly active HOAt ester. The active intermediate ester reacts quickly with an amino component to give the condensation product and HOAt. The liberated HOAt functions as a catalyst for the coupling. The coupling procedure was successfully applied to an efficient synthesis of Alamethicin F30, a linear peptide consisting of 19 α-amino acids including eight Aib residues and a C-terminal α,α-aminoalcohol, phenylalaninol (Pheol). The stepwise coupling of Aib residues using CIP/HOAt proceeded nearly quantitatively in CH_2Cl_2 or DMF at 25 °C within 60 min to yield the necessary fragments without any difficulty.[9]

More complex and hindered α,α-dialkylamino acids can also be coupled with CIP/HOAt. CIP-mediated activation was applied for the difficult coupling of three adjacent α-methylcysteine residues. The coupling reaction proceeded at 25 °C in a 60% yield (eq 2).[10]

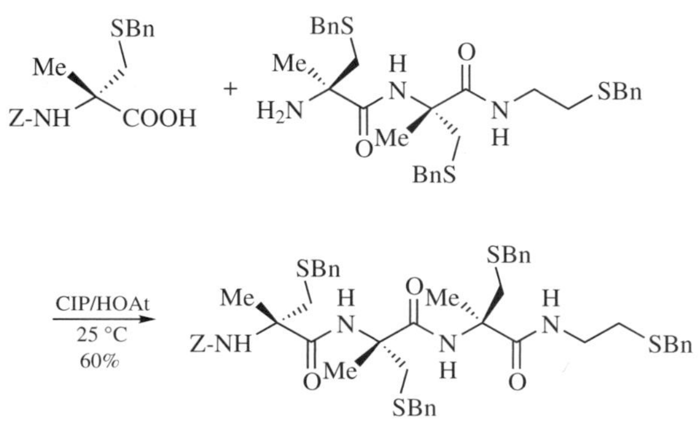

(2)

The procedure was the key step in the convergent syntheses of the polythiazoline alkaloids, mirabazole B (**3**),[11] mirabazole C (**4**),[10] and thiangazole(**5**).[12]

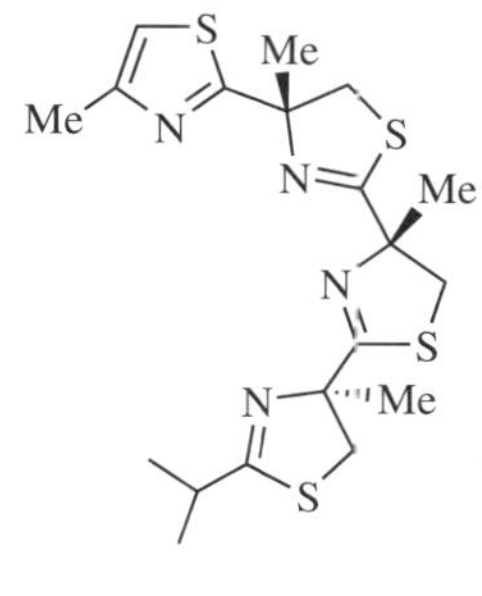

Mirabazole B **3**

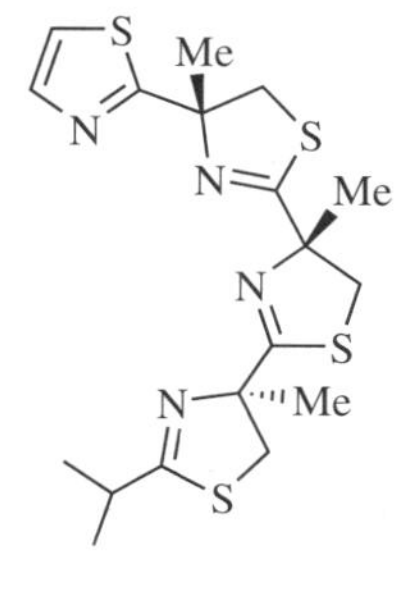

Mirabazole C **4**

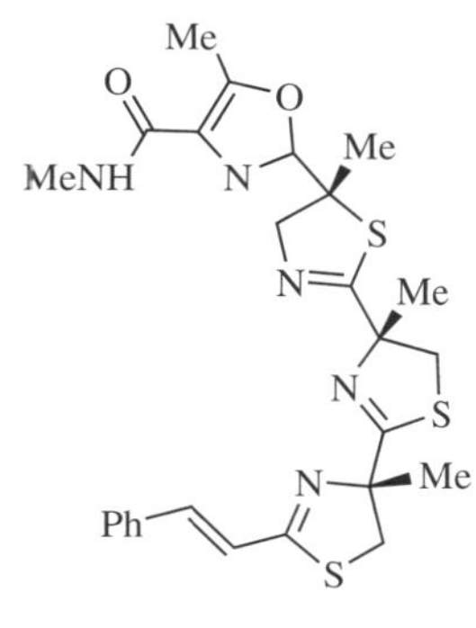

Thiangazole **5**

Coupling of *N*-methylamino Acid. An *N*-methylamino acid is incorporated with difficulty using conventional peptide coupling reagents because of its steric hindrance and marked tendency to racemize.[13] CIP/HOAt was found to mediate the solution-phase coupling of the *N*-methylamino acid at 25 °C in CH_2Cl_2 in quantitative yield. The most hindered coupling between two *N*-methylated residues proceeds within 3 h, to give a 97% yield of the product with no detectable racemization: the corresponding D-isomer was produced in less than 0.5% yield (eq 3).[14]

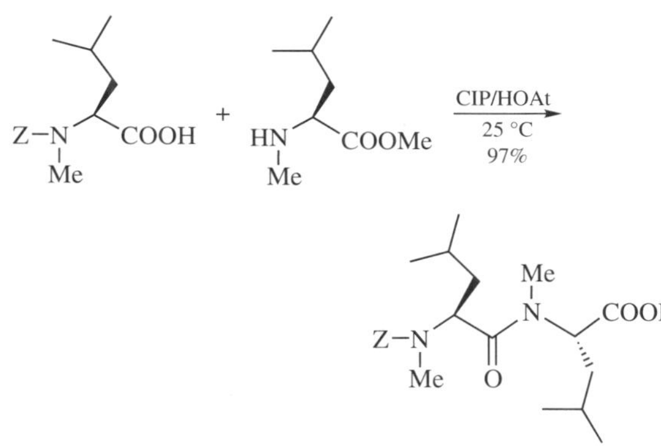

(3)

In the coupling of Z-MeVal-OH with MeAla-NHNH-Boc in solution phase, an efficient conventional reagent, BOP-Cl,[15] resulted in an 87% yield after an 18 h reaction time providing <0.5% isomerization, whereas the 1,3-dicyclohexylcarbodiimide (DCC)/DMAP procedure resulted in a 72% yield after a 24 h reaction, with 37% isomerization.[16]

The CIP-mediated activation procedure can be applied to the coupling of *N*-methylamino acids on solid support.[14,17] Quantitative-to-moderate coupling yields were obtained by a single coupling reaction (1–3 h) at 25 °C with practically no racemization. The coupling between two *N*-methylated amino acids proceeded on solid support in 53% yield after a 3 h single coupling reaction (eq 4). The content of the D-isomer was less than 2%. As a reaction solvent, the use of DMF resulted in better yields than CH_2Cl_2.

Solid phase coupling using CIP/HOAt was successfully applied to a convergent synthesis of Dolastatin 15, a cytostatic depsipeptide isolated from a marine mollusk. A tetrapeptide fragment was obtained by cleavage of **6**, which was prepared by the successive CIP-mediated coupling of *N*-methylamino acids on a 2-chlorotrityl polystyrene resin. The overall yield based on starting resin was 50%.[14]

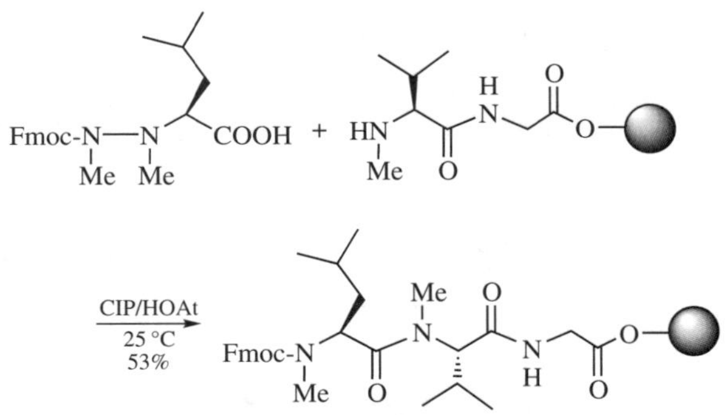

(4)

Fragment Coupling. The reaction efficiency of fragment coupling is generally low, compared with the coupling of a single amino acid. A greater fragment size results in a lower coupling efficiency. Fragment coupling is also prone to racemization via the oxazolone, which is formed during the activation of the C-terminus, i.e. during the activation of the *N*-acylamino acid.[18] These difficulties can be avoided by taking advantage of the high reactivity of a coupling reagent combined with a judicious choice of coupling position.

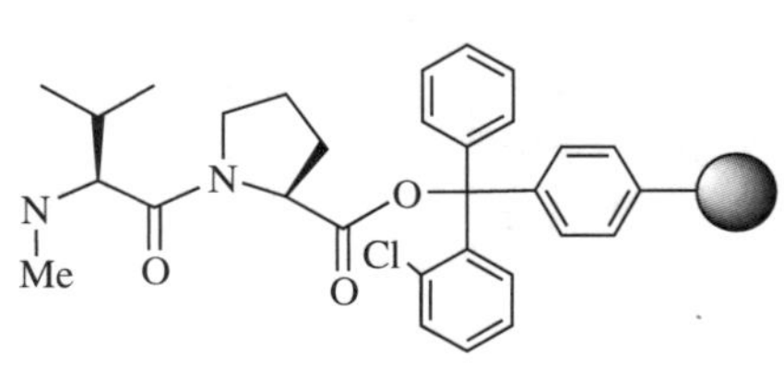

6

CIP/HOAt was employed for the coupling of an *N*-terminal 11-residue fragment with a Gly residue at its C-terminus, and a C-terminal nine-residue fragment to yield protected Alamethicin F-30. The coupling reaction proceeded at 25 °C within 60 min to yield 66% of the homogeneous product after purification.[19] The same coupling procedure was used in the final coupling reaction in the synthesis of Dolastatin 15 (**7**). The coupling reaction at the Pro residue proceeded at 25 °C within 2 h to yield Dolastatin 15 in 89% yield.[14]

CIP/HOAt can also be applied to the coupling of fragments with an optically active α-amino acid at the C-terminus. In the synthesis of Trichovirin I4A (**8**) using CIP/HOAt, coupling efficiency and the rate of racemization were found to be greatly dependent on the coupling site. Fragment coupling between the Ala and Val residues proceeded within 60 min at 25 °C in 81% yield with no detectable racemization, whereas fragment coupling between the more hindered Val and Aib residues resulted in a lower yield (69%) of the epimerized product.[19]

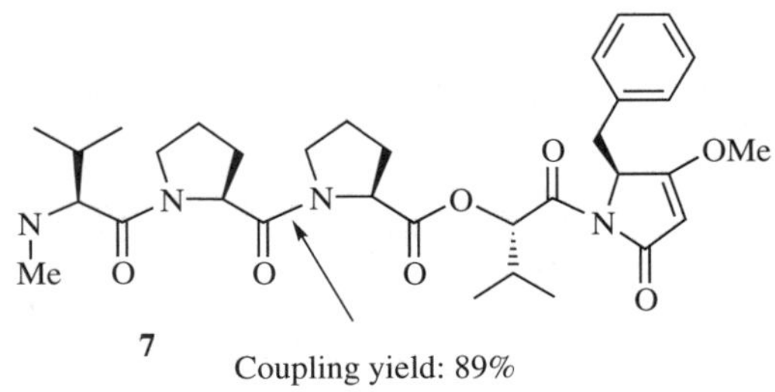

7

Macrocyclization. Head-to-tail cyclization represents a type of intramolecular fragment coupling and an efficient coupling reagent required for this transformation. A low reaction rate, as the result of an improper choice of the coupling reagent or the coupling site, facilitates extensive side reactions such as dimerization, oligomerization, and racemization.

Ac-Aib-Asn-Leu-Aib-Pro-Ala-Val-Aib-Pro-Aib-Leu-Aib-Pro-Leuol

8

Coupling yield: 81% Yield of epimerized product: 69%

CIP/HOAt was employed for the intramolecular cyclization to yield the key intermediate **9** in the synthesis of MEN11420, a glycosylated bicyclic peptide which has potent tachykinin antagonistic activity. CIP-mediated cyclization at the site between Dap(Boc) and Phe proceeded in DMF within 15 min to yield the cyclized monomeric product containing 6% of the D-Phe isomer. In contrast, cyclization at the more hindered site between the glycosylated-Asn and Leu proceeded in the presence of an excess amount of CIP (9 equiv) within 15 min to yield the monomeric product contaminated with 33% of the D-Leu isomer.[20]

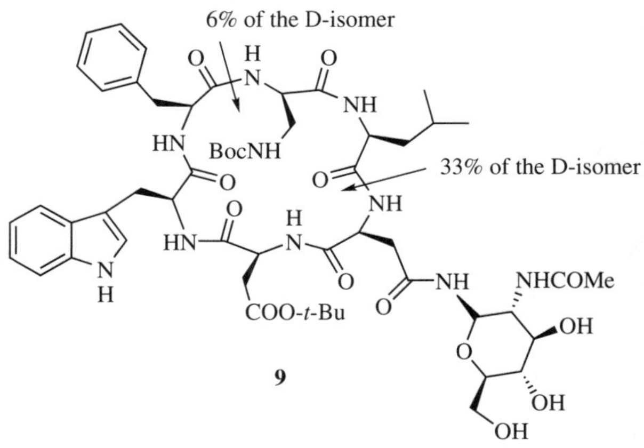

9

Esterification. CIP can be used for the esterification of a carboxylic acid to a hindered alcohol, such as a hindered secondary alcohol and a resin-bound alcohol. The CIP-mediated esterification of an Fmoc amino acid to a 4-alkoxybenzylalcohol resin was conducted at 25 °C within 60 min using 3 equiv of the Fmoc amino acid and 6 equiv of CIP in pyridine/CH_2Cl_2 (1:1). The reaction proceeded faster, with the same or lower level of racemization than the anchoring reaction using DIPCDI or PyBrop. For example, Fmoc-Lys(Boc)-OH was anchored to the alcohol resin in 74% yield accompanied by an 0.8% formation of the corresponding D-isomer, whereas a 57% yield using DIPCDI and a 33% yield using PyBrop was obtained.[21]

The esterification of Boc-Pro-OH with a secondary alcohol was achieved, after a 10 h reaction at 25 °C, using CIP in the presence of an additive (eq 5).[14] The additive, DMAP, gave better results than HOAt, giving the desired product in 92% yield.

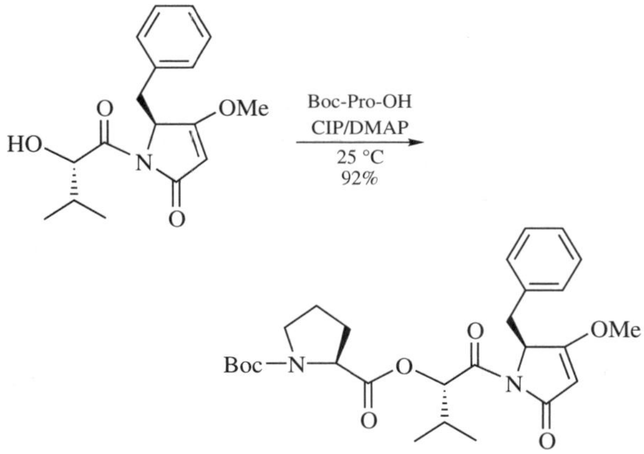

(5)

CIP can also be used for the introduction of a sterically hindered 2-azidomethylbenzoyl (AZMB) group into the 3′-hydroxy function of nucleosides (eq 6). The esterification reaction proceeded within 2 h at 25 °C to yield the desired product in 66% isolated yield.[22] An excess amount of CIP (2.5 equiv) was necessary for a smooth reaction because of the steric hindrance involved.

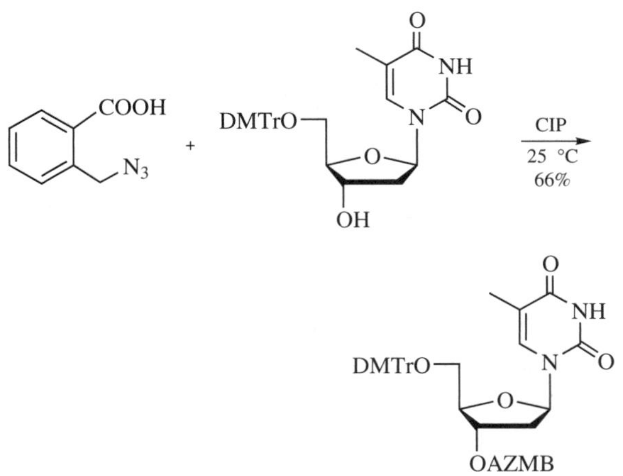

(6)

Related Reagents. 3-(Diethoxyphosphoryloxy)-1,2,3-benzotriazin-4(3*H*)-one (DEPBT);[23] ethyl 1-hydroxy-1*H*-1,2,3-triazole-4-carboxylate (HOCt);[24] 5-(1*H*-benzotriazol-1-yloxy)-3,4-dihydro-1-methyl 2*H*-pyrrolium hexachloroantimonate (BDMP);[25] 2-bromo-3-ethyl-4-methyl thiazolium tetrafluoroborate (BEMT).[26]

1. (a) Humphery, J. M.; Chamberlin, R., *Chem. Rev.* **1997**, *97*, 2243. (b) Albericio, F.; Carpino, L. A., *Methods Enzymol.* **1997**, *289*, 104.
2. Carpino, L. A., *J. Am. Chem. Soc.* **1993**, *115*, 4397.
3. Hassner, A.; Krepski, L. R.; Alexanian, V., *Tetrahedron* **1978**, *34*, 2069.
4. Spencer, J. R.; Antonenko, V. V.; Delaet, N. G. J.; Goodman, M., *Int. J. Peptide Protein Res.* **1992**, *40*, 282.
5. Akaji, K.; Kuriyama, N.; Kiso, Y., *Tetrahedron Lett.* **1994**, *35*, 3315.
6. Frerot, E.; Coste, J.; Pantaloni, A.; Dufour, M. N.; Jouin, P., *Tetrahedron* **1991**, *47*, 259.
7. Frerot, E.; Coste, J.; Poncet, J.; Jouin, P.; Castro, B., *Tetrahedron Lett.* **1992**, *33*, 2815.
8. (a) Perlow, D. S.; Erb, J. M.; Gould, N. P.; Tung R. D.; Freidinger, R. M.; Williams, P. D.; Veber. D. F., *J. Org. Chem.* **1992**, *57*, 4394. (b) Paquet, A.; Chen, F. M.; Benoiton, N. L., *Can. J. Chem.* **1984**, *62*, 1335.
9. Akaji, K.; Tamai, Y.; Kiso, Y., *Tetrahedron Lett.* **1995**, *36*, 9341.
10. Akaji, K.; Kuriyama, N.; Kiso, Y., *J. Org. Chem.* **1996**, *61*, 3350.
11. Kuriyama, N.; Akaji, K.; Kiso, Y., *Tetrahedron* **1997**, *53*, 8323.
12. Akaji, K.; Kiso, Y., *Tetrahedron* **1999**, *55*, 10685.
13. Coste, J.; Frerot, E.; Jouin, P., *J. Org. Chem.* **1994**, *59*, 2437.
14. Akaji, K.; Hayashi, Y. ; Kiso, Y ; Kuriyama, N., *J. Org. Chem.* **1999**, *64*, 405.
15. Diago-Meseguer, J.; Palomo-Coll, A. L.; Fernandez-Lizarbe, J. R.; Zugaza-Bilbao, A., *Synthesis* **1980**, 547.
16. Ward, D. E. ; Lazny, R.; Pedras, M. S., *Tetrahedron Lett.* **1997**, *38*, 339.
17. Raman, P.; Stokes, S. S.; Angell, Y. M.; Flentke, G. R.; Rich, D. H., *J. Org. Chem.* **1998**, *63*, 5734.
18. (a) Carpino, L. A.; Ionescu, D.; El-Faham, A., *J. Org. Chem.* **1996**, *61*, 2460. (b) Carpino, L. A.; El-Faham, A., *Tetrahedron* **1999**, *55*, 6813.
19. Akaji, K.; Tamai, Y.; Kiso, Y., *Tetrahedron* **1997**, *53*, 567.
20. Akaji, K.; Aimoto, S., *Tetrahedron* **2001**, *57*, 1749.
21. Akaji, K.; Kuriyama, N.; Kimura, T.; Fujiwara, Y.; Kiso, Y., *Tetrahedron Lett.* **1992**, *33*, 3177.
22. Wada, T.; Ohkubo, A.; Mochizuki, A.; Sekine, M., *Tetrahedron Lett.* **2001**, *42*, 1069.
23. Li, H.; Jiang, X.; Ye, Y.; Fan, C.; Pomoff, T.; Goodman, M., *Org. Lett.* **1999**, *1*, 91.
24. Robertson, N.; Jiang, L.; Ramage, R., *Tetrahedron* **1999**, *55*, 2713.
25. Li, P.; Xu, J.-C., *Tetrahedron* **2000**, *56*, 4437.
26. Li, P. ; Xu, J.-C., *J. Org. Chem.* **2000**, *65*, 2951.

Kenichi Akaji
Osaka University, Osaka, Japan

2-Chloro-1-methylpyridinium Iodide[1]

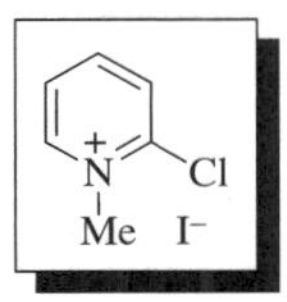

[14338-32-0] C_6H_7ClIN (MW 255.49)

(activation of carboxylic acids;[1] formation of esters[2] and amides;[3] lactonization;[4] ketene formation;[5] β-lactam synthesis;[6,7] carbodiimide synthesis[8])

Alternate Names: Mukaiyama's reagent.
Physical Data: mp 204–206 °C.
Form Supplied in: commercially available yellow solid.
Preparative Method: by reaction between 2-chloropyridine and Iodomethane in acetone at reflux.[9]

Purification: recrystallization from acetone.
Handling, Storage, and Precautions: hygroscopic.

Activation of Carboxylic Acids: Ester Formation. 2-Chloro-1-methylpyridinium iodide (**1**) reacts with a mixture of a carboxylic acid and an alcohol, in the presence of two equivalents of base, to form an ester (eq 1).[2] The pyridinium salt (**2**) is formed rapidly by displacement of chloride from (**1**) by the carboxylate; subsequent reaction with the alcohol results in formation of the ester, along with 1-methyl-2-pyridone (**3**). A variety of solvents may be employed, but yields are highest in dichloromethane or pyridine. Tri-n-butylamine or Triethylamine are often used as base. The co-product (**3**) is insoluble in dichloromethane and so precipitates from this solvent. Good results are obtained even for hindered carboxylic acids and alcohols.

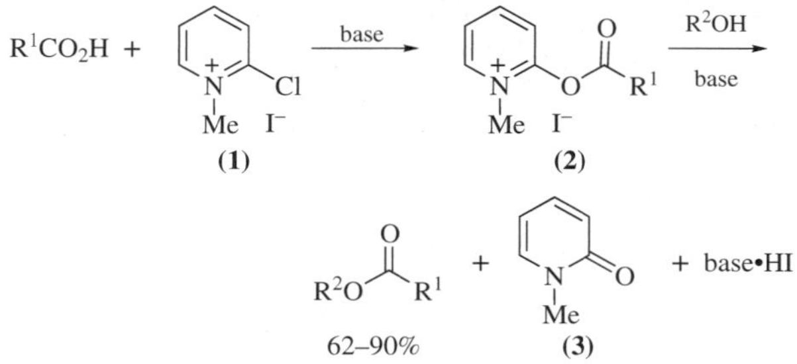
(1)

Amide Synthesis. In a similar way, amides can be prepared efficiently by treatment of an equimolar amount of a carboxylic acid and an amine with (**1**) in the presence of base (eq 2).[3]

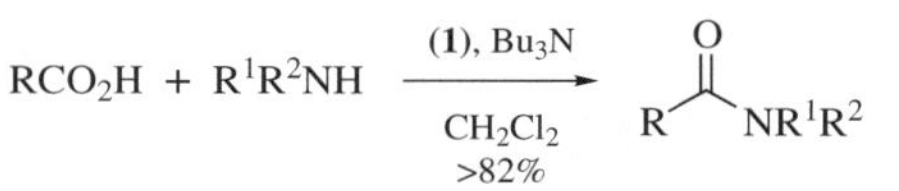
(2)

Lactone Formation. The ester formation method has found widespread use in the synthesis of lactones by cyclization of hydroxy acids (eq 3).[4] The cyclization process is believed to be entropically favored because of the close proximity of all reactants to a central pyridinium salt (**4**). Optimum conditions for the process involve addition over 8 h of a solution of the hydroxy acid (0.0125 M) and triethylamine (8 equiv) in dichloromethane or acetonitrile to a solution of (**1**) (4 equiv; 0.04 M) in the same solvent at reflux. After a further 30 min at reflux, evaporation of the solvent and chromatography affords the lactone. Results of a study of the formation of lactones of various ring sizes are shown in Table 1. Good results are obtained for ring sizes 7 and$\geq$ 12, but attempts to form 8- or 9-membered lactones result in substantial dimerization.

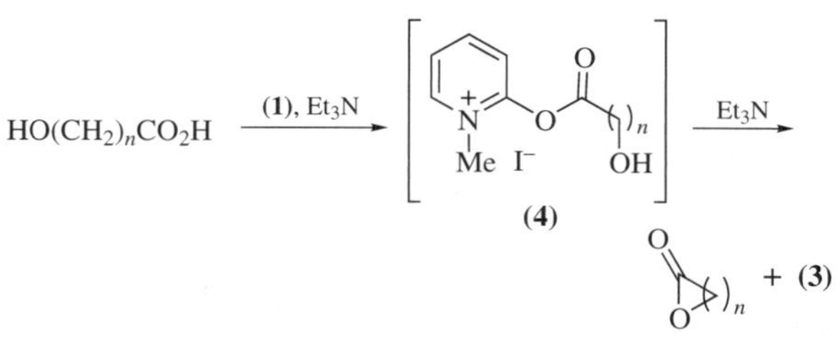
(3)

Cyclization of sensitive *trans*-γ-hydroxy acids to give strained *trans*-bicyclic γ-lactones of the type (**5**) is often problematic due to dehydration of the tertiary alcohol. A study showed (**1**) to be the reagent of choice for this transformation (eq 4).[10] Other reagents, such as 1,3-Dicyclohexylcarbodiimide/ Pyridine or 2,2′-Dipyridyl Disulfide/ Triphenylphosphine, give far lower yields.

Table 1 Lactone Formation by Cyclization of Hydroxy Acids with Reagent (**1**)

Ring size	% Yield lactone	% Yield dimer
7	89	0
8	0	93
9	13	34
12	61	24
13	69	14
16	84	3

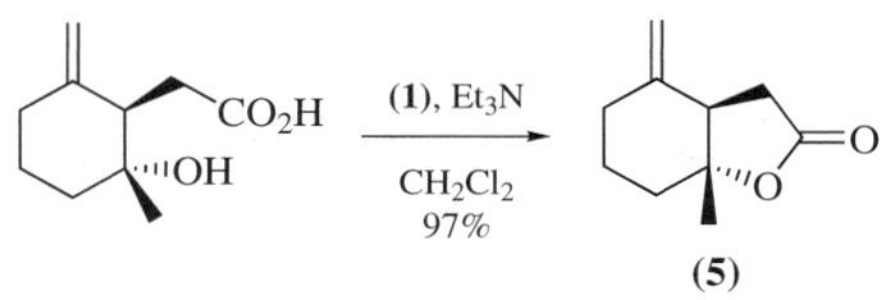
(4)

Ketene Formation. There is evidence that the above esterification procedure may occur at least partly via formation of the ketene from the carboxylic acid.[5] If a carboxylic acid contains a suitably placed alkene, the resulting ketene can undergo intramolecular [2 + 2] cycloaddition (eq 5).[5] Benzene is the best solvent for this reaction, giving yields comparable to those obtained by ketene formation from the corresponding acid chloride.

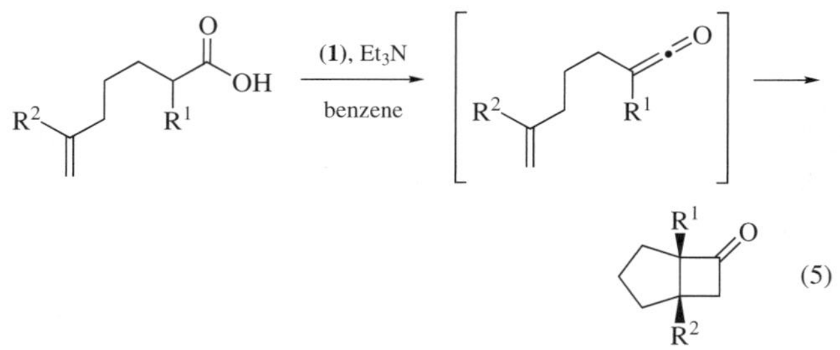
(5)

β-Lactam Synthesis. Dehydration of β-amino acids to give β-lactams can be effected in high yield using (**1**) in CH_2Cl_2 or MeCN (eq 6).[6]

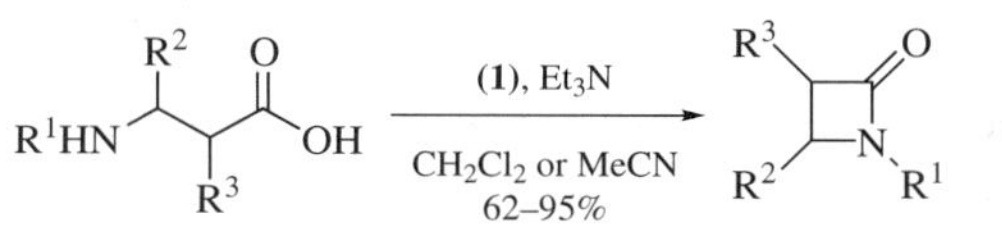
(6)

β-Lactams can also be prepared by the reaction of carboxylic acids, activated by (**1**), with imines; best results are obtained with Tripropylamine as base in CH_2Cl_2 at reflux (eq 7).[7]

Carbodiimides from Thioureas. Conversion of *N,N*′-disubstituted thioureas into carbodiimides can be effected by treatment with (**1**) and 2 equiv of base (eq 8).[8]

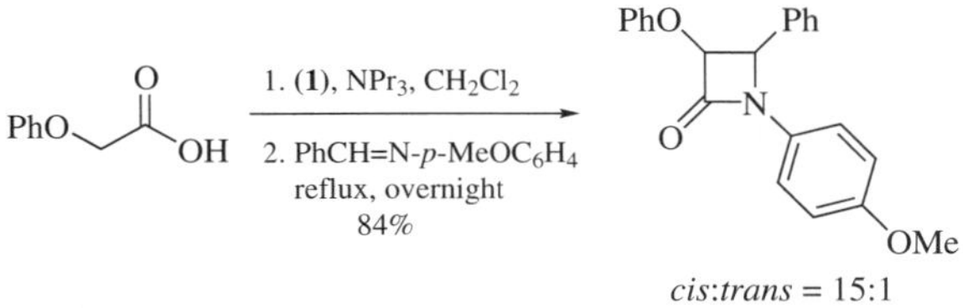

(7)

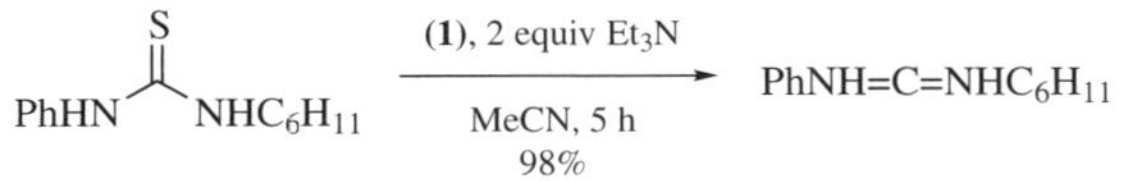

(8)

1. Mukaiyama, T. *Angew. Chem., Int. Ed. Engl.* **1979**, *18*, 707.
2. Mukaiyama, T.; Usui, M.; Shimada, E.; Saigo, K. *Chem. Lett.* **1975**, 1045.
3. Bald, E.; Saigo, K.; Mukaiyama, T. *Chem. Lett.* **1975**, 1163.
4. Mukaiyama, T.; Usui, M.; Saigo, K. *Chem. Lett.* **1976**, 49.
5. Funk, R. L.; Abelman, M. M.; Jellison, K. M. *Synlett* **1989**, 36.
6. Huang, H.; Iwasawa, N.; Mukaiyama, T. *Chem. Lett.* **1984**, 1465.
7. Georg, G. I.; Mashava, P. M.; Guan, X. *Tetrahedron Lett.* **1991**, *32*, 581.
8. Shibanuma, T.; Shiono, M.; Mukaiyama, T. *Chem. Lett.* **1977**, 575.
9. Amin, S. G.; Glazer, R. D.; Manhas, M. S. *Synthesis* **1979**, 210.
10. Strekowski, L.; Visnick, M.; Battiste, M. A. *Synthesis* **1983**, 493.

Alan Armstrong
University of Bath, UK

2-Chloro-4,4,5,5-tetramethyl-1,3,2-dioxaphospholane

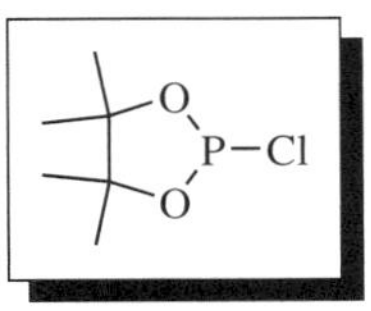

[14812-59-0] $C_6H_{12}ClO_2P$ (MW 182.58)

(reagent used for the phosphitylation of alcohols and other heteroatomic nucleophiles, resulting in the formation of useful glycosyl donors and ligands)

Physical Data: bp 81.5–82 °C/13 mm Hg; *d* 1.149.

Solubility: soluble in $CHCl_3$, ether, THF, and most organic solvents.

Form Supplied in: colorless liquid, available from Aldrich Chemical Company, Inc.

Preparative Methods: the reagent is prepared via the reaction of 2,3-dimethyl-butane-2,3-diol (pinacol) and phosphorous(III) chloride in diethyl ether as solvent with pyridine as base. Other bases such as triethylamine have been used, and yields are moderate.[1]

Handling, Storage, and Precautions: the reagent is very sensitive to moisture and thus all procedures should be performed under an inert atmosphere. Storage in a sealed container under nitrogen is necessary and, if stored as a solution, the addition of 3Å molecular sieves is recommended.

Phosphitylation. The basic function of 2-chloro-4,4,5,5-tetramethyl-1,3,2-dioxaphospholane is the derivatization of various heteroatom nucleophiles as the phosphite esters and, by subsequent oxidation, as the corresponding phosphates.[2] In addition to the phosphitylation of a broad range of alcohols,[2,3] phosphite derivatives of sulfur,[4] nitrogen,[2,5] and selenium[6] based nucleophiles, as well as of phenols,[7] have been formed with this reagent. Reaction with hydroxylamines[8] in the presence of base results in rearrangement of the initial phosphite to the corresponding phosphoramide.[9] Interesting mixed anhydrides of P(III) with P(V), which serve as precursors to phosphoryl radicals, have been formed by the phosphitylation of phosphoric acid derivatives with 2-chloro-4,4,5,5-tetramethyl-1,3,2-dioxaphospholane.[10]

Use in Glycosylation. Reaction of anomeric hemiacetals with 2-chloro-4,4,5,5-tetramethyl-1,3,2-dioxaphospholane results in the formation of glycosyl phosphites, which function as highly reactive glycosyl donors in the presence of trimethylsilyl trifluoromethanesulfonate.[11] The reactivity of glycosyl phosphites is a function of the substituents, with the pinacolyl derivatives being less active than the corresponding dialkyl substances. This difference in reactivity is manifest in the greater stability of the pinacolyl derivatives, facilitating their isolation,[12] and has enabled a one-pot synthesis of a trisaccharide as shown in eq 1.[13]

Use in the Formation of Ligands for Transition Metal-Catalyzed Processes. Ligands derived from the reaction of alcohols and vicinal diols with 2-chloro-4,4,5,5-tetramethyl-1,3,2-dioxaphospholane have found considerable application in hydrogenations, hydroformylations, and isoprene dimerization processes catalyzed by transition metals, including Ni, Rh, and Pt.[2,14]

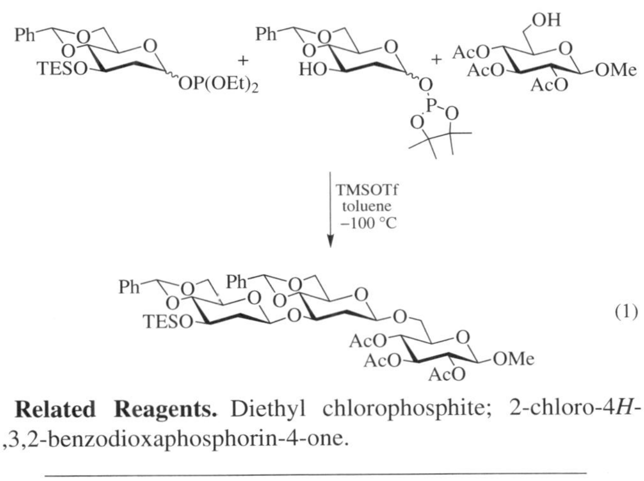

(1)

Related Reagents. Diethyl chlorophosphite; 2-chloro-4*H*-1,3,2-benzodioxaphosphorin-4-one.

1. Arbusov, A. E.; Azanovskaya, M. M., *Izv. Akad. Nauk. SSSR, Ser. Khim.* **1949**, *473*, 478.
2. Denney, D. Z.; Chen, G. Y.; Denney, D. B., *J. Am. Chem. Soc.* **1969**, *91*, 6838.
3. van Leeuwen, P. W. N. M.; Roobeek, C. F., *Tetrahedron* **1981**, *37*, 1973.

4. Bone, S.; Trippett, S.; Whittle, P. J., *J. Chem. Soc., Perkin Trans. 1* **1974**, 2125.
5. Nifant'ev, E. E.; Negrebetskii, V. V.; Grachev, M. K., *Zh. Obshch. Khim.* **1991**, *61*, 1581.
6. Johnson, M. P.; Trippett, S., *J. Chem. Soc., Perkin Trans. 1* **1981**, 3074.
7. Jiang, Z.-H.; Argyropoulos, D. S.; Granata, A., *Magn. Reson. Chem.* **1995**, *33*, 375.
8. Bailly, T.; Burgada, R., *Phosphorus, Sulfur, Silicon, Rel. Elements* **1991**, *63*, 33.
9. Angelov, C. M.; Dahl, O., *Tetrahedron Lett.* **1983**, *24*, 1643.
10. Morita, I.; Kunimoto, K.; Tsuda, M.; Tada, S.; Kise, M.; Kimura, K., *Chem. Pharm. Bull.* **1987**, *35*, 4144.
11. Koeller, K. M.; Wong, C. H., *Chem. Rev.* **2000**, *100*, 4465.
12. Guo, Y.; Sulikowski, G. A., *J. Am. Chem. Soc.* **1998**, *120*, 1392.
13. Pongdee, R.; Wu, B.; Sulikowski, G. A., *Org. Lett.* **2001**, *3*, 3523.
14. Kadyrov, R.; Heller, D.; Selke, R., *Tetrahedron: Asymmetry* **1998**, *9*, 329.

John Picione
University of Illinois at Chicago, IL, USA

1-Chloro-*N,N*,2-trimethylpropenylamine[1]

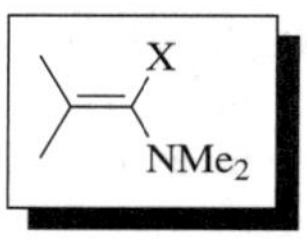

(X = Cl)
[26189-59-3] $C_6H_{12}ClN$ (MW 133.62)
(X = F)
[65560-29-4] $C_6H_{12}FN$ (MW 117.17)
(X = Br)
[73630-93-0] $C_6H_{12}BrN$ (MW 178.07)
(X = I)
[65560-41-0] $C_6H_{12}IN$ (MW 225.07)

(mild halogenation of alcohols and acids under neutral conditions;[2] conversion of *N*-protected amino acids into peptides without racemization;[3] coupling of acids and allylic alcohols with organometallics;[9] [2 + 2] cycloaddition synthons, more reactive than dimethylketene[1b,4,5])

Alternate Name: tetramethyl-α-chloroenamine; TMCE

Physical Data: X = Cl: bp 129–130 °C (760 mmHg), *d* 1.01 g cm^{-3}; X = F: bp 89–92 °C (760 mmHg); X = Br: bp 42–45 °C (15 mmHg); X = I: bp 61–63 °C (9 mmHg), *d* 1.55 g cm^{-3}.

Solubility: very sol most organic solvents; reacts instantaneously with H_2O, alcohols, and protic solvents.

Preparative Methods: 1-chloro-*N,N*,2-trimethylpropenylamine is conveniently prepared by the reaction of *N,N*,2-trimethylpropanamide with Phosgene followed by dehydrochlorination of the intermediate α-chloroiminium chloride (eq 1).[1a,6] Recent unpublished results of our laboratory have shown that Phosphorus Oxychloride or di- or triphosgene can also be used for the preparation of TMCE.

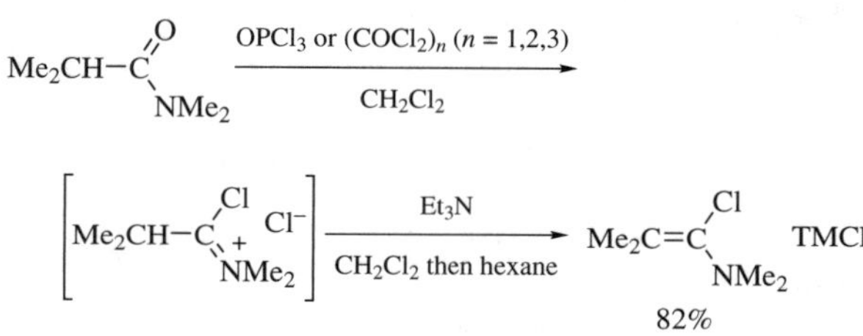

(1)

The corresponding 1-fluoro-, 1-bromo- and 1-iodo-*N,N*,2-trimethylpropenylamines (tetramethyl-α-fluoro-, -bromo-, and -iodoenamines; TMFE, TMBE, and TMIE) are readily prepared from TMCE by halogen exchange (eq 2).[1a,7]

Analysis of Reagent Purity: IR, ^{1}H NMR, elemental analysis.[6]

Handling, Storage, and Precautions: 1-chloro-, 1-fluoro-, and 1-bromo-*N,N*,2-propenylamines are thermally stable. They must be transferred in the absence of moisture and stored under nitrogen in sealed tubes. In spite of these precautions, light precipitates sometimes may form. 1-Iodo-*N,N*,2-trimethylpropenylamine is less stable and should be used when freshly prepared.

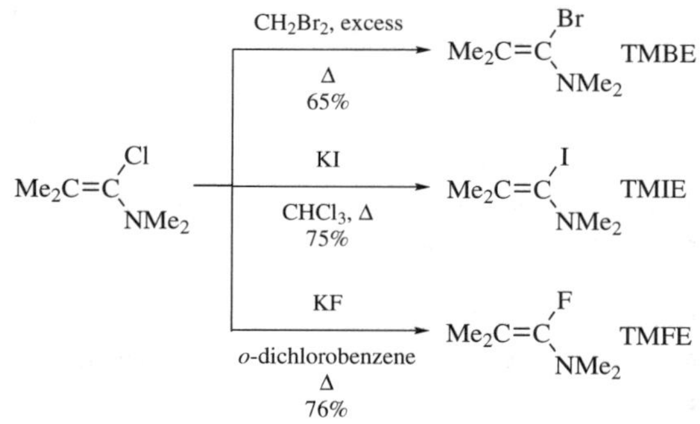

(2)

Halogenation Reactions. 1-Chloro-, 1-bromo- and 1-iodo-*N,N*,2-trimethylpropenylamines are mild reagents which transform alcohols into the corresponding halides (eqs 3–6).[2b] The reaction proceeds under neutral conditions, thus allowing the presence of acid-sensitive functional groups. The reaction occurs with inversion of configuration (eq 4). Secondary allylic and propargylic alcohols give some rearranged halides (eqs 5 and 6), but their formation seems to occur after the halogenation step.

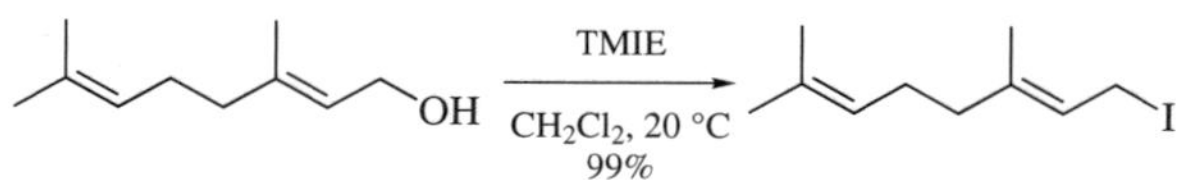

(3)

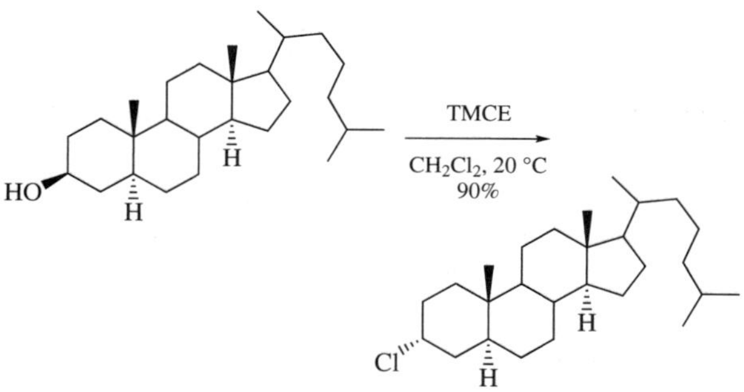

(4)

100% inversion

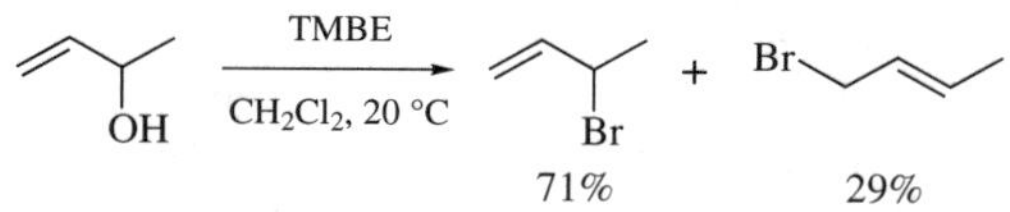

(5)

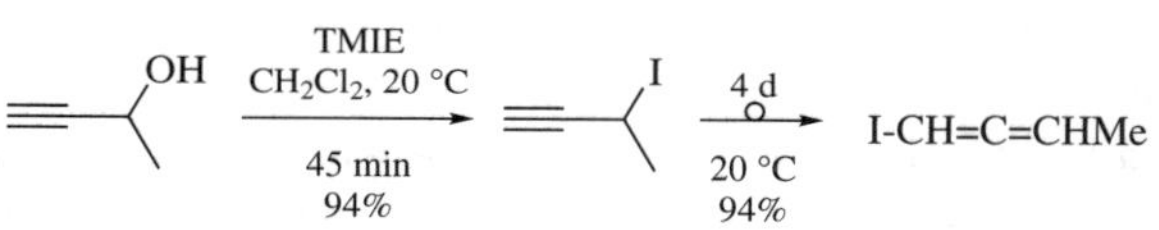

(6)

The halogenation reaction has been applied to the modification of various furanose and pyranose hemiacetals (eqs 7 and 8).[8]

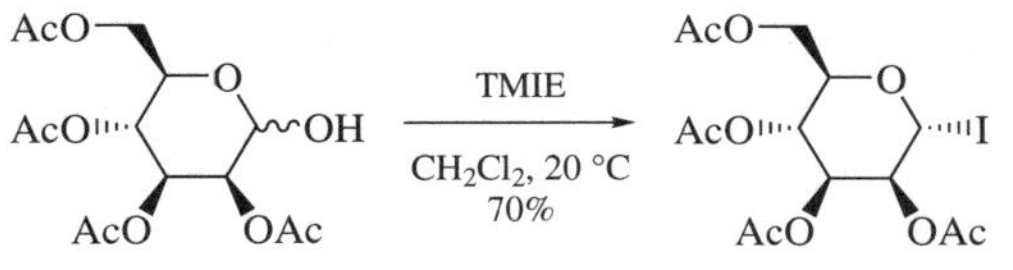
(7)

100% α

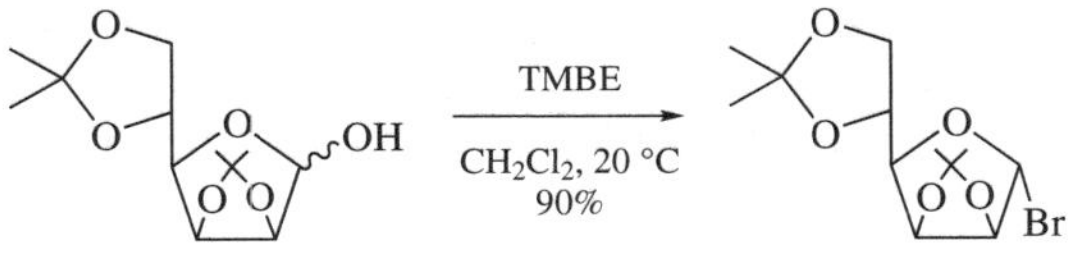
(8)

100% α

The reaction of 1-fluoro-*N,N*,2-trimethylpropenylamine with alcohols usually gives a mixture of products.[2b] In some cases a bulkier derivative has been successfully used to transform alcohols into fluorides (eqs 9 and 10).[2b,8]

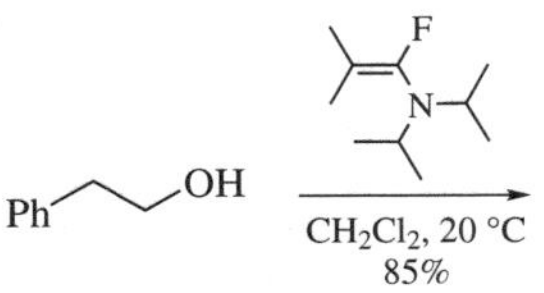
(9)

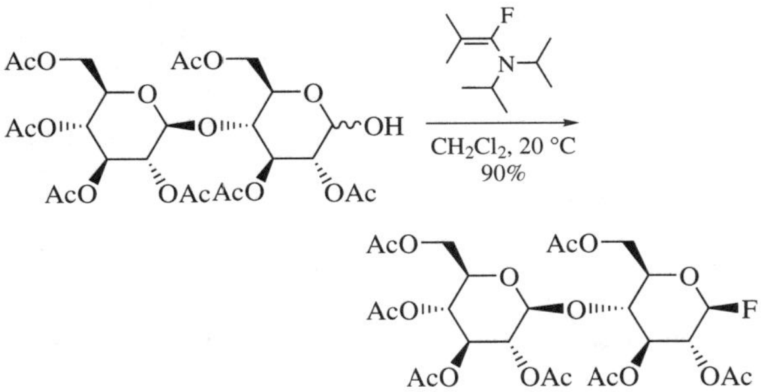
(10)

100% β

All the 1-halo-*N,N*,2-trimethylpropenylamines rapidly convert acids into acid halides (eqs 11 and 12).[2a] Since the only byproduct is the relatively inert *N,N*,2-trimethylpropanamide, the isolation of the acid halide will often be unnecessary. This method has been applied in peptide synthesis; no racemization was observed (eq 13).[3]

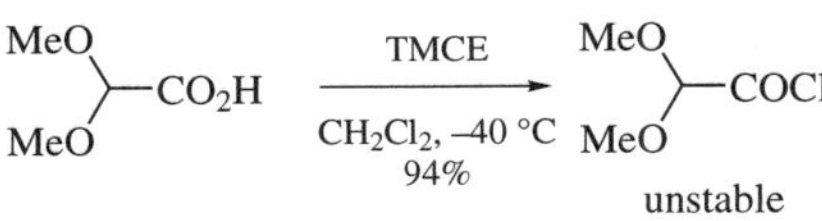
(11)

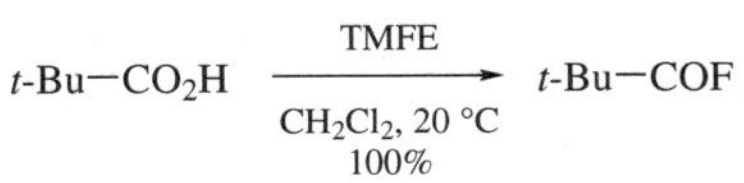
(12)

Coupling Reactions. 1-Chloro-*N,N*,2-trimethylpropenylamine is an excellent reagent for the coupling of carboxylic acids (eq 14) or allylic alcohols (eq 15) with organometallic compounds.[9]

[2 + 2] Cycloadditions. TMCE is a source of tetramethylketeniminium chloride.[1a] This highly electrophilic intermediate can be trapped by Schiff bases to generate azetidiniminium salts which readily hydrolyze to β-lactams (eq 16).[1b,5a] This has been developed into a general method to produce β-lactams from tertiary amides (eqs 17 and 18).[5c,5d]

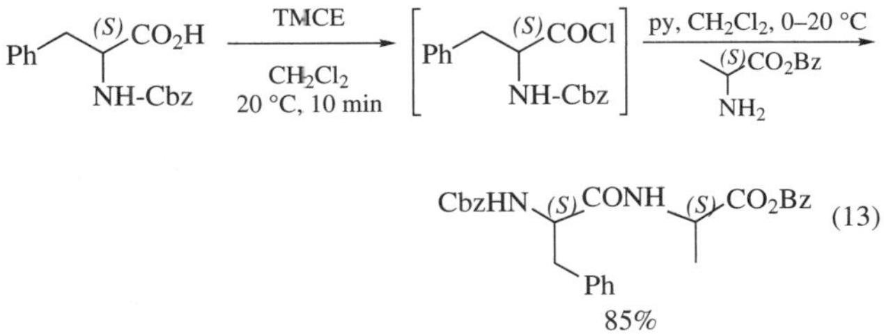
(13)

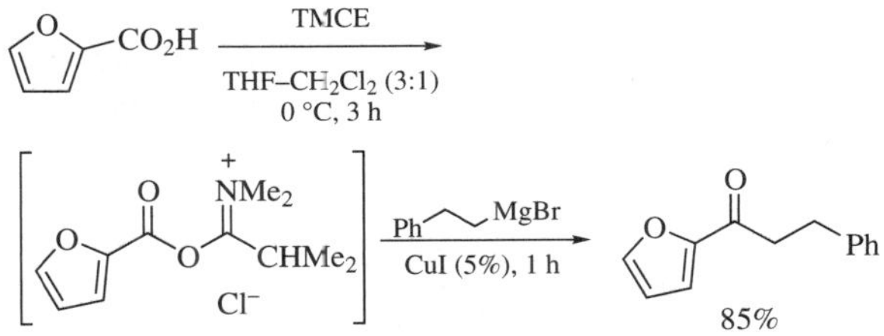
(14)

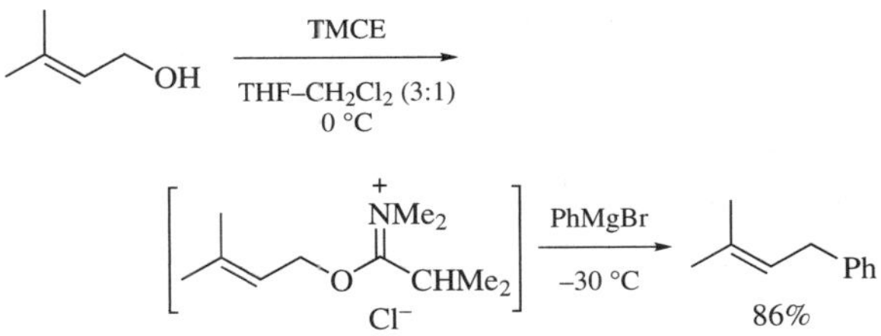
(15)

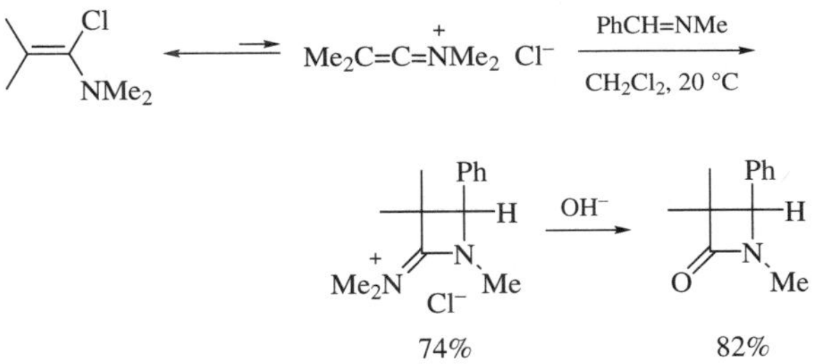
(16)

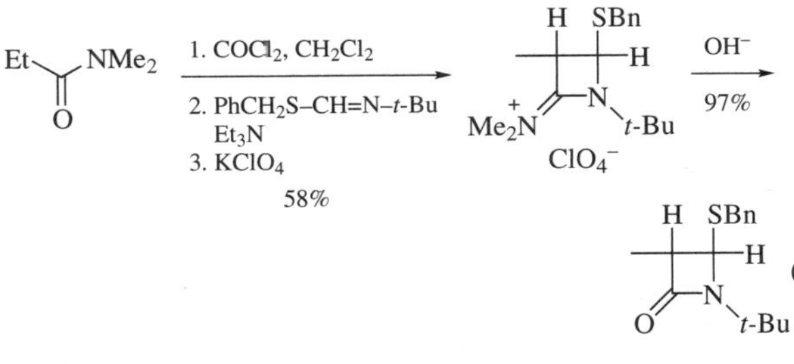
(17)

With the less nucleophilic carbon–carbon multiple bonds, cycloaddition will only occur in the presence of a Lewis acid which displaces the equilibrium toward the cumulene form (eq 19).[1a,4] $AgBF_4$, $ZnCl_2$, $AlCl_3$, and $SnCl_4$ have been successfully used, but the combination of TMCE with Zinc Chloride is the most convenient in many cases.[4b] Thus tetramethylketeniminium salts, which are most often generated in situ, behave as highly reactive equivalents of dimethylketene for the preparation of cyclobutanones and cyclobutenones. Indeed, the cyclobutane- and cyclobuteneiminium salts intermediates are readily hydrolyzed (eqs 20–23).

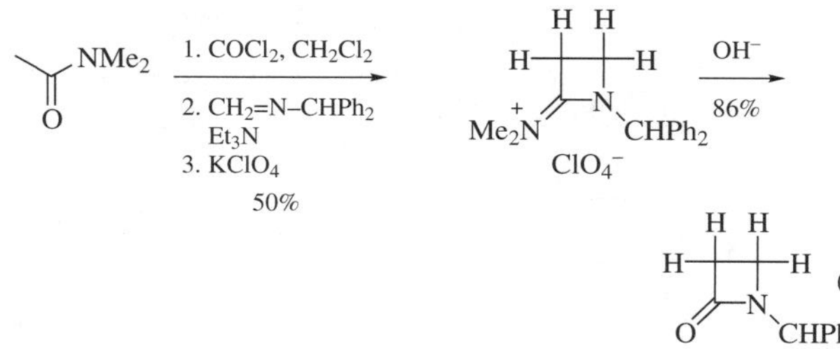

(18)

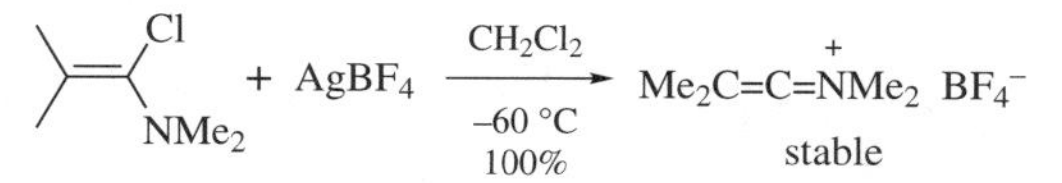

(19)

The reaction has been extended to other 2-substituted 1-chloro-*N,N*-dialkylvinylamines.[10] An asymmetric version of the cycloaddition has been described (eq 24).[11]

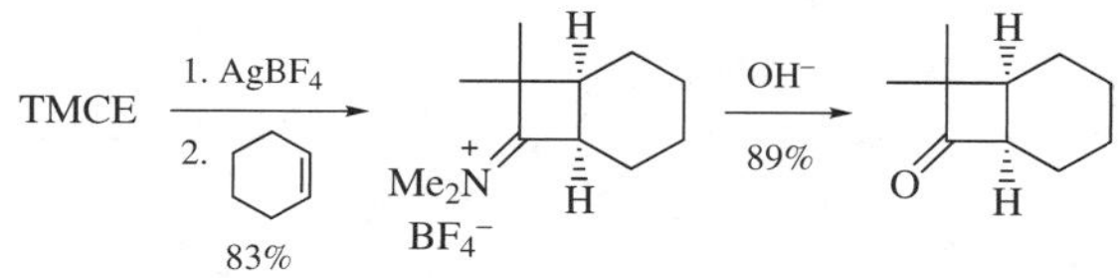

(20)

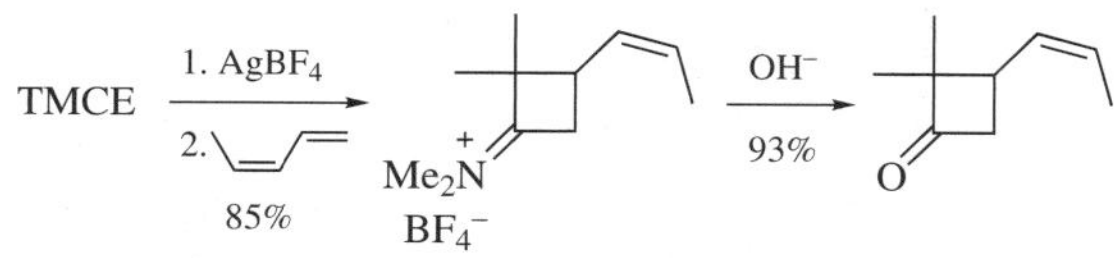

(21)

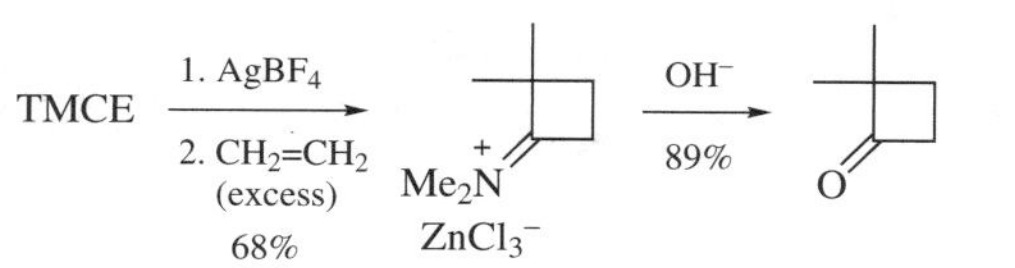

(22)

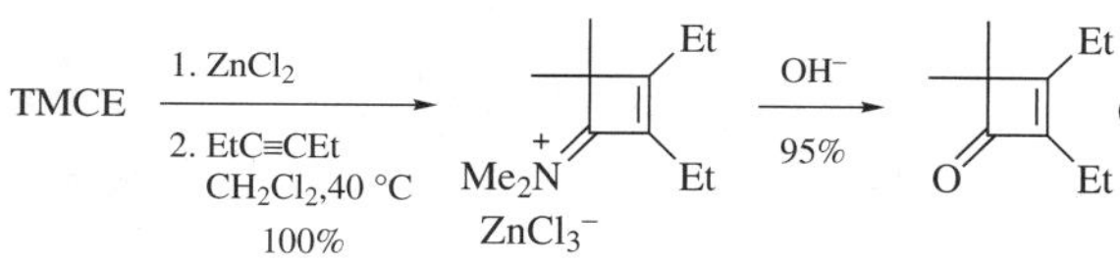

(23)

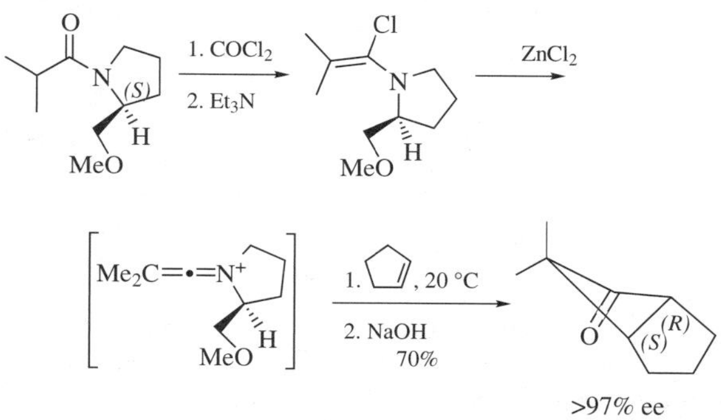

(24)

Miscellaneous Reactions. Reaction of 1-chloro-*N,N*,2-trialkylpropenylamines with azide ion gives 2-amino-1-azirines; these three-membered ring amidines are useful intermediates for the synthesis of pyridines.[12] 1-Chloro-*N,N*,2-trialkylpropenylamines react readily with electron-rich aromatics in electrophilic substitutions, without the need of an acid catalyst.[13] The corresponding α-metalated vinylamines give nucleophilic aminoalkenylation reactions.[14] TMCE reacts with potassium cyanide to yield 1-cyano-*N,N*,2-trimethylpropenylamine, a synthetic intermediate for the preparation of α-diketones.[15] TMCE is smoothly converted to *N,N*-dimethylmethacrylamide in the presence of Pyridine N-Oxide and Triethylamine; this reaction has been applied successfully to the preparation of various α,β-unsaturated amides.[16] 1-Phosphonato enamines have been obtained by treatment of TMCE with trialkyl phosphites.[17] The reaction of metal carbonyl anions with TMCE has been described; this leads to the formation of new transition metal organometallic compounds with novel structural features.[18]

---

1. (a) Ghosez, L.; Marchand-Brynaert, J. In *Advances in Organic Chemistry*; Raphael, R. A.; Taylor, E. C.; Wynberg, H., Eds.; Interscience: New York, 1976; Part 1, pp 421–523. (b) Ghosez, L.; O'Donnell, M. J. *Pericyclic Reactions*; Marchand, A. P.; Lehr, R. C., Eds.; Academic: New York, 1977; Vol. 2, Chapter 2. (c) Ghosez, L. *Medicinal Chemistry V*; Elsevier: Amsterdam, 1977; pp 363–385. (d) Ghosez, L. *Organic Synthesis Today and Tomorrow*; Trost, B. M.; Hutchinson, C. R., Eds.; Pergamon: Oxford, 1981; pp 145–162. (e) Ghosez, L. *New Synthetic Methodology and Functionally Interesting Compounds*; Yoshida, Z., Ed.; Kodansha: Tokyo, 1986; pp 99–117.
2. (a) Devos, A.; Rémion, J.; Frisque-Hesbain, A.-M.; Colens, A.; Ghosez, L. *Chem. Commun./J. Chem. Soc., Chem. Commun.* **1979**, 1180. (b) Munyemana, F.; Frisque-Hesbain, A.-M.; Devos, A.; Ghosez, L. *Tetrahedron Lett.* **1989**, *30*, 3077. (c) Schmidt, U.; Werner, J. *Chem. Commun./J. Chem. Soc., Chem. Commun.* **1986**, 996.
3. (a) Schmidt, U.; Lieberknecht, A.; Griesser, H.; Utz, R.; Beuttler, T.; Bartkowiak, F. *Synthesis* **1986**, 361. (b) Schmidt, U.; Kroner, M.; Beutler, U. *Synthesis* **1988**, 475.
4. (a) Marchand-Brynaert, J.; Ghosez, L. *J. Am. Chem. Soc.* **1972**, *94*, 2870. (b) Sidani, A.; Marchand-Brynaert, J.; Ghosez, L. *Angew. Chem., Int. Ed. Engl.* **1974**, *13*, 267. (c) Hoornaert, C.; Hesbain-Frisque, A.-M.; Ghosez, L. *Angew. Chem., Int. Ed. Engl.* **1975**, *14*, 569. (d) Heine, H.-G.; Hartmann, W. *Angew. Cheml* **1981**, *93*, 805. (e) Heine, H.-G.; Hartmann, W. *Synthesis* **1981**, 706.
5. (a) De Poortere, M.; Marchand-Brynaert, J.; Ghosez, L. *Angew. Chem., Int. Ed. Engl.* **1974**, *13*, 267. (b) Marchand-Brynaert, J.; Moya-Portuguez, M.; Lesuisse, D.; Ghosez, L. *Chem. Commun./J. Chem. Soc., Chem. Commun.* **1980**, 173. (c) Ghosez, L.; Bogdan, S.; Cérésiat, M.; Frydrych, C.; Marchand-Brynaert, J.; Moya-Portuguez, M.; Huber, I. *Pure Appl. Chem.* **1987**, *59*, 393. (d) Marchand-Brynaert, J.; Moya-Portuguez, M.; Huber, I.; Ghosez, L. *Chem. Commun./J. Chem. Soc., Chem. Commun.* **1983**, 818.
6. (a) Haveaux, B.; Dekoker, A.; Rens, M.; Sidani, A. R.; Toye, J.; Ghosez, L. *Q. Rev., Chem. Soc.* **1979**, *59*, 26. (b) Ghosez, L.; Koch, I.Swiss Pat. 681 623A, 1993.
7. (a) Colens, A.; Demuylder, M.; Téchy, B.; Ghosez, L. *Nouv. J. Chim.* **1977**, *1*, 369. (b) Colens, A.; Ghosez, L. *Nouv. J. Chim.* **1977**, *1*, 371.
8. (a) Ernst, B.; Winkler, T. *Tetrahedron Lett.* **1989**, *30*, 3773. (b) Ernst, B.; De Mesmaeker, A.; Hoffmann, P.; Ernst, B. *Tetrahedron Lett.* **1989**, *30*, 3773. (c) Ernst, B.; De Mesmaeker, A.; Wagner, B.; Winkler, T. *Tetrahedron Lett.* **1990**, *31*, 6167. (d) Ernst, B.Eur. Pat. Appl. 373 118A2, 1990.
9. (a) Fujisawa, T.; Mori, T.; Higuchi, K.; Sato, T. *Chem. Lett.* **1983**, 1791. (b) Fujisawa, T.; Iida, S.; Yukizaki, H.; Sato, T. *Tetrahedron Lett.* **1983**, *24*, 5745. (c) Fujisawa, T.; Umezu, K.; Sato, T. *Chem. Lett.* **1985**, 1453. (d) Fujisawa, T.; Sato, T. *Q. Rev., Chem. Soc.* **1988**, *66*, 116.
10. (a) Falmagne, J.-B.; Escudero, J.; Taleb-Sahraoui, S.; Ghosez, L. *Angew. Chem., Int. Ed. Engl.* **1981**, *20*, 879. (b) Schmit, C.; Sahraoui-Taleb, S.; Differding, E.; Dehasse-De Lombaert, C.-G.; Ghosez, L.

Tetrahedron Lett. **1984**, *25*, 5043. (c) Génicot, C.; Gobeaux, B.; Ghosez, L. *Tetrahedron Lett.* **1991**, *31*, 3827.

11. (a) Houge, C.; Frisque-Hesbain, A.-M.; Mockel, A.; Ghosez, L.; Declercq, J. P.; Germain, G.; Van Meerssche, M. *J. Am. Chem. Soc.* **1982**, *104*, 2920. (b) Saimoto, H.; Houge, C.; Hesbain-Frisque, A.-M.; Mockel, A.; Ghosez, L. *Tetrahedron Lett.* **1983**, *24*, 2251. (c) Génicot, C.; Ghosez, L. *Tetrahedron Lett.* **1992**, *33*, 7357. (d) Ghosez, L.; Génicot, C.; Gouverneur, V. *Pure Appl. Chem.* **1992**, *64*, 1849.
12. (a) Rens, M.; Ghosez, L. *Tetrahedron Lett.* **1970**, 3765. (b) Vittorelli, P.; Heimgartner, H.; Schmid, H.; Hoet, P.; Ghosez, L. *Tetrahedron* **1974**, *30*, 3737. (c) Demoulin, A.; Gorissen, H.; Hesbain-Frisque, A.-M.; Ghosez, L. *J. Am. Chem. Soc.*, **1975**, *97*, 4409.
13. Marchand-Brynaert, J.; Ghosez, L. *J. Am. Chem. Soc.* **1972**, *94*, 2869.
14. Wiaux-Zamar, C.; Dejonghe, J.-P.; Ghosez, L.; Normant, J.-F.; Villieras, J. *Angew. Chem., Int. Ed. Engl.* **1976**, *15*, 371.
15. Toye, J.; Ghosez, L. *J. Am. Chem. Soc.* **1975**, *97*, 2276.
16. Da Costa, R.; Gillard, M.; Falmagne, J.-B.; Ghosez, L. *J. Am. Chem. Soc.* **1979**, *101*, 4381.
17. Ahlbrecht, H.; Farnung, W. *Synthesis* **1977**, 336.
18. (a) King, R. B.; Hodges, K. C. *J. Am. Chem. Soc.* **1974**, *96*, 1263. (b) King, R. B.; Hodges, K. C. *J. Am. Chem. Soc.* **1975**, *97*, 2702.

Léon Ghosez & Jacqueline Marchand-Brynaert
Université Catholique de Louvain, Louvain-la-Neuve, Belgium

Chlorotripyrrolidinophosphonium Hexafluorophosphate

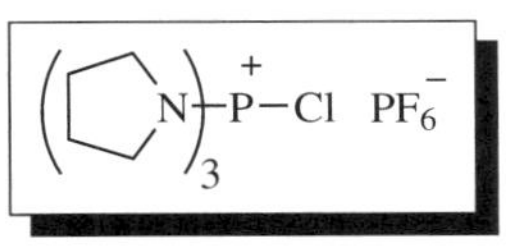

[133894-48-1] $C_{12}H_{24}ClF_6N_3P_2$ (MW 421)

(reagent for carboxylic acid activation, coupling agent in peptide synthesis)

Alternate Name: PyCloP.
Physical Data: mp 150–151 °C.[1]
Solubility: soluble in CH_2Cl_2, $CHCl_3$, DMF, DMSO, NMP, acetone.
Form Supplied in: colorless crystals.
Analysis of Reagent Purity: ^{31}P NMR, HPLC.
Preparative Methods: addition of chlorine to a diethyl ether solution of tripyrrolidinophosphine, then addition of KPF_6 in water.[2] Addition of tripyrrolidinophosphine oxide to a DCM solution of phosphoryl trichloride, then addition of KPF_6 in water.[1–3]
Purification: recrystallization (CH_2Cl_2/Et_2O).
Handling, Storage, and Precautions: the toxicological properties of this reagent have not been investigated and no special instructions for storage and handling are mentioned in the literature.

Introduction. Treatment of a carboxylic acid with PyCloP in the presence of base produces the unstable acylphosphonium intermediate. This species directly reacts with a nucleophile or decomposes into a different activated form (e.g. carboxylic anhydride). This reagent gives results that are comparable to those obtained with **PyBroP** (bromotripyrrolidinophosphonium hexafluorophosphate).[1]

Reagent for Peptide Synthesis. Compared with hydroxybenzotriazol-based activators like BOP, PyBOP, HBTU, or DCC/HOBT, which produce moderately activated benzotriazolyl esters from carboxylic acids, PyCloP is highly recommended for coupling *N*-methylated amino acids, as illustrated by the difficult coupling of NMeVal (eq 1),[1,4] and used in the total synthesis of natural peptoids.[5–7] PyCloP activation of Boc-amino acids can, however, give moderate yields due to the formation of *N*-carboxyanhydride (NCA).[1,8]

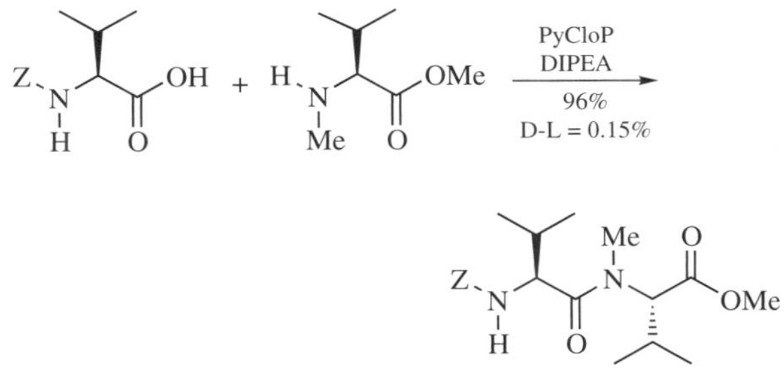

(1)

Activation of a Monothiocarboxylic Acid. Using different phosphorus reagents for activation of a monothiocarboxylic acid, the observed oxygen/sulfur selectivities correlate with the hardness/softness of the reaction center. Compared with the soft leaving bromine atom in PyBroP (S selective) or the hard oxygen atom in PyBOP (O selective), the PyCloP is borderline in O/S (25:47) selectivity (eq 2).[3]

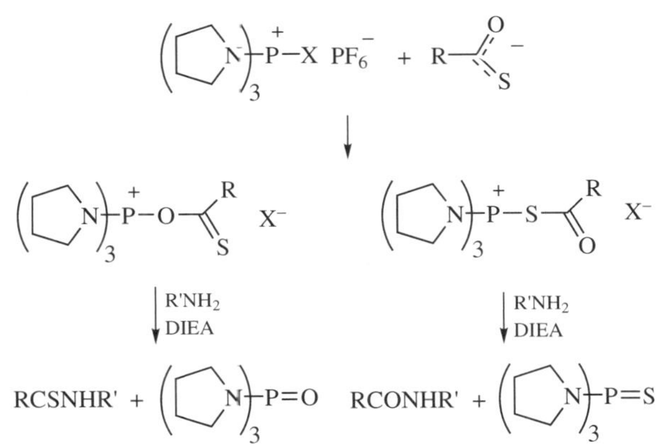

(2)

X=OBt(71%); Br(4%); Cl(25%) X=OBt(8%); Br(73%); Cl(47%)

Preparation of Coupling Reagents for Peptide Synthesis. PyBOP, a related phosphonium coupling reagent for peptide synthesis, has been prepared from PyCloP,[2] and several analogs of PyBOP have also been prepared as reagents for thioacylation of monothiocarboxylic acid[3] (eq 3).

Related Reagents. PyBroP, bromotripyrrolidinophosphonium hexafluorophosphate; PyBOP, (1*H*-benzotriazol-1-yl) oxytripyr rolidinophosphonium hexafluorophosphate; PyClU, 1,1,3,3-bis-(tetramethylene)chlorouronium hexafluorophosphate; TpyClU, 1,1,3,3-bis(tetramethylene)chlorouronium tetrafluoroborate; CIP, 2-chloro-1,3-dimethylimidazolidium hexafluorophosphate; PyAOP,(7-azabenzotriazol-1-yl)oxytripyrrolidinophosphonium hexafluorophosphate; PyPOP, (pentafluorophenyl) oxytripyrrolidinophosphonium hexafluorophosphate; PyNOP, (6-nitrobenzotriazol-1)-yloxytripyrrolidinophosphonium hexafluorophosphate; PyFOP, [(6-trifluoromethyl)benzotriazol-1]-yloxytripyrrolidinophosphonium hexafluorophosphate; PyDOP, (3,4-dihydro-4-oxo-1,2,3-benzotriazin-3-yl)oxytri pyrrolidinophosphonium hexafluorophosphate.

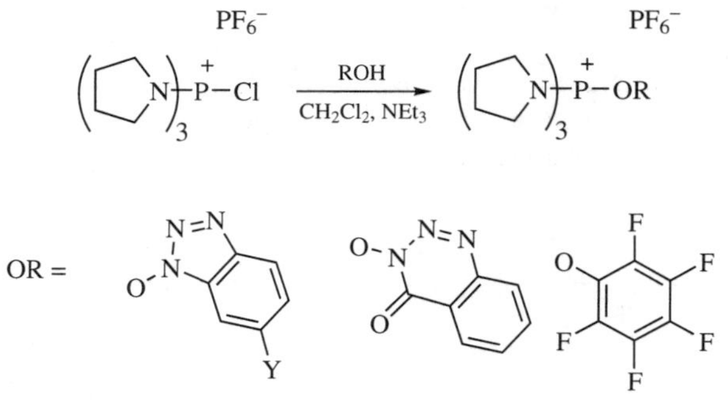

(3)

Y = H, PyBOP (75%)
NO2, PyNOP (68%)
CF3, PyFOP (92%)

PyDOP (94%) PyPOP (92%)

1. Coste, J.; Frérot, E.; Jouin, P., *J. Org. Chem.* **1994**, *59*, 2437.
2. Castro, B.; Coste, J., I Patent WO 90/10009, 1990.
3. Hoeg-Jensen, T.; Olsen, C. E.; Holm, A., *J. Org. Chem.* **1994**, *59*, 1257.
4. Coste, J.; Frérot, E.; Jouin, P.; Castro, B., *Tetrahedron Lett.* **1991**, *32*, 1967.
5. Calmes, M.; Cavelier-Frontin, F.; Jacquier, R.; Mercadier, J.-L.; Sabil, S.; Verducci, J.; Quiot, J.-M.; Vey, A., *Int. J. Peptide Protein Res.* **1993**, *41*, 528.
6. Patino, N.; Frérot, E.; Galéotti, N.; Poncet, J.; Coste, J.; Dufour, M.-N.; Jouin, P., *Tetrahedron* **1992**, *48*, 4115.
7. Poncet, J.; Busquet, M.; Roux, F.; Pierré, A.; Atassi, G.; Jouin, P., *J. Med. Chem.* **1998**, *41*, 1524.
8. Frérot, E.; Coste, J.; Poncet, J.; Jouin, P.; Castro, B., *Tetrahedron Lett.* **1992**, *33*, 2815.

Jacques Coste & Patrick Jouin
CNRS-UPR9023, Montpellier, France

2,4,6-Collidine

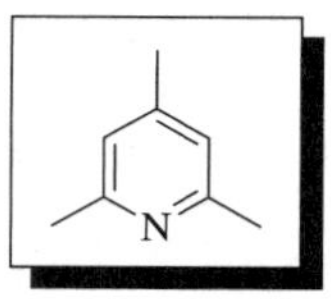

[108-75-8] $C_8H_{11}N$ (MW 121.18)

(most useful in dehydrohalogenation reactions;[1] used for basic properties as reaction solvent or additive)

Alternate Names: 2,4,6-trimethylpyridine.
Physical Data: bp 170 °C; mp −44.5 °C; *d* 0.913 g cm^{-3} at 20 °C. Isomers of the title reagent include 2,3,4-collidine (*[2233-29-6]*; bp 192–193 °C), 2,3,5-collidine (*[695-98-7]*; bp 182– 183 °C/739 mmHg), 2,3,6-collidine (*[1462-84-6]*; bp 176– 178 °C), 2,4,5-collidine (*[1122-39-0]*; bp 165–168 °C), and 3,4, 5-collidine (*[20579-43-5]*; bp 205–207 °C).[2–4]
Solubility: sol 3.5 g/100 mL H_2O (20 °C) and 20.8 g/100 mL H_2O (6 °C); miscible with ether; sol methanol, ethanol, chloroform, benzene, toluene, dilute acids.[3]
Form Supplied in: liquid.
Analysis of Reagent Purity: GLC.
Purification: by distillation.
Handling, Storage, and Precautions: protect from light; moisture sensitive; incompatible with strong oxidizing agents.[5]

Dehydrohalogenation Reactions. Numerous examples in the literature describe the utility of collidines to aid in dehydrohalogenation reactions. One example (eq 1) is the synthesis of 2-methyl-2-cyclohexenone (**2**) from 2-chloro-2-methylcyclohexanone (**1**).[6] Another example involves the dehydrobromination of tetrahydrofuran (**3**) to diene (**4**) en route from linalool to karahanaenone (62%, overall, eq 2).[7] Similarly, dehydrobromination of stigmasteryl acetate (**5**) affords 7-dehydrostigmasteryl acetate (**6**) in a refluxing mixture of collidine and mesitylene (eq 3).[1]

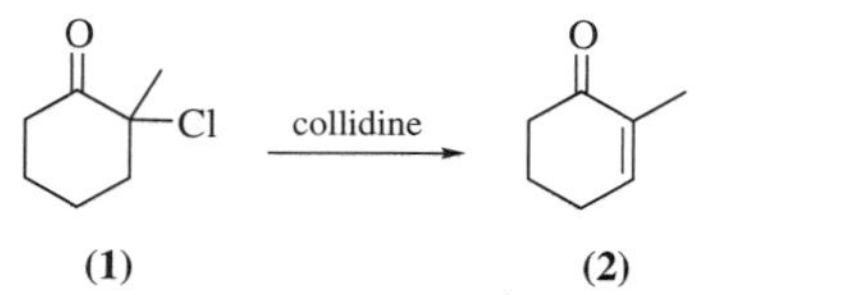

(1)

(1) **(2)** (45–49% from 2-methylcyclohexanone)

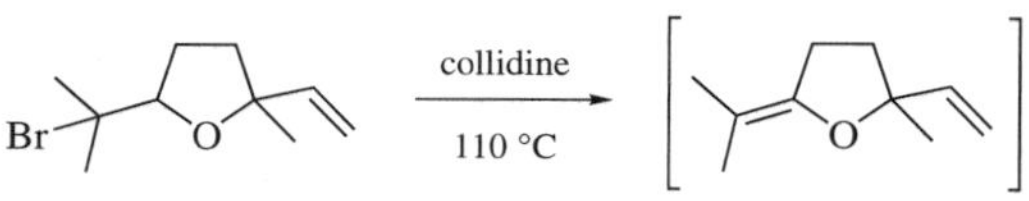

(2)

(3) **(4)**

A key step in a synthesis of pyridocarbazoles used collidine in THF for a dehydrochlorination. A hindered base such as collidine was required to avoid cleavage of the lactone ring, the preferred reaction using a base such as *N,N*-dimethylaniline (eq 4).[8]

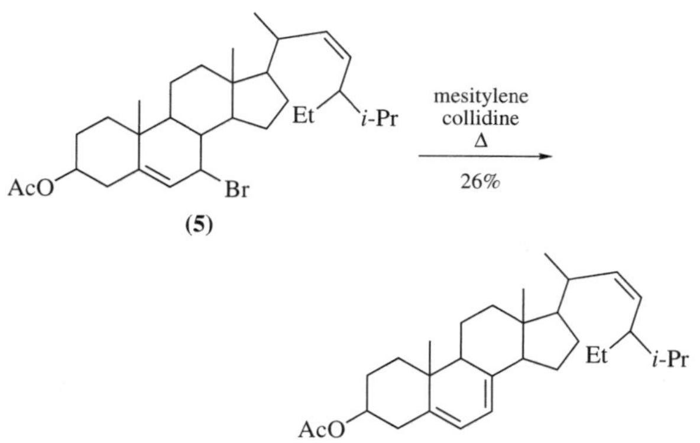

(3)

(5) **(6)**

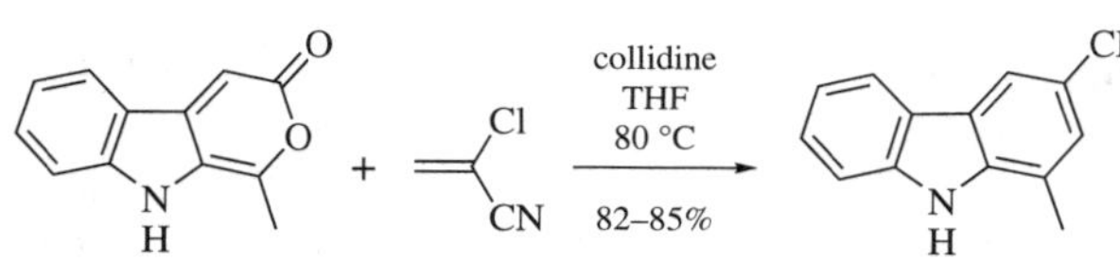

(4)

2,4,6-Collidine is used as solvent in the preparation of 2-benzylcyclopentanone (**8**) from 2-benzyl-2-methoxycarbonylcyclopentane (**7**) (eq 5).[9]

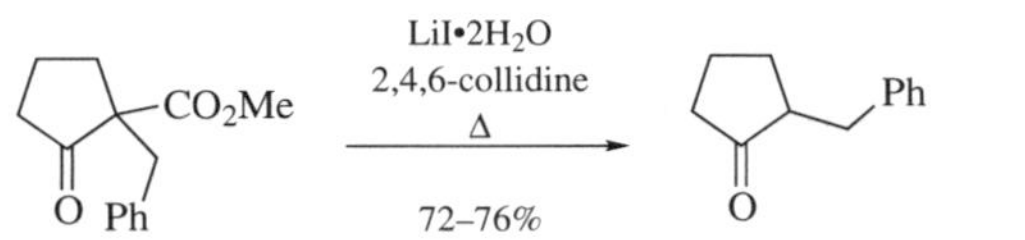

(**7**) (**8**) (5)

Cleavage of alkyl aryl ethers (72–76%) occurs in high yield with Lithium Iodide with 2,4,6-collidine as solvent.[10] In the absence of a solvent, high reaction temperatures are required.

Collidine as a Base Catalyst. Collidine is useful as a base in a variety of reactions. Sometimes it doubles as reaction base and solvent. A mixture of monobromo isomers is obtained in very high yield when 1-methylcyclopropyl-3-methyl-3-butenylcarbinol (**9**) is treated with Phosphorus(III) Bromide and collidine (eq 6).[11] The bromo compounds shown underwent further reaction with Zinc Bromide to form *trans*-1-bromo-3,7-dimethylocta-3,7-diene in 85–90% yield.

Collidine is effective as a proton scavenger in the Koenigs–Knorr synthesis of disaccharide glycosides. Silver(I) Trifluoromethanesulfonate and collidine are paired as a useful promoter of glycosylation reactions of alcohols with glycosyl halides. For example, the reaction of hepta-*O*-acetyl-α-D-cellobiosyl bromide (**10**) with 8-ethoxycarbonyloctanol (**11**) in the presence of silver triflate and collidine afforded a 78% yield of the corresponding 1,2-orthoacetate derivative (**12**) (eq 7).[12] When a similar reaction was carried out with *N,N,N′,N′*-tetramethylurea in place of collidine, a different reaction path was followed to give isomeric 1,2-*trans*-glycosides.

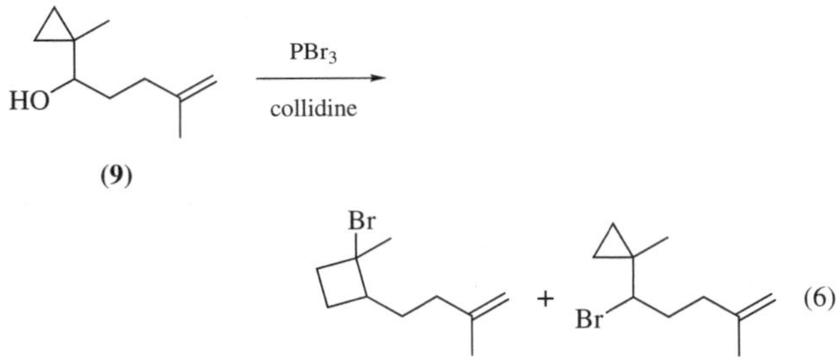

(**9**) (6)

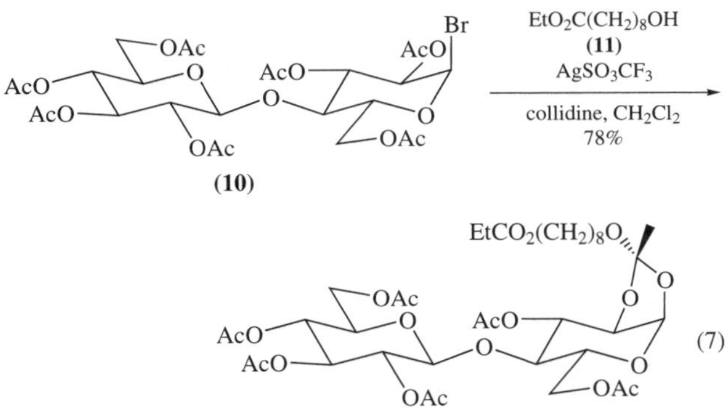

(**10**) (**12**) (7)

Collidine is also used as the base for glycosidation reactions promoted by Tin(II) Trifluoromethanesulfonate. Starting with glucose derivative (**13**) and protected sugar derivatives, 1,2-*trans*-β-D-disaccharides have been prepared by this method (eq 8)[13]

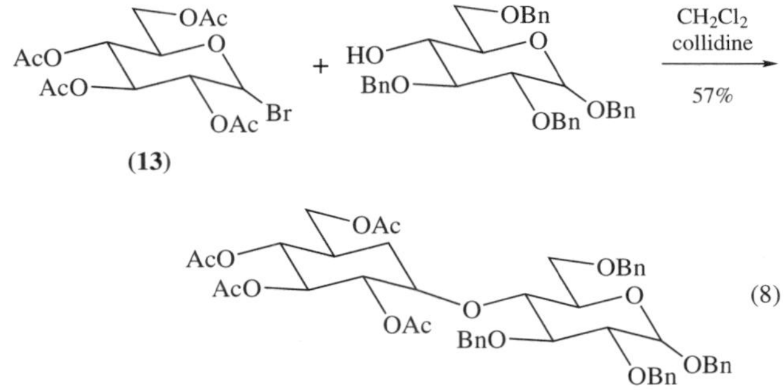

(**13**) (8)

Collidine behaves as a nucleophilic catalyst in reactions such as the hydrolysis of picryl chloride. The reaction proceeds by formation of intermediate *N*-picryl-2,4,6-trimethylpyridinium ion (**14**) (eq 9).[14]

Collidine acts as a sterically hindered base in tritylation reactions of very weakly acidic compounds. The tritylation reaction proceeds in high yield for both acetone and acetonitrile, which serve as both reactant and solvent in the syntheses of 1,1,1-triphenyl-3-butanone (**15**) and 1,1,1-triphenylpropionitrile (**16**), respectively (eqs 10 and 11).[15] This tritylation is a general reaction, applicable to more strongly acidic compounds such as diethyl malonate and nitromethane. In the nitromethane reaction, a low yield of the desired product is obtained due to side reactions.

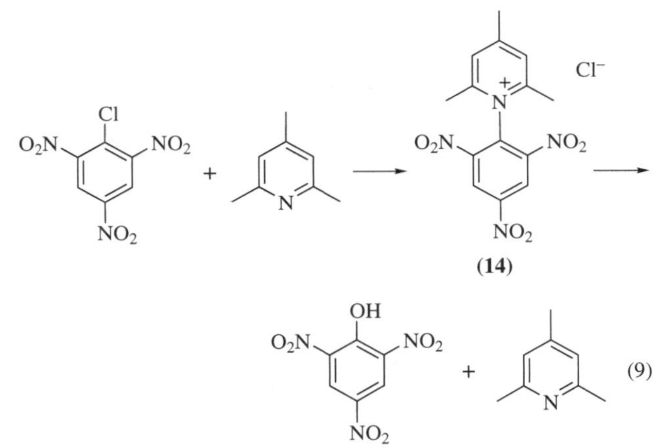

(**14**) (9)

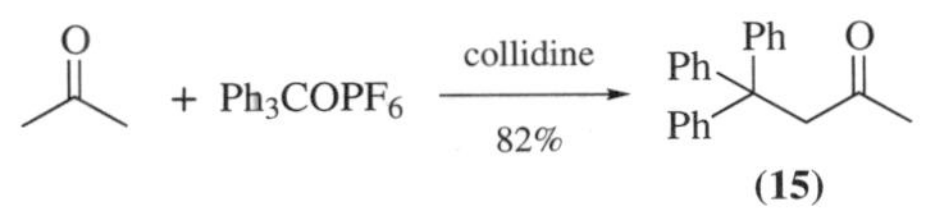

(**15**) (10)

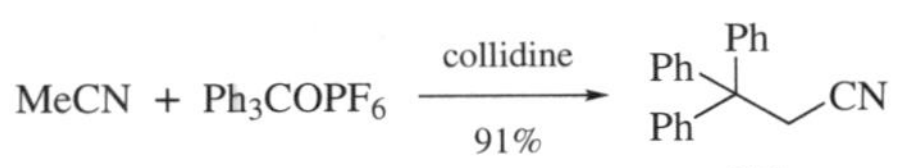

(**16**) (11)

1. Kircher, H. W.; Rosenstein, F. U. *J. Org. Chem.* **1973**, *38*, 2259.
2. Goe, G. L.In *Kirk-Othmer Encyclopedia of Chemical Technology*; Wiley:New York, **1982**; Vol. 19, p 454.
3. *The Merck Index*, 11th ed.; Budavari, S., Ed.; Merck:Rahway, NJ, 1989; p 1529.

4. *Dictionary of Organic Compounds*, 5th ed.; Buckingham, J., Ed.; Chapman & Hall:London, 1982; p 5592.
5. *The Sigma Aldrich Library of Regulatory & Safety Data*; Aldrich: Milwaukee, WI, 1993; Vol 2, p 2503.
6. Warnhoff, E. W.; Martin, D. G.; Johnson, W. S. *Org. Synth., Coll. Vol.* **1963**, *4*, 162.
7. Demole, E.; Enggist, P. *Helv. Chim. Acta* **1971**, *54*, 456.
8. Narasimhan, N. S.; Gokhale, S. M. *Chem. Commun./J. Chem. Soc., Chem. Commun.* **1985**, 86.
9. Elsinger, F. *Org. Synth., Coll. Vol.* **1973**, *5*, 76.
10. Harrison, I. T. *Chem. Commun./J. Chem. Soc., Chem. Commun.* **1969**, 616.
11. Brady, S. F.; Ilton, M. A.; Johnson, W. S. *J. Am. Chem. Soc.* **1968**, *90*, 2882.
12. Banoub, J.; Bundle, D. R. *Can. J. Chem.* **1979**, *57*, 2091.
13. Lubineau, A.; Malleron, A. *Tetrahedron Lett.* **1985**, *26*, 1713.
14. de Rossi, R. H.; deVargas, E. B. *J. Org. Chem.* **1979**, *44*, 4100.
15. Bidan, G.; Cauquis, G.; Genies, M. *Tetrahedron* **1979**, *35*, 177.

Angela R. Sherman
Reilly Industries, Indianapolis, IN, USA

2-Cyanoethyl Diisopropylchlorophosphoramidite

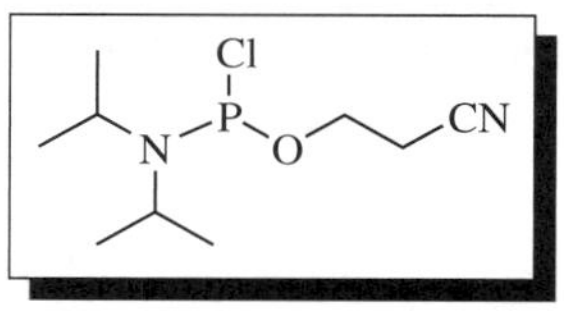

[89992-70-1] $C_9H_{18}ClN_2OP$ (MW 236.68)

(reagent widely used for the phosphinylation of the 3′-hydroxy function of modified and unmodified nucleosides, and of various reporter groups for incorporation into oligonucleotides)

Alternate Name: 2-cyanoethyl *N,N*-diisopropylphosphonamidic chloride; chloro-(2-cyanoethoxy)-*N,N*-diisopropylaminophosphine; chloro(diisopropylamino)-β-cyanoethoxyphosphine; 2-cyanoethyl *N,N*-diisopropylphosphoramidochloridite.
Physical Data: bp 103–104 °C (0.08 mmHg; d 1.061 g cm^{-3}).
Solubility: soluble in most organic solvents.
Form Supplied in: colorless liquid.
Analysis of Reagent Purity: ^{31}P NMR, ^{1}H NMR, EI-MS.
Preparative Methods: the reagent is prepared from the reaction of *N,N*-diisopropylamine or its *N*-trimethylsilyl derivative with (2-cyanoethoxy)dichlorophosphine in anhydrous diethyl ether and isolated by vacuum distillation.
Purification: vacuum distillation.
Handling, Storage, and Precautions: the reagent is moisture-sensitive and corrosive. It should therefore be stored in a dry environment, in a tightly closed container at −20 °C. Caution should be exercised when vacuum-distilling the reagent as it may explode if heated at a temperature (internal) exceeding 140 °C. Distillation should be carried out only if a vacuum of less than 0.1 mmHg can be generated.
Disposal: the reagent should be neutralized at a temperature below −20 °C by slowly adding an excess of ethanolic potassium hydroxide. The deactivated material can then be discarded in a chemical waste container.

Phosphinylation of Modified Nucleosides for Incorporation into Oligonucleotides. Although 2-cyanoethyl diisopropylchlorophosphoramidite has been used extensively for the phosphinylation of unmodified nucleosides toward solid-phase oligonucleotide synthesis,[1] its application to the preparation of modified nucleoside phosphoramidites for incorporation into oligonucleotides has recently become the focus of intense scrutiny. For example, 2-cyanoethyl diisopropylchlorophosphoramidite is used to phosphinylate 1,N^6-ethano-2′-deoxyadenosine.[2] Insertion of the modified nucleoside phosphoramidite (**1**) at a specific location in an oligodeoxyribonucleotide is reported as a means to gather information about the repair, mutagenesis, and structural properties of the lesion. Studies such as these should lead to a better understanding of the cytotoxic and carcinogenic mechanisms of 1,3-bis(2-chloroethyl)nitrosourea, which has been frequently used in cancer treatment; 1,N^6-ethano-2′-deoxyadenosine is one of the many adducts generated from the reaction of the nitrosourea with DNA.

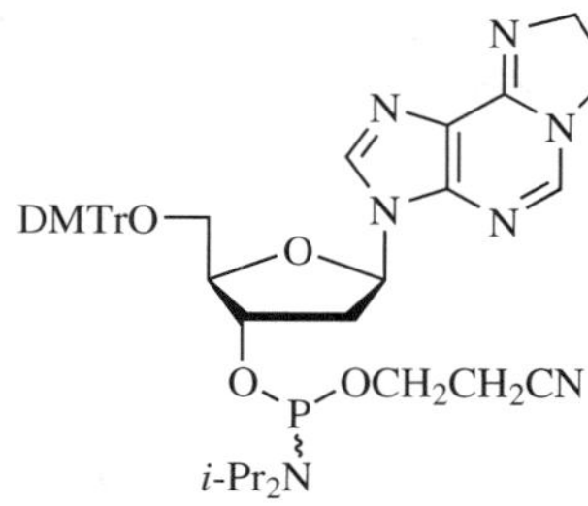

1, DMTr = 4,4′-dimethoxytrityl

It should be noted that 3,N^4-etheno-2′-deoxycytidine and 3,N^4-ethano-2′-deoxycytidine have also been converted to the corresponding phosphoramidites upon reaction with 2-cyanoethyl diisopropylchlorophosphoramidite.[3] The incorporation of these modified nucleoside phosphoramidites into oligonucleotides at predetermined sites provides additional models for studying the cytotoxic mechanisms of 1,3-bis(2-chloroethyl)nitrosourea, vinyl chloride, and other α,β-unsaturated compounds. In this context, polycyclic aromatic hydrocarbons are metabolized to bay-region diol epoxides, which are thought to exert their carcinogenic and other genotoxic effects predominantly through alkylation of DNA.[4] In order to investigate the biochemical events associated with cancer induction subsequent to DNA-adduct formation, site-specific integration of a bay-region diol epoxide adduct into an oligonucleotide may indicate whether a preference for specific mismatches opposite the adduct is achieved during replication of the adducted DNA. Thus, reaction of 2-cyanoethyl diisopropylchlorophosphoramidite with a tetrahydrophenanthrene-2′-deoxyadenosine derivative affords the modified deoxyribonucleoside phosphoramidite derivative (**2**) for incorporation into a nonanucleotide at predetermined sites.[4]

Similar syntheses of oligonucleotides containing both *cis*[5] and *trans*[6] ring-opened adducts of 3,4-epoxy-1,2,3,4-tetrahydrophenanthrene with 2′-deoxyadenosine, and that of an analogous *trans* adduct with 2′-deoxyguanosine[7] have been reported.

Oligonucleotides modified with a specific chemical functionality, such as a transition or lanthanide metal complex, are of widespread interest for analytical and therapeutic applications, and for mechanistic studies. The use of solid-phase DNA synthesis methodologies for site-specific labeling oligonucleotides with a transition metal complex has led to the preparation of a novel ruthenium-nucleoside phosphoramidite. Specifically, phosphinylation of a $Ru(bpy)_3^{2+}$-labeled 2′-deoxyuridine derivative with 2-cyanoethyl diisopropylchlorophosphoramidite produces the ruthenium-labeled phosphoramidite (**3**), which was inserted at selected sites in hexadecanucleotides.[8] These ruthenium-modified oligonucleotides form stable luminescent DNA duplexes at ambient temperature and are amenable to photophysical studies.[8]

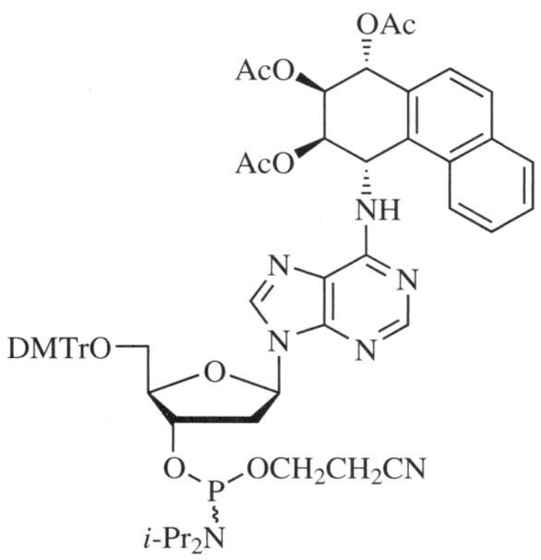

2

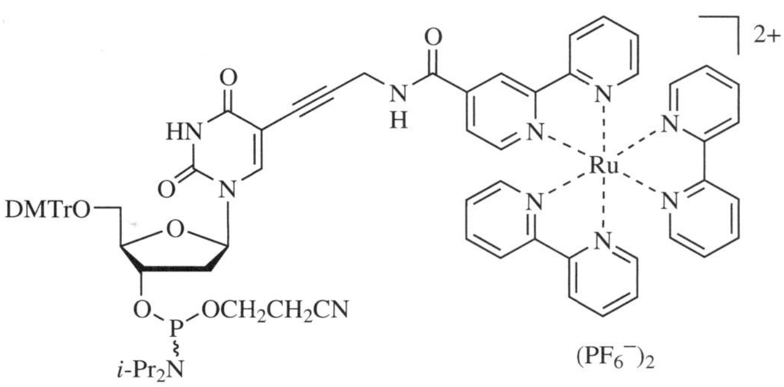

3

Oligonucleotides composed of carbohydrate-modified nucleosides have been extensively investigated in an attempt to identify those oligonucleotides that would demonstrate an increased affinity for complementary DNA/RNA oligomers while maintaining sequence specificity in hybridization experiments. In this regard, the incorporation of two anthraquinonylmethyl groups into oligonucleotides at specific carbohydrate residues has been accomplished by solid-phase synthesis.[9] Thus, condensation of 2-cyanoethyl diisopropylchlorophosphoramidite with 5′-*O*-(4,4′-dimethoxytrityl)-2′-*O*-(2-anthraquinonylmethyl) uridine gives the modified nucleoside phosphoramidite (**4**) required for oligonucleotide synthesis. An oligonucleotide having two 2′-*O*-anthraquinonylmethyl groups at sites separated by four nucleotides exhibits the highest affinity for both unmodified complementary DNA ($\Delta T_m = +5.7\,^\circ C$ per modification) and RNA ($\Delta T_m = +3.3\,^\circ C$ per modification) oligonucleotides without altering binding specificity.[9]

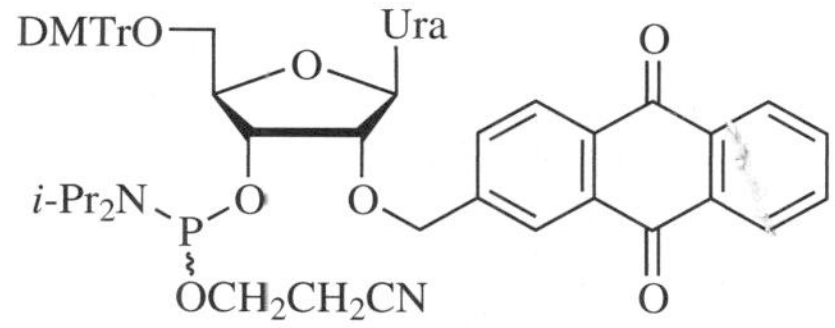

4

'Locked nucleic acids' (LNA) composed of 2′-*O*,4′-*C*-methylene bicyclonucleosides are efficiently prepared from the phosphoramidite derivative of these bicyclonucleosides.[10] 2-Cyanoethyl diisopropylchlorophosphoramidite has been efficient, as a phosphinylating agent, in the synthesis of the modified phosphoramidites (**5**).

LNA display unprecedented binding affinity for unmodified complementary DNA and RNA oligonucleotides; ΔT_m values of +3 to +5 °C per modification and +4 to +8 °C per modification were determined for LNA-DNA and LNA-RNA hybrids, respectively. LNA also demonstrate excellent stability against 3′-exonucleases.[11]

Replacement of the five-membered furanose ring of natural DNA by a cyclohexene ring leads to cyclohexene nucleic acids (CeNA) as a novel class of structurally modified nucleic acids. Cyclohexenyl nucleosides[12] are converted to their corresponding phosphoramidites such as **6** upon reaction with 2-cyanoethyl diisopropylchlorophosphoramidite, and then incorporated into oligonucleotides according to standard solid-phase synthesis.[13]

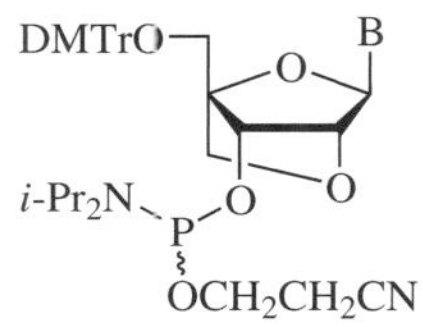

5 B = uracil-1-yl
= 4-*N*-benzoylcytosin-1-yl
= 6-*N*-benzoyladenin-9-yl
= 2-*N*-isobutyrylguanin-9-yl

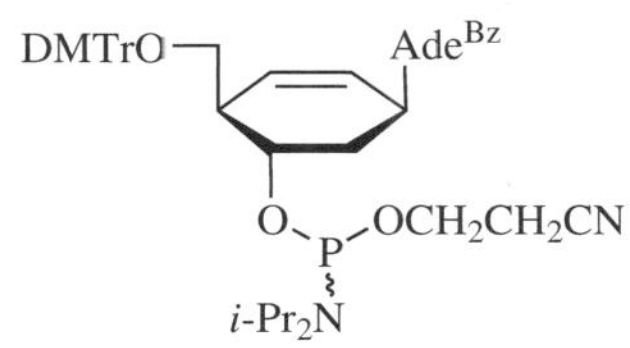

6

The integration of cyclohexenyladenosine into oligonucleotides results only in a small decrease of the stability of DNA-DNA hybrids; a ΔT_m/mod of −0.5 °C is observed. However, when cyclohexenyladenosine-modified DNA oligonucleotides are mixed with complementary unmodified RNA oligomers, the stability of these DNA-RNA hybrids increases; the ΔT_m/mod ranges from +0.8 to +1.7 °C depending on the incorporation site of cyclohexenyladenosine. CeNA are stable against fetal calf serum 3′-exonuclease relative to unmodified DNA and, remarkably, CeNA/RNA hybrids are able to activate *Escherichia coli* RNase H, thereby resulting in cleavage of the RNA strand of each hybrid.[13] Thus, CeNA represent a promising class of oligonucleotide analogs for evaluation in antisense therapeutic applications.

2-Cyanoethyl diisopropylchlorophosphoramidite has also been applied to the synthesis of cyclohexanyl nucleoside phosphoramidites[14] and hexitol nucleoside phosphoramidites[15] toward the preparation of cyclohexanyl nucleic acids (CAN) and hexitol nucleic acids (HNA), respectively. HNA are completely stable to 3′-exonucleases and form stable duplexes with complementary DNA and RNA oligonucleotides.[15,16] It is worth mentioning that, almost a decade ago, hexopyranosyl nucleoside phosphoramidites were prepared using 2-cyanoethyl diisopropylchlorophosphoramidite, and inserted in oligonucleotides to evaluate base-pairing properties and enzymatic stability of this still maturing class of carbohydrate-modified oligonucleotides.[17,18]

Acyclic nucleoside phosphoramidites such as **7** and **8** are prepared by phosphinylation of the parent acyclic nucleosides with 2-cyanoethyl diisopropylchlorophosphoramidite.[19]

Incorporation of these phosphoramidites into oligonucleotides at the 3′-terminus results in significant stabilization of the modified oligonucleotides to 3′-exonucleases like snake venom phosphodiesterase.[19] 2-Cyanoethyl diisopropylchlorophosphoramidite has similarly been employed in the synthesis of glyceronucleoside phosphoramidites[20] and silicon-containing acyclic nucleoside phosphoramidites[21] for integration into oligonucleotides. The stability of DNA-DNA hybrids incorporating acyclic nucleosides decreases by 9–15 °C for each flexible nucleoside incorporated into a strand,[20] and thus casts doubt about the usefulness of flexible oligonucleotide analogs in antisense experiments or as diagnostic probes.

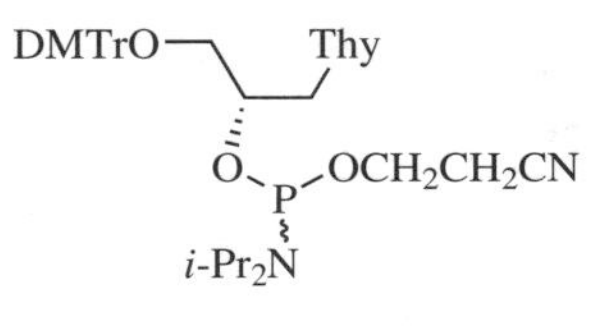

7

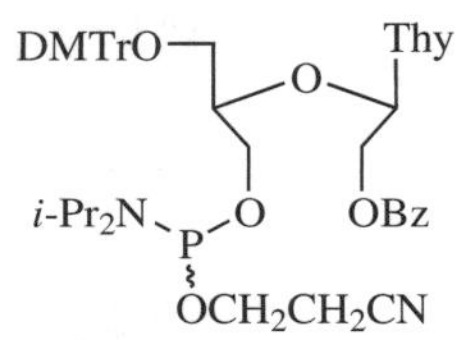

8

An efficient approach to the synthesis of oligonucleotides containing methylene acetal linkages serves as a model to the preparation of therapeutic oligonucleotides containing modified internucleoside linkages.[22] Oligonucleotides such as these are not only resistant to serum nucleases but are designed to facilitate cellular uptake by decreasing the highly charged nature of native phosphodiester groups. The incorporation of nucleoside dimers containing (3′–5′)-methylene linkages into oligonucleotides is achieved by converting the dimers to phosphoramidite derivatives comparable to **9** by treatment with 2-cyanoethyl diisopropylchlorophosphoramidite followed by conventional solid-phase synthesis.[22]

2-Cyanoethyl diisopropylchlorophosphoramidite has found an important application in the synthesis of photolabile 5′-*O*-(α-methyl-6-nitropiperonyloxycarbonyl)-*N*-acyl-2′-deoxyribonucleoside phosphoramidites (**10**).[23] These phosphoramidites are used in the preparation of densely packed oligonucleotides on microarrays. The precise removal of selected 5′-*O*-photolabile protecting groups can be achieved through properly designed photolithographic masks.[23]

For similar applications, the synthesis of photolabile 5′-*O*-(3″,5″-dimethoxybenzoin)carbonyl-*N*-acyl-2′-deoxyribonucleoside phosphoramidites[24,25] and 5′/3′-*O*-[2-(2-nitrophenyl) ethoxycarbonyl]-*N*-protected-2′-deoxyribonucleoside phosphoramidites[26,27] has been accomplished using 2-cyanoethyl diisopropylchlorophosphoramidite.

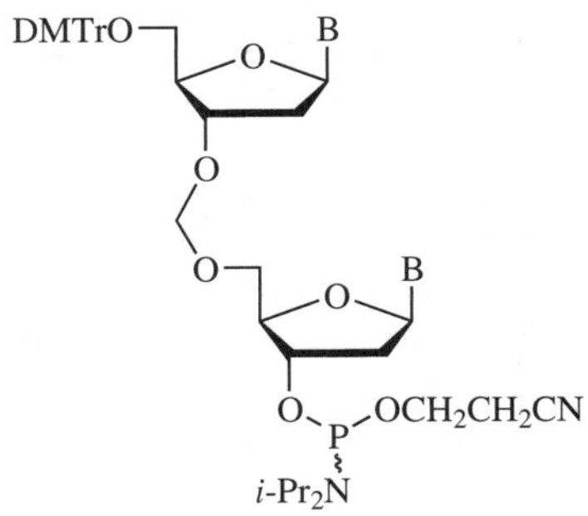

9, B = 3-*N*-benzoylthymin-1-yl

Addition of Functional Groups and Ligands to the 5′-Terminus of Oligonucleotides. Chemical methods have been developed for the 5′-phosphitylation of fully protected oligodeoxyribonucleotides. Phosphinylation of 2-(4,4′-dimethoxytrityloxyethylsulfonyl)ethanol with 2-cyanoethyl diisopropylchlorophosphoramidite affords the phosphoramidite derivative **11**.[28]

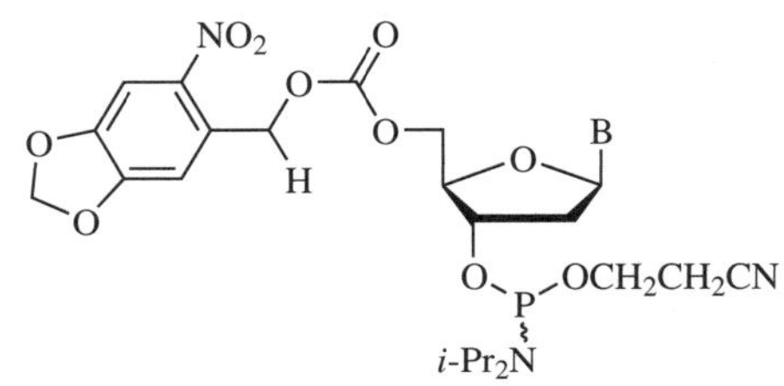

10

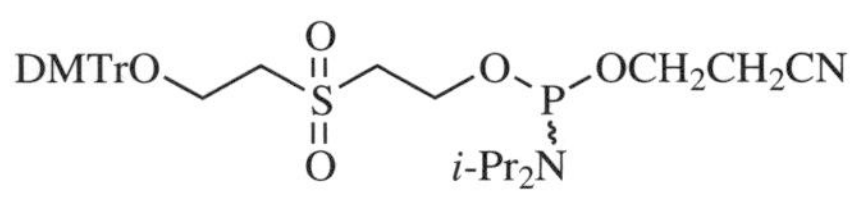

11

The reagent is used on automated DNA synthesizers to functionalize synthetic oligonucleotides with either a 5′-phosphate or a 5′-phosphorothioate group. The coupling efficiency of the reagent can be monitored upon release of the 4,4′-dimethoxytrityl cation under acidic conditions.[28] Other functional groups have been added to the 5′-terminus of oligonucleotides using a similar strategy. For example, the reaction of 2-cyanoethyl diisopropylchlorophosphoramidite with 2-(*N*-trifluoroacetyl) aminoethanol,[29] 2-[*N*-(9-fluorenylmethoxycarbonyl)amino]ethanol,[30] 6-(*N*-monomethoxytrityl)aminohexanol,[31] 3-(tritylthio)propanol,[31] 6-(tritylthio)hexanol[31,32a] and 2-[2-(2-acetylthioethoxy)ethoxy] ethanol,[32b] gives the parental phosphoramidite derivatives required for subsequent addition of a 5′-amino or a 5′-mercapto function to oligonucleotides. The conjugation of 5′-aminoalkylated or 5′-mercaptoalkylated oligonucleotides with a variety of ligands has been extensively reviewed in the literature.[33,34] In addition, phosphinylation of 3-(4,4′-dimethoxytrityloxy)propan-1-ol,[35] 18-(4,4′-dimethoxytrityloxy)-3,6,9,12,15,18-hexaoxaoctadecan-1-ol,[36] and *N*-[6-(4,4′-dimethoxytrityloxy)hexyl]-*N*′-(6-hydroxyhexyl)terephthalamide,[37] 6-(3-hydroxypropyl)-N^1,N^5, N^{10}, N^{14}-tetraphenoxyacetylspermine,[38] 1,3-*O*-bis(*N*-[3-*O*-(4,4′-dimethoxytrityloxy)propyl] carbamoyl-2,2-bis[*N*-(3-hydroxypropyl) carbamoyl]oxymethyl)-1,3-propanediol,[39] and *N*-Fmoc-tyramine[40] with 2-cyanoethyl diisopropylchlorophosphoramidite has enabled the integration of the resulting phosphoramidites into oligonucleotides. These modified oligonucleotides are useful in assessing DNA triplex formation,[37–39] the mechanism of ligation effected by the FLP recombinase and mammalian topoisomerase I,[40] the quality of oligonucleotides on glass supports for biosensor development,[36] and the folding and binding of oligonucleotides in solution.[37]

Condensation of 1,6-bis(4,4′-dimethoxytrityloxy)hexan-2-ol,[41] 2,2-bis(4,4′-dimethoxytrityloxymethyl)propan-1-ol,[42] and tris-2, 2,2-[3-(4,4′-dimethoxytrityloxy) propyloxymethyl]ethanol[43] with 2-cyanoethyl diisopropylchlorophosphoramidite gives the phosphoramidites (**12–14**), which can be incorporated into oligonucleotides to allow attachment of multiple ligands[41,42] or generate oligonucleotide dendrimers.[43]

Many ligands functionalized with linkers have been converted to phosphoramidites upon reaction with 2-cyanoethyl diisopropylchlorophosphoramidite, and then added singly or repeatedly to the 5′-terminus of oligonucleotides to generate conjugates with specific properties. These ligands include acridine (**15**),[44a] psoralen (**16**),[44] 2,4-dinitrophenyl (**17**),[45] biotin (**18**),[46–49] phosphotyrosine (**19**),[48] mannopyranosyl (**20**),[42] phenanthrodihydrodioxin (**21**),[50] and vitamin E (**22**)[51] derivatives.

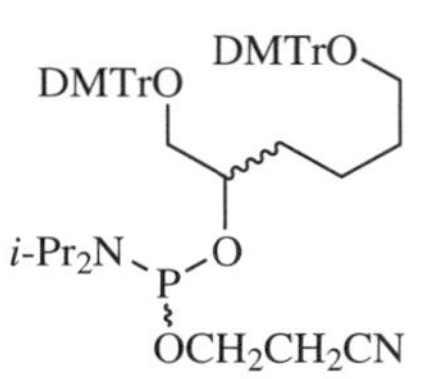

12

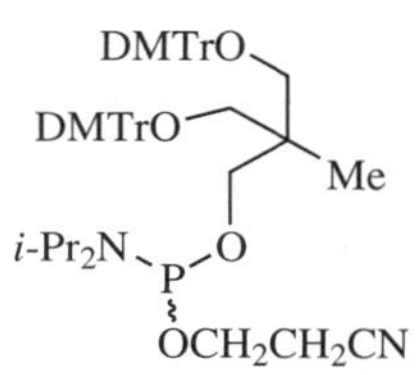

13

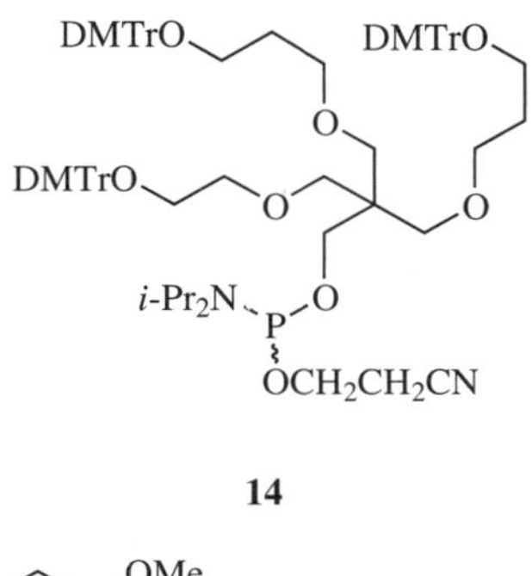

14

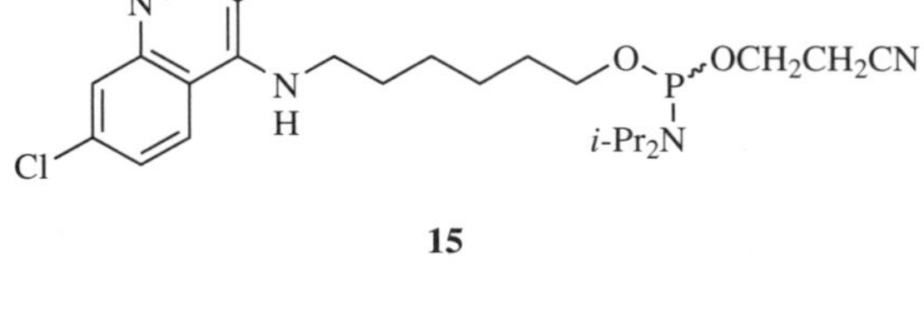

15

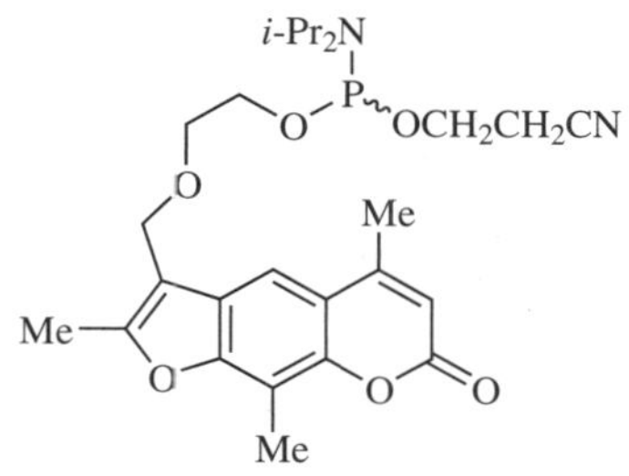

16

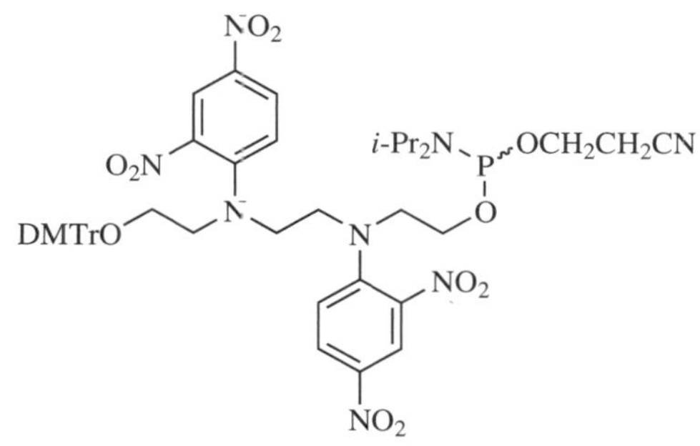

17

Boron-rich Oligomers and other Derivatives. Several carboranyl alcohol derivatives, such as **23**,[52] have been converted to the corresponding phosphoramidite derivatives by treatment with 2-cyanoethyl diisopropylchlorophosphoramidite toward the preparation of boron-rich oligophosphates for the treatment of cancer by boron neutron capture therapy.

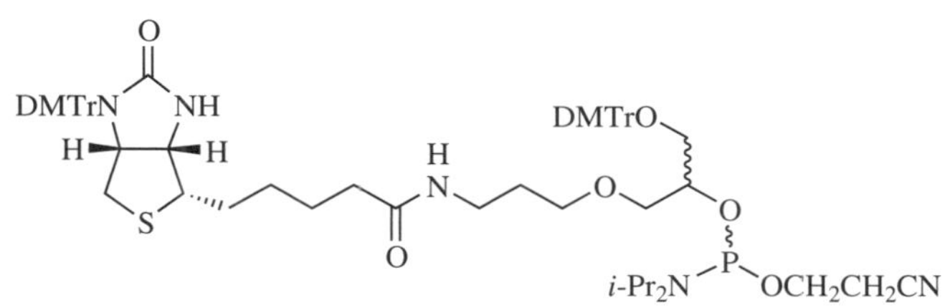

18

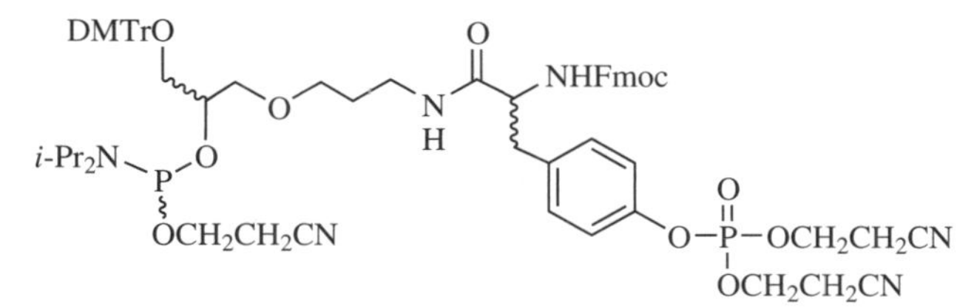

19

An analogous approach to the synthesis of boron-rich oligophosphates uses *S*-alkyl derivatives of mercaptoundecahydrododecaborate as starting materials for phosphoramidite monomers.[53]

2-Cyanoethyl diisopropylchlorophosphoramidite has been particularly useful in the preparation of phosphoramidites toward the synthesis of cyclic sphingosine 1,3-phosphate,[54] bis-phosphatidylcholine lipids,[55] lysophosphatidylnucleosides,[56] glycosyl phospholipids of 5-fluorouridine and arabinofuranosylcytidine,[57] lactosyl thiophosphate diester derivatives of arabinofuranosylcytidine and arabinofuranosyladenosine,[58] D- and L-*myo*-inositol 1-phosphorothioate,[59] and *myo*-inositol 1,4,5-trisphosphorothioate.[60]

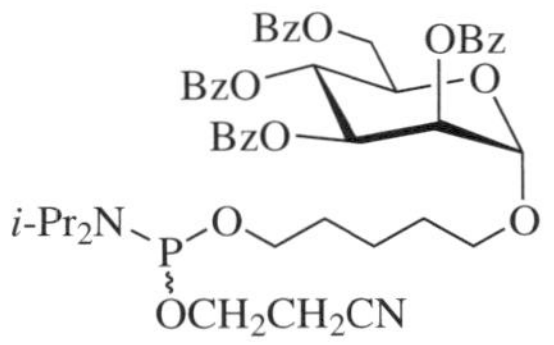

20

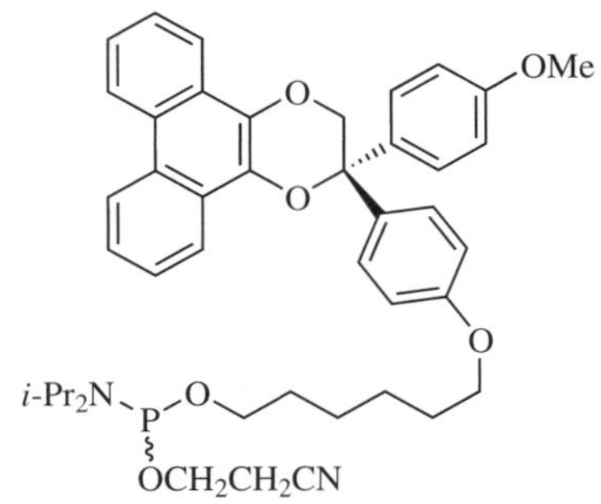

21

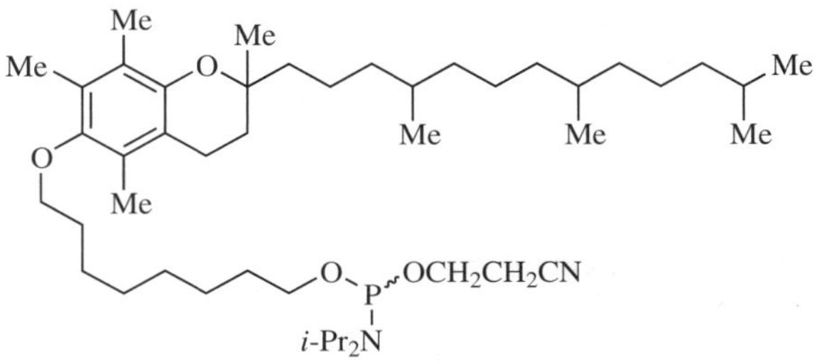

22

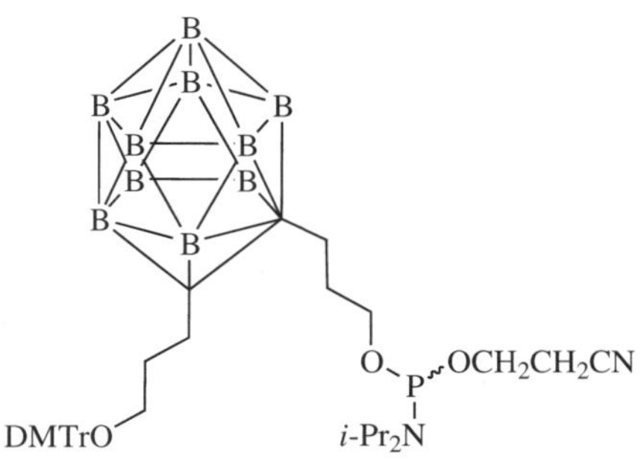

23

Related Reagents. 2-Cyanoethyl tetraisopropylphosphorodiamidite.

1. Sinha, N. D.; Biernat, J.; McManus, J.; Köster, H., *Nucleic Acids Res.* **1984**, *12*, 4539.
2. Maruenda, H.; Chenna, A.; Liem, L.-K.; Singer, B., *J. Org. Chem.* **1998**, *63*, 4385.
3. Zhang, W.; Rieger, R.; Iden, C.; Johnson, F., *Chem. Res. Toxicol.* **1995**, *8*, 148.
4. Lakshman, M. K.; Sayer, J. M.; Yagi, H.; Jerina, D. M., *J. Org. Chem.* **1992**, *57*, 4585.
5. Lakshman, M. K.; Sayer, J. M.; Jerina, D. M., *J. Org. Chem.* **1992**, *57*, 3438.
6. Lakshman, M. K.; Sayer, J. M.; Jerina, D. M., *J. Am. Chem. Soc.* **1991**, *113*, 6589.
7. Zajc, B.; Lakshman, M. K.; Sayer, J. M.; Jerina, D. M., *Tetrahedron Lett.* **1992**, *33*, 3409.
8. Khan, S. I.; Beilstein, A. E.; Grinstaff, M. W., *Inorg. Chem.* **1999**, *38*, 418.
9. Yamana, K.; Mitsui, T.; Yoshioka, J.; Isuno, T.; Nakano, H., *Bioconjugate Chem.* **1996**, *7*, 715.
10. Koshkin, A. A.; Singh, S. K.; Nielsen, P.; Rajwanshi, V. K.; Kumar, R.; Meldgaard, M.; Olsen, C. E.; Wengel, J., *Tetrahedron* **1998**, *54*, 3607.
11. Singh, S. K.; Nielsen, P.; Koshkin, A. A.; Wengel, J., *J. Chem. Soc., Chem. Commun.* **1998**, 455.
12. Wang, J.; Herdewijn, P., *J. Org. Chem.* **1999**, *64*, 7820.
13. Wang, J.; Verbeure, B.; Luyten, I.; Lescrinier, E.; Froeyen, M.; Hendrix, C.; Rosemeyer, H.; Seela, F.; Van Aerschot, A.; Herdewijn, P., *J. Am. Chem. Soc.* **2000**, *122*, 8595.
14. Maurinsh, Y.; Rosemeyer, H.; Esnouf, R.; Medvedovici, A.; Wang, J.; Ceulemans, G.; Lescrinier, E.; Hendrix, C.; Busson, R.; Sandra, P.; Seela, F.; Van Aerschot, A.; Herdewijn, P., *Chem. Eur. J.* **1999**, *5*, 2139.
15. Hendrix, C.; Rosemeyer, H.; Verheggen, I.; Seela, F.; Van Aerschot, A.; Herdewijn, P., *Chem. Eur. J.* **1997**, *3*, 110.
16. Hendrix, C.; Rosemeyer, H.; De Bouvere, B.; Van Aerschot, A.; Seela, F.; Herdewijn, P., *Chem. Eur. J.* **1997**, *3*, 1513.
17. Augustyns, K.; Vandendriessche, F.; Van Aerschot, A.; Busson, R.; Urbanke, C.; Herdewijn, P., *Nucleic Acids Res.* **1992**, *20*, 4711.
18. Augustyns, K.; Godard, G.; Hendrix, C.; Van Aerschot, A.; Rozenski, J.; Saison-Behmoaras, T.; Herdewijn, P., *Nucleic Acids Res.* **1993**, *21*, 4670.
19. Nielsen, P.; Dreiøe, L. H.; Wengel, J., *Bioorg. Med. Chem.* **1995**, *3*, 19.
20. Schneider, K. C.; Benner, S. A., *J. Am. Chem. Soc.* **1990**, *112*, 453.
21. Thibon, J.; Latxague, L.; Déléris, G., *J. Org. Chem.* **1997**, *62*, 4635.
22. Veeneman, G. H.; Van Der Marel, G. A.; Van Den Elst, H.; Van Boom, J. H., *Tetrahedron* **1991**, *47*, 1547.
23. Caviani Pease, A.; Solas, D.; Sullivan, E. J.; Cronin, M. T.; Holmes, C. P.; Fodor, S. P. A., *Proc. Natl. Acad. Sci. USA* **1994**, *91*, 5022.
24. Pirrung, M. C.; Fallon, L.; McGall, G., *J. Org. Chem.* **1998**, *63*, 241.
25. Pirrung, M. C.; Bradley, J.-C., *J. Org. Chem.* **1995**, *60*, 6270.
26. Giegrich, H.; Eisele-Bühler, S.; Hermann, C.; Kvasyuk, E.; Charubala, R.; Pfleiderer, W., *Nucleosides Nucleotides* **1998**, *17*, 1987.
27. Pirrung, M. C.; Wang, L.; Montague-Smith, M. P., *Org. Letters* **2001**, *3*, 1105.
28. Horn, T.; Urdea, M. S., *Tetrahedron Lett.* **1986**, *27*, 4705.
29. Coull, J. M.; Weith, H. L.; Bischoff, R., *Tetrahedron Lett.* **1986**, *27*, 3991.
30. Agrawal, S.; Christodoulou, C.; Gait, M. J., *Nucleic Acids Res.* **1986**, *14*, 6227.
31. Sinha, N. D.; Cook, R. M., *Nucleic Acids Res.* **1988**, *16*, 2659.
32. (a) Ede, N. J.; Tregear, G. M.; Haralambidis, J., *Bioconjugate Chem.* **1994**, *5*, 373. (b) Asseline, U.; Thuong, N. T., In *Current Protocols in Nucleic Acid Chemistry*, Beaucage, S. L.; Bergstrom, D. E.; Glick, G. D.; Jones, R. A., Eds. Wiley: New York; 2001, 4.9.1.
33. Beaucage, S. L.; Iyer, R. P., *Tetrahedron* **1993**, *49*, 1925.
34. Beaucage, S. L., In *Comprehensive Natural Product Chemistry, Volume 7*, Barton, D.; Nakanishi, K.; Meth-Cohn, O.; Kool, E. T., Eds. Elsevier: Amsterdam, 1999; 153.
35. Ono, A.; Chen, C.-N.; Kan, L.-S., *Biochemistry* **1991**, *30*, 9914.
36. Sojka, B.; Piunno, P. A. E.; Wust, C. C.; Krull, U. J., *Anal. Chim. Acta* **1999**, *395*, 273.
37. Salunkhe, M.; Wu, T.; Letsinger, R. L., *J. Am. Chem. Soc.* **1992**, *114*, 8768.
38. Sund, C.; Puri, N.; Chattopadhyaya, J., *Nucleosides Nucleotides* **1997**, *16*, 755.
39. Ueno, Y.; Takeba, M.; Mikawa, M.; Matsuda, A., *J. Org. Chem.* **1999**, *64*, 1211.
40. Panigrahi, G.; Zhao, B.-P.; Krepinsky, J. J.; Sadowski, P. D., *J. Am. Chem. Soc.* **1996**, *118*, 12004.

41. Teigelkamp, S.; Ebel, S.; Will, D. W.; Brown, T.; Beggs, J. D., *Nucleic Acids Res.* **1993**, *21*, 4651.
42. Wijsman, E. R.; Filippov, D.; Valentijn, A. R. P. M.; van der Marel, G. A.; van Boom, J. H., *Rec. Trav. Chim. Pays-Bas* **1996**, *115*, 397.
43. Shchepinov, M. S.; Udalova, I. A.; Bridgman, A. J.; Southern, E. M., *Nucleic Acids Res.* **1997**, *25*, 4447.
44. (a) Asseline, U.; Thuong, N. T., In *Current Protocols in Nucleic Acid Chemistry*, Beaucage, S. L.; Bergstrom, D. E.; Glick, G. D.; Jones, R. A., Eds. Wiley: New York, 2000; 4.3.1. (b) Pieles, U.; Englisch, U., *Nucleic Acids Res.* **1989**, *17*, 285.
45. Will, D. W.; Pritchard, C. E.; Brown, T., *Carbohydr. Res.* **1991**, *216*, 315.
46. Pon, R. T., *Tetrahedron Lett.* **1991**, *32*, 1715.
47. Olejnik, J.; Krzymanska-Olejnik, E.; Rothschild, K. J., *Nucleic Acids Res.* **1996**, *24*, 361.
48. Misiura, K.; Durrant, I.; Evans, M. R.; Gait, M. J., *Nucleic Acids Res.* **1990**, *18*, 4345.
49. Neuner, P., *Bioorg. Med. Chem. Lett.* **1996**, *6*, 147.
50. Bendinskas, K. G.; Harsch, A.; Wilson, R. M.; Midden, W. R., *Bioconjugate Chem.* **1998**, *9*, 555.
51. Will, D. W.; Brown, T., *Tetrahedron Lett.* **1992**, *33*, 2729.
52. Drechsel, K.; Lee, C. S.; Leung, E. W.; Kane, R. R.; Hawthorne, M. F., *Tetrahedron Lett.* **1994**, *35*, 6217.
53. Perleberg, O.; Gabel, D., In *Advances in Neutron Capture Therapy. Volume II, Chemistry and Biology*, Larsson, B.; Crawford, J.; Weinreich, R., Eds. Elsevier: Amsterdam, 1997, 119.
54. Qiao, L.; Kozikowski, A. P., *Tetrahedron Lett.* **1998**, *39*, 8959.
55. Klotz, P.; Cuccia, L. A.; Mohamed, N.; Just, G.; Lennox, R. B., *J. Chem. Soc. Chem. Commun.* **1994**, 2043.
56. Chillemi, R.; Russo, D.; Sciuto, S., *J. Org. Chem.* **1998**, *63*, 3224.
57. Le Bec, C.; Huynh-Dinh, T., *Tetrahedron Lett.* **1991**, *32*, 6553.
58. Cai, T.-W.; Min, J.-M.; Zhang, L.-H., *Carbohydr. Res.* **1997**, *303*, 113.
59. Baker, G. R.; Billington, D. C.; Gani, D., *Tetrahedron* **1991**, *47*, 3895.
60. Cooke, A. M.; Gigg, R.; Potter, B. V. L., *J. Chem. Soc., Chem. Commun.* **1987**, 1525.

Serge L. Beaucage
Food and Drug Administration, Bethesda, Maryland, USA

2-Cyanoethylphosphoric Acid[1]

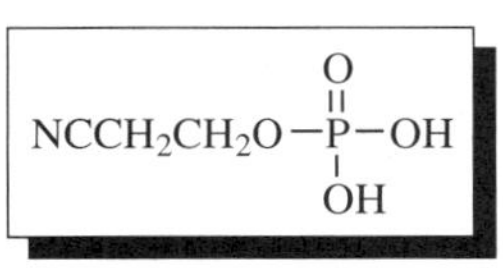

[2212-88-6] $C_3H_6O_4NP$ (MW 151.06)

(used in the synthesis of phosphoric acid esters of carbohydrates, nucleosides, nucleotides, and other molecules)

Physical Data: mp 165 °C;[1] IR: 2500, 2210, 1623, 1548 cm^{-1};[1] ^1H NMR (CF_3CO_2H): δ 5.64 (*dt*, 2H, $J = 5.5$, 6.5 Hz), 7.12 (*t*, 2H, $J = 5.5$ Hz);[2] ^{31}P NMR: δ 1.2 ($J_{PH} = 725$ Hz).[3]

Form Supplied in: as barium salt.

Handling, Storage, and Precautions: a stock solution of 2-cyanoethylphosphoric acid is prepared by dissolving an exact quantity of the barium salt in water and by passage of this solution through a column of Dowex-50W (H^+ form). The eluate is evaporated in vacuo resulting in a thick syrup that is dried by addition and evaporation of pyridine. The resulting residue is dissolved in pyridine and the solution is transferred (using pyridine washes) to a volumetric flask and diluted with pyridine to give a final concentration of 1.0 mmol/mL.[1]

Preparation of Barium Salt of 2-Cyanoethylphosphate.[4] To a solution of freshly distilled phosphorus oxychloride in anhydrous pyridine at −10 °C is added hydroxyacrylonitrile. After the addition, the mixture is stirred for 1 h at the same temperature and the suspension is slowly poured into a mixture of water, pyridine, and ice. A solution of barium acetate is then added and the resulting suspension is kept at room temperature for 2 h and filtered. Ethanol is then added with stirring and the white suspension is refrigerated overnight. The product is collected by filtration, washed repeatedly with 50% aqueous ethanol and finally with absolute ethanol. The material is air-dried initially and then dried in vacuo over phosphorus pentoxide for 16 h at room temperature (eq 1).

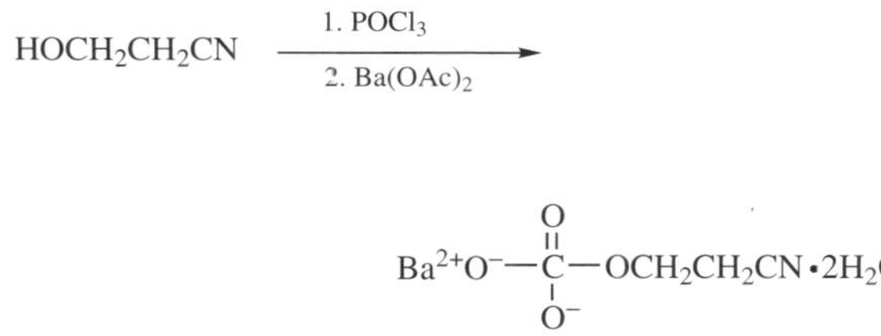

(1)

Synthesis of 3′-Phosphate and 5′-Phosphate Esters of Nucleosides.[1,5,6–8] The reagent is useful for the synthesis of phosphomonoesters of alcohols, particularly, of nucleosides. It is coupled to the alcoholic component commonly with pyridine using *N,N*-dicyclohexylcarbodiimide (DCC) as a condensing agent (eq 2). In case of thymidine (eq 3), selective phosphorylation of the primary alcohol group can be achieved under controlled conditions. However, phosphorylation at the 3′-position requires protection of the 5′-hydroxyl group. Acid hydrolysis eliminates

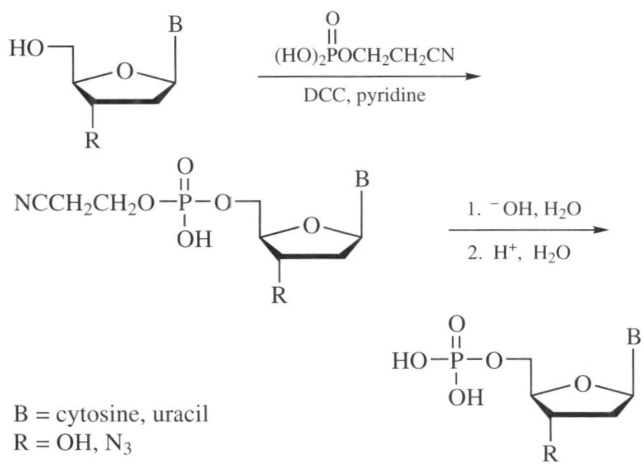

(2)

the protective trityl group. The cyanoethyl phosphate group has the advantage of being completely stable towards water, but it can be hydrolyzed under very mild alkaline conditions.

Synthesis of Steroidal Phosphate Esters.[9] In the presence of DCC, 2-cyanoethylphosphoric acid can be used for the direct phosphorylation of hydroxylated steroids (eq 4). Selected examples include triamcinolone acetonide, hydrocortisone, dexamethasone, prednisolone, and D-homotriamicinolone acetonide.

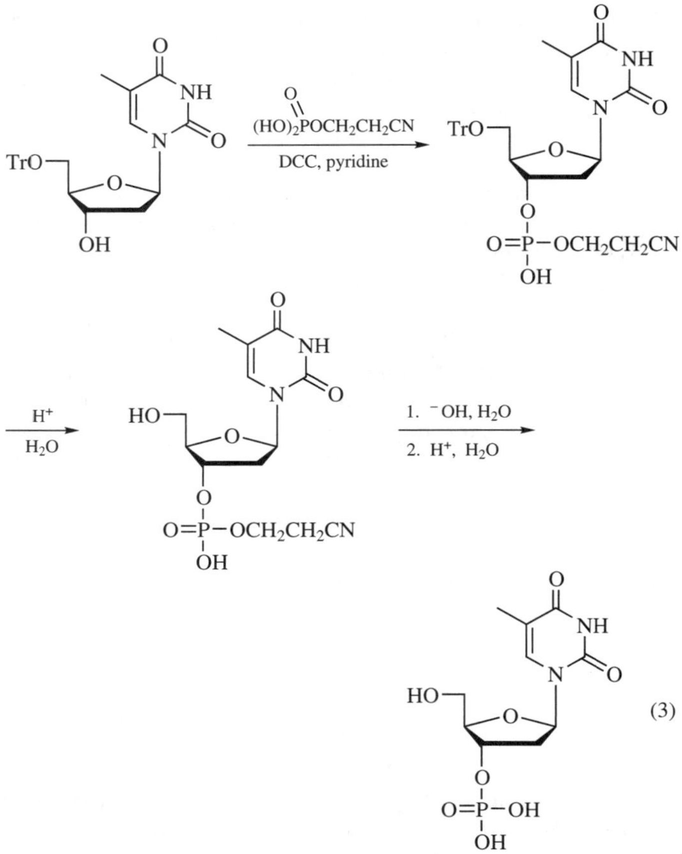

(3)

Phosphorylation of Carbohydrates (Synthesis of Sucrose6′-phosphate).[10] Sucrose 6′-phosphate is synthesized by phosphorylation of sucrose heptaacetate by means of cyanoethylphosphoric acid and DCC in pyridine (eq 5). The product of

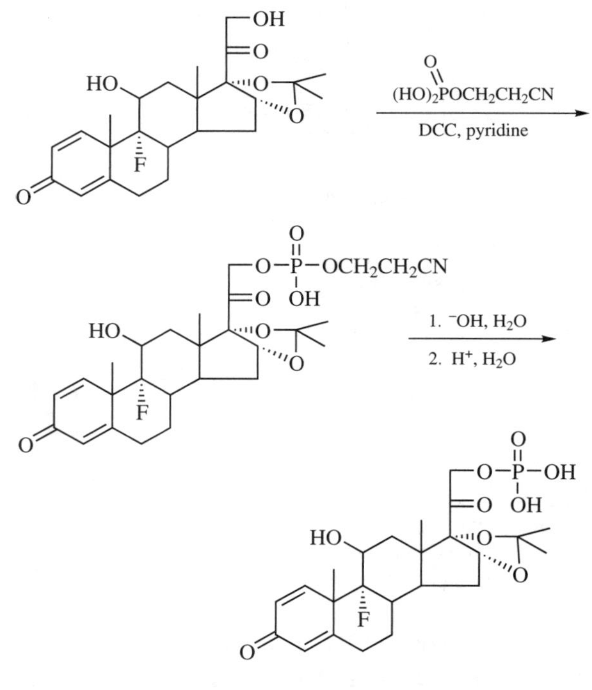

(4)

phosphorylation, the phosphodiester, is treated with barium hydroxide to remove the protecting groups and to hydrolyze excess reagent. Sucrose 6′-phosphate is isolated as the barium salt.

Phosphorylation of Non-natural Dinucleotides.[11–13] Both natural and non-natural dinucleotides can also be phosphorylated at the primary 5′-hydroxyl group with 2-cyanoethylphosphoric acid under the normal conditions to give, after deprotection with ammonium hydroxide, the dinucleotide 5′-phosphate. This is illustrated in eq 6 for an HIV integrase inhibitor.

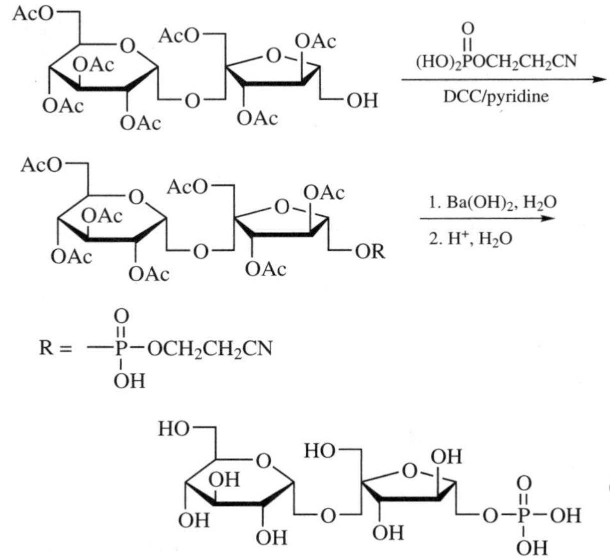

(5)

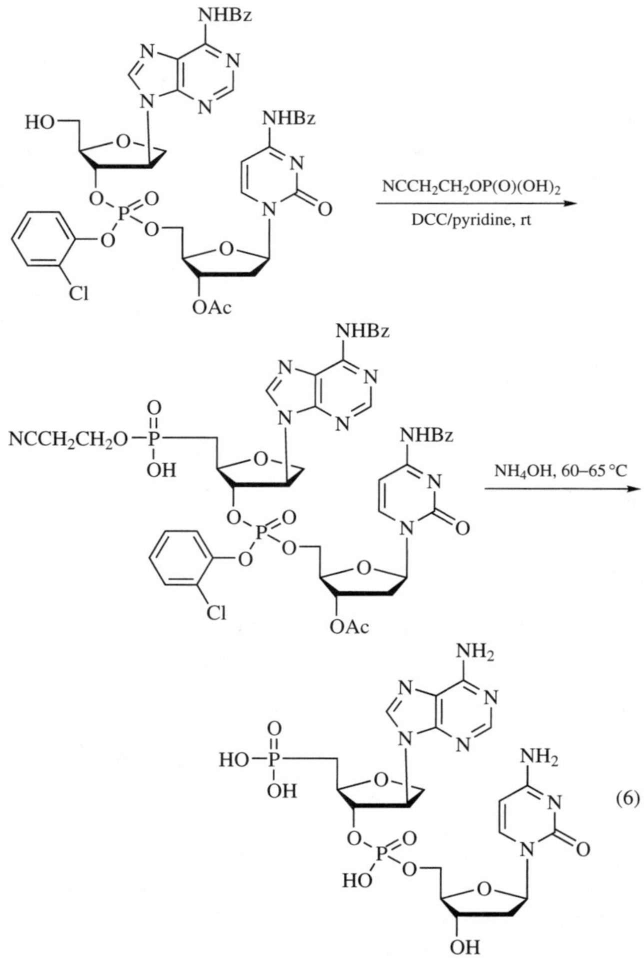

(6)

1. Tener, G. M., *J. Am. Chem. Soc.* **1961**, *83*, 159.
2. Khwaja, T. A.; Reese, C. B.; Stewart, J. C. M., *J. Chem. Soc. (C)* **1970**, 2092.
3. Markowska, A.; Olejnik, J.; Michalski, J., *Bulletin de L'academie Polonaise des Sciences, Sénie des Sciences chimiques* **1979**, *27*, 115.
4. Clayton, S. J.; Rose, J. D.; Fitzgibbon, Jr, W. E., In *Synthetic Procedures in Nucleic Acid Chemistry*; Zorbach, W. W.; Tipson, R. S., Eds.; Wiley: New York, 1968, Vol. 1, pp 523–524.
5. Moffatt, J. G., *J. Am. Chem. Soc.* **1963**, *85*, 1118.
6. Torrence, P. F.; Witkop, B., In *Nucleic Acid Chemistry*, Part 2; Townsend, L. B.; Tipson, R. S., Eds.; Wiley-Interscience: New York, 1978, pp 977–988.
7. Tener, G. M., *Biochem. Prep.* **1962**, *9*, 5.
8. Hong, C. I.; Nechaev, A.; Kirisits, A. J.; Vig, R.; West, R.; Manouilov, K. K.; Chu, C. K., *J. Med. Chem.* **1996**, *39*, 1771.
9. Brownfield, R. B.; Shultz, W., *Steroids* **1963**, *2*, 597.
10. Buchanan, J. G.; Cummerson, D. A.; Turner, D. M., *Carbohydrate Research* **1972**, *21*, 283.
11. Taktakishvili, M.; Neamati, N.; Pommier, Y.; Pal, S.; Nair, V., *J. Am. Chem. Soc.* **2000**, *122*, 5671.
12. Taktakishvili, M.; Neamati, N.; Pommier, Y.; Nair, V., *Bioorg. Med. Chem. Lett.* **2000**, *10*, 249.
13. Taktakishvili, M.; Neamati, N.; Pommier, Y.; Nair, V., *Bioorg. Med. Chem. Lett.* **2001**, *11*, 1433.

Vasu Nair & Byung I. Seo
University of Georgia, Athens, Georgia, USA

2-Cyanoethyl Tetraisopropylphosphorodiamidite

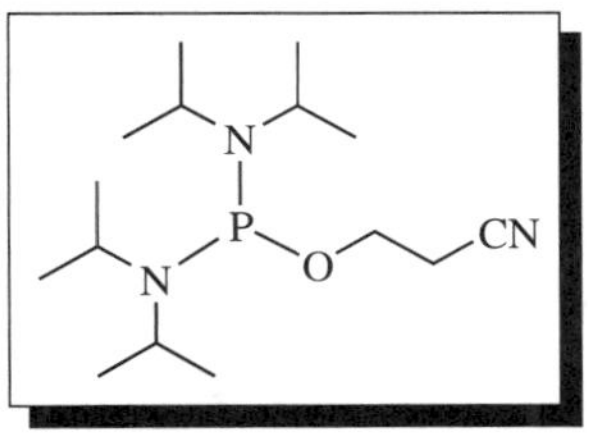

[102691-36-1] $C_9H_{18}ClN_2OP$ (MW 236.68)

(reagent widely used for the phosphinylation of functional and reporter groups, ligands, and 3′-hydroxyls of modified and unmodified nucleosides for incorporation into oligonucleotides)

Alternate Name: bis(*N,N*-diisopropylamino)-2-cyanoethoxyphosphine; (2-cyanoethoxy) bis(*N,N*-diisopropylamino)phosphine, 2-cyanoethyl *N,N,N′,N′*-tetraisopropylphosphordiamidite.
Physical Data: bp 105–107 °C/0.4 mm Hg; d 1.04 g cm^{-3}.
Solubility: soluble in most organic solvents.
Form Supplied in: colorless liquid.
Analysis of Reagent Purity: ^{31}P NMR; ^{1}H NMR; elemental analysis.
Preparative Methods: the reagent is prepared from the reaction of (2-cyanoethoxy)dichlorophosphine with *N,N*-diisopropylamine in anhydrous diethyl ether and isolated by vacuum distillation.
Purification: can be purified by vacuum distillation.
Handling, Storage, and Precautions: the reagent can be handled without decomposition under normal laboratory conditions and can be stored for months at −20 °C. Caution should be exercised when vacuum-distilling the reagent as it may explode if heated at a temperature (internal) exceeding 140 °C. Distillation should be carried out only if a vacuum of less than 0.5 mm Hg can be generated.
Disposal: the reagent should first be carefully destroyed by slow addition to a stirred ethanolic solution of potassium hydroxide. The resulting solution can then be discarded in an appropriate chemical waste container.

Introduction. 2-Cyanoethyl tetraisopropylphosphorodiamidite is inherently more stable to hydrolysis than 2-cyanoethyl diisopropylchlorophosphoramidite and is therefore easier to handle under normal laboratory conditions. Specifically, 2-cyanoethyl tetraisopropylphosphorodiamidite is activated upon treatment with less than a molar equivalent of 1*H*-tetrazole or its diisopropylammonium salt[1] in the preparation of nucleoside phosphoramidite monomers for automated solid-phase DNA/RNA oligonucleotide syntheses. In this regard, 2-cyanoethyl tetraisopropylphosphorodiamidite has been particularly useful in producing deoxyribonucleoside phosphoramidites in situ for direct use in solid-phase synthesis of DNA oligonucleotides.[2]

Since it is generally accepted that 2-cyanoethyl tetraisopropylphosphorodiamidite is the reagent of choice for the preparation of unmodified nucleoside phosphoramidites, its use has rapidly spread to the synthesis of modified nucleoside phosphoramidites for incorporation into oligonucleotides. These phosphoramidites typically include those with altered nucleobases or carbohydrates and those with dinucleosides coupled through modified internucleoside linkages. Considering the biomedical significance of modified oligonucleotides, the preparation of phosphoramidite derivatives required for the synthesis of these biomolecules will be discussed in this article.

The addition of functional and reporter groups, and ligands to the 5′-terminus of oligonucleotides or at predetermined sites, via phosphoramidite synthons is an attractive approach towards providing diagnostic and/or biologically relevant properties to nucleic acids. The usefulness of 2-cyanoethyl tetraisopropylphosphorodiamidite in these applications will also be the subject of this review.

Finally, the utilization of 2-cyanoethyl tetraisopropylphosphorodiamidite in the syntheses of phospholipids and phosphoinositols will be reported to underscore the phosphinylation of molecules other than nucleosides and nucleotides.

This article is not comprehensive, as its coverage of the scientific literature is mostly limited to the past seven years. Consequently, the readers are urged to check each literature reference for the additional contributions that have regrettably been omitted only to minimize the voluminous nature of this treatise. Comprehensive reports on the use of 2-cyanoethyl tetraisopropylphosphorodiamidite and 2-cyanoethyl diisopropylchlorophosphoramidite in the synthesis of various phosphoramidite derivatives prior to 1996 have already been published.[3]

It should be understood that all the phosphoramidites described in this article have been prepared using exclusively 2-cyanoethyl tetraisopropylphosphorodiamidite even though the usage of this phosphinylating reagent is not always explicitly stated to prevent much too frequent reiterations.

Phosphinylation of Nucleobase-modified Nucleosides. Oligodeoxyribonucleotides containing structurally defined adducts of carcinogens, drugs, and other xenobiotics have been over the years the focus of interest for structural and biological studies. In this context, an improved methodology has been developed for the preparation of DNA oligonucleotides carrying adducts on guanine at *N*-2.[4] The method entails the incorporation of the modified deoxyribonucleoside phosphoramidite **1** into an oligonucleotide chain using solid-phase techniques. The treatment of the partially deprotected oligonucleotide with (*R*)-2-amino-2-phenylethanol followed by complete deprotection led to the desired N^2-styrene oxide-adducted oligomer.[4]

The same strategy has been used to incorporate a N^2-[*S*-(2-ethyl)]glutathione-adducted 2′-deoxyguanosine residue into synthetic DNA oligonucleotides, and enable in vitro site-specific misincorporation studies opposite the lesion to characterize the mutagenesis caused by ethylene dibromide.[5]

The deoxyribonucleoside phosphoramidite **2** has alternatively been prepared through the functionalization of N^2-phenoxyacetyl-5′-*O*-(4,4′-dimethoxytrityl)-2′-deoxyguanosine with *N*-(9-fluorenylmethoxycarbonyl)-*S*-(2-hydroxyethyl)glutathione dimethyl ester and phosphinylation effected by 2-cyanoethyl tetraisopropylphosphorodiamidite.[5] Integration of **2** into synthetic oligonucleotides allowed additional in vitro site-specific misincorporation studies, involving guanine residues derivatized with *S*-(2-hydroxyethyl)glutathione at *O*-6 to provide further insight on the mutagenic mechanisms of ethylene dibromide.[5]

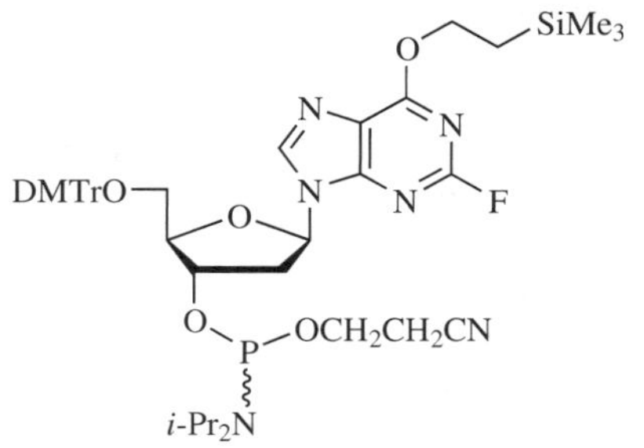

1, DMTr = 4,4′-dimethoxytrityl

In related studies, the incorporation of phosphoramidite **3** into synthetic oligonucleotides produced G→A and G→T mutations (<1% each) in misincorporation studies performed on *E. coli*.[6] The coupling efficiency of **3** was ~50% at the site of addition in a 19-mer DNA oligonucleotide.[6]

Nitrous acid is a mutagenic substance converting the exocyclic amino groups of DNA to carbonyl groups and forming interstrand cross-links in DNA duplexes. There is a considerable interest in these reactions due to the dietary and environmental exposure of humans to oxides of nitrogen which can initiate these reactions.[7] Nitrous acid-induced DNA interstrand cross-links are formed preferentially between the two proximal guanine residues through a common *N*-2 atom. The synthesis of a nitrous acid interstrand cross-linked DNA lesion would be useful to determine its structural characteristics and investigate the repair of such a lesion. The phosphoramidite **4** was synthesized and then used to generate a self-complementary 12-mer cross-linked DNA duplex on a heavily loaded solid support. Upon deprotection and purification by denaturing polyacrylamide gel electrophoresis, the cross-linked DNA duplex was obtained in an overall yield of 10%.[7] NMR analysis of the duplex indicates the formation of at least ten stable base pairs in the 12-mer cross-linked duplex.

5′,8-Cyclo-2′-deoxyadenosine is one of the main radiation-induced decomposition products of 2′-deoxyadenosine. The formation of this product results from the intramolecular attack of the *C*-5′ carbon radical produced by the action of •OH radicals on the *C*-8, *N*-7 double bond of the purine. 5′,8-Cyclodeoxyadenosine is considered as a tandem DNA lesion because both 2-deoxyribose and adenine are being altered.[8] In order to assess the conformational changes and biochemical features associated with the formation of 5′,8-cyclodeoxyadenosine in DNA, insertion of this double lesion at defined sites into oligonucleotides (14-mers and 22-mers) has been achieved using phosphoramidite **5**. While the coupling time for **5** was increased by a factor of 2, relative to that of standard phosphoramidites, the coupling time for an unmodified phosphoramidite monomer, next to the modification, was extended even more (330 s) to achieve a coupling efficiency better than 85%. The phosphoramidite **5** was obtained in nine steps from N^6-benzoyl-2′-deoxyadenosine.

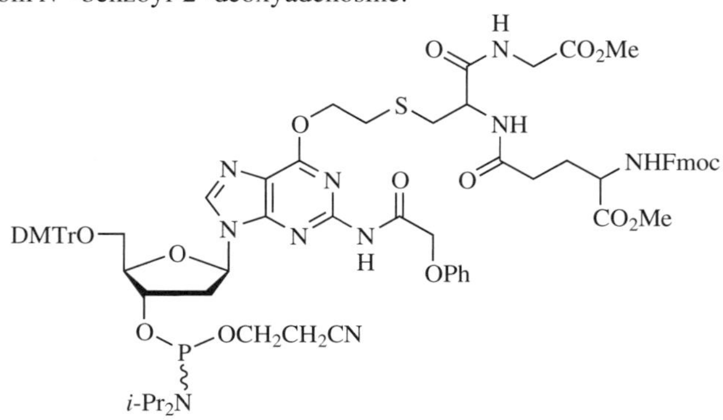

2, Fmoc = 9-fluorenylmethoxycarbonyl

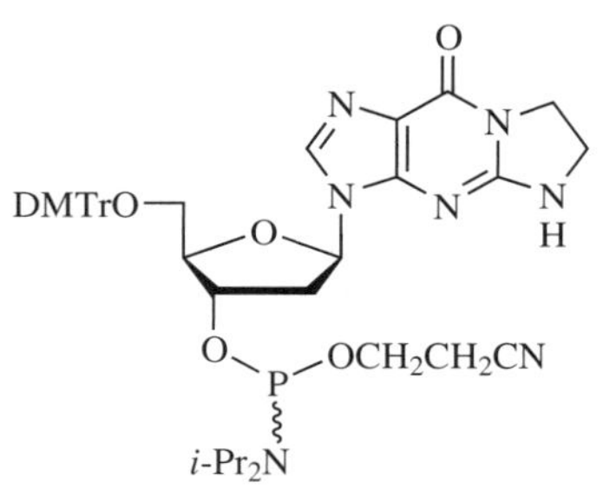

3

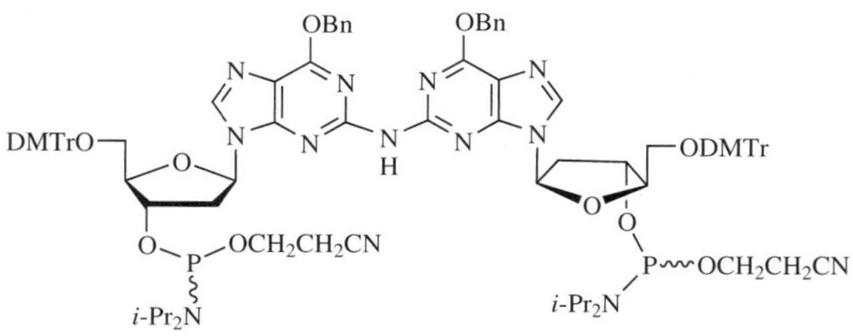

4, Bn = benzyl

Even though the thermostability of the DNA duplex containing the modified nucleoside is lower than that of an identical unmodified DNA duplex ($\Delta T_m = -6\,°C$), oligonucleotides containing 5′,8-cyclodeoxyadenosine residues are suitable for mutagenesis and DNA repair studies.[8]

1-(2-Deoxy-β-D-*erythro*-pentofuranosyl)cyanuric acid has also been shown to be formed upon exposure of DNA to ionizing radiation and excited photosensitizers.[9] The biological and structural significance of such a modified nucleobase in DNA was assessed by its chemical synthesis and incorporation into oligonucleotides at selected sites. The cyanuric acid nucleoside phosphoramidite **6** was prepared in four steps from 2-deoxy-3,5-di-*O*-*p*-toluoyl-D-*erythro*-pentofuranosyl chloride and 2,4,6-trihydroxy-1,3,5-triazine. The coupling time for **6** was extended to 120 s to ensure a coupling efficiency of more than 90%.

The presence of a cyanuric acid nucleoside in a 14-mer was found destabilizing opposite to cytosine in a double-stranded DNA complex, when compared with the parent unmodified DNA duplex ($\Delta T_m = -15\,^\circ C$). Such a destabilization is attributed to a lack of hydrogen bonding with the complementary cytosine base. In vitro replication experiments, using the Klenow fragment of *E. coli* DNA polymerase I, show that the cyanuric acid nucleoside did not stop the transcription of the primed 22-mer. Both dAMP and, to a lesser extent, dGMP are incorporated opposite to the cyanuric acid nucleoside. Thus, cyanuric acid nucleobases exhibit miscoding potential during replication leading to mutagenic transversions.[9]

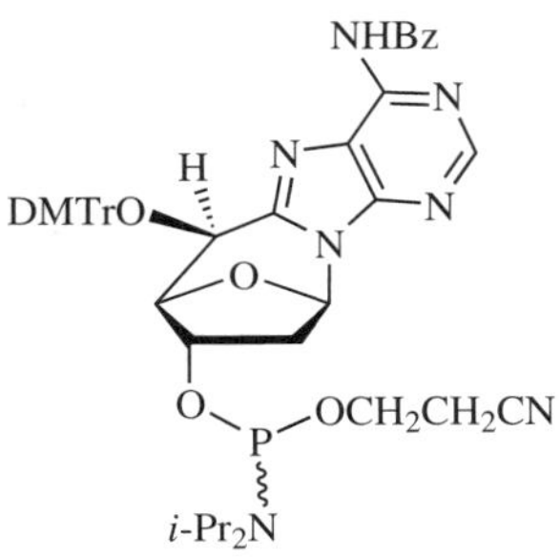

5, Bz = benzoyl

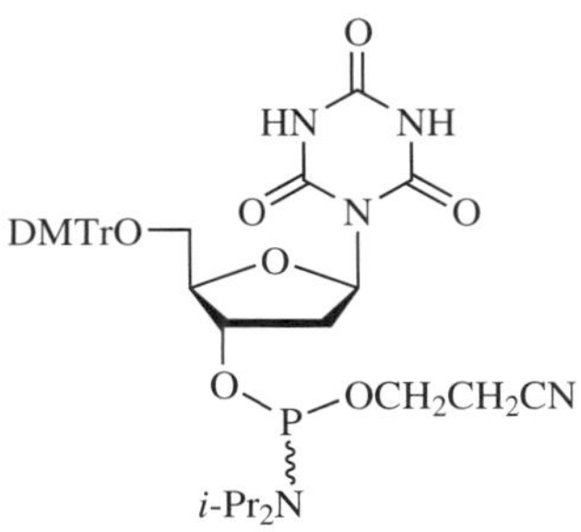

6

5-Formyldeoxyuridine is a new type of oxidized thymine lesion formed in DNA by ionizing radiation and photosensitization with, for example, 2-methyl-1,4-naphthoquinone. To better understand the genetic effects of this oxidative lesion, site-specific incorporation of 5-formyl deoxyuridine into oligonucleotides is required. Given the instability of 5-formyldeoxyuridine towards basic conditions, the phosphoramidite **7** has been prepared to generate 5-(1,2-dihydroxyethyl)uracil-containing oligonucleotide, which can then be readily oxidized by sodium periodate to 5-formyluracil-containing oligonucleotides.[10] The phosphoramidite **7** was prepared in seven steps from 5-iodo-2′-deoxyuridine, and was incorporated into DNA with a coupling efficiency higher than 97%.

The thermostability of DNA hybrids containing one 5-formyluracil residue is moderately reduced ($\Delta T_m = -2.2\,^\circ C$ to $-5.8\,^\circ C$) when compared with that of the same unmodified DNA duplex.[10] Thus, the incorporation of a 5-formyluracil residue does not significantly perturb the structure of duplex DNA. However, the discovery of 5-formyluracil DNA glycolase activity in cell-free extracts of human, mouse, and rat tissues confirms the potential mutagenic nature of 5-formyluracil lesions and the necessity for repairing such DNA damages. Indeed, incubation of double-stranded DNA containing 5-formyluracil with cell-free extracts from HeLa cells resulted in an efficient cleavage of the oligomer at the 5-formyluracil site.[10]

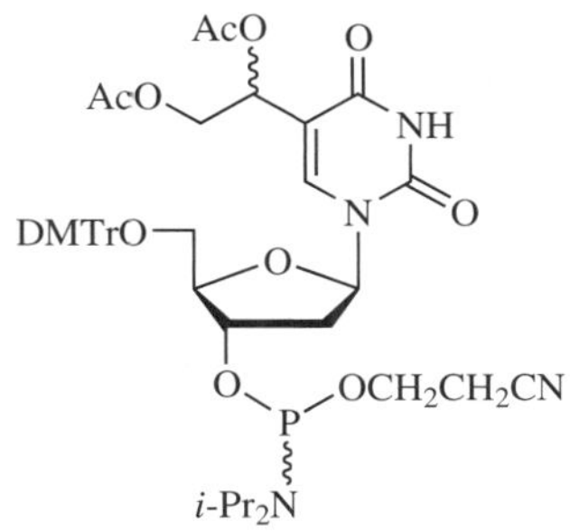

7

In a somewhat related study, synthesis of the ribonucleoside phosphoramidite **8** has been achieved to enable integration of a single 6-formylcytidine into a DNA duplex as a replacement for thymidine in the *Myb* binding sequence d(CAGT̲T).[11] The phosphoramidite **8** was prepared in several steps from O^2,2′-cyclouridine and incorporated into a 23-mer DNA oligonucleotide using solid-phase technology. After ammonolysis and purification, the oligomer was treated with sodium periodate to convert the 6-(1,2-dihydroxy)ethyl-2′-*O*-methyl cytidine residue to its 6-formyl derivative. The binding affinity of the modified duplex toward the DNA-binding domain of *c-Myb* is decreased significantly when compared with that of the unmodified duplex.[11] This experiment suggests that, contrary to expectations, no basic amino acid residue of *c-Myb* reacted with the formyl group in the DNA-binding domain.

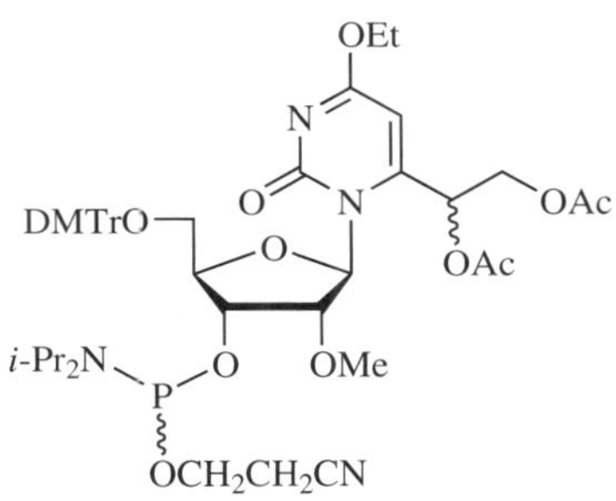

8

The synthesis and radioiodination of a stannylated oligodeoxyribonucleotide were performed to evaluate a γ-ray emitting ligand for thrombus imaging in vivo.[12] The modified deoxyribonucleoside phosphoramidite **9** was synthesized in three steps from 5-iodo-2′-deoxyuridine, and was efficiently added to the 5′-terminus of the thrombin-binding aptamer d(XGGTTGGTGTGGTTGG), in which X represents the stannylated nucleoside residue.

The oxidation of the phosphite triester with 3-chloroperoxybenzoic acid is necessary to prevent premature destannylation. 5′-*O*-Fmoc protection is also required to avoid exposure of the oligonucleotide to acid conditions, which would lead to destannylation.[12] Radioiodination of purified stannylated oligonucleotide provided ^{123}I-labeling yields up to 97% and high specific activity (15 000 Ci/mmol). The binding affinity of the resulting iodovinyl aptamer ($K_D = 36$ nM) for human α-thrombin is similar to that of the unmodified aptamer ($K_D = 45$ nM).[12]

The deoxyribonucleotide d(TGGGAG) functionalized with a 3,4-dibenzyloxybenzyl group at the 5′-terminus exhibited high anti-HIV-1 activity and low cytoxicity.[13] Modification of the 3′-end with a 2-hydroxyethyl phosphate group further enhanced anti-

HIV-1 activity, as well as stability towards nucleases in humans. The modified oligonucleotide presumably acts as an aptamer and binds to the gp120 protein of HIV-1, thereby inhibiting cellular absorption and entry of the virus.[13] Since this kind of short guanine-rich oligonucleotide forms a parallel quadruplex, synthesis of the N^2-methyl-2′-deoxyguanosine phosphoramidite (**10**) was performed to evaluate N^2-methylation on G-quadruplex stability.[13] Incorporation of three contiguous N^2-methyl-2′-deoxyguanosines into d(T GGGAG) produced the most stable parallel G-quadruplex with a twofold increase in anti-HIV activity.[13] N^2-Alkylation of guanine residues is therefore a promising way to improve the stability of G-quadruplexes.

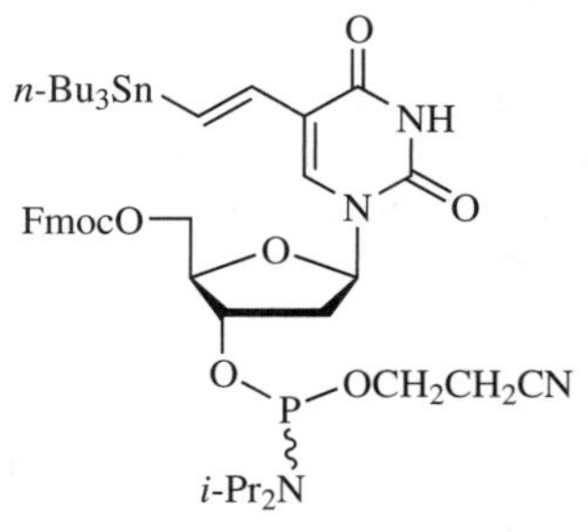

9

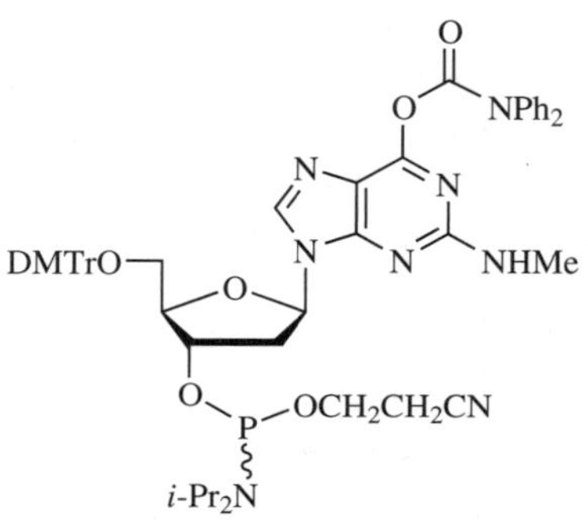

10

The synthesis of several *C*-nucleosides has been undertaken to explore the biophysical and biochemical properties of DNA modified with these nucleosides. This led to a number of studies addressing the roles hydrogen-bonding, base-stacking, and dipolar interactions may have on both duplex stability, and the ability of oligonucleotides to serve as templates. One functional group, which would induce a pronounced directional dipole and potential for hydrogen-bonding specificity, is the aromatic *N*-oxide. It was envisioned that the nucleoside analogs functionalized with a *N*-oxide at a structural position analogous to *O*-2 of pyrimidines would provide information on the importance of this contact in DNA polymerase catalysis. To test this hypothesis, a thiazole *C*-nucleoside analog was produced when applying the Hantzsch cyclization method to a 2′-deoxy-β-D-ribofuranosyl cyanide derivative.[14] Peroxide oxidation of the *C*-nucleoside afforded its *N*-oxide, which was subsequently converted to the phosphoramidite **11** upon reaction with 2-cyanoethyl tetraisopropylphosphorodiamidite.

Incorporation of **11** into DNA oligonucleotides was successful although significant deoxygenation of the *N*-oxide group occurred during DNA synthesis. The *N*-oxide of **11** may be reacting with the phosphorus of tetrazole-activated *N*-oxide phosphoramidites, thereby decreasing the concentration of active **11** and causing lower coupling efficiency.[14] The thermodynamic and enzymatic behavior of these modified oligonucleotides are currently under investigation.

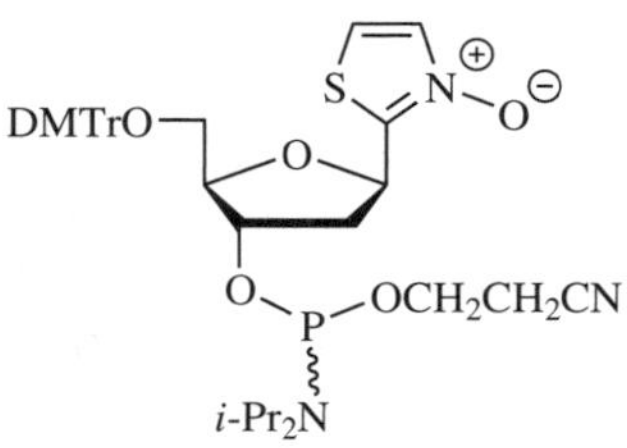

11

The existence of protein–DNA complexes in which DNA is dramatically bent and/or twisted, relative to canonical double helices, spurs interest in the energy cost of these distortions and motivates further the study of DNA dynamics.[15] A key issue in such studies is whether sequence-dependent differences in DNA flexibility contribute to the selectivity with which DNA-deforming proteins bind to distinct DNA sequences. Electron paramagnetic resonance (EPR) spectroscopy and spin-labeled nucleobases have been used to monitor DNA dynamics. However, motions of the spin-labeled nucleobases alone cannot be representative of the dynamics of the nucleic acid itself. To address the limitations of such EPR-based models, the phosphoramidite **12** has been prepared and inserted at a middle position in an undecaoligonucleotide, opposite to a 2-aminopurinyl residue in a DNA duplex.[15]

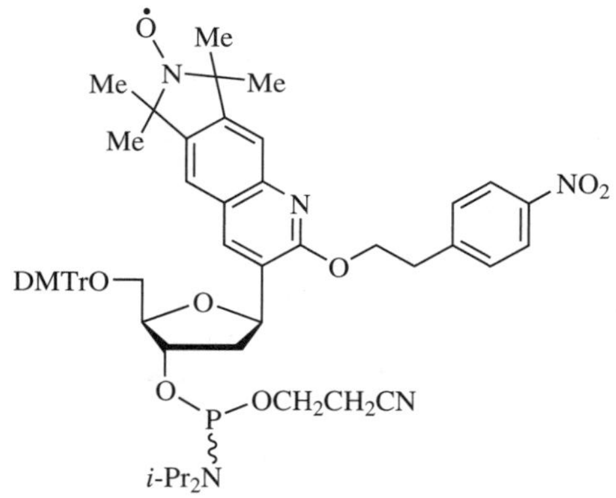

12

This study was prompted as a result of the incorporation of 2-hydroxyquinoline deoxyriboside phosphoramidite (**13**) into DNA oligonucleotides, which led to adequate thermal stability of several DNA duplexes when a modified nucleobase was opposite to 2-aminopurine ($\Delta T_m = -1\,^\circ C$ to $-3\,^\circ C$ relative to an unmodified A/T base pair).[16] On the basis of circular dichroism analyses, spin-labeled DNA duplexes adopt the conformation characteristic of right-handed double helices. EPR spectra of spin-labeled DNAs, prepared by the use of **12**, provide the first glimpse of nucleic acid dynamics uncompromised by independent spin probe motion.[15]

Ligandosides are nucleoside mimics in which the heterocyclic base is replaced by a strong chelator. These modified nucleosides are designed to place charged-metal complexes at the center of the DNA helix. The ligandoside phosphoramidite **14** has been prepared in a multistep synthesis using the lithium salt of 5-methyl-2,2′-bipyridine and 3,5-di-*O*-benzyl-2-deoxy-D-ribofuranose as starting materials.[17]

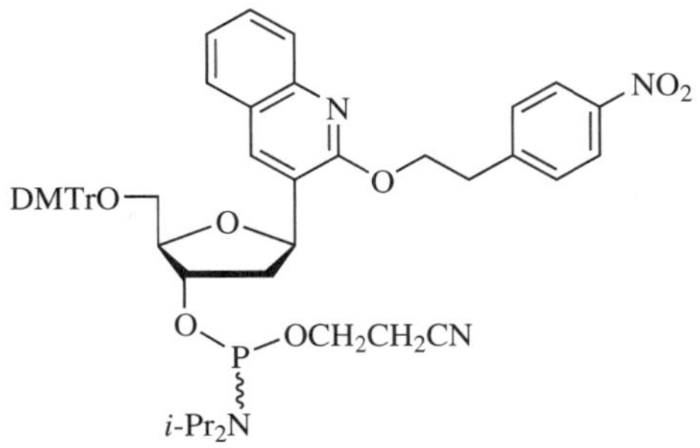

13

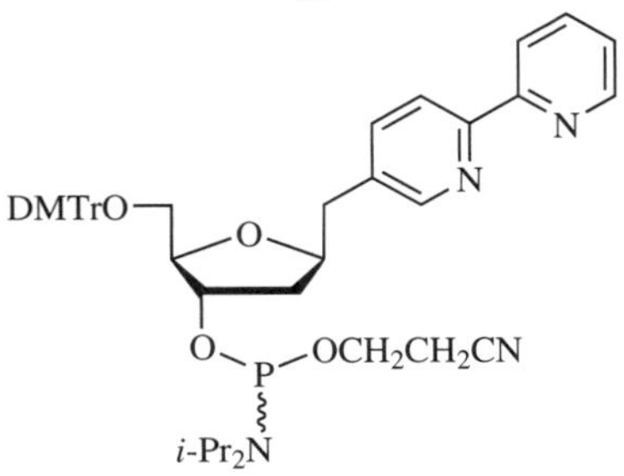

14

The phosphoramidite **14** was incorporated into a self-complementary DNA oligonucleotide (11-mer) at a single site using solid-phase DNA synthesis. Thermal denaturation experiments revealed a T_m of 56 °C for the modified duplex. Upon addition of 1.0 equiv of $Cu(OAc)_2$, the T_m of the ligandoside-containing duplex increased to 64 °C. This enhanced thermostability is consistent with an interstrand, intraduplex teracoordinated complex formation between ligandoside residues.[17] This is a unique approach to metal-mediated interstrand binding or reversible interstrand cross-linking.

There is much interest in long-range electron transfer through DNA duplex. However, lack of information on the precise location of an electron acceptor in a DNA duplex prevents accurate determination of the rate constants of distance-dependent electron transfer reactions.[18] The incorporation of an electron-accepting group at a well-defined position and geometry into a DNA duplex would, therefore, be desirable. In this context, synthesis of the deoxyribonucleoside phosphoramidite **15** has been accomplished in five steps from 3′,5′-di-*O*-(*tert*-butyldimethylsilyl)-2′-deoxyuridine and inserted as shown (X) in d(TAXTATGGTT). Preliminary studies indicate that *p*-cyano-substituted 5-benzoyldeoxyuridine has a unique property as an electron-accepting nucleobase, which led to photoinduced DNA cleavage at the 5′-G of the GG doublet via an electron transfer process.[18]

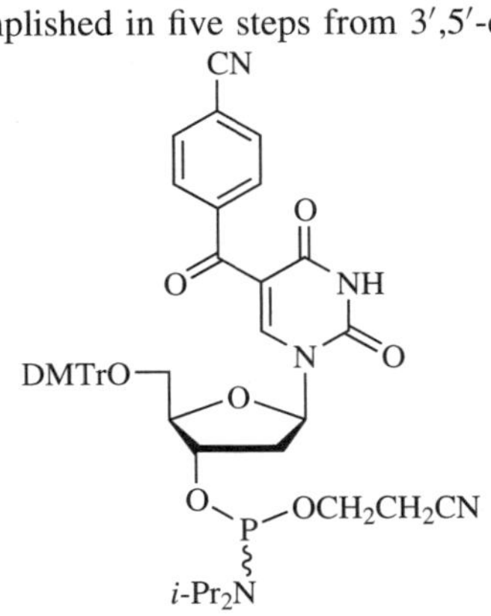

15

The π-stacking of DNA nucleobases plays an important role in duplex stabilization and in long-range charge transport through the DNA duplex. Efficiency of charge transport is strongly affected by π-stacking in DNA structures, such as A-type DNA, bulge-containing DNA sequences, and mismatched DNA. Furthermore, expansion of the π-stacking area lowers the ionization potentials of stacked nucleobases. To further study the extension of π-conjugate systems on lowering oxidation potentials and stabilizing DNA duplexes by increased π-stacking, 7-vinyl-7-deazaguanine deoxyribonucleoside phosphoramidite (**16**) was obtained in four steps from 7-iodo-7-deaza-2′-deoxyguanosine.[19]

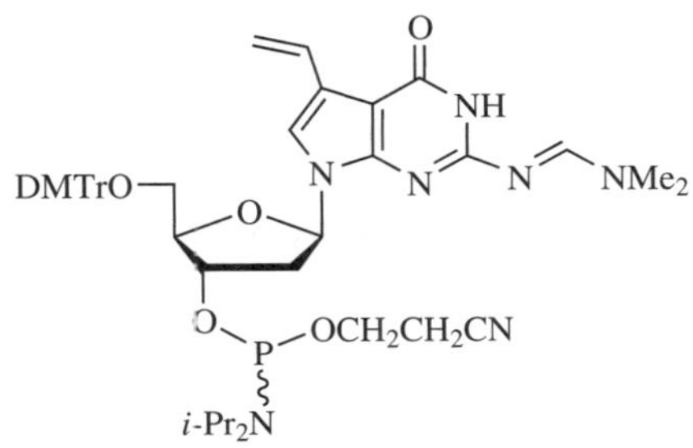

16

Incorporation of **16** into 5′-d(TTTGGTTGTTT) generated a duplex with a complementary DNA strand exhibiting a denaturation temperature higher ($\Delta T_m = +2.1$ °C) than that of the corresponding duplex containing native dG or 7-deaza dG.[19] Stabilization of the duplex is presumably due to the favorable enthalpy change caused by an increased π-stacking between the vinyl group and the 5′-flanking sequence. 7-Vinyl-7-deazaguanine would therefore likely serve as an electron-donating nucleobase for DNA-mediated electron transfer chemistry.

The synthesis of 7-propynyl-, 7-iodo-, and 7-cyano-7-deaza-2-amino-2′-deoxyadenosines from 2-deoxy-3,5-di-*O*-*p*-toluoyl-α-D-*erythro*-pentofuranosyl chloride and appropriate purine nucleobases has been reported along with that of the respective phosphoramidites shown in **17–19**.[20]

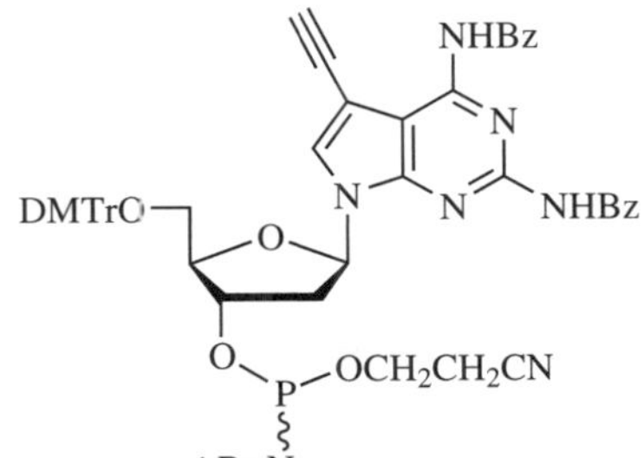

17

These phosphoramidites were incorporated into DNA oligonucleotides with coupling efficiencies exceeding 90%. Hybridization of the modified oligonucleotides with complementary RNA strands resulted in increased thermostability of the corresponding

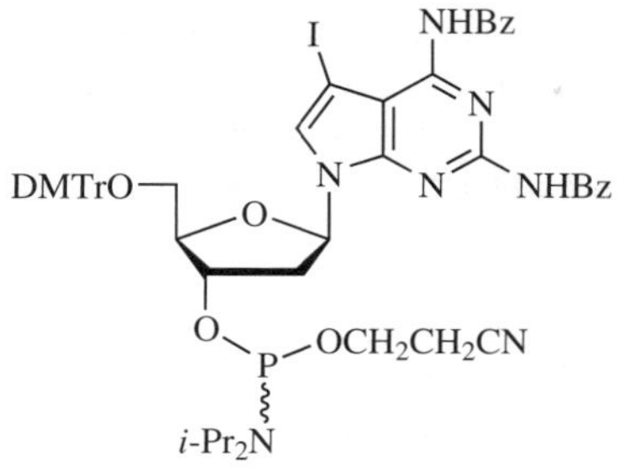

18

heteroduplexes [$\Delta T_m = +3\,^\circ$C to $+4\,^\circ$C/modification or $+0.3\,^\circ$C to $+1.5\,^\circ$C/modification for single or multiple incorporation(s), respectively] relative to unmodified DNA oligonucleotides.[20]

Aside from the traditional Watson–Crick base-pairing, nucleobases can also recognize their complements via Hoogsteen-type base-pairing. The stability of such type of base-pairing in a triplex structure can be improved through appropriately designed nucleobase modifications. Specifically, the deoxyribonucleoside phosphoramidite **20** has been prepared in eight steps beginning with 2′-deoxyadenosine, and integrated either singly or repeatedly into oligonucleotide d(GCTAAGAGAAAGATCG) using a DNA synthesizer.[21] Triplex structures were formed by mixing the modified oligonucleotides with d(CGATCTTTCTCTTAGC) and d(TTCTCTTTCT).

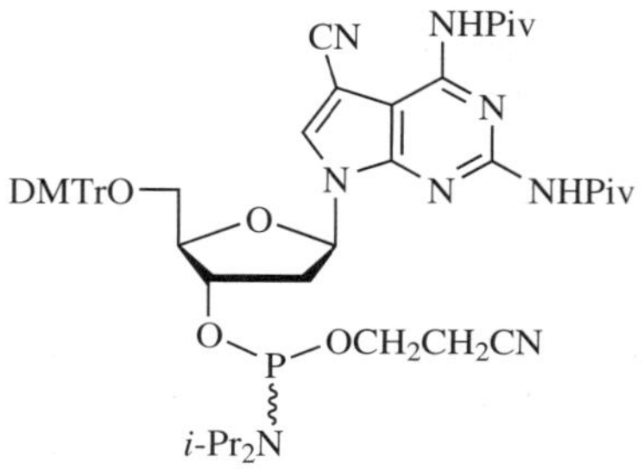

19, Piv = trimethylacetyl

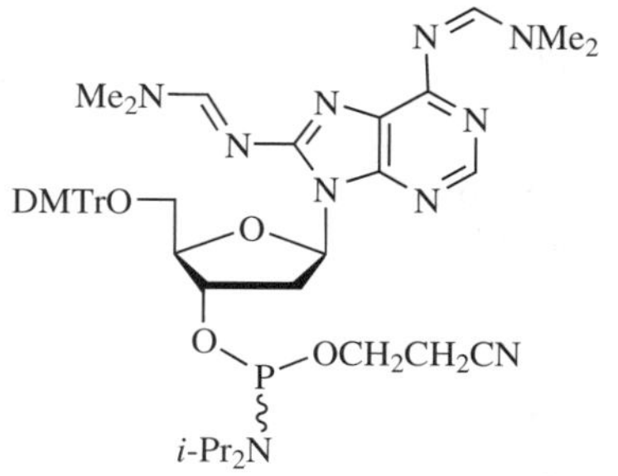

20

Interestingly, the thermostability of these triplexes increased with the number of 8-amino-2′-deoxyadenosine residues in the DNA duplex [5′-d(GCTAAGAGAAAGATCG)/3′-d(CGATTCTCTTTCTAGC)]. Indeed, the temperature at which the third strand of a triplex dissociated from the DNA duplex containing three 8-amino-2′-deoxyadenosine residues at pH 6.0, is 24.5 °C higher than that recorded for the unmodified triplex under identical conditions.[21] Thus, incorporation of 8-amino-2′-deoxyadenosine into DNA may provide a useful tool for stabilizing nucleic acid structures of biological importance.

A myriad of modified DNA oligonucleotides have been synthesized to improve their binding affinities toward complementary DNA or RNA targets. Oligonucleotides functionalized with cationic groups seem an obvious choice for accomodating interactions between two polyanionic entities. In fact, endogenous and synthetic polyamines are known to stabilize the DNA against thermal denaturation, presumably because of the formation of a *trans*-strand network of ammonium hydrogen bonds involving *O*-2 of pyrimidines and *N*-3 of purines in the minor groove. In an attempt to improve the affinity of oligonucleotides for their DNA or RNA complements, the polyamine spermine was covalently linked to an oligonucleotide through the *C*-2 position of a purine, which bisects the minor groove in the duplex DNA. The sperminyl deoxyribonucleoside phosphoramidite **21** was prepared from a 2-fluoro-2′-deoxyinosine intermediate and was found less reactive than unmodified phosphoramidites, possibly because of the steric hindrance created by the sperminyl group. Best coupling results were obtained with two coupling steps of 10 min each using a 0.09 M solution of **21** in acetonitrile.[22]

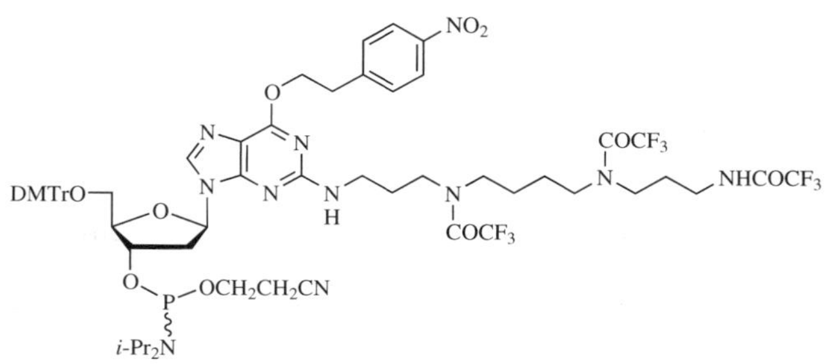

21

The phosphoramidite **21** was incorporated into an oligonucleotide (22-mer) at up to six different sites. Duplex formation with a complementary DNA strand led to hybrids with increased thermostability ($\Delta T_m = +3.0 \pm 0.2\,^\circ$C/mod) when compared with that of the unmodified duplex.[22] On the basis of a reasonable extrapolation, a duplex composed of an oligonucleotide (22-mer) bearing seven spermine residues is expected to be as stable as genomic DNA, and would be capable of sequence-selective recognition of double-stranded DNA through strand invasion.[22] Modified oligonucleotides such as these may find wide applications in - diagnosis and therapy.

The covalent attachment of natural/unnatural polycationic polyamines to oligonucleotides may, in addition to decreasing the net charge of such conjugates, result in enhanced nuclease resistance and cell membrane permeability. Given that insertion of **21** in oligonucleotides led to the formation of DNA duplexes with improved thermostability properties, synthesis of the sperminyl deoxyribonucleoside phosphoramidite **22** has also been described.[23] Incorporation of **22** into synthetic oligonucleotides at predetermined positions was performed to determine whether these oligonucleotides might lead to useful antisense drugs.

Contrary to the expectations, the T_m's of DNA complexes containing one or two 5-methyl-4-*N*-(sperminyl)cytosine residues are lower, to the extent of 5 °C per modification, than that of the unmodified duplex.[23] These unanticipated findings cast doubts on the efficacy of this specific type of modified oligonucleotides in therapeutic settings.

The studies on oligonucleotides containing deoxyribonucleotides of 3-nitropyrrole or 4-nitroazole and 5- or 6-nitroindole have demonstrated that these can function as primers for both dideoxy-sequencing reactions and PCR amplifications.[24] The deoxyribonucleoside phosphoramidites of 3-nitropyrrole (**23**), 4-nitropyrazole (**24**), 4-nitroimidazole (**25**), and 5-nitroindole (**26**) were incorporated into oligonucleotide d(CGC<u>X</u>AATT<u>Y</u>GCG) (where X corresponds to the modified nucleosides and Y to A, C, G, or T) to provide insight into the significance of electronic structure and heterocyclic size toward the development of more effective universal spacers for nucleic acid duplexes. The phosphoramidites **23–26** were synthesized in four steps starting with 2-deoxy-3,5-di-*O*-*p*-toluoyl-α-D-*erythro*-pentofuranosyl chloride and the appropriate nitroazole nucleobases.

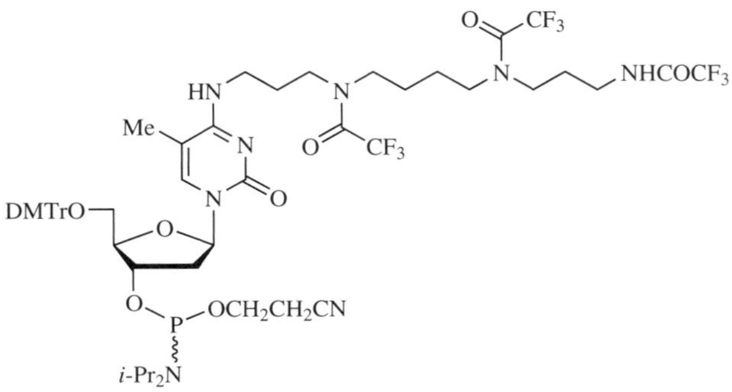

22

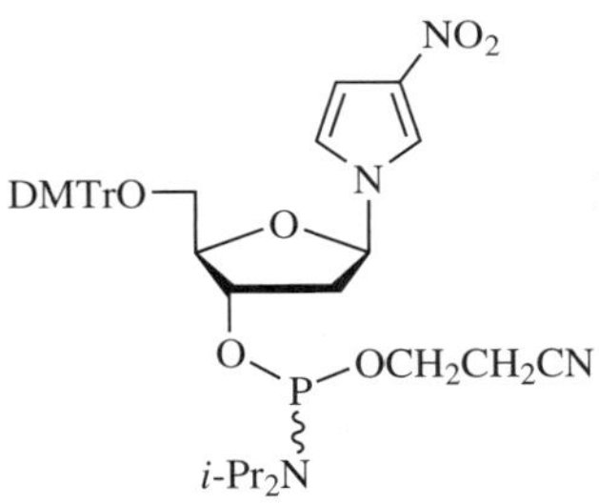

23

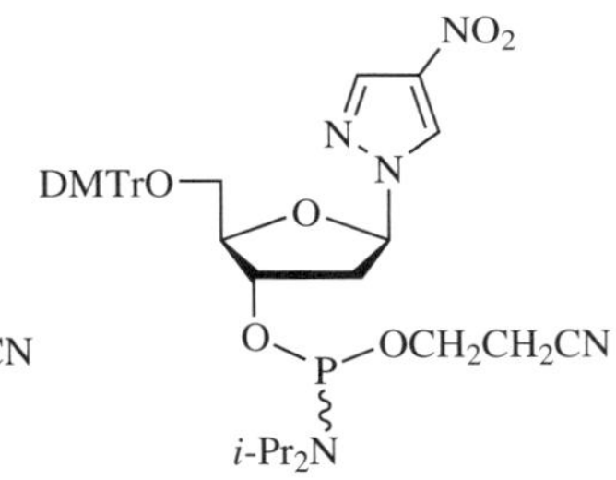

24

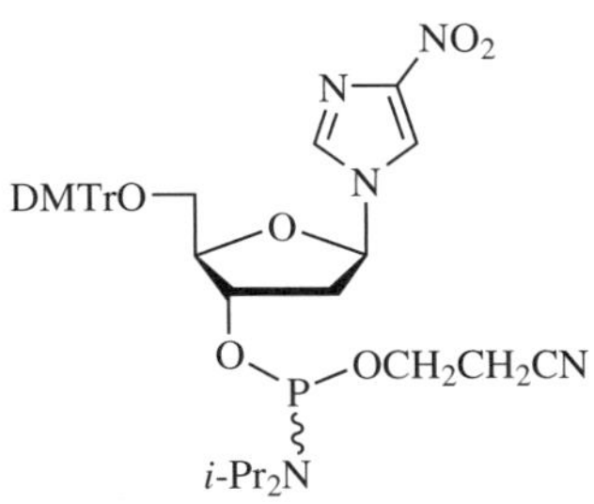

25

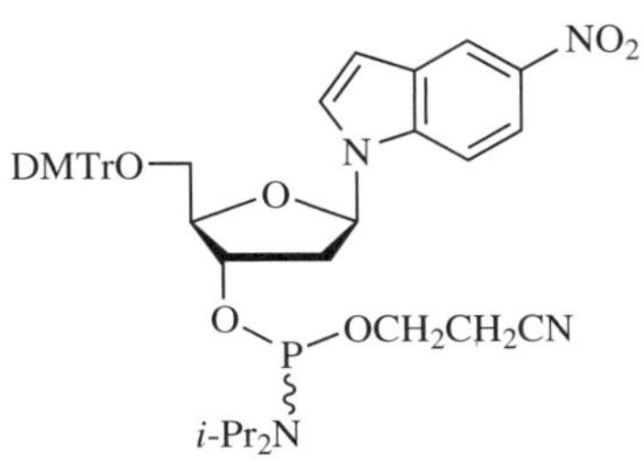

26

The coupling yields of these phosphoramidites ranged from 96% to 98%.[24] The comparison of T_m and thermodynamic data indicates that 3-nitropyrrole and 5-nitroindole are non-discriminating base-pairing partners. Addition of a ring nitrogen, capable of functioning as a hydrogen bond acceptor, decreases the non-discriminatory character of the nitroazole base. 4-Nitroimidazole shows specific base-pairing with G, whereas 4-nitropyrazole shows some preference for pairing with A.[24]

The synthesis of DNA oligomers containing modified nucleobases may be carried out using the 'convertible' nucleoside phosphoramidite **27**. The phosphoramidite was prepared in five steps from thymidine and trifluoroacetyl nitrate, and integrated into oligonucleotides without the loss of coupling efficiency under conditions similar to those employed for standard phosphoramidites.[25]

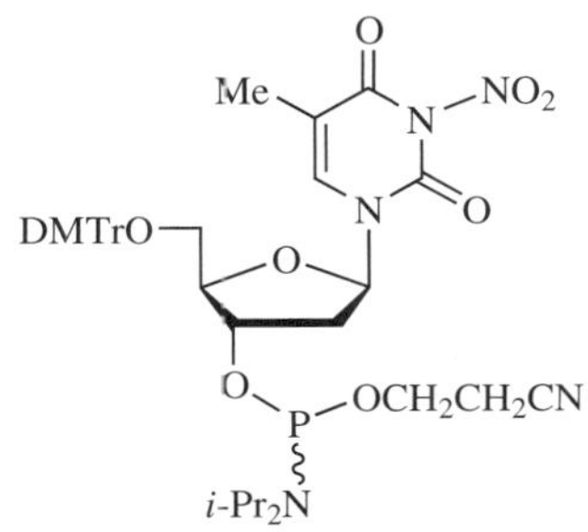

27

The treatment of the modified oligonucleotides still attached to the solid support with primary amines, diaminoalkanes, and aminoalcohols resulted in the conversion of N^3-nitrothymine residues to the corresponding N^3-alkylated thymines. The reaction mechanism is thought to proceed through ring-opening of the nucleobase upon nucleophilic attack at *C*-4 by primary amines followed by ring-closure and departure of H_2NNO_2.[25]

An aldehyde function is often used to conjugate biopolymers to other molecules by the process of reductive amination or formation of adducts with hydroxylamines, hydrazines, and semicarbazides. In this regard, a convenient method for incorporating an aldehyde function into an oligonucleotide at a specific location has been reported.[26] The method uses the formylindole phosphoramidite **28**, prepared in four steps from 3-formylindole and 2-deoxy-3,5-di-*O*-toluoyl-D-*erythro*-pentofuranosyl chloride.

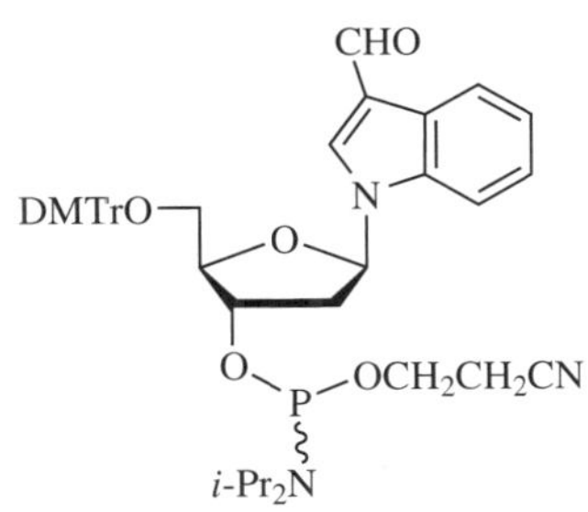

28

Incorporation of **28** into d(GCGATG<u>X</u>GTAGCG) afforded, after mixing with d(CGCTAC<u>N</u>CATCGC) (where <u>N</u> = C, G, T, and A), DNA duplexes that were less stable ($\Delta T_m = -7\,°C$ to $-10\,°C$) than the one with a native A-T base pair, regardless of the base opposite to X.[26] The formylindole residue was then reacted with ferrocenecarbohydrazide to give the corresponding ferrocenecarbohydrazone adduct **29**. Incidentally, oligonucleotides functionalized with ferrocene are described as excellent electrochemical probes.[26] The insertion of an aldehyde function in oligonucleotides is undoubtedly an attractive method for post-synthetic and site-selective attachment of reporter groups and ligands to DNA.

Pteridine *N*-1 and *N*-8 nucleosides are structurally related to the naturally occurring purine and pyrimidine nucleosides.[27] A striking feature of the pteridines is their strong fluorescence, which can be valuable in labeling oligonucleotides at specific sites. Synthesis of pteridine phosphoramidite **30** has been achieved in several steps from 6-methyl-2-methylthio-8-(2-deoxy-3,5-di-*O*-*p*-toluoyl-β-D-ribofuranosyl)-3*H*,8*H*-pteridine-4,7-dione. This new fluorescent phosphoramidite for oligonucleotide synthesis can be considered as a substitute for 2′-deoxyguanosine.[27]

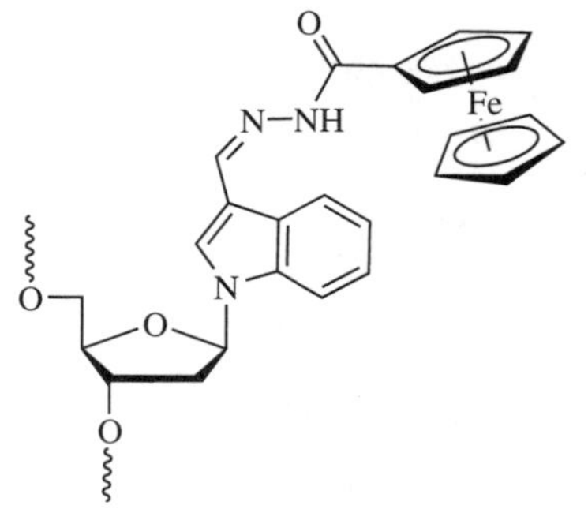

29

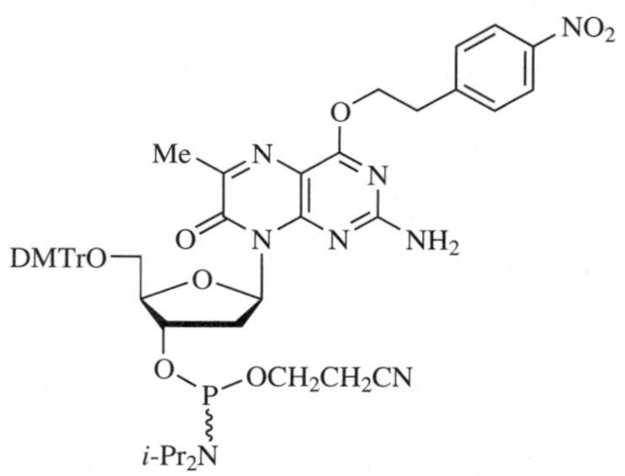

30

2-Cyanoethyl tetraisopropylphosphorodiamidite has not only been useful in the phosphinylation of nucleobase-modified nucleosides, it has also been commonly used in the phosphinylation of nucleosides with altered carbohydrates. This topic will be the focus of the following section of this article.

Phosphinylation of Carbohydrate-modified Nucleosides. The application of oligonucleotides as antisense inhibitors of gene expression has generated much interest in the discovery of new therapeutic drugs. This technology is based on the premises that an antisense oligonucleotide must: (i) hybridize with the target RNA sequence with a high degree of specificity; (ii) have sufficient cellular penetration and distribution; and (iii) have sufficient resistance to an environment rich in nucleases. In this regard, the synthesis of α-2′-deoxyribonucleoside phosphoramidites (**31a–31d**) has been reported toward the preparation of alternating α,β-oligodeoxyribonucleotides with 3′→3′ and 5′→5′ internucleotidic linkages (*alt*DNA) for potential antisense applications.[28] *alt*DNA oligonucleotides were designed assuming that oligonucleotides having alternating 3′→3′ and 5′→5′ internucleotide linkages may not be hydrolyzed by nucleases as rapidly as native oligomers.

When tested, *alt*DNA oligonucleotides were found less resistant to snake venom phosphodiesterase and calf spleen phosphodiesterases than phosphorothioated β-oligodeoxyribonucleotides, but were considerably more stable than native DNA oligomers under similar conditions. Hybridization studies conducted with *alt*DNA oligonucleotides and complementary unmodified DNA or RNA strands indicated that *alt*DNA–DNA duplexes are less thermostable than unmodified DNA–DNA duplexes ($\Delta T_m =$ − 1.1 °C/modification) and more stable than *alt*DNA–RNA complexes.[28] The reduced thermostability of *alt*DNA–RNA complexes may result from the limited ability of *alt*DNA to establish base-pairing within A-type helices. One can argue that a greater flexibility of *alt*DNA nucleobases may compensate for the unnatural arrangement of the internucleotidic phosphodiester motifs of these oligomers, which may detrimentally affect base-pair formation. In order to evaluate this postulate, the deoxyribonucleoside phosphoramidites shown in **32a**, **32b–35a**, **35b** were prepared,[29–31] and incorporated into oligonucleotides with defined internucleotidic motifs at centrally located sites.[31]

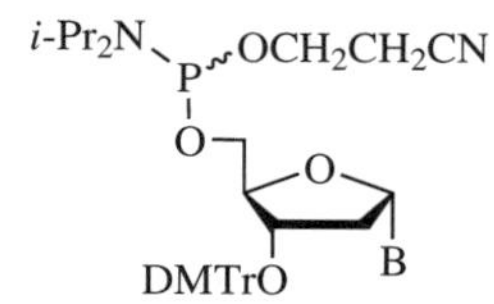

31a, B = thymin-1-yl
31b, B = 4-*N*-benzoylcytosin-1-yl
31c, B = 6-*N*-benzoyladenin-9-yl
31d, B = 2-*N*-isobutyrylguanin-9-yl

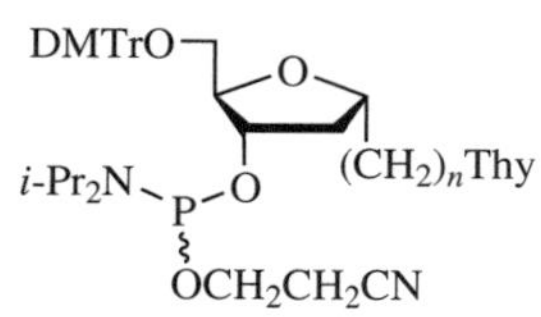

32a, *n* = 2
33a, *n* = 1

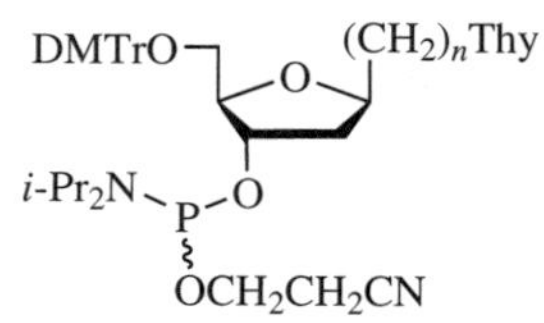

32b, *n* = 2
33b, *n* = 1

Insertion of **32a**, **32b** and **34a**, **34b** according to either a 3′→5′ or a 3′→3′ internucleotidic polarity, produced oligonucleotides exhibiting a slightly higher affinity for their unmodified DNA complements than for the corresponding RNA complements. Conversely, the incorporation of **35a** into 5′-d(TCGACACCCAATTCTGAAAATGGA), according to a 3′→3′ orientation, generated, for the first time, an *alt*DNA oligonucleotide displaying a greater affinity for its complementary unmodified RNA sequence ($\Delta T_m =$ −6 °C) than for the corresponding DNA sequence ($\Delta T_m = -10$ °C). This hybrid was, however, thermodynamically less stable than the duplex having **31a** similarly incorporated ($\Delta T_m = -3$ °C).[31] Thus, the nucleobase flexibility imparted by the integration of **35a** into an oligonucleotide presumably contributed to entropic factors that adversely affected the thermostability of the modified DNA–RNA hybrid to a greater extent than that observed with the incorporation of **31a** under identical conditions.[31] Because of poor affinity toward RNA targets, *alt*DNA oligonucleotides are not suitable for therapeutic applications.

Phosphorothioated oligonucleotides incorporating 2′-fluoro-2′-deoxyribonucleosides have been assembled from the deoxyribonucleoside phosphoramidites as shown in **36a–36d**,[32] and have demonstrated significant resistance to nucleases.

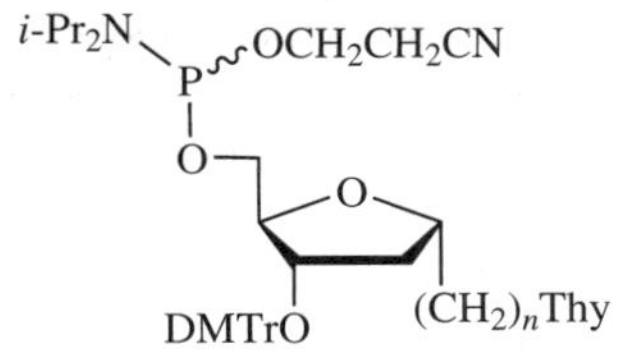

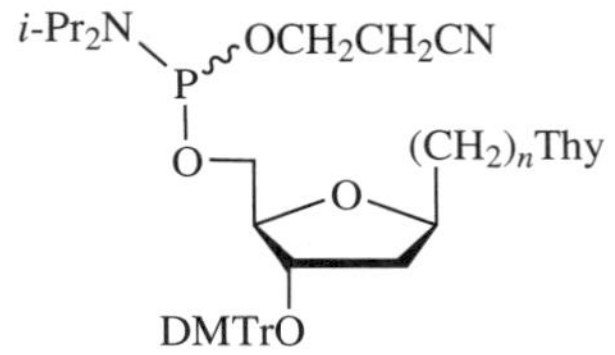

34a, $n = 2$
35a, $n = 1$

34b, $n = 2$
35b, $n = 1$

Thy = thymin-1-yl

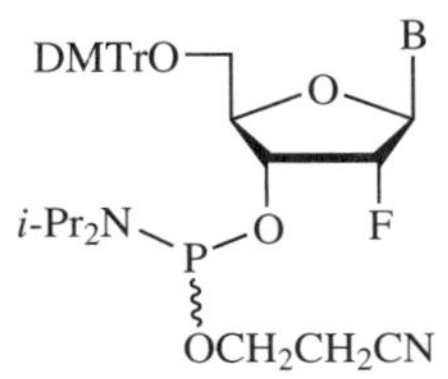

36a, B = uracil-1-yl
36b, B = 4-*N*-benzoylcytosin-1-yl
36c, B = 6-*N*-benzoyladenin-9-yl
36d, B = 2-*N*-isobutyrylguanin-9-yl

These modified oligonucleotides exhibited high-binding affinity for complementary RNAs, and CD spectra of the resulting duplexes revealed A-type helices. However, these duplexes were not substrates for HeLa RNase H.[32] Such a limitation was circumvented by flanking an unmodified DNA segment capable of activating RNase H with phosphorothioated 2′-fluoro DNA segments on each side of the native DNA gap. The resulting chimeric oligonucleotides elicited regioselective RNase H degradation of the complementary RNA strand of the modified DNA–RNA heteroduplexes.[32] This RNase H-mediated mechanism is generally viewed as a major effector of antisense activity.

Incorporation of an intercalating agent into an oligodeoxyribonucleotide has the potential to enhance its binding affinity, increase its hydrophobicity, and improve its resistance to nuclease hydrolysis. Site-specific intercalation of 2′-*O*-(anthraquinone-2-methyl)adenosine (rA*) into the palindromic dodecanucleotide d(CGCrA*CATGTGCG) has been achieved through the use of phosphoramidite **37**, which was prepared in four steps from 2′,3′-

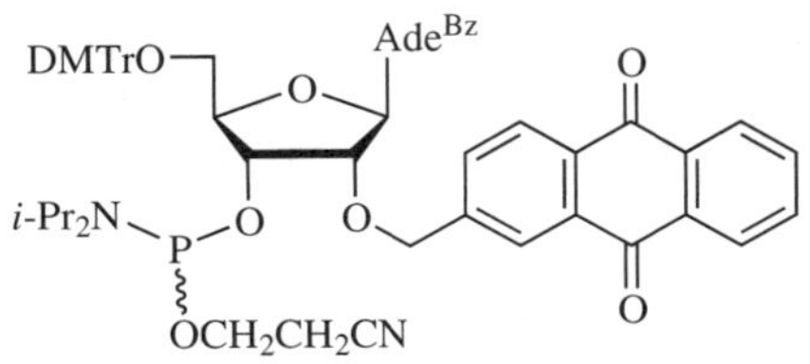

37, Ade = adenin-9-yl

(dibutylstannylene)adenosine.[33] A coupling reaction of 10 min was necessary for this phosphoramidite to effect a coupling yield better than 90%.

Thermal denaturation measurements established that intercalation of the anthraquinone ring into the self-complementary B-type helix stabilized the duplex by 3 °C to 3.5 °C relative to the unmodified duplex.[33] In contrast to these findings, incorporation of 2′-*O*-phenethyl ribonucleoside phosphoramidite (**38**) at the same location into the same self-complementary dodecanucleotide produced significant duplex destabilization ($\Delta T_m = -12\,°C$).[34]

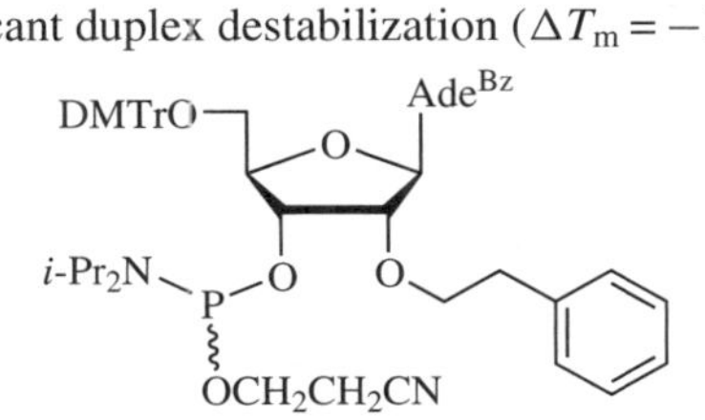

38

On the basis of NMR studies, the plane of the phenyl ring lies in the major groove of the DNA duplex perpendicular to the stacked bases.[34] Such an unusual orientation of an aromatic group into the major groove of DNA deserves consideration when designing oligonucleotides containing major groove-binding molecules.

Nonanucleotides containing a centrally located 2′-*O*-(1-pyrenylmethyl)uridine nucleoside were prepared using the ribonucleoside phosphoramidite **39**, which was obtained in 80% yield from 5′-*O*-(4,4′-dimethoxytrityl)-2′-*O*-(1-pyrenylmethyl)uridine, 2-cyanoethyl tetraisopropylphosphorodiamidite, 1*H*-tetrazole, and diisopropylamine.[35]

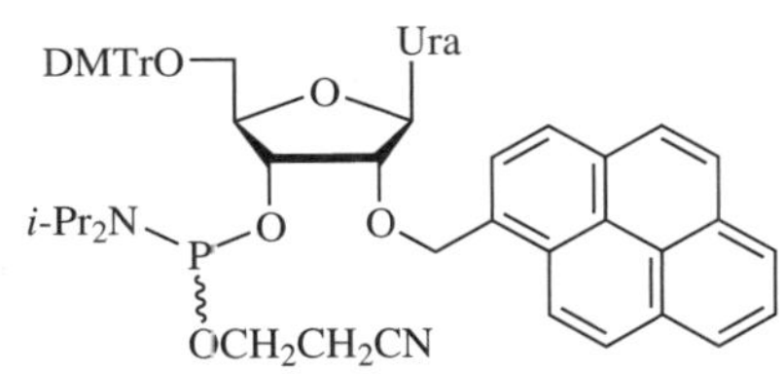

39, Ura = uracil-1-yl

The coupling efficiency of **39** in solid-phase oligonucleotide synthesis is ~90%. Thermal denaturation studies indicate that pyrene-modified oligonucleotides can bind to both DNA and RNA complements in aqueous solution. Typically, pyrene-modified oligonucleotides exhibit higher affinity for complementary DNA sequences than for RNA sequences.[35] However, addition of complementary RNAs to pyrene-modified oligonucleotides results in enhancement of the pyrene monomer emission, the extent of which depends on the nucleobase 3′ to the pyrene modification. Specifically, a pyrenylated oligonucleotide containing a deoxycytidine adjacent to the modification shows a ~250-fold increase in fluorescence upon binding to a complementary RNA sequence. These findings may find applications in ribozyme studies and in the monitoring of in vitro transcription experiments.[35]

Oligonucleotides conjugated with various ligands are in demand for applications in bioorganic chemistry, molecular biology, and biotechnology. The 2′-hydroxy group of ribonucleosides is an ideal site for attachment of ligands to nucleic acids. While the conjugation of ligands is usually accomplished through 2′-*O*-alkylation of nucleosides, the reaction proceeds with notoriously poor yields. A convenient method for the preparation of uridin-2′-yl carbamate derivatives offers an attractive alternative to 2′-*O*-alkylation for the incorporation of ligands into oligonucleotides.[36] Specifically, the reaction of 3′,5′-*O*-(tetraisopropyldisiloxane-1,3-diyl)uridine with 1,1′-carbonyldiimidazole followed by treatment with primary or secondary amines gives 2′-carbamates in high yields. Conversion of the resulting carbamates to the phosphoramidites **40** via 2-cyanoethyl tetraisopropylphosphorodiamidite was effected in three steps without 2′-3′-migration of the carbamate groups.[36]

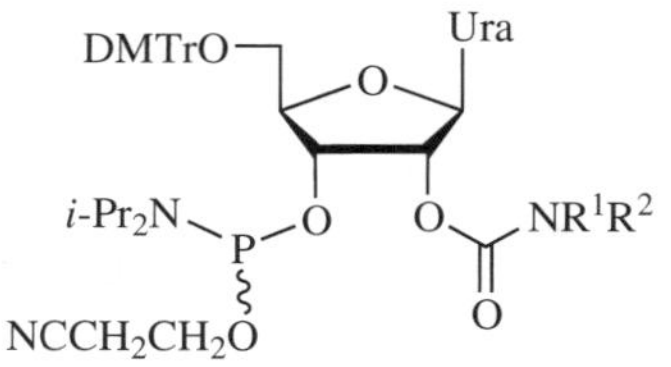

40, NR^1R^2 = primary or secondary amines

The phosphoramidites shown in **40** were used in solid-phase DNA synthesis using a coupling time of 10 min, allowing incorporation of 2′-*O*-(pyren-1-ylmethylcarbamoyl)uridine residues into DNA oligonucleotides. The pyrene modification was able to stabilize a duplex with a mismatched deoxycytidine directly opposite to the carbamate-modified nucleoside. Moreover, the fluorophore showed a considerable fluorescence increase (up to 30-fold) when pyrenylated oligonucleotides were mixed with complementary RNA sequences. This approach to fluorescence enhancement may therefore find useful applications in the detection of RNA.[36]

The utility of 2-cyanoethyl tetraisopropylphosphorodiamidite as a bifunctional phosphitylating reagent has been demonstrated in the preparation of a novel 2′,3′-cyclic phosphate derived from 2′-homouridine (eq 1).[37]

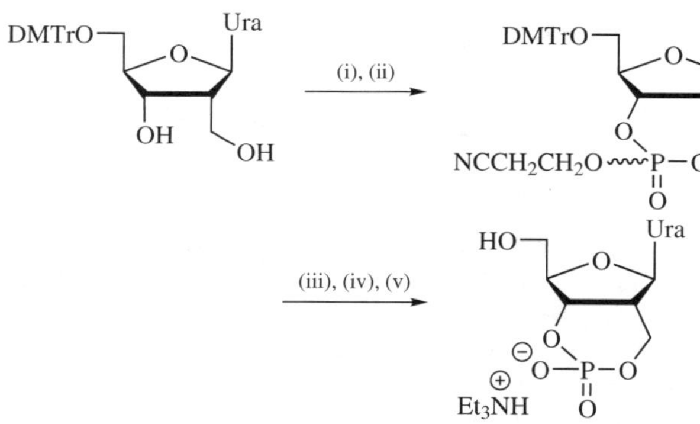

(1)

Conditions:

(i) $(i\text{-}Pr_2N)_2POCH_2CH_2CN$; (ii) I_2/pyridine/H_2O;

(iii) Et_3N/CH_2Cl_2 (1:1); (iv) AcOH/H_2O (4:1);

(v) DEAE Sephadex Chromatography.

This is the first example of a cyclic phosphate prepared from 2′-homouridine, which will serve as a model in evaluating the hydrolytic behavior of oligonucleotides containing 2′-homouridine entities. In this context, 2′-homouridine phosphoramidite (**41**) has been prepared and used in the synthesis of 2′-deoxy-2′-α-*C*-(hydroxymethyl)uridylyl-3′,5′-uridine.[38] This dinucleotide was found 1840-fold more stable to base hydrolysis than the unmodified dinucleotide UpU.

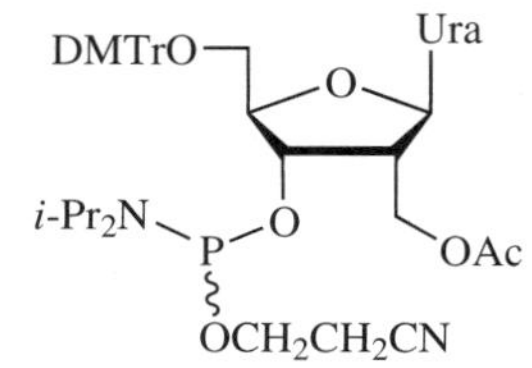

41

The modified dinucleotide was untouched by ribonuclease A over a period of 2 days, whereas UpU was completely hydrolyzed within 30 min under the same conditions.[38] These findings suggest that longer oligomers containing 2′-homouridine residues shall exhibit increased resistance to hydrolytic cleavage and may potentially be valuable as antisense oligonucleotides.

In an effort to develop therapeutic oligonucleotides that would regulate gene expression, the nucleoside phosphoramidites shown in **42–44** were synthesized in several steps from uridine and inserted at predetermined sites in oligonucleotides (16- or 18-mers) using standard solid-phase technology.[39]

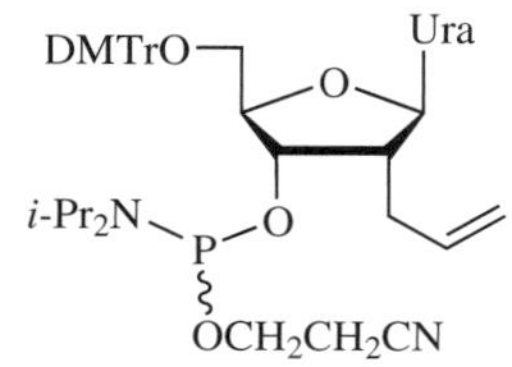

42

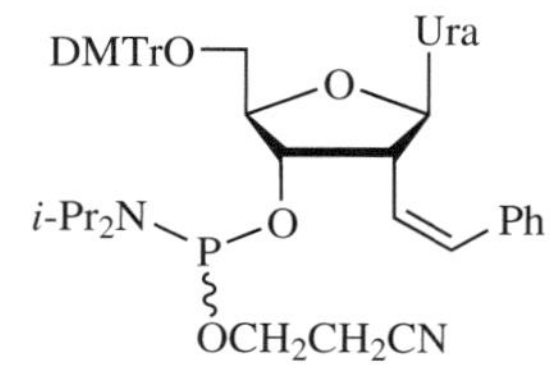

43

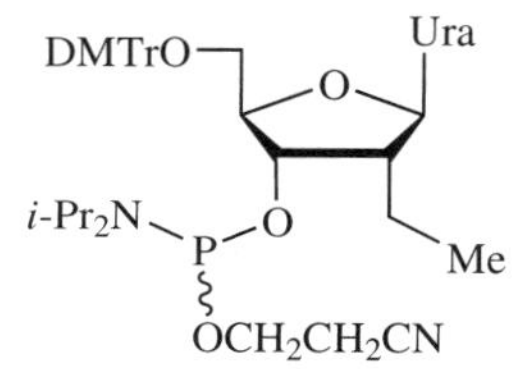

44

These modified oligonucleotides significantly destabilized duplexes formed with RNA complements. Denaturation temperatures of −3.1 °C to −7.3 °C per modification were determined for these hybrids relative to those recorded for the same unmodified duplexes. These results contrast with those reported for the 2′-*O*-alkyl ribonucleotide analogs, which are known to stabilize duplex formation with complementary RNAs. Steric, dipolar, and/or conformational factors may account for differences in duplex stability between 2′-*C*-alkyl and 2′-*O*-alkyl modified heteroduplexes.[39] On the basis of these findings, 2′-*C*-alkylated oligonucleotides are not recommended for therapeutic antisense applications.

The difluoromethyleneuridine phosphoramidites shown in **45** and **46** have been prepared from 5′-*O*-DMTr-2′- or 3′-*O*-TBDMS uridine derivatives for incorporation into oligonucleotides. It was expected that these modified oligonucleotides would exhibit increased stability against nucleases given the conformational restrictions imposed by the difluoromethylene groups on ribose puckering.[40]

Thus, 5′-*O*-DMTr-3′-*O*-TBDMS uridine was converted to the corresponding 2′-keto derivative when reacted with pyridinium dichromate. Treatment of the 2′-keto nucleoside intermediate with bromodifluoromethyl[tris(dimethylamino)]phosphonium bromide and zinc gave the 2′-difluoromethylene-3′-*O*-TBDMS uridine derivative. After desilylation and phosphinylation with 2-cyanoethyl tetraisopropylphosphorodiamidite/diisopropylammonium tetrazolide, the phosphoramidite **45** was isolated in good yields.[40] Syntheses of oligonucleotides using either **45** or **46** remain, however, to be reported.

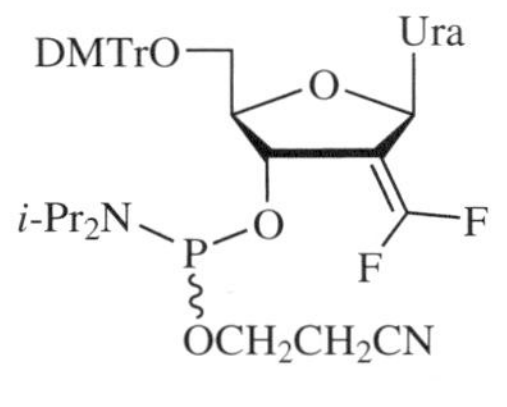

45

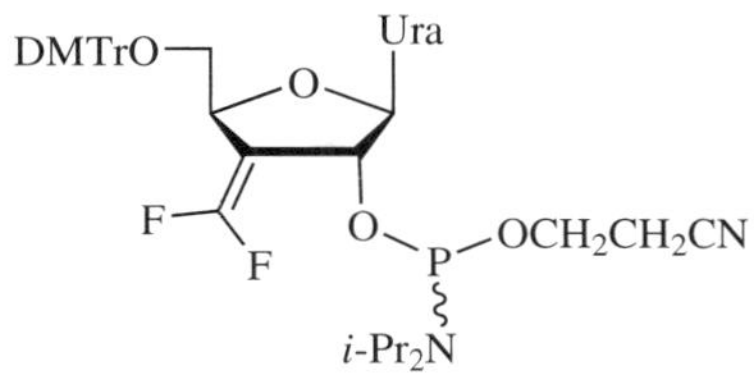

46

Of interest, 3′-deoxyriboadenosine phosphoramidite (**47**) has been employed in the synthesis of 2′→5′phosphorothioated 3′-deoxyriboadenylate trimers (dA_3) to define the requirements for binding to, and activating, the 2′→5′oligoadenylate (2-5A)-dependent endoribonuclease (RNase L) and to evaluate the role of phosphorothioate chirality on the biological activity.[41]

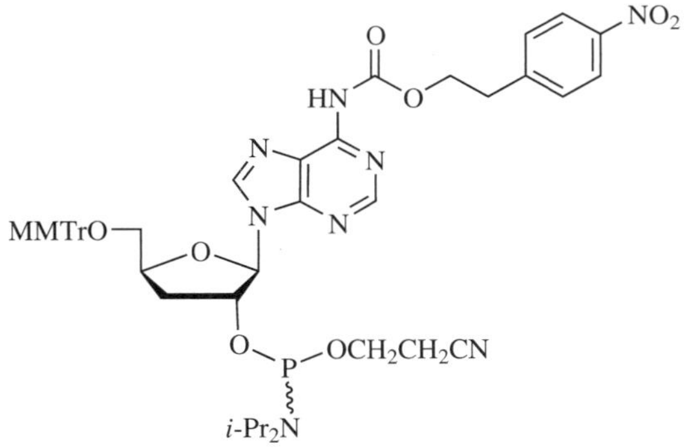

47, MMTr = 4-methoxytrityl

It was found that (2′→5′)-3′-dA_3 with either R_P- or S_P-(2′→5′)-internucleotide linkages can bind to, but cannot activate, RNase L. Interestingly, the 5′-monophosphate of [R_P,R_P]-(2′→5′)-3′-dA_3 and [S_P,R_P]-(2′→5′)-3′-dA_3 phosphorothioates can bind to and activate RNase L, whereas the 5′-monophosphate of [R_P,S_P]-(2′→5′)-3′-dA_3 and [S_P,S_P]-(2′→5′)-3′-dA_3 phosphorothioates can bind to but cannot activate RNase L.[41] While phosphorothioated analogs of 2-5A (5′-monophosphates and 5′-triphosphates) are effective inhibitors of HIV-1 reverse transcriptase and HIV-1 replication, the 5′-monophosphate of thioated (2′→5′)-3′-dA_3 is also a potent inhibitor of HIV-1 replication when encapsulated in liposomes coated with anti-CD3 antibodies.[41]

One major problem associated with the therapeutic application of 2′→5′oligoadenylates is the sensitivity of these oligonucleotides to nucleases leading to rapid hydrolysis. Chemical modification of 2′→5′oligoadenylates involving replacement of the 5′- or 3′(2′)-hydroxyl with an amino group is known to improve the stability of these oligonucleotides to nucleases. To further assess the magnitude of this enhanced resistance to enzymatic hydrolysis, 5′-aminated- and 5′-amidated-2′-ribonucleoside phosphoramidites (**48a**, **48b**) were prepared in several steps starting with 2′,3′-*O*-isopropylidene adenosine and added to the 5′-terminus of a suitably protected 2′→5′adenylate dimer. After deprotection, the 5′-aminated and 5′-amidated 2′→5′-linked adenylate trimers were, respectively, found 3-fold and 15-fold more stable to snake venom phosphodiesterase than unmodified (2′→5′)-ApApA.[42] The mechanism through which a terminal 5′-amino or 5′-amido function provides 2′→5′polyadenylic acids with increased stability to nucleases is still unclear.

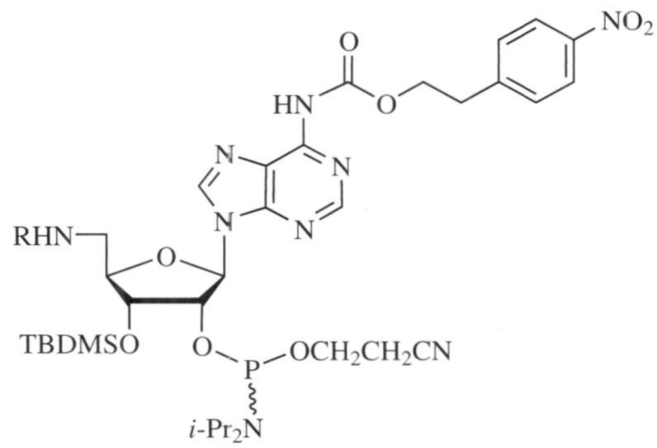

48a, R = 2-(4-nitrophenyl)ethoxycarbonyl; TBDMS = *tert*-butyldimethylsilyl
48b, R = palmitoyl

Modified nucleotides have been discovered at position 64 of eukaryotic cytoplasmic initiator methionine tRNAs and have been identified as the β-anomers of *O*-ribosyl(1″→2′)-5″-phosphates of adenosine and guanosine, respectively.[43] The 2′-*O*-phosphoribosylation is thought to be a general feature of cytoplasmic initiator methionine tRNAs in plants and fungi, which may control initiation vs elongation function. In order to carry out further studies in this area, the synthesis and structure of 2′-*O*-ribosyl nucleic acid derivatives have been reported. 2-Cyanoethyl tetraisopropylphosphorodiamidite was particularly useful in the synthesis of the phosphoramidites shown in **49a–49d**, which eventually led to the preparation of 2′-*O*-ribosyl dinucleotide (**50**).[43]

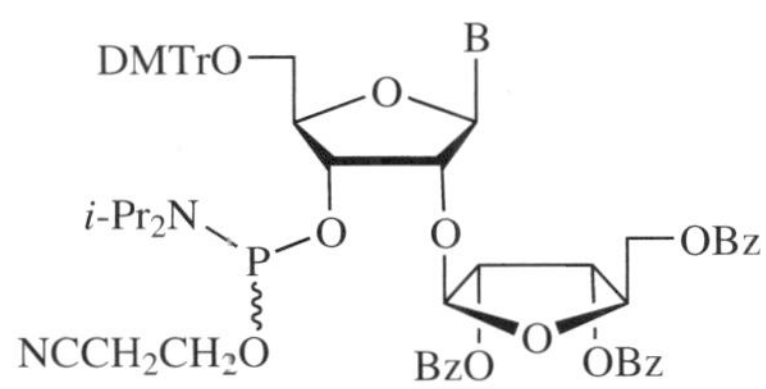

49a, B = 6-*N*-benzoyladenin-9-yl
49b, B = 4-*N*-benzoylcytosin-1-yl
49c, B = 2-*N*-isobutyrylguanin-9-yl
49d, B = uracil-1-yl

The synthesis of 2-amino-2-deoxyarabinonucleoside phosphoramidites (**51a–51e**) has been accomplished in several steps from native nucleosides. The incorporation of **51a–51e** into oligonucleotides is anticipated to impart them with better stability against exonucleases and improved hybridization properties.[44]

The enzymatic stability and binding properties of the oligodeoxyribonucleotide analogs (17-mers) that have been prepared via the phosphoramidite derivative of 1-[2,3-dideoxy-5-*O*-(4,4′-dimethoxytrityl)-3-*C*-hydroxymethyl-β-D-*threo*-pentofuranosyl]thymine (**52**) have been reported.[45] Specifically, an oligomer having three modifications in the middle of its sequence forms a duplex with its DNA complement that is less thermostable ($\Delta T_m = -3.0\,^{\circ}$C/modification) than the corresponding unmodified DNA duplex. In addition, end-modified oligonucleotides exhibited a 10-fold increase in stability to snake venom phosphodiesterase.[45] The phosphoramidite **52** was prepared in four steps from 5′-*O*-(4,4′-dimethoxytrityl) thymidine and its coupling efficiency was greater than 98%.[45]

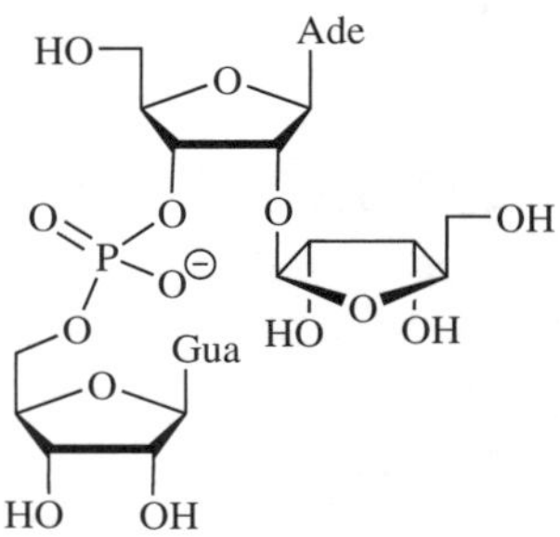

50, Gua = guanin-9-yl

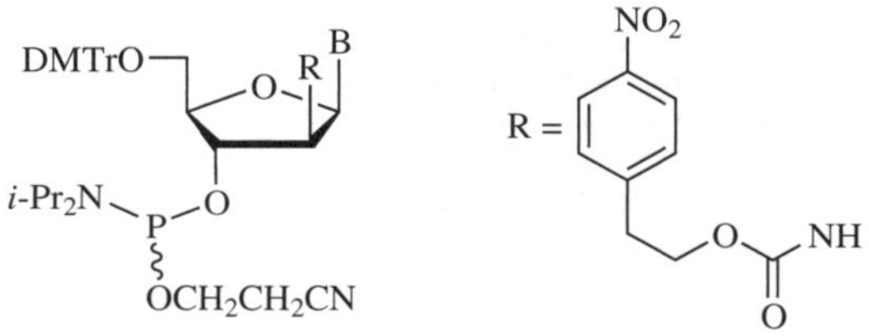

51a, B = uracil-1-yl
51b, B = thymin-1-yl
51c, B = N^4-[2-(4-nitrophenyl)ethoxycarbonyl]cytosin-1-yl
51d, B = N^6-[2-(4-nitrophenyl)ethoxycarbonyl]adenin-9-yl
51e, B = N^2-[2-(4-nitrophenyl)ethoxycarbonyl]-O^6-[2-(4-nitrophenyl)ethyl]guanin-9-yl

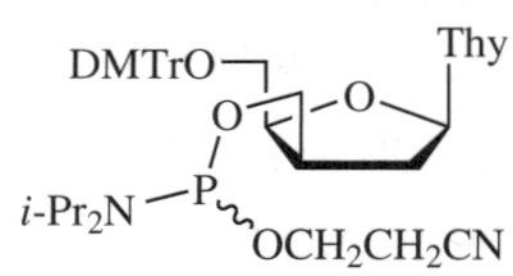

52

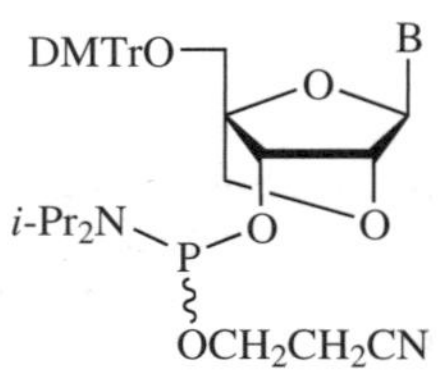

53, B = thymin-1-yl
54, B = H

Since the binding affinity of natural pyrimidine oligonucleotides towards dsDNA is relatively low under physiological conditions, the development of novel DNA and RNA analogs to foster the practical use of antigene technologies has become the object of extensive research efforts. In spite of their importance, there is only limited information on the triplex-forming properties of conformationally restricted or locked oligonucleotides. To better understand the triplex-forming ability and sequence specificity of oligonucleotides containing only one 2′-*O*,4′-*C*-methylene-bridged nucleoside, the phosphoramidites **53** and **54** have been synthesized and inserted in the triplex-forming oligonucleotide 5′-d(TTTTTCTXTCTCTCT).[46]

The integration of **53** into the triplex-forming oligonucleotide resulted in a significant increase in the triplex stability ($\Delta T_m = +10\,^{\circ}$C) when compared with that of the unmodified triplex. Conversely, the incorporation of **54** into the same oligonucleotide led to a large decrease in the triplex stability ($\Delta T_m = -37\,^{\circ}$C) to commensurate with mismatched triplex-forming oligonucleotide.[46] Thus, incorporation of 2′-*O*,4′-*C*-methylene-bridged nucleosides into triplex-forming oligonucleotides is a promising approach in the production of functional antigene biomolecules.

The potential therapeutic significance of triple helix-forming oligonucleotides, as inhibitors of DNA transcription, provides the impetus to design oligonucleotides that should target duplex sites other than just purine-rich ones. Along these lines, the deoxyribonucleoside phosphoramidites **55a**, **55b** and **56a**, **56b** were produced, and were found stable to the conditions used in the solid-phase oligonucleotide synthesis and oligonucleotide deprotection. A series of triplex-forming oligonucleotides with stretched phosphodiester linkages was assembled with stepwise phosphoramidite coupling efficiencies ranging from 96% to 98%.[47]

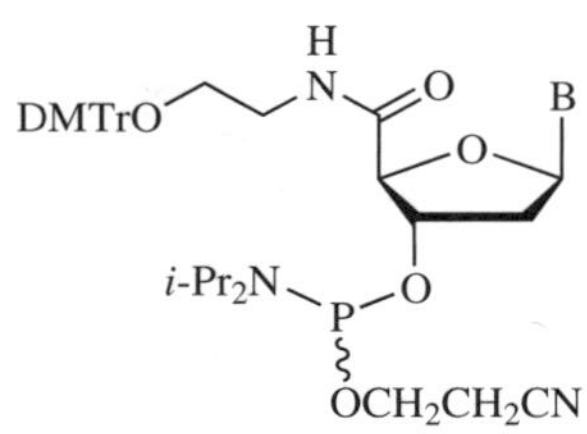

55

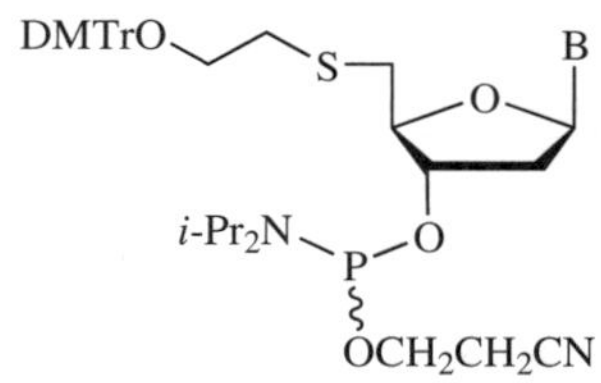

56

55a, **56a**; B = thymin-1-yl
55b, **56b**; B = 2-*N*-isobutyrylguanin-9-yl

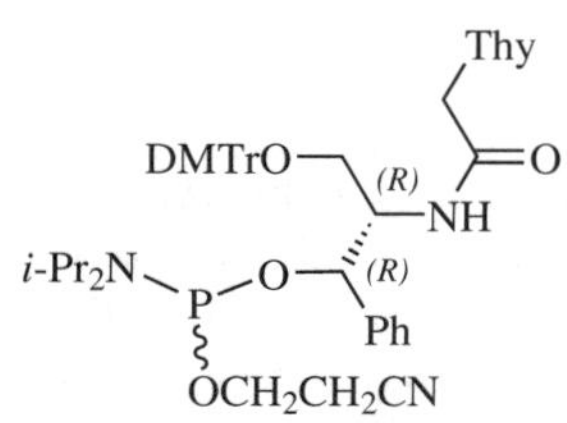

57

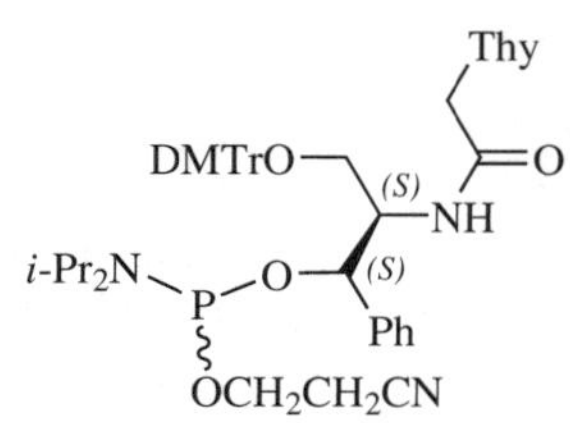

58

In a quest for modified oligonucleotides with superior duplex and triplex stabilities, the acyclic phosphoramidites **57** and **58** were incorporated at predetermined sites into the oligonucleotides.[48] These phosphoramidites were prepared in four steps from (1*R*,2*R*)- and (1*S*,2*S*)-2-amino-1-phenyl-1,3-propanediol. Triplex-forming oligonucleotides (18-mers) harboring either a (*R*,*R*)- or (*S*,*S*)-modification at the 3′-terminus generated more stable triplexes than those having the modification at the 5′-terminus. However, these modifications destabilized triplex formation by 3–8 °C relative to the unmodified DNA triplexes. Increasing the number of modifications toward the 3′-terminus further destabilized triplex formation ($\Delta T_m = -13\,°C$ to $-17\,°C$). The most destabilizing triplex-forming oligonucleotides are the ones carrying a centrally located modification ($\Delta T_m = -17\,°C$ to $-22\,°C$). Interestingly, triplex-forming oligonucleotides with (*R*,*R*)-modifications produced less stable triplexes than those with (*S*,*S*)-modifications.[48]

Further studies demonstrate that replacing the phenyl group in **57** and **58** with a *p*-nitrophenyl or a *p*-(benzoylamido)phenyl group slightly decreased the stability of the corresponding DNA duplexes and triplexes, but provided opportunities for fluorescent labeling in the case of the *p*-amidophenyl substitution.[49] A finding of importance is that the placement of these chiral acyclic nucleosides at the 3′-end of oligonucleotides provided resistance against snake venom phosphodiesterase that is at least 30-fold greater than what was measured with unmodified DNA oligonucleotides.[49]

Acyclic analogs of nucleosides are of interest in terms of their chemical and pharmacological properties. While a number of acyclic nucleoside derivatives are effective antiviral agents, the synthesis of isosteric glyceronucleosides is rather aimed at the discovery of potential antisense or antigene therapeutics with improved stability to nucleases and binding affinity toward complementary DNA or RNA strands. Acyclic nucleoside phosphoramidites in **59** and **60** have been synthesized using the methyl ester of bis-hydroxymethylphosphinic acid and N^1-(2-hydroxyethyl)thymine or N^1-(2-aminoethyl)thymine as starting materials, and 2-cyanoethyl tetraisopropylphosphorodiamidite as the phosphinylating reagent.[50]

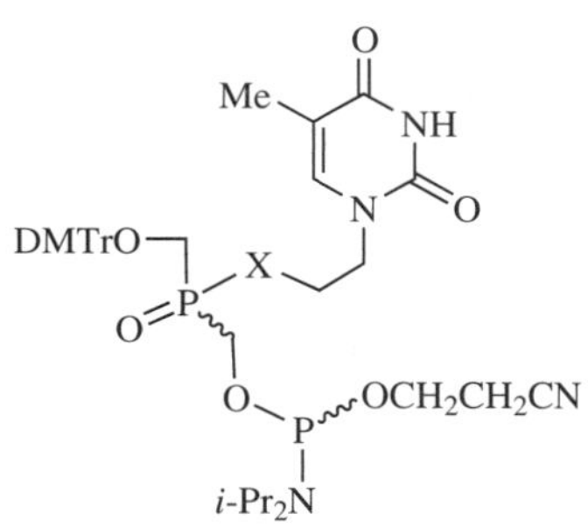

59, X = O
60, X = NH

The phosphoramidite **59** was integrated into the hexanucleotide d(TTTTTT), whereas the phosphoramidite **60** was incorporated into the trinucleotide d(ATC) via automated solid-phase synthesis.[50] The coupling time for **59** and **60** was extended to 600 s to provide a coupling efficiency of 95–97%. While the trinucleotide generated from **60** was stable to the basic conditions required for oligonucleotide deprotection, the hexanucleotide prepared from **59** was not. The oligothymidylate was isolated only when oxalyl-long chain alkylamine controlled-pore glass was used as the solid support, since the release of the hexanucleotide from this support and removal of the 2-cyanoethyl phosphate protecting groups could be achieved under very mild conditions using 1% aqueous triethylamine for 10 min at ambient temperature.[50] Application of **59** and **60** to the synthesis of longer oligonucleotides is ongoing along with the studies aimed at determining the hybridization properties and nuclease resistance of these oligonucleotide analogs.

Nucleoproteins are formed during replication of the genetic material of a number of viruses. In these DNA- or RNA-protein complexes, a covalent phosphodiester linkage is established between the 5′-OH of a nucleic acid and the hydroxyl group of a serine, threonine, or tyrosine residue of a protein. To facilitate structural studies or biomedical applications of nucleopeptides, the deoxyribonucleoside phosphoramidites in **61–63** have been prepared from suitably protected nucleosides, 2-cyanoethyl tetraisopropylphosphorodiamidite, and 1*H*-tetrazole.[51]

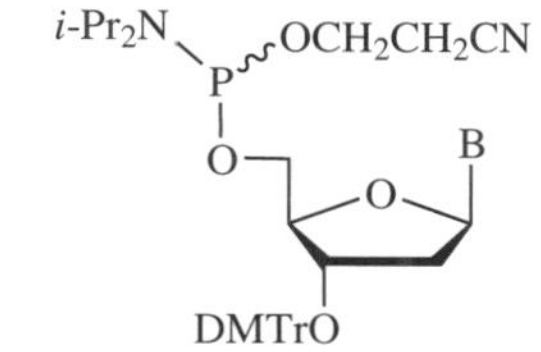

61, B = thymin-1-yl
62, B = 4-*N*-isobutyrylcytosin-1-yl
63, B = 6-*N*-(dimethylaminomethylene)adenin-9-yl

These phosphoramidites are used to initiate DNA synthesis from the serinyl hydroxy function of a resin-bound peptide upon activation with 5-(*o*-nitrophenyl)-1*H*-tetrazole. Average coupling yields were 95%. Following mild basic deprotection and purification by anion-exchange chromatography, the nucleopeptide Phac-Ser(p5′-CATCAT)-Gly-Asp-OH was isolated in 33% yield.[51]

Alternatively, the deoxyribonucleoside phosphoramidite **64**, which was obtained in four steps from thymidine, was instrumental in the assembly of peptide-DNA hybrids by generating a terminal 5′-amino group on the protected DNA oligonucleotides still linked to the solid support.[52]

The commercially available *N*-α-Fmoc-protected amino acids, uronium activating agent, hydroxybenzotriazole, and Hünig's base were used to initiate peptide synthesis from the terminal 5′-aminated thymidine of solid-phase-linked oligonucleotides. While the incorporation of the first amino acid was achieved with a coupling efficiency of 50–90%, much higher coupling yields (>90%) were obtained for subsequent amino acid additions.[52] This synthetic approach to nucleopeptides may likely lead to the identification of antisense inhibitors with improved bioavailability and target selectivity.

The following section of this paper will address the usefulness of 2-cyanoethyl tetraisopropylphosphorodiamidite in the preparation of dinucleoside phosphoramidites joined through modified internucleosidic links.

Phosphinylation of Dinucleosides Coupled Through Modified Internucleoside Linkages. Oligonucleotides with modified internucleoside phosphate linkages have recently emerged as an antisense strategy to regulate cellular and viral gene expression. In an attempt to overcome the instability of unmodified DNA oligonucleotides to exo- and endonucleases, dimers in which the natural phosphodiester linkage is replaced with amides were prepared for incorporation into oligonucleotides. Specifically, the dinucleoside phosphoramidites in **65–71** were synthesized and inserted in oligonucleotides to assess their hybridization properties with complementary RNA strands.[53–55]

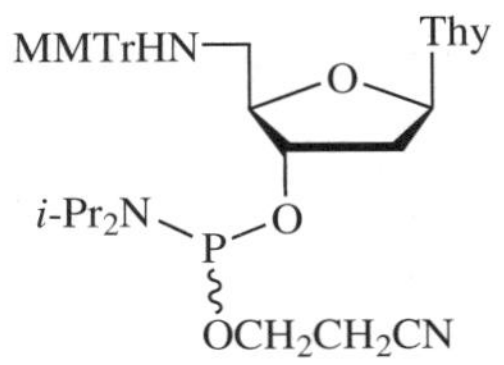

64

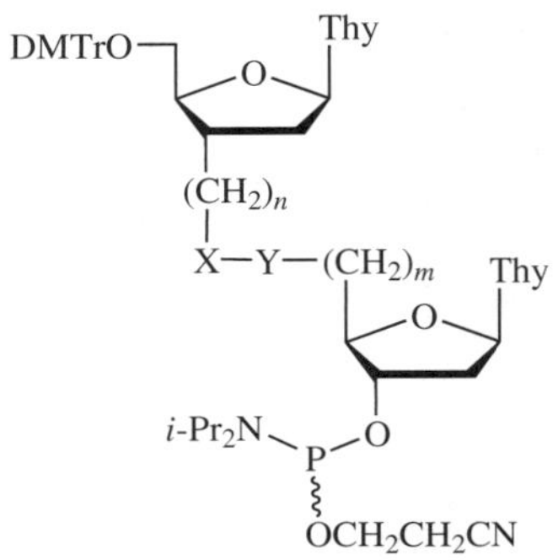

65; $n = 1$, X = CO, Y = NH, $m = 1$
66; $n = 1$, X = NH, Y = CO, $m = 1$
67; $n = 0$, X = NH, Y = CO, $m = 2$
68; $n = 2$, X = NH, Y = CO, $m = 0$
69; $n = 2$, X = CO, Y = NH, $m = 1$
70; $n = 1$, X = NH, Y = CO, $m = 0$
71; $n = 0$, X = CO, Y = NH, $m = 1$

When compared with unmodified oligonucleotides, oligomers constructed from the incorporation of **65** or **66** exhibit an average ΔT_m/modification of +0.4 °C or 0 °C, respectively. Insertion of **69** in oligonucleotides is better accommodated in a duplex structure (ΔT_m/modification = −1.1 °C) than those oligonucleotides prepared from the integration of **70** ($\Delta T_m = -1.8$ °C/modification) or **71** ($\Delta T_m = -5.0$ °C/modification), which have shorter backbones.[54] The single or multiple incorporation(s) of **67** or **68** into oligonucleotides also lead(s) to less thermostable hybrids ($\Delta T_m =$ −2.2 °C to −3.9 °C/modification[55] or −0.5 °C to −2.8 °C/modification,[53] respectively) with complementary RNA oligonucleotides than those obtained with unmodified oligonucleotides. Noteworthy, however, is the addition of **68** to the 3′-terminus of an oligonucleotide, which results in a 5- to 6-fold increase in stability against the nucleolytic activity of 10% fetal calf serum over that of an unmodified oligonucleotide.[53]

A thymidine dinucleoside phosphoramidite (**72**), bearing an internucleosidic carbamate linkage, has been prepared for single or multiple insertion(s) in oligonucleotides.[56]

The hybridization of these modified oligonucleotides with complementary RNA oligonucleotides generated hybrids displaying ΔT_m's ranging from −1.1 °C/modification, for a single incorporation of **72**, to −3.9 °C/modification for multiple incorporations of the dimeric phosphoramidite. Oligonucleotides having a carbamate modification located at the 3′-terminus show a 4-fold increase in resistance to hydrolysis catalyzed by 10% fetal calf serum relative to that determined for the parent unmodified oligonucleotides.[56]

The incorporation of dimeric phosphoramidite **73** functionalized with an internucleosidic urea linkage into oligonucleotides resulted in the formation of hybrids with complementary RNA strands that are significantly less thermostable ($\Delta T_m = -2.9$ °C to −3.4 °C/modification) than identical unmodified duplexes.[57]

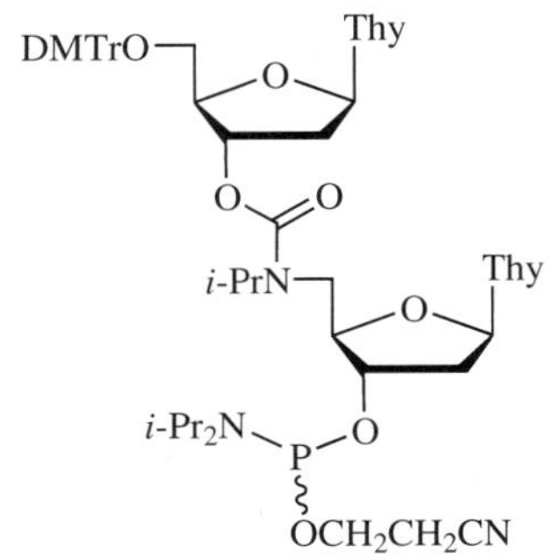

72

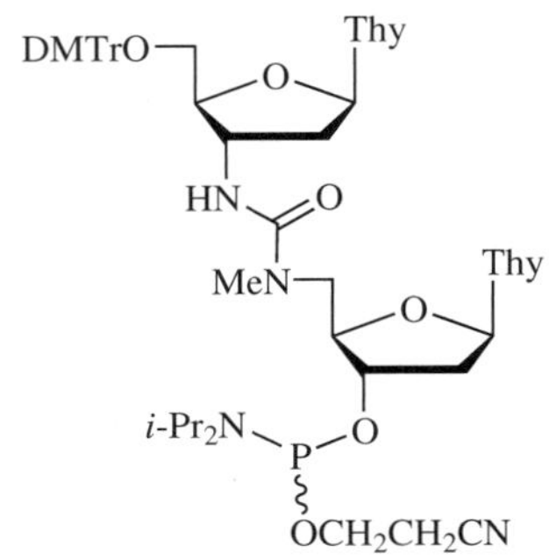

73

However, oligonucleotides carrying an internucleosidic urea function near the 3′-terminus exhibited a 15-fold increase in stability against the hydrolytic activity of 10% fetal calf serum relative to the unmodified oligonucleotides.[57] Since the replacement of native phosphodiester linkages with either carbamates or ureas significantly affects the binding affinities of the modified oligonucleotides toward RNA, these internucleosidic linkages are therefore unsuitable for antisense applications.[56,57]

Following synthesis, the dinucleoside carboxamide phosphoramidite **74** has been integrated into oligonucleotides with a coupling efficiency similar to that of the commercial phosphoramidites (∼99%).[58]

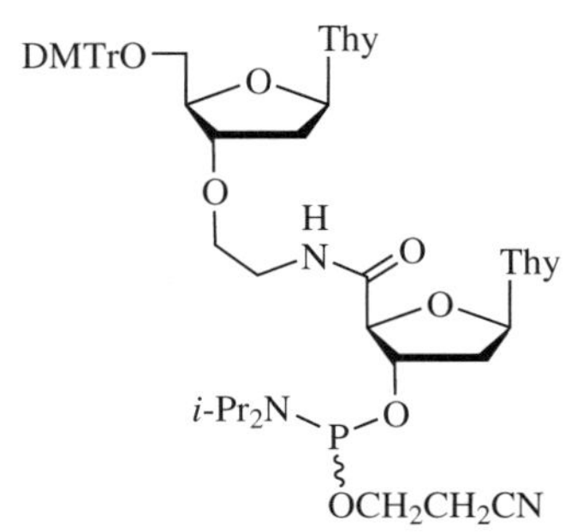

74

Modified oligonucleotides having one or two modifications in the middle of a 17-mer or a 22-mer show only moderate reduction of duplex stability ($\Delta T_m \sim -2$ °C/modification) when hybridized with complementary DNA strands.[58] The use of these modified oligonucleotides in the primer extension studies causes specific termination of DNA polymerization catalyzed by *E. coli* DNA polymerase I, T4 DNA polymerase, and AMV reverse transcriptase or Sequenase.[58] Incorporation of **74** into oligonucleotides causes complete inhibition of exonuclease III near the modification site, and the treatment of a modified DNA duplex with S1 nuclease does not show detectable internal cutting sites, which would have been indicative of looped-out or bubble-type structures.[58]

The methylene(methylimino) (MMI) modification for antisense oligonucleotides entails replacement of the *C*-3′ oxygen atom and phosphodiester function of natural DNA with a methylene and a *N*-methylhydroxylamino group, respectively. In order to evaluate this promising modification, a novel diribonucleoside-MMI phosphoramidite (**75**) has been constructed in several steps from 1,2-*O*-isopropylidene-D-xylose.[59]

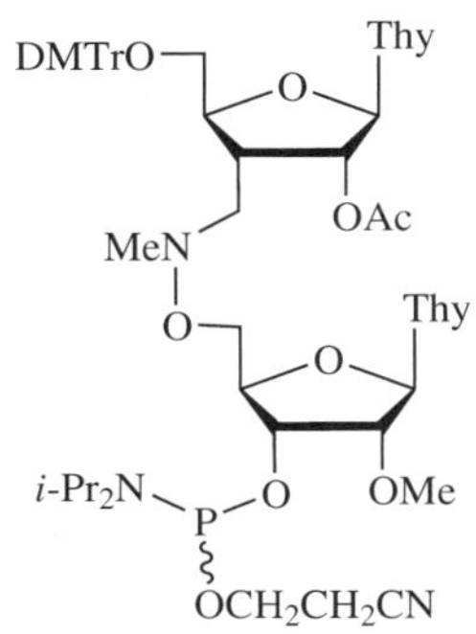

75

The phosphoramidite **75** was incorporated at selected sites into DNA sequences; an extended coupling reaction time (15 min) was necessary to ensure high coupling efficiency of the phosphoramidite. The modified oligonucleotides were mixed with complementary RNA oligonucleotides in order to perform thermostability studies on the resulting heteroduplexes. These studies showed a substantial increase in the stability of each duplex ($\Delta T_m = +2.3\,^{\circ}C$ to $+3.8\,^{\circ}C$/modification) relative to that measured for the corresponding unmodified heteroduplexes.[59] While MMI linkages provide considerable nuclease stability to antisense oligonucleotides, the pharmacology of these modified oligonucleotides is currently being investigated.

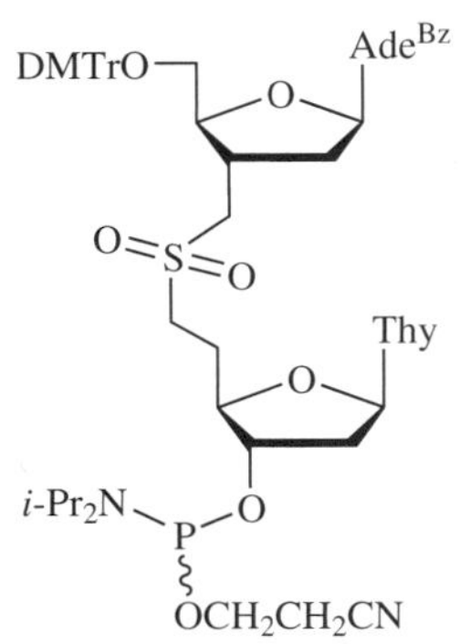

76

Many proteins (such as restriction enzyme), transcription factors, and other regulatory elements bind DNA and distort it by bending and/or unwinding. DNA oligonucleotide analogs that are predistorted by modified internucleosidic linkages should therefore act as general tools for blocking a variety of protein-nucleic acid interactions. To test this hypothesis, the dinucleoside phosphoramidite **76** has been prepared and inserted, as indicated, in oligonucleotides d(ACCAGAATTCGGATCCAGA$_{SO2}$TATCG CCA) and d(ACCAGAATTCGGATCCAGATA$_{SO2}$TCGCCA) via standard automated solid-phase oligonucleotide synthesis.[60] The underlined sequences correspond to the *Eco*RV recognition site.

It was found that DNA duplexes containing the modified oligonucleotides inhibit cleavage of a self-complementary *Eco*RV substrate. The DNA duplex having the recognition sequence GATA$_{SO2}$TC is the most potent competitive inhibitor of *Eco*RV with an inhibition constant of 25 pM.[60] The tight binding of this duplex to *Eco*RV suggests that the distortion induced by the dimethylene sulfone modification allows the modified duplex to fit in the active site of the enzyme without consuming binding energy to distort the structure from its ground-state conformation.[60]

Dinucleosides functionalized with internucleosidic sulfamide and sulfamate linkages have been synthesized and converted to the phosphoramidites **77** and **78**, respectively, for potential incorporation into oligonucleotides.[61,62]

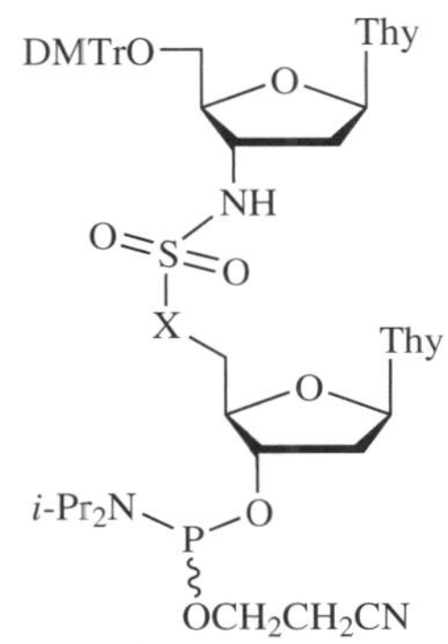

77, X = NH
78, X = O

The conformational analysis of these modifications using NMR spectroscopy shows a shift from a predominantly southern (*C2′*-endo) to northern (*C3′*-endo) ribose conformation. The effect of temperature changes on ribose conformation indicates that dinucleosides with an internucleosidic 3′-*N*-sulfamate linkage have a greater propensity to base-stack intramolecularly.[61] Since an increase in northern ribose conformation and intramolecular base-stacking are known to contribute to more stable heteroduplexes with RNA, the incorporation of **77** and **78** into oligonucleotides may lead to promising candidates for therapeutic applications. The integration of **77** or **78** into d(GCGT$_{10}$GCG) at one and up to five centrally located sites was achieved using solid-phase DNA synthesis protocols. A 15-min coupling time was, however, required to produce a coupling efficiency better than 98%.[62] Hybridization of deprotected oligonucleotides with complementary DNA or RNA strands reveals that the sulfamide modification destabilizes the resulting duplexes (ΔT_m of ca. $-3.0\,^{\circ}C$ per modification at low salt concentration). Conversely, 3′-*N*-sulfamate-modified oligonucleotides have little effect on duplex stability and even stabilize DNA duplexes at low salt concentrations.[62] These findings suggest that 3′-*N*-sulfamate-modified oligonucleotides are worthy of further development for applications as antisense or antigene therapeutic agents.

The synthesis of the dinucleoside phosphoramidites **79a**, **79b** led to the preparation of α-oligonucleotides.[63] Specifically, the modified α-oligonucleotide 5′-d(TCTTAA*CCCACA), which was prepared by inserting **79b** or **79a**, formed a complex with a complementary DNA oligonucleotide which is more or less thermostable (ΔT_m = + 1.5 °C or −0.5 °C, respectively) than that obtained with an unmodified α-oligonucleotide. However, the *P*-chirality of either modified α-oligonuclotide had a much lesser differential effect on the affinity of each oligomer for its RNA target. On the basis of these observations, stereoselective synthesis of **79a** or **79b** is not necessary for efficient binding of α-oligonuclotides carrying these modifications with complementary RNAs.

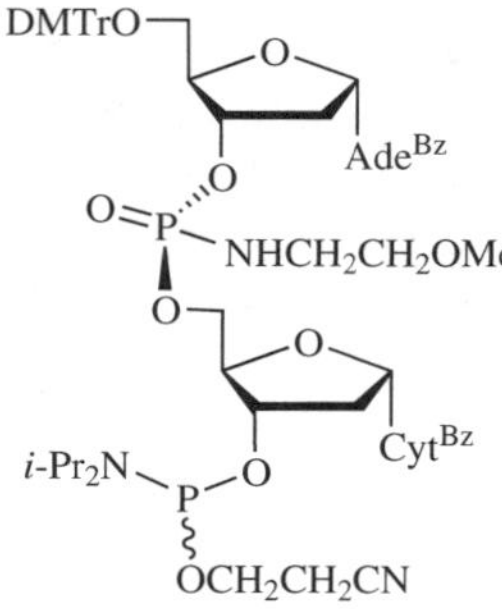

79a, Cyt = cytosin-1-yl

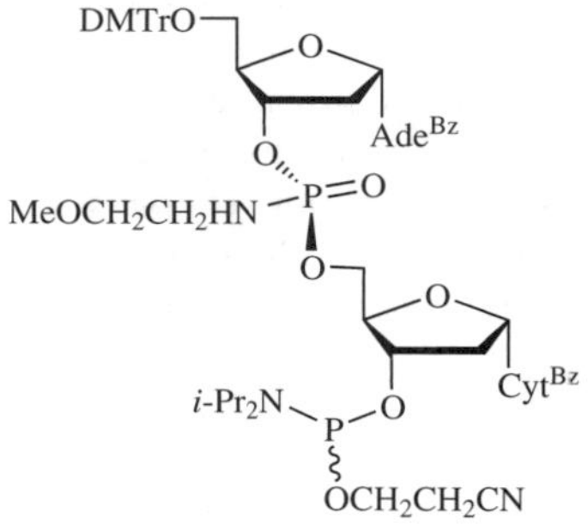

79b

Another class of dinucleoside phosphoramidite (**80**) for single or multiple incorporation(s) into dodecathymidylates has been reported in the literature.[64] The coupling efficiency of **80** was similar to that of regular deoxyribonucleoside phosphoramidites.

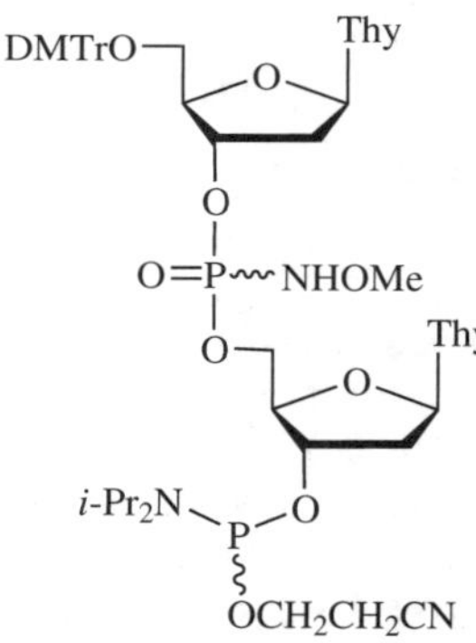

80

Given the sensitivity of phosphoramidate linkages to concentrated ammonium hydroxide, an oxalyl linker bridging the oligonucleotide chain to the controlled-pore glass (CPG) support was used instead of the conventional succinyl linker. This oxalyl-functionalized CPG support allowed the release of oligothymidylates under very mild basic conditions.[64] The phosphoramidate-modified oligothymidylates formed complexes with poly dA and poly rA which were less thermostable (ΔT_m = −1.5 °C/modification and −3.4 °C/modification, respectively) than those formed with an unmodified oligothymidylate. The lower thermostability of poly rA hybrids is indicative of an appreciable distortion of the DNA–RNA complexes relative to that of an unmodified heteroduplex. It is, however, relevant to note that an oligothymidylate carrying five phosphoramidate modifications is 200-, 70-, and 800-fold more resistant to S1 nuclease, snake venom phosphodiesterase, and calf spleen phosphodiesterase, respectively, than the parent unmodified oligonucleotide.[64] Because of the relatively low affinity of phosphoramidate-modified oligothymidylates for poly rA, it is likely that this class of phosphoramidate-modified oligonucleotides will be of limited value in therapeutic applications.

The investigations related to the replacement of both bridging oxygens of a DNA dinucleotide with methylene groups have been reported.[65] The details of these studies involve the preparation of dinucleoside phosphoramidite **81** and its insertion at the predetermined sites in oligonucleotides.

The hybridization experiments demonstrate that ethylphosphinate-modified oligonucleotides exhibit better binding affinity toward RNA (ΔT_m = −2.2 °C/modification) than DNA (ΔT_m = −5.5 °C/modification), and that diastereomeric ethylphosphinate modifications are not more destabilizing than achiral phosphinate modifications.[65] Furthermore, the stability of oligonucleotides bearing two 3′-terminal ethylphosphinates to hydrolysis catalyzed by 10% fetal calf serum is 48-fold greater than that of wild-type DNA.[65] Given these properties, ethylphosphinate-modified oligonucleotides have potential as antisense oligonucleotides. In this context, synthesis of the dinucleoside phosphoramidite **82** from a diastereomeric thymidyl-3′-*H*-phosphinate intermediate and 5′-azidothymidine has been documented.[66]

The R_P-ethylphosphonamidate modification created by the incorporation of **82** into oligonucleotides was found to enhance binding with complementary RNAs to the extent of +0.9 °C per modification. To further investigate the properties of this duplex-stabilizing modification, the phosphoramidite **82** will be incorporated into a broader array of DNA sequences beyond those containing adjacent thymidines.[66]

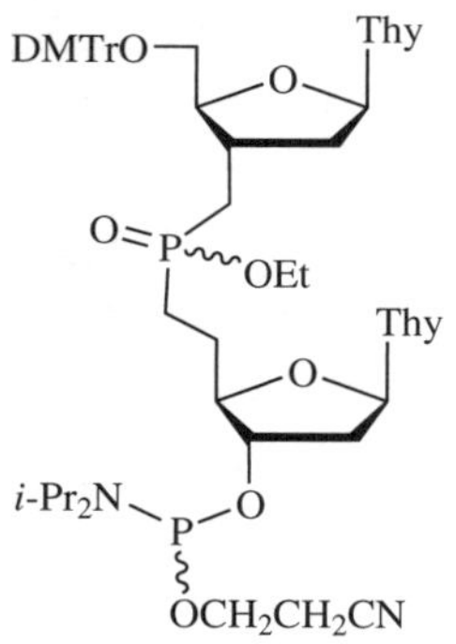

81

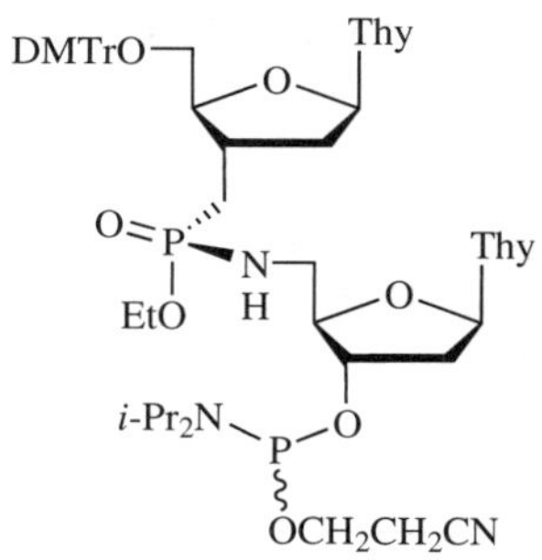

82

The internucleosidic α-hydroxybenzylphosphonate group has been identified as a lipophilic pseudo-phosphate modification and has been proposed for the synthesis of pro-oligonucleotides. Thus, rearrangement and/or hydrolytic cleavage of the α-hydroxybenzylphosphonate groups of pro-oligonucleotides should generate, when needed, oligonucleotides with native internucleosidic phosphodiester linkages. In order to test this pro-oligonucleotide strategy, the dinucleoside phosphoramidite **83** has been prepared in several steps starting with a 5′-*O*-protected thymidyl-3′-*H*-phosphonate, and used in the solid-phase synthesis of a 3′-modified dT_{15} oligonucleotide.[67]

The stability of the modified oligonucleotides against 3′-exonucleases, such as snake venom phosphodiesterase, is enhanced by a factor of ~7.[67] Strong electron-withdrawing groups in the aromatic ring of the internucleosidic phosphonate group were shown to induce rearrangement of the phosphonate to a benzyl phosphotriester, whereas electron-donating groups triggered a direct cleavage reaction to generate the desired phosphodiester function.[67] These findings underscore the development of lipophilic pro-oligonucleotides as means to facilitate the delivery of therapeutic oligonucleotide drugs across the cell membranes.

The phosphorothioated dinucleoside phosphoramidite **84** has been used in the solid-phase synthesis of 'perthioated' oligodeoxyribonucleotides which were the first generation of antisense drugs to reach the clinics.[68]

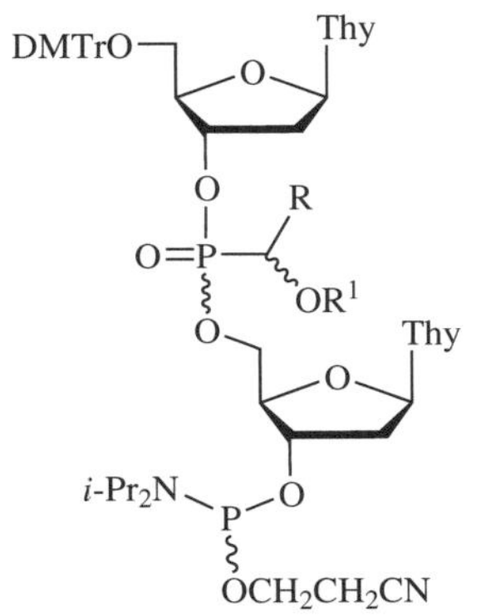

83, R = 2-nitrophenyl or phenyl;
R^1 = trihexylsilyl

This approach to oligonucleotide analogs syntheses has been proposed to reduce the content of both (n-1)-mers and phosphate linkages, which have been encountered when using monomeric nucleoside phosphoramidites in the synthesis of phosphorothioated oligonucleotides. The dinucleoside phosphoramidite **84** is prepared by the reaction of 3′-*O*-levulinyl thymidine or

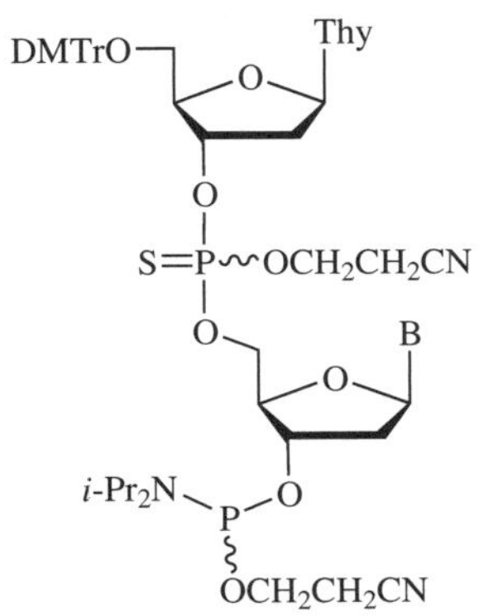

84, B = Thy or Cyt^{Bz}

N^4-benzoyl-3′-*O*-levulinyl-2′-deoxycytidine with commercial thymidine CED-phosphoramidite and 1*H*-tetrazole, followed by sulfurization of the newly formed phosphite triester with 3*H*-1,2-benzodithiol-3-one 1,1-dioxide. Upon hydrazinolysis of the 3′-*O*-levulinyl group, the phosphorothioated dinucleotide is phosphinylated with 2-cyanoethyl tetraisopropylphosphorodiamidite and 1*H*-tetrazole to give **84**. When applied to the solid-phase synthesis of perthioated dT_{19} or $d(TC)_9T$, the use of **84** is as efficient as monomeric phosphoramidites and further results in a better than 70% reduction of the (n-1)-mer population and a ~50% reduction in the content of phosphate linkages.[68] However, a combinatorial library of up to 16 phosphorothioated dinucleoside phosphoramidites may be required to synthesize any DNA sequence for therapeutic applications. The question of whether the benefits realized from the use of phosphorothioated dinucleoside phosphoramidites outweigh the labor necessary to construct a combinatorial library of dimeric phosphoramidites remains a matter of debate.

One major application of 2-cyanoethyl tetraisopropylphosphorodiamidite is the phosphinylation of functional groups for incorporation into oligonucleotides. This application will be emphasized in the following section of this article.

Phosphinylation of Functional Groups Including Ligands and Reporter Groups. Given the potential of synthetic oligonucleotides as therapeutic agents, the development of methods that would facilitate the purification of these biomolecules is of interest. The most commonly used method for the purification of oligonucleotides consists of reverse-phase (RP) HPLC separation of full-length 5′-dimethoxytrityl-containing oligonucleotides from all non-lipophilic components. This method has, however, two disadvantages: (i) it requires a post-purification acidic detritylation step and (ii) loss of some full-length oligonucleotides occurs during the purification process due to unwanted detritylation. To circumvent these shortcomings, a lipophilic phosphoramidite **85** was obtained from the reaction of n-octanol with 2-cyanoethyl tetraisopropylphosphorodiamidite and diisopropylammonium tetrazolide, and was used as a 'capping' reagent during solid-phase oligonucleotide synthesis.[69] The lipophilicity, thus imparted to the capped failure sequences, greatly facilitates RP-HPLC separation of these truncated oligomers from 'trityl-off' full-length oligonucleotides, and improves recovery of the desired oligonucleotide products.[69]

A terminal thiophosphate group provides a convenient site for regiospecific conjugation of synthetic oligonucleotides to a variety of ligands carrying an electrophilic group. An attractive approach to terminal phosphorylation and thiophosphorylation of synthetic oligonucleotides entails the use of phosphoramidite **86**,[70] which was prepared in two steps from 2,2-diethoxycarbonyl-1,3-propanediol.[70] Addition of **86** to the 5′-terminus of oligonucleotides is achieved by following a solid-phase synthesis protocol that includes a sulfurization step. Oligonucleotides are subsequently deprotected and released from the support upon treatment with concentrated NH_4OH at 55 °C. The presence of a terminal phosphorothioate group in any oligonucleotides is confirmed by an *S*-alkylation reaction with fluorescent iodoacetamide derivatives.[70]

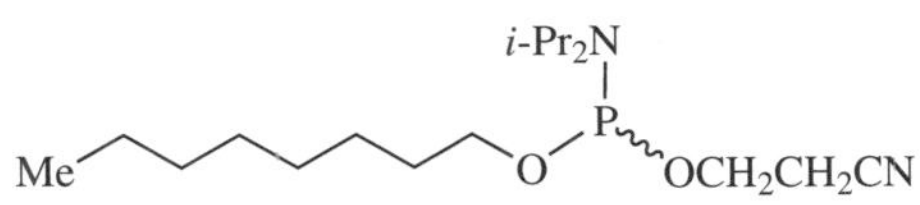

85

The potential of modified oligonucleotides in analytical and therapeutic applications has spurred the development of new methods for linking reporter groups to oligonucleotides. Noteworthy is the synthesis of the tritylated aminooxyphosphoramidite derivative **87** from 6-bromohexan-1-ol and 2-cyanoethyl tetraisopropylphosphorodiamidite for final phosphinylation.[71] Introduction of phosphoramidite **87** to the 5′-terminus of synthethic oligonucleotides proved very efficient. This functionalization method allows 'trityl-on' oligonucleotide purification, and subsequent mild acidic cleavage of the trityl group affords a reactive aminooxy-oligonucleotide without by-product formation. This modified oligonucleotide is rapidly and quantitatively derivatized with a reporter group bearing an aldehyde function.[71]

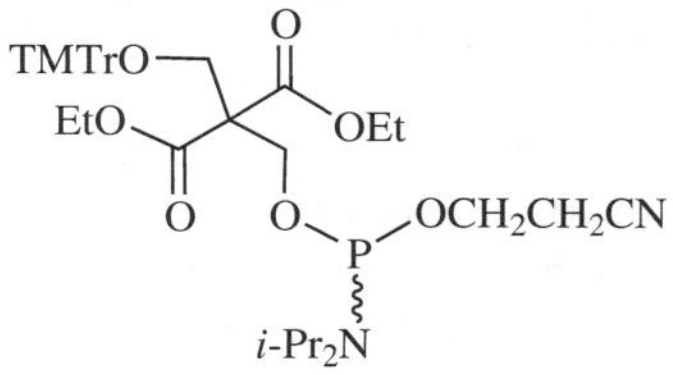

86, TMTr = 4,4',4''-trimethoxytrityl

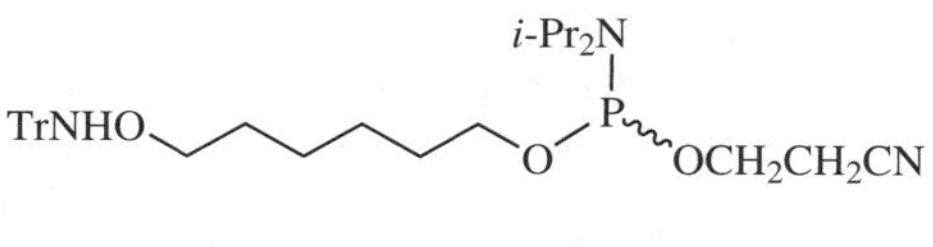

87, Tr = triphenylmethyl

A versatile method for labeling or chemically modifying oligonucleotides involves the phosphinylation of non-nucleosidic, chirally defined 2-deoxy-β-D-ribosyl derivatives with 2-cyanoethyl tetraisopropylphosphorodiamidite to generate the phosphoramidites shown in **88**. These phosphoramidites incorporate biotin, cholesterol, and protected amino groups into oligonucleotides via automated solid-phase methods.[72] Because of the 2-deoxyribosyl spacer, the enantiomeric, anomeric, conformational, and internucleotidic properties of the modified oligonucleotides are conserved when compared with those of the corresponding unmodified oligonucleotides.

Phosphinylation of 1-*O*-(4,4′dimethoxytrityl)-2-[*N*-(9-fluorenylmethoxycarbonyl)-4-aminobutyl]-1,3-propanediol, 1-*O*-(4,4′-dimethoxytrityl)-2-(*N*-biotinoyl-4-aminobutyl)-1,3-propanediol, 1-*O*-(4,4′-dimethoxytrityl)-2-[*N*-(dipivaloyl-5-fluoresceinthiocarbamoyl)-4-aminobutyl]-1,3-propanediol, and 1-*O*-(4,4-dimethoxytrityl)-2-[*N*-9-(6-chloro-2-methoxyacridinyl)aminobutyl]-1,3-propanediol with 2-cyanoethyl tetraisopropylphosphorodiamidite provides flexible phosphoramidite derivatives for multiple insertions of biotin, fluorescein, acridine, and aliphatic amine groups in oligonucleotides.[73] Biotinylated and fluorescently labeled oligonucleotides have found invaluable applications in molecular biology techniques such as DNA sequencing and the polymerase chain reaction (PCR).

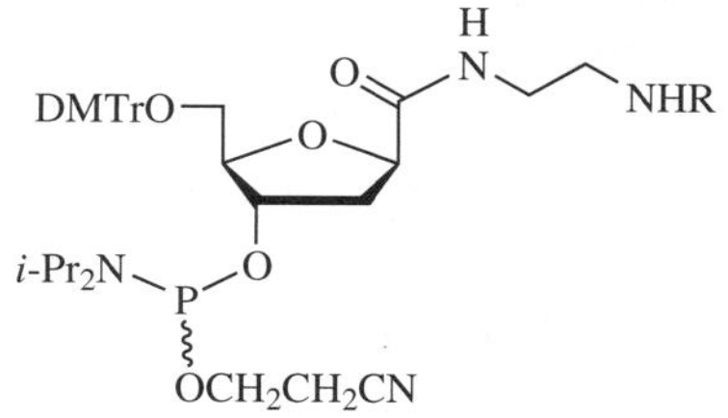

88, R = Fmoc, trifluoroacetyl, biotinyl, or cholesteryl

The Diels–Alder cycloaddition between a diene and a dienophile remains among the most useful carbon-carbon bond-forming reactions available to synthetic organic chemists. The reaction is also amenable to oligonucleotide bioconjugation as its efficiency compares favorably with that of established nucleic acid bioconjugation techniques. In this regard, the synthesis of phosphoramidites **89** and **90** has enabled the incorporation of diene modifications into synthetic oligonucleotides.[74] Optimal coupling yields were obtained when using 15 equiv of **89** or **90** over two successive 15-min coupling steps.[74] Clean and efficient Diels–Alder cycloadditions of diene-modified oligonucleotides with 10 equiv of biotin maleimide were optimally conducted at 40 °C and at pH 5.5. Under these conditions, biotinylation of oligonucleotides was complete within 30 min.[74] The Diels–Alder reaction is therefore commendable for the preparation of oligonucleotide conjugates under mild conditions in aqueous solution.

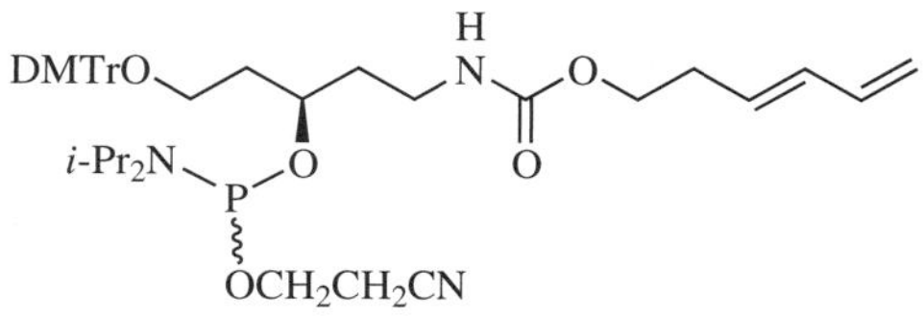

89

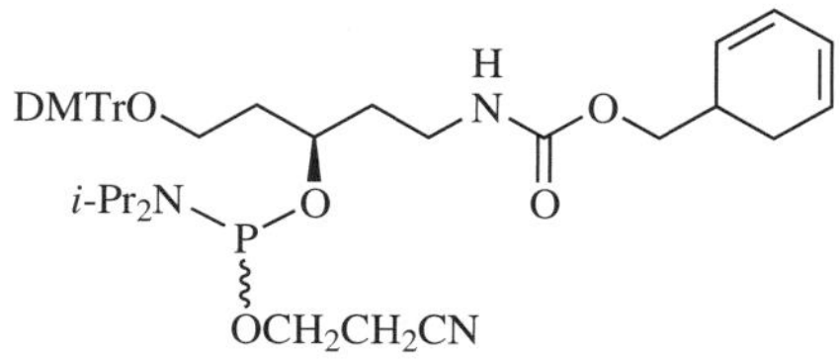

90

2-Cyanoethyl tetraisopropylphosphorodiamidite has also been instrumental in the preparation of non-nucleosidic phosphoramidites shown in **91** for oligonucleotide labeling.[75] Addition of **91** to the 5′-terminus of oligonucleotides, using solid-phase technology, led to polyaminated or polydansylated oligonucleotides along with their phosphorothioate analogs. Addition of up to six dansyl groups to the 5′-terminus of oligonucleotides had no or only minor effects on their hybridization properties.[75]

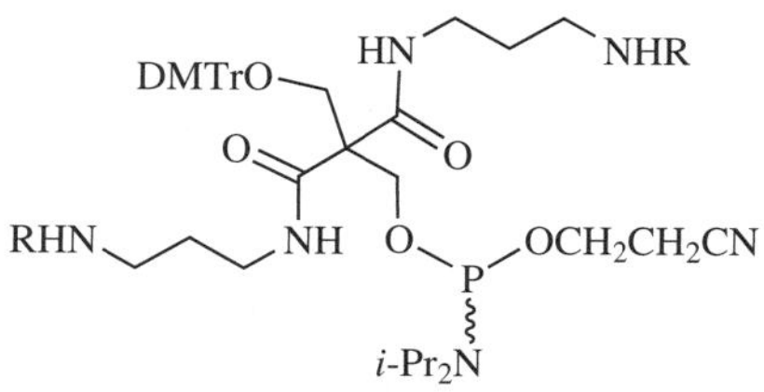

91, R = Fmoc, trifluoroacetyl, or dansyl

The condensation of the non-nucleosidic phosphoramidites **92** and **93** with the 5′-terminus of oligonucleotides or with the 1′-OH of any incorporated 3′-deoxypsiconucleoside residues is particularly useful for post-synthesis attachment of reporter groups[76] and ligands[77] to oligonucleotides. These phosphoramidites were prepared from the reaction of 2,2,2-trifluoroethyl glycolate with 2-cyanoethyl tetraisopropylphosphorodiamidite and 1*H*-tetrazole. Typically, when **92** is added to the 5′-terminus of synthetic oligonucleotides, subsequent reaction with aqueous diamine affords 5′-aminoalkylated oligonucleotides which may then be condensed with electrophiles, such as dabsyl chloride, to complete the labeling reaction.[76] Alternatively, addition of **93** to the selected oligonucleotidic sites followed by the reaction with 1-(3-aminopropyl)imidazole or histamine generated imidazole-tethered oligodeoxyribonucleotides, which have been tested for their ability to hydrolyze complementary RNA sequences in the presence of Zn^{+2} ions.[77]

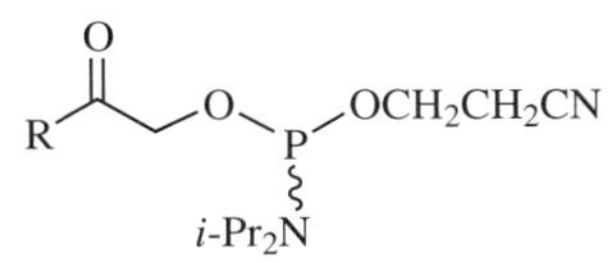

92, R = CF_3CH_2O
93, R = $PhCH_2S$

2-Cyanoethyl tetraisopropylphosphorodiamidite has similarly been applied to the synthesis of imidazolyl-linked phosphoramidite synthons towards the preparation of oligonucleotides functionalized with terminal imidazole groups for the cleavage of RNA in a sequence-specific manner.[78] Along these lines, the preparation of phosphoramidite **94** from benzyl 4-{*N*-[2-[(4,4′-dimethoxytrityloxy)methyl]-2-(hydroxymethyl)propionyl]amino}butylate and 2-cyanoethyl tetraisopropylphosphorodiamidite has been achieved, and **94** was successfully incorporated at predetermined sites into oligonucleotides with a coupling efficiency of 98%.[79]

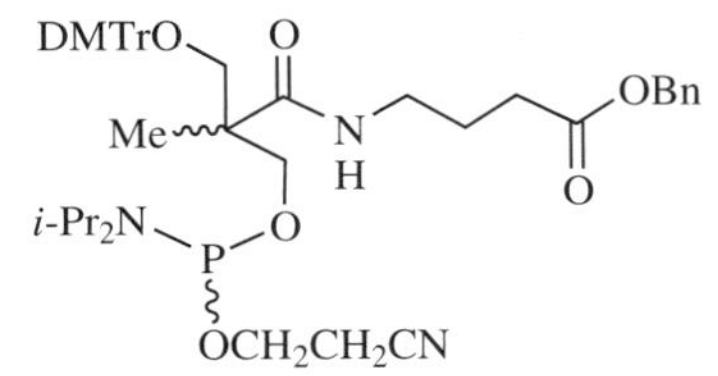

94

Aminolysis of the benzyl ester effected by triethylenetetramine gives oligoaminated oligonucleotides which after complete deprotection and hybridization with complementary RNA targets hydrolyze the RNA strands at precisely defined sites without affecting much the stability of the duplexes.[79] Such artificial cleaving enzymes may become valuable tools for various biomedical applications.

Structurally related to **94** is the azobenzene phosphoramidite derivative **95**, which is prepared from the phosphinylation of suitably protected *m*-amidoazobenzene alcohol with 2-cyanoethyl tetraisopropylphosphorodiamidite.[80] Incorporation of **95** into a synthetic oligonucleotide provided control over the formation and dissociation of its duplex with a complementary DNA strand via photoinduced *cis*→*trans* isomerization of the *m*-azobenzene group.

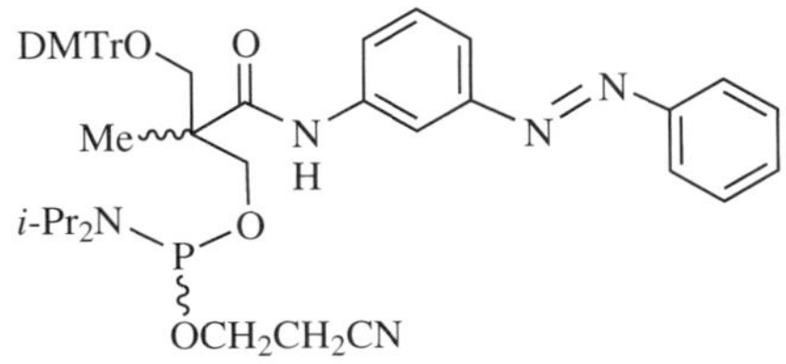

95

An isostere of **95**, the *p*-substituted azobenzene phosphoramidite, has also been integrated into an oligonucleotide and has been effective in controlling DNA duplex formation upon photoirradiation.[81] The *p*-substituted azobenzene modifications were, however, much more sensitive to the thermal *cis*→*trans* isomerization than the *m*-substituted azobenzene lesions.[80] Thus, the insertion of **95** in oligonucleotides is more appropriate for strictly photoregulating DNA duplex formation.

Immobilization of DNA on solid surfaces has wide application in areas such as diagnostics and biosensors. Covalent ligand immobilization via photoreactive species (photoprobes) offers the possibility of selective activation and high reactivity in an orientation-specific manner. In order to perform covalent anthraquinone-mediated DNA immobilization on surfaces such as polystyrene, polycarbonate, polypropylene, teflon, and silylated glass, the anthraquinone phosphoramidite **96** was prepared using 2-(hydroxymethyl)anthraquinone, 2-cyanoethyl tetraisopropylphosphorodiamidite, and 1*H*-tetrazole.[82]

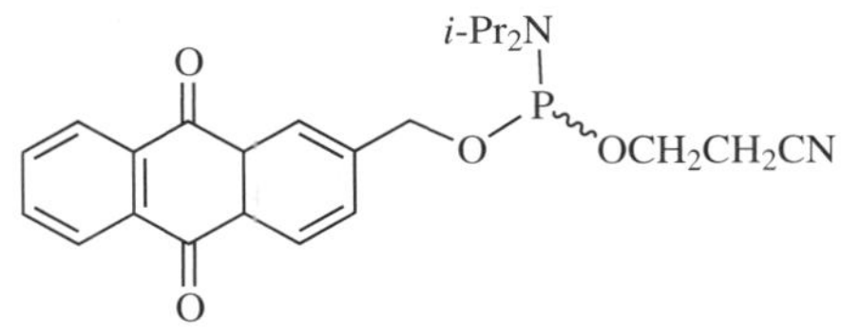

96

Addition of **96** to the 5′-terminus of solid-phase-linked oligonucleotides is accomplished using a coupling time of 5 min. Anthraquinone-conjugated oligonucleotides were then used as PCR primers and, following amplification, the amplicons were immobilized on a solid surface. However, when the primers are immobilized on a solid support prior to PCR amplification, solid-phase PCR can be performed to generate amplicons directly on the solid support.[82] The generality of this immobilization technique permits the immobilization of PCR primers on various polymers suitable for PCR platforms.

The vicinal diol-containing phosphoramidite, [5,6-bis(benzoyloxy)-1-hexyloxy](2-cyanoethoxy)-*N*,*N*-diisopropylaminophosphine, was obtained from the reaction of 5,6-di-*O*-benzoyl-1,5,6-hexanetriol with 2-cyanoethyl tetraisopropylphosphorodiamidite and diisopropylammonium tetrazolide, and used to functionalize the 5′-terminus of synthetic $d(CA)_{25}$.[83] Following oligonucleotide deprotection, the 5′-terminal vicinal diol is oxidized to an aldehyde and conjugated to keyhole limpet hemocyanin (KLH) to provide, upon reductive amination, an oligonucleotide-KLH conjugate as a model for the treatment of systemic lupus erythematosus.[83]

Along similar lines, the conjugation of synthetic oligonucleotides to bovine serum albumin (BSA) is achieved by first adding the tritylated non-nucleosidic phosphoramidite **97** to the 5′-terminus of oligonucleotides via solid-phase synthesis protocols.[84]

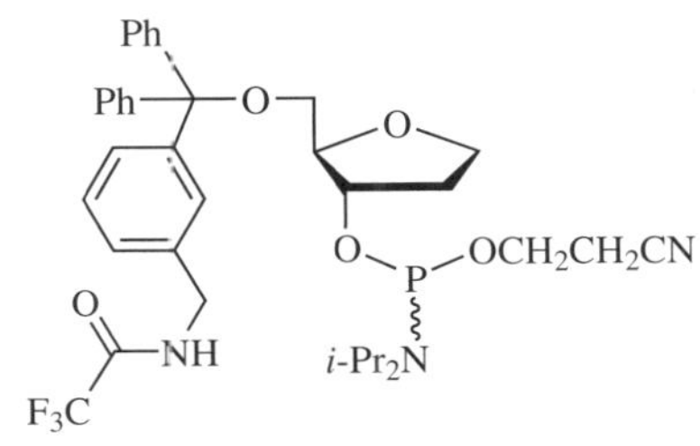

97

After oligonucleotide deprotection under basic conditions, the 5′-aminated trityl group is acylated by treatment with the

N-hydroxysuccinimidyl ester of *p*-benzoylbenzoic acid to generate a photoactivable trityl group and permit its photocrosslinking with BSA. When a slide projector is employed as an irradiation device, the BSA-oligonucleotide conjugate is produced to the extent of 40 mol %.[84] These findings pave the way to new approaches in cell-targeting and intracellular release of synthetic DNA oligonucleotides.

A generally applicable strategy for peptide-oligonucleotide conjugation consists of separate solid-phase assembly of peptide and oligonucleotide sequences and their subsequent solution-phase or solid-phase chemoselective coupling. To further assess this strategy, the cysteine phosphoramidites **98** and **99** have been produced from the commercially available *N*-α-Fmoc-*S*-protected cysteine pentafluorophenyl esters and appropriate amino alcohols. The resulting alcohols were then phosphinylated with 2-cyanoethyl tetraisopropylphosphorodiamidite and diisopropylammonium tetrazolide to give the desired phosphoramidites.[85] These were added to the 5′-terminus of solid-phase-linked oligonucleotides with a coupling efficiency of >97%.

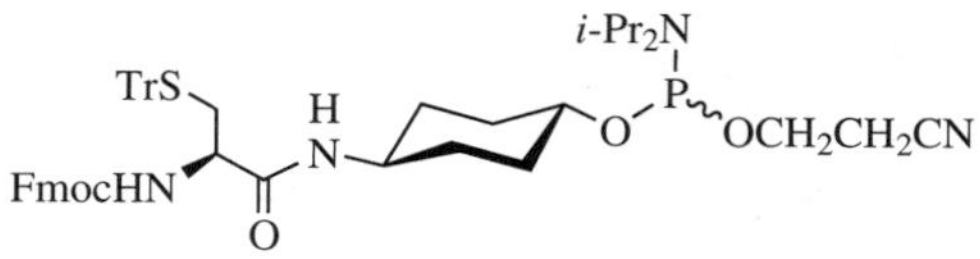

98

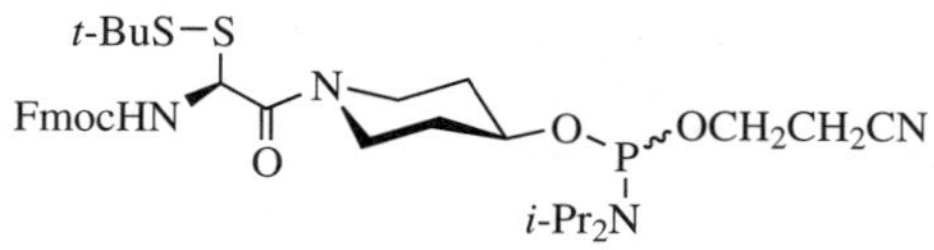

99

Subsequent to deprotection and removal of *S*-protecting group, a 5′-cysteine oligonucleotide was successfully conjugated with an ε-(*N*-maleimido)hexanoyl-PKKKRKV peptide in aqueous solution.[85] Alternatively, 5′-cysteine oligonucleotides generated by the use of **98** can also be conjugated with *N*-terminal thioester-functionalized peptides. Several peptide-oligonucleotide conjugates have been prepared using this 'native ligation' method.[86] 5′-Cysteine-substituted oligonucleotides should also be useful for conjugation with fluorescent markers and a variety of ligands.

Condensation of *trans*-4-[*N,N*-di-(2-methoxyoxalamidoethyl)-aminoethyl]-aminooxalamido-1-cyclohexanol with 2-cyanoethyl tetraisopropylphosphorodiamidite gives the phosphoramidite **100**.[87] Addition of **100** to the 5′-terminus of oligonucleotides is

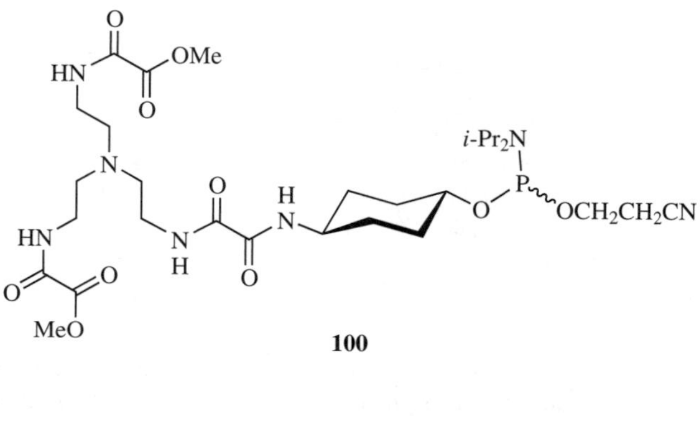

100

achieved by conventional solid-phase techniques using a 3-min coupling time to ensure high coupling yields. Treatment of the modified oligonucleotides with aqueous polyamine affords 5′-polyaminoalkylated oligonucleotides,[87] which are amenable to further functionalization with reporter groups and ligands.

The synthesis of the phosphoramidite **101** has been accomplished in four steps from hydroxydodecanoic acid to modify oligo- and polynucleotides for incorporation of reporter groups such as biotin.[88]

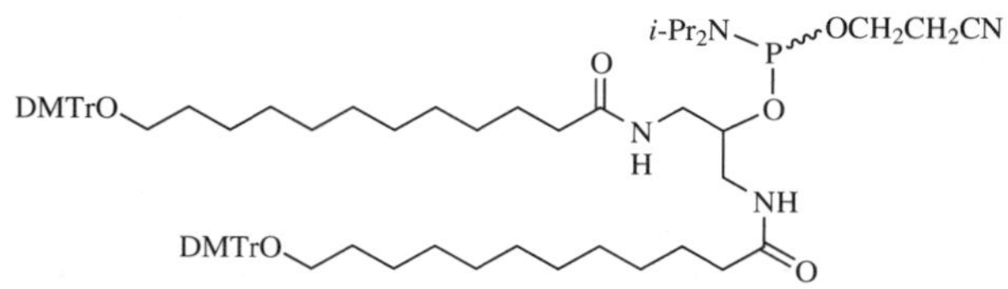

101

In addition, the biotinyl phosphoramidite **102** was prepared through a multistep synthesis from ε-aminohexanoic acid. The final phosphinylation was done under conditions similar to those used in the synthesis of **101**.[88] However, unlike **101**, which can be used after being dissolved for a week in acetonitrile, **102** must be employed within 2–3 days after being solubilized. The use of **102** or structurally related biotinyl phosphoramidite derivatives[89] permits the multiple addition of biotin to the 5′-terminus of oligonucleotides or its incorporation into oligonucleotides at any site during solid-phase synthesis.

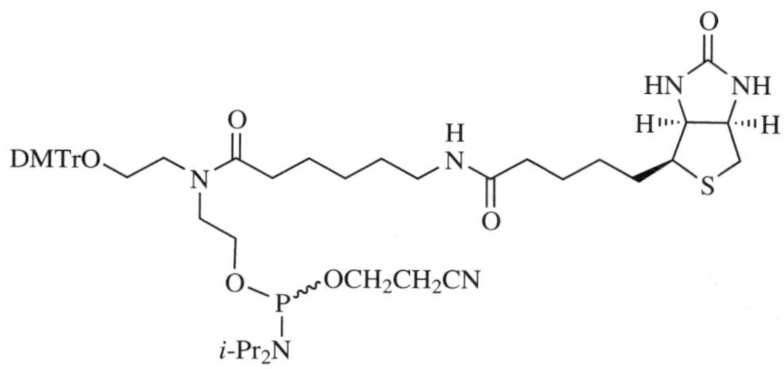

102

The functionalization of a solid-phase-bound oligothymidylate at the 5′-terminus has also been achieved via phosphoramidite **103**, which was obtained from the phosphinylation of 11-(phthalimidooxy)-3,6,9-trioxaundecan-1-ol with 2-cyanoethyl tetraisopropylphosphorodiamidite and 1*H*-tetrazole.[90]

Treatment of the functionalized oligonucleotide with hydrazinium acetate in pyridine removes the phthaloyl group within 30 min without releasing the oligonucleotide from the solid support. The 5′-aminooxyalkylated oligonucleotide was reacted with pyrenecarbaldehyde or *cis*-retinal and then with concentrated NH_4OH to release the desired 5′-labeled oligothymidylate from the solid support and demonstrate the simplicity of the labeling method.[90]

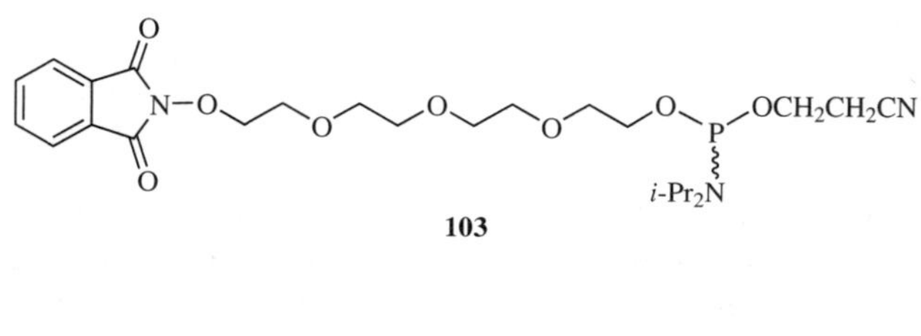

103

In an effort to further facilitate the modification of oligonucleotides with reporter groups for applications in molecular biology and as diagnostic tools, the non-nucleosidic phosphoramidite **104** has been produced from cyclohexyl-4-amino-1,1-dimethanol.[91] The incorporation of biotin, fluorescein, or amino groups into oligonucleotides at any site was achieved efficiently using the solid-phase DNA synthesis technology. It is postulated that the rigid cyclohexyl ring keeps the reporter group away from the oligonucleotide chain and results in higher coupling efficiency.[91]

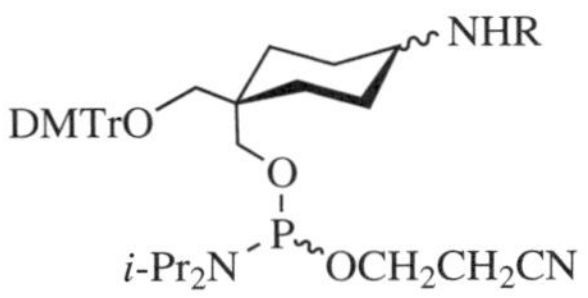

104, R = Biotinyl, fluoresceinyl, or 5-trifluoroacetamido-1-pentyl

A chemically stable acridinyl phosphoramidite derivative has been proposed for labeling synthetic oligonucleotides.[92,93] This fluorescent dye is also intercalating and, hence, stabilizes the complexes of primers and templates. The advantage in using the acridinyl phosphoramidite **105**, as opposed to 9-aminoacridine derivatives, is the stability of the dye to the conditions used for oligonucleotide synthesis and deprotection. The phosphoramidite **105** was obtained in four steps from 6-chloro-2-methoxy-9-methylacridine.[92] Short dye-labeled oligonucleotides have been used as primers in template-controlled polymerization reactions of RNA.[93]

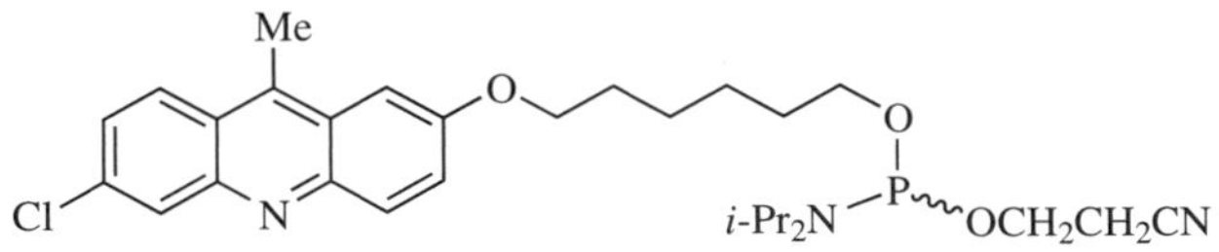

105

Fluorescent intercalators covalently attached to oligonucleotides stabilize duplexes formed with complementary DNA strands and provide excellent models for fluorescent energy transfer experiments. The methidium phosphoramidite **106** was obtained from 2,7-diamino-9-(*p*-carboxyphenyl)-10-methylphenanthridinium chloride, a flexible hydrophilic linker, and 2-cyanoethyl tetraisopropylphosphorodiamidite.[94] The synthesis of 5′-d(CCAT G*GCTAT), carrying an internal methidium insert (*), was achieved by site-specific incorporation of **106** using traditional solid-phase methodology. Hybridization of the modified oligonucleotide with complementary 5′-d(ATAGCCATGG) resulted in a duplex that is more stable ($\Delta T_m = +8.1\,°C$) than the unmodified duplex.[94] Modified oligonucleotides such as these may therefore serve as sensitive diagnostic probes.

The development of high-performance oligonucleotide labeling methods that are robust enough for automated solid-phase oligonucleotide synthesis still presents a challenge. To confront such adversity, the tetramethyl rhodamine (Tamra) phosphoramidite (**107**) was prepared from 5′-carboxyTamra, 4-hydroxypiperidine, and 2-cyanoethyl tetraisopropylphosphorodiamidite.[95] Condensation of **107** with the 5′-hydroxyls of solid-phase-linked DNA oligonucleotides proceeded with coupling efficiencies exceeding 95%; a post-coupling acetylation step was required to improve the quality of fully deprotected DNA conjugates.[95] This approach to 5′-labeling DNA with fluorophores is recommended over coupling 5′-aminoalkylated DNA oligonucleotides with active esters.

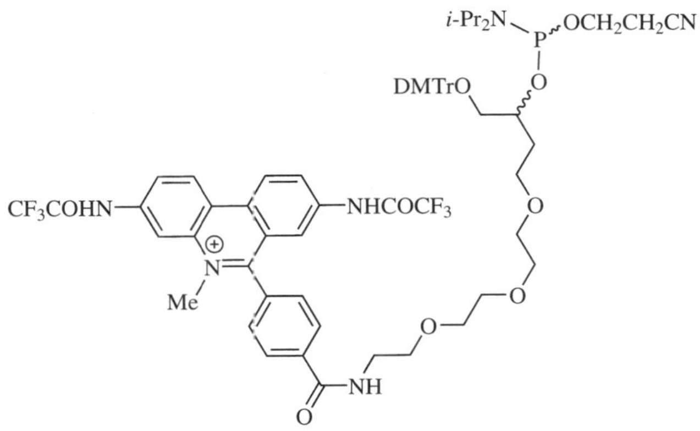

106

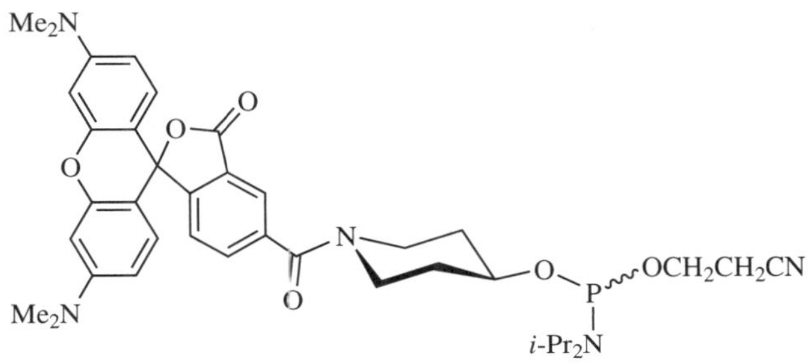

107

The interactions of nucleic acids with covalently bound pyrene residues have been studied by various spectroscopic techniques.[96] Pyrene is able to intercalate into dsDNA and stack with terminal nucleic bases. The pyrene fluorophore has a longer lifetime of the excited state and can form excimers, thereby rendering the molecular probe sensitive to changes in the microenvironment. Thus, the attachment of two or more pyrene residues to oligonucleotides produced excimeric fluorescence and led to changes in fluorescence intensity and excimer-to-monomer ratios, which should be taken into consideration in the detection of duplex formation.[96] To further study the interaction of pyrene residues with dsDNA, the pyrenylated phosphoramidite derivative **108** has been prepared in five steps from pentafluorophenyl 1-pyreneacetate.[96]

The phosphoramidite **108** has enabled the incorporation of pyrene at predetermined positions into an oligonucleotide. Integration of one or two pyrenes into a DNA duplex at terminal dangling positions increases the T_m of the duplex by 2.5 °C to 3.5 °C, regardless of which end (5′- or 3′-) of the strand the pyrenes are attached to.[96] The effect of solvent polarity on the fluorescence of a pentapyrenylated oligonucleotide is noteworthy as a 10-fold increase of excimer emission with increasing methanol concentration resulted from the aggregation of the pyrenylated oligonucleotide to form micelle-like structures.[96] In this regard, multipyrenylated oligonucleotides may potentially be useful in membrane studies.

The activation mechanism of anti-HIV nucleoside analogs in vivo involves phosphorylation by cellular kinases to the corresponding 2′-deoxyribonucleoside mono-, di-, and triphosphates. Then, the biologically active triphosphate is incorporated into the growing viral DNA chain causing chain termination. In this context, the R_P isomers of α-*P*-boronated deoxyribonucleoside di- and triphosphates have been found as better substrates than their natural counterparts for kinases and viral polymerases. Indeed, the α-(R_P)-boranodiphosphate of AZT is typically a 10-fold better substrate for diphosphate kinase than AZT diphosphate, whereas the α-(R_P) boranotriphosphate of AZT has an incorporation efficiency with HIV-RT that is 9-fold better than AZT triphosphate. In order to generate nucleoside boranophosphate prodrugs with enhanced bioavailability and cellular uptake properties, the preparation of L-tyrosinyl phosphoramidite (**109**) from suitably protected L-tyrosine and 2-cyanoethyl tetraisopropylphosphorodiamidite has been performed.[97] Condensation of **109** with AZT or 3′-*O*-acetyl-5-fluoro-2′-deoxyuridine and 1*H*-tetrazole afforded, after boronation and deprotection, the boranophosphate **110** or **111**, respectively, as a mixture of diastereomers.[97] These prodrugs are expected to be intracellularly hydrolyzed by enzymes to an amino acid and bioactive boronated nucleotides. The potential utility of **110** and **111**, as carriers of ^{10}B in boron capture therapy, may lead to promising antiviral and anticancer applications.

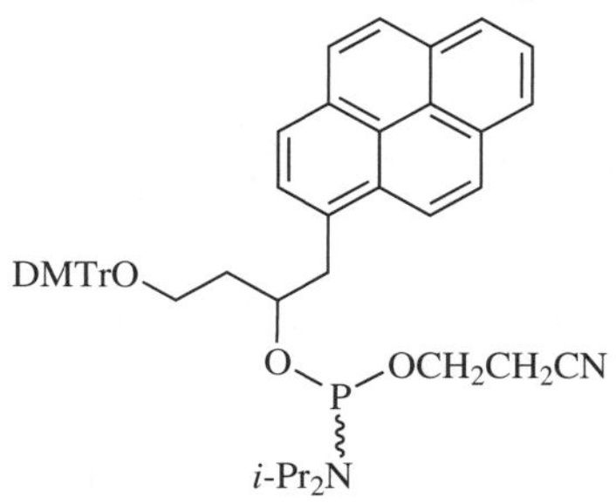

108

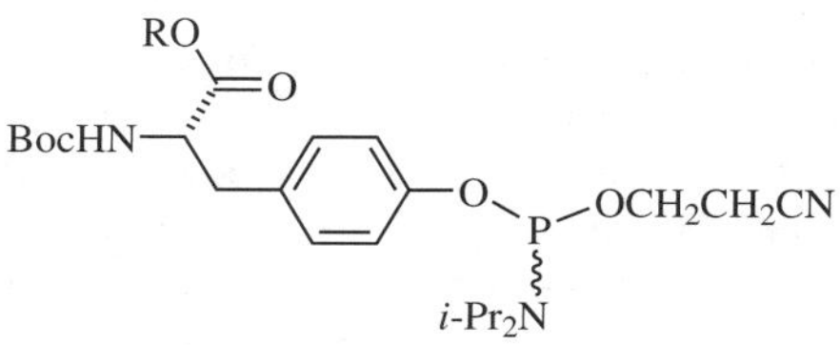

109, R = *p*-nitrobenzyl,
Boc = *tert*-butyloxycarbonyl

Several 5′-monophosphate derivatives of 2′,3′-didehydro-3′-deoxythymidine (D4T) were synthesized as potential lipophilic prodrugs of D4T. Specifically, the cholesteryl phosphoramidite **112** was prepared from the corresponding alcohol and 2-cyanoethyl tetraisopropylphosphorodiamidite, and then reacted with D4T/ 1*H*-tetrazole to give, after iodine oxidation and deprotection, the desired 5′-cholesteryl D4T phosphate.[98]

The synthesis of bis-(5′-D4T) phosphate from the reaction of D4T with 2-cyanoethyl tetraisopropylphosphorodiamidite and 1*H*-tetrazole was also described. Both bis-(5′-D4T) phosphate and 5′-cholesteryl D4T phosphate were tested for antiviral activity in vitro, and both phosphodiesters inhibited HIV replication in CEM-13 cells more efficiently than D4T. Moreover, the low toxicity of bis-(5′-D4T) phosphate brings its selectivity index to the same level as that of AZT, which is 10-fold higher than D4T.[98]

The insertion of transition metal complexes in oligonucleotides has been the focus of intense scrutiny, as the transition metals have been employed in electron transfer processes and in the cleavage of DNA or RNA through hydrolytic or free radical chemistry. To further evaluate the properties conferred to DNA by transition metals, the phosphoramidite derivative **113** has been engineered to enable incorporation of 2,2′-bipyridine ligands into oligodeoxyribonucleotides and subsequent formation of ruthenium–DNA complexes.[99] The solid-phase DNA synthesis using **113** proceeded without difficulty. The reaction of the modified oligonucleotides with *cis*-dichloro bis(2,2′-bipyridyl)Ru(II) gave ruthenium-containing DNA oligonucleotide complexes exhibiting UV and fluorescence properties consistent with those of simple tris-2,2′-bipyridyl Ru(II) complexes. Moreover, oligonucleotides functionalized with a ruthenium complex form DNA duplexes and triplexes that are more stable than those obtained from oligonucleotides tethered with tri(ethylene glycol or hexaethylene glycol) linkers.[99]

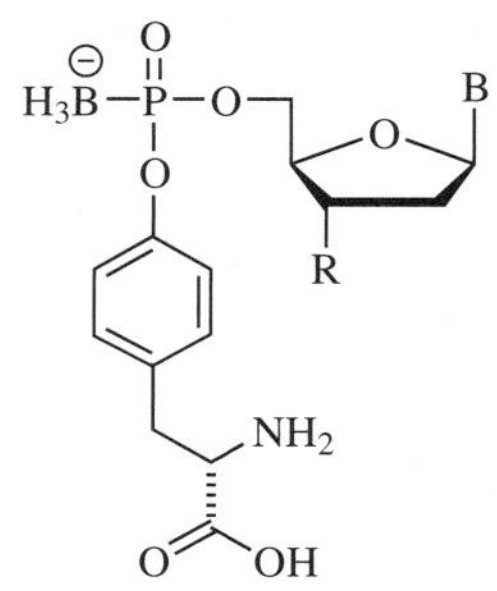

110, R = N_3, B = thymin-1-yl
111, R = OH, B = 5-fluorouracil-1-yl

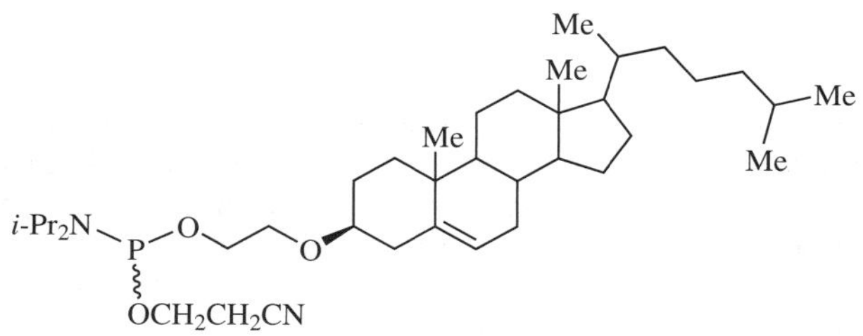

112

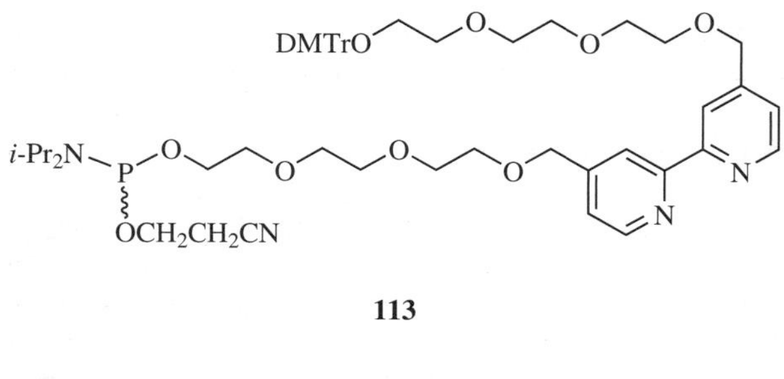

113

The monodeoxynucleotide-tethered platinum(IV) complex (**115**) was obtained from the reaction of nucleoside phosphoramidite **114** with dichloro[*N*-(2-hydroxymethyl) ethylenediamine]platinum(II) and 5-methylthio-1*H*-tetrazole (eq 2).[100,101]

The platinum(IV) complex (**115**) is the first nucleotide conjugate showing in vitro antitumor activity, and selectivity among the leukemia and ovarian cancer cell lines studied.[101] Thus, oligonucleotide–platinum complex conjugates may inhibit the expression of, for example, *erb*B-2 in a sequence-specific manner. This oncogene is especially a good target since it is overexpressed in 25–30% of primary breast and ovarian cancer patients.[101]

The dabcyl phosphoramide derivative **116** has been exploited in the automated preparation of molecular beacons,[102] which are particularly useful in the detection of specific nucleic acid sequences in hybridization-based assays. Incorporation of **116** into fluorescein- and rhodamine-containing beacons demonstrated that the dabcyl group quenched the fluorescence of fluorescein dyes at 22 °C much more efficiently than that of rhodamine dyes.[103] Moreover, since the fluorescence of fluorescein- and rhodamine-containing beacons increased 15- to 30-fold and 3-fold, respectively, at 73 °C,[103] molecular beacons carrying both dabcyl and fluorescein groups are recommended for the sensitive detection of single-stranded nucleic acids.

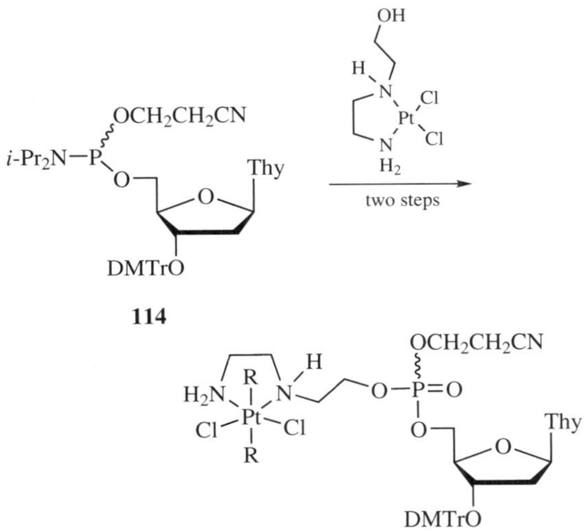

114

(2)

115, R = OAc

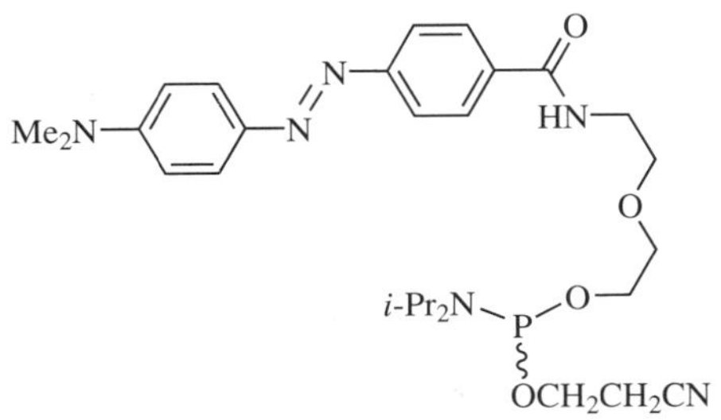

116

The synthesis and screening of oligonucleotide combinatorial libraries has become a useful tool in the search for novel lead sequences against a variety of cellular and viral targets. Propane-2,3-diol phosphoramidite monomers (**117**) have been prepared to provide diverse oligomerizing functionalities in addition to the heterocyclic adenine, guanine, thymine, and cytosine bases present in oligonucleotide combinatorial libraries.[104]

Novel heterocyclic, aromatic, and alkyl groups shall provide additional interactions with biological targets. The conformational flexibility of propane phosphodiesters relative to deoxyribose phosphodiesters will determine the tertiary structure of combinatorial oligomers. Unique molecular shapes have been postulated to be responsible for the biological activity of novel pharmacophores derived from combinatorial libraries.[104]

2-Cyanoethyl tetraisopropylphosphorodiamidite has not only been employed for the phosphinylation of modified nucleosides/nucleotides, reporter groups, and ligands for incorporation into oligonucleotides, it has also found application as a phosphinylating reagent for non-nucleosidic biomolecules. This application will be addressed in the following and the last section of this article.

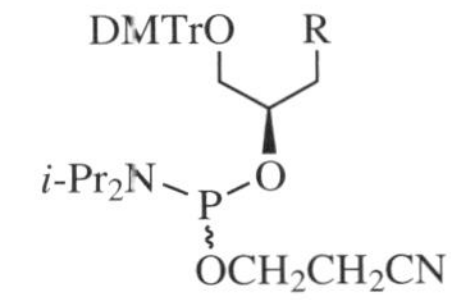

117

R = Thymin-1-yl; cytosin-1-yl; adenin-9-yl; guanin-9-yl; 4-benzoyl cytosin-1-yl; 6-benzoyladenin-9-yl; 2-isobutyrylguanin-9-yl; imidazol-1-yl; carbazol-9-yl; 2-amino-6-chloropurin-1-yl; phenyl; isopropyl; allyl; octyl.

Phosphinylation of Non-nucleosidic Biomolecules. Phosphosphingolipids play an important role as membrane constituents and D-*erythro*-ceramide 1-phosphoinositols have been recognized as being active in the glycosylphosphoinositols (GPI)-mediated anchoring of proteins. In an effort to synthesize these biomolecules, the ceramide phosphoramidite **118** has been assembled and coupled with a properly protected and functionalized D-*myo*-inositol to give, after oxidation and rounds of deprotection, the ceramide derivative of 2-amino-2-deoxy-D-glucopyranosyl-α-(1→6)-D-*myo*-inositol-1-phosphate (**119**).[105]

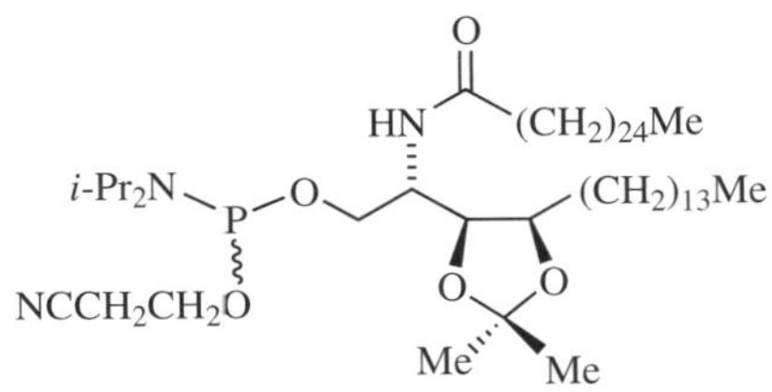

118

Given the role of GPI-anchored proteins in cell-cell recognition and as enzymes, the challenging chemical synthesis of structurally homogenous GPI anchors was attempted. In this regard, the ceramide phosphoramidite **118** was further used in the total synthesis of the ceramide-containing GPI anchor of yeast (**120**).[106]

Artificial glycoconjugate polymers carrying biologically active carbohydrates constitute a new class of biomimetic compounds. They have been widely used in cell cultivations, tumor diagnosis, and in the detection and trapping of viruses and toxins. Such a widespread utility can be attributed to ubiquitous lectin-like receptors on germ and mammalian cell surfaces, which recognize certain types of carbohydrate chains in glycopeptides and glycolipids. In this context, phosphocholine-containing glycoglycerolipids GGPL-1 are the main cell membrane glycolipids of *Mycoplasma fermentans*. These microorganisms are suspected of exhibiting pathogenic roles in the progression of HIV diseases and rheumatoid arthritis. Findings such as these prompted the development of a synthetic pathway to artificial glycoconjugate polymers carrying GGPL-1 analogs from D-glucose. Specifically, the glucopyranosyl phosphoramidite intermediate **121** was prepared in situ from *p*-nitrophenyl 2,3,4-tri-*O*-benzyl-α-D-glucopyranoside and 2-cyanoethyl tetraisopropylphosphorodiamidite.[107] A phosphocholine group was subsequently generated at *O*-6 and the 1-*O*-*p*-nitrophenyl group was converted to a *p*-*N*-methacrylamidophenyl group for radical polymerization.[107] This synthetic approach may become a general method for the synthesis of other glycoconjugate polymers carrying 6-*O*-phosphocholine glycosides as GGPL analogs.

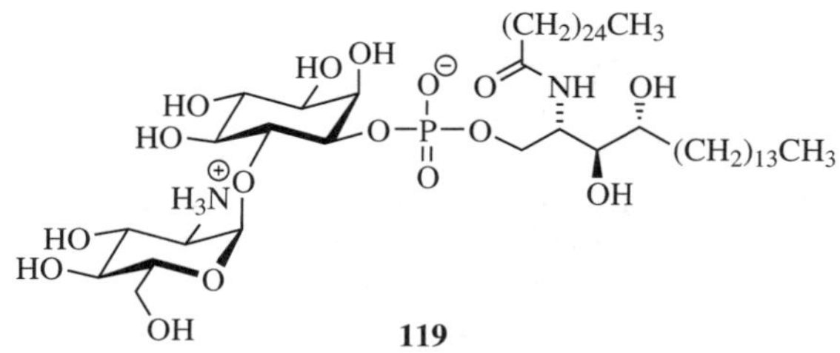

119

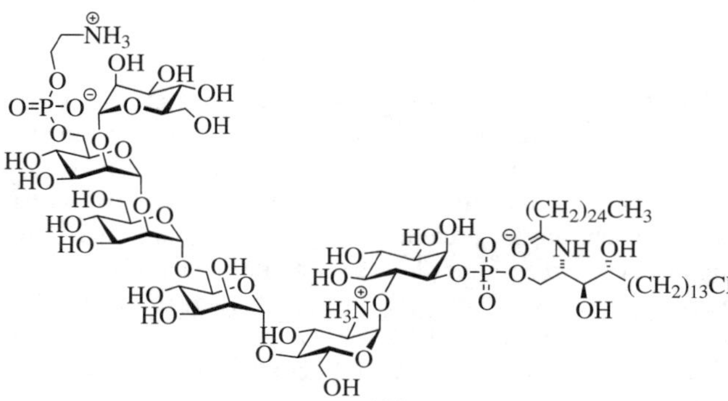

120

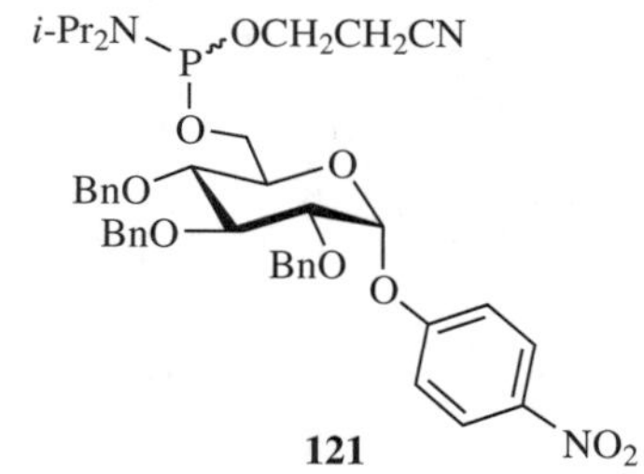

121

Phospholipids, such as phosphatidylcholine and phosphatidylserine, are used extensively in the formation of liposomes which have an enormous potential as drug delivery systems. Liposomes have also been used as model membrane systems to study cell-cell or liposome-cell interactions and the molecular mechanisms of biological membrane fusion. To further support these investigations, the *N*-Boc-L-serine benzhydryl ester phosphoramidite (**122**)was prepared from *N*-Boc-L-serine benzhydryl ester and 2-cyanoethyl tetraisopropylphosphorodiamidite.[108]

The condensation of **122** with 1,2-*O*-diacyl-*sn*-glycerol gave, after oxidation, deprotection, and purification, 1,2-*O*-diacyl-*sn*-glycerophosphatidylserine for the preparation of liposomes with different phosphatidylcholine/phosphatidylserine ratios but with identical fatty acyl compositions.[108]

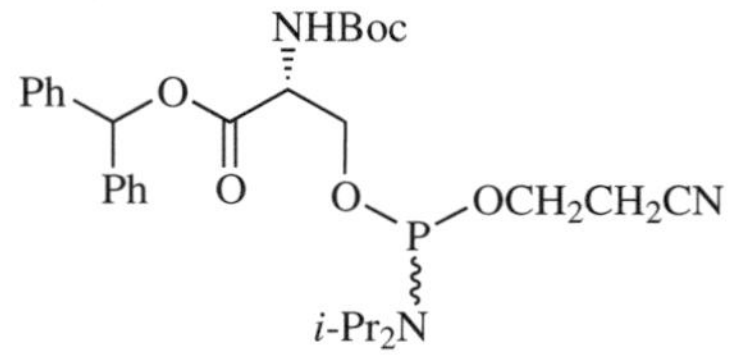

122

L-α-Phosphatidyl-D-*myo*-inositol 3,4,5-trisphosphate (PIP_3) and L-α-phosphatidyl-D-*myo*-inositol 4,5-bisphosphate (PIP_2) play a variety of essential roles in biological processes. Indeed, PIP_2 is particularly important in the recruitment of signaling proteins to membranes, as mediated by the pleckstrin homology domains. A synthetic route was developed for the preparation of *O*-(3-aminopropyl)phosphotriester analogs of PIP_2 and PIP_3.[109] Typically, phosphinylation of optically pure 1,2-*O*-diacyl-*sn*-glycerol with 2-cyanoethyl tetraisopropylphosphorodiamidite afforded the phosphoramidite **123**, which was then coupled with protected inositol precursors to generate the desired 3-aminopropyl phosphotriesters as shown in **124a** and **124b** after reductive deprotection.[109]

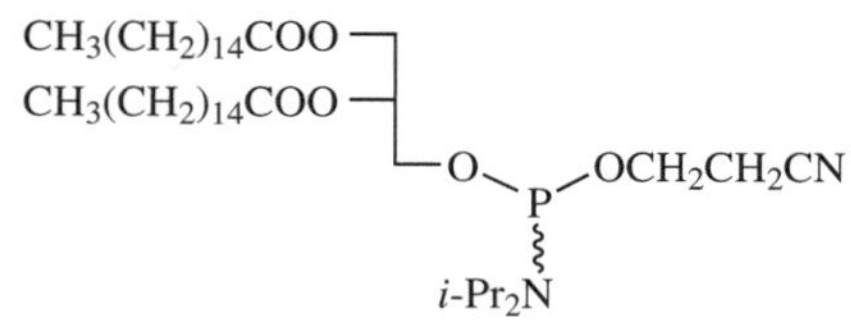

123

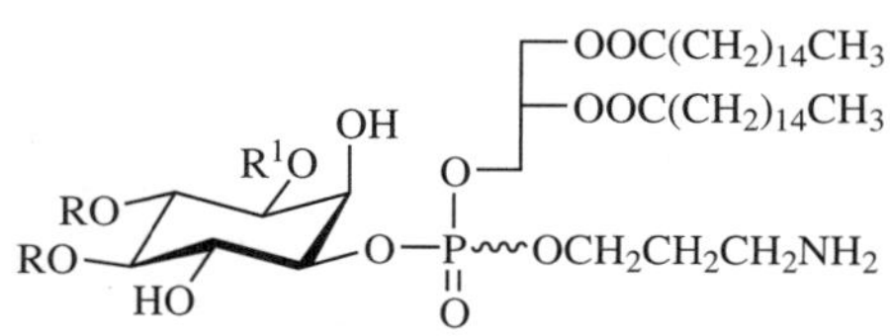

124a, $R = R^1 = Na_2PO_3$

124b, $R = Na_2PO_3$; $R^1 = H$

These phosphotriesters were designed to permit the attachment of a reporter group at the surface of the bilayer and in the vicinity of the phosphorylated inositol. This reporter group may interact with membrane-proximal regions of PIP_2- and PIP_3-binding proteins recruited to membrane sites by electrostatic interactions between the phosphates of the phospholipid and basic regions of the proteins.[109] These affinity probes may therefore provide an opportunity to examine the interactions of PIP_2 and PIP_3 with proteins at the surface of the membranes.

In recent years, considerable efforts have been devoted to improve the therapeutic potential of the nucleoside analogs by the use of nucleotide prodrugs. Many strategies have been considered to mask or reduce the negative charges of 5′-nucleoside monophosphates. One of these strategies entails the use of β-thioglucopyranosidylethyl groups as phosphate masking groups, since it is anticipated that nucleotide prodrugs functionalized with these groups would eventually revert to the corresponding 5′-nucleoside monophosphates through a glucosidase-mediated activation mechanism. To test this strategy, the phosphoramidite **125** was produced by the reaction of 2-hydroxyethyl 2,3,4,6-tetra-*O*-acetyl-1-α-D-thioglucopyranoside with 2-cyanoethyl tetraisopropylphosphorodiamidite, 1*H*-tetrazole, and diisopropylamine.[110] Condensation of AZT with **125** and 1*H*-tetrazole led, after oxidation and a number of additional steps, to the AZT prodrug (**126**).

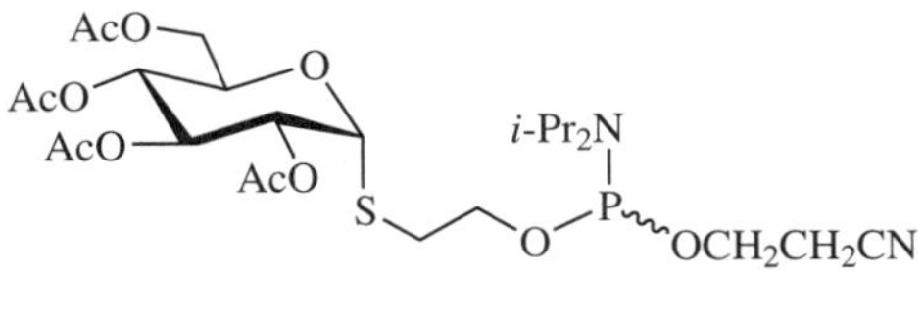

125

The stability of pronucleotide **126** in several media including culture medium is relatively low and may be related to a chemical decomposition mechanism involving the sulfur atom as the participating neighboring group.[110] The half-life of **126** in any media is nonetheless sufficient for comparative in vitro anti-HIV activity testing.

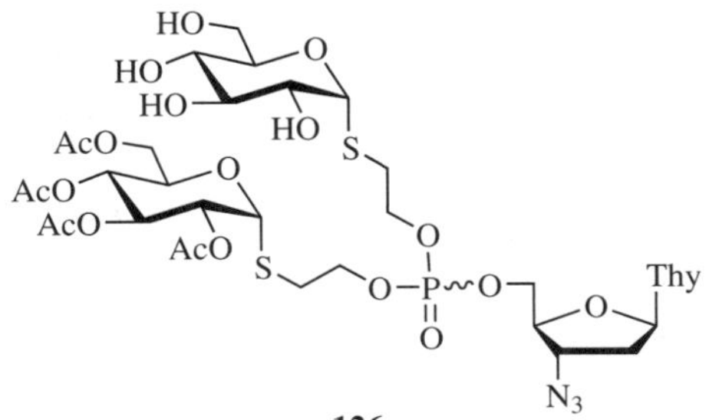

126

A new class of DNA-carbohydrate conjugates, termed nucleoglycoconjugates, has been reported to have promising biomedical applications. The mono- and disaccharide phosphoramidite derivatives (**127–129**) were then generated for internal or terminal incorporation into oligonucleotides using solid-phase chemistry.[111] The phosphoramidite **127** was prepared in nine steps from the 4′,6′-benzylidene derivative of methyl-β-D-lactoside, whereas the phosphoramidite **128** was obtained in four steps from methyl-β-D-glucopyranoside. The phosphoramidite **129** was in turn produced in several steps from glucose pentaacetate.[111]

The coupling efficiency of the disaccharide phosphoramidite **127** is greater than 95%. However, extension of the oligonucleotide chain from the inserted disaccharide proceeded with a coupling yield of ~70%. Comparatively, the monosaccharide phosphoramidites **128** and **129** reacted with the terminal hydroxyls of DNA oligonucleotides with a coupling efficiency of >90%.[111] The nucleoglycoconjugates formed stable duplexes with complementary DNA oligonucleotides even though the presence of a disaccharide, near a middle position in a 29 nt duplex structure, moderately destabilized the hybrid ($\Delta T_m = -8\,°C$) relative to the parent unmodified duplex. The nucleoglycoconjugates are substrates for common DNA-manipulating enzymes, such as kinases and polymerases, and can be utilized as primers in template-directed extension reactions.[111] In consideration of the prevalence of cell surface proteins (lectins) that recognize carbohydrate epitopes, these DNA-glycoconjugates may also be useful in increasing the specificity and cellular uptake of therapeutic oligonucleotides.

While this article covers only a fraction of the applications of 2-cyanoethyl tetraisopropylphosphorodiamidite as a phospinylating reagent, the diversity of such applications toward nucleic acids and others biomolecules is a testimony of the impact phosphoramidite derivatives still have on the biomedical sciences.

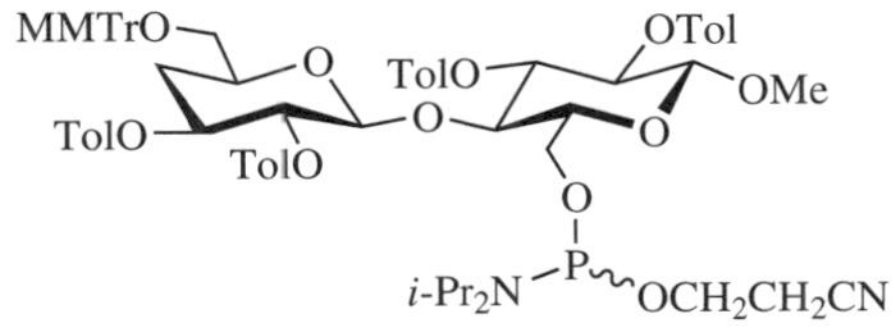

127, Tol = toluoyl

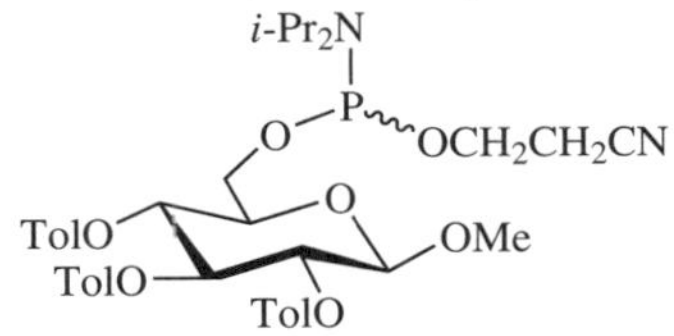

128

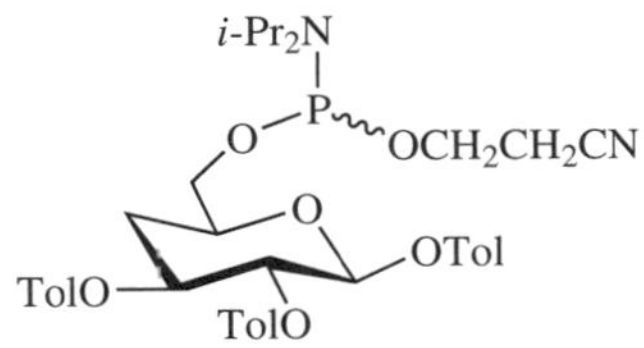

129

Related Reagents. 2-Cyanoethyl diisopropylchlorophosphoramidite.

1. Caruthers, M. H.; Barone, A. D.; Beaucage, S. L.; Dodds, D. R.; Fisher, E. F.; McBride, L. J.; Matteucci, M.; Stabinsky, Z.; Tang, J.-Y., In *Methods in Enzymology*; Wu, R.; Grossman, L., Eds.; Academic Press: San Diego, 1987, Vol. 154, pp 287–313.
2. Nielsen, J.; Taagaard. M.; Marugg, J. E.; van Boom, J. H.; Dahl, O.; *Nucleic Acids Res.* **1986**, *14*, 7391.
3. (a) Beaucage, S. L.; Iyer, R. P., *Tetrahedron* **1992**, *48*, 2223. (b) Beaucage, S. L.; Iyer, R. P., *Tetrahedron* **1993**, *49*, 1925. (c) Beaucage, S. L.; Iyer, R. P., *Tetrahedron* **1993**, *49*, 6123. (d) Beaucage, S. L.; Iyer, R. P., *Tetrahedron* **1993**, *49*, 10441. (e) Iyer, R. P.; Beaucage, S. L., In *Comprehensive Natural Product Chemistry*; Barton, D.; Nakanishi, K.; Meth-Cohn, O.; Kool, E. T., Eds.; Elsevier: Amsterdam, 1999, p 105. (f) Beaucage, S. L., In *Comprehensive Natural Product Chemistry*; Barton, D.; Nakanishi, K.; Meth-Cohn, O.; Kool, E. T., Eds.; Elsevier: Amsterdam, 1999, p 153.
4. DeCorte, B. L.; Tsarouhtsis, D.; Kuchimanchi, S.; Cooper, M. D.; Horton, P.; Harris, C. M.; Harris, T. M., *Chem. Res. Toxicol.* **1996**, *9*, 630.
5. Kim, M.-S.; Guengerich, F. P., *Chem. Res. Toxicol.* **1997**, *10*, 1133.
6. Langouët, S.; Mican, A. N.; Müller, M.; Fink, S. P.; Marnett, L. J.; Muhle, S. A.; Guengerich, F. P., *Biochemistry* **1998**, *37*, 5184.
7. Harwood, E. A.; Sigurdsson, S. T.; Edfeldt, N. B. F.; Reid, B. R.; Hopkins, P. B., *J. Am. Chem. Soc.* **1999**, *121*, 5081.
8. Romieu, A.; Gasparutto, D.; Molko, D.; Cadet, J., *J. Org. Chem.* **1998**, *63*, 5245.
9. Gasparutto, D.; Da Cruz, S.; Bourdat, A.-G.; Jaquinod, M.; Cadet, J., *Chem. Res. Toxicol.* **1999**, *12*, 630.
10. Sugiyama, H.; Matsuda, S.; Kino, K.; Zhang, Q.-M.; Yonei, S.; Saito, I., *Tetrahedron Lett.* **1996**, *37*, 9067.
11. Kittaka, A.; Kuze, T.; Amano, M.; Tanaka, H.; Miyasaka, T.; Hirose, K.; Yoshida, T.; Sarai, A.; Yasukawa, T.; Ishii, S., *Nucleosides Nucleotides* **1999**, *18*, 2769.

12. Dougan, H.; Hobbs, J. B.; Weitz, J. I.; Lyster, D. M., *Nucleic Acids Res.* **1997**, *25*, 2897.
13. Koizumi, M.; Akahori, K.; Ohmine, T.; Tsutsumi, S.; Sone, J.; Kosaka, T.; Kaneko, M.; Kimura, S.; Shimada, K., *Bioorg. Med. Chem. Lett.* **2000**, *10*, 2213.
14. Miller, T. J.; Farquar, H. D.; Sheybani, A.; Tallini, C. E.; Saurage, A. S.; Fronczek, F. R.; Hammer, R. P., *Org. Lett.* **2002**, *4*, 877.
15. Miller, T. R.; Alley, S. C.; Reese, A. W.; Solomon, M. S.; McCallister, W. V.; Mailer, C.; Robinson, B. H.; Hopkins, P. B., *J. Am. Chem. Soc.* **1995**, *117*, 9377.
16. Solomon, M. S.; Hopkins, P. B., *J. Org. Chem.* **1993**, *58*, 2232.
17. Weizman, H.; Tor, Y., *J. Am. Chem. Soc.* **2001**, *123*, 3375.
18. Nakatani, K.; Fujisawa, K.; Dohno, C.; Nakamura, T.; Saito, I., *Tetrahedron Lett.* **1998**, *39*, 5995.
19. Okamoto, A.; Taiji, T.; Tanaka, K.; Saito, I., *Tetrahedron Lett.* **2000**, *41*, 10035.
20. Balow, G.; Mohan, V.; Lesnik, E. A.; Johnston, J. F.; Monia, B. P.; Acevedo, O. L., *Nucleic Acids Res.* **1998**, *26*, 3350.
21. Kawai, K.; Saito, I.; Sugiyama, H., *Tetrahedron Lett.* **1998**, *39*, 5221.
22. Potier, P.; Abdennaji, A.; Behr, J.-P., *Chem. Eur. J.* **2000**, *6*, 4188.
23. Prakash, T. P.; Barawkar, D. A.; Kumar, V.; Ganesh, K. N., *Bioorg. Med. Chem. Lett.* **1994**, *4*, 1733.
24. Bergstrom, D. E.; Zhang, P.; Johnson, W. T., *Nucleic Acids Res.* **1997**, *25*, 1935.
25. Gorchs, O.; Hernández, M.; Garriga, L.; Pedroso, E.; Grandas, A.; Farràs J., *Org. Lett.* **2002**, *4*, 1827.
26. Okamoto, A.; Tainaka, K.; Saito, I., *Tetrahedron Lett.* **2002**, *43*, 4581.
27. Melguizo, M.; Gottlieb, M.; Charubala, R.; Pfleiderer, W., *Nucleosides Nucleotides* **1998**, *17*, 175.
28. Koga, M.; Wilk, A.; Moore, M. F.; Scremin, C. L.; Zhou, L.; Beaucage, S. L., *J. Org. Chem.* **1995**, *60*, 1520.
29. Scremin, C. L.; Boal, J. H.; Wilk, A.; Phillips, L. R.; Zhou, L.; Beaucage, S. L., *Tetrahedron Lett.* **1995**, *36*, 8953.
30. Scremin, C. L.; Boal, J. H.; Wilk, A.; Phillips, L. R.; Beaucage, S. L., *Bioorg. Med. Chem. Lett.* **1996**, *6*, 207.
31. Boal, J. H.; Wilk, A.; Scremin, C. L.; Gray, G. N.; Phillips, L. R.; Beaucage, S. L., *J. Org. Chem.* **1996**, *61*, 8617.
32. Kawasaki, A. M.; Casper, M. D.; Freier, S. M.; Lesnik, E. A.; Zounes, M. C.; Cummins, L. L.; Gonzalez, C.; Cook, P. D., *J. Med. Chem.* **1993**, *36*, 831.
33. Deshmukh, H. M.; Joglekar, S. P.; Broom, A. D., *Bioconjugate Chem.* **1995**, *6*, 578.
34. Deshmukh, H. M.; Broom, A. D., *Bioconjugate Chem.* **1996**, *7*, 108.
35. Yamana, K.; Iwase, R.; Furutani, S.; Tsuchida, H.; Zako, H.; Yamaoka, T.; Murakami, A., *Nucleic Acids Res.* **1999**, *27*, 2837.
36. Korshun, V. A.; Stetsenko, D.; Gait, M. J., *J. Chem. Soc., Perkin Trans 1* **2002**, 1092.
37. Pavey, J. B. J.; Cosstick, R.; O'Neil, I. A., *Tetrahedron Lett.* **1998**, *39*, 8923.
38. Pavey, J. B. J.; Lawrence, A. J.; Potter, A. J.; Cosstick, R.; O'Neil, I. A., *Tetrahedron Lett.* **1998**, *39*, 6967.
39. De Mesmaeker, A.; Lebreton, J.; Hoffmann, P.; Freier, S. M., *Synlett* **1993**, 677.
40. Serafinowski, P. J.; Barnes, C. L., *Tetrahedron* **1996**, *52*, 7929.
41. Sobol, R. W.; Charubala, R.; Pfleiderer, W.; Suhadolnik, R. J., *Nucleic Acids Res.* **1993**, *21*, 2437.
42. Kvasyuk, E. I.; Kulak, T. I.; Mikhailopulo, I. A.; Charubala, R.; Pfleiderer, W., *Helv. Chim. Acta* **1995**, *78*, 1777.
43. Markiewicz, W. T.; Niewczyk, A.; Gdaniec, Z.; Adamiak, D. A.; Dauter, Z.; Rypniewski, W.; Chmielewski, M., *Nucleosides Nucleotides* **1998**, *17*, 411.
44. Walcher, G.; Pfleiderer, W., *Helv. Chim. Acta* **1996**, *79*, 1067.
45. Svendsen, M. L.; Wengel, J.; Dahl, O.; Kirpekar, F.; Roepstorff, P., *Tetrahedron* **1993**, *49*, 11341.
46. Obika, S.; Hari, Y.; Sugimoto, T.; Sekiguchi, M.; Imanishi, T., *Tetrahedron Lett.* **2000**, *41*, 8923.
47. Rao, T. S.; Jayaraman, K.; Revankar, G. R., *Tetrahedron Lett.* **1993**, *34*, 6189.
48. Rana, V. S.; Kumar, V. A.; Ganesh, K. N., *Bioorg. Med. Chem. Lett.* **1997**, *7*, 2837.
49. Rana, V. S.; Kumar, V. A.; Ganesh, K. N., *Tetrahedron* **2001**, *57*, 1311.
50. Nawrot, B.; Michalak, O.; Nowak, M.; Okruszek, A.; Dera, M.; Stec, W. J., *Tetrahedron Lett.* **2002**, *43*, 5397.
51. Robles, J.; Pedroso, E.; Grandas, A., *Nucleic Acids Res.* **1995**, *23*, 4151.
52. Tetzlaff, C. N.; Schwope, I.; Bleczinski, C. F.; Steinberg, J. A.; Richert, C., *Tetrahedron Lett.* **1998**, *39*, 4215.
53. De Mesmaeker, A.; Lebreton, J.; Waldner, A.; Fritsch, V.; Wolf, R. M.; Freier, S. M., *Synlett* **1993**, 733.
54. De Mesmaeker, A.; Jouanno, C.; Wolf, R. M.; Wendeborn, S., *Bioorg. Med. Chem. Lett.* **1997**, *7*, 447.
55. Lebreton, J.; De Mesmaeker, A.; Waldner, A.; Fritsch, V.; Wolf, R. M.; Freier, S. M., *Tetrahedron Lett.* **1993**, *34*, 6383.
56. Waldner, A.; De Mesmaeker, A.; Lebreton, J., *Bioorg. Med. Chem. Lett.* **1994**, *4*, 405.
57. Waldner, A.; De Mesmaeker, A.; Lebreton, J.; Fritsch, V.; Wolf, R. M., *Synlett* **1994**, 57.
58. Chur, A.; Holst, B.; Dahl, O.; Valentin-Hansen, P.; Pedersen, E. B., *Nucleic Acids Res.* **1993**, *21*, 5179.
59. Prhavc, M.; Just, G.; Bhat, B.; Cook, P. D.; Manoharan, M., *Tetrahedron Lett.* **2000**, *41*, 9967.
60. Blättler, M. O.; Wenz, C.; Pingoud, A.; Benner, S. A., *J. Am. Chem. Soc.* **1998**, *120*, 2674.
61. Micklefield, J.; Fettes, K. J., *Tetrahedron* **1998**, *54*, 2129.
62. Fettes, K. J.; Howard, N.; Hickman, D. T.; Adah, S.; Player, M. R.; Torrence, P. F.; Micklefield, J., *J. Chem. Soc., Perkin Trans. 1* **2002**, 485.
63. Laurent, A.; Naval, M.; Debart, F.; Vasseur, J.-J.; Rayner, B., *Nucleic Acids Res.* **1999**, *27*, 4151.
64. Peyrottes, S.; Vasseur, J. J.; Imbach, J. L.; Rayner, B., *Nucleosides Nucleotides* **1994**, *13*, 2135.
65. Collingwood, S. P.; Baxter, A. D., *Synlett* **1995**, 703.
66. Fairhurst, R. A.; Collingwood, S. P.; Lambert, D., *Synlett* **2001**, 473.
67. Mauritz, R. P.; Meier, C.; Uhlmann, E., *Nucleosides Nucleotides* **1997**, *16*, 1209.
68. Krotz, A. H.; Klopchin, P.; Cole, D. L.; Ravikumar, V. T., *Bioorg. Med. Chem. Lett.* **1997**, *7*, 73.
69. Natt, F.; Häner, R., *Tetrahedron* **1997**, *53*, 9629.
70. Guzaev, A. P.; Manoharan, M., *Tetrahedron Lett.* **2001**, *42*, 4769.
71. Defrancq, E.; Lhomme, J., *Bioorg. Med. Chem. Lett.* **2001**, *11*, 931.
72. Smith, T. H.; Kent, M. A.; Muthini, S.; Boone, S. J.; Nelson, P. S., *Nucleosides Nucleotides* **1996**, *15*, 1581.
73. Nelson, P. S.; Kent, M.; Muthini, S., *Nucleic Acids Res.* **1992**, *20*, 6253.
74. Hill, K. W.; Taunton-Rigby, J.; Carter, J. D.; Kropp, E.; Vagle, K.; Pieken, W.; McGee, D. P. C.; Husar, G. M.; Leuck, M.; Anziano, D. J.; Sebesta, D. P., *J. Org. Chem.* **2001**, *66*, 5332.
75. Guzaev, A.; Salo, H.; Azhayev, A.; Lönnberg, H., *Bioconjugate Chem.* **1996**, *7*, 240.
76. Hovinen, J.; Guzaev, A.; Azhayev, A.; Lönnberg, H., *J. Chem. Soc., Perkin Trans. 1* **1994**, 2745.
77. Hovinen, J.; Guzaev, A.; Azhayeva, E.; Azhayev, A.; Lönnberg, H., *J. Org. Chem.* **1995**, *60*, 2205.
78. Polushin, N. N.; Chen, B.; Anderson, L. W.; Cohen, J. S., *J. Org. Chem.* **1993**, *58*, 4606.
79. Endo, M.; Azuma, Y.; Saga, Y.; Kuzuya, A.; Kawai, G.; Komiyama, M., *J. Org. Chem.* **1997**, *62*, 846.

80. Asanuma, H.; Liang, X.; Komiyama, M., *Tetrahedron Lett.* **2000**, *41*, 1055.
81. Asanuma, H.; Ito, T.; Komiyama, M., *Tetrahedron Lett.* **1998**, *39*, 9015.
82. Koch, T.; Jacobsen, N.; Fensholdt, J.; Boas, U.; Fenger, M.; Jakobsen, M. H., *Bioconjugate Chem.* **2000**, *11*, 474.
83. Jones, D. S.; Hachmann, J. P.; Osgood, S. A.; Hayag, M. S.; Barstad, P. A.; Iverson, G. M.; Coutts, S. M., *Bioconjugate Chem.* **1994**, *5*, 390.
84. Bidaine, A.; Berens, C.; Sonveaux, E., *Bioorg. Med. Chem. Lett.* **1996**, *6*, 1167.
85. Stetsenko, D. A.; Gait, M. J., *Nucleos. Nucleot. Nucleic Acids* **2000**, *19*, 1751.
86. Stetsenko, D. A.; Gait, M. J., *J. Org. Chem.* **2000**, *65*, 4900.
87. Polushin, N. N., *Nucleic Acids Res.* **2000**, *28*, 3125.
88. Korshun, V. A.; Pestov, N. B.; Nozhevnikova, E. V.; Prokhorenko, I. A.; Gontarev, S. V.; Berlin, Y. A., *Synth. Commun.* **1996**, *26*, 2531.
89. Zhao, Z.; Ackroyd, J., *Nucleosides Nucleotides* **1999**, *18*, 1231.
90. Salo, H.; Virta, P.; Hakala, H.; Prakash, T. P.; Kawasaki, A. M.; Manoharan, M.; Lönnberg, H., *Bioconjugate Chem.* **1999**, *10*, 815.
91. Su, S.-H.; Iyer, R. S.; Aggarwal, S. K.; Kalra, K. L., *Bioorg. Med. Chem. Lett.* **1997**, *7*, 1639.
92. Schütz, K.; Kurz, M.; Göbel, M. W., *Tetrahedron Lett.* **1995**, *36*, 8407.
93. Kurz, M.; Göbel, K.; Hartel, C.; Göbel, M. W., *Helv. Chim. Acta* **1998**, *81*, 1156.
94. Timofeev, E. N.; Smirnov, I. P.; Hoff, L. A.; Tishchenko, E. I.; Mirzabekov, A. D.; Florentiev, V. L., *Tetrahedron Lett.* **1996**, *37*, 8467.
95. Lyttle, M. H.; Carter, T. G.; Dick, D. J.; Cook, R. M., *J. Org. Chem.* **2000**, *65*, 9033.
96. Korshun, V. A.; Balakin, K. V.; Proskurina, T. S.; Mikhalev, I. I.; Malakhov, A. D.; Berlin, Y. A., *Nucleosides Nucleotides* **1999**, *18*, 2661.
97. Li, P.; Shaw, B. R., *Org. Lett.* **2002**, *4*, 2009.
98. Sergheraert, C.; Pierlot, C.; Tartar, A.; Henin, Y.; Lemaitre, M., *J. Med. Chem.* **1993**, *36*, 826.
99. Wiederholt, K.; McLaughlin, L. W., *Nucleic Acids Res.* **1999**, *27*, 2487.
100. Ren, S.; Cai, L.; Segal, B. M., *J. Chem. Soc., Dalton Trans.* **1999**, 1413.
101. Cai, L.; Lim, K.; Ren, S.; Cadena, R. S.; Beck, W. T., *J. Med. Chem.* **2001**, *44*, 2959.
102. Tyagi, S.; Kramer, F. R., *Nature Biotechnology* **1996**, *14*, 303.
103. Mullah, B.; Livak, K., *Nucleosides Nucleotides* **1999**, *18*, 1311.
104. Acevedo, O. L.; Andrews, R. S., *Tetrahedron Lett.* **1996**, *37*, 3931.
105. Kratzer, B.; Mayer, T. G.; Schmidt, R. R., *Eur. J. Org. Chem.* **1998**, 291.
106. Mayer, T. G.; Schmidt, R. R., *Eur. J. Org. Chem.* **1999**, 1153.
107. Nishida, Y.; Takamori, Y.; Matsuda, K.; Ohrui, H.; Yamada, T.; Kobayashi, K., *J. Carbohydr. Chem.* **1999**, *18*, 985.
108. Morillo, M.; Luisa Sagristá, M.; Africa de Madariaga, M.; Eritja, R., *Lipids* **1996**, *31*, 541.
109. Gu, Q.-M.; Prestwich, G. D., *J. Org. Chem.* **1996**, *61*, 8642.
110. Schlienger, N.; Périgaud, C.; Gosselin, G.; Imbach, J.-L., *J. Org. Chem.* **1997**, *62*, 7216.
111. Sheppard, T. L.; Wong, C.-H.; Joyce, G. F., *Angew. Chem., Int. Ed. Engl.* **2000**, *39*, 3660.

Serge L. Beaucage
Food and Drug Administration, Bethesda, MD, USA

Cyanuric Fluoride

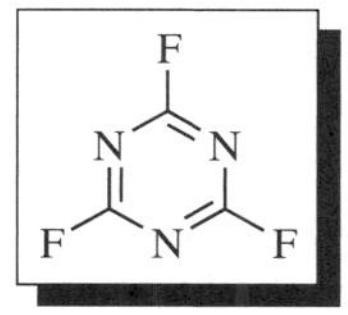

[675-14-9] $C_3F_3N_3$ (MW 135.05)

(reagent used for conversion of carboxylic acids to acid fluorides; deoxygenation of alkyl sulfoxides; and activation of alkenyl boronic acids for 1,4-addition reactions)

Alternate Name: 2,4,6-trifluoro-1,3,5-triazine, cyanuryl fluoride.
Physical Data: mp −38 °C; bp 72–74 °C; d 1.574 g mL^{-1}.
Form Supplied in: colorless liquid.
Purification: the commercial reagent is generally used as received.
Handling, Storage, and Precautions: highly toxic; very toxic by inhalation, in contact with skin, and if swallowed. Causes burns. Readily absorbed through the skin. Lachrymator. In case of accident or if you feel ill, seek medical advice immediately (show the label when possible). In case of contact with eyes, rinse immediately with plenty of water and seek medical advice. Do not breath vapor. Wear suitable protective clothing, gloves, and face/eye protection. Refrigerate. Moisture sensitive; keep container tightly closed.
Disposal: dissolve or mix with a combustible solvent and incinerate.

Acid Fluoride Synthesis. Cyanuric fluoride first appeared as a reagent for the mild and direct conversion of carboxylic acids into acid fluorides. This method represented a significant advancement for the preparation of acid fluorides.[1] Previously reported methods include treatment of acid chlorides with potassium fluoride,[2] potassium hydrogen fluoride,[3] or potassium fluorosulfinate.[4] Acid fluorides were also prepared from carboxylic acids using sulfur tetrafluoride,[5] diethyl(2-chloro-1,1,2-trifluoroethyl)amine,[6] thionyl fluoride,[7] benzoyl fluoride,[8] and selenium tetrafluoride.[9] These methods are less direct and are often incompatible with other functional groups in the substrate.

In a typical procedure,[1] a solution of 0.4 equiv of cyanuric fluoride in acetonitrile is added to a mixture of 1.0 equiv each of a carboxylic acid and pyridine in acetonitrile at room temperature. After being stirred for 1 h, the reaction is subjected to a standard aqueous work-up which then provides the desired acid fluoride in generally high yields for a variety of carboxylic acid starting materials (eq 1). Each mole equivalent of cyanuric fluoride reacts to give three moles of acyl fluoride.

The ability to prepare acid fluorides under these mild conditions opened up a number of synthetic applications. Perhaps the most significant application to date is the increased use of acid fluorides as activated intermediates in peptide bond formation.[10,11] The use of Fmoc amino acid chlorides as activated N-protected peptide building blocks has been limited by the propensity of *t*-butyl ester- or ether-protected side chains (e.g. serine, threonine, glutamate, aspartate) to cyclize onto the acid chloride with the loss of *t*-butyl chloride. Carpino et al. have shown that the corresponding amino acid fluorides are not subject to this same side reaction.[10] The general reactivity pattern for acid fluorides compared with acid chlorides is that they are less sensitive to uncharged oxygen nucleophiles such as water or methanol but display similar reactivity to charged or nitrogen nucleophiles. The Fmoc amino acid fluorides were prepared from the corresponding Fmoc amino acids under similar conditions to those described above and were isolated as white crystalline solids. A number of the exemplified amino acids contained protected acid and amide side chains (eq 2 and 3).

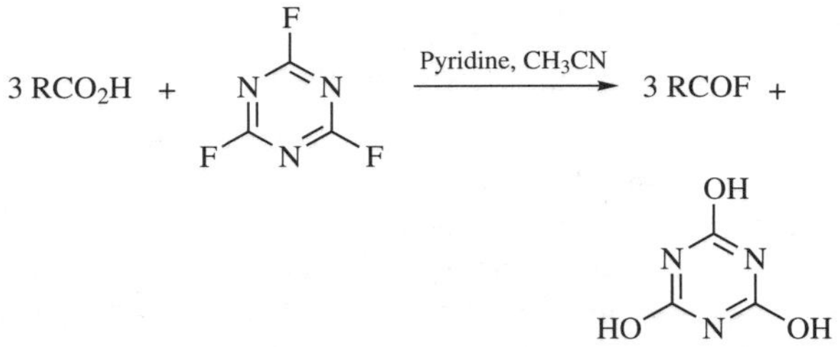

(1)

Acid	Yield (%)
Acetic acid	70
Propiolic acid	85
Cyclopentylcarboxylic acid	54
Cyclohexylcarboxylic acid	95
Cinnamic acid	95
Benzoic acid	97

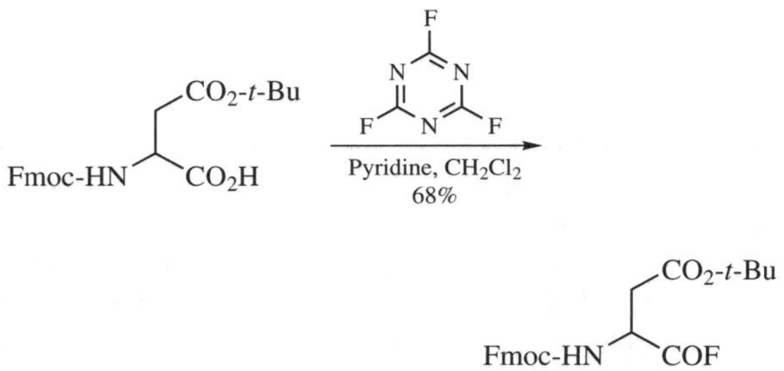

(2)

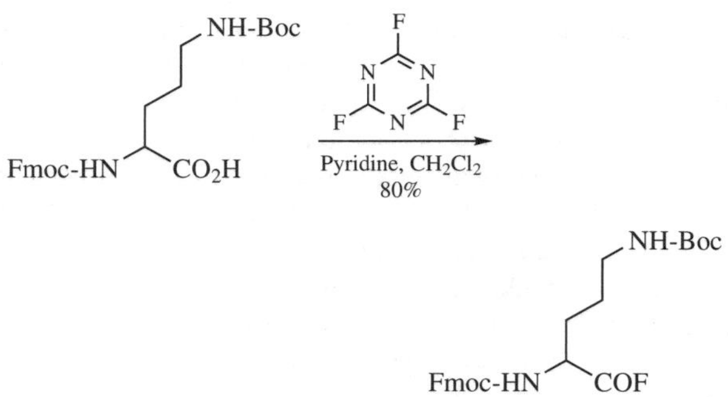

(3)

Acid fluorides have the additional advantage that they are more reactive in peptide coupling reactions of hindered amino acid derivatives. This is due at least in part to the small size of the fluoride leaving group.[12] Hindered acid fluorides like Fmoc-Aib-F can be prepared in the usual way with cyanuric fluoride.

Reduction of Acids to Alcohols. The mild conditions required for the conversion of carboxylic acids to acid fluorides has resulted in a method for the reduction of carboxylic acids to alcohols by way of the acid fluorides.[13] A variety of carboxylic acids were converted to acid fluorides under standard conditions. A quick aqueous work-up was followed by $NaBH_4$ reduction (eq 4). This mild reaction sequence has the distinct advantage over $NaBH_4$-I_2 in that it is compatible with N-protected amino acids. The resulting alcohols are obtained in generally high yields with no observable loss in enantiomeric purity.

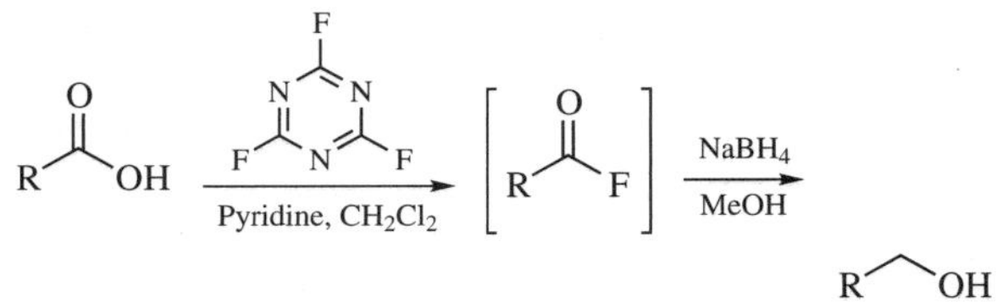

(4)

R	Yield (%)
Ph	95
Bn	94
$-(CH_2)_8Me$	92
Cbz-Phe	81
Fmoc-Leu	90

Other Uses. A recent use for cyanuric fluoride is the activation of alkenyl boronic acids for the 1,4 addition of α,β-unsaturated ketones.[14,15] Hara et al. have shown that alkenyl boronic acids when treated with cyanuric fluoride undergo 1,4 addition to unsaturated ketones.[14] Presumably, this reaction takes place via the alkenyl fluoro alkoxy borane intermediate followed by coordination of the carbonyl oxygen to the boron and intramolecular carbon-carbon bond formation (eq 5). Selective 1,4 additions to α,β,α′, β′-unsaturated ketones could be achieved by introducing substituents on one of the double bonds (eq 6).[15]

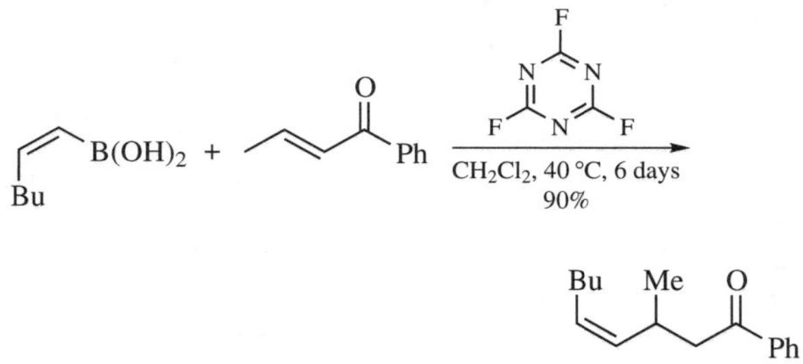

(5)

Cyanuric fluoride has also been used as a reagent for the selective deoxygenation of sulfoxides (eq 7 and 8).[16] Stirring the sulfoxide with cyanuric fluoride in dioxane followed by aqueous work-up gave good yields of the corresponding sulfides.

Finally, cyanuric fluoride has been used in solid-phase synthesis as a method for activating a carboxylic acid functionality on solid support for loading a nucleophile onto a resin.[17,18] For example, in 1998 Böhm et al. demonstrated carboxypolystyrene activation as the acyl fluoride followed by resin loading with a primary amine linker (eq 9).[17] In 1999, Sams et al. employed a similar method to activate substituted benzoic acids loaded onto Wang resin with *N*-methylmorpholine (NMM) in dichloromethane. The resulting acyl fluorides were coupled with derivatives of *N*-hydroxyamidine in the presence of *N*-methylmorpholine (eq 10).[18]

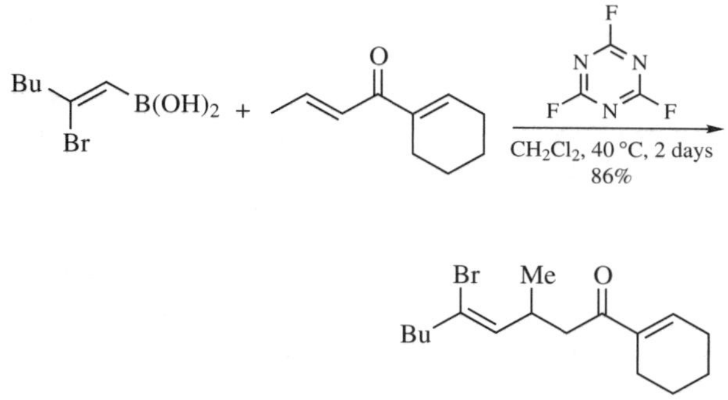

(6)

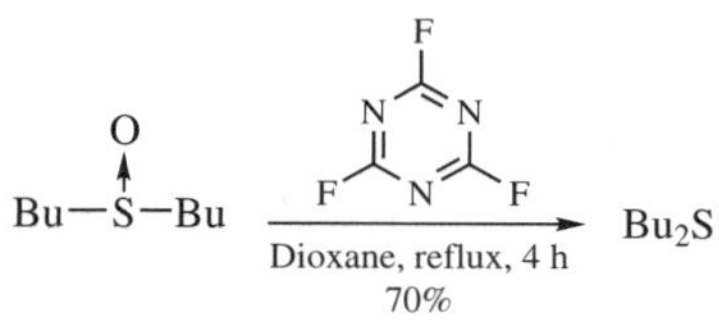

(7)

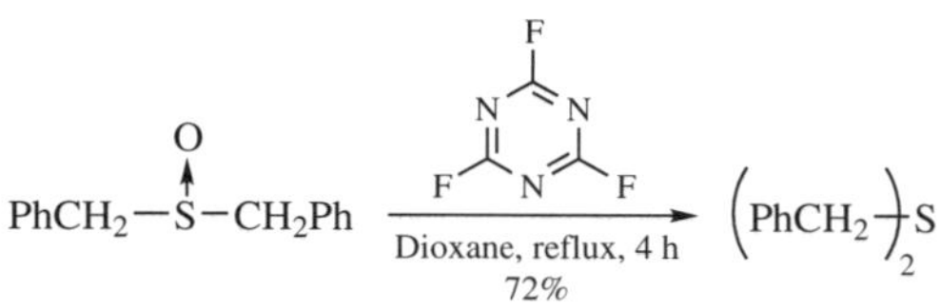

(8)

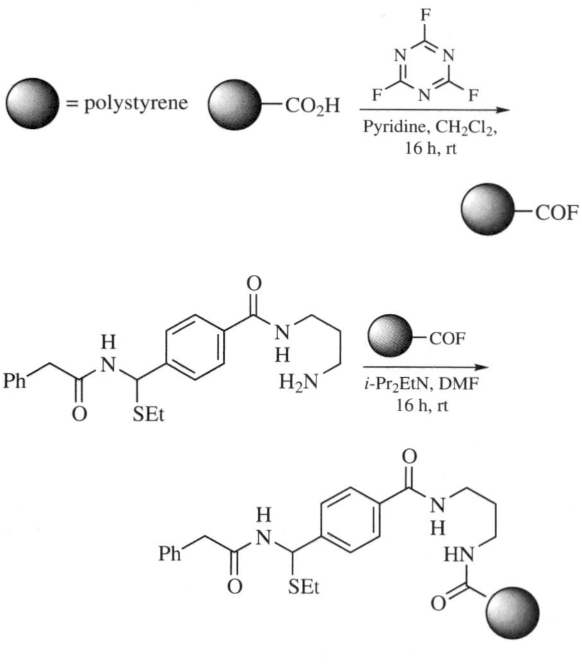

(9)

1. Olah, G. A.; Nojima, M.; Kerekes, I., *Synthesis* **1973**, 487.
2. Pittman, A. G.; Sharp, D. L., *J. Org. Chem.* **1966**, *31*, 2316.
3. Olah, G. A.; Kuhn, S.; Beke, S., *Chem. Ber.* **1956**, *89*, 862.
4. Seel, F.; Lange, J., *Chem. Ber.* **1958**, *91*, 2553.
5. Hasek, W. R.; Smith, W. C.; Engelhardt, V. A., *J. Am. Chem. Soc.* **1960**, *82*, 543.

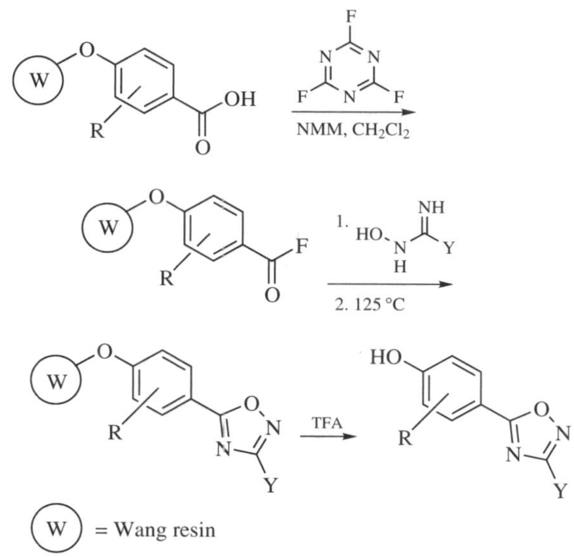

(10)

6. Yarovenko, N. N.; Radsha, M. A., *J. Gen. Chem. USSR (Engl. Transl.)* **1959**, *29*, 2125.
7. Roberts, H. L., British Patent 908177 (1963) (*Chem. Abstr.* **1963**, *58*, 5579.)
8. Olah, G. A.; Kuhn, S. J., *Org. Synth.* **1965**, *45*, 3.
9. Olah, G. A.; Nojima, M.; Kerekes, I., *J. Am. Chem. Soc.* **1974**, *96*, 925.
10. Carpino, L. A.; Sadat-Aalaee, D.; Chao, H. G.; DeSelms, R. H.; *J. Am. Chem. Soc.* **1990**, *112*, 9651.
11. Carpino, L. A.; Beyermann, M.; Wenschuh, H.; Bienert, M., *Acc. Chem. Res.* **1996**, *29*, 268.
12. Wenschuh, H.; Beyermann, M.; Krause, E.; Brudel, M.; Winter, R.; Schümann, M.; Carpino, L. A.; Bienert, M., *J. Org. Chem.* **1994**, *59*, 3275.
13. Kokotos, G.; Noula, C., *J. Org. Chem.* **1996**,*61*, 6994.
14. Hara, S.; Shudoh, H.; Ishimura, S.; Suzuki, A., *Bull. Chem. Soc. Jpn.* **1998**, *71*, 2403.
15. Hara, S.; Ishimura, S.; Suzuki, A., *Synlett* **1996**, 993.
16. Olah, G. A.; Fung, A. P.; Gupta, B. G. B.; Narang, S. C., *Synthesis* **1980**, 221.
17. Böhm, G.; Dowden, J.; Rice, D. C.; Burgess, I.; Pilard, J.-F.; Guilbert, B.; Haxton, A.; Hunter, R. C.; Turner, N. J.; Flitsch, S. L., *Tetrahedron Lett.* **1998**, *39*, 3819.
18. Sams, C. K.; Lau, J., *Tetrahedron Lett.* **1999**, *40*, 9359.

David A. Barda
Eli Lilly and Company, Indianapolis, IN, USA

Cyclohexyl Chloroformate

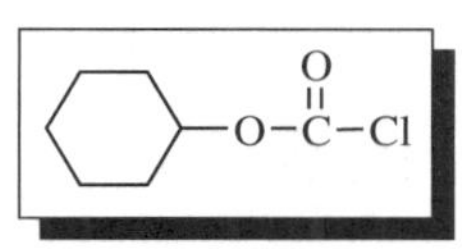

[13248-54-9] $C_7H_{11}ClO_2$ (MW 162.62)

(cyclohexyloxycarbonyl (Hoc); amine and indole protecting group for peptide synthesis)

Alternate Names: carbonochloridic acid cyclohexyl ester; cyclohexyloxycarbonyl chloride; chloroformic acid cyclohexyl ester.
Analysis of Reagent Purity: ^{1}H NMR.
Form Supplied in: colorless liquid; not commercially available.
Solubility: soluble in most organic solvents.
Handling, Storage, and Precautions: decomposes to CO_2 and cyclohexyl chloride on prolonged storage. Store in a refrigerator. Bottles may develop pressure and should be cooled before opening. Use in a fume hood. Irritating to skin, eyes, mucous membranes.
Physical Data: bp 38–44 °C 2 mmHg;[1] n_D^{20} 1.4587.[1]
Preparative Methods: a 1 L round-bottomed flask equipped with a delivery tube extending to the bottom of the flask, a pressure-equalizing addition funnel and a cold-finger condenser cooled with dry ice-acetone were placed in an ice bath. The exit tube of the condenser was connected to a $CaCl_2$ drying tube and suitable trap for phosgene and HCl. Phosgene was introduced into the flask until 109 g (1.1 mol) has been condensed. The reaction flask is stirred while 100 g (1 mol) of cyclohexyl alcohol was added through the dropping funnel at a rate to keep the mixture in gentle reflux for 1 h. The flask was allowed to stand at room temperature with stirring for 2 h. The solution was then concentrated under reduced pressure at room temperature to remove HCl and the excess phosgene. The residue may be used without further purification.
Note: Phosgene is a highly toxic gas; use in a fume hood.

Introduction. The cyclohexyloxycarbonyl (Hoc) group was originally developed by McKay and Albertson as acid labile N^{α}-amino protection used for peptide synthesis.[2] The reagent used for the introduction of the Hoc group, cyclohexyl chlorofomate, is easily prepared from inexpensive starting materials, cyclohexyl alcohol, and phosgene (eq 1).

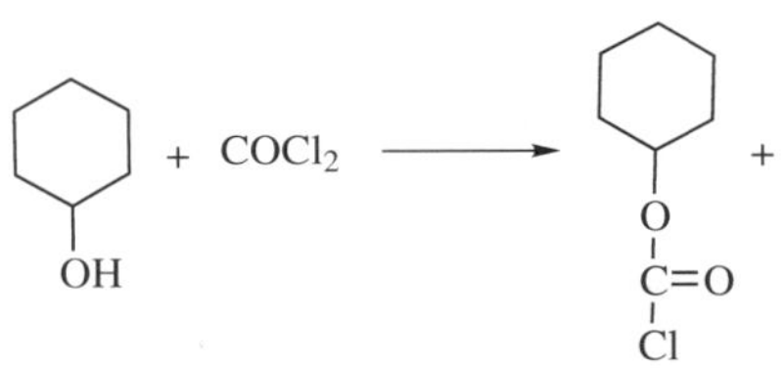

(1)

In elongating peptide chains properly, there must be a highly compatibility between the protecting groups for N^{α}-amino function and amino acid side-chain functional groups. The N^{α}-Hoc group is readily cleaved by strong acids such as hydrogen bromide (HBr) in acetic acid[2] or liquid hydrogen fluoride (HF),[3] although most side-chain protecting groups are also removed under these conditions. When compared with other N^{α}-amino protecting groups commonly used for *t*-butyloxycarbonyl (Boc)- or benzyloxycarbonyl (Z)-mediated peptide synthesis, however, the Boc and the Z groups can be selectively removed by much milder conditions such that the side-chain protecting groups remain intact. Therefore, the Hoc group has not gained the practical application for N^{α}-amino protection since it offers no advantage over the Boc and Z groups. However, recently the use of Hoc group has been used as the N^{in}-protecting group for the synthesis of tryptophan (Trp)-containing peptides[4] and also as another orthogonal amino-protecting group in Boc chemistry.[5]

Hoc Group as a Base-Resistant N^{in}-protecting group for Trp. The Hoc group, which is compatible with Boc-mediated peptide synthesis, was introduced to the indole moiety of Trp to avoid indole oxidation and alkylation during repetitive acid treatment such as with trifluoroacetic acid (TFA).[4,6] The Hoc group was introduced by acylation of Boc-Trp-OPac (Pac = phenacyl) or Boc-Trp-OBzl using cyclohexyl chloroformate in dichloromethane in the presence of pulverized NaOH and a catalytic amount of tetra-*n*-butyl ammonium hydrogensulfate (TBAHS) at room temperature for 1 h. Removal of the Pac or Bzl ester by using zinc dust in acetic acid or catalytic hydrogenation gave the Trp derivative (eq 2).

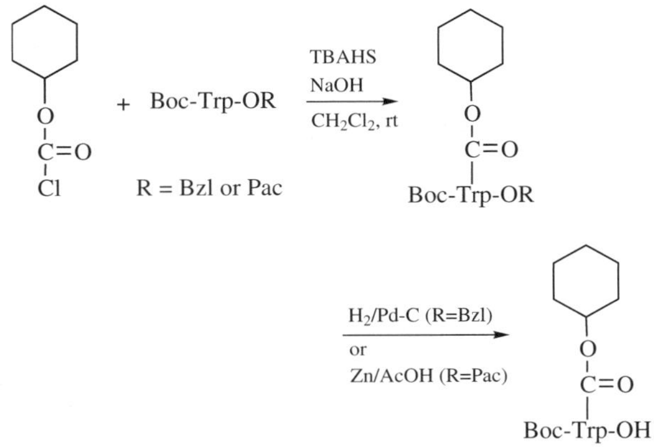

(2)

In contrast with the widely employed formyl protecting group for Trp,[7] the Hoc group is readily cleaved by HF without resorting to the use of thiols and is also far more stable under conditions using basic nucleophiles (e.g. morpholine and piperidine)[4] (Table 1).

Table 1 Stability of the side-chain protecting groups for Trp in acid and base. Determined by HPLC after treatment of Boc-Trp(X) under acidic (500 eq) or basic (100 eq) conditions (rt, 18 h)

	Loss of protecting group X (%)	
X=	Hoc	For
TFA	0	0
10% DIEA/DMF	0	28
10% piperidine/DMF	0	100
20% morpholine/DMF	0	100

When synthesizing Trp-containing peptides by the Boc-mediated peptide synthesis, therefore, the N^{in}-Hoc group is applicable not only to the standard solid-phase peptide synthesis but also to the preparation of protected peptide segments on base labile linkers such as an *N*-[9-(hydroxymethyl)-2-fluorenyl]succinamic acid (HMFS) linker.[8] The HMFS linker is designed to release fully protected segments with a free α-carboxyl group by treatment with 20% morpholine in DMF. The Hoc group on Trp is completely stable under basic conditions for detachment of the segments from the HMFS resin. The protected segments thus obtained are used for the subsequent segment condensation reaction performed using either solution or solid phase method. Employing this procedure based on performing segment assembly on the solid support and segment condensation in solution, the synthesis of muscarinic toxin 1 (a 66-residue peptide)[9] and green fluorescent protein (a 238-residue protein)[10] was reported. With accumulation of experiments employing the N^{in}-Hoc group, however, it became apparent that modification of Trp associated with use of this group produces two *trans*-isomers of 2-[2′-(*p*-cresol)]-2,3-dihydrotryptophan (**1**) during the HF reaction in the presence of *p*-cresol as a scavenger.[11]

This side reaction for Trp proved to be inherent to the carbamate-type N^{in}-protecting groups (e.g. Hoc, Z and 2,4-dimethylpent-3-yloxycarbonyl[12]). The use of thiols such as butane-1,4-dithiol during the HF reaction was found to almost completely avoid this side reaction, although the Hoc group could be removed by HF without any thiol.

Hoc Group as Another Orthogonal Amino-Protecting Group in Boc Chemistry. In Boc-mediated peptide synthesis, the Boc group is used as N^{α} protection in combination with benzyl (Bzl)/cyclohexyl (Hex)-based side-chain protecting groups. Thus, the orthogonality of protecting scheme consists in the difference in reaction rate when acidolysis is used to remove different classes of protecting groups: generally TFA and HF are employed for removal of the Boc and the side-chain protecting groups, respectively. Regarding removal of the side-chain protecting groups, Bzl-based groups can be cleaved by 1M trimethylsilyl trifluoromethanesulphonate (TMSOTf)-thioanisole/TFA mixture[13] to which Hex-based groups are stable. Therefore, free amino and/or carboxyl groups protected by Bzl-based groups are available prior to cleaving Hex-based protecting groups. By taking advantage of the difference in their removability under acidic conditions, a possibility for the assembly of cyclic lactam peptides was demonstrated.[5]

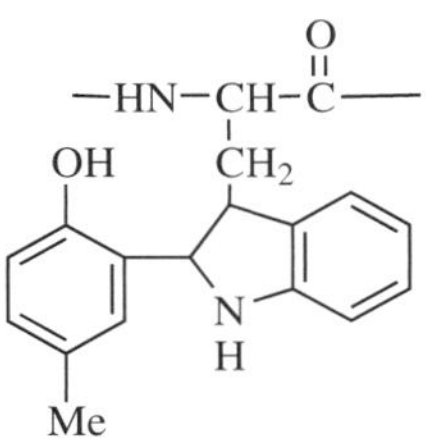

1

Related Reagents. 2-(*t*-Butoxycarbonyloxyimino)-2-phenylacetonitrile (BOC-ON), *t*-butylchloroformate (Boc-Cl), benzyl chloroformate.

1. Fourneau, M. E.; Montaigne, M.; Puyal, J., *J. Anales Soc. Espanol Fis. Quim.* **1920**, *73*, 323.
2. McKay, F. C.; Albertson, N. F., *J. Am. Chem. Soc.* **1957**, *79*, 4686.
3. Sakakibara, S.; Shimonishi, Y.; Kishida, Y.; Okada, M.; Sugihara, H., *Bull. Chem. Soc. Jpn* **1967**, *40*, 2164.
4. Nishiuchi, Y.; Nishio, H.; Inui, T.; Kimura, T.; Sakakibara, S., *Tetrahedron Lett.* **1996**, *37*, 7529.
5. Mező, G.; Mihala, N.; Kóczán, G.; Hudecz, F., *Tetrahedron* **1998**, *54*, 6757.
6. Jaeger, E.; Thamm, P.; Knof, S.; Wünsch, E.; Löw, M.; Kisfaludy, L., *Hoppe-Seyler's Z. Physiol. Chem.* **1978**, *359*, 1617.
7. Tam, J. P.; Heath, W. F.; Merrifield, R. B., *J. Am. Chem. Soc.* **1983**, *105*, 6442.
8. Rabanal, F.; Giralt, E.; Albericio, F. *Tetrahedron* **1995**, *51*, 1449.
9. Nishiuchi, Y.; Nishio, H.; Inui, T.; Bódi, J.; Kimura, T., *J. Peptide Sci.* **2000**, *6*, 84.
10. Nishiuchi, Y.; Inui, T.; Nishio, H.; Bódi, J.; Kimura, T.; Tsuji, F.I.; Sakakibara, S., *Proc. Natl. Acad. Sci. USA* **1998**, *95*, 13549.
11. Nishio, H.; Nishiuchi, Y.; Inui, T.; Yoshizawa-Kumagaye, K.; Kimura, T., *Tetrahedron Lett.* **2000**, *41*, 6839.
12. Karlström, A.; Undèn, A., *Chem. Commun.* **1996**, 1471.
13. Fujii, N.; Otaka, A.; Ikemura, O.; Akaji, K.; Funakoshi, S.; Hayashi, Y.; Kuroda, Y.; Yajima, H. J., *Chem. Soc., Chem. Commun.* **1987**, 274.

Yuji Nishiuchi
Peptide Institute Inc., Osaka, Japan

1-Cyclohexyl-3-(2-morpholinoethyl) carbodiimide[1]

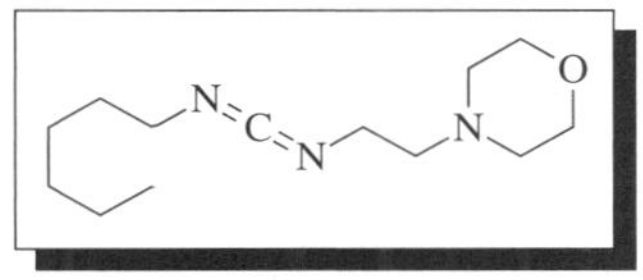

[15580-20-8] $C_{13}H_{23}N_3O$ (MW 237.34)

(coupling agent for peptide synthesis; used to make sulfate esters from alcohols)

Physical Data: bp 145 °C/0.2 mmHg.
Solubility: sol DMF, CH_2Cl_2, dioxane, xylene.
Preparative Methods: from cyclohexyl isothiocyanate and *N*-(2-aminoethyl)morpholine (67% yield);[2] by dehydration of the corresponding urea with p-Toluenesulfonyl Chloride, Potassium Carbonate, and Benzyltriethylammonium Chloride (91% yield);[3] or by the reaction of cyclohexyl isocyanate, 2-(4-morpholinyl)ethyl diethylphosphoramidate, and potassium carbonate (82% yield).[4] The corresponding *N*-metho-*p*-toluenesulfonate salt is commercially available (see 1-Cyclohexyl-3-(2-morpholinoethyl)carbodiimide Metho-p-toluenesulfonate).

Peptide Synthesis. Amide bonds may be formed using carbodiimides but workup procedures are sometimes complicated by the presence of the urea byproduct. 1-Cyclohexyl-3-(2-morpholinoethyl)carbodiimide (**1**) carries a tertiary amine and has a urea derivative that is soluble in dilute acid, making product separation far easier. It has been used to couple glycine ethyl ester hydrochloride and phthaloyl-L-phenylalanine to give phthaloyl-L-phenylalanylglycine ethyl ester in 81% yield (eq 1).[2]

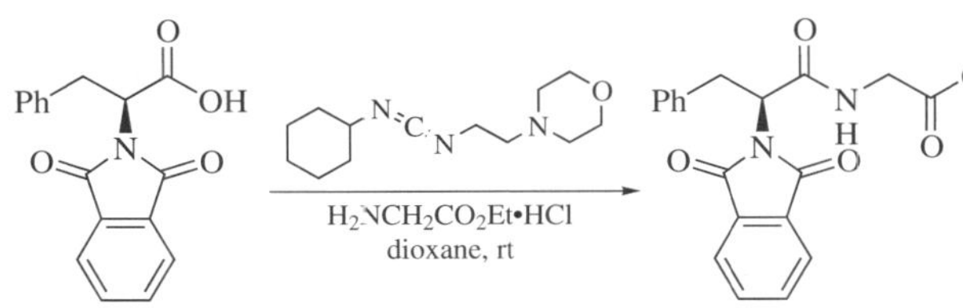

(1)

(**1**) has been found to be superior to 1,3-Dicyclohexylcarbodiimide (DCC) for the final cyclization step in the synthesis of a large-ring peptide lactone related to the antibiotic vernamycin B_{α}, with methylene chloride being the solvent of choice.[5] Of all the other coupling reagents examined, only N,N′-Carbonyldiimidazole gave any cyclization product, although in lower yields.

Preparation of Sulfate Esters. 1-Cyclohexyl-3-(2-morpholinoethyl)carbodiimide has been used to mediate the sulfation of

1-hexadecanol to give the corresponding monoalkyl sulfate in 56% yield (eq 2).[6] DCC is superior in this reaction, giving the monoalkyl sulfate in 92% yield.

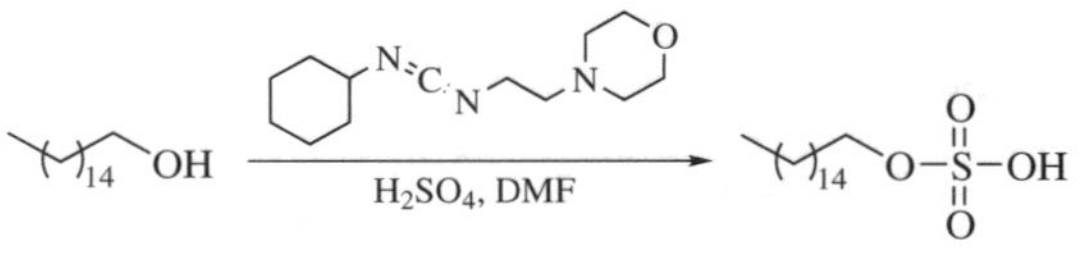

(2)

1. For general reviews on the use of carbodiimides, see (a) Kurzer, F.; Douraghi-Zadeh, K. *Chem. Rev.* **1967**, *67*, 107. (b) Williams, A.; Ibrahim, I. T. *Chem. Rev.* **1981**, *81*, 589. (c) Mikolajczyk, M.; Kielbasiński, P. *Tetrahedron* **1981**, *37*, 233.
2. Sheehan, J. C.; Hlavka, J. J. *J. Org. Chem.* **1956**, *21*, 439.
3. Jászay, Z. M.; Petneházy, I.; Tϕke, L.; Szajáni, B. *Synthesis* **1987**, 520.
4. Jászay, Z. M.; Petneházy, I.; Tϕke, L.; Szajáni, B. *Synthesis* **1988**, 397.
5. Ondetti, M. A.; Thomas, P. L. *J. Am. Chem. Soc.* **1965**, *87*, 4373.
6. Mumma, R. O.; Hoiberg, C. P. *J. Chem. Eng. Data* **1971**, *16*, 492.

Andrew M. Griffin
University of Bristol, UK

1-Cyclohexyl-3-(2-morpholinoethyl) carbodiimide Metho-*p*-toluenesulfonate[1]

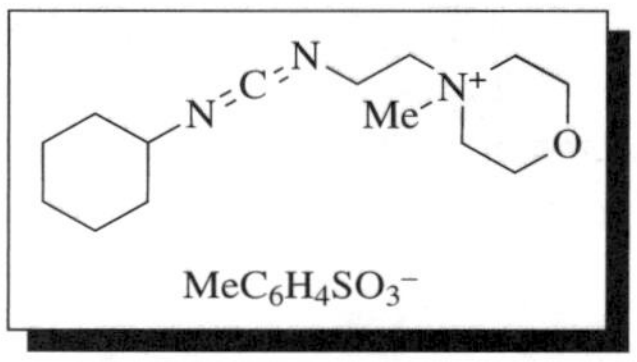

[2491-17-0]	$C_{21}H_{33}N_3O_4S$	(MW 423.58)

(used extensively in the activation of alcohols and carboxylic acids)

Alternate Name: morpho-CDI.
Physical Data: mp 113–115 °C.
Solubility: sol DMF, MeCN.
Form Supplied in: commercially available as a colorless solid.
Preparative Methods: prepared by quaternization of the corresponding carbodiimide (see 1-Cyclohexyl-3-(2-morpholinoethyl)carbodiimide) with methyl *p*-toluenesulfonate. The carbodiimide is obtained by dehydration of the thiourea.[2]
Handling, Storage, and Precautions: assumed to be harmful by ingestion and skin contact.

Peptide Synthesis.[2] Phthaloyl-L-phenylalanyl-L-leucine ethyl ester has been prepared in 80% yield by the 1-cyclohexyl-3-(2-morpholinoethyl)carbodiimide metho-*p*-toluenesulfonate (**1**) mediated coupling of L-leucine ethyl ester hydrochloride to phthaloyl-L-phenylalanine (eq 1). Any excess of reagent and the corresponding urea were easily separated from the product by washing with aqueous solutions. The ease of separation of byproduct is the main advantage of this reagent over 1,3-Dicyclohexylcarbodiimide(DCC).

Macrocyclic Lactonization.[3] Lactonization of secoacid (**2**) was effected with (**1**), and subsequent cleavage of the methoxymethyl ether results in the antibiotic milbemycin β_3 (**3**) in 85% overall yield (eq 2).

Cyclodehydrations of *N*-Acyl-α-amino Acids.[4] 2-Oxazolin-5-ones are conveniently prepared by dehydration of *N*-acyl-α-amino acids with (**1**) (e.g. eq 3).

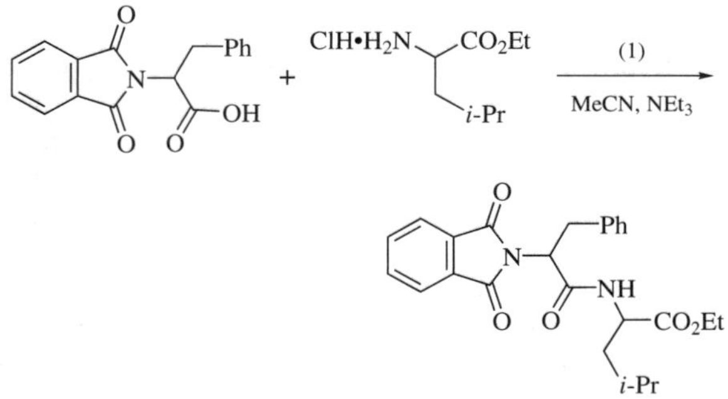

(1)

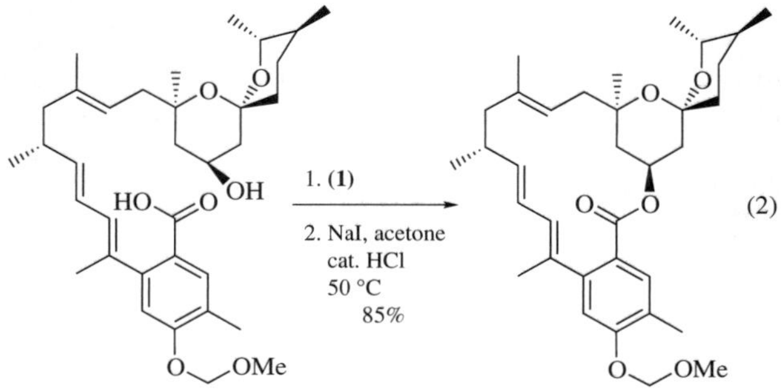

(2) (3)

(2)

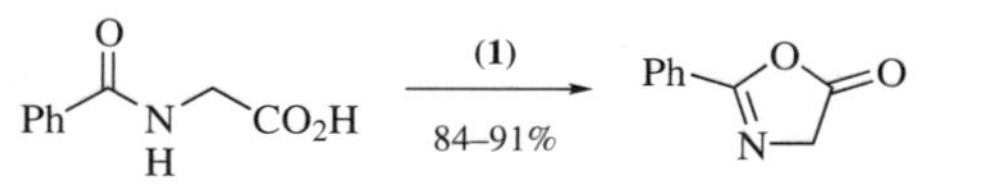

(3)

Modified Moffat-Type Oxidations.[5] (**1**) used in conjunction with DMSO has been used to oxidize alcohols to ketones, as exemplified in the total synthesis of kopsanone (eq 4). This is another example where the advantage of the water solubility of(**1**) and the corresponding urea makes the use of this reagent more attractive than DCC as the diimide component.

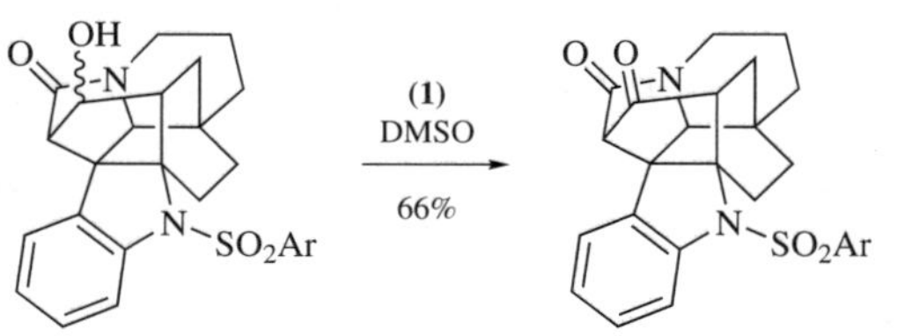

(4)

Dehydration of α-Hydroxymethyl Lactones.[6] α-Hydroxymethyl lactones can be dehydrated to the corresponding α-methylene lactone with the use of (**1**) in the presence of Copper(II) Chloride (eq 5). This procedure has been successfully demonstrated in the total synthesis of isolobophytolide.

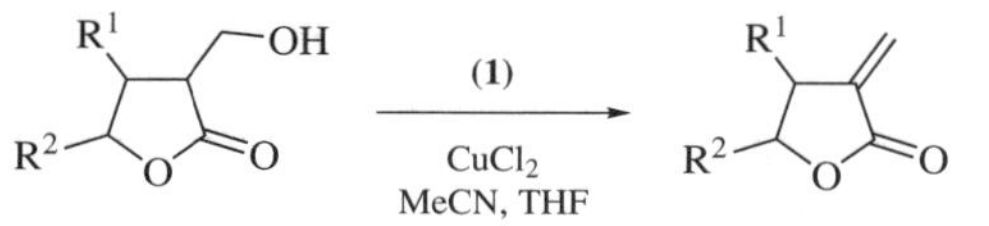

(5)

1. For general reviews on carbodiimides, see (a) Kurzer, F.; Douraghi-Zadeh, K. *Chem. Rev.* **1967**, *67*, 107; (b) Williams, A.; Ibrahim, I. T. *Chem. Rev.* **1981**, *81*, 589; (c) Mikolajczyk, M.; Kielbasiński, P. *Tetrahedron* **1981**, *37*, 233.
2. Sheehan, J. C.; Hlavka, J. J. *J. Org. Chem.* **1956**, *21*, 439.
3. Williams, D. R.; Barner, B. A.; Nishitani, K.; Phillips, J. G. *J. Am. Chem. Soc.* **1982**, *104*, 4708.
4. Fieser, M. *Fieser & Fieser* **1984**, *11*, 151.
5. Magnus, P.; Gallagher, T.; Brown, P.; Huffman, J. C. *J. Am. Chem. Soc.* **1984**, *106*, 2105.
6. Marshall, J. A.; Andrews, R. C.; Lebioda, L. *J. Org. Chem.* **1987**, *52*, 2378.

Peter Szeto
University of Bristol, UK

17*H*-Cyclopenta[1, 2-1,:3,4-1′] diphenanthren-17-yl Chloroformate

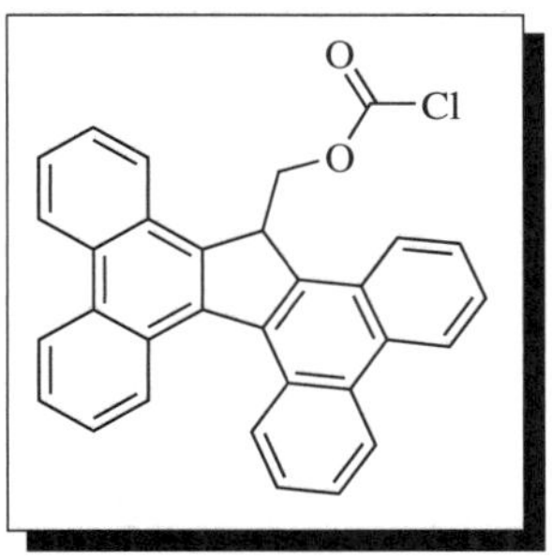

[139746-34-2] $C_{31}H_{19}ClO_2$ (MW 458.94)

(base labile protecting group for primary amines, used in peptide synthesis for affinity purification)

Alternate Name: (17-tetrabenzo[*a,c,g,i*]fluorenyl)methyl chloroformate, Tbfmoc-Cl.
Physical Data: mp 194–195 °C.
Solubility: soluble in organic solvents (CH_2Cl_2, THF, dioxane, toluene, and DMF).
Preparative Methods: preparation involves addition of triphosgene to the corresponding alcohol prepared in four steps from bromophenanthrene.[1,2]
Handling, Storage, and Precautions: Tbfmoc-Cl is an acid chloride, and should be protected from moisture and heat. The reagent can evolve CO_2 upon prolonged storage; therefore bottles should be opened carefully. Polycyclic aromatic hydrocarbons are toxic and should be handled with caution.

Introduction. The Tbfmoc group is an N^{α}-amino protecting group; it has been designed to facilitate the elimination of truncated peptides and other impurities accumulated during solid phase peptide synthesis (SPPS) from the desired product. It is structurally similar to 9-fluorenylmethyl chloroformate (Fmoc-Cl) and is therefore incorporated and cleaved by standard procedures in SPPS. Four additional aromatic rings have been fused onto the basic Fmoc-Cl structure. These extra rings are able to form strong non-covalent affinity interactions with porous graphitized carbon (PGC), which is a strong hydrophobic adsorbent. This property allows the selective retention of Tbfmoc derivatives on PGC.

Incorporation of the Tbfmoc Protecting Group. The Tbfmoc group has been introduced either directly with Tbfmoc-Cl, by sonication in the presence of diisopropylethylamine, to the *N*-termini of resin bound peptides[1] (eq 1); or with Tbfmoc-Gly-OH as the terminal amino acid in solid phase peptide synthesis[3,4] (eq 2).

H-Leu-Ile-Phe-Ala-Gly-Resin —(Tbfmoc-Cl / DIEA / CH_2Cl_2, sonication; quantitative)→

Tbfmoc-Leu-Ile-Phe-Ala-Gly-Resin (1)

H-Ser-Met-Val-Leu-Ser-Resin —(Tbfmoc-Gly-OH, DIC /dioxane; 80%)→

Tbfmoc-Gly-Ser-Met-Val-Leu-Ser-Resin (2)

After incorporation of the Tbfmoc group, the Tbfmoc peptides can be cleaved from the resin by standard methods (eq 3).

Purification and Deprotection of Tbfmoc-peptides. Two procedures for purification have been developed. In the first, a solution of the crude Tbfmoc peptides in 70% aqueous acetonitrile is applied to a short column of PGC. Under these conditions, the Tbfmoc peptides are totally retained, whereas the acetylated truncated peptide impurities are eluted.

Deprotection of the Tbfmoc group and concomitant release of the purified peptide is effected by washing the column with 1–20% piperidine/70% aqueous acetonitrile (eq 3). The second procedure involves direct addition of PGC to a solution of the crude peptide in 70% aqueous acetonitrile, whereby the Tbfmoc peptide is adsorbed onto the surface of the PGC and the impurities remain in the solution. The impurities are removed by filtration and the peptide released from the PGC by a suitable base treatment (10% piperidine/70% aqueous acetonitrile). During the process of deprotection, the protecting group residues[2] are retained on the PGC.

Tbfmoc-Gly-Ser-Met-Val-Leu-Ser-Resin —(1. TFA/anisole/EtSMe; 2. adsorption on PGC; 3. 20% piperidine; 85%)→

H-Gly-Ser-Met-Val-Leu-Ser-OH (3)

The Tbfmoc group also allows the simplification of peptide purification by RP–HPLC. The highly hydrophobic Tbfmoc peptides elute significantly later than the hydrophilic *N*-acetylated truncated peptides. Furthermore, unambiguous identification of the Tbfmoc peptides is facilitated by the specific UV absorbance of the Tbfmoc group at 364 nm and its fluorescence at 424 nm.

These purification methodologies have been applied to the synthesis of large peptides[2,3] and peptide libraries.[5]

1. Brown, A. R.; Irving, S. L.; Ramage, R., *Tetrahedron Lett.* **1993**, *34*, 7129.
2. Brown, A. R.; Irving, S. L.; Ramage, R.; Raphy, G., *Tetrahedron* **1995**, *51*, 11815.
3. Ramage, R.; Raphy, G., *Tetrahedron Lett.* **1992**, *33*, 385.
4. U.K. Pat. Application: GB 2 251 242 A (to Ramage, R.; Raphy, G.) (1992).
5. Ramage, R.; Swenson, H. R.; Shaw, K. T., *Tetrahedron Lett.* **1998**, *39*, 8715.

Stéphane Gastaldi
Université Aix-Marseille III, UMR 6517, France

α-Cyclopropylcyclopropanemethanol

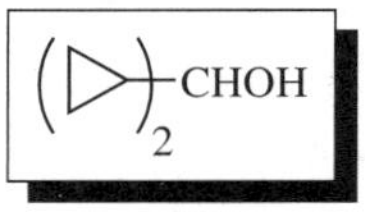

[14300-33-5] $C_7H_{12}O$ (MW 112.17)

(reagent for carboxylic acid protection[1])

Alternate Name: dicyclopropylmethanol, dicyclopropylcarbinol, Dcpm-OH.
Physical Data: bp 66–68 °C (12 mmHg),[2] 48 °C (3 mmHg).[3]
Solubility: soluble in all common organic solvents.
Analysis of Reagent Purity: ^{1}H NMR, elemental analysis.
Preparative Methods: the title reagent can be prepared by the reduction of dicyclopropyl ketone by means of $LiAlH_4$[2,3] or $NaBH_4$.[4]
Handling, Storage, and Precautions: stable under normal conditions but suffers ring opening in the presence of acids.[5] Some Fmoc amino acid Dcpm esters suffer deblocking to the free acid on TLC or column chromatography over silica gel.[1]

Carboxylic Acid Protection. The Dcpm group was recommended for protection of the carboxyl function in cases where selectivity in deblocking relative to *O-tert*-butyl or *N*-trityl protection was needed.[1] An example involves the rapid solution assembly of **1** and its conversion to **2** by treatment with 1% trifluoroacetic acid in methylene chloride.

Fmoc-Ile-Thr(*tert*-Bu)-Arg(Pmc)-Gln(Trt)-Arg(Pmc)-Tyr(*tert*-Bu)-ODcpm

1

Fmoc-Ile-Thr(*tert*-Bu)-Arg(Pmc)-Gln(Trt)-Arg(Pmc)-Tyr(*tert*-Bu)-OH

2

A general method has been given for the synthesis of amino acid Dcpm esters via the acid chloride or acid fluoride in the presence of pyridine.[1]

Homoallylic Rearrangement of Dcpm-OH. As is the case for most secondary and tertiary cyclopropylmethyl alcohols, Dcpm-OH undergoes ring opening with conversion to the homoallyl alcohol on treatment with an acidic reagent (eq 1).[5–8] The *trans* product is favored. Ring opening is avoided in the presence of a good nucleophile such as a mercaptan or azide ion.

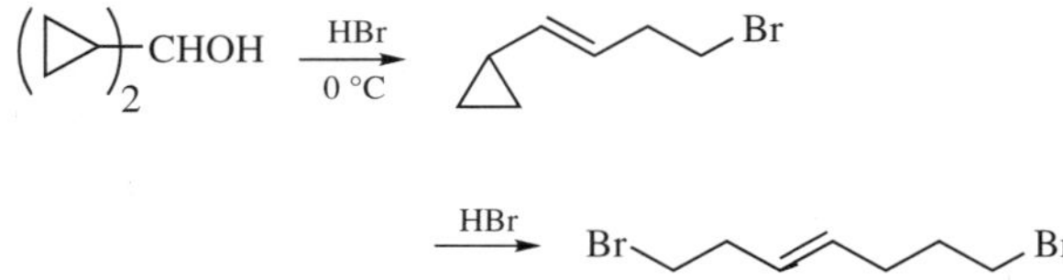

(1)

Conversion to Ring-Retained Sulfides. Dcpm protection of mercaptans is possible via reaction of the alkoxide of Dcpm-OH in the presence of *N*,*N*-methylphenylaminotriphenylphosphonium iodide (eq 2).[9]

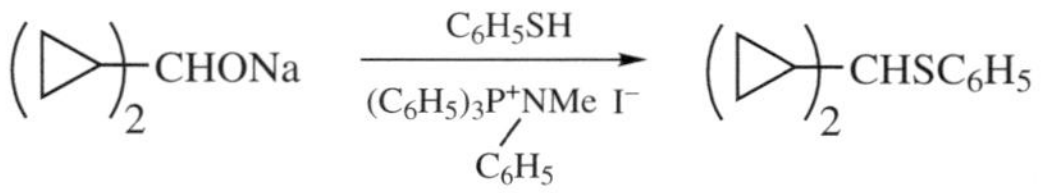

(2)

Conversion to Dcpm-NH$_2$. Dcpm-OH is a convenient source of the corresponding amine, Dcpm-NH$_2$, useful as a less-sensitive amide protectant than the Dmcp analog.[1] Reaction with NaN_3/CCl_3CO_2H to give the azide is followed by reduction to the amine (eq 3).[1]

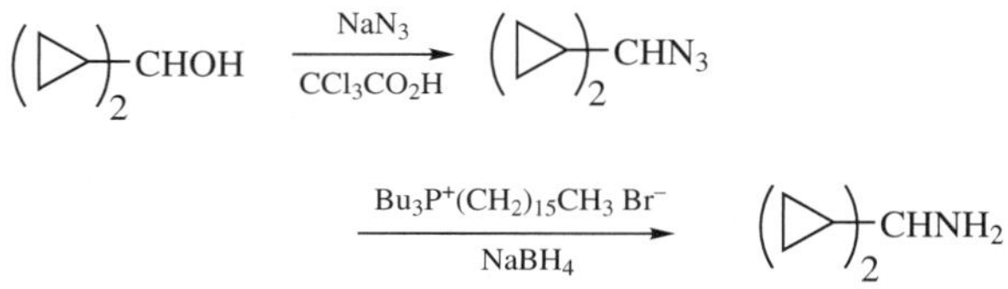

(3)

1. Carpino, L. A.; Chao, H.-G.; Ghassemi, S.; Mansour, E. M. E.; Riemer, C.; Warrass, R.; Sadat-Aalaee, D.; Truran, G. A.; Imazumi, H.; El-Faham, A.; Ionescu, D.; Ismail, M.; Kowaleski, T. L.; Han, C.-H.; Wenschuh, H.; Beyermann, M.; Bienert, M.; Shroff, H.; Albericio, F.; Triolo, S. A.; Sole, N.; Kates, S. A., *J. Org. Chem.* **1995**, *60*, 7718.
2. Hanack, M.; Eggensperger, H., *Liebigs Ann. Chem.* **1963**, *663*, 31.
3. Hart, H.; Curtis, O. E., Jr., *J. Am. Chem. Soc.* **1956**, *78*, 112.
4. Neckers, D. C.; Schaap, A. P., *J. Org. Chem.* **1967**, *32*, 22.
5. Sarel, S.; Yovell, J.; Sarel-Imber, M., *Angew. Chem., Int. Ed. Engl.* **1968**, *7*, 577.
6. McCormick, J. P.; Barton, D. L., *J. Chem. Soc., Chem. Commun.* **1975**, 303.
7. Argenti, L.; Bellina, F.; Carpita, A.; Dell'Amico, N.; Rossi, R., *Synth. Commun.* **1994**, *24*, 3167.
8. Matveeva, E. D.; Kvasha, M. P.; Kurts, A. L., *Russ. J. Org. Chem.* **1996**, *32*, 17.
9. Tanigawa, Y.; Kanamaru, H.; Murahashi, S.-I., *Tetrahedron Lett.*, **1975**, *16*, 4655.

Louis A. Carpino
University of Massachusetts, Amherst, Massachusetts, USA

D

Dibenzyl *N,N*-Diisopropylphosphoramidite[1]

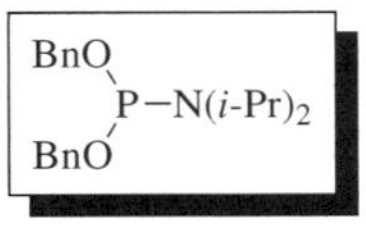

[108549-23-1] $C_{20}H_{28}NO_2P$ (MW 345.42)

(phosphitylating reagent for alcohols; the resulting alkyl dibenzyl phosphite is transformed to the phosphoric monoesters by oxidation and subsequent debenzylation)[2]

Solubility: insoluble in cold H_2O; soluble in CH_2Cl_2, THF, acetonitrile.
Analysis of Reagent Purity: NMR: δ_P ($CDCl_3$) +148.2.
Preparative Methods: there are several synthetic methods,[2–4] among which the reaction of benzyl alcohol with *N,N*-diisopropylphosphoramidous dichloride in the presence of *t*-amine in diethyl ether, THF, or dioxane is convenient.[5]
Purification: chromatography on silica gel.
Handling, Storage, and Precautions: can be stored in the absence of moisture and air for more than several months. A stock solution can be also prepared.

Original Commentary

Shoichiro Ozaki & Yutaka Watanabe
Ehime University, Matsuyama, Japan

Phosphoric Monoester Synthesis. Dibenzyl phosphoramidites such as dibenzyl *N,N*-diisopropylphosphoramidite (**1**) (see also Dibenzyl *N,N*-Diethylphosphoramidite) react quickly with various alcohols in the presence of 1*H*-tetrazole at room temperature to afford the phosphorus triesters which are oxidized in situ with an oxidant such as m-Chloroperbenzoic Acid, t-Butyl Hydroperoxide, or 30% Hydrogen Peroxide to give the corresponding alkyl dibenzyl phosphoric triesters in excellent yields. The triesters are deprotected by simple and clean catalytic hydrogenolysis on a Pd catalyst such as Palladium on Carbon or Palladium(II) Hydroxide. The usefulness of the sequence has been demonstrated by many examples which involve synthesis of phosphate derivatives of amino acids,[2,6] peptides,[7] sugars,[8] and cyclitols (eq 1).[4]

Dibenzyl phosphoric triesters are prepared alternatively by using Dibenzyl Phosphorochloridate, tetrabenzyl pyrophosphate, or dibenzyl phosphorofluoridate. See also cyclic homologue of (**1**), *o*-Xylylene *N,N*-Diethylphosphoramidite. These amidite reagents are more effective than P^V reagents, especially for phosphorylation of sterically hindered or base-sensitive alcohols, due to their high reactivity under milder conditions.

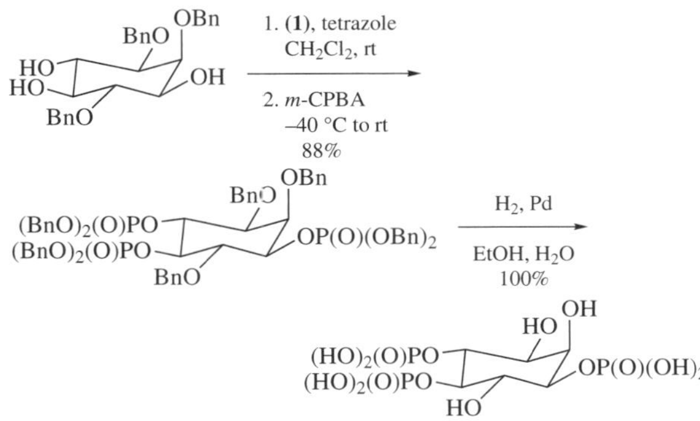

(1)

First Update

Lisheng Cai
National Institute of Mental Health, NIH, Bethesda, MD, USA

Phosphoric Monoester Synthesis. It has been demonstrated that phenols may be phosphorylated in the presence of carboxylic acids with this reagent (eq 2).[9,10]

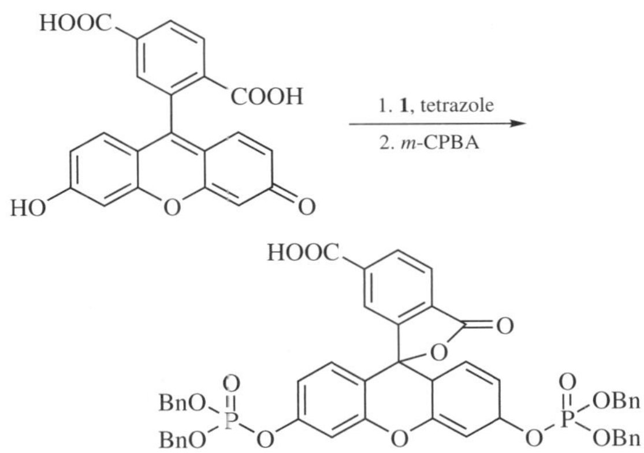

(2)

The synthesis of phosphopeptides may, in principal, be addressed by two strategies. In the first one, the phosphorylated amino acid building blocks are prepared[11] and then incorporated into the stepwise mode of solid phase synthesis. Alternatively, the preassembled peptide can be phosphorylated if the residue to be phosphorylated is incorporated as the side-chain in an unprotected form (eq 3). This latter approach is referred to as global phosphorylation.[12,13]

The oxidation of P(III) phosphite by elemental sulfur affords phosphorothioate esters, which upon hydrolysis give phosphorothioate salts (eq 4). These phosphorothioates may provide novel properties not accessible for their oxoates.[14]

Other phosphorothioate esters have been prepared with analogous reagent combinations (eq 5).[15]

Dibenzyl *N,N*-diisopropylphosphoramidite exhibits high selectivity for the phosphorylation of OH versus other acidic functional groups, such as amides (eq 6)[16] and amines (eq 7).[17]

Following phosphitylation of an alcohol, the oxidative transformation to the phosphate may be achieved in such a manner as to cleave one of the benzyl ester groups and permit a second coupling reaction. For example, phosphitylation of diacyl glycerol with dibenzyl *N,N*-diisopropylphosphoramidite gave a dibenzyl diacylglyceryl phosphite. This compound,

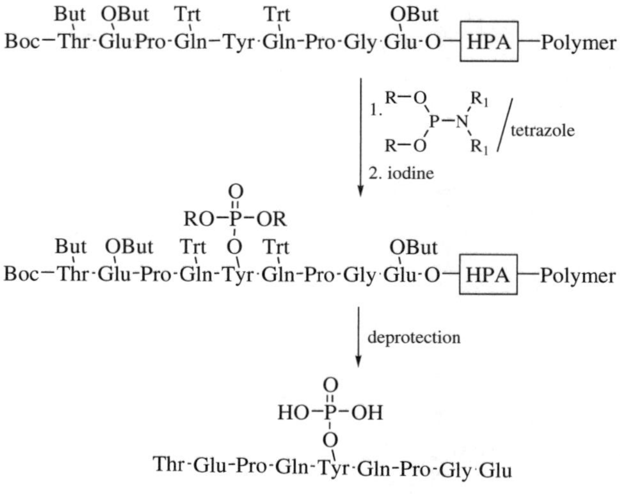

(3)

Tyr(*P*)[527]]p60[src] (523-531)

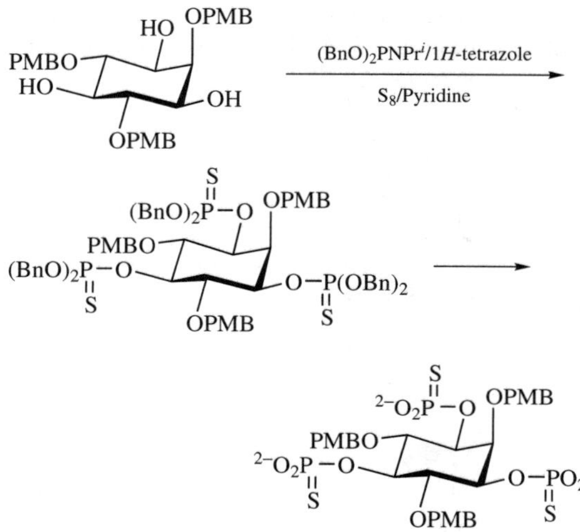

(4)

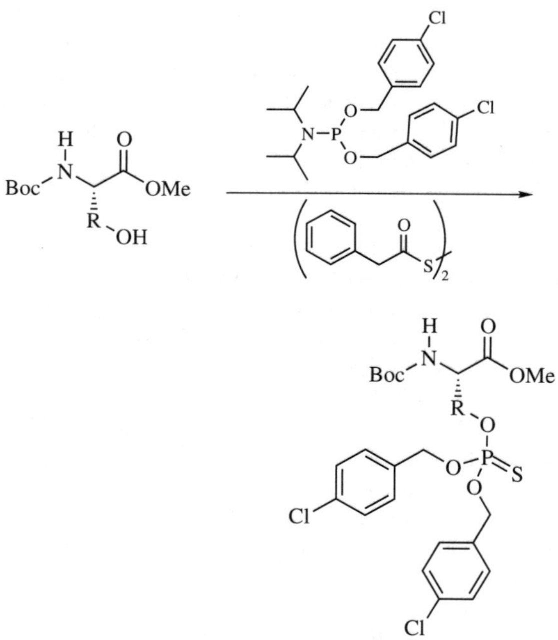

(5)

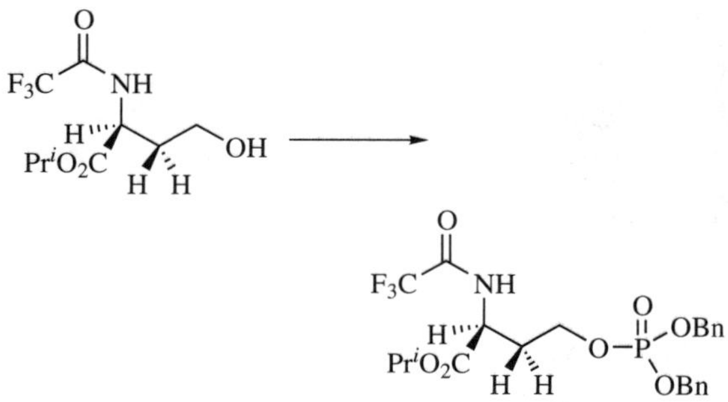

(6)

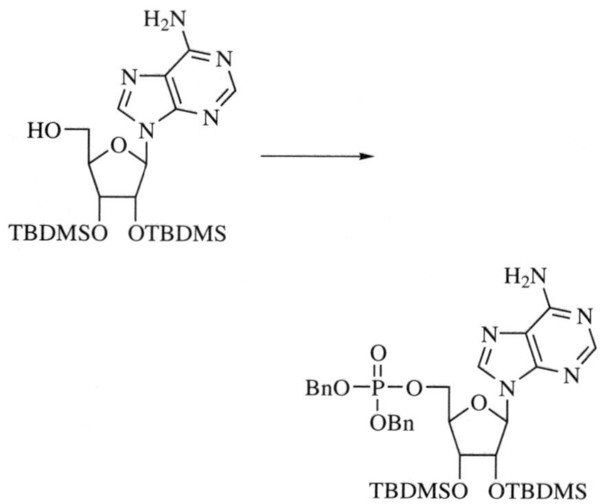

(7)

on activation with pyridinium tribromide in the presence of lutidine, was then oxidatively coupled to dicyclohexylidene *myo*-inositol, with cleavage of a benzyl group, to give a triphosphate carrying the inosityl, diacylglyceryl, and benzyl moieties (eq 8).[18] Interestingly, when the same sequence was conducted with *o*-xylylene *N,N*-diethylphosphoramidite, the final product incorporated the xylylene residue in the form of an *o*-bromomethylbenzyl ester.

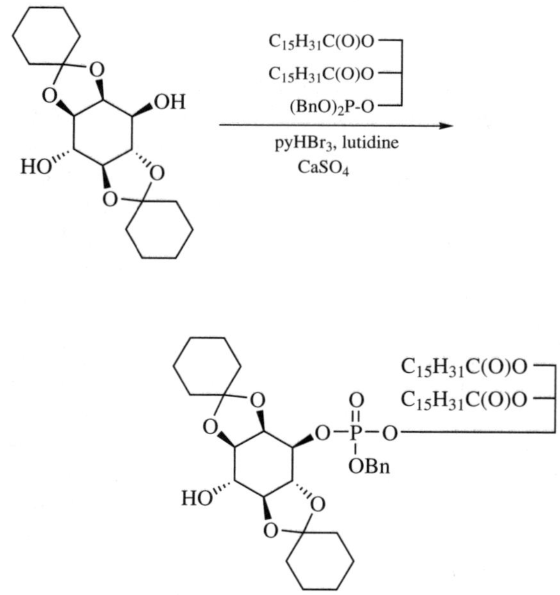

(8)

Different combinations of reagents have been used to substitute the original one. For example, 5-phenyltetrazole and other substituted tetrazoles have been used as promoter in place of tetrazole itself, and magnesium monoperoxyphthalate hexahydrate (MMPP) has served as oxidant.[19,20] Interestingly, in some special situations, $Et_2NP(O\text{-}t\text{-}Bu)_2$, $(i\text{-}Pr)_2NP(O\text{-}t\text{-}Bu)_2$, and $(i\text{-}Pr)_2NP(OCH_2CH_2OSO_2CH_3)_2$ have been employed to synthesize the corresponding phosphoric monoesters.[21–24] Alternatively, $(i\text{-}Pr_2N)_2P(OCH_2CH_2CN)$ has been used to generate the corresponding triester, which could be hydrolyzed under basic conditions.[25,26] $(i\text{-}Pr_2N)_2P(OAll)$ was used analogously to synthesize the triester. Removal of the allyl group was accomplished with $(Ph_3P)_4Pd/n\text{-}BuNH_2$.[27] $(EtO)_2PCl$[28] and $(CF_3CH_2O)_2PCl$[29] were also used in the synthesis of triesters, permitting subsequent partial hydrolysis under basic conditions. The use of $(BnO)_2P(O)H$ in the analogous transformation has also been described (eq 9).[30]

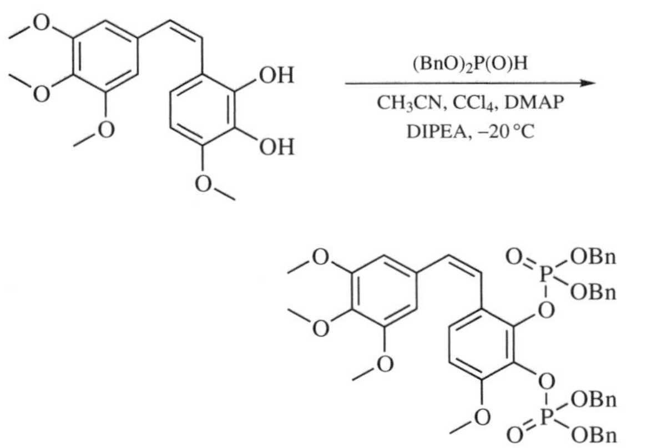

(9)

***N*-Phosphoryl Amino Acids.** *N*-Phosphoryl amino acids have been synthesized through a phosphoramidite amine-exchange reaction (eq 10).[31]

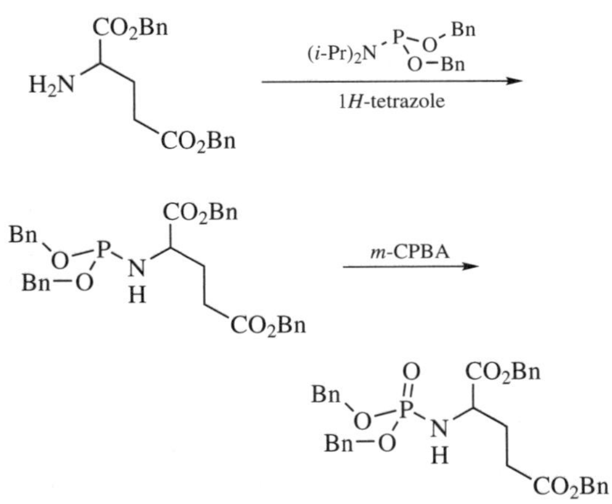

(10)

Related Reagents. Di-(*tert*-butyl) *N,N*-diethylphosphoramidite; di-(*tert*-butyl) *N,N*-diisopropylphosphoramidite; di-[2-(methanesulfonyloxy)ethyl] *N,N*-diisopropylphosphoramidite; diallyl *N,N*-diisopropylphosphoramidite; diethyl chlorophosphite; dibenzyl phosphite; xylylene *N,N*-diethylphosphoramidite.

1. Beaucage, S. L.; Iyer, R. P., *Tetrahedron* **1992**, *48*, 2223.
2. Bannwarth, W.; Trzeciak, A., *Helv. Chim. Acta* **1987**, *70*, 175.
3. Smirnova, L. I.; Malenkovskaya, M. A.; Predvoditelev, D. A.; Nifant'ev, É. E., *Zh. Org. Khim.* **1980**, *16*, 1011.
4. Yu, K.-L.; Fraser-Reid, B., *Tetrahedron Lett.* **1988**, *29*, 979.
5. Fukase, K.; Kamikawa, T.; Iwai, Y.; Shiba, T.; Rietschel, E. T.; Kusumoto, S., *Bull. Chem. Soc. Jpn.* **1991**, *64*, 3267.
6. de Bont, H. B. A.; Veeneman, G. H.; van Boom, J. H.; Liskamp, R. M. J., *Recl. Trav. Chim. Pays-Bas* **1987**, *106*, 641.
7. Perich, J. W.; Johns, R. B., *Tetrahedron Lett.* **1988**, *29*, 2369.
8. Ichikawa, Y.; Sim, M. M.; Wong, C.-H., *J. Org. Chem.* **1992**, *57*, 2943.
9. Takakusa, H.; Kikuchi, K.; Urano, Y.; Kojima, H.; Nagano, T., *Chem. Eur. J.* **2003**, *9*, 1479–1485.
10. Davidson, J. P.; Martin, S. F., *Tetrahedron Lett.* **2000**, *41*, 9459–9464.
11. Kitas, E.; Kueng, E.; Bannwarth, W., *Int. J. Pept. Prot. Res.* **1994**, *43*, 146–153.
12. Bannwarth, W.; Küng, E.; Vorherr, T., *Bioorg. Med. Chem. Lett.* **1996**, *6*, 2141–2146.
13. Bannwarth, W.; Kitas, E. A., *Helv. Chim. Acta* **1992**, *75*, 707–714.
14. Lampe, D.; Liu, C.; Potter, B. V. L., *J. Med. Chem.* **1994**, *37*, 907–912.
15. De Bont, D. B. A.; Moree, W. J.; van Boom, J. H.; Liskamp, R. M. J., *J. Org. Chem.* **1993**, *58*, 1309–1317.
16. Barclay, F.; Chrystal, E.; Gani, D., *Chem. Commun.* **1994**, 815–816.
17. Hartgers, W. A.; De Boer, R. F.; Wanner, M. J.; Koomen, G. J., *Nucleosides & Nucleotides* **1992**, *11*, 1325–1332.
18. Watanabe, Y.; Munetsugu, H.; Hayashi, M., *Chem. Lett.* **2002**, 292–293.
19. Yin, N.; Marshall, R. L.; Matheson, S.; Savage, P. B., *J. Am. Chem. Soc.* **2003**, *125*, 2426–2435.
20. Graham, S. M.; Pope, S. C., *Org. Lett.* **1999**, *1*, 733–736.
21. Xu, B.; Stephens, A.; Kirschenheuter, G.; Greslin, A. F.; Cheng, X.; Sennelo, J.; Cattaneo, M.; Zighetti, M. L.; Chen, A.; Kim, S.-A.; Kim, H. S.; Bischofberger, N.; Cook, G.; Jacobson, K. A., *J. Med. Chem.* **2002**, *45*, 5694–5709.
22. Lindberg, J.; Ekeroth, J.; Konradsson, P., *J. Org. Chem.* **2002**, *67*, 194–199.
23. Kim, H. S.; Barak, D.; Harden, T. K.; Boyer, J. L.; Jacobson, K. A., *J. Med. Chem.* **2001**, *44*, 3092–3108.
24. de Kort, M.; Luijendijk, J.; van der Marel, G. A.; van Boom, J. H., *Eur. J. Org. Chem.* **2000**, 3085–3092.
25. Matsui, T.; Kondo, T.; Nishita, Y.; Itadani, S.; Tsuruta, H.; Fujita, S.; Omawari, N.; Sakai, M.; Nakazawa, S.; Ogata, A.; Mori, H.; Kamoshima, W.; Terai, K.; Ohno, H.; Obata, T.; Nakai, H.; Toda, M., *Bioorg. Med. Chem.* **2002**, *10*, 3787–3805.
26. Matsui, T.; Kondo, T.; Nishita, Y.; Itadani, S.; Nakatani, S.; Omawari, N.; Sakai, M.; Nakazawa, S.; Ogata, A.; Mori, H.; Terai, K.; Kamoshima, W.; Ohno, H.; Obata, T.; Nakai, H.; Toda, M., *Bioorg. Med. Chem.* **2002**, *10*, 3757–3786.
27. Zhang, Y.; Fussel, S.; Reimer, U.; Schutkowski, M.; Fischer, G., *Biochemistry* **2002**, *41*, 11868–11877.
28. Chung, S.-K.; Kwon, Y.-U.; Shin, J.-H.; Chang, Y.-T.; Lee, C.; Shin, B.-G.; Kim, K.-C.; Kim, M.-J., *J. Org. Chem.* **2002**, *67*, 5626–5637.
29. Muller, T.; Schmidt, R. R., *Liebigs Ann./Recueil* **1997**, 1907–1914.
30. Pettit, G. R.; Lippert, J. W., *Anti-Cancer Drug Design* **2000**, *15*, 203–216.
31. Chow, C. P.; Berkman, C. E., *Tetrahedron Lett.* **1998**, *39*, 7471–7474.

Dibenzylphosphite

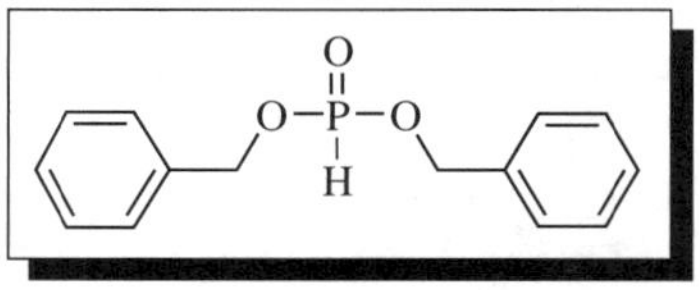

1

[17176-77-1] $C_{14}H_{15}O_3P$ (MW 262.244)

(reagent mainly used for phosphorylation and as a phosphonylating agent)

Alternate Name: dibenzyl phosphonate.
Physical Data: bp 110–120 °C at 0.01 mm Hg, 165 °C at 0.1 mm Hg; 1.187 g cm^{-3}.
Form Supplied in: colorless liquid; widely available.
Analysis of Reagent Purity: ^{31}P NMR δ 7.9 ppm ($^1J_{PH}$ = 713 Hz).
Purification: vacuum distillation at 0.01 mm Hg.
Handling, Storage, and Precautions: may explode during distillation, decomposes violently at 160 °C.

Phosphorylation of Alcohols and Amines. Dibenzyl *H*-phosphonate (**1**) in combination with carbon tetrachloride and triethyl amine was used for the first time for the phosphorylation of amines[1] in the Todd laboratory. These reaction conditions, which involve in situ generation of dialkyl phosphorochloridate from the corresponding dialkyl *H*-phosphonate in the presence of CCl_4 and tertiary amine, are now referred to as the Atherton-Todd reaction. Although dibenzyl *H*-phosphonates can be converted into dibenzyl phosphorochloridate using various oxidizing agents, e.g., chlorine[2] and sulfuryl chloride,[3] which can be prepared separately prior to the reaction with nucleophiles, most of the contemporary applications of these reagents make use of generation of the reactive phosphorylating species in situ under the Atherton-Todd conditions.[1,4–6] In a typical experiment (eq 1),[1] dibenzyl *H*-phosphonate (**1**) is dissolved in carbon tetrachloride and treated with a slight excess of amine. This produces overnight the corresponding dibenzyl phosphoramidates in high yields.[1,3,7,8]

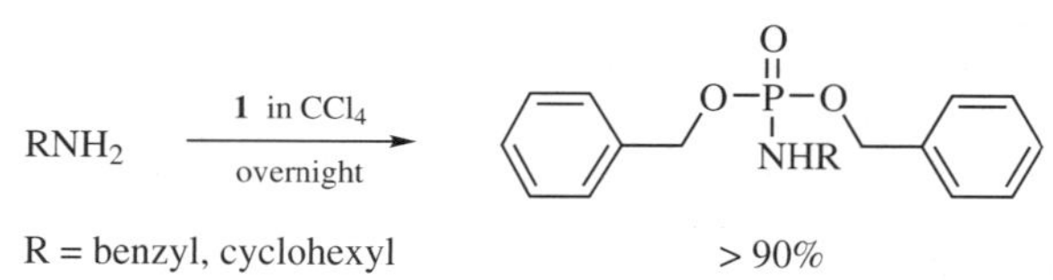

(1)

Phosphorylation of alcohols with this reagent system is usually less efficient. Atherton and Todd reported[9] on a successful phosphorylation of ethanol using (**1**) and trichlorobromomethane in the presence of tertiary base (note the in situ generation of dibenzyl phosphorobromidate), but the reaction failed when carbon tetrachloride was used as an oxidant. The low efficiency of this transformation was probably due to the generally lower reactivity of alcohols in S_N2(P) reactions and the lack of a nucleophilic catalyst in the reaction mixture. Indeed, when these reactions are carried out in pyridine, efficient phosphorylation of alcohols can be achieved, as exemplified by the synthesis of various 5′-phosphorylated nucleosides[10] or phospholipid derivatives.[11] Alternatively, phosphorylation of alcohols using dibenzyl *H*-phosphonate can be carried out in high yield under solid-liquid phase transfer catalysis (S-L PTC), and these conditions have been successfully applied to the phosphorylation of various carbohydrates.[12] The main advantage of this S-L PTC procedure is that it does not require anhydrous reaction conditions and permits easy handling of the reactants.

Recently, the efficient phosphorylation of phenols under modified Atherton-Todd conditions, involving the use of acetonitrile as a solvent and catalytic amounts of 4-*N,N*-dimethylaminopyridine (DMAP), was reported.[4] In this protocol (eq 2), a phenol to be phosphorylated is dissolved in acetonitrile and treated with 1.45 equiv of (**1**) in the presence of 5 equiv of CCl_4, 2.1 equiv of *N,N*-diisopropylethylamine, and 0.1 equiv of DMAP.[4] This affords within 45 min the phosphorylated phenols in ca. 90% yields. A distinctive feature of these reaction conditions is that they permit selective phosphorylation of phenols in the presence of aliphatic hydroxylic functions.[4]

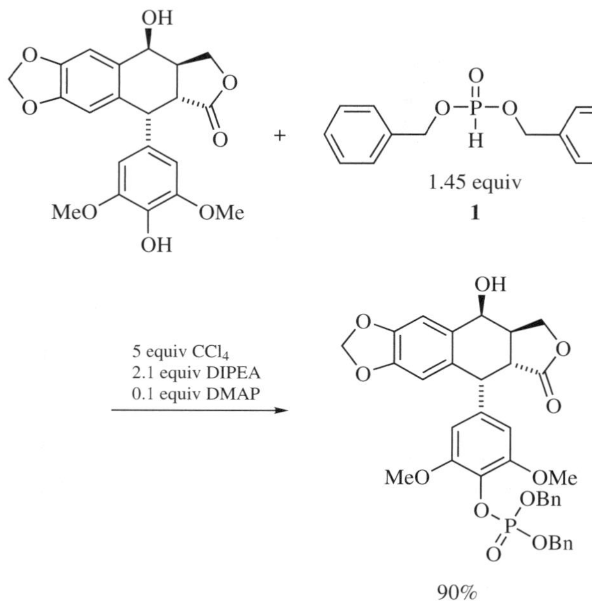

(2)

Dibenzyl phosphoramidates, produced from phosphorylation of amines, or dibenzyl phosphotriesters, obtained from phosphorylation of alcohols, can be converted either into the corresponding monoesters (via the removal of both benzyl groups) or into diesters, (via a selective removal of only one benzyl group). The conversion into the phosphomonoesters can be effected most conveniently via hydrogenolysis at normal pressure over Pd/C catalyst, with removal of both benzyl groups.[1,4–6] For the selective removal of one benzyl group from the phosphotriesters to produce the corresponding diesters (which are important synthetic intermediates), hydrolytic conditions (e.g. aqueous ammonia[13] or acidic conditions[14]), or dealkylation with soft nucleophiles (e.g. iodide ion[15,16]) or with tertiary amines (e.g. DABCO in toluene[8]) can be employed.

Reactions of Dibenzyl H-Phosphonate with Electrophiles. In the presence of base, dibenzyl *H*-phosphonate reacts with various electrophiles; examples include aldehydes,[17] ketones,[18] imines,[19] aziridines,[20] and alkyl halides.[21,22] In this way, α-hydroxyalkylphosphonates,[17,18] α-aminoalkylphosphonates,[19] β-aminoalkylphosphonates,[20] and other C-phosphonate derivatives[21,22] can be obtained.

Aldehydes and reactive carbonyl compounds undergo smooth addition to dibenzyl H-phosphonate (**1**). In a typical reaction (eq 3), an aldehyde is treated with 1.5 equiv of dibenzyl H-phosphonate in the presence of a catalytic amount of triethylamine to afford the desired α-hydroxyphosphonate in high yields.[17]

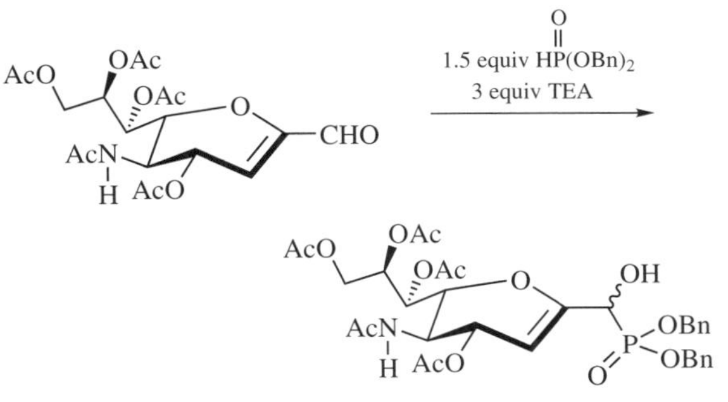

(3)

100%, *R*:*S*=4:1

α-Bromo ketones[16] react rapidly with (**1**) without added base to produce the corresponding β-bromo-α-hydroxyphosphonates, that eliminates of HBr under basic conditions to afford the corresponding enol phosphates.[16] Ketones with a nonactivated carbonyl group react less efficiently. However, carrying out this type of reaction under heterogeneous conditions in the presence of γ-alumina-potassium fluoride was claimed to overcome this problem.[18]

Imines usually react with H-phosphonate diesters under thermal conditions to produce the corresponding α-aminoalkylphosphonates.[23] Recently, an improved procedure for the preparation of these compounds using **1** (or other H-phosphonates) under sonochemical conditions was reported.[19] Another set of important synthetic intermediates, the N-hydroxyaminomethylphosphonates, can be obtained in a base catalyzed reaction of N-substituted hydroxylamines, paraformaldehyde, and (**1**).[24] This reaction works well also for other aldehydes and ketones.[24]

Opening of an N-acylated or N-sulfonylated aziridine ring with phosphorus nucleophiles leads to the corresponding β-aminoalkylphosphonate derivatives.[20] The dibenzyl esters of these compounds can be obtained by using dibenzyl H-phosphonate (**1**), which prior to the reaction with aziridine derivatives is usually converted into its tervalent sodium salt.[20] Also, reactions with carbon electrophiles (e.g., alkylation of H-phosphonates) are more efficient when sodium salt of (**1**), generated in situ using NaH, is employed.[21,22]

Other Applications. Dibenzyl H-phosphonate (**1**) was instrumental in the synthesis of a novel type of triphosphate analogs having P-N-P-N-P skeleton, the diimidotriphosphoric acids, which involved as a key step phosphorylation of o-xylenediamine.[8,25]

Dibenzyl H-phosphonate (**1**) also serves as a starting material for the preparation of benzyl H-phosphonate monoesters,[26] which can be used in the synthesis of asymmetric H-phosphonate diesters.

Related Reagents. Dibenzyl N,N-diisopropyl phosphoramidite; dibenzyl N,N-diethyl phosphoramidite; dibenzyl phosphorochloridate; dibenzyl phosphate; diethyl chlorophosphite; diethyl phosphorochloridate; dimethyl chlorophosphite; 2-chloro-4,4,5,5-tetramethyl-1,3,2-dioxaphospholane; H-pyrophosphonic acid; 2-chloro-4H-1,3,2-benzodioxaphosphorinan-4-one.

1. Atherton, F. R.; Openshaw, H. T.; Todd, A. R., *J. Chem. Soc.* **1945**, 660.
2. Atherton, F. R.; Openshaw, H. T.; Todd, A. R., *J. Chem. Soc.* **1945**, 382.
3. Li, S.-O., *J. Am. Chem. Soc.* **1952**, *74*, 5959.
4. Silverberg, L. J.; Dillon, J. L.; Vemishetti, P., *Tetrahedron Lett.* **1969**, *37*, 771.
5. Harvey, W. E.; Michalski, J.; Todd, A., *J. Chem. Soc.* **1951**, 2271.
6. Schick, A.; Kolter, T.; Giannis, A.; Sandhoff, K., *Tetrahedron* **1969**, *52*, 2945.
7. Li, S.-O.; Eakin, R. E., *J. Am. Chem. Soc.* **1955**, *77*, 1866.
8. Saady, M.; Lebeau, L.; Mioskowski, C., *Tetrahedron Lett.* **1995**, *36*, 4237.
9. Atherton, R. F.; Todd, A. R., *J. Chem. Soc.* **1947**, 674.
10. Michelson, A. M.; Todd, A. R., *J. Chem. Soc.* **1949**, 2476.
11. Brown, D. M.; Higson, H. M., *J. Chem. Soc.* **1957**, 2034.
12. Ding, Y. X.; Zhou, S. F., *Phosphor. Sulfur Silicon* **1994**, *92*, 219.
13. Hayes, D. H.; Michelson, A. M.; Todd, A., *J. Chem. Soc.* **1955**, 808.
14. Baddiley, J.; Todd, A., *J. Chem. Soc.* **1947**, 648.
15. Baddiley, J.; Buchanan, J. G.; Fawcett, C. P., *J. Chem. Soc.* **1959**, 2192.
16. Cramer, F.; Voges, D., *Chem. Ber.* **1959**, *92*, 952.
17. Amann, F.; Schaub, C.; Müller, B.; Schmidt, R. R., *Chem. Eur. J.* **1998**, *4*, 1106.
18. Texier-Boullet, F.; Lequitte, M., *Tetrahedron Lett.* **1986**, *27*, 3515.
19. Hubert, C.; Oussaid, B.; Etemad-Moghadam, G.; Koenig, M.; Garrigues, B., *Synthesis* **1994**, 51.
20. Stamm, H.; Gerster, G.; Baumann, T., *Chem. Ber.* **1983**, *116*, 2936.
21. Bookser, B. C.; Kasibhatla, S. R.; Appleman, J. R.; Erion, M. D., *J. Med. Chem.* **2000**, *43*, 1495.
22. Cushman, M.; Yang, D.; Mihalic, J. T.; Chen, J.; Gerhardt, S.; Huber, R.; Fischer, M.; Kis, K.; Bacher, A., *J. Org. Chem.* **2002**, *67*, 6871.
23. Redmore, D., *Topics in Phosphorus Chemistry* **1976**, *8*, 515.
24. Pastor, S. D.; Ravichandran, R.; Meuwly, R., *Tetrahedron* **1992**, *48*, 2911.
25. Saady, M.; Valleix, A.; Lebeau, L.; Mioskowski, C., *J. Org. Chem.* **1995**, *60*, 3685.
26. Kers, A.; Kers, I.; Stawinski, J.; Sobkowski, M.; Kraszewski, A., *Synthesis* **1995**, 427.

Martin Kullberg & Jacek Stawinski
Stockholm University, Stockholm, Sweden

2,7-Dibromo-9H-fluoren-9-ylmethyl Chloroformate

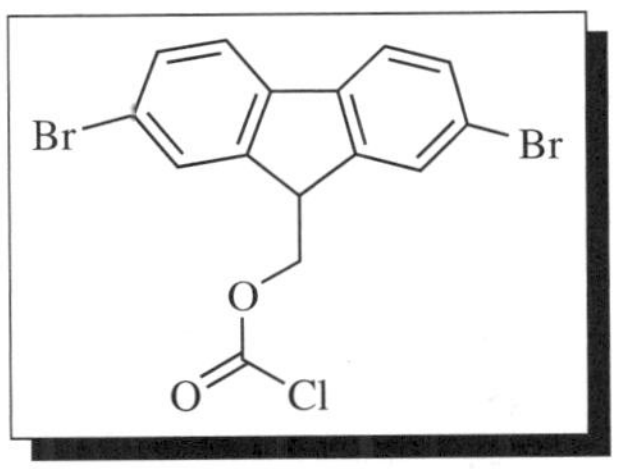

[74808-82-5] $C_{15}H_9Br_2ClO_2$ (MW 416.49)

(reagent used to introduce the 2,7-dibromo-9-fluorenylmethyl carbamate protecting group for amines)

Alternate Name: carbonochloridic acid, (2,7-dibromo-9*H*-fluoren-9-yl)methyl ester.
Physical Data: mp 124–126 °C (white crystals).[1]
Solubility: soluble in chloroform, dichloromethane, ether, and THF.
Analysis of Reagent Purity: ^{1}H NMR, elemental analysis.[1]
Preparative Methods: the reagent is prepared directly in 37% yield (eq 1) by the reaction, at room temperature, of bromine with (9-fluorenyl)methyl chloroformate in the presence of iron (III) chloride.[1] Alternatively, the title reagent may be directly prepared from (2,7-dibromo-9-fluorenyl)methanol by treatment with phosgene.[1] The alcohol is obtained by bromination of 9-fluorenylmethanol[2] with NBS in acetic acid in the presence of HBr in 50% yield (eq 2).[2]

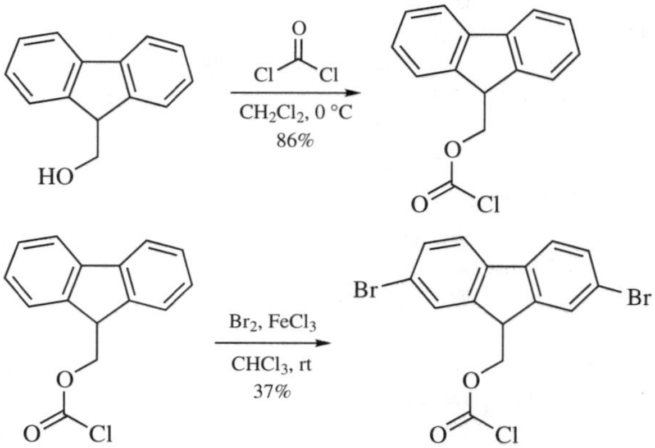

(1)

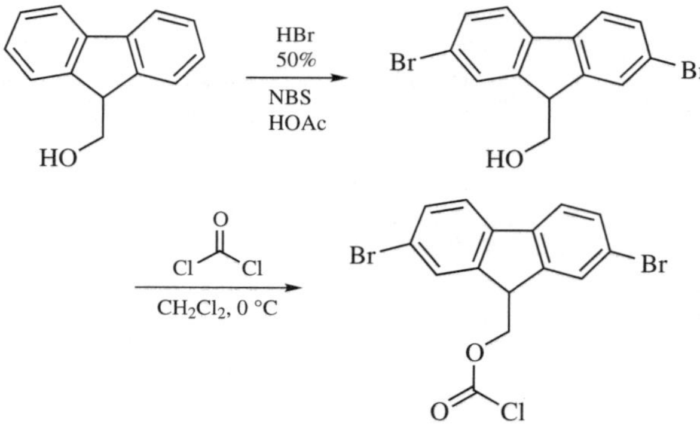

(2)

Purification: trituration with ligroin (bp 30–60 °C), filtration, and recrystallization from a 2:1 mixture of ligroin (bp 60–70 °C) and ligroin (bp 88–98 °C).[1]
Handling, Storage, and Precautions: very sensitive to hydrolysis. Should be stored in closed vessels under an inert atmosphere in the refrigerator; use of the recrystallized reagent is recommended; should be used in a well-ventilated fumehood.

Protection Deprotection Reactions. Most common amino-protecting groups are deblocked under various acidic conditions.[3] In contrast, the popularity of the 9-fluorenylmethyloxycarbonyl group (Fmoc),[4,5] in solution-phase[6] and solid-phase[7] peptide synthesis, arises from the stability of Fmoc-carbamates toward acids and catalytic hydrogenation but they readily cleave under mild basic, non-aqueous conditions. The title reagent (2,7-dibromo-9-fluorenyl)methyl chloroformate was prepared with a view to obtain a more base labile system which is otherwise stable and introduced in high yields. Thus, (2,7-dibromo-9-fluorenyl)methyl *N*-(*p*-chlorophenyl)carbamate was prepared by treatment of (2,7-dibromo-9-fluorenyl)methyl chloroformate with *p*-chloroaniline (eq 3).

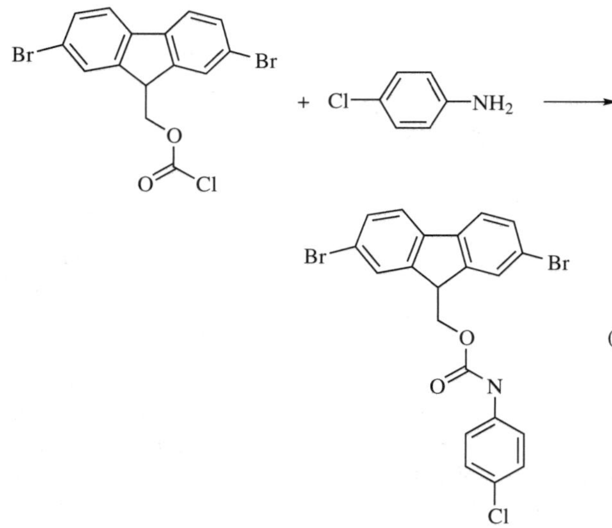

(3)

The same compound was obtained in 74% yield by treatment of (2,7-dibromo-9-fluorenyl)methanol with *p*-chlorophenyl isocyanate in the presence of triethylamine (eq 4).

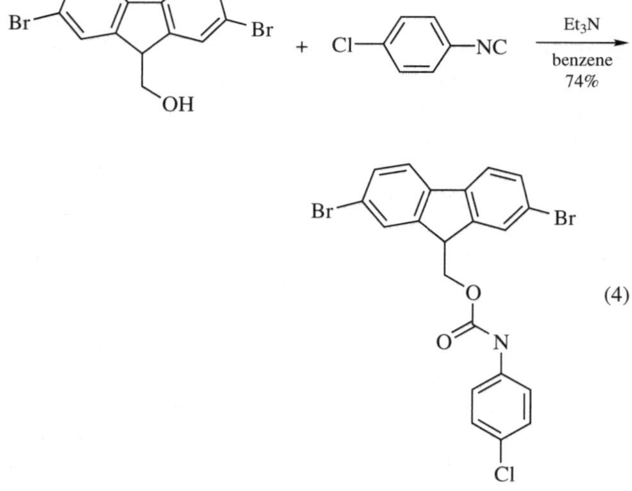

In a comparison of the labilities of substituted Fmoc anilates, it was shown that 2,7-dihalo substitution along with a solution of piperidine in toluene greatly accelerates base-catalyzed deblocking of 2,7-dibromo-9-fluorenylmethyl carbanilate. The consumption of the urethane was complete in about 45 min in the case of the dibromo derivative (eq 5), whereas for the unsubstituted case the formation of *p*-chloroaniline was only apparent after 12 h, with 5.5 days required for the completion of the whole reaction (eq 6). In these two examples, the secondary amine adds to the initially formed dibenzofulvene to give the corresponding adducts in good yields.

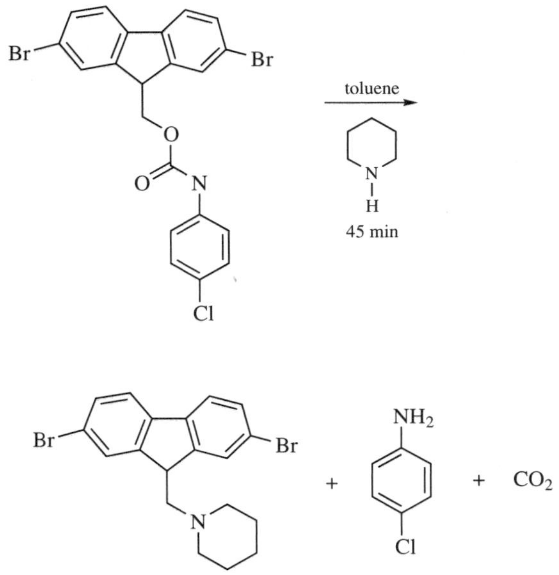

(5)

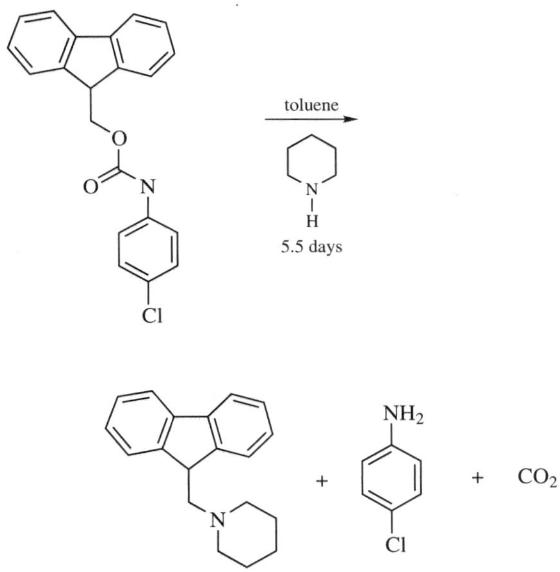

(6)

Similar tests in pyridine as both solvent and deblocking agent gave evidence for the appearance of *p*-chloroaniline immediately after mixing, e.g., in the case of the 2,7-dibromo derivative with complete consumption of urethane in 24 h (eq 7). In the case of the parent Fmoc system, the first evidence of dibenzofulvene[8] formation was only after 10 h, and the starting material was still present after 8 days (eq 8).[9] Although no mechanistic studies were carried out, the reaction probably represents an ElcB-type elimination process activated by the aromaticity effect of the dibenzocyclopentadienyl anion.

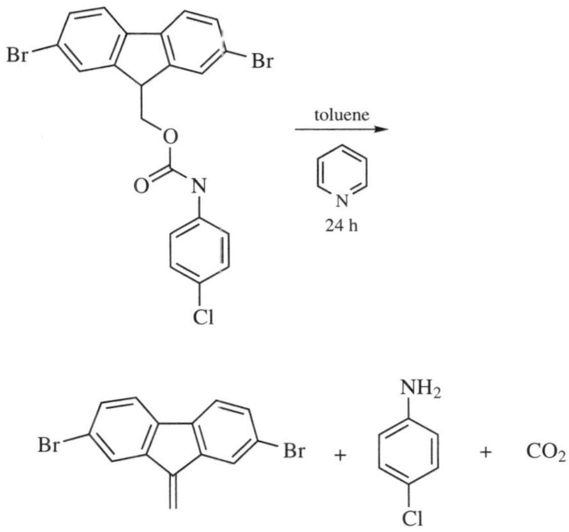

(7)

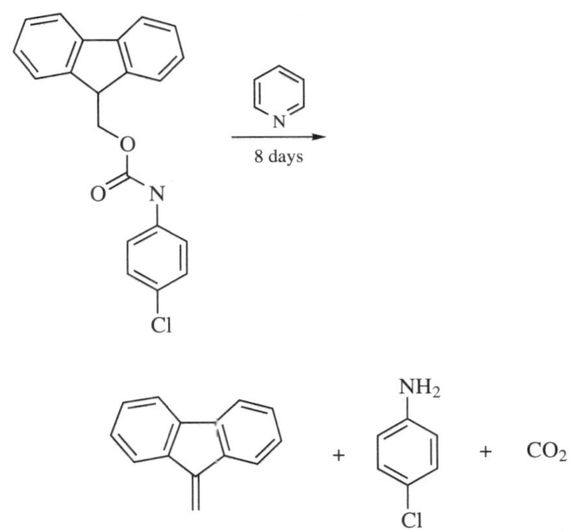

(8)

Related Reagents. 2,7-Dichloro-9-fluorene;[1,9] (2,7-dichloro-9-fluorenyl)methanol;[1] (2,7-dichloro-9-fluorenyl)methyl chloroformate;[1] (9-fluorenyl)methyl chloroformate;[4] (2,7-dichloro-9-fluorenyl)methyl acetate;[1] (2,7-dichloro-9-fluorenyl)methyl *N*-(*p*-chlorophenyl)carbamate.[1]

1. Carpino, L. A., *J. Org. Chem.* **1980**, *45*, 4250.
2. Brown, W. G.; Bluestein, B. A., *J. Am. Chem. Soc.* **1943**, *65*, 1082
3. (a) Schöder, E.; Lübke, K., *The Peptides*; Academic Press: New York, 1965, Vol. 1, pp. 3–51. (b) Wolman, Y., *The Chemistry of the Amino Group*; Patai, S., Ed.; Wiley-Interscience: New York, 1968, Chap. 11.
4. Carpino, L. A.; Han, G. Y., *J. Am. Chem. Soc.* **1970**, *92*, 5748
5. (a) Carpino, L. A.; Han, G. Y., *J. Org. Chem.* **1972**, *37*, 3404; (b) Carpino, L. A.; Han, G. Y., *ibid.* **1973**, *38*, 4218.
6. (a) Bodanszky, A.; Bodanszky, M.; Chandramouli, N.; Kwei, J. Z.; Martinez, J.; Tolle, J. C., *J. Org. Chem.* **1980**, *45*, 72. (b) Bodanszky, M.; Bodanszky, A.; Tolle, J. C.; Bednarek, M. A., In *Chemical Synthesis and Sequencing of Peptides and Proteins*; Liu, T. Y.; Schechter, A. N.; Heinrikson, R. L.; Condliffe, P. G., Eds.; Elsevier: Amsterdam, 1981, p. 163. (c) Kisfaludy, L.; Schön, I., *Synthesis* **1983**, 325. (d) Ten Kortenaar, P. B. W.; Krüse, J.; Hemminga, M. A.; Tesser, G. I., *Int. J. Pept. Protein Res.* **1986**, *27*, 401. (e) Toth, G. K.; Penke, B.; Zarandi, M.; Kovacs, K., *Int. J. Pept. Protein Res.* **1985**, *26*, 630. (f) Kessler, H.; Kuhn, M.; Loeschner, T., *Liebigs Ann. Chem.* **1986**, 21.
7. (a) Meienhofer, J.; Waki, M.; Heimer, E. P.; Lambros, T. J.; Makofske, R. C.; Chang, C.-D., *Int. J. Pept. Protein Res.* **1979**, *13*, 35. (b) Atherton, E.; Gait, M. J.; Sheppard, R. C.; Williams, B. J., *Bioorg. Chem.* **1979**, *8*, 351. (c) Atherton, E.; Sheppard, R. C.; Ward, P., *J. Chem. Soc., Perkin*

Trans. 1 **1985**, 2065. (d) Scanlon, D. B.; Eefting, M. A.; Lloyd, C. J.; Burges, A. W.; Simpson, R. J., *J. Chem. Soc., Chem. Commun.* **1987**, 516.

8. Sieglitz, A.; Jassoy, H., *Ber. Dtsch. Chem. Ges.* **1922**, *55*, 2032.
9. Hodgkinson, W. R.; Matthews, F. E., *J. Chem. Soc.* **1883**, *43*, 170.

Marco de la Mora
University of Illinois, Chicago, IL, USA

Dibromomethylbenzene

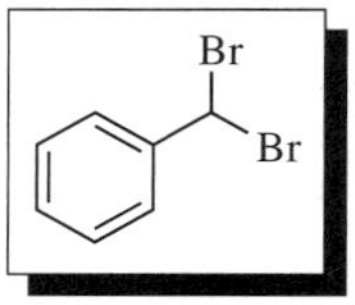

[618-31-5] $C_7H_6Br_2$ (MW 249.93)

(reagent used for the introduction of a benzylidene group)

Alternate Name: benzal bromide.
Physical Data: bp 154–156 °C at 31 mm Hg, *d* 1.845 g cm^{-3}.
Form Supplied in: colorless liquid, widely available.
Handling, Storage, and Precautions: stable at room temperature in closed containers under normal storage and handling conditions. Benzal bromide causes burns and is toxic by inhalation, in contact with the skin, and if swallowed. The reagent should be handled using suitable protective equipment in a well ventilated fume hood. This compound is incompatible with strong bases and strong oxidizing agents.

Nucleophilic Substitution Reactions. Benzal bromide is converted to benzaldehyde by the action of water at elevated temperatures, a reaction that has been used for the preparation of ^{18}O labeled benzaldehyde.[1] The same transformation has also been achieved by heating benzal bromide in DMSO[2] or DMF[3] (via initial iminium salt formation involving nucleophilic attack by dimethylamine, formed by DMF decomposition, followed by hydrolysis). Many of the reported reactions involving nucleophilic displacement of bromide from benzal bromide are broadly comparable to reactions of nucleophilic species with benzaldehyde. Heating primary amines with benzal bromide in the presence of potassium carbonate in acetonitrile gives rise to the corresponding imines.[4] Benzal bromide may also be used for the synthesis of benzylidene acetals, a commonly used protecting group for 1,2- and 1,3-diols. While acetal formation using benzaldehyde usually requires acid catalysis,[5] the equivalent reaction using benzal bromide is carried out under basic conditions. The formation of benzylidene acetals by reaction of a 1,3-diol with benzal bromide was first carried out using potassium *t*-butoxide in refluxing benzene/*t*-butanol.[6] However, the more usual conditions for acetal formation involve refluxing a mixture of the diol and benzal bromide in anhydrous pyridine (eq 1).[7,8] This methodology can be used for the selective protection of molecules containing 1,2-[9] and 1,3-diols.[10–14] Variants of this methodology include the addition of triethylamine[15] or triethylamine and DMAP[16] to the reaction mixture. It has also been shown that the reaction times required for the formation of benzylidene acetals can be dramatically reduced by microwave heating of a substrate diol with benzal bromide in acetonitrile in the presence of polymer supported DMAP as the HBr scavenger (eq 2).[17]

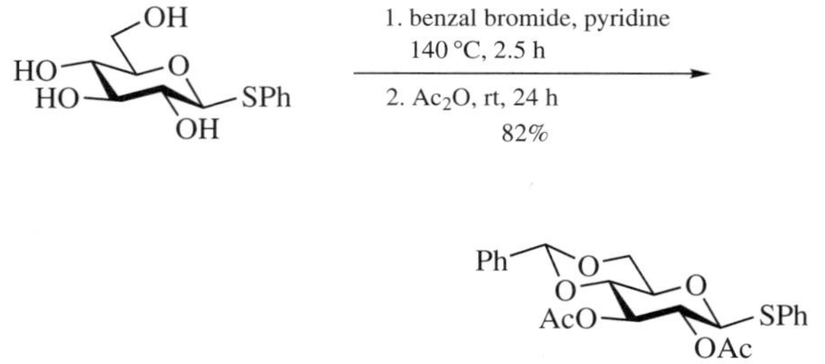

(1)

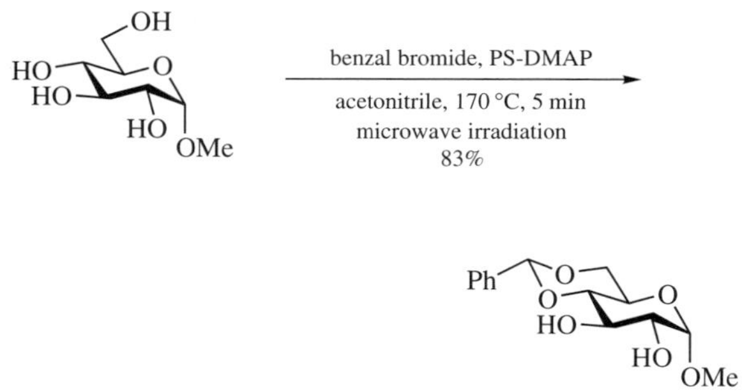

(2)

Benzal bromide has also been used for the formation of other benzylidene heterocyclic compounds of structure similar to benzylidene acetals. This includes substituted 2-phenyl-[1,3]diselenanes,[18] 2-phenyl-[1,3]thiazetidines,[19] and 2-phenyl-[1,3]thiazinanes (eq 3).[20]

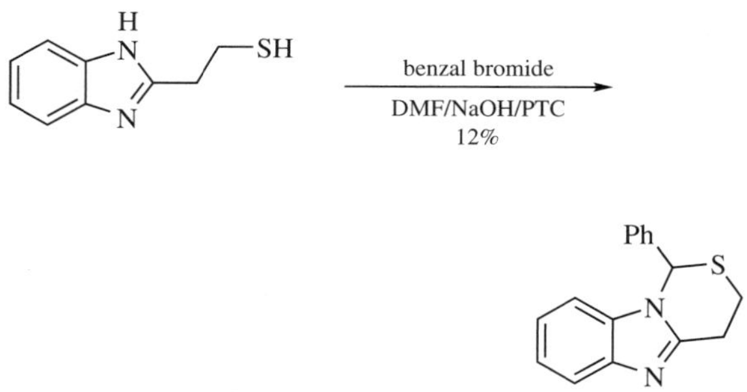

(3)

Cyclopropane Formation and Related Reactions. Benzal bromide can be transformed into two distinct carbenoid species dependent upon the reaction conditions employed. The phenylbromo carbenoid species "PhCBr" is generated from benzal bromide by the action of potassium *t*-butoxide and can be reacted in situ with a variety of alkenes to produce 1-phenyl-1-bromocyclopropanes (eq 4).[21,22] For reactions where both *syn*- and *anti*-cyclopropanation products are formed, a preference for the cyclopropane with the bromine in the *syn*-position is usually observed.

The generation of metallated species from benzal bromide by metal-halogen exchange gives rise to phenyl carbenoids. Carbenoid generation has been carried out by the reaction of benzal bromide with methyllithium,[23] *n*-butyllithium,[23,24] magnesium,[25] or electrochemical reduction using a zinc anode.[26] The carbenoids formed react with alkenes to form 1-phenylcyclopropanes with the *syn*-cyclopropane being formed as the major isomer (eq 5). The cyclopropane products are usually accompanied by varying quantities of stilbene from dimerization of the carbenoids or their precursors and low yields of the cyclopropane are usually obtained.

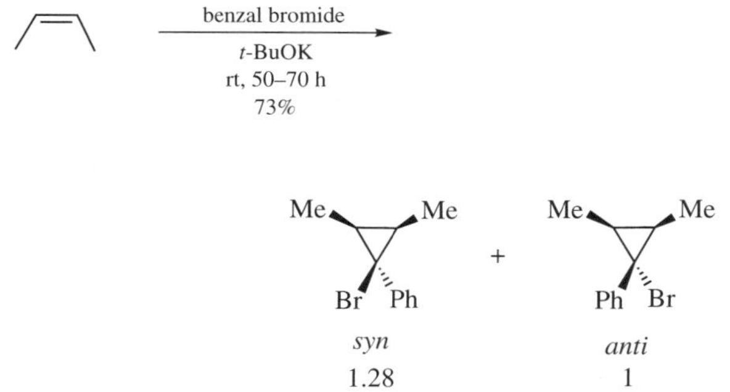

(4)

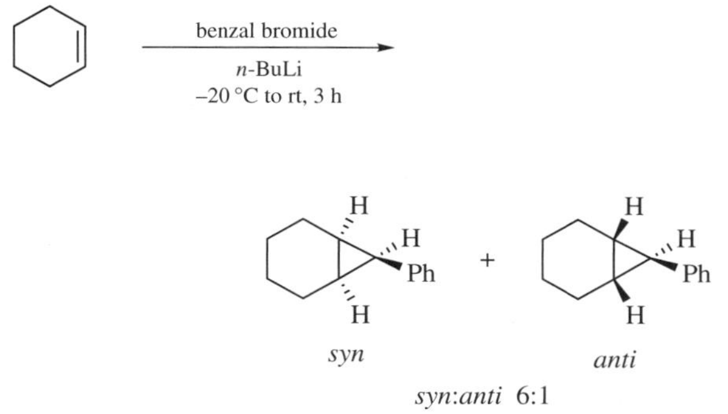

(5)

A related reaction involves the preparation of aziridines by magnesium activation of benzal bromide in the presence of benzaldehyde imines under ultrasonic irradiation (eq 6).[25] The major reaction pathway is again thought to involve the phenyl carbenoid species and leads to *cis*-aziridines as major isomers.[27] Stilbenes, as well as other by-products, are produced in the reaction and good yields are only produced when a large excess of the imine is used.

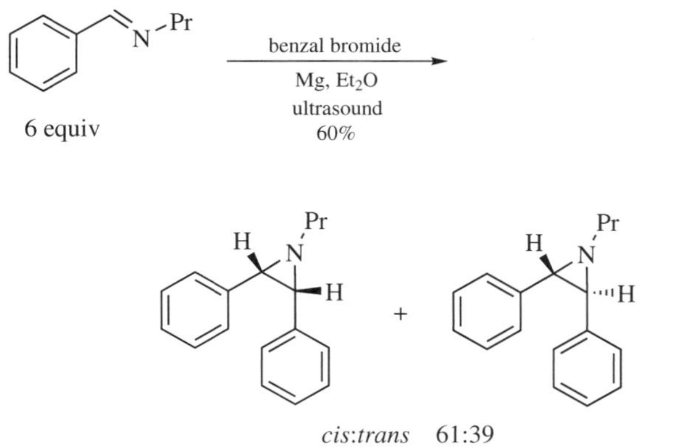

(6)

Formation of Benzylidene Derivatives. Stilbene is formed by the reductive homocoupling of benzal bromide using a variety of metal catalysts including metallic nickel,[28,29] palladium(0) complexes,[30,31] and cobalt(I) complexes.[32,33] In every case, there is a strong preference for the formation of the (*E*)-isomer. Unsymmetrical stilbenes can be formed by the reaction of benzal bromide with benzaldehydes in the presence of lithium (eq 7).[34] The reaction pathway is thought to involve the addition of bromobenzyl lithium to the aldehyde, followed by epoxide formation, and subsequent in situ reduction to the stilbene.

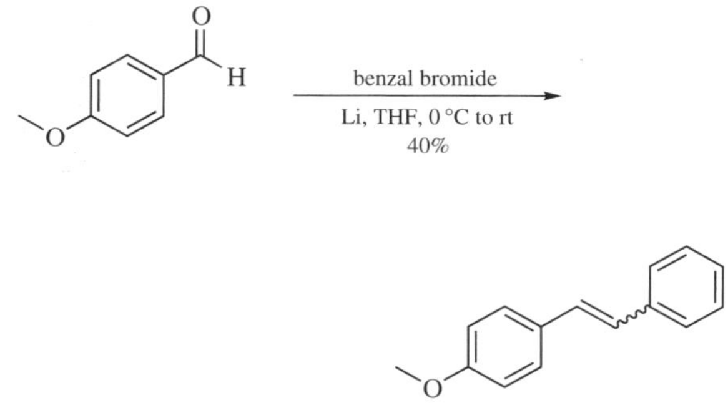

(7)

α-Lithio-α,α-dibromotoluene, generated by the reaction of LDA and benzal bromide at low temperature, has been shown to react with alkyl iodides in the presence of HMPT to form the corresponding α-bromostyrene derivatives (eq 8).[35]

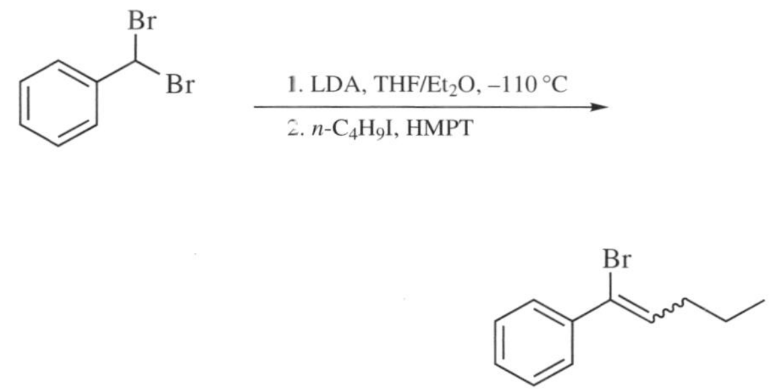

(8)

Benzylidene phthalides may be synthesized by the action of benzal bromide on phthalides in the presence of zinc dust and titanium tetrachloride (eq 9).[36]

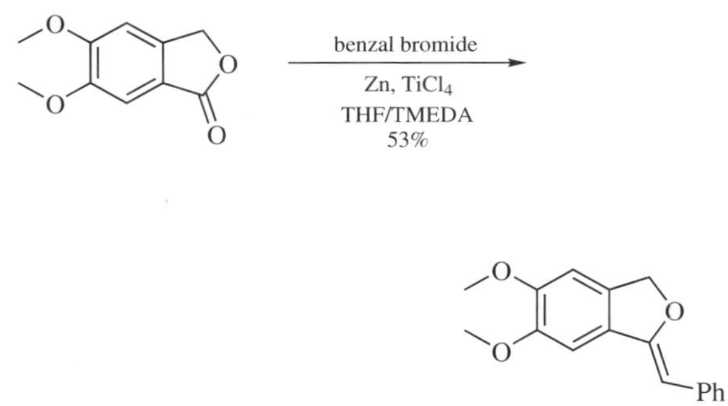

(9)

A similar reaction using chromium(II) chloride to promote the reaction of benzal bromide with an aldehyde in a Takai-type reaction has also been reported (eq 10).[37]

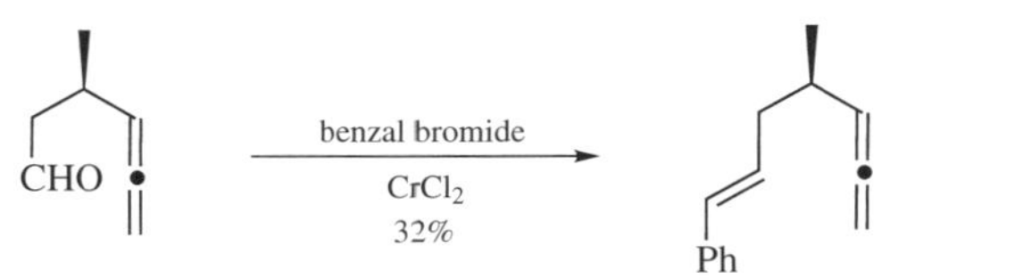

(10)

Related Reagents. Benzaldehyde; benzal chloride; benzaldehyde dimethyl acetal; 1,2-bis(phenyl)sulfonylethylene.

1. Brown, R. F. C.; Browne, N. R.; Eastwood, F. W., *Aust. J. Chem.* **1983**, *36*, 2355.

2. Li, W.; Li, J.; DeVincentis, D.; Mansour, T. S., *Tetrahedron Lett.* **2004**, *45*, 1071.
3. Bankston, D., *Synthesis* **2004**, 283.
4. Weibel, N.; Charbonniere, L. J.; Ziessel, R. F., *J. Org. Chem.* **2002**, *67*, 7876.
5. Kocienski, P. J., *Protecting Groups*; Thieme: New York, 1994, 96.
6. Bagget, N.; Duxbury, J. M.; Foster, A. B.; Webber, J. M., *Carbohydr. Res.* **1965**, *1*, 22.
7. Garegg, P. J.; Maron, L.; Swahn, C-G, *Acta. Chem. Scand.* **1972**, 518.
8. Garegg, P. J.; Swahnm, C-G, *Acta. Chem. Scand.* **1972**, 3895.
9. Weigel, T. M.; Liu, H. W., *Tetrahedron Lett.* **1988**, *29*, 4221.
10. RajanBabu, T. V., *J. Org. Chem.* **1988**, *53*, 4522.
11. Motawia, M. S.; Olsen, C. E.; Enevoldsen, K.; Marcussen, J.; Moeller, B. L., *Carbohydr. Res.* **1995**, *277*, 109.
12. Motawia, M. S.; Larsen, K.; Olsen, C. E.; Moller, B. L., *Synthesis* **2000**, 1547.
13. Damager, I.; Olsen, C. E.; Moller, B. L.; Motawia, M. S., *Synthesis* **2002**, 418.
14. Hoffmann, B.; Bernet, B.; Vasella, A., *Helv. Chim. Acta.* **2002**, *85*, 265.
15. Bowers, S. G.; Coe, D. M.; Boons, G-J, *J. Org. Chem.* **1998**, *63*, 4570.
16. Adinolfi, M.; Barone, G.; De Napoli, L.; Iadonisi, A.; Piccialli, G., *Tetrahedron Lett.* **1996**, *37*, 5007.
17. Soderberg, E.; Westman, J.; Oscarson, S., *J. Carbohydr. Chem.* **2001**, *20*, 397.
18. Kaneno, D.; Zhang, J.; Iwaoka, M.; Tomoda, S., *Heteroatom Chem.* **2001**, *12*, 358.
19. Segawa, J.; Kitano, M.; Kazuno, K.; Matsuoka, M.; Shirahase, I.; Ozaki, M.; Matsuda, M.; Tomii, Y.; Kise, M., *J. Med. Chem.* **1992**, *35*, 4727.
20. Chimirri, A.; Monforte, A. M.; Monforte, P.; Nicolo, F.; Rao, A.; Zappala, M., *Heterocycles* **2000**, *53*, 613.
21. Moss, R. A.; Gerstl R., *Tetrahedron* **1966**, *22*, 2687.
22. Applequist, D. E.; Nickel, G. W., *J. Org. Chem.* **1979**, *44*, 321.
23. Closs, G. L.; Moss, R. A., *J. Am. Chem. Soc.* **1964**, *86*, 4042.
24. Savino, T. G.; Kanakarajan, K.; Platz, M. S., *J. Org. Chem.* **1986**, *51*, 1305.
25. Oudeyer, S.; Aaziz, A.; Leonel, E.; Paugam, J. P.; Nedelec, J-Y., *Synlett* **2003**, 485.
26. Durandetti, S.; Sibille, S.; Perichon, J., *J. Org. Chem.* **1991**, *56*, 3255.
27. Biscoe, M. R.; Fry, A. J., *Tetrahedron Lett.* **2001**, *42*, 2759.
28. Inaba, S.; Matsumoto, H.; Rieke, R. D., *Tetrahedron Lett.* **1982**, *23*, 4215.
29. Inaba, S.; Matsumoto, H.; Rieke, R. D., *J. Org. Chem.* **1984**, *49*, 2093.
30. Grushin, V. V.; Alper, H., *Organometallics* **1993**, *12*, 3846.
31. Mataka, S.; Liu, G-B; Tashiro, M., *Synthesis* **1995**, 133.
32. Yamada, Y.; Momose, D., *Chem. Lett.* **1981**, 1277.
33. Momose, D.; Iguchi, K.; Sugiyama, T.; Yamada, Y., *Chem. Pharm. Bull.* **1984**, *32*, 1840.
34. Pasha, M. A.; Ravindranath, B., *Indian J. Chem.*; Sect. B **1985**, 24B: 781.
35. Villieras, J.; Bacquet, C.; Normant, J. F., *Bull. Soc. Chim. Fr.* **1975**, 1797.
36. Meegalla, S. K.; Rodrigo, R., *Synthesis* **1989**, 942.
37. Matsubara, S.; Toda, N.; Kobata, M.; Utimoto, K., *Synlett* **2000**, 987.

Nicholas J. Lawrence & Benjamin Greedy
University of South Florida, Tampa, USA

Dibromomethyl Methyl Ether

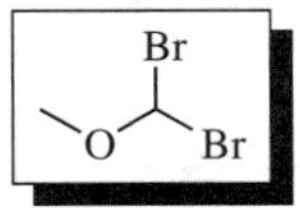

[3492-44-2] $C_2H_4Br_2O$ (MW 203.86)

(brominating reagent, reagent to effect conversions at the anomeric center of carbohydrates)

Physical Data: bp 32–33 °C/13 mm Hg.
Solubility: soluble in most common organic solvents; reacts vigorously with alcohols.
Form Supplied in: commercially not available.
Preparative Methods: from methyl formate and catechyl phosphorus tribromide.[1]
Purification: distillation.
Handling, Storage, and Precautions: dibromomethyl methyl ether is very reactive and moisture sensitive; redistilled product can be kept in flame-sealed glass tubes at −10 °C for several months.[2] It is a lachrymator and a suspect carcinogen. Handle in well-vented fume hood. Conversions effected by DBMME normally require presence of a Lewis acid. For this purpose, $ZnBr_2$, $ZnCl_2$, and BF_3 etherate have been used most frequently.

Bromination.

Direct Brominative Deoxygenation of Ketones

Dibromomethyl methyl ether has been used to convert ketones to alkenyl bromides. This is not a trivial reaction, as treatment of ketones with the most obvious reagent to effect the conversion, phosphorus pentabromide, normally gives α-bromo ketones. Ketones of general formula **1** can be converted to products **3**, via **2**, by treatment with dibromomethyl methyl ether (eq 1).[3]

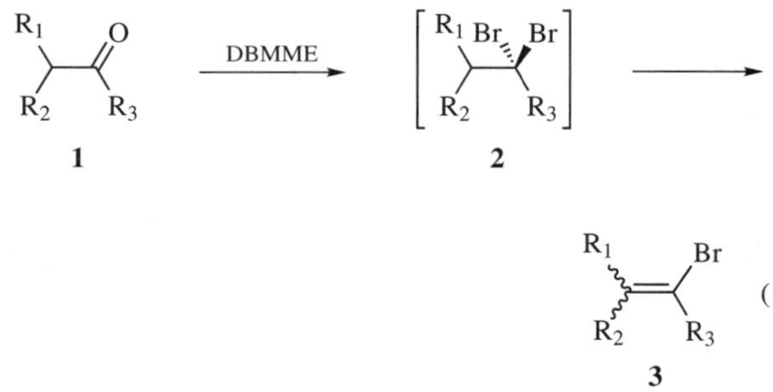

(1)

Conversion of esters to alkyl halides

Acetylated alkanols react with DBMME-$ZnBr_2$ to give the corresponding alkyl bromides and methyl formate (eq 2).[4]

$$\text{R—OAc} \xrightarrow[\text{ZnBr}_2]{\text{DBMME}} \text{R—Br} \qquad (2)$$

n-Butyl acetate reacts with DBMME (Table 1) in the presence of $ZnBr_2$, to give butyl bromide after 1.5 h. While benzyl acetate also reacts rapidly, no reaction takes place in the absence of $ZnBr_2$.

Table 1 Reaction rates of alkyl acetates with DBMME

R	Time
t-Butyl	<2 min
Benzyl	<2 min
s-Butyl	1 h
Cyclohexyl	10 h
Isobutyl	~2 weeks

Secondary acetates react in a similar way. Thus, *cis*- (**4**) and *trans*-1,2-diacetoxycyclohexane (**5**) react with the DBMME-$ZnBr_2$ reagent to give *trans*-2-bromo-cyclohexyl acetate (**6**). Further treatment of **6** with the reagent gives (eq 3) *trans*-1,2-dibromocyclohexane (**7**).

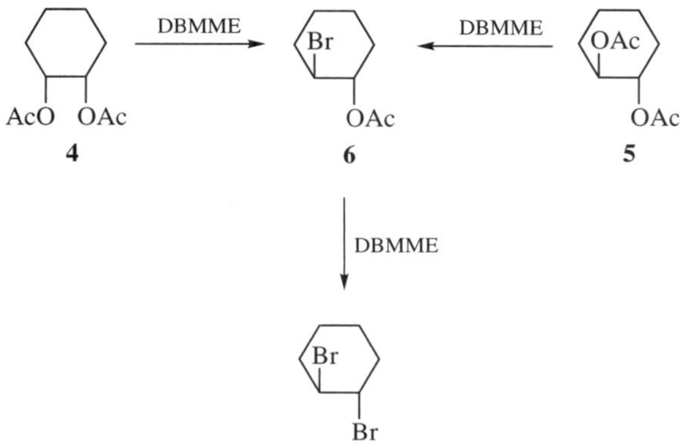

4 **6** **5** **7** (3)

3-Acetoxy-2-bromo-tetrahydropyrane (**8**), when treated with the DBMME-$ZnBr_2$ reagent, is readily converted (eq 4) into 2,3-dibromotetrahydropyrane (**9**), while 3-acetoxytetrahydropyrane (**10**) is stable towards the reagent (eq 5).

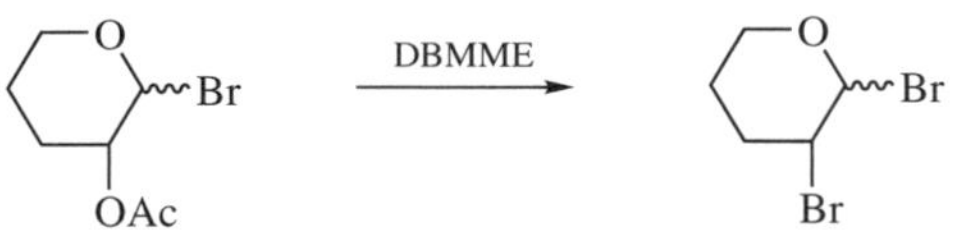

8 **9** (4)

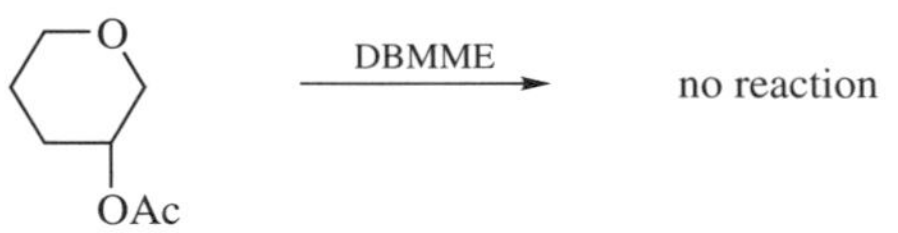

10 (5)

Preparation of carboxylic acid bromides

Carboxylic acids can be conveniently converted to the corresponding acyl bromides by simple treatment with DBMME (eq 6).[5] Examples in the literature are limited to conversions of protected aldonic and glycaric acids, but other carboxylic acids can be expected to react in the same way.

$$\text{R-COOH} \xrightarrow{\text{DBMME}} \text{R-COBr} + \text{HBr} + \text{HCOOCH}_3 \quad (6)$$

Halides derived from per-*O*-acetylated D-arabonic, -galactonic, -gluconic, -cellobionic, and mucic acids were thus obtained.

The work-up procedure comprises simple evaporation of volatiles, whereupon crystalline materials were usually obtained in high yields.

Preparation of sulfinyl bromides

Both dichloromethyl and dibromomethyl ether halogenate sulfinate anions yielding the corresponding sulfonyl halides. For example, treatment of sodium *p*-toluenesulfinate with DBMME in refluxing benzene gave tosyl bromide in 58.8% yield.[6]

Conversions at the Anomeric Center of Carbohydrates.

As its chlorinated counterpart[7,8] DBMME is a useful and versatile reagent in carbohydrate chemistry. Among other useful applications, it can be used to convert many glycosyl derivatives, including glycosides and oligosaccharides, into their respective glycosyl bromides.

Simple cleavage of anomeric derivatives

In the presence of a catalytic amount of freshly fused $ZnCl_2$, fully acetylated[5,9] or acetylated 1-OH-free[5] sugars, irrespective of their anomeric configuration, are converted to the corresponding thermodynamically more stable glycosyl bromides. In this way, for example, peracetylated methyl α-glucuronate (**11**), β-glucose (**13**), and 2,3,4,6-tetra-*O*-acetyl-α,β-D-glucopyranose (**15**) are converted to the corresponding α-glycosyl bromides **12**, **14**, and **16**, respectively (eqs 7–9).

11 **12** (7)

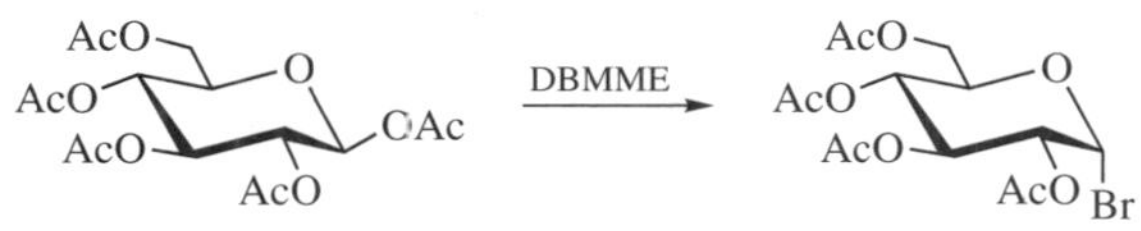

13 **14** (8)

15 **16** (9)

Alkyl and phenyl glycosides of monosaccharides and oligosaccharides can be converted in a similar way.[10] Thus, for example (eqs 10 and 11), benzyl tetra-*O*-acetyl-β-D-glucopyranoside (**17**) and methyl phenyl hepta-*O*-acetyl-β-lactoside (**19**) give the corresponding bromides **18** and **20**.

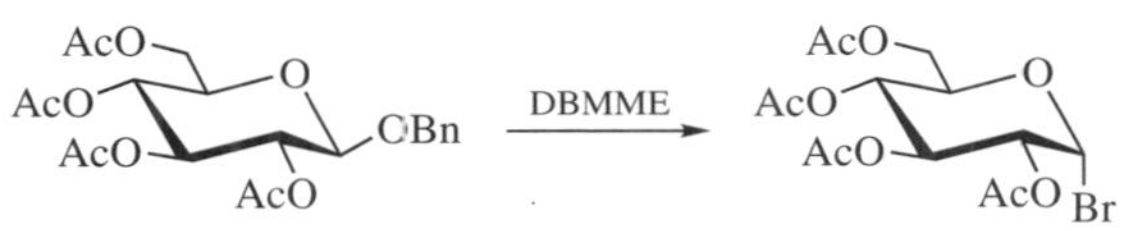

17 **18** (10)

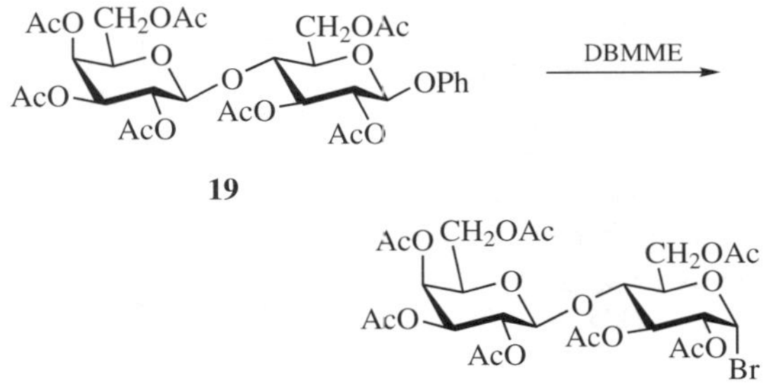

19 20 (11)

An important application of the conversions just described is selective cleavage of the linkage between oligosaccharides and aglycons in flavonoid glycosides.[11,12] For example (eq 12), rutinose can be released in this way from the peracetylated flavonoid rutine (**21**) in the form of its glycosyl bromide (**22**).

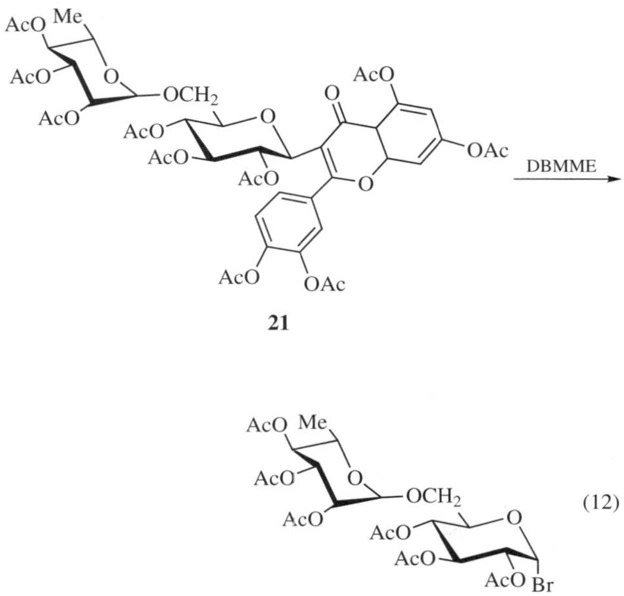

21 22 (12)

In analogy with a similar conversion effected with dichloromethyl methyl ether,[13] when boron trifluoride etherate is used as catalyst 1,2-*trans-O*-acetyl derivatives can be converted to the thermodynamically less stable 1,2-trans glycosyl bromides, which are not readily obtainable by other methods. Accordingly (eqs 13 and 14), methyl β-glucuronate[9] (**23**) and 4-*O*-methyl-β-glucuronate[14] derivatives (**24**) give β-bromides **25** and **26**, respectively.

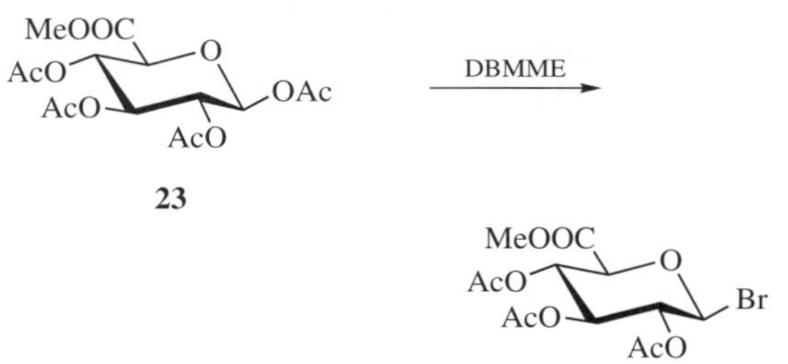

23 25 (13)

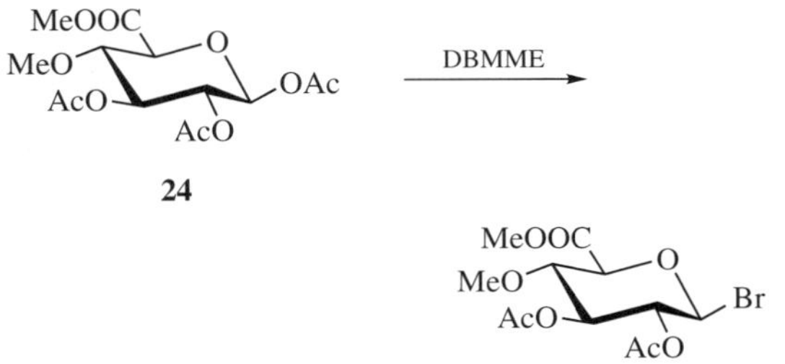

24 26 (14)

A noteworthy advantage of DBMME over other reagents capable of converting various glycosyl derivatives of sugars to glycosyl bromides is that it works with derivatives bearing acid labile alkylidene groups. For example (eqs 15 and 16), the isopropylidene[15] and cyclohexylidene[16] groups survive mild treatment with the DBMME-$ZnCl_2$ reagent. In this way glycosyl bromides **28** and **30** could be prepared in high yield from methyl glycosides **27** and **29**, respectively.

27 28 (15)

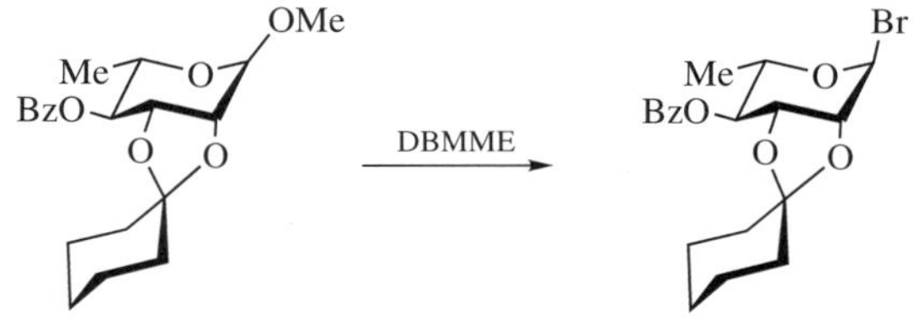

29 30 (16)

Cleavage of 1,6-anhydro sugars

Opening of the anhydro ring in 2,3,4-tri-*O*-acetyl-1,6-anhydro-β-D-glucopyranose (**31**) and formation of the glycosyl bromide (**32**) is accompanied by formylation of the primary hydroxyl group (eq 17).[17]

31 32 (17)

Formation of deoxybromo sugars from the corresponding glycosyl bromides

As can be surmised from some observations described above, primary products of treatment of pentofuranosides with the DBMME-$ZnBr_2$ reagent are the corresponding glycosyl bromides. Prolonged treatment with the reagent results in further transformations to form 2-bromo-2-deoxy derivatives, and the conversion provides an easy access to this useful class of carbohydrate derivatives. Thus (eq 18), for example,[18] methyl 2,3,5-tri-*O*-acetyl-β-Darabinofuranoside (**33**) can be converted to 3,5-di-*O*-acetyl-2-bromo-2-deoxy-α-D-arabinofuranosyl bromide (**35**), via the 2-*O*-acetyl compound **34**. Similarly (eq 19),[18] 2,3,5-tri-*O*-acetyl-1-*O*-benzoyl-β-D-xylofuranose (**36**) gives, eventually, 3,5-di-*O*-acetyl-2-bromo-2-deoxy-α-D-xylofuranosyl bromide (**37**).

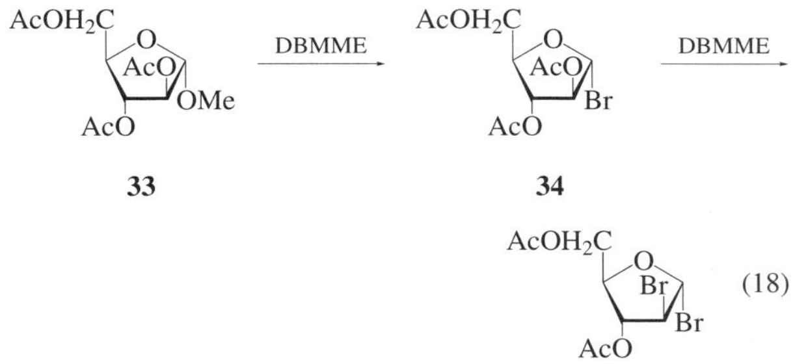

33 **34** **35** (18)

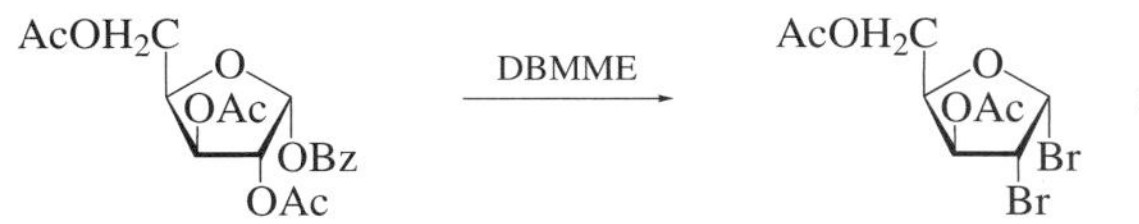

36 **37** (19)

Generally, similar reaction of pyranosides[19,20] is preparatively less attractive, but the product of further treatment with DBMME of glycosyl bromide **39**, obtained from the corresponding 1-*O*-benzoate **38** gave the 2-deoxy-bromosugar **40** as the main product (eq 20).

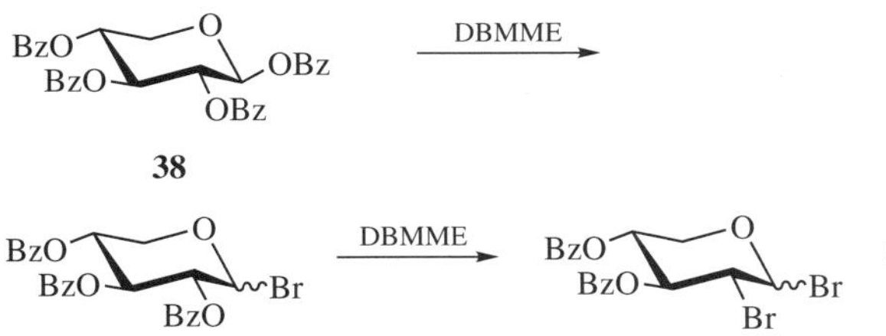

38 **39** **40** (20)

Related Reagents. Dichloromethyl methyl ether.

1. Gross, H.; and Karsch, U., *J. pract. Chem.* **1965**, *29*, 315–3318.
2. Kováč, P.; unpublished results.
3. von Roman, U.; Ruhdorfer, J.; Knorr, R., *Synthesis* **1993**, 985–992.
4. Bock, K.; Pedersen, C.; Rasmussen, P., *Acta Chem. Scand., Series B: Organic Chemistry and Biochemistry* **1976**, *B30*, 172–176.
5. Farkas, I.; Menyhart, M.; Bognar, R.; Gross, H., *Chem. Ber.* **1965**, *98*, 1419–1426.
6. Schank, K.; Schroeder, F., *Phosphorus and Sulfur and the Related Elements* **1976**, *1*, 307–308.
7. Gross, H.; Farkas, I.; Bognár, R., *Z. Chem.* **1978**, *18*, 201–210.
8. Kováč P., In *Synthesis of Glycosyl Halides for Oligosaccharide Synthesis Using Dihalogenomethyl Methyl Ethers*; Khan, S. H.; O'Neill, R. A., Eds.; Harwood Academic Publishers: Amsterdam, 1996, pp. 55–81.
9. Kováč, P.; Farkas, I.; Mihalov, V.; Palovcik, R.; Bognar, R., *J. Carbohydrates, Nucleosides, Nucleosides* **1976**, *3*, 57–69.
10. Farkas Szabo, I.; Farkas, I.; Bognar, R.; Gross, H., *Acta Chimica Academiae Scientiarum Hungaricae* **1970**, *64*, 67–80.
11. Bognar, R.; Farkas-Szabo, I.; Farkas, I.; Gross, H., *Carbohydr. Res.* **1967**, *5*, 241–243.
12. Koeppen, B. H., *Carbohydrate Research* **1969**, *10*, 105–112.
13. Farkas, I.; Szabo, I. F.; Bognar, R.; Anderle, D., *Carbohydr. Res.* **1976**, *48*, 136–138.
14. Kováč, P.; Palovcik, R., *Carbohydr. Res.* **1977**, *56*, 399–403.
15. Thiem, J.; Gerken, M.; Bock, K., *Justus Liebigs Ann. Chem.* **1983**, 462–470.
16. Iversen, T.; Bundle, D. R., *J. Org. Chem.* **1981**, *46*, 5389–5393.
17. Bognar, R.; Farkas, I.; Menyhart, M.; Gross, H.; Paulsen, H., *Carbohydr. Res.* **1968**, *6*, 404–413.
18. Bock, K.; Pedersen, C.; Rasmussen, P., *Acta Chem. Scand.* **1975**, *29 B*, 185–190.
19. Bock, K.; Pedersen, C.; Rasmussen, P., *J. Chem. Soc., Perkin Trans. 1* **1973**, 1456–1461.
20. Bock, K.; Pedersen, C.; Thiem, J., *Carbohydrate Research* **1979**, *73*, 85–91.

Pavol Kováč
National Institutes of Health, LMC, Bethesda, MD, USA.

Di-*t*-butyl Dicarbonate

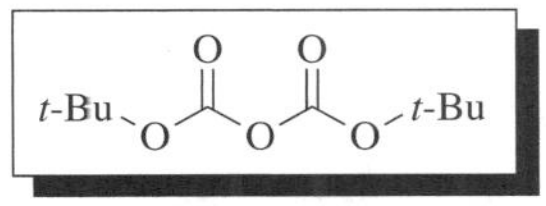

[24424-99-5] $C_{10}H_{18}O_5$ (MW 218.25)

(efficient *t*-butoxycarbonylating agent for amines,[1,2] phenols, and thiols; *t*-butoxycarbonylation of alcohols, amides,[3] lactams, carbamates[4] and NH-pyrroles[5] occurs in the presence of 4-dimethylaminopyridine)

Alternate Names: di-*t*-butyl pyrocarbonate; Boc anhydride; Boc_2O.

Physical Data: mp 22–24 °C; bp 56–57 °C/0.5 mm Hg; fp 37 °C; *d* 0.950 g cm^{-3}; n_D^{20} 1.4103.

Solubility: insol cold H_2O; sol most organic solvents such as decalin, toluene, CCl_4, THF, dioxane, alcohols, acetone, MeCN, DMF.

Form Supplied in: colorless liquid; widely available.

Analysis of Reagent Purity: IR 1810 and 1765 cm^{-1}; NMR 1.50 (CCl_4).

Preparative Method: prepared by the action of Phosgene on potassium *t*-butyl carbonate, followed by conversion of the obtained di-*t*-butyl tricarbonate under the action of basic catalysts such as 1,4-Diazabicyclo[2.2.2]octane.[6]

Handling, Storage, and Precautions: the liquid is flammable and must be stored in a refrigerator in the absence of moisture. Do not heat above 80 °C.

***t*-Butoxycabonylation of Amines.** Owing to the instability of the corresponding chloride (t-Butyl Chloroformate), the Boc_2O reagent is widely used for the introduction of the *t*-butoxycarbonyl amino protecting group (Boc). This group is easily removed under moderately acidic conditions such as treatment with Trifluoroacetic Acid, or by thermolysis. The Boc–NHR group is not hydrolyzed under basic conditions and is inert to many other nucleophiles.[1] Therefore the use of this amino protecting group is not limited to amino acid and peptide synthesis; it has been extended to the synthesis of amino sugars, alkaloids, etc. The great importance of the *t*-butyl carbamate function in peptide synthesis is not only due to the ease of its cleavage, but also to the fact that the oxazolones formed from the activated *N*-alkoxycarbonyl amino acids do not usually epimerize (racemize) during coupling with amines.[7]

Boc_2O reacts smoothly and rapidly with amino compounds in organic solvents, or aqueous organic solvent mixtures, to form pure derivatives in high yields, the only byproducts being the innocuous and easily removable *t*-butanol and carbon dioxide (eq 1). Innumerable examples are available.[8–11] Hydroxylamine slightly accelerates the reaction through the in situ formation of the Boc–ONH_2 intermediate.[12] The reaction is also accelerated by sonication of the amine salts.[13]

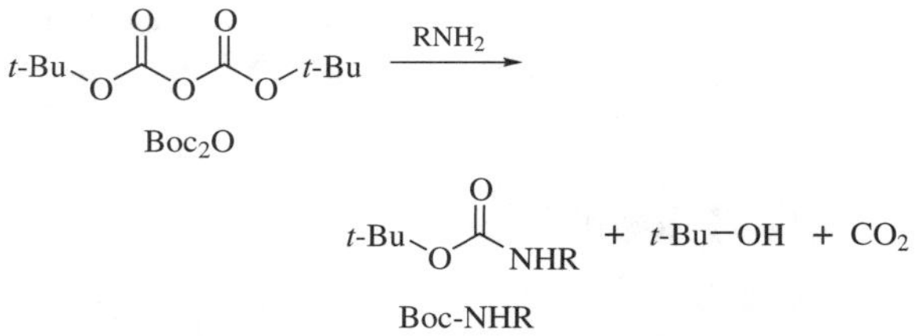
(1)

Aliphatic,[14] alicyclic,[15] aromatic,[16,17] and heterocyclic amines[18,19] have been *t*-butoxycarbonylated under a variety of conditions. Anilines are sometimes heated at reflux in THF to obtain the corresponding Boc derivatives in good yields. However, *t*-butoxycarbonylation also occurs at rt,[17] and 4-aminobenzoic acid reacts in DMF/aq NaOH at rt.[20] The Boc–NHR substituent is an *ortho*-directing group in heteroatom-facilitated lithiations of aromatic and heterocyclic amines.[16,18,19,21] Reaction of the resultant carbanions (eq 2), as well as that of the analogous benzylic carbanions, with electrophiles is a powerful method for the synthesis of substituted anilines and heterocycles.[16,22]

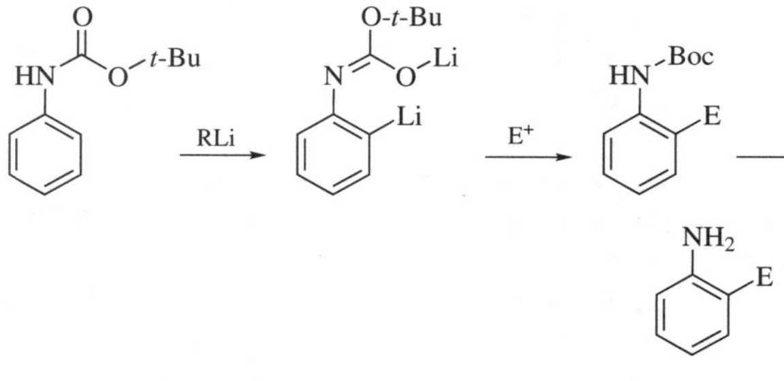
(2)

Asymmetric deprotonation of Boc-pyrrolidine with s-Butyllithium/ (-)-Sparteine gives a chiral organolithium reagent, which undergoes reaction with electrophiles to give enantiomerically enriched products.[23] The sequence of equatorial α-lithiation followed by an electrophilic substitution of a series of *N*-Boc cyclic amines provides amines which are substituted adjacent to the nitrogen atom.[15]

A suspension of 4-dimethylaminopyridinium fluoroborate in ethyl acetate reacts with Boc_2O to give the corresponding 1-*t*-butoxycarbonyl-4-dimethylaminopyridinium salt, which is an efficient water-soluble agent for the *t*-butoxycarbonylation of amines (eq 3).[24]

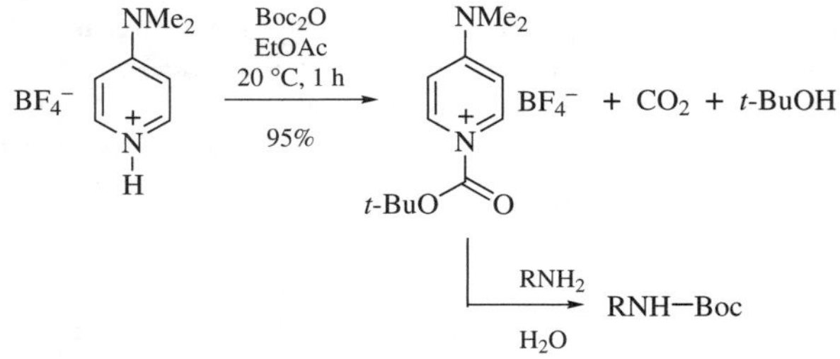
(3)

Interesting chemo- and regioselective alkoxycarbonylation of polyfunctional substrates, e.g. diamines, amino alcohols, and amino acids, can be achieved under mild conditions. Treatment of α,ω-alkanediamines (6–10 molar excess) with the reagent in dioxane or acetone as solvent leads to the corresponding mono-Boc diamines in 75–90% yields (eq 4).[25,26]

N^{ε}-Boc-L-lysine has been selectively prepared from the copper complex of lysine.[27] This complex, as well as that of N^{δ}-Boc-L-ornithine, has been directly benzyloxycarbonylated to give the corresponding α-Cbz-ε-Boc amino acids (eq 5).[28]

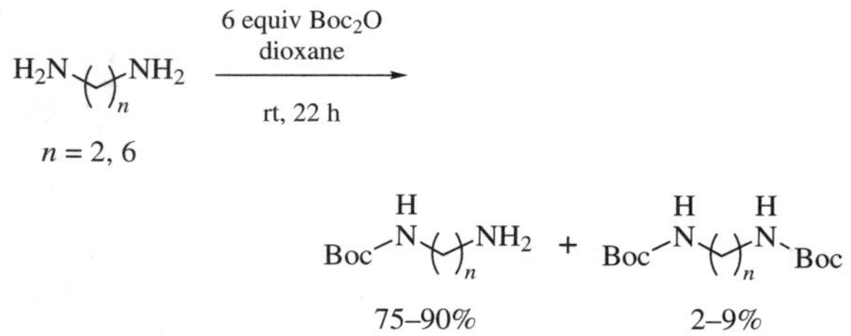
(4)

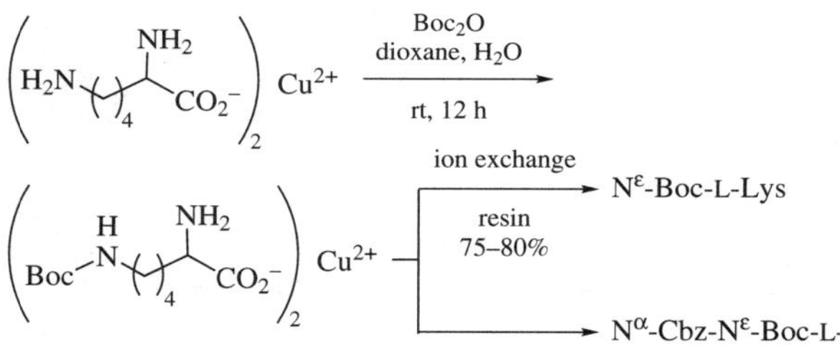
(5)

In CH_2Cl_2 at low temperature, selective *t*-butoxycarbonylation of a β-amino- in the presence of an α-amino-ester function has been recently realized (eq 6).[29]

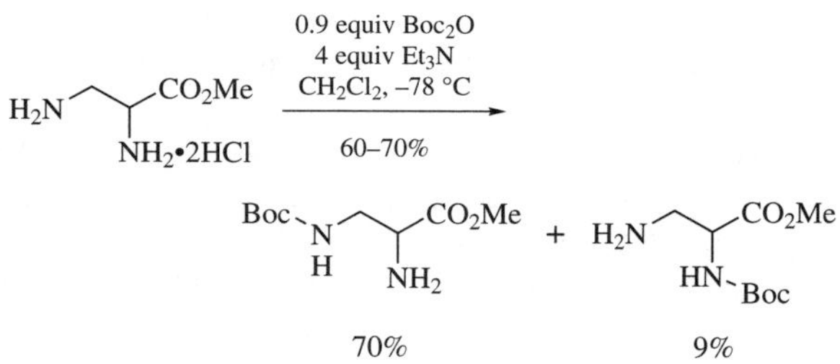
(6)

The substitution of a secondary amino group in the presence of a primary amine has been performed by an indirect route: formation of the labile Schiff base of the primary amine and then *t*-butoxycarbonylation of the secondary nitrogen (eq 7).[30] Selective protection of polyamines may also be achieved by an indirect route in the presence of DMAP (see below).

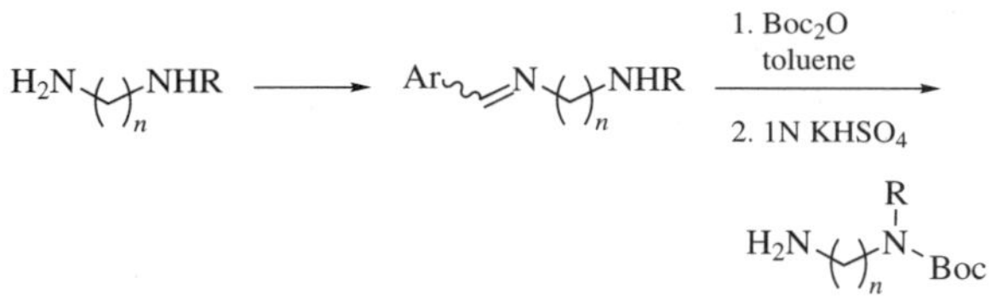

(7)

The Boc_2O reagent has been widely used for the *N*-protection of amino acids and peptides. When a carboxylic acid function is present in the molecule, 1 equiv of base (NaOH, Et_3N, Triton B) is necessary (eq 8).[10,11,31]

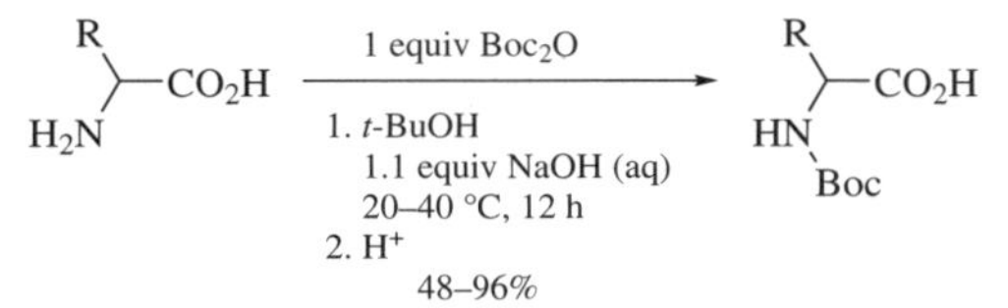

(8)

To avoid loss of the *O*-labeling, trialkylammonium salts of amino acids have been *t*-butoxycarbonylated in dry methanol or DMF.[32] *N-t*-butoxycarbonylation of an alkaline labile and hindered 2,2-dimethylthiazolidine-4-carboxylic acid has been performed in acetonitrile (eq 9).[33]

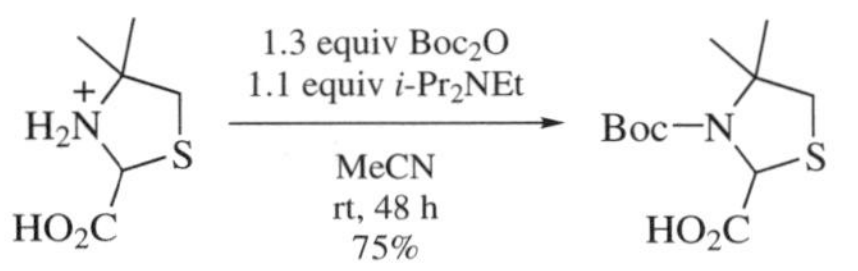

(9)

In the presence of pyridine, a free carboxylic function reacts with the reagent to give a mixed anhydride (see below). This reaction also occurs during the acylation of an amino acid salt with Boc_2O and is responsible for the formation of small amounts of dipeptide byproducts by aminolysis of the mixed anhydride intermediate (eq 10).[34]

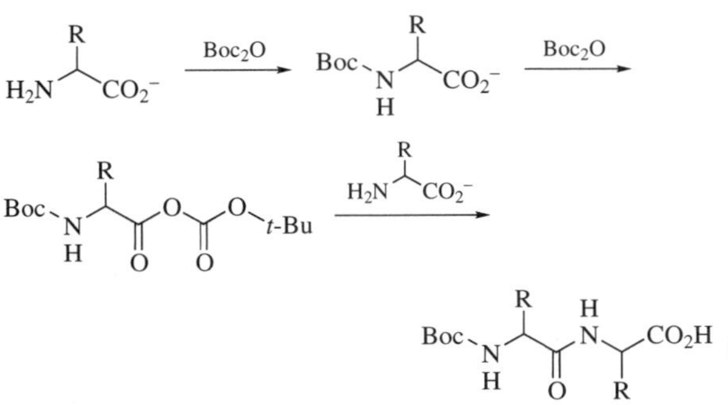

(10)

t-Butoxycarbonylation of phenolic amino acids in aqueous Potassium Hydroxide gives the *N,O*-bis-Boc derivatives.[35] Under phase-transfer conditions, selective *N-t*-butoxycarbonylation of tyrosine ethyl ester has been observed (see below). The SH group of cysteine,[36] the imidazole of histidine,[37] and even the guanidine function of arginine[38] have been acylated. Boc_2O has been used for the chemical modification of histidine residues of proteins.[39] For the *t*-butoxycarbonylation of insulin, Boc_2O is much more reactive than the corresponding succinimidyl and *p*-nitrophenyl esters.[40] Bamberger ring-cleavage of the imidazole nucleus slowly occurs in the presence of Potassium t-Butoxide.[41] Selective *N*-protection of amino alcohols is also easily performed (eq 11).[42]

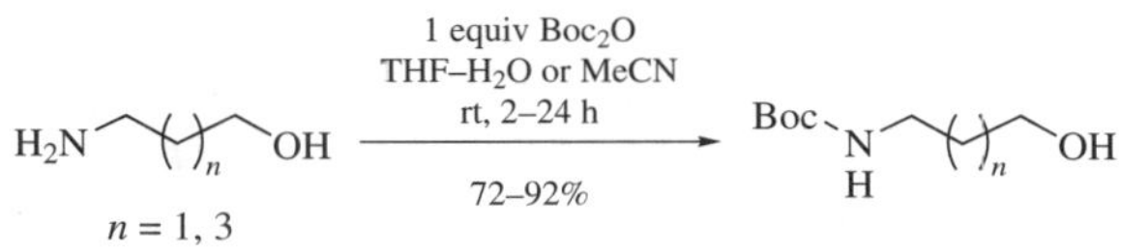

(11)

At pH 10.3, hydroxylamine itself gives the *O*-substituted product,[12] whereas treatment of *N*-alkylhydroxylamines in dioxane furnishes the *N*-Boc derivatives (eq 12).[43] The *O*-Boc hydroxylamine[12,44] as well as *O*-Boc oximes[45] are themselves efficient *t*-butoxycarbonylating agents.

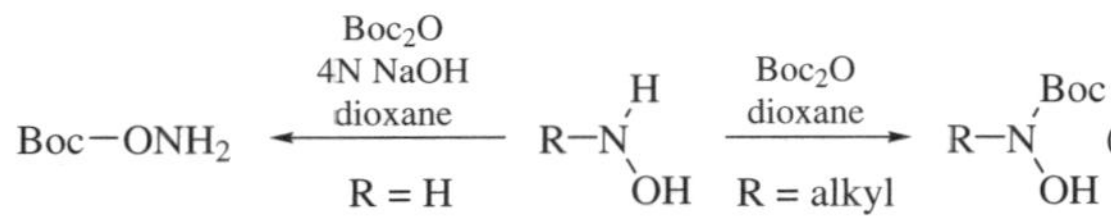

(12)

Hydrazines and hydrazides are acylated by the reagent.[46] The mono-Boc derivatives of hydrazine or phenylhydrazine, obtained in isopropanol, react further with Boc_2O in benzene or dioxane to give the $(Boc–NR)_2$ diacylated derivatives. Regioselective N^{α}-protection of ethyl hydrazinoacetate is performed at 0 °C and the product used for the synthesis of hydrazino peptides (eq 13).[47]

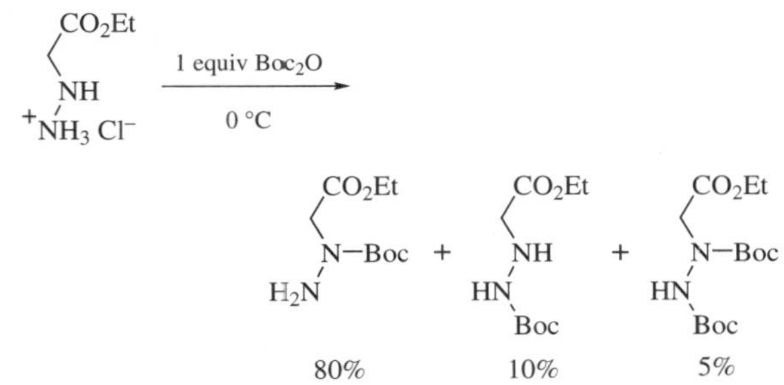

(13)

Boc-amines can also be prepared directly from azides by hydrogenation in the presence of Boc_2O (eq 14).[48]

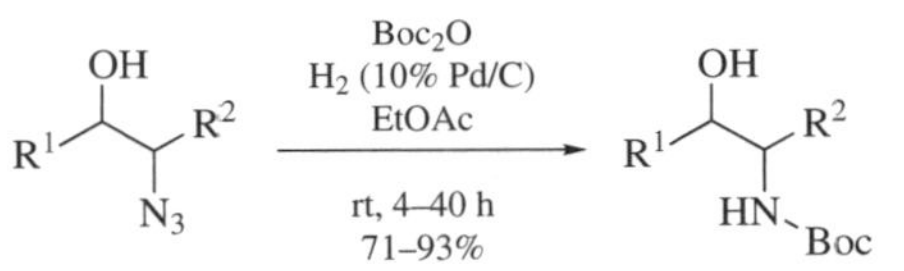

(14)

The alkylation of Boc-amines, particularly that of the more acidic Boc-anilines[17] and the selective alkylation of *N*-Boc amino acids,[49–51] has been done by using an alkylating agent and Sodium Hydride or Silver(II) Oxide as a base. For the *N*-perbenzylation of *N*-Boc peptides a P4-phosphazene base (see Phosphazene Base P_4-t-Bu) has been used at −100 °C.[52]

Exchange of Amino-Protecting Groups. *N*-Benzyloxycarbonyl (Cbz) and 9-fluorenylmethyloxycarbonyl (Fmoc) amino protecting groups are also widely used in peptide synthesis.[2] The choice of protecting groups, and hence of the conditions for their cleavage (Cbz: hydrogenolysis or HBr/HOAc; Fmoc: piperidine or F^-)[1] plays a key role in the planning of a synthesis. For instance, orthogonal protection (Boc/Cbz or Fmoc/Boc or Cbz) is of crucial importance in the case of the trifunctional amino acids. For a given peptide segment, chemoselective transformation of an amino protecting group into another one under neutral conditions may be very useful for the synthesis.

In the presence of a slight excess of Boc_2O, conversion of a benzyl carbamate into a *t*-butyl carbamate has been achieved by catalytic hydrogenation,[53] or by the use of cyclohexadiene as an hydrogen donor (eq 15).[54] The catalytic transfer hydrogenation is the more selective: benzyl and BOM ethers are stable under the reaction conditions. The Cbz → Boc transformation can also be performed using triethylsilane and a palladium catalyst.[53] The replacement of a Fmoc group by a Boc group has recently been achieved by the in situ cleavage of the Fmoc group with Potassium Fluoride in DMF (eq 15).[55]

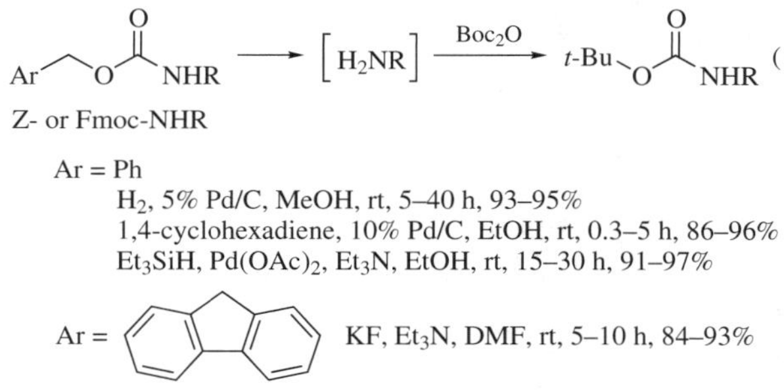

(15)

***t*-Butoxycarbonylation of phenols, Alcohols, Enols, and Thiols** In alkaline media the nucleophilic phenolate salts are easily *t*-butoxycarbonylated.[35,56] Acylation of *p*-dimethylsulfoniophenol gives a sulfonium salt which is a new water-soluble *t*-butoxycarbonylating agent for amino acids.[57] Under anhydrous conditions, thiolate salts are also substituted in good yields using the Boc_2O reagent.[56] Under phase-transfer conditions, *t*-butoxycarbonylation of phenols as well as enols, thiols, or alcohols can be realized. For phenols, enols, and thiols, K_2CO_3 (powder)/18-crown-6/THF are the best experimental conditions. Biphasic conditions (Bu_4NHSO_4/CH_2Cl_2/aq. NaOH) are also effective with sluggish alcoholic groups.[56] Using 1 equiv of Boc_2O, selective *N-t*-butoxycarbonylation of tyrosine ethyl ester can be achieved. *N,S*-Bis(Boc)-cysteine ethyl ester has also been prepared.[56] The *t*-butoxycarbonylation of alcohols to give alkyl *t*-butyl carbonates is efficiently catalyzed by DMAP.[58]

***t*-Butoxycarbonylation of Amides, Lactams, and carbamates.** In the presence of 4-Dimethylaminopyridine, Boc_2O is a strong acylating agent able to react with the weakly nucleophilic amide and carbamate groups under anhydrous conditions. The reaction probably involves a 1-*t*-butoxycarbonyl-4-dimethylaminopyridinium ion intermediate (eq 3).[24] This enhanced reactivity has led to numerous interesting synthetic applications. A wide range of amides, lactams, and urethanes have been converted to the corresponding Boc derivatives using Boc_2O as reagent and DMAP as catalyst.

Secondary amides and lactams react with Boc_2O in the presence of both Triethylamine and DMAP in CH_2Cl_2 to give stable *N-t*-butoxycarbonylated products (eq 16).[3] In the reaction of disubstituted pyrrolidones, the yields are much higher using THF as solvent than with dichloromethane.[59]

Exhaustive acylation of a series of secondary amides has also been performed with Boc_2O in dry acetonitrile at ambient temperature using DMAP as catalyst. The method has some steric limitations. For instance, it is retarded in the case of the *ortho*-substituted benzamides. The reaction has been extended to the acylation of carbamates and other amide type functions: sulfonamides, sulfenamides, and phosphinamides.[60] β-Lactams are also *N*-protected in good yields (eq 17).[61]

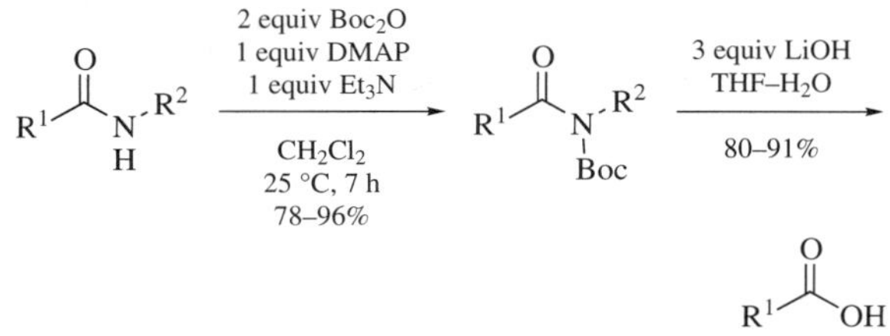

(16)

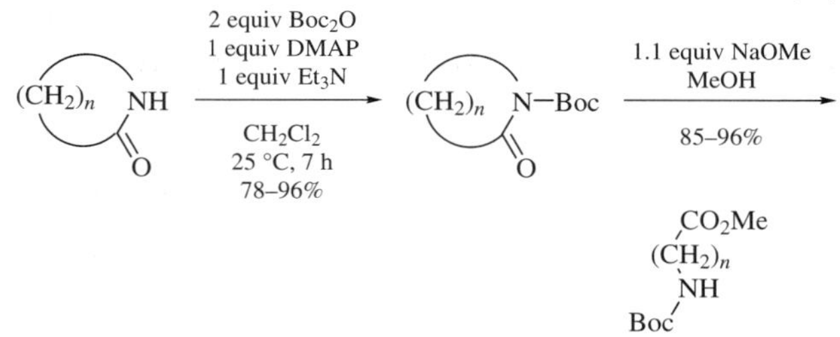

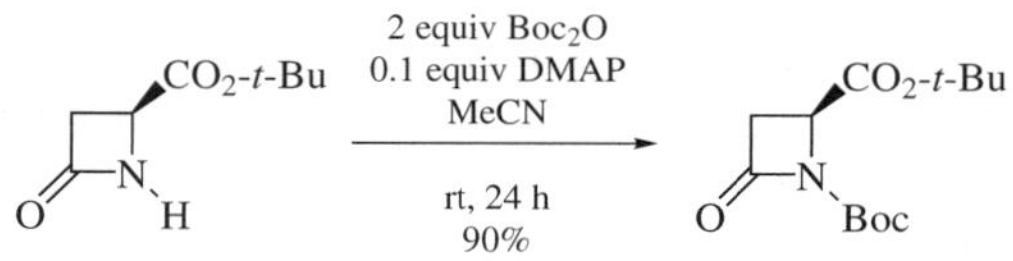

(17)

Primary carboxamides react with 2 equiv of Boc_2O and 1 equiv of DMAP to give isolable *N*-acylimidodicarbonates which are acylating agents for amines (eq 18).[62] Formamide itself furnishes the unstable di-*t*-butyl *N*-formylimidodicarbonate (eq 19).[63]

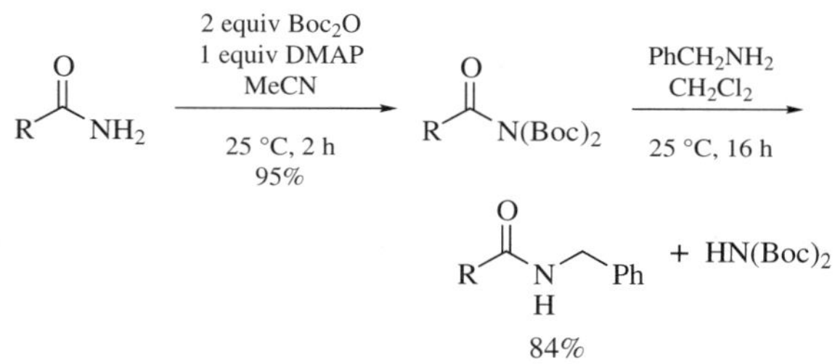

(18)

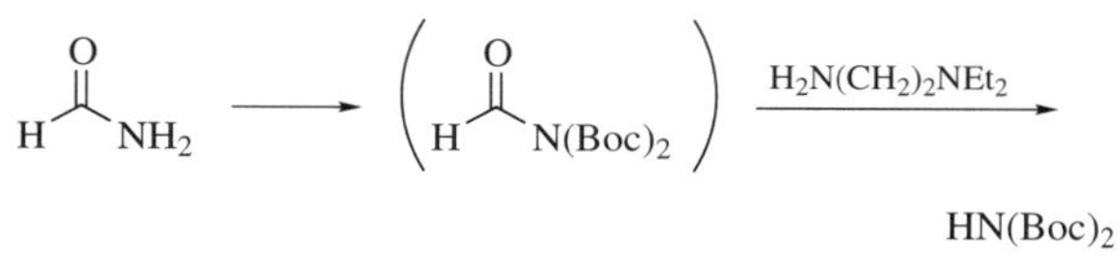

(19)

Grieco has shown that the regioselective hydrolysis or methanolysis of *N-t*-butoxycarbonyl derivatives of amides and lactams affords the corresponding carboxylic acids or methyl esters, respectively (eq 16).[3] 1,1,3,3-Tetramethylguanidine efficiently catalyzes the methanolic cleavage.[64] Grieco's method has been applied to a formal synthesis of gabaculine, in which Lithium Hydroxide treatment induces both sulfoxide elimination and lactam cleavage (eq 20).[65]

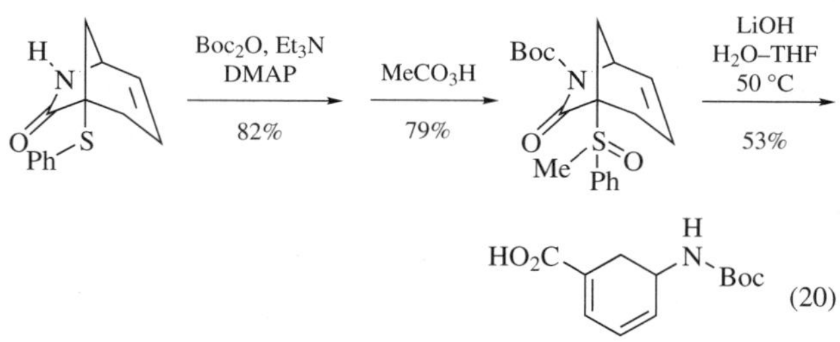

(20)

Other regioselective attacks at the acyl function are known. Aminolysis of substituted pyrrolidones gives the corresponding *N*-Boc γ-amino acid amides.[59] The *N*-Boc five- to eight-membered lactams are cleaved with Grignard reagents at the endocyclic carbonyl group to give ω-Boc-amino ketones,[66–68] which can be deprotected with TFA and cyclized with NaOH to yield cyclic imines.[67] Regioselective ring opening of the *N*-Boc-pyroglutamate ethyl ester with the lithium enolates of carboxylic esters, dithiane anion, alcohols, amines, and thiols occurs without epimerization.[69–71] Selective cleavage of cyclic carbamates gives *N*-protected amino alcohols (eq 21). In this reaction, methyl ester substituents are not transformed into acids by using Cesium Carbonate as a base.[72]

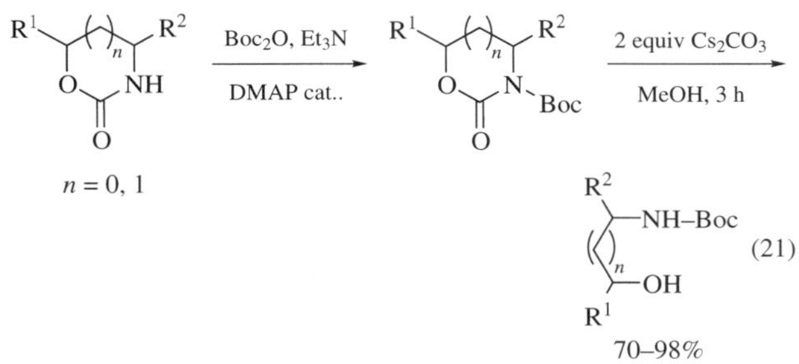

The selective deacylation of *N*-Boc carboxamides has also been achieved by aminolysis to give *N*-Boc amines. 2-Diethylaminoethylamine in acetonitrile is particularly convenient, but other conditions such as hydrazine in methanol can also be used.[64,73] Aminolysis of di-*t*-butyl *N*-formylimidodicarbonate affords di-*t*-butyl imidodicarbonate, $HN(Boc)_2$ (eq 19).[63] This compound, as well as the unsymmetrical benzyl *t*-butyl imidocarbonate, has been used as a substitute for Phthalimide in the Gabriel and Mitsunobu reactions to give primary amines. Compared to the classical method, which requires the hydrazinolysis of the phthaloyl protecting group, the main advantage of these novel Gabriel reagents is the much milder and rather specific conditions for the final deprotection step.[74] The lithium salt of $HN(Boc)_2$ has also been used in a palladium-catalyzed allylic amination.[75] Labeled di-*t*-butyl imidodicarbonate leads to ^{13}C and ^{15}N glycine derivatives.[76] Alkylation with triflates prepared from α-hydroxy acid esters gives ^{15}N-labeled chiral Boc-amino acids suitable for the synthesis of labeled peptides.[77]

t-Butoxycarbonylation of benzyl or allyl *N*-Boc or *N*-Cbz amino acid esters using Boc_2O/DMAP gives *N*-bis(alkoxycarbonyl) amino acid esters. Specific cleavage of the allyl protecting group with Chlorotris(triphenylphosphine)rhodium(I), or hydrogenolysis of the benzyl ester function, respectively, gives N^{α}-benzyloxycarbonyl, N^{α}-*t*-butoxycarbonyl or N^{α}-di-*t*-butoxycarbonyl amino acids (eq 22).[78,79]

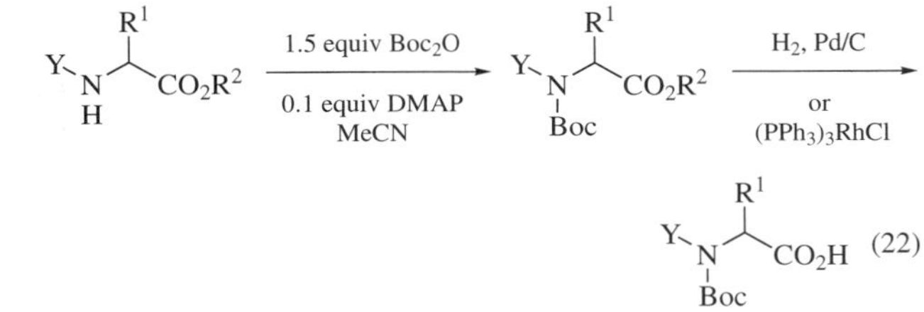

where H_2, Pd/C is used, R^2 = Bn, Y = Boc
where $(PPh_3)_3RhCl$ is used, R^2 = allyl, Y = Cbz or Boc

The di-*t*-butoxycarbonyl amino acids have been applied to solution peptide synthesis using the 1,3-Dicyclohexylcarbodiimide or *p*-nitrophenyl ester method in DMF for coupling[79]

Reaction of a bis(alkoxycarbonyl)amino acid pyridinium salt with cyanuric fluoride at −30 °C gives a *N*-bis(alkoxycarbonyl)amino acid fluoride (eq 23). Owing to the high electronegativity and small size of the fluorine atom, this derivative is an efficient acylating agent for amines and pyrrole anions.[80] Use of the Vilsmeier reagent instead of cyanuric fluoride leads to a chemically and optically pure *N*-Boc or *N*-Cbz amino acid *N*-carboxy anhydride (Boc- or Cbz-NCA) without the use of phosgene and the unstable *t*-butyl chlorocarbonate (eq 23).[80] Peptide synthesis with these urethane protected amino acid *N*-carboxy anhydrides (U-NCAs) is a very clean reaction which liberates only CO_2 as a byproduct.[81]

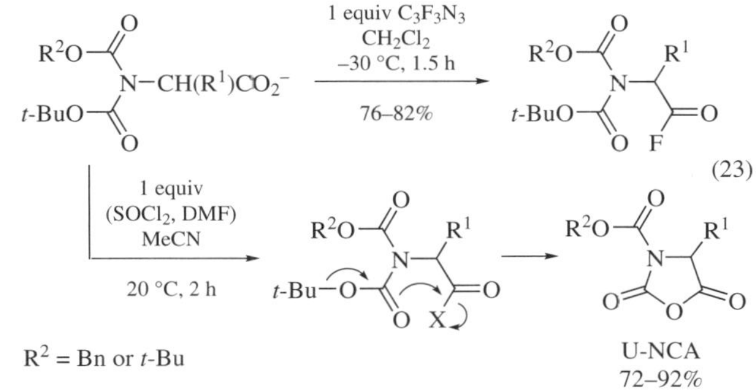

Treatment with $Boc_2O/DMAP_{cat}$ allows exhaustive *N*-*t*-butoxycarbonylation of some peptides.[82] The *t*-butoxycarbonylation of the internal amide bonds favors the cyclization of the peptides.[83] Interestingly, one Boc group can be selectively removed from a bis(Boc)-amine[75,76] with a slight excess of trifluoroacetic acid. Use of the $Boc_2O/DMAP_{cat}$/MeCN conditions also allows selective protection of mixed primary–secondary diamines. *t*-Butoxycarbonylation of a *N,N'*-bis(benzyloxycarbonyl)diamine followed by hydrogenolysis of the Cbz groups furnishes the *t*-butoxycarbonylated derivative of the primary amino function (eq 24).[84] Analogous strategies have been used for the selective indirect acylation of polyamines such as spermidine.[85,86]

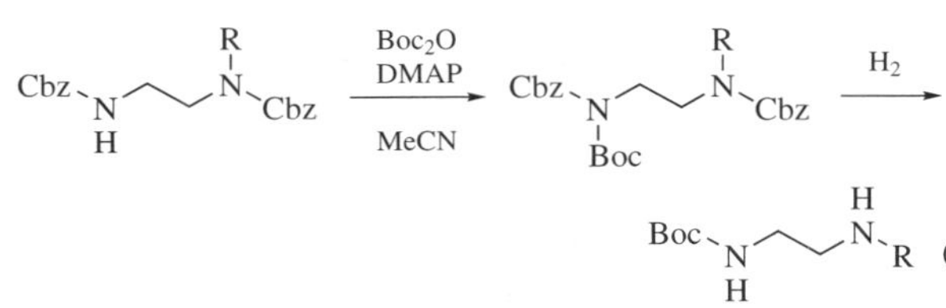

The *t*-butoxycarbonylation of the sodium salt of thiourea does not require DMAP catalysis. The resulting *N,N'*-bis(*t*-butoxycarbonyl)thiourea reacts with amines to give bis(*t*-butoxycarbonyl)guanidines (eq 25).[87]

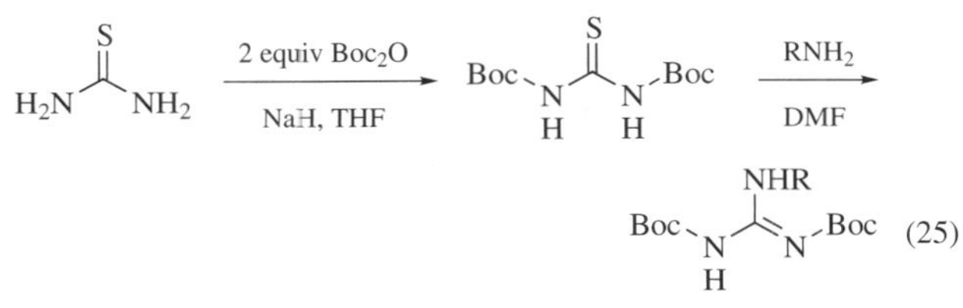

Sequential diacylation of 1-guanidinylpyrazole gives the *N,N'*-bis(Boc)-guanidinylpyrazole, which is also an efficient reagent for the conversion of amines to protected monosubstituted guanidines and therefore for the synthesis of arginine-containing peptides from their ornithine counterparts (eq 26).[88]

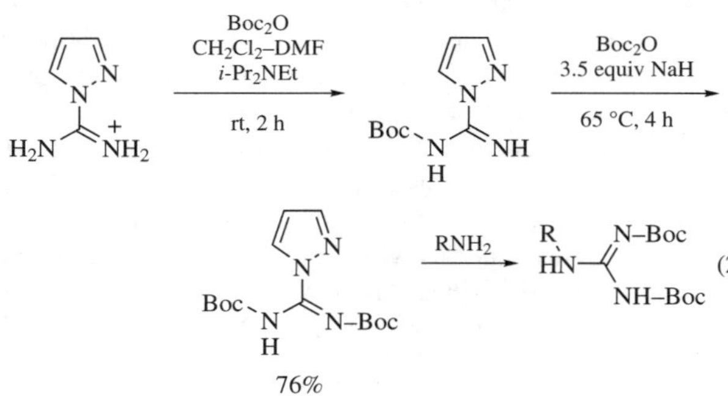

(26)

***N*-*t*-Butoxycarbonylation of Pyrroles and Indoles.** A series of substituted pyrroles and indoles[89–92] react smoothly with Boc_2O in the presence of catalytic amounts of DMAP in CH_2Cl_2 or acetonitrile to give the corresponding *N*-Boc pyrrole or indole derivatives. The *N*-*t*-butoxycarbonylation allows direct lithiation α to the indole and pyrrole nitrogen at low temperature. Alternatively, lithium–halogen exchange of *N*-Boc-2-bromopyrroles generates the *N*-protected 2-lithiopyrroles (eq 27).[91] After reaction with an electrophile, the Boc protecting group is easily removed either by methoxide ion in methanol or by thermolysis.[90,93]

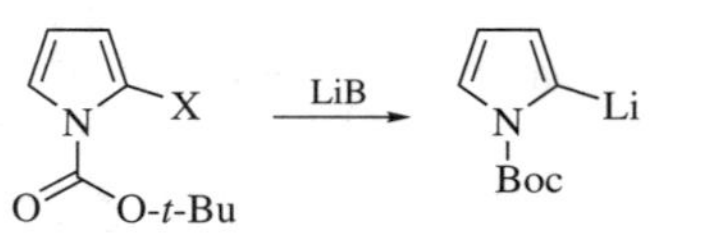

(27)

X = H, LiB = lithium tetramethylpiperidide
X = Br, LiB = BuLi

In the reaction with Boc_2O/DMAP, N^{α}-Boc-tryptophan methyl ester affords the N^{α},N^{in}-di-Boc derivative in almost quantitative yield. This product has been used in peptide synthesis. Moreover, selective cleavage of the N^{α}-Boc group is possible in the presence of the N^{in}-Boc function by using 2.7 N HCl in dioxane.[94]

Activation of the Carboxyl Group. Formation of Esters and Anhydrides. In the presence of Pyridine in aprotic solvents, the carboxyl function of a carboxylic acid or a *N*-Boc-amino acid is activated with the Boc_2O reagent. The resulting mixed anhydride can react further with one molecule of the starting acid to give a symmetrical anhydride. At an equimolar ratio of the acid and the Boc_2O reagent a mixture of the mixed and the symmetrical anhydrides is formed. If 2 equiv of acid are used, the symmetrical anhydride becomes the main reaction product. Addition of certain nucleophilic agents to the Boc_2O/carboxylic acid/pyridine mixture leads to the acylation of the nucleophile (eq 28).[58]

From a preparative point of view, it is essential that Boc_2O reacts much more rapidly with the carboxylate function than with the nucleophile. The reaction has been applied to the synthesis of aryl and alkyl esters and anilides such as *o*-nitrophenyl, benzyl, or trifluoroethyl esters and *p*-nitroanilides of *N*-protected amino acids. The formation of esters derived from secondary alcohols (benzhydryl, menthyl, or cholesteryl esters) is accelerated by addition of DMAP. Depsipeptides are obtained in good yields. MAP is a necessary catalyst for the synthesis of *t*-butyl esters and the esterification of *N*-protected amino acids occurs without epimerization.[58]

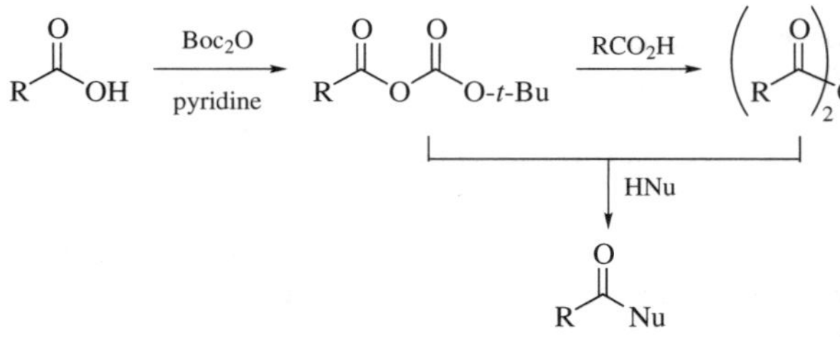

(28)

t-Butoxycarbonylation of Carbanions. Preferential *C*-*t*-butoxycarbonylation of diethyl acetamidomalonate occurs by treatment with $Boc_2O/DMAP_{cat}$ in acetonitrile (eq 29).[60] From the reaction of the potassium salt of diphenylphosphinyl isocyanide and Boc_2O in CH_2Cl_2 at −70 °C, the corresponding Wittig–Horner reagent, *t*-butyl (diphenylphosphinyl)isocyanoacetate, is formed in near quantitative yield.[95] Similarly, treatment of α-lithioalkylphosphonates with Boc_2O gives *C*-*t*-butoxycarbonylated enolates which react with aldehydes to give substituted *t*-butyl acrylates.[96] The lithium salt of the Schiff base derived from the aminomethyldiphenylphosphine oxide reacts with Boc_2O at low temperature to give the imine of *t*-butyl α-(diphenylphosphinoyl)glycinate.[97]

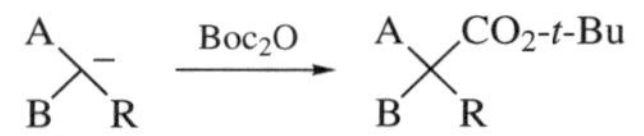

(29)

A = B = CO_2Et		R = NHAc	MeCN rt	93%
A = Ph_2PO	B = N=C	R = H	CH_2Cl_2 −70 °C	92%
A = $(Et_2O)_2PO$	B = alkyl	R = H	THF −78 °C	quant.
A = Ph_2PO	B = N=CPh	R = H	THF −60 °C	80%

t-Butoxycarbonylation of Imines In previous examples, the acylation occurred either at the amino function[30] or at the carbanionic center[97] and not at the imine function. However, the reaction of a nucleophilic imine possessing an electron-releasing substituent (Ar = *p*-$MeOC_6H_4$) with Boc_2O in ethanol proceeds through the intermediacy of an *N*-alkoxycarbonyl iminium ion to give an α-ethoxy carbamate (eq 30). As the trapping of the iminium ion by the alcohol is a reversible process, the α-ethoxy *t*-butyl carbamate is a precursor of the electrophilic *N*-Boc iminium ion in an intermolecular condensation with propargyltrimethylsilane.[98]

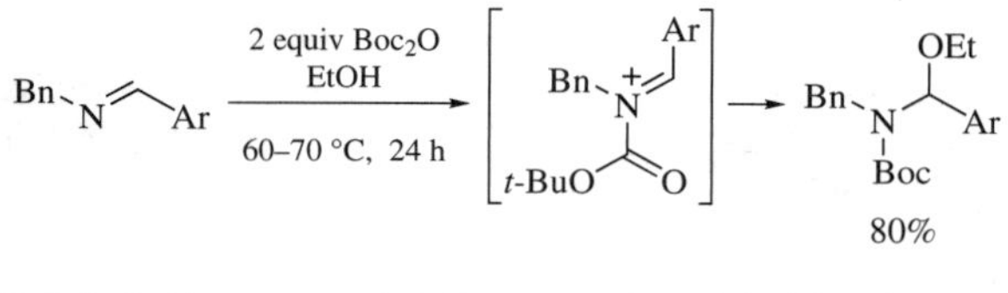

(30)

Related Reagents. 1-(t-Butoxycarbonyl)imidazole; 1-N-(t-Butoxycarbonyl)-1 H-benzotriazole 3-N-Oxide; 2-(t-Buto xycarbonyloxyimino)-2-phenylacetonitrile; N-(t-Butoxycarbonyloxy)phthalimide; N-(t-Butoxycarbonyloxy)succinimide; 1-t-Butoxycarbonyl-1,2,4-triazole; t-Butyl Azidoformate; t-Butyl Chloroformate .

1. Greene, T. W.; Wuts, G. M. *Protective Groups in Organic Synthesis*2nd ed.; Wiley:New York, **1991**; p 327.

2. Bodansky, M. *Principles of Peptide Chemistry*; Springer; New York, **1984**; p 99.
3. Flynn, D. L.; Zelle, R. E.; Grieco, P. A. *J. Org. Chem.* **1983**, *48*, 2424.
4. Gunnarsson, K.; Grehn, L.; Ragnarsson, U. *Angew. Chem., Int. Ed. Engl.* **1988**, *27*, 400.
5. Grehn, L.; Ragnarsson, U. *Angew. Chem., Int. Ed. Engl.* **1984**, *23*, 296.
6. Pope, B. M.; Yamamoto, Y.; Tarbell, D. S. *Org. Synth.* **1977**, *57*, 45.
7. Benoiton, N. L.; Lee, Y. C.; Steinaur, R.; Chen, F. M. F. *Int. J. Pept. Protein Res.* **1992**, *40*, 559.
8. Pozdnev, V. F. *Khim. Prir. Soedin.* **1971**, 384 (*Chem. Abstr.* **1971**, *75*, 110 568x) .
9. Pope, B. M.; Yamamoto, Y.; Tarbell, D. S. *Proc. Natl. Acad. Sci. USA* **1972**, *69*, 730.
10. Moroder, L.; Hallett, A.; Wünsch, E.; Keller, O.; Wersin, G. *Z. Physiol. Chem.* **1976**, *357*, 1652.
11. Keller, O.; Keller, W. E.; van Look, G.; Wersin, G. *Org. Synth.* **1985**, *63*, 160.
12. Harris, R. B.; Wilson, I. B. *Tetrahedron Lett.* **1983**, *24*, 231.
13. Einhorn, J.; Einhorn, C.; Luche, J. L. *Synlett* **1991**, 37.
14. Guzman, A.; Quintero, C.; Muchowski, J. M. *Can. J. Chem.* **1991**, *69*, 2059.
15. Beak, P.; Lee, W. K. *J. Org. Chem.* **1993**, *58*, 1109.
16. Muchowski, J. M.; Venuti, M. C. *J. Org. Chem.* **1980**, *45*, 4798.
17. Della Croce, P.; La Rosa, C.; Ritieni, A. *J. Chem. Res. (S)* **1988**, 346.
18. Fishwick, C. W. G.; Storr, R. C.; Manley, P. W. *Chem. Commun./J. Chem. Soc., Chem. Commun.* **1984**, 1304.
19. Fiakpui, C. Y.; Knaus, E. E. *Can. J. Chem.* **1987**, *65*, 1158.
20. Hofmann, K.; Finn, F.; Kiso, Y. *J. Am. Chem. Soc.* **1978**, *100*, 3585.
21. Stanetty, P.; Koller, H.; Mihovilovic, M. *J. Org. Chem.* **1992**, *57*, 6833.
22. Clark, R. D.; Muchowski, J. M.; Fisher, L. E.; Flippin, L. A.; Rebke, D. B.; Souchet, M. *Synthesis* **1991**, 871.
23. Kerrick, S. T.; Beak, P. *J. Am. Chem. Soc.* **1991**, *113*, 9708.
24. Guibé-Jampel, E.; Wakselman, M. *Synthesis* **1977**, 772.
25. Shai, Y.; Kirk, K. L.; Channing, M. A.; Dunn, B. B.; Lesniak, M. A.; Eastman, R. C.; Finn, R. D.; Roth, J.; Jakobsen, K. A. *Biochemistry* **1989**, *28*, 4801.
26. Krapcho, A. P.; Kuell, C. S. *Synth. Commun.* **1990**, *20*, 2559.
27. Scott, J. W.; Parker, D.; Parrish, D. R. *Synth. Commun.* **1981**, *11*, 303.
28. Masukiewicz, E.; Rzeszotarska, B.; Szczerbaniewicz, J. *Org. Prep. Proced. Int.* **1992**, *24*, 191.
29. Egbertson, M. S.; Homnick, C. F.; Hartman, G. D. *Synth. Commun.* **1993**, *23*, 703.
30. Prugh, J. D.; Birchenough, L. A.; Egberston, M. S. *Synth. Commun.* **1992**, *22*, 2357.
31. Perseo, G.; Piani, S.; De Castiglione, R. *Int. J. Pept. Protein Res.* **1983**, *21*, 227.
32. Ponnusamy, E.; Fotatar, U.; Spisni, A.; Fiat, D. *Tetrahedron Lett.* **1986**, 48.
33. Kemp, D. S.; Carey, R. I. *J. Org. Chem.* **1989**, *54*, 3640.
34. Chen, F. M. F.; Benoiton, N. L. *Can. J. Chem.* **1987**, *65*, 1224.
35. Pozdnev, V. F. *Khim. Prir. Soedin.* **1982**, *129* (*Chem. Abstr.* **1983**, *97*, 92 706p) .
36. Pozdnev, V. F. *Bioorg. Khim.* **1977**, *3*, 1604 (*Chem. Abstr.* **1978**, *88*, 62 595y) .
37. Pozdnev, V. F. *Zh. Obshch. Khim.* **1977**, *48*, 476 (*Chem. Abstr.* **1978**, *89*, 24 739m) .
38. Pozdnev, V. F. *Bioorg. Khim.* **1986**, *12*, 1013.
39. Harris, R. B.; Wilson, I. B. *J. Biol. Chem.* **1983**, *258*, 1357.
40. Losse, G.; Naumann, W.; Raddatz, H. *Z. Chem.* **1983**, *23*, 22.
41. Altman, Shoef, N.; Wilchek, M.; Warshawsky, A. *Chem. Commun./J. Chem. Soc., Chem. Commun.* **1985**, 1133.
42. Lee, B. H.; Miller, M. J. *J. Org. Chem.* **1983**, *38*, 24.
43. Baldwin, J. E.; Adlington, R. M.; Birch, J. B. *Tetrahedron Lett.* **1985**, *26*, 5931.
44. Harris, R. B.; Wilson, I. B. *Int. J. Pept. Protein Res.* **1984**, *23*, 55.
45. Itoh, H.; Hagiwara, D.; Kamiya, T. *Bull. Chem. Soc. Jpn.* **1977**, *50*, 718.
46. Pozdnev, V. F. *Zh. Org. Khim.* **1977**, *13*, 2531 (*Chem. Abstr.* **1978**, *88*, 89 063k) .
47. Lecoq, A.; Marraud, M.; Aubry, A. *Tetrahedron Lett.* **1991**, *32*, 2765.
48. Saito, S.; Nakajima, H.; Inaba, M.; Moriwake, T. *Tetrahedron Lett.* **1989**, *30*, 837.
49. Cheung, S. T.; Benoiton, N. L. *Can. J. Chem.* **1977**, *55*, 906.
50. Hansen, D. W.; Pilipauskas, D. *J. Org. Chem.* **1985**, *50*, 945.
51. Rich, D. H.; Dhaon, M. K.; Dunlap, B.; Miller, S. P. F. *J. Med. Chem.* **1986**, *29*, 978.
52. Pietzonka, T.; Seebach, D. *Angew. Chem., Int. Ed. Engl.* **1992**, *31*, 1481.
53. Sakatani, M.; Hori, K.; Ohfune, Y. *Tetrahedron Lett.* **1988**, *29*, 2983.
54. Bajwa, J. S. *Tetrahedron Lett.* **1992**, *33*, 2955.
55. Li, W.-R.; Jiang, J.; Joullié, M. M. *Tetrahedron Lett.* **1993**, *34*, 1413.
56. Houlihan, F.; Bouchard, F.; Fréchet, J. M. F.; Wilson, C. G. *Can. J. Chem.* **1985**, *63*, 153.
57. Aruze, I.; Okai, H.; Kouge, K.; Yamamoto, Y.; Koizumi, T. *Chem. Express* **1988**, *3*, 45.
58. Pozdnev, V. F. *Int. J. Pept. Protein Res.* **1992**, *40*, 407.
59. Klutchko, S.; O'Brien, P.; Hodges, J. C. *Synth. Commun.* **1989**, *19*, 2573.
60. Grehn, L.; Gunnarsson, K.; Ragnarsson, U. *Acta Chem. Scand.* **1986**, *B40*, 745.
61. Baldwin, J. E.; Adlington, R. M.; Godfrey, C. R. A.; Gollins, D. W.; Smith, M; L.Russel, A. *Synlett* **1993**, 51.
62. Davidsen, S. K.; May, P. D.; Summers, J. B. *J. Org. Chem.* **1991**, *56*, 5482.
63. Grehn, L.; Ragnarsson, U. *Synthesis* **1987**, 275.
64. Grehn, L.; Gunnarsson, K.; Ragnarsson, U. *Acta Chem. Scand.* **1987**, *B41*, 18.
65. Hiemstra, H.; Klaver, W. J.; Speckamp, W. N. *Tetrahedron Lett.* **1986**, *27*, 1411.
66. Ohta, T.; Hosoi, A.; Kimura, T.; Nozoe, S. *Chem. Lett.* **1987**, 2091.
67. Giovannini, A.; Savcia, D.; Umani-Ronchi, A. *J. Org. Chem.* **1989**, *54*, 228.
68. Klutchko, S.; Hamby, J. M.; Reilly, M.; Taylor, M. D.; Hodges, J. C. *Synth. Commun.* **1993**, *23*, 971.
69. Ohta, T.; Kimura, T.; Sato, N.; Nozoe, S. *Tetrahedron Lett.* **1988**, *29*, 4303.
70. Ezquerra, J.; de Mendoza, J.; Pedregal, C.; Ramírez, C. *Tetrahedron Lett.* **1992**, *33*, 5589.
71. Molina, M. T.; del Valle, C.; Escribano, A. M.; Ezquerra, J.; Pedregal, C. *Tetrahedron* **1993**, *49*, 3801.
72. Ishizuka, T.; Kunieda, T. *Tetrahedron Lett.* **1987**, *28*, 4185.
73. Grehn, L.; Gunnarsson, K.; Ragnarsson, U. *Chem. Commun./J. Chem. Soc., Chem. Commun.* **1985**, 1317.
74. Ragnarsson, U.; Grehn, L. *Acc. Chem. Res.* **1991**, *24*, 285.
75. Connell, R. D.; Rein, T.; B.; Helquist, P. *J. Org. Chem.* **1988**, *53*, 3845.
76. Grehn, L.; Bondesson, U.; Pehk, T.; Ragnarsson, U. *Chem. Commun./J. Chem. Soc., Chem. Commun.* **1992**, 1332.
77. Degerbeck, F.; Fransson, B.; Grehn, L.; Ragnarsson, U. *J. Chem. Soc., Perkin Trans. 1* **1993**, 11.
78. Gunnarsson, K.; Grehn, L.; Ragnarsson, U. *Angew. Chem., Int. Ed. Engl.* **1988**, *27*, 400.
79. Gunnarsson, K.; Ragnarsson, U. *Acta Chem. Scand.* **1990**, *44*, 944.

80. (a) Savrda, J.; Wakselman, M. *Chem. Commun./J. Chem. Soc., Chem. Commun.* **1992**, 812. (b) Carpino, L. A.; Mansour, E.-S. M. E.; El-Faham, A. *J. Org. Chem.* **1993**, *58*, 4162.
81. Fuller, W. D.; Cohen, M. P.; Shabankaresh, M.; Blair, R. K.; Goodman, M.; Naider, F. R. *J. Am. Chem. Soc.* **1990**, *112*, 7414.
82. Grehn, L.; Ragnarsson, U. *Angew. Chem., Int. Ed. Engl.* **1985**, *24*, 400.
83. Cavelier-Frontin, F.; Achmad, S.; Verducci, J.; Pépe, G.; Jacquier, R. *J. Mol. Struct. (Theochem)* **1993**, *105*, 125.
84. Almeida, M. L. S.; Grehn, L.; Ragnarsson, U. *Chem. Commun./J. Chem. Soc., Chem. Commun.* **1987**, 1250.
85. Almeida, M. L. S.; Grehn, L.; Ragnarsson, U. *J. Chem. Soc., Perkin Trans. 1* **1988**, 1905.
86. Araújo, M. J. S. M. P.; Ragnarsson, U.; Almeida, M. L. S.; Amaral Trigo, M. J. S. A. *J. Chem. Res. (S)* **1992**, 110.
87. Iwanowicz, E. J.; Poss, M. A.; Lin, J. *Synth. Commun.* **1993**, *23*, 1443.
88. Bernatowicz, M. S.; Wu, Y.; Matsueda, G. R. *Tetrahedron Lett.* **1993**, *34*, 3389.
89. Grehn, L.; Ragnarsson, U. *Angew. Chem., Int. Ed. Engl.* **1984**, *23*, 296.
90. Rawal, V. H.; Cava, M. P. *Tetrahedron Lett.* **1985**, *26*, 6141.
91. Chen, W.; Cava, M. P. *Tetrahedron Lett.* **1987**, *28*, 6025.
92. Woolridge, E. M.; Rokita, S. E. *Tetrahedron Lett.* **1989**, *30*, 6117.
93. Leroy, J.; Cantacuzene, D.; Wakselman, C. *Synthesis* **1982**, 313.
94. Franzén, H.; Grehn, L.; Ragnarsson, U. *Chem. Commun./J. Chem. Soc., Chem. Commun.* **1984**, 1699.
95. Rachoń, J.; Schöllkopf, U. *Justus Liebigs Ann. Chem./Liebigs Ann. Chem.* **1981**, 99.
96. Teulade, M. P.; Savignac, P.; About-Jaudet, E.; Colignon, N. *Synth. Commun.* **1989**, *19*, 71.
97. van Es, J. J. G. S.; Jaarveld, K.; van der Gen, A. *J. Org. Chem.* **1990**, *55*, 4063.
98. Esch, P. M.; Hiemstra, H.; Speckamp, W. N. *Tetrahedron* **1992**, *48*, 3445.

Michel Wakselman
CNRS-CERCOA, Thiais, France

2,7-Di-*t*-Butyl-9-fluorenylmethyl Chloroformate[1]

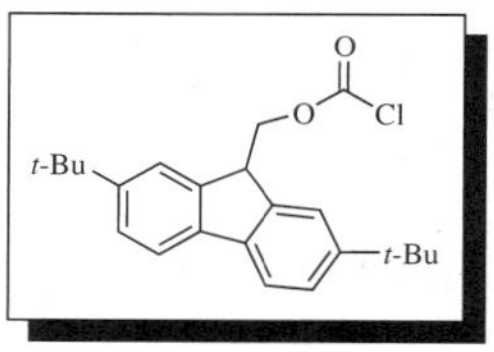

[287381-46-8] $C_{23}H_{27}ClO_2$ (MW 370.92)

(base-labile protecting group for primary and secondary amines with the resulting carbamates having improved solubility in organic solvents over those of the Fmoc protecting group)

Alternate Name: 2,7-di-*tert*-butyl-9-fluorenylmethoxycarbonyl chloride; carbonochloridic acid, [2,7-bis(1,1-dimethylethyl)-9*H*-fluoren-9-yl]methyl ester; Fmoc*-Cl.

Physical Data: mp 63–65 °C.

Solubility: soluble in organics (CH_2Cl_2, $CHCl_3$, pentane, etc.); reacts with alcohols, amines, and water.

Form Supplied in: 2,7-di-*tert*-butyl-9-fluorenylmethanol chloroformate and its immediate synthetic precursor (2,7-di-*tert*-butyl-9-fluorenylmethanol) are available from Aldrich, with the immediate synthetic precursor priced considerably more economically.

Analysis of Reagent Purity: ^1H NMR, IR, elemental analysis.

Preparative Methods: the title reagent can be prepared in three steps from fluorene[1] by Friedel-Crafts alkylation with $(CH_3)_3CCl$ and $FeCl_3$ to give 2,7-di-*tert*-butyl-9-fluorene, hydroxymethylation with *n*-BuLi and paraformaldehyde to give 2,7-di-*tert*-butyl-9-fluorenylmethanol, and conversion to the chloroformate with phosgene.

Purification: low-temperature (−78 °C) recrystallization from pentane.

Handling, Storage, and Precautions: treat with precautions typical of nonvolatile acid chlorides and chloroformates—avoid skin and eye contact; avoid breathing fumes, which may contain HCl, by working in a fume hood; open bottles carefully, as pressure may develop; protect from moisture; reactions with nucleophiles, such as amines, may be exothermic and should be performed with adequate dilution and cooling.

2,7-Di-*tert*-butyl-9-fluorenylmethyl chloroformate (Fmoc*-Cl) is useful as a base-labile protecting group for primary and secondary amines for applications in which the 9-fluorenylmethyloxycarbonyl (Fmoc) protecting group[2] would generally be used but is precluded by the low solubility of the Fmoc carbamates.[3] It may be coupled with amines in the presence of a suitable base, such as aqueous sodium carbonate, and the resulting carbamate may be subjected to various synthetic transformations (eq 1).[1] At an appropriate point, the Fmoc* protecting group may be cleaved with a suitable base, such as **piperidine**, to liberate the amine group (eq 2).[1] As such, 2,7-di-*tert*-butyl-9-fluorenylmethyl chloroformate may be viewed as an alternative to **9-fluorenylmethyl chloroformate**. An added benefit of Fmoc* is that the piperidine adduct that forms upon deprotection is highly lipophilic, allowing the removal of this adduct by extraction of a DMF solution of the deprotected amine with several portions of hexanes.

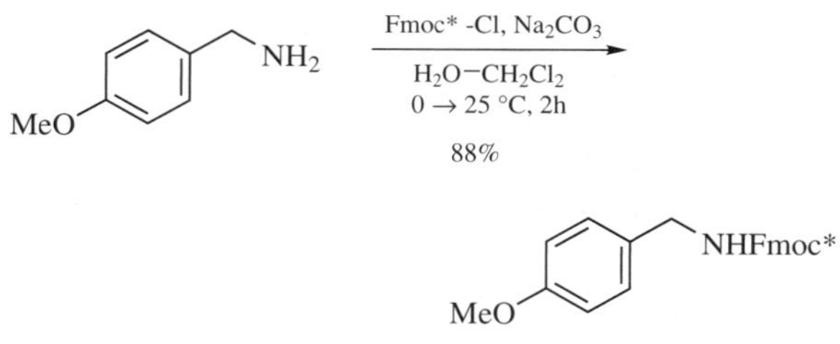

(1)

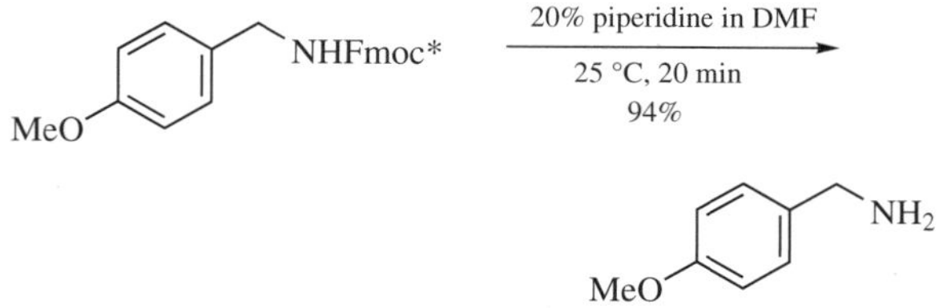

(2)

The Fmoc-protecting group has emerged as one of the two most important α-amino protecting groups in peptide synthesis. Although Fmoc-protected versions of the 20 proteinogenic amino acids with good solubilities in organic solvents are available, the growing importance of combinatorial and peptidomimetic chemistry has created a demand for many other protected amino acids.

In some cases, the Fmoc-protected versions of these amino acids or their synthetic precursors have poor solubility in organic solvents. The Fmoc*-protecting group provides what may be the only practical method of preparing and incorporating these amino acids in Fmoc-based peptide synthesis. Such is the situation with the unnatural amino acid Hao (5-HO_2CCONH-2-MeO-C_6H_3-$CONHNH_2$).[4] Hao is readily prepared as its Fmoc*-protected derivative (Fmoc*-Hao) by the condensation of Fmoc*-hydrazine with 5-methoxy-2-nitrobenzoyl chloride, reduction of the resulting Fmoc*-protected nitrobenzoic hydrazide to the corresponding aminobenzoic hydrazide, followed by condensation with ethyl oxalyl chloride and saponification of the ethyl ester group (eq 3). Fmoc*-Hao is readily incorporated into standard solid-phase peptide synthesis, as illustrated by the preparation of *i*-PrCO-Phe-Hao-Val-NHBu (eq 4).

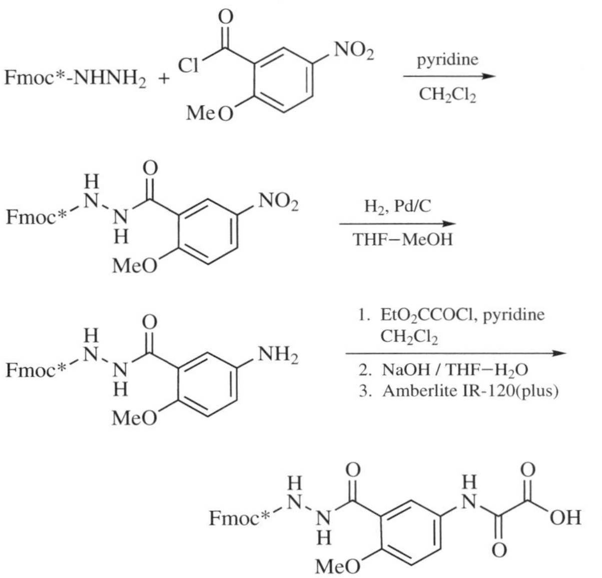

(Fmoc*-Hao) (3)

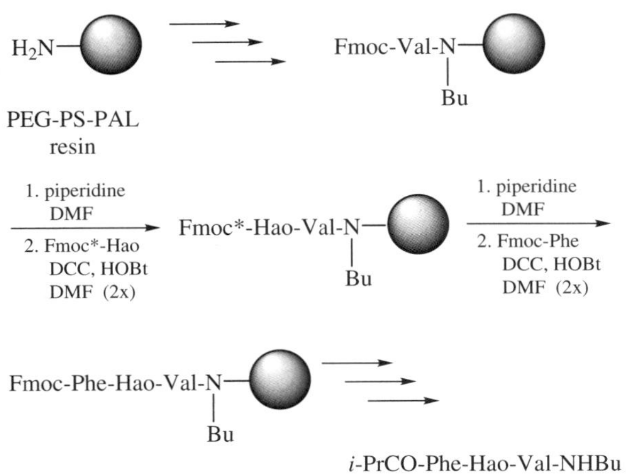

(4)

Improvements in solubility of Fmoc*-protected amines of up to one-to-two orders of magnitude over their Fmoc-protected counterparts have been reported.[1] Thus, the Fmoc-protected nitrobenzoic hydrazide 5-NO_2-2-MeO-C_6H_3-CONHNH-Fmoc (**1**) exhibits extremely limited solubility in $CHCl_3$ (1 mg mL^{-1}) and limited solubility in THF (8 mg mL^{-1}), while the corresponding Fmoc*-protected compound (5-NO_2-2-MeO-C_6H_3-CONHNH-Fmoc*) (**2**) is freely soluble in both of these solvents (>100 mg mL^{-1}). The limited solubility of the Fmoc-protected compound precludes its use in the synthesis of Fmoc-Hao by a route analogous to that shown in eq 3, and the use of the Fmoc*-protected compound is essential.

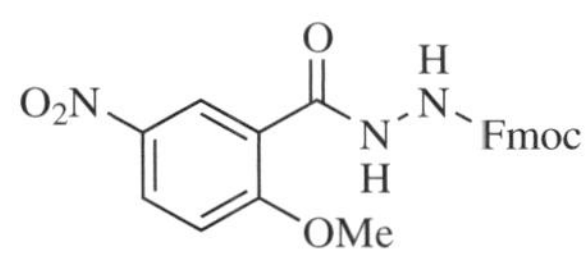

1

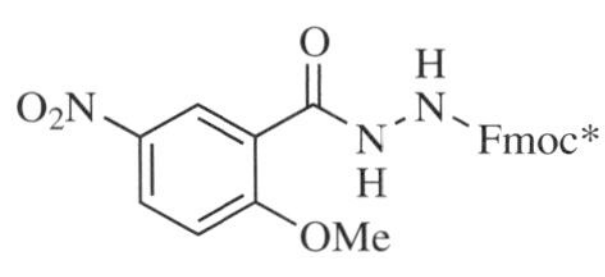

2

An alternative to the Fmoc*-protecting group, which bears trimethylsilyl groups in place of the two *tert*-butyl groups, has been reported to have even better solubility properties than Fmoc*.[5] The removal of the 2,7-bis(trimethylsilyl)fluorenylmethyl protecting group with amine bases proceeds in a rate almost comparable to that of the removal of Fmoc, while the removal of Fmoc* has been reported to proceed at either a nearly comparable rate to that of Fmoc[1] or somewhat more slowly.[5] A major drawback of the 2,7-bis(trimethylsilyl)fluorenylmethyl protecting group is that 2,7-bis(trimethylsilyl)fluorenylmethyl chloroformate is not commercially available and its preparation from fluorene requires seven steps (as opposed to three for Fmoc*-Cl).[5,6,7]

Related Reagents. 9-Fluorenylmethyl chloroformate (Fmoc-Cl); 2,7-bis(trimethylsilyl)-9-fluorenylmethyl chloroformate.[5]

1. Stigers, K. D.; Koutroulis, M. R.; Chung, D. M.; Nowick, J. S., *J. Org. Chem.* **2000**, *65*, 3858.
2. Carpino, L. A.; Han, G. Y., *J. Org. Chem.* **1972**, *37*, 3404.
3. Kocienski, P. J., *Protecting Groups*. Thieme, New York, 1994, p 202.
4. Nowick, J. S.; Chung, D. M.; Maitra, K.; Maitra, S.; Stigers, K. D.; Sun, Y., *J. Am. Chem. Soc.* **2000**, *122*, 7654.
5. Carpino, L. A.; Wu, A.-C., *J. Org. Chem.* **2000**, *65*, 9238.
6. Eaborn, C.; Shaw, R. A., *J. Chem. Soc.* **1955**, 1420.
7. Smith, W. K.; Hardin, J. N.; Rabideau, P. W., *J. Org. Chem.* **1990**, *55*, 5301.

James S. Nowick
University of California, Irvine, California, USA

2,6-Di-*t*-butylmethylpyridine

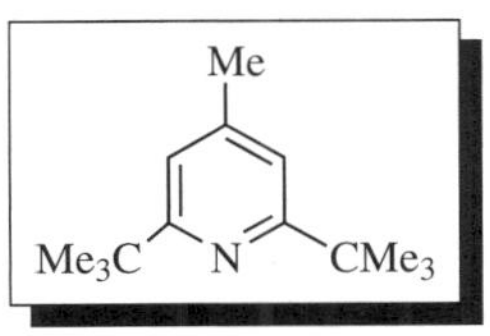

[38222-83-2] $C_{14}H_{23}N$ (MW 205.34)

(sterically hindered non-nucleophilic Brønsted base that may be protonated by Brønsted acids but does not react with Lewis acids)

Alternate Name: 2,6-bis(1,1-dimethylethyl)-4-methylpyridine, DTBMP.

Physical Data: mp 34 °C; bp 223 °C/760 Torr, 150 °C/95 Torr; n_D^{20} 1.4763; Fp 84 °C; enthalpy of vaporization: 45 kJ/mol; vapor pressure 0.087 Torr at 25 °C; pK_a 4.41 (50% EtOH).

Solubility: sparingly soluble in H_2O at pH >7, slightly soluble at pH 4, and soluble at pH 1; soluble in EtOH, AcOH, Et_2O, Me_2CO.

Form Supplied in: white solid at temperatures lower than 30 °C.

Purification: chromatography on activated alumina, recrystallization from cold pentane, sublimation, or distillation under reduced pressure.

Handling, Storage, and Precautions: may be stored indefinitely; irritant, potentially toxic, handle with care. Use in a fume hood.

Salts: trifluoromethanesulfonate, mp 203 °C; hexachloroplatinate mp 213–214 °C (dec.).

Syntheses. The diacylation of isobutene with pivaloyl derivatives (by means of the Balaban-Nenitzescu-Praill reaction [1]) affords 2,6-di-*tert*-butyl-4-methylpyrylium salts that are readily converted by ammonia in a separate step into 2,6-di-*tert*-butyl-4-methylpyridine (eq 1). The first hint of this reaction was the serendipitous discovery that the above pyrylium tetrachloroaluminate resulted from *tert*-butyl chloride and carbon monoxide at high pressure in the presence of anhydrous aluminum chloride.[2] Strong Lewis acids decarbonylate the pivaloyl cation, but this reaction becomes reversed at higher pressure and this explains the assembly of five fragments, namely one *iso*butene molecule and two (CO + *t*-Bu$^+$) synthons, to form the pyrylium cation.[3]

$$Me_3CX \rightleftharpoons Me_3C^+ X^- \rightleftharpoons Me_2C{=}CH_2 + HX$$

$$Me_3C^+ + CO \rightleftharpoons Me_3C{-}CO^+$$

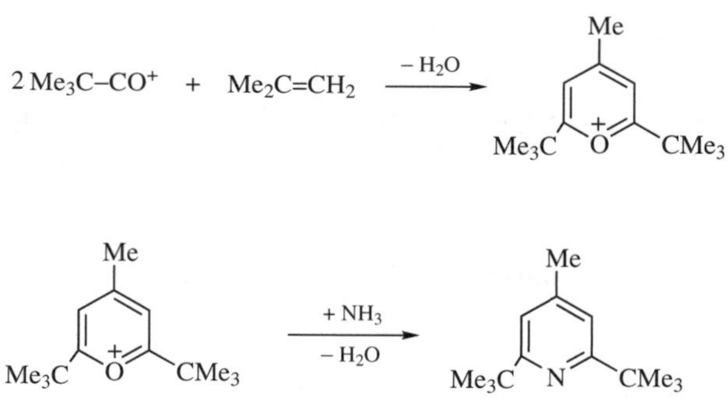

(1)

Subsequently, it was shown that tin tetrachloride, which is a milder Lewis acid, did not cause decarbonylation of pivaloyl chloride.[4] A synthetic procedure was developed for obtaining the above pyrylium salt and pyridine from *tert*-butyl chloride, pivaloyl chloride, and tin tetrachloride.[5] For isolating and purifying the pyrylium salt, one can replace the tetrachloroaluminate anion by perchlorate,[4,5] or better (in order to avoid handling potentially explosive perchlorates) the tetrafluoroborate (mp 199 °C) or the less soluble hexafluorophosphate (mp 259 °C).[6]

Instead of a mild Lewis acid, one may use a strong Brønsted acid (trifluoromethanesulfonic acid) as shown by Anderson and Stang (eq 2).[7,8] The procedure is carried out using a 5 L three-necked flask in a hood under a nitrogen atmosphere, cooling the reflux condenser with isopropanol and dry ice, and introducing during 2–3 min 109 mL (187.5 g, 1.25 mol) of triflic acid into a stirred mixture of 300 g (2.5 mol) of pivaloyl chloride and 46 g (0.62 mol) of anhydrous *tert*-butanol preheated to 85 °C. After heating at around 100 °C for 10 min, the mixture is allowed to spontaneously cool to 50 °C, then chilled to −10 °C. After adding 1 L of cold diethyl ether, the resulting crude pyrylium triflate is filtered off, washed with diethyl ether, and air-dried (53–62% yield, mp 168–169 °C when pure). Then this is converted by ammonia into DTBMP in 95% yield by suspending the crude pyrylium triflate (about 120 g) in 2 L of 95% ethanol, cooling to −60 °C, and adding 1 L of concentrated (d 0.90) aqueous ammonium hydroxide also cooled at −60 °C. After stirring for 30 min at −60 °C and at −40 °C for two additional hrs, the clear solution is allowed to reach room temperature. Aqueous 10% sodium hydroxide (2 L) is added and the solution is extracted four times with a total of 2 L of pentane. After washing the combined extracts with 100 mL of saturated aqueous sodium chloride, the pentane is rotavaporated and the oily product is taken into 150 mL of pentane and column-chromatographed over 300 g of activated alumina, eluting with 2 L of pentane, and finally evaporating the solvent on a rotary evaporator.

$$Me_3COH + 4Me_3CCOCl + F_3CSO_3H \xrightarrow[-\,2Me_3CCO_2H]{-4\,HCl}$$

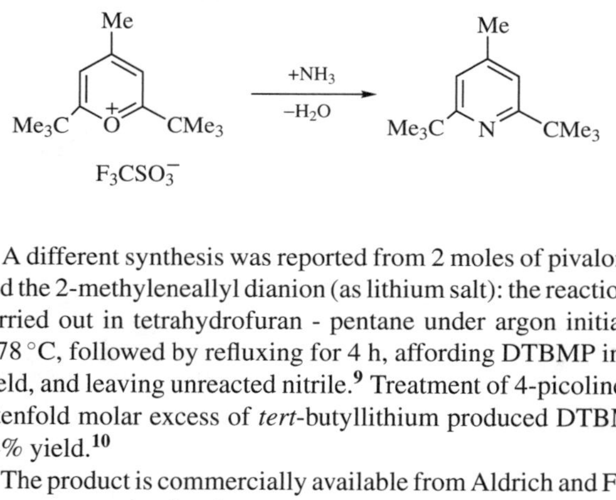

(2)

A different synthesis was reported from 2 moles of pivalonitrile and the 2-methyleneallyl dianion (as lithium salt): the reaction was carried out in tetrahydrofuran - pentane under argon initially at −78 °C, followed by refluxing for 4 h, affording DTBMP in 30% yield, and leaving unreacted nitrile.[9] Treatment of 4-picoline with a tenfold molar excess of *tert*-butyllithium produced DTBMP in 44% yield.[10]

The product is commercially available from Aldrich and Fisher-Acros companies, but for a very pure compound the traces of basic impurities have to be removed by sublimation.[11]

Reactions. Like di-*tert*-butylpyridine, this 4-methyl congener, which is prepared by a simpler reaction and is therefore less expensive, has its heteroatom sterically shielded, and is thus a non-nucleophilic base.[12] It cannot be quaternized by methyl iodide at normal pressure, but the *N*-methyl-derivative can be obtained (as fluorosulfate salt) with methyl fluorosulfate at high pressure.[13,14] Also, the *N*-amino-derivative was synthesized,[15] and the *N*-oxide was prepared from the corresponding pyrylium salt and hydroxylamine in 60% yield.[16] 2,6-Di-*tert*-butyl-*N*-fluoro-4-methylpyridinium salts with various anions were prepared in 5% yield by Umemoto et al. from DTBMP and fluorine in the presence of sodium fluoride; the corresponding N-fluoro triflate has mp 158–159 °C.[17]

Along with 2,6-di-*tert*-butylpyridine and 2,6-di-*tert*-butyl-4-methylpyridine, other related heterocyclic non-nucleophilic bases (2,4,6-tri-*tert*-butylpyrimidine[18] and the commercially available 2,4,6-tri-*tert*-butylpyridine) have been shown to share the same usefulness in the synthesis of enolic triflates and in technologically important polymerizations. A related compound is also a weakly nucleophilic base: 4-ethyl-2,6-diisopropyl-3,5-dimethylpyridine.[19] This compound can also be synthesized by alkene diacylation (with hexafluorophosphoric acid, isobutyric anhydride, and 3-ethylpentanol) but without interference of decarbonylation reactions; owing to a double Janus effect, the two *iso*propyl groups buttressed by the neighboring groups are twisted such that they mimic *tert*-butyl groups towards nucleophiles approaching the nitrogen heteroatom.

An interesting and still unexplained observation was that the bromination of β-substituted 2-butenoic esters, e.g., methyl (*Z*)-4-*tert*-butyldiphenylsilyloxy-3-methyl-2-butenoate, followed by potassium *tert*-butoxide-induced dehydrobromination, afforded α-bromoderivatives with reversed stereoselectivities in the presence of DTBMP (formation of *E*-bromides) and *N,N*-diisopropylethylamine (formation of *Z*-bromides, as useful intermediates in the synthesis of taxol). Sterically nonhindered bases such as 2,4,6-trimethylpyridine, or no base at all during bromination, led to the formation of an *E/Z* mixture in approximately equal amounts.[20]

In the presence of DTBMP, the mixed anhydride of trifluoroacetic and triflic acids (trifluoroacetyl triflate) is an electrophilic trifluoroacetylation agent for alcohols, amines, and enolizable ketones (eq 3).[21]

$$\mathrm{F_3C{-}CO{-}O{-}Tf} \;+\; \mathrm{ROH} \longrightarrow \mathrm{F_3C{-}CO{-}O{-}R} \;+\; \mathrm{TfOH} \qquad (3)$$

Non-Nucleophilic Base for Preparing Vinylic Triflates. Stang and Treptow advocated the use of DTBMP and TBMP as the key reagent allowing the high-yield synthesis of primary vinyl triflates from enolizable aldehydes (eq 4) or enolizable ketones.[22] Other tertiary bases that have no steric hindrance to quaternization such as pyridine or triethylamine give lower yields because with triflic anhydride they form dark-colored, often heterogeneous mixtures, containing sulfonylammonium triflates.

$$\mathrm{R_2CH{-}CHO} \xrightarrow[\text{DTBMP}]{(\mathrm{F_3CSO_2})_2\mathrm{O}} \mathrm{R_2C{=}CH{-}O{-}SO_2\,CF_3} \qquad (4)$$

Such vinyl triflates form new C–C bonds via palladium-catalyzed coupling with organostannanes yielding conjugated dienes (eq 5).[23]

An interesting application of this Stille cross coupling using a vinyl triflate whose synthesis was assisted by DTBMP was the synthesis of a potassium channel activator.[24] Alternatively, the triflate group may be removed reductively in DMF solution on treatment with $Pd(OAc)_2$ and Ph_3P in the presence of formic acid; thus, cholest-4-en-3-one affords cholesta-3,5-diene in a reaction that is a formal two-step deoxygenation of a ketonic enol, yielding the corresponding alkene.[25]

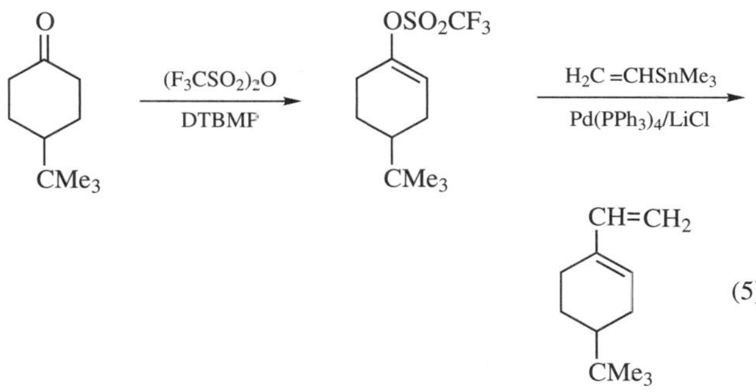

(5)

Similarly, in the presence of DTBMP, 1-(ethynyl)-vinyl triflates can be prepared in good yields from triflic anhydride and 1-acyl-2-trimethylsilyl-alkynes, followed by desilylation with excess potassium fluoride dihydrate in methanol (eq 6).[26]

$$\mathrm{R_2CH{-}CO{-}C{\equiv}C{-}SiMe_3} \xrightarrow[\text{DTBMP} \\ \text{2. KF}]{\text{1. }(\mathrm{F_3CSO_2})_2\mathrm{O}} \mathrm{R_2C{=}C(O{-}SO_2{-}CF_3){-}C{\equiv}CH} \qquad (6)$$

By using DTBMP as proton sponge (proton trap), silylvinyl triflates may be obtained from enolizable trimethylsilyl ketones,[27,28] and alkyl trimethylsilylbutadiynyl ketones react with triflic anhydride yielding 1-trimethylsilyl-5-alkene-1,3-diyn-5-yl triflates that may be desilylated subsequently to 1-(butadiynyl)-vinyl triflates.[27] A comprehensive review is available on syntheses and applications of perfluoroalkanesulfonic esters (including triflates and nonaflates).[28] Interesting vinylic nonafluorobutanesulfonates (eq 7) may also be prepared from enolizable ketones and nonafluorobutanesulfonic anhydride using DTBMP.[29]

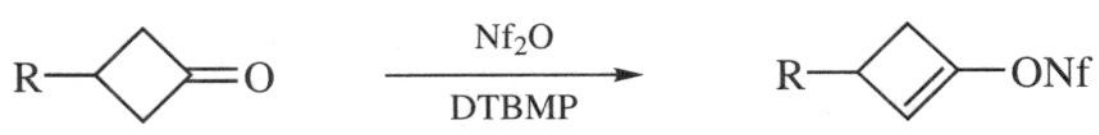

(7)

The methyl group of DTBMP is acidic enough to react with *sec*-BuLi at −78 °C in THF;[30] reaction of the resulting lithium derivative with chloromethylated polystyrene beads in cyclohexane followed by washing with benzene/methanol afforded a polymer-bound non-nucleophilic resin that was used in vinyl triflate syntheses.[31a] The use of such polymer-supported versions of DTBMP provides optimal conditions for preparing vinyl triflates from enolizable aldehydes or ketones. These reactions involving intermediately formed *gem*-bis(triflates) that decompose thermally to vinyl triflates include chlorinated solvents ($CHCl_3$ or $ClCH_2CH_2Cl$) and temperatures in the range 50–70 °C.[31b]

Non-Nucleophilic Base for Glycosylation Reactions. One of the most important applications of DTBMP is in allowing the synthesis of complicated polysaccharides. Unlike 2,4,6-collidine, which is alkylated to an *N*-substituted collidinium salt by primary alkyl triflates, DTBMP permits the methylation of unprotected hydroxy groups of carbohydrates under mild conditions with methyl triflate (MeOTf) without any anomerization (eq 8).[32,33]

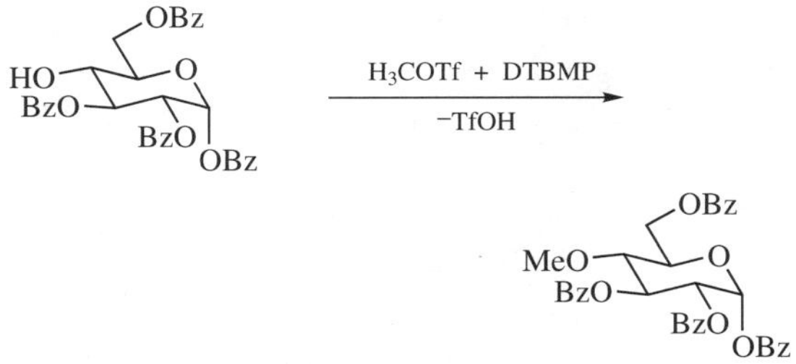

(8)

Unlike pyridine or collidine, DTBMP allowed the isolation of the *O*-triflate formed from triflic anhydride and 1,2,3,4-tetra-*O*-acetyl-β-D-glucopyranose; then the triflate group can be substituted by halogens.[34,35] For Hal = Cl, Br, and I yields were in the range 85–97%, but for F yields were only 27% (eq 9).

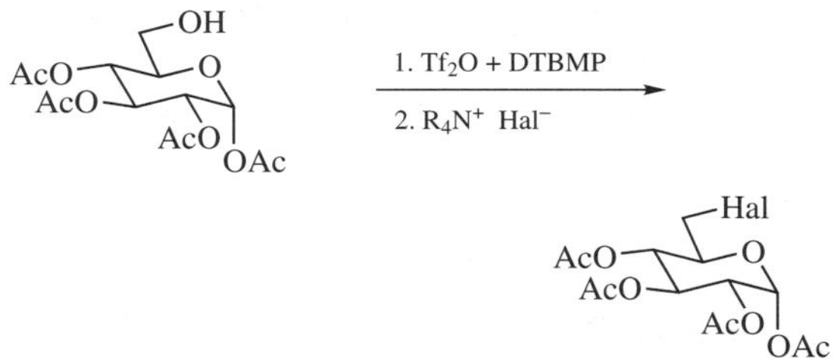

(9)

It was shown that DTBMP can assist in the cyclization of carbohydrates having vicinal hydroxy and triflate groups (eq 10).[36]

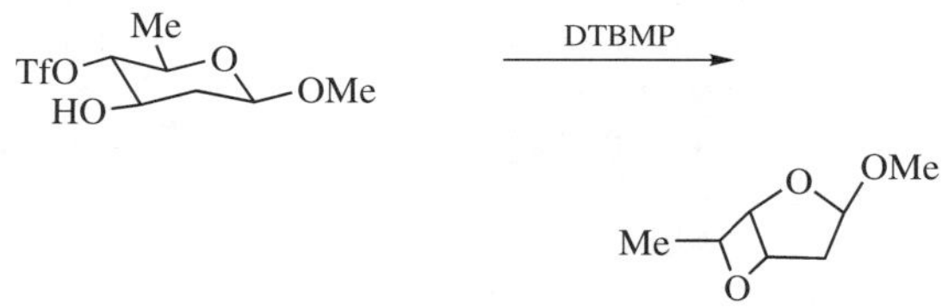

(10)

Kahne et al. have shown that anomeric sulfoxides act as glycosyl donors to alcoholic or phenolic hydroxy groups, even when they are sterically hindered.[37] As DTBMP and boron trifluoride do not form an amine-borane complex, it is possible to employ a combination of these reagents in the neighboring group directed glycosylation of acid sensitive substrates (eq 11). The DTBMP serves to buffer the reaction mixture while the BF_3 acts to promote the rearrangement of intermediate orthoesters to the glycosides. In this manner, Kahne and coworkers were able to glycosylate the phenolic vancomycin aglycons using the sulfoxide glycosylation method, thus laying the groundwork for investigating how the carbohydrate moieties of glycopeptides may be varied.[38]

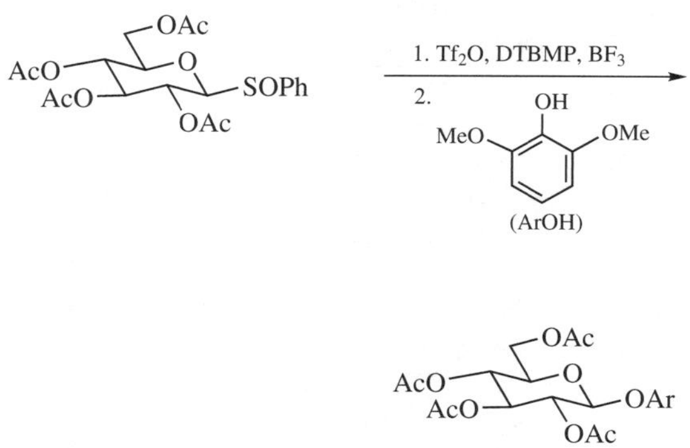

(11)

With 4,6-*O*-benzylidene protected mannosyl sulfoxides, the method provides a highly stereoselective entry into the normally difficult to access β-mannosides (eq 12). Crich et al. investigated in detail this DTBMP-assisted distereoselective synthesis and showed it to involve the in situ formation of α-mannosyl triflates. Using ^{13}C-labeled monosaccharides and various types of NMR spectra (^{1}H, ^{13}C, ^{19}F) the same group was able to demonstrate the formation of intermediate dioxalenium ions in the presence of DTBMP from protected glycosyl donors.[39–41]

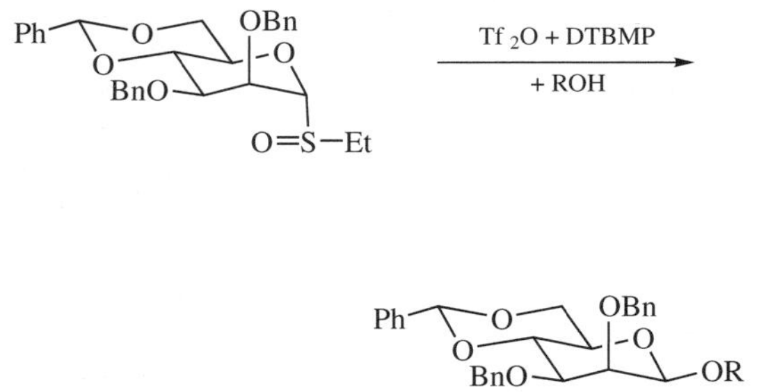

(12)

Other applications of DTBMP in the synthesis of glycosylated natural products or bioactive compounds have been reported. Thus, syntheses of the caloporoside disaccharide,[42] of inhibitors of sialyltransferases,[43] of kojidextrins,[44] of phosphatidyl-inositol-kinases,[45] of Amadori compounds,[46] trisaccharides,[47] and even hexasaccharides were accomplished using by a variety of *O*- and *C*-glycoside forming reactions in the presence of DTBMP.[48] Nicolaou et al. also made use of DTBMP-buffered glycosylations in their total synthesis of lipopolysaccharides involved in the assimilation of atmospheric nitrogen by root nodules of *Rhizobium melitoti*.[49]

Non-Nucleophilic Base for Living Cationic Polymerizations. It is well known that the room-temperature cationic polymerization of alkenes such as isobutene initiated by Brønsted or Lewis acids yields low molecular weight (MW) polymers, and that only low-temperature polymerization (at or below −78 °C) affords high MW polymers. The explanation is that for growing polymer cationic species, the C–H bond breaking (proton transfer) has higher activation energy than C–C bond formation (polymer growth). Whereas the ratio between reaction rate constants for transfer to monomer (k_{trM}) and polymerization (k_p) for styrene is about 0.01 at −28 °C, it becomes a hundred times lower at −60 °C. It was only after 1980 that observations of linear growth of MW with the amount of polymer formed, and of narrowing polydispersity during the course of the polymerization, led to the hypothesis that *living polymers* could also occur in cationic (not only anionic) polymerizations, despite the short lifetime of carbocationic species.

At room temperature, bimolecular rate constants (k_2 in $L mol^{-1} s^{-1}$) between pyridine derivatives and 1,2,2-tris(*para*-anisyl)vinyl cations were measured by the flash photolysis technique in acetonitrile:[50] for sterically nonhindered pyridines, $k_2 \approx 3 \times 10^7$; for DTBMP, $k_2 = 8.7 \times 10^4$; and for 2,6-di-*tert*-butylpyridine, $k_2 = 3.4 \times 10^4$. Specific rate constants were also determined for the polymerization of indene initiated by cumyl chloride and tin tetrachloride.[51]

Ring-opening homopolymerizations and copolymerizations of tetrahydrofuran and 7-oxabicyclo[2.2.1]heptane were carried out with or without DTBMP in the presence of triflic anhydride or PhCOF + SbF_5 as initiator.[52] The diad/triad structure of block copolymers was determined by ^{1}H- and ^{13}C-NMR, evidencing the higher reactivity of the more strained bicyclic comonomer. In the presence of DTBMP, lengths of homopolymer sequences were ten times higher than in its absence.[53] All possible repeating irreducible sequences in linear binary, tertiary, quaternary, and quinquenary copolymers and in stereoregular homopolymers were enumerated.[54]

The DTBMP-controlled polymerization of styrene initiated by $F_3B.OEt_2$ with aq. Bu_4NOH resulted in a narrow molecular weight distribution.[55] It is possible to distinguish between various modes of cationic polymerization: when initiated by Lewis acids, adding DTBMP has no effect; when initiated by Brønsted acids, adding an equivalent amount of DTBMP completely inhibits the polymerization; when both types of acids contribute, adding DTBMP quenches only the protonic species involved in the formation of chain carriers, directly or through cocatalysis.[56] The elucidation of the apparent 'negative activation energies' in carbocationic polymerizations with the help of DTBMP and other additives was reviewed.[57–60]

Biological Activity. It was reported that a QSAR study based on molecular topology indicated that 2,6-di-*tert*-butylpyridine and its 4-methyl congener should have hypolipemic activity;[61] pharmacological tests carried out on rats using the Triton WR-1339 test confirmed that these compounds decreased the cholesterol level more than the reference drug Clofibrate.[62]

Related Reagents. 2,6-Di-*tert*-butylpyridine; 2,4,6-tri-*tert*-butylpyridine; 2,4,6-tri-*tert*-butylpyrimidine.

---

1. Hassner, A.; Stummer, C., *Organic Syntheses Based on Name Reactions*, 2nd Ed.; Elsevier: Amsterdam, **2002**, p. 17.
2. Balaban, A. T.; Nenitzescu, C. D., *Liebigs Ann. Chem.* **1959**, *625*, 66.
3. Balaban, A. T., *Chem. Intelligencer* **1998**, *4*, 55.
4. Balaban, A. T.; Nenitzescu, C. D., *Liebigs Ann. Chem.* **1959**, *625*, 74.
5. Balaban, A. T., *Org. Prep. Proced. Int.* **1977**, *9*, 125.
6. Balaban, T. S.; Balaban, A. T., *In Science of Synthesis. Houben-Weyl Methods of Molecular Transformations*; Thomas, E. J., Ed.; Georg Thieme Verlag: Stuttgart, 2003, Vol. 14, pp 11–200.
7. Anderson, A. G.; Stang, P. J., *J. Org. Chem.* **1976**, *41*, 3034.
8. Anderson, A. G.; Stang, P., J., *Org. Synth.* **1981**, *60*, 34.
9. Bates, R. B.; Gordon III, B.; Keller, P. C.; Rund, J. V.; Mills, N. S., *J. Org. Chem.* **1980**, *45*, 168–169.
10. Deutsch, E.; Cheung, N. K. V., *J. Org. Chem.* **1973**, *38*, 1123.
11. Chardin, A.; Laurence, C.; Berthelot, M., *J. Chem. Res. (S)* **1996**, 332.
12. (a) Brown, H. C.; Kanner, B., *J. Am. Chem. Soc.* **1953**, *75*, 3865. (b) **1966**, *88*, 986.
13. Okamoto, Y.; Shimakawa, Y., *Tetrahedron Lett.* **1966**, 317.
14. Hou, C. J.; Okamoto, Y., *J. Org. Chem.* **1982**, *47*, 1977.
15. Gawinecki, R.; Karner, B., *Monatsh. Chem.* **1994**, *125*, 35.
16. Weber, H.; Rohn, T., *Z. Naturforsch.* **1990**, *45b*, 701.
17. Umemoto, T.; Harasawa, K.; Tomizawa, G.; Kawada, K.; Tomita, K., *Bull. Chem. Soc. Jpn.* **1991**, *64*, 1081.
18. Crich, D.; Smith, M; Yao, Q.; Picione, J., *Synthesis* **2001**, 323.
19. Balaban, A. T.; Ghiviriga, I.; Czerwinski, E. W.; De. P.; Faust, R., *J. Org. Chem.* **2004**, *69*, 536.
20. Nakada, M.; Kojima, E.; Ichinose, H., *Synth. Commun.* **2000**, *30*, 863.
21. Forbus Jr., T. R.; Taylor, S. L.; Martin, J. C., *J. Org. Chem.* **1987**, *52*, 4156.
22. Stang, P. J.; Treptow, W., *Synthesis* **1980**, 283.
23. Scott, W. J.; Crisp, G. T.; Stille, J. K., *Org. Synth.* **1990**, *68*, 116.
24. Yoo, S.; Suh, J. H.; Lee, S. J.; Jeong, N., *Bioorg. Med. Chem. Lett.* **1992**, *2*, 381.
25. Cacchi, S.; Morera, E ; Ortar, G., *Org. Synth.* **1990**, *68*, 138.
26. Stang, P. J.; Fisk, T. E., *Synthesis* **1979**, 438.
27. Stang, P. J.; Ladika, M., *Synthesis* **1981**, 29.
28. Stang, P. J.; Hanack, M.; Subramanian, L. R., *Synthesis* **1982**, 85.
29. Hanack, M.; Auchter, G., *J. Am. Chem. Soc.* **1985**, *107*, 5238.
30. Wright, M. E.; Yamada, L. Y., *Synth. Commun.* **1988**, *18*, 1449.
31. (a) Wright, M. E.; Pulley, S. R., *J. Org. Chem.* **1987**, *52*, 1623, 5036. (b) **1989**, *54*, 2886.
32. Arnarp, J.; Kenne, L.; Lindberg, B.; Lönngren, J., *Carbohydr. Res.* **1975**, *44*, C5.
33. Arnarp, J.; Lönngren, J., *Acta Chem. Scand.* **1978**, *B32*, 465.
34. Ambrose, M. G.; Binkley, R. W., *J. Org. Chem.* **1983**, *48*, 674.
35. Binkley, R. W., *Modern Carbohydrate Chemistry*; Dekker: New York, 1988.
36. (a) Binkley, R. W., *J. Org. Chem.* **1992**, *57*, 2353. (b) Binkley, R. W. *J. Carbohyr. Chem.* **1985**, *4*, 227.
37. Kahne, D.; Walker, S.; Cheng, Y.; Van Engen, D., *J. Am. Chem. Soc.* **1989**, *111*, 6881.
38. Thompson, C.; Ge, M.; Kahne, D., *J. Am. Chem. Soc.* **1999**, *121*, 1237.
39. Crich, D.; Sun, S., *Tetrahedron* **1998**, *54*, 8321.
40. Crich, D.; Sun, S., *J. Am. Chem. Soc.* **1997**, *119*, 11217.
41. Crich, D.; Cai, W., *J. Org. Chem.* **1999**, *64*, 4926.
42. Crich, D.; Barba, G. R., *Tetrahedron Lett.* **1998**, *39*, 9339.
43. Cheng, X.; Khan, N.; Mootoo, D. R., *J. Org. Chem.* **2000**, *65*, 2544.
44. Pozsgay, V.; Dubois, E. R.; Pannell, L., *J. Org. Chem.* **1997**, *62*, 2832.
45. Watanabe, Y.; Nakatomi, M., *Tetrahedron* **1999**, *55*, 9743.
46. Lopez, M. G.; Gruenwedel, D. W., *Carbohydr. Res.* **1991**, *212*, 37.
47. La Ferla, B.; Lay, L.; Poletti, L.; Russo, G.; Panza, L., *J. Carbohydr. Chem.* **2000**, *19*, 331.
48. Nitz, M.; Bundle, D. R., *J. Org. Chem.* **2001**, *66*, 8411.
49. Nicolaou, K. C.; Bockovich, N. J.; Carcanague, D. R.; Hummel, C. W.; Evan, L. F., *J. Am. Chem. Soc.* **1992**, *114*, 8701.
50. Dembinski, A.; Yagci. Y.; Schnabel, W., *Polym. Comm.* **1993**, *34*, 3738.
51. Givehchi, M.; Tardi, M.; Polton, A.; Sigwalt, P., *Macromolecules* **2000**, *33*, 9512.
52. Pretula, J.; Kaluzynski, K.; Szymanski, R.; Penczek, S., *Macromolecules* **1996**, *29*, 6700.
53. (a) Sigwalt, P., *Macromol. Symp.* **1998**, *132*, 127. (b) *Makromol. Chem., Macromol. Symp.* **1991**, *47*, 179.
54. Balaban, A. T.; Artemi, C., *Macromol. Chem.* **1988**, *189*, 863.
55. Satoh, K.; Nakashima, J.; Kamigaito, M.; Sawamoto, M., *Macromolecules* **2001**, *34*, 396.
56. Moulis, J. M.; Collomb, J.; Gandini, A.; Cheradame, H., *Polymer Bull.* **1980**, *3*, 197.
57. Matyjaszewski, K., *Polymer Internat.* **1994**, *35*, 1.
58. *Cationic Polymerization. Fundamentals and Applications.* ACS Symp. Series 665, Faust, R.; Shaffer, T. D., Eds.; Washington, DC, 1997.
59. *Cationic Polymerization*, Matyjaszewski, K., Ed.; Marcel Dekker: New York, 1996.
60. Kennedy, J. P.; Ivan, B., *Designed Polymers by Carbocationic Macromolecular Engineering*; Hanser: München, 1992.
61. Cercos-del-Poso, R. A.; Perez-Gimenez, F.; Salabert-Salvador, M. T.; Garcia-March, F. J., *J. Chem. Inf. Comput. Sci.* **2000**, *140*, 178.
62. Cercos-del-Poso, R. A.; Perez-Gimenez, F.; Salabert-Salvador, M. T.; Garcia-March, F. J.; Murcia-Soler, M., *Quant. Struct. Act. Rel.* **1999**, *18*, 464.

Alexandru T. Balaban
Texas A&M University at Galveston, Galveston TX, USA.

Di-*n*-butyltin Oxide[1,2]

Bu_2SnO

[818-08-6] $C_8H_{18}OSn$ (MW 248.92)

(converts diols to *O*-stannylene acetals[2] which undergo regioselective acylation, alkylation, oxidation, and condensation with activated carboxylic acid derivatives; activates α,ω-hydroxy acids and α,ω-amino acids for lactone and lactam formation, respectively;[3] mediates epoxidation of allylic alcohols by *t*-butyl hydroperoxide;[4] catalyzes $TMSN_3$ addition to nitrile;[5] useful as a source of tin for Otera transesterification[6] and 'organotin phosphate condensate'[7] catalysts)

Alternate Name: DBTO.
Physical Data: mp >300 °C.
Solubility: insol most organic solvents with which it does not react. However solvents like toluene, benzene, and methanol are routinely used for reactions of hydroxylic substrates.
Form Supplied in: white powder.
Handling, Storage, and Precautions: toxic and irritant;[8,9] use in a fume hood.

Original Commentary

T. V. (Babu) RajanBabu
The Ohio State University, Columbus, OH, USA

Introduction. Di-*n*-butyltin oxide, because of its biocidal and antifouling properties, is one of the most widely used industrial chemicals.[9] It is the final product of hydrolysis of dialkyltin halides (eq 1) and exists as an amorphous polymeric solid insoluble in nonreacting solvents. A reticulated network (**1**) consisting of four-membered rings of alternating tin and oxygen has been suggested

$$R_2SnX_2 \longrightarrow R_2Sn(X)OSn(X)R_2 \longrightarrow R_2(X)SnOSnR_2(OH) \longrightarrow R_2SnO \quad (1)$$

as a possible structure for this compound.[10] The interaction of diorganotin oxides with various biomolecules like amino acids, peptides, carbohydrates, and nucleic acid components has been reviewed[9] and is beyond the scope of this article.

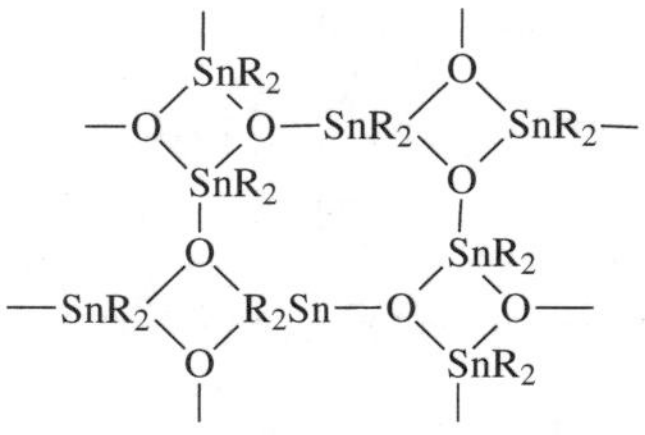

(**1**)

Formation and Reactions of Dibutylstannylenes. By far the largest application of DBTO in organic chemistry is in the generation and reactions of stannylenes from polyhydroxy compounds. Dibutylstannylenes are prepared in nearly quantitative yields by heating stoichiometric amounts of DBTO and the polyol in a solvent such as benzene or toluene with concomitant removal of water (eq 2).[11] Compounds (**2**)[12] and (**3**)[13] are other prototypical stannylenes that have been prepared by this method. Formation and reactions of 1,6-stannylenes from monosaccharides have also been reported recently.[14] While the stannylene from ethylene glycol (eq 2) is an infinite coordination chain polymer with hexacoordinated Sn,[15] the stannylenes from cyclohexane-1,2-diols and the methyl 4,6-*O*-benzylidene-α-D-glucopyranoside (**3**)[16] are dimeric (e.g. **4**).

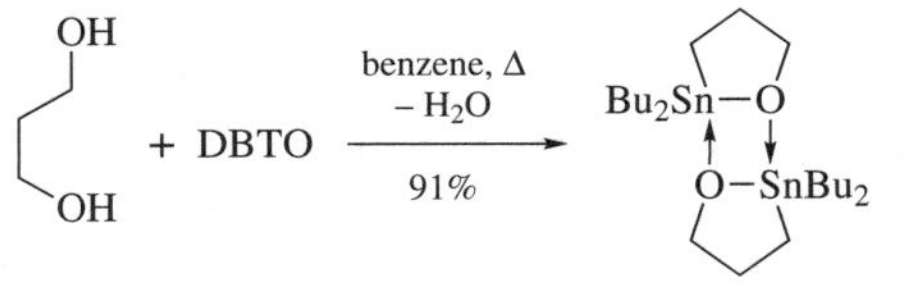

(2)

mp 223–226 °C

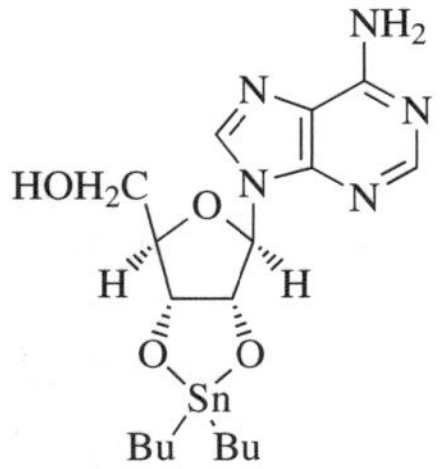

(**2**)

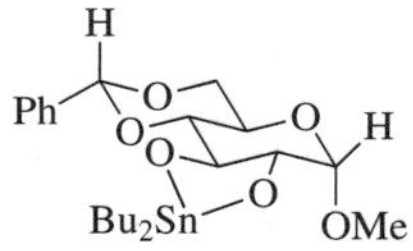

(**3**)

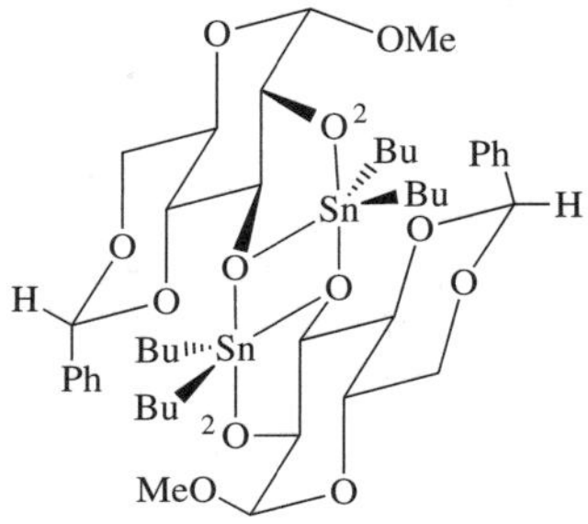

(**4**)

Regioselective Alkylation, Acylation, and Sulfonylation. Synthetic applications of stannylenes have followed the elegant studies of Moffatt[17] and Ogawa,[18] who showed that the inherent differences in the nucleophilicities of carbohydrate hydroxys can be amplified by the formation of trialkyltin ethers. Several selective acylations and alkylations could thus be accomplished. Further it was noted that while acylation proceeds without any catalyst, alkylation is a sluggish reaction and needs assistance from tetrabutylammonium halides (eq 3).[19]

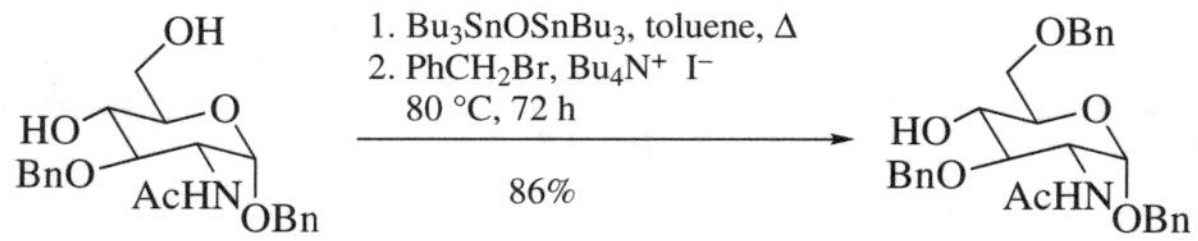

(3)

Reactions of stannylene (**2**) are typical and are shown in eq 4.[2]

The stannylene (**3**) reacts with benzoyl chloride or tosyl chloride to give the corresponding 2-*O*-derivative (eq 5). The alkylation is less selective.[20] Use of nonpolar solvents and tetrabutylammonium halide has been a significant improvement (eq 6) in this alkylation procedure.[21] Several other examples of selective acylation and alkylations are given by David and Hanessian.[2] An intramolecular ether formation using DBTO has been used for the synthesis of octosyl acid.[22] Monoacylation of dimethyl L-tartrate has been carried out using this procedure (eq 7).[23] Two highly readable accounts of how the anomer composition and stoichiometry of reagents affect the regiochemistry of acylation[24] and tosylation[25] have recently appeared.

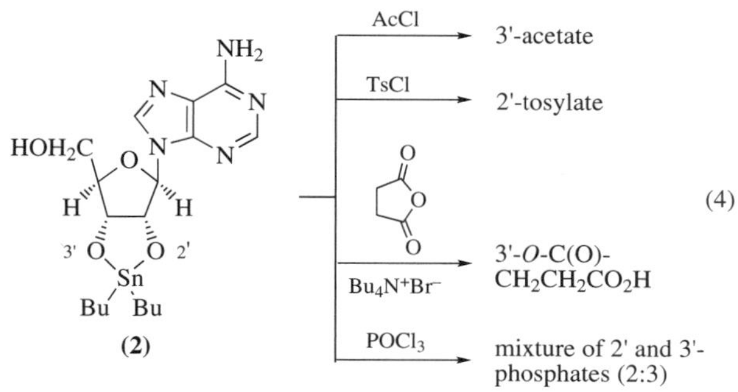

(2) (4)

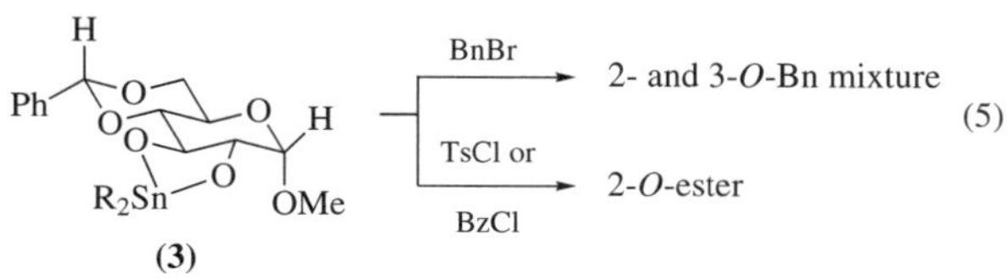

(3) (5)

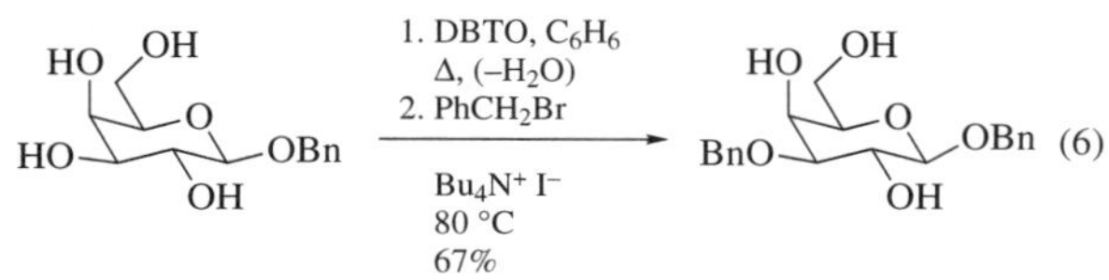

(6)

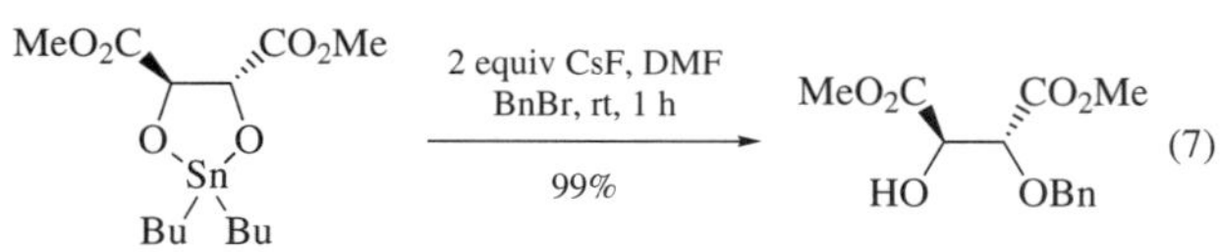

(7)

(also reactions with *t*-BuCOCl and TBDMS-Cl)

The structure of the dimeric complex (**4**) has been used to explain the observed preferential reactivity of the C-2 hydroxy groups.[2] It is argued that the divalent oxygens that occupy the axial position within the tin cordination sphere (i.e. this case, the C-2 oxygens) are not only more nucleophilic, but also are more accessible sterically. It is believed that the orientational preference of the hydroxy group is related to its acidity, and thus by forming a Sn complex with the more acidic hydroxy group in the axial position, its reactivity towards nucleophiles is enhanced. The stannylene procedure can be used to reverse the stereochemistry of the classical monoacylation regiochemistry observed in some unsymmetrical diols.[26]

Regioselective Oxidation. Following the original discovery by David and Thieffrey,[27] several stannylenes have been oxidized by bromine to acyloins (eq 8). Hanessian employed this reaction for the synthesis of a densely functionalized intermediate which was converted into (+)-spectinomycin.[28] A detailed study of monooxidation of unprotected carbohydrates has also appeared.[29] Stannylenes are oxidized the same way as diols, with Sodium Periodate and Lead(IV) Acetate.[2]

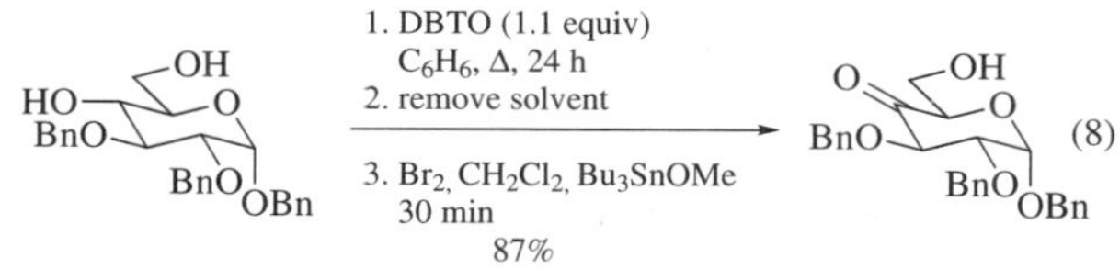

(8)

Esterification Catalyst. In the presence of DBTO, alcohols react with carboxylic acids to give esters[30] and the reaction can be extended to make lactones and lactams (eqs 9 and 10).[3] It should be noted that the stereochemistry of the alcohol center is unaffected. A 'template-driven' extrusion process, in which an intermediate where the hydroxy and the carboxyl groups are brought into close proximity by tin oxide, has been proposed for this reaction.

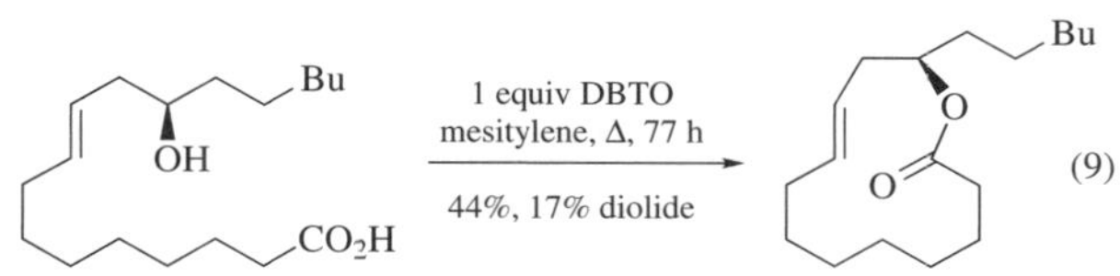

(9)

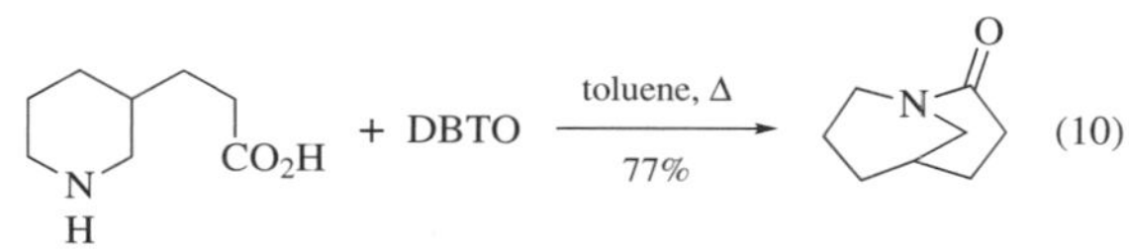

(10)

Stannylenes derived from diols react with bifunctional carboxylic acid derivatives to give macrocyclic polyolides of various sizes.[31,32] This approach has been used for the synthesis of (+)-dicrotaline (eq 11).[33]

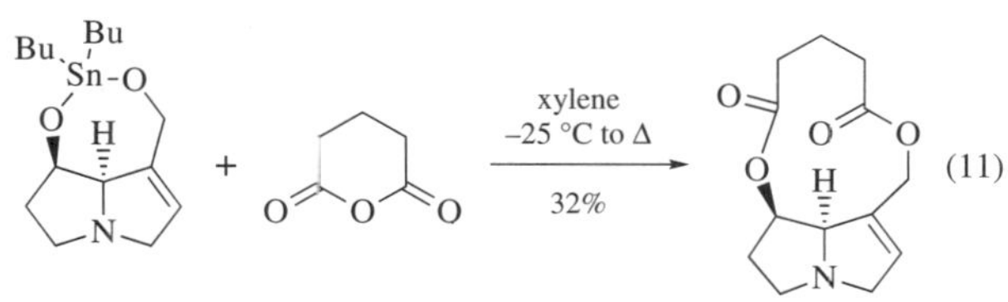

(11)

Dibutyltin oxide is the best source of tin for the Otera transesterification catalysts (eq 12),[6] which have found wide use in organic synthesis. DBTO has also been used as a catalyst for hydrolysis of amides in very sensitive molecules, where other procedures have failed.[34]

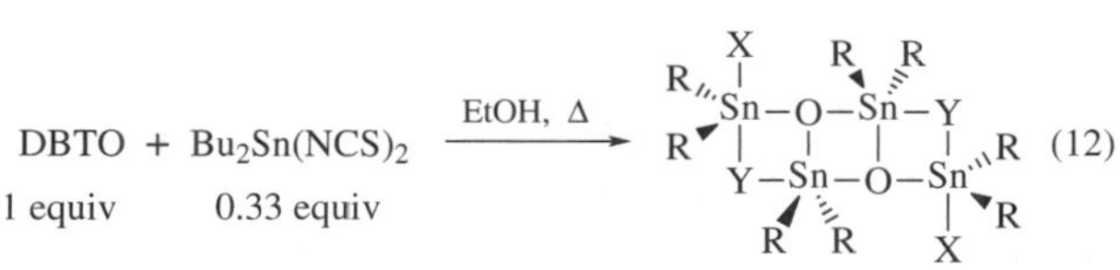

(12)

R = Bu, Y = NCS, X = OH

Catalyst in Oxidation Reactions. DBTO has been used as a catalyst in Fe^{III}-mediated oxidation of thiols to disulfides, even though Tri-n-butyl(methoxy)stannane seems to be better suited for this purpose.[35] Epoxidation of terminal alkenes in a two-phase system (chloroform–water) containing H_2O_2/ammonium molybdate/DBTO has also been reported.[36] A combination of DBTO and t-Butyl Hydroperoxide oxidizes allylic alcohols with moderate regio- and stereoselectivity.[4] Tri- and tetrasubstituted double bonds are most easily oxidized and the selectivities are comparable to those of the corresponding Vanadyl Bis(acetylacetonate) mediated reactions.

Miscelleneous Applications. The method of choice for the regioselective opening of benzylic and tertiary epoxides with alcohols (eq 13) appears to be a reaction mediated by 'organotin phosphate condensate (OPC)', which is readily prepared from DBTO and tributyl phosphate.[7] DBTO catalyzes rearrangement of 3-hydroxy-2-oxo carboxylic acid esters (eq 14), a reaction reminiscent of a similar one mediated by the enzyme reductoisomerase.[37] DBTO has been used as a catalyst for the addition of Azidotrimethylsilane to nitriles[5] for the production of tetrazoles (eq 15).

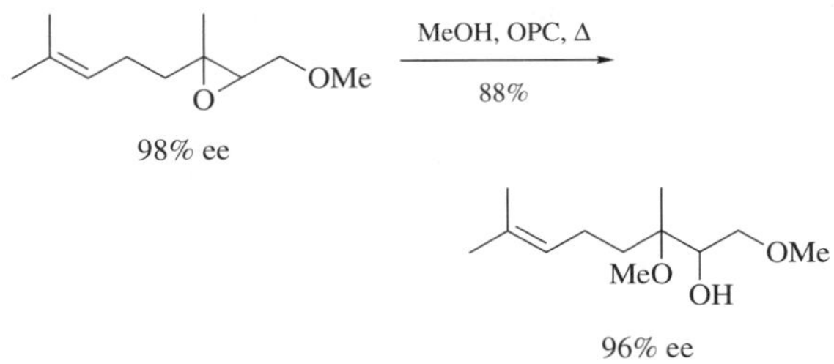

(13)

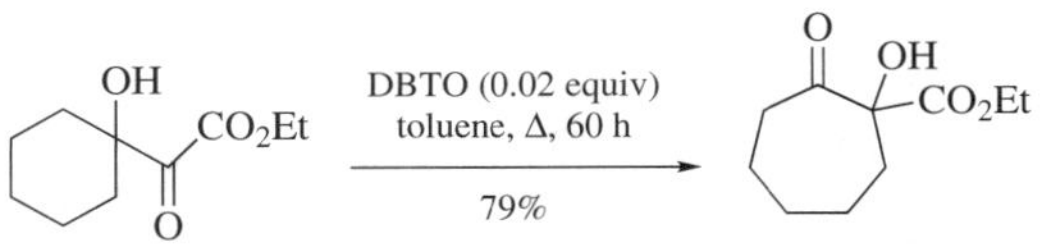

(14)

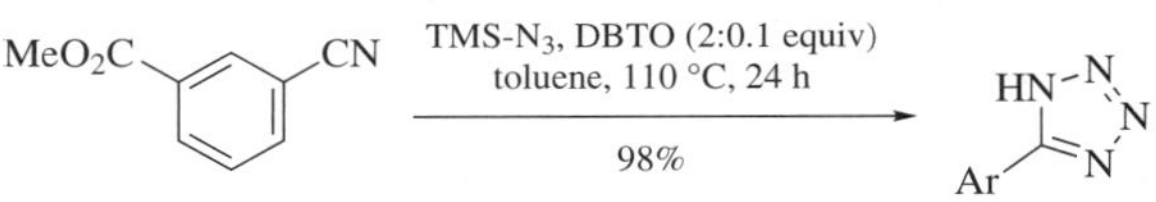

(15)

First Update

Junzo Otera
Okayama University of Science, Okayama, Japan

Selective Protection of Diols. *O*-Dibutylstannylene acetals obtained by reaction of DBTO with diols continue to play versatile roles in organic synthesis, particularly in sugar chemistry. The necessity to use a stoichiometric amount of DBTO, one of disadvantages of this protocol, has been circumvented by microwave irradiation. Mono-benzoylation of diols can be performed with a substoichiometric or catalytic amount of DBTO (eq 16).[38] Irradiation of a mixture of DBTO and diol in toluene followed by addition of benzoyl chloride furnishes the desired monobenzoates smoothly. On the other hand, tosylation of adenosine is effected by use of 10 mol % of DBTO even without microwave irradiation (eq 17).[39]

Benzoylation of carbohydrates such as galactosides, glucosides, mannosides, and lactosides with an excess of DBTO leads to unusual regiochemistry (eq 18).[40] For example, reaction of galactoside **1** with 3 equiv of DBTO by heating azeotropically in toluene-benzene (1:1) followed by treating with 3 equiv of benzoyl chloride at 100 °C furnishes 3,4,6-*O*-tribenzoyl derivative **2** in 89% yield instead of the 2,3,6-derivative which is expected to form because the 2-OH position is normally more reactive than the 4-OH position in a β-galactoside series, even in the tin complex case. This protocol provides building blocks that usually require many steps with traditional synthesis.

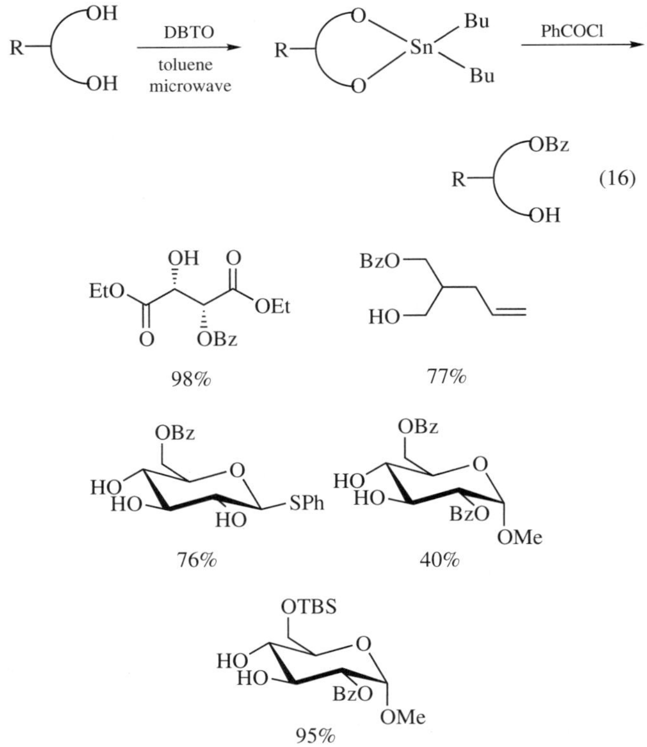

(16)

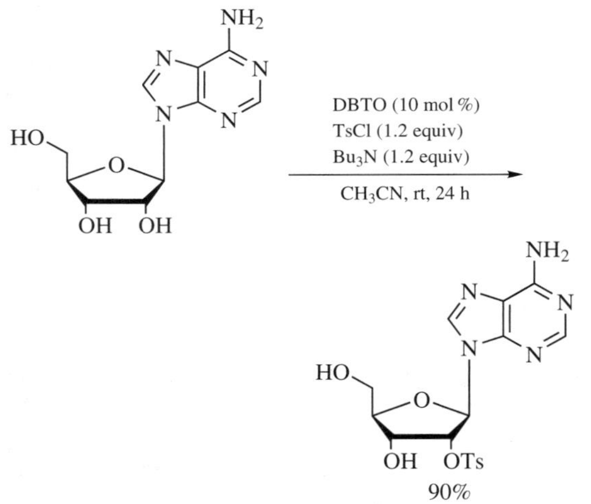

(17)

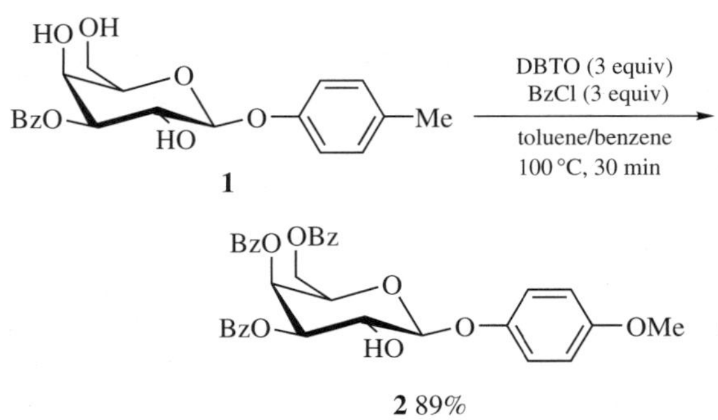

(18)

The stannylene method is also effective for etherification or esterification of polyhydroxyl compounds such as pentaerythritol, pentitiols, hexitols, etc. Chemical transformations of these compounds are generally difficult on account of their poor solubility, but exposure of the polyols to DBTO under forced conditions provides soluble stannylenes which easily undergo further reactions regioselectively under normal conditions.[41,42]

Synthesis of Heterocyclic Compounds. Stannylene intermediates are in situ converted to heterocyclic compounds. As shown in eq 19, *syn*-1,2-diols like tartrate esters and stilbene diol derivatives can be transformed to 2-oxazolidinones in one-pot.[43,44] The stannylenes obtained from the diols and DBTO is treated with isothiocyanates to give iminocarbonates, which, upon in situ exposure to halide anion, are converted to 2-oxazolidinones. The same class of compounds are obtained when 1,2-aminoalcohols are treated with DBTO under CO_2 atmosphere (5 MPa) at 180 °C (eq 20).[45]

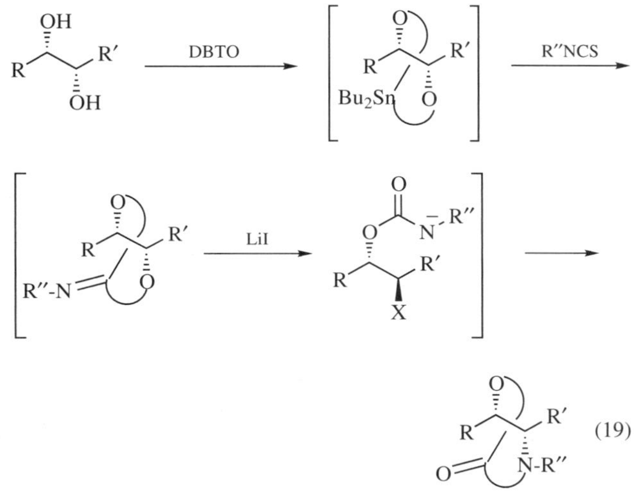

(19)

R = $CO_2{}^iPr$, CH_3, Ph; R′ = $CO_2{}^iPr$, CO_2Et, *p*-tolyl, Ph

R″ = Bz, Ts, Ph, Et, Bn

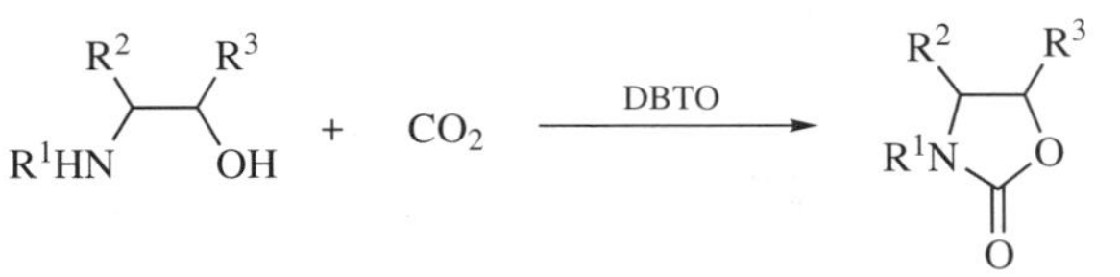

(20)

R^1 = Me, Et, H; R^2 = Me, H; R^3 = Me, H

Glycosylation. Glycosylation with high regio- and stereoselectivities is realized via stannylene intermediates. β-(1→4)-Linked disaccharides, **3** and **4**, as well as β-(1→6)-linked disaccharide **5** are obtained from 1,2-*O*-stannylene acetals in reasonable yields (eq 21).[46] β-(1→3)-Linked disaccharide **6** is also synthesized from stannylene-activated α-L-rhamnopyranoside and per-*O*-pivaloylglucuronyl bromide by promotion with Ag-silica alumina (eq 22).[47] The use of the *O*-pivaroyl protection is crucial for the smooth reaction. When the bezoyl group is employed instead, the benzoyl carbonyl is acylated to give intersugar orthoesters. The stereocontrolled construction of 1,1′-disaccharides and 1,1′:1′,2-trisaccharides is feasible as well (eq 23).[48,49] Reaction of stannylene **7** with trichloroacetimidate **8** in the presence of TMSOTf furnishes 1β,1′α-disaccharide **9**. Furthermore, exposure of the same stannylene to glycosyl fluoride **10** (2.2 equiv) leads to trisaccharide **11**, in which all three glycosydic bonds are formed stereoselectively. This technology was applied to total synthesis of everninomicin 13,384-1.

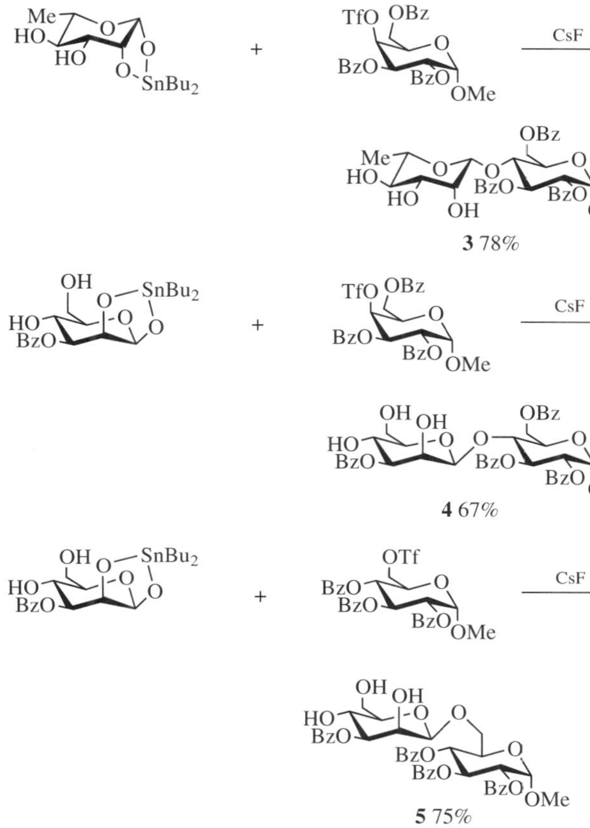

(21)

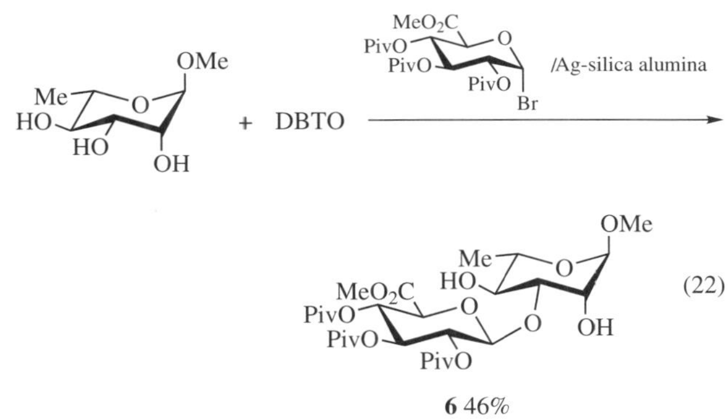

(22)

Applications in Polymer Chemistry. When DBTO is heated with ethylene glycol azeotropically in dry toluene, the corresponding stannylene is obtained as a monomer (**12a**)/dimer (**12b**) mixture (eq 24).[50] This mixture catalyzes ring-opening polymerization of 1,5-dioxepan-2-one. Analogously, spirocyclic stannylenes **13** are formed from pentaerythritol ethoxylates and DBTO (eq 25).[51] These compounds catalyze ring-opening polymerization of L-lactide. DBTO is effective for degradation of polycarbonates. For example, heating poly(bisphenol A carbonate) and DBTO in a 10:1 ratio at 210 °C leads to stannylene **14** (eq 26).[52] This tin-containg macrocycle is highly reactive and allows the synthesis of various telechelic oligocarbonates. Thus, treatment of **14** with acid chlorides furnishes oligocarbonates with two ester end groups.

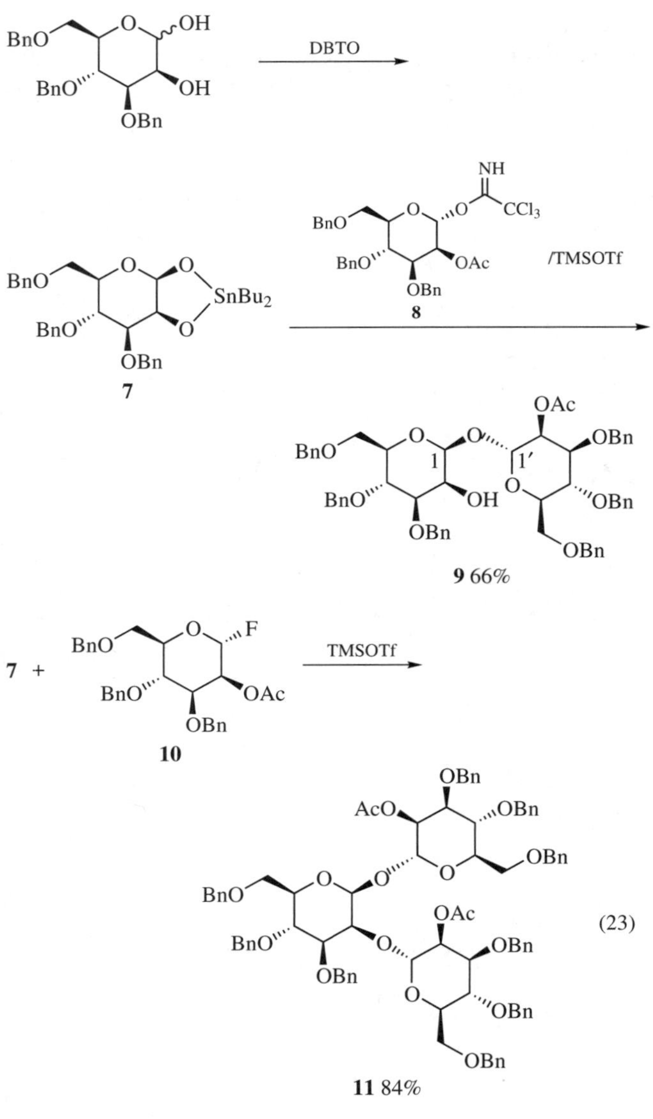

(23)

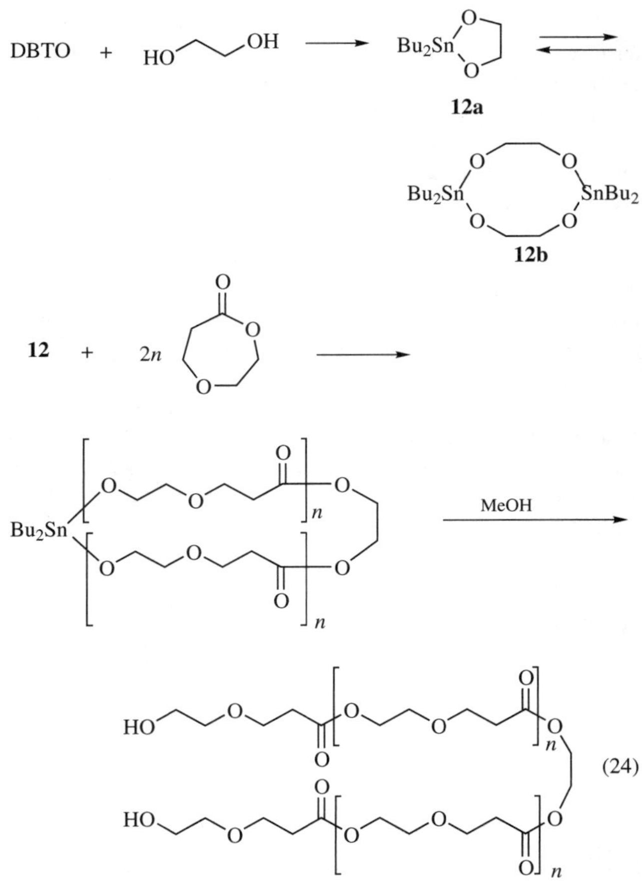

(24)

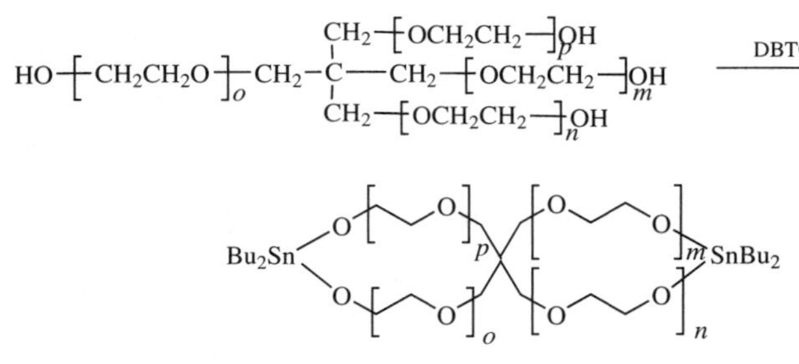

(25)

13

1. Pereyre, M.; Quintard, J. P.; Rahm, A. *Tin in Organic Synthesis*; Butterworth: London, 1987.
2. David, S.; Hanessian, S., *Tetrahedron* **1985**, *41*, 643.
3. Steliou, K.; Poupart, M. A., *J. Am. Chem. Soc.* **1983**, *105*, 7130. See also: Steliou, K.; Szczygielska-Nowosielska, A.; Favre, A.; Poupart, M. A.; Hanessian, S., *J. Am. Chem. Soc.* **1980**, *102*, 7578.
4. Kanemoto, S.; Nonaka, T.; Oshima, K.; Utimoto, K.; Nozaki, H., *Tetrahedron Lett.* **1986**, *27*, 3387. For the possible structure of the reagent see: Davies, A. G.; Grahan, I. F., *Chem. Ind. (London)* **1963**, 1622.
5. Wittenberger, S. J.; Donner, B. G., *J. Org. Chem.* **1993**, *58*, 4139.
6. Otera, J.; Dan-oh, N.; Nozaki, H., *J. Org. Chem.* **1991**, *56*, 5307.
7. Otera, J.; Niibo, Y.; Nozaki, H., *Tetrahedron* **1991**, *47*, 7625.
8. Selwyn, M. J. In *Chemistry of Tin*; Harrison, P. G., Ed.; Chapman and Hall: New York, 1989; p. 359.
9. Molloy, K. C. In *The Chemistry of the Metal–Carbon Bond*; Hartley, F. R., Ed.; Wiley: Chichester, 1989; Vol. 5, p. 465.
10. Davies, A. J.; Smith, P. J. In *Comprehensive Organometallic Chemistry*; Wilkinson, G., Ed.; Pergamon: Oxford, 1982; Vol. 2, p. 519.
11. Considine, W. J., *J. Organomet. Chem.* **1966**, *5*, 263. This compound was first described in the patent literature: Ramsden, H. E.; Banks, C. K., U.S. Patent 2 789 994, **1957**.
12. Wagner, D.; Verheyden, J. P. H.; Moffatt, J. G., *J. Org. Chem.* **1974**, *39*, 24.
13. David, S.; Thieffry, A., *C.R. Hebd. Seances Acad. Sci., Ser. C* **1974**, *279*, 1045.
14. Köpper, S.; Brandenburg, A., *Justus Liebigs Ann. Chem./Liebigs Ann. Chem.* **1992**, 933.
15. Davies, A. G.; Price, A. J.; Dawes, H. M.; Hursthouse, M. B., *J. Chem. Soc., Dalton Trans.* **1986**, 297. See also Ref. 10.

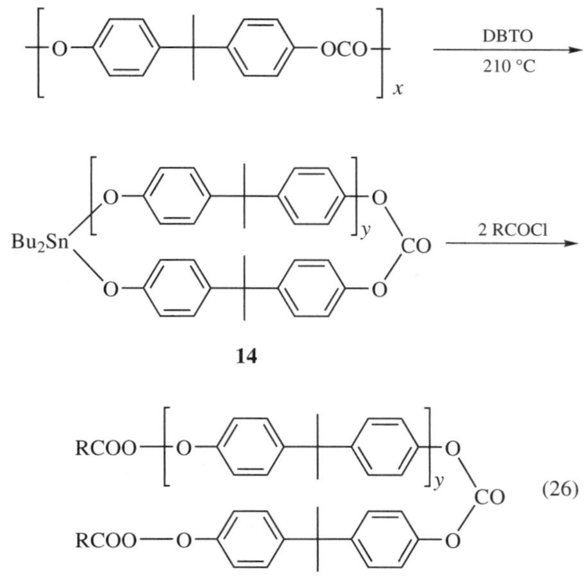

14

(26)

R = $ClCH_2$, CH_2=$CH(CH_2)_8$, *p*-BrC_6H_4, *p*-$NO_2C_6H_4$
PhCH=CH, etc.

16. David, S.; Pascard, C.; Cesario, M., *Nouv. J. Chim.* **1979**, *3*, 63.
17. Jenkins, I. D.; Verheyden, J. P. H.; Moffatt, J. G., *J. Am. Chem. Soc.* **1971**, *93*, 4323.
18. Ogawa, T.; Matsui, M., *Carbohydr. Res.* **1977**, *56*,C1. See also: Ogawa, T.; Matsui, M., *Tetrahedron* **1981**, *37*, 2363.
19. Veyrières, A., *J. Chem. Soc., Perkin Trans. 1* **1981**, 1626. See also: David, S.; Thieffry, A.; Veyrières, A., *J. Chem. Soc., Perkin Trans. 1* **1981**, 1796.
20. Ogawa, T.; Kaburagi, T., *Carbohydr. Res.* **1982**, *53*, 1033. See also: Nashed, M. A.; Anderson, L., *Tetrahedron Lett.* **1976**, 3503. For a recent reference dealing with related chemistry in partially protected carbohydrates see: Tsuda, Y.; Nishimura, M.; Kobayashi, T.; Sato, Y.; Kanemitzu, K., *Chem. Pharm. Bull.* **1991**, *39*, 2883.
21. David, S.; Thieffry, A.; Veyrières, A., *J. Chem. Soc., Perkin Trans. 1* **1981**, 1796.
22. Danishefsky, S. J.; Hungate, R.; Schulte, G., *J. Am. Chem. Soc.* **1988**, *110*, 7434.
23. Nagashima, N.; Ohno, M., *Chem. Lett.* **1987**, 141. For an application in the inositol area see: Yu, K.-L.; Fraser-Reid, B., *Tetrahedron Lett.* **1988**, *29*, 979.
24. Helm, R. F.; Ralph, J.; Anderson, L., *J. Org. Chem.* **1991**, *56*, 7015.
25. Tsuda, Y.; Nishimura, M.; Kobayashi, T.; Sato, Y.; Kanemitzu, K., *Chem. Pharm. Bull.* **1991**, *39*, 2883.
26. Ricci, A.; Roelens, S.; Vannuchi, A., *Chem. Commun./J. Chem. Soc., Chem. Commun.* **1985**, 1457.
27. David, S.; Thieffry, A., *J. Chem. Soc., Perkin Trans. 1* **1979**, 1568. For such oxidations on unprotected carbohydrates see: Tsuda, Y.; Hanajima, M.; Matsuhira, N.; Okuno, Y.; Kanemitzu, K., *Chem. Pharm. Bull.* **1989**, *37*, 2344.
28. Hanessian, S.; Roy, R., *J. Am. Chem. Soc.* **1979**, *101*, 5839.
29. Tsuda, Y.; Hanajima, M.; Matsuhira, N.; Okuno, Y.; Kanemitzu, K., *Chem. Pharm. Bull.* **1989**, *37*, 2344.
30. Habib, O. M. O.; Malek, J., *Collect. Czech. Chem. Commun.* **1976**, *41*, 2724.
31. Shanzer, A., *Acc. Chem. Res.* **1983**, *16*, 60.
32. Bredenkamp, M. W.; Flowers, H. M.; Holzapfel, C. W., *Ber. Dtsch. Chem. Ges./Chem. Ber.* **1992**, *125*, 1159.
33. Niwa, H.; Okamoto, O.; Ishiwata, H.; Kuroda, A.; Uosaki, Y.; Yamada, K., *Bull. Chem. Soc. Jpn.* **1988**, *61*, 3017.
34. Chwang, T. L.; Nemec, J.; Welch, A. D., *J. Carbohydr., Nucleosides Nucleotides* **1980**, *7*, 159.
35. Tsuneo, S.; Otera, J.; Nozaki, H., *Tetrahedron Lett.* **1990**, *31*, 3591.
36. Kamiyama, T.; Inoue, M.; Kashiwagi, H.; Enomoto, S., *Bull. Chem. Soc. Jpn.* **1990**, *63*, 1559 and references cited therein.
37. Crout, D. H. G.; Rathone, D. L., *Chem. Commun./J. Chem. Soc., Chem. Commun.* **1987**, 290.
38. Morcuende, A.; Valverde, S.; Herrodón, B., *Synlett* **1994**, 89–91.
39. Kawana, M.; Tsujimoto, M.; Takahashi, S., *Carbohydr. Chem.* **2000**, *19*, 67–78.
40. Zhang, Z.; Wong, C. H., *Tetrahedron* **2002**, *58*, 6513–6519.
41. David, S., *Carbohydr. Res.* **2001**, *331*, 327–329.
42. Halila, S.; Benazza, M.; Demailly, G., *J. Carbohydr. Chem.* **2001**, *20*, 467–483.
43. Cho, G. Y.; Ko, S. Y., *J. Org. Chem.* **1999**, *64*, 8745–8747.
44. Cho, G. Y.; Choi, J. Y.; Ko, S. Y., *Bull. Korean Chem. Soc.* **2002**, *23*, 11–12.
45. Tominaga, K.; Sasaki, Y., *Synlett* **2002**, 307–309.
46. Hodosi, G.; Kovác, P., *J. Am. Chem. Soc.* **1997**, *119*, 2335–2336.
47. Kaji, E.; Harita, N., *Tetrahedron Lett.* **2000**, *41*, 53–56.
48. Nicolaou, K. C.; van Delft, F. L.; Conley, S. R.; Mitchell, H. J.; Jin, Z.; Rodrigues, R. M., *J. Am. Chem. Soc.* **1997**, *119*, 9057–9058.
49. Nicolaou, K. C.; Fylaktakidou, K. C.; Mitchell, H. J.; van Delft, F. L.; Rodrigues, R. M.; Conley, S. R.; Jin, Z., *Chem. Eur. J.* **2000**, *6*, 3166–3185.
50. Stridsberg, K.; Albertsson, A. C., *J. Polymer Sci: Part A: Polymer Chem.* **1999**, *37*, 3407–3417.
51. Finne, A.; Albertsson, A. C., *Biomacromolecules* **2002**, *3*, 684–690.
52. Krichedorf, H. R.; Probst, N.; Langank, D.; Schwarz, G., *Macromol. Rapid Commun.* **2002**, *22*, 750–755.

Dichlorobis(cyclopentadienyl)hafnium–Silver(I) Salts[1]

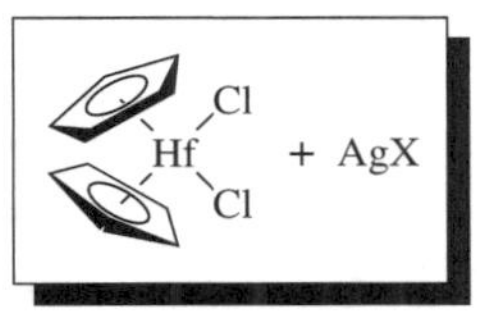

(Cp_2HfCl_2)		
[12116-66-4]	$C_{10}H_{10}Cl_2Hf$	(MW 379.59)
($AgClO_4$)		
[7783-93-9]	$AgClO_4$	(MW 207.32)
($AgOSO_2CF_3$; AgOTf)		
[2923-28-6]	$CAgF_3O_3S$	(MW 256.93)
($AgBF_4$)		
[14104-20-2]	$AgBF_4$	(MW 194.68)
($AgPF_6$)		
[26042-63-7]	AgF_6P	(MW 252.83)

(can activate glycosyl fluorides for *O*-glycosidation;[2] can convert aryl *O*-glycosides to the corresponding aryl *C*-glycosides[3])

Physical Data: see entries for Dichlorobis(cyclopentadienyl) hafnium and individual silver salts.

Handling, Storage, and Precautions: the reagents are combined just before use by adding the Ag salt to a solution of Cp_2HfCl_2. $AgClO_4$ leads to the best reactivity, but has a potential hazard of explosion.[4] AgOTf is recommended as a safe alternative with slightly decreased reactivity. Other silver salts (e.g.

$AgBF_4$, $AgPF_6$, $AgAsF_6$) are less effective, but sometimes offer unique stereochemical preferences.[2,3] Molecular sieves 4Å are often used for removal of water in small scale runs, but are not necessarily essential.

Glycosyl fluorides are widely used in glycoside synthesis, because of their stability, ease of handling and availability of specific activation methods.[5] Although Tin(II) Chloride–$AgClO_4$[5a] is most often utilized, the extremely high reactivity offered by Cp_2HfCl_2–$AgClO_4$[2] renders it the promoter of choice for the glycosidation of sterically demanding substrates. Intermediate cationic hafnocene complexes are believed to be the origin of the high reactivity, which led to the finding that the use of the combination in 1:2 ratio leads to an even more increased reactivity (eq 1).[2d] The efficiency of this new protocol is clearly demonstrated in the synthesis of oligo- or even polysaccharides.[6e]

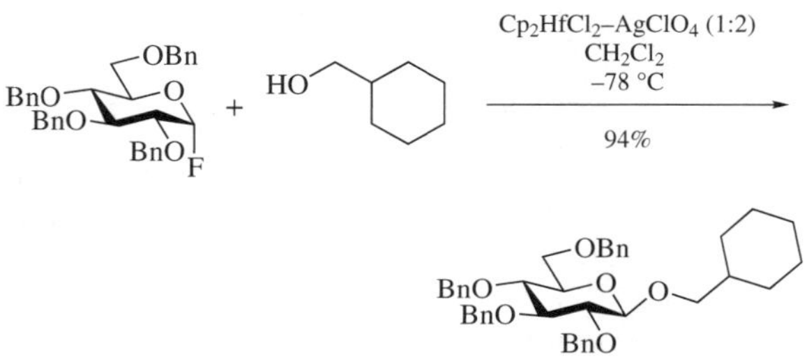

(1)

α:β = 1:13

A number of useful applications are reported for the synthesis of various classes of glycoconjugates ranging from macrolides, e.g. (**1**),[2a,b,c] to oligosaccharide structures embedded in glycolipids.[6] One of the highlights in the latter context is the block synthesis of sialyl Le^X related oligosaccharides via the coupling of the glycosyl fluoride (**2**).[6a,b] Applicability to glycosidation of amino sugars is another notable feature of this activation method.[2b,c,6]

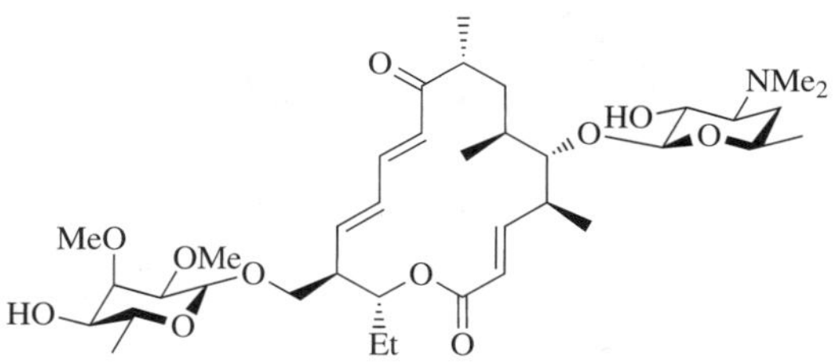

(**1**)

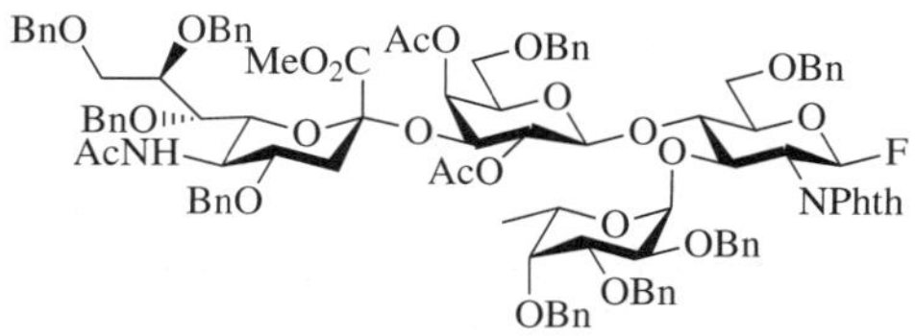

(**2**)

A Lewis acid-mediated synthetic reaction of aryl *C*-glycosides is reported, which involves (1) the initial formation of aryl *O*-glycosides at low temperature, and (2) its in situ rearrangement to the corresponding aryl *C*-glycosides (eq 2).[3] A particularly useful feature is that the C–C bond is formed regioselectively at the position *ortho* to the phenol.

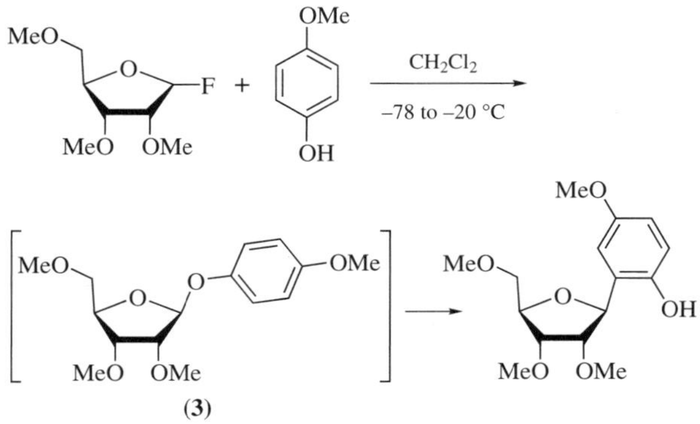

(2)

Cp_2HfCl_2–$AgClO_4$, 79% α:β = 1:9
BF_3•OEt_2, 50% α:β = 1:1 + (**3**) 41%

Both processes can be promoted by general Lewis acids, but Cp_2HfCl_2–$AgClO_4$ is prominent in terms of the reactivity and stereoselectivity. These features are well illustrated by their application in the total syntheses of vineomycinone B_2 methyl ester (**4**)[7] and gilvocarcin M (**5**).[8] Donor solvents, such as Et_2O or MeCN, retard the second step, thereby providing a useful synthetic method for aryl *O*-glycosides.[3b]

Cp_2HfCl_2–$AgClO_4$ is useful in nucleoside synthesis via the coupling of glycosyl fluorides and bis(trimethylsilyl)uracil (eq 3).[9]

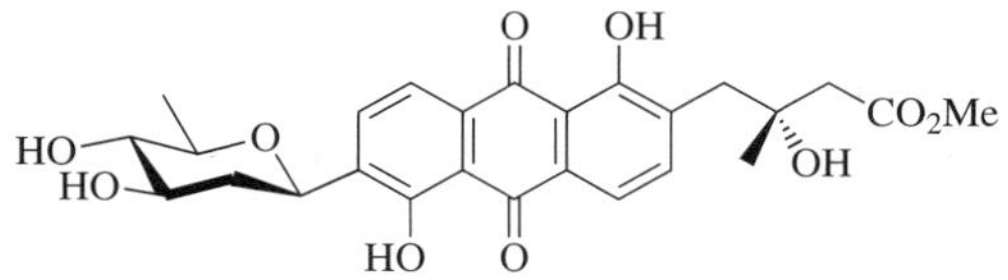

(**4**)

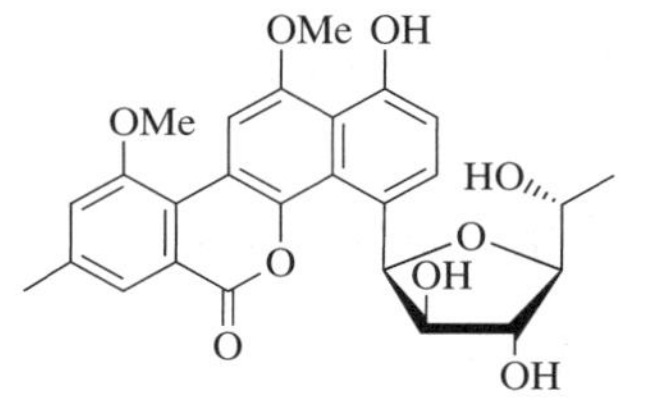

(**5**)

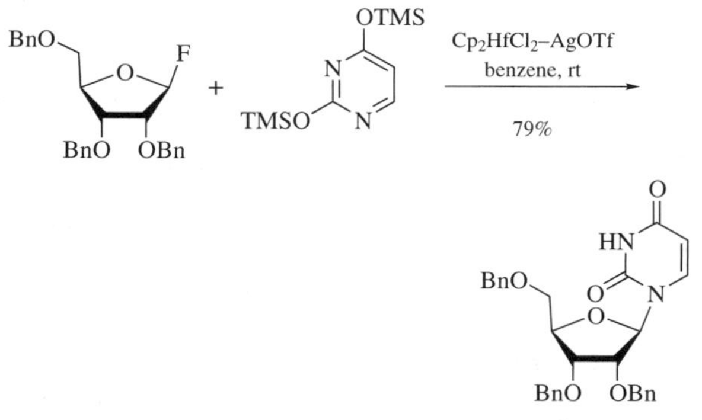

(3)

α:β = 5:95

Related Reagents. Dichlorobis(cyclopentadienyl)hafnium; Dichlorobis(cyclopentadienyl)zirconium–Silver(I) Salts.

---

1. Suzuki, K.; Matsumoto, T. *J. Synth. Org. Chem. Jpn.* **1993**, *51*, 718.
2. (a) Matsumoto, T.; Maeta, H.; Suzuki, K.; Tsuchihashi, G. *Tetrahedron Lett.* **1988**, *29*, 3567. (b) Suzuki, K.; Maeta, H.; Matsumoto, T.; Tsuchihashi, G. *Tetrahedron Lett.* **1988**, *29*, 3571. (c) Matsumoto, T.; Maeta, H.; Suzuki, K.; Tsuchihashi, G. *Tetrahedron Lett.* **1988**, *29*, 3575. (d) Suzuki, K.; Maeta, H.; Matsumoto, T. *Tetrahedron Lett.* **1989**, *30*, 4853.
3. (a) Matsumoto, T.; Katsuki, M.; Suzuki, K. *Tetrahedron Lett.* **1988**, *29*, 6935. (b) Matsumoto, T.; Katsuki, M.; Suzuki, K. *Chem. Lett.* **1989**, 437. (c) Matsumoto, T.; Hosoya, T.; Suzuki, K. *Tetrahedron Lett.* **1990**, *31*, 4629. (d) Matsumoto, T.; Hosoya, T.; Suzuki, K. *Synlett* **1991**, 709.
4. Brinkley, Jr., S. R. *J. Am. Chem. Soc.* **1940**, *62*, 3524.
5. (a) Mukaiyama, T.; Murai, Y.; Shoda, S. *Chem. Lett.* **1981**, 431 ($SnCl_2$–$AgClO_4$) . (b) Hashimoto, S.; Hayashi, M.; Noyori, R. *Tetrahedron Lett.* **1984**, *25*, 1379 (TMSOTf and SiF_4) . (c) Nicolaou, K. C.; Chucholowski, A.; Dolle, R. E.; Randall, J. L. *Chem. Commun./J. Chem. Soc., Chem. Commun.* **1984**, 1155 ($BF_3{\cdot}OEt_2$) . (d) Kunz, H.; Sager, W. *Helv. Chim. Acta* **1985**, *68*, 283 ($BF_3{\cdot}OEt_2$) . (e) Kreuzer, M.; Thiem, J. *Carbohydr. Res.* **1986**, *149*, 347 (TiF_4) . (f) Kobayashi, S.; Koide, K.; Ohno, M. *Tetrahedron Lett.* **1990**, *31*, 2435 (Me_2GaX) . (g) Wessel, H. P. *Tetrahedron Lett.* **1990**, *31*, 6863 (Tf_2O) . (h) Maeta, H.; Matsumoto, T.; Suzuki, K. *Carbohydr. Res.* **1993**, *249*, 49 (Bu_2SnCl_2–$2AgClO_4$) .
6. (a) Nicolaou, K. C.; Caulfield, T. J.; Kataoka, H.; Stylianides, N. A. *J. Am. Chem. Soc.* **1990**, *112*, 3693. (b) Nicolaou, K. C.; Hummel, C. W.; Iwabuchi, Y. *J. Am. Chem. Soc.* **1992**, *114*, 3126. (c) Matsuzaki, Y.; Ito, Y.; Ogawa, T. *Tetrahedron Lett.* **1992**, *33*, 6343. (d) Nicolaou, K. C.; Bockovich, N. J.; Carcanague, D. R.; Hummel, C. W.; Even, L. F. *J. Am. Chem. Soc.* **1992**, *114*, 8701. (e) Matsuzaki, Y.; Ito, Y.; Nakahara, Y.; Ogawa, T. *Tetrahedron Lett.* **1993**, *34*, 1061.
7. (a) Matsumoto, T.; Katsuki, M.; Jona, H.; Suzuki, K. *Tetrahedron Lett.* **1989**, *30*, 6185. (b) Matsumoto, T.; Katsuki, M.; Jona, H.; Suzuki, K. *J. Am. Chem. Soc.* **1991**, *113*, 6982.
8. Matsumoto, T.; Hosoya, T.; Suzuki, K. *J. Am. Chem. Soc.* **1992**, *114*, 3568.
9. Matheu, M. I.; Echarri, R.; Castillón, S. *Tetrahedron Lett.* **1992**, *33*, 1093.

Keisuke Suzuki & Takashi Matsumoto
Keio University, Yokohama, Japan

Dichloromethyl Methyl Ether

[4885-02-3] $C_2H_4Cl_2O$ (MW 114.96)

(carbon monoxide equivalent in reactions with boranes and borinates; formylating reagent for aromatic compounds and vinylsilanes; formation of α-keto acid chlorides)

Alternate Name: dichloromethoxymethane; DCME.
Physical Data: bp 85 °C; *d* 1.271 g cm^{-3}.
Solubility: sol in most common organic solvents.
Form Supplied in: commercially available in 98% purity.
Preparative Methods: from Methyl Formate and Phosphorus(V) Chloride.[1]
Purification: by distillation.
Handling, Storage, and Precautions: dichloromethyl methyl ether is moisture sensitive, a lachrymator, and a suspect carcinogen. Handle in a fume hood.

Original Commentary

Kevin E. Henegar
The Upjohn Company, Kalamazoo, MI, USA

Carbon Monoxide Equivalent. Dichloromethyl methyl ether functions as a carbon monoxide equivalent in the reaction with boranes to yield tertiary alcohols. The reaction works well with both symmetrical and unsymmetrical boranes and for boranes containing tertiary alkyl groups (eq 1).[2,3] Hindered lithium alkoxides are the preferred bases.

$$Bu_3B + MeOCHCl_2 \xrightarrow[\text{2. } H_2O_2,\text{ NaOH; } 94\%]{\text{1. LiOCEt}_3} Bu_3COH \quad (1)$$

A similar reaction occurs with borinate esters to yield ketones.[4] This reaction has been used in a facile synthesis of bicyclo[3.3.0]nonan-9-one from 9-Borabicyclo[3.3.1]nonane (eq 2).[5] The insertion reaction proceeds with retention of configuration of the migrating alkyl groups and provides a route to optically active ketones from chiral borinates and boronates (eq 3).[6] This reaction is applicable to the enantioselective synthesis of α-alkynyl ketones[7] and alkenones,[8] and as in the reaction of boranes with dichloromethyl methyl ether, hindered lithium alkoxides are the preferred bases. Borinates also undergo an elimination–rearrangement reaction to yield alkenes on treatment with dichloromethyl methyl ether and heat or Silver(I) Nitrate (eq 4).[9]

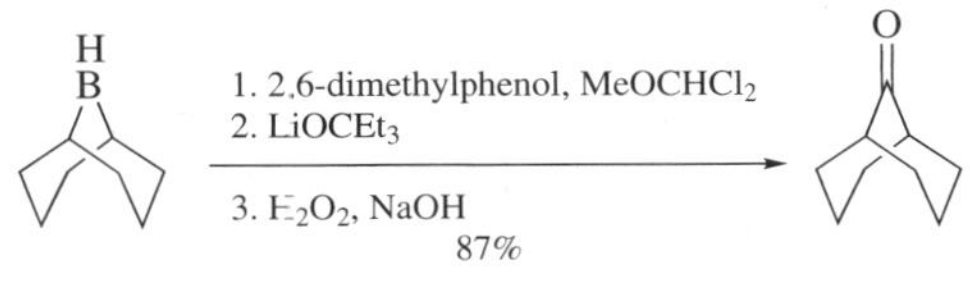

(2)

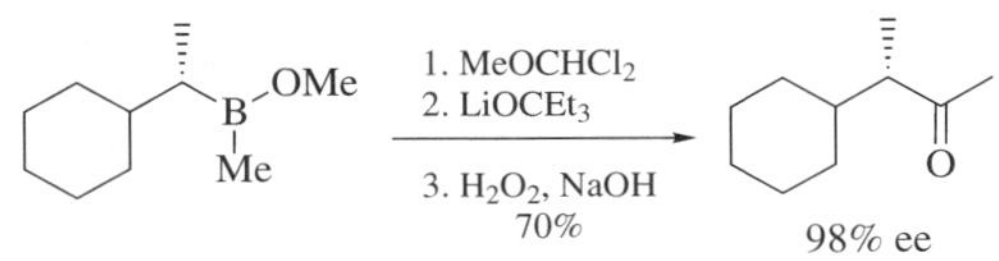

(3)

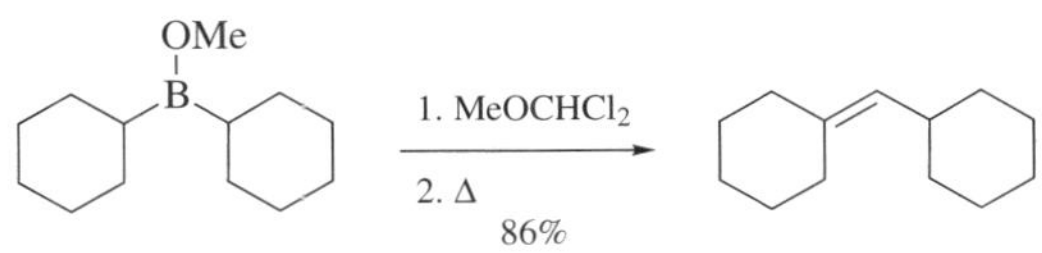

(4)

Formylation of Aromatic Compounds. Dichloromethyl methyl ether has been used extensively to formylate aromatic compounds.[10–12] The reaction works well for the formylation of arenes and heterocycles such as indoles and pyrroles. Titanium(IV) Chloride has generally been used as the Lewis acid but an improved procedure recommending Tin(IV) Chloride has been reported.[13] A typical example is shown in eq 5.[14]

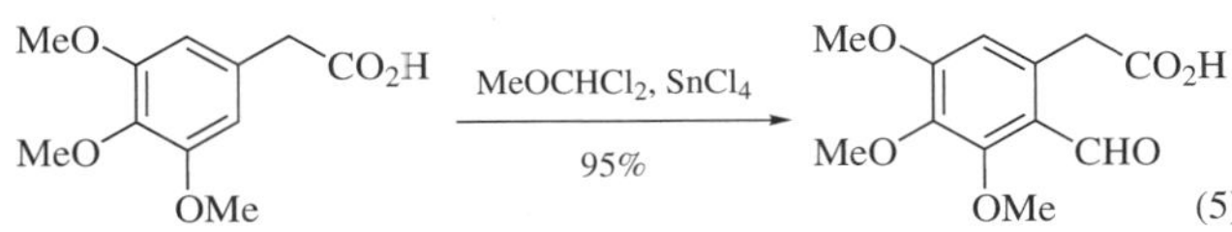

(5)

Formylation of Vinylsilanes. Vinylsilanes are formylated by reaction with dichloromethyl methyl ether, providing a route to α,β-unsaturated aldehydes (eq 6).[15,16] This reaction is also works well for the synthesis of dienals from 1-trimethylsilyl-1,3-dienes.[17]

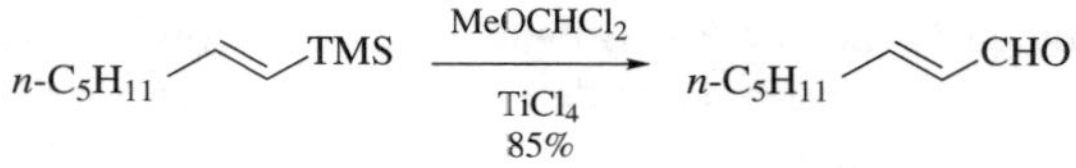

(6)

α-Keto Acid Chlorides. Dichloromethyl methyl ether is used in the preparation of α-keto acid chlorides. Conventional reagents such as oxalyl chloride or phosphorus halides give poor results (eq 7).[18]

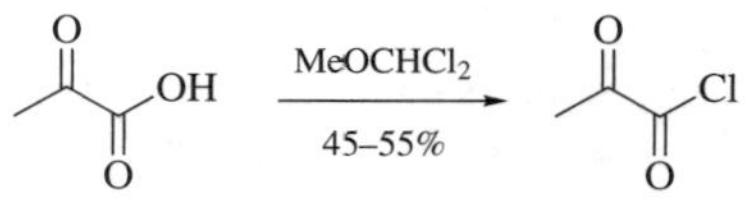

(7)

First Update

Leticia Quintero & Rosa L. Meza-León
Universidad Autónoma de Puebla, Puebla, México

Carbon Monoxide Equivalent. In an asymmetric variant on the conversion of cyclic boranes to the corresponding ketones, it was demonstrated that hydroboration of 1-allylcyclohexane with isopinocampheyldichloroborne, trimethylsilane, and lithium aluminum hydride followed by removal of the auxiliary with acetaldehyde and exposure to dichloromethyl methyl ether, then oxidative work-up afforded *trans*-decalinone in >99% ee and de (eq 8).[19]

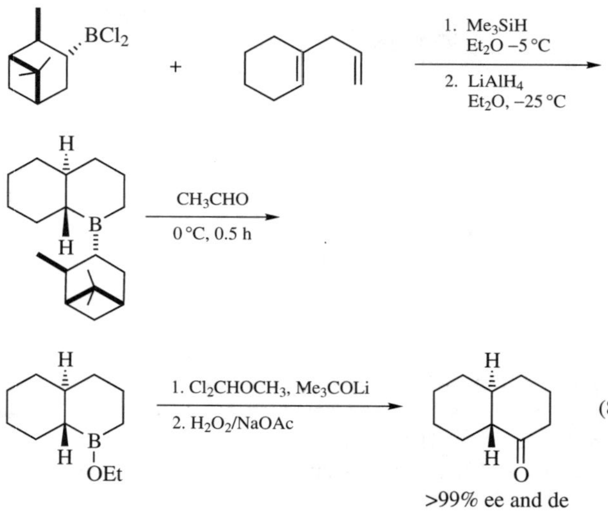

(8)

Formation of Orthoesters. The interaction of stoichiometric quantities of glycidol with dichloromethyl methyl ether in the presence of 2 equiv of NEt_3 in diethyl ether leads to formation of diglycidyloxy methoxy methanes (eq 9). When equimolar quantities of the reagents in benzene are employed along with 1 equiv of NEt_3, 4-chloromethyl-2-methoxy-1,3-dioxolane is the major product (eq 9).[20]

Preparation of Glycosyl Chlorides. Dichloromethyl methyl ether finds wide application in the formation of glycosyl chlorides from a range of anomeric derivatives, and is often to be considered the reagent of choice for this transformation.[21] A selection of illustrative examples is given in eqs 10–18.[22–30]

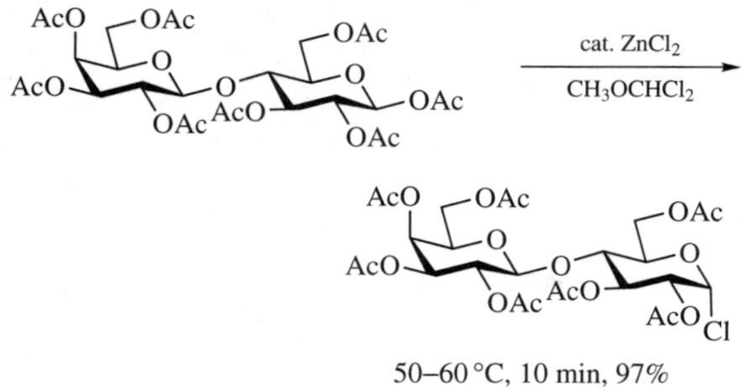

(10)

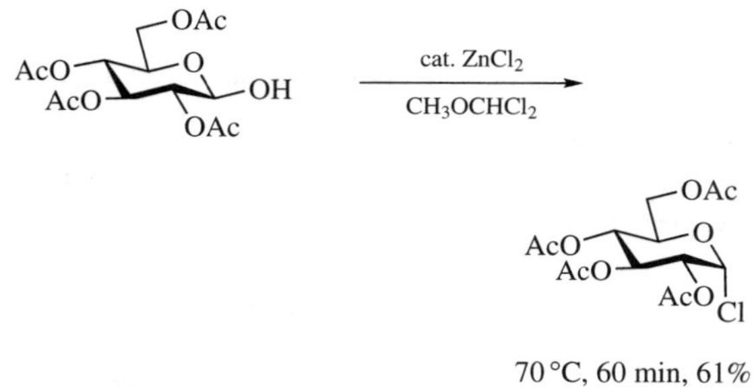

(11)

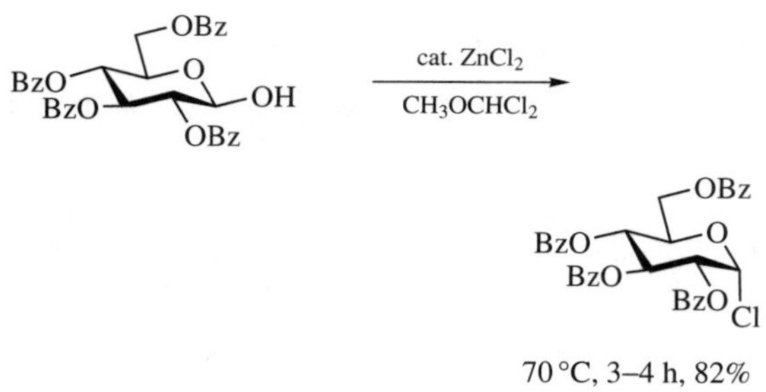

(12)

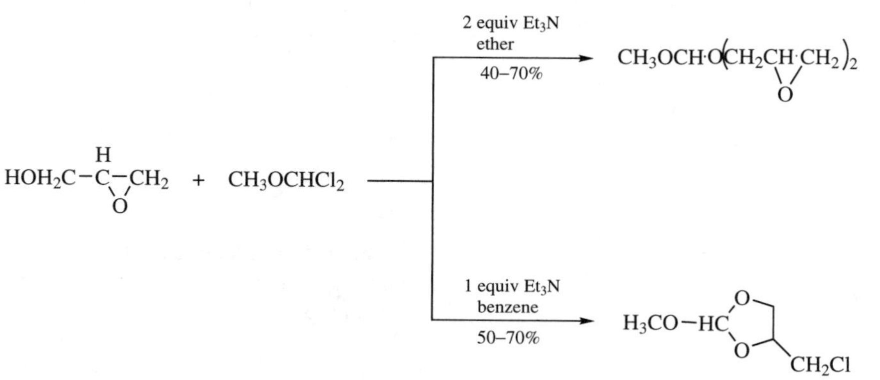

(9)

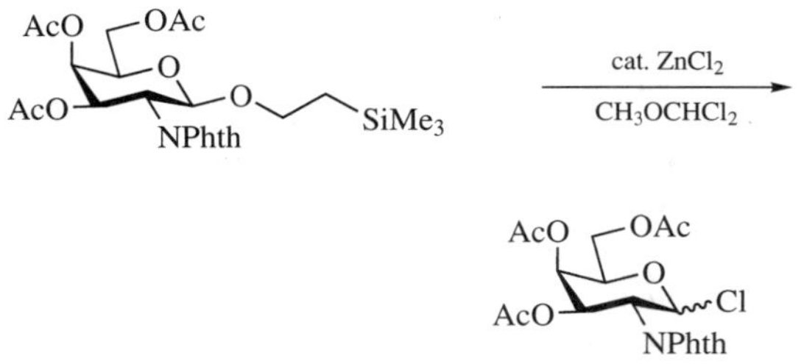

(13)

rt, 2 h, 99%, α:β = 88:12

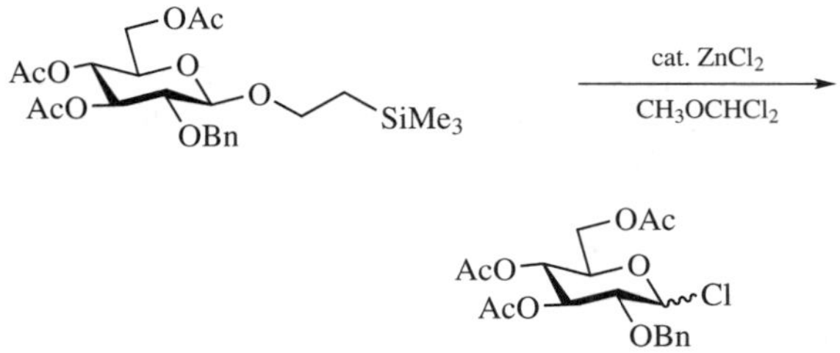

(14)

rt, 3 h, 98% α:β = 92:8

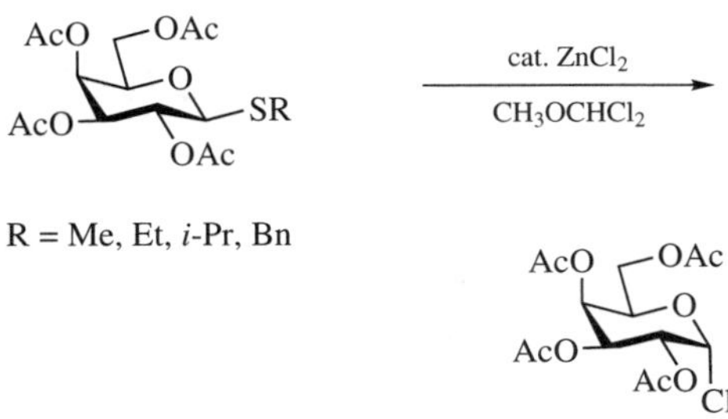

R = Me, Et, *i*-Pr, Bn

(15)

70–75 °C, 1 h, 61–68%

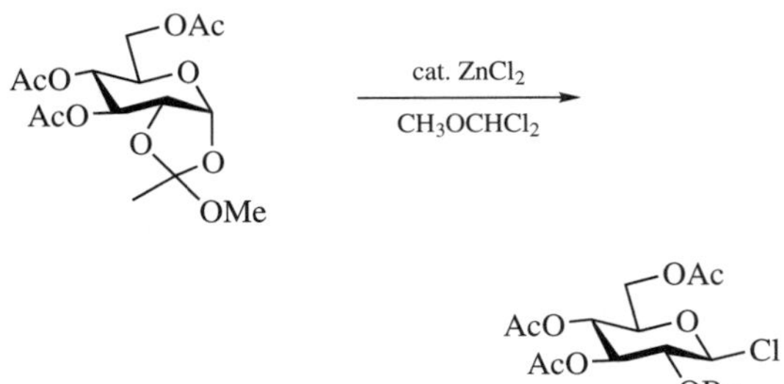

(16)

45 °C, 1 h, 60%

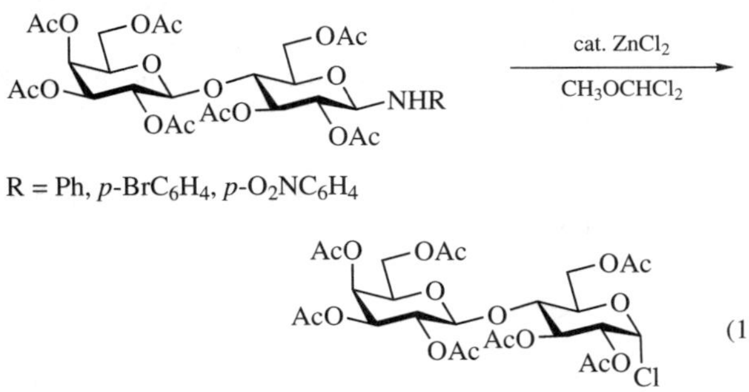

R = Ph, *p*-BrC$_6$H$_4$, *p*-O$_2$NC$_6$H$_4$

(17)

65–75 °C, 30–60 min, 41–57%

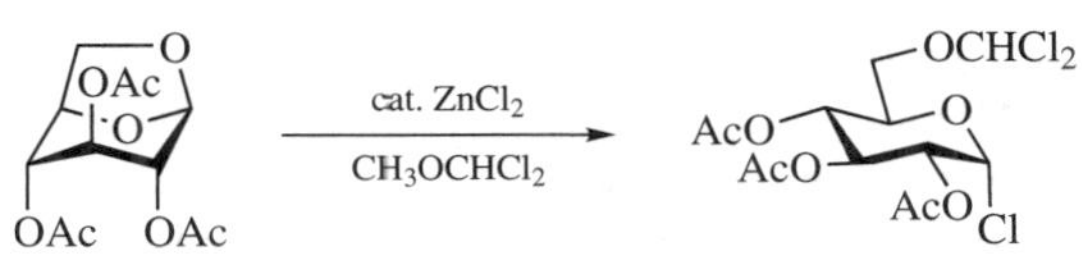

(18)

75–80 °C, 60 min, 70%

Under the usual conditions (eqs 10–18), heating with zinc chloride catalysis, the more stable axial chloride is typically, but not always (eq 16), the predominant product. When the reaction is conducted at room temperature using boron trifluoride as catalyst, ester-protected substrates provide the less stable equatorial chlorides in a stereospecific reaction that proceeds with neighboring group participation (eqs 19–21).[31–33]

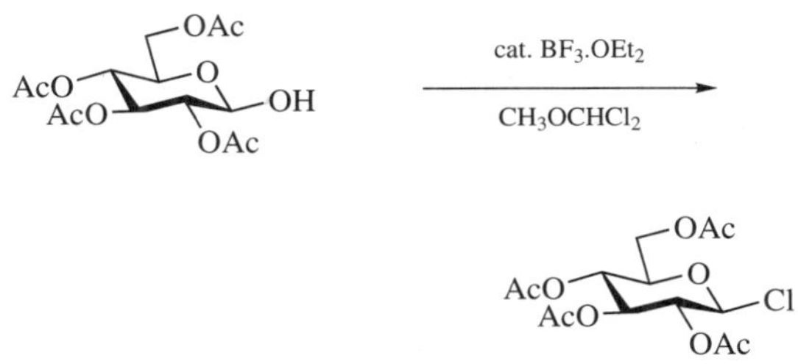

(19)

20 °C, 75 min, 89%

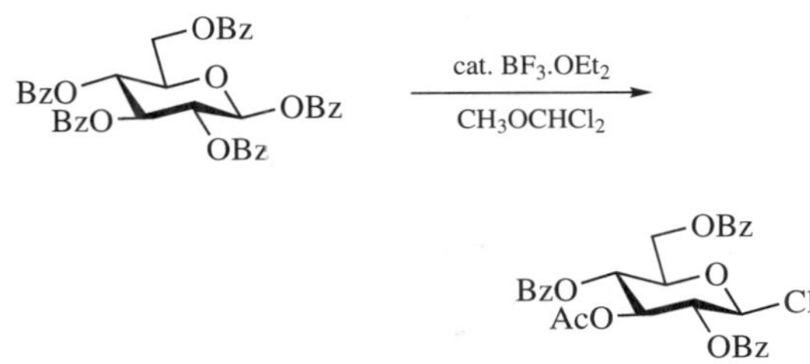

(20)

20 °C, 75 min, 91%

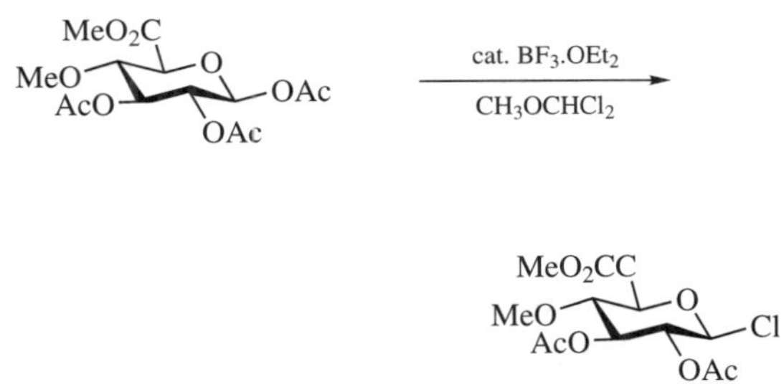

(21)

40 °C, 60 min, 82%

Related Reagents. Carbon Monoxide; *N*-Dichloromethylene-*N,N*-dimethyliminium Chloride; Potassium Cyanide. dibromomethyl methyl ether.

1. Gross, H.; Rieche, A.; Hoft, E.; Beyer, E., *Org. Synth., Coll. Vol.* **1973**, *5*, 365.
2. Brown, H. C.; Carlson, B. A., *J. Org. Chem.* **1973**, *38*, 2422.
3. Brown, H. C.; Katz J.-J.; Carlson, B. A., *J. Org. Chem.* **1973**, *38*, 3968.
4. Carlson, B. A.; Brown, H. C., *J. Am. Chem. Soc.* **1973**, *95*, 6876.
5. (a) Carlson, B. A.; Brown, H. C., *Synthesis* **1973**, 776. (b) Carlson, B. A.; Brown, H. C., *Org. Synth., Coll. Vol.* **1988**, *6*, 137.
6. (a) Brown, H. C.; Srebnik, M.; Bakshi, R. K.; Cole, T. E., *J. Am. Chem. Soc.* **1987**, *109*, 5420. (b) Rangaishenvi, M. V.; Singaram, B.; Brown, H. C., *J. Org. Chem.* **1991**, *56*, 3286.

7. Brown, H. C.; Gupta, A. K.; Vara Prasad, J. V. N.; Srebnik, M., *J. Org. Chem.* **1988**, *53*, 1391.
8. Brown, H. C.; Mahindroo, V. K., *Synlett* **1992**, 626.
9. (a) Brown, H. C.; Katz, J.-J.; Carlson, B. A., *J. Org. Chem.* **1974**, *39*, 2817. (b) Brown, H. C.; Katz, J.-J.; Carlson, B. A., *J. Org. Chem.* **1975**, *40*, 813.
10. Rieche, A.; Gross, H.; Hoft, E., *Ber. Dtsch. Chem. Ges./Chem. Ber.* **1960**, *93*, 88.
11. Rieche, A.; Gross, H.; Hoft, E., *Org. Synth., Coll. Vol.* **1973**, *5*, 49.
12. Gross, H.; Rieche, A.; Matthey, G., *Ber. Dtsch. Chem. Ges./Chem. Ber.* **1963**, *96*, 308.
13. Lewin, A. H.; Parker, S. R.; Fleming, N. B.; Carroll, F. I., *Org. Prep. Proced. Int.* **1978**, *10*, 201.
14. McBride, B. J.; Garst, M. E., *Tetrahedron* **1993**, *49*, 2839.
15. Yamamoto, K.; Ninokawa, O.; Tsuji, J., *Synthesis* **1977**, 721.
16. Yamamoto, K.; Yoshitake, J.; Qui, N. T.; Tsuji, J., *Chem. Lett.* **1978**, 859.
17. Andreini, B. P.; Carpita, A.; Rossi, R.; Scamuzzi, B., *Tetrahedron* **1989**, *45*, 5621.
18. Ottenheijm, H. C. J.; de Man, J. H. M., *Synthesis* **1975**, 163.
19. Brown, H. C.; Mahindroo, V. K.; Dhokte, U. P., *J. Org. Chem.* **1996**, *61*, 1906.
20. Bredikhin, A. A.; Lazarev, S. N., *Mendeleev Com.* **1998**, 81.
21. Kovác P. in *Modern Methods in Carbohydrate Synthesis*, Khan, S. H.; O'Neill, R. A., Eds.; Harwood Academic Press, Amsterdam, 1996, pp55.
22. Rieche, A.; Gross, H., *Chem Ber.* **1959**, *92*, 83.
23. Gross, H.; Farkas, I., *Chem Ber.* **1960**, *93*, 95.
24. Farkas, I.; Menyhárt, M.; Bognár, R.; Gross, H., *Chem Ber.* **1965**, *98*, 1419.
25. Kartha, K. P. R.; Jennings, H. J., *Tetrahedron Lett.* **1990**, *31*, 2537.
26. Jansson, K.; Noori, G.; Magnusson, G., *J. Org. Chem.* **1990**, *55*, 3181.
27. Farkas, I.; Bognár, R.; Menyhart, M. M.; Tarnai, A. K.; Bihari, M.; Tamás, J., *Acta Chim. Sci. Hung.* **1975**, *84*, 325.
28. Farkas, I.; Dinya, Z.; Szabó, I. F.; Bognár, R., *Carbohydr. Res.* **1972**, *21*, 331.
29. Menyhárt, M. M.; Bognár, R.; Farkas, I., *Carbohydr. Res.* **1976**, *48*, 139.
30. Bognár, R.; Farkas, I.; Menyhárt, M.; Paulsen, H., *Carbohydr. Res.* **1968**, *6*, 404.
31. Bock, K.; Pedersen, C.; Rasmussen, P., *J. Chem. Soc., Perkin Trans. 1* **1973**, 1456.
32. Farkas, I.; Szabó, I. F.; Bognár, R.; Anderle, D., *Carbohydr. Res.* **1976**, *48*, 136.
33. Kovác, P.; Farkas, I.; Mihálov, V.; Palovcík, R.; Bognár, R., *J. Carbohydr. Nucleosides, Nucleotides* **1976**, *3*, 57.

1,3-Dichloro-1,1,3,3-tetraisopropyldisiloxane

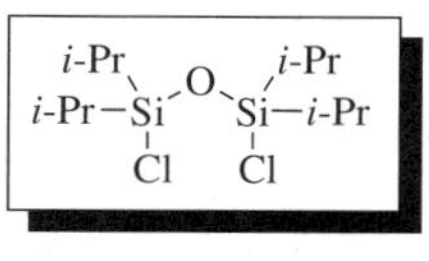

[69304-37-6] $C_{12}H_{28}Cl_2OSi_2$ (MW 315.43)

(simultaneous protection of the 3′- and 5′-hydroxyl functions of furanose nucleosides;[2,3] protecting group for carbohydrates;[4,5] protecting group reagent for open-chain polyhydroxy compounds;[6] preparation of cyclic bridged peptides[7])

Alternate Name: TIPSCl.
Physical Data: bp 70 °C/0.5 mmHg; *d* 0.986 g cm^{-3}.
Solubility: unaffected by water, 0.3M TsOH in dioxane, 10% TFA in chloroform, 5M NH_3 in dioxane–water (4:1), or isobutylamine–methanol (1:9), or tertiary amines (Et_3N, pyridine).
Form Supplied in: liquid; commercially available.
Preparative Method: can be prepared starting from Trichlorosilane by treatment with *i*-PrMgCl, H_2O, and Acetyl Chloride.[1]
Handling, Storage, and Precautions: the liquid is corrosive and moisture sensitive.

Protection of the 3′- and 5′-Hydroxy Functions. The reagent design was based upon the fact that triisopropylsilyl chloride reacts 1000 times faster with primary alcohols than secondary alcohols. Thus in the case of 3′,5′-dihydroxy nucleosides the reagent initially silylates at the 5′-position. This is then followed by intramolecular reaction with the secondary alcohol at the 3′-position to give the doubly protected derivative (eq 1).[2,3,8]

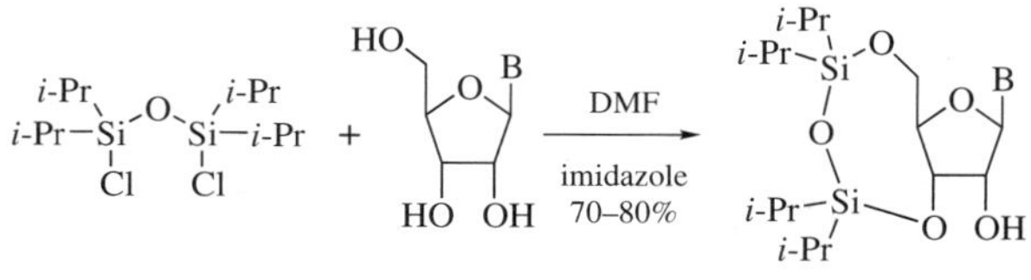

(1)

Rendering this protecting group of further utility is the fact that, in DMF under acidic conditions, silyl migration occurs affording the 2′,3′-disilyl derivative (eq 2).[4] This migration has been employed in the synthesis of carbohydrates (eq 3).[5,9]

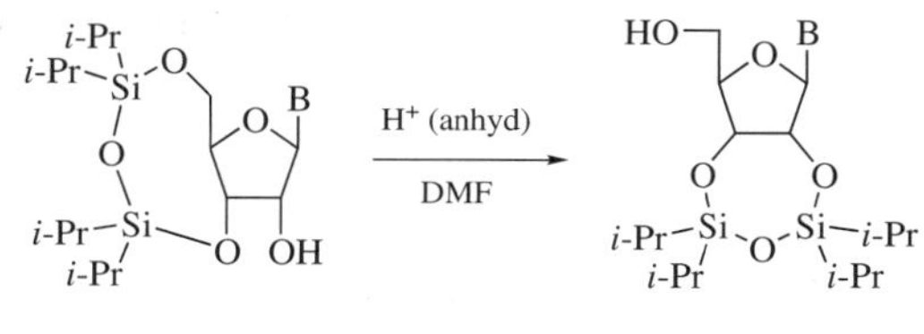

(2)

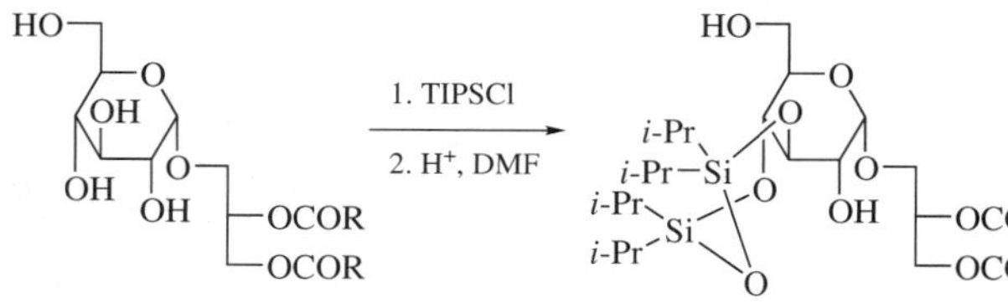

(3)

Protection of Open-Chain Polyhydroxy Compounds. The utility of 1,3-dichloro-1,1,3,3-tetraisopropyldisiloxane was extended to include open-chain polyhydroxy compounds. Glycerol reacted to form the eight-membered ring (eq 4) while the D-erythropentose derivative gave only the seven-membered ring (eq 5).[6] This strategy was used as a key step in the synthesis of the steroid 19-norcanrenone.[10]

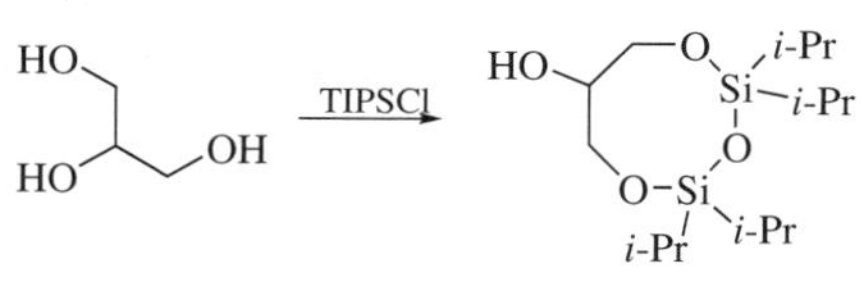

(4)

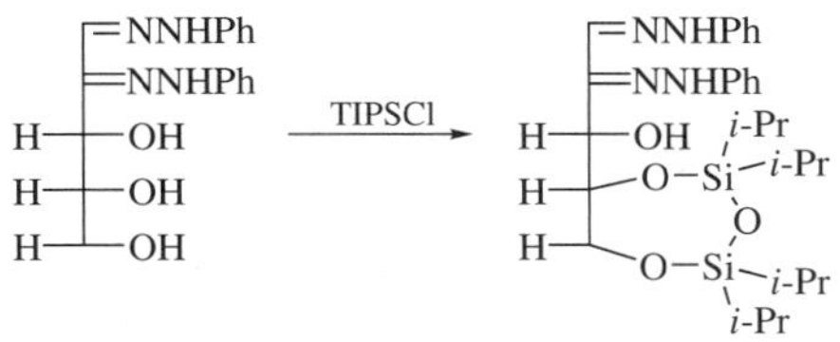

(5)

Preparation of Cyclic Bridged Peptides. As part of a study aimed at improving the transport of peptides by derivatization with silyl reagents, the feasibility of restricting the conformational mobility of the enkephalin peptide backbone by bridging two serine residues with a disiloxane bridge was investigated. TIPSCl was chosen due to its specificity for primary alcohols as well as its ability to bridge secondary alcohols if present in the correct spatial environment.[7] Thus the solution phase syntheses of protected [Ser^2,Ser^5]- and [D-Ser^2,Ser^5]enkephalins were carried out followed by the reaction of the two free hydroxy groups in each with TIPSCl to afford the bridged disiloxane structures.

1. Zhang, H. X.; Guibe, F.; Balavoine, G. *Synth. Commun.* **1987**, *17*, 1299.
2. (a) Markiewicz, W. *J. Chem. Res. (S)* **1979**, 24. (b) Markiewicz, W. *J. Chem. Res. (M)* **1979**, 181.
3. Markiewicz, W.; Padyukova, N. Sh.; Samek, Z.; Smrt, J. *Collect. Czech. Chem. Commun.* **1980**, *45*, 1860.
4. Verdegaal, C. H. M. N.; Jansse, P. L.; de Rooij, J. F. M.; van Boom, J. H. *Tetrahedron Lett.* **1980**, *21*, 1571.
5. van Boeckel, C. A. A.; van Boom, J. H. *Tetrahedron Lett.* **1980**, *21*, 3705.
6. Markiewicz, W.Samek, Z; Smrt, J. *Tetrahedron Lett.* **1980**, *21*, 4523.
7. Davies, J. S.; Tremer, E. J. *J. Chem. Soc., Perkin Trans. 1* **1987**, 1107.
8. (a) Schaumberg, J. P.; Hokanson, G. C.; French, J. C.; Smal, E.; Baker, D. C. *J. Org. Chem.* **1985**, *50*, 1651. (b) Tatsuoka, T.; Imao, K.; Suzuki, K. H **1986**, *24*, 617. (c) Robins, M. J.; Wilson, J. S.; Hansske, F. *J. Am. Chem. Soc.* **1983**, *105*, 4059. (d) Robins, M. J.; Wilson, J. S.; Sawyer, L.; James, M. N. G. *Can. J. Chem.* **1983**, *61*, 1911. (e) Hagen, M. D.; Scalfi-Happ, C.; Happ, E.; Chladek, S. *J. Org. Chem.* **1988**, *53*, 5040.
9. (a) Ziegler, T.; Eckhardt, E.; Neumann, K.; Birault, V. *Synthesis* **1992**, 1013. (b) Thiem, J.; Duckstein, V.; Prahst, A.; Matzke, M. *Justus Liebigs Ann. Chem./Liebigs Ann. Chem.* **1987**, 289. (c) Oltvoort, J. J.; Kloosterman, M.; van Boom, J. H. *Recl. Trav. Chim. Pays-Bas* **1983**, *102*, 501.
10. Nemoto, H.; Fujita, S.; Nagai, M.; Fukumoto, K.; Kametani, T. *J. Am. Chem. Soc.* **1988**, *110*, 2931.

Joel Slade
Ciba-Geigy Corp., Summit, NJ, USA

4,5-Dicyanoimidazole

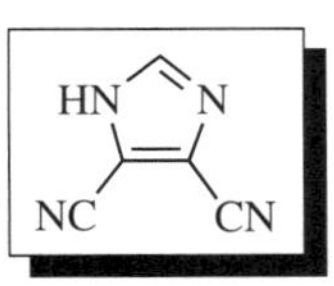

[1122-28-7] $C_5H_2N_4$ (MW 118.10)

(reagent used for activation of phosphoramidites; oligonucleotide synthesis; nucleophilic catalyst)

Physical Data: mp 168–175 °C (95% pure); 175–176 °C (99% pure).
Solubility: soluble in acetonitrile, organic solvents.
Form Supplied in: off-white to tan powder (95% purity); widely available. White, crystalline solid (99% purity); available from Glen Research. A 0.25 M solution in acetonitrile; available from Proligo, LLC and Glen Research.
Purification: decolorized with charcoal and crystallized from hot water.
Handling, Storage, and Precautions: 4,5-dicyanoimidazole is classified as an irritant and contact with skin and mucus membranes or exposure by inhalation or by swallowing can be harmful. Solutions of 4,5-dicyanoimidazole in acetonitrile are hygroscopic and should be stored under an inert atmosphere. Incompatible with oxidizing agents.

Activation of Nucleoside Phosphoramidites During Solid-Phase Oligonucleotide Synthesis. The key step in oligonucleotide chain elongation during solid-phase synthesis by the phosphoramidite method[1] (eq 1) is the nucleophilic activation of the

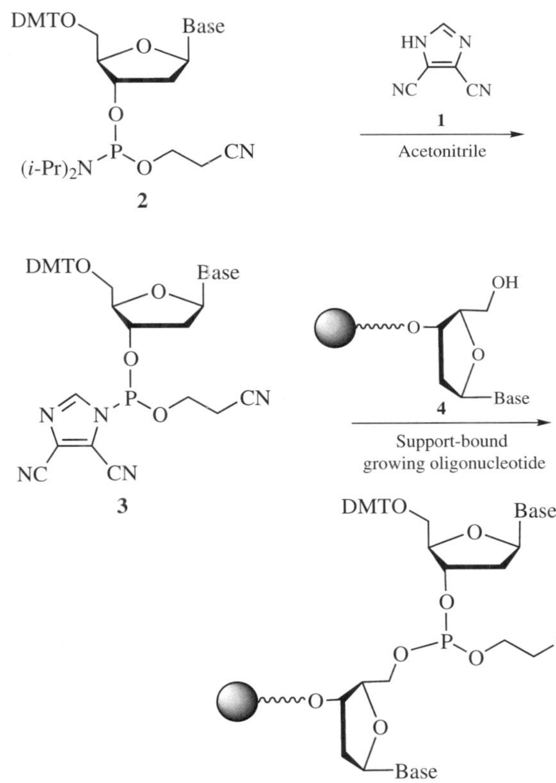

(1)

phosphoramidite nucleotide synthon **2** by a weak acid, which enables phosphitylating the 5′-hydroxy group of the support-bound oligonucleotide chain. The mechanism of this process is known to involve formation of intermediate **3** via initial phosphorus protonation by the acidic activator, phosphorus-to-nitrogen proton transfer, followed by rate limiting nucleophilic substitution of the diisopropylamine ligand by the conjugate base of the activator.[2] Intermediate **3** reacts rapidly with the hydroxy group of the support-bound oligonucleotide **4**, affording phosphite triester **5**. The synthesis cycle is then completed by capping unreacted hydroxy groups and oxidation of the phosphite triester to the corresponding phosphate triester. Although automated oligonucleotide synthesis was developed and widely commercialized employing 1*H*-tetrazole (pKa = 4.8) as a 0.45 M solution in acetonitrile as the activator for the coupling step, it has been shown that acetonitrile solutions (0.25–1.0 M) of highly pure (99+%), crystalline 4,5-dicyanoimidazole (**1**) have many advantages as high performance activators for oligonucleotide synthesis.[3] When substituted for 1*H*-tetrazole, the less acidic 4,5-dicyanoimidazole (pKa = 5.2) has been shown to afford fewer chain elongated by-products resulting from undesired 5′-DMT ether cleavage during the coupling step; this constitutes a key advantage of the reagent. Furthermore, solution-phase model studies revealed rates of phosphoramidite activation and coupling by DCI were at least doubled relative to 1*H*-tetrazole (eq 2). The decreased reaction acidity and improved kinetics of phosphoramidite couplings catalyzed by **1** are particularly significant when less reactive 2′-modified nucleoside phosphoramidites are involved. Such monomers require longer coupling times and, therefore, suffer prolonged exposure to the weak acid activator. This has been shown to result in the generation of impurities that are one or more monomeric units longer than the desired oligomeric product and difficult to separate. Improved purities of crude oligonucleotide prepared from syntheses employing 2′-OTBS and 2′-trifluoroacetamido monomers were reported using **1** as the activator.[3]

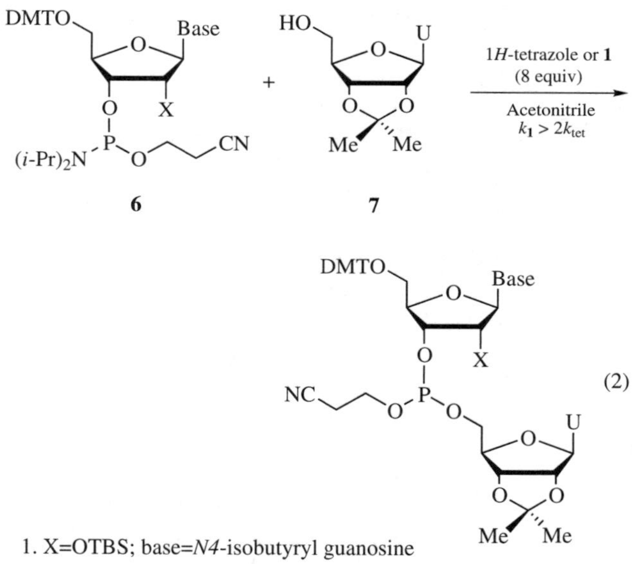

6 **7** **8**

1. X=OTBS; base=*N4*-isobutyryl guanosine
2. X=OMe; base=uridine
3. X=NHCF$_3$; base=uridine
4. X=*N*-pthalimido; base=uridine
5. DMT = Dimethoxytrityl

An additional advantage is that 4,5-dicyanoimidazole displays improved solubility in acetonitrile relative to tetrazole (up to 1.1 M compared to saturation at 0.5 M for tetrazole). This has proven to be particularly significant as oligonucleotide synthesis scales are being engineered ever lower[4] and, consequently, synthesizer fluid delivery systems are correspondingly more sensitive to particulates in the reagent solutions that are employed. A final practical advantage of 4,5-dicyanoimidazole over 1*H*-tetrazole is the comparative safety. 1*H*-tetrazole is classified as an explosive and the shipping and storage of this activator have become increasingly complicated by regulatory agencies.[5]

Synthesis of Nucleoside Phosphoramidites. The efficient preparation of 5′-*O*-dimethoxytrityl thymidine phosphoramidite (**11**) from the corresponding 5′-DMT nucleoside **9**, reagent **10**, and **1** has been described (eq 3).[3] This protocol serves as an alternative to methods employing 1*H*-tetrazole under these conditions or chloro 2-cyanoethyl diisopropylphosphoramidite and diisopropylethyl amine.[6]

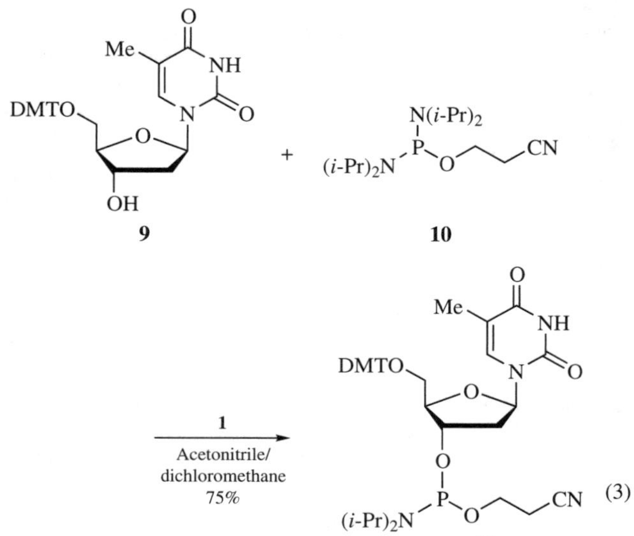

Precursor to 2-Bromo-4,5-dicyanoimidazole. 4,5-Dicyanoimidazole is a precursor to 2-bromo-4,5-dicyanoimidazole (**12**; eq 4).[7] This brominated derivative was employed by Xin and Just as an activator for the diastereoselective formation of phosphite triester **14** from chiral, cyclic nucleoside phosphoramidite **13** and methanol (eq 5).[8] The authors also evaluated several other heterocyclic activators, including 1*H*-tetrazole, 3-nitroimidazole, benzotriazole, 5-chlorobenzotriazole, 4,5-dichloroimidazole, 4,5-dicyanoimidazole, 2-nitroimidazole, and tribromoimidazole. Interestingly, they found that **1** displayed fastest reaction kinetics in the activation of the phosphoramidite, while 2-bromo-4,5-dicyanoimidazole afforded the highest diastereoselectivity (50:1 vs 12:1 for **1** and 1:1 for 1H-tetrazole). 2-Bromo-4,5-dicyanoimidazole is also a versatile precursor to various imidazole heterocycles.[9]

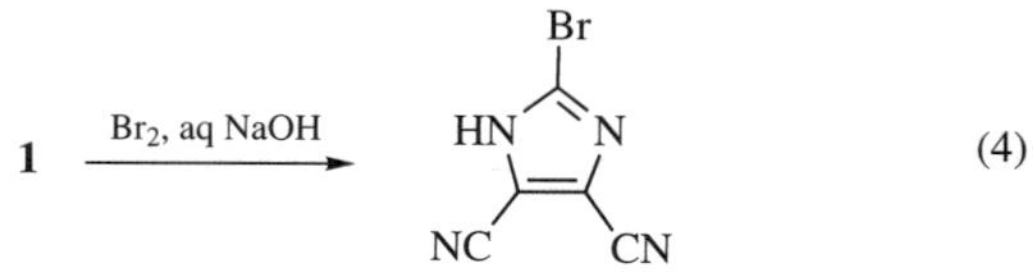

12

Synthesis of Novel Nucleosides. 4,5-Dicyanoimidazole has been used for the preparation of novel nucleoside analogues. Hosmane and co-workers[10] reported the synthesis of ring expanded purine analogue **16** formed by condensation of **1** with guanidine (**15**) that is aromatic and predicted to be flat by molecular modeling (eq 6). 4,5-Dicyanoimidazole was also used by this group in the glycosidation process using Vorbruggen conditions. The resultant *N*-glycoside (**18**) was condensed with **15** to give the nucleoside analogue **19** (eq 7).

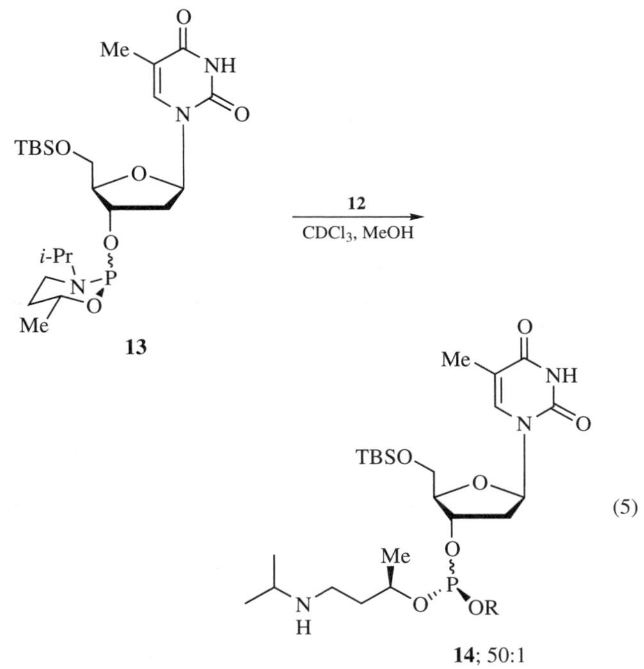

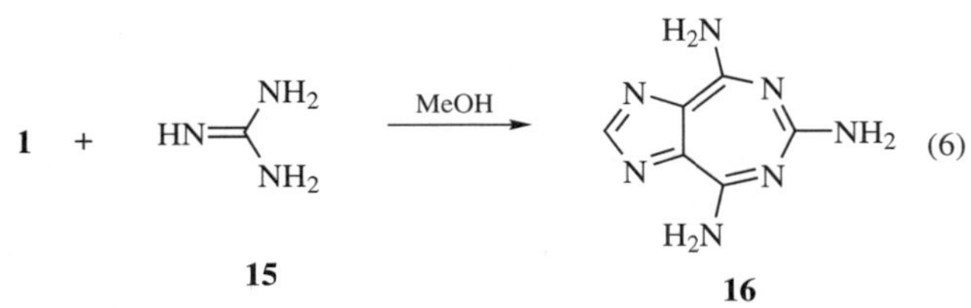

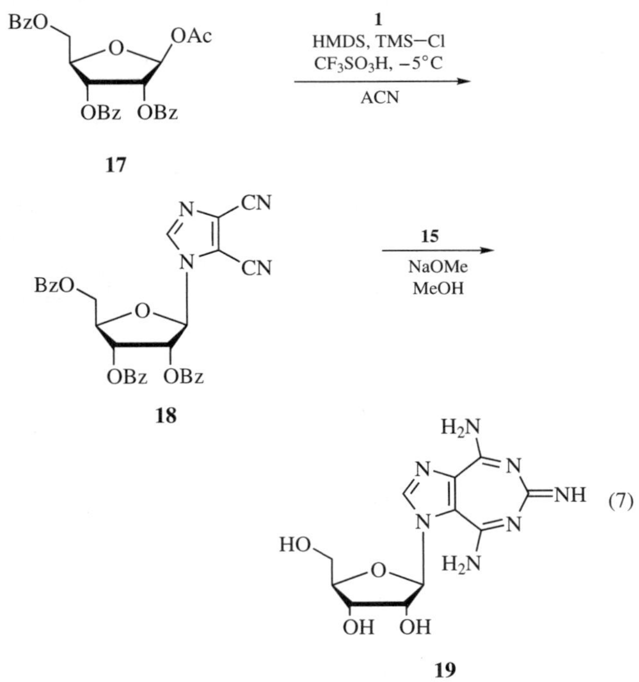

Ferris and co-workers synthesized a ribopyranose nucleoside derivative of 4,5-dicyanoimidazole by acid catalyzed glycosylation with 1,2,3,4-tetra-*O*-acetyl-β-D-ribopyranose (eq 8).[11] As expected, the β-anomer was formed preferentially.

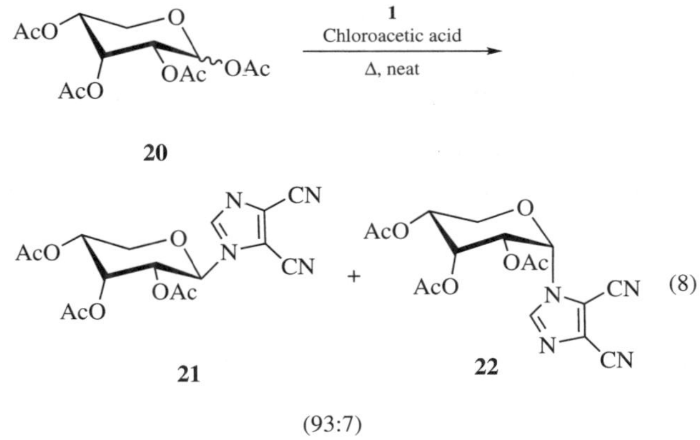

Synthesis of 1-alkyl Dicyanoimidazole Derivatives. The pyrrolic nitrogen of 4,5-dicyanoimidazole has been derivatized via alkylation using Et_3N,[12] NaH,[13] K_2CO_3[10] or $NaHCO_3$[14] as the base with electrophiles including BnCl, $(CH_3)_2SO_4$ or CH_3OCH_2Cl (eq 9). 4,5-Dicyanoimidazole was also *N*-alkylated by chiral alcohols under Mitsunobu conditions, affording enantiomerically pure *N*-alkylimidazoles that have interesting pharmaceutical properties.[15] Imidazole itself did not react under Mitsunobu conditions, however, 4,5-dicyanoimidazole gave a single enantiomer in alkylation reactions of enantiomerically pure alcohols (except for benzylic alcohols, where the Mitsunobu reaction may proceed with both retention and inversion). 4,5-Dicyanoimidazole can also be alkylated by heating with a variety of trialkyl orthoformates as well as orthoacetates and tetraethyl orthocarbonate.[16] The imidazole nitrogen can be liberated by employing standard *N*-deprotection methods[17] such as LiCl/NMP/Δ when R = Me,[12] acid treatment when $R = CH_2OCH_3$[12] and Na/NH_3[13] or $Pd(OH)_2$–C/H_2[10] when $R = CH_2Ph$.

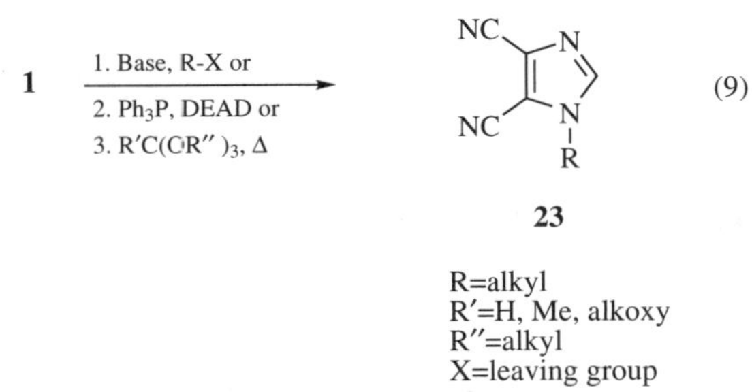

Related Reagents. 1*H*-Tetrazole [*288-94-8*];[1] 5-ethylthio-1*H*-tetrazole [*89797-68-2*];[18] 5-(4-nitrophenyl)-1*H*-tetrazole [*16687-60-8*];[19] 5-methylthio-1*H*-tetrazole [*255377-50-5*];[19] benzimidazolium triflate [*99257-95-1*].[20]

1. (a) Beaucage, S. L.; Caruthers, M. H., *Tetrahedron Lett.* **1981**, *22*, 1859. (b) Sinha, N. D.; Biernat, J.; McManus, J.; Koester, H., *Nucleic Acids Res.* **1984**, *12*, 4539.

2. (a) Dahl, B. H.; Nielson, J.; Dahl, O., *Nucleic Acids Res.* **1987**, *15*, 1729. (b) Berner, S.; Muehlegger, K.; Seliger, H., *Nucleic Acids Res.* **1989**, *17*, 853.
3. Vargeese, C.; Carter, J.; Yegge, J.; Krivjansky, S.; Settle, A.; Kropp, E.; Peterson, K.; Pieken, W., *Nucleic Acids Research* **1998**, *26*, 1046.
4. For example, see: Hughes, T. R.; Mao, M.; Jones, A. R.; Burchard, J.; Marton, M. J.; Shannon, K. W.; Lefkowitz, S. M.; Ziman, M.; Schelter, J. M.; Meyer, M. R.; Kobayashi, S.; Davis, C.; Dai, H.; He, Y. D.; Stephaniants, S. B.; Cavet, G.; Walker, W. L.; West, A.; Coffey, E.; Shoemaker, D. D.; Stoughton, R.; Blanchard, A. P.; Friend, S. H.; Linsley, P. S., *Nature Biotechnol.* **2001**, *19*, 342.
5. *International Air Transport Association, Dangerous Goods Regulations*, 42nd ed., effective January 1, 2001.
6. (a) Nielsen, J.; Taagaard, M.; Marugg, J. E.; van Boom, J. H.; Dahl, O., *Nucleic Acids Res.* **1986**, *14*, 7391. (b) Beijer, B.; Sulston, I.; Sproat, B. S.; Rider, P.; Lamond, A. I.; Neuner, P., *Nucleic Acids Res.* **1990**, *18*, 5143.
7. Apen, P. G.; Rasmussen, P. G., *Heterocycles* **1989**, *29*, 1325.
8. Xin, Z.; Just, G., *Tetrahedron Lett.* **1996**, *37*, 969.
9. (a) Coad, E. C. (1994) Synthesis and Characterization of Materials Derived from 4,5-Dicyanoimidazoles, Ph.D. thesis dissertation from the University of Michigan. (b) See also, reference 7.
10. Wang, L.; Bhan, A.; Hosmane, R. S., *Nucleosides & Nucleotides* **1994**, *13*, 2307.
11. Ferris, J. P.; Devadas, B.; Huang, C.-H.; Ren, W.-Y., *J. Org. Chem.* **1985**, *50*, 747.
12. Apen, P. G.; Rasmussen, P. G., *J. Am. Chem. Soc.* **1991**, *113*, 6178.
13. Matthews, D. P.; McCarthy, J. R.; Whitten, J. P.; Kastner, P. R.; Barney, C. L.; Marshall, F. N.; Ertel, M. A.; Burkhard, T.; Shea, P. J.; Kariya, T., *J. Med. Chem.* **1990**, *33*, 317.
14. O'Connell, J. F.; Parquette, J.; Yelle, W. E.; Wang, W.; Rapoport, H., *Synthesis* **1988**, 767.
15. Botta, M.; Summa, V.; Trapassi, G.; Monteagudo, E.; Corelli, F., *Tetrahedron: Asymmetry* **1994**, *5*, 181.
16. Johnson, S. J., *Synthesis* **1991**, 75.
17. Greene, T.W.; Wuts, P. G. M., *Protective Groups in Organic Synthesis*, 3rd ed.; Wiley: New York, 1999.
18. Polushin, N. N., *Nucleic Acids Res.* **2000**, *28*, 3125.
19. (a) Nurminen, E. J.; Mattinen, J. K.; Lonnberg, H., *J. Chem. Soc., Perkin Trans. 2* **1999**, *11*, 2551. (b) Froehler, B. C.; Matteucci, M. D., *Tetrahedron Lett.* **1983**, *24*, 3171.
20. Hayakawa, Y.; Kataoka, M.; Noyori, R., *J. Org. Chem.* **1996**, *61*, 7996.

David P. Sebesta & Kurt Vagle
Proligo, LLC, Boulder, CO, USA

1,3-Dicyclohexylcarbodiimide[1]

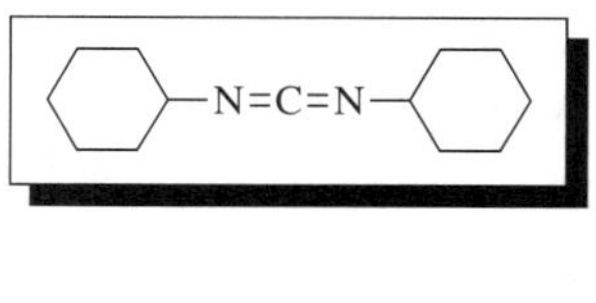

[538-75-0] $C_{13}H_{22}N_2$ (MW 206.33)

(powerful dehydrating agent commonly used for the preparation of amides,[2] esters,[3] and anhydrides;[4] used with DMSO for the mild oxidation of alcohols to ketones;[5] used in the dehydrative conversion of primary amides to nitriles,[6] β-hydroxy ketones to α,β-unsaturated ketones,[7] and can effect the stereochemical inversion of secondary alcohols[8])

Alternate Name: DCC.
Physical Data: mp 34–35 °C; bp 122–124 °C.
Solubility: highly sol dichloromethane, THF, acetonitrile, DMF.
Form Supplied in: opalescent solid; widely available.
Handling, Storage, and Precautions: is an acute skin irritant in susceptible individuals. Because of its low melting point, it is conveniently handled as a liquid by gentle warming of the reagent container. It should be handled with gloves in a fume hood, and stored under anhydrous conditions.

Amide Formation. Since the initial reports,[9,10] DCC has become the most common reagent in peptide synthesis[2,11,12] and in other amide bond-forming reactions of primary and secondary amines with carboxylic acids. The mechanism is considered to be well understood.[2,5]

Typically, DCC (1.1 equiv) is added to a concentrated solution (0.1–1.0 M) of the carboxylic acid (1.0 equiv), amine (1.0 equiv), and catalyst (when used) in methylene chloride or acetonitrile at 0 °C. The hydrated DCC adduct, dicyclohexylurea (DCU), quickly precipitates and the reaction is generally complete within 1 h at rt. The solvents THF and DMF can be used, but are reported to reduce reaction rates and encourage the formation of the *N*-acylurea side product, as well as increasing racemization in chiral carboxylic acids.[13–16] If the amine is initially present as the salt (i.e. amine hydrochloride), it may be neutralized by adding 1 equiv of Diisopropylethylamine prior to adding DCC; however, the addition of tertiary amines (particularly Triethylamine) can facilitate *N*-acylurea formation and racemization.[2] Racemization occurs via the formation of an oxazalone intermediate.[2] The addition of coupling agents (acylation catalysts) such as 1-Hydroxybenzotriazole (HOBt),[17] 1-hydroxy-7-azabenzotriazole (HOAt),[18] N-Hydroxysuccinimide (HOSu),[19] and 3-hydroxy-3,4-dihydro-1,2,3-benzotriazin-4-one[20] can generally ameliorate both problems. These additives are required for the coupling of sterically hindered components, or when the amine is weakly nucleophilic.

Another potential problem with DCC is that at the completion of the reaction some DCU remains in solution with the product, necessitating additional purification. Water-soluble carbodiimide derivatives such as 1-Cyclohexyl-3-(2-morpholinoethyl)carbodiimide Metho-p-toluenesulfonate[21] and 1-Ethyl-3-(3′-dimethylaminopropyl)carbodiimide Hydrochloride (EDCI)[22] obviate this problem, as they are removed by a simple extraction. Many newer coupling agents have been developed for peptide synthesis and other acylation reactions. These include Benzotriazol-1-yloxytris(dimethylamino)phosphonium Hexafluorophosphate (BOP),[23] O-Benzotriazol-1-yl-N,N,N′, N′-tetramethyluronium Hexafluorophosphate (HBTU),[24] Bis(2-oxo-3-oxazolidinyl)phosphinic Chloride[25] (BOP-Cl), and (1*H*-1,2,3-benzotriazol-1-yloxy)tris(pyrrolidino)phosphonium hexafluorophosphate (PyBOP).[26] In addition to linear and polymeric amides, lactams of various ring sizes have been synthesized using these methods (eq 1).[27]

(1)

$R = (CH_2)_n$, $n = 0–6$

Ester and Thioester Formation. These reactions occur through the same *O*-acylurea or anhydride active intermediate as in the amide coupling reactions, and the discussion of associated problems applies here as well. In general, alkyl and (particularly) aryl thiols can be efficiently coupled to carboxylic acids using DCC.[28] Reactions of primary and secondary alcohols proceed reliably, but require the presence of an acylation catalyst. This is usually 4-Dimethylaminopyridine (DMAP),[29,30] (see also 1,3-Dicyclohexylcarbodiimide–4-Dimethylaminopyridine), but others have been used including 4-pyrrolidinopyridine[31] and pyridine (solvent) with catalytic p-Toluenesulfonic Acid.[32] The acylation of more hindered alcohols often results in reduced yields; however, even *t*-butanol can be acylated, providing a useful route to *t*-butyl esters.[3,31] Various other carbodiimide derivatives have also been used in the preparation of esters.[33,34] As with amides, which are not limited to intermolecular reactions, a wide variety of lactones can also be synthesized.[35,36]

Anhydride Formation. Among other anhydride-forming reagents, including Acetic Anhydride, Trifluoroacetic Anhydride, and Phosphorus(V) Oxide, DCC is one of the simplest, mildest, and most effective reagents for the preparation of symmetrical anhydrides,[4,37] including formic anhydride,[38] which is useful in the preparation of formamides (eq 2).

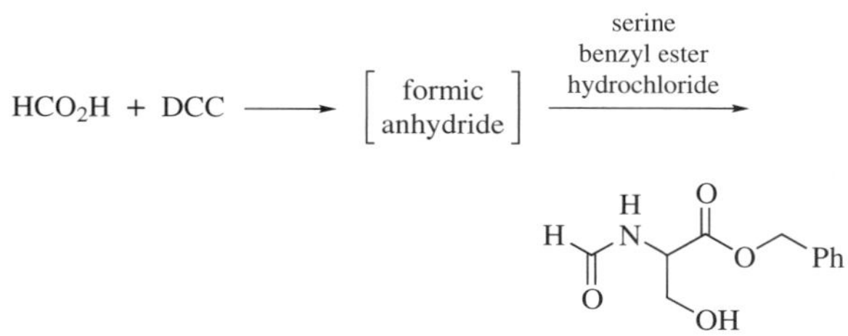

(2)

Anhydride formation is associated with each of the reaction types involving carboxylic acids. The anhydride is often the reactive species, to the extent that it competes with the *O*-acylurea or covalent catalyst adduct (see discussion of amide bond formation reactions).

Sulfinate and Phosphate Esters. Anhydrides of sulfinic and sulfonic acids[39,40] and phosphate monoesters can be prepared using DCC. Further reaction with alcohols or phenols gives the corresponding sulfinates[41] and phosphate diesters.[42,43] Pyrophosphates can be prepared from mono- and disubstituted phosphate esters.[44] Reaction of activated sulfinates with amines gives the corresponding sulfinamides.[45]

Oxidation of Alcohols to Aldehydes and Ketones: Moffatt Oxidation. The oxidation of primary and secondary alcohols[46] to aldehydes and ketones, respectively, can be carried out by reaction with DMSO activated by DCC (see Dimethyl Sulfoxide–Dicyclohexylcarbodiimide).[5,47,48] In comparison to the many metal-mediated oxidative reactions, the Moffatt oxidation is carried out under very mild conditions, and has found widespread use for reactions in the presence of sensitive functional groups (eq 3).[49] In addition, over-oxidation of aldehydes to form carboxylic acids is not observed. Under typical reaction conditions, a solution of the alcohol (1 equiv), DCC (3 equiv), and a proton source (pyridinium trifluoroacetate) (0.5 equiv) is stirred in DMSO or DMSO/benzene overnight at rt. After quenching the reaction with aqueous acetic acid and removing DCU by filtration, the product can be isolated by extraction. In modifications of this reaction,[46] DCC can be replaced by other DMSO-activating agents including acetic anhydride (see Dimethyl Sulfoxide–Acetic Anhydride),[50] trifluoroacetic anhydride (see Dimethyl Sulfoxide–Trifluoroacetic Anhydride),[51] and oxalyl chloride (see Dimethyl Sulfoxide–Oxalyl Chloride).[52,53]

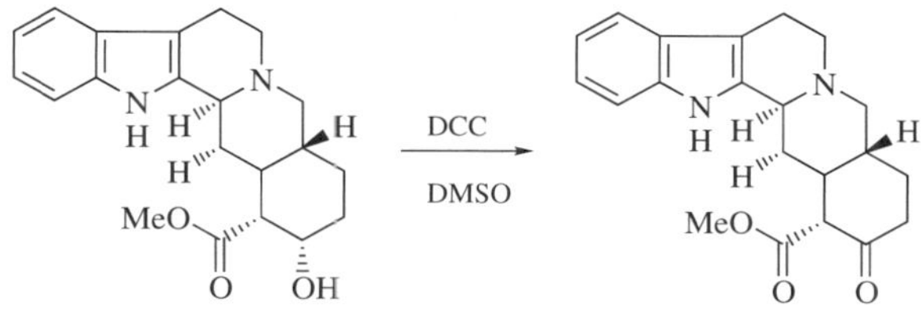

(3)

Dehydrative-Type Couplings. Because of the power of DCC as a dehydrating agent, it has found many uses in a variety of other dehydrative coupling reactions. These include the reaction of primary amines with DCC and Carbon Dioxide or Carbon Disulfide to form ureas (eq 4)[54] or isothiocyanates (eq 5),[55] respectively.

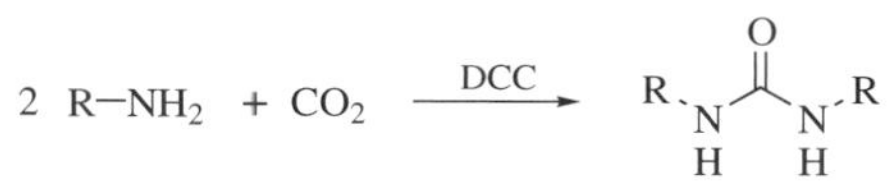

(4)

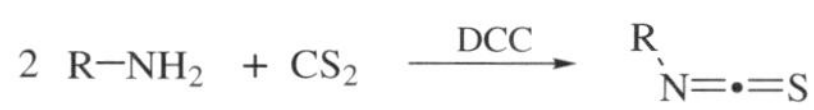

(5)

Symmetrical peroxides can be synthesized from benzoic acid derivatives and hydrogen peroxide (eq 6).[56] Unsymmetrical peroxides can be synthesized from the corresponding carboxylic acid and peroxy acid.[56]

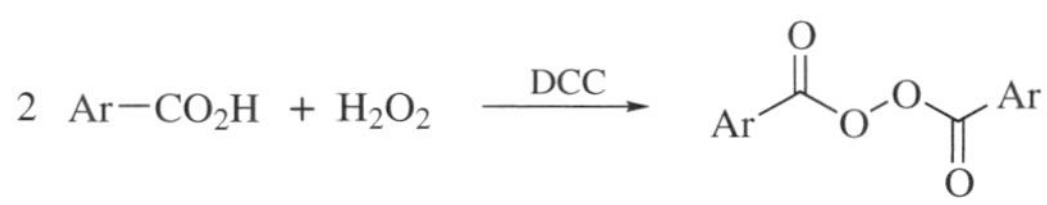

(6)

Aromatic and aliphatic alcohols and thiophenols can be coupled to form ethers (eq 7)[57,58] and thioethers (eq 8),[59,60] respectively, using DCC.

(7)

(8)

In cases where sensitivity of the compound precludes the formation of an acid chloride, α-diazo ketones can be synthesized by the DCC-mediated coupling of diazomethane to carboxylic acids (eq 9).[61,62]

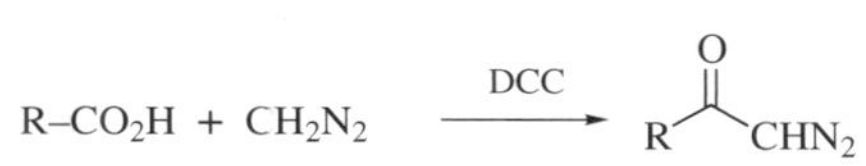

(9)

Dehydration to Alkenes, Epoxides, Nitriles, and Ketenes. β-Hydroxy ketones and β-hydroxy esters can be dehydrated, using DCC, to α,β-unsaturated ketones (eq 10)[7,63] and esters (eq 11),[64] respectively. Cyclopropanes can be synthesized by the 'dehydration' of γ-hydroxy ketones (eq 12).[65]

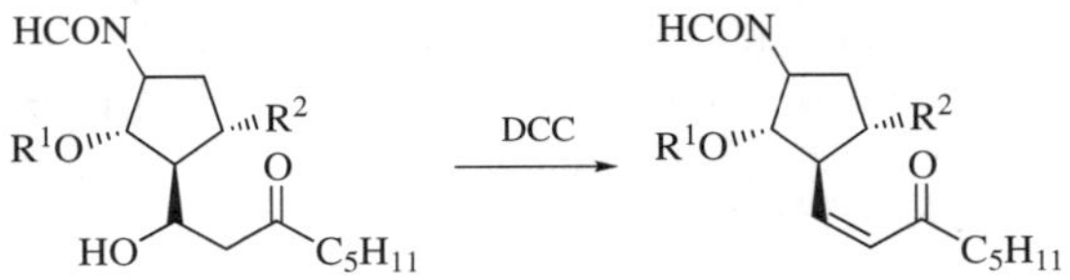

(10)

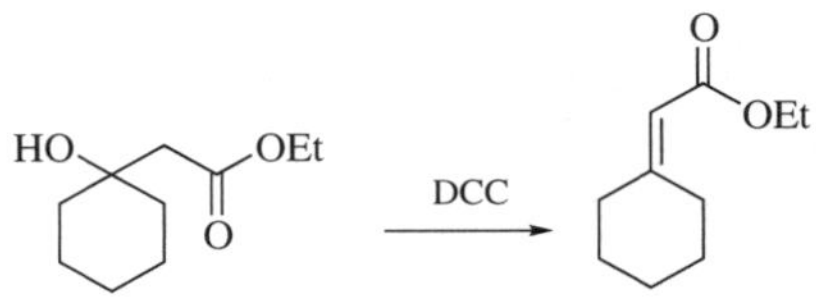

(11)

(12)

Other dehydration reactions include the conversion of primary amides to nitriles,[6] aldehydes to nitriles (via the hydroxylamine) (eq 13),[66,67] and carboxylic acids to ketenes (eq 14).[68]

$$\text{R–CHO} + \text{HONH}_2\text{•HCl} \xrightarrow[\text{CuSO}_4]{\text{DCC}} \text{R–≡N} \qquad (13)$$

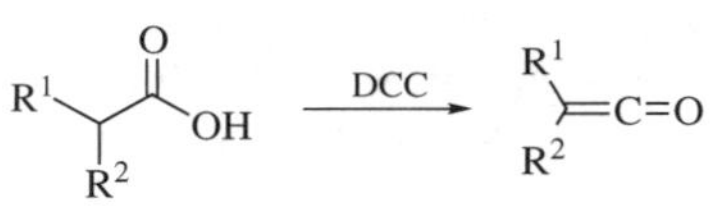

(14)

Dehydroxylation of Alcohols. Alcohols and phenols can be dehydroxylated to the corresponding alkane by hydrogenation of the *O*-acylurea adduct formed from the reaction of the alcohol with DCC (eq 15).[69,70]

$$\text{R–OH} \xrightarrow{\text{DCC}} \left[\begin{matrix} O\text{-acyl} \\ \text{urea} \end{matrix}\right] \xrightarrow[\text{Pd/C}]{\text{H}_2} \text{R–H} \qquad (15)$$

Inversion of Secondary Alcohols. Secondary alcohols can be stereochemically inverted by formylation (or esterification) with DCC, followed by saponification (eq 16).[8]

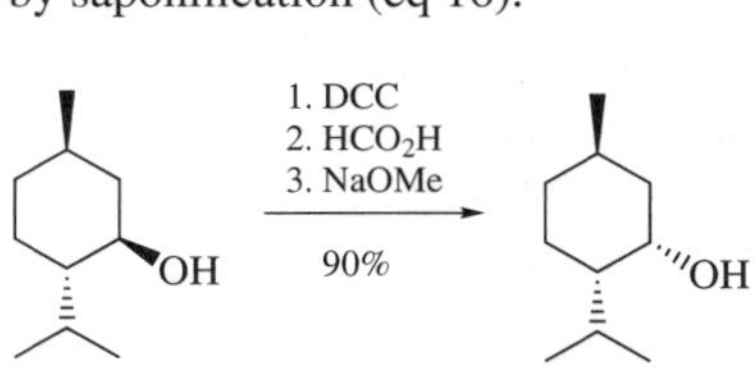

(16)

>99% de

Heterocyclization Reactions. DCC has frequently been used both as a reagent and as a reactant in the synthesis of heterocycles.[1] For example, DCC-mediated cyclodesulfurative annulation reactions[71] have been used to synthesize guanosine-type nucleotide analogs (eq 17).[72]

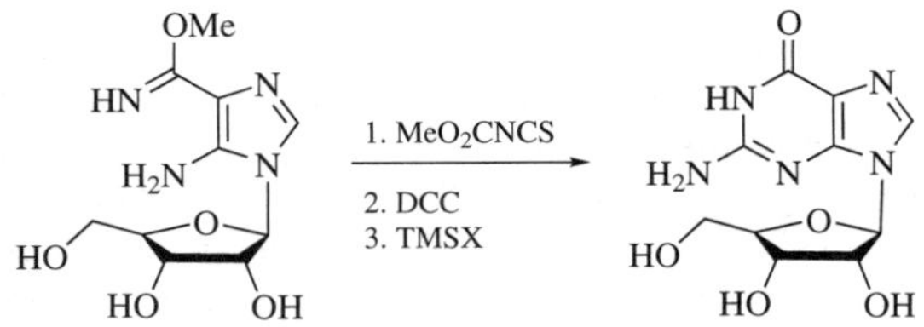

(17)

1. (a) Mikolajczyk, M.; Kielbasinski, P. *Tetrahedron* **1981**, *37*, 233. (b) Williams, A.; Ibrahim, I. T. *Chem. Rev.* **1981**, *81*, 589. (c) Kurzer, F.; Douraghi-Zadeh, K. *Chem. Rev.* **1967**, *67*, 107.
2. Bodanszky, M. *Peptide Chemistry: A Practical Textbook*; Springer:New York, **1988**.
3. Neises, B.; Steglich, W. *Org. Synth.* **1985**, *63*, 183.
4. Chen, F. M. F.; Kuroda, K.; Benoiton, N. L. *Synthesis* **1978**, 928.
5. Moffatt, J. G. *J. Org. Chem.* **1971**, *36*, 1909.
6. Ressler, C.; Ratzkin, H. *J. Org. Chem.* **1961**, *26*, 3356.
7. Corey, E. J.; Andersen, N. H.; Carlson, R., M.; Paust, J.; Vedejs, E.; Vlattas, I.; Winter, R. E. K. *J. Am. Chem. Soc.* **1968**, *90*, 3245.
8. Kaulen, J. *Angew. Chem., Int. Ed. Engl.* **1987**, *26*, 773.
9. Sheehan, J. C.; Hess, G. P. *J. Am. Chem. Soc.* **1955**, *77*, 1067.
10. Khorana, H. G. *Chem. Ind. (London)* **1955**, 1087.
11. Hudson, D. *J. Org. Chem.* **1988**, *53*, 617.
12. Wang, S. S.; Tam, J. P.; Wang, B. S. H.; Merrifield, R. B. *Int. J. Pept. Protein Res.* **1981**, *19*, 459.
13. Balcom, B. J.; Petersen, N. O. *J. Org. Chem.* **1989**, *54*, 1922.
14. DeTar, D. F.; Silverstein, R. *J. Am. Chem. Soc.* **1966**, *88*, 1013.
15. DeTar, D. F.; Silverstein, R. *J. Am. Chem. Soc.* **1966**, *88*, 1020.
16. Mironova, D. F.; Dvorko, G. F.; Skuratovskaya, T. N. *Ukr. Khim. Zh. (Russ. Ed.)* **1969**, *35*, 726.
17. König, W.; Geiger, R. *Ber. Dtsch. Chem. Ges./Chem. Ber.* **1970**, *103*, 788.
18. Carpino, L. A. *J. Am. Chem. Soc.* **1993**, *115*, 4397.
19. Wünsch, E.; Drees, F. *Ber. Dtsch. Chem. Ges./Chem. Ber.* **1966**, *99*, 110.
20. König, W.; Geiger, R. *Ber. Dtsch. Chem. Ges./Chem. Ber.* **1970**, *103*, 2034.
21. Sheehan, J. C.; Hlavka, J. J. *J. Org. Chem.* **1956**, *21*, 439.
22. Sheehan, J. C.; Cruickshank, P. A.; Boshart, G. L. *J. Org. Chem.* **1961**, *26*, 2525.
23. Castro, B.; Dormoy, J. R.; Evin, G.; Selve, C. *Tetrahedron Lett.* **1975**, 1219.
24. Knorr, R.; Trzeciak, A.; Bannwarth, W.; Gillessen, D. *Tetrahedron Lett.* **1989**, *30*, 1927.
25. Cabré, J.; Palomo, A. L. *Synthesis* **1984**, 413.
26. Frérot, E.; Coste, J.; Pantaloni, A.; Dufour, M.; Jouin, P. *Tetrahedron Lett.* **1991**, *47*, 259.
27. Tanner, D.; Somfai, P. *Tetrahedron* **1988**, *44*, 613.
28. Grunwell, J. R.; Foerst, D. L. *Synth. Commun.* **1976**, *6*, 453.
29. Smith, R. J.; Capaldi, R. A.; Muchmore, D.; Dahlquist, F. *Biochemistry* **1978**, *17*, 3719.
30. Neises, B.; Steglich, W. *Angew. Chem., Int. Ed. Engl.* **1978**, *17*, 522.
31. Hassner, A.; Alexanian, V. *Tetrahedron Lett.* **1978**, 4475.
32. Holmberg, K.; Hansen, B. *Acta Chem. Scand.* **1979**, *33*, 410.
33. Dhaon, M. K.; Olsen, R. K.; Ramasamy, K. *J. Org. Chem.* **1982**, *47*, 1962.

34. Ohta, S.; Shimabayashi, A.; Aono, M.; Okamoto, M. *Synthesis* **1982**, 833.
35. Johnson, W. S.; Bauer, V. J.; Margrave, J. L.; Frisch, M. A.; Dreger, L. H.; Hubbard, W. N. *J. Am. Chem. Soc.* **1961**, *83*, 606.
36. Woodward, R. B.; Bader, F. E.; Bickel, H.; Frey, A. J.; Kierstead, R. W. *Tetrahedron* **1958**, *2*, 1.
37. Rammler, D. H.; Khorana, H. G. *J. Am. Chem. Soc.* **1963**, *85*, 1997.
38. Chen, F. M. F.; Benoiton, N. L. *Synthesis* **1979**, 709.
39. Samuel, D.; Silver, B. L. *J. Am. Chem. Soc.* **1963**, *85*, 1197.
40. Khorana, H. G. *Can. J. Chem.* **1953**, *31*, 585.
41. Mijaji, Y.; Minoto, H.; Kobayashi, M. *Bull. Chem. Soc. Jpn.* **1971**, *44*, 862.
42. Kampe, W. *Ber. Dtsch. Chem. Ges./Chem. Ber.* **1965**, *98*, 1031.
43. Pal, B. C.; Schmidt, D. G.; Farrelly, J. G. *Nucleic Acid Chem.* **1978**, *2*, 963.
44. Khorana, H. G.; Todd, A. R. *J. Chem. Soc* **1953**, 2257.
45. Furukawa, M.; Okawara, T. *Synthesis* **1976**, 339.
46. Tidwell, T. T. *Org. React.* **1990**, *39*, Chapter 3.
47. Pfitzner, K. E.; Moffatt, J. G. *J. Am. Chem. Soc.* **1965**, *87*, 5661.
48. Moffatt, J. G. In *Oxidation*; Augustine, R. L. andTrecker, D. J., Ed.; Dekker:New York, **1971**; Vol. *2*, Chapter 1.
49. Albright, J. D.; Goldman, L. *J. Org. Chem.* **1965**, *30*, 1107.
50. Albright, J. D.; Goldman, L. *J. Am. Chem. Soc.* **1967**, *89*, 2416.
51. Huang, S. L.; Omura, K.; Swern, D. *Synthesis* **1978**, 297.
52. Omura, K.; Swern, D. *Tetrahedron* **1978**, *34*, 1651.
53. Marx, M.; Tidwell, T. T. *J. Org. Chem.* **1984**, *49*, 788.
54. Ogura, H.; Takeda, K.; Tokue, R.; Kobayashi, T. *Synthesis* **1978**, 394.
55. Jochims, J. C.; Seeliger, A. *Angew. Chem., Int. Ed. Engl.* **1967**, *6*, 174.
56. Greene, F. D.; Kazan, J. *J. Org. Chem.* **1963**, *28*, 2168.
57. Vowinkel, E. *Ber. Dtsch. Chem. Ges./Chem. Ber.* **1966**, *99*, 42.
58. Vowinkel, E. *Ber. Dtsch. Chem. Ges./Chem. Ber.* **1966**, *99*, 1499.
59. Vowinkel, E.; Wolff, C. *Ber. Dtsch. Chem. Ges./Chem. Ber.* **1974**, *107*, 496.
60. Vowinkel, E. *Synthesis* **1974**, 430.
61. Penke, B.; Czombos, J.; Baláspiri, L.; Petres, J.; Kovács, K. *Helv. Chim. Acta* **1970**, *53*, 1057.
62. Hodson, D.; Holt, G.; Wall, D. K. *J. Chem. Soc. (C)* **1970**, 971.
63. Alexandre, C.; Rouessac, F. *Bull. Soc. Claim. Fr., Part 2* **1971**, 1837.
64. Alexandre, C.; Rouessac, F. *C.R. Hebd. Seances Acad. Sci., Ser. C* **1972**, *274*, 1585.
65. Alexandre, C.; Rouessac, F. *Tetrahedron Lett.* **1970**, 1011.
66. Vowinkel, E.; Bartel, J. *Ber. Dtsch. Chem. Ges./Chem. Ber.* **1974**, *107*, 1221.
67. Vowinkel, E. *Angew. Chem., Int. Ed. Engl.* **1974**, *13*, 351.
68. Olah, G. A.; Wu, A.; Farooq, O. *Synthesis* **1989**, 568.
69. Vowinkel, E.; Wolff, C. *Ber. Dtsch. Chem. Ges./Chem. Ber.* **1974**, *107*, 907.
70. Vowinkel, E.; Büthe, I. *Ber. Dtsch. Chem. Ges./Chem. Ber.* **1974**, *107*, 1353.
71. Mohsen, A.; Omar, M. E. M. E. O.; Habib, N. S.; Aboulwafa, O. M. *Synthesis* **1977**, 864.
72. Groziak, M. P.; Chern, J.; Townsend, L. B. *J. Org. Chem.* **1986**, *51*, 1065.

Jeffrey S. Albert & Andrew D. Hamilton
University of Pittsburgh, PA, USA

1,3-Dicyclohexylcarbodiimide-4-dimethylaminopyridine[1]

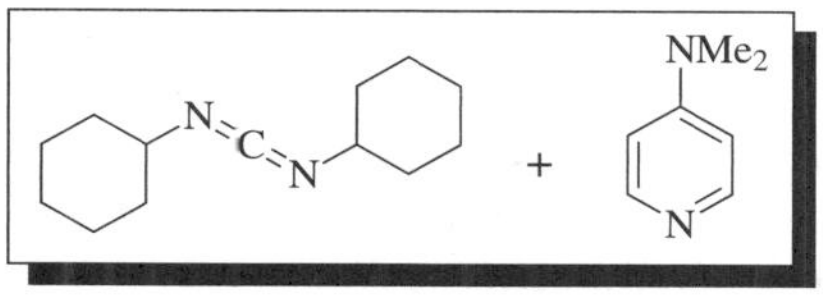

(DCC)		
[538-75-0]	$C_{13}H_{22}N_2$	(MW 206.33)
(DMAP)		
[1122-58-3]	$C_7H_{10}N_2$	(MW 122.17)

(powerful dehydrating mixture used for the preparation of esters,[2] lactones,[3,4] and amides from carboxylic acids and alcohols or amines)

Alternate Name: DCC/DMAP.
Physical Data: DCC: mp 34–35 °C; bp 122–124 °C. DMAP: mp 108–110 °C.
Solubility: DCC: highly sol dichloromethane, THF, acetonitrile, DMF. DMAP: sol dichloromethane, THF, acetonitrile, DMF, ethyl acetate.
Form Supplied in: DCC: opalescent solid; widely available. DMAP: yellow to tan crystals; widely available.
Handling, Storage, and Precautions: DCC is an acute skin irritant in susceptible individuals. Because of its low melting point, it is conveniently handled as a liquid by gentle warming of the reagent container. It should be handled with gloves in a fume hood, and stored under anhydrous conditions. DMAP causes skin and eye burns, and is readily absorbed through the skin. It should be handled with gloves in a fume hood and stored under dry conditions.

Ester Synthesis. The combination of the powerful dehydrating agent 1,3-Dicyclohexylcarbodiimide (DCC) and the nucleophilic acylation catalyst 4-Dimethylaminopyridine (DMAP)[1] is widely used in the preparation of esters from carboxylic acids and alcohols. The coupling of carboxylic acids with phenols, thiophenols, and thiols can generally be carried out in the absence of any acylation catalyst,[5] but reactions with alcohols do require a catalyst. DMAP is most commonly used, although other related derivatives such as 4-pyrrolidinopyridine (PPY) have in some cases been more effective.[6] These reactions are quite general, although yields typically decrease as steric bulk increases. *t*-Butyl esters can frequently be synthesized in yields greater than 80%.[6–8] To facilitate purification of the product, water-soluble carbodiimides such as 1-Ethyl-3-(3′-dimethylaminopropyl)carbodiimide Hydrochloride (EDCI)[9] can be used in place of DCC.

Under typical conditions, DCC (1.1 equiv) is added to a concentrated solution (0.1–1.0 M) of the carboxylic acid (1.0 equiv), alcohol (1.0–3.0 equiv), and DMAP (0.01–0.10 equiv) in dichloromethane or dichloro methane/DMF at 0 °C. After 1 h the solution is allowed to warm to rt and stirring is continued for an additional 1–5 h. After removal of the precipitated dicyclohexylurea (DCU) by filtration, the product can be recovered by extraction.

Because DMAP and its related derivatives (in conjunction with a coupling agent) are such powerful acylating agents, racemization has frequently been observed in reactions with chiral carboxylic acids, particularly with amino acids. For example, in the synthesis of the *t*-butyl ester of Boc–Asp(OBz)–CO_2H, greater than 60% racemization was observed[2] using EDCI with DMAP (0.5 equiv). The degree of racemization is highly dependent upon the components, and generally increases with steric bulk. Racemization can be suppressed by using low temperatures, short reaction times, and a minimum of DMAP.

The role of DMAP in these reactions is two-fold. First, the pyridine nitrogen can undergo nucleophilic attack upon the *O*-acyl active ester, forming the *N*-acylpyridinium. This charge-separated adduct facilitates electrophilic attack of the nucleophile and increases the leaving group ability. The second role of DMAP is its function as a base (pK_a9.70). As such, it can assist in the deprotonation of the nucleophile. If DMAP is used in stoichiometric amounts, it can be used to neutralize acids formed if the activating agent is an anhydride rather than a carbodiimide.

Lactones. DCC/DMAP has been frequently used in the synthesis of lactones of various ring sizes. The reaction conditions and considerations are essentially the same as for ester formation. In several cases, however, the catalyst was effective only when used as the hydrochloride or trifluoroacetate salt.[4,10,11]

Amides. Although amide bond formation reactions can often be carried out without any acyl-transfer catalyst, the use of DMAP can dramatically increase the reaction rate. For example, the rate of reaction between *m*-chloroaniline and benzoyl chloride was increased by a factor of nearly 6×10^3 when pyridine was replaced by DMAP as the catalyst.[12] However, DMAP is not frequently used for this purpose[13] because of the potential for racemization with chiral carboxylic acids. In these cases, less powerful acyl-transfer catalysts are used.

1. Höfle, G.; Steglich, W.; Vorbrüggen, H. *Angew. Chem., Int. Ed. Engl.* **1978**, *17*, 569.
2. Dhaon, M. K.; Olsen, R. K.; Ramasamy, K. *J. Org. Chem.* **1982**, *47*, 1962.
3. Johnson, W. S.; Bauer, V. J.; Margrave, J. L.; Frisch, M. A.; Dreger, L. H.; Hubbard, W. N. *J. Am. Chem. Soc.* **1961**, *83*, 606.
4. Stork, G.; Rychnovsky, S. D. *J. Am. Chem. Soc.* **1987**, *109*, 1565.
5. Grunwell, J. R.; Foerst, D. L. *Synth. Commun.* **1976**, *6*, 453.
6. Hassner, A.; Alexanian, V. *Tetrahedron Lett.* **1978**, 4475.
7. Neises, B.; Steglich, W. *Org. Synth.* **1985**, *63*, 183.
8. Neises, B.; Steglich, W. *Angew. Chem., Int. Ed. Engl.* **1978**, *17*, 522.
9. Sheehan, J. C.; Cruickshank, P. A.; Boshart, G. L. *J. Org. Chem.* **1961**, *26*, 2525.
10. Keck, G. E.; Boden, E. P.; Wiley, M. R. *J. Org. Chem.* **1989**, *54*, 896.
11. Bowden, E. P.; Keck, G. E. *J. Org. Chem.* **1985**, *50*, 2394.
12. Litvinenko, L. M.; Kirichenko, A. I. *Dokl. Akad. Nauk SSSR* **1967**, *176*, 97 (*Chem. Abstr.* **1968**, *68*, 68 325u) .
13. Wang, S. S.; Tam, J. P.; Wang, B. S. H.; Merrifield, R. B. *Int. J. Pept. Protein Res.* **1981**, *19*, 459.

Jeffrey S. Albert & Andrew D. Hamilton
University of Pittsburgh, PA, USA

3-(Diethoxyphosphoryloxy)-1,2,3-benzotriazin-4(3*H*)-one (DEPBT)

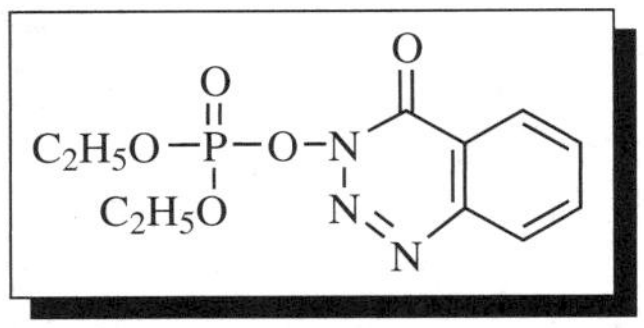

[165534-43-0] $C_{11}H_{14}N_3O_5P$ (MW 299.22)

(coupling reagent for peptide synthesis[1])

Physical Data: mp 72–74 °C.
Solubility: soluble in ethyl acetate, dichloromethane, *N,N*-dimethylformamide, and most organic solvents.
Form Supplied in: white solid.
Analysis of Reagent Purity: IR (KBr), 1712 cm^{-1} (C=O), 1302 cm^{-1} (P=O). ^{1}H-NMR ($CDCl_3$) δ 1.48 (t, 6H), 4.53 (m, 4H), 7.84–8.40(m, 4H). ^{31}P-NMR ($CDCl_3$) δ 32 (89% H_3PO_4).
Purification: recrystallization from ethyl acetate and petroleum ether.
Handling, Storage, and Precautions: DEPBT is very stable. It has a shelf life of several months at room temperature.

Peptide Coupling Reagent. The crystalline phosphate reagent mediates peptide bond formation with a remarkable resistance to racemization. It has been used as a coupling reagent both in solution and in solid-phase peptide synthesis. In solution, equimolar amounts of *N*-protected amino acid, α-carboxyl-protected amino acid, DEPBT, and 2 equiv base (diisopropylethylamine or triethylamine) in *N,N*-dimethylformamide, tetrahydrofuran or acetonitrile react at room temperature for 2–6 h to yield desired peptide in good yields.

A number of peptide coupling reactions were carried out using DEPBT.[1,2] It is worthwhile noting that it is not necessary to protect the hydroxy group of the amino component (such as serine, threonine, and tyrosine). The side reaction of asparagine dehydration, easily caused by DCC, was not observed with DEPBT-mediated peptide coupling. It has also been demonstrated that DEPBT can be used as the coupling reagent in solid-phase peptide synthesis.[3]

Comparative studies were carried out to assess racemization with DEPBT, the commonly used phosphonium reagents (PyBrop and BOP) and the uronium reagents (HATU and HBTU).[4] In the study, an activated carboxyl intermediate was allowed to stand in the presence of tertiary amine for a short time (4–60 min). The intermediate was then allowed to react with benzylamine, and the resulting ratio of enantiomers was determined by chiral HPLC. To maximize sensitivity, experiments were carried out using Boc-Ser(Bzl)-OH as the carboxyl component, since it has the highest susceptibility toward racemization. In Table 1, the degree of racemization for DEPBT and other commonly used coupling reagents are compared. The results clearly indicate that DEPBT is superior to other typical phosphonium and uronium coupling reagents in terms of inhibition of racemization.

Table 1 Comparative studies of racemization during in situ activation

Boc-Ser(Bzl)-OH —(1.0 equiv or 2.0 equiv activator; 2.0 equiv DIEA, CH_2Cl_2, 20 °C)→ —Delay→ —Benzylamine→ measure L:D ratio by chiral HPLC

Activators	Delay time (min)	L:D ratio	Yield (%)
PyBrop (1.0 equiv)	4	65:35	81
HATU (1.0 equiv)	4	84:16	91
HBTU (1.0 equiv)	15	79:21	>99
BOP (1.0 equiv)	15	85:15	95
DEPBT (1.0 equiv)	60[a]	95:5	70
DEPBT (2.0 equiv)	60[a]	96:4	>99

[a]Even with a long delay time very little racemization was observed.

Peptide Fragment Condensation. The high efficiency of DEPBT and its utility in the synthesis of complex natural products have been demonstrated.[5,6] In the total synthesis of (−)-tamandarin B, DEPBT was employed to couple the side chain to the macrocycle.[5] Various coupling reagents were used to investigate the effectiveness for this transformation. DEPBT stood out for its resistance to racemization of sensitive substrates and its high yields in amide bond formation. Another excellent example of utilizing DEPBT for amide bond formation can be seen in Boger's synthesis of the glycopeptide antibiotic teicoplanin aglycon.[6] The coupling reagent DEPBT was used in a number of the phenylglycine-related amide bond formations. As with the analogous substrates in the vancomycin series, these coupling reactions occasionally led to low conversions or significant racemization, depending on the reagents or reaction conditions employed. However, the use of DEPBT (0 °C, THF), uniformly gave the best results, providing good to superb conversions, and most importantly, no detectable racemization. One of the greatest challenges in the synthesis of teicoplanin aglycon arises from the epimerization of the $C_2{}^3$ center of the teicoplanin (1% aqueous $NaHCO_3$, 80 °C) to its more stable *R*-configuration under mildly basic conditions.[7,8] In Boger's synthesis, the amide bond formation between $C_2{}^3$ and $C_2{}^4$ was achieved by using DEPBT (THF, $NaHCO_3$, 0 °C, 2 h, 68–83%) in excellent yield, and without detectable epimerization (eq 1). This is very notable since this coupling involves the use of an activated phenylglycine derivative prone to racemization. In contrast to DEPBT, typical reagents (EDC, PyBOP, HATU) and conditions, including those used for the vancomycin synthesis, provided near 1:1 mixtures of the $C_2{}^3$ diastereomers.[9]

Peptide Cyclization.[3,10,11,12] Two cyclopentapeptides, c(Pro-Tyr-Leu-Ala-Gly) and c(Ala-Tyr-Leu-Ala-Gly), isolated from a Chinese medicinal herb called *Stellaria yunnanensis Franch (M)* were used to evaluate the cyclization efficiency of DEPBT, BOP, EDCI, HBTU, and DPPA. The linear peptides H-Pro-Tyr-Leu-Ala-Gly-OH and H-Ala-Tyr-Leu-Ala-Gly-OH were dissolved in DMF (10^{-3}M) separately. Four equivalents of coupling reagent was added. Triethylamine was added to maintain the pH between 8 and 9. The reaction was allowed to proceed for 3–5 days. The yields of the cyclization reactions are shown in Table 2. It is readily seen that DEPBT-mediated cyclization gave the highest yields.

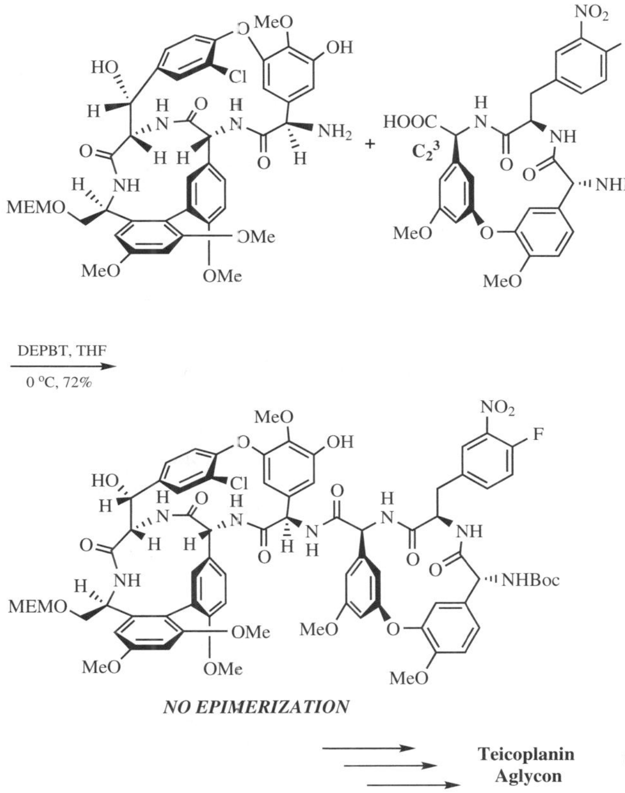

(1)

Table 2 Synthesis of *N*-protected peptido-alcohols using DEPBT

P-AAOH + NH_2-CH(R)-OH —(DEPBT, Solvent[a])→ P-AANH-CH(R)-OH

Reaction product	Carboxyl component	Amino component	Yield (%)
Z-Ala-NHCH(C_2H_5)CH_2OH	Z-Ala	DL-2-aminobutanol	58
Z-Phe-NHCH_2CH(OH)CH_3	Z-Phe	DL-1-amino-2-propanol	43
Z-Phe-NHCH(CH_3)CH_2OH	Z-Phe	D-Alaol	51
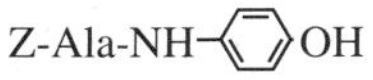	Z-Ala	*p*-aminophenol	76
Z-Ala-NHCH(CH_2OH)CH(C_6H_5)OH	Z-Ala	(1*S*,2*S*)-(+)-2-amino-1-phenyl-1,3-propanediol	80
Z-Tyr(OBzl)-Thr-ol	Z-Tyr(OBzl)	Throl	76
Fmoc-Cys(Acm)-Thr-ol	Fmoc-Cys(Acm)	Throl	68
Z-Tyr(OBzl)-Ser-ol	Z-Tyr(OBzl)	Serol	77
Fmoc-Lys(Boc)-Thr-ol	Fmoc-Lys(Boc)	Throl	69
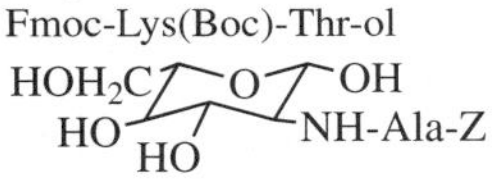	Z-Ala	Glucosamine hydrochloride	47

[a]Solvent selected depending on solubility. Solvents used were : Anhydrous CH_2Cl_2, THF, and DMF or mixtures of CH_2Cl_2, THF, DMF, and C_5H_5N and H_2O.

Table 3 The yields of peptide cyclizations of the compounds H-Pro-Tyr-Leu-Ala-Gly-OH and H-Ala-Tyr-Leu-Ala-Gly-OH using various coupling reagents

Cyclopeptide	Coupling reagent	Yield (%)
c(Pro-Tyr-Leu-Ala-Gly)	DEPBT	54
	BOP	35
	EDC	25
	HBTU	40
	DPPA	52
c(Ala-Tyr-Leu-Ala-Gly)	DEPBT	52
	TBTU	45
	DCC	15

Reagent to Synthesize Peptido-Alcohols and *N*-Glycopeptide. When DEPBT is used for peptide coupling, it is not necessary to protect the hydroxy group of the amino component. Therefore, DEPBT can be used to synthesize peptido-alcohols. The use of DEPBT has been reported for the synthesis of a series of peptido-alcohols and a glycopeptide derivative from *N*-protected amino acids and different amine components with one or more unprotected hydroxyl groups (Table 3).[13] When DEPBT is used as the coupling reagent, the results clearly show that the carboxyl component selectively reacts with the amine group of the amino alcohol substrate to form the amide bond. The hydroxyl group in the amino component did not react (i.e. no formation of the ester was observed). This is a unique property of DEPBT, since the hydroxyl group of the amino alcohol must be protected when DCC or EDC are used in the synthesis of peptido-alcohols.

Related Reagents. *O*-Benzotriazol-1-yl-*N,N,N′,N′*-tetramethyluronium hexafluorophosphate (HBTU); (1*H*-1,2,3-benzotriazol-1-yloxy) tris(pyrrolidino) phosphonium hexafluorophosphate (PyBOP); benzotriazol-1-yl-oxy tris(dimethylamino) phosphonium hexafluorophosphate (BOP); 1,3-dicyclohexylcarbodiimide (DCC); diethyl Phosphorocyanidate (DEPC); 1-ethyl-3-(3′-dimethylaminopropyl) carbodiimide hydrochloride (EDCI); and diphenylphosphoryl azide (DPPA).

1. Fan, C. X.; Hao, X. L.; Ye, Y. H. *Synth. Commun.* **1996**, *26*, 1455.
2. Ye, Y. H.; Fan, C. X.; Zhang, D. Y.; Xie, H. B.; Tian, G. L. *Chem. J. Chin. Univ.* **1997**, *18*, 1086.
3. Xie, H. B.; Tian, G. L.; Ye, Y. H. *Synth. Commun.* **2000**, *30*(23), 4233.
4. Li, H. T.; Jiang, X. H.; Ye, Y. H.; Fan, C. X.; Romoff, T.; Goodman, M. *Org. Lett.* **1999**, *1*, 91.
5. Joullie, M. M.; Portonovo, P.; Liang, B.; Richard, D. J. *Tetrahedron Lett.* **2000**, *41*, 9373.
6. Boger, D. L.; Kim, S. H.; Mori, Y.; Weng, J. H.; Rogel, O.; Castle, S. L.; McAtee, J. J. *J. Am. Chem. Soc,* **2001**, *123*, 1862.
7. Barna, J. C. J.; Williams, D. H.; Strazzolini, P.; Malabarba, A.; Leung, T. W. C. *J. Antibiot.* **1984**, *37*,1204.
8. Boger, D. L.; Weng, J. H.; Miyazaki, S.; McAtee, J. J.; Castle, S. L.; Kim, S. H.; Mori, Y.; Rogel, O.; Strittmatter, H.; Jin, Q. *J. Am. Chem.Soc.* **2000**, *122*, 10047.
9. Boger, D. L.; Miyazaki, S.; Kim, S. H.; Wu, J. H.; Castle, S. L.; Loiseleur, O.; Jin, Q. *J. Am. Chem. Soc.* **1999**, *121*, 10004.
10. Tang, Y. C.; Gao, X. M.; Tian, G. L.; Ye, Y. H. *Chem. Lett.* **2000**, *7*, 826.
11. Li, H. T.; Jiang, X. H.; Goodman M. J. *Peptide Sci.* **2001**, *7*, 82.
12. Li, H. T.; Jiang, X. H.; Howell, S. B.; Goodman M. J. *Peptide Sci.* **2000**, *6*, 26.
13. Liu, P.; Sun, B. Y.; Chen, X. H.; Tian, G. L.; Ye, Y. H. *Synth. Commun.* **2002**, *32(3)*, 473–480.

Haitao Li, Yun-hua Ye & Murray Goodman
Pfizer Global Research & Development, La Jolla Laboratories, CA, USA
Peking University, Peking, P.R. China
University of California, San Diego, CA, USA

Diethyl Chlorophosphite

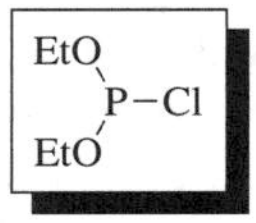

[589-57-1] $C_4H_{10}ClO_2P$ (MW 156.55)

(synthesis of allenyl phosphonates,[11] β-ketophosphonates;[8,12] phosphorylation of alcohols;[15] coupling agent[19])

Alternate Name: diethyl phosphorochloridite; diethyl chlorophosphonite; ethyl phosphorochloridite.

Physical Data: bp 153–155 °C/760 mmHg, 56–57.5 °C/30 mmHg; *d* 1.082 g cm^{-3}; fp 1 °C.

Solubility: soluble in benzene, THF, diethyl ether, methylene chloride, acetonitrile.

Form Supplied in: colorless liquid.

Preparative Methods: the reaction of Phosphorus(III) Chloride with Triethyl Phosphite,[1] anhydrous ethanol,[2] or Sodium Ethoxide[3] has been employed, although diethyl chlorophosphite is commercially available.

Purification: vacuum distillation.

Handling, Storage, and Precautions: the neat liquid is flammable and combustion produces toxic byproducts. Caution should be taken when handling the material due to its toxicity. Bottles of diethyl chlorophosphite should be sealed and stored under N_2 to prevent contact with moisture, which produces HCl vapor. Use in a fume hood.

Original Commentary

Eric J. Stoner
Abbott Laboratories, North Chicago, IL, USA

Peter P. Giannousis
Ciba-Geigy Corporation, Summit, NJ, USA

General Notes. Diethyl chlorophosphite is a reactive electrophile for the introduction of phosphorus into organic compounds. It is known to react readily with a variety of nucleophiles including amines,[4] amides,[5] alcohols (see below), and various organometallic derivatives including Grignard reagents,[6] enolates, and Reformatsky reagents.[7] Oxidation of the initial phosphorus(III) species to a phosphorus(V) species, typically from phosphite to phosphate or phosphonite to phosphonate, is accomplished with a number of reagents including air (O_2),[8] Iodine,[9] Ozone,[5] and t-Butyl Hydroperoxide.[10]

Arbuzov-Type Rearrangements Leading to Allenyl and Allyl Phosphonates. Reaction of diethyl chlorophosphite with propargyl alcohols in the presence of base results in the rapid formation of allenylphosphonates (eq 1).[11] This general method is believed to proceed through a cyclic transition state with rearrangement occurring at or below rt in many cases. Substitution greatly affects the rate of rearrangement. Allenylphosphonates react with a variety of nucleophiles, including alkoxides and secondary amines, to form β-ketophosphonates after hydrolysis of the intermediate enamine or enol ether (eq 2).[12]

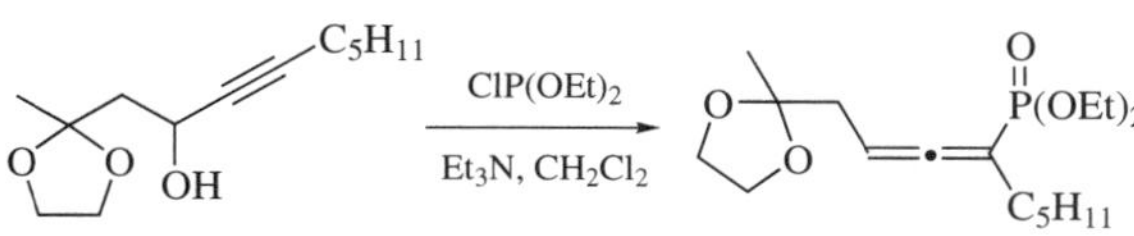

(1)

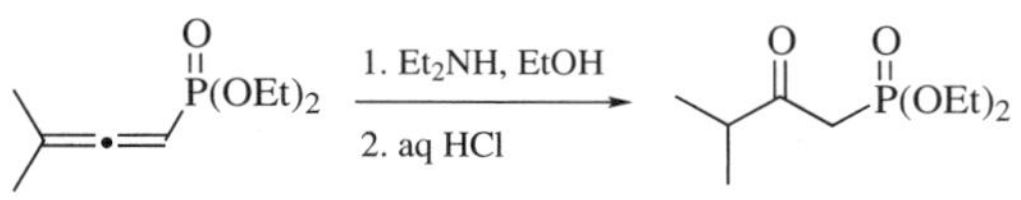

(2)

Allyl alcohols can be similarly treated to afford allylphosphonates where the selectivity in alkene geometry is dependent upon substitution (eq 3).[13] Rearrangement of the intermediate diethyl allyl phosphites takes place at 70–100 °C. A related rearrangement involves the conversion of 2,3-butadienyl phosphites (homoallenyl phosphites) to 1,3-butadienyl-2-phosphonates.

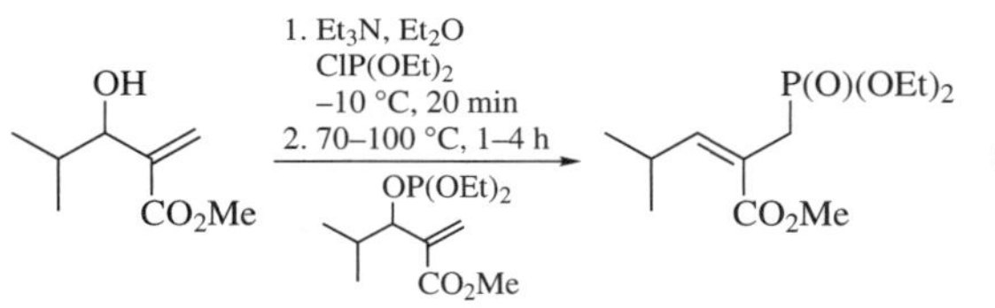

(3)

Phosphorylation of Alcohols. Protein phosphorylation is now recognized as a critical reaction in the control of cell metabolism.[14] Diethyl chlorophosphite has been shown to be a more effective and reactive phosphorus electrophile than diphenyl chlorophosphonate.[15] In the case of *N*-protected serine derivatives (eq 4), an excess of the phosphite is required due to the competitive reaction with the carboxylic acid.[9] Oxidation to the phosphate takes place under a variety of conditions already described.

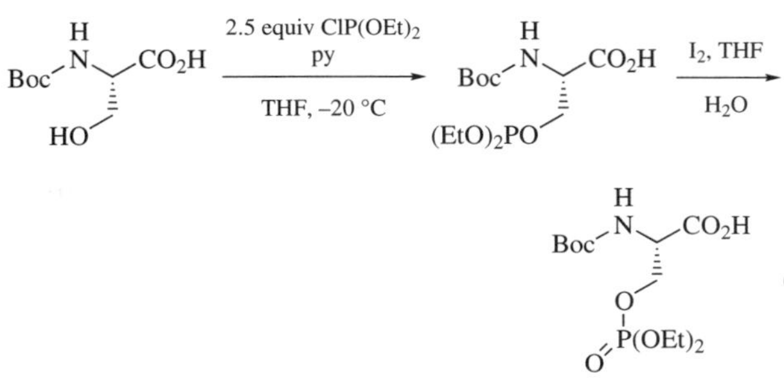

(4)

Organophosphates are also available from alcohols using diethyl chlorophosphite. In compounds containing multiple alcohol functionalities, such as derivatives of inositol (eq 5), chlorophosphites are preferred over more traditional phosphorus(V) reagents which suffer from not only lower relative reactivity, but also the tendency of the intermediate phosphate triesters to form cyclic phosphates or undergo acid-induced rearrangements.[16] Phosphate triesters can be dealkylated to the phosphates with Bromotrimethylsilane.[10]

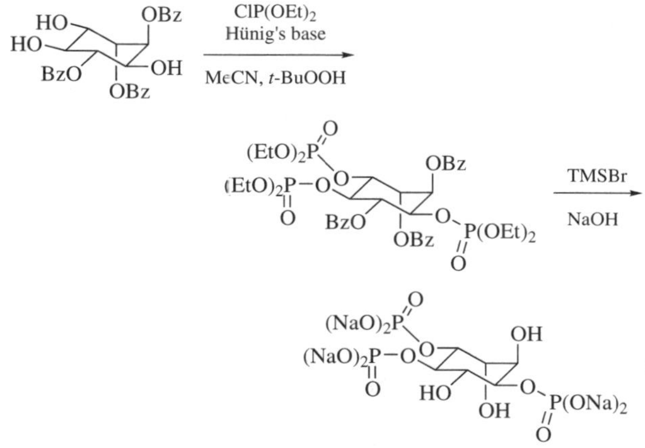

(5)

Diethyl Chlorophosphite as a Coupling Agent. Schmidt and Martin have demonstrated the use of phosphite as a leaving group in the sialylation of sugars with catalytic Trimethylsilyl Trifluoromethanesulfonate. Their procedure allows for higher yields than could be obtained using other methods for the formation of the glycosidic bond.[17] Diethyl chlorophosphite activation of amines allows for the formation of peptides.[18] In certain cases, diethyl anilinophosphite (formed by the reaction of the chlorophosphite with aniline) has also been used as the phosphite source. The researchers found that the corresponding phosphorus(V) reagents do not work as well with this coupling protocol. Reaction of 3′-(diethyl phosphonite) thymidine derivatives with a 5′-azidothymidine leads to the formation of the dinucleotide at room temperature with extrusion of N_2 (eq 6). This coupling procedure is easily accomplished and high yielding.[19]

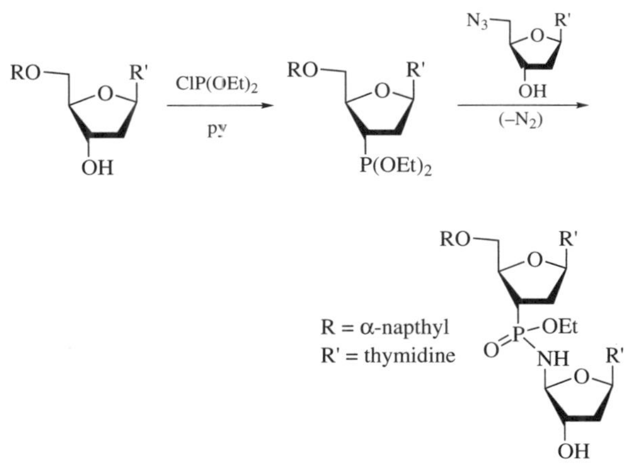

(6)

Preparation of Horner–Wadsworth–Emmons Alkenation Reagents. Treatment of an α-hydroxy ketone with diethyl chlorophosphite in the presence of a Lewis acid results in a rapid rearrangement to the corresponding β-ketophosphonate (eq 7).[20] Treatment of an α-hydroxy ester results in the formation of an H-phosphonate diester instead of the expected rearrangement product. Horner–Wadsworth–Emmons condensation reagents[21] can also be prepared by the reaction of ketone or ester enolates with diethyl chlorophosphite followed by oxidation of the resultant phosphite (eq 8).[8] This methodology offers several advantages to the Michaelis–Arbuzov reaction,[22] including application to cyclic ketones and availability of starting materials. The competitive formation of enol phosphates by this method is not a problem as it is with phosphorylating agents.[23] Reaction of diethyl chlorophosphite with phosphoryl-stabilized anions has been reported with the products leading to 1,3-butadienylphosphonites.[24]

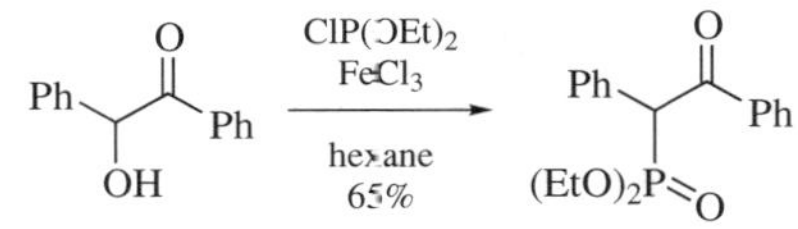

(7)

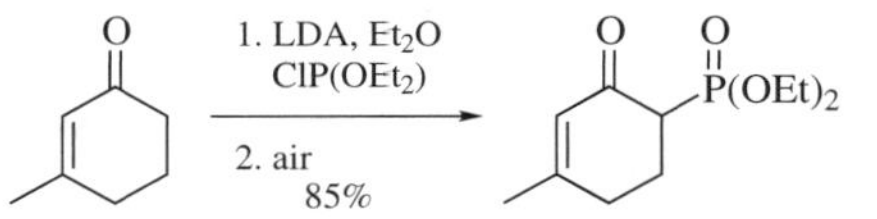

(8)

First Update

Lisheng Cai
National Institute of Mental Health, Bethesda, MD, USA

As Electrophile. Diethyl chlorophosphite can function as an electrophile for organometallic compounds to synthesize phosphates (eqs 9 and 10).[25–27] The new organophosphites can be used as ligands in organometallic chemistry.

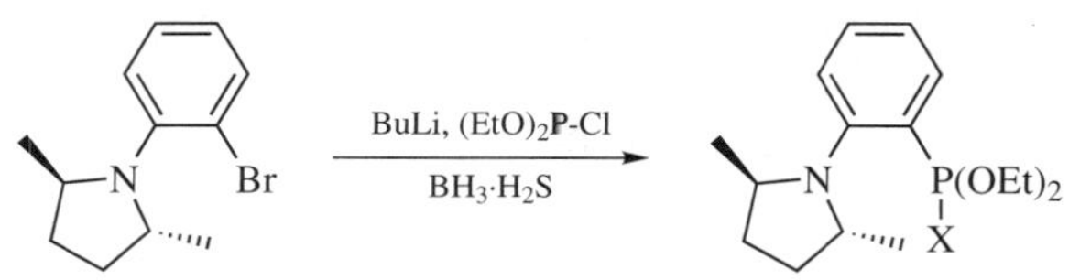

(9)

X = BH_3

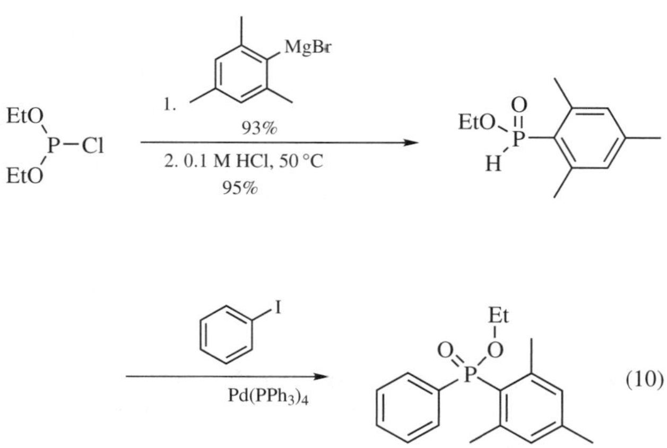

(10)

Diethyl chlorophosphite can also function as electrophile for carbanions to synthesize phosphonic esters,[28,29] which may have important biological interest or may be synthons in Wittig–Horner and other reactions (eqs 11 and 12).

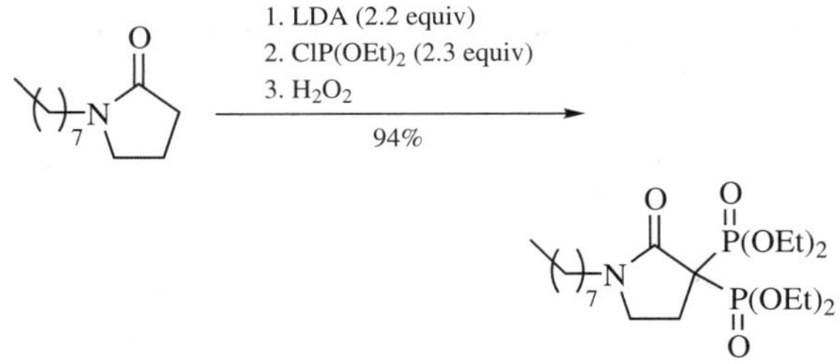

(11)

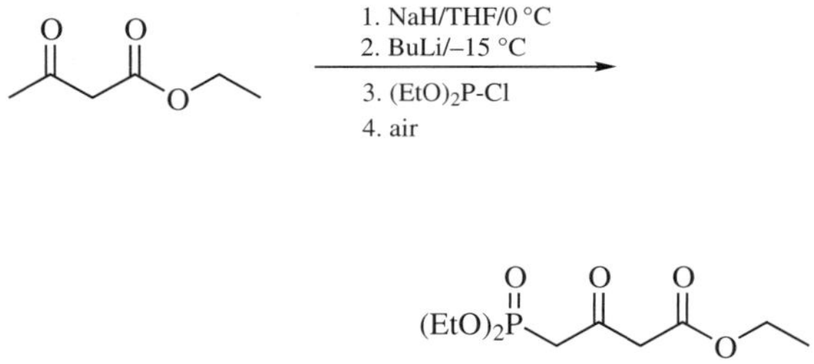

(12)

Regioselectivity of Attack by Ambident Nucleophiles

Oxygen vs Carbon Selectivity

Enolates attack diethyl chlorophosphite via O or C depending on the degree of delocalization and the counterion (eqs 13–15).[30–32]

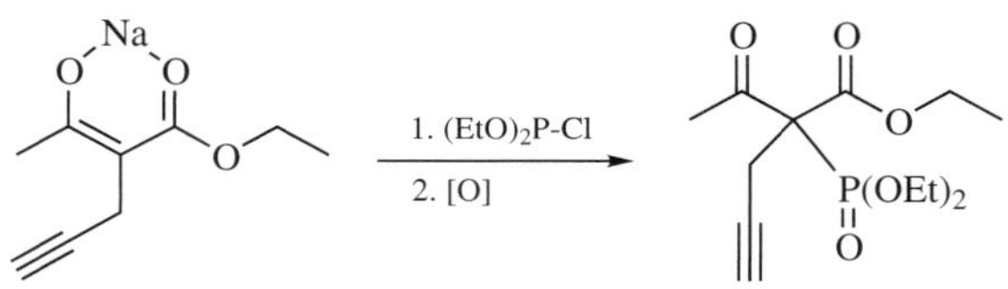

(13)

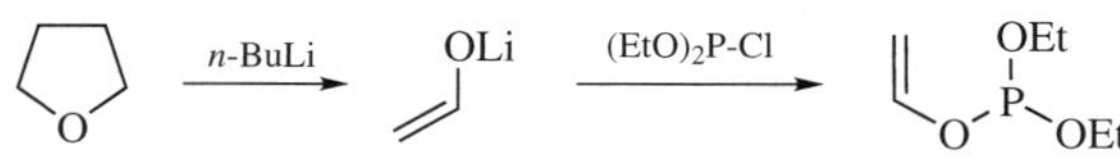

(14)

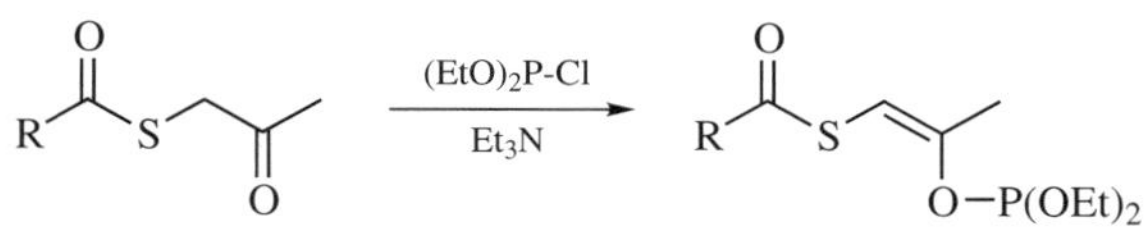

(15)

With silyl enolates attack takes place preferentially at oxygen (eq 16).[33]

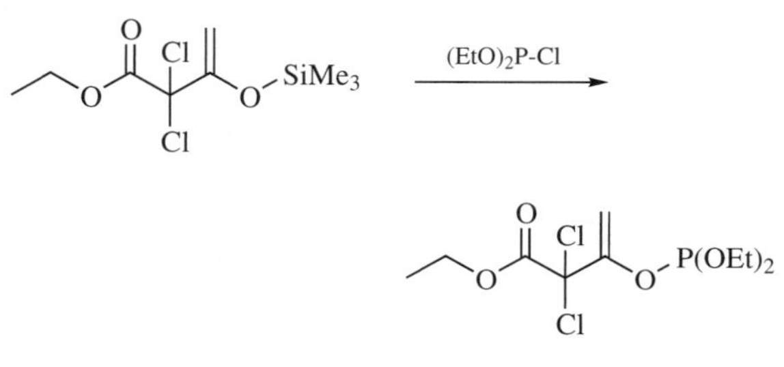

(16)

NH_2 vs OH Selectivity

Diethyl chlorophosphite always prefers OH over NH_2 (eq 17).[34] However, the corresponding phosphorus(V) species prefers the NH_2 instead (eq 18).[34]

$$HO-(CH_2)_n\text{-}NH_2 \xrightarrow[\text{[O]}]{(EtO)_2P\text{-}Cl/NEt_3} (EtO)_2P(=O)-O-(CH_2)_n\text{-}NH_2 \quad (17)$$

$$HO-(CH_2)_n\text{-}NH_2 \xrightarrow[CCl_4/NEt_3]{(EtO)_2P\text{-}OH} HO-(CH_2)_n-NH-P(=O)(OEt)_2 \quad (18)$$

Synthesis of α-phosphono Esters and Lactones. The trapping of ester and lactone enolates with diethyl chlorophosphite provides a convenient alternative to the traditional Arbuzov synthesis of α-phosphono esters and lactones (eq 19).[35]

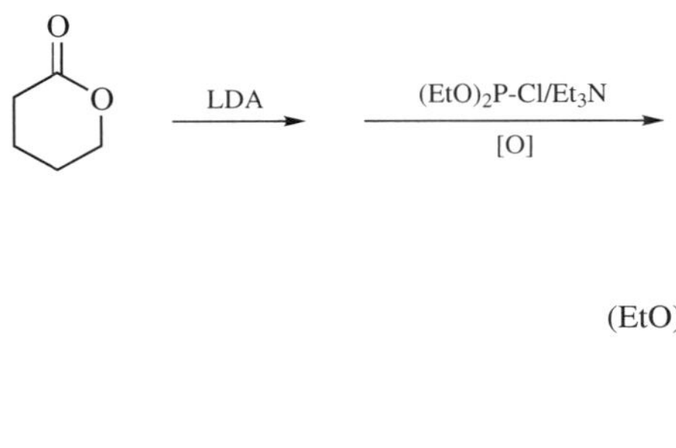
(19)

Enamines similarly provide α-phosphono ketones when exposed to diethyl chlorophosphite (eq 20).[36]

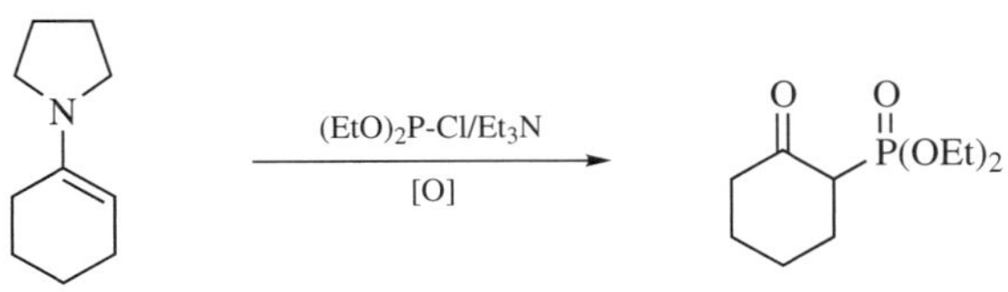
(20)

Phosphorotropic Rearrangement. Under appropriate conditions, newly formed phosphites may undergo rearrangement to form C-phosphorylated products (eq 21).[37]

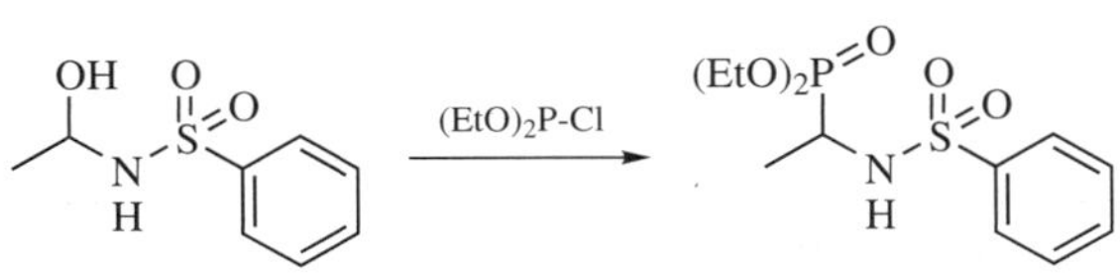
(21)

Available hemiacetals and hemithioacetals of α-halo-β-oxo aldehydes were reacted with diethyl chlorophosphite in the presence of an organic base to obtain previously unknown α-hetero-substituted phosphites, which undergo intramolecular Perkow reaction to give unsaturated thioethers and phosphorochloridic acid diethyl ester (eq 22).[38]

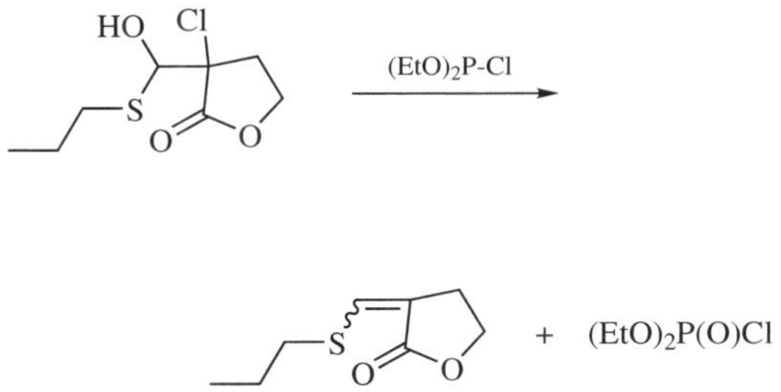
(22)

The half amidal has also been used in an analogous reaction, but the intermediate does not undergo the intramolecular elimination (eq 23).[39]

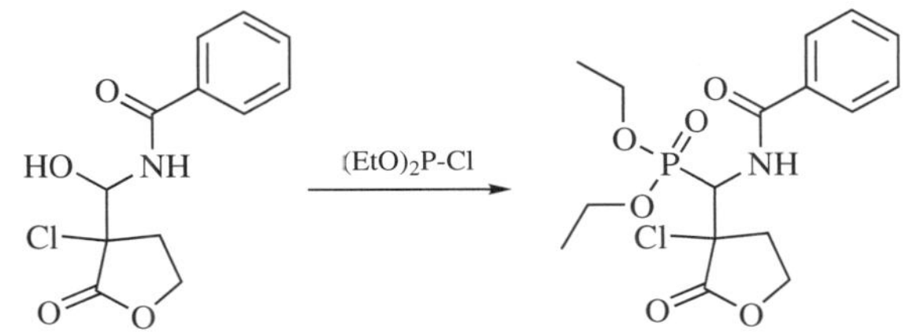
(23)

The in situ reaction of amines, aldehydes, and diethyl chlorophosphite has been used in the synthesis of 1-aminoalkylphosphonates (eq 24).[40] Presumably, this reaction follows a pathway analogous to that of eq 23.

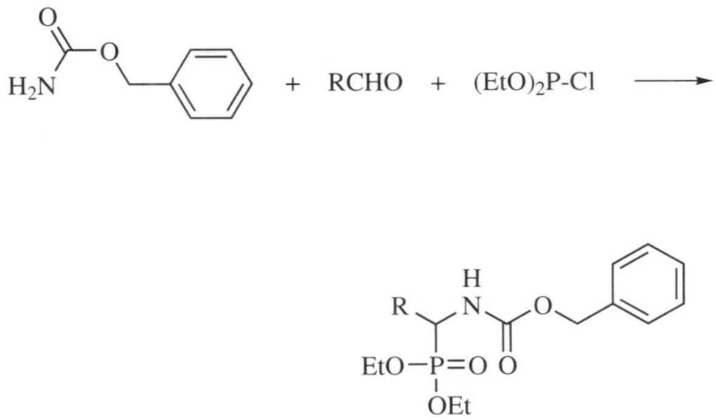
(24)

Reaction with Imines. Diethyl chlorophosphite can also serve as a nucleophile. Its reaction with imines belongs to this category. Normally, further reactions always ensue, such as cyclization with alcohols to form oxazaphosphorinanes (eqs 25 and 26).[41,42]

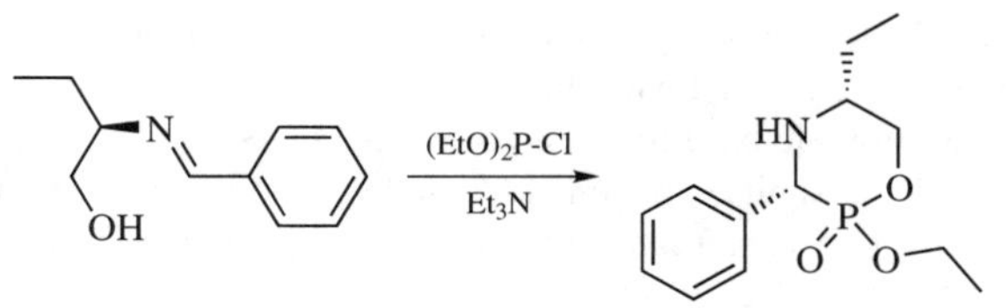

(25)

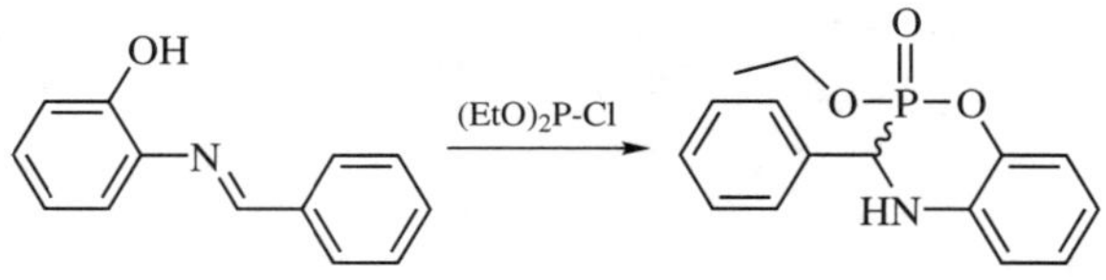

(26)

The reaction of diethyl chlorophosphite with dimethyl ketazine is a typical nucleophilic attack by diethyl chlorophosphite (eq 27).[43] However, with the cyclic analog, a 2,3-dihydropyrazine, the converse is true and diethyl chlorophosphite serves as an electrophile (eq 28).[44]

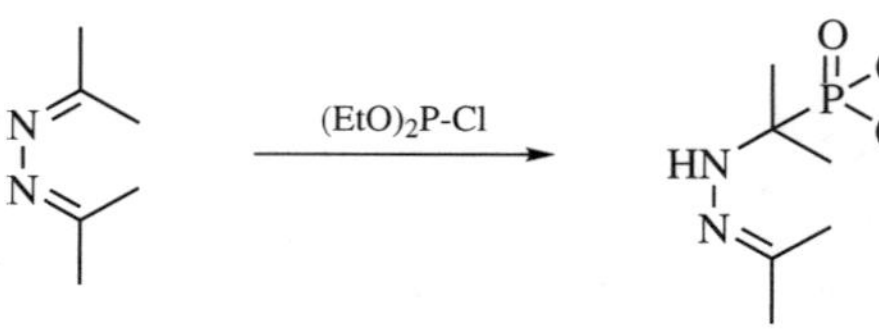

(27)

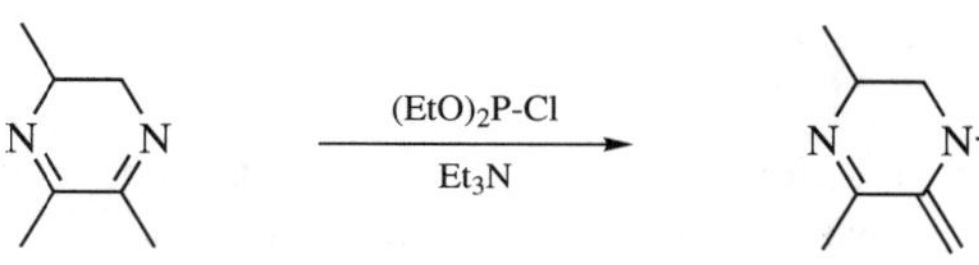

(28)

Nucleophilic Addition. Nucleophilic addition of diethyl chlorophosphite to electron-deficient double bond gives Michael addition type products.[45] In the course of these additions, the phosphorus center is oxidized to phosphorus(V). The addition may be coupled to subsequent reactions such as the elimination of chloride (eq 29).[46]

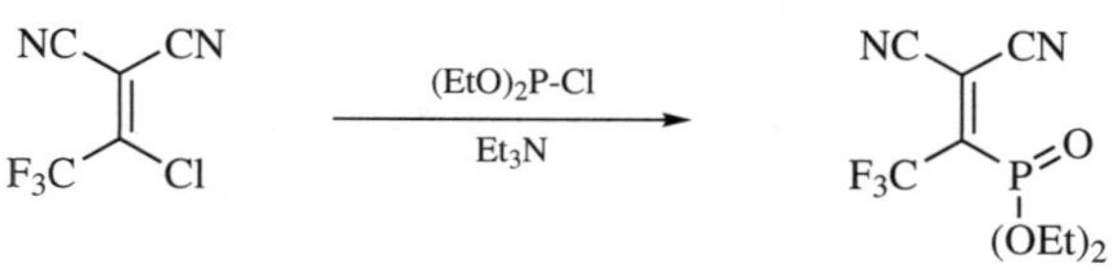

(29)

37–77%

Diethyl Chlorophosphite as a Biphilic Agent. The intrinsic oxophilicity of diethyl chlorophosphite renders it capable of acting either as an electrophile or as a nucleophile toward a variety of oxygen functional groups. This makes it a versatile reagent capable of carrying out a number of interesting reductions (eq 30) and oxidations (eq 31).[47] Further examples include the reduction of pyridine *N*-oxides to pyridines, that of sulfoxides to sulfides, and the opening of epoxides to chlorohydrins.[47]

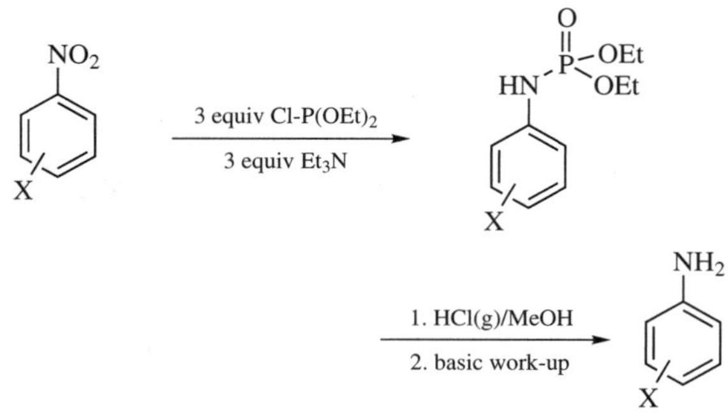

(30)

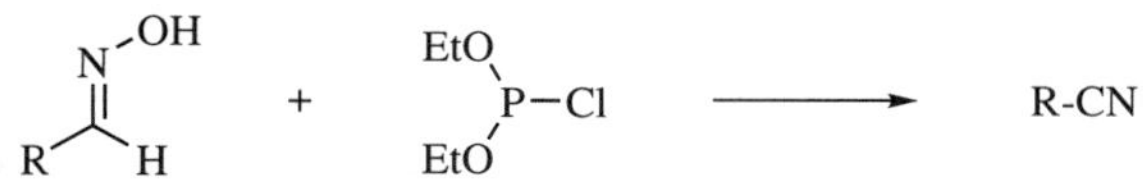

(31)

Horner–Mark [2,3]-Sigmatropic Rearrangement (Arbuzov-type Rearrangement Leading to Allenyl Phosphonates). This reaction class has been extended recently to the synthesis of complex organic compounds (eq 32).[48]

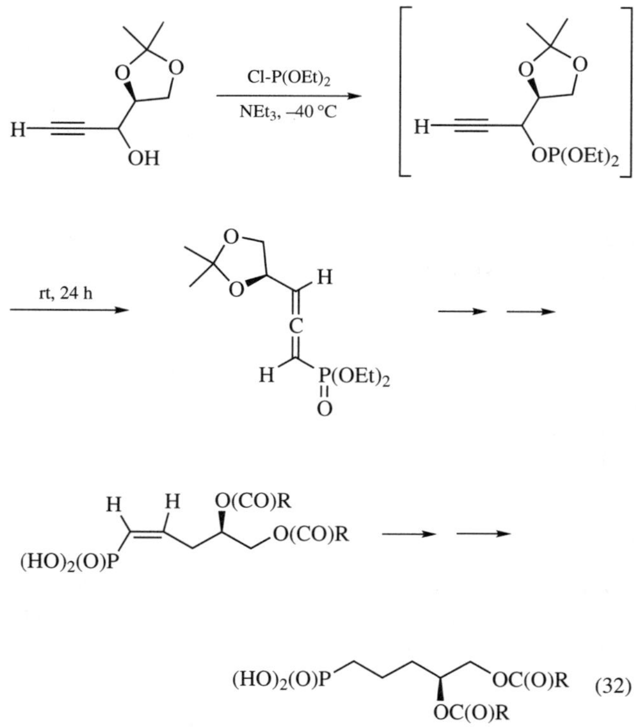

(32)

This reaction has been used to synthesize the otherwise difficult to access 2-(diethylphosphono)-2*H*-azirines (eq 33).[49]

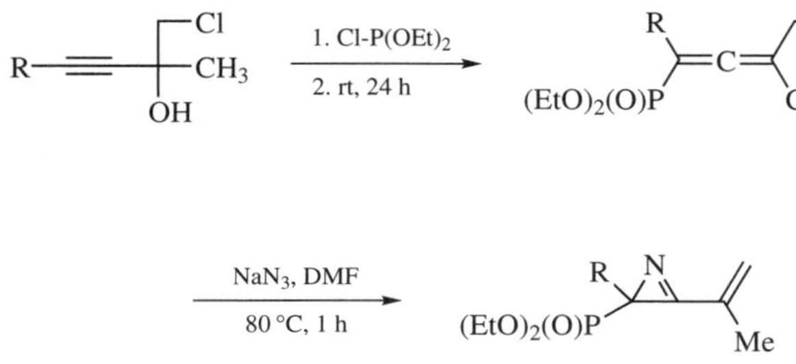

(33)

Extension of the Horner–Mark [2,3]-sigmatropic rearrangement to diynes has been reported (eq 34).[50] The reaction with diyne diols has been organized into a cascade sequence resulting in the formation of cyclobutenes (eq 35).[51] Depending on the functionality present, allenyl phosphonates, generated from propargyl alcohols by the Horner–Mark rearrangement, undergo a number of interesting intra- and intermolecular reactions leading to cyclic[52] and polycyclic (eq 36)[53] compounds, and cumulenes (eq 37).[54] It is noteworthy that the complete sequence is compatible with the presence of arenechromium tricarbonyl complexes (eq 36).[53]

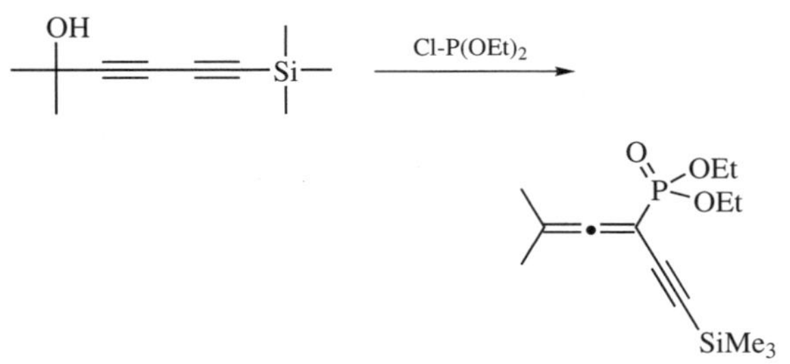

(34)

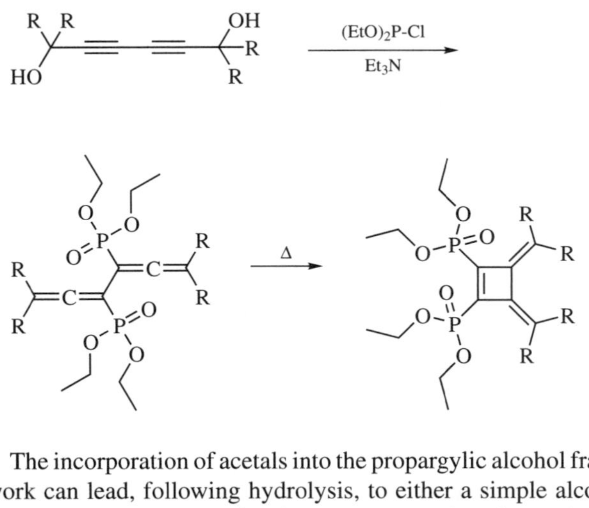

(35)

The incorporation of acetals into the propargylic alcohol framework can lead, following hydrolysis, to either a simple alcohol-functionalized allenyl phosphonates or to phosphorus heterocycles depending on the substituent pattern and steric environment (eqs 38 and 39).[55–57]

[4+1] Cycloaddition. The [4+1] cycloaddition of diethyl chlorophosphite to a number of heterodienes has been reported (eq 40).[58,59]

Certain non-conjugated dienes such as norbornadiene also undergo cycloaddition with diethyl chlorophosphite (eq 41).[60] The adamantanyl framework may be accessed by a cascade sequence where the ability of the reagent to activate a ketone is of advantage (eq 42).[61]

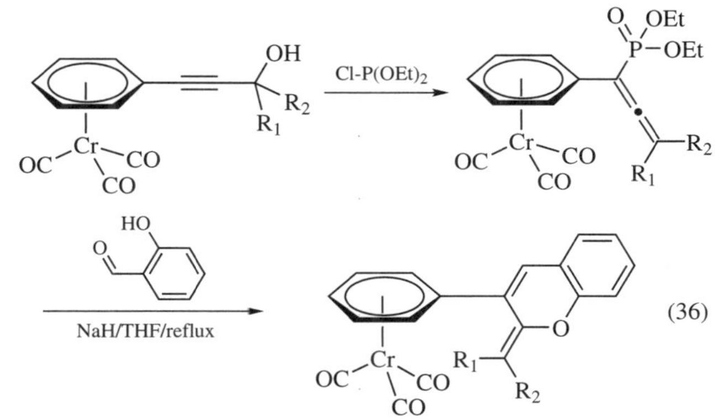

(36)

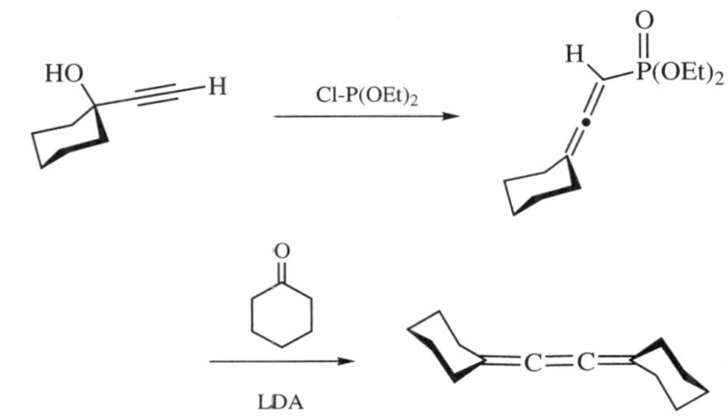

(37)

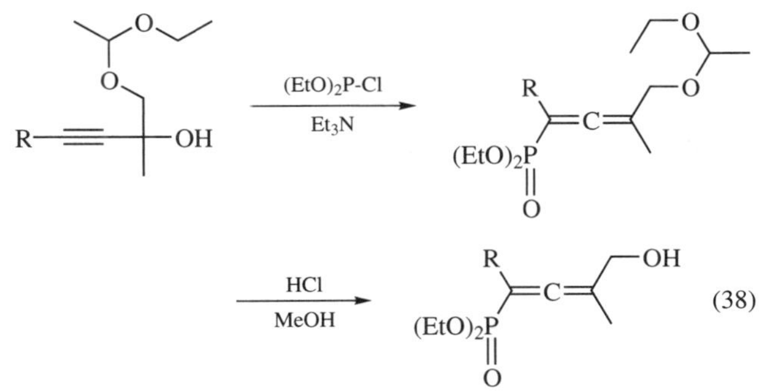

(38)

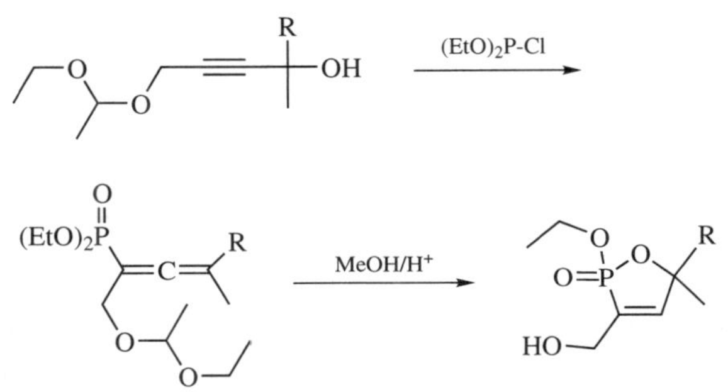

(39)

Use in Carbohydrate and Nucleotide Chemistry. Phosphite esters prepared from diethyl chlorophosphite and anomeric hydroxyl groups have been used as glycosylating agents for the synthesis of 2-deoxy-β-oligosaccharides (eq 43)[62] as well as a host of other glycosidic bonds.[63]

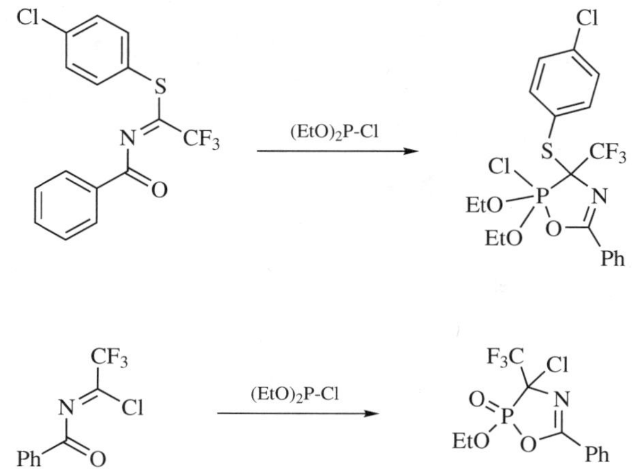

(40)

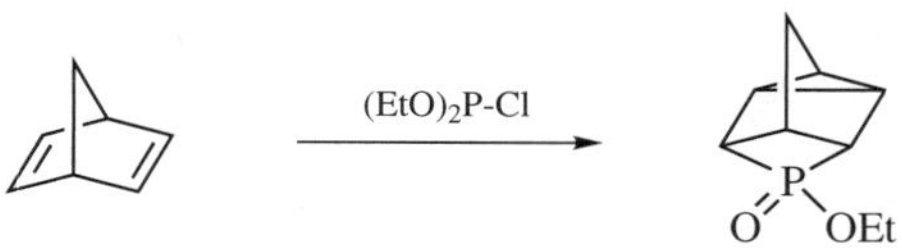

(41)

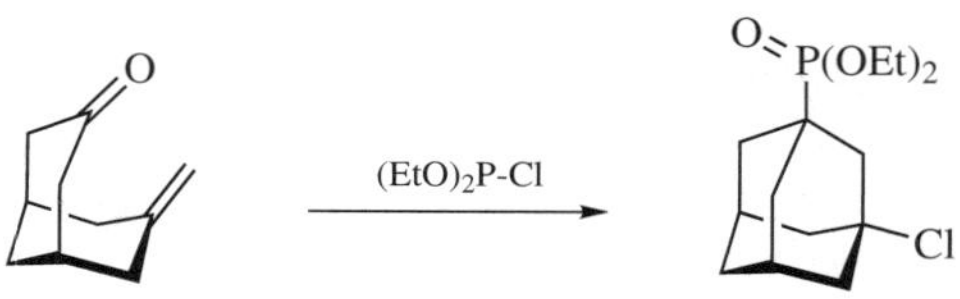

(42)

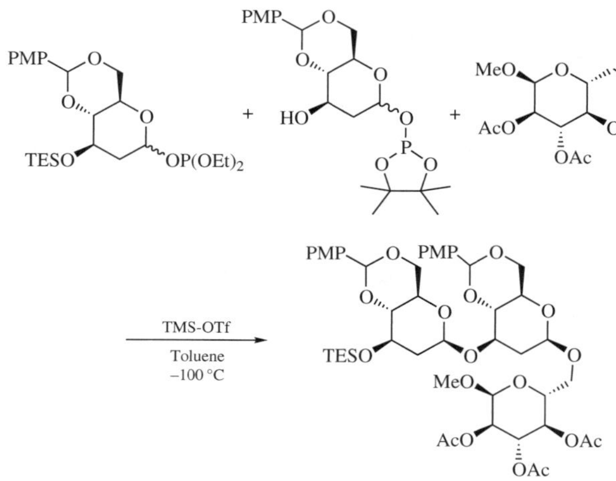

(43)

50% yield

The reaction of orthoesters derived from nucleosides with diethyl chlorophosphite affords interesting phosphonate intermediates which can be rearranged on treatment with sodium azide to give deoxy azido nucleotides (eq 44).[64,65]

Radical Reactions. Phosphitylation of oximes with diethyl chlorophosphite leads to convenient precursors to generate iminyl radicals for use in radical cyclizations. In the example shown (eq 45), the radical chain reaction is initiated by attack of a phenylselenyl radical on P(III) with subsequent cleavage of the N-O bond and valence expansion to P(VI).[66]

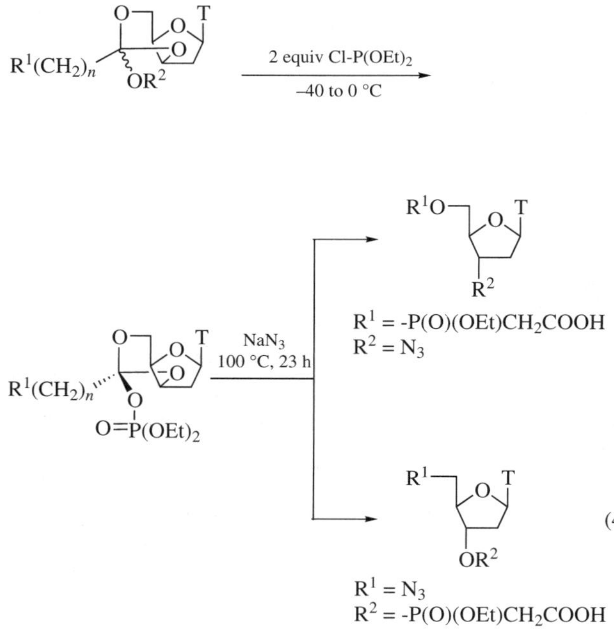

(44)

Reaction with Selenol, Thiol, and Silyl Compound. Selenols (eq 46) and thiols react with diethyl chlorophosphite to give selenophosphonates and thiophosphonates, respectively.[67,68] Trimethylsilylated thiols, on the other hand, enable the reaction to be stopped at the level of diethyl thiophosphite (eq 47).[69]

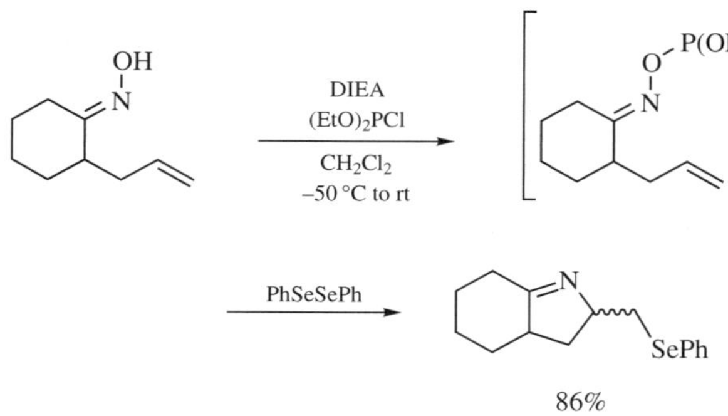

(45)

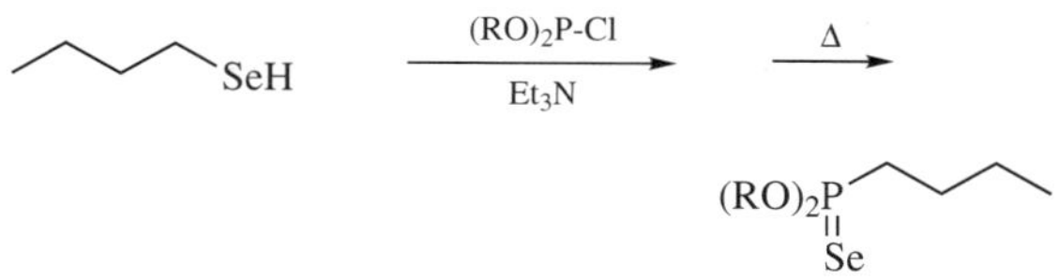

(46)

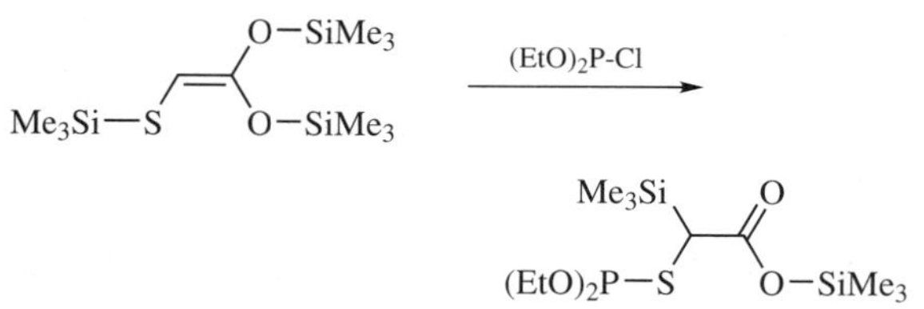

(47)

1. Cook, H. G.; Ilett, J. D.; Saunders, B. C.; Stacey, G. J.; Watson, H. G.; Wilding, I. G. E.; Woodcock, S. J., *J. Chem. Soc., Perkin Trans. 2* **1949**, 2921.

2. Michalski, J.; Modro, T.; Zwierzak, A., *J. Chem. Soc., Perkin Trans. 2* **1961**, 4904.
3. Arbusow, A. E.; Arbusow, B. A., *Ber. Dtsch. Chem. Ges./Chem. Ber.* **1932**, *65*, 195.
4. Afarinka, K.; Cadogan, J. I.; Rees, C. W., *Chem. Commun./J. Chem. Soc., Chem. Commun.* **1992**, 285.
5. Kawonobe, W.; Yamaguchi, K.; Nakahama, S.; Yamazaki, N., *Chem. Lett.* **1982**, 825.
6. (a) Krasil'nikova, E. A.; Razumov, A. I.; Nevzoravo, O. L., *J. Gen. Chem. USSR (Engl. Transl.)* **1977**, *47*, 1095 (*Chem. Abstr.* **1977**, *87*, 85 097). (b) Collins, D. J.; Drygala, P. F.; Swan, J. M., *Aust. J. Chem.* **1983**, *36*, 2517.
7. Malenko, D. M.; Gololobov, Y. G., *J. Gen. Chem. USSR (Engl. Transl.)* **1976**, *46*, 2291 (*Chem. Abstr.* **1976**, *84*, 43 785).
8. Lee, K.; Wiemer, D. F., *J. Org. Chem.* **1991**, *56*, 5556.
9. Perich, J. W.; Alewood, P. F.; Johns, R. B., *Synthesis* **1986**, 572.
10. Liu, C.; Nahorski, S. R.; Potter, B. V. L., *Carbohydr. Res.* **1992**, *234*, 107.
11. Mark, V., In *Mechanisms of Molecular Migrations*; Thyagarajan, B. S. Ed.; Wiley: New York, 1971, Vol 2, p 319.
12. (a) Altenbach, H.-J.; Korff, R., *Tetrahedron Lett.* **1981**, *22*, 5175. (b) Altenbach, H.-J.; Korff, R., *Angew. Chem., Int. Ed. Engl.* **1982**, *21*, 371.
13. Janecki, T.; Bodalski, R., *Synthesis* **1990**, 799.
14. Cohen, P., *Nature* **1982**, *296*, 613.
15. (a) Alewood, P. F.; Perich, J. W.; Johns, R. B., *Synth. Commun.* **1982**, *12*, 821. (b) Perich, J. W.; Johns, R. B., *Aust. J. Chem.* **1990**, *43*, 1609.
16. Meek, J. L.; Davidson, F.; Hobbs, F. W., *J. Am. Chem. Soc.* **1988**, *110*, 2317.
17. Martin, T. J.; Schmidt, R. R., *Tetrahedron Lett.* **1992**, *33*, 6123.
18. Anderson, G. W.; Blodinger, J.; Young, R. W.; Welcher, A. D., *J. Am. Chem. Soc.* **1952**, *74*, 5304.
19. Letsinger, R. L.; Heavner, G. A., *Tetrahedron Lett.* **1975**, *2*, 147.
20. Roussis, V.; Wiemer, D. F., *J. Org. Chem.* **1989**, *54*, 627.
21. Wadsworth, W. S., *Org. React.* **1984**, *25*, Chapter 2.
22. Arbusow, B. E., *Pure Appl. Chem.* **1964**, *9*, 307.
23. (a) Hammond, G. B.; Calogeropoulou, T.; Wiemer, D. F., *Tetrahedron Lett.* **1986**, *27*, 4265. (b) Sampson, P.; Hammond, G. B.; Wiemer, D. F., *J. Org. Chem.* **1986**, *51*, 4342.
24. Teulade, M.-P.; Savignac, P., *Tetrahedron Lett.* **1989**, *30*, 6327.
25. Vasse, J.-L.; Stranne, R.; Zalubovskis, R.; Gayet, C.; Moberg, C., *J. Org. Chem.* **2003**, *68*, 3258–3270.
26. Roch-Neirey, C.; Le Bris, N.; Laurent, P.; Clément, J.-C.; des Abbayes, H., *Tetrahedron Lett.* **2001**, *42*, 643–645.
27. Schuman, M.; Lopez, X.; Karplus, M.; Gouverneur, V., *Tetrahedron* **2001**, *57*, 10299–10307.
28. Du, Y. M.; Jung, K. Y.; Wiemer, D. F., *Tetrahedron Lett.* **2002**, *43*, 8665–8668.
29. du Pisani, C.; Schneider, D. F.; Venter, P. C. R., *Synth. Commun.* **2002**, *32*, 305–314.
30. Vizer, S. A.; Erzhanov, K. B.; Manchuk, Z. N., *Russ. J. Org. Chem.* **2000**, *36*, 1072–1074.
31. Johnston, K. F.; Padias, A. B.; Hall, H. K., *Polym. Bull.* **2000**, *45*, 359–364.
32. Burilov, A. R.; Nikolaeva, I. L.; Pudovik, M. A., *Zh. Obshch. Khim.* **1994**, *64*, 597–601.
33. Malenko, D. M.; Repina, L. A.; Simurova, N. V.; Sinitsa, A. D., *Zh. Obshch. Khim.* **1992**, *62*, 1179–1181.
34. Yin, Y.; Chen, Y.; Zhao, Y.; Li, L., *Phosphorus, Sulfur, Silicon Rel. Elements* **1993**, *82*, 67–72.
35. Lee, K.; Wiemer, D. F., *Phosphorus, Sulfur, Silicon Rel. Elements* **1993**, *75*, 87–90.
36. Barkallah, S.; Boukraa, M.; Zantour, H.; Baccar, B., *Phosphorus, Sulfur, Silicon Rel. Elements* **1996**, *108*, 51–56.
37. Guseinov, F. I.; Burangulova, R. N., *Russ. Chem. Bull.* **2002**, *51*, 897–898.
38. Guseinov, F. I.; Burangulova, R. N., *Russ. J. Gen. Chem.* **2002**, *72*, 46–48.
39. Guseinov, F. I.; Burangulova, R. N., *Chem. Heterocycl. Compds.* **1997**, *33*, 1040–1042.
40. Xu, J.; Ma, Y.; Duan, L. F., *Heteroatom Chem.* **2000**, *11*, 417–421.
41. Dimukhametov, M. N.; Bajandina, E. V.; Davydova, E. Y.; Dobrynin, A. B.; Gubaidullin, A. T.; Litvinov, I. A.; Alfonsov, V. A., *Mendeleev Commun.* **2001**, 196–197.
42. Touil, S.; Zantour, H., *Phosphorus Sulfur, Silicon Rel. Elements* **2001**, *175*, 183–192.
43. Gryaznov, P. I.; Kurochkina, S. N.; Musin, R. Z.; Pudovik, A. N.; Kibardin, A. M.; Arbuzov, A. E., *Zh. Obshch. Khim.* **1996**, *66*, 383–385.
44. Lodochnikov, V. G.; Gryaznova, T. V.; Gryaznov, P. I.; Musin, R. Z.; Kibardin, A. M., *Russ. J. Gen. Chem.* **1999**, *69*, 1883–1887.
45. Kolomnikova, G. D.; Krylova, T. O.; Chernoglazova, I. V.; Petrovskii, P. V.; Gololobov, Y. G., *Izv. Akad. Nauk, Ser. Khim.* **1993**, 1245–1247.
46. Bal'on, Y. G.; Kozhushko, B. N.; Paliichuk, Yu. A.; Shokol, V. A., *Zh. Obshch. Khim.* **1992**, *62*, 2530–2539.
47. Jie, Z.; Rammoorty, V.; Fischer, B., *J. Org. Chem.* **2002**, *67*, 711–719.
48. Brel, V. K.; Stang, P. J., *Eur. J. Org. Chem.* **2003**, 224–229.
49. Brel, V. K., *Synthesis* **2002**, 1829–1832.
50. Saalfrank, R. W.; Haubner, M.; Deutscher, C.; Bauer, W., *Eur. J. Org. Chem.* **1999**, 2367–2372.
51. Cai, B.-Z.; Blackburn, G. M., *Synth. Commun.* **1997**, *27*, 3943–3949.
52. Brel, V. K., *Synthesis* **2001**, 1539–1545.
53. Müller, T. J. J.; Ansorge, M., *Tetrahedron* **1998**, *54*, 1457–1470.
54. Pollack, S. K.; Fiseha, A.; Narayanswamy, B., *Macromolecules* **1997**, *30*, 5265–5270.
55. Brel, V. K., *Synth. Commun.* **1999**, *29*, 3869–3880.
56. Brel, V. K., *Synthesis* **1999**, 463–466.
57. Brel, V. K., *Synthesis* **1998**, 710–712.
58. Onys'ko, P. P.; Kolodka, T. V.; Sinitsa, A. D., *Zh. Obshch. Khim.* **1995**, *65*, 948–954.
59. Onys'ko, P. P.; Kolodka, T. V.; Kudryavtsev, A. A.; Sinitsa, A. D., *Zh. Obshch. Khim.* **1993**, *63*, 1562–1566.
60. Roussis, V.; Baenziger, N. C.; Wiemer, D. F., *J. Heterocycl. Chem.* **1996**, *33*, 979–981.
61. Yurchenko, R. I.; Martynyuk, E. G.; Yurchenko, A. G., *Zh. Obshch. Khim.* **1995**, *65*, 876–878.
62. Pongdee, R.; Wu, B.; Sulikowski, G. A., *Org. Lett.* **2001**, *3*, 3523–3525.
63. Kondo, H.; Aoki, S.; Ichikawa, Y.; Halcomb, R. L.; Ritzen, H.; Wong, C.-H., *J. Org. Chem.* **1994**, *59*, 864–877.
64. Endova, M.; Masojidkova, M.; Budesinsky, M.; Rosenberg, I., *Tetrahedron* **1998**, *54*, 11187–11208.
65. Endova, M.; Masojidkova, M.; Rosenberg, I., *Coll. Czech. Chem. Commun.* **1996**, *61*, S167–S170.
66. Lin, X.; Artman, G. D.; Stien, D.; Weinreb, S. M., *Tetrahedron* **2001**, *57*, 8779–8791.
67. Mel'nik, Ya. I.; Mel'nik, G. F., *Russ. J. Gen. Chem.* **1999**, *69*, 74–76.
68. Burilov, A. R.; Cherepachkin, D. V.; Nikolaeva, I. L.; Pudovik, M. A., *Izv. Akad. Nauk. Ser Khim.* **1995**, 353–357.
69. Burilov, A. R.; Cherepashkin, D. V.; Pudovik, M. A., *Zh. Obshch. Khim.* **1994**, *64*, 1621–1624.

Diethyl Phosphorochloridate[1]

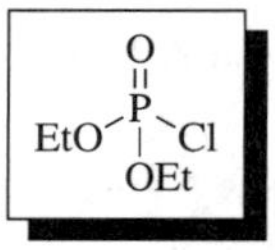

[814-49-3] $C_4H_{10}ClO_3P$ (MW 172.55)

(highly electrophilic phosphorylating reagent easily installed on anionic carbon,[2] oxygen,[3] nitrogen,[4] and sulfur;[5] ketones are converted to enol phosphates which can be reduced to alkenes[6] or alkanes,[7] or coupled with organometallic reagents to form substituted alkenes;[8] enol phosphates can be tranformed into β-keto phosphonates,[9] useful for Horner–Emmons homologation, or into terminal alkynes;[10] used to convert carboxylic acids to other carboxylic derivatives[11])

Alternate Name: diethyl chlorophosphate.
Physical Data: bp 60 °C/2 mmHg; *d* 1.194 g cm^{-3}.
Handling, Storage, and Precautions: highly toxic, corrosive. Use in a fume hood.

Phosphorylation. Diethyl phosphorochloridate is highly electrophilic and can be cleanly reacted at an anionic center provided that prior metalation is regio- and chemoselective. For example, phenols,[12] thiophenols,[5] and anilines[4] can be phosphorylated under basic conditions (eq 1). The phosphorylated compound can be isolated, or treated further with bases, resulting in orthometalation followed by a facile 1,3-phosphorus migration from the heteroatom to carbon. If carbon phosphorylation is desired, simply treating thiophenol with 2 equiv of n-Butyllithium followed by diethyl phosphorochloridate gave a moderate yield (58%).[5]

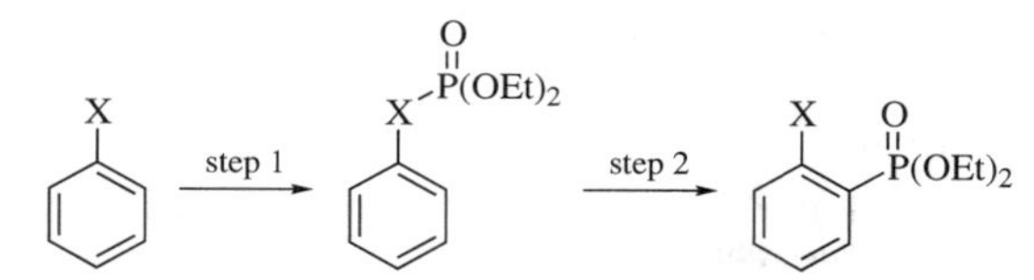

(1)

(1) **(2)** **(3)**

X	Step 1 (% yield **(2)**)	Step 2 (% yield **(3)**)
OH	NaH, THF; $(EtO)_2P(O)Cl$ (50–85%)	LDA, THF −78 °C to rt (90%)
SH	NaH, THF, rt; $(EtO)_2P(O)Cl$ (63%)	LDA, THF −78 to 0 °C (16%)
HNMe	$(EtO)_2P(O)Cl$, Et_3N, THF (97%)	LDA, THF −78 to rt (63%)

Synthetic preparation of oligonucleotides in a cost-efficient manner can be complicated by low chemoselectivity during phosphorylation, suggesting the need for prior nitrogen protection. It has been determined that despite the higher acidity of guanosine and thymidine over a hydroxy, addition of 1 equiv of the phosphorylating agent provided only the phosphates (eq 2).[3] Adenosine and cytidine bases were compatible as well.

Phosphorylation of substituted 1,4-dihydropyridine dianions occurred selectively (eq 3).[2b] Other reports documented selective phosphorylation at C-5 of thiophenes[2a] and the *ortho* position of various substituted benzenes.[13] It has been noted that certain heterocycles demonstrated alternative pathways, such as nucleophilic attack at carbon rather than phosphorus (eq 4).[14]

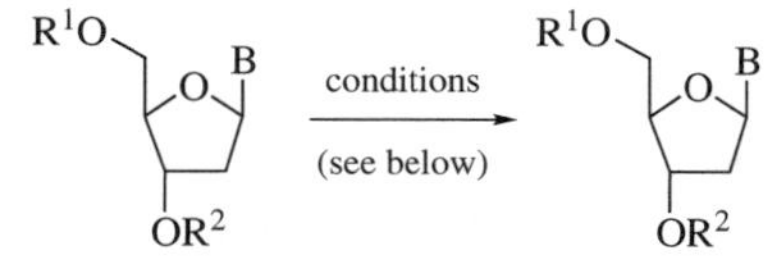

(4) **(5)** (2)

(4) R[1]	R[2]	B	Conditions	(5) R[1]	R[2]	% Yield
TBS	H	guanosine	*t*-BuMgCl (2 equiv) THF, rt $(EtO)_2P(O)Cl$	TBS	$P(O)(OEt)_2$	94
H	TBS	thymidine	*t*-BuLi, (2 equiv) THF, −50 °C $(EtO)_2P(O)Cl$	$P(O)(OEt)_2$	H	85

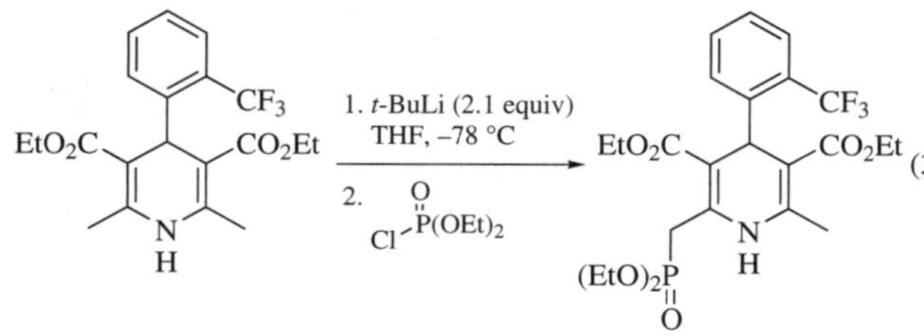

(3)

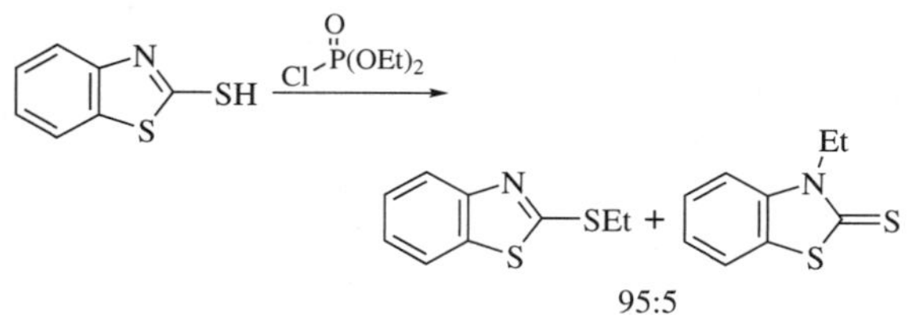

(4)

Trapping of ester, lactone, or ketone enolates results in rapid *O*-phosphorylation, thus providing the enol phosphate (eq 5).[9] Further treatment with base resulted in facile rearrangement to the β-ketophosphonate (eq 5).[15] However, regioisomeric β-ketophosphonates can often be observed from a regiochemically pure enol phosphate (eq 6).[9a] Since the classical Arbuzov reaction is limited to primary alkyl iodides (competing Perkov reaction is known with secondary halides), an alternative method was sought. Installation of an alkene blocking group sufficed (eq 7).[9a] Dienol phosphates have also been prepared and the authors note that Lithium 2,2,6,6-Tetramethylpiperidide was the preferred base for regiochemical control.[16]

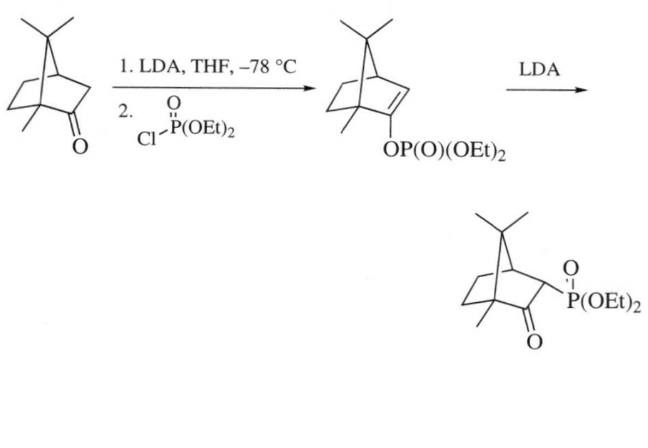

(5)

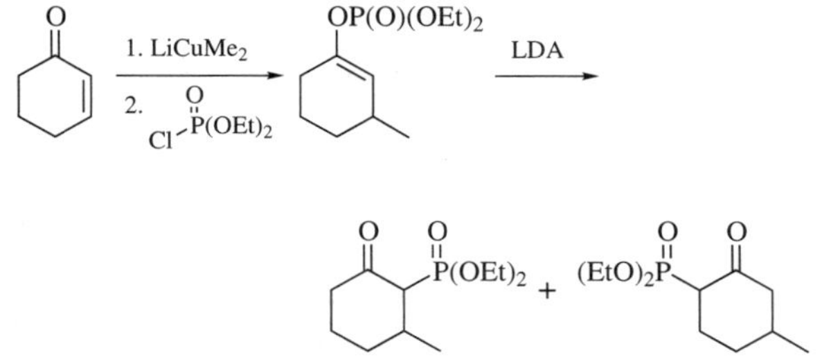

(6)

64:36

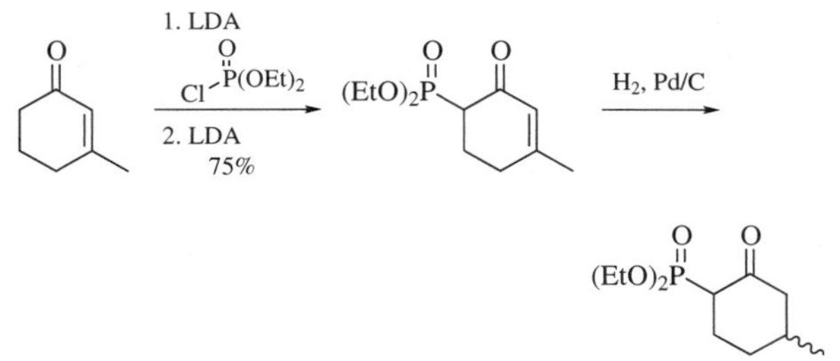

(7)

β-Ketophosphonates are valuable intermediates in the realm of Horner–Emmons alkenation methodology. Acyclic variants are difficult to obtain from enol phosphates due to competing alkyne and allene formation. One solution utilized the dianion derived from α-bromo ketones and trapping with diethyl phosphorochloridate;[17] however, only moderate yields of β-keto phosphonates were reported. The most efficient procedure utilizes the anion derived from dialkyl methylphosphonate, addition to an aldehyde, followed by oxidation (eq 8).[18]

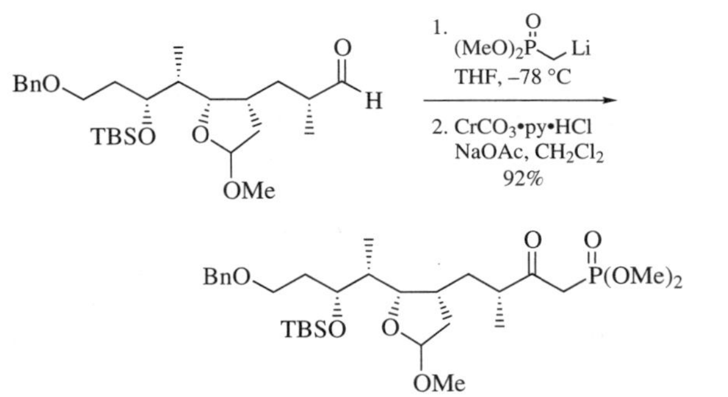

(8)

Interconversion of Carboxylic Acid Derivatives. Diethyl phosphorochloridate is useful for activation of carboxylic acids toward nucleophilic attack. Subsequent treatment of the phosphate ester with thallium sulfides produced thiol esters.[19] Variations[11] on this theme included prior formation of a heterocyclic phosphonate followed by treatment with alcohols, amines, or thiols, thus providing a racemization-free method to prepare esters, amides, and thiol esters, respectively (eq 9).[11b]

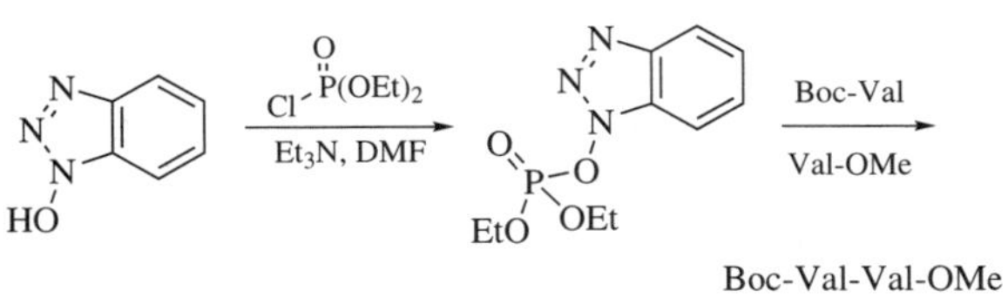

(9)

Amines can be readily transformed to an alcohol surrogate (eq 10)[20] and phenols to anilines.[21]

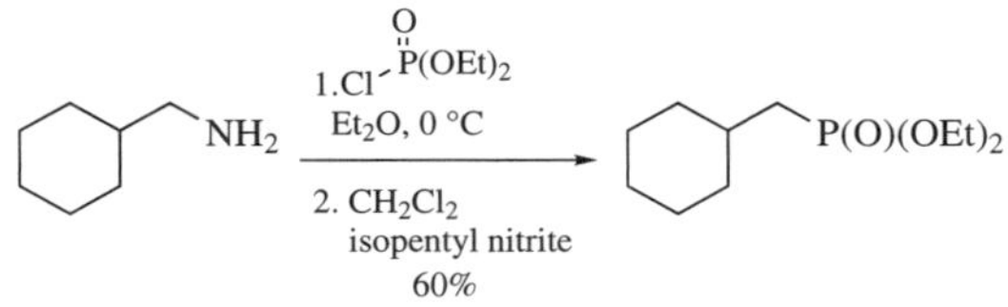

(10)

Deoxygenation of Phenols and Ketones. Excision of oxygen from a molecule is often encountered in a synthetic sequence. Phenols are readily deoxygenated by formation of the phosphate followed by reduction under dissolving metal conditions.[22] It has been noted that the Birch conditions result in low yields, whereas an alternative method utilizing activated Titanium metal is superior (eq 11).[23] Under this protocol, enol phosphates are efficiently reduced as well.[6b] Conversion of an enone to a regiochemically defined alkene was accomplished via 1,4-reduction, enolate trapping, and reduction of the trapped enol phosphate (eq 12).[6a] Enol phosphates can be fully reduced to alkanes by hydrogenation of palladium catalysts[7] or converted to vinyl iodides when treated with Iodotrimethylsilane.[24]

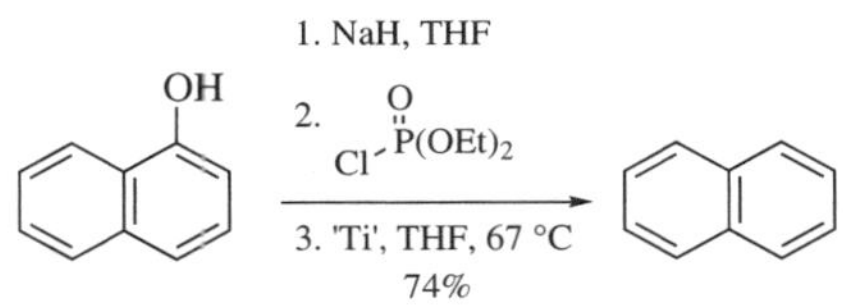

(11)

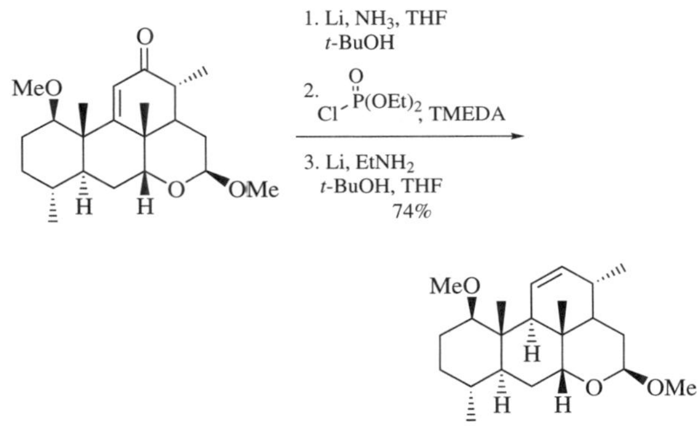

(12)

Synthesis of Alkenes and Alkynes. Enol phosphates are smoothly transformed into substituted alkenes when treated with organometallic reagents. If the enol phosphate was derived from a β-keto ester, cuprate reagents are generally reactive enough to encourage conjugate addition–phosphate elimination (eq 13).[25] In the event that this coupling fails, the combination of Pd^0 and Trimethylaluminum results in regio- and stereospecific methylation.[8b] Substrates lacking an ester moiety on the enol phosphate can be alkylated with Grignard reagents under nickel catalysis (eq 14).[8a]

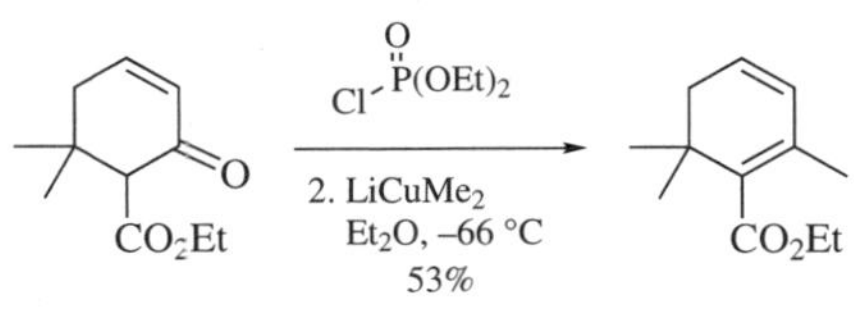

(13)

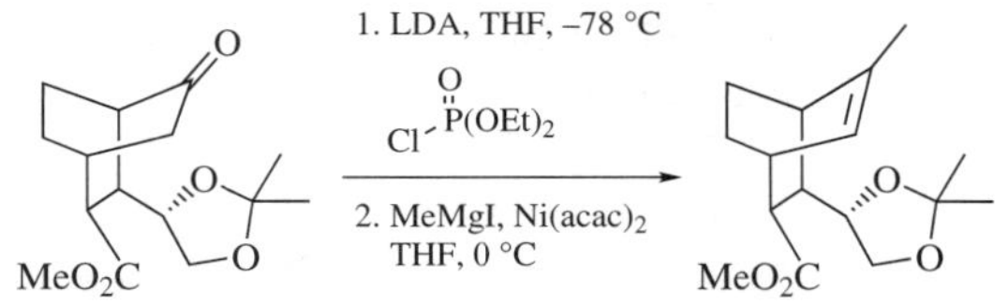

(14)

One of the most useful applications of β-keto phosphonates is the Horner–Emmons alkenation procedure (eq 15).[18] Variations of this theme have employed β-imino[26] and β-sulfonyl phosphates.[27]

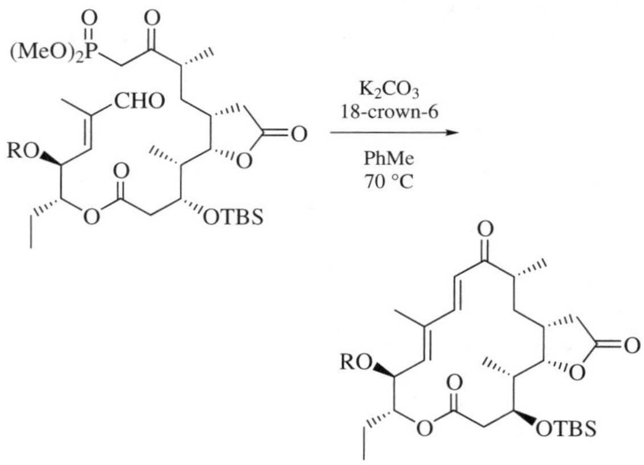

(15)

Alkynes are available from methyl ketones[10] by elimination of the enol phosphate (eq 16).[28] When the ketone contains α-branching, lithium tetramethylpiperidide has been recommended to circumvent allene formation.

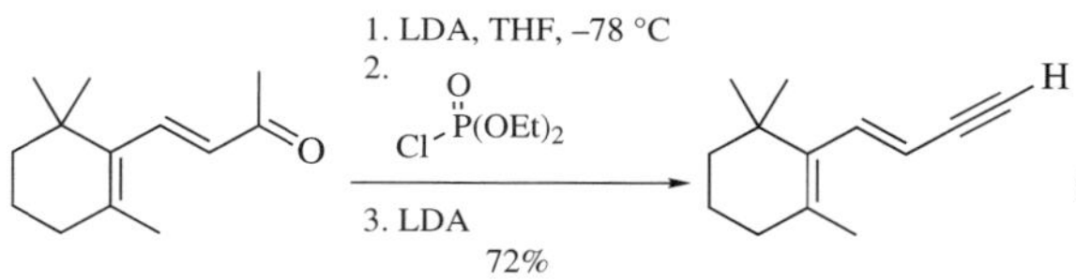

(16)

Miscellaneous. Allylic phosphates have found applications in π-allyl palladium chemistry. It has been demonstrated that allylic phosphates undergo oxidative addition more readily than the corresponding acetate, such that chemoselectivity could be achieved when these functionalities were present in the same molecule (eq 17).[29] Finally, β-keto phosphonates were coupled with epoxides to provide useful yields of spirocyclopropanes (eq 18).[30]

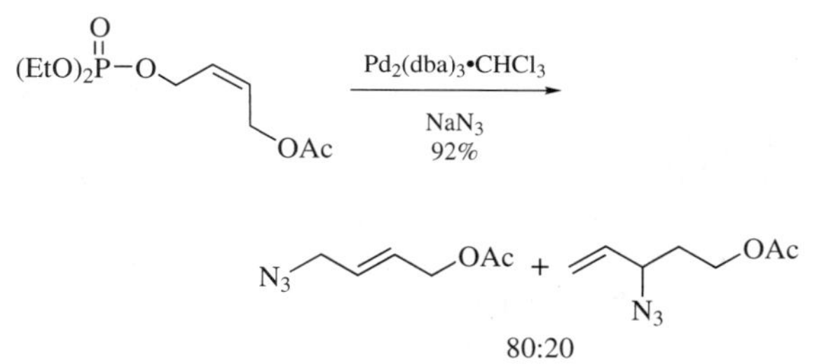

(17)

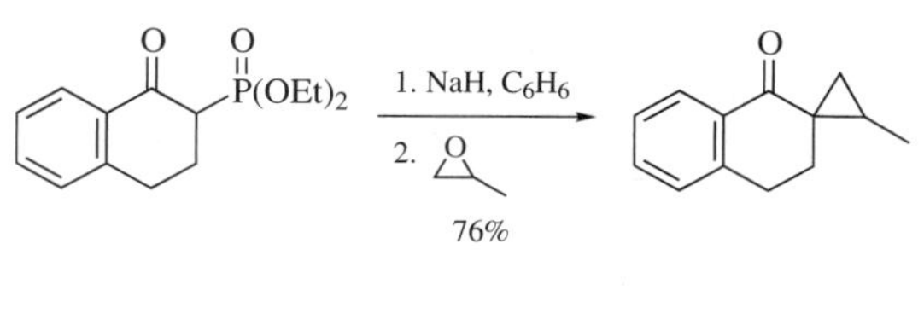

(18)

1. Koh, Y. J.; Oh, D. Y. *Synth. Commun.* **1993**, *23*, 1771.
2. (a) Graham, S. L.; Scholz, T. H. *J. Org. Chem.* **1991**, *56*, 4260. (b) Poindexter, G. S.; Licause, J. F.; Dolan, P. L.; Foley, M. A.; Combs, C. M. *J. Org. Chem.* **1993**, *58*, 3811.
3. (a) Hayakawa, Y.; Aso, Y. *Tetrahedron Lett.* **1983**, *24*, 1165. (b) Uchiyama, M.; Aso, Y.; Noyori, R.; Hayakawa, Y. *J. Org. Chem.* **1993**, *58*, 373.
4. Jardine, A. M.; Vather, S. M. *J. Org. Chem.* **1988**, *53*, 3983.
5. Masson, S.; Saint-Clair, J.-F.; Saquet, M. *Synthesis* **1993**, 485.
6. (a) Grieco, P. A.; Nargund, R. P.; Parker, D. T. *J. Am. Chem. Soc.* **1989**, *111*, 6287. (b) Welch, S. C.; Walters, M. E. *J. Org. Chem.* **1978**, *43*, 2715.
7. Jung, A.; Engel, R. *J. Org. Chem.* **1975**, *40*, 3652.
8. (a) Iwashima, M.; Nagaoka, H.; Kobayashi, K.; Yamada, Y. *Tetrahedron Lett.* **1992**, *33*, 81. (b) Asao, K.; Iio, H.; Tokoroyama, T. *Synthesis* **1990**, 382.
9. (a) Gloer, K. B.; Calogeropoulou, T.; Jackson, J. A.; Wiemer, D. F. *J. Org. Chem.* **1990**, *55*, 2842. (b) Jackson, J. A.; Hammond, G. B.; Wiemer, D. F. *J. Org. Chem.* **1989**, *54*, 4750.
10. Negishi, E.; King, A. O.; Tour, J. M. *Org. Synth.* **1986**, *64*, 44.
11. (a) Mikaye, M.; Kirisawa, M.; Tokutake, N. *Chem. Lett.* **1985**, 123. (b) Kim, S.; Chang, H.; Ko, Y. K. *Tetrahedron Lett.* **1985**, *26*, 1341.
12. Casteel, D. A.; Peri, S. P. *Synthesis* **1991**, 691.
13. Takenaka, H.; Hayase, Y. *Heterocycles* **1989**, *29*, 1185.
14. Rani, B. R.; Bhalerao, U. T.; Rahman, M. F. *Synth. Commun.* **1990**, *20*, 3045.
15. Hammond, G. B.; Calogeropoulou, T.; Wiemer, D. F. *Tetrahedron Lett.* **1986**, *27*, 4265.
16. Blotny, G.; Pollack, R. M. *Synthesis* **1988**, 109.
17. Sampson, P.; Hammond, G. B.; Wiemer, D. F. *J. Org. Chem.* **1986**, *51*, 4342.
18. Nicolaou, K. C.; Pavia, M. R.; Seitz, S. P. *J. Am. Chem. Soc.* **1982**, *104*, 2027.
19. Masamune, S.; Kamata, S.; Diakur, J.; Sugihara, Y.; Bates, G. S. *Can. J. Chem.* **1975**, *53*, 3693.
20. Nikolaides, N.; Ganem, B. *Tetrahedron Lett.* **1990**, *31*, 1113.
21. Rossi, R. A.; Bunnett, J. F. *J. Org. Chem.* **1972**, *37*, 3570.
22. Kenner, G. W.; Williams, N. J. *J. Chem. Soc* **1955**, 522.
23. Welch, S. C.; Walters, M. E. *J. Org. Chem.* **1978**, *43*, 4797.
24. Lee, K.; Wiemer, D. F. *Tetrahedron Lett.* **1993**, *34*, 2433.
25. Moorhoff, C. M.; Schneider, D. F. *Tetrahedron Lett.* **1987**, *28*, 4721.
26. (a) Molin, H.; Pring, B. G. *Tetrahedron Lett.* **1985**, *26*, 677. (b) Meyers, A. I.; Shipman, M. *J. Org. Chem.* **1991**, *56*, 7098. (c) Highet, R. J.; Jones, T. H. *J. Org. Chem.* **1992**, *57*, 4038.
27. Musicki, B.; Widlanski, T. S. *Tetrahedron Lett.* **1991**, *32*, 1267.
28. Okuda, Y.; Morizawa, Y.; Oshima, K.; Nokaki, H. *Tetrahedron Lett.* **1984**, *25*, 2483.
29. Murahashi, S.-I.; Taniguchi, Y.; Imada, Y.; Tanigawa, Y. *J. Org. Chem.* **1989**, *54*, 3292.
30. Jacks, T. E.; Nibbe, H.; Wiemer, D. F. *J. Org. Chem.* **1993**, *58*, 4584.

Jonathan R. Young
University of Wisconsin, Madison, WI, USA

Diethyl Phosphorocyanidate (DEPC)[1]

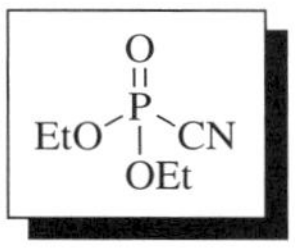

[2942-58-7] $C_5H_{10}NO_3P$ (MW 163.11)

(synthesis of amides,[3] esters,[2] α-aminonitriles,[12] aryl thiocyanates,[15] and cyanophosphates[17])

Alternate Names: diethyl cyanophosphonate; diethylphosphoryl cyanide; diethoxyphosphoryl cyanide; diethyl cyanophosphate; DEPC.

Physical Data: bp 93–96 °C/14 mmHg; fp 80 °C; *d* 1.075 g cm^{-3}.

Solubility: sol DMF, THF, Et_2O, toluene.

Form Supplied in: pale yellow liquid; 93% (technical grade).

Preparative Method: can be readily prepared from Triethyl Phosphite and Cyanogen Bromide.[2]

Purification: in most cases, commercial samples are used without purification. However, freshly prepared DEPC contains an impurity of diethyl phosporoisocyanidate. Addition of anhydrous Cobalt(II) Chloride to a sample of crude DEPC, followed by dilution with ether forms a green precipitate which is removed by filtration; distillation of the filtrate gives pure DEPC.[2]

Handling, Storage, and Precautions: is very toxic and corrosive. It must be handled in a fume hood. The reagent is moisture sensitive and should be stored under nitrogen in a refrigerator.

Synthesis of Amides. DEPC is an efficient and useful reagent for amide formation and is applicable to racemization-free peptide synthesis. Amides of various types can be obtained by the simple mixing of carboxylic acids and amines with DEPC in the presence of Triethylamine (eq 1).[3] Both aromatic and aliphatic acids easily react with aromatic and aliphatic amines. The reaction is rapid, clean, and usually free of contamination from side products.

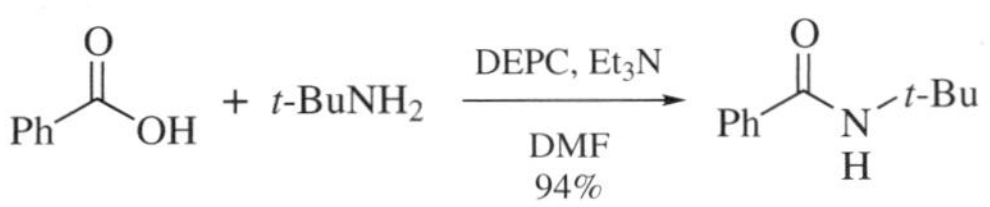

(1)

Studies of selected coupling methods for the attachment of amino acid derivatives to *cis-* and *trans*-4-aminocyclohexanecarboxylic acid have shown DEPC to be the reagent of choice as coupling agent.[4] A study directed towards racemization during peptide coupling of benzyloxycarbonyl-L-phenylalanyl-L-valine with L-proline *t*-butyl ester revealed that, among different coupling agents, DEPC coupling afforded the least racemization.[5]

Following their application as effective peptide coupling agents in the solution phase, both DEPC and Diphenyl Phosphorazidate (DPPA)[6] have been shown to be useful in solid-phase peptide synthesis.[7,8] Using these reagents, synthesis of porcine motilin, comprising 22 amino acids, was successfully achieved.[7]

Synthesis of Esters and Thioesters. Reaction of DEPC with carboxylic acids in the presence of triethylamine gives transient acyl cyanides, which in the presence of alcohols[2] or thiols[9] results in the formation of the corresponding esters or thiolesters (eq 2).

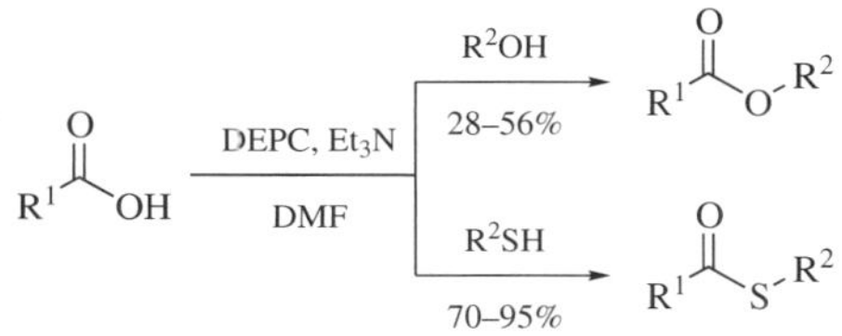

(2)

Although esterification yields are not generally satisfactory, the advantage of this method is that the reaction proceeds under mild and almost neutral conditions. The method can be efficiently applied to the synthesis of thiolesters in high yields with little racemization, as demonstrated by the formation of the ethylthiol ester of benzoyl-L-leucine.[9]

***C*-Acylation.** A direct *C*-acylation of active methylene compounds can be achieved using a carboxylic acid and DEPC in the presence of base under exceptionally mild conditions (eq 3).[10,11] The reactions are best carried out in DMF solution, though hexane, toluene, ether, or THF may be used. Triethylamine, 1,8-Diazabicyclo[5.4.0]undec-7-ene, *N,N,N′, N′*-Tetramethylethylenediamine, Sodium Hydride, or Potassium Carbonate may be used as a base.

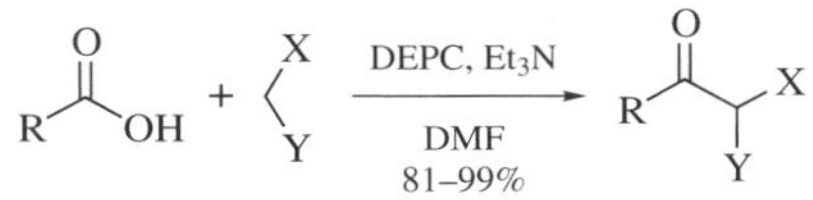

(3)

X = CN, NO_2, CO_2R
Y = H, CN, CO_2R

Synthesis of α-Aminonitriles. α-Aminonitriles can be efficiently prepared from carbonyl compounds with DEPC and amines under mild conditions, in a modified Strecker synthesis.[12] Since the nitrile functionality of α-aminonitrile can be easily converted to the acid, this constitutes a convenient method for the preparation of α-amino acids. The superiority of this method over the classical Strecker synthesis of α-aminonitriles was demonstrated by the reaction of 4-cholesten-3-one (**1**) with DEPC and Pyrrolidine, which afforded the α-aminonitrile (**2**) in 62% yield (eq 4). The classical Strecker conditions using a mixture of KCN and pyrrolidine hydrochloride in aqueous THF did not yield any α-aminonitriles.

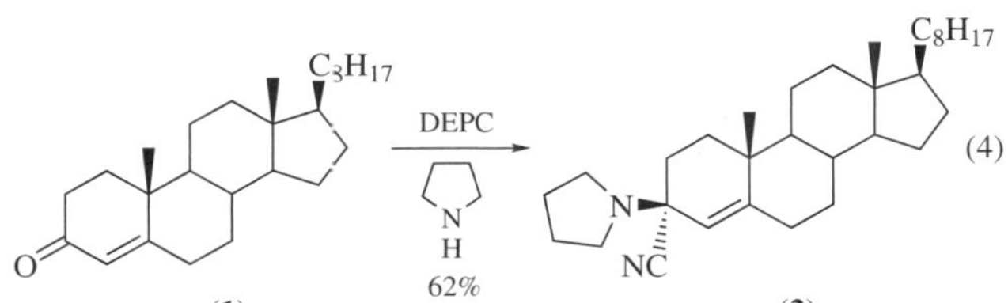

(**1**) (**2**) (4)

Modified Reissert–Henze Reaction. The α-cyanation of aromatic amine oxides (Reissert–Henze reaction) is usually carried out using benzoyl chloride and potassium cyanide.[13] Alternatively, α-cyano-substituted aromatic nitrogen heterocycles can be prepared by reaction of their corresponding *N*-oxide with DEPC in the presence of triethylamine (eq 5).[14]

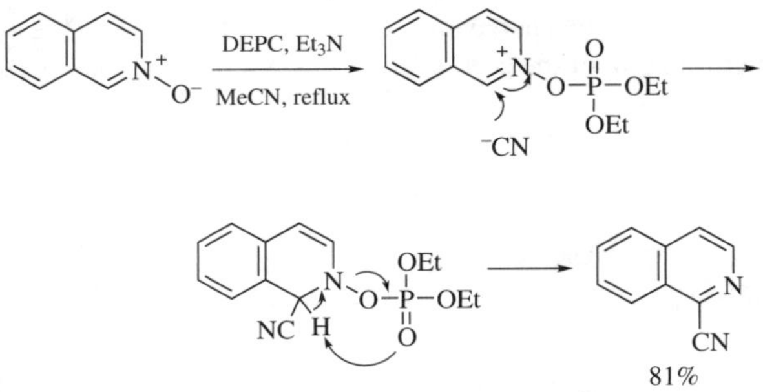

(5)

Synthesis of Aryl Thiocyanates. Aryl thiocyanates can be prepared in one step from the sodium salt of the arylsulfinic acid by treatment with DEPC in refluxing THF. The thiocyanates are obtained in good yields and even hindered substrates like adamantane give the corresponding thiocyanate in moderate yield (eq 6).[15]

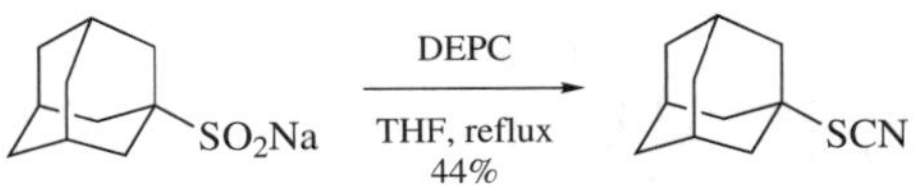

(6)

Epoxidation of Alkenes. A mixture of DEPC and Hydrogen Peroxide smoothly epoxidizes alkenes at rt in the presence of 2-Hydroxypyridine or 1,2,4-Triazole.[16] Methylene chloride is reported to be the solvent of choice. Various alkenes were epoxidized in 60–93% yields. *trans*-Stilbene and 1-dodecene, which are less susceptible to electrophilic attack by peroxycarboxylic acids, smoothly underwent this DEPC/H_2O_2 reaction.

Synthesis of Cyanophosphates as Versatile Intermediates. Reaction of aldehydes or ketones with DEPC in the presence of a catalytic amount of Lithium Diisopropylamide (LDA) in THF gives cyanophosphates in high yield (eq 7).[17] Various cyclic and aromatic ketones react with DEPC to give cyanophosphates. Even conjugated or hindered ketones such as α-tetralone or benzophenone, which are known not to give cyanohydrins under the usual reaction conditions, react smoothly with DEPC to give cyanophosphates.

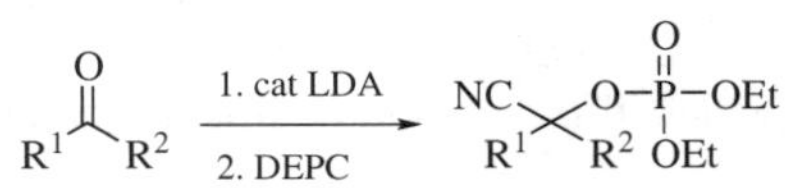

(7)

Reductive Cyanation of Ketones and Aldehydes. Saturated and unsaturated carbonyl compounds react with DEPC and Lithium Cyanide to give intermediate cyanophosphates which can be reduced with Samarium(II) Iodide in the presence of *t*-butanol to give nitriles (eq 8).[18]

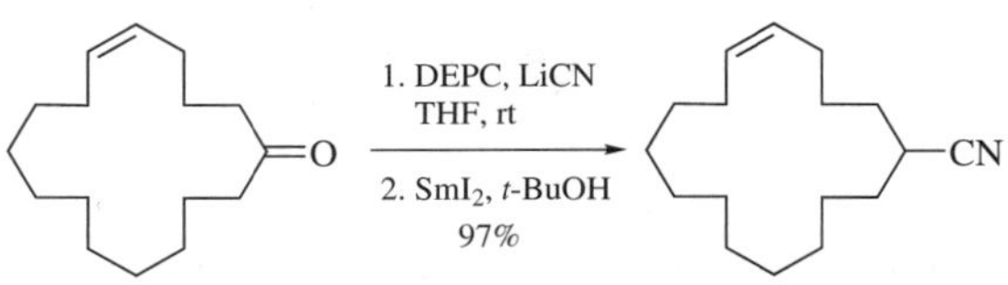

(8)

α,β-Unsaturated Nitriles. Cyanophosphates prepared by the reaction of aromatic ketones in the presence of lithium cyanide can be readily converted into α,β-unsaturated nitriles by treatment with Boron Trifluoride Etherate at rt (eq 9).[19] This reaction was used to convert 4-isobutylacetophenone to α-cyanostyrene, which upon hydrogenation followed by hydrolysis gives ibuprofen in 70% overall yield (eq 10).[19a]

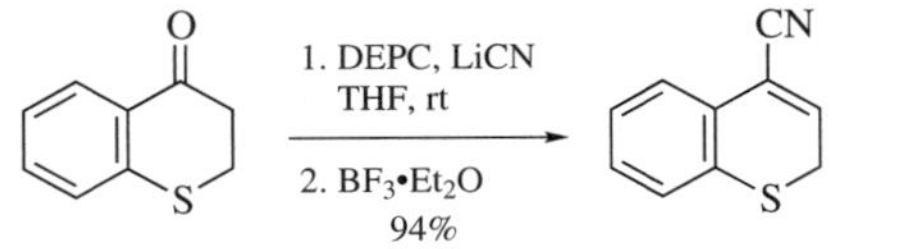

(9)

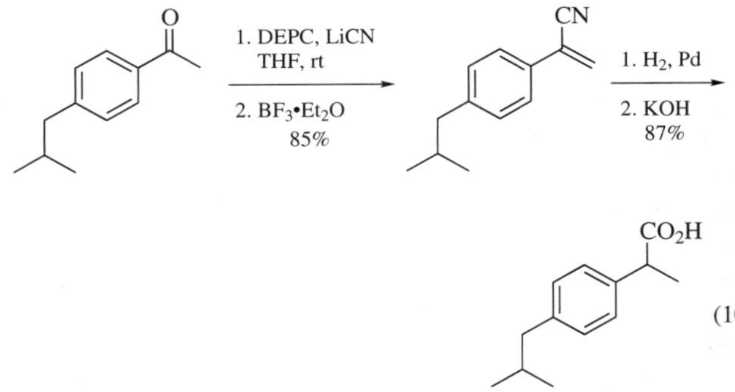

(10)

Deoxygenation of α,β-Unsaturated Ketones. Cyanophosphates, prepared from α,β-unsaturated ketones or aromatic ketones by treatment with DEPC and lithium cyanide, can be reductively cleaved with Lithium in liquid ammonia to give the corresponding methylene compounds (eq 11).[20] In the case of α,β-enones, this procedure is superior to the more commonly employed thioacetal and desulfurization deoxygenation method. This method for α,β-enones is simple and the products are obtained in excellent yields without isomerization or reduction of the double bond.

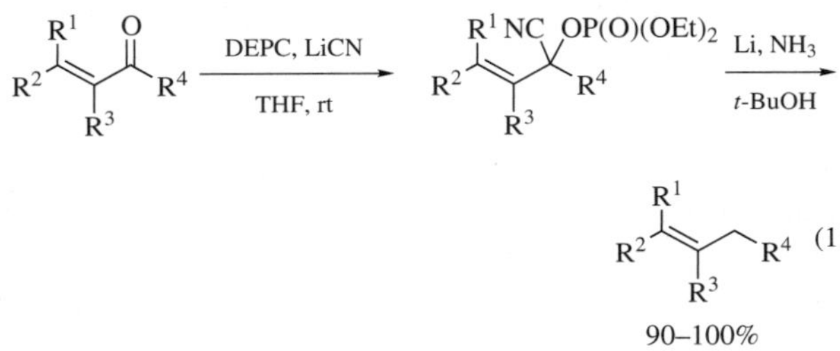

(11)

1. Saunders, B. C.; Stacey, G. T.; Wild, F.; Wilding, I. G. E. *J. Chem. Soc* **1948**, 699.
2. Shioiri, T.; Yokoyama, Y.; Kasai, Y.; Yamada, S. *Tetrahedron* **1976**, *32*, 2211.
3. Yamada, S.; Kasai, Y.; Shioiri, T. *Tetrahedron Lett.* **1973**, 1595.
4. Chen, W.-Y.; Olsen, R. K. *J. Org. Chem.* **1975**, *40*, 350.
5. Takuma, S.; Hamada, Y.; Shioiri, T. *Chem. Pharm. Bull.* **1982**, *30*, 3147.
6. Hamada, Y.; Rishi, S.; Shioiri, T.; Yamada, S. *Chem. Pharm. Bull.* **1977**, *25*, 224.
7. Yamada, S.; Ikota, N.; Shioiri, T.; Tachibana, S. *J. Am. Chem. Soc.* **1975**, *97*, 7174.
8. (a) Ikota, N.; Shioiri, T.; Yamada, S. *Chem. Pharm. Bull.* **1980**, *28*, 3064. (b) Ikota, N.; Shioiri, T.; Yamada, S.; Tachibana, S. *Chem. Pharm. Bull.* **1980**, *28*, 3347.

9. (a) Yamada, S.; Yokoyama, Y.; Shioiri, T. *J. Org. Chem.* **1974**, *39*, 3302. (b) Yokoyama, Y.; Shioiri, T.; Yamada, S. *Chem. Pharm. Bull.* **1997**, *25*, 2423.
10. Shioiri, T.; Hamada, Y. *J. Org. Chem.* **1978**, *43*, 3631.
11. (a) Kato, N.; Hamada, Y.; Shioiri, T. *Chem. Pharm. Bull.* **1984**, *32*, 1679. (b) Kato, N.; Hamada, Y.; Shioiri, T. *Chem. Pharm. Bull.* **1984**, *32*, 3323.
12. Harusawa, S.; Hamada, Y.; Shioiri, T. *Tetrahedron Lett.* **1979**, 4663.
13. Henze, M. *Ber. Dtsch. Chem. Ges./Chem. Ber.* **1936**, *69*, 1566.
14. Harusawa, S.; Hamada, Y.; Shioiri, T. *Heterocycles* **1981**, *15*, 981.
15. Harusawa, S.; Shioiri, T. *Tetrahedron Lett.* **1982**, *23*, 447.
16. Mizuno, A.; Hamada, Y.; Shioiri, T. *Chem. Pharm. Bull.* **1981**, *29*, 1774.
17. Harusawa, S.; Yoneda, R.; Kurihara, T.; Hamada, Y.; Shioiri, T. *Chem. Pharm. Bull.* **1983**, *31*, 2932.
18. (a) Yoneda, R.; Harusawa, S.; Kurihara, T. *Tetrahedron Lett.* **1989**, *30*, 3681. (b) Yoneda, R.; Harusawa, S.; Kurihara, T. *J. Org. Chem.* **1989**, *56*, 1827.
19. (a) Harusawa, S.; Yoneda, R.; Kurihara, T.; Hamada, Y.; Shioiri, T. *Tetrahedron Lett.* **1984**, *25*, 427. (b) Yoneda, R.; Terada, T.; Harusawa, S.; Kurihara, T. *Heterocycles* **1985**, *23*, 557. (c) Kurihara, T.; Terada, T.; Satoda, S.; Yoneda, R. *Chem. Pharm. Bull.* **1986**, *34*, 2786.
20. Yoneda, R.; Osaki, H.; Harusawa, S.; Kurihara, T. *Chem. Pharm. Bull.* **1989**, *37*, 2817.

Hemantkumar H. Patel
Abbott Laboratories, North Chicago, IL, USA

3,4-Dihydro-2*H*-pyrido[1,2-*a*]pyrimidin-2-one

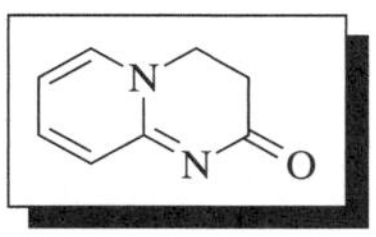

[5439-14-5] $C_8H_8N_2O$ (MW 148.16))

(synthesis of esters and amides from carboxylic acids; glycosylation)

Physical Data: mp 155–156 °C; mp (HCl salt) 284–285.5 °C (dec); mp (HBr salt) 299–300 °C (dec); mp (HI salt) 263–264 °C (dec).

Solubility: sol MeOH, EtOH, H_2O.

Form Supplied in: both the free base and salts are obtained as colorless solids.

Analysis of Reagent Purity: the reagent has a betaine-like structure and behaves as a salt in terms of the reagent's solubility. The literature reports various melting points values for both the anhydrous and hydrated forms of the reagent but the conditions used to determine melting point have been shown to be critical and care must be exercised if this property is to be used as a measure of reagent purity.[1]

Preparative Methods: is prepared by reaction of 2-aminopyridine with either Acrylic Acid itself or an equivalent (2-bromo- or 2-iodopropionic acid, Methyl Acrylate, ethyl 2-bromopropionate, β-Propiolactone).[1,2]

Purification: the free base and salts are readily recrystallized from MeOH–Et_2O.[1]

Handling, Storage, and Precautions: the reagent may be dried over P_2O_5 in vacuo (135 °C/17 mmHg).

Introduction. The principal synthetic application of 3,4-dihydro-2*H*-pyrido[1,2-*a*]pyrimidin-2-one (**1**) is an acid scavenger; it is an example of a 'proton sponge' (see 1,8-Bis(dimethylamino) naphthalene). The reagent has been used in the esterification and amination of alkyl and aryl carboxylic acids and also in glycosylation of 2-amino-2-deoxy sugars.

Synthesis of Esters from Carboxylic Acids.[3] Esterification (eq 1) proceeds by activation of the carboxyl component using a 2-halopyridinium salt (**2**; X = Cl, Y = I or X = F, Y = OTs) and (**1**) (2.4 equiv) in the presence of an alcohol. The best solvents for this process are CH_2Cl_2 or MeCN, but a range of other solvents may also be used.

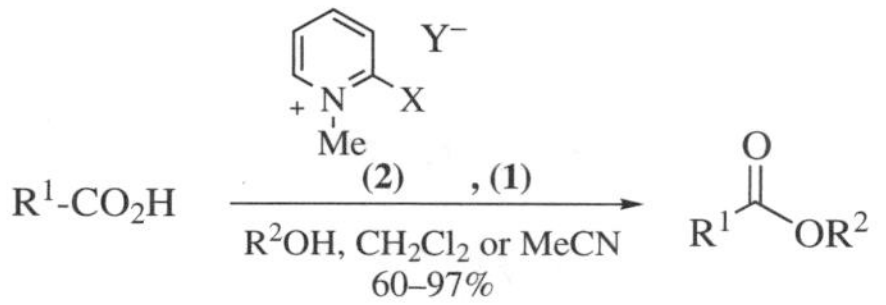

(1)

Synthesis of Amides from Carboxylic Acids.[4] In the corresponding amidation process (eq 2) there is a requirement for the

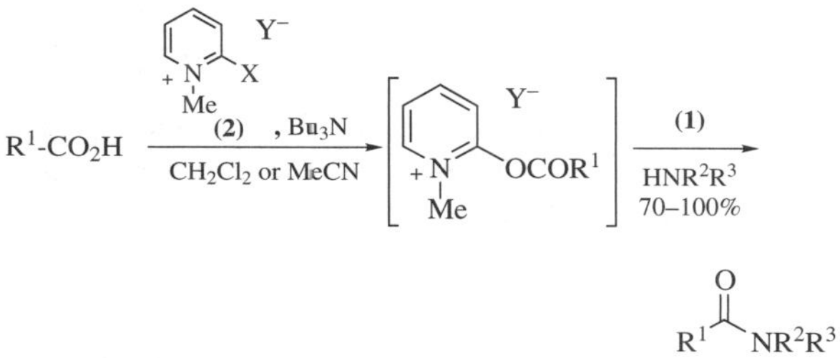

(2)

addition of 1 equiv of a tertiary amine (Tri-n-butylamine) to ensure efficient utilization of the amine component (R^2R^3NH). Once again, carboxyl activation is achieved using a 2-halopyridinium salt (**2**) but, unlike the esterification reaction, amidation is best carried out as a two-step one-pot process.

Yields for both the esterification (eq 1) and amidation (eq 2) reactions are generally high, but the advantages associated with use of (**1**) over other reagents for these types of transformation have yet to be clearly defined.

Glycosylation.[5] 3,4-Dihydro-2*H*-pyrido[1,2-*a*]pyrimidin-2-one has been used in conjunction with Silver(I) Trifluoromethanesulfonate as a reagent combination to mediate the glycosylation of *N*-protected 2-amino-2-deoxy sugars using a glycosyl bromide as the glycosyl donor (eq 3).

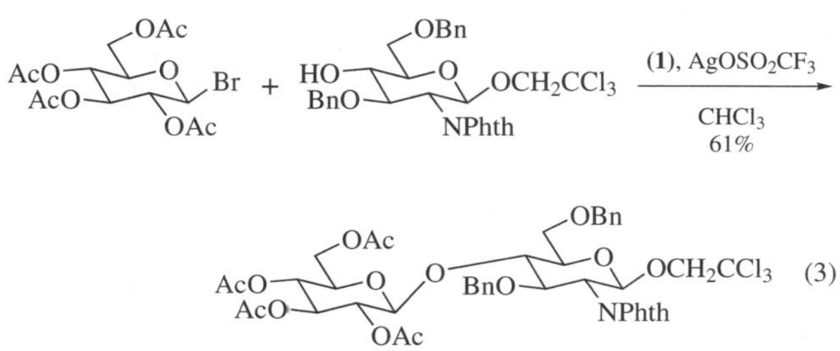

(3)

1. Hurd, C. D.; Hayao, S. *J. Am. Chem. Soc.* **1955**, *77*, 117.
2. (a) Magidson, O. Y.; Elina, A. S. *J. Gen. Chem. USSR (Engl. Transl.)* **1946**, *16*, 1933 (*Chem. Abstr.* **1947**, *41*, 6219) . (b) Adams, R.; Pachter, I. J. *J. Am. Chem. Soc.* **1952**, *74*, 4906, 5491. (c) Krishnan, M.; *Proc. Indian Acad. Sci., Sect. A* **1955**, *42*, 289. (d) Baltrusis, R.; Maciulis, A.; Purenas, A. *Lietuvos TSR Moksln Akad. Darbai. Ser. B* **1962**, 125 (*Chem. Abstr.* **1963**, *58*, 6827a) . (e) Biniecki, S.; Modrzejewska, W. *Acta Pol. Pharm.* **1984**, *41*, 607 (*Chem. Abstr.* **1986**, *104*, 50 843p) .
3. Mukaiyama, T.; Toda, H.; Kobayashi, S. *Chem. Lett.* **1976**, 13.
4. Mukaiyama, T.; Aikawa, Y.; Kobayashi, S. *Chem. Lett.* **1976**, 57.
5. Ogawa, T.; Nakabayashi, S.; Sasajima, K. *C.R. Hebd. Seances Acad. Sci.*, **1981**, *96*, 29.

Timothy Gallagher
University of Bristol, UK

1,3-Diisopropylcarbodiimide[1]

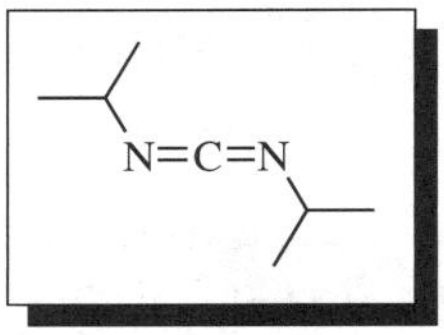

[693-13-0] $C_7H_{14}N_2$ (MW 126.20)

(coupling agent used for the preparation of amides, peptides, ureas, heterocycles, and unsymmetrical carbodiimides; used with DMSO to perform a modified Moffatt-type oxidation; activator in some polymerization reactions)

Alternate Name: DIC, DIPC, DIPCD, DIPCDI, *N,N′*-methanetetraylbis-2-propanamine, *N,N′*-diisopropylcarbodiimide, diisopropylcarbodiimide, 2-propanamine, *N,N′*-methanetetraylbis-(9ci).
Physical Data: bp 145–148 °C; fp 33.88 °C; density 0.806.
Form Supplied in: colorless liquid; widely available.
Purification: can be purified by distillation.
Handling, Storage, and Precautions: severe eye[2] and skin irritant; may cause allergic respiratory and skin reactions in certain individuals; handle in a well-ventilated chemical fume hood with goggles, gloves, and a lab coat; store in a cool dry place.

Introduction. The first carbodiimide synthesized was 1,3-diethylcarbodiimide in 1893,[3] followed by several other carbodiimides in 1938[4] and diisopropylcarbodiimide (DIC) in 1941.[5] It was observed that most of the carbodiimides either decomposed or polymerized except for dicyclohexylcarbodiimide (DCC) and DIC. Due to their greater stability and availability, DIC and DCC are now two of the most popular carbodiimides.

The mechanism of carbodiimide-mediated coupling reactions was first proposed in 1955[6] and involved attack of the carbodiimide by a carboxylic acid to form the *O*-acylisourea (**1**) intermediate, followed by nucleophilic attack on the activated acid to form the product **2** and urea by-product **3** (eq 1). However, this mechanism did not account for the significant amount of epimerization that is sometimes observed in the product. Epimerization can be avoided if the nucleophile attacks the *O*-acylisourea before it undergoes an intramolecular cyclization to form the 5(4*H*)-oxazolone (**4**) or attack by another acid to form the symmetrical anhydride **5**.[7] Both **4** and **5** can racemize to give the undesired epimers. The pathways to epimerization are shown in eq 2.

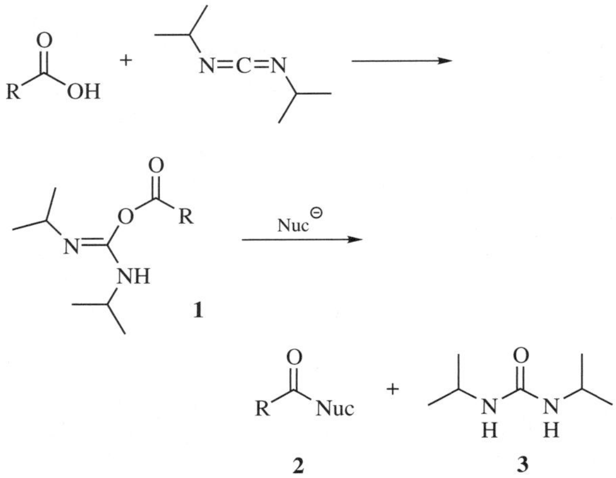

(1)

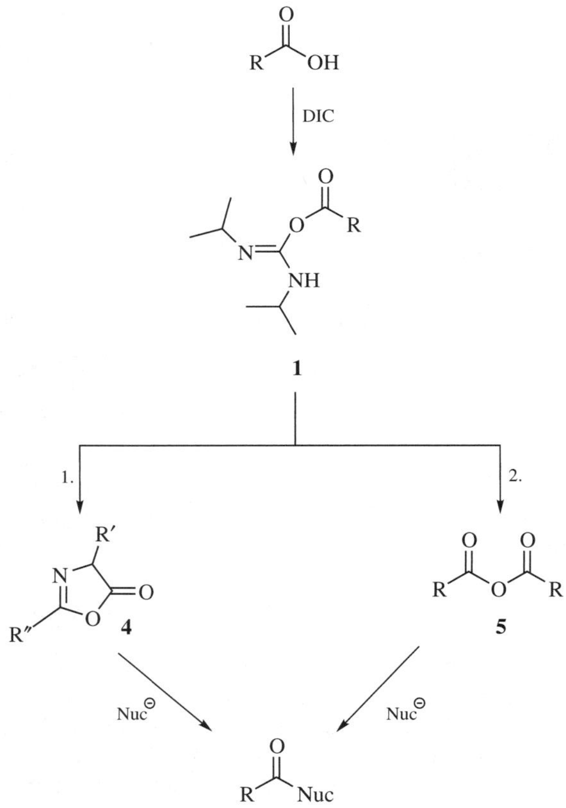

(2)

1. R = CH(R')NHCOR"
2. additional RCOOH

Several additives have been developed to facilitate DIC coupling reactions and minimize epimerization of the activated acids.[8] The effect of such additives is attributed to their ability to intercept the *O*-acylisourea, or to react with the 5(4*H*)-oxazolones and symmetrical anhydrides before they can racemize. Some of the common additives include *N*-hydroxysuccinimide (NHS),[9]

1-hydroxybenzotriazole (HOBt),[10–12] 1-hydroxy-7-azabenzotriazole (HOAt),[13] and 1-hydroxy-1*H*-1,2,3-triazole-4-carboxylate (HOCt).[14,15] Interestingly, DIC/HOAt coupling systems have been shown to be better than DIC/HOBt systems,[13,16,17] and HOCt has been proven superior to both HOAt and HOBt, but unfortunately it is not commercially available.[14,15] Also, the amount of time allowed for generation of the active ester can affect the extent of racemization; in some cases, only 5 min of preactivation retains the maximum amount of sterochemistry.[17]

The choice of solvent can also affect the extent of racemization. DIC-mediated couplings performed in methylene chloride (CH_2Cl_2) exhibit very little racemization, but the reactions are rather slow.[16] Conversely, dimethylformamide (DMF) has been shown to increase the reaction rate, but also increases the extent of racemization.[16] Hence, maximized reaction rates with minimal racemization of the product have been achieved with mixtures of DMF and methylene chloride.[16,17] In addition to optimizing the solvent systems for DIC coupling reactions, other reagents can be added to accelerate the couplings. Some examples include the bases dimethylaminopyridine (DMAP),[16,18,19] diisopropylethylamine (DIPEA),[17,18] *N*-methylmorpholine (NMM), and 2,4,6-trimethylpyridine (TMP), the latter of which was shown to be the most effective.[17]

Amide Bond and Peptide Formation. Carbodiimides are invaluable in the synthesis of peptides, thus peptide bond formation has come to be their principle use.[20] DIC is a very effective agent for the activation of carboxylic acids toward nitrogen nucleophiles. In fact, in a comparative study of carbodiimides, DIC was shown to be a better coupling agent in solution phase than EDC–HCl, EDC–methohexafluorophosphate, and EDC–methiodide.[15] One example of this utility is the acylation of oxazolidinones with a series of substituted phenylacetic acids in the presence of DIC and DMAP to give the substituted oxazolidinones in 61–77% (eq 3). The methodology was also applied to the coupling of several acryloyl derivatives to oxazolidinones to form α,β-unsaturated acylated derivatives in 62–84% (eq 4).[21]

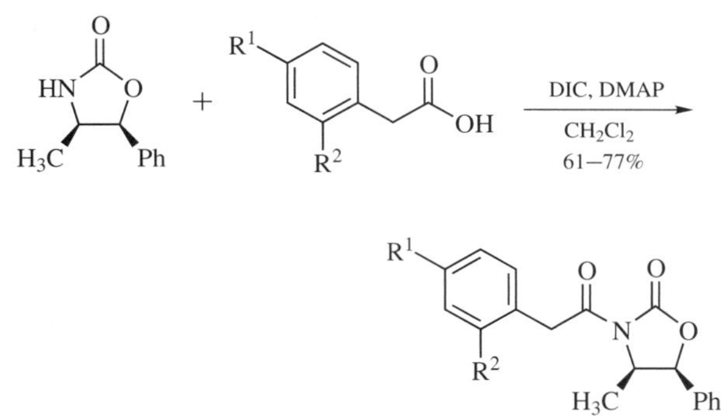

(3)

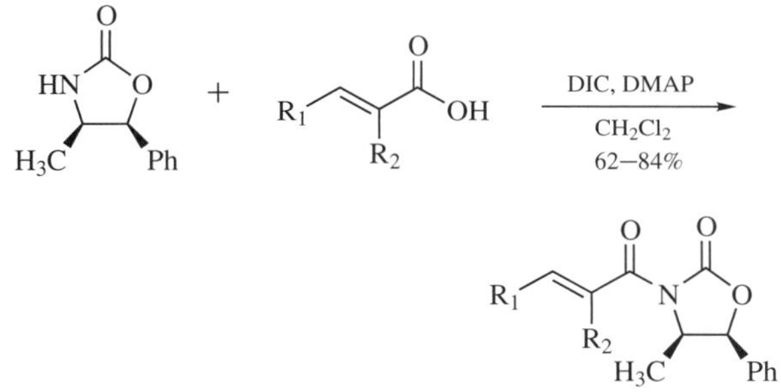

(4)

A DIC/HOBt coupling system has been employed to synthesize 2-amino-1,4-naphthoquinone derivatives from aminoquinones and *N*-Boc-amino acids, however the yields were moderate to poor (37–54%) (eq 5).[22]

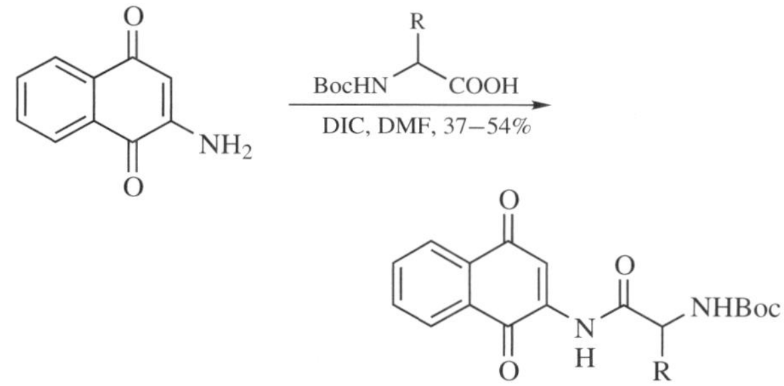

(5)

DCC was the first carbodiimide to be used in solid-phase synthesis,[23] but complications and low yields resulting from the dicyclohexyl urea precipitate quickly made it a less attractive choice. DIC, as a coupling agent, is more amenable to solid-phase peptide synthesis than DCC and EDC for several reasons. DIC is commercially available as a liquid, the diisopropyl urea is easier to remove than the DCC by-product urea, and DIC is less costly than EDC. Other carbodiimides such as *N-tert*-butyl-*N′*-methylcarbodiimide,[24] *N-tert*-butyl-*N′*-ethyl-carbodiimide,[25] and *N*-cyclohexyl-*N′*isopropylcarbodiimide[26] have found some success in solving the problem of by-product solubility, but are comparatively time consuming and costly to prepare. DIC has been used successfully to attach linkers via ester bonds on to alcohol Wang resins,[19,27] and on to solid supports such as Fmoc-PAL-Nle-MBHA-polystyrene[18] and Tentagel[16] via amide bonds. These couplings are sometimes facilitated by microwave radiation.[27] Once the linkers are attached to the resins, DIC has again been used to couple various compounds by amide[28,29] or ester linkages.[19,30] Interestingly, one study of coupling leucine-enkephalin to a polyethylene-COOH solid support found that DIC worked best, followed by EDC and then DCC.[31]

Isourea and Urea Formation. *O*-Alkylisoureas serve as valuable alkylating reagents and are readily prepared from the reaction of an alkyl alcohol with a carbodiimide in the presence of catalytic CuCl (eq 6).[32] Attempts to apply this methodology to the hydroxyl group of serine and threonine derivatives in the presence of DIC gave only the dehydro amino acids, arising from β-elimination of the corresponding isoureas (eq 7).[33] The *N*-protected dehydro amino acids of serine, cysteine, and threonine esters were prepared this way in 65–82% yield. The cycloaddition of carbodiimides with oxiranes to form the corresponding 2-oxazolidinimines has been demonstrated in the presence of organotin halide-Lewis base catalysts. An attempt to apply this transformation to DIC and propylene oxide, however, resulted in formation of the isourea (eq 8).[34]

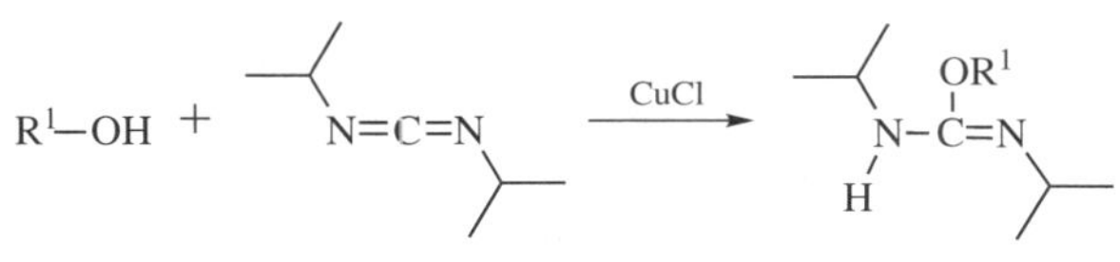

(6)

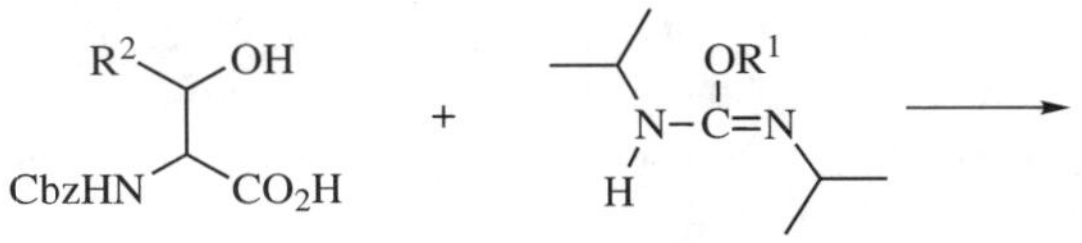

$R^1 = CH_3, CH_2Ph$
$R^2 = H, CH_3$

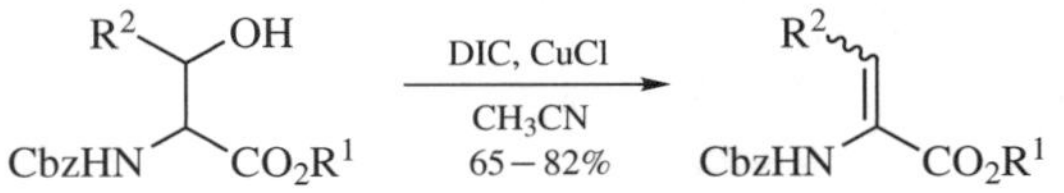

(7)

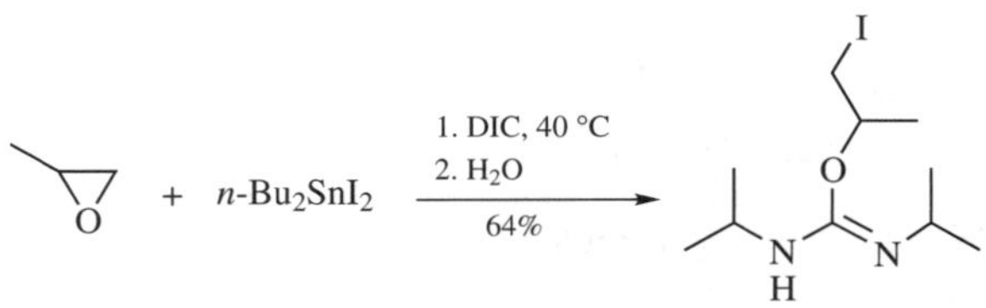

(8)

Ureas are not only the by-products of carbodiimide mediated coupling reactions, but they can also serve as valuable intermediates in small molecule synthesis. One example of this is the treatment of DIC with dilute acetic acid to form the diisopropylurea, which is then converted into *N,N′*-diisopropylbarbituric acid.[35] Another example is the three component coupling of methyl hydrogen maleate, an alcohol or amine, and DIC to form *N*-carbamoylaspartic acid derivatives (**6**) (eq 9).[36]

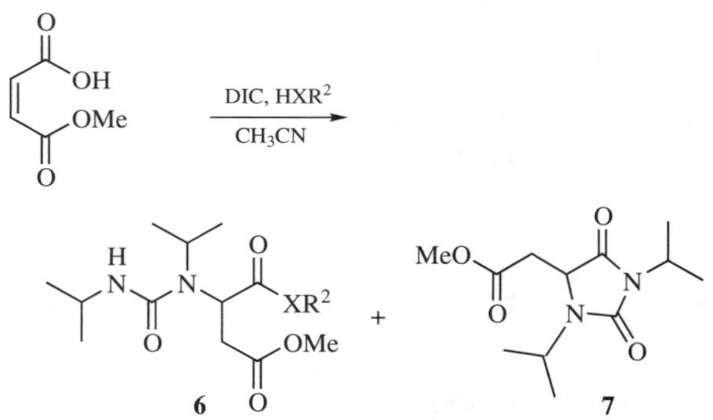

(9)

XR^2	**6**	**7**
BnNH	71%	8%
Et_2N	55%	0%
MeO	60%	17%
i-PrO	65%	10%

Modified Moffatt-Type Oxidation. The Moffatt oxidation has been employed to convert primary and secondary hydroxyl groups to aldehydes and ketones. This is traditionally accomplished by treatment of the alcohol with 3 equiv of DCC and 0.5 equiv of an acid of intermediate strength in DMSO.[37,38] DIC is a viable substitute for DCC in this conversion and has the advantage of producing an organic soluble *N,N′*-dialkylurea by-product. One example is the oxidation of the primary alcohol of 2′,3′-*O*-cyclohexylideneuridine to the corresponding aldehyde in 65% yield with DIC in a solution of dichloroacetic acid in DMSO (eq 10).[39] Another example is the oxidation of a primary alcohol in a derivatized thiazole to the corresponding aldehyde with DMSO and DIC.[40] The mild reaction conditions that tolerate most conventional protecting groups, and the specific oxidation of primary hydroxyls to aldehydes make this oxidation methodology very attractive for sensitive substrates.[39] Related oxidative methods include using DMSO and acetic anhydride,[41] phosphorus pentoxide[42] or the sulfur trioxide-pyridine complex.[43]

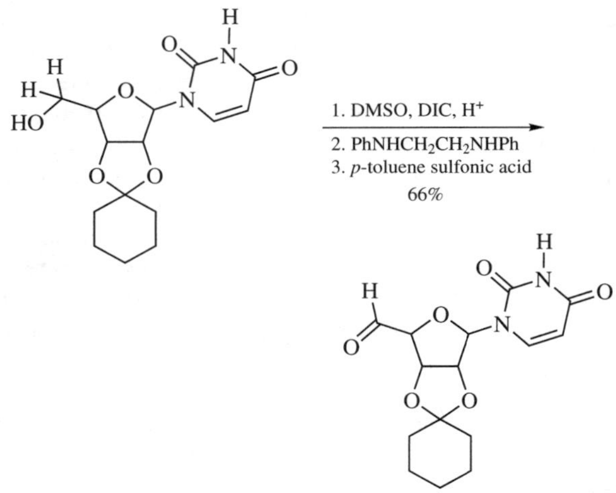

(10)

DIC-Derived Heterocycles. Several instances exist in which DIC reacts with various compounds and is incorporated into the heterocyclic product. One example is the cyclization of *O*-thiobenzoquinone methide with DIC to form 3,4-dihydro-2*H*-1,3-benzothiazin-2-imine (eq 11).[44] Another example is the generation of spirocyclohexadienones from the reaction of DIC with 4-hydroxyphenyl acetic acid (eq 12).[45] A [3 + 2] cycloaddition of heteroallenes to DIC can also occur to generate a 4,5-dihydro-1*H*-tetrazolium antimonate salt.[46] Additionally, barbituric acid derivitives,[47] β-lactones,[48] 4,5-dihydro-5-imino-1*H*-1,2,4-triazolium salts,[49] and oxizinoquinolines[50] have been synthesized by the incorporation of DIC.

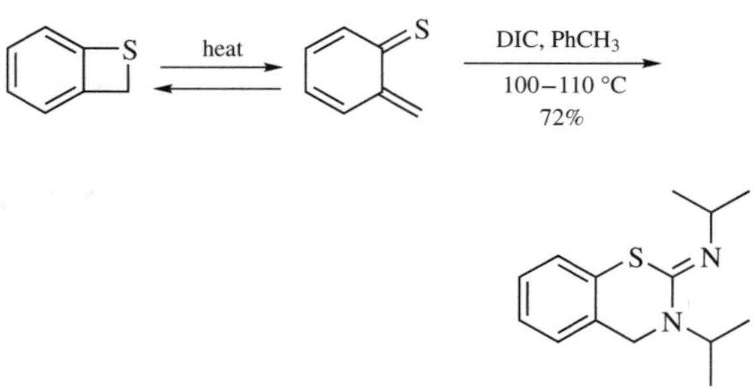

(11)

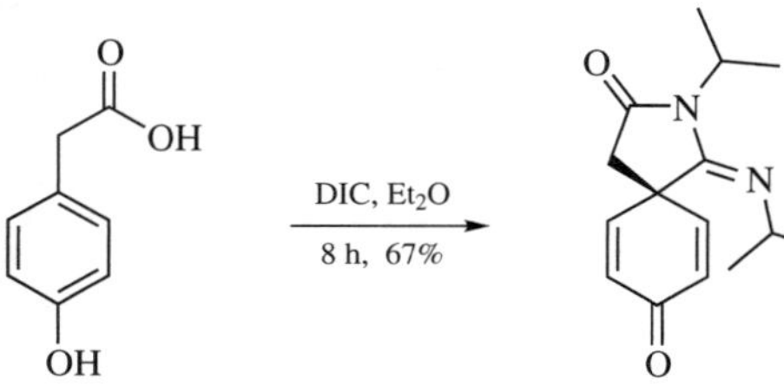

(12)

Additional Reactions

Preparation of Unsymmetrical Carbodiimides

The demand for unsymmetrical carbodiimides arises when portions of, or the whole carbodiimide framework is incorporated into the product, as seen in the cases of some heterocycle formations. Hence, there are several methods for synthesizing unsymmetrical carbodiimides from DIC. One is to heat DIC in toluene to 120 °C with another symmetrical carbodiimide, such as di(*p*-tolyl)carbodiimide, in the presence of 10 mole % iminophosphorane.[51] A similar method involves the use of a vanadium oxo or a vanadium imido complex to catalyze the reaction.[52] The best results from this method involved the formation of *N,N′*-cyclohexylisopropylcarbodiimide in 69% yield using $V(O)(acac)_2$ as the catalyst and the two corresponding carbodiimides.

Synthesis of Hydroxamic Acids

Although reactions of *N*-unsubstituted hydroxamic acids with carbodiimides initiate Lossen rearrangements,[53] hydroxamic acids reportedly have been formed from the corresponding free carboxylic acids using DIC and hydroxylamine hydrochloride (eq 13).[54,55] However, more efficient syntheses of hydroxamic acids have been developed, such as carbodiimide mediated couplings of acids and *O*-benzylhydroxylamine.[56]

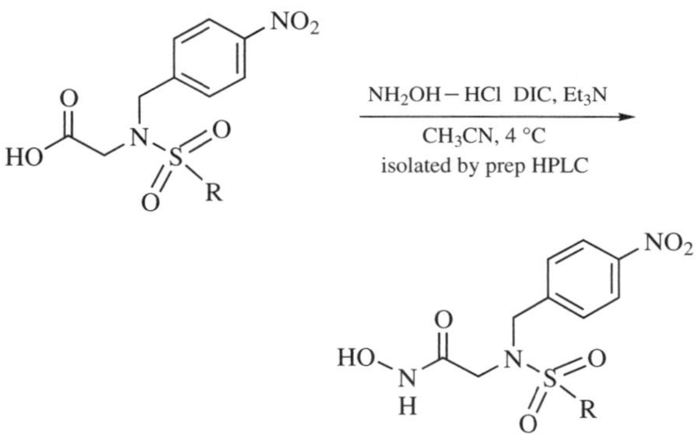

(13)

Polymers

DIC has been shown to be an effective activator in the polymerization of caprolactam.[57,58] DCC and CO_2 were also effective activators; the only notable difference between the three was the morphology and particle size of the polycaproamide products. DCC generated the largest particle sizes and CO_2 gave the smallest.

Phosphorus Chemistry

DIC has also been applied to phosphorus chemistry. Examples include a mechanistic study of the reaction of DIC with various phosphorus acids,[59] the DIC-mediated transfer of a phosphate between hydroxyl groups of a sugar,[60] and a DIC coupling of a modified diphosphate.[61]

Related Reagents. Polyhexamethylenecarbodiimide; 1-cyclo-hexyl-3-(2-morpholinoethyl)-carbodiimide; 1-cyclohexyl-3-(2-morpholinoethyl)-carbodiimide metho-*p*-toluenesulfonate (morpho-CDI); 1-ethyl-3-(3′-dimethylaminopropyl)-carbodiimide hydrochloride (EDC); 1,3-dicyclohexylcarbodiimide (DCC); 1,3-dicyclohexylcarbodiimide-copper(I) chloride; 1,3-dicyclohexylcarbodiimide-4-dimethylaminopyridine.

---

1. (a) Khorana, H. G., *Chem. Rev.* **1953**, *53*, 145. (b) Neumann, W.; Fisher, P., *Angew. Chem. Int. Ed. Engl.* **1962**, *1*, 621. (c) Kurzer, F.; Douraghi-Zadeh, K., *Chem. Rev.* **1967**, *67*, 107. (d) Mikolajczyk, M.; Kielbasinski, P., *Tetrahedron* **1981**, *37*, 233.
2. Hayes, B. B.; Gerber, P. C.; Griffey, S. S.; Meade, B. J., *Drug Chem. Toxicol.* **1998**, *21*, 195.
3. Chancel, F., *Compt. Rend.* **1893**, *116*, 330.
4. Schmidt, E.; Hitzler, F.; Lahde, E., *Chem. Ber.* **1938**, *71*, **1933**.
5. Schmidt, E.; Striewsky, W., *Chem. Ber.* **1941**, *74*, 1285.
6. Khorana, H. G., *Chem. Industry* **1955**, 1087.
7. Slebioda, M., *Polish, J., Chem.* **1994**, *68*, 957.
8. Weber, P. J. A.; Bader, J. E.; Folkers, G.; Beck-Sickinger, A. G., *Bioorganic & Med. Chem. Lett.* **1998**, *8*, 597.
9. Weygand, F.; Hoffmann, D.; Wünsch, E., *Z Naturforsch., Teil B* **1966**, *21*, 426.
10. König, W.; Geiger, R., *Chem. Ber.* **1970**, *103*, 788.
11. König, W.; Geiger, R., *Chem. Ber.* **1970**, *103*, 2024.
12. König, W.; Geiger, R., *Chem. Ber.* **1970**, *103*, 2034.
13. Carpino, L. A.; El-Faham, A.; Minor, C. A.; Albericio, F., *J. Chem. Soc. Chem. Commun.* **1994**, 201.
14. Robertson, N.; Jiang, L.; Ramage, R., *Tetrahedron* **1999**, *55*, 2713.
15. Jiang, L.; Davison, A.; Tennant, G.; Ramage, R., *Tetrahedron* **1998**, *54*, 14233.
16. Carpino, L. A.; El-Faham, A., *Tetrahedron* **1999**, *55*, 6813.
17. Han, Y.; Albericio, F.; Barany, G., *J. Org. Chem.* **1997**, *62*, 4307.
18. Angell, Y. M.; García-Echeverría, C.; Rich, D. H., *Tetrahedron Lett.* **1994**, *35*, 5981.
19. Kuisle, O.; Quiñoá, E.; Riguera, R., *J. Org. Chem.* **1999**, *64*, 8063.
20. Sheehan, J. C.; Hess, G. P., *J. Am. Chem. Soc.* **1955**, *77*, 1067.
21. Grahm, J. M.; Shireman, B. T.; Maddux, T. M.; Brandstedt, C. M.; Zeller, W. E., *Synthetic Communications* **2000**, *30*, 1221.
22. Pachatouridis, C.; Couladouros, E. A.; Papageorgio, V. P.; Liakopoulou-Kyriakides, M., *Lett. Pept. Sci* **1998**, *5*, 259.
23. Shenoy, N. R.; Bailey, J. M.; Shively, J. E., *Protein Sci.* **1992**, *1*, 58.
24. Sarantakis, D.; Teichman, J.; Lien, E. L.; Fenichel, R. L., *Biochem. Biophys. Res. Commun.* **1976**, *73*, 336.
25. Tartar, A.; Gesquiere, J. C., *J. Org. Chem.* **1979**, *44*, 5000.
26. Izdebski, J.; Bondaruk, J.; Gumulka, S. W.; Krzascik, P., *Int. J. Peptide Protein Res.* **1989**, *33*, 77.
27. Stadler, A.; Kappe, C. O., *Tetrahedron* **2001**, *57*, 3915.
28. Page, P.; Bradley, M.; Walters, I.; Teague, S., *J. Org. Chem.* **1999**, *64*, 794.
29. Haver, A. C.; Smith, D. D., *Tetrahedron. Lett.* **1993**, *34*, 2239.
30. Li, W.; Czarnik, A. W.; Lillig, J.; Xiao, X., *J. Comb. Chem.* **2000**, *2*, 224.
31. Izebski, J.; Bondaruk, J.; Kunce, D.; Chung, N. N.; Schiller, P. W., *Polish, J., Chem.* **1994**, *68*, 109.
32. Mathias, L. J., *Synthesis* **1979**, 561.
33. Miller, M. J., *J. Org. Chem.* **1980**, *45*, 3131.
34. Shibata, I.; Baba, A.; Iwasaki, H.; Matsuda, H., *J. Org. Chem.* **1986**, *51*, 2177.
35. Bohanon, T. M.; Caruso, P. L.; Denzinger, S.; Fink, R.; Möbius, D.; Paulus, W.; Preece, J. A.; Ringsdorf, H.; Schollmeyer, D., *Langmuir* **1999**, *15*, 174.
36. Kishikawa, K.; Sankhavasi, W.; Yoshizaki, K.; Kohmoto, S.; Yamamoto, M.; Yamada, K., *J. Chem. Soc., Perkin Trans. 1* **1994**, 1205.
37. Pfitzner, K. E.; Moffatt, J. G., *J. Am. Chem. Soc.* **1965**, *87*, 5661.
38. Pfitzner, K. E.; Moffatt, J. G., *J. Am. Chem. Soc.* **1965**, *87*, 5670.
39. Jones, G. H.; Moffatt, J. G., In *Methods in Carbohydrate Chemistry*, Whistler, R. L.; BeMiller, J. N., Eds.; Academic Press: New York, **1972**, p. 315.
40. Hara, T.; Sheehan, J. C., *Heterocycles* **1981**, *16*, 1295.

41. Albright, J. D.; Goldman, L., *J. Am. Chem. Soc.* **1967**, *89*, 2416.
42. Onodera, K.; Hirano, S.; Kashimura, N., *Carbohyd. Res.* **1968**, *6*, 276.
43. Parikh, J. R.; von Doering, W. E., *J. Am. Chem. Soc.* **1967**, *89*, 5505.
44. Gröschl, D.; Niedermann, H. P.; Meier, H., *Chem. Ber.* **1994**, *127*, 955.
45. Gubernator, K.; Hofeditz, W.; Plieninger, H., *Chem. Ber.* **1980**, *113*, 669.
46. Wirschun, W.; Winkler, M.; Lutz, K.; Jochims, J. C., *J. Chem. Soc., Perkin Trans. 1* **1998**, 1755.
47. Buckingham, D. A.; Clark, C. R.; McKeown, R. H.; Wong, O., *J. Am. Chem. Soc.* **1987**, *109*, 466.
48. Geffken, D.; Zinner, G., *Chem. Ber.* **1975**, *108*, 3730.
49. Wang, Q.; Amer, A.; Troll, C.; Fischer, H.; Jochims, J. C., *Chem. Ber.* **1993**, *126*, 2519.
50. March, L. C.; Romanchick, W. A.; Bajwa, G. S.; Joullie, M. M., *J. Med. Chem.* **1973**, *16*, 337.
51. Bell, S. A.; Geib, S. J.; Meyer, T. Y., *J. Chem . Soc. Chem. Commun.* **2000**, 1375.
52. Birdwhistell, K. R.; Lanza, J.; Pasos, J., *J. Organomet. Chem.* **1999**, *584*, 200.
53. Hoare, D. G.; Olson, A.; Koshland, D. E., *J. Am. Chem. Soc.* **1968**, *90*, 1638.
54. Scozzafava, A.; Supuran, C. T., *Bioorganic & Med. Chem.* **2000**, *8*, 637.
55. Scozzafava, A.; Supuran, C. T., *J. Med. Chem.* **2000**, *43*, 1858.
56. Apfel, C.; Banner, D. W.; Bur, D.; Dietz, M.; Hirata, T.; Hubschwerlen, C.; Locher, H.; Page, M. G. P.; Pirson, W.; Rossé, G.; Specklin, J. L., *J. Med. Chem.* **2000**, *43*, 2324.
57. Dan, F.; Vasiliu-Oprea, C., *Colloid Polym. Sci.* **1998**, *276*, 483.
58. Dan, F.; Vasiliu-Oprea, C., *J. Appl. Polym. Sci.* **1998**, *67*, 231.
59. Mikolajczyk, M.; Kielbasinski, P.; Basinski, W., *J. Org. Chem.* **1984**, *49*, 899.
60. Walsh, J. P.; Caldwell, K. K.; Majerus, P. W., *Proc. Natl. Ac. Sci. USA* **1991**, *88*, 9184.
61. (a) Lesiak, K.; Watanabe, K. A.; Pankiewicz, K. W., *Nucleosides and Nucleotides* **1998**, *17*, **1857**. (b) Lesiak, K.; Watanabe, K. A.; George, J.; Pankiewicz, K. W., *J. Org. Chem.* **1998**, *63*, **1906**. (c) Lesiak, K.; Watanabe, K. A.; Majumdar, A.; Powell, J.; Seidman, M.; Vanderveen, K.; Goldstein, B. M.; Pankiewicz, K. W., *J. Med. Chem.* **1998**, *41*, 618.

George P. Nora, Clara Schous, Susanna M. Troyer & Marvin J. Miller
University of Notre Dame, IN, USA

2-(Dimethoxymethyl)naphthalene

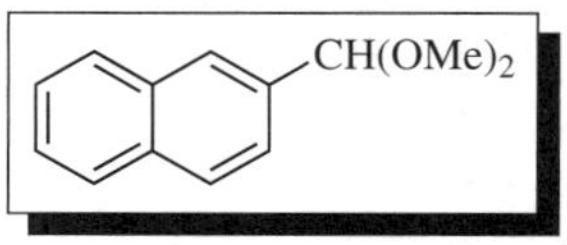

[77196-31-7] $C_{13}H_{14}O_2$ (MW 202.25)

(reagent used for the protection of diols as the versatile 2-naphthylmethylene acetal)

Alternate Name: 2-naphthaldehyde dimethylacetal.
Physical Data: bp 141–143 °C/2 mm Hg; *d* 1.10 g cm^{-3}.
Solubility: soluble in alcohol, ether, and most organic solvents.
Form Supplied in: colorless clear liquid. The title reagent is not commercially available, it can be prepared from 2-naphthaldehyde which is commercially available.
Preparative Methods: trimethyl orthoformate and Amberlite IR 120 (H^+) resin were added to a solution of 2-naphthaldehyde in methanol and stirred overnight at room temperature. The mixture was filtered, neutralized with aq $NaHCO_3$, and evaporated. The residue was dissolved in dichloromethane, washed with water, dried over $MgSO_4$, and evaporated in vacuum. The residue is a yellowish clear liquid, pure enough for synthesis (>95%). Vacuum distillation can be applied for further purification which gives a colorless liquid. A preparative method without solvent has also been described.[1]
Handling, Storage, and Precautions: should be kept in a cool and dry place.

Protection of Diols. Dioxane and dioxolane-type 2-naphthylmethylene acetals are easily prepared from diols by acid-catalyzed transacetalization with 2-(dimethoxymethyl)naphthalene in DMF.[2–4] 2-(Diisobutoxymethyl)naphthalene can also be used in an acid-catalyzed acetal exchange reaction for the synthesis of acetals.[5,6] Sugar hexopyranosides yield the dioxane-type 4,6-*O*-2-naphthylmethylene acetals exclusively in the thermodynamically stable stereoisomeric form, which is with the 2-naphthyl group in the equatorial position (eq 1).[2,3,5]

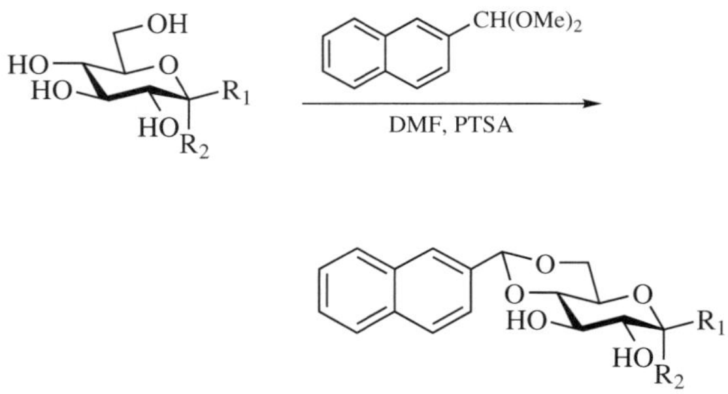

(1)

1, $R_1 = H$, $R_2 = OMe$, 91%
2, $R_1 = OAll$, $R_2 = H$, 87%
3, $R_1 = OPMP$, $R_2 = H$, 92%

The reaction of 2-(dimethoxymethyl)naphthalene with *cis* axial/equatorial hydroxyls of pyranosides gives diastereomeric dioxolane-type 2-naphthylmethylene acetals, in almost 1:1 ratio (eq 2), except for the 1,6-anhydro-β-D-manno- and -galactopyranoses which give predominantly the *endo*-naphthyl isomers.[4] In the presence of 2-naphthylmethylene acetals, alkylation, acylation, and silylation can be achieved with very high yields.[2–7]

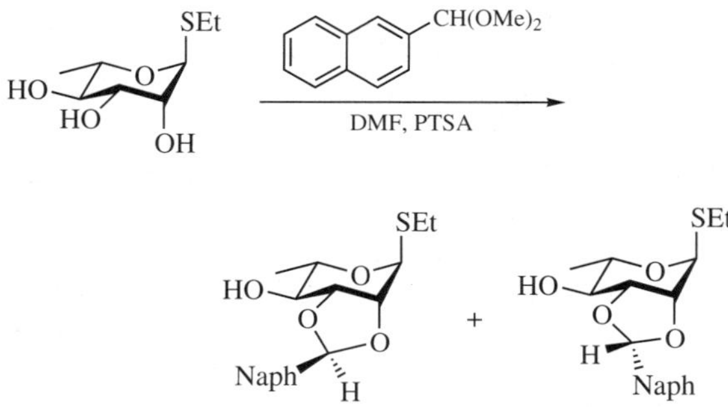

(2)

exo (41%) *endo* (47%)

As with the benzylidene and *p*-methoxybenzylidene groups (see *p*-methoxybenzaldehyde), one of the two *C-O* bonds in this acetal can be cleaved selectively. The *p*-methoxybenzyl- and the 2-naphthylmethyl ethers (see 2-naphthylmethyl bromide) can be removed selectively in the presence of benzyl ethers. The advantage of the 2-naphthylmethylene acetals and the 2-naphthylmethyl ethers over the *p*-methoxybenzylidene acetals and the *p*-methoxybenzyl ethers is their higher stability under acidic conditions. The 2-naphthylmethylene acetal group can be removed by using 0.4 equiv of DDQ at room temperature.

Partial Selective Hydrogenolysis of the Acetal. In the field of carbohydrates, partial reductive cleavage of the dioxane-type acetals is possible under mild conditions with high regioselectivity. The proper choice of the reagents leads to the formation of the appropriate NAP ethers.[2,3,5–7] The use of $AlH_3(3LiAlH_4–AlCl_3)$,[2,3] DIBAL-H in toluene–dichloromethane, or $Et_3SiH–TfOH$[7] results in the formation of the 4-ONAP/6-OH derivatives, whereas $BH_3.NMe_3–AlCl_3$[2,3,5] in THF, $NaCNBH_3–AlCl_3$,[2,3] or $Et_3SiH–PhBCl_2$[7] produces the 6-ONAP/4-OH derivatives (eq 3).

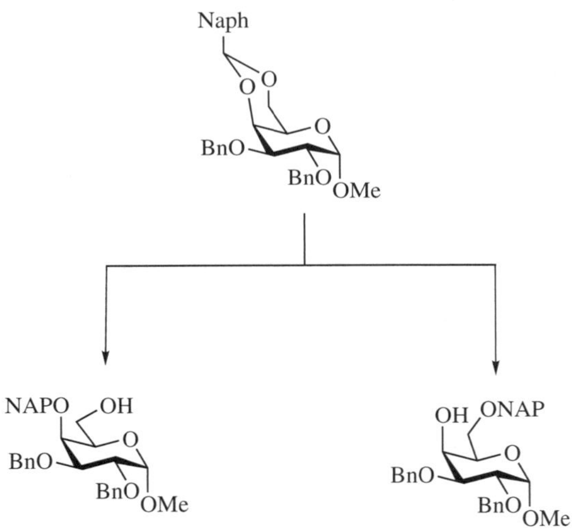

(3)

85% ($3LiAlH_4–AlCl_3$, Et_2O–DCM)
86% (Et_3SiH–TfOH, 4 Å MS, DCM)

64% ($Me_3N.BH_3–AlCl_3$, THF)
71% ($Et_3SiH–PhBCl_2$, 4 Å MS, DCM)

In the case of dioxolane-type acetals of hexopyranosides, the direction of the acetal cleavage is determined by the configuration of the acetalic carbon atom. The *endo*-naphthyl isomers furnish axial NAP-ethers, and the *exo*-naphthyl isomers give equatorial NAP-ethers using either AlH_3 ($3LiAlH_4–AlCl_3$) or $BH_3.NMe_3–AlCl_3$ in THF (eq 4).

Reaction with an *N*-Nucleophile. Reaction of 2-(dimethoxymethyl)naphthalene with benzylidene sulfamide results in an unsymmetrical 1,2-diaryl ethanediamine derivative via the formation of a 2-naphthylmethylene imine followed by a reductive cross-coupling and a hydrolytic step (eq 5).[8]

Related Reagents. *p*-Methoxybenzaldehyde; α,α-dimethoxy toluene.

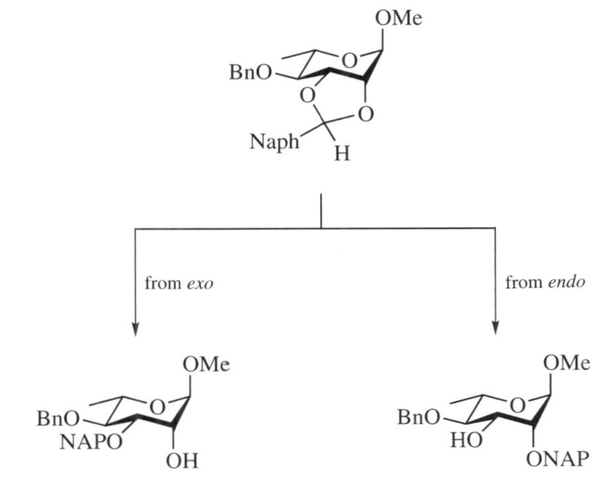

(4)

93% ($3LiAlH_4–AlCl_3$, Et_2O–DCM)
91% ($Me_3N.BH_3–AlCl_3$, THF)

92% ($3LiAlH_4–AlCl_3$, Et_2O–DCM)

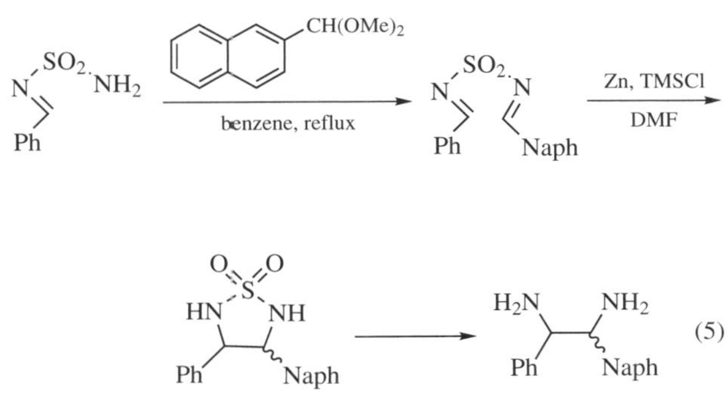

(5)

cis/*trans*: 1/1 (75%)

58%

1. Kimura, M.; Kuboki, A.; Sugai, T., *Tetrahedron: Asymmetry* **2002**, *13*, 1059.
2. Lipták, A.; Borbás, A.; Jánossy, L.; Szilágyi, L., *Tetrahedron Lett.* **2000**, *41*, 4949.
3. Borbás, A.; Szabó, Z. B.; Szilágyi, L.; Bényei, A.; Lipták, A., *Tetrahedron* **2002**, *58*, 5723.
4. Borbás, A.; Szabó, Z. B.; Szilágyi, L.; Bényei, A.; Lipták, A., *Carbohydr. Res.* **2002**, *337*, 1941.
5. Matsuoka, K.; Nishimura, S. I.; Lee, Y. C., *Tetrahedron: Asymmetry* **1994**, *5*, 2335.
6. Matsuoka, K.; Nishimura, S. I.; Lee, Y. C., *Carbohydr. Res.* **1995**, *276*, 31.
7. Wright, J. A.; Yu, J.; Spencer, J. B., *Tetrahedron Lett.* **2001**, *42*, 4033.
8. Pansare, S. V.; Malusare, M. G., *Tetrahedron Lett.* **1996**, *37*, 2859.

Anikó Borbás, Zoltán B. Szabó, Magdolna Csávás & András Lipták
HAS-DU Research Group for Carbohydrates, Debrecen, Hungary

2,2-Dimethoxypropane

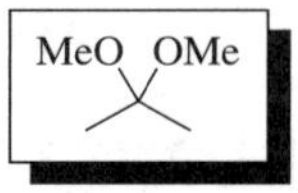

[77-76-9] $C_5H_{12}O_2$ (MW 104.15)

(reagent used in the preparation of acetals,[1,2] acetonides,[3] isopropylidene derivatives of sugars[4–6] and nucleosides,[7] methyl esters of amino acids,[8] and enol ethers;[9] undergoes the aldol condensation with enol silyl ethers;[10] used in the preparation of hypophosphite esters[11] and dimethylboronate esters[12])

Physical Data: bp 83 °C; *d* 0.847 g cm^{-3}.
Form Supplied in: liquid; widely available.
Handling, Storage, and Precautions: flammable liquid; irritant.

Formation of Acetals. The reagent can be used to prepare a variety of acetals via transacetalization. This involves interchanging the alkoxy or ketone groups in an acidic medium. Both symmetrical [$R^1R^2C(OR^3)_2$] and mixed [$R^1R^2C(OR^3)OR^4$] acetals can be obtained, depending upon the choice of conditions.[1]

The three equilibria shown in eqs 1–3 are established when 2,2-dimethoxypropane, an alcohol, and a catalytic amount of acid are mixed. Distillation of the methanol formed shifts the equilibrium far in the direction of the new acetals. Methanol and 2,2-dimethoxypropane form a binary azeotrope; however, this can be broken by the addition of a hydrocarbon solvent such as hexane or benzene.

(1)

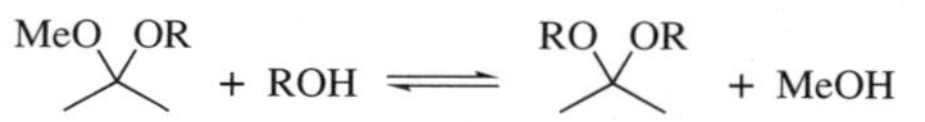

(2)

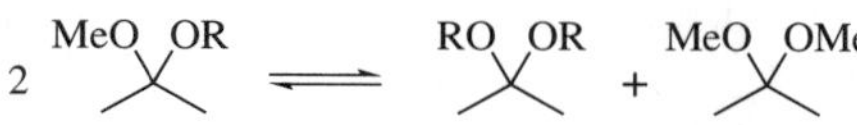

(3)

Methyl acetals of other ketones can be obtained by acidifying a mixture of the ketone, methanol, and 2,2-dimethoxypropane and removing the acetone formed by distillation. In this case, the function of the 2,2-dimethoxypropane is to react with the water formed to give methanol and acetone (eqs 4 and 5). This is evidenced by the fact that the reaction is very slow in the absence of alcohol or water, but fast when methanol is present. The rate of the reaction is also directly related to the alcohol concentration.[1]

(4)

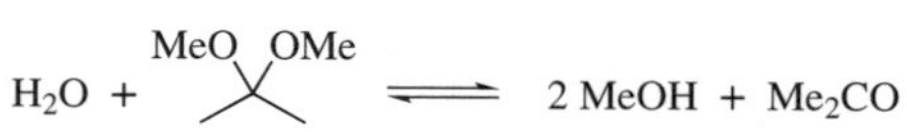

(5)

One undesirable side reaction is the formation of enol ethers via the acid-catalyzed loss of one molecule of alcohol from the acetal. This occurs even at relatively low temperatures (50 °C). In order to suppress this reaction, the procedure was modified to include a neutralization with base prior to workup.

A recent example of the use of 2,2-dimethoxypropane in the above capacity involves the reaction of an aldehyde or ketone with o-Nitrobenzyl Alcohol to afford the bis-*o*-nitrobenzyl acetal derivative.[2] This protecting group has the advantage that it is photoremovable in high yield by irradiation at 350 nm.

Acetonide Formation. The reagent has been used to great advantage as a protecting group for the labile dihydroxyacetone side chain of the corticosteroids. Cyclic acetal formation between the 17α,21-diol grouping of the prednisolone side chain and 2,2-dimethoxypropane in an acid-catalyzed exchange reaction gives 17α,21-isopropylidenedioxypregnane in good yield (eq 6).[3]

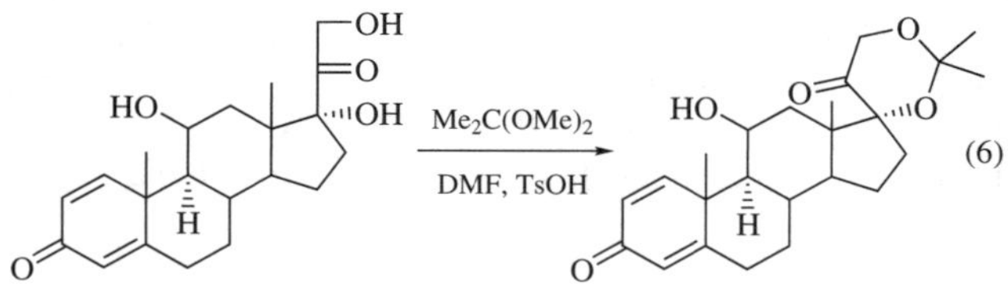

(6)

The base stability of the isopropylidenedioxy function makes it useful for the preparation of C-21 modified cortical hormones whose formation requires strongly basic conditions. The acetonide is easily cleaved with aqueous acetic or formic acid in the presence of heat.

Isopropylidene Derivatives of Sugars and Nucleosides. A mixture of 2,2-dimethoxypropane, DMF, and a catalytic amount of p-Toluenesulfonic Acid constitutes an efficient acetonating agent capable of protecting vicinal, *trans*-diequatorial hydroxy groups and of forming *N*-acetyl-2,2-dimethyloxazolidines from vicinal hydroxy and acetamido groups.[4] Reaction of methyl α-D-glucopyranoside under the above conditions gives recovered starting material, methyl 4,6-*O*-isopropylidene-α-D-glucoside, and a trace amount of the diisopropylidene glucoside (eq 7). Similarly, deoxyinosamine, upon treatment with the same reagent mixture, affords as the major product the diisopropylidene derivative (eq 8).

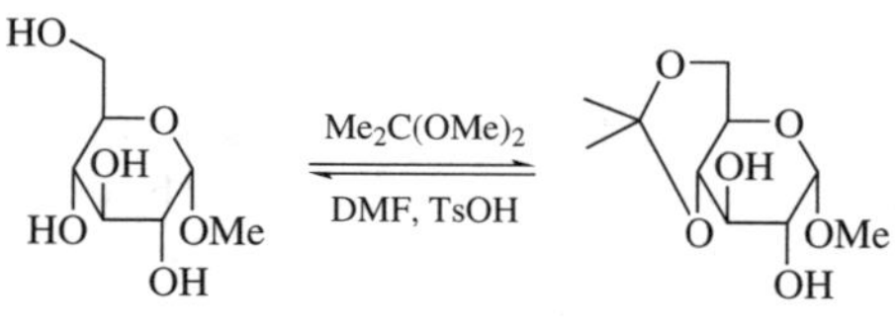

(7)

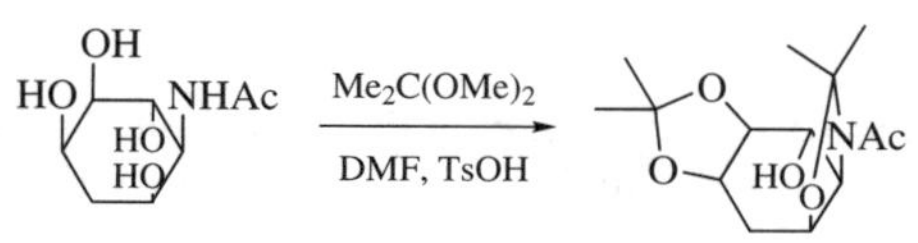

(8)

A further improvement to the existing procedure involves the substitution of Tin(II) Chloride for the *p*-toluenesulfonic acid and 1,2-dimethoxyethane for DMF.[5a] Studies involving D-mannitol have shown that the tin(II) chloride does not act as a Lewis acid catalyst or as a source of HCl (by reaction with the hydroxy groups) since the use of zinc chloride under identical conditions gives no reaction. In this particular case, it was subsequently demonstrated that no catalyst is required and that isopropylidenation occurs under completely neutral conditions, presumably via acetal exchange.[5b] Tin(II) chloride has been used to catalyze isopropylidene formation in other carbohydrate systems.[6]

The ability of 2,2-dimethoxypropane to act as a water scavenger has resulted in its use (in conjunction with acetone and an acid catalyst) in the protection of the 2′,3′-hydroxy groups of purine and pyrimidine nucleosides.[7] Here again, the formation of acetone and methanol serves to drive the equilibrium towards the product.

Esterification of Amino Acids. Methyl ester hydrochlorides of amino acids have been prepared in which 2,2-dimethoxypropane has been used as the source of the methoxy group, the reaction solvent, and the reagent for the removal of the water formed by virtue of the hydrolysis of the acetal to methanol and acetone.[8a] This procedure has also been used to prepare the methyl esters of fatty acids.[8b]

Enol Ethers. Treatment of testosterone acetate with 2,2-dimethoxypropane/DMF/TsOH converts this material into the enol ether (eq 9).[9] In this case, enol ether formation is attributed to loss of water directly from the unsaturated hemiacetal.

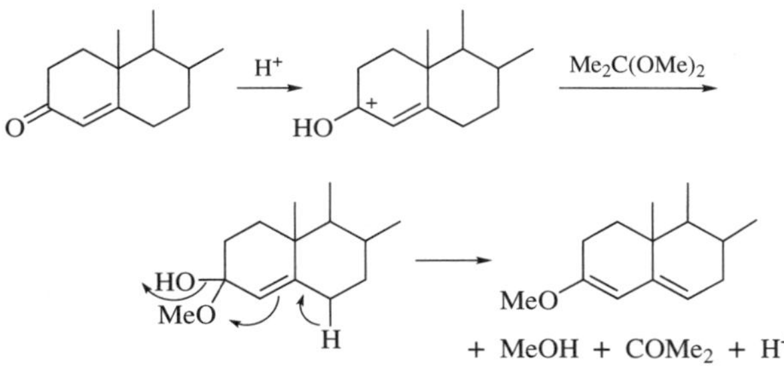

(9)

Aldol Condensation with Enol Silyl Ethers. The first example of an aldol condensation between an unactivated enolate and an electrophilically activated carbonyl substrate was accomplished using enol silyl ethers and acetals or orthoesters in conjunction with a catalytic amount of Trimethylsilyl Trifluoromethanesulfonate. The reaction proceeds very readily at −78 °C and affords the *erythro*-aldol product with high stereoselectivity.[10] The use of 2,2-dimethoxypropane in this reaction affords β-methoxy ketones (eq 10).

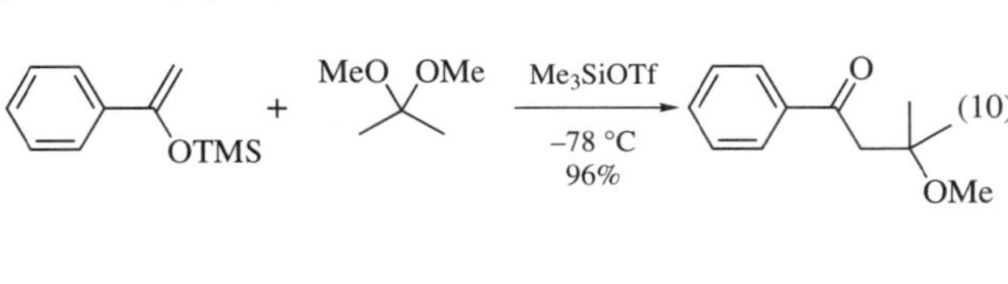

(10)

Hypophosphite Esters. Methyl hypophosphite has been prepared by the reaction between Hypophosphorous Acid and 2,2-dimethoxypropane (eq 11).[11] However, the product reacts further with the acetone formed to give 1-hydroxy-1,1-dimethylmethylphosphinate (eq 12).

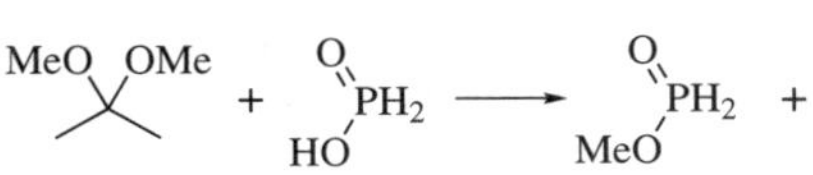

+ MeOH + Me$_2$CO (11)

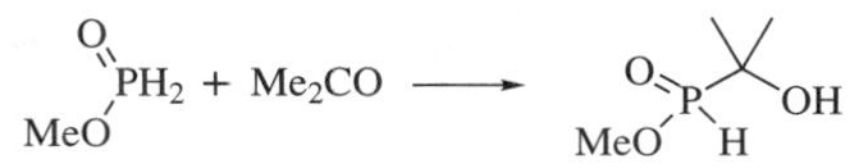

(12)

Dimethyl Boronate Esters. Dimethyl esters of arylboronic acids have been prepared by heating a mixture of the acid, 2,2-dimethoxypropane, and Zinc Chloride (catalyst).[12]

---

1. (a) Lorette, N. B.; Howard, W. L. *J. Org. Chem.* **1960**, *25*, 521, 525. (b) Brown, B. R.; MacBride, J. A. H. *J. Chem. Soc* **1964**, 3822.
2. Gravel, D; Murray, S.; Ladouceur, G. *Chem. Commun./J. Chem. Soc., Chem. Commun.* **1985**, 1828.
3. (a) Tanabe, M.; Bigley, B. *J. Am. Chem. Soc.* **1961**, *83*, 756. (b) Robinson, C. H.; Finckenor, L. E.; Tiberi, R.; Olivetto, E. P. *J. Org. Chem.* **1961**, *26*, 2863.
4. (a) deBelder, A. N. *Adv. Carbohydr. Chem. Biochem* **1965**, *20*, 219. (b) deBelder, A. N., *Adv. Carbohydr. Chem. Biochem.* **1977**, *34*, 179. (c) Evans, M. E.; Parrish, F. W.; Long, L. *Carbohydr. Res* **1967**, *3*, 453. (d) Hasegawa, A.; Fletcher, H. G. *Carbohydr. Res.* **1973**, *29*, 209, 223. (e) Hasegawa, A.; Nakajima, M. *Carbohydr. Res.* **1973**, *29*, 239.
5. (a) Chittenden, G. *Carbohydr. Res.* **1980**, *84*, 350. (b) Chittenden, G. *Carbohydr. Res.* **1980**, *87*, 219.
6. (a) Vekemans, J. A. J. M.; deBruyn, R. G. M.; Caris, R. C. H. M.; Kokx, A. J. P. M.; Konings, J. J. H. G.; Godefroi, E. F.; Chittenden, G. *J. Org. Chem.* **1987**, *52*, 1093. (b) Chittenden, G. *Recl. Trav. Chim. Pays-Bas* **1988**, *107*, 455.
7. (a) Hampton, A. *J. Am. Chem. Soc.* **1961**, *83*, 3640. (b) Moffatt, J. G. *J. Am. Chem. Soc.* **1963**, *85*, 1118. (c) Hampton, A. *J. Am. Chem. Soc.* **1965**, *87*, 4654.
8. (a) Rachele, J. R. *J. Org. Chem.* **1963**, *28*, 2898. (b) Radin, N. S.; Hajra, A. K.; Akahori, Y. *J. Lipid Res* **1960**, *1*, 250.
9. Nussbaum, A. L.; Yuan, E.; Dincer, D.; Olivetto, E. P. *J. Org. Chem.* **1961**, *26*, 3925.
10. Murata, S.; Suzuki, M.; Noyori, R. *J. Am. Chem. Soc.* **1980**, *102*, 3248.
11. Fitch, S. J. *J. Am. Chem. Soc.* **1964**, *86*, 61.
12. Matteson, D. S.; Kramer, E. *J. Am. Chem. Soc.* **1968**, *90*, 7261.

Joel Slade
Ciba-Geigy Corporation, Summit, NJ, USA

4,4′-Dimethoxytrityl Chloride

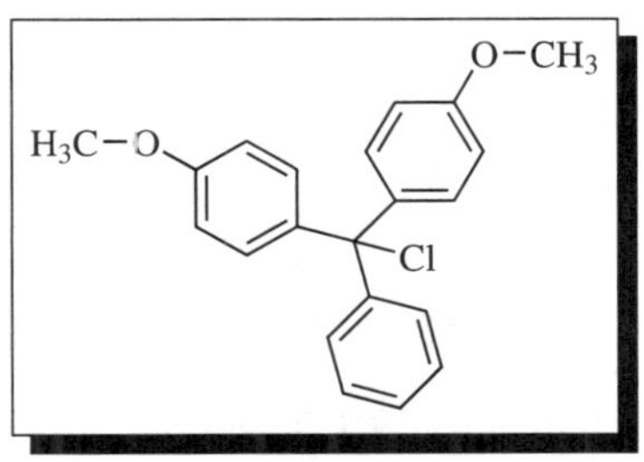

[40615-36-9] $C_{21}H_{19}ClO_2$ (MW 338.83)

(reagent used for the protection of hydroxyl, amino, and sulfhydryl groups[1])

Alternate Name: DMTr-Cl, 4,4′-dimethoxytriphenylmethyl chloride or di-*p*-anisylphenylmethyl chloride.
Physical Data: mp 119–123 °C.

Solubility: readily soluble in most organic solvents, mainly used in methylene chloride, THF, and pyridine.

Form Supplied in: pink powder; widely available.

Handling, Storage, and Precautions: Use in a well-ventilated hood with chemical safety goggles and rubber gloves. Do not breathe dust. Avoid prolonged or repeated exposure. Wash thoroughly after handling. It is an irritant. Keep tightly closed and refrigerate. Incompatible with strong oxidizing agents as well as strong bases. Toxic decomposition product is hydrogen chloride gas.

Selective Protection of Alcohols in Carbohydrate Chemistry. In the search for protecting groups of alcohols for the use in the synthesis of polynucleotides, Khorana and co-workers[2] first prepared the 5′-*O*-dimethoxytrityl (DMTr) ethers of uridine and adenosine derivatives.[2a] These ethers proved to be labile enough to be removed with acid treatment without causing depurination that had been a problem with standard trityl ethers. Since the introduction by Khorana, the DMTr ether has been a standard protecting group in oligonucleotide synthesis for over 40 years and continues to be used in standard protocols today. For instance, Khorana and co-workers produced 5′-*O*-di-*p*-anisyldiphenylmethyluridine from uridine using 4,4′-dimethoxytrityl chloride in pyridine in 78% yield (eq 1).[2a] The DMTr group is removed in 15 min in 80% acetic acid at 27 °C. By comparison, trityl ethers require 48 h for complete deprotection under these conditions, whereas the monomethoxytrityl (MMTr) group requires approximately 2 h for deprotection under these conditions.

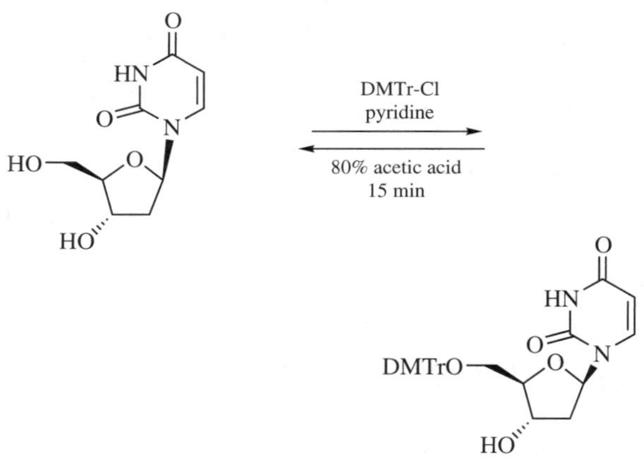

(1)

78% yield

The DMTr group has allowed for the large scale synthesis of oligonucleotides by the modified triester method. It is usually the first protecting group introduced to the base pair. The large scale preparation of 5′-*O*-dimethoxytritylthymidine is described in a review of the modified triester method by Narang et al.[3] Three other useful intermediates for phosphoramidite synthesis are the amidines derived from deoxyadenosine, deoxyguanosine, and deoxycytidine developed by Caruthers. To prepare for phosphoramidite coupling, the primary hydroxyl group of 2-*N*-(1-(dimethylamino)ethylidene)-2′-deoxyguanosine was treated with DMTr-Cl in pyridine to afford an 81% yield of the DMTr ether (eq 2).[4]

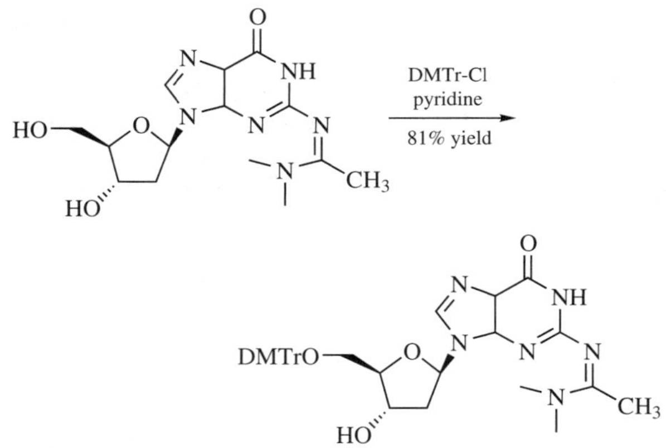

(2)

This protecting group has also been utilized with the corresponding ribose sugars. For instance, *N*-6-dimethylaminomethylene-adenosine has been treated with DMTr-Cl in a mixture of DMF and pyridine to afford the DMTr ether in high yield (eq 3).[5]

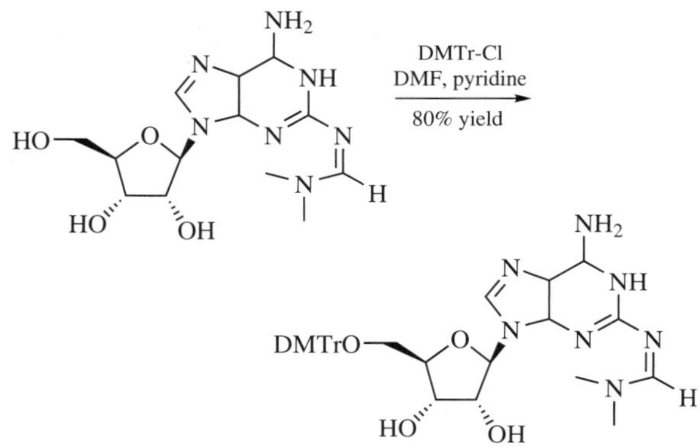

(3)

More recently, it has been shown that the amine containing nucleosides can be tritylated in the presence of the free amine. The 5′-*O*-selective dimethoxytritylation of 2′-deoxyguanosine was conducted using three different methods—either by the use of DMTr-Cl in the presence of a mixture of imidazolium mesylate and diisopropylethylamine, a mixture of imidazole and diisopropylethylammonium mesylate or a 1:1:1 mixture of imidazole, diisopropylamine, and methanesulfonic acid in DMF at room temperature (eq 4).[6]

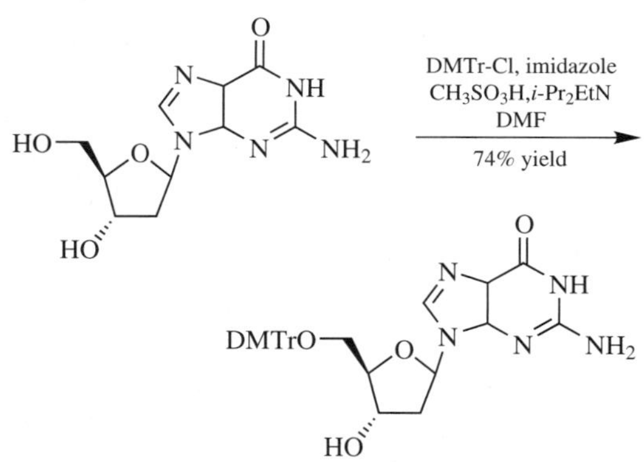

(4)

The DMTr group is compatible with the disulfide linkage and has been used in the synthesis of a C-5-modified deoxyuridine for the preparation of an *n*-butyl-DNA disulfide conjugate. The requisite phosphoramidite was prepared by a two-step sequence of protection of the 5′-alcohol as the dimethoxytrityl ether and functionalization of the 3′-hydroxyl as the phosphoramidite (eq 5).[7]

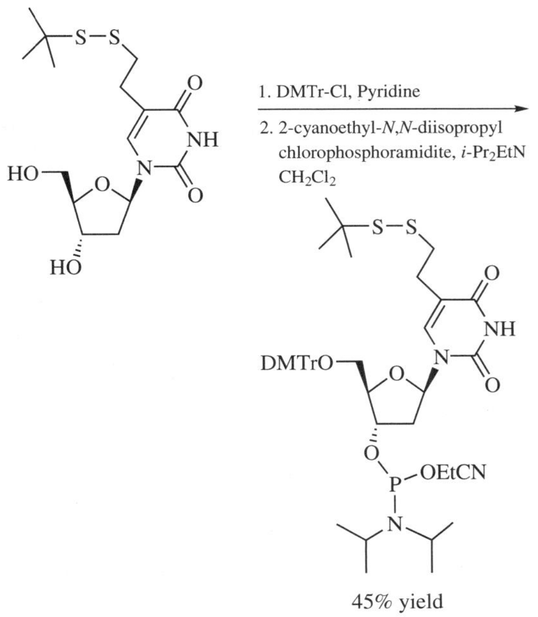

(5)

Thionucleic acids have been prepared for use in oligonucleotide synthesis, thus the dimethoxytrityl ethers containing sulfur heterocycles are readily prepared.[8,9] The DMTr ether of *S*-(2-cyanoethyl)-4-thio-2′-deoxyuridine was synthesized in high yield under standard conditions (eq 6).[9]

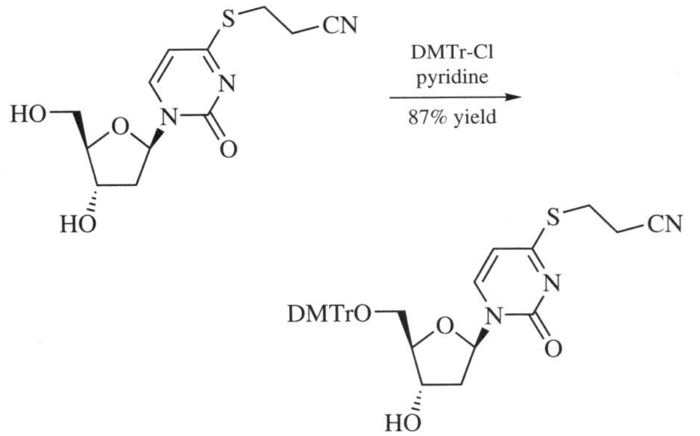

(6)

An interesting recent report on process method for the capture and reuse of the 4,4′-dimethoxytriphenymethyl group during manufacturing of oligonucleotides was disclosed.[10]

Simple sugar derivatives are easily monotritylated at the primary hydroxyl. For instance, allyl β-D-galactopyranoside was treated with DMTr-Cl in pyridine at room temperature to give the DMTr ether in 82% yield (eq 7).[11]

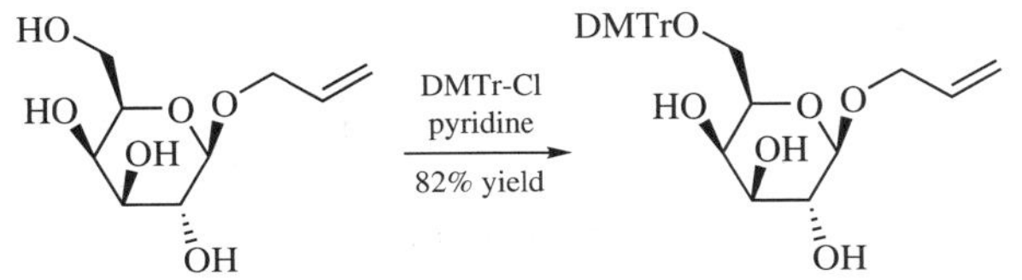

(7)

Selective Protection of Cyclic Alcohols. During the synthesis of (−)-shikimate-3-phosphate the secondary alcohol at C-3 of methyl shikimate was selectively protected through the use of an *O*-stannylene acetal (eq 8).[12]

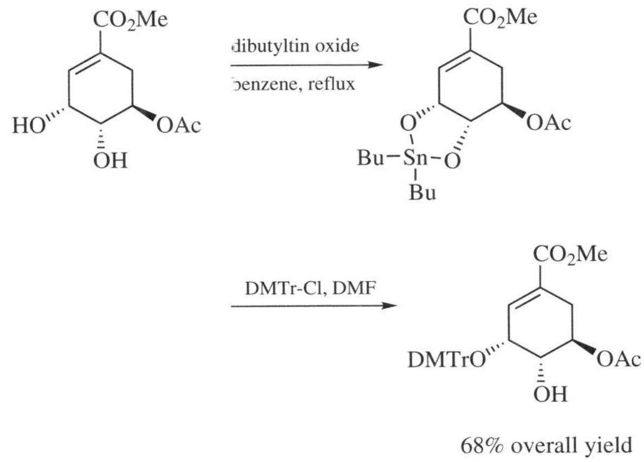

(8)

Selective Protection of Primary Alcohols. In simple acyclic systems, primary hydroxyl groups can be selectively protected over secondary alcohols. For instance, (*R*)-3-allyloxy-propane-1,2-diol was converted to its DMTr ether in good yield under standard conditions (eq 9).[13]

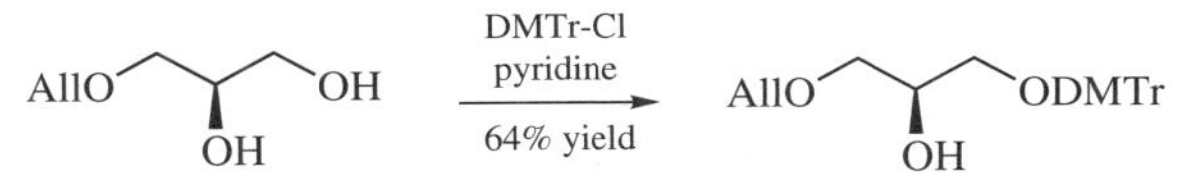

(9)

For the partial synthesis of the antifeedant (−)-specionin, the primary hydroxyl group (a (−)-catalpol derivative) was selectively protected as its DMTr ether (eq 10).[14]

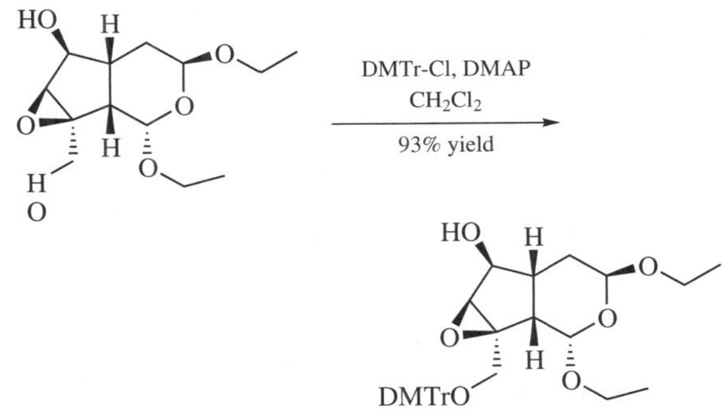

(10)

The DMTr group can be used to protect a primary hydroxyl group in the presence of a carboxylic acid. Thus, 10-hydroxydecanoic acid was converted to its DMTr ether under standard conditions in 69% yield.[15] Additionally, the primary hydroxyl of a glycine pyrimidinyl derivative was protected as its DMTr ether in the presence of a carboxylic acid using DMTr-Cl, triethylamine as base, and DMF and methylene chloride as solvent (eq 11).[16]

Protection of Nitrogen-Containing Functional Groups. In nucleoside chemistry, protection of a primary amino group in a 5′-amino-2′,5′-dideoxyadenosine derivative was used for the synthesis of oligonucleotides with modified primary structure (eq 12).[17]

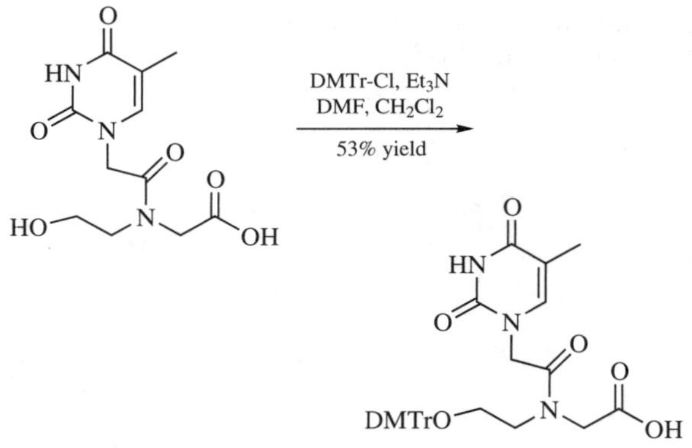

(11)

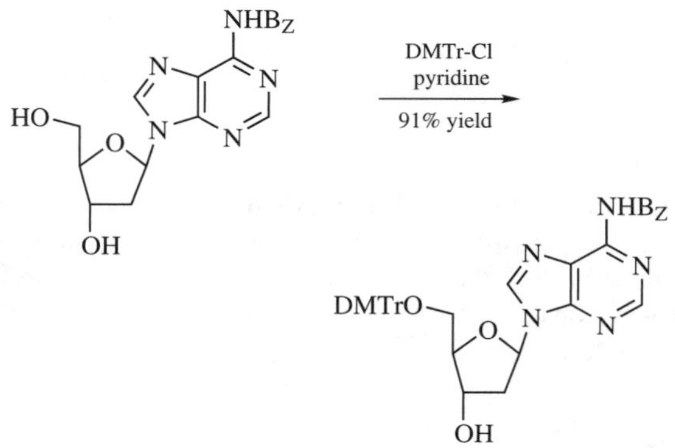

(12)

An imidazole nitrogen can be protected with a DMTr. 1*H*-imidazole-4-acetic acid ethyl ester was treated with DMTr-Cl and triethylamine in methylene chloride to give the DMTr derivative in good yield (eq 13).[18]

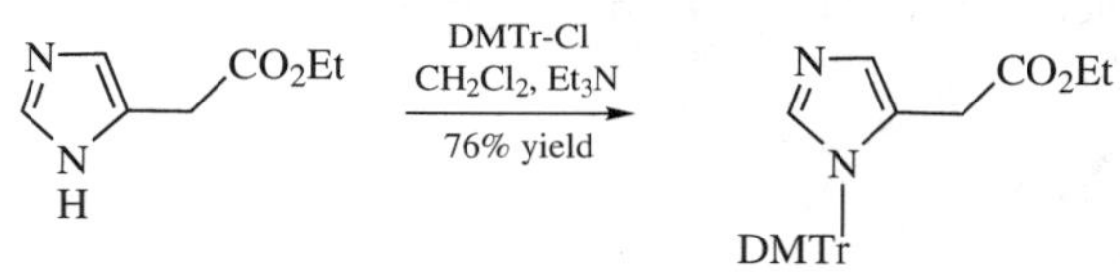

(13)

One of the urea nitrogens of a biotin derivative has been converted into its DMTr derivative to allow for incorporation of the biotin moiety into oligonucleotides via the phosphoramidite method.[19]

Protection of Sulfhydryl Groups. A primary sulfhydryl group of a uridine derivative was protected as its DMTr thioether, by treatment with DMTr-Cl and triethylamine and then selective removal of the DMTr group from oxygen was carried out with 80% aqueous acetic acid (rt, 10 min) (eq 14).[20] The DMTr group can selectively be cleaved from the thiol by treatment with silver nitrate.

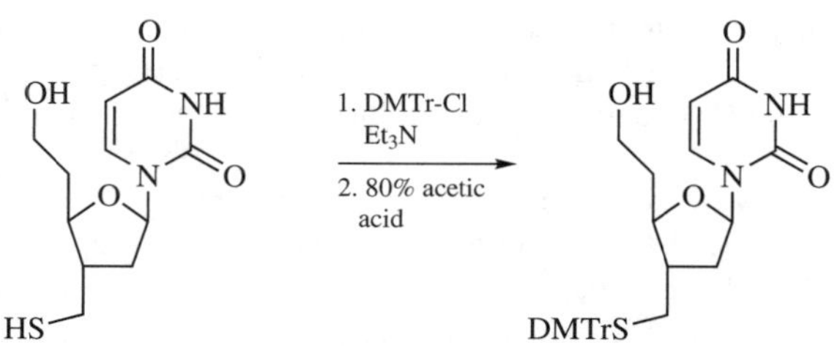

(14)

Related Reagents. 4-Methoxytrityl chloride; trityl chloride.

1. Greene, T. W.; Wuts, P. G. M., *Protective Groups in Organic Synthesis*; Wiley: New York, 1999; p 105.
2. (a) Smith M.; Rammler, D. H.; Goldberg, I. H.; Khorana, H. G., *J. Am. Chem. Soc.* **1962**, *84*, 430. (b) Khorana, H. G., *Pure Appl. Chem.* **1968**, *17*, 349. (c) For a review of the use of various trityl groups in nucleotide synthesis, see Beaucage, S. L.; Iyer, R. P., *Tetrahedron* **1992**, *48*, 2223.
3. Narang, S. A.; Brousseau, R.; Hsiung, H. M.; Michniewicz, J. J., In *Methods in Enzymology*; Dennis, M. G.; Dennis, E. A., Eds.; Academic: New York, **1980**; Vol. 62, p. 610.
4. (a) Mcbride, L. J.; Kierzek, R.; Beaucage, S.; Caruthers, M. H., *J. Am. Chem. Soc.* **1986**, *108*, 2040. (b) Jones, R. A.; Kung, P. P., *Tetrahedron Lett.* **1992**, *33*, 5869.
5. Xi, C.; Zhang, J. D.; Zhang, L. H., *Synthesis* **1989**, *5*, 383.
6. Kataoka, M.; Hayakawa, Y., *J. Org. Chem.* **1999**, *64*, 6087.
7. Goodwin, J. T.; Glick, G. D., *Tetrahedron Lett.* **1993**, *34*, 5549.
8. (a) Clivio, P.; Fourrey, J. L., *J. Org. Chem.* **1994**, *59*, 7273. (b) Clivio, P.; Fourrey, J. L.; Gasche, J.; Favre, A., *Tetrahedron Lett.* **1992**, *33*, 69.
9. Coleman, R. S.; Kesicki, E. A., *J. Am. Chem. Soc.* **1994**, *116*, 11636.
10. Guo, Z.; Pfundheller, H. M.; Sanghvi, Y. S., *Org. Proc. Res. & Develop.* **1998**, *2*, 415.
11. Nakahara, Y.; Ogawa, T., *Carbohydr. Res.* **1989**, *194*, 95.
12. Chahoua, L.; Baltas, M.; Gorrichon, L.; Tisnes, P.; Zedde, C., *J. Org. Chem.* **1992**, *57*, 5798.
13. Toepfer, A.; Kretzchmar, G.; Bartnik, E., *Tetrahedron Lett.* **1995**, *36*, 9161.
14. Van der Eycken, E.; Janssens, A.; Vandewale, M., *Tetrahedron Lett.* **1987**, *28*, 3519.
15. Truffert, J. C.; Asseline, U.; Brack, A.; Thuong, N. T., *Tetrahedron Lett.* **1996**, *52*, 3005.
16. Breipohl, G.; Will, D. W.; Peyman, A.; Uhlmann, E., *Tetrahedron* **1997**, *53*, 14671.
17. Hennningeld, K. A.; Arslan, T.; Hecht, S. M., *J. Am. Chem. Soc.* **1996**, *118*, 11701.
18. Bashkin, J. K.; Gard, J. K.; Modak, A. S., *J. Org. Chem.* **1990**, *55*, 5125.
19. Pon, R. T., *Tetrahedron Lett.* **1991**, *32*, 1715.
20. Huang, Z.; Benner, S. A., *Synlett* **1993**, *1*, 83.

David J. Madar
Abbott Laboratories, Abbott Park, Illinois

N-[(Dimethylamino)-1*H*-1,2,3-triazolo[4,5-*b*]pyridin-1-ylmethylene]-*N*-methylmethanaminium Hexafluorophosphate *N*-Oxide

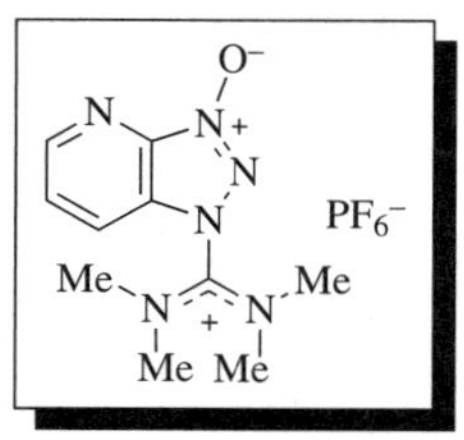

[148893-10-1] $C_{10}H_{15}F_6N_6OP$ (MW 380.28)

(coupling reagent for peptide synthesis[1–4])

Alternate Name: HATU.
Physical Data: mp: sample decomposes at temperatures above 180 °C.
Solubility: sol DMF (0.6 mol L^{-1}).
Form Supplied in: white solid; commercially available.
Purification: by crystallization from a mixture of MeCN and CH_2Cl_2.
Handling, Storage, and Precautions: is very stable, not hygroscopic, and can be stored indefinitely at 0 °C. HATU solutions in DMF (0.5 M) can be stored in an inert atmosphere for weeks (after 4 weeks, purity is 85 %). Syntheses of peptides carried out with freshly prepared and 3-week old solutions show crude products of similar quality.[5] Violent decomposition can occur when dried at elevated temperature.

General Considerations. HATU is the 7-azabenzotriazolyl analog of O-Benzotriazol-1-yl-N,N,N′, N′-tetramethyluronium Hexafluorophosphate (HBTU).[6] X-ray structure determinations of both HATU and HBTU have revealed that the solid-state structures differ from the formulations commonly presented in the literature.[7] The solid-state structure is not the *N,N,N′,N′*-tetramethyluronium salt but rather the guanidinium *N*-oxide.

HATU and its pyrrolidinium derivative, 1-(1-pyrrolidinyl-1*H*-1,2,3-triazolo[4,5-*b*]pyridin-1-ylmethylene)pyrrolidinium hexafluorophosphate *N*-oxide (HAPyU), have been used as condensing reagents for the preparation of peptides by both solution and solid-phase techniques.[1–3] In solution, reaction of α-amino protected amino acids or dipeptides with equimolar amounts of α-carboxyl protected amino acid hydrochlorides or amino amides, and base (2 equiv) [Diisopropylethylamine (DIPEA), *N*-methylmorpholine (NMM), 2,4,6-Collidine, or 1,8-Bis(dimethylamino)naphthalene (Proton Sponge, PS)] in DMF proceeds cleanly, yielding the corresponding peptides in good yields. The level of epimerization in HATU-mediated couplings was reduced compared to HBTU-promoted acylations (eq 1) (Table 1).[1,4,8]

$$RCO_2H + R'NH_2 \xrightarrow[\text{DIPEA}]{\text{HATU}} RCONHR' \quad (1)$$

The results, collected in Table 1, clearly indicated that (i) HATU is superior to HBTU (entries 1 vs. 2, 3 vs. 4, 5 vs. 6, 7 vs. 9, and 12 vs. 13); (ii) HAPyU is similar to HATU, although in some cases HAPyU can give slightly better results (entries 7 vs. 10); and (iii) the use of collidine considerably reduces the level of epimerization (entries 7 vs. 8, 12 vs. 14, 13 vs. 15, and 16 vs. 17).

HATU activation has been adapted for automated stepwise solid-phase peptide synthesis for both fluorenylmethyloxycarbonyl (Fmoc) and *t*-butyloxycarbonyl (Boc) strategies. For the former, the fragment 65–74 (H–Val–Gln–Ala–Ala–Ile–Asp–Tyr–Ile–Asn–Gly–NH_2) of the acyl carrier protein was chosen as a model to compare the performance of HATU and HBTU. Results from these studies indicated that uronium salt coupling is superior to carbodiimide and pentafluorophenyl ester-mediated methods, and the aza derivatives are more efficient than HBTU.[3]

HATU has shown special utility for the synthesis of peptides containing hindered amino acids. Thus the solid-phase assembly of the model pentapeptide H–Tyr–Aib–Aib–Phe–Leu–NH_2 was compared via three different coupling methods, HATU, HBTU, and Fmoc-amino acid fluorides, the last named being known as one of the most efficient reagents for couplings involving consecutive Aib units.[9] After 2 h coupling using 4 equiv of both Fmoc–Aib–OH and Fmoc–Tyr(*t*-Bu)–OH, as well as HXTU plus 8 equiv of DIPEA (all remaining couplings were carried out for 30 min), the pentapeptide was obtained with a purity of 94% for the HATU synthesis, and only 43% for HBTU. The synthesis conducted under similar conditions with Fmoc–Aib–F gave a purity of 96%, indicating that the reactivity of HATU is much closer to that of fluorides than to that of HBTU.[3]

The use of HATU for the solid-phase coupling of *N*-methyl amino acids has also been shown to be beneficial if compared with HBTU, as found during the syntheses of fragments of cyclosporin (CsA): H–Val–MeLeu–Ala–NH_2 (HATU, >99%; HBTU, 87%);[10] H–MeLeu–MeLeu–Ala–NH_2 (HATU, >99% HBTU, 34%);[10] H–D-Ala–MeLeu–MeLeu–MeVal–Phe–Val–OH (HATU, 85%; HBTU, 8%).[3]

Table 1 Racemization during HATU/HBTU Couplings in Solution

Entry	RCOOH	R′NH_2 or R′NH_2·HCl	Conditions	DL-isomer (%)
1	Z–Phg–OH	H–Val–OMe·HCl	HATU, PS (3 equiv), 7 h	<1–2
2	Z–Phg–OH	H–Val–OMe·HCl	HBTU, PS (3 equiv), 7 h	4.0
3	Bn–Val–OH	H–Val–OMe·HCl	HATU, NMM (3 equiv), 3 h	28.3
4	Bn–Val–OH	H–Val–OMe·HCl	HBTU, NMM (3 equiv), 3.5 h	46.8
5	Z–Phe–Val–OH	H–Val–OMe·HCl	HATU, DIPEA (3 equiv), 3.5 h	<1–2
6	Z–Phe–Val–OH	H–Val–OMe·HCl	HBTU, DIPEA (3 equiv), 4 h	4.1
7	Z–Gly–Phe–OH	H–Pro–NH_2	HATU, DIPEA (2 equiv)	1.6
8	Z–Gly–Phe–OH	H–Pro–NH_2	HATU, collidine (2 equiv)	0.8
9	Z–Gly–Phe–OH	H–Pro–NH_2	HBTU, DIPEA (2 equiv)	3.9
10	Z–Gly–Phe–OH	H–Pro–NH_2	HAPyU, DIPEA (2 equiv)	1.3
11	Z–Gly–Phe–OH	H–Pro–NH_2	HAPyU, collidine (2 equiv)	0.8
12	Fmoc–Phe–Ser(*t*-Bu)–OH	H–Pro–NH_2	HATU, DIPEA (2 equiv), 3 h	4.0
13	Fmoc–Phe–Ser(*t*-Bu)–OH	H–Pro–NH_2	HBTU, DIPEA (2 equiv), 3 h	17
14	Fmoc–Phe–Ser(*t*-Bu)–OH	H–Pro–NH_2	HATU, collidine (2 equiv), 3 h	<0.1
15	Fmoc–Phe–Ser(*t*-Bu)–OH	H–Pro–NH_2	HBTU, collidine (2 equiv), 3 hp	12
16	Fmoc–Phe–Ser(*t*-Bu)–OH	H–Pro–NH_2	HAPyU, DIPEA (2 equiv), 3 h	5.0
17	Fmoc–Phe–Ser(*t*-Bu)–OH	H–Pro–NH_2	HAPyU, collidine (2 equiv), 3 h	<0.1

Various protected peptide segments have been coupled by solid-phase techniques using HATU-mediated coupling methods.[11] Thus the synthesis of the *N*-terminal domain of γ-zein protein has been reported in good yield by eight successive segment couplings of the same protected hexapeptide to give the desired sequence using HATU and 3-Hydroxy-3 H-1,2,3-triazolo[4,5-b]pyridine (HOAt) in the presence of DIPEA.[12]

Less racemization is observed for solid-phase segment condensations in the presence of HATU relative to HBTU. The coupling of Fmoc–Phe–Ser(*t*-Bu)–OH onto H–Pro–resin was accompanied by 5% and 15% racemization, respectively, with HATU and HBTU in the presence of DIPEA. Replacement of DIPEA with collidine reduced the racemization level to 3%, confirming the effects observed for coupling reactions carried out in solution.[4]

HATU and HAPyU have been used for the cyclization of linear peptides in both solution and solid-phase modes. Cyclization in solution with HAPyU of a linear hexapeptide constructed exclusively from L-amino acids gave a better yield and less racemization at the *C*-terminal residue compared to other uronium or phosphonium salts.[13] HATU has also been shown to be the method of choice for cyclizations conducted while the peptide still remains anchored to the polymeric support. In the preparation of a library of 1296 cyclic pentapeptides, cyclization was complete after 1 h as judged by the ninhydrin test.[14]

In the absence of the carboxylic component, HATU, as is the case for HBTU, reacts with amino groups leading to the formation of tetramethylguanidinium derivatives.[5,15] In syntheses conducted in solution, an excess of either HATU or amino component should be avoided. In both solution and solid-phase strategies, the order of addition of reagents is critical. HATU should be delivered to the carboxylic component for preactivation prior to addition of the amine.

In addition, formation of the guanidinium derivative can also occur during slow reactions, such as the preparation of cyclic peptides, where both amino and carboxylic components are of necessity present together in equimolar amounts. An excess of the uronium salt can block the amino group.[16] For the synthesis of cyclic peptides, the phosphonium derivatives 7-azabenzotriazol-1-yloxytris(pyrrolidino)phosphonium hexafluorophosphate (PyAOP) and 7-azabenzotriazol-1-yloxytris(dimethylamino)phosphonium hexafluorophosphate (AOP), which are also commercially available, may be more useful.

1. Carpino, L. A. *J. Am. Chem. Soc.* **1993**, *115*, 4397.
2. Carpino, L. A.; El-Faham, A.; Minor, C. A.; Albericio, F. *J. Chem. Soc., Dalton Trans.* **1994**, 201.
3. Carpino, L. A.; El-Faham, A. *J. Org. Chem.* **1994**, *59*, 695.
4. Carpino, L. A.; El-Faham, A.; Albericio, F. *Tetrahedron Lett.* **1994**, *35*, 2279.
5. Kates, S. A.; Minor, C. A.; Shroff, H.; Haaseth, R. C.; Triolo, S. A.; El-Faham, A.; Carpino, L. A.; Albericio, F. In *Peptides 1994: Proceedings of the Twenty-Third European Peptide Symposium*; Maia, H. L. S., Ed.; ESCOM:Leiden, **1995**; pp 248–249.
6. (a) Dourtoglou, V.; Ziegler, J. C.; Gross, B. *Tetrahedron Lett.* **1978**, 1269. (b) Dourtoglou, V.; Gross, B.; Lambropoulou, V.; Zioudrou, C. *Synthesis* **1984**, 572. (c) Knorr, R.; Trzeciak, A.; Bannwarth, W.; Gillessen, D. *Tetrahedron Lett.* **1989**, *30*, 1927.
7. Abdelmoty, I.; Albericio, F.; Carpino, L. A.; Foxman, B. F.; Kates, S. A. *Lett. Peptide Sci.* **1994**, *1*, 57.
8. Albericio, F.; Abdelmoty, I.; Bofill, J. M.; Carpino, L. A.; El-Faham, A.; Foxman, B. F.; Gairí, M.; Griffin, G. W.; Kates, S. A.; Lloyd-Williams, P.; Scarmoutzos, L. M.; Shroff, H.; Triolo, S. A.; Wenschuh, H. In *Peptides 1994: Proceedings of the Twenty-Third European Peptide Symposium*; Maia, H. L. S., Ed.; ESCOM:Leiden, **1995**; pp 23–25.
9. Wenschuh, H.; Beyermann, M.; Krause, E.; Brudel, M.; Winter, R.; Schümann, M.; Carpino, L. A.; Bienert, M. *J. Org. Chem.* **1994**, *59*, 3275.
10. Angell, Y. M.; García-Echeverría, C.; Rich, D. H. *Tetrahedron Lett.* **1994**, *35*, 5981.
11. Giralt, E.; Albericio, F.; Dalcol, I.; Gairí, M.; Lloyd-Williams, P.; Rabanal, F. In *Innovation and Perspectives in Solid-Phase Peptide Synthesis: Biological & Biomedical Applications*; Epton, R., Ed.; Mayflower Worldwide:West Midlands, UK, **1994**; pp 51–60.
12. Dalcol, I.; Rabanal, F.; Albericio, F.; Pons, M.; Geli, M.; Torrent, M.; Ludevid, M. D.; Giralt, E. In *Peptides 1994: Proceedings of the Twenty-Third European Peptide Symposium*; Maia, H. L. S., Ed.; ESCOM:Leiden, **1995**; pp 60–61.
13. Ehrlich, A.; Rothemund, S.; Brudel, M.; Beyermann, M.; Carpino, L. A.; Bienert, M. *Tetrahedron Lett.* **1993**, *34*, 4781.
14. Darlak, K.; Romanovskis, P.; Spatola, A. F. In *Proceedings of the Thirteenth American Peptide Symposium*; Hodges, R. S., Smith, J. A., Eds.; ESCOM:Leiden, **1994**; pp 981–983.
15. Gausepohl, H.; Pieles, U.; Frank, R. W. In *Proceedings of the Twelfth American Peptide Symposium*; Smith, J. A.; Rivier, J. E., Eds.; ESCOM:Leiden, **1992**; pp 523–524 (*Chem. Abstr.* **1992**, *117*, 171 981j).
16. Arttamangkul, S.; Murray, T. F.; DeLander, G. E.; Aldrich, J. V. *J. Med. Chem.* **1995**, *38*, 2410.

Fernando Albericio
University of Barcelona, Spain
Steven A. Kates
PerSeptive Biosystems, Framingham, MA, USA
Louis A. Carpino
University of Massachusetts, Amherst, MA, USA

α,α-Dimethyl Cyclopropanemethanamine

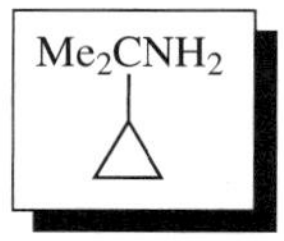

[172947-13-6] $C_6H_{13}N$ (MW 99.2)

(reagent for amide protection[1])

Alternate Name: dimethylcyclopropylmethylamine, dimethylcyclopropylcarbinylamine, Dmcp-NH_2.
Physical Data: hydrochloride, mp 223–225 °C (decomposes),[2] *p*-toluenesulfonic acid salt, mp 168–170 °C.[1]
Solubility: soluble in water and common organic solvents.
Analysis of Reagent Purity: ^{1}H NMR, elemental analysis.
Preparative Methods: the title reagent can be prepared by hydrolysis of dimethylcyclopropylcarbinyl-*N*-(dimethylcyclopropylcarbinyl) carbamate by means of concentrated hydrochloric acid[2] or, more conveniently, by reduction of the azide via $LiAlH_4$ and conversion to the *p*-toluenesulfonic acid salt.[1]

Handling, Storage, and Precautions: stored in its salt form, the reagent is relatively safe and stable.

Reagent for Side Chain Amide Protection. The Dmcp group was recommended for protection of the amide function of Fmoc-protected asparagine and glutamine. Syntheses are carried out via prior synthesis of the Z-analog (eq 1).[1]

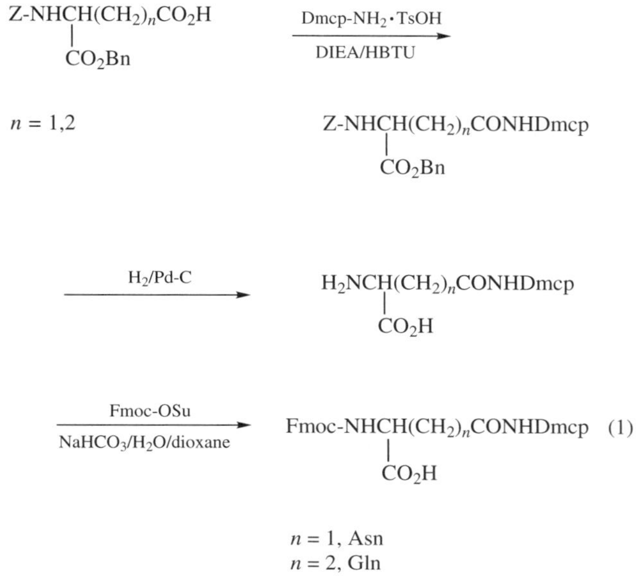

The corresponding benzo[b]thiophenesulfone-2-methyloxycarbonyl derivatives, Bsmoc-Asn(Dmcp)-OH and Bsmoc-Gln (Dmcp)-OH were made similarly for use in Bsmoc rapid solution syntheses of short peptides.[3]

Reagent for C-Terminal Amide Protection. The Dmcp group enhances the solubility of a growing peptide during rapid solution assembly of short peptide amides. An example is the assembly of protected pentapeptide **1** in 76% yield (eq 2).[1]

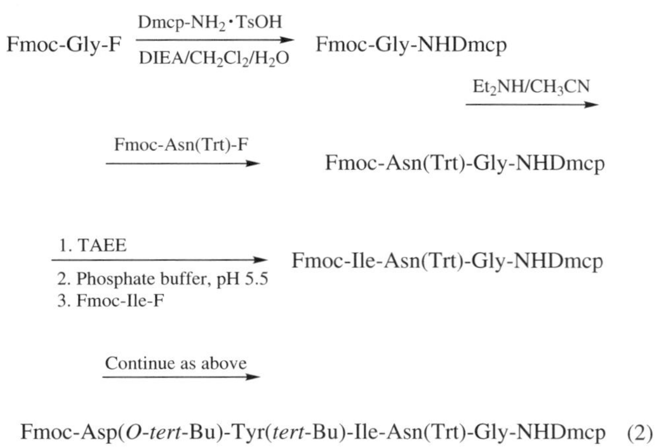

1

Octapeptide (**2**) was made similarly via Bsmoc protection and either HATU or acid fluoride coupling.[3]

pGlu-Leu-Thr-Phe-Thr-Pro-Asn-Trp-NH_2

2

Related Reagents. Dicyclopropylmethylamine[4,5] shows similar properties but is deblocked by acidic reagents somewhat less readily than the Dmcp group.[1] Dmcp-NH_2·HCl, mp 263–265 °C;[4] 250 °C (decomposes).[1]

1. Carpino, L. A.; Chao, H.-G.; Ghassemi, S.; Mansour, E. M. E.; Riemer, C.; Warrass, R.; Sadat-Aalaee, D.; Truran, G. A.; Imazumi, H.; El-Faham, A.; Ionescu, D.; Ismail, M.; Kowaleski, T. L.; Han, C. H.; Wenschuh, H.; Beyermann, M.; Bienert, M.; Shroff, H.; Albericio, F.; Triolo, S. A.; Sole, N. A.; Kates, S. A., *J. Org. Chem.* **1995**, *60*, 7718.
2. Timberlake, J. W.; Martin, J. C., *J. Org. Chem.* **1968**, *33*, 4054.
3. Carpino, L. A.; Ismail, M.; Truran, G. A.; Mansour, E. M. E.; Iguchi, S.; Ionescu, D.; El-Faham, A.; Riemer, C.; Warrass, R., *J. Org. Chem.* **1999**, *64*, 4324.
4. Corrodi, H., *Helv. Chim. Acta* **1963**, *46*, 1059.
5. Hanack, M.; Eggensperger, H., *Liebigs Ann. Chem.* **1963**, *663*, 31.

Louis A. Carpino
University of Massachusetts, Amherst, Massachusetts, USA

Dimethyl(methylthio)sulfonium Tetrafluoroborate[1]

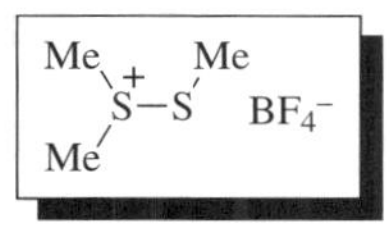

[5799-67-7] $C_3H_9BF_4S_2$ (MW 196.07)

(electrophilic sulfenylation reagent capable of reacting with nucleophilic atoms;[2] reacts with electron-rich alkenes to promote addition reactions,[3] cyclizations;[4] activates dithioacetals,[5] trithioorthoesters,[6] and thioglycosides[7] for carbon–carbon or carbon–heteroatom bond forming reactions)

Alternate Name: DMTSF.
Physical Data: mp 89–92 °C.
Solubility: soluble in MeCN, $MeNO_2$; insoluble in ether, pet ether; reacts with H_2O and other protic solvents.
Form Supplied in: cream-colored to white solid; available from various suppliers.
Preparative Method: DMTSF can be conveniently prepared from Dimethyl Disulfide and Trimethyloxonium Tetrafluoroborate (Me_3O^+ BF_4^-) in MeCN.[8] Precipitation of the product results after addition of ether and isolation requires filtration under anhydrous conditions.
Handling, Storage, and Precautions: DMTSF must be stored under anhydrous conditions, preferably in a well sealed bottle under nitrogen. Exposure to moisture results in a stench and the reagent is corrosive. Use in a fume hood.

Original Commentary

Edward J. Adams
E. I. DuPont de Nemours and Co., Newark, DE, USA

Methylsulfenylation of Nucleophilic Atoms. DMTSF reacts with Potassium Cyanide to yield MeSCN and with Phenylmagnesium Bromide to yield thioanisole.[8] Extensive investigations addressed the reactivity of DMTSF with nucleophilic agents.[9] DMTSF does not react with phenol at rt, but reacts slowly at 60 °C to yield the methylthiophenols (*ortho*/*para* = 0.22). The reaction with anisole also requires a temperature of 60 °C and results in *p*-methylthioanisole and *p*-methylthiophenol, a consequence of ether cleavage under the reaction conditions. Unlike phenol, dimethylaniline reacts rapidly with DMTSF at rt, yielding *p*-methylthio-*N,N*-dimethylaniline in 60% yield (eq 1). Despite the presence of an electron-withdrawing group, *p*-nitro-*N,N*-dimethylaniline also reacts rapidly with DMTSF to yield *p*-nitro-*N,N*-dimethyl-*N*-methylthioanilinium tetrafluoroborate.

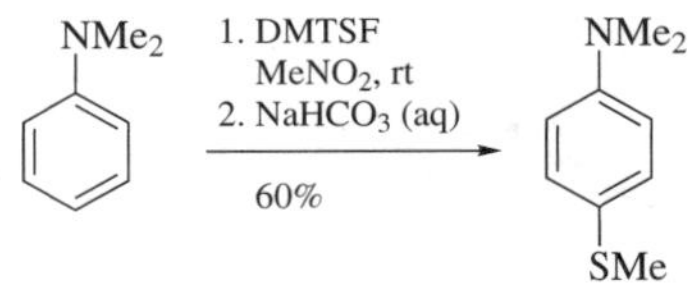
(1)

DMTSF reacts rapidly at rt with pyridine to yield the hydrolytically unstable methylthiopyridinium ion, which failed to methylate the MeSMe byproduct. When 4-cyanopyridine was used, an equilibrium mixture was detected by NMR.

Triphenyl Phosphite reacts rapidly with DMTSF at rt to yield the methylthiophosphonium salt which, unlike the pyridinium analog, methylates the MeSMe byproduct to yield $Me_3S^+ BF_4^-$. This salt is removed by precipitation with ether and chromatography provides triphenyl thionophosphate in an 86% yield (eq 2). This technique complements other methods of thionophosphate synthesis from phosphites.[10]

$$(PhO)_3P \xrightarrow[86\%]{\text{DMTSF, } MeNO_2\text{, rt}} (PhO)_3P{=}S \qquad (2)$$

Synthesis of mixed disulfides involving a methylthio unit may utilize DMTSF. Thiols and thiolates react rapidly with the reagent to yield mixed disulfides. Unfortunately, disproportionation occurs, resulting in a mixture of products.[9] Subsequent investigations suggest that the addition of a sterically hindered base (eq 3),[11] or the use of a 2-(trimethylsilyl)ethanethiol derived component (eq 4),[12] permit high yields of unsymmetrical disulfides.

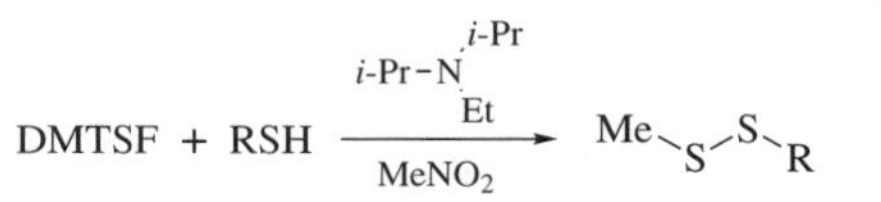
(3)

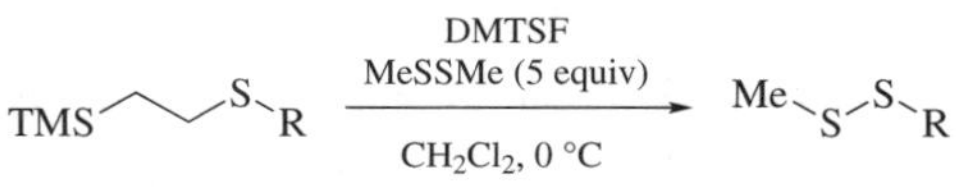
(4)

In systems having a tertiary amine and a sulfide unit, DMTSF demonstrates a chemoselectivity for sulfur (eq 5).[13]

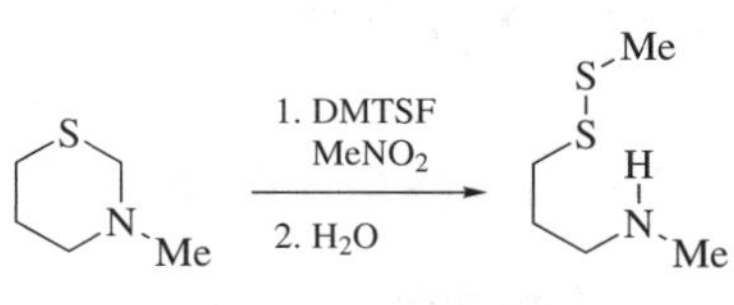
(5)

Allylic sulfides react with DMTSF chemoselectively at the sulfur atom, generating thiosulfonium ions. This ion formation is reversible and the allylic thiosulfonium ion is capable of a [2,3]-sigmatropic rearrangement before subsequent electrophilic addition to the alkene.[14]

Methylsulfenylation of Alkenes. A key application of DMTSF in organic synthesis involves electrophilic addition to alkenes and subsequent addition of a nucleophile to the bridged sulfonium ion. Support for this mechanism is provided by the *anti* addition products observed.[15]

Carbon Nucleophiles. Most nucleophilic carbon sources are too basic for application to these reaction conditions. Alkynylalanate complexes demonstrate an exception (eq 6).[16]

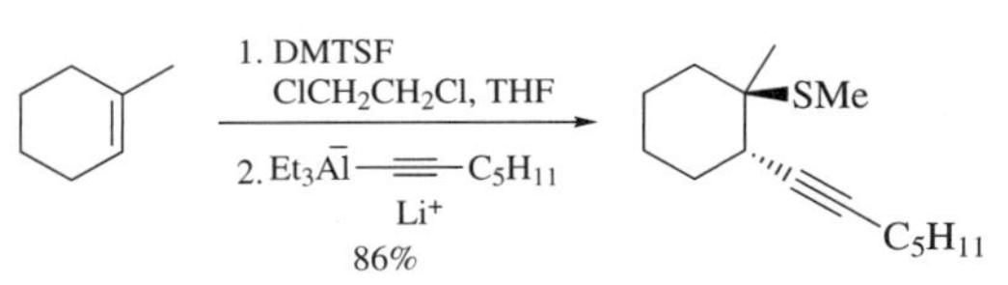
(6)

Nitrogen Nucleophiles. Nitrogen nucleophiles, specifically NH_3, amines, azide, and nitrite, resulted in high yields of alkene addition products (eq 7).[15a] In the absence of additional nucleophiles, the resulting sulfonium salt reacted with a tertiary amine to yield the demethylated product or the ammonium salt with a retention of configuration (eq 8).[15b]

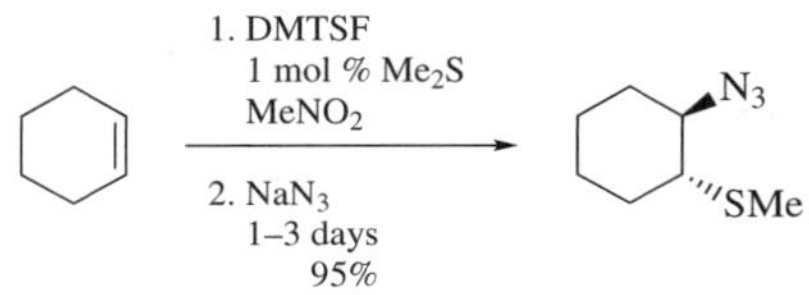
(7)

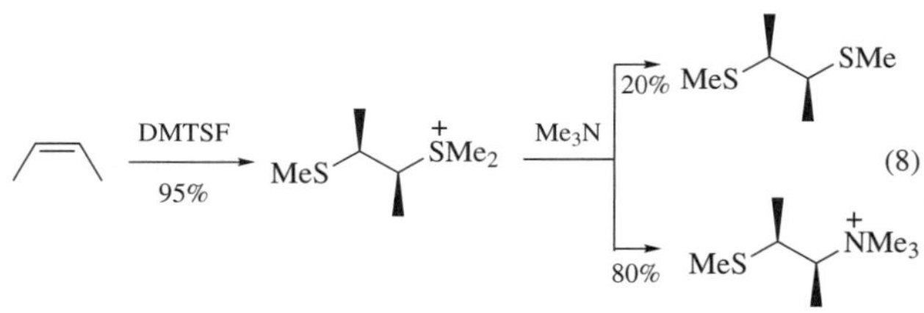
(8)

Oxygen Nucleophiles. Successful application of oxygen nucleophiles includes anhydrous KOAc (anti-Markovnikov product), $H_2O/CaCO_3$ (Markovnikov product), and DMSO which, upon treatment with Diisopropylethylamine, yields α-methyl thioketones (eq 9).[17]

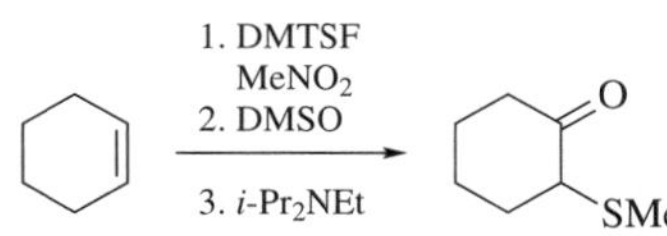
(9)

Electrophilic ring closure demonstrates another utility of DMTSF. Examples of sulfenyl etherification and sulfenyl lactonization have been reported (eq 10).[18]

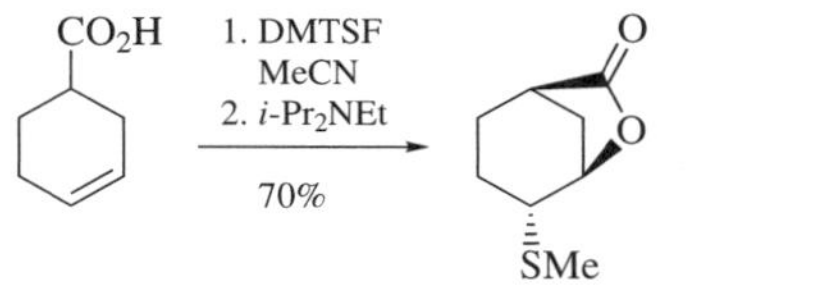

(10)

Phosphorus Nucleophiles. Triphenylphosphine is an effective nucleophile for reacting with the bridged sulfonium ion in high yield. The phosphonium salts generated eliminate MeSH upon reaction with base, such as 1,8-Diazabicyclo[5.4.0]undec-7-ene, resulting in a synthesis of vinylphosphonium salts. If NaOH is used, a vinylphosphine oxide is generated.[19]

Other Nucleophiles. DMTSF and $Et_3N{\cdot}3HF$ react with alkenes to yield products formally of a MeSF addition. Yields are in excess of 80% and the reaction proceeds rapidly.[20] Homoallylic sulfides cyclized in the presence of DMTSF. The resulting sulfonium salt was successfully demethylated with Me_3N to yield 3-methylthiotetrahydrothiophenes.[21]

Dithioacetal, Trithioorthoester, and Thioglycoside Activation. DMTSF provides a chemoselective reagent for dithioacetal activation in the presence of electron-rich alkenes.[22] This bond-forming procedure complements the $TiCl_4$ mediated reaction between acetals and silyl enol ethers demonstrated by Mukaiyama (see also Titanium(IV) Chloride and Trimethylsilyl Trifluoromethanesulfonate). The generation of ketone precursors through dithioacetal carbanion (acyl anion equivalents) chemistry further amplifies the significance of DMTSF activation.

Dithioacetal Activation. DMTSF chemoselectively reacts with dithioacetals in the presence of a vinylsilane[22] and in the presence of a silyl enol ether (eq 11), resulting in cyclization.[23]

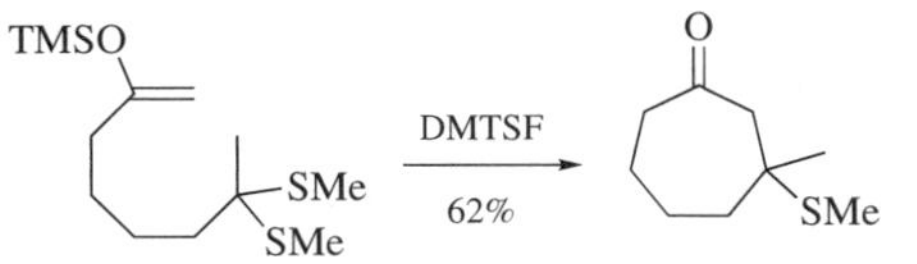

(11)

Many carbon nucleophiles tend to be too basic for successful addition to the carbocationic center provided by DMTSF activation of dithioacetals and usually result in vinyl sulfide formation. In addition, the Me_2S generated competes for the electrophilic center. Allylstannanes overcome these constraints and successfully generate new carbon–carbon bonds (eq 12).[24] DMTSF activation of a thioacetal resulted in an electrophilic cyclization at the 3-position of a 3-substituted indole.[25]

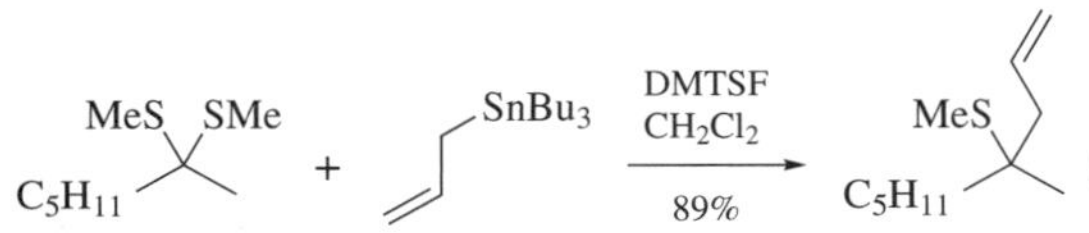

(12)

Trithioorthoester Activation. Electron-rich aromatic rings undergo electrophilic aromatic substitution with Tris(phenylthio) methane in the presence of DMTSF. Subsequent hydrolysis results in an aldehyde and a net 'electrophilic formylation'.[6] Intramolecular reaction between a tris(phenylthio)methane unit and an alcohol represents an approach to lactone formation which utilizes the chemoselectivity of DMTSF.[26]

Thioglycoside Activation. A thioglycoside was successfully converted into a glycosyl fluoride with DMTSF.[7]

First Update

Stefan Oscarson
Stockholm University, Stockholm, Sweden

Methylsulfenylation of Nucleophilic Atoms.

Methylsulfenylation of Metal Complexes. Metal complexes have been sulfenylated by the use of DMTSF. The methylsulfenyl group is either inserted into a metal-metal bond or added to a metal-carbon multiple bond (double or triple bond) (eq 13).[27,28]

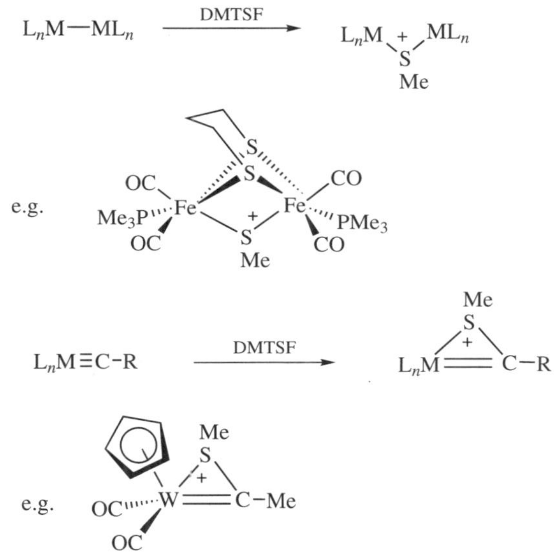

(13)

Methylsulfenylation of Sulfides. DMTSF can be used to remove several common thiol protecting groups to produce asymmetric methyl disulfides. Examples of such protecting groups are trimethylsilylethyl, trityl, benzyl, *p*-methoxybenzyl, and acetamidomethyl groups. If two protected thiol groups are present in the same molecule, disulfide bridge formation occurs.[29]

DMTSF reacts with flavan-3-ol-based benzyl sulfides with activation of the C_4–S bond towards substitution by carbon nucleophiles, allowing the formation of the interflavanyl bond in procyanidins (eq 14).[30]

Methylsulfenylation of Alkenes. Azidosulfenylation of various substituted cyclopentenes with DMTSF in conjunction with azides[15] takes place in moderate yield with preferential methylsulfenylation, *syn* to the ring substituent(s) (eq 15).[31]

Cyclopentenes were also activated (with *anti*–directed methylsulfenylation) by DMTSF in the presence of nucleoside base anions yielding nucleoside analogs that could be converted into the antiviral agent Carbovir and related substances (eq 16).[32]

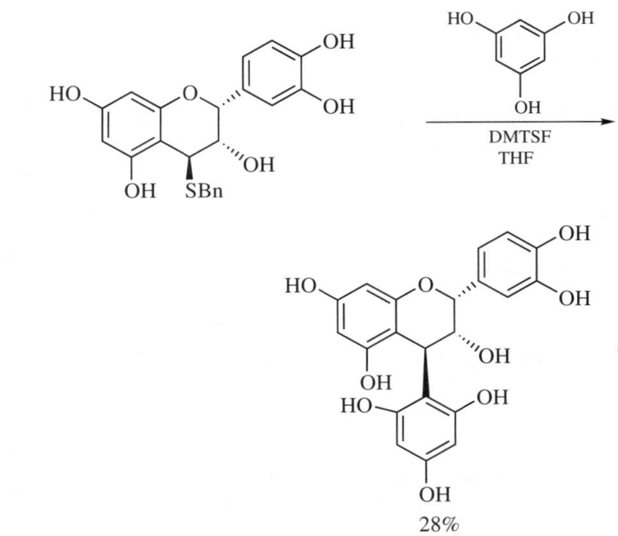

(14)

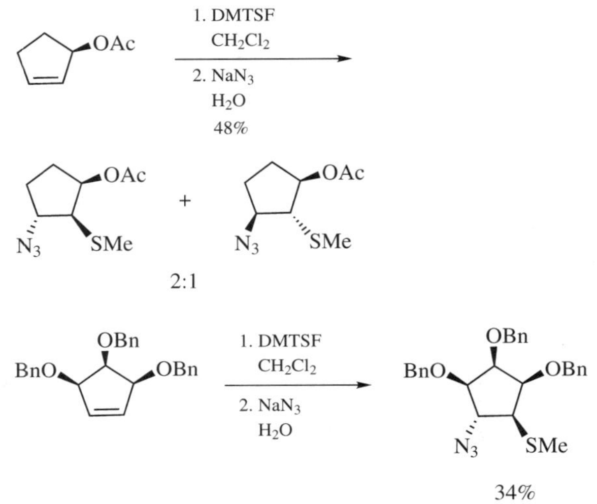

(15)

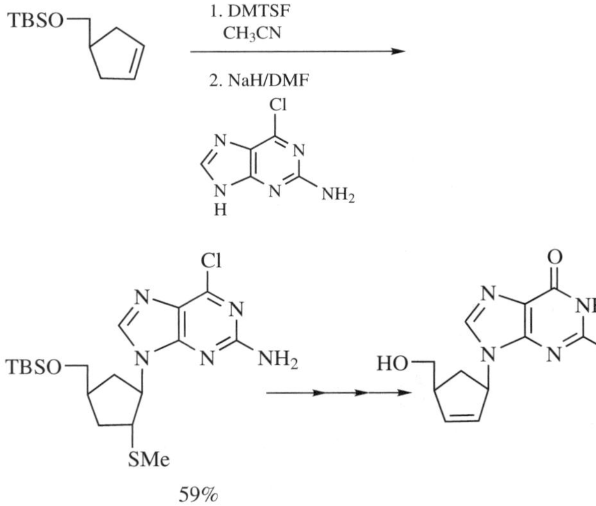

(16)

DMTSF-treatment of a dihydropyridine afforded an iminium ion, which was attacked intramolecularly by an indole moiety to form a key intermediate (albeit in a low yield) in a total synthesis of *N*-methylvitsine (eq 17).[33]

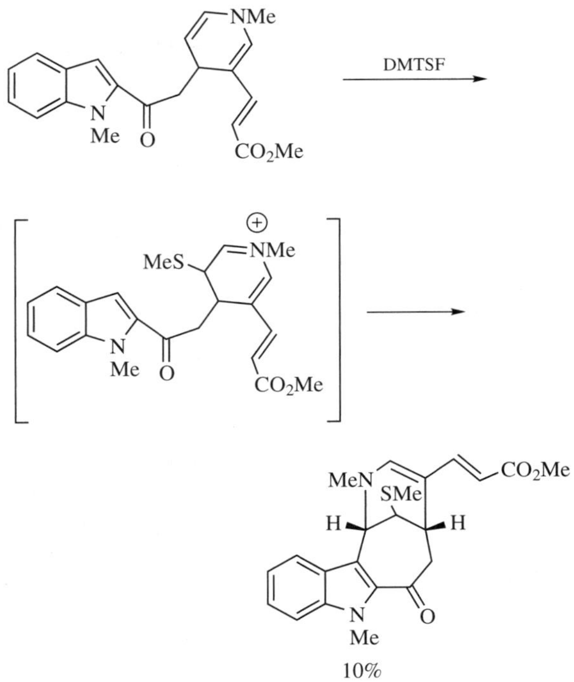

(17)

A further example of the use of DMTSF and $Et_3N{\cdot}3HF$[20] to produce Markovnikov-oriented fluoro methylthio ethers of methallyl chloride has been communicated (eq 18).[34]

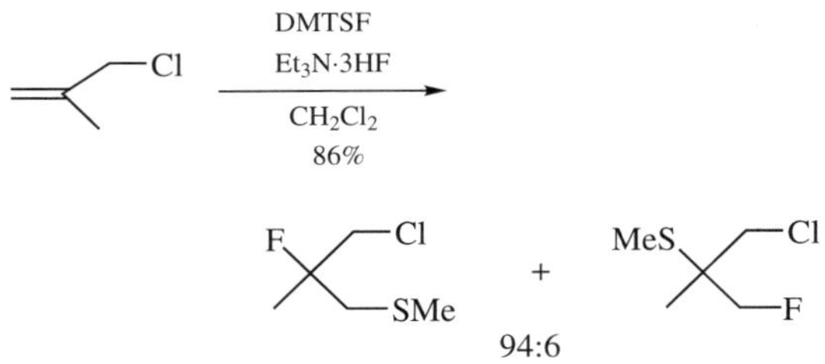

(18)

Addition of hydroxybenzaldehydes to DMTSF-activated cycloolefins were described as a first step in a library synthesis of methylsulfanylcycloalkoxyethers (eq 19).[35]

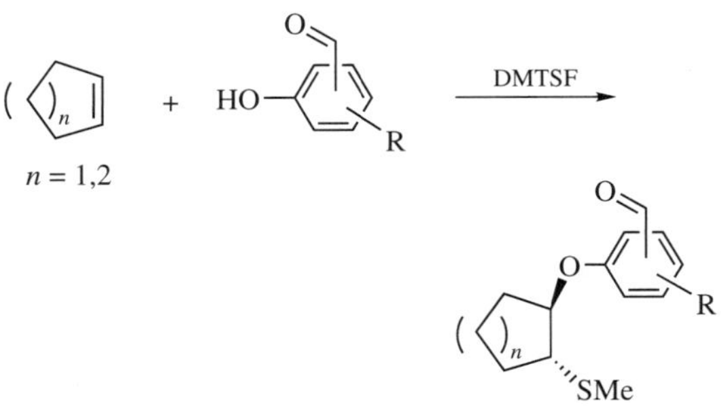

(19)

Activation of Thioacetals Including Thioglycosides.

Dithioacetals. DMTSF-treatment of substituted 2-acyloxy-4-oxo-dithioacetals conveniently yields substituted 2-thiomethylfuran derivates in high yield (eq 20).[36] The reaction is rather substituent-independent and accepts both lactones, lactams, and carbamates, with the formation of bicyclic products.[37]

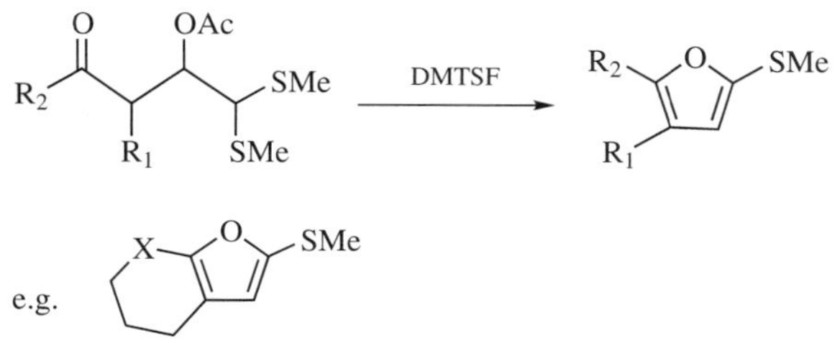

(20)

X = O, NMe, NTs, NCO*t*Bu

The presence of alkenes is also tolerated, which enables consecutive cycloaddition reactions leading to bi- or tricyclic targets (eq 21).[38]

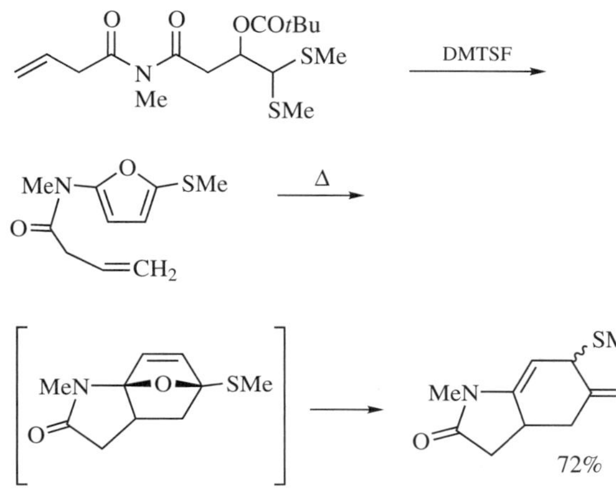

(21)

In systems bearing secondary amide based chains, the nitrogen atom may act as nucleophile resulting in Mannich-type intermediates. These can then be trapped by suitable electron rich aromatic rings to give azatricyclic ring systems (eq 22).[39]

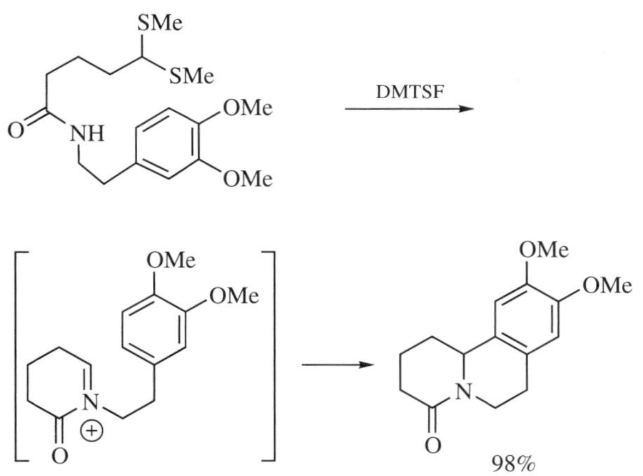

(22)

Alkenes can also be used as nucleophiles in this type of reaction (eq 23).[40]

A further example of DMTSF activation of a dithioacetal followed by cyclization onto an olefin, this time with a subsequent pinacol rearrangement, was reported as part of a total synthesis of Shahamin K (eq 24).[41]

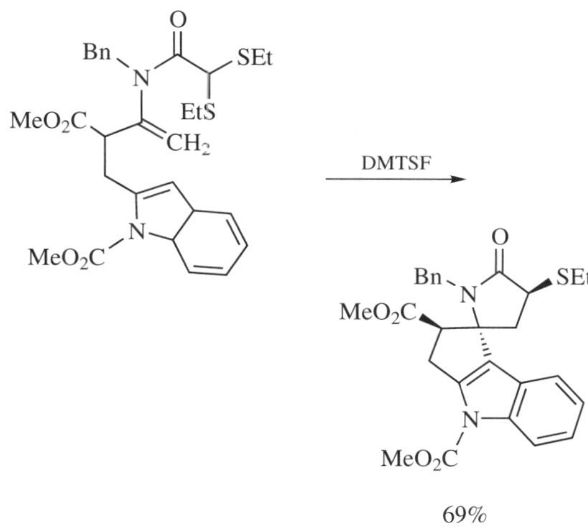

(23)

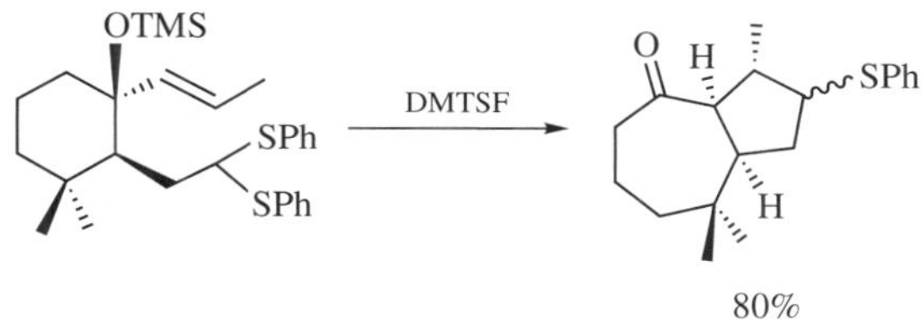

(24)

***S,N*-Acetals.** Cyclic amidothietanes undergo ring opening to form asymmetric methyl disulfides when treated with DMTSF (eq 25).[42]

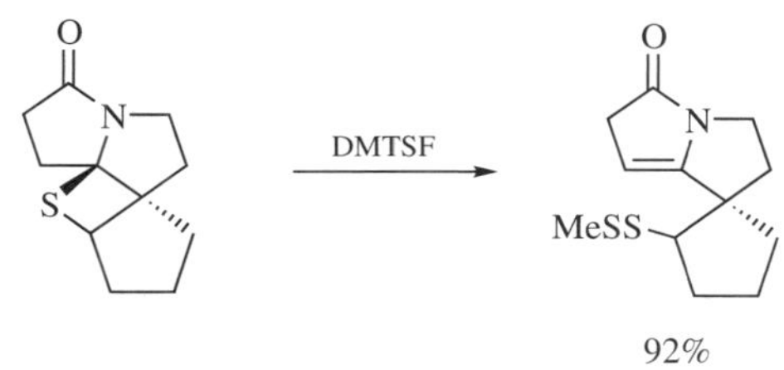

(25)

***S,O*-Acetals Including Thioglycosides.** Activation of *S,O*-acetals with DMTSF in the presence of intramolecular silyl enol ethers affects cyclization (eq 26). As expected the activation is completely selective for the thiol group.[43]

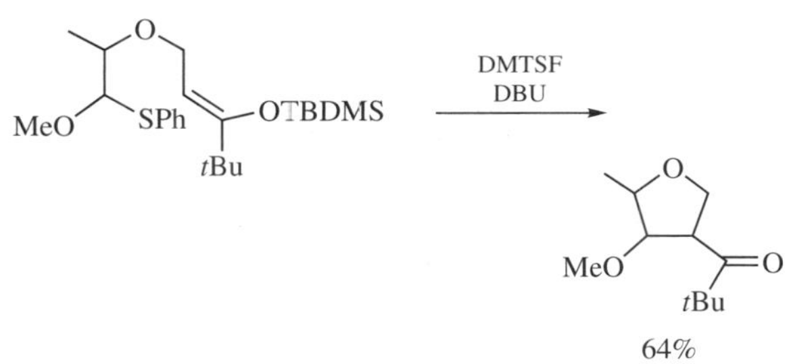

(26)

DMTSF has been used infrequently in the activation of thioglycosides with external acceptors,[7,44] for example, 2-deoxy donors[45] and in the release of solid phase-attached thioglycosides.[46] Usually the triflate salt, i.e. DMTST, is used, since normally the triflate is a better counterion in glycosylation reactions.[47] DMTSF has mainly been used in the context of synthesis of nucleosides and nucleotide analogs, using an internal glycosylation methodology. A pyrimidine base was attached to a hydroxyl group (most often 5-OH of a ribose derivative) via alkylation. Activation of the thioglycoside by DMTSF was followed by an intramolecular attack of a pyrimidine ring nitrogen to afford a bicyclic *N*-glycoside pyridinium intermediate, which was hydrolyzed with base to yield the desired nucleoside (eq 27).[48] Opening of the intermediate with different sulfur and nitrogen nucleophiles afforded 5′-substituted analogs.[49] Many variants of both the precursor thioglycoside and pyrimidine base have been used to produce different target analogs.[50]

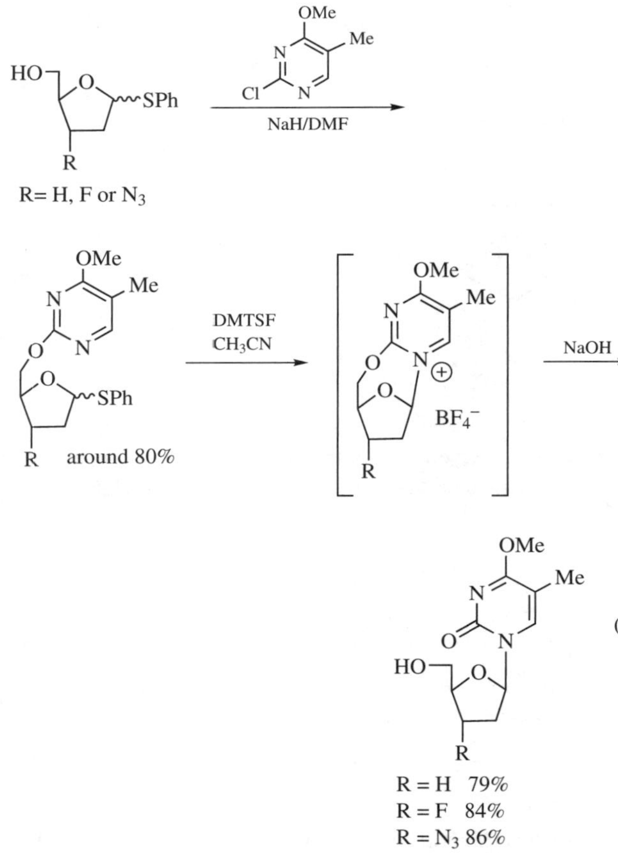

(27)

When a 2′,3′-unsaturated pentafuranose thioglycoside is utilized allylic displacement is observed and the product is the 3′-isomeric nucleoside (eq 28).[51]

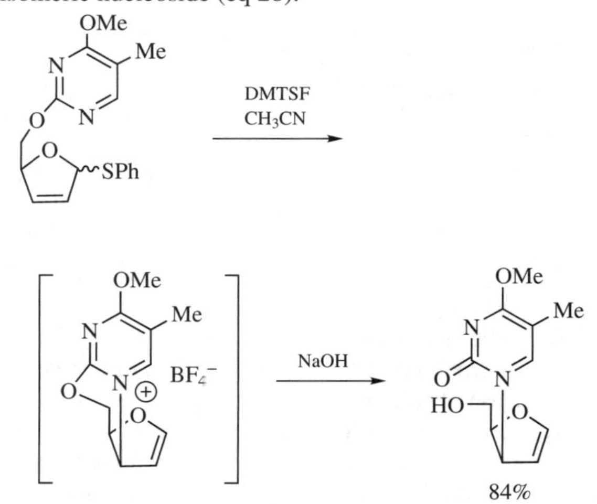

(28)

Similar processes, both direct and allylic nucleophilic displacements, are observed when hexopyranose thioglycosides bearing a pyrimidine ring on the 6-OH group are activated with DMTS (eq 29).[52]

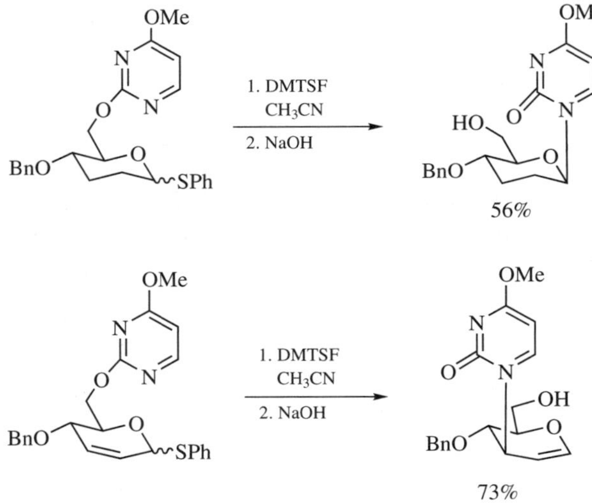

(29)

Nucleosides have also been synthesized starting from open-chain dithioacetals in a three-step procedure involving introduction of the base, deprotection, and cyclization. In the first (dithioacetal) and last step (*N,S*-acetal) DMTSF was used as reagent (eq 30).[53]

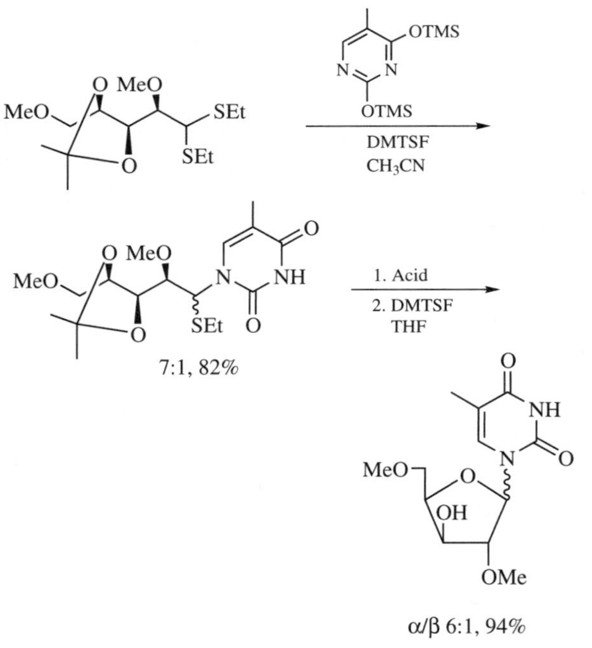

(30)

Related Reagents. Dimethyl(methylthio)sulfonium trifluoromethanesulfonate (DMTST); *N*-iodosuccinimide; *N*-iodosaccharin; 1-benzenesulfinyl piperidine; diphenyl sulfoxide.

---

1. (a) Meerwein, H.; Zenner, K.-F.; Gipp, R., *Justus Liebigs Ann. Chem./Liebigs Ann. Chem.* **1965**, *688*, 67. (b) Alternate reagent:

dimethyl(thiomethyl)sulfonium 2,4,6-trinitrobenzenesulfonate: Helmkamp, G. K.; Cassey, H. N.; Olsen, B. A.; Pettitt, D. J., *J. Org. Chem.* **1965**, *30*, 933. (c) Review: Horito, S.; Hashimoto, H., *Yuki Gosei Kagaku Kyokai Shi* **1990**, *48*, 1028 (, *Chem. Abstr.* **1991**, *114*, 121 385k).

2. Minato, H.; Miura, T.; Kobayashi, M., *Chem. Lett.* **1975**, 701.
3. Trost, B. M.; Shibata, T.; Martin, S. J., *J. Am. Chem. Soc.* **1982**, *104*, 3228.
4. (a) Kline, M. L.; Beutow, N.; Kim, J. K.; Caserio, M. C., *J. Org. Chem.* **1979**, *44*, 1904. (b) O'Malley, G. J.; Cava, M. P., *Tetrahedron Lett.* **1985**, *26*, 6159.
5. Trost, B. M.; Murayama, E., *J. Am. Chem. Soc.* **1981**, *103*, 6529.
6. Smith, R. A. J.; Bin Manas, A. R., *Synthesis* **1984**, 166.
7. Blomberg, L.; Norberg, T., *J. Carbohydr. Chem.* **1992**, *11*, 751.
8. Meerwein, H.; Zenner, K.-F.; Gipp, R., *Justus Liebigs Ann. Chem./Liebigs Ann. Chem.* **1965**, *688*, 67.
9. Minato, H.; Miura, T.; Kobayashi, M., *Chem. Lett.* **1975**, 701.
10. (a) Lawesson's Reagent: Zabirov, N. G.; Cherkasov, R. A.; Khalikov, I. S.; Pudovik, A. N., *Zh. Obshch. Khim.* **1986**, *56*, 2673 (, *Chem. Abstr.* **1987**, *107*, 236 846j) . (b) Sulfur: Bruzik, K. S.; Salamonczyk, G.; Stec, W. J., *J. Org. Chem.* **1986**, *51*, 2368. (c) Dibenzoyl disulfide: Schönberg reaction: Kamer, P. C. J.; Roelenm, H. C. P. F.; van den Elst, G. A.; van der Marel, G. A.; van Boom, J. H., *Tetrahedron Lett.* **1989**, *30*, 6757. (d) Tetraethylthiuram disulfide: Vu, H.; Hirschbein, B. L., *Tetrahedron Lett.* **1991**, *32*, 3005. (e) 3*H*-1, 2-Benzodithiole-3-one 1, 1-dioxide: Iyer, R. P.; Egan, W.; Regan, J. B.; Beaucage, S. L., *J. Am. Chem. Soc.* **1990**, *112*, 1253.
11. Dubs, P.; Stüssi, R., *Helv. Chim. Acta* **1976**, *59*, 1307.
12. Anderson, M. B.; Ranasinghe, M. G.; Palmer, J. T.; Fuchs, P. L., *J. Org. Chem.* **1988**, *53*, 3125.
13. Kim, J. K.; Souma, Y.; Beutow, N.; Ibbeson, C.; Caserio, M. C., *J. Org. Chem.* **1989**, *54*, 1714.
14. Kim, J. K.; Caserio, M. C., *Tetrahedron Lett.* **1981**, *22*, 4159.
15. (a) Trost, B. M.; Shibata, T., *J. Am. Chem. Soc.* **1982**, *104*, 3225. (b) Caserio, M. C.; Kim, J. K., *J. Am. Chem. Soc.* **1982**, *104*, 3231.
16. Trost, B. M.; Martin, S. F., *J. Am. Chem. Soc.* **1984**, *106*, 4263.
17. Trost, B. M.; Shibata, T.; Martin, S. J., *J. Am. Chem. Soc.* **1982**, *104*, 3228.
18. O'Malley, G. J.; Cava, M. P., *Tetrahedron Lett.* **1985**, *26*, 6159.
19. (a) Okuma, K.; Koike, T.; Yamamoto, S.-i.; Yonekura, K.; Ohta, H., *Chem. Lett.* **1989**, 1953. (b) Okuma, K.; Koike, T.; Yamamoto, S.-i.; Takeuchi, H.; Yonekura, K.; Ono, M.; Ohta, H., *Bull. Chem. Soc. Jpn.* **1992**, *65*, 2375.
20. (a) Haufe, G.; Alvernhe, G.; Anker, D.; Laurent, A.; Saluzzo, C., *Tetrahedron Lett.* **1988**, *29*, 2311. (b) Haufe, G.; Alvernhe, G.; Anker, D.; Laurent, A.; Saluzzo, C., *J. Org. Chem.* **1992**, *57*, 714.
21. Kline, M. L.; Beutow, N.; Kim, J. K.; Caserio, M. C., *J. Org. Chem.* **1979**, *44*, 1904.
22. Trost, B. M.; Murayama, E., *J. Am. Chem. Soc.* **1981**, *103*, 6529.
23. Braish, T. F.; Saddler, J. C.; Fuchs, P. L., *J. Org. Chem.* **1988**, *53*, 3647.
24. Trost, B. M.; Sato, T., *J. Am. Chem. Soc.* **1985**, *107*, 719.
25. Amat, M.; Alvarez, M.; Bonjoch, J.; Casamitjana, N.; Gràcia, J.; Lavilla, R.; Gracía, X.; Bosch, J., *Tetrahedron Lett.* **1990**, *31*, 3453.
26. Posner, G. H.; Asirvatham, E.; Hamill, T. G.; Webb, K. S., *J. Org. Chem.* **1990**, *55*, 2132.
27. Kreissl, F. R.; Schuett, W.; Stegmair, C. M.; Ullrich, N.; Keller, H.; Ostermeier, J.; Herdtweck, E., *Chem Ber.* **1993**, *126*, 1609.
28. Georgakaki, I. P.; Miller, M. L.; Darensbourg, M. Y., *Inorg. Chem.* **2003**, *42*, 2489.
29. Bishop, P.; Jones, C.; Chmielewski, J., *Tetrahedron Lett.* **1993**, *34*, 4469.
30. Steynberg, P. J.; Nel, R. J. J.; Van Rensburg, H.; Bezuidenhoudt, B. C. B.; Ferreira, D., *Tetrahedron* **1998**, *54*, 8153.
31. Clark, M. A.; Goering, B. K.; Li, J.; Ganem, B., *J. Org. Chem.* **2000**, *65*, 4058.
32. Jung, M. E.; Rhee, H., *Tetrahedron Lett.* **1993**, *34*, 4449.
33. Bennasar, M. L.; Vidal, B.; Bosch, J., *Chem. Commun.* **1995**, 125.
34. Haufe, G.; Wessel, U.; Schulze, K.; Alvernhe, G., *J. Fluorine Chem.* **1995**, *74*, 283.
35. Krueger, C. A.; Chiu, B. C.; Walters, M. A. Abstracts of Papers, *224th ACS National Meeting* **2002**, ORGN-151.
36. Padwa, A.; Ginn, J. D.; McClure, M. S., *Org. Lett.* **1999**, *1*, 1559.
37. Padwa, A.; Eidell, C. K.; Ginn, J. D.; McClure, M. S., *J. Org. Chem.* **2002**, *67*, 1595.
38. (a) Ginn, J. D.; Lynch, S. M.; Padwa, A., *Tetrahedron Lett.* **2000**, *41*, 9387. (b) Padwa, A.; Ginn, J. D.; Bur, S. K.; Eidell, C. K.; Lynch, S. M., *J. Org. Chem.* **2002**, *67*, 3412. (c) Padwa, A.; Eidell, C. K.; Lynch, S. M., *Heterocycles* **2002**, *58*, 227.
39. (a) Padwa, A.; Waterson, A. G., *Tetrahedron Lett.* **1998**, *39*, 8585. (b) Padwa, A.; Waterson, A. G., *J. Org. Chem.* **2000**, *65*, 235.
40. Padwa, A.; Heidelbaugh, T. M.; Kuethe, J. T.; McClure, M. S.; Wang, Q., *J. Org. Chem.* **2002**, *67*, 5928.
41. Lebsack, A. D.; Overman, L. E.; Valentekovich, R. J., *J. Am. Chem. Soc.* **2001**, *123*, 4851.
42. (a) Padwa, A.; Jacquez, M. N.; Schmidt, A., *Org. Lett.* **2001**, *3*, 1781. (b) Padwa, A.; Jacquez, M. N.; Schmidt, A., *J. Org. Chem.* **2004**, *69*, 33.
43. Craig, D.; King, N. P.; Shaw, A. N., *Tetrahedron Lett.* **1997**, *38*, 8599.
44. Aaberg, P. M. l.; Blomberg, L.; Loenn, H.; Norberg, T., *J. Carbohydr. Chem.* **1994**, *13*, 141.
45. Lear, M. J.; Yoshimura, F.; Hirama, M., *Angew. Chem., Int. Ed.* **2001**, *40*, 946.
46. Rademann, J.; Schmidt, R. R., *Tetrahedron Lett.* **1996**, *37*, 3989.
47. Garegg, P. J., *Adv. Carbohydr. Chem. Biochem.* **1997**, *52*, 179.
48. (a) Sujino, K.; Sugimura, H., *Tetrahedron Lett.* **1994**, *35*, 1883. (b) Sujino, K.; Sugimura, H., *Chem. Lett.* **1993**, 1187.
49. Sugimura, H.; Katoh, Y., *Chem. Lett.* **1999**, 361.
50. (a) Sugimura, H.; Motegi, M.; Sujino, K., *Nucleosides & Nucleotides* **1995**, *14*, 413. (b) Yokomatsu, T.; Shimizu, T.; Sada, T.; Shibuya, S., *Heterocycles* **1999**, *50*, 21. (c) Sugimura, H.; Sujino, K., *Yuki Gosei Kagaku Kyokaishi* **2001**, *59*, 879.
51. Sujino, K.; Sugimura, H., *Chem. Commun.* **1994**, 2541.
52. (a) Sugimura, H., *Nucleosides, Nucleotides & Nucleic Acids* **2000**, *19*, 629. (b) Sugimura, H., *Nucleic Acid Res. Suppl.* **2003**, 21–22.
53. Guindon, Y.; Gagnon, M.; Thumin, I.; Chapdelaine, D.; Jung, G.; Guerin, B., *Org. Lett.* **2002**, *4*, 241.

Dimethyl(methylthio)sulfonium Trifluoromethanesulfonate

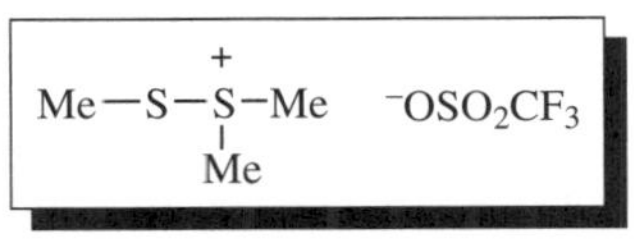

[85055-60-3] $C_3H_9S_2^+\ CF_3O_3S^-$ (MW 258.30)

(methylsulfenylating agent; activates thioglycosides and other types of glycosyl donors for glycoside synthesis[1–3])

Alternate Name: dimethyl(methylthio)sulfonium triflate, dimethylthiomethyl sulfonium trifluoromethanesulfonate, DMTST.

Physical Data: mp 28–36 °C;[4] 53–55 °C.[5] Variation in reported melting points is probably due to the hygroscopic nature of the reagent.

Solubility: soluble in CH_2Cl_2, MeCN, $MeNO_2$; sparingly soluble in diethyl ether, benzene, toluene; insoluble in petroleum ether.
Form Supplied in: white crystals; not commercially available.
Analysis of Reagent Purity: ^1H NMR.
Purification: recrystallization from dichloromethane–diethyl ether.
Preparative Methods: readily prepared by alkylation of methyl disulfide with methyl trifluoromethanesulfonate in dichloromethane[1,4,5] or neat[6] (eq 1). The reagent crystallizes from dichloromethane by the addition of diethyl ether. Alternatively, the dichloromethane solution can be used directly for most applications.

$$\text{Me—S—S—Me} + \text{MeOSO}_2\text{CF}_3 \longrightarrow \text{Me—S—}\overset{+}{\underset{\text{Me}}{\text{S}}}\text{—Me}\ {}^{-}\text{OSO}_2\text{CF}_3 \quad (1)$$

Handling, Storage, and Precautions: the reagent is hygroscopic, it liquifies in air. Store in a closed vessel or over a strong drying agent, preferably, in the refrigerator. A 1 M solution in dichloromethane, which partially crystallizes on storage in the refrigerator, can be stored for prolonged periods of time.[1] In cases of doubts the use of freshly prepared DMTST is recommended.

Introduction. The sulfur–sulfur bond in alkylated disulfides [dialkyl(alkylthio)sulfonium salts] is susceptible to nucleophiles and thus alkylated disulfides are alkylsulfenylating agents, or potential sources of alkylsulfenyl (RS^+) ions.[7,8] This, in combination with the excellent leaving property of the triflate group, led to the introduction of dimethyl(methylthio)sulfonium triflate (DMTST) as a synthetic reagent for activation of thioglycosides in glycosylation reactions,[1–3] which is still its primary use. Other applications of the reagent (acetalization, addition to unsaturated compounds) are also based on its alkylsulfenylating capability.

Glycosylation.

Syntheses of O-glycosides from thioglycosides

Thioglycosides have the advantage over most glycosylating agents that they are stable, are easy to functionalize, can serve both as glycosyl acceptors and glycosyl donors, and are therefore widely used in oligosaccharide syntheses.[9,10] DMTST, a highly thiophilic reagent, activates thioglycosides for glycosylations. It has been used in a large number of oligosaccharide syntheses for the construction of diverse glycosidic and interglycosidic linkages.

Activation of thioglycosides by DMTST takes place by sulfenylation of the thioglycoside sulfur. The resulting positively charged intermediate then can react directly or, more probably, through the common intermediates of glycosylation reactions (oxocarbonium ions, acyloxonium ions, glycosyl triflates) with alcohols (or other nucleophiles) to give glycosides (eq 2).

In the presence of neighboring group participating substituents the glycosylations lead to 1,2-*trans*-glycosides, as shown in the synthesis of the cellobioside derivative (eq 3).[1]

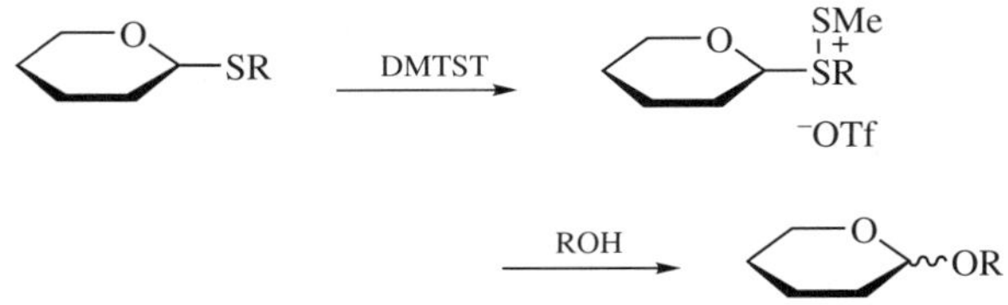

(2)

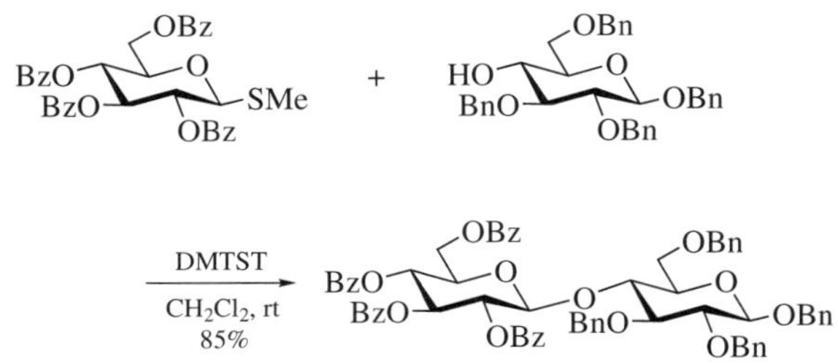

(3)

Reactions are usually performed at or below room temperature in solvents such as dichloromethane or acetonitrile with a 3–5-fold excess of the reagent. The reaction mixture may be buffered with 2,6-di-*tert*-butyl-4-methylpyridine or other sterically hindered bases if required. DMTST has been used to activate a variety of different monosaccharides including thioglycoside derivatives of D-glucose,[1,2,11] D-galactose,[1,12–14] D-mannose,[15] D-fructofuranose,[16] amino sugars,[1,12,17] uronic acids[15,18] and heptoses.[19] Di-,[13,14,17] tri-,[20,21] tetra-,[22] and even pentasaccharide[23] thioglycosides have also been activated by this reagent. The most frequently used neighboring group participating protecting groups are *O*-acetyl, *O*-pivaloyl and *O*-benzoyl; for amino functions *N*-phthaloyl[1,12,17] and *N*-trichloroethoxycarbonyl[24,25] are the most frequently used ones. Reactions with *O*-benzoyl-protected thioglycosides tend to give higher yields and cleaner reactions than those of *O*-acetyl protected ones[1,18]. The thioglycosides commonly used in conjunction with DMTST have simple thioalkyl (SMe, SEt) and thioaryl (SPh, STol) groups; heterocyclic thioglycosides have been used occasionally.[26] A wide range of alcohols, including mono- and oligosaccharides, as well as peptides,[24,25,27] has been used as glycosyl acceptors. It should be noted that although neighboring group participation is generally used to govern the formation of *trans*-glycosides, several examples of stereoselective synthesis of *trans*-glycosides without neighboring group participation in DMTST-promoted glycosylations are also known.[19,28,29]

DMTST-promoted glycosylations with thioglycosides having non-participating substituents result in anomeric mixtures of *O*-glycosides,[2] but very often useful, or even complete, stereoselectivity can be achieved. Methods increasing the stereoselectivity include the addition of bromide ion to the glycosylation mixture,[2] changing the solvent,[2] tuning the reactivity of the glycosyl donors and acceptors by the proper choice of the protecting groups, and intramolecular aglycon delivery. Addition of **tetra-n-butylammonium bromide** to the glycosylation mixture results in the in situ formation of a glycosyl bromide, which then reacts in a halide ion-catalyzed glycosylation reaction[30] to yield 1,2-*cis*-glycosides. Performing the reaction in acetonitrile at low temperature changes the stereoselectivity due to the formation of nitrilium ion intermediates.[2] Application of this principle to the biologically important sugar *N*-acetylneuraminic acid, resulted in a very useful synthesis of α-glycosides of *N*-acetylneuraminic acid[31,32] (eq 4), previously accessible only with great difficulty.

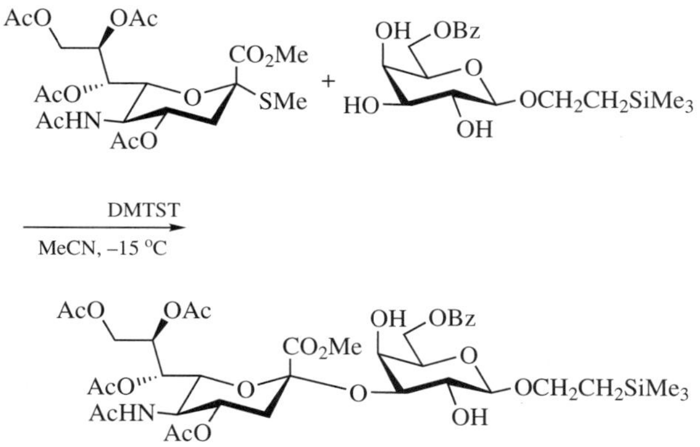

(4)

Ethereal solvents and solvents of low polarity such as benzene also shift the stereoselectivity towards *cis*-glycosides. Although DMTST is poorly soluble in diethyl ether or benzene, its solubility is sufficient for reactions to proceed.[33] A large number of stereoselective α-L-fucosylations have been performed this way, as for example in the total synthesis of the biologically important sialyl Lewis x oligosaccharide.[14] Internal aglycon delivery, i.e. performing the glycosylation on a tethered derivative of the glycosyl donor and acceptor, was found to be a valuable tool for the synthesis of β-D-fructofuranosides.[34,35] Whereas common glycosylations with D-fructofuranosyl donors give a preponderance of α-D-fructoside,[16] reactions of tethered derivatives gave exclusively the β-D-fructofuranosides. In an analogous manner DMTST-promoted reaction of the intramolecularly locked fructofuranosyl donor in eq 5 resulted in the first stereoselective, high-yielding synthesis of sucrose[36] (eq 5).

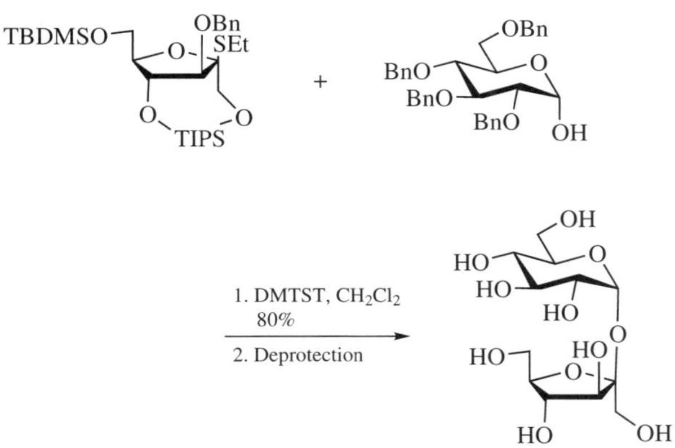

(5)

Compared with other promoters (**methyl trifluoromethanesulfonate** and the **N-iodosuccinimide trifluoromethanesulfonic acid** [NIS-TfOH] system) frequently used in glycosylation reactions with thioglycoside donors, DMTST is more efficient[3,27] than methyl triflate, but it is less powerful than NIS–TfOH. DMTST has the advantage that side-reactions, such as alkylation of the acceptor (which can be encountered in methyl triflate-promoted reactions[1,37]), formation of glycosyl-succinimides,[16,38,39] or iodination of certain protecting groups[40] can be avoided. The related compound, **dimethyl(methylthio)sulfonium tetrafluoroborate**, which was introduced as a glycosylation promoter at the same time as DMTST,[1] also found application in oligosaccharide synthesis.[41] However, DMTSF-promoted glycosylations are likely to proceed through glycosyl fluoride intermediates, as glycosyl fluorides have been isolated in reactions of thioglycosides with the fluoroborate salt performed in the absence of glycosyl acceptors.[42]

The reactivity differences of thioglycosides depending on their substituents in the carbohydrate unit ('armed-disarmed' donors[43,44]) or in the thioaglycon ('latent-active' donors[45–47]) permit the use of thioglycosides both as glycosyl donors and glycosyl acceptors in the same reaction. Glycosylation of an unreactive thioglycoside with a more reactive one takes place without self-condensation of the acceptor. This was elegantly exploited in the programmable one-pot oligosaccharide synthesis developed by Wong.[38,39] DMTST has found application in one-pot oligosaccharide synthesis, where its use is especially beneficial when the reactivities of glycosyl donors are relatively high.[38,39] Chain termination by the accumulation of glycosylsuccinimide byproducts formed in NIS-promoted reactions is avoided in DMTST-promoted reactions. Among the applications of DMTST in oligosaccharide synthesis, its use in the solid-phase synthesis of large and structurally complex oligosaccharides[11,21,48] and in the combinatorial synthesis of oligosaccharide libraries[6,38,39] deserves particular attention.

Syntheses of O-Glycosides from Compounds other than Thioglycosides

Other types of compounds having a thio functionality at the anomeric center can be activated by DMTST. Anomeric *S*-xanthates[49,50] (eq 6) and dithiocarbamates[51] have also been used as glycosyl donors in combination with the reagent. Isopropenyl glycosides,[52] though not having sulfur at the anomeric leaving group, could also be activated for glycoside synthesis with DMTST.

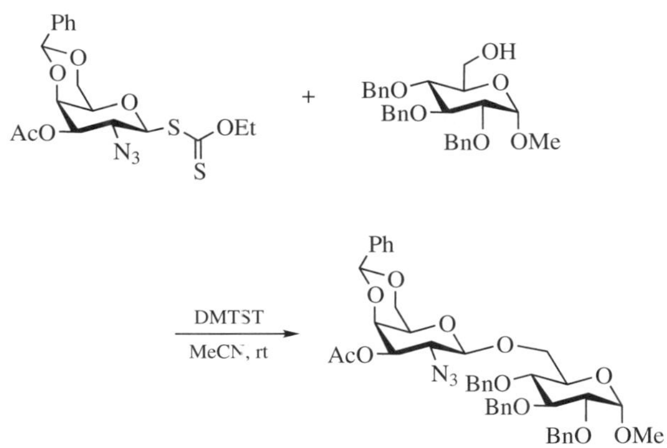

(6)

Syntheses of other types of Compounds from Thioglycosides

DMTST promotes the reaction of thioglycosides with nucleophiles other than alcohols. Glycosylation of *H*-phosphonate monoesters followed by oxidation of the phosphonate diester was found to be a viable alternative to glycosyl phosphodiesters not accessible by other methods (eq 7).[53] Reaction of thioglycosides with pyrimidine bases gave nucleoside analogs,[54–56] reaction with water constitutes a mild hydrolysis of thioglycosides to the hemiacetal[57] and reaction with alcohols in the presence of excess base was used for the synthesis of orthoesters.[58] The formation of glycosyl bromides from thioglycosides with the aid of DMTST and bromide ions[2] was discussed earlier.

Acetalization. Cyclic acetals of pyruvic acid are not easily prepared by using commonly available acetalization methods. Cyclic pyruvic acid acetals of carbohydrates were prepared by DMTST-promoted transacetalization reaction between methyl pyruvate diphenyl dithioacetal and carbohydrate diols (eq 8).[59]

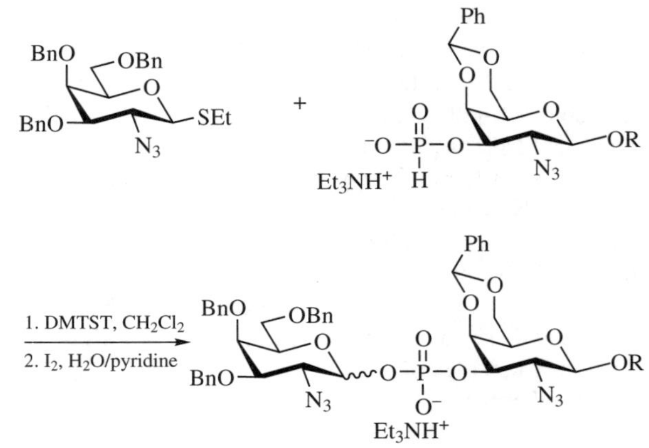

(7)

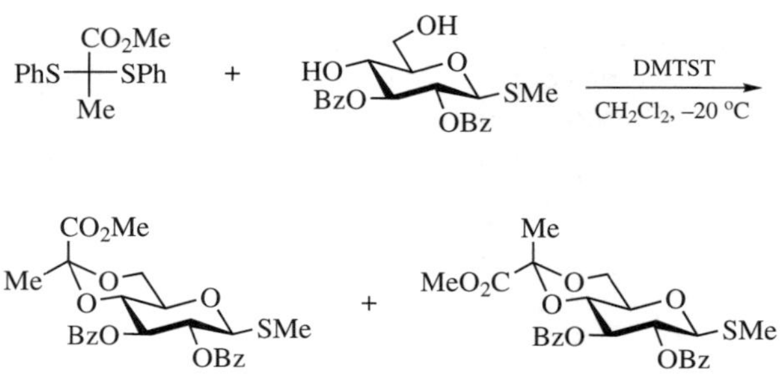

(8)

Addition to Double Bonds. Compared with other alkylsulfenylating agents, addition of dimethyl(methylthio)sulfonium trifluoromethanesulfonate to double bonds is less frequently used. Treatment of olefins with the reagent followed by the addition of triphenylphosphine gave 2-methylthioalkylphosphonium salts in high yields, which could be converted into vinylphosphonium salts or vinyl phosphine oxides (eq 9).[5]

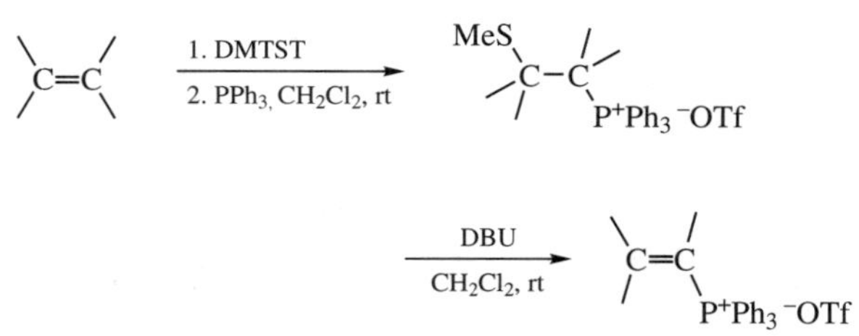

(9)

Addition of the reagent to double bonds[60,61] limits the use of unsaturated protecting groups (e.g. allyl, allyloxycarbonyl) in dimethyl(methylthio)sulfonium triflate-promoted glycosylation reactions.

Related Reagents. Dimethyl(methylthio)sulfonium fluoroborate (DMTSF); dimethyl(methylthio)sulfonium tetrafluoroborate, methylsulfenyl trifluoromethanesulfonate, methyl trifluoromethanesulfonate, *N*-iodosuccinimide (with triflic acid).

1. Fügedi, P.; Garegg, P. J., *Carbohydr. Res.* **1986**, *149*, C9.
2. Andersson, F.; Fügedi, P.; Garegg, P. J.; Nashed, M., *Tetrahedron Lett.* **1986**, *27*, 3919.
3. Andersson, F.; Birberg, W.; Fügedi, P.; Garegg, P. J.; Nashed, M.; Pilotti, Å., *ACS Symp. Ser.* **1989**, *386*, 117.
4. Ravenscroft, M.; Roberts, R. M. G.; Tillett, J. G., *J. Chem. Soc., Perkin Trans. 2* **1982**, 1569.
5. Okuma, K.; Koike, T.; Yamamoto, S.; Takeuchi, H.; Yonekura, K.; Ono, M.; Ohta, H., *Bull. Chem. Soc. Japan* **1992**, *65*, 2375.
6. Wong, C. H.; Ye, X. S.; Zhang, Z. Y., *J. Am. Chem. Soc.* **1998**, *120*, 7137.
7. Helmkamp, G. K.; Olsen, B. A.; Pettitt, D. J., *J. Org. Chem.* **1965**, *30*, 676.
8. Smallcombe, S. H.; Caserio, M. C., *J. Am. Chem. Soc.* **1971**, *93*, 5826.
9. Fügedi, P.; Garegg, P. J.; Lönn, H.; Norberg, T., *Glycoconjugate J.* **1987**, *4*, 97.
10. Garegg, P. J., *Adv. Carbohydr. Chem. Biochem.* **1997**, *52*,179.
11. Nicolaou, K. C.; Winssinger, N.; Pastor, J.; DeRoose, F., *J. Am. Chem. Soc.* **1997**, *119*, 449.
12. Kiso, M.; Katagiri, H.; Furui, H.; Hasegawa, A., *J. Carbohydr. Chem.* **1992**, *11*, 627.
13. Kameyama, A.; Ishida, H.; Kiso, M.; Hasegawa, A., *Carbohydr. Res.* **1990**, *200*, 269.
14. Kameyama, A.; Ishida, H.; Kiso, M.; Hasegawa, A., *Carbohydr. Res.* **1991**, *209*, C1.
15. Garegg, P. J.; Olsson, L.; Oscarson, S., *Bioorg. Med. Chem.* **1996**, *4*, 1867.
16. Krog-Jensen, C.; Oscarson, S., *J. Org. Chem.* **1996**, *61*, 1234.
17. Nilsson, M.; Norberg, T., *Carbohydr. Res.* **1988**, *183*, 71.
18. Garegg, P. J.; Olsson, L.; Oscarson, S., *J. Org. Chem.* **1995**, *60*, 2200.
19. Ekelöf, K.; Oscarson, S., *J. Carbohydr. Chem.* **1995**, *14*, 299.
20. Ishida, H. K.; Ishida, H.; Kiso, M.; Hasegawa, A., *Tetrahedron: Asymmetry* **1994**, *5*, 2493.
21. Nicolaou, K. C.; Watanabe, N.; Li, J.; Pastor, J.; Winssinger, N., *Angew. Chem., Int. Ed.* **1998**, *37*, 1559.
22. Chernyak, A.; Oscarson, S.; Turek, D., *Carbohydr. Res.* **2000**, *329*, 309.
23. Ruda, K.; Lindberg, J.; Garegg, P. J.; Oscarson, S.; Konradsson, P., *J. Am. Chem. Soc.* **2000**, *122*, 11067.
24. Schultz, M.; Kunz, H., *Tetrahedron Lett.* **1992**, *33*, 5319.
25. Schultz, M.; Kunz, H., *Tetrahedron: Asymmetry* **1993**, *4*, 1205.
26. Takeda, K.; Tsuboyama, K.; Torii, K.; Furuhata, K.; Sato, N.; Ogura, H., *Carbohydr. Res.* **1990**, *203*, 57.
27. Paulsen, H.; Rauwald, W.; Weichert, U., *Liebigs Ann. Chem.* **1988**, 75.
28. Hällgren, C.; Hindsgaul, O., *Carbohydr. Res.* **1994**, *260*, 63.
29. Bernlind, C.; Oscarson, S., *Carbohydr. Res.* **1997**, *297*, 251.
30. Lemieux, R. U.; Hendriks, K. B.; Stick, R. V.; James, K., *J. Am. Chem. Soc.* **1975**, *97*, 4056.
31. Murase, T.; Ishida, H.; Kiso, M.; Hasegawa, A., *Carbohydr. Res.* **1988**, *184*, C1.
32. Kanie, O.; Kiso, M.; Hasegawa, A., *J. Carbohydr. Chem.* **1988**, *7*, 501.
33. Garegg, P. J.; Oscarson, S.; Ritzén, H.; Szönyi, M., *Carbohydr. Res.* **1992**, *228*, 121.
34. Krog-Jensen, C.; Oscarson, S., *J. Org. Chem.* **1996**, *61*, 4512.
35. Krog-Jensen, C.; Oscarson, S., *J. Org. Chem.* **1998**, *63*, 1780.
36. Oscarson, S.; Sehgelmeble, F. W., *J. Am. Chem. Soc.* **2000**, *122*, 8869.
37. Bhattacharya, S. K.; Danishefsky, S. J., *J. Org. Chem.* **2000**, *65*, 144.
38. Zhang, Z. H.; Ollmann, I. R.; Ye, X. S.; Wischnat, R.; Baasov, T.; Wong, C. H., *J. Am. Chem. Soc.* **1999**, *121*, 734.
39. Ye, X. S.; Wong, C. H., *J. Org. Chem.* **2000**, *65*, 2410.
40. Mehta, S.; Whitfield, D. M., *Tetrahedron* **2000**, *56*, 6415.
41. Åberg, P. M.; Blomberg, L.; Lönn, H.; Norberg, T., *J. Carbohydr. Chem.* **1994**, *13*, 141.

42. Blomberg, L.; Norberg, T., *J. Carbohydr. Chem.* **1992**, *11*, 751.
43. Veeneman, G. H.; van Boom, J. H., *Tetrahedron Lett.* **1990**, *31*, 275.
44. Veeneman, G. H.; van Leeuwen, S. H.; van Boom, J. H., *Tetrahedron Lett.* **1990**, *31*, 1331.
45. Roy, R.; Andersson, F. O.; Letellier, M., *Tetrahedron Lett.* **1992**, *33*, 6053.
46. Cao, S. D.; Hernández-Matéo, F.; Roy, R., *J. Carbohydr. Chem.* **1998**, *17*, 609.
47. Sliedregt, L. A. J. M.; Zegelaar-Jaarsveld, K.; van der Marel, G. A.; van Boom, J. H., *Synlett* **1993**, 335.
48. Kanemitsu, T.; Kanie, O.; Wong, C. H., *Angew. Chem., Int. Ed.* **1998**, *37*, 3415.
49. Marra, A.; Gauffeny, F.; Sinaÿ, P., *Tetrahedron* **1991**, *47*, 5149.
50. Marra, A.; Sinaÿ, P., *Carbohydr. Res.* **1990**, *195*, 303.
51. Fügedi, P.; Garegg, P. J.; Oscarson, S.; Rosen, G.; Silwanis, B. A., *Carbohydr. Res.* **1991**, *211*, 157.
52. Chenault, H. K.; Castro, A.; Chafin, L. F.; Yang, J., *J. Org. Chem.* **1996**, *61*, 5024.
53. Garegg, P. J.; Hansson, J.; Helland, A. C.; Oscarson, S., *Tetrahedron Lett.* **1999**, *40*, 3049.
54. Sugimura, H.; Muramoto, I.; Nakamura, T.; Osumi, K., *Chem. Lett.* **1993**, 169.
55. Sugimura, H.; Sujino, K., Jpn. Kokai Tokkyo Koho JP 10 182,691 (*Chem. Abstr.* **1998**, *129*, 149179e).
56. Sugimura, H., Jpn. Kokai Tokkyo Koho JP 10 237,090 (*Chem. Abstr.* **1998**, *129*, 245422z).
57. Hasegawa, A.; Murase, T.; Ogawa M.; Ishida, H.; Kiso, M., *J. Carbohydr. Chem.* **1990**, *9*, 415.
58. Hällgren, C., *J. Carbohydr. Chem.* **1992**, *11*, 527.
59. Lipták, A.; Szabó, L., *Carbohydr. Res.* **1988**, *184*, C5–C8.
60. Chernyak, A. Y.; Antonov, K. V.; Kochetkov, N. K., *Bioorg. Khim.* **1989**, *15*, 1113 (*Chem. Abstr.* **1990**, *112*, 158760p).
61. Kunz, H.; Wernig, P.; Schultz, M., *Synlett* **1990**, 631.

Péter Fügedi
Chemical Research Center, Hungarian Academy of Sciences
Budapest, Hungary

Dimethylphosphinothioyl Chloride

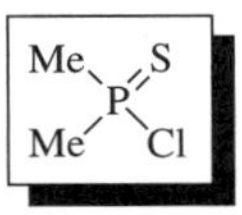

[993-12-4] C_2H_6ClPS (MW 128.57))

(reagent for protection in peptide[1,2] and carbohydrate[3,4] synthesis)

Alternate Name: Mpt-Cl.
Physical Data: mp 24–25 °C; bp 107–107.5 °C/16 mmHg; bp 69–70 °C/11 mmHg;[5] n_D^{20} 1.546.
Form Supplied in: liquid; commercially available.
Preparative Method: chlorination of tetramethyldiphosphine disulfide[7] with Sulfuryl Chloride.[1,6]
Handling, Storage, and Precautions: stable for several months in a refrigerator.[1] Should be handled in the fume hood.

Peptide Synthesis. The dimethylphosphinothioyl (Mpt) group has been used to protect *N*-amino acids. It is more readily removed than *t*-butoxycarbonyl group, making it a convenient reagent for solid phase synthesis (eq 1).[1]

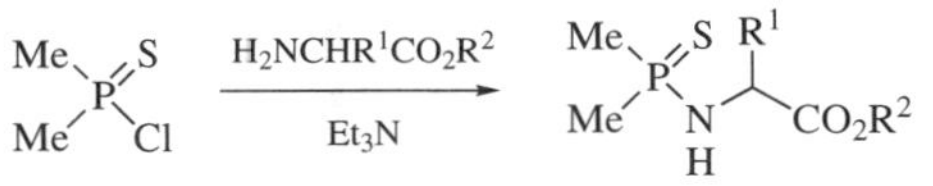

(1)

This method has been effectively demonstrated in the synthesis of the opioid agonist L-Leu-enkaphalin (L-Tyr-Gly-Gly-L-Phe-L-Leu).

Mpt-Cl also reacts rapidly with *N*-protected amino acids in the presence of Triethylamine to afford Mpt mixed anhydrides (Mpt-MA). These mixed anhydrides can be isolated by silica-gel column chromatography and stored without decomposition in a freezer. Cbz-Ser-Mpt-MA has been isolated in 17% yield, without protection of the side-chain hydroxy group (eq 2).[8,9] As a consequence, these active Mpt-MAs have been used for peptide synthesis in methanol or ethanol rather than the usual dipolar aprotic solvents (HMPA, DMF, or DMSO). Synthesis of a heptapeptide (**1**) has been completed by this method.[2]

Cbz–Tyr–Leu–Asn–Ser–Ile–Asn–NH_2

(**1**)

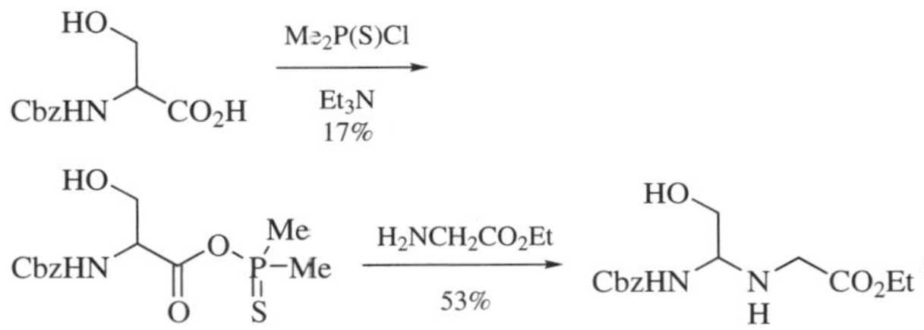

(2)

Carbohydrate Protection/Synthesis. Hydroxy groups of carbohydrates can be protected using Mpt-Cl in the presence of 1,8-Diazabicyclo[5.4.0]undec-7-ene and catalytic amounts of 4-Dimethylaminopyridine. The Mpt group is stable to acidic hydrolysis conditions that cleave trityl and isopropylidene (acetonide) groups, DBU/MeOH, Bu_4NF, Bu_3SnH, Grignard reagents, and cat. NaOMe/MeOH. The Mpt group does not migrate as do acyl groups (carboxylate esters) under acidic and basic conditions.[4]

Mpt-Cl has been used in the stereoselective synthesis of B-glucopyranomides.[3]

---

1. Veki, M.; Inazu, T.; Ikeda, S. *Bull. Chem. Soc. Jpn.* **1979**, *52*, 2424.
2. Veki, M.; Santo, T.; Sasaya, Jun.; Ikeda, S.; Oyamada, H. *Bull. Chem. Soc. Jpn.* **1988**, *61*, 3653.
3. Yamanoi, T.; Inazu, T. *Noguchi Kenkyusho Jiho* **1988**, *31*, 49 (*Chem. Abstr.* **1989**, *111*, 7684v) .
4. Inazu, T.; Yamanoi, T. *Noguchi Kenkyusho Jiho* **1988**, *31*, 43 (*Chem. Abstr.* **1989**, *111*, 7685w) .
5. *Dictionary of Organophosphorus Compounds*; Chapman & Hall: New York, **1988**; p 327.
6. Maier, L. *Ber. Dtsch. Chem. Ges./Chem. Ber.* **1961**, *94*, 3051.
7. Parshall, G. W. *Org. Synth., Coll. Vol.* **1973**, *4*, 1016.
8. Veki, M.; Okazaki, K.; Sano, Y.; Ikeda, S.Pept., Proc. Eur. Peptide Symp. **1984**, 1814.
9. Veki, M.; Inazu, T. *Chem. Lett.* **1982**, 45.

Harjinder S. Bansal
Zeneca Agrochemicals, Bracknell, UK

Diphenylphosphinic Chloride

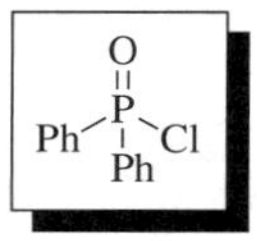

[1499-21-4] $C_{12}H_{10}ClOP$ (MW 236.64)

(agent for acid activation,[1] alkylphosphine oxide formation,[2] and amine protection[3])

Alternate Name: chlorodiphenylphosphine oxide.
Physical Data: bp 222 °C/16 mmHg; *d* 1.240 g cm^{-3}.
Form Supplied in: colorless liquid; widely available.
Handling, Storage, and Precautions: handle only in a chemical fume hood, wear chemical-resistant gloves and safety goggles, do not breathe vapor, and avoid contact with eyes or skin. Store in a 'flammable storage' cabinet under nitrogen. Contact with water generates hydrogen chloride gas.

Acid Activation. Chlorodiphenylphosphine oxide has been utilized for the activation of acids via in situ formation of the diphenylphosphinic (DPP) mixed anhydrides. These anhydrides are superior to carbon-based mixed anhydrides because they do not suffer from disproportionation to symmetrical anhydrides.[1] Also, nucleophiles prefer to attack at carbon rather than phosphorus, solving the regioselectivity problem associated with carbon-based mixed anhydrides.[4] Additionally, diphenylphosphinic mixed anhydrides are more electrophilic. Finally, diphenylphosphinic mixed anhydrides form rapidly allowing shorter activation times.

Diphenylphosphinic mixed anhydrides have been utilized to form peptide bonds.[5] Peptides are easier to isolate by this method than by employing 1,3-Dicyclohexylcarbodiimide. These anhydrides are the method of choice for the formation of amides of 2-alkenoic acids (eq 1).[6] Carbodiimide and acyl carbonate

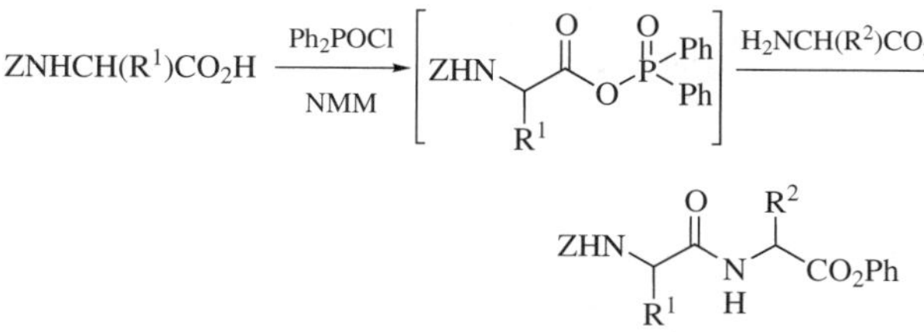

(1)

methods proved to be inferior. Primary amines result in better yields than secondary amines. This activation protocol can be employed to form thiol esters.[7] β-Amino acids are readily converted to β-lactams with chlorodiphenylphosphine oxide (eq 2).[8]

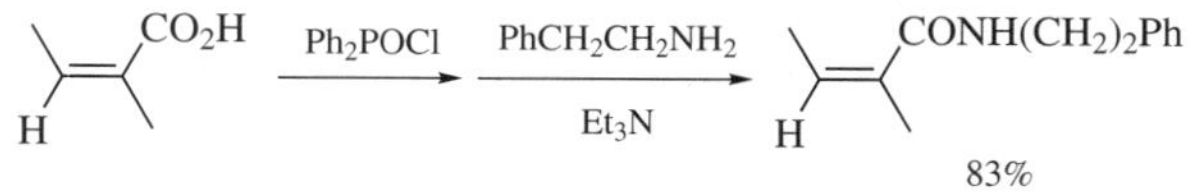

(2)

Secondary amines work best. This activation protocol has been utilized to convert acids to amines via a Curtius rearrangement (eq 3).[9] Phenols have been generated from diene acids, presumably via base-induced elimination of diphenylphosphinic acid from the mixed anhydrides to form ketenes which spontaneously cyclize (eq 4).[10] Acids have been converted to ketones via activation followed by reaction with organometallic reagents (eq 5).[11]

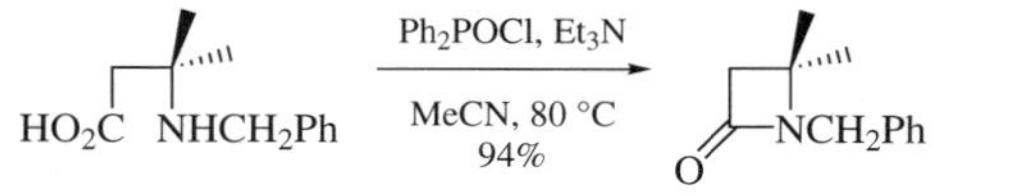

(3)

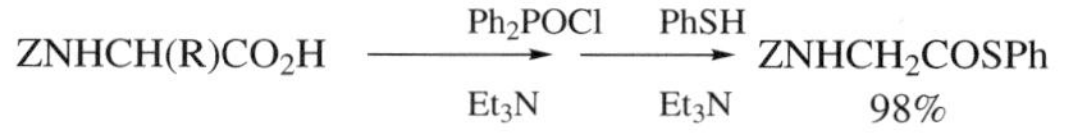

(4)

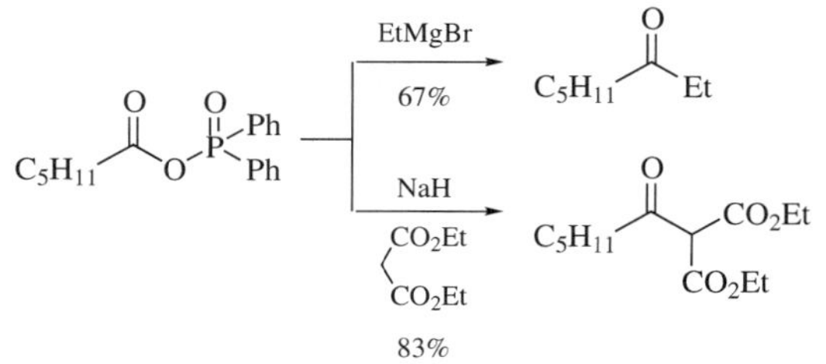

(5)

Phosphine Oxides. This reagent reacts with organometallic reagents to form phosphine oxides.[2] Alkyl phosphonates have been converted to diphenylvinylphosphine oxides via trapping of the α-lithioalkylphosphonate with diphenylchlorophosphine oxide, deprotonation, and reaction with aldehydes (eq 6).[12] Organometallic reagents have been converted to 1,3-dienes utilizing $Ph_2P(O)Cl$ and carbonyl compounds (eq 7).[13]

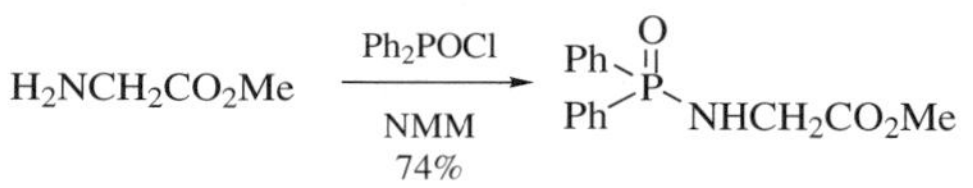

(6)

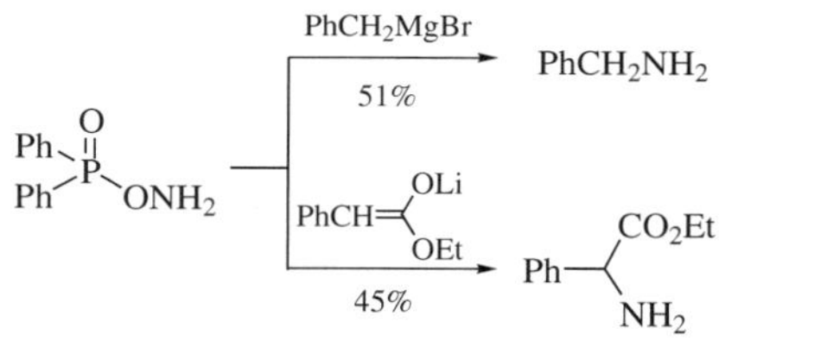

(7)

Thiol Esters and Ketones. The Dpp mixed anhydrides react with thiols to yield thiol esters (eq 8)[7] and undergo carbon-acylation reactions with both Grignard reagents and diethyl sodiomalonate in good yields (eq 9).[14]

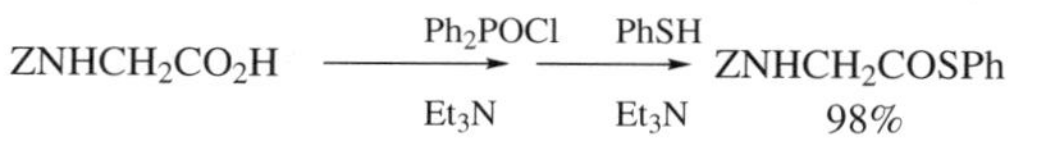

(8)

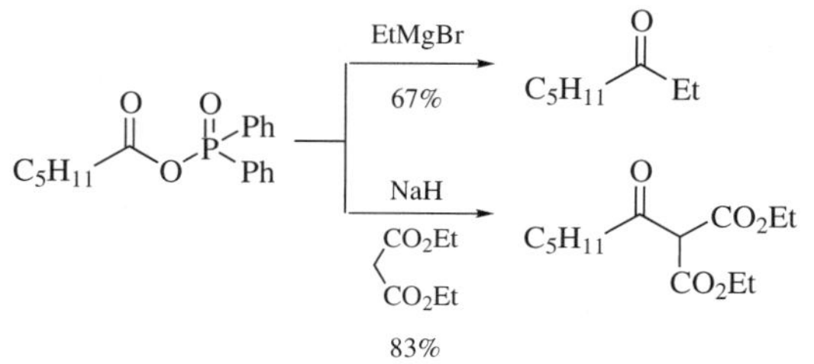

(9)

Protection of Amino Groups. In the presence of *N*-methylmorpholine, Ph_2POCl reacts with amino acid esters in CH_2Cl_2 at 0 °C in 1.5–2 h to give *N*-diphenylphosphinyl (Dpp) amino acid esters (65–80%), which are then hydrolyzed under mild alkaline conditions to give *N*-Dpp amino acids (eq 10).[3,4] Since Ph_2POCl hydrolyzes extremely quickly under aqueous conditions, it is not possible to prepare *N*-Dpp amino acids directly.

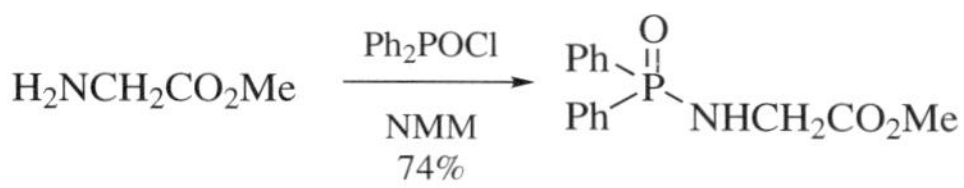

(10)

The Dpp derivatives are slightly more acid-labile than the Boc derivatives and are cleaved by HOAc–HCOOH–H_2O (7:1:2) (24 h), 1 M HCl in H_2O–dioxane (2:1) (3 h), TFA–CH_2Cl_2 (1:1) (40 min), and 95% TFA (10 min). Selective cleavage of the Dpp group in the presence of a *t*-butyl ester and a phenyl ester is possible.

Amination. Ph_2POCl reacts with Hydroxylamine to yield *O*-(diphenylphosphinyl)hydroxylamine.[15] This reagent aminates stabilized carbanions and Grignard reagents in moderate yields (eq 11).[16] Electrophilic *N*-amination of imide sodium salts is also possible using this reagent (eq 12).[17]

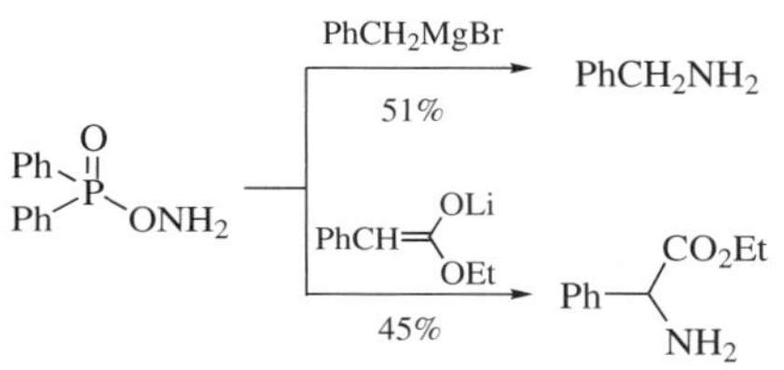

(11)

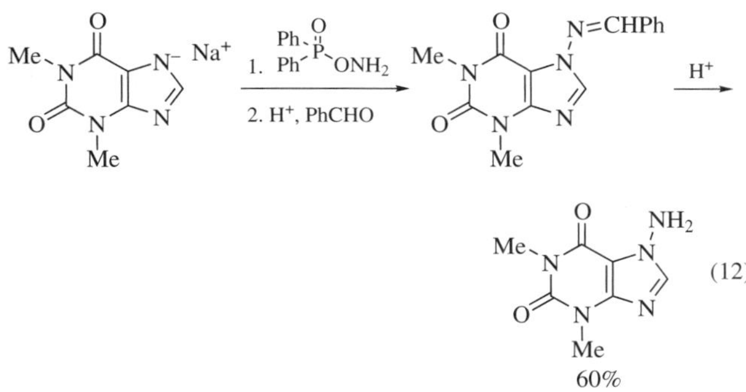

(12)

Diphenylphosphinamide[18] and 1-diphenylphosphinyl-2,2-dimethylaziridine,[19] which are derived from Ph_2POCl, react with Grignard reagents to give high yields of primary amines and moderate yields of α,α-dimethylarylalkylamines, respectively.

1. Jackson, A. G.; Kenner, G. W.; Moore, G. A.; Ramage, R.; Thorpe, W. D. *Tetrahedron Lett.* **1976**, 3627.
2. Takaya, H.; Mashima, K.; Koyano, K.; Yagi, M.; Kumobayashi, H.; Taketomi, T.; Akutagawa, S.; Noyori, R. *J. Org. Chem.* **1986**, *51*, 629.
3. (a) Kenner, G. W.; Moore, G. A.; Ramage, R. *Tetrahedron Lett.* **1976**, 3623. (b) Ramage, R.; Atrash, B.; Hopton, D.; Parrott, M. J. *J. Chem. Soc., Perkin Trans. 1* **1985**, 1217.
4. Ramage, R.; Atrash, B.; Hopton, D.; Parrott, M. J. *J. Chem. Soc., Perkin Trans. 1* **1985**, 1617.
5. Ramage, R.; Hopton, D.; Parrott, M. J.; Richardson, R. S.; Kenner, G. W.; Moore, G. A. *J. Chem. Soc., Perkin Trans. 1* **1985**, 461.
6. Bernasconi, S.; Comini, A.; Corbella, A.; Gariboldi, P.; Sisti, M. *Synthesis* **1980**, 385.
7. Horiki, K. *Synth. Commun.* **1977**, *7*, 251.
8. Kim, S.; Lee, P. H.; Lee, T. A. *Chem. Commun./J. Chem. Soc., Chem. Commun.* **1988**, 1242.
9. (a) Armstrong, V. W.; Coulton, S.; Ramage, R. *Tetrahedron Lett.* **1976**, 4311. (b) Ramage, R.; Armstrong, V. W.; Coulton, S. *Tetrahedron* **1981**, *37*, Supplement No. 1, 157.
10. Clinch, K.; Marquez, C. J.; Parrott, M. J.; Ramage, R. *Tetrahedron* **1989**, *45*, 239.
11. Soai, K.; Ookawa, A. *Chem. Commun./J. Chem. Soc., Chem. Commun.* **1986**, 412.
12. (a) Savignac, P.; Teulade, M.-P.; Aboujaoude, E. E.; Collignon, N. *Synth. Commun.* **1987**, *17*, 1559. (b) Aboujaoude, E. E.; Lietje, S.; Collignon, N. *Tetrahedron Lett.* **1985**, *26*, 4435.
13. Davidson, A. H.; Warren, S. *Chem. Commun./J. Chem. Soc., Chem. Commun.* **1975**, 148.
14. Kende, A. S.; Scholz, D.; Schneider, J. *Synth. Commun.* **1978**, 59.
15. (a) Harger, M. J. P. *Chem. Commun./J. Chem. Soc., Chem. Commun.* **1979**, 768. (b) Harger, M. J. P. *J. Chem. Soc., Perkin Trans. 1* **1981**, 3284.
16. (a) Colvin, E. W.; Kirby, G. W.; Wilson, A. C. *Tetrahedron Lett.* **1982**, *23*, 3835. (b) Boche, G.; Bernheim, M.; Schrott, W. *Tetrahedron Lett.* **1982**, *23*, 5399.
17. Klotzer, W.; Stadlwieser, J.; Raneburger, J. *Org. Synth.* **1985**, *64*, 96.
18. Zwierzak, A.; Slusarska, E. *Synthesis* **1979**, 691.
19. Buchowiecki, W.; Grosman-Zjawiona, Z.; Zjawiony, J. *Tetrahedron Lett.* **1985**, *26*, 1245.

David N. Deaton
Glaxo Research Institute, Research Triangle Park, NC, USA

Sunggak Kim
Korea Advanced Institute of Science and Technology, , Taejon, Korea

Diphenyl Phosphonate

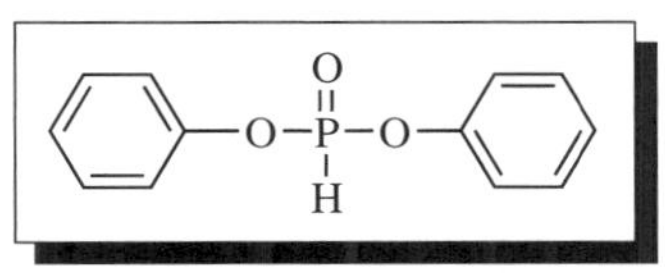

[4712-55-4] $C_{12}H_{11}O_3P$ (MW 234.20)

(phosphonylating agent; used in the synthesis of H-phosphonate monoesters, H-phosphonothioate monoesters, phosphate analogs, cyclic phosphates, P–C bond formation, aminoalkylphosphonates, and enamines, mild coupling agent for synthesis of esters and amides)

Alternate Name: diphenyl H-phosphonate; diphenyl phosphite, DPHP.

Physical Data: mp 12 °C (25 °C);[1a] bp 100 °C/0.008 mmHg, 218–219 °C/26 mmHg; n_D^{25} 1.5564–1.5570; δ ^{31}P 1.9 ($^1J_{PH}$ 756 Hz).

Solubility: soluble in most aprotic organic solvents. In protic solvents under basic conditions, it undergoes decomposition (transesterification). Readily hydrolyzes in the presence of bases to a monoester.

Form Supplied in: colourless liquid, purity ca. 90%, widely available. Contains variable amounts of phenol (10–15%) and sometimes triphenyl phosphite.

Analysis of Reagent Purity: ^{31}P NMR [in acetonitrile, $\delta_P = 2.1$ ppm, $^1J_{PH} = 741$ Hz (d)]. Possible contaminations: triphenyl phosphite ($\delta_P = 129.7$ ppm) and phenyl H-phosphonate monoester [$\delta_P = -0.95$ ppm, $^1J_{PH} = 633$ Hz (d)].

Purification: vacuum distillation, although for most synthetic applications the reagent can be used as supplied. The amount of phenol contaminant (based on ^{1}H NMR analysis) can be significantly decreased (if so desired) by heating the reagent at 130 °C at 20 mm Hg.[2]

Handling, Storage, and Precautions: store in a cool, dry place. Keep tightly closed. Upon frequent contacts with atmospheric moisture the reagent slowly hydrolyzes. The bottles are usually labeled with precautionary notes: stench, highly toxic, vesicant, readily absorbed through skin, etc. It may affect liver, kidneys, nerves, although no toxicological data are available.

Original Commentary

Michael J. Gallagher
University of New South Wales, Kensington, NSW, Australia

Amide and Ester Coupling. In the presence of tertiary amines, preferably Pyridine or Imidazole, the ester acts as an efficient coupling agent in the formation of peptides (eq 1).[1,3] Ureas and thioureas may be similarly prepared from Carbon Dioxide or Carbon Disulfide (eq 2).[4]

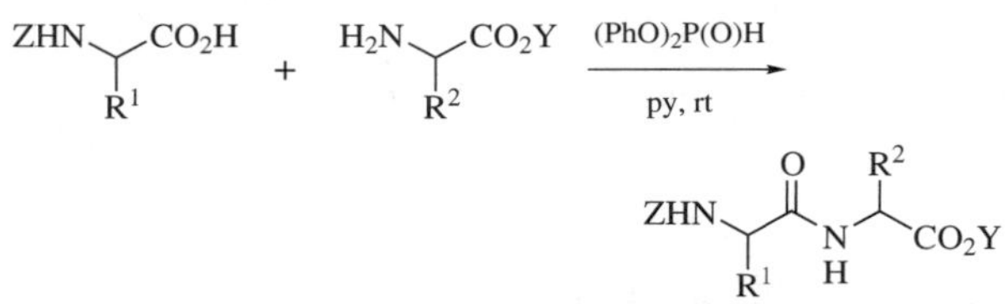

(1)

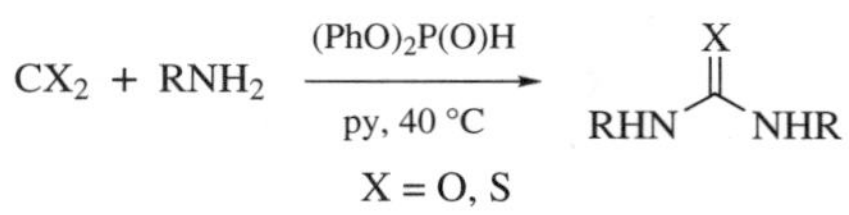

(2)

X = O, S

The diester undergoes easy transesterification in two distinct stages, allowing the preparation of unsymmetrical diesters (eq 3).[5]

$$(PhO)_2P(O)H \xrightarrow[\text{py, rt}]{R^1OH} (PhO)(R^1O)P(O)H \xrightarrow{R^2OH} (R^1O)(R^2O)P(O)H \quad (3)$$

R^1 = primary, secondary, tertiary; R^2 = primary, secondary

The use of stronger bases than pyridine, e.g. Triethylamine, causes rapid disproportionation to $(PhO)_3P$ and $PhOPHO_2^-$.[2c,5]

First Update

Jacek Stawinski
Stockholm University, Stockholm, Sweden

Adam Kraszewski
Institute of Bioorganic Chemistry, PAS, Poznan, Poland

General Comments. Diphenyl H-phosphonate (DPHP) is very susceptible to the substitution at the phosphorus centre. In pyridine, in the presence of water, the reagent undergoes instantaneous hydrolysis to phenyl H-phosphonate monoester. Under anhydrous basic conditions, the reagent disproportionates to triphenyl phosphite and phenyl H-phosphonate monoester.[5,6] The disproportionation may occur as a competing reaction pathway in all transformations involving DPHP and its rate is highly dependent on the strength of the base used.

Synthesis of H-phosphonate Monoesters. DPHP in pyridine undergoes rapid transesterification with various hydroxylic compounds to produce the corresponding monophenyl H-phosphonate derivatives together with variable amounts of symmetrical H-phosphonate diesters (eq 4).[7] Such mixtures can be converted into H-phosphonate monoesters upon mild basic hydrolysis (aqueous ammonia). This protocol is most suitable for the preparation of simple alkyl H-phosphonate derivatives.[7]

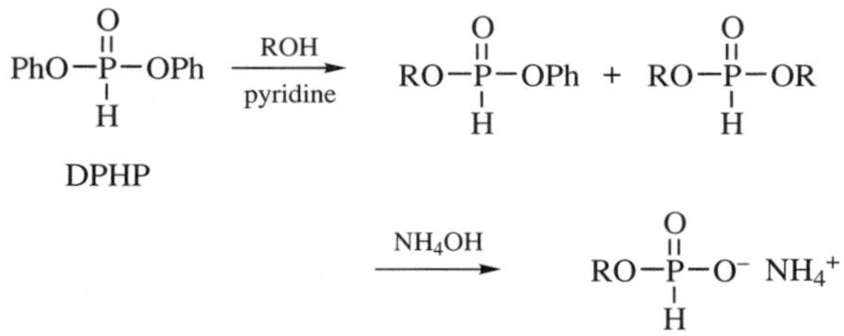

(4)

The reaction works well also for more complex hydroxylic components (e.g., nucleosides, eq 5).[8] To minimize the formation of symmetrical H-phosphonate diesters, in these instances (2–5)-fold excess of DPHP is usually used.

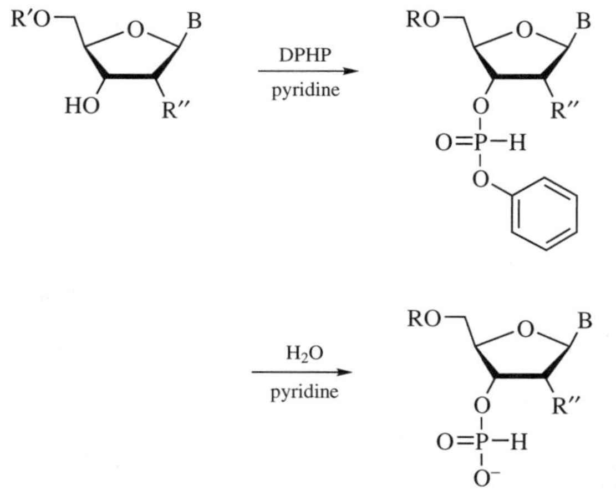

(5)

R′ = Protecting group
R″ = H or OTBDMS, OMe
B = Heterocyclic base

Transesterification with amino alcohols produces exclusively compounds with P–O bonds.[9] Due to this O- vs N-chemoselectivity, the method seems to be well suited for the phosphonylation of compounds bearing unprotected amino functions.[10]

Synthesis of H-phosphonothioate and H-phosphonodithioate Monoesters. Phenyl alkyl H-phosphonate diesters, generated in situ in the reaction of alcohols with DPHP, are useful substrates in the synthesis of alkyl H-phosphonomonothioate and alkyl H-phosphonodithioate monoesters. When treated with H_2S in the presence of pyridine these intermediates produce the corresponding alkyl H-phosphonodithioates.[11,12] However, when sulfhydrolysis is carried out in the presence of trimethylsilyl chloride and triethylamine, the products of this reaction are alkyl H-phosphonomonothioates[11,12] (eq 6).

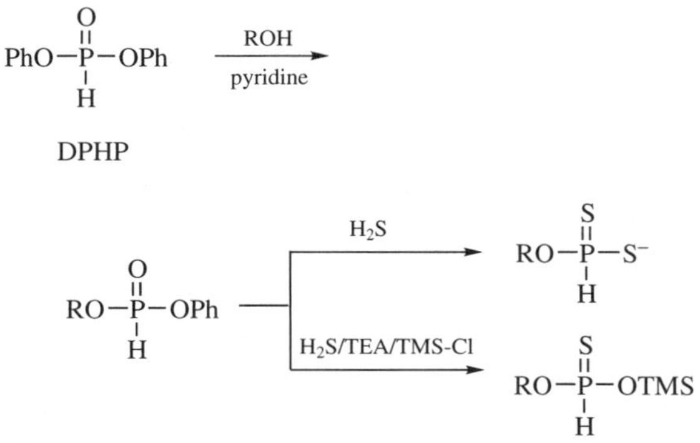

(6)

TEA = Triethylamine
TMS = Trimethylsilyl moiety

These protocols are also applicable to nucleosides and other hydroxylic compounds, and afford the corresponding H-phosphonomono and H-phosphonodithioate monoesters in practically quantitative yields.

Synthesis of Cyclic Phosphates and their Analogs. DPHP and equimolar amount of ethylene glycol in pyridine afford cyclic H-phosphonate diester as a sole product.[13] Similarly, 5′-O-protected ribonucleosides bearing a free 2′, 3′-cis-diol functionality generate rapidly in the reaction with DPHP in pyridine the corresponding 2′,3′-cyclic H-phosphonates, which upon sulfurization with elemental sulfur provide in high yields nucleoside 2′,3′-cyclic phosphorothioates[14] (eq 7).

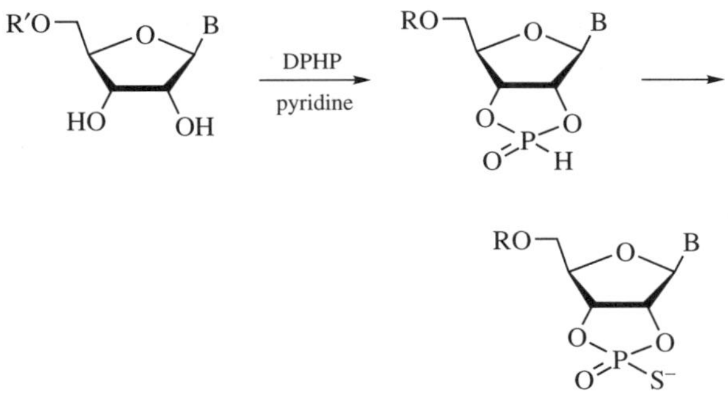

(7)

R′ = Protecting group
B = Heterocyclic base

In analogous reactions when selenium, instead of elemental sulfur, is used as an oxidant cyclic phosphoroselenoates are produced.[15]

Formation of P–C Bonds. Diesters of phosphonic acid add easily to various imines (the Kabachnik–Field reaction)[16] to produce the corresponding α-aminoalkyl-C-phosphonates (eq 8).

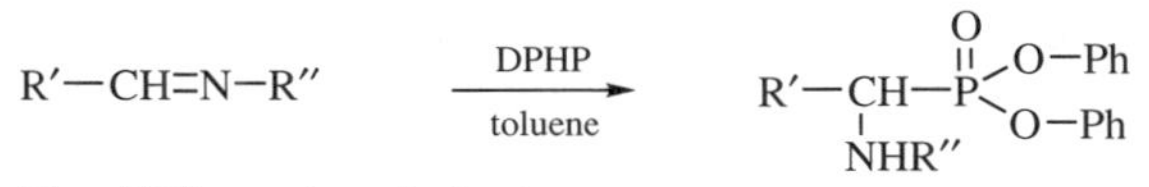

(8)

R′ and R″ = aryl or alkyl groups

The reaction with DPHP occurs at an ambient temperature affording usually crystalline α-aminophosphonates,[17,18] from which unprotected aminophosphonic acids can be obtained.[19] In the presence of base, α-anions can be generated from aminoalkylphosphonates. These may react with various carbonyl compounds (in a modified Horner–Wittig reaction) to form enamines, which upon hydrolysis afford carbonyl derivatives.[18,20] Recently, this methodology was used as an alternative approach to a Suzuki or Stille cross-coupling for the construction of the diaryl pyridine scaffold in the synthesis of cyclooxygenase (COX) inhibitors.[21]

DPHP also reacts easily with acrylate derivatives (in a Michael addition) to produce the corresponding C-phosphonates. This type of reaction, followed by a reductive elimination of diphenyl silyl phosphite from the resulting C-phosphonate (effected by DBU and trimethylsilyl triflate) was used as a crucial step in the synthesis of carbapenem antibiotics.[22]

DPHP was found to be a superior phosphorus nucleophile in a Michaelis–Becker reaction involving tertiary alkyl halides or substrates with pseudohalide character. In contrast to dialkyl H-phosphonates, which usually favour X-philic substitution, the reactions with DPHP in acetonitrile in the presence of DBU afford the desired C-phosphonates in high yields.[23]

---

1. Stelzel, P., *Methoden Org. Chem. (Houben-Weyl)* **1974**, *15/2*, 246.
2. (a) Houalla, D.; Wolf, R., *Chem. Rev.* **1958**, *247*, 482. (b) Walsh, E. N., *J. Am. Chem. Soc.* **1959**, *81*, 3023. (c) Yamazaki, N.; Iguchi, T.; Higashi, F., *Chem. Lett.* **1977**, 185.
3. (a) Yamazaki, N.; Higushi, F., *Tetrahedron Lett.* **1972**, 5047. (b) Yamazaki, N.; Higashi, F., *Tetrahedron Lett.* **1974**, 1323.
4. Yamazaki, N.; Higashi, F.; Iguchi, T., *Tetrahedron Lett.* **1974**, 1191.
5. Gallagher, M. J.; Garbutt, R.; Liu Yuan Hua; Gum Hee Lee, *Phosphorus Sulfur* **1993**, 201.
6. Kers, A.; Kers, I.; Stawinski, J.; Sobkowski, M.; Kraszewski, A., *Tetrahedron* **1996**, *52*, 9931.
7. Kers, A.; Kers, I.; Stawinski, J.; Sobkowski, M.; Kraszewski, A., *Synthesis* **1995**, 427.
8. Jankowska, J.; Sobkowski, M.; Stawinski, J.; Kraszewski, A., *Tetrahedron Lett.* **1994**, *35*, 3355.
9. Sobkowski, M.; Stawinski, J.; Kraszewski, A., *Coll. Czech. Chem. Commun.* **1996**, *61*, S238.
10. (a) Wada, T.; Honda, F.; Sato, Y.; Sekine, M., *Tetrahedron Lett.* **1999**, *40*, 915. (b) Robles, J.; Grandas, A.; Pedroso, E., *Tetrahedron* **2001**, *57*, 179.
11. Kers, I.; Kers, A.; Stawinski, J.; Kraszewski, A., *Tetrahedron Lett.* **1999**, *40*, 3945.
12. Cieslak, J.; Jankowska, J.; Stawinski, J.; Kraszewski, A., *J. Org. Chem.* **2000**, *65*, 7049.
13. Sobkowski, M.; Wenska, M.; Kraszewski, A.; Stawinski, J., *Nucleoside & Nucleotides* **2000**, *19*, 1487.
14. Jankowska, J.; Wenska, M.; Popenda, M.; Stawinski, J.; Kraszewski, A., *Tetrahedron Lett.* **2000**, *41*, 2227.
15. Cieslak, J.; Sobkowski, M.; Jankowska, J.; Wenska, M.; Szymczak, M.; Imiolczyk, B.; Zagorowska, I.; Shugar, D.; Stawinski, J.; Kraszewski, A., *Acta Biochim. Pol.* **2001**, *48*, 429.
16. Fields, S. C., *Tetrahedron* **1999**, *55*, 12237.

17. Smith, S. J.; Zimmer, H.; Fluck, E.; Fischer, P., *Phosphorus Sulfur* **1988**, *35*, 105.
18. Zimmer, H.; Bercz, P. J., *J. L. Ann. Chem.* **1965**, *686*, 107.
19. (a) Lewkowski, J.; Rzezniczak, M.; Skowronski, R., *Organic Preparations Procedures International* **2000**, *32*, 453. (b) Lewkowski, J.; Rzezniczak, M.; Skowronski, R., *Heteroatom Chemistry* **2000**, *11*, 144. (c) Boduszek, B., *Phosphorus, Sulfur Silicon Relat. Elem.* **1996**, *113*, 209. (d) Soroka, M.; Zygmunt, J., *Synthesis* **1988**, 370.
20. Bunting, J. W.; Stefanidis, D., *J. Am. Chem. Soc.* **1988**, *110*, 4008.
21. Davies, I. W.; Marcoux, J.-F.; Corley, E. G.; Journet, M.; Cai, D.-W.; Palucki, M.; Wu, J.; Larsen, R. D.; Rossen, K.; Pye, P. J.; DiMichele, L.; Dormer, P.; Reider, P. J., *J. Org. Chem.* **2000**, *65*, 8415.
22. Hakimelahi, G. H.; Moosavi-Movahedi, A. A.; Tsay, S.-C.; Tsai, F.-Y.; Wright, J. D.; Dudev, T.; Hakimelahi, S.; Lim, C., *J. Med. Chem.* **2000**, *43*, 3632.
23. Kers, A.; Stawinski, J.; Dembkowski, L.; Kraszewski, A., *Tetrahedron* **1997**, *53*, 12691.

Diphenyl Phosphorazidate

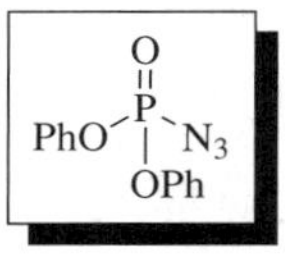

[26386-88-9] $C_{12}H_{10}N_3O_3P$ (MW 275.22)

(peptide synthesis;[1] synthesis of α-aryl carboxylic acids;[4] synthesis of thiol esters;[7,8] stereospecific conversion of alcohols to azides;[9] ring contraction of cycloalkanones;[13] decarbonylation of aldehydes[26])

Alternate Names: DPPA; *O,O*-diphenylphosphoryl azide.

Physical Data: bp 157 °C/0.17 mmHg; *d* 1.277 g cm^{-3}; fp >110 °C.

Solubility: sol toluene, THF, DMF, *t*-butyl alcohol.

Form Supplied in: colorless liquid; commercially available in kilogram quantities.

Preparative Methods: by the reaction of Diphenyl Phosphorochloridate with a slight excess of Sodium Azide in acetone at rt (85–90% yield).[1]

Purification: vacuum distillation (bath temp <200 °C, dec >200 °C)

Handling, Storage, and Precautions: handle and store under nitrogen. Refrigerate. Decomposes at >200 °C and may produce toxic fumes of phosphorus oxides and/or phosphine. Toxic, irritant. Harmful if swallowed, inhaled, or absorbed through the skin. Use in fume hood. Incompatible with oxidizing agents and acids. Hazardous combustion or decomposition products include carbon monoxide, carbon dioxide, and nitrogen oxides. Prolonged contact with moisture may produce explosive hydrogen azide (HN_3).

General. Diphenyl phosphorazidate is a readily available, nonexplosive, and relatively stable azide widely used as a reagent in peptide synthesis,[1–3] and as a versatile reagent in a wide array of organic transformations. DPPA has been successfully utilized in the synthesis of α-amino acids[4] and α-aryl carboxylic acids;[5,6] direct preparation of thiol esters from carboxylic acids and thiols;[7,8] the stereospecific preparation of alkyl azides;[9] and the phosphorylation of alcohols and amines.[10] The application of DPPA in a modified Curtius[11] reaction permits a simple one-step conversion of carboxylic acids to urethanes under mild reaction conditions. DPPA acts as a nitrene source,[12] and can undergo 1,3-dipolar cycloaddition reactions.[5,13] The Curtius degradation[14] of carboxylic acids in the presence of *t*-butanol gives the Boc-protected amine directly (eq 1).

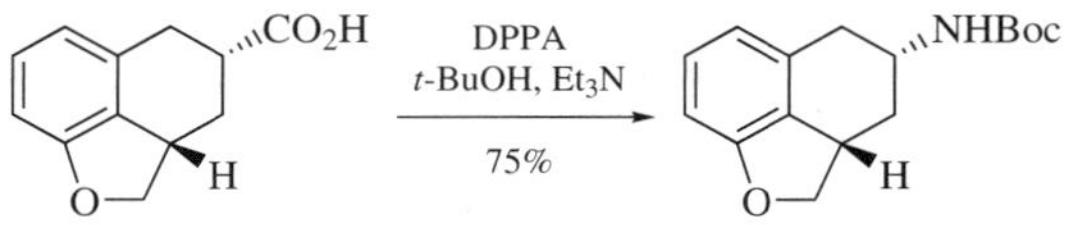

(1)

DPPA has been utilized in the synthesis of 1,4-dinitrocubane by Eaton and co-workers.[15] Refluxing cubane-1,4-dicarboxylic acid with DPPA and Triethylamine in *t*-butanol forms the intermediate *t*-butyl carbamate in nearly quantitative yield. This method avoids the formation of the explosive diacyl azide.

Peptide Synthesis. Urethanes are readily prepared by refluxing equimolar mixtures of DPPA, a carboxylic acid, an alcohol, and triethylamine. The reaction involves transfer of the azide group from DPPA to the carboxylic acid. The resulting acyl azide subsequently undergoes a Curtius rearrangement. This reaction has been successfully applied in the area of peptide synthesis.[1] The coupling of acylamino acids or peptides with amino acid esters or peptide esters in the presence of base proceeds in high yields without racemization, and is compatible with a variety of functional groups. The reaction of malonic acid half-esters results in the corresponding α-amino acid derivatives[4] (eq 2). It should be noted that addition of the alcohol at the beginning of the reaction results in esterification.

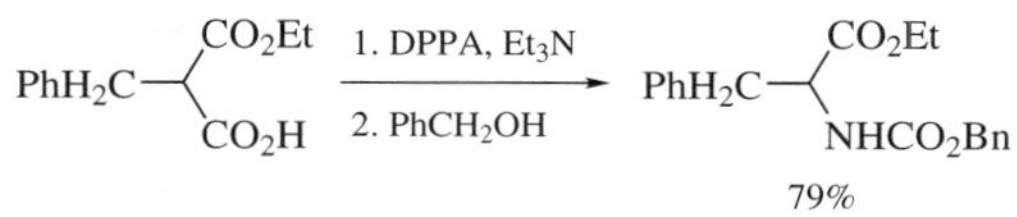

(2)

α-Aryl Carboxylic Acids. Alkyl aryl ketones are converted into the corresponding α-aryl alkanoic acids[4] via a three-step sequence. Yields of >90% are possible if the synthesis is carried out in a one-flask procedure. The sequence includes a 1,3-dipolar cycloaddition of DPPA to the corresponding enamines (eq 3).

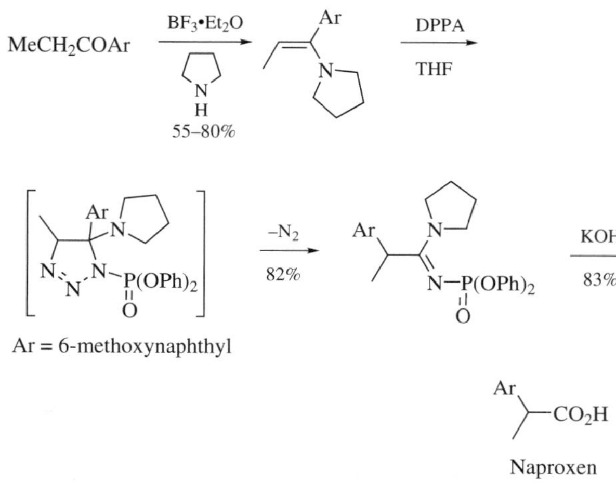

(3)

Although Thallium(III) Nitrate has also been used in similar oxidative rearrangements,[16] the DPPA method exhibits higher functional specificity, uses less toxic reagents, and gives preparatively useful yields. Naproxen, a nonsterodial anti-inflammatory therapeutic, has been prepared by this route.[6]

Stereospecific Conversion of Alcohols to Azides. Reaction of an alcohol with DPPA, Triphenylphosphine, and Diethyl Azodicarboxylate forms the corresponding azides in 60–90% yields.[9] The stereospecific nature of this reaction permits the conversion of Δ^5-sterols such as 3β-cholestanol exclusively to the 3α-cholestanyl azide in 75% yield. This synthesis is clearly superior to the alcohol → tosylate → azide route which is longer and also prone to competing elimination reactions.

Ring Contraction Reactions. DPPA undergoes 1,3-dipolar cycloadditions to the enamines of cyclic ketones.[13] The resulting Δ^2-triazolines, which are not isolated, undergo loss of nitrogen to form ring-contracted products, which, on hydrolysis, yield cycloalkanoic acids. In the case of six- to eight-membered cycloalkanones, overall yields as high as 75% have been reported (eq 4).

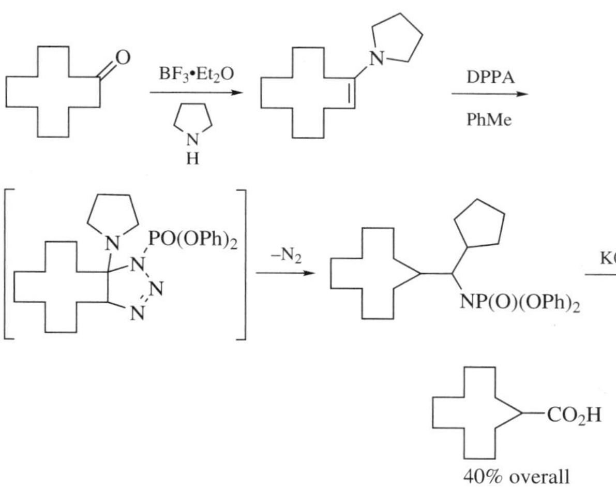

(4)

Amination of Aromatic and Heteroaromatic Organometallics. Reaction of organic azides with Grignard and lithium compounds gives 1,3-disubstituted triazenes,[17] which are readily converted to amines by reductive[18,19] or hydrolytic[20,21] workup. These methods are limited to either the aromatic Grignard[18,20] or lithium[21] compounds, and have not been very successful with heteroaromatic organometallics. Aromatic and heteroaromatic organometallics (Grignard and lithium compounds) are aminated in good yields in a one-pot process by treatment with DPPA and reduction of the resulting phosphoryltriazenes with Sodium Bis(2-methoxyethoxy) aluminum Hydride (eq 5).[22]

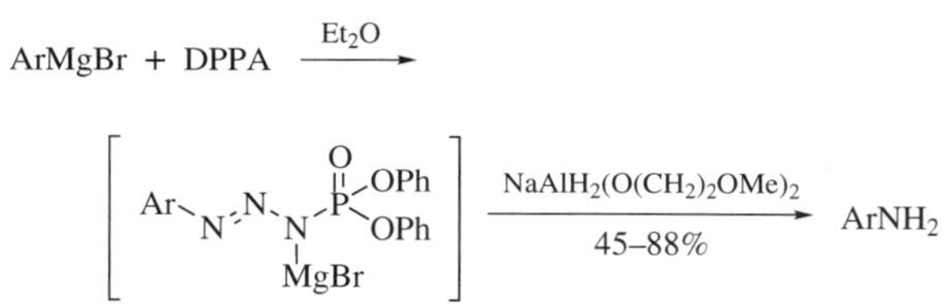

(5)

Metal-Catalyzed Decarbonylation of Primary Aldehydes. The decarbonylation of primary aldehydes under catalytic conditions is difficult, requiring high temperatures or involving radical mechanisms;[23] the rt decarbonylations[24,25] using stoichiometric Chlorotris(triphenylphosphine)rhodium(I) have had limited practicality due to the high cost of the reagent. Recently, a high-yielding, rt decarbonylation of primary aldehydes using catalytic amounts (5 mol %) of $Rh(PPh_3)_3Cl$ and stoichiometric amounts of DPPA, with minimal formation of alkene byproducts, has been developed[26] (eq 6).

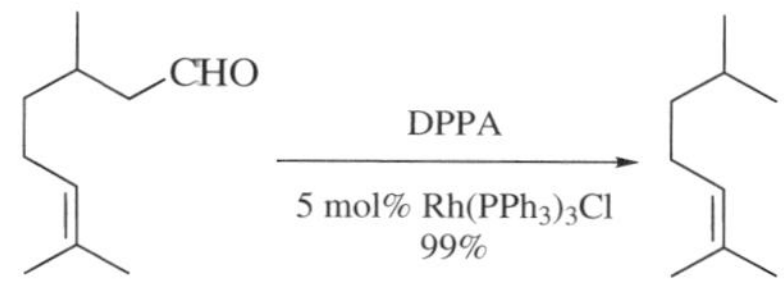

(6)

Synthesis of Macrocyclic Lactams. Macrocyclic lactams are conventionally prepared by the reaction of dicarboxylic acid chlorides with diamines, a method which is effective with simple acyl chlorides. Activation of the carboxylic groups in order to overcome the drawbacks of high dilution and the low yields and purity encountered with larger acyl chlorides has seen only limited success.[27] The superior activating ability of DPPA has recently been demonstrated.[28] The reactions of dicarboxylic acids and diamines in the presence of DPPA form macrobicyclic lactams in yields as high as 82%. By comparison, in a control experiment using conventional high-dilution techniques, the corresponding acyl chlorides cyclized with the diamine to form the lactam in only 24% yield.

DPPA has been used for the direct *C*-acylation of methyl isocyanoacetate with carboxylic acids to give 4-methoxycarbonyloxazoles.[29] L-Daunosamine, the glycone component of anticancer anthracycline antibiotics, has been synthesized from L-lactic acid in 9 steps with a 24% overall yield,[30] where a key step in the sequence is the direct *C*-acylation of methyl isocyanoacetate with the lithium salt of the lactate ester using diphenyl phosphorazidate (eq 7).

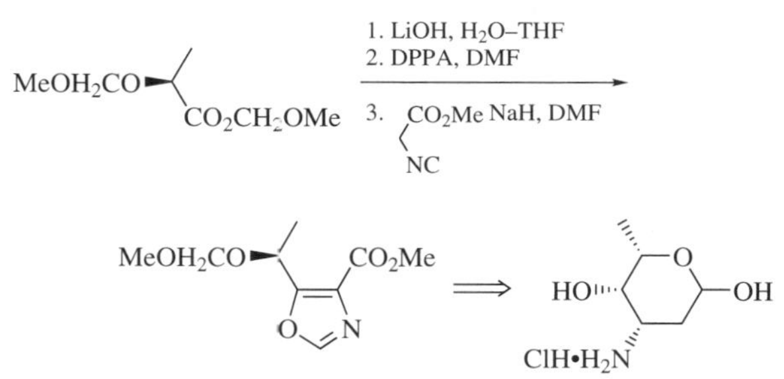

(7)

1. Shioiri, T.; Ninomiya, K.; Yamada, S. *J. Am. Chem. Soc.* **1972**, *94*, 6203.
2. Shioiri, T.; Yamada, S. *Chem. Pharm. Bull.* **1974**, *22*, 849.
3. Yamada, S.; Ikota, N.; Shioiri, T.; Tachibana, S. *J. Am. Chem. Soc.* **1975**, *97*, 7174.
4. Yamada, S.; Ninomiya, K.; Shioiri, T. *Tetrahedron Lett.* **1973**, 2343.
5. Shioiri, T.; Kawai, N. *J. Org. Chem.* **1978**, *43*, 2936.
6. Riegel, J.; Madox, M. L.; Harrison, I. T. *J. Med. Chem.* **1974**, *17*, 377.
7. Yamada, S.; Yokoyama, Y.; Shioiri, T. *J. Org. Chem.* **1974**, *39*, 3302.

8. Yokoyama, Y.; Shioiri, T.; Yamada, S. *Chem. Pharm. Bull.* **1977**, *25*, 2423.
9. Lal, B.; Pramanik, B.; Manhas, M. S.; Bose, A. K. *Tetrahedron Lett.* **1977**, 1977.
10. Cremlyn, R. J. W. *Aust. J. Chem.* **1973**, *26*, 1591.
11. (a) Ninomiya, K.; Shioiri, T.; Yamada, S. *Tetrahedron* **1974**, *30*, 2151. (b) Ninomiya, K.; Shioiri, T.; Yamada, S. *Chem. Pharm. Bull.* **1974**, *22*, 1398.
12. Breslow, R.; Feiring, A.; Herman, F. *J. Am. Chem. Soc.* **1974**, *96*, 5937.
13. Yamada, S.; Hamada, Y.; Ninomiya, K.; Shioiri, T. *Tetrahedron Lett.* **1976**, 4749.
14. Haefliger, W.; Klöppner, E. *Helv. Chim. Acta* **1982**, *65*, 1837.
15. Eaton, P. E.; Ravi Shanker, B. K. *J. Org. Chem.* **1984**, *49*, 185.
16. Taylor, E. C.; Chiang, C.-S.; McKillop, A.; White, J. F. *J. Am. Chem. Soc.* **1976**, *98*, 6750.
17. Dimroth, R. *Ber. Dtsch. Chem. Ges./Chem. Ber.* **1903**, *36*, 909.
18. Smith, P. A. S.; Rowe, C. D.; Bruner, L. B. *J. Org. Chem.* **1969**, *34*, 3430.
19. Reed, J. N.; Snieckus, V. *Tetrahedron Lett.* **1983**, *24*, 3795.
20. Trost, B. M.; Pearson, W. H. *J. Am. Chem. Soc.* **1981**, *103*, 2483.
21. Hassner, A.; Munger, P.; Belinka, B. A. *Tetrahedron Lett.* **1982**, *23*, 699.
22. Mori, S.; Aoyama, T.; Shioiri, T. *Tetrahedron Lett.* **1984**, *25*, 429.
23. Domazetis, G.; Tarpey, B.; Dolphin, D.; James, B. R. *Chem. Commun./J. Chem. Soc., Chem. Commun.* **1980**, 939.
24. Tsuji, J.; Ohno, K. *Tetrahedron Lett.* **1965**, 3969.
25. Osborn, J. A.; Jardine, F. H.; Young, J. F.; Wilkinson, G. *J. Chem. Soc. (A)* **1966**, 1711.
26. O'Connor J. M.; Ma J. *J. Org. Chem.* **1992**, *57*, 5075.
27. Cazaux, L.; Duriez, M. C.; Picard, C.; Moieties, P. *Tetrahedron Lett.* **1989**, *30*, 1369.
28. Qian, L.; Sun, Z.; Deffo, T.; Mertes, K. B. *Tetrahedron Lett.* **1990**, *31*, 6469.
29. Hamada, Y.; Shioiri, T. *Tetrahedron Lett.* **1982**, *23*, 235, 1226.
30. Hamada, Y.; Kawai, A.; Shioiri, T. *Tetrahedron Lett.* **1984**, *25*, 5409.

Albert V. Thomas
Abbott Laboratories, North Chicago, IL, USA

Diphenyl Sulfoxide

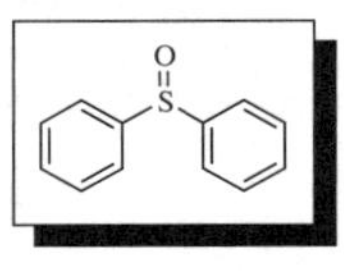

[945-51-7] $C_{12}H_{10}OS$ (MW 202.27)

(reagent used as an oxidant and hydroxyl activator in combination with a variety of electrophilic reagents)

Physical Data: mp 71 °C.
Solubility: soluble in most organic solvents.
Form Supplied in: widely available as white solid.

Original Commentary

David Y. Gin
University of Illinois, Urbana, IL, USA

Oxidation of Alcohols. Diphenyl sulfoxide has been employed as an oxidant in conjunction with molybdenum (VI) and osmium (VIII) catalysts for the conversion of alcohols to carbonyl compounds. Diphenyl sulfoxide in combination with catalytic quantities of $MoO_2(acac)_2$ oxidizes alcohols to ketones or aldehydes.[1] Higher yields are obtained with allylic or benzylic alcohols. Catalytic OsO_4 in association with Ph_2SO can oxidize primary and secondary alcohols to aldehydes and ketones in the presence of catalytic quantities of *p*-toluenesulfonic acid at elevated temperatures (eq 1).[2]

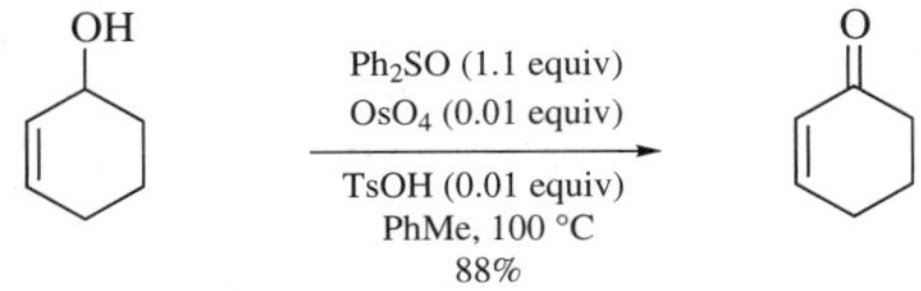

(1)

Oxidation of Sulfides and Disulfides. A mild method for the oxidation of dialkyl and monoaryl sulfides to sulfoxides employs $Re(O)Cl_3(PPh_3)_2$ as a catalyst and diphenyl sulfoxide as an oxidant.[3] Over-oxidation to the corresponding sulfone is not observed. The same reagent system can be applied to the efficient oxidation of acyclic symmetrical disulfides to the corresponding thiosulfonates. Cyclic five- and six-membered disulfides afford the corresponding thiosulfinate products with this reagent system (eq 2).[4] The oxidative coupling of *S*-protected cysteine to cystine can be accomplished with Ph_2SO and $MeSiCl_3$ in TFA (eq 3).[5,6] *S*-Protective groups such as acetamidomethyl, benzyl, *p*-methoxybenzyl, *p*-methylbenzyl, and *t*-butyl thioethers are susceptible to the deprotective dimerization process. Other silicon electrophiles, such as TMSCl, $SiCl_4$, and TMSOTf can also be used. This method has been applied to regioselective disulfide formation with complex peptide fragments.[7,8]

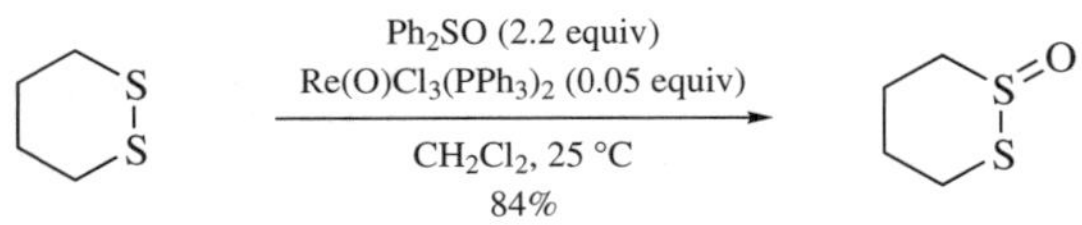

(2)

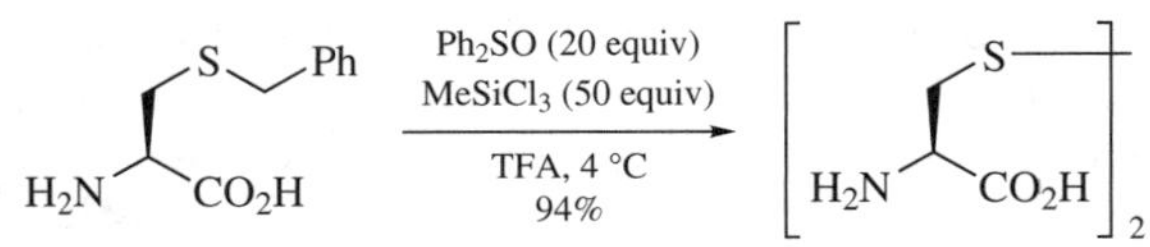

(3)

Electrophilic Aromatic Substitution. When activated with electrophilic reagents, such as trifluoroacetic anhydride (TFAA), diphenyl sulfoxide is converted into an acyloxysulfonium ion, which can effect electrophilic aromatic substitution. When this mode of sulfoxide activation is performed in the presence of simple aromatic substrates, such as *o*-xylene, the corresponding triarylsulfonium salt is formed in good yield (eq 4).[9] The $Ph_2SO{\cdot}TFAA$ reagent combination reacts similarly with more complex substrates, such as indoles, affording the 3-indolylsulfonium salts.[10]

Oxidative Glycosylation. Glycal substrates undergo rapid electrophilic activation with the reagent combination of Ph_2SO

and Tf_2O. In the presence of excess Ph_2SO, transfer of the sulfoxide oxygen to the C_2-position of the glycal occurs. Following the addition of an appropriate nucleophile, C_2-hydroxy glycoside products are obtained (eq 5).[11] This sulfoxide-mediated oxygen transfer reaction was shown to proceed through formation of a 1,2-anhydropyranoside intermediate.[12]

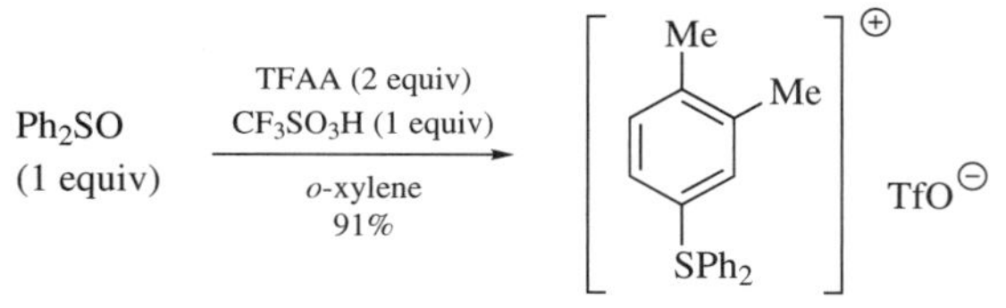

(4)

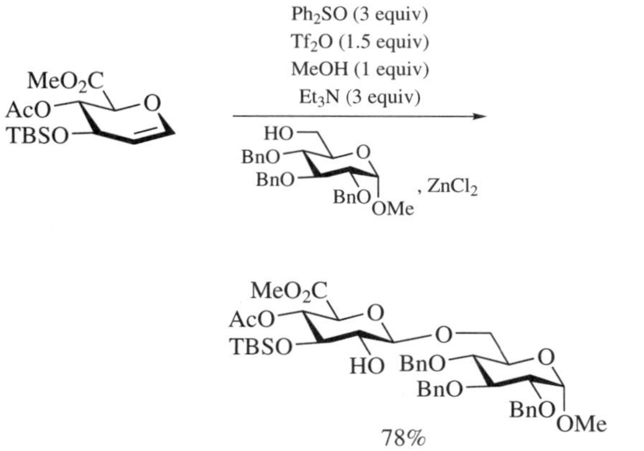

(5)

Dehydrative Glycosylation. The reagent combination of Ph_2SO and Tf_2O is known to rapidly activate the C1-hemiacetal functionality of carbohydrates.[13] Formation of an intermediate glycosyl oxosulfonium species allows for glycosidic bond formation with a wide array of nucleophiles to afford a variety of glycoconjugates (eq 6).[14]

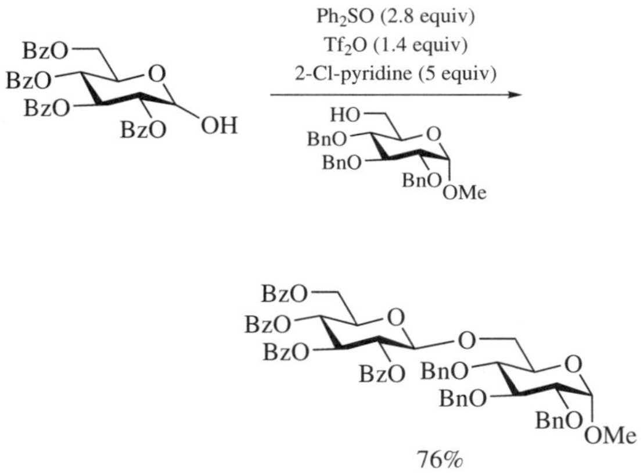

(6)

Other Applications. Mixtures of diphenyl sulfoxide and acid chlorides have been shown to act as chlorinating agents toward alkenes.[15] Diphenyl sulfoxide as well as other dialkyl sulfoxides have also been employed in the solvent extraction separation of various metals.[16]

First Update

Abhisek Banerjee
University of Illinois at Chicago, Chicago, IL, USA

C-Benzylation. The diphenyl sulfoxide and Tf_2O combination, used in conjunction with benzyl alcohol, can be exploited as an efficient one-pot *C*-benzylating system for various sodium enolates (eq 7).[17]

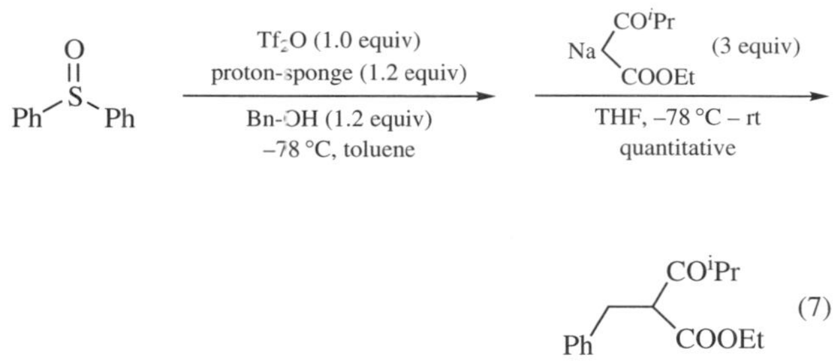

(7)

Predominant formation of the *C*-benzylated product can be understood by considering the soft carbon nucleophile attacking the benzyl carbon of the oxosulfonium trifluoromethane sulfonate, with diphenyl sulfoxide acting as a soft leaving group. The reaction also proceeds smoothly with various other substituted benzyl alcohols with moderate to good yields. Sodium enolates derived from esters, α-cyano esters, α-aromatic and aliphatic ketones can also be benzylated with consistent high yields.

Synthesis of 2-Arylaziridines. An effective and convenient method for the synthesis of 2-arylaziridines from styrene derivatives and primary amines involves the following two-step protocol: preparation of 2-arylethenyl(diphenyl)sulfonium salts from styrenes and the diphenyl sulfoxide-Tf_2O reagent combination at low temperature and subsequent treatment with primary amines (eq 8).[18]

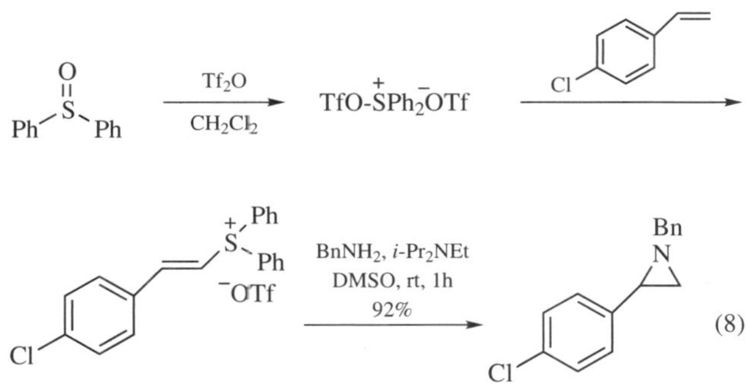

(8)

tert-Butylamine can replace Hünig's base in this chemistry and simplifies the work-up procedure because of its low boiling point (46 °C). The reaction time highly depends on steric bulk of the primary amine, but can be shortened by heating the reaction mixture.

Aziridination of *trans*- and *cis*-β-methylstyrene with benzylamine gave two diastereomeric aziridines with almost the same yields and diastereomeric ratio (1:1), suggesting that the stereochemistry of aziridines is not influenced by the olefin configuration. A one step protocol can also be applied in case of unstable 2-arylethenyl(diphenyl)sulfonium salts.

Synthesis of α-Imidostyrenes. α-Imidostyrenes, useful precursors to vinylamines can be synthesized by introducing an imido group to the α-position of styrenes with the diphenyl sulfoxide-Tf_2O reagent combination.[19] Styrenes with a hydrogen atom at the α-position react with diphenyl(trifluoromethanesulfonyloxy)sulfonium triflate to form diphenylstyrylsulfonium triflates which, in turn are converted into the corresponding α-imidostyrenes on treatment with sodium or potassium salts of cyclic imides (eq 9).

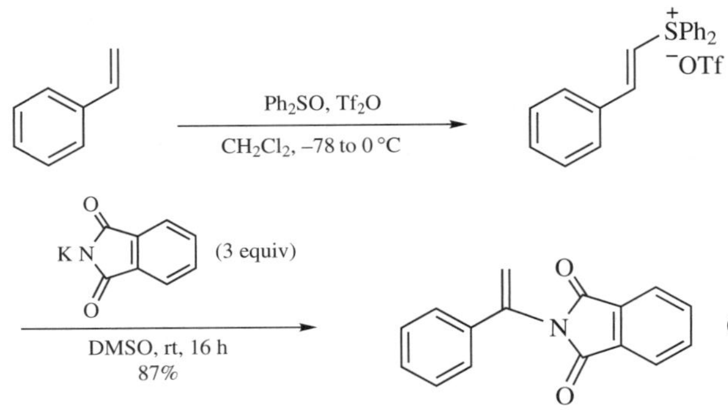

(9)

Several styrene derivatives can be used, but the presence of an electron donating group in the benzene ring diminishes the yield. The reaction time is highly dependent on the steric bulk of the cyclic imide which should be used in excess as it is in part quenched by triflic acid generated during the reaction.

Synthesis of β,γ-Unsaturated Amines. 1,1-Disubstituted alkenes can be converted into the corresponding β,γ-unsaturated amines, valuable synthons in medicinal chemistry, and versatile synthetic intermediates, via a three-step protocol with moderate to good yields. The initial formation of α,β-unsaturated diphenylsulfonium triflates from alkenes, such as α-methylstyrene, and the combination of diphenyl sulfoxide and Tf_2O, is followed by double bond migration to form β,γ-unsaturated sulfonium triflates. Subsequent nucleophilic substitution with primary or secondary amines such as *tert*-butylamine leads to the corresponding β,γ-unsaturated amines (eq 10).[20]

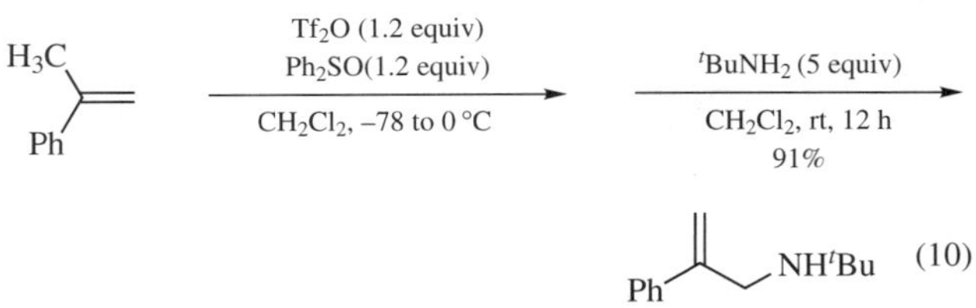

(10)

Steric hindrance at the γ-position of the alkene inhibits smooth double bond migration. For example, in case of 2-phenyl-3-methyl-1-butene, consumption of the corresponding sulfonium salt was incomplete even after prolonged reaction time or heating, and yielded only 45% of the desired product. Trisubstituted alkenes also generate the desired products, however yields are low.

Synthesis of Alkylidene 1,2-Oxasiletanes. The regiospecific oxidation of siliranes to 1,2-oxasiletanes can be accomplished in moderate yield with diphenyl sulfoxide.[21] When 1,1-dimesityl-2*Z*-neopentylidenesilirane was heated with diphenyl sulfoxide in benzene, 2,2-dimesityl-4*Z*-neopentylidene-1,2-oxasiletane was formed in 70% yield (eq 11). In contrast, use of cyclohexyldimethylamine *N*-oxide instead of diphenyl sulfoxide gave the isomeric 1,2-oxasiletane in 44% yield along with mesitol in 40% yield (eq 12).

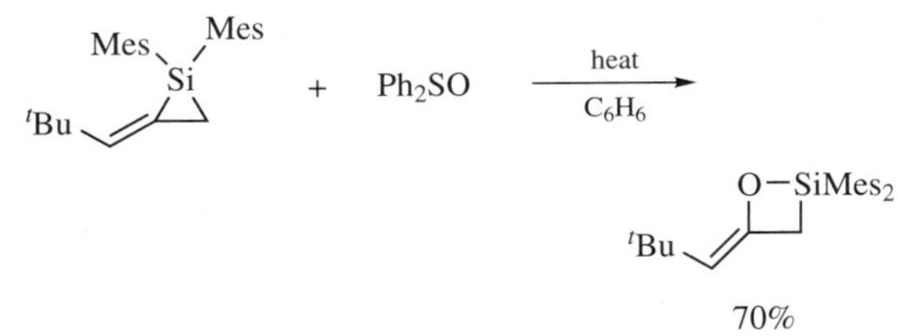

(11)

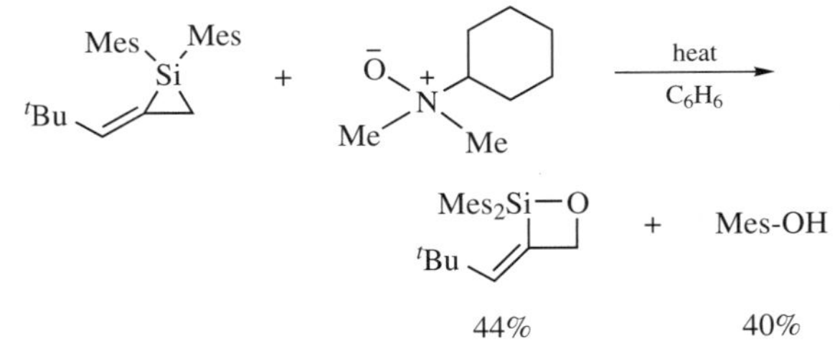

(12)

Interestingly, when the electrophilicity of the sulfur atom was lowered by introducing a *p*-methoxy group to the benzene ring of diphenyl sulfoxide, the regioselectivity dropped significantly. The nucleophilicity of oxygen atom should also be considered as no reaction was observed with *p*-nitrophenyl phenyl sulfoxide. Both oxasiletanes were thermally stable and did not undergo retro [2+2] reaction to give silanone and olefin.

Chemoselective Glycosylation. Diphenyl sulfoxide in combination with Tf_2O provides a very potent thiophilic glycosylation promoter system, which is capable of activating highly disarmed thioglycosides and mechanistically works in a similar manner to benzenesulfinyl piperidine (BSP)/Tf_2O.[22] Glycosylation of 4,6-*O*-benzylidene, 4,6-*O*-phenylboronate, and 4,6-*O*-polystrylboronate protected galacto-, gluco-, and mannopyranosyl thioglycosides was investigated under BSP/Tf_2O and diphenyl sulfoxide/Tf_2O conditions. In all three cases both methods gave comparable yields and selectivities.[23]

However, the BSP/Tf_2O promoter system failed to effectuate the condensation of *S*-phenyl-2-azido-3-*O*-benzyl-4,6-*O*-benzylidene thiomannoside (**1**) with a variety of acceptors, a task that was accomplished in high yield with the diphenyl sulfoxide/Tf_2O promoter system (eq 13).

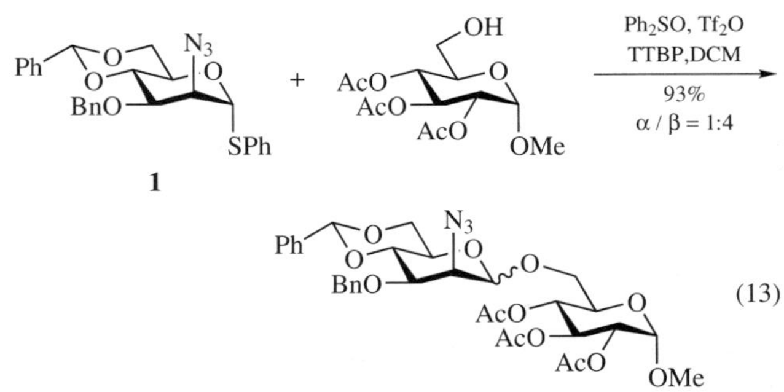

(13)

Another disarmed donor **2** was activated with diphenyl sulfoxide and Tf_2O and coupled with relatively unreactive acceptor **3** with excellent yield (eq 14).[24]

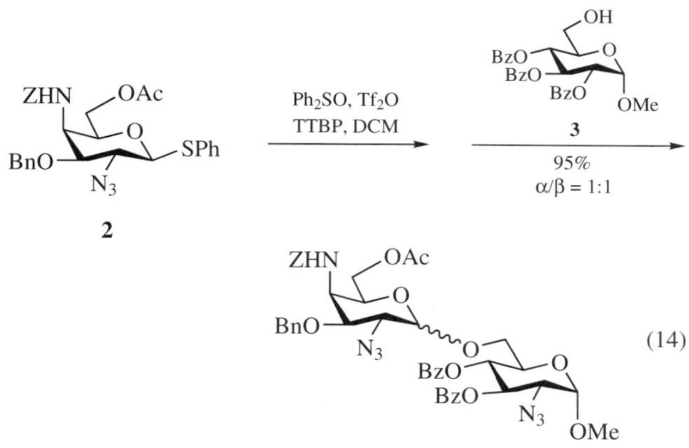

(14)

This difference in reactivity between the BSP/Tf$_2$O and diphenyl sulfoxide/Tf$_2$O reagents can be exploited in chemoselective oligosaccharide synthesis.[22,25]

Such an application is depicted in the assembly of trisaccharide **8** (eq 15). Activation of disarmed galactose donor **4** with BSP/Tf$_2$O followed by treatment with highly disarmed thioglycoside acceptor **5** and quenching with (EtO)$_3$P afforded dimer **6** with 64% yield. Subsequent diphenyl sulfoxide/Tf$_2$O-mediated activation and condensation with **7** furnished the trisaccharide **8** in acceptable yield (61%).[25]

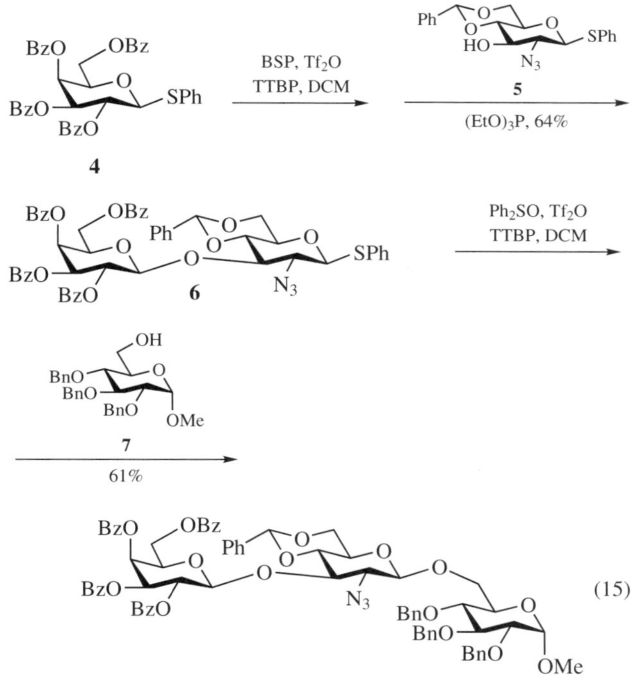

(15)

8

Dehydrative Sialylation. The recently developed dehydrative glycosylation of hemiacetal donors using the diphenyl sulfoxide-Tf$_2$O reagent combination[13] can also be employed in the preparation of the more challenging sialyl conjugates, in which anomeric bond formation occurs at the C2-ketal carbon. The anomeric selectivity however, favors the non-natural β-glycosidic linkage (eq 16).[26]

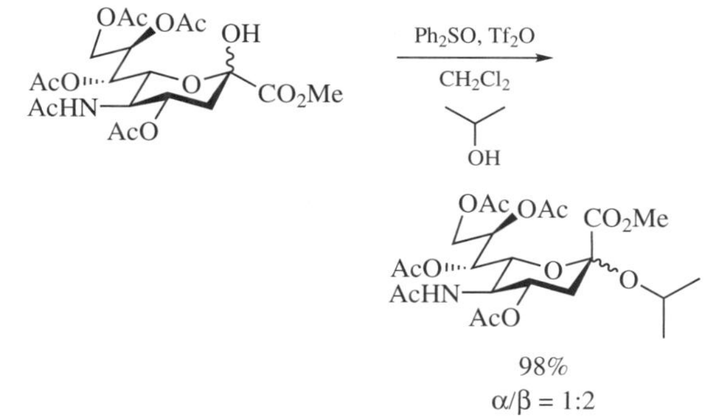

(16)

To augment the naturally-occurring, thermodynamically-disfavored α-selective sialyl conjugate, the C1-*N*,*N*-dimethyl glycolamide protective group can be employed. Although the proportion of α-sialoside product increases with this group, the reaction was still not sufficiently α-selective (eq 17).

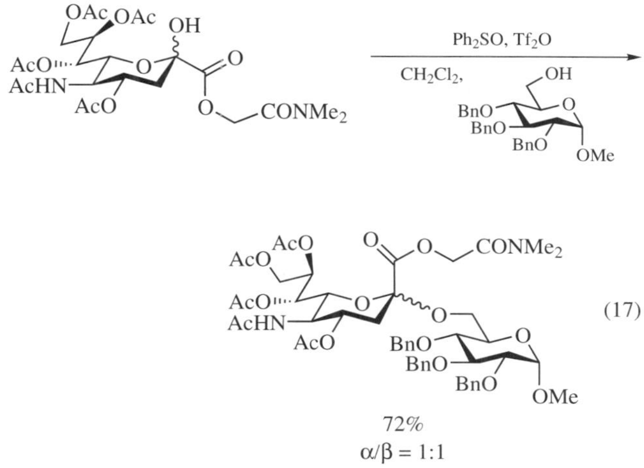

(17)

The problem can be circumvented by using an electron-deficient diaryl sulfoxide. For example, (*p*-nitrophenyl)(phenyl) sulfoxide-Tf$_2$O mediated dehydrative sialylation is highly stereoselective and favors the naturally occurring α-sialyl conjugate (eq 18).

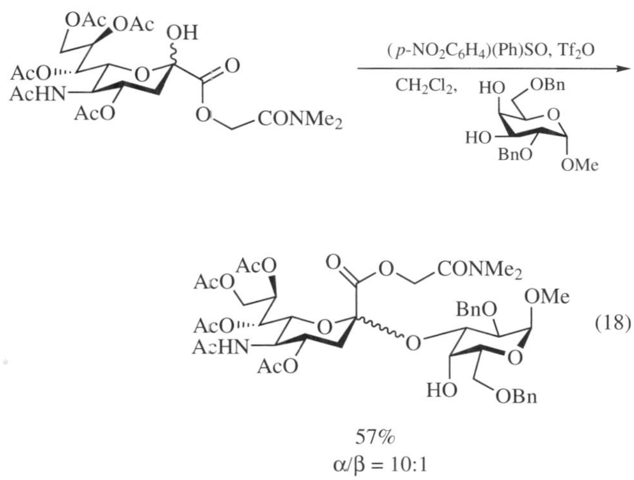

(18)

Iterative Dehyrdative Glycosylation. The dehydrative glycosidic coupling method using diphenyl sulfoxide-Tf$_2$O reagent combination[13] has been demonstrated to bring about selective activation of an armed hemiacetal in the presence of a disarmed one.[27] The scope of this chemistry is illustrated in the synthesis of a 1,4-α-linked tetrasaccharide **10** by two routes (eq 19). Route **a** shows the applicability of the iterative protocol whereas route **b** highlights the adaptability of this strategy to the block synthesis of oligosaccharides.

Iterative Glycosylation of 1-Hydroxy and 1-Thiodonors. The diphenyl sulfoxide-Tf_2O mediated dehydrative condensation of 1-hydroxyl donors with thioglycosides affords in good yield the thiodisaccharides, which in turn can be activated by the same activator system to furnish trisaccharides.[28] This sequential glycosylation is exemplified by the synthesis of trisaccharide **11** (eq 20). The synthesis can be conducted in a step-wise procedure (64 and 69%) as well as in a one-pot procedure (80%) with a much improved yield. Thioglycuronic acids can also be used both as donor and acceptor in the aforementioned coupling procedure toward acidic di- and trisaccharides.[29]

Related Reagents 1-Benzenesulfinyl piperidine; methoxyphenyl benzenethiosulfinate; benzothianthrene oxide; *N*-iodosuccinimide; dimethyl(methylthio)sulfonium tetrafluoroborate.

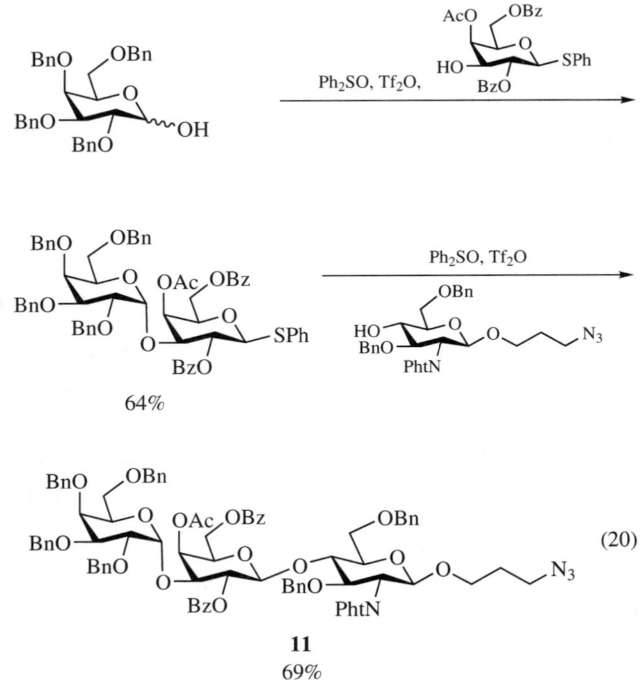

(20)

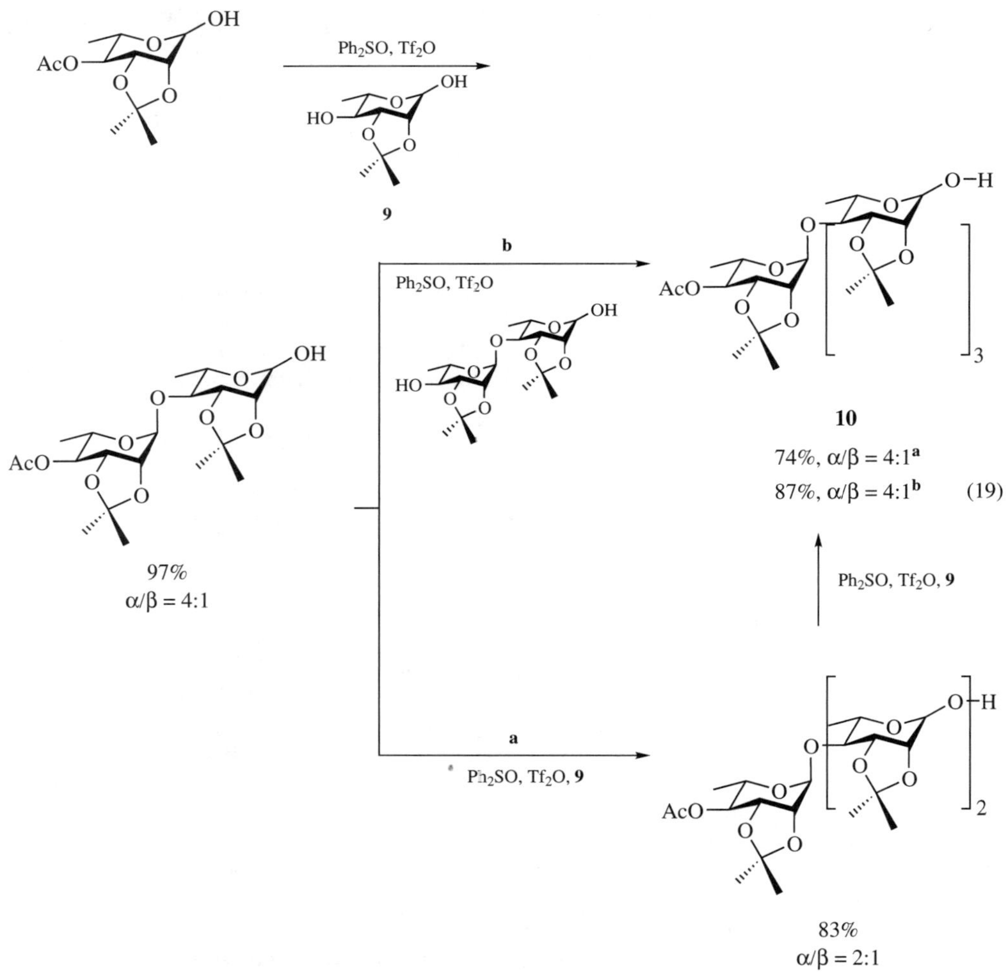

(19)

1. Lorber, C. Y.; Pauls, I.; Osborn, J. A., *Bull. Soc. Chim. Fr.* **1996**, *133*, 755.
2. Coleman, K. S.; Bedel, L. J. L.; Osborn, J. A., *C. R. Acad. Sci. Paris, IIC* **2000**, *3*, 765.
3. Arterburn, J. B.; Nelson, S. L., *J. Org. Chem.* **1996**, *61*, 2260.
4. Arterburn, J. B.; Perry, M. C.; Nelson, S. L.; Dible, B. R.; Holguin, M. S., *J. Am. Chem. Soc.* **1997**, *119*, 9309.
5. Koide, T.; Otaka, A.; Suzuki, H.; Fujii, N., *Synlett* **1991**, 345.
6. Akaji, K.; Tatsumi, T.; Yoshida, M.; Kimura, T.; Fujiwara, Y.; Kiso, Y., *J. Chem. Soc., Chem. Commun.* **1991**, 167.
7. Akaji, K.; Tatsumi, T.; Yoshida, M.; Kimura, T.; Fujiwara, Y.; Kiso, Y., *J. Am. Chem. Soc.* **1992**, *114*, 4137.
8. Akaji, K.; Fujino, K.; Tatsumi, T.; Kiso, Y., *J. Am. Chem. Soc.* **1993**, *115*, 11384.
9. Endo, Y.; Shudo, K.; Okamoto, T., *Chem. Pharm. Bull.* **1981**, *29*, 3753.
10. Hartke, K.; Strangemann, D., *Heterocycles* **1986**, *24*, 2399.
11. Di Bussolo, V.; Kim, Y.-J.; Gin, D. Y., *J. Am. Chem. Soc.* **1998**, *120*, 13515.
12. Honda, E.; Gin, D. Y., *J. Am. Chem. Soc.* **2002**, *124*, 7343.
13. Garcia, B. A.; Poole, J. L.; Gin, D. Y., *J. Am. Chem. Soc.* **1997**, *119*, 7597.
14. Garcia, B. A.; Gin, D. Y., *J. Am. Chem. Soc.* **2000**, *122*, 4269.
15. Nagai, U.; Sammes, P. G., *J. Chem. Soc. C* **1967**, 2230.
16. Reddy, A. S.; Reddy, B. R., *Can. J. Chem.* **1982**, *60*, 40.
17. Takuwa, T.; Onishi, Y. J.; Matsuo, J.-i.; Mukaiyama, T., *Chem. Lett.* **2004**, *33*, 8.
18. Matsuo, J.-i.; Yamanaka, H.; Kawana, A.; Mukaiyama, T., *Chem. Lett.* **2003**, *32*, 392.
19. Yamanaka, H.; Mukaiyama, T., *Chem. Lett.* **2003**, *32*, 1192.
20. Yamanaka, H.; Matsuo, J.-i.; Kawana, A.; Mukaiyama, T., *Chem. Lett.* **2003**, *32*, 626.
21. Saso, H.; Yoshida, H.; Ando, W., *Tetrahedron Lett.* **1988**, *29*, 4747.
22. Codée, J. D. C.; Litjens, R. E. J. N.; den Heeten, R.; Overkleeft, H. S.; van Boom, J. H.; van der Marel, G. A., *Org. Lett.* **2003**, *5*, 1519.
23. Crich, D.; de la Mora, M.; Vinod, A. U., *J. Org. Chem.* **2003**, *68*, 8142.
24. van den Bos, L. J.; Codée, J. D. C.; van Boom, J. H.; Overkleeft, H. S.; van der Marel, G. A., *Org. Biomol. Chem.* **2003**, *1*, 4160.
25. Codée, J. D. C.; van den Bos, L. J.; Litjens, R. E. J. N.; Overkleeft, H. S.; van Boeckel, C. A. A.; van Boom, J. H.; van der Marel, G. A., *Tetrahedron* **2004**, *60*, 1057.
26. Haberman, J. M.; Gin, D. Y., *Org. Lett.* **2003**, *5*, 2539.
27. Nguyen, H. M.; Poole, J. L.; Gin, D. Y., *Angew. Chem. Int. Ed.* **2001**, *40*, 414.
28. Codée, J. D. C.; van den Bos, L. J.; Litjens, R. E. J. N.; Overkleeft, H. S.; van Boom, J. H.; van der Marel, G. A., *Org. Lett.* **2003**, *5*, 1947.
29. van den Bos, L. J.; Codée, J. D. C.; van der Toorn, J. C.; Boltje, T. J.; van Boom, J. H.; Overkleeft, H. S.; van der Marel, G. A., *Org. Lett.* **2004**, *6*, 2165.

Di-2-pyridyl Carbonate[1]

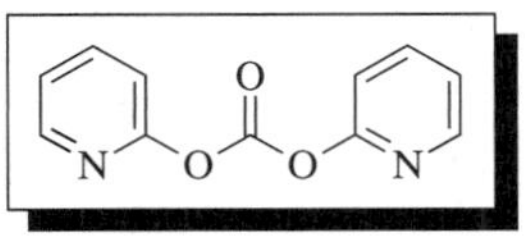

[1659-31-0] $C_{11}H_8N_2O_3$ (MW 216.21))

(dehydration agent; preparation of esters,[3] amides,[4] nitriles;[5] alkoxycarbonylation[6] or protection[7] of amines; carbonyl transfer[8])

Alternate Name: 2-DPC.
Physical Data: mp 76–78 °C.
Solubility: sol CH_2Cl_2.
Form Supplied in: white solid.
Preparative Methods: from Phosgene and 2-Hydroxypyridine in the presence of Pyridine in THF[9] or Triethylamine in CH_2Cl_2 (eq 1).[3,7] Phosgene can also be replaced with triphosgene (Bis(trichloromethyl) Carbonate).[2] The usual yield is about 80%.

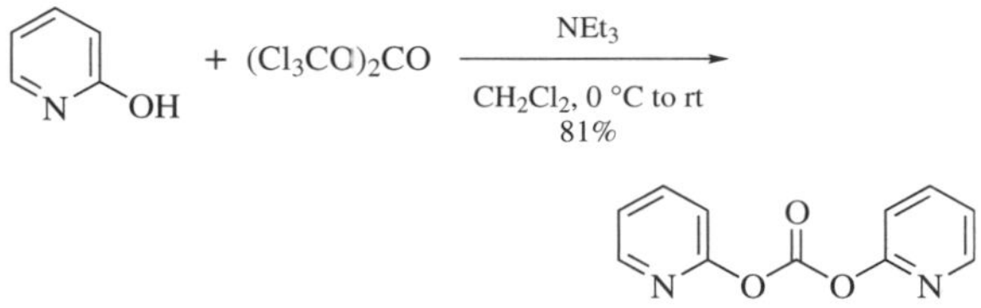

(1)

Handling, Storage, and Precautions: quite stable and can be stored under nitrogen at rt for several months without any noticeable decomposition.

Introduction. The advantages of using 2-DPC are the following: 2-hydroxypyridine is a neutral compound and can be removed completely from the reaction mixture by the usual aqueous workup. The 2-pyridyl esters are also highly reactive.[10]

As a Dehydrating Agent. 2-DPC can be used for the direct esterification of carboxylic acids in good yield and under mild conditions.[1,3] The reaction can be carried out with equimolar amounts of acids, alcohols, and 2-DPC in the presence of 0.1 equiv 4-Dimethylaminopyridine (DMAP) in CH_2Cl_2 at rt. The reaction of diphenylacetic acid and benzyl alcohol gave the corresponding ester in 90% yield (eq 2).

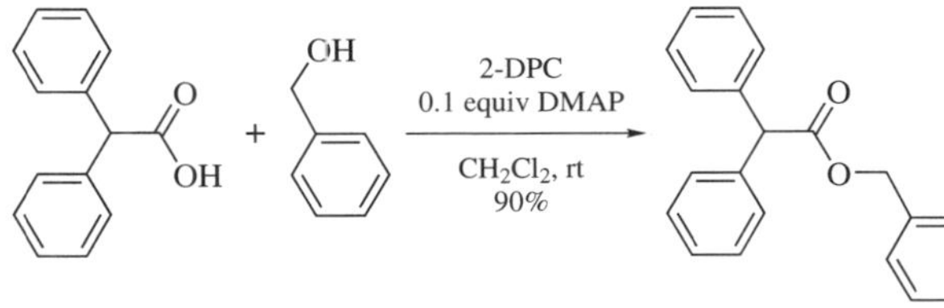

(2)

Typical yields are as follows: methyl diphenylacetate, 95%; benzyl octanoate, 94%; isopropyl phenylacetate, 90%. With aromatic acids and hindered carboxylic acids the esterification is not very satisfactory. The yields drop to 68% for ethyl benzoate or 42% for benzyl pivalate. In these cases the corresponding 2-pyridyl benzoate and pivalate are isolated.

For the preparation of amides a stepwise procedure must be used.[4,11] Formation of activated pyridyl esters and subsequent treatment with amines provide the carboxylic amides. This reaction gave benzyloctanamide in 80% and cyclohexylbenzamide in 85% yield.[1]

2-DPC can also be used to form nitriles via dehydration of the corresponding aldoximes. The reaction was carried out in refluxing toluene or in presence of 0.1 equiv of 1,8-Diazabicyclo[5.4.0]undec-7-ene in CH_2Cl_2 at rt.[5] The last procedure gave benzonitrile in 83% yield.

Alkoxycarbonylation of Amines. 2-DPC is useful for the *N*-protection of amino acids via an activated carbonate.[7] The mixed carbonate is obtained by the reaction of 2-DPC with an equimolar amount of *t*-butyl alcohol in the presence of 0.1 equiv DMAP in CH_2Cl_2. The isolated carbonate obtained in 80% yield was added to a solution of L-proline and Triethylamine in DMF–water (eq 3). After crystallization the yield of *N*-Boc-proline was 98%. Some other typical yields were as follows: *N*-Boc-Ala, 84%; *N*-Boc-Leu, 95%.

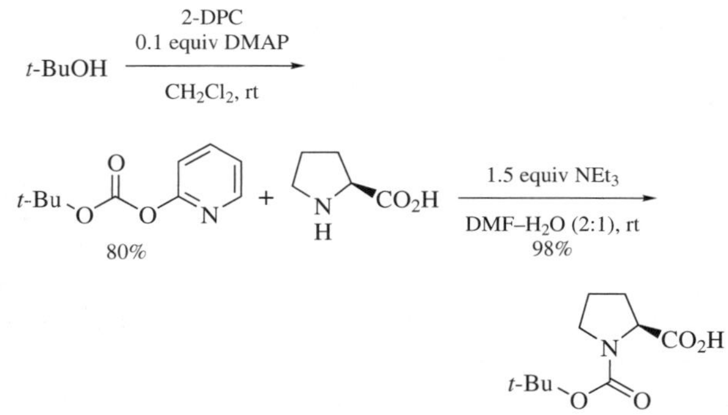
(3)

The same strategy was used to synthesize functionalized carbamates.[2] The reaction between (+)-menthol and 2-DPC in the presence of triethylamine (CH_2Cl_2) gave the mixed carbonate which was used directly, after classical workup. The reaction of this carbonate and the amine gave the carbamate in 81% yield (eq 4). Amines produced in situ by reduction of the azide can also be transformed into the corresponding carbamate.[6]

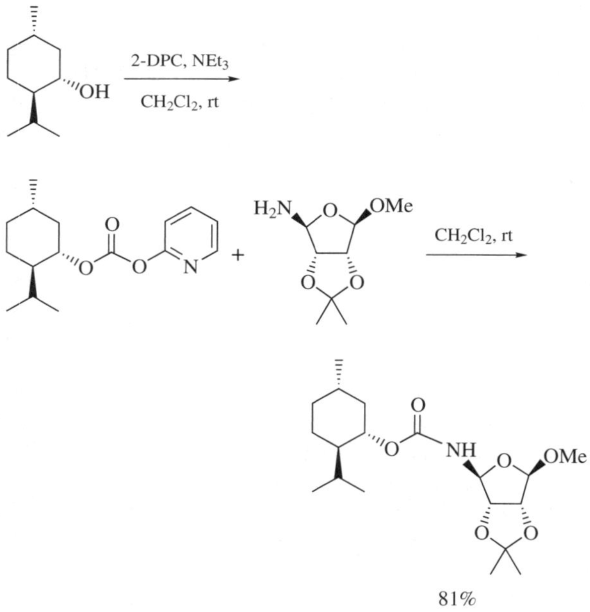
(4)

Carbonyl Transfer Reaction. 2-DPC is useful for the preparation of cyclic carbonates or 2-oxazolidones.[1,8] The reaction with diols is carried out in refluxing toluene. The reaction of ethylene glycol gave an 85% yield of the corresponding carbonate. In the presence of DMAP in CH_2Cl_2 at rt, 1,3-butanediol gave the cyclic carbonate in 85% yield. The reaction of 2-DPC with β-amino alcohols in CH_2Cl_2 at rt gave good yields of the corresponding 2-oxazolidone: 1-aminopropan-2-ol, 87%; 1-phenyl-2-aminoethanol, 93%; 2-amino-4-nitrophenol, 96% (eq 5).

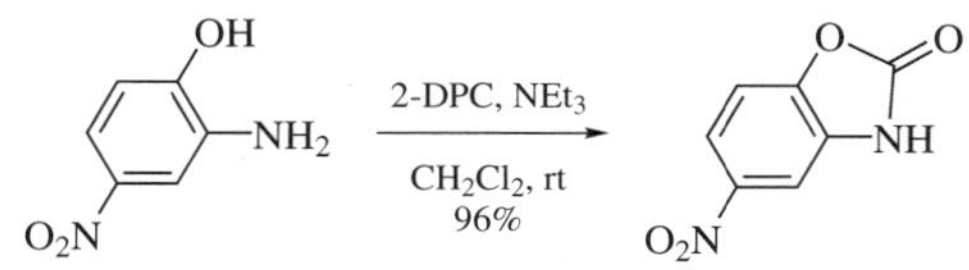
(5)

Related Reagents. 1-(t-Butoxycarbonyl)imidazole; 1-N-(t-Butoxycarbonyl)-1 H-benzotriazole 3-N-Oxide; 2-(t-Butoxycarbonyloxyimino)-2-phenylacetonitrile; t-Butyl Chloroformate; Di-t-butyl Dicarbonate

1. Kim, S. *Org. Prep. Proced. Int.* **1988**, *20*, 145.
2. Ghosh, A. K.; Duong, T. T.; McKee, S. P. *Tetrahedron Lett.* **1991**, *32*, 4251.
3. Kim, S.; Lee, J. I.; Ko, Y. K. *Tetrahedron Lett.* **1984**, *25*, 4943.
4. Kim, S.; Ko, Y. K. *Chem. Commun./J. Chem. Soc., Chem. Commun.* **1985**, 473.
5. Kim, S.; Ko, Y. K. *Bull. Korean Chem. Soc.* **1987**, *8*, 50.
6. Ghosh, A. K.; McKee, S. P.; Duong, T. T.; Thomson, W. J. *Chem. Commun./J. Chem. Soc., Chem. Commun.* **1992**, 1308.
7. Kim, S.; Lee, J. I.; Yi, K. Y. *Bull. Chem. Soc. Jpn.* **1985**, *58*, 3570.
8. Kim, S.; Ko, Y. K. *Heterocycles* **1986**, *24*, 1625.
9. Kampe, W. *Angew. Chem., Int. Ed. Engl.* **1963**, *2*, 479.
10. Dutta, A. S.; Morley, J. S. *J. Chem. Soc. (C)* **1971**, 2896.
11. Kim, S.; Ko, Y. K. *Bull. Korean Chem. Soc.* **1985**, *6*, 175.

Joseph Schoepfer
University of Neuchâtel, Switzerland

2,2′-Dipyridyl Disulfide[1]

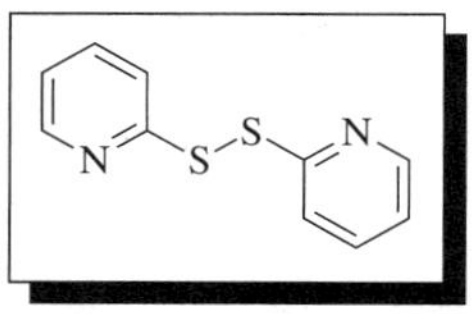

[2127-03-9] $C_{10}H_8N_2S_2$ (MW 220.34)

(macrolactonization of ω-hydroxy acids,[2] acylation of pyrroles,[3] couplings at anomeric centers;[4] lactamization, active ester formation, disulfide formation, thioacetal formation)

Alternate Name: 2,2′-dithiopyridine.
Physical Data: mp 57–58 °C.
Solubility: benzene, THF, CH_2Cl_2, MeCN, DMF.
Form Supplied in: white (colorless) crystalline solid.
Preparative Methods: although 2,2′-dipyridyl disulfide is commercially available, it can be readily prepared from 2-pyridinethione by oxidation with a number of reagents.[5]
Purification: recrystallization from hexane at 30 mL g^{-1} concentration.[6]
Handling, Storage, and Precautions: the solid is irritating to eyes, mucus membranes, and respiratory tract, and physical contact should be avoided. Combustion produces toxic by-products including nitrogen oxides. Cold storage in air-tight containers is recommended. Use in a fume hood.

Original Commentary

Eric J. Stoner
Abbott laboratories, North Chicago, IL, USA

Macrolactonization.[7] When a carboxylic acid is treated with 2,2′-dipyridyl disulfide in the presence of Triphenylphosphine, the corresponding 2-pyridinethiol ester is formed.[8] Corey and Nicolaou have developed an efficient method for the synthesis of macrocyclic lactones based on these 2-pyridinethiol esters.[2] When an ω-hydroxy thiolester is heated in refluxing xylene under high dilution conditions (10^{-5} M, typically accomplished with syringe pump techniques), macrolactonization occurs, liberating triphenylphosphine oxide and pyridinethione. The reaction is quite general and is believed to proceed by a 'double activation' mechanism in which the basic 2-pyridinethiol ester simultaneously activates both the hydroxy and the carboxylic acid moieties with a single proton transfer. It has been shown that the cyclization rate is not affected by the presence of acids, bases, or any of the possible reaction contaminants.[9]

This method is mild, highly efficient for the preparation of medium to large rings (7–16,[2] 12–21[9]), and has been applied to the synthesis of a number of important macrocyclic targets including monensin,[10] brefeldin A (eq 1) and erythronolide B,[11] (±)-11-hydroxy-*trans*-8-dodecenoic acid lactone,[12] (±)-vermiculine,[13] enterobactin,[14] and prostaglandins.[10,15]

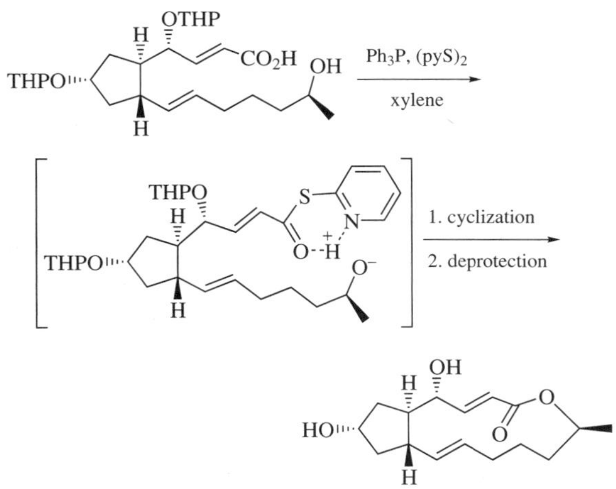

Brefeldin A (1)

In an effort to develop an even milder lactonization protocol, other diaryl disulfides were explored, the most promising being imidazole derivatives (**1**) and (**2**).[16] The formation of dilides (dimers of macrocyclic lactones) is occasionally a problem, necessitating the use of other cyclization methods.[7,17] A variation of Corey's method using silver salts has been developed.[6,18]

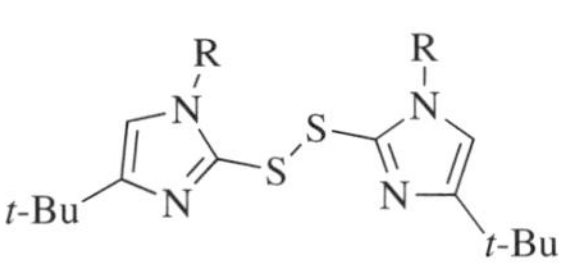

(**1**) R = Me
(**2**) R = *i*-Pr

If an ω-aminopyridinethiol ester is used, macrolactamization occurs.[19] A method for the synthesis of β-lactams from β-amino acids has also been developed where the use of MeCN has been found to be critical (eq 2).[20] The intermolecular version of this reaction is a powerful peptide coupling method.[21,22]

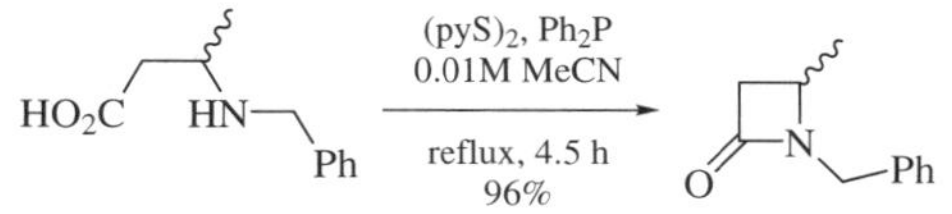

(2)

Related Reactions of 2-Pyridinethiol Esters. In the absence of an internal nucleophile, the thiopyridyl esters generated by the reaction of carboxylic acids with 2,2′-dipyridyl disulfide and triphenylphosphine exhibit other reactivities. Thiolesters are potent acylating agents, reacting with Grignard reagents to yield ketones instead of tertiary alcohols.[23,3] This activity has been exploited for the synthesis of 2-acylpyrroles using pyrrylmagnesium salts as the nucleophilic species. Nicolaou and co-workers have exploited 2-pyridinethiol esters in the synthesis of a number of complex 2-acylpyrroles (eq 3).[24] In certain troublesome cases, the use of stoichiometric Copper(I) Iodide has been found to significantly improve yields.[25]

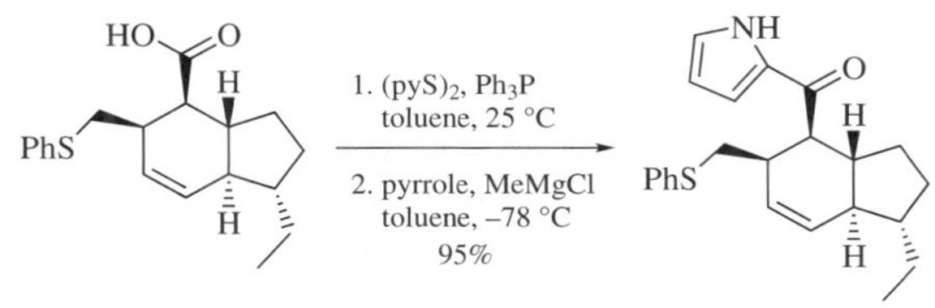

(3)

2-Pyridinethiol esters are readily reduced with Sodium Borohydride in the presence of isopropanol.[26] Activation of 2-pyridinethiol esters with Iodomethane allows for mild trapping with alcohols or benzenethiol, yielding esters or thiolesters.[27] In these cases the use of iodomethane avoids the need for thiophilic silver or mercury salts and allows for transthiolesterification.

Disaccharide Formation. Treatment of glycoside (**3**) with 2,2′-dipyridyl disulfide and Tri-n-bulylphoshine in CH_2Cl_2 rapidly yields thiopyridyl derivative (**4**) as a mixture of anomers. Activation of (**4**) with iodomethane, followed by treatment with glycoside acceptor (**5**), affords the disaccharide fragment of avermectin (**6**),[4] exclusively α-linked in 78% yield (eq 4).[28] Although other methods are available, this protocol offers advantages in practicality and stereoselectivity.

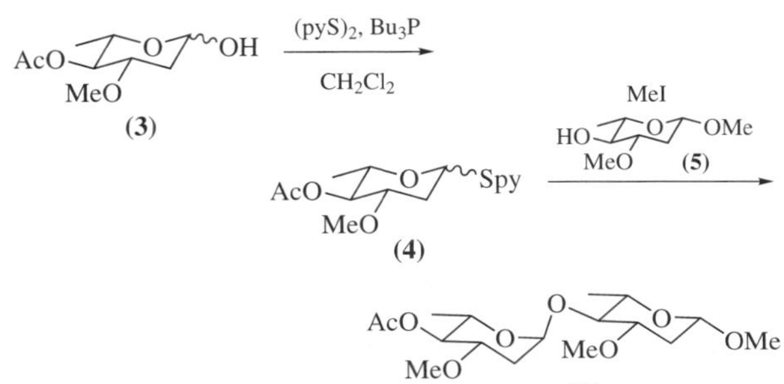

(4)

Carbon–carbon bonds can be similarly formed at the anomeric center of carbohydrates. Oxonium ions, formed by the treatment of 1-(2′-thiopyridyl)glycosides with Silver (I) Trifluoromethanesulfonate, can be trapped with silyl enol ethers, silyl ketene acetals, and reactive aromatic compounds, where the stereoselectivity of the addition is determined by solvent and nucleophile choice.[29] The intramolecular version of this process has also been examined (eq 5).[30] Similarly, bicyclic piperazinediones are available by the intramolecular trapping of iminium ions, generated from the appropriate thiopyridyl derivatives with $PhHgClO_4$ (eq 6).[31]

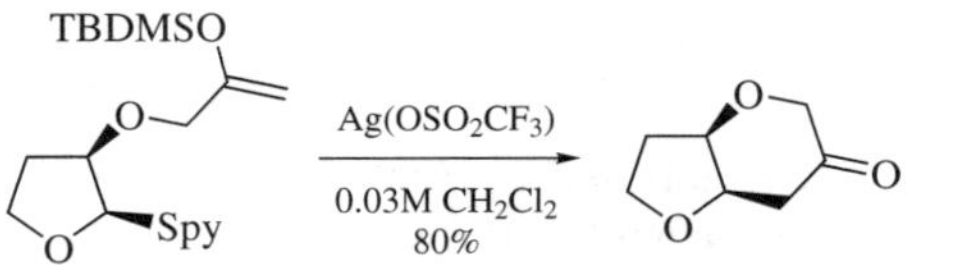

(5)

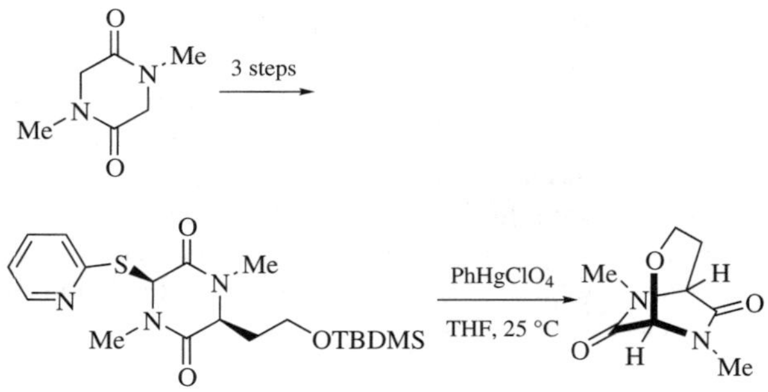

(6)

Alkene Formation. The elimination of pyridinethione or 2-pyridylsulfenic acid forms the basis of a number of alkene syntheses. For example, treatment of dithioacetal anions with 2,2′-dipyridyl disulfide affords the α-thiopyridyl derivative, which undergoes elimination at below room temperature to give ketene dithioacetals.[32] Ester enolates have been similarly treated, although m-Chloroperbenzoic Acid is first used to generate the sulfoxide which reacts further to yield an α,β-unsaturated ester.[33] This methodology has been applied to the synthesis of methyl dehydrojasmonate (**7**) (eq 7).[34]

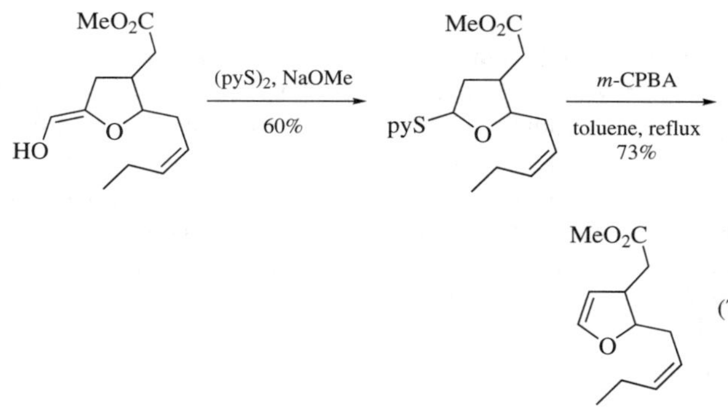

(7)

(**7**)

Other Reactions. Treatment of an active hydroxy compound with 2,2′-dipyridyl disulfide and *n*-Bu_3P yields the corresponding thiopyridyl derivative. This methodology has been applied to the preparation of 5-arylthio-5′-deoxyribonucleosides (eq 8).[35] Monophosphate esters [$ROP(O)(OH)_2$] will react similarly to form the activated triphenylphosphonium adduct, which, in the absence of an added external nucleophile, dimerizes yielding a pyrophosphate.[36] *N*-Methylimidazole has been found to catalyze this transformation. The addition of alcohols or amines, however, traps the phosphoryloxyphosphonium salt as the mixed diphosphate ester or mixed ester/amide, respectively (eq 9).[37] Chlorotrimethylsilane and $(pyS)_2$ have also been reported to facilitate the oxidation of phosphites to phosphates.[38]

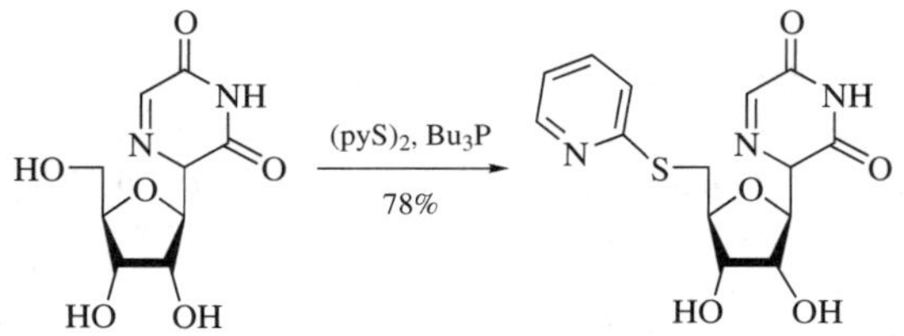

(8)

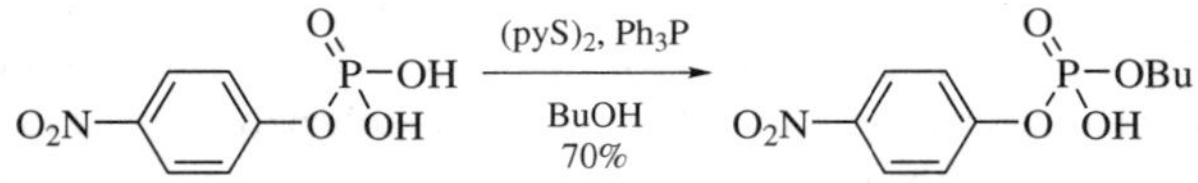

(9)

General Reactions of Diaryl Disulfides[1]**.** 2,2′-Dipyridyl disulfide reacts with a variety of carbon nucleophiles, yielding the corresponding thiopyridyl derivatives. These include phosphonate stabilized anions,[39] bromouridine derivatives,[40] indole anions,[41] and heterocyclic stabilized anions.[42]

First Update

Guillermo Negrón
Universidad Autónoma Metropolitanana, México DF, México
Rubén Gaviño
Universidad Nacional Autónoma de México, Ciudad Universitaria, México DF, México

Lactam and Lactone Formation. A number of β-amino and β-hydroxyacids have been converted to the corresponding β-lactams[43–52] and β-lactones[53] by means of the dipyridyl disulfide/triphenylphosphine couple, as exemplified in eqs 10–14.[46,47,49,50,53]

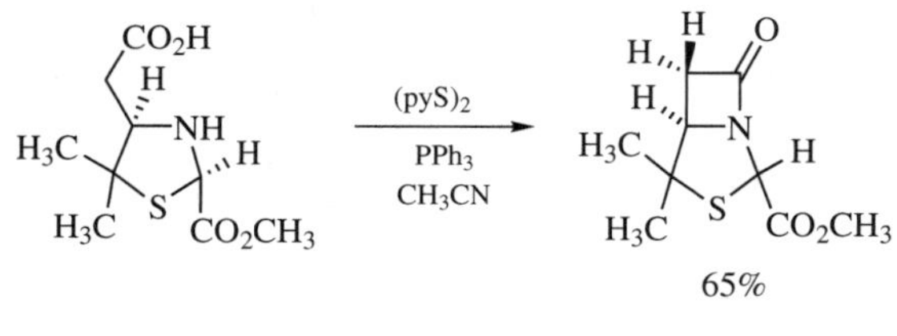

(10)

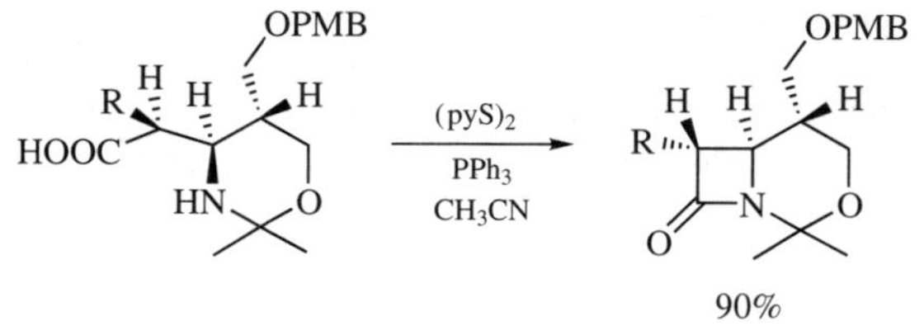

(11)

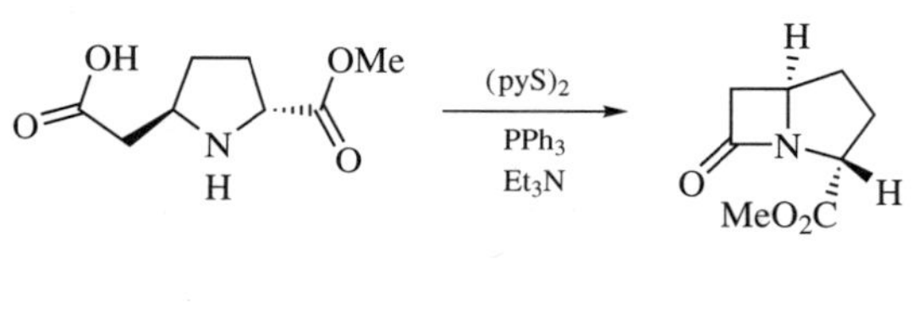

(12)

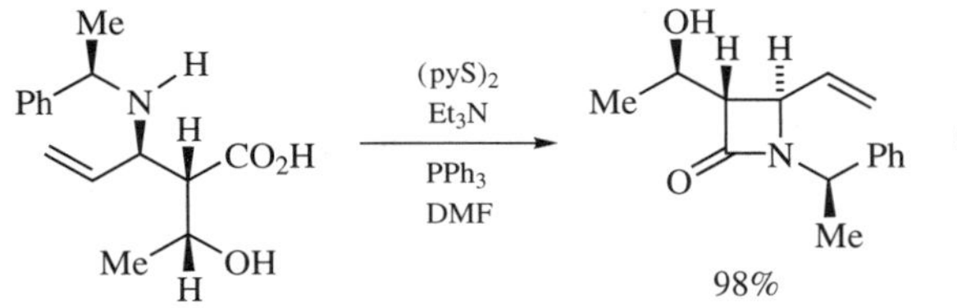

(13)

98%

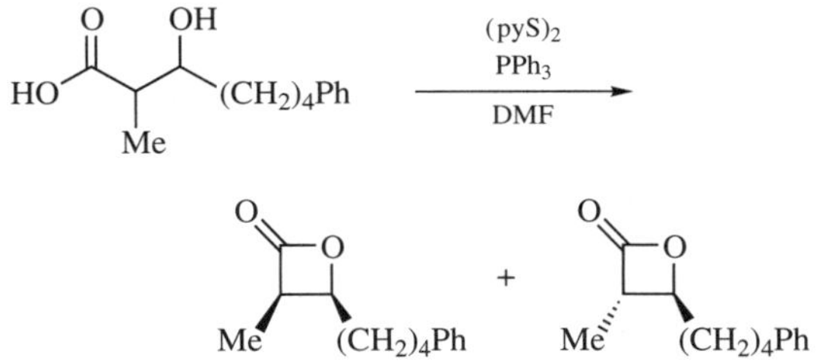

(14)

66% equimolar mixture

γ-Lactams are also readily accessible from the seco-derivatives with the dipyridyl disulfide/triphenyl phosphine couple (eq 15).[54]

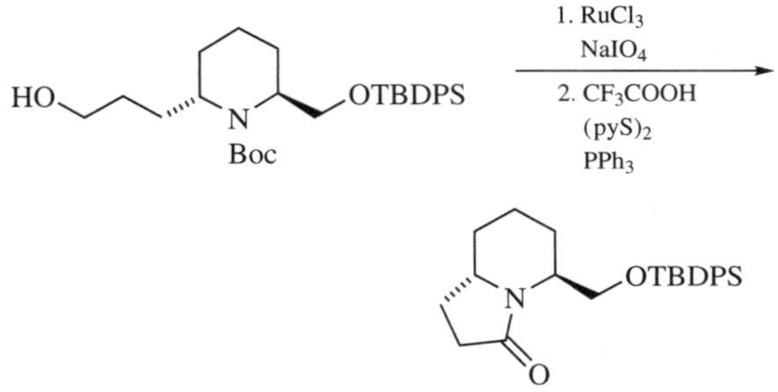

(15)

63%

Macrolactonization. The use of dipyridyl disulfide in conjunction with triphenylphosphine continues to be an important method of macrolactonization both alone (eqs 16 and 17)[55,56] and in the presence of silver salts (eqs 18 and 19)[57,58] or copper salts (eq 20).[59]

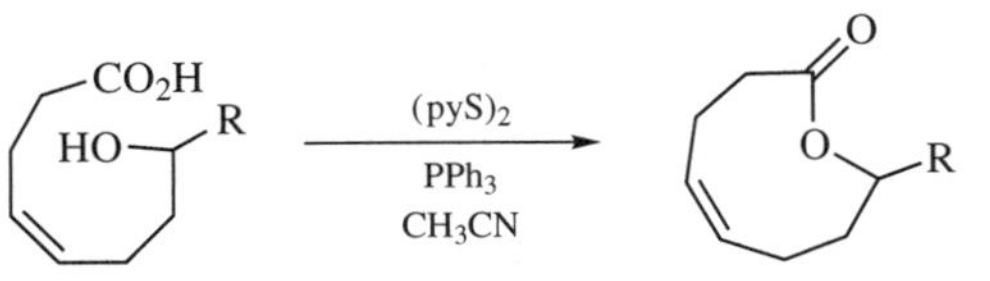

(16)

78%

Formation and Application of 2-Pyridylthiol Esters. 2-Pyridylthiol esters continue to enjoy widespread use as active esters in organic synthesis.[60–66] Some interesting examples of their formation, beyond the bounds of traditional synthesis are given in eqs 21 and 22.[67,68]

Formation of Monothioacetals and Thioglycosides. The reaction of hemiacetals with dipyridyl disulfide and a phosphine, leading to the formation of monothioacetals and thioglycosides has been applied in the arabinofuranose series (eq 23)[69] and to increasingly complex substrates (eq 24).[70] Surprisingly, in the case of eq 24, the product was isolated not in the form of the *S*-pyridyl monothioacetal but as the SH compound.[70]

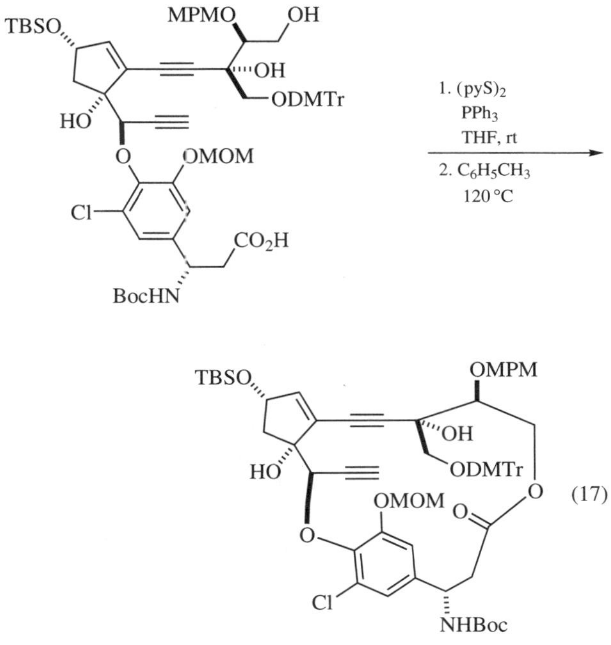

(17)

29%

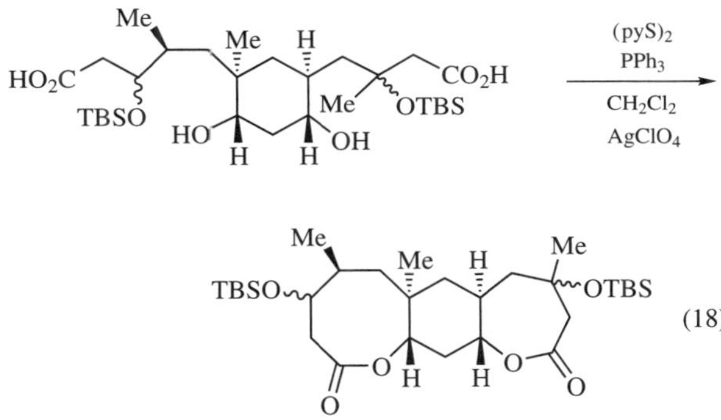

(18)

76%

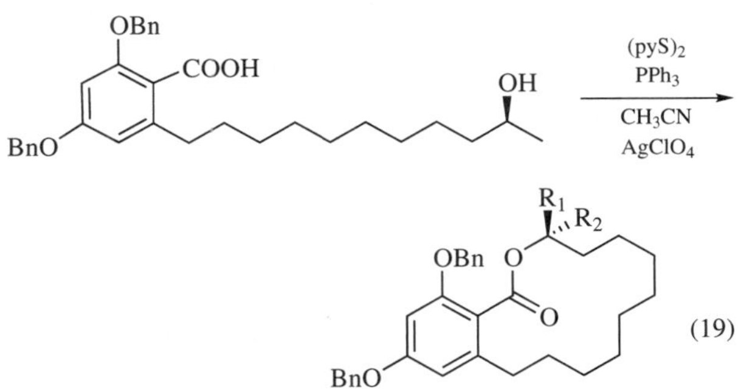

(19)

R_1 = CH3, R_2 = H (*S*) 30%
R_1 = H, R_2 = CH3 (*R*) 49%

In an interesting departure from the usual clean formation of the *S*-2-pyridyl sulfide,[71] Craig and co-workers observed competing formation of the tautomeric aminal when the dipyridyl disulfide/tributylphosphine couple was applied to a five-membered cyclic hemiacetal (eq 25).[72]

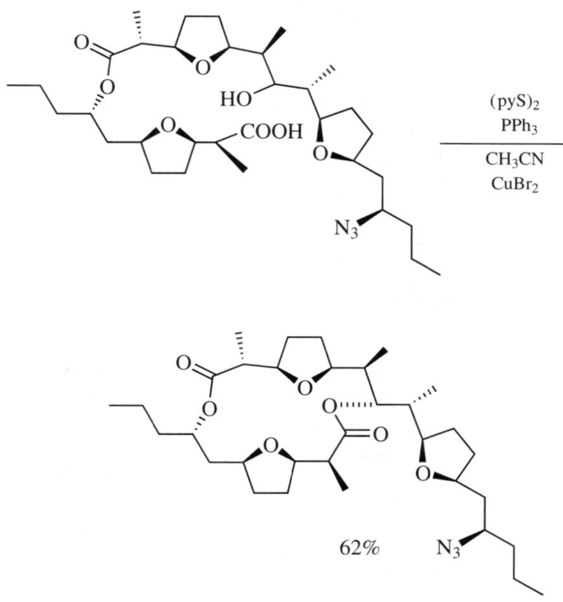

(20)

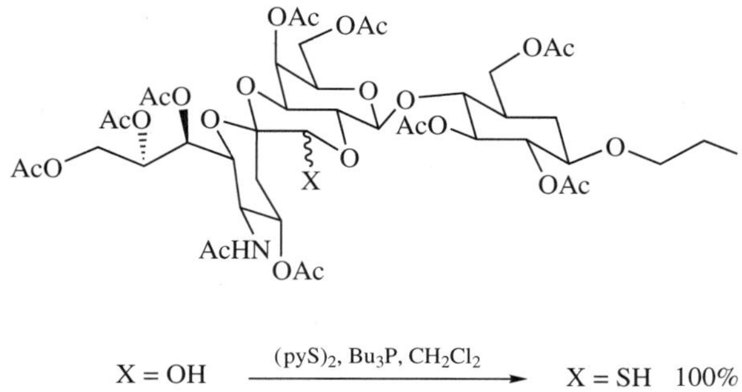

(24)

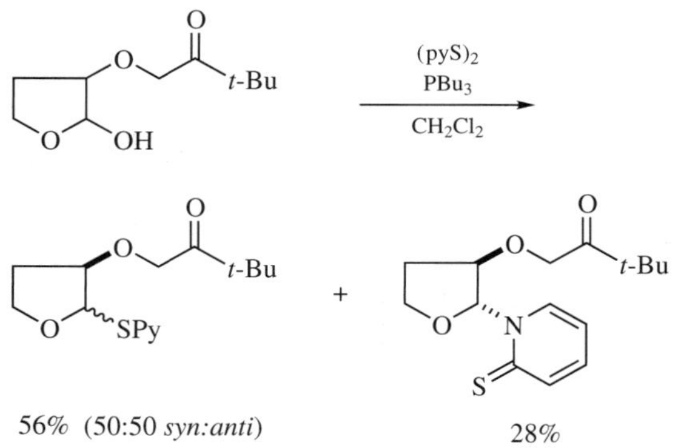

(25)

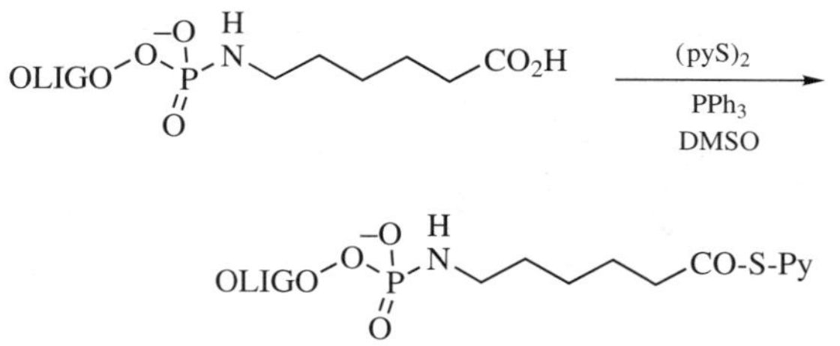

(21)

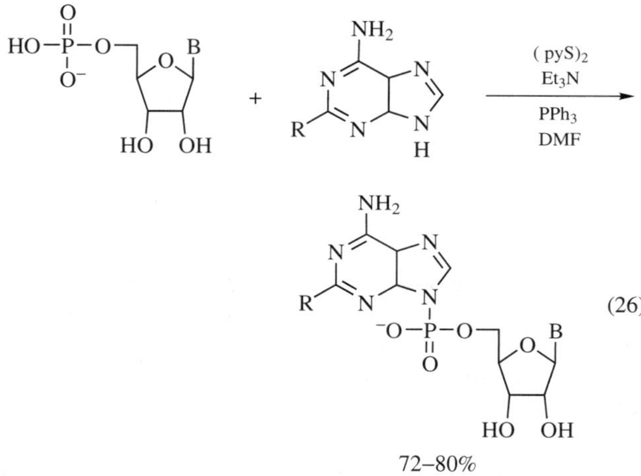

(26)

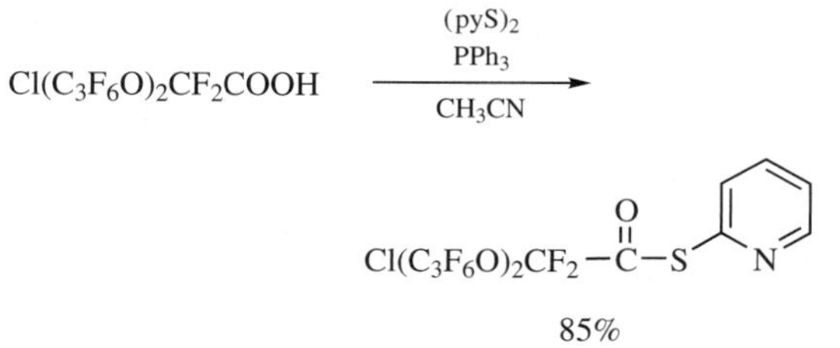

(22)

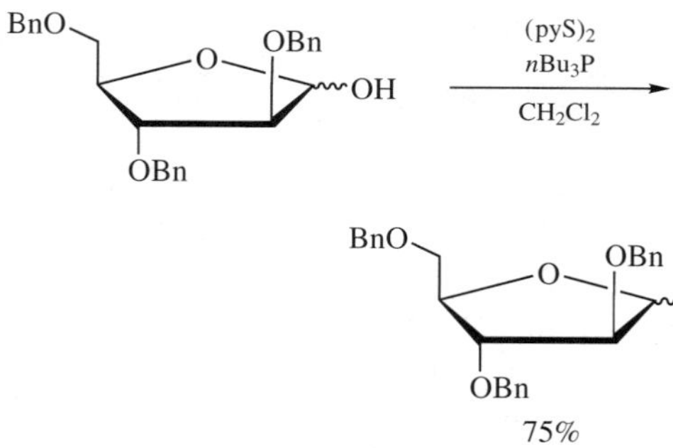

(23)

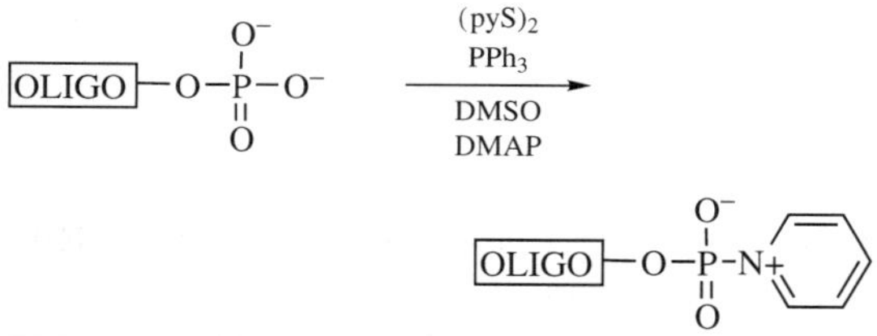

(27)

OLIGO = 5′-TTT-TC*T-TTT-C*C*C*-C*C*C*-T-3′

C* = 5-methylcytosine

Phosphoramide Formation. In the field of nucleotide chemistry the dipyridyl disulfide/phosphine combination provides a useful means of forming phosphoramides. The reaction has been applied to simple nucleosides (eq 26)[73] and in more complex oligomers, using a range of amines as nucleophiles (eq 27).[74]

Interestingly, using this chemistry it is even possible to modify internal phosphoramidate groups (np) in an oligonucleotide (eq 28) as well as the more common internal phosphate (p) and terminal N3′-P5′-phosphoramidate bond.[75]

Formation of 2-Pyridyl Alkyl or Aryl Disulfides. One of the most important applications of 2,2′-dipyridyl disulfide is in the

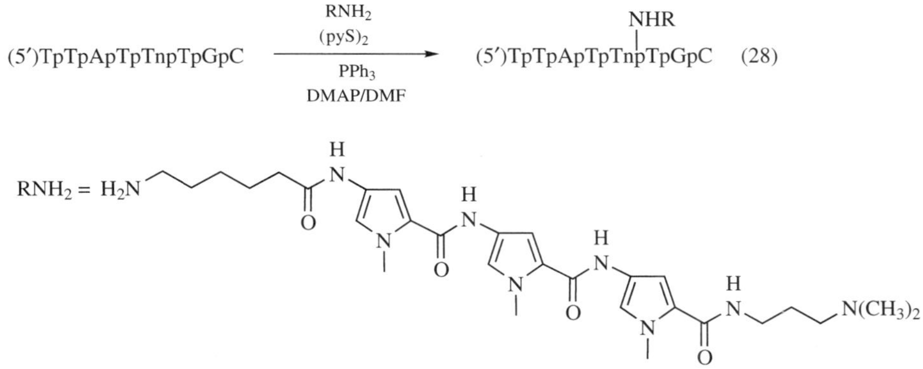

formation of mixed alkyl or aryl 2-pyridyl disulfides. These reactions, which proceed cleanly owing to the liberated 2-mercaptopyridine existing mainly in the form of the thiopyridone tautomer, are extremely important as further reaction of the alkyl pyridyl sulfide with a second thiol enables clean formation of mixed disulfides, again with liberation of thiopyridone. Such reaction sequences are widely employed in the formation of bioconjugates but also find application wherever mixed disulfides are required. The chemistry, which functions in protic solvents, has been applied to a range of simple substrates and is tolerant of many standard functional groups, as illustrated in eqs 29–32.[76–79]

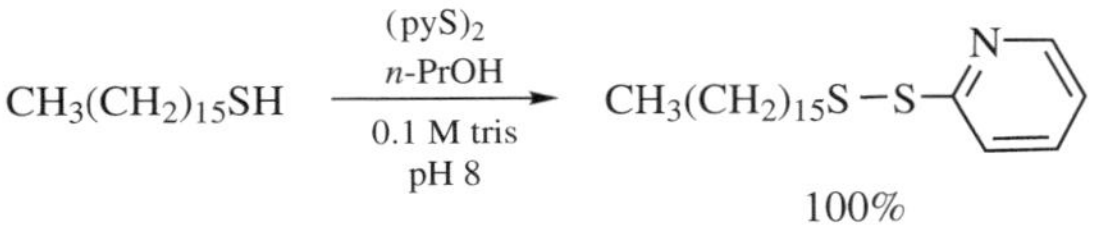

(29)

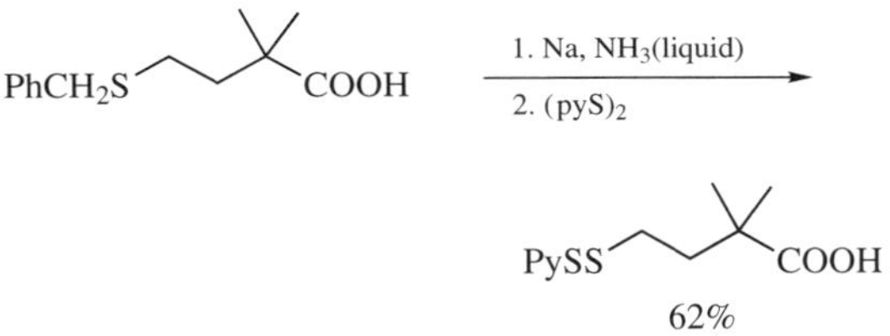

(30)

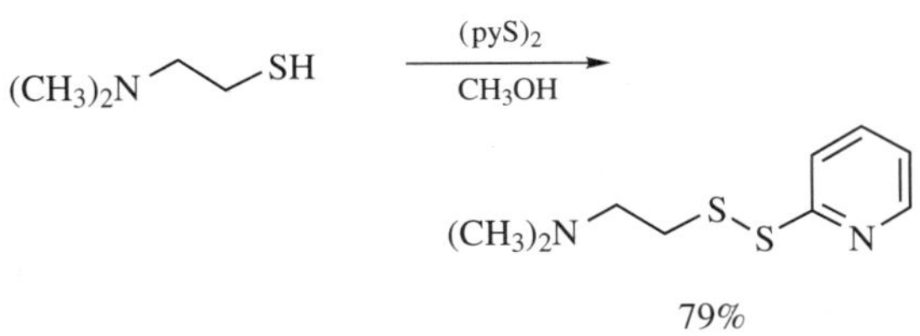

(31)

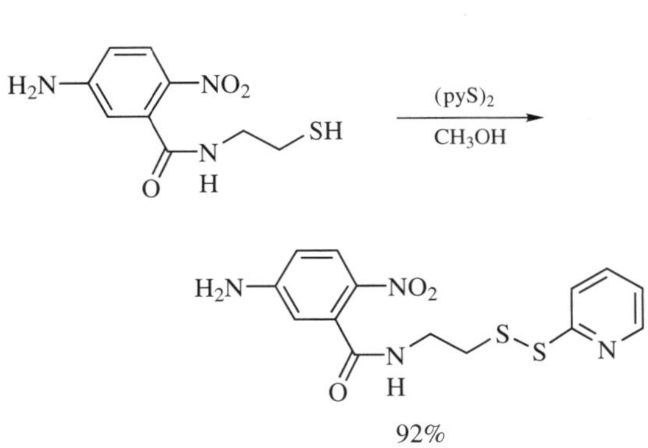

(32)

With excess thiol, the symmetrical disulfide is formed (eq 33).[80] The chemistry is ideally suited to the functionalization of the thiol group in cysteine,[81] in cysteine-containing peptides (eq 34)[82] and related substances (eq 35).[83]

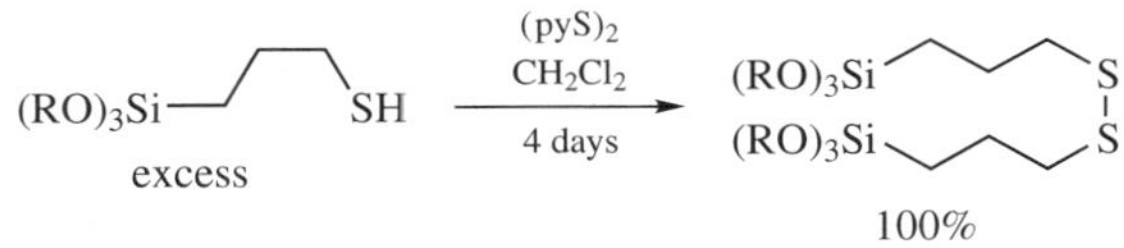

(33)

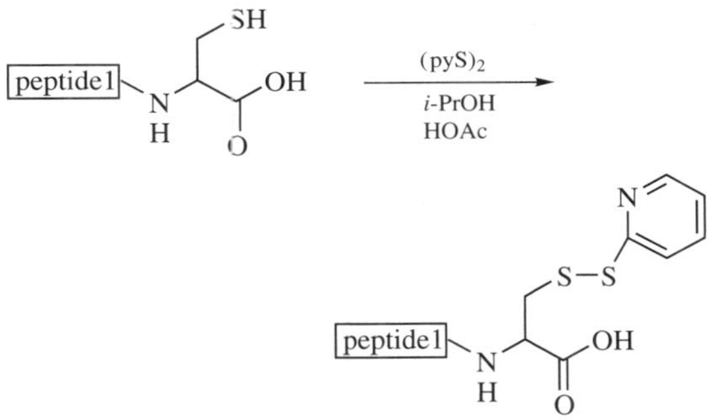

(34)

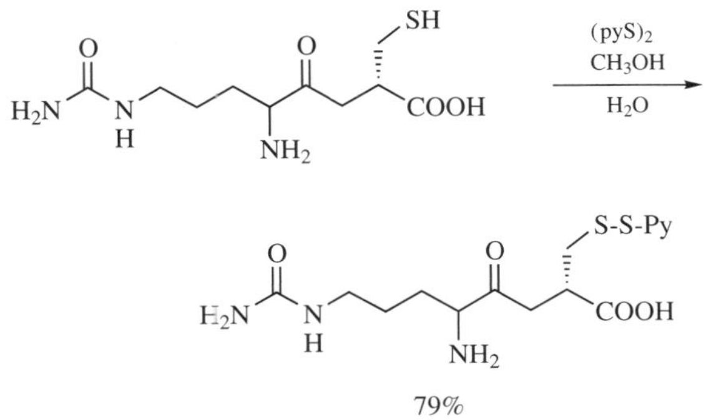

(35)

Still making use of the cysteine thiol, the reaction has been applied, with substoichiometric dipyridyl disulfide to the preparation of a small library of peptides dendrimers containing a central disulfide linkage (eq 36).[84]

This chemistry has also found application in the nucleoside field (eq 37),[85] in the preparation of folic acid (eq 38),[86] and biotinylated synthons for affinity labeling (eq 39).[87]

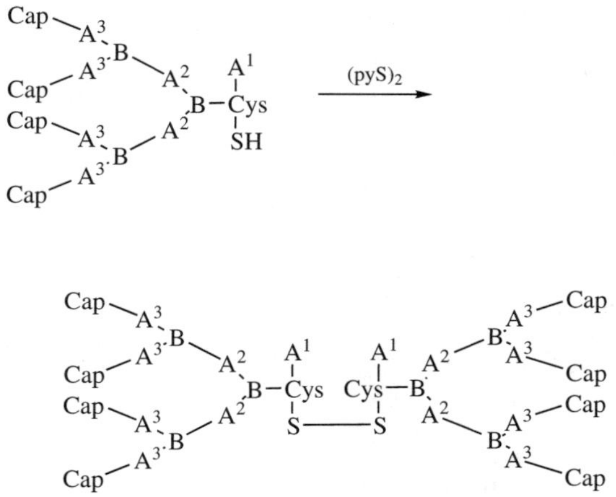

(36)

A^1, A^2, A^3 = His, Asp, Ser

B =

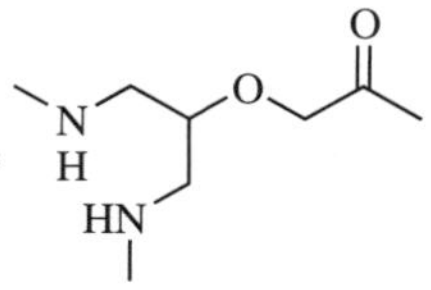

Cap =

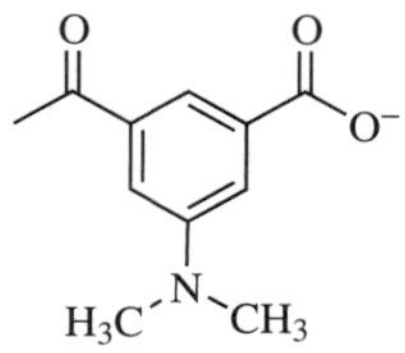

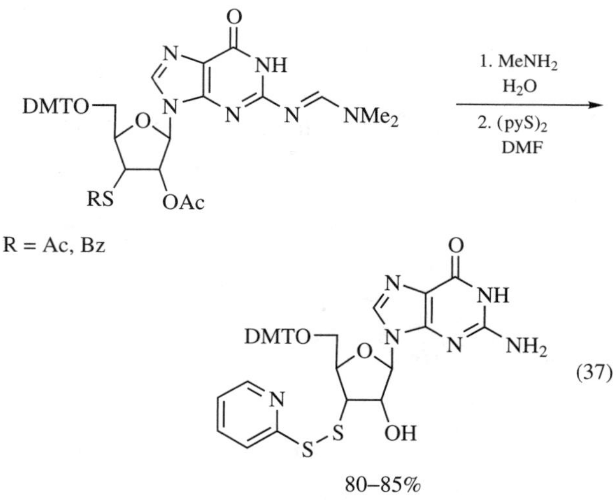

(37)

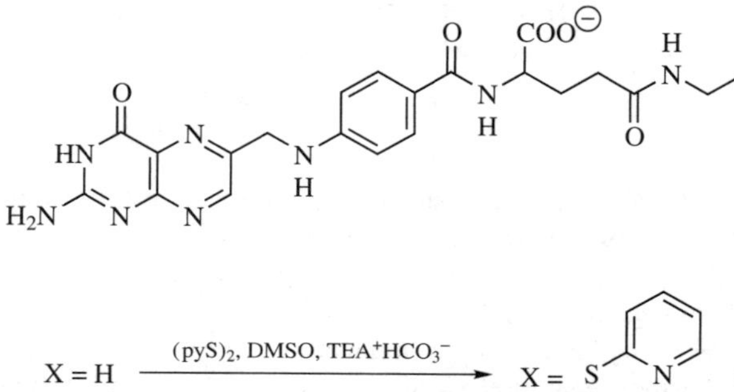

(38)

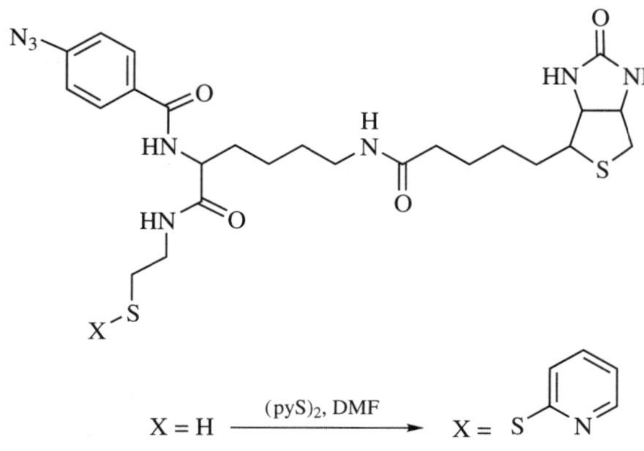

(39)

Other Reactions. Reaction of 2,2′-dipyridyl disulfide with iodine generates, in situ, a potential sufenylating agent. This chemistry has been applied on the solid phase to the modification of thiazoles (eq 40).[88]

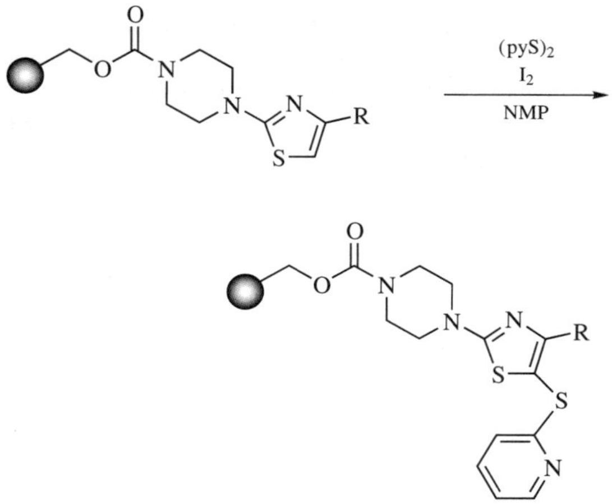

(40)

1. (a) Capozzi, G.; Modena, G., *The Chemistry of the Thiol Group*; Patai, S.; Ed.; Wiley: New York, 1974; Part 2, pp 785–839. (b) Jocelyn, P. C., *The Chemistry of the SH Group*; Academic: New York, 1972.
2. Corey, E. J.; Nicolaou, K. C., *J. Am. Chem. Soc.* **1974**, *96*, 5614.
3. For other recent methods for the synthesis of 2-acylpyrroles, see: (a) Review: Patterson, J. M., *Synthesis* **1976**, 281. (b) Kozikowski, A. P.; Ames, A., *J. Am. Chem. Soc.* **1980**, *102*, 860. (c) Martinez, G. R.; Grieco, P. A.; Srinivasan, C. V., *J. Org. Chem.* **1981**, *46*, 3760. (d) Edwards, M. P.; Ley, S. V.; Lister, S. G.; Palmer, B. D., *Chem. Commun./J. Chem. Soc., Chem. Commun.* **1983**, 630.
4. Springer, J. P.; Arison, B. H.; Hirshfield, J. M.; Hoogsteen, K., *J. Am. Chem. Soc.* **1981**, *103*, 4221.
5. (a) NaOH/KI_3: Marckwald, W.; Klemm, W.; Trabert, H., *Ber. Dtsch. Chem. Ges./Chem. Ber* **1900**, *33*, 1556. (b) Nickel peroxide: Nakagawa, K.; Shiba, S.; Horikawa, M.; Sato, K.; Nakamura, H.; Harada, N.; Harada, F., *Synth. Commun.* **1980**, *10*, 305. (c) I_2: McAllan, D. T.; Cullum, T. V.; Dean, R. A.; Fidler, F. A., *J. Am. Chem. Soc.* **1951**, *73*, 3627. (d) $Zn(BiO_3)_2$: Firouzabadi, H.; Mohammadpour-Baltork, I., *Bull. Chem. Soc. Jpn.* **1992**, *65*, 1131. (e) $(NH_4)_2Ce(NO_3)_6$ (CAN): Dhar, D. N.; Bag, A. K., *Indian J. Chem., Sect.B* **1984**, *23B*, 974. (f) Bromodimethylsulfonium bromide: Olah, G. A.; Arvanaghi, M.; Vankar, Y. D., *Synthesis* **1979**, 721. (g) $NaBO_3$: McKillop, A.; Koyunçu, D., *Tetrahedron Lett.* **1990**, *31*, 5007. (h) Diethyl bromomalonate: Kato, E.; Oya, M.; Iso, T.; Iwao, J.-I., *Chem. Pharm. Bull.* **1986**, *34*, 486. (i) $FeCl_3$–Bu_3SnOMe: Sato, T.; Otera, J.; Nozaki, H., *Tetrahedron Lett.* **1990**, *31*, 3591.

6. Thalmann, A.; Oertle, K.; Gerlach, H., *Org. Synth., Coll. Vol.* **1990**, *7*, 470.
7. Reviews: (a) Nicolaou, K. C., *Tetrahedron* **1977**, *33*, 683. (b) Back, T. G., *Tetrahedron* **1977**, *33*, 3041. (c) Ogliaruso, M. A.; Wolfe, J. A. In *Synthesis of Lactones and Lactams*; Patai, S.; Rappaport, Z.; Eds.; Wiley: New York, 1993.
8. Mukaiyama, T., *Synth. Commun.* **1972**, *2*, 243.
9. Corey, E. J.; Brunelle, D. J.; Stork, P. J., *Tetrahedron Lett.* **1976**, 3405.
10. Corey, E. J.; Nicolaou, K. C.; Melvin, L. S., *J. Am. Chem. Soc.* **1975**, *97*, 653.
11. Corey, E. J.; Nicolaou, K. C.; Melvin, L. S., *J. Am. Chem. Soc.* **1975**, *97*, 654.
12. Corey, E. J.; Ulrich, P.; Fitzpatrick, J. M., *J. Am. Chem. Soc.* **1976**, *98*, 222.
13. Corey, E. J.; Nicolaou, K. C.; Toru, T., *J. Am. Chem. Soc.* **1975**, *97*, 2287.
14. Corey, E. J.; Bhattacharyya, S., *Tetrahedron Lett.* **1977**, 3919.
15. Bundy, G. L.; Peterson, D. C.; Cornette, J. C.; Miller, W. L.; Spilman, C. H.; Wilks, J. W., *J. Med. Chem.* **1983**, *26*, 1089.
16. Corey, E. J.; Brunelle, D. J., *Tetrahedron Lett.* **1976**, 3409.
17. (a) Karim, M. R.; Sampson, P., *J. Org. Chem.* **1990**, *55*, 598. (b) Justus, K.; Steglich, W., *Tetrahedron Lett.* **1991**, *32*, 5781.
18. Gerlach, H.; Thalmann, A., *Helv. Chim. Acta* **1974**, *57*, 2661.
19. Bai, D.; Shi, Y., *Tetrahedron Lett.* **1992**, *33*, 943.
20. (a) Ohno, M.; Kobayashi, S.; Iimori, T.; Wang. Y.-F.; Izawa, T., *J. Am. Chem. Soc.* **1981**, *103*, 2405. (b) Ohno, M.; Kobayashi, S.; Iimori, T.; Wang. Y.-F.; Izawa, T., *J. Am. Chem. Soc.* **1981**, *103*, 2406.
21. Reviews: (a) Klausner, Y. S.; Bodanszky, M., *Synthesis* **1972**, 453. (b) Mukaiyama, T., *Angew. Chem., Int. Ed. Engl.* **1976**, *15*, 94.
22. (a) Mukaiyama, T.; Matsueda, R.; Suzuki, M., *Tetrahedron Lett.* **1970**, 1901. (b) Mukaiyama, T.; Matsueda, R.; Maruyama, H., *Bull. Chem. Soc. Jpn.* **1970**, *43*, 1271.
23. (a) Araki, M.; Sakata, S.; Takei, H.; Mukaiyama, T., *Bull. Chem. Soc. Jpn.* **1974**, *47*, 1777. (b) Loader, C. E.; Anderson, H. J., *Can. J. Chem.* **1971**, *49*, 45.
24. Nicolaou, K. C.; Claremon, D. A.; Papahatjis, D. P., *Tetrahedron Lett.* **1981**, *22*, 4647.
25. Nakahara, Y.; Fujita, A.; Ogawa, T., *Agric. Biol. Chem.* **1985**, *49*, 1491.
26. Nicolaou, K. C.; Maligres, P.; Suzuki, T.; Wendeborn, S. V.; Dai, W.-M.; Chadha, R. K., *J. Am. Chem. Soc.* **1992**, *114*, 8890.
27. Ravi, D.; Mereyala, H. B., *Tetrahedron Lett.* **1989**, *30*, 6089.
28. (a) Mereyala, H. B.; Kulkarni, V. R.; Ravi, D.; Sharma, G. V. M.; Rao, B. V.; Reddy, G. B., *Tetrahedron* **1992**, *48*, 545. (b) Ravi, D.; Kulkarni, V. R.; Mereyala, H. B., *Tetrahedron Lett.* **1989**, *30*, 4287. (c) for mechanistic details see: Mereyala, H. B.; Reddy, G. V., *Tetrahedron* **1991**, *47*, 6435.
29. Williams, R. M.; Stewart, A. O., *Tetrahedron Lett.* **1983**, *24*, 2715.
30. Craig, D.; Munasinghe, V. R. N., *Tetrahedron Lett.* **1992**, *33*, 663.
31. Williams, R. M.; Dung, J.-S.; Josey, J.; Armstrong, R. W.; Meyers, H., *J. Am. Chem. Soc.* **1983**, *105*, 3214.
32. Nagao, Y.; Seno, K.; Fujita, E., *Tetrahedron Lett.* **1979**, 4403.
33. Nagao, Y.; Takao, S.; Miyasaka, T.; Fujita, E., *Chem. Commun./J. Chem. Soc., Chem. Commun.* **1981**, 286.
34. Dubs, P.; Stüssi, R., *Helv. Chim. Acta* **1978**, *61*, 998.
35. (a) Nakagawa, I.; Hata. T., *Tetrahedron Lett.* **1975**, 1409. (b) Nakagawa, I.; Aki, K.; Hata, T., *J. Chem. Soc., Perkin Trans. 1* **1983**, 1315.
36. Kanavorioti, A.; Lu, J.; Rosenbach, M. T.; Hurley, T. B., *Tetrahedron Lett.* **1991**, *32*, 6065.
37. Mukaiyama, T.; Hashimoto, M., *Bull. Chem. Soc. Jpn.* **1971**, *44*, 196.
38. Hata, T.; Sekine, M., *Tetrahedron Lett.* **1974**, 3943.
39. Ebertino, F. H.; Degenhart, C. R.; Jamieson, L. A.; Burdsall, D. C., *Heterocycles* **1990**, *30*, 855.
40. Hirota, K.; Tomishi, T.; Maki, Y., *Heterocycles* **1987**, *26*, 3089.
41. Atkinson, J. G.; Hanel, P.; Girard, Y., *Synthesis* **1988**, 480.
42. Mirazaei, Y. R.; Simpson, B. M.; Triggle, D. J.; Natale, N. R., *J. Org. Chem.* **1992**, *57*, 6271.
43. Seki, M.; Miyake, T.; Yamanaka, T.; Ohmizu, H., *Synlett* **1996**, 455.
44. Seki, M.; Furutani, T.; Miyake, T.; Yamanaka, T.; Ohmusi, H., *Tetrahedron: Asymmetry* **1996**, *7*, 1241.
45. Collis, M. P.; Perlmutter, P., *Tetrahedron: Asymmetry* **1996**, *7*, 2117.
46. Davies, S. G.; Fenwick, D. R., *Chem. Commun.* **1997**, 565.
47. Maison, W.; Kosten. M.; Charpy, A.; Kintscher-Langenhagen, J.; Schlemminger, I.; Lützen, A.; Westerhoff, O.; Martens, J., *Eur. J. Org. Chem.* **1999**, 2433.
48. Ham, W. H.; Oh, C. Y.; Lee, Y. S.; Jeong, J. H., *J. Org. Chem.* **2000**, *65*, 8372.
49. Guanti, G.; Moro, A.; Narisano, E., *Tetrahedron Lett.* **2000**, *41*, 3203.
50. Tanaka, H.; Sakagami, H.; Ogasawara, K., *Tetrahedron Lett.* **2002**, *43*, 93.
51. Chippindale, A. M.; Davies, S. G.; Iwamoto, K.; Parkin, R. M.; Smethurst, C. A. P.; Smith, A. D.; Rodríguez-Solla, H., *Tetrahedron* **2003**, *59*, 3253.
52. Whisler, M. C.; Beak. P., *J. Org. Chem.* **2003**, *68*, 1207.
53. Groneberg, R. D.; Burns, C. J.; Morrissette, M. M.; Ullrich, J. W.; Morris, R. L.; Darnbrough, S.; Djuric, S. W.; Condon, S. M.; McGeehan, G. M.; Labaudiniere, R.; Neuenschwander, K.; Scotese, A. C.; Kline, JA., *J. Med. Chem.* **1999**, *42*, 541.
54. Shimizu, M.; Nemoto, H.; Kakuda, H.; Takahata, H., *Heterocycles* **2003**, *59*, 245.
55. McNaughton-Smith, G.; Taylor, R. J. K., *Tetrahedron* **1996**, *52*, 2113.
56. Sasaki, T.; Inoue, M.; Hirama, M., *Tetrahedron Lett.* **2001**, *42*, 5299.
57. Nicolaou, K. C.; Burnage, M. E.; McGarry, D. G.; Shi, S.; Somers, P. K.; Wallace, P. A.; Chu, X. J.; Agrios, K. A.; Gunzner, J. L.; Yang, Z., *Chem. Eur. J.* **1999**, *5*, 599.
58. Bracher, F.; Krau J., *Eur. J. Org. Chem.* **2001**, *24*, 4701.
59. Kang, S. H.; Jeon, J. W.; Hwang, Y. S.; Lee, S. B., *Angew, Chem. Int., Ed.* **2002**, *41*, 1392.
60. Zhuang, W.; Hazell, R. G.; Jørgensen, K. A., *Chem. Commun.* **2001**, 1240.
61. Annunziata, R.; Benaglia, M.; Cinquini, M.; Cozzi, F., *Tetrahedron: Asymmetry* **1999**, *10*, 4841.
62. Tsunoda, T.; Nishii, T.; Yoshizuka, M.; Yamasaki, C.; Suzuki, T.; Itô, S., *Tetrahedron Lett.* **2000**, *41*, 7667.
63. Couladouros, E. A.; Strongilos, A. T., *Eur. J. Org. Chem.* **2002**, 3341.
64. Di Fabio, R.; Conti, N.; De Magistris, E.; Feriani, A, Provera, S.; Sabbatini, F. M.; Reggiani, A.; Rovatti, L.; Barnaby, R. J., *J. Med. Chem.* **1999**, *42*, 3486.
65. Kanoh, N.; Ishihara, J.; Yamamoto, Y.; Murai, A., *Synthesis* **2000**, 1878.
66. Di Fabio, R.; Alvaro. G.; Bertani, B.; Giacobbe, S., *Can. J. Chem.* **2000**, *78*, 809.
67. Bergamin, M.; Da Ros, T.; Spalluto, G.; Boutorine, A.; Prato, M., *Chem. Commun.* **2001**, 17.
68. Campestrini, S.; Lora, G.; Tonellato, U., *Tetrahedron Lett.* **2001**, *42*, 7045.
69. Chang, G. X.; Lowary, T. L., *Org. Lett.* **2000**, *2*, 1505.
70. Tietze, L. F.; Keim, H.; Janßen, C. O.; Tappertzhofen, C.; Olschimke, J., *Chem.-A Eur. J.* **2000**, *6*, 2801.
71. Craig, D.; Pennington, M. W.; Warner, P., *Tetrahedron* **1999**, *55*, 13495.
72. Craig, D.; Munasinghe, V. R. N.; Tierney, J. P.; White, A. J. P.; Williams, D. J.; Williamson, C., *Tetrahedron* **1999**, *55*, 15025.
73. Prabahar, K. J.; Ferris, J. P., *J. Am .Chem. Soc.* **1997**, *119*, 4330.
74. Da Ros, T.; Bergamin, M.; Vázquez, E.; Spalluto, G.; Baiti, B.; Moro, S.; Boutorine, A.; Prato, M., *Eur., J. Org. Chem.* **2002**, 405.
75. Sinyakov, A. N.; Boutorine, A. S.; Novopashina, D. S.; Halby, L.; Ryabinin, V. A., *Russ. J. Bioorg. Chem.* **2003**, *29*, 88.
76. Wilkinson, T. A.; Yin, J.; Pidgeon, C.; Post, C. B., *J. Peptide Res.* **2000**, *55*, 140.

77. Vrudhula, V. M.; MacMaster, J. F.; Li, Z.; Kerr, D. E.; Senter, P. D., *Bioorg. Med. Chem. Lett.* **2002**, *12*, 3591.
78. Uragami, M.; Miyake, Y.; Tokutake, N.; Zhang, L.; Regen, S. L., *Langmuir* **2000**, *16*, 8010.
79. Kempin, U.; Hennig, L.; Knoll, D.; Welzel, P.; Müller, D.; Markus, A.; van Heijenoort, J., *Tetrahedron* **1997**, *53*, 17669.
80. Dufaud, V.; Davis, M. E., *J. Am. Chem. Soc.* **2003**, *125*, 9403.
81. Zhang, Q.; Kelly, J. W., *Biochemistry* **2003**, *42*, 8756.
82. Schulz, A.; Adermann, K.; Eulitz, M.; Feller, S. M.; Kardinal, C., *Tetrahedron* **2000**, *56*, 3889.
83. Sengle, G.; Jenne, A.; Arora, P. S.; Seeling, B.; Nowick, J. S.; Jäschke, A.; Famulok, M., *Bioorg. Med. Chem.* **2000**, *8*, 1317.
84. Esposito, A.; Delort, E.; Lagnoux, D.; Djojo, F.; Reymond, J. L., *Angew. Chem. Int. Ed.* **2003**, *42*, 1381.
85. Matulic-Adamic, J.; Beigelman, L., *Helv. Chim. Acta.* **1999**, *82*, 2141.
86. Matulic-Adamic, J.; Sanseverino, M.; Beigelman, L., *Tetrahedron Lett.* **2002**, *43*, 4439.
87. Alley, S. C.; Ishmael, F. T.; Jones, A. D.; Benkovic, S. J., *J. Am. Chem. Soc.* **2000**, *122*, 6126.
88. Grimstrup, M.; Zaragoza, F., *Eur. J. Org. Chem.* **2002**, 2953.

2,2′-Dipyridyl Disulfide-*N,N*′-Dioxide

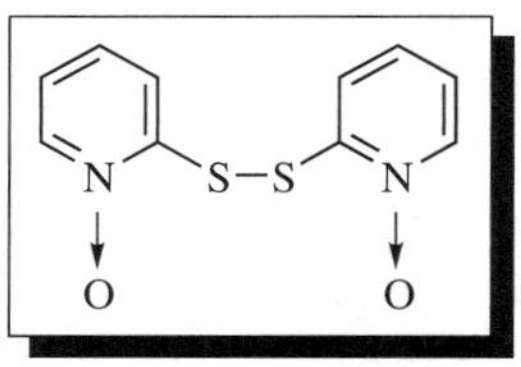

[7275-43-6] $C_{10}H_8N_2O_2S_2$ (MW 252.32)

(reagent used for the preparation of *O*-acyl thiohydroxamates (Barton esters), which are the precursors of mainly sp^3 carbon-centered radicals formed via decarboxylation of acyloxyl radicals)

Physical Data: mp 205–206 °C[1] (white crystalline solid).

Analysis of Reagent Purity: ^{1}H-NMR ($CDCl_3$, TMS): $\delta = 8.28$ (2H, d), 7.58 (2H, d), 7.22 (4H, m). ^{13}C-NMR: $\delta = 148.5$, 138.7, 126.6 (d), 122.5 (d), 121.9 (d).

Preparative Methods: generally, 2,2′-dipyridyl disulfide-*N,N*′-dioxide is obtained by the oxidation of *N*-hydroxypyridine-2-thione (2-mercaptopyridine-1-oxide, thiohydroxamic acid). To a suspension of 42.3 g of *N*-hydroxypyridine-2-thione in water (300 mL) was added 30% hydrogen peroxide (35 mL). The reaction is slightly exothermic. The reaction mixture was stirred for 1 h with the temperature maintained at 45 °C. A white solid, 34.6 g (81%), was collected by filtration (mp 200–201 °C). Recrystallization from methanol gave white crystals (mp 205–206 °C).[1a,2] Other preparative methods include irradiation of *N*-hydroxypyridine-2-thione with a tungsten lamp (500 W) in benzene which affords a mixture of 2,2′-dipyridyl disulfide, 2,2′-dipyridyl disulfide-*N,N*′-dioxide, and its mono-*N*-oxide (eq 1).[3a] The mono-*N*-oxide is rather unstable and undergoes disproportionation to form dipyridyl disulfide and 2,2′-dipyridyl disulfide-*N,N*′-dioxide.[3] Photolysis (350 nm) of *N*-hydroxypyridine-2-thione with a mercury lamp provides 2,2′-dipyridyl disulfide (7%), 2,2′-dipyridyl disulfide-*N,N*′-dioxide (38%), and its mono-*N*-oxide (15%), together with 2-pyridinesulfonic acid (22%).[3]

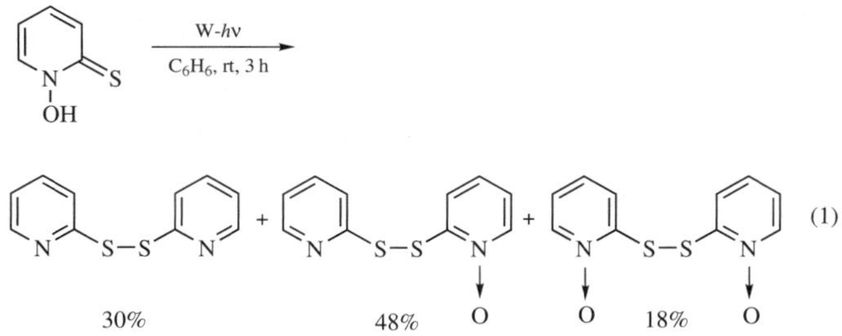

(1)

Handling, Storage, and Precautions: air-stable solid.

Preparation of Disulfides. Unsymmetrical alkyl and aryl pyridyl disulfide-*N*-oxides are prepared by the reaction of 2,2′-dipyridyl disulfide-*N,N*′-dioxide with various thiols in good yields (eq 2).[2] *N*-hydroxypyridine-2-thione is a by-product of this reaction.

Alkyl and aryl pyridyl disulfide-*N*-oxides are also effective sulfenylating agents. Thus, treatment under prolonged reaction time of 2,2′-dipyridyl disulfide-*N,N*′-dioxide with 2 equiv of thiol generates symmetrical disulfides and *N*-hydroxy-2-thiopyridone as shown in eq 3.[2]

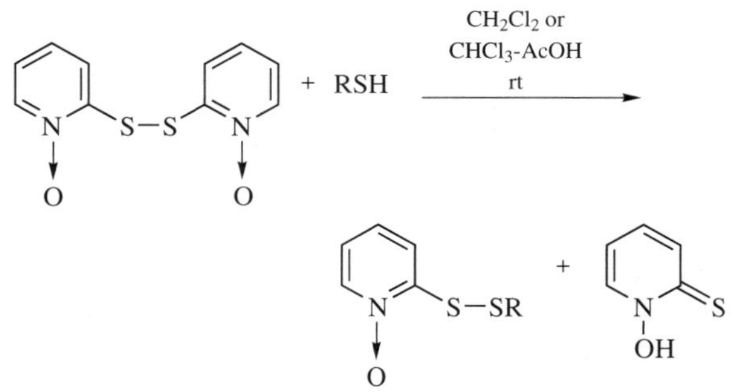

(2)

R	Time (h)	Yield (%)
n-C_3H_7	30	100
CH_2CO_2H	12	98
$CH_2CH(NHAc)CO_2H$	30	96

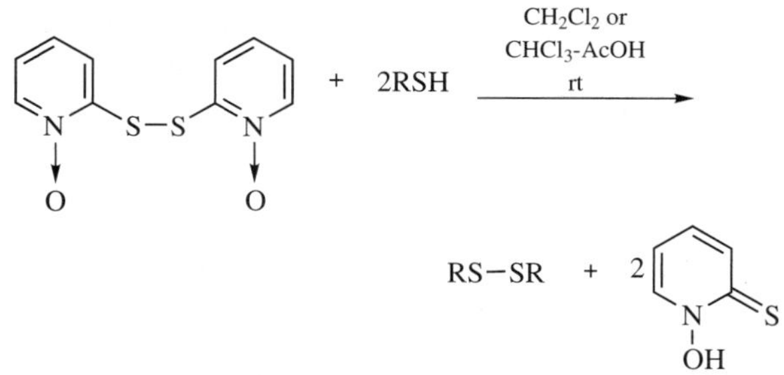

(3)

R	Time (h)	Yield (%)
CH_2CO_2H	18	100
C_6H_5	54	92

Preparation of O-Acyl Thiohydroxamates (Barton Esters). *O*-Acyl esters (Barton esters) of *N*-hydroxypyridine-2-thione are excellent precursors of carbon-centered radicals through the rapid

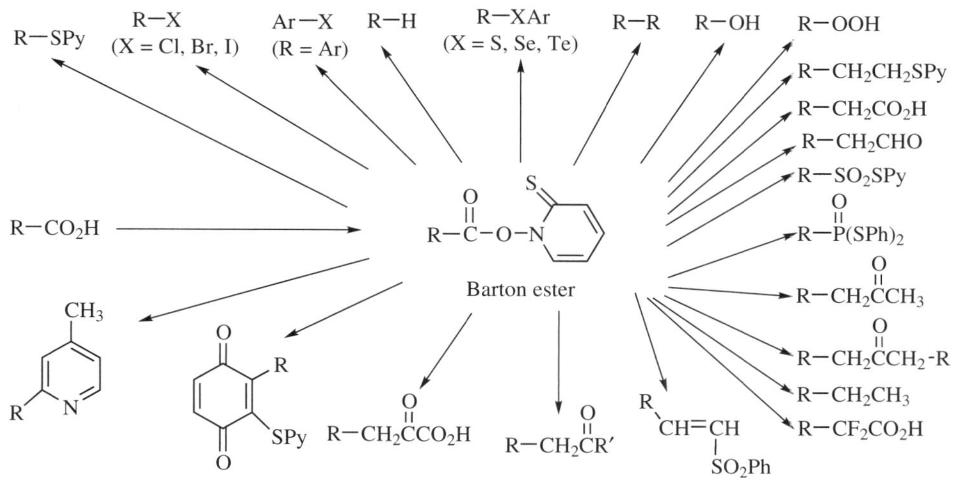

Figure 1 Functional group introduction and C-C bond formation using Barton esters

decarboxylation of the acyloxyl radicals formed, under mild reaction conditions (photolysis with a tungsten lamp at room temperature, or refluxing in benzene or toluene). Carbon-centered radicals formed in this manner can be used for various types of functional group conversions and C-C bond formations (Figure 1).[4]

N-Hydroxypyridine-2-thione derivatives such as carbamates and carbonates are also precursors of nitrogen- and oxygen-centered radicals. Generally, Barton esters are prepared from (a) acid chlorides with *N*-hydroxypyridine-2-thione in the presence of a base, (b) carboxylic acids and *N*-hydroxypyridine-2-thione with DCC, (c) carboxylic acids and *N*-hydroxypyridine-2-thione with isobutyl chloroformate via mixed anhydrides, (d) carboxylic acids with a salt prepared from sodium salt of *N*-hydroxypyridine-2-thione and phosgene, in the presence of a base. These methods are applicable to a wide variety of carboxylic acids including those derived from peptides and nucleosides. However, for very sensitive carboxylic acids, reductive condensation with 2,2′-dipyridyl disulfide-*N,N*′-dioxide and a phosphine is effective for the preparation of Barton esters. A particular advantage of this reagent combination is the polar nature of the reagents and by-products which enables the Barton ester to be rapidly eluted from a silica gel column, thereby minimizing decomposition. Accordingly, treatment of primary, secondary, and tertiary carboxylic acids with 2,2′-dipyridyl disulfide-*N,N*′-dioxide and tributylphosphine under dark conditions at room temperature provides the corresponding *O*-acyl esters of *N*-hydroxy-2-thiopyridone in high yields (eq 4).[5]

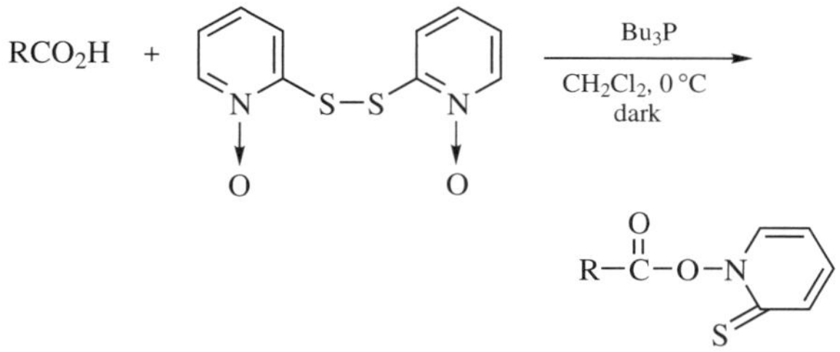

(4)

R	Yield (%)
2-phenylethyl	92
cyclohexyl	86
1-adamantyl	84

Examples of carboxylic acids containing sensitive functional groups that have been successfully converted to Barton esters in this manner are given in eqs 5–7.[6]

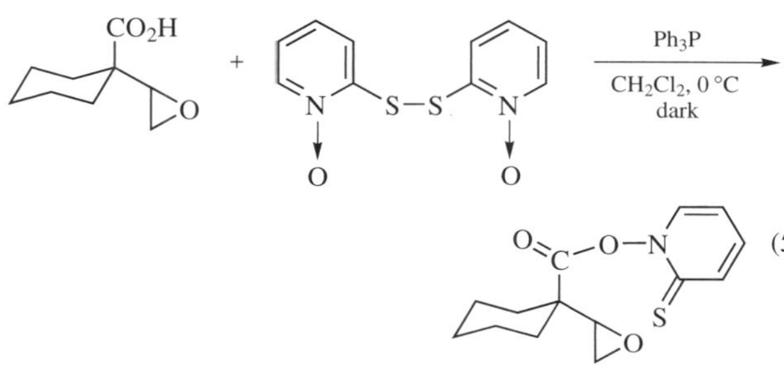

(5)

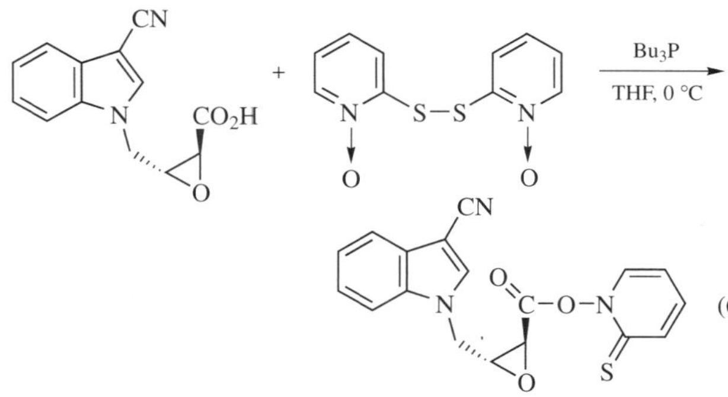

(6)

Barton esters prepared with 2,2′-dipyridyl disulfide-*N,N*′-dioxide and tributylphosphine have been used for the determination of the rate constants for the *5-exo-trig* ring-closure of *N*-allyl amide radicals and for the ring-opening of cyclopropylmethyl radicals (eq 8).[7]

The natural product tyromycin was prepared in a good yield by the photolytic treatment of the double Barton ester of a dicarboxylic acid with a tungsten lamp (500 W) in the presence of citranic anhydride (eq 9).[8]

In connection with studies on the mechanism of DNA degradation by anti-cancer agents such as bleomycin and enediyne antibiotics, model systems with Barton esters bearing a phosphatoxy bleomycin and enediyne antibiotics, as well as model systems

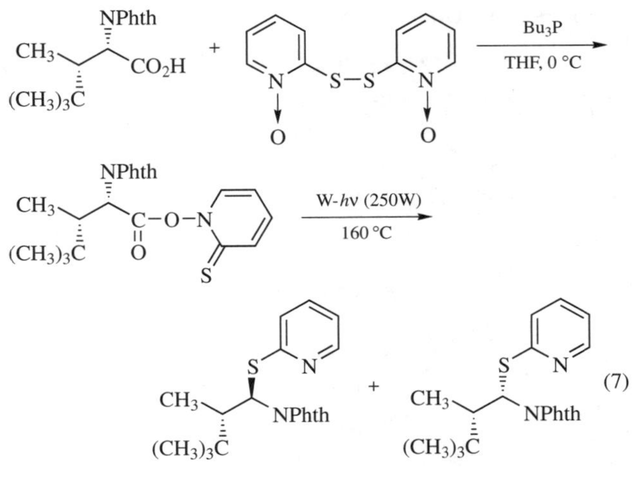

53% (5.5:1)

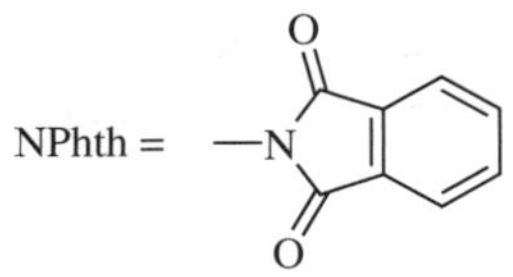

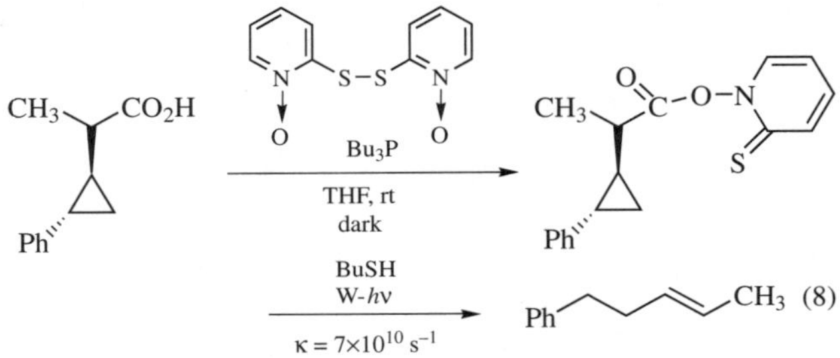

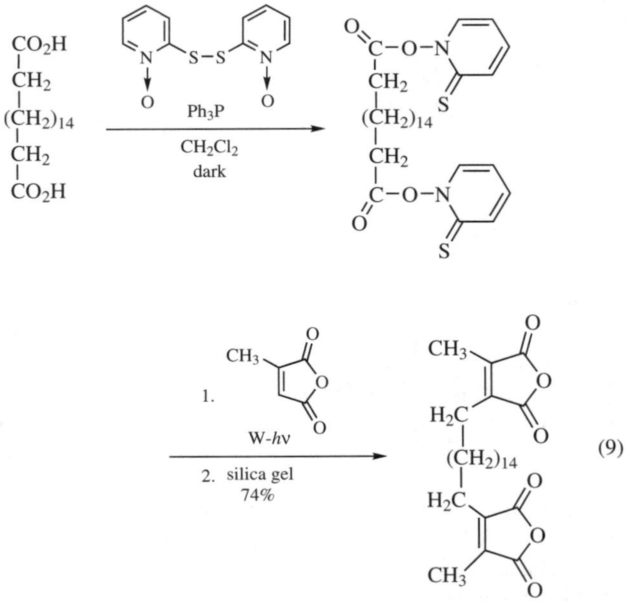

Tyromycin

with Barton esters bearing a phosphatoxy group and related functional groups at the β-position, have been investigated in detail.[9]

In this chemistry, the Barton esters were again prepared with 2,2′-dipyridyl disulfide-*N,N′*-dioxide and tributylphosphine (eq 10).

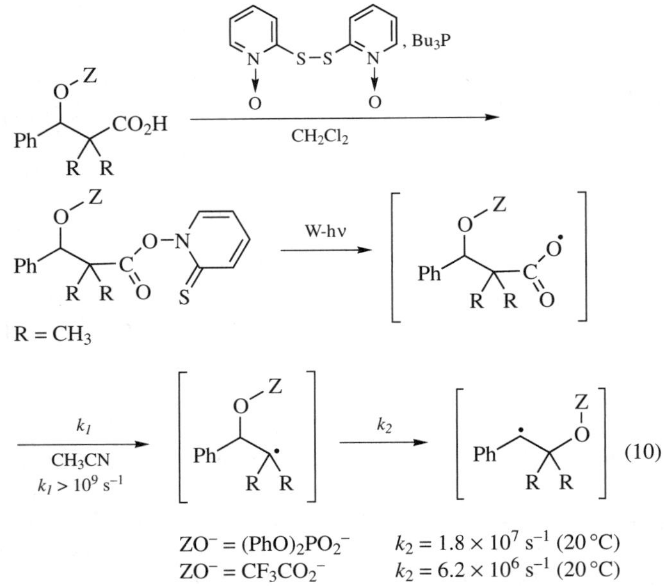

Finally, Barton esters formed in this manner react in situ with benzenesulfenamides at low temperatures to give amides with no racemization. The by-product of this reaction is an unsymmetrical pyridyl disulfide-*N*-oxide. This reaction is exemplified by the treatment of amino acids with 2,2′-dipyridyl disulfide-*N,N′*-dioxide and tributylphosphine, followed by the addition of a sulfenamide, which provides racemization-free peptides (eq 11).[10]

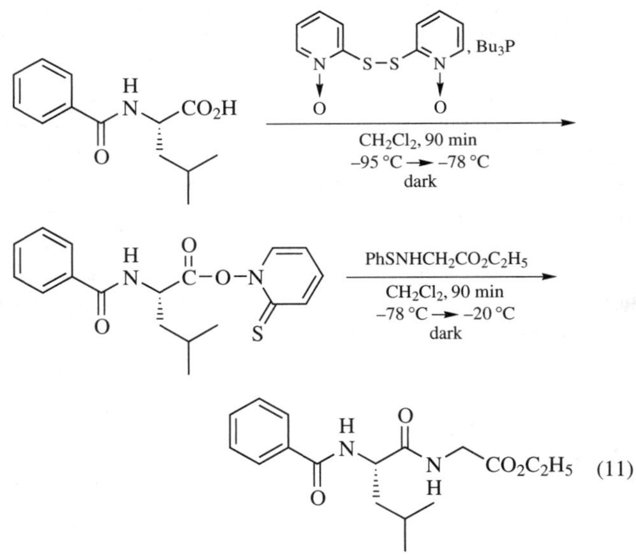

90% (96% ee)

Related Reagents. 2,2′-Dipyridyl disulfide (a reductive condensation reagent system with triphenylphosphine); *N*-hydroxypyridine-2-thione.

1. (a) U.S. Pat. 2,742,476 (to Bernstein, J.; Losee, K. A.) (1956) (*CA*. **1956**, *50*, 89530). (b) U.S. Pat. 3,700,676 (to Damico, R. A.) (1972) (*CA*. **1973**, *78*, 43278z).
2. Barton, D. H. R.; Chen, C.; Wall, G. M., *Tetrahedron* **1991**, *47*, 6127.
3. (a) Boivin, J.; Crépon, E.; Zard, S. Z., *Tetrahedron Lett.* **1990**, *31*, 6869. (b) Barton, D. H. R.; Jaszberenyi, J. C.; Morrell, A. I., *Tetrahedron Lett.* **1991**, *32*, 311. (c) Adam, W.; Grimm, G. N.; Marquardt, S.; Saha-Möller, C. R., *J. Am. Chem. Soc.* **1999**, *121*, 1179.
4. (a) Crich, D., *Aldrichimica Acta* **1987**, *20*, 35. (b) Barton, D. H. R., *Aldrichimica Acta* **1990**, *23*, 3.
5. Barton, D. H. R.; Samadi, M., *Tetrahedron* **1992**, *48*, 7083.
6. (a) Ziegler, F. E.; Harran, P. G., *Tetrahedron Lett.* **1993**, *34*, 4505. (b) Damm, W.; Hoffmann, U.; Macko, L.; Neuburger, M.; Zehnder, M.; Giese, B., *Tetrahedron* **1994**, *50*, 7029. (c) Krishnamurthy, V.; Rawal, V. H., *J. Org. Chem.* **1997**, *62*, 1572.
7. (a) Musa, O. M.; Horner, J. H.; Newcomb, M., *J. Org. Chem.* **1999**, *64*, 1022. (b) Choi, S.-Y.; Toy, P. H.; Newcomb, M., *J. Org. Chem.* **1998**, *63*, 8609.
8. Poigny, S.; Guyot, M.; Samadi, M., *J. Org. Chem.* **1998**, *63*, 1342.
9. (a) Choi, S.-Y.; Crich, D.; Horner, J. H.; Huang, X.; Newcomb, M.; Whitted, P. O., *Tetrahedron* **1999**, *55*, 3317. (b) Newcomb, M.; Horner, J. H.; Whitted, P. O.; Crich, D.; Huang, X.; Yao, Q.; Zipse, H., *J. Am. Chem. Soc.* **1999**, *121*, 10685. (c) Newcomb, M.; Miranda, N.; Huang, X.; Crich, D., *J. Am. Chem. Soc.* **2000**, *122*, 6128. (d) Crich, D.; Huang, X.; Newcomb, M., *J. Org. Chem.* **2000**, *65*, 523. (e) Bales, B. C.; Horner, J. H.; Huang, X.; Newcomb, M.; Crich, D.; Greenberg, M. M., *J. Am. Chem. Soc.* **2001**, *123*, 3623. (f) Horner, J. H.; Taxil, E.; Newcomb, M., *J. Am. Chem. Soc.* **2002**, *124*, 5402.
10. Barton, D. H. R.; Ferreira, J. A., *Tetrahedron* **1996**, *52*, 9367.

Hideo Togo
Chiba University, Chiba, Japan

E

N-Ethoxycarbonyl Phthalimide

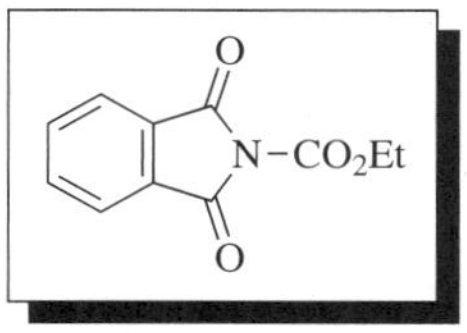

[22509-74-6] $C_{11}H_9NO_4$ (MW 219.20)

(phthaloylating agent selective for amines;[1] selective protection of amino acids[2] and amino alcohols[3])

Physical Data: mp 90–92 °C.
Solubility: sol DMF, benzene, THF, ether, chloroform; partially sol water, ethanol, isopropyl alcohol; insol petroleum ether.
Form Supplied in: crystalline (colorless).
Analysis of Reagent Purity: commercial material is >95% pure.
Preparative Methods: addition of Ethyl Chloroformate to a cold solution of Phthalimide and Triethylamine in DMF;[4] alkylation of potassium phthalimide with ethyl chloroformate.[5]
Purification: commercial product can be purified by recrystallization from ethanol, although some loss of material occurs due to ring opening. Synthetic material should be purified by the method of McArthur.[4] A 2 M solution of the crude material in chloroform is cooled overnight in a refrigerator and filtered to remove phthalimide. The solution is then concentrated to approximately 70% of the volume and an equal volume of petroleum ether is added. The crystals that form are collected and dried to give the desired *N*-ethoxycarbonylphthalimide.
Handling, Storage, and Precautions: no special handling conditions or precautions are required.

Phthaloylation of Amines. The phthalimide protecting group has long been utilized for amines in organic synthesis.[6] While this group can be introduced under a variety of conditions such as phthalic anhydride or Phthaloyl Dichloride, these generally require drastic conditions and are not typically selective. The use of *N*-ethoxycarbonylphthalimide allows the phthaloylation of primary amines under mild conditions (eq 1).[7] Primary amines can be

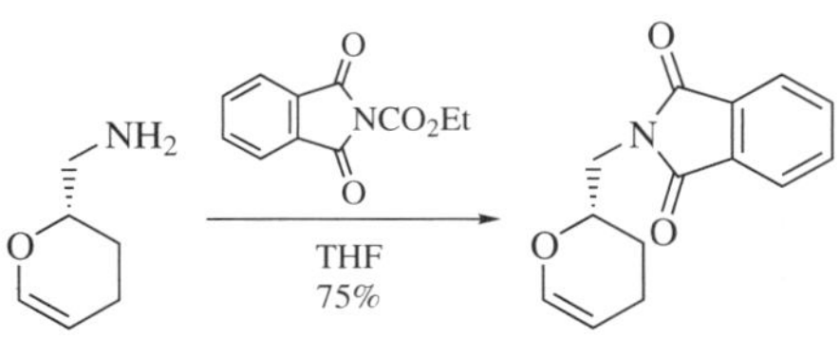

(1)

selectively protected in the presence of secondary amines (eq 2),[8] as well as amides and carbamates (eq 3).[9] The advantage of the phthalimido protecting group are its stability to both acid and mild base as well as the well known methods for selective removal with Hydrazine (eq 4)[10] or Phenylhydrazine (eq 5).[11]

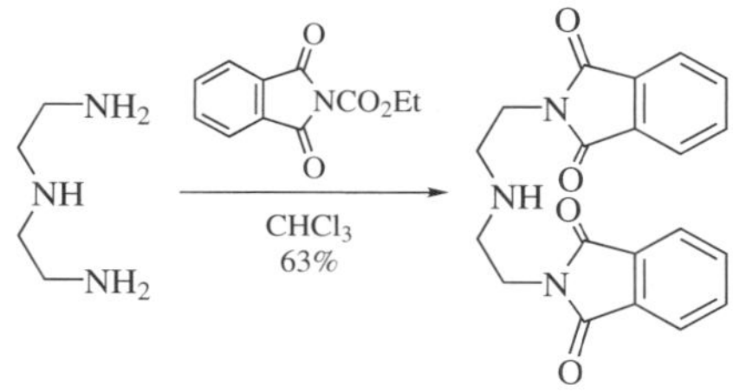

(2)

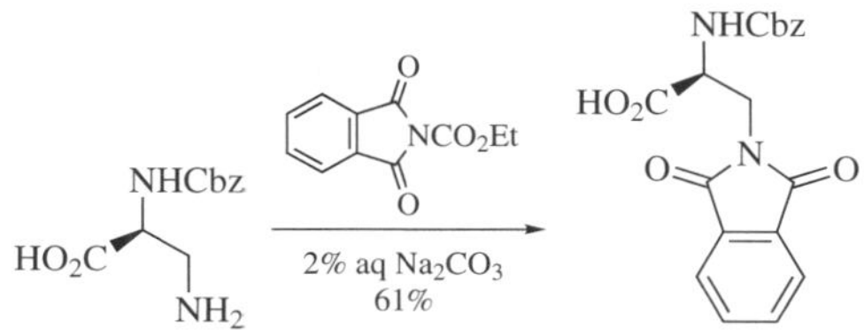

(3)

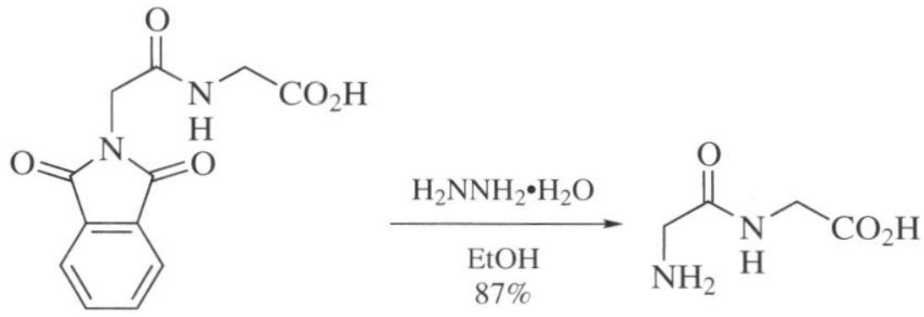

(4)

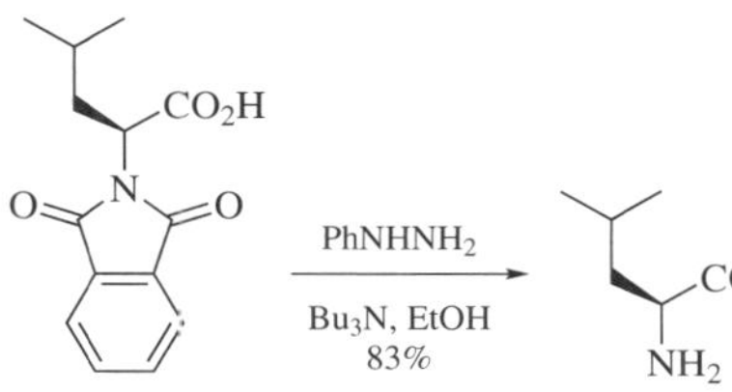

(5)

Selective Phthaloylation of Amino Acids. In his attempt to develop a method for the selective amine protection of amino acids, Nefkens introduced *N*-ethoxycarbonylphthalimide.[5] He reported the protection of a variety of amino acids without racemization (eq 6).[2] These phthaloyl derivatives are typically highly crystalline, thereby facilitating purification. The phthaloylation of hindered amino acids such as valine could be achieved by initial deprotonation followed by protection in ethyl acetate (eq 7).[12] Recently, McArthur has shown that the aqueous conditions initially developed by Nefkens lead to the formation of undesired side products.[4] Heating *N*-ethoxycarbonylphthalimide in ethanol generates the corresponding carbamate (eq 8).[4] He has developed an alternative method for the phthaloylation of amino acids in THF that occurs in exceptional yields without racemization (eq 9).[1] These protections can be performed on amino acids with a variety of sensitive functionalities such as hydroxyl groups (eq 10)[2] or halogens without reaction at these centers (eq 11).[13]

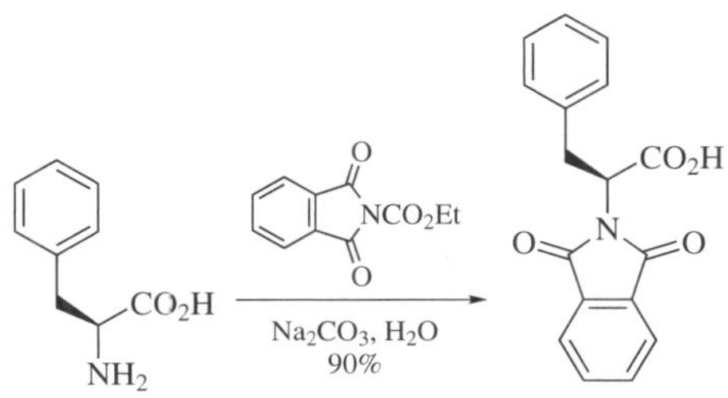

(6)

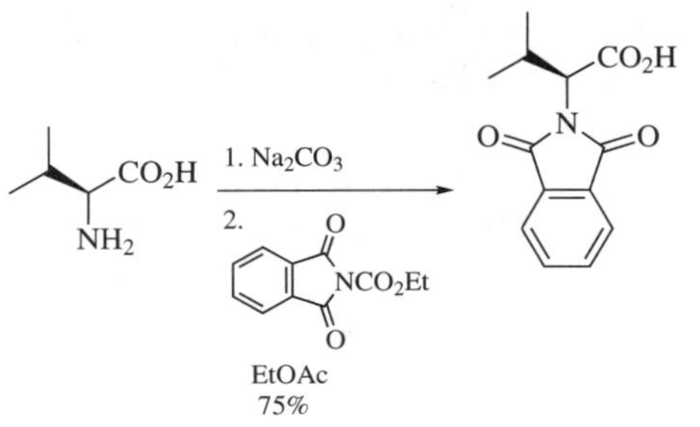

(7)

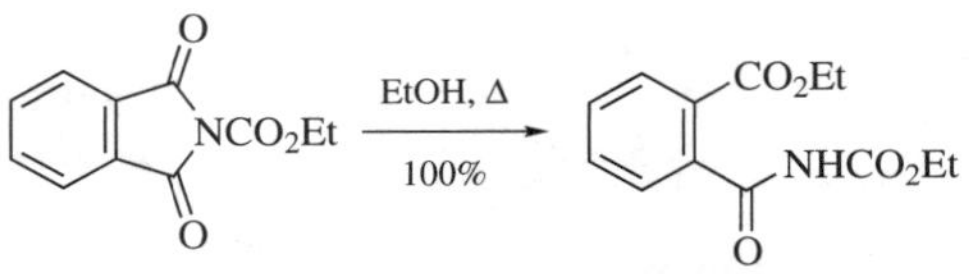

(8)

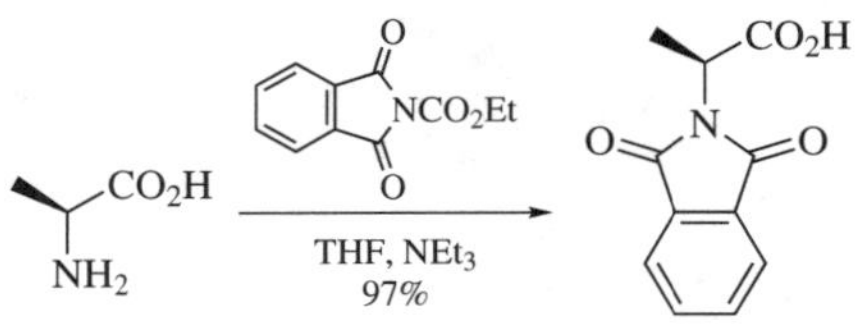

(9)

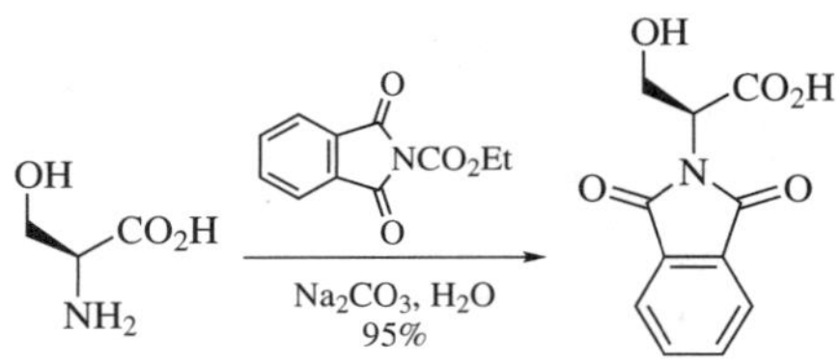

(10)

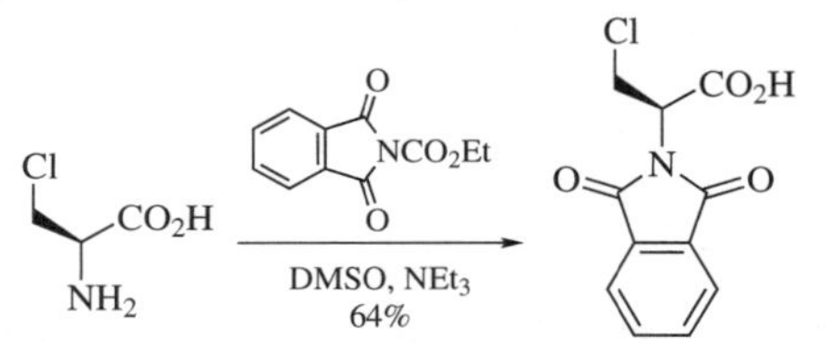

(11)

Phthaloylation of Amino Alcohols. The use of *N*-ethoxycarbonylphthalimide to protect amino alcohols has also been demonstrated (eq 12).[3]

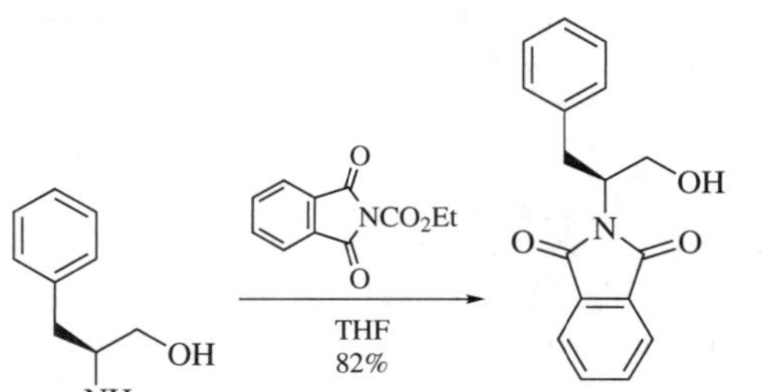

(12)

---

1. McArthur, C. R.; Worster, P. M.; Okon, A. U. *Synth. Commun.* **1983**, *13*, 393.
2. Nefkens, G. H. L.; Tesser, G. I.; Nivard, R. J. F. *Recl. Trav. Chim. Pays-Bas* **1960**, *79*, 688.
3. McArthur, C. R.; Worster, P. M.; Jiang, J.-L.; Leznoff, C. C. *Can. J. Chem.* **1982**, *60*, 1836.
4. Worster, P. M.; Leznoff, C. C.; McArthur, C. R. *J. Org. Chem.* **1980**, *45*, 174.
5. Nefkens, G. H. L. *Nature (London)* **1960**, *185*, 309.
6. Greene, T. W.; Wuts, P. G. M. *Protective Groups in Organic Synthesis*; Wiley:New York, **1991**; pp 358–362.
7. Krstenansky, J. L.; delRosario-Chow, M.; Currie, B. L. *J. Heterocycl. Chem.* **1992**, *29*, 707.
8. Sosnovsky, G.; Lukszo, J. *Z. Naturforsch., Tell B* **1986**, *41b*, 122.
9. Funabashi, Y.; Tsubotani, S.; Koyama, K.; Katayama, N.; Harada, S. *Tetrahedron* **1993**, *49*, 13.
10. Greenstein, J. P.; Winitz, M. *Chemistry of the Amino Acids*; Wiley:New York, **1961**; Vol. 2, p.1249.
11. Schumann, I.; Boissonnas, R. A. *Helv. Chim. Acta* **1952**, *35*, 2235.
12. Adriaens, P.; Meesschaert, B.; Janssen, G.; Dumon, L.; Eyssen, H. *Recl. Trav. Chim. Pays-Bas* **1978**, *97*, 260.
13. Clarke, S.; Hider, R. C.; John, D. I. *J. Chem. Soc., Perkin Trans. 1* **1973**, 230.

James W. Leahy
University of California, Berkeley, CA, USA

2-Ethoxy-1-ethoxycarbonyl-1,2-dihydroquinoline

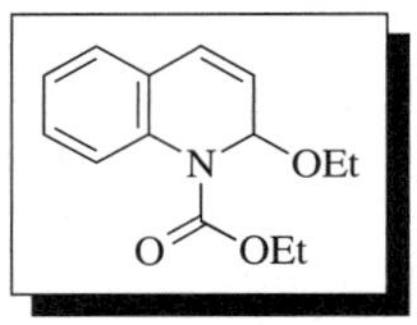

[16357-59-8] $C_{14}H_{17}NO_3$ (MW 247.29)

(activating agent used in peptide synthesis)

Alternate Names: ethyl 1,2-dihydro-2-ethoxy-1-quinolinecarboxylate; EEDQ.
Physical Data: mp 66–67 °C; bp 125–128 °C/0.1 mmHg.
Solubility: sol water, alcohol.
Form Supplied in: white powder; widely available and can be easily prepared.[1]

Peptide Synthesis. In peptide synthesis, peptide bonds are created from the activated carboxylic group of a N^{α}-protected amino acid that is reacted with the nucleophilic free amino group of another amino acid (eq 1).

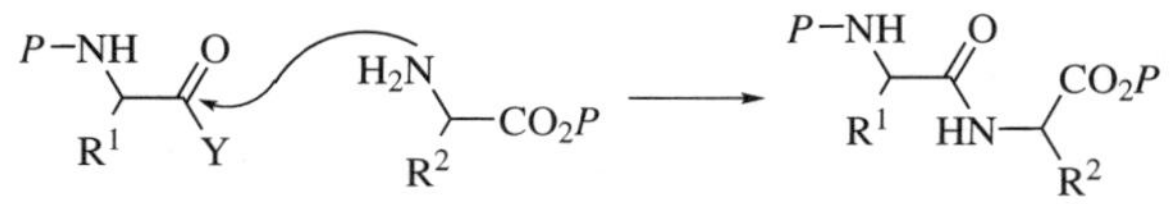

(1)

Activation of the carboxylic acid can be performed according to several strategies, leading mainly to two kinds of intermediates: active esters and acid anhydrides, which can be either symmetrical anhydrides or mixed anhydrides. EEDQ is a reagent used in peptide synthesis for the mixed carbonic anhydride strategy. It allows a slow formation of unsymmetrical anhydrides under very mild conditions (eq 2).[2]

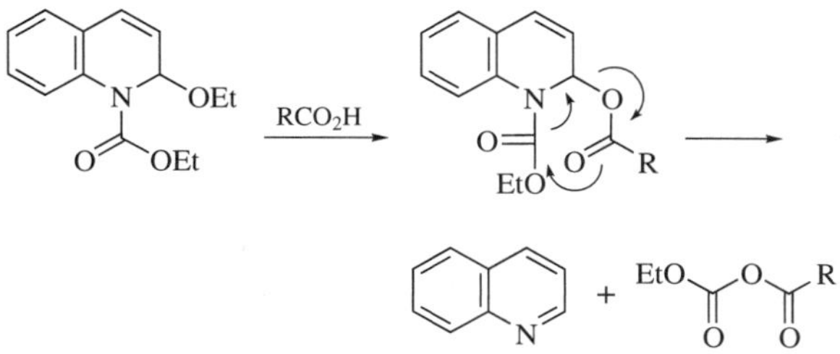

(2)

Faster reaction of the mixed anhydride with nucleophilic species reduces the extent of side reactions that occur with high concentration of the anhydride (eq 3).[3]

$$\text{Boc–Cys(Bn)–OH} + \text{H–Tyr(Bn)–Gln–Asn–Cys(Bn)–OBn} \xrightarrow[\text{NEt}_3]{\text{EEDQ}} \text{Boc–Cys(Bn)–Tyr(Bn)–Gln–Asn–Cys(Bn)–OBn} \quad (3)$$

The tendency to racemization is limited to a small extent. Nevertheless, wrong-side attack, leading to the formation of urethane, is not suppressed (eq 4).[4]

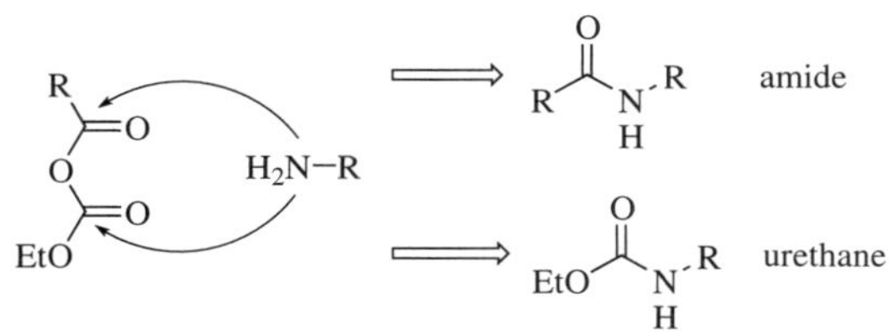

(4)

As best results had been obtained with mixed anhydrides of Isobutyl Chloroformate, probably because the isobutyl group decreases the amount of urethane products, IIDQ (**1**), an analog of EEDQ, has been proposed for use in peptide synthesis.[5,6]

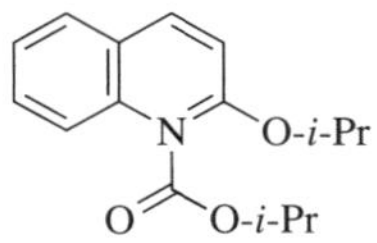

(**1**)

As another alternative, EEDQ has been linked to a polymeric support, providing a supported coupling reagent in high yield (eq 5).[7]

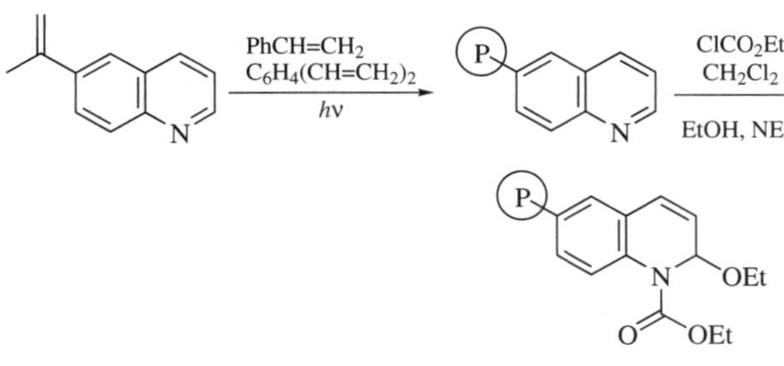

(5)

Coupling reactions mediated by this reagent occur cleanly in high yield and with a low degree of racemization, comparable with that observed with EEDQ itself.[8] Nevertheless, the use of EEDQ in peptide synthesis is rather limited, other coupling reagents being very much more popular. Examples are: the carbodiimides, mainly 1,3-Dicyclohexylcarbodiimide and diisopropylcarbodiimide; the family of BOP reagents such as Benzotriazol-1-yloxytris (dimethylamino)phosphonium Hexafluorophosphate; uroniums such as O-Benzotriazol-1-yl-N,N,N′, N′-tetramethyluronium Hexafluorophosphate (HBTU), or urethane *N*-protected carbonic anhydrides (UNCA), and active esters.

1. Fieser, M.; Fieser, L. F. *Fieser & Fieser* **1969**, *2*, 191.
2. Belleau, B.; Malek, G *J. Am. Chem. Soc.* **1968**, *90*, 1651.
3. Mühlemann, M.; Titov, M., I.; Scweyzer, R.; Rudinger, J. *Helv. Chim. Acta* **1972**, *55*, 2854.
4. Lombardino, J. G.; Anderson, S. L.; Norris, C. P. *J. Heterocycl. Chem.* **1978**, *15*, 655.
5. Kiso, Y.; Kai, Y.; Yajima, H. *Chem. Pharm. Bull.* **1973**, *90*, 1651.
6. Bodanszky, M.; Bodanszky, A. *The Practice of Peptide Synthesis*; Springer:Berlin, **1984**; p 148.
7. Brown, J.; Williams, R. E. *Can. J. Chem.* **1971**, *49*, 3765.
8. Lauren, D. R.; Williams, R. E. *Tetrahedron Lett.* **1972**, 2665.

Jean-Claude Gesquière
Institut Pasteur, Lille, France

N-Ethylbenzisoxazolium Tetrafluoroborate

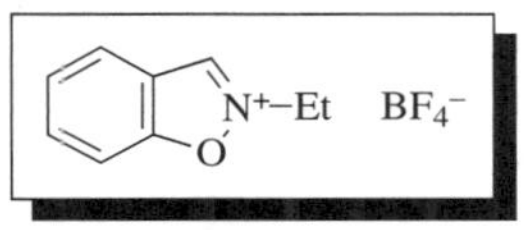

[4611-62-5] $C_9H_{10}BF_4NO$ (MW 234.99)

(peptide-coupling reagent;[1] reacts with a wide range of nucleophiles to give *o*-hydroxy-*N*-ethylbenzamido derivatives[2])

Physical Data: mp 109.5–110.2 °C.
Solubility: sol acetonitrile; slightly sol dichloromethane; sol aqueous acid (unstable above pH 4).
Form Supplied in: crystalline solid; commercially available.
Preparative Methods: ethylation of benzisoxazole with purified Triethyloxonium Tetrafluoroborate.[2]
Handling, Storage, and Precautions: the salt is reported not to be hygroscopic, although it etches glass on long exposure to moist air; should be protected from light; reported to be an irritant.

Peptide Coupling. *N*-Ethylbenzisoxazolium tetrafluoroborate (EBF) is used to couple amines and carboxylic acids in a two-stage procedure. Firstly, EBF reacts with the carboxylic acid to form an active phenolic ester. Secondly, the active ester reacts with the amine to form the peptide linkage (eq 1).[3] The procedure for the formation of the active ester employs a carefully buffered aqueous solution of the Cbz-protected aminoacid (or oligopeptide), to which an equal volume of ethyl acetate or dichloromethane is added, followed by 1.1 equiv of finely powdered EBF. The acylsalicylamide ester is isolated from the organic solvent and may be recrystallized. This product is stable to storage. This procedure does not suffer when used with the sterically demanding valine or proline residues. The active ester cleanly reacts with peptide amines in dipolar aprotic solvents, yielding the amide and *N*-ethylsalicylamide, which can be removed with alkali or carbon tetrachloride. The second step is, however, very slow, taking ca. 48 h at room temperature (roughly 20 times slower than the rate of *p*-nitrophenyl ester couplings[4]), and suffers from unwanted side-reactions (Brenner rearrangements[5]) in the presence of strongly basic amines such as triethylamine. More seriously, significant amounts (up to 1–2%) of epimerization may occur.[6]

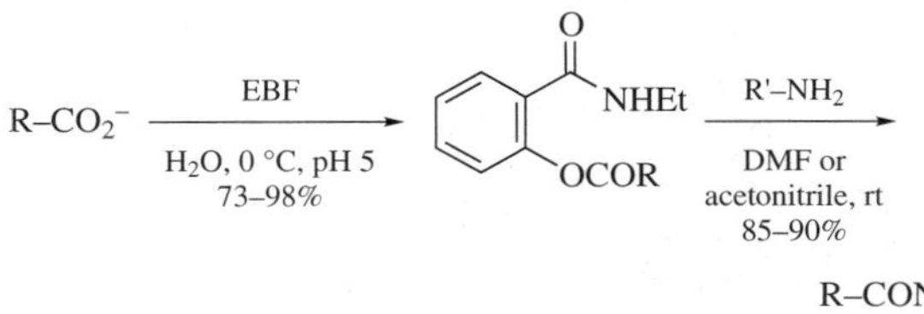

(1)

EBF has been used to prepare peptide cyclodimers in yields which compare favorably with other procedures.[1] For this method, intermediate Cbz-protected acylsalicylamide esters were hydrogenolyzed to form the free amines which then yielded the cyclodimers (eqs 2 and 3).

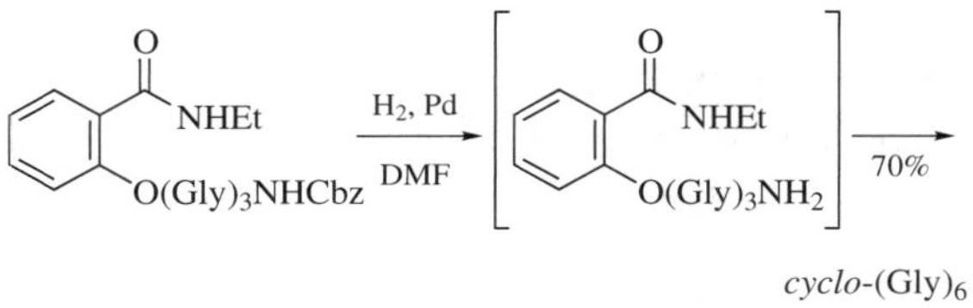

(2)

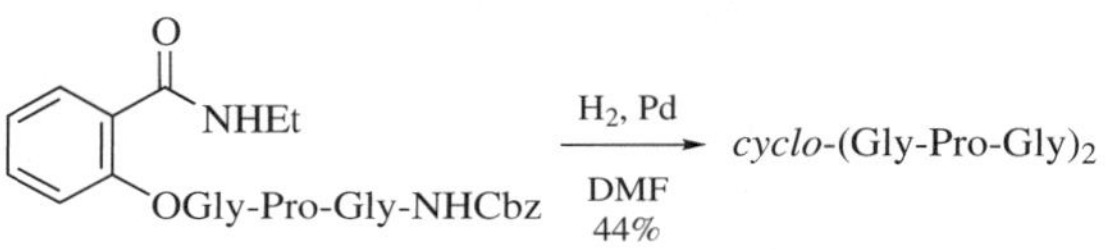

(3)

Peptide cyclotrimers have also been prepared in this manner.[7,8] After deblocking at nitrogen, a highly dilute solution of the resulting free amine in dry pyridine was stirred for 65 h at room temperature to give the cyclotrimer in an impressive 90% yield (eq 4).

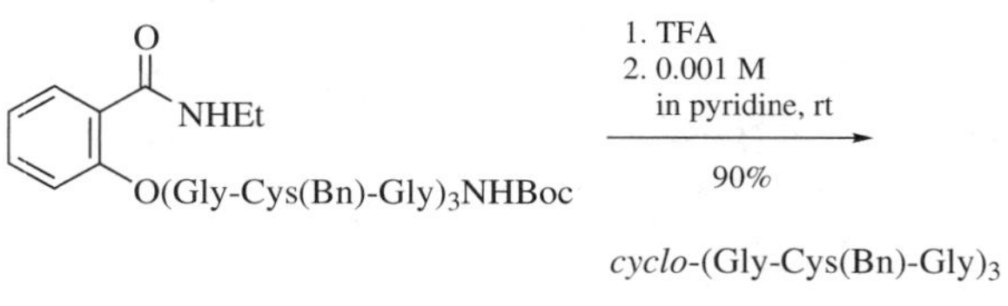

(4)

Detailed mechanistic studies indicate that the *N*-ethylbenzisoxazolium cation readily undergoes base-catalyzed elimination to form a transitory *N*-ethylbenzoketoketenimine (eq 5).[9]

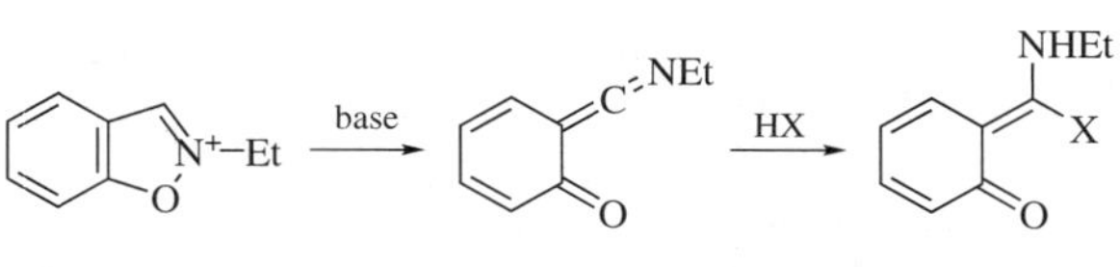

(5)

In the case of reactions with carboxylic acids ($X = RCO_2$), the initial addition product then rapidly rearranges to the active ester (eq 6).

(6)

Several hydroxy-,[10] nitro-,[3] and chloro-substituted[11] *N*-ethylbenzisoxazolium cations have been studied in attempts to vary the reactivity of the intermediate active ester (see 2-Ethyl-7-hydroxybenzisoxazolium Tetrafluoroborate).

Reactions with Nucleophiles. The *N*-ethylbenzisoxazolium cation reacts rapidly in aqueous solution with a range of nucleophiles, yielding *o*-hydroxy-*N*-ethylbenzimido derivatives. Hydrolysis proceeds smoothly in mild acid, but at pH above 7, polymeric material is obtained (eq 7).[2]

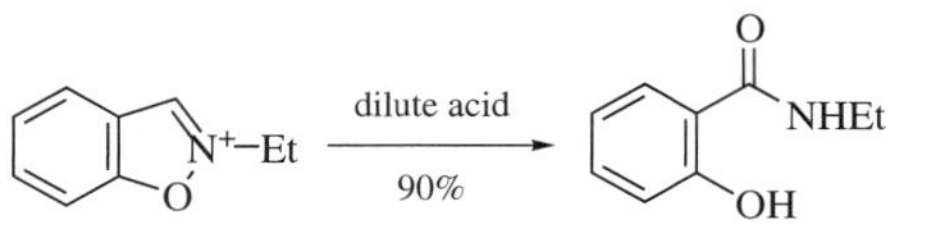

(7)

Treatment of EBF with aqueous solutions of simple nucleophiles (HS^-, F^-, CN^-) or with methanolic methoxide gives high yields of the benzimido derivatives (eq 8).

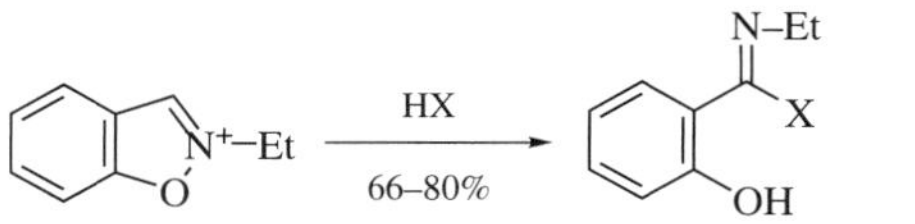

(8)

Reaction of EBF with cyanate, thiocyanate, and thiourea generates initial adducts which can undergo internal rearrangements. For example, the adduct of cyanate to EBF yields a cyclic urethane (eq 9); reaction with thiocyanate yields the analogous thiourethane.[2]

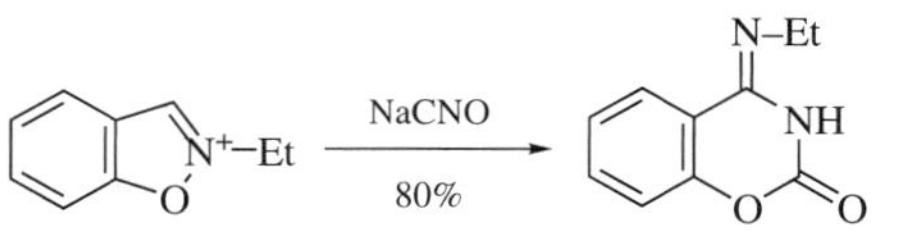

(9)

1. Rajappa, S.; Akerkar, A. S. *Tetrahedron Lett.* **1966**, *25*, 2893.
2. Kemp, D. S.; Woodward, R. B. *Tetrahedron* **1965**, *21*, 3019.
3. Kemp, D. S.; Wang, S.-W.; Mollan, R. C.; Hsia, S.-L.; Confalone, P. N. *Tetrahedron* **1974**, *30*, 3677.
4. Kemp, D. S.; Wang, S.-W. *J. Am. Chem. Soc.* **1967**, *89*, 2745.
5. Kemp, D. S.; Duclos, J. M.; Bernstein, Z.; Welch, W. M. *J. Org. Chem.* **1971**, *36*, 157.
6. Kemp, D. S.; Wang, S.-W.; Busby, G., III; Hugel, G. *J. Am. Chem. Soc.* **1970**, *92*, 1043.
7. Kemp, D. S.; McNamara, P. *Tetrahedron Lett.* **1981**, *22*, 4571.
8. Kemp, D. S.; McNamara, P. *J. Org. Chem.* **1985**, *50*, 5834.
9. Kemp, D. S. *Tetrahedron* **1967**, *23*, 2001.
10. Kemp, D. S.; Chien, S.-W. *J. Am. Chem. Soc.* **1967**, *89*, 2743.
11. Rajappa, S.; Akerkar, A. S. *Chem. Commun./J. Chem. Soc., Chem. Commun.* **1966**, 826.

Ross P. McGeary
University of Cambridge, UK

1-Ethyl-3-(3′-dimethylaminopropyl) carbodiimide hydrochloride[1]

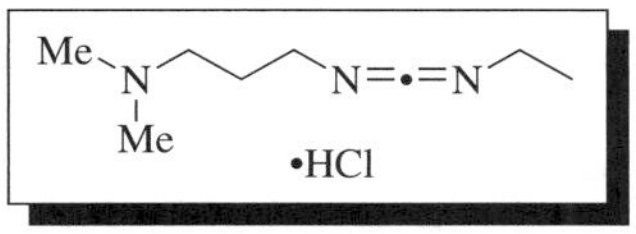

[25952-53-8] $C_8H_{18}ClN_3$ (MW 191.74)
(base)
[1892-57-5]
(.MeI)
[22572-40-3]

(peptide coupling reagent;[1b,2] amide formation;[3] ester formation;[4] protein modification;[5] mild oxidations of primary alcohols[6])

Alternate Name: 1-(3-dimethylaminopropyl)-3-ethylcarbodiimide; water-soluble carbodiimide; EDC; EDCI.
Physical Data: free base is an oil, bp 47–48 °C/0.27 mmHg; HCl salt is a white powder, mp 111–113 °C; MeI salt mp 97–99 °C.
Solubility: sol H_2O, CH_2Cl_2, DMF, THF.
Form Supplied in: commercially available as HCl salt and as methiodide salt that are white solids. Reagents are >98% pure; main impurity is the urea that can form upon exposure to moisture.
Analysis of Reagent Purity: IR: 2150 cm^{-1} (N=C=N stretch); urea has C=O stretch near 1600–1700 cm^{-1}.
Purification: recrystallization from CH_2Cl_2/ether.
Handling, Storage, and Precautions: EDC is moisture-sensitive; store under N_2 in a cool dry place. It is incompatible with strong oxidizers and strong acids. EDC is a skin irritant and a contact allergen; therefore avoid exposure to skin and eyes.

Peptide Coupling Reagent. This carbodiimide (EDC) reacts very similarly to 1,3-Dicyclohexylcarbodiimide and other carbodiimides. The advantage EDC has over DCC is that the urea produced is water soluble and, therefore, is easily extracted. Dicyclohexylurea is only sparingly soluble in many solvents and is removed by filtration which may not be as effective as extraction. A typical example of EDC for peptide coupling is shown in (eq 1).[2]

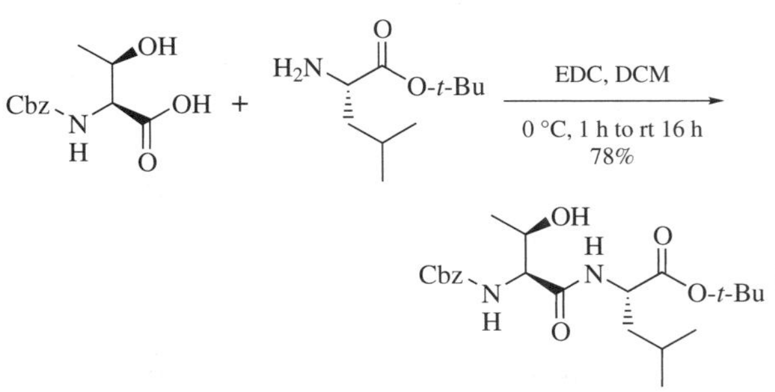

(1)

The problems associated with carbodiimide couplings are mainly from the *O*-acylisourea intermediate (**1**) having poor selectivity for specific nucleophiles. This intermediate can rearrange to an *N*-acylurea (**2**), resulting in contamination of the product and low yields. Additionally it can rearrange to 5(4*H*)-oxazolones (**3**) that tautomerize readily, resulting in racemization. Using low dielectric solvents like CH_2Cl_2 or additives to trap (**1**) minimizes these side reactions by favoring intermolecular nucleophilic attack on (**1**).[1b]

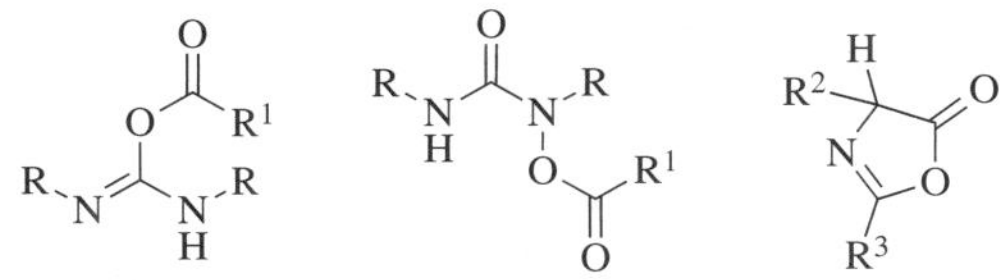

(1) **(2)** **(3)**

R = carbodiimide chain; R^1 = amino acid; R^2 = amino acid side chain; R^3 = NH protecting group

When low dielectric constant solvents are used, the carboxylic acid tends to dimerize, thus promoting symmetrical anhydride formation. This intermediate is stable enough to give good yields of the desired product. High dielectric solvents such as DMF retard acylation of amino acids, and *N*-acylurea can be a major byproduct.[1b] Unfortunately, many starting materials require polar solvents for dissolution.

Addition of trapping agents such as N-Hydroxysuccinimide[7] and 1-Hydroxybenzotriazole (HOBt)[8] reduce the extent of many side reactions, especially *N*-acylurea formation. Also, racemization is suppressed when these additives are present. The latter reagent eliminates the intramolecular dehydration of ω-amides of asparagine and glutamine that occurs with carbodiimides (eq 2).[8,9]

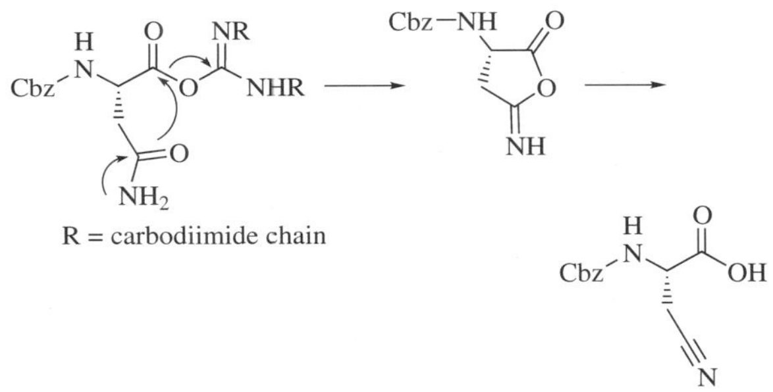

(2)

With EDC the addition of Copper(II) Chloride suppresses racemization to <0.1% compared to 0.4% with HOBt under ideal conditions.[10] Combinations of HOBt and $CuCl_2$ are also useful.[10b] The suggested stoichiometry for the highest optical purity and yield is 2 equiv of HOBt and 0.25–0.5 equiv of $CuCl_2$ in DMF.

A practical example of EDC's utility is the solution synthesis of human epidermal growth factor, a 53-residue protein.[11] This included several couplings of fragments ranging from three to six residues in length. All couplings were performed with EDC/HOBt in DMF or NMP. At the completion of the synthesis, no racemized material could be detected by HPLC.

Amide Formation. Formylated amino acids and peptides are prepared in high yields by forming the acid anhydride (eq 3).[3] These products are pure without requiring chromatography or recrystallization.

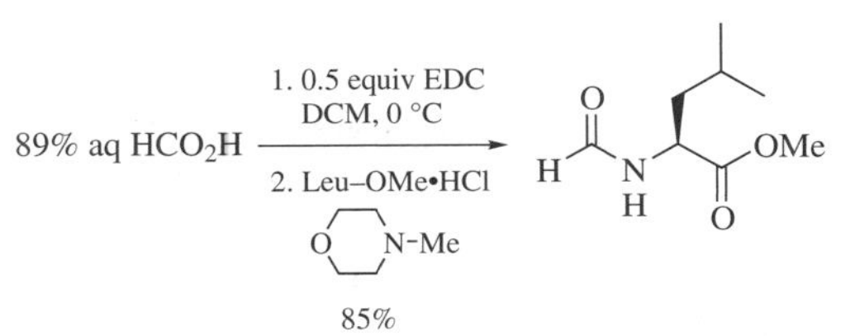

(3)

Another method for formylation of amines is with *p*-nitrophenyl formate, which usually gives products in high yield. However, removing the last traces of the *p*-nitrophenol is difficult.[12]

Treating carbon dioxide and amines with EDC gives symmetrical ureas (eq 4).[13] DCC with CO_2 at ambient pressure works equally well.

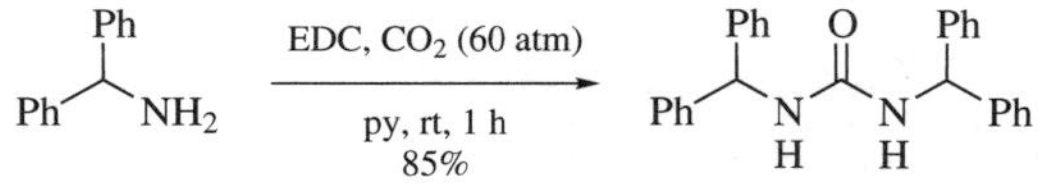

(4)

EDC facilitates the synthesis of xanthine analogs by condensing a diamine with water-soluble acids (eq 5).[14] The use of water as the solvent precludes the use of DCC in this case.

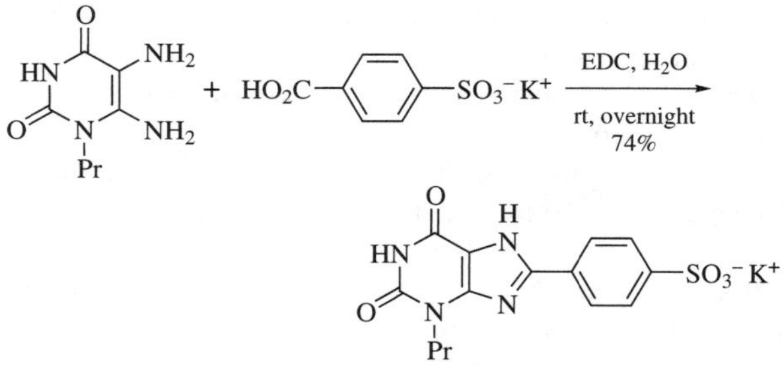

(5)

Ester Formation. Esters of *N*-protected amino acids are prepared in high yield with EDC and 4-Dimethylaminopyridine (eq 6).[4] DMAP causes extensive racemization if not used in a catalytic amount.[15] However, when esterifying the α-carboxyl of β- and γ-benzyl esters of aspartyl and glutamyl derivatives, extensive racemization was observed even with DMAP present in catalytic amounts. It is postulated that the side-chain esters contribute in some fashion to the lability of the α-H.

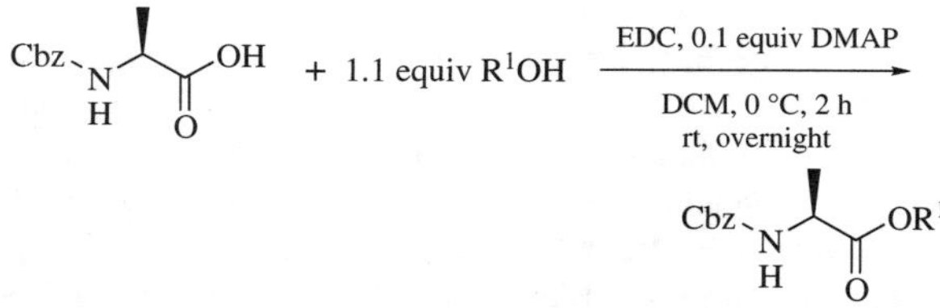

(6)

R^1 = *t*-Bu, 76%; Me, 96%; CH_2CCl_3, 87%

Carbodiimides including EDC are also useful in preparing active esters such as the *p*-nitrophenyl, pentafluorophenyl, and *N*-hydroxysuccinimide esters.[1b] Numerous additional methods have been reported.[16]

Protein Modification. Carboxyl groups in proteins react with EDC, resulting in an activated group that can be trapped with a nucleophile such as glycine methyl ester.[5] The nucleophile can also be amino groups in the protein, causing cross-linking. Applications for carboxylate modification include determining which groups are buried versus exposed, and mechanistic studies for determining which carboxyl groups are essential for an enzyme's activity.

EDC is used extensively in cross-linking proteins to solid supports for affinity chromatography.[17]

Oxidation of Primary Alcohols. One example of EDC substituting for DCC in the Pfitzner–Moffatt oxidation is shown in (eq 7).[6] The use of DCC in this example resulted in a difficult purification of the product and lower yields.

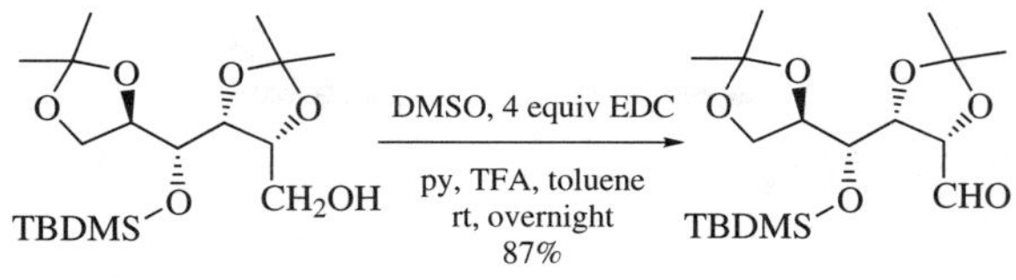

(7)

Miscellaneous Reactions. EDC is a useful water scavenger as well. An example of this utility is given in (eq 8).[18] The carbodiimide is required in this reaction, but DCC does not work well in this transformation.

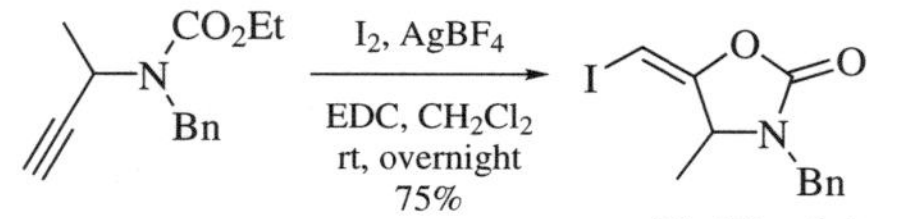

(*E*):(*Z*) > 9:1

(8)

Another example of EDC having a marked improvement over DCC is the formation of cyanoguanidines (eq 9).[19] This was originally attempted with DCC, which gives poor yields even after extended reaction times. It is thought the positively charged nitrogen in EDC facilitates the C–S bond cleavage and since DCC lacks this atom the reaction goes slowly.

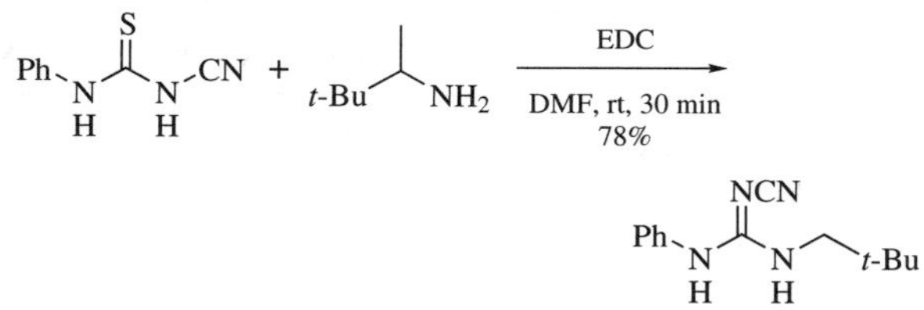

(9)

Pyrroles are formed when acetylenes undergo cycloadditions with *N*-acylamino acids in the presence of EDC.[20] An *N*-alkenylmunchnone azomethine ylide is generated in situ and is trapped by the dipolaraphile. Loss of CO_2 yields the pyrrole (eq 10).

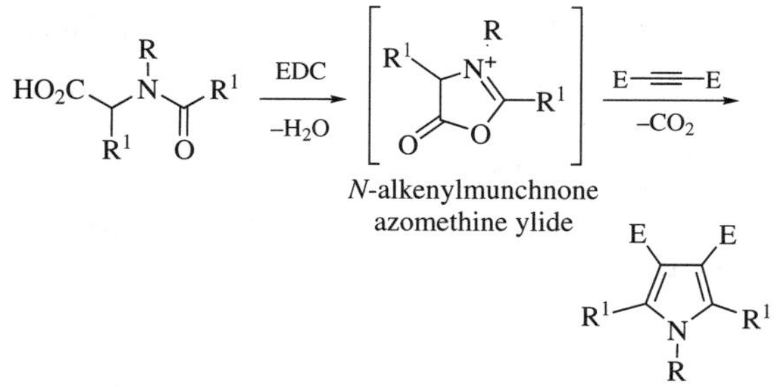

(10)

Related Reagents. Benzotriazol-1-yloxytris(dimethylamino) phosphonium Hexafluorophosphate; *N,N*′-Carbonyldiimidazole; 1-Cyclohexyl-3-(2-morpholinoethyl)carbodiimide; 1,3-Dicyclohexylcarbodiimide; Diphenyl Phosphorazidate; Di-p-tolylcarbodiimide; *N*-Ethyl-5-phenylisoxazolium-3′-sulfonate; 1-Hydroxybenzotriazole; Isobutene; Isobutyl Chloroformate; 1,1′-Thionylimidazole.

1. (a) Kurzer, F.; Douraghi-Zader, K. *Chem. Rev.* **1967**, *67*, 107. (b) Rich, D. H.; Singh, J. *The Peptides: Analysis, Synthesis, Biology*; Academic: New York, 1979; Vol. 1, pp 241–261.
2. Sheehan, J. C.; Ledis, S. L. *J. Am. Chem. Soc.* **1973**, *95*, 875.
3. Chen, F. M. F.; Benoiton, N. L. *Synthesis* **1979**, 709.
4. Dhaon, M. K.; Olsen, R. K.; Ramasamy, K. *J. Org. Chem.* **1982**, *47*, 1962.
5. Carraway, K. L.; Koshland, Jr., D. E., Jr. *Methods Enzymol.* **1972**, *25*, 616.
6. Ramage, R.; MacLeod, A. M.; Rose, G. W. *Tetrahedron* **1991**, *47*, 5625.
7. Wuensch, E.; Drees, F. *Ber. Dtsch. Chem. Ges./Chem. Ber.* **1966**, *99*, 110.
8. Konig, W.; Geiger, R. *Ber. Dtsch. Chem. Ges./Chem. Ber.* **1970**, *103*, 788.
9. Gish, D. T.; Katsoyannis, P. G.; Hess, G. P.; Stedman, R. J. *J. Am. Chem. Soc.* **1956**, *78*, 5954.
10. (a) Miyazawa, T.; Otomatsu, T.; Yamada, T.; Kuwata, S. *Tetrahedron Lett.* **1984**, *25*, 771. (b) Miyazawa, T.; Otomatsu, T.; Fukui, Y.; Yamada, T.; Kuwata, S. *Chem. Commun./J. Chem. Soc., Chem. Commun.* **1988**, 419.
11. Hagiwara, D.; Neya, M.; Miyazaki, Y.; Hemmi, K.; Hashimoto, M. *Chem. Commun./J. Chem. Soc., Chem. Commun.* **1984**, 1676.
12. Okawa, K.; Hase, S. *Bull. Chem. Soc. Jpn.* **1963**, *36*, 754.
13. Ogura, H.; Takeda, K.; Tokue, R.; Kobayashi, T. *Synthesis* **1978**, 394.
14. Shamim, M. T.; Ukena, D.; Padgett, W. L.; Daly, J. W. *J. Med. Chem.* **1989**, *32*, 1231.
15. Atherton, E.; Benoiton, N. L.; Brown, E.; Sheppard, R. C.; Williams, B. J. *Chem. Commun./J. Chem. Soc., Chem. Commun.* **1981**, 336.
16. Greene, T. W. *Protective Groups in Organic Synthesis*, Wiley: New York, 1981; pp 154–180.
17. Keeton, T. K.; Krutzsch, H.; Lovenberg, W. *Science* **1981**, *211*, 586.
18. Tam, T. F.; Thomas, E.; Kruntz, A. *Tetrahedron Lett.* **1987**, *28*, 1127.
19. Atwal, K. S.; Ahmed, S. Z.; O'Reilly, B. C. *Tetrahedron Lett.* **1989**, *30*, 7313.
20. Anderson, W. A.; Heider, A. R. *Synth. Commun.* **1986**, 357.

Richard S. Pottorf
University of Bristol, UK

Peter Szeto
Marion Merrell Dow Research Institute, Cincinnati, OH, USA

N-Ethyl-5-phenylisoxazolium-3′-sulfonate[1] (Woodward's Reagent K)

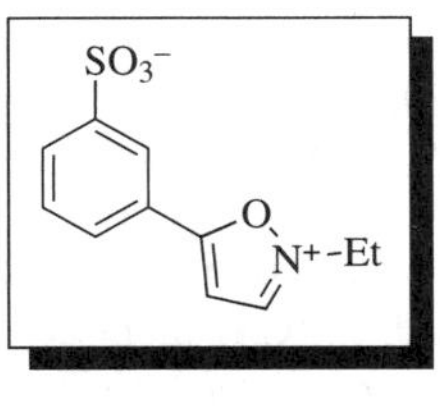

[4156-16-5] $C_{11}H_{11}NO_4S$ (MW 253.30)

(coupling reagent for peptide synthesis;[2] protein modification;[3] reduction of carboxylic acids into alcohols[4])

Alternate Name: Woodward's reagent K; NEPIS.

Physical Data: mp 206–208 °C (dec); UV λ_{max} (ε) 283 nm (22 500) in 0.1 N HCl.

Solubility: insol MeCN, $MeNO_2$, DMF, CH_2Cl_2; sol H_2O.

Form Supplied in: crystalline, anhydrous zwitterion; widely available as 95% pure that may contain some of the *para* sulfonate derivative.

Analysis of Reagent Purity: material that is sufficiently pure has mp 206–208 °C.

Purification: dissolve in aqueous 1 N HCl and reprecipitate by the slow addition of acetone. After filtration and drying the reagent is a fluffy solid.

Handling, Storage, and Precautions: stable to light and temperature and nonhygroscopic. No extra precautions need to be observed.

Peptide Coupling Reagent. Woodward's reagent K reacts with an *N*-protected amino acid, under mild conditions, in the presence of a tertiary amine to form an enol ester intermediate, e.g. (**1**). This enol ester acylates an amino acid ester or peptide to afford the elongated peptide derivative (eq 1). The isoxazolium salt is not soluble in organic solvents, but as it reacts with a carboxyl group in the presence of a tertiary amine the reaction mixture clears.

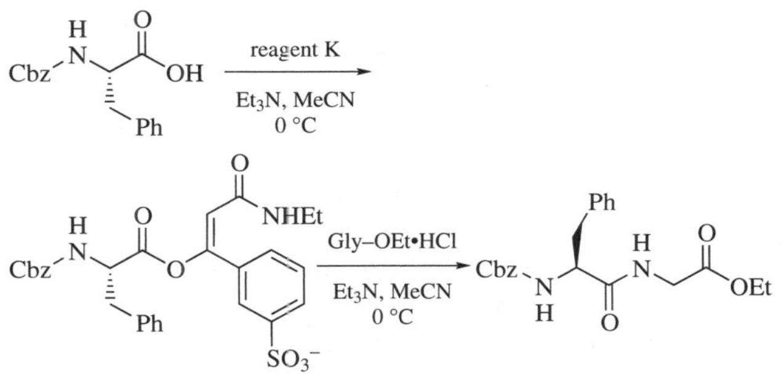

(**1**) (1)

Examples of di- and tripeptide couplings abound with yields typically >80%; minimal purification steps are required since the byproducts are water soluble (eq 2).[2]

$$\text{Cbz-Gly} + \text{Gly-Tyr-OMe} \xrightarrow[\text{Et}_3\text{N, MeCN, 0 °C, 84\%}]{\text{reagent K}} \text{Cbz-Gly-Gly-Tyr-OMe} \quad (2)$$

The recommended coupling procedure is to activate the carboxyl group in $MeNO_2$ or MeCN with vigorous stirring for 10 min at rt or 1 h at 0 °C. Then the amino ester, dissolved in $MeNO_2$ or MeCN, is added and the coupling allowed to go at rt for several hours.

Good yields are obtained in coupling asparagine and glutamine. With some coupling reagents the carboxamide of these residues can undergo an intramolecular dehydration when the α-carboxylate is activated to give a nitrile residue (**2**) in the peptide.[5] However, this byproduct has not been observed with reagent K.

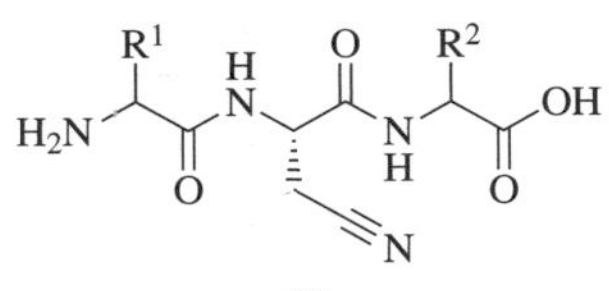

(**2**)

An additional advantage of reagent K is that when serine, threonine, etc., are activated their hydroxyl groups do not require protection.[6] For instance, the hydroxyproline tripeptide in eq 3 was obtained in good yield after crystallization.[2]

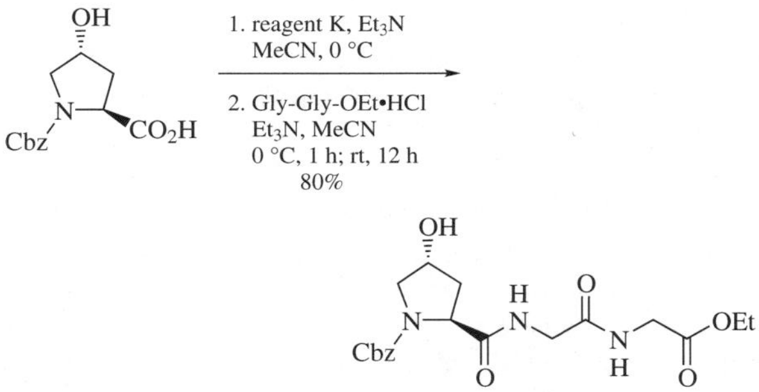

(3)

Several studies[7] on the extent of racemization with peptide coupling reagents have been published. This generally is a problem when performing a fragment condensation, but normally not a problem when the amino group is protected as a carbamate. For best control of the extent of racemization, accurate measurement of the tertiary amine is advised.[7b] The extent of racemization observed with reagent K is just a few percent for acylated amino acids,[7c] which is no worse than most coupling reagents except for the azide coupling procedure. There are some disadvantages of this reagent in peptide couplings. The most serious one is the limitation of solvents. Acetonitrile and nitromethane are the best for high yield, high optical purity products. However, many peptides are not readily soluble in these solvents. The next disadvantage is that the enol ester intermediate can rearrange to the imide side-product (**3**) which is difficult to remove.[1] Another problem with reagent K is its incompatibility with solid-phase peptide synthesis because of its poor solubility in organic solvents and relatively slow activation of carboxyl groups.[8]

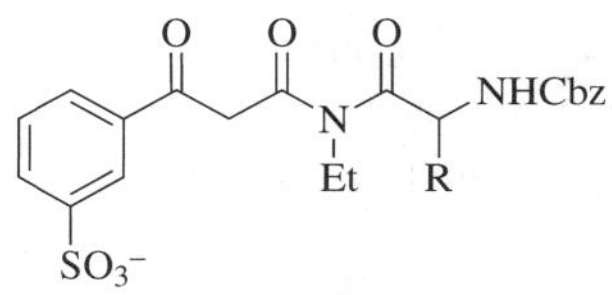

(**3**)

A good comparison of reagent K with other reagents is in the synthesis of a protected derivative of oxytocin (eq 4).[9] This product was further elaborated to give oxytocin. The results are compared with those for 1,3-Dicyclohexylcarbodiimide and N,N′-Carbonyldiimidazole in Table 1.

Cbz-Cys(Bn)-Tyr-Ile-Gln-Asn + Cys(Bn)-Pro-Leu-Gly-NH_2 $\xrightarrow[\text{DMP, 0 °C; 85\%}]{\text{reagent K}}$

Cbz-Cys(Bn)-Tyr-Ile-Gln-Asn-Cys(Bn)-Pro-Leu-Gly-NH_2 (4)

Table 1 Comparison of Coupling Agents for Protected Oxytocin

Coupling method	Yield (%)	Mp (°C)	Biological activity[a]
DCC	80	225–230	81
CDI	70.5	225–230	50
Reagent K	85	240–242	134

[a]Units mg^{-1} in avian depressor activity for deprotected and oxidized nonapeptide.

An additional utility for reagent K is peptide cyclization.[10] The activation and cyclization steps are performed separately; therefore each can be carried out under the appropriate dilution conditions. This is advantageous when the carboxyl group activates slowly since it is in concentrated solution.

The isoxazolium salt has been used in the synthesis of other difficult peptide structures. A peptidoglycan has been obtained in good yield by this procedure (eq 5).[11]

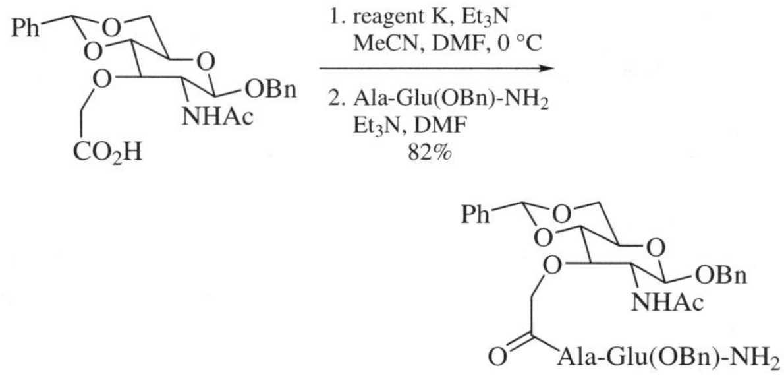

(5)

Additionally, this reagent has been used repetitively in the synthesis of a steroid–peptide adduct with a free hydroxyl group (eq 6).[12] Deprotection and another coupling with Cbz–Arg(N^G-nitro) gave the dipeptide–steroid adduct in good yield. In both couplings the steroid hydroxyl group was not acylated.

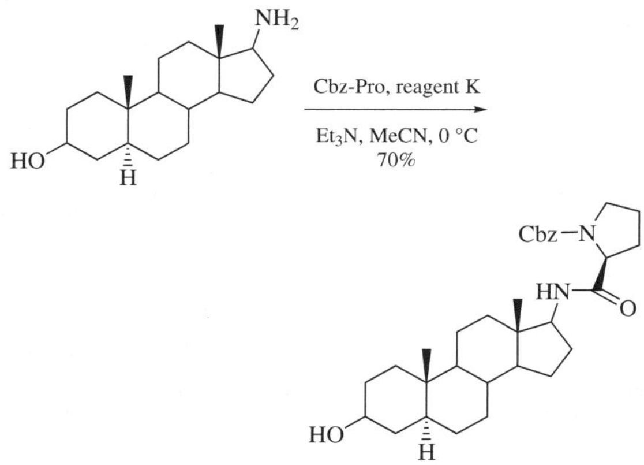

(6)

Protein Modification. Since this reagent is water soluble and is reactive under aqueous conditions below pH 4.75, it can specifically modify carboxyl groups in proteins.[3,13] This has proven useful in determining essential carboxyl groups in various enzymes. The enol ester is stable enough to isolate the modified protein and 'tag' with a nucleophile for eventual identification of the modified amino acid. Other reagents such as the water-soluble carbodiimides are also used extensively in these mechanistic studies. However, an advantage with reagent K is that the enol ester formed has a UV absorbance at 340 nm ($E_{340} = 7000\ M^{-1}\ cm^{-1}$);[14] thus the number of carboxyl groups modified can be quantified.

Reduction of Carboxylic Acids to Alcohols. The conversion of carboxylic acids to alcohols occurs in a two-step process under mild conditions using reagent K with yields for simple alcohols ranging from 50–100%.[4] The reduction of a dipeptide (eq 7) suggests a potential utility in modifying carboxyl groups in proteins.

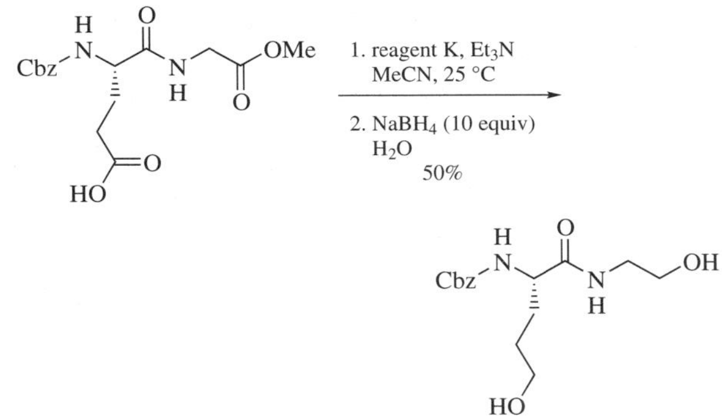

(7)

Another method[15] to obtain alcohols is to reduce the mixed carbonic acid anhydride prepared with Ethyl Chloroformate with an excess of Sodium Borohydride. This method uses less costly reagents; thus it is more useful for synthetic transformations.

Miscellaneous Reactions. Pantothenic acid has been coupled to a long chain amine with reagent K (eq 8).[16] This is another example of the nonreactivity of hydroxyl groups with the enol ester intermediate. Attempts to prepare this amide through the mixed anhydride procedure give low yields.

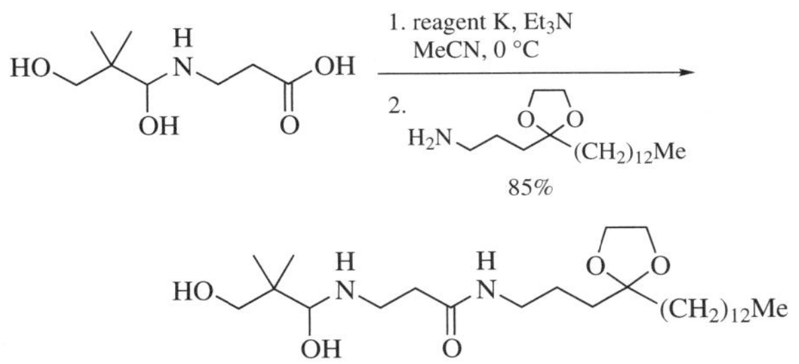

(8)

Reagent K may have utility in sequencing peptides and proteins from the *C*-terminus (eq 9).[17] The 2-thiohydantoin produced is used to analyze a peptide sequence upon cleavage of the thiohydantoin from the peptide. This mild procedure gives good yields with amino acids that have sensitive side chains such as serine and lysine. The original Acetic Anhydride procedure[18] for *C*-terminal sequencing does not work well with these amino acids; therefore this procedure is an improvement. Another activation reagent, hydroxysuccinimide esters of Fmoc-amino acids, fails to give the thiohydantoin.[17]

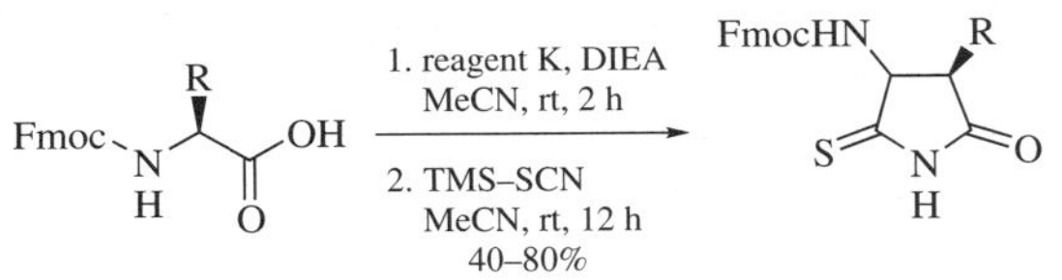

(9)

R = variety of amino acid side chains

Related Reagents. Benzotriazol-1-yloxytris(dimethylamino)phosphonium Hexafluorophosphate; 1-Cyclohexyl-3-(2-morpholinoethyl)carbodiimide; 1-Ethyl-3-(3′-dimethylaminopropyl)carbodiimide Hydrochloride; Isobutyl Chloroformate; *p*-Nitrophenol; Pentafluorophenol; 1,1′-Thionylimidazole

1. Woodward, R. B.; Olofson, R. A.; Mayer, H. *Tetrahedron* **1969**, *22 (S8)*, 321.
2. Woodward, R. B.; Olofson, R. A. *Org. Synth.* **1977**, *56*, 88.
3. Bodlaender, P.; Feinstein, G.; Shaw, E. *Biochemistry* **1969**, *8*, 4941.
4. Hall, P. L.; Perfetti, R. B. *J. Org. Chem.* **1974**, *39*, 111.
5. Gish, D. T.; Katsoyannis, P. G.; Hess, G. P.; Stedman, R. J. *J. Am. Chem. Soc.* **1956**, *78*, 5954.
6. Klausner, Y. S.; Bodanszky, M. *Synthesis* **1972**, 453.
7. (a) Kemp, D. S. *The Peptides: Analysis, Synthesis and Biology*; Academic: New York, 1979; Vol. 1, pp 315–383. (b) Woodward, R. B.; Woodman, D. J. *J. Org. Chem.* **1969**, *34*, 2742. (c) Kemp, D. S.; Wang, S. W.; Busby, G., III; Hugel, G. *J. Am. Chem. Soc.* **1970**, *92*, 1043.
8. Hudson, D. *J. Org. Chem.* **1988**, *53*, 617.
9. Fosker, A. P.; Law, H. D. *J. Chem. Soc* **1965**, 4922.
10. Blaha, K.; Rudinger, J. *Collect. Czech. Chem. Commun.* **1965**, *30*, 3325.
11. Merser, C.; Sinay, P.; Adam, A. *Biochem. Biophys. Res. Commun.* **1975**, *66*, 1316.
12. (a) Pettit, G. R.; Gupta, A. K. D.; Smith, R. L. *Can. J. Chem.* **1966**, *44*, 2023. (b) Pettit, G. R.; Smith, R. L.; Klinger, H. *J. Org. Chem.* **1967**, *10*, 145.
13. Baker, A. J.; Weber, B. H. *J. Biol. Chem.* **1974**, *249*, 5452.
14. Sinha, U.; Brewer, J. M. *Anal. Biochem.* **1985**, *151*, 327.
15. Ramsamy, K.; Olsen, R. K.; Emery, T. *Synthesis* **1982**, 42.
16. Wagner, A. P.; Retey, J. *Eur. J. Biochem.* **1991**, *195*, 699.
17. Boyd, V. L.; Hawke, D. H.; Geiser, T. G. *Tetrahedron Lett.* **1990**, *31*, 3849.
18. Cromwell, L. D.; Stark, G. R. *Biochemistry* **1969**, *8*, 4735.

Richard S. Pottorf
Marion Merrell Dow Research Institute, Cincinnati, OH, USA

5-Ethylthio-1*H*-tetrazole

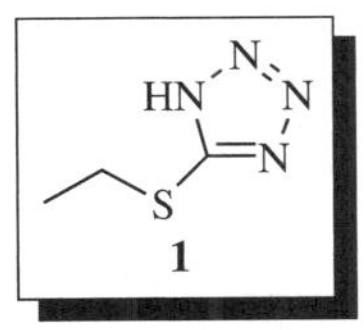

[89797-68-2] $C_3H_6N_4S$ (MW 130.17)

(reagent used a powerful activator in DNA and RNA syntheses)

Physical Data: mp 86–88 °C.[1,2]
Solubility: soluble in CH_3CN, DMF.
Purification: recrystallized from toluene.
Analysis of Reagent Purity: IR, UV, elemental analysis.[2]
Preparative Methods: the title reagent **1** is prepared from the corresponding alkyl thiocyanate and sodium azide through a [2+3] cycloaddition under phase transfer catalytic conditions (eq 1).[1] The ratio of the solvent (water to toluene) as well as the reaction temperature are important factors in obtaining preparative yields of the alkylthiotetrazoles. 5-Ethylthio-1*H*-tetrazole has also been prepared using DMF as solvent albeit in much lower yield.[2]

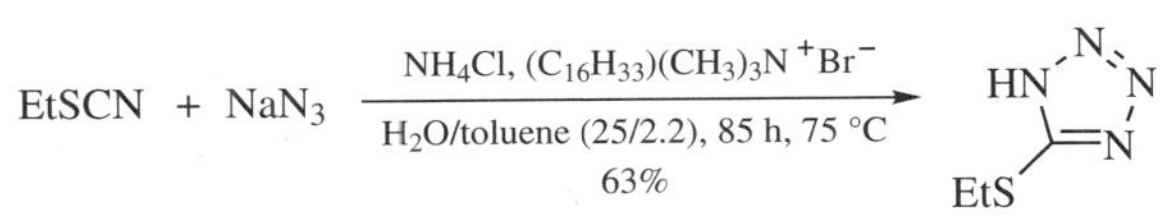

(1)

1

Handling, Storage, and Precautions: stable reagent; stored under argon, may be toxic.

Introduction. Tetrazoles have widespread applications in medicinal chemistry as well as in energetic materials.[3] 5-Alkylthiotetrazoles, especially 5-ethylthio-1*H*-tetrazole (**1**), serve as powerful activators in DNA and RNA syntheses.[4,5] The title compound is a better activator than the corresponding 5-alkyl or 5-aryltetrazoles due to the presence of the alkylthio group which renders the tetrazole more acidic.[2] **1** is considerably more soluble in acetonitrile than simple alkyltetrazoles, and the higher concentration results in a better activator performance.[4,5]

Selective Phosphitylation of the Primary Hydroxyl Group of Nucleosides. 5-Ethylthio-1*H*-tetrazole has been used for the selective 5′-phosphitylation of nucleosides.[6] The commonly accepted mechanism involves protonation of a phosphoramidite by the weakly acidic tetrazole, followed by nucleophilic attack by tetrazolide ion giving a tetrazoyl phosphoramidite (**2**) (eq 2). The nucleophilic attack of the 5′-hydroxyl group then produces the phosphite triester, which is then carried through further transformations (eq 3).[6,7]

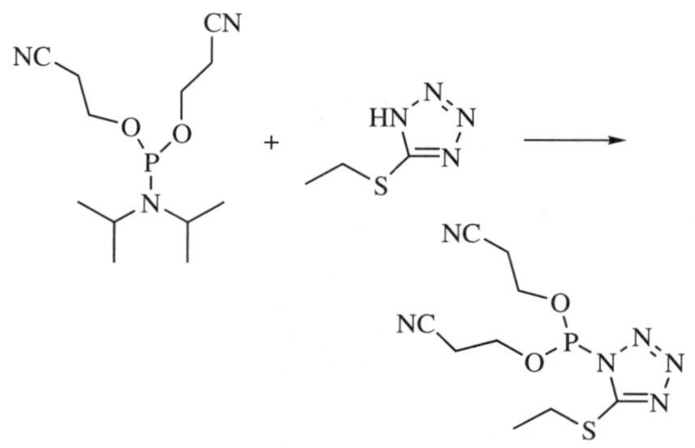

(2)

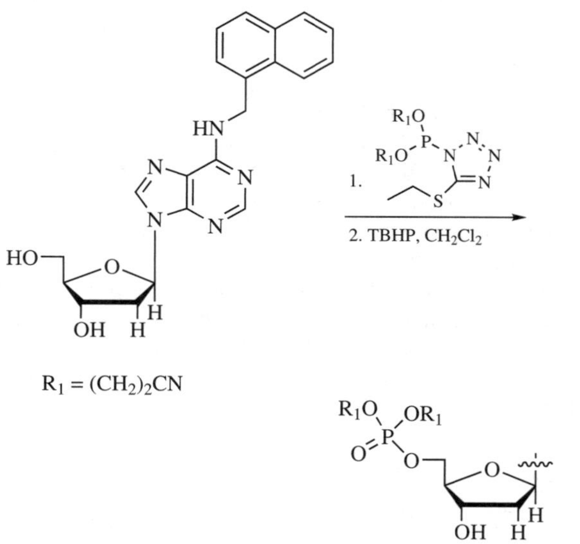

(3)

Catalyst in Oligonucleotide Derivatization. New terminus modifiers, bearing a phosphoramidite moiety and a methoxyoxalamido (MOX) precursor group, have been coupled onto the 5′-end of a synthetic oligodeoxyribonucleotide in the last step of an automated synthesis to form the MOX precursor oligonucleotide (eq 4).[8] It is observed that the coupling time is greatly reduced in the presence of a powerful catalyst such as 5-ethylthio-1*H*-tetrazole. The MOX groups can then be post-synthetically derivatized with an appropriate primary amine to construct a 5′-modified oligonucleotide.[8,9]

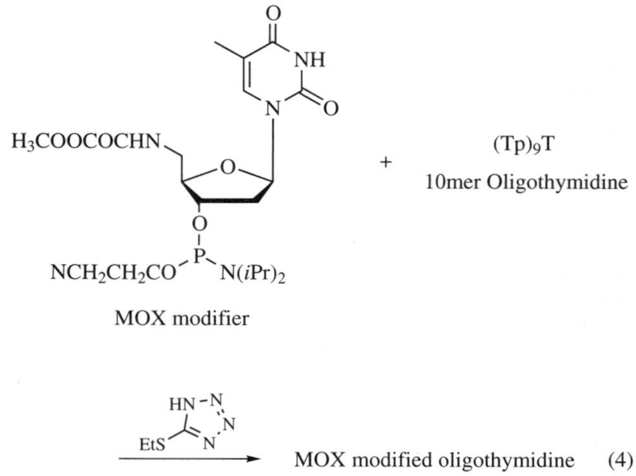

(4)

Activator in the Synthesis of RNA. 5-Ethylthio-1*H*-tetrazole has been used as an activator in the automated synthesis of a 14mer RNA sequence (5′>GAGCCUGGGAGCUC<3′). The key step in this automated synthesis by the phosphoramidite approach is the tetrazole-mediated coupling of a 5′-*O*-DMT-nucleoside-3′-*O*-phosphoramidite with the 5′-hydroxyl group of the growing chain anchored to a polymeric support. It was found that 5-ethylthio-1*H*-tetrazole as an activator gave a superior product with sixfold excess of phosphoramidite and 15 min coupling time.[10] These concentrations and coupling times are much less than those required with 1*H*-tetrazole as an activator.[11]

1,3-Addition Reactions of Azoles. 1,3-Addition reactions of azoles with nitrile oxides and nitrile imines have been carried out by dissolving equimolar amounts of the azole and the 1,3-dipole precursor (hydroximoyl chloride or hydrazonoyl halide) in an appropriate solvent at room temperature in the presence of excess triethylamine (eq 5).[12] The additions yielded mixtures of 1- and 2-substituted tetrazoles with the ratio depending mainly on the substituent in the 5-position of the tetrazole ring (eqs 5 and 6).[12]

On thermolysis in a suitable sovent, these 1,2- adducts (eq 5) and 1,3-adducts (eq 6) afforded almost quantitative yields of substituted heterocycles.

Decomposition at Higher Temperatures. It has been demonstrated that *S*-substituted mercaptotetrazoles, upon heating slightly above their melting points, yield hydrazoic acid and the corresponding organic thiocyanates (eq 7).[2]

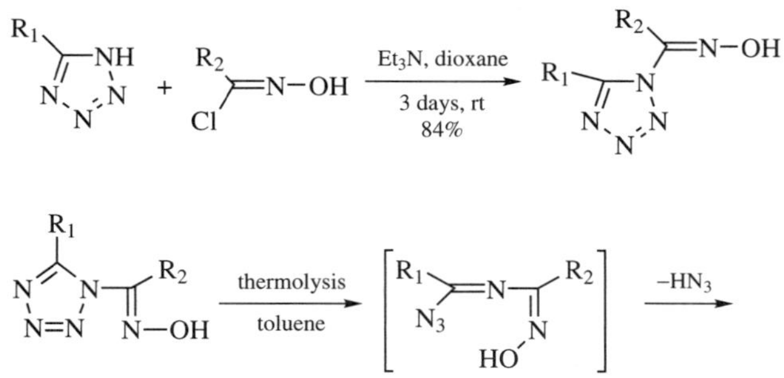

R_1 = SEt
R_2 = 2,4,6-$(CH_3)_3$-C_6H_2

(5)

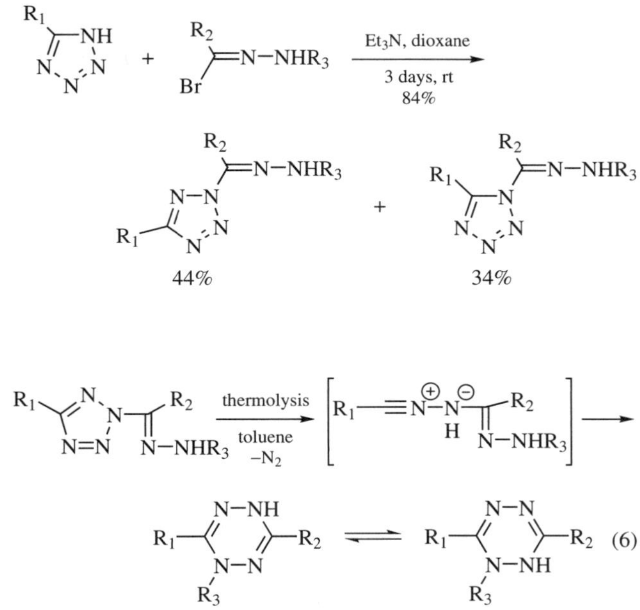

(6)

R_1= SEt
R_2 = *t*-Bu
R_3 = 4-NO_2-C_6H_4

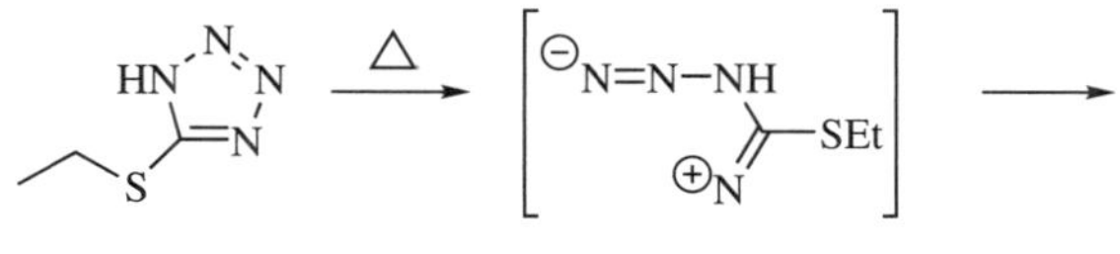

HN_3 + EtSCN (7)

Related Reagents. 1*H*-Tetrazole;[13] 5-methylthio- 1*H*-tetrazole;[14] 5-(*p*-nitrophenyl)-1*H*-tetrazole;[15] 5-butylthio-1*H*-tetrazole.[1]

---

1. LeBlanc, B. W.; Jursic, B. S., *Synth. Commun.* **1998**, *28*, 3591–3599.
2. Leiber, E.; Enkoji, T., *J. Org. Chem.* **1961**, *26*, 4472–4479.
3. Butler, R. N., In *Comprehensive Heterocyclic Chemistry*; Katritzky, A. R.; Rees, C. W.; Scriven, E. F. V.; Ed.; Pergamon Press: Oxford, 1988, Vol. 5, p 791.
4. Wincott, F.; Direnzo, A.; Shaffer, C.; Grimm, C.; Tracz, D.; Workman, C.; Sweedler, D.; Gonzalez, C.; Scaringe, S.; Usman, N., *Nucleic Acids Res.* **1995**, *23*, 2677–2684.
5. Tsou, D.; Hampel, A.; Andrus, A.; Vinayak, R., *Nucleosides Nucleotides* **1995**, *14*, 1481–1492.
6. Kennedy, K. J.; Bressi, J. C.; Gelb, M. H., *Bioorg. Med. Chem. Lett.* **2001**, *11*, 95–98.
7. Graham, S. M.; Pope, S. C., *Org. Lett.* **1999**, *1*, 733–736.
8. Polushin, N. N., *Nucleic Acids Research* **2000**, *28*, 3125–3133.
9. Polushin, N. N., *Nucleosides, Nucleotides & Nucleic Acids* **2001**, *20*, 973–976.
10. Sproat, B.; Colonna, F.; Mullah, B.; Tsou, D.; Andrus, A.; Hampel, A.; Vinayak, R., *Nucleosides Nucleotides* **1995**, *14*, 255–273.
11. Vinayak, R.; Colonna, F.; Tsou, D.; Mullah, B.; Andrus, A.; Sproat, B., *Nucleic Acids Symposium Series* **1994**, *31*, 165–166.
12. Plenkiewicz, J.; Zdrojewski, T., *Bull. Soc. Chim. Belges* **1987**, *96*, 675–709.
13. Claramunt, R. M.; Lopez, C.; Sanz, D.; Alkorta, I.; Elguero, J., *Heterocycles* **2001**, *55*, 2109–2122.
14. Wright, P.; Lloyd, D.; Rapp, W.; Andrus, A., *Tetrahedron Lett.* **1993**, *34*, 3373–3376.
15. Froehler, B. C.; Matteucci, M. D., *Tetrahedron Lett.* **1983**, *24*, 3171–3174.

Santhosh Neelamkavil
University of Pennsylvania, Philadelphia, PA

F

9-Fluorenylmethyl Chloroformate[1]

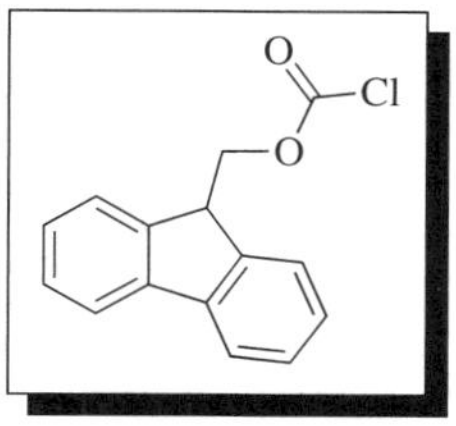

[28920-43-6] $C_{15}H_{11}ClO_2$ (MW 258.71)

(base-labile protecting group for primary and secondary amines;[3] used in peptide synthesis;[5] used for derivatization of amines and amino acids prior to HPLC analysis and fluorescence detection[13])

Alternate Names: 9-fluorenylmethoxycarbonyl chloride; FMOC-Cl.
Physical Data: mp 62–64 °C.
Solubility: sol organics (CH_2Cl_2, THF, dioxane); reacts with alcohols, amines, and water.
Form Supplied in: white crystalline solid; widely available.
Handling, Storage, and Precautions: FMOC-Cl is an acid chloride, and should be protected from moisture and heat. The reagent can evolve CO_2 (pressure) upon prolonged storage; therefore bottles should be opened carefully. FMOC-Cl is harmful if swallowed, inhaled, or absorbed through skin, is corrosive, and is a strong lachrymator. Inhalation may be fatal. Use in a fume hood.

N-Terminus Blocking in Peptide Synthesis. Solid-phase peptide synthesis based on base-labile FMOC protection[1] is rapidly augmenting and/or replacing older methods based on Boc and Z (Cbz) protection. The major advantage offered by the use of FMOC protection is that repetitive treatment with HF or CF_3CO_2H is no longer required for the deprotection steps. Most FMOC-protected amino acids are commercially available, although the search for orthogonal side-chain functional group protection continues.[2] FMOC protection is currently the method of choice for the synthesis of glycopeptides which carry acid-labile *O*-glycosidic bonds (eq 1).[3,5b,7]

FMOC protection of amino acids is accomplished with Shotten–Bauman conditions (e.g. $NaHCO_3$/dioxane/H_2O or $NaHCO_3$/DMF), or with anhydrous conditions (e.g. pyridine/CH_2Cl_2) at room temperature or below (eq 1).[3a,4] A typical synthetic protocol for solid-phase peptide synthesis would involve coupling (eq 2), followed by deprotection with 20–30% Piperidine in DMF (eq 3), although other amines (e.g. 2% DBU, 10% Et_2NH, or 50% morpholine in DMF) have been used. Many workers no longer use the chloroformate ester FMOC-Cl (**1**), preferring the more shelf-stable carbonate FMOC-OSu (**2**) *[82911-69-1]*,[5] or FMOC-pfp (**3**) *[88744-04-1]*. When used in the standard way,[4] FMOC-Cl (**1**) has been shown to promote the formation of 'FMOC-dipeptides' in 3–7% yield, presumably via the mixed anhydride intermediate.[5a]

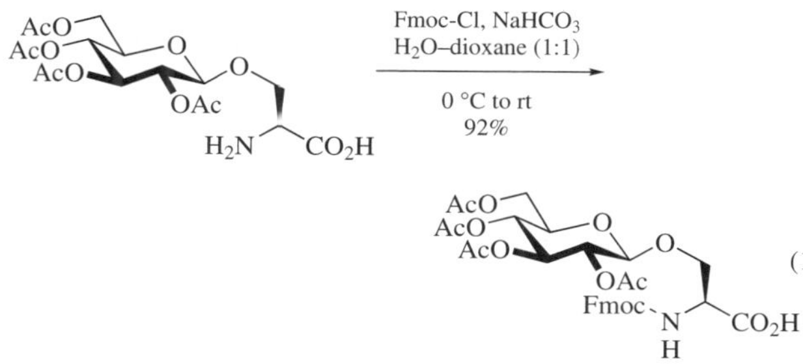

(1)

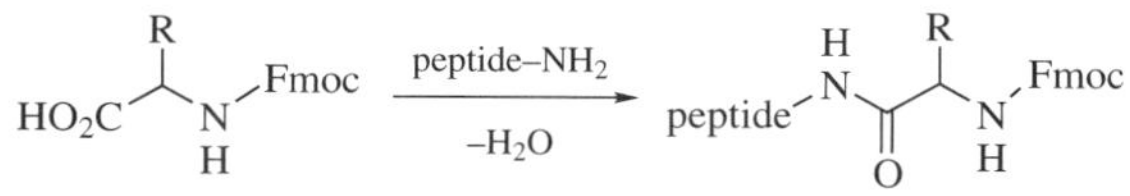

(2)

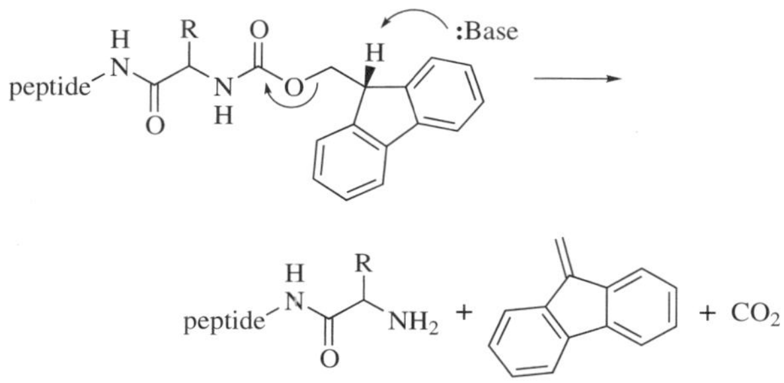

(3)

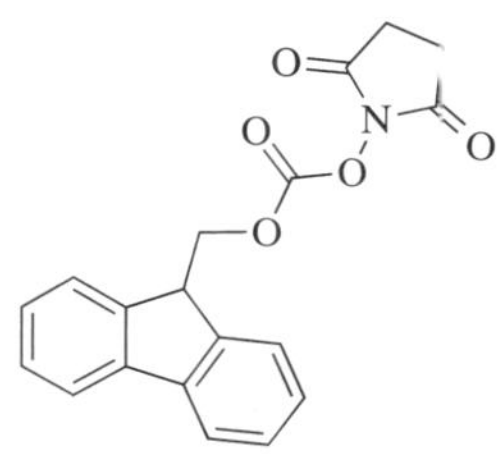

(2)

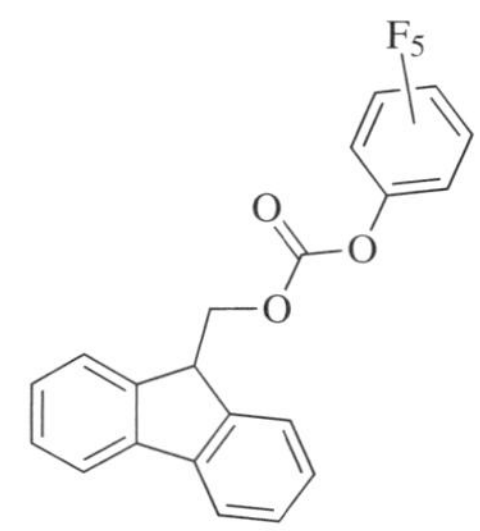

(3)

Use of FMOC-OSu (**2**) provides *N*-protected amino acids in comparable yields without the formation of the dipeptide impurities.[5] FMOC-pfp (**3**) has been used to prepare either FMOC-protected amino acids, or the corresponding pfp esters in one pot. Advantages of this reagent include supression of the dipeptide byproducts, as well as the production of the activated pfp esters using the same pentafluorophenol released during the *N*-protection step (eq 4).[6] This same *N*-protection/*C*-activation procedure has been used to produce serine glycosides (eq 5),[3b] and for the solid-phase synthesis of human intestinal mucin fragments (eq 6).[7] The use of FMOC-N_3 (**4**) is also claimed to reduce dipeptide formation during the protection step,[8] but the storage or use of potentially explosive oxycarbonyl azides cannot be recommended by this reviewer. Alternatively, the azide (**4**) may be generated in situ from (**1**).[4]

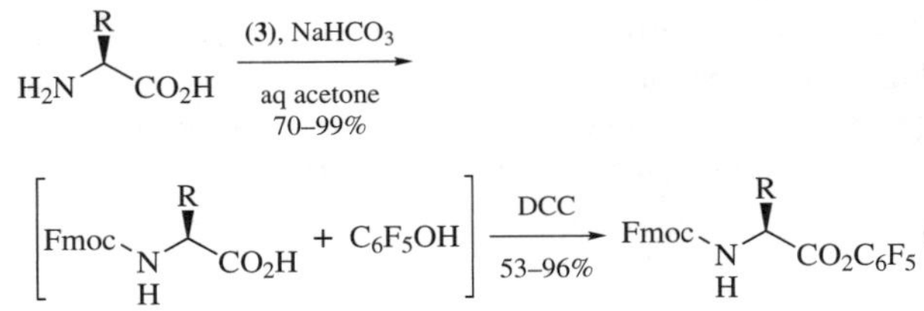

(4)

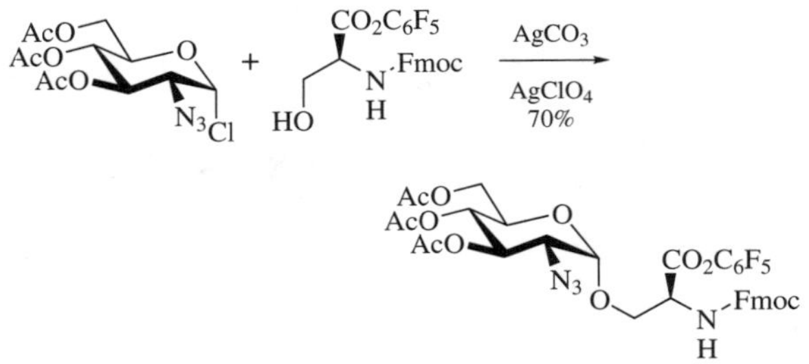

(5)

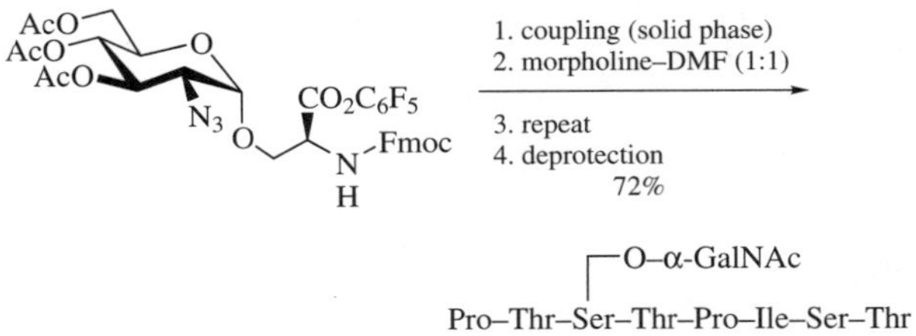

(6)

Fluorenylmethyl esters have been used for *C*-terminus protection in combination with FMOC *N*-terminus protection for solution-phase coupling of protected amino acids and peptide fragments. This permits simultaneous deprotection of both termini with Et_2NH/DMF (eq 7).[9]

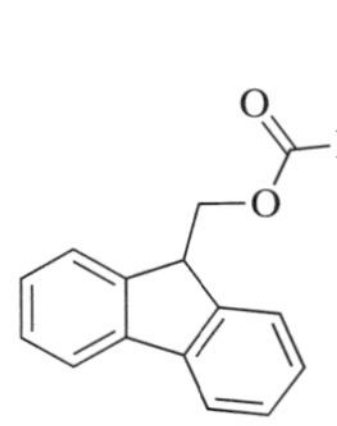

(4)

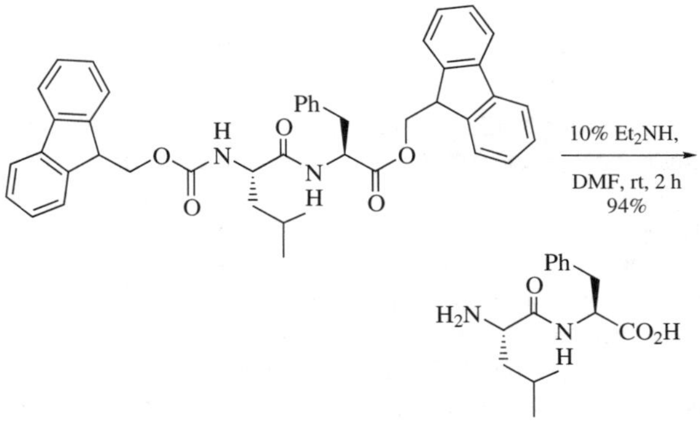

(7)

Amine and Hydroxyl Protection for Nucleosides, Carbohydrates, and Natural Products. FMOC has been used to protect the amino groups on various nucleoside bases (adenosine, cytosine, guanine, and their 2-deoxy derivatives) for solid-phase oligonucleotide synthesis in a one-pot procedure.[10] Chlorotrimethylsilane in dry pyridine was used for transient protection of the 2′,3′,5′- or 3′,5′-hydroxyls, followed by the addition of FMOC-Cl, and hydrolytic workup of the TMS ethers to produce crystalline FMOC-protected nucleosides and 2-deoxynucleosides in good yield (eq 8). The FMOC group was removed from the heterocyclic bases with excess Triethylamine in dry pyridine.

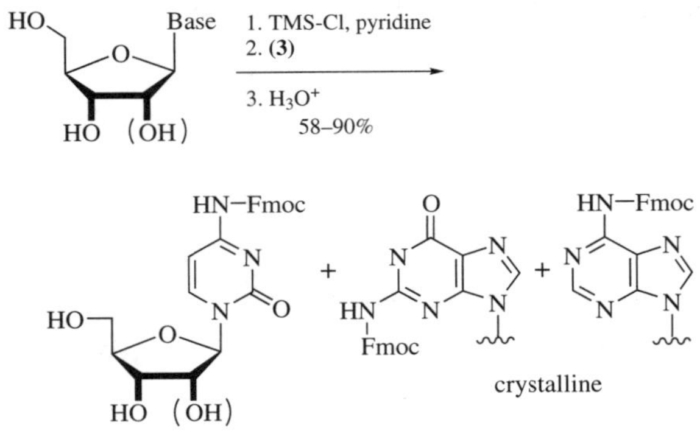

(8)

FMOC-Cl has been used for 3′,5′-hydroxyl protection of deoxyribose for the introduction of modified bases such as azacytidine using the classical Vorbrüggen reaction.[11] While the carbonate protecting groups do not alter the stereoselectivity of the *N*-glycoside formation, the FMOC groups can be removed under much milder conditions (Et_3N/pyridine) than the typical 4-methylbenzoyl group (eq 9).

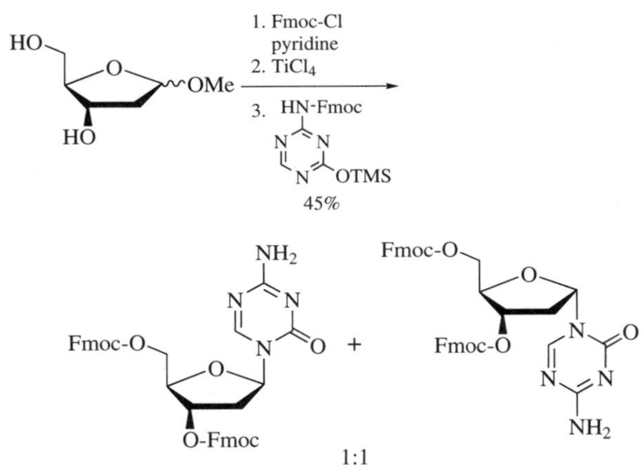

(9)

Perhaps the most impressive use of the FMOC group in the synthesis of natural products is provided by the key tetrasaccharide intermediate to calicheamicin.[12] The robust nature of this particular FMOC group is amply demonstrated by the fact that it survives 15 discrete reaction steps, including exposure to cat. $NaH/HOCH_2CH_2OH$, Br_2, EtN_3, K-Selectride, i-Bu_2AlH, and NaSMe (eq 10). Clearly, the use of FMOC protection is not limited to peptide synthesis.

FMOC Derivatization for Fluorescence Detection and Analysis. Derivatization of amino acids and other primary and secondary amines with (**1**) for HPLC analysis has been described. The chief advantage of this methodology is the high sensitivity due to the intense fluorescence of the fluorenyl moiety, but the FMOC also reduces polarity for faster elution from the column. Amino acids have been detected in concentrations as low as 26 femtomolar using a precolumn derivatization scheme.[13] FMOC derivatization of *lyso*-gangliosides in a two-phase Et_2O–H_2O system followed by chromatographic purification with fluoresence detection has also been reported.[14] Optically active 1-(9-fluorenyl)ethyl chloroformate (FLEC®) (**5**) is commercially available in both (+) [*107474-79-3*] and (−) forms for the analysis of racemates.

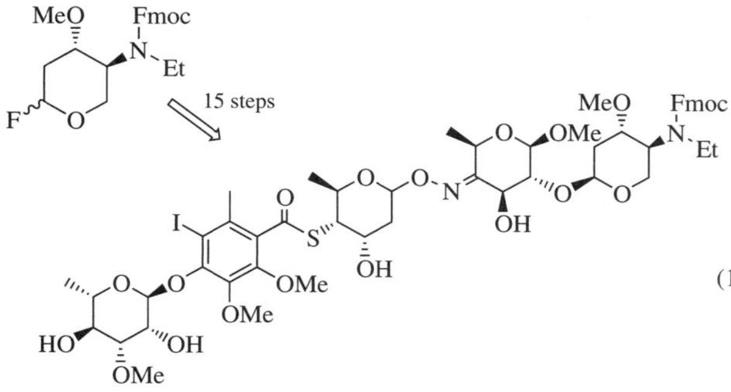

(10)

1. For illustrative procedures using FMOC and other protection schemes, see *The Practice of Peptide Synthesis*; Bodanszky, M.; Bodanszky, A., Eds.; Springer:Berlin, **1984**.
2. (a) Histidine: Fischer, P. M. et al. *Int. J. Pept. Protein Res.* **1992**, *40*, 19. (b) Glutamate:Handa, B. K.; Keech, E. B. K. *Int. J. Pept. Protein Res.* **1992**, *40*, 66.
3. (a) Polt, R.; Szabo, L.; Treiberg, J.; Li, Y.; Hruby, V. J. *J. Am. Chem. Soc.* **1992**, *114*, 10249. (b) Lüning, B.; Norberg, T.; Tejbrant, J. *Chem. Commun./J. Chem. Soc., Chem. Commun.* **1989**, 1267.
4. Carpino, L. A.; Han, G. Y. *J. Org. Chem.* **1972**, *37*, 3404.
5. (a) Lapatsanis, L.; Milias, G.; Froussios, K.; Kolovos, M. *Synthesis* **1983**, 671. (b) Bardaji, E.; Torres, J. L.; Clapes, P.; Albericio, F.; Barany, G.; Rodriguez, R. E.; Sacristin, M. P.; Valencio, G. *J. Chem. Soc., Perkin Trans. 1* **1991**, 1755. For an alternate synthesis of FMOC-OSu (**2**) from fluorenylmethyl alcohol and phosgene: (c) Ten Kortenaar, P. B. W. et al. *Int. J. Pept. Protein Res.* **1986**, *27*, 398.
6. Schön, I.; Kisfalady, L. *Synthesis* **1986**, 303.
7. (a) Jansson, A. M.; Meldal, M.; Bock, K. *Tetrahedron Lett.* **1990**, *31*, 6991. (b) Bielfeldt, T.; Peters, S.; Meldal, M.; Paulsen, H.; Bock, K. *Angew. Chem., Int. Ed. Engl.* **1992**, *31*, 857.
8. Tessier, M. et al. *Int. J.Pept. Protein Res.* **1983**, *22*, 125.
9. (a) Bednarek, M. A.; Bodanszky, M. *Int. J. Pept. Protein Res.* **1983**, *21*, 196. (b) Bodanszky, M. et al. *Int. J. Pept. Protein Res.* **1981**, *17*, 444.
10. Heikkila, J.; Chattopadhyaya, J. *Acta Chem. Scand.* **1983**, *B37*, 263.
11. Ben-Hattar, J.; Jiricny, J. *J. Org. Chem.* **1986**, *51*, 3211.
12. Nicolaou, K. C.; Groneborg, R. D.; Miyazaki, T.; Stylianides, N.; Schulze, T. J.; Stahl, W. *J. Am. Chem. Soc.* **1990**, *112*, 8193.
13. (a) Einarsson, S.; Folestad, S.; Josefsson, B. *Anal. Chem.* **1986**, *58*, 1638. (b) Einarsson, S. *J. Chromatogr.* **1985**, *348*, 213. (c) Veuthey, J.-L. *J. Chromatogr.* **1990**, *515*, 385. (d) For a review:Betner, I.; Foeldi, P. In *Modern Methods in Protein Chemistry*; Tschesche, H., Ed.de Gruyter:Berlin, **1988**; *3*, p 227.
14. Neuenhofer, S.; Schwarzmann, G.; Egge, H.; Sandhoff, K. *Biochemistry* **1985**, *24*, 525.

Robin L. Polt
University of Arizona, Tucson, AZ, USA

N-(9-Fluorenylmethyloxycarbonyl) oxybenzotriazole (Fmoc-HOBT)

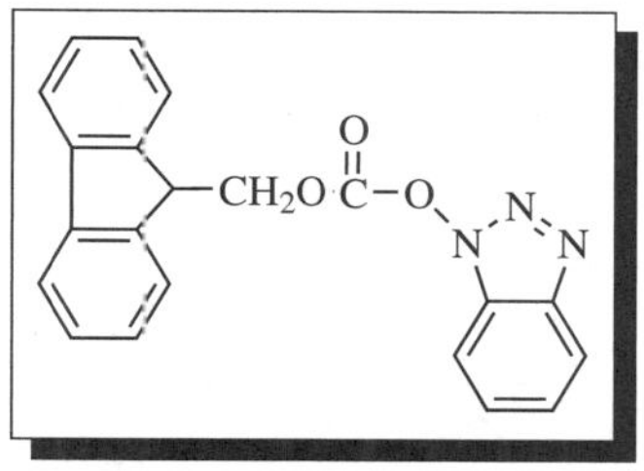

[*82911-71-5*] $C_{21}H_{15}N_3O_3$ (MW 357.4)

(reagent for the preparation of 9-fluorenylmethoxycarbonyl (Fmoc) derivatives of amines and amino acids)

Alternate Name: 3-oxy-benzotriazolinium-1-carboxylic acid 9*H*-fluoren-9-ylmethyl ester; benzotriazol-1-yl-9-fluorenylmethyl carbonate; Fmoc-OBt.
Physical Data: white solid; mp 180 °C.[1]
Solubility: soluble in common organic solvents (i.e., diethyl ether, THF, and CH_2Cl_2).
Form Supplied in: solid.
Analysis of Reagent Purity: melting point.
Preparative Methods: it can be prepared from 9-fluorenylmethyl chloroformate and dicyclohexylammonium salt of 1-hydroxybenzotriazole. This protocol provided the product in 83% yield.[1]
Purification: recrystallization from chloroform–ether.[1]
Handling, Storage, and Precautions: irritating to eyes, respiratory system, and skin. Stored at 2–8 °C.

Protection of Amines and Amino Acids. 9-Fluorenylmethyl 1-benzotriazolyl carbonate (Fmoc-OBt) has been used for the conversion of amines and amino acids to their 9-fluorenylmethyloxycarbonyl (Fmoc) derivatives (eq 1).[1]

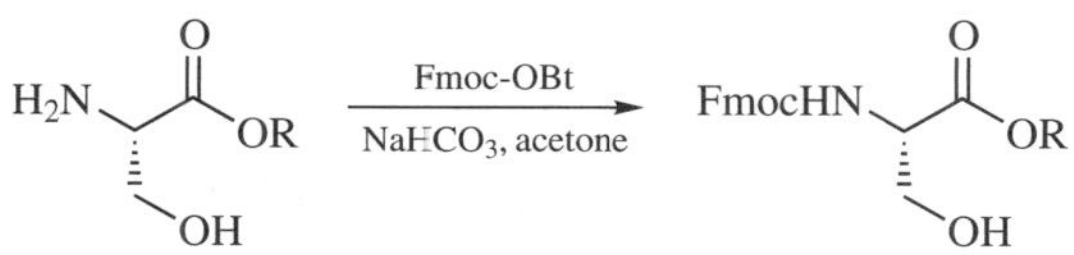

(1)

R = H; 64%
R = Me; 89%

Although the structure of this reagent was orginally formulated as the HOBt-ester, a subsequent X-ray diffraction study has revealed that, as least in the solid state, it is in zwitterionic HOBt-amide form [Fmoc-(*N*)OBt] (eq 2).[2] The structure of Fmoc-OBt in solution has not been established.[3]

The treatment of primary amines with Fmoc-OBt in the presence of a base, such as triethylamine, provides the corresponding Fmoc derivatives in good to excellent yield. This mild reagent tolerates a variety of functional groups including alcohols,[1] carboxylic acids,[1,4] carboxylate esters,[1] and phosphate esters. For example, a series of (phosphonomethyl)-D,L-phenylalanine amino acids, some bearing a free hydroxy group, have been derivatized with Fmoc-OBt (eq 3).[5]

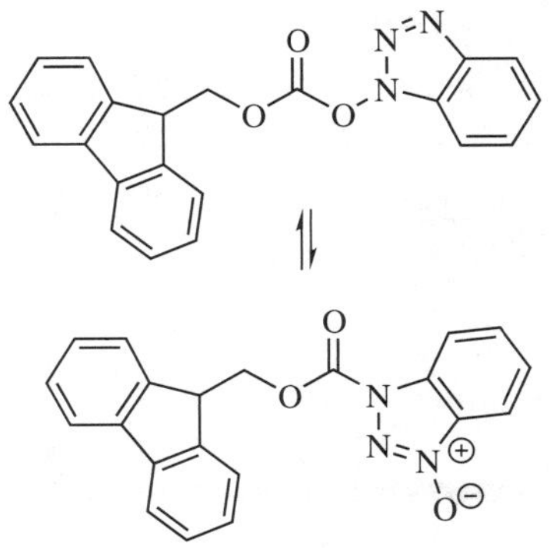

(2)

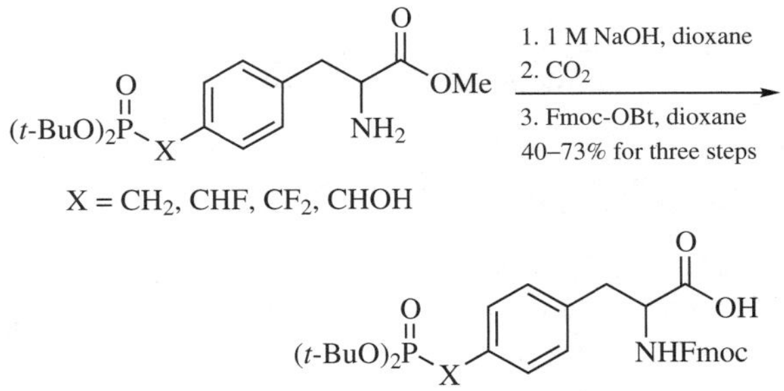

(3)

During the synthesis of the 1-azabicyclo[3.2.0]hexane core of the azinomycins, it was found that a Cbz group could not be removed in the presence of vinyl bromide (eq 4). Accordingly, the benzyl carbamate was removed by catalytic hydrogenation prior to the introduction of vinyl bromide, and the resulting aziridine was protected using Fmoc-OBt. Addition of bromine, dehydrobromination, and removal of the Fmoc group under standard base-catalyzed conditions then furnished the sensitive aziridino[1,2-*a*]pyrrolidine system, the product of addition–elimination (eq 5).[6]

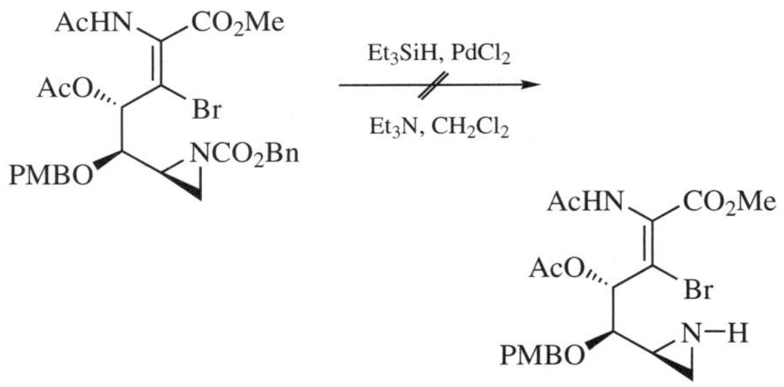

(4)

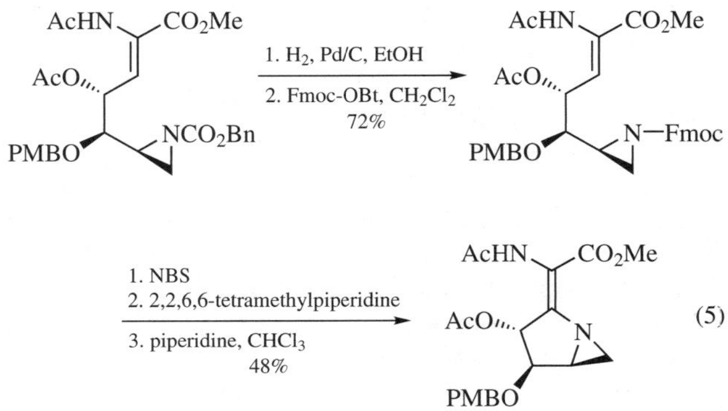

(5)

Fmoc-OBt has been employed in the protection of a nucleoside-bound primary amine during the synthesis of a modified M1 ribozyme (eq 6).[7]

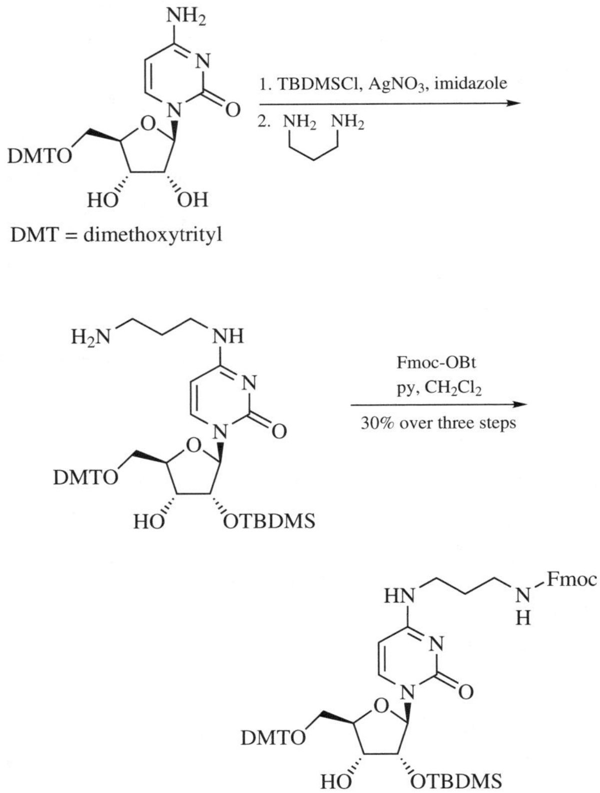

(6)

A solid phase analog of Fmoc-OBt has also been developed. It is readily prepared by the treatment of polymer-bound 1-hydroxy-benzotriazole (PB-HOBT) with Fmoc-Cl in this presence of pyridine.[8] This reagent provides Fmoc-protected primary amines, secondary amines, and amino acids in good to excellent yields without purification. Aromatic and sterically congested amines gave lower yields.

Related Reagents. 9-Fluorenylmethyl chloroformate; 9-fluorenylmethyl succinimidyl carbonate.

---

1. Paquet, A., *Can. J. Chem.* **1982**, *60*, 976.
2. Crisma, M.; Valle, G.; Moretto, V.; Formaggio, F.; Toniolo, C.; Albericio, F., *Lett. Pept. Sci.* **1998**, *5*, 247.
3. Singh, J.; Fox, R.; Wong, M.; Kissick, T. P.; Moniot, J. L.; Gougoutas, J. Z.; Malley, M. F.; Kocy, O., *J. Org. Chem.* **1988**, *53*, 205.
4. Sigler, G. F.; Fuller, W. D.; Chaturvedi, N. C.; Goodman, M.; Verlander, M., *Biopolymers* **1983**, *22*, 2157.
5. Burke, T. R., Jr; Smyth, M. S.; Nomizu, M.; Otaka, A.; Roller, P. R., *J. Org. Chem.* **1993**, *58*, 1336.
6. Coleman, R. S.; Kong, J. S., *J. Am. Chem. Soc.* **1998**, *120*, 3538.
7. Richardson, P. L.; Gross, M. L.; Light-Wahl, K. J.; Smith, R. D.; Schepartz, A., *Bioorg. Med. Chem. Lett.* **1994**, *4*, 2133.
8. Dendrinos, K. G.; Kalivretenos, A. G., *J. Chem. Soc., Perkin Trans. 1* **1998**, 1463.

Duncan J. Wardrop & Edward G. Bowen
University of Illinois at Chicago, Chicago, IL

N-(9-Fluorenylmethyloxycarbonyl) oxysuccinimide

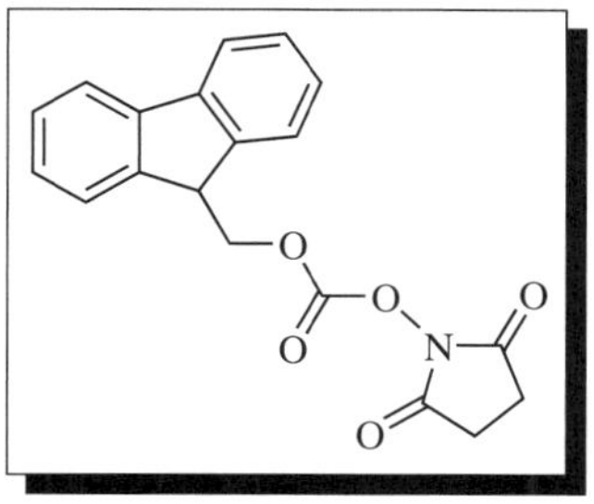

[82911-69-1] $C_{19}H_{15}NO_5$ (MW 337.34)

Alternate Name: Fmoc-OSu.
Physical Data: mp 151 °C.
Solubility: soluble in organic solvents such THF, dioxane, DMF, CH_2Cl_2 as well as alcohols. Please see also the section on Fmoc-Cl in this encyclopedia; Fmoc-OSu undergoes many of the same reactions as Fmoc-Cl, but at a somewhat slower rate than Fmoc-Cl.
Form Supplied in: white solid.
Analysis of Reagent Purity: HPLC and NMR.
Preparative Methods: the title reagent can be prepared by reaction of fluorenylmethanol with phosgene and *N*-hydroxysuccinimide,[1] by treating Fmoc-Cl with *N*-hydroxysuccinimide[2] or by reacting 9-fluorenylmethanol with succinimidyl chloroformate.[3]
Purification: recrystallization from chloroform/ether.
Handling, Storage, and Precautions: store dry.

The 9-fluorenylmethoxycarbonyl protecting group for amines was introduced in 1970 by Carpino.[4] It is exceptionally stable towards acid; thus, carboxylic acids that contain Fmoc-groups can be converted to acid chlorides with thionyl chloride[5] or *tert*-butyl esters using H_2SO_4 and isobutene.[6] In addition, Fmoc-groups are unaffected by HBr in HOAc or CF_3COOH and, therefore, allow the selective deprotection of Cbz and Boc groups.

The traditional Fmoc protection of amino acids uses Fmoc-Cl in a Schotten–Baumann-type procedure.[4,7] It was later found that this procedure can lead to the formation of significant amounts of Fmoc oligopeptides that can be difficult to remove.[2]

Two methods have been developed that allow the synthesis of Fmoc amino acids free of any corresponding Fmoc oligopeptides: one procedure uses Fmoc-OSu in mixtures of water and organic solvent together with a base,[1,2,3] whereas the other consists of silylation of an amino acid with TMS-Cl and a tertiary amine base in an organic solvent followed by treatment with Fmoc-Cl.[8] Concepts from each of these methods have recently been combined into a procedure that affords Fmoc amino acids in high purity under neutral, anhydrous conditions.[9] In this procedure a pertrimethylsilyl amino acid (**2**) is formed with *N*-methyl-*N*-(trimethylsilyl)trifluoroacetamide; **2** is then treated with Fmoc-OSu, followed by methanol (eq 1).

For large scale Fmoc amino acid preparations, the less expensive N,O-bis(trimethylsilyl)acetamide (BSA) may be employed. This is illustrated with the scale-up of a photolinker **3** for combinatorial chemistry, which was achieved using this anhydrous method (eq 2).[9]

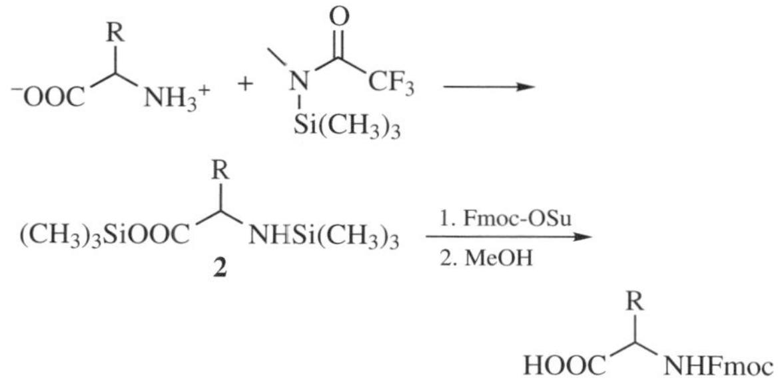

(1)

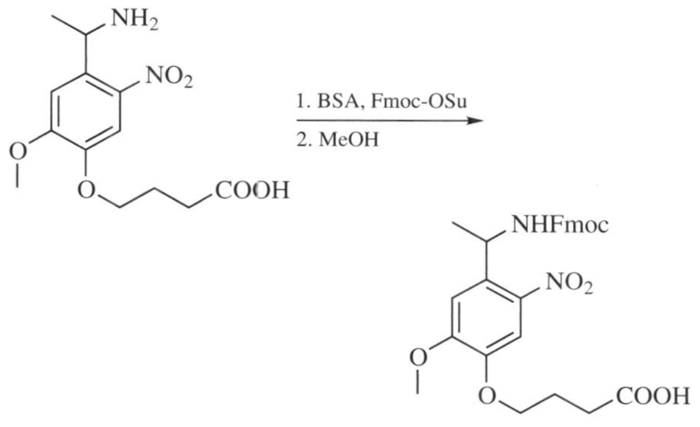

(2)

3

Other less reactive Fmoc-reagents in addition to Fmoc-OSu have been used to introduce the Fmoc-group, such as Fmoc-benzotriazole (**4**)[3] and Fmoc-pentafluorophenyl (**5**).[10]

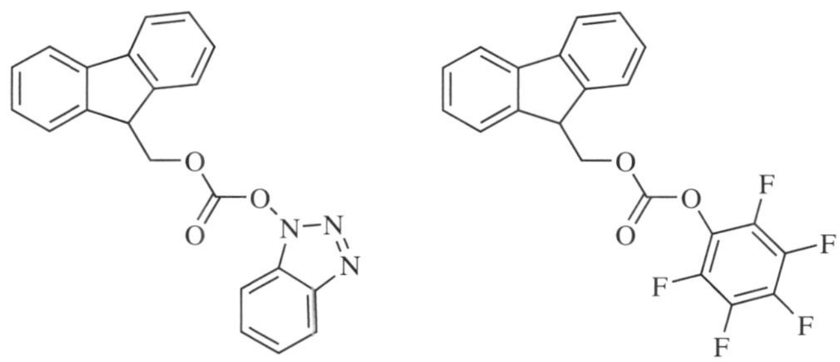

4 **5**

The use of the Fmoc protection group in peptide synthesis has been extensively reviewed.[11]

The Fmoc-group is susceptible to hydrogenolysis but at a slower rate than *O*-benzyl systems. One can, therefore, expect some selectivity as has been shown in a one-pot conversion of benzyl carbamates (**6**) into fluorenylmethyl carbamates (**7**) by hydrogenation with a poisoned catalyst (eq 3).[12]

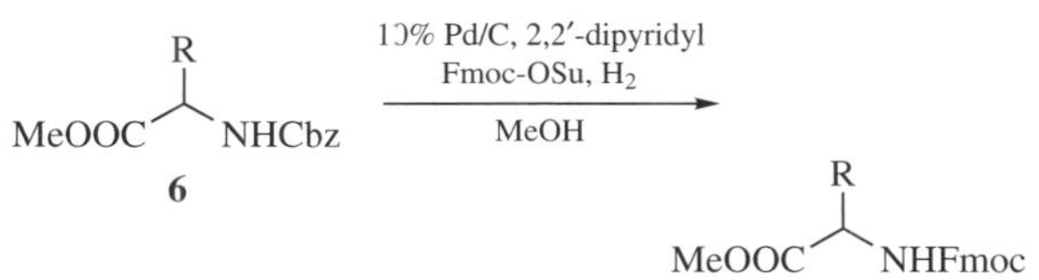

(3)

7

Cleavage of the Fmoc-group is achieved via a rapid nonhydrolytic elimination mechanism by treatment of the Fmoc-compound **8** with an amine base such as ammonia, piperidine, morpholine, triethylamine, and others, usually in polar solvents such as DMF, NMP or acetonitrile.[13] Other bases that have been used include TBAF[14] and DBU.[15] Generally a large excess of amine base is used because it acts as a scavenger to trap the liberated dibenzofulvene (**9**) to form **10** (eq 4).[16]

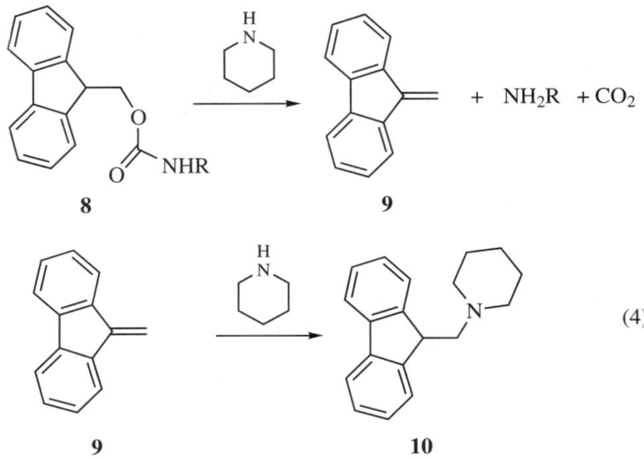

(4)

A drawback of the Fmoc-group is that Fmoc-protected compounds frequently show low solubility in common organic solvents. In addition, the by-products of the deprotection step can often only be removed via a chromatographic step. This has recently led to the development of Fmoc* (**11**) as a more soluble alternative to Fmoc.[17]

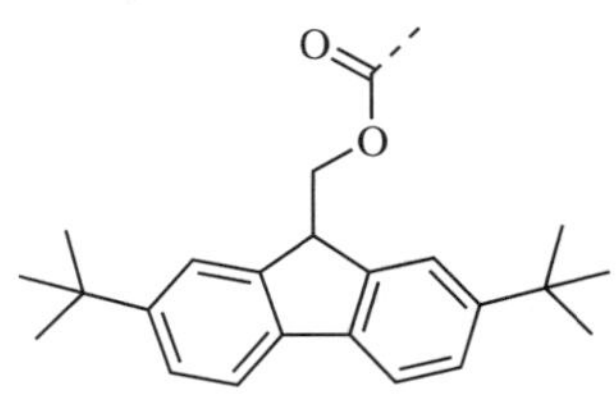

11

An added benefit of **11** is that the piperidine adduct that forms upon deprotection with piperidine in DMF is highly lipophilic and can be removed from a DMF solution of the deprotected amine by extraction with hexane.

Another convenient and scaleable procedure for removing the Fmoc-group in solution uses catalytic DBU in the presence of a polymer-supported thiol scavenger (**12**) (eq 5). The resin is commercially available and can be recycled.[16]

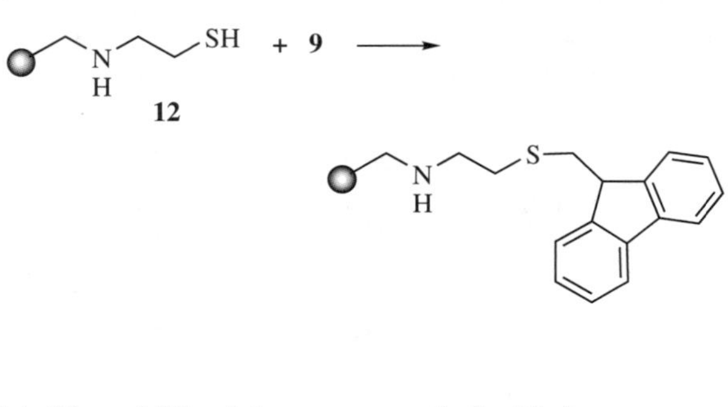

(5)

It was recently demonstrated that Fmoc-OSu can be converted into 2-nitro Fmoc-OSu (**13**) in one step by treatment with nitric acid (eq 6).[18] The nitro compound **13** can then be used as a photocleavable protecting group.

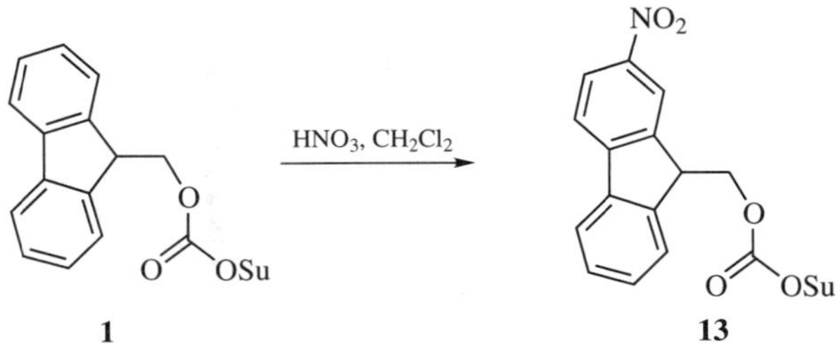

(6)

Related Reagents. Fmoc-Cl.

1. Ten Kortenaar, P. B. W.; Van Dijk, B. G.; Peeters, J. M.; Raaben, B. J.; Adams, P. J. H. M.; Tesser, G. I., *Int. J. Pept. Prot. Res.* **1986**, *27*, 398.
2. Sigler, G. F.; Fuller, W. D.; Chaturvedi, N.; Goodman, M.; Verlander, M., *Biopolymers* **1983**, *22*, 2157.
3. Paquet, A., *Can. J. Chem.* **1982**, *60*, 976.
4. Carpino, L. A.; Han, G. Y., *J. Am. Chem. Soc.* **1970**, *92*, 5748.
5. Carpino, L. A.; Cohen, B. J.; Stephens, K. E.; Sadat-Aalaee, S. Y.; Tien, J.-H.; Langridge, D. C., *J. Org. Chem.* **1986**, *51*, 3732.
6. Kelly, R. C.; Gebhard, I.; Wicnienski, N., *J. Org. Chem.* **1986**, *51*, 4590.
7. Chang, C. D.; Waki, M.; Ahmad, M.; Meienhofer, J.; Lundell, E. O.; Haug, J., *Int. J. Pept. Prot. Res.* **1980**, *15*, 59.
8. Bolin, D. R.; Sytwu, I. I.; Humiec, F.; Meienhofer, J., *Int. J. Pept. Prot. Res.* **1989**, *33*, 353.
9. Raillard, S. P.; Mann, A. D.; Baer, T. A., *Org. Prep. Proc. Int.* **1998**, *30*, 183.
10. Schön, I.; Kisfaludy, L., *Synthesis* **1986**, 303.
11. Udenfriend, S.; Meienhofer, J., *The Peptides*; Academic: San Diego, 1987; Vol. 9, p 1.
12. Dzubeck, V.; Schneider, J., *Tetrahedron. Lett.* **2000**, *41*, 9953.
13. Greene, T.; Wuts, P. G. M., *Protective Groups in Organic Synthesis*, 3rd ed.; Wiley: New York, **1999**, p 506.
14. Schmidt, U.; Mundinger, K.; Mangold, R.; Lieberknecht, A., *J. Chem. Soc. Chem. Comm.* **1990**, 1216.
15. Ueki, M.; Amemiya, M., *Tetrahedron. Lett.* **1987**, *28*, 6617.
16. Sheppeck, J. E., Jr; Kar, H.; Hong, H., *Tetrahedron. Lett.* **2000**, *41*, 5329.
17. Stigers, K. D.; Koutroulis, M. R.; Chung, D. M.; Nowick, J. S., *J. Org. Chem.* **2000**, *65*, 3858.
18. Henkel, B.; Bayer, E., *Synthesis* **2000**, 1211.

Stephen P. Raillard
Affymax Research Institute, Santa Clara, CA, USA

O-(9-Fluorenylmethyloxycarbonyl) Pentafluorophenol

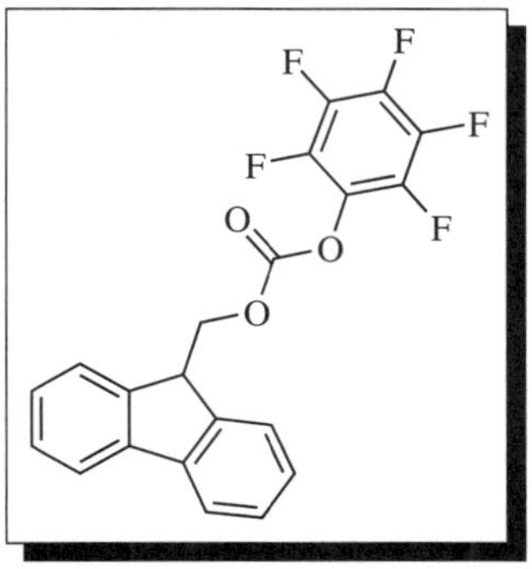

[88744-04-1] $C_{21}H_{11}F_5O_3$ (MW 406.305)

(used for the preparation of *N*-Fmoc amino acids and their pentafluorophenyl esters;[1,2] used in peptide synthesis;[3] used as a fluorescent labeling agent for amino acids[4])

Alternate Name: *O*-(9-fluorenylmethoxycarbonyl)pentafluorophenol; Fmoc-pfp.
Physical Data: mp 84–86 °C.
Solubility: soluble in organic solvents such as CH_2Cl_2, THF, acetone; slightly soluble in *n*-hexane
Form Supplied in: widely available as white crystalline solid.
Handling, Storage, and Precautions: bottles may develop pressure upon prolonged storage and should be opened carefully. Should be kept away from moisture and heat.

***N*-Terminal Protection in Peptide Synthesis.** The 9-fluorenylmethoxycarbonyl (Fmoc) group is widely used for the protection of amino acids in peptide synthesis.[1] It is stable to strong acidic conditions but can be cleaved very rapidly with mild bases, such as 20% piperidine in DMF. The combination of the Fmoc *N*-terminal protecting group and the appropriate side-chain protections has become the standard method for the solid-phase synthesis of peptides,[2b,5] and the method is applied in the preparation of combinatorial libraries (eq 1).[5a,5c]

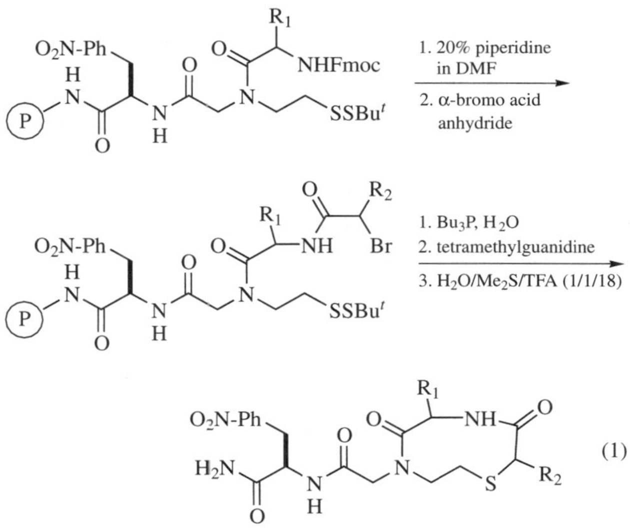

(1)

(P) = {4-[(2',4'-dimethoxyphenyl)aminomethyl] phenoxy}acetic acid derivatized resin

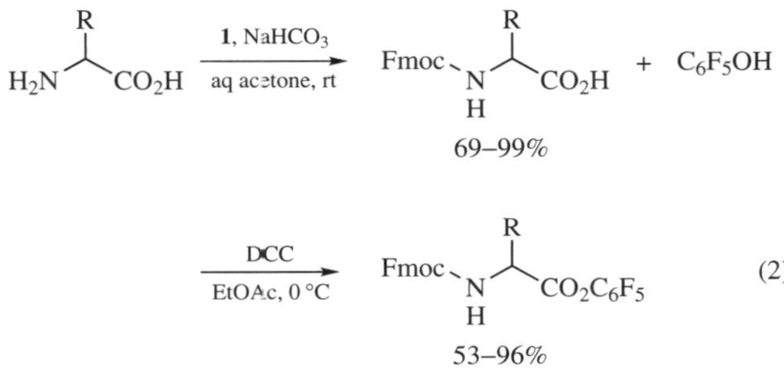

(2)

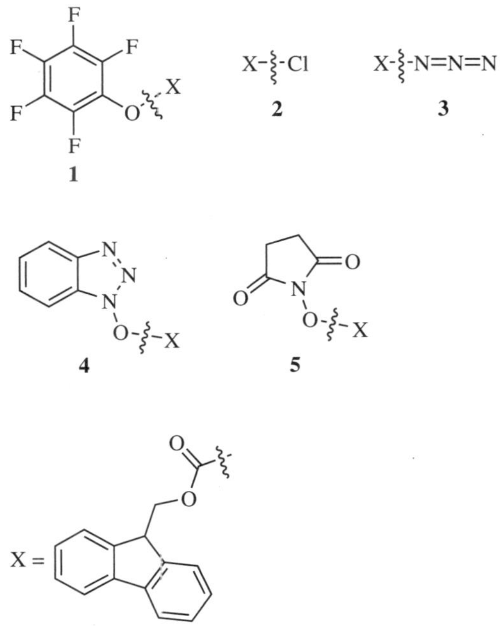

Many reagents have been developed for the incorporation of the Fmoc group. The widely used chloroformate ester Fmoc-Cl (**2**) has been applied in the peptide synthesis for two decades.[6] However, this chloroformate is relatively unstable and has been shown to promote the formation of undesired 'Fmoc-dipeptides' in significant amounts (3–20%), presumably via mixed anhydride intermediates.[7] The corresponding azide derivative Fmoc-N_3 (**3**) is claimed to reduce dipeptide formation,[8] but is rarely used in peptide synthesis due to its potential explosive nature as well as storage problems. Fmoc-OBt (**4**),[9] Fmoc-OSu (**5**),[7,8,10] and Fmoc-pfp (**1**)[2] have been developed to avoid these problems and are used preferentially by many workers. These reagents provide the *N*-protected amino acids in comparable yields without the formation of the dipeptide by-products.[2,7,9,10]

Fmoc-pfp is of particular interest because Fmoc-protected amino acids as well as the corresponding pfp esters can be prepared in one pot. Fmoc protection of amino acids using this reagent is carried out in $NaHCO_3$/water/acetone (eqs 2[2a] and 4[2b]) or under anhydrous conditions (e.g., base/DMF)[2c] at room temperature. The anhydrous conditions were used to prepare a new *O*-phosphotyrosine isosteric building block for incorporation into SH2-binding peptides (eq 3).[2c] The corresponding pfp esters are obtained by adding a coupling reagent (such as DCC) to the ethyl acetate extracts of the reaction mixture (eqs 2[2a] and 4[2b]) or directly from the amino acids by using excess Fmoc-pfp under anhydrous conditions (eq 5).[2d] This *N*-protection/*C*-activation procedure is particularly useful in glycopeptide assembly,[3] such as in the synthesis of 2-acetamido-2-deoxy-β-D-glucopyranose *O*-glycopeptides (eqs 6 and 7).[3e] The advantage of this method is that the pfp ester is stable under the conditions of glycosidic bond formation; it therefore acts as both a *C*-terminal protecting group and activated group for peptide bond coupling.

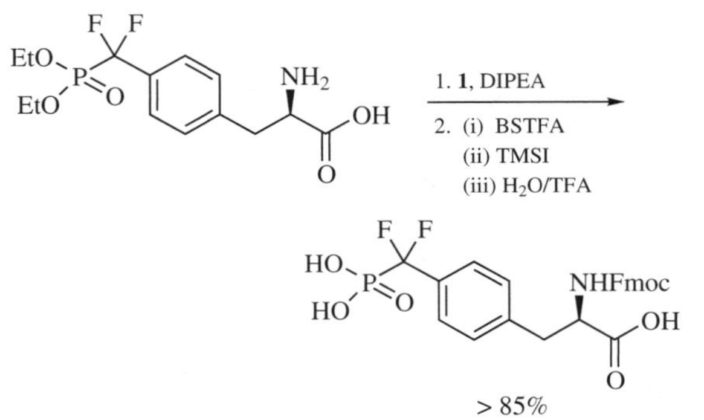

(3)

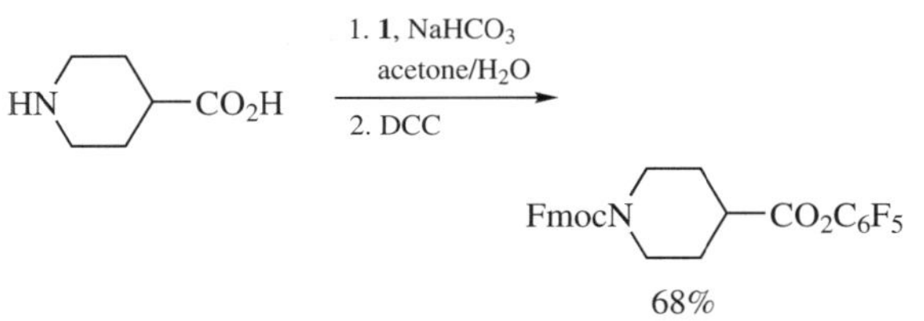

(4)

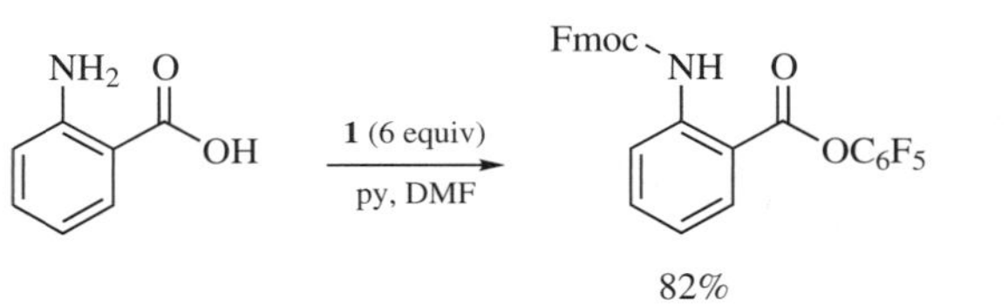

(5)

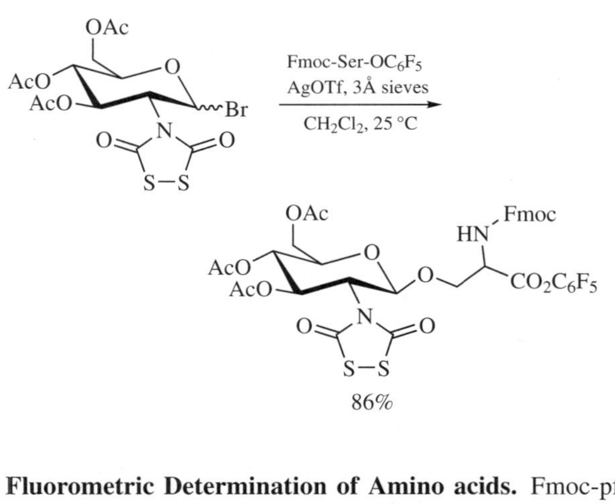

(6)

Fluorometric Determination of Amino acids. Fmoc-pfp is used in the derivatization of amino acids for fluorimetric detection. For example, the derivatization and fluorometric determination of D-penicillamine, a very useful hydrolytic product of penicillin, has been described.[4] This method is both very specific and ultrasensitive. The protection of the amino group was best performed in the range of pH 6.5–7.0 using borate buffer (eq 8); Fmoc-pfp being the preferred protecting reagent because Fmoc-Cl reacts with water to yield the corresponding fluorescent alcohol.[11]

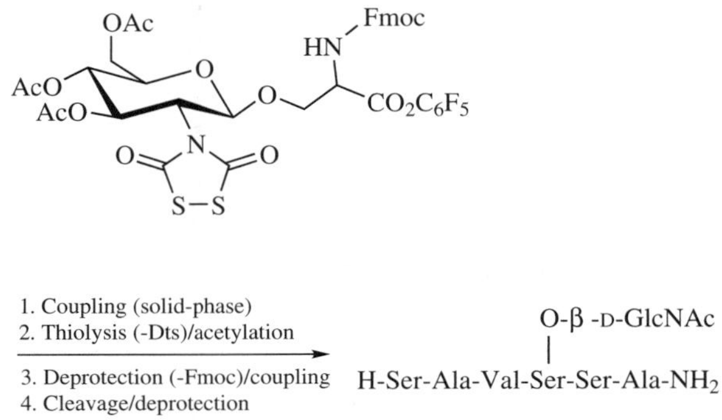

(7)

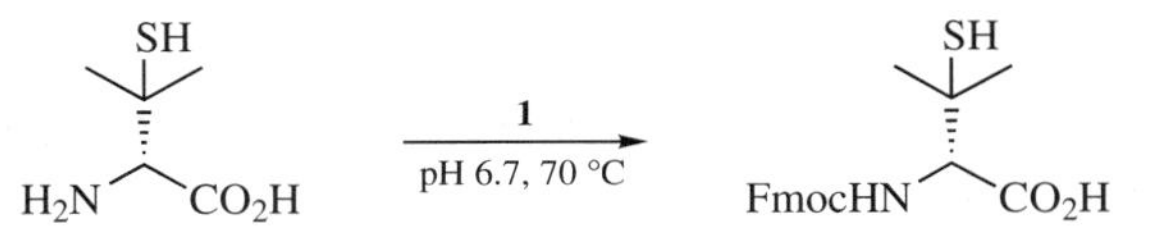

(8)

Related Polymer-supported Reagents. Recently, polymer-supported reagents for the Fmoc-protection of amino acids have been developed because of easy recycling of reagents and the simplification of product isolation.[12] These reagents are generated from polystyrene-bound HOBt (**6**),[12a] HOSu (**7**),[12b] **8**,[12c] and Fmoc-Cl. The Fmoc-protection reaction is performed under similar conditions to those used with Fmoc-pfp. The Fmoc-protected amino acids are isolated in comparable yields and the residual polymer-supported succinimide or hydroxybenzotriazole generated after the protection reaction can be separated by simple filtration and reused.

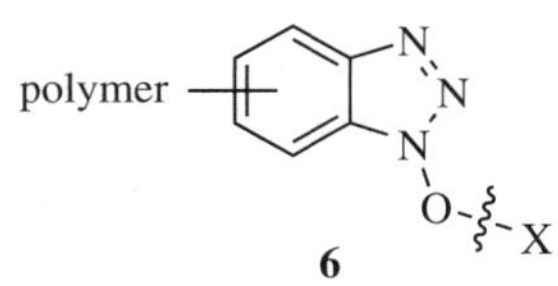

6

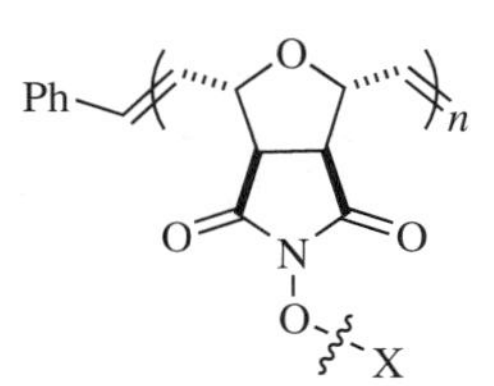

7

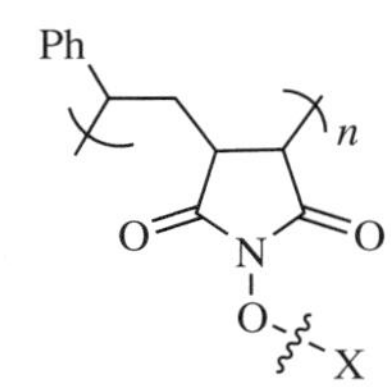

8

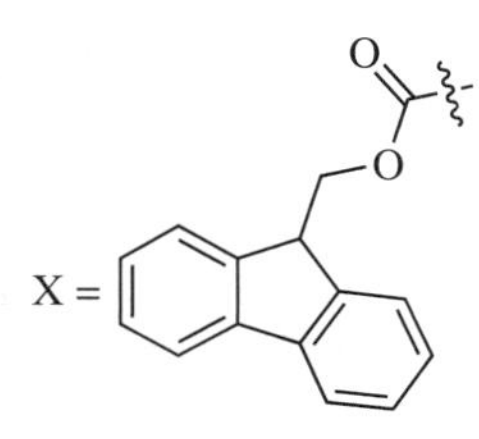

1. (a) *The Practice of Peptide Synthesis*; Bodanszky, M.; Bodanszky, A., Eds, 2nd ed.; Springer-Verlag: Berlin, 1994. (b) *Principles of Peptide Synthesis*; Bodaszky, M., Ed., 2nd ed.; Springer-Verlag: Berlin, 1993.
2. (a) Schön, I.; Kisfaludy, L., *Synthesis* **1986**, 303. (b) Zych, A. J.; Iverson, B. L., *J. Am. Chem. Soc.* **2000**, *122*, 8898. (c) Gordeev, M. F.; Patel, D. V.; Barker, P. L.; Gordon, E. M., *Tetrahedron Lett.* **1994**, *35*, 7585. (d) Gayo, L. M.; Suto, M. J., *Tetrahedron Lett.* **1996**, *37*, 4915.
3. (a) Jansson, A. M.; Meldal, M.; Bock, K., *Tetrahedron Lett.* **1990**, *31*, 6991. (b) Peters, S.; Bielfeldt, T.; Meldal, M.; Bock, K.; Paulsen, H., *Tetrahedron Lett.* **1992**, *33*, 6445. (c) Paulsen, H.; Bielfeldt, T.; Peters, S.; Meldal, M.; Bock, K., *Liebigs. Ann. Chem.* **1994**, 369. (d) Paulsen, H.; Peters, S.; Bielfeldt, T.; Meldal, M.; Bock, K., *Carbohydr. Res.* **1995**, *268*, 17. (e) Jensen, K. J.; Hansen, P. R.; Venugopal, D.; Barany, G.; *J. Am. Chem. Soc.* **1996**, *118*, 3148.
4. Byeon, S. Y.; Choi, J. K.; Yoo, G. S., *Yakche Hakhoechi* **1990**, *20*, 73 (*Chem. Abstr.* **1991**, *114*, 150276).
5. (a) Virgilio, A. A.; Ellman, J. A., *J. Am. Chem. Soc.* **1994**, *116*, 11580. (b) Atherton, E.; Logan, C. J.; Sheppard, R. C., *J. Chem Soc., Perkin Trans.* **1981**, 538. (c) Gordon, E. M.; Barrett, R. W.; Dower, W. J.; Foder, S. P. A.; Gallop, M. A., *J. Med. Chem.* **1994**, *37*, 1385.
6. (a) Polt, R.; Szabo, L.; Treiberg, J.; Li, Y.; Hruby, V. J., *J. Am. Chem. Soc.* **1992**, *114*, 10249. (b) Lüning, B.; Norberg, T.; Tejbrant, J., *J. Chem. Soc., Chem. Commun.* **1989**, 1267.
7. Lapatsanis, L.; Milias, G.; Froussios, K.; Kolovos, M., *Synthesis* **1983**, 671.
8. (a) Carpino, L. A.; Han, G. Y., *J. Org. Chem.* **1972**, *37*, 3404. (b) Tessier, M.; Albericio, F.; Pedroso, E.; Grandas, A.; Eritja, R.; Giralt, E.; Granier, C.; Van Rietschoten, J., *Int. J. Pept. Protein Res.* **1983**, *22*,125.
9. Paquet, A., *Can. J. Chem.* **1982**, *60*, 976.
10. (a) Ten Kortenaar, P. B. W.; van Dijk, B. G.; Peeters, J. M.; Raaben, B. J.; Adams, P. J. H. M.; Tesser, G. I., *Int. J. Pept. Protein. Res.* **1986**, *27*, 398. (b) Bardaji, E.; Torres, J. L.; Clapes, P.; Albericio, F.; Barany, G.; Rodriguez, R. E.; Sacristin, M. P.; Valencio, G., *J. Chem Soc., Perkin Trans. 1.* **1991**, 1755.
11. Einarsson, S.; Josefsson, B.; Lagerkvist, S., *J. Chromatogr.* **1983**, *282*, 609.
12. (a) Dendrinos, K. G.; Kalivretenos, A. G., *J. Chem Soc., Perkin Trans. 1* **1998**, 1463. (b) Barrett, A. G. M.; Cramp, S. M.; Roberts, R. S.; Zecri, F. J., *Org. Lett.* **2000**, *2*, 261. (c) Chinchilla, R.; Dodsworth, D. J.; Nájera, C.; Soriano, J. M., *Tetrahedron Lett.* **2001**, *42*, 7579.

Xiaolin Hao
Tularik Inc., South San Francisco, CA, USA

H

Hexafluorophenylpropanol

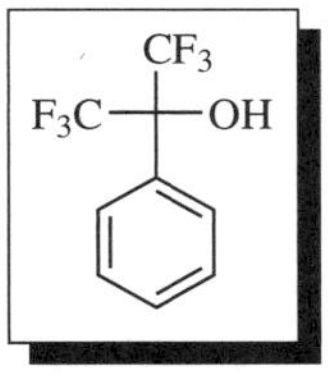

[718-64-9] $C_9H_6F_6O$ (MW 244.13)

(reagent used for the protection of primary alcohols; as a precursor to sulfuranes and other hypervalent substances)

Alternate Name: 2-phenyl-hexafluoro-2-propanol; α,α-bis(trifluoromethyl)benzyl alcohol, hexafluorophenylisopropanol.
Physical Data: bp 160 °C; *d* 1.450 g cm^{-3}; fp 60 °C.
Solubility: soluble in most organic solvents.
Preparative Methods: from benzene and hexafluoroacetone in the presence of various Lewis acids in a Friedel-Crafts reaction,[1] or by nucleophilic trifluoromethylation of α,α,α-trifluoroacetophenone.[2]

Protecting Group for Primary Alcohols. Hexafluoro-2-phenyl-2-propanol serves as the acidic component and nucleophile in Mitsunobu reactions with primary and secondary alcohols, leading to the formation of hexafluorophenylpropyl (HIP) ethers.[3–5] As is typical for the Mitsunobu reaction, primary alcohols are functionalized more rapidly and efficiently than secondary alcohols; tertiary alcohols are recovered unchanged. This provides a convenient means of selective protection of primary alcohols in polyols (eq 1).[3b] HIP ethers are stable across a broad pH range and

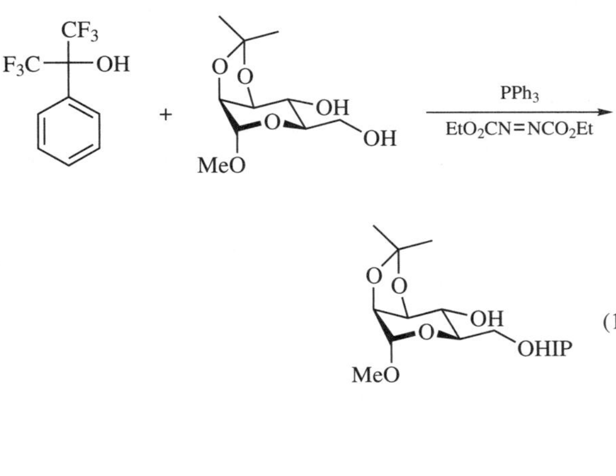

(1)

are resistant to oxidants such as PCC, NBS, nucleophiles such as MeLi, N_2H_4, Lewis acids, $BF_3.Et_2O$, and many reducing agents including Zn/HCl, Na (Hg), H_2 (1atm)/Pd-C, and $LiAlH_4$.[3b] The chemical stability of HIP ethers enables the selective removal of numerous other protecting groups in their presence, including trityl, tetrahydropyranyl, benzyl, and silyl ethers, and esters.[3b] In addition, HIP ethers are also amenable to mass spectrometry and gas chromatography with the fluorine atoms rendering them suitable for electron capture detection.[3b]

The sodium salt of 2-phenylhexafluoro-2-propanol may also be alkylated in DMF solution by reaction with such powerful electrophiles as dimethyl sulfate, trifluoroethyl trifluoromethanesulfonate, and 1,1,2,2,3,3-hexahydroperfluoroundecyl iodide, giving the corresponding HIP ethers.[4]

In common with trifluoroethyl ethers,[6] HIP ethers are cleaved with stoichiometric lithium naphthalenide at −78 °C in THF; many common functional groups including amides, carboxylic acids, isolated olefins, and acetylenes are stable under these conditions.[3b] Esters, epoxides, ketones and halides, however, do not resist the lithium naphthalenide cleavage conditions.[3b]

Use as a Supporting Ligand for Hypervalent Atoms and High Oxidation State Metals. An important use of 2-phenylhexafluoro-2-propanol is as a readily exchangeable, supporting ligand for diphenylbis(1,1,1,3,3,3-hexafluoro-2-phenyl-2-propoxy) sulfurane, the Martin sulfurane (eq 2),[7] with its myriad uses in organic synthesis.

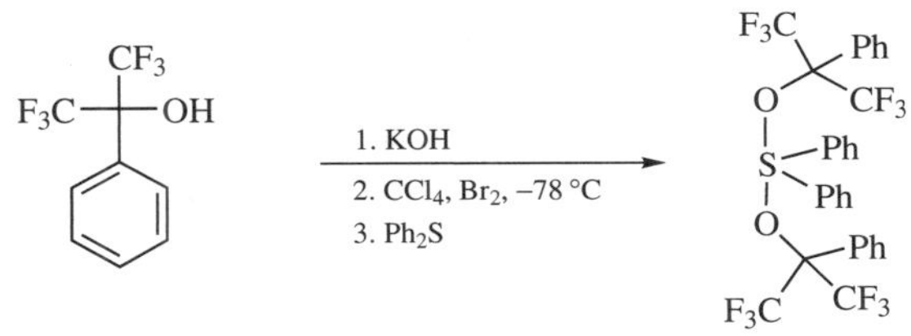

(2)

Metallation of 2-phenyl-hexafluoro-2-propanol with *sec*-butyllithium affords the *o*-lithiated alkoxide which can be quenched with trichloroalkyl silanes, thionyl chloride, and phosphorus oxychloride leading to a series of hypervalent silicon, sulfur, and phosphorus derivatives stabilized by the bidentate ligand (eq 3).[8]

With a catalytic amount of rhodium on carbon, hexafluorophenylpropanol is converted to 2-cyclohexyl-perfluoro-2-propanol (eq 4), an alcohol with the potential to act as stabilizing ligand for high oxidation state metals.[9]

Reaction with Cyanogen Chloride. On reaction of the sodium or potassium salt of hexafluorophenylpropanol with cyanogen chloride, the initial cyanate ester undergoes rearrangement to give *o*-hexafluoroisopropylphenyl carbamate, a reaction which is facilitated by the electron-withdrawing effect of the two trifluoromethyl groups (eq 5).[10]

Related Reagents. Perfluoro-*tert*-butanol; triphenylmethyl chloride; diphenylbis(1,1,1,3,3,3-hexafluoro-2-phenyl-2-propoxy) sulfurane.

---

1. (a) Sepioł, J.; Soulen, R. L., *J. Fluorine Chem.* **1984**, *24*, 61. (b) Krespan, C. G., *J. Org. Chem.* **1979**, *44*, 4924. (c) Farah, B. S.; Gilbert, E. E.; Sibilia, J. P., *J. Org. Chem.* **1965**, *30*, 998.

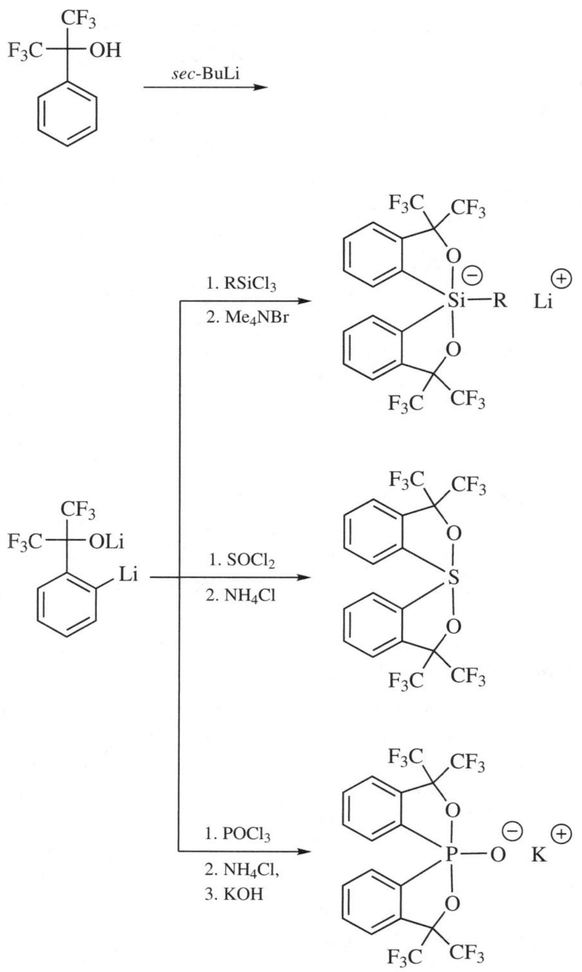

(3)

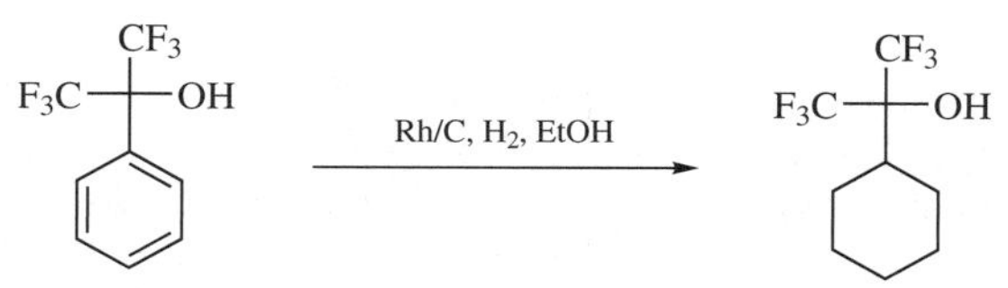

(4)

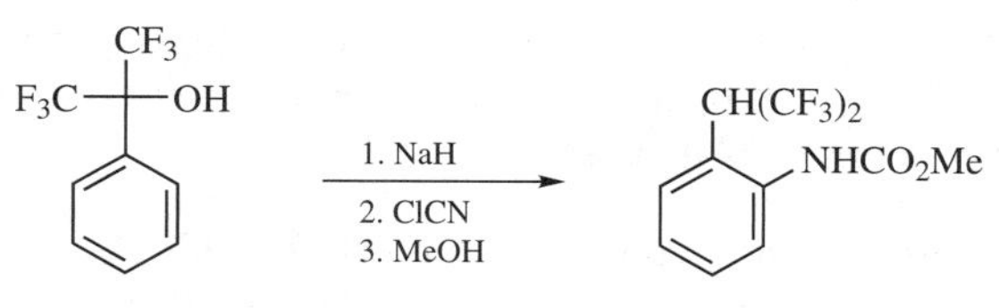

(5)

2. Hartkopf, W.; De Meijere, A., *Angew. Chem. Int. Ed. Engl.* **1982**, *21*, 443.
3. (a) Falck, J. R.; Yu, J.; Cho, H.-S., *Tetrahedron Lett.* **1994**, *35*, 5997. (b) Cho, H.-S.; Yu, J.; Falck, J. R., *J. Am. Chem. Soc.***1994**, *116*, 8354.
4. Rábai, J.; Szabó, D.; Borbás, E. K.; Kövesi, I.; Kövesdi, I.; Csámpai, A.; Gömöry, A.; Pashinik, V. E.; Shermolovich, Y. G.; *J. Fluorine Chem.* **2002**, *114*, 199.
5. Nga, T. T. T.; Ménage, C.; Bégué, J. P., Bonnet-Delpon, D.; Gantier, J. C.; Pradines, B.; Doury, J. C.; Thac, T. D., *J. Med. Chem.* **1998**, *41*, 4101.
6. Sargent, G. D., *J. Am. Chem. Soc.* **1971**, *93*, 5268.
7. (a) Martin, J. C.; Arhart, R. J., *J. Am. Chem. Soc.* **1971**, *93*, 4327. (b) Martin, J. C.; Arhart, R. J.; Franz, J. A.; Perozzi, E. F.; Kaplan, L. J., *Org. Synth. Coll. Vol 6.* **1988**, 163; (c) Kaplan, L. J.; Martin, J. C., *J. Am. Chem. Soc.* **1973**, *95*, 793 (d) Perozzi, E. F.; Martin, J. C., *J. Org Chem.* **1977**, *42*, 3222. (e) Martin, J. C.; Arhart, R. J., *J. Am. Chem. Soc.* **1971**, *93*, 2339.
8. (a) Perozzi, E. F.; Martin, J. C., *J. Am. Chem. Soc*, **1979**, *101*, 1591. (b) Farnham, W. B.; Harlow, R. L., *J. Am. Chem. Soc.* **1981**, *103*, 4608.
9. Barbarich, T. J.; Nolan, B. G.; Tsujioka, S.; Miller, S. M.; Anderson, O. P.; Strauss, S. H., *J. Fluorine Chem.* **2001**, *112*, 335.
10. Krespan, C. G., *Org. Chem.* **1969**, *34*, 1278.

Zongmin Dai
Protez Pharmaceuticals, Inc., Chicago, IL, USA

2,5-Hexanedione

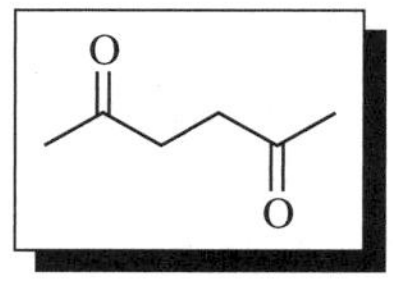

[110-13-4] $C_6H_{10}O_2$ (MW 114.14)

(reagent used for the protection of amino groups in amino sugars and nucleosides; for the synthesis of five-membered heterocycles, in particular 2,5-dimethyl aminopyrroles, indane-type and benzannulated systems).

Alternate Name: acetonylacetone, diacetonyl, 1,2-diacetylethane, and 2,5-diketohexane.
Physical Data: clear to yellow liquid, bp 188–191 °C.
Solubility: soluble in most organic solvents. Miscible with water.
Analysis of Reagent Purity: ^{1}H NMR,[1] IR,[2] Mass spectrum (commercially available).[3]
Purification: distillation through a fractionating column (bp 78–79 °C /15 mm Hg).
Handling, Storage, and Precautions: shelf-stable, oral rat LD_{50}: 2076 mg/kg, NFPA ratings: health-1, flammability-2, reactivity-0.

Paal–Knorr Synthesis and Related Pyrrole Systems. 2,5-Hexanedione is a useful reagent in an array of organic transformations, particularly in the synthesis of heterocyclic systems. The enolizable 1,4-dicarbonyl moiety allows condensation to form five-membered heterocycles with various heteroatom functionalities in a typical Paal–Knorr[4] fashion. As part of a general investigation of the scope and limitations of the Paal–Knorr synthesis of 2,5-dimethylpyrroles by condensation of 2,5-hexanedione with primary amines, Buu-Hoi and Xuong have prepared a number of pyrrole derivatives presenting potential pharmacological interest. *N,N*-Diethyl-l,3-diaminopropane reacted vigorously with hexanedione to give l-α-(γ-diethylaminopropyl)-2,5-dimethylpyrrole (**1**) in excellent yield, while the secondary amine *N,N*-diethyl-1,4-diaminopentane reacted only sluggishly under the same conditions to afford l-β-(ω-diethylaminopentyl)-2,5-dimethylpyrrole (**2**).[5]

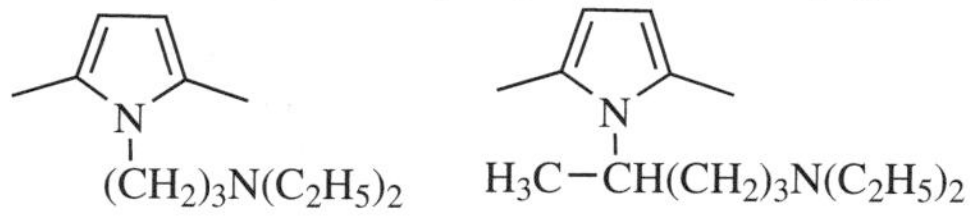

1 **2**

It is noteworthy that 2,5-dimethylpyrrole formation can also be effected in benzene by azeotropic distillation (eq 1).[6]

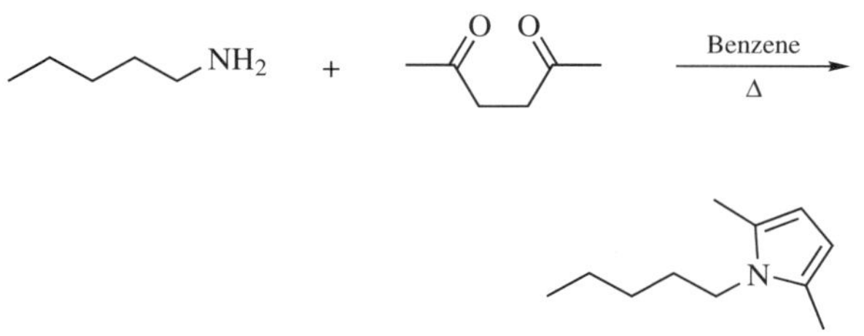

(1)

A facile synthesis of substituted pyrroles from 2,5-hexanedione and amines over montmorillonite KSF clay has been accomplished in excellent yield. Thus, reacting 2,5-hexanedione with $PhNH_2$ gave *N*-phenyl-2,5-dimethylpyrrole in 96% yield (eq 2). Other amines such as 1-naphthylamine and 2-pyridinamine gave similar products.[7]

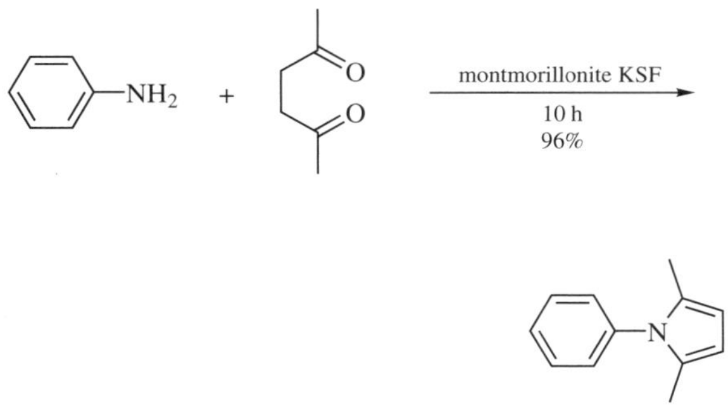

(2)

As a Protecting Group for 1° Amines. The use of 2,5-dimethylpyrrole as an amino protecting group in an efficient synthesis of 5-amino-3-{[*N*-methyl-pyrrolidin-2(*R*)-yl]methyl}indole, the precursor for the synthesis of CP-135,807, has been reported by Macor et al. In this system, the pyrrole protection satisfies the following criteria required: (a) non-ionizable (i.e., no NH in the protecting group); (b) stable to strong base (i.e., alkylmagnesium bromide); (c) stable to strong reducing agents (i.e., $LiAlH_4$); (d) relatively non-nucleophilic (i.e., low reactivity to acid chlorides); and (e) removable under conditions in which a 5-aminoindole derivative was stable (i.e., neutral or basic conditions) (eq 3).[8]

The amino groups in aminodeoxy sugars can be protected as 2,5-dimethylpyrroles, which are compatible with many protecting group manipulations and sets of glycosylation conditions. The dimethylpyrrole protecting group is readily installed by treatment of an amine with 2,5-hexanedione in the presence of triethylamine in methanol. When glucosamine hydrochloride was reacted under these conditions, followed by *O*-acetylation with acetic anhydride and pyridine, the fully protected sugar (**3**) was obtained in a yield of 51% (eq 4). This could then be easily converted to various acetylated glycosyl donors.[9]

Enhanced reactivity has been found for the pyrrole protected trichloroacetimidate donor (**4**) which upon TMSOTf-mediated glycosylation gave exclusive formation of β-glycosides (eq 5).[9]

Although the pyrrole group is highly electron rich and somewhat acid sensitive, the yields of glycosylations were generally high for both primary and secondary alcohols. The high β-selectivity is also remarkable since the C-2 dimethylpyrrole moiety is electronically a non-participating functionality. Probably, the α-face of the glycoside is sterically shielded, forcing the glycosyl acceptor to approach from the β-face. The post-glycosylation deprotection can be effected by treatment with hydroxylamine hydrochloride.[9]

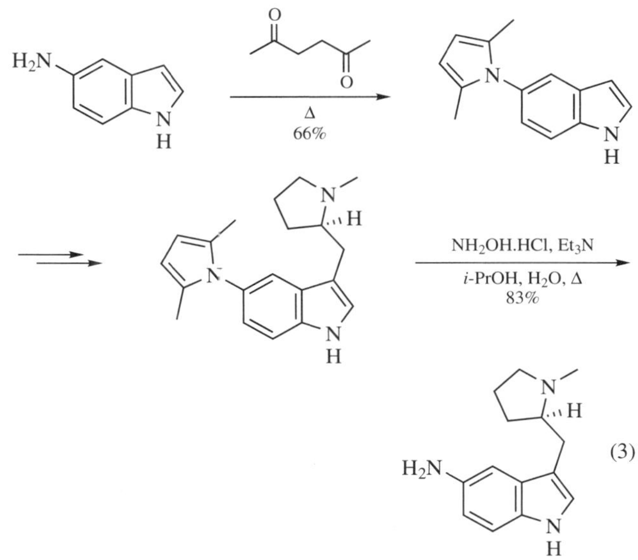

(3)

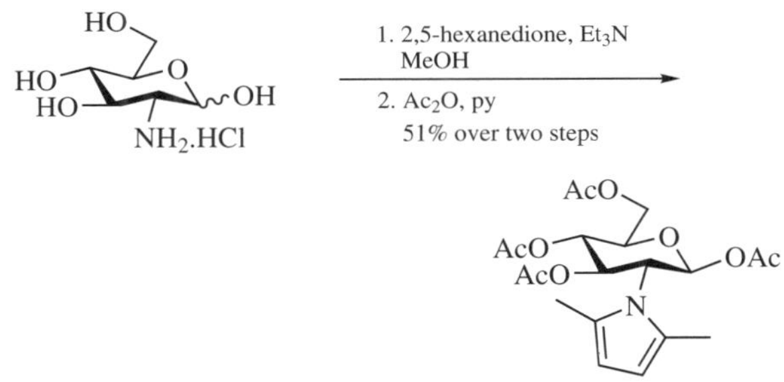

(4)

3

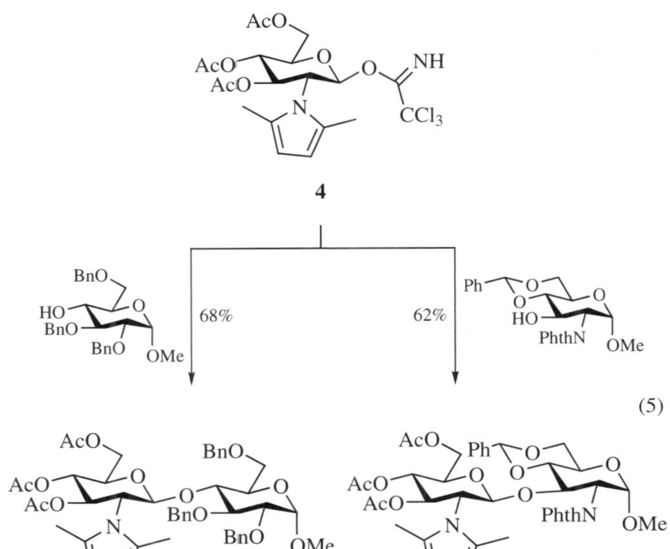

(5)

Protection of the amino group of both adenine and guanine nucleosides was achieved by heating the substrates in 2,5-hexanedione. The resulting 2,5-dimethylpyrrole adducts were stable toward bases but were hydrolyzed by TFA/H_2O (9:1) with regeneration of the amino function (eq 6).[10]

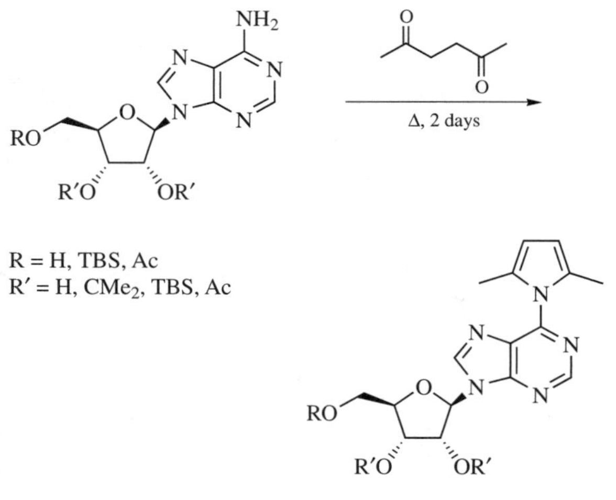

(6)

Synthesis of 2,5-Dimethylfuran, Thiophene, and Selenophene. An improvement of the sealed tube method of Paal[11] for the synthesis of 2,5-dimethylthiophene was reported by Campaigne and Foye. When acetonylacetone was added dropwise to a boiling mixture of phosphorus pentasulfide in dry tetralin and the crude 2,5-dimethylthiophene rapidly distilled, the product was obtained in 70% yield. When phosphorus pentoxide was substituted for the pentasulfide in this preparation, 2,5-dimethylfuran was obtained in 53% yield.[12] P_4Se_{10} prepared from red amorphous Se was sufficiently reactive toward 1,4-diketones to effect their cyclization to 2,5-disubstituted selenophenes in moderate yields. Thus, 2,5-hexanedione was cyclized in 38% yield to give 2,5-dimethylselenophene (eq 7).[11,13]

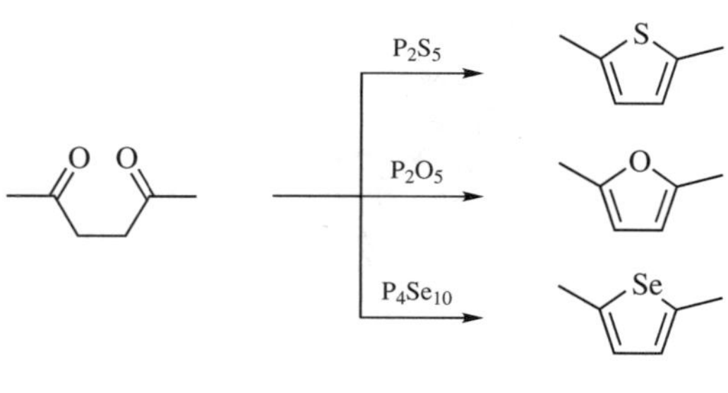

(7)

Indane-Type and Benzannulated Systems. Indane-type compounds are obtained when 2,5-hexanedione and catechol are subjected to condensation in the presence of an acid catalyst (eq 8).[14]

With pyrogallol and hydroquinone, 2,5-hexanedione undergoes cyclization followed by aromatization. Pyrogallol reacted with one molecule of hexanedione, whereas hydroquinone underwent a double cyclization-aromatization sequence with two molecules of dione. In the first case, 1,2,3-tri-hydroxy-5,8-dimethyl naphthalene was produced (eq 9), while in the second case 1,4,5,8-tetramethyl anthraquinone was formed.[15]

Carbazole derivatives are obtained by the reaction of indoles with 2,5-hexanedione in the presence of Lewis acids, thus 5,6-methylenedioxyindole and the title reagent were condensed in the presence of BF_3-etherate in refluxing dioxane to give 6,9-dimethyl-5*H*-[1,3]dioxolo[4,5-*b*]carbazole (eq 10).[16]

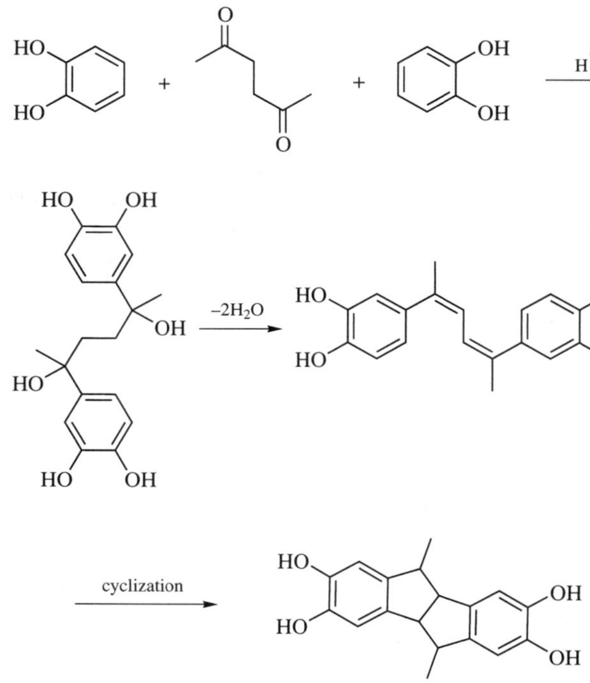

(8)

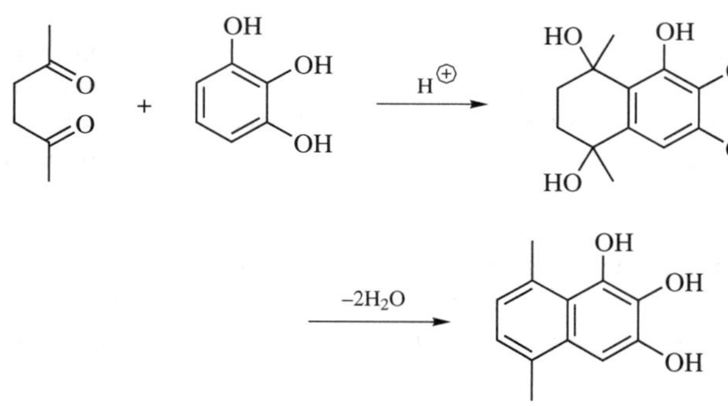

(9)

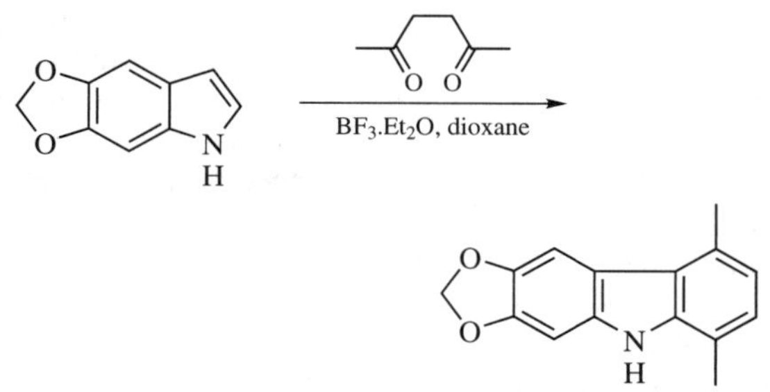

(10)

Treatment of 2,5-hexanedione with cyclopentadiene in methanol in the presence of 2.5 equiv of sodium methoxide at ambient temperature gave exclusively 4,7-dimethylindene, which was isolated in 84% yield. Under these conditions, the intramolecular condensation step starting from intermediate **5** is apparently faster than the introduction of a second fulvene unit. In contrast, bis(fulvene) was obtained and isolated in 89% yield when 2,5-hexanedione was treated with 2 mol equiv of cyclopentadiene in MeOH at 0 °C in the presence of 1.5 equiv of pyrrolidine (eq 11).[17]

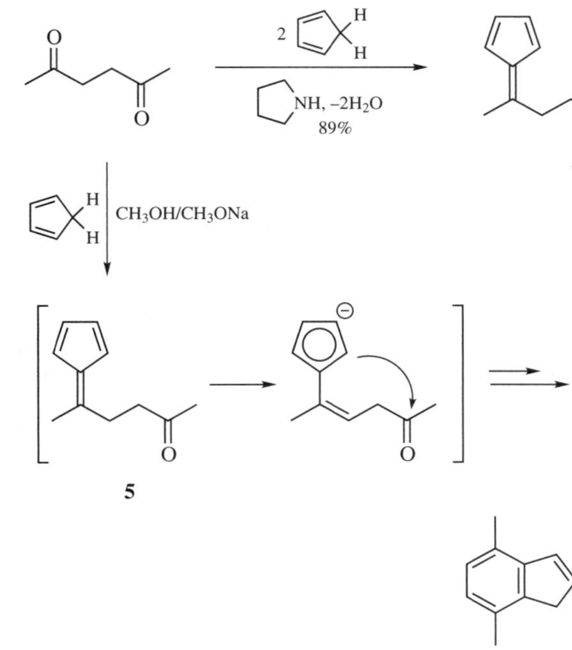

(11)

Miscellaneous Reactions. 2,5-Hexanedione was evaluated as a substrate in the Friedländer reaction with 2-aminopyridine-3-carbaldehyde. Under the typical Friedländer conditions (NaOH, EtOH), little or none of **6** is observed, with isomers **7** and **8** being formed in a 70:30 ratio, respectively. In contrast, the TABO (1,3,3-trimethyl-6-azabicyclo[3.2.1]octane) catalyzed Friedländer provides bis(naphthyridine) (**6**) as the predominant product (eq 12).[18]

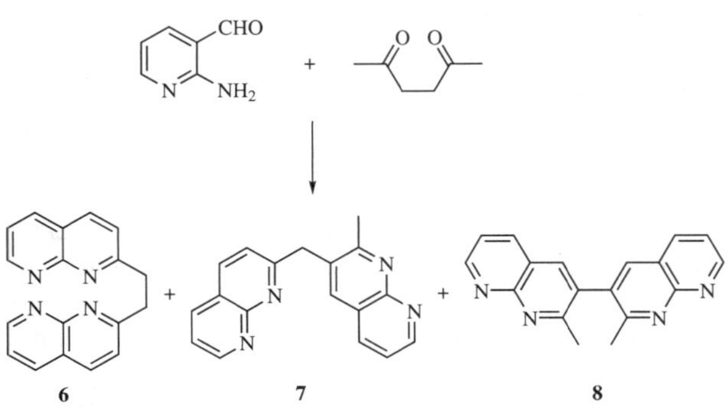

(12)

Conditions	Ratio (**6**:**7**:**8**)	
NaOH, EtOH	<1:70:30	–
1.1 equiv TABO, 0.05 equiv H_2SO_4, EtOH, 85 °C	95:05:00	68% assay yield 46% isolated yield (direct crystallization)

The intramolecular condensation of 2,5-hexanedione with an immobilized acid catalyst (Nafion-H or Amberlite IR-120) gave mainly 2,5-dimethylfuran with the intramolecular aldol product, 3-methyl-2-cyclopentenone, only being formed in low yield (eq 13).[19]

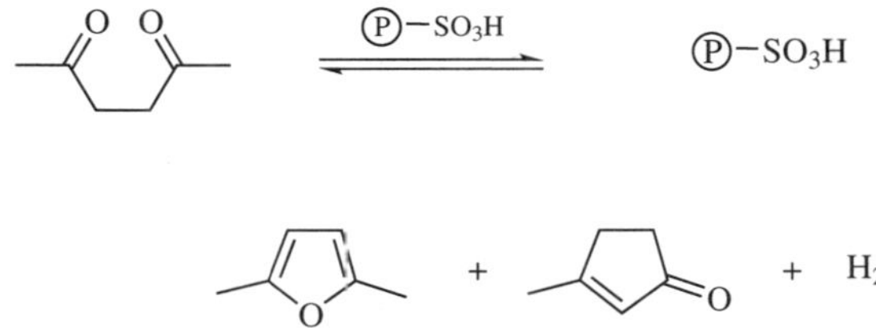

(13)

Conditions:
2,5-Hexanedione (27 mmol), Ⓟ−SO_3H (0.54 mmol), 12 h

Catalyst	Temperature (°C)	Yield (%)	
		Furan	Cyclopentenone
Nafion-H	100	30	0
	150	29	2
	180	30	2

In contrast, the cyclopentenone is prepared using basic catalysts in bilayer solvents comprising H_2O and aromatic hydrocarbons (eq 14).[20]

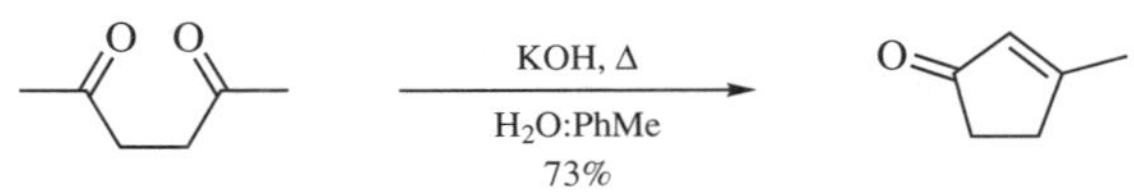

(14)

Samarium diiodide-mediated pinacolization of 1,4-diketones has been applied to the synthesis of a variety of *cis*-selective cyclobutane-1,2-diols, with good yields being obtained. 2,5-Hexanedione afforded 1,2-dimethylcyclobutane-1,2-diol with high *cis*-selectivity (eq 15).[21]

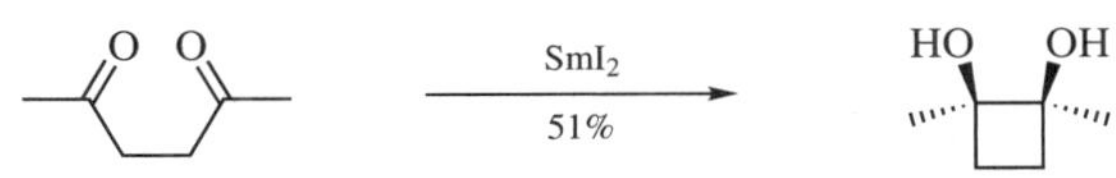

(15)

SnF_2-mediated cyclocondensation of 3-iodo-2-[(trimethylsilyl)methyl]propene (**9**) with 2,5-hexanedione gave 91% of the bicyclic ether (**10**). This process involves reaction of **9** with SnF_2 to generate a trimethylenemethane dianionic synthon (eq 16).[22]

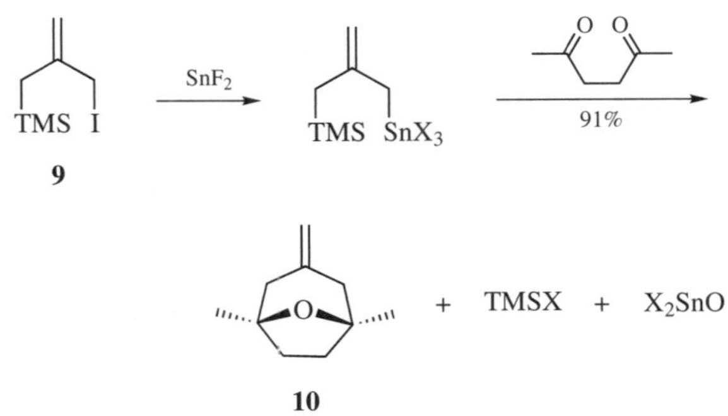

(16)

Related Reagents. Various substituted 1,4-dicarbonyl compounds.

1. Kornblum, N.; Wade, P. A., *J. Org. Chem.* **1973**, *38*, 1418–1420.
2. 'The Sadtler Standard Spectra, Standard Infrared,' Sadtler Research Laboratories, Philadelphia, Spectrum No. 121, 1970.

3. 'Catalog of Mass Spectral Data,' American Petroleum Institute Research Project No. 44, Texas A&M University, Texas, 1970.
4. (a) Paal, C., *Chem Ber.* **1885**, *18*, 367–371. (b) Knorr, L., *Chem Ber.* **1885**, *18*, 299–311. (c) Young, D. M.; Allen, C. F. H., *Org. Synth., Coll. Vol.* **1943**, *2*, 219–220.
5. Buu-Hoi, N. P.; Xuong, N. D., *J. Org. Chem.* **1955**, *20*, 850–854.
6. Bishop, W. S., *J. Am. Chem. Soc.* **1945**, *67*, 2261–2262.
7. Samajdar, S.; Becker, F. F.; Banik, B. K., *Heterocycles* **2001**, *55*, 1019–1022.
8. Macor, J. E.; Chenard, B. L.; Post, R. J., *J. Org. Chem.* **1994**, *59*, 7496–7498.
9. Bowers, S. G.; Coe, D. M.; Boons, G.-J., *J. Org. Chem.* **1998**, *63*, 4570–4571.
10. Nowak, I.; Robins, M. J., *Org. Lett.* **2003**, *5*, 3345–3348.
11. Paal, C., *Chem. Ber.* **1885**, *18*, 2252–2254.
12. Campaigne, E.; Foye, W. O., *J. Org. Chem.* **1952**, *17*, 1405–1412.
13. Gronowitz, S.; Konar, A., *Chem. Scr.* **1977**, *12*, 11–14.
14. Niederl, J. B.; Nagel, R. H., *J. Am. Chem. Soc.* **1940**, *62*, 3070–3072.
15. Niederl, J. B.; Nagel, R. H., *J. Am. Chem. Soc.* **1941**, *63*, 307–308.
16. Guthrie, R. W.; Brossi, A.; Mennona, F. A.; Mullin, J. G.; Kierstead, R. W.; Grunberg, E., *J. Med. Chem.* **1975**, *18*, 755–760.
17. Erker, G.; Psiorz, C.; Froehlich, R.; Grehl, M.; Krueger, C.; Noe, R.; Nolte, M., *Tetrahedron* **1995**, *51*, 4347–4358.
18. Dormer, P. G.; Eng, K. K.; Farr, R. N.; Humphrey, G. R.; McWilliams, J. C.; Reider, P. J.; Sager, J. W.; Volante, R. P., *J. Org. Chem.* **2003**, *68*, 467–477.
19. Pittman, C. U., Jr; Liang, Y. F., *J. Org. Chem.* **1980**, *45*, 5048–5052.
20. Jpn. Kokai Tokkyo Koho Pat. 10130192 (to Kato, K.; Tsukamoto, K.) (1998).
21. Hoffmann, H. M. R.; Muennich, I.; Nowitzki, O.; Stucke, H.; Williams, D. J., *Tetrahedron* **1996**, *52*, 11783–11798.
22. Molander, G. A.; Shubert, D. C., *J. Am. Chem. Soc.* **1987**, *109*, 6877–6878.

A. U. Vinod
University of Illinois at Chicago, Chicago, IL, USA

N-Hydroxybenzotriazole

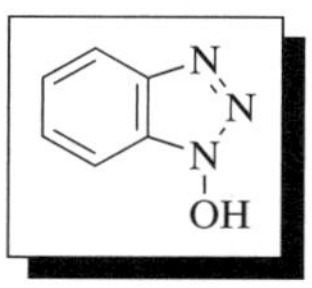

[2592-95-2] $C_6H_5N_3O$ (MW 135.14)
(hydrate)
[123333-53-9]

(peptide synthesis; nucleotide synthesis)

Alternate Names: HOBT; HOBt.
Physical Data: mp 155–160 °C; exact melting point depends on the amount of water of hydration present.
Form Supplied in: white solid usually containing 12–17% water of hydration; widely available commercially.
Analysis of Reagent Purity: should be a white solid; if it becomes discolored, it is advisable to purify.
Purification: recrystallize from either water or aqueous ethanol.
Handling, Storage, and Precautions: should be stored in the dark; avoid contact with strong acids, oxidizing agents, and reducing agents; heating above 180 °C causes rapid exothermic decomposition; toxicity not fully investigated so should be treated with caution.

Peptide Coupling. 1-Hydroxybenzotriazole is most widely used in reactions involving the coupling of amino acid units to give peptides.[1] In this context it has been used mainly as an 'additive' to a coupling reaction, although there are also examples of HOBT being incorporated into the coupling reagent itself. The most common use of 1-hydroxybenzotriazole in peptide synthesis is in conjunction with a carbodiimide such as 1,3-Dicyclohexylcarbodiimide (DCC). Although it is quite possible to couple amino acids using DCC alone, it is found that the addition of 1-hydroxybenzotriazole to the reaction system results in improved reaction rates and suppressed epimerization of the chiral centres present in the peptide (eq 1).[2] This reaction appears to

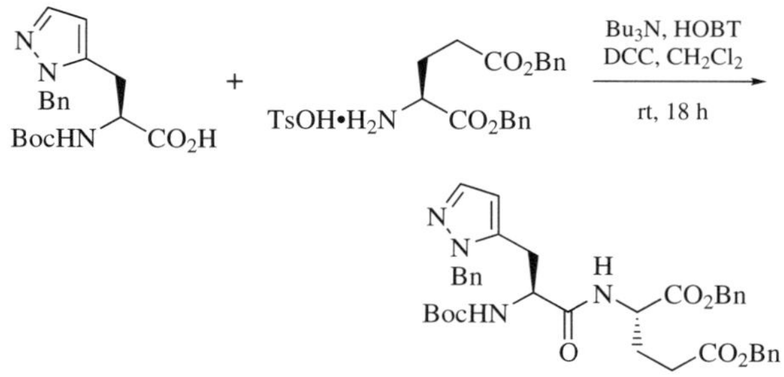

(1)

< 0.1% DL
(1.8% DL in absence of HOBT)

proceed via DCC-mediated formation of a hydroxybenzotriazole ester intermediate which then reacts with the amino function of a second amino acid to give the coupled product.

This coupling protocol is not limited to the synthesis of small peptides in solution, but can also be used in the solid-phase preparation of larger peptides, and for the coupling of larger peptide fragments. It should be noted, however, that there are some drawbacks to this method; most notably, it is sometimes difficult to purify the peptide due to contamination with dicyclohexylurea and the method is unsatisfactory for the coupling of *N*-methylated amino acids.[3] The problem of purification can often be overcome by using an alternative carbodiimide, e.g. ones that give water-soluble byproducts can be used, such as EDC·MeI (see 1-Ethyl-3-(3′-dimethylaminopropyl)carbodiimide Hydrochloride).[4] Polymer-bound variants of HOBT have also been used to overcome purification problems,[5] since the urea byproduct can be removed from the polymer-bound active ester intermediate (**1**) by washing with chloroform–isopropyl alcohol mixtures. Once the urea has been removed, the active ester can be reacted with the second peptide fragment to give the product (eq 2), regenerating the polymer-bound HOBT. While this method appears quite successful for the preparation of small peptide fragments, it is clearly unsuitable for the preparation of larger peptides by solid-phase synthesis.

Although the addition of HOBT to the carbodiimide-mediated coupling of amino acids has been shown to suppress epimerization of the chiral centres present, there are still a number of cases where the level of epimerization is unsatisfactory. One method for further suppressing epimerization is to add Copper(II) Chloride to the reaction system.[6] In this case it is important that the correct stochiometry is determined, since the copper(II) chloride not only suppresses epimerization, but it also slows down the reaction rate and reduces the overall yield (eq 3).

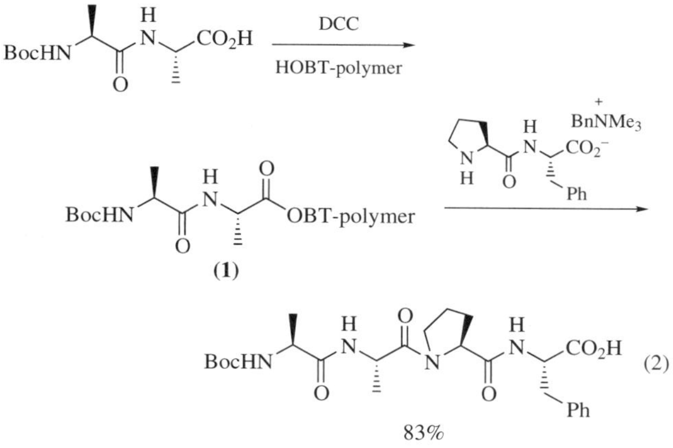

The problem of epimerization during carbodiimide-mediated coupling of amino acids can also be overcome by the use of an alternative coupling agent. Once such system that still employs HOBT as an additive involves the use of bis(2-oxo-3-oxazolidinyl) phosphinic chloride (BOP-Cl) and *N*-methylmorpholine (NMM) (eq 4).[7] See also Bis(2-oxo-3-oxazolidinyl)phosphinic Chloride.

HOBT has also been used as an additive in other reactions involving amino acid derivatives.[8] It would appear that most reaction systems that are capable of generating a hydroxybenzotriazole ester intermediate will give successful coupling. One such system involves the use of amino acid derived trichlorophenyl esters.[9] These can be readily reacted with a second amino acid unit in the presence of HOBT to give the coupled product with little or no racemization (eq 5). *p*-Nitrophenyl and pentachlorophenyl esters can also be employed with similar success. This reaction again is thought to proceed via the hydroxybenzotriazole ester intermediate and it has been shown that use of the potassium salt of HOBT in conjunction with a crown ether can lead to substantial rate increase,[10] presumably by enhancing the rate of formation of the active ester intermediate.

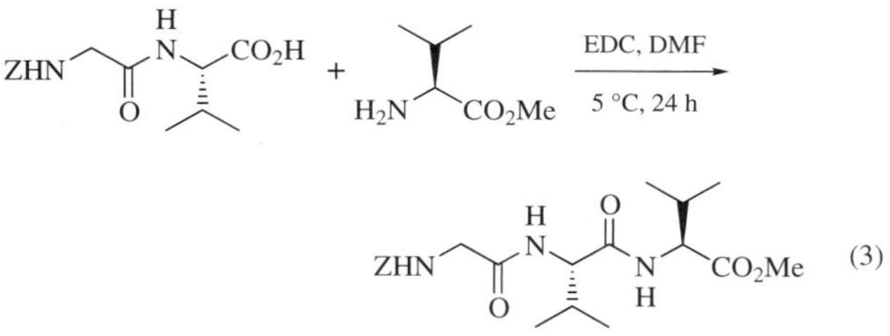

HOBT	$CuCl_2$	% DL	% Yield
0 equiv	0 equiv	38	22
1 equiv	0 equiv	1.3	96
1 equiv	0.25 equiv	<0.1	79

HOBT has also been used in conjunction with amino acid chlorides to facilitate peptide bond formation. This has proved of particular use in the case of FMOC-protected amino acid chlorides in solid-phase peptide synthesis where direct coupling is rather slow and suffers from competing oxazolone formation.[11]

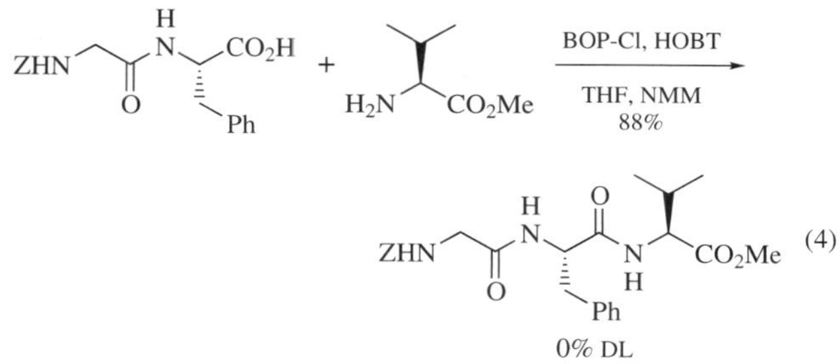

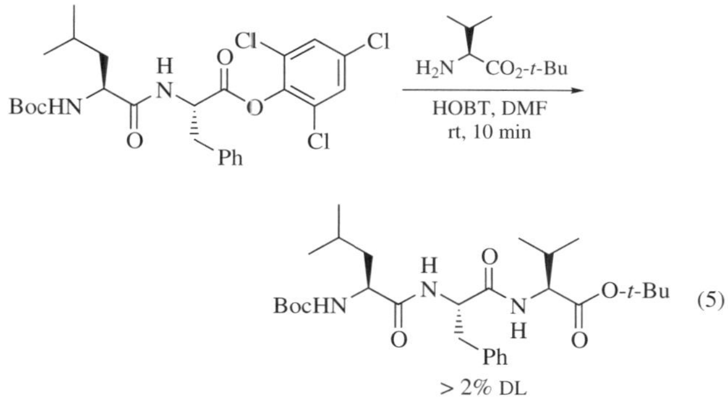

As mentioned earlier, HOBT has also been incorporated into the peptide coupling reagent itself and one example of such a reagent is benzotriazol-1-yloxytris(dimethylamino)phosphonium chloride (BOP).[12] BOP is prepared by the reaction of HOBT with Hexamethylphosphorous Triamide in the presence of carbon tetrachloride, and is very effective in the coupling of amino acids (eq 6). See also Benzotriazol-1-yloxytris(dimethylamino) phosphonium Hexafluorophosphate (also known as BOP).

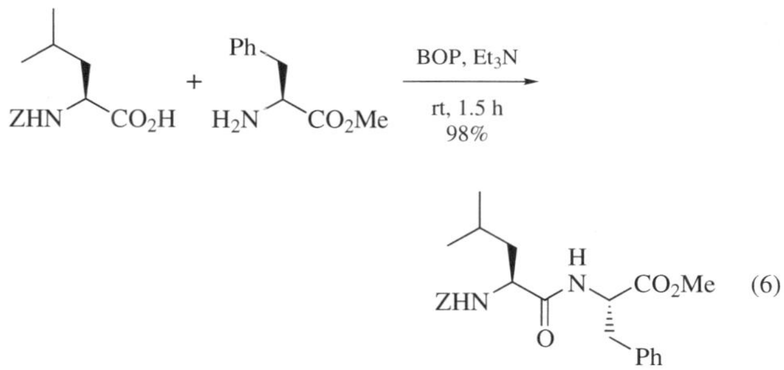

A second peptide coupling reagent that incorporates the HOBT unit is O-Benzotriazol-1-yl-N,N,N′, N′-tetramethyluronium Hexafluorophosphate (HBTU),[13] which can be prepared by the reaction of tetramethylurea with oxalyl chloride followed by treatment with HOBT and Potassium Hexafluorophosphate. HBTU can then be used as a direct coupling agent as outlined in eq 7.

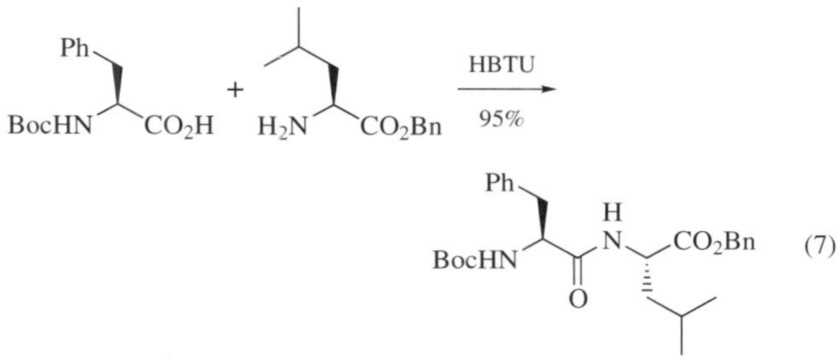

Nucleotide Synthesis. Phosphotriester derivatives of HOBT have been employed in the synthesis of nucleotides.[14] The phosphotriesters are readily formed by reaction of HOBT with the corresponding aryl phosphorodichloridates in the presence of pyridine and can then be used to couple nucleosides as outlined in eq 8. This reaction takes advantage of the fact that a differentially protected nucleoside will react rapidly with the reagent, displacing one molecule of HOBT to form an intermediate phosphotriester (**2**), but at this stage reaction with a second nucleoside molecule is extremely slow due to steric hindrance. Consequently the intermediate can then be reacted with a second nucleoside, the more reactive primary hydroxy displacing the second molecule of HOBT and giving the coupled product. This method for coupling nucleosides has been applied to both solution and solid-phase synthesis of a variety of RNA fragments.

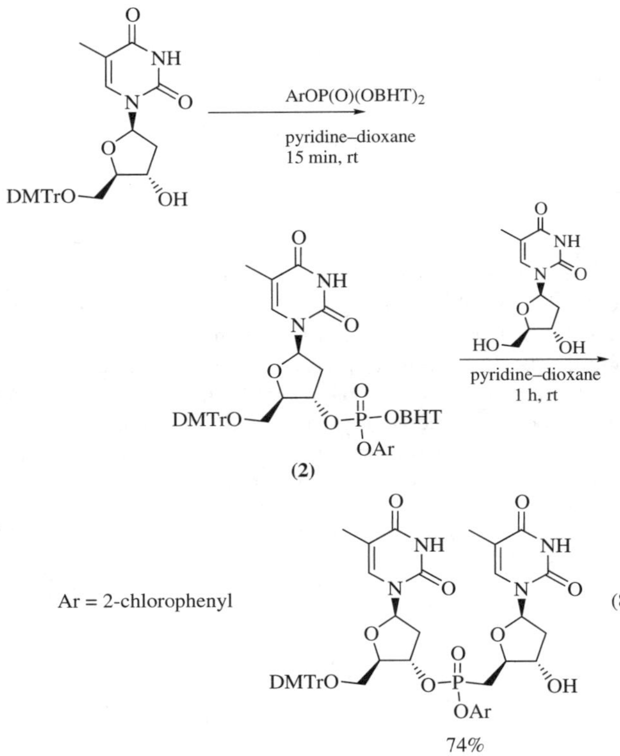

(8)

Other Applications. HOBT has also been employed as a catalyst for the conversion of isoamides into maleimides (eq 9).

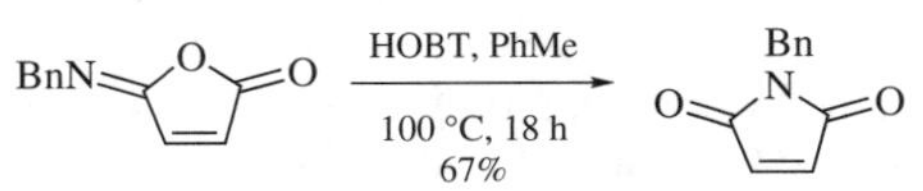

(9)

1. For recent general reviews on peptide synthesis involving the use of HOBT, see: Bodanszki, M. *Int. J. Pept. Protein Res.* **1985**, *25*, 449; Kent, S. B. H.; *Annu. Rev. Biochem.* **1988**, *57*, 957; Bodanszky, M. *J. Protein Chem.* **1989**, *8*, 461; Fields, G. B.; Nobel, R. L. *Int. J. Pept. Protein Res.* **1990**, *35*, 161.
2. König, W.; Geiger, R. *Ber. Dtsch. Chem. Ges./Chem. Ber.* **1970**, *103*, 788; König, W.; Geiger, R. *Ber. Dtsch. Chem. Ges./Chem. Ber.* **1970**, *103*, 2024; König, W.; Geiger, R. *Ber. Dtsch. Chem. Ges./Chem. Ber.* **1970**, *103*, 2034; Windridge, G. C.; Jorgensen, E. C. *J. Am. Chem. Soc.* **1971**, *93*, 6318; Nagaraj, R.; Balaram, P. *Tetrahedron* **1981**, *37*, 2001; Benoiton, N. L.; Kuroda, K. *Int. J. Pept. Protein Res.* **1981**, *17*, 197; Chen, S. T.; Wu, S. H.; Wang, K. T. *Synthesis* **1989**, 37; Dardoize, F.; Goasdoué, C.; Goasdoué, N.; Laborit, H. M.; Topall, G. *Tetrahedron* **1989**, *45*, 7783; Bennoiton, N. L.; Lee, Y. C.; Steinaur, R.; Chen, F. M. F. *Int. J. Pept. Protein Res.* **1992**, *40*, 559; Bennoiton, N. L.; Lee, Y. C.; Chen, F. M. F. *Int. J. Pept. Protein Res.* **1993**, *41*, 587.
3. Coste, J.; Frérot, E.; Jouin, P.; Castro, B. *Tetrahedron Lett.* **1991**, *32*, 1967.
4. Kimura, T.; Takai, M.; Masui, Y.; Morikawa, T.; Sakakibara, S. *Biopolymers* **1981**, *20*, 1823; Hagiwara, D.; Neya, M.; Miyazaki, Y.; Hemmi, K.; Hashimoto, M. *Chem. Commun./J. Chem. Soc., Chem. Commun.* **1984**, 1676.
5. Berrada, A.; Cavelier, F.; Jacquier, R.; Verducci, J. *BSF (1)* **1989**, 511; Grigor'ev, E. I.; Zhil'tsov, O. S. *J. Org. Chem. USSR (Engl. Transl.)* **1989**, *25*, 1774; Chen, S. T.; Chang, C. H.; Wang, K. T. *J. Chem. Res. (S)* **1991**, 206.
6. Miyazawa, T.; Otomatsu, T.; Fukui, Y.; Yamada, T.; Kuwata, S. *Chem. Commun./J. Chem. Soc., Chem. Commun.* **1988**, 419; Miyazawa, T.; Otomatsu, T.; Fukui, Y.; Yamada, T.; Kuwata, S. *Int. J. Pept. Protein Res.* **1992**, *39*, 237; Miyazawa, T.; Otomatsu, T.; Fukui, Y.; Yamada, T.; Kuwata, S. *Int. J. Pept. Protein Res.* **1992**, *39*, 308.
7. van der Auwera, C.; van Damme, S.; Anteunis, M. J. O. *Int. J. Pept. Protein Res.* **1987**, *29*, 464.
8. Knorr, R.; Trzeciak, A.; Bannwarth, W.; Gillessen, D. *Tetrahedron Lett.* **1989**, *30*, 1927.
9. König, W.; Geiger, R. *Ber. Dtsch. Chem. Ges./Chem. Ber.* **1973**, *106*, 3626.
10. Horiki, K.; Murakami, A. *Heterocycles* **1989**, *28*, 615.
11. Carpino, L. A.; Chao, H. G.; Beyermann, M.; Bienert, M. *J. Org. Chem.* **1991**, *56*, 2635; Sivanandaiah, K. M.; Babu, V. V. S.; Renukeshwar, C. *Int. J. Pept. Protein Res.* **1992**, *39*, 201.
12. Castro, B.; Dormoy, J. R.; Evin, G.; Selve, C. *Tetrahedron Lett.* **1975**, 1219.
13. Dourtoglou, V.; Gross, B.; Lambropoulou, V.; Zioudrou, C. *Synthesis* **1984**, 572.
14. van der Marel, G.; van Boeckel, C. A. A.; Wille, G.; van Boom, J. H. *Tetrahedron Lett.* **1981**, *22*, 3887; Marugg, J. E.; Tromp, M.; Jhurani, P.; Hoyng, C. F.; van der Marel, G. A.; van Boom, J. H. *Tetrahedron* **1984**, *40*, 73; Gottikh, M.; Ivanovskaya, M.; Shabarova, Z. *Bioorg. Khim.* **1988**, *14*, 500; Hirao, I.; Miura, K. *Chem. Lett.* **1989**, 1799; Colonna, F. P.; Scremin, C. L.; Bonora, G. M. *Tetrahedron Lett.* **1991**, *32*, 3251.

Barry Lygo
Salford University, UK

2-Hydroxyethyl Phenyl Sulfone

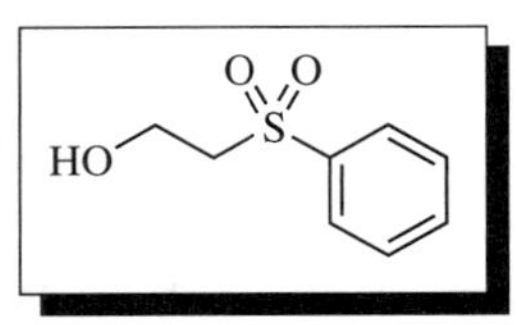

[20611-21-6] $C_8H_{10}O_3S$ (MW 186.2)

(protection of carboxylic acids;[1] preparation of a chloroformate reagent for protecting alcohols and amines;[2] protection of phosphates and phosphonates;[3] preparation of a phosphitylation reagent for nucleotide synthesis;[4] 2-ethanol carbanion equivalent;[5] preparation of a triflate reagent for amine quaternization[6])

Alternate Name: 2-(phenylsulfonyl)ethanol.
Physical Data: bp 177 °C/2 mmHg.

Solubility: soluble in organic solvents.
Form Supplied in: colorless liquid.
Analysis of Reagent Purity: NMR, HPLC.
Purification: distillation.
Handling, Storage, and Precautions: irritant, handle with gloves in a fume hood.

Protection of Carboxylic Acids. 2-Hydroxyethyl phenyl sulfone (**1**) is a useful reagent for the protection of carboxylic acids.[1] Thus, treatment of the protected muramic acid derivative (**2**) with **1** and DCC afforded the 2-(phenylsulfonyl)ethyl ester (**3**) in excellent yield (eq 1).[1a]

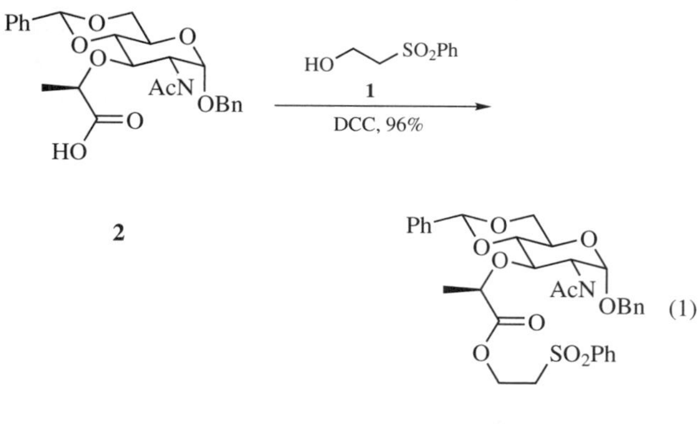

2 **3**

In another example, the unsaturated acid **4** was protected in the same manner to afford ester **5** (eq 2).[1b,c]

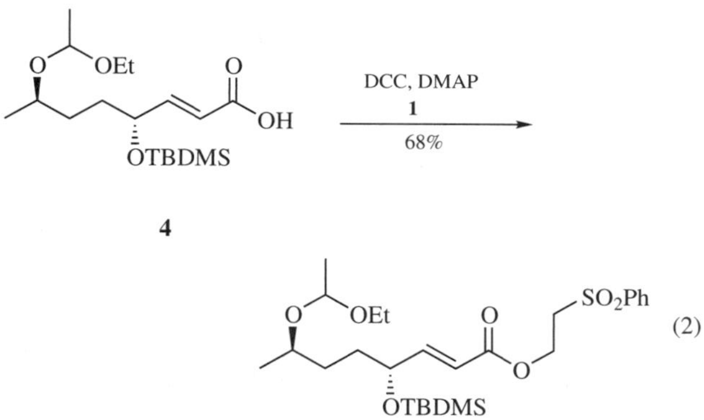

4 **5**

The 2-(phenylsulfonyl)ethyl ester is stable to acidic conditions [tetrahydropyranyl (THP), *tert*-butoxycarbonyl (BOC), and ethoxyethyl (EE) groups can be selectively removed], and to oxidation (peroxide) but can be readily removed by treatment with DBU. Note that in both examples cited above, the 2-(phenylsulfonyl)ethyl ester was cleaved without causing epimerization of the acid. The closely related reagent, 2-(p-toluenesulfonyl)ethanol, has also been used as a carboxylic acid protecting group.[1d,e] The two reagents work equally well for this purpose and both are commercially available.

Preparation of a Chloroformate Reagent (6) for the Protection of Alcohols and Amines. Reaction of **1** with phosgene affords the chloroformate (**6**) (eq 3),[2a] a useful reagent for the protection of alcohols[2a,b] and amines.[2c,d]

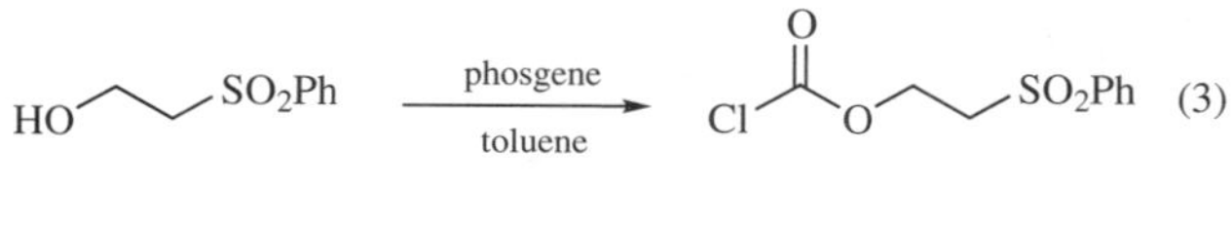

1 **6**

Chloroformate (**6**) reacts with alcohols in pyridine at room temperature to afford carbonates **7**.[2a] The 2-phenylsulfonylethoxycarbonyl (PSEC) group is stable to the conditions required to cleave tetrahydropyranyl (THP) ethers, 2-trimethylsilylethoxy carbonates (TEOC), and levulinate and benzoate esters.[2a,b] The alcohol can be regenerated when desired by treatment with triethylamine in pyridine (eq 4).

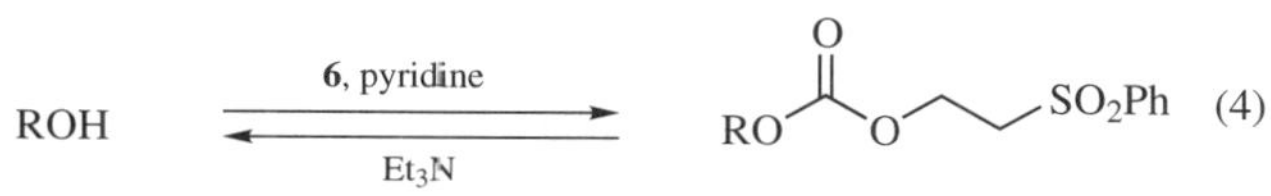

7

Reagent **6** has also been used to protect amines, generating carbamates which are readily cleaved under mild basic conditions.[2c,d] In one case, the carbamate resulting from reaction of **6** with daunomycin was proposed as a potential prodrug for daunomycin.[2d] As mentioned above for carboxylic acid protection, similar chemistry has been reported using 2-(*p*-toluenesulfonyl)ethanol instead of **1**.

Protection of Phosphates and Phosphonates. 2-Hydroxyethyl phenyl sulfone (**1**) has been used to differentially protect phosphate esters for nucleotide synthesis.[3a,b] Thus, reaction of the diphosphate (**8**) with **1** and 1-(mesitylene-2-sulfonyl)-3-nitro-1,2,4-triazole (MSNT) in pyridine afforded the differentially protected phosphate **9** in very good yield (eq 5).[3a] The 2-phenylsulfonylethyl phosphate can be cleaved under mild conditions by treatment with triethylamine.

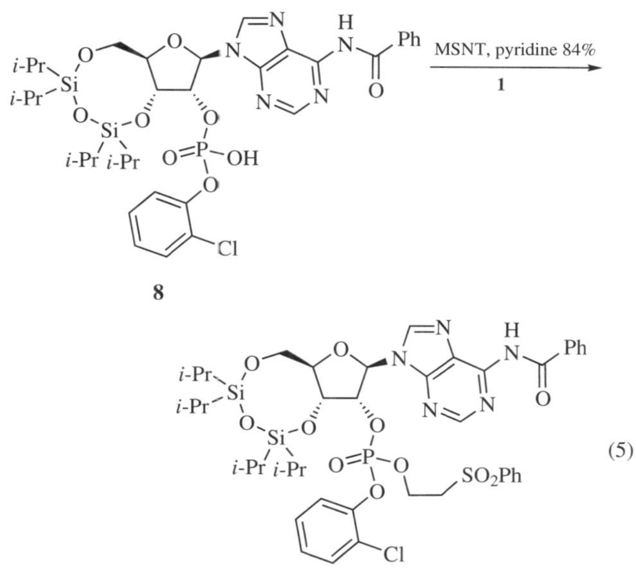

8 **9**

Alcohol **1** has also been used to protect a phosphonic acid (**10**) as part of a synthesis of cyclic nucleotides (eq 6).[3c]

Preparation of a Phosphitylation Reagent (12) for Nucleotide Synthesis. The phosphitylation reagent (**12**), useful for nucleotide synthesis, is readily prepared by reaction of **1** with phosphorus trichloride in acetonitrile (eq 7).[4a,b]

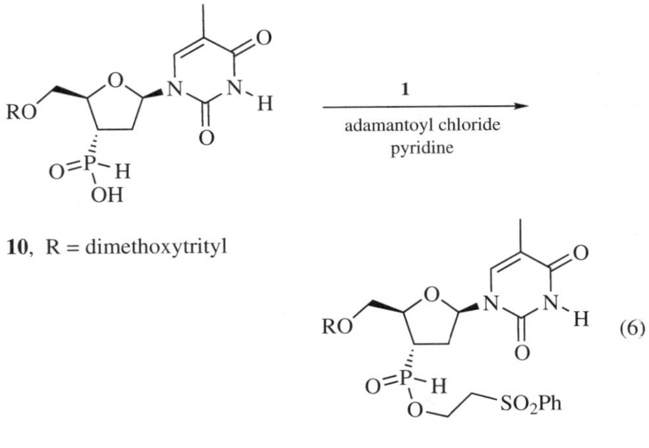

10, R = dimethoxytrityl

11

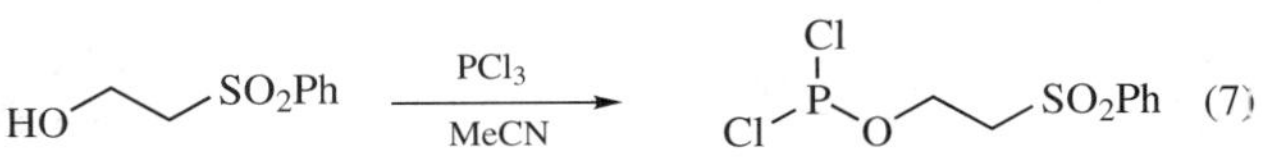

1 **12**

Reagent **12** is an easily handled solid which is stable for at least one year when stored in a freezer.[4b]

Use as a 2-Ethanol Carbanion Equivalent. Treatment of **1** with more than 2 equiv of butyllithium followed by treatment of the resulting dianion with 1 equiv of an electrophile and removal (either reductively or by elimination) of the phenylsulfonyl group results in a net addition of a 2-ethanol carbanion to the electrophile.[5] Thus, reaction of **1** with 2.2 equiv of butyllithium followed by quenching of the dianion with benzyl chloride afforded intermediate **13** in very good yield (eq 8).[5a] Desulfurization of **13** with Raney nickel afforded 3-phenylpropanol (**14**) in good yield.

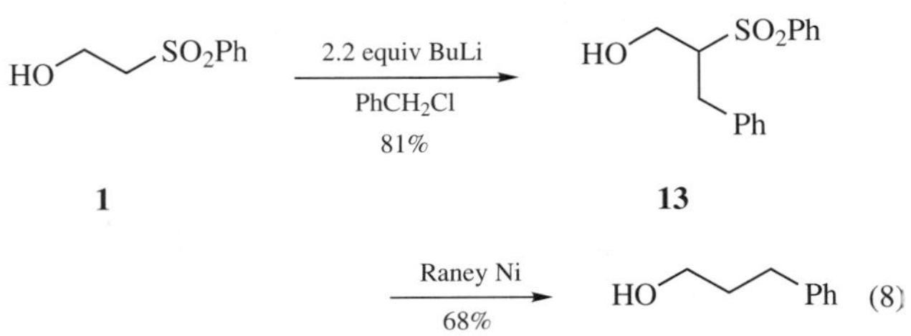

1 **13** **14**

The initial alkylation product can be subjected to a second alkylation to afford a more highly substituted product. For example, alkylation of **13** with allyl bromide provides the di-substituted analog **15** (eq 9).[5c] These bis-alkylated products have been elaborated further to furanones[5b] and furans.[5c]

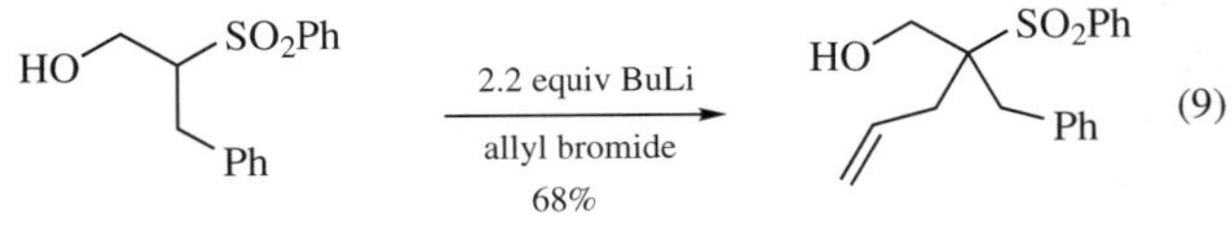

13 **15**

Preparation of a Triflate Reagent (16) for Nitrogen Quaternization. Reaction of **1** with triflic anhydride in pyridine affords the triflate (**16**) in excellent yield (86%).[6] Triflate (**16**) reacted with oxazole (**17**) to afford the quaternized oxazole (**18**) (eq 10). Subsequently, **18** was reduced with sodium borohydride to afford a dihydro-oxazole with a phenylsulfonylethyl substituent on the tertiary nitrogen. Thus, **16** was used to activate the oxazole ring to chemistry which generated a protected secondary amine.

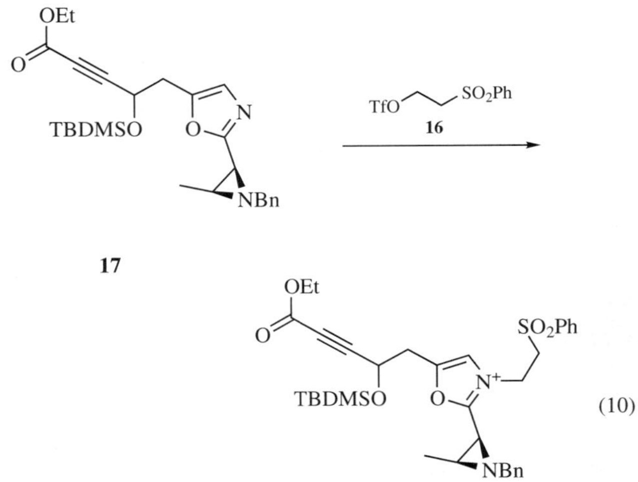

17

18

1. (a) Hitchcock, S. A.; Eid, C. N.; Aikins, J. A.; Zia-Ebrahimi, M.; Blaszczak, L. C., *J. Am. Chem. Soc.* **1998**, *120*, 1916. (b) Dommerholt, F. J.; Thijs, L.; Zwanenburg, B., *Tetrahedron Lett.* **1991**, *32*, 1495. (c) Dommerholt, F. J.; Thijs, L.; Zwanenburg, B., *Tetrahedron Lett.* **1991**, *32*, 1499. (d) Roush, W. R.; Blizzard, T. A., *J. Org. Chem.* **1984**, *49*, 4332. (e) Colvin, E. W.; Purcell, T. A.; Raphael, R. A., *J. Chem. Soc. Chem. Comm.* **1972**, 1031.
2. (a) Balgobin, N.; Josehpson, S.; Chattopadhyaya, J. B., *Tetrahedron Lett.* **1981**, *22*, 3667. (b) Josephson, S.; Balgobin, N.; Chattopadhyaya, J., *Tetrahedron Lett.* **1981**, *22*, 4537. (c) Nyilas, A.; Foldesi, A.; Chattopadhyaya, J., *Nucleosides Nucleotides* **1988**, *7*, 787. (d) Maligres, P. E.; Nicolaou, K. C.; Wrasidlo, W., *Bioorg. Med. Chem. Lett.* **1993**, *3*, 1051.
3. (a) Zhou, X.; Remaud, G.; Chattopadhyaya, J., *Tetrahedron* **1988**, *44*, 6471. (b) Sund, C.; Foldesi, A.; Yamakage, S.; Agback, P.; Chattopadhyaya, J., *Tetrahedron* **1991**, *47*, 6305. (c) Zeng, F.; Jones, R. A., *Nucleosides Nucleotides* **1996**, *15*, 1679.
4. (a) Balgobin, N.; Chattopadhyaya, J., *Acta. Chem. Scand. Ser. B.* **1985**, *39*, 883. (b) Claesen, C. A. A.; Segers, R. P. A. M.; Tesser, G. I., *Rec. Trav. Chim. Pays. Bas.* **1985**, *104*, 119.
5. (a) Julia, M.; Uguen, D.; Callipolitis, A., *Bull. Soc. Chim. Fr.* **1976**, 519. (b) Tanaka, K.; Ootake, K.; Imai, K.; Tanaka, N.; Aritsune, K. *Chem. Lett.*

1983, 633. (c) Jung, J. H.; Lee, J. W.; Oh, D. Y., *Tetrahedron Lett.* **1995**, *36*, 923.

6. Vedejs, E.; Piotrowski, D. W.; Tucci, F. C., *J. Org. Chem.* **2000**, *65*, 5498.

T. A. Blizzard
Merck Research Laboratories, Rahway, New Jersey, USA

N-Hydroxysuccinimide

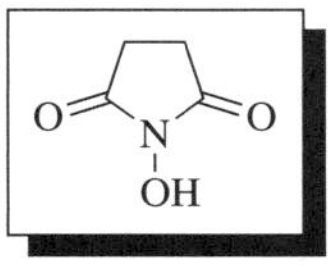

[6066-32-6] $C_4H_5NO_3$ (MW 115.10)

(activating agent for carboxylic acids in amide synthesis and related coupling reactions; acyl transfer reagent)

Alternate Name: HOSu.
Physical Data: mp 96–98 °C (99–100 °C).
Solubility: sol H_2O, DMF, alcohols, EtOAc; insol cold ether.
Form Supplied in: colorless crystalline solid, widely available.
Preparative Methods: by heating Succinic Anhydride with Hydroxylamine or, better, hydroxylamine hydrochloride followed by crystallizations from ether, 1-butanol, and finally EtOAc.[1]
Purification: can be crystallized from 1-butanol, EtOAc, or EtOH– EtOAc.

Peptide Bond Formation. The most important use of *N*-hydroxysuccinimide is in the formation of (isolable) activated derivatives of N^{α}-protected α-amino acids, which subsequently undergo generally smooth coupling reactions with amino esters. Typically, 1,3-Dicyclohexylcarbodiimide (DCC) is used as the initial coupling agent (the HOSu–DCC method) (eq 1).[1–3] *N*-Hydroxyphthalimide can be employed in much the same way, but a distinct advantage in the case of HOSu is its high water solubility, which facilitates product purification.

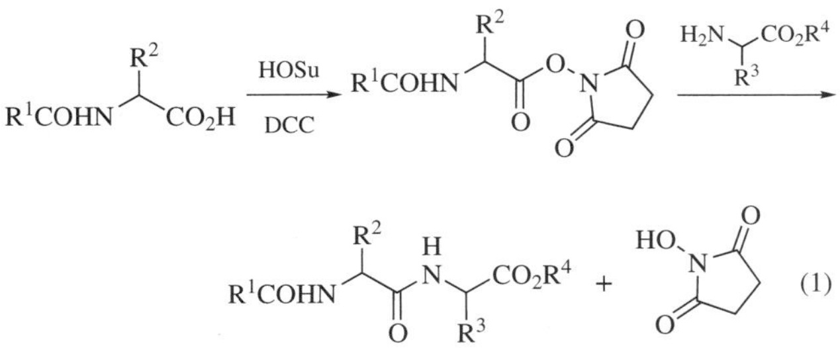

In some cases, up to 3% racemization has been observed when using the HOSu–DCC method, which renders it unsuitable for some types of peptide synthesis.[4,5] Esters derived from HOSu are relatively reactive, as shown by a kinetic study of their rates of hydrolysis in aqueous buffers[6] although, in some examples, *endo*-*N*-hydroxy-5-norbornene-2,3-dicarboximide has proven superior to HOSu in peptide bond formation.[7] Of course, the HOSu–DCC method is not limited to peptide synthesis and is well suited to the preparation of amides in general, both from ammonia and primary amines (e.g. eq 2).[8]

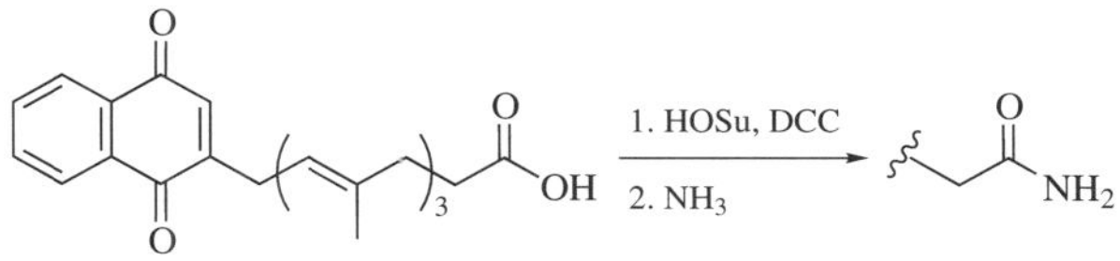

(2)

In order to avoid the need for DCC, a number of activated derivatives of *N*-hydroxysuccinimide itself have been developed which react directly with, for example, an α-amino acid to provide the required activated α-amino acid esters. These include the commercially available carbonate (**1**), a stable crystalline solid obtained from HOSu and Trichloromethyl Chloroformate or from *O*-trimethylsilyloxysuccinimide and Phosgene,[9] and the related phosphate (**2**) derived from HOSu and Diphenyl Phosphorochloridate under Schotten–Baumann conditions.[10] Another alternative (also commercially available) is the oxalate (**3**), formed from HOSu and Oxalyl Chloride.[11] In general, these intermediates react rapidly with an N^{α}-protected α-amino acid, usually in the presence of a mild base such as pyridine, to provide excellent yields of the required hydroxysuccinimide esters, generally with less racemization than is sometimes associated with the HOSu–DCC method. It is also possible to carry out the entire process of peptide synthesis in one pot using these reagents.

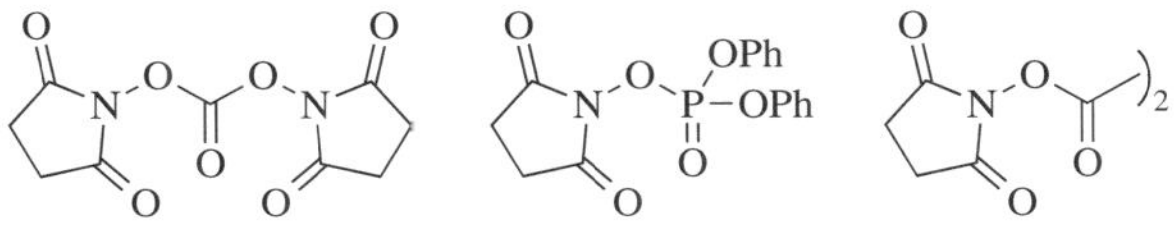

(**1**) (**2**) (**3**)

A rather different approach to peptide bond formation and amide synthesis in general is to treat a mixture of a carboxylic acid and an amine with an isocyanide (2-Morpholinoethyl Isocyanide is especially suitable), which effectively acts as a dehydrating agent (eq 3). The procedure can result in extensive racemization of both reactants and products which may be supressed by the addition of HOSu; presumably an HOSu ester is the penultimate intermediate.[12] The addition of HOSu also decreases racemization in polypeptide synthesis when 'Bates reagent' $\{[(Me_2N)_3P^+]_2O\,(BF_4^-)_2\}$ is used as the coupling agent.[13]

$$R^1CO_2H + H_2NR^2 + R^3NC \longrightarrow R^1CONHR^2 + R^3NHCHO \quad (3)$$

A rather unusual coupling reaction between HOSu and glyoxylic acid tosylhydrazone leads to succinimidyl diazoacetate (eq 4); again, the leaving ability of the HOSu residue renders this a useful compound for effecting the direct transfer of a diazoacetyl function to amines, phenols, and peptides.[14]

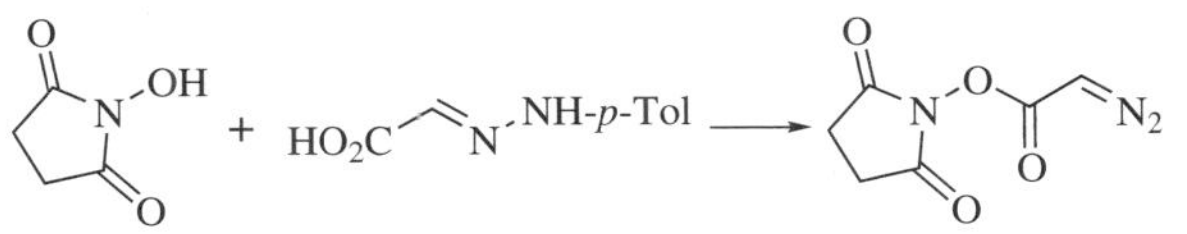

(4)

Modified Barton–McCombie intermediates are best prepared from pentafluorophenyl chlorothionoformate and a catalytic quantity (15–20%) of HOSu in refluxing benzene rather than by using 4-Dimethylaminopyridine, the rather more conventional catalyst of acyl group transfer (eq 5).[15]

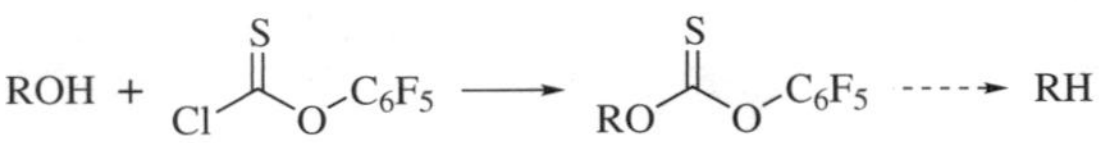

(5)

1. Anderson, G. W.; Zimmerman, J. E.; Callahan, F. M. *J. Am. Chem. Soc.* **1964**, *86*, 1839; Wegler, R.; Grewe, F.; Mehlhose, K.U.S. Patent 2 816 111, **1957***Chem. Abstr.* **1958**, *52*, 6405i) .
2. For a review, see: Klausner, Y. S.; Bodansky, M. *Synthesis* **1972**, 453.
3. For examples, see: Wunsch, E.; Drees, F. *Ber. Dtsch. Chem. Ges./Chem. Ber.* **1966**, *99*, 110; Manesis, N. J.; Goodman, M. *J. Org. Chem.* **1987**, *52*, 5331; Mukaiyama, T.; Goto, K.; Matsuda, R.; Ueki, M. *Tetrahedron Lett.* **1970**, 1901; Bosshard, H. R.; Schechter, I.; Beger, A. *Helv. Chim. Acta* **1973**, *56*, 717.
4. Kemp, D. S.; Trangle, M.; Trangle, K. *Tetrahedron Lett.* **1974**, 2695.
5. For a review of side reactions in peptide synthesis, see: Martinez, J. *Synthesis* **1981**, 333.
6. Cline, G. W.; Hanna, S. B. *J. Org. Chem.* **1988**, *53*, 3583.
7. Fujino, M.; Kobayashi, S.; Obayashi, M.; Fukuda, T.; Shinagawa, S.; Nishimura, O. *Chem. Pharm. Bull.* **1974**, *22*, 1857.
8. Terao, S.; Shiraishi, M.; Kato, K.; Ohkawa, S.; Ashida, Y.; Maki, Y. *J. Chem. Soc., Perkin Trans. 1* **1982**, 2909.
9. Ogura, H.; Kobayashi, T.; Shimizu, K.; Kawabe, K.; Takeda, K. *Tetrahedron Lett.* **1979**, 4745.
10. Ogura, H.; Nagai, S.; Takeda, K. *Tetrahedron Lett.* **1980**, *21*, 1467.
11. Takeda, K.; Sawada, I.; Suzuki, A.; Ogura, H. *Tetrahedron Lett.* **1983**, *24*, 4451.
12. Wackerle, L. *Synthesis* **1979**, 197; Marquarding, D.; Aignar, H. Ger. Offen. 2 942 606, **1979** (*Chem. Abstr.* **1981**, *95*, 62 731j) .
13. Bates, A. J.; Galpin, I. J.; Hallett, A.; Hudson, D.; Kenner, G. W.; Ramage, R.; Sheppard, R. C. *Helv. Chim. Acta* **1975**, *58*, 688.
14. Ouihia, A.; Rene, L.; Guilhem, J.; Pascard, C.; Badet, B. *J. Org. Chem.* **1993**, *58*, 1641.
15. Barton, D. H. R.; Jaszberenyi, J. Cs. *Tetrahedron Lett.* **1989**, *30*, 2619. For an application, see:Gervay, J.; Danishefsky, S. *J. Org. Chem.* **1991**, *56*, 548.

David W. Knight
Nottingham University, UK

1-Hydroxy-1,2,3-triazole-4-carboxylate

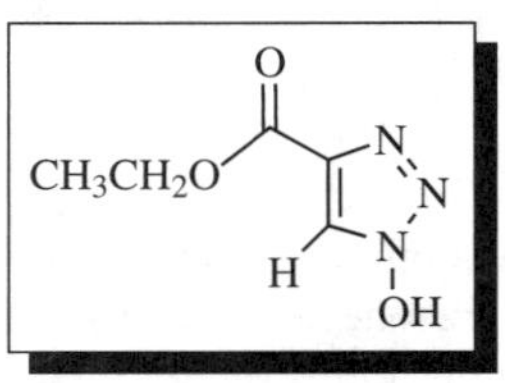

$C_5H_7N_3O_3$ (MW 157.05)

(used together with DIC as coupling reagents for peptide synthesis)

Alternate Name: HOCT.
Physical Data: mp110–112 °C.[1]
Solubility: soluble in H_2O, alcohol, DCM, DMF, and most organic solvents.
Form Supplied in: colorless crystals, available from Hecheng Tech. China.
Analysis of Reagent Purity: IR, ^{1}H NMR, elemental analysis, and X-ray crystallography.
Preparative Methods: the title reagent can be prepared by the reaction of ethyl 2-diazoethanoate with dimethylformamidinium chloride, the Vilsmeier reagent.[1] The product obtained reacted with hydroxylamine to give the diazo oxime, which then cyclizes in chloroform at room temperature to give the product.
Purification: re-crystallized from ethyl acetate and dried over P_2O_5 under vacuum.
Handling, Storage, and Precautions: very stable at room temperature. If moisture is to be avoided in the peptide coupling reaction, it is recommended that the reagent be stored in a desiccator over drying reagents.

Formation of Amide Bonds—Activation of Carboxyl Group. The principal reaction in peptide synthesis is acylation of the α-amino group of an amino acid by the carboxyl group of a second amino acid to form an amide bond with the assistance of coupling reagents, namely through the activation of a carboxyl group (eq 1).[2,3]

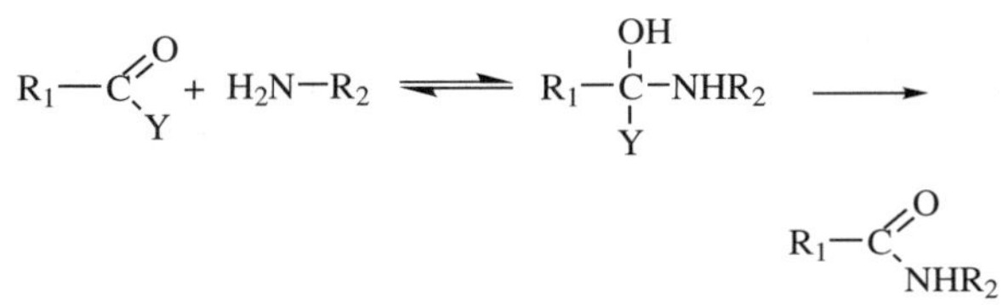

(1)

The design of HOCt was based on the well-established reagent HOBt (eq 2).[4,5] A series of imidazoles and triazoles were studied[6] which led to the selection of HOCt as the optimal reagent due to the fact that it minimizes steric hinderance in the N–OH region and it is more acidic, which means –OCt is a better leaving group.[1] In a number of comparative syntheses, HOCt/DIC proved to be greatly superior to HOBt/DIC and was effective for the incorporation of Aib into peptide sequences. The most marked beneficial effect was observed in the synthesis of long chain peptides (> 100 amino acids).[7]

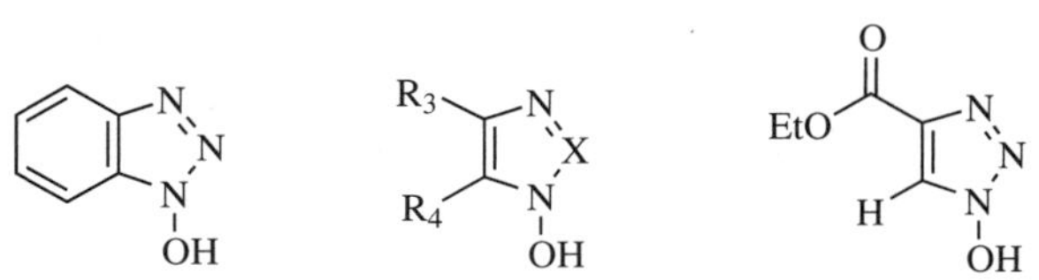

(2)

HOBt X= CR or N HOCt

No racemization was found using HOCt/DIC as coupling reagents under normal coupling conditions except for His.[2] The racemization could be reduced if activation was carried out at low temperature (0 °C, sonic bath) or for a short time. Excess of HOCt (> 3 equiv) also reduced the racemization to a negligible level (< 1%).[1,8]

On-line Monitoring of Peptide Synthesis. Other advantages of HOCt including the fact that it has very little absorbance at

302 nm, a key feature to allow on-line monitoring of peptide coupling using Fmoc chemistry reference. Before the development of HOCt, peptide synthesis could only be monitored, after each coupling and deprotection cycle, by the estimation of the Fmoc-adduct in the deprotection mixture since all coupling reagents based on HOBt have some absorbance at 302 nm, a wavelength at which the Fmoc-group is examined. Although this monitoring method provides useful information regarding the process of each coupling step, no intervention in the coupling step could be made until the next synthesis. This is most undesirable when a long peptide chain assembly is approaching its conclusion, or when expensive building blocks are to be incorporated in the synthesis. On-line monitoring through the examination of the coupling mixture before the subsequent deprotection of N^{α}-protecting group in the cycle would be of great advantage under these circumstances. Since HOCt has no absorption at this wavelength, on-line monitoring could be incorporated wherever required. This technology has greatly increased the chance of success especially in the synthesis of long chain peptides or proteins.[1]

Fragment Coupling. Although stepwise solid-phase peptide synthesis (SPPS) has been greatly improved by the development of new coupling reagents and methods, fragment coupling provides an alternative convergent route towards the synthesis of large peptides or proteins. With the increase in the size of synthetic targets, a marriage of stepwise and convergent strategies should be a better approach. Used in situ HOCt converts a peptidyl azide into the more thermally stable active ester, which allows the coupling reaction to be carried out at room temperature in higher yield (eq 3). This is termed transfer active ester condensation (TAEC).[9] It was also found that the excess HOCt used greatly enhanced the solubility of peptide fragments, another advantage in applying HOCt as coupling reagent. Another coupling reagent HOAt[10] is almost as efficient as HOCt with regard to fragment coupling, however HOBt is not as useful for this methodology.[9]

$$\sim\!\mathrm{CONHNH_2} \xrightarrow[\mathrm{HOCt/rt}]{t\text{-}\mathrm{BuONO}} \left[\sim\!\mathrm{CON_3} \longrightarrow \sim\!\mathrm{COOCt}\right] \xrightarrow{\mathrm{H_2N}\sim} \sim\!\mathrm{CONH}\!\sim \quad (3)$$

C-terminal Modification. In a very similar way to the fragment coupling, many *C*-terminal modifications can be achieved through the combination of azide method and in situ formation of the active ester (eq 4).[11]

$$\sim\!\mathrm{CONHNH_2} \xrightarrow[\mathrm{HOCt/rt}]{t\text{-}\mathrm{BuONO}} \left[\sim\!\mathrm{CON_3} \longrightarrow \sim\!\mathrm{COOCt}\right] \xrightarrow{\mathrm{HNu}} \sim\!\mathrm{CONu} \quad (4)$$

$\mathrm{HNu} = \mathrm{NH_2OH}$; $\mathrm{NH_2NHCONH_2}$; $\mathrm{HN(OCH_3)CH_3}$; HOEt; $\mathrm{HOCH_2CF_3}$; $\mathrm{HSCH_2C_6H_5}$

Related Reagents. 1-Hydroxybenzotriazole (HOBt); 3-hydroxy-3*H*-1,2,3-trizolo[4,5-b]-pyridine (HOAt); 2-(1*H*-benzotriazole-1-yl)-1,1,3,3-tetramethyluronium tetrafluoroborate (TBTU, CAS # 132705-51-2), *O*-(benzotriazol-1-yl)-1,1,3,3-tetramethyluronium hexafluorophosphate (HBTU), (7-azabenzotriazol-1-yl)-1,1,3,3-tetramethyluronium tetrafluorophosphate (TATU), *O*-7-(azabenzotriazol-1-yl)-1,1,3,3-tetramethyluronium hexafluorophosphate (HATU).

1. Jiang, L.; Davison, A.; Tennant, G.; Ramage, R., *Tetrahedron* **1998**, *54*, 14233.
2. Ramage, R.; Biggin, G. W.; Brown, A.; Comer, A.; Davison, A.; Draffin, L.; Jiang, L.; Morton, G.; et al., In *Innovation and Perspective in Solid Phase Synthesis and Combinatorial Libraries*, Epton, R., Ed.; Mayflower Scientific Limited: Birmingham, 1996, Vol. 1.
3. (a) Jones, J. H., In *Peptides,* Academic Press: New York, 1979, Vol. 1, 65–104. (b) Bodanszky, M., In *Peptide Chemistry, a Practical Textbook*; Springer-Verlag: London, 1988.
4. (a) Konig, W.; Geiger, R., *Chem. Ber.* **1970**, *103*, 788. (b) Konig, W.; Geiger, R., *Chem. Ber.* **1970**, *103*, 2024.
5. (a) Knorr, R.; Trzeciak, A.; Banwarth, W.; Gillessen, D., *Tetrahedron Lett.* **1989**, *30*, 1927. (b) Knorr, R.; Trzeciak, A.; Banwarth, W.; Gillessen, D., In *Peptides* 1990; Giralt, E., Andreu, D., Eds, ESCOM, Leiden: The Netherlands, 1991, p 62.
6. Davison, A.; *PhD Thesis*, Edinburgh University, UK, 1993.
7. Robertson, N.; Ramage, R., *J. Chem. Soc. Perk I* **1999**, *8*, 1015.
8. Robertson, N.; Jiang, L.; Ramage R., *Tetrahedron* **1999**, *55*, 2713.
9. Wang, P.; Layfield, R.; Landon, M.; Mayer, R. J.; Ramage, R., *Tetrahedron Lett.* **1998**, *39*, 8711.
10. (a) Carpino, L. A.; El-Faham, A.; Minor, C. A.; Albericio, F., *J. Chem. Soc. Chem. Commun.* **1994**, 201. (b) Carpino, L. A.; El-Faham, A.; Albericio, F., *Tetrahedron Lett.* **1994**, *35*, 2279.
11. Wang, P.; Shaw, K.; Whigham, B., Ramage, R.; *Tetrahedron Lett.* **1998**, *39*, 8719.

Lu Jiang & Robert Ramage
Edinburgh University, Edinburgh, UK

I

Iodine Azide[1]

IN_3

[14696-82-3] IN (MW 168.93)

(additions to multiple bonds;[1–4] regioselective and stereoselective introduction of the azide function[1–3] and other nitrogen functions including amines; synthesis of aziridines[5] and azirines;[6] radical substitution active hydrogen)

Physical Data: dark crystals;[7] explosive when neat. In crystalline form, IN_3 has a polymeric structure in the form of zigzag chains.[8] In gas phase it was found to have a monomeric structure as determined by diffraction analysis.[9]
Solubility: soluble in most organic solvents.
Form Supplied in: not available commercially; prepared in situ in solution.
Preparative Methods: the first reported preparation[7] from AgN_3 and I_2 is not recommended. IN_3 can be prepared in situ from Sodium Azide and Iodine Monochloride preferably in MeCN,[2] DMF, ether, or CH_2Cl_2.[2,3] It can also be prepared from TlN_3 and ICl[3c] in MeCN or CH_2Cl_2,[3] or from N-Iodosuccinimide and Hydrazoic Acid in CH_2Cl_2.[3] The following is a typical procedure for addition to alkenes:[2] to a slurry of 15.0 g (0.25 mol) of NaN_3 in 100 mL of MeCN at −10 to 0 °C is added, over 10–20 min, 18.3 g (0.113 mol) of ICl neat or in MeCN. After 10–20 min of further stirring, 0.1 mol of alkene is added neat or in MeCN and the mixture is warmed to 20 °C and stirred for 8–24 h. The red–brown slurry is poured into 250 mL of water and extracted three times with 85 mL portions of ether. The combined ether fractions are washed with 5% aq sodium thiosulfate until colorless (see precautions), then with water, and dried. Removal of the solvent in vacuo is followed by chromatography (neutral alumina) to afford the iodo azide adduct in 80–90% yield. In 1999, Kirschning et al., described a stable electrophilic polymer-bound reagent that synthetically behaves like iodine azide.[10] By reacting the polystyrene-bound iodide **1** with (diacetoxyiodo)benzene in dichloromethane at room temperature and subsequently with trimethysilylazide resin **3** was generated. They also obtain **3** by direct azido transfer after treatment of iodide (**1**) with (diazido)benzene (eq 1). This polymer-bound iodine azide resolved the problems arising from the explosive character of iodine azide.
Handling, Storage, and Precautions: solutions of IN_3 in MeCN can be stored in the refrigerator for 1 d. Explosions have been reported on evaporation of solutions of IN_3 to dryness. Therefore, after reaction with multiple bonds, the mixture should be washed with sodium bisulfite until colorless to destroy excess IN_3 reagent. Violent reaction can also occur on addition of IN_3 to alkenes containing low-valent S-functions. An explosion shield should be used when handling this reagent. Operations should be conducted in a fume hood.

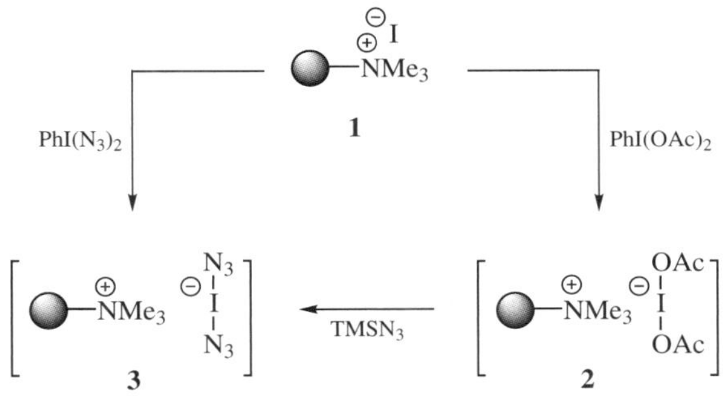

(1)

Original Commentary

Alfred Hassner
Bar-Ilan University, Ramat-Gan, Israel

Additions to Alkenes. Addition of IN_3 to alkenes (see also RB257) usually proceeds via a three-membered ring iodonium species, resulting from attack of I^+ on the double bond. The reaction is highly stereoselective (*anti*) and regioselective. Thus *cis*- and *trans*-2-butene give stereospecifically the *threo* and *erythro* adducts, respectively. The azide–carbon bond usually forms at the more substituted carbon or the one better able to support a positive charge in the transition state. *trans*-1-Phenylpropene gives, stereo- and regiospecifically, *erythro*-1-azido-2-iodo-1-phenylpropane (eq 2). *Anti* elimination of HI from the adduct leads to a vinyl azide.[4a] However, steric hindrance can influence the position of attack by azide ion. Thus *t*-butylethylene affords the terminal azide (eq 3).[2b]

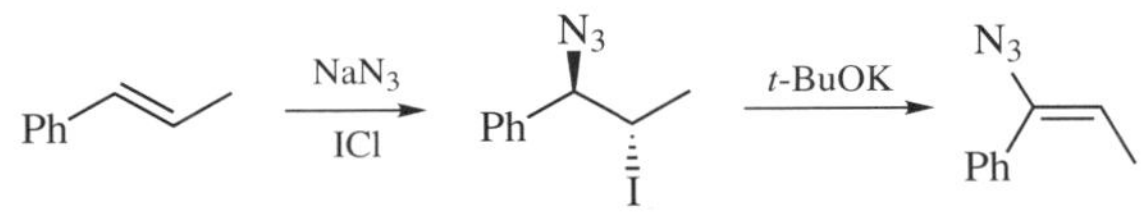

(2)

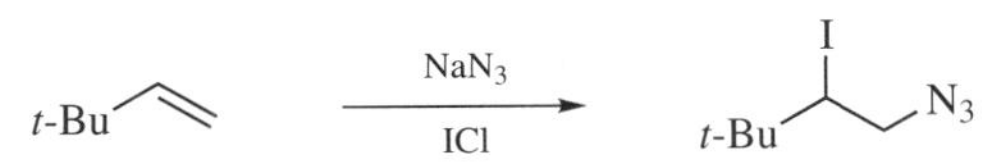

(3)

The structure of the adduct with phenylcyclohexene has been questioned[11a] but it has been proven to be the one expected (eq 4).[11b,c] With 2-cholestene the *trans* diaxial adduct, 2β-azido-3α-iodocholestane, results from the α-iodonium species,[1,2] but products from β-attack on other steroid alkenes are also known.[12a] In rare cases, *syn* addition has been reported,[9] for instance to strained cyclobutenes.[12b,c]

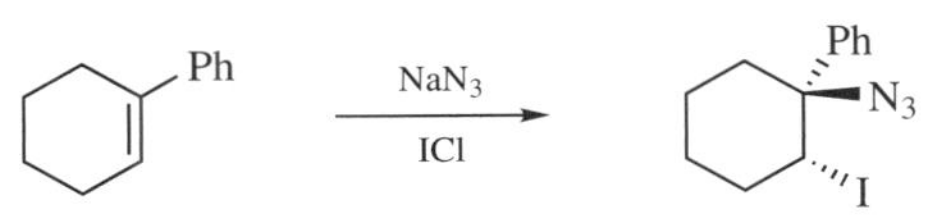

(4)

Addition of IN_3 to conjugated ketones can lead to a mixture of regiosomers, sometimes depending on the presence of oxygen or on the method of generation of the reagent.[3,4] In such cases the addition can proceed via a radical mechanism,[3,13] in which the azide radical rather than I^+ is the attacking species and different regioisomers can result.[3] For instance, cyclopentenone affords 2-iodo-2-cyclopentenone, while cyclohexenone gives a mixture of products.[3e]

Enantioselective addition of IN_3 to conjugated amides derived from sultams or oxazolidones as chiral auxiliaries has been reported.[14] Although diastereoselection only averages 34% (eq 5), crystallization raises this to 94%.

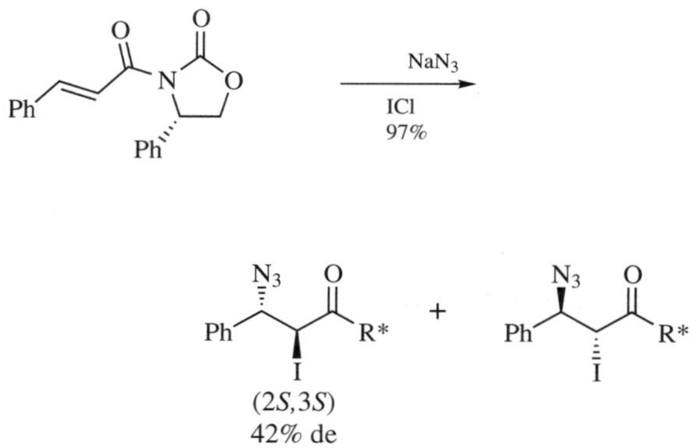

(5)

Enol ethers, as expected, give α-azido-β-iodo ethers,[15a] but benzofuran and some indoles give a mixture of *cis*- and *trans*-2,3-diazido products.[15b] Allenes can give mono- or bis-adducts.[16] When the monoadduct is an allyl azide, then [3,3] rearrangements are observed (eq 6). When the monoadduct is a vinyl azide, addition of a second molecule of IN_3 is faster than the first addition and a bis-adduct results. The cyclopropane group in norcarane is opened by IN_3 to give a mixture of products.[17]

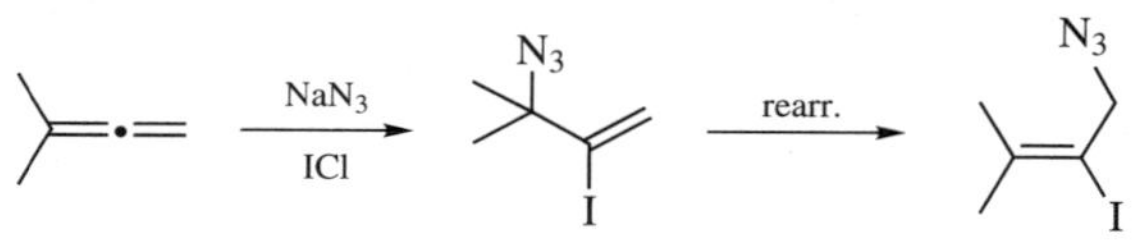

(6)

Synthesis of Aziridines. IN_3 addition to alkenes followed by Lithium Aluminum Hydride (or Diborane) reduction is a convenient method ('Hassner aziridine synthesis') for the preparation of aziridines.[6] This method has been applied to the synthesis of monoaziridine derivatives of terpenes such as squalene, farnesol, and geraniol (eq 7).[18]

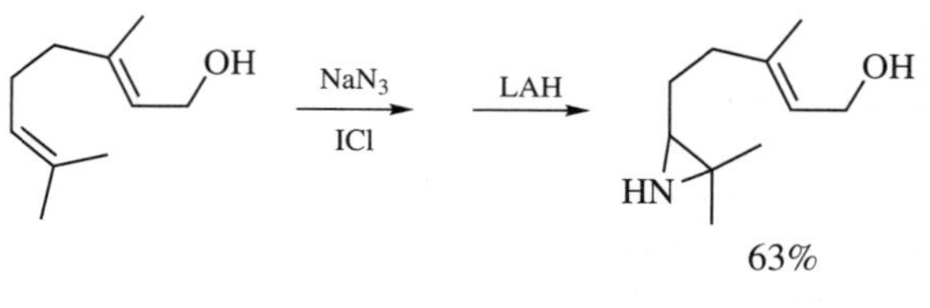

(7)

With *anti*-oriented iodo and azide functions, LAH leads to a large amount of IN_3 elimination.[6] Many other reagents can be used for the reduction of IN_3 adducts to aziridines,[6,19] including Hydrazine–Raney Nickel, $Me_2NH{\cdot}BH_3$, and phosphines or phosphites; the last reagents lead to *N*-phosphorylated aziridines.[19a] When alkyl- or aryldichloroborane is used, it provides stereospecific entry into *N*-substituted aziridines (eq 8).[19d]

(8)

Selective reduction of the adducts has also been achieved, as in the formation of β-amino acids from acrylic esters.[20] Rearrangements during IN_3 additions have been observed in rearrangement-prone systems such as methylenenorbornene, benzonorbornadiene, the pinenes, and tritylethylene.[21] In α- and β-pinene, rearrangement is accompanied by trapping with MeCN followed by N_3^- to afford tetrazoles.[22]

Synthesis of Vinyl Azides and Azirines. Various bases, especially Potassium t-Butoxide, cause stereospecific elimination of HI from the adducts and the resulting vinyl azides serve as useful intermediates for the synthesis of azirines (eq 9),[6,23] ketones,[24] and cyanoketenes,[25] and for the amination of aromatic compounds.[26]

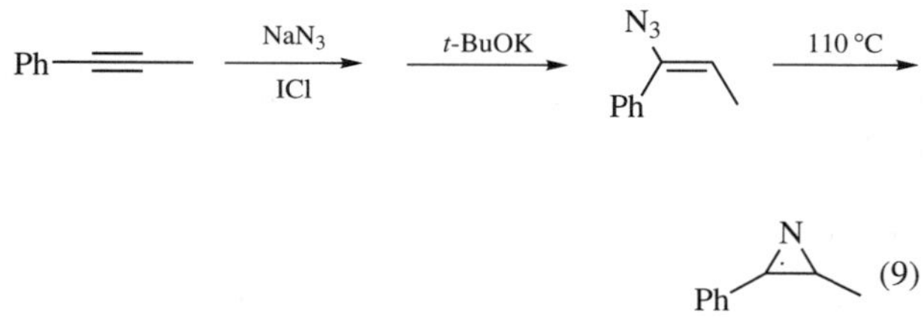

(9)

Miscellaneous Reactions. Addition of IN_3 to methylphenylacetylene leads to 2-azido-1-iodo-1-phenylpropene.[27] IN_3 also adds to isocyanides to produce iodotetrazoles.[2] Dimethyl thioacetals react with IN_3 to afford an α-azido thioether which can undergo Schmidt rearrangement in the presence of Trifluoroacetic Acid.[28]

First Update

Lavinia Marinescu & Mikael Bols
Aarhus University, Aarhus, Denmark

Addition to [1.1.1]Propellane. Addition of IN_3 to [1.1.1]propellane, the smallest tricyclic bridgehead compound, proceeds across the 1,3 positions, providing 3-iodobicyclo [1.1.1]pentyl azide which was expected to supply 1-bicyclo[1.1.1]pentylamine by reduction.[29] The final product from the reduction of 3-iodobicyclo[1.1.1]pentyl azide with lithium aluminum hydride in THF was 3-chloro-3-methylbutylamine hydrochloride due to a series of rearrangements (eq 10).

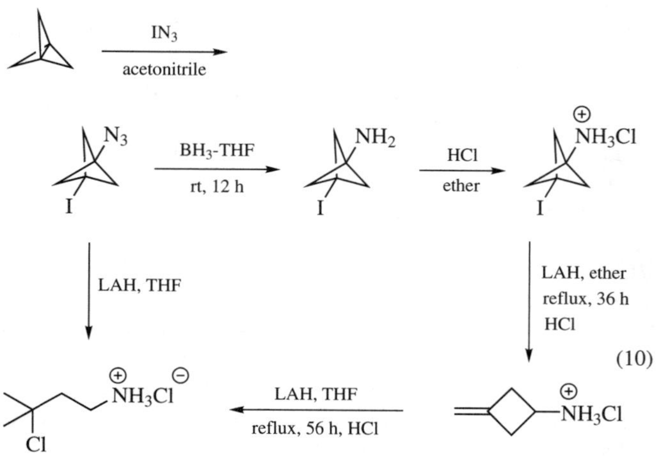

(10)

Radical Substitutions of IN_3. Iodine azide, well known for its propensity for ionic reactions, can also undergo bond homolysis. Radical substitutions with iodine azide can be carried out on C–H bonds with low bond dissociation energies, such as benzyl ethers (eq 11), to give α-azido ethers, benzal acetals, and aldehydes. The azidyl radical is formed by oxidation of azide ion and is an electrophilic species that is related to halogen and pseudohalogen radicals.[30,31]

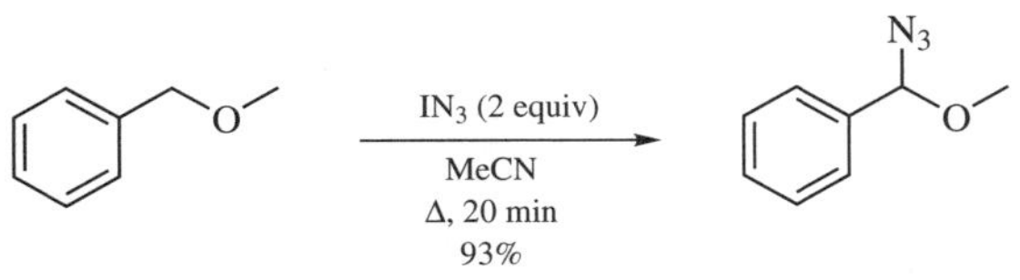

(11)

Radical Substitutions to Benzyl Ethers. Benzyl ethers react with IN_3 in refluxing acetonitrile to give the corresponding α-azido ethers.[32] A related reaction which may involve radicals has been reported by Magnus et al., who were able to directly azidonate in the α-position of amines using $PhIO/TMSN_3$.[33]

Substitution of the activated α-hydrogen in benzyl ethers by azido groups using IN_3 was highly efficient and tolerates other functionalities such as ester (eq 12), epoxide (eq 13), or acetal units (eq 14).

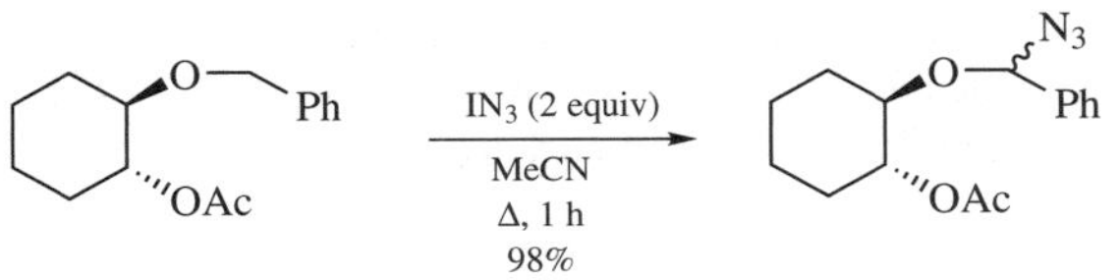

(12)

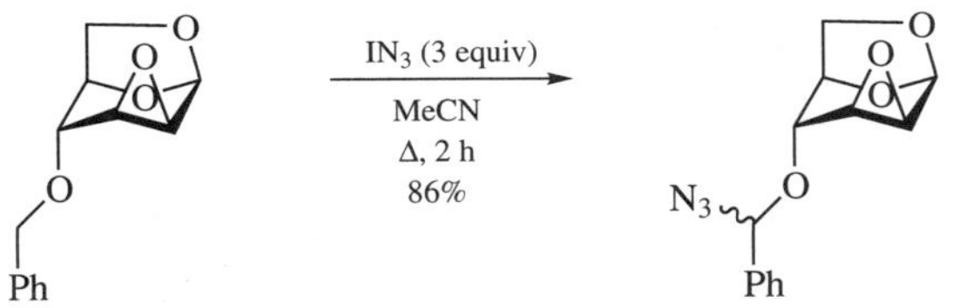

(13)

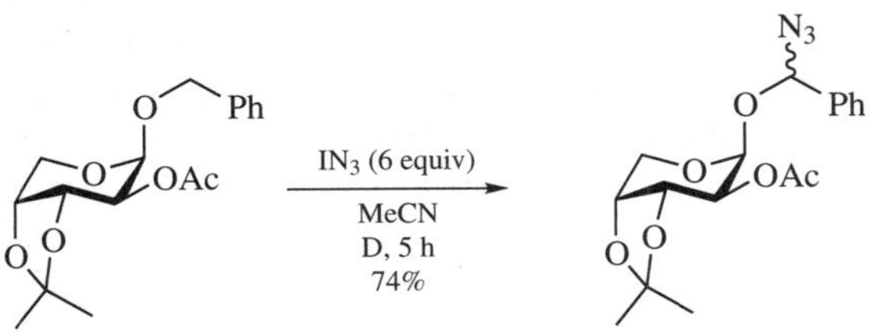

(14)

The suggested mechanism involves first a homolysis of IN_3 resulting in an azide radical that abstracts a hydrogen atom from the substrate to form a benzylic radical. This radical is oxidized to a benzylic cation which can react with azide ion to give the product. Support for the radical mechanism was found by running the reaction in the presence of the radical trap *N-tert*-butyl-α-phenylnitrone, which led to recovery of the starting materials.

Radical Substitutions and Benzylidene Acetal Opening Reactions. Benzylidene acetals are transformed to α-azidobenzoates by this reagent[34] in an azido-Hanessian type reaction. The Hanessian reaction[35] involves the conversion of a benzylidene acetal into a 1,2- or 1,3-bromobenzoate using *N*-bromosuccinimide: this very useful transformation allows the change of a protection group into a reactive functionality. When a nitrogen-containing target is desired, the reaction of benzal acetals with IN_3 can be used and a step is saved compared with the Hanessian type transformation.

The charactersitics of these reactions are reflected in the following example. Following a nucleophilic substitution, the azide prefers the primary position (eq 15). Electron-donating groups have an enhancing effect, resulting in a faster reaction (eq 16) while electron-withdrawing substituents give a sluggish reaction and a poor yield (eq 17).

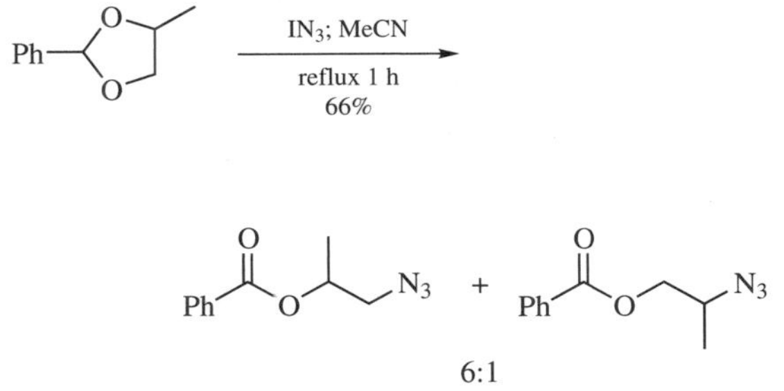

(15)

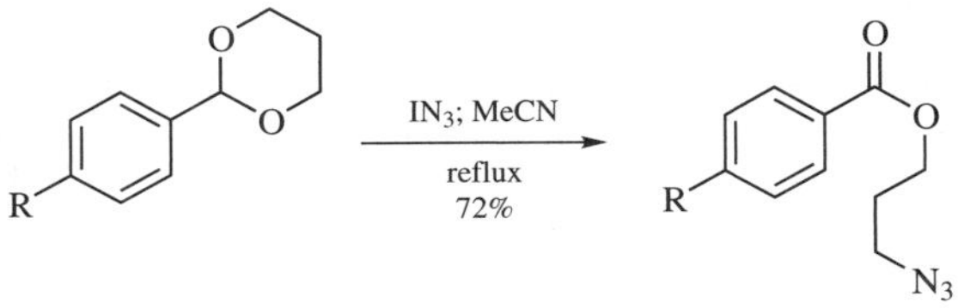

(16)

R = H, OMe

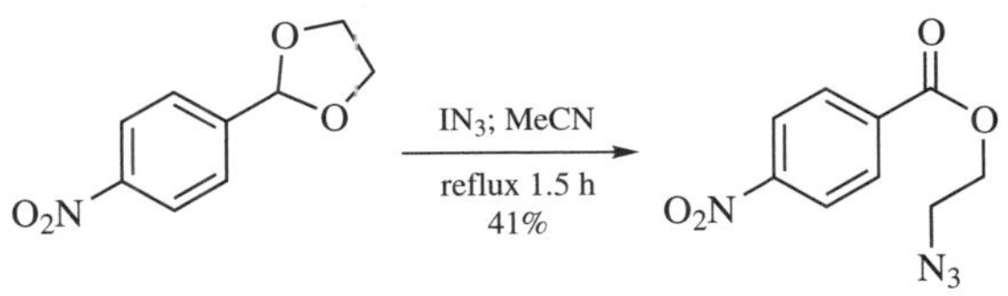

(17)

The mechanism involves abstraction of the benzylic hydrogen in the first step, resulting in a benzylic radical, which is subsequently oxidized to a phenyl dioxolanium ion by an iodonium ion. The cation is finally cleaved with N_3^- to give the product (eq 18).

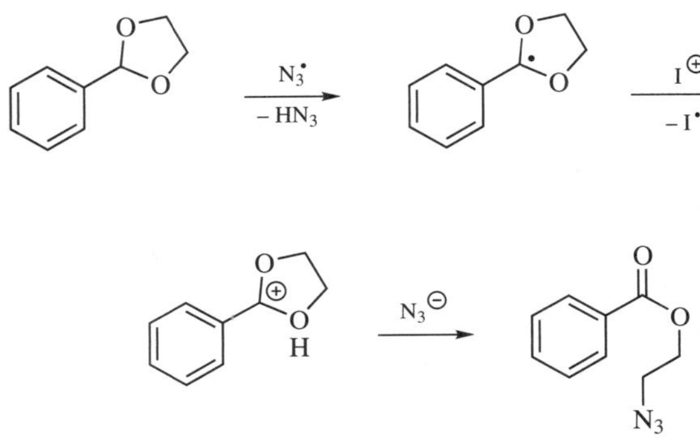

(18)

Radical Azidonation of Aldehydes. With a C–H bond dissociation energy of 89 kcal mol^{-1}, aldehydes can easily undergo radical substitution reactions by treatment with IN_3 at room temperature, to give acyl azides (eq 19).[36] The transformation of aldehydes to acyl azides has previously been reported with $CrO_3/TMSN_3$[37] or $MnO_2/SiCl_4/NaN_3$[38] via different mechanisms.

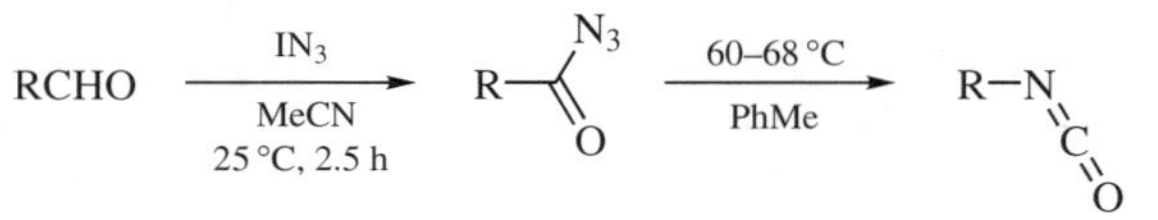
(19)

The results of the reaction of some aliphatic and aromatic aldehydes with IN_3 in acetonitrile at 25 °C are given in Table 1. By heating the acyl azide in toluene[39] it undergoes Curtius rearrangement to form the isocyanate in high yield.

Table 1 Reaction of Aldehydes with Iodine Azide at room temperature and subsequent Curtius rearrangement[a]

	Yield(%)	
R	(Acyl azide)	(Isocyanate)
$PhCH_2CH_2$	82	97
$CH_3(CH_2)_8$	Quant.	98
$CH_3(CH_2)_6$	86	99

[a]Yield determined by NMR.

Because acyl azides and isocyanates are very reactive compounds that cannot be isolated by chromatography, it is more convenient to carry out the azidonation reactions at reflux, which results in spontaneous Curtius rearrangement and formation of carbamoyl azide by addition of azide to the isocyanate (eq 20). The results are summarized in Table 2.

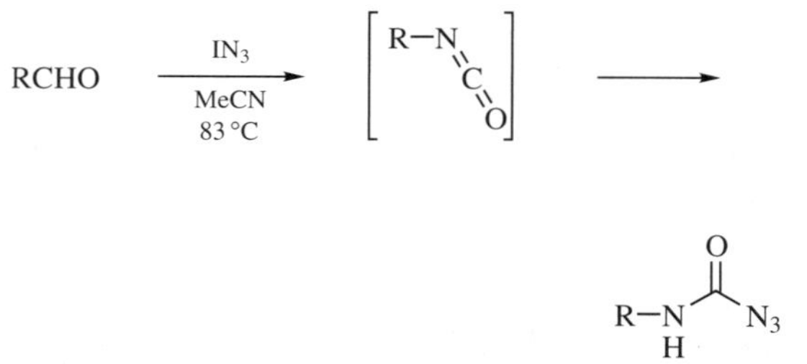
(20)

Table 2 Reaction of Aldehydes with Iodine Azide (2 equiv of ICl and 3 equiv of NaN_3 in MeCN) at Reflux

		Yield [a]
R	Time (h)	(Carbamoyl azide) (%)
$CH_3(CH_2)_6$	4	86
$CH_3(CH_2)_8$	4	84
$PhCH_2CH_2$	4	73
Ph	4	97
p-MeOPh	4	70
p-MePh	4	96

[a]Yields are of carbamoyl azide after chromatography.

The reaction can be performed with BrN_3 formed in situ in acetonitrile from NBS and sodium azide or with NIS/NaN_3 but the yields are lower; cyclohexyl carboxaldehyde gave a 42% yield of azidonation product using BrN_3.

Since carbamoyl azides[40] can be converted to amines, carbamates, or ureas, the reaction is a valuable and simple route from aldehydes to amines or their derivatives.

The following reaction course was suggested: homolysis of the weak iodine-azide bond (eq 21) with formation of iodine atom which abstracts the aldehydic hydrogen atom (eq 22). The carbon-centered radical reacts with iodine azide to produce an acyl azide and an iodine atom (eq 23).

$$I{-}N_3 \longrightarrow I^\bullet + N_3^\bullet \quad (21)$$

(22)

(23)

Carbamoyl azides were not formed when the reaction was carried out in presence of the radical trap *N-tert*-butyl-α-phenylnitrone which was taken as support for a radical reaction mechanism. Furthermore, the IN_3 reaction of cyclohexyl carboxaldehyde in EtOH leads to complete conversion with carbamoyl azide being the major product; no cyclohexyl carboxylic acid ethyl ester was observed, so no ionic intermediates appear to be involved in the reaction.

Substitution of the α-Azido Group with Grignard Reagents. In 1998, Magnus et al. reported the substitution of α-azidoamine with phenylmagnesium bromide in THF (eq 24).[41]

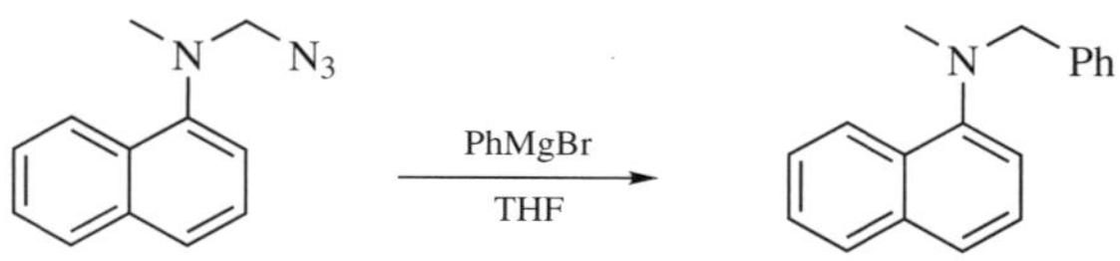
(24)

A similar substitution reaction was applied to α-azidobenzyl ethers using aromatic, allylic and aliphatic Grignard reagents.[42] The reaction of methylmagnesium bromide with α-azidobenzyl methyl ether when performed in THF failed. However, the reactivity of the Grignard reagent was increased using toluene as solvent[43,44] and the 1-phenylethyl methyl ether was obtained after 20 min at room temperature with 90% yield, see eq 25 and Table 3.

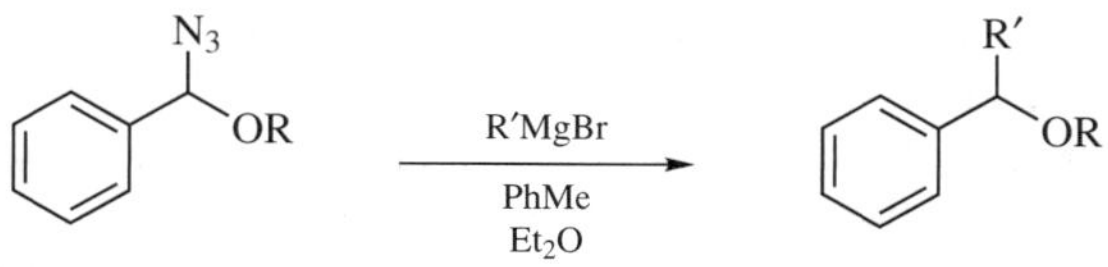
(25)

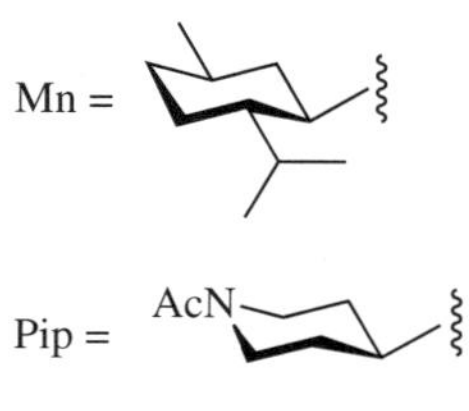

Table 3 The results from the reaction shown in eq 25

R	R′	Time (min)	Yield (%)
Me	Me	20	90
Me	Ph	20	89
Me	Allyl	90	85
Et	Me	20	89
Et	Ph	20	87
Mn	Me	30	82
Pip	Me	10	94
Pip	n-C_6H_{13}	10	90

This reaction was used in a combinatorial way to obtain libraries of compounds which can be tested as bioactive compounds. Thus a four-component Grignard reagent was made by reaction of a mixture of MeI, PhBr, $CH_3CH_2CH_2Br$, and $CH_2{=}CHCH_2Br$ with excess of magnesium and then added to a solution of α-azidobenzyl methyl ether in toluene-ether. A mixture of the corresponding α-substituted ethers was obtained in ratio 3:4:3:4 (eq 26).

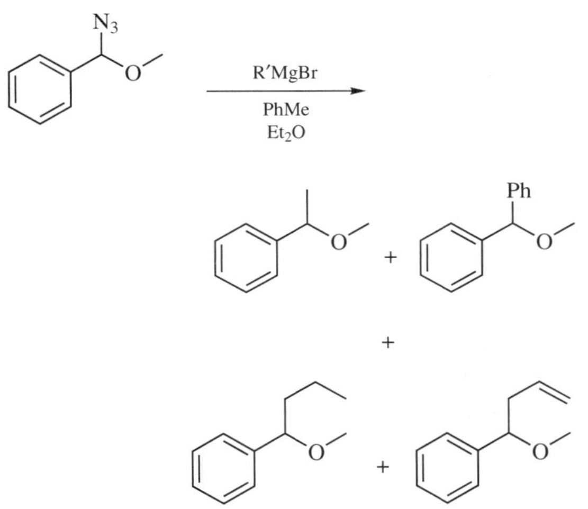

(26)

Similar libraries using eight-component Grignard reagents have been made and analyzed by mass spectra and ^{1}H NMR.

Related Reagents. bromine azide; *N*-bromosuccinimide; *N*-iodosuccinimide; trimethylsilyl azide.

1. Hassner, A., *Ann. Chim. (Rome)* **1971**, *4*, 9.
2. (a) Hassner, A.; Levy, L. A., *J. Am. Chem. Soc.* **1965**, *87*, 4203. (b) Fowler, F. W.; Hassner, A.; Levy, L. A., *J. Am. Chem. Soc.* **1967**, *89*, 2077. (c) Hassner, A.; Fowler, F. W., *J. Org. Chem.* **1968**, *33*, 2686.
3. (a) L'abbé, G.; Hassner, A., *J. Org. Chem.* **1971**, *36*, 258. (b) Cambie, R. C.; Rutledge, P. S.; Smith-Palmer, T.; Woodgate, P. D., *J. Chem. Soc., Perkin Trans. 1* **1977**, 2250. (c) Cambie, R. C.; Jurlina, J. L.; Rutledge, P. S.; Woodgate, P. D., *J. Chem. Soc., Perkin Trans. 1* **1982**, 315. (d) Cambie, R. C.; Jurlina, J. L.; Rutledge, P. S.; Swedlund, B. E.; Woodgate, P. D., *J. Chem. Soc., Perkin Trans. 1* **1982**, 327. (e) Ponsold, K.; Wuderwald, M., *Z. Chem.* **1979**, *19*, 25. (f) McIntosh, J. M.; *Can. J. Chem.* **1971**, *49*, 3045.
4. (a) L'abbé, G.; Hassner, A., *Angew. Chem., Int. Ed. Engl.* **1971**, *10*, 98. (b) Hassner, A., *Methoden Org. Chem. (Houben-weyl)* **1990**, *10/2*, 1243. (c) Hassner, A.; Isbister, R. J.; Friederang, A., *Tetrahedron Lett.* **1969**, 2939.
5. Hassner, A.; Matthews, G. J.; Fowler, F. W., *J. Am. Chem. Soc.* **1969**, *91*, 5046.
6. Hassner, A.; Fowler, F. W., *J. Am. Chem. Soc.* **1968**, *90*, 2869.
7. (a) Hantzsch, A., *Ber. Dtsch. Chem. Ges./Chem. Ber.* **1900**, *33*, 522. (b) Dehnike, K., *Angew. Chem., Int. Ed. Engl.* **1976**, *15*, 553.
8. Tornieporth-Oetting, I. C.; Klapötke, T. M., *Angew. Chem. Int. Ed. Engl.* **1995**, *34*, 511–520.
9. Hargittai, M.; Molnár, J.; Klapötke, T. M.; Tornieporth-Oetting, I. C.; Kolonits, M.; Hargittai, I., *J. Phys. Chem.* **1994**, *98*, 10095.
10. Kirschning, A.; Monenschein, H.; Schmeck, C., *Angew. Chem. Int. Ed.* **1999**, *38(17)*, 2594–2696.
11. (a) Sivasubramanian, S.; Aravind, S.; Kumarasingh, L. T.; Arumugam, N., *J. Org. Chem.* **1986**, *51*, 1985. (b) Hassner, A.; Dehaen, W., *J. Org. Chem.* **1990**, *55*, 2243. (c) Crotti, P.; Chini, M.; Uccello-Barretta, G.; Macchia, F., *J. Org. Chem.* **1989**, *54*, 4525.
12. (a) Carlon, F. E.; Draper, R. W., *J. Chem. Soc., Perkin Trans. 1* **1983**, 2793. (b) Mehta, G.; Dutta, P. K.; Pandey, P. N., *Tetrahedron Lett.* **1975**, 445. (c) Sasaki, K.; Kanematsu, K.; Kondo, A., *Tetrahedron* **1975**, *31*, 2215. (d) Cambie, R. C.; Hayward, R. C.; Rutledge, P. S.; Smith-Palmer, T.; Swedlund, B. E.; Woodgate, P. D., *J. Chem. Soc., Perkin Trans. 1* **1979**, 180.
13. Hassner, A.; Boerwinkle, F., *J. Am. Chem. Soc.* **1968**, *90*, 216.
14. Lee, P.-C.; Wu, C.-C.; Cheng, M.-C.; Wang, Y.; Wu, M.-J., *J. Chin. Chem. Soc.* **1992**, *39*, 87.
15. (a) Ghosez, L.; Sainte, F.; Rivera, M; Bernard-Henriet, C.; Gouverneur, V., *Recl. Trav. Chim Pays-Bas* **1986**, *105*, 456. (b) Tamura, Y.; Chun, M. W.; Ohno, K.; Kwon, S.; Ikeda, M., *Chem. Pharm. Bull.* **1978**, *26*, 2874. (c) Tamura, Y.; Chun, M. W.; Kwon, S.; Bayomi, S. M.; Okada, T.; Ikeda, M., *Chem. Pharm. Bull.* **1978**, *26*, 3515.
16. Hassner, A.; Keogh, J., *J. Org. Chem.* **1986**, *51*, 2767.
17. Cambie, R. C.; Dixon, G.; Rutledge, P. S.; Woodgate, P. D., *J. Chem. Soc., Perkin Trans. 1* **1982**, 961.
18. (a) Parish, E. J.; Nes. W. D., *Synth. Commun.* **1988**, *18*, 221. (b) Avruch, L.; Oehlschlager, A. C., *Synthesis* **1973**, 622.
19. (a) Hassner, A.; Galle, J. E., *J. Am. Chem. Soc.* **1970**, *92*, 3733. (b) Sehgal, R. K.; Almassian, B.; Rosenbaum, D. P.; Zadrozny, R.; Sengupta, S. K., *J. Med. Chem.* **1987**, *30*, 1626. (c) Campbell, M. M.; Abbas, N.; Sainsbury, M., *Tetrahedron* **1985**, *41*, 5637. (d) Levy, A. B.; Brown, H. C., *J. Am. Chem. Soc.* **1973**, *95*, 4067.
20. Wasserman, H. H.; Brunner, R. K.; Buynak, J. D.; Carter, C. G.; Oku, T.; Robinson, R. P., *J. Am. Chem. Soc.* **1985**, *107*, 519.
21. (a) Hassner, A.; Teeter, J. S., *J. Org. Chem.* **1970**, *35*, 3397. (b) Hassner, A.; Teeter, J. S., *J. Org. Chem.* **1971**, *36*, 2176.
22. Ranganathan, S.; Ranganathan, D.; Mehrotra, A. K., *Tetrahedron Lett.* **1973**, 2265.
23. (a) Hassner, A.; Fowler, F. W., *Tetrahedron Lett.* **1967**, 1545. (b) Padwa, A.; Blacklock, T.; Tremper, A., *Org. Synth.* **1977**, *57*, 83.
24. (a) Belinka, B. A., Jr.; Hassner, A., *J. Org. Chem.* **1979**, *44*, 4712. (b) Hassner, A.; Belinka, B. A., Jr., *J. Am. Chem. Soc.* **1980**, *102*, 6185.
25. Moore, H. W.; Decker, O. H. W., *Chem. Rev.* **1986**, *86*, 821.
26. Hassner, A.; Munger, P.; Belinka, B. A., Jr., *Tetrahedron Lett.* **1982**, *23*, 699.
27. Hassner, A.; Isbister, R. J.; Friederang, A., *Tetrahedron Lett.* **1969**, 2939.
28. Trost, B. M.; Vaultier, M.; Santiago, M. L., *J. Am. Chem. Soc.* **1980**, *102*, 7929.
29. Hossain Md, T.; Timberlake, J. W., *J. Org. Chem.* **2001**, *66*, 4409–4412.
30. Worketin, M. S.; Wagner, B. D.; Lusztyk, J.; Wayner, D. D. M., *J. Am. Chem. Soc.* **1995**, *117*, 119–126.
31. Fontana, F.; Minisci, F.; Yan, Y. M.; Zhao, L., *Tetrahedron Lett.* **1993**, *34*, 2517–2520.
32. Viuf, C.; Bols, M., *Angew. Chem. Int. Ed.* **2001**, *40*, 623–625.
33. Magnus, P.; Lacour, J.; Weber, W., *J. Am. Chem. Soc.* **1993**, *115*, 9347–9348.

34. Baruah, M.; Bols, M., *Synlett.* **2002**, *7*, 1111–1112.
35. Hanessian, S.; Plessas, N. R., *J. Org. Chem.* **1969**, *34*, 1045–1053.
36. Marinescu, L.; Thinggaard, J.; Thomsen Ib, B.; Bols, M., *J. Org. Chem.* **2003**, *68*, 9453–9455.
37. Lee, J. G.; Kwak, K. H., *Tetrahedron Lett.* **1992**, *33*, 3165–3166.
38. Elmorsy, S. S., *Tetrahedron Lett.* **1995**, *36*, 1341–1342.
39. Allen, C. F. H.; Bell, A., *Organic Synthesis*; Wiley: New York, 1955, Vol. III, pp. 846–847.
40. Lieber, E.; Minnis, R. L.; Rao, C. N. R., *Chem. Rev.* **1965**, *65*, 377–384.
41. Magnus, P.; Lacour, J.; Weber, W., *Synthesis*, **1998**, 547–551.
42. Baruah, M.; Bols, M., *J. Chem. Soc., Perkin Trans.* , **2002** 509–512.
43. Ishikawa, H.; Ikeda, S.; Mukaiyama, T., *Chem. Lett.* **1975**, 1051–1054.
44. Ishikawa, H.; Mukaiyama, T., *Chem. Lett.* **1975**, 305–306.

Iodine Monobromide

BrI

[7789-33-5] BrI (MW 206.80)

(diastereoselective cyclizations of homoallylic carbonates;[1] electrophilic addition to alkenes;[2] α-bromination of steroidal aldehydes and ketones;[3,4] cleavage of carbon–metal bonds[5])

Physical Data: mp 40 °C; bp 116 °C (dec); *d* 4.416 g cm^{-3}.
Solubility: sol alcohol, ether, CS_2, acetic acid, acetone, CH_2Cl_2, MeCN, DME, toluene (but with limited solubility below −80 °C); soluble in H_2O, but hydrolyzes to HBr and IOH.
Form Supplied in: brownish-black crystals or very hard, black solid; widely available (usually 97–98% purity).
Handling, Storage, and Precautions: corrosive solid and vapor; readily absorbed through skin and mucous membranes; use in a fume hood and wear gloves and appropriate protective clothing; air-, moisture-, and light-sensitive; should be stored refrigerated (under N_2) in a tightly sealed amber bottle. Forms explosive mixtures with potassium and sodium, and reacts extremely exothermically with phosphorus and tin.

Original Commentary

Ronald G. Brisbois & Randall A. Wanke
Hamline University, St. Paul, MN, USA

Electrophilic Cyclization of Homoallylic Carbonates.[1] Iodine monobromide provides very good diastereoselectivity in electrophilic cyclizations of homoallylic carbonates (eq 1). Isolated yields of α-iodo cyclic carbonates are excellent, and these products can be easily and efficiently converted into the corresponding epoxides (eq 2). Optimal diastereoselectivity is obtained from reactions performed in toluene at −80 to −85 °C. Use of Iodine/MeCN[6] or Iodine Monochloride/CH_2Cl_2 yields much lower diastereoselectivity than iodine monobromide under similar conditions. Results of an equilibration experiment strongly indicate that the diastereoselectivity in iodine monobromide-induced cyclization stems from kinetic control.

Electrophilic Addition to Alkenes.[2] Iodine monobromide addition across alkenes is very facile, yielding vicinal bromoiodoalkanes. In general, addition occurs with high *anti* stereospecificity, and mechanistic studies have implicated an iodonium ion intermediate in the overall reaction sequence. The regiochemistry of iodine monobromide addition is very sensitive to the initial alkene structure, such that anti-Markovnikov products are possible (Table 1).

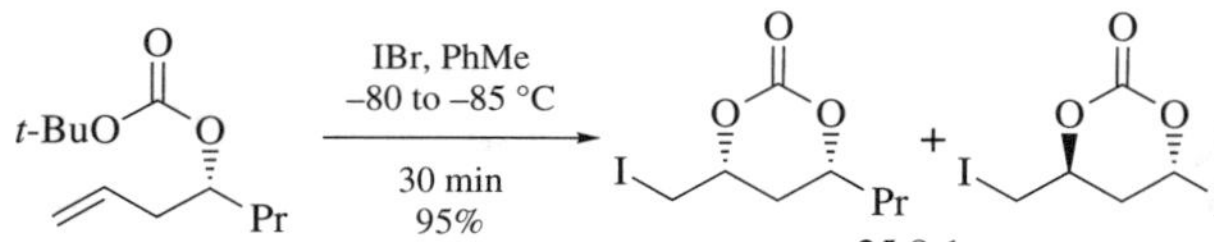

25.8:1 (1)

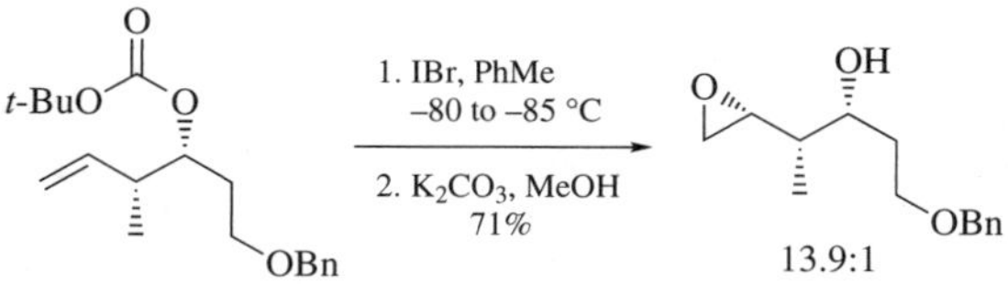

13.9:1 (2)

Table 1 Regiochemistry of IBr Addition to Alkenes

Alkene	% Markovnikov	% anti-Markovnikov
MeCH=CH$_2$	65	35
ErCH=CH$_2$	45	55
i-PrCH=CH$_2$	15	85
t-BuCH=CH$_2$	0	100
PhCH=CH$_2$	100	0

α-Bromination of Steroidal Aldehydes and Ketones. Iodine monobromide is an effective monobrominating reagent for steroidal aldehydes (eq 3).[3] The ratio of diastereomeric products is in part a function of the reaction site's proximity to an asymmetric center, with β-orientation favored for the entering bromine. In an analogous application, methyl 3-oxo-5β-cholanate has been monobrominated (IBr/$MeCO_2H$; 50% yield) to produce the 2β-bromo ketone.[4]

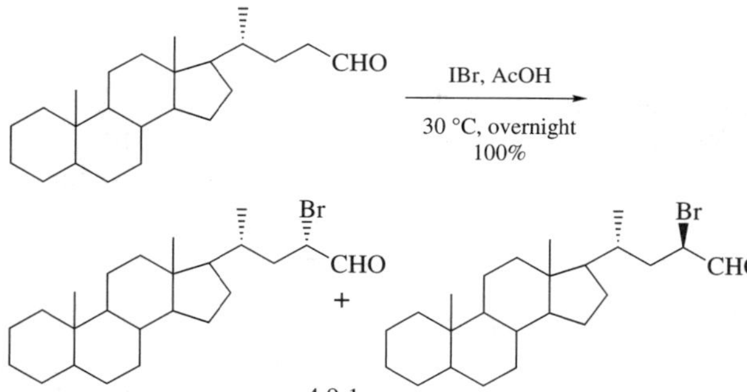

4.9:1 (3)

Bond Cleavage Reactions.[5] Cleavage of carbon–metal bonds by iodine monobromide typically produces the organic iodide and the metal bromide. Cleavage of organotin,[5a–c] organoarsenic,[5d] organoantimony,[5d] organogermanium,[5e] and organolead[5e] bonds has been investigated (see also Iodine Monochloride for use in cleaving carbon–metal bonds).

Other Applications. Iodine monobromide has been used for brominating phenol and aryl ether derivatives.[7] Ethers and esters have been cleaved by catalytic iodine monobromide in the presence of Bromotrimethylsilane.[8] Aryl acetals can be directly oxidized to the corresponding esters in moderate yield.[9]

First Update

Robert A. Field & Balaram Mukhopadhyay
University of East Anglia, Norwich, UK

Electrophilic Cyclization. Iodine monobromide continues to be exploited in stereoselective cyclization reactions *en route* to natural products, such as calyculin[10] and halichondrin B.[11] It can also be employed in diastereoselective isothiourea iodocyclization reactions, which proceed smoothly using IBr at low temperature. Advanced intermediates in the synthesis of manzacidin A, amongst other examples, have been obtained using this procedure (eq 4).[12]

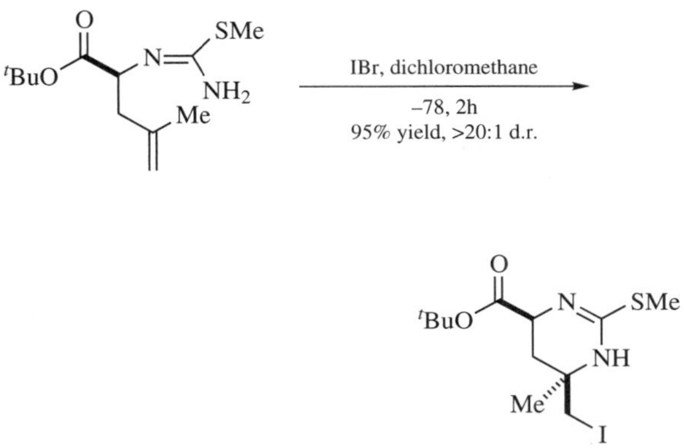

(4)

Electrophilic Addition to Alkenes. Trihalide-based ionic liquids, produced by mixing equimolar amounts of IBr and [bmim][Br], prove to be effective reagent-solvents for stereoselective iodination of alkenes and alkynes.[13] Yields of iodobromo adducts range from very good to almost quantitative for a range of substrates (eq 5).

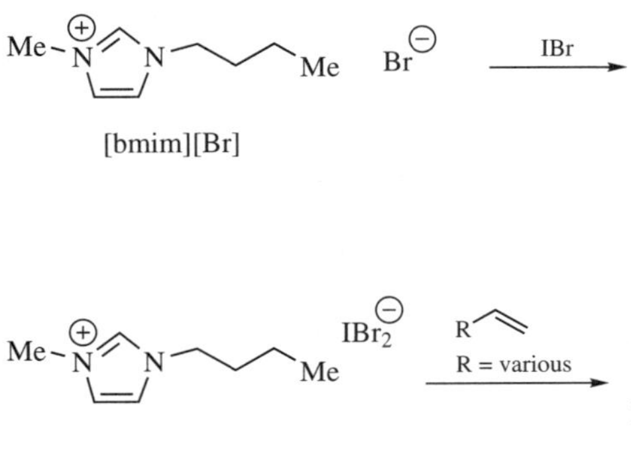

(5)

Formation of Glycosyl Bromides. Iodine monobromide is used for the synthesis of glycosyl bromides from various alkyl and aryl thioglycosides (eq 6).[14] The reaction of IBr with thioglycosides is compatible with many of the common protecting groups, such as isopropylidene acetals, cyclohexane 1,2-diacetals, 4-methoxybenzyl and benzyl ethers, phthalimido groups, etc. Compounds containing intersugar glycosidic linkages form glycosyl bromides at their reducing terminal only, without affecting existing glycosidic linkages.

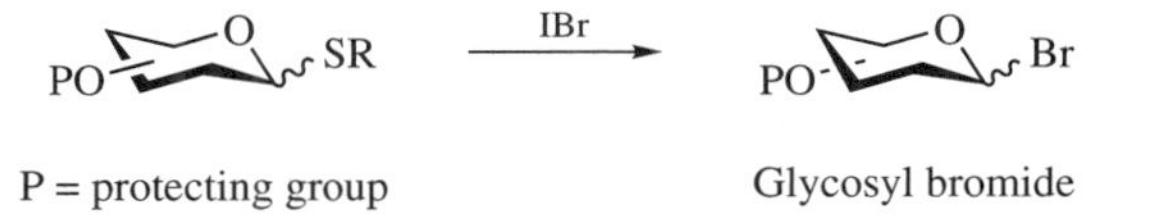

(6)

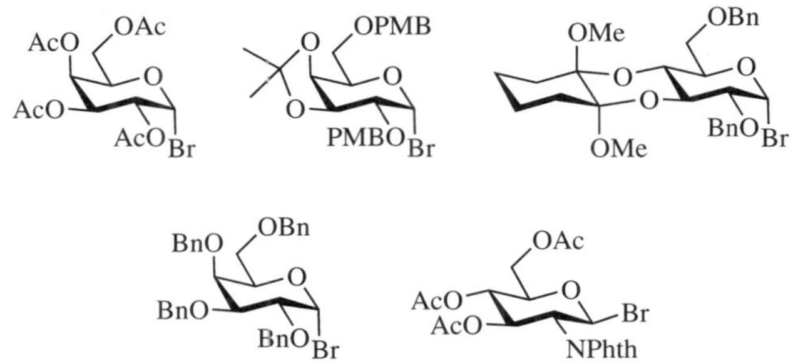

Glycosylation Chemistry. Iodine monobromide is an efficient reagent for *O*-glycosylation using "disarmed" thioglycoside donors.[14,15] It provides excellent 1,2-*trans* stereoselectivity and it is compatible with most common protecting groups and existing *O*-glycosides. However, issues can arise from the Lewis acidity of IBr, which can effect acyl migration. In addition, competing glycosyl bromide formation may have an impact on reaction outcome.[15] These problems can largely be overcome through the use of IBr in conjunction with silver triflate. Studies on interhalogen/silver trifluoromethanesulfonate (IX/AgOTf) promoted glycosylations reactions identified differences in the sensitivity of the formed oxocarbenium ions, e.g., from compounds with or without participating groups) toward halide nucleophiles.[16] These differences can be explained using the Hard-Soft-Acid-Base theory. By applying this theory silylation yields were increased for a model reaction from ~40% using ICl to 74% using IBr. By increasing the amount of AgOTf from 1 to 1.5 equiv (with respect to IBr) the yield in the model reaction improved to 89%, with a move from xanthate to thiophenyl glycoside donor further increasing the yield of the model reaction to 97% (eq 7).

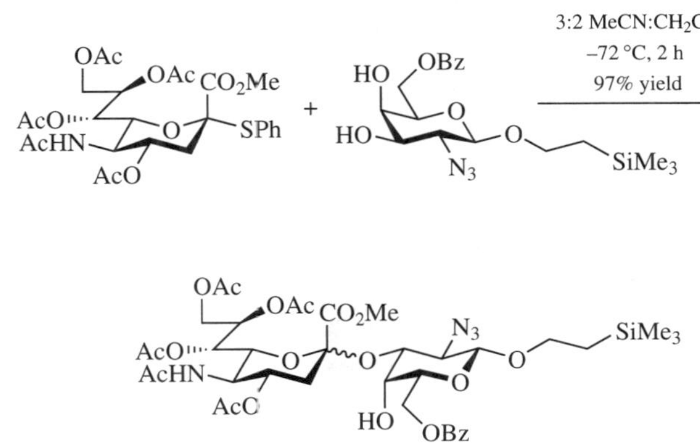

(7)

Other Applications. Iodine monobromide has been used as a mild reagent for selective deprotection of *O-tert*-butyldimethylsilyl ethers of simple alcohols, carbohydrates, and nucleosides.[17] Acid labile functionalities such as acetals, *O-p*-methoxybenzyl ethers, etc., as well as base labile esters and amides, are stable in the presence of iodine monobromide. Among several solvents used for the reaction, methanol gives the best yields, which may indicate that desilylation is a result of in situ generation of HBr.

Radical addition of iodine monobromide to chlorotrifluoroethylene forms a useful intermediate in the synthesis of 4,5,5-trifluoro-4-ene-pentanol, a useful monomer for fluoropolymer synthesis.[18]

Related Reagents. Iodine monochloride; bromine N-iodosuccinimide; N-bromosuccinimide, N-iodosaccharin; iodonium dicollidine perchlorate

1. Duan, J. J.-W.; Smith, A. B., III., *J. Org. Chem.* **1993**, *58*, 3703.
2. Schmid, G. H.; Garratt, D. G. In *The Chemistry of Doubly-Bonded Functional Groups*; Patai, S., Ed.; Wiley: New York, 1977; Part 2, pp 799–801.
3. Yanuka, Y.; Katz, R.; Sarel, S., *Chem. Commun./J. Chem. Soc., Chem. Commun.* **1968**, 849.
4. Yanuka, Y.; Halperin, G., *J. Org. Chem.* **1974**, *39*, 3047.
5. (a) Singhal, K., *Indian J. Chem., Sect. A* **1993**, *32A*, 542. (b) Bhattacharya, S. N.; Husain, I., *Indian J. Chem., Sect. A* **1981**, *20A*, 1119. (c) Folarami, A.; McLean, R. A. N.; Wadibia, N., *J. Organomet. chem.* **1974**, *73*, 59. (d) Raizada, M. S.; Nigam, H. L., *J. Inorg. Nucl. Chem.* **1986**, *63*, 722. (e) Bhattacharya, S. N.; Raj, P.; Srivastava, R. C., *J. Organomet. chem.* **1976**, *105*, 45.
6. Bartlett, P. A.; Meadows, J. D.; Brown, E. G.; Morimoto, A.; Jernstedt, K. K., *J. Org. Chem.* **1982**, *47*, 4013.
7. (a) Rao, M. D. P.; Ahmad, M.; Kanungo, T. K., *Indian J. Chem., Sect. A* **1983**, *22A*, 699. (b) Militzer, M., *J. Am. Chem. Soc.* **1938**, *60*, 256.
8. Friedrich, E. C.; DeLucca, G., *J. Org. Chem.* **1983**, *48*, 1678.
9. Alphonse, I.; Xavier, N.; Arulraj, S. J., *Indian J. Chem., Sect. B* **1987**, *26B*, 736.
10. Smith, A. B., III; Friestad, G. K.; Barbosa, J.; Bertounesque, E.; Hull, K. G.; Iwashima, M.; Qiu, Y.; Salvatore, B. A.; Spoors, P. G.; Duan, J. J.-W., *J. Am. Chem. Soc.* **1999**, *121*, 10468.
11. Austad, B. C.; Hart, A. C.; Burke, S. D., *Tetrahedron* **2002**, *58*, 2011.
12. Woo, J. C. S.; MacKay, D. B., *Tetrahedron Lett.* **2003**, *44*, 2881.
13. Bortolini, O.; Bottai, M.; Chiappe, C.; Conte, V.; Pieraccini, D., *Green Chem.* **2002**, *4*, 621.
14. Kartha, K. P. R.; Field, R. A., *Tetrahedron Lett.* **1997**, *38*, 8233.
15. Kartha, K. P. R.; Cura, P.; Aloui, M.; Readman, S. K.; Rutherford, T. J.; Field, R. A., *Tetrahedron Asymmetry* **2000**, *11*, 581.
16. Meijer, A.; Ellervik, U., *J. Org. Chem.* **2002**, *67*, 7407.
17. Kartha, K. P. R.; Field, R. A., *Synlett.* **1999**, *3*, 311.
18. Améduri, B.; Boutevin, B.; Kostov, G. K.; Petrova, P., *J. Fluorine Chem.* **1999**, *93*, 117.

Iodine Monochloride[1]

ICl

[7790-99-0] ClI (MW 162.35)

(iodinating reagent for aromatic compounds;[2] chloroiodinating reagent for unsaturated compounds;[3] cleaves carbon–metal bonds[4])

Physical Data: mp 27.2 °C (α-form; black needles, ruby-red by transmitted light); mp 13.9 °C (β-form; black platelets, brownish-red by transmitted light); bp 97 °C (β-form, partial decomposition); *d* 3.10 g cm^{-3}.

Solubility: soluble in alcohol, ether, CS_2, acetic acid, acetone, pyridine; soluble in H_2O, but hydrolyzes to HCl and IOH.

Form Supplied in: low melting black or brownish-red solid; widely available (usually 97–98% purity).

Handling, Storage, and Precautions: corrosive solid and vapor; readily absorbed through skin and mucous membranes; use in a fume hood and wear gloves and appropriate protective clothing; air-, moisture-, and light-sensitive; should be stored refrigerated (under N_2) in a tightly sealed amber bottle. Explodes on contact with potassium metal; mixtures with sodium metal can explode if impacted. Reaction with PCl_3 is extremely exothermic.

Original Commentary

Ronald G. Brisbois & Randall A. Wanke
Hamline University, St. Paul, MN, USA

Iodination of Aromatic Compounds. As a source of electrophilic iodine, iodine monochloride is an effective reagent for iodination of a wide variety of aromatic compounds,[2] including heterocyclic aromatic compounds (eq 1).[5] In a related mode of reactivity, uracil bases and protected nucleosides can be iodinated in excellent yield (eq 2).[6] See also Iodine–Silver(I) Fluoride and Trifluoroacetyl Hypoiodite for iodination of aromatic compounds.

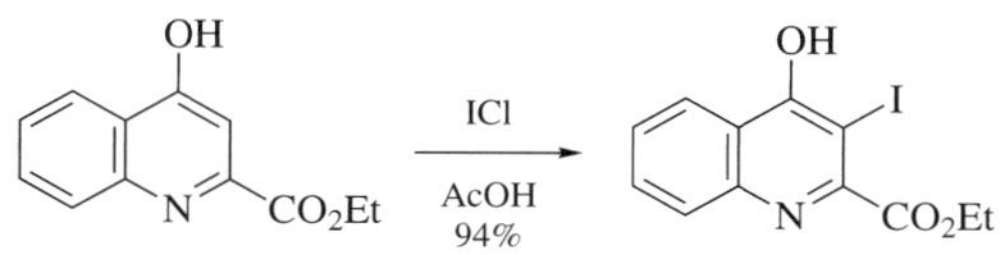

(1)

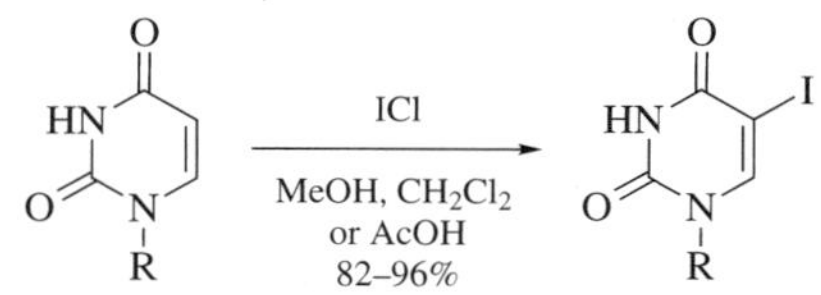

(2)

Chloroiodination of Unsaturated Compounds.[3] Formation of vicinal chloroiodoalkanes via iodine monochloride addition to alkenes generally occurs with high *anti* stereospecificity, implicating the intermediacy of an iodonium ion in the reaction sequence. Unsymmetrical alkenes typically undergo predominantly Markovnikov addition (e.g. reaction of 1-hexene produces a 66:34 mixture of 2-chloro-1-iodohexane and 1-chloro-2-iodohexane in 75% overall yield[7]). Regioselectivity and stereoselectivity in additions to alkynes are very dependent on the alkyne structure.

Bond Cleavage Reactions.[4] Except in the case of C–Fe and C–Co bonds, cleavage of carbon–metal bonds by iodine monochloride specifically produces the organic iodide and the metal chloride. Organosilicon[8] (eq 3), organoboron[9] (eq 4), and organoaluminum[10] (eq 5) bond cleavage reactions are probably of greatest utility to synthetic organic chemists.

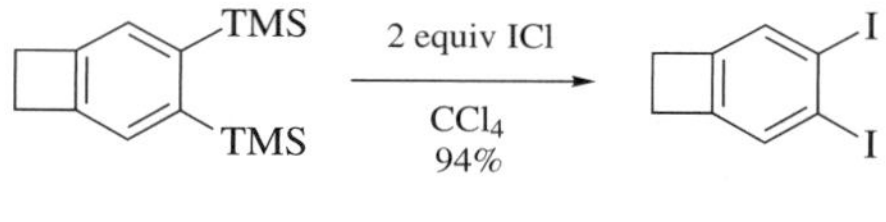

(3)

Other Applications. Iodine monochloride can function as a chlorinating agent because of its tendency to disproportionate to molecular iodine and molecular chlorine in nonpolar solvents.[11] Oxidation of dioxolanes to esters has been reported to proceed in good to excellent yield.[12] Also, a variety of conditions have been reported for the in situ generation of iodine monochloride, including t-Butyl Hypochlorite/Iodine,[13] Antimony(V) Chloride/I_2,[14] and Iodine–Copper(II) Chloride.[15]

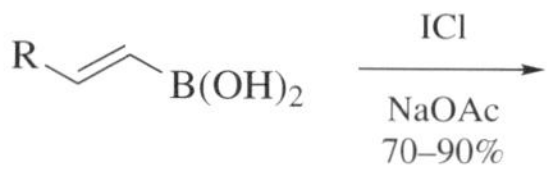

(4)

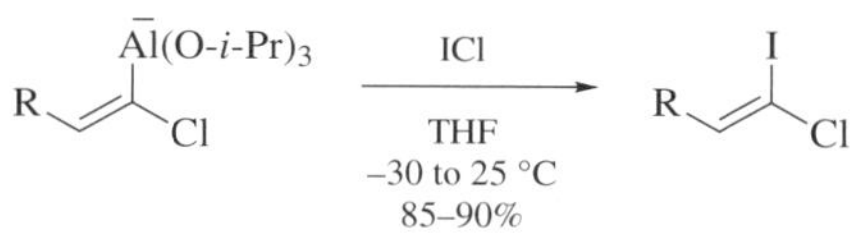

(5)

First Update

Keith A. Stubbs & Robert V. Stick
The University of Western Australia, Crawley, WA, Australia

Iodination of Aromatic Compounds. This source of electrophilic iodine has been extended to the diiodination of dialkoxy-substitiuted aromatic derivatives,[16] important precursors in the synthesis of phenylene vinylene polymers.[17] The use of hydroxylic solvents such as methanol (eq 6) was found to give higher yields than aprotic solvents such as acetonitrile. A major by-product in aprotic solvents is molecular iodine.

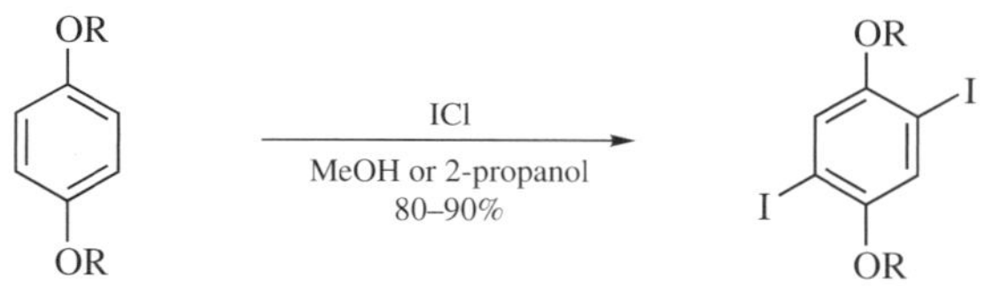

(6)

Chloroiodination of Unsaturated Compounds. Addition to acetylenic ketones gives mixed stereoisomers of the resulting dihaloalkenes.[18] The ratio of these alkenes has been controlled by the addition of base (eq 7).

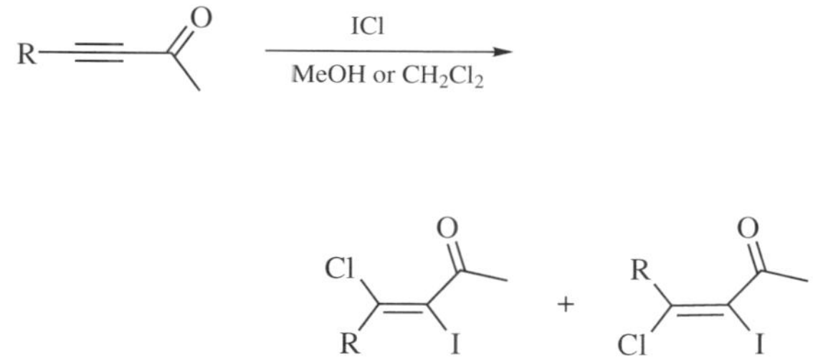

(7)

1 **2**

Base	Yields (%)	
	1	2
no base	25–89	11–75
Pyridine	72–98	2–28

Diiodination of Unsaturated Compounds. The use of ICl has been extended to the synthesis of diiodoalkenes,[19] which are further used in reactions such as the formation of carbon-carbon bonds and are the precursors for Suzuki and Heck coupling reactions.[20] Treatment of an alkyne with ICl/NaI at 0 °C produces almost exclusively the thermodynamically more stable *E*-diiodide (eq 8). In contrast, at −78 °C the *Z*-isomer was obtained (eq 9).[21]

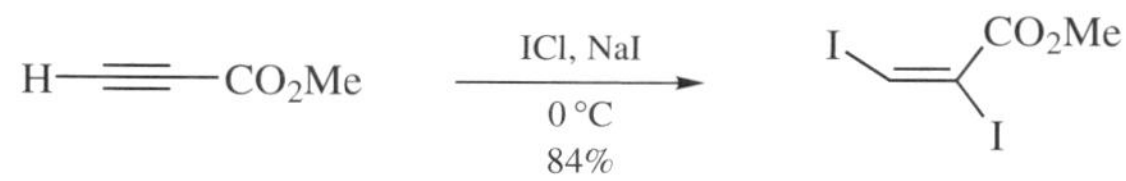

(8)

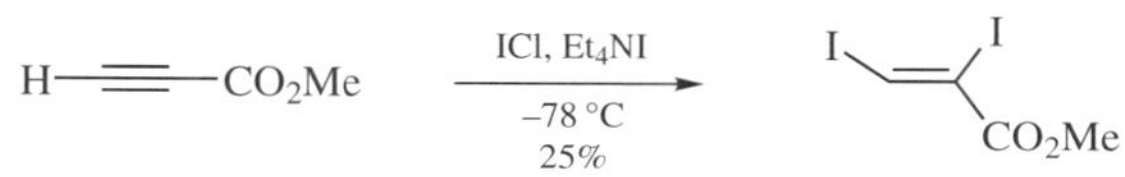

(9)

Cyclization Reactions. Treatment of ω-acetylenic α-iodo cycloalkanones with $AlCl_3$/ICl gave the corresponding spirobicyclic ketones in good yields (eq 10).[22] Even though ICl is probably regenerated in the reaction,[23] it was found that the reagent must be present in greater than stoichiometric amounts to ensure good yields.

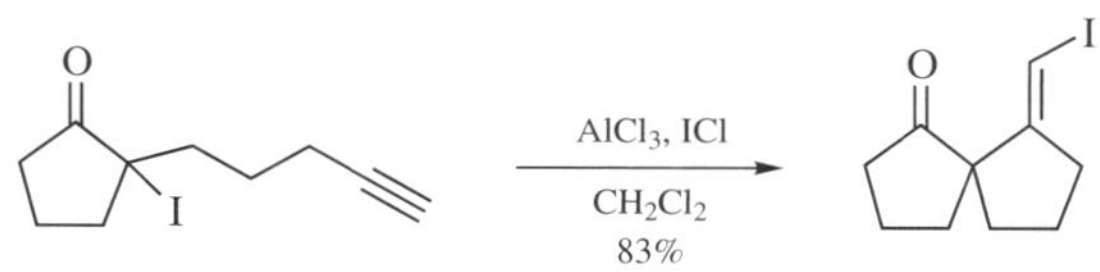

(10)

Iodination of Carbohydrates. Halogenation of sugars represents an important class of reaction in carbohydrate chemistry. The deoxygenated product can lead to various other derivatives, such as amino deoxy and anhydro sugars.[24,25] Although not as common as iodine in the halogenation reaction, ICl represents a mild and efficient way to introduce a halogen. Activation of an unsaturated carbohydrate using diborane, followed by treatment of the intermediate with ICl/NaOAc, yielded the corresponding iodo derivative in a good yield (eq 11).[26] The versatility of ICl is such that it can be used in the presence of sensitive protecting groups.

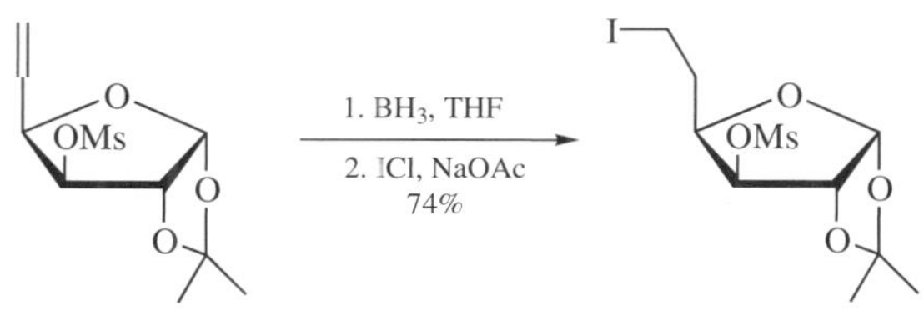

(11)

The iodination of glycals is also an important reaction and can be used to prepare useful precursors for aminoglycoside synthesis.[27] Addition of ICl and NaN_3 to tri-*O*-benzyl-D-glucal gives the azide in an excellent yield (eq 12).[28]

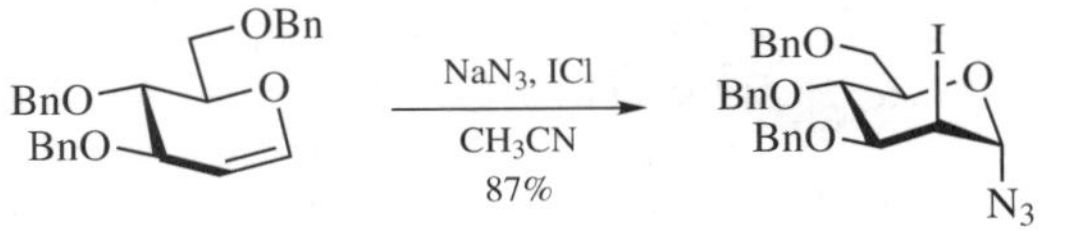

(12)

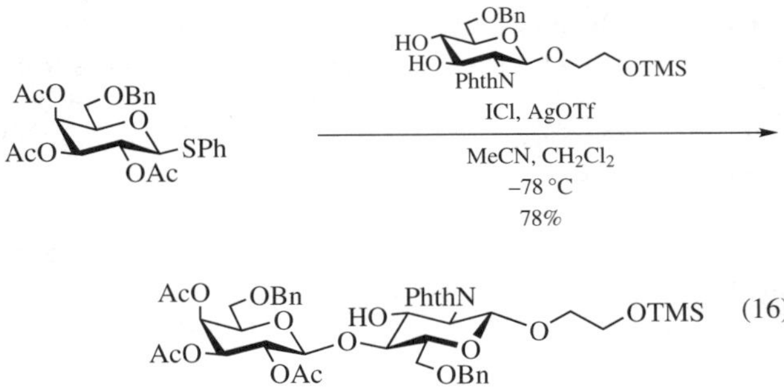

(16)

Glycosylation Promoter. The promotion of glycoside bond formation is an extremely important part of carbohydrate chemistry; iodine monochloride is primarily used in the activation of thioglycoside donors.[29] Treatment of 'disarmed' methylthio D-galactosides with ICl in a variety of solvents gave anomeric mixtures of D-galactosyl chlorides (another type of glycosyl donor), with the β-chloride being the predominant anomer (eq 13). The participating group at C2 is thought to underlie the stereoselectivity of this process.[30] On the other hand, treatment of an 'armed' donor with ICl gave exclusively the α-anomer (eq 14).[30] These glycosyl chlorides have been used in situ to prepare new glycosides,[31] for example D-glucuronates.[32]

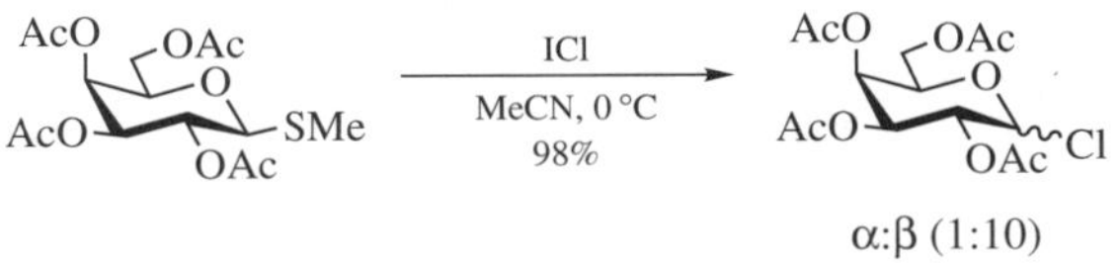

(13)

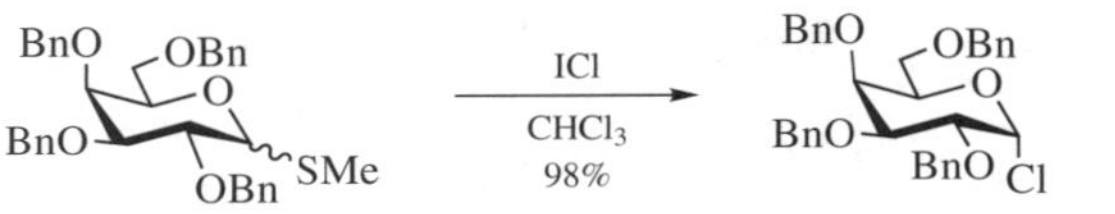

(14)

The use of the ICl/AgOTf couple has the advantage over other thioglycoside activators such as MeSBr/AgOTf[33] and NIS/TfOH[34,35] of being less toxic and of not generating succinimide, which can be difficult to remove from the product during purification, respectively. The ICl/AgOTf couple in various mixtures of MeCN and CH_2Cl_2, with the appropriate starting materials, results in good yields of products (eqs 15 and 16).[36]

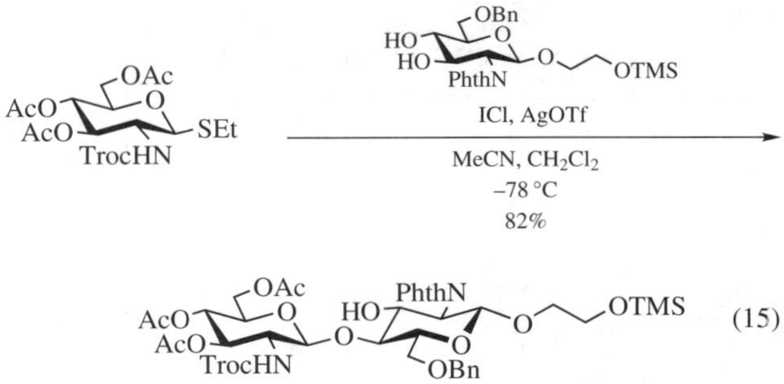

(15)

ICl/AgOTf has also been used in the glycosylation of sialic acids,[37] a somewhat complex set of molecules of considerable importance in glycobiology, but for which there are few effective methods of glycosylation (eq 17).[38]

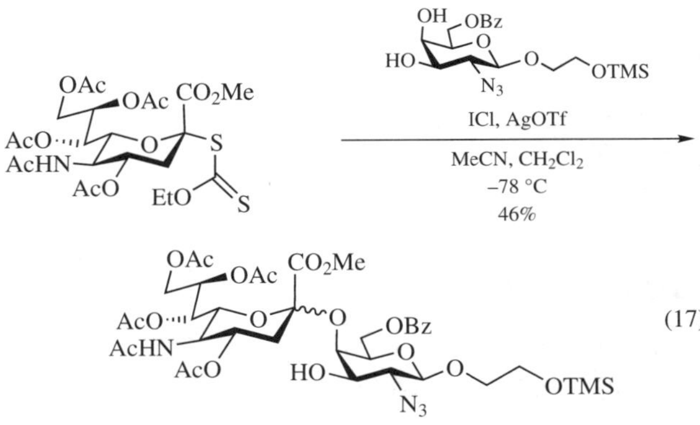

(17)

Other Applications. Various polyfluoroethylenes have been treated with ICl[39–41] to yield polyhalogenated derivatives (eq 18). Similarly, ICl has been used with highly fluorinated epoxides in the presence of nickel again to give polyhalogenated derivatives (eq 19).[42]

$$F_2C{=}CH_2 \xrightarrow[20\text{–}140\ ^\circ C]{ICl} \underset{99\text{–}100\%}{ClF_2C{-}CH_2I} + \underset{0\text{–}1\%}{IF_2C{-}CH_2Cl} \quad (18)$$

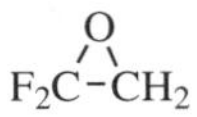

$$\xrightarrow[Ni]{ICl} CF_2I_2 + CF_2ICl + CF_2Cl_2 + CF_3COF \quad (19)$$

The reagent has also been used in a reaction with substituted cyclopropanes to give mixtures of the ring-opened products (eq 20).[43]

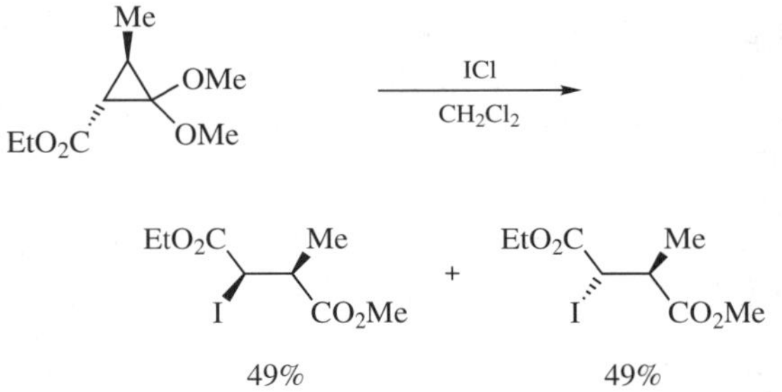

(20)

Even though ICl is regarded as more of an iodinating than chlorinating reagent, instances of the latter have been noted. Thus, when treated with an imidazole nucleoside, ICl gave a good yield of the chlorinated product (eq 21).[44]

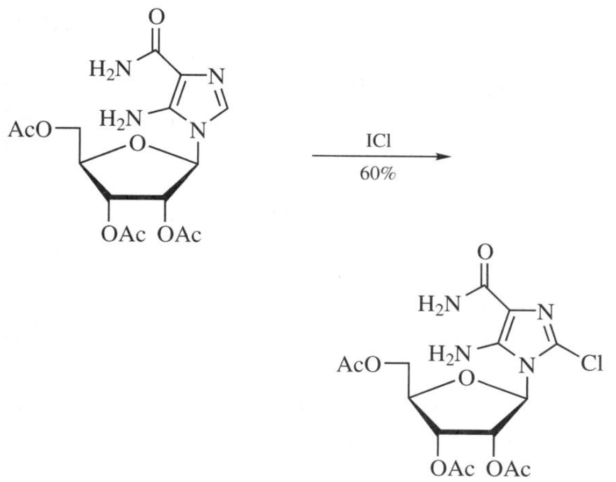

(21)

Treatment of polycyclic aromatic derivatives with ICl also gave the chloro derivatives in good yields (eq 22).[45]

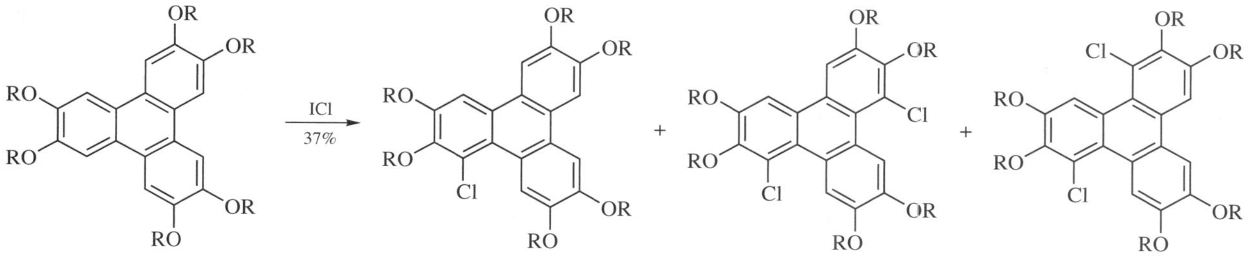

(22)

R = alkyl

Related Reagents. Silver trifluoromethanesulfonate; *N*-iodosuccinimide; trifluoromethanesulfonic acid; iodine; iodine monobromide.

1. McCleland, C. W., In *Synthetic Reagents*; Pizey, J. S., Ed.; Wiley: New York, 1983; Vol. 5, pp 85–164.
2. (a) See Ref. 1, pp 88–100. (b) For a general review of aromatic iodination, see Merkushev, E. B., *Synthesis* **1988**, 923.
3. (a) See Ref. 1, pp 101–120. (b) Schmid, G. H.; Garratt, D. G., In *The Chemistry of Doubly-Bonded Functional Groups*; Patai, S., Ed.; Wiley: New York, 1977; Part 2, pp 796–799.
4. See Ref. 1, pp 130–143.
5. Surrey, A. R.; Cutler, R. A., *J. Am. Chem. Soc.* **1946**, *68*, 2570.
6. Robins, M. J.; Barr, P. J.; Giziewicz, J., *Can. J. Chem.* **1982**, *60*, 554.
7. Ingold, C. K.; Smith, H. G., *J. Chem. Soc.* **1931**, 2742.
8. Hillard, R. L., III; Vollhardt, K. P. C., *J. Am. Chem. Soc.* **1977**, *99*, 4058.
9. Kabalka, G. W.; Gooch, E. E.; Hsue, H. C., *Synth. Commun.* **1981**, *11*, 247.
10. Zweifel, G.; Lewis, W.; On, H. P., *J. Am. Chem. Soc.* **1979**, *101*, 5101.
11. See Ref. 1, pp 121–129.
12. (a) See Ref. 1, pp 143–147. (b) Mingotaud, A.-F.; Florentin, D.; Marquet, A., *Synth. Commun.* **1992**, *22*, 2401.
13. Akhtar, M.; Barton, D. H. R., *J. Am. Chem. Soc.* **1964**, *86*, 1528.
14. Uemura, S.; Onoe, A.; Okano, M., *Bull. Chem. Soc. Jpn.* **1974**, *47*, 147.
15. Baird, W. C., Jr; Surridge, J. H.; Buza, M., *J. Org. Chem.* **1971**, *36*, 3324.
16. Wariishi, K.; Morishima, S.; Inagaki, Y., *Org. Proc. Res. Dev.* **2003**, *7*, 98.
17. Manecke, G.; Zerpner, D., *Makromolekulare Chemie* **1969**, *129*, 183.
18. Heasley, V. L.; Buczala, D. M.; Chappell, A. E.; Hill, D. J.; Whisenand, J. M.; Shellhamer, D. F., *J. Org. Chem.* **2002**, *67*, 2183.
19. Hénaff, N.; Stewart, S. K.; Whiting, A., *Tetrahedron Lett.* **1997**, *38*, 4525.
20. Hunt, A. R.; Stewart, S. K.; Whiting, A., *Tetrahedron Lett.* **1993**, *34*, 3599.
21. Henaff, N.; Whiting, A., *J. Chem. Soc., Perkin Trans. 1* **2000**, 395.
22. Sha, C.-K.; Lee, F.-L.; Lin, H.-H., *Chem. Commun.* **2001**, 39.
23. Matsubara, S.; Tsuboniwa, N.; Morizawa, Y.; Oshima, K.; Zozaki, H., *Bull. Chem. Soc. Jpn.* **1984**, *57*, 3242.
24. Barnett, J. E. G., *Adv. Carbohydr. Chem.* **1967**, *22*, 177.
25. Hanessian, S., *Adv. Chem. Ser.* **1968**, *74*, 159.
26. Hall, L. D.; Neeser, J., *Can. J. Chem.* **1982**, *60*, 2082.
27. Danishefsky, S. J.; Bilodeau, M. T., *Angew. Chem., Int. Ed. Engl.* **1996**, *35*, 1380.
28. Boschi, A.; Chiappe, C.; De Rubertis, A.; Ruarre, M. F., *J. Org. Chem.* **2000**, *65*, 8470.
29. Kartha, K. P. R.; Aloui, M.; Field, R. A., *Tetrahedron Lett.* **1996**, *37*, 5175.
30. Kartha, K. P. R.; Cura, P.; Aloui, M.; Readman, S. K.; Rutherford, T. J.; Field, R. A., *Tetrahedron: Asymmetry* **2000**, *11*, 581.
31. Cura, P.; Aloui, M.; Kartha, K. P. R.; Field, R. A., *Synlett* **2000**, 1279.
32. Stachulski, A. V., *Tetrahedron Lett.* **2001**, *42*, 6611.
33. Dasgupta, F.; Garegg, P. J., *Carbohydr. Res.* **1988**, *177*, C13.
34. Veeneman, G. H.; van Leeuwen, S. H.; van Boom, J. H., *Tetrahedron Lett.* **1990**, *31*, 1331.
35. Buskas, T.; Konradsson, P.; Oscarsson, S.; Ohberg, L., *J. Carbohydr. Chem.* **2001**, *20*, 569.

36. Ercegovic, T.; Meijer, A.; Magnusson, G.; Ellervik, U., *Org. Lett.* **2001**, *3*, 913.
37. Meijer, A.; Ellervik, U., *J. Org. Chem.* **2002**, *67*, 7407.
38. Boons, G. J.; Demchenko, A. V., *Chem. Rev.* **2000**, *100*, 4539.
39. Bissel, E. R., *J. Org. Chem.* **1964**, *29*, 252.
40. Haszeldine, R. N., *J. Chem. Soc.* **1955**, 4291.
41. Kharroubi, M.; Manseri, A.; Ameduri, B.; Boutevin, B., *J. Fluorine Chem.* **2000**, *103*, 145.
42. Yang, Z.-Y., *J. Am. Chem. Soc.* **1996**, *118*, 8140.
43. Piccialli, V.; Graziano, M. L.; Iesce, M. R.; Cermola, F., *Tetrahedron Lett.* **2002**, *43*, 8067.
44. Srivastava, P. C.; Hasan, A.; Pratap, R.; Bhakuni, D. S., *Nucleosides & Nucleotides* **1989**, *8*, 1281.
45. Boden, N.; Bushby, R. J.; Cammidge, A. N.; Headdock, G., *Tetrahedron Lett.* **1995**, *36*, 8685.

p-Iodobenzyl Bromide

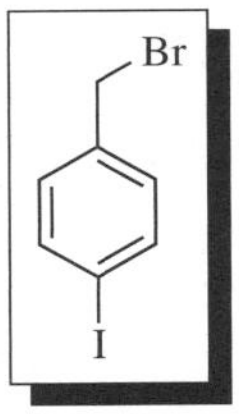

[16004-15-2] C_7H_6BrI (MW 296.94)

(reagent used as a stable protecting group removable by Pd-catalyzed amination followed by acid treatment)

Alternate Name: α-bromo-4-iodotoluene; 1-bromomethyl-4-iodo-benzene.

Physical Data: mp 78–82 °C, bp 115–120 °C (at 0.6 Torr).

Solubility: soluble in ether and most organic solvents; insoluble in water.

Form Supplied in: light beige crystalline solid.

Analysis of Reagent Purity: ^{1}H NMR, IR, and elemental analysis.

Purification: crystallize from ethanol.

Preparative Methods: bromination of 4-iodotoluene with NBS in CBr_4 and catalytic benzoyl peroxide.[1] Bromination of 4-iodobenzyl alcohol with CBr_4, triphenyl phosphine, and CH_2Cl_2.[2] Commercially available.

Disposal: contact a licensed professional waste disposal service to dispose of this material. Dissolve or mix with a combustible solvent and burn in a chemical incinerator equipped with an afterburner and scrubber.

Handling, Storage, and Precautions: corrosive, lachrymator, moisture sensitive, stored under nitrogen, causes burns, irritant to eyes, respiratory system, and skin.

***p*-Iodobenzyl Bromide (IBB) as a Hydroxyl Protecting Group.** *p*-Iodobenzyl (PIB) ethers are readily obtained by the reaction of a hydroxyl group with *p*-iodobenzyl bromide.[3] Once installed, the *p*-iodobenzyl ethers are stable to a variety of reaction conditions used to remove common protecting groups such as esters, silyl ethers, and *p*-methoxybenzyl (PMB) ethers.[3] The selective removal of the PIB ether can be accomplished under mild conditions via a two-step reaction. First, the iodide is coupled with *N*-methyl aniline using NaO*t*-Bu, $Pd_2(dba)_3$ (1 mol % Pd), and 1-(*N,N*-dimethylamino)-1′-(dicyclohexylphosphino)biphenyl as a ligand to form a labile substituted arylamine. This procedure is followed by brief exposure to Lewis acids, protic acids or oxidants to obtain the free hydroxyl (eq 1).[3]

p-Iodobenzyl ethers, as well as other halobenzyl ethers, compare favorably to other substituted benzyl ethers as protecting groups in terms of stability and orthogonality of removal in the presence of commonly used protecting groups. The widely used PMB ethers are acid labile, and although this reactivity can be exploited to selectively remove them in the presence of unsubstituted benzyl ethers, it greatly reduces their synthetic utility.[3] Other *p*-substituted benzyl ethers containing nitro, acetyl, and trialkylsilyl substituents have been reported;[4,5] however, their removal requires treatment with Zn/AcOH, base or fluoride, rendering them incompatible with the commonly used ester, silyl or PMB protecting groups. However, PIB ethers and related compounds (e.g. *p*-bromobenzyl, *p*-chlorobenzyl, *o*-bromobenzyl, and *o*-chlorobenzyl ethers) are fully orthogonal to the commonly used protecting groups and are as stable as unsubstituted benzyl ethers.

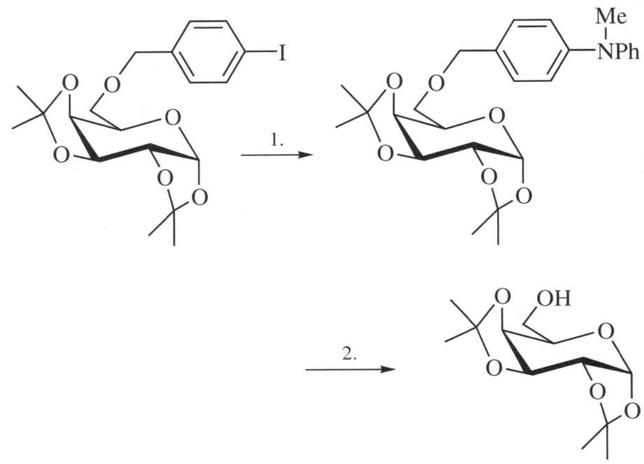

(1)

Conditions:

1. $Pd_2(dba)_3$, NaOt-Bu, PhN(H)Me, 1-(*N,N*-dimethylamino)-1′-(dicyclohexylphosphino)biphenyl, rt, 4 h.
2. Cl_2HCCO_2H

Halobenzyl ether protecting groups introduce three more degrees of orthogonality to a synthetic scheme since they can be sequentially removed. *p*-Iodobenzyl ethers can be selectively removed in the presence of other halobenzyl ethers such as *p*-chlorobenzyl and *p*-bromobenzyl ethers as seen in eq 2.[3] Similarly, *p*-bromobenzyl ethers can be removed in the presence of *p*-chlorobenzyl ethers (eq 3).[3]

The selectivity in the deprotection process can be explained through analysis of the Pd-catalyzed amination of the substituted aryl halides (eq 4). The rate of oxidative addition (step i) of the aryl halides to Pd(0) follows the general rule I>Br>Cl>F,[6] thus setting the basis for the observed selectivity.[7] The Pd–N bond formation (step ii) and the reductive amination processes (step iii) are fast and proceed at a similar rate for all reactions when substituted anilines are used as nucleophiles.[6]

Related Reagents. *p*-Bromobenzyl bromide; *p*-chlorobenzyl chloride; benzyl bromide; *p*-methoxybenzyl bromide.

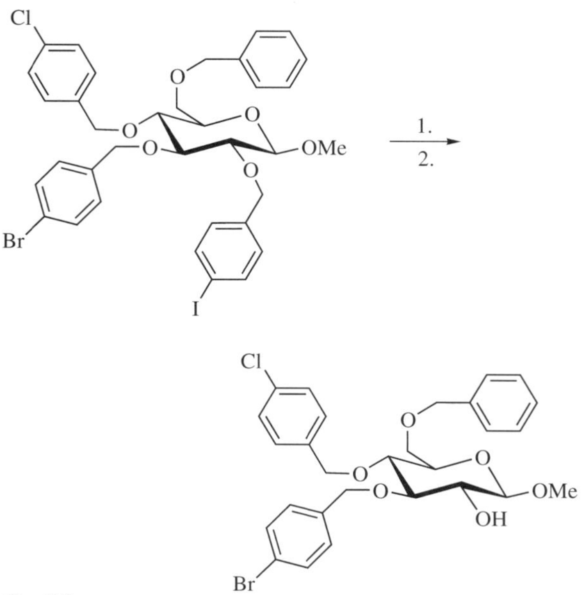

(2)

Conditions:

1. $Pd_2(dba)_3$, NaO*t*-Bu, PhN(H)Me, 1-(*N,N*-dimethylamino)-1′-(dicyclohexylphosphino)biphenyl, rt, 4 h.
2. Cl_2HCCO_2H, 84%.

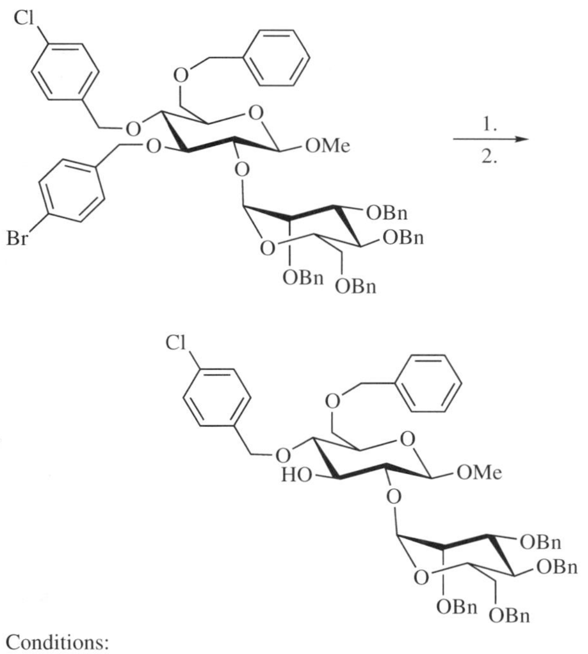

(3)

Conditions:

1. $Pd_2(dba)_3$, NaO*t*-Bu, PhN(H)Me, (*o*-biphenyl)P(*t*-Bu)$_2$, 80 °C.
2. $SnCl_4$, 82%.

1. Patel, H. K.; Kilburn, J. D.; Langley, G. J.; Edwards, P. D.; Mitchell, T.; Southgate, R., *Tetrahedron Lett.* **1994**, *35*, 481.
2. Sucholeiki, I., *Med. Chem. Res.* **1995**, *5*, 618.
3. Plante, O. J.; Buchwald, S. L.; Seeberger, P. H., *J. Am. Chem. Soc.* **2000**, *122*, 7148.
4. Greene, T. W.; Wuts, P. G., *Protective Groups in Organic Synthesis*; John Wiley & Sons, Inc.: New York, **1999**; pp 86–95.

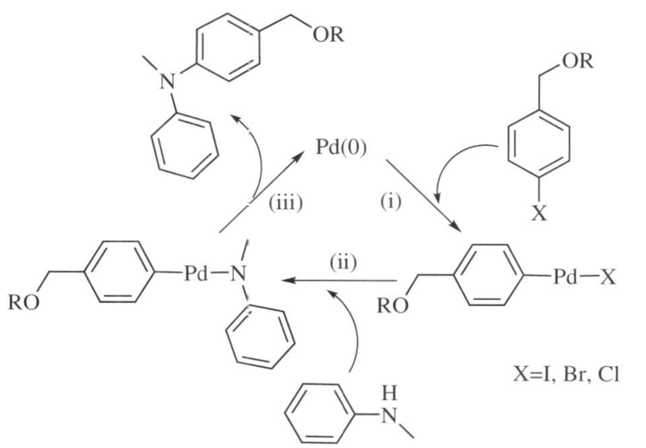

(4)

5. Jobron, L.; Hindsgaul, O., *J. Am. Chem. Soc.* **1999**, *121*, 5835.
6. Wolf, J. P.; Tomori, H.; Sadighi, J. P.; Yin, J.; Buchwald, S. L., *J. Org. Chem.* **2000**, *65*, 1158.
7. Collman, J. P.; Hegedus, L. S.; Norton, J. R.; Finke, R. G., *Principles and Applications of Organotransition Metal Chemistry*; University Science Books: Mill Valley, 1987; p 306–320.

Luis G. Melean & Peter H. Seeberger

Massachusetts Institute of Technology, Cambridge, MA, USA

Iodonium Di-*sym*-collidine Perchlorate

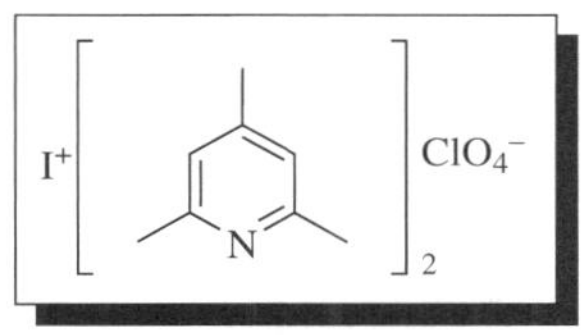

[69417-67-0] $C_{16}H_{22}ClIN_2O_4$ (MW 468.75)

(very reactive electrophile, superior source of I^+,[1] useful in the synthesis of *cis*-β-hydroxy amines,[2] activates glycosides for glycosylation;[5,6] can be used for iodolactonization[12,13] and vicinal *cis*-diol[14] preparation)

Alternate Name: IDCP.
Solubility: soluble in chloroform; insoluble in ether.
Form Supplied in: fine colorless crystalline powder.
Drying: see Bromonium Di-sym-collidine Perchlorate.
Handling, Storage, and Precautions: see Bromonium Di-sym-collidine Perchlorate.

Original Commentary

Tapan Ray

Sandoz Research Institute, East Hanover, NJ, USA

***cis*-Oxyamination.**[2] IDCP (**1**) is useful in the synthesis of *cis*-hydroxyamino sugars, e.g. methyl *N*-acetylristosaminide has been obtained from an oxazoline which can be made by the reaction of a trichloromethyl imidate with IDCP. The imidate can be prepared by reaction of the corresponding allylic alcohol with Trichloroacetonitrile in presence of Sodium Hydride (eq 1).

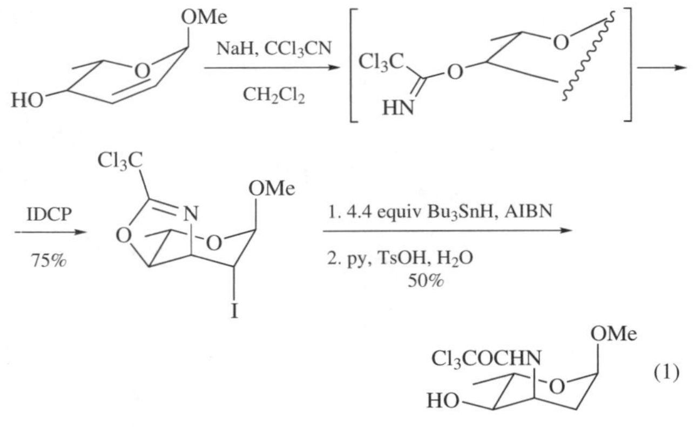

(1)

The synthesis of methyl α,L-garosaminide,[3] a key component of aminocyclitol antibiotics, is complicated by the presence of a *cis*-hydroxyamino group and by the tertiary character of the hydroxy group. The problems have been resolved by use of the allylic epoxide as starting material. This epoxide was converted in three steps into an allylic amine. Treatment of iodonium salt gave the iodooxazolidinone in 82% yield. The product was reduced and the ethoxy ethyl group was removed and finally converted to the desired product by hydrolysis (eq 2).

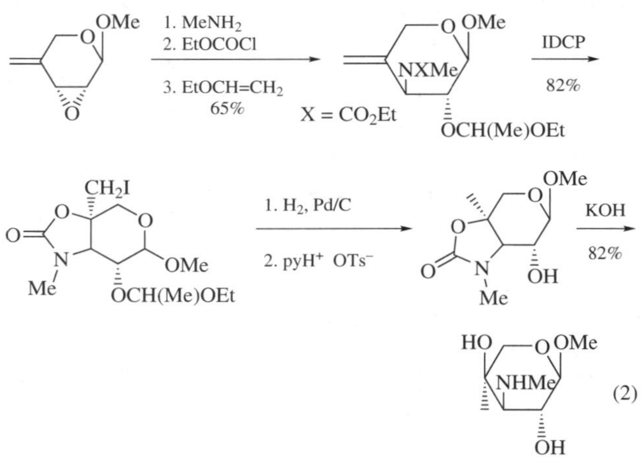

(2)

Similar methodology can convert an internal allylic amine into a *cis*-β-hydroxyamine, as illustrated in a synthesis of holacosamin, a component of some glycosteroids (eq 3).[4]

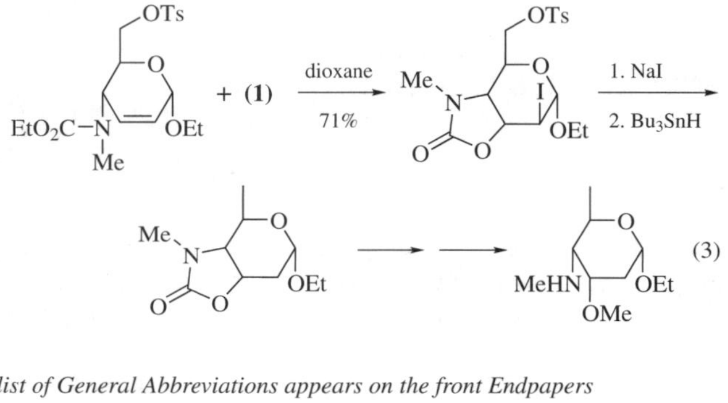

(3)

α-Linked Disaccharides.[5] The reagent functions as a superior source of I^+, probably because of the nonnucleophilic counterion. Thus a pyranoid diene reacts with IDCP to give a planar ion to which an alcohol adds in a 1,4-sense to give an α-glycoside. Thus tetraacetylfructose reacts to give an α-disaccharide in 45% yield (eq 4). The β-isomer is not detected.

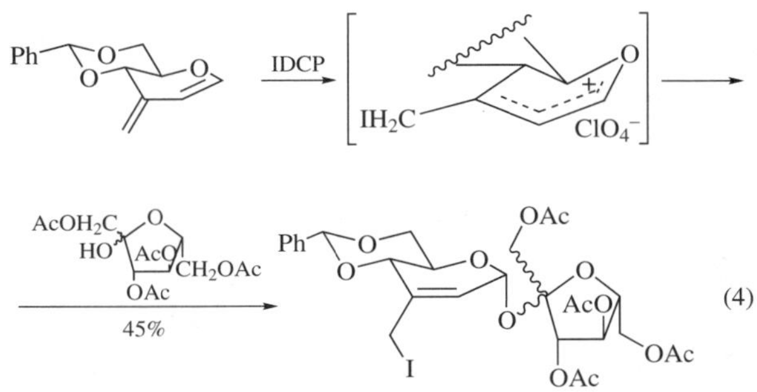

(4)

Examples are known where this reagent has activated pent-4-enyl glycosides for glycosylation,[6] but α- and β-glycosides were obtained in various proportions regardless of the nature of donors or acceptors (primary or secondary hydroxyl groups). Glycosylations of 1,2:5,6-di-*O*-isopropylidene-α-D-galactopyranose and methyl 2,3,4-tri-*O*-benzyl-α-D-glucopyranoside were investigated with pent-4-enyl-2,3,4-tri-*O*-benzyl-β-D-glucopyranoside derivatives in which 6-OH was protected by a benzyl, a trityl, or a TBDMS group in order to assess the effect of the bulk of the 6-substituents. The presence of a bulky 6-substituent (a) increases significantly the proportion of the α-product; (b) decreases the yield when a secondary hydroxy group is glycosylated, but the effect is less or opposite when a primary hydroxy group is involved; (c) lowers the increase in yield of the α-product when a primary hydroxy group is glycosylated; (d) gives a much better yield of the α-anomer when there is a higher proportion of ether in the solvent.

Chemospecific glycosidation of partially benzoylated thioglycosides ('disarmed' acceptors) with perbenzylated thioglycosides ('armed' donors) can be realized in the presence of the promotor IDCP (eq 5).[7]

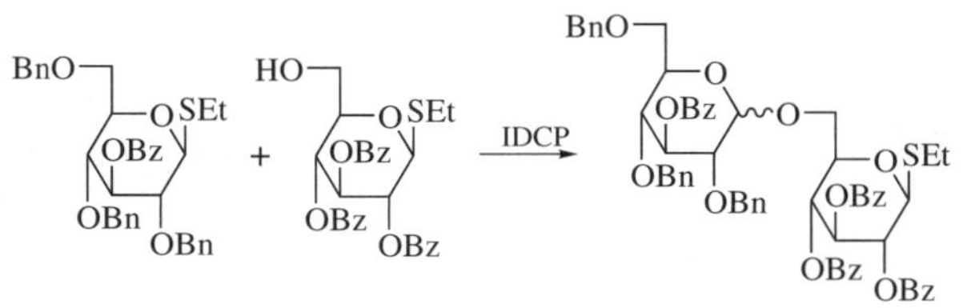

(5)

Reaction of IDCP with unsaturated alcohols and carboxylic acids in dichloromethane at ambient temperature has afforded three- to seven-membered ring iodoethers and four- to seven-membered ring iodolactones, respectively, in moderate yields and generally with high regioselectivity. The reaction has particular utility for synthesis of 2-(1-iodoalkyl)oxiranes and -oxetanes.[11] Glycosylation has also been achieved by electrophile-induced activation of anomeric *O*-glycosyl *N*-allyl carbamates (eq 6).[8]

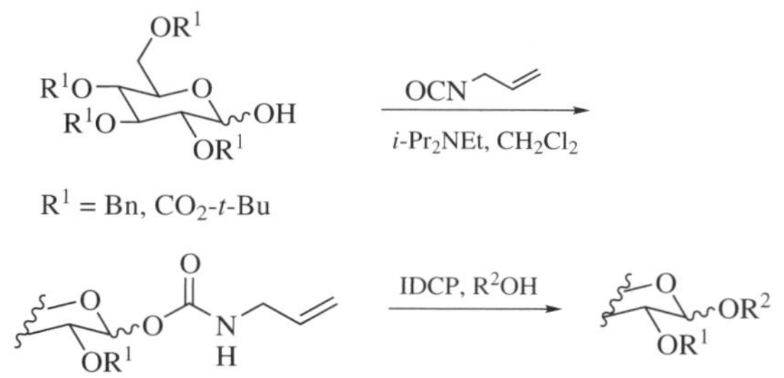

(6)

Treatment of fully benzylated 1-methylene-D-glucose with IDCP gives easy access to 1-iodoheptuloses or a 1-iodomethylene derivative.[9] The latter compound is, in turn, further amenable to similar IDCP-mediated addition reaction. In the synthesis of ciclamycin O the required trisaccharide glycol was assembled by substituent-directed iodinative coupling of glycals as shown in (eq 7).[10]

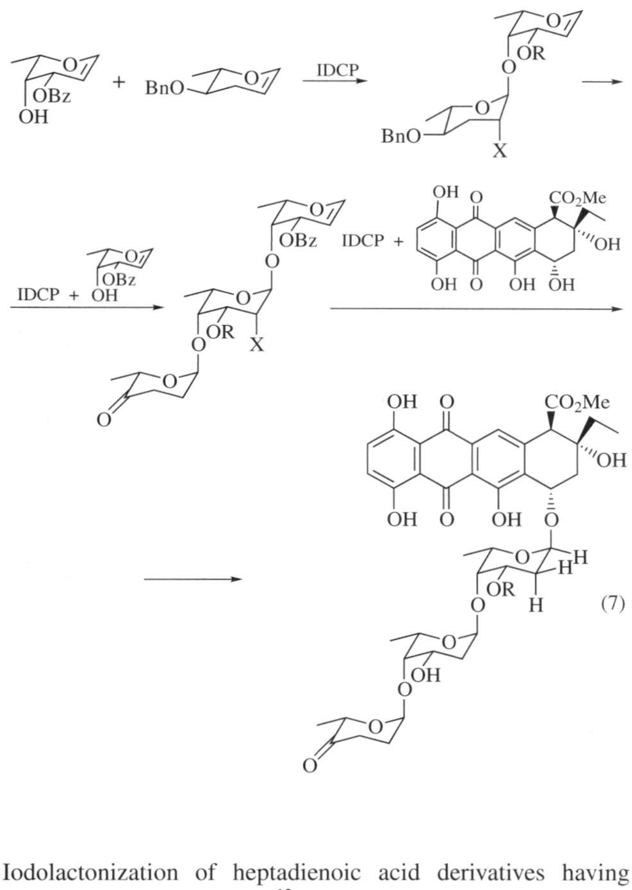

(7)

Iodolactonization of heptadienoic acid derivatives having oxazolidin-2-ones or a sultam[12] as chiral auxiliary gave the chiral iodolactone with moderate to excellent enantioselectivity.[13]

Vicinal *cis*-Diols.[14] An allylic alcohol is converted into its primary urethane derivative, which is then subjected to iodonium ion induced cyclization to give a single iodocarbonate. The carbonate is then deiodinated reductively and hydrolyzed to afford the vicinal diol.

First Update

Alexei V. Demchenko
University of Missouri, St. Louis, MO, USA

Iodosulfonamidation. Danishefsky has developed a glycal iodosulfonamidation approach to the diastereoselective synthesis of *N*-, *O*-, or *S*-linked glycosides. According to this technique, D-glycal derivatives are treated with the combination of IDCP and benzenesulfonamide in the presence of molecular sieves (4Å) in dichloromethane to afford 2-iodo-1-sulfonamido derivatives with the α-D-manno configuration.[15] These derivatives could be efficiently applied in glycoside synthesis by reaction with an appropriate nucleophile in the presence of AgOTf and a strong base (lithium tetramethylpiperidide). This technique was found to be suitable for the synthesis of a range of *O*-linked di- and oligosaccharides, of 1-azido derivatives,[16] and of *S*-ethyl thioglycosides (eq 8).[15] The 1-azido derivatives were subsequently transformed into *N*-linked glycopeptides.[16] This method was extended to the synthesis of a variety of oligosaccharides and glycoconjugates.[17]

Acetal Cleavage and Cyclic Ether Formation. The activation of a suitably placed alkene in the side chain of a glycoside with IDCP leads to cleavage of the acetal and formation of a cyclic iodomethylether. This methodology, initially designed for acetal cleavage under neutral conditions, was extended to a number of other applications (eq 9). Thus, it was demonstrated that the

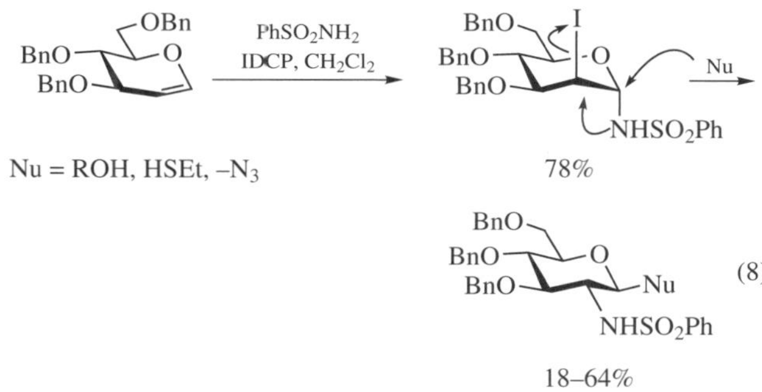

(8)

intermediate oxacarbenium ion could be quenched intermolecularly with an incoming nucleophile, for example, water.[18] The product of this trapping reaction could be either reduced to obtain a polyol derivative or subjected to the isopropylidene acetal cleavage.[19] Alternatively, the key oxacarbenium ion intermediate could be trapped intramolecularly by a further alkene in a highly diastereoselective C-C bond forming process leading to a tetrahydropyran ring, juxtaposed with an iodomethyltetrahydrofuran.[20] The latter could then be reductively cleaved to give overall a pyranose derivative,[20] thereby demonstrating the versatility of the approach and its potential applicability to the synthesis of a range of natural compounds (eq 9).[21]

Synthesis of C-aryl Glycosides. A protected *O*-pentenyl glycoside was used as a glycosyl donor in an IDCP-mediated intramolecular C-arylation.[22] It was noted that the resulting product was kinetically favored, however, the thermodynamic *trans*-fused product could be obtained by subsequent treatment with a Lewis acid (BF_3-Et_2O) (eq 10).

Electrophilic Additions. In electrophilic additions to alkenes, it was reported that higher diastereoselectivities could be obtained

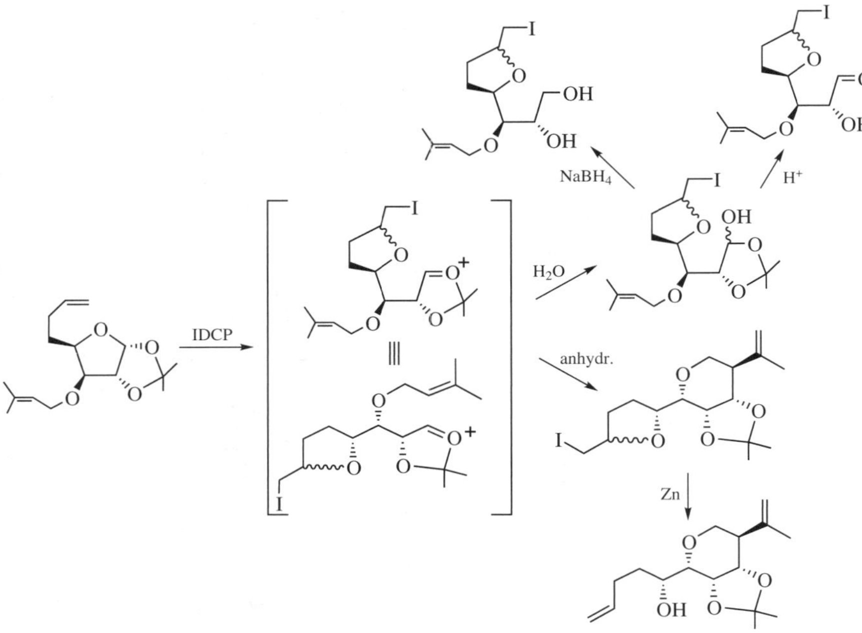

(9)

when bulky electrophilic reagents were used (IDCP or PhSeBr) (eq 11). This was rationalized via steric interactions in the transition state.[23] The steric effect of the neighboring group R in these transformations was also highlighted.

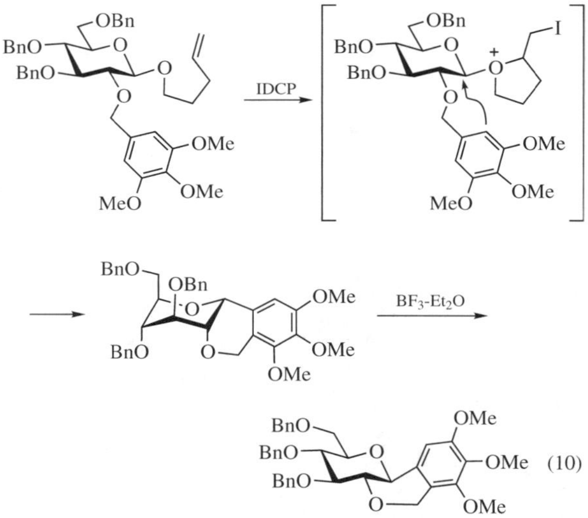

(10)

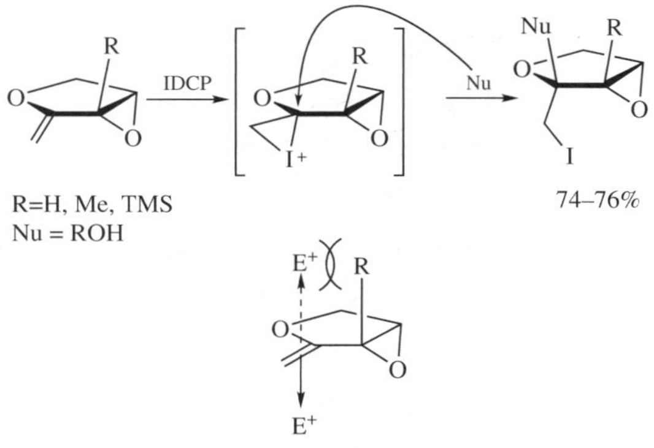

(11)

1. Lemieux, R. U.; Morgan, A. R., *Can. J. Chem.* **1965**, *43*, 2190.
2. Pauls, H. W.; Fraser-Reid, B., *J. Org. Chem.* **1983**, *48*, 1392.
3. Pauls, H. W.; Fraser-Reid, B., *J. Am. Chem. Soc.* **1980**, *102*, 3956.
4. George, M.; Fraser-Reid, B., *Tetrahedron Lett.* **1981**, *22*, 4635.
5. Fraser-Reid, B.; Iley, D. E., *Can. J. Chem.* **1979**, *57*, 645.
6. Houdier, S.; Voltero, P. J. A., *Carbohydr. Res.* **1992**, *232*, 349.
7. (a) Veeneman, G. H.; Van Boom, J. H., *Tetrahedron Lett.* **1990**, *31*, 275. (b) Veeneman G. H.; Van Leeuwen, S. H.; Van Boom, J. H., *Tetrahedron Lett.* **1990**, *31*, 1331.
8. Kuns, H.; Simmer, J., *Tetrahedron Lett.* **1993**, *34*, 2907.
9. Noort, D.; Veeneman, G. H.; Boons, G. P. H.; Vander Marel, G. A.; Mulder, G. J.; Van Boom, J. H., *Synlett* **1990**, 205.
10. Susuki, K.; Sulikowski, G. A.; Friesen, R. W.; Danishefsky, S. J., *J. Am. Chem. Soc.* **1990**, *112*, 8895.
11. Evans, R. D.; Magee, J. W.; Schauble, J. H., *Synthesis* **1988**, 862.
12. Oppolser, W.; Chapuis, C.; Bernardielli, G., *Helv. Chim. Acta* **1984**, *67*, 1397.
13. Yokomatsu, T.; Iwasawa, H.; Shibuya, S., *Chem. Commun.* **1992**, 510.
14. Pauls, H. W.; Fraser-Reid, B., *J. Carbohydr. Chem.* **1985**, *4*, 1.
15. Griffith, D. A.; Danishefsky, S. J., *J. Am. Chem. Soc.* **1990**, *112*, 5811.
16. McDonald, F. E.; Danishefsky, S. J., *J. Org. Chem.* **1992**, *57*, 7001.
17. (a) Roberge, J. Y.; Beebe, X.; Danishefsky, S. J., *J. Am. Chem. Soc.* **1998**, *120*, 3915. (b) Kwon, O.; Danishefsky, S. J., *J. Am. Chem. Soc.* **1998**, *120*, 1588.
18. Elvey, S. P.; Mootoo, D. R., *J. Am. Chem. Soc.* **1992**, *114*, 9685.
19. Shan, W.; Wilson, P.; Liang, W.; Mootoo, D. R., *J. Org. Chem.* **1994**, *59*, 7986.
20. Khan, N.; Xiao, H.; Zhang, B.; Cheng, X.; Mootoo, D. R., *Tetrahedron* **1999**, *55*, 8303.
21. Zhao, H.; Hans, S.; Cheng, X.; Mootoo, D. R., *J. Org. Chem.* **2001**, *66*, 1761.
22. Rousseau, C.; Martin, O. R., *Tetrahedron: Asymmetry* **2000**, *11*, 409.
23. Dalla, V.; Pale, P., *Tetrahedron Lett.* **1996**, *37*, 2777.

Iodonium Di-*sym*-collidine Trifluoromethanesulfonate

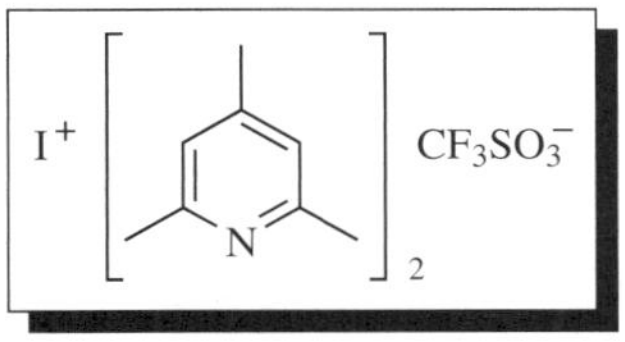

[132799-16-7] $C_{17}H_{22}F_3IN_2O_3S$ (MW 518.33)

(convenient source of iodonium ion, effective in activation of glycosyl donors such as thioglycosides and *n*-pentenyl orthoesters, iodination of *exo*-glycals, and acetal formation from vinyl ethers)

Alternate Name: IDCT.
Physical Data: crystalline solid.
Solubility: soluble in dichloromethane and dichloroethane; insoluble in ether.
Preparative Methods: prepared[1] from the reaction of AgOTf with collidine in dichloromethane, followed by the addition of I_2. The reaction mixture was filtered through Celite and crystallization was induced by the addition of ether. The resulting solid was filtered and dried in vacuo before use.

Glycosylation Using Thioglycosides. In an effort to design new vaccines for tuberculosis, Van Boom and co-workers synthesized aminoethyl disaccharide and trisaccharide analogs of a *Mycobacterium tuberculosis* glycolipid.[1] A key feature of the synthesis was the "armed/disarmed" glycosylation between ethyl 2,3,4,-tri-*O*-methyl-1-thio-α-L-fucopyranoside (armed) and a protected thioethyl rhamnoside alcohol (disarmed). Activation of the armed thioglycoside with iodonium di-*sym*-collidine perchlorate (IDCP) furnished the protected disaccharide in 60% yield, while IDCT proved more effective, giving the product in a 73% yield (eq 1).[1] In other cases, however, IDCT has been shown to be an ineffective thioglycoside promoter.[2]

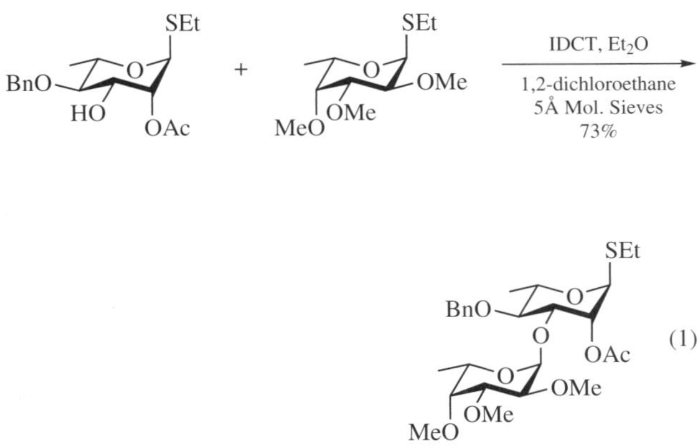

(1)

Glycosylation Using *n*-Pentenyl 1,2-Orthoesters (NPOEs). During studies comparing *n*-pentenyl glycosides (NPGs) and NPOEs as glycosylating agents, Fraser-Reid and co-workers[3] noted that the product formation from NPOEs depended heavily on the promoter and coupling partners. Acidic conditions such as *N*-bromosuccinimide (NBS) and triethylsilyl triflate (TESOTf) favored rearrangement to glycosides while slightly basic reagents such as IDCP and IDCT typically favored *trans*-orthoesterification. Thus, reaction of an NPOE with IDCT in the presence of *iso*-propyl alcohol afforded an *iso*-propyl orthoester in 76% yield (eq 2).

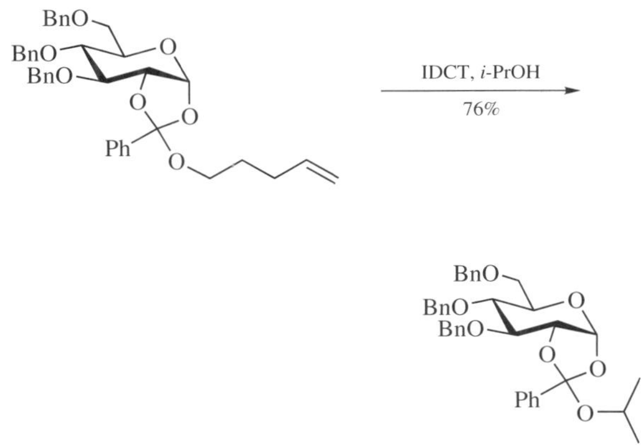

(2)

In contrast, when the same NPOE was reacted with a disaccharide alcohol and IDCT, the trisaccharide product was obtained in 67% yield with no evidence of *trans*-orthoesterification. These results indicate that the size of the glycosyl acceptor influences the outcome of the reaction (eq 3).[3]

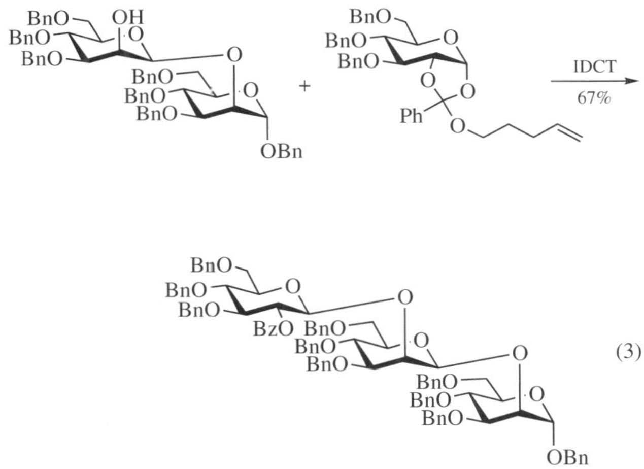

(3)

Iodination of *Exo*-Glycals. IDCT-mediated iodination of 1-*exo*-methylene pyranoses was found to proceed in a stereoselective fashion to give the (*Z*)-iodo-*exo*-glycal in moderate to good yields. It was proposed that the specificity of the reaction was the result of steric repulsion of the large benzyloxy substituent at C-2 of the pyranoid ring and the iodine atom in the pre-elimination intermediate (eq 4).[4] Additional examples of the iodination of hexofuranosyl-based *exo*-glycals have been reported.[5]

Intramolecular Glycosylation via Vinyl Ethers. Studies on intramolecular aglycon delivery (IAD) performed by Fairbanks and co-workers[6] used 2-*O*-allylic ethers that were first isomerized to the corresponding vinyl ether. The required acetal was formed by treatment with *N*-iodosuccinimide (NIS) and the acceptor alcohol. Various conditions were then employed to effect the glycosylation to afford β-manno-linked disaccharides and α-gluco-linked disaccharides. A limitation of the acetal-forming step with NIS was the formation of a succinimide-containing by-product in some instances. The use of iodinium ion sources with counterions less nucleophilic than succinimide were then explored and it was discovered that this by-product could be avoided by the use of IDCT, which was prepared in-situ (eq 5).

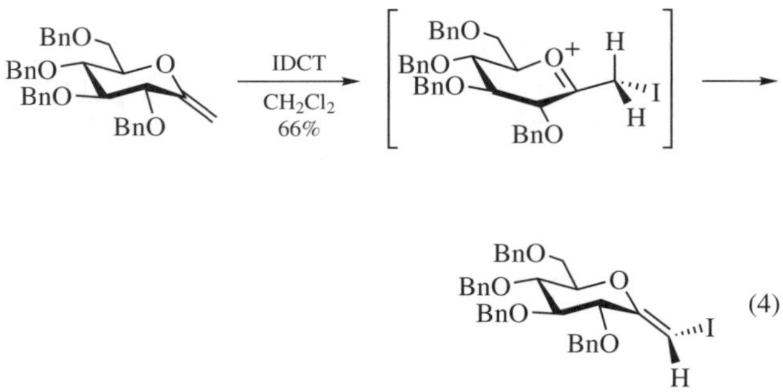

(4)

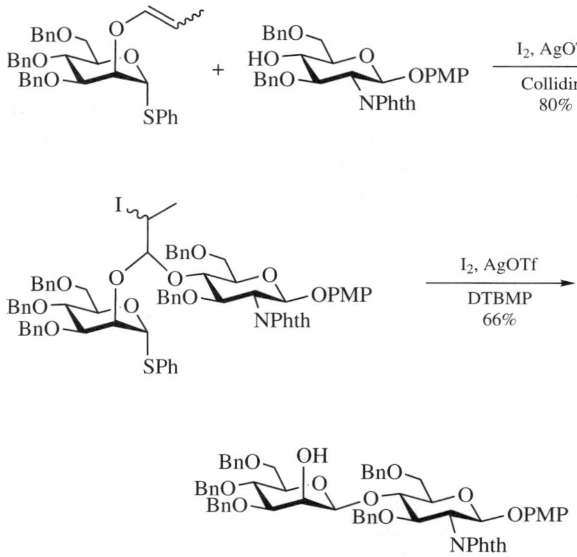

(5)

Related Reagents. Iodonium di-*sym*-collidine perchlorate; *N*-iodosuccinimide; *N*-iodosaccharin; dimethyl(methylthio)sulfonium tetrafluoroborate; dimethyl(methylthio)sulfonium trifluoromethanesulfonate; diphenyl sulfoxide; 1-benzenesulfinyl piperidine.

1. Veeneman, G. H.; Van Leeuwen, S. H.; Zuurmond, H.; Van Boom, J. H., *J. Carboydr. Chem.* **1990**, *9*, 783–796.
2. Broddefalk, J.; Forsgren, M.; Sethson, I.; Kihlberg, J., *J. Org. Chem.* **1999**, *64*, 8948–8953.
3. Mach, M.; Schlueter, U.; Mathew, F.; Fraser-Reid, B.; Hazen, K. C., *Tetrahedron* **2002**, *58*, 7345–7354.
4. Gomez, A. M.; Pedregosa, A.; Valverde, S.; Lopez, J. C., *Tetrahedron Lett.* **2003**, *44*, 6111–6116.
5. Gomez, A. M.; Danelon, G. O.; Pedregosa, A.; Valverde, S.; Lopez, J. C., *Chem. Commun.* **2002**, 2024–2025.
6. Aloui, M.; Chambers, D. J.; Cumpstey, I.; Fairbanks, A. J.; Redgrave, A. J.; Seward, C. M. P., *Chem. Eur. J.* **2002**, *8*, 2608–2621.

Robert S. McGavin & Todd L. Lowary
University of Alberta, Alberta, Canada

2-(2-Iodophenyl)ethanethiol

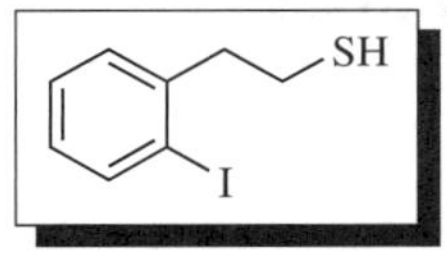

[177748-62-8] C_8H_9IS (MW 264.13)

(reagent used to generate acyl radicals from thioesters)

Solubility: soluble in hexane, ether, chloroform, and most organic solvents.
Form Supplied in: colorless oil, not commercially available.
Handling, Storage, and Precautions: should be kept in a refrigerator.

Preparation. 2-(2-Iodophenyl)ethanethiol (**1**) can be readily prepared in 70% overall yield from 2-(2-bromophenyl)ethanol (eq 1).[1] Protection of the OH group as a TMS ether followed by a halogen/lithium exchange and subsequent iodination of the resulting aryllithium affords the iodoalcohol after hydrolytic work-up. The alcohol is converted to the thiol by Mitsunobu thiolesterification and DIBALH reduction of the thiolester.

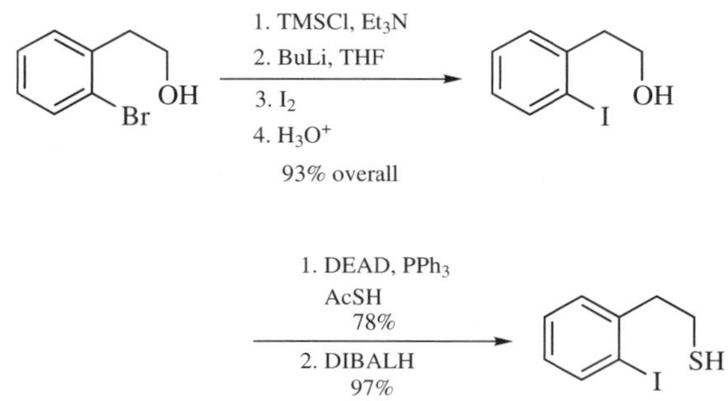

(1)

1

Generation of Acyl Radicals from Thioesters. Acyl radicals are valuable reactive intermediates which undergo a number of synthetically useful transformations such as intra- or intermolecular carbon-carbon bond forming reactions.[2] Traditionally, acyl radicals are generated from acyl selenides in stannane-mediated chain sequences. Thiolesters are attractive alternatives of acyl selenides in terms of their enhanced stability and ease of preparation. However, direct homolytic dissociation of *S*-2-naphthyl and other thiolesters as means of direct entry into acyl radicals was shown to be less efficient both photochemically, due to low quantum yields,[3]and in conjunction with the standard stannanes and silanes.[4] Thiolesters derived from **1**, readily accessible by the coupling of **1** and a carboxylic acid or the corresponding acid chloride by standard means, are excellent precursors of acyl radicals when tri-*n*-butylstannane (Bu_3SnH) and tris(trimethylsilyl)silane (TMS_3SiH) are used as the chain transfer agents.[1] In this process, the lack of reactivity of standard thiolesters is overcome by two efficient free radical reactions, namely, iodine abstraction by tributyltin and tris(trimethylsilyl)silyl radicals from an aryl iodine and intramolecular homolytic attack on the sulfur by the aryl radical, with release of the acyl radical (eq 2).

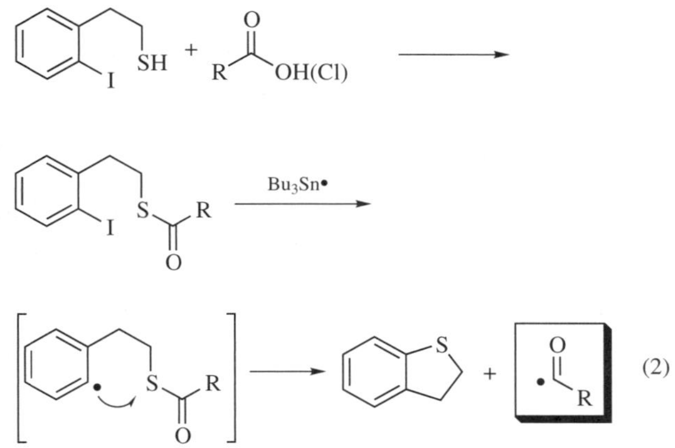

Acyl radicals generated from thiolesters, derived from **1**, participate smoothly in various chain sequences including radical cyclization and decarbonylation.[1] For example, thiolester **2** was shown to react with Bu_3SnH in the presence of AIBN in refluxing benzene to give the cyclic ketone **3** in excellent yield (eq 3). This strategy was used to generate aryl- and alkyl-derived azidoacyl radicals in a new approach to the synthesis of lactams as demonstrated by the example in eq 4.[5]

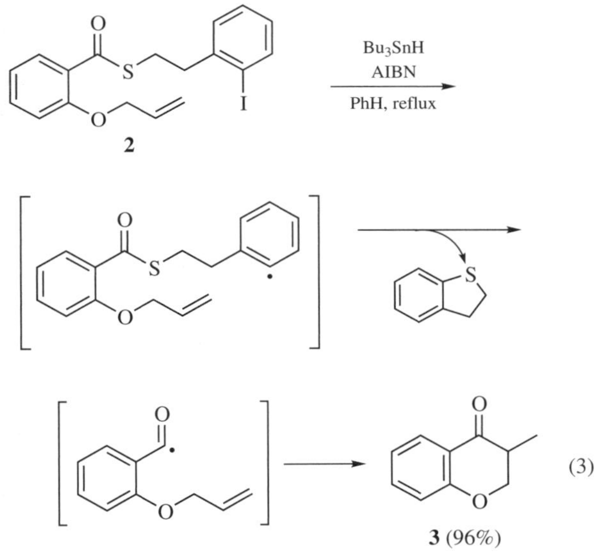

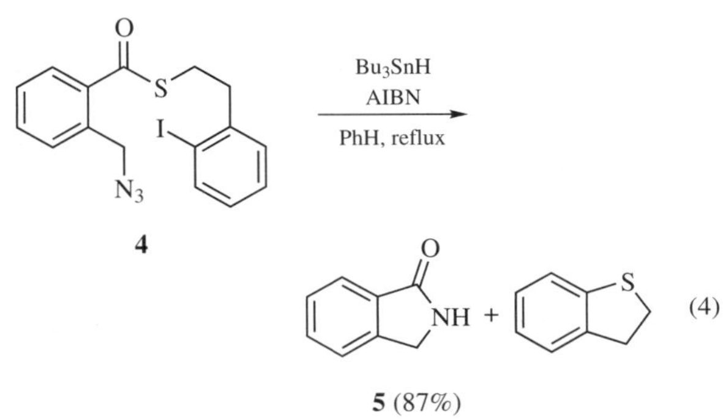

The chemistry has been applied to the generation of a nucleotide C4′ radical via decarbonylation of the acyl radical (eq 5).[6] Pyranoside C6 radicals can also be generated through the sequential

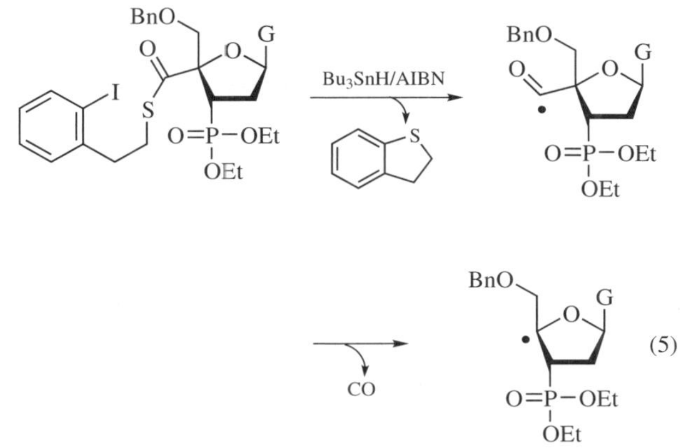

decarbonylation of the initially formed acyl radicals and ring-opening of the resulting benzylidene radicals in the context of carbohydrate synthesis (eq 6).[7]

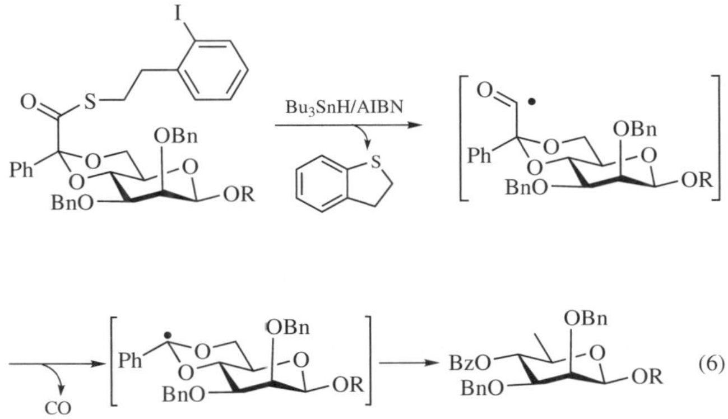

Application of the system to α,β-unsaturated carboxylic acids enables the generation of α-ketenyl radicals, which permit numerous interesting tandem cyclizations as, for example, in eq 7.[8]

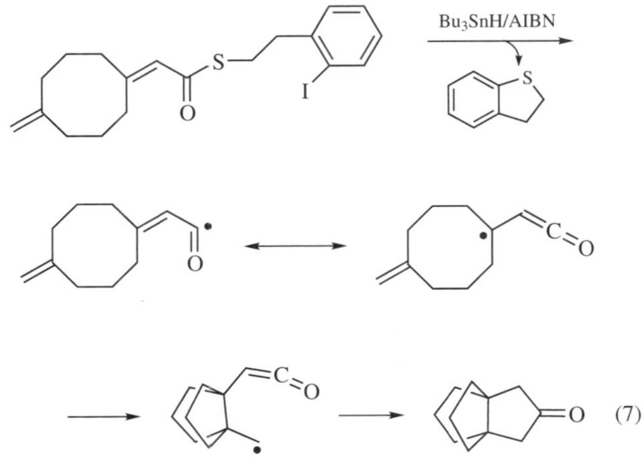

Related Reagents. 2-(2-Iodophenyl)-2-methylpropanethiol;[1] 2-(2-bromophenyl)ethanethiol;[1] 2′-iodo-2-biphenylthiol;[9] *tert*-butyl [2-(2-mercaptoethyl)phenyl]carbamate.[10]

1. Crich, D.; Yao, Q., *J. Org. Chem.* **1996**, *61*, 3566.
2. Chatgilialoglu, C.; Crich, D.; Ryu, I.; Komatsu, M., *Chem. Rev.* **1999**, *99*, 1991.

3. Penn, J. H.; Liu, F., *J. Org. Chem.* **1994**, *59*, 2608.
4. (a) Crich, D.; Eustace, K. A.; Fortt, S. M.; Ritchie, T. J., *Tetrahedron* **1990**, *46*, 2135. (b) Boger, D. L.; Mathvink, R. J., *J. Org. Chem.* **1988**, *53*, 3377.
5. Benati, L.; Leardini, R.; Minozzi, M.; Nanni, D.; Spagnolo, P.; Strazzari, S.; Zanardi, G., *Org. Lett.* **2002**, *4*, 3079.
6. Crich, D.; Suk, D.-H.; Hao, X., *Tetrahedron* **2002**, *58*, 5789.
7. Crich, D.; Yao, Q., *Org. Lett.* **2003**, *5*, 2189.
8. Pattenden, G.; De Boeck, B., *Tetrahedron Lett.* **1998**, *39*, 6975.
9. Ooi, T.; Furuya, M.; Sakai, D.; Maruoka, K., *Adv. Synth. Catal.* **2001**, *343*, 166.
10. Crich, D.; Hao, X., *J. Org. Chem.* **1997**, *62*, 5982.

Qingjia Yao
University of Illinois at Chicago, Chicago, IL, USA

N-Iodosaccharin

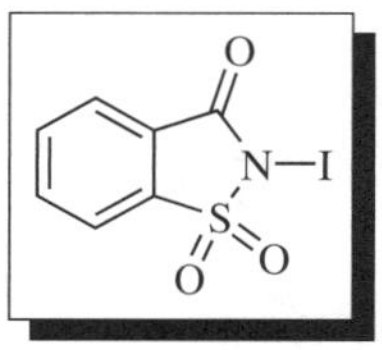

[86340-94-5] $C_7H_4INO_3S$ (MW 309.08)

(reagent used in the iodination of alkenic compounds and activated aromatics, and also acts as an activator for thiophenylglycosides)

Alternate Name: NISac.
Physical Data: mp 206–208 °C.
Solubility: soluble in acetone, acetonitrile, dichloroethane, and other polar organic solvents. Not soluble in water.
Form Supplied in: white or pale yellow crystals; as a crystallohydrate with 1 equiv of water; not currently commercially available.
Analysis of Reagent Purity: iodometric titration, IR, mp, NMR, elemental analysis.
Preparative Methods: silver nitrate (8.50 g, 50 mmol) was dissolved in water (50 mL), heated to 80 °C, and a solution of the sodium salt of saccharin (10.5 g, 51 mmol) in water (50 mL) was added dropwise. The resulting white precipitate was filtered, washed with water and acetone, and dried in air to give the silver salt of saccharin (14.3 g, 99%). The dry silver salt (5.80 g, 20 mmol) and iodine (5.21 g, 20.5 mmol) were stirred in acetone (50 mL) at rt in the dark. After 5 h, the precipitated AgI was filtered, and the filtrate was concentrated in vacuo to give crude NISac (6.30 g, 96%). The crude product can be used for most purposes, or can be further purified.
Purification: crude NISac is dissolved in THF and precipitated by addition of pentane; the white precipitate is filtered and dried under high vacuum in the dark.
Handling, Storage, and Precautions: can be stored indefinitely in crystalline form in a freezer in the dark. Decomposes on prolonged exposure to light. Toxicity is unknown.

Introduction. *N*-Haloamides have long been known as a class of moderately electrophilic halogenating reagents. The most commonly used reagents of this class are those derived from succinimide (see, e.g., *N*-iodosuccinimide, *N*-bromosuccinimide). However, a comparison of the pK_a values of succinimide (pK_a = 9.62) and saccharin (pK_a = 1.30) implies that NISac should be a more potent source of I^+ than its succinimide cousin.[1]

Iodination of Alkenic Compounds. In the presence of NISac, alkenes are transformed to the corresponding adducts of iodine and nucleophile (eq 1).[2] Additions follow the Markovnikov rule with high regioselectivity. Reaction with disymmetric alkenes results in the formation of a single diastereomer as a result of exclusive *anti* addition.

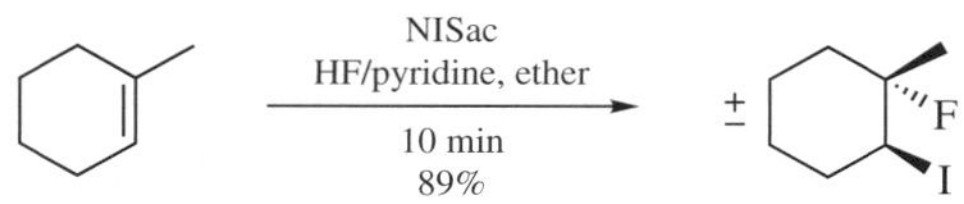

(1)

In comparison with *N*-iodosuccinimide (NIS), iodination of cyclohexene by NISac was found to be 500 times faster.[2]

Iodination of Activated Aromatics. NISac is effective in the iodination of activated aromatics such as phenols and anilines.[2] While NIS alone has been shown to halogenate similar substrates in low yields,[3] an acid co-reactant is more commonly used.[4,5] Electrophilic aromatic substitution proceeds with the expected *ortho*/*para* regiochemistry; iodination of deactivated aniline and anisole derivatives results in mono-iodination (eq 2), while phenol derivatives undergo di-iodination.[2] It is noteworthy that NISac is tolerant of hydroxyl groups and aldehydes under these conditions.[2]

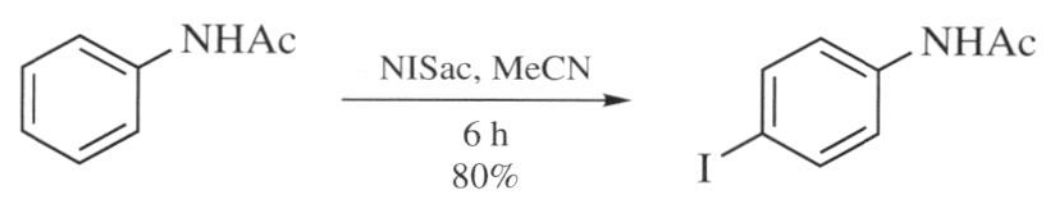

(2)

NISac as an Activator for Thiophenylglycosides. The use of NIS in the activation of thioglycosides is well known.[6,7] However, thioglycoside activation with NIS alone is usually slow and, commonly, co-reagents are used to increase its reactivity.[7,8] Thiophenylglycosides can be activated by use of NISac alone,[1] exemplifying its increased potency relative to that of NIS. Disarmed[9] thiophenyl glucosides do not undergo any reaction, but glycosylation of a fully armed thiophenyl glucoside resulted in rapid formation of the desired methyl glycoside as well as minor products resulting from the nucleophilic attack by saccharin (eq 3). Under identical conditions, glycosylation of a fully armed thiomethyl glucoside is slow, a direct contradiction to the usual expectation that anomeric SMe is easier to activate than anomeric SPh.[10] It is proposed that NISac can act as a chemoselective glycosylation reagent as a potent species for the selective activation of armed thiophenyl glycosides.[1]

Solvent Incorporation During NISac-Mediated Glycosylation. Solvent participation in glycosylation reactions has been well documented.[11,12] However, solvent incorporation during glycosylation reactions is much less common,[13,14] but has been observed for the case of NISac activation of a thiophenyl glucoside in either acetone or cyclohexanone.[15] Glycosylation of a fully armed[9] thiophenyl glucoside in the presence of NISac with methanol in acetone gave an anomerically pure α-linked mixed acetal as the sole product (eq 4).

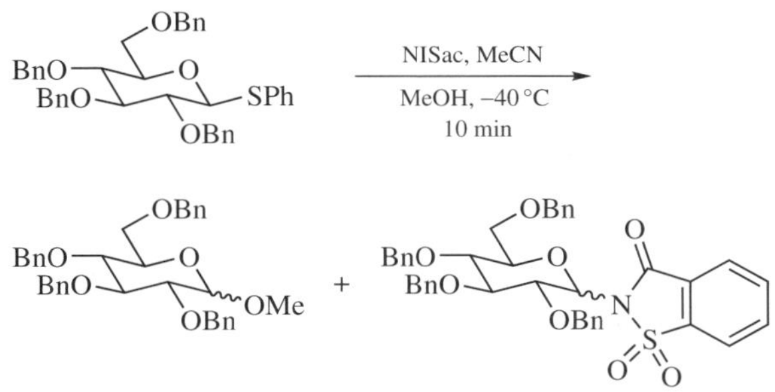

70%, α:β = 1:6 15% (3)

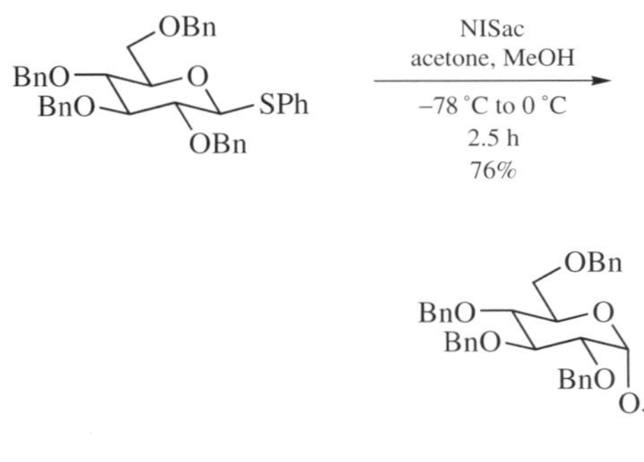

(4)

Related Reagents. *N*-Iodosuccinimide; *N*-bromosuccinimide; iodine.

1. Aloui, M.; Fairbanks, A. J., *Synlett* **2001**, 797–799.
2. Dolenc, D., *Synlett* **2000**, 544–546.
3. Nicolaou, K. C.; Ramanjulu, J. M.; Natarajan, S.; Bräse, S.; Li, H.; Boddy, C. N. C.; Rübsam, F., *Chem. Commun.* **1997**, 1899–1900.
4. Castanet, A.-S.; Colobert, F.; Broutin, P.-E., *Tetrahedron Lett.* **2002**, *43*, 5047–5048.
5. Boger, D. L.; Brunette, S. R.; Garbaccio, R. M., *J. Org. Chem.* **2001**, *66*, 5163–5173.
6. *Best Synthetic Methods: Carbohydrates*; Osborn, H. M. I., Ed.; Elsevier: New York, 2003, 128–137.
7. Veeneman, G. H.; van Leeuwen, S. H.; van Boom, J. H., *Tetrahedron Lett.* **1990**, *31*, 1331–1334.
8. Konradsson, P.; Udodong, U. E.; Fraser-Reid, B., *Tetrahedron Lett.* **1990**, *31*, 4313–4316.
9. Mootoo, D. R.; Konradsson, P.; Udodong, U. E.; Fraser-Reid, B., *J. Am. Chem. Soc.* **1988**, *110*, 5583–5584.
10. Zuurmond, H. M.; van der Laan, S. C.; van der Marel, G. A.; van Boom, J. H., *Carbohydr. Res.* **1991**, *215*, c1-c3.
11. Schmidt, R. R.; Behrendt, M.; Toepfer, A., *Synlett* **1990**, 694–696.
12. Pougny, J.-R.; Sinaÿ, P., *Tetrahedron Lett.* **1976**, *17*, 4073–4076.
13. Somsák, L.; Kovács, L.; Gyóllai, V.; Ósz, E., *Chem. Commun.* **1999**, 591–592.
14. Ratcliffe, A. J.; Fraser-Reid, B., *J. Chem. Soc., Perkin Trans. 1* **1990**, 747–750.
15. Aloui, M.; Fairbanks, A. J., *Chem. Commun.* **2001**, 1406–1407.

David J. Chambers & Antony J. Fairbanks
Dyson Perrins Laboratory, Oxford, UK

N-Iodosuccinimide

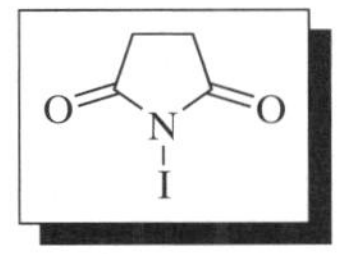

[516-12-1] $C_4H_4INO_2$ (MW 224.99)

(electrophilic iodination of arenes, alkenes, and alkynes; activation of glycosyl donors)

Alternate Name: NIS; 1-iodo-2,5-pyrrolidinedione.
Physical Data: mp 193–199 °C (dec).
Solubility: soluble in dioxane, THF, MeCN; insoluble in ether, CCl_4.
Form Supplied in: white powder; widely available, but quite expensive.
Preparative Methods: reaction of *N*-silversuccinimide with I_2 in anhydrous dioxane, removal of AgI by fitration, and addition of CCl_4 to promote crystallization.[1]
Purification: recrystallization from dioxane–CCl_4 at −20 °C.
Handling, Storage, and Precautions: stored at 0 °C under nitrogen, and protected from light and moisture to avoid decomposition; an irritating solid and precautions should be taken to avoid inhalation of the powder.

Original Commentary

Scott C. Virgil
Massachusetts Institute of Technology, Cambridge, MA, USA

Introduction. *N*-Iodosuccinimide is an efficient reagent for the electrophilic iodination of organic compounds for which the reactivity of I_2 is often insufficient. A variety of functional groups are electrophilically iodinated using NIS with similar effectiveness to the corresponding brominations using N-Bromosuccinimide (NBS).[2] Electrophilic attack on soft nucleophiles (e.g. as in dithiane deprotection) is achieved in the presence of other sensitive groups with optimum selectivity by iodination using NIS under mild conditions (see also 1,3-Diiodo-5,5-dimethylhydantoin).[3] For other iodinations a more powerful source of iodonium ion can be generated by the combination of NIS with protic acids including AcOH, TsOH, and TfOH.[4]

Reaction with Organometallic Derivatives. Similarly, the lesser reactivity of NIS compared with NBS allows it to be used for the synthesis of vinyl iodides by iodination of vinylaluminum intermediates in the presence of reactive alkenic groups.[5] NIS has also been found to be an excellent reagent for the oxidative cyclization of dipeptide dianions to β-lactams (eq 1).[6]

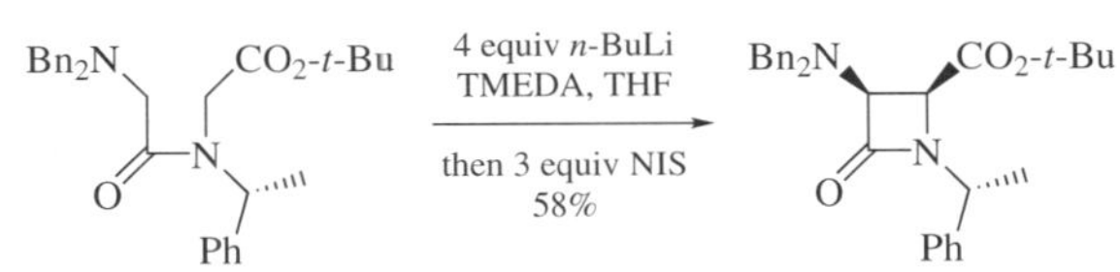

(1)

Iodination of Alkenic Compounds. Iodolactonization and related reactions with alkenes can be achieved using NIS in aprotic solvents. In addition, iodoesters may also be prepared by the reaction of alkenes with NIS in the presence of carboxylic acids (eq 2),[7] and allylic trichloroacetimidates may be cyclized to 5-iododihydro-1,3-oxazines (eq 3).[8]

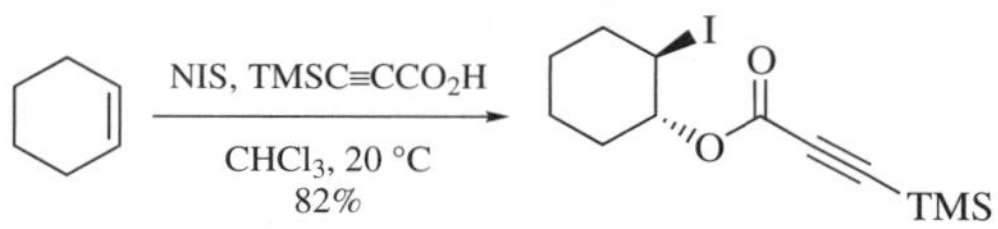

(2)

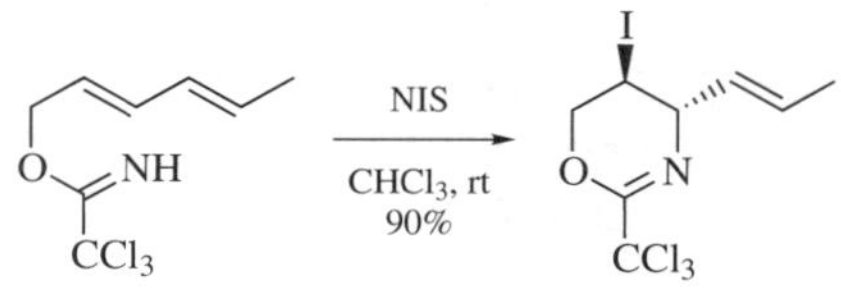

(3)

In combination with protic acid, NIS has been used for the preparation of α-iodo enones and β,β-dihalo enones (eqs 4 and 5; HTIB = [Hydroxy(tosyloxy)iodo]benzene).[9] More reactive alkenes such as enol acetates react with NIS to afford α-iodo ketones.[10] The reaction of enol ethers with NIS in the presence of an alcohol, affording iodoacetals, allows the synthesis of acetals by dehalogenation or mixed ketene acetals by elimination.[11]

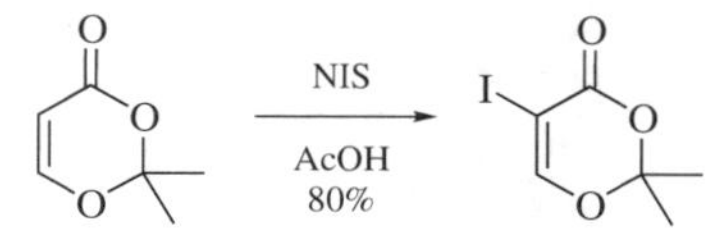

(4)

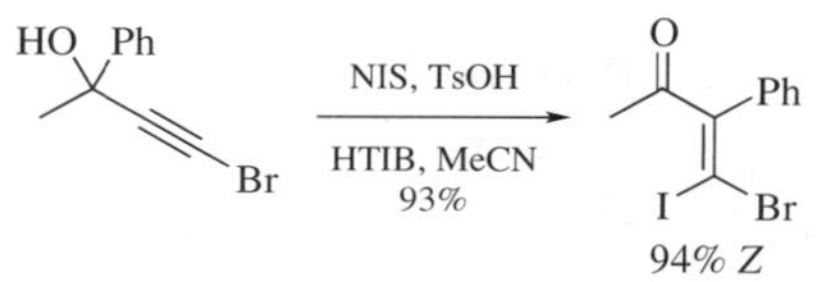

(5)

Activation of Glycosyl Donors.[12] NIS is a useful reagent for the coupling of glycals to alcohols (eq 6)[13] or other carbohydrates,[14] affording 2-iodoglycosides which are readily dehalogenated to 2-deoxysugars, or can serve as precursors to 2-aminosugars via the epoxide intermediate.[15] Selective activation of glycosyl thioethers for coupling with carboxylic acids and heterocyclic compounds is achieved under mild conditions by reaction with NIS and Trifluoromethanesulfonic Acid (eq 7).[16] In addition, the combination of NIS and TfOH is also the reagent of choice for the activation of 4-pentenylglycosides.[17]

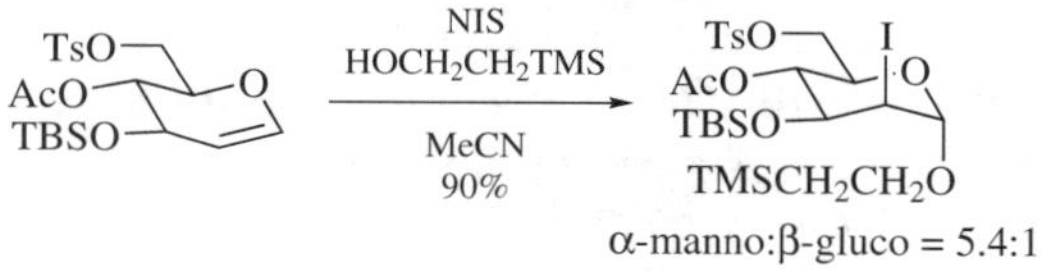

(6)

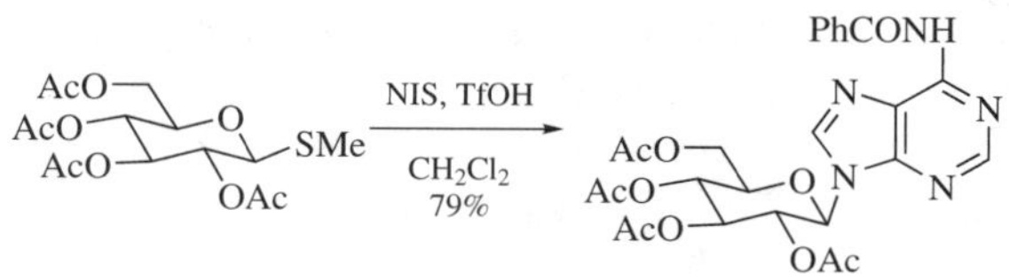

(7)

Oxidation Reactions. Hypoiodite intermediates may be generated from the reaction of simple alcohols with NIS. When conducted under photochemical irradiation, the products of Barton-type or fragmentation reactions of alkoxyl radical intermediates may be obtained.[18] Aldehydes are oxidized to methyl esters via hemiacetal intermediates by reaction with NIS in methanol at rt.[19] However, such conditions are not effective for the oxidation of simple alcohols. The combination of NIS and Tetra-n-butylammonium Iodide in dichloromethane has been developed for the oxidation of a variety of alcohols to the corresponding carbonyl compounds (eq 8).[20] This reagent system is most widely used for the oxidation of lactols to lactones, in which near-quantitative yields are generally obtained under mild conditions (eq 9).[21]

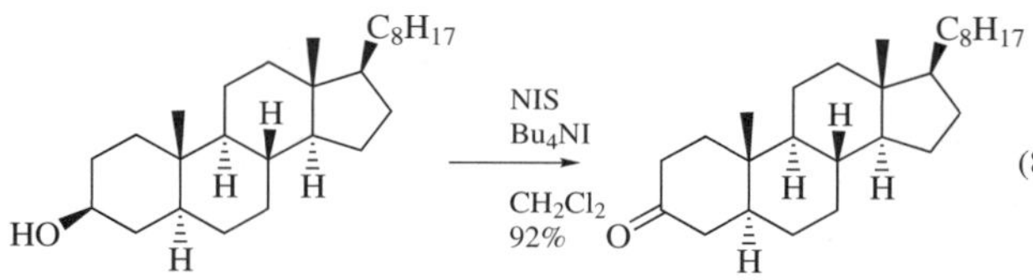

(8)

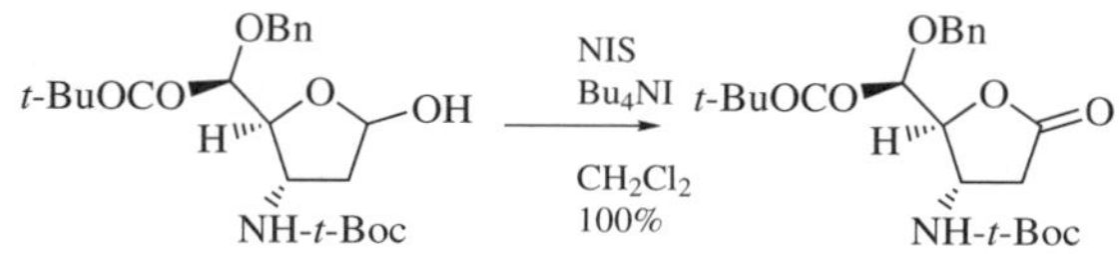

(9)

First Update

Ying Zeng & Fanzuo Kong
Research Center for Eco-Environmental Sciences, Academia Sinica, Beijing, China

Reaction with Thiols. In combination with triphenylphosphine (PPh_3), NIS is used as an iodide donor to convert alkyl thiols to alkyl iodides under mild conditions (eq 10),[22] analogous to the conversion of alcohols to alkyl iodides.

$$CH_3(CH_2)_{11}SH \xrightarrow[CH_2Cl_2]{PPh_3/NIS} \underset{70\%}{CH_3(CH_2)_{11}I} \quad (10)$$

Reaction with Carborane Compounds. Reaction of $Na_2[B_9H_9]$ with NIS gives the monoiodinated anion $[1\text{-}IB_9H_8]^{2-}$, while treatment of $[Me_3NH][nido\text{-}7,8\text{-}C_2B_9H_{12}]$ with NIS affords the corresponding $[Me_3NH][nido\text{-}9,11\text{-}I_2\text{-}7,8\text{-}C_2B_9H_{10}]$.[23,24] Reaction of sandwich type carborane $CpCl_2Ta(2,3\text{-}Et_2C_2B_4H_4)$ with NIS generates B(5)-monoiodo, B(4,5)-diiodo, and in some cases B(4,5,6)-triiodo complexes.[25,26] These compounds, in turn, react with Grignard reagents to give the B-alkyl and B-aryl derivatives.

Application in Glycosylations. NIS is employed to achieve both tethering and thioglycoside activation allowing the stereoselective synthesis of 1,2-*cis*-glycosides either in a one- or two-step procedure. Thus, treatment of 2-*O*-alkenyl protected thioglycosides with NIS and a variety of alcohols in THF produces ketal intermediates, which allow the synthesis of 1,2-*cis*-glycosidic linkages (eq 11).[27] 1,2-*cis*-Glycosidic linkages are similarly obtained from 2-*O*-alkenyl protected glycosyl fluoride donors.[28]

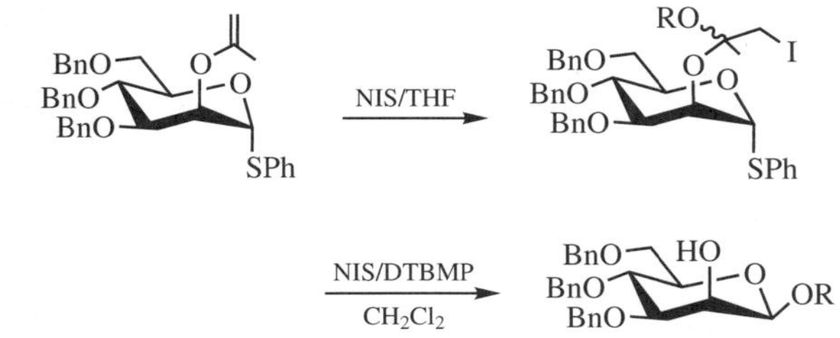

(11)

A new method for the deprotection of benzyl ethers with NIS through a hypoiodite intermediate has been developed (eq 12).[29]

(12)

R=Bn 75%

Applications in Fluorination. Iodofluorinated compounds are obtained using the so-called 'three-component reaction' [a substrate, an electrophile (NIS), and a nucleophile (fluorination agent)] from alkenes or alkynes in a regio- and stereoselective way (eq 13).[30–32]

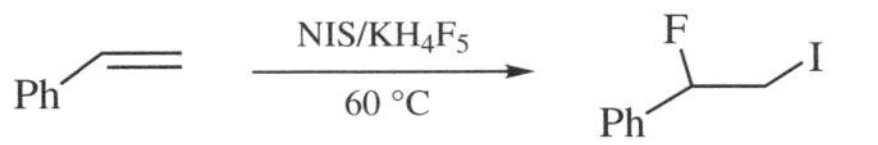

(13)

80%

A reagent consisting of NIS and 70% HF-pyridine converts xanthates R-OCS_2Me (R=alkyl) to R-OCF_3 (R = primary) or R-F (R = benzylic, secondary, tertiary) by an oxidative desulfurization-fluorination reaction.[33] The C-S bonds of the dithiocarbamates are directly replaced by C-F bonds using NIS and tetrabutylammonium dihydrogentrifluoride ($TBAH_2F_3$) (eq 14).[34]

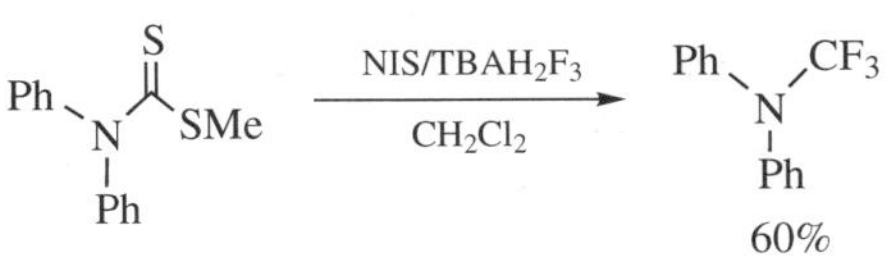

(14)

60%

Reactions with Alkyne and Alkene Compounds. NIS is an excellent iodinating reagent that can smoothly convert vinyl silanes and vinyl stannanes to the corresponding vinyl iodides (eqs 15 and 16).[35,36] It also enables the mild conversion of several types of alkenyl boronic acids to geometrically pure alkenyl iodides (eq 17).[37]

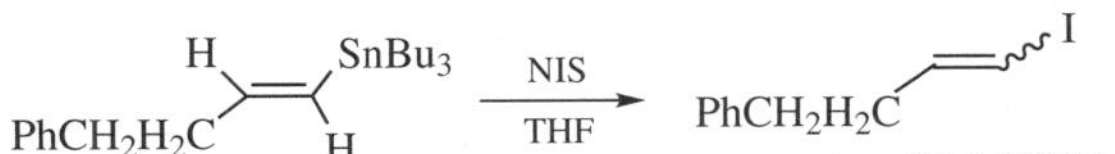

(15)

trans:*cis* = 13:1 (83%)

(16)

90%

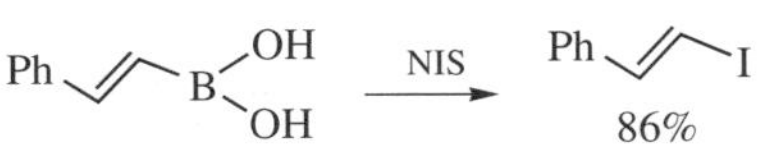

(17)

86%

Alkenyl carboxylic acids and propiolic acids are converted to the corresponding iodoalkenes and iodoalkynes in excellent yields with NIS (eq 18).[38,39]

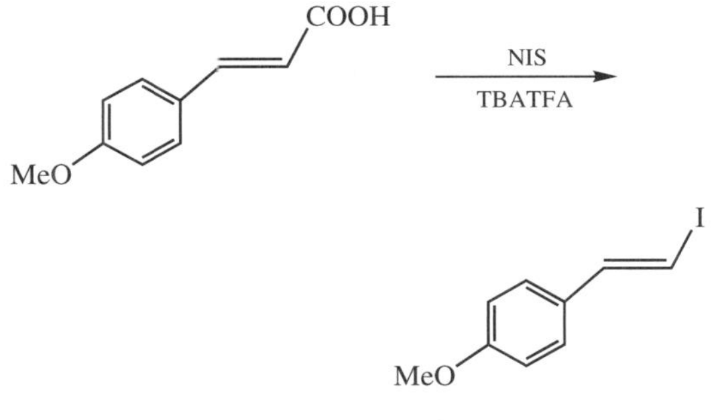

(18)

74%

NIS reacts with carbonyl-conjugated alkynes in the presence of protic acid to give diiodo ketoesters, diiodo diketones (eq 19), or diiodo acetals.[40]

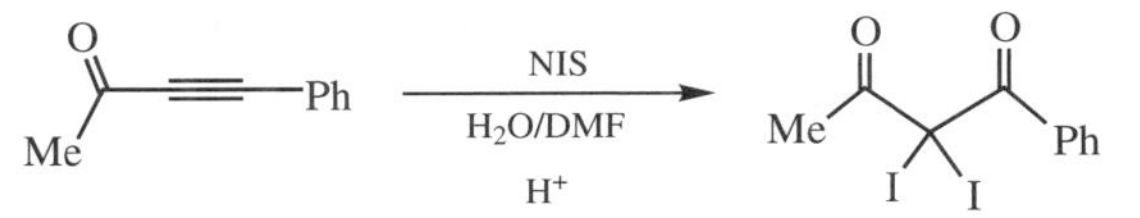

(19)

95%

Applications in Retro-Mannich Reactions. Treatment of unprotected gramine with NIS results in smooth conversion to 3-iodoindole (eq 20). 3,4-Disubstituted indoles are obtained by combination of the retro-Mannich process with the directed ortho metalation reaction (DoM) or the Negishi cross-coupling protocol.[41]

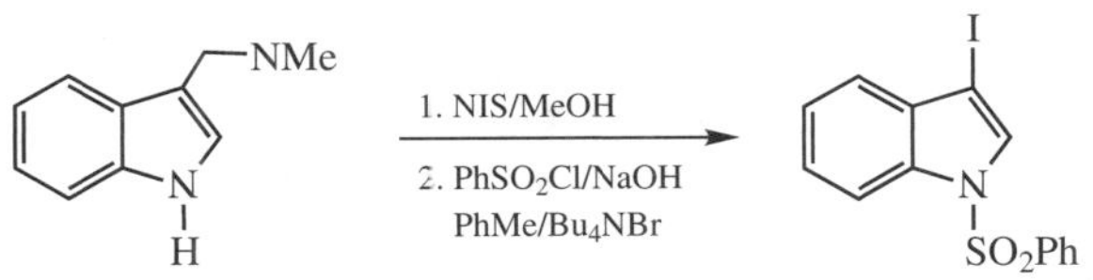

(20)

78%

Reaction with Stabilized Phosphoranes. Iodo enol lactones are obtained from reaction of the corresponding ω-carboxy keto phosphoranes with NIS (eq 21).[42]

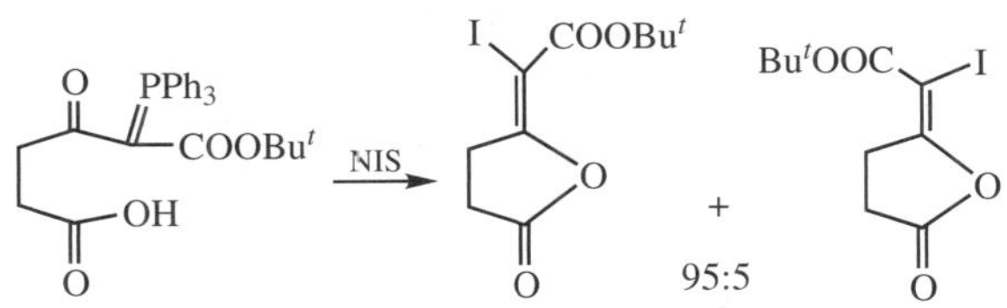

(21)

95:5
88%

Other Reactions. Treatment of 4-halogenocyclohexa-2,5-dienones with 4-hydroperoxycyclohexa-2,5-dienones in the presence of NIS gives substituted bis(4-oxocyclohexa-2,5-dienyl) peroxides (eq 22), some of which have not been prepared by the classical method.[43]

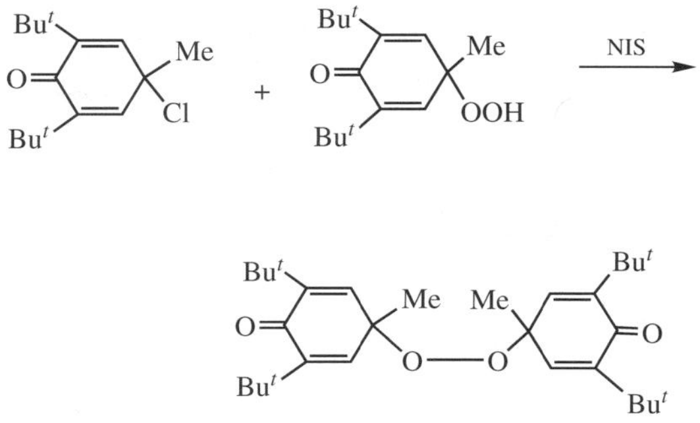

(22)

74%

NIS is used as a selective *N*-demethylating agent under mild conditions. These conditions have been successfully applied to the synthesis of metabolites of potent antibacterial agents within the macrolide and ketolide classes (eq 23).[44]

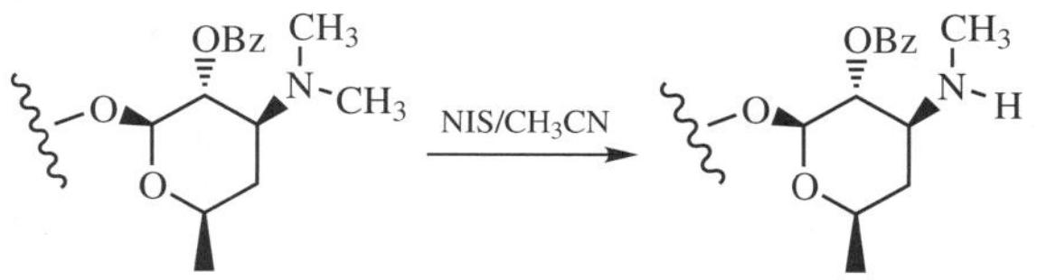

(23)

1. Benson, W. R.; McBee, E. T.; Rand, L., *Org. Synth., Coll. Vol.* **1973**, *5*, 663.
2. NIS is generally not effective in the range of radical halogenations possible for NBS. See also: Taneja, S. C.; Dhar, K. L.; Atal, C. K., *J. Org. Chem.* **1978**, *43*, 997.
3. Burri, K. F.; Cardone, R. A.; Chen, W. Y.; Rosen, P., *J. Am. Chem. Soc.* **1978**, *100*, 7069.
4. Olah, G. A.; Wang, Q.; Sandford, G.; Prakash, G. K. S., *J. Org. Chem.* **1993**, *58*, 3194.
5. Marshall, J. A.; Lebreton, J.; DeHoff, B. S.; Jenson, T. M., *Tetrahedron Lett.* **1987**, *28*, 723.
6. Kawabata, T.; Minami, T.; Hiyama, T., *J. Org. Chem.* **1992**, *57*, 1864.
7. (a) Adinolfi, M.; Parrilli, M.; Barone, G.; Laonigro, G.; Mangoni, L., *Tetrahedron Lett.* **1976**, 3661. (b) Haaima, G.; Weavers, R. T., *Tetrahedron Lett.* **1988**, *29*, 1085.
8. Bongini, A.; Cardillo, G.; Orena, M.; Sandri, S.; Tomasini, C., *J. Org. Chem.* **1986**, *51*, 4905.
9. (a) Iwaoka, T.; Murohashi, T.; Katagiri, N.; Sato, M.; Kaneko, C., *J. Chem. Soc., Perkin Trans. 1* **1992**, 1393. (b) Angara, G. J.; Bovonsombat, P.; McNelis, E., *Tetrahedron Lett.* **1992**, *33*, 2285.
10. Djerassi, C.; Grossman, J.; Thomas, G. H., *J. Am. Chem. Soc.* **1955**, *77*, 3826.
11. Middleton, D. S.; Simpkins, N. S., *Synth. Commun.* **1989**, *19*, 21.
12. (a) Theim, J.; Karl, H.; Schwentner, J., *Synthesis* **1978**, 696. (b) Theim, J., In *Trends in Synthetic Carbohydrate Chemistry*; Horton, D.; Hawkins, L. D.; McGarvey, G. J., Eds.; ACS: Washington, 1989; Vol. 386, Chapter 8.
13. Sebesta, D. P.; Roush, W. R., *J. Org. Chem.* **1992**, *57*, 4799.
14. Horton, D.; Priebe, W.; Sznaidman, M., *Carbohydr. Res.* **1990**, *205*, 71.
15. Friesen, R. W.; Danishefsky, S. J., *Tetrahedron* **1990**, *46*, 103.
16. (a) Veeneman, G. H.; van Leeuwen, S. H.; van Boom, J. H., *Tetrahedron Lett.* **1990**, *31*, 1331. (b) Knapp, S.; Shieh, W.-C., *Tetrahedron Lett.* **1991**, *32*, 3627. (c) Konradsson, P.; Udodong, U. E.; Fraser-Reid, B., *Tetrahedron Lett.* **1990**, *31*, 4313.
17. (a) Konradsson, P.; Mootoo, D. R.; McDevitt, R. E.; Fraser-Reid, B., *Chem. Commun.* **1990**, 270. (b) Ratcliffe, A. J.; Konradsson, P.; Fraser-Reid, B., *J. Am. Chem. Soc.* **1990**, *112*, 5665.
18. (a) McDonald, C. E.; Beebe, T. R.; Beard, M.; McMillen, D.; Selski, D., *Tetrahedron Lett.* **1989**, *30*, 4791. (b) McDonald, C. E.; Holcomb, H.; Leathers, T.; Ampadu-Nyarko, F.; Frommer, J., Jr., *Tetrahedron Lett.* **1990**, *31*, 6283.
19. McDonald, C.; Holcomb, H.; Kennedy, K.; Kirkpatrick, E.; Leathers, T.; Vanemon, P., *J. Org. Chem.* **1989**, *54*, 1213.
20. Hanessian, S.; Wong, D. H.; Therien, M., *Synthesis* **1981**, 394.
21. Hamada, Y.; Kawai, A.; Matsui, T.; Hara, O.; Shioiri, T., *Tetrahedron* **1990**, *46*, 4823.
22. Iranpoor, N.; Firouzabadi, H.; Aghapour, G., *Synlett* **2001**, 1176–1178.
23. Siegburg, K.; Preetz, W., *Inorg. Chem.* **2000**, *39*, 3280–3282.
24. Santos, E. C.; Pinkerton, A. B.; Kinkead, S. A.; Hurlburt, P. K.; Jasper, S. A.; Sellers, C. W.; Huffman, J. C.; Todd, L. J., *Polyhedron* **2000**, *19*, 1777–1781.
25. Stockman, K. E.; Boring, E. A.; Sabat, M.; Finn, M. G.; Grimes, R. N., *Organometallics* **2000**, *19*, 2200–2207.
26. Stockman, K. E.; Garrett, D. L.; Grimes, R. N., *Organometallics* **1995**, *14*, 4661–4667.
27. Ennis, S. C.; Fairbanks, A. J.; Slinn, C. A.; Tennant-Eyles, R. J.; Yeates, H. S., *Tetrahedron* **2001**, *57*, 4221–4230.
28. Cumpstey, I.; Fairbanks, A. J.; Redgrave, A. J., *Org. Lett.* **2001**, *3*, 2371–2374.
29. Madsen, J.; Bols, M., *Angrew. Chem. Int. Ed.* **1998**, *37*, 3177–3178.
30. Tamura, M.; Shibakami, M.; Sekiya, A., *Synthesis* **1995**, *5*, 515–517.
31. Shellhamer, D. F.; Horney, M. J.; Pettus, B. J.; Petttus, T. L.; Stringer, J. M.; Heasley, V. L.; Syvret, R. G.; Dobrolsky, J. M., *J. Org. Chem.* **1999**, *64*, 1094–1098.
32. Shellhamer, D. F.; Jones, B. C.; Pettus, B. J.; Petttus, T. L.; Stringer, J. M.; Heasley, V. L., *J. Fluorine Chem.* **1998**, *88*, 37–39.
33. Kanie, K.; Tanaka, Y., Shimizu, M.; Kuroboshi, M.; Hiyama, T., *Chem. Commun.* **1997**, 309–310.
34. Kanie, K.; Mizuno, K.; Kuroboshi, M.; Hiyama, T., *Bull. Chem. Soc. Jpn.* **1998**, *71*, 1973–1991.
35. Stamos, D. P.; Taylor, A. G.; Kishi, Y., *Tetrahedron Lett.* **1996**, *37*, 8647–8650.
36. Boden, C. D. J.; Pattenden, G.; Ye, T., *J. Chem. Soc., Perkin Trans. 1* **1996**, 2417–2419.
37. Petasis, N. A.; Zavialov, L. A., *Tetrahedron Lett.* **1996**, *37*, 567–570.
38. Naskar, D.; Chowdhury, S.; Roy, S., *Tetrahedron Lett.* **1998**, *39*, 699–702.
39. Naskar, D.; Roy S., *J. Org. Chem.* **1999**, *64*, 6896–6897.
40. Heasley, V. L.;. Shellhamer, D. F.; Chappell, A. E.; Cox, J. M.; Hill, D. J.; McGovern, S. L.; Eden, C. C.; Kissel, C. L., *J. Org. Chem.* **1998**, *63*, 4433–4437.
41. Chauder, B.; Larkin, A.; Snieckus, V., *Org. Lett.* **2002**, *4*, 815–817.
42. Kayser, M. M.; Zhu, J.; Hooper, D. L., *Can. J. Chem.* **1997**, *75*, 1322–1330.
43. Omura, K., *J. Org. Chem.* **1997**, *62*, 8790–8793.
44. Stenmark, H. G.; Brazzale, A.; Ma, Z., *J. Org. Chem.* **2000**, *65*, 3875–3876.

Isobutyl Chloroformate

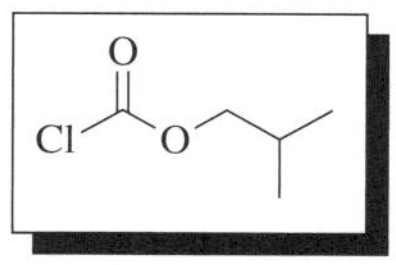

[543-27-1] $C_5H_9ClO_2$ (MW 136.59)

(used for making mixed anhydrides which serve as active intermediates for peptide synthesis;[1–3] used to block the 5′-hydroxy function of deoxyribosides in oligonucleotide synthesis;[4] condensing reagent[5])

Physical Data: bp 128.8 °C; *d* 1.04 g cm^{-3}.
Solubility: miscible with benzene, chloroform, ether.
Form Supplied in: clear colorless liquid.
Handling, Storage, and Precautions: gradually decomposed by water and alcohol; highly toxic; vapor irritates eyes and mucous membranes; stable on storage.[6,7] Use in a fume hood.

Peptide Reagent.[1,3] Protected amino acids, such as the *N*-benzyloxycarbonyl (Cbz) derivatives, are brought into solution in toluene or THF by addition of enough Triethylamine to form the salt, and isobutyl chloroformate is added at 0 °C to produce the mixed anhydride. A solution of an amino acid ester (or peptide ester) to be *N*-acylated is added in an inert solvent and the mixture is allowed to come to rt. Evolution of carbon dioxide begins immediately (eq 1). The isobutyl chloroformate is preferred to lower esters for preparation of peptides of moderate or high molecular weight;[6,7] ethyl chloroformate is preferred for synthesis of dipeptides.

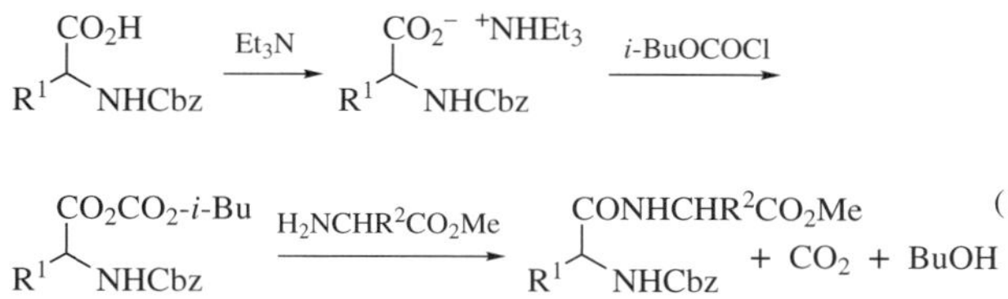
(1)

Boc-Gly-Gly-OEt and Boc-Gly-Gly-Gly-OEt have been successfully synthesized in reasonable yields (77% and 88%, respectively) via mixed anhydride derivatives from isobutyl chloroformate.[8] Previously it was reported[9] that considerable quantities of diacylated side products are formed in mixed anhydride syntheses with glycine when ethyl chloroformate is used.

5′-Hydroxy Function Blocking Agent for Deoxyribosides. The reagent has been used to block the 5′-hydroxy function of deoxyribosides in oligonucleotide synthesis (eq 2).[4] The reagent shows good selectivity for the 5′-hydroxy group, as is evident in the high yield of (**2**). The blocking group is removed smoothly by alkaline hydrolysis to give (**4**).

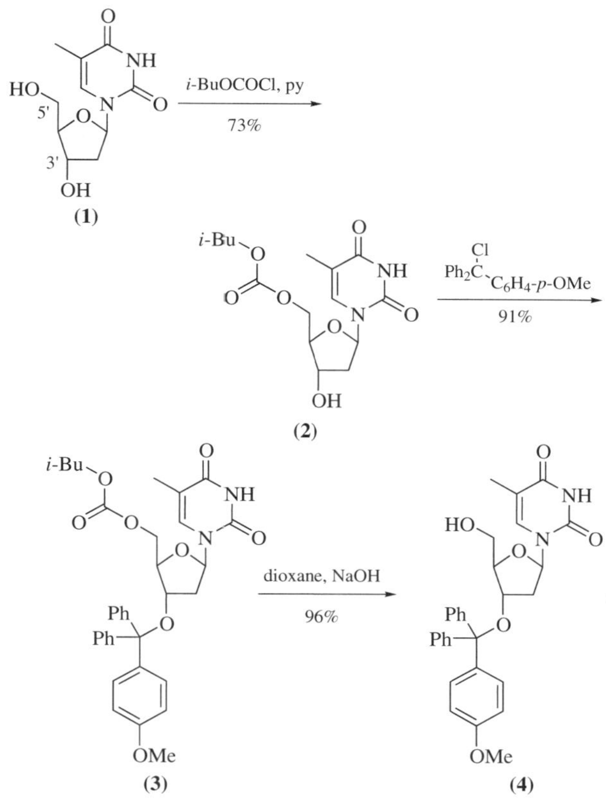
(2)

***N*-Protection.**[10] In the synthesis of enantiomerically pure symmetric vicinal diamines involving chirality transfer from a single chiral source, 1,2-diphenylethane-1,2-diamine, hydrolysis of the bisacetamides to the corresponding vicinal diamines was attempted using a number of amide hydrolysis procedures. However, the various base- and acid-catalyzed procedures did not result in a clean hydrolysis of the acetamide groups. This was achieved by converting amines into biscarbamate derivatives, followed by treatment with 30% HBr in acetic acid (eq 3).

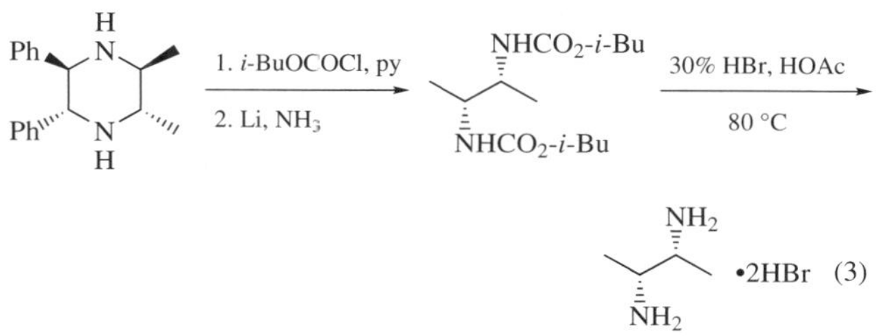

Cyclodehydration.[5] In the synthesis of 1,4-dihydropyran[3,4-*b*]indolones, the reagent was used along with triethylamine for cyclodehydration of hydroxymethylindoleacetic acid to produce the lactone (eq 4).

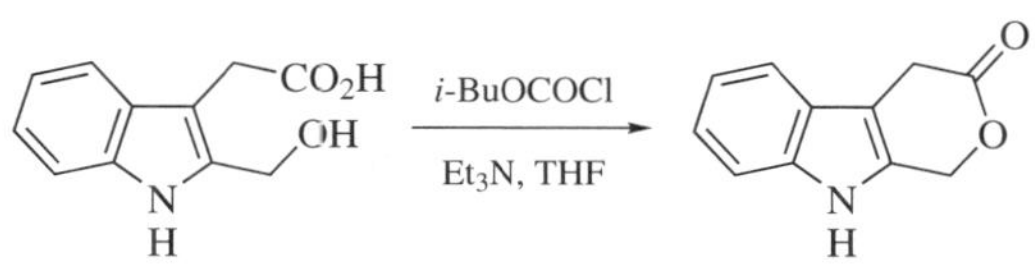
(4)

Regioselective Arylation.[11] Arylation in the 4-position in pyridines results from 1:1 adduct formation between an aryltriisopropoxytitanium reagent and an *N*-isobutyloxycarbonylpyridinium salt, and after successive 2,3-Dichloro-5,6-dicyano-1,4-benzoquinone dehydrogenation and cleavage of the 1-substituent (eq 5).

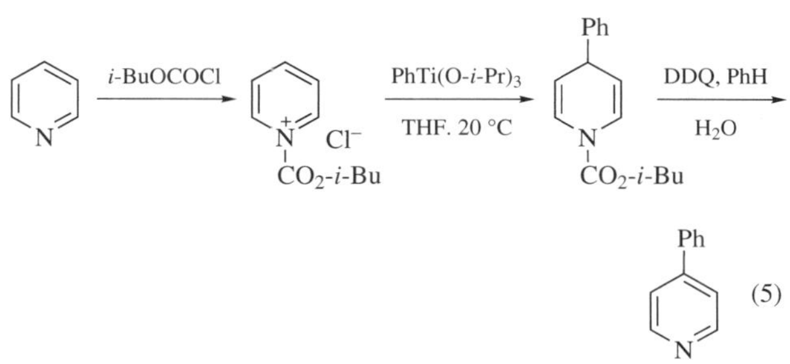

Isobutyl Sulfonate.[12] With Silver(I) p-Toluenesulfonate, alkyl chloroformate forms sulfonic carbonic anhydride which when heated liberates carbon dioxide, producing alkyl sulfonate. In order to find out the mechanism of the decomposition of the mixed anhydride, reaction of isobutyl chloroformate and silver *p*-toluenesulfonate was investigated. If the free alkyl cation, i.e. isobutyl carbonium ion, is formed, the rearrangement to *s*-butyl and *t*-butyl carbonium ion would be expected, finally resulting in the formation of *s*-butyl *p*-toluenesulfonate. *t-B*utyl *p*-toluenesulfonate is unstable and is not expected to survive under the reaction conditions. In fact, decomposition gave a mixture of isobutyl *p*-toluenesulfonate and *s*-butyl *p*-toluenesulfonate in a 0.8:1 mol ratio, with a quantity of isobutene and free *p*-toluenesulfonic acid. Isobutene was considered to be derived from the *t*-butyl cation.

1. Boissonnas, R. A. *Helv. Chim. Acta* **1951**, *34*, 874; *Helv. Chim. Acta* **1952**, *35*, 2229 and2237.
2. Vaughan, J. R. *J. Am. Chem. Soc.* **1951**, *73*, 3547; *J. Am. Chem. Soc.* **1952**, *74*, 676 and6137; *J. Am. Chem. Soc.* **1953**, *75*, 5556; *J. Am. Chem. Soc.* **1954**, *76*, 2474.
3. Wieland, T.; Bernhard, H. *Justus Liebigs Ann. Chem./Liebigs Ann. Chem.* **1951**, *572*, 190.
4. Ogilvie, K. K.; Letsinger, R. L. *J. Org. Chem.* **1967**, *32*, 2365.
5. Fray, E. B.; Moody, C. J.; Shah, P. *Tetrahedron* **1993**, *49*, 439.
6. Vaughan, J. R.; Osato, R. L. *J. Am. Chem. Soc.* **1952**, *74*, 676.
7. Anderson, G. W.; Zimmerman, J. E.; Callahan, F. M. *J. Am. Chem. Soc.* **1966**, *88*, 1336.ibid **1967**, *89*, 5012.
8. Hoffmann, J. A.; Tilak, M. A. *Org. Prep. Proced. Int.* **1975**, *7*, 215.
9. Kopple, K. D.; Renick, K. J. *J. Org. Chem.* **1974**, *39*, 660.
10. Nantz, M. H.; Lee, D. A.; Bender, D. M.; Roohi, A. H. *J. Org. Chem.* **1992**, *57*, 6653.
11. Gundersen, L.; Rise, F.; Undheim, K. *Tetrahedron* **1992**, *48*, 5647.
12. Yamamoto, A.; Kobayashi, M. *Bull. Chem. Soc. Jpn.* **1966**, *39*, 1283.

Tapan Ray
Sandoz Pharmaceuticals, East Hanover, NJ, USA

Isopropenyl Chloroformate

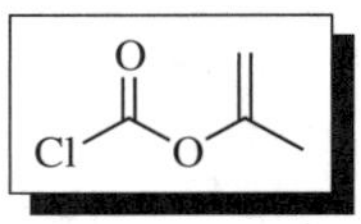

[57933-83-2] $C_4H_5ClO_2$ (MW 120.54)

(reagent used as an amino protecting group and in the coupling of amino acids)

Alternate Name: chlorocarbonic acid isopropenyl ester; carbonochloridic acid, 1-methylethenyl ester; IPCF.

Physical Data: bp 93 °C/746 mmHg; d 1.1007; Fp 9 °C; RI 1.415/589 nm; specific gravity 1.103.

Solubility: soluble in most organic solvents.

Form Supplied in: colorless liquid, stabilized with 50 to 100 ppm BHT; widely available.

Preparative Methods: can be generated by the addition of diacetonylmercury to excess phosgene in CH_2Cl_2, followed by distillation.

Handling, Storage, and Precautions: isopropenyl chloroformate is highly flammable; precautions should be taken against static discharge.[1] Toxic on inhalation; lachrymator; causes burns; should be handled in fume hood. It is moisture sensitive. The flash point for isopropenyl chloroformate is 9 °C, thus it should be stored below −4 °C, and vented periodically. Hazardous decomposition products include hydrogen chloride gas, carbon dioxide, and carbon monoxide.

General Utility. Isopropenyl chloroformate is an extremely versatile chloroformate for carboxylic acid activation via a mixed anhydride. IPCF was first synthesized in 1978, as one in a series of enol chloroformates, to launch an investigation into the potential utility of vinyl chloroformate derivatives.[2] Since then, IPCF has been used most commonly in esterification and peptide coupling reactions, but effectiveness has also been demonstrated with other nucleophiles and in oxazolone and glycosidic bond formations. Equation 1 illustrates some of the transformations detailed in this review. Additionally, halogenated derivatives of IPCF have been investigated for synthetic utility more recently, but they fall beyond the scope of this review.[3]

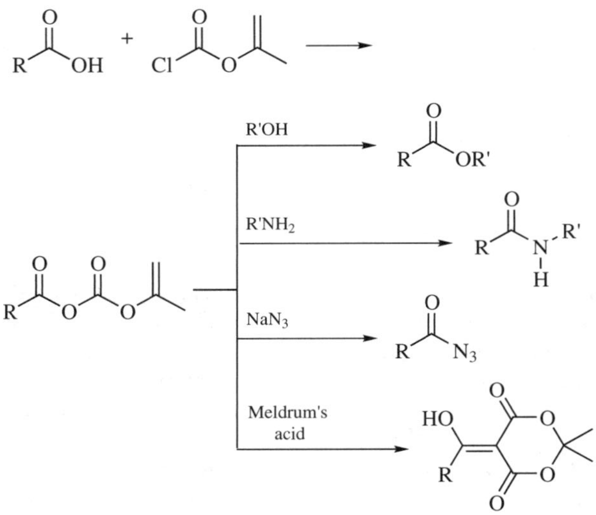

(1)

Hydroxyl Nucleophiles

Esterification of N-protected Amino Acids

Esters of *N*-protected α-amino acids are commonly prepared from alkyl halides or one of many other methods for activation of the acid toward nucleophilic attack. The use of IPCF allows for the mild, one-pot esterification of *N*-protected α-amino acids.[4,5] The initial reaction between IPCF and the amino acid substrate results in a mixed anhydride (eq 2), as seen with other alkyl chloroformates. Nucleophilic attack of an alcohol on the activated species yields the corresponding ester with acetone and carbon dioxide as by-products (eq 3). Triethylamine, pyridine, and/or catalytic dimethylaminopyridine (DMAP) are used as bases for this transformation. Interestingly, when this transformation is pursued with other alkyl chloroformates in the presence of DMAP, the DMAP-mediated activation of the intermediate mixed anhydride results in the liberation of the corresponding alkyl alcohol (originating from the chloroformate reagent); the alcohol then serves to displace the DMAP resulting in an ester. This reaction has been taken advantage of for the formation of esters,[6] but it is limited to commercially available or chemically accessible chloroformates. In contrast, when the isopropenyl-derived mixed anhydride is displaced by DMAP, acetone is liberated, which does not serve as a viable nucleophile, so the addition of 1 equiv of a separate alcohol can lead exclusively to the corresponding ester. In this sense, IPCF is seen as a more versatile chloroformate for acid activation.

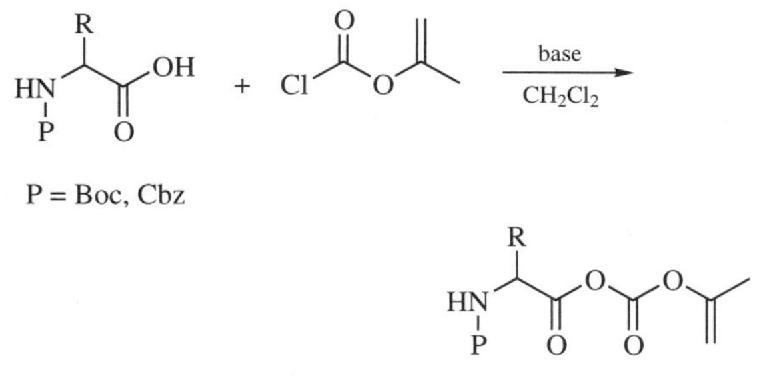

(2)

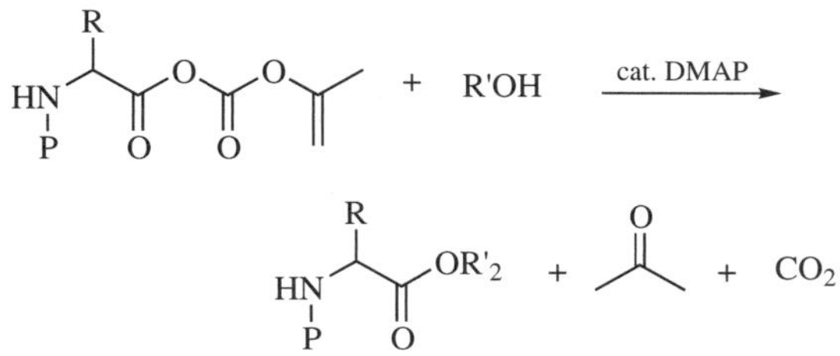

(3)

Primary, secondary, and tertiary alcohols have all been shown to act as effective nucleophiles in the esterification.[4,5] As an example, IPCF has been utilized as a coupling reagent to gain access to a variety of 1,3-diacylamino-2-(*O*-acyl)-propanes by acylation of the secondary alcohol (eq 4).[7] IPCF has also been used to prepare amino esters for the synthesis of the peptidic antibiotic valinomycin.[5]

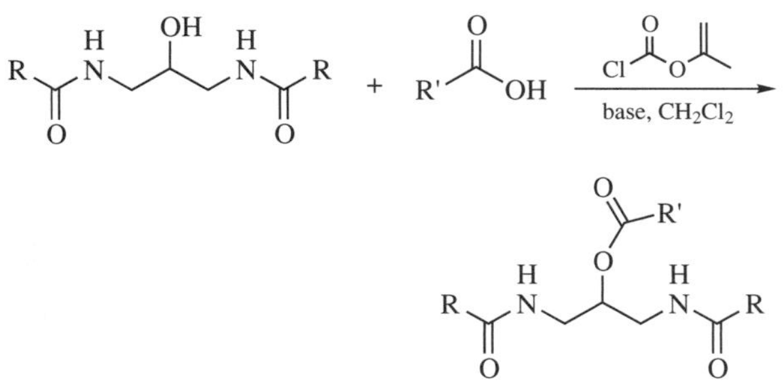

(4)

For similar reactivity principles, amino acids with unprotected hydroxyl or amine side chains, such as serine and threonine, were not successfully esterified at the *C*-terminus with IPCF. The IPCF-mediated esterification can be accomplished without racemization of the adjacent stereocenter, provided that a carbamate protecting group is installed on the amino terminus. Non-carbamate protecting groups (e.g., *N*-acyl) have been shown to result in significant racemization.[4,5]

Mono-esterification of Di-acids. As an extension of the methods for esterification of *N*-protected amino acids, Jouin et al.,[8] reported another application of IPCF for the mono-esterification of aspartic (n = 1) and glutamic (n = 2) di-acids (eq 5). Previous reports for the mono-esterification of these acids involved opening of an internal mixed anhydride that had been prepared from acetic anhydride[9] or carbodiimide.[10] The chloroformate-activated mono-esterifications were carried out in the presence of 1 equiv of chloroformate, 2 equiv of base, and/or catalytic DMAP. The reaction proceeded through a mixture of regioisomers of the activated anhydrides (**3a, 3b, 4a, 4b**), followed by decomposition to form a heterocyclic mixed anhydride (**5,6**). This active species was then opened by an alcohol to give a mixture of mono-esterified products (**7a, 7b, 8a, 8b**). As evidence for the proposed reaction pathway, the internal anhydride **5** was isolated, and it's reactivity toward alcohols verified.

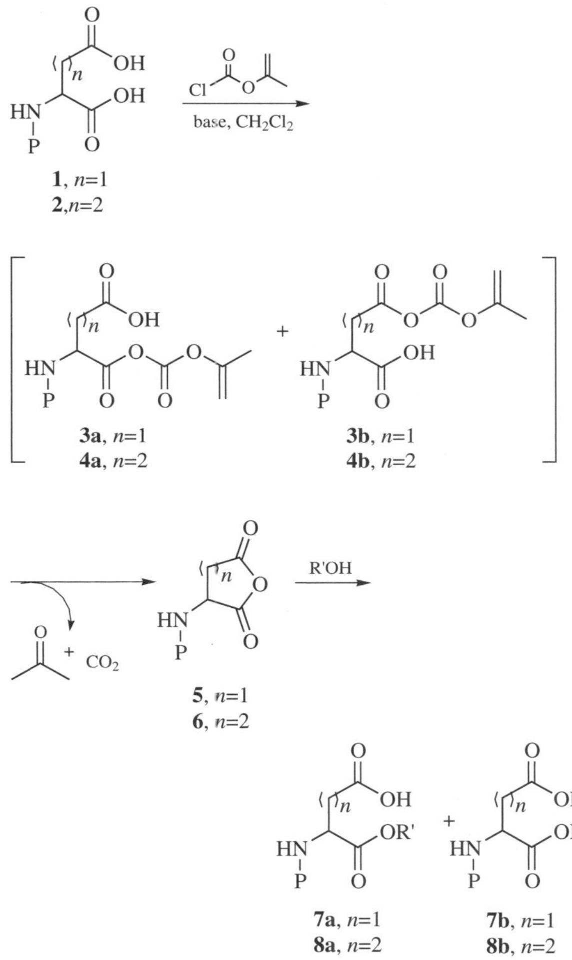

(5)

Other alkyl chloroformates were also investigated in this mono-esterification reaction and gave comparable reaction times and yields to IPCF. However, as observed in the esterification of amino acids, the reactions with various chloroformates proceed similarly, but are limited by the availability of the alkyl chloroformate. It was further demonstrated that by using IPCF access to lipophilic mono-esters and other unusual esters are possible.[8] The general regioselectivity, under a variety of conditions, favored the α-ester (**7a, 8a**). The one exception was when glutamic acid was reacted in the presence of DMAP; then the γ-mono-ester was favored. Parallel studies were done with dicyclohexylcarbodiimide (DCC) as the activating agent, and the regioselectivities under the same reaction conditions were identical to those observed with the chloroformates.

Lactonization. Isopropenyl chloroformate has also been applied to lactonization reactions via a mixed anhydride activation. One example is an IPCF-mediated lactonization as the key final step in the synthesis of a cyclic analog of the antimitotic agent Dolastatin 10 (eq 6). This conversion was accomplished with IPCF and DMAP in 26% yield.[11]

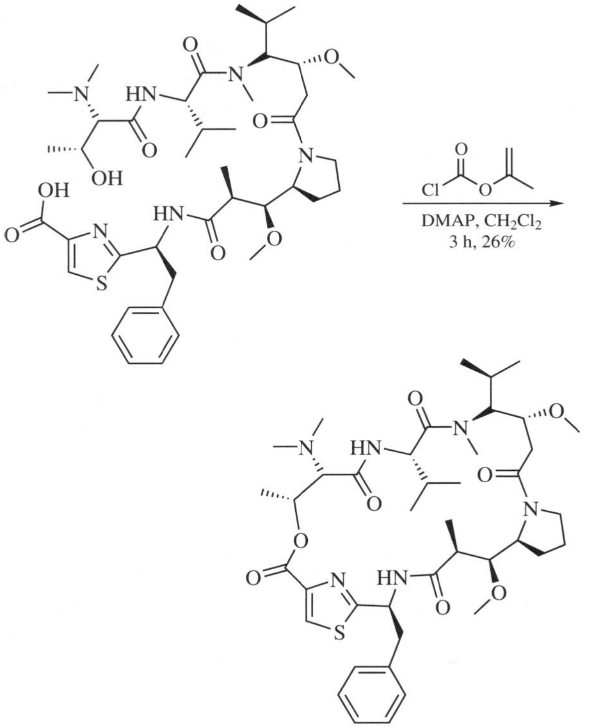

(6)

Another example is the lactonization to form the key bicyclic lactone bond in the synthesis of an active analog of *clasto*-lactacystin β-lactone (eq 7). This was accomplished using IPCF and triethylamine in a 75% to 79% yield.[12]

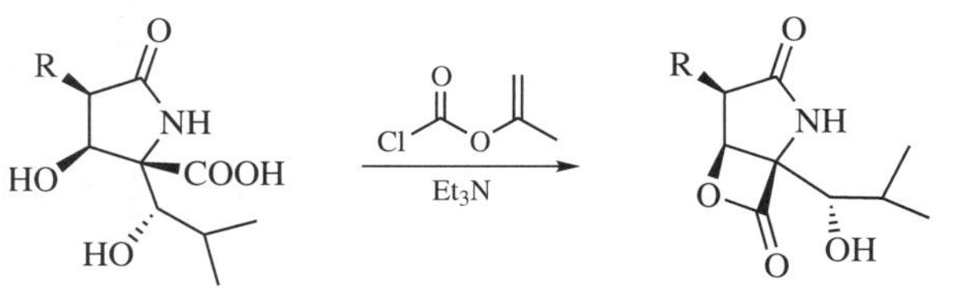

(7)

R = *n*-Pr, 75%
R = Me, 79%

Amine Nucleophiles. Peptide synthesis by aminolysis of a mixed anhydride dates back to the early 1950's.[13] Mixed anhydrides were prepared from the reaction of an *N*-protected amino acid with an alkyl chloroformate at low temperature in anhydrous solvent. In 1984, IPCF was used for activation of the carboxylic acid in dipeptide synthesis.[14] IPCF was deemed a superior choice over classical alkyl chloroformates because it reacted under a wider range of temperature conditions. In particular, the reagent allowed successful activation at ambient temperatures. As with the classical chloroformates, the use of IPCF is reserved for step by step coupling reactions in the liquid phase. IPCF is specifically recommended for the coupling of bulky non-functional residues.

The simplified mechanism for peptide synthesis involves two steps (eq 8). First, formation of a mixed anhydride results from a reaction between IPCF and an *N*-protected amino acid in the presence of an equivalent of base. Second, aminolysis of the anhydride by an unprotected amine yields the desired dipeptide (pathway a). A degenerative pathway also exists however, resulting from aminolysis at the carbonate moiety of the anhydride (pathway b).[15] Thus, carbamate formation leads to loss of the *C*-protected amino acid, and regeneration of the starting *N*-protected amino acid. This pathway is a well-known obstacle to peptide synthesis via mixed anhydrides.[16] Reaction conditions that minimize the carbamate product, or REMA conditions (repetitive excess of mixed anhydride),[17] have been studied. Jaouadi et al. adapted these conditions to the use of IPCF and were able to maximize the desired dipeptide yield to 96% while minimizing the carbamate product to 1.8%. The general REMA method involves using approximately 1.5 equiv of the mixed anhydride to 1 equiv of the free amine.

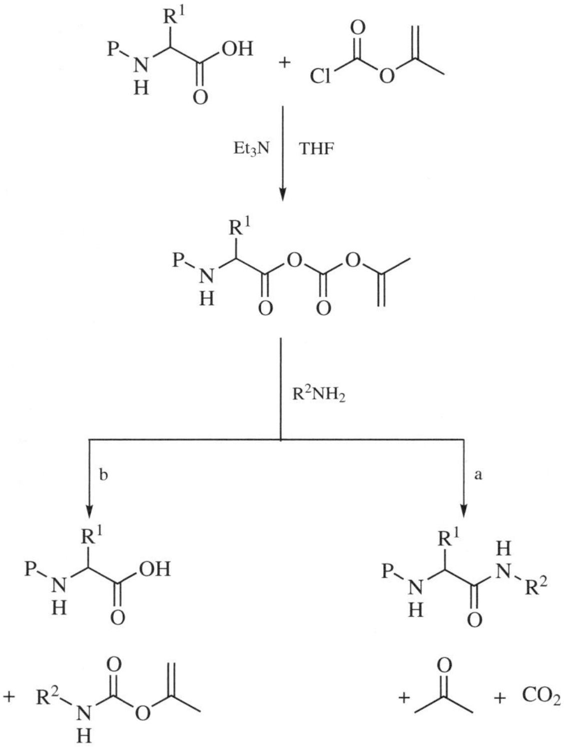

(8)

P = Boc, Cbz

An additional application of IPCF under the REMA conditions involves the synthesis of protected oligopeptides tuftsine (a tetrapeptide) and bradykinine (a nonapeptide).[18] The successful synthesis of the oligopeptides at ambient temperature with a minimization of the urethane product demonstrated the competitiveness of the IPCF-derived mixed anhydride method for the synthesis of peptides. Additionally, monitoring of the reaction intermediates revealed the absence of an oxazolone species, thus eliminating a potential racemization pathway.

In addition to the synthesis of peptides, other kinds of amides can be prepared from the reaction of a free acid with IPCF. One example is the use of IPCF to form the amide bond in the synthesis of a series of phosphodiesterase type-IV inhibitors.[19] The IPCF-derived mixed anhydrides of a variety of quinolone-3-carboxylic acids were treated with various primary amines to form the corresponding amides (eq 9). A rather narrow range of nitrogen and oxygen containing functionalities were tolerated at the R^3 position.

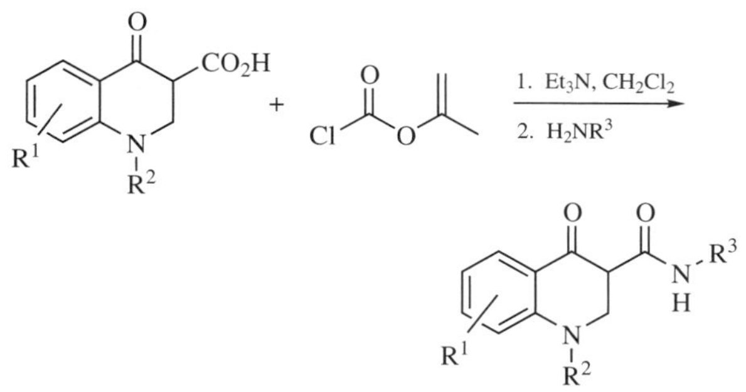

(9)

Activated Methylene Nucleophiles. IPCF has been utilized in a novel synthesis of a series of γ-amino-β-keto-esters and δ-amino-β-keto-esters involving a two-carbon extension of α- and β-amino acids, respectively.[20] The reactions proceed through an IPCF-derived mixed anhydride of an *N*-protected amino acid, followed by condensation with the activated methylene of Meldrum's acid to form the desired chain-extended intermediate (eq 10). Upon refluxing with 2 equiv of an alcohol in benzene,

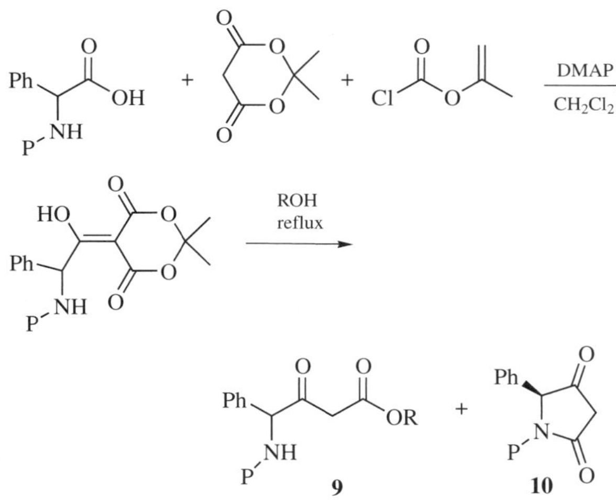

(10)

P	R	9:10	Yield (%)
Boc	Et	1:2	90
Boc	^{i}Pr	2:5	78
Boc	^{t}Bu	1:3	83
Cbz	Et	1:2	81

the corresponding esters (**9**) are formed. One disadvantage in the formation of the γ-amino-β-keto esters from α-amino acids is the generation of a major undesired product—the tetramic acid derived from cyclization of the chain extended intermediate **10**. However, this unwanted product is not observed in the synthesis of the δ-amino-β-keto-esters. In general, this transformation is efficient, high yielding, and tolerant of Boc, Cbz, and phthalimido *N*-protecting groups and affords a variety of γ-amino-β-keto-esters and δ-amino-β-keto-esters.

Oxazolone Formation. 2-Alkoxy-5(4*H*)-oxazolones can undergo aminolysis without racemization to give chirally homogeneous peptides. 2-Alkoxy-5(4*H*)-oxazolones can be regarded as the ideal activated form of the original acids as nothing is liberated during their aminolysis, and they couple efficiently and stereoselectively with amino acid esters.[21] In fact, a common activated form of some symmetrical amino acid anhydrides, mixed anhydrides, and Fmoc-amino acid chlorides is the oxazolone.[21] A general method, therefore, for the preparation of these oxazolones from mixed anhydrides was desired.

The ease of oxazolone formation is dependent on the stability of the alkyl group on the carbonate moiety of the anhydride. Simple alkyl groups such as methyl, ethyl, and methylsulfonylethyl cyclize more readily, but the heterocycle is also more reactive toward the resulting alcohol to form the corresponding ester. Liberated secondary alcohols are less likely to open the oxazolone, but the parent mixed anhydrides also cyclize much slower. Alternatively, the IPCF-derived mixed anhydrides of *N*-alkoxy carbonyl amino acids cyclize to the oxazolones faster than the corresponding ethyl and isopropyl anhydrides, and the by-products of cyclization (acetone and carbon dioxide) are not reactive toward the product. The IPCF mixed anhydrides are prepared from the desired protected acids and IPCF, and after isolation they are allowed to sit in chloroform or DMF for 0.5–24 h (eq 11). The resultant oxazolones are formed in moderate to high yields, isolated and used as valuable intermediates in peptide couplings.

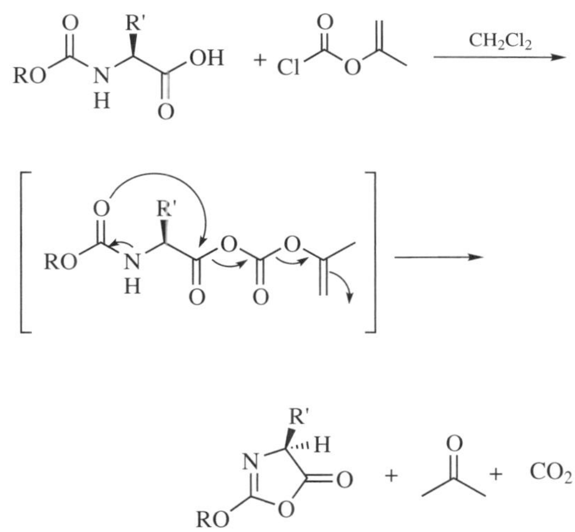

(11)

R = PhCH$_2$, (CH$_3$)$_3$C, 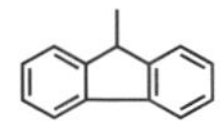

Azide Nucleophiles. In the same way that IPCF activates carboxylic acids for nucleophilic attack by hydroxyl and amine functionalities, acids can also be activated for attack by azide. An example lies in the conversion of amido acids into pseudo-peptides containing a *cis*-2-amino-cyclopropane carboxylate unit (eq 12).[22] The activation of the carboxylic acid toward nucleophilic attack by azide, followed by a Curtius rearrangement[5,23] to form the desired isocyanate is a key step in the synthesis of several of the desired peptides.

Glycosidic Linkage Formation. Isopropenyl glycosides, a novel class of glycosyl donor, can undergo glycosylation reactions with primary and secondary carbohydrate alcohols to form disaccharides with good β-selectivity.[24] The isopropenyl glycosides arise from treatment of the corresponding acetates with the Tebbe reagent. The Tebbe procedure can be too harsh, however, for some substrates, so the need for an alternative method was apparent. Therefore, a route to disaccharides utilizing isopropenyl glycosyl carbonates (**11**) as glycosylating agents was developed.[24] Isopropenyl glycosyl carbonates react with primary and secondary carbohydrate acceptors to give disaccharides in high yield and with good β-selectivity (eqs 13 and 14). The isopropenyl glycosyl carbonates are prepared in nearly quantitative yield by treatment of the hemiacetal with IPCF and pyridine in methylene chloride at 0 °C for 1 h.

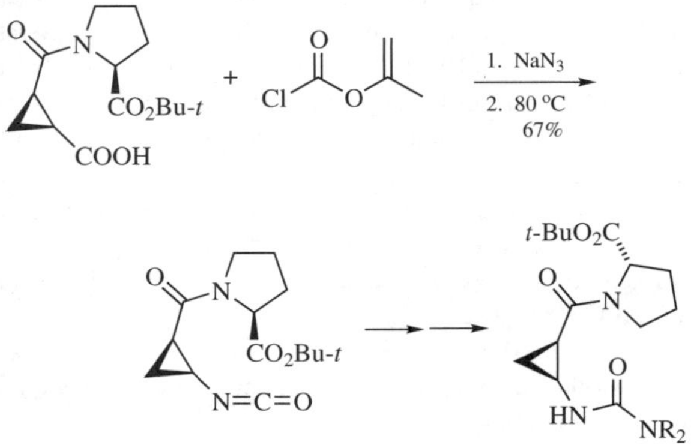

(12)

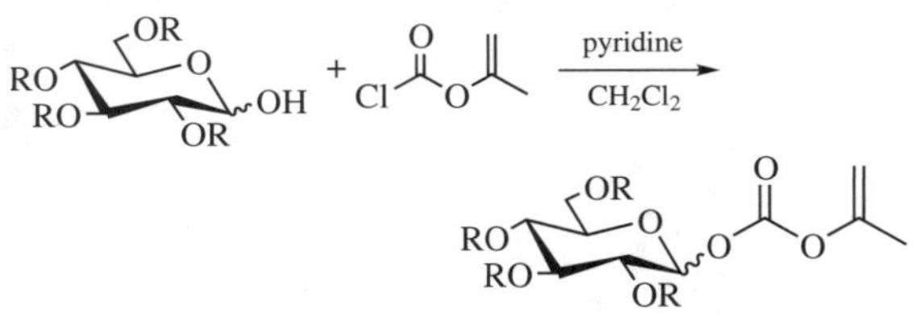

(13)

11

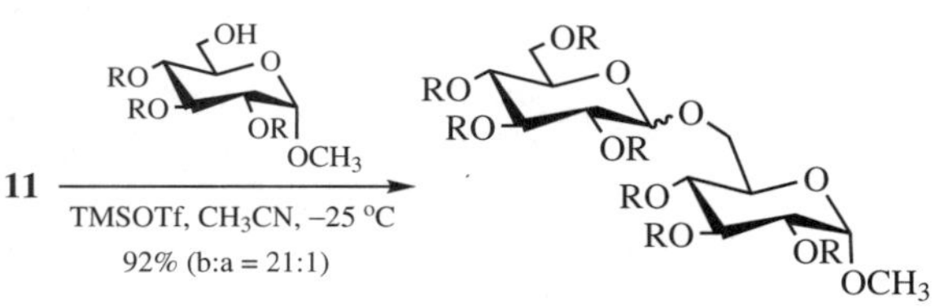

(14)

Related Reagents. Vinyl chloroformate; allyl isopropenyl dicarbonate; aryl isopropenyl dicarbonate; 2,2-dichlorovinyl chloroformate.

1. Website: www.msdssolutions.com; search for isopropenyl chloroformate.
2. Olofson, R. A.; Bauman, B. A.; Wancowicz, D. J., *J. Org. Chem.* **1978**, *43*, 752.
3. Bowman, M. P.; Senet, J. P. G.; Malfroot, T.; Olofson, R. A., *J. Org. Chem.* **1990**, *55*, 5982.
4. Jouin, P.; Castro, B.; Zeggef, C.; Pantaloni, A.; Senet, J. P.; Lecolier, S.; Sennyey, G., *Tetrahedron Lett.* **1987**, *28*, 1661.
5. Zeggaf, C.; Poncet, J.; Jouin, P.; Dufour, M.-N.; Castro, B., *Tetrahedron* **1989**, *45*, 5039.
6. Dhaon, M. K.; Olsen, R. K.; Ramasamy, K., *J. Org. Chem.* **1982**, *47*, 1962.
7. Mergen, F.; Lambert, D. M.; Poupaert, J. H.; Bidaine, A.; Dumont, P., *Chem. Phys. Lipids.* **1991**, *59*, 267.
8. Jouin, P.; Castro, B.; Zeggaf, C.; Pantaloni, A.; Senet, J. P.; Lecolier, S.; Sennyey, G., *Tetrahedron Lett.* **1987**, *28*, 1665.
9. Riniker, B.; Schwyzer, R., *Helv. Chim. Acta.* **1964**, *47*, 2357.
10. Schroeder, E.; Klieger, E., *Justus Liebigs Ann. Chem.* **1964**, *673*, 208.
11. Poncet, J.; Hortala, L.; Busquet, M.; Guéritte-Voegelein, F.; Thoret, S.; Pierré, A.; Atassi, G.; Jouin, P., *Bioorg. Med. Chem. Lett.* **1998**, *8*, 2855.
12. Soucy, F.; Grenier, L.; Behnke, M. L.; Destree, A. T.; McCormack, T. A.; Adams, J.; Plamondon, L., *J. Am. Chem. Soc.* **1999**, *121*, 9967.
13. (a) Wieland, T.; Bernhard, H., *Justus Liebigs Ann. Chem.* **1951**, *572*, 190. (b) Boissonnas, R. A., *Helv. Chim. Acta.* **1951**, *34*, 874. (c) Vaughan, J. R., *J. Am. Chem. Soc.* **1951**, *73*, 3547.
14. Jaouadi, M.; Selve, C.; Dormoy, J. R.; Castro, B., *Bull. Soc. Chim. Fr.* **1984**, *9–10*, 409.
15. Chen, F. M. F.; Benoiton, N. L., *Can. J. Chem.* **1987**, *65*, 619.
16. (a) Battersby, A. R.; Robinson, J. C., *J. Chem. Soc.* **1955**, 259. (b) Albertson, N. F., *Org. React.* **1962**, *12*, 157.
17. Van Zon, A.; Beyerman, H. C., *Helv. Chim. Acta.* **1976**, *59*, 1112.
18. Jaouadi, M.; Selve, C.; Dormoy, J. R.; Castro, B., *Bull. Soc. Chim. Fr.* **1988**, *5*, 870.
19. Beasley, S. C.; Cooper, N.; Gowers, L.; Gregory, J. P.; Haughan, A. F.; Hellewell, P. G.; Macari, D.; Miotla, J.; Montana, J. G.; Morgan, T.; Naylor, R.; Runcie, K. A.; Tuladhar, B.; Warneck, J. B. H., *Bioorg. & Med. Chem. Lett.* **1998**, *8*, 2629.
20. Li, B.; Franck, R. W., *Bioorg. & Med. Chem. Lett.* **1999**, *9*, 2629.
21. Benoiton, N. L.; Chen, M. F. M., *Int. J. Peptide Protein Res.* **1993**, *42*, 455.
22. Jones, I.; Jones, W.; North, M.; Teijeira, M.; Uriarte, E., *Tetrahedron Lett.* **1997**, *38*, 889.
23. (a) Canonne, P.; Akssira, M.; Dahdouh, A.; Kasmi, H.; Boumzebra, M., *Tetrahedron* **1993**, *49*, 1985. (b) Furuta, K.; Hayashi, S.; Miwa, Y.; Yamamoto, H., *Tetrahedron Lett.* **1987**, *28*, 5841. (c) Arai, Y.; Kawanami, S.; Koizumi, T., *J. Chem. Soc., Perkin Trans. 1* **1991**, 2969.
24. Marra, A.; Esnault, J.; Veyriéres, A.; Sinay, P., *J. Am. Chem. Soc.* **1992**, *114*, 6354.

Clara E. Schous, Weiquang Huang, Alicia Thatcher & Marvin J. Miller
University of Notre Dame, IN, USA

L

Levulinic Anhydride

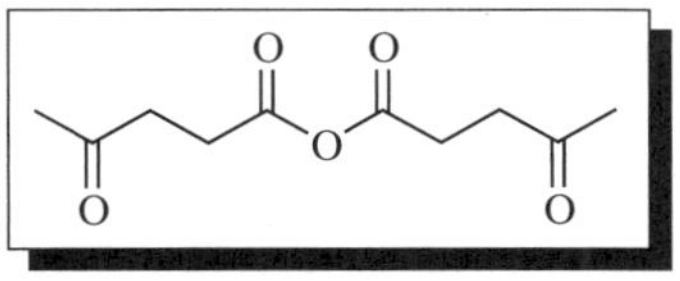

[40608-06-8] $C_{10}H_{14}O_5$ (MW 214.22)

(reagent used to protect hydroxyl groups and some amino groups)

Alternate Name: 4-oxopentanoic anhydride.
Physical Data: bp 330.6 ± 27.0 °C; fp 145.4 ± 42.8 °C.
Solubility: soluble in most organic solvents.
Form Supplied in: colorless liquid, not commercially available
Preparative Methods: Levulinic anhydride can be obtained in quantitative yield by reaction of levulinic acid (20 mmol) with dicyclohexylcarbodiimide (10 mmol) in 65 mL of diethyl ether for 5 h followed by filtration to remove the solid side products and evaporation of the solvent under reduced pressure.[1]
Handling, Storage, and Precautions: irritating; sensitive to moisture. Store under a nitrogen atmosphere in a cool, dry place.

The levulinoyl group has been widely utilized to protect hydroxyl groups temporarily in the synthesis of a variety of structures, especially those having multiple hydroxyl groups such as carbohydrates,[2–5] nucleotides,[6–9] and steroids,[10] due to that fact that it can be selectively removed under very mild conditions without affecting other ester linkages and functional groups. The carbonyl group of levulinoyl group has also been used as a molecular handle to couple with other molecules via ligation methods in biological studies.[11,12]

Levulinates are normally prepared by the reaction of levulinic anhydride with substrate alcohols in pyridine (eq 1), whereas levulinic chloride is not usually used for this purpose as it mainly leads to unstable pseudo esters of levulinic acid.[13,14] Levulinic anhydride was first introduced as a levulinating reagent in early 1970s.[1,15] A typical procedure for its introduction involves dropwise addition of excess levulinic anhydride to a solution of the substrate alcohol dissolved in the mixture of an organic solvent and pyridine [often with 4-dimethylaminopyridine (DMAP) as a catalyst]. The solution is stirred at room temperature under an inert atmosphere until the reaction is complete as shown by TLC. Ice water is added to quench the reaction, and the mixture is extracted with chloroform. The organic layer is washed with brine, dried over sodium sulfate, and finally concentrated in vacuo to afford the expected levulinate which may be purified further by standard methods.

Primary alcohols can be selectively acylated with levulinic anhydride in the presence of secondary hydroxyl groups.[16,17] Regioselectivities range from moderate to very good, depending on the properties and structures of the alcohols (eq 2).

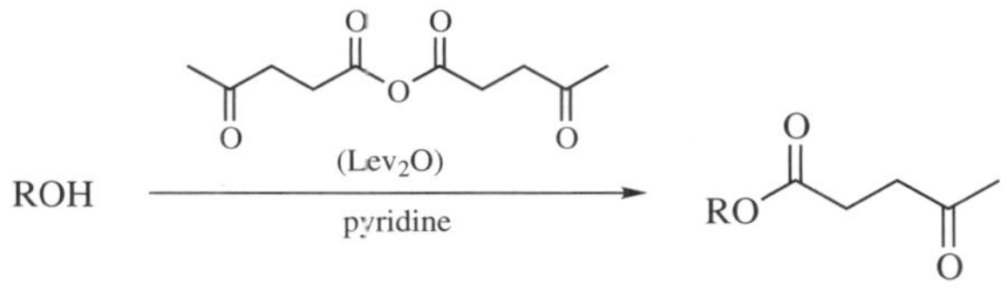

(1)

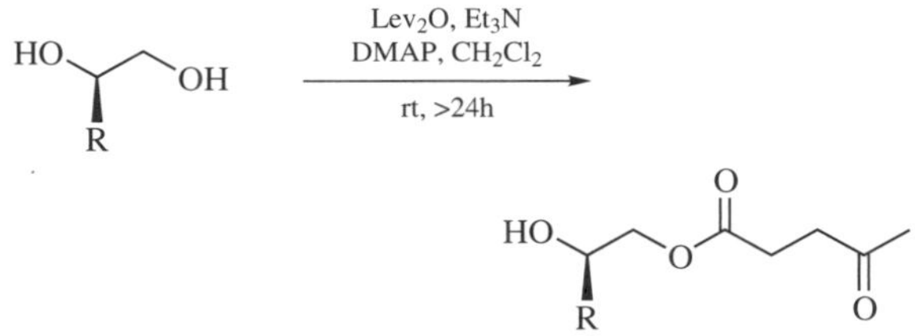

(2)

R = Me, Ph, $(CH_2)_2COOMe$
Yield 60–80%

Like most esters, levulinates are rather stable to acid, but not to base. However, the levulinoyl group can be readily and selectively removed under very mild conditions without affecting other acyl or base labile groups. Several methods have been developed in this respect. For example, Hassner et al.[1] recommend the use of sodium borohydride ($NaBH_4$) to remove the levulinoyl group (eq 3); the alkoxide anion formed by $NaBH_4$ reduction of the carbonyl group attacks the ester bond to give a γ-lactone with the release of the alcohol. Ono's method is based upon a similar mechanism but uses sulfite ion as the nucleophile (eq 4).[18] This deprotection can be carried out under almost neutral conditions.

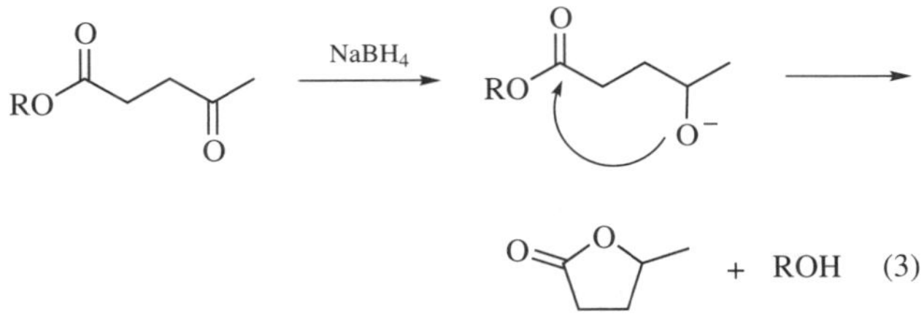

(3)

The most common method for the removal of levulinoyl group, developed by van Boom and Burgers,[6,17] employs hydrazine as the nucleophile (eq 5). The reaction, which completes in minutes, has no effect on other esters, and the products are isolated in quantitative yield. Typically, a levulinate is dissolved in the solution of hydrazine monohydrate (0.5 M) in pyridine and acetic acid (4:1) at room temperature under an inert atmosphere. After the starting material is completely consumed (usually in just a few minutes) the mixture is diluted with chloroform and then washed with water and brine. The organic layer is dried over sodium sulfate and concentrated in vacuo. The residue is purified by flash chromatography or recrystallization to afford the expected alcohol.

The convenient attachment and removal of levulinoyl group make it a very useful protecting group in oligosaccharide and oligonucleotide synthesis,[2–9] including solid-phase synthesis.[19,20] It has been extensively employed to differentiate one specific hydroxyl group from other hydroxyl groups in a building block or an intermediate. Moreover, the mild conditions used to remove levulinoyl group make it especially useful in the synthesis of carbohydrates containing sensitive functional groups.[5,21,22]

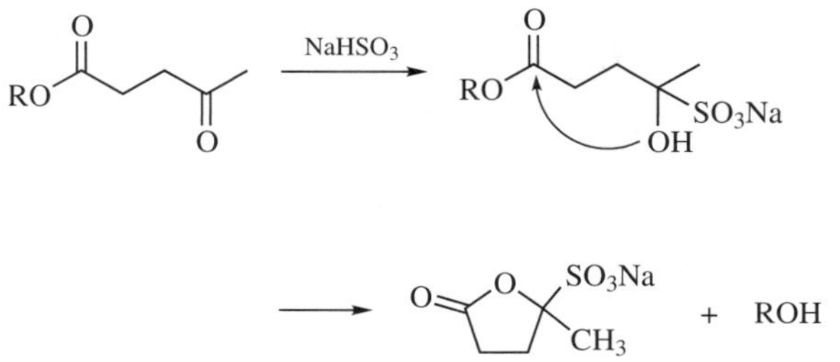

(4)

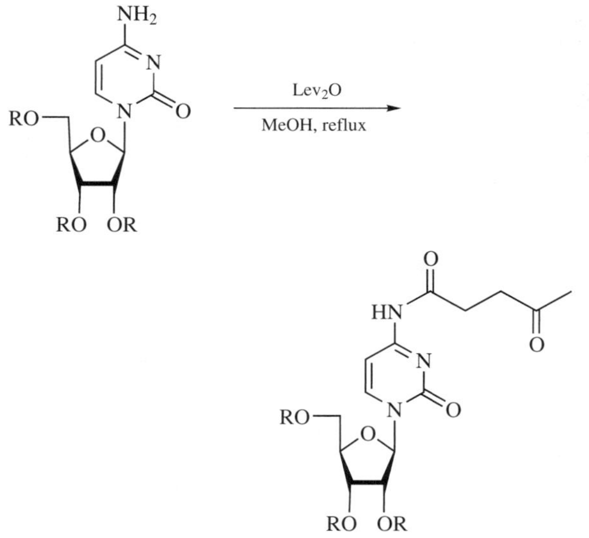

(6)

The levulinoyl group also has been used to protect amino groups in the synthesis of nucleotides.[23] The acylation of the amino groups of cystidine, adenosine, and guanosine has been studied. The reaction requires harsher conditions, e.g., refluxing with levulinic anhydride in methanol (eq 6), than for the acylation reactions of alcohols. The *N*-levulinoyl group is stable to all general derivatization techniques employed in nucleotide synthesis, while it can be readily removed by the standard hydrazine method in 5–15 min.

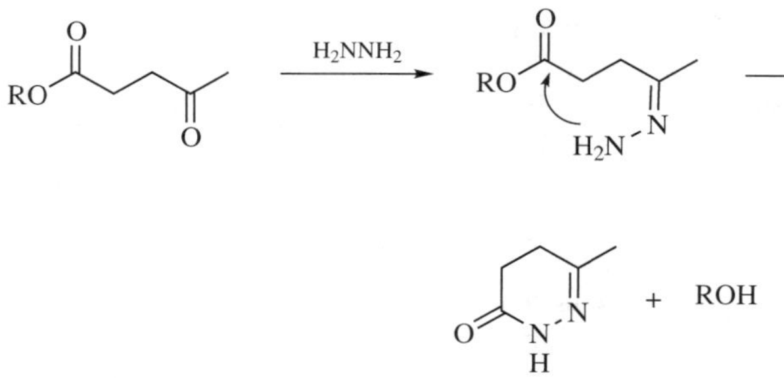

(5)

The presence of a carbonyl group in levulinoyl group facilitates the attachment of molecules to the levulinated structures via selective coupling reactions, such as condensation with amines and reductive aminations. This property is very useful in the preparation of bioconjugates. For instance, by using *N*-levulinoyl-D-mannosamine as an artificial substrate of sialoglycoconjugate biosynthetic machineries, Bertozzi and co-workers[11] metabolically delivered levulinoyl groups onto cell surfaces. The hydrazine-functionalized biological epitopes were then covalently attached to cells by selective reactions with the delivered carbonyl groups (eq 7). This strategy can be used to bioengineer cells for various purposes.

1. Hassner, A.; Strand, G.; Rubinstein, M.; Patchornik, A., *J. Am. Chem. Soc.* **1975**, *97*, 1614.
2. Koeners, H. J.; Verhoeven, J.; Van Boom, J. H., *Recl. Trav. Chim. Pays-Bas* **1981**, *100*, 65.
3. Lergenmüller, M.; Ito, Y.; Ogawa, T., *Tetrahedron* **1998**, *54*, 1381.
4. Tamura, J.; Nishihara, J., *J. Org. Chem.* **2001**, *66*, 3074.
5. Ito, H.; Ishida, H.; Collins, B. E.; Fromholt, S. E.; Schnaar, R. L.; Kiso, M., *Carbohydr. Res.* **2003**, *338*, 1621.
6. Van Boom, J. H.; Burgers, P. M. J., *Tetrahedron Lett.* **1976**, 4875.
7. Grøtli, M.; Eritja, R.; Sproat, B., *Tetrahedron* **1997**, *53*, 11317.
8. Sørensen, M. D.; Meldgaard, M.; Rajwanshi, V. K.; Wengel, J., *Bioorg. Med. Chem. Lett.* **2000**, *10*, 1853.
9. Kamaike, K.; Namiki, T.; Kawashima, E., *Nucleosides, Nucleotides & Nucleic Acids* **2003**, *22*, 469.
10. Verdegaal, C. H. M.; Jansse, P. L.; De Rooij, J. F. M.; Van Boom, J. H., *Tetrahedron Lett.* **1980**, *21*, 1571.
11. Mahal, L. K.; Yarema, K. L.; Bertozzi, C. R., *Science* **1997**, *276*, 1125.
12. Bertozzi, C. R.; Kiessling, L. L., *Science* **2001**; *291*, 2357.
13. Newman, M. S.; Gill, N.; Darre, B., *J. Org. Chem.* **1966**, *31*, 2713.
14. Fife, T. H., *J. Am. Chem. Soc.* **1965**, *87*, 271.
15. Guthrie, R. D.; Lucas, T. J.; Khan, R., *Carbohydr. Res.* **1974**, *33*, 391.
16. Hall, D. G.; Schultz, P. G., *Tetrahedron Lett.* **1997**, *38*, 7825.
17. Van Boom, J. H.; Burgers, P. M. J., *Recl. Trav. Chim. Pays-Bas* **1978**, *97*, 73.
18. Ono, M.; Itoh, I., *Chem. Lett.* **1988**, 585.

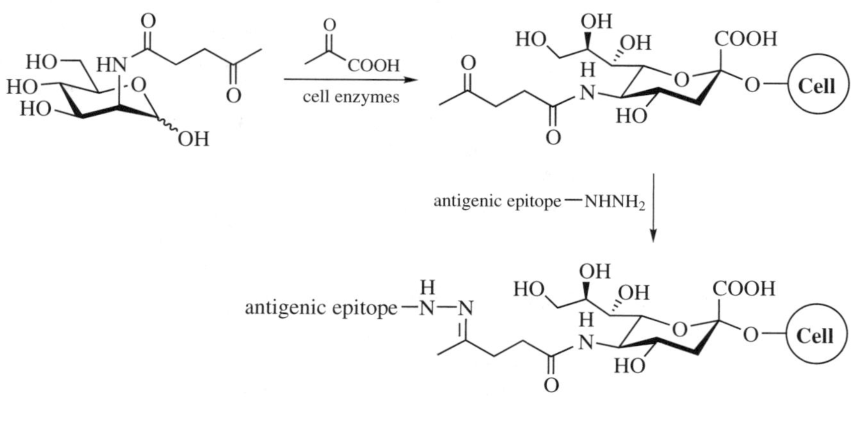

(7)

19. Urdea, M. S.; Horn, T., *Tetrahedron Lett.* **1986**, *27*, 2933.

20. Melean, L. G.; Love, K. R.; Seeberger, P. H., *Carbohydr. Res.* **2002**, *337*, 1893.

21. Nakano, T.; Ito, Y.; Ogawa, T., *Tetrahedron Lett.* **1991**, *32*, 1569.

22. Goto, F.; Ogawa, T., *Tetrahedron Lett.* **1992**, *33*, 5099.

23. Ogilvie, K. K.; Nemer, M. J.; Hakimelahi, G. H.; Proba, Z. A.; Lucas, M., *Tetrahedron Lett.* **1982**, *23*, 2615.

Zhongwu Guo & Jie Xue
Case Western Reserve University, Cleveland, OH, USA

M

6-Mercapto-4-dibenzofuranol

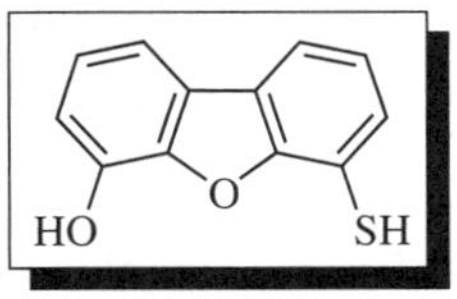

[101697-54-5] $C_{12}H_8O_2S$ (MW 216.26)

(peptide synthesis by prior thiol capture; solid phase synthesis of polypeptides)

Alternate Name: 4-hydroxy-6-mercaptodibenzofuran.

Introduction. 6-Mercapto-4-dibenzofuranol is a reagent developed specifically[1] as an optimal spacing element to facilitate the intramolecular *O,N* acyl transfer during the second stage of the amide-bond-forming protocol known as the "thiol capture strategy." As a result, it plays a special role in a relatively new solid phase polypeptide synthesis strategy.

Peptide Synthesis by Prior Thiol Capture. A coupling method has been designed to be used at the late stages of fragment condensation syntheses of large peptides that contain cysteine residues spaced 5–20 amino acid residues apart.[1,2] The method takes advantage of an N-terminal cysteine residue. The rationale of the synthesis[3] calls for a template that puts the participating fragments in the right proximity, with the right orientation. The fundamental premise[4] is that a rationally designed system that displays the appropriate distance and orientation of the two reaction centers can induce rapid intramolecular acyl transfer to a weakly basic (p*K*a ca. 7.0) cysteine amino function from a very weakly activated phenyl ester across a span of at least nine atoms. After examination of several templates,[4,5] 6-mercapto-4-dibenzofuranol emerged as the appropriate choice. The thiol capture strategy involves three stages described in eq 1.[1,2]

Step A (eq 1)—the disulfide bond formation between a peptide fragment derivatized at the C-terminus as 6-mercaptodibenzofuranyl ester and a peptide fragment bearing an *S*-sulfenylthiocarbonate (Scm) activated cysteine residue at its N-terminus, occurs cleanly and very rapidly at submillimolar concentrations in protic solvents or solvent mixtures. The intramolecular acyl transfer (step B, eq 1) is highly efficient (effective molarity > 1 M),[5,6] to the effect that, for most cases, a very weakly activated phenolic ester can be employed and many quite reactive side-chain functions in R and R′ can be left unprotected.[1] These high reactivity criteria are met best in the protein-solubilizing solvent hexafluoro-2-propanol (HFIP) and its mixtures with water and acetonitrile.[2]

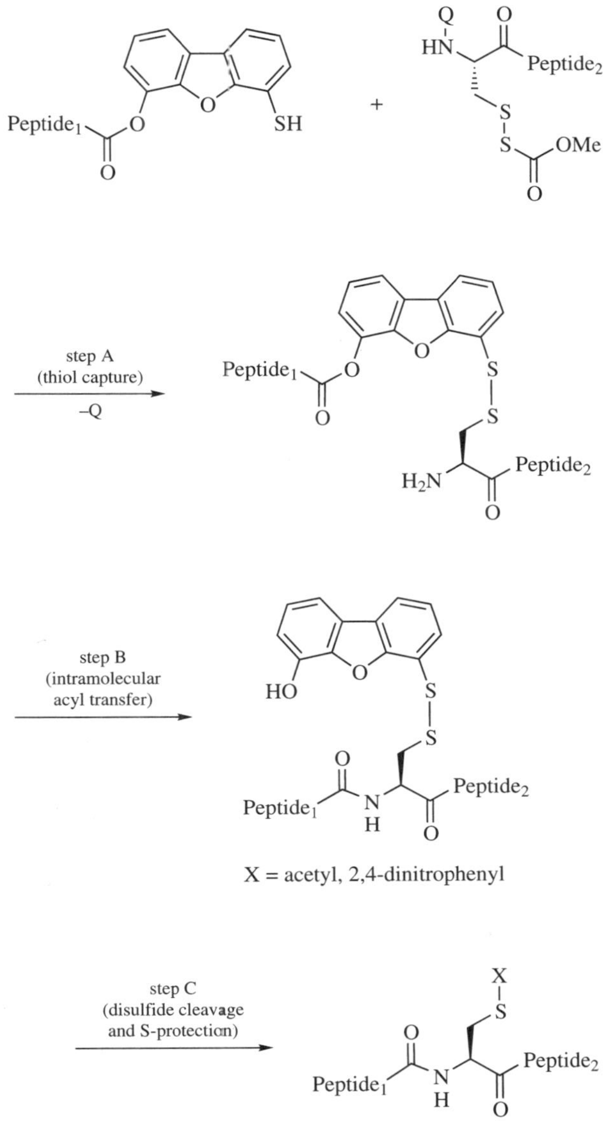

(1)

The use of a phenolic ester as anchoring group limits the choice of the protective groups. An appropriate choice is *tert*-butyloxycarbonyl (Boc, trifluoroacetic acid cleavable) for N^{α}-blocking of the amino acid esterified with the phenol and functionalized benzyl groups for side-chain functions.[1] An alternative (preferred in many applications) is *tert*-butyl derived side-chain blocking groups and 2-p-biphenylyl-2-propyloxycarbonyl (Bpoc) for N^{α}-protection.[2]

The resin-bound fragment preparation starts with the chlorine displacement from the chloromethylene function on chloromethylated polystyrene by the cesium salt of *Z*-L-Cys(Trt)-OH. The resulting benzyl ester is treated with methoxycarbonyl sulfenyl chloride (Scm-Cl) to form the active intermediate sulfenyl thiocarbonate which is allowed to react with 4-acyloxy-6-mercaptodibenzofuran, preferably in dichloromethane–hexafluoro-2-propanol, with or without catalytic triethylamine to afford the unsymmetrical disulfide (eq 2).[1,7]

An alternative to acetamidomethyl (Acm) and trityl (Trt) for the temporary protection for cysteine is 2,2-dimethylthiazolidine-4-carboxylic acid (H-Dmt-OH, used as the anhydride derived from Boc-Dmt-OH or as Boc-Dmt-OPfp).[8]

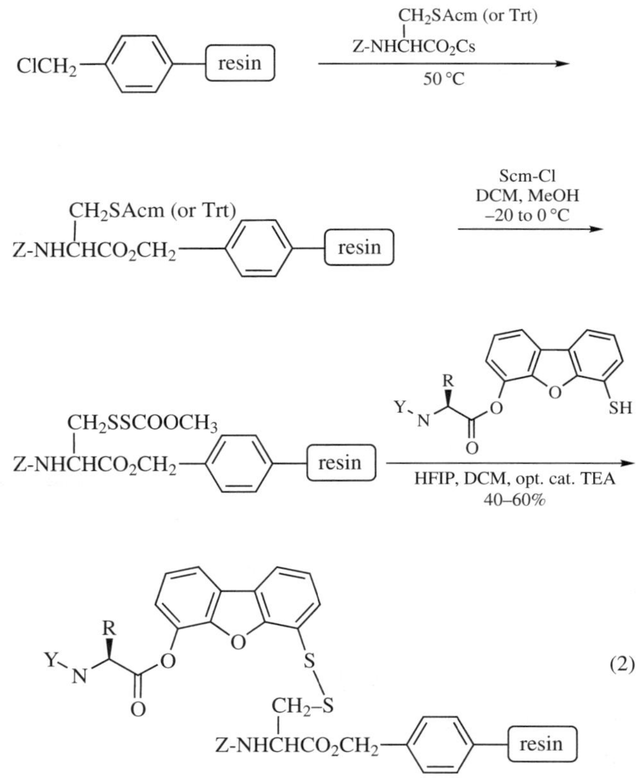

(2)

It is often simpler to modify the sequence in eq 2 to the one in which the resin-bound sulfenyl thiocarbonate intermediate is first reacted with 4-hydroxy-6-mercaptodibenzofuran. The free phenol is then acylated by 3–4 equiv of a symmetrical anhydride derived from an N^{α}-carbamate-blocked amino acid in the presence of di-isopropylethylamine in dichloromethane (ca. 10–30 min reaction time at 0 °C) (eq 3).[1]

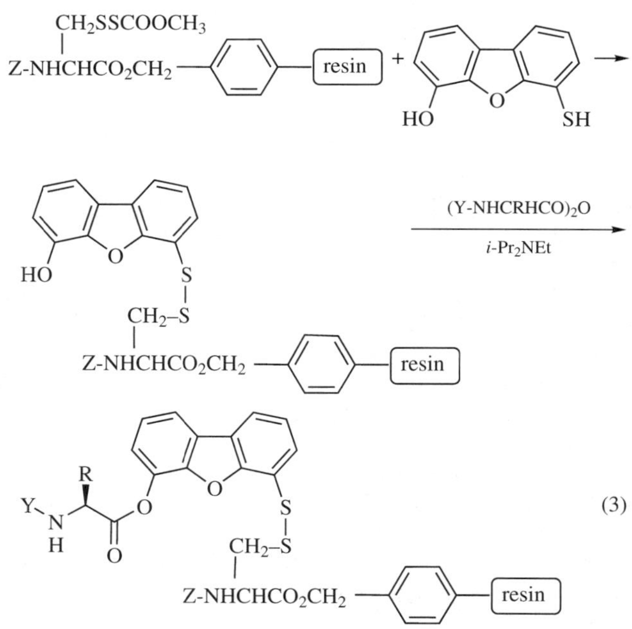

(3)

Y = Boc, Bpoc

A study of the effect of the steric bulk of the side-chain substituent of the acylating agent (R in eq 3) during the second stage of the process (step B, eq 1) revealed that the acyl transfer rates are anomalously low in the case of Pro and especially Val. Also, Asp shows evidence of intramolecular general base catalysis by the neighboring carboxylate group. For Ala, Leu, Lys(Z), Asn, and Arg(9-anthracenesulfonyl) the half-life of the acyl transfer lies between 2 and 4 h when running the reactions in DMSO at room temperature.[9] Racemization during the thiol capture protocol appears to be insignificant.[10]

The reductive cleavage of the disulfide bond (step C, eq 1) can be performed cleanly and with high reaction rates (almost instantaneously) by treatment with 1 equiv of tributylphosphine or other trialkylphosphines in dioxane-water at room temperature. An acidic medium (TFA) is required.[1]

Solid Phase Synthesis of Polypeptides. A reaction sequence has been designed for the solid-phase synthesis of a polypeptide esterified at the C-terminus with the phenol functionality in 4-hydroxy-6-mercaptodibenzofuran and resin bound with a disulfide bond (eq 4).[1] There are certain liabilities associated with the method (the lability of the disulfide bond, that of the phenyl ester, cyclization with formation of diketopiperazides after the amine is liberated in the second chain elongation step, etc.). A detailed protocol has been designed to circumvent these issues.[1,11]

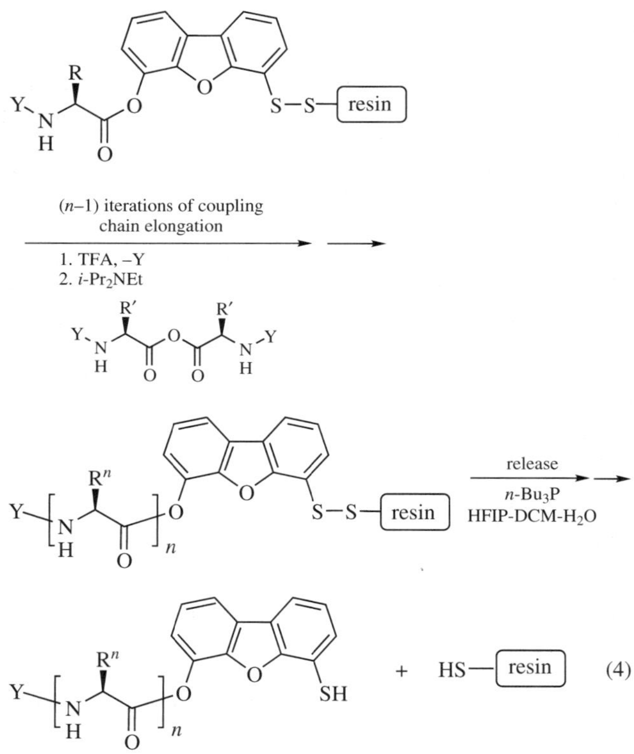

(4)

Related Reagents. 4-Methoxy-*a*-[{[(4-methylphenyl)methyl]thio}methyl]benzenemethanamine.

1. Kemp, D. S.; Galakatos, N. G., *J. Org. Chem.* **1986**, *51*, 1821.
2. Fotouhi, N.; Galakatos, N. G.; Kemp, D. S., *J. Org. Chem.* **1989**, *54*, 2803.
3. Kemp, D. S., *Biopolymers* **1981**, *20*, 1793.
4. Kemp, D. S.; Galakatos, N. G.; Bowen, B.; Tan, K., *J. Org. Chem.* **1986**, *51*, 1829.

5. Kemp, D. S.; Carey, R. I.; Dewan, J. C.; Galakatos, N. G.; Kerkman, D.; Leung, S.–L., *J. Org. Chem.* **1989**, *54*, 1589.
6. Kemp, D. S.; Carey, R. I., *J. Org. Chem.* **1993**, *58*, 2216.
7. Brois, S. J.; Pilot, J. F.; Barnum, H. W., *J. Am. Chem. Soc.* **1970**, *92*, 7629.
8. Kemp, D. S.; Carey, R. I., *J. Org. Chem.* **1989**, *54*, 3640.
9. Kemp, D. S.; Galakatos, N. G.; Dranginis, S.; Ashton, C.; Fotouhi, N.; Curran, T. P., *J. Org. Chem.* **1986**, *51*, 3320.
10. McBride, B. J.; Kemp, D. S., *Tetrahedron Lett.* **1987**, *28*, 3435.
11. Fotouhi, N.; Kemp, D. S., *Int. J. Peptide Protein Res.* **1993**, *41*, 153.

Livia A. Enache
deCode Genetics, Woodridge, IL, USA

Mercuric Bromide

$HgBr_2$

[7789-47-1] Br_2Hg (MW 360.40)

(reagent used in glycosylation, organomercurial formation, Friedel–Crafts type reactions)

Alternate Name: mercuric bromide.
Physical Data: mp 236 °C, bp 322 °C, d 6.05 g cm^{-3}.
Solubility: soluble in H_2O (0.61 g/100 g H_2O at 25 °C), methanol, ethanol, HCl, HBr, and alkali bromide solutions; slightly soluble in chloroform.
Form Supplied in: white crystals or crystalline powder.
Preparative Methods: usually prepared from the elements.
Handling, Storage, and Precautions: highly toxic; absorbed through skin; protect from light.

Promoter for Glycosylation. Mercury(II) bromide is a mild Lewis acid promoter for glycosylations employing glycosyl halides, most commonly glycosyl bromides, as donors. $HgBr_2$ is weaker than silver trifluoromethanesulfonate but stronger than tetralkylammonium bromide, which are other common promoters for glycosyl bromides.[1] $HgBr_2$, often utilized together with $Hg(CN)_2$ or HgO, is efficient in activating a wide range of glycosyl bromides including pyranosides (eqs 1–3),[2,3] furanosides (eq 4),[4]

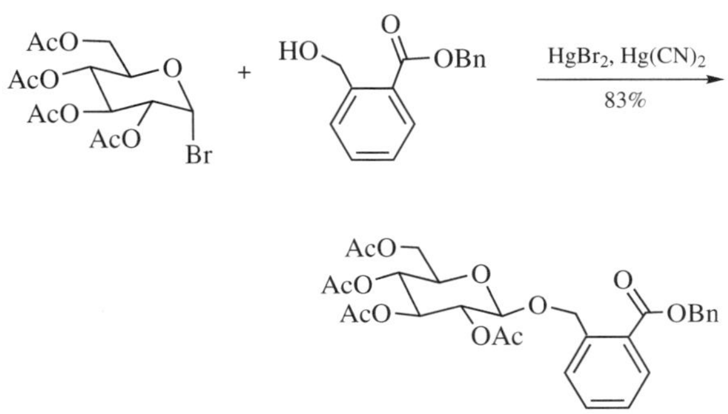

(1)

and pyranosylonic acids (eq 5)[5] to form glycosides with various glycosyl acceptors such as alkyl alcohols (eqs 1 and 3), carbohydrate alcohols (eqs 2 and 5), and a protected pyrimidine (eq 4). In certain cases, ortho ester intermediates were isolated, which were subsequently isomerized to glycosides with trimethylsilyl trifluoromethanesulfonate (TMSOTf) (eq 5).

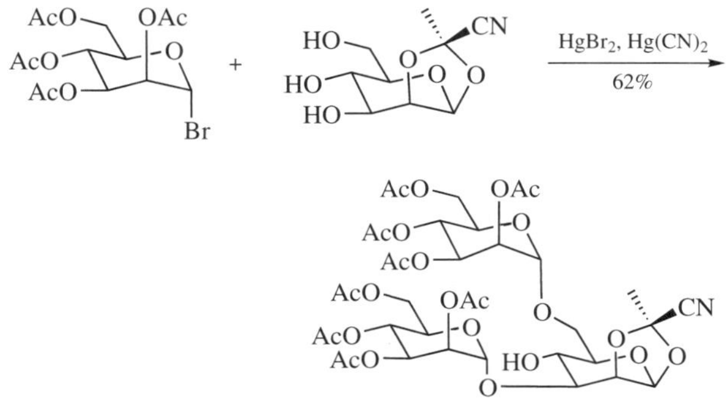

(2)

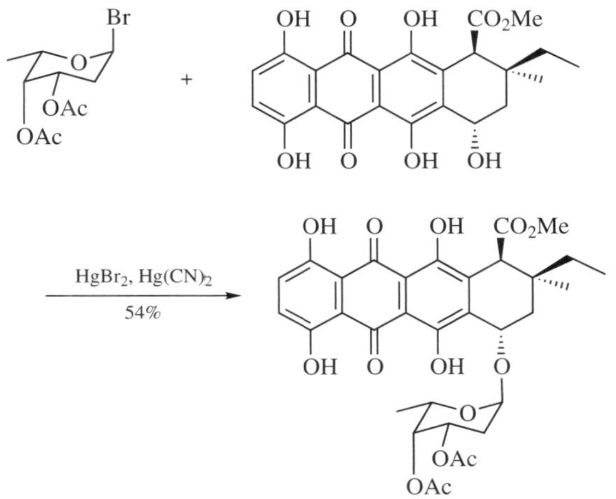

(3)

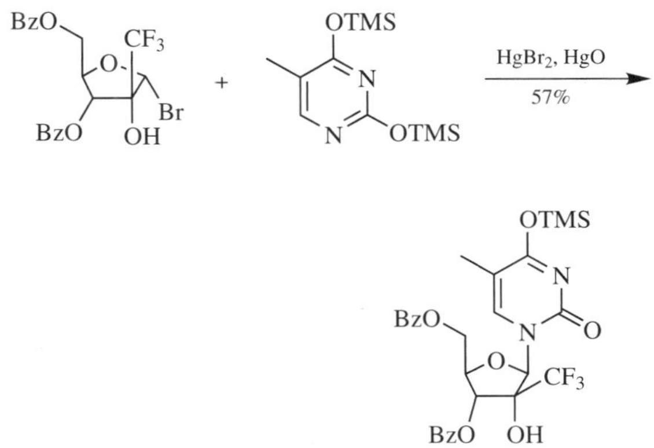

(4)

Glycosyl chlorides can be activated by $HgBr_2$ in a way similar to glycosyl bromides.[6,7] Treatment of 2-acetamido-3,4,6-tri-*O*-acetyl-2-deoxy-α-D-glucopyranosyl chloride and a galactosamine derivative with $HgBr_2$ and $Hg(CN)_2$ led to a disaccharide in 48% yield (eq 6) without the formation of an oxazoline side product, which is often a major product with glycosyl donors bearing acetamido moiety at the C-2 position.[6]

Besides glycosyl halides, glycosyl esters and glycosyl orthoesters are also effective glycosyl donors with $HgBr_2$ as the promoter. 2,3,4-Tri-*O*-acetyl-β-L-fucopyranosyl cyanide was obtained from the corresponding glycosyl acetate in 95% yield catalyzed by 0.1 equiv $HgBr_2$, whose use was more advantageous than that of $BF_3 \cdot Et_2O$ due to its mildness and the absence of a cyanoethylidene side product (eq 7).[8] Selenoglycosides (eq 8) and thioglycosides[3,9] were synthesized from glycosyl orthoesters with $HgBr_2$.

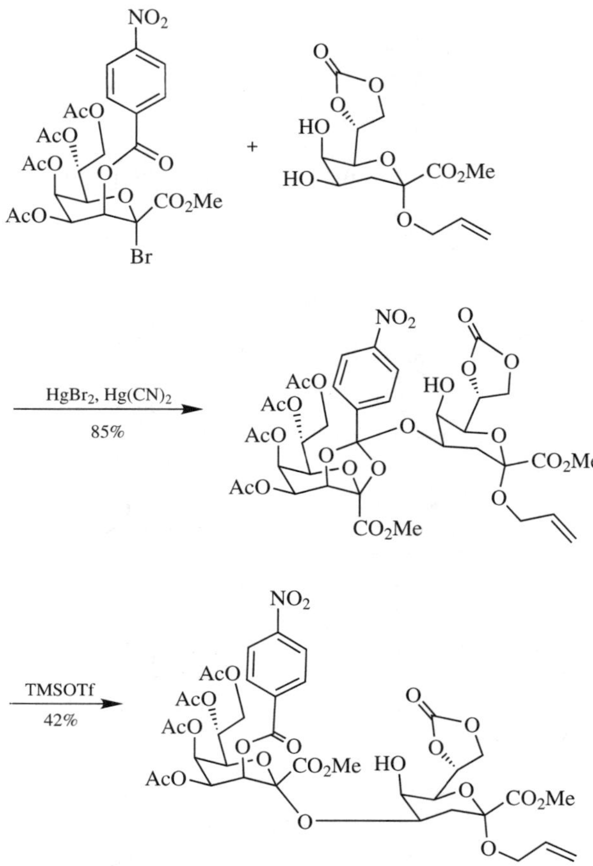
(5)

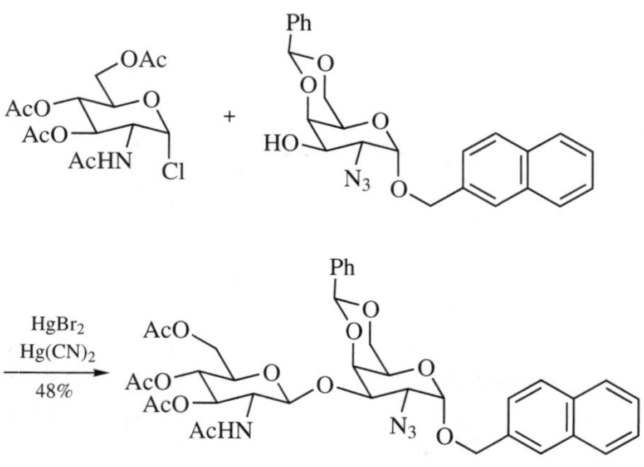
(6)

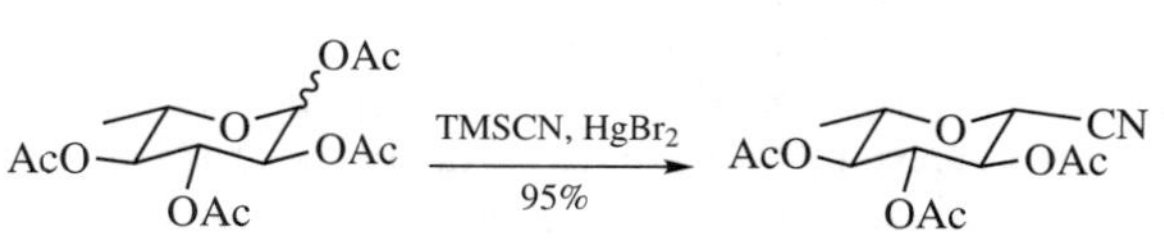
(7)

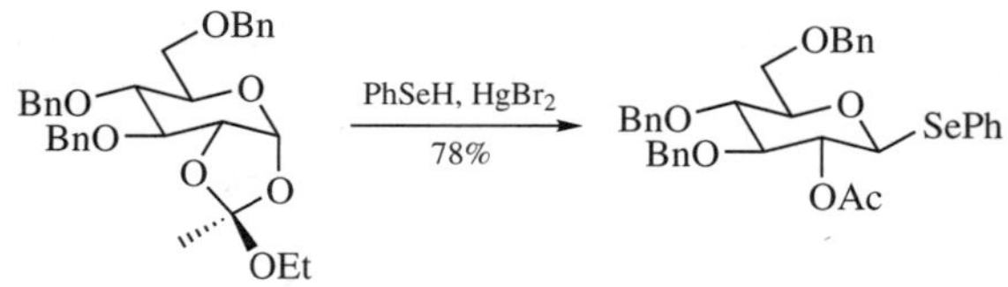
(8)

The stereochemistry of the glycosylation product is generally determined by the identity of the protective group at the C-2 position. In the presence of a participating group (e.g., acetyl), the 1,2-*trans* glycoside is usually predominantly formed (eqs 1, 2, 6, and 7). In the 2-deoxy series, the thermodynamically more favored axial glycosides are preferentially generated due to the anomeric effect (eq 3).

Organomercurial Formation. Organomercurials can be synthesized through transmetalation of other organometallics such as organolithiums and Grignard reagents with $HgBr_2$.[10]

Conversion of Aromatic Amines into Bromides. $HgBr_2$ has been utilized in transforming aromatic amines into aromatic bromides in moderate yields through pyrolysis of the double salt formed from the diazonium salt and $HgBr_2$ (eq 9).[11]

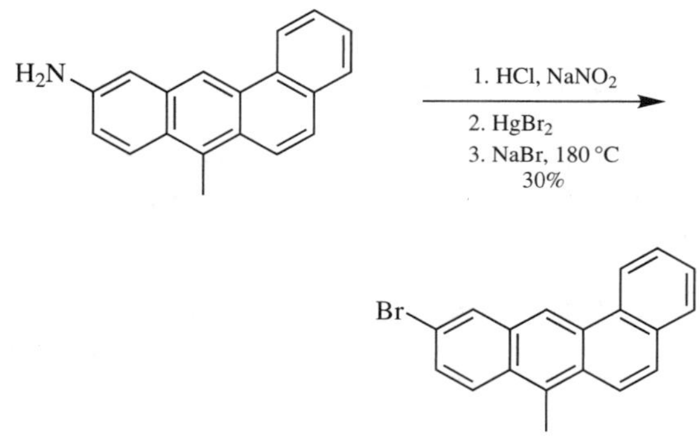
(9)

Friedel–Crafts Type Carbamoylation, Alkylation, and Aroylation. Due to the high affinity of selenium atoms towards soft metal atoms, $HgBr_2$ was efficacious in promoting carbomoylation of aromatic compounds with bis(*N,N*-diethylcarbamoyl) diselenide (eq 10).[12] In a similar manner, aroylation and alkylation were also successfully performed. The regioselectivities of these reactions were determined by the substituents on the aromatic ring as in typical Friedel–Crafts reactions. Other Lewis acids including $FeCl_3$, $AlBr_3$, $Hg(OAc)_2$, and Hg_2Cl_2 were not effective.

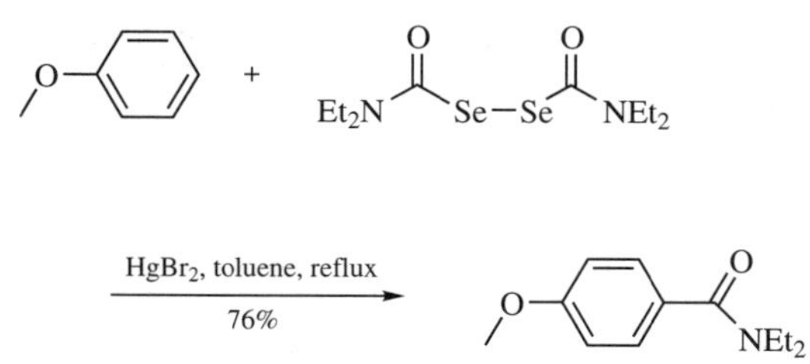
(10)

1. Bowden, T.; Garegg, P. J.; Maloisel, J.-L.; Konradsson, P., *Isr. J. Chem.* **2000**, *40*, 271–277.
2. (a) Kim, K. S.; Kim, J. H.; Lee, Y. J.; Lee, Y. J.; Park, J., *J. Am. Chem. Soc.* **2001**, *123*, 8477–8481. (b) Klaffke, W.; Pudlo, P.; Springer, D.; Thiem, J., *Liebigs Ann. Chem.* **1991**, 509–512.

3. Backinowsky, L. V.; Abronina, P. I.; Shashkov, A. S.; Grachev, A. A.; Kochetkov, N. K.; Nepogodiev, S. A.; Stoddart, J. F., *Chem. Eur. J.* **2002**, *8*, 4412–4423.
4. Li, N.-S.; Tang, X.-Q.; Piccirilli, J. A., *Org. Lett.* **2001**, *3*, 1025–1028.
5. Wimmer, N.; Brade, H.; Kosma, P., *Carbohydr. Res.* **2000**, *329*, 549–560.
6. Sarkar, A. K.; Brown, J. R.; Esko, J. D., *Carbohydr. Res.* **2000**, *329*, 287–300.
7. (a) Tsui, D. S.; Gorin, P. A. J., *Carbohydr. Res.* **1986**, *156*, 1–8. (b) Murakata, C.; Ogawa, T., *Carbohydr. Res.* **1992**, *235*, 95–114.
8. Phiasivongsa, P.; Gallagher, J.; Chen, C.-N.; Jones, P. R.; Samoshin, V. V.; Gross, P. H., *Org. Lett.* **2002**, *4*, 4587–4590.
9. (a) Aloui, M.; Chambers, D. J.; Cumpstey, I.; Fairbanks, A. J.; Redgrave, A. J.; Seward, C. M. P., *Chem. Eur. J.* **2002**, *8*, 2608–2621. (b) Baeschlin, D. K.; Green, L. G.; Hahn, M. G.; Hinzen, B.; Ince, S. J.; Ley, S. V., *Tetrahedron: Asymmetry* **2000**, *11*, 173–197. (c) Bamhaoud, T.; Sanchez, S.; Prandi, J., *Chem. Commun.* **2000**, 659–660.
10. (a) Trécourt, F.; Breton, G.; Bonnet, V.; Mongin, F.; Marsais, F.; Quéguiner, G., *Tetrahedron* **2000**, *56*, 1349–1360. (b) Rot, N.; Bickelhaupt, F., *Organometallics* **1997**, *16*, 5027–5031. (c) Kočovský, P.; Dunn, V.; Grech, J. M.; Šrogl, J.; Mitchell, W. L., *Tetrahedron Lett.* **1996**, *37*, 5585–5588. (d) Seetz, J. W. F. L.; Schat, G.; Akkerman, O. S.; Bickelhaupt, F., *J. Am. Chem. Soc.* **1982**, *104*, 6848–6849.
11. (a) Newman, M. S.; Hussain, N. S., *J. Org. Chem.* **1982**, *47*, 2837–2840. (b) Brown, K. J.; Berry, M. S.; Murdoch, J. R., *J. Org. Chem.* **1985**, *50*, 4345–4349. (c) Dauben, W. G.; Saegebarth, K. A., *J. Am. Chem. Soc.* **1951**, *73*, 1853–1854.
12. Fujiwara, S.-I.; Ogawa, A.; Kambe, N.; Ryu, I.; Sonoda, N., *Tetrahedron Lett.* **1988**, *29*, 6121–6124.

Xuefei Huang
University of Toledo, Toledo, OH, USA

Mercuric Cyanide

$Hg(CN)_2$

[592-04-1] C_2HgN_2 (MW 252.63)

(glycosidation reagent[1])

Alternate Name: mercuric cyanide.
Physical Data: mp 320 °C dec; *d* 3.996 g mL^{-3}.
Solubility: insoluble in benzene; soluble in H_2O, alcohol, NH_3; slightly soluble in ether.
Form Supplied in: colorless, odorless, tetragonal crystals or white powder.
Handling, Storage, and Precautions: acute poison. Exposure to all mercury compounds is to be strictly avoided. Releases toxic Hg fumes when heated to decomposition. Protect from light.

Original Commentary

Pavel Kočovský
University of Leicester, UK

Mercury(II) cyanide is sometimes superior to Ag_2CO_3 in the formation of glycosides from bromo sugars with isolated yields up to 90%; $MeNO_2$ or CH_2Cl_2 seem to be the best solvents.[1] Evomonoside, a steroidal cardioactive glycoside, has been prepared from digitoxigenin in 44% yield;[2] attempted synthesis with Ag_2CO_3 failed because of an extensive elimination of the 14β-OH of the aglycon.[2] Nucleosides can be obtained from pyrimidine bases and protected halo sugars by heating with $Hg(CN)_2$ in an inert solvent (toluene, MeCN, or $MeNO_2$).[3]

First Update

Guijun Wang & Vibha Sharma
University of New Orleans, LA, USA

***O*-Glycoside Formation.** Mercury (II) cyanide has been used extensively in glycosylation reactions involving protected bromo or chloro sugars and different nucleophiles. Sometimes mercury (II) bromide is used as a co-catalyst. Alcohols are common nucleophiles and a typical application is the formation of interglycosidic linkages.[4–28]

Beside alcohols, acyl ketene aminals have also been reported as acceptors in *O*-glycosylations using mercury cyanide as a catalyst.[29,30] These reactions are highly stereoselective, typically giving *E*-enol ethers (eq 1).

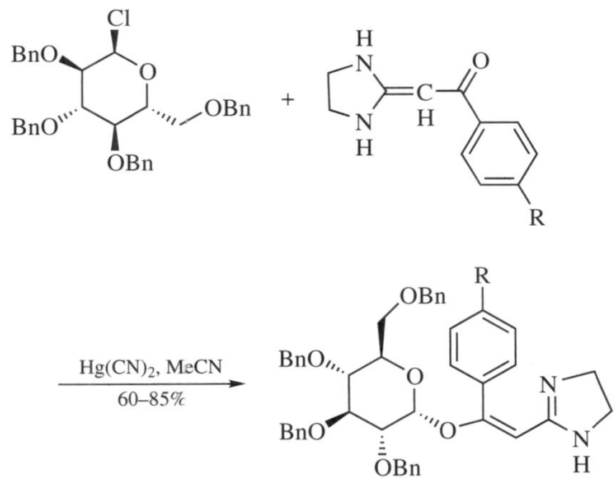

(1)

***N*-Glycoside Formation.** Nitrogen can be introduced to a sugar moiety by glycosylation of amines or nitrogen containing heterocycles.[31–35] *N*-Glycosylation has been used in the synthesis of liposidomycin analogs,[31] benzodiazepine nucleosides,[32] nitroimidazole glycosides (eq 2),[33] pyrazole glycosides,[34] and glycosylated amino acids.[35]

***C*-Glycoside Formation.** *C*-Glycosides can be formed from halosugars by using mercury cyanide as mediator.[36–38] The cyano group can subsequently be converted to a carboxyl group. Glycamino acids, a family of amino acids in which a carbohydrate ring separates the carboxylic acid from the amino group, can be synthesized by this route (eq 3).[6] A cyano-glycosylation accompanied by ring-opening of steroidal spirostanols has also been reported (eq 4).[39] Direct displacement of a halo group by a cyano group was also observed.[40]

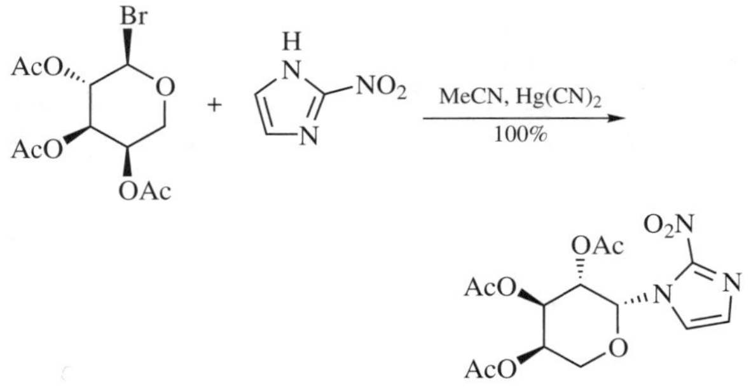

(2)

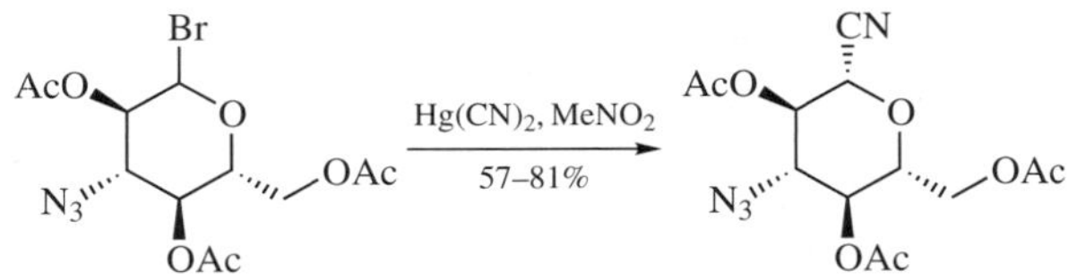

(3)

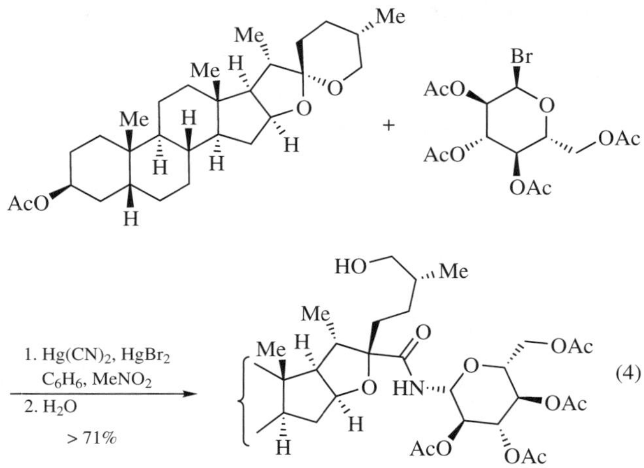

(4)

Other Reactions. $Hg(CN)_2$was used as a catalyst in the hydrolysis of *N*-(phenylthiomethyl)oxazolidinone to the corresponding enantiopure α-hydroxy aldehydes, with yields in the range 67–78%.[41] It has also been used as a catalyst for silylation reactions.[42] One other application of mercury cyanide is as a catalyst in Grignard type reactions.[43] Ketones with C-C multiple bonds at the δ-position can form cyclized products under the conditions indicated in eq 5.

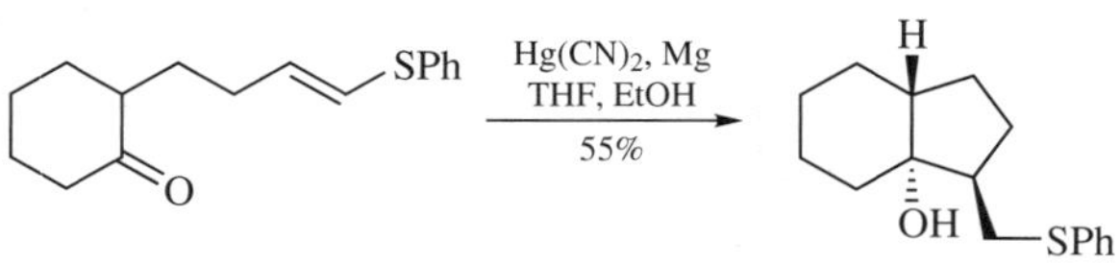

(5)

1. (a) Helferich, B.; Weis, K., *Ber. Dtsch. Chem. Ges./Chem. Ber.* **1956**, *89*, 314. (b) Helferich, B.; Steinpreis, R., *Ber. Dtsch. Chem. Ges./Chem. Ber.* **1958**, *91*, 1794.
2. Zorbach, W. W.; Valiaveedan, G. D.; Kashelikar, D. V., *J. Org. Chem.* **1962**, *27*, 1766.
3. (a) Rogers, G. T.; Ulbricht, T. L. V., *Chem. Commun.* **1969**, 508. (b) Rogers, G. T.; Ulbricht, T. L. V., *J. Chem. Soc. (C)* **1969**, 2450.
4. Vega-Pérez, J. M.; Candela, J. I.; Blanco, E.; Iglesias-Guerra, F., *Tetrahedron: Asymmetry* **2002**, *13*, 2471.
5. Backinowsky, L. V.; Abronina, P. I.; Shashkov, A. S.; Grachev, A. A.; Kochetkov, N. K.; Nepogodiev, S. A.; Stoddart, J. F., *Chem. Eur. J.* **2002**, *8*, 4412.
6. Garcia-Barrientos, A.; Garcia-Lopez, J. J.; Isac-Garcia, J.; Ortega-Caballero, F.; Uriel, C.; Vargas-Berenguel, A.; Santoyo-Gonzalez, F., *Synthesis* **2001**, 1057.
7. Kim, K. S.; Kim, J. H.; Lee, Y. J.; Lee, Y. J.; Park, J., *J. Am. Chem. Soc.* **2001**, *123*, 8477.
8. Wimmer, N.; Brade, H.; Kosma, P., *Carbohydr. Res.* **2000**, *329*, 549.
9. Roy, R.; Baek, M.-G.; Rittenhouse-Olson, K., *J. Am. Chem. Soc.* **2001**, *123*, 1809.
10. Ross, A. J.; Ivanova, I. A.; Ferguson, M. A. J.; Nikolaev, A. V., *J. Chem. Soc., Perkin. Trans. 1* **2001**, 72.
11. Zamani, K.; Rama, N. H.; Iqbal, R., *J. Heterocyclic Chem.* **2000**, *37*, 1651.
12. Khatuntseva, E. A.; Ustuzhanina, N. E.; Zatonskii, G. V.; Shashkov, A. S.; Usov, A. I.; Nifant'ev, N. E., *J. Carbohydr. Chem.* **2000**, *19*, 1151.
13. Vega-Pérez, J. M.; Candela, J. I.; Romero, I.; Blanco, E.; Iglesias-Guerra, F., *Eur. J. Org. Chem.* **2000**, *23*, 3949.
14. Nicolaou, K. C.; Mitchell, H. J.; Jain, N. F.; Bando, T.; Hughes, R.; Winssinger, N.; Natarajan, S.; Koumbis, A. E., *Chem. Eur. J.* **1999**, *5*, 2648.
15. Xia, J.; Srikrishnan, T.; Alderfer, J. L.; Jain, R. K.; Piskorz, C. F.; Matta, K. L., *Carbohydr. Res.* **2000**, *329*, 561.
16. Chernyak, A.; Oscarson, S.; Turek, D., *Carbohydr. Res.* **2000**, *329*, 309.
17. Utille, J.-P.; Priem, B., *Carbohyd. Res.* **2000**, *329*, 431.
18. Sarkar, A. K.; Brown, J. R.; Esko, J. D., *Carbohydr. Res.* **2000**, *329*, 287.
19. Li, H.; Li, Q.; Cai, M. S.; Li, Z. J., *Carbohydr. Res.* **2000**, *328*, 611.
20. Bai, Y.; Boons, G. J.; Burton, A.; Johnson, M.; Haller, M., *J. Carbohydr. Chem.* **2000**, *19*, 939.
21. Kramer, S.; Nolting, B.; Ott, A. J.; Vogel, C., *J. Carbohydr. Chem.* **2000**, *19*, 891.
22. Ellervik, U.; Magnusson, G., *Carbohydr. Res.* **1996**, *280*, 251.
23. Nakahara, Y.; Shibayama, S.; Nakahara, Y.; Ogawa, T., *Carbohydr. Res.* **1996**, *280*, 67.
24. Xia, J.; Alderfer, J. L.; Piskorz, C. F.; Matta, K. L., *Chem. Eur. J.* **2000**, *6*, 3442.
25. Adamczyk, M.; Reddy, R. E.; Rege, S. D., *Synth. Commun.* **2000**, *30*, 3281.
26. Miethchen, R.; Rentsch, D.; Frank, M.; Lipták, A., *Carbohydr. Res.* **1996**, *281*, 61.
27. Snider, B. B.; Vo, N. H.; O'Neil, S. V.; Foxman, B. M., *J. Am. Chem. Soc.* **1996**, *118*, 7644.
28. Qiu, D.; Gandhi, S. S.; Koganty, R. R., *Tetrahedron Lett.* **1996**, *37*, 595.
29. Chen, X. M.; Li, Z. J.; Huang, Z. T., *Carbohydr. Res.* **2000**, *328*, 253.
30. Li, Z. J.; Wang, L. B.; Huang, Z. T., *Carbohydr. Res.* **1996**, *295*, 77.
31. Dini, C.; Didier-Laurent, S.; Drochon, N.; Feteanu, S.; Guillot, J. C.; Monti, F.; Uridat, E.; Zhang, J.; Aszodi, J., *Bioorg. Med. Chem. Lett.* **2002**, *12*, 1209.
32. Ahmad, R.; Zia-Ul-Haq, M.; Hameed, S.; Nadeem, H.; Duddeck, H., *J. Chem. Soc. Pakistan* **2000**, *22*, 302–308.
33. Kumar, P.; Wiebe, L. I.; Atrazheva, E.; Tandon, M., *Tetrahedron Lett.* **2001**, *42*, 2077.
34. Liberek, B.; Smiatacz, Z., *J. Carbohydr. Chem.* **2000**, *19*, 1259.
35. Andersch, J.; Hennig, L.; Wilde, H., *Carbohydr. Res.* **2000**, *329*, 693.
36. Suhara, Y.; Yamaguchi, Y.; Collins, B.; Schnaar, R. L.; Yanagishita, M.; Hildreth, J. E. K.; Shimada, I.; Ichikawa, Y., *Bioorg. Med. Chem.* **2002**, *10*, 1999.
37. Stöckle, M.; Voll, G.; Günther, R.; Lohof, E.; Locardi, E.; Gruner, S.; Kessler, H., *Org. Lett.* **2002**, *4*, 2501.

38. Somsák, L.; Nagy, V., *Tetrahedron: Asymmetry* **2000**, *11*, 1719.

39. Tobari, A.; Miyamae, H.; Nagasawa, A.; Koyanagi, J.; Kawase, M.; Saito, S., *Heterocycles* **2000**, *53*, 1745.

40. Unterhalt, B.; Jostingmeier, R.; Sanatgar, A., *Pharmazie* **1997**, *52*, 186.

41. Gawley, R. E.; Campagna, S. A.; Santiago, M.; Ren, T., *Tetrahedron: Asymmetry* **2002**, *13*, 29.

42. Lin, S.; Danishefsky, S. J., *Angew. Chem., Int. Ed. Engl.* **2001**, *40*, 1967.

43. Lee, G. H.; Ha, S. J.; Yoon, I. K.; Pak, C. S., *Tetrahedron Lett.* **1999**, *40*, 2581.

Mesitylenesulfonyl Chloride

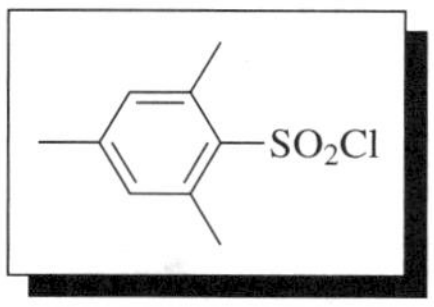

[773-64-8] $C_9H_{11}ClO_2S$ (MW 218.72)

(preparation of sulfonates;[1] condensing agent in nucleotide synthesis[2])

Alternate Name: 2,4,6-trimethylbenzenesulfonyl chloride.
Physical Data: mp 55–57 °C.
Solubility: sol organic solvents.
Form Supplied in: white crystalline solid; widely available.
Purification: can be recrystallized from hexane or pentane.
Handling, Storage, and Precautions: corrosive; moisture sensitive; handle and store under nitrogen.

Sulfonamide Formation. Reaction of mesitylenesulfonyl chloride with amines in the presence of Pyridine or Triethylamine yields the corresponding sulfonamides (eq 1),[3] which have been used as intermediates in synthesis.

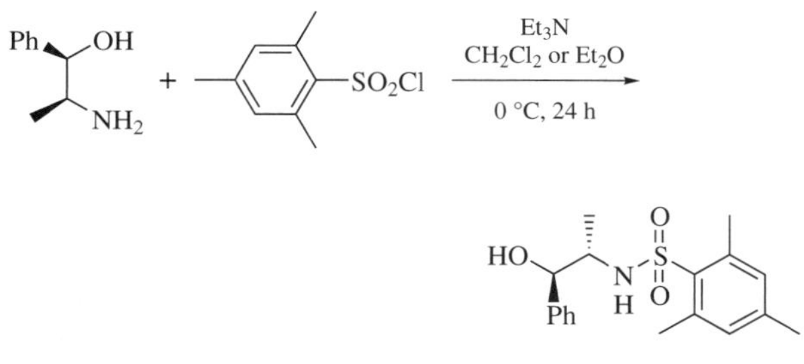

(1)

The mesityl group has been used as a protecting group for various amines including amino acids (eq 2)[4] and peptides.[5–7]

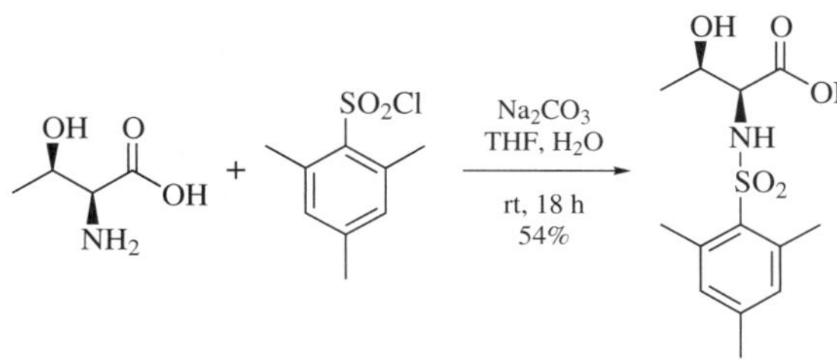

(2)

The mesitylsulfonyl-protected amino acids can be deblocked by treatment with Hydrogen Bromide and Acetic Acid. The stability of the sulfonamide towards both acidic and basic conditions makes it a useful protecting group for peptide synthesis. The mesitylenesulfonyl group has been used to protect the indole group of tryptophan[6] and the guanidine group of arginine.[7] These mesityl derivatives can be cleaved with Trifluoromethanesulfonic Acid or Methanesulfonic Acid. Partial cleavage of the guanidinomesityl group occurs when treated with HBr/HOAc. The mesitylenesulfonyl group has also been used as a blocking group for other indoles. Boteju et al. reported the formation of the indole sulfonamide by treatment with n-Butyllithium and mesitylenesulfonyl chloride, but the sulfonamide group was not subsequently removed (eq 3).[8]

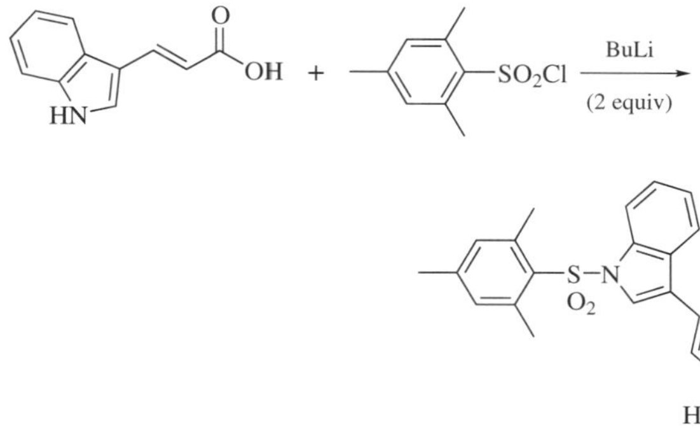

(3)

Sulfonate Formation. Treatment of alcohols with mesitylenesulfonyl chloride yields the corresponding sulfonates. Mesitylenesulfonyl chloride is particularly useful for the selective sulfonation of polyhydroxylic systems such as carbohydrates (eq 4).[1,9] It is

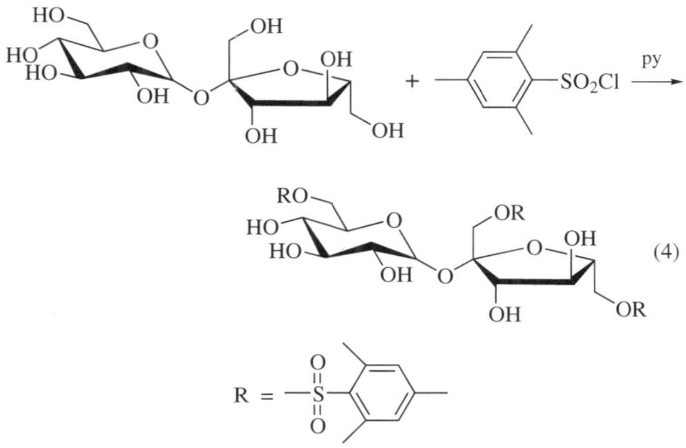

(4)

more selective than p-Toluenesulfonyl Chloride, which has been frequently used but gives mixtures of products. Unfortunately, the mesitylenesulfonates are not as reactive as the corresponding tosylates. Numerous 1′-derivatives of sucrose have been synthesized via mesitylenesulfonyl derivatives.[10]

The selectivity of sulfonation has also been applied to other polyhydroxylic systems such as α- and β-cyclodextrins[11,12] and nucleotides.[13] Monosulfonated products have been reported from the treatment of diols with mesitylenesulfonyl chloride (eq 5).[14,15]

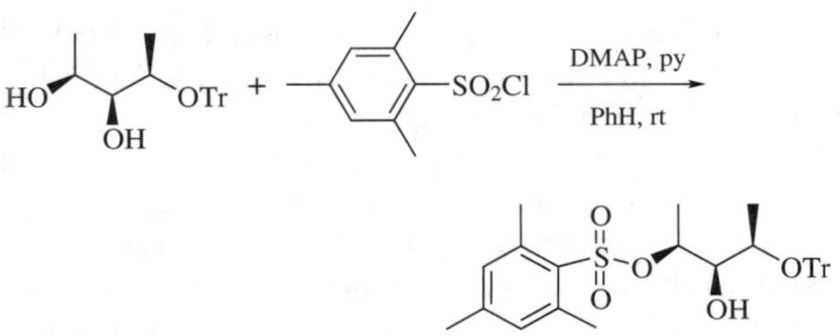

(5)

Mesitylenesulfonyl derivatives which have specific biological activities, such as 2,3-diaziridinyl-1,4-napthoquinone sulfonates[16] and 1-arenesulfonyloxy-2-alkanones,[17] have also been prepared.

Nucleotide Synthesis. Mesitylenesulfonyl chloride has been used as a condensing agent in the diester approach to nucleotide synthesis (eq 6).[2] However, it is not effective in the triester approach due to extensive sulfonation of the 5′-hydroxy group.[18] Other condensing agents include 1,3-Dicyclohexylcarbodiimide (DCC), benzenesulfonyltriazole (BST), (1-mesitylyl-2-sulfonyl)-3-nitro-1,2,4-triazole (MSNT), Mesitylsulfonyl-1 H-1,2,4-triazole (MST), *p*-nitrobenzenesulfonyltriazole (*p*-NBST), *p*-toluenesulfonyltriazole (TST), and triisopropylbenzenesulfonyl chloride (TPS).

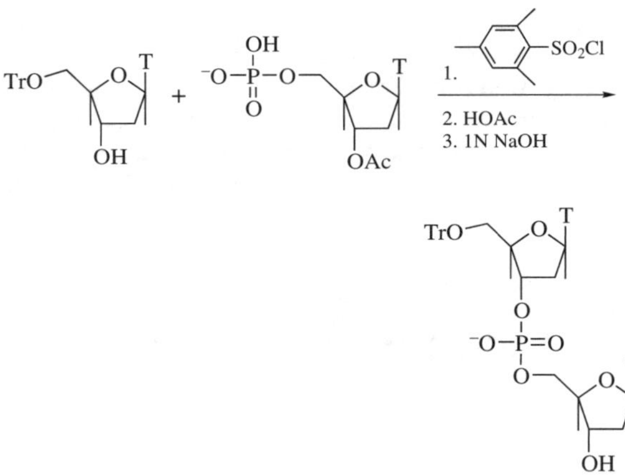

(6)

Condensation Reactions. Mesitylenesulfonyl chloride has also been employed as a condensing agent for the preparation of esters in the synthesis of nonactin (eq 7),[19] *N*-methylmaysenine,[20] and maysine.[21]

Sulfinate Ester Synthesis. Sulfinate esters of menthol can be prepared via reaction of mesitylenesulfonyl chloride (as well as other sulfonyl chlorides) with menthol using Trimethyl Phosphite (eq 8).[22] Sulfinate esters of menthol are used for the preparation of optically active sulfoxides. See also 10-Camphorsulfonyl Chloride[1] and p-Toluenesulfonyl Chloride.

Sulfonylalkyne Synthesis. Sulfonylalkynes have been synthesized (eq 9) and subsequently reacted with *N*-(1-alkynyl)anilines to give 2-anilino-5-sulfinylfurans.[23]

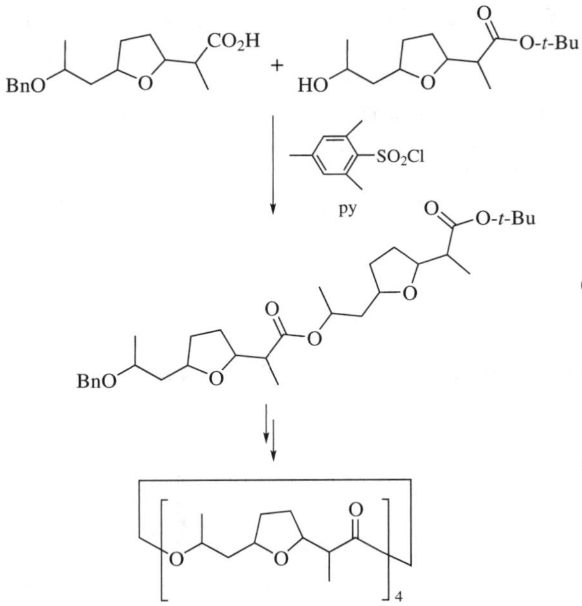

(7)

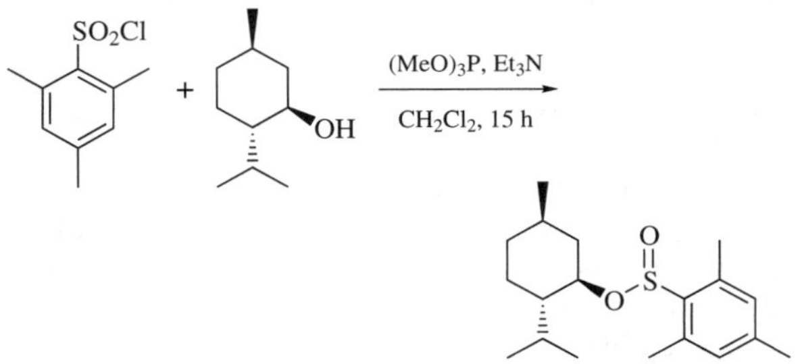

(8)

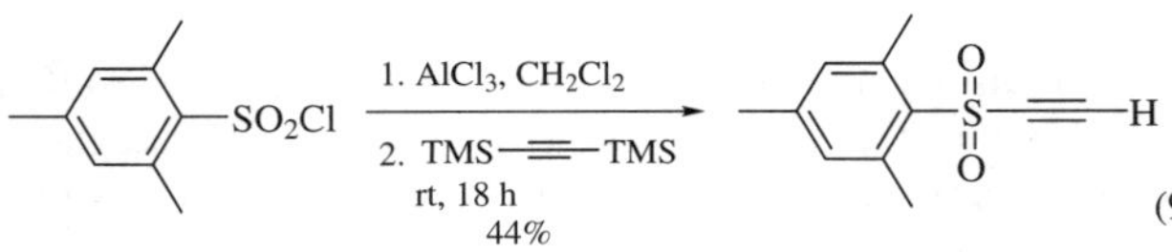

(9)

1. Guthrie, R. D.; Thang, S. *Aust. J. Chem.* **1987**, *40*, 2133.
2. (a) Narang, S. A.; Khorana, H. G. *J. Am. Chem. Soc.* **1965**, *87*, 2981. (b) *Fieser & Fieser* **1967**, *1*, 661.
3. (a) Reetz, M. T.; Kukenhohner, T.; Weinig, P. *Tetrahedron Lett.* **1986**, *27*, 5711. (b) Pavlidis, V. H.; Chan, E. D.; Pennington, L.; McParland, M.; Whitehead, M. *Synth. Commun.* **1988**, *18*, 1615. (c) Hoppe, I.; Hoffmann, H.; Gartner, I.; Krettek, T.; Hoppe, D. *Synthesis* **1991**, 1157. (d) Takahashi, H.; Takahashi, K.; Ohno, M.; Yoshioka, M.; Kobayashi, S. *Tetrahedron* **1992**, *48*, 5691.
4. Roemmele, R. C.; Rapoport, H. *J. Org. Chem.* **1988**, 2367.
5. (a) Corey, E. J.; Weigel, L. O.; Floyd, D.; Bock, M. G. *J. Am. Chem. Soc.* **1978**, *100*, 2916. (b) *Fieser & Fieser* **1980**, *8*, 318.
6. Fuji, N.; Futaki, S.; Yasumura, K.; Yajima, H. *Chem. Pharm. Bull.* **1984**, 2660.
7. (a) Yajima, H.; Takeyama, M.; Kanaki, J.; Mitani, K. *Chem. Commun./J. Chem. Soc.* **1978**, 482. (b) Yajima, H.; Takeyama, M.; Kanaki, J.; Nishimura, O.; Fujino, M. *Chem. Pharm. Bull.* **1978**, *26*, 3752.
8. Boteju, L. W.; Wegner, K.; Hruby, V. J. *Tetrahedron Lett.* **1992**, 7291.
9. Gilchrist, T. L.; Rees, C. W.; Stanton, E. *Chem. Commun./J. Chem. Soc.* **1971**, 801.

10. Hough, L.; Phadnis, S. P.; Tarelli, E. *Carbohydr. Res.* **1975**, *44*, C12. (b) Ball, D. H.; Bisett, F. H.; Chalk, R. C. *Carbohydr. Res.* **1977**, *55*, 149. (c) Guthrie, R. D.; Jenkins, I. D.; Watters, J. J. *Aust. J. Chem.* **1980**, *33*, 2487.
11. Fujita, K.; Matsunaga, A.; Imoto, T. *J. Am. Chem. Soc.* **1984**, *106*, 5740.
12. (a) Fujita, K.; Yamamura, H.; Imoto, T. *J. Org. Chem.* **1985**, *50*, 4393. (b) Fujita, K.; Ishizu, T.; Minamiura, N.; Yamamoto, T. *Chem. Lett.* **1991**, 1889. (c) Fujita, K.; Ishizu, T.; Obe, K.; Minamiura, N.; Yamamoto, T. *J. Org. Chem.* **1992**, *57*, 5606.
13. (a) Ueda, T.; Usui, H.; Shuto, S.; Inoue, H. *Chem. Pharm. Bull.* **1984**, *32*, 3410. (b) Tanimura, H.; Sekine, M.; Hata, T. *Tetrahedron Lett.* **1986**, *27*, 4047. (c) Xu, Y. Z.; Zheng, Q.; Swann, P. F. *Tetrahedron Lett.* **1991**, *24*, 2817. (d) Xu, Y. Z.; Zheng, Q.; Swann, P. F. *Tetrahedron* **1992**, 1729.
14. Nakata, T.; Fukui, M.; Ohtsuka, H.; Oishi, T. *Tetrahedron Lett.* **1983**, *24*, 2661.
15. Rama Rao, A. V.; Gaitonde, A.; Prahlada Rao, S. *Indian J. Chem., Sect. B* **1992**, *31B*, 641.
16. Lin, T. S.; Xu, S. P.; Zhu, L. Y.; Cosby, L. A.; Sartorelli, A. C. *J. Med. Chem.* **1989**, *32*, 1467.
17. Ogawa, K.; Terada, T.; Muranaka, Y.; Hamakawa, T.; Hashimoto, S.; Fujii, S. *Chem. Pharm. Bull.* **1986**, *34*, 3252.
18. Katagiri, N.; Itakura, K.; Narang, S. A. *J. Am. Chem. Soc.* **1975**, *97*, 7332.
19. (a) Gerlach, H.; Oertle, K.; Thalmann, A.; Servi, S. *Helv. Chim. Acta* **1975**, *58*, 2036. (b) *Fieser & Fieser* **1977**, *6*, 625.
20. Corey, E. J.; Weigel, L. O.; Floyd, D.; Bock, M. G. *J. Am. Chem. Soc.* **1978**, *100*, 2916.
21. Kitamura, M.; Isobe, M.; Ichikawa, Y.; Goto, T. *J. Org. Chem.* **1984**, *49*, 3517.
22. Sharpless, K. B.; Klunder, J. M. *J. Org. Chem.* **1987**, *52*, 2598.
23. Kosack, S.; Himbert, G. *Ber. Dtsch. Chem. Ges./Chem. Ber.* **1987**, *120*, 71.

Valerie Vaillancourt & Michele M. Cudahy
The Upjohn Co., Kalamazoo, MI, USA

1-(Mesitylene-2-sulfonyl)-3-nitro-1,2,4-triazole (MSNT)

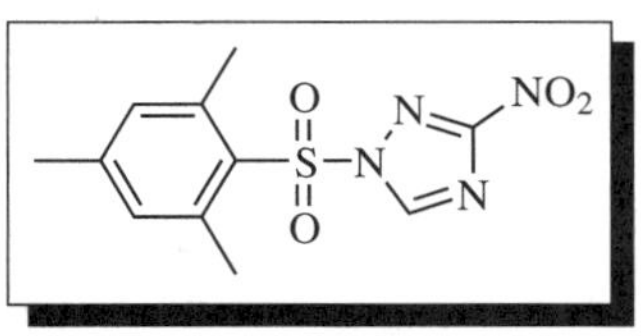

[74257-00-4] $C_{11}H_{12}N_4O_4S$ (MW 296.30)

(coupling agent for both solution and solid phase synthesis of oligonucleotides using the phosphotriester method)

Physical Data: mp 132–134 °C.

Solubility: insoluble in H_2O; soluble in CH_2Cl_2, $CHCl_3$, DMF, and pyridine.

Form Supplied in: crystalline solid; commercially available.

Analysis of Reagent Purity: the reagent may be checked by mp or analyzed by ^{1}H NMR.

Preparative Methods: this reagent can be prepared by the reaction of 3-nitro-1,2,4-triazole (1 equiv) and mesitylsulfonyl chloride (1 equiv) in the presence of triethylamine (1 equiv) in dioxane at 0 °C for 1 h. The product is isolated in 92% yield after crystallization from benzene.[1]

Purification: recrystallization from benzene.

Handling, Storage, and Precautions: this reagent may be an irritant. Avoid contact with the skin. It should be handled under nitrogen and stored below −20 °C.

Synthesis of Oligodeoxyribonucleotides by the Phosphotriester Method. 1-(Mesitylene-2-sulfonyl)-3-nitro-1,2,4-triazole-(MSNT) has been widely used as a condensing agent for the synthesis of oligodeoxynucleotides[2,3] by the phosphotriester method.[4]

For example, MSNT readily promotes the coupling of the triethylammonium salt of 5′-*O*-(*p*-chlorophenoxyacetyl)thymidine 3′-(2-chlorophenyl)phosphate and 3′-*O*-methoxytetrahydropyranylthymidine inpyridine (eq 1).[2]

MSNT has been employed as the condensing agent in a continuous-flow, solid-phase synthesis of oligodeoxyribonucleotides using the phosphotriester method (eq 2).[5]

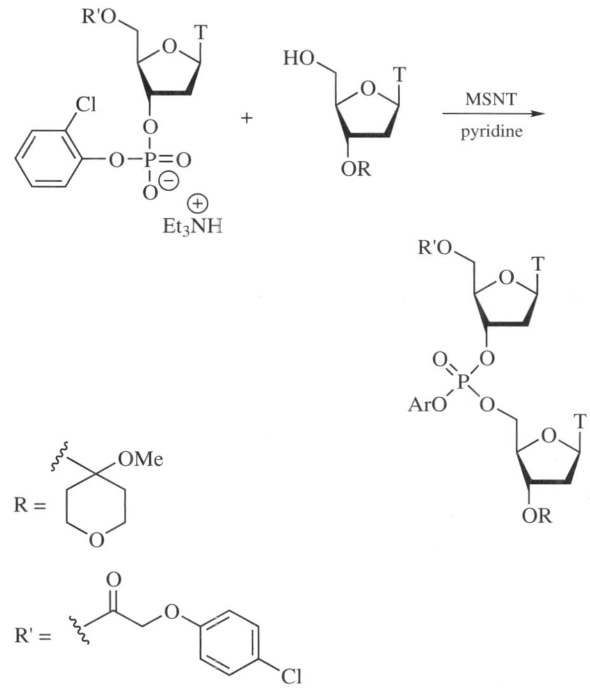

(1)

An additional advantage of using MSNT as the condensing agent is that it is not always necessary to protect the 3′-hydroxy group of the nucleoside. For example, in the presence of MSNT, unprotected thymidine can be coupled to nucleotide **1** regioselectively to give the corresponding 3′-5′ dinucleotide in good yield (eq 3).[3]

The time required for MSNT-mediated couplings can be considerably shortened by the addition of *N*-methylimidazole (NMI). For example, the time for the monomer (as the triethylammonium salt of the phosphodiester) addition in a solid-phase synthesis of oligodeoxynucleotides performed on controlled pore glass can be reduced to less than 10 min when MSNT is used in combination with NMI.[6] A large-scale soluble polymer PEG-supported oligodeoxynucleotide synthesis has been developed using MSNT/NMI as the coupling agent as shown in eq 4 for the synthesis of d(TpT).[7] This synthesis can be extended to higher oligomers such as d(TAGCGCTA).

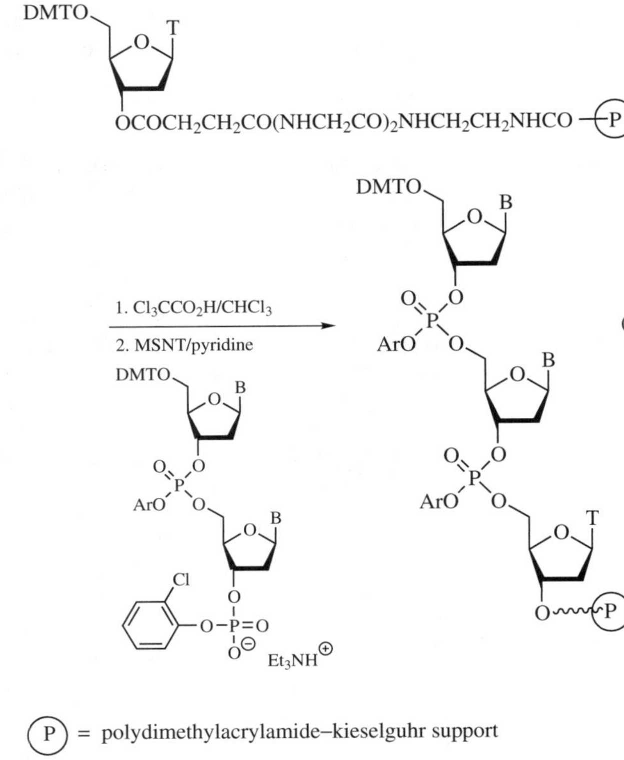

(2)

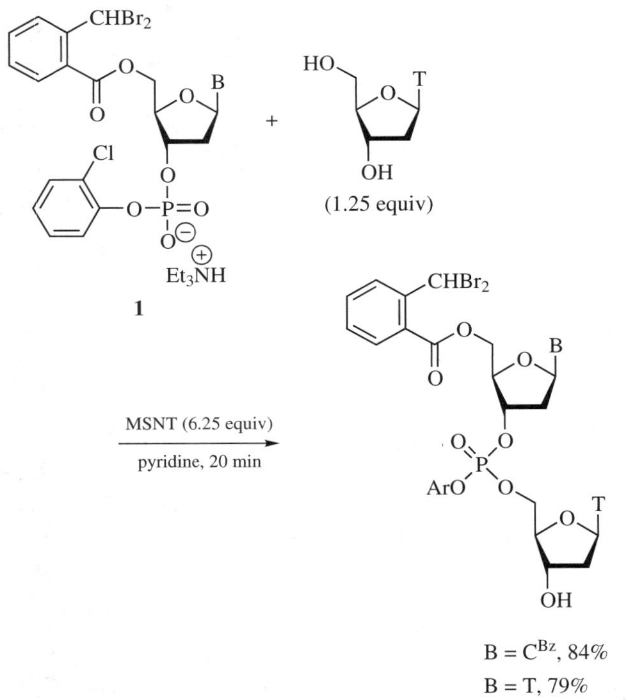

(3)

Synthesis of Oligoribonucleotides by the Phosphotriester Method. The use of MSNT as the condensing agent has also been extended to both the solution and the solid-phase synthesis of oligoribonucleotides. For example, the uridine-derived mononucleotide **2** couples to 2′-*O*-methoxytetrahydropyranyl-4-*N*-benzoylcytidine (**3**) in the presence of MSNT in pyridine in just 20 min to give the partially protected dinucleotide phosphate in 79% yield (eq 5).[1] Cyclic oligoribonucleotides can also be synthesized by the phosphotriester method using MSNT as the condensing agent.[8]

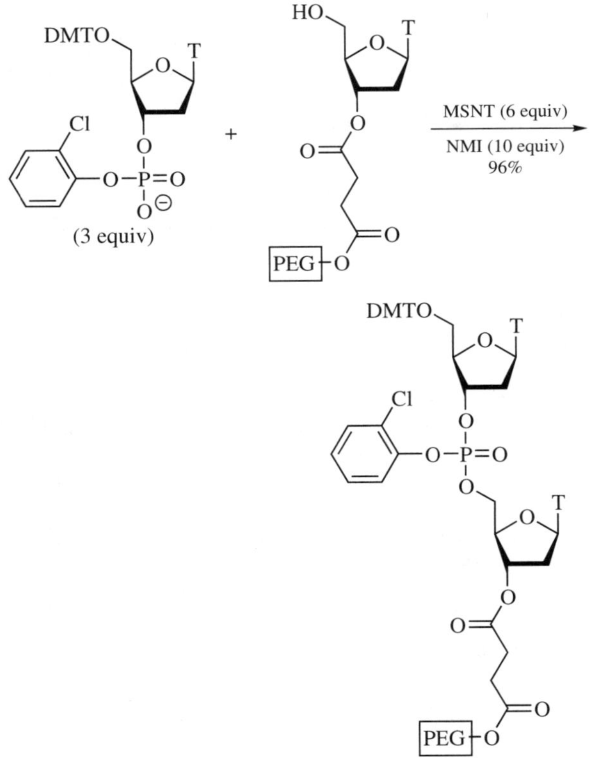

(4)

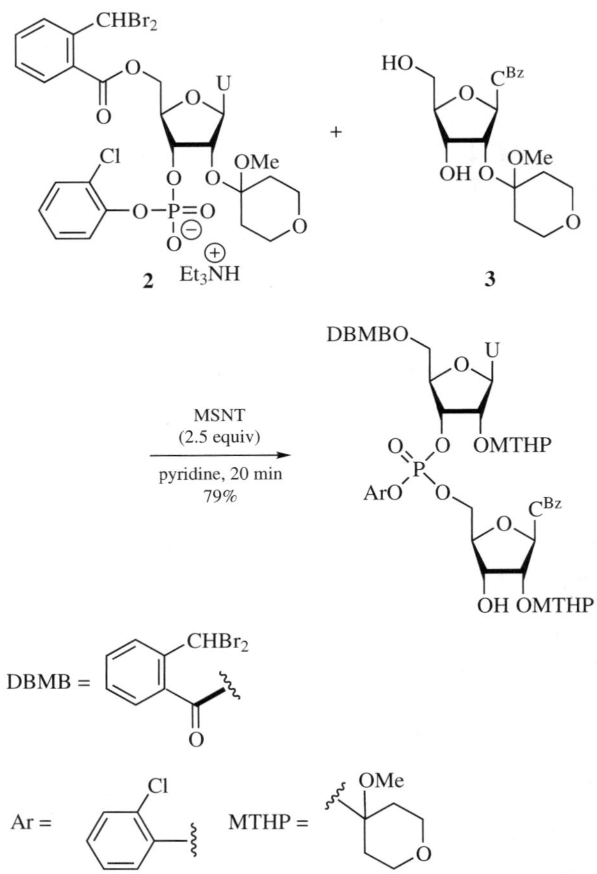

(5)

MSNT has been used for the solid-phase synthesis of oligonucleotides on cross-linked polystyrene (eq 6).[9] Best coupling result can be obtained by adding NMI as a catalyst, which not only increases the average condensation yield but also allows the dinucleotide phosphate to be coupled to the 5′-OH group of the growing oligonucleotide chain on the solid support.

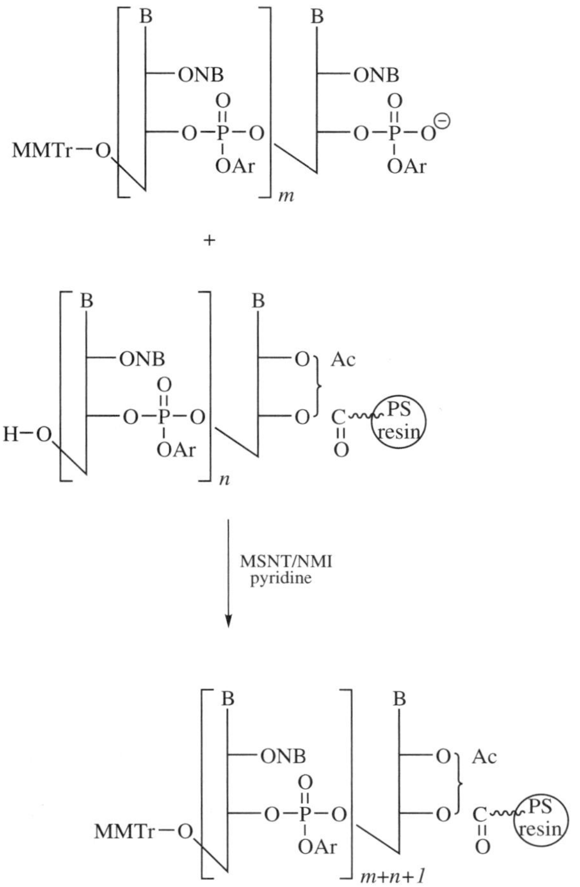

(6)

NB = *o*-nitrobenzyl
Ar = *o*-chlorophenyl; *m* = 0, 1

Formation of Nucleoside Phosphorothiolates. MSNT also promotes the formation of nucleoside phosphorothiolate triesters.[10,11] As shown in eq 7, the coupling of nucleoside phosphorothiolate **4** and nucleoside **5** is rapid in the presence of MSNT, giving the dinucleoside phosphorothiolate in excellent yield.[10]

Esterification, Amide Bond Formation, and Peptide Synthesis. In the presence of NMI, MSNT can be used as a coupling agent for amide bond formation[12] both in solution and in solid-phase peptide synthesis. The dipeptide Phe-Ala is formed very rapidly in 84.3% yield with virtually no racemization using MSNT as the coupling agent in the presence of NMI (eq 8).[12] The combination of MSNT and NMI can also be used for peptide formation on solid supports.[13]

In the presence of NMI, MSNT can be used as an efficient coupling agent for ester bond formation. For example, Fmoc-amino acids (Fmoc-AA) can be anchored to hydroxyl-functionalised solid supports using MSNT as the coupling agent in the presence of NMI.[14] Best results are obtained with the ratio Fmoc-AA:MSNT:NMI = 1:1:0.75. The MSNT:NMI method proved to be of general utility and gives better results than the conventional symmetrical anhydride/DMAP method. This protocol has been applied to the anchoring of Fmoc-Ser(*t*-Bu)OH onto ArgoGel resin modified with a fluorine-labeled alcohol linker (eq 9).[15]

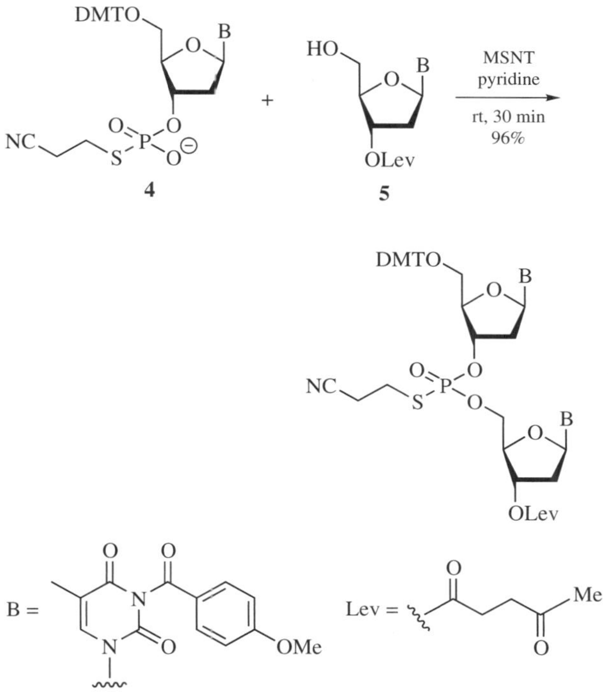

(7)

N-Boc-Phe-OH + H_2N-Ala-OMe $\xrightarrow[\text{DMF, 1 min; 84.3\%}]{\text{MSNT, NMI}}$ *N*-Boc-Phe-Ala-OMe (8)

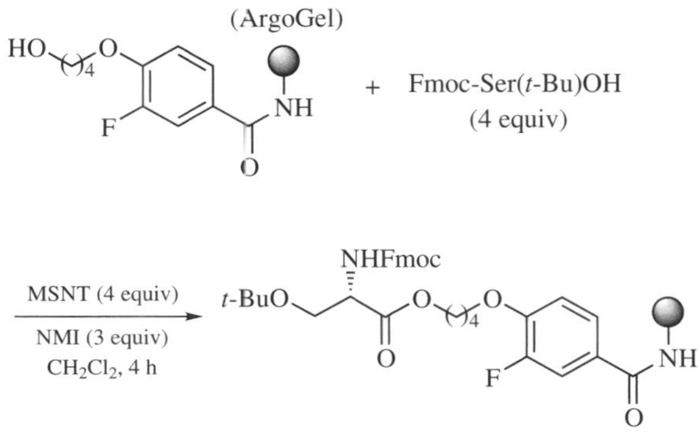

(9)

70% for the first coupling;
100% after a second cycle

Aminoacylation of the 5′-hydroxyl group of partially protected oligouridylic acids on controlled pore glass support is achieved with high yield by using MSNT as the condensing agent in the presence of Hunig's base (eq 10).[16] The same reactions using other coupling methods, including DCC/HOBT, DCC/DMAP, or PPh_3/DIAD, either failed or gave much lower conversion.

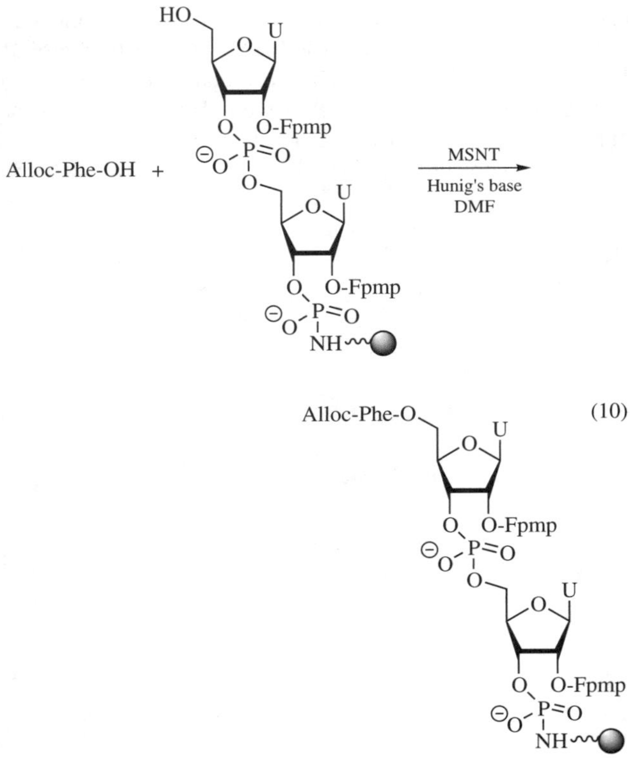

(10)

Fpmp = 1-(2-fluorophenyl)-4-methoxypiperidin-4-yl

1. Jones, S. S.; Rayner, B.; Reese, C. B.; Ubasawa, A.; Ubasawa, M., *Tetrahedron* **1980**, *36*, 3075–3085.
2. Reese, C. B.; Titmas, R. C.; Yau, L., *Tetrahedron Lett.* **1978**, *19*, 2727–2730.
3. Chattopadhyaya, J. B.; Reese, C. B., *Tetrahedron Lett.* **1979**, *20*, 5059–5062.
4. Reese, C. B., *Tetrahedron* **1978**, *34*, 3143–3179.
5. Gait, M. J.; Matthes, H. W. D.; Singh, M.; Titmas, R. C., *J. Chem. Soc., Chem. Commun.* **1982**, 37–40.
6. Sproat, B. S.; Bannwarth, W., *Tetrahedron Lett.* **1983**, *24*, 5771–5774.
7. Bonora, G. M.; Scremin, C. L.; Colonna, F. P.; Garbesi, A., *Nucleosides Nucleotides* **1991**, *10*, 269–273.
8. De Napoli, L.; Galeone, A.; Mayol, L.; Messere, A.; Piccialli, G.; Santacroce, C., *J. Chem. Soc., Perkin Trans. 1* **1993**, 747–749.
9. Tanaka, T.; Orita, M.; Uesugi, S.; Ikehara, M., *Tetrahedron* **1988**, *44*, 4331–4338.
10. Reese, C. B.; Song, Q.; Rao, M. V.; Beckett, I., *Nucleosides Nucleotides* **1998**, *17*, 451–470.
11. Reese, C. B.; Song, Q., *Nucleic Acids Res.* **1999**, *27*, 2672–2681.
12. Bannwarth, W., *Chimia* **1987**, *41*, 302–316.
13. Jorba, X.; Albericio, F.; Grandas, A.; Bannwarth, W.; Giralt, E., *Tetrahedron Lett.* **1990**, *31*, 1915–1918.
14. Blankemeyer-Menge, B.; Nimtz, M.; Frank, R., *Tetrahedron Lett.* **1990**, *31*, 1701–1704.
15. Mogemark, M.; Elofsson, M.; Kihlberg, J., *Org. Lett.* **2001**, *3*, 1463–1466.
16. Tetzlaff, C. N.; Richert, C., *Tetrahedron Lett.* **2001**, *42*, 5681–5684.

Qingwei Yao
Northern Illinois University, DeKalb, IL, USA

Mesitylsulfonyl-1*H*-1,2,4-triazole

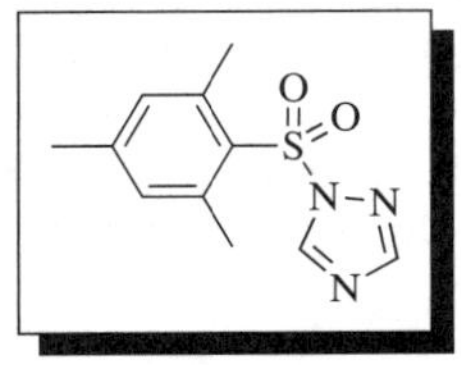

[54230-59-0] $C_{11}H_{13}N_3O_2S$ (MW 251.34)

(reagent for phosphate activation[1] and nucleotide condensation[2])

Alternate Name: MST.
Physical Data: mp 135 °C.
Form Supplied in: solid; widely available from commercial sources.
Preparative Method: readily prepared by the reaction of equimolar ratios of Mesitylenesulfonyl Chloride, 1,2,4-Triazole, and Triethylamine in a chloroform solution (eq 1).[3]

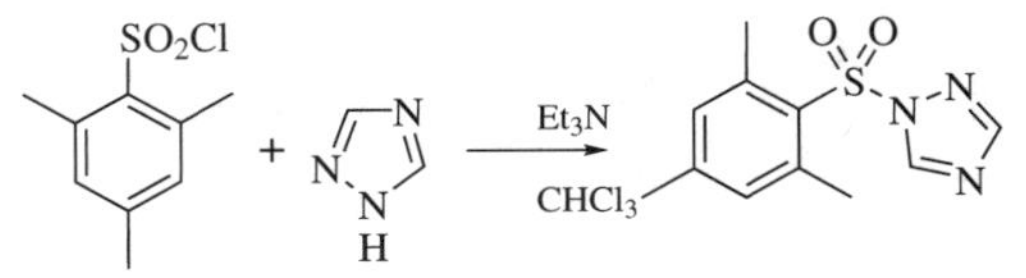

(1)

Purification: recrystallization from benzene.

Phosphate Activator. MST was shown to be an effective catalyst for the phosphorylation of a 5′-protected mononucleoside.[1] MST was stirred with *p*-chlorophenyl phosphorodichloridate and triethylamine and the resulting agent was used directly for the phosphorylation of 5′-protected nucleosides in 70–85% yields.

Oligonucleotide Synthesis. As originally reported, the triester approach to the formation of oligonucleotides involved using triisopropylbenzenesulfonyl chloride (TPS) as a condensing agent for a 5′–3′ internucleotide condensation. Since that time, many modifications of this procedure have been reported, including (i) using a stepwise phosphorylation then condensation procedure,[2] (ii) synthesizing the oligonucleotide in the 3′–5′ rather than the 5′–3′ direction,[2] (iii) varying the protecting groups used in these transformations, and (iv) changing the condensing agent. MST is one of the more recently reported condensing agents for the synthesis of oligonucleotides through the formation of 3′,5′-internucleotide bonds via the triester approach. The initial report of the use of MST was in the synthesis of a fully protected dinucleotide (eq 2).[3]

A follow-up full paper by the same authors extended this methodology to the synthesis of hexanucleotides by an iterative phosphorylation and condensation sequence.[1] This paper also kinetically compared this reagent to some of the previously used condensing agents. It was reported that although the reaction using MST was significantly slower than that using TPS, the reaction mixture was much cleaner and the yields were much higher. This was especially true for synthesis of nucleotides that contained purine units, which were formed only in low yields when

TPS was employed as the condensing agent, perhaps due to the liberation of HCl from the reagent. MST was also employed as the condensing reagent in the synthesis of 3′,5′-bisphosphorylated oligonucleotides.[4]

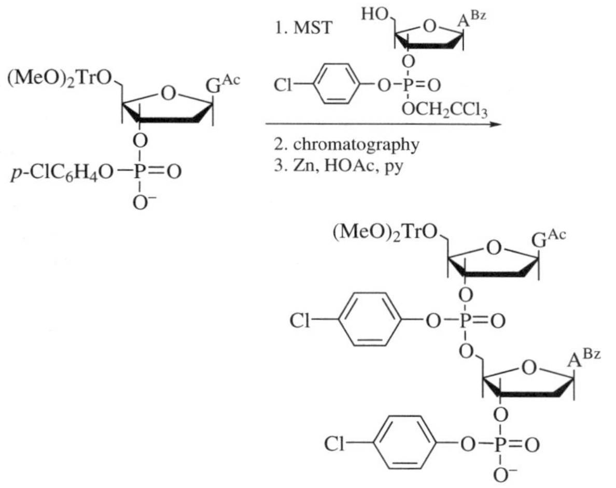

(2)

Oligoribonucleotide Synthesis. Subsequent work showed that this reagent was useful in the synthesis of the less stable oligoribonucleotides as well as deoxyribonucleotides.[5] In the same iterative manner as described above for deoxyribonucleotide synthesis, the tetrameric ribonucleotide GpApGpC was formed(eq 3).

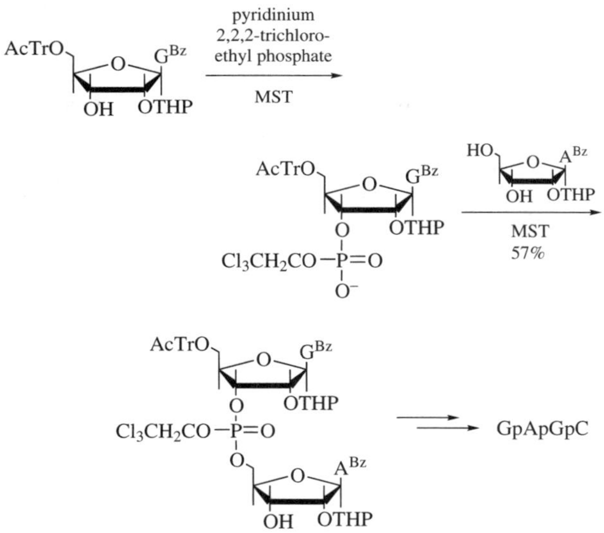

(3)

Yields were reported to be significantly improved over previously described methods that employed TPS as the condensing agent, especially in the coupling of purine residues. Note again that MST can also be used in the phosphorylation step of the condensation sequence. MST was also used as a condensing agent for the synthesis of 2′(3′)-*O*-aminoacyl triribonucleoside diphosphates.[6]

1. Katagiri, N.; Itakura, K.; Narang, S. A. *J. Am. Chem. Soc.* **1975**, *97*, 7332.
2. Itakura, K.; Katagiri, N.; Bahl, C. P.; Wightman, R. H.; Narang, S. A. *J. Am. Chem. Soc.* **1975**, *97*, 7327.
3. Katagiri, N.; Itakura, K.; Narang, S. A. *J. Chem. Soc. (C)* **1974**, 325.
4. Ikehara, N.; Oshie, K.; Hasegawa, A.; Ohtsuka, E. *Nucleic Acids Res.* **1981**, *9*, 2003.
5. England, T. E.; Nielson, T. *Can. J. Chem.* **1976**, *54*, 1714.
6. Kumar, G.; Chladek, S. *Tetrahedron Lett.* **1981**, *22*, 827.

Valerie Vaillancourt & Michele M. Cudahy
The Upjohn Co., Kalamazoo, MI, USA

p-Methoxybenzyl Chloride

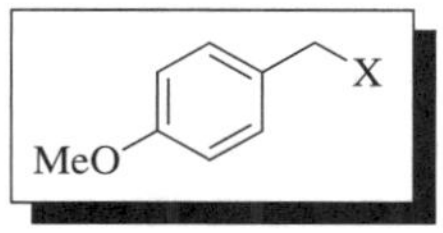

(X = Cl)		
[824-98-6]	C_8H_9ClO	(MW 156.62)
(X = Br)		
[2746-25-0]	C_8H_9BrO	(MW 201.07)

(protection of alcohols,[3] thiols,[42] phenols,[29] amides,[55] amines,[50] and carboxylic acids[31])

Alternate Name: MPMCl.
Physical Data: X = Cl: bp 117–118 °C/14 mmHg; *d* 1.155 g cm^{-3}.
Solubility: sol all organic solvents.
Form Supplied in: the chloride is commercially available. The bromide is not commercially available and is prepared immediately before use because of its instability.[2]
Preparative Methods: to prepare the bromide, an ether solution of the alcohol is stirred with concd HBr, the phases are separated, and the organic phase is washed with saturated aq. NaBr, dried over K_2CO_3, and concentrated.
Handling, Storage, and Precautions: the chloride is stored over K_2CO_3 to stabilize it; the bromide is stored in the freezer to prevent polymerization, which occurs within a few days at rt.[2]

Protection of Alcohols. The inherent stability of the MPM ether, coupled with a large repertoire of methods for its removal under mild conditions that do not normally effect other functional groups, makes it a particularly effective derivative for the protection of alcohols. The most common method for its introduction is by the Williamson ether synthesis. A number of bases can be used to generate the alkoxide, but Sodium Hydride[3] in DMF (eq 1) or THF[4] (eq 2) is the most common. Other bases such as n-Butyllithium, [5] Potassium Methylsulfinylmethylide (dimsylpotassium)[6] (eq 3), and Sodium Hydroxide under phase-transfer conditions[7] are also used. From these results, it is clear that protection can be achieved without interference from Payne rearrangement, and considerable selectivity can be obtained. In the ribose case, selectivity is probably achieved because of the increased acidity of the 2′-hydroxy group. The additive Tetra-n-butylammonium Iodide[8] is used for in situ preparation of the highly reactive *p*-methoxybenzyl iodide, thus improving the protection of very hindered alcohols. Selective monoprotection of diols is readily occasioned with *O*-stannylene acetals.[9]

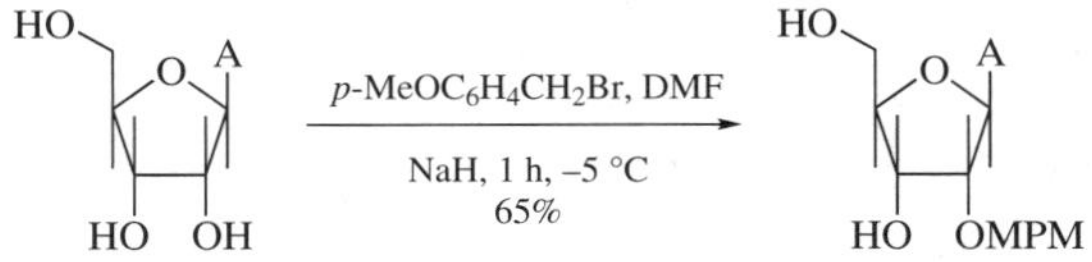

(1)

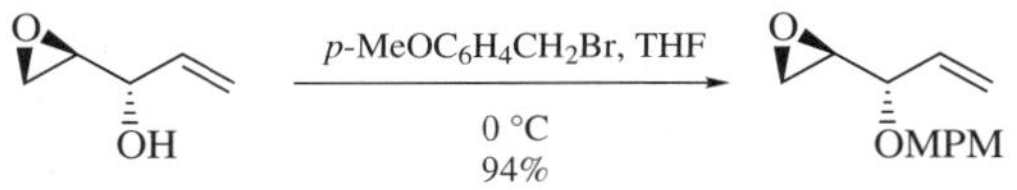

(2)

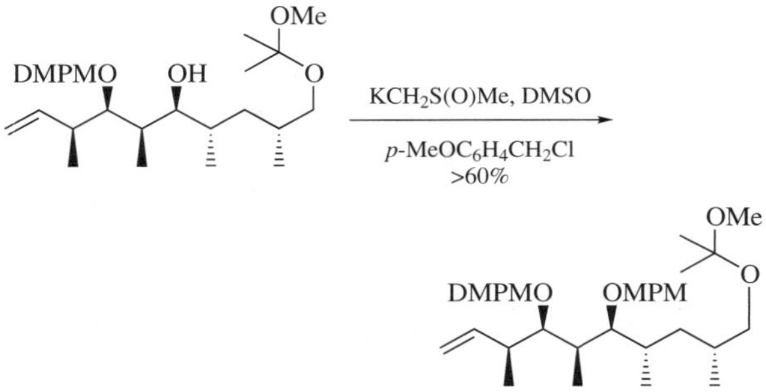

(3)

Cleavage of the MPM group is usually achieved oxidatively with 2,3-Dichloro-5,6-dicyano-1,4-benzoquinone.[10–12] When DDQ is used, overoxidation of the alcohol occasionally occurs,[13,14] but for the most part this is not a problem. In general, selective cleavage of the MPM group is achieved in molecules containing benzyl groups, but more forcing conditions will result in benzyl cleavage.[15] Cerium(IV) Ammonium Nitrate,[16] trityl tetrafluoroborate,[17] photolysis,[18] electrolytic oxidation,[19] dimethylboron bromide,[20] Boron Trichloride,[21] Iodotrimethylsilane,[22] Tin(II) Chloride/ Chlorotrimethylsilane/anisole,[23] hydrogenolysis,[24] and ozonolysis[25] have also been used to cleave this derivative. Selectivity for the *O*-MPM is achieved over an amide MPM derivative with DDQ.[26]

In the presence of neighboring hydroxyl groups, the DDQ oxidation of MPM ethers leads to cyclic acetals, thus affording a level of selectivity which may not be possible to achieve otherwise.[27] In a synthesis of erythronolide (eq 4), the critical oxidative

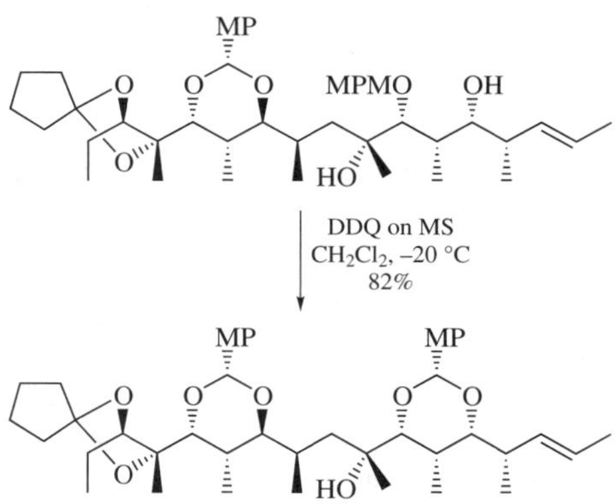

(4)

cyclization had to be performed with freshly recrystallized DDQ supported on molecular sieves in order to prevent acid-catalyzed isomerization of the acetal from its initial kinetic disposition to the more thermodynamically favorable orientation.[28]

Protection of Phenols. Protection of phenols is achieved using *p*-methoxybenzyl chloride ($Bu_4N^+I^-$, K_2CO_3) in acetone,[29] or the benzyl bromide (*i*-Pr_2NEt) in CH_2Cl_2.[30] Deprotection is occasioned with Trifluoroacetic Acid[29] or 10-Camphorsulfonic Acid/$Me_2C(OMe)_2$.[30]

Protection of Carboxylic Acids. *p*-Methoxybenzyl esters have been prepared from the Ag^I salt of amino acids and the benzyl halide (Et_3N, $CHCl_3$, 25 °C, 24 h, 60% yield),[31] and from cephalosporin precursors and the benzyl alcohol [$Me_2NCH(OCH_2$-*t*-$Bu)_2$, CH_2Cl_2, 90% yield],[32] and from the benzyl halide ($NaHCO_3$, DMF).[33] They are also prepared by activation of the acid with isopropenyl oxychloroformate ($MeOC_6H_4CH_2OH$, DMAP, 0 °C, CH_2Cl_2, 91%).[34] They are cleaved by acidic hydrolysis (CF_3CO_2H/ PhOMe, 25 °C, 3 min, 98% yield;[35] HCO_2H, 22 °C, 1 h, 81% yield),[31] by heating with phenol,[36] by treatment with $AlCl_3$ (anisole, CH_2Cl_2 or $MeNO_2$, −50 °C; $NaHCO_3$, −50 °C, 73–95% yield),[37,38] or by $CF_3CO_2H/B(OTf)_3$.[39]

Protection of Thiols. The protection of thiols as *S*-*p*-methoxybenzyl thioethers is occasioned with MPMCl and a base such as NH_3,[40] KOH (MeOH),[41] and NaH (THF, 60 °C, 1 h).[42] The reduction of disulfides with Lithium Aluminum Hydride affords an intermediate thiolate that can be trapped with MPMCl to give the thioether.[43] The cleavage of *S*-*p*-methoxybenzyl thioethers is effected with $Hg(OAc)_2/CF_3CO_2H$ (0 °C, 10–30 min), $Hg(OCOCF_3)_2$/aq AcOH (20 °C, 2–3 h, followed by H_2S or $HSCH_2CH_2OH$, 100% yield),[44,45] $Hg(OCOCF_3)_2/CF_3CO_2H$/ anisole,[46] CF_3CO_2H/reflux,[40] anhydrous HF/anisole (25 °C, 1 h, quant),[47] $(4\text{-}BrC_6H_4)_3NH^+\ SbCl_6^-$ (75%),[48] HCO_2H (5 °C, 30 min, 45%).[42] During the synthesis of peptides which contain 4-methoxybenzyl-protected cysteine residues, sulfoxide formation may occur (eq 5). These sulfoxides, when treated with HF/anisole, form thiophenyl ethers that cannot be deprotected; therefore the peptides should be subjected to a reduction step prior to deprotection.[49]

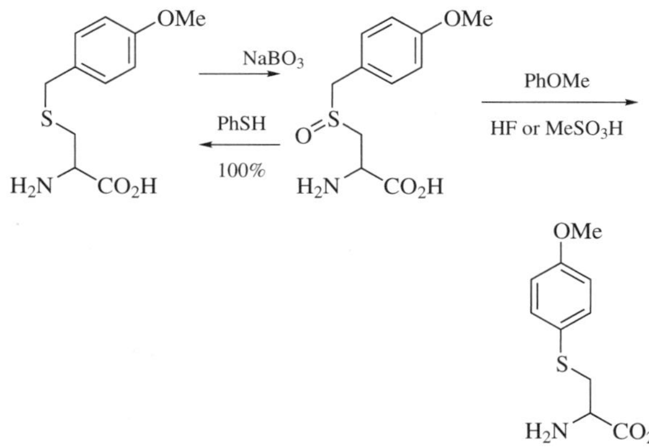

(5)

Protection of Amines. The use of the MPM group to protect amines has not seen many applications, but a few examples are available in the literature. The MPM group can be introduced on an aromatic amine with MPMBr (DMF, KI, K_2CO_3, 92%),[50] or by reductive amination (anisaldehyde, PhH, H_2O; $NaBH_4$, MeOH, 92%),[51] but for the most part the MPM group acts as a carrier for the nitrogen to be introduced into a molecule. MPM amines can be cleaved by hydrogenolysis,[52] oxidation to the imine followed by hydrolysis,[53] or acylation with MeCHClOCOCl (CH_2Cl_2, −7 °C, 5 min; MeOH, reflux, 89%).[54]

Protection of the Amide Nitrogen. Of the few methods for protection of the amide NH, the MPM group has found success where other methods have not been entirely successful.[55] The group can be introduced using methods similar to those used for alcohols: MPMCl/NaH/DMF,[56] MPMBr/NaH,[55] MPMCl (DBU, MeCN, 60 °C, 5 h, 92%),[57] MPMCl/Ag_2O.[58]

The most common method for removal of the amide MPM group is by oxidation with DDQ[59] or CAN.[60] With CAN, the benzoyl imide is sometimes a byproduct of the reaction (eq 6), but this is easily cleaved by hydrolysis.[61] The amide MPM group is also cleaved by solvolysis in TFA[51] or $AlCl_3$/anisole.[62] Metalation with t-Butyllithium followed by reaction with oxygen (62%) has been used successfully (eq 7).[63]

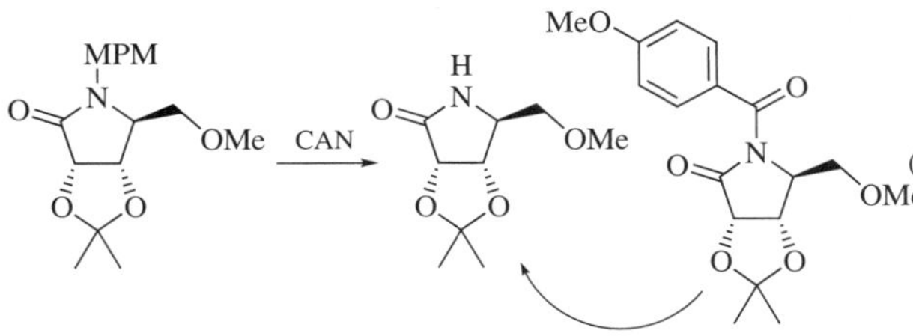

(6)

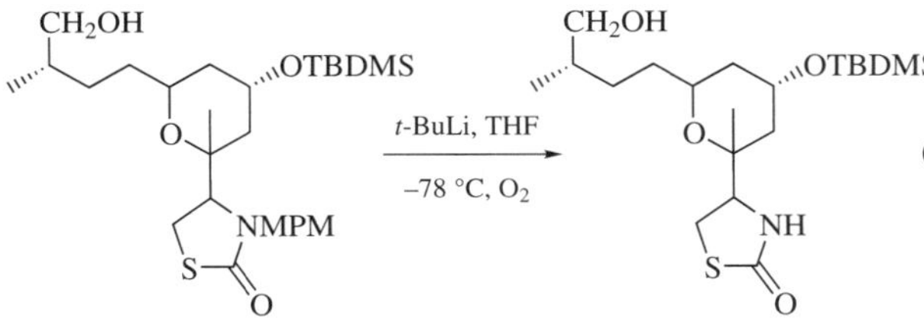

(7)

Protection of Phosphates. There is one example where the MPM group is used for phosphate protection. It is cleaved with 48% aq HF/MeCN/H_2O (eq 8).[64]

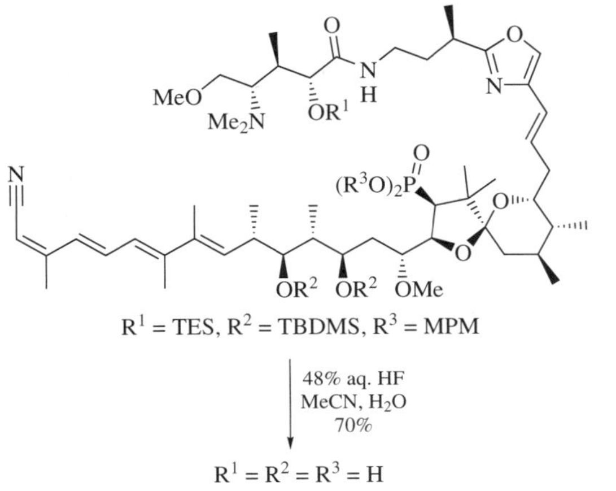

(8)

Related Reagents. Benzyl Bromide; Benzyl Chloride; Benzyl Iodide; 3,4-Dimethoxybenzyl Bromide; (p-Methoxybenzyloxy)methyl Chloride; 4-Methoxybenzyl 2,2,2-Trichloroacetimidate.

---

1. For a review of protective groups see:Greene, T. W.; Wuts, P. G. M. *Protective Groups in Organic Synthesis*, 2nd ed.; Wiley:New York, **1991**.
2. Ruder, S. M.; Ronald, R. C. *Tetrahedron Lett.* **1987**, *28*, 135.
3. Takaku, H.; Kamaike, K.; Tsuchiya, H. *J. Org. Chem.* **1984**, *49*, 51.
4. Schreiber, S. L.; Smith, D. B. *J. Org. Chem.* **1989**, *54*, 9.
5. Hoshi, H.; Ohnuma, T.; Aburaki, S.; Konishi, M.; Oki, T. *Tetrahedron Lett.* **1993**, *34*, 1047.
6. Nakajima, N.; Hamada, T.; Tanaka, T.; Oikawa, Y.; Yonemitsu, O. *J. Am. Chem. Soc.* **1986**, *108*, 4645.
7. Garegg, P. J.; Oscarson, S.; Ritzen, H. *Carbohydr. Res.* **1988**, *181*, 89.
8. (a) Mootoo, D. R.; Fraser Reid, B. *Tetrahedron* **1990**, *46*, 185. (b) Trost, B. M.; Hipskind, P. A.; Chung, J. Y. L.; Chan, C. *Angew. Chem., Int. Ed. Engl.* **1989**, *28*, 1502.
9. (a) Nagashima, N.; Ohno, M. *Chem. Lett.* **1987**, 141. (b) Verduyn, R.; Elle, C. J. J.; Dreef, C. E.; Van der Marel, G. A.; van Boom, J. H. *Recl. Trav. Chim. Pays-Bas* **1990**, *109*, 591. (c) Danishefsky, S.; Lee, J. Y. *J. Am. Chem. Soc.* **1989**, *111*, 4829.
10. (a) Horita, K.; Yoshioka, T.; Tanaka, T.; Oikawa, Y.; Yonemitsu, O. *Tetrahedron* **1986**, *42*, 3021. (b) Tanaka, T.; Oikawa, Y.; Hamada, T.; Yonemitsu, O. *Tetrahedron Lett.* **1986**, *27*, 3651.
11. Oikawa, Y.; Tanaka, T.; Horita, K.; Yonemitsu, O. *Tetrahedron Lett.* **1984**, *25*, 5397.
12. Oikawa, Y.; Yoshioka, T.; Yonemitsu, O. *Tetrahedron Lett.* **1982**, *23*, 885.
13. Nagaoka, H.; Baba, A.; Yamada, Y. *Tetrahedron Lett.* **1991**, *32*, 6741.
14. Trost, B. M.; Chung, J. Y. L. *J. Am. Chem. Soc.* **1985**, *107*, 4586.
15. (a) Horita, K.; Nagato, S.; Oikawa, Y.; Yonemitsu, O. *Tetrahedron Lett.* **1987**, *28*, 3253. (b) Ikemoto, N.; Schreiber, S. L. *J. Am. Chem. Soc.* **1992**, *114*, 2525.
16. (a) Classon, B.; Garegg, P. J.; Samuelsson, B. *Acta Chem. Scand. (B)* **1984**, *B38*, 419. (b) Johansson, R.; Samuelsson, B. *J. Chem. Soc., Perkin Trans. 1* **1984**, 2371.
17. Nakajima, N.; Abe, R.; Yonemitsu, O. *Chem. Pharm. Bull.* **1988**, *36*, 1988.
18. Pandey, G.; Krishna, A. *Synth. Commun.* **1988**, *18*, 2309.
19. (a) Schmidt, W.; Steckhan, E. *Angew. Chem., Int. Ed. Engl.* **1979**, *18*, 801. (b) Weinreb, S. M.; Epling, G. A.; Comi, R.; Reitano, M. *J. Org. Chem.* **1975**, *40*, 1356.
20. Hebert, N.; Beck, A.; Lennox, R. B.; Just, G. *J. Org. Chem.* **1992**, *57*, 1777.
21. Vanhessche, K.; Bello, C. G.; Vandewalle, M. *Synlett* **1991**, 921.
22. Gordon, D. M.; Danishefsky, S. J. *J. Am. Chem. Soc.* **1992**, *114*, 695.
23. Akiyama, T.; Shima, H.; Ozaki, S. *Synlett* **1992**, 415.
24. Hikota, M.; Tone, H.; Horita, K.; Yonemitsu, O. *J. Org. Chem.* **1990**, *55*, 7.
25. Hirama, M.; Shimizu, M. *Synth. Commun.* **1983**, *13*, 781.
26. Hamada, Y.; Tanada, Y.; Ykokawa, F.; Siori, T. *Tetrahedron Lett.* **1991**, *32*, 5983.
27. Oikawa, Y.; Yoshioka, T.; Yonemitsu, O. *Tetrahedron Lett.* **1982**, *23*, 889.
28. Stürmer, R.; Ritter, K.; Hoffmann, R. W. *Angew. Chem., Int. Ed. Engl.* **1993**, *32*, 101.
29. White, J. D.; Amedio, J. C., Jr. *J. Org. Chem.* **1989**, *54*, 736.
30. Nagaoka, H.; Schmid, G.; Iio, H.; Kishi, Y. *Tetrahedron Lett.* **1981**, *22*, 899.
31. Stelakatos, G. C.; Argyropoulos, N. *J. Chem. Soc. (C)* **1970**, 964.
32. Webber, J. A.; Van Heyningen, E. M.; Vasileff, R. T. *J. Am. Chem. Soc.* **1969**, *91*, 5674.

33. Baldwin, J. E.; Flinn, A. *Tetrahedron Lett.* **1987**, *28*, 3605.
34. Jouin, P.; Castro, B.; Zeggaf, C.; Pantaloni, A.; Senet, J. P.; Lecolier, S.; Sennyey, G. *Tetrahedron Lett.* **1987**, *28*, 1661.
35. Stewart, F. H. C. *Aust. J. Chem.* **1968**, *21*, 2543.
36. Tanaka, H.; Taniguchi, M.; Kameyama, Y.; Torii, S.; Sasaoka, M.; Shiroi, T.; Kikuchi, R.; Kawahara, I.; Shimabayashi, A.; Nagao, S. *Tetrahedron Lett.* **1990**, *31*, 6661.
37. Ohtani, M.; Watanabe, F.; Narisada, M. *J. Org. Chem.* **1984**, *49*, 5271.
38. Tsuji, T.; Kataoka, T.; Yoshioka, M.; Sendo, Y.; Nishitani, Y.; Hirai, S.; Maeda, T.; Nagata, W. *Tetrahedron Lett.* **1979**, 2793.
39. Young, S. D.; Tamburini, P. P. *J. Am. Chem. Soc.* **1989**, *111*, 1933.
40. Akabori, S.; Sakakibara, S.; Shimonishi, Y.; Nobuhara, Y. *Bull. Chem. Soc. Jpn.* **1964**, *37*, 433.
41. Chambers, M. S.; Thomas, E. J. *Chem. Commun./J. Chem. Soc.* **1989**, 23.
42. Taylor, A. W.; Dean, D. K. *Tetrahedron Lett.* **1988**, *29*, 1845.
43. Baidoo, K. E.; Lever, S. Z. *Tetrahedron Lett.* **1990**, *31*, 5701.
44. Nishimura, O.; Kitada, C.; Fujino, M. *Chem. Pharm. Bull.* **1978**, *26*, 1576.
45. Gordon, E. M.; Godfrey, J. D.; Delaney, N. G.; Asaad, M. M.; Von Langen, D.; Cushman, D. W. *J. Med. Chem.* **1988**, *31*, 2199.
46. Holler, T. P.; Spaltenstein, A.; Turner, E.; Klevit, R. E.; Shapiro, B. M.; Hopkins, P. B. *J. Org. Chem.* **1987**, *52*, 4420.
47. Sakakibara, S.; Shimonishi, Y. *Bull. Chem. Soc. Jpn.* **1965**, *38*, 1412.
48. Platten, M.; Steckhan, E. *Justus Liebigs Ann. Chem./Liebigs Ann. Chem.* **1984**, 1563.
49. Funakoshi, S.; Fujii, N.; Akaji, K.; Irie, H.; Yajima, H. *Chem. Pharm. Bull.* **1979**, *27*, 2151.
50. Tamato, M.; Takeuchi, Y.; Ikeda, Y. *Heterocycles* **1987**, *26*, 191.
51. Smith, A. B., III; Rano, T. A.; Chida, N.; Sulikowski, G. A. *J. Org. Chem.* **1990**, *55*, 1136.
52. Rowley, M. *Tetrahedron* **1992**, *48*, 3557.
53. Maruyama, K.; Kusukawa, T.; Higuchi, Y.; Nishinaga, A. *Chem. Lett.* **1991**, 1093.
54. Yang, B. V.; O'Rourke, D.; Li, J. *Synlett* **1993**, 195.
55. Smith, A. B., III; Leahy, J. W.; Noda, I.; Remiszewski, S. W.; Liverton, N. J.; Zibuck, R. *J. Am. Chem. Soc.* **1992**, *114*, 2995.
56. Williams, R. M.; Maruyama, L. K. *J. Org. Chem.* **1987**, *52*, 4044.
57. Akiyama, T. *Bull. Chem. Soc. Jpn.* **1990**, *63*, 3356.
58. Takahashi, Y.; Yamshita, H.; Kobayashi, S.; Ohno, M. *Chem. Pharm. Bull.* **1986**, *34*, 2732.
59. Smith, A. B., III; Rano, T. A.; Chida, N.; Sulikowski, G. A.; Wood, J. L. *J. Am. Chem. Soc.* **1992**, *114*, 8008.
60. Williams, R. M.; Sabol, M. R.; Kim, H.-d.; Kwast, A. *J. Am. Chem. Soc.* **1991**, *113*, 6621.
61. Smith, A. B., III; Salvatore, B. A.; Hull, K. G.; Duan, J.J.-W. *Tetrahedron Lett.* **1991**, *32*, 4859.
62. Akiyama, T.; Kumegawa, M.; Takesue, Y.; Nishimoto, H.; Ozaki, S. *Chem. Lett.* **1990**, 339.
63. Smith, A. B., III; Noda, I.; Remiszewski, S. W.; Liverton, N. J.; Zibuck, R. *J. Org. Chem.* **1990**, *55*, 3977.
64. Evans, D. A.; Gage, J. R.; Leighton, J. L. *J. Am. Chem. Soc.* **1992**, *114*, 9434.

Peter G. M. Wuts
The Upjohn Co., Kalamazoo, MI, USA

Methoxyphenyl Benzenethiosulfinate (MPBT)

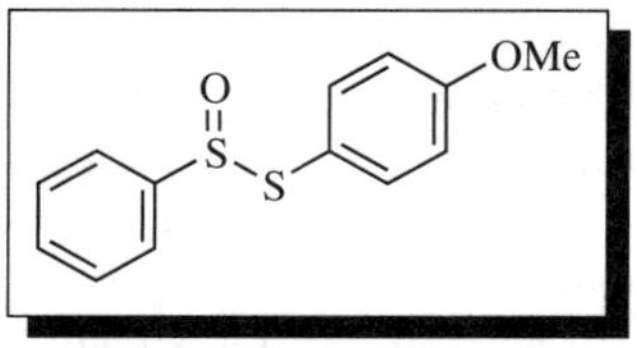

[26974-26-5] $C_{13}H_{12}O_2S_2$ (MW 264.36)

(reagent used for the activation of thioglycosides[1])

Physical Data: pale yellow crystals, mp 77–78 °C; mass spectrum.[2]
Solubility: soluble in most organic solvents.
Analysis of Reagent Purity: ^{1}H and ^{13}C NMR.[1]
Preparative Methods: the reagent is prepared from benzenesulfinyl chloride and *p*-methoxythiophenol (eq 1).[1]

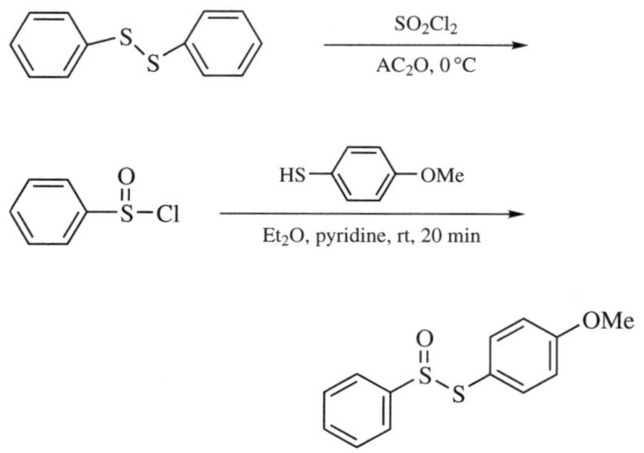

(1)

Purification: recrystallization from ether-petroleum ether (bp 40–60 °C).
Handling, Storage, and Precautions: crystalline, odorless, and stable on the laboratory bench at room temperature when stored in standard amber bottles.

***S*-(4-Methoxyphenyl) Benzenethiosulfinate (MPBT)/Trifluoromethanesulfonic Anhydride (Tf_2O): A Powerful Activating System for Thioglycosides and for the Formation of Glycosidic Linkages.** The combination of MPBT with Tf_2O forms a powerful, metal free, thiophile that can readily activate thioglycosides via glycosyl triflates in a matter of minutes at −60 °C in dichloromethane, in the presence of 2,6-di-*tert*-butyl-4-methylpyridine (DTBMP).[2] The glycosyl triflates are rapidly and cleanly converted to glycosides upon treatment with alcohols. This method has allowed access to β-mannosides (eq 2) and α-glucosides in good yield and selectivity.

Generally, the combination of MPBT/Tf_2O cannot activate disarmed thioglycosides at low temperature, but in some case when the more electron-donating *p*-methoxyphenylthio group is used as the anomeric function, good yield and stereoselectivity can still be achieved (eq 3).[3]

This limitation of activation has been overcome by the combination of 1-benzenesulfinyl piperidine (BSP) and Tf_2O.[4]

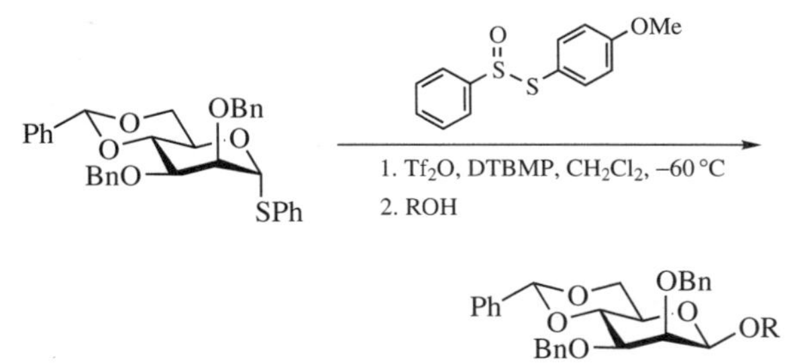

(2)

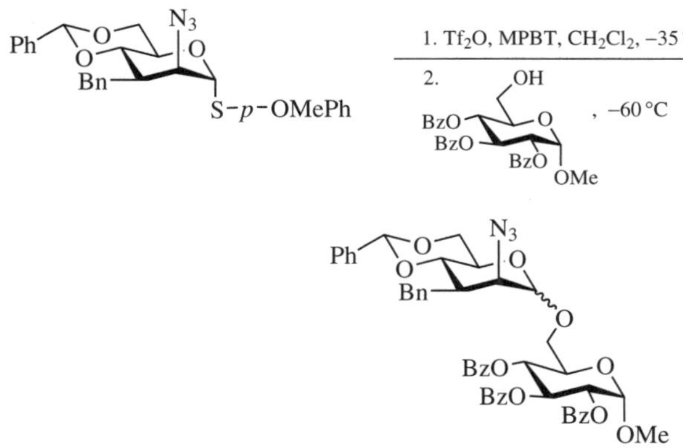

(3)

yield: 87%
ratio: α/β = 1/4

Other Reactions. In some glycosylation reactons, stoichiometric amount of *S*-(4-methoxyphenyl)benzenethiosulfinate is reported to function as an acid scavenger to avoid the undesired removal of PMB protection that prevailed on activation by Tf_2O alone (eq 4).[5]

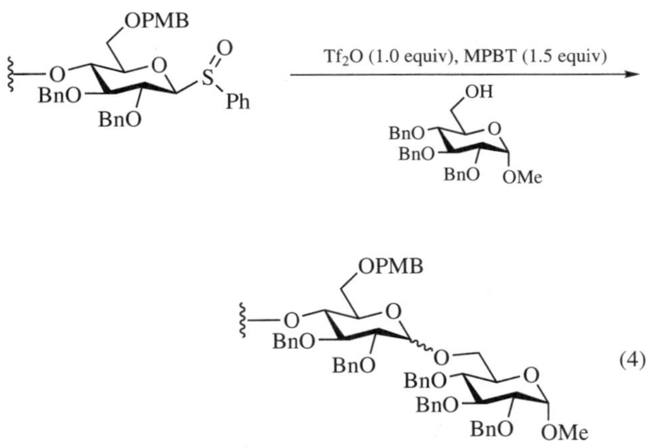

(4)

yield: 58%
ratio: α/β = 1/4

It is reported that the mixture of disulfide and thiosulfinate, such as *S*-(4-methoxyphenyl) benzenethiosulfinate, can be used as an antioxdant for the retardation of autoxidation of olefins.[6]

Related Reagents. 1-Benzenesulfinyl piperidine; dimethyl (methylthio)sulfonium tetrafluoroborate; *N*-iodosuccinimide; diphenyl sulfoxide.

1. Crich, D.; Smith, M., *Org. Lett.* **2000**, *2*, 4067–4069.
2. Kozuka, S.; Takahashi, H.; Oae, S., *Bull. Chem. Soc. Jpn.* **1970**, *43*, 129–134.
3. Litjens, R. E. J. N.; Leeuwenburgh, M. A.; Marel, G. A.; Boom, J. H., *Tetrahedron Lett.* **2001**, *42*, 8693–8696.
4. Crich, D.; Smith, M., *J. Am. Chem. Soc.* **2001**, *123*, 9015–9020.
5. Ferguson, J.; Marzabadi, C., *Tetrahedron Lett.* **2003**, *44*, 3573–3577.
6. Cole, E. R., *Phosohorus Sulfur Silicon* **1991**, *63*, 147–159.

Wenju Li
University of Illinois at Chicago, IL, USA

1-(4-Methoxyphenyl)-2-(4′-methylbenzylthio)ethylamine[1]

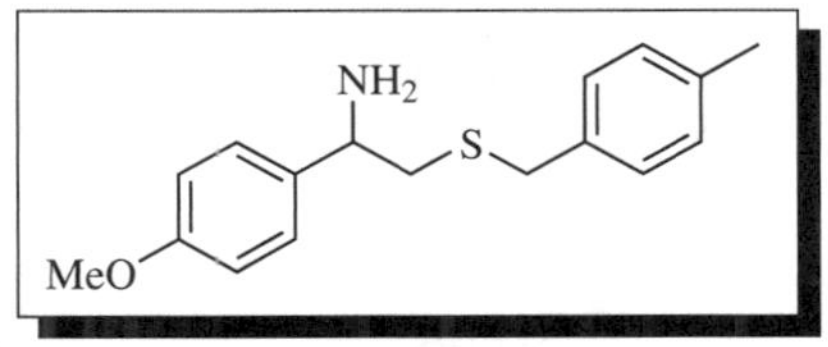

[403739-55-9] $C_{17}H_{21}NOS$ (MW 287.42)

(auxiliary used in the native chemical ligation of peptides and proteins)

Form Supplied in: not commercially available.

Preparative Methods: 1-(4-methoxyphenyl)-2-(4′-methylbenzylthio)ethylamine **3** was prepared in three steps from *p*-methoxy bromoacetophenone **1**. Reaction with a suitable resin-bound bromoacetyl peptide,[2] followed by *S*-deprotection gives the peptide **4**, derivatized with the N^{α}-(1-(4-methoxyphenyl)-2-mercaptoethyl) auxiliary (eq 1).

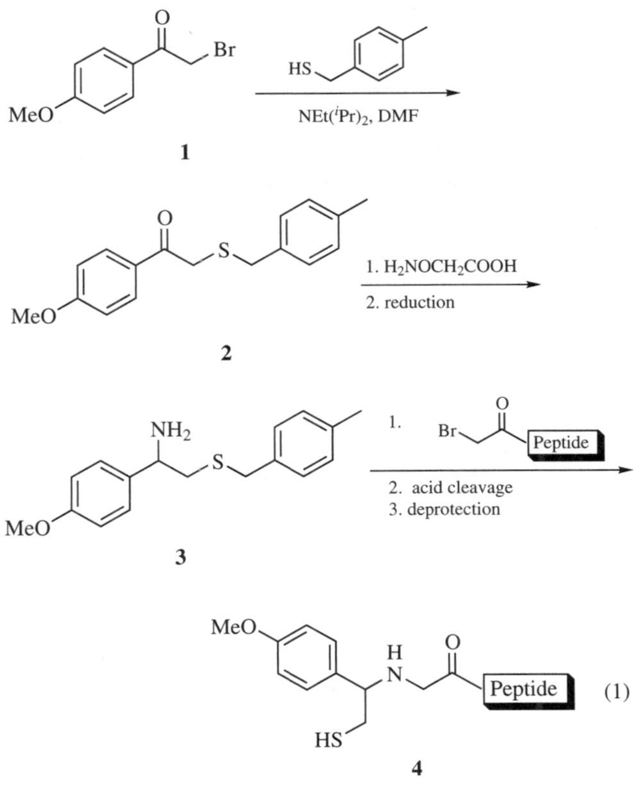

(1)

The title compound **3** has been used in native chemical ligation (NCL) for the chemical synthesis of proteins.[3–7] NCL selectively couples unprotected peptide segments to form larger polypeptides and proteins (up to ~300 amino acid residues). The reaction occurs in aqueous systems at pH 7.0–7.2, and is chemoselective between an *N*-terminal cysteine peptide and a thioester. The thioester undergoes dynamic thiol exchange with the cysteine functionality (usually catalyzed by an aryl or alkyl thiol), which then rearranges "irreversibly" through a five-membered transition state to the native amide bond (eq 2). The chemoselectivity of this reaction arises from: 1. the protonation of other nucleophiles in the peptide at pH 7 (disfavoring undesirable acylation, *e.g.*, at the γ-amino group of lysine residues); 2. the facile thioester exchange between COSR and the cysteine SH; and 3. the rapid rearrangement of the resulting α-amino thioester. This method has been combined with molecular biology techniques such as protein design and expression to expand the scope of proteins that can be synthesized and modified.[8,9]

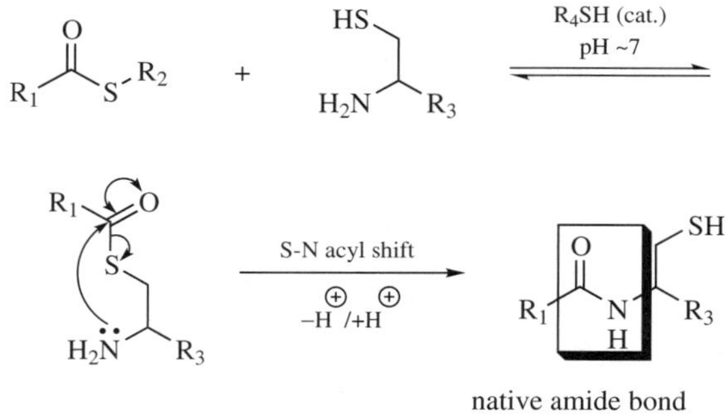

(2)

One of the major limitations of the thioester-cysteine mediated ligation is the necessity for cysteine residues in the target molecule. Similar ligation strategies have been developed that exploit alternative acyl shifts between other functionality, *e.g.*, the imidazole of histidine residues.[5] However, the most general of these alternative strategies involves the use of removable ligation auxiliaries such as **3**, which enable disconnection at a glycine residue. These auxiliaries provide an aryl or alkyl thiol α- or β- to an amine for thioester exchange and rearrangement to the amide. The use of auxiliary **3** offers several advantages over earlier systems, including rapid ligation rates and ease of removal post-ligation. The effectiveness of this auxiliary was first assessed using a model peptide of sequence N^{α}-(Aux)GlySerTyrArgPheLeu.[1] This was successfully ligated with a number of thioester peptides, as illustrated below (eq 3), (Table 1), with post-ligation auxiliary removal achieved using either 95% HF/5% *p*-cresol or trifluoroacetic acid (TFA)/bromotrimethyl silane (TMSBr).[10,11]

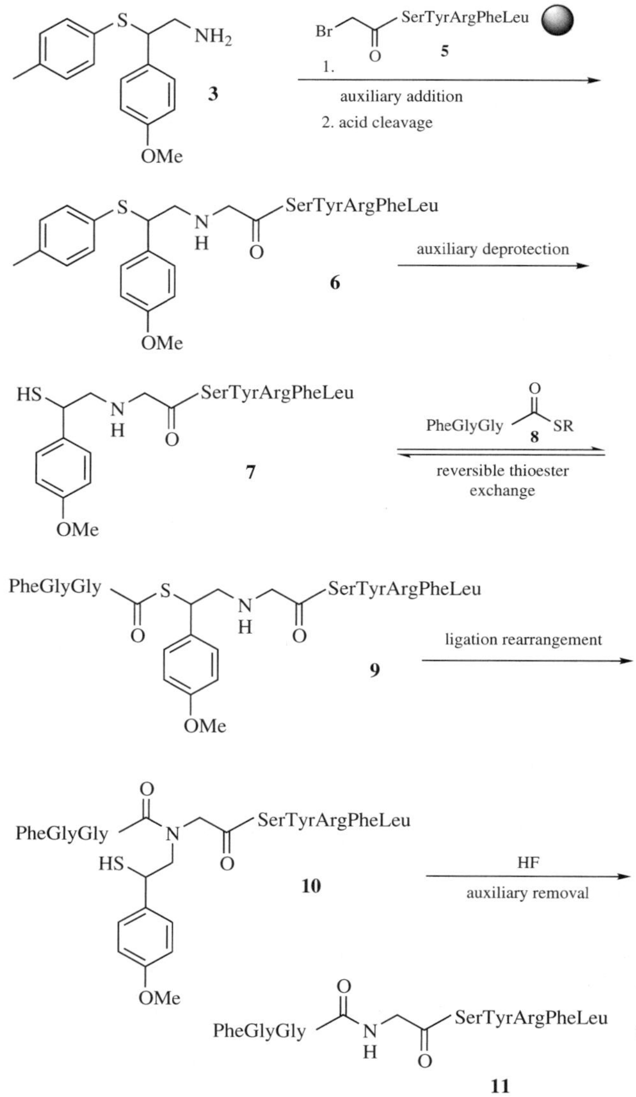

(3)

Table 1 Ligation studies with auxiliary **3**

Peptide-αCO-SR	Reaction Time (h)[a]	Ligation Yield(%)[b]	Auxiliary Removal
Phe-Gly-Gly	16	>98	HF
Mouse Larc 1-31 (Ala)	40	92	HF
MCP1 1–35 (Lys)	40	76	TFA/TMSBr

[a] Ligation conditions: 5 mM N^{α}-(Aux) peptide in 6 M guanidinium phosphate buffer pH 7, 25 °C, 2% thiophenol
[b] Based on consumption of the N^{α}-(Aux) peptide. Thioester peptides were used in 1.5-fold excess.

Use of 1-(4-Methoxyphenyl)-2-(4′-methylbenzylthio)ethylamine. Auxiliary **3** has been applied to the formation of cyclic peptides[12] and in the synthesis of cytochrome b562.[13]

Cyclic Peptides.[12] Cyclic peptides are known to possess greater metabolic stability, biological activity, and receptor selectivity than their linear counterparts.[14] Consequently, much effort has been directed towards their synthesis, via cyclic lactam or disulfide bridge formation for example.[15–18] However, these methods suffer from problems such as the necessity of protecting group manipulation and the use of organic solvents in lactam formation and the metabolic instability of disulfide bonds. Classical NCL, making use of cysteine residues, has recently been applied to surmount these problems, resulting in the synthesis of a variety of cyclic peptides.[19] Earlier incarnations of NCL auxiliaries,[20] albeit with the reported problems associated with these reagents, have also been employed in the synthesis of cyclic peptides. The 1-(4-methoxyphenyl)-2-(4′-methylbenzylthio)ethylamine auxiliary **3** has been used to form an 11-amino acid cyclic peptide, the linear analog of which is derived from an anti HIV-1 lead compound. The peptide is amenable neither to Xxx-Cys nor Gly-Gly cyclative coupling, necessitating the use of the auxiliary **3**. Linear peptide synthesis was achieved using standard solid phase peptide synthesis techniques with a thioester linkage to the solid support, with the auxiliary attached to the *N*-terminus via the bromide as discussed above (eq 4). To effect cyclization, **9** was dissolved in freshly degassed sodium phosphate (0.2 M) pH 7.5, thiophenol was added (0.1% v/v) and the pH adjusted to 7.4. The reaction was complete within 2 h (by RP-HPLC) and purification was achieved with preparative RP-HPLC (C_{18} column, gradient 20–90% B over 35 min; A: 0.1% aq TFA, B: acetonitrile + 0.1% TFA). The auxiliary was removed using 95% HF, 5% *p*-cresol at 0 °C for 1 h to give the cyclic peptide in 22.5% yield from the linear peptide.

Cytochrome b562 synthesis.[13] The total chemical synthesis of cytochrome b562, and that of an axial ligand analog, [SeMet^7]cyt b562, was achieved using native chemical ligation of two unprotected peptide segments via the 1-(4-methoxyphenyl)-2-(4′-methylbenzylthio)ethylamine auxiliary approach.

Automated solid phase peptide synthesis was used to generate three peptide fragments, with the auxiliary introduced as an on-resin modification of the amino terminus of peptide cyt b562 (64-106) [G(Aux)FDILVGQIDDALKLANEGKVK-EAQAAAEQL KTTRNAYHQKYR]. Two further peptide sequences were then synthesised, representing cyt b562(1-63) wild-type and [SeMet^7] cyt b562(1-63), using a thioester generating resin to form the peptides required for native chemical ligation [ADLEDN*X*ETLNDN LKVIEKADNAAQVKDALTKMRAAALDAQKATPPKLEDK SPDSPEMKDFRH(COSR), where *X* = Met or SeMet].

This synthesis exemplifies the specific advantages provided by the use of ligation auxiliaries. Analysis of the protein structure predicted that modification of the sequence to include Cys as a replacement for Gly^{64} would create unfavorable steric interactions and that the resulting protein would not fold. Attempts to use D-Cys in place of L-Cys (thus circumventing the steric problems) still disrupted folding of the protein (presumably due to destabilization of the helix by the D-amino acid). Thus the ligation auxiliary **3** was employed, allowing ligation of the peptides while still leaving the native Gly at position 64. Native chemical ligation was conducted at room temperature, with peptides dissolved in 6 M guanidine.HCl buffered at pH 7 with 0.3 M sodium phosphate, with 0.5–2% thiophenol as a catalyst. Following purification by semipreparative HPLC, treatment with HF for 1 h at 0 °C cleaved the auxiliary, providing the native polypeptide. Protein folding was then accomplished using a heme solution, giving the native protein cyt b562 and [SeMet^7]cyt b562, as determined by electrospray MS, CD spectra, and cyclic voltammetry.

Potential Use of Ligation Auxiliaries Outside of Peptide and Protein Synthesis. Auxiliaries such as **3** have greatly improved the generality of native chemical ligation. They also have the potential to expand the scope of chemical ligation beyond protein and peptide synthesis, as in the ligation of fluorophores and affinity tags to biomolecules via cysteine.[21,22] The use of auxiliary-mediated ligation could further generalize the conjugation of non-peptide molecules in a mild, chemoselective fashion. This chemistry provides a tool for the synthetic chemist, and may prove particularly useful as a method for the selective formation of amide

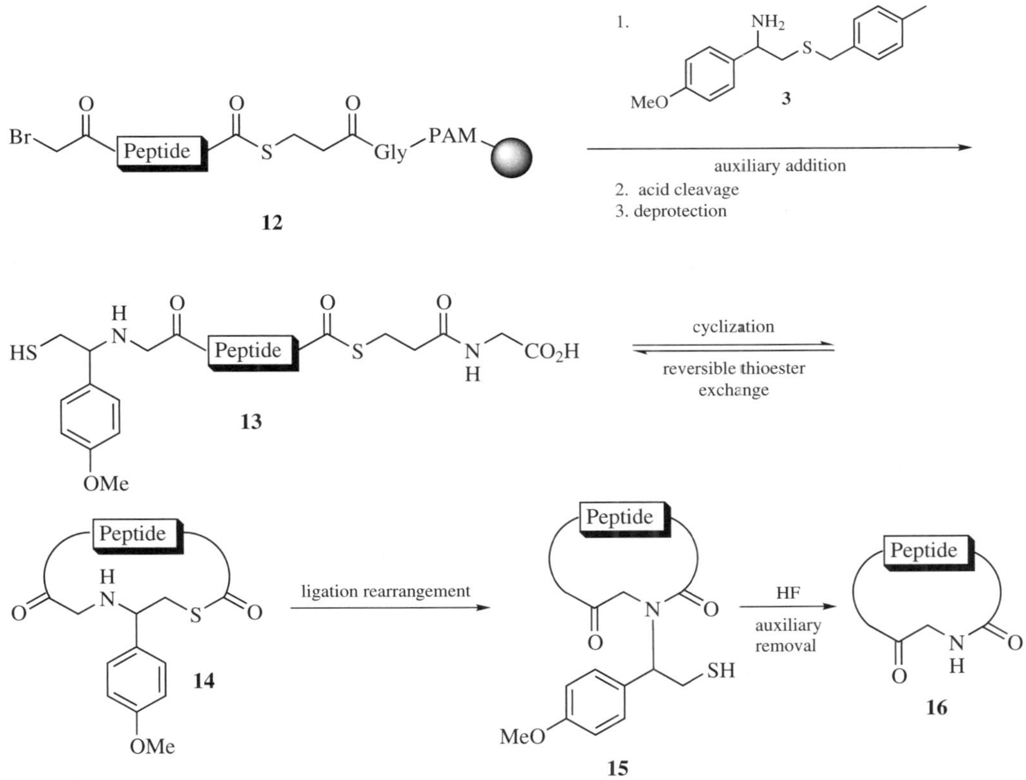

(4)

bonds in the presence of other amines with minimal use of protecting groups.

The majority of these auxiliaries have not yet found widespread application in the synthesis of complex peptides, despite the fact that cysteine is a relatively rare amino acid, constituting

Related Reagents. *S*-(4-Methylbenzyl)-2-(aminooxy)ethanethiol: This auxiliary, which is easily removed following ligation by N-O bond cleavage with Zn in acetic acid, suffers from a slower cyclization step due the six-membered transition state.[20]

S-[(3-Carboxy-4-nitrophenyl)thio]-2-aminoethanethiol: The five-membered transition state permitted by this auxiliary allows rapid ligation rearrangement, offering a solution to the problem of the above reagent, however, this auxiliary is no longer removable after ligation.[20]

2-Mercaptobenzaldehyde: Condensation of this auxiliary with a terminal amine gives an acid stable benzyl amine, but leaves an acid labile benzylamide after cyclization. This method again suffers due to a six-membered transition state.[24]

4,5-Dimethoxy-2-tritylthiobenzylamine,[25] 4,5,6-Trimethoxy-(2-4′-methylbenzylthio)-benzylamine:[26] These groups are also reductively coupled to terminal amines, with the methoxy groups serving to increase aryl ring electron density and hence ligation efficiency and acid lability.

1-(2-Nitrophenyl)-2-tritylthioethylamine: Additition of the nitro group to the benzylamine allows this reagent to take advantage of the rapid ligation rates of **3**; post ligation removal is achieved by simple photolysis.[27]

1-Hydroxyl-1-(4-methoxyphenyl)-2-*tert*-butylthioethane: This intermediate, formed from in situ from the corresponding alcohol, allows the attachment of the 2-mercaptobenzylamine-type auxiliary to a terminal amine, thereby extending the number of amino acids to which this moiety may be coupled.[28]

1. (a) Botti, P.; Carrasco, M. R.; Kent, S. B. H., *Tetrahedron Lett.* **2001**, *42*, 1831. (b) Tchertchian, S.; Hartley, O.; Botti, P., *J. Org. Chem.* **2004**, *in press*.
2. Zuckermann, R. N.; Kerr, J. M.; Kent, S. B. H.; Moos, W. H., *J. Am. Chem. Soc.* **1992**, *114*, 10646.
3. Casi, G.; Hilvert, D., *Current Opinion in Structural Biology* **2003**, *13*, 589.
4. Kent, S., *J. Pept. Sci.* **2003**, *9*, 574.
5. Ma, Y.; Zhao, Y. F., *Chinese Sci. Bull.* **2003**, *48*, 1.
6. Tam, J. P.; Xu, J.; Eom, K. D., *Biopolymers* **2001**, *60*, 194.
7. Dawson, P. E.; Muir, T. W.; Clark-Lewis, I.; Kent, S. B. H., *Science* **1994**, *266*, 776.
8. Muir, T. W.; Sondhi, D.; Cole, P. A., *Proc. Natl. Acad. Sci. USA* **1998**, *95*, 6705.
9. David, R.; Richter, M. P. O.; Sickinger, A. G., *Eur. J. Biochem.* **2004**, *271*, 663.
10. Guo, L.; Funakoshi, S.; Fujii, N.; Yajima, H., *Chem. Pharm. Bull.* **1988**, *36*, 4989.
11. Yajima, H.; Fujii, N.; Funakoshi, S.; Watanabe, T.; Murayama, E.; Otaka, A., *Tetrahedron* **1988**, *44*, 805.
12. Cardona, V. M. F.; Hartley, O.; Botti, P., *J. Pept. Res.* **2003**, *61*, 152.
13. Low, D. W.; Hill, M. G.; Carrasco, M. R.; Kent, S. B. H.; Botti, P., *Proc. Natl. Acad. Sci. USA* **2001**, *98*, 6554.
14. Hruby, V. J., *Life Sciences* **1982**, *31*, 189.
15. Chorev, M.; Roubini, E.; McKee, R. L.; Gibbons, S. W.; Goldman, M. E.; Caulfield, M. P.; Rosenblatt, M., *Biochemistry* **1991**, *30*, 5968.
16. Botti, P.; Pallin, T. D.; Tam, J. P., *J. Am. Chem. Soc.* **1996**, *118*, 10018.
17. Zhang, L. S.; Tam, J. P., *J. Am. Chem. Soc.* **1997**, *119*, 2363.
18. Camarero, J. A.; Pavel, J.; Muir, T. W., *Angew. Chem. Int. Ed., Engl.* **1998**, *37*, 347.
19. Shao, Y.; Lu, W.; Kent, S. B. H., *Tetrahedron Lett.* **1998**, *39*, 3911.
20. Canne, L. E.; Bark, S. J.; Kent, S. B. H., *J. Am. Chem. Soc.* **1996**, *118*, 5891.
21. Wood, R. J.; Pascoe, D. D.; Brown, Z. K.; Medlicott, E. M.; Kriek, M.; Neylon, C.; Roach, P. L., *Bioconj. Chem.* **2004**, *15*, 366.
22. Lue, R. Y. P.; Chen, G. Y. J.; Hu, Y.; Zhu, Q.; Yao, S. Q., *J. Am. Chem. Soc.* **2004**, *126*, 1055.
23. McCaldon, P.; Argos, P., *Proteins-Struct. Funct. Genetics* **1988**, *4*, 99.
24. Offer, J.; Dawson, P. E., *Org. Lett.* **2000**, *2*, 23.
25. Vizzavona, J.; Dick, F.; Vorherr, T., *Bioorg. Med. Chem. Lett.* **2002**, *12*, 1963.
26. Offer, J.; Boddy, C. N. C.; Dawson, P. E., *J. Am. Chem. Soc.* **2002**, *124*, 4642.
27. Kawakami, T.; Aimoto, S., *Tetrahedron Lett.* **2003**, *44*, 6059.
28. Clive, D. L. J.; Hisaindee, S.; Coltart, D. M., *J. Org. Chem.* **2003**, *68*, 9247.

Thomas K. Hutton
University of Illinois at Chicago, Chicago, IL, USA

Laurie L. Parker
University of Chicago, Chicago, IL, USA

2-Methoxypropene[1]

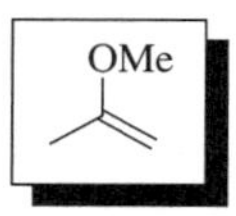

[116-11-0] C_4H_8O (MW 72.12)

(electron-rich alkene used in the monoprotection of aliphatic,[2] allylic,[3] and propargylic[4] alcohols, often followed by Claisen[3b–e] or other[2a,4c] rearrangements; protection of peroxides,[5] cyanohydrins,[6] α-hydroxy ketones,[2b] and phenols;[7] condensation with amines;[8] protection of 1,2-[9,10] and 1,3-diols,[10,11] sometimes followed by Prins[9f,g] rearrangements; protection of 1,2-dithiols[10a] and α-hydroxy carbamates;[12] participant in pericyclic reactions;[13] formation of 2-methoxyallyl halides[14] and substituted furans[4b])

Alternate Names: 2-MP; isopropenyl methyl ether.
Physical Data: bp 34–36 °C; n_D^{20} 1.3820; *d* 0.753 g cm^{-3}; flash point −29 °C.
Form Supplied in: colorless liquid; commercially available.
Preparative Method: prepared in high yield from Succinic Anhydride, 2,2-Dimethoxypropane, benzoic acid, and Pyridine.[15]
Handling, Storage, and Precautions: flammable liquid; lightsens itive; should be refrigerated.

Monoprotection of Alcohols. 2-Methoxypropene is used as a protective group for aliphatic,[2,11d] allylic,[3] and propargylic[4] alcohols, masking them as their mixed acetals (eq 1). Deprotection can be accomplished by stirring in MeOH over ion exchange resin,[4a] by reaction in methanol with catalytic Acetyl Chloride,[11d] by Potassium Carbonate in methanol,[2c] or by 20% Acetic Acid.[3a] A general advantage of this acetal, which undergoes hydrolysis at approximately 10^3 times the rate of THP,[2d,3a] is that it does not confer an additional diastereomeric center to the protected substrate. Access to allyl vinyl ethers for subsequent Claisen rearrangements[3b–e] is illustrated in eq 2.

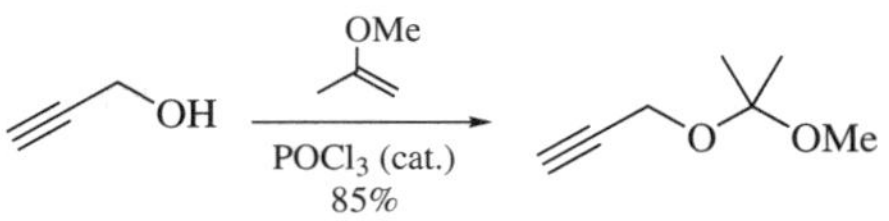
(1)

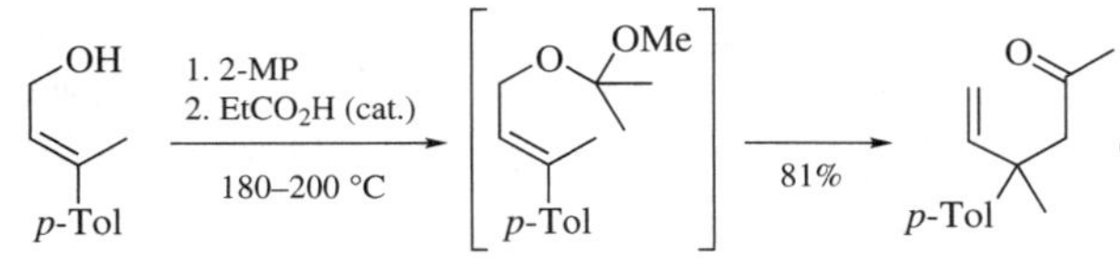
(2)

A useful entry into β-keto allenes is provided by the reaction of 2-MP with tertiary propargylic alcohols.[4c] Base-catalyzed rearrangement of the allenes affords conjugated dienones (eq 3).

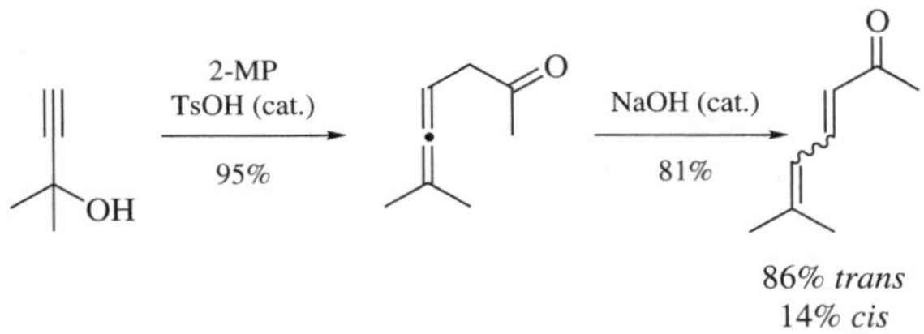
(3)

Stable allylic peroxyacetals have been prepared by reacting 2-MP with hydroperoxides (eq 4).[5] Organomercury functionality is tolerated in this reaction.[5b] Cyanohydrins,[6] α-hydroxy ketones,[2b] and phenols[7] are similarly protected.

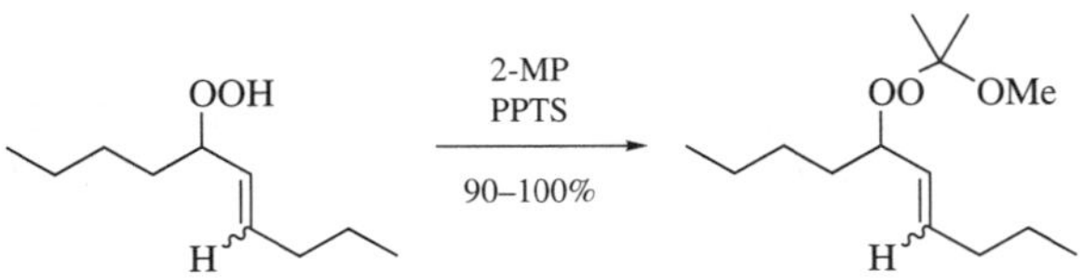
(4)

Condensation with Amines. Imine formation with loss of methanol occurs under acidic conditions, as illustrated in eq 5.[8]

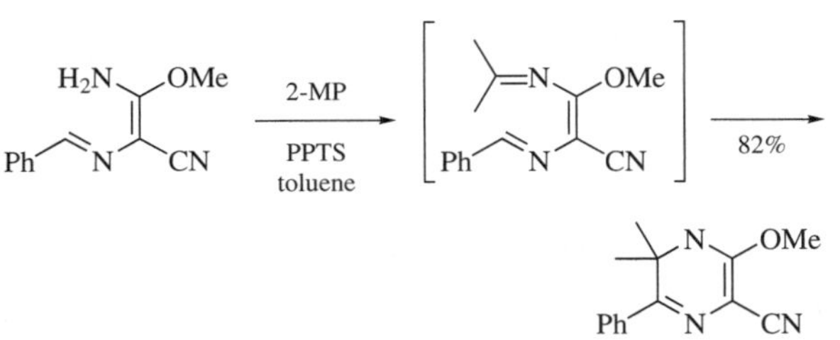
(5)

Protection of 1,2- and 1,3-Diols and 1,3-Dithiols. When another hydroxy group is in close proximity to the initially protected alcohol functionality, acetonide formation occurs.[9–11] The kinetic product is preferentially formed.[10b,c,11c] As anomeric hydroxy groups do not usually[11] participate, this reagent provides complementary selectivity to the more traditional acid/acetone acetonation protocol. Control of the amount of 2-MP can result in varying degrees of protection (eq 6).[10b]

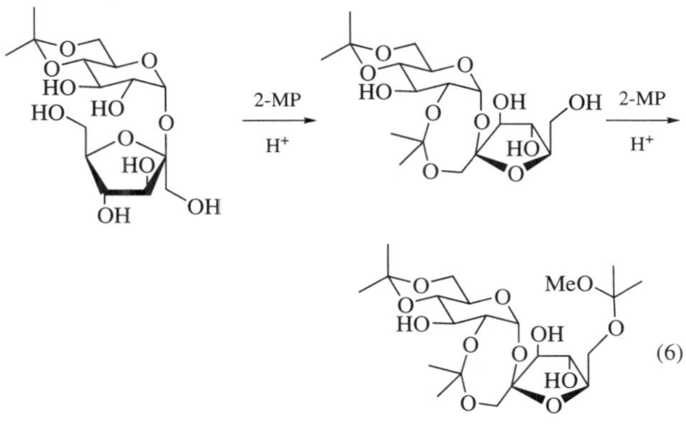
(6)

Acetonide formation using n-Butyltrichlorostannane with 2-MP[10a] is quite general for 1,2- and 1,3-diols, as well as for 1,3-dithiols. Monoacetonide formation occurs with 1,2,3- and 1,2,4-triols. Yields of the acetonides (e.g. **1**–**5**) show improvement over conventional procedures.[10a]

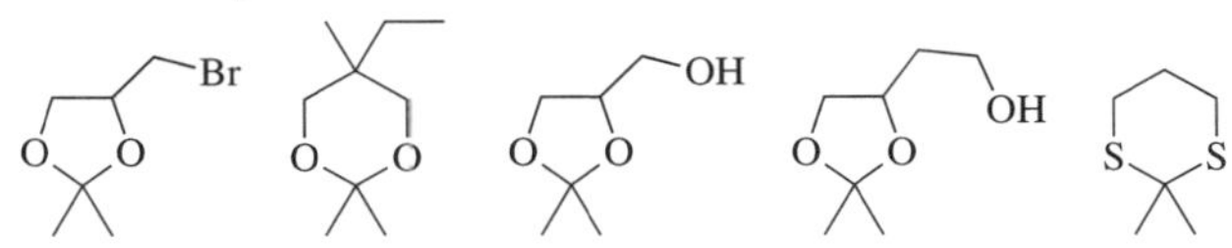
(**1**) 83% (**2**) 94% (**3**) 77% (**4**) 82% (**5**) 98%

Protection of α-Hydroxy Carbamates. Aminal formation from α-hydroxy carbamates also occurs using 2-MP, as illustrated in eq 7.[12]

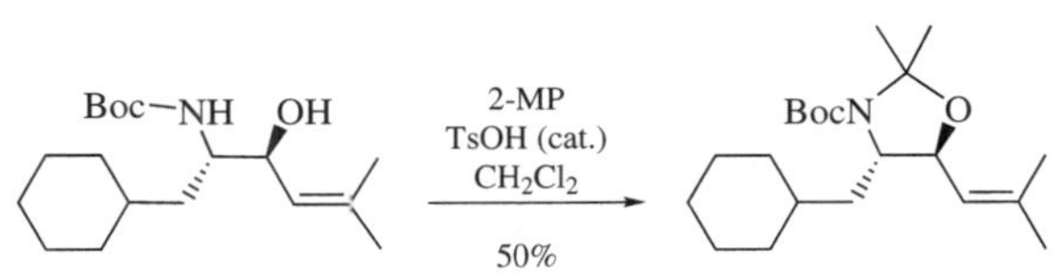
(7)

Participation in Pericyclic Reactions. 2-MP is an active participant in pericyclic reactions. A few examples include a substitution-dependent ene reaction with an alkynyl Fischer carbene complex (eq 8),[13c] a [2 + 3] reaction with 2-amino-1,4-naphthoquinone (eq 9),[13f] a solvent-dependent photochemical cycloaddition with *o*-naphthoquinones,[13j] and reaction with 2-nitro-1,3-dienes to provide α,β-unsaturated 1,4-diones (eq 10).[13g]

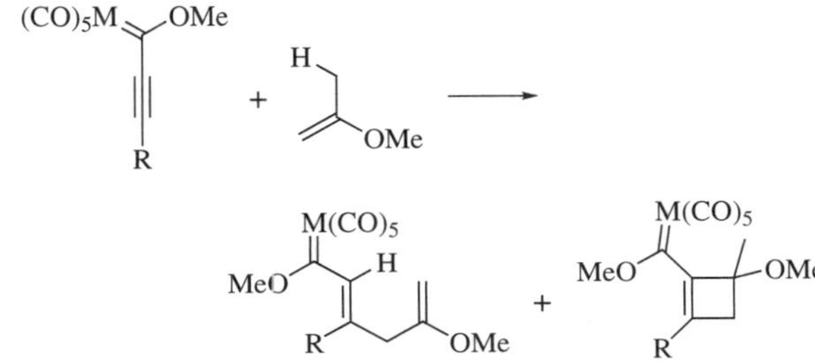
(8)

M = W, Cr	R = TMS	30%	<2%
	R = Me	<2%	63%

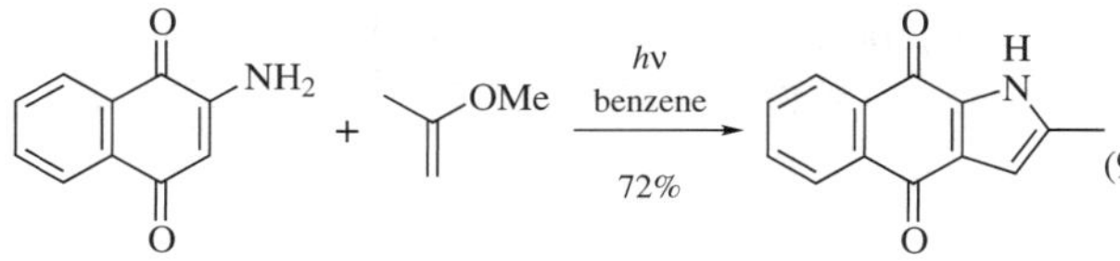

(9)

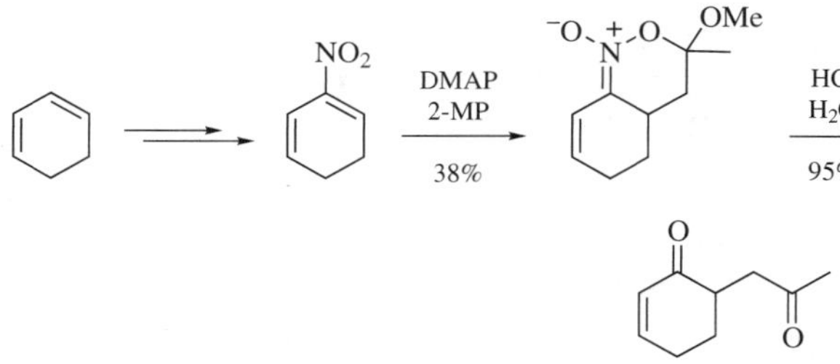

(10)

Reaction with N-Halosuccinimides. 2-Methoxyallyl and -vinyl halides are formed by reaction of 2-MP with *N*-halosuccinimides.[14] Also, treatment of propargylic alcohols with 2-MP and N-Bromosuccinimide results in the formation of bromo acetals, which are subsequently cyclized to furans in good yield (eq 11).[4b]

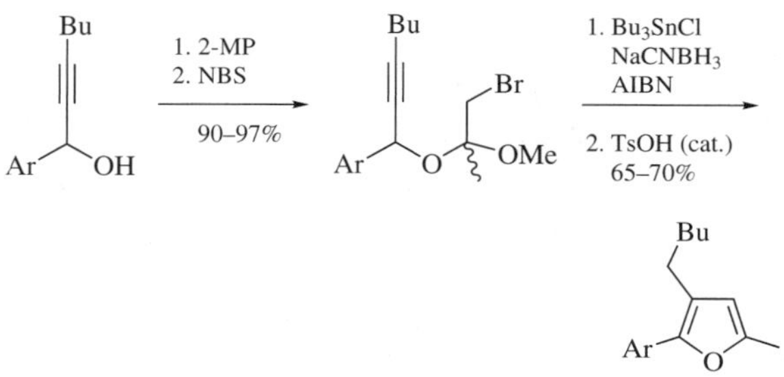

(11)

Related Reagents. 2,2-Dimethoxypropane.

1. *Fieser & Fieser* **1969**, *2*, 230; **1975**, *5*, 360; **1981**, *9*, 304; **1984**, *11*, 329; **1986**, *12*, 291.
2. (a) Kang, S. H.; Hwang, T. S.; Kim, W. J.; Lim, J. K. *Tetrahedron Lett.* **1991**, *32*, 4015. (b) Wilson, T. M.; Kocienski, P.; Jarowicki, K.; Isaac, K.; Faller, A.; Campbell, S. F.; Bordner, J. *Tetrahedron* **1990**, *46*, 1757. (c) Tietze, L. F.; Lögers, M. *Justus Liebigs Ann. Chem./Liebigs Ann. Chem.* **1990**, 261. (d) Reese, C. B.; Saffhill, R.; Sulston, J. E. *J. Am. Chem. Soc.* **1967**, *89*, 3366.
3. (a) Kluge, A. F.; Untch, K. G.; Fried, J. H. *J. Am. Chem. Soc.* **1972**, *94*, 7827. (b) Srikrishna, A.; Krishnan, K.; Vankateswarlu, S. *Chem. Commun./J. Chem. Soc.* **1993**, 143. (c) Srikrishna, A.; Krishnan, K. *Indian J. Chem., Sect. B* **1990**, *29B*, 879. (d) Gajewski, J. J.; Gee, K. R.; Jurayj, J. *J. Org. Chem.* **1990**, *55*, 1813. (e) Saucy, G.; Marbet, R. *Helv. Chim. Acta* **1967**, *50*, 2091.
4. (a) Lampilas, M.; Lett, R. *Tetrahedron Lett.* **1992**, *33*, 773. (b) Srikrishna, A.; Sundarababu, G. *Tetrahedron* **1990**, *46*, 7901. (c) Saucy, G.; Marbet, R. *Helv. Chim. Acta* **1967**, *50*, 1158.
5. (a) Dussault, P.; Sahli, A.; Westermeyer, T. *J. Org. Chem.* **1993**, *58*, 5469. (b) Bloodworth, A. J.; Cooksey, C. J.; Korkodilos, D. *Chem. Commun./J. Chem. Soc.* **1992**, 926. (c) Dussault, P.; Sahli, A. *Tetrahedron Lett.* **1990**, *31*, 5117.
6. Zanderbergen, P.; Van den Nieuwendijk, A. M. C. H.; Brussee, J.; van der Gen, A. *Tetrahedron* **1992**, *48*, 3977.
7. Gibson, C. P.; Bem, D. S.; Falloon, S. B.; Hitchens, T. K.; Cortopassi, J. E. *Organometallics* **1992**, *11*, 1742.
8. Freeman, F.; Kim, D. S. H. L. *J. Org. Chem.* **1992**, *57*, 550.
9. (a) Kozikowski, A. P.; Ognyanov, V. I.; Fauq, A. H.; Nahorski, S. R.; Wilcox, R. A. *J. Am. Chem. Soc.* **1993**, *115*, 4429. (b) van der Klein, P. A. M.; Filemon, W.; Boons, G. J. P. H.; Veeneman, G. H.; van der Marel, G. A.; van Boom, J. H. *Tetrahedron* **1992**, *48*, 4649. (c) Barbat, J.; Gelas, J.; Horton, D. *Carbohydr. Res.* **1991**, *219*, 115. (d) Shing, T. K. M.; Tang, Y. *Tetrahedron* **1991**, *47*, 4571. (e) Fujisawa, T.; Takemura, I.; Ukaji, Y. *Tetrahedron Lett.* **1990**, *31*, 5479. (f) Hopkins, M. H.; Overman, L. E.; Rishton, G. M. *J. Am. Chem. Soc.* **1991**, *113*, 5354. (g) Brown, M. J.; Harrison, T.; Herrington, P. M.; Hopkins, M. H.; Hutchinson, K. D.; Mishra, P.; Overman, L. E. *J. Am. Chem. Soc.* **1991**, *113*, 5365.
10. (a) Carofiglio, T.; Marton, D.; Stivanello, D.; Tagliavini, G. *Main Group Chem.* **1992**, *15*, 247. (b) Fanton, E.; Gelas, J.; Horton, D. *J. Org. Chem.* **1981**, *46*, 4057. (c) Gelas, J.; Horton, D. *Heterocycles* **1981**, *16*, 1587.
11. (a) Guanti, G.; Banfi, L.; Narisano, E.; Thea, S. *Synlett* **1992**, 311. (b) Srikrishna, A.; Krishnan, K.; Yelamaggad, C. V. *Tetrahedron* **1992**, *48*, 9725. (c) Jung, M. E.; Clevenger, G. L. *Tetrahedron Lett.* **1991**, *32*, 6089. (d) Thomé, M. A.; Giudicelli, M. B.; Picq, D.; Anker, D. *J. Carbohydr. Chem.* **1991**, *10*, 923.
12. Melnick, M. J.; Bisaha, S. N.; Gammill, R. B. *Tetrahedron Lett.* **1990**, *31*, 961.
13. (a) Deaton, M. V.; Ciufolini, M. A. *Tetrahedron Lett.* **1993**, *34*, 2409. (b) Beak, P.; Song, Z.; Resek, J. E. *J. Org. Chem.* **1992**, *57*, 944. (c) Faron, K. L.; Wulff, W. D. *J. Am. Chem. Soc.* **1990**, *112*, 6419. (d) Kobayashi, K.; Suzuki, M.; Suginome, H. *J. Org. Chem.* **1992**, *57*, 599. (e) Fannes, C.; Meerpoel, L.; Toppet, S.; Hoornaert, G. *Synthesis* **1992**, 705. (f) Kobayashi, K.; Takeuchi, H.; Seko, S.; Suginome, H. *Helv. Chim. Acta* **1991**, *74*, 1091. (g) Bäckvall, J.-E.; Karlsson, U.; Chinchilla, R. *Tetrahedron Lett.* **1991**, *32*, 5607. (h) Coleman, R. S.; Grant, E. B. *Tetrahedron Lett.* **1990**, *31*, 3677. (i) Gupta, R. B.; Franck, R. W. *Synlett* **1990**, 355. (j) Takuwa, A.; Sumikawa, M. *Chem. Lett.* **1989**, 9.
14. (a) Jacobson, R. M.; Raths, R. A.; McDonald, J. H., III *J. Org. Chem.* **1977**, *42*, 2545. (b) Greenwood, G.; Hoffmann, H. M. R. *J. Org. Chem.* **1972**, *37*, 611.
15. Newman, M. S.; Vander Zwan, M. C. *J. Org. Chem.* **1973**, *38*, 2910.

K. Sinclair Whitaker
Wayne State University, Detroit, MI, USA

D. Todd Whitaker
Detroit County Day School, Beverly Hills, MI, USA

4-Methoxytriphenylmethyl Chloride

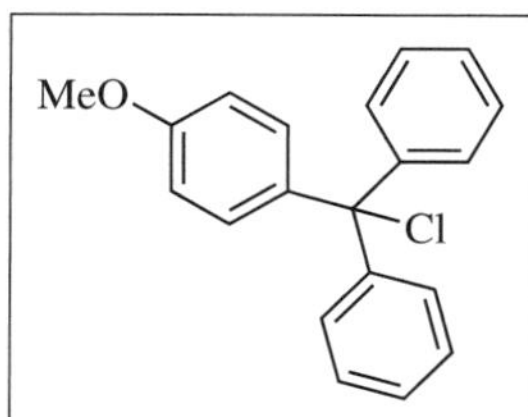

[14470-28-1] $C_{20}H_{17}ClO$ (MW 308.81)

(reagent used for the protection of hydroxyl and amino groups[1])

Alternate Name: MMTr-Cl, monomethoxytrityl chloride, *p*-anisylchlorodiphenylmethane.
Physical Data: mp 122–124 °C.
Solubility: readily soluble in most organic solvents, mainly used in methylene chloride, THF, and pyridine.

Form Supplied in: yellow to orange powder; widely available

Handling, Storage, and Precautions: use in a fume hood, with chemical safety goggles and rubber gloves. Do not breathe dust. Avoid prolonged or repeated exposure. Wash thoroughly after handling. Irritant. Keep tightly closed and refrigerate. Incompatible with strong oxidizing agents as well as strong bases. Toxic decomposition product is hydrogen chloride gas.

Selective Protection of Alcohols in Carbohydrate Chemistry. In a search for protecting groups of alcohols for the use in the synthesis of polynucleotides, Khorana[2] first prepared the 5′-*O*-monomethoxytrityl (5′-*O*-MMTr) ethers of uridine and adenosine derivatives.[2a] These ethers proved to be labile enough to be removed with acid treatment without causing depurination which had been a problem with standard trityl ethers.[2c] Since the introduction by Khorana almost 40 years ago, the MMTr ether (and to a greater extent the dimethoxytrityl ether) has been a standard protecting group in oligonucleotide synthesis and continues to be used in standard protocols today.[2] For instance, Khorana et. al. produced 5′-*O*-mono-*p*-methoxytritylthymidine from thymidine using 4-methoxytritylchloride in pyridine in 82% yield (eq 1).[3] The MMTr group is removed in 1.5 h in 80% acetic acid at 27 °C. The MMTr group is removed 24 times faster under these conditions than a trityl ether, and eight times slower than the dimethoxytrityl group.[2]

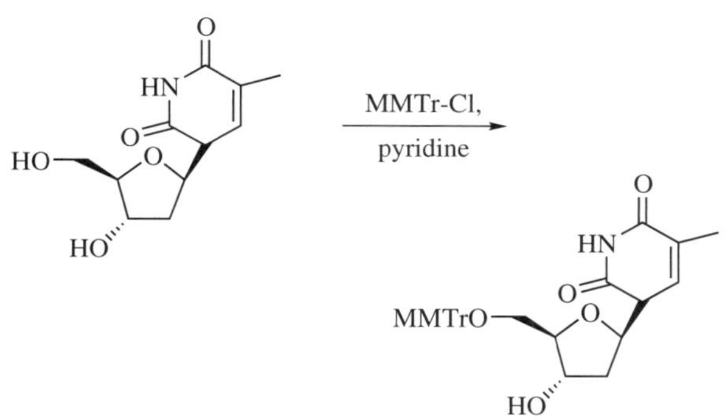

(1)

82% yield

As another example of selective protection, the primary hydroxyl group of 2,3-*O*-isopropylidene-β-D-ribofruanose was protected as its MMTr ether in high yield under similar conditions (eq 2).[4]

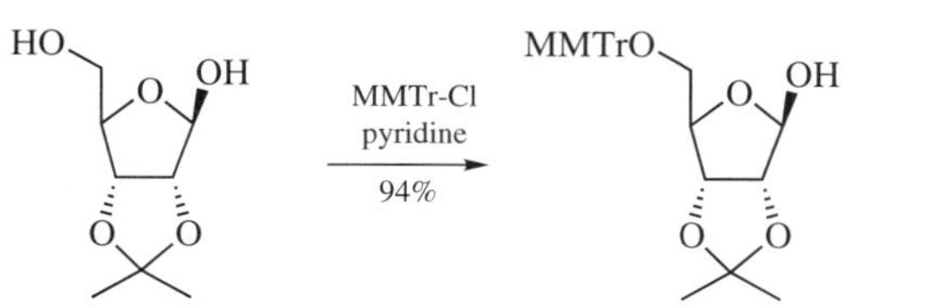

(2)

The primary hydroxyl groups of glucitol, mannitol, and galactitol were selectively protected as their MMTr ethers in high yield for use in the synthesis of some conduritol derivatives (eq 3).[5]

For the synthesis of some inhibitors of the D-fructose transporter GLUT5, the selective protection of the primary hydroxyl group of methyl α-D-fructofuranose was utilized (eq 4).[6]

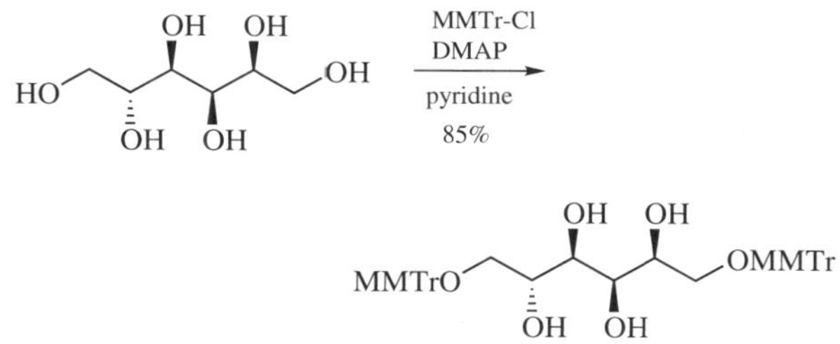

(3)

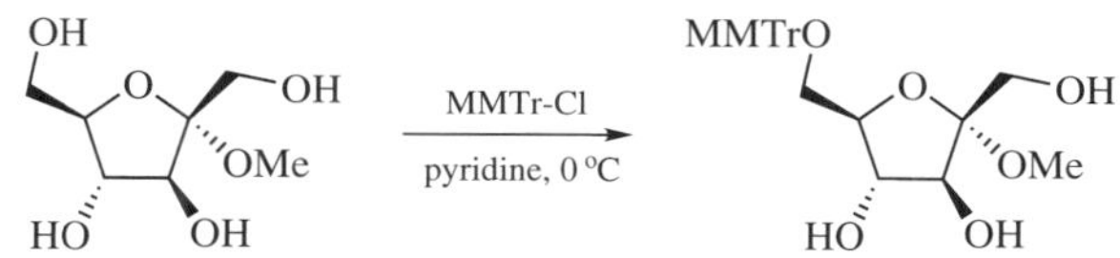

(4)

In the synthesis of some interesting 4-purinylpyrrolidine nucleosides, the selective protection of an *N*-Boc-D-prolinol derivative gave a key intermediate (eq 5).[7]

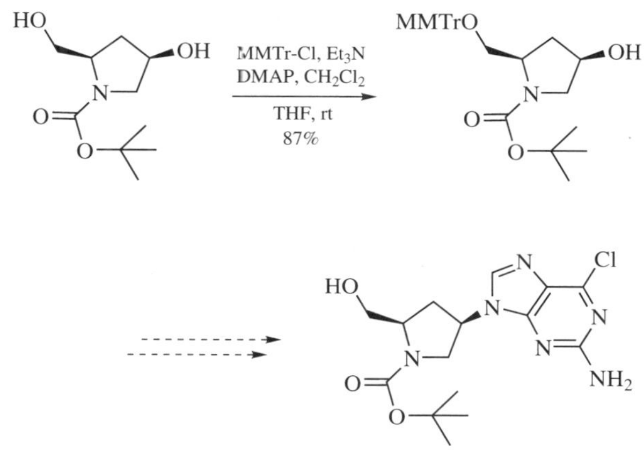

(5)

During the synthesis of D-altritol nucleoside building blocks for oligonucleotide synthesis the D-*altro*-hexitol was tritylated in DMF with silver nitrate as a catalyst (eq 6).[8]

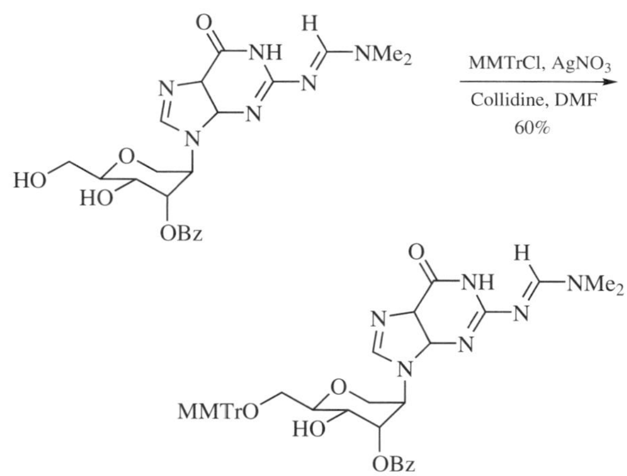

(6)

Selective Protection of Hydroxyl Groups in Acyclic Systems. The primary hydroxyl of the diol in eq 7 was selectively protected over the secondary allylic alcohol in good yield.[9]

For the synthesis of an abasic linker for the use in some DNA-peptide chimeras, the MMTr ether of a glycine derivative was prepared (eq 8).[10]

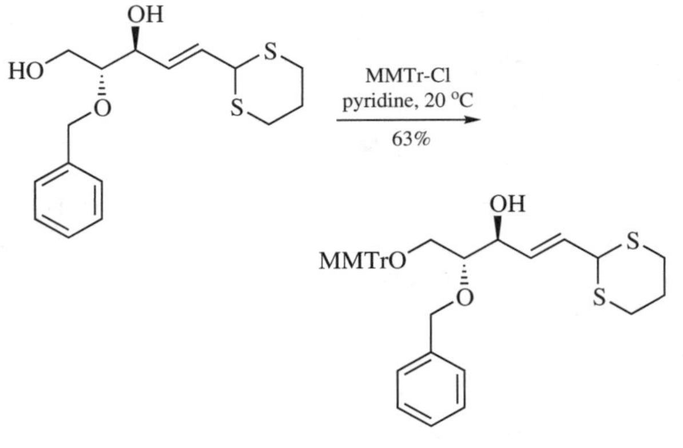

(7)

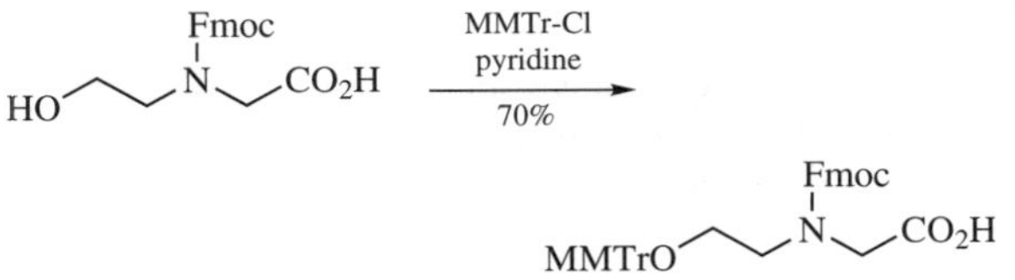

(8)

Protection of the Amino Group. The MMTr group has been used in a few cases as an amino protecting group. MMTr-protected amines react with acid chlorides and chloroformates above −40 °C in the presence of base, but only react with activated esters, carbonates, and alkylating agents under vigorous conditions. The amine group on the side chain of lysine was protected as its MMTr derivative by a two step procedure starting from Fmoc-lysine (eq 9).[11] Silylation of the amino acid derivative with TMS-Cl, followed by treatment of the transiently bis-silyalted compound with MMTr-Cl gave the MMTr amine in 91% yield. The MMTr-protected lysine was then incorporated into a doxorubicin analog.[11]

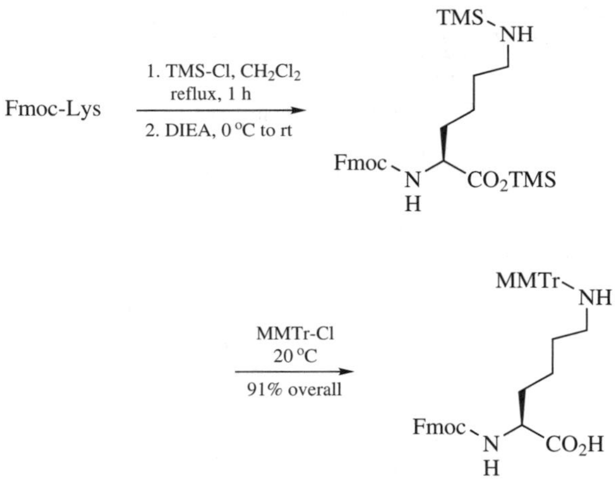

(9)

Protection of the primary amine of 3-aminopropanol allowed for the hydroxyl group to be coupled to a proline derivative in an EDC coupling (eq 10).[12]

MMTr amines have been utilized in the synthesis of oligonucleotides. The amino unit of an adenosine derivative was protected as its MMTr amine and then incorporated into a short DNA strand to probe the structure of the topoisomerase I-DNA binary complex (eq 11).[13]

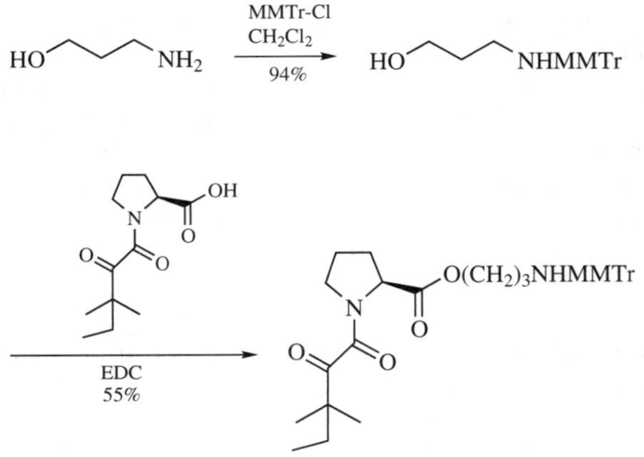

(10)

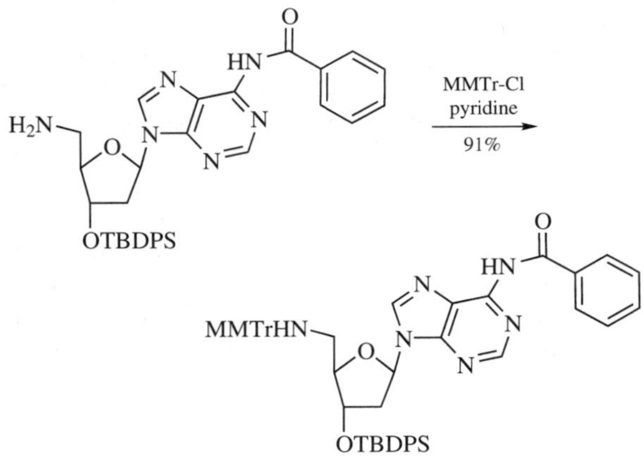

(11)

The MMTr amine was used for the synthesis of the azirdine portion of the antitumor antibiotic azinomycin. A serine derivative was treated with MMTr chloride and triethylamine at −10 °C in methylene chloride, followed by tosylation and cyclization by heating in the presence of triethylamine (eq 12).[14] This MMTr-protected aziridine proved to be remarkably stable in a variety of conditions.

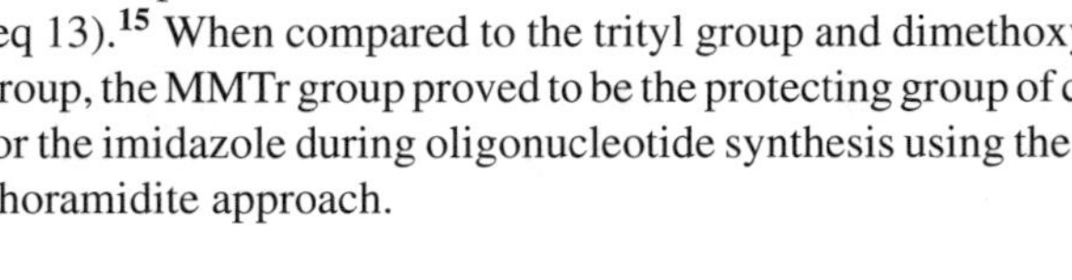

(12)

The MMTr group has been used successfully as a protecting on an imidazole nitrogen. The imidazole ring of 4-imidazoleacetic acid was protected as a MMTr ether under standard conditions (eq 13).[15] When compared to the trityl group and dimethoxytrityl group, the MMTr group proved to be the protecting group of choice for the imidazole during oligonucleotide synthesis using the phosphoramidite approach.

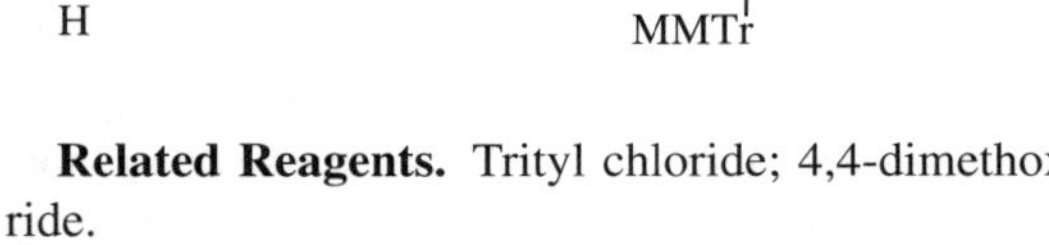

(13)

Related Reagents. Trityl chloride; 4,4-dimethoxytrityl chloride.

1. Greene, T. W.; Wuts, P. G. M., *Protecting Groups in Organic Synthesis*, Wiley: New York, 1999, 105–106.
2. (a) Smith, M.; Rammler, D. H.; Goldberg, I. H.; Khorana, H. G., *J. Am. Chem. Soc.* **1962**, *84*, 430. (b) Khorana, H. G., *Pure Appl. Chem.* **1968**, *17*, 349. (c) For a review of the use of various trityl protecting groups in nucleotide synthesis, see: Beaucage, S. L.; Iyer, R. P., *Tetrahedron* **1992**, *48*, 2223.
3. Schaller, H.; Weimann, G.; Lerch, B.; Khorana, H.G., *J. Am. Chem. Soc.* **1963**, *85*, 3821.
4. Mandal, S. B.; Achari, B., *Synth. Commun.* **1993**, *23*, 1239.
5. Ackermann, L.; El Tom, D.; Furstner, A., *Tetrahedron* **2000**, *56*, 2195.
6. Tatibouet, A.; Yang, J.; Morin, C.; Holman, G. D., *Bio. Med. Chem. Lett.* **2000**, 8, 1825.
7. Vince, R.; Peterson, M. L., *J. Med. Chem.* **1991**, *34*, 2787.
8. Allart, B.; Busson, R.; Rozenski, J.; Aerschot, A. V.; Herdewijn, P., *Tetrahedron* **1999**, *55*, 6527.
9. Hassan, H. A. M.; Abdel, M. M., *Synth. Commun.* **2000**, *30*, 201.
10. Greiner, B.; Breipohl, G.; Uhlmann, E., *Helv. Chim. Acta* **1999**, *82*, 2151.
11. Dubowchik, G. M.; Radia, S., *Tetrahedron Lett.* **1997**, *38*, 5257.
12. Dubowchik, G. M.; Ditta, J. L.; Herbst, J. J.; Bollini, S.; Vinitsky, A., *Bio. Med. Chem. Lett.* **2000**, *10*, 559.
13. Henningfeld, K. A.; Arslan, T.; Hecht, S. M., *J. Am. Chem. Soc.* **1996**, *118*, 11701.
14. Moran, E. J.; Tellew, J. E.; Zhao, Z.; Armstrong, R. W., *J. Org. Chem.* **1993**, *58*, 7848.
15. Polushin, N. N.; Chen, B.-C.; Anderson, L. W.; Cohen, J. S., *J. Org. Chem.* **1993**, *58*, 4606.

David J. Madar
Abbott Laboratories, Abbott Park, Illinois, USA

N-Methylcarbazole

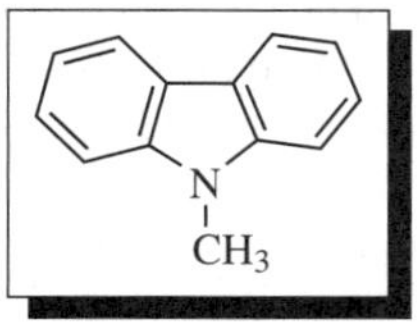

[1484-12-4] $C_{13}H_{11}N$ (MW 181.23)

(reagent used for the deoxygenation of benzoyl derivatives,[1–14] the decarboxylation of *N*-acyloxyphthalimides,[15] and reduction of aryl sulfoxides[16] under photoinduced electron-transfer conditions)

Physical Data: mp 90–92 °C.
Solubility: soluble in benzene, CH_2Cl_2, $CHCl_3$, THF, and alcohols.
Form Supplied in: solid, commercially available.
Purification: recrystallization from ethanol.

Photoinduced Electron-Transfer (PET) Deoxygenation of Benzoate Esters. Saito first reported that secondary benzoates could be deoxygenated under photochemical conditions using 9-methylcarbazole (MCZ) as the photosensitizer (eq 1).[1] The *m*-trifluoromethylbenzoyl derivative provided the highest yield and shortest reaction times. This reaction is of value as an alternative to the Barton-McCombie and related reactions.[17] The advantage of the PET deoxygenation is that it avoids the use of toxic tin hydrides and benzoyl derivatives are easily prepared in high yield.

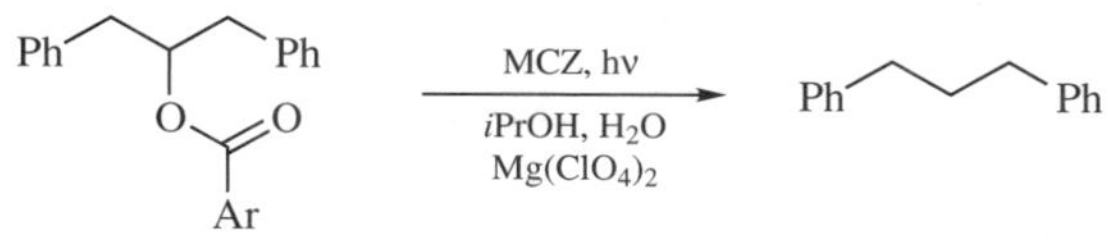

(1)

Ar = C_6H_5, *p*-ClC_6H_4, *m*-$CF_3C_6H_4$

The mechanism (eq 2) involves an excited state electron-transfer from MCZ radical cation to the benzoate to give the MCZ/benzoate radical anion pair; salts such as magnesium perchlorate promote solvent cage escape and therefore, can accelerate the reaction. β-Scission of the intermediate benzoate radical gives the deoxygenated radical. Hydrogen atom transfer from the solvent (i.e., 2-propanol or THF) provides the deoxygenated product. Oxidation of the solvent radical by the radical cation of MCZ regenerates the donor. In principle, the donor could be used in substoichiometric amounts since it is regenerated, however in practice 1 equiv of MCZ is necessary for the PET deoxygenation.

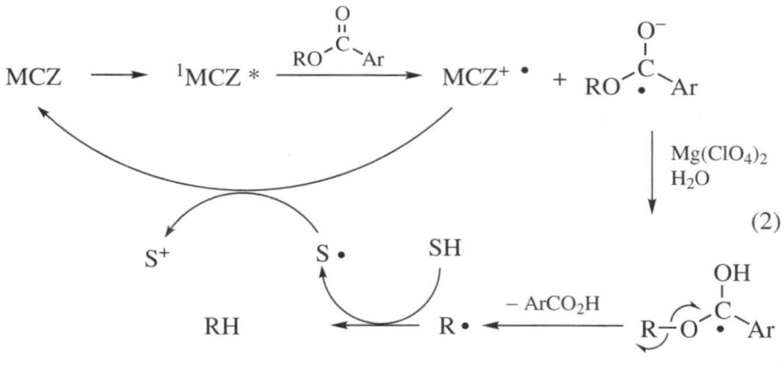

(2)

The PET deoxygenation of benzoates and *m*-trifluoromethylbenzoates has been particularly useful for the synthesis of deoxynucleosides,[3,5,6,9,10] deoxycarbohydrates,[11,12,14] and their analogs. Notable examples used the *m*-trifluormethylbenzoyl group at the 2-position of a protected ribose in order to direct the glycosylation reaction and serve as a deoxygenation precursor (eq 3).[5,9] This example demonstrated that a *m*-trifluoromethylbenzoate can be selectively deoxygenated in the presence of a benzoate, although prolonged irradiation gave the dideoxynucleoside. The PET deoxygenation has also been used to generate nucleoside radicals for mechanistic studies.[19–21]

Radical cyclizations have been initiated by the PET deoxygenation of a *m*-trifluorobenzoate in THF.[2,8,22] In the example (eq 4), 1,4-cyclohexadiene was used as the hydrogen atom donor. This method of generating a carbon-centered free radical was also used to initiate a cyclopropylcarbinyl ring opening (eq 5).[4]

3,6-Dimethyl-9-ethylcarbazole has been developed as an improved photosensitizer for the PET deoxygenation.[13,23,24] This reagent is typically used at the 10 mol % level and is reported to be more reactive than 9-methylcarbazole. The methyl groups at the 3- and 6-positions improve the stability of the carbazole radical cation intermediate. Although not commercially available, 3,6-dimethyl-9-ethylcarbazole is easily synthesized.[18]

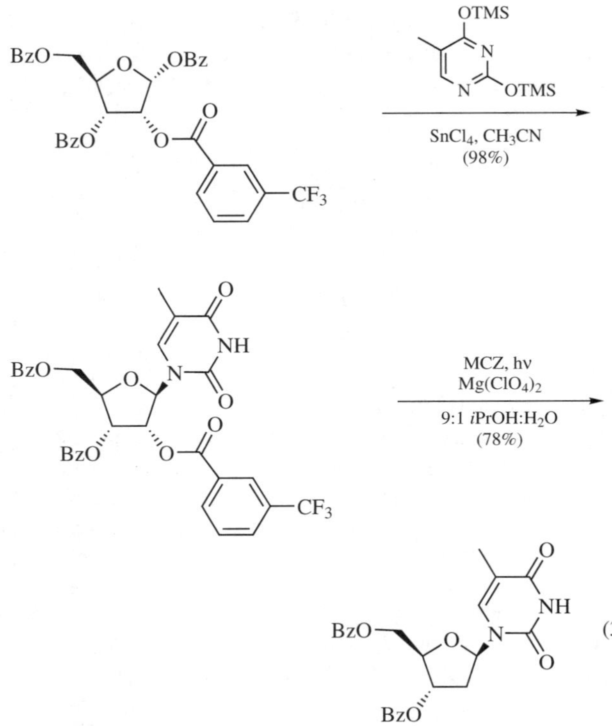

(3)

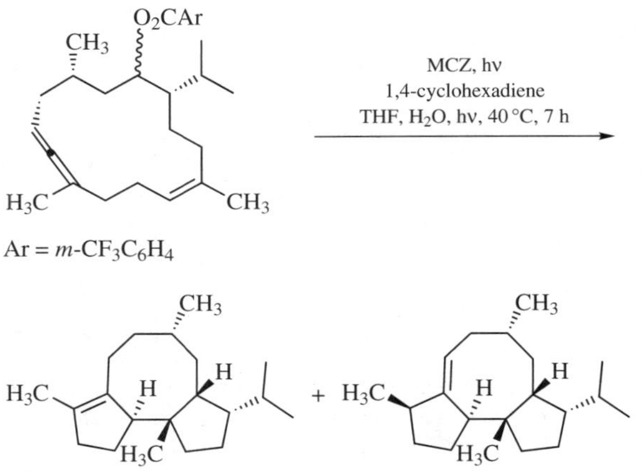

(4)

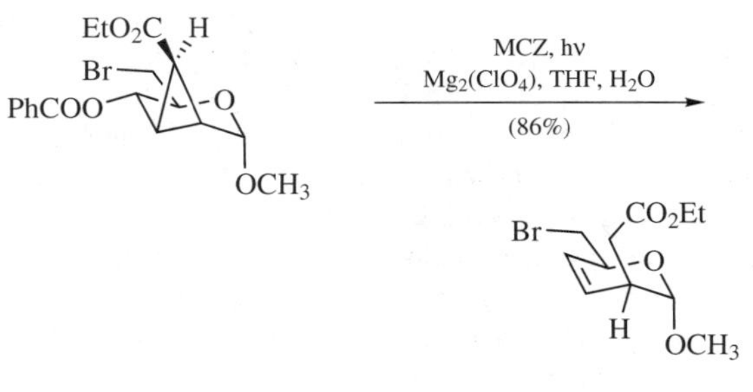

(5)

The functional group compatibility of the PET deoxygenation has not been extensively explored. Two reported examples in which the reaction did not give the desired products are shown in eqs 6 and 7. The attempted 2′-deoxygenation of 6-chloropurine ribonucleoside gave dehalogenation instead, although the deoxygenation of a 5-fluorouridine derivative was successful.[24] Attempted PET deoxygenation in the presence of a tertiary amine lead to decomposition of the substrate (eq 7).[25] It was hypothesized that the failure of this reaction may be due to interference of the tertiary amine, which can be oxidized to the aminium radical cation under photochemical conditions.

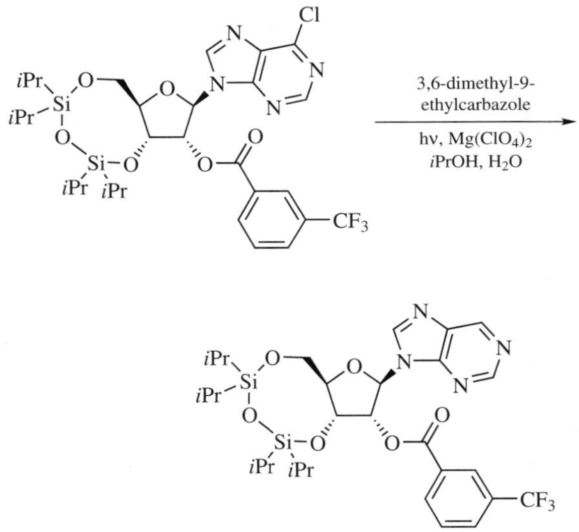

(6)

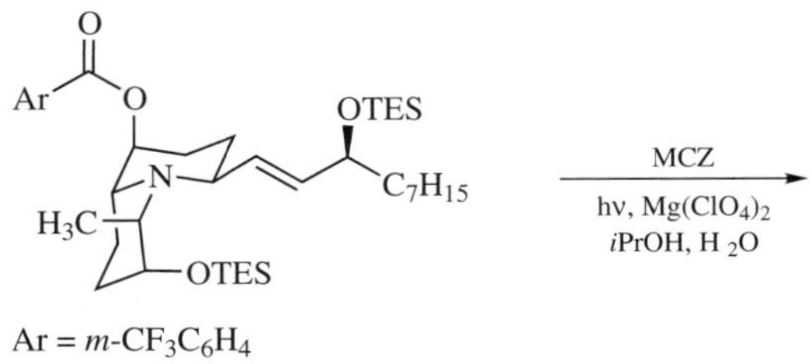

Decomposition (7)

Photosensitized Decarboxylation. 9-Methylcarbazole and 9-phenylcarbazole were reported to be effective photosensitizers for the decarboylation of *N*-acyloxyphthalimides (eq 8).[15] High yields of reduced products were achieved with simple substrates. However, 1,6-bis(dimethylamino)pyrene (BDMAP) was judged to be a superior reagent because of its large extinction coefficient in the visible light region. Unfortunately, this reagent is not commercially available. The photosensitized decarboxylation was also shown to be effective for a radical cyclization reaction if THF was used as the solvent.

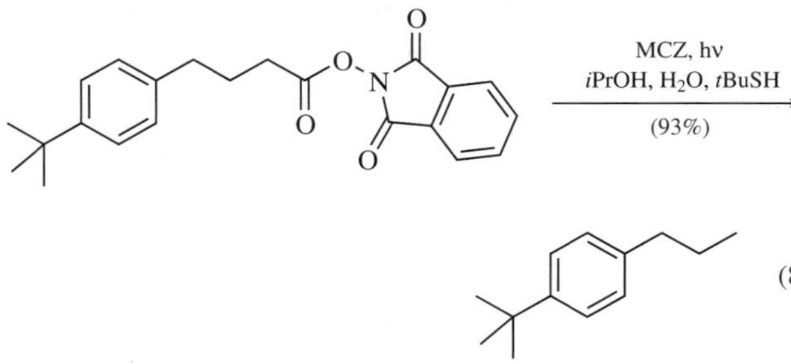

(8)

Photoreduction of Aromatic Sulfoxides to Sulfides. Aromatic sulfoxides can be reduced under photochemical conditions using 9-methylcarbazole as the photosensitizer in methanol.[16] The corresponding sulfides were produced in quantitative yield. Aliphatic sulfoxides, however, were resistant to photoreduction under these conditions.

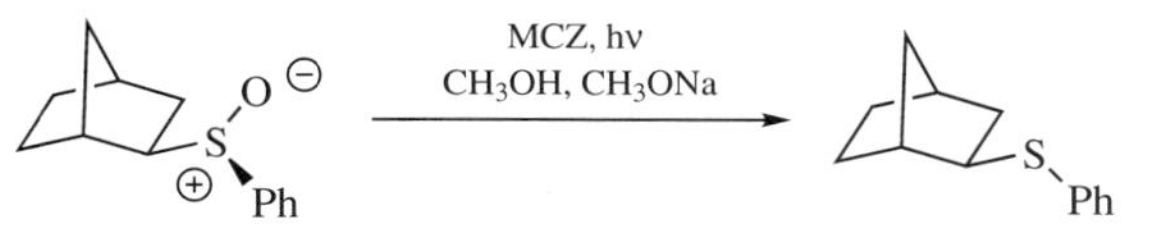

(9)

Related Reagents 3,6-Dimethyl-9-ethylcarbazole; 1,6-bis(dimethylamino)pyrene.

1. Saito, I.; Ikehira, H.; Kasatani, R.; Watanabe, M.; Matsuura, T., *J. Am. Chem. Soc.* **1986**, *108*, 3115.
2. Suzuki, M.; Koyano, H.; Noyori, R., *J. Org. Chem.* **1987**, *52*, 5583.
3. Almond, M. R.; Collins, J. L.; Reitter, B. E.; Rideout, J. L.; Freeman, G. A.; St Clair, M. H., *Tetrahedron Lett.* **1991**, *32*, 5745.
4. Clive, D. L. J.; Daigneault, S., *J. Org. Chem.* **1991**, *56*, 3801.
5. Huang, Z.; Schneider, K. C.; Benner, S. A., *J. Org. Chem.* **1991**, *56*, 3869.
6. Barvian, M. R.; Greenberg, M. M., *Tetrahedron Lett.* **1992**, *33*, 6057.
7. Ireland, R. E.; Armstrong, J. D.; Lebreton, J.; Meissner, R. S.; Rizzacasa, M. A., *J. Am. Chem. Soc.* **1993**, *115*, 7152.
8. Myers, A. G.; Condroski, K. R., *J. Am. Chem. Soc.* **1995**, *117*, 3057.
9. Park, M.; Rizzo, C. J., *J. Org. Chem.* **1996**, *61*, 6092.
10. Jung, P. M. J.; Burger, A.; Biellmann, J. F., *J. Org. Chem.* **1997**, *62*, 8309.
11. Rubira, M. J.; Perez-Perez, M. J.; Balzarini, J.; Camarasa, M. J., *Synlett* **1998**, 177.
12. Riedel, S.; Donnerstag, A.; Hennig, L.; Welzel, P.; Richter, J.; Hobert, K.; Muller, D.; Heijenoort, J., *Tetrahedron* **1999**, *55*, 1921.
13. Burgdorf, L. T.; Carell, T., *Chem. Eur. J.* **2002**, *8*, 293.
14. Chiocconi, A.; Marino, C.; Otal, E.; de Lederkremer, R. M., *Carbohydr. Res.* **2002**, *337*, 2119.
15. Okada, K.; Okamoto, K.; Oda, M., *J. Am. Chem. Soc.* **1988**, *110*, 8736.
16. Cubbage, J. W.; Tetzlaff, T. A.; Groundwater, H.; McCulla, R. D.; Nag, M.; Jenks, W. S., *J. Org. Chem.* **2001**, *66*, 8621.
17. McCombie, S. W. In Comprehensive Organic Synthesis; Trost, B. M., Fleming, I., Eds.; Pergamon Press:Oxford, **1991**; Vol. *8*, pp 811.
18. Park, M.; Buck, J. R.; Rizzo, C. J., *Tetrahedron* **1998**, *54*, 12707.
19. Matray, T. J.; Greenberg, M. M., *J. Am. Chem. Soc.* **1994**, *116*, 6931.
20. Barvian, M. R.; Barkley, R. M.; Greenberg, M. M., *J. Am. Chem. Soc.* **1995**, *117*, 4894.
21. Greenberg, M. M.; Barvian, M. R.; Cook, G. P.; Goodman, B. K.; Matray, T. J.; Tronche, C.; Venkatesan, H., *J. Am. Chem. Soc.* **1997**, *119*, 1828.
22. Myers, A. G.; Condroski, K. R., *J. Am. Chem. Soc.* **1993**, *117*, 7926.
23. Prudhomme, D. R.; Wang, Z.; Rizzo, C. J., *J. Org. Chem.* **1997**, *62*, 8257.
24. Wang, Z.; Prudhomme, D. R.; Buck, J. R.; Park, M.; Rizzo, C. J., *J. Org. Chem.* **2000**, *65*, 5969.
25. Ha, J. D.; Cha, J. K., *J. Am. Chem. Soc.* **1999**, *121*, 10012.

Carmelo J. Rizzo

Vanderbilt University, Nashville, TN, USA

Methyl Fluorosulfonate and Methyl Trifluoromethanesulfonate

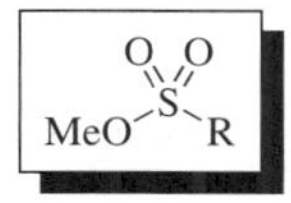

($R = CF_3$)		
[333-27-7]	$C_2H_3F_3O_3S$	(MW 164.12)
($R = F$)		
[421-20-5]	CH_3FO_3S	(MW 114.11)

(powerful methylating agents[1,2])

Alternate Name: $R = CF_3$, methyl triflate; $R = F$, Magic Methyl.

Physical Data: methyl triflate: bp 99 °C, mp −64 °C, *d* 1.50 g cm^{-3}. Methyl fluorosulfonate: bp 92 °C, mp −92.5 °C, *d* 1.45 g cm^{-3}.

Solubility: both reagents are miscible with all organic solvents, but react with many. They are only sl sol water, but hydrolyze rapidly as they dissolve. Useful inert solvents are CH_2Cl_2, SO_2, sulfolane, nitromethane, Me_2SO_4, and Me_3PO_4.

Form Supplied in: methyl triflate (MeOTf) is available as a colorless liquid. Methyl fluorosulfonate ($MeOSO_2F$) was formerly available as Magic Methyl™ but has been withdrawn (see below).

Analysis of Reagent Purity: MeOTf gives a singlet in ^1H NMR at δ 4.18, with ^{13}C absorption at δ 61.60 and 119.32 (q), and a ^{19}F shift of 75.4 ppm. $MeOSO_2F$ absorbs at δ 4.19 in ^1H NMR (J_{HF} ; 0.4 Hz or less), with ^{13}C absorption at δ 62.45, and a ^{19}F shift of −31.2 ppm. ^{33}S and ^{17}O NMR data have been reported for both compounds.[9]

Preparative Methods: both reagents are prepared[3,4] by distilling an equimolar mixture of the corresponding acid with Dimethyl Sulfate in an all-glass apparatus with a short Vigreux column. They may be dried by standing over fused K_2CO_3 and redistillation. Trifluoromethanesulfonic Acid and Flurosulfuric Acid are both available, and are comparably priced.

Handling, Storage, and Precautions: both reagents are **extremely hazardous**. All possible precautions should be taken to avoid inhalation or absorption through the skin. A fatality has occurred with $MeOSO_2F$ through inhalation of the vapors leading to pulmonary edema.[5] Dexamethasone isonicotinate (Auxiloson® spray) has been recommended as a first aid in the treatment of such pulmonary irritation.[5] The oral LD_{50} of $MeOSO_2F$ is 112 mg/kg in mice, and an LC_{50} for 1 h exposure for rats between 5 and 6 ppm has been reported; severe eye irritation was noted.[6] It is very unlikely[7] that MeOTf is less dangerous, but no data on toxicity have been reported. Both materials are extremely destructive to the tissue of the mucous membranes and upper respiratory tract, eyes, and skin. Inhalation may be fatal as a result of spasm, inflammation, and edema of the larynx and bronchi, chemical pneumonitis, and pulmonary edema. Use in a fume hood.

The reagents are stable in glass when dry, but storage of $MeOSO_2F$ in bottles with ground glass joints should be avoided as these slowly become fused. It is reported[8] that $MeOSO_2F$ after storage over CaH_2 for two weeks contained 17% Me_2SO_4.

Original Commentary

Roger W. Alder & Justin G. E. Phillips
University of Bristol, UK

General Reactivity. MeOTf and $MeOSO_2F$ are two of the most powerful reagents for methylation and are more reactive by a factor of $\sim 10^4$ than Iodomethane and Me_2SO_4.[1,2] The only reagents which are substantially more powerful are Methyl Fluoride–Antimony(V) Fluoride and related reagents[10] and the dimethylhalonium ions; these reagents pose more severe handling problems. The reactivity of MeOTf and $MeOSO_2F$ is not effectively enhanced by addition of Lewis acids. Thus addition of Antimony(V) Chloride to $MeOSO_2F$ and Dimethyl Sulfone led to complexation of the sulfone rather than methylation;[11] it is also reported that addition of $SbCl_5$ leads to formation of MeCl.[12]

The qualitative reactivity of some methylating agents toward a range of functional groups is shown in tabular form (Table 1).[1] Few quantitative data are available, but MeOTf and $MeOSO_2F$ are only a little less reactive than the trimethyloxonium ion, with the dimethoxycarbenium ion probably somewhat more reactive again. Substitution rates with various nucleophiles are reported to be in the order: $Me_3O^+ > MeOTf > MeOSO_2F > MeOClO_3$, with rate ratios of 109:23:8:6:1.0 for reaction with acetonitrile at 0 °C.[13,14] Methyl perchlorate, an explosion hazard, therefore never offers a practical advantage. Our experience is that the relative rates for MeOTf and $MeOSO_2F$ rarely differ by more than a factor of 2–5.

It is noteworthy that neither MeOTf nor $MeOSO_2F$ shows any reactivity on their own in Friedel–Crafts methylation reactions even with highly reactive substrates (Me_3O^+ ion is similar). Reaction has been observed in the presence of protic or Lewis acids.[15] Olah has recently suggested that this reactivity is associated with generation of superelectrophiles by further protonation.[16] $MeOSO_2F$ has been reported to react as a methylsulfonylating agent towards phenol and anisole,[17] but this alternative reactivity is usually not significant.

The alkylation reactions of MeOTf and $MeOSO_2F$ are discussed according to the atom alkylated. To avoid unnecessary repetition, reactions can be assumed to use MeOTf unless $MeOSO_2F$ is specified. Historically, $MeOSO_2F$ was mainly used up until about 1980, but MeOTf has since become the reagent of choice. In the great majority of cases these reagents can probably be used interchangeably.

Alkylation at Nitrogen. Most amines react violently with MeOTf or $MeOSO_2F$, and only those with severe steric hindrance or conjugated with strong electron-withdrawing groups really require the use of these reagents. Of derivatives with sp^3 nitrogen, Diisopropylamine, *N*,2,2,6,6-pentamethylpiperidine, and 1,8-Bis(dimethylamino)naphthalene (Proton Sponge™) can all be quaternized by $MeOSO_2F$, and *N*,*N*,2,6-tetramethylaniline reacts on heating.[1] However, it has been reported that *i*-Pr_3N does not react with $MeOSO_2F$.[19] 2,6-Lutidine reacts exothermically with $MeOSO_2F$ at room temperature, and 2,6-dimethoxycarbonylpyridine can be quantitatively quaternized.[1] A range of [*n*](2,6)-pyridinophanes have been methylated with $MeOSO_2F$, with *n* as small as 6 (eq 1).[20]

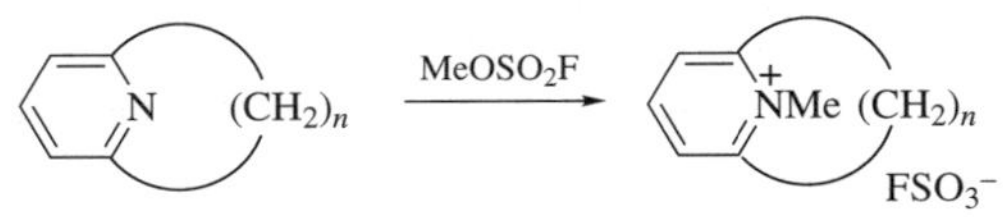

(1)

2,6-Di-t-butylpyridine does not react at normal pressure, and this or 2,6-di-*t*-butyl-4-methylpyridine (synthesis[21]), are often used in applications which require base (see below). Note that 2,6-di-*t*-butylpyridine can be alkylated under high pressure with $MeOSO_2F$ to give >90% of the methylation product when water is carefully excluded.[22]

Simple imines, such as benzylideneaniline, react readily. The *N*-methylation of imine and amidine derivatives of amino acid esters, followed by hydrolysis, has been used as a method for the preparation of *N*-alkylated amino acids with minimal racemization.[23] A convenient synthesis of *N*-methyl nitrones has been developed by alkylation of OTMS oximes with MeOTf, followed by treatment with fluoride ion.[24]

Besides the pyridine derivatives already mentioned, most heterocyclic nitrogens can be alkylated. Aspects of the quaternization

Table 1 Reactivity[a] of Methylating Agents towards Functional Groups

	Approx. pK_a[18]	MeI	Me_2SO_4	$MeOSO_2F$ (MeOTf)	Me_3O^+	$HC(OMe)_2^+$	Me_2Cl^+
Me_3N	9	+	+	++	++	++	++
Pyridine	5	+	+	++	++	++	++
Me_2S	1	+	+	++	++	++	++
Me_2SO	0	+	+	+	+	+	++
$HCONMe_2$	0	−	+	+	+	+	++
MeCN	−10	−	−	+	+	+	+
Me_2O	−5	−	−	+	(+)	+	+
PhCHO	−6	−	−	+	−	+	+
R_2CO	−6	−	−	±	−	±	+
Lactone	−7	−	−	+	−	+	+
Ester	−7	−	−	(+)	−	+	+
MeI	?	−	−	(+)	−	(+)	+
$MeNO_2$	−10	−	−	−	−	−	+
Me_2SO_2	−10	−	−	−	−	−	+

[a]Key: ++ methylation occurs and may be strongly exothermic; + methylation occurs; − little reaction; (+) equilibrium transfer reaction; ± reaction is complex.

of heteroaromatics have been recently reviewed.[25] With respect to the limits of reactivity, it is interesting that dimethylation of 2-phenyl-4,6-dimethylpyrimidine could only be achieved with Trimethyloxonium Tetrafluoroborate; $MeOSO_2F$ only gave monoalkylation.[26] 2-Benzoylbenzothiazole can be *N*-alkylated with $MeOSO_2F$ (but not with MeI) and then acts as an active acylating agent.[27] Formation of an *N*-methylindole from an *N*-ethoxycarbonylindole has been achieved with $MeOSO_2F$.[28]

A number of alkylation products from MeOTf and heterocycles have been advocated as useful intermediates. Thus treatment of 2-substituted thiazoles with MeOTf in acetonitrile, followed by reduction of the salt formed with Sodium Borohydride/CuO in CH_2Cl_2, leads to aldehydes.[29] 1-(Benzenesulfonyl)-3-methylimidazolium and 1-(*p*-toluenesulfonyl)-3-methylimidazolium triflates have been proposed as efficient reagents for the preparation of aryl sulfonamides and aryl sulfonates.[30] MeOTf alkylates 2,5-oxazoles to give salts which can be reduced by $PhSiH_3$/CsF to give 4-oxazolines, and these provide a route to stabilized azomethine ylides.[31]

A novel synthesis of 2-aryl-4-piperidones by Mannich cyclization of imino acetals, initiated by methylation of the imine, has been described.[32] MeOTf has been used in the generation of a munchnone for cycloaddition.[33] Finally, methylation of 1-lithio-2-*n*-butyl-1,2-dihydropyridine with MeOTf gives 2-butyl-5-methylpyridine in 42% yield.[34]

Nitriles, with sp hybridized nitrogen, are unreactive to MeI or Me_2SO_4, but are readily methylated by $MeOSO_2F$,[1] MeOTf,[35] or oxonium ions. Nitrilium salts have been shown to have a number of useful applications. The reduction of nitrilium salts by $NaBH_4$ in alcohols leads first to iminoethers and subsequently to amines.[36] Reduction of *N*-alkylnitrilium ions by organosilicon hydrides gives *n*-alkylaldimines, and thus provides a route to aldehydes from nitriles.[37] Nitrilium triflate salts have been shown to be useful reagents for the synthesis of ketones and ketenimines by electrophilic substitution of reactive aromatics, and also provide good routes to amidinium, imidate, and thioimidate salts.[35] Reaction with 2-amino alcohols gives oxazolidines.[38] Synthesis of either 5-substituted 1-methyl-1*H*-tetrazoles or 3,5-disubstituted 1,4-dimethyltriazolium salts from *N*-methylnitrilium triflate salts can be controlled in reactions with $(Me_2N)_2C{=}NH_2^+N_3^-$.[39] Reaction of nitrilium ions with alkyl azides gives 1,2,3-trisubstituted tetrazolium salts.[40] These can be deprotonated to highly reactive 2-methylenetetrazoles.[41]

Amides are alkylated largely on oxygen, as expected (see below), although some *N*-alkylation can be seen by NMR.[1] *N*-Alkylation is more apparent with carbamates (see the section on ambident nucleophiles). *N,N*-Dimethylmethanesulfonamides can be *N*-alkylated ($MeOSO_2F$) to provide salts which are effective reagents for mesylation.[42] *N,N*-Dimethylsulfamate esters react with $MeOSO_2F$ to give trimethylammoniumsulfate esters, which rapidly give methyl esters unless the O-group is aryl, showing that $Me_3N^+SO_3^-$ is a very powerful leaving group.[43] $Me_3N^+SO_2OPh$ reacts with nucleophiles at either the sulfur or a methyl carbon atom.[44]

Azo compounds, which do not react with methyl iodide, can be *N*-methylated by $MeOSO_2F$.[45]

A steroidal oxaziridine was converted ($MeOSO_2F$) to an oxaziridinium salt which showed oxidizing properties.[46]

Alkylation at Oxygen. Most neutral functionalities with lone pairs on oxygen are not alkylated by MeI or Me_2SO_4 but do react with MeOTf or $MeOSO_2F$, although not all such reactions are preparatively useful.[1,2] Ethers react reversibly, and the ultimate product depends on the conditions. Thus good yields of the oxonium ion can be obtained from reaction of THF with stoichiometric amounts of MeOTf, but the use of catalytic amounts leads to polymerization. Cationic ring opening polymerization, initiated by MeOTf and $MeOSO_2F$ among other reagents, has been extensively investigated and recently reviewed.[47]

Reaction of $MeOSO_2F$ with 2-methoxyethyl carboxylates gives 2-alkyl-1,3-dioxolanium ions.[1] The reaction of these ions with trialkylalkynylborate anions provides versatile and direct routes to (*Z*)-α,β-unsaturated ketones (eq 2). Specifically protected 1,3-diketones and other ketonic species can also be prepared from the intermediates.[48]

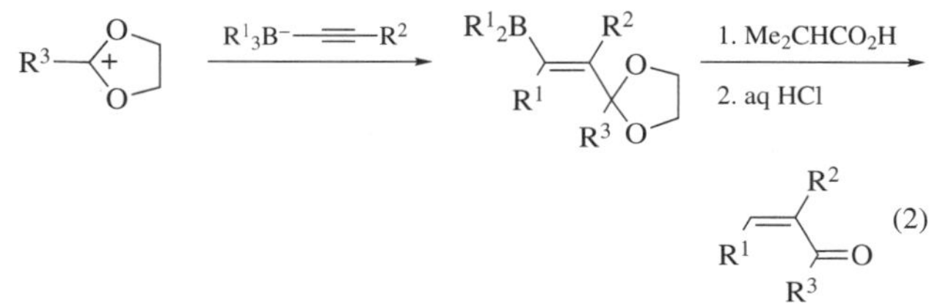

(2)

Almost all carbonyl functions can be methylated. Enolizable aldehydes and ketones usually lead to complex mixtures, probably because of deprotonation to enol ethers, followed by reaction of these with the electrophilic species in the reaction mixture. Nonenolizable aldehydes and ketones give methoxycarbenium ions cleanly, and the relative thermodynamic stabilities of these have been assessed via pairwise equilibrations.[49] Most esters only generate low equilibrium concentrations of dialkoxycarbenium ions, but lactones are readily alkylated.[1]

Amides, carbamates, and ureas are rapidly alkylated, usually on carbonyl oxygen (see the section on ambident nucleophiles). Alkylation of amides with MeOT5f in CH_2Cl_2 followed by reduction of the salts provides a route for the selective reduction of amides; esters, nitriles, acetals, and double bonds are left unaffected by this procedure.[50] Alkylation of isoindolin-1-ones and subsequent deprotonation can provide routes to methoxyisoindoles.[51]

Alcohols can be converted to methyl ethers by the use of MeOTf + 2,6-di-*t*-butylpyridine or 2,6-di-*t*-butyl-4-methylpyridine. This procedure was initially developed in the carbohydrate field.[52] Me_3PO_4 provides a good polar solvent for this process.[53] A recent application, in the synthesis directed at lonomycin, was to methylation of the complex alcohol (**1**) without causing retroaldol cleaveage.[54]

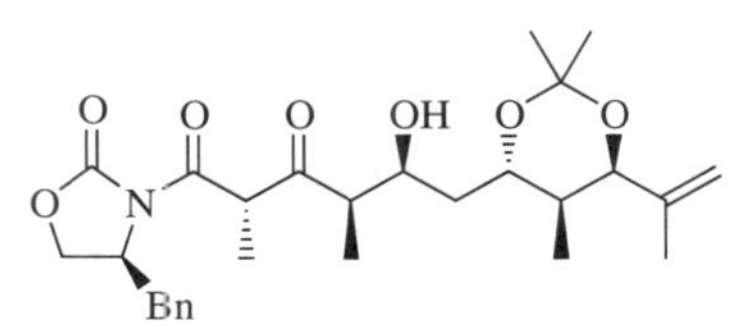

(**1**)

MeOTf has been reported to effect complete methylation of inositol polyphosphates, P–OMe groups being formed as well.[55]

MeOTf will effect *O*-alkylation at the anomeric center, but the stereochemistry is affected by the presence of a crown ether.[56] Reaction of Meisenheimer complexes with $MeOSO_2F$ can lead to capture of the anion as a nitronate ester.[57]

Alkylation at Sulfur and Selenium. Dialkyl and most arylalkyl sulfides are readily converted into sulfonium salts.[1] Cyclic sulfides, especially dithiolanes, react somewhat faster.[58] Reaction of disulfides with MeOTf gives Me_2SSMe^+ OTf^-; this salt, with Triphenylphosphine, reacts with alkenes in a stereo- and regioselective fashion and the products can be converted into vinylphosphonium salts.[59] Thione groups are also alkylated even when electronegative groups are present. Thus 4,5-bis(trifluoromethyl)-1,3-dithiolane-2-thione was converted into a methylated salt.[60] Sulfoxides are *O*-alkylated,[1] and formation of various oxa- and azasulfonium ions has been reported, up to and including triazasulfonium salts.[61] Alkylation of R_2SO with $MeOSO_2F$, followed by reduction with Sodium Cyanoborohydride, leads to sulfides.[62]

Reaction of MeOTf with the product from P_2S_5/Na_2CO_3 ($Na_2P_4S_{10}O$) gave a useful electrophilic thionation reagent.[63] $MeOSO_2F$ has been used in the conversion of thiols to reactive sulfenating agents (eq 3).[64]

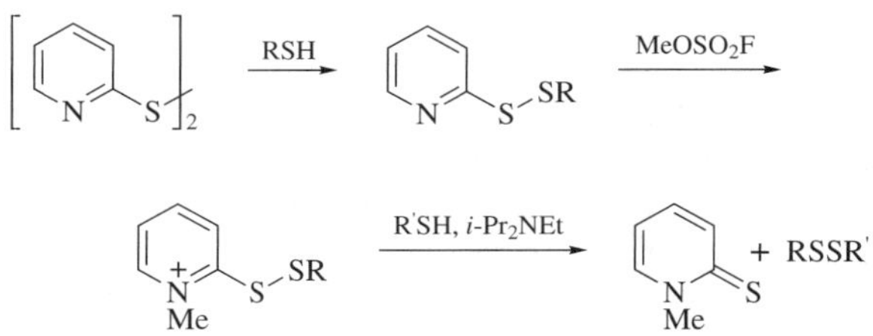

(3)

Methylation of dithioacetals by $MeOSO_2F$, followed by reaction with various nucleophiles, has been used for the removal of this protecting group or its conversion into other protecting groups, e.g. acetals (eq 4).[65–68]

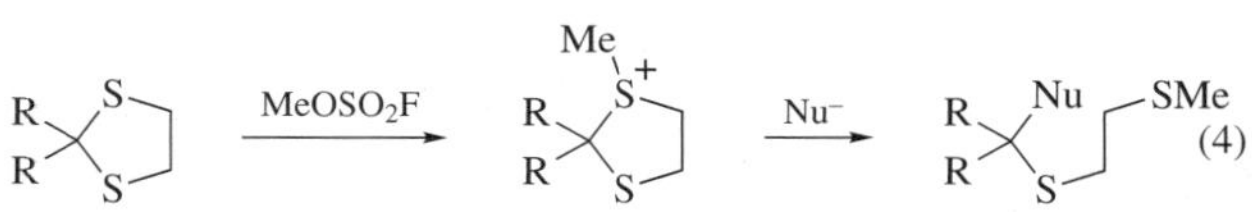

(4)

The reaction of the sulfonium intermediate with alcohols leads to their protection as hemithioacetals.[69] Treatment of thioglycosides with MeOTf gives an efficient glycosylating agent,[70] and pyruvic acetal formation from a pyruvyl thioacetal has been achieved in a reaction catalyzed by MeOTf (amongst other electrophiles).[71]

Alkenes can be prepared by methylation of selenides by $MeOSO_2F$ and treatment of the selenonium ions formed with Potassium t-Butoxide.[72]

Alkylation at Phosphorus. Phosphines and phosphites undergo easy quaternization. Thus methylation of tris(2,6-dimethylphenoxy)phosphine with MeOTf, followed by treatment of the product with sodium 2,6-dimethylphenoxide, gave methyltetrakis (2,6-dimethylphenoxy)phosphane.[73] Methoxyphosphonium triflates are relatively stable intermediates in Arbuzov reactions.[74] Phosphine oxides and sulfides are alkylated. *S*-Methylation of chiral phosphine sulfides, followed by treatment with Hexamethylphosphorous Triamide, has been advocated as a general synthesis of optically active phosphines.[75]

Ambident Nucleophiles. Amides and related functional groups can be alkylated on oxygen or nitrogen and, as has been noted already, alkylation on carbonyl oxygen normally predominates. In the case of carbamates, *O*-alkylation by $MeOSO_2F$ can be faster, but *N*-alkylation predominates at equilibrium.[76] It has been noted that methylation of secondary amides and thioamides occurs at the protonless heteroatom in the major tautomer.[77] The ionic products of these reactions can be deprotonated to give synthetically useful products, e.g. imidates,[78] but excess $MeOSO_2F$ should be removed before treatment with base.[77]

Reaction of most enolates with MeOTf or $MeOSO_2F$ is always likely to be kinetically controlled. There does not appear to have been a definitive study, but *O*-alkylation is the normal outcome. *O*-Alkylation of a bicyclodecatrienone by $MeOSO_2F$ is enhanced by the use of polar solvents like HMPA.[79] *O*-Alkylation of enolates of appropriate cyclohexadienones by MeOTf has been used to generate various 3a*H*-indenes.[80] A ketene acetal is formed by exclusive *O*-alkylation of the sodium enolate of isopropyl bis(pentachlorophenyl)acetate by MeOTf.[81]

Alkylations at Carbon. In an important recent development, primary α-alkylation of carbonyl compounds under nonbasic conditions has been achieved (eq 5) by alkylation of silyl enol ethers with MeOTf and other primary alkyl triflates, catalyzed by Methylaluminum Bis(4-bromo-2,6-di-t-butylphenoxide (MABR).[82]

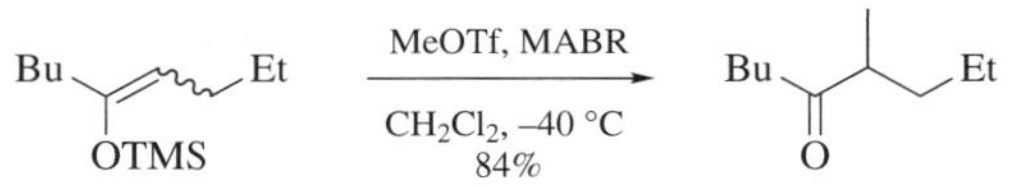

(5)

Related Reagents. Dimethoxycarbenium Tetrafluoroborate; Dimethyliodonium Hexafluoroantimonate; *O*-Methyldibenzofuranium Tetrafluoroborate; Triethyloxonium Tetrafluoroborate.

First Update

Lijun Huang & Xuefei Huang
University of Toledo, Toledo, OH, USA

Alkylation at Nitrogen. MeOTf can readily react with many types of nitrogen containing compounds under mild conditions. Recent developments of MeOTf mediated *N*-alkylation focused on the utility of *N*-quaternary intermediates.

N-Methylation confers a positive charge on nitrogen atoms, thus converting them into good leaving groups. *N,N'*-Sulfuryldiimidazole was activated towards displacement by *N*-methylation with MeOTf. The imidazolium group of the resulting triflate salt was then displaced with a variety of amines and anilines to afford unsymmetrical sulfonylureas in excellent yield (eq 6).[83] Quaternization of an *N*-(2-pyridyl)tosylamide in a protected amino alcohol by MeOTf, followed by base treatment provided a mild method for removing the pyridine moiety without epimerization (eq 7).[84] The reaction of a sulfonylated bicyclic amine with excess MeOTf followed by treatment with sodium amalgam led to an unsaturated cyclic amine in 71% yield (eq 8).[85] Intramolecular regioselective epoxidation of a less electron-rich alkene in a nonconjugated diene was achieved by the oxaziridinium salts formed from oxaziridines and MeOTf (eq 9).[86] MeOTf can also be used for solid phase synthesis. An aminoxy functionalized resin was utilized in traceless solid phase syntheses of amines. Basic cleavage following MeOTf alkylation of the nitrogen afforded tertiary methylamines in high purity.[87,88]

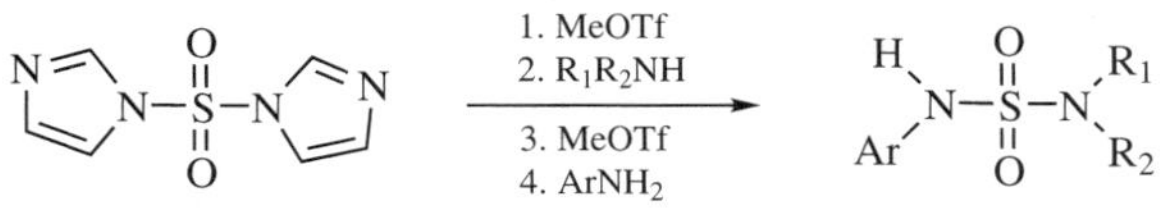

(6)

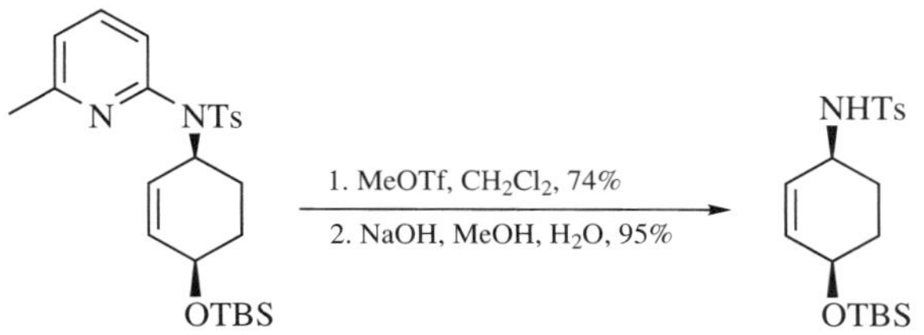

(7)

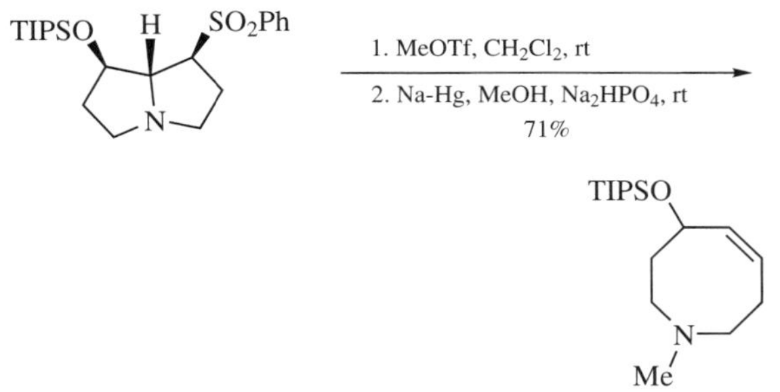

(8)

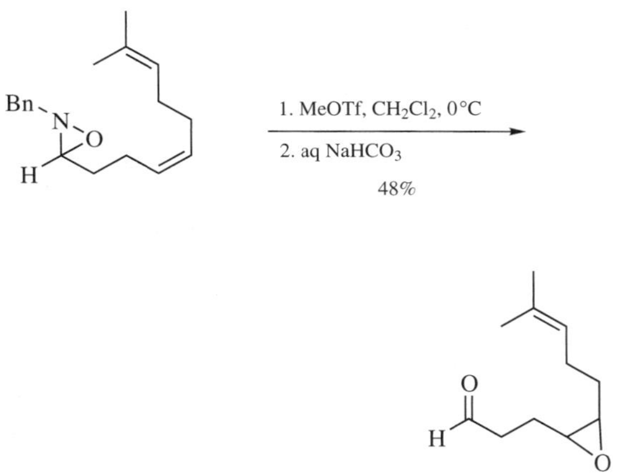

(9)

Quaternization of the nitrogen atom in a heterocyclic system can activate the ring towards addition reactions. An oxazoline moiety was converted to an aldehyde in 97% overall yield by alkylation of oxazoline with MeOTf, reduction with sodium borohydride, and hydrolysis of the resulting aminal (eq 10).[89] Alternatively, treatment of an *N*-methylated oxazolium triflate with a Grignard reagent followed by aqueous acid produced a ketone.[90] Reaction of alkenyl oxazoles with MeOTf induced spontaneous intramolecular [4 + 2] cycloaddition at room temperature leading to a hydroindole or a hydroisoquinoline after reduction by sodium borohydride (eq 11).[91]

MeOTf, as an initiator, can promote cationic ring-opening polymerization of a six-membered cyclic pseudourea via *N*-methylation, producing poly(*N*-carbamoylimino) trimethylene (eq 12). The reaction with alkyl halides as promoters gave a polymer mainly consisting of 1,3-diazin-2-one-1,3-diylalkylene unit.[92]

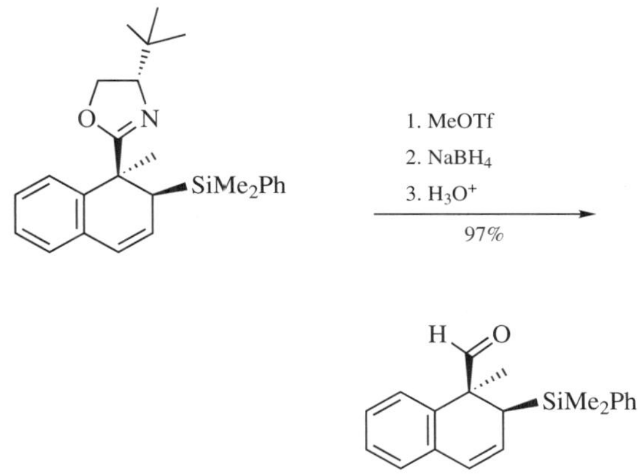

(10)

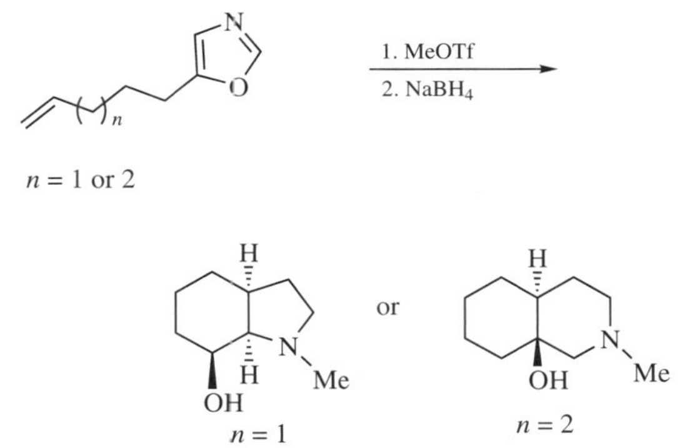

(11)

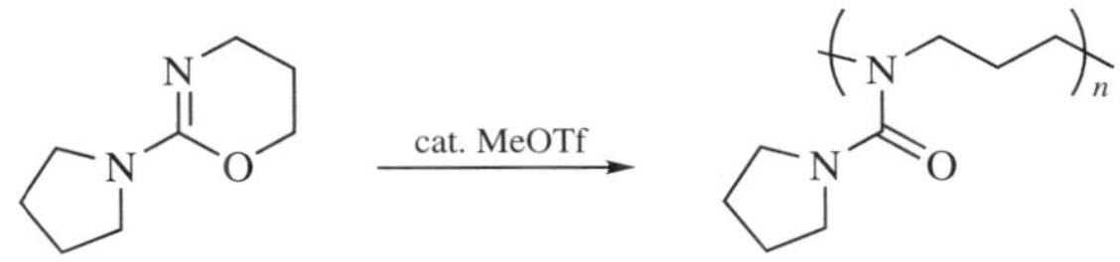

(12)

Due to its superior reactivity, ^{11}C labeled MeOTf has been introduced as a highly reactive alternative to ^{11}C labeled methyl iodide in preparation of ^{11}C radiopharmaceuticals through methylation of amines.[93]

Alkylation at Oxygen. Alcohols can be methylated by MeOTf in the presence of sterically hidered bases such as di-*tert*-butyl-4-methyl-pyridine (DTBMP),[94] or with the aid of organolithium reagents.[95] A secondary alcohol was converted to methyl ether in high yield using MeOTf and DTBMP without affecting a *tert*-butyldiphenyl silyl (TBDPS) ether (eq 13).[96] Significant TBDPS migration to the secondary hydroxyl group was observed when more basic conditions (NaH with MeI) were employed. *O*-Methylation of tylosin derivatives with excess MeOTf and DTBMP gave the 3-methyl ether in 49% yield in the presence of adjacent acetoxy groups (eq 14). Reactions in other solvents, and the use of MeI-Ag_2O, led to migration of the acetyl group or partial deacetylation to give a complex reaction mixture.[97] Sterically hindered alcohols also could be methylated with MeOTf, although they required more forcing conditions (*n*BuLi, MeOTf) (eq 15).[98,99]

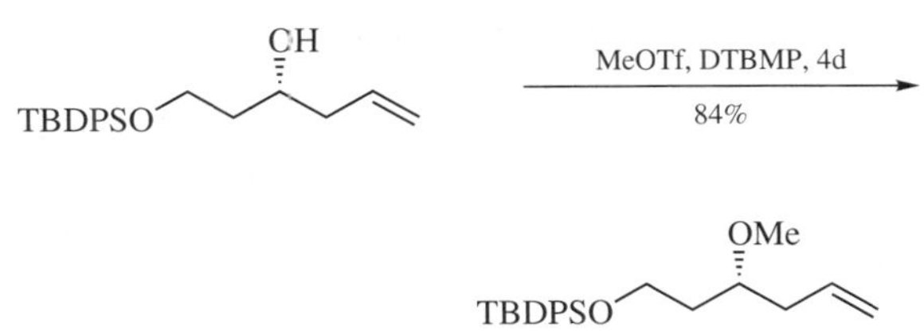

(13)

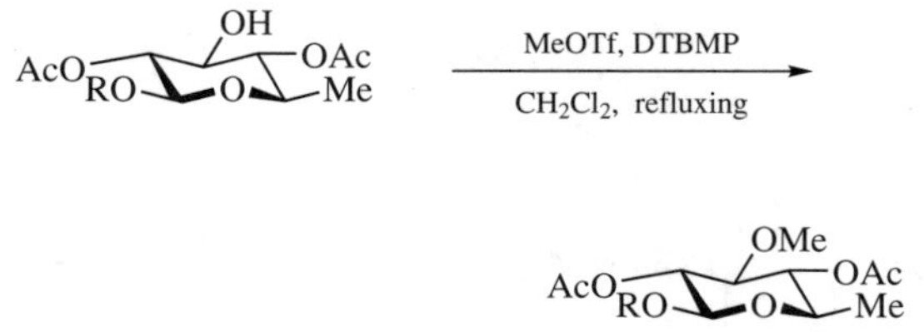

(14)

R = mycaminosyltylonolide alygcon

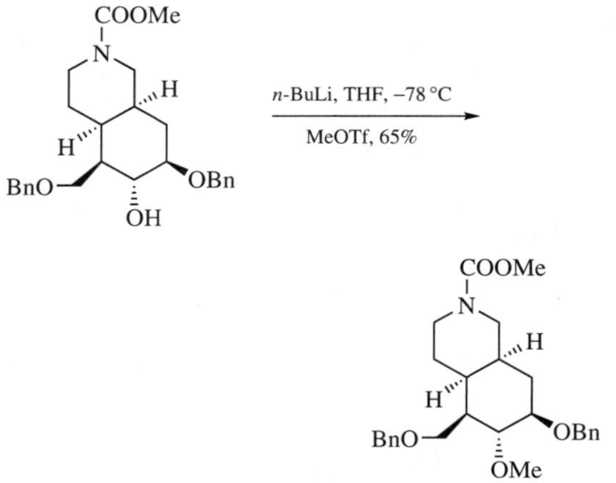

(15)

MeOTf has been reported to effect the dealkoxylation of a perfluoroalkyltrialkoxyboronate to generate the corresponding boronic ester.[100] Conversely, an alkenyl boroxycarbene complex was reacted with MeOTf to remove the borane-based chiral auxiliary yielding a Fischer carbene complex.[101] *tert*-Amide substituted oxetanes rearranged in anhydrous nitrobenzene at 150 °C with a catalytic amount of MeOTf to produce ester-substituted azetidines (eq 16).[102] Other acids such as boron trifluoride etherate, trifluoromethanesulfonic acid, and benzylthiolanium hexafluoroantimonate led to low yields of the desired azetidines.

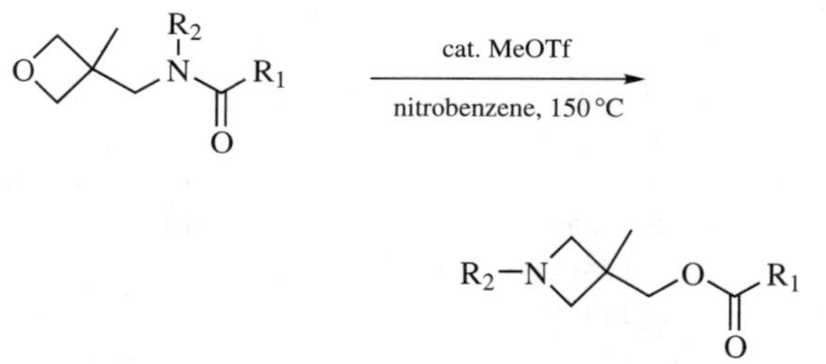

(16)

Alkylation at Carbon. MeOTf can serve as a carbanion alkylating reagent giving one carbon extended homologs. Methylation of the sodium salt of a terminal alkyne with MeOTf afforded a propynyl intermediate in excellent yield.[103] It has been reported that alkylation of ketone enolates with MeOTf in ether or hexane proceeded in high yields without the exchange products, that occurred when methyl halides or methyl sulfate were used.[104] Lithiation of 2-methyloxazoles with lithium diethylamide followed by treatment with MeOTf resulted in the selective formation of 2-ethyl oxazoles.[105] The usage of lithium diethylamide was crucial to achieve high selectivity, as the conjugated acid, diethylamine, led to formation of the most stable carbanion. The nature of the electrophile and substrate was found to greatly affect the selectivities of alkylation.[106]

Methylation by MeOTf can also be carried out asymmetrically. Treatment of α-lithio benzyl 2-quinolyl sulfide with MeOTf in the presence of a bis(oxazoline) as an external chiral ligand yielded chiral 1-phenylethanethio quinolyl sulfide (eq 17).[107] Lithiation adjacent to nitrogen of chiral 3,4-substituted piperidines and pyrrolidines followed by diastereoselective substitution with MeOTf offered general routes to 2,4,5- and 2,4,5,6-substituted piperidines as well as 2,3,4- and 2,3,4,5-substituted pyrrolidines.[108]

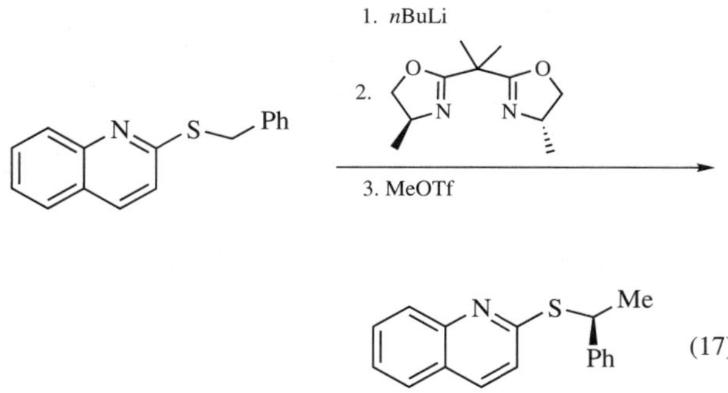

(17)

85% yield, 78% ee

Alkylation at Sulfur and Selenium. Thioamides can be converted to the corresponding thioimidates via alkylation with MeOTf, which is very useful for introducing different functional groups at the thiocarbonyl carbon. Direct conversion of a thioamide to the amidrazone by heating with hydrazine was not successful, but the thioimidate readily reacted with hydrazine at room temperature to afford the desired product (eq 18).[109] It has been reported that thioiminium salts generated in situ from thioamides and MeOTf can be used to construct carbon-carbon bonds by sequential reaction with organolithium and Grignard reagents. A variety of tertiary propargylamines were conveniently synthesized by this protocol (eq 19).[110] Treatment of selenoamides with MeOTf, followed by reaction with lithium acetylides furnished β-methylselenenyl α, β-unsaturated ketones in high yield. These reaction proceed with high stereoselectivity to give exclusively *Z*-isomers.[111]

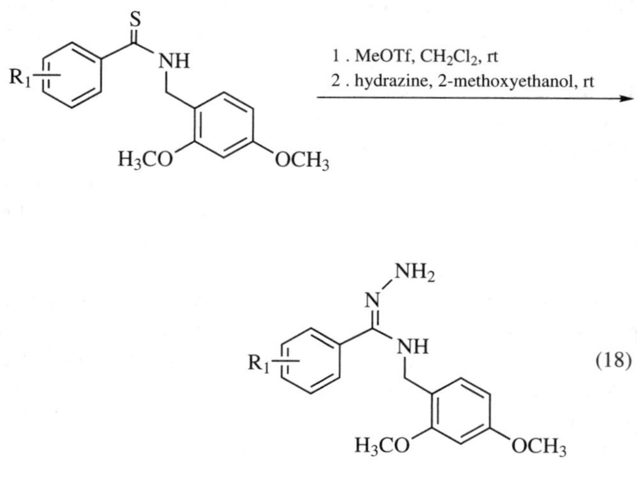

(18)

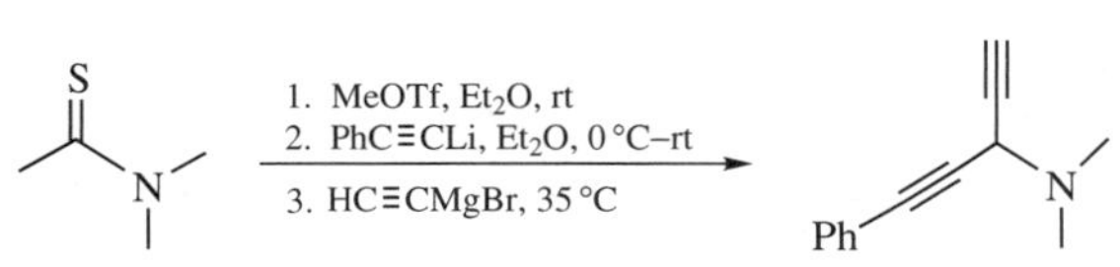

(19)

Thioethers and selenoethers can be displaced with various nucleophiles after methylation by MeOTf. This has been used extensively in the carbohydrate synthesis field (see the section on glycosylation). A phenylselenide moiety was substituted stereospecifically by an alkyl stannane following MeOTf activation in the construction of a C–C bond between two quaternary carbon centers (eq 20).[112]

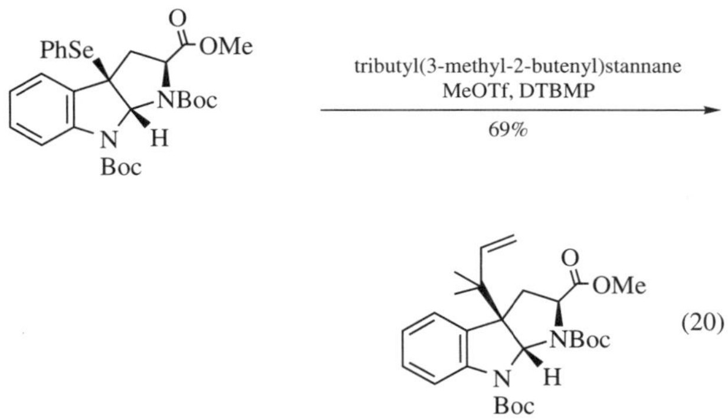

(20)

Alkylation of Metal Complexes. MeOTf has been reported to alkylate transition metal complexes on either the metal center or the ligands. The reaction of η^2-acroleinpalladium complexes with MeOTf in toluene at room temperature quantitatively gave a palladium complex with a η^3-methoxyallyl structure, which has been proposed as an important intermediate in several catalytic reactions (eq 21).[113] The replacement of MeOTf with MeI led to oxidative addition, forming methyl metal complexes with concomitant dissociation of the acrolein. The sequential addition of alkyllithium reagents and MeOTf to a series of cyclomanganated 2-phenylpyridine derivatives afforded metallospiralenes.[114,115] *N*-Rhenaimine β-lactams underwent clean demetalation with MeOTf to yield the *N*-methyl-β-lactams.[116] Novel square-planar, terdentate, aryl-substituted pyridine-diimine Rh(I) and Ir(I) triflate complexes were synthesized by treatment of the corresponding chloro complexes with MeOTf to produce a methylated Rh(III) or Ir (III) complex followed by reduction.[117]

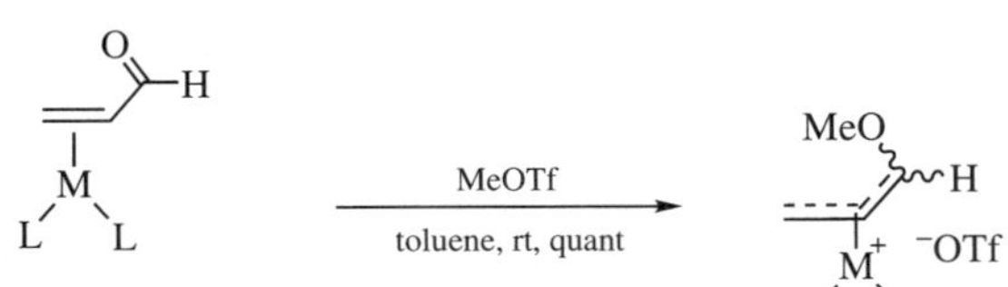

(21)

M=Pd or Pt
L=PPh_3 or dppf

Alkylation at Phosphorus. Bis(trifluoromethyl)phosphines were methylated by MeOTf, yielding the corresponding phosphonium salts, which can be deprotonated using MeN=P(NEt_2)$_3$ leading to the phosphorus ylides.[118] An amino-phosphino carbene was transformed through *P*-methylation with MeOTf into an amino-phosphonio carbene, which underwent nucleophilic substitution reactions at the carbene center (eq 22). Various carbenes can be synthesized starting from a single precursor, allowing facile adjustment of electronic and steric effects at the carbene, which can facilitate the study of the structure/catalytic activity relationship for carbine-metal complexes.[119]

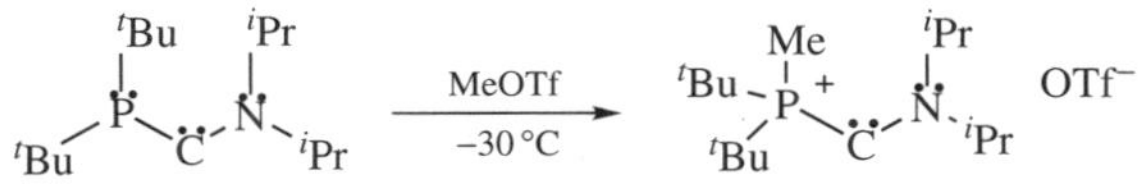

(22)

Phosphine oxides can be efficiently alkylated with MeOTf. Optically active *P*-chirogenic phosphine oxides were reduced with inversion of configuration at the phosphorus atom to yield chiral phosphines by sequential treatment with MeOTf and $LiAlH_4$.[120]

Promoter for Glycosylation. MeOTf is a moderately reactive promoter for thioglycosyl donors, most commonly *S*-methyl or ethyl thioglycosides.[121] MeOTf is less potent than the widely used *N*-iodosuccinimide (NIS) and triflic acid (TfOH)/silver triflate system, therefore an excess of reagent is often required for complete activation. MeOTf can activate a wide array of thioglycosides including glucosamine,[122] mannose,[123] and glucose[124] derivatives with ether-protected thioglycosides being more reactive than the corresponding esters. Complex oligosaccharides can be assembled using MeOTf promoted thioglycoside glycosylation.[125] *O*-Methylation of the glycosyl acceptor hydroxyl group has been observed as a side reaction in the glycosylation of an unreactive thioglycoside with a reactive acceptor. Treatment of phenyl 2,3-*O*-thionocarbonyl-1-thio-α-L-rhamnopyranosides with MeOTf resulted in 1,2-migration and concurrent glycosylation (eq 23).[126] Besides intermolecular glycosylation, MeOTf can also be used to promote the intramolecular aglycon delivery forming the challenging β-mannosyl linkage (eq 24).[125,127]

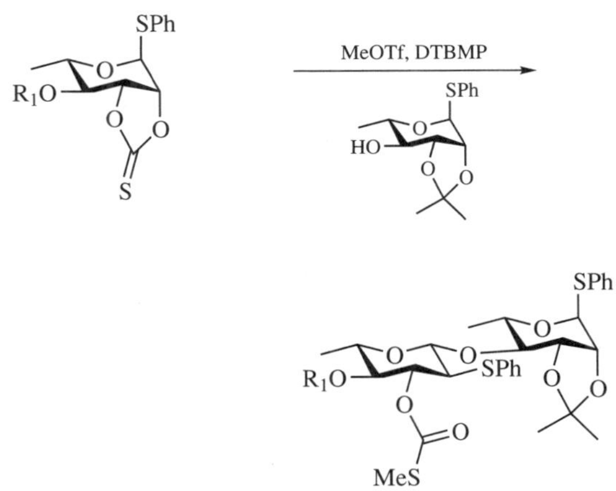

(23)

83%

Due to the high reactivity of MeOTf towards the sulfur atom, MeOTf can chemoseletively activate thioglycosides in the presence of an alkene. This has allowed the development of a chemoselective glycosylation protocol in which a thioglycosyl donor is selectively glycosylated with a glycosyl acceptor bearing an *O*-pentenyl moiety at its anomeric center (eq 25). The resulting

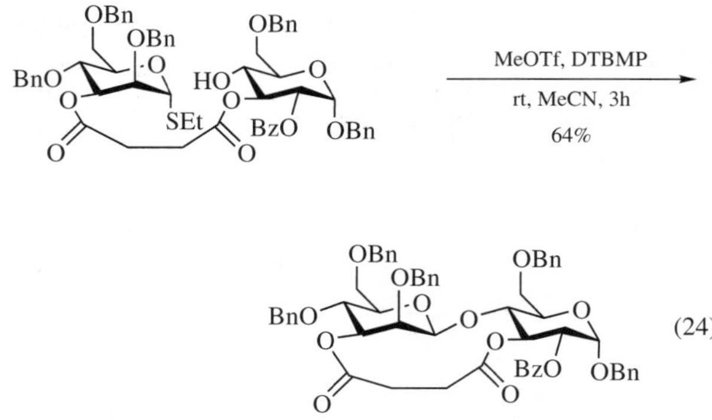

(24)

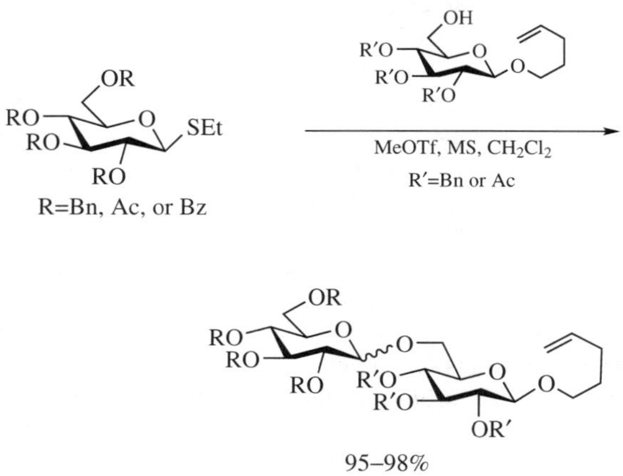

(25)

oligosaccharide can be further activated enabling rapid synthesis of complex oligosaccharides.[128] In contrast, the usage of other thiophilic promoter systems such as NIS/TfOH and iodonium (di-γ-collidine) perchlorate led to low yields of the desired oligosaccharide due to competing reaction with the alkene.

In addition to thioglycosides, MeOTf can promote the activation of 3-methoxypyridyl *O*-unprotected glycosyl donors,[129] and of *S*-benzoxazolyl glycosides for highly stereoselective 1,2-*cis*-glycosylations.[130]

Due to the possibility of side reactions of acceptors with MeOTf and safety considerations, a more specific promoter, dimethyl (methylthio)sulfonium triflate (DMTST), can be utilized instead of MeOTf; it is produced by mixing MeOTf with dimethyl disulfide.[59]

Related Reagents. Methyl iodide; methyl sulfate; 1-benzenesulfinyl piperidine; diphenyl sulfoxide; dimethyl(methylthio) sulfonium tetrafluoroborate.

1. Alder, R. W., *Chem. Ind. (London)* **1973**, 983.
2. (a) Stang, P. J.; Hanack, M.; Subramanian, L. R., *Synthesis* **1982**, 85. (b) For other reviews of trifluoromethanesulfonates, see: Howells, R. D.; McCown, J. D., *Chem. Rev.* **1977**, *77*, 69 and Stang, P. J.; White, M. R., *Aldrichim. Acta* **1983**, *16*, 15.
3. Ahmed, M. G.; Alder, R. W.; James, G. H.; Sinnott, M. L.; Whiting, M. C., *Chem. Commun./J. Chem. soc., Chem. Commun.* **1968**, 1533.
4. Beard, C. D.; Baum, K.; Grakauskas, V., *J. Org. Chem.* **1973**, *38*, 3673.
5. (a) van der Ham, D. M. W.; van der Meer, D., *Chem. Ind. (London)* **1976**, 782. (b) van der Ham, D. M. W.; van der Meer, D., *Chem. Eng. News* **1976**, *54* (36), 5.
6. Hite, M.; Rinehart, W.; Braun, W.; Peck, H., *Am. Ind. Hyg. Assoc.* **1979**, *40*, 600 (*Chem. Abstr.* **1979**, *91*, 84 580d).
7. Alder, R. W.; Sinnott, M. L.; Whiting, M. C.; Evans, D. A.; *Chem. Eng. News* **1976**, *54* (36), 56.
8. Hase, T.; Kivikari, R., *Synth. Commun.* **1979**, *9*, 107.
9. Barbarella, G.; Chatigilialoglu, C.; Rossini, S.; Tugnoli, V., *J. Magn. Reson.* **1986**, *70*, 204.
10. Christie, J. J.; Lewis, E. S.; Casserly, E. F., *J. Org. Chem.* **1983**, *48*, 2531.
11. Minato, H.; Yamaguchi, K.; Miura, T.; Kobayashi, M., *Chem. Lett.* **1976**, 593.
12. Binder, G. E.; Schmidt, A., *Z. Anorg. Allg. Chem.* **1980**, *467*, 197 (*Chem. Abstr.* **1980**, *93*, 196 905m).
13. Kevill, D. N.; Lin, G. M. L., *Tetrahedron Lett.* **1978**, 949.
14. (a) Lewis, E. S.; Vanderpool, S., *J. Am. Chem. Soc.* **1977**, *99*, 1946; (b) Lewis, E. S.; Vanderpool, S., *J. Am. Chem. Soc.* **1978**, *100*, 6421; (c) Lewis, E. S.; Kukes, S.; Slater, C. D., *J. Am. Chem. Soc.* **1980**, *102*, 303.
15. (a) Olah, G. A.; Nishimura, J., *J. Am. Chem. Soc.* **1974**, *96*, 2214; (b) Booth, B. L.; Haszeldine, R. N.; Laali, K., *J. Chem. Soc., Perkin Trans. 1* **1980**, 2887.
16. Olah, G. A., *Angew. Chem., Int. Ed. Engl.* **1993**, *32*, 767.
17. Kametani, T.; Takanashi, K.; Ogasawara, K., *Synthesis* **1972**, 473.
18. Arnett, E. M., *Prog. Phys. Org. Chem.* **1963**, *1*, 223.
19. Wieland, G.; Simchen, G., *Justus Liebigs Ann. Chem./Liebigs Ann. Chem.* **1985**, 2178 (*Chem. Abstr.* **1986**, *104*, 33 742n).
20. Weber, H.; Pant, J.; Wunderlich, H., *Ber. Dtsch. Chem. Ges./Chem. Ber.* **1985**, *118*, 4259 (*Chem. Abstr.* **1986**, *104*, 129 776c).
21. Anderson, A. G.; Stang, P. J., *Org. Synth.* **1981**, *60*, 34.
22. Hou, C. J.; Okamoto, Y., *J. Org. Chem.* **1982**, *47*, 1977.
23. O'Donnell, M. J.; Bruder, W. A.; Daugherty, B. W.; Liu, D.; Wojciechowski, K., *Tetrahedron Lett.* **1984**, *25*, 3651.
24. Le Bel, N. A.; Balasubramanian, N., *Tetrahedron Lett.* **1985**, *26*, 4331.
25. Gallo, R.; Roussel, C.; Berg, U., *Adv. Heterocycl. Chem.* **1988**, *43*, 173.
26. Douglass, J. E.; Bumgarner, D. L., *J. Heterocycl. Chem.* **1981**, *18*, 417.
27. Chikashita, H.; Ishihara, M.; Takigawa, K.; Itoh, K., *Bull. Chem. Soc. Jpn.* **1991**, *64*, 3256.
28. Kametani, T.; Suzuki, T.; Ogasawara, K., *Chem. Pharm. Bull.* **1972**, *20*, 2057.
29. Dondoni, A.; Marra, A.; Perrone, D., *J. Org. Chem.* **1993**, *58*, 275.
30. O'Connell, J. F.; Rapoport, H., *J. Org. Chem.* **1992**, *57*, 4775.
31. Vedejs, E.; Grissom, J. W., *J. Am. Chem. Soc.* **1988**, *110*, 3238.
32. Bosch, J.; Rubiralta, M.; Moral, M.; Valls, M., *J. Heterocycl. Chem.* **1983**, *20*, 595.
33. Hershenson, F. M.; Pavia, M. R., *Synthesis* **1988**, 999.
34. Knaus, E. E.; Ondrus, T. A.; Giam, C. S., *J. Heterocycl. Chem.* **1976**, *13*, 789.
35. Booth, B. L.; Jibodu, K. O.; Proenca, M. F. J. R. P., *Chem. Commun./J. Chem. soc., Chem. Commun.* **1980**, 1151.
36. Borch, R. F., *Chem. Commun./J. Chem. soc., Chem. Commun.* **1968**, 442.
37. Fry, J. L.; Ott, R. A., *J. Org. Chem.* **1981**, *46*, 602.
38. Booth, B. L.; Jibodu, K. O.; Proenca, M. F. J. R. P., *J. Chem. Soc., Perkin Trans. 1* **1983**, 1067.
39. Amer, M. I. K.; Booth, B. L., *J. Chem. Res. (S)* **1993**, 4.
40. Carboni, B.; Carrie, R., *Tetrahedron* **1984**, *40*, 4115.
41. Quast, H.; Hergenroether, T., *Justus Liebigs Ann. Chem./Liebigs Ann. Chem.* **1992**, 581 (*Chem. Abstr.* **1992**, *117*, 48 426e).
42. King, J. F.; du Manior, J. R., *J. Am. Chem. Soc.* **1975**, *97*, 2566.
43. King, J. F.; Lee, T. M., *Chem. Commun./J. Chem. soc., Chem. Commun.* **1978**, 48.

44. King, J. F.; Lee, T. M. L., *Can. J. Chem.* **1981**, *59*, 356.
45. Ferguson, A. N., *Tetrahedron Lett.* **1973**, 2889.
46. Milliet, P.; Picot, A.; Lusinchi, X., *Tetrahedron Lett.* **1976**, 1573, 1577.
47. Penczek, S.; Kubisa, P., In *Ring Opening Polymerization*; Brunelle, D. J., Ed.; Hanser, 1993; Chapter 2.
48. Pelter, A.; Colclough, M. E., *Tetrahedron Lett.* **1986**, *27*, 1935.
49. Quirk, R. P.; Gambill, C. R.; Thyvelikakath, G. X., *J. Org. Chem.* **1981**, *46*, 3181.
50. Tsay, S. C.; Robl, J. A.; Hwu, J. R., *J. Chem. Soc., Perkin Trans. 1* **1990**, 757.
51. Kreher, R. P.; Hennige, H.; Konrad, M.; Uhrig, J.; Clemens, A., *Z. Naturforsch., Teil B* **1991**, *46*, 809 (*Chem. Abstr.* **1991**, *115*, 92 001v) and references therein.
52. Arnarp, J.; Kenne, L.; Lindbreg, B.; Lönngren, J., *Carbohydr. Res.* **1975**, *44*, C5; Arnarp, J.; Lönngren, J., *Acta Chem. Scand.* **1978**, *32B*, 465. Berry, J. M.; Hall, L. D. J., *Carbohydr. Res.* **1976**, *47*, 307. Gilleron, M.; Fournie, J. J.; Pougny, J. R.; Puzo, G., *J. Carbohydr. Chem.* **1988**, *7*, 733.
53. Prehm, P., *Carbohydr. Res.* **1980**, *78*, 372.
54. Evans, D. A.; Sheppard, G. S., *J. Org. Chem.* **1990**, *55*, 5192.
55. Goldman, H. D.; Hsu, F. F.; Sherman, W. R., *Biomed. Environ. Mass Spectrom.* **1990**, *19*, 771.
56. Schmidt, R. R.; Moering, U.; Reichrath, M., *Ber. Dtsch. Chem. Ges./Chem. Ber.* **1982**, *115*, 39 (*Chem. Abstr.* **1982**, *96*, 123 134m).
57. Drozd, V. N.; Grandberg, N. V.; Udachin, Y. M., *Zh. Org. Khim.* **1982**, *18*, 1249 (*Chem. Abstr.* **1982**, *97*, 109 904b).
58. Roberts, R. M. G.; Tillett, J. G.; Ravenscroft, M., *J. Chem. Soc., Perkin Trans. 2* **1982**, 1569.
59. Okuma, K.; Koike, T.; Yamamoto, S.; Yonekura, K.; Ohta, H., *Chem. Lett.* **1989**, 1953.
60. Frasch, M.; Mono, S.; Pritzkow, H.; Sundermeyer, W., *Ber. Dtsch. Chem. Ges./Chem. Ber.* **1993**, *126*, 273.
61. Minato, H.; Okuma, K.; Kobayashi, M., *J. Org. Chem.* **1978**, *43*, 652 and references therein.
62. Durst, H. D.; Zubrick, J. W.; Kieczyowski G. R., *Tetrahedron Lett.* **1974**, 1777.
63. Brillon, D., *Synth. Commun.* **1990**, *20*, 3085.
64. Barton, D. H. R.; Hesse, R. H.; O'Sullivan, A. C.; Pechet, M. M., *J. Org. Chem.* **1991**, *56*, 6697.
65. Fetizon, M.; Jurion, M., *Chem. Commun./J. Chem. soc., Chem. Commun.* **1972**, 382.
66. Ho, T. L.; Wong, C. M., *Synthesis* **1972**, 561.
67. Marshall, J. A.; Seitz, D. E., *J. Org. Chem.* **1975**, *40*, 534.
68. Corey, E. J.; Hase, T., *Tetrahedron Lett.* **1975**, 3267.
69. Hase, T. A.; Kivikari, R., *Synth. Commun.* **1979**, *9*, 107.
70. Lonn, H., *Carbohydr. Res.* **1985**, *139*, 105, 115.
71. Liptak, A.; Szabo, L., *Carbohydr. Res.* **1988**, *184*, C5.
72. Halazy, S.; Krief, A., *Tetrahedron Lett.* **1979**, 4233.
73. Szele, I.; Kubisen, S. J.; Westheimer, F. H., *J. Am. Chem. Soc.* **1976**, *98*, 3533.
74. (a) Colle, K. S.; Lewis, E. S., *J. Org. Chem.* **1978**, *43*, 571; (b) Lewis, E. S.; Hamp, D., *J. Org. Chem.* **1983**, *48*, 2025.
75. Omelanczuk, J.; Mikolajczyk, M., *Tetrahedron Lett.* **1984**, *25*, 2493.
76. Ahmed, M. G.; Alder, R. W., *Chem. Commun./J. Chem. soc., Chem. Commun.* **1969**, 1389.
77. Beak, P.; Lee, J.; McKinnie, B. G., *J. Org. Chem.* **1978**, *43*, 1367.
78. Julia, S.; Ryan, R. J., *C. R. Hebd. Seances Acad. Sci., Ser. C* **1972**, *274*, 1207 (*Chem. Abstr.* **1972**, *77*, 5657u).
79. Press, J. B.; Shechter, H., *Tetrahedron Lett.* **1972**, 2677.
80. Foster, S. J.; Rees, C. W.; Williams, D. J., *J. Chem. Soc., Perkin Trans. 1* **1985**, 711 and references therein.
81. O'Neill, P.; Hegarty, A. F., *J. Org. Chem.* **1987**, *52*, 2113.
82. Maruoka, K.; Sato, J.; Yamamoto, H., *J. Am. Chem. Soc.* **1992**, *114*, 4422.
83. Beaudoin, S.; Kinsey, K. E.; Burns, J. F., *J. Org. Chem.* **2003**, *68*, 115.
84. Yamamoto, Y.; Yamamoto, H., *J. Am. Chem. Soc.* **2004**, *126*, 4128.
85. Iradier, F.; Gomez-Arrayas, R.; Carretero, J. C., *Org. Lett.* **2001**, *3*, 2957.
86. Armstrong, A.; Draffan, A. G., *Synlett* **1998**, 646.
87. Gustafsson, M.; Olsson, R.; Andersson, C.-M., *Tetrahedron Lett.* **2002**, *42*, 133.
88. Blaney, P.; Grigg, R ; Rankovic, Z.; Thoroughgood, M., *Tetrahedron Lett.* **2000**, *41*, 6639.
89. Degnan, A. P.; Meyers, A. I., *J. Org. Chem.* **2000**, *65*, 3503.
90. Florio, S.; Capriati, V.; Di Martino, S.; Abbotto, A., *Eur. J. Org. Chem.* **1999**, 409.
91. Wenkert, D.; Chen, T.-F.; Ramachandran, K.; Valasinas, L.; Weng, L. I.; McPhail, A. T., *Org. Lett.* **2001**, *3*, 2301.
92. Miyamoto, M.; Aoi, K.; Saegusa, T., *J. Polym. Sci. Part A: Polym. Chem.* **1997**, *35*, 933.
93. Någren, K.; Sutinen, E.; Jyrkkiö, S., *J. Label. Compd. Radiopharm.* **2000**, *43*, 1013.
94. Kozmin, S. A., *Org. Lett.* **2001**, *3*, 755.
95. Christoph, G.; Hoppe, D., *Org. Lett.* **2002**, *4*, 2189.
96. Smith, A. B., III; Minbiole, K. P.; Verhoest, P. R.; Schelhaas, M., *J. Am. Chem. Soc.* **2001**, *123*, 10942.
97. Gotoh, Y.; Saitoh, H.; Miyake, T., *Carbohydr. Res.* **1998**, *309*, 45.
98. Sparks, S. M.; Shea. K. J., *Org. Lett.* **2001**, *3*, 2265.
99. Baxter, E. W.; Labaree, D.; Ammon, H. L.; Mariano, P. S., *J. Am. Chem. Soc.* **1990**, *112*, 7682.
100. Kolomeitsev, A. A.; Kadyrov, A. A.; Szczepkowska-Sztolcman, J.; Milewska, M.; Koroniak, H.; Bissky, G.; Barten, J. A.; Röschenthaler, G.-V., *Tetrahedron Lett.* **2003**, *44*, 8273.
101. Barluenga, J.; Dieguez, A.; Rodriguez, F.; Florez, J.; Fananas, F. J., *J. Am. Chem. Soc.* **2002**, *124*, 9056.
102. Kanoh, S.; Nishimura, T.; Kita, Y.; Ogawa, H.; Motoi, M.; Takani, M.; Tanaka, T., *J. Org. Chem.* **2000**, *65*, 2253.
103. Brenneman, J. B.; Martin, S. F., *Org. Lett.* **2004**, *6*, 1329.
104. Bates, R. B.; Taylor, S. R., *J. Org. Chem.* **1993**, *58*, 4469.
105. Evans, D. A.; Cee, V. J.; Smith, T. E.; Santiago, K. J., *Org. Lett.* **1999**, *1*, 87.
106. Smith, T. E.; Mourad, M. S.; Velander, A. J., *Heterocycles* **2002**, *57*, 1211.
107. Nakamura, S.; Furutani, A.; Toru, T., *Eur. J. Org. Chem.* **2002**: 1690.
108. Johnson, T. A.; Jang, D. O.; Slafer, B. W.; Curtis, M. D.; Beak, P., *J. Am. Chem. Soc.* **2001**, *124*, 11689.
109. Larsen, S. D.; DiPaolo, B. A., *Org. Lett.* **2001**, *3*, 3341.
110. Murai, T.; Mutoh, Y.; Ohta, Y.; Murakami, M., *J. Am. Chem. Soc.* **2004**, *126*, 5968.
111. Murai, T.; Mutoh, Y.; Kato, S., *Org. Lett.* **2001**, *3*, 1993.
112. Schiavi, B. M.; Richard, D. J.; Joullie, M. M., *J. Org. Chem.* **2002**, *67*, 620.
113. Morita, M.; Inoue, K.; Ogoshi, S.; Kurosawa, H., *Organometallics* **2003**, *22*, 5468.
114. Djukic, J.- P.; Michon, C.; Maisse-François, A.; Allagapen, R.; Pfeffer, M.; Dötz K. H.; De Cian, A.; Fischer, J., *Chem. Eur. J.* **2000**, *6*, 1064.
115. Djukic, J.-P.; Maisse-Francois, A.; Pfeffer, M.; Dötz K. H.; De Cian, A.; Fischer, J., *Organometallics* **2000**, *19*, 5484.
116. Hevia, E.; Perez, J.; Riera, V.; Miguel, D.; Campomanes, P.; Menendez, M. I.; Sordo, T. L.; Garcia-Granda, S., *J. Am. Chem. Soc.* **2003**, *125*, 3706.
117. Nückel, S.; Burger, P., *Organometallics* **2001**, *20*, 4345.

118. Dieckbreder, U.; Röschenthaler, G.-V.; Kolomeitsev, A. A., *Heteroatom Chem.* **2002**, *13*, 650.
119. Merceron-Saffon, N.; Baceiredo, A.; Gornitzka, H.; Bertrand, G., *Science* **2003**, *301*, 1223.
120. Imamoto, T.; Kikuchi, S.-i.; Miura, T.; Wada, Y., *Org. Lett.* **2001**, *3*, 87.
121. Lönn, H., *Carbohydr. Res.* **1987**, *139*, 105.
122. Hansson, J.; Garegg, P. J.; Oscarson, S., *J. Org. Chem.* **2001**, *66*, 6234.
123. Takatani, M.; Nakano, J.; Arai, M. A.; Ishiwata, A.; Ohta, H.; Ito, Y., *Tetrahedron Lett.* **2004**, *45*, 3929.
124. Nilsson, S.; Lönn, H.; Norberg, T., *Glycocojugate J.* **1991**, *8*, 9.
125. Matsuo, I.; Wada, M.; Manabe, S.; Yamaguchi, Y.; Otake, K.; Kato, K.; Ito, Y., *J. Am. Chem. Soc.* **2003**, *125*, 3402.
126. Yang, Z.; Cao, H.; Hu, J.; Shan, R.; Yu, B., *Tetrahedron* **2003**, *59*, 249.
127. Ziegler, T.; Lemanski, G., *Angew. Chem. Int. Ed.* **1998**, *37*, 3129.
128. Demchenko, A. V.; De Meo, C., *Tetrahedron Lett.* **2002**, *43*, 8819.
129. Hanessian, S.; Lou, B., *Chem. Rev.* **2000**, *100*, 4443.
130. Demchenko, A. V.; Malysheva, N. N.; De Meo, C., *Org. Lett.* **2003**, *5*, 455.

2-Methyl-1-(1-methylethyl)propyl Chloroformate(Doc-Cl)

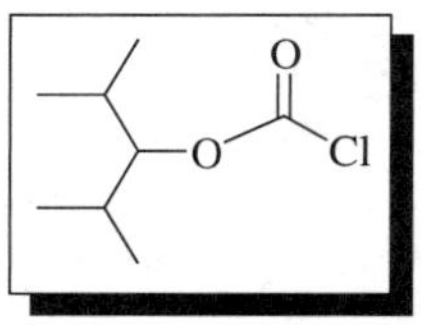

[30250-57-8] $C_8H_{15}ClO_2$ (MW 178.5)

(reagent for the introduction of the Doc protecting group for phenols and amines in peptide synthesis)

Alternate Name: 2,4-dimethyl-3-pentyloxycarbonyl chloride (DocCl).
Physical Data: bp 90–100 °C,[1] 33–35 °C / 6 cm Hg,[2] n_D^{20} 1.4225.[1]
Solubility: soluble in chloroform, dichloromethane, ether, THF, and toluene.
Analysis of Reagent Purity: IR,[1] elemental analysis.[1]
Preparative Methods: prepared by reaction of 2,4-dimethyl-3-pentanol with phosgene at −78 °C,[1] rt,[2] or 0 °C.[3]
Purification: can be purified by distillation.[1,2]
Handling, Storage, and Precautions: gradually decomposed by water and alcohol. Stable on storage under argon atmosphere, used in a fumehood.

Anhydride Formation. 2-Methyl-1-(1-methylethyl)propyl chloroformate (DocCl) reacts with benzoic acid in the presence of a base to form a stable mixed anhydride (eq 1), which may be decomposed thermally with extrusion of carbon dioxide at 170 °C to afford 2,4-dimethylpentyl benzoate (eq 2).[1,4]

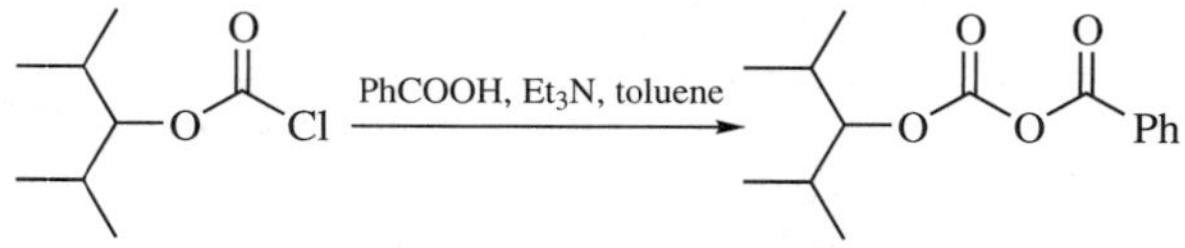 (1)

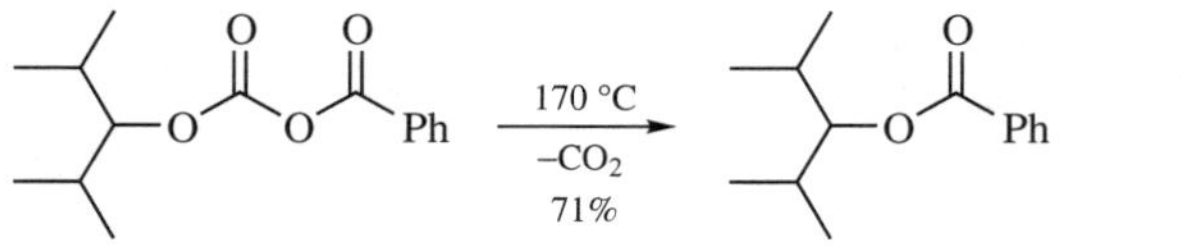 (2)

Acylations. The mixed anhydride derived from Doc chloride with benzoic acid has been employed for the acylation of alcohols and carboxylic acids. The yields, as in the case of the esterification of ethanol, however, are low; and the major product is 2,4-dimethyl-3-pentyl benzoate (57%) (eq 3).[1,4]

When DMAP is employed as catalyst, extrusion of carbon dioxide from the mixed anhydride becomes the predominant pathway, which therefore constitutes a preparative method for the esterification of carboxylic acids (eq 4).[5]

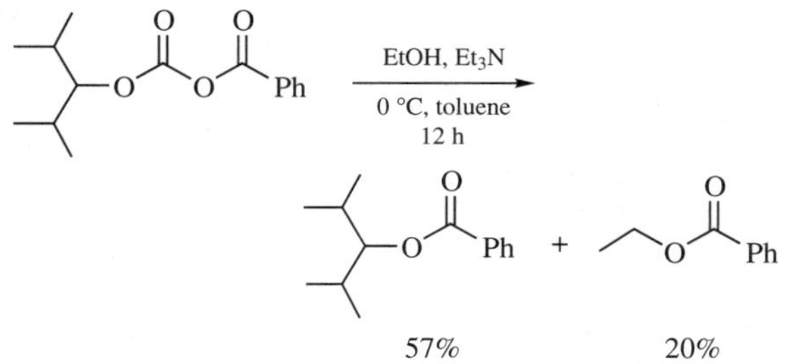 (3)

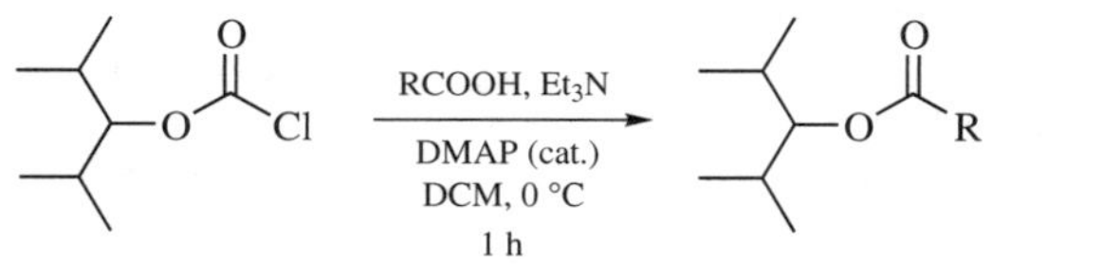 (4)

R	Yield (%)
$H_3C(CH_2)_6$-	96
$(H_3C)_2CH$-	93
$(H_5C_6)_2CH$-	98

Protection of Phenols. The 2-methyl-1-(1-methylethyl) propylcarbonyl moiety is a convenient protecting group for phenols in the synthesis of peptides (eq 5).[6]

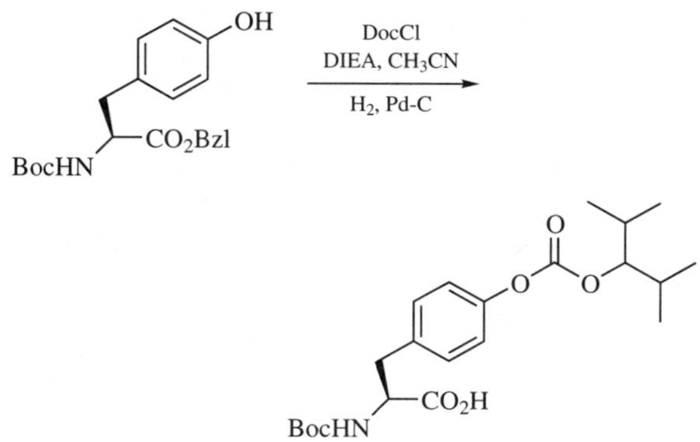 (5)

The *O*-Doc tyrosine protecting group is very resistant to nucleophilic attack with a half-life of 8 h in 20% piperidine–DMF, which compares very favorably to that of the *O*-2-BrZ group (30 s) under the same conditions.[6] Studies on the stability of the Doc-protected phenolic group of tyrosine in 50% TFA–DCM revealed the *O*-Doc group to be somewhat less stable than the *O*-2-BrZ group, showing 1.2% cleavage after 0.33 days, while the latter remained unchanged (eq 6).[6] However, the acidolytic cleavage of the *O*-Doc group suffers neither from peptide alkylation nor from protecting group migration, both of which are problematic with the *O*-2-Br-Z system. The slight increase in lability is easily offset by these desirable characteristics.[6]

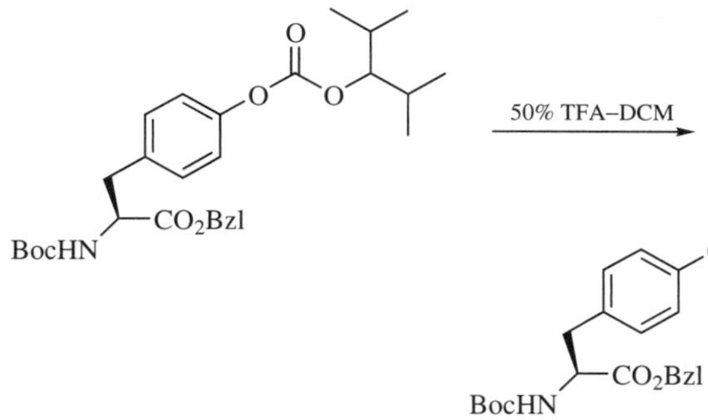

(6)

Protection of Amines. The imidazole nitrogen of histidine could be selectively protected for peptide synthesis using Doc chloride in the presence of a base (eq 7).[3,7]

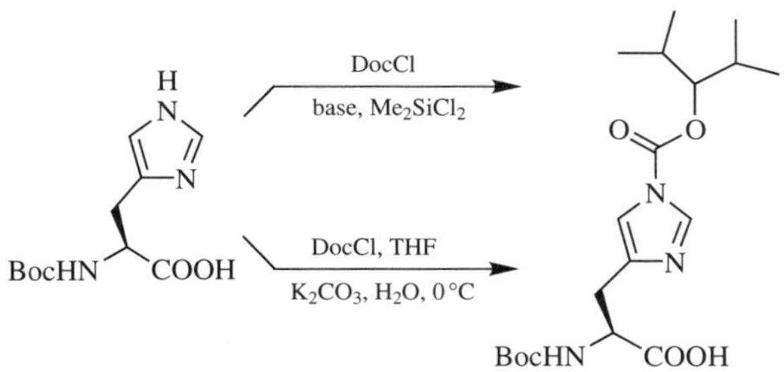

(7)

The indole nitrogen of tryptophan[8] was efficiently protected with the 2,4-dimethyl-3-pentyloxycarbonyl group in presence of DMAP or ^{t}BuOK. This protection was slow with DMAP as base and catalyst but no racemization was observed, while derivatization with ^{t}BuOK was very rapid with <2% racemization (eq 8).[8]

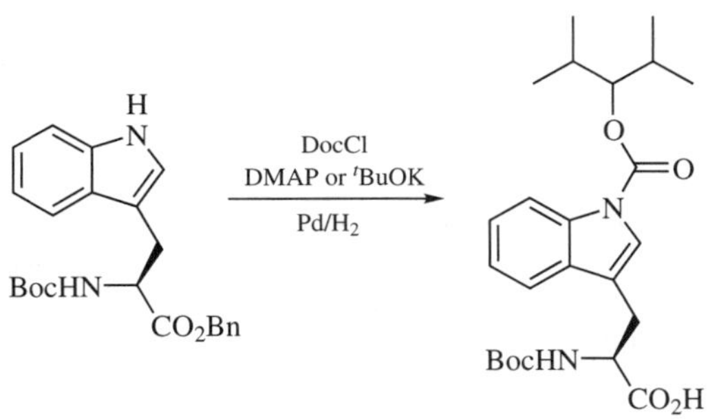

(8)

The *N*-Doc histidine group is very stable to nucleophiles compared to the related protecting groups. This is readily seen from a comparison of the cleavage of *N*-Doc, *N*-Dnp, and *N*-2-Adoc groups of histidine in 20% piperidine–DMF when the respective half-lives were 84 h, 7.3 min, and 12 min. Despite the stability of the *N*-Doc group to nucleophiles, it can be cleaved with a $t_{1/2}$ of 5 min in 5% hydrazine–DMF; conditions to which the *N*-Dnp and *N*-2-Adoc groups were resistant (eq 9).[7]

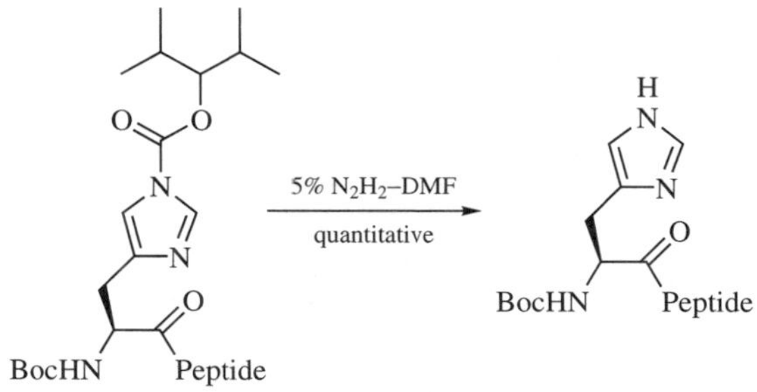

(9)

The *N*-Doc tryptophan group is remarkably stable towards 20% and 5% hydrazine. However, the Doc group was completely removed with 10% *p*-cresol–HF at 0 °C in 1 h and also with TFMSA–thioanisole–EDT–TFA (2:2:1:20) in 10 min at rt (eq 10). On the other hand, it was calculated that only 5% of the Doc group was cleaved in 50% TFA–DCM at rt in 7 days.[8]

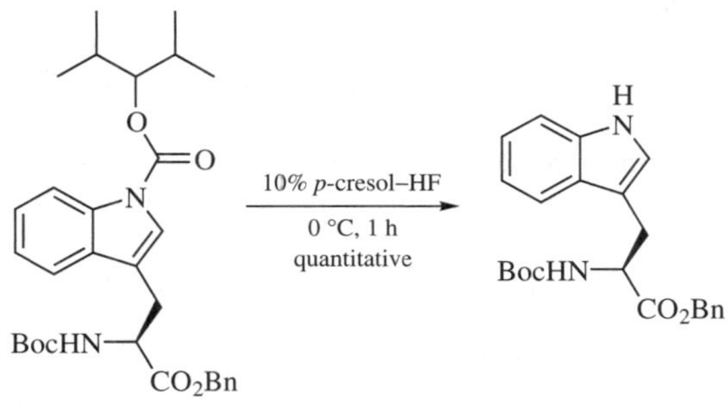

(10)

Related Reagents. 2-Adamantyl chloroformate (2-Adoc); 2-bromobenzyl chloroformate (2-BrZ); dinitrophenyl chloroformate (Dnp); *N-tert*-butyl chloroformate; isobutyl chloroformate; 9-fluorenylmethyl chloroformate (9-Fmoc).

1. Tarbell, D. S.; Price, J. A., *J. Org. Chem.* **1957**, *22*, 245–250.
2. W. O. patent 9526975 A1. (to Unden, A.) (October 1995).
3. W. O. Patent 2001025257 A2. (to Deslongchamps, P.; Dory, Y.; Berthiaume, G.; Quellet, L.; Lan, R.) (April 2001).
4. Tarbell, D. S.; Longosz, E. J., *J. Org. Chem.* **1959**, *24*, 774–778.
5. (a) Kim, S.; Kim, Y. C.; Lee, J. I., *Tetrahedron Lett.* **1983**, *24*, 3365–3368. (b) Kim, S.; Lee, J. I.; Kim, Y. C., *J. Org. Chem.* **1985**, *50*, 560–565.
6. Rosenthal, K.; Karlstrom, A.; Unden, A., *Tetrahedron Lett.* **1997**, *38*, 1075–1078.
7. Karlstrom, A.; Unden, A., *Chem. Commun.* **1996**, 959–960.
8. Karlstrom, A.; Unden, A., *Chem. Commun.* **1996**, 1471–1472.

Sochanchingwung Rumthao
University of Illinois at Chicago, Chicago, IL, USA

N-Methylmorpholine

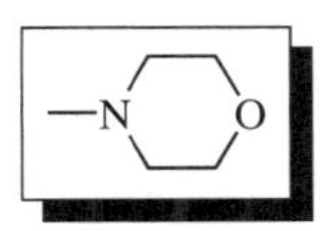

[109-02-4] $C_5H_{11}NO$ (MW 101.15)

(tertiary amine base used in dihydroxylation, dehydrohalogenation, and substitution reactions)

Alternate Name: NMM.
Physical Data: mp −66 °C; bp 116–117 °C; d = 0.920 g cm^{-3}.
Solubility: soluble in most organic solvents.
Form Supplied in: liquid.
Purification: dried by refluxing with BaO or sodium, then fractionally distilled through a helices-packed column.[1]
Handling, Storage, and Precautions: corrosive and flammable liquid. Should be stored under an inert gas blanket. Can be absorbed through skin and cause moderate to severe irritation of the eyes, skin, and mucous membranes. Use in a fume hood.

Introduction. *N*-Methylmorpholine is a widely used amine in synthetic organic chemistry. Due to its lower p*K*a (7.61) it is a frequent replacement for triethylamine (p*K*a 10.65) and diisopropylethylamine (p*K*a 10.61) in base sensitive reactions such as peptide couplings. *N*-Methylmorpholine also finds wide use in eliminations as well as in substitution and addition reactions.

Use as a Base. NMM is very often the base of choice in peptide coupling reactions given its inherently low basicity and high steric encumbrance, which lead to lower degrees of racemization during activation of the acid component.[2] Studies comparing tertiary amine bases used in the presence of *N,N′*-dicyclohexylcarbodiimide uronium, or phosphonium based coupling reagents have shown NMM to be superior to diisopropylethylamine.[3] In other instances, carboxylic acids activated as anhydrides or alkyloxycarbonic anhydrides by reaction with the corresponding acyl chloride or chloroformate in the presence of NMM, can undergo a variety of reactions mostly involving nucleophilic substitution by amines. Other nucleophiles like lithiated carbamates,[4] organolithiums,[5] diazomethane,[6] hydroxylamines,[7] and sulfides[8] also participate in similar reactions.

In a solid phase synthesis developed to generate a variety of tertiary amides, the by-product arising from deprotonation of the acidic proton α to the nitrogen could be effectively suppressed by switching bases from Et_3N to NMM[9] during the formation of the acyl ammonium intermediate (eq 1).

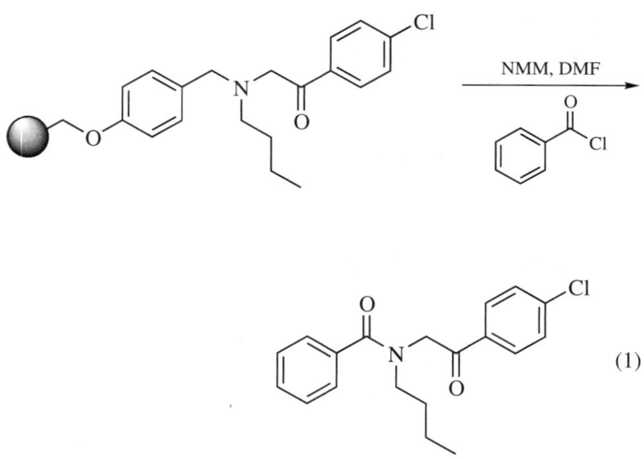

(1)

NMM acts as an effective base in dehydrogenation and dehydrohalogenation reactions. For example, in the presence of $Pd(OTf)_2$, α,β unsaturated aldehydes[10] are generated by dehydrogenation through an oxo-π-allylpalladium complex (eq 2).

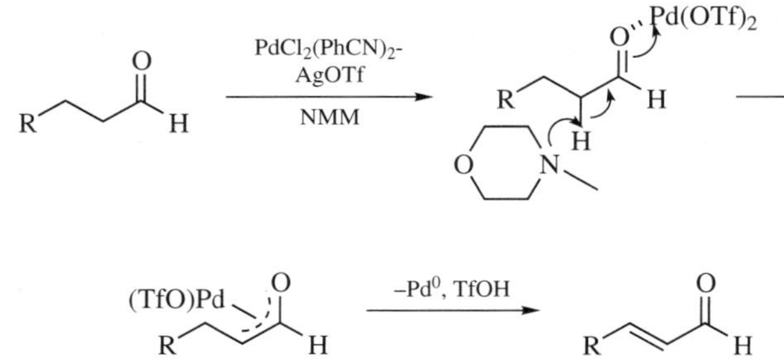

(2)

R = alkyl, benzyl

Dehydrohalogenation of an acid chloride with NMM proceeds via the acylammonium intermediate and generates a ketene. The resulting highly reactive ketene can undergo [2 + 2] cycloadditions[11] with functionalized imines to form β-lactams in good yield and high selectivity using this process (eq 3).

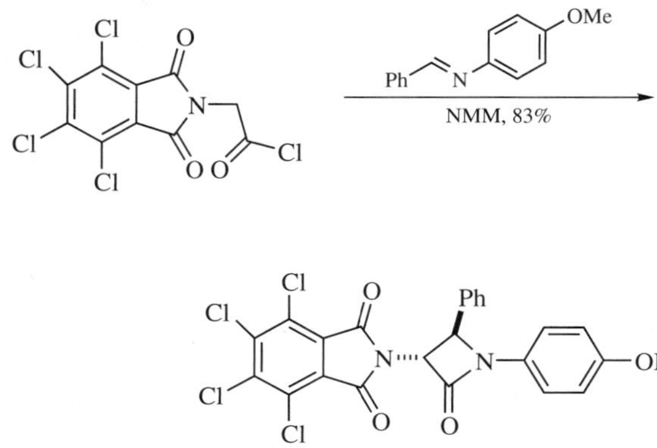

(3)

67%, along with 16% of *cis*-product

Subjecting alkyl or aryl carboxamides to $TiCl_4$ and NMM effects dehydration to form nitriles[12] in good yields (65–100%), with NMM acting as a proton scavenger (eq 4).

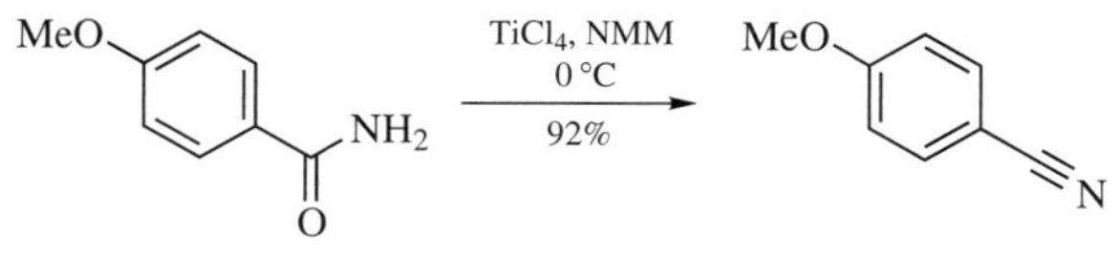

(4)

In a methodology used to prepare a diverse array of α-amino acid derivatives, NMM was used effectively in the Knovenagel condensation, in the presence of $TiCl_4$, to give the corresponding α,β-unsaturated nitroacetates in excellent yields (eq 5).[13]

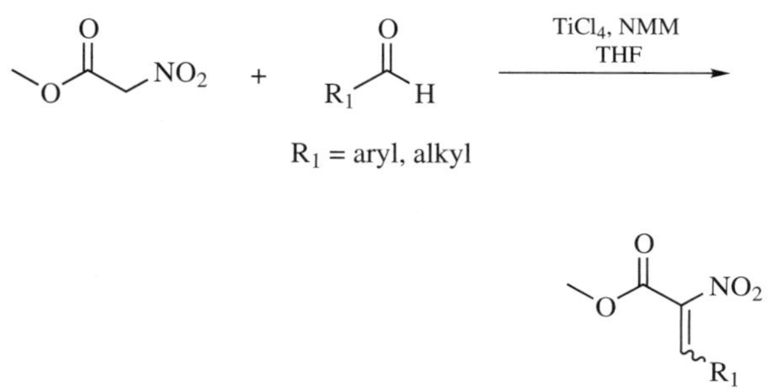

R_1 = aryl, alkyl

(5)

In a synthesis of ketomethylene isosteres, the reduced basicity of NMM enabled suppression of racemization that took place when Et_3N was employed as a base in a Horner-Emmons reaction with an amino acid derived phosphonate (eq 6).[14]

NMM is a proton scavenger for generation of the intermediate hydroxamate in the Barton *O*-acyl 2-thiopyridylhydroxamate fragmentation reaction.[15–17] The radical subsequently generated on photolytic fragmentation of the hydroxamate can be stereoselectively trapped by a radical acceptor[15] (eq 7) or quenched with a hydrogen radical to give the decarboxylated product.[16,17]

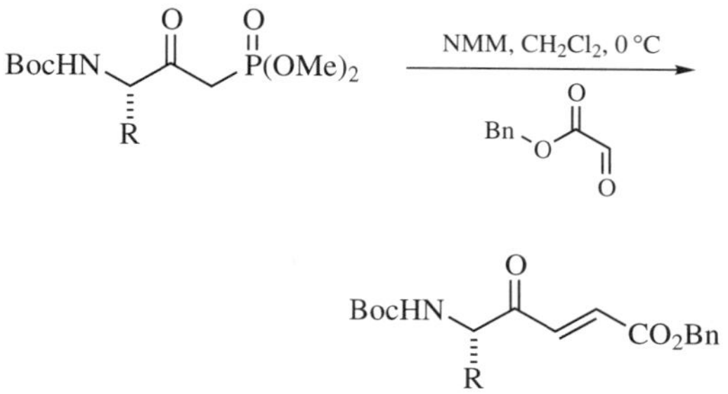

(6)

R = *t*-Bu, *i*-Pr, *i*-Bu
Bn, Me

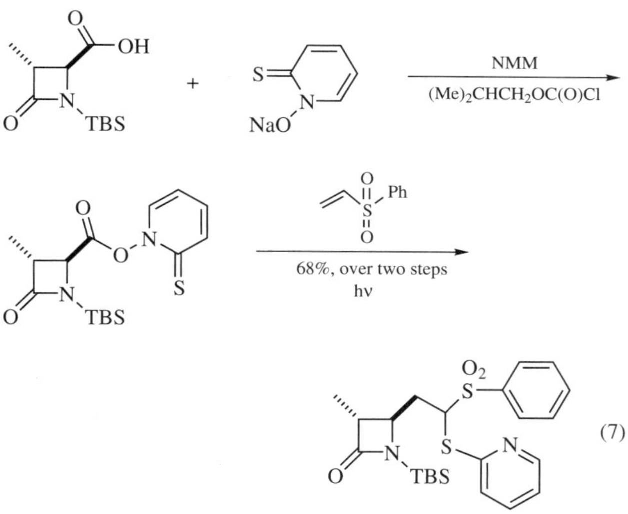

(7)

During the conversion of acyloins to ketones using a TBAF-thiol system,[18] NMM is crucial in capturing hydrogen fluoride formed in situ, thereby suppressing formation of unwanted by-products (eq 8).

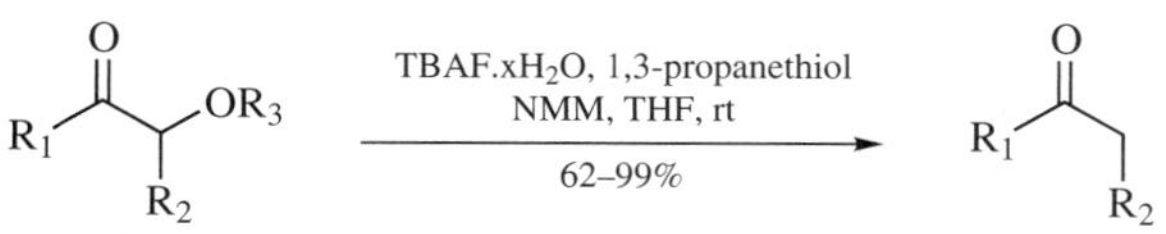

(8)

R_1 = alkyl, aryl
R_2 = alkyl, aryl, H
R_3 = alkyl, aryl

N-Methylmorpholine also acts as a base in Pd(0) catalyzed phosphinylation reactions[19–21] such as used for the preparation of phosphinylmethylphosphonates (eq 9).[20]

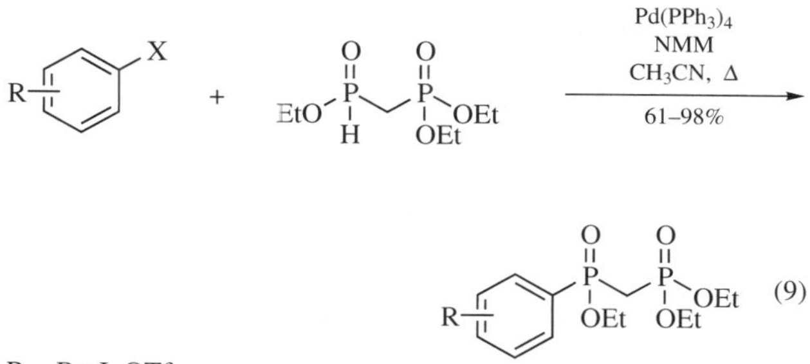

(9)

R = Br, I, OTf
X = H, *p*-CO_2Me, *p*-NO_2, Br

Conjugate additions of β-dicarbonyls, a widely used technique to generate carbon-carbon bonds, both chirally[22] and achirally,[23] have often benefited from the use of NMM as cocatalyst when other more basic tertiary amines[22] proved to be detrimental to the reaction (eq 10).

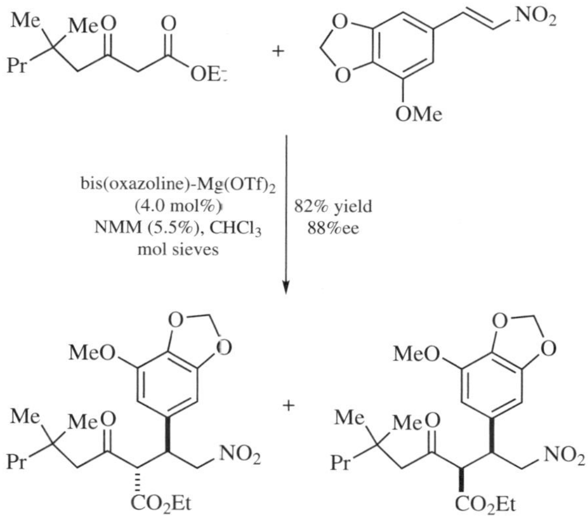

(10)

major product

[3 + 2] Cycloadditions of pyrazolidinone-based dipoles with thioxoacetates, generated in situ by NMM-catalyzed cleavage of the *N*-(alkoxycarbonylmethylthio) phthalimides, yielded the pyrazolidinone core common to various antibacterial agents (eq 11).[24]

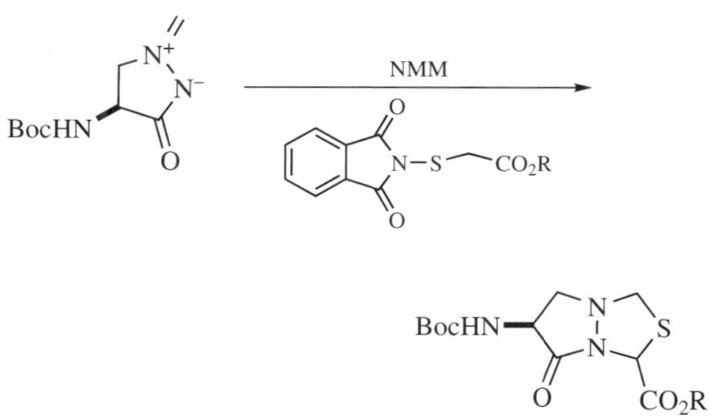

(11)

cis:*trans* = 3:1

R = Me, *t*-Bu, HC(Ph)$_2$

Amines can be protected as carbamates by reaction with chloroformates using NMM as a base.[25] Similarly, NMM has also been used to trap the hydrogen halide formed in situ during the formation of phosphonamides,[26] sulfonamides,[27] and sulfenamides[28] from amines, sulfonates from alcohols[29] and acids,[30] esters[31] and carbonates[32] from alcohols, and thiations,[33] formation of cyclic ureas with trichloromethyl chloroformate,[34] formation of amides,[35] substituted amines,[36] and selenoesters.[37]

Use as Nucleophile. When substituted triazine-based coupling reagents are used,[38] NMM serves both to activate the reagent and liberate the nucleophilic amine from its salt (eq 12).

In the field of protecting group chemistry, NMM has a variety of uses. For example, in the presence of dithiothreitol, NMM rapidly removes the *S*-[*N'*-methyl-*N'*-phenylcarbamoyl)sulfenyl (Snm) protecting group of cysteine.[39] Also, using NMM under high pressure, a variety of esters can be hydrolyzed cleanly to the corresponding carboxylic acids[40] with complete selectivity over carbamates (eq 13). In the absence of NMM, higher temperatures are required for this reaction resulting in lower selectivity.

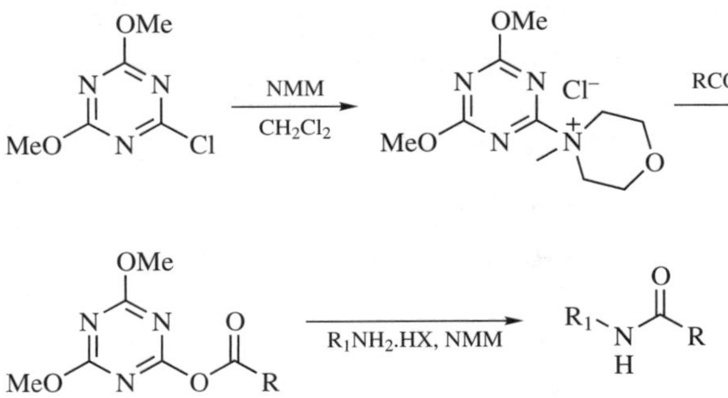

(12)

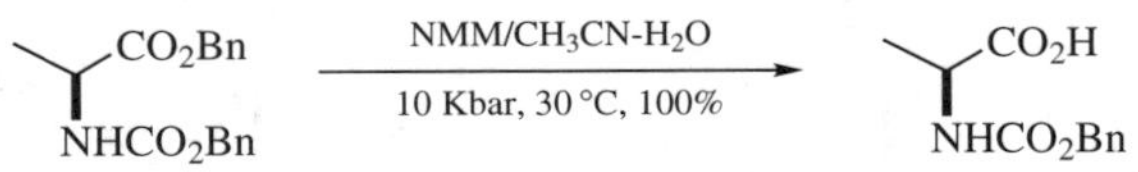

(13)

Vinyl ethers were prepared utilizing the propensity of 1,3-dicarbonyl compounds to undergo *O*-conjugate addition to reactive acetylenic esters[23] in the presence of NMM (eq 14). Alcohols also undergo similar reactions in excellent yields.[23b,c]

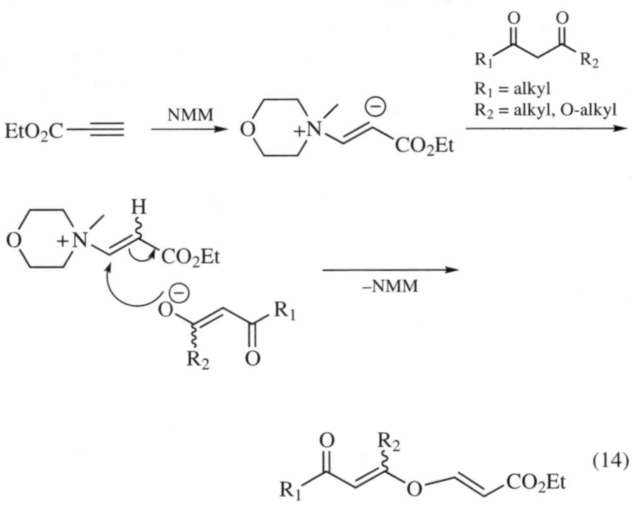

(14)

Osmium catalyzed dihydroxylation of olefins is one of the most reliable methods to produce 1,2 diols. Although *N*-methylmorpholine *N*-oxide is often used in this reaction, there are cases where NMM in the presence of a co-oxidant, have led to better regioselectivity (eq 15).[41]

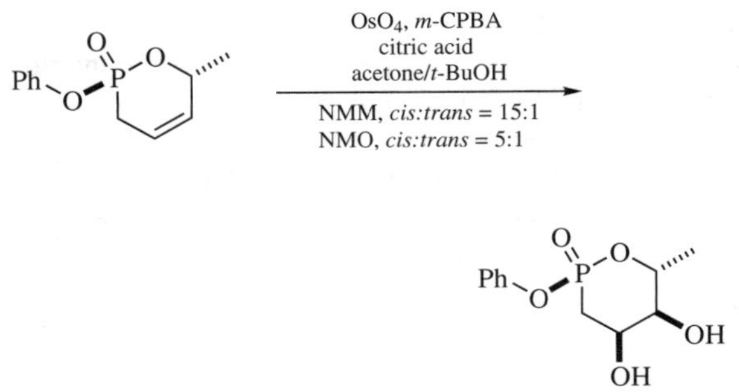

(15)

Aldehydes have been degraded to next lower carbon homolog, via the formation of the corresponding enol, when hydroxylated in the presence of NMM.[42] The mild reaction conditions tolerate different functional groups and do not epimerize α-substituted aldehydes (eq 16).

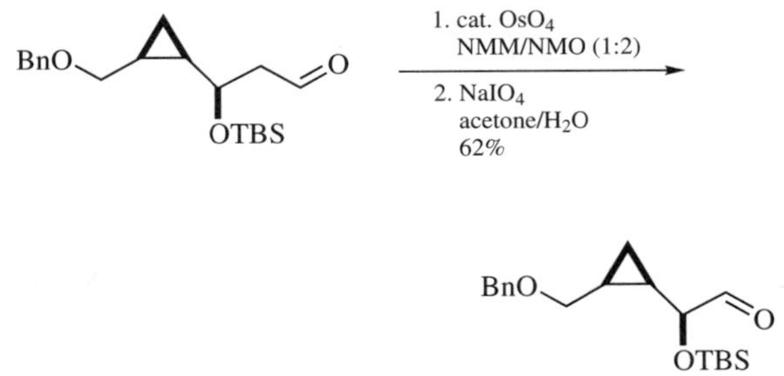

(16)

In a recent development,[43] replacement of NMO with NMM in the presence of H_2O_2 using a mixed heterogenous catalyst (Pd/Os/W) system gave chiral diols in higher yields and ee's than the pre-existing Kobayashi system (eq 17).[43] Other teriary amines (Et_3N and Me_3N) led to poorer yields and ee's.

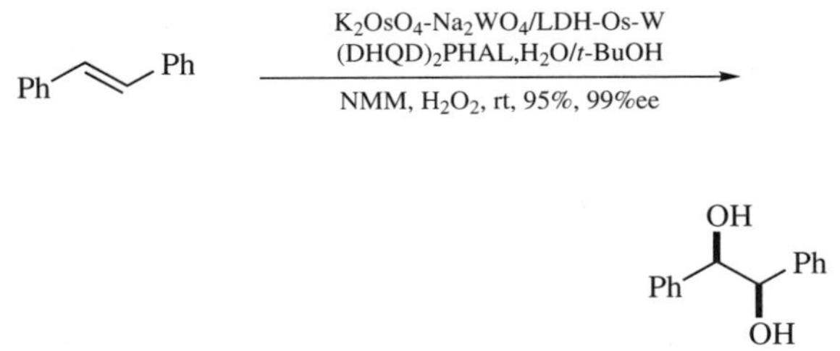

(17)

In another example, involving the synthesis of asymmetrical trisulfanes, NMM was the catalyst of choice where Et_3N led to disproportionation reactions yielding both symmetrical and asymmetrical polysulfanes. However, in some cases involving highly reactive systems, aniline gave superior results (eq 18).[44]

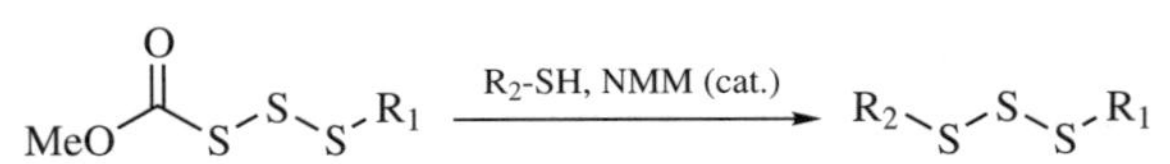

(18)

There are several instances in which NMM can form quaternary salts by displacement of a suitable leaving group.[45,46] Demethylation of the resulting quaternary salt by a halide ion to give *N*-substituted morpholine has also been observed (eq 19).[46]

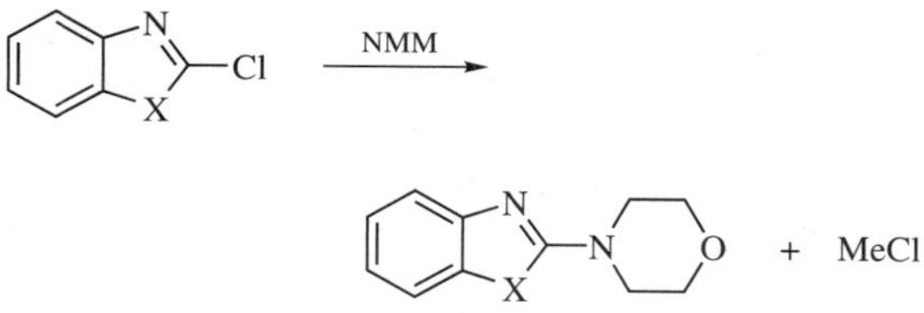

(19)

X = O,S

1. Perrin, D. D.; Armarego, W. L. F., *Purification of Laboratory Chemicals*, 4th Ed., Butterworth-Heinemann, 1996; pp 270.
2. (a) Bodanszky, M.; Bodanszky, A., *Chem. Commun.* **1967**, 591. (b) Anderson, G. W.; Zimmerman, J. E.; Callahan, F. M., *J. Am. Chem. Soc.* **1967**, *89*, 5012. (c) Slebioda, M.; St-Amand, M. A.; Chen, F. M. F.; Benoiton, N. L., *Can. J. Chem.* **1988**, *66*, 2540.
3. Carpino, L. A.; El-Faham, A., *J. Org. Chem.* **1994**, *59*, 695.
4. Liu, W.-Q.; Olszowy, C.; Bischoff, L.; Garbay, C., *Tetrahedron Lett.* **2002**, *43*, 1417.
5. Cupps, T. L.; Boutin, R. H.; Rapoport, H., *J. Org. Chem.* **1985**, *50*, 3972.
6. Walker, B.; Brown, M. F.; Lynas, J. F.; Martin, S. L.; McDowell, A.; Badet, B.; Hill, A. J., *Biorg. Med. Chem. Lett.* **2000**, *10*, 2795.
7. Nakonieczna, L.; Milewska, M.; Kolasa, T.; Chimiak, A., *Synthesis* **1985**, 929.
8. Kates, M. J.; Schauble, J. H., *J. Heterocycl. Chem.* **1995**, *32*, 971.
9. Lee, H. B.; Balasubramanian, S., *Org. Lett.* **2000**, *2*, 323.
10. Mukaiyama, T.; Oshima, M.; Nakatsuka, T., *Chem. Lett.* **1983**, 1207.
11. (a) Bose, A. K.; Manhas, M. S.; Ghosh, M.; Shah, M.; Raju, V. S.; Bari, S. S.; Newaz, S. N.; Banik, B. K.; Chaudhary, A. G.; Barakat, K. J., *J. Org. Chem.* **1991**, *56*, 6968. (b) Bose, A. K.; Jayaraman, M.; Okawa, A.; Bari, S. S.; Robb, E. W.; Manhas, M. S., *Tetrahedron Lett.* **1996**, *37*, 6989.
12. (a) Lehnert, W., *Tetrahedron Lett.* **1971**, *12*,1501. (b) Mukaiyama, T., *Angew. Chem., In.t Ed. Engl.* **1977**, *16*, 817.
13. Fornicola, F. S.; Oblinger, E.; Montgomery, J., *J. Org. Chem.* **1998**, *63*, 3528.
14. Deziel, R.; Plante, R.; Caron, V.; Grenier, L.; Llinas-Brunet, M.; Duceppe, J.-S.; Malenfant, E.; Moss, N., *J. Org. Chem.* **1996**, *61*, 2901.
15. Sumi, K.; Di Fabio, R.; Hanessian, S., *Tetrahedron Lett.* **1992**, *33*, 749.
16. Linz, G.; Weetman, J.; Hady, A. F. A.; Helmchen, G., *Tetrahedron Lett.* **1989**, *30*, 5599.
17. Barton, D. H. R.; Herve, Y.; Potier, P.; Thierry, J., *J. Chem. Soc., Chem. Commun.* **1984**, 1298.
18. Ueki, M.; Okamura, A.; Yamaguchi, J.-i., *Tetrahedron Lett.* **1995**, *36*, 7467.
19. Lei, H.; Stoakes, M. S.; Schwabacher, A. W., *Synthesis* **1992**, 1255.
20. Luke, G. P.; Shakespeare, W. C., *Synth. Commun.* **2002**, *32*, 2951.
21. Petrakis, K. S.; Nagabhushan, T. L., *J. Am. Chem. Soc.* **1987**, *109*, 2831.
22. Barnes, D. M.; Ji, J.; Fickes, M. G.; Fitzgerald, M. A.; King, S. A.; Morton, H. E.; Plagge, F. A.; Preskill, M.; Wagaw, S. H.; Wittenberger, S. J.; Zhang, J., *J. Am. Chem. Soc.* **2002**, *124*, 13097.
23. (a) Tae, J.; Kim, K.-O., *Tetrahedron Lett.* **2003**, *44*, 2125.(b) Hori, N.; Matsukura, H.; Matsuo, G.; Nakata, T, *Tetrahedron* **2002**, *58*, 1853. (c) Berlin, S.; Ericsson, C.; Engman, L., *Org. Lett.* **2002**, *4*, 3.
24. (a) Shepherd, T. A.; Jungheim, L. N., *Tetrahedron Lett.* **1988**, *29*, 5061–5064. (b) Jungheim, L. N., *Tetrahedron Lett.* **1989**, *30*, 1889.
25. St-Denis, Y.; Augelli-Szafran, C. E.; Bachand, B.; Berryman, K. A.; DiMaio, J.; Doherty, A. M.; Edmunds, J. J.; Leblond, L.; Lévesque, S.; Narasimhan, L. S.; Penvose-Yi, J. R.; Rubin, J. R.; Tarazi, M.; Winocour, P. D.; Siddiqui, M. A., *Biorg. Med. Chem. Lett.* **1998**, *8*, 3193.
26. Sawa, M.; Kurokawa, K.; Inoue, Y.; Kondo, H.; Yoshino, K., *Biorg. Med. Chem. Lett.* **2003**, *13*, 2021.
27. Ho, J. Z.; Gibson, T. S.; Semple, J. E., *Biorg. Med. Chem. Lett.* **2002**, *12*, 743. (b) Papageorgiou, G.; Corrie, J. E. T., *Tetrahedron* **1999**, *55*, 237.
28. Decicco, C. P.; Seng, J. L.; Kennedy, K. E.; Covington, M. B.; Welch, P. K.; Arner, E. C.; Magolda, R. L.; Nelson, D. J., *Biorg. Med. Chem. Lett.* **1997**, *7*, 2331.
29. Sin, H. S.; Kwon, Y. J.; Han, H. S.; Park, S. H.; Park, M. S.; Rho, Y. S.; Um, S. J., *Bull. Korean Chem. Soc.* **2002**, *23*, 1806.
30. Watanabe, Y.; Oka, A.; Shimizu, Y.; Ozaki, S., *Tetrahedron Lett.* **1990**, *31*, 2613.
31. Obushak, N. D.; Matiichuk, V. S.; Martyak, R. L., *Chem. Heterocyc. Compd.* **2001**, *37*, 909.
32. Park, K.-H.; Cox, L. J., *Tetrahedron Lett.* **2002**, *43*, 3899.
33. Mosier, G. G. J.; Lawton, R. G., *J. Org. Chem.* **1995**, *60*, 6953.
34. Kukla, M. J.; Breslin, H. J.; Pauwels, R.; Fedde, C. L.; Miranda, M.; Scott, M. K.; Sherrill, R. G.; Raeymaekers, A.; Van Gelder, J.; Andries, K.; Janssen, M. A. C.; De Clerq, E.; Janssen, P. A. J., *J. Med. Chem.* **1991**, *34*, 746.
35. (a) Krueger, E. B.; Hopkins, T. P.; Keaney, M. T.; Walters, M. A.; Boldi, A. M., *J. Comb. Chem.* **2002**, *4*, 229. (b) Johnson, E. P.; Cantrell, W. R. J.; Jenson, T. M.; Miller, S. A.; Parker, D. J.; Reel, N. M.; Sylvester, L. G.; Szendroi, R. J.; Vargas, K. J.; Xu, J.; Carlson, J. A., *Org. Proc. Res. Dev.* **1998**, *2*, 238.
36. Delarue, S.; Girault, S ; Ali, F. D.; Maes, L.; Grellier, P.; Sergheraert, C., *Chem. Pharm. Bull.* **2001**, *49*, 933.
37. Brooks, G.; Hunt, E., *J. Chem. Soc., Perkin Trans. 1* **1983**, 2513.
38. (a) Bandgar, B. P.; Pandit, S. S., *Tetrahedron Lett.* **2003**, *44*, 2331. (b) Augé, F.; Horneback, W.; Decarme, M.; Laronze, J.-Y., *Biorg. Med. Chem. Lett.* **2003**, *13*, 1783. (c) Shönfeld, F.; Troschütz, R., *Heterocycles* **2001**, *55*,1679.
39. Schroll, A. L.; Barany, G., *J. Org. Chem.* **1989**, *54*, 244.
40. Yamamoto, Y.; Furuta, T.; Matsuo, J.; Kuruta, T., *J. Org. Chem.* **1991**, *56*, 5737.
41. Stoianova, D. S.; Hanson, P. R., *Org. Lett.* **2001**, *3*, 3285.
42. Belotti, D.; Andreatta, G.; Pradaux, F.; Bouzbouz, S.; Cossy, J., *Tetrahedron Lett.* **2003**, *44*, 3613.
43. Choudary, B. M.; Chowdari, N. S.; Madhi, S.; Kantam, M. L., *J. Org. Chem.* **2003**, *68*, 1736. (b) Kobayashi, S.; Ishida, T.; Akiyama, R., *Org. Lett.* **2001**, *3*, 2649.
44. Mott, A. W.; Barany, G., *Synthesis* **1984**, *8*, 657.
45. Maslov, M. A.; Morozova, N. G.; Serebrennikova, G. A., *Russ. Chem. Bull.* **2002**, *51*,1931.
46. Khalaf, A. I.; Alvarez, R. G.; Suckling, C. J.; Waigh, R. D., *Tetrahedron* **2000**, *56*, 8567.

Mousumi Sannigrahi & Angela D. Kerekes
Schering Plough Research Institute, Kenilworth, USA

N-Methylmorpholine Polystyrene (PS-NMM)

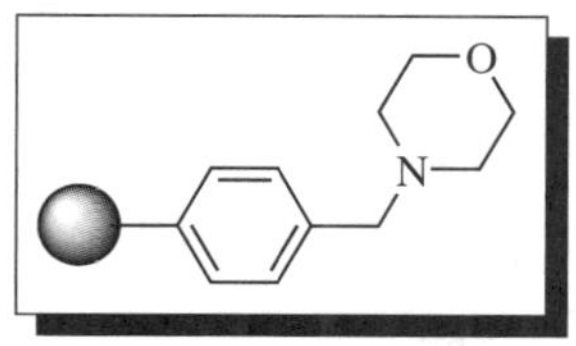

[01-64-0171]

(polymer-supported tertiary amine base used as an acid scavenger)

Alternate Name: morpholinomethyl polystyrene, PS-NMM.
Solvent Compatibility: soluble in DMF, THF, DCM, DCE, and toluene.

Form Supplied in: copoly(styrene-2% DVB), 200–400 mesh. Available from NovaBiochem, Polymer Laboratories, Rapp Polymere, Aldrich, and Matrix Technologies.
Preparative Methods: treatment of Merrifield resin with morpholine (more than 200 mol %) in DMF at 65 °C for 6 h.[1]
Loading: typical > 3.2 mmol g^{-1} as determined by elemental analysis.
Typical Reaction Conditions: 3 equivalents relative to an electrophile.

Introduction. PS-NMM, a polymer-supported equivalent of *N*-methylmorpholine (NMM), became the most popular tertiary amine base for scavenging acids as compared to the other polymer-supported trialkylamines such as piperidine (PS-piperidine) and diisopropylethyl amine (PS-DIEA).[2–5] It has been widely used to scavenge acids produced in the synthesis of many useful compounds, such as amides, sulfonamides, carbamates, ureas, thioureas, and pyrazoles.[1–3,6–12] The resin is usually employed with PS-trisamine and PS-isocyanate as scavengers for electrophiles and nucleophiles, respectively.

Synthesis of Amides, Sulfonamides, and Carbamates. The solution-phase synthesis of amides, sulfonamides, and carbamates by condensation reactions of amines typically involves very clean and efficient procedures.[2] The ease with which these compounds can be prepared, and the commercial availability of a large number of subunits used in their synthesis, makes them a logical choice for parallel solution-phase synthesis and purification.[2] In all cases, one of the reagents, such as acid chloride, sulfonyl chloride, or chloroformate, is used in excess and the unused portion remains in solution, contaminating the product. The hydrochloric acid or sulfonic acid generated in the process can be removed with PS-NMM, followed by addition of PS-trisamine and PS-isocyanate to remove the remaining excess of acid chloride, sulfonyl chloride, or chloroformate by simple filtration (eqs 1–4).[1,2,6–10] This method of purification of organic reactions is called the polymer-supported quench (PSQ) method.[1]

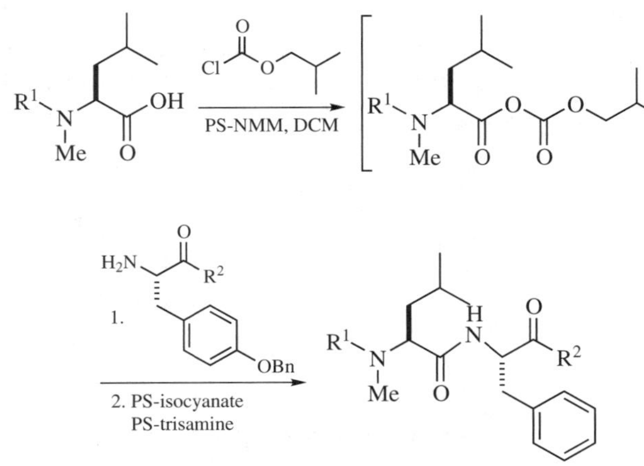

The first equation illustrating the application of PS-NMM as an acid scavenger presents the synthesis of a series of *N,N*-dialkyldipeptidylamines which are important *N*-type VSCC (Voltage Sensitive Calcium Channel) blockers (eq 1).[6] The acid was activated with a chloroformate in the presence of PS-NMM, which scavenged the HCl generated in the process. The remaining excess of reagents was removed with PS-trisamine and PS-isocyanate. Thirty examples were described in this study.

In another typical application for amide synthesis, a set of secondary amines (10 examples) were condensed with acid chlorides (80 examples) to give the corresponding amides (80 examples) in the presence of PS-NMM (eq 2).[7,8] The compounds were obtained in 39–75% yield and the purity was greater than 80% in all cases.

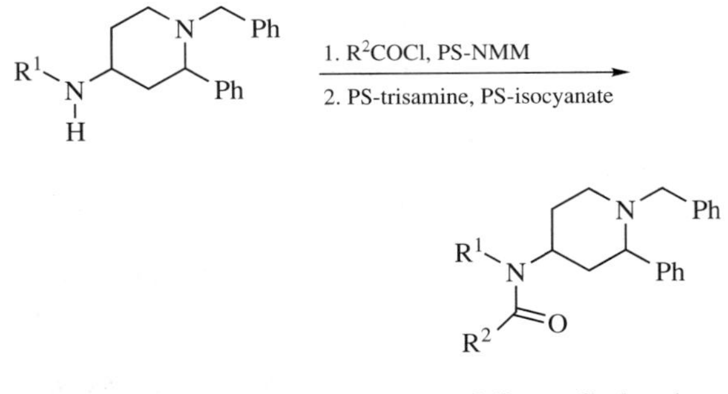

(2)

The PSQ method can also be successfully applied to the synthesis of sulfonamides (eq 3). PS-NMM is employed as a base during the reaction to scavenge the sulfonic acid by-product. It should be noted that it is not necessary to remove PS-NMM before addition of the quenching resin PS-trisamine.[1,2] The yield of the obtained sulfonamide was 65% and the purity assessed by HPLC was 99%.

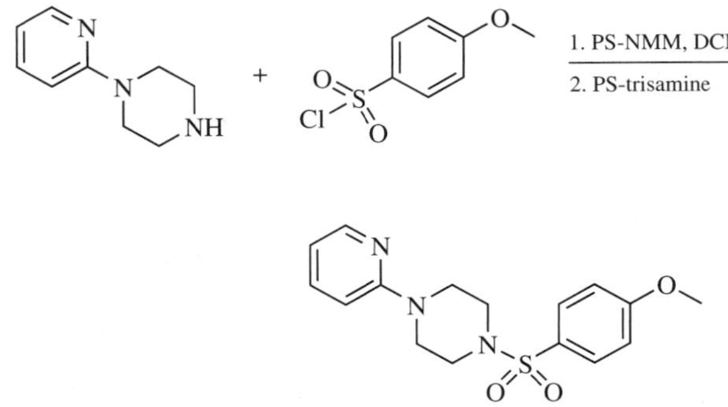

(3)

A representative example of solution-phase parallel synthesis of *N*-(pyrrolidinylmethyl)hydroxamic acids to target fucosyltransferases is presented in eq 4.[9] The reaction was conducted in the presence of acid scavenger PS-NMM, and the excess of the

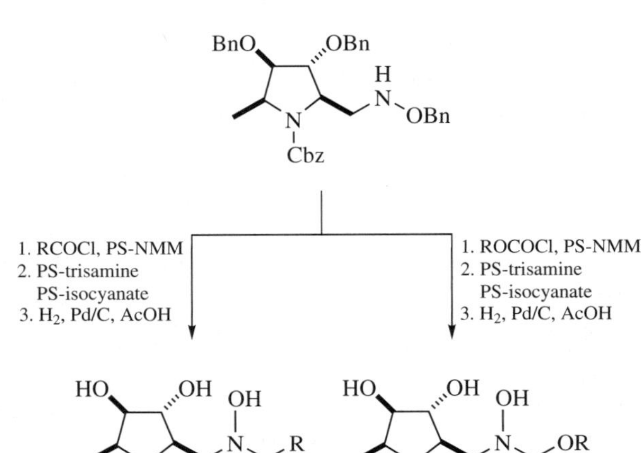

(4)

acid chloride and chloroformate were removed with PS-trisamine and PS-isocyanate to give the desired compounds. Four examples were described. The compounds were obtained in >80% yield and >80% purity.

Alkylation of Amines. PS-NMM can also be used as a scavenger of other electrophiles than protons. This example describes the use of PS-NMM as a scavenger of an alkylating agent by quaternary ammonium salt formation (eq 5).[10] The resulting compound was isolated after addition of PS-NMM and used for further modifications without purification.

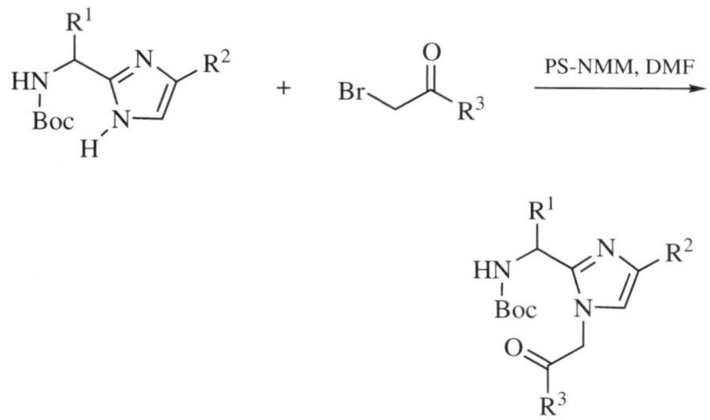

(5)

Application of Other Resin-supported Morpholine Derivatives. Morpholine can be linked to the polystyrene backbone by other means. For example, it was linked through a propylene sulfonamide moiety (Figure 1).[1,4] This resin is more stable than the typical PS-NMM derivatives in the presence of chloroformates, e.g., no cleavage of the morpholine unit from the resin was observed in the presence of methyl chloroformate (DCM, 16 h), whereas the treatment of morpholinomethyl polystyrene under similar conditions led to 90% cleavage at the benzylic anchor point. This resin can also be used in a similar fashion as the typical PS-NMM to efficiently scavenge acid by-products in the synthesis of amides, sulfonamides, and carbamates.[1,4]

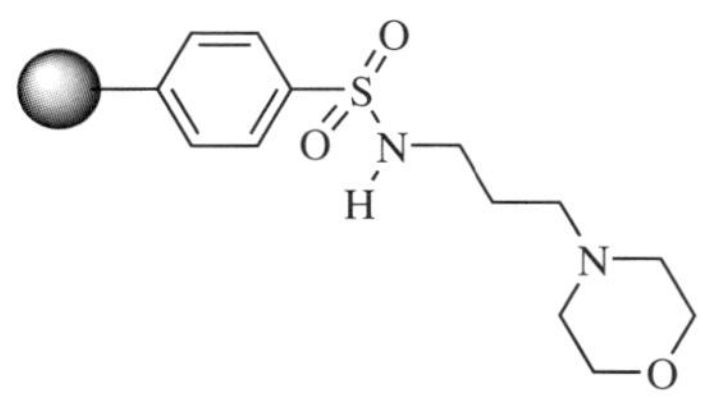

Figure 1 Structure of 3-(morpholino)propyl polystyrenesulfonamide[1,4]

A series of morpholine-containing copolymeric resins were described in a patent.[3] These basic resins (Figure 2) can be used as scavengers for acid chlorides, acid anhydrides, activated esters, alkyl chlorides, sulfonates, boronic acids, diazoalkanes, epoxides, α-haloketones, imidazoles, isocyanates, isothiocyanates, Meerwein reagents, phosphoryl chloride, silyl chloride, triflate, alcohols, primary and secondary alkyl amines, carbanions, carboxylic acids, phenols, silanols, thiols, aryl bromides, aryl iodides, aryl triflates, vinyl bromides, vinyl iodides, and vinyl triflates.[3,11]

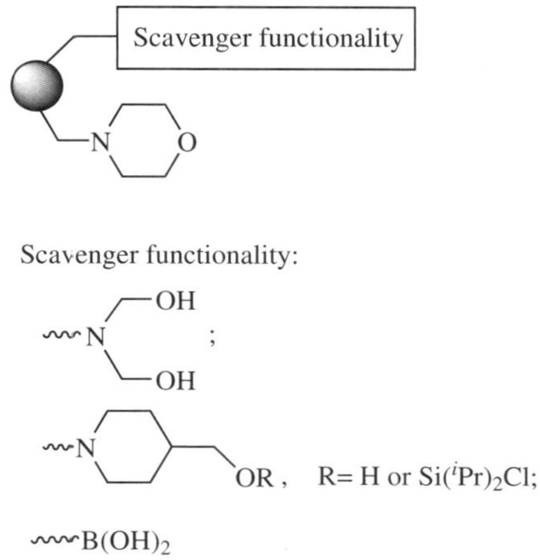

Figure 2 Structures of morpholine-containing copolymeric resins (ref 3)

Related Reagents. PS-piperidine;[2,4] PS-DIEA;[3,10] PS-dimethylamine.[8,13–18]

---

1. Booth, R. J.; Hodges, J. C., *J. Am. Chem. Soc.* **1997**, *119*, 4882.
2. Booth, R. J.; Hodges, J. C., *Acc. Chem. Res.* **1999**, *32*, 18.
3. U.S. Pat. WO97/42230 (to Bolton, G. L.; Booth, R. J.; Creswell, M. W.; Hodges, J. C.; Warmus, J. S.; Wilson, M. W.; Kennedy, R. M.) (1997).
4. Conti, P.; Demont, D.; Cals, J.; Ottenheijm, H. C. J.; Leysen, D., *Tetrahedron Lett.* **1997**, *38*, 2915.
5. Gooding, O. W.; Bansal, R. P., *Synth. Comm.* **1995**, *25*, 1155.
6. Ryder, T. R.; Hu, L.-Y.; Rafferty, M. F.; Millerman, E.; Szoke, B. G.; Tarczy-Hornoch, K., *Bioorg. Med. Chem. Lett.* **1999**, *9*, 1813.
7. Creswell, M. W.; Bolton, G. L.; Hodges, J. C.; Meppen, M., *Tetrahedron* **1998**, *54*, 3983.
8. Eames, J.; Watkinson, M., *Eur. J. Org. Chem.* **2001**, 1213.
9. Takayanagi, M.; Flessner, T.; Wong, Ch.-H., *J. Org. Chem.* **2000**, *65*, 3811.
10. Contour-Galcéra, M.-O.; Poitout, L.; Moinet, Ch.; Morgan, B.; Gordon, T.; Roubert, P.; Thurieau, Ch., *Bioorg. Med. Chem. Lett.* **2001**, *11*, 741.
11. Ley, S. V.; Baxendale, I. R.; Bream, R. N.; Jackson, P. S.; Leach, A. G.; Longbottom, D. A.; Nesi, M.; Scott, J. S.; Storer, R. I.; Taylor, S. J., *J. Chem. Soc., Perkin Trans. 1* **2000**, 3815.
12. Kaldor, S. W.; Siegel, M. G.; Fritz, J. E.; Dressman, B. A.; Hahn, P. J., *Tetrahedron Lett.* **1996**, *37*, 7193.
13. Bauer, U.; Egner, B. J.; Nilsson, I.; Berghult, M., *Tetrahedron Lett.* **2000**, *41*, 2713.
14. Flynn, D. L.; Crich, J. Z.; Devraj, R. V.; Hockerman, S. L.; Parlow, J. J.; South, M. S.; Woodward, S., *J. Am. Chem. Soc.* **1997**, *119*, 4874.
15. Kaldor, S. W.; Fritz, J. E.; Tang, J.; McKinney, E. R., *Bioorg. Med. Chem. Lett.* **1996**, *6*, 3041.
16. Kirschning, A.; Monenschein, H.; Wittenberg, R., *Angew. Chem. Int. Ed.* **2001**, *40*, 650.
17. Lee, A.; Ellman, J. A., *Org. Lett.* **2001**, *3*, 3707.
18. Shinkai, S.; Tsuji, H.; Hara, Y.; Manabe, O., *Bull. Chem. Soc. Jpn.* **1981**, *54*, 631.

Dennis G. Hall & Agnieszka Ulaczyk Lesanko
University of Alberta, Edmonton, Canada

2-Methylsulfonyl-3-phenyl-1-prop-2-enyl Chloroformate

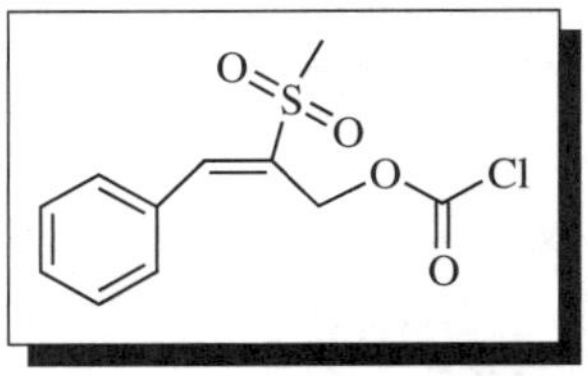

[252668-49-8] $C_{11}H_{11}ClO_4S$ (MW 274.72)

(reagent used for the introduction of a protecting group for amines)

Alternate Name: Mspoc-Cl.
Physical Data: mp 118–120 °C, colorless crystals.[1]
Solubility: soluble in CH_2Cl_2 and $CHCl_3$; insoluble in hexanes.
Purification: recrystallized from CH_2Cl_2/hexanes.
Analysis of Reagent Purity: ^{1}H NMR, IR, elemental analysis.[1]
Preparative Methods: the title reagent is prepared by the reaction of 2-(methylsulfonyl)-3-phenyl-2-propenyl alcohol and phosgene (eq 1).[1] The alcohol is obtained from the corresponding allylic bromide[2] by formate-catalyzed hydrolysis.[3] The synthesis of the bromide involves the $Cu(OAc)_2$-catalyzed addition of methanesulfonyl chloride to β-methylstyrene followed by elimination of hydrogen chloride and subsequent free radical bromination of the methyl group (eq 1).[2]

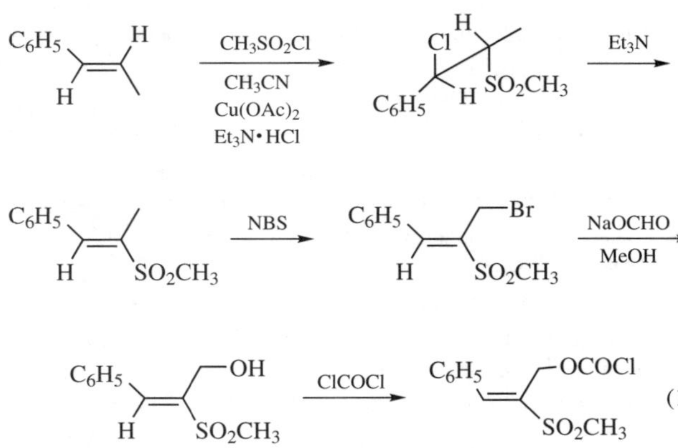

(1)

Handling, Storage, and Precautions: relatively stable reagent, but moisture sensitive and best when freshly recrystallized before use. Used in a well-ventilated fume hood.

Protecting Group for Amines. 2-Methylsulfonyl-3-phenyl-1-prop-2-enyl chloroformate (Mspoc-Cl) has been introduced as an effective amino protecting group, particularly useful for peptide synthesis. This group is more stable than the 2-*tert*-butylsulfonyl-2-propenoxy carbonyl (Bspoc) residue,[4] and avoids the problem of premature deprotection seen with Bspoc protected amines in peptide synthesis.[4] The Mspoc group can be introduced by simply treating the amine with the chloroformate in the presence of a base (eq 2).[1]

Mspoc-Cl has also been transformed into the *N*-hydroxycarbonate (Mspoc-OSu), which subsequently may be used for the protection of amino acids (eq 3).[1]

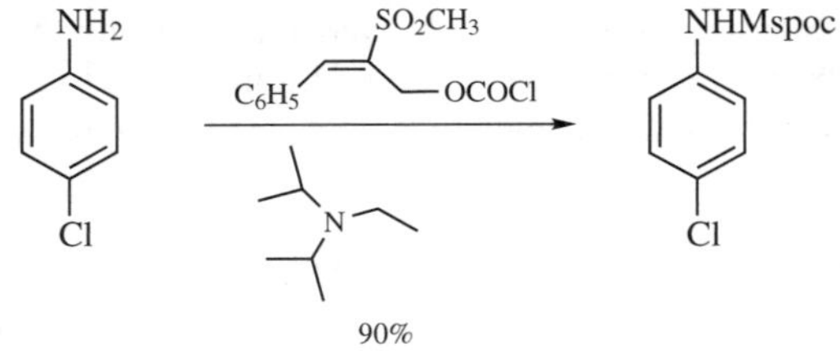

(2)

Various *N*-Mspoc protected amino acids have been synthesized using the Mspoc-OSu reagent (eq 4).[1]

The Mspoc protected amino acids can be converted to their respective acyl fluorides[5,6] by treatment with cyanuric fluoride and pyridine (eq 5). The Mspoc acyl fluorides are obtained in crystalline form,[1] in contrast to the undefined foams or amorphous products that are typically obtained with other protecting groups.

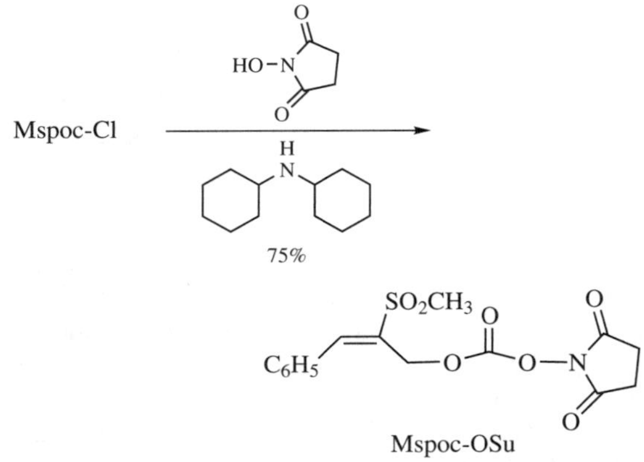

(3)

$$\text{R-NH-CH}_2\text{-CO}_2\text{H} \xrightarrow[\text{NaHCO}_3\text{, acetone}]{\text{Mspoc-OSu}} \text{R-N(Mspoc)-CH}_2\text{-CO}_2\text{H} \quad (4)$$

R = H, 84%
R = Bn, 82%

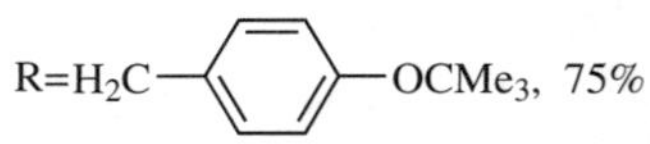

R=H_2C–C$_6H_4$–OCMe$_3$, 75%

$$\text{R-N(Mspoc)-CH}_2\text{-CO}_2\text{H} \xrightarrow[\text{pyridine}]{\text{cyanuric fluoride}} \text{R-N(Mspoc)-CH}_2\text{-COF} \quad (5)$$

R = H, 82%
R = Bn, 81%

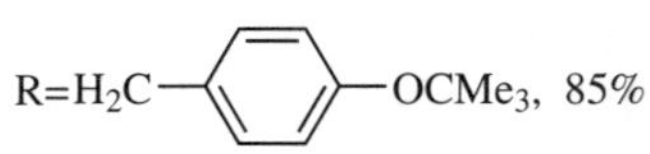

R=H_2C–C$_6H_4$–OCMe$_3$, 85%

Deprotection of the Mspoc Group. The key deprotection method for this group involves the addition of a nucleophilic reagent to the α,β-unsaturated sulfone system with the consequent ejection of the carbamate anion. The major advantages of such a process over systems for which deprotection involves a classic β-elimination process [e.g., the 2-(methylsulfonyl)ethoxycarbonyl (Msc)[7] or 9-fluorenylmethyloxycarbonyl (Fmoc)[8] systems] are

that (a) lower concentrations of piperidine or weaker bases can be used for deprotection, thus minimizing the base-catalyzed side reactions, and (b) application to the technique of rapid continuous synthesis is greatly improved. Piperidine or pyrrolidine can be used to deprotect the *N*-Mspoc amino compounds and the deprotection is found to be both more rapid and specific with the latter (eq 6).[1]

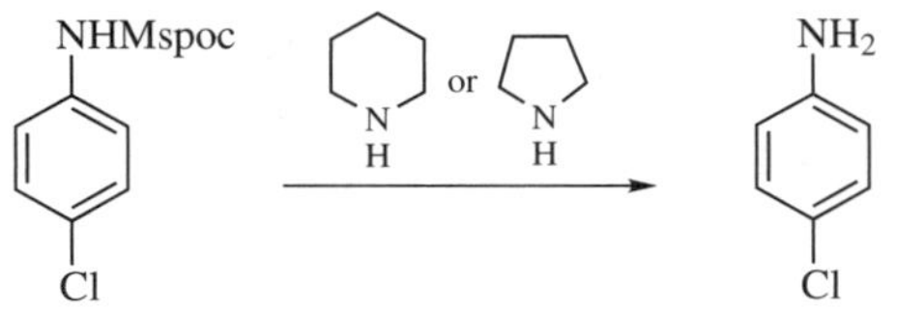

(6)

According to ^{1}H NMR analysis the initial Michael-like adduct, formed upon treatment of *N*-Mspoc amines with piperidine, exists in equilibrium with its regioisomer (eq 7).

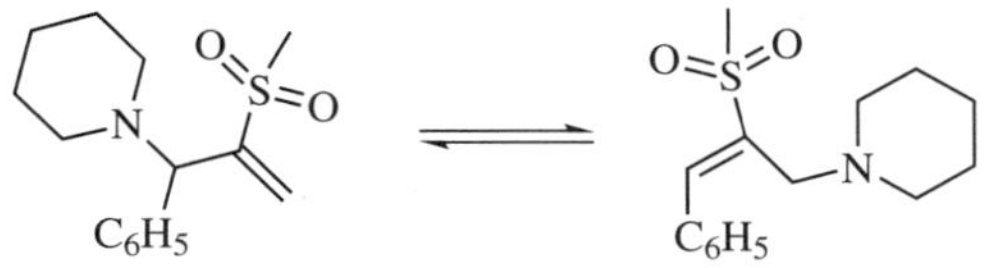

(7)

The Mspoc group is also sensitive to the base-catalyzed mercaptan deprotection. For example, 3-mercapto-1,2-propanediol rapidly deprotects the Mspoc derivative of *p*-chloroaniline in the presence of diisopropylethylamine to give an equimolar mixture of the amine and the thio adduct (eq 8).[1]

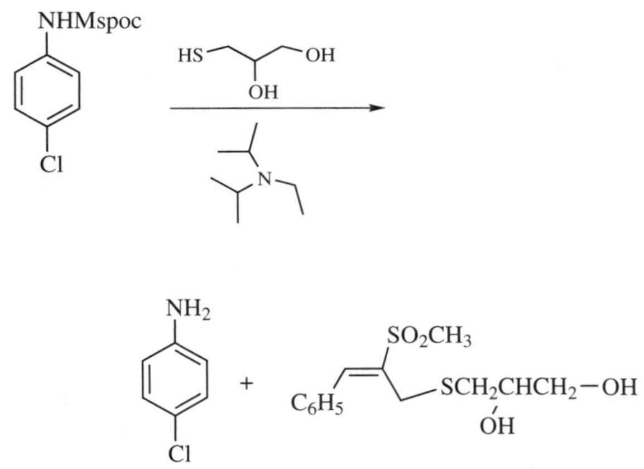

(8)

Related Reagents. 2-(*tert*-Butylsulfonyl)-2-propenyl chloroformate (Bspoc-Cl);[4] 1,1-dioxobenzo[b]thiophene-2-ylmethyl chloroformate (Bsmoc-Cl);[9] 9-fluorenylmethyl chloroformate (Fmoc-Cl);[8] 2-(methylsulfonyl)ethyl chloroformate (Msc-Cl).[7]

1. Carpino, L. A.; Mansour, E. M. E., *J. Org. Chem.* **1999**, *64*, 8399–8401.
2. Doomes, E.; Overton, B. M., *J. Org. Chem.* **1987**, *52*, 1544–1548.
3. Harris, M.; Bull, M. J., *Synth. Commun.* **1985**, *15*, 1225–1228.
4. Carpino, L. A.; Philbin, M., *J. Org. Chem.* **1999**, *64*, 4315–4323.
5. Carpino, L. A.; Sadat-Aalaee, D.; Chao H.-G.; DeSelms, R. H., *J. Am. Chem. Soc.* **1990**, *112*, 9651–9652.
6. Carpino, L. A.; Mansour, E. M. E.; Sadat-Aalaee, D., *J. Org. Chem.* **1991**, *56*, 2611–2614.
7. Tesser, G. I.; Balvert-Geers, I. C., *Int. J. Pept. Protien Res.* **1975**, *7*, 295–305.
8. Carpino, L. A., *Acc. Chem. Res.* **1987**, *20*, 401–407.
9. Carpino, L. A.; Philbin, M.; Ismail, M.; Truran, G. A.; Mansour, E. M. E.; Iguchi, S.; Ionescu, D.; El-Faham, A.; Riemer, C.; Warrass, R.; Weiss, M. S., *J. Am. Chem. Soc.* **1997**, *119*, 9915–9916.

Santhosh Neelamkavil
University of Illinois at Chicago, IL, USA

5-Methylthio-1*H*-tetrazole

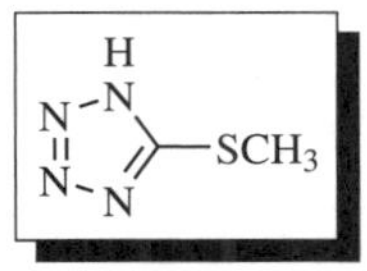

[29515-99-9] $C_2H_4N_4S$ (MW 116.15)

(coupling agent for the synthesis of oligonucleotides using the phosphoramidite method)

Physical Data: mp 149–151 °C.
Solubility: partially soluble in water; soluble in DMF and CH_3CN.
Form Supplied in: crystalline solid.
Preparative Methods: this reagent can be readily prepared from methyl thiocyanate and sodium azide in the presence of ammonium chloride in anhydrous DMF.[1,2] Alternatively, it can be prepared by the reaction of methyl thiocyanate and sodium azide in water with zinc bromide as the catalyst.[3]
Purification: recrystallize from water or dioxane.
Handling, Storage, and Precautions: this reagent should be handled with precaution. It may be irritating to the eyes, the respiratory system, and harmful to the skin. Avoid contact with strong heat and metal salts since tetrazoles may be explosive. It should be stored in the dark in a desiccator.

Activation of Phosphoramidites in Nucleotide Synthesis. Similar to 5-(4-nitrophenyl)-1*H*-tetrazole, 5-methylthio-1*H*-tetrazole (**1**) is a more efficient activator of phosphoramidites than 1*H*-tetrazole in the synthesis of oligonucleotides.[4] This reagent gives 2% higher average yield per coupling cycle than 1*H*-tetrazole as well as higher purity oligonucleotide product. Compared to 5-(4-nitrophenyl)-1*H*-tetrazole, which has poor solubility in acetonitrile (<0.12 M),[5] 5-methylthio-1*H*-tetrazole has a greater solubility (>2.0 M in acetonitrile) and is more suitable for application in large-scale synthesis of oligonucleotides on high loading support. Thus, the TentaGel-supported nucleoside (**2**) (loading 137–193 μmoles per gram) is readily coupled to the phosphoramidite **3** (eq 1) with the total synthesis time for four complete cycles (coupling, capping, oxidation, and detritylation) being only 45 min. The average stepwise yield for the synthesis of an 18mer oligonucleotide 5′TCACAGTCTGAT CTCGAC3′has been estimated to be 96.6% on a 1-mmole scale synthesis.

The use of 5-methylthio-1*H*-tetrazole as the coupling agent also allows the primary hydroxyl group in unprotected carbohydrates and nucleosides to be selectively phosphitylated (eq 2).[6] When the same reactions are performed with 1*H*-tetrazole as the activator, much worse selectivity is observed, resulting in diminished yields.

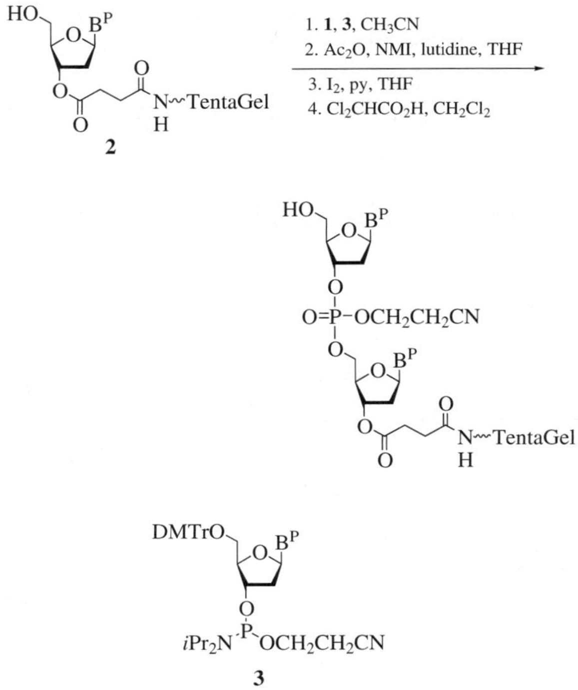

(1)

B^P (a general protected base) =A^{bz}, G^{ibu}, C^{bz}, and T

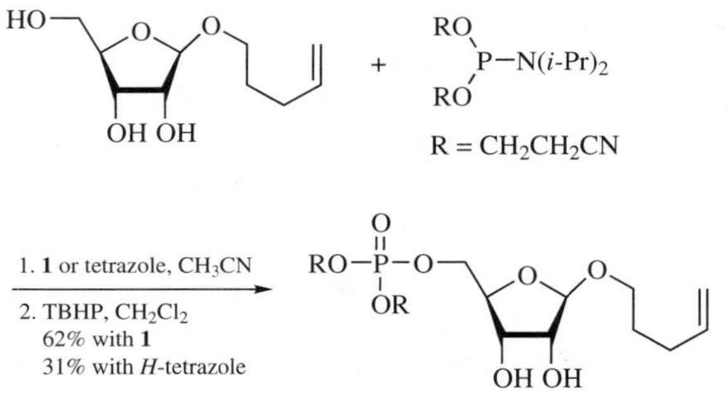

(2)

Other Applications. 5-Methylthio-1*H*-tetrazole has also been used for the synthesis of phosphoramidite and phosphite triester containing platinum complexes (eqs 3 and 4).[7]

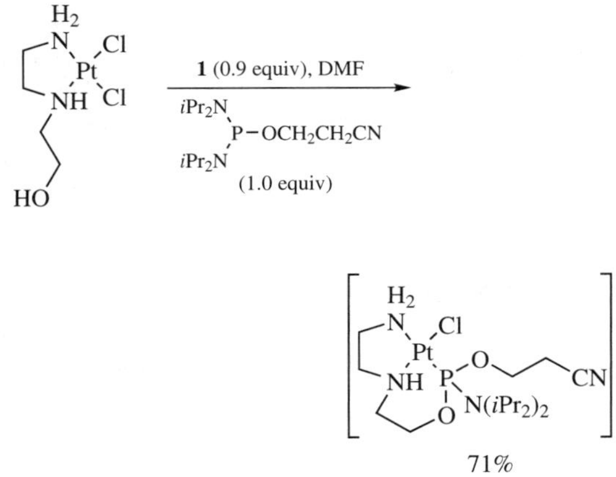

(3)

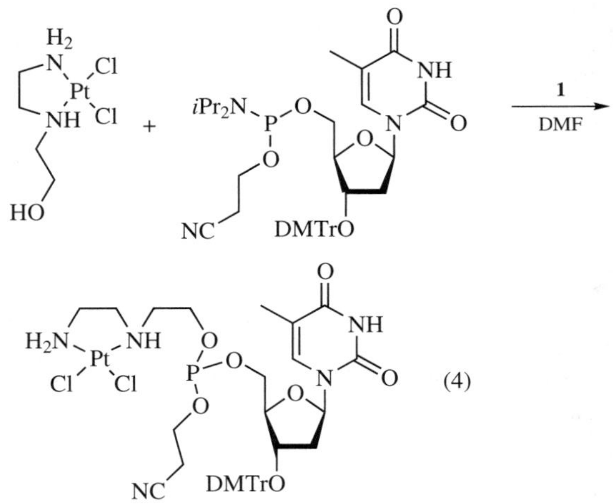

(4)

Related Reagents. 5-(4-Nitrophenyl)-1*H*-tetrazole.

1. Finnegan, W. G.; Henry, R. A.; Lofquist, R., *J. Am. Chem. Soc.* **1958**, *80*, 3908–3911.
2. Lieber, E.; Enkoji, T., *J. Am. Chem. Soc.* **1961**, *26*, 4472–4479.
3. Demko, Z. P.; Sharpless, K. B., *J. Org. Chem.* **2001**, *66*, 7945–7950.
4. Wright, P.; Lloyd, D.; Rapp, W.; Andrus, A., *Tetrahedron Lett.* **1993**, *34*, 3373–3376.
5. Froehler, B. C.; Matteucci, M. D., *Tetrahedron Lett.* **1983**, *24*, 3171–3174.
6. Graham, S. M.; Pope, S. C., *Org. Lett.* **1999**, *1*, 733–736.
7. Ren, S.; Cai, L.; Segal, B. M., *J. Chem. Soc., Dalton Trans.: Inorg. Chem.* **1999**, 1413–1422.

Qingwei Yao
Northern Illinois University, DeKalb, IL, USA

N

2-Naphthylmethyl Bromide

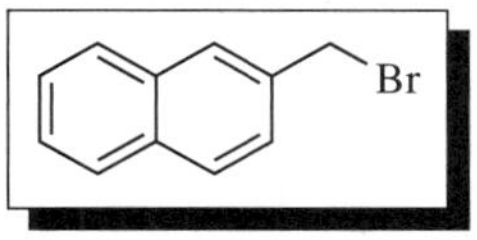

[939-26-4] $C_{11}H_9Br$ (MW 221.10)

(2-naphthylmethylating agent for heteroatomic functional groups and a widely applicable protecting group for alcohols, thiols, and amides)

Alternate Name: 2-(bromomethyl)naphthalene, NAPBr.
Physical Data: mp 51–54 °C; bp 213 °C/100 mm Hg; Fp > 230 °F (110 °C).
Solubility: soluble in nearly all organic solvents.
Form Supplied in: commercially available, 96% pure, slightly yellowish solid.
Handling, Storage, and Precautions: irritating, lachrymatory agent. Should be kept in a cool, dry place.

Protection of Alcohols. 2-Naphthylmethylation of alcohols finds widespread application in the field of organic chemistry,[1] especially in the carbohydrate field.[2–7] Similar to benzylation, this reagent can be applied under a variety of basic conditions. The classical method involves generating the corresponding alkoxides from alcohols with a metal hydride (NaH, KH), and subsequently allowing them to react with 2-(bromomethyl)naphthalene.[8] In order to increase the reactivity of the alkylating agent, a catalytic amount of Bu_4NI[9] may be added to the reaction mixture, which generates a more reactive substance (NAPI). An example of 2-naphthylmethylation using KOH as the base, in the presence of 18-crown-6 catalyst in THF, has been reported.[10] In the field of carbohydrate chemistry, particularly in oligosaccharide synthesis, selective alkylation of an equatorial alcoholic group in the presence of an axial alcoholic OH via stannylene acetal is very important (eq 1, see dibutyltin oxide).[11]

When a free diol is heated with Bu_2SnO in toluene, a five- (dioxolane-type) or a six- (dioxane-type) membered stannylene acetal ring forms, with five-membered rings being possible between either *cis* vicinal hydroxyl groups or their *trans* diequatorial counterparts.[12] On treatment of the formed dioxolane-type tin-acetal compound with NAPBr, selective alkylation of the equatorial hydroxyl ensues,[11] while dioxane-type tin-acetals undergo alkylation of the primary OH group (eq 2).[2] If the stannylene acetal is formed between *trans* functional groups, the group of higher reactivity is alkylated. The rate of the reaction can be enhanced by the addition of Bu_4NI and/or CsF, the latter increases the reactivity of the stannylene acetal ring.[12]

2-Naphthylmethylation of alcoholic OH groups can also be performed by the reductive ring opening of 2-naphthylmethylene acetals formed between vicinal (or neighboring)[6,7,13,14] hydroxyl groups and a transacetalizing agent (see dimethoxymethyl naphthalene). In this reaction one of the hydroxyl groups is liberated, while the other remains protected as a NAP ether.

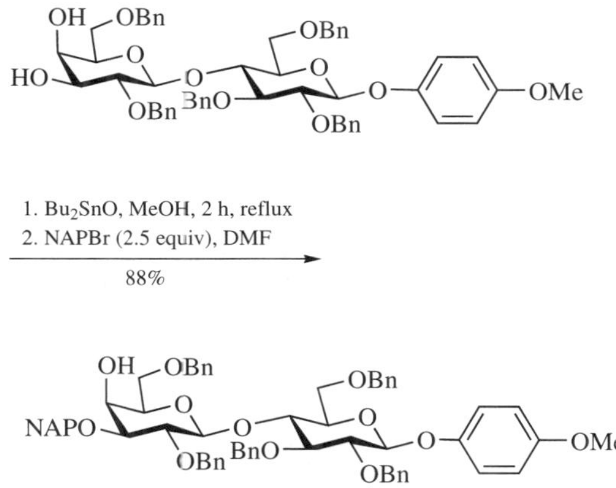

(1)

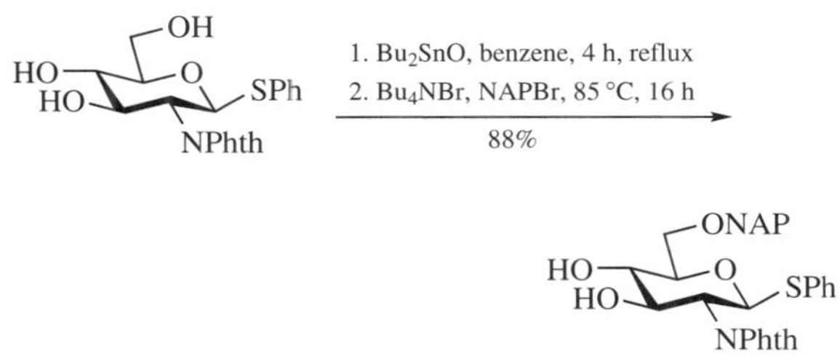

(2)

Reactions with *N*-, *S*-, and *C*-Nucleophiles. 2-(Bromomethyl) naphthalene readily reacts with the three types of nucleophiles listed above. In the case of *N*-nucleophiles, both amines[15,16] and amides[17,18] can be transformed into their *N*-alkylated derivatives in the presence of a suitable base (e.g., $NaHCO_3$, K_2CO_3, Et_3N) and solvent (e.g., CH_3CN, H_2O). Thus, amines of all order; *N*-alkyl, *N,N*-dialkyl amides; and *N*-alkyl imides[19] can be obtained (eq 3).

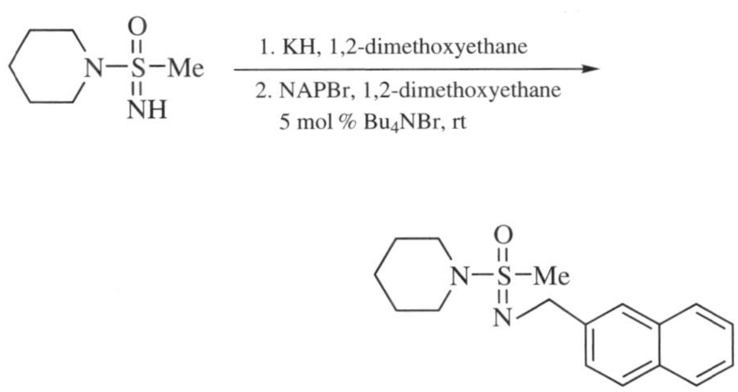

(3)

Similar to *N*-nucleophiles, the NAPBr reagent reacts with *S*-nucleophiles with good yield. With the correct choice of reaction conditions (solvent, base), alkylation of thiols,[20] thiones,[21] and thiohydantoins[22] may also be achieved (eq 4).

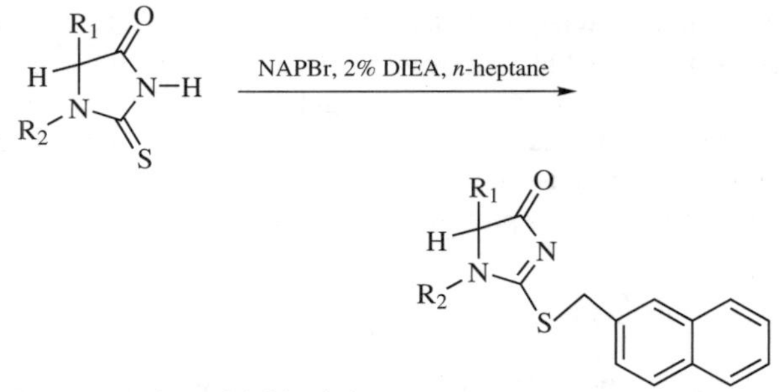

(4)

R_1= any amino acid side chain
R_2= polypeptide chain

The situation is almost the same in the case of *C*-nucleophiles. A good yielding reaction takes place between NAPBr and species having either a negatively charged[23] or polarized[24] *C*-atom. NAPBr itself can also be converted into an organometallic compound[25] by reaction with Zn or Mg.

Properties of *O*-2-Naphthylmethyl Ethers. The NAP ether finds a widespread application as an orthogonal protecting group in carbohydrate chemistry because of its stability under acidic (in the presence of 4% trifluoroacetic acid and 80% acetic acid, conditions of glycosylation) and basic conditions.[9] On the other hand, both hydrogenolytic and oxidative methods are known for its deprotection. 2-Naphthylmethyl ethers can be selectively removed in the presence of a benzyl ether by catalytic hydrogenolysis.[1,10] The NAP ether can be also cleaved under oxidative conditions using cerium ammonium nitrate (CAN),[14] DDQ[6,7,10,11,13,26] (see also 3,4-dichloro-5,6-dicyano-1,4-benzoquinone), or dimethyl dioxirane[27] in the presence of benzyl ethers (the last used only at lower temperatures). Equation 5 shows an interesting example of the consecutive deprotection of the two most frequently used ether-type protecting groups in addition to benzyl, namely the *p*-methoxybenzyl and the 2-naphthylmethyl group.

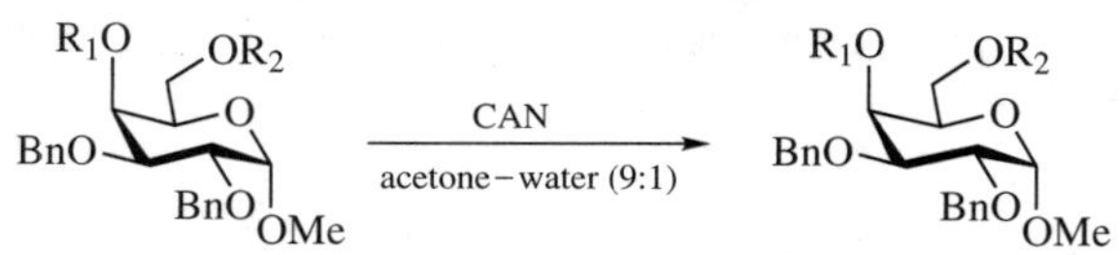

1; R_1 = PMB, R_2 = NAP
2; R_1 = NAP, R_2 = PMB

3; R_1 = H, R_2 = NAP (83%)
4; R_1 = NAP, R_2 = H (72%)

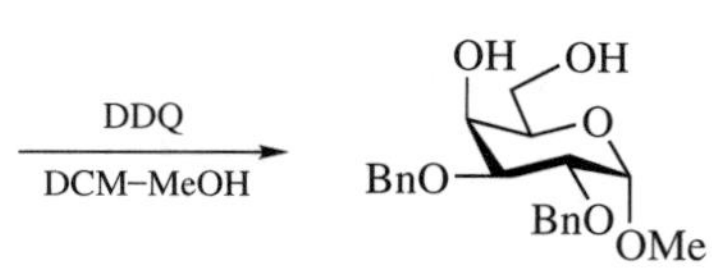

(5)

82%

2-(Bromomethyl)naphthalene can be used not only for the protection of OH groups, but also for the *N* and *S* analogs of the hydroxyl group. Deprotection of NAP from a *N*-containing compound requires almost the same conditions developed for the *O*-derivatives.[28] It is also possible to use the 2-naphthylmethyl group for the protection of the carboxylic acid function.[29]

Related Reagents. Benzyl bromide; *p*-methoxybenzyl chloride.

1. Gaunt, M. J.; Yu, J.; Spencer, J. B., *J. Org. Chem.* **1998**, *63*, 4172. (b) Nardi, D.; Cappelletti, R.; Catto, A.; Leonardi, A.; Tajana, A.; Veronese, M., *Arzneim. Forsch.* **1981**, *31*, 2123.
2. Xia, J.; Piskorz, C. F.; Alderfer, J. L.; Locke, R. D.; Matta, K. L., *Tetrahedron Lett.* **2000**, *41*, 2773.
3. Xia, J.; Alderfer, J. L.; Piskorz, C. F.; Matta, K. L., *Chem. Eur. J.* **2000**, *6*, 3442.
4. Xia, J.; Srikrishnan, T.; Alderfer, J. L.; Jain, R. K.; Piskorz, C. F.; Matta, K. L., *Carbohydr. Res.* **2000**, *329*, 561.
5. Xia, J.; Alderfer, J. L.; Piskorz, C. F.; Matta, K. L., *Chem. Eur. J.* **2001**, *7*, 356.
6. Borbás, A.; Szabó, Z. B.; Szilágyi, L.; Bényei, A.; Lipták, A., *Carbohydr. Res.* **2002**, *337*, 1941.
7. Borbás, A.; Szabó, Z. B.; Szilágyi, L.; Bényei, A.; Lipták, A., *Tetrahedron* **2002**, *58*, 5723.
8. Kang, J.; Lim, G. J.; Yoon, S. K.; Kim, M. Y., *J. Org. Chem.* **1995**, *60*, 564.
9. Xia, J.; Abbas, S. A.; Locke, R. D.; Piskorz, C. F.; Alderfer, J. L.; Matta, K. L., *Tetrahedron Lett.* **2000**, *41*, 169.
10. Liao, W.; Locke, R. D.; Matta, K. L., *Chem. Commun.* **2000**, 369.
11. Borbás, A.; Csávás, M.; Szilágyi, L.; Lipták, A., *J. Carbohydr. Chem.* **2004**, in press.
12. Grindley, T. B., *Adv. Carbohydr. Chem. Biochem.* **1998**, *53*, 17.
13. Lipták, A.; Borbás, A.; Jánossy, L.; Szilágyi, L., *Tetrahedron Lett.* **2000**, *41*, 4949.
14. Wright, J. A.; Yu, J.; Spencer, J. B., *Tetrahedron Lett.* **2001**, *42*, 4033.
15. Chapman, N. B.; James, J. W., *J. Chem. Soc. Abstr.* **1953**, 1865.
16. O'Sullivan, M. C.; Zhou, Q.; Li, Z.; Durham, T. B.; Rattendi, D.; Lane, S.; Bacchi, C. J., *Bioorg. Med. Chem.* **1997**, *5*, 2145.
17. Wu, S.; Lee, S.; Beak, P., *J. Am. Chem. Soc.* **1996**, *118*, 715.
18. Mellor, S. L.; Chan, W. C., *Chem. Commun.* **1997**, *20*, 2005.
19. Johnson, C. R.; Lavergne, O. M., *J. Org. Chem.* **1993**, *58*, 1922.
20. Paul, B.; Korytnyk, W., *Carbohydr. Res.* **1984**, *126*, 27.
21. Shearer, B. G.; Oplinger, J. A.; Lee, S., *Tetrahedron. Lett.* **1997**, *38*, 179.
22. Samyn, B.; Hardeman, K.; Van der Eycken, J.; Van Beeumen, J., *Anal. Chem.* **2000**, *72*, 1389.
23. (a) Mazón, P.; Chinchilla, R.; Nájera, C.; Guillena, G.; Kreiter, R.; Klein Gebbink, R. J. M.; van Koten, G., *Tetrahedron: Asymmetry* **2002**, *13*, 2181. (b) Zhou, C.; Shao, Y.; Gibbs, R. A., *Bioorg. Med. Chem. Lett.* **2002**, *12*, 1417.
24. (a) Balema, V. P.; Wiench, J. W.; Pruski, M.; Pecharsky, V. K., *J. Am. Chem. Soc.* **2002**, *124*, 6244. (b) Thornton, T. A.; Ross, G. A.; Patil, D.; Mukaida, K.; Warwick, J. O.; Wollsey, N. F.; Bartak, D. E., *J. Am. Chem. Soc.* **1989**, *111*, 2434.
25. (a) Aitken, R. A.; Hodgson, P. K. G.; Morrison, J. J.; Oyewale, A. O., *J. Chem. Soc., Perkin Trans. 1* **2002**, *3*, 402. (b) Katritzky, A. R.; Odens, H. H.; Voronkov, M. V., *J. Org. Chem.* **2000**, *65*, 1886.
26. Csávás, M.; Borbás, A.; Szilágyi, L.; Lipták, A., *Synlett* **2002**, 887.
27. Csuk, R.; Dörr, P., *Tetrahedron* **1994**, *50*, 9983.
28. Godin, G.; Compain, P.; Martin, O. R., *Synlett* **2003**, 2065–2067.
29. Gaunt, M. J.; Boschetti, C. E.; Yu, J.; Spencer, J. B., *Tetrahedron Lett.* **1999**, *40*, 1803.

Magdolna Csávás, Zoltán B. Szabó, Anikó Borbás & András Lipták
University of Debrecen, Debrecen, Hungary

p-Nitrobenzenesulfonyl-4-nitroimidazole

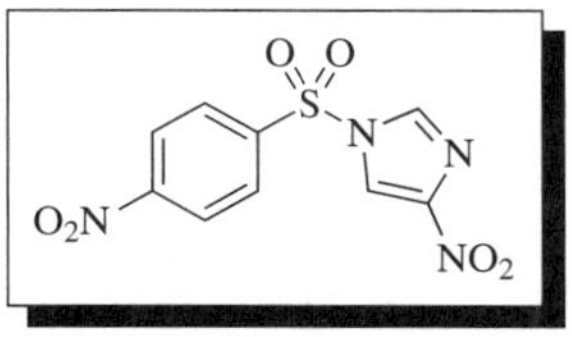

[81006-79-3] $C_9H_6N_4O_6S$ (MW 298.26)

(condensing agent for the triester coupling of nucleotide fragments[1])

Alternate Name: p-NBSNI.
Physical Data: mp (sealed tube) 179–185 °C (dec); UV (dioxane) 253 nm.
Solubility: sol pyridine, DMSO; slightly sol dioxane, ethyl acetate.
Form Supplied in: white needles; not available commercially; may be contaminated with p-nitrobenzenesulfonic acid.
Analysis of Reagent Purity: ^1H NMR (60 MHz, DMSO-d_6) δ 8.45 (s, 4H); 8.55 (d, J = 2 Hz, 1H); 9.10 (d, J = 2 Hz, 1H); sulfonic acid impurities are easily detected as higher field signals.
Preparative Method: by treatment of 4-nitroimidazole (1.025 equiv) and p-nitrobenzenesulfonyl chloride (1.0 equiv) with Triethylamine (1.1 equiv) in dry dioxane at 0 °C.[1]
Purification: recrystallization from ethyl acetate.
Handling, Storage, and Precautions: stable to air and moisture in screw cap bottles at rt for over six months.

Condensing Agent. p-Nitrobenzenesulfonyl-4-nitroimidazole has been used as a condensing agent in the triester synthesis of oligonucleotides. The reagent activates the phosphodiester towards attack by the incoming 5′-hydroxyl through formation of a mixed phosphoric–sulfonic anhydride. Typically, the suitably protected nucleotide fragments, in pyridine solution, are treated with p-NBSNI (2.5 equiv) and stirred at rt for approximately 15 h. After workup, purification is achieved through two chromatographic separations over silica gel, affording the desired polynucleotide in 60–83% yield (eq 1).[1]

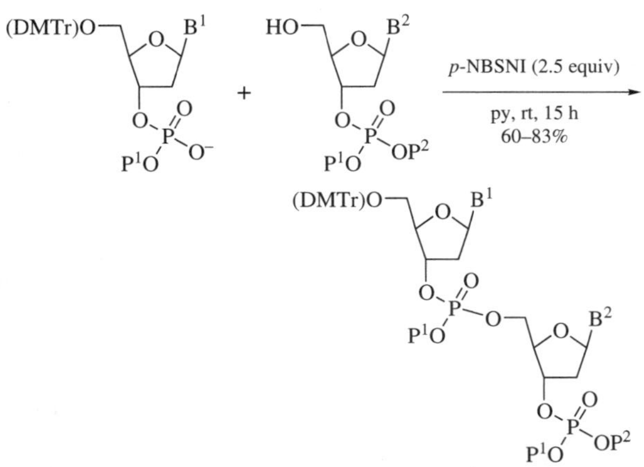

(1)

DMTr = 5′-O-dimethoxytrityl; P^1 = p-ClC_6H_4; P^2 = cyanoethyl

p-Nitrobenzenesulfonyl-4-nitroimidazole holds the advantage over older reagents, such as triisopropylbenzenesulfonyl chloride, of giving fewer side products resulting from sulfonation of the free 5′-hydroxyl. The long reaction times necessary are a disadvantage and the condensing agents 2,4,6-triisopropylbenzenesulfonyl-1,2,3,4-tetrazole (TPSTe) and 1-(mesitylenesulfonyl)-3-nitro-1,2,4-triazole (MSNT) are preferable.[2]

---

1. Leach, C. A.; Tamm, C.; Waldmeier, F. *Helv. Chim. Acta* **1981**, *64*, 2515.
2. Ikehara, M.; Ohtsuka, E.; Uesugi, S.; Tanaka, T. In *Chemistry of Nucleosides and Nucleotides*; Townsend, L. B., Ed.; Plenum:New York, **1988**; *1*, Chapter 3.

Matthew P. Braun & Carl R. Johnson
Wayne State University, Detroit, MI, USA

o-Nitrobenzyl Bromide

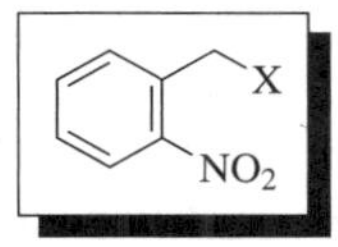

(R = Br)
[3958-60-9] $C_7H_6BrNO_2$ (MW 216.04)
(R = Cl)
[612-23-7] $C_7H_6ClNO_2$ (MW 171.58)

(introduction of the 2-nitrobenzyl protecting group which can be cleaved selectively by photolysis[1–4])

Alternate Name: 2-nitrobenzyl bromide.
Physical Data: (X = Br) mp 45–48 °C. (X = Cl) mp 46–48 °C, bp 127–133 °C/10 mmHg.
Solubility: sol most solvents; alkylations are frequently carried out in polar solvents such as DMF.
Form Supplied in: commercially available from many sources.
Handling, Storage, and Precautions: 2-nitrobenzyl halides are lachrymators and should be used in a well-ventilated fume hood.

The 2-Nitrobenzyl Group as a Protecting Group. The 2-nitrobenzyl group is one of an increasing number of functions that have been employed as photolabile protecting groups in organic synthesis,[1–4] and photolytic cleavage usefully adds to the range of deprotecting strategies in the armory of the organic chemist. One of the first examples of such a protecting group was in the use of 2-nitrobenzyl esters as protected carboxylic acids.[5] UV irradiation of the esters yields the free acid and initially 2-nitrosobenzaldehyde (the mechanism of the rearrangement has been studied extensively[1–4]). The benzaldehyde is further converted to azobenzene-2,2′-dicarboxylic acid which then acts as an internal light filter and so severely hampers the reaction. Benzylic substituents remove this problem and derivatives such as the α-phenyl-2-nitrobenzyl esters give much better yields (eq 1). The 2-nitrobenzyl unit has also been incorporated into a wide variety of photolabile polymer supports for solid phase peptide synthesis.[6] Polystyrene supported 2-nitrobenzyl bromide has been used as a resin for anchoring the C-terminus amino acid residue to the solid phase as its 2-nitrobenzyl ester.[7] Phosphate esters incorporating the photolabile 2-nitrobenzyl group have also been synthesized.[8]

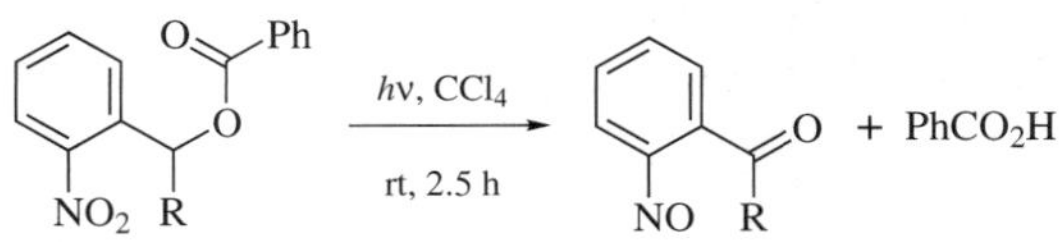

(1)

R = H, 17%; Ph, 90%

2-Nitrobenzyl halides have been used extensively in the field of oligonucleotide synthesis[9–15] as a protecting group (eqs 2 and 3) for the 2′-OH of uridine,[9] adenosine,[10] cytidine,[10] and guanosine.[11] Treatment of the nucleoside with a strong base such as Bis(tri-n-butyltin) Oxide[9] or Sodium Hydride[10,11] followed by 2-nitrobenzyl bromide gives the expected ethers, though in very poor yields (ca. 25–30%). The benzylic protons of nitrobenzyl halides are quite acidic and so decomposition of the halides is significant under the strongly basic conditions. To bypass this problem some workers have used 2-nitrophenyldiazomethane instead of the substituted benzyl halide,[16] but the increased yields are accompanied by poorer selectivity for the 2′-OH over the 3′-OH and introduce the problem of product separation. Another method, which has been used for making 4-nitrobenzyl ethers of simple alcohols, is to use *p*-nitrobenzyl trifluoromethanesulfonate derived from the corresponding alcohol.[17]

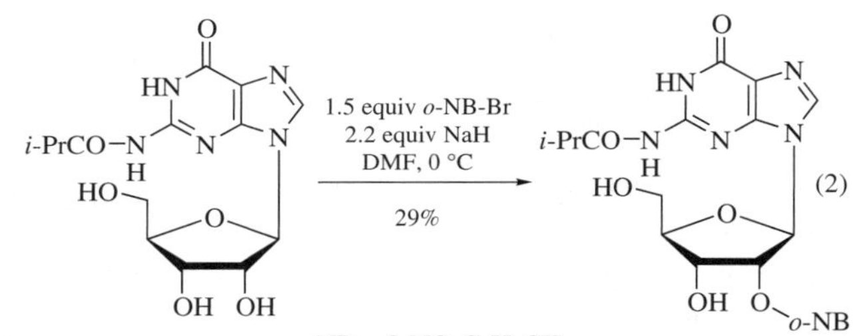

(2)

o-NB = $2\text{-}NO_2C_6H_4CH_2–$

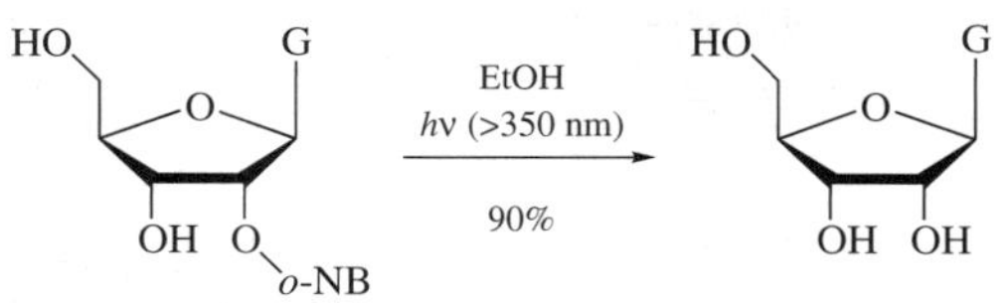

(3)

In the etherification of an alcohol using 2-nitrobenzyl bromide, Ireland[18] used BaO/ Barium Hydroxide instead of the stronger bases mentioned above (eq 4). The reaction is still capricious, and investigation of silver-assisted alkylations resulted in failure. Deprotection near the end of the synthesis though is effected smoothly and cleanly with UV irradiation (eq 5). For the protection of phenolic hydroxyl groups both Silver(II) Oxide in benzene[19] and a Potassium Carbonate/ Cesium Carbonate mixture[20] (eq 6) have been used successfully.

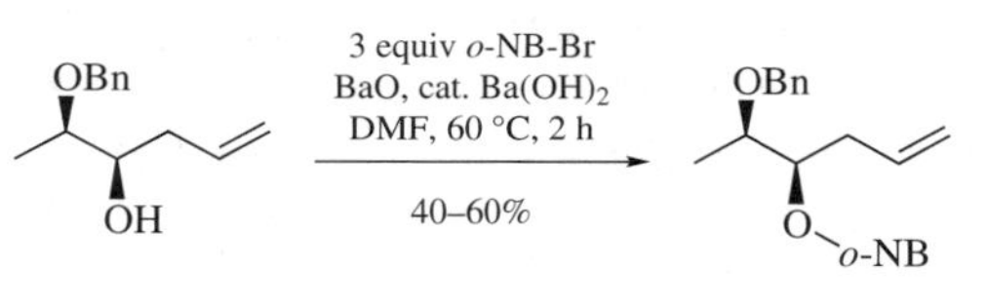

(4)

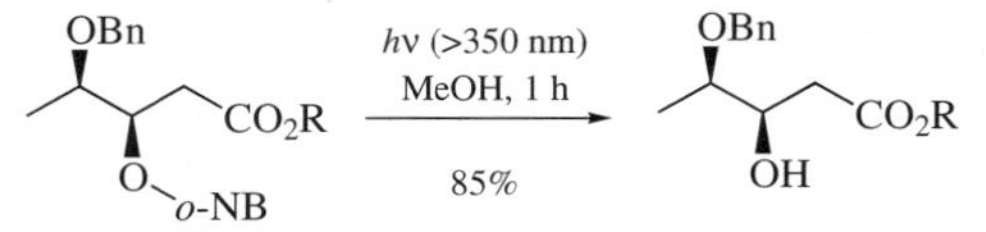

(5)

R = complex spiroacetal

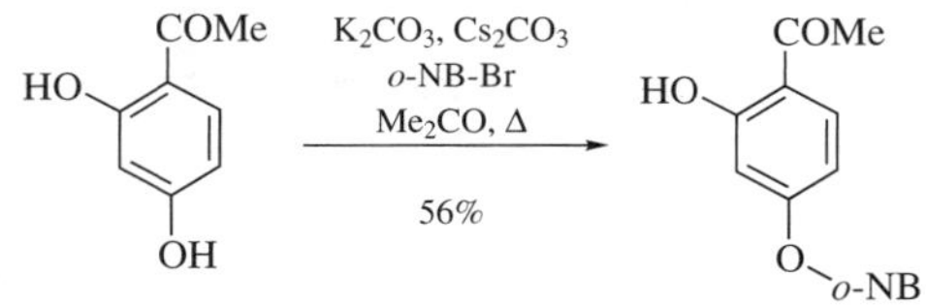

(6)

The photolabile 2-nitrobenzyl group has been employed in carbohydrate chemistry[21] for the protection of anomeric hydroxyl groups, and has also been used for other acetals.[22] These derivatives, synthesized from 2-nitrobenzyl alcohol, are particularly useful as they impart greater crystallinity to the compounds and aid in purification.

The 2-nitrobenzyl group has been used to protect heteroatoms other than oxygen. The sulfur atom in the phosphorothioate group of some nucleotides has been protected as the *S*-(2-nitrobenzyl) phosphorothioate (eq 7), and with this protection the nucleotides can be successfully coupled and deprotected to give di- and trinucleotides.[23]

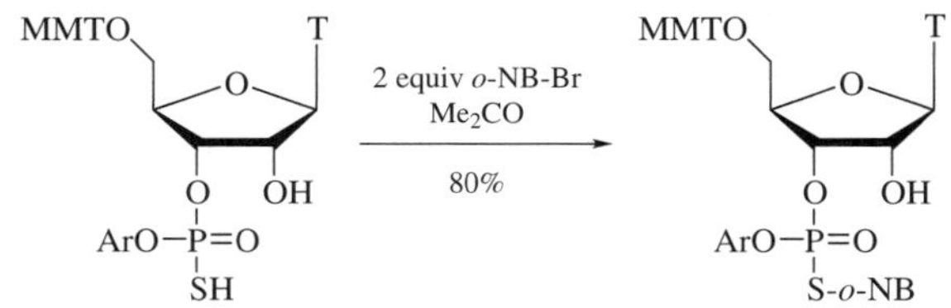

(7)

Ar = $p\text{-}NO_2C_6H_4–$

The N-1 atom in pyrazolidin-3-ones has been protected using 2-nitrobenzyl bromide in DMF (eq 8),[24] as has the N^{im}-3 atom in histidine (eqs 9 and 10).[25]

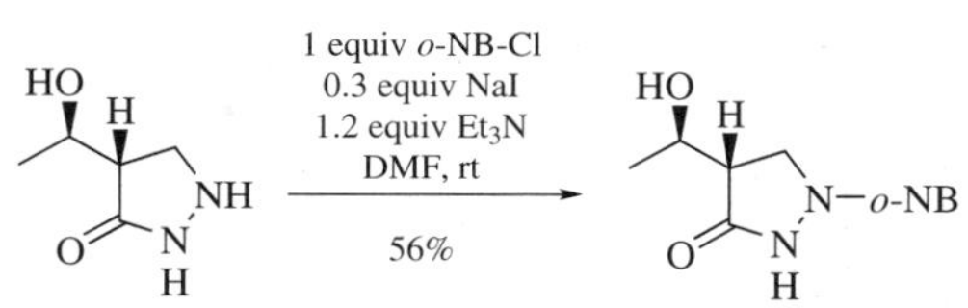

(8)

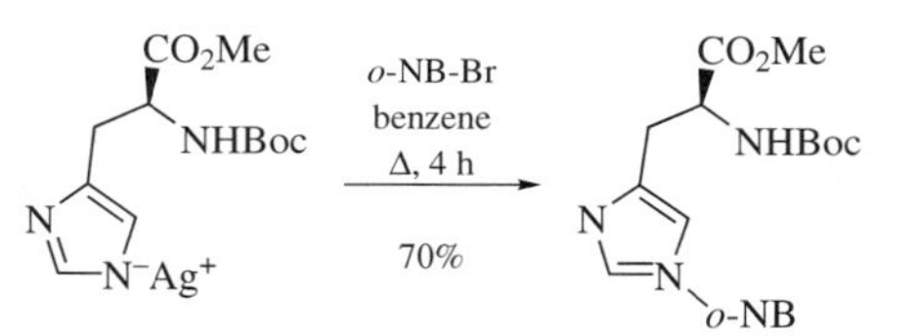

(9)

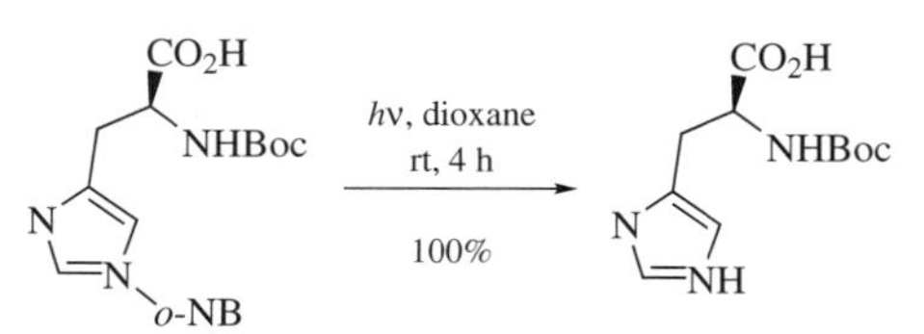

(10)

In Williams' synthesis of (±)-aplysiatoxin, he used a *N*-(2-nitrobenzyl) protected glycine unit.[26] The 2-nitrobenzyl group has been used to *N*-protect other biologically important compounds such as 5-fluorouracil, 2′-deoxy-5-fluorouridine (eq 11),[27] and purine bases.[28] Glycine, an important inhibitory neurotransmitter, has been *N*-derivatized with the 2-nitrobenzyl group so that the biologically active compound can be released with UV photolysis as desired.[29] This use of photolabile protecting groups for biologically active molecules has been reviewed recently.[30]

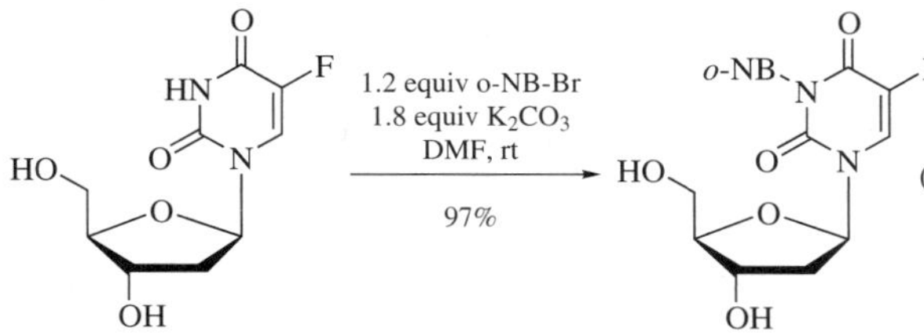

(11)

The 2-nitrobenzyl substituent has also found a use in the protection of amino groups (eq 12) as urethanes using 2-nitrobenzyloxycarbonyl chloride, giving a photolabile equivalent to the benzyloxycarbonyl (Z) group.[31]

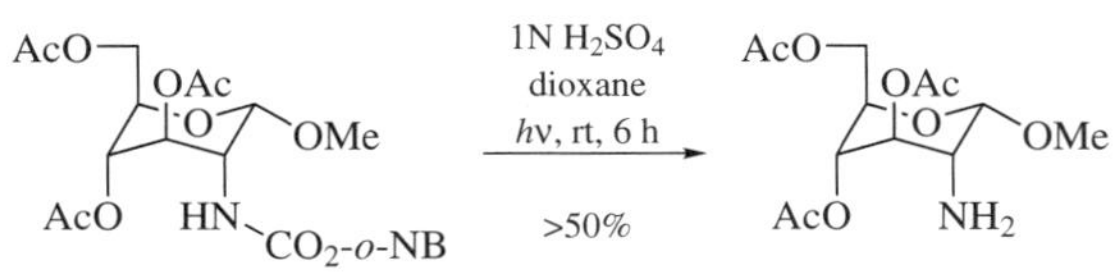

(12)

Related Reagents. o-Nitrobenzoyl Chloride; o-Nitrobenzyl Alcohol.

1. Amit, B.; Zehavi, U.; Patchornik, A. *Isr. J. Chem.* **1974**, *12*, 103.
2. Pillai, V. N. R. *Synthesis* **1980**, 1.
3. Pillai, V. N. R. In *Organic Photochemistry*; Padwa, A., Ed.; Dekker:New York, **1987**; *9*, p 225.
4. Zehavi, U. *Adv. Carbohydr. Chem. Biochem.* **1988**, *46*, 179.
5. (a) Barltrop, J. A.; Plant, J. A.; Schofield, P. *Chem. Commun./J. Chem. Soc., Chem. Commun.* **1966**, 822. (b) Patchornik, A.; Amit, B.; Woodward, R. B. *J. Am. Chem. Soc.* **1970**, *92*, 6333.
6. Ajayaghosh, A.; Pillai, V. N. R. *J. Org. Chem.* **1990**, *55*, 2826 and refs therein.
7. Ajayaghosh, A.; Pillai, V. N. R. *Tetrahedron* **1988**, *44*, 6661.
8. (a) Rubinstein, M.; Patchornik, A. *Tetrahedron* **1975**, *31*, 2107. (b) Givens, R. S.; Keuper, L. W. *Chem. Rev.* **1993**, *93*, 55.
9. Ohtsuka, E.; Tanaka, S.; Ikehara, M. *Nucleic Acids Res.* **1974**, *1*, 1351.
10. Ohtsuka, E.; Tanaka, S.; Ikehara, M. *Chem. Pharm. Bull.* **1977**, *25*, 949.
11. Ohtsuka, E.; Tanaka, S.; Ikehara, M. *Synthesis* **1977**, 453.
12. Tanaka, T.; Orita, M.; Uesugi, S.; Ikehara, M. *Tetrahedron* **1988**, *44*, 4331 and refs therein.
13. Cullis, P. M. *Chem. Commun./J. Chem. Soc., Chem. Commun.* **1984**, 1510.
14. Hayes, J. A.; Brunden, M. J.; Gilham, P. T.; Gough, G. R. *Tetrahedron Lett.* **1985**, *26*, 2407.
15. Gough, G. R.; Brunden, M. J.; Gilham, P. T. *Tetrahedron Lett.* **1981**, *22*, 4177.
16. Bartholomew, D. G.; Broom, A. D. *Chem. Commun./J. Chem. Soc., Chem. Commun.* **1975**, 38.
17. Fukase, K.; Tanaka, H.; Torii, S.; Kusumoto, S. *Tetrahedron Lett.* **1990**, *31*, 389.
18. Ireland, R. E.; Thaisrivongs, S.; Dussault, P. H. *J. Am. Chem. Soc.* **1988**, *110*, 5768.
19. Krafft, G. A.; Sutton, W. R.; Cummings, R. T. *J. Am. Chem. Soc.* **1988**, *110*, 301.
20. Kees, K. L.; Musser, J. H.; Chang, J.; Skowronek, M.; Lewis, A. J. *J. Med. Chem.* **1986**, *29*, 2329.
21. (a) Zehavi, U.; Amit, B.; Patchornik, A. *J. Org. Chem.* **1972**, *37*, 2281. (b) Zehavi, U.; Patchornik, A. *J. Org. Chem.* **1972**, *37*, 2285.
22. Winnik, F. M.; Carver, J. P.; Krepinsky, J. J. *J. Org. Chem.* **1982**, *47*, 2701.
23. Lesnikowski, Z. J.; Jaworska, M. M. *Tetrahedron Lett.* **1989**, *30*, 3821.
24. Perri, S. T.; Slater, S. C.; Toske, S. G.; White, J. D. *J. Org. Chem.* **1990**, *55*, 6037.
25. Kalbag, S. M.; Roeske, R. W. *J. Am. Chem. Soc.* **1975**, *97*, 440.
26. Miknis, G. F.; Williams, R. M. *J. Am. Chem. Soc.* **1993**, *115*, 536.
27. Lin, T.-S.; Wang, L.; Antonini, I.; Cosby, L. A.; Shiba, D. A.; Kirkpatrick, D. L.; Sartorelli, A. C. *J. Med. Chem.* **1986**, *29*, 84.
28. Kelley, J. L.; Linn, J. A.; Selway, J. W. T. *J. Med. Chem.* **1989**, *32*, 1757.
29. Billington, A. P.; Walstrom, K. M.; Ramesh, D.; Guzikowski, A. P.; Carpenter, B. K.; Hess, G. P. *Biochemistry* **1992**, *31*, 5500.
30. McCray, J. A.; Trentham, D. R. *Annu. Rev. Biophys. Chem.* **1989**, *18*, 239.
31. (a) Amit, B.; Zehavi, U.; Patchornik, A. *J. Org. Chem.* **1974**, *39*, 192. (b) Cameron, J. F.; Frechet, J. M. J. *J. Am. Chem. Soc.* **1991**, *113*, 4303. (c) Cummings, R. T.; Krafft, G. A. *Tetrahedron Lett.* **1988**, *29*, 65.

Andrew N. Boa & Paul R. Jenkins
Leicester University, UK

p-Nitrobenzyl Bromide

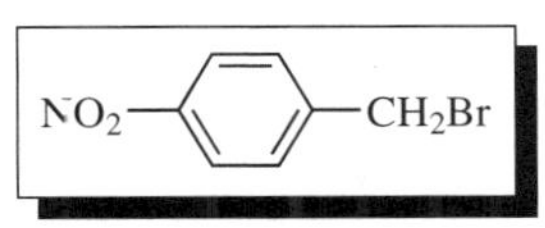

[100-11-8] $C_7H_6BrNO_2$ (MW 216.04)

(benzylating agent for a variety of heteroatomic functional groups as well as carbon nucleophiles)

Alternate Name: 4-nitrobenzyl bromide, α-bromo-4-nitrotoluene, 1-(bromomethyl)-4-nitrobenzene, PNB-Br.

Physical Data: yellow solid; mp 98–99 °C.[1]

Solubility: soluble in common organic solvents (i.e., CH_2Cl_2, cyclohexane, and benzene).

Form Supplied in: widely available commerically.

Analysis of Reagent Purity: melting point.

Preparative Methods: *p*-nitrobenzyl bromide is most readily prepared through bromination of *p*-nitrotoluene.[1,2] Other preparative methods involve bromination of *p*-nitrobenzyl alcohol[3] and nitration of benzyl bromide.[4]

Purification: *p*-nitrobenzyl bromide (280 g) is dissolved in hot petroleum ether (ligroin) (3 L), boiled with decolorizing carbon (10 g), and filtered with suction. After cooling the filtrate in an ice bath, the resulting crystals are collected on a Büchner funnel, pressed, and washed with two 25 mL portions of cold petroleum ether to provide *p*-nitrobenzyl bromide as pale yellow crystals (250 g).[1]

Handling, Storage, and Precautions: *p*-nitrobenzyl bromide is a corrosive lachrymator, and accordingly should be dispensed and used in a well-ventilated fumehood.

Introduction. *p*-Nitrobenzyl bromide (PNB-Br) is a powerful alkylating agent, which undergoes reaction with a wide range of nucleophiles. Previously, it has been employed to identify carboxylic acids[5] and phenols[6] through conversion to their respective *p*-nitrobenzyl esters and ethers. These derivatives are often crystalline solids with sharp melting points. Most contemporary applications of *p*-nitrobenzyl bromide, however, involve its use as a reagent for the introduction of the *p*-nitrobenzyl (PNB) protecting group to alcohols, thiols, and carboxylic acids. *p*-Nitrobenzyl ethers, thioethers, esters, and carbamates (PNZ) are useful protecting groups, which exhibit enhanced stability toward acidic and oxidizing conditions over the corresponding benzyl-based groups. In addition to the crystallinity which PNB substituents often confer, this family of protecting groups are readily cleaved in the presence of a number of protecting groups which are sensitive to hydrogenation, including allyl, benzyl, and *p*-methoxybenzyl ethers and benzyloxycarbonyl (Cbz) groups.

Protection of Alcohols. Introduction of *p*-nitrobenzyl groups by means of the Williamson ether synthesis, using strong bases (NaOH, NaH, or $Ba(OH)_2$) and polar solvents such as DMF and THF, is not practical since both *p*-nitrobenzyl bromide and the resulting ethers undergo rapid decomposition under these conditions. *p*-Nitrobenzyl ethers can, however, be prepared in good to excellent yield by treatment of alcohols with *p*-nitrobenzyl bromide and silver oxide (Ag_2O) in non-polar solvents, such as CH_2Cl_2, cyclohexane, or benzene (eq 1).[7] This method is not compatible with the more polar solvents DMF and THF.

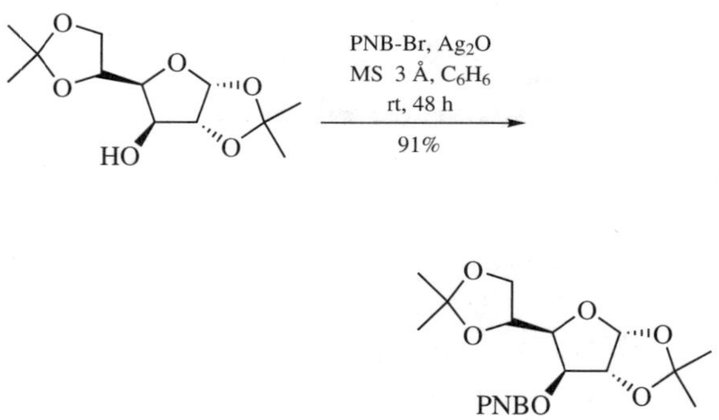

(1)

p-Nitrobenzylation of hydroxyl groups can also be accomplished in high yield using *p*-nitrobenzyl triflate, generated in situ by the reaction of *p*-nitrobenzyl bromide with silver triflate in the presence of 2,4,6-collidine.[7] Other effective methods for the formation of PNB ethers include *O*-monoalkylation of *O*-stannylene acetals with *p*-nitrobenzyl bromide in the presence of cesium fluoride (eq 2)[8] and reductive cleavage of *p*-nitrobenzylidene acetals using $Zn(BH_4)_2$ and trimethylsilyl chloride.[9]

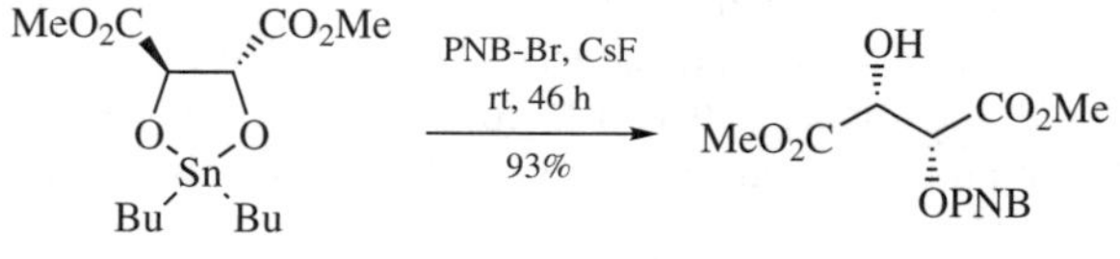

(2)

In common with benzyl ethers, *p*-nitrobenzyl ethers can be cleaved in one step by catalytic hydrogenolysis.[7] In this respect, *p*-nitrobenzyl groups offer few advantages, other than their enhanced crystallinity. However, *p*-nitrobenzyl ethers are valuable since they can be selectively removed in the presence of a wide range of other protecting groups including allyl, benzyl, and *p*-methoxybenzyl ethers, as well as phosphate esters, acetate esters, and Cbz groups. Deprotection is achieved through a two-step sequence involving reduction of the nitro group to form the more labile *p*-aminobenzyl ether, which is then cleaved under oxidative conditions to provide the free hydroxyl group (Figure 1). Since reduction of aromatic nitro substituents often proceeds significantly faster than hydrogenolysis of benzyl ethers and related protecting groups, selective removal of PNB ethers is possible.

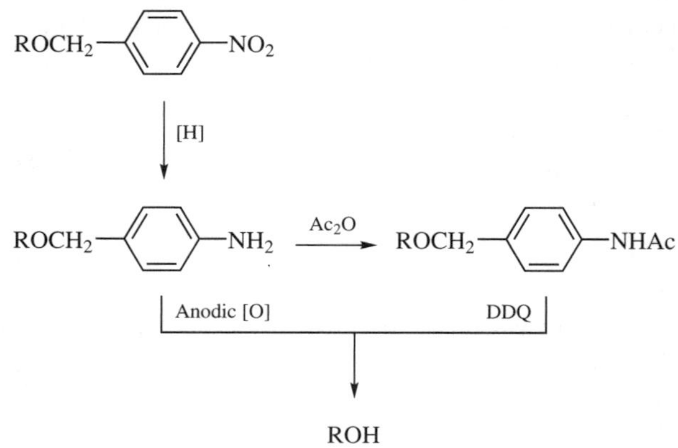

Figure 1 Two-step cleavage of *p*-nitrobenzyl ethers

Chemoselective reduction of the nitro group can be accomplished by catalytic hydrogenation in the presence of Pd-black, using a variety of hydrogen donors (H_2, HCO_2NH_4, HCO_2H, 1,4-cyclohexadiene) (eq 3),[7] or alternatively, using zinc-copper couple (Zn-Cu) in the presence of acetylacetone (eq 4).[10] A number of methods are also available for cleavage of the intermediate *p*-aminobenzyl ether, including anodic oxidation using sulfuric acid as the supporting electrolyte (4.2–10 F mol^{-1}, dioxane, H_2O, H_2SO_4).[7,11] Employing these conditions, *p*-aminobenzyl ethers can be removed in the presence of both benzyl and *p*-methoxybenzyl (PMB) ethers (eq 3).[7] In contrast, PMB ethers can be removed in the presence of PNB ethers through oxidation with DDQ.

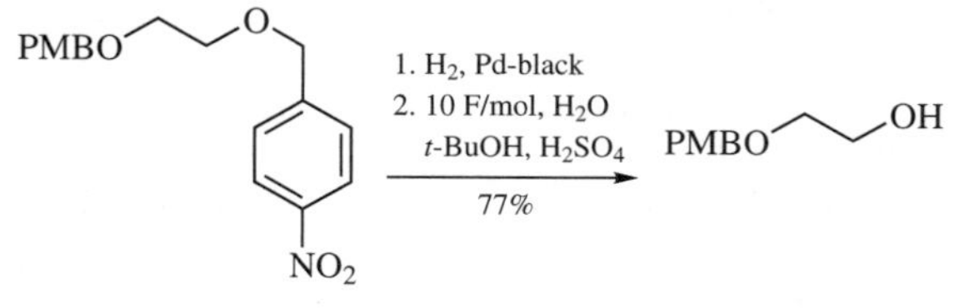

(3)

In analogy with *p*-methoxybenzyl ethers, *p*-aminobenzyl ethers can also be cleaved through oxidation with 2,3-dichloro-5,6-dicyano-1,4-benzoquinone (DDQ) (Figure 1). In this case, yields are improved by first converting the *p*-aminobenzyl ether to the corresponding acetamido derivative (eq 4).[7]

Protection of Carboxylic Acids. *p*-Nitrobenzyl esters are readily prepared through alkylation of carboxylic acids with *p*-

nitrobenzyl bromide in the presence of a base, typically triethylamine (eq 5).[12]

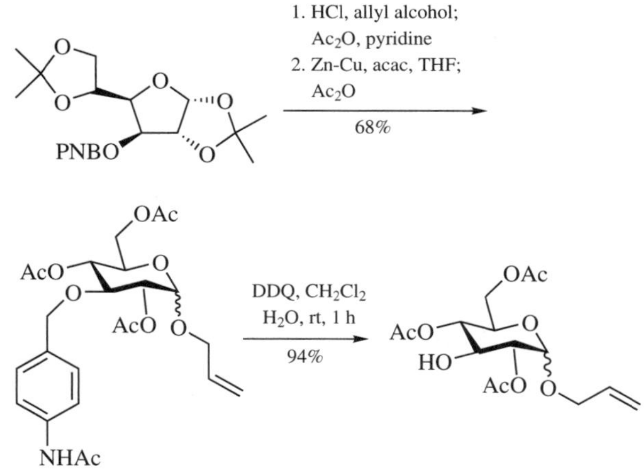

(4)

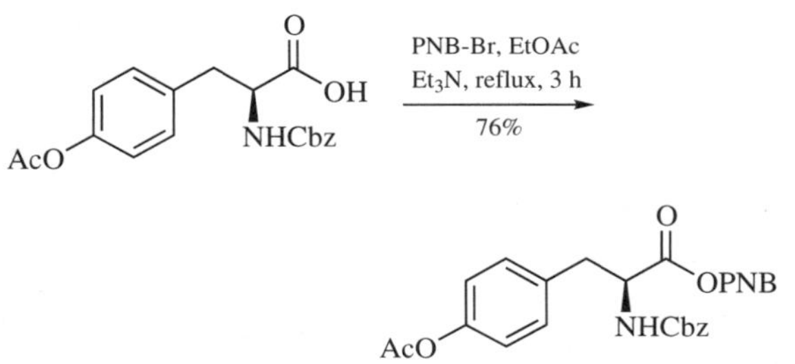

(5)

In cases where competitive alkylation is a problem or where protection of an *N*-acyl amino acid or peptide is to be undertaken, alternate methods are recommended, including treatment of sodium or trialkylammonium carboxylate salts with *p*-nitrobenzyl tosylate (prepared by the reaction of *p*-nitrobenzyl bromide and silver tosylate),[13] or alternatively, Fischer esterification with *p*-nitrobenzyl alcohol (benzenesulfonic acid, CCl_4, reflux, 69–100%).[14]

p-Nitrobenzyl esters display greater stability towards acidic conditions than conventional benzyl esters (*p*-nitrobenzyl esters are approximately 50 times as stable as benzyl esters to HBr in acetic acid), yet they can be removed by reductive hydrolysis under mild conditions. It is for these reasons that PNB esters have been recommended for the side-chain protection of aspartic acid and glutamic acid residues during both liquid-phase[12,15,16] and solid-phase peptide syntheses.[17,18] The enhanced stability of PMB esters towards acidic hydrolysis facilitates the preferential removal of Cbz and Boc groups from the peptide chain with HBr. Benzyl esters, on the other hand, are not completely stable towards these strongly acidic conditions; and their use can give rise to a cumulative loss of side-chain protection, and an increase in the possibility of branching in the peptide chain.

A wide variety of mild conditions are available for the deprotection of *p*-nitrobenzyl esters under reducing conditions, such as Zn/AcOH,[16] Zn/HCl,[19] or sodium dithionite ($Na_2S_2O_4$) (eq 6).[20,21]

The cleavage of PNB esters using chemical reducing agents is complicated, however, by the formation of a yellow by-product, which is believed to result from the polymerization and/or hydrolysis of *p*-benzoquinoneimine (Figure 2).[20] This unstable compound arises through 1,6-elimination of the *p*-aminobenzyl esters generated upon reduction of the *p*-nitro group. Removal of this contaminant from the final peptide product has proven to be troublesome, particularly during solid-phase synthesis.

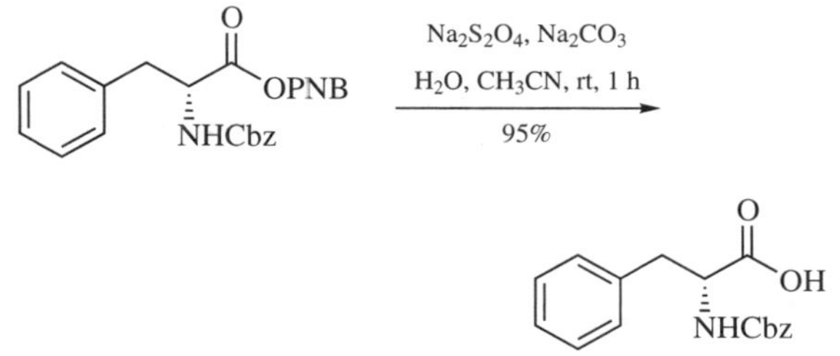

(6)

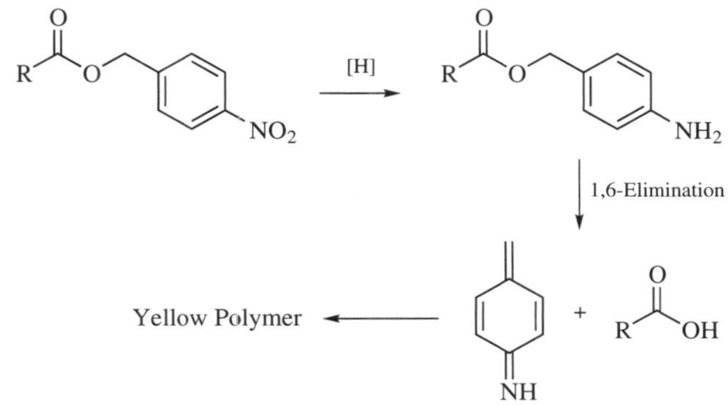

Figure 2 Reductive cleavage of *p*-nitrobenzyl esters

An effective solution to this problem involves carrying out the deprotection under mildly acidic conditions, using a solution of stannous chloride ($SnCl_2$), phenol, and acetic acid in DMF. The yellow contaminant generated during the deprotection step can then be scavenged by treatment of the polymer-bound peptide with benzene sulfinic acid in DMF (eq 7).[18,22]

As an alternative to chemical reduction, tetrabutylammonium fluoride (TBAF, THF, 15–30 min, rt) can be used to deprotect *p*-nitrobenzyl esters. The major by-product in this case is *p*-nitrobenzyl alcohol, which can be readily removed by extraction; no yellow by-products are formed. Notably, this mild reagent is compatible with a range of other common protecting groups, including benzyl, *t*-butyl, and cyclohexyl esters, as well as *t*-butyloxycarbonyl (Boc) and Cbz groups.[23]

p-Nitrobenzyl esters can also be unmasked through catalytic hydrogenolysis in the presence of Pd-C[15,18,21] or Adams catalyst (PtO_2).[24] This approach has been successfully employed in the final step of the Merck synthesis of the fragile β-lactam antibiotic (+)-thienamycin (eq 8).[24]

Catalytic hydrogenolysis of PNB esters can, however, be problematic with sulfur-containing substrates such as penicillins and cephalosporins. In such instances, alkaline hydrolysis with aqueous sodium sulfide ($Na_2S{\cdot}9H_2O$) offers a simple and rapid alternative, which is compatible even with highly sensitive substrates such as azetidinone antibiotics (eq 9).[25] The deprotection of 3-cephem PNB esters under these conditions is accompanied by isomerization of the Δ^3 double bond to the Δ^2-position (eq 10).[25] The 4-nitrobenzyl esters of 3-cephem derivatives have been removed by hydrogenolysis, using ammonium formate or phosphinic acid as the catalytic transfer agents in the presence of 10% Pd-C.[26] Δ^3-Δ^2 rearrangement is not observed under these conditions.

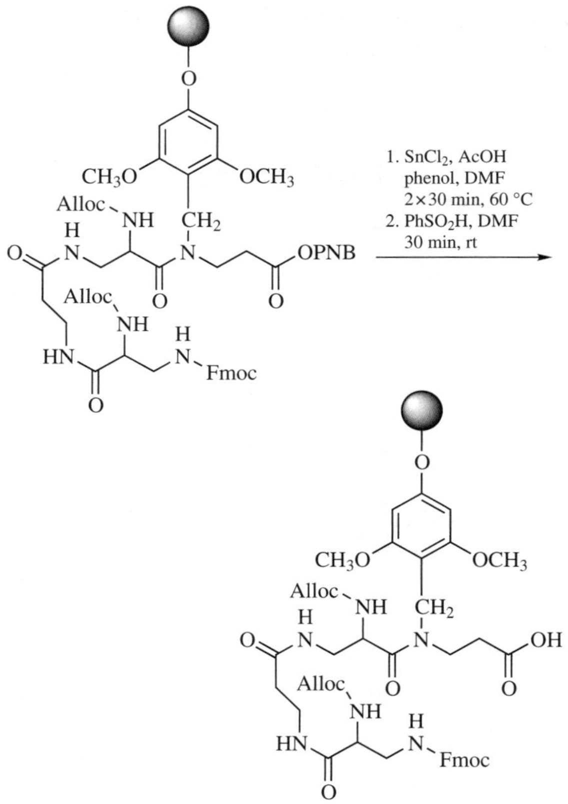

(7)

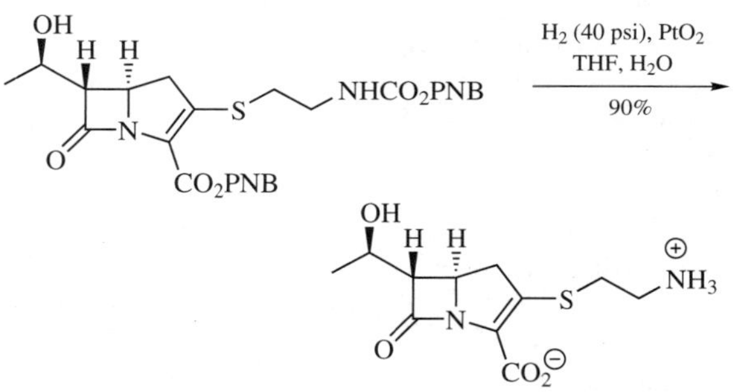

(8)

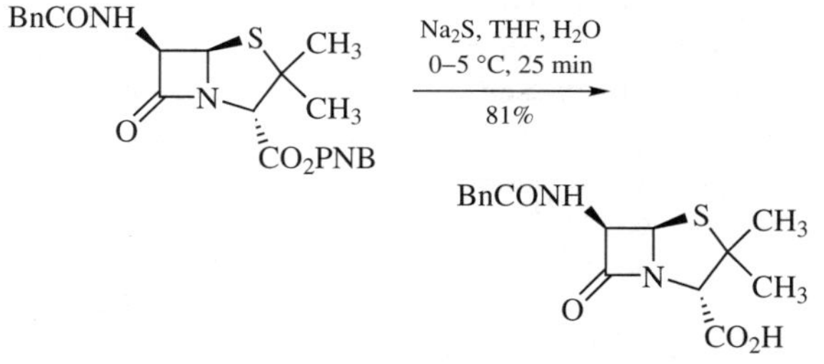

(9)

Lithium iodide (LiI) promotes the dealkylation of *N*-acyl amino *p*-nitrobenzyl esters under very mild conditions.[27] While this reagent has traditionally been limited to the cleavage of methyl esters, coordination of the lithium ion by the ester carbonyl and the amide carbonyl at the γ-position serves to activate this type of substrate. Notably, *p*-nitrobenzyl esters of 3-cephem derivatives undergo dealkylation without Δ^3-Δ^2 rearrangement to provide 3-cephem acids in excellent yield (eq 11).[27]

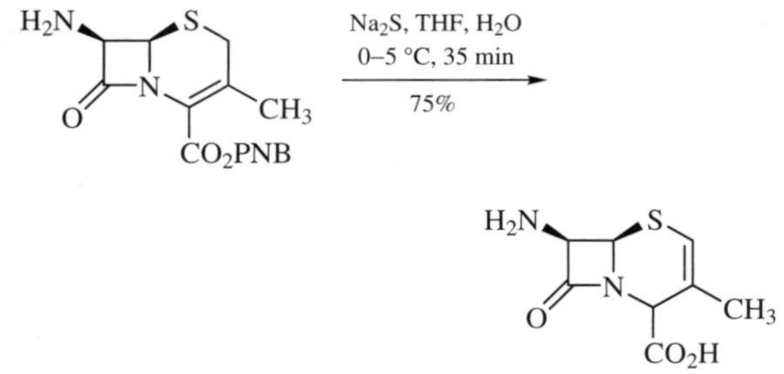

(10)

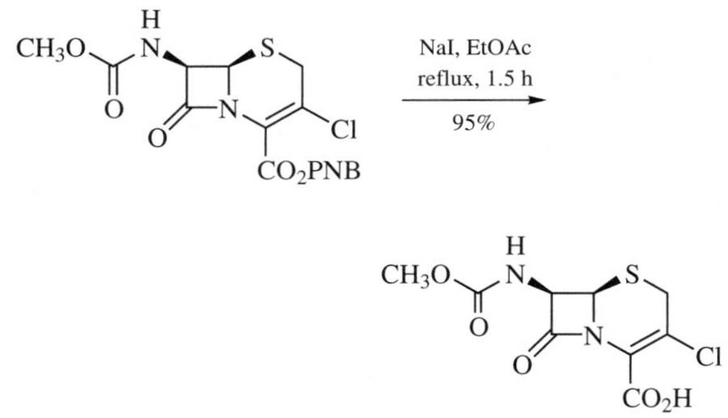

(11)

Miscellaneous. In addition to its role in protecting group chemistry, *p*-nitrobenzyl bromide has also been utilized to alkylate a diverse range of nucleophiles, including thiolates (eq 12),[28,29] sulfides (thietanes) (eq 13),[30] amines (eq 14),[31,32] benzodiazepinones,[33] pyridines (eq 15),[34] phosphines,[35] and ketone enolates (eq 16).[36]

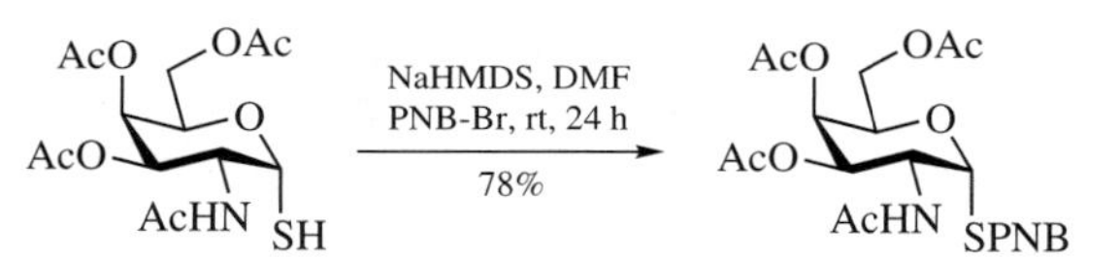

(12)

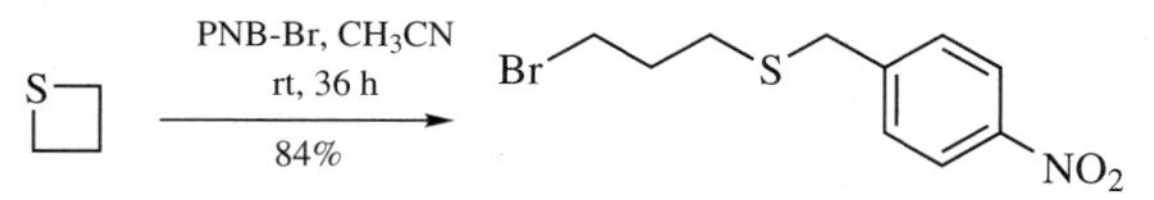

(13)

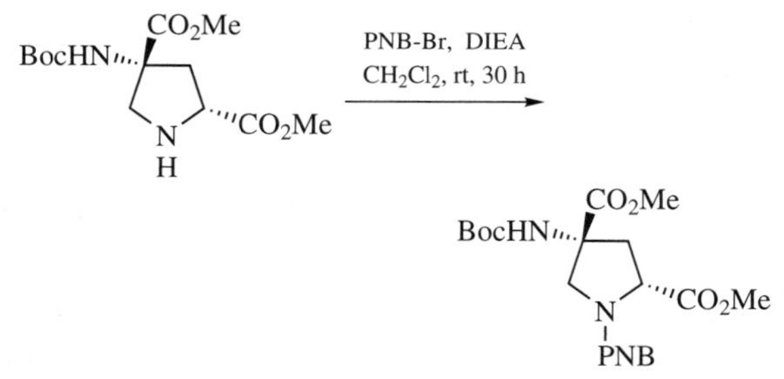

(14)

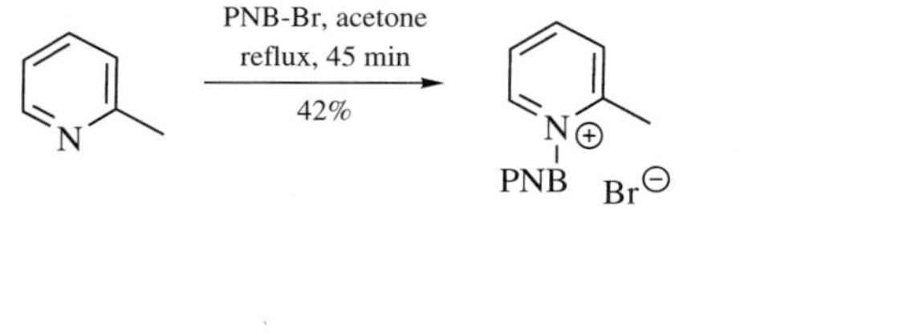

(15)

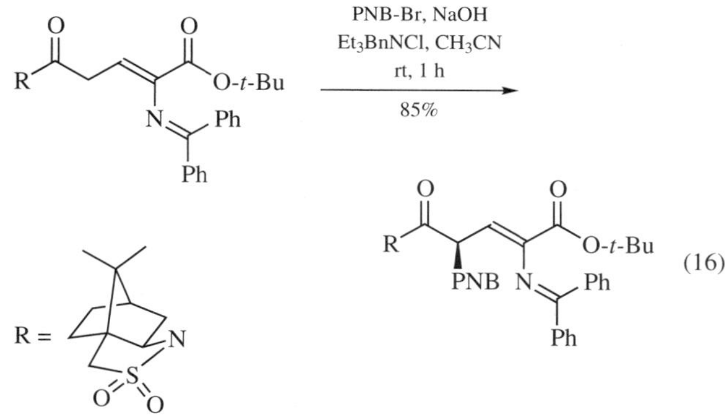

Related Reagents. *p*-Nitrobenzyl chloride; *p*-nitrobenzyl chloroformate; *O*-nitrobenzyl bromide; *O*-nitrobenzyl chloride.

1. Coleman, G. H.; Honeywell, G. E., *Org. Synth., Coll. Vol.* **1943**, *2*, 443.
2. Brewster, J. F., *J. Am. Chem. Soc.* **1918**, *40*, 406.
3. Norris, J. F.; Watt, M.; Thomas, R., *J. Am. Chem. Soc.* **1916**, *38*, 1071.
4. Moureu, C.; Brown, R. L., *Bull. Soc. Chim. Fr.* **1921**, *29*, 1006.
5. Reid, E. E., *J. Am. Chem. Soc.* **1917**, *39*, 124.
6. Lyman, J. A.; Reid, E. E., *J. Am. Chem. Soc.* **1920**, *42*, 615.
7. Fukase, K.; Tanaka, H.; Torii, S.; Kusumoto, S., *Tetrahedron Lett.* **1990**, *31*, 389.
8. Nagashima, N.; Ohno, M., *Chem. Lett.* **1987**, 141.
9. Kotsuki, H.; Ushio, Y.; Yoshimura, N.; Ochi, M., *J. Org. Chem.* **1987**, *52*, 2594.
10. Fukase, K.; Hase, S.; Ikenaka, T.; Kusumoto, S., *Bull. Chem. Soc. Jpn.* **1992**, *65*, 436.
11. Mairanovsky, V. G., *Angew. Chem., Int. Ed. Engl.* **1976**, *15*, 281.
12. Schwyzer, R.; Sieber, P., *Helv. Chim. Acta* **1959**, *42*, 972.
13. Theodoropoulos, D.; Tsangaris, J., *J. Org. Chem.* **1964**, *29*, 2272.
14. Shields, J. E.; McGregor, W. H.; Carpenter, F. H., *J. Org. Chem.* **1961**, *26*, 1491.
15. Schwarz, H.; Arakawa, K., *J. Am. Chem. Soc.* **1959**, *81*, 5691.
16. Suzuki, K.; Endo, N., *Chem. Pharm. Bull.* **1978**, *26*, 2269.
17. Prestidge, R. L.; Harding, D. R. K.; Hancock, W. S., *J. Org. Chem* **1976**, *41*, 2579.
18. Hocker, M. D.; Caldwell, C. G.; Macsata, R. W.; Lyttle, M. H., *Pept. Res.* **1995**, *8*, 310.
19. Chauvette, R. R.; Pennington, P. A., *J. Med. Chem.* **1975**, *18*, 403.
20. Guibe-Jampel, E.; Wakselman, M., *Synth. Commun.* **1982**, *12*, 219.
21. Perich, J. W.; Alewood, P. F.; Johns, R. B., *Aust. J. Chem.* **1991**, *44*, 233.
22. Royo, M.; Farrera-Sinfreu, J.; Solé, L.; Albericio, F., *Tetrahedron Lett.* **2002**, *43*, 2029.
23. (a) Namikoshi, M.; Kundu, B.; Rinehart, K. L., *J. Org. Chem.* **1991**, *56*, 5464. (b) Ueki, M.; Kai, K.; Amemiya, M.; Horino, H.; Oyamada, H., *J. Chem. Soc., Chem. Commun.* **1988**, 414.
24. (a) Melillo, D. G.; Shinkai, I.; Liu, T.; Ryan, K.; Sletzinger, M., *Tetrahedron Lett.* **1980**, *21*, 2783. (b) Salzmann, T. N.; Ratcliffe, R. W.; Christensen, B. G.; Bouffard, F. A., *J. Am. Chem. Soc.* **1980**, *102*, 6161.
25. Lammert, S. R.; Ellis, A. I.; Chauvette, R. R.; Kukolja, S., *J. Org. Chem.* **1978**, *43*, 1243.
26. Albanese, D.; Leone, M.; Penso, M.; Seminati, M.; Zenoni, M., *Tetrahedron Lett.* **1998**, *39*, 2405.
27. Fischer, J. W.; Trinkle, K. L., *Tetrahedron Lett.* **1994**, *35*, 2505.
28. Knapp, S.; Myers, D. S., *J. Org. Chem.* **2002**, *67*, 2995.
29. Matta, K. L.; Girotra, R. N.; Barlow, J. J., *Carbohydr. Res.* **1975**, *43*, 101.
30. Palmer, D. C.; Taylor, E. C., *J. Org. Chem.* **1986**, *51*, 846.
31. Mukhopadhyaya, J. K ; Kozikowski, A. P.; Grajkowska, E.; Pshenichkin, S.; Wroblewski, J. T., *Bioorg. Med. Chem. Lett.* **2001**, *11*, 1919.
32. Singh, P.; Arora, G., *Indian J. Chem., Sect. B* **1986**, *25B*, 1034.
33. Lee, J.; Gauthier, D.; Rivero, R. A., *J. Org. Chem.* **1999**, *64*, 3060.
34. Alvarez-Builla, J.; Gonzalez Trigo, G.; Ezquerra, J.; Fombella, M. E., *J. Heterocycl. Chem.* **1985**, *22*, 681.
35. Cremer, S. E.; Cowles, J. M.; Farr, F. R.; Hwang, H. O.; Kremer, P. W.; Peterson, A. C.; Gray, G. A., *J. Org. Chem.* **1992**, *57*, 511.
36. Alvarez-Ibarra, C.; Csákÿ, A. G.; Gómez de la Oliva, C.; Rodríguez, E., *Tetrahedron Lett.* **2001**, *42*, 2129.

Duncan J. Wardrop & Chad L. Landrie
University of Illinois at Chicago, Chicago, IL, USA

p-Nitrophenol

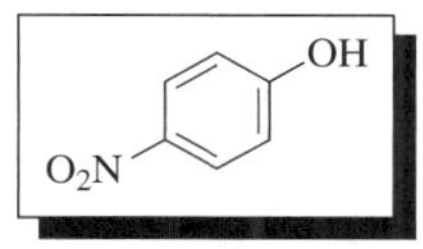

[100-02-7] $C_6H_5NO_3$ (MW 139.12)

(preparation of active esters for peptide synthesis[1])

Physical Data: mp 114 °C; pK_a 7.15 (at 25 °C); dimorphous.
Solubility: moderately sol cold water; sol alcohol, chloroform, ether.
Form Supplied in: generally obtained as a mixture of the two crystalline forms.
Purification: recrystallization from toluene gives the metastable α-form if performed at >63 °C; the β-form is obtained as yellow prisms at <63 °C.
Handling, Storage, and Precautions: the α-form is light stable, whereas the yellow β-form gradually turns red in light. Moderately toxic by inhalation and skin absorption.

Preparation of Active Esters for Peptide Synthesis. *p*-Nitrophenyl esters, prepared by 1,3-Dicyclohexylcarbodiimide-mediated coupling between *N*-protected amino acids and *p*-nitrophenol, are among the most useful activated esters for peptide synthesis (eq 1).[1] They are readily purified by recrystallization, often from

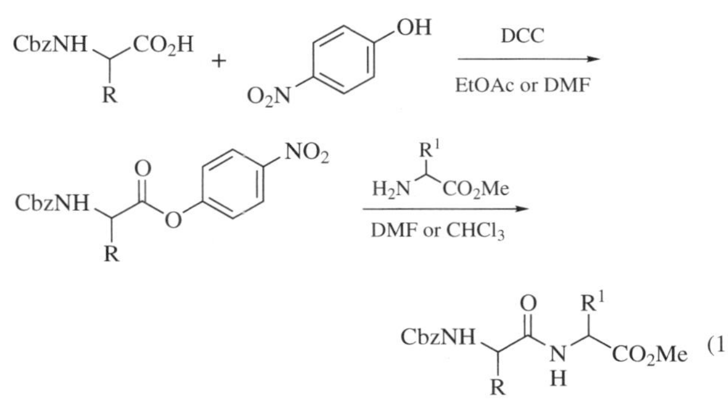

alcoholic solvents with which they do not react. Aminolysis occurs at a reasonable rate, generally at room temperature without a catalyst. The *p*-nitrophenol generated as a byproduct in this step is easily removed. *o*-Nitrophenyl esters can be used similarly (see o-Nitrophenol); a study of the rates of aminolysis of Boc-L-leucine nitrophenyl esters showed the *o*-isomer to react 4–9 times faster than the *p*-isomer.[2]

1. Bodanszky, M.; Du Vigneaud, V. *J. Am. Chem. Soc.* **1959**, *81*, 5688; Bodanszky, M.; Meienhofer, J.; Du Vigneaud, V. *J. Am. Chem. Soc.* **1960**, *82*, 3195.
2. Bodanszky, M.; Bath, R. J. *Chem. Commun./J. Chem. Soc., Chem. Commun.* **1969**, 1259.

Alan Armstrong
University of Bath, UK

5-(4-Nitrophenyl)-1*H*-tetrazole

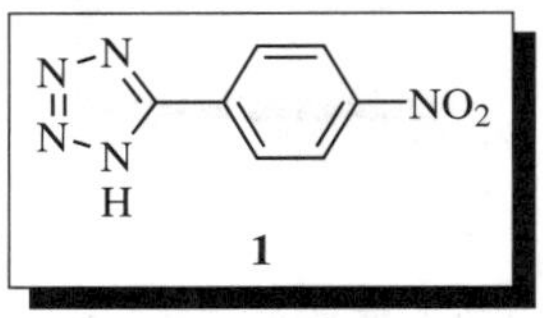

1

[16687-60-8] $C_7H_5N_5O_2$ (MW 191.15)

(reagent used as activator for both solution- and solid-phase synthesis of oligonucleotides using the phosphoramidite method)

Physical Data: mp 225–226 °C.
Solubility: insoluble in H_2O; soluble in DMSO; this reagent is only slightly soluble in the most commonly used solvent for oligonucleotide synthesis, i.e., acetonitrile (< 0.12 M),[3] but is more soluble in THF (up to 0.8 M).[4]
Form Supplied in: white or slightly colored powder; commercially available.
Analysis of Reagent Purity: the reagent may be checked by mp or analyzed by standard techniques.
Preparative Methods: the reagent can be readily prepared from 4-nitrobenzonitrile and sodium azide in the presence of ammonium chloride in anhydrous DMF.[1] It can also be prepared by the reaction of 4-nitrobenzonitrile and sodium azide in water using zinc bromide as the catalyst.[2]
Purification: recrystallization from either ethanol, isopropanol, or ethanol/water.
Handling, Storage, and Precautions: this reagent should be handled with caution. It may be irritating to the eyes and the respiratory system, and harmful to the skin. Avoid contact with strong heat and metal salts, since tetrazoles may be explosive. It may be stored in the dark in a desiccator.

Nucleotide Synthesis. 5-(4-Nitrophenyl)-1*H*-tetrazole (**1**) is primarily used as a highly efficient coupling agent for the synthesis of oligonucleotides by the phosphoramidite method.[3,4] The enhanced acidity of **1** relative to tetrazole[5] makes it a superior activator of phosphoramidites in the synthesis of oligonucleotides. Kinetic studies[3] on the coupling of the 5′-hydroxyl group of silica gel–supported deoxythymidine (**2**) and phosphoramidite (**3**) with different tetrazole derivatives reveals the following reactivity: **1** > 5-(3-nitrophenyl)-1*H*-tetrazole > 5-(3-bromophenyl)-1*H*-tetrazole > 1*H*-tetrazole (eq 1). Despite its enhanced acidity, this reagent does not cause significant detritylation when an equimolar solution of **1** and 5′-*O*-DMTr-thymidine in CH_3CN is stored for 24 h.[3]

Use of **1** as the activator in the synthesis of oligonucleotides via the phosphoramidite method can also be achieved in a catalytic manner (eq 2).[6] This is particularly useful for a large scale synthesis of a short oligonucleotide. Dideoxyribonucleoside phosphates are prepared on a multigram scale in 92–99% yields through the reaction of nucleoside 3′-*N*,*N*-diethylphosphoramidites (1.05 equiv) and 5′-*O*-free nucleosides (1.00 equiv) with **1** (0.05 equiv) in the presence of molecular sieves 13X as the amine scavenger, followed by TMSOTf-catalyzed oxidation with bis(trimethylsilyl)peroxide. This catalytic approach is also effective for the synthesis of longer deoxyribonucleotides such as d($^{5'}$CTACCTGT$^{3'}$).

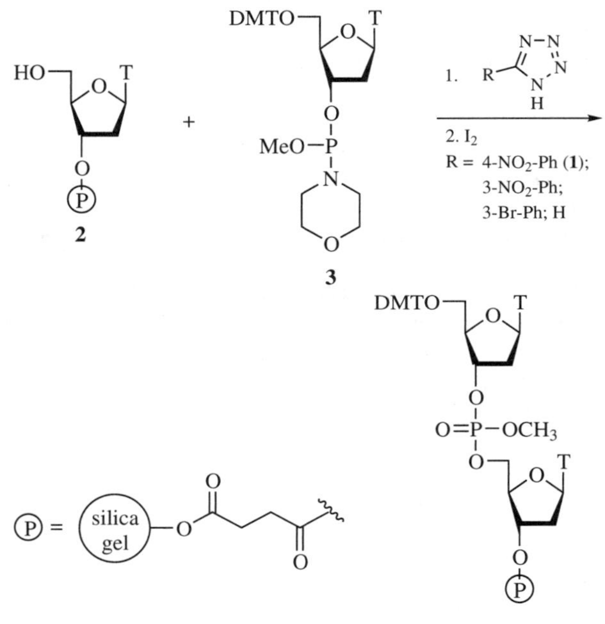

(1)

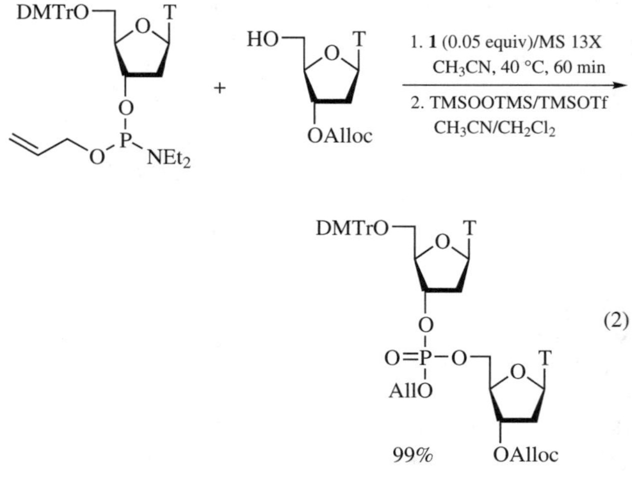

(2)

The enhanced activity of **1** compared to that of 1*H*-tetrazole is more pronounced in the synthesis of the 3′-thio analogs of deoxynucleotides.[7,8] Thus, **1** promotes the coupling of 5′-*O*-MMTr-thymidine-3′-*S*-thiophosphoramidite and 3′-*O*-acetylthymidine, yielding the corresponding dinucleotide in 75% yield after oxidation of the thiophosphite intermediate (eq 3).[7] In contrast, use of 1*H*-tetrazole as the activator yields no product. Oligodecadeoxyribonucleotides containing a 3′-thio analog of the thymidine 3′-phosphate such as d(TpTpTspTpT)[7] and d(GCACGTspTGCACG)[8] have been synthesized on solid supports using **1** as the coupling agent.

Synthesis of oligoribonucleotides by the phosphoramidite method can also be carried out using **1** as the activator. As shown in eq 4, the coupling of phosphoramidite (**4**) with the 5′-hydroxyl group of nucleoside **5** attached to functionalized controlled pore glass (CPG) takes only 15 min.[9] The coupling efficiency has been estimated to average about 93% for each cycle. Using this methodology, the 3′-terminal decaribonucleoside nonaphosphate and nonadecanucleoside octadecaphosphate sequences of yeast *r*-RNA^{Ala} can be readily synthesized. Similarly, the heptaribonucleotides (UACUAAC) and undecaribonucleotides (GUAUGUUAAUA) have been synthesized[10] on controlled pore glass beads using **1** as the activator for the nucleoside-phosphoramidite coupling.

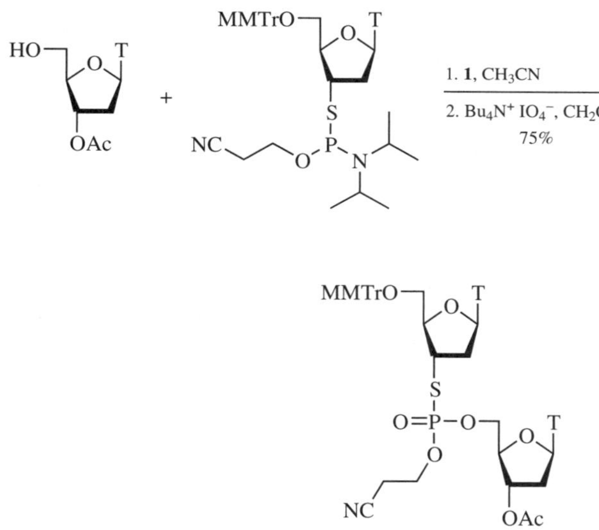

(3)

5′-*O*-protected ribonucleoside 2′,3′-cyclic phosphoramidites, such as **6**, can also be activated by **1** during the solid-phase synthesis of diribonucleotides UpA and ApA (eq 5).[11] Again, the more acidic tetrazole **1** activates the phosphoramidite more efficiently than other azoles.

It has been shown that when **1** is used as the phosphoramidite activator, the coupling efficiency could be further improved by the presence of a catalytic amount of DMAP. Thus, the homo-oligomer $(Up)_8U$ has been synthesized[12] on controlled pore glass support with an average yield of 96.4% for each coupling step using **1** as the activator in the presence of catalytic DMAP. The average coupling yield dropped to 91.9% when **1** was used alone as the activator.

Either 3′-thioribonucleoside 3′-*S*-phosphoramidites[13] or 2′-thioribonucleoside 2′-*S*-phosphoramidites such as **7** (eq 6)[14] can be used to install 3′-phosphothiolate or 2′-phosphothiolate linkages into oligonucleotides using **1** as the activator in the formation of the phosphite intermediates.

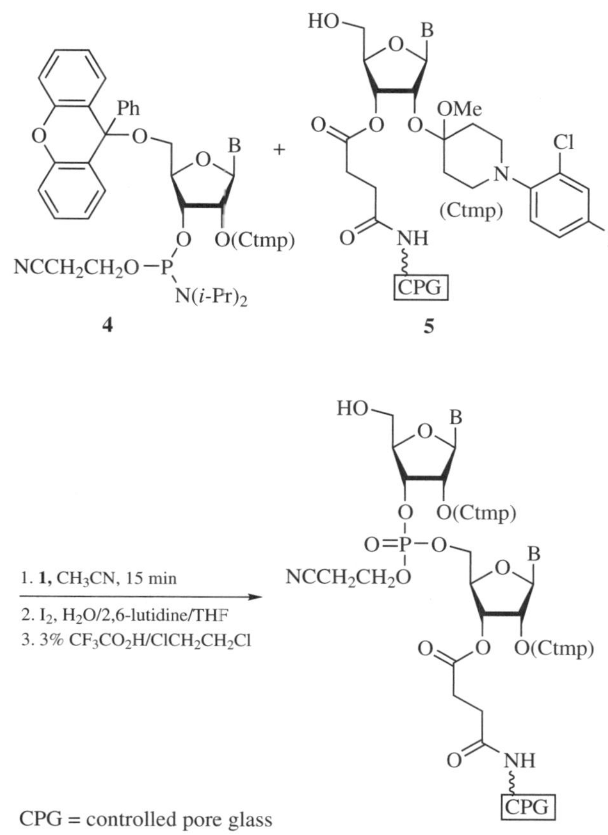

(4)

CPG = controlled pore glass

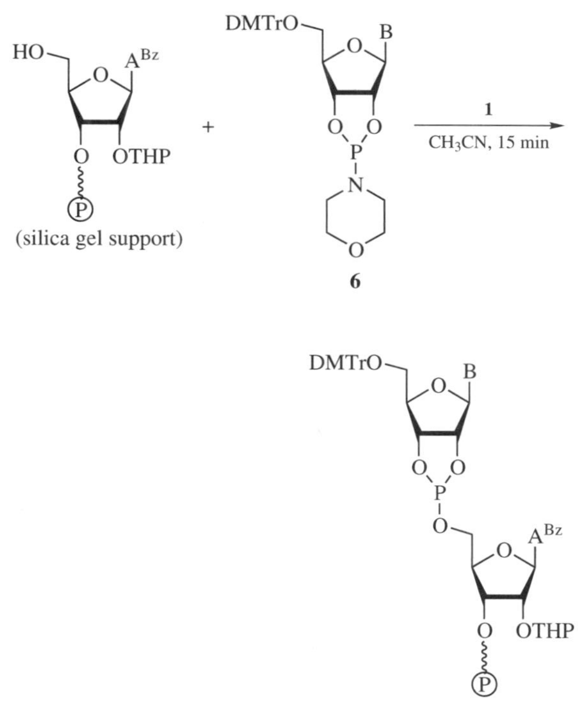

(5)

Selective phosphitylation of a primary hydroxyl group in unprotected carbohydrates and nucleosides can be achieved by the formation of the phosphite intermediates using the corresponding phosphoramidites and **1** as the activator (eq 7).[15] The same reaction performed with tetrazole as the activator is less selective.

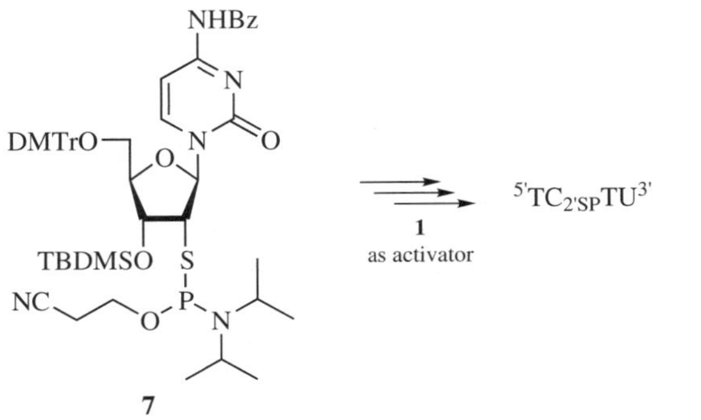

(6)

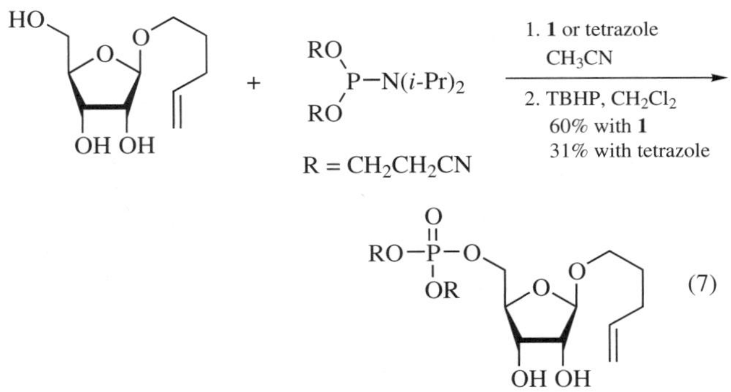

(7)

Formation of N-Acyl Phosphoramidate Linkage. The syntheses of new aminoacyl-adenylate analogs having an *N*-acyl phosphoramidate linkage can be achieved by condensing the amide functionality of an amino acid derivative such as **8** with a phosphoramidite in presence of **1** as the activator (eq 8).[16,17]

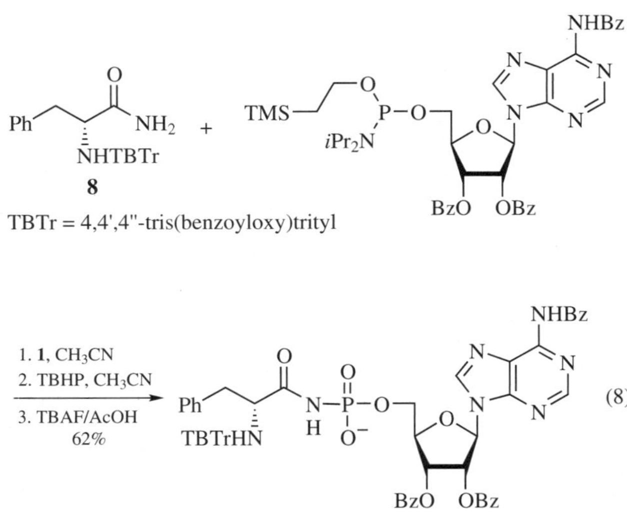

(8)

Syntheses of 2,5-Disubstituted 1,3,4-Oxadiazoles. 1,3,4-Oxadiazoles are synthesized by the reaction of tetrazole (**1**) with either an acid in the presence of DCC (eq 9)[18] or with an acid anhydride.[19]

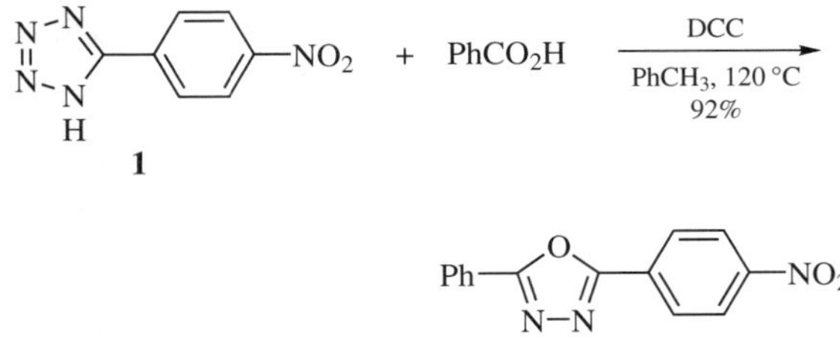

(9)

1. Finnegan, W. G.; Henry, R. A.; Lofquist, R., *J. Am. Chem. Soc.* **1958**, *80*, 3908–3911.
2. Demko, Z. P.; Sharpless, K. B., *J. Org. Chem.* **2001**, *66*, 7945–7950.
3. Froehler, B. C.; Matteucci, M. D., *Tetrahedron Lett.* **1983**, *24*, 3171–3174.
4. Montserrat, F. X.; Grandas, A.; Eritja, R.; Pedroso, E., *Tetrahedron* **1994**, *50*, 2617–2622.
5. Beaucage, S. L.; Caruthers, M. H., *Tetrahedron Lett.* **1981**, *22*, 1859–1862.
6. Hayakawa, Y.; Kataoka, M., *J. Am. Chem. Soc.* **1997**, *119*, 11758–11762.
7. Cosstick, R.; Vyle, J. S., *Tetrahedron Lett.* **1989**, *30*, 4693–4696.
8. Murata, T.; Iwai, S., Ohtsuka, E., *Heterocycles* **1992**, *33*, 529–531.
9. Rao, T. S.; Reese, C. B.; Serafinowska, H. T.; Takaku, H.; Zappia, G., *Tetrahedron Lett.* **1987**, *28*, 4897–4900.
10. Tanaka, T.; Tamatsukuri, S.; Ikehara, M., *Nucleic Acids Res.* **1986**, *14*, 6265–6279.
11. Ozaki, H.; Nakamoto, A.; Shimidzu, T., *Chem. Lett.* **1990**, *6*, 959–962.
12. Pon, R. T., *Tetrahedron Lett.* **1987**, *28*, 3643–3646.
13. Sun, S.; Yoshida, A.; Piccirilli, J. A., *RNA* **1997**, *3*, 1352–1363.
14. Hamm, M. L.; Piccirilli, J. A., *J. Org. Chem.* **1999**, *64*, 5700–5704.
15. Graham, S. M.; Pope, S. C., *Org. Lett.* **1999**, *1*, 733–736.
16. Moriguchi, T.; Yanagi, T.; Wada, T.; Sekine, M., *Tetrahedron Lett.* **1998**, *39*, 3725–3728.
17. Moriguchi, T.; Yanagi, T.; Kunimori, M.; Wada, T.; Sekine, M., *J. Org. Chem.* **2000**, *65*, 8229–8238.
18. Jursic, B. S.; Zdravkovski, Z., *Synth. Commun.* **1994**, *24*, 1575–1582.
19. Povazanec, F.; Kovac, J.; Svoboda, J., *Collect. Czech. Chem. Commun.* **1980**, *45*, 1299–1300.

Qingwei Yao
Northern Illinois University, DeKalb, IL, USA

Nitrosonium Tetrafluoroborate[1]

$NO^+BF_4^-$

[14635-75-7] BF_4NO (MW 116.82)

(efficient nitrosating and diazotizing agent;[1] mild oxidant;[2] hydride- and halide-abstracting agent[2])

Physical Data: *d* 2.185 g cm^{-3}.[3]
Solubility: soluble in acetonitrile.
Form Supplied in: colorless crystalline compound; commercially available.
Purification: commercial $NO^+BF_4^-$ frequently contains some $NO_2^+BF_4^-$; NO_2^+ can be removed from $NO^+BF_4^-$ by washing with dry benzene; sublimation (200–250 °C/0.01 mmHg) can be used if highly pure $NO^+BF_4^-$ is needed.[3]

Handling, Storage, and Precautions: due to its hygroscopic nature, $NO^+BF_4^-$ should be stored and used under anhyd conditions; use in a fume hood.

Original Commentary

George A. Olah, G. K. Surya Prakash,
Qi Wang & Xing-ya Li
University of Southern California, Los Angeles, CA, USA

Nitrosation. $NO^+BF_4^-$ is a highly efficient nitrosating agent, and has been used widely for synthetic purposes. Alcohols and secondary amines react readily with $NO^+BF_4^-$ to form alkyl nitrites and nitrosamines, respectively, in high yields (eqs 1 and 2).[4] Nitrosation of less reactive *N*-alkylamides and sulfonamides can also be achieved without difficulty, where other nitrosating agents are not effective.[5,6] Reaction of $NO^+BF_4^-$ with pyridine yields *N*-nitrosopyridinium tetrafluoroborate.[7] Aziridines react with $NO^+BF_4^-$ to yield nitrosoaziridines. Extrusion of N_2O from the products under thermal decomposition conditions led to formation of carbon–carbon double bonds.[8] This reaction has been employed for the synthesis of *cis*-diazabis-σ-homobenzene.[8a]

$$\text{ROH} + NO^+BF_4^- \longrightarrow \text{RONO} \qquad (1)$$

$$R^1R^2NH + NO^+BF_4^- \longrightarrow R^1R^2NNO \qquad (2)$$

R^1 = alkyl; R^2 = alkyl, acyl

Diazotization and Related Reactions. When primary amines react with $NO^+BF_4^-$, diazonium tetrafluoroborates are generated.[1] This reaction has been extensively studied with arylamines, since it is particularly useful for the isolation of the corresponding diazonium tetrafluoroborates (eq 3).[9] Decomposition of these salts produces fluoroarenes.[10] Isocyanates also react with $NO^+BF_4^-$ to yield diazonium salts.[2a,11] Primary amides can be converted to acids by $NO^+BF_4^-$ via diazotization and subsequent hydrolysis.[12]

$$ArNH_2 \xrightarrow{NO^+BF_4^-} ArN_2^+BF_4^- \longrightarrow ArF \qquad (3)$$

Treatment of alkyl azides with $NO^+BF_4^-$ led to the formation of carbenium ions, which can be intercepted with various nucleophiles such as fluoride, hydroxide, carboxides, and nitriles.[13] The reaction using nitriles has been utilized to prepare dihydroisoquinolines and oxazoles. Vinyl azides react with $NO^+BF_4^-$ to form 1,2,5-oxadiazoles and/or 1,2,4-oxadiazoles, depending on the structures of the substrates.[14] When 1,2-dialkylvinyl azides are used, 1,2,5-oxadiazoles were obtained as the only heterocyclic compounds in 75–80% yields.

Various substituted hydrazines react with $NO^+BF_4^-$ to form azides.[15]

Oxidation. $NO^+BF_4^-$ is a mild oxidant, with NO/NO^+ having 1.5 V standard oxidation–reduction potential in acetonitrile.[16] It has been widely used as a single electron transfer oxidant. Many arene- and sulfur-containing radical cations have been prepared by treatment of corresponding precursors with $NO^+BF_4^-$ (eq 4).[17,18] The easy removal of the side-product NO makes the procedure especially convenient. These radical cations have played a significant role in the development of organic ferromagnets and conductors.

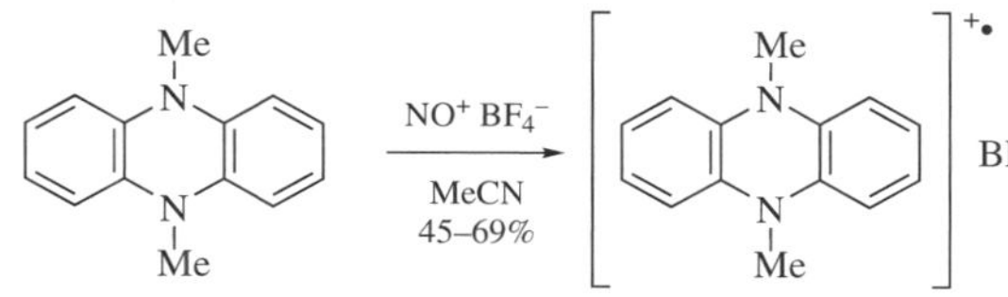

(4)

The oxidative property of $NO^+BF_4^-$ has also been synthetically utilized to regenerate carbonyl groups from masked derivatives such as 1,3-dithioacetals, oximes, and dimethylhydrazones.[19]

The same principle was applied to stereospecific construction of *O*-glycosidic linkages.[20] Treatment of readily available methyl *S*-glycosides with an equimolar amount of $NO^+BF_4^-$ in dry methylene chloride at temperatures between 0 and 25 °C gives high yields of *O*-glycosides in the presence of a hydroxylic component (eq 5). The stereospecificity and lack of complications compared with earlier methods make this procedure very attractive.

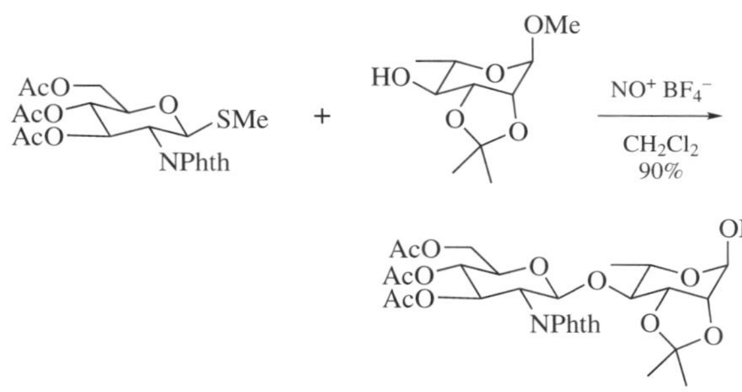

(5)

Noteworthy is the oxidative iodination of aromatic compounds with $NO^+BF_4^-$ as catalyst.[21] A wide range of arenes were iodinated with the system $I^-/O_2/NO^+BF_4^-$.[21b] This method is not as effective as I_2/Ag^+, but much more efficient than the use of other oxidative iodination systems such as I_2/Ce^{IV} salts and $I_2/(NH_4)_2S_2O_8$. The more important feature of this transformation is the suppression of side-chain oxidation of alkylarenes.[21c]

Electrophilic Addition to Unsaturated Carbon–Carbon Bonds. As an electrophilic agent, $NO^+BF_4^-$ is able to initiate electrophilic addition to unsaturated carbon–carbon bonds. Reaction of alkenes with $NO^+BF_4^-$ in the presence of acetonitrile yields imidazoles after reduction (eq 6).[22a] Changing acetonitrile to dimethyl sulfide led to formation of nitrososulfonium tetrafluoroborate (in dimer form).[22b] The dimer can be depolymerized, tautomerized, and hydrolyzed to give oxosulfonium salt, which can be utilized synthetically. $NO^+BF_4^-$ was reported to dimerize certain alkenes and alkynes.[22c,d] In the latter case, the true identity of the reagent that initiates the reaction (whether it is NO^+ or NO_2^+) is unclear. $NO_2^+BF_4^-$ is known to induce the same reaction.[22d]

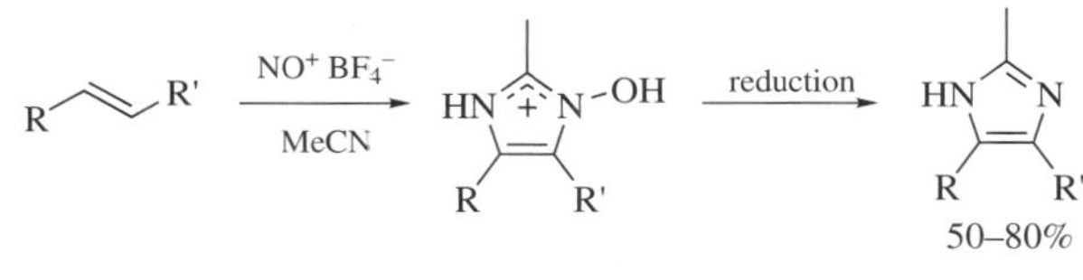

(6)

NO^+ can also insert into the cyclopropane ring, offering a new route to 2-isoxazolines.[23]

Hydride and Halide Abstraction. $NO^+BF_4^-$ is able to abstract a hydride from a tertiary C–H bond, generating a carbocation. The carbocationic intermediate can be trapped by a fluoride or alkyl nitriles, forming alkyl fluorides or *N*-alkylamides respectively.[24] Silanes also react with $NO^+BF_4^-$ to form fluorosilanes.[25]

In a similar fashion, NO^+ can also abstract a halogen atom from alkyl halides. In the presence of alkyl nitriles, *N*-alkylamides were obtained as a consequence of the Ritter reaction.[26]

Addition to Allenylsilanes. Allenylsilanes react with nitrosonium tetrafluoroborate in acetonitrile at −30 °C to form 4-trialkylsilylisoxazole derivatives in good yield.[27] This method provides a regiospecific [3 + 2] annulation route to unsymmetrically substituted isoxazoles. The silylisoxazole products can be further elaborated by electrophilic substitution in the same operation (eqs 7 and 8). In these [3 + 2] annulations, $NOBF_4$ is distinctly superior to other nitrosating agents such as Nitrosyl Chloride and Nitrosylsulfuric Acid.

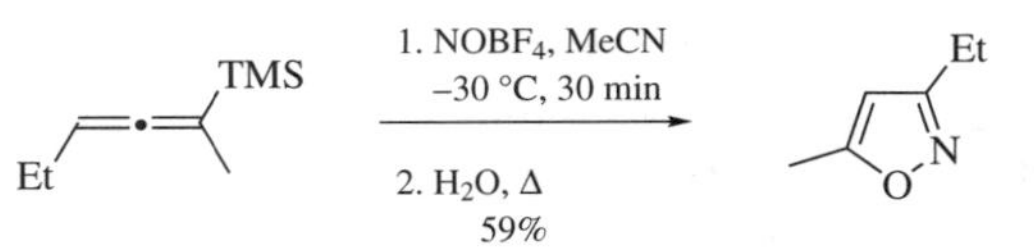

(7)

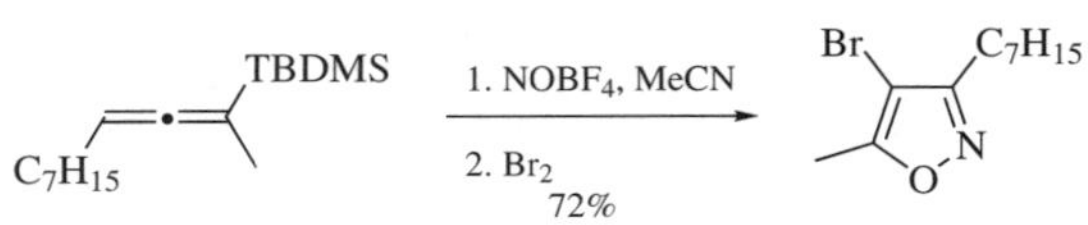

(8)

First Update

G. K. Surya Prakash & Jinbo Hu
University of Southern California, Los Angeles, CA, USA

Electrophilic Aromatic Nitrosations. $NO^+BF_4^-$ can be used for the direct electrophilic nitrosation of electron-rich aromatic compounds (such as anisoles and polymethylbenzenes) in good yields under mild conditions in which the conventional procedure (based on the nitrite with strong acid) is ineffective (eqs 9 and 10).[28] The reaction rates of nitrosation are much slower than those of aromatic nitration, which is ascribed to a rate-limiting deprotonation of the reversibly formed Wheland intermediate $[Ar^+(H)NO]$. Moreover, nitrosoaromatic products are better electron donors than the aromatic precursors from which they are derived.[28] The highly disparate rates of electrophilic aromatic nitrosation and nitration have also been explained by a charge-transfer mechanism.[29]

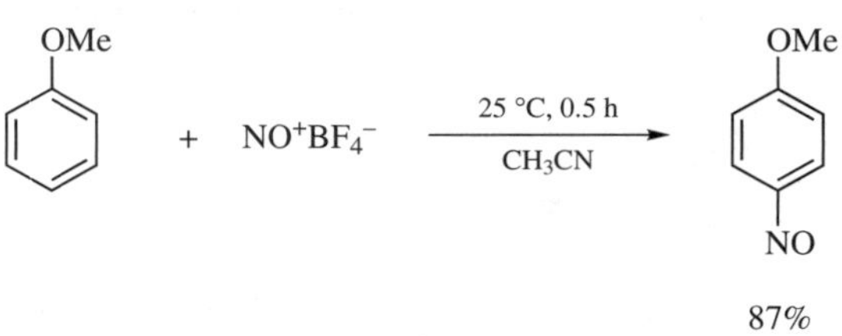

(9)

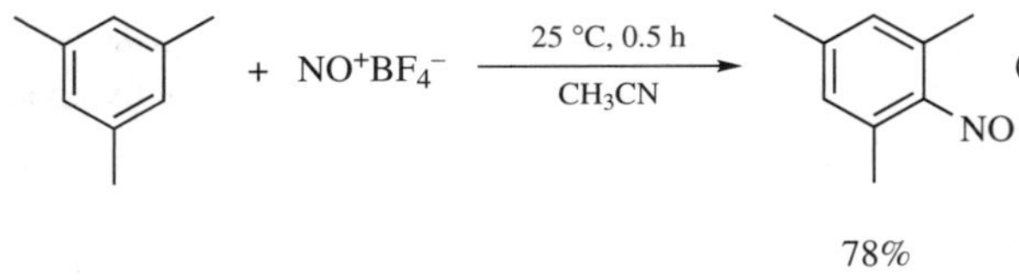

(10)

Selective Fluorinations with $NO^+BF_4^-$/Pyridinium Poly (Hydrogen Fluoride). $NO^+BF_4^-$/pyridinium poly(hydrogen fluoride) (PPHF with 60 wt % HF) is a useful nucleophilic fluorinating agent, which can be employed for the selective fluorination of ketoximes,[30] diarylacetylenes,[31] aryl sulfides, and aryl dithiolane derivatives[32] to give the corresponding fluorinated products under mild reaction conditions (eqs 11–14). The mechanism of these fluorinations is based on the electrophilic activation by nitrosonium ion (NO^+) and/or acid (H^+) on the substrates, followed by nucleophilic fluorination with fluoride ions (F^-).[30–32] The $NO^+BF_4^-$/PPHF method provides a convenient and mild alternative way to the conventional fluorination with SF_4, SeF_4, DAST, or elemental fluorine (F_2).

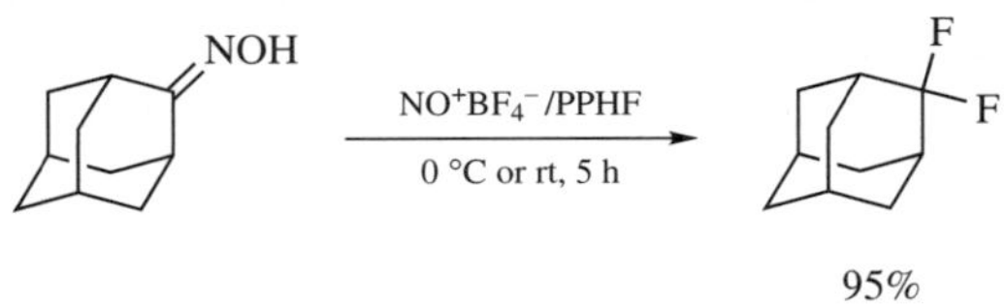

(11)

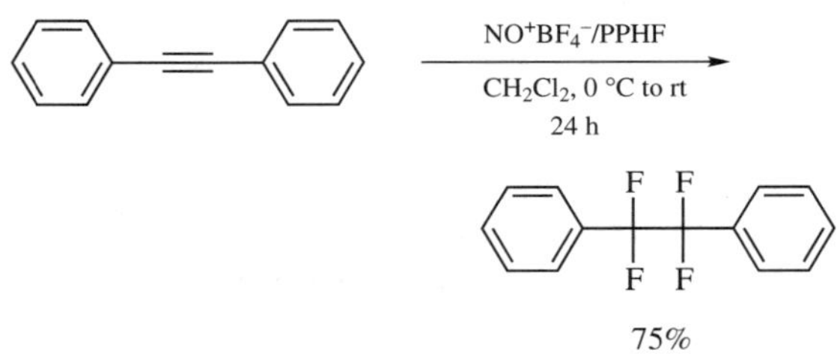

(12)

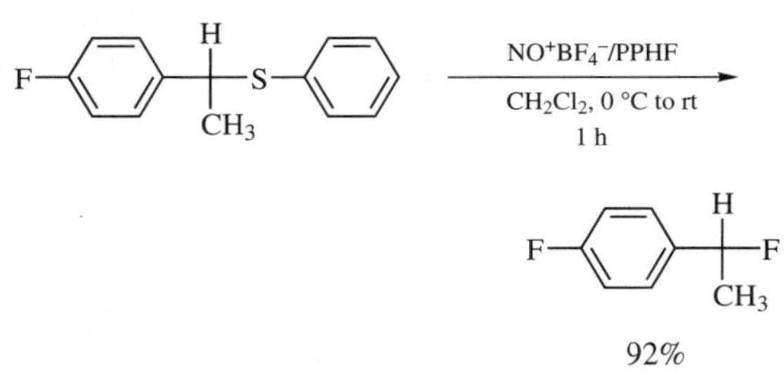

(13)

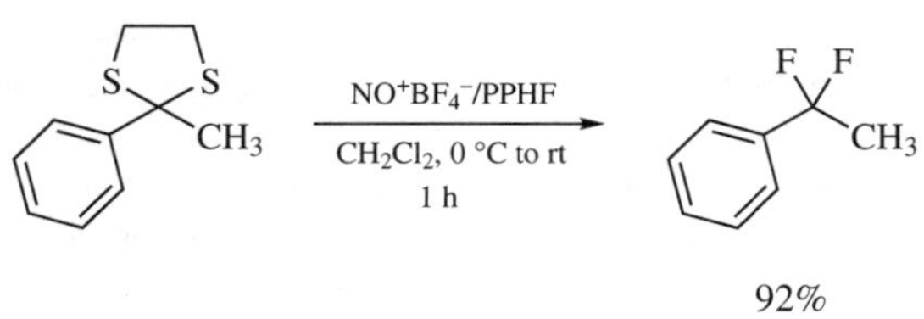

(14)

Oxidations of Alcohols and Related Compounds. Although many alcohols react readily with $NO^+BF_4^-$ to form alkyl nitrites,[4] benzylic alcohols can be oxidized by $NO^+BF_4^-$ under mild conditions to give satisfactory yields of aldehydes or ketones (eq 15).[33] This is ascribed to the benzylic hydride abstraction by $NO^+BF_4^-$

to generate a carbocation that in fact is a protonated carbonyl compound.[33] *exo*-Bicyclo[2.2.1]-heptan-2-ol can also be oxidized with $NO^+BF_4^-$ to give bicyclo[2.2.1]heptan-2-one.[34] Simple aliphatic alcohols do not undergo clean oxidation with $NO^+BF_4^-$, however, their *O*-tributylstannyl and *O*-trimethylsilyl derivatives can be readily oxidized by $NO^+BF_4^-$ to give corresponding carbonyl compounds in good yields (eq 16).[33]

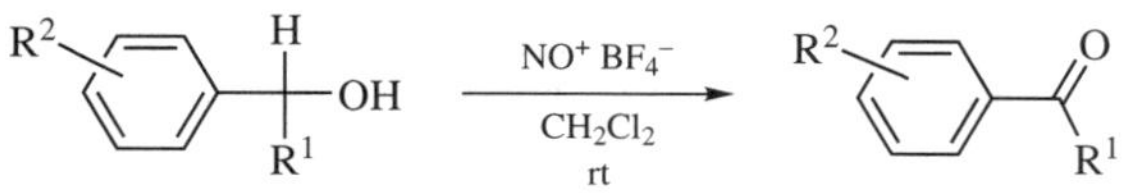
(15)

R^1 = H, alkyl, aryl
R^2 = H, F, alkyl

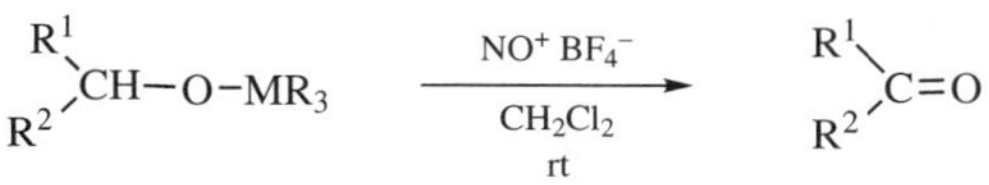
(16)

R^1, R^2 = H, alkyl, aryl, cyclic
MR_3 = $-Sn(n\text{-}C_4H_9)_3$, $-Si(CH_3)_3$

Additions to Allylic Esters. $NO^+BF_4^-$ can also undergo electrophilic addition to allylic esters such as allyl benzoate, allyl phenylacetate, allyl 3-phenylpropinoate, crotyl esters, 4-bromo-2-butenyl esters, and 1,4-bis(aroyloxy)-2-butenes in acetonitrile at – 23 °C to give the corresponding 2-isoxazolines (eq 17).[35]

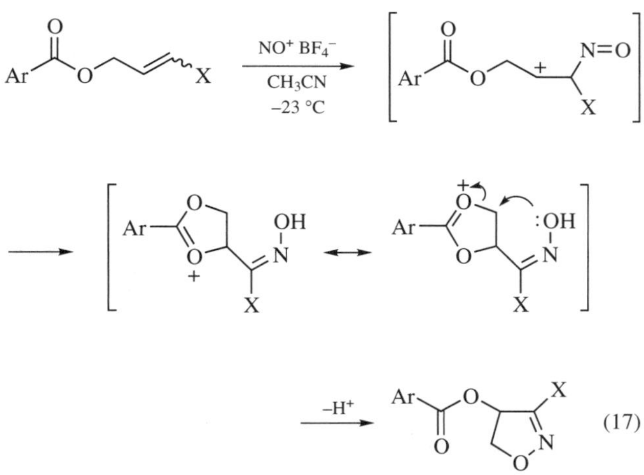
(17)

Ar = aryl; X = H, CH_3, CH_2O_2CAr

Disproportion Reactions of Disulfides. A catalytic amount (0.005 equiv) of $NO^+BF_4^-$ can promote the disproportion reactions of disulfides in acetonitrile at room temperature (eq 18).[36] A possible mechanism involves cleavage of the disulfide S-S bond following attack by NO^+ to afford the *S*-nitrosothiol and a sulfenium cation. The sulfenium cation can then attack another molecule of disulfide to give an unsymmetrical disulfide and a new sulfenium cation, which propagates the reaction and drives it to the statistically most probable 1:2:1 mixture of three disulfides (eq 18).[36]

$$\text{Et-S-S-Et} + \text{Bn-S-S-Bn} \xrightarrow[\text{CH}_3\text{CN, 1 h}]{\text{NO}^+\text{BF}_4^-\ (0.005\ \text{equiv})}$$

$$1/2\ \text{Et-S-S-Et} + \underset{48\%}{\text{Et-S-S-Bn}} + 1/2\ \text{Bn-S-S-Bn} \quad (18)$$

1. (a) Williams, D. L H., *Nitrosation*; Cambridge University Press: Cambridge, 1988.
2. (a) Olah, G. A., *Aldrichim. Acta* **1979**, *12*, 43. (b) Olah, G. A., *Acc. Chem. Res.* **1980**, *13*, 330.
3. Balz, G.; Mailander, E., *Z. Anorg. Allg. Chem.* **1934**, *217*, 161.
4. (a) Olah, G.; Noszko, L.; Kuhn, S.; Szelke, M., *Ber. Dtsch. Chem. Ges.* **1956**, *89*, 2374. (b) Nagasawa, H. T.; Fraser, P. S.; Yuzon, D. L., *J. Med. Chem.* **1973**, *16*, 583. (c) Weib, R.; Wagner, K. G.; Hertel, M., *Ber. Dtsch. Chem. Ges.* **1984**, *117*, 1965.
5. (a) Olah, G. A.; Olah, J. A., In *Friedel–Crafts and Related Reactions*; Olah, G. A., Ed.; Interscience: New York; Vol 3, p 1267. (b) Romea, P.; Urpi, F.; Vilarrasa, J., *J. Org. Chem.* **1989**, *54*, 3209.
6. Simpson, J. M., *Synthesis* **1979**, 100.
7. Olah, G. A.; Olah, J. A.; Overchuk, N. A., *J. Org. Chem.* **1965**, *30*, 3373.
8. (a) Prinzbach, H.; Breuninger, M.; Gallenkamp, B.; Schwesinger, R.; Hunkler, D., *Angew. Chem., Int. Ed. Engl.* **1975**, *14*, 348. (b) Prinzbach, H.; Schwesinger, R.; Breuninger, M.; Gallenkamp, B.; Hunkler, D., *Angew. Chem., Int. Ed. Engl.* **1975**, *14*, 347. (c) Schafer-Ridder, M.; Brocker, U.; Vogel, E., *Angew. Chem., Int. Ed. Engl.* **1976**, *15*, 228.
9. (a) Wannagat, U.; Hohlstein, G., *Ber. Dtsch. Chem. Ges.* **1955**, *88*, 1839. (b) Olah, G. A.; Tolgyesi, W. S., *J. Org. Chem.* **1961**, *26*, 2319.
10. (a) Olah, G. A.; Li, X. Y., In *Synthetic Fluorine Chemistry*; Olah, G. A.; Chambers, R. D.; Prakash, G. K. S., Eds.; Wiley: New York, 1992. (b) Milner, D. J., *Synth. Commun.* **1992**, *22*, 73. (c) Jovanovic, M. V., *Heterocycles* **1985**, *23*, 1969.
11. (a) Olah, G. A.; Friedman, N.; Bollinger, J. M.; Lukas, J., *J. Am. Chem. Soc.* **1966**, *88*, 5328. (b) Bott, K., *Angew. Chem., Int. Ed. Engl.* **1965**, *4*, 148. (c) Bott, K., *Ber. Dtsch. Chem. Ges.* **1975**, *108*, 402. (d) Bachman, G. B.; Michaelowicz, W., *J. Org. Chem.* **1958**, *23*, 1800.
12. Olah, G. A.; Olah, J. A., *J. Org. Chem.* **1965**, *30*, 2386.
13. (a) Doyle, M. P.; Wierenga, W., *J. Am. Chem. Soc.* **1972**, *94*, 3896. (b) *ibid.* **1972**, *94*, 3901. (c) McGirk, R. H.; White, E. H., *J. Am. Chem. Soc.* **1973**, *95*, 3804. (d) Doyle, M. P.; Spoelhof, G. D.; Zaleta, M. A., *J. Heterocycl. Chem.* **1975**, *12*, 263. (e) Doyle, M. P.; Hedstrand, D. M.; Busman, S. C.; Alexander, D., *J. Am. Chem. Soc.* **1975**, *97*, 5554. (f) Doyle, M. P.; Whitefleet, J. L.; Zaleta, M. A., *Tetrahedron Lett.* **1975**, 4201. (g) Owen, G. R.; Verheyden, J. P. H.; Moffatt, J. G., *J. Org. Chem.* **1976**, *41*, 3010. (h) Doyle, M. P.; Whitefleet, J. L.; Bosch, R., *J. Org. Chem.* **1979**, *44*, 2923.
14. Thakore, A. N.; Buchshriber, J.; Oehlschlager, A. C., *Can. J. Chem.* **1973**, *51*, 2406.
15. Pozsgay, V.; Jennings, H. J., *Tetrahedron Lett.* **1987**, *28*, 5091.
16. Morkovnik, A. S.; Dobaeva, N. M.; Okhlobystin, O. Y., *Zh. Obshch. Khim.* **1989**, *59*, 921.
17. (a) Bandlish, B. K.; Shine, H. J., *J. Org. Chem.* **1977**, *42*, 561. (b) Inozemtsev, A. N.; Koshechko, V. G.; Pokhodenko, V. D., *Zh. Obshch. Khim.* **1984**, *54*, 1160. (c) Koshechko, V. G.; Inozemtsev, A. N.; Pokhodenko, V. D., *Zh. Org. Khim.* **1983**, *19*, 751.
18. (a) Musker, W. K.; Roush, P. B., *J. Am. Chem. Soc.* **1976**, *98*, 6745. (b) Doyle, M. P.; Hedstrand, D. M., *Chem. Commun.* **1977**, 643. (c) Musker, W. K.; Hirschon, A. S.; Doi, J. T., *J. Am. Chem. Soc.* **1978**, *100*, 7754. (d) Bock, H.; Brahler, G.; Henkel, U.; Schlecker, R.; Seebach, D., *Ber. Dtsch. Chem. Ges.* **1980**, *113*, 289. (e) Blankespoor, R. L.; Doyle, M. P.; Hedstrand, D, M.; Tamblyn, W. H.; Van Dyke, D. A., *J. Am. Chem. Soc.* **1981**, *103*, 7096. (f) Blankespoor, R. L.; Doyle, M. P.;

Smith, D. J.; Van Dyke, D. A.; Waldyke, M. J., *J. Org. Chem.* **1983**, *48*, 1176. (g) Boduszek, B.; Shine, H. J., *J. Org. Chem.* **1988**, *53*, 5142. (h) Gompper, R.; Polborn, K.; Strassberger, W., *Chem. Commun.* **1989**, 1346. (i) Tsubouchi, A.; Matsumura, N.; Inoue, H., *Chem. Commun.* **1991**, 520.

19. (a) Olah, G. A.; Ho, T. L., *Synthesis* **1976**, 610. (b) Olah, G. A.; Narang, S. C.; Salem, G. F.; Gupta, B. G. B., *Synthesis* **1979**, 273.
20. (a) Jennings, H. J.; Pozsgay, V., *J. Org. Chem.* **1987**, *52*, 4635. (b) Jennings, H. J.; Pozsgay, V., *J. Carbohydr. Chem.* **1990**, *9*, 333. (c) Liptak, A.; Szabo, L., *Carbohydr. Res.* **1988**, *185*, c5.
21. (a) Makhon'kov, D. I.; Cheprakov, A. V.; Rodkin, M. A.; Beletskaya, I. P., *Zh. Org. Khim.* **1988**, *24*, 241. (b) Radner, F., *J. Org. Chem.* **1988**, *53*, 3548. (c) Galli, C., *J. Org. Chem.* **1987**, *56*, 3238.
22. (a) Scheinbaum, M. L.; Dines, M. B., *Tetrahedron Lett.* **1971**, 2205. (b) Chow, Y. L.; Iwai, K., *J. Chem. Soc. Perkin Trans. 2* **1980**, 931. (c) Lee, G. H.; Lee, J. M. Jeong, W. B.; Kim, K., *Tetrahedron Lett.* **1988**, *29*, 4437. (d) Brittelli, D. R.; Boswell, G. A., Jr., *J. Org. Chem.* **1981**, *46*, 312.
23. (a) Mizuno, K.; Ichinose, N.; Tamai, T.; Otsuji, Y., *J. Org. Chem.* **1992**, *57*, 4669. (b) Ichinose, N.; Mizuno, K.; Otsuji, Y., *Chem. Lett.* **1989**, 457.
24. (a) Olah, G. A.; Shih, J. G.; Singh, B. P.; Gupta, B. G. B., *J. Org. Chem.* **1983**, *48*, 3356. (b) Olah, G. A.; Gupta, B. G. B., *J. Org. Chem.* **1980**, *45*, 3532.
25. Prakash, G. K. S.; Wang, Q.; Li, X. Y.; Olah, G. A., *Nouv. J. Chim.* **1990**, *14*, 791.
26. Olah, G. A.; Gupta, B. G. B; Narang, S. C., *Synthesis* **1979**, 274.
27. Danheiser, R. L.; Becker, D. A., *Heterocycles* **1987**, *25*, 277.
28. Bosch, E.; Kochi, J. K., *J. Org. Chem.* **1994**, *59*, 5573.
29. Gwaltney, S. R.; Rosokha, S. V.; Head-Gordon, M.; Kochi, J. K., *J. Am. Chem. Soc.* **2002**, *125*, 3273.
30. York, C.; Prakash, G. K. S., Wang, Q.; Olah, G. A., *Synlett* **1994**, 425.
31. York, C.; Prakash, G. K. S.; Olah, G. A., *J. Org. Chem.* **1994**, *59*, 6493.
32. York, C.; Prakash, G. K. S.; Olah, G. A., *Tetrahedron* **1996**, *52*, 9.
33. Olah, G. A.; Ho, T.-L., *Synthesis* **1976**, 609.
34. Olah, G. A.; Ramaiah, P., *J. Org. Chem.* **1993**, *58*, 4639.
35. Chang, R.; Kim, K., *Tetrahedron Lett.* **1999**, *40*, 6773.
36. Itoh, T.; Tsutsumi, N.; Ohsawa, A., *Bioorg. Med. Chem. Lett.* **1999**, *9*, 2161.

2-(5-Norbornene-2,3-dicarboxamido)-1,1,3,3-Tetramethyluronium Tetrafluoroborate

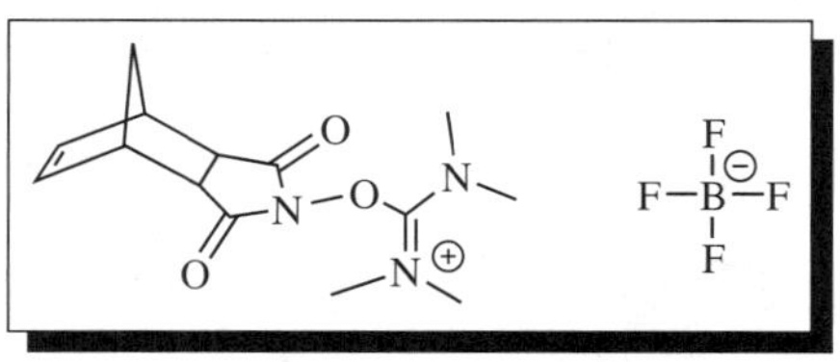

[125700-73-4] $C_{14}H_{20}BF_4N_3O_3$ (MW 365.13)

(reagent used in low racemization peptide coupling reactions)

Alternate Name: *O*-(5-norbornene-2,3-dicarboximido)-*N*,*N*,*N*′,*N*′-tetramethyluronium tetrafluoroborate, TNTU.
Physical Data: mp 214–216 °C.
Solubility: soluble in acetonitrile, water, dichloromethane, and methanol.
Form Supplied in: white powder; widely available.
Handling, Storage, and Precautions: the toxicological properties of this substance have not yet been yet fully investigated. It may be harmful if inhaled, ingested, or absorbed by the skin. Incompatible with strong oxidizing agents, and sensitive to moisture and light.

General. 2-(5-Norbornene-2,3-dicarboximido)-1,1,3,3-tetramethyluronium tetrafluoroborate (TNTU) belongs to a class of uronium coupling reagents which has become very popular in solid-phase peptide couplings[1–3] and oligonucleotide[4,5] synthesis. Their popularity over more traditional coupling reagents is due in part to their increased stability, non-hygroscopic character, and ease of handling. Moreover, couplings proceed in high yield and with low racemization.

Peptide Couplings. The coupling of the dipeptide Z-glycine-phenylalanine with methyl valine (eq 1) with TNTU as coupling reagent proceeded smoothly, but 28% of the D-isomer was identified by RP-HPLC.[6] Addition of 1-hydroxybenzotriazole (HOBT) to the reaction mixture reduced the amount of racemization to the extent that only 0.5% of the D-isomer was formed.

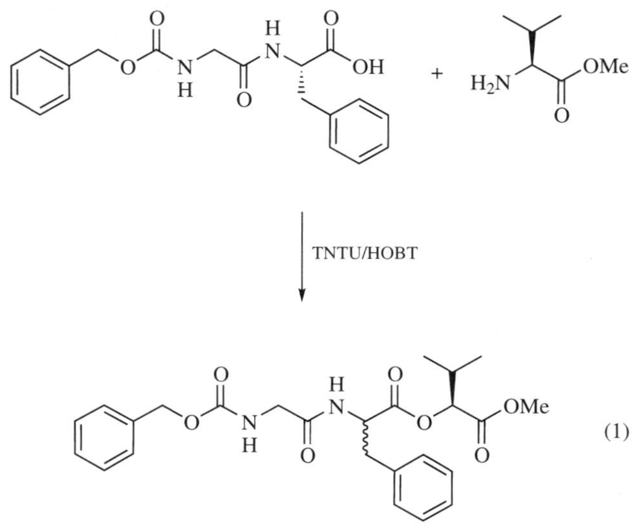

(1)

Oligonucleotide Synthesis. TNTU has been investigated for its use as a coupling reagent in esterification reactions for the preparation of immobilized oligonucleotides in solid-phase automatic synthesis.[7] Automated couplings of nucleoside 3′-*O*-succinates to a hydroxyl-derivatized solid-phase support (OH-CPG), utilizing the primary ester linkage (eq 2), were performed using a series of uronium coupling reagents. TNTU gave a nucleoside loading yield of 2 μmol/g which, however, is 15 times lower than that obtained with the related 7-azabenzotriazolyl derivative (HATU) in a direct comparison experiment.

Esterification. The use of TNTU, along with that of various related uranium salts, has been investigated for the esterifcation of Fmoc-protected amino acids. The reagent, however, was not as effective for this purpose as the *N*-carboxyanhyrides or bromotrispyrrolidinophosphonium hexafluorophosphate (pyBroP).[8]

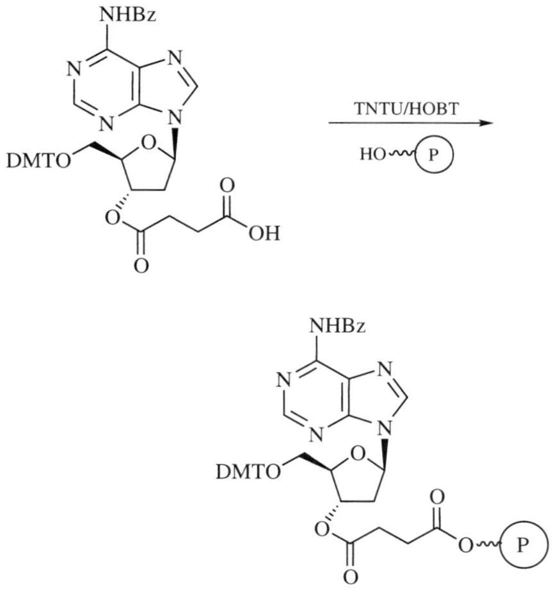

(2)

Related Reagents. 2-(1*H*-Benzotriazol-1-yl)-1,1,3,3-tetramethyluronium hexafluorophosphate (HBTU); 2-(7-azabenzotriazol-1-yl)-1,1,3,3-tetramethyluronium hexafluorophosphate (HATU); 2-(7-azabenzotriazol-1-yl)-1,1,3,3-tetramethyluronium tetrafluoroborate (TATU); 2-(1*H*-benzotriazol-1-yl)-1,1,3,3-bis(pentamethylene)uronium hexafluorophosphate (HBPipU); 2-(1*H*-benzotriazol-1-yl)-1,1,3,3-bis(tetramethylene)uronium hexafluorophosphate (HBPyU).

1. Albericio, F.; Bofill, J. M.; El-Faham, A.; Kates, S. A., *J. Org. Chem.* **1998**, *63*, 9678.
2. Fields, C. G.; Lloyd, D. H.; Macdonald, R. L.; Oteson, K. M.; Noble, R. L., *Peptide Res.* **1991**, *4*, 95.
3. Pon, R. T.; Yu, S., *Tetrahedron Lett.* **1997**, *38*, 3331.
4. Bergmann, F.; Bannwarth, W., *Tetrahedron Lett.* **1995**, *36*, 1839.
5. Finn, P. J.; Gibson, N. J.; Fallon, R.; Hamilton, A.; Brown, T., *Nucleic Acids Res.* **1996**, *24*, 3357.
6. Knorr, R.; Trzeciak, A.; Bannwarth, W.; Gillessen, D., *Tetrahedron Lett.* **1989**, *30*, 1927.
7. Pon, R. T.; Yu, S.; Sanghvi, Y. S., *Bioconjugate Chem.* **1999**, *10*, 1051.
8. Davies, J. S.; Hewe, J ; Le Breton, M., *J. Chem. Soc., Perkin Trans. 2* **1995**, 2335.

Duncan Batty
Celltech R&D Ltd, Cambridge, UK

O

Octenediol

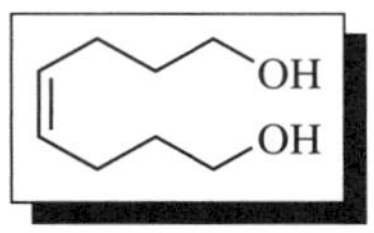

[56506-08-2]
[62422-45-1] $C_8H_{16}O_2$ (MW 144.2114)

(reagent used as cleavable linker for solid-phase synthesis of oligosaccharides and as a polyfunctional source of (*Z*)-olefin in the synthesis of natural products)

Alternate Name: (*Z*)-4-octene-1,8-diol.
Physical Data: clear oil, bp 78 °C (at 0.001 Torr).
Solubility: soluble in most organic solvents.
Form Supplied in: clear oil, not commercially available.
Analysis of Reagent Purity: ^{1}H NMR, bp, IR, elemental analysis, and mass spectrometry.
Preparative Methods: preparation of octenediol involves selective oxidative cleavage of 1,5-cyclooctadiene (commercially available) followed by reduction to diol (eq 1). *m*-CPBA epoxidation followed by oxidation with periodic acid to give 1,8-octenedial and subsequent reduction by sodium borohydride to yield the diol.[1,2] Other methods include selective ozonolysis to the dialdehyde followed by sodium borohydride reduction.[3–5]

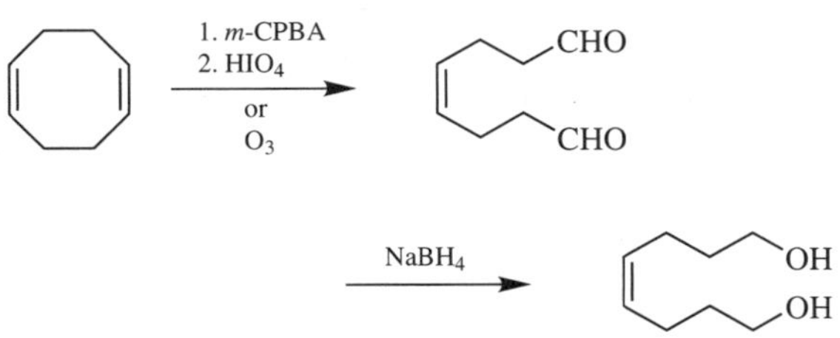

(1)

Purification: distillation (90–95 °C, 0.02 Torr) or silica gel flash chromatography.
Handling, Storage, and Precautions: store neat, refrigerate, inert atmosphere.
Disposal: dispose of reagent through licensed professional waste disposal service.

Octenediol as Cleavable Linker for Solid-phase Synthesis of Oligosaccharides. Octenediol has been shown to be a versatile linker for coupling carbohydrates to a polymeric solid-support for oligosaccharide synthesis.[6,7] Quantifiable functionalization of a solid-support by octenediol may be achieved through the mono-dimethoxytritylation of octenediol followed by nucleophilic addition to an appropriate polymeric support (e.g., Merrifield's resin, chloromethylated polystyrene cross-linked with 1% divinylbenzene). Loading may be determined by colorimetric assay of the removal of the dimethoxytrityl group with dichloroacetic acid (eq 2).[8]

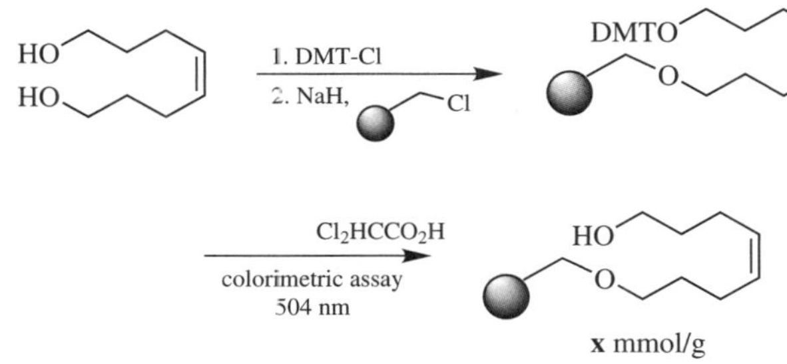

(2)

The free hydroxyl group of support-bound octenediol serves as a nucleophilic acceptor for a range of glycosyl donors, including glycosyl trichloroacetimidates and phosphates (eq 3). Octenediol linker based solid-phase oligosaccharide synthesis (SOS), including automated SOS, has demonstrated its utility in the synthesis of a wide array of complex type oligosaccharides.[6,7,9,10]

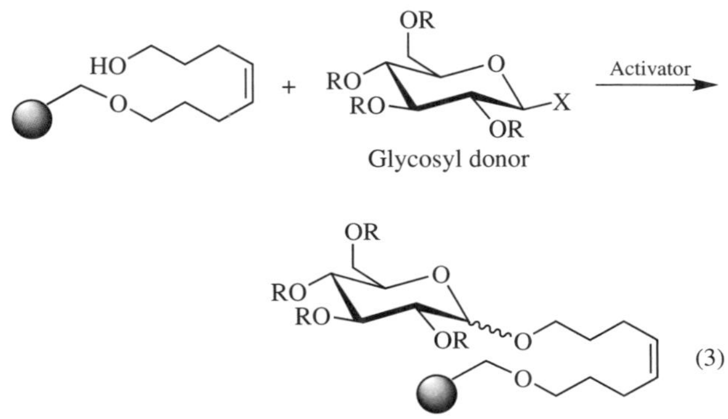

(3)

Cleavage of linker bound oligosaccharides may be achieved through several routes, depending on the demands of a particular synthesis (eq 4). Olefin cross-metathesis of the bound

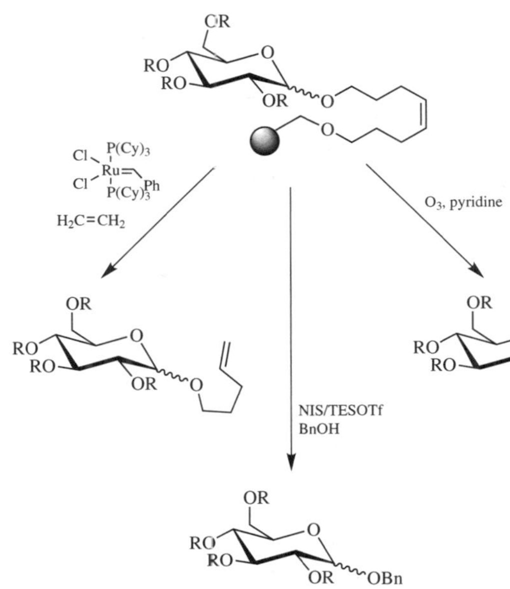

(4)

carbohydrate with Grubbs' catalyst under an atmosphere of ethylene furnishes the free pentenyl glycoside. In addition to serving as a glycosyl donor, the pentenyl glycoside provides a reactive handle through the terminal olefin for further functionalization.[11,12] Cleavage of oligosaccharide from the support can also be achieved by treatment with *N*-iodosuccinimide (NIS) and TESOTf in the presence of benzyl alcohol to provide the benzyl glycoside. Additionally, ozonolysis furnishes the aldehyde functionalized glycoside, suitable for the formation of neoglycoconjugates.

The internal olefin of the octenediol linker must be masked for transformations on the solid-support that require strongly electrophilic conditions, such as the *N*-iodosuccinimide/TMSOTf activation of a thioethyl glycosyl donor. This can be achieved through dibromination. Reductive debromination restores the double bond, permitting cleavage from the solid-support (eq 5).[13]

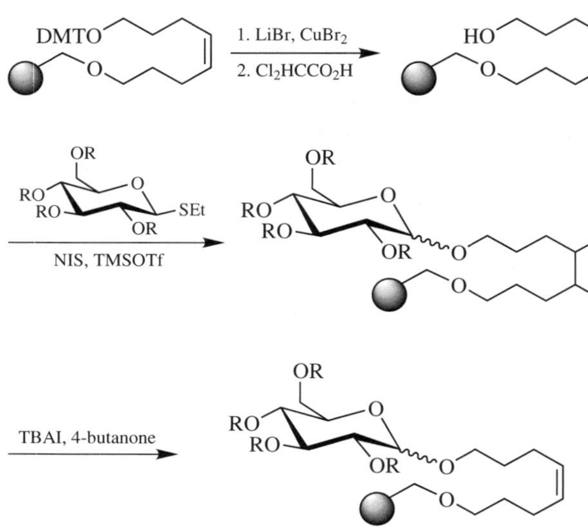

(5)

Octenediol as Polyfunctional Carbon Skeleton for Natural Product Synthesis. Access to polyfunctional reagents containing an internal olefin with the required *E*/*Z* configuration can be valuable in the synthesis of natural products. The utility of pheromones for pest control precipitated interest in the stereospecific synthesis of sex attractants for a number of insect species. Early syntheses of the sex pheromone for the gypsy moth (*Porthetria dispar*), (*Z*)-7,8-epoxy-2-methyl-octadecane (disparlure), were based on the fixed double bond (*Z*)-geometry of (*Z*)-4-octene-1,8-diol. Stereochemical control is afforded through the partial ozonolysis and reduction of readily available (*Z*,*Z*)-1,5-cyclooctadiene. Subsequent conversion of the diol to the dibromide, followed by sequential alkyl Grignard additions, and epoxidation with monoperoxyphthalic acid or *m*-CPBA afforded the racemic epoxide product (eq 6).[3–5]

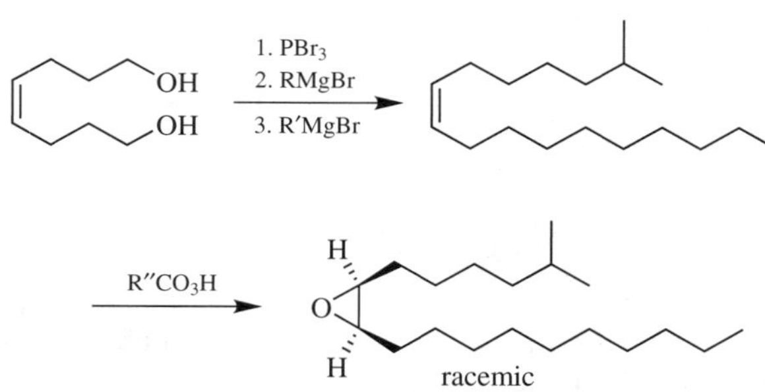

(6)

Octenediol has also been accessed through the partial epoxidation of 1,5-cyclooctadiene, followed by oxidative cleavage of the epoxide to the dialdehyde with periodic acid. Reduction of the dialdehyde by sodium borohydride gives free (*Z*)-4-octene-1,8-diol. The stereocontrol afforded by octenediol was used in the study of the enantiomers of the marine lipid demospongic acid, i.e., 22-methyl-5,9-octacosadienoic acid (eq 7).[2]

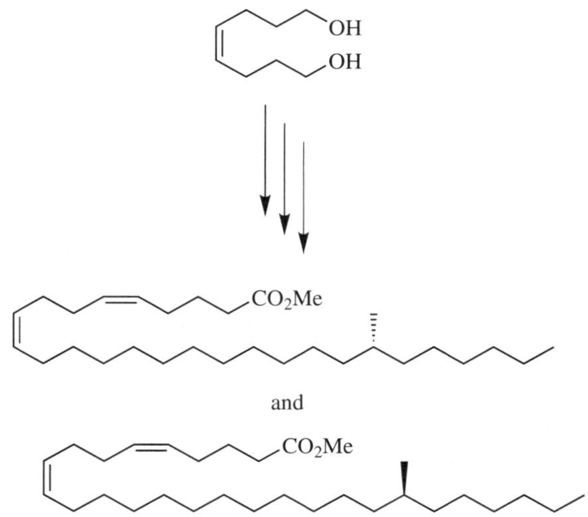

(7)

Related Reagents. 1,5-Cyclooctadiene.

1. Nagarkatti, J. P.; Ashley, K. R., *Tetrahedron Lett.* **1973**, *14*, 4599.
2. Raederstorff, D.; Shu, A. Y. L.; Thompson, J. E.; Djerassi, C., *J. Org. Chem.* **1987**, *52*, 2337.
3. Tolstikov, G. A.; Odinokov, V. N.; Galeeva, R. I.; Bakeeva, R. S., *Tetrahedron Lett.* **1978**, *19*, 1857.
4. Tolstikov, G. A.; Odinokov, V. N.; Galeeva, R. I.; Bakeeva, R. S.; Rafikov, S. R., *Dok. Chem. (Engl. Transl.)* **1978**, *239*, 174.
5. Odinokov, V. N.; Galeeva, R. I.; Bakeeva, R. S.; Grishova, R. S.; Tolstikov, G. A., *J. Org. Chem USSR (Engl. Transl.)* **1979**, *15*, 1252.
6. Andrade, R. B.; Plante, O. J.; Melean, L. G.; Seeberger, P. H., *Org. Lett.* **1999**, *1*, 1811.
7. Plante, O. J.; Palmacci, E. R.; Seeberger, P. H., *Science* **2001**, *291*, 1523.
8. Pon, R. T.; Agrawal, S. (Eds.); *Methods in Molecular Biology 20: Protocols for Oligonucleotides and Analogues*; Humana Press: Totowa, 1993, p. 467.
9. Hewitt, M. C.; Seeberger, P. H., *J. Org. Chem.* **2001**, *66*, 4233.
10. Hewitt, M. C.; Seeberger, P. H., *Org. Lett.* **2001**, *3*, 3699.
11. Fraser-Reid, B.; Udodong, U. E.; Wu, Z.; Ottosson, H.; Merritt, J. R.; Rao, C. S.; Roberts, C.; Madsen R., *Synlett* **1992**, 927.
12. Buskas, T.; Soderberg, E.; Konradsson, P.; Fraser-Reid, B., *J. Org. Chem.* **2000**, *65*, 958.
13. Melean, L. G.; Haase, W.-C.; Seeberger, P. H., *Tetrahedron Lett.* **2000**, *41*, 4329.

Daniel M. Ratner & Peter H. Seeberger
Massachusetts Institute of Technology, Cambridge, MA, USA

S-(1-Oxido-2-pyridinyl)-1,1,3,3-tetramethylthiouronium Hexafluorophosphate (HOTT)

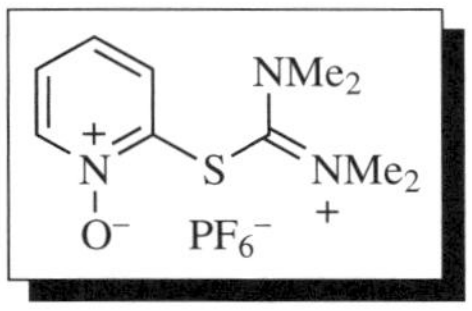

[212333-72-7] $C_{10}H_{16}N_3OSPF_6$ (MW 371.28)

(reagent used to convert carboxylic acids into Barton esters and for peptide coupling/amidation)

Physical Data: mp 115–116 °C.
Solubility: acetonitrile, CH_2Cl_2, THF (moderate solubility).
Form Supplied in: white solid.
Purification: recrystallized from CH_2Cl_2.
Preparative Methods: the title reagent can be prepared by slowly adding Et_3N to a dry CH_2Cl_2 solution of 2-mercaptopyridine-*N*-oxide and tetramethylchloroformamidinium hexafluorophosphate.[1] After removal of CH_2Cl_2 from the reaction mixture, the resulting solid mass is pulverized, washed with $CHCl_3$, and then filtered to give a white solid of sufficient purity to be used in subsequent reactions.[2]
Handling, Storage, and Precautions: it is advised to protect the solid from prolonged exposure to light since 2-mercaptopyridine *N*-oxide and related compounds can be light-sensitive.

Barton Esterification: Reductive Decarboxylation. *O*-Acyl thiohydroxamates or Barton esters are useful precursors of carbon-centered radicals via thermolysis or photolysis.[3] Several different methods are available for converting carboxylic acids into Barton esters (eq 1).[4] These reactions generally proceed via the attack of a 2-mercaptopyridine-*N*-oxide salt on an activated carboxylic acid that has either been preformed (acid chloride, mixed anhydride) or generated in situ (with 1,3-dicyclohexylcarbodiimide or tri-*n*-butylphosphine + 2,2′-dithiodipyridine-1,1′-dioxide). However, HOTT has the distinct advantages of (1) being easy to prepare and handle without the need for any special precautions, (2) facilitates efficient Barton esterification of carboxylic acids, and (3) simplifies subsequent work-up and purifications by avoiding the need to remove by-products like 1,3-dicyclohexylurea.

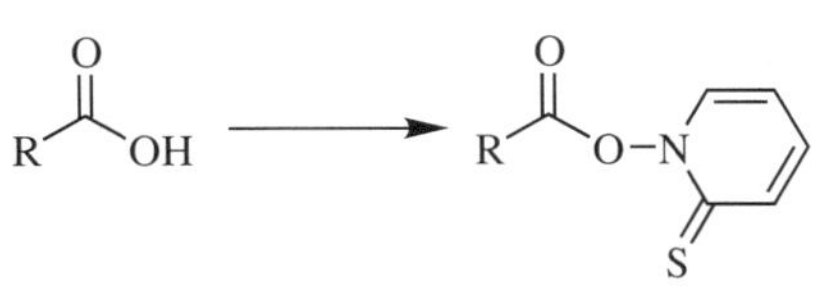

(1)

The HOTT reagent has been shown to significantly improve the yields of reductive decarboxylations of 2,3:4,6-di-*O*-isopropylidene-2-keto-L-gulonic acid (eq 2) and peracetylated *N*-acetylneuraminic acid (eq 3).[2] In both cases, the yield of reduced product nearly doubled when HOTT was used to esterify these hindered carboxylic acids.

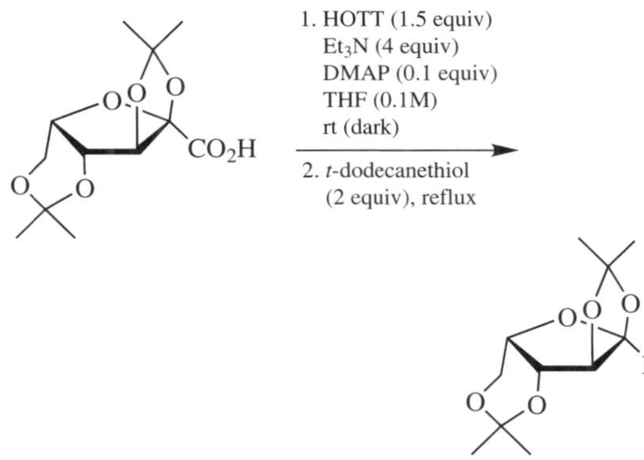

78% yield
(42% via acid chloride) (2)

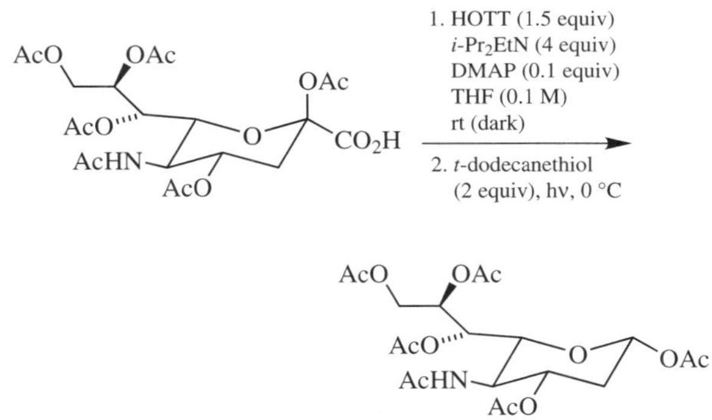

(3)

50% yield
(27% via carbodiimide)

Barton Esterification: Oxidative Decarboxylation. HOTT-mediated Barton esterification was coupled to oxidative decarboxylation in a synthesis of the sesquiterpene (+)-culmorin (eq 4).[5] Use of the HOTT reagent was clearly superior with this hindered substrate when compared with the acid chloride method.

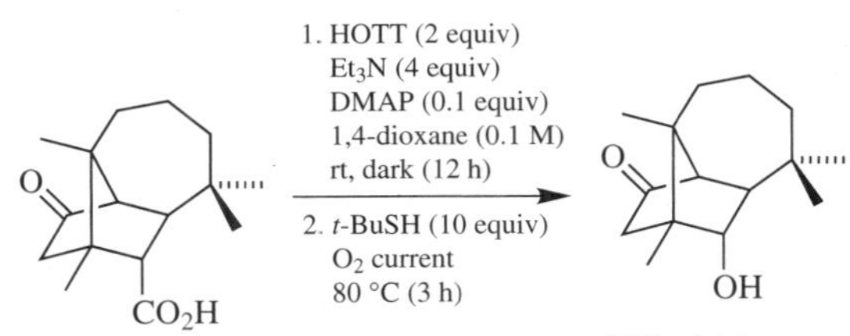

82% yield
(48% via acid chloride) (4)

Barton Esterification: Radical Addition. One of the best examples illustrating the benefits of HOTT for this transformation is shown in eq 5. The Barton esterification of this very hindered acid was followed by IR spectroscopy by monitoring the disappearance of the carbonyl stretch of the acid (1740 cm^{-1}) and the appearance of the carbonyl stretch of the Barton ester (1810 cm^{-1}). Barton esterification using HOTT was complete within 20 min, whereas over 4 h was required when using the combination of DCC and 2-mercaptopyridine-*N*-oxide.

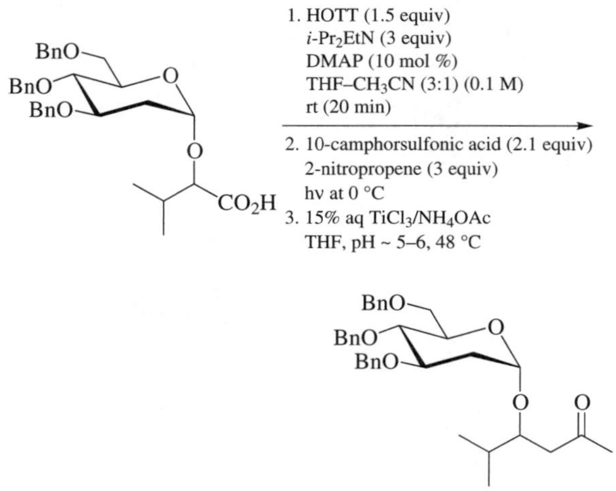

(5)

73% yield
(47% via carbodiimide)

HOTT was used to effect Barton esterification in a novel approach to 1, 3, 5, ... (2*n*+1) polyols based on iterative stereocontrolled homologation of chiral hydroxyalkyl radicals (eq 6).[6]

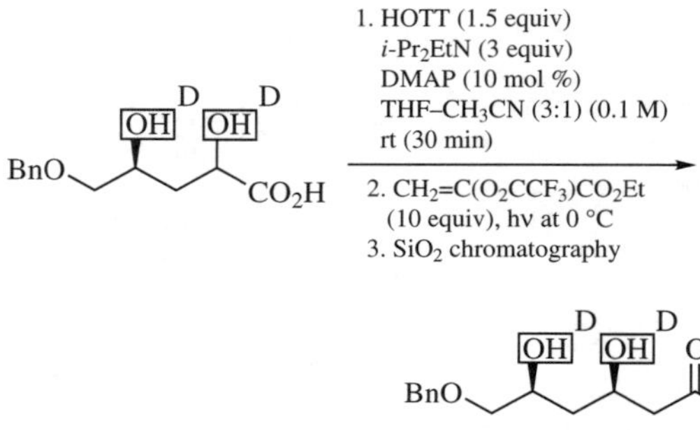

(6)

80% yield, ds = 9:1

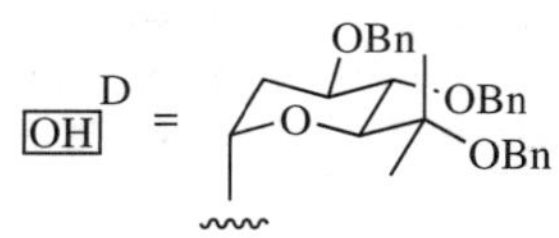

General notes: As with most Barton esterifications, the reaction should be performed in the dark and under anhydrous conditions.

Peptide Coupling. The HOTT reagent, as well as the corresponding tetrafluoroborate salt (TOTT), have also been reported to be inexpensive alternatives to uronium- and phosphonium-based peptide coupling reagents (eq 7).[7] Yields were generally on the same order as those observed with standard peptide coupling reagents. An advantage of these reagents—at least in some instances—may be a reduced propensity of the *N*-protected amino acid component to racemize during the coupling reaction.

Synthesis of Primary Amides. Carboxylic acids can be converted to their primary amides using HOTT (or TOTT) as the coupling agent (eq 8 and 9).[8] The reaction conditions are very mild and do not adversely affect other functionality prone to nucleophilic attack by ammonia. A simple extractive work-up is sufficient to obtain the primary amides in pure form.

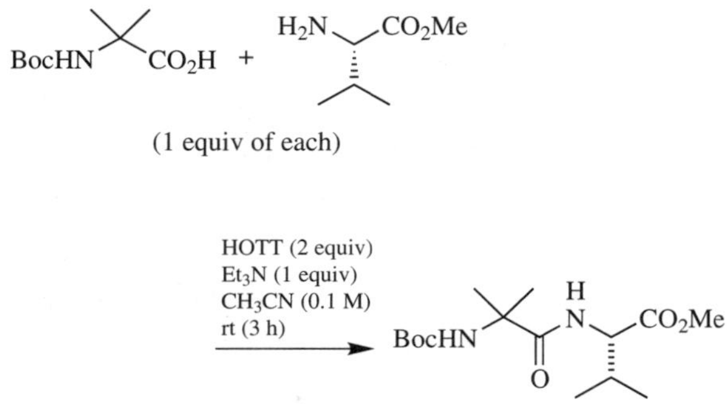

(7)

58% yield

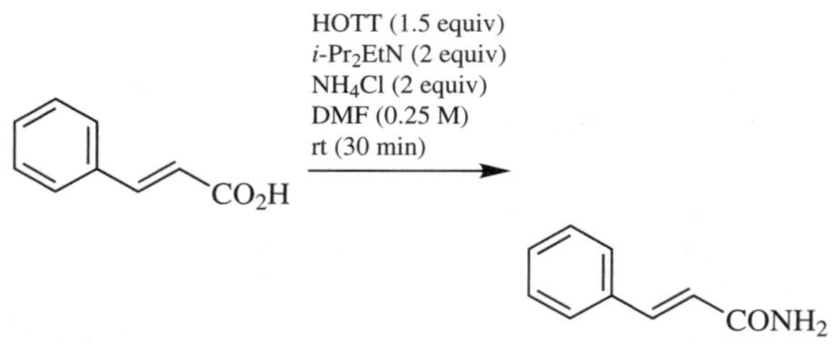

(8)

93% yield

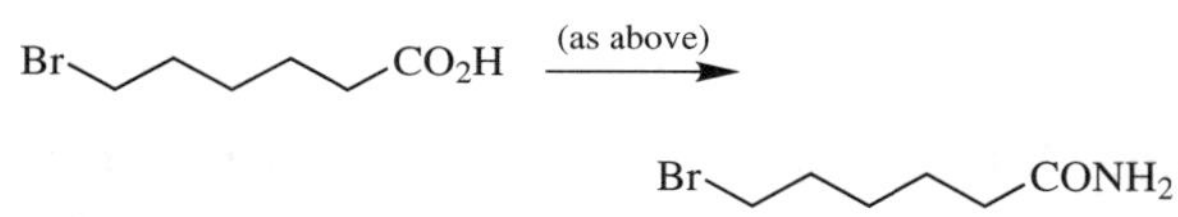

(9)

93% yield

1. Dourtoglou, V.; Gross, B.; Lambropoulou, V.; Zioudrou, C., *Synthesis* **1984**, 572.
2. Garner, P.; Anderson, J. T.; Dey, S.; Youngs, W. J.; Galat, K., *J. Org. Chem.* **1998**, *63*, 5732.
3. Crich, D., *Aldrichimica Acta* **1987**, *20*, 35.
4. Barton, D. H. R.; Samadi, M., *Tetrahedron* **1992**, *48*, 7083.
5. Takasu, K.; Mizutani, S.; Noguchi, M.; Ihara, M., *J. Org. Chem.* **2000**, *65*, 4112.
6. Garner, P.; Anderson, J. A., *Org. Lett.* **1999**, *1*, 1057.
7. Bailén, M. A.; Chinchilla, R.; Dodsworth, D. J.; Nájera, C., *J. Org. Chem.* **1999**, *64*, 8936.
8. Bailén, M. A.; Chinchilla, R.; Dodsworth, D. J.; Nájera, C., *Tetrahedron Lett.* **2000**, *41*, 9809.

James T. Anderson, Subhakar Dey & Philip Garner
Case Western Reserve University, Cleveland, OH, USA

P

Pentafluorophenol

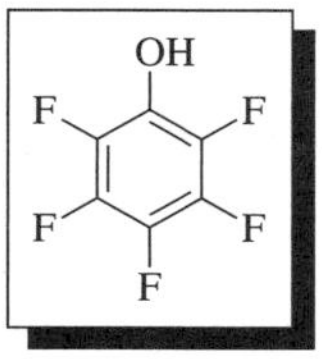

[771-61-9] C_6HF_5O (MW 184.07)

(preparation of pentafluorophenyl active esters for peptide synthesis;[1] formylating agent; preparation of aromatic fluoro derivatives)

Physical Data: white solid, mp 34–36 °C; bp 143 °C; fp >72 °C; n_D 1.4263.
Solubility: sol most organic solvents, particularly aprotic ones.
Form Supplied in: available commercially.
Handling, Storage, and Precautions: toxic and irritant. Harmful if swallowed, inhaled, or absorbed through the skin. Incompatible with strong oxidizing agents, bases, acid chlorides, and anhydrides. Keep away from heat and naked flames: combustible solid/liquid. Store in a cool, dry place.

Use in Peptide Synthesis.

Solution Techniques. Simple alkyl esters of amino acids (which may be protected) undergo aminolysis at a rate too slow for peptide bond synthesis. Phenyl esters are more reactive and, if electronegative substituents are present on the ring, may undergo aminolysis at rates similar to those of anhydrides. The overriding requirements in peptide bond formation are efficiency and freedom from side reactions. Esters of pentafluorophenol have found wide application in the synthesis of peptides (cf. Pentachlorophenol).[1a–c] Pentafluorophenol (**1**) serves as a coupling reagent, i.e. a compound added to a mixture of carboxyl and amino components to promote condensation. However it is very difficult to find efficient coupling reagents which do not give undesirable side reactions. Kinetic studies have shown the superiority of pentafluorophenyl esters compared to other coupling reagents with respect to speed of coupling (and thus lowering or eliminating undesirable reactions, e.g. α-hydrogen abstraction leading to racemization).[2a,b] Relative rates are OPFP >> OPCP > ONp, corresponding to 111:3.4:1, where PFP = pentafluorophenyl; PCP = pentachlorophenyl; Np = nitrophenyl.

Pentafluorophenyl esters have been used in both solution and solid phase peptide syntheses, the difference being that solid phase synthesis involves linking the peptide to a solid support (see below), whilst solution techniques involve just one phase. There are many examples of pentafluorophenyl esters being used in peptide synthesis,[1a,3a–e] e.g. the peptide sequence *t*-Boc–Phe–Pro–Pro–Phe–Phe–Val–Pro–Pro–Ala–Phe–OMe[3a] and the alkaloid peptide (**2**)[3b] have been synthesized by these methods.

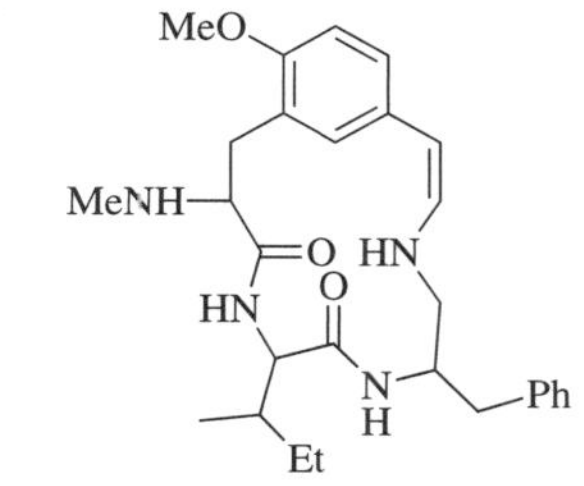

(2)

More recently, the base labile 9-fluorenylmethoxycarbonyl (Fmoc) group has become well established in peptide synthesis for the protection of amino functionalities.[4,5] The problem associated with this group is cleavage of the Fmoc group by the free amino component to be acylated, which results in double acylations. The principal method of avoiding this now is to use the Fmoc group in association with highly reactive pentafluorophenyl esters to greatly increase the rate of coupling compared with the undesirable reactions. Formation of these esters is shown in eq 1.[4]

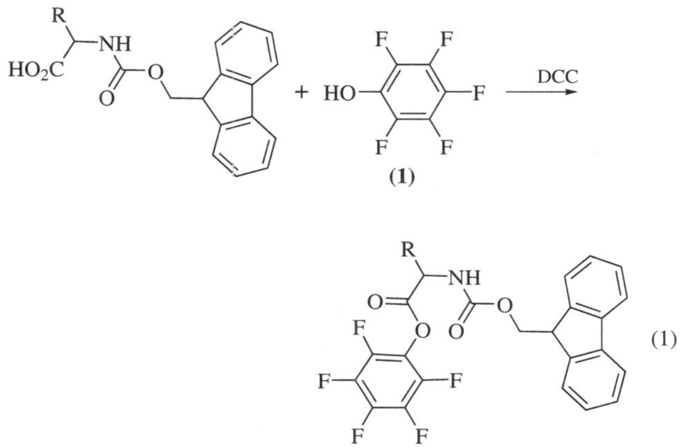

(1)

For example, this methodology was first used in the synthesis of Fmoc-glycyl-L-tryptophyl-L-leucyl-L-aspartyl-L-phenylalaninamide (Fmoc–Gly–Trp–Leu–Asp–Phe–NH_2) with the isolation of all pentafluorophenyl ester intermediates in excellent yields (75–99%); the coupled amino acids were subsequently prepared stepwise, again in very good yields, with the minimum of undesirable side reactions.[4]

This method has since been modified and improved further by utilizing 9-fluorenylmethyl pentafluorophenyl carbonate (eq 2).[5a]

There is a newer alternative method of synthesizing Fmoc-protected pentafluorophenyl ester amino acids employing pentafluorophenyl trifluoroacetate (**3**).[5b]

Table 1 gives some typical examples of the *N*-9-Fluorenylmethoxycarbonyl (Fmoc) amino acids and their pentafluorophenyl (PFP) esters which have been prepared by these methods.[4,6,7] The physical data are in agreement with either preparative method, and with early preparations of these compounds.[4,5a,b]

Solid Phase Peptide Synthesis (SPPS). Pentafluorophenyl esters have been used in both solution and solid phase peptide synthesis. So far, only solution methods have been considered. The

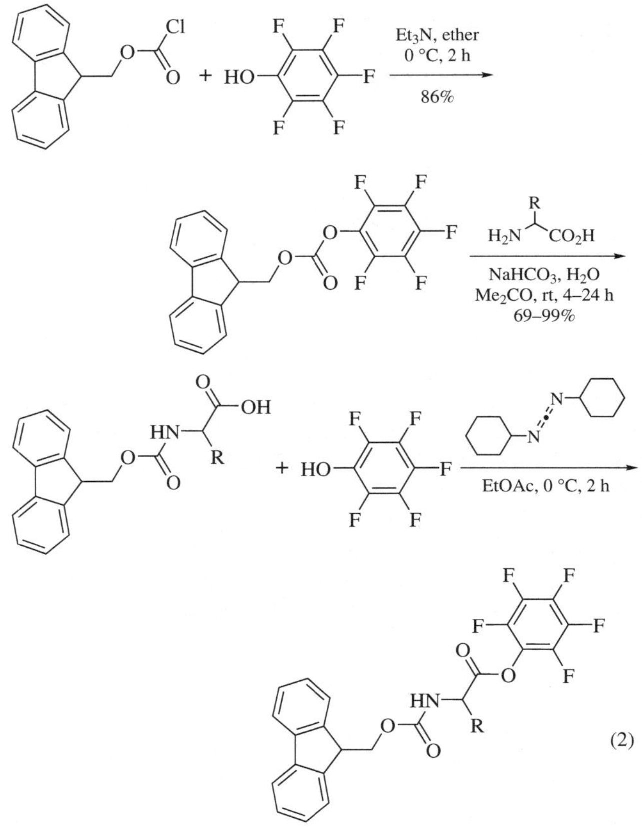

(2)

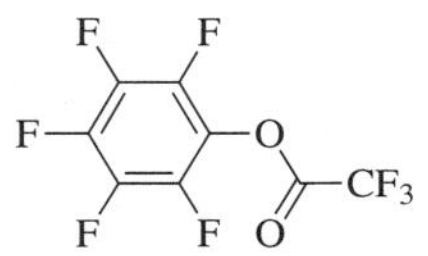

(3)

Table 1 Examples of Fmoc-Protected Amino Acids and their Pentafluorophenyl (PFP) Esters

Amino acid	Fmoc derivative mp (°C)	Fmoc derivative yield (%)	Fmoc amino acid PFP ester mp (°C)	Fmoc amino acid PFP ester yield (%)	Ref.
Gly	171–172	84	158–160	72	4, 6
Ala	141–142	93	173–175	74	4, 6
Leu	151–152	83	113–114	78	4, 6
Val	143–144	90	121–123	77	4, 7
Pro	114–115	82	125–127	89	4, 7
Phe	181–182	86	152–154	90	4, 6
Trp	163–165	89	184–185	85	4, 6
Tyr	98–107	70	152–156	70	4
Ser	90–94	78	124–126	83	4, 7
Ser (*t*-Bu)	–	–	67–71	67	4, 7

principal of SPPS is incorporation of single amino acid residues into peptide chains built on insoluble polymeric supports. The famous Merrifield paper[8] started the tradition of SPPS using *t*-butoxycarbonyl (Boc) amino acids activated with 1,3-Dicyclohexylcarbodiimide, although this has now been superseded with the introduction of the Fmoc group in polystyrene- and polyamide-based SPPS. The theory of SPPS will not be discussed here. Needless to say, the use of pentafluorophenyl esters in SPPS was inevitable and was first reported by Atherton in 1985.[9] Fmoc–amino acid pentafluorophenyl active esters were used in the polyamide solid phase series in polar DMF solutions in the presence of 1-Hydroxybenzotriazole as a catalyst.

The preparation of the decapeptide H–Val–Gln–Ala–Ile–Asp–Tyr–Ile–Asn–Gly–OH (**4**) will be used as an example of pentafluorophenyl esters in SPPS.[9,10] This is a good example to use since it contains normal, sterically hindered, and functionalized amino acids. All Fmoc–pentafluorophenyl esters were prepared by the method of Kisfaludy and Schön already described.[4,5] Beaded polymethylacrylamide gel resin was used as the solid phase with fivefold excesses of the pentafluorophenyl esters dissolved in a minimum of DMF. The resin was first functionalized with a reference norleucine residue and then with an acid-labile linkage agent (**5**).

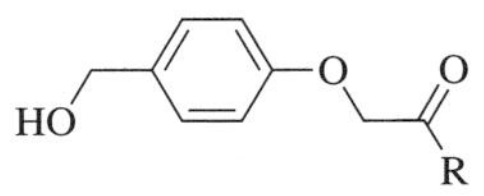

(**5**) R = 2,4,5-$Cl_3C_6H_2O$

Esterification of the first residue, glycine, to the free OH used the symmetrical anhydride/ 4-Dimethylaminopyridine procedure.[11] The subsequent peptides were all coupled using standard techniques like those already described above and are fully documented. Slightly varying conditions and catalysts were required for some of the amino acids.[10] After addition of the last amino acid, the final Fmoc group was cleaved and the complete decapeptide detached from the support resin using 95% aqueous Trifluoroacetic Acid. Side chain protecting groups were also cleaved simultaneously. Crude peptide analysis was achieved by HPLC and showed the samples obtained by this method to be extremely pure.

More recently, SPPS has also been employed in glycopeptide synthesis.[12] There is an excellent review of active esters in SPPS, including pentafluorophenyl esters.[13]

A New and Efficient Derivative of Pentafluorophenol.[14] Pentafluorophenol has found many applications in both solution and solid phase peptide syntheses. Diphenylphosphinic mixed anhydrides are also well known in peptide chemistry. A new and efficient reagent has been prepared from these precursors: pentafluorophenyl diphenylphosphinate (FDPP; **6**) which can be used directly as a coupling reagent without side reactions. FDPP is prepared from mixing equimolar amounts of Diphenylphosphinic Chloride, pentafluorophenol (**1**), and Imidazole in CH_2Cl_2 at room temperature (eq 3).[14]

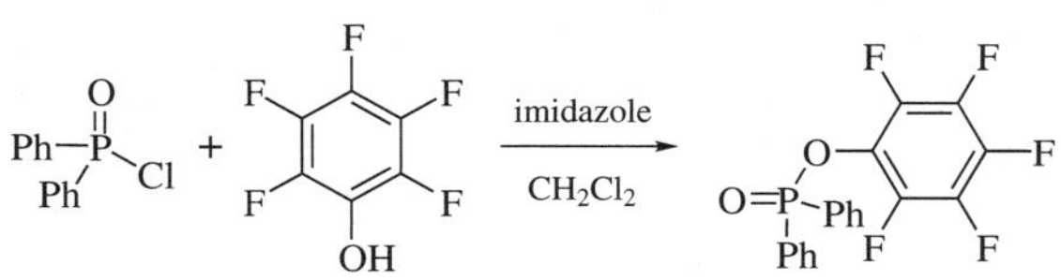

(**1**) (**6**) (3)

FDPP may be kept refrigerated for several months and is used as an efficient coupling reagent, by simply mixing the carbonyl and amine components with a tertiary amine and FDPP in an organic solvent, of which DMF was found to be the most efficient (eq 4).

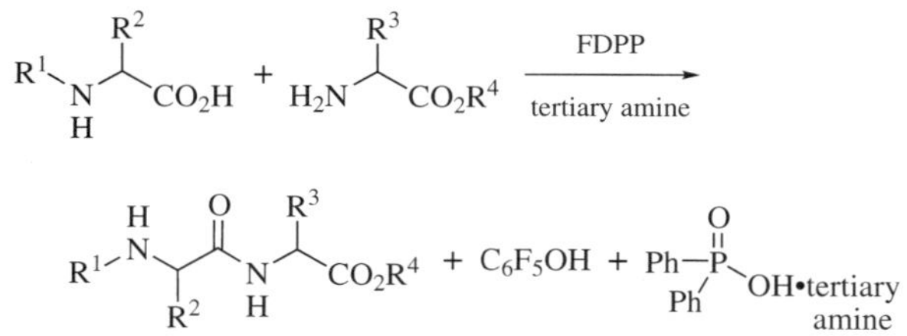

(4)

FDPP has been used in both solid and solution peptide syntheses and analysis has shown it to have the lowest rate of racemization of any of the common coupling reagents.[14]

Comparison of Pentafluorophenyl Active Esters with Pentachlorophenyl Esters.[15] It is interesting to compare these two compounds with respect to active ester formation since they are very closely related and both have been employed in both liquid and solid phase peptide syntheses (see Pentachlorophenol). Pentachlorophenyl esters generally have higher melting points, which can be of advantage if crystallization is difficult or a compound is of low melting point, e.g. for serine and threonine the oily pentafluorophenyl esters are usually replaced by the pentachlorophenyl esters.

As already mentioned, pentafluorophenyl derivatives are much more reactive than pentachlorophenyl ones. The reason for this is still not clear. Many calculations have been performed on the net atomic energies of these esters and show that there is no significant difference between the environment of the electrophilic center in OPFP esters compared to OPCP ones. Analysis of the carbonyl stretching frequencies of these compounds is quite revealing (Table 2), i.e. there is no significant difference in the carbonyl stretching frequencies of these functionalized esters. This, plus additional evidence from MM calculations and NMR studies, suggests that electronic factors do not play a significant role in determining the reactivity of these esters.

Table 2 IR Stretching Frequencies of Carbonyl Groups in Pentachlorophenyl (OPCP) and Pentafluorophenyl (OPFP) Active Esters[a]

Amino acid	OPCP	OPFP	$\Delta\nu$
Z–Gly	1794	1800	6
Boc–Val	1782	1788	6
Boc–Phe	1784	1790	6
Boc–Tyr(*t*-Bu)	1786	1794	8
Z–Cys(Bn)	1784	1794	10

[a] In $CHCl_3$; in cm^{-1}.

There is however, an important steric effect, as would be expected given the differing sizes of the halogens involved (van der Waals atomic radii = 1.75 Å for Cl and 1.47 Å for F). X-ray crystal analysis is interesting in that it shows there is no intramolecular interaction in OPFP esters and the structure is an extended conformation. The main interaction is the strong dipole–dipole of a negative fluorine to a positive carbon of another PFP ring in a neighboring molecule. CO···NH interactions appear to be only minor in comparison and the carbonyl group and PFP ring are almost perpendicular. Studies have shown that the same conformation applies whether in solid or solution.

One of the explanations offered for the high reactivity of the OPFP ester is neighboring group participation of an *ortho* fluorine atom in accelerating the collapse of a tetrahedral intermediate transition state, i.e. anchimeric acceleration. This would explain the low solvent effect on aminolysis rates, the high second-order rate constant, and unaffected third-order rate constant. Pentahalogen esters fall midway between quinol-8-yl esters, where there is definite participation, and nitrophenol esters where there is none (from studies on the model system of phenyl acetates and piperidine).[15] The PFP ester shows greater interaction than the PCP ester. There are other possible explanations: this could still be explained in steric terms, or by the high solvation of the leaving group (pentafluorophenol is much more soluble and reactive in aprotic solvents than pentachlorophenol).

Formylation Reactions.[16] The problem with Acetic Formic Anhydride, the most commonly used formylating agent, is undesirable side reactions, particularly if there are acid sensitive functionalities in the compound. One of the best alternatives is to use pentafluorophenyl formate (**7**), an easily prepared and stable compound which reacts with N-nucleophiles in minutes at room temperature to yield the *N*-formyl derivative (eq 5) (Table 3). Significantly, no reaction occurs with alcohols, thiols, or sterically hindered amines.

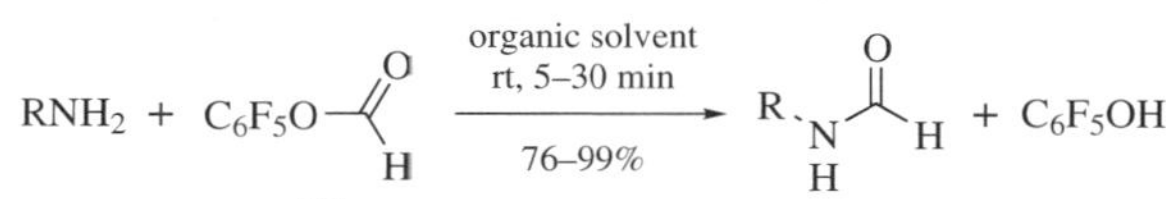

(**7**) (5)

Table 3 Examples of *N*-Formylamines Prepared from Primary Amines and Pentafluorophenyl Esters

Amine	Solvent	Yield (%)	mp (°C)	R_f
Aniline	$CHCl_3$	90	48	0.6[a]
2-Amino-biphenyl	$CHCl_3$	99	67–70	0.6[a]
Benzylamine	$CHCl_3$	90	62–63	0.35[a]
H–Met–Leu–Phe–OMe	$CHCl_3$	93	135–136	0.4[b]

[a] Chloroform–methanol (9:1).
[b] Ethyl acetate.

Preparation of Aromatic Fluoro Derivatives. Reaction of (**1**) with VF_2, VF_4, Vanadyl Trifluoride, or Xenon(II) Fluoride in HF, $CFCl_3$, or MeCN at −30 to 20 °C yields mixtures of perfluoro-2,5-cyclohexadien-1-one together with its 2- and 4-pentafluorophenoxy derivatives and/or perfluoro-6-phenoxy-2,4-cyclohexadien-1-one, along with traces of perfluoro-2-cyclohexen-1-one or perfluoro-*p*-benzoquinone. Using Antimony(V) Fluoride or NbF_4, however, only gives stable complexes or pentafluorophenolates.[17]

Pentafluorophenol (**1**) reacts with Phosphorus(V) Chloride in the presence of a base, triethylamine or γ-collidine, to yield perfluoropentaphenoxyphosphorane (**8**) plus perfluorotriphenyl phosphate (**9**) and perfluorodiphenyl ether (**10**) (eq 6).[18]

$$PCl_5 + C_6F_5OH \xrightarrow[\text{or } \gamma\text{-collidine}]{NEt_3} \underset{(\mathbf{8})}{(C_6F_5O)_5P} + \underset{(\mathbf{9})}{(C_6F_5O)_3P{=}O} + \underset{(\mathbf{10})}{(C_6F_5)_2O} \quad (6)$$

This reaction has also been used to prepare pentafluorothiophenol phosphorus derivatives. These compounds have been utilized in ^{31}P, ^{19}F, and ^{1}H variable temperature NMR studies on intramolecular ligand rearrangement processes in phosphorus compounds.[18]

The preparation and reactions of Polyfluoroaromatics, including pentafluorophenol derivatives, have been studied by various Russian workers, who have also evaluated the pentafluorophenoxy group as a leaving group (eq 7).[19a]

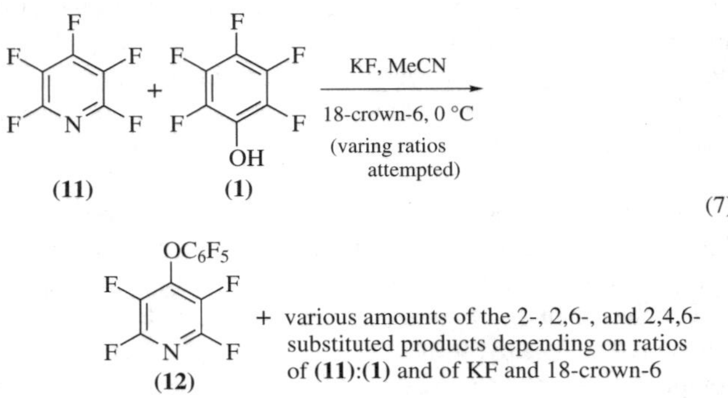

(7)

The main product of the reaction was the 4-substituted tetrafluoropyridine, although varying yields of the 2-, 2,6- and 2,4,6-pentafluorophenoxypyridines were obtained depending on the ratios of (**11**):(**1**) and also on the levels of Potassium Fluoride, 18-Crown-6, the reaction temperature, and solvent. This reaction has been studied in comparison with 4-$NO_2C_6H_4$ (which is of similar basicity) and in competition with F^- anions.

More recently it has been shown that the substitution reaction of pentafluoropyridine with (**1**) in the presence of Cesium Fluoride/graphite gave much better yields of (**12**), but still with traces of the other products. The catalyst activity was found to increase in the order: NaF < KF < RbF < CsF and was similar for the free fluorides and their graphite-supported analogs.[19b]

1. For examples of early work on pentafluorophenyl esters in peptide synthesis, see: (a) Kisfaludy, L.; Roberts, J. E.; Johnson, R. H.; Mayers, G. L.; Kovacs, J. *J. Org. Chem.* **1970**, *35*, 3563 and references therein. (b) Kisfaludy, L.; Schön, I.; Szirtes, T.; Nyéki, O.; Lów, M. *Tetrahedron Lett.* **1974**, 1785 and references therein. (c) Kovacs, J.; Kisfaludy, L.; Ceprini, M. Q. *J. Am. Chem. Soc.* **1967**, *89*, 183.
2. (a) For a kinetic study of coupling rates and competing reactions, see:Gross, E.; Meienhofer, J. In *The Peptides–Analysis, Synthesis and Biology*, 2nd ed.; Academic:New York, **1980**; p 519; (b) see also Ref. 1 (c) .
3. There are many examples of solution peptide synthesis utilizing pentafluorophenyl esters. A few notable examples are cited here: (a) Kisfaludy, L.; Schön, I.; Szirtes, T.; Nyéki, O.; Low, M. *Tetrahedron Lett.* **1974**, 1785. (b) Schmidt, U.; Schanbacher, U. *Justus Liebigs Ann. Chem./Liebigs Ann. Chem.* **1984**, *6*, 1205. See also Ref. [1] (a) . (c) Schmidt, U.; Lieberknecht, A.; Griesser, H.; Haeusler, J. *Justus Liebigs Ann. Chem./Liebigs Ann. Chem.* **1982**, *12*, 2153. (d) Karel'skii, V. N.; Krysin, E. P.; Antonov, A. A.; Rostovskaya, G. E. *Khim. Prir. Soedin.* **1982**, *1*, 96 (*Chem. Abstr.* **1982**, *97*, 92 730s) . (e) Polevaya, L. K.; Vegners, R.; Ars, G.; Grinsteine, I.; Cipens, G. *Latv. PSR Zinat. Akad. Vestis, Khim. Ser.* **1981**, *4*, 469 (*Chem. Abstr.* **1982**, *96*, 7059s) .
4. Kisfaludy, L.; Schön, I. *Synthesis* **1983**, 325.
5. (a) Kisfaludy, L.; Schön, I. *Synthesis* **1986**, 303. (b) Green, M.; Berman, J. *Tetrahedron Lett.* **1990**, *31*, 5851.
6. Meienhofer, J.; Waki, M.; Heimer, E. P.; Lambros, T. J.; Makofske, R. C.; Chang, C-D. *Int. J. Pept. Protein Res.* **1979**, *13*, 35.
7. Chang, C-D.; Waki, M.; Ahmad, M.; Meienhofer, J.; Lundell, E. O.; Haug, J. D. *Int. J. Pept. Protein Res.* **1980**, *15*, 59.
8. Merrifield, R. B. *J. Am. Chem. Soc.* **1963**, *85*, 2149.
9. Atherton, E.; Sheppard, R. C. *Chem. Commun./J. Chem. Soc., Chem. Commun.* **1985**, 165.
10. Atherton, E.; Cameron, L. R.; Sheppard, R. C. *Tetrahedron* **1988**, *44*, 843.
11. Atherton, E.; Logan, C. J.; Sheppard, R. C. *J. Chem. Soc., Perkin Trans. 1* **1981**, 538.
12. For the use of pentafluorophenyl esters in SPPS of glycopeptides, see: Meldal, M.; Jensen, K. J. *Chem. Commun./J. Chem. Soc., Chem. Commun.* **1990**, 483 and references therein.
13. For a review of active esters in SPPS, see: Bodanszky, M.; Bednarek, M. A. *J. Protein Chem.* **1989**, *8*, 461.
14. Chen, S.; Jiecheng, X. *Tetrahedron Lett.* **1991**, *32*, 6711.
15. Kisfaludy, L.; Low, M.; Argay, G.; Czugler, M.; Komives, T.; Sohar, P.; Darvas, F. In *Peptides (Proceedings 14th European Peptide Symposium)* ; Loffet, A., Ed.; Editions de l'Université de Bruxelles:Bruxelles, **1976**; p 55 and references therein.
16. Kisfaludy, L.; Ötvos, Jr., L. *Synthesis* **1987**, 510.
17. Avramenko, A. A.; Bardin, V. V.; Karelin, A. L.; Krasil'nikov, V. A.; Tushin, P. P.; Furin, G. G.; Yakobson, G. G. *Zh. Org. Khim.* **1985**, *21*, 822 (*Chem. Abstr.* **1985**, *103*, 141 551n) .
18. Denny, D. B.; Denney, D. Z.; Liu L-T. *Phosphorus Sulfur/Phosphorus Sulfur Silicon* **1982**, *13*, 1.
19. (a) Aksenov, V. V.; Vlasov, V. M.; Yakobson, G. G. *J. Fluorine Chem.* **1982**, *20*, 439. (b) Aksenov, V. V.; Vlasov, V. M.; Danilkin, V. I.; Naumova, O. Y.; Rodionov, P. P.; Chertok, V. S.; Shnitko, G. N.; Yakobson, G. G. *Izv. Akad. Nauk SSSR, Ser. Khim.* **1984**, *9*, 2158 (*Chem. Abstr.* **1985**, *102*, 131 881k) .

Keith Jones
King's College London, UK

Pentafluorophenyl Diphenylphosphinate (FDPP)

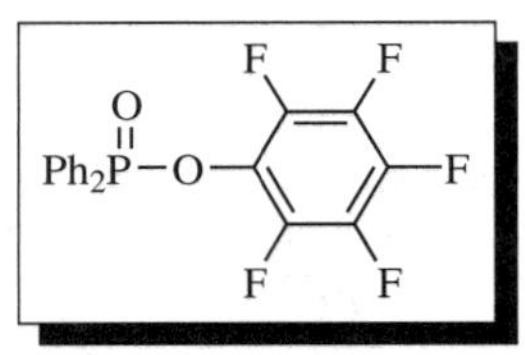

[138687-69-1] $C_{18}H_{10}F_5PO_2$ (MW 384.25)

(reagent used as carboxyl activating group and coupling agent in a wide variety of amide bond formations)

Physical Data: mp 47–50 °C.
Solubility: soluble in most organic solvents.
Form Supplied in: white powder and chunks.
Handling, Storage, and Precautions: decomposition occurs upon prolonged exposure to moisture in air. Storage in a cool dry place extends the life of the reagent. The toxicological properties have not been thoroughly investigated; however, hazardous combustion products may include, but are not limited to, carbon monoxide, carbon dioxide, hydrogen fluoride, and phosphorus oxides. FDPP is incompatible with strong oxidizing agents.
Preparative Methods: FDPP can be conveniently prepared by mixing equimolar amounts of diphenylphosphinic chloride, pentafluorophenol, and imidazole in anhydrous CH_2Cl_2 at room

temperature (eq 1).[1] There are slight procedural variations in the literature as to order of addition and reaction duration.[2,3] It has been found that using an excess of the diphenylphosphinic chloride (1.5–2 equiv) to drive the reaction to completion simplifies the purification, as the residual hydrolyzed reagent adheres tightly to the silica gel column and is, therefore, easier to separate than residual pentafluorophenol.[4]

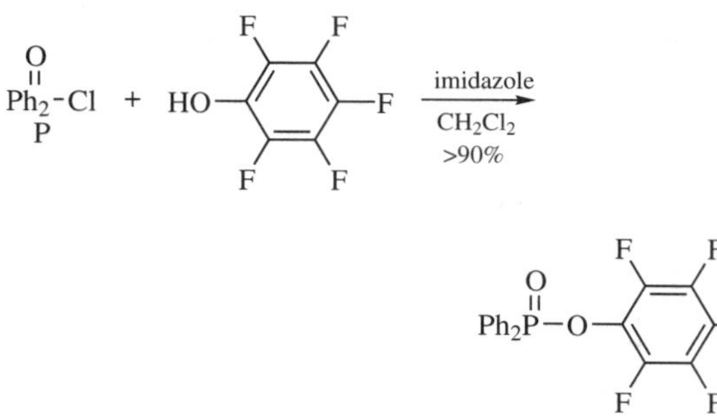

(1)

Purification: silica gel chromatography (short column to minimize decomposition) eluting with 20% ethyl acetate in petroleum ether followed by drying under reduced pressure for 2 days over phosphorus pentoxide.

Coupling to Generate Linear Molecules. When Chen and Xu reported the new reagent, FDPP, in 1991,[1] the use of diphenylphosphinic mixed anhydrides in peptide chemistry was known,[5,6] as was the use of pentafluorophenyl esters.[7–11] However, the need to preactivate amino acids or to prepare active esters was unattractive. FDPP can be used directly as a coupling reagent without preactivation.[1] Chen and Xu demonstrated the utility of the reagent by preparing several small oligopeptides in solution and by preparing Leu-enkephalin via solid-phase synthesis.[1] Their solution-phase protocol called for equimolar amounts of the carboxyl and amine hydrochloride components, 3 equiv of Hünig's base, and 1.2 equiv of FDPP in DMF at room temperature for 1.5 h. The yields of isolated products were all greater than 80%. Using HPLC[12] (for the coupling product of Z-Gly-Phe-OH and Val-OMe•HCl) and the Young test[13] (for the coupling product of Bz-Leu-OH and Gly-OEt•HCl), they observed almost no racemization with FDPP; whereas, pentafluorophenyl diphenylphosphate (FDP)[14,15], 2-(1*H*-benzotriazol-1-yl)-1,1,3,3-tetramethyluronium hexafluorophosphate (HBTU), benzotriazol-1-yl-oxytris(dimethylamino)phosphonium hexafluorophosphate/1-hydroxybenzatriazole (BOP/HOBt), and dicyclohexylcarbodiimide (DCC) all showed significant levels of racemization. In the solid phase protocol, they employed the chloromethyl resin and carried out the reactions with 3 equiv of Boc-amino acid, FDPP, and 9 equiv of Hünig's base. This method provided extremely high yields and proved superior to the DCC method because no insoluble by-product was generated.

The use of FDPP in the formation of a variety of dipeptides has been studied in comparison to BOP, HBTU, and isobutyl chloroformate (IBCF).[3] Product yields, HPLC diastereomeric yields,[12] and optical rotations were measured for each example. In general, FDPP did give high yields and good optical purity. When Fmoc was used as the nitrogen protecting group, significantly lower yields and lower optical purity resulted. FDPP gave excellent yields for the formation of proline methyl ester dipeptides, outperforming HBTU, BOP, and IBCF; however, these reagents provided better optical purity than FDPP. Overall, BOP was the most successful coupling reagent for the dipeptides studied.

The search for an economical synthesis of the β-hydroxyproline fragment found in the cyclopeptide alkaloids nummularine F and mauritine was the impetus for another comparative study of carboxyl activating reagents (eq 2).[16] Relatively expensive isopropenyl chloroformate (IPCF) was used previously for this reaction. It was shown that the more affordable FDPP and BOP-Cl both possessed reactivity comparable to IPCF. In addition, the reactivity of FDPP in this reaction was shown to be greater than that of BOP, pentafluorophenol/DCC, phosgene, diphenylphosphinic chloride, DCC, *N*,*N*-carbodiimidazole, 2,4,6-trichlorobenzoyl chloride, pentafluorophenyl trifluoroacetate, and oxalyl chloride.

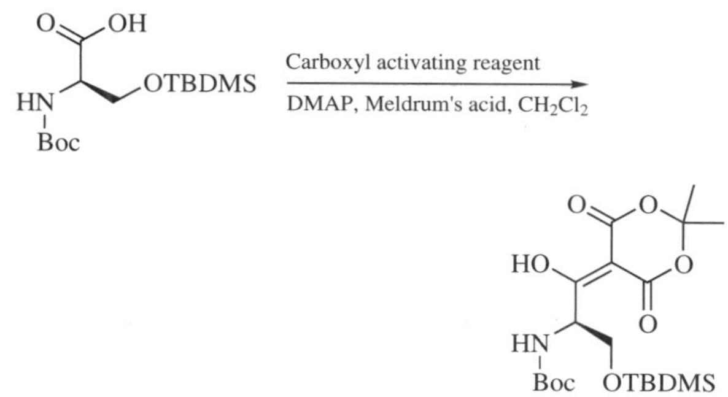

(2)

In the cryptophycin synthesis, FDPP was employed for the coupling shown in eq 3.[2] The amine component was used as its TFA salt, and when DCC/HOBt or BOP-Cl was used, the desired product was contaminated with the trifluoroacetamide to varying degrees. The problem could be circumvented by exchanging the counterion for tosylate, but this replacement incurred significantly longer reaction times. FDPP gave a comparable yield of product uncontaminated by the trifluoroacetamide.

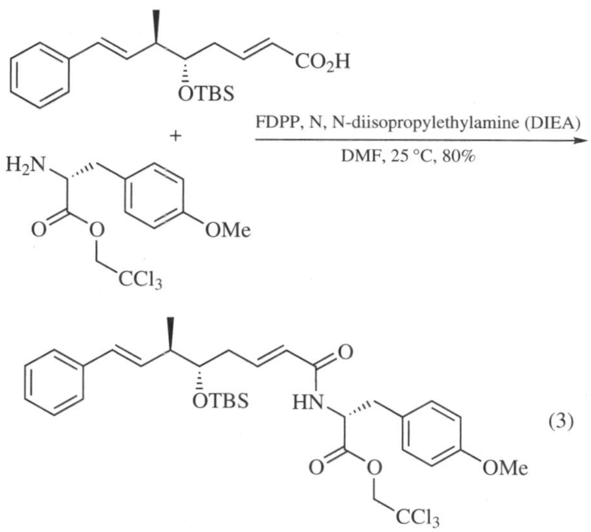

(3)

FDPP has also been employed in the coupling of *N*-protected amino acids to substituted γ-thiapyrone amines (eq 4).[17] When α-amino acids were used, the yields varied from 48–61% over a range of substitution patterns. The authors observed no signs of racemization in the ^{1}H and ^{13}C NMR spectra. Using Fmoc-protected β-amino acids, yields ranged from 39–46% for different substitution patterns, and longer reaction times (7–9 h) were required. Peptoid carboxyl components gave even lower yields at the same extended reaction time. It is unclear whether the low yields, with β-amino acids and peptoids, are due to the nature of these molecules or to the Fmoc protecting group, which is known to give lower coupling yields when used with FDPP.[3]

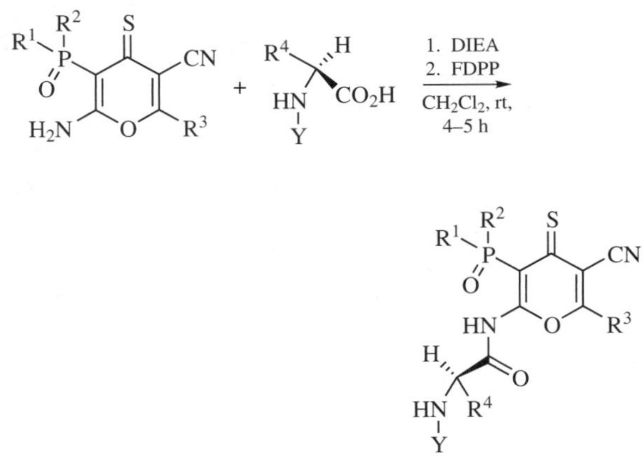

(4)

One way of immobilizing stationary phases onto silica gel involves amide bond formation; wherein the coupling partners are stationary phases derivatized with a free carboxylic acid and silica gel derivatized with an amino group. Low yields and racemization of chiral stationary phases often prove problematic, so Yang and co-workers investigated a variety of conditions and coupling reagents for this reaction.[18] While model reactions in solution phase gave generally high yields, the extent of racemization encountered with FDPP was greater than that found when using diisopropylcarbodiimide/1-hydroxy-7-azabenzotriazole (DIC/HOAt), DIC/HOBt, diphenylphosphoryl azide (DPPA), 2-ethoxy-1-ethoxycarbonyl-1,2-dihydroquinoline (EEDQ), *O*-(7-azabenzotriazol-1-yl)-*N*,*N*,*N'*,*N'*-tetramethyluronium hexafluorophosphate (HATU), and benzotriazolyloxy-tris[pyrrolidino]-phosphonium hexafluorophosphate (PyBop). However, for aminopropylsilica gel couplings (eq 5), yields were generally

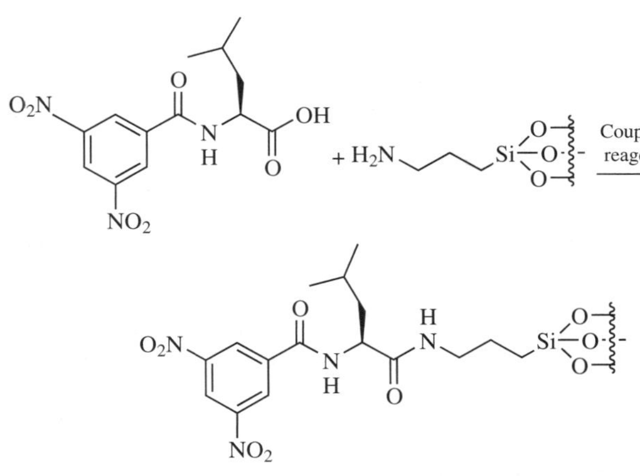

(5)

much lower, and the extent of racemization was much greater with all reagents except EEDQ. In this particular coupling, FDPP performed similarly to DIC/HOAt, DIC/HOBt, DPPA, HATU, and PyBop.

The use of FDPP has also been investigated for the thioacylation of iminoacid residues with monothioacids (eq 6).[19] FDPP gave a good yield of coupled product (81%) but a very low ratio of thio to oxo product. It was found that, for this reaction, coupling reagents with a leaving group bearing a delocalized charge, such as $PfpO^-$, had inherently lower selectivity for thioamide formation. Polar solvents also lowered the selectivity, and FDPP couplings are usually performed in DMF.

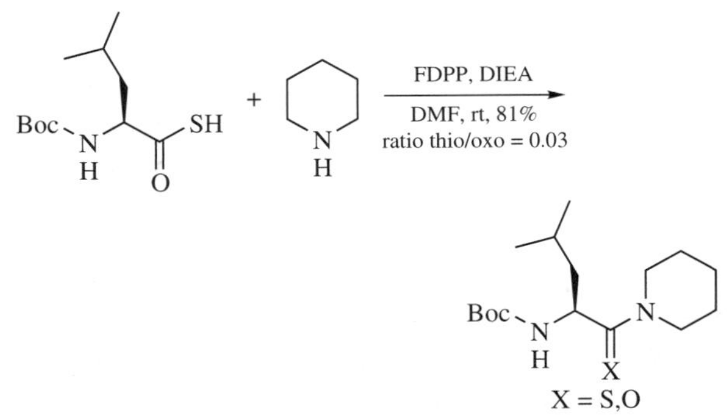

(6)

FDPP has also shown utility in the coupling of *N*-protected amino acid residues to linear diamino substrates bearing a free secondary hydroxyl group. The desired hydroxy amides were obtained in good yields for testing as HIV-1 protease inhibitors.[20]

Macrocyclization. While it is not always the reagent of choice, FDPP often provides excellent yields in macrocyclizations over a wide variety of substrates. In Deng and co-workers' 1994 synthesis of the macrocyclic pentapeptide cyclotheonamide B, FDPP was used to accomplish the macrolactamization in >80 % yield (eq 7).[21]

In their 1996 approach to the same molecule,[22] the WSCI/HOBt-mediated macrolactamization, which had been performed in 42% yield on an analog that had a vinylogous D-tyrosine in place of the vinylogous L-tyrosine, yielded only 16% for the desired substrate. DPPA and diethyl phosphorocyanidate (DEPC) gave only 8% and 12%, respectively, even after an extended reaction time of 3.5 days. The pentafluorophenyl ester also gave only 8% cyclized product. However, cyclization with FDPP gave 43% yield in only 4.5 h, and optimization increased the yield to 57% (eq 8).

Deng and co-workers had similar success using FDPP for the macrolactamization en route to alterobactin A, a siderophore from an open ocean bacterium.[23,24] In this case, FDPP in the presence of NMM in DMF provided a 53% yield of cyclized product despite the relative steric congestion about the amino terminus (eq 9).

During the synthesis of new didemnin B analogs for investigation of the structure activity relationship of these cyclodepsipeptides,[25] improvements were made to the original[26] macrocycle synthesis. Among these modifications was the use of FDPP for the cyclization step. Previously, the best yields (42%) had been attained with DPPA after 72 h at 0 °C.[26] The product of dimerization was always a major contaminant despite the use of very dilute conditions. The use of FDPP at room temperature for 3–4 h accomplished the cyclization in 68% yield (eq 10).[3,25]

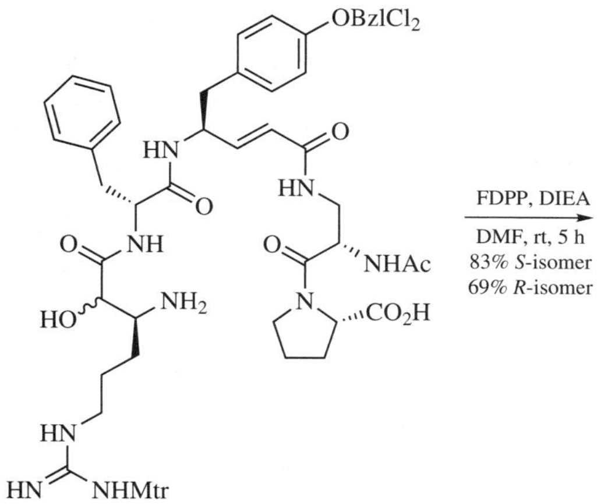

Mtr = 2,3,6-trimethyl-4-methoxy benzenesulfonyl

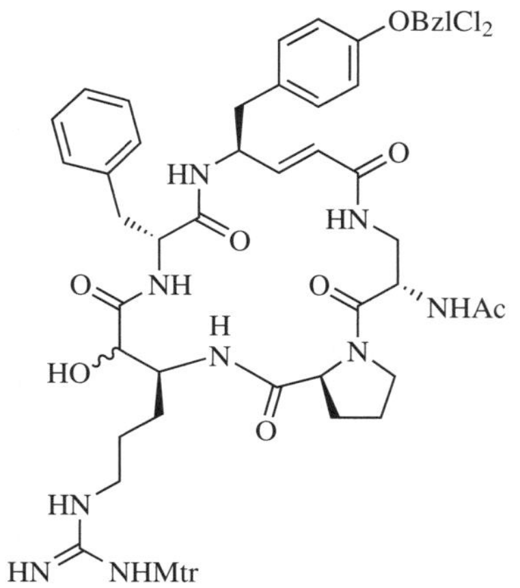

(7)

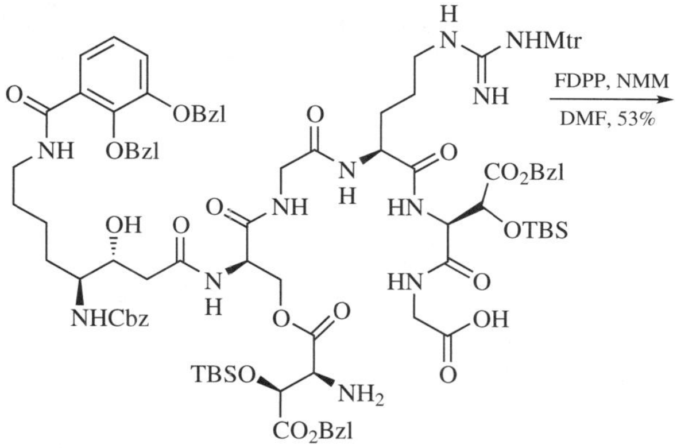

Mtr = 2,3,6-trimethyl-4-methoxy benzenesulfonyl

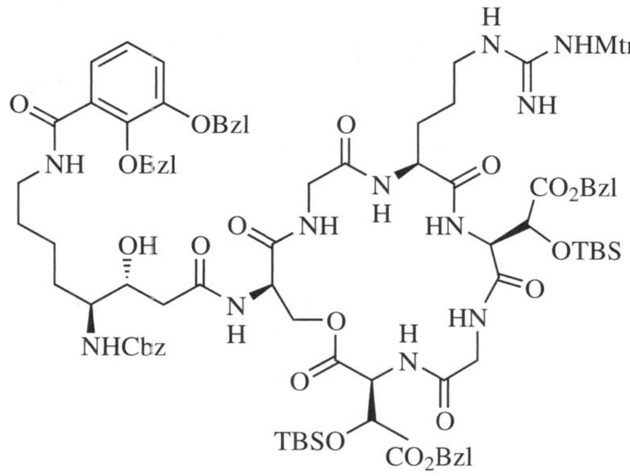

(9)

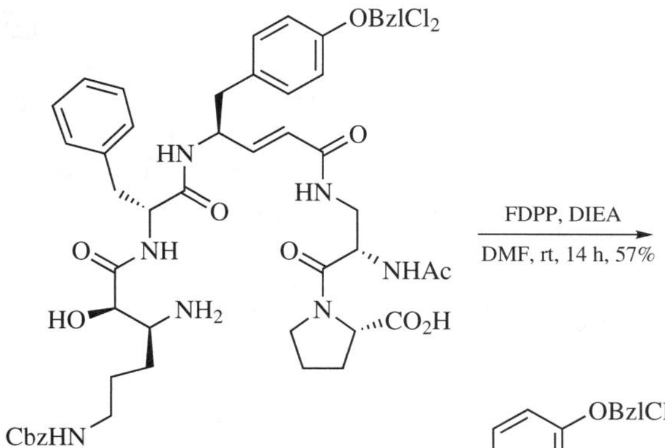

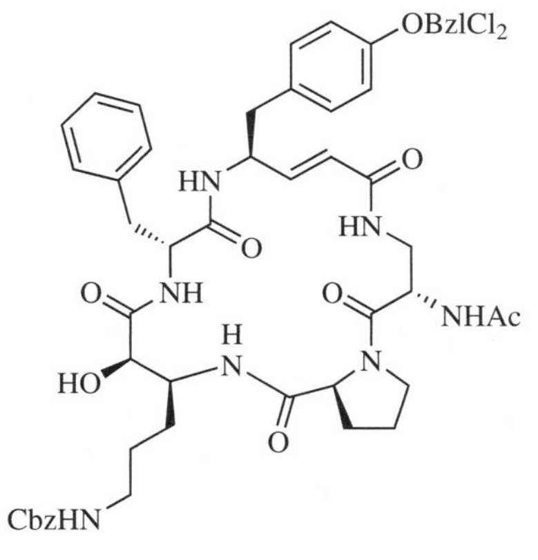

(8)

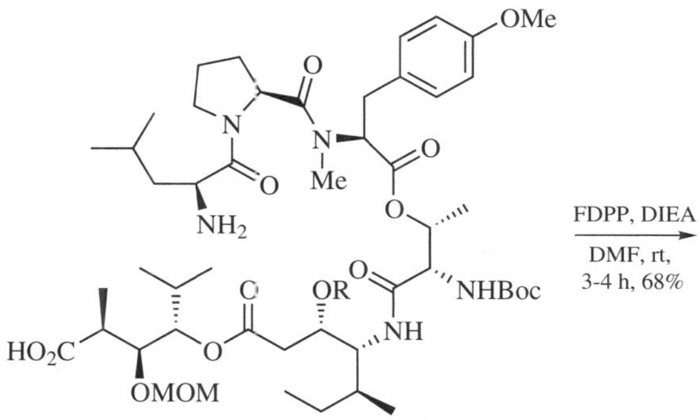

R = Si(i-Pr)$_3$

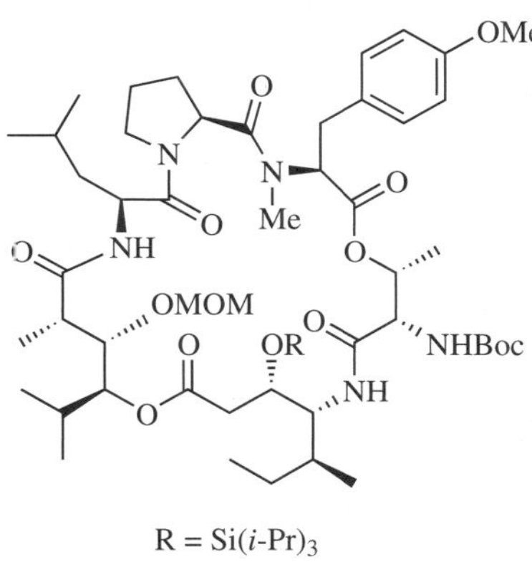

(10)

R = Si(i-Pr)$_3$

In a 1995 synthesis of the cryptophycins, FDPP was used for the macrolactamization reactions, giving a 62% yield of cryptophycin D (eq 11).[2,27,28]

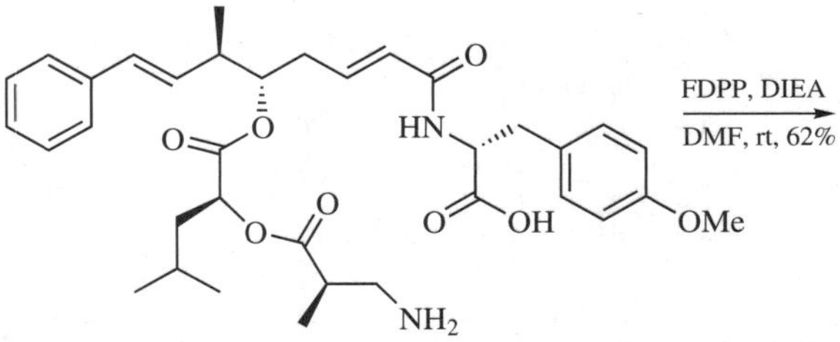

Cryptophycin D (11)

The same yield was achieved for a similar cyclization to give a product that proved not to be identical with cryptophycin C, leading to a revision of the structure of this natural product. The following year, a chemoenzymatic approach to the molecule was published by a different group in an effort to make the synthesis amenable to gram-scale preparation.[29] A more hindered site of cyclization was chosen, and FDPP cyclized in >70% yield in each case examined. A representative example is shown in eq 12.

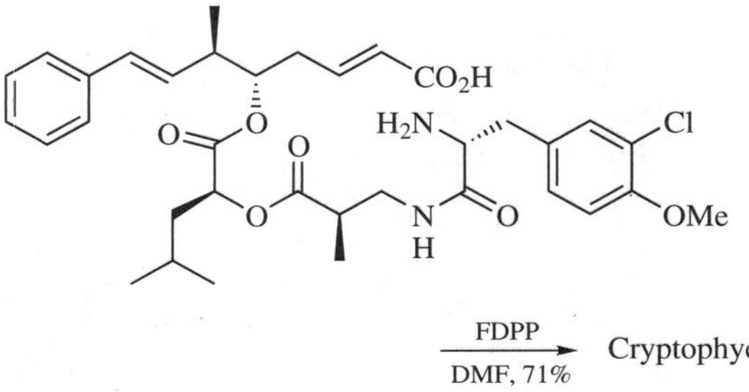

Cryptophycin C (12)

In the synthesis of astin G,[30] FDPP was useful for the coupling of the dipeptide and tripeptide fragments to form the linear precursor, giving the same yield (80%) as BOP. However, cyclization to form the 16-membered ring was unsuccessful with both FDPP and DPPA. HBTU proved to be the best reagent for this difficult cyclization, which proceeded in only 16% yield (eq 13).

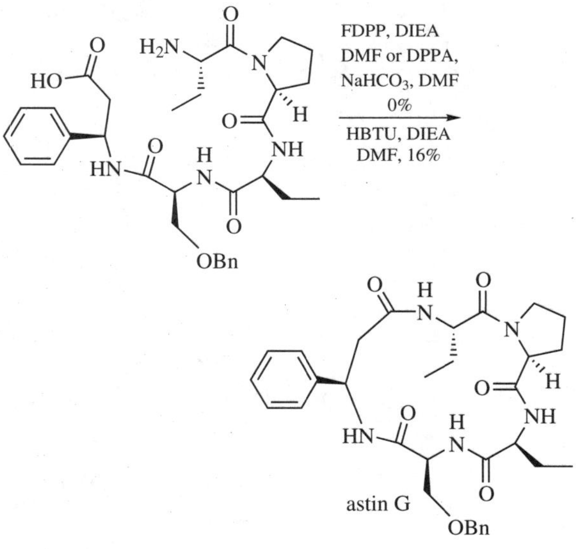

(13)

Cyclooligomerization. A less common use for FDPP is the cyclooligomerization reaction. Bertram et al. found that cyclic peptides containing thiazole moieties could be efficiently constructed in one step from the monomeric unit (eq 14).[31] They found that FDPP gave consistently high yields of cyclic products. The optimum procedure employed 3 equiv of Hünig's base and 1.5 equiv of FDPP added to a suspension of the monomer at a concentration of about 50 mM in acetonitrile. The reaction duration was approximately 18 h at room temperature. The yields were consistently >80% under these conditions, and the major products were the cyclic trimer and tetramer, in a ratio of 5:2.

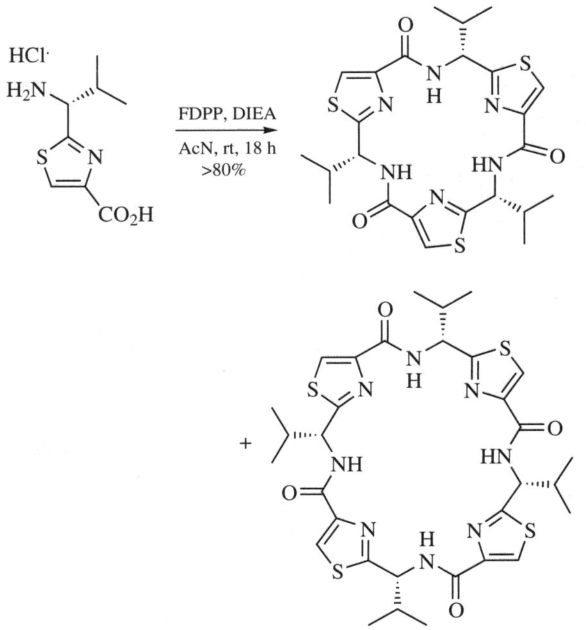

(14)

Couplings on Microarrays. In a recent adaptation of DNA microarray technology, Korbel et al. have used chiral fluorescent probes to measure the ee of tens of thousands of samples on reaction microarrays.[32] Their protocol employed an FDPP-mediated coupling of an equimolar mixture of two pseudoenantiomeric fluorescent reporters to the immobilized amino acid samples. During the coupling, the ee information of the sample becomes coded as a ratio of the fluorescent probes that is observable upon measurement of fluorescent emission at the appropriate excitation wavelength for each fluorophore.

Related Reagents. DCC, EDAC, IPCF, DPPA, BOP, BOP-Cl, HBTU, HATU, DEPBT.

1. Chen, S.; Xu, J., *Tetrahedron Lett.* **1991**, *32*, 6711–6714.
2. Barrow, R. A.; Hemscheidt, T.; Liang, J.; Paik, S.; Moore, R. E.; Tius, M. A., *J. Am. Chem. Soc.* **1995**, *117*, 2479–2490.
3. Dudash, J. J.; Jiang, J.; Mayer, S. C.; Joullié, M. M., *Synth. Commun.* **1993**, *23*, 349–356.
4. Ding, X.; Joullié, M. M., unpublished results.
5. Jackson, A. G.; Kenner, G. W.; Moore, G. A.; Ramage, R.; Thorpe, W. D., *Tetrahedron Lett.* **1976**, *40*, 3627.
6. Ramage, R.; Hopton, D.; Parrott, M. J.; Richardson, R. S.; Kenner, G. W.; Moore, G. A., *J. Chem. Soc., Perkin Trans. I* **1985**, 461.
7. Kovacs, J.; Mayers, G. L.; Johnson, R. H.; Cover, R. E.; Ghatak, U. R., *J. Chem. Soc., Chem. Commun.* **1970**, 53.
8. Kovacs, J.; Kisfaludy, L.; Ceprini, M. Q., *J. Am. Chem. Soc.* **1967**, *89*, 183.

9. Kisfaludy, L.; Schon, I. *Synthesis* **1983**, 325.
10. Atherton, E.; Sheppard, R. C., *J. Chem. Soc., Chem. Commun.* **1985**, 165.
11. Atherton, E.; Cameron, L. R.; Sheppard, R. C., *Tetrahedron* **1988**, *44*, 843.
12. Van der Auwera, C.; Van Damme, S.; Anteunis, M. J. O., *Int. J. Peptide Protein Res.* **1987**, *29*, 464.
13. Williams, M. W.; Young, G. T., *J. Chem. Soc.* **1963**, 881.
14. Kabachnik, M. I.; Zakharov, L. S.; Kudryavtsev, I. Y.; Kharchenko, A. P.; Tataurov, G. P.; Korobeinikova, L. I. *Izv. Akad. Nauk SSSR, Ser. Khim.* **1979**, *4*, 896.
15. Chen, S. Q.; Xu, J. C., *Chinese J. Chem.* **1995**, *13*, 175.
16. Jiang, J.; Li, W.-R.; Przeslawski, R. M.; Joullié, M. M., *Tetrahedron Lett.* **1993**, *34*, 6705.
17. Hahn, D. U.; Neidlein, R., *Heterocycles* **1999**, *50*, 1041.
18. Yang, A.; Gehring, A. P.; Li, T., *J. Chromatogr., A* **2000**, *878*, 165.
19. Le, H.-T.; Gallard, J.-F.; Mayer, M.; Guittet, E.; Michelot, R., *Bioorg. Med. Chem.* **1996**, *4*, 2201.
20. Konig, S.; Ugi, I.; Schramm, H. J., *Arch. Pharm.* **1995**, *328*, 699.
21. Deng, J.; Hamada, Y.; Shioiri, T.; Matsunaga, S.; Fusetani, N., *Angew. Chem., Int. Ed. Engl.* **1994**, *33*, 1729.
22. Deng, J.; Hamada, Y.; Shioiri, T., *Tetrahedron Lett.* **1996**, *37*, 2261.
23. Deng, J.; Hamada, Y.; Shioiri, T., *J. Am. Chem. Soc.* **1995**, *117*, 7824.
24. Deng, J.; Hamada, Y.; Shioiri, T. *Pept. Chem.* **1996**, *Volume date 1995*, 109.
25. Mayer, S. C.; Ramanjulu, J.; Vera, M. D.; Pfizenmayer, A. J.; Joullié, M. M., *J. Org. Chem.* **1994**, *59*, 5192.
26. Li, W.-R.; Ewing, W. R.; Harris, B. D.; Joullié, M. M., *J. Am. Chem. Soc.* **1990**, *112*, 7659.
27. Shih, C., PCT Int. Appl. WO 9723211 A1, 1997.
28. Aikins, J. A.; Briggs, B. S.; Zhang, T. Y.; Zmijewski Jr, M. J., PCT Int. Appl., 2000023429, 2000.
29. Salamonczyk, G. M.; Han, K.; Guo, Z.-W.; Sih, C. J., *J. Org. Chem.* **1996**, *61*, 6893.
30. Schumacher, K. K.; Hauze, D. B.; Jiang, J.; Szewczyk, J.; Reddy, R. E.; Davis, F. A.; Joullié, M. M., *Tetrahedron Lett.* **1999**, *40*, 455.
31. Bertram, A.; Hannam, J. S.; Jolliffe, K. A.; González-López de Turiso, F.; Pattenden, G., *Synlett* **1999**, 1723.
32. Korbel, G. A.; Lalic, G.; Shair, M. D., *J. Am. Chem. Soc.* **2001**, *123*, 361.

Michael S. Leonard & Madeleine M. Joullié
University of Pennsylvania, Philadelphia, PA, USA

4-Pentenoic Acid

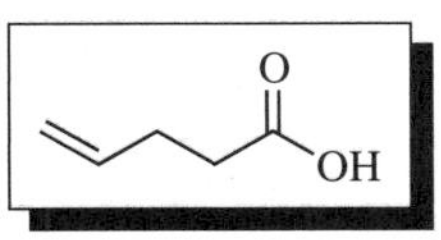

[591-80-0] $C_5H_8O_2$ (MW 100.12)

(reagent used for protection of amines and activation of certain alcohols[1])

Physical Data: mp −22.5 °C; bp 83–84 °C/12 mmHg; d 0.981 g cm^{-3}.
Solubility: soluble in alcohol, diethyl ether, and most organic solvents.
Form Supplied in: colorless liquid; widely available.
Purification: fractional distillation using a Vigreux column.
Handling, Storage, and Precautions: corrosive organic acid with an acrid odor. Use in a well-ventilated fume hood with gloves, eye protection, and protective clothing to avoid contact. Moderately toxic by ingestion. LD_{50} (rat, oral): 470 mg/kg.

Amides and Esters of 4-Pentenoic Acid. Amides and esters of 4-pentenoic acid will undergo ring-closure to form five-membered lactones in the presence of electrophiles. In the case of 4-pentenoic amides, this lactonization liberates the amine; hence, the 4-pentenoyl group serves as an amine-protecting group.[1] In the case of 4-pentenoic esters the lactonization cleaves the alkyl ester bond. Hereby the 4-pentenoic ester can serve as a leaving group for the corresponding alcohol.[2] This alcohol, however, has to be allylic or activated as a lactol for the lactonization to occur.

4-Pentenoic amides and esters can be prepared from the corresponding amine or alcohol by **1,3-dicyclohexylcarbodiimide** coupling with 4-pentenoic acid.[2,3] Alternative procedures involve reaction of the amine or the alcohol with 4-pentenoic anhydride or 4-pentenoyl chloride in the presence of a tertiary amine.[1,4] Both 4-pentenoic anhydride and 4-pentenoyl chloride are commercially available, but can also be easily prepared from the parent acid.[1,4] In addition, 4-pentenoyl chloride can be generated in situ by allowing 4-pentenoic acid to react with 1-amino-1-chloro-*N,N*,2-trimethyl-1-propene.[5] In a special case, cyanomethyl 4-pentenoate has been used for enzymatic resolution of amines by forming the 4-pentenoic amide from only one of the enantiomers (eq 1).[6]

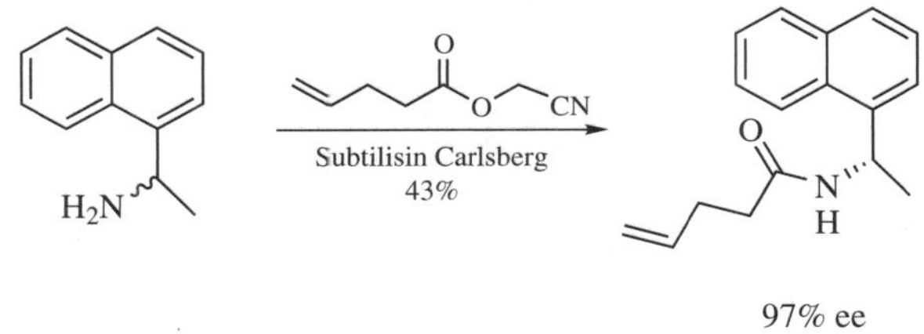

(1)

Deprotection of 4-Pentenoic Amides. 4-Pentenoic amides can be cleaved and the free amine regenerated by treatment with 3 equiv of **iodine** in THF–H_2O mixture (eq 2).[1] The deprotection

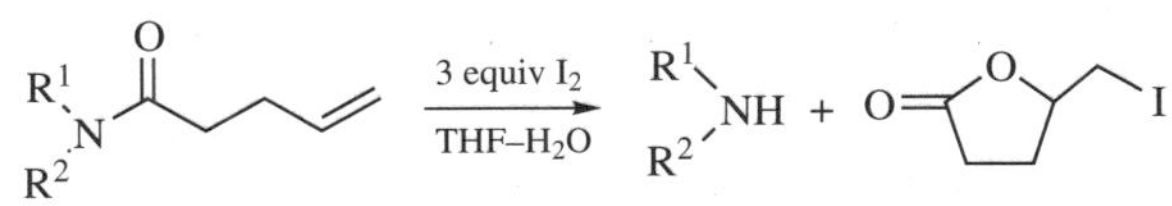

(2)

is usually complete within a few minutes and proceeds through an intermediate iminolactone which undergoes hydrolysis to the free amine and lactone. Although an oxidizing medium is used for the deprotection, a large variety of other oxidizable functionalities are stable to these conditions including allyl and *para*-methoxybenzyl ethers as well as alkyl sulfides.[1] Most acid and base labile functional groups are also stable to the deprotection conditions.[1] Even a 4-pentenyl glycoside is not affected by the pentenoyl deprotection (eq 3).[1] This has been used for preparation of 4-pentenyl 2-azido-2-deoxy-glucopyranosides which are useful glycosyl donors in oligosaccharide synthesis.[7] Notably, the 4-pentenoyl group is orthogonal to several other amine-protecting groups including BOC and phthalimide which is very useful in synthetic strategies involving several amino groups.[8]

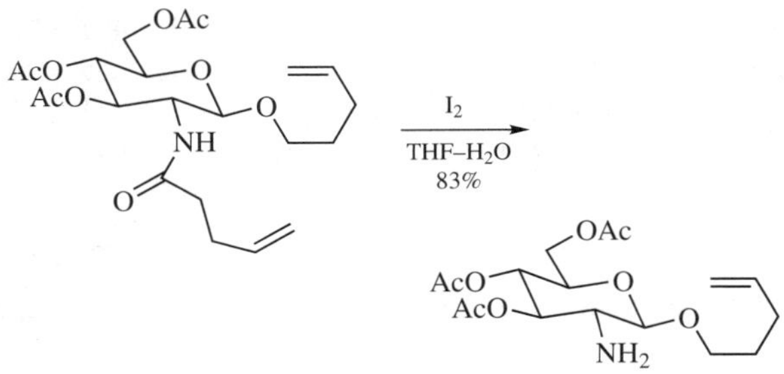

(3)

Deprotection of 4-pentenoic amides can also be achieved on solid phase.[9] This has been exploited for preparation of oligonucleotides where the 4-pentenoyl group serves as a protective group for the amino group in adenine, cytosine, and guanine.[10] The 4-pentenoyl protected nucleosides are coupled together on solid phase using the phosphoramidite or the *H*-phosphonate method.[10] Before cleaving the oligonucleotide from the solid support the 4-pentenoyl groups are removed with 2% iodine in pyridine–MeOH.[10]

Besides applications in carbohydrate and nucleoside chemistry the 4-pentenoyl group can also be used for protection of amino acids.[1,3] No epimerization of the amino acids is observed in the deprotections. In an interesting study valine benzyl ester was protected with the 4-pentenoyl group as well as derivatives containing alkyl substituents in the 2, 3, 4, or 5 positions (eq 4).[3] The purpose was to investigate the effect of alkyl substitution on the cleavage reaction. In the event, it turned out that the parent acid ($R^1 = R^2 = R^3 = R^4 = H$) was the best amine-protecting group.[3]

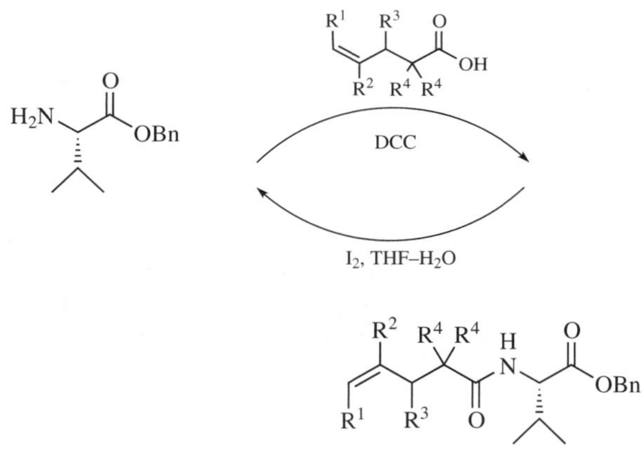

(4)

Ring-Closure of 4-Pentenoic Esters. Glycopyranoses which are protected at all positions except the anomeric center can be esterified with 4-pentenoic acid to form glycopyranosyl 4-pentenoates.[11] These can be activated with electrophiles for glycosylations in the same way as the corresponding 4-pentenyl glycosides.[12] Using **iodonium di-sym-collidine perchlorate** (IDCP) or *N*-**iodosuccinimide/trifluoromethanesulfonic acid** as the promoter the 4-pentenoate at the anomeric center undergoes the iodolactonization process and glycosylation of an added alcohol can be achieved (eq 5).[11]

Glycosyl 4-pentenoates can also be prepared directly from 2-(trimethylsilyl)ethyl glycosides by treatment with 4-pentenoic anhydride and **boron trifluoride etherate** (eq 6).[13]

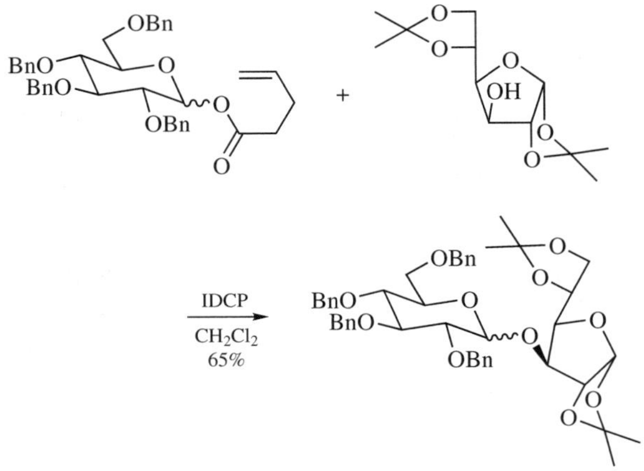

(5)

$\alpha/\beta = 3{:}1$

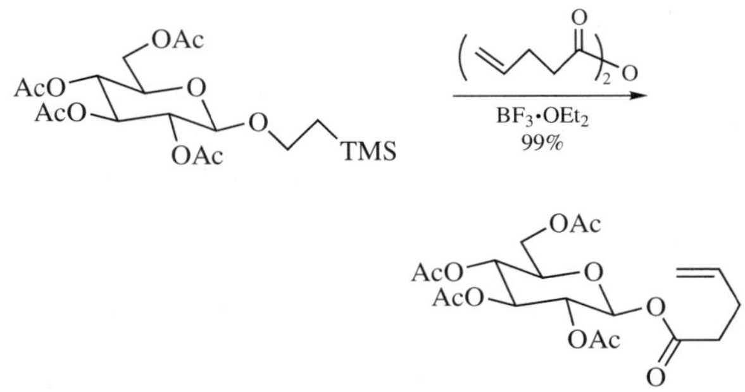

(6)

The 4-pentenoic esters can also participate in another glycosylation reaction involving a Ferrier rearrangement under neutral conditions.[2] In this case, the 3-hydroxy group in a glycal is esterified with 4-pentenoic acid. The glycal 3-pentenoate is then treated with IDCP and an alcohol to effect allylic rearrangement to the corresponding alkyl hex-2-enopyranoside (eq 7).[2]

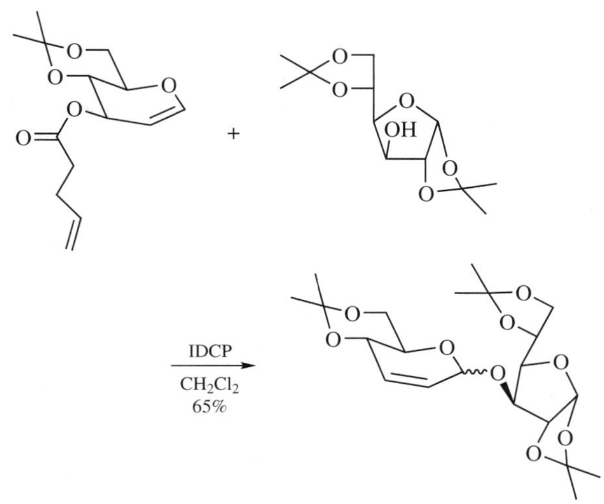

(7)

$\alpha/\beta = 6.7{:}1$

Finally, it should be noted that 4-pentenoic esters only undergo lactonization with electrophiles if they are attached at the anomeric center or at the 3-position in glycals. When the 4-pentenoic ester is placed at other positions in sugars, simple addition to the double bond will occur upon treatment with iodine and ROH.

Related Reagents. Benzyl chloroformate; di-*t*-butyl dicarbonate; tetrachlorophthalic anhydride.

1. Madsen, R.; Roberts, C.; Fraser-Reid, B., *J. Org. Chem.* **1995**, *60*, 7920.
2. López, J. C.; Gómez, A. M.; Valverde, S.; Fraser-Reid, B., *J. Org. Chem.* **1995**, *60*, 3851.
3. Lodder, M.; Golovine, S.; Laikhter, A. L.; Karginov, V. A.; Hecht, S. M., *J. Org. Chem.* **1998**, *63*, 794.
4. Kudo, H.; Sanda, F.; Endo, T., *Macromolecules* **1999**, *32*, 8370.
5. Fürstner, A.; Langemann, K., *J. Am. Chem. Soc.* **1997**, *119*, 9130.
6. Takayama, S.; Moree, W. J.; Wong, C.-H., *Tetrahedron Lett.* **1996**, *37*, 6287.
7. Olsson, L.; Jia, Z. J.; Fraser-Reid, B., *J. Org. Chem.* **1998**, *63*, 3790.
8. Debenham, J.; Rodebaugh, R.; Fraser-Reid, B., *Liebigs Ann./Recueil* **1997**, 791.
9. Carrington, S.; Renault, J.; Tomasi, S.; Corbel, J.-C.; Uriac, P.; Blagbrough, I. S., *Chem. Commun.* **1999**, 1341.
10. (a) Iyer, R. P.; Yu, D.; Habus, I.; Ho, N.-H.; Johnson, S.; Devlin, T.; Jiang, Z.; Zhou, W.; Xie, J.; Agrawal, S., *Tetrahedron* **1997**, *53*, 2731. (b) Iyer, R. P.; Yu, D.; Jiang, Z.; Agrawal, S., *Tetrahedron* **1996**, *52*, 14419. (c) Iyer, R. P.; Yu, D.; Ho, N.-H.; Devlin, T.; Agrawal, S., *J. Org. Chem.* **1995**, *60*, 8132.
11. (a) López, J. C.; Fraser-Reid, B., *J. Chem. Soc., Chem. Commun.* **1991**, 159. (b) Kunz, H.; Wernig, P.; Schultz, M., *Synlett* **1990**, 631.
12. Fraser-Reid, B.; Udodong, U. E.; Wu, Z.; Ottosson, H.; Merritt, J. R.; Rao, C. S.; Roberts, C.; Madsen, R., *Synlett* **1992**, 927.
13. Ellervik, U.; Magnusson, G., *Acta Chem. Scand.* **1993**, *47*, 826.

Robert Madsen & Bert Fraser-Reid
NPG Research Institute, Durham, North Carolina, USA

4-Pentenol

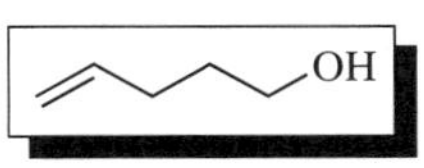

[821-09-0] $C_5H_{10}O$ (MW 86.13)

(reagent used to prepare synthetic intermediates containing the 4-pentenyloxy moiety which can be activated for use as a leaving group by exposure to electrophiles)

Physical Data: bp 139 °C, bp_3 63 °C, $bp_{0.45}$ 37–39 °C; density 0.858 g cm^{-3}; $\eta_D{}^{25}$ 1.4290.

Solubility: 5.7% by weight in water at 25 °C,[1] soluble in common organic solvents.

Form Supplied in: colorless liquid; widely available.

Purification: can be purified by distillation.

Handling, Storage, and Precautions: keep containers tightly closed, away from heat and open flame. Store in a cool dry place. Avoid contact with eyes, skin, and clothing; avoid breathing the vapor. Wash thoroughly after handling. The toxological properties of this compound have not been thoroughly investigated.

Use of 4-Pentenyloxy Groups as Glycosyl Leaving Groups. 4-Pentenyloxy groups are among the most effective glycosyl leaving groups for the synthesis of oligosaccharides,[2–6] and 4-pentenyl glycosides have recently been shown to be useful as linker arm precursors.[7–10] 4-Pentenyl glycosides can be obtained by any of the normal methods used for preparing glycosides from simple alcohols of which the most common ones are (i) the acid-catalyzed reaction of the unprotected sugar with 4-penten-1-ol which also acts as the solvent (Fischer glycosidation) (eq 1) and (ii) the reaction of peracylated glycosyl bromides with 4-penten-1-ol and a silver salt (Koenings–Knorr glycosidation) (eq 2).[5]

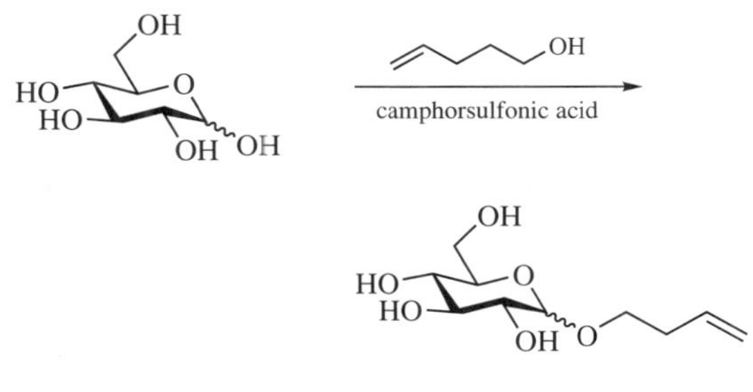

α:β (6:1) (1)

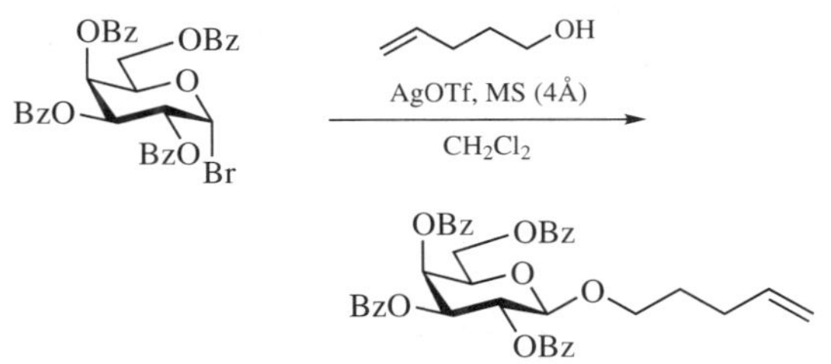

(2)

They have also been formed from the glycosides linked to solid supports by means of an 4-octene-1,8-diol linker by cross metathesis with ethylene.[11]

The 4-pentenyloxy group is activated as a leaving group by reaction with electrophiles, initially *N*-bromosuccinimide (NBS),[12] as shown in Figure 1.

Choosing[12] NBS as the activator for 4-pentenyl glycosides led to the synthetically useful discovery that 4-pentenyl glycosides with the remaining hydroxyls protected as esters react much more slowly than those with the hydroxyls protected by ethers. Ether-protected 4-pentenyl glycosides and ester-protected 4-pentenyl glycosides were termed 'armed' and 'disarmed' glycosyl donors, respectively. Iodonium dicollidine perchlorate (IDCP) only activates 'armed' glycosyl donors, not 'disarmed' glycosyl donors (eq 3),[5,13,14] allowing the development of strategies for oligosaccharide synthesis.[3,15,16]

More reactive activators, such as *N*-iodosuccinimide–triflic acid[5] and *N*-iodosuccinimide–triethylsilyl triflate,[2,5] cause all 4-pentenyl glycosides to be activated. It was recently shown that *N*-iodosuccinimide–ytterbium(III) triflate[17] is effective as an activator as it does not cleave acid-sensitive functional groups. 4-Pentenyloxy glycosides can be protected as dibromides by reaction with bromine and excess bromide ion in dichloromethane;[14] the alkene can be regenerated with zinc, samarium(II) iodide, or sodium iodide.[18]

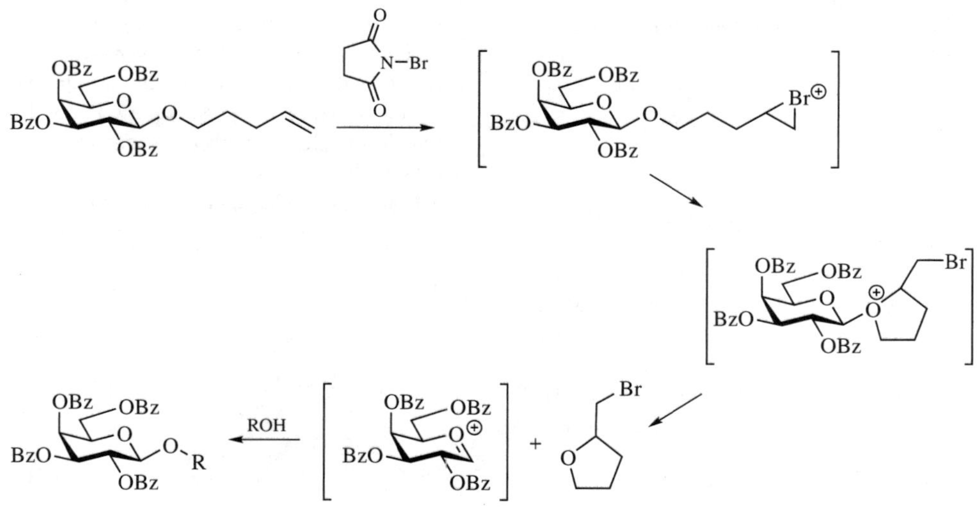

Figure 1 Activation of a 4-pentenyl glycoside

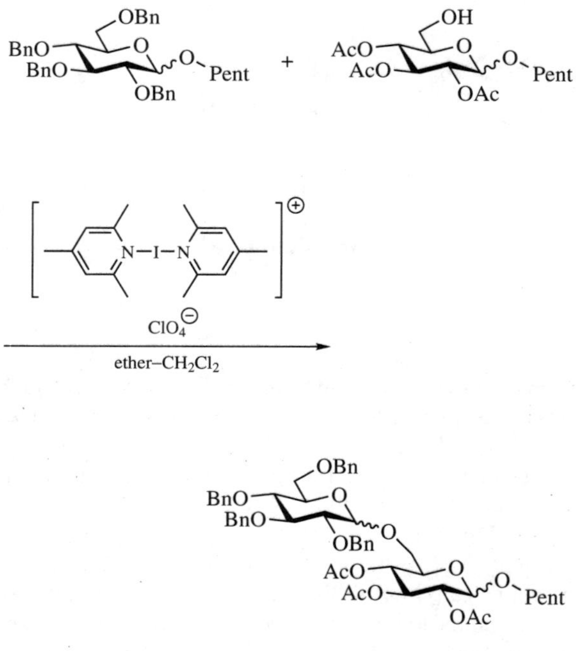

(3)

62%

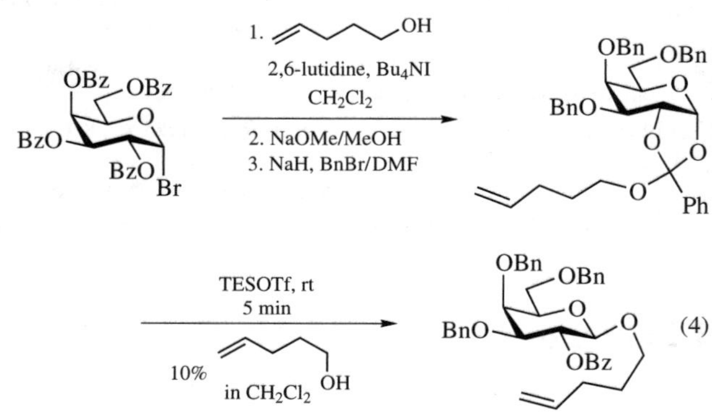

(4)

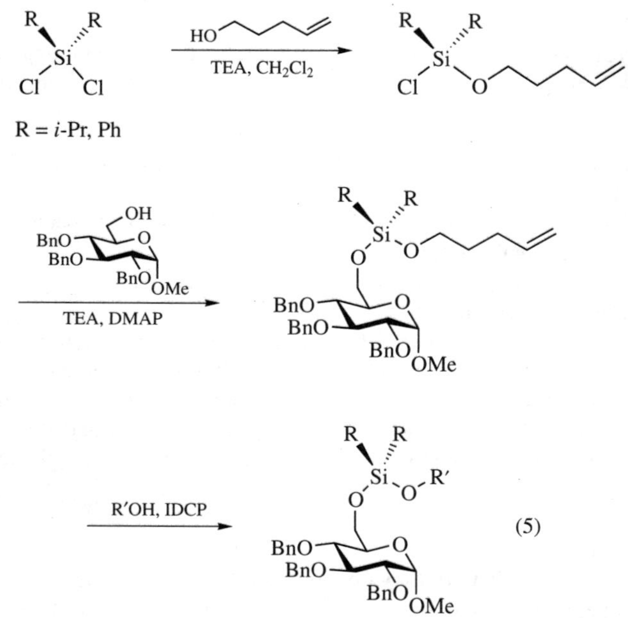

(5)

4-Pentenyloxy glycosyl orthoesters also serve as glycosyl donors and are often convenient intermediates in preparing differentially protected 4-pentenyl glycosides (eq 4).[2,19]

Silaketals. Silaketals are useful intermediates for the formation of bonds between molecules bearing hydroxyl groups using a variety of reactions.[20] Following earlier methodology,[21] Beau and co-workers[22] showed that 4-pentenyl chlorosilyl ethers are useful intermediates for the formation of silaketals (eqs 5 and 6).

Silaketals, such as the product of eq 6, have been used to make β-mannopyranosides[23] and, via radial reactions with pyridylsulfonyl glycosides, C-glycosides with *cis*-stereochemistry.[24]

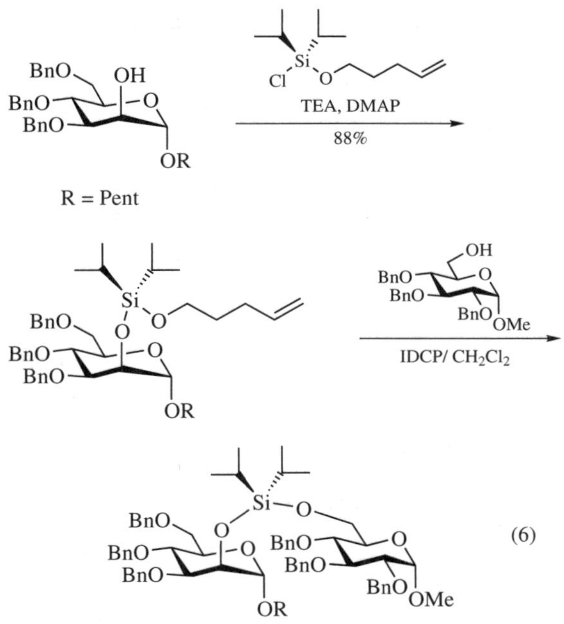

(6)

Conversion into Useful Synthons. 4-Penten-1-ol has been converted into two potentially useful synthons. The protection of the hydroxyl group as a tetrahydropyranyl ether (THP) through the reaction with dihydropyran,[25] followed by ozonolysis afforded 4-tetrahydropyranyloxybutanal (eq 7).[26]

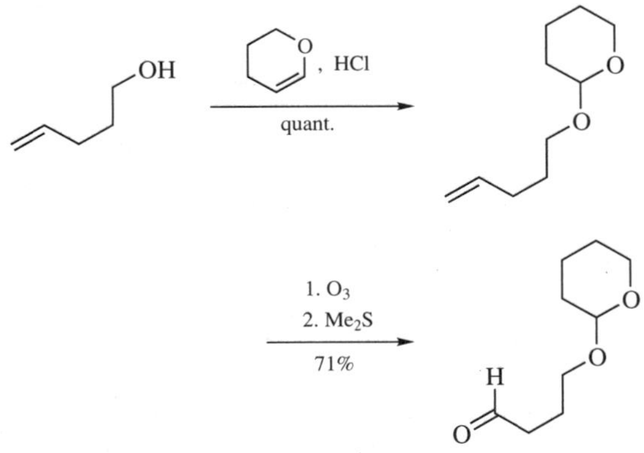

(7)

This synthon was used in a Wittig reaction during the synthesis of the diterpene, biflora-4,10(19),15-triene.[26]

4-Penten-1-ol has been epoxidized in good yields using an oxaziridinium salt. This result along with those obtained using *m*-CPBA are summarized in eq 8.[27]

Dipent-4-enyl Acetals. Dipent-4-enyl acetals were prepared in order to create reagents for acetal transfer under close to neutral conditions.[28] 4-Penten-1-ol was found to react with different aldehydes and ketones in hexane at 0 °C to give the corresponding acetals (the acetal transfer reagents) in good to excellent yields (eq 9).[29]

These products were found to transfer acetals in excellent yields to a variety of carbohydrate substrates containing acid-sensitive groups including silyl ethers, trityl ethers, benzylidene acetals, double bonds, and 1,6-anhydro rings. Of the substrates tested, only orthoesters were too acid-sensitive to give good yields (eq 10).[28]

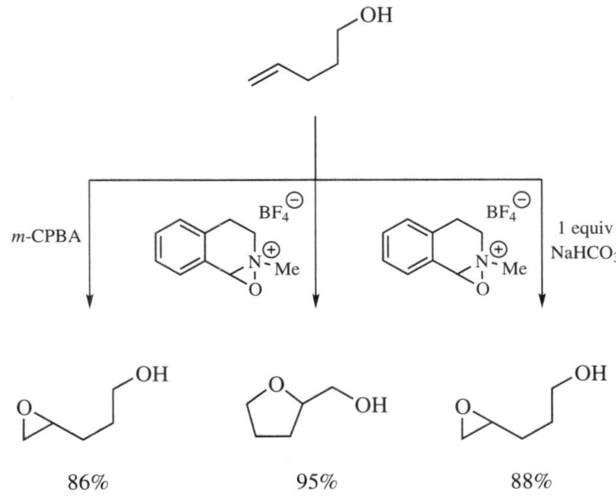

(8)

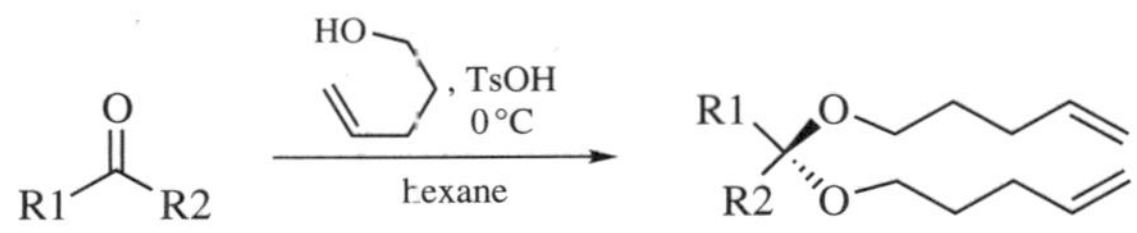

(9)

R1	R2	Yield (%)
Ph	H	76
4-MeOPh	H	95
Me	Me	84
$-(CH_2)_5-$		82

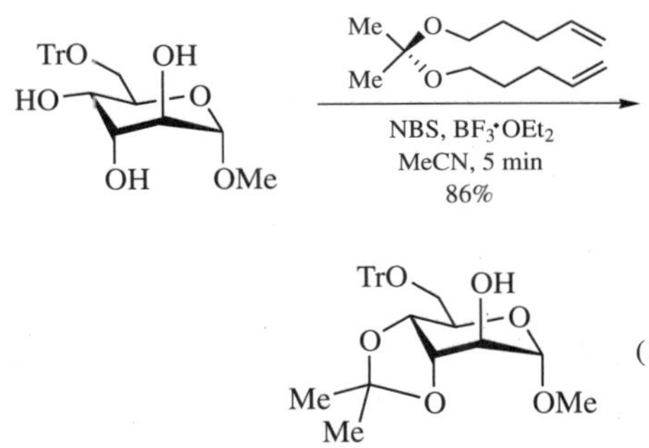

(10)

Other Reactions. 3-Methyl-2-hydroxycyclopentenone was reacted with 4-penten-1-ol and an acid to give enol ether which underwent intramolecular photocycloaddition (eq 11).[30]

In a convenient one-pot two-step procedure, alcohols (of which 4-penten-1-ol was one example) were converted into alkyl aryl sulfones via activation with NBS and PPh_3 (eq 12).[30]

Under Mitsunobu[31,32] conditions, 4-penten-1-ol was found to react with tetrachlorophthalimide[33] or dimethylmaleimide.[34] These products undergo formal [5 + 2] photochemical intramolecular cycloadditions (eq 13).[34]

The alkene group of 4-penten-1-ol serves as a radical acceptor in the synthesis of an α-GalNAc thioconjugate by reaction of excess alkene at reflux in chloroform with glycosyl thiol and trace azo(bis)isobutyronitrile as initiator.[35] It also acts as a terminal alkene in a cross metathesis reaction with acrylonitrile using an active ruthenium catalyst (eq 14).[36]

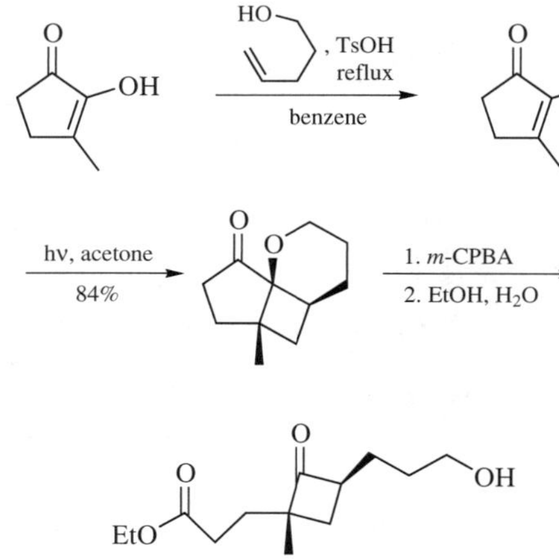

(11)

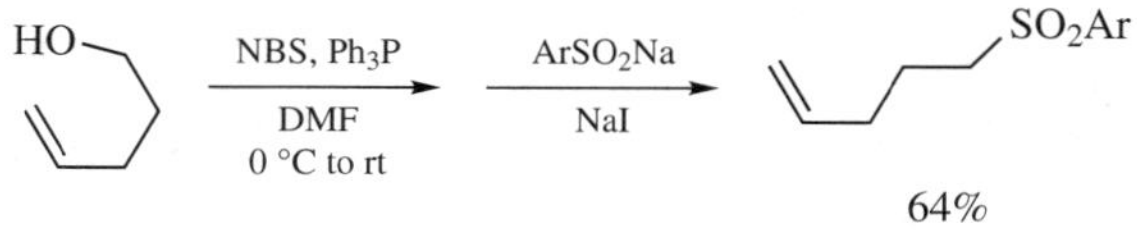

(12)

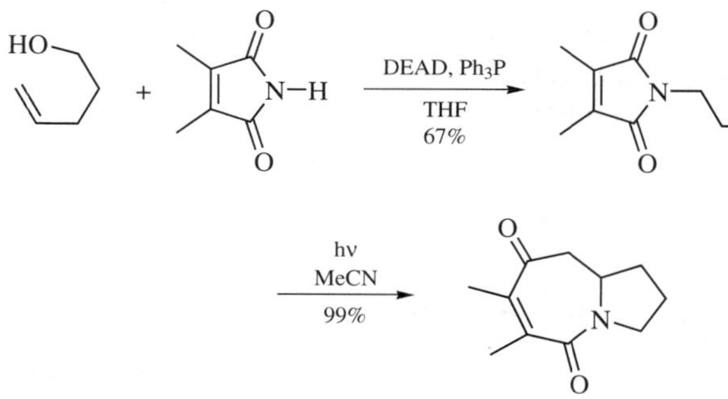

(13)

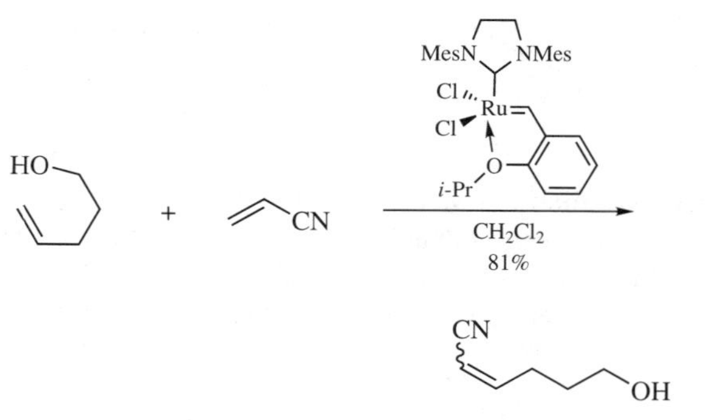

(14)

Z:*E* (3:1)

Related Reagents. Other glycosyl leaving groups, such as trichloroacetimidates, alkylthio groups, arylthiogroups, sulfoxides, phosphates; other alkenols.

1. Ginnings, P. M.; Herring, E.; Coltrane, D., *J Am. Chem. Soc.* **1939**, *61*, 807.
2. Allen, J. G.; Fraser-Reid, B., *J. Am. Chem. Soc.* **1999**, *121*, 468.
3. Arasappan, A.; Fraser-Reid, B., *J. Org. Chem.* **1996**, *61*, 2401.
4. Debenham, J. S.; Rodebaugh, R.; Fraser-Reid, B., *J. Org. Chem.* **1997**, *62*, 4591.
5. Fraser-Reid, B.; Udodong, U. E.; Wu, Z.; Ottosson, H.; Merritt, J. R.; Rao, C. S.; Roberts, C.; Madsen, R., *Synlett* **1992**, 927.
6. Fraser-Reid, B.; Merritt, J. R.; Handlon, A. L.; Andrews, C. W., *Pure Appl. Chem.* **1993**, *65*, 779.
7. Buskas, T.; Soderberg, E.; Konradsson, P.; Fraser-Reid, B., *J. Org. Chem.* **2000**, *65*, 958.
8. Allen, J. R.; Danishefsky, S. J., *J. Am. Chem. Soc.* **1999**, *121*, 10875.
9. Allen, J. R.; Harris, C. R.; Danishefsky, S. J., *J. Am. Chem. Soc.* **2001**, *123*, 1890.
10. Ratner, D. M.; Plante, O. J.; Seeberger, P. H., *Eur. J. Org. Chem.* **2002**, 826.
11. Andrade, R. B.; Plante, O. J.; Melean, L. G.; Seeberger, P. H., *Org. Lett.* **1999**, *1*, 1811.
12. Mootoo, D. R.; Date, V.; Fraser-Reid, B., *J. Am. Chem. Soc.* **1988**, *110*, 2662.
13. Mootoo, D. R.; Konradsson, P.; Udodong, U.; Fraser-Reid, B., *J. Am. Chem. Soc.* **1988**, *110*, 5583.
14. Fraser-Reid, B.; Wu, Z.; Udodong, U. E.; Ottosson, H., *J. Org. Chem.* **1990**, *55*, 6068.
15. Merritt, J. R.; Naisang, E.; Fraser-Reid, B., *J. Org. Chem.* **1994**, *59*, 4443.
16. Rodebaugh, R.; Debenham, J. S.; Fraser-Reid, B., *J. Carbohydr. Chem.* **1997**, *16*, 1407.
17. Jayaprakash, K. N.; Radhakrishnan, K. V.; Fraser-Reid, B., *Tetrahedron Lett.* **2002**, *43*, 6953.
18. Merritt, J. R.; Debenham, J. S.; Fraser-Reid, B., *J. Carbohydr. Chem.* **1996**, *15*, 65.
19. Mach, M.; Schlueter, U.; Mathew, F.; Fraser-Reid, B.; Hazen, K. C., *Tetrahedron* **2002**, *58*, 7345.
20. Fensterbank, L.; Malacria, M.; Sieburth, S. M., *Synthesis* **1997**, 813.
21. Gillard, J. W.; Fortin, R.; Morton, H. E.; Yoakim, C.; Quesnelle, C. A.; Daignault, S.; Guindon, Y., *J. Org. Chem.* **1988**, *53*, 2602.
22. Colombier, C.; Skrydstrup, T.; Beau, J. M., *Tetrahedron Lett.* **1994**, *35*, 8167.
23. Stork, G.; Kim, G., *J. Am. Chem. Soc.* **1992**, *114*, 1087.
24. Skrydstrup, T.; Mazeas, D.; Elmouchir, M.; Doisneau, G.; Riche, C.; Chiaroni, A.; Beau, J. M., *Chem. Eur. J.* **1997**, *3*, 1342.
25. Woods, G. F.; Kramer, D. N., *J. Am. Chem. Soc.* **1947**, *69*, 2246.
26. Parker, K. A.; Farmar, J. G., *J. Org. Chem.* **1986**, *51*, 4023.
27. Lusinchi, X.; Hanquet, G., *Tetrahedron* **1997**, *53*, 13727.
28. Madsen, R.; Fraser-Reid, B., *J. Chem. Soc., Chem. Commun.* **1994**, 749.
29. Madsen, R.; Fraser-Reid, B., *J. Org. Chem.* **1995**, *60*, 772.
30. Pirrung, M. C.; Chang, V. K.; DeAmicis, C. V., *J. Am. Chem. Soc.* **1989**, *111*, 5824.
31. Mitsunobu, O., *Synthesis* **1981**, 1.
32. Hughes, D. L., *Org. Prep. Proced. Int.* **1996**, *28*, 127.
33. Jia, Z. J.; Kelberlau, S.; Olsson, L.; Anilkumar, G.; Fraser-Reid, B., *Synlett* **1999**, 565.
34. Booker-Milburn, K. I.; Anson, C. E.; Clissold, C.; Costin, N. J.; Dainty, R. F.; Murray, M.; Patel, D.; Sharpe, A., *Eur. J. Org. Chem.* **2001**, 1473.
35. Knapp, S.; Myers, D. S., *J. Org. Chem.* **2002**, *67*, 2995.
36. Randl, S.; Gessler, S.; Wakamatsu, H.; Blechert, S., *Synlett* **2001**, 430.

Hussein Al-Mughaid & T. Bruce Grindley
Dalhousie University, Halifax, NS, Canada

Phenyl-Bis(thiophenyl)sulfonium Hexachloroantimonate

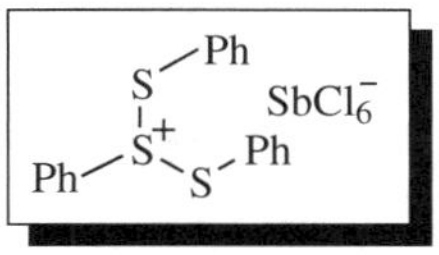

$C_{18}H_{15}S_3SbCl_6$ (MW 661.99)

(reagent used as phenylsulfenium cation transfer agent in Ad_E reactions)

Physical Data: colorless solid, stable for at least 12 h at −78 °C, decomposes above −20 °C.
Solubility: soluble in SO_2, CH_2Cl_2, CH_3NO_2; unsoluble in Et_2O.
Form Supplied in: generated in situ from available reagents.
Purification: no purification is required.
Handling, Storage, and Precautions: moisture sensitive, can be stored under nitrogen in the solution of CH_2Cl_2 or SO_2 at −78 °C indefinitely or at −20 °C for at least 24 h.

Preparation of the Reagent. The title reagent can be easily prepared using either of the alternative procedures represented below by the eqs 1 or 2:

$$3(PhS)_2 + 3SbCl_6 \rightarrow 2[PhSS^+(Ph)SPhSbCl_6^-] + SbCl_3 \quad (1)$$

$$PhSCl + (PhS)_2 + SbCl_6 \rightarrow [PhSS^+(Ph)SPhSbCl_6^-] \quad (2)$$

In both cases the reaction is carried out by mixing equimolar amounts of the indicated reagents in CH_2Cl_2 solution at −78 °C. The reaction is complete within 1 h (^{13}C NMR monitoring data) to give a quantitative yield of the reagent as 0.25–0.5 M solution ready for the further use.[1–3]

Generation of Episulfonium salts upon Interaction with Alkenes. Phenyl-bis(phenylthio)sulfonium hexachloroantimonate as well as other aryl-bis(arylthio)sulfonium salts (vide infra) are the reagents of choice for the preparation of the episulfonium salts from various alkenes under mild conditions. In fact Ad_E reaction of this salt with representative alkenes such as *cis*-cyclooctene, norbornene, tetramethylethylene, cyclohexene, 1-methylcyclohexene, and styrene proceeded readily at −60 °C to give the respective episulfonium hexachloroantimonates in good to excellent yields (eqs 3–8).[1,2]

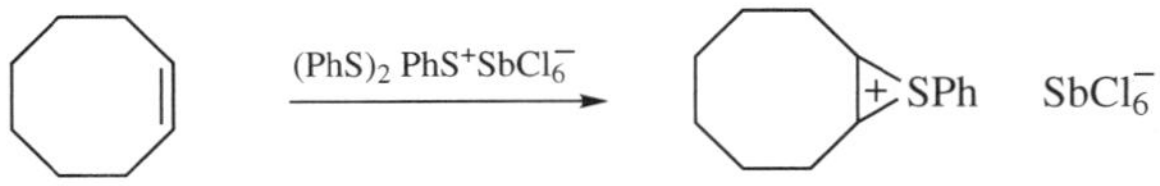
(3)

(4)

(5)

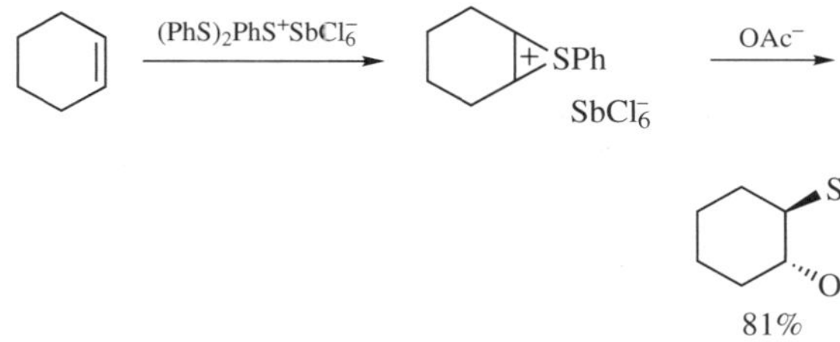
(6)

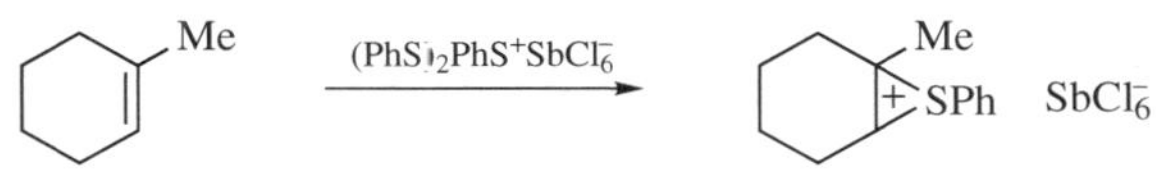
(7)

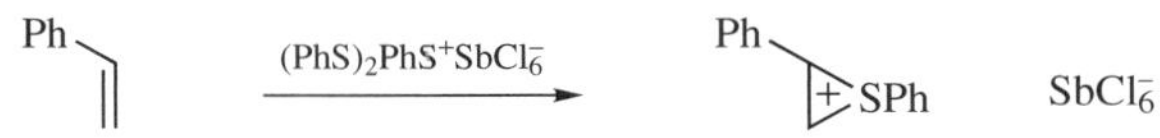
(8)

S-Phenylepisulfonium salts formed in the reactions with norbornene and cyclooctene (eqs 3 and 4) were isolated as stable solids in nearly quantitative yields and were fully characterized. For other examples, complete conversion of the starting alkenes into the respective episulfonium ions was ascertained by ^1H and ^{13}C NMR studies of the reaction mixtures and/or by the high yield isolation of the conventional products formed upon the quenching of these intermediates with nucleophiles such as AcO^- (e.g., eq 6).

Electrophilic Reagent in the Synthesis of 2-Deoxy-β-glycosides from Glycals. The title reagent was successfully employed as the coupling agent to achieve an efficient preparation of the series of 2-deoxy-β-glycosides from readily available protected glycals in accordance with the general eq 9.[4]

The glycosylation reactions could be carried out in CH_2Cl_2 at −60 °C either by the addition of the sulfonium salt to the equimolar solution of aglycone and glycal or by the addition of glycal to the solution of aglycone and sulfonium salt. The reaction is rather general in scope and is applicable for quite a number of various protected glycals. The list of aglycones successfully used in glycosilation with this reagent includes aliphatic alcohols, stannyl ether derivatives of phenols or diisopropylidenegalactose. The products are usually formed in good yields. Typically, the reaction proceeds with good to excellent diastereofacial selectivity and furnishes β-glycosides as the major product arising from below-plane attack of electrophile (PhS^+) followed by the trans-opening of episulfonium ion-like intermediate by aglycone. The ratio of β-glycoside/α-glycoside varies within the limits 2:1 and 11.5:1 depending on the nature of aglycone and glycals. The only exception to this generalization is the glycal bearing an axial alkoxy group at C-3 which gave predominantly α-glycoside (α:β = 10:1). The elaborated procedure seems to be more general and reliable than previously reported methods based upon the utilization of preformed phenylsulfenate ethers of the desired aglycone or selenium electrophiles. It is recommended as a practical route for the synthesis of various 2-deoxy-β-glycosides.

Related Reagents. A number of other aryl-bis(arylthio)sulfonium salts, $ArS^+(SAr)_2SbCl_6^-$, where Ar = 2- or 4-methylphenyl, 3,4-dimethylphenyl, 2,4,6-trimethylphenyl, 4-methoxyphenyl, 4-chlorophenyl, and 2-naphtyl had been also prepared and studied as reagents both for the preparation of *S*-arylepisulfonium salts or in the synthesis of 2-deoxyglycosides. It was established that the stability of episulfonium salts may vary dramatically depending on the nature of *S*-aryl substituent. Variation of the same parameter might also affect the steric outcome of the glycosylation reaction.

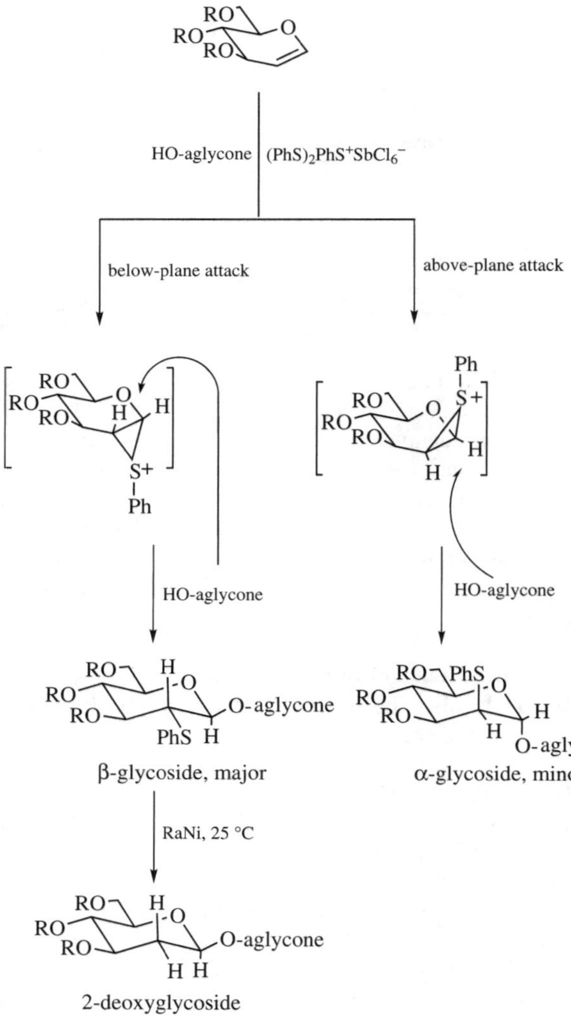

(9)

1. Gybin, A. S.; Smit, W. A.; Bogdanov, V. S.; Krimer, M. Z.; Kalyan, Y. B., *Tetrahedron Lett.* **1980**, *21*, 383.
2. Bogdanov, V. S.; Gybin, A. S.; Cherepanova, E. G.; Smit, W. A., *Izv. Acad. Nauk SSSR, Ser. Khim.* **1981**, 2681.
3. Gybin, A. S.; Smit, W. A.; Cherepanova, E. G.; Kalyan, Y. B.; Krimer, M. Z.; Lemeshev, A. N., *Izv. Acad. Nauk SSSR, Ser. Khim.* **1982**, 387.
4. Grewal, G.; Kaila, N.; Franck, R. W., *J. Org. Chem.* **1992**, *57*, 2084.

William A. Smit
Russian Academy of Sciences, Moscow, Russia

Phenyl Chlorothionocarbonate

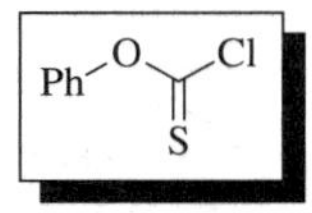

[1005-56-7] C_7H_5ClOS (MW 172.64)

(forms thionocarbonate ester derivatives of alcohols which can be deoxygenated with tin hydride reagents;[1–5] converts ribonucleosides to 2′-deoxynucleosides;[2] provides allylic thionocarbonates which undergo [3,3]-sigmatropic shifts;[6–9] provides precursors for radical bond-forming reactions;[10,11] reagent for thioacylation[12–16])

Physical Data: bp 81–83 °C/6 mmHg; fp 81 °C; *d* 1.248 g cm^{-3}.
Solubility: sol chloroform, THF.
Form Supplied in: colorless liquid.
Handling, Storage, and Precautions: corrosive; moisture sensitive; should be stored in airtight containers which exclude moisture; incompatible with alcohol solvents.

Deoxygenation of Secondary Alcohols. Reaction of secondary alcohols with the reagent in the presence of Pyridine or 4-Dimethylaminopyridine (DMAP) provides thiocarbonate ester derivatives which can be reduced to alkanes using Tri-n-butylstannane (eq 1).[1] The advantage of this method is the ability to deoxygenate alcohols via radical intermediates and thereby avoid problems associated with ionic reaction conditions (i.e. carbonium ion rearrangements, reduction of other functional groups). This method is particularly useful for the conversion of ribonucleosides to 2′-deoxynucleosides. For example, adenosine can be converted to 2′-deoxyadenosine in 78% overall yield by initial protection of the 3′- and 5′-hydroxyl groups as a cyclic disiloxane, thiocarbonylation, reductive cleavage, and then final deprotection using a fluoride source (eq 2).[2] Treatment of 2′-bromo-3′-phenoxythiocarbonyl nucleosides with tributyltin hydride affords 2′,3′-didehydro-2′,3′-dideoxy nucleosides via radical β-elimination.[3]

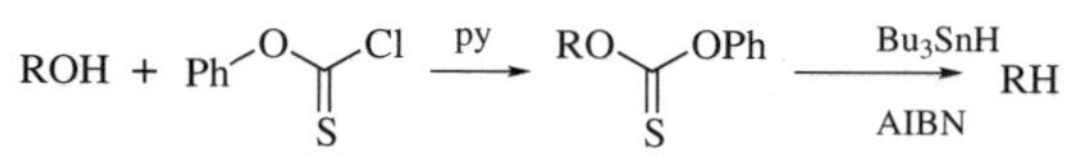

(1)

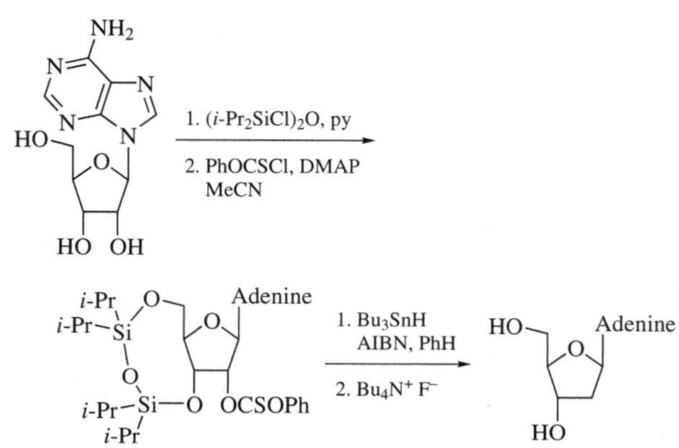

(2)

Synthetic intermediates can be selectively deoxygenated without reduction of other functional groups such as esters, ketones, and oxime ethers (eq 3),[4] as well as epoxides, acetate esters, and alkenes.[5]

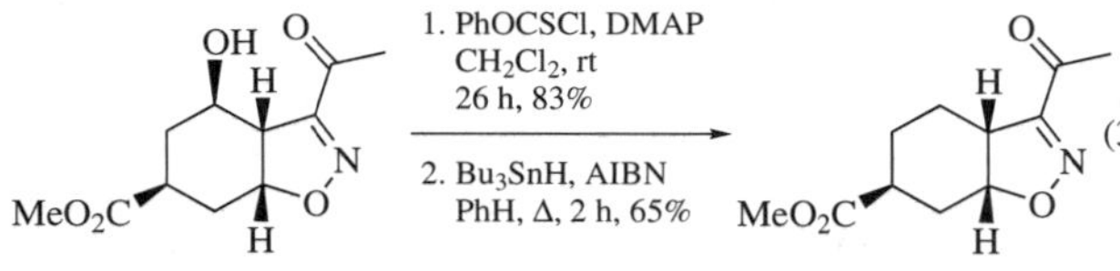

(3)

Sigmatropic Rearrangements of Allylic Thionocarbonates. The reaction of allyl alcohols with aryl chlorothionocarbonates affords *S*-allyl aryl thiocarbonates by [3,3]-sigmatropic rearrangement via the intermediate thionocarbonate esters.[6–9] For example, treatment of 2-methyl-1-penten-3-ol with the reagent in pyridine at −20 °C affords phenyl 2-methyl-2-pentenyl thionocarbonate in 67% yield (*E*:*Z* = 96.5:3.5).[7] This type of rearrangement, coupled with tin hydride mediated reduction of the phenyl thiocarbonate ester product, was used as a key step in the synthesis of isocarbacyclin (eq 4).[8] Rearrangement of cyclic thionocarbonates contained in eight-membered rings or smaller provides two-atom ring enlarged thiocarbonates having (*Z*) double bond geometry (eq 5).[9] Depending on the system, the cyclic thiocarbonates are obtained by either treatment of the diol monothionocarbonate with base or by reaction of the diol with 1,1′-thiocarbonyldi-2,2′-pyridone. Cyclic thionocarbonates of ring size nine or larger afford ring expanded products with exclusive (*E*) double bond geometry in modest yields.

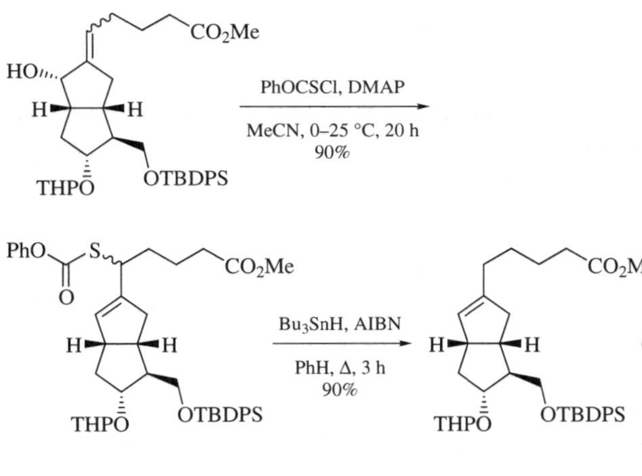

(4)

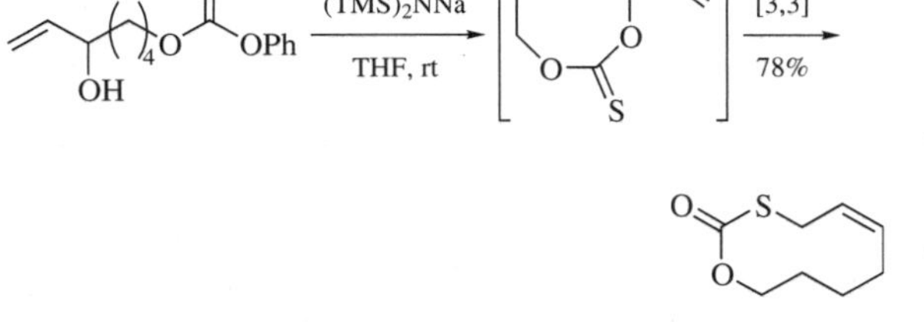

(5)

Radical Coupling and Cyclization Reactions. Phenyl thionocarbonate esters derived from alcohols serve as efficient precursors for the generation of radical intermediates which can be used for the formation of new carbon–carbon bonds. For example, a 4-thionocarbonate ester derived from L-lyxose undergoes a stereoselective allylation upon photolysis in toluene in the presence of 2.0 equiv of Allyltributylstannane (eq 6).[10] Photochemical initiation is preferable to chemical initiation using Azobisisobutyronitrile which results in the formation of side products at the expense of the desired product. The allylation product was used further in a total synthesis of pseudomic acid C.

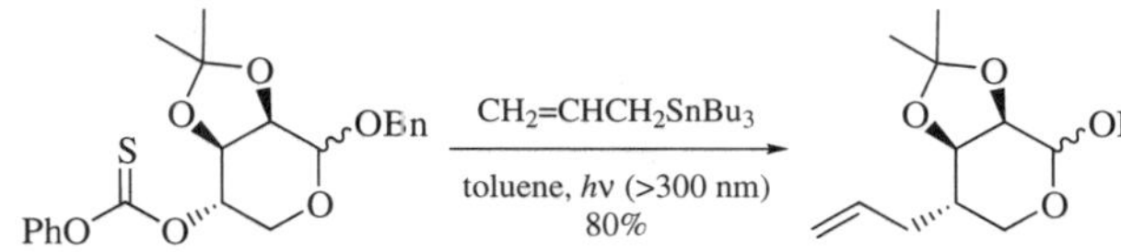

(6)

Oxime ethers derived from hydroxy aldehydes, upon conversion to their phenyl thionocarbonate esters, undergo radical cyclizations resulting in the formation of carbocycles.[11] For example, an oxime ether obtained from D-glucose is converted into its phenyl thionocarbonate ester at C-5 and, upon heating in benzene in the presence of tributyltin hydride, affords cyclopentanes in 93% yield as a 62:38 mixture of two diastereomers (eq 7). In general, only low to modest stereoselectivity between the newly formed stereocenters is observed in a number of substrates examined.

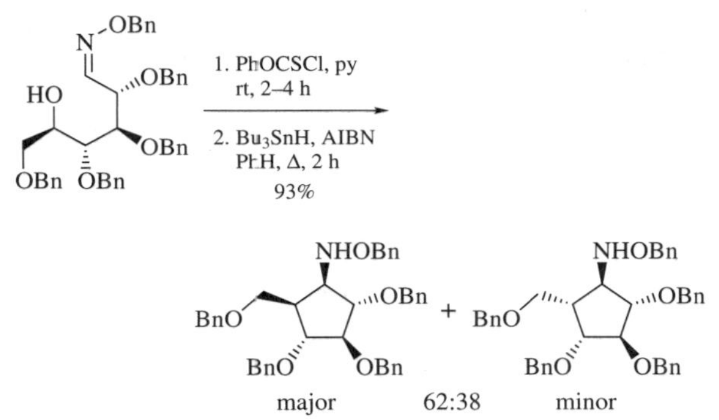

(7)

Thioacylation Reagent. The regioselective thioacylation of unprotected carbohydrates via the agency of Di-n-butyltin Oxide and the reagent has been investigated.[12] Glycopyranosides having a *cis*-diol arrangement, e.g. galactose, form cyclic thiocarbonates which can be either converted into dihydropyranosides using the Corey elimination procedure or deoxygenated to a mixture of 3- and 4-deoxyglycopyranosides (eq 8). Methyl D-glucopyranoside is monothioacylated with the regioselectivity dependent upon the configuration at the anomeric carbon; the α-epimer gives 83% of 2-thionocarbonate and the β-epimer gives 85% of 6-thionocarbonate. Further treatment with tributyltin hydride affords the corresponding deoxyglucose derivatives in high yield.

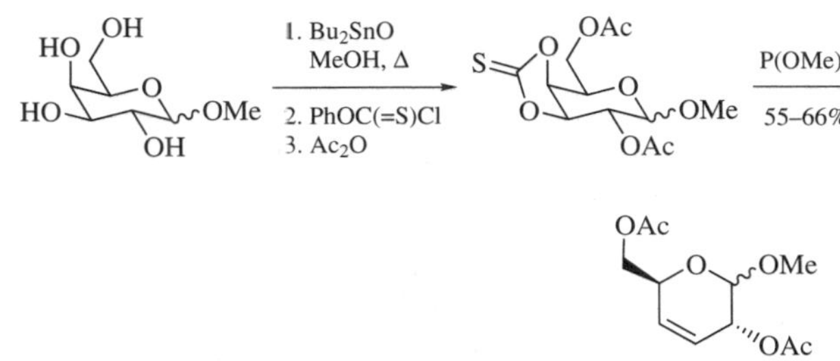

(8)

In a key step leading to a synthesis of saxitoxin, radical fragmentation of a pyrazolidine ring followed by intramolecular thioacylation afforded the ring expanded tetrahydropyrimidine intermediate (eq 9).[13] The thionocarbamate activation of the pyrazolidine N-H was found to be necessary to effect the desired transformation.

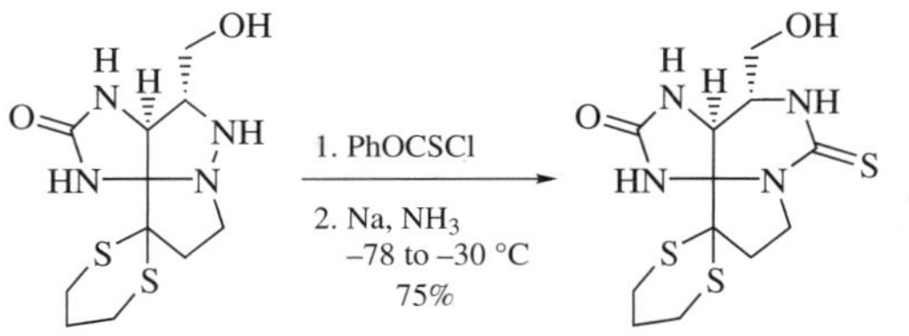

(9)

The intramolecular thioacylation of an ester enolate was used for the synthesis of 2-alkylthiopenem carboxylic acid derivatives.[14] Sequential acylations have led to the synthesis of zwitterionic pyrazole-5-thiones from acyclic precursors,[15] whereas 2-ethoxyoxazolidines react with the reagent to afford the products of N-acylation.[16]

Heating of O-phenyl thionocarbonates of pyrrolidine and piperidine-2-ethanols in acetonitrile gives a ring expanded azepine or an octahydroazocine accompanied by the pyrrolidine and piperidine O-phenyl ethers (eq 10).[17] These products arise via the internal expulsion of carbonyl sulfide, leading to formation of an azetidinium intermediate followed by nucleophilic ring opening with phenoxide ion.

Related Reagents. Carbon Disulfide; p-Tolyl Chlorothionocarbonate.

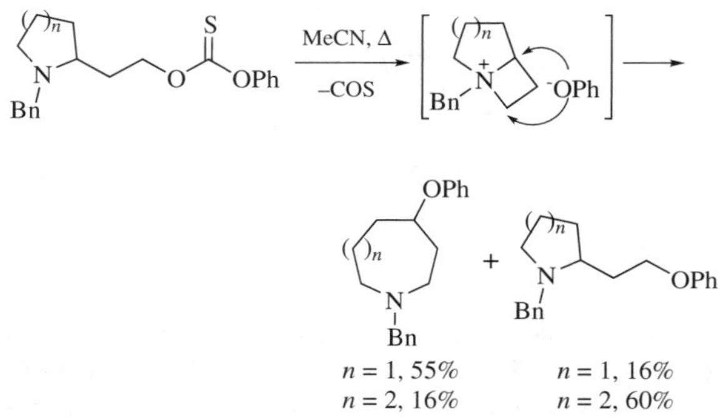

(10)

1. Barton, D. H. R.; Subramanian, R. *J. Chem. Soc., Perkin Trans. 1* **1977**, 1718.
2. Robins, M. J.; Wilson, J. S. *J. Am. Chem. Soc.* **1981**, *103*, 932; Robins, M. J.; Wilson, J. S.; Hansske, F. *J. Am. Chem. Soc.* **1983**, *105*, 4059.
3. Serafinowski, P. *Synthesis* **1990**, 411.
4. Martin, S. F.; Dappen, M. S.; Dupre, B.; Murphy, C. J. *J. Org. Chem.* **1987**, *52*, 3706.
5. Schuda, P. F.; Potlock, S. J.; Wannemacher, Jr., R. W. *J. Nat. Prod.* **1984**, *47*, 514.
6. Garmaise, D. L.; Uchiyama, A.; McKay, A. F. *J. Org. Chem.* **1962**, *27*, 4509.
7. Faulkner, D. J.; Petersen, M. R. *J. Am. Chem. Soc.* **1973**, *95*, 553.
8. Torisawa, Y.; Okabe, H.; Ikegami, S. *Chem. Commun./J. Chem. Soc., Chem. Commun.* **1984**, 1602.
9. Harusawa, S.; Osaki, H.; Kurokawa, T.; Fujii, H.; Yoneda, R.; Kurihara, T. *Chem. Pharm. Bull.* **1991**, *39*, 1659.
10. Keck, G. E.; Kachensky, D. F.; Enholm, E. J. *J. Org. Chem.* **1985**, *50*, 4317.
11. Bartlett, P. A.; McLaren, K. L.; Ting, P. C. *J. Am. Chem. Soc.* **1988**, *110*, 1633.
12. Haque, M. E.; Kikuchi, T.; Kanemitsu, K.; Tsuda, Y. *Chem. Pharm. Bull.* **1987**, *35*, 1016.
13. Jacobi, P. A.; Martinelli, M. J.; Polanc, S. *J. Am. Chem. Soc.* **1984**, *106*, 5594.
14. Leanza, W. J.; DiNinno, F.; Muthard, D. A.; Wilkering, R. R.; Wildonger, K. J.; Ratcliffe, R. W.:Christensen, B. G. *Tetrahedron* **1983**, *39*, 2505.
15. Grohe, K.; Heitzer, H.; Wendisch, D. *Justus Liebigs Ann. Chem./Liebigs Ann. Chem.* **1982**, 1602.
16. Widera, R.; Muehlstaedt, M. *J. Prakt. Chem.* **1982**, *324*, 1005.
17. Sakanoue, S.; Harusawa, S.; Yamazaki, N.; Yoneda, R.; Kurihara, T. *Chem. Pharm. Bull.* **1990**, *38*, 2981.

Eric D. Edstrom
Utah State University, Logan, UT, USA

1,2-Phenylene Phosphorochloridate

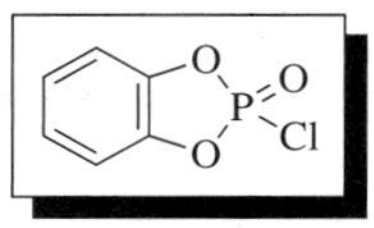

[1449-17-8] $C_6H_4ClO_3P$ (MW 190.52)

(phosphorylating agent for primary alcohols,[1,2] particularly in nucleoside chemistry[3])

Physical Data: bp 120 °C/9 mmHg; 91 °C/0.9 mmHg.
Solubility: sol most organic solvents; react with alcohols.
Form Supplied in: liquid; commercially available.
Analysis of Reagent Purity: FT-IR.[11]
Preparative Methods: reaction of Catechol with Phosphorus(V) Chloride gives the intermediate (**2**) which on warming with Acetic Anhydride decomposes giving acetyl chloride and 1,2-phenylene phosphorochloridate (**1**) in 88% yield (eq 1).[1]

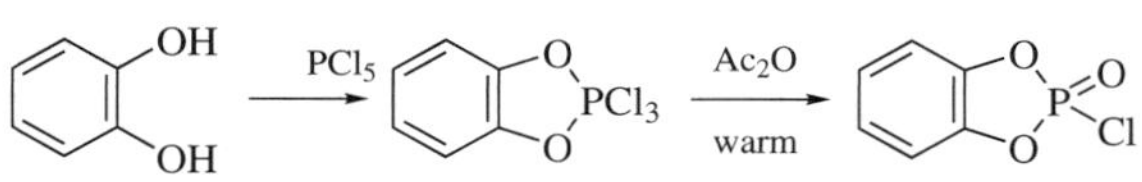

(**2**) (**1**) (1)

Purification: redistil under reduced pressure.
Handling, Storage, and Precautions: will keep indefinitely in a sealed tube. Harmful and irritant; moisture sensitive. Incompatible with alcohols and water.[4] Use in a fume hood.

Phosphorylation. 1,2-Phenylene phosphorochloridate is a convenient and efficient phosphorylating agent. Reaction with primary alcohols in the presence of base (Pyridine) gives the phosphate ester (**3**). This is readily hydrolysed to the o-hydroxyphenyl ester (**4**), which on treatment with Bromine in neutral aqueous buffer gives the monophosphate ester (**5**) in good yield (eq 2).[2,5]

Adams' catalyst[6] has been used as an alternative to bromine in the last step and the use of other oxidative conditions involving periodic acid and Lead(IV) Acetate has also been reported.[3] Unsaturated alcohols are not compatible with some of the reaction conditions used to cleave (**4**).

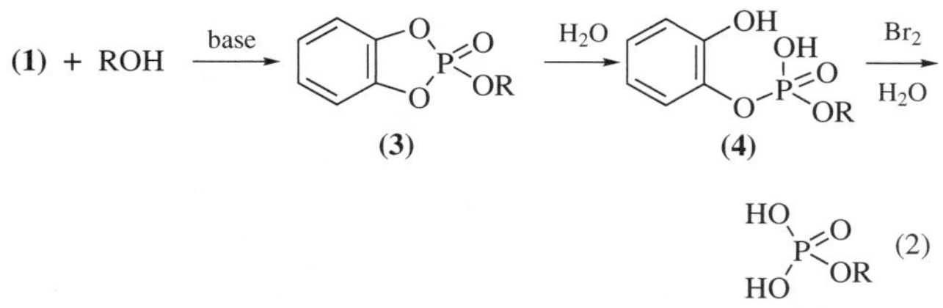

(3) (4) (5) (2)

The reagent is particularly useful for the phosphorylation of nucleotides (eq 3).[3] Reaction of (**1**) with 2′,3′-*O*-isopropylidene ribonucleosides (**6**) is carried out at room temperature in the presence of 2,6-Lutidine for 3 hours. Addition of water hydrolyses the intermediate phosphotriesters and gives the 5′-*o*-hydroxyphenyl phosphates (**7**) in good yields. This functionality is stable to the acidic conditions required for removal of the isopropylidene group and subsequent treatment with an excess of bromine water gives the 5′-phosphates (**8**) as the sole nucleotide products. For example, with B = adenine, the overall yield for the sequence was 78%.[3] Phosphorylation of a 2′-deoxyribonucleotide using this procedure was successful.[3]

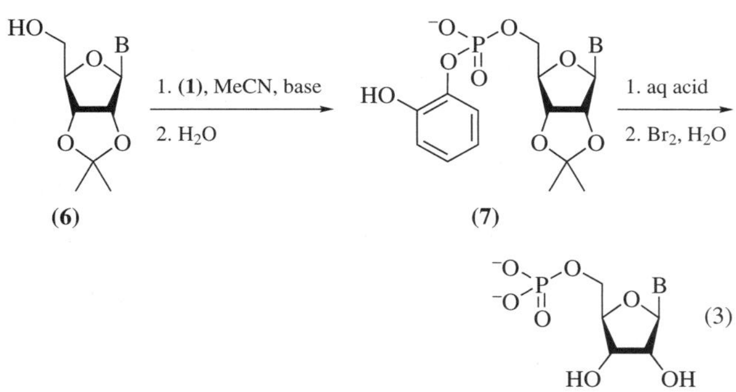

(6) (7) (8) (3)

Phosphorylation of Hydrazines. The reaction of 3-trifluoromethylphenylhydrazine with (**1**) gives the ring-opened monohydrazide (**9**) in 66% yield (eq 4).[7]

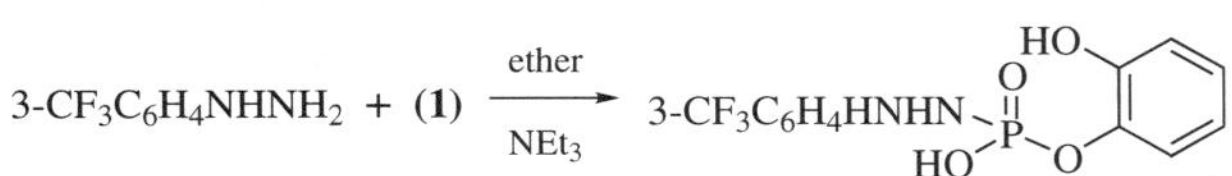

(9) (4)

Spirophosphoranes. Reaction of (**1**) with 1,2-diols, 2-hydroxy acids, or 2-keto acids leads to the formation of spirophosphoranes (**10**) which are in equilibrium with the phosphotriesters (**11**) (eq 5).[8] The use of (**1**) in the synthesis of a hexacovalent phosphate ester has also been reported.[9]

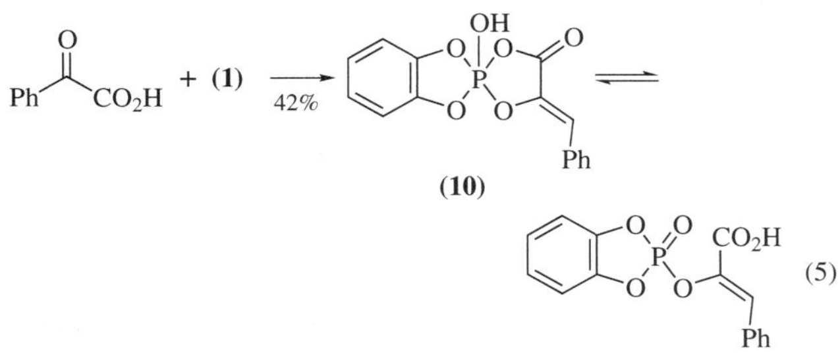

(10) (11) (5)

Reaction with Grignard Reagents. Reaction of (**1**) with a range of aryl Grignard reagents leads to the formation of *o*-hydroxyphenyl phosphinates (**12**) (eq 6).[10]

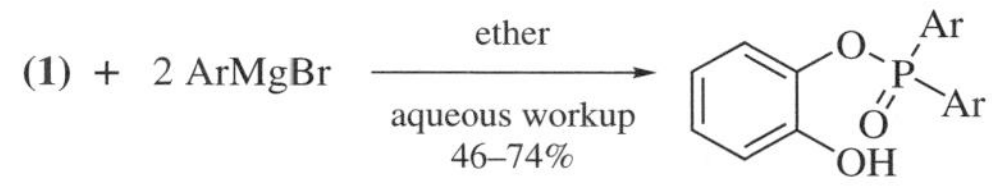

(12) (6)

Ar = Ph, 1-naphthyl, 4-$MeOC_6H_4$, 2,4,6-$Me_3C_6H_2$, 2-thienyl

1. Reich, W. S. *Nature* **1946**, *157*, 133; Lora Tomayo, M.; Calderón, J. *An. Real. Soc. Esp. Fis. Quim.* **1950**, *46B*, 475 (*Chem. Abstr.* **1951**, *45*, 7046e) .
2. Khwaja, T. A.; Reese, C. B. *J. Am. Chem. Soc.* **1966**, *88*, 3446.
3. Khwaja, T. A.; Reese, C. B.*Tetrahedron* **1971**, *27*, 6189.
4. *The Sigma-Aldrich Library of Chemical Safety Data*, 2nd ed.; Lenga, R. E., Ed.; Sigma-Aldrich:Milwaukee, WI, **1987**; Vol. *2*, p 2776B.
5. Khwaja, T. A.; Reese, C. B.; Stewart, J. C. M. *J. Chem. Soc. (C)* **1970**, 2092.
6. Calderón, J. *An. Real Soc. Esp. Fis. Quim.* **1957**, *53B*69 (*Chem. Abstr.* **1957**, *51*, 11 273f) ; Calderón, J.; Moreno, G. *An. Real. Soc. Esp. Fis. Quim.* **1960**, *56B*, 1960; Calama, M. A.; Calderón, J. *An. Real Soc. Esp. Fis. Quim.* **1966**, *62B*, 1015.
7. Gusar, N. I.; Rendina, L. V.; Shurubura, A. K. *J. Gen. Chem. USSR (Engl. Transl.)* **1989**, *59*, 486.
8. Munoz, A.; Garrigues, B.; Koenig, M. *Tetrahedron* **1980**, *36*, 2467; Munoz, A.; Lamandé, L. *Phosphorus Sulfur/Phosphorus Sulfur Silicon* **1988**, *35*, 195.
9. Gloede, J.; Gross, H. *Tetrahedron Lett.* **1976**, *17*, 917.
10. Gloede, J.; Hauser, A.; Ramm, M. *Phosphorus Sulfur/Phosphorus Sulfur Silicon* **1992**, *66*, 245.
11. Pouchert, C. J. *The Aldrich Library of FT-IR Spectra*; Aldrich: Milwaukee, WI, **1989**; Vol. *1*, p 556D.

Keith Jones
King's College London, UK

Phenyl *N*-Phenylphosphoramidochloridate

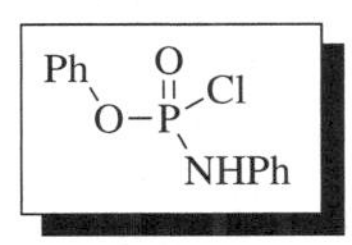

[51766-21-3] $C_{12}H_{11}ClNO_2P$ (MW 267.66))

(phosphorylating agent in oligonucleotide synthesis via the phosphotriester approach; for the conversion of acids to their symmetrical anhydrides and for the coupling of acids and amines to give amides; two important derivatives are also considered)

Physical Data: mp 132–134 °C.
Solubility: sol org solvents, e.g. acetone, dichloromethane.
Form Supplied in: white solid; widely available commercially.
Analysis of Reagent Purity: FT-IR;[22] R_f (pure) = 0.63 (chloroform– acetone (5:1) on silica).
Drying: is moisture sensitive; dry in vacuo ($\leq 10^{-2}$ mmHg) over P_2O_5 at rt.
Preparative Methods: obtainable from phenyl phosphorodichloridate (**2**) and aniline (eq 1).[1]

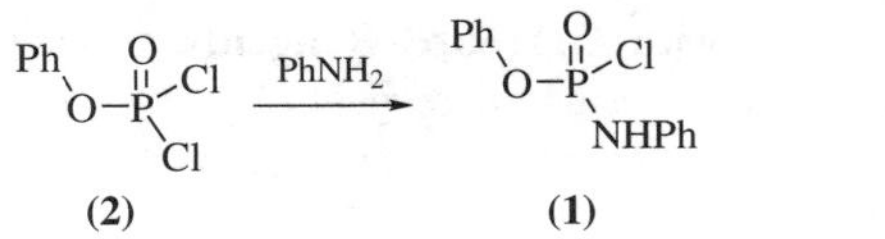
(1)

Handling, Storage, and Precautions: irritant and moisture sensitive. May be harmful by inhalation (irritating to mucous membranes and upper respiratory tract), ingestion, or skin absorption. Incompatible with strong oxidizing agents and strong bases.

Phosphorylating Agent. Phenyl *N*-phenylphosphoramidochloridate (**1**) reacts readily at 0 °C in dry pyridine with equimolar amounts of a corresponding alcohol or phenol to give the *O*-alkyl-*O*-phenyl-*N*-phenylphosphoramides (eq 2).[2]

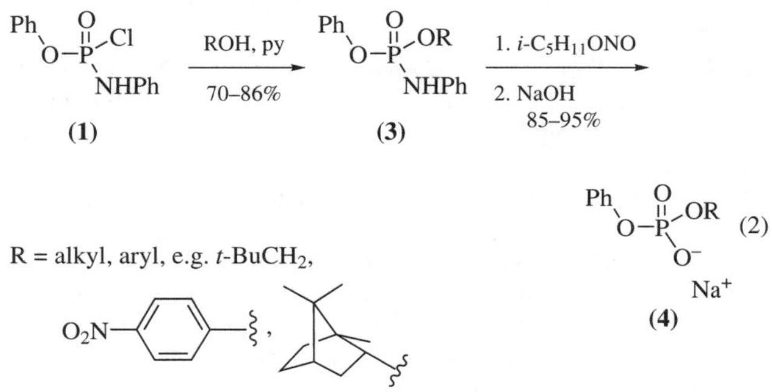
(2)

The compound (**3**) is a useful intermediate for the synthesis of mixed diesters of phosphoric acid (**4**). There are several methods of cleaving the N–P bond, of which the method of Ikehara is the most convenient.[3] Reaction of (**3**) with a 40 molar excess of 3-methylbutyl nitrite in acetic acid–pyridine (1:1 v/v) is successful in removing the aniline protecting group from the phosphoryl moiety in almost quantitative yields, giving the *O*-alkyl/aryl-*O*-phenyl hydrogen phosphate.

Given the good yields and simplicity of the process, this procedure has been used to prepare intermediates for oligonucleotide synthesis, e.g. in the synthesis of 5′-*O*-monomethoxytritylthymidine 3′-phenylphosphate (**7**) (eq 3). Phosphorylation of

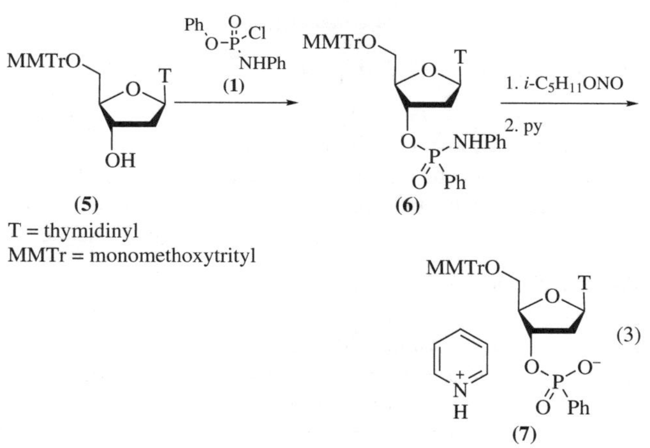
(3)

5′-*O*-monomethoxytritylthymidine (**5**) gave the protected nucleoside phosphate (**6**) in 98% yield which, after removal of the aniline protecting group and workup, gave the pyridinium salt in 85% yield.

This chemistry has additionally been utilized in the preparation of nucleoside phosphorothioates.[4] Reaction of (**6**) with NaH to remove the *P*-anilido group followed by CS_2 gave the corresponding 5′-monomethoxytritylthymidine 3′-*O*-phenylphosphorothioate (as a mixture of diastereomers, **8**) which was converted to the *S*-methyl ester (**9**) (eq 4). Owing to the chiral nature of pentavalent phosphorus, the compound (**9**) could exist as a mixture of diastereomers; however, the above reaction was the first example of an asymmetric synthesis of a phosphorothioate. There are now more examples of stereospecific synthesis of nucleoside derivatives utilizing this type of chemistry.[5]

There are numerous reviews on phosphorylation methods in biological molecules, some of which include aryl phosphoramidochloridates.[6]

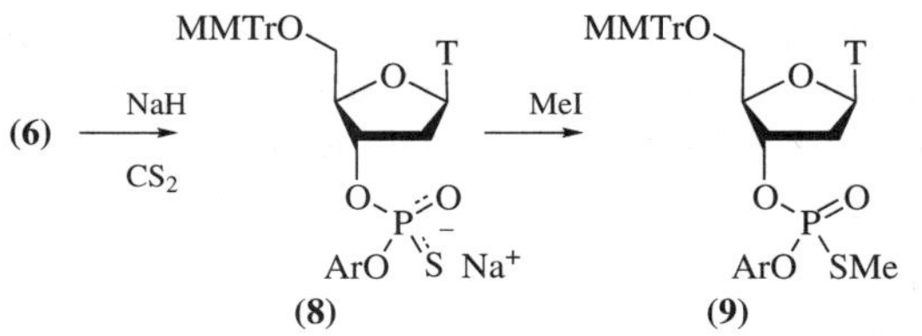
(4)

Activation of Carboxylic Acids.

Formation of Symmetrical Anhydrides. Symmetrical carboxylic anhydrides are useful reagents for peptide or amide synthesis.[7,8] Unfortunately, most methods require transformation of the acid first into the acid chloride or the use of condensation agents, e.g. 1,3-Dicyclohexylcarbodiimide (DCC), which can lead to separation problems of the product anhydride from the hydrated derivatives of the condensation reagents. The use of Thionyl Chloride/ Pyridine to form the acid chloride is often complicated by racemization with amino acid derivatives.

However, carboxylic acids are converted almost quantitatively to their corresponding symmetrical anhydride by treatment with 1 equiv of Triethylamine or 1-Ethylpiperidine and 0.5 equiv of phenyl *N*-phenylphosphoramidochloridate (or Diphenyl Phosphorochloridate) in acetone or dichloromethane, with the product anhydride (**10**) being obtained by filtration or from evaporation of the solvent after washing with water.[9] The resulting crude products are very nearly pure and the phenyl *N*-phenylphosphoramidochloridate can be regenerated as the aqueous solutions yield the phosphoramidic acid on acidification, which can be converted to (**1**) by reaction with Phosphorus(V) Chloride. The proposed pathway for this is through the mixed anhydride (**11**), which then reacts with a second molecule of acid. The only limitation of this method is the reaction of some of the anhydrides with water, e.g. as observed with trichloroacetic acid.

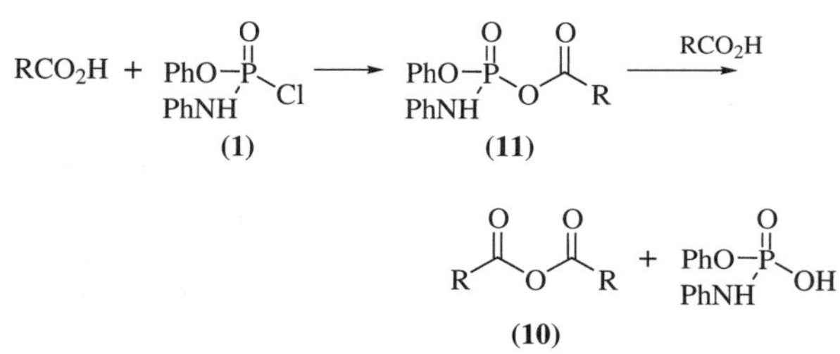
(5)

The Synthesis of Polyanhydrides by Dehydrative Coupling. The synthesis of polyanhydrides was first reported in 1909[10] and led on to a series of aliphatic polyanhydrides being prepared in the 1930s, as potential materials for the textile industry. One method for the formation of such polyanhydrides, particularly where sensitive monomers are involved, is to find suitably powerful dehydrating agents that function in mild reaction conditions. Organophosphorus compounds appear to be one such group of compounds. Phenyl N-phenylphosphoramidochloridate was found to be one of the most successful dehydrating agents.[11] The proposed mechanism of polyanhydride formation is shown in eq 6.

Formation of Carboxamides. There are many reagents for the conversion of carboxylic acids to carboxamides.[8,12–14] Activating agents such as DCC[15] can lead to purification problems, whilst involvement of the acid chloride can lead to racemization; there are also examples using phosphorus containing reagents, e.g. Diethyl Phosphorocyanidate,[13,16] but again these reagents only react well in certain systems. Other methods involve long reaction times, high temperatures, or low yields.

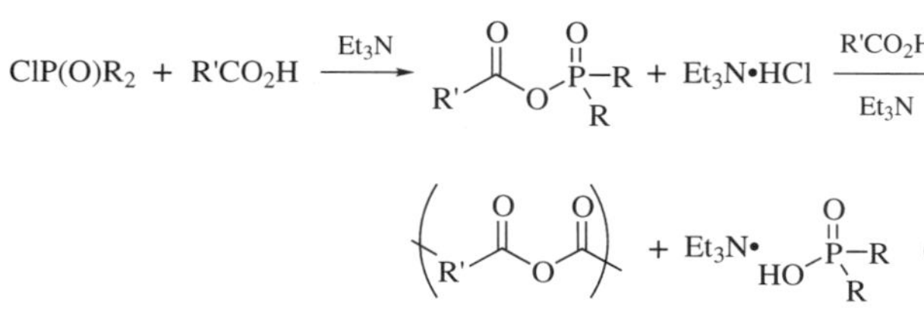

(6)

However, the use of phenyl N-phenylphosphoramidochloridate as a condensing agent for carboxylic acids and amines yields carboxamides in a one-step method (eq 7).[17] The amide (**12**) is prepared from the carboxylic acid, 2 equiv of triethylamine, 1 equiv of amine, and 1 equiv of reagent (**1**) and isolated by filtration or evaporation of solvent and washing with water. Once again, a mixed anhydride intermediate (**13**) is presumed, which results from nucleophilic attack by the carboxylate anion on the phosphorus atom with elimination of a chloride ion; a further nucleophilic attack by the amine on the carbonyl group of (**13**) and breakdown of the complex yields the amide (**12**) and phenyl N-phenylphosphoramidate (**14**) which can be recycled to (**1**) by reaction with phosphorus pentachloride.

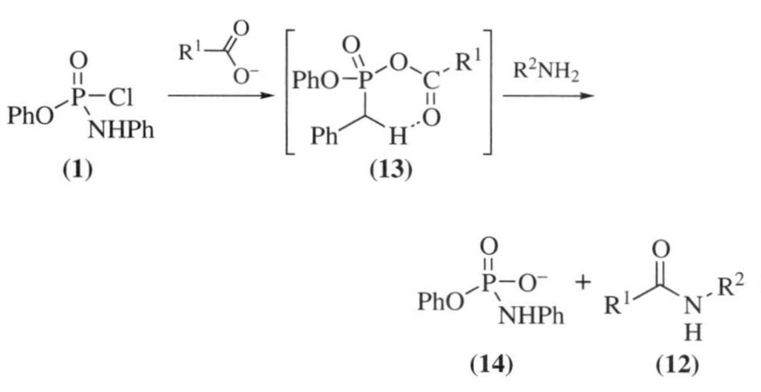

(7)

A distinct order of reactivity is observed in these reactions with (**1**): carboxylate anion ≥ aliphatic amines >> aromatic amines. This has led to slight variations in the practical procedures for preparing the amides but these are well documented in the literature.[17] The first route involves a one-step procedure; the second involves reaction of (**1**) with the carboxylate anion and subsequent treatment with the amine and is the preferred method for converting alkanamines to the amides since the first one-step method produces large amounts of phenyl phosphorodiamidates as well as the amide, by nucleophilic attack of (**1**) by the amine. The final method prevents the formation of phenyl phosphorodiamidates as the carboxylic acid first reacts with 0.5 equiv of (**1**) to form its anhydride, which is then converted to the amide in almost quantitative yield (this is largely utilizing the anhydride-formation chemistry described above) (eq 8).

Generally, phenyl N-phenylphosphoramidochloridate and other very similar related organophosphorus reagents e.g. phosphorus oxychloride, phenyl dichlorophosphate, diphenylchlorophosphate and N,N′-bis(2-oxo-3-oxazolinyl)phosphorodiamidic chloride have been utilised in recent years not only to form symmetrical carboxylic acid anhydrides but also in esterification and thiol esterification reactions; although phenyl N-phenylphosphoramidochloridate is not so efficient for the latter two reactions compared to its analogs but is probably the best choice for forming the anhydride.[18]

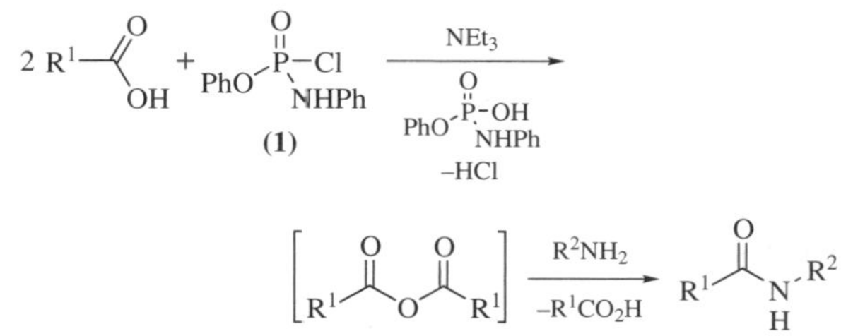

(8)

Two Important Derivatives of Phenyl N-Phenylphosphoramidochloridate.

Phenyl N-Phenylphosphoramidoazidate. This compound (**15**) is a white solid, which is easily prepared from (**1**) (eq 9).[19] It is obtained in excellent yields and is unusual for an organophosphorus compound in that it shows little sensitivity to moisture at room temperature for several months.

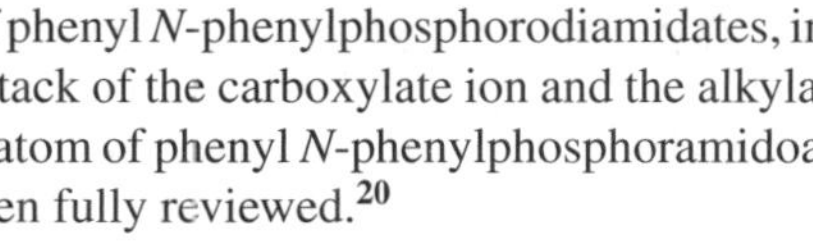

(9)

It is an excellent reagent for the preparation of N,N′-diarylureas. Reaction of a carboxylic acid, primary amine, and (**15**) in acetonitrile leads to the disubstituted urea (**16**) (eq 10). There is an additional competing reaction to the formation of the diarylurea: formation of phenyl N-phenylphosphorodiamidates, involving nucleophilic attack of the carboxylate ion and the alkylamine on the phosphorus atom of phenyl N-phenylphosphoramidoazidate. This work has been fully reviewed.[20]

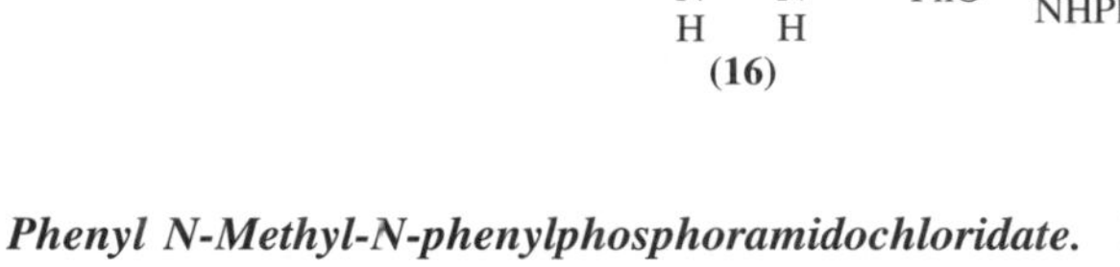

(10)

Phenyl N-Methyl-N-phenylphosphoramidochloridate. Like (**1**), phenyl N-methyl-N-phenylphosphoramidochloridate (**17**) reacts under analogous conditions to give anhydrides and carboxamides (see above). However, phenyl *N*-methyl-*N*-phenylphosphoramidochloridate undergoes one additional reaction which is not observed to any significant degree in (**1**). It is used for the conversion of imines to α-substituted β-lactams (eq 11).[21] The reaction is believed not to work for (**1**) as it requires an excess of triethylamine in the presence of a less nucleophilic imine, which is unfavorable to a mixed anhydride intermediate. The proposed pathway for this reaction is shown in eq 11, and has been performed on a wide variety of imines, proceeding in good yields (44–78%).

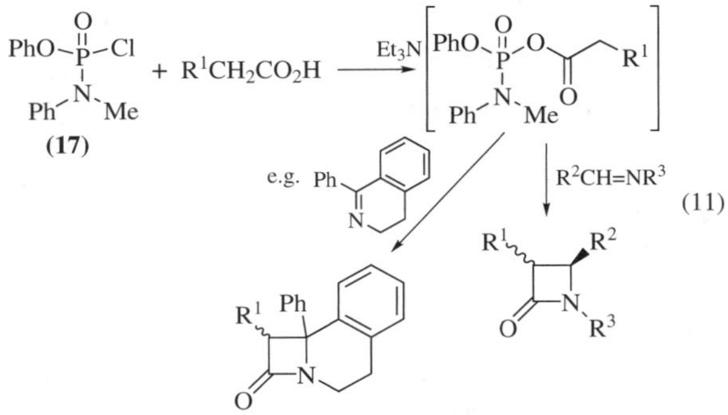

1. Cremlyn, R. W. J.; Kishore, D. N. *J. Chem. Soc., Perkin Trans. 1* **1972**, 585.
2. Zielinski, W. S.; Lesnikowski, Z. J. *Synthesis* **1976**, 185.
3. Ikehara, M.; Uesugi, S.; Fukui, T. *Chem. Pharm. Bull.* **1967**, *15*, 440.
4. Zielinski, W. S.; Lesnikowski, Z. J.; Stec, W. J. *Chem. Commun./J. Chem. Soc., Chem. Commun.* **1976**, 772.
5. Lesnikowski, Z. J.; Niewiarowski, W.; Zielinski, W. S.; Stec, W. J. *Tetrahedron* **1984**, *40*, 15.
6. Slotin, L. A. *Synthesis* **1977**, 737 and references therein.
7. Chen, F. M. F.; Koroda, K.; Benoiton, N. L. *Synthesis* **1978**, 928.
8. Jones, J. H. In *The Peptides*Gross, E.; Meienhofer, J., Eds.; Academic:New York, **1979**; p 65.
9. Mestres, R.; Palomo, C. *Synthesis* **1981**, 219.
10. Bucher, J. E.; Slade, W. C. *J. Am. Chem. Soc.* **1909**, *31*, 1319.
11. Leong, K. W.; Simonte, V.; Langer, R. *Macromolecules* **1987**, *20*, 705.
12. Ogliaruso, M. A.; Wolfe, J. F. In *The Chemistry of Acid Derivatives*Wiley:New York, **1979**; Part 1, p 474.
13. Haslam, E., *Chem. Ind. (London)* **1979**, 610.
14. Haslam, E. *Tetrahedron* **1980**, *36*, 2409.
15. Mikolajczyk, M.; Kielbasinski, P. *Tetrahedron* **1981**, *37*, 233.
16. Shiori, T.; Yokoyama, Y.; Kasai, Y.; Yamada, S. *Tetrahedron* **1976**, *32*, 2213.
17. Mestres, R.; Palomo, C. *Synthesis* **1982**, 288.
18. Arrieta, A.; Garcia, T.; Lago, J. M.; Palomo, C. *Synth. Commun.* **1983**, *13*, 471.
19. Arrieta, A.; Palomo, C. *Tetrahedron Lett.* **1981**, *22*, 1729.
20. Arrieta, A.; Palomo, C. *Bull. Soc. Claim. Fr., Part 2* **1982**, 7.
21. Shridhar, D. R.; Bhagat, R.; Narayana, V. L.; Awasthi, A. K.; Reddy, G. J. *Synthesis* **1984**, 846.
22. Pouchert, C. J. *Aldrich Library of FT-IR Spectra*; Aldrich:Milwaukee, WI, **1989**; *1*, p 556C.

Adrian P. Dobbs
King's College London, UK

Phenyl Phosphorodi(1-imidazolate)

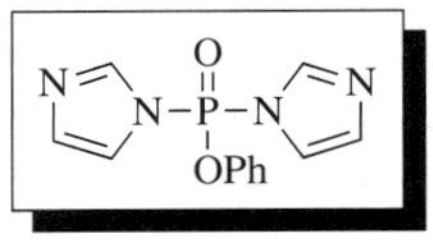

[15706-68-0] $C_{12}H_{11}N_4O_2P$ (MW 274.24)

(used in phosphorylation,[1] as a peptide coupling reagent,[1] and in the dehydration of aldoximes to nitriles[2])

Alternate Name: phenyl diimidazol-1-ylphosphinate.
Physical Data: an exceedingly hygroscopic white solid, mp 96–97 °C[2] (mp 90–92 °C also reported[1]).
Solubility: sol acetonitrile, dioxane, THF; relatively insol benzene, ether.
Form Supplied in: not commercially available.
Preparative Methods: this reagent was first prepared in 81% yield by reaction of 4 equiv of imidazole with 1 equiv of phenyl phosphorodichloridate in benzene at 60 °C.[4] The use of 2 equiv of imidazole and 2 equiv of triethylamine as base has been reported to give a yield of 98% of (**1**) (eq 1).[2]

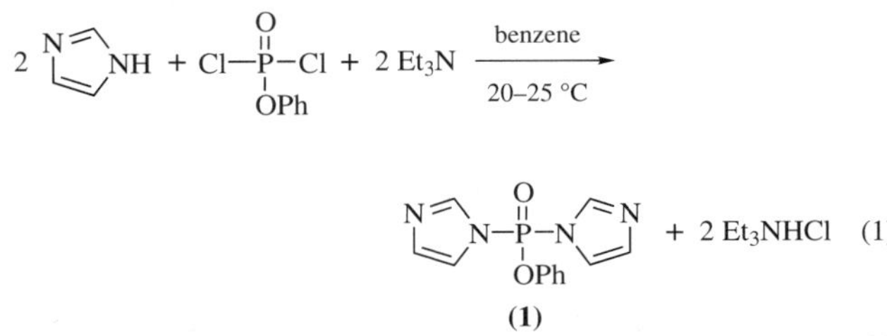

Purification: recrystallization from benzene, filtration under a water-free atmosphere, and drying under vacuum.
Handling, Storage, and Precautions: owing to its hygroscopic nature, exclusion of moisture in handling and use is essential; should be stored in a sealed tube at −20 °C.[3]

Phosphorylation. The title reagent (**1**) reacts with amines, alcohols, and water to give amides and esters of phenyl phosphate (eqs 2–4).[1] The monoimidazole derivatives are formed rapidly with a much slower reaction to give the disubstitution products.[1]

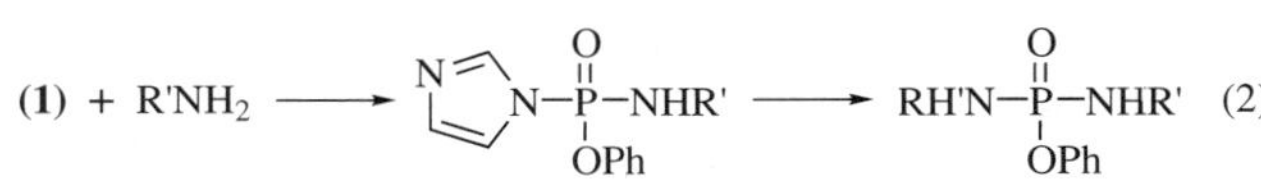

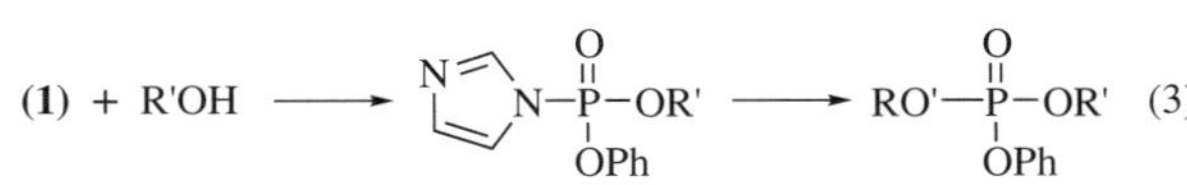

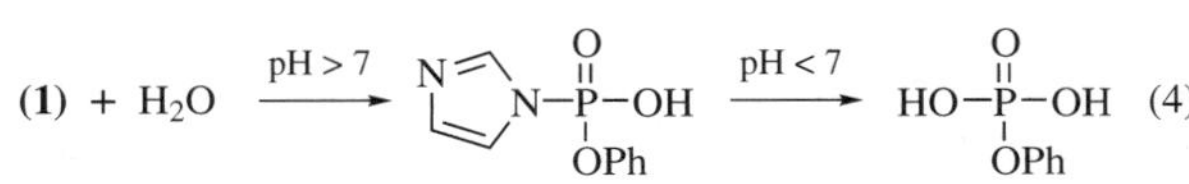

This behavior can be used to synthesize mixed diphosphates by treatment with a small excess of alcohol followed by hydrolysis of the monoimidazole derivative. In this manner the monophenyl ester of 2′,3′-isopropylideneadenosine 5′-phosphate (**2**) has been synthesized in 53.5% yield (eq 5).[1]

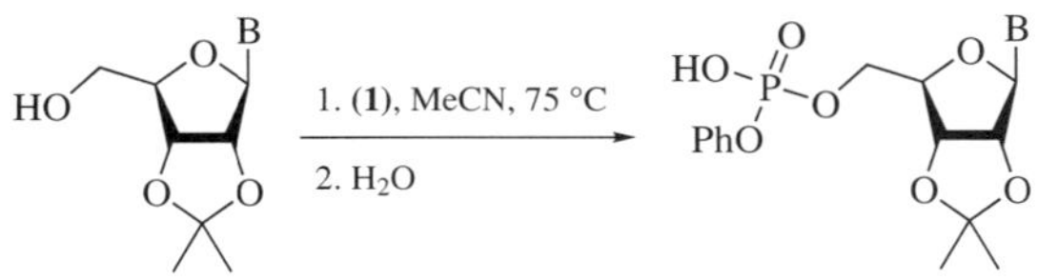

(5)

B = adenine (**2**)

Peptide Synthesis. Reagent (**1**) has been used as a coupling reagent in the synthesis of benzyloxycarbonylalanylalanine (**3**).[1] Initial treatment of benzyloxycarbonylalanine with (**1**) in acetonitrile at 45 °C for 12 h followed by reaction with an excess of an aqueous solution (pH 8–9) of alanine gave the protected dipeptide in 85% yield (eq 6). No reference to the stereochemistry of the starting materials or product was made.

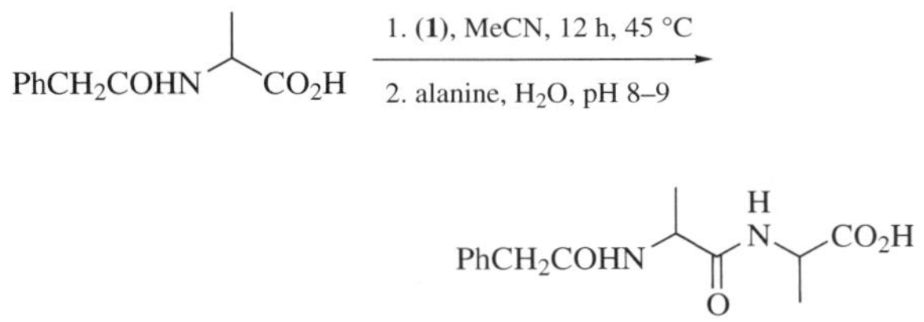

(6)

(**3**)

Acylimidazolides have been isolated from the reaction of (**1**) with carboxylic acids[1] and may be involved in the above reaction.

Dehydration of Aldoximes to Nitriles. Reagent (**1**) has been shown to be excellent for the dehydration of aldoximes of both aromatic and aliphatic aldehydes.[2] This is accomplished by reacting the aldoxime with (**1**) in dioxane at 20 °C for 20 h (eq 7).[2]

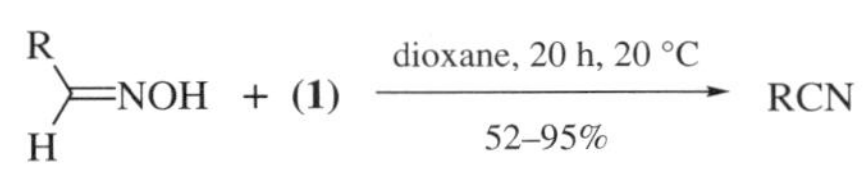

(7)

R = pentyl, hexyl, cyclohexyl, undecyl, cinnamyl, 4-methoxyphenyl, 4-chlorophenyl, 2,4-dichlorophenyl, 4-nitrophenyl, 2-pyridyl

Reagent (**1**) is superior in terms of yield to a variety of phosphorochloridates, including Phosphorus Oxychloride. For example, in the case of 2-pyridyl aldoxime, with (**1**) a yield of 52% was obtained, whereas with phosphorus oxychloride the yield was 33% and with phenyl phosphorodichloridate the yield was 47%.[2] The reaction is thought to proceed via an intermediate formed by *O*-phosphorylation of the aldoxime.[2] The use of spin-labeled versions of (**1**) has been explored, in which the phenyl is replaced with the 2,2,6,6-tetramethylpiperidyl *N*-oxide group.[3,5]

1. Cramer, F.; Schaller, H. *Ber. Dtsch. Chem. Ges./Chem. Ber.* **1961**, *94*, 1634.
2. Konieczny, M.; Sosnovsky, G. *Z. Naturforsch., Tell B* **1978**, *33b*, 1033.
3. Konieczny, M.; Sosnovsky, G. *Synthesis* **1976**, 537.
4. Cramer, F.; Schaller, H.; Staab, H. A. *Ber. Dtsch. Chem. Ges./Chem. Ber.* **1961**, *94*, 1612.
5. Konieczny, M.; Sosnovsky, G. *Z. Naturforsch., Tell B* **1977**, *32b*, 1179.

Keith Jones
King's College London, UK

2-(Phenylseleno)-1*H*-isoindole-1,3(2*H*)-dione

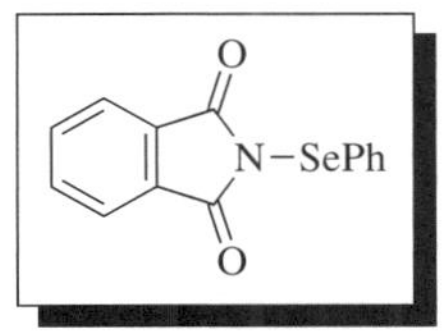

[71098-88-9] $C_{14}H_9NO_2Se$ (MW 302.20)

(reagent for electrophilic phenylselenenylation of alkenes, dienes, enols, and other nucleophiles)

Alternate Name: NPSP.
Physical Data: mp 171–175 °C (dec).
Form Supplied in: colorless solid.
Preparative Methods: prepared in 87% yield by the reaction of Benzeneselenenyl Chloride with potassium Phthalimide in hexane.[1]
Handling, Storage, and Precautions: indefinitely stable when stored at −20 °C in the dark. Use in a fume hood.

Original Commentary

Tom Livinghouse
Montana State University, Bozeman, MT, USA

Selenenylation. *N*-Phenylselenophthalimide is a very mild reagent for electrophilic selenenylation and reacts with a wide variety of organic substrates to provide products bearing the PhSe group. The generalized stepwise conversion of an alkene to a functionalized selenide with this reagent is shown in eq 1.

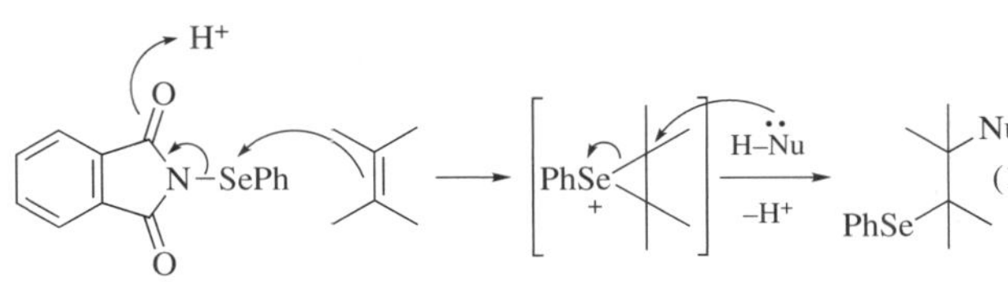

(1)

Various alkenes have been found to react in the intermolecular sense with nucleophiles in condensation reactions mediated by NPSP. Representative examples are given in eqs 2 and 3.[1–3]

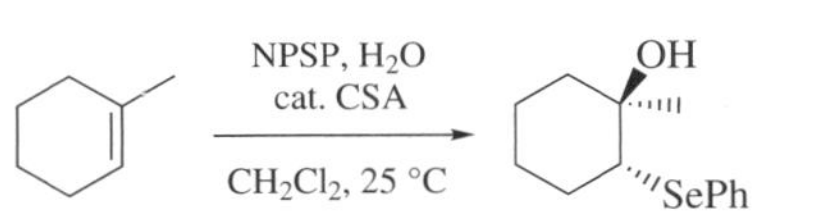

(2)

Dialkenes react with NPSP in the presence of water in a closely related manner to provide cyclic ethers with the net incorporation of two PhSe moieties (eq 4).[3]

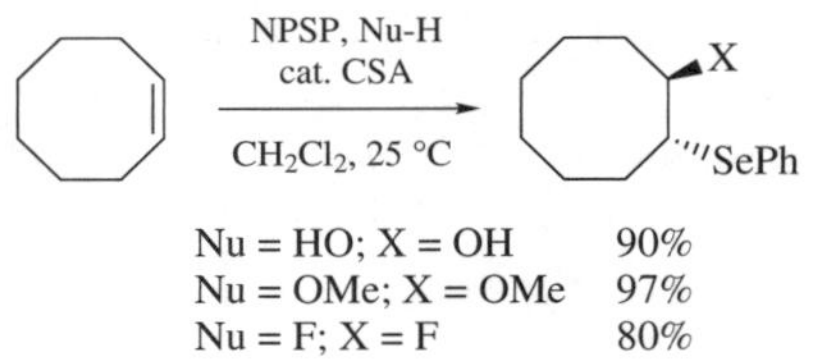

(3)

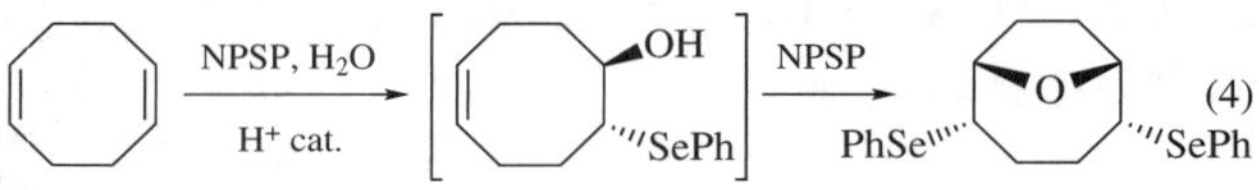

(4)

NPSP is a very efficient reagent for effecting the cyclization of unsaturated substrates containing internal nucleophiles and has been used for macrocyclizations (eqs 5–10)[3] as well as cyclizations of homogeranic acid derivatives.[4] In addition, NPSP has been used to effect annulations leading to carbohydrates.[5]

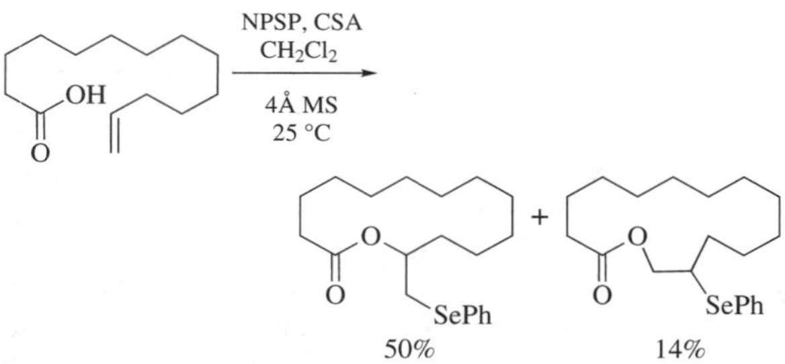

(5)

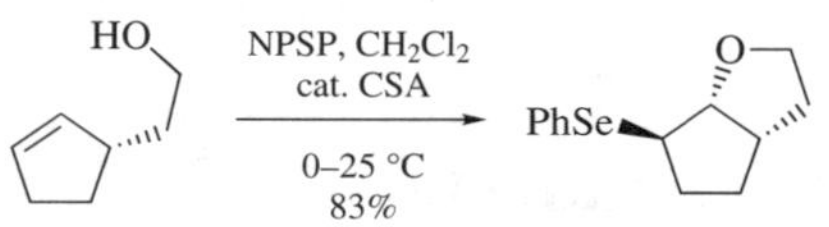

(6)

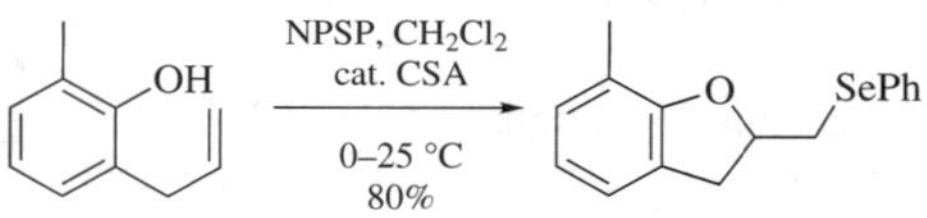

(7)

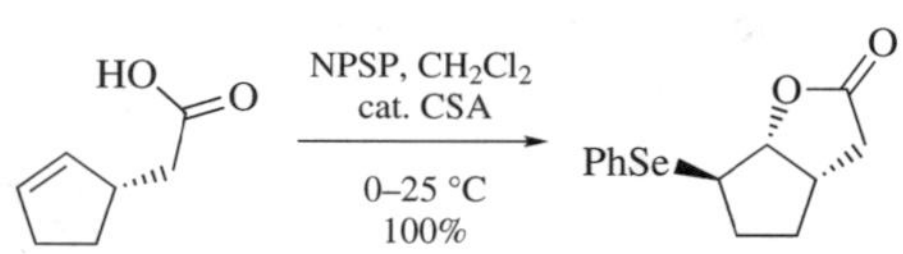

(8)

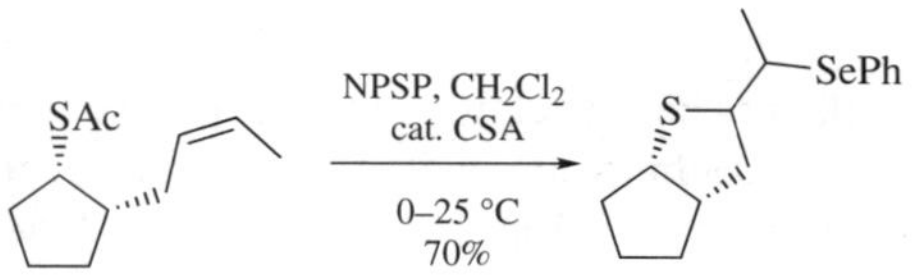

(9)

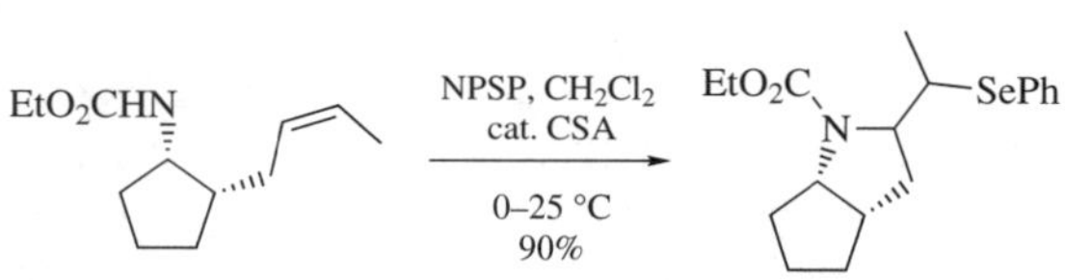

(10)

In several instances the use of NPSP has been found advantageous when compared to the more classical reagents Benzeneselenenyl Bromide or Benzeneselenenyl Chloride. In this connection the latter reagents have been reported to be useful for the selenenylative cyclization of unsaturated alcohols,[6] carboxylic acids,[7] thio compounds,[8] and urethanes.[9]

Selenolactonization of alkynoic acids also proceeds efficiently in the presence of NPSP (eq 11).[10]

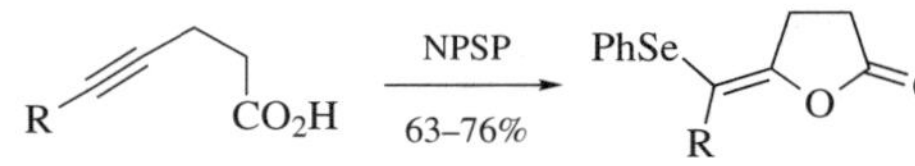

(11)

In addition, suitably disposed β-keto ester moieties can be induced to stereoselectively cyclize onto alkenes in the presence of NPSP (eq 12).[11]

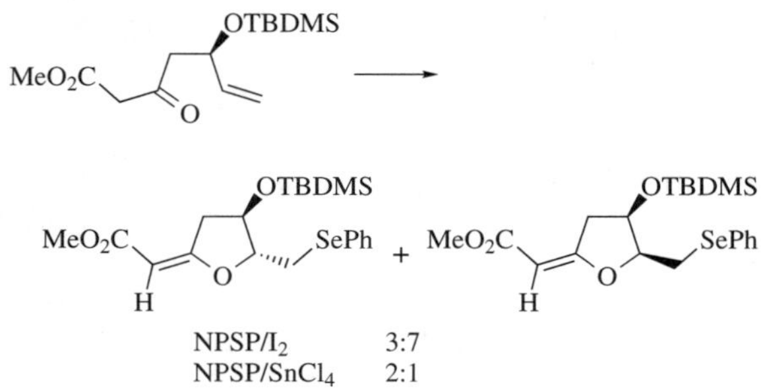

(12)

Heterocyclizations leading to oxazolidine derivatives have been reported to proceed efficiently via the selenenylation of alkenyl amides by NPSP (eq 13).[12] Danishefsky and collaborators have reported an interesting phenylselenenylation–radical functionalization sequence using NPSP[13] and have also used this reagent in the synthesis of mitomycin congeners.[14]

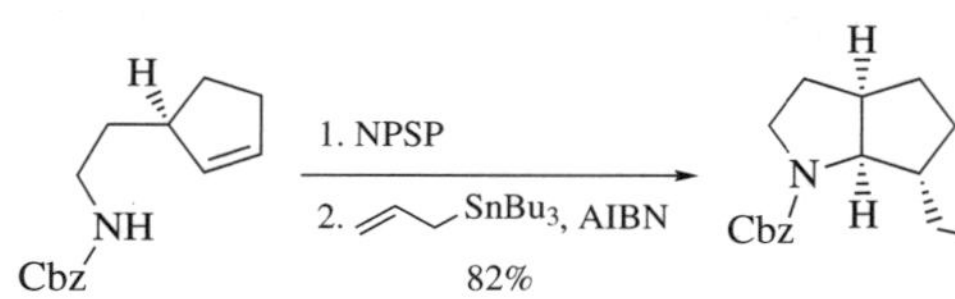

(13)

Ley and co-workers have published extensively on the use of NPSP in the presence of Lewis acids to effect carbocyclizations (e.g. eq 14).[15] In addition, these authors have utilized NPSP for the synthesis of spiroacetals.[16] For the majority of the preceding transformations, NPSP has been found the superior reagent both in terms of chemical efficiency as well as the mild nature of the reaction conditions that are typically employed.[17]

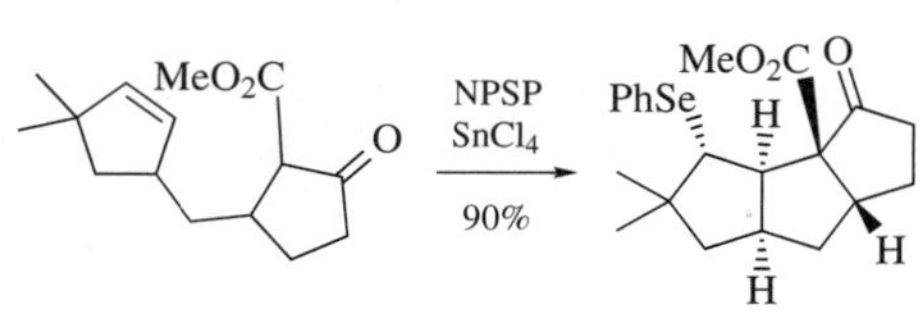

(14)

Alcohols and carboxylic acids have been found to condense with NPSP in the presence of Tri-n-butylphosphine to provide the corresponding selenoethers (eq 15) and acyl phenyl selenides (eq 16), respectively.[18–20]

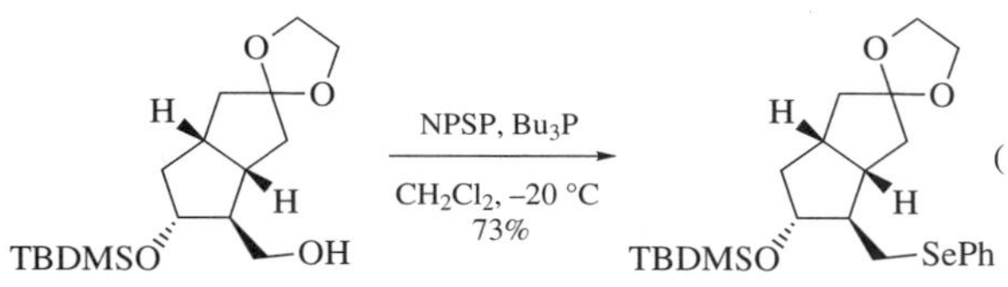
(15)

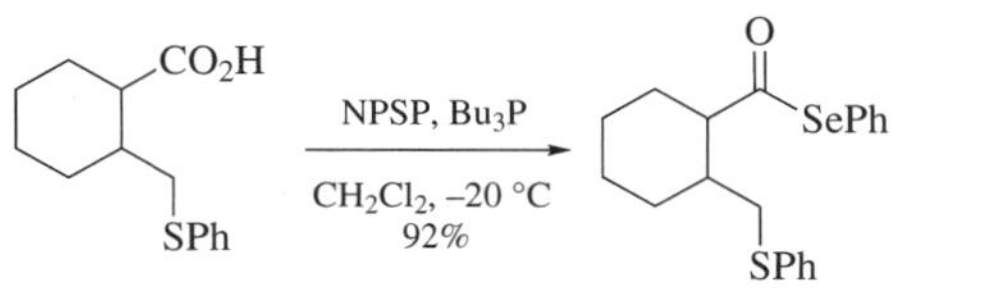
(16)

Sequential hydrozirconation–phenylselenenylation of conjugated enynes has provided a convenient route to stereodefined dienyl selenides (eq 17).[21]

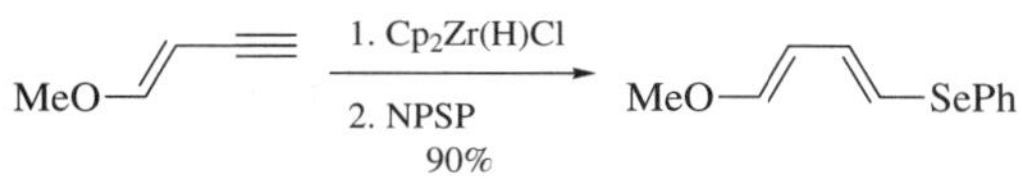
(17)

Related Reagents. Benzeneselenenyl Bromide; Benzeneselenenyl Chloride; Benzeneselenenyl Trifluoromethanesulfonate; Benzenesulfenyl Chloride; *N,N*-Diethylbenzeneselenenamide; 2-Pyridineselenenyl Bromide.

First Update

Thomas G. Back *University of Calgary, Calgary, Alberta, Canada*

Introduction. Two recent monographs[22] and a review article[23] provided detailed reports on the additions and cyclizations promoted by electrophilic selenium reagents, including NPSP. Further examples of oxyselenenylations[24] and fluoroselenenylations[25a] of alkenes, as exemplified by eqs 2 and 3, as well as the fluoroselenenylation of alkynes[25b] with this reagent have appeared. Additional examples of cyclizations of unsaturated alcohols,[26] β-dicarbonyl compounds,[27] and thioesters[28] (as in eqs 6, 7, 9, and 14) have been reported, along with new instances where NPSP was utilized for the conversion of alcohols to selenides[29] (as in eq 15) and carboxylic acids to selenoesters[30] (as in eq 16). Several new applications of NPSP, as well as recent variations of the above processes, are described below.

Electrophilic Selenenylation. Treatment of enol ethers with NPSP in the presence of an alcohol or water, and usually an acid catalyst, afforded the corresponding β-(phenylseleno)acetal[31a,b] or hemiacetal[32] (e.g., eq 18).[31b] Similarly, 2-cholestene produced vicinal phenylseleno carbamates and cyanamides via eqs 19 and 20, respectively.[33] The combination of NPSP with azidotrimethylsilane provides a useful reagent for the 1,2-addition of phenylseleno and azide groups to alkenes,[34] as in the example shown in eq 21.[34a]

Electrophilic Cyclization. Electrophilic selenenylations are accompanied by cyclization when internal nucleophiles are present. In a variation of eq 4, 1,4-cyclooctadiene reacted with NPSP and cyanamide to afford aza-bridged regioisomers (eq 22).[33] A catalytic variation of the usual lactonization of unsaturated carboxylic acids (as in eq 8) was observed when NPSP was used in the presence of the stoichiometric oxidant ammonium persulfate. This resulted in selenoxide elimination of the initially cyclized product and generated other electrophilic selenium species that continued the cyclization process (eq 23).[35] In a reaction reminiscent of eq 14, the cyclization of an unsaturated enol acetate was effected with NPSP in the presence of tin tetrachloride (eq 24).[36] The cyclization of tryptophan derivatives to the corresponding pyrroloindoles was reported simultaneously by two groups.[37] An example is shown in eq 25.[37a] NPSP was also effective in promoting the intramolecular alkylation of a phenol to the corresponding 4-substituted dienone (eq 26),[38] presumably via an initially formed seleniranium ion intermediate.

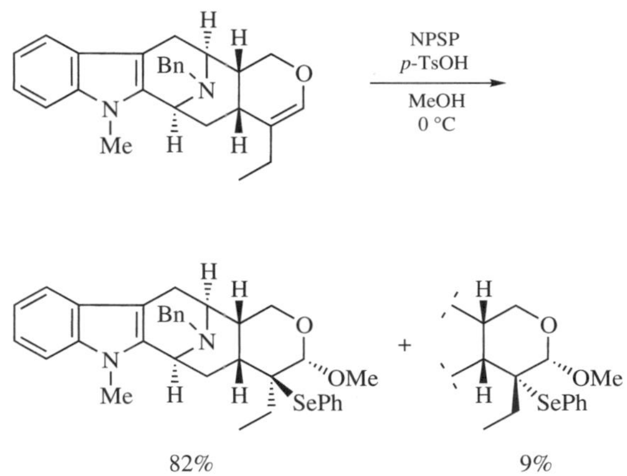
(18)

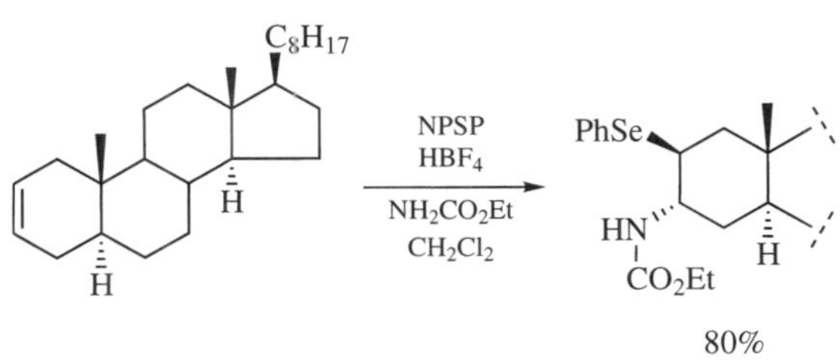
(19)

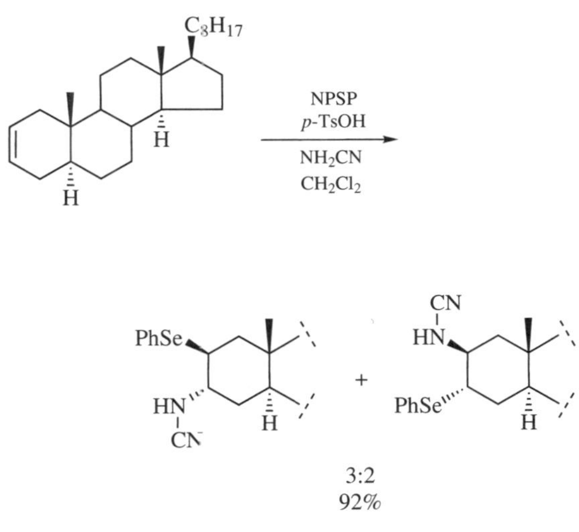
(20)

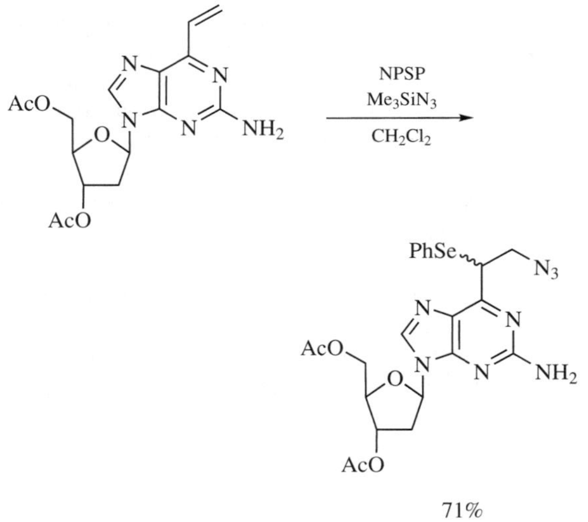

(21)

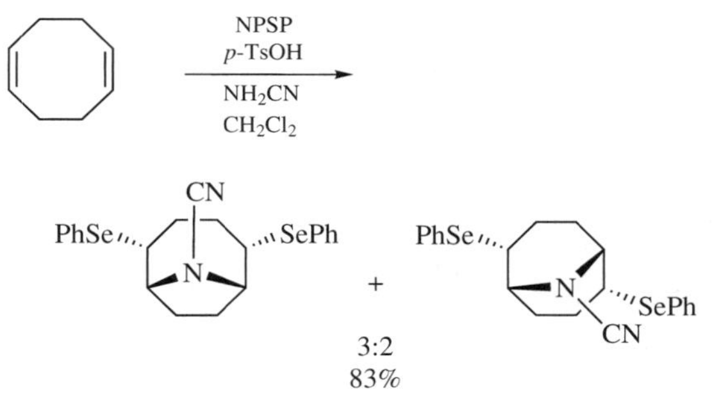

(22)

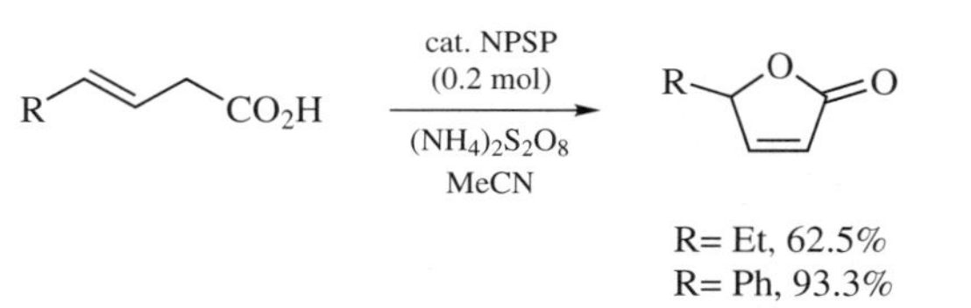

(23)

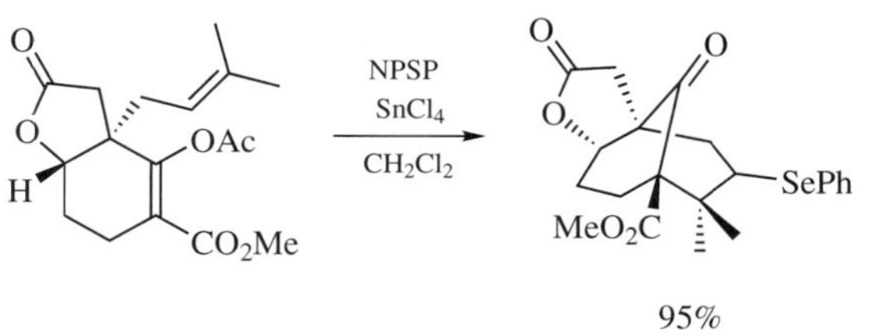

(24)

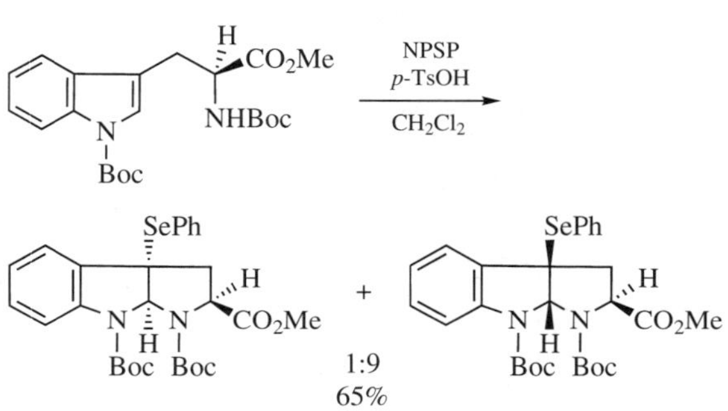

(25)

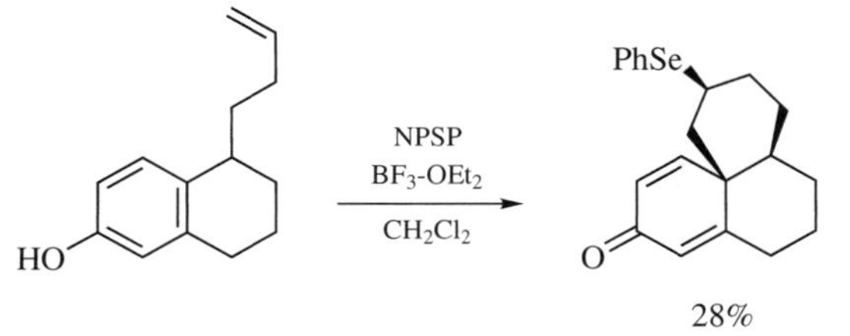

(26)

α-Selenenylation of Carbonyl Compounds. Carbonyl compounds undergo α-selenenylation via their enols or enolates when treated with NPSP. Ketones afford α-phenylseleno derivatives at their more substituted α-position under acidic conditions (e.g., eq 27).[39] The further selenenylation of α-phenylseleno ketones with NPSP under basic conditions produces diselenoacetals (eq 28).[40] Enolates derived from lactones[41] and β-dicarbonyl compounds[42] have been selenenylated with NPSP, while an enol silyl ether was reported to furnish a mixture of the corresponding α-phenylseleno ketone and enone in modest yield when treated with NPSP and trimethylsilyl triflate[43] (eqs 29–31, respectively).

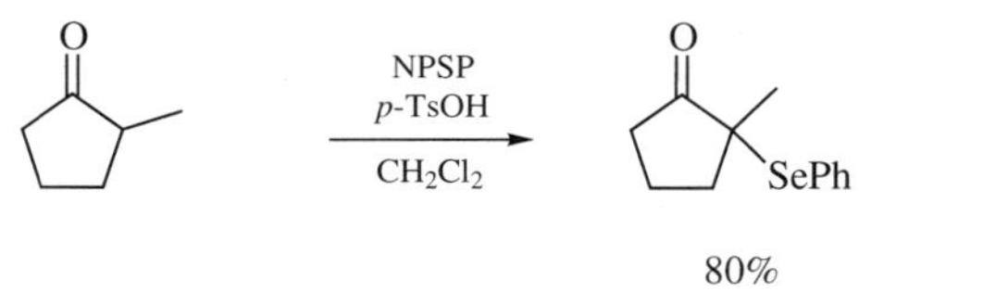

(27)

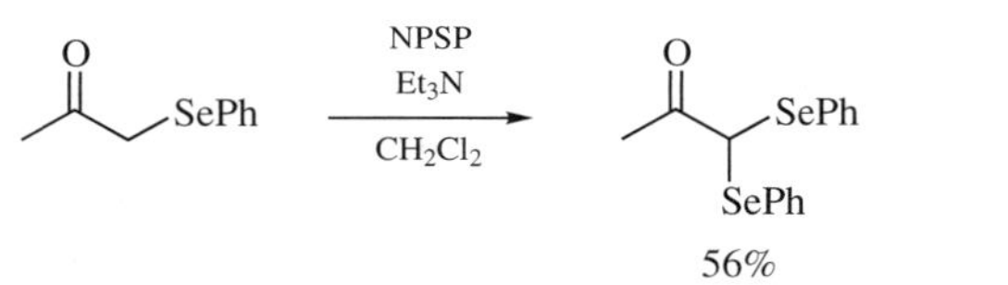

(28)

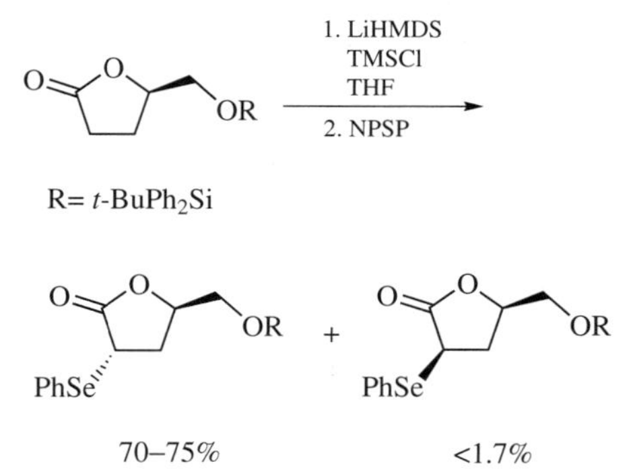

(29)

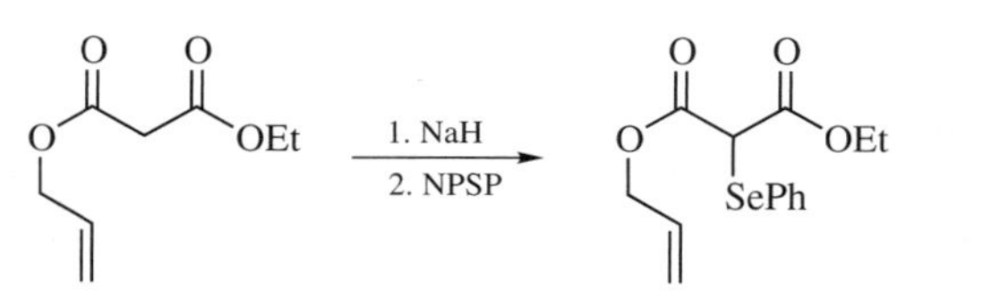

(30)

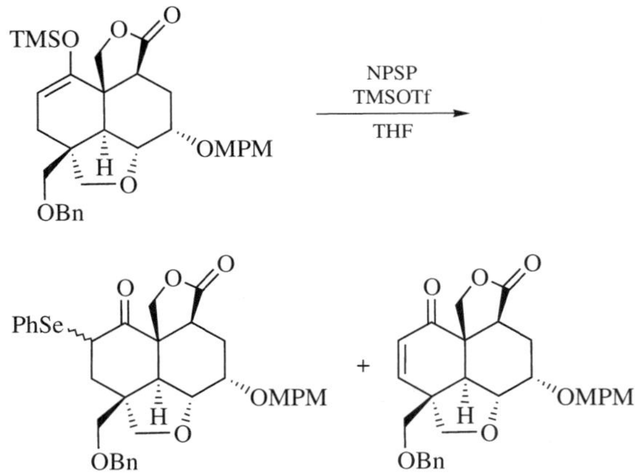

(31)

18% 28%

Other Reactions. The reaction of 5-hexenylstannanes, such as shown in eq 32, with NPSP and a Lewis acid afforded the corresponding cyclized selenides.[44] A similar selenenylation–destannylation process was used to generate an *o*-quinodimethane intermediate from *o*-(vinyl)benzylstannane (eq 33).[45] The intermediate was trapped in situ by various dienophiles to produce Diels–Alder cycloadducts. Finally, the combination of NPSP with trimethylsilyl triflate and cyclohexanol was employed to activate a thioglycoside and convert it to its cyclohexyl glycoside (eq 34).[46]

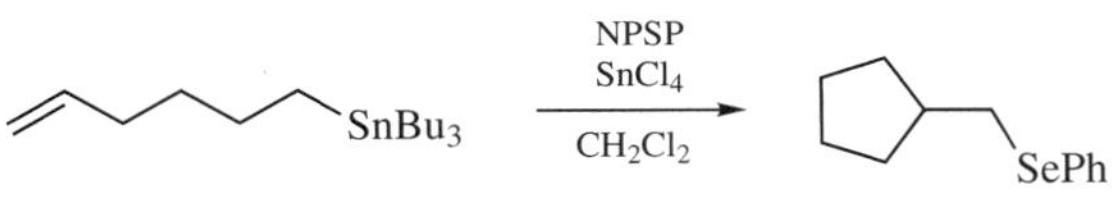

(32)

99%

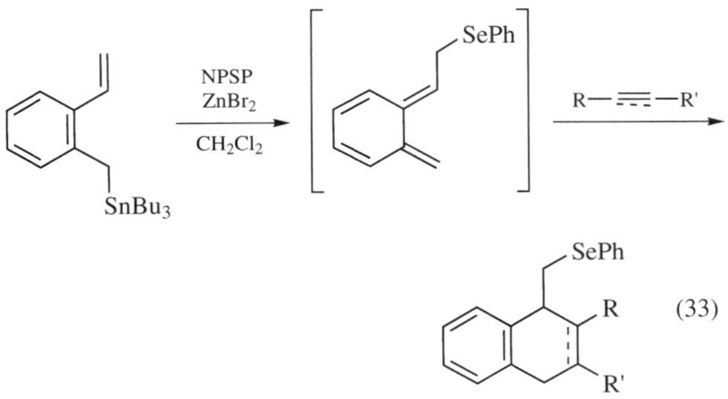

(33)

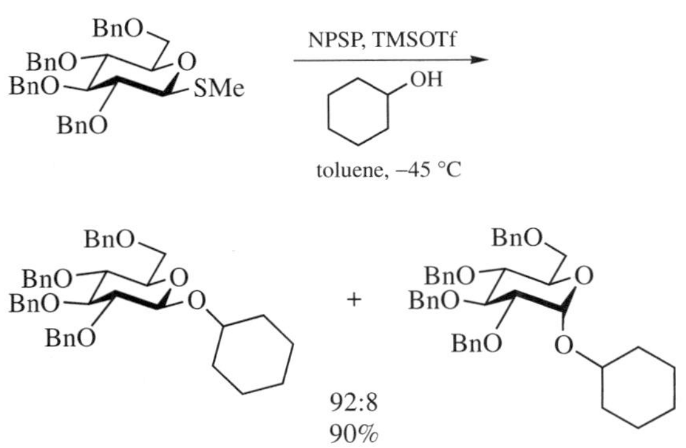

(34)

1. Nicolaou, K. C.; Petasis, N. A.; Claremon, D. A., *Tetrahedron* **1985**, *41*, 4835.
2. Saluzzo, C.; Alvernhe, G.; Anker, D., *Tetrahedron Lett.* **1990**, *31*, 663.
3. Nicolaou, K. C.; Claremon, D. A.; Barnette, W. E.; Seitz, S. P., *J. Am. Chem. Soc.* **1979**, *101*, 3704.
4. (a) Rouessac, A.; Rouessac, F.; Zamarlik, H., *Tetrahedron Lett.* **1981**, *22*, 2641. (b) Rouessac, F.; Zamarlik, H., *Tetrahedron Lett.* **1981**, *22*, 2643.
5. Suzuki, K.; Mukaiyama, T., *Chem. Lett.* **1982**, 683.
6. (a) Nicolaou, K. C.; Magolda, R. L.; Sipio, W. J.; Barnette, W. D.; Lysenko, Z.; Joullie, M. M., *J. Am. Chem. Soc.* **1980**, *102*, 3784. (b) Nicolaou, K. C.; Lysenko, Z., *Tetrahedron Lett.* **1977**, 1257. (c) Clive, D. L. J.; Chittattu, G.; Wong, C. K., *Can. J. Chem.* **1977**, *55*, 3894. (d) Clive, D. L. J.; Chittattu, G.; Curtis, N. J.; Kiel, W. A.; Wong, C. K., *Chem. Commun.* **1977**, 725.
7. (a) Nicolaou, K. C.; Seitz, S. P.; Sipio, W. J.; Blount, J. F., *J. Am. Chem. Soc.* **1979**, *101*, 3884. (b) Nicolaou, K. C.; Lysenko, Z., *J. Am. Chem. Soc.* **1977**, *99*, 3185. (c) Clive, D. L. J.; Chittattu, G., *Chem. Commun.* **1977**, 484. (d) Clive, D. L. J.; Russell, C. G.; Chittattu, G.; Singh, A., *Tetrahedron* **1980**, *36*, 1399.
8. Nicolaou, K. C.; Barnette, W. E.; Magolda, R. L., *J. Am. Chem. Soc.* **1978**, *100*, 2567.
9. (a) Clive, D. L. J.; Farina, V.; Singh, A.; Wong, C. K.; Kiel, W. A.; Menchen, S. M., *J. Org. Chem.* **1980**, *45*, 2120. (b) Clive, D. L. J.; Wong, C. K.; Kiel, W. A.; Menchen, S. M., *Chem. Commun.* **1978**, 379.
10. Toru, T.; Fujita, S.; Saito, M.; Maekawa, E., *J. Chem. Soc., Perkin Trans. 1* **1986**, 1999.
11. Brussani, G.; Ley, S. V.; Wright, J. L.; Williams, D. J., *J. Chem. Soc., Perkin Trans. 1* **1986**. 303.
12. Chrétien, F.; Chapleur, Y., *J. Org. Chem.* **1988**, *53*, 3615.
13. Webb, R. R., II; Danishefsky, S., *Tetrahedron Lett.* **1983**, *24*, 1357.
14. Danishefsky, S.; Berman, E. M.; Ciufolini, M.; Etheredge, S. J.; Segmuller, B. E., *J. Am. Chem. Soc.* **1985**, *107*, 3891.
15. (a) Jackson, W. P.; Ley, S. V.; Morton, J. A., *Chem. Commun.* **1980**, 1028. (b) Jackson, W. P.; Ley, S. V.; Whittle, A. J., *Chem. Commun.* **1980**, 1173. (c) Jackson, W. P., Ley, S. V.; Morton, J. A., *Tetrahedron Lett.* **1981**, *22*, 2601. (d) Ley, S. V.; Lygo, B.; Molines, H.; Morton, J. A., *Chem. Commun.* **1982**, 1251. (e) Ley, S. V.; Murray, P. J.; Palmer, B. D., *Tetrahedron* **1985**, *41*, 4765. (f) Ley, S. V.; Murray, P. J., *Chem. Commun.* **1982**, 1252.
16. Ley, S. V.; Lygo, B., *Tetrahedron Lett.* **1982**, *23*, 4625.
17. For a comprehensive review, see: Harring, S. R.; Edstrom, E. D.; Livinghouse, T., In *Advances in Heterocyclic Natural Product Synthesis*; Pearson, W. H., Ed.; JAI: Greenwich, CT, 1992; Vol. 2, pp 299–376.
18. (a) Grieco, P. A.; Gilman, S.; Nishizawa, M., *J. Org. Chem.* **1976**, *41*, 1485.
(b) Nicolaou, K. C.; Petasis, N. A., *Selenium in Natural Products Synthesis*; CIS: Philadelphia, PA, 1984.
19. (a) Ley, S. V.; Simpkins, N. S.; Whittle, A. J., *Chem. Commun.* **1981**, 1001 and **1983**, 503. (b) Ley, S. V.; Neuhaus, D.; Simpkins, N. S.; Whittle, A. J., *J. Chem. Soc., Perkin Trans. 1* **1982**, 2157.
20. Grieco, P. A.; Jaw, J. Y.; Claremon, D. A.; Nicolaou, K. C., *J. Org. Chem.* **1981**, *46*, 1215.
21. Fryzuk, M. D.; Bates, G. S.; Stone, C., *J. Org. Chem.* **1991**, *56*, 7201.
22. (a) Beaulieu, P. L., Déziel, R., In *Organoselenium Chemistry – A Practical Approach*; Back, T. G., Ed.; Oxford University Press: Oxford, 1999, Chap 3. (b) Tiecco, M., In *Organoselenium Chemistry*; Wirth, T., Ed.; Topics in Current Chemistry, Vol 208, Springer-Verlag: Berlin, 2000, pp 7–54.
23. Petragnani, N.; Stefani, H. A.; Valduga, C. J., *Tetrahedron* **2001**, *57*, 1411–1448.
24. (a) Kim, K. S.; Park, J. I.; Moon, H. K.; Yi, H., *J. Chem. Soc., Chem. Commun.* **1998**, 1945–1946. (b) Kim, K. S.; Choi, S. O.; Park, J. M.; Lee, Y. J.; Kim, J. H., *Tetrahedron: Asymmetry* **2001**, *12*, 2649–2655. (c) Kim, K. S.; Park, J. I.; Ding, P., *Tetrahedron Lett.* **1998**, *39*,

6471–6474. (d) Haughan, A. F.; Sweeney, J. B., *Tetrahedron: Asymmetry* **1994**, *5*, 177–180. (e) Bange, J.; Haughan, A. F.; Sweeney, J. B., *Tetrahedron Lett.* **1994**, *35*, 1405–1408. (f) Haughan, A. F.; Knight, J. R.; Sweeney, J. B., *Tetrahedron Lett.* **1994**, *35*, 1781–1784. (g) Bhandari, P.; Crombie, L.; Kilbee, G. W.; Pegg, S. J.; Proudfoot, G.; Rossiter, J.; Sanders, M.; Whiting, D. A., *J. Chem. Soc., Perkin Trans. 1* **1992**, 851–863.

25. (a) Saluzzo, C.; La Spina, A.-M.; Picq, D.; Alvernhe, G.; Anker, D.; Wolf, D.; Haufe, G., *Bull. Soc. Chim. Fr.* **1994**, *131*, 831–843. (b) Saluzzo, C.; Alvernhe, G.; Anker, D.; Haufe, G., *Tetrahedron Lett.* **1990**, *31*, 2127–2130.
26. (a) Rosselli, S.; Bruno, M.; Pibiri, I.; Piozzi, F., *Eur. J. Org. Chem.* **2002**, 4169–4173. (b) Bravo, F.; Viso, A.; Castillon, S., *J. Org. Chem.* **2003**, *68*, 1172–1175. (c) Iwasaki, K.; Nakatani, M.; Inoue, M.; Katoh, T., *Tetrahedron Lett.* **2002**, *43*, 7937–7940. (d) Allin, S. M.; Baird, R. D.; Lins, R. J., *Tetrahedron Lett.* **2002**, *43*, 4195–4197. (e) Kühnert, S. M., Maier, M. E., *Org. Lett.* **2002**, *4*, 643–646. (f) Germay, O.; Kumar, N.; Thomas, E. J., *Tetrahedron Lett.* **2001**, *42*, 4969–4974. (g) Bravo, F.; Castillón, S., *Eur. J. Org. Chem.* **2001**, 507–516. (h) Bravo, F.; Kassou, M.; Castillón, S., *Tetrahedron Lett.* **1999**, *40*, 1187–1190. (i) Greeves, N.; Lee, W.-M., *Tetrahedron Lett.* **1997**, *38*, 6449–6452. (j) Greeves, N.; Lee, W.-M., Barkley, J. V., *Tetrahedron Lett.* **1997**, *38*, 6453–6456. (k) Colins, C. C.; Cronin, M. F.; Moynihan, H. A.; McCarthy, D. G., *J. Chem. Soc., Perkin Trans. 1* **1997**, 1267–1269. (l) Höfle, G.; O'Sullivan, A. C.; Rihs, G.; Sutter, M.; Winkler, T., *Tetrahedron* **1995**, *51*, 3159–3174. (m) Paquette, L. A.; Ezquerra, J.; He, W., *J. Org. Chem.* **1995**, *60*, 1435–1447. (n) Carman, R. M.; Handley, P. N., *Aust. J. Chem.* **2001**, *54*, 769–776.
27. Blaney, W.M.; Cuñat, A. C.; Ley, S. V.; Montgomery, F. J.; Simmonds, M. S. J., *Tetrahedron Lett.* **1994**, *35*, 4861–4864.
28. Jana, G.; Viso, A.; Díaz, Y.; Castillón, S., *Eur. J. Org. Chem.* **2003**, 209–216.
29. (a) Prévost, N.; Shipman, M., *Tetrahedron* **2002**, *58*, 7165–7175. (b) Chiba, A.; Aoyama, T.; Suzuki, R.; Eguchi, T.; Oshima, T.; Kakinuma, K., *J. Org. Chem.* **1999**, *64*, 6159–6165. (c) Yamauchi, N.; Terachi, T.; Eguchi, T.; Kakinuma, K., *Tetrahedron* **1994**, *50*, 4125–4136.
30. (a) Allin, S. M.; Barton, W. R. S.; Bowman, W. R.; McInally, T., *Tetrahedron Lett.* **2001**, *42*, 7887–7890. (b) Chatgilialoglu, C.; Ferreri, C.; Bazzanini, R.; Guerra, M.; Choi, S.-Y.; Emanuel, C. J.; Horner, J. H.; Newcomb, M., *J. Am. Chem. Soc.* **2000**, *122*, 9525–9533. (c) Harrington-Frost, N. M.; Pattenden, G., *Synlett* **1999**, 1917–1918. (d) Crich, D.; Eustace, K. A.; Fortt, S. M.; Ritchie, T. J., *Tetrahedron* **1990**, *46*, 2135–2148.
31. (a) Kim, C. U.; Luh, B. Y.; Martin, J. C., *J. Org. Chem.* **1991**, *56*, 2642–2647. (b) Yu, P.; Wang, T.; Li, J.; Cook, J. M., *J. Org. Chem.* **2000**, *65*, 3173–3191.
32. Gan, T.; Cook, J. M., *J. Org. Chem.* **1998**, *63*, 1478–1483.
33. Francisco, C. G.; Hernández, R.; Léon, E. I.; Salazar, J. A.; Suárez, E., *J. Chem. Soc., Perkin Trans. 1* **1990**, 2417–2427.
34. (a) Czernecki, S.; Hoang, A.; Valéry, J.-M., *Tetrahedron Lett.* **1996**, *37*, 8857–8860. (b) Czernecki, S.; Randriamandimby, D., *Tetrahedron Lett.* **1993**, *34*, 7915–7916. (c) Chelain, E.; Czernecki, S.; Chmielewski, M.; Kaluza, Z., *J. Carbohydr. Chem.* **1996**, *15*, 571–579. (d) Giuliano, R. M.; Duarte, F., *Synlett* **1992**, 419–421.
35. Fujita, K.; Taka, H.; Oishi, A.; Ikeda, Y.; Taguchi, Y.; Fujie, K.; Saeki, T.; Sakuma, M., *Synlett* **2000**, 1509–1511.
36. Nicolaou, K.C.; Pfefferkorn, J. A.; Kim, S.; Wei, H. X., *J. Am. Chem. Soc.* **1999**, *121*, 4724–4725.
37. (a) Crich, D.; Huang, X., *J. Org. Chem.* **1999**, *64*, 7218–7223. (b) Depew, K. M.; Marsden, S. P.; Zatorska, D.; Zatorski, A.; Bornmann, W. G.; Danishefsky, S. J., *J. Am. Chem. Soc.* **1999**, *121*, 11953–11963.
38. Hares, O.; Hobbs-Mallyon, D.; Whiting, D. A., *J. Chem. Soc., Perkin Trans. 1* **1993**, 1481–1492.
39. Cossy, J.; Furet, N., *Tetrahedron Lett.* **1993**, *34*, 7755–7756.
40. Ponthieux, S.; Outurquin, F.; Paulmier, C., *Tetrahedron Lett.* **1995**, *36*, 6453–6456.
41. Chen, S.-H.; Li, X.; Li, J.; Niu, C.; Carmichael, E.; Doyle, T. W., *J. Org. Chem.* **1997**, *62*, 3449–3452.
42. Byers, J. H.; Gleason, T. G.; Knight, K. S., *J. Chem. Soc., Chem. Commun.* **1991**, 354–356.
43. Kanoh, N.; Ishihara, J.; Yamamoto, Y.; Murai, A., *Synthesis* **2000**, 1878–1893.
44. Herndon, J. W.; Harp, J. J., *Tetrahedron Lett.* **1992**, *33*, 6243–6246.
45. Woo, S. H., *Tetrahedron Lett.* **1993**, *34*, 7587–7590.
46. Shimizu, H.; Ito, Y.; Ogawa, T., *Synlett* **1994**, 535–536.

(2-Phenylsulfonyl)ethoxy Dichlorophosphine

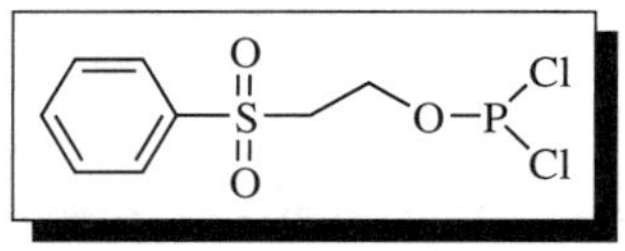

[98288-60-9] $C_8H_9Cl_2O_3PS$ (MW 287.10)

(reagent used for the synthesis of nucleotides by phosphoramidite method)

Physical Data: mp 32–35 °C;[1] ^{1}H NMR;[1] ^{31}P NMR;[1] stable for at least one year when stored at −18 °C.[1]

Preparative Methods: **1** can be prepared from phenylsulfonylethanol with an excess of phosphorus trichloride in acetonitrile solution without the use of a base (eq 1).

$$PhSO_2CH_2CH_2OH \xrightarrow[\text{rt, 10 min; 97\%}]{PCl_3\text{ (7 equiv), }CH_3CN} PhSO_2CH_2CH_2OPCl_2 \quad (1)$$

Nucleotide Synthesis by the Phosphoramidite Method. (2-Phenylsulfonyl)ethoxy dichlorophosphine is readily converted into the corresponding crystalline phosphoromorpholido chloridite (**1**) on exposure to *N*-trimethylsilylmorpholine (eq 2).

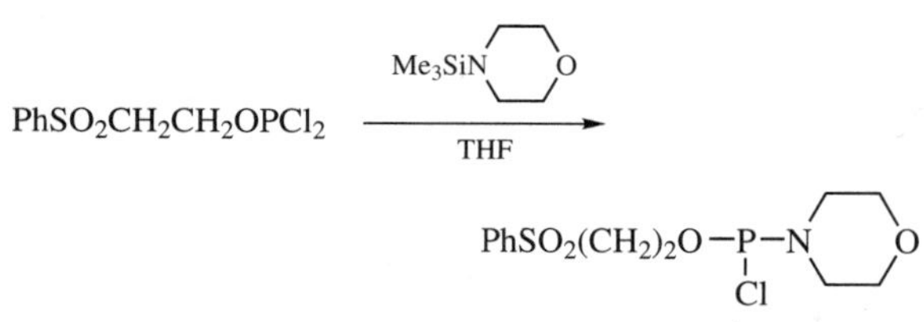

(2)

1

This substance then reacts with 5′-*O*-protected nucleosides to give the 3′-*O*-phosphoromorpholodites which are typically solids and may be purified by column chromatography. For example, the reaction with 5′-*O*-(dimethoxytrityl)thymidine gave corresponding phosphoromorpholodite (eq 3). **1** has been used to synthesize the thymidine decamer on a solid support using 1-hydroxybenzotriazole as the promoter and the usual combination of iodine and water to effect oxidation to the phosphate after coupling.[2]

(2-Phenylsulfonyl)ethyl phosphites and phosphates are stable in the presence of acids, towards acid, hydrogenolysis, and tertiary amines in aprotic solvents, enabling the orthogonal removal of numerous other protecting groups.

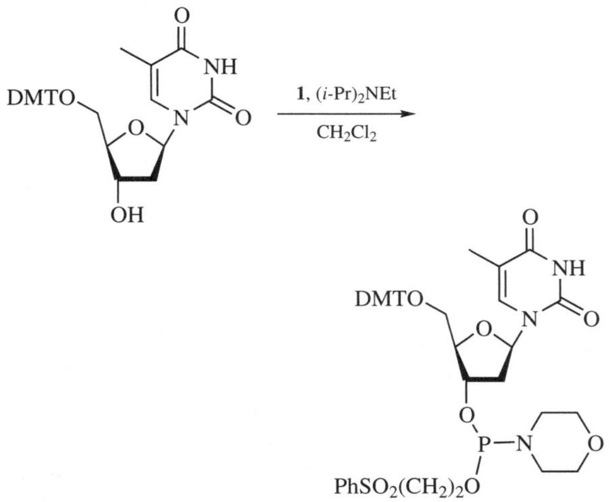

(3)

Following coupling and oxidation to the phosphate oxidation level, deprotection is achieved rapidly and cleanly by β-elimination in the presence of the mixture of dioxane/methanol/4 N NaOH(14/5/1) in 5 min (eq 4).[1,3]

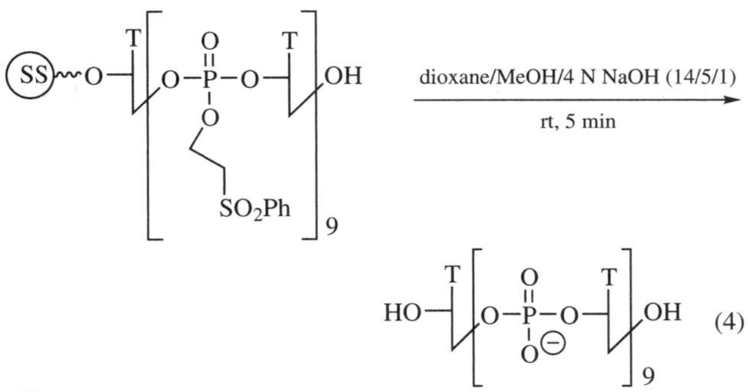

(4)

(SS) = solid support, T = thymine

1. Claesen, C. A. A.; Segers, R. P. A. M.; Tesser, G. I., *Recl. Trav. Chim. Pays-Bas* **1985**, *104*, 119.
2. Claesen, C. A. A.; Segers, R. P. A. M.; Tesser, G. I., *Recl. Trav. Chim. Pays-Bas* **1985**, *104*, 209.
3. (a) Claesen, C.; Tesser, G. I.; Dreef, C. E.; Marugg, J. E.; van der Marel, G. A.; van Boom, J. H., *Tetrahedron Lett.* **1984**, *25*, 1307. (b) Beld, A.; Claesen, C. A. A.; Roersma, E. S.; Schippers, W. J. M.; Keizer, L. M.; Tesser, G. I., *Recl. Trav. Chim. Pays-Bas* **1984**, *103*, 196.

Jae-Taeg Hwang
LG Life Sciences, Daejon, Korea

2-Phenylthioethanol

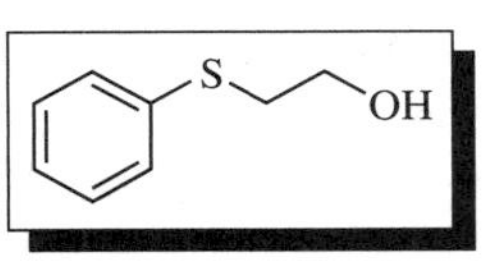

[699-12-7] $C_8H_{10}OS$ (MW 154.23)

(reagent used as a protecting agent in nucleotide synthesis)

Alternate Name: 2-phenylmercaptoethanol.
Physical Data: bp 115–116 °C/ 2 mm Hg, density 1.14 g cm^{-3}; mp 57–58 °C (as 4-nitrobenzoate ester).
Form Supplied in: stench liquid; commercially available.
Preparative Methods: prepared by the reaction of thiophenol with ethylene chlorohydrin in 20% NaOH aqueous solution.[1]
Handling, Storage, and Precautions: toxicity and other health hazards are not reported.

Phosphate Protecting Group in Oligonucleotide Synthesis. Protection of phosphates is particularly important in the synthesis of oligonucleotides. The commonly used protecting group is 2-cyanoethyl, which is introduced to a phosphate with 2-cyanoethanol and DCC/pyridine and can be cleaved by aqueous ammonia. In automated oligonucleotide synthesis, 2-cyanoethyl repeatedly protects and deprotects the 3′-phosphate at each condensation step in high yields. When condensation starts with the 5′-phosphate nucleotide, the phosphate protecting group must resist all the condensation processes, including deprotection of 3′-OH and amino groups. 2-Phenylthioethanol is employed for such a protection.[2] Protection of the deoxymononucleotide 5′-phosphate is accomplished by treating its pyridinium salt with 2-phenylthioethanol and DCC. The protection can also be performed with triisopropylbenzenesulfonyl chloride (TPS). Condensation with a second nucleotide and succeeding condensations do not touch the protected 5′-phosphate of the terminal nucleotide (eq 1). The phosphate protected by 2-phenylthioethanol is stable under 1 M aqueous NaOH, concentrated ammonium hydroxide at 50 °C, and aqueous HCl at pH 2 at room temperature.

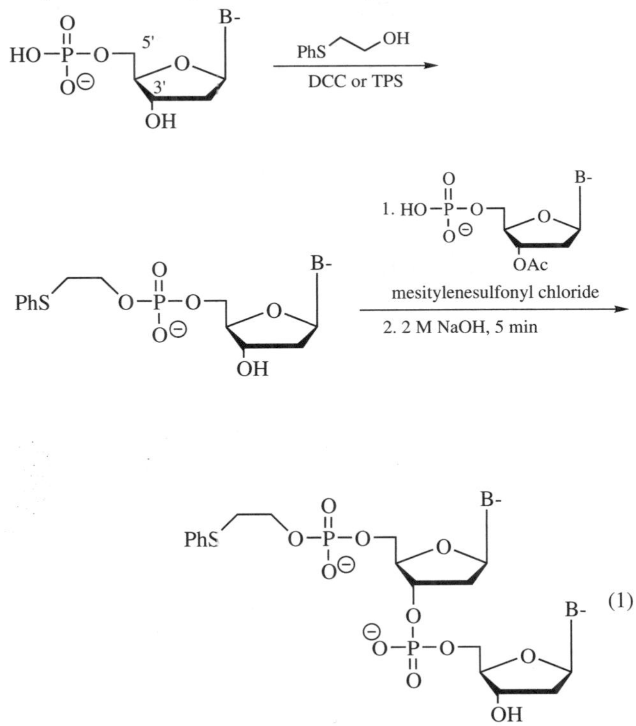

(1)

Deprotection of the 5′-phosphate is easily accomplished after oxidative treatment. Oxidation with sodium metaperiodate in water (or in 0.1 M triethylammonium bicarbonate) gives the corresponding sulfoxide derivative. The free nucleotide is liberated by treatment with 2 M aqueous sodium hydroxide at room temperature (eq 2). Alternatively, the sulfide is oxidized to sulfone with

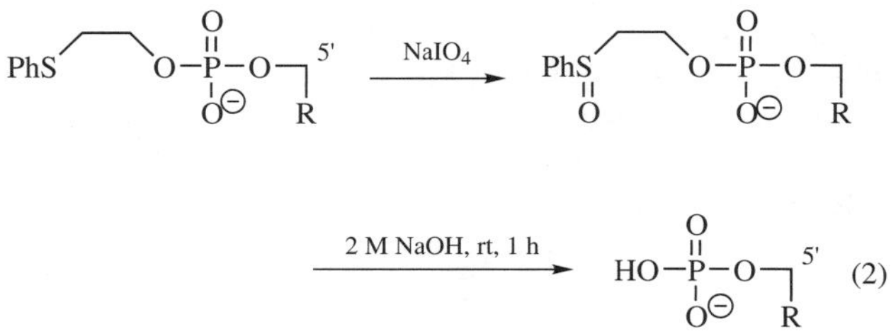

NCS in 0.1 M triethylammonium bicarbonate buffer at room temperature. In this case, deprotection can be achieved successfully under milder conditions involving 1 M aqueous sodium hydroxide at 0 °C for 5 min (eq 3).[3]

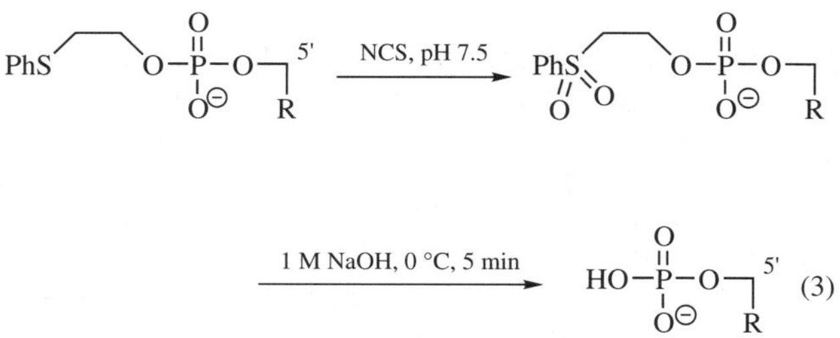

1. Ford-Moore, A. H.; Peters, R. A.; Wakelin, R. W., *J. Chem. Soc.* **1949**, 1754.
2. (a) Narang, S. A.; Bhanot, O. S.; Goodchild, J.; Wightman, R. H.; Dheer, S. K., *J. Am. Chem. Soc.* **1972**, *94*, 6183. (b) Wightman, R. H.; Narang, S. A.; Itakura, K., *Can. J. Chem.* **1972**, *50*, 456. (c) Narang, S. A.; Itakura, K.; Bahl, C. P.; Katagiri, N., *J. Am. Chem. Soc.* **1974**, *96*, 7074.
3. Agarwal, K. L.; Fridkin, M.; Jay, E.; Khorana, H. G., *J. Am. Chem. Soc.* **1973**, *95*, 2020.

Takashi Sugimura
Himeji Institute of Technology, Hyogo, Japan

N-Phenyltrifluoroacetimidoyl Chloride

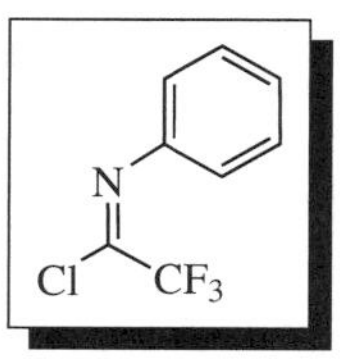

[61881-19-4] $C_8H_5ClF_3N$ (MW 207.58)

(reagent used as a two-carbon building block for synthesis of compounds containing a trifluoromethyl residue and used for preparation of glycosyl trifluoroacetimidates as glycosy donors)

Physical Data: bp 55–56 °C (11 Torr).

Preparative Methods: to a mixture of Ph_3P (51.8 g, 0.20 mol), Et_3N (11 mL, 0.080 mol), and CCl_4 (32 mL, 0.33 mol) at 0 °C was added trifluoroacetic acid (5 mL, 0.065 mol). After stirring for 10 min, a solution of aniline (7.3 mL, 0.080 mol) in CCl_4 (31 mL) was added, and the mixture was then refluxed for 4 h. After cooling down to rt, the solvent was decanted. The residual solids (Ph_3PO, Ph_3P, and $Et_3N \cdot HCl$) were washed several times with hexane. The combined solvents were concentrated under reduced pressure, and the residue was distilled to afford *N*-phenyltrifluoroacetimidoyl chloride (9.6 g, 72%) as a pale yellow liquid.[1]

Handling, Storage, and Precautions: N-phenyltrifluoroacetimidoyl chloride is thermally rather stable, but is slowly hydrolyzed to the amide upon exposure to moisture. Storage under a nitrogen atmosphere and refrigeration is recommended.

Addition of Nucleophiles. *N*-Aryltrifluoroacetimidoyl chlorides react with a variety of nucleophiles readily, providing compounds bearing an *N*-aryltrifluoroethanimine residue.[2] Grignard addition gives smoothly the trifluoromethyl ketimines.[3] Addition of lithium enolates[4,5] or active methylene compounds (in the presence of bases)[3,6] is followed by a tautomerization, providing the corresponding enamines in good yields. A chiral sulfinylmethyl lithium reacts at −78 °C to afford the sulfinylimine (eq 1).[7] In a similar manner, treatment of an alkyl phosphonate with lithium diisopropylamide followed by *N*-phenyltrifluoroacetimidoyl chloride provides a mixture of the iminophosphonate and enamine, which take part in a Horner-Emmons olefination on addition of aldehydes giving 2-trifluoromethyl-1-phenyl-1-azadienes and, on hydrolysis, α,β-unsaturated trifluoromethyl ketones.[8] Benzoin-type condensation with aroyl anions gives trifluoroacetimidoyl aryl ketones.[9] In addition, Friedel-Crafts reaction with 2,4-di-*tert*-butyl-phenol affords 2,4-di-*tert*-butyl-6-(2,2,2-trifluoro-1-phenylimino-ethyl)-phenol in good yield.[10]

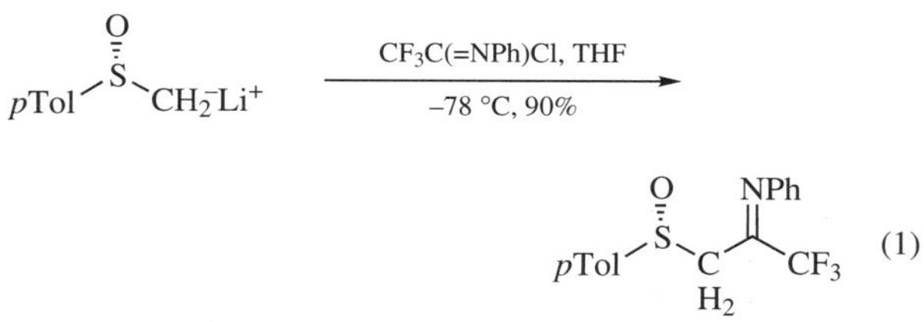

Addition with nitrogen nucleophiles, such as phenylamines,[11] hydrazines,[12] urethanes,[12] acetone hydrazone,[13] and azide,[14] provides the corresponding trifluoroethanimidic derivatives, which are versatile precursors for the preparation of various polynitrogen heterocycles. Hydroxy-acetic acid esters react in the presence of NEt_3 at 50 °C to afford the imino ethers in good yields, which undergo a Wittig-type rearrangement to γ-fluorinated-α-hydroxy-β- iminoesters.[15] Thiols react more easily than the alcohols.[16,17] Arbuzov-type reaction with trialkyl phosphites gives trifluoroethylphosphonates in excellent yields.[18] In addition, *N*-aryltrifluoroacetimidoyl iodides are prepared by replacement of the imidoyl chloride with NaI in acetone quantitatively.[1]

Metal Mediated Additions. *N*-Aryltrifluoroacetimidoyl iodides, rather than the chlorides, undergo palladium-catalyzed carbonylation[19] and Heck-type alkenylation/alkynlation.[20] Trifluoroacetimidoyl zinc derivatives are generated from the iodides in a Zn/Al-DMF-HMPA system, and react with aldehydes at rt to give, after tautomerization, the α-amino-ketones.[21] The iodides are also exchangeable with lithium (BuLi, THF, <-60 °C) to produce cleanly the *N*-aryltrifluoroacetimidoyl lithium species, which react readily with aldehydes, ketones, and acyl chlorides. Substituents on the *N*-aryl residue exert a significant effect on the stability of the lithium species.[22] *N*-Aryltrifluoroacetimidoyl chlorides react with trimethylsilyl lithium in the presence of CuCN to afford the trifluoroacetimidoyl trimethylsilanes, which behave also as trifluoroacetimidoyl anion equivalents under the action of fluoride ion.[23] Rh(I) complexes catalyze the coupling cyclization with alkynes to afford 2-trifluoromethyl quinolines in good yields (eq 2) and regioselectivity (for unsymmetrical alkynes).[24] A versatile bis-silylated difluoroenamine, which behaves as a dianion equivalent and undergoes sequential reactions with two kinds of electrophiles, is generated in a Mg/Me_3SiCl system (eq 3).[25]

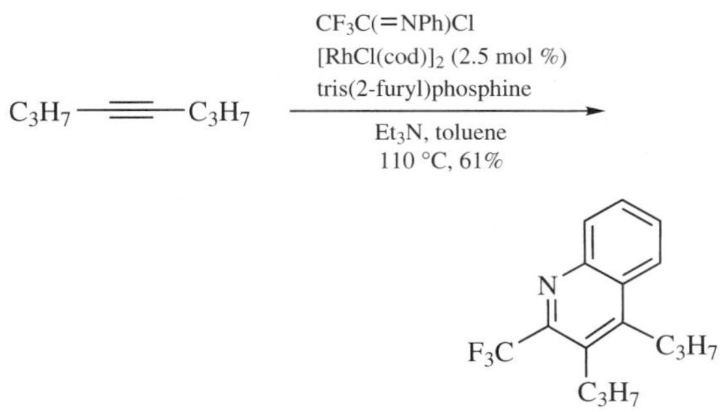

(2)

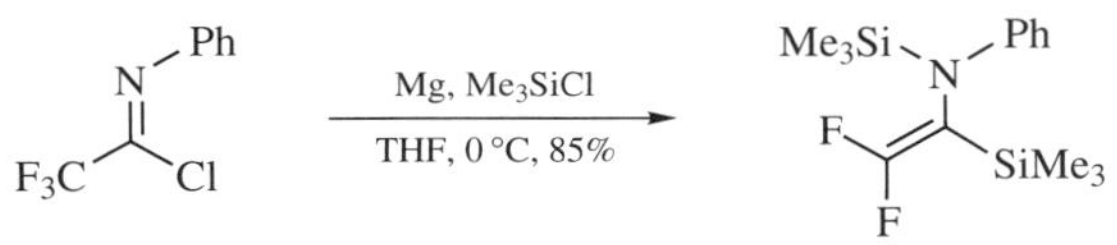

(3)

Precursor to Imidoyl Radicals. *N*-Aryltrifluoroacetimidoyl iodides,[26] tellurides,[27] and azo-derivatives,[28] readily prepared from the chlorides, produce the corresponding imidoyl radicals (under the promotion of tin radical, photolysis, or thermolysis), which react with alkynes intermolecularly or intramolecularly.

Preparation of Glycosyl Trifluoroacetimidates. Reaction with 1-OH sugars in the presence of K_2CO_3 in acetone at rt affords the corresponding glycosyl trifluoroacetimidates in excellent yields in favor of the α anomers (eq 4).[29–34] An alternative set of reaction conditions uses diisopropylethylamine as a base and CH_2Cl_2 as solvent.[35] The resulting *N*-phenyltrifluoroacetimidates are excellent donors for glycosylation reactions, coupling with a variety of hydroxyl groups (including those of steroids,[29,30] triterpenes,[31,32] flavonoids,[33] and saccharides[29–34]) to provide the corresponding glycoconjugates and oligosaccharides in satisfactory yields. Normally, 0.05 molar equivalents of TMSOTf are employed to drive the glycosylation to completion (rt, 4 Å MS, CH_2Cl_2).[29–32] $BF_3 \cdot OEt_2$,[33] I_2/Et_3SiH,[35] $Yb(OTf)_3$,[36] and acid-washed molecular sieves[37] are also effective promoters. Compared to the widely employed glycosyl trichloroacetimidates, *N*-phenyltrifluoroacetimidates are slightly more stable,[38] which facilitates purification and storage. In addition, the 2-*O*-trifluoroacetimidate of methyl *N*-acetylneuraminic acid (Neu5Ac) has been successfully prepared (eq 5), while the corresponding trichloroacetimidate is not accessible, and behaves as one of the most effective donors for direct construction of the important but formidable equatorial 2-α-ketosidic sialyl linkages.[39]

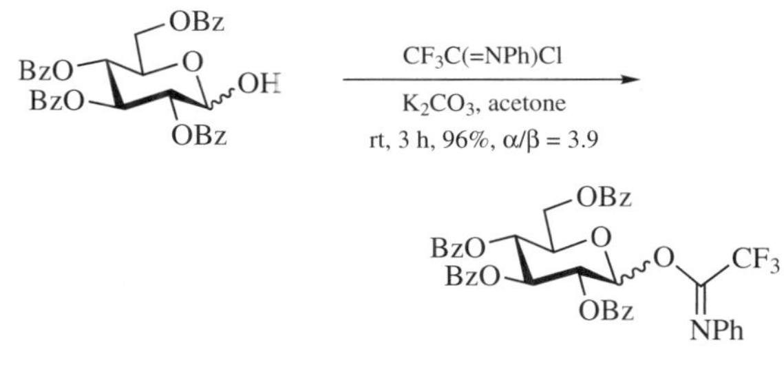

(4)

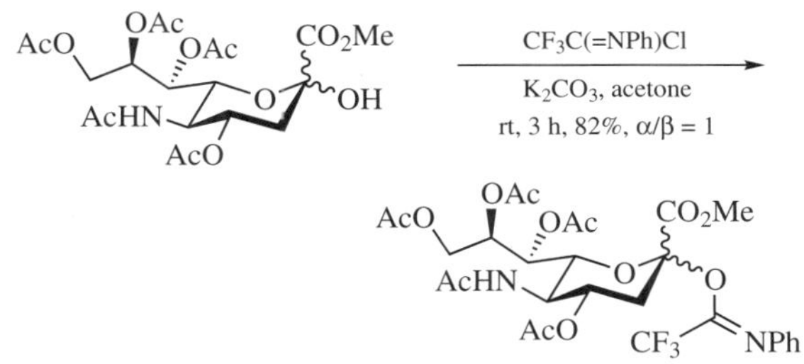

(5)

Related Reagents. *N*-Aryltrifluoroacetimidoyl chlorides; *N*-aryltrifluoroacetimidoyl bromides and iodides; *N*-arylperfluoroalkylimidoyl halides; trichloroacetonitrile.

---

1. Tamura, K.; Mizukami, H.; Maeda, K.; Watanabe, H.; Uneyama, K., *J. Org. Chem.* **1993**, *58*, 32.
2. Uneyama, K., *J. Fluorine Chem.* **1999**, *97*, 11.
3. Uneyama, K.; Morimoto, O.; Yamashita, F., *Tetrahedron Lett.* **1989**, *30*, 4821.
4. Ohkura, H.; Handa, M.; Katagiri, T.; Uneyama, K., *J. Org. Chem.* **2002**, *67*, 2692.
5. Fustero, S.; Pina, B.; de la Torre, M. G.; Navarro, A.; Ramirez de Arellano, C.; Simon, A., *Org. Lett.* **1999**, *1*, 977.
6. Huang, W.; Yuan, C., *Synthesis* **1996**, 511.
7. Bravo, P.; Cavicchio, G.; Crucianello, M.; Markovsky, A. L.; Volonterio, A.; Zanda, M., *Synlett* **1996**, 887.
8. Huang, W.; Yuan, C., *J. Chem. Soc., Perkin Trans. 1* **1995**, 741.
9. Huang, W.; Yuan, C.; Wang, Z., *J. Fluorine Chem.* **1995**, *74*, 247.
10. Reinartz, S.; Mason, A. F.; Lobkovsky, E. B.; Coates, G.W., *Organometallics* **2003**, *22*, 2542.
11. Uneyama, K.; Kobayashi, M., *J. Org. Chem.* **1994**, *59*, 3003.
12. Uneyama, K.; Yamashita, F.; Sugimoto, K.; Morimoto, O., *Tetrahedron Lett.* **1990**, *31*, 2717.
13. Uneyama, K.; Sugimoto, K., *J. Org. Chem.* **1992**, *57*, 6014.
14. Xiao, J.; Zhang, X.; Wang, D.; Yuan, C., *J. Fluorine Chem.* **1999**, *99*, 83.
15. Uneyama, K.; Hao, J.; Amii, H., *Tetrahedron Lett.* **1998**, *39*, 4079.
16. Uneyama, K.; Ohkura, H.; Hao, J.; Amii, H., *J. Org. Chem.* **2001**, *66*, 1026.
17. Kim, S.; Kavali, R., *Tetrahedron Lett.* **2002**, *43*, 7189.
18. Yuan, C.; Zhang, Y.; Luo, W.; Yao, Z., *Heteroatom. Chem.* **1998**, *9*, 139.
19. Watanabe, H.; Hashizume, Y.; Uneyama, K., *Tetrahedron Lett.* **1992**, *33*, 4333.
20. Uneyama, K.; Watanabe, H., *Tetrahedron Lett.* **1991**, *32*, 1459.
21. Tamura, J.; Yan, F. Y.; Sakai, T.; Uneyama, K., *Bull. Chem. Soc. Jpn.* **1994**, *67*, 300.
22. Watanabe, H.; Yan, F.; Sakai, T.; Uneyama, K., *J. Org. Chem.* **1994**, *59*, 758.
23. Uneyama, K.; Noritake, C.; Sadamune, K., *J. Org. Chem.* **1996**, *61*, 6055.

24. Amii, H.; Kishikawa, Y.; Uneyama, K., *Org. Lett.* **2001**, *3*, 1109.
25. Kobayashi, T.; Nakagawa, T.; Amii, H.; Uneyama, K., *Org. Lett.* **2003**, *5*, 4297.
26. Dan-oh, Y.; Matta, H.; Uemura, J.; Watanabe, H.; Uneyama, K., *Bull. Chem. Soc. Jpn.* **1995**, *68*, 1497.
27. Ueda, Y.; Watanabe, H.; Uemura, J.; Uneyama, K.; *Tetrahedron Lett.* **1993**, *34*, 7933.
28. Kadota, K.; Dan-oh, Y.; Uneyama, K., *J. Org. Chem.* **1996**, *61*, 2364.
29. Yu, B.; Tao, H., *Tetrahedron Lett.* **2001**, *42*, 2405.
30. Yu, B.; Tao H., *J. Org. Chem.* **2002**, *67*, 9099.
31. Sun, J.; Han, X.; Yu, B., *Carbohydr. Res.* **2003**, *338*, 827.
32. Peng, W.; Sun, J.; Lin, F.; Han, X.; Yu, B., *Synlett* **2004**, 259.
33. Li, M.; Han, X.; Yu, B., *J. Org. Chem.* **2003**, *68*, 6842.
34. Zhang, Z.; Yu, B., *J. Org. Chem.* **2003**, *68*, 6309.
35. Adinolfi, M.; Barone, G.; Iadonisi, A.; Schiattarella, M., *Synlett* **2002**, 269.
36. Adinolfi, M.; Barone, G.; Iadonisi, A.; Schiattarella, M., *Tetrahedron Lett.* **2002**, *43*, 5573.
37. Adinolfi, M.; Barone, G.; Iadonisi, A.; Schiattarella, M., *Org. Lett.* **2003**, *5*, 987.
38. Nakajima, N.; Saito, M.; Kudo, M.; Ubukata, M., *Tetrahedron* **2002**, *58*, 3579.
39. Cai, S.; Yu, B., *Org. Lett.* **2003**, *5*, 3827.

Biao Yu
Shanghai Institute of Organic Chemistry, Chinese Academy of Sciences, Shanghai, China

Phosphorus Oxychloride

$POCl_3$

[10025-87-3] Cl_3OP (MW 153.32)

(formylation of aromatic rings[1,2] (Vilsmeier–Haack reaction); phosphorylating agent;[3] dehydrating agent for amides; halogenation of alcohols, phenols, and heterocycles)

Alternate Name: phosphoryl chloride.
Physical Data: mp 1 °C; bp 106 °C; *d* 1.675 g mL^{-1}; *n* 1.461.
Solubility: sol THF, MeCN, CH_2Cl_2, many other solvents.
Form Supplied in: colorless, fuming liquid; commercially available.
Purification: by distillation in vacuo.[4]
Handling, Storage, and Precautions: toxic and corrosive; reacts vigorously with alcohols and water, liberating HCl, phosphoric acid, and heat. Protect from water. Use in a fume hood.

Chloromethyleneiminium Ions.[5–8] Phosphorus oxychloride (**1**) is a strong Lewis acid that is widely used in synthesis. Of particular importance are reactions of (**1**) with substituted amides, most often N,N-Dimethylformamide or dimethylacetamide (DMA). These reactions lead to the formation of chloromethyleneiminium salts (**2**) (eq 1). These salts (Vilsmeier reagents, see Dimethylchloromethyleneammonium Chloride) are highly versatile intermediates[6] and are involved in numerous important reactions, including the Vilsmeier–Haack[2,6] and Bischler–Napieralski[9] reactions. Formation of (**2**) can be achieved using Thionyl Chloride or Phosgene, although there have been reports of differences in reactivity between $POCl_3$/DMF, $SOCl_2$/DMF, and $COCl_2$/DMF systems.[7,10,11]

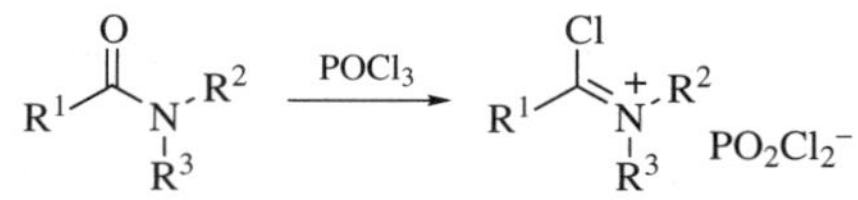

(**2**) (1)

Formylation of Aromatic Rings.[2,5,6,8,11] The Vilsmeier reagent attacks electron-rich aromatic systems to form arylmethyleneiminium ions which liberate a formylated aromatic compound upon hydrolysis (eq 2). Thio- and selenoaldehydes can be prepared by hydrolysis in the presence of Sodium Hydrogen Sulfide[6,12] or Sodium Hydrogen Selenide.[13] A wide range of aromatic systems[6,11] can be formylated in this fashion, including benzene derivatives, polyaromatic hydrocarbons (eq 3), and azulene.[11] Substitution occurs at relatively electron-rich positions.

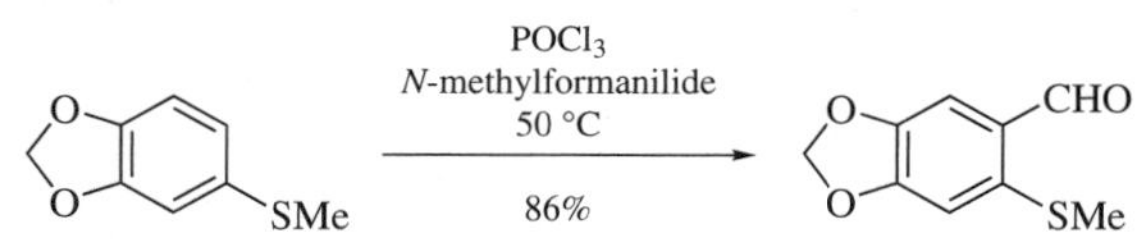

(2)

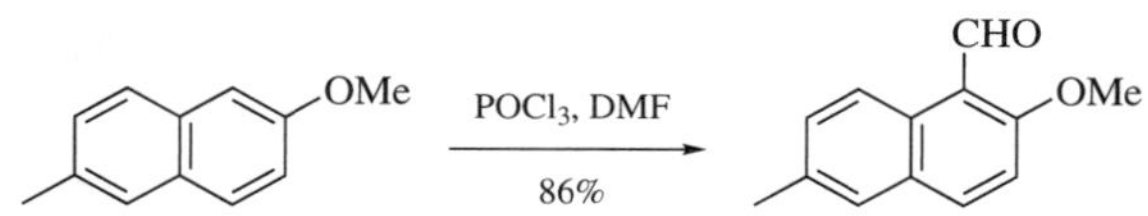

(3)

Formylation of Heterocycles.[6,11,14] Many heterocycles are readily formylated by $POCl_3$/DMF, including pyrroles, thiophenes, furans, indoles, quinolines, pteridines, and purines.[6] Reaction at the 2- and 5-positions of pyrroles, furans, and thiophenes is preferred, and indoles undergo attack at the 3-position (eq 4).[6]

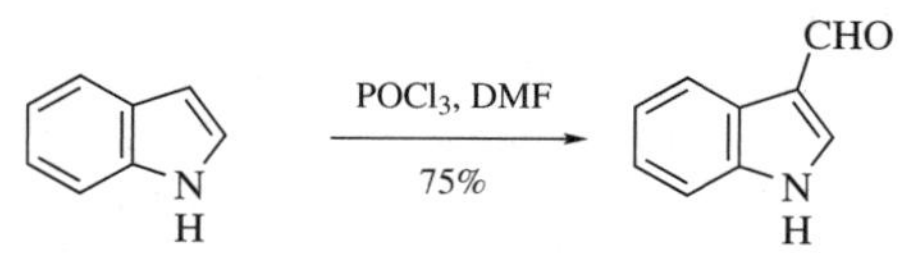

(4)

Formylation of Alkenes.[10,11] Both activated and unactivated alkenes undergo electrophilic attack by (**2**).[5] Reaction at terminal positions is preferred.[6] Styrene derivatives give cinnamaldehydes; enamines[6] and aryl polyenes[15,16] react at terminal positions (eq 5).

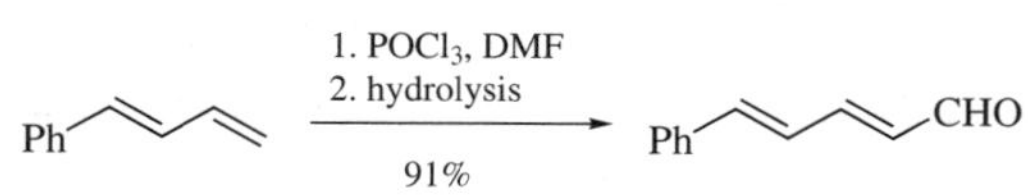

(5)

Bischler–Napieralski Reaction. A widely used method for cyclization of *N*-β-phenylethyl amides to form dihydroisoquinolines and isoquinolines (eq 6)[17–19] is the Bischler–Napieralski reaction.[9] A nitrilium ion intermediate has been implicated in this reaction.[20,21]

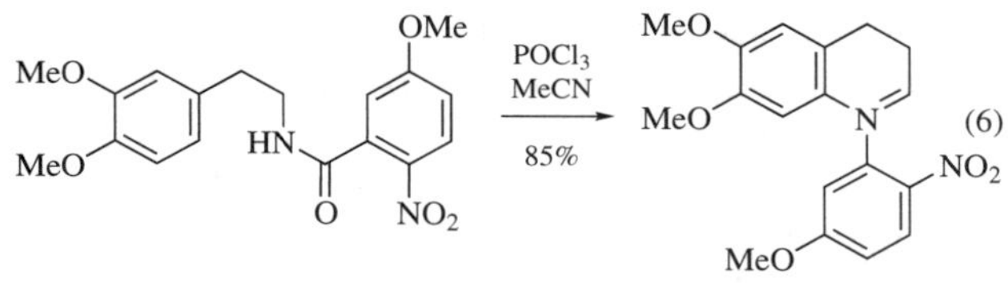

(6)

Fragmentation of the nitrilium ion interferes with the cyclization[20,21] of 1,2-diphenylethane derivatives, leading to formation of stilbenes rather than cyclization products. This problem has been overcome by using Oxalyl Chloride instead of (**1**).[22] Fragmentation (von Braun reaction or retro-Ritter reaction) occurs whenever a highly stable cation can be formed, such as benzyl[20] or *t*-butyl (eq 7).[21,23,24]

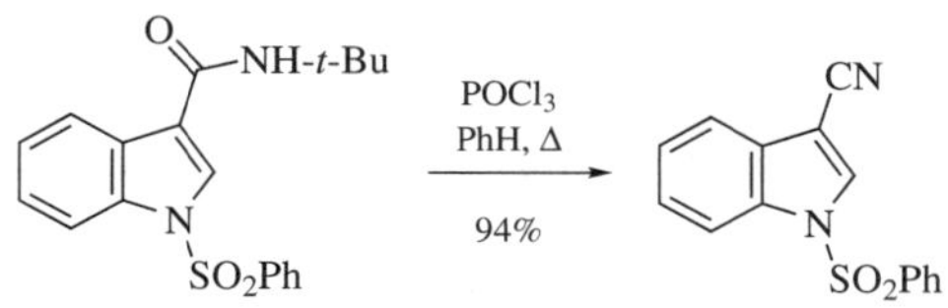

(7)

Treatment of tertiary amides with $POCl_3$/DMF (**2**) results in formation of β-dimethylamino α,β-unsaturated amides. The highly electrophilic iminium ion (**3**) is formed in this reaction, and in the presence of an alkene this undergoes cyclization (eq 8).[14,25]

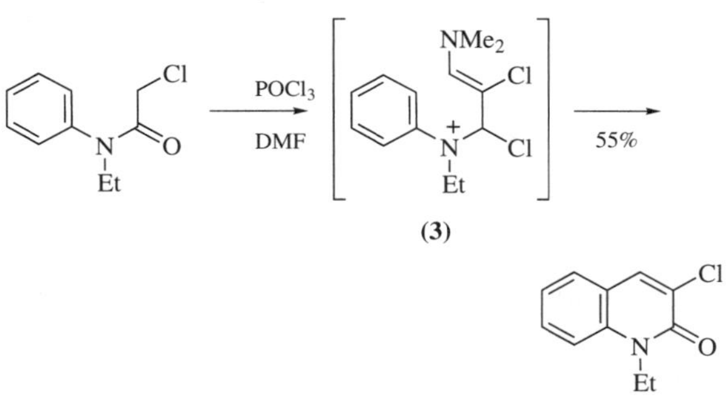

(**3**)

(8)

Excess $POCl_3$ and DMF can lead to formylation of the initial heterocycle (eq 9). A series of α-substituted acrylonitriles has been prepared by dehydration of α,β-unsaturated amides (eq 10). $POCl_3$/DMF and $SOCl_2$/DMF give comparable results.[26]

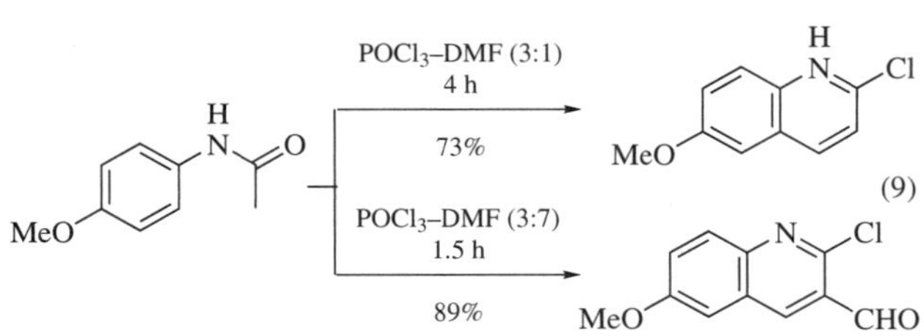

(9)

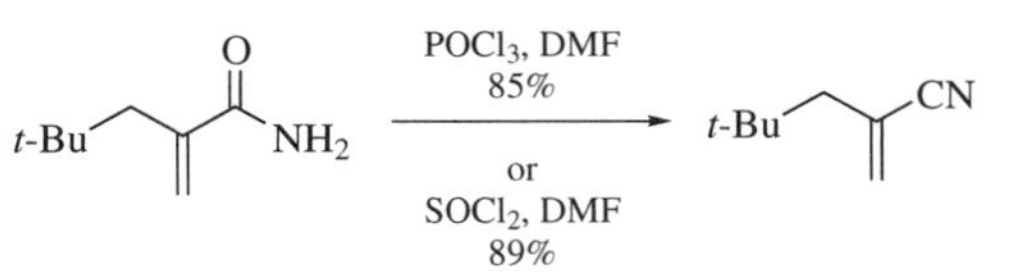

(10)

Chloroaldehydes from Carbonyl Compounds.[7] The enol tautomers of ketones react with $POCl_3$/DMF to form β-chloroacrolein derivatives (eq 11). The regiochemistry is determined both by the stability of the enol as well as by steric factors. Nonsymmetrical ketones often give a mixture of regioisomers, and most ketones produce a mixture of (*E*) and (*Z*) isomers of the product.[27] Aldehydes will undergo this type of reaction, although there are relatively few reports.

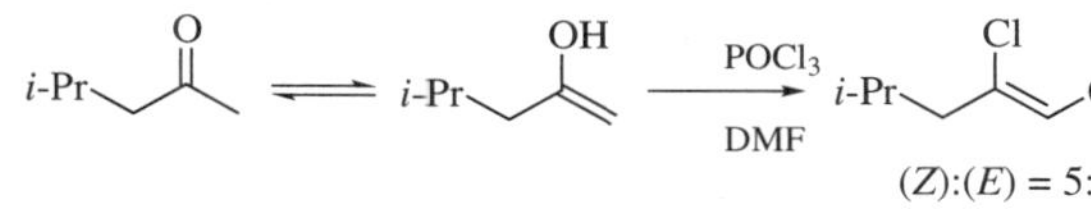

(11)

Reaction of $POCl_3$/DMF with carboxylic acids results in vinamidinium ions (eq 12).[28,29] These ions react with a variety of nucleophiles to produce pyrazoles, oxazoles, pyrimidines, diazepines, quinolines, quinolizines, and vinylogous sulfonamides.[30]

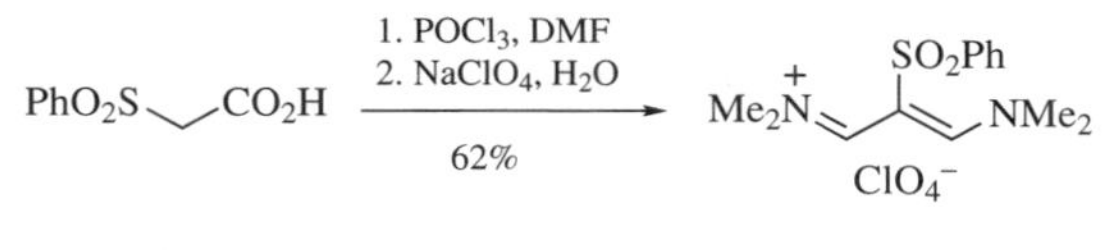

(12)

The Vilsmeier reagent (**2**) reacts with Grignard reagents as well as with alkylzinc and alkylaluminum reagents to form tertiary amines.[31] It also reacts with heteroatom nucleophiles, including thiols,[32] alcohols, and amines,[8] as well as with nucleophilic amines containing other functional groups.[5,33,34] Aromatic alcohols react with $POCl_3$/DMF to form aryl formates in 50–80% yield,[35] but aliphatic alcohols are more efficiently formylated with Benzoyl Chloride/DMF.[36] Homoallylic alcohols such as (**4**) produce biphenyls in 30–98% yield[16] through initial dehydration[6] (eq 13) followed by electrophilic attack on the resulting aryl diene.

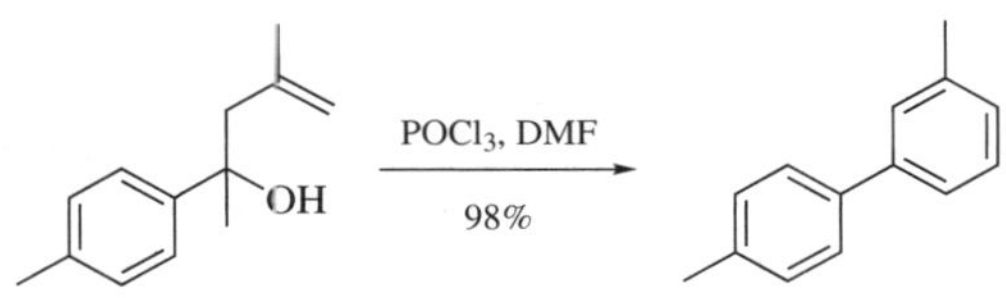

(**4**)

(13)

Halogenation of Alcohols. The combination of $POCl_3$ and DMF can be used to halogenate primary, secondary, and tertiary aliphatic alcohols (eq 14),[37] whereas the reaction of primary alcohols and $POCl_3$ without DMF or DMA will generally lead to the formation of trialkyl phosphates.

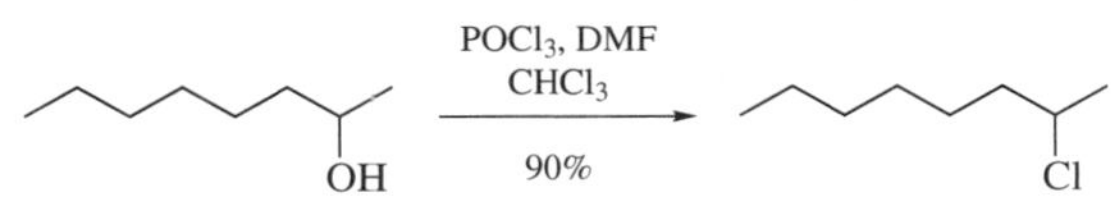

(14)

A comparison of $POCl_3$ with SO_2Cl_2, PCl_3, MsCl, and $SOCl_2$ indicated that optimum conditions for the conversion of (**5**) to (**6**) (eq 15) are 1.1 equiv $POCl_3$ in DMF at 0 °C.[38]

(**5**) (**6**)

(15)

Halogenation of Heterocycles. Phosphorus oxychloride is widely used in the chlorination of heterocycles. In general, heterocyclic ketones (eq 16) or alcohols (eq 17) react readily with (**1**), including pteridines,[39] purines,[40,41] and others.

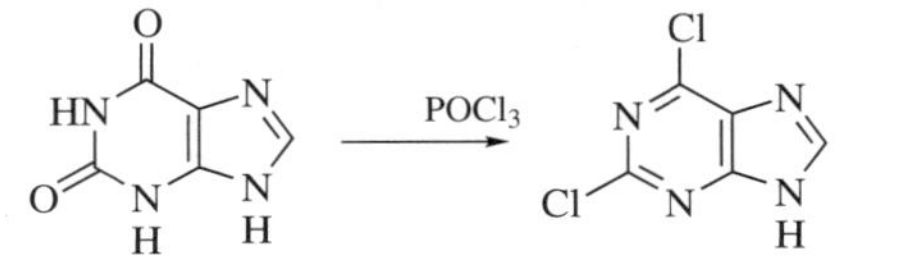

(16)

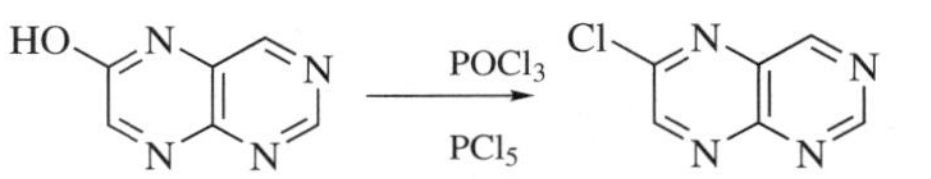

(17)

In the chlorination of the isoxazole (**7**), it was noted that freshly distilled $POCl_3$ is ineffective but that an older sample leads to the formation of the chloride (**8**). Further investigation showed that a mixture of $POCl_3$, acid (H_3PO_4), and an additional chloride source (pyridinium chloride) leads to reliable conversion of (**7**) to (**8**) (eq 18).[42]

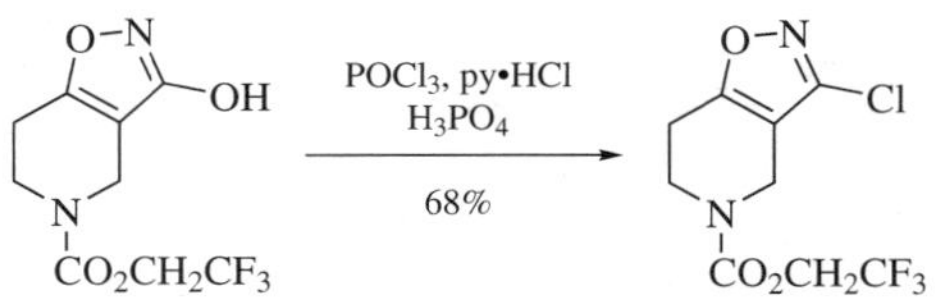

(18)

(**7**) (**8**)

Phosphorylation.[3,43] Phosphorus oxychloride reacts with alcohols, amines, and thiols, resulting in phosphorylation of these functional groups. Trimethyl phosphate is a particularly effective solvent,[44] and tertiary amine bases are generally used as well. Treatment of primary alcohols with $POCl_3$ results in the formation of phosphonyl dichloride intermediates which, in the presence of excess alcohol, convert to symmetrical trialkyl phosphates.[10] It is generally possible to isolate aryl phosphorodichloridates when a two-fold excess of (**1**) is used and $AlCl_3$, KCl, or pyridine is used as a catalyst.[45,46] Secondary and tertiary alcohols tend to form alkyl chlorides and phosphoric acid.

Phosphorus oxychloride is very useful in the preparation of nucleoside 5′-phosphates (eq 19) given appropriate mixtures of $POCl_3$, pyridine, and water.[44,47] The primary hydroxy (5′) is sufficiently reactive for there to be minimal formation of cyclic phosphates at the 2,3-positions of nucleosides. The intermediate phosphorodichloridate can be converted to the corresponding triphosphate by treatment with inorganic phosphate in a convenient one-pot fashion.

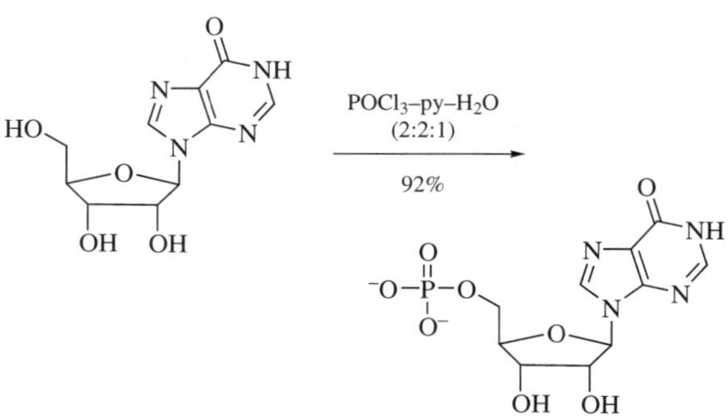

(19)

Under anhydrous conditions, $POCl_3$ can be used to produce cyclic dialkyl phosphonyl chlorides from diols, and these can be hydrolyzed with water to the cyclic phosphates or with an alcohol to trialkyl phosphates. Treatment of forskolin (**9**) with $POCl_3$ leads to the formation of the cyclic phosphoryl chloride (**10**) in 55% yield as a mixture of two stereoisomers at phosphorus (eq 20).[48] A comparison of several phosphorylating agents showed that $POCl_3$, $(MeOC_6H_4)_2POCl$, and 2-chloro-2-oxo-1,3,2-dioxaphospholane all give the analogous products in virtually identical yield.[48]

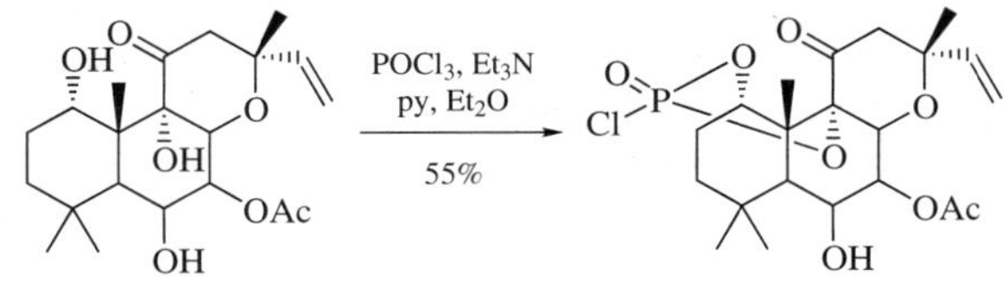

(20)

(**9**) (**10**)

Similar results can be obtained with primary amines,[49] although it is possible to obtain diamidophosphorochloridates by reaction of $POCl_3$ with substoichiometric amounts of amine.[50] Phosphorus oxychloride reacts more readily with primary amines than with primary alcohols, so that it is possible to prepare phosphoramides.[51] Alkoxides, however, react more rapidly than amines so it is possible to phosphorylate hydroxyls in the presence of amines. Again, like the reaction of $POCl_3$ with diols, diamines react with $POCl_3$ to form diazaphospholidines (eq 21).[52]

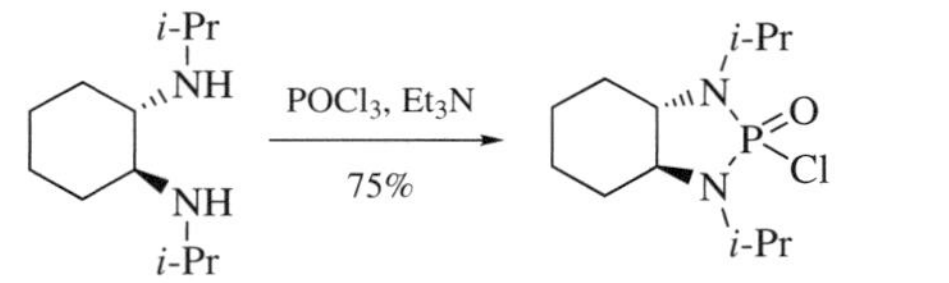

(21)

Carboxylic anhydrides can be prepared by treatment of carboxylic acids with $POCl_3$,[32] allowing the formation of both esters[32] and amides, although other methods are more common.[53] Dichloroacetyl chloride produces dichloroketene when treated with $POCl_3$ and Zinc/Copper Couple.[54]

Dehydration of Amides. Unsubstituted amides undergo dehydration upon treatment with $POCl_3$. This reaction can also be performed with P_2O_5, $SOCl_2$, or other reagents, and in a study of racemization at the α-position it was determined that dehydration with $POCl_3$ leads to more (albeit little) epimerization than dehydration with P_2O_5 or $SOCl_2$ (eq 22).[55]

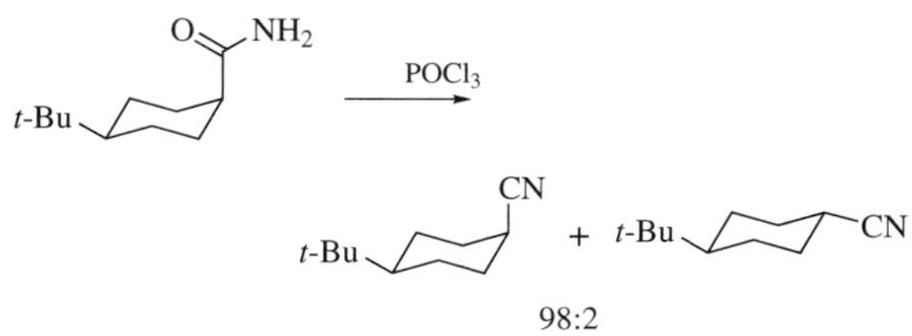

(22)

Dehydration of alkyl and aryl *N*-formyl compounds with $POCl_3$ is one of the more general routes to alkyl and aryl isocyanides (eq 23).[56–59] A base, typically an amine base or Potassium t-Butoxide, is also required. The method is simple and effective, although less useful for small, volatile isocyanides than other techniques.[60] The Bischler–Napieralski reaction can occur preferentially to isocyanide formation.[17]

Dehydration of Alcohols. The combination of $POCl_3$ and Pyridine is an effective dehydrating agent for alcohols (eq 24)[61] and cyanohydrins.[62] The stereochemistry of elimination is *anti*, although the regioselectivity is often not high, particularly for tertiary alcohols.[63]

The combination of phosphorus oxychloride and Tin(II) Chloride reduces halohydrins to alkenes (eq 25). The elimination proceeds in an *anti* fashion.[64]

Beckmann Rearrangement. Treatment of ketoximes with $POCl_3$ induces Beckmann rearrangement to form amides (eq 26).[65] Numerous other Lewis acids can be used for this transformation.[66,67]

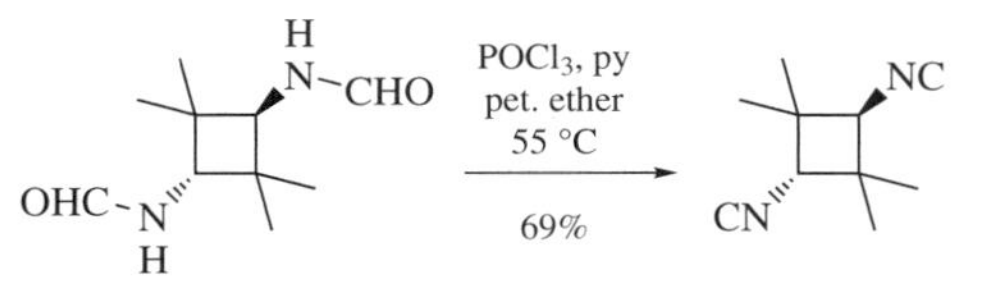

(23)

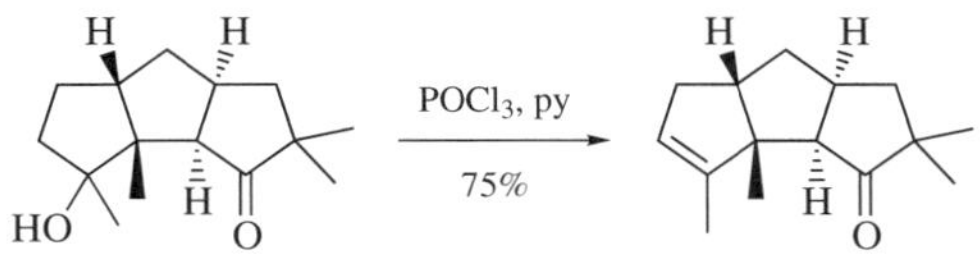

(24)

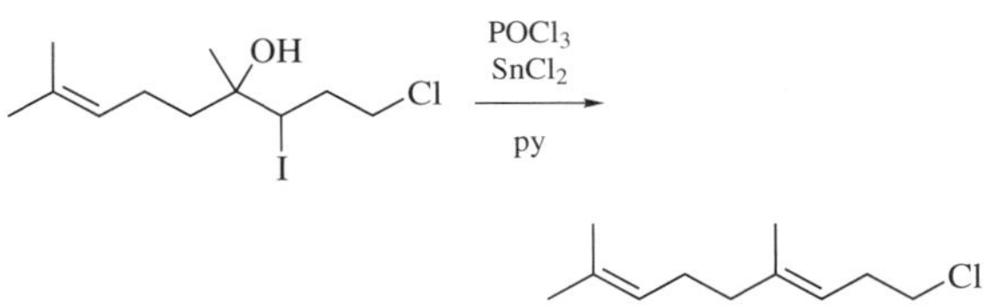

(25)

86% from the epoxide precursor

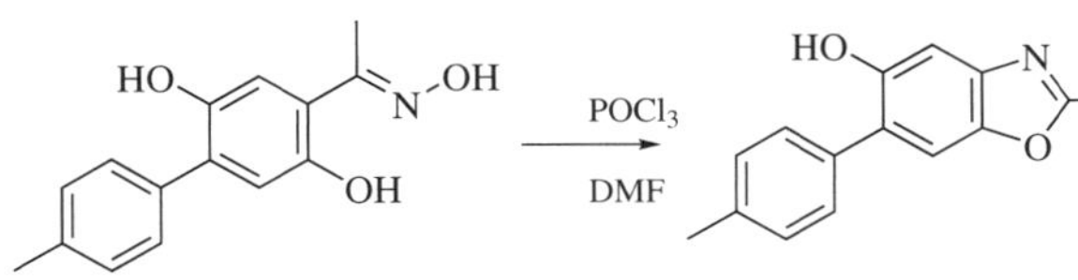

(26)

Related Reagents. Bis(trichloromethyl) Carbonate; Phosphorus(V) Oxide; Phosphorus Oxychloride–Zinc(II) Chloride; Thionyl Chloride; *p*-Tolyl Vinyl Sulfoxide; Trichloromethyl Chloroformate .

1. *Fieser & Fieser* **1967**, *1*, 876.
2. Seshadri, S. *J. Sci. Ind. Res.* **1973**, *32*, 128.
3. Hayakawa, H. *Comprehensive Organic Synthesis* **1991**, *6*, 601.
4. Perrin, D. D.; Armarego, W. L. F. *Purification of Laboratory Chemicals*, 3rd ed.; Pergamon:Oxford, **1988**.
5. Meth-Cohn, O.; Stanforth, S. P. *Comprehensive Organic Synthesis* **1991**, *2*, 777.
6. Jutz, C.In *Iminium Salts in Organic Chemistry*; Böhme, H.; Viehe, H. G.; Eds.; Wiley:New York, **1976**; Vol. *9*, pp 225–342.
7. Marson, C. M. *Tetrahedron* **1992**, *48*, 3659.
8. Kantlenher, W. *Comprehensive Organic Synthesis* **1991**, *6*, 485.
9. Hilger, C. S.; Fugmann, B.; Steglich, W. *Tetrahedron Lett.* **1985**, *26*, 5975.
10. Burn, D. *Chem. Ind. (London)* **1973**, 870.
11. Khimii, U. *Russ. Chem. Rev. (Engl. Transl.)* **1960**, *29*, 599.
12. Lin, Y.; Lang, S. A. *J. Org. Chem.* **1980**, *45*, 4857.
13. Reid, D. H.; Webster, R. G.; McKenzie, S. *J. Chem. Soc., Perkin Trans. 1* **1979**, 2334.
14. Meth-Cohn, O. *Heterocycles* **1993**, *35*, 539.
15. Jutz, C.; Heinicke, R. *Ber. Dtsch. Chem. Ges./Chem. Ber.* **1969**, *103*, 623.
16. Suresh Chandler Rao, M. S. C.; Krishna Rao, G. S. K. *Synthesis* **1987**, 231.
17. Badia, D.; Carrillo, L.; Dominguez, E.; Cameno, A. G.; Martinez de Marigorta, E.; Vincente, T. *J. Heterocycl. Chem.* **1990**, *27*, 1287.
18. Kametani, T.; Fukumoto, K.In *Isoquinolines*; G. Grethe, Ed.; Wiley: New York, **1981**; Vol. *31*, Part 1, pp 142–160.
19. Knabe, J.; Krause, W.; Powilleit, H.; Sierocks, K. *Pharmazie* **1970**, *25*, 313.
20. Fodor, G.; Gal, J.; Phillips, B. A. *Angew. Chem., Int. Ed. Engl.* **1972**, *11*, 919.
21. Fodor, G.; Nagubandi, S. *Tetrahedron* **1980**, *36*, 1279.
22. Larsen, R. D.; Reamer, R. A.; Corley, E. G.; Davis, P.; Grabowski, E. J. J.; Reider, P. J.; Shinkai, I. *J. Org. Chem.* **1991**, *56*, 6034.
23. Ketcha, D. M.; Gribble, G. W. *J. Org. Chem.* **1985**, *50*, 5451.
24. Perni, R. B.; Gribble, G. W. *Org. Prep. Proced. Int.* **1983**, *15*, 297.
25. Meth-Cohn, O.; Tarnowski, B. *Adv. Heterocycl. Chem.* **1982**, *31*, 207.
26. Bargar, T. M.; Riley, C. M. *Synth. Commun.* **1980**, *10*, 479.
27. Schellhorn, H.; Hauptmann, S.; Frischleder, H. *Z. Chem.* **1973**, *13*, 97.
28. Gupton, J. T.; Riesinger, S. W.; Shah, A. S.; Bevirt, K. M. *J. Org. Chem.* **1991**, *56*, 976.
29. Gupton, J. T.; Gall, J. E.; Riesinger, S. W.; Smith, S. Q.; Bevirt, K. M.; Sikorski, J. A.; Dahl, M. L.; Arnold, Z. *J. Heterocycl. Chem.* **1991**, *28*, 1281.
30. McNab, H.; Lloyd, D. *Angew. Chem., Int. Ed. Engl.* **1976**, *15*, 459.
31. Mesnard, D.; Miginiac, L. *J. Organomet. Chem.* **1989**, *373*, 1.
32. Arrieta, A.; Garcia, T.; Lago, J. M.; Palomo, C. *Synth. Commun.* **1983**, *13*, 471.
33. Harris, L. N. *Synthesis* **1981**, 907.
34. Zelenin, K. N.; Khrustalev, V. A.; Sergutina, V. P. *Chem. Abstr.* **1980**, *93*, 70 910.
35. Morimura, S.; Horiuchi, H.; Muruyama, K. *Bull. Chem. Soc. Jpn.* **1977**, *50*, 2189.
36. Barluenga, J.; Campos, P. J.; Gonzalez-Nuñez, E.; Asensio, G. *Synthesis* **1985**, 426.
37. Yoshihara, M.; Eda, T.; Sakaki, K.; Maeshima, T. *Synthesis* **1980**, 746.
38. Sanda, K.; Rigal, L.; Delmas, M.; Gaset, A. *Synthesis* **1992**, *6*, 541.
39. Albert, A.; Clark, J. *J. Chem. Soc.* **1964**, 1666.
40. Golovchinskaya, E. S. *Russ. Chem. Rev. (Engl. Transl.)* **1974**, *43*, 1089.
41. Robins, R. K.; Revankar, G. R.; O'Brien, D. E.; Springer, R. H.; Novinson, T.; Albert, A.; Senga, K.; Miller, J. P.; Streeter, D. G. *J. Heterocycl. Chem.* **1985**, *22*, 601.
42. Andersen, K.; Begtrup, M. *Acta Chem. Scand.* **1992**, *46*, 1130.
43. Edmunson, R. S.In *Comprehensive Organic Chemistry*; Barton, D. H. R.; Ollis, W. D., Eds.; Pergamon:Oxford, **1979**; Vol. *2*, pp 1262–1263.
44. Yoshikawa, M.; Kato, T.; Takenishi, T. *Bull. Chem. Soc. Jpn.* **1969**, *42*, 3505.
45. Owen, G. R.; Rees, C. B.; Ransom, C. J.; van Boom, J. H.; Herscheid, J. D. H. *Synthesis* **1974**, 704.
46. Taguchi, Y.; Mushika, Y. *Tetrahedron Lett.* **1975**, 1913.
47. Sowa, T.; Ouchi, S. *Bull. Chem. Soc. Jpn.* **1975**, 2084.
48. Lal, B.; Gangopadhyay, A. K. *J. Chem. Soc., Perkin Trans. 1* **1992**, *15*, 1993.
49. Edmunson, R. S.In *Comprehensive Organic Chemistry*; Barton, D. H. R.; Ollis, W. D., Eds.; Pergamon:Oxford, **1979**; Vol. *2*, pp 1262–1265.

50. Ireland, R. E.; O'Neil, T. H.; Tolman, G. L. *Org. Synth.* **1983**, *61*, 116.
51. Crans, D. C.; Whitesides, G. M. *J. Am. Chem. Soc.* **1985**, *107*, 7008.
52. Alexakis, A.; Mutti, S.; Mangeney, P. *J. Org. Chem.* **1992**, *57*, 1224.
53. Mulzer, J. *Comprehensive Organic Synthesis* **1991**, *6*, 323.
54. Hassner, A.; Dillon, J. *Synthesis* **1979**, 689.
55. Rickborn, B.; Jensen, F. R. *J. Org. Chem.* **1962**, *27*, 4608.
56. Sandler, S. R.; Karo, W.In *Organic Functional Group Preparations*, 2nd ed.; Academic:San Diego, **1989**; Vol. *3*, pp 207–238.
57. Ugi, I.; Meyr, R.; Lipinski, M.; Bodesheim, F.; Rosendahl, F. *Org. Synth., Coll. Vol.* **1973**, *5*, 300.
58. van Leusen, D.; van Leusen, A. M. *Recl. Trav. Chim. Pays-Bas* **1992**, *47*, 1249.
59. van Leusen, A. M.; Boerma, G. J. M.; Helmholdt, R. B.; Siderius, H.; Strating, J. *Tetrahedron Lett.* **1972**, 2367.
60. Höfle, G.; Lange, B. *Org. Synth.* **1983**, *61*, 14.
61. Mehta, G.; Murthy, A. N.; Reddy, D. S.; Reddy, A. V. *J. Am. Chem. Soc.* **1986**, *108*, 3443.
62. Oda, M.; Yamauro, A.; Watabe, T. *Chem. Lett.* **1979**, 1427.
63. Giner, J.-L.; Margot, C.; Djerassi, C. *J. Org. Chem.* **1989**, *54*, 369.
64. Cornforth, J. W.; Cornforth, R. H.; Mathew, K. K. *J. Chem. Soc.* **1959**, 2539.
65. Fujita, S.; Kotauna, K.; Inagaki, Y. *Synthesis* **1982**, 68.
66. Gawley, R. E. *Org. React.* **1988**, *35*, 1.
67. Donaruma, L. G.; Heldt, W. Z. *Org. React.* **1960**, *11*, 1.

Mark S. Meier
University of Kentucky, Lexington, KY, USA

Suzanne M. Ruder
Virginia Commonwealth University, Richmond, VA, USA

Phthalic Anhydride

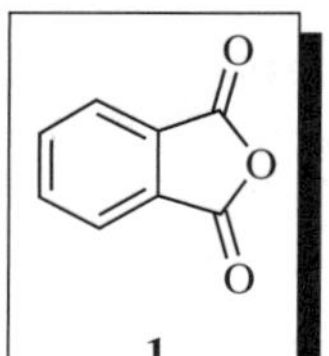

1

[85-44-9] $C_8H_4O_3$ (MW 148.12)

(reagent used in acylation of amines, alcohols; Friedel-Craft acylation; olefination reactions; oxidation reactions; and dehydration reactions)

Alternate Name: 1,3-isobenzofurandione, 1,2-benzenedicarboxylic acid anhydride, 1,3-dioxophthalan, 1,3-phthalandione; phthalic acid anhydride; 1,2-benzenedicarboxylic anhydride; phthalandione; 1,3-dihydro-1,3-dioxoisobenzofuran.

Physical Data: mp 130–131 °C, bp 284 °C, d 1.53, white flaky crystals.

Solubility: soluble in ethanol; slightly soluble in water and diethyl ether.

Introduction. Phthalic anhydride (**1**) is synthesized commercially by catalytic oxidation of *o*-xylene or naphthalene using V_2O_5, MoO_3, V_2O_5-TiO_2, V_2O_5-Al_2O_3, or V_2O_5-SnO_2. On a laboratory scale, it can be obtained from phthalic acid or its esters using various conditions.[1–3] Phthalic anhydride is widely used as a plastizicer; as hardener of resins; and in the synthesis of polyesters, phenolphthalein, and other phthaleins, dyes, chlorinated products, pharmaceutical intermediates, and insecticides. It is also a common laboratory reagent and is used for the protection of amines, acylation of hydroxyls, dehydration, enol- and spirolactone synthesis, and *N*-, *S*-oxide formation.

Acylation of Amines. Phthalic anhydride is a convenient acylating reagent and reacts with primary amines to give phthalimides, which are widely used as amino protecting groups especially in carbohydrate chemistry.[4] They can be introduced under mild conditions by reaction of phthalic anhydride with an amino group in chloroform at elevated temperatures to give the products in high yields.[5] Reaction of phthalic anhydride in the presence of a base such as K_2CO_3, NEt_3, or pyridine in DMF followed by ring-closure with Ac_2O also gives phthalimides in high yields (eq 1).[4,6–8] An attractive solvent-free procedure involved microwave irradiation on $TaCl_5$-silica gel.[9] Alternative reagents for the introduction of phthalimides include *o*-(MeOOC)C_6H_4COCl,[10,11] 3-(dicyanomethylidene) phthalide,[12] phthaloyl dichloride,[13] and *N*-ethoxycarbonylphthalimide;[14] the latter reagent is particularly useful for the protection of α-amino groups of amino acids.[15]

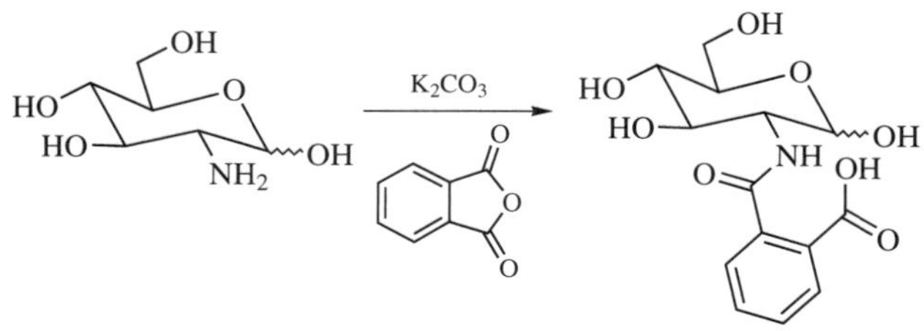

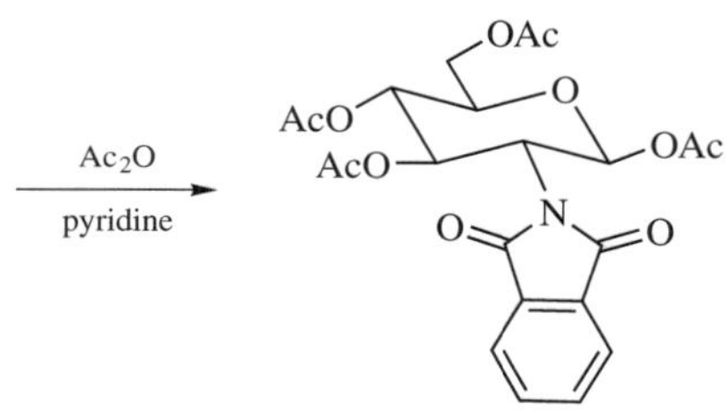

(1)

Phthalimides are stable to a wide range of reaction conditions including acids such as HBr/AcOH; oxidative reagents such as H_2O_2, O_3, and Jones oxidation; and mild transesterification conditions used, for example, to cleave acetyl ester protecting groups.[16] Care has to be taken during benzylation reactions using NaH and BnBr in DMF; these transformations can more reliably be performed when THF is used as a solvent.

The most common procedures for cleavage of phthalimido moieties to give primary amines include the treatment with large excesses of hydrazine,[17,18] $NaBH_4$,[19,20] butylamine,[21] hydroxylamine,[22] ethylenediamine,[23] or alkyldiamines immobilized on polystyrene beads.[24] These procedures can, however, affect base-sensitive functionalities and therefore several modified phthalimides have been reported which can be cleaved under milder conditions. For example, the tetrachlorophthaloyl group can be introduced under similar conditions to the phthaloyl group, but can be cleaved by treatment with ethylene diamine in MeCN/THF/EtOH at 60 °C.[25,26]

Phthalimido derivatives of 2-amino-2-deoxy sugars are often used in oligosaccharide assembly.[27,28] This protecting group is compatible with many glycosylation procedures involving glycosyl donors such as glycosyl bromides and fluorides, trichloroacetimidates, thioglycosides, and pentenyl glycosides. They often give excellent yields and direct the glycosylation to the formation of 1,2-*trans* glycosides by neighboring group participation. Phthalimides are also commonly used in natural product, peptide, and peptidomimetic synthesis.[29–31] For example, this protecting group has been used in the synthesis of a peptidomimetic in anticipation of rather harsh conditions used in the synthetic sequence (eq 2).[32]

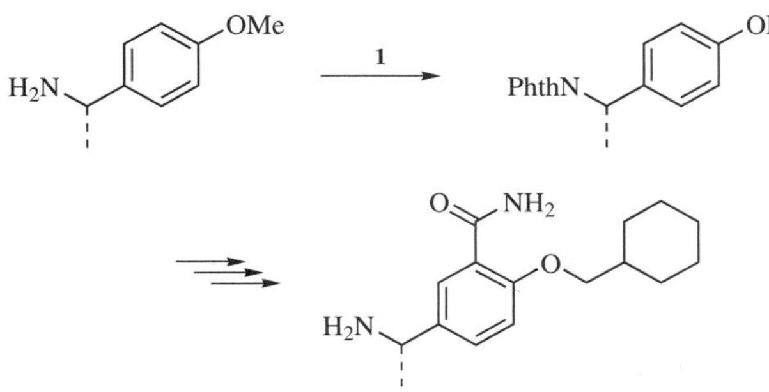

(2)

Acylation of Alcohols and the Formation of Polymers. Phthalic anhydride reacts with primary, secondary, and tertiary alcohols to form phthalate esters. These esters are extensively used in polymerization reactions with ethylene glycol and glycerol to give polyesters. Since ethylene glycol has no secondary hydroxyls, reaction with phthalic anhydride initially gives a half-ester which reacts further to give a diester. Repeating these reaction steps leads to the formation of linear polyesters (eq 3). When glycerol and phthalic anhydride are condensed at ~150 °C, only the primary hydroxyls are involved in the acylation, resulting in a long linear polyester. However, at higher temperatures, a second stage polymerization occurs with the secondary hydroxyls producing a cross-linked polymer. The level of cross-linking is determined by the reaction conditions, and the resulting cross-linked polyesters are more rigid than their linear counterparts prepared from ethylene glycol.

Phthalic anhydride has been used for the resolution of racemic alcohols. In this process, phthalic anhydride is reacted with a racemic alcohol to give a half-ester (hydrogen phthalate). Enantiomeric hydrogen phthalates could be resolved by first formation of a diastereoisomeric salt by addition of a chiral base followed by selective crystallization. This method was applied to the resolution of a key intermediate in the chemical synthesis of optically pure D-biotin.[33] Thus, an intermediate (±)-alcohol was reacted with **1** in pyridine to form a corresponding (±)-phthalic acid mono-ester. The (±)-esters were resolved by salt formation with (+)-α-methylbenzylamine followed by recrystallization from methanol. After extraction, the optically pure salt was treated with H_2SO_4, followed by hydrolysis of the phthalate ester with NaOH to give an alcohol in 90% enantiomeric excess (eq 4).

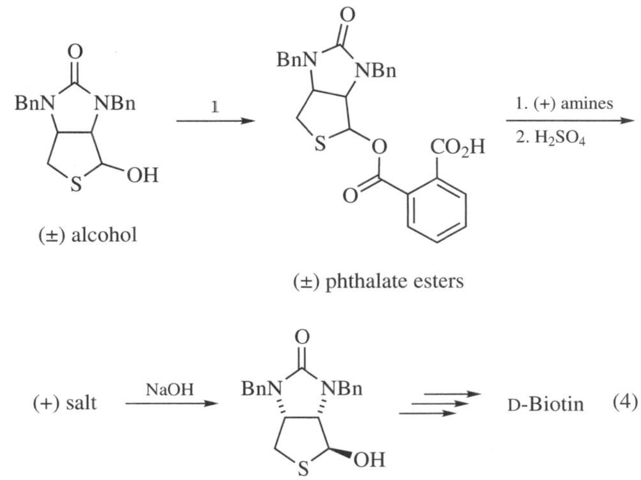

(4)

***C*-Acylations: Friedel-Crafts Acylations.** Phthalic anhydride has been used as an electrophilic reagent in electrophilic aromatic substitutions (Friedel-Crafts acylations). It reacts with a Lewis acid (such as $AlCl_3$) at elevated temperatures (190 °C) to give a reactive acylium ion, which reacts with aromatic compounds to give substituted products that have a carboxylic acid side chain. Activation of the latter functionality will allow a second Friedel-Craft acylation introducing a new ring system. Friedel-Crafts acylations with phthalic anhydride have been used for the synthesis of tetracycline (anthracycline) antibiotics such as Daunorubicin and Adriamycin and their analogs.[34] These syntheses involve the reaction of phthalic anhydride with 2,3-dimethylhydroquinone in a melt of $AlCl_3$/NaCl at 190 °C followed by a tandem double electrophilic aromatic substitution to give a tricyclic quinone intermediate (eq 5).[35]

Organometallic Reagents: *C*-Acylation. Phthalic anhydride reacts with Grignard reagents to give acylated derivatives and this reaction has, for example, been used for the synthesis of benz(α)anthracene derivatives (eq 6).[36] Reaction of dibromomagnesio Grignard reagents with phthalic anhydride led to the formation of spirolactones. For example, the reaction with 1,4-di(bromomagnesio)pentane in THF formed an aromatic spirolactone as a single diastereoisomer (eq 7).[37] This reaction could be extended to several other cyclic anhydrides and lactones.

Phthalic anhydride is a useful reagent for the synthesis of *o*-benzoylbenzoic acid derivatives, and its reaction with

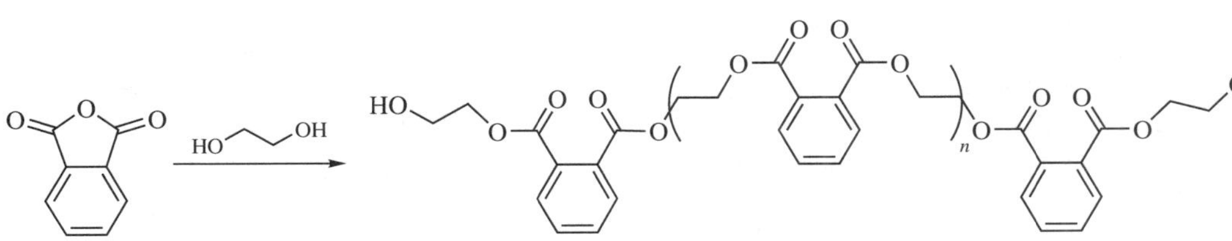

(3)

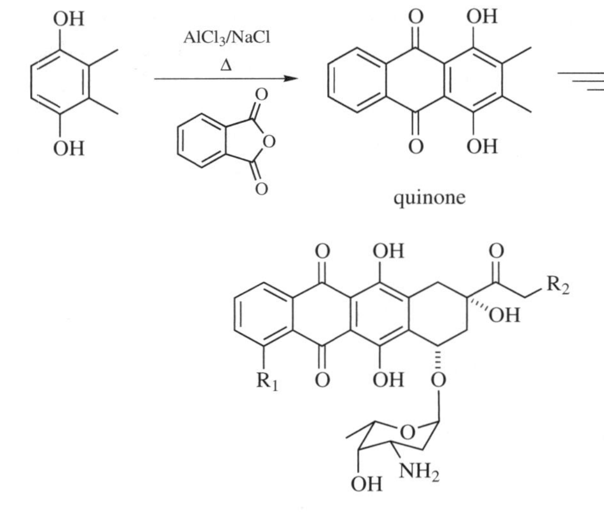

(5)

Daunorubicin/Adriamycin

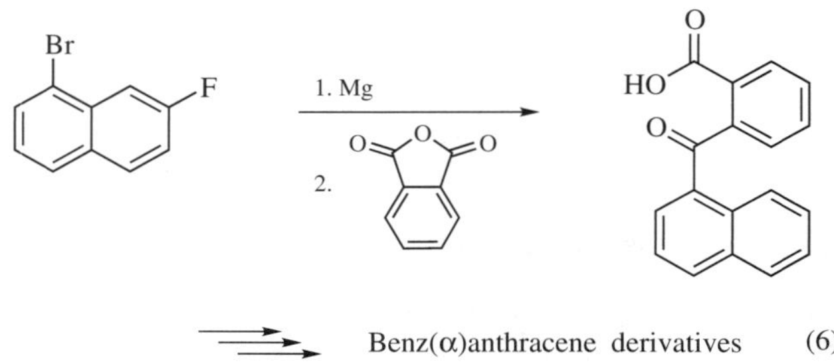

Benz(α)anthracene derivatives (6)

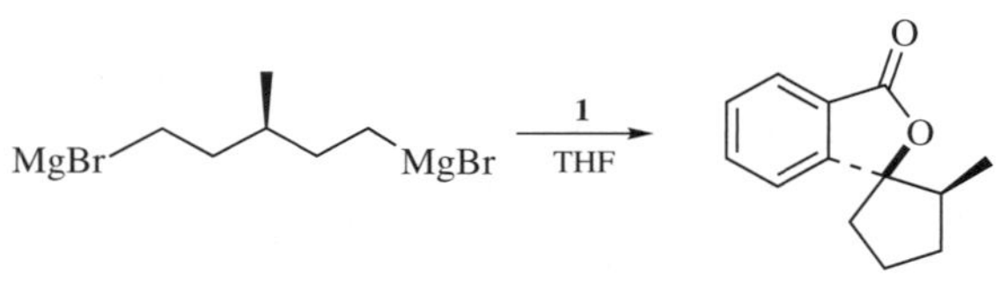

(7)

cis-spirolactone

aryllithium reagents has been examined as a route to these compounds. When equimolar quantities of phthalic anhydride and phenyl lithium were added at −78 °C, a bi-substituted lactone was formed in 78% yield (eq 8). When the addition was reversed, the lactone was obtained as a minor compound and the required product, *o*-benzoylbenzoic acid, was obtained in a yield of 35%. The yield of *o*-benzoylbenzoic acid could be further increased by employing 2 equiv of phthalic anhydride.[38] In a similar fashion, derivatives of benz[b]carbazoles[39] were conveniently prepared in modest yields by the reaction of 2,3-dilithio-*N*-methylindole, generated from 2,3-diiodo-*N*-methylindole, with phthalic anhydride (eq 9).

Olefination Reactions. Enol-lactones are precursors for the synthesis of several biologically important compounds. These precursors can be synthesized by the reaction of carbonyl-stabilized ylides with phthalic anhydride. The initial condensation leads to the formation of acyclic adducts that cyclize to produce an enol-lactone (eq 10).[40,41] These condensations show a surprisingly high stereoselectivity in the favor of a *trans*-double bond. It has been proposed that the electron-rich system interacts with the electron-deficient aromatic ring to give preferentially the *trans*-adduct. The fact that trans selectivity was shown to increase with the electron-donating capacity of the X group provided support for the proposed transition state.

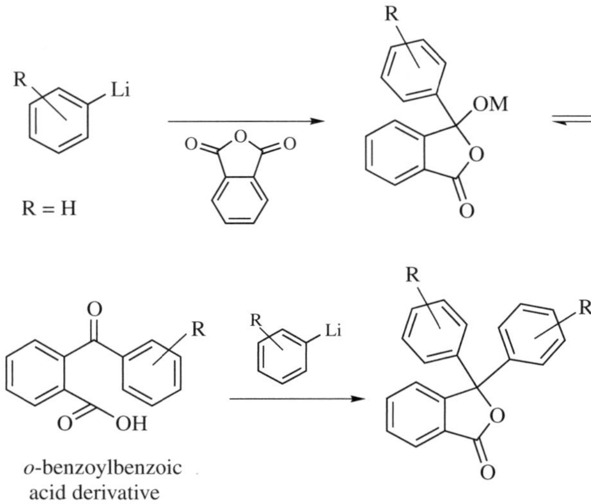

(8)

lactone

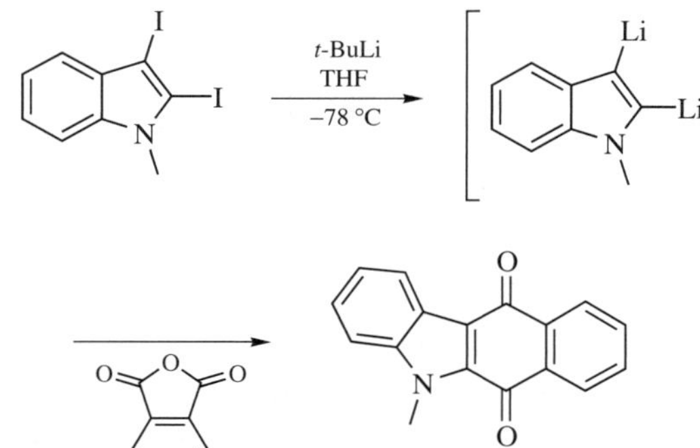

(9)

benzo[b]carbazoles

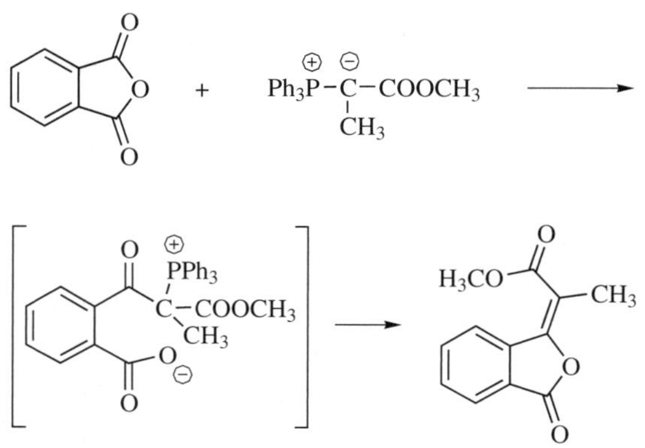

(10)

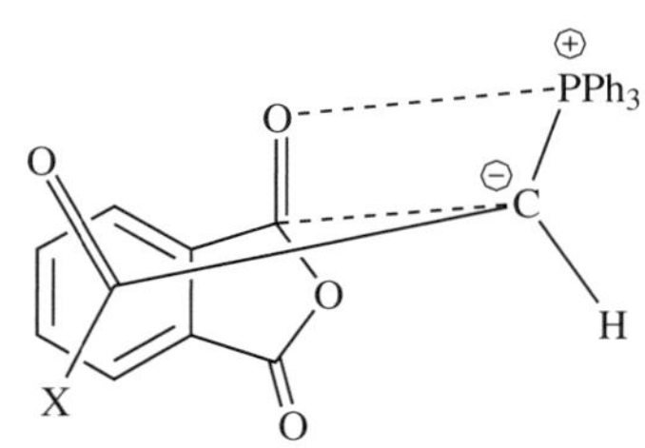

Substituted benzalphthalides are intermediates for the synthesis of compounds of medicinal importance, and these have been obtained by the condensation of phthalic anhydride and a substituted phenylacetic acid to give the corresponding benzalphthalide (eq 11).[42]

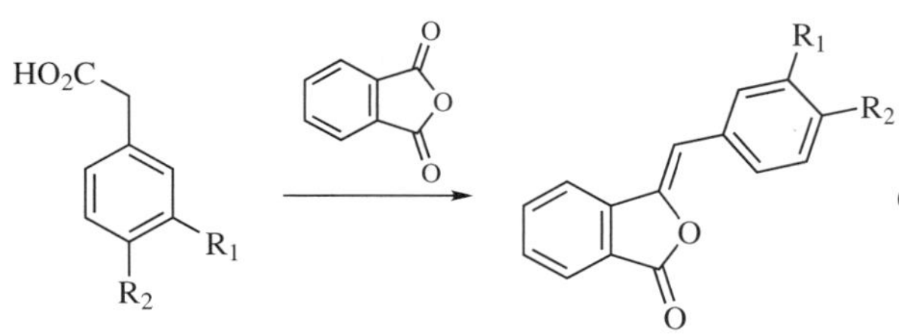

(11)

Oxidation Reactions. Phthalic anhydride in combination with hydrogen peroxide/urea (UHP) has been used as a mild and safe oxidation reagent for the conversion of sulfides into sulfoxides (eq 12). Other reagents that are commonly used for this transformation, such as hydrogen peroxide, *m*-CPBA, and sodium metaperiodate, may cause over-oxidation to sulfones. Furthermore, these methods require severe reaction conditions that are not always satisfactory for oxidation of sulfides with bulky groups or highly electron-withdrawing substituents. The combination of UHP and phthalic anhydride acts, however, as a safe and efficient oxidation reagent for the large-scale preparation of a wide range of sulfoxides.[43]

A combination of H_2O_2/urea and phthalic anhydride also acts as a useful reagent for mild and selective *N*-oxidation of *N*-heteroaromatics.[44] For example, when *m*-CPBA in $CHCl_3$ was used for the *N*-oxidation of 4,4′-bipyridine, the *N,N′*-dioxide was obtained in low yield. However, the *N,N′*-dioxide was obtained as a single compound in excellent yield (95%) when an excess of UHP and phthalic anhydride in CH_2Cl_2 was employed (eq 12).[44]

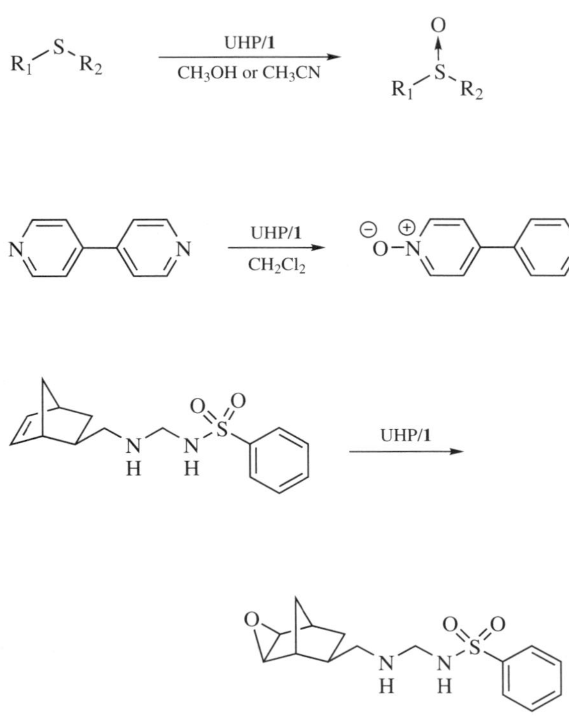

(12)

A mixture of phthalic anhydride and UHP has also been used for the synthesis of epoxides. This mixture of reagents generates in situ monoperphthalic acid (phthalic monoperoxy acid). This peroxyacid oxidizes alkenes to epoxides. Stereoselective epoxidation of the sulfonylureas of norbornene has been achieved by using this approach (eq 12).[45] Monoperphthalic acid has also been used for the epoxidation of β-carotene to epoxycartenoids, which are believed to reduce the risk of cancer.[46,47]

Dehydration Reactions. Many functionalities such as alcohols, oximes, and aldehydes are readily dehydrated upon treatment with phthalic anhydride.

Alcohols are easily dehydrated to alkenes by the action of phathalic anhydride and benzenesulfonic acid.[48] The dehydration could be accomplished without the acid, but prolonged reaction times were required.

Direct conversion of oximes into nitriles can also be performed with reagents such as trimethylsilyl iodide, oxalyl diimidazole, and carbon monoxide/water in the presence Rh carbonyl cluster using enzymatic methods. Phthalic anhydride can convert aldoximes to nitriles rapidly and efficiently under the microwave conditions (MW).[49] When benzaldoxime and phthalic anhydride dissolved in acetonitrile were heated in a microwave for few minutes, benzonitrile was obtained in almost quantitative yield. This method offers a mild, efficient, and inexpensive conversion of aldoximes to nitriles (eq 13).

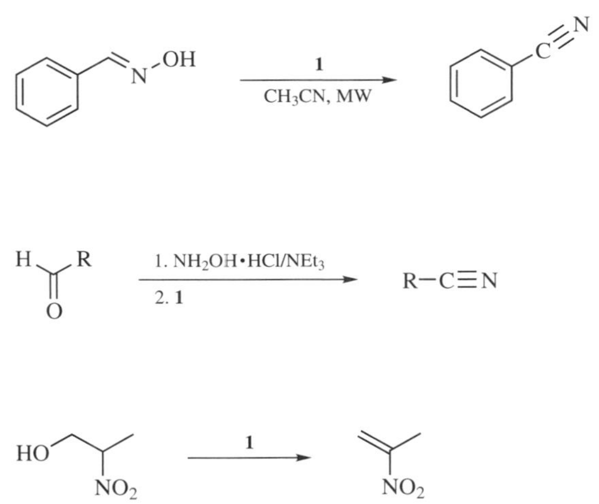

(13)

Aldehydes can be converted into nitriles either by conversion into aldoximes followed by dehydration as described above, or through a direct conversion by treatment with reagents such as trimethylsilyl azide, triazidochloro silane, hydroxylamine/polyphosporic acid, and hydroxylamine hydrochloride/selenium dioxide. Most of these methods either use harsh conditions and tedious work up, or are low yielding. A one-pot two-step conversion of aldehydes into nitriles using hydroxylamine and phthalic anhydride is simple, efficient, and high yielding. This conversion is achieved when hydroxylamine hydrochloride, NEt_3, and aldehyde are stirred in acetonitrile for 30 min followed by the addition of phthalic anhydride to dehydrate the intermediate aldoxime (eq 13).[50] This conversion can also be achieved by microwave heating under solvent-free conditions to give products in high yields.[51]

Phthalic anhydride is a useful reagent for the synthesis of various 2-nitro-olefins. 2-Nitro-propanol is readily dehydrated by phthalic anhydride to form 2-nitro-propene (eq 13).[52] The nitroolefins thus obtained are important alkylating reagents.

---

1. Kim, J.; Jang, D. O., *Synth. Commun.* **2001**, *31*, 395.
2. Shulman, G. P.; Bennett, J. H.; Botteron, D. G., *J. Org. Chem.* **1962**, *27*, 3923.
3. McMaster, L.; Ahmann, F. F., *J. Am. Chem. Soc.* **1928**, *50*, 145.
4. Boons, G. J., In *Organic Synthesis with Carbohydrates*; Boons, G. J., Ed.; Blackie Academic & Professional: London, 2000, pp. 155–174.
5. Sasaki, T.; Minamoto, K.; Itoh, H., *J. Org. Chem.* **1978**, *43*, 2320.
6. Lemieux, R. U.; Takeda, T.; Chung, B. Y., *ACS Symp. Ser.* **1976**, *39*, 90.
7. Akiya, S.; Osawa, T., *Chem. Pharm. Bull.* **1960**, *8*, 583.
8. Baker, B. R.; Joseph, J. P.; Schaub, R. E.; Williams, J. H., *J. Org. Chem.* **1954**, *19*, 1786.
9. Chandrasekhar, S.; Takhi, M.; Uma, G., *Tetrahedron Lett.* **1997**, *38*, 8089.
10. Hoogwater, D. A.; Reinhoudt, D. N.; Lie, T. S.; Gunneweg, J. J.; Beyerman, H. C., *Recl. Trav. Chim. Pays-Bas* **1973**, *92*, 819.
11. Kelly, R. C.; Schletter, I.; Stein, S. J.; Wierenga, W., *J. Am. Chem. Soc.* **1979**, *101*, 1054.
12. Moore, J. A.; Kim, J. H., *Tetrahedron Lett.* **1991**, *32*, 3449.
13. Nicolaou, K. C., *Angew. Chem., Int. Ed. Engl.* **1993**, *32*, 1377.
14. Nefkins, G. H. L.; Tesser, G. I.; Nivard, R. J. F., *Recl. Trav. Chim. Pays-Bas* **1960**, *79*, 688.
15. Wade, P. A.; Singh, S. M.; Pillay, M. K., *Tetrahedron* **1984**, *40*, 601.
16. Kocienski, P. J., *Protecting Groups*; Georg Thieme Verlag: Stuttgart, 1994.
17. Sakai, K.; Nakahara, Y.; Ogawa, T., *Tetrahedron Lett.* **1990**, *31*, 3035.
18. Kahn, M. N., *J. Org. Chem.* **1995**, *60*, 4536.
19. Dasgupta, F.; Garegg, P. J., *J. Carbohydr. Chem.* **1988**, *7*, 701.
20. Osby, J. O.; Martin, M. G.; Ganem, B., *Tetrahedron Lett.* **1984**, *25*, 2093.
21. Durette, P. L.; Meitzner, E. P.; Shen, T. Y., *Tetrahedron Lett.* **1979**, *20*, 4013.
22. Mootoo, D. R.; Fraser-Reid, B., *Tetrahedron Lett.* **1989**, *30*, 2363.
23. Kanie, O. C.; Crawley, S. C.; Palcic, M. M.; Hindsgaul, O., *Bioorg. Med. Chem.* **1993**, *2*, 1231.
24. Stangier, P.; Hindsgaul, O., *Synlett* **1996**, 179.
25. Debenham, J.; Rodebaugh, R.; Fraser-Reid, B., *Liebigs Ann./Recl* **1997**, 791.
26. Shimizu, H.; Ito, Y.; Matsuzaki, Y.; Iijima, H.; Ogawa, T., *Biosci., Biotechnol., Biochem.* **1996**, *60*, 73.
27. Boons, G.-J., *Contemp. Org. Synth.* **1996**, *3*, 173.
28. Boons, G. J., *Tetrahedron* **1996**, *52*, 1095.
29. Wasserman, H. H.; Robinson, R. P.; Carter, C. G., *J. Am. Chem. Soc.* **1983**, *105*, 1697.
30. Smith, A. L.; Hwang, C. K.; Pitsinos, E.; Scarlato, G. R.; Nicolaou, K. C., *J. Am. Chem. Soc.* **1992**, *114*, 3134.
31. Kamiya, T.; Hashimoto, M.; Nakaguchi, O.; Oku, T., *Tetrahedron* **1979**, *35*, 323.
32. Luke, G. P.; Holt. D. A., *Tetrahedron: Assymetry* **1999**, *10*, 4393.
33. Lu, Y. X.; Huang, Y. M.; Huang, Y. F.; Lu, Y., *Synth. Commun.* **2002**, *32*, 781.
34. Kerdesky, F. A. J.; Ardecky, R. J.; Lakshmikantham, M. V.; Cava, M. P., *J. Am. Chem. Soc.* **1981**, *103*, 1992.
35. Nilsson, J. L. G.; Sievertsson, H.; Selander, H., *Acta Pharm. Suec.* **1968**, *5*, 215.
36. Newman, M. S.; Tuncay, A., *J. Org. Chem.* **1980**, *45*, 348.
37. Canonne, P.; Boulanger, R.; Bernatchez, M., *Tetrahedron Lett.* **1987**, *28*, 4997.
38. Parham, W. E.; Piccirilli, R. M., *J. Org. Chem.* **1976**, *41*, 1268.
39. Liu, Y.; Gribble, G. W., *Tetrahedron Lett.* **2001**, *42*, 2949.
40. Kayser, M. M.; Hatt, K. L., *Can. J. Chem.* **1992**, *70*, 1985.
41. Manning, C.; McClory, M. R.; McCullough, J. J., *J. Org. Chem.* **1981**, *46*, 919.
42. del Olmo, E.; Armas, M. G.; Lopez-Perez, J. L.; Munoz, V.; Deharo, E.; Feliciano, A. S., *Bioorg. Med. Chem. Lett.* **2000**, *11*, 2123.
43. Balicki, R.; Kaczmarek, L.; Nantka-Namirski, P., *Liebigs Ann. Chem.* **1992**, 883.
44. Bergeron, R. J.; Wiegand, J.; Weimar, W. R.; Vinson, J. R. T.; Bussenius, J.; Yao, G. W.; McManis, J. S., *J. Med. Chem* **1999**, *42*, 95.
45. Kasyan, A. O.; Tarabara, I. N.; Golodaeva, E. A.; Kasyan, L. I., *Russ. J. Org. Chem.* **2001**, *37*, 1659.
46. Karrer, P.; Jucker, E., *Helv. Chim. Acta* **1945**, *28*, 427.
47. Barua, A. B., *Biochem. J.* **1999**, *339*, 359.
48. Waldmann, H.; Petru, F., *Chem. Ber.* **1950**, *83*, 287.
49. Hekmatshoar, R.; Beheshtiha, Y. S.; Kheirkhah, M.; Faridbod, F., *Monatshefte Fur Chemie* **2002**, *133*, 669.
50. Wang, E. C.; Lin, G. J., *Tetrahedron Lett.* **1998**, *39*, 4047.
51. Veverkova, E.; Toma, S., *Synth. Commun.* **2000**, *30*, 3109.
52. Miyashita, M.; Yanami, T.; Yoshikoshi, A., *Org. Synth. Coll.* **1990**, *7*, 414.

Balaji Santhanam & Geert-Jan Boons
University of Georgia, Athens, GA 30602

Polyhexamethylenecarbodiimide[1]

$-[(CH_2)_6N{=}C{=}N]_n-$

[31667-72-8] $(C_7H_{12}N_2)_n$ (MW n(124.21)

(insoluble carbodiimide condensation reagent)

Solubility: insol CH_2Cl_2.

Preparative Method: by catalytic decarboxylation of 1,6-diisocyanatohexane, followed by several purification steps.[1]

Peptide Bond Formation.[1] In what may be termed a 'reverse Merrifield' approach, an *N*-protected amino acid is reacted with a sizable molar excess of the polymeric carbodiimide, suspended in CH_2Cl_2, and an equimolar amount of a *C*-protected amino acid (side chain reactive groups of the amino acid derivatives also appropriately protected). After removal of the catalyst by filtration, and washing with aqueous acid and base, evaporation of solvent leaves the substantially pure coupled material in good yields (82–93%).

The coupling and workup procedures are similar to those used in the classical 1,3-Dicyclohexylcarbodiimide coupling. Both this and the water-soluble carbodiimide 1-Ethyl-3-(3′-dimethylaminopropyl)carbodiimide Hydrochloride, which requires no filtration step, are commercially available, and perform these same reactions with equal facility. They are, moreover, required in only stoichiometric proportions, and are accordingly very widely used, in distinct contrast to this polymeric reagent.

Other insoluble polymeric carbodiimides[2] have also been reported. Another,[3] which is soluble in typical reaction solvents

(CH_2Cl_2, DMF, DMSO), but subsequently precipitated by addition of ether, is also known. None of these appear to have found any great popularity either.

1. Wolman, Y.; Kivity, S.; Frankel, M. *Chem. Commun./J. Chem. Soc., Chem. Commun.* **1967**, 629.
2. (a) Akelah, A. *Synthesis* **1981**, 413. (b) Weinshenker, N. M.; Shen, C. M.; Wong, J. Y. *Org. Synth., Coll. Vol.* **1988**, *6*, 951. (c) Ito, H.; Takamatsu, N.; Ichikizaki, I. *Chem. Lett.* **1975**, 577.
3. Mutter, M. *Tetrahedron Lett.* **1978**, 2843.

Peter Ham
SmithKline Beecham Pharmaceuticals, Harlow, UK

Polystyrene-bound Dimethylaminopyridine (PS-DMAP)

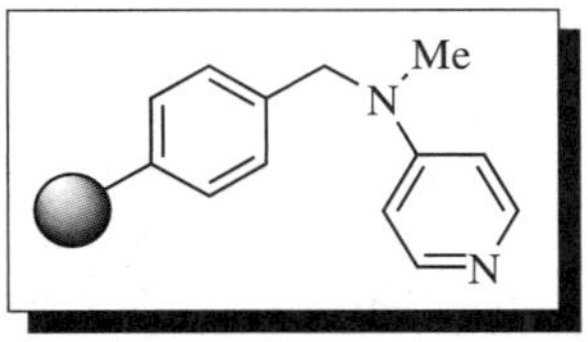

[82942-26-5]

(*N*-benzylmethylaminopyridine polymer bound; *N*-methyl-*N*-(4-pyridyl)aminomethyl-polystyrene cross-linked with divinylbenzene; reagent used as recyclable catalyst for acylation reactions)

Form Supplied in: 2–4% cross-linked poly(styrene-*co*-divinylbenzene) as resin type; commercially available.

Analysis of Reagent Purity: typical loading 2.5–3.0 mmol N g^{-1} (dry).

Preparative Methods: several syntheses for typical polymeric forms of (dimethylamino)pyridine have been published: (i) Condensation reaction of laurylated poly(ethyleneimine) with 3-[*N*-methyl-*N*-(4-pyridyl)amino]propionic acid.[1] (ii) Substitution reaction of polystyrene bearing (methylamino)methyl substituents with 4-chloropyridine.[2] (iii) Copolymerization of 4-(*N*-methyl-*N*-*para*-vinylbenzylamino)pyridine, styrene and commercially available divinylbenzene (DVB).[3] (iv) Chemical modification of halogenated bead polymers such as chloromethylstyrene copolymer with 4-(*N*-methylamino)pyridine.[4]

Introduction. Polystyrene-bound dimethylaminopyridine (PS-DMAP) is a polymer-bound equivalent of 4-(dimethylamino) pyridine (DMAP)[5] which may be used as a catalyst for acylation and related reactions. PS-DMAP can be recovered easily from the reaction mixture and is recyclable. Therefore, PS-DMAP can be used in flow systems such as packed or fluidized bed reactors.

Acylations. The application of PS-DMAP as a catalyst for the acylation of hydroxy groups with acid anhydride has been investigated, in particular for acetic anhydride. Catalytic PS-DMAP accelerates the acylation of sluggish nucleophiles, e.g. tertiary alcohols. The catalytic activity of PS-DMAP for the acylation of linalool (eq 1)[3,6] and 1-methylcyclohexanol[4,7,8] with acetic anhydride has been studied.

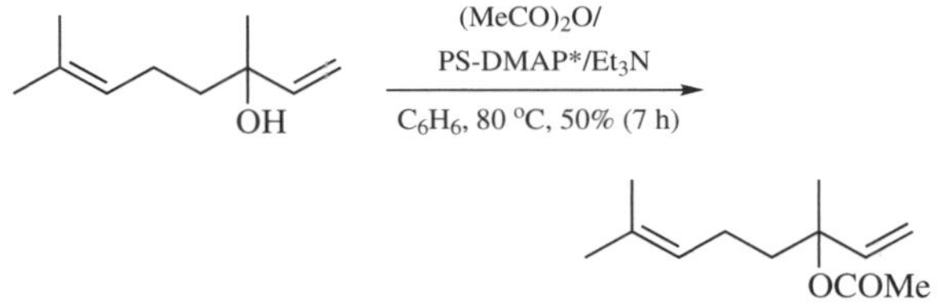

(1)

*PS-DMAP (0.2 molar equiv based on linalool)

Although the acetylation is hardly catalyzed by triethylamine alone, it proceeds easily in the presence of PS-DMAP.[3] The activity of PS-DMAP, cross-linked with 2 mol % divinylbenzene (DVB), is slightly lower than that of DMAP. In this reaction, the catalyst is recovered by filtration, washed with aqueous sodium hydroxide and acetone, and the recovered PS-DMAP can be reused without a significant reduction in catalytic activity.

The reaction mechanism suggests that the catalytic activity of DMAP moiety is due to both the increased electron-donating ability of the ring nitrogen and the increased stability of the acetylpyridinium species.[1a,3,6]

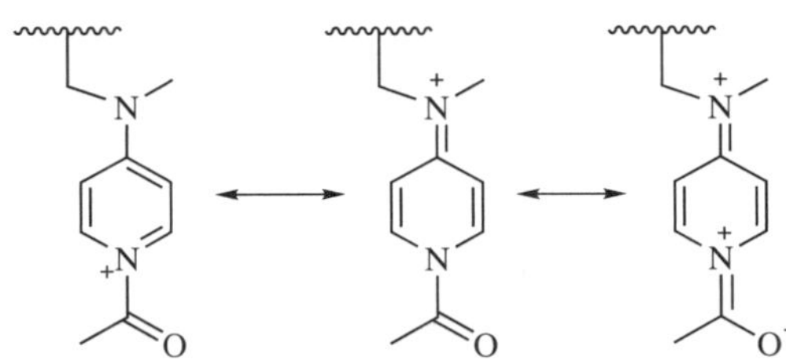

PS-DMAP can also be used for *O*-acetylation of enols (eq 2), *O*-formylation (eq 3), *O*-silylation (eq 4), *N*-acetylation (eq 5), and *O*-tritylation (eq 6) in the same manner as DMAP under homogeneous conditions.[4a] Generally, the yields in the two-phase system are somewhat lower than those in the homogeneous environment.

$$Me(CH_2)_5CHO + (MeCO)_2O \xrightarrow[\text{DMAP:60\%}]{\text{PS-DMAP: 59\%}} Me(CH_2)_4CH{=}CHOCOMe \quad (2)$$

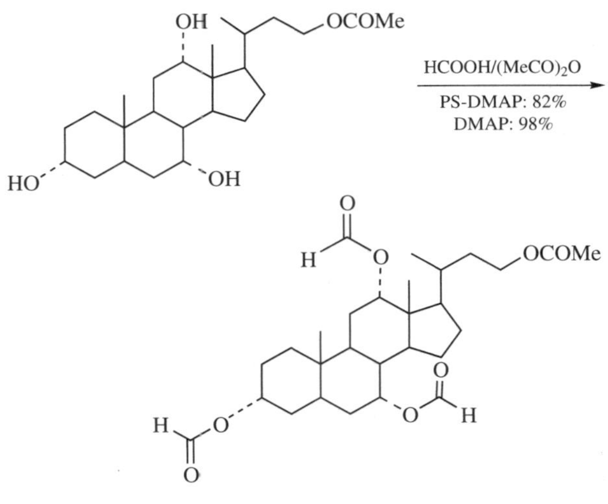

(3)

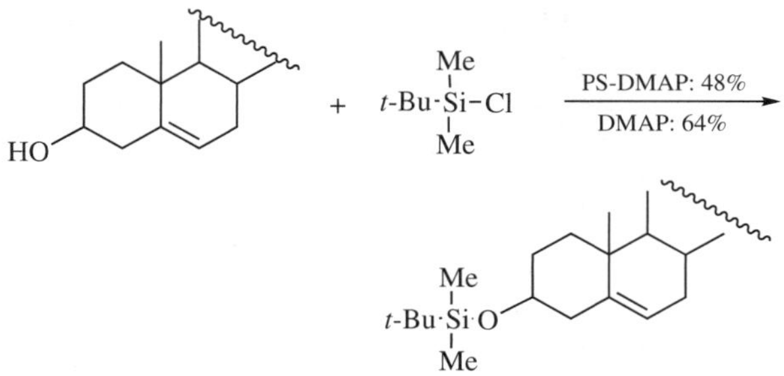

(4)

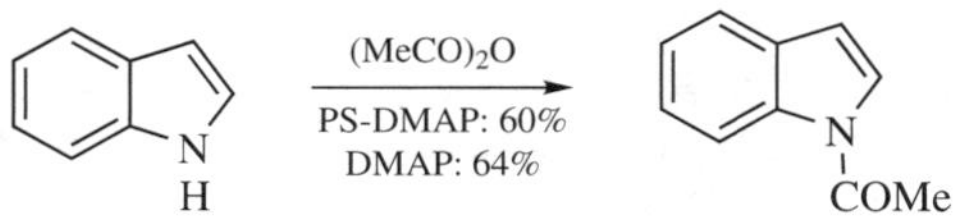

(5)

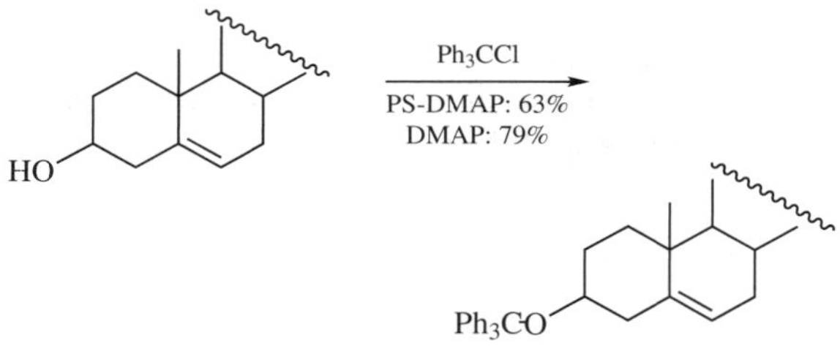

(6)

Esterifications. PS-DMAP beads act as a catalyst for ester syntheses by the dicyclohexylcarbodiimide (DCC) method (eq 7)[2], in the transesterification of *para*-nitrophenyl active ester (eq 8)[4a] and in the esterification with dialkyl dicarbonate (eqs 9 and 10).[9]

(7)

PS-DMAP: 1 h, 61%; 3 h, 59%; 1 day, 65%
DMAP: 0.5 h, 100%

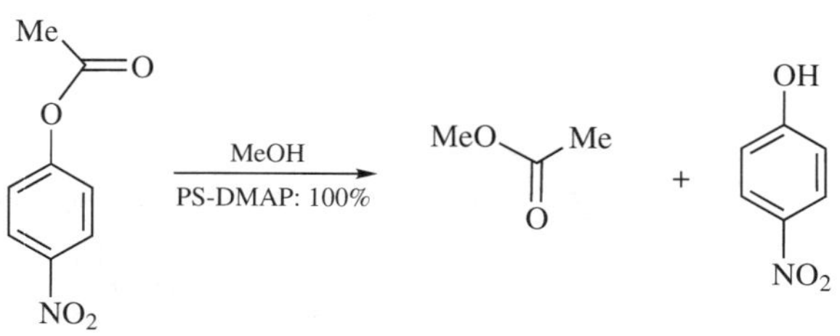

(8)

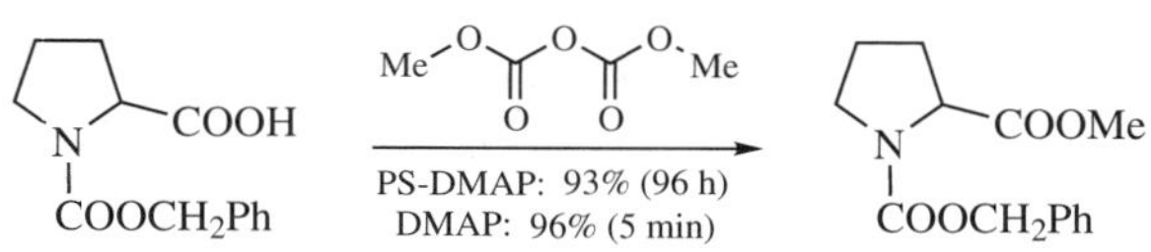

(9)

In the esterification of benzoic acid with MeOH by the DCC method, the process catalyzed by the polymer-bound reagent is relatively slow and the yield is lower than with DMAP under homogeneous conditions (eq 7). Transesterification of *para*-nitrophenyl active ester with methanol proceeds in quantitative yield.

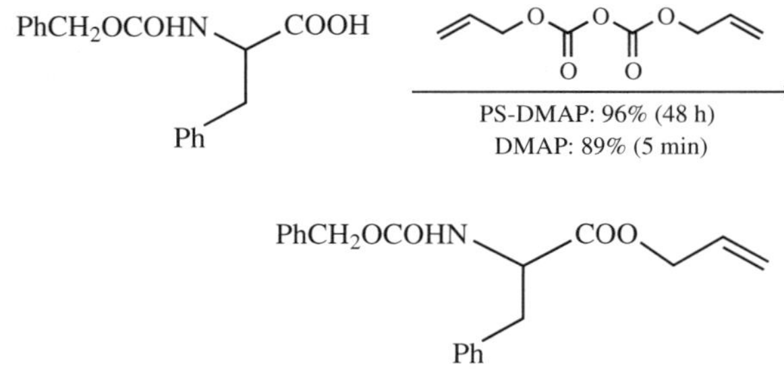

(10)

In the esterification of *N*-protected amino acids using dimethyl dicarbonate (eq 9) and diallyl dicarbonate (eq 10) as reagents, the use of PS-DMAP leads to product formation in satisfactory yield and optical purity, although a longer reaction time is required than in the case of DMAP catalyst.[9]

Application of PS-DMAP as a Storable Acylium Repository for Amide, Ester and Thioester Formations.[10,11] PS-DMAP may also be used as a stoichiometric polymer-bound acylium repository for *N*-substituted acyl pyridinium salts derived from acid chlorides, and a variety of acyl derivatives, including esters, amides, and thioesters can be synthesized in this way. The electrophilic component is reacted with PS-DMAP and an acyl pyridinium salt forms which is then reacted with various nucleophiles such as alcohols, amines, and thiols without addition of a tertiary amines base (eq 11, Table 1). This condensation reaction can also be used for sulfonamide and phosphonamide syntheses.

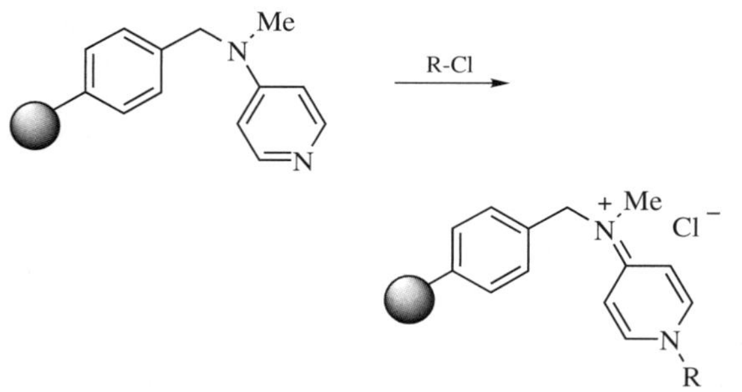

(11)

Nu-H: R'-NH$_2$ ⟶ R-CONH-R'
R'-OH ⟶ R-COO-R'
R'-SH ⟶ R-COS-R'

Table 1 Synthesis of acyl derivatives using PS-DMAP as an acylium repository

R	Nu-H	Product	Yield (%)
PhCO	HOC$_6$H$_4$-*p*-NO$_2$	PhCOOC$_6$H$_4$-*p*-NO$_2$	100
PhCO	HSC$_6$H$_4$-*p*-NO$_2$	PhCOSC$_6$H$_4$-*p*-NO$_2$	100
PhCO	Menthol	Menthol benzoate	30
PhCO	PhCOOH	Benzoic anhydride	92
PhCO	MeCOOH	MeCOOCOPh	54
p-MeC$_6$H$_4$SO$_2$	NH$_2$CH$_2$Ph	*p*-Tosyl-NHCH$_2$Ph	100

The key to this approach is the ability to purify the polymer-bound acylium salt by solvent washes. When the nucleophile is used as the limiting reagent, the product is isolated in high purity by filtration, with the excess electrophile remaining bound to the resin.

PS-DMAP and isobutyl chloroformate act as a polymeric activating agent for carboxylic acids to form the mixed carbonic-carboxylic anhydride, which may then acylate nucleophiles after polymer removal (eq 12).[10,11]

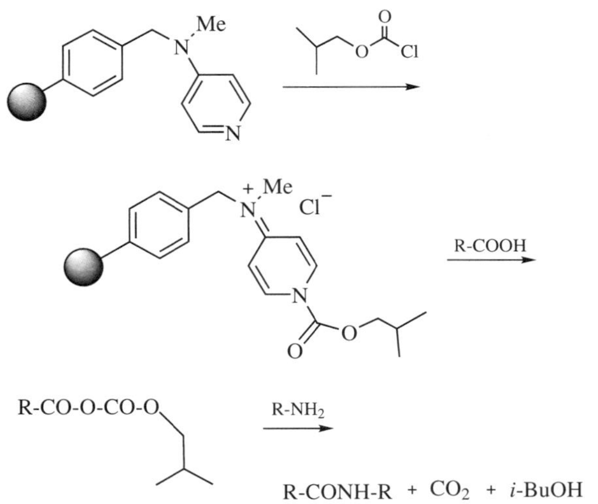

(12)

Related Reagents. 4-(Dimethylamino)pyridine (DMAP).

1. (a) Hierl, M. A.; Gamson, E. P.; Klotz, I. M., *J. Am. Chem. Soc.* **1979**, *101*, 6020. (b) Delaney, E. J.; Wood, L. E.; Klotz, I. M., *J. Am. Chem. Soc.* **1982**, *104*, 799.
2. Shinkai, S.; Tsuji, H.; Hara, Y.; Manabe, O., *Bull. Chem. Soc. Jpn.* **1981**, *54*, 631.
3. Tomoi, M.; Akada, Y.; Kakiuchi, H., *Macromol. Chem., Rapid Commun.* **1982**, *3*, 537.
4. (a) Menger, F. M.; McCann, D. J., *J. Org. Chem.* **1985**, *50*, 3928. (b) Deratani, A.; Darling, G. D.; Horak, D.; Fréchet, J. M. J., *Macromolecules* **1987**, *20*, 767.
5. Höfle, G.; Steglich, W.; Vorbrüggen, H., *Angew. Chem., Int. Ed. Engl.* **1978**, *17*, 569.
6. Tomoi, M.; Goto, M.; Kakiuchi, H., *J. Polym. Sci. Part A, Polym. Chem.* **1987**, *25*, 77.
7. Keay, J. G.; Scriven, E. F. V., *Chem. Ind.* **1994**, *53*, 339.
8. Guendouz, F.; Jacquier, R.; Verducci, J., *Tetrahedron* **1988**, *44*, 7095.
9. Takeda, K.; Akiyama, A.; Nakamura, H.; Takizawa, S.; Mizuno, Y.; Takayanagi, H.; Harigaya, Y., *Synthesis* **1994**, 1063.
10. Shai, Y.; Jacobson, K. A.; Patchornik, A., *J. Am. Chem. Soc.* **1985**, *107*, 4249.
11. Patchornik, A., *Chemtech.* **1987**, *17*, 58.

Kazuyoshi Takeda
Ebara Research Co. Ltd., Fujisawa-shi, Japan

Polystyrene-bound 1-Hydroxybenzotriazole (PS-HOBT)

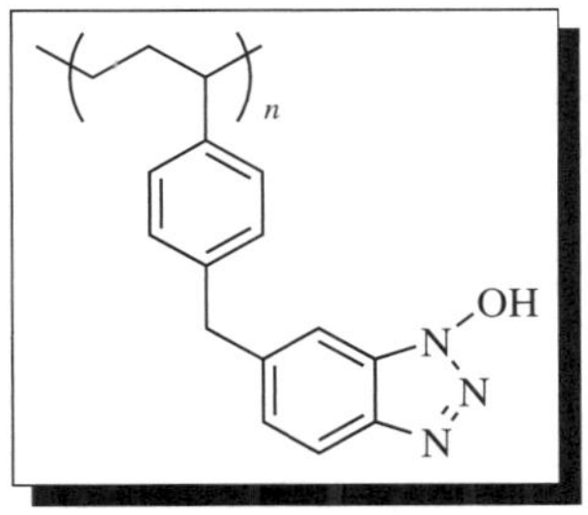

[2592-95-2] $(C_{15}H_{13}N_3O)_n$ (MW 251.28)

(polystyrene-divinylbenzene (ca. 1 mol %) copolymer resin; reagent used for peptide coupling)

Alternate Name: PS-HOBt, PS-HOBT.
Solubility: insoluble in all solvents; polymer swells in DMF, CH_2Cl_2, and CH_3CN.
Form Supplied in: brown powder; available as 1-hydroxybenzotriazole-6-carboxamidomethyl and 6-sulfonamidomethyl hydrochloric acid.
Analysis of Reagent Activity: the number of active sites on the resin is determined by formation of the polymer-bound acetate ester by treatment of 2 g of polymer with acetic anhydride or acetyl chloride (4 mmol) in CH_2Cl_2 (10 mL) in the presence of pyridine (0.5 mL). After 30 min stirring at room temperature, the polymer was filtered off and washed with DMF (2×20 mL), CH_2Cl_2 (3×20 mL), and dry ether (3×20 mL). Subsequent treatment with isopropylamine (large excess) yielded *N*-isopropylacetamide and recovered polymer.[3] Optimal capacity between 0.2 and 1.5 mmol g^{-1}.[4]
Preparative Methods: prepared from copolystyrene/1 mol % divinylbenzene (100–200 or 200–400 mesh) and 4-chloro-3-nitrobenzyl alcohol using aluminium trichloride at 90 °C, followed by treatment with hydrazine at 100 °C, and final cyclization with hydrochloric acid and neutralization.[1]
Handling, Storage, and Precautions: very stable, not hygroscopic and can be stored for more than 2 years under ambient conditions.[2]

Applications as Peptide Coupling Reagent. Polystyrene-bound 1-hydroxybenzotriazole was initially employed for the preparation of immobilized active esters by reaction with carboxylic acids using dicyclohexylcarbodiimide (DCC) as coupling reagent.[1,5–12] These polymeric active esters can be used even in continuous peptide synthesis.[6] This method is considered to be a hybrid of solution- and solid-phase peptide synthesis, and is especially useful when there are several repeating units of the same amino acid. In general, couplings take place in DMF, with triethylamine or diisopropylethylamine as bases, in high yields within 10–30 min. PS-HOBt can be filtered and reused without appreciable loss of activity. Active PS-HOBt esters have mechanical stability, good swelling properties, and high reactivity.[1] The polymeric esters have been used in the synthesis of several short peptides[1,6,8,11] and two blocked segments Boc-Asp(O-*t*-Bu)-Leu-Lys(2Cbz)-Glu

(OBzl)-Lys(2Cbz)-Lys(2Cbz)-OH and Boc-Glu(OBzl)-Val-Val-Glu(OBzl)-Glu(OBzl)-Ala-Glu(OBzl)-AsnOBzl for the convergent synthesis of the acidic *C*-terminal segment of thymosin α_1 (15–28)[7] and thymosin-like peptides.[9]

The racemization-free coupling of PS-HOBt esters can be carried out directly in the absence of an amine with benzyltrimethylammonium salts derived from amino acids or peptides using chloroform as solvent.[10]

A soluble (CH_2Cl_2, DMF, H_2O) polymeric HOBt (**1**) bearing a polyethylene glycol chain (PEG, MW 4000) has been prepared with PEG-NH_2 and 1-hydroxybenzotriazole-5-carboxylic acid using DCC as a coupling reagent.[13] The corresponding esters were prepared by reaction with an excess of carboxylic acid anhydrides and used for the synthesis of the model tetrapeptide Z-Leu-Ala-Gly-Val-O-*t*-Bu.[14]

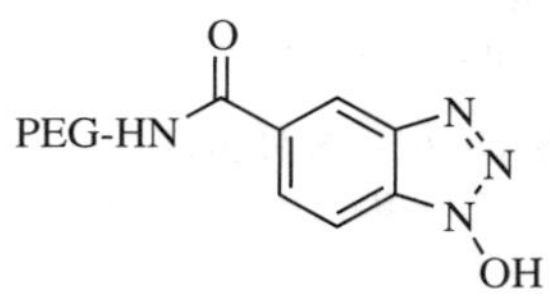

1

Amidation Reactions. In the preparation of a library of amides derived from primary and secondary amines, PS-HOBt esters immobilized on macroporous polystyrene beads (100–200 mesh) gave high yields in a variety of solvents even in the presence of water (eq 1).[3]

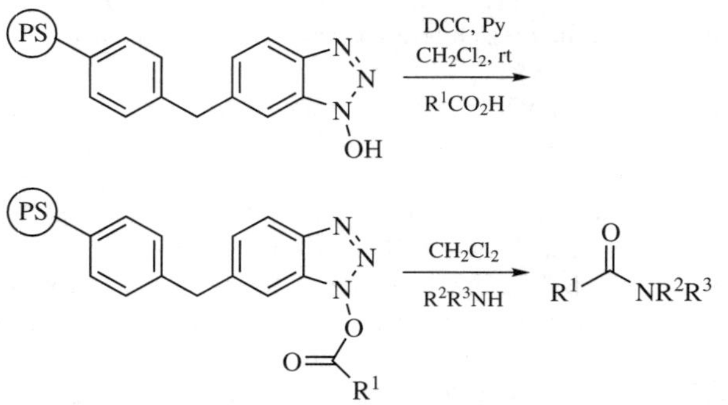

(1)

The aminomethylated polystyrene 6-sulfonamide-1-hydroxybenzotriazole (**2**) has been used for the parallel synthesis of amide libraries.[15] The higher acidity conferred by the sulfonyl moiety increases the efficiency of this polymeric HOBt, which has been recently commercialized as its hydrochloride salt with a substitution level of 0.8–1 mmol g^{-1}.

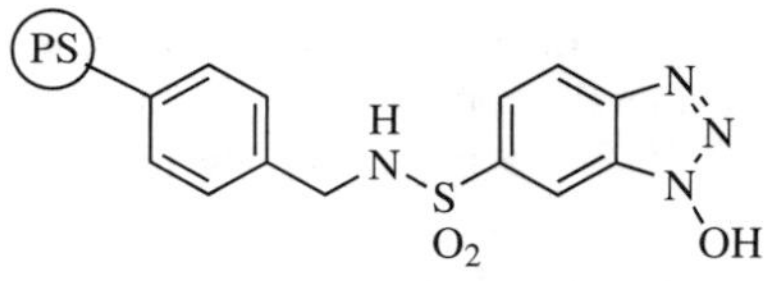

2

A recent convergent synthesis of sildenafil, a potent and selective inhibitor of the enzyme phosphodiesterase, is based on polymer-supported reagents.[16] The coupling of aminopyrazole with a benzoic acid derivative was performed in THF using a polystyrene-supported ester of a commercially available 6-carboxamidomethyl-1-hydroxybenzotriazole prepared from the acid and PyBrOP (eq 2). A versatile fluorescence labeling method is based on amidation reactions with the ester derived from HOBt-6-carboxamidomethyl polystyrene and 4-acetamido-1,8-naphthalimide-*N*-caproic acid.[17] In the synthesis of a library of α,α-disubstituted-α-acylaminoketones, esters of HOBt-6-carboxamidomethyl polystyrene have been coupled with different α-aminoketones.[18]

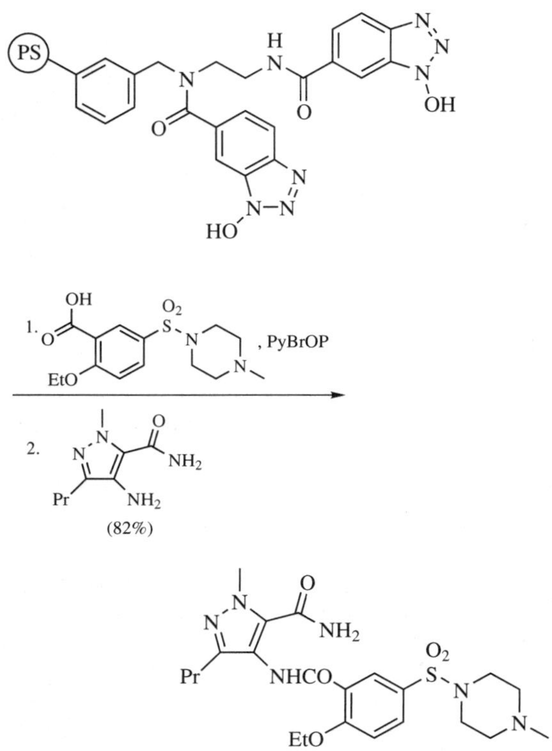

(2)

Medium ring lactams (7-, 9-, 11-, and 13-membered rings) can be prepared in moderate yields by intramolecular cyclization of PS-HOBt esters derived from the corresponding substituted amino acids (eq 3).[19]

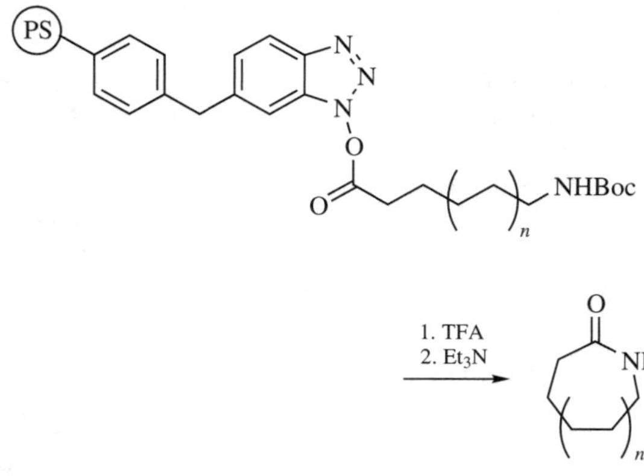

(3)

Boc-L-Alanine and di-Boc-L-histidine PS-HOBt esters have been coupled with *N*-cetyl *N′*,*N′*-dimethylethylenediamine for the preparation of amino acid functionalized surfactants.[20]

Other Applications. *N*-Hydroxysuccinimide esters, commonly used in solid-phase peptide synthesis, have been prepared in a highly pure form from the corresponding polystyrene-bound HOBt esters of carboxylic acids or α-amino acids (eq 4).[2] The utility of this method, in contrast to polymer-bound EDC, is that the PS-HOBt can be recycled and the HOSu esters do not need additional purification.

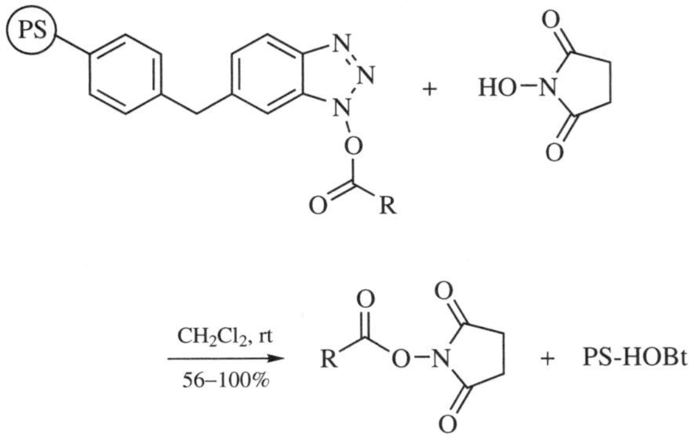

(4)

Protecting group transfer reagents derived from PS-HOBt and fluorenylmethoxycarbonyl (Fmoc), benzyloxycarbonyl (Cbz), and *t*-butoxycarbonyl (Boc) have been synthetized by the reaction of the corresponding chloroformates in the presence of pyridine, and from Boc_2O.[21] These immobilized carbonates (**3**) (activity: 0.25 mmol g^{-1}) were used for the protection of primary and secondary aliphatic amines in good yields and pure form, and only the Boc derivatives were obtained in poor yields. The reaction can be performed in CH_2Cl_2 or in DMF–H_2O (3:1) according to amine solubility.

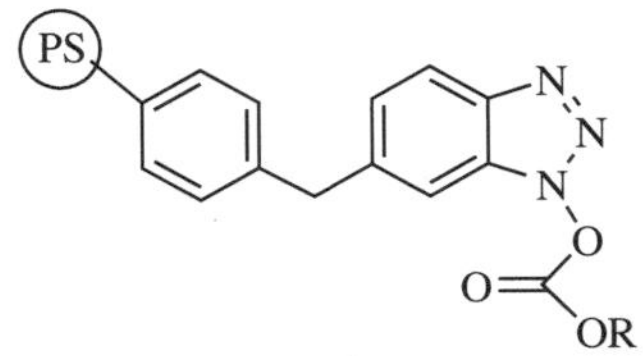

3

R = *t*-Bu, $PhCH_2$,
9-fluorenylmethyl

PS-HOBt and the 3,5-dinitrophenol derived ester have been used for the derivatization of aliphatic amines, aminoalcohols, and amino acids and for the HPLC analysis of amphetamine.[22]

The polymer-bound TBTU (PS-TBTU, **4**) has been prepared[23] from polystyrene-bound 1-hydroxybenzotriazole (PS-HOBt) and has been used for di and tripeptide synthesis, using pyridine as base, even in wet acetonitrile. Yields are similar or slightly lower than those obtained using TBTU, but racemization ratios are lower than those obtained using HATU.

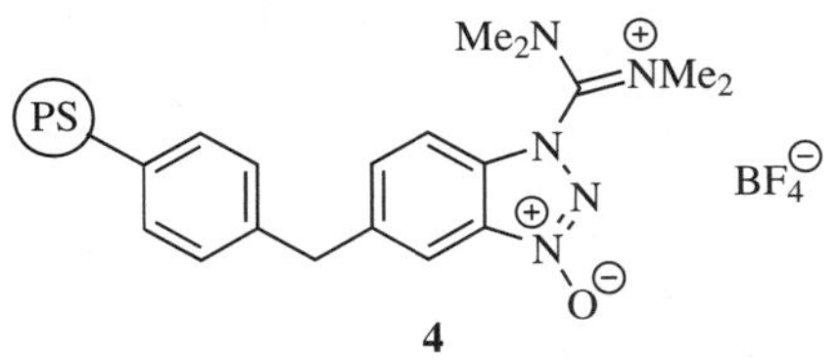

4

Reductive cleavage of the hydroxy group of PS-HOBt with either PCl_3 or SmI_2 afforded the corresponding polymer-supported benzotriazole used as traceless linker for solid-phase synthesis of amines. This PS-Bt reacts with aldehydes and amines to give Mannich-type adducts which, after cleavage with Grignard reagents, provide homologated amines.[24]

1. Kalir, R.; Warshawsky, A.; Fridkin, M.; Patchornik, A., *Eur. J. Biochem.* **1975**, *59*, 55.
2. Dendrinos, K. G.; Kalivretenos, A. G., *Tetrahedron Lett.* **1998**, *39*, 1321.
3. Dendrinos, K. G.; Jeong, J.; Huang, W.; Kalivretenos, A. G., *Chem. Commun.* **1998**, 499.
4. Berrada, A.; Cavelier, F.; Jacquier, R.; Verducci, J., *Bull. Soc. Chim. Fr.* **1989**, 511.
5. Heusel, G.; Bovermann, G.; Göhring, W.; Jung, G., *Angew. Chem, Int. Ed. Engl.* **1977**, *16*, 642.
6. Stern, M.; Kalir, R.; Patchornik, A.; Warshawsky, A.; Fridkin, M., *Solid-Phase Biochem.* **1977**, *2*, 131.
7. Mokotoff, M.; Patchornik, A., *Int. J. Peptide Res.* **1983**, *21*, 145.
8. Sheh, L.; Mokotoff, M.; Abraham, D. J., *Int. J. Peptide Protein Res.* **1987**, *29*, 509.
9. Mokotoff, M.; Zhao, M.; Roth, S. M.; Shelley, J. A.; Slavoski, J. N.; Kouttab, N. M., *J. Med. Chem.* **1990**, *33*, 354.
10. Chen, S. T.; Chang, C. H.; Wang, K. T., *J. Chem. Res. (S)* **1991**, 206.
11. Grigor'ev, E. I.; Zhil'tsov, O. S., *Zh. Org. Khim.* **1989**, *25*, 1963 (*Chem. Abstr.* **1990**, *112*, 199064).
12. (a) Albericio, F.; Carpino, L. A., *Methods Enzymol.* **1997**, *289*, 104. (b) Humphrey, J. M.; Chamberlin, A. R., *Chem. Rev.* **1997**, *97*, 2243. (c) Albericio, F.; Kates, S. A., In *Solid-Phase Synthesis, A Practical Guide*; Kates, S. A.; Albericio, F., Eds; Marcel Dekker: New York, 2000, p 275. (d) Albericio, F.; Chinchilla, R.; Dodsworth, D. J.; Nájera, C., *Org. Prep. & Proced. Int.* **2001**, *33*, 203.
13. Mutter, M., *Tetrahedron Lett.* **1978**, 2839.
14. Mutter, M., *Tetrahedron Lett.* **1978**, 2843.
15. Pop, I. E.; Déprez, B. P.; Tartar, A. L., *J. Org. Chem.* **1997**, *62*, 2594.
16. Baxendale, I. R.; Ley, S. V., *Biorg. Med. Chem. Lett.* **2000**, *10*, 1983.
17. Chang, Y. T.; Schultz, P. G., *Biorg. Med. Chem. Lett.* **1999**, *9*, 2479.
18. García, J.; Nicolàs, E.; Albericio, F.; Michelotti, E. L.; Tice, C. M., *Tetrahedron Lett.* **2002**, *43*, 7495.
19. Huang, W.; Kalivretenos, A. G., *Tetrahedron Lett.* **1995**, *36*, 9113.
20. Moss, R. A.; Lukas, T. J.; Nahas, R. C., *Tetrahedron Lett.* **1977**, 3851.
21. Dendrinos, K. G.; Kalivretenos, A. G., *J. Chem. Soc., Perkin Trans. 1* **1998**, 1463.
22. Bourgue, A. J.; Krull, I. S., *J. Chromatogr.* **1991**, *537*, 123.
23. (a) Chinchilla, R.; Dodsworth, D. J.; Nájera, C.; Soriano, J. M., *Tetrahedron Lett.* **2000**, *41*, 2463. (b) Nájera, C., In *Electronic Encyclopedia of Reagents for Organic Synthesis*; Paquette, L. A., Ed.; Wiley: Chichester.
24. Schiemann, K.; Showalter, H. D. H., *J. Org. Chem.* **1999**, *64*, 4972.

Carmen Nájera
Universidad de Alicante, Spain

Polymer-bound *N*-Hydroxysuccinimide

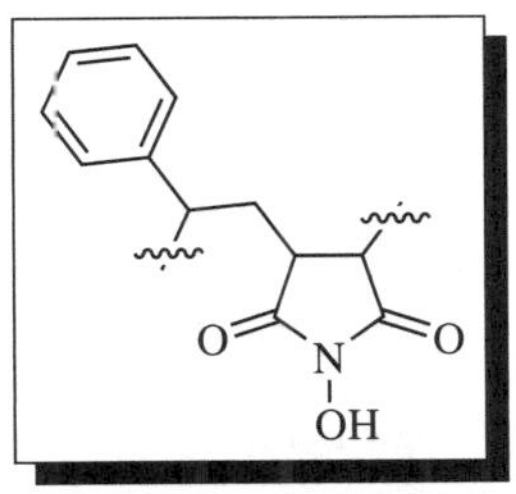

Physical Data: white solid; OH pKa = 6.0. A commercial preparation of styrene-maleic anhydride copolymer was reported to have an average molecular weight of 56000 Da.[1]

Solubility: such a polymer is soluble in DMF but precipitates in methanol. Polymers of larger average molecular weights are insoluble in all solvents.

Analysis of Reagent Purity: IR (KBr). OH stretch: 3500–2700 cm^{-1}(strong); C=O stretch: 1780 cm^{-1}(medium), 1705 cm^{-1} (strong), 1655 cm^{-1}(shoulder); others: 1230 cm^{-1}(shoulder), 1060 cm^{-1}(medium). Elemental analysis indicated that anhydride content could be as high as 50 mol %. The information has been reproduced by permission of Wiley.

Loading: upon acetylation of this polymer with acetic anhydride in pyridine, a loading of 3.28 mmol g^{-1} was obtained.

Handling, Storage, and Precautions: the reagent is non-toxic and non-hydroscopic; however, it is best to store the reagent under desiccated, inert atmosphere. The reagent is stable indefinitely at room temperature.

Various Uses. Polymer bound *N*-hydroxysuccinimide (**1**) has been reported as the reagent for resin capture-and-release amide bond synthesis methodology.[2] This method of solid-phase synthesis is distinct from the method in which solid-phase linkers are used to stablize a wide range of chemical conditions. Capture-and-release methodology is a synthetic strategy of convenience for the loading and activation of a reagent (usually a carboxylic acid) followed by reaction with a nucleophile. This results in the release of the coupled product (Figure 1). The capture and activation of the acyl moiety is usually stable enough for storage of the dry resin at room temperature, but in most cases it is not considered sufficiently stable for multiple synthetic steps on solid phase. For multi-step solid-phase synthesis, linkers cleavable only under specific conditions (e.g., Rink linker treated with strong acid) are advised.

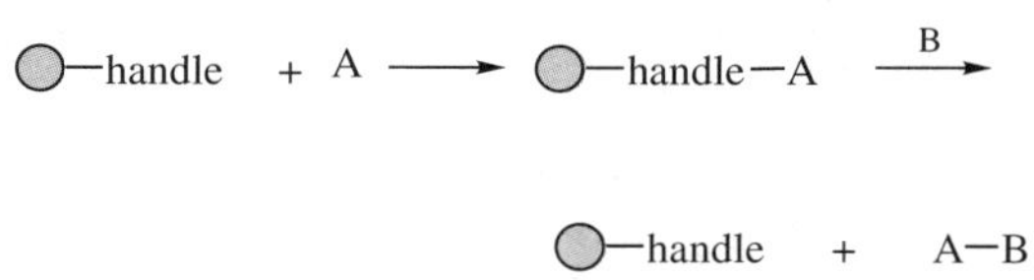

Figure 1 The capture-and-release concept applied to small molecule synthesis using solid-phase polymers containing functional handles

Of the capture-and-release resins that have been reported, *N*-hydroxysuccinimide is appealing for its comparable ease of preparation and reactivity to the analogous solution-phase protocols. Coupling of amino acid residues with *N*-hydroxysuccinimide activation is one of the several carboxylic acid activation methods that minimizes racemization.

The loading of activated *N*-hydroxysuccinimide esters to this resin is effected by treatment with excess carboxylic acid under acylating conditions (e.g., excess acetic anhydride) giving resin **3**. The resin is filtered, washed, and dried. An aliquot of the resin is then used for titration with the reactive amine **4** (e.g., *p*-methoxybenzyl amine) as a means of quantification of resin loading. When the resin is swollen in an inert solvent such as DMF or DMC and treated with some sub-equivalent of reactive nucleophile, the captured acyl group is transferred to the nucleophile giving product **5** (eq 1). Treatment of the nucleophile with excess acylating resin drives the reaction to completion, but does not create excess reaction impurities. Purification and isolation of the desired acylated product is accomplished by filtration and evaporation of the solvent. The ease of preparation, storage, handling, and reaction of solid-phase reagents is particularly appealing for high-throughput synthesis. This reagent has also found some use in the coupling of small peptide fragments.

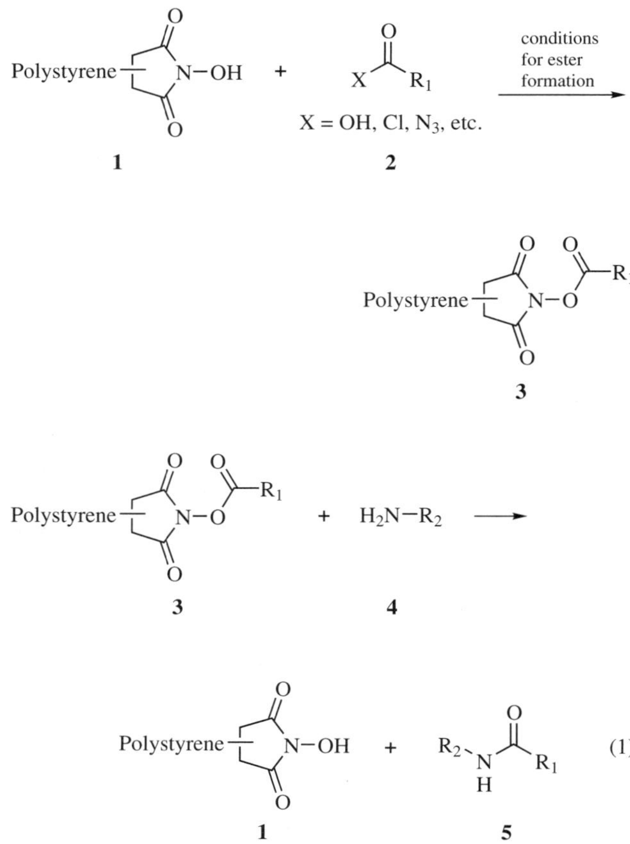

(1)

Primary amine nucleophiles generally give high yields. With decreased nucleophilicity, acylation yields are lower. Enhanced reaction rates and yields have been achieved by increasing the number of equivalents of acylating resin, the reaction temperature, and the reaction time, as well as by adding a base. The reactivity of *N*-hydroxysuccinimide polymers derivatized with acetyl and benzoyl groups were measured against a variety of amine nucleophiles (Table 1). Yields of approximately 80% for primary alkyl amines were obtained. Amines of lesser reactivity (e.g., aniline) resulted in lower product yields.

Preparative Methods. The co-polymer of *N*-benzyloxymaleimide with styrene and divinylbenzene has been reported to contain as much as 47 mol % of imide functionality (eq 2).[3] *N*-acetoxymaleimide co-polymer preparation is similarly reported by the same authors. The benzoyl and acetyl groups were removed by treatment of the polymer with 20% HBr in acetic acid, resulting in the formation of the *N*-hydroxysuccinimide resin. After treatment of the *N*-hydroxysuccinimide resin with acetic anhydride, resins containing 3.2 mmol g^{-1} acetylating functionality were obtained.

Table 1 Solid-phase acylating efficacy: crude product yields by treatment of *N*-acyloxyimide polymers (**3**, R=CH_3 and R=Ph)[a]

	Crude yield (%	
Amine	Acetyl polymer IIIa	Benzoyl polymer IIIb
Benzylamine	81	55
Cyclohexylamine	85	91
Aniline	25	10
p-Toluidine	66	15
Ethyl glycinate	84	85
Ethyl DL-leucinate	90	59

[a]Reaction conditions: 0.4 molar solutions of amine were treated with 1.55 equiv of acetyl polymer **3** (R=CH_3) and 1.31 equiv of benzoyl polymer **3** (R=Ph) (reproduced by permission of Wiley).

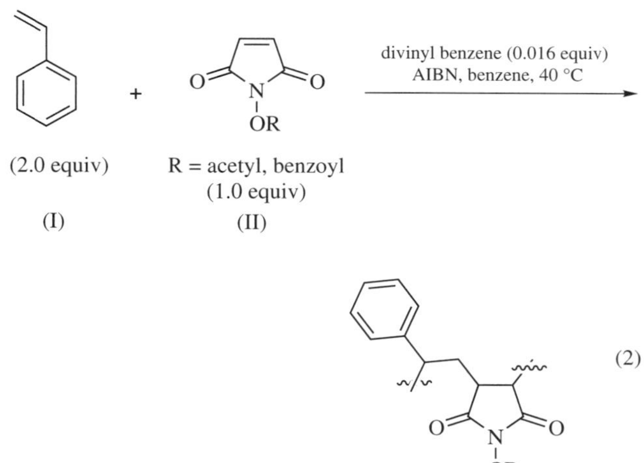

(2.0 equiv) R = acetyl, benzoyl (1.0 equiv)

(I) (II)

(IIIa / IIIb) (2)

Related Reagents

2-Polystryene-5-hydroxy-tetrahydro-cyclopenta[1,2-c;3,4-c′] dipyrrole-1,3,4,6-tetraone as an Alternative to Polymer-bound N-hydroxysuccinimide

Polymer-bound *N*-hydroxysuccinimide, useful for solid-phase organic synthesis, is normally not available from commercial sources. Although relatively convenient to prepare, the details of measuring and assuring functional group loading as well as consistent polymer particle sizing have prompted several groups to prepare alternative reagents from the commercially available polymers, specifically manufactured for solid-phase organic synthesis. These alternative reagents are comparable with respect to the scope and details of their reactivity profile for simple capture-and-release chemistry.

Roche scientist have reported the condensation of commercially available 5-hydroxy-tetrahydro-2-oxa-5-aza-cyclopenta[a] pentalene-1,3 (**6**) with commercially available aminomethyl polystyrene resin, resulting in the formation of *N*-hydroxysuccinimidyl active ester linker (**7**), followed by treatment with hydroxylamine hydrochloride to form 2-polystryene-5-hydroxy-tetrahydro-cyclopenta[1,2-c;3,4-c′]difuran-1, 3, 4,6-tetraone **8** (eq 3).[4] The infrared absorption (KBr) gives us the following information about the compound: OH stretch : 3460 cm^{-1}(strong); C=O stretch : 1700 cm^{-1}(strong), 1684 cm^{-1}(shoulder).

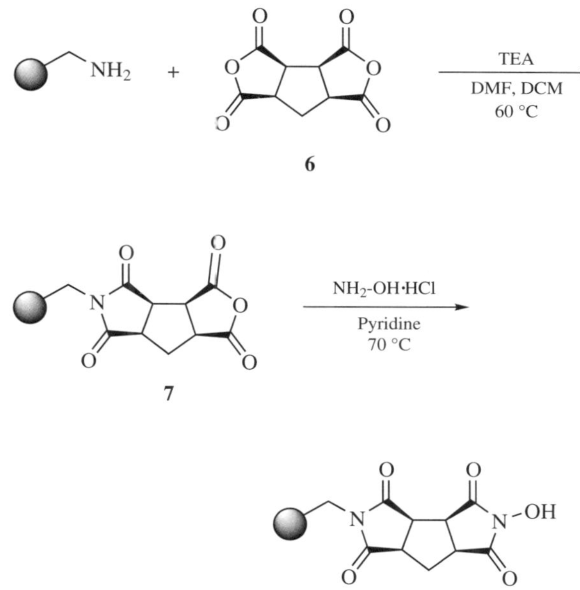

6 **7** **8** (3)

The reactivity profile of various activated esters and nucleophiles is shown in Table 2. Some decrease in reaction yields, product purity, and reaction rate is noted for less nucleophilic amines (e.g., aniline, data not shown). Slower reactions can be accelerated thermally to some extent.

1-Hydroxy-3-polystryene-sulfanyl-pyrrolidine-2,5-dione

Two groups have independently reported on the formation of thiol ether linked *N*-hydroxysuccimide resins.[5,6] These resins have been prepared from commercially available thiol-containing solid-phase synthesis resin materials reacted with *N*-oxymaleimide precursors. The synthesis of 1-hydroxy-3-(4-polystyrene-benzylsulfanyl)-pyrrolidine-2,5-dione (**9**), as shown in eq 4, is according to the method reported in ref 10. The applications of capture-and-release chemistry methods to a diverse set of small-molecule primary amine substrates were also reported. *N*-hydroxysuccinimide residue bound to polymeric resin in this manner was used for column-free fluorescent-labeling and biotin-labeling of various primary amines.[5]

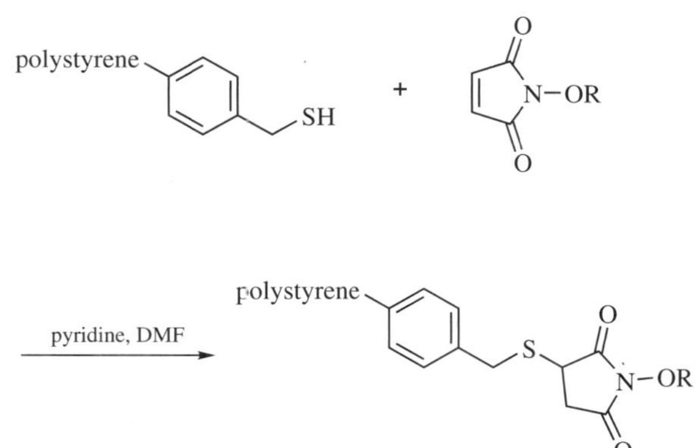

9 (4)

R = H or diverse acyl reagents

Finally, in several reports amide bond couplings have been described using various polymer-bound reagents, such as containing nitrophenol,[7] HOBt (1-hydroxybenzotriazole),[8] carbodiimides,[9]

Table 2 *N*-Acylation reactions using *N*-hydroxysuccinimidyl active ester resin (reproduced by permission of Elsevier Science)[a,b]

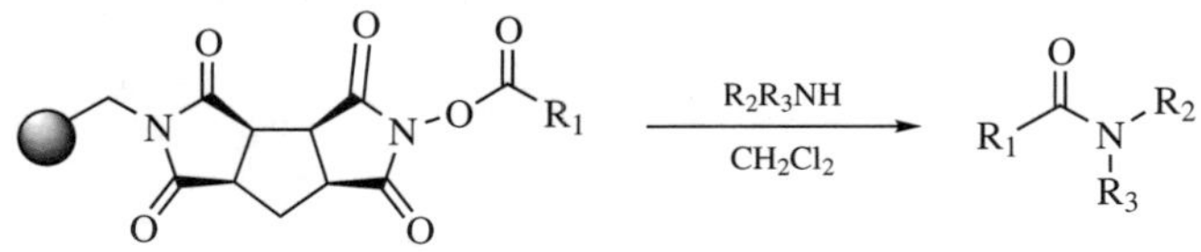

Amines (R_2R_3NH) / Carboxylic acids (R_1-COOH)	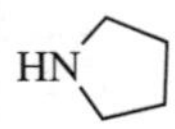	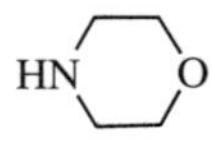	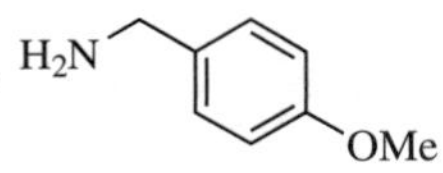	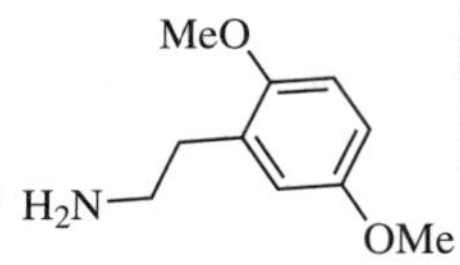
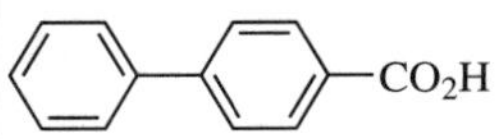	99%[a] (81%[b])	100% (73%)	97% (73%)	99% (68%)
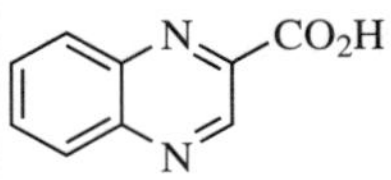	100% (78%)	100% (81%)	86% (84%)	98% (74%)
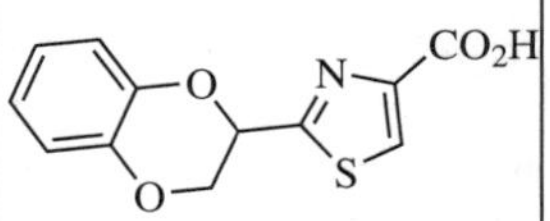	100% (93%)	100% (90%)	97% (79%)	98% (92%)
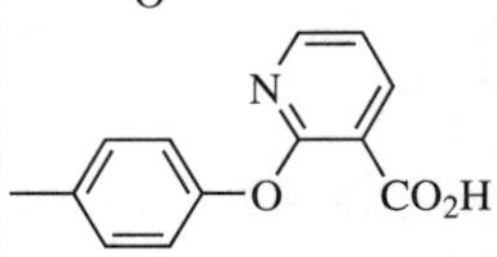	99% (72%)	100% (77%)	99% (86%)	100% (77%)

[a]Purity for all table entries was determined by LC/MS–ELSD.

[b]Yield for all table entries was isolated yield based on the amine nucleophile as the limiting reagent.

and Kaiser oxime resin.[10] Compared to other active ester reagents, *N*-hydroxysuccinimidyl esters tend to be more stable, easy to store, and give clean products in solution reactions.

---

1. Akiyama, M.; Narita, M.; Okawara, M., *J. Polym. Sci.* **1969**, *7*, 1299–1306.
2. Kirschning, A.; Monenschein, H.; Wittenberg, R., *Chem. Eur. J.* **2000**, *6*, 4445–4450.
3. Narita, M.; Teramoto, T.; Okawara, M., *Bull. Soc. Chem. Jpn.* **1972**, *45*, 3149–3155.
4. Shao, H.; Zhang, Q.; Goodnow, R.; Chen, L.; Tam, S., *Tetrahedron Lett.* **2000**, *41*, 4257–4260.
5. Katoh, M.; Sodeoka, M., *Bioorg. Med. Chem. Lett.* **1999**, *9*, 881–884.
6. Adamczyk, M.; Fishpaugh, J. R.; Mattingly, P. G., *Tetrahedron Lett.* **1999**, *40*, 463–466.
7. (a) Fridkin, M.; Hazum, E.; Kalir, R.; Rotman, M.; Koch, Y., *J. Solid-Phase Biochem.* **1977**, *2*, 175. (b) Carpino, L. A.; Cohen, B. J.; Lin, Y. Z.; Stephens, Jr, K. E.; Triolo, S. A., *J. Org. Chem.* **1990**, *55*, 251–259. (c) Cohen, B. J.; Karoly-Hafeli, H.; Patchornik, A., *J. Org. Chem.* **1984**, *49*, 922–924. (d) Reichwein, J. F.; Liskamp, M. J., *Tetrahedron Lett.* **1998**, *39*, 1243–1246. (e) Chang, Y.-T.; Schultz, P. G., *Bioorg. & Med. Chem. Lett.* **1999**, *9*, 2479–2482. (f) Kim, K.; Le, K., *Synlett* **1999**, 1957–1959.
8. (a) Kalir, R.; Warshawsky, A.; Fridkin, M.; Patchornic, A., *Eur. J. Biochem.* **1975**, *59*, 55. (b) Mokotoff, M.; Patchornic, A., *Int. J. Pept. Protein. Res.* **1983**, *21*, 145. (c) Mokotoff, M.; Zhao, M.; Roth, S. M.; Konttab, N. M.; Slavoski, J. N.; Shelley, J. A., *J. Med. Chem.* **1990**, *3*, 354. (d) Pop, I. E.; Deprez, B. P.; Tartar, A. L., *J. Org. Chem.* **1997**, *62*, 2594–2603. (e) Dendrinos, K. G.; Kalivretenos, A. G., *Tetrahedron Lett.* **1998**, *39*, 1321–1324. (f) Dendrinos, K.; Jeong, J.; Huang, W.; Kalivretenos, A. G., *Chem. Commun.* **1998**, 499–500. (g) Schiemann, K.; Showalter, H. D. H., *J. Org. Chem.* **1999**, *64*, 4972–4975.
9. (a) Weinshenker, N. M.; Shen, C. M., *Tetrahedron Lett.* **1972**, 3281. (b) Desai, M. C.; Stephens Straminello, L. M., *Tetrahedron Lett.* **1993**, *34*, 7685.
10. (a) Smith, R. A.; Bobko, M. A.; Lee, W., *Bioorg. & Med. Chem. Lett.* **1998**, *8*, 2369–2374. (b) Golebiowski, A.; Klopfenstein, S., *Tetrahedron Lett.* **1998**, *39*, 3397–3400. (c) Scialdone, M. A.; Shuey, S. W.; Soper, P.; Hamuro, Y.; Burns, D. M., *J. Org. Chem.* **1998**, *63*, 4802–4807.

Robert A. Goodnow, Jr & Hui Shao
Hoffmann-La Roche, Nutley, New Jersey USA

Polystyrene-1-ethyl-3-(3′-dimethylaminopropyl)-carbodiimide Hydrochloride (PS-EDCI)

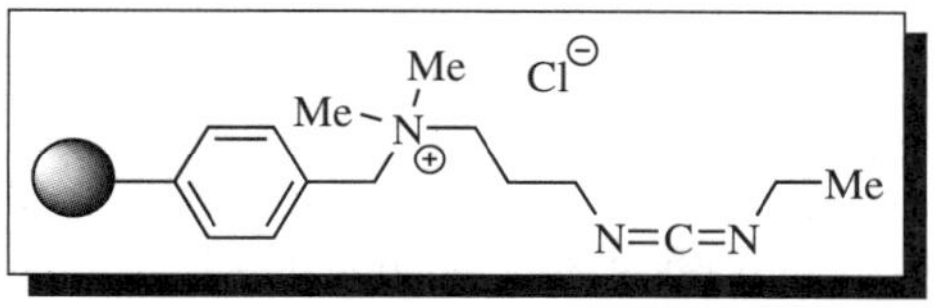

[99270-53-8]

(used as a peptide coupling reagent; used in acylsulfonamide libraries, disulfide analog synthesis, heterocyclic carboxamides via a CMR/R approach, formation of Mosher amides, preparation of hapten active esters, thiol ester formation, cyclodehydration reactions)

Alternate Name: polymer-bound (or polymer-tethered) 1-(3-dimethylaminopropyl)-3-ethylcarbodiimide; polymer-bound (P)-EDCI; P-EDC; polymeric-EDC; solid-supported (SS)-EDAC; polymer-bound EDAC.

Solubility: reaction solvents include chloroform ($CHCl_3$), methylene chloride (CH_2Cl_2), dimethylformamide (DMF), or acetonitrile (MeCN). In certain instances, *t*-butanol (*t*-BuOH) has been added as a co-solvent in order to dissolve insoluble amine or carboxylic acid components. Other solvents, such as tetrahydrofuran (THF) or diethyl ether (Et_2O), have been shown to slow down the reaction substantially.

Preparative Methods: PS-EDCI (**1**) is prepared from commercially available EDCI, which may be bound via alkylation utilizing the nucleophilic property of the dimethylamino group of the reagent, to the commercially available chloromethylated polystyrene-divinylbenzene. Suitable polymers have a 200–400 mesh size, 1–2% divinylbenzene cross-linking, and a level of chloromethylation of 0.8 mmol of Cl per g of resin. The preparation of PS-EDCI is accomplished in one step by heating EDCI (1.5 equiv) with chloromethylated polystyrene-divinylbenzene and a suitable solvent such as dimethylformamide (DMF, 100 °C) or acetonitrile (MeCN, reflux) for 15 h (i.e., overnight). A long reflux period and the use of 50% excess EDCI are to ensure completeness of the reaction (i.e., in the quaternization, accessible chloromethyl groups are all consumed). The functionalized polymer is obtained upon cooling and subsequent filtration. Washing with THF followed by diethyl ether and drying (P_2O_5) gives the PS-EDCI reagent product in pure form. The reagent can then be stored at 0 °C without loss of activity for several months.[1]

Handling, Storage, and Precautions: the beige colored, free floating resin can be stored without significant loss of reactivity at 0 °C for several months.

Peptide Coupling Reagent. The synthesis and utility of PS-EDCI (**1**), a polymer-supported 1-ethyl-3-(3′-dimethylaminopropyl)-carbodiimide, was reported in 1973.[1] PS-EDCI is a dehydrative solid-phase coupling reagent which mediates the formation of amides through the coupling of amines with carboxylic acids. The reagent has alternate names, such as polymer-bound EDC (P-EDC) (i.e., original designation) and solid-supported EDAC.

PS-EDCI comprises an insoluble polymer which has a pendant carbodiimide-functionalized side-chain terminating with the formula shown in **2**, containing a positively charged quaternary nitrogen and a chloride counteranion.[2] However, it has been reported that in a more general sense, side-chains as shown in **3**, where *n* can range from 1–5 and the counteranion Y^- could be bromide, iodide, mesylate, or trifluoroacetate, have also been found to be useful.

PS-EDCI has wide applicability as a coupling reagent in amide bond forming reactions. As described by Desai and Stramiello,[1] the coupling of an amine with a carboxylic acid is effected in chloroform at room temperature with PS-EDCI, and after the reaction is completed (0.5–48 h), the product is filtered and rinsed with chloroform, followed by the evaporation of chloroform. The process yields the essentially pure desired product. Chloroform or any other suitable solvent [e.g., methylene chloride, dimethylformamide (DMF), or acetonitrile] is preferred over solvents such as ether or THF which slow down the reaction. Up to 25% *tert*-butanol can be added to aid in solubilizing the amine or carboxylic acid if needed. The amount of PS-EDCI used should contain a total molar amount of carbodiimide groups equal to or greater than the desired moles of amide, and equivalent amounts of amine and carboxylic acid should be added. A greater molar amount of carboxylic acid relative to the amine (i.e., 1.2 to 1) could be used. Excess carboxylic acid and urea byproducts, if any, remain attached to the polymer. When PS-EDCI is the coupling reagent in the formation of peptides, where the reactants are α-amino acids or peptide fragments containing more than one amino acid residue, amino blocking groups [e.g., *t*-butoxycarbonyl (*t*-BOC), benzyloxycarbonyl (Cbz), formyl, acetyl, trifluoroacetyl, or toluenesulfonyl] are used to control the reaction and they decide which amino group will react; alkoxy groups derived from esterification of alcohols (e.g., methanol, ethanol, propanol, *t*-butanol, and benzyl alcohol) are carboxylic acid blocking groups.[2]

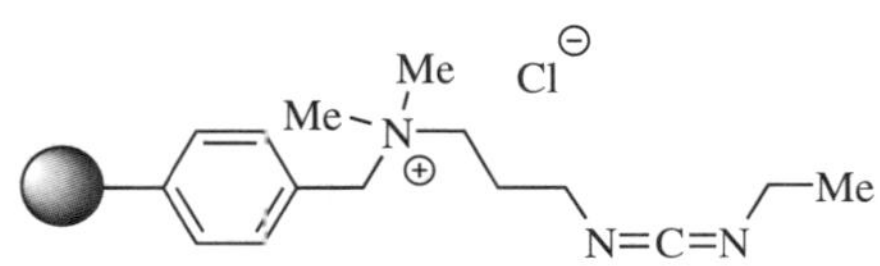

1

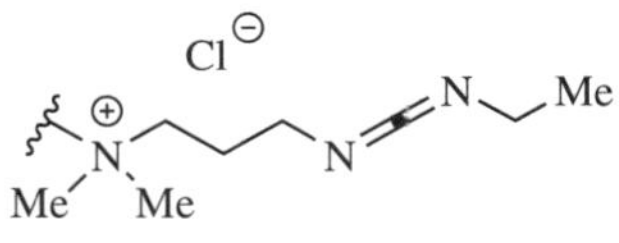

2

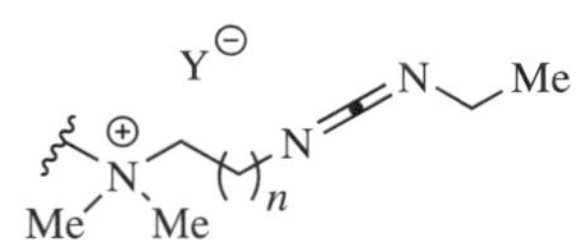

3a–3e

In mediating the formation of an amide, PS-EDCI is not limited to any particular amine or carboxylic acid. Examples of classes of amines that can be used include arylalkyl amines (e.g., benzylamine, 2-phenylethylamine, and 3-phenylpropylamine), aliphatic (acyclic) amines (e.g., methylamine, ethylamine, *n*-propylamine, *n*-butylamine, *n*-pentylamine, *n*-octylamine, and decylamine), alicyclic amines (e.g., cyclohexylamine, cyclopentylamine, cyclobutylamine, and cyclopropylamine), and heterocyclic amines within which the amino nitrogen is a saturated ring atom (e.g., pyrrolidine, piperidine, morpholine, and piperazine). A broad range of carboxylic acids can be used as well, including aromatic acids with the acid group attached to an aromatic ring (e.g., benzoic acid, the phenylbenzoic acids, and the naphthoic acids), alkyl acids (e.g., acetic acid, propionic acid, and butyric acid), arylalkyl acids (e.g., phenylacetic acid and phenylpropionic acid), aromatic heterocyclic acids (e.g., pyridine carboxylic acids and benzimidazole carboxylic acids), and non-aromatic heterocyclic acids (e.g., piperidine carboxylic acids, piperazine carboxylic acids, pyrrolidine carboxylic acids, and quinuclidine carboxylic acids).[2]

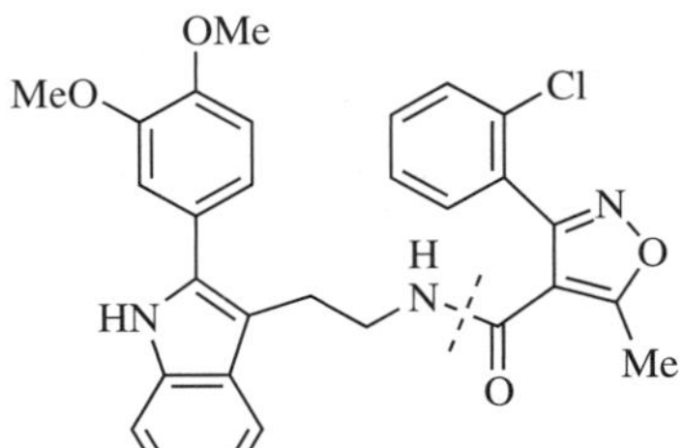

81%

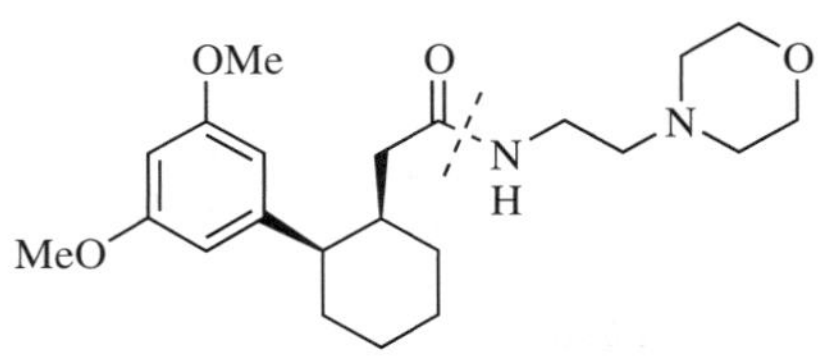

quantitative

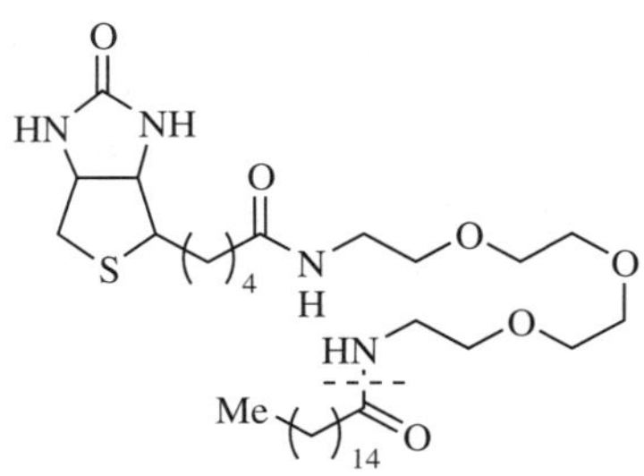

73%

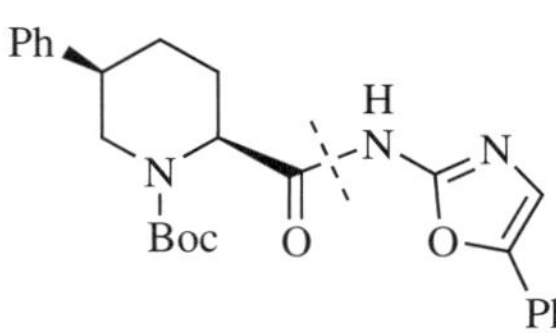

83%

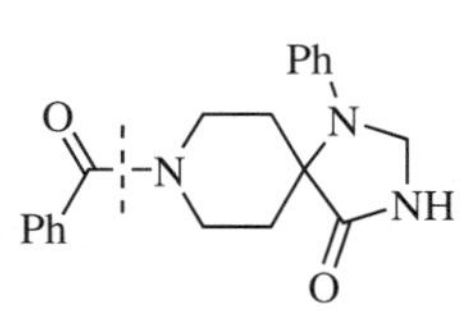

97%

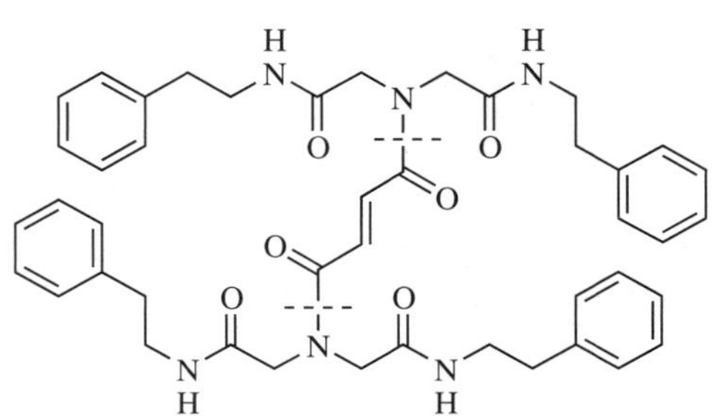

10%

Due to the general effectiveness of coupling amines and carboxylic acids, PS-EDCI has been exploited for the quick production of various high throughput lead discovery libraries shown below.[1–7]

Acylsulfonamide Libraries. PS-EDCI usage in the preparation of small molecule libraries offers advantages over traditional solid-phase organic synthesis (SPOS), since the starting materials need not possess a suitable functional group for attachment to the solid support, and in some instances the polymer-bound EDCI reagent is used in conjunction with SPOS. Since PS-EDCI and other polymer-supported reagents do not interact, more than one of these reagents may be used in a reaction. A case in point was the reported use of PS-EDCI as the dehydrating agent in the preparation of acylsulfonamide libraries of benzoic acid (eq 1).[8] In these reactions, benzoic acid reacts with 4-isopropylbenzenesulfonamide to yield acylsulfonamide. PS-EDCI serves as the dehydrating agent and Amberlyst-15 (A-15), a polymer-bound sulfonic acid, is used to protonate acylsufonamide and to remove DMAP from the solution as its corresponding salt.

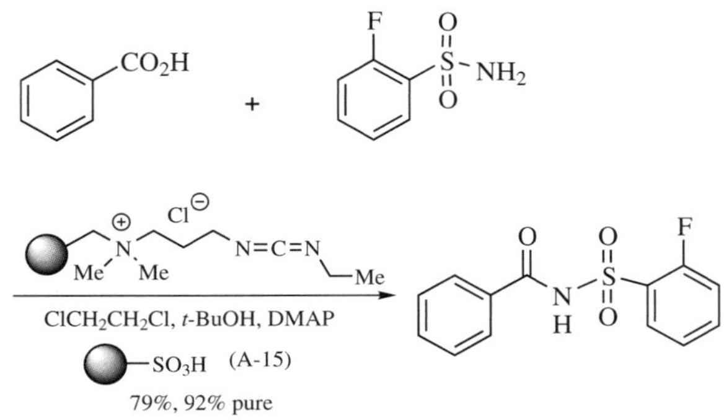

(1)

Disulfide Analog Synthesis. Similarly, the synthesis of disulfide substrates for the enzyme trypanothione reductase was reported in a superior fashion upon employing PS-EDCI, when compared to the normal solution-phase preparations in terms of simplicity, efficiency, and yields (eq 2).[9] The generation of the disulfide included the activation of the carboxyl group of the substrate prepared from glycine and benzylchloroformate in the presence of 3-*N*,*N*-dimethylaminopropylamine, and after filtration and removal of the benzyloxycarbonyl (Cbz) group by catalytic hydrogenation, the resulting product was coupled with a cystine derivative in the presence of PS-EDCI.

Heterocyclic Carboxamides via a CMR/R Approach. The dehydrating power of PS-EDCI was also utilized with the fundamental principles of complementary molecular reactivity and molecular recognition (CMR/R) to synthesize carboxamides with different heterocyclic motifs, thus enabling rapid lead generation and refinement (eq 3).[10] As a dehydrating agent, PS-EDCI facilitated the conversion of heterocyclic carboxylic acids into their respective anhydrides which were then reacted with substituted aniline derivatives to furnish carboxamide products in high yields and purities.

Formation of Mosher Amides. The use of PS-EDCI in a novel method for the synthesis of Mosher amides in deuterochloroform

($CDCl_3$) was reported (eq 4).[11] The desired amides were afforded in high yields through the coupling of an amine with Mosher's acid [α-methoxy-α-trifluoromethyl-phenylacetic acid] mediated by PS-EDCI.

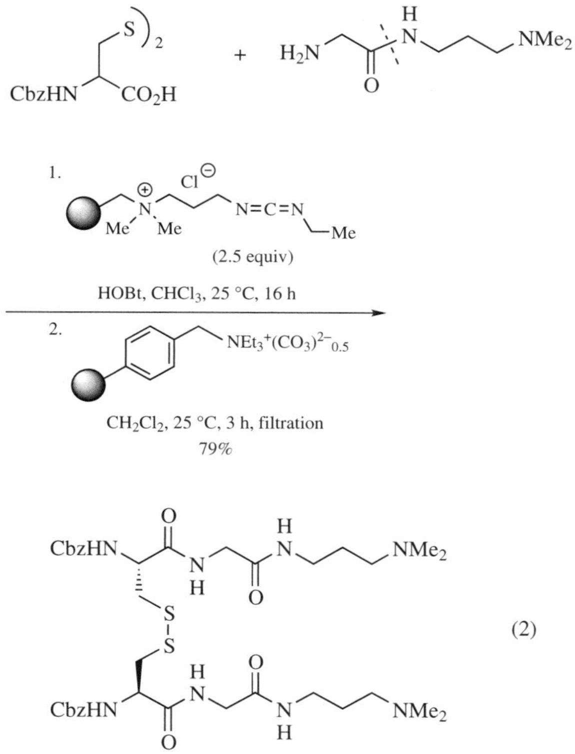

(2)

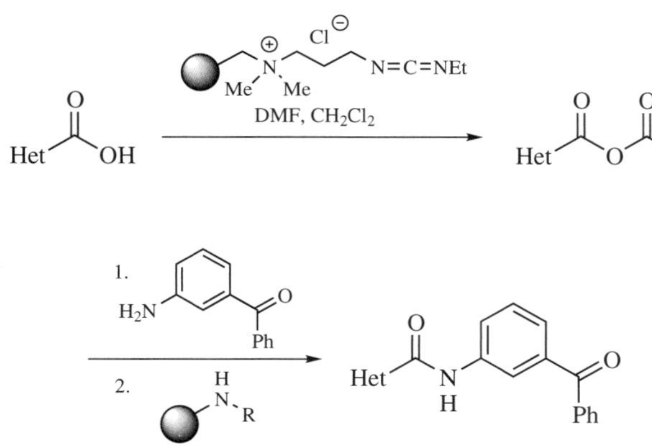

(3)

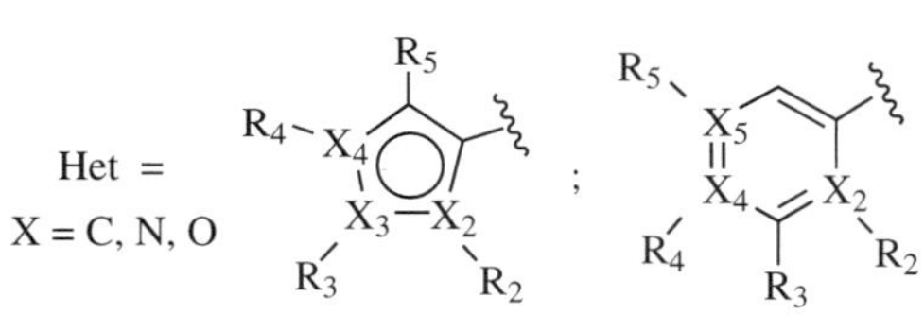

Preparation of Hapten Active Esters. The preparation of bioconjugates, such as immunogens and enzyme conjugates, by direct coupling of carboxylic acid haptens to proteins using EDCI (not polymer-supported) results in undesired side reactions, such as cross-linking of the protein, *N*-acylurea formation

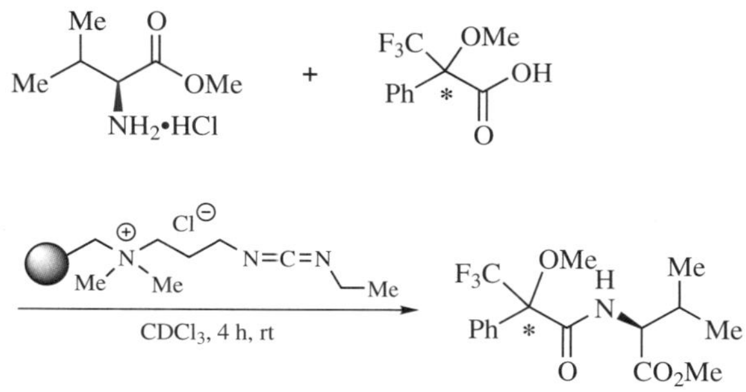

(4)

Configuration	Yield (%)
R	90
S	95
Racemic	80

on the protein, and formation of urea by products for which removal is not trivial. A practical method for the synthesis of pure hapten active esters for bioconjugation, which overcomes these disadvantages, uses polymer-supported EDCI to produce *N*-hydroxysuccinimidyl and pentafluorophenyl active esters (eqs 5 and 6).[12] PS-EDCI was applied to activate carboxylic acids which were reacted with either *N*-hydroxysuccinimide (HOSu) or pentafluorophenol (PFP-OH) as the limiting reagent.

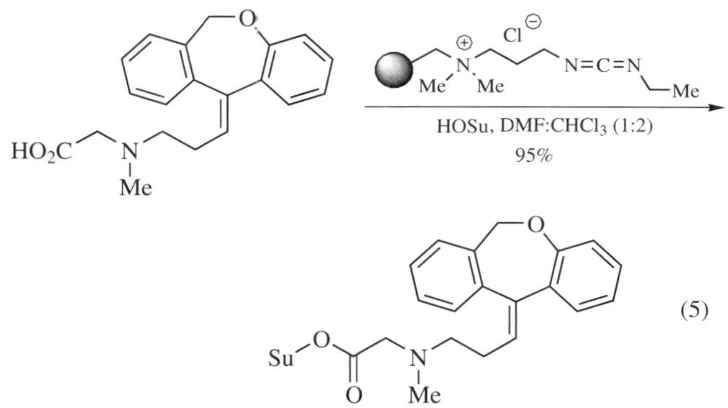

(5)

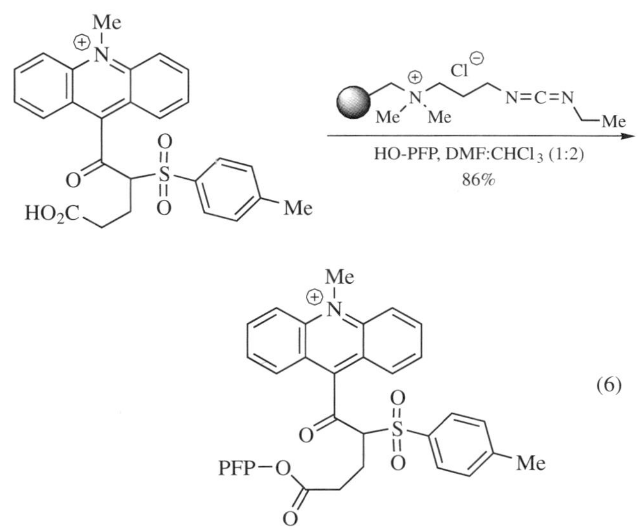

(6)

Thiol Ester Formation. Thiol esters were prepared in high yield and purity through the coupling of a thiol with a carboxylic acid mediated by PS-EDCI (eqs 7–9).[13] In general, the formation of the primary and aryl thiol esters was found to be quite facile compared to the secondary or tertiary systems. High conversions, however, can be established by either employing excess carboxylic acid (in the case of expensive thiols) or excess thiol (in reactions where the thiol component is volatile).

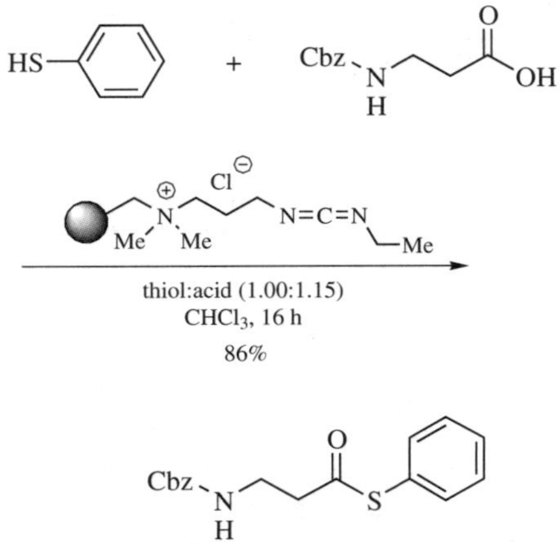

(7)

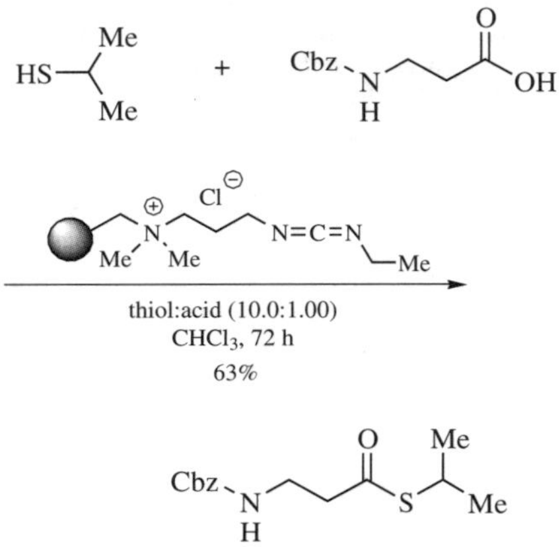

(8)

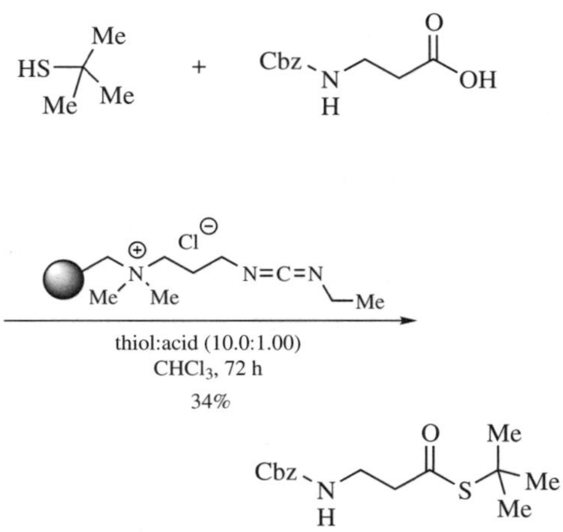

(9)

Cyclodehydration Reactions. Due to the emphasis on high-throughput production of small molecule libraries, the multi-step synthesis of various benzoxazinone and benzoxatine scaffolds concluding with ring closure via a cyclization/dehydration method mediated by PS-EDCI was adapted to a fully automated process.[14,15] PS-EDCI sequestered any uncyclized acid as imino-anhydride adducts (removed from the reaction mixture by filtration), and thus served a dual role as a reagent-sequestering resin. For example, formation of 2-aminobenzoxazines from anthranilic acids and isocyanates was achieved in a two-step protocol: urea formation and cyclodehydration (eq 10).[14]

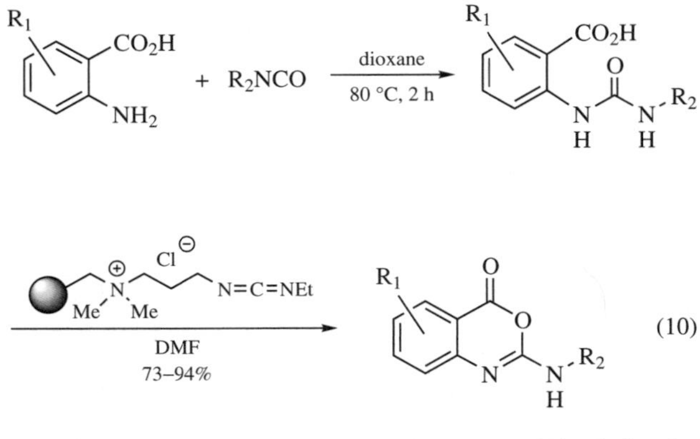

(10)

Related Reagents. Reagents related to PS-EDCI include other polymer-supported reagents, such as *N*-cyclohexylcarbodiimide-*N*′-propyloxymethyl polystyrene (PS-CDI) (this reagent is commercially available from Argonaut Technologies Inc. and Novabiochem) and polymer-supported 1-(3-pyrrolidinylpropyl)-3-ethylcarbodiimide (PS-EPC),[3a] as well as the non-bounded amide coupling agents including EDCI and 1,3-dicyclohexylcarbodiimide (DCC). The use of EDCI for amide coupling reactions requires that the urea and any *N*-acylurea produced be separated from products and extracted into an acidic aqueous phase. This is facilitated by the fact that the urea produced is water soluble. EDCI may, therefore, be preferable to the amide coupling reagent 1,3-dicyclohexylcarbodiimide (DCC) for which dicyclohexylurea is only slightly soluble and must be removed by filtration which is possibly less effective than extraction. Additionally, any *N*-acylurea of carbodiimide couplings can contaminate the product and cause low yields. Furthermore, rearrangement of this coupling intermediate can, upon tautomerization, result in racemization. The addition of trapping agents, such as 1-hydroxybenzotriazole (HOBt), reduces *N*-acylurea formation and suppresses racemization. The HOBt may then be scavenged with MP-carbonate or PS-trisamine resin. PS-EDCI behaves similar to EDCI in solution, but the by products would remain on the polymer, permitting the product to be simply isolated by filtration followed by evaporation of the filtrate and thereby avoiding the aqueous work-up.

1. Desai, M. C.; Stephens Stramiello, L. M., *Tetrahedron Lett.* **1993**, *34*, 7685.
2. European Pat. 623589 (to Desai, M. C.) (Pfizer Inc., 1994).
3. (a) Kaldor, S. W.; Siegel, M. G., *Curr. Opin. Chem. Biol.* **1997**, *1*, 101. (b) Baxter, A. D., *Curr. Opin. Chem. Biol.* **1997**, *1*, 79. (c) Drewry, D. H.; Coe, D. M.; Poon, S., *Med. Res. Rev.* **1999**, *19*, 97. (d) Kirschning, A.; Monenschein, H.; Wittenberg, R., *Angew. Chem. Int. Ed.* **2001**, *40*, 650.
4. Boger, D. L.; Goldberg, J.; Jiang, W.; Chai, W.; Ducray, P.; Lee, J. K.; Ozer, R. S.; Andersson, C.-M., *Bioorg. Med. Chem.* **1998**, *6*, 1347.
5. McReynolds, K. D.; Bhat, A.; Conboy, J. C.; Saavedra, S. S.; Gervay-Hague, J., *Bioorg. Med. Chem. Lett.* **2002**, *10*, 625.
6. Krantz, A.; Song, Y.; DeNagel, D.; Hartmann, C.; Bridon, D., *J. Drug Target* **1999**, *7*, 113.
7. Mortezaei, R.; Ida, S.; Campbell, D. A., *Mol. Divers.* **1999**, *4*, 143.
8. Sturino, C. F.; Labelle, M., *Tetrahedron Lett.* **1998**, *39*, 5891.

9. Chibale, K.; Chipeleme, A.; Warren, S., *Tetrahedron Lett.* **2002**, *43*, 1587.
10. Parlow, J. J.; Mischke, D. A.; Woodard, S. S., *J. Org. Chem.* **1997**, *62*, 5908.
11. Adamczyk, M.; Fishpaugh, J. R., *Tetrahedron Lett.* **1996**, *37*, 7171.
12. Adamczyk, M.; Fishpaugh, J. R.; Mattingly, P. G., *Tetrahedron Lett.* **1995**, *36*, 8345.
13. Adamczyk, M.; Fishpaugh, J. R., *Tetrahedron Lett.* **1996**, *37*, 4305.
14. Buckman, B. O.; Morrissey, M. M.; Mohan, R., *Tetrahedron Lett.* **1998**, *39*, 1487.
15. (a) Parlow, J. J.; Flynn, D. L., *Tetrahedron* **1998**, *54*, 4013. (b) Flynn, D. L.; Devraj, R. V.; Naing, W.; Parlow, J. J.; Weidner, J. J.; Yang, S., *Med. Chem. Res.* **1998**, *8*, 219.

Michael M. Miller & Dale L. Boger
The Scripps Research Institute, La Jolla, CA, USA

Polystyrylboronic Acid[1]

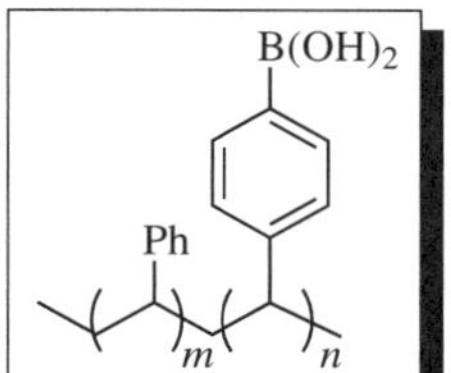

1

[52384-35-7]

(used as selective, recoverable protective agent for diols; in the solid phase synthesis of di- and trisaccharides; as a torsionally disarming protective group in glycosidic bond formation)

Alternate Name: PSBA.

Solubility: used as an insoluble resin. Suitable solvents include pyridine, DMF, and DCM.

Form Supplied in: white solid; not commercially available; usually prepared crosslinked with 2% divinylbenzene.

Analysis of Reagent Purity: boron loading can be determined from diol binding capacity, elemental analysis, or via digestion followed by ICP boron analysis.

Preparative Methods: the most common method for the preparation of PSBA involves the treatment of commercially available brominated polystyrene with *n*-butyllithium, followed by a trialkyl borate such as trimethyl borate. Acid hydrolysis of the intermediate dialkyl boronate gives PSBA (eq 1).[2–4]

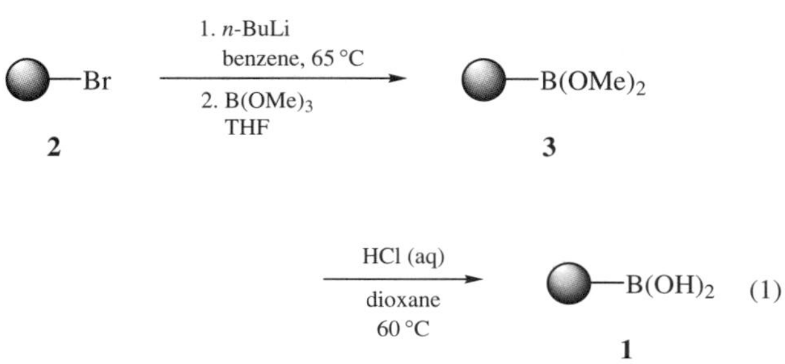

2 **3** **1**

 = 2% crosslinked polystyrene

Handling, Storage, and Precautions: PSBA does not appear to have any special handling requirements. It is quite stable and readily recycled.

Diol Protection in the Selective Derivatization of Monosaccharides. Phenylboronic acid (PBA) has been long established as a labile protective agent of polyols, especially 1,3-diols, with the protection and deprotection steps occurring under relatively mild conditions.[1,5] In 1970s, it was shown that polystyrylboronic acid (PSBA, **1**) can be used in a similar way, with protection and deprotection occurring under even milder conditions to those used with PBA, and that higher yields of the desired derivatized products are usually obtained. This is demonstrated by the selective mono-acylation of methyl α-D-*xylo*pyranoside (**4**).[3] Using PSBA, 3-acetyl pyranoside (**6**) was obtained in 87% yield (eq 2), whereas the solution phase alternative using PBA gives 42% yield. Of note is the simple and efficient deprotection procedure which involves the room temperature treatment of the resin with aqueous acetone.

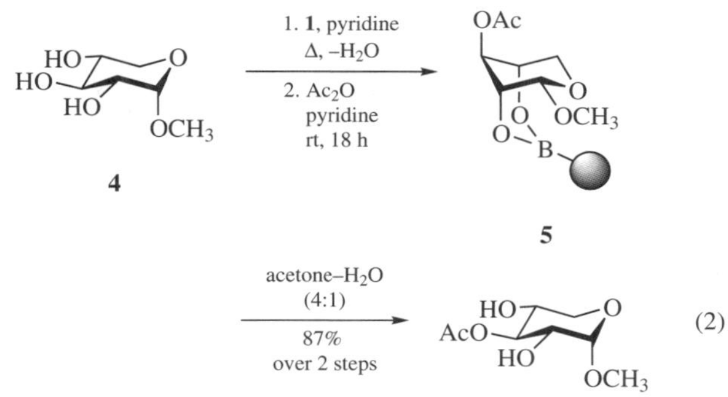

4 **5** **6**

A similar strategy can be used to introduce pyrimidine groups to monosaccharides.[6] Condensation of α-D-glucopyranoside **7** with PSBA, followed by deprotonation with NaH, treatment with 2-methylsulfonyl-4,6-dimethylpyrimidine, and then hydrolysis of the intermediate boronate esters gives two products, a monopyrimidinyl glucopyranoside (**8**) and a di-pyrimidinyl glucopyranoside (**9**), in a combined yield of 80% (eq 3). This is a major improvement upon the reaction with the unprotected pyranoside, which produces at least seven products.

Di- and Trisaccharide Synthesis. PSBA can be used to prepare di- and trisaccharides on the solid phase. For example, immobilization of glycosyl acceptor **10** as 4,6-*O*-boronate **11** leaves the 3-OH available for glycosylation with thioglycosyl donors. β-Disaccharide **13** can be prepared in anomerically pure form in 95% yield using this approach (eq 4).[7] Trisaccharides can be prepared in a similar way. In cases where the α:β selectivity of the first glycosylation reaction is much lower, a 'loading-release-loading' strategy can be used to obtain trisaccharides with high anomeric purity. For example, glycosylation of the immobilized cyclic boronate **11** with an alternative thioglycosyl donor **14**, followed by Fmoc removal and cleavage from the resin gives a disaccharide with an α:β ratio of 1.6:1.0. Isolation of α-anomer **15**, followed by reattachment to PSBA allows the formation of a second glycosidic bond in high anomeric selectivity such that trisaccharide **17** can be readily obtained in anomerically pure form (eq 5).[8]

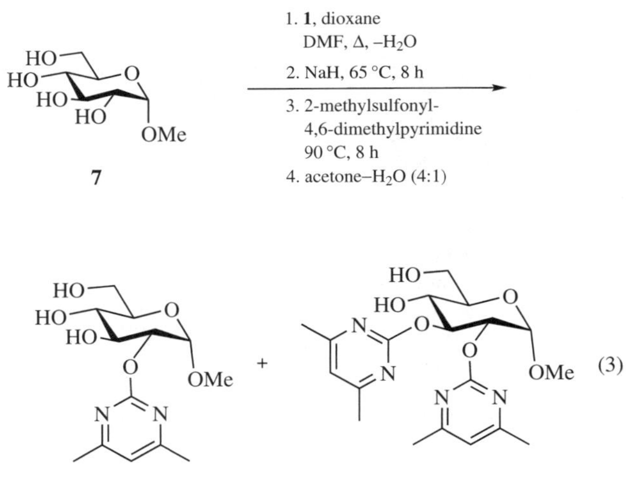

7 **8** 65% **9** 15%

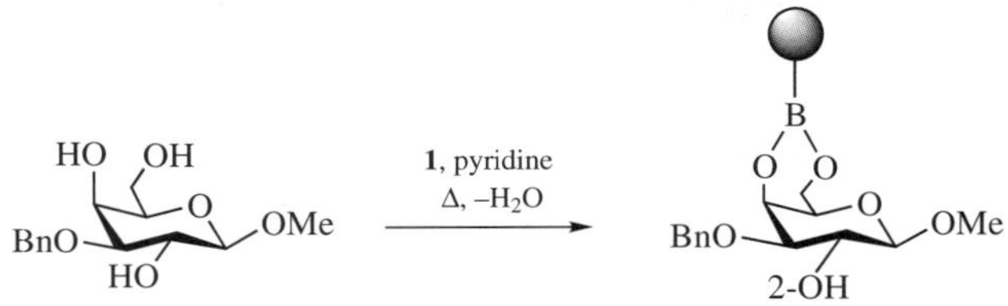

10 **11**

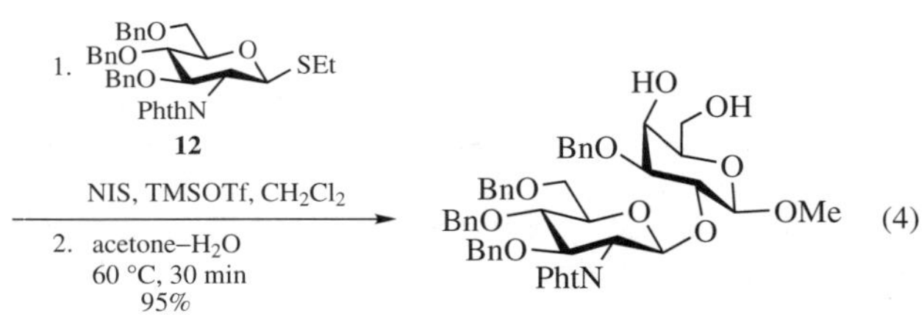

13

Immobilization of the thioglycosyl donor on the solid phase, rather than the acceptor, can also lead to efficient disaccharide formation (eq 6).[7]

The torsionally disarming ability of protecting groups such as 4,6-*O*-benzylidene, which encourages mannosylation reactions of carbohydrate derivatives to proceed with high selectivity for formation of the β-glycoside, has also been demonstrated for the corresponding 4,6-*O*-boronates. Protection of the 4- and 6-hydroxy groups of thioglycosyl donor **21** with either PBA or PSBA appears to destabilize the corresponding 1-oxacarbenium ion intermediate enough to significantly limit S_N1 reactions with α-mannopyranosyl triflate **22**, thus providing direct and facile access to the normally difficult to obtain β-mannosides.[4] This is demonstrated by the preparation of β-mannoside **24** from **22** and the 1,2; 5,6-*O*-di-acetonide of glucofuranose (**23**) in high yield and high β:α selectivity (eq 7). In this particular case, hydrolysis from the resin is accompanied by cleavage of the 5,6-acetonide group.

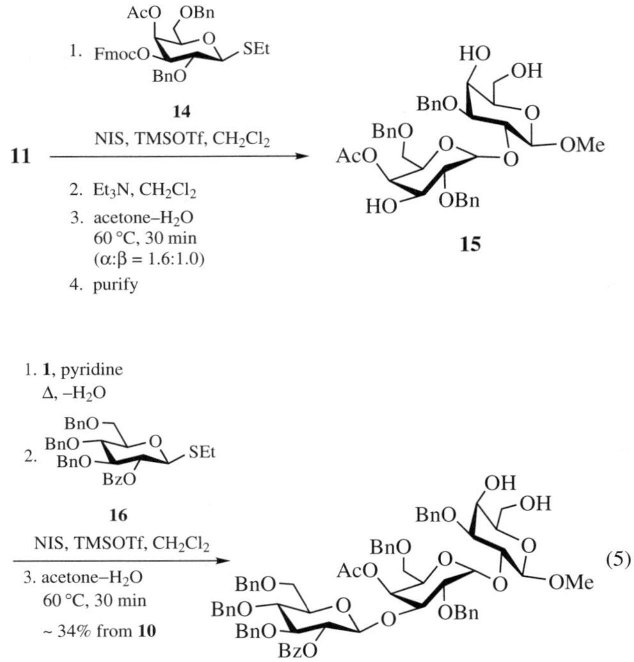

17

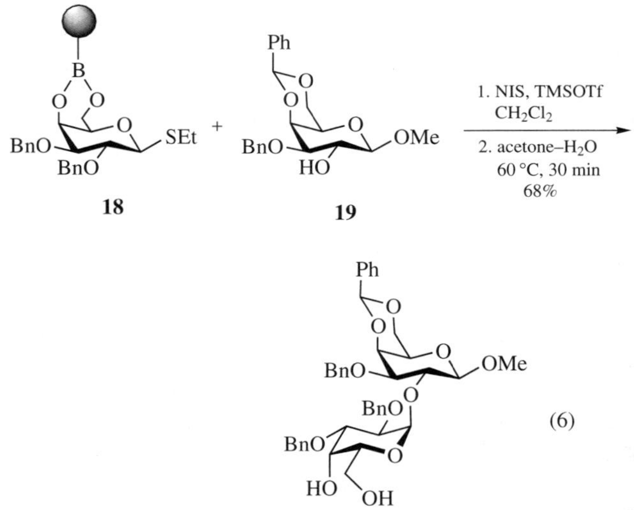

18 **19** **20**

Related Reagents. Phenylboronic acid (PBA).[1,5]

1. Duggan, P. J.; Tyndall, E. M., *J. Chem. Soc., Perkin Trans. 1* **2002**, 1325–1339.
2. Frechet, J. M. J.; Farrall, M. J., *J. Org. Chem.* **1976**, *41*, 3877–3882.
3. Frechet, J. M. J.; Nuyens, N. J.; Seymour, E., *J. Am. Chem. Soc.* **1979**, *101*, 432–436.
4. Crich, D.; Smith, M., *J. Am. Chem. Soc.* **2002**, *124*, 8867–8869.

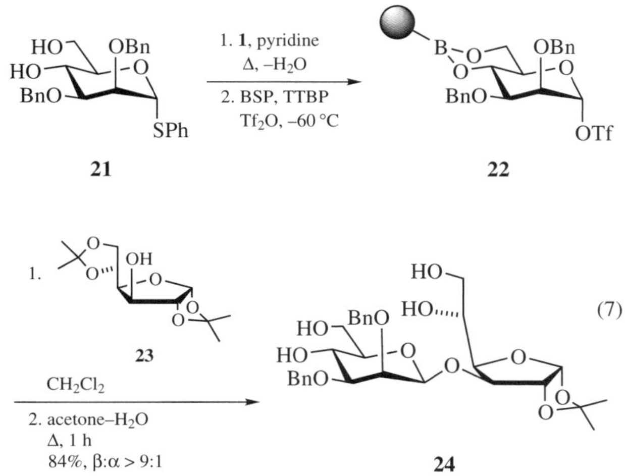

(7)

5. Ferrier, R. J., *Adv. Carbohydr. Chem. Biochem.* **1978**, *35*, 31–80.

6. Liao, Y.; Li, Z., *Synth. Commun.* **1998**, *28*, 3539–3547.

7. (a) Belogi, G.; Zhu, T.; Boons, G.-J., *Tetrahedron Lett.* **2000**, *41*, 6965–6968. (b) Crich, D.; de la Mora, M.; Vinod, A. U., *J. Org. Chem.* **2003**, *68*, 8142–8148.

8. Belogi, G.; Zhu, T.; Boons, G.-J., *Tetrahedron Lett.* **2000**, *41*, 6969–6972.

Peter J. Duggan
Monash University, Clayton, Victoria, Australia

Potassium Ethyl Xanthate

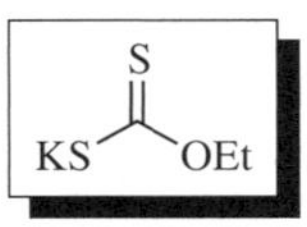

[140-89-6] $C_3H_5KOS_2$ (MW 160.30)

(preparation of thiols, sulfides, and disulfides; heterocycle formation; reaction with alkyl halides leads to the formation of alkyl xanthates which are versatile radical precursors; glycosidation; esterification)

Alternate Name: potassium xanthate, potassium xanthogenate, potassium *O*-xanthogenate, carbonodithioic acid *O*-ethyl ester potassium salt, *O*-ethyl *S*-potassium dithiocarbonate.

Physical Data: mp 222 °C;[1] density $1.558^{21.5}$.

Solubility: very soluble in cold water; decomposes in hot water; soluble in ethanol; insoluble in ether.

Form Supplied in: pale yellow solid.

Preparative Methods: commercially available. Potassium ethyl xanthate and its congeners can be prepared by the reaction of the corresponding potassium alkoxide with carbon disulfide.[2,3] Methods of purification of the commercial compound,[4] as well as for the preparation of extremely pure xanthate derivatives have been reported.[5]

Handling, Storage, and Precautions: reasonably stable when stored in a dark place, protected from air and moisture.

Substitution Reactions: Preparation of Thiols, Sulfides, and Heterocycle Formation. *O*-Alkyl dithiocarbonate salts are strong and soft nucleophiles, which readily undergo S_N2 substitution with alkyl halides or sulfonates.[6,7] This reaction has been used for the preparation of aliphatic thiols.[8,9] The advantage of this xanthate, in comparison to sodium sulfide, or sodium hydrogen sulfide, is that thioethers are not formed and the preparation does not require the use of gaseous hydrogen sulfide. The conversion of the intermediate xanthates into mercaptans can be accomplished by hydrolysis under basic conditions; the best reagent for this transformation is 1,2-diaminoethane,[10] although aqueous ammonia,[11] ethanolamine,[12] or morpholine[13] have also been used (eq 1). Alternatively, xanthate intermediates can be reduced to thiols with lithium aluminum hydride.[14] Use of a (−)-menthol derived xanthate allowed for the preparation of optically pure benzylic mercaptans; the intermediary diastereomeric xanthates could be separated by fractional crystallization (eq 2).[13]

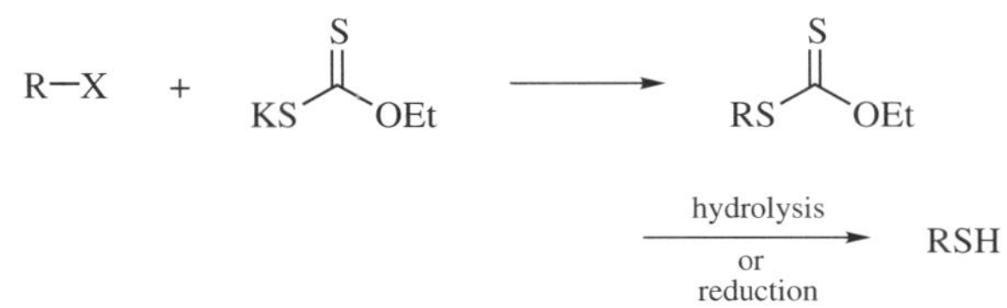

(1)

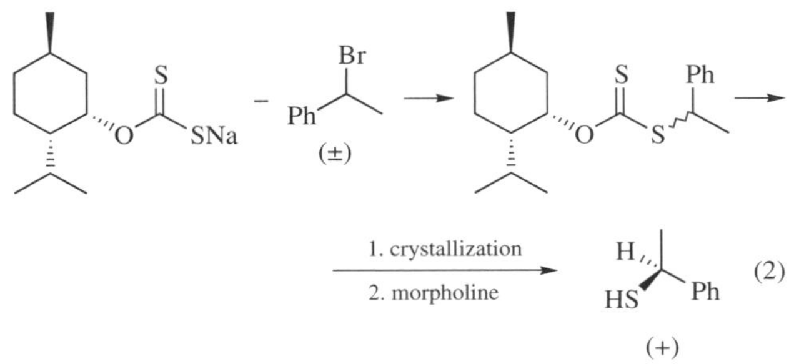

(2)

Under basic conditions thiols are particularly sensitive towards oxidation. Substituting the *tert*-butyl group for ethyl in the xanthate reagent obviates the need for a basic hydrolysis: the decomposition of the intermediate is performed in situ, without its isolation, or the addition of base (eq 3).[15]

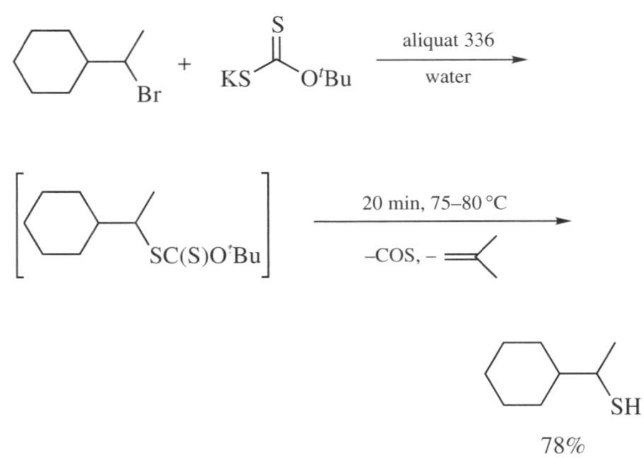

(3)

The dithiocarbonate method also allows for a direct, one pot preparation of unsymmetrical sufides from alkyl halides, or sulfonates (eq 4).[16]

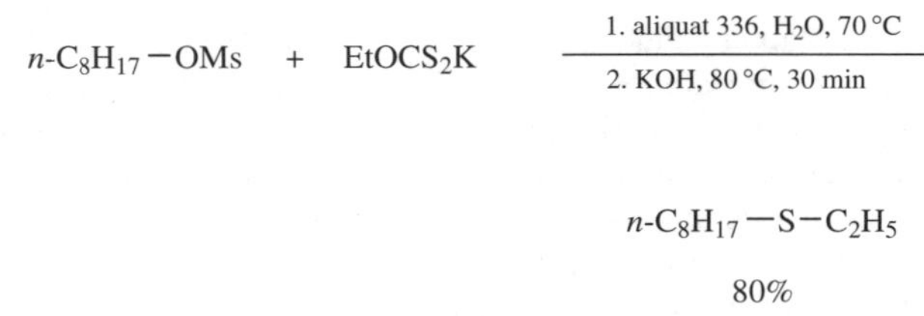

(4)

S_NAr substitution with aryl halides, or diazonium salts, has been used for the introduction of sulfur functionalities into aromatic rings, in the presence of groups such as methoxyl,[17] hydroxyl,[18] or iodide[19] in the aromatic nucleus, or in the side chain. Thus, the reaction of the *m*-toluidine derived diazonium salt with potassium *O*-ethyl xanthate affords the *S*-aryl xanthate, which was hydrolyzed to the corresponding *m*-thiocresol (eq 5).[20] The presence of trace amounts of nickel salts has been shown to enhance the reaction rate considerably,[21] completely eliminating the risk of explosion of the intermediate diazonium xanthate. This type of transformation can also be accomplished with thiourea, albeit in inferior yields. When the reaction is performed with two equivalents of aryl halide, the diaryl sulfide is obtained in good yield (eq 6).[2] The use of sodium sulfide instead of the xanthate led to a mixture of products.[22]

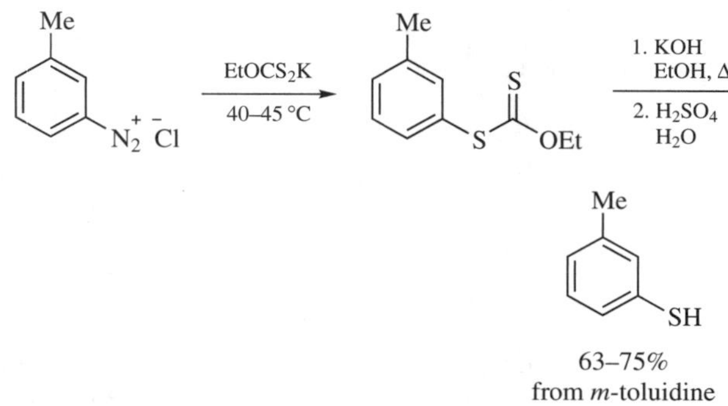

(5)

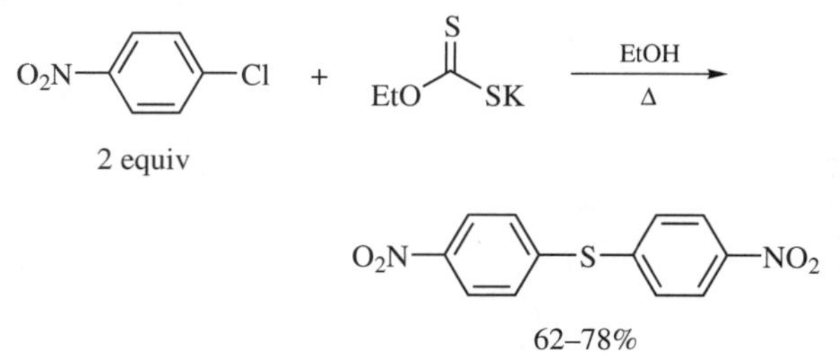

(6)

Potassium *O*-ethyl xanthate can act as a double electrophilic reagent, thus providing a convenient synthetic equivalent of thiophosgene. This type of reactivity, however, is restricted to fairly reactive substrates, such as *o*-phenylenediamine which is readily transformed into 2-mercaptobenzimidazole (eq 7).[23] 1,2-Aminophenols react similarly providing fused oxazole derivatives. Upon alkylation the sulfur moiety can be easily substituted by nitrogen nucleophiles, which opens up a useful entry into 2-amino substituted benzoxazoles and their congeners—an important class of heterocyclic compounds (eq 8).[24]

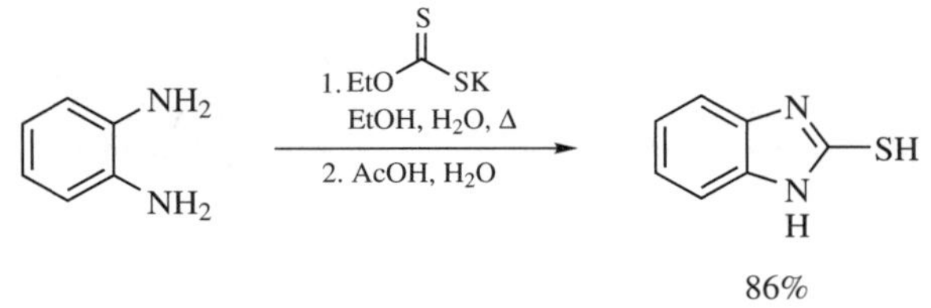

(7)

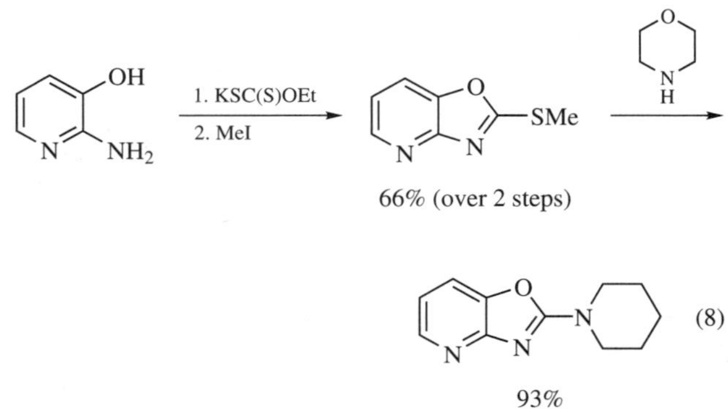

(8)

Vicinal aminoalkyl hydrogensulfates can be converted with potassium ethyl xanthate into thiazolidine-2-thiones.[25] Adducts obtained by radical addition of 2,4-difluorophenacyl xanthate to alkenes (vide infra) can afford both tetrahydro-1-benzothiepine, or thieno[2,3-b]-benzothiopyran systems, depending on the reaction conditions.[26] Upon treatment with cold, concentrated sulfuric acid, the xanthate derivative, obtained from 3-chloro-2-oxo-1,4-dithiane by nucleophilic substitution, undergoes cyclization into 1,4-dithiine; the latter compound can be reductively dimerized into bis(ethylenedithio)tetrathiafulvalene (eq 9).[27]

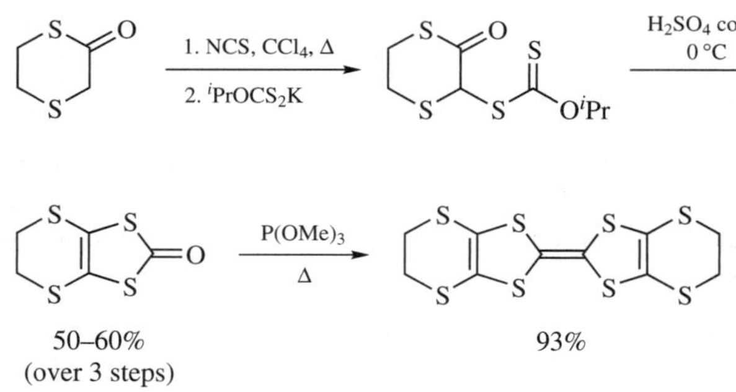

(9)

α-Halo oximes, prepared from the corresponding ketones, can be converted into xanthates with potassium ethyl xanthate. On treatment with zinc chloride these compounds cyclize into *N*-hydroxythiazole-2(3*H*)-thiones—a highly useful class of radical precursors (eq 10).[28,29]

Radical Reactions. *S*-Alkyl *O*-ethyl dithiocarbonates are a synthetically useful, versatile source of alkyl radicals.[30,31,32,33] The principle of radical formation (eq 11) relies on a reversible addition/fragmentation process, known as a group-transfer.[30] In order to secure the regioselectivity of the fragmentation step, •R^1 should be a stabilized radical. This method of radical generation is particularly well suited for the reactions of stabilized, electrophilic, generally less reactive radicals, which can not be generated in a synthetically useful manner using reagents such as tributyltin hydride or *N*-hydroxypyridine-2-thione. The experimental conditions are relatively simple and involve heating of the solution of reactant(s) in an inert solvent (1,2-dichloroethane, toluene, cyclohexane, etc) in the presence of a peroxide initiator (most often dilauroyl peroxide, DLP) under an inert atmosphere. Alternatively, photolytic initiation can be used.

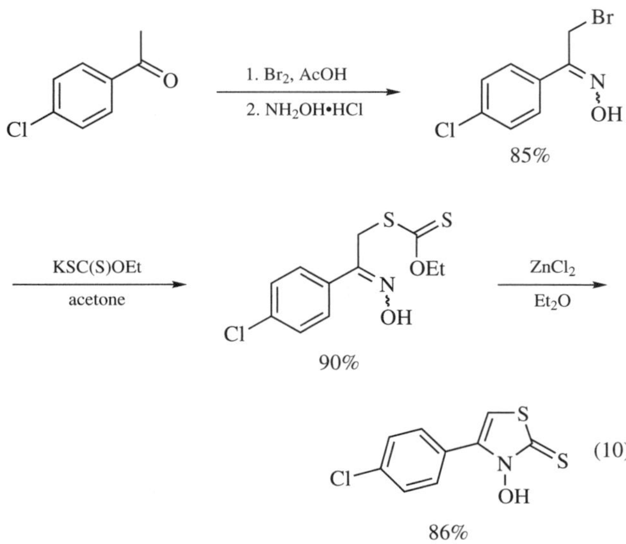

(10)

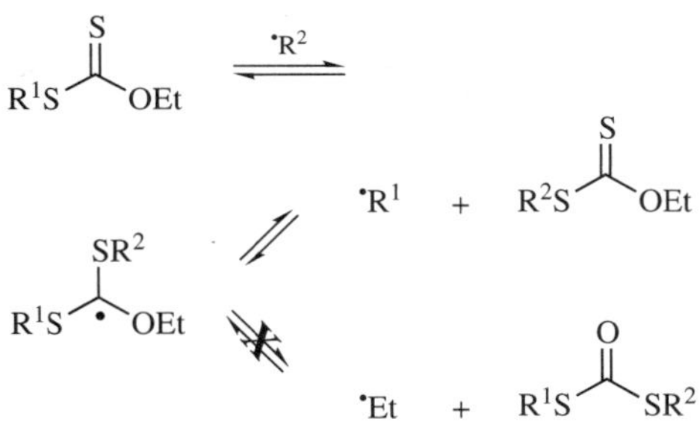

(11)

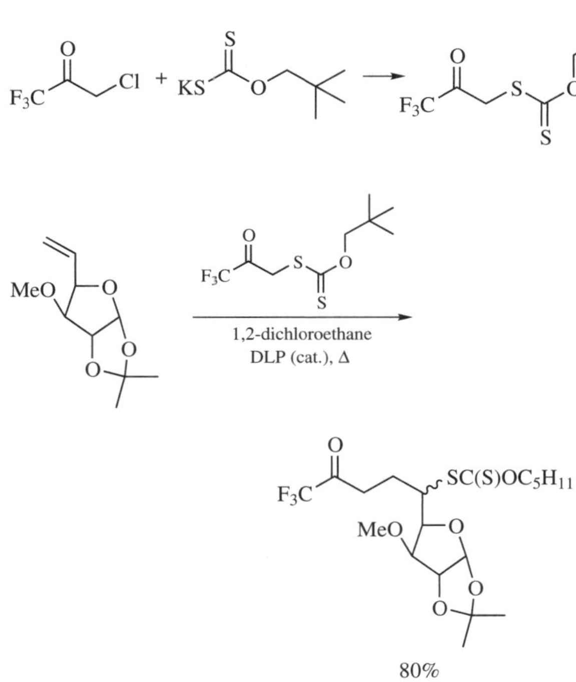

(12)

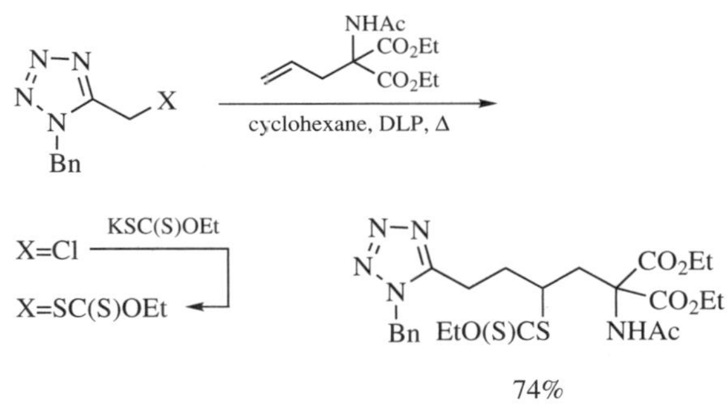

(13)

The xanthate method has been successfully used for carbon-carbon bond forming reactions. Intermolecular additions were mostly performed with electron-rich alkenes (although examples of additions to electron-deficient alkenes have also been reported);[34,35,36] examples of such reactions are provided in eqs 12[37] and 13.[38] It is experimentally convenient that these reactions are performed at relatively high concentrations (0.5 M): due to the degenerate nature of the xanthate transfer step, there is no need for high dilution, or slow addition techniques, as frequently required with other methods of radical generation.

The broad tolerance of functional groups enabled the introduction of an activated carboxylic group into pleuromutilin, without protection of the hydroxyl functionalities; this approach proved suitable for the creation of libraries of pleuromutilin derivatives (eq 14).[39]

Xanthate radical precursors can also be prepared by a hetero-Michael addition of potassium ethyl xanthate to enones. Subsequent intermolecular addition of the corresponding radicals is an efficient approach for the creation of quaternary carbon centres and cyclic systems. The product of the radical addition to vinyl acetate is a protected aldehyde, which may take part in further transformations (eq 15).[40] Intermolecular additions of xanthate derived radicals onto small, strained cycloalkenes have been reported.[41,42]

Intermolecular additions to aromatic rings are generally difficult to achieve under free radical conditions. This transformation has been successfully accomplished with a series of heterocycles by means of the xanthate methodology (eqs 16 and 17).[43] In order for the intermediate adduct radicals to rearomatize, a full equivalent of a peroxide initiator is required.

The absence of competing reactions enables the xanthate method to be used effectively for slower radical processes such as 8-*endo*- (eq 18),[44] or 6-*exo*-cyclizations with the formation of bridged bicyclic systems (eq 19).[45]

Intramolecular addition onto an aromatic nucleus also works reasonably well, and provides an expedient entry into various carbo- and heteropolycyclic aromatic systems. Five- (eq 20),[46] six- (eq 21),[47] and even seven-membered rings (eq 22)[48] can be annulated to the aromatic core; the cyclization precursors, in turn, can also be prepared by radical addition, as shown in the examples.

Domino reactions have been successfully performed with xanthate precursors. Tandem cyclization (6-*exo*/6-*exo*) was exploited in a very efficient manner for the total synthesis of matrine (eq 23).[49] Performing the reaction in refluxing isopropanol allowed for the reductive removal of the xanthate group from the tetracyclic product. A combination of intermolecular addition and cyclization results in the annulation of a cyclopentane ring (eq 24).[50]

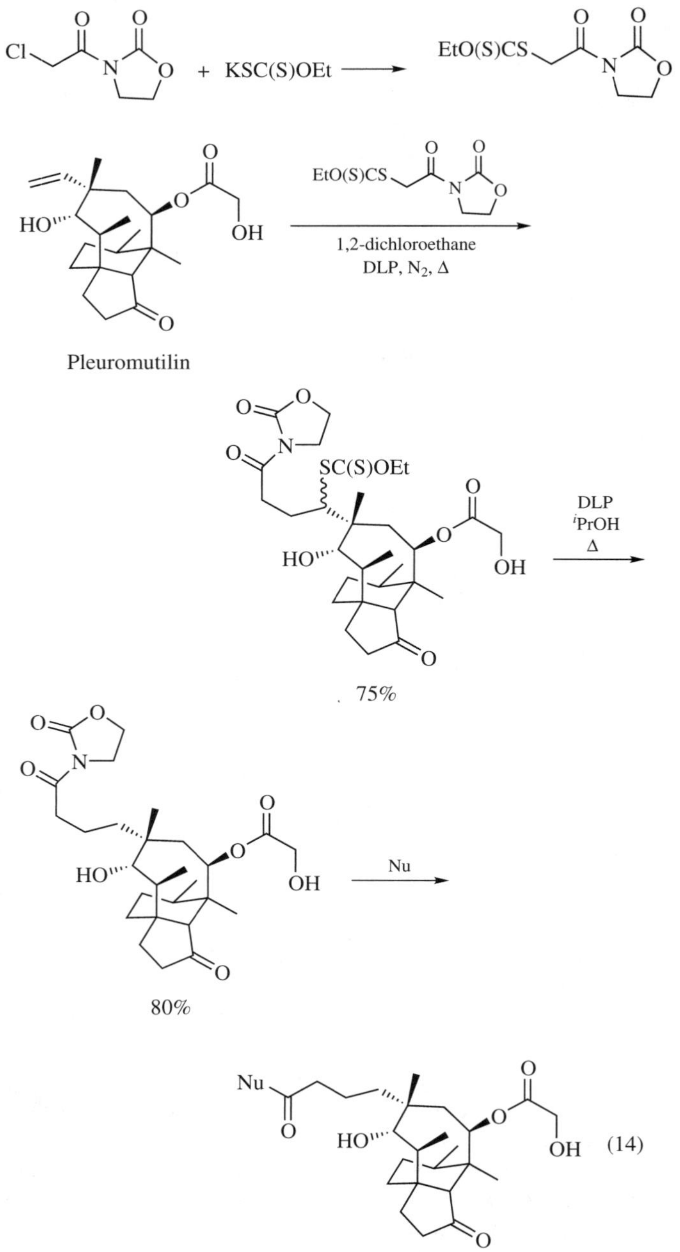

(14)

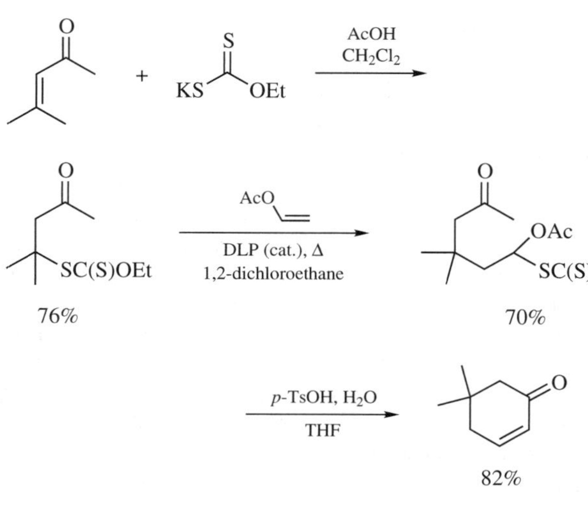

(15)

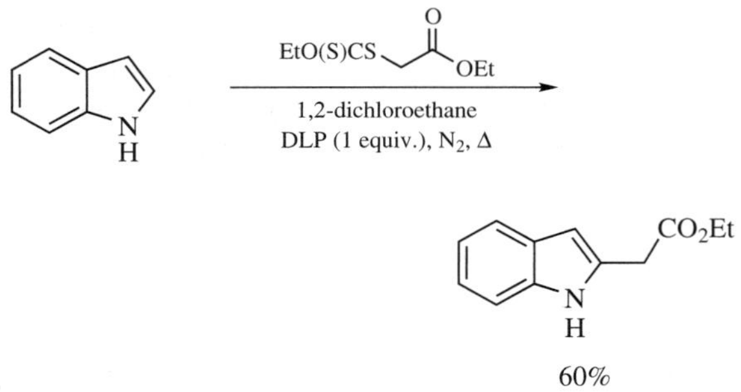

(16)

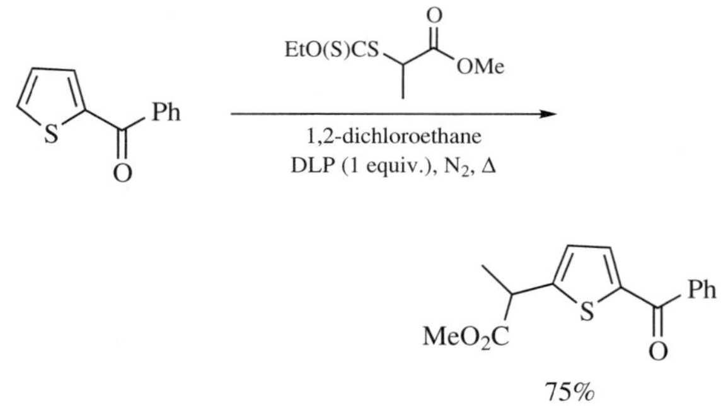

(17)

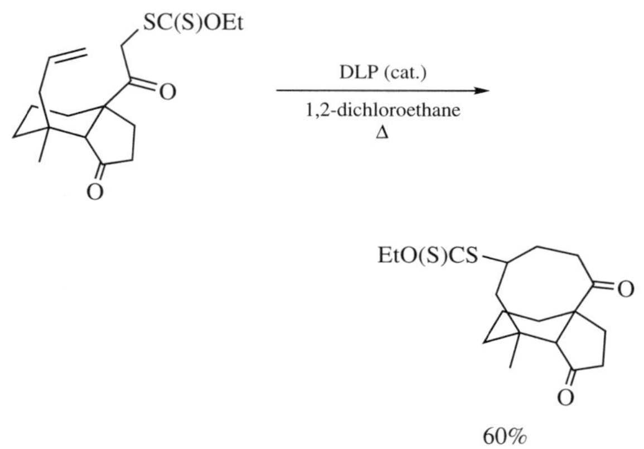

(18)

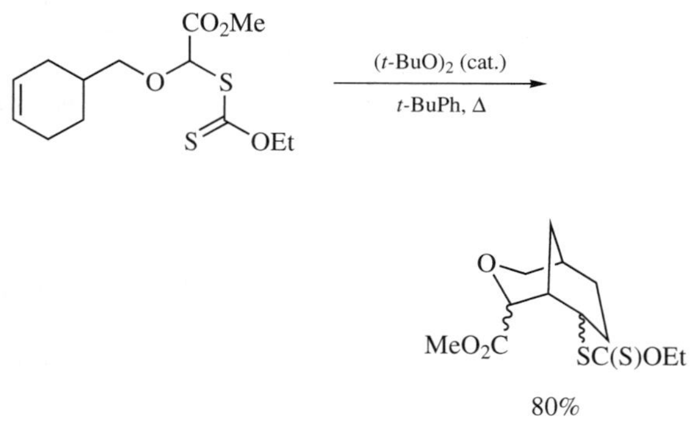

(19)

The introduction of allyl group into organic compounds is a very important reaction, owing to a multitude of methods for subsequent synthetic transformations of this structural unit. Allyltributylstannane is generally considered as the most efficient reagent for the radical allylation.[51] However, the toxicity of organotin compounds prevents its use in medicinal chemistry, or in large scale reactions. The use of xanthates as radical precursors, in combination with allyl sufones, allows for a metal-free, environment friendly allylation procedure. In addition to a simple allyl group, 2-chloro-, or 2-methyl allyl derivatives can also be introduced (eq 25).[52] Domino reactions are also possible, with cyclization preceeding the allylation reaction (eq 26).[52]

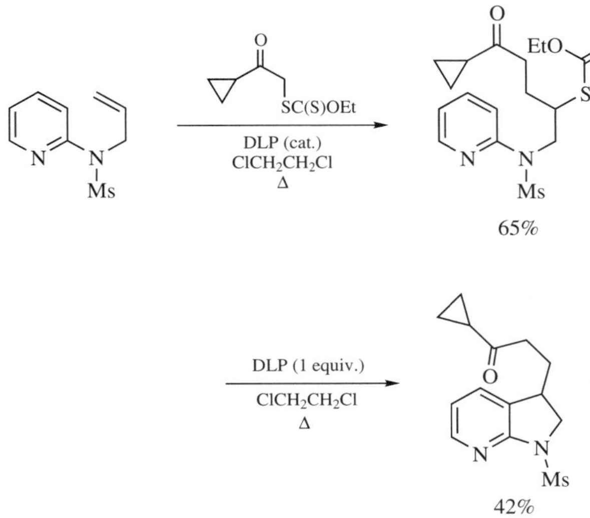

(20)

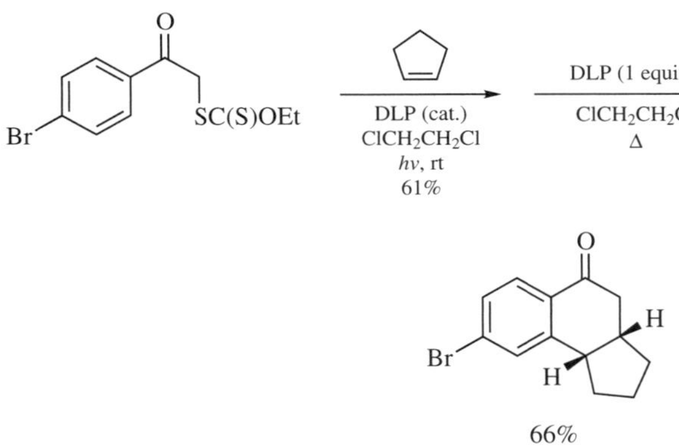

(21)

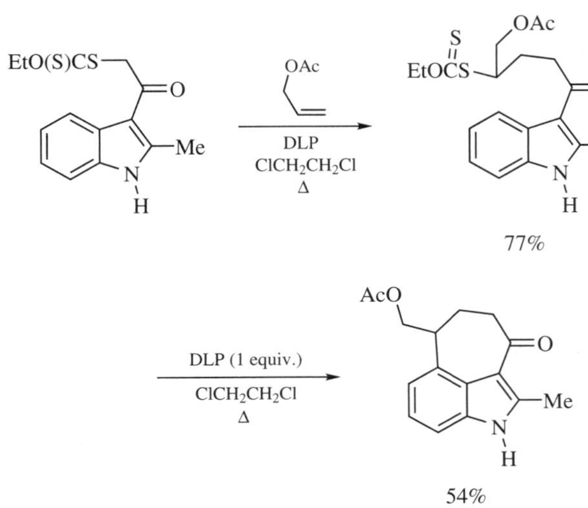

(22)

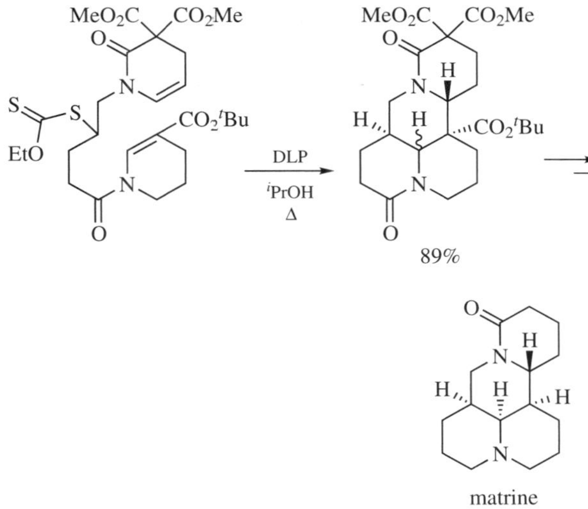

(23)

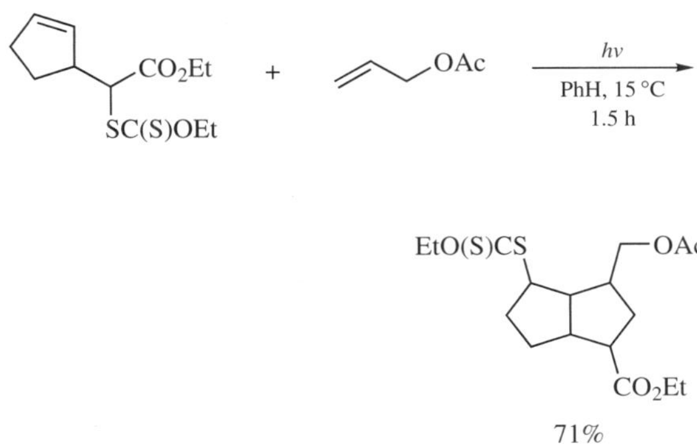

(24)

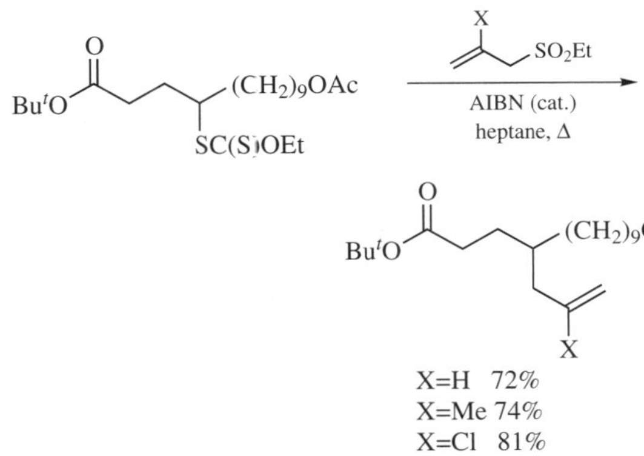

(25)

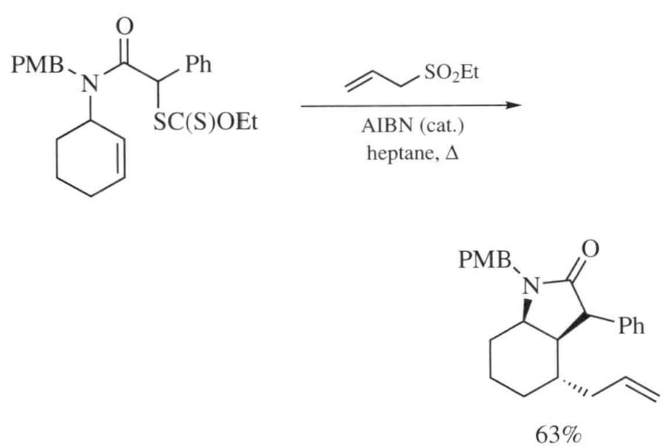

(26)

With 2,2-dichlorophenylsulfone (eq 27),[53] or 2-phenylsulfone (eq 28),[54] the analogous addition/elimination sequence affords vinylated derivatives. One-carbon homologation of xanthates with methanesulfonyl oxime ether, via an analogous tandem reaction, has also been described.[55]

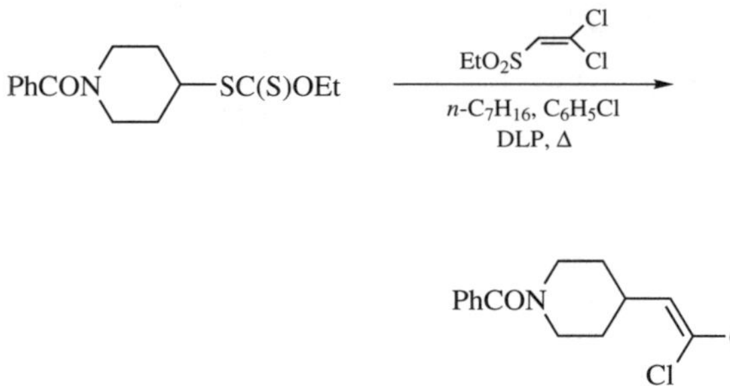

(27)

76%

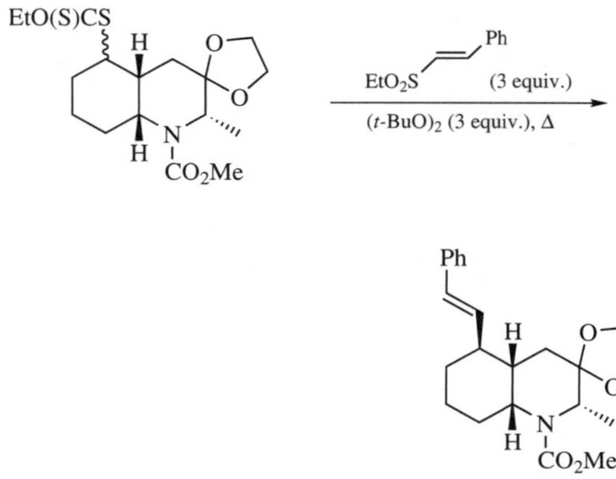

(28)

75–80%

Acyl halides react with potassium ethyl xanthate to give *S*-acyl xanthates which are synthetically useful precursors of acyl radicals (it should be noted that even a slight excess of the potassium *O*-ethyl xanthate induces the decomposition of the initially formed *S*-acyl xanthate, by an ionic chain reaction).[56] Radical generation occurs upon irradiation with visible light, and the resulting acyl radicals can be trapped inter- or intramolecularly (eq 29).[57] Treatment of homoallylic alcohols with phosgene, followed by potassium ethyl xanthate, affords *S*-alkoxycarbonyl xanthates; the cyclization of the corresponding alkoxycarbonyl radicals gives rise to butanolides with a xanthate group in the β-position. These intermediates readily undergo elimination (base-catalyzed or thermally-induced), which allows for an efficient synthesis of unsaturated lactones (eqs 30 and 31).[58]

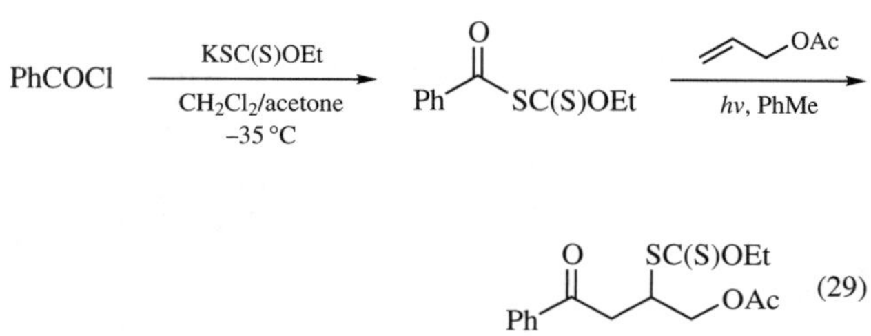

(29)

60%

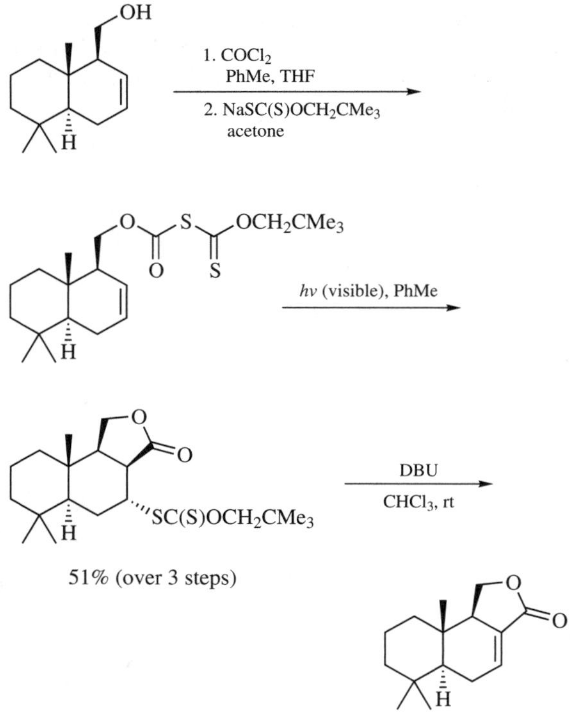

(30)

cinnamolide, 80%

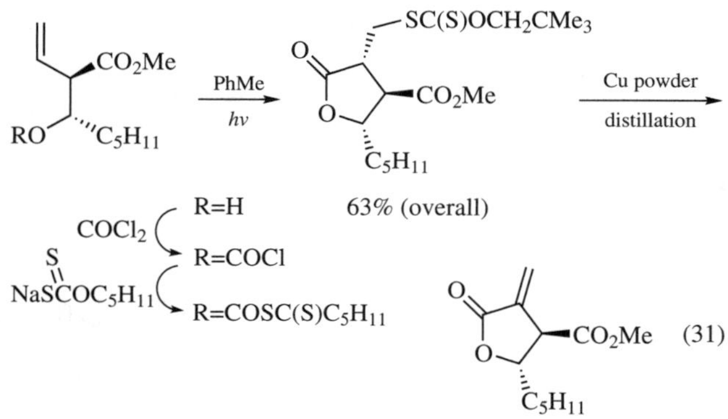

(31)

methylenolactocin, 62%

A significant advantage of the xanthate method of radical generation, as compared to the tributyltin hydride method, is itsnonreductive nature, *i.e.*, the xanthate functionality remains in the final products and is available for further functional transformations. When the reduced products are required, refluxing the *S*-alkyl xanthate in *iso*-propanol, in the presence of DLP, is a convenient, environment friendly alternative for the reductive removal of the xanthate group (eq 32).[59] Hypophosphorous acid/triethylamine, or diethyl phosphite, can also be used as reductants for this purpose and are especially suitable for sensitive substrates (eq 32).[60] On treatment with ethanesulfonyl azide in hot chlorobenzene, alkyl xanthates are smoothly converted into azides (eq 32),[61] a reaction that can be implemented in domino sequences such as carboazidation.[62] Exchange of bromo substituent for xanthate can be accomplished under free radical conditions, in the reaction with ethyl 2-bromo-2-methylpropanoate (eq 32).[63] Finally, the xanthate group can be ionically transformed into the sulfonium salt, which can act as a leaving group (eq 32).[64]

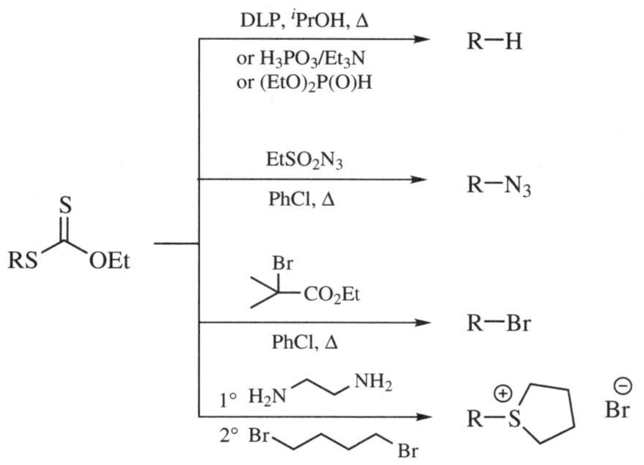

(32)

Reaction of potassium ethyl xanthate, or preferably potassium neopentyl xanthate, with triphenyltin chloride affords triphenylstannyl xanthate (eq 33) — a useful reagent which combines the versatility of tin hydride with the nonreductive character of the xanthate group transfer method (eq 34).[65]

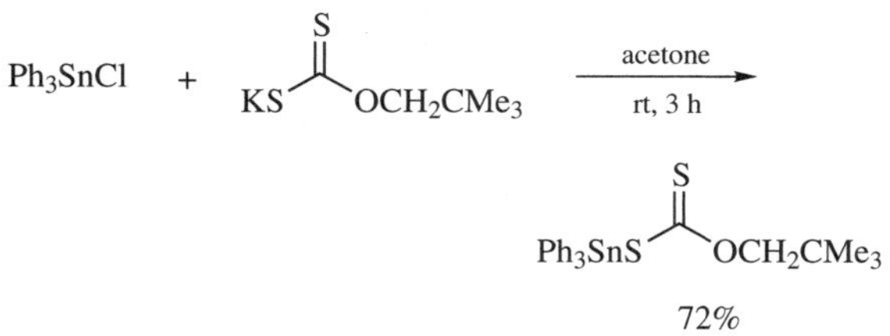

(33)

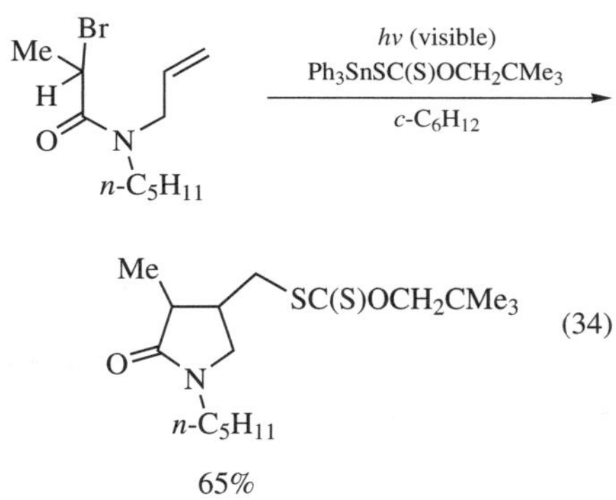

(34)

Polymer-supported Synthesis. Xanthates are excellent initiators for a living polymerization of radicophilic alkenes. The ease of introduction of xanthate functionality into organic compounds contributes to the applicability of this method for the preparation of polymeric supports "tailored" according to the substrate structure and the reaction type. Submitting a bile acid xanthate derivative to the typical conditions in the presence of excess styrene resulted in the formation of a soluble polymer, suitable for the polymer-supported synthesis (eq 35).[66] After the polymerization is complete, the xanthate group can be removed oxidatively, or can be converted into a thiol and used for the placement of a label. While maintaining all the advantages of solution chemistry, this type of reagent can be precipitated with methanol, which allows for a rapid purification of synthetic intermediates. This type of polymeric support has been shown to be superior to Wang resin for free radical additions.[67] The method shows promise for applications in parallel synthesis.

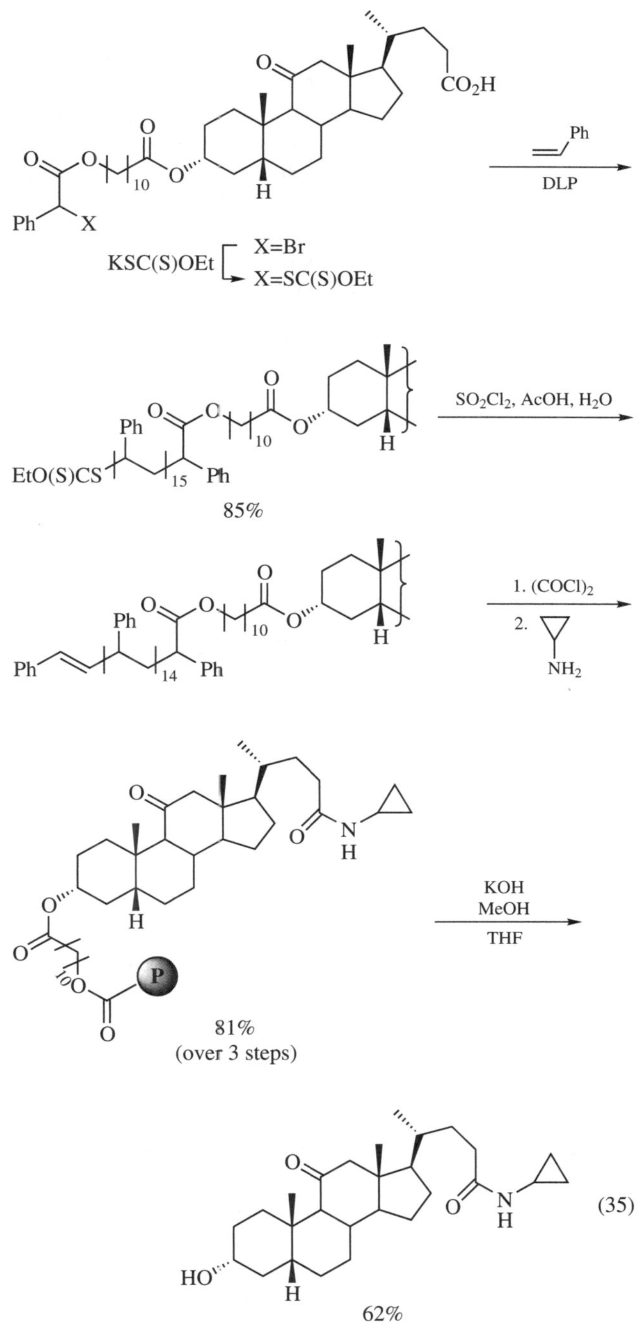

(35)

Glycosylation Reactions. Anomeric xanthates, which can be obtained from the reaction of glycosyl halides with potassium ethyl xanthate, are valuable glycosyl donors,[68,69] as well as precursors of thioglycosides.[70,71] Owing to the excellent nucleophilic properties of xanthates (vide supra), the substitution works well even with anomeric nitrates which, in turn, can be obtained by azidonitration of the corresponding glycals.[72] Notably, attempts to substitute the nitrate group with other sulfur nucleophiles, such as thiophenoxide or thioacetate anion, led only to denitration of the precursors.[73] The anomeric xanthates thus obtained can be efficiently coupled with suitable sugar derivatives to give the corresponding glycosides in good yields (eq 36).[73] The glycosylation procedure also works well with *S*-furanosyl xanthates,[74] which can be obtained from the corresponding furanoses.[75] An alternative to potassium ethylxanthate, for the preparation of anomeric xanthates in both pyranose and furanose series, which relies on the use of dixanthogen and tributylphosphine has been reported.[76] An *N*-acetylneuraminic acid derived xanthate[77] has been used for highly efficient sialylation of various substrates, in the presence of various glycosylation promoters, including dimethyl(methylthio) sulfonium triflate,[78] methylsulfenyl triflate,[79,80] phenylsulfenyl triflate,[81] and iodonium species[82] (eq 37).[81] The xanthate-based sialylation was the method of choice

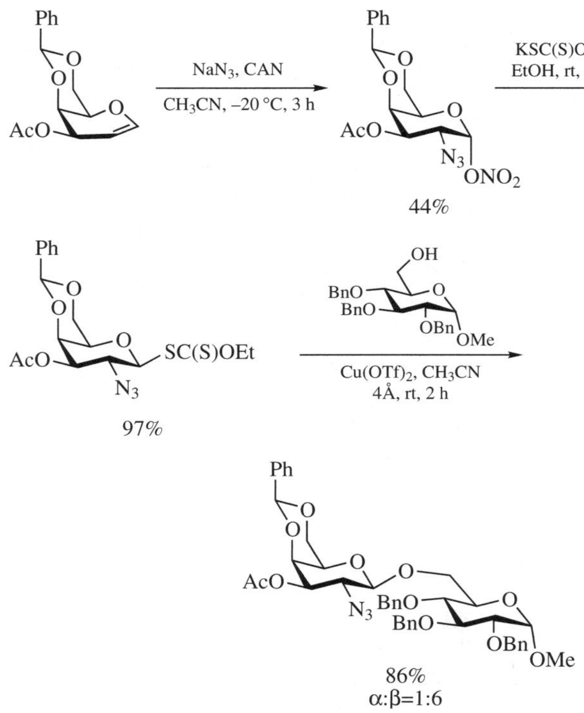

(36)

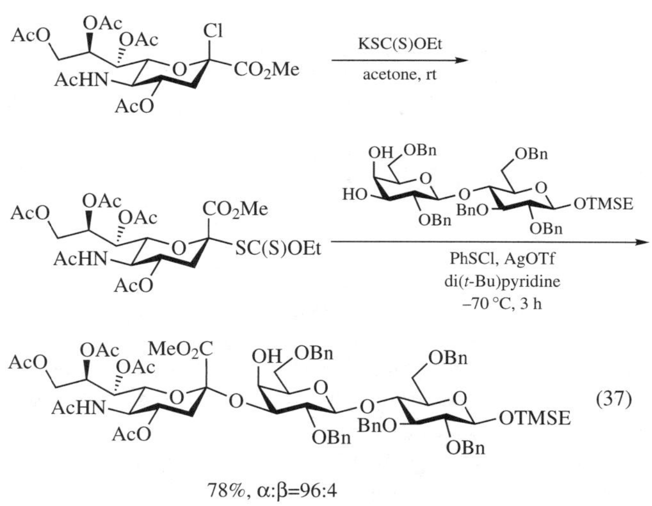

(37)

in the synthesis of rather complex oligosaccharides,[83] and some isotopically labeled gangliosides.[84] Its successful application in the synthesis of sialylated building blocks for glycopeptide libraries has also been reported.[85]

Sigmatropic Rearrangements. Propargylic xanthates, readily obtainable from the corresponding propargylic halides or sulfonates, rearrange on heating into cyclic betaines (eq 38). The latter species react with electron-deficient alkenes by a tandem addition/elimination mechanism, which results in the formation of the products of a formal [3+2] annulation reaction (eq 39).[86]

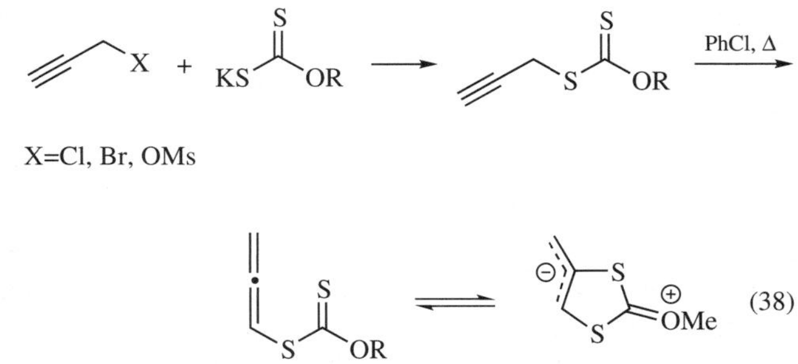

(38)

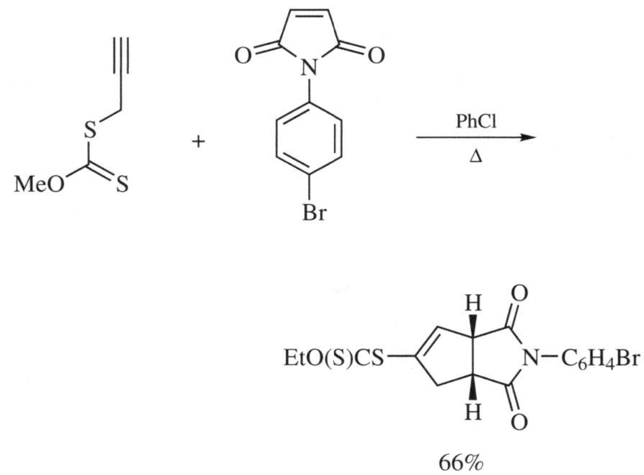

(39)

Propargylic xanthates, which possess a leaving group at the distal propargylic position, give rise to dienes which are highly reactive partners in the Diels-Alder reaction (eq 40).[87] Other reactions of the betaine intermediates include Michael addition,[88] Knoevenagel condensation,[89] and acylation.[90]

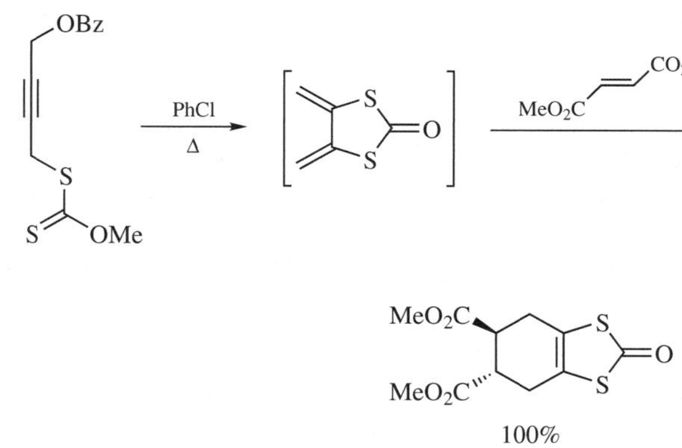

(40)

O-Alkyl *S*-propargyl xanthate derived betaines are capable of deprotonating carboxylic acids and transferring the *O*-alkyl group to the carboxylic anion. Thus, heating a carboxylic acid with propargyl *S*-alkyl xanthate is an excellent method of esterification under essentially neutral conditions, which was successfully applied to a variety of sensitive substrates (eq 41).[91] Esterification with secondary alcohol derived xanthates occurs with the inversion of configuration.

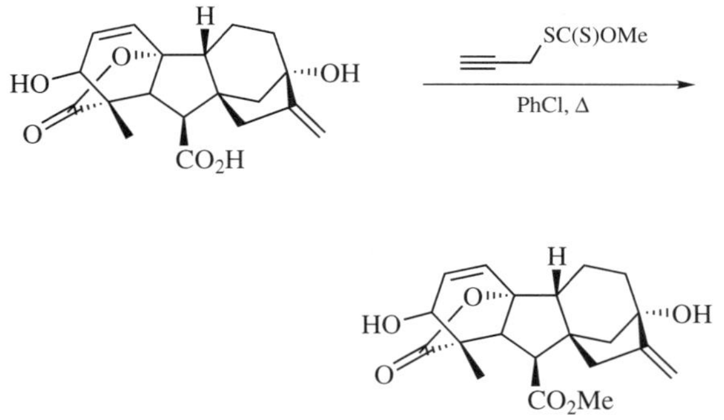

(41)

78%

Related Reagents. Carbon disulfide; sodium sulphide; sodium hydrogen sulphide; thiourea; thiophosgene.

1. Drawert, F.; Reuter K.-H.; Born, F., *Chem. Ber.* **1960**, *93*, 3056–3065.
2. Price, C. C.; Stacy, G. W., *Org. Synth. Coll. Vol.* **1955**, *3*, 667–668.
3. Rao, S. R., *Xanthates and Related Compounds*, Marcel Dekker Inc.: New York, 1971.
4. DeWit, C.C.; Roper, E.E., *J. Am. Chem. Soc.* **1932**, *54*, 444–455.
5. Stueber, D.; Patterson, D.; Mayne, C. L.; Orendt, A. M.; Grant, D. M.; Parry, R. W., *Inorg. Chem.* **2001**, *40*, 1902–1911.
6. Kleinpeter, E.; Pihlaja, K., In *Comprehensive Organic Functional Group Transformations*; Katritzky, A. R.; Meth-Cohn, O.; Rees, C. W.; Gilchrist, T. L., Eds.; Oxford: 1995, Vol. 6, pp 545–552.
7. Degani, I.; Fochi, R., *Synthesis* **1978**, 365–368.
8. Wardell, J. L. In *The Chemistry of the Thiol Group*; Patai, S., Ed.; Wiley-Interscience: New York, 1975, pp 163–269.
9. Schöberl, A.; Wagner, A. In *Houben-Weyl Methoden der Organischen Chemie*; Müller, E., Ed. Georg Thieme Verlag: Stuttgart, 1955, Vol. 9, pp 12–13.
10. Mori, K.; Nakamura, Y., *J. Org. Chem.* **1969**, *34*, 4170–4172.
11. Lehr, H.; Karlan, S.; Goldberg, M. W., *J. Med. Chem.* **1963**, *6*, 136–141.
12. Taguchi, T.; Kiyoshima, Y.; Komori, O.; Mori, M., *Tetrahedron Lett.* **1969**, *10*, 3631–3634.
13. Isola, M.; Ciuffarin, E.; Sagramora, L., *Synthesis* **1976**, 326–329.
14. Djerassi, C.; Gorman, M.; Markley, F. X.; Oldenbergm, E.B., *J. Am. Chem. Soc.* **1955**, *77*, 568–571.
15. Degani, I.; Fochi, R.; Santi, M., *Synthesis* **1977**, 873–874.
16. Degani, I.; Fochi, R.; Regondi, V., *Synthesis* **1979**, 178–181.
17. Offer, J.; Boddy, C. N. C.; Dawson, P. E., *J. Am. Chem. Soc.* **2002**, *124*, 4642–4646.
18. Jaeger, D. A.; Wang, J., *J. Org. Chem.* **1993**, *58*, 6745–6755.
19. Xiao, W.-J.; Alper, H., *J. Org. Chem.* **1999**, *64*, 9646–9652.
20. Tarbell, D. S.; Fukushima, D. K., *Org. Synth. Coll. Vol.* **1955**, *3*, 809–811.
21. *Org. Synth. Coll. Vol.* **1973**, *5*, 1050.
22. Kehrmann, F.; Baner, E., *Chem. Ber.* **1896**, *29*, 2362–2367.
23. VanAllan, J. A.; Deacon, B. D., *Org. Synth. Coll. Vol.* **1963**, *4*, 569–570.
24. Chu-Moyer, M. Y.; Berger, R., *J. Org. Chem.* **1995**, *60*, 5721–5725.
25. Dewey, C. S.; Bafford, R. A., *J. Org. Chem.* **1965**, *30*, 495–500.
26. Boivin, J.; Boutillier, P.; Zard, S. Z., *Tetrahedron Lett.* **1999**, *40*, 2529–2532.
27. Larsen, J.; Lenoir, C., *Org. Synth. Coll. Vol.* **1998**, *9*, 72–76.
28. Hartung, J.; Schwartz, M., *Org. Synth.* **2002**, *79*, 228–235.
29. Barton D. H. R.; Crich, D.; Kretzschmar, G., *J. Chem. Soc. Perkin Trans. 1* **1986**, 39–53.
30. Zard, S. Z., *Radical Reaction in Organic Synthesis*, Oxford University Press: Oxford, 2003, pp 135–175.
31. Zard, S. Z., In *Radicals in Organic Synthesis*; Renaud, P.; Sibi, M. Eds.; Wiley-WCH: Weinheim, 2001, Vol. 1, pp 90–108.
32. Quiclet-Sire, B.; Zard, S. Z., *Phosphorus, Sulfur Silicon Relat. Elem.* **1999**, 137–154.
33. Zard, S. Z., *Angew. Chem. Int. Ed. Engl.* **1997**, *36*, 673–685.
34. Denieul, M. P.; Quiclet-Sire, B.; Zard, S. Z., *Tetrahedron Lett.* **1996**, *37*, 5495–5498.
35. Delduc, P.; Tailhan, C.; Zard, S. Z., *J. Chem. Soc., Chem. Commun.* **1988**, 308–310.
36. Udding, J. H.; Hiemstra, H.; Speckamp, W. N., *J. Org. Chem.* **1994**, *59*, 3721–3725.
37. Denieul, M. P.; Quiclet-Sire, B.; Zard, S. Z., *J. Chem. Soc., Chem. Commun.* **1996**, 2511–2512.
38. Biadatti, T.; Quiclet-Sire, B.; Saunier, J. B.; Zard, S. Z., *Tetrahedron Lett.* **1998**, *39*, 19–22.
39. Bacque, E.; Pautrat, F.; Zard, S. Z., *J. Chem. Soc., Chem. Commun.* **2002**, 2312–2313.
40. Binot, G.; Quiclet-Sire, B.; Saleh, T.; Zard, S. Z., *Synlett* **2003**, 382–386.
41. Ferjancic, Z.; Cekovic, Z.; Saicic, R. N., *Tetrahedron Lett.* **2000**, *41*, 2979–2982.
42. Legrand, N.; Quiclet-Sire, B.; Zard, S. Z., *Tetrahedron Lett.* **2000**, *41*, 9815–9818.
43. Osornio, Y. M.; Cruz-Almanza, R.; Jimenez-Montano, V.; Miranda, L. D., *J. Chem. Soc., Chem. Commun.* **2003**, 2316–2317.
44. Bacque, E.; Pautrat, F.; Zard, S. Z., *Org. Lett.* **2003**, *5*, 325–328.
45. Udding, J. H.; Giesselink, J. P. M.; Hiemstra, H.; Speckamp, W. N., *J. Org. Chem.* **1994**, *59*, 6671–6682.
46. Ly, T. M.; Quiclet-Sire, B.; Sortais, B.; Zard, S. Z., *Tetrahedron Lett.* **1999**, *40*, 2533–2536.
47. Liard, A.; Quiclet-Sire, B.; Saicic, R. N.; Zard, S. Z., *Tetrahedron Lett.* **1997**, *38*, 1759–1762.
48. Kaoudi, T.; Quiclet-Sire, B.; Seguin, S.; Zard, S. Z., *Angew. Chem. Int. Ed. Engl.* **2000**, *39*, 731–733.
49. Boiteau, L.; Boivin, J.; Liard, A.; Quiclet-Sire, B.; Zard, S. Z., *Angew. Chem. Int. Ed. Engl.* **1998**, *37*, 1128–1131.
50. Maslak, V.; Cekovic, Z.; Saicic, R. N., *Synlett* **1998**, 1435–1437.
51. Rosenstein, I. J., In *Radicals in Organic Synthesis*, Renaud, P.; Sibi, M. Eds. Wiley-WCH: Weinheim, 2001, Vol. 1, pp 50–71.
52. Quiclet-Sire, B.; Seguin, S.; Zard, S. Z., *Angew. Chem. Int. Ed. Engl.* **1998**, *37*, 2864–2866.
53. Bertrand, F.; Quiclet-Sire, B.; Zard, S. Z., *Angew. Chem. Int. Ed. Engl.* **1999**, *38*, 1943–1946.
54. Kalai, C.; Tate, E.; Zard, S. Z., *J. Chem. Soc., Chem. Commun.* **2002**, 1430–1431.
55. Kim, S.; Song, H. J.; Choi, T. L.; Yoon, J. Y., *Angew. Chem. Int. Ed. Engl.* **2001**, *40*, 2524–2526.
56. Barton, D. H. R.; George, M. V.; Tomoeda M. *J. Chem. Soc.* **1962**, 1967–1974.
57. Delduc, P.; Tailhan, C.; Zard, S. Z., *J. Chem. Soc., Chem. Commun.* **1988**, 308–310.
58. Forbes, J. E.; Saicic, R. N.; Zard, S. Z., *Tetrahedron* **1999**, *55*, 3791–3802.
59. Liard, A.; Quiclet-Sire, B.; Zard, S. Z., *Tetrahedron Lett.* **1996**, *37*, 5877–5880.
60. Boivin, J.; Jrad, R.; Juge, S.; Nguyen, V. T., *Org. Lett.* **2003**, *5*, 1645–1648.
61. Ollivier, C.; Renaud, P., *J. Am. Chem. Soc.* **2000**, *122*, 6496–6497.
62. Renaud, P.; Ollivier, C ; Panchaud, P., *Angew. Chem. Int. Ed. Engl.* **2002**, *41*, 3460–3462.

63. Barbier, F.; Pautrat, F.; Quiclet-Sire, B.; Zard, S. Z., *Synlett* **2002**, 811–813.
64. Boivin, J.; Pothier, J.; Ramos, L.; Zard, S. Z., *Tetrahedron Lett.* **1999**, *40*, 9239–9240.
65. Boivin, J.; Camara, J.; Zard, S. Z., *J. Am. Chem. Soc.* **1992**, *114*, 7909–7910.
66. Quiclet-Sire, B.; Wilczewska, A.; Zard, S. Z., *Tetrahedron Lett.* **2000**, *41*, 5673–5677.
67. Dublanchet, A. C.; Lusinchi, M.; Zard, S. Z., *Tetrahedron* **2002**, *58*, 5715–5721.
68. Sinaÿ, P., *Pure. Appl. Chem.* **1991**, *63*, 519–528.
69. Sinaÿ P., *Phosphorus, Sulfur, Silicon Relat. Elem.* **1994**, *95&96*, 89–102.
70. Sakata, M.; Haga, M.; Tejima, S., *Carbohyd. Res.* **1970**, *13*, 379–390.
71. Ponpipom, M. M.; Bugianesi, R. L.; Blake, T. J., *J. Med. Chem.* **1987**, *30*, 705–710.
72. Marra, A.; Shun, L. K. S.; Gauffeny, F.; Sinaÿ, P., *Synlett* 1990, 445–448.
73. Marra, A.; Gauffeny, F.; Sinaÿ, P., *Tetrahedron* **1991**, *47*, 5149–5160.
74. Bogusiak, J.; Szeja, W., *Carbohyd. Res.* **1996**, *295*, 235–243.
75. Bogusiak, J.; Wandzik, I.; Szeja, W., *Carbohydrate Letters* **1996**, *1*, 411–416.
76. Gueyrard, D.; Tatibouet, A.; Gareau, Y.; Rollin, P., *Org. Lett.* **1999**, *1*, 521–522.
77. Marra, A.; Sinaÿ, P., *Carbohyd. Res.* **1989**, *187*, 35–42.
78. Marra, A.; Sinaÿ, P., *Carbohyd. Res.* **1990**, *195*, 303–308.
79. Birberg, W.; Lönn, H., *Tetrahedron Lett.* **1991**, *32*, 7453–7456.
80. Lönn, H.; Stenvall, K., *Tetrahedron Lett.* **1992**, *33*, 115–116.
81. Martichonok, V.; Whitesides, G. M., *J. Org. Chem.* **1996**, *61*, 1702–1706.
82. Meijer, A.; Ellervik, U., *J. Org. Chem.* **2002**, *67*, 7407–7412.
83. Tietze, L. F.; Janssen, C. O.; Gewert, J. A., *Eur. J. Org. Chem.* **1998**, 1887–1894.
84. Tietze, L. F.; Gretzke, D, *Eur. J. Org. Chem.* **1998**, 1895–1899.
85. Halkes, K. M.; St. Hilaire, P. M.; Jansson, A. M.; Gotfredsen, C. H.; Meldal, M., *J. Chem. Soc., Perkin Trans. 1* **2000**, 2127–2133.
86. Boivin, J.; Tailhan, C.; Zard, S. Z., *J. Am. Chem. Soc.* **1991**, *113*, 5874–5876.
87. Boivin, J.; Tailhan, C.; Zard, S. Z., *Tetrahedron Lett.* **1992**, *33*, 7853–7856.
88. Poelert, M.; Roger, W.; Zard, S. Z., *J. Chem. Soc., Chem. Commun.* **1996**, 743–744.
89. Boivin, J.; Henriet, E. B.; Zard, S. Z., *Tetrahedron Lett.* **1995**, *36*, 5171–5174.
90. Poelert, M.; Zard, S. Z., *Synlett* **1995**, 325–326.
91. Boivin, J.; Henriet, E.; Zard, S. Z., *J. Am. Chem. Soc.* **1994**, *116*, 9739–9740.

Radomir N. Saičić
University of Belgrade, Serbia and Montenegro

N-[(Propargyloxycarbonyl)oxy] maleimide-styrene Copolymer (Proc-resin)

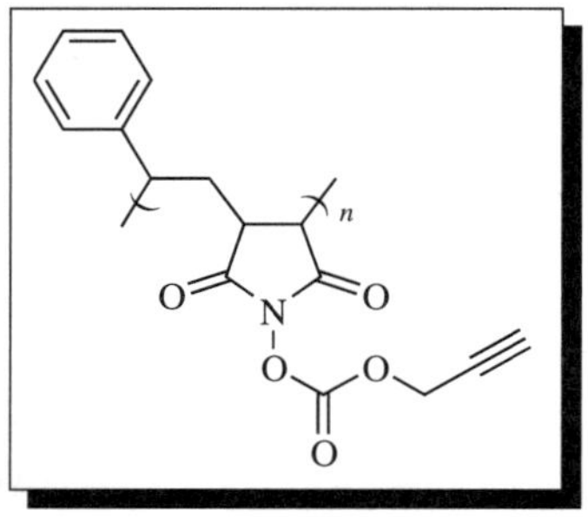

$C_{16}H_{15}NO_5$ (MW *n* (301.29))

(reagent used for the *N*-protection of amines and amino acids with the propargyloxycarbonyl (Proc or Poc) group)

Alternate Name: Proc-P-OSu.
Solubility: soluble in DMSO and DMF; partially soluble in acetone, CH_2Cl_2, THF, CH_3CN, and dioxane; insoluble in toluene, water, $CHCl_3$, AcOEt, and hexane.
Form Supplied in: white powder.
Preparative Methods: prepared by reaction of alternating *N*-hydroxymaleimide-styrene copolymer (P-HOSu)[1] with propargyl chloroformate in the presence of triethylamine, in dichloromethane at room temperature, followed by precipitation after addition of toluene.[2]
Handling, Storage, and Precautions: very stable, not hygroscopic, remains active after extended periods of time of storage under ambient conditions.

Introduction. The propargyloxycarbonyl (Proc or Poc) function has recently been reported as an alternative to the allyloxycarbonyl (Alloc or Aloc) group for the protection of amines.[3] The latter group is specially useful in peptide synthesis. Similar to Alloc, the Proc group can also resist acidic and mild basic conditions, but can be cleaved under neutral conditions with benzylammonium tetrathiomolybdate under ultrasonic irradiation,[3a] with Amberlite-bound tetrathiomolybdate,[3d] $Co_2(CO)_8$,[3b] and with various other metal reagents.[3c]

The Proc protection of the amino group has been achieved by reaction with propargyl chloroformate,[3a–c] a moisture-sensitive reagent, which can contain traces of very toxic phosgene,[4] which can induce the formation of protected dipeptide by-products of amino acid substrates.[5] Therefore, more stable reagents for the introduction of the Proc group have been developed, such as propargyl pentafluorophenyl carbonate (Proc-OPfp).[6] The *N*-[(propargyloxycarbonyl)oxy]maleimide-styrene copolymer (Proc-P-OSu)[2] can be considered as a Proc-containing copolymeric *N*-hydroxysuccinimide, thus representing a polymeric version of Proc-OSu, a compound closely related to the known Alloc-protecting reagent (Alloc-OSu).[7]

***N*-Protection of Amines and Amino Acids.**[2] Primary amines and amino acid esters reacted with a suspension of Proc-P-OSu (**1**) (1 equiv) in dicholoromethane as a solvent and at room temperature during 1 day, affording the corresponding *N*-Proc-protected compounds (eq 1). When the *N*-protection reaction was carried out on free α-amino acids, a solution of the amino acid and K_2CO_3 in water was added to the suspension of **1** in dichloromethane. When the protection reaction was completed, the polymeric *N*-hydroxysuccinimide (P-HOSu) (**2**) residue was removed by filtration,[1] thus allowing its recycling for the preparation of new Proc-P-OSu.[2]

The resulting yields of the Proc-protected amino acids are usually lower with Proc-P-OSu than with nonpolymeric Proc-OSu,[2] although they can be increased when the Proc-protection takes place on the amino acid ester instead of the free amino acid (eq 2). However, the use of Proc-P-OSu allows the easy separation of the polymeric *N*-hydroxysuccinimide residue from the reaction mixture by filtration.

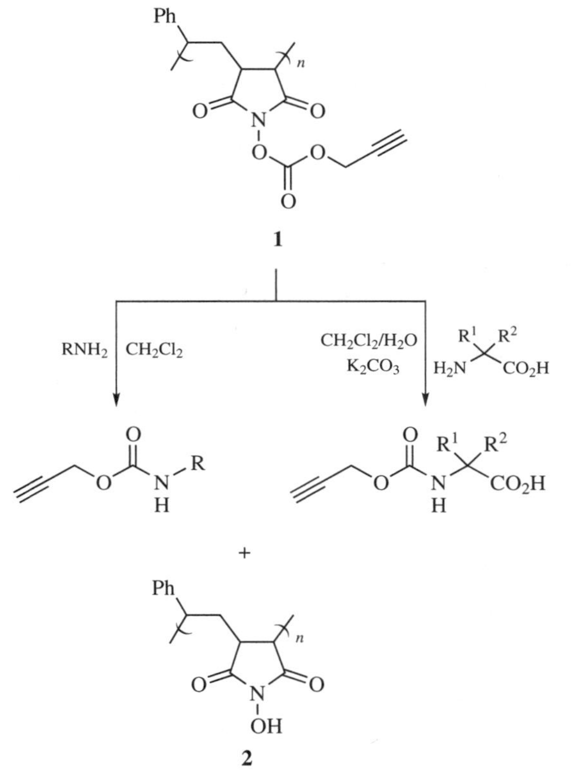

(1)

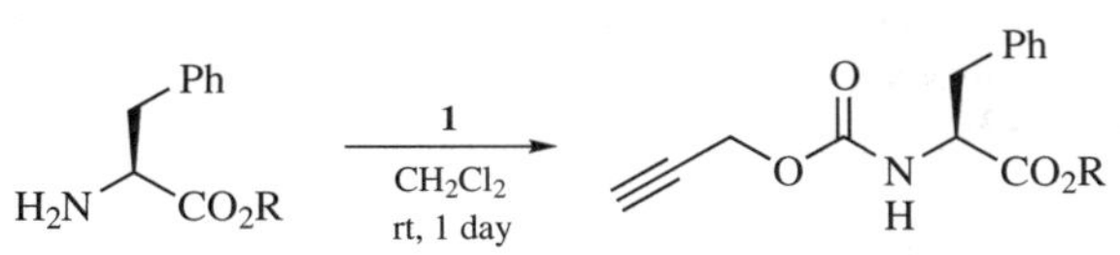

(2)

R = H, 71%
R = Bn, 87%

1. Chinchilla, R.; Dodsworth, D. J.; Nájera, C.; Soriano, J. M., *Tetrahedron Lett.* **2001**, *42*, 4487.
2. Chinchilla, R.; Dodsworth, D. J.; Nájera, C.; Soriano, J. M., *Synlett* **2003**, 809.
3. (a) Sinha, S.; Ilankumaran, P.; Chandrasekaran, S., *Tetrahedron Lett.* **1999**, *40*, 771. (b) Fukase, Y.; Fukase, K.; Kusumoto, S., *Tetrahedron Lett.* **1999**, *40*, 1169. (c) Fukase, Y.; Fukase, K.; Kusumoto, S., *Pept. Sci.* **1999**, *36*, 93. (d) Bhat, R. G.; Sinha, S.; Chandrasekaran, S., *Chem. Commun.* **2002**, 812.
4. Herrinton, P. M., In *Encyclopedia of Reagents for Organic Synthesis*; Paquette, L. A., Ed.; Wiley: Chichester, **1995**, p 95.
5. Albericio, F., *Biopolymers (Peptide Science)* **2000**, *55*, 123.
6. Bhat, R. G.; Kérourédan, E.; Porhiel, E.; Chandrasekaran, S., *Tetrahedron Lett.* **2002**, *43*, 2467.
7. Blaakmeer, J.; Tijsse-Klasen, T.; Tesser, G. I., *Int. J. Pept. Protein Res.* **1991**, *37*, 556.

Carmen Nájera & Rafael Chinchilla
Universidad de Alicante, Alicante, Spain

Pyridine-2-thiol

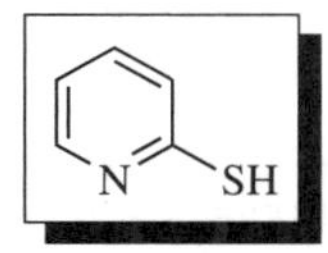

[2637-34-5] C_5H_5NS (MW 111.18)

(reagent for the synthesis of thiolesters which acylate nucleophiles[1] or yield ketones with organometallic reagents;[2] utilized in protection and deprotection chemistry;[3,4] glycoside coupling agent;[5,6] component of a selective reducing agent;[7] precursor of electrophilic sulfur reagents;[8,9] applied to dienophile synthesis[10])

Physical Data: mp 128–130 °C.
Solubility: sol in most organic solvents; soluble in aqueous base.
Form Supplied in: crystalline solid, available from various suppliers.
Analysis of Reagent Purity: mp measurement is indicative of purity; values below 120 °C suggest an impure sample.
Purification: recrystallization from benzene.
Handling, Storage, and Precautions: demonstrates good stability if stored in a well sealed vial under an inert atmosphere at room temperature. On continued exposure to air the reagent becomes contaminated with moisture and disulfide. 2-Pyridinethiol is irritating to eyes and skin and imparts a stench.

Original Commentary

Edward J. Adams
E. I. DuPont de Nemours and Co., Newark, DE, USA

Thiolester Synthesis and Utility. Thiolesters represent a class of acylating agents.

Synthesis

Pyridine-based thiolester generation utilizes carboxylic acids and 2-pyridinethiol under 1,3-Dicyclohexylcarbodiimide coupling conditions (EtOAc, 0 °C).[11] Quantitative conversions of the carboxylic acid to the 2-pyridyl thiolester involve premixing 2-pyridinethiol and Phosgene before introduction of the acid.[12] A mixed anhydride method (Trichloroacetyl Chloride, Triethylamine) of acid activation successfully generates the thiolester[13] as does use of an acyl chloride.[14] Other methods of acid activation provide routes to the 2-pyridyl thiolesters.[15] Alternate methods are available from 2,2′-Dipyridyl Disulfide.[16] 2-Pyridyl thiolesters may also be generated from ketenes[17] using the tin(II) sulfides obtained from 2-pyridinethiol.[18] This technique provides a tin(II) enolate which reacts with aldehydes with *syn* selectivity and provides a 3-hydroxy thiolester.

Acylation of Nucleophiles

2-Pyridyl thiolesters and 2-pyridyl esters provide a coupling technique in peptide synthesis.[11a] Macrolactonization with a thiolester has been demonstrated.[16] Alkylation of the 2-pyridylthiolester with nonthiophilic agents such as Triethyloxonium Tetrafluoroborate provides successful acylation of phenols, amines, and carboxylic acids.[1] Using 2 equiv of methylenetriphenylphosphorane, 2-pyridyl thiolesters provide a route to stabilized ylides.[11d]

Ketone Generation from a Carboxylic Acid Derivative

An historical quest for organic chemists has been the generation of ketones from carboxylic acid derivatives. Carbon nucleophiles select the intermediate ketone in preference to the acid derivative, resulting in alcohol products. The stabilized intermediate provided by Grignard addition to a 2-pyridyl thiolester provides isolation of ketones from a carboxylic acid derivative.[2] Successful application of this technique to complex molecules requires a tolerance for other functional groups. Ketone formation has been accomplished in the presence of imide,[19] *t*-butoxycarbonyl, and ester[11b,13] moieties. A comparison of acid activation agents used for ketone formation from Grignard reagents examined 2-pyridyl thiolesters.[20] Investigations of ketone formation using alkynylmetals suggest that *N*-alkoxyamides may be better substrates for these organometallic reagents[21] (see N,O-Dimethylhydroxylamine).

Protecting Group Chemistry. 2-Pyridinethiol provides reagents for facile protection of amines as the benzyloxycarbonyl or *t*-butoxycarbonyl protected species.[3,22] 2-Pyridinethiol protects the imide function of uridine and guanosine units in nucleotide synthesis and is displaced by nucleophiles such as amines.[23] Deprotection of thiols[24] and the azetidinone nitrogen[4] is accomplished with 2-pyridinethiol as the reagent of choice.

Glycoside Synthesis. Thioglycosides generated from 2-pyridinethiol provide important synthetic intermediates in glycoside chemistry. The synthesis of the thioglycosides utilizes an acid[5,6,25] or Lewis acid catalyst.[26] Silyl enol ethers react with the 2-pyridylthioglycosides generating *C*-glycosides. Using Silver(I) Trifluoromethanesulfonate (silver triflate) activation, the electrophilic glycoside center accomplishes electrophilic aromatic substitution of electron-rich aromatic rings.[5] *N*-Alkylation of the 2-pyridyl thioglycoside provides the activation for facile glycosidic linkage formation.[6]

Selective Reducing Agent. 2-Pyridinethiol, ***Tin(II) Chloride***, and Et_3N produce a complex that provides selective reduction of azides to primary amines, primary and secondary nitro alkanes to oximes, and tertiary nitro alkanes and nitro aromatics to hydroxylamines.[7] This complex tolerates ketone, sulfoxide, sulfone, nitrile, and ester functional groups. Applications to macrolactamization have been demonstrated.[27]

Synthesis of Electrophilic Agents. The chloramine generated from ***Morpholine*** reacts with 2-pyridinethiol to generate an electrophilic sulfenylation agent.[8] This reagent provides a mixed disulfide containing a 2-pyridyl unit (see also 2-Pyridinesulfonyl Chloride and 2-Pyridinesulfenyl Bromide) which is *N*-alkylated by nonthiophilic agents such as Triethyloxonium Tetrafluoroborate. The resulting salt is readily displaced to generate mixed disulfides (from thiols, thioamides, dithiocarbamates, thiocyanate), sulfenamides (from amines), or sulfides (from β-dicarbonyl nucleophiles).

2-Pyridinethiol is transformed into methyl 2-pyridinesulfinate using MeOH and N-Bromosuccinimide. Enolate sulfinylation with this reagent, followed by thermal elimination, provides a formal dehydrogenation to an enone with (*E*) selectivity. Generating this reagent from 2-pyridinethiol provides an internal buffering during the elimination step.[9]

Application to Diels–Alder Chemistry. Conjugate addition of 2-pyridinethiol to menthol propiolate results in a *cis*-alkene which undergoes a diastereoselective oxidation of sulfur using m-Chloroperbenzoic Acid to yield a highly reactive dienophile. Subsequent application to cycloaddition chemistry has been demonstrated.[10,28]

Other Applications. 2-Pyridinethiol photochemically adds to alkenes to yield a bis(2-pyridylethyl) disulfide.[29] 2-Pyridinesulfenylation of 2,5-piperazinediones provides an opportunity for carbon–carbon bond formation using silver triflate activation in the presence of a silyl enol ether.[30] Under nickel catalysis, 2-pyridinethiol provided a 2-pyridyl unit in an aryl–aryl coupling with subsequent loss of sulfur.[31]

First Update

Troels Skrydstrup & Karl B. Lindsay
Aarhus University, Aarhus, Denmark

Thiolester Synthesis and Utility. The preparation of α,β-unsaturated 2-pyridinethiol esters is particularly challenging due to competing conjugate addition of the soft thiolate nucleophile. In these cases, a thiouronium-based reagent, *S*-(2-pyridinyl)-1,1,3,3-tetramethylthiouronium hexafluorophosphate, has been developed for the synthesis of the thioester under non-nucleophilic conditions.[32]

Stereoselective Synthesis of β-Lactams and β-Lactones. The short enantioselective synthesis of highly functionalized β-lactams is highly relevant in medicinal chemistry. 2-Pyridyl thioesters may be reacted with $TiCl_4$ or $SnCl_4$ in the presence of NEt_3 to give titanium and tin enolates, respectively.[33] These enolates may then be condensed with imines to give β-lactams.[33] The use of chiral *N*-substituted imines imparts excellent diastereoselectivity.[34] Similarly, $BCl_3{\cdot}SMe_2$ generated boron enolates give β-lactams, and in these cases chiral amino alcohols [e.g. (1*S*,2*R*)-*N*-methyl ephedrine] can be employed to impart enantioselectivity via ligation to the boron-enolate intermediate (eq 1).[35] Other Lewis acids such as $AlBr_3$ or $EtAlCl_2$ are also effective (eq 2).[36] Silyl ketene thioacetals may be used in place of enolates with various Lewis acids,[37] and a catalytic approach employing $Yb(OTf)_3$ is available.[38]

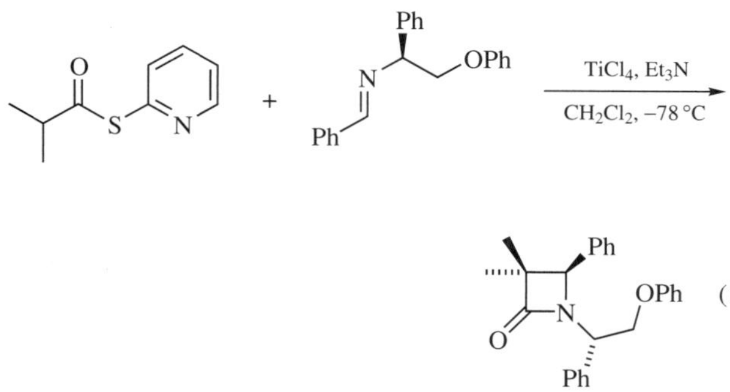

80%
d:r = 96:4
(1)

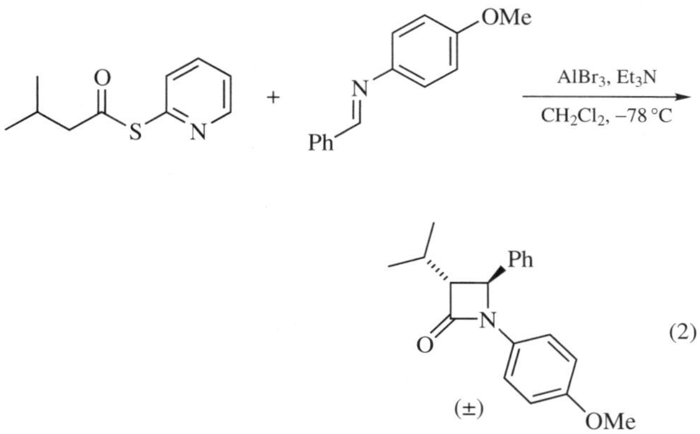

81%
trans:cis = 95:5
(2)

O-Silyl ketene-*O,S*-acetals derived from 2-pyridyl thioesters readily undergo $ZnCl_2$-promoted [2+2] cycloadditions (known also as a tandem Mukaiyama aldol lactonization) with aldehydes to provide *trans*-β-lactones with good diastereomeric control (eq 3).[39,40] Best results were obtained with aldehydes free of an α-alkyl substituent. The Lewis acid $SnCl_4$ gives the complementary *cis*-lactones.[41]

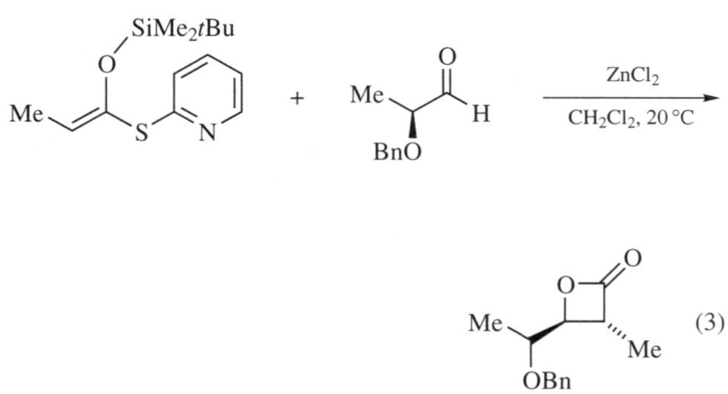

69%
trans:cis > 19:1
syn:anti = 20:1
(3)

Synthesis of *C*-Glycosides. *C*-Glycosides represent stable mimics of *O*-glycosides in which the interglycosidic linkage is replaced by a methylene group. The introduction of 2-pyridinethiol at the anomeric position of monosaccharides followed by oxidation to their corresponding sulfone generates a class of exceptional *C*-glycoside donors.[42,43] Coupling of such sulfones with aldehydes in the presence of samarium diiodide at room temperature affords 1,2-*trans*-*C*-glycosides via a glycosyl samarium(III) intermediate (eq 4).[44,45] In this way, a variety of *C*-glycosides has been prepared.

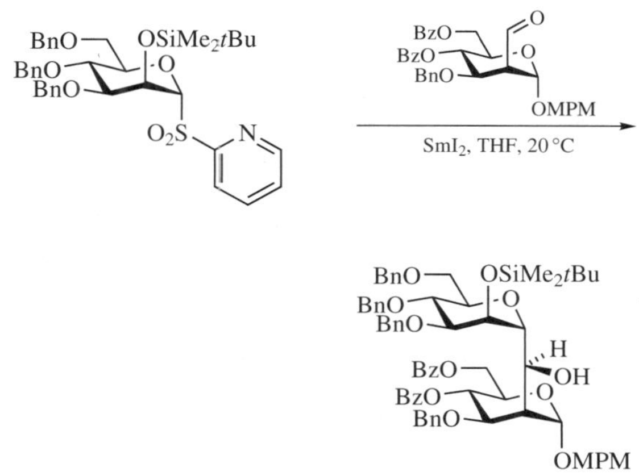

85%
(4)

Synthesis of 1,2-*cis*-*C*-glycosides was achieved via a radical cyclization reaction with glycosyl pyridyl sulfones bearing a silicon-tethered alkene or alkyne at the C2-hydroxyl group (eq 5).[46] The pyridyl group is essential for these mild carbon-carbon bond forming reactions as the corresponding glycosyl phenyl sulfones are inert under the same reaction conditions. Other heterocycles tested were less productive.

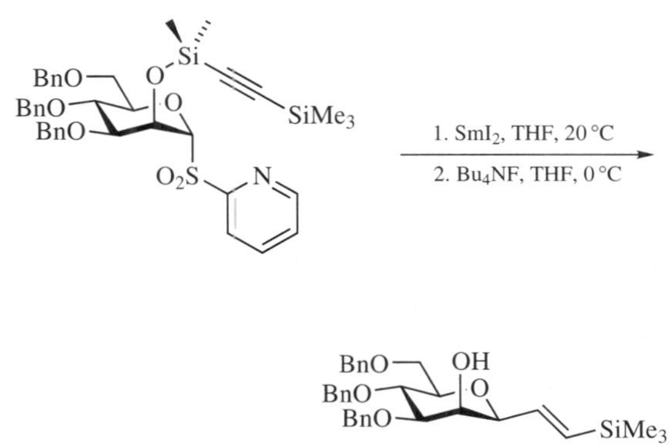

61% (*E*:*Z* = 10:1)
(5)

Benzyloxyalkylations. A variant to the *C*-glycosylation protocol is represented by the benzyloxyalkylation of carbonyls with benzyloxyalkyl pyridyl sulfones.[47] Such compounds are prepared by the treatment of 2-pyridinethiol with benzyloxymethyl chloride in the presence of base. Oxidation to the sulfone and *C*-alkylation of this compound with BuLi and an alkyl halide provided a series of *C*-substituted benzyloxymethyl pyridyl sulfones. Treatment of these compounds with carbonyls and samarium diiodide afforded instantaneously the monoprotected 1,2-diol in good yields (eq 6).

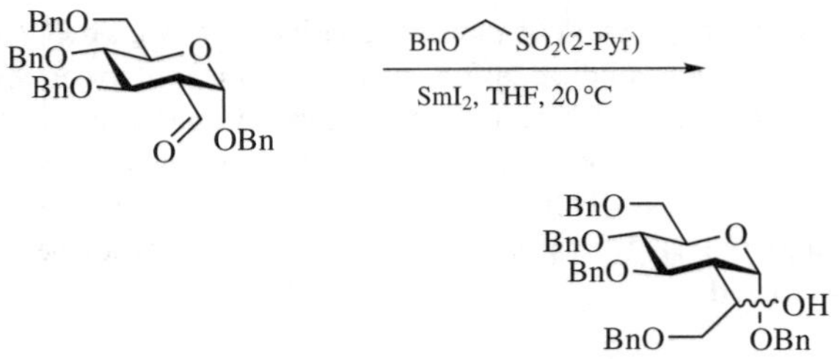

(6)

86% (d:r = 1:1)

***C*-Alkylation of Glycine Units.** Introduction of 2-pyridinethiol onto the α-carbon of glycine units in small peptides generates a series of precursors for the introduction of new side chains on peptides.[48,49] These compounds were prepared by two routes; 1. bromination of glycine residues with NBS followed by displacement with the thiolate ion,[48] and 2. oxidative degradation of serine residues with $Pb(OAc)_4$ and subsequent treatment with the same thiol.[49] Subjection of these sulfide-containing peptides with a carbonyl compound and samarium diiodide at −78 °C to 20 °C in the presence of NiI_2 (1%) provided in good yields a series of new peptides possessing nonproteinogenic side-chains (eq 7). An intramolecular version has also been provided for the preparation of a tricyclic β-lactam.[50]

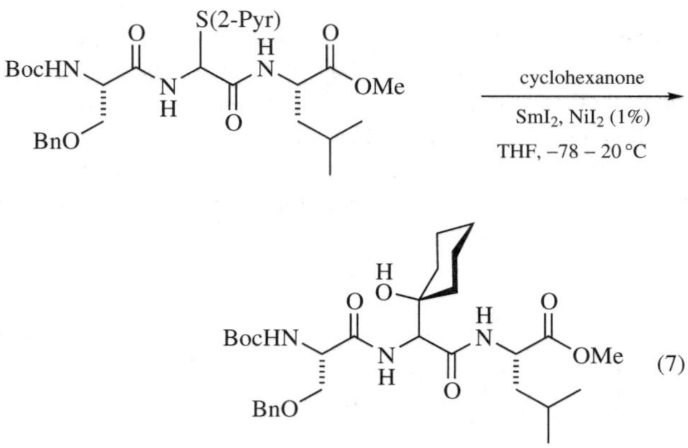

(7)

90% (d:r = 1:1)

Chiral Formal Anion Equivalents. Unsymmetrical dithioacetals such as *t*-butylthio(2-pyridylthio)methane can easily be prepared via the sequential additions of each thiol to bromochloromethane in the presence of base. The enantioselective reaction of the lithiated dithioacetal with aldehydes proceeded well with good diastereoselectivities and enantiomeric excesses in the presence of chiral bisoxazolines (eq 8).[51] Coupling products were subjected to a three step protocol allowing for the formation of chiral diols.

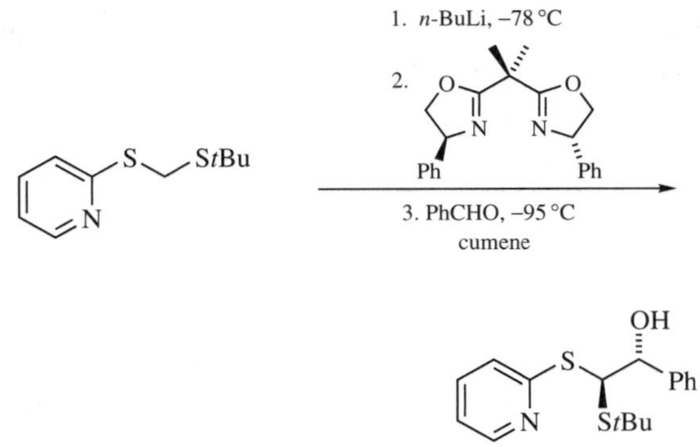

(8)

93%
anti:*syn* = 86:14
ee = 85%

Other Applications. 2-Pyridyl thioethers may be oxidized to the corresponding 2-pyridyl sulfates, and then used in Julia olefination reactions with aldehydes.[52]

Related Reagents. *S*-(2-Pyridinyl)-1,1,3,3-tetramethylthiouronium hexafluorophosphate; dipyridyl disulfide; *N*-hydroxypyridine-2-thione; thiphenol.

1. Sakakibara, T.; Watabe, Y.; Yamada, M.; Sudoh, R., *Bull. Chem. Soc. Jpn.* **1988**, *61*, 247.
2. Mukaiyama, T.; Araki, M.; Takei, H., *J. Am. Chem. Soc.* **1973**, *95*, 4763.
3. Kim, S.; Lee, J. I.; Yi, K. Y., *Bull. Chem. Soc. Jpn.* **1985**, *58*, 3570.
4. Shah, N. V.; Cama, L. D., *Heterocycles* **1987**, *25*, 221.
5. Stewart, A. O.; Williams, R. M., *J. Am. Chem. Soc.* **1985**, *107*, 4289.
6. Mereyala, H. B.; Ravi, D., *Tetrahedron Lett.* **1991**, *32*, 7317.
7. Bartra, M.; Romea, P.; Urpí, F.; Vilarasa, J., *Tetrahedron* **1990**, *46*, 587.
8. Barton, D. H. R.; Hesse, R. H.; O'Sullivan, A. C.; Pechet, M. M., *J. Org. Chem.* **1991**, *56*, 6697, 6702.
9. Trost, B. M.; Parquette, J. R., *J. Org. Chem.* **1993**, *58*, 1579.
10. Arai, Y.; Hayashi, Y.; Yamamoto, M.; Takayama, H.; Koizumi, T., *J. Chem. Soc., Perkin Trans. 1* **1988**, 3133.
11. (a) Lloyd, K.; Young, G. T., *J. Chem. Soc. (C)* **1971**, 2890. (b) Conrow, R.; Portoghese, P. S., *J. Org. Chem.* **1986**, *51*, 938. (c) Almquist, R. G.; Chao, W.-R.; Judd, A. K.; Mitoma, C.; Rossi, D. J.; Panasevich, R. E.; Matthews, R. J., *J. Med. Chem.* **1988**, *31*, 561. (d) Savage, I.; Thomas, E. J., *Chem. Commun./J. Chem. Soc., Chem. Commun.* **1989**, 717.
12. Corey, E. J.; Clark, D. A., *Tetrahedron Lett.* **1979**, 2875.
13. Kato, E.; Yamamoto, K.; Kawashima, Y.; Watanabe, T.; Oya, M.; Iso, T.; Iwao, J.-I., *Chem. Pharm. Bull.* **1985**, *33*, 4836.
14. Ueno, S.; Asakawa, S.; Imoto, E., *Nippon Kagaku Kaishi* **1968**, *89*, 101.
15. Kitagawa, T.; Kuroda, H.; Iida, K.; Ito, M.; Nakamura, M., *Chem. Pharm. Bull.* **1989**, *37*, 3225, and references cited within.
16. Corey, E. J.; Nicolaou, K. C., *J. Am. Chem. Soc.* **1974**, *96*, 5614.
17. Mukaiyama, T.; Yamasaki, N.; Stevens, R. W.; Murakami, M., *Chem. Lett.* **1986**, 213.
18. (a) Harrison, P. G.; Stobart, S. R., *J. Chem. Soc., Daltron Trans.* **1973**, 940. (b) Harrison, P. G.; Stobart, S. R., *Inorg. Chim. Acta* **1973**, *7*, 306.
19. Jennings-White, C.; Almquist, R. G., *Tetrahedron Lett.* **1982**, *23*, 2533.
20. Araki, M.; Sakata, S.; Takei, H.; Mukaiyama, T., *Bull. Chem. Soc. Jpn.* **1974**, *47*, 1777.
21. Cupps, T. L.; Boutin, R. H.; Rapoport, H., *J. Org. Chem.* **1985**, *50*, 3972.
22. Romani, S.; Moroder, L.; Bovermann, G.; Wünsch, E., *Synthesis* **1985**, 738.
23. Zhou, X.-X.; Welch, C. J.; Chattopadhyaya, J., *Acta. Chem. Scand.* **1986**, *40B*, 806.
24. Schroll, A. L.; Barany, G., *J. Org. Chem.* **1989**, *54*, 244.
25. Mereyala, H. B., *Carbohydr. Res.* **1987**, *168*, 136.
26. Gurjar, M. K.; Dhar, T. G. M., *J. Carbohydr. Chem.* **1987**, *6*, 313.
27. (a) Bartra, M.; Bou, V.; Garcia, J.; Urpi, F.; Vilarrasa, J., *Chem. Commun./J. Chem. Soc., Chem. Commun.* **1988**, 270. (b) Bartra, M.; Urpi, F.; Vilarrasa, J., *Tetrahedron Lett.* **1992**, *33*, 3669.
28. (a) Takayama, H.; Hayashi, K.; Takeuchi, Y.; Koizumi, T., *Heterocycles* **1986**, *24*, 2137. (b) Takayama, H.; Iyobe, A.; Koizumi, T., *Chem. Commun./J. Chem. Soc., Chem. Commun.* **1986**, 771.
29. Sato, E.; Hasebe, M.; Nishio, T.; Ikeda, Y.; Kanaoka, Y., *Justus Liebigs Ann. Chem./Liebigs Ann. Chem.* **1988**, 733.

30. Williams, R. M.; Armstrong, R. W.; Maruyama, L. K.; Dung, J.-S.; Anderson, O. P., *J. Am. Chem. Soc.* **1985**, *107*, 3246.
31. Sugimura, H.; Okamura, H.; Miura, M.; Yoshida, M.; Takei, H., *Nippon Kagaku Kaishi* **1985**, 416 (*Chem. Abstr.* **1986**, *104*, 108 625n).
32. Scardovi, N.; Garner, P. P.; Protasiewicz, J. D., *Org. Lett.* **2003**, *5*, 1633.
33. Annunziata, R.; Benaglia, M.; Cinquini, M.; Cozzi, F.; Raimondi, L., *Tetrahedron* **1994**, *50*, 5821.
34. Annunziata, R.; Benaglia, M.; Cinquini, M.; Cozzi, F.; Raimondi, L., *Tetrahedron* **1994**, *50*, 9471.
35. Annunziata, R.; Benaglia, M.; Cinquini, M.; Cozzi, F.; Molteni, V.; Raimondi, L., *Tetrahedron* **1995**, *51*, 8941.
36. Annunziata, R.; Benaglia, M.; Cinquini, M.; Cozzi, F.; Martini, O.; Molteni, V., *Tetrahedron* **1996**, *52*, 2583.
37. Annunziata, R.; Cinquini, M.; Cozzi, F.; Molteni, V.; Schupp, O., *Tetrahedron* **1996**, *52*, 2573.
38. Annunziata, R.; Cinquini, M.; Cozzi, F.; Molteni, V.; Schupp, O., *J. Org. Chem.* **1996**, *61*, 8293.
39. Yang, H. W.; Zhao, C.; Romo, D., *Tetrahedron* **1997**, *53*, 16471.
40. Yang, H. W.; Romo, D., *J. Org. Chem.* **1998**, *63*, 1344.
41. Wang, Y.; Zhao, C.; Romo, D., *Organic Letters* **1999**, *1*, 1197.
42. Skrydstrup, T.; Jarreton, O.; Mazéas, D.; Urban, D.; Beau, J. M., *Chem. Eur. J.* **1998**, *4*, 655.
43. Urban, D.; Skrydstrup, T.; Beau, J. M., *J. Org. Chem.* **1998**, *63*, 2507.
44. Jarreton, O.; Skrydstrup, T.; Espinosa, J. F.; Jiménez-Barbero, J.; Beau, J. M., *Chem. Eur. J.* **1999**, *5*, 430.
45. Mikkelsen, L. M.; Krintel, S. L.; Jiménez-Barbero, J.; Skrydstrup, T., *J. Org. Chem.* **2002**, *67*, 6297.
46. Skrydstrup, T.; Mazéas, D.; Elmouchir, M.; Doisneau, G.; Riche, C.; Chiaroni, A.; Beau, J. M., *Chem. Eur. J.* **1997**, *3*, 1342.
47. Skrydstrup, T.; Jespersen, T. M.; Beau, J. M.; Bols, M., *J. Chem. Soc., Chem.Commun.* **1996**, 515.
48. Ricci, M.; Blakskjær, P.; Skrydstrup, T., *J. Am. Chem. Soc.* **2000**, *122*, 12413.
49. Paulitz, C.; Steglich, W., *J. Org. Chem.* **1997**, *62*, 8474.
50. Jacobsen, M. F.; Turks, M.; Hazell, R.; Skrydstrup. T., *J. Org. Chem.* **2002**, *67*, 2411.
51. Nakamura, S.; Ito, Y.; Wang, L.; Toru, T., *J. Org. Chem.* **2004**, *69*, 1581.
52. Charette, A. B.; Berthelette, C.; St-Martin, D., *Tetrahedron Lett.* **2001**, *42*, 5149.

H-Pyrophosphonic Acid

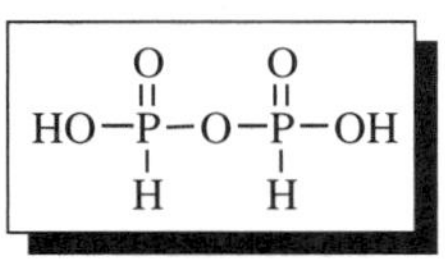

[36465-90-4] $H_4O_5P_2$ (MW 145.98)

(reagent used in the synthesis of *H*-phosphonate monoesters; phosphonylation of oligonucleotides)

Physical Data: mp 36–39 °C.[1]

Preparative Methods: the pyridinium salt of the title reagent is prepared in situ by reaction in pyridine phosphonic acid with a condensing agent [usually pivaloyl chloride (**2**) or 5,5-dimethyl-2-oxo-2-chloro-1,3,2-dioxaphosphorinane (**3**)].[2]

Handling, Storage, and Precautions: coloration after 2 to 3 days at 30 °C; decomposition on standing in a vacuum desiccator, accelerated by light; deliquescent in air; hydrolyzed rapidly by dilute HCl, KOH, or NaOH or more slowly by dilute $NaHCO_3$ solution; stable in I_2 + $NaHCO_3$ solution; oxidized in Br_2 + $NaHCO_3$ solution.[3]

Original Commentary

Jacek Stawinski
Stockholm University, Stockholm, Sweden
Mats Thelin
AstraZeneca, Södertälje, Sweden

General Comments. By reacting phosphonic acid in pyridine with 0.5 equiv. of a condensing agent, a pyridinium salt of the title reagent is formed almost exclusively (eq 1).[2]

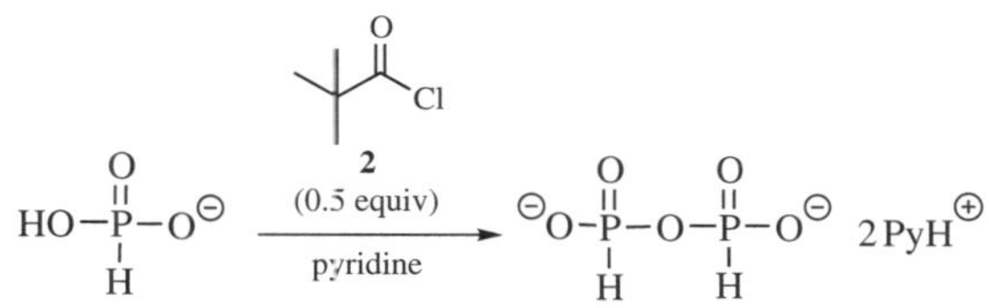

1 (1)

Reagent **1** is stable and can be stored for months as a stock solution in pyridine. Since the formation of **1** is rapid, the reagent may also be generated in situ in the presence of an alcohol to be phosphonylated. One should avoid strongly basic conditions (e.g., presence of triethylamine) as they significantly slow down the phosphonylation reaction.

Synthesis of *H*-phosphonate Monoesters. *H*-Pyrophosphonic acid in the form of the pyridinium salt (**1**) is a mild phosphonylating reagent that reacts with hydroxylic compounds in pyridine to produce the corresponding *H*-phosphonate monoesters. To speed up the reaction, a 5 mol equivalent excess of the reagent is usually used for the phosphonylation. Reagent **1** turned out to be very efficient for the preparation of 5′-*O*-dimethoxytrityldeoxyribonucleoside 3′-*H*-phosphonate monoesters which could be obtained in high yields (86–92%) after silica gel chromatography (eq 2).[2]

Since the condensing agent is completely consumed in the activation process, its chemical nature is usually of minor importance for the yield of the produced *H*-phosphonate monoesters.

α-Anomers of naturally occurring deoxyribonucleosides also undergo smooth phosphonylation with **1** producing the corresponding α-nucleoside 3′-*H*-phophonates in good yields (70–80%).[4]

The reaction is not confined to nucleosides, as other hydroxylic compounds can also be efficiently phosphonylated with this reagent. For example, using reagent **1**, *O*-(benzyl-L-phenyl lactate) *H*-phosphonate was obtained in 96% yield from benzyl-L-phenyl lactate(eq 3).[5]

Due to the low reactivity of **1**, the reagent seems to be less suited for the preparation of sterically hindered *H*-phosphonate monoesters (e.g., for the 3′-*O*-phosphonylation of ribonucleosides with bulky 2′-OH protecting groups).

Other Applications. Reagent **1** can also be used for efficient 5′-*O*-phosphonylation of oligonucleotides attached to a solid support. For example, the reaction of 5′-OH unprotected oligonucleotides bound to a glass support with **1** produced in ca. 2 h, a yield of > 90% of the corresponding 5′-*O*-phosphonylated oligonucleotides.[6]

The disodium salt of *H*-pyrophosphonic acid transfers under aqueous conditions an *H*-phosphonate group to nucleotides[7] and sugar phosphates,[8] producing the corresponding phosphonate-phosphate mixed anhydrides.

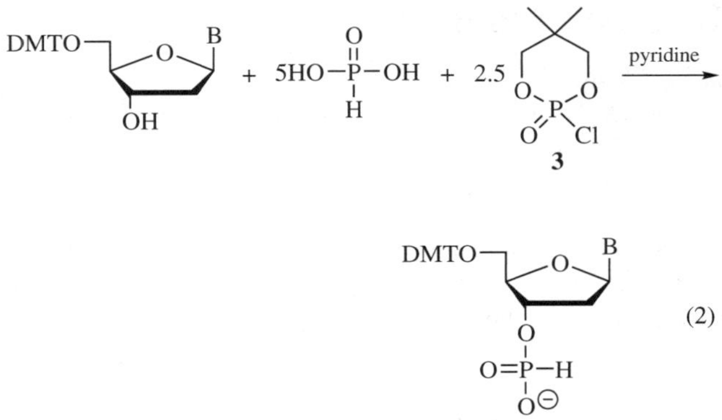

(2)

C^{Bz} = N^4-benzoylcytosin-1-yl
A^{Bz} = *N*-benzoyladenin-9-yl
G^{iBu} = *N*-isobutyrylguanin-9-yl

B= T; 92%
B=C^{Bz}; 86%
B=A^{Bz}; 89%
B=G^{iBu}; 88%

DMT = 4,4'-dimethoxytrityl
T = thymin-1-yl

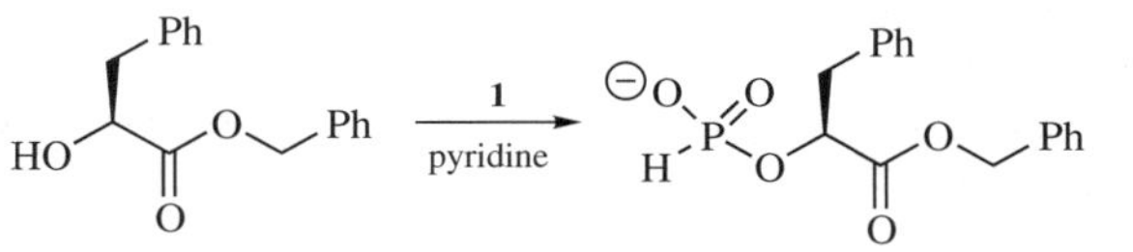

(3)

First Update

Dae-Hwan Suk
University of Illinois at Chicago, Chicago, IL, USA

Preparative Methods. *H*-Pyrophosphonic acid can be prepared from phosphorous acid and excess phosphorus trichloride in a sealed tube at 25–100 °C.[1a] Reaction of phosphorus tribromide and phosphorous acid (molar ratio, 1:5) also leads to the formation of *H*-pyrophosphonic acid as demonstrated by NMR spectroscopy. The equilibrium is influenced by the high solubility of HBr in the reaction medium.[9] The reagent can be formed on hydrolysis of phosphorus trichloride in ether,[10] and has also been prepared by violent agitation of a mixture of phosphorous acid and phosphorus trichloride or phosphorus tribromide in a current of carbon dioxide at about 35 °C.[1a] The disodium salt of *H*-pyrophosphonic acid, $Na_2P_2H_2O_5$, can be prepared from the disodium salt of hypophosphoric acid, $Na_2H_2P_2O_6$, by pyrolysis at 230 °C.[11,12]

Synthesis of *H*-Phosphonate Monoesters. *H*-pyrophosphonic acid provides a simple, general, and efficient method for the preparation of *H*-phosphonate monoesters of various hydroxylic compounds. It is effectively unreactive toward sterically hindered hydroxyl functions and can be applied in the presence of unprotected amino groups.[13]

It has been demonstrated that *H*-pyrophosphonic anhydride, generated in situ with triphosgene, can be used to prepare *H*-phosphonates of ribonucleosides selectively methylated on *O*-2′ (eq 4).[14]

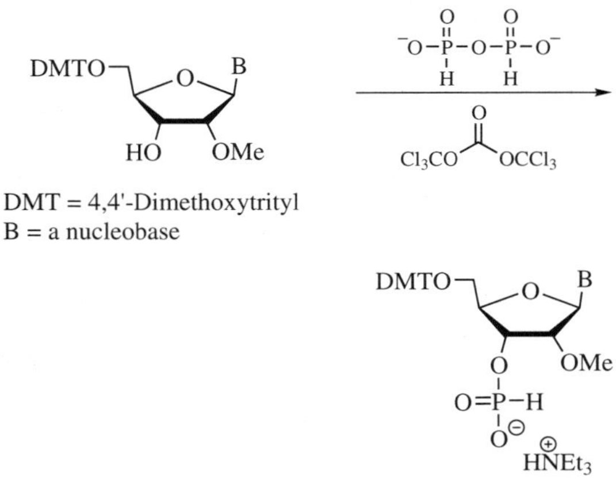

DMT = 4,4'-Dimethoxytrityl
B = a nucleobase

(4)

Phosphonylation of AMP. AMP has been phosphonylated with *H*-pyrophosphonate in aqueous solution (80 °C, pH 5.5–6). A mixture of the 2′- and 3′-phosphonylated products was obtained. While competing attack on the base was not a problem, reaction with the 5′-*O*-phosphate group was a complicating factor.

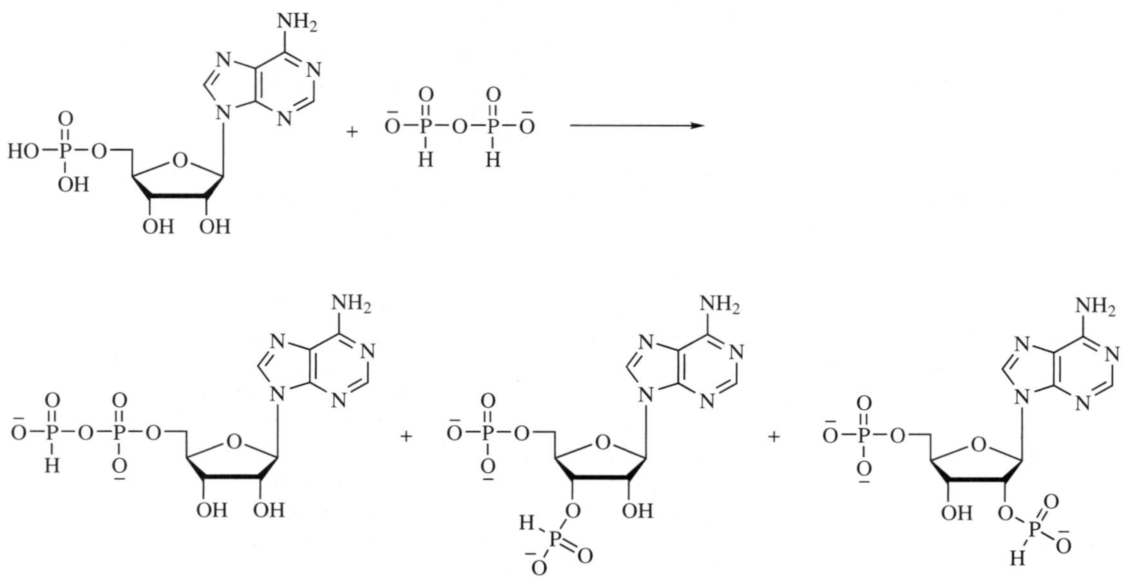

(5)

Under the same conditions, phosphonylation of dAMP did not occur leading to the suggestion that the phosphonylation of AMP proceeds in two steps: (1) formation of hydrogen bond between OH group and phosphorylating agent and (2) subsequent nucleophilic attack of the adjacent OH group at the phosphorus (eq 5).[7]

Related Reagents. Salicylchlorophosphite; PCl_3/imidazole; diphenyl *H*-phosphonate.

1. (a) Schwarzmann, E., van Wazer, J. R., *J. Inorg. Nuclear Chem.* **1960**, *14*, 296. (b) Baudler, M., *Z. Naturforsch.* **1957**, *12b*, 347.
2. Stawinski, J.; Thelin, M., *J. Chem. Soc., Perkin Trans. 2* **1990**, 849–853.
3. Ohashi, S., *Top. Phosphorus Chem.* **1964**, *1*, 113.
4. Laurent, A.; Naval, M.; Debart, F.; Vasseur, J.-J.; Rayner, B., *Nucl. Acids Res.* **1999**, *27*, 4151–4159.
5. Hill, J. M.; Lowe, G., *J. Chem. Soc., Perkin Trans. 1* **1995**, 2001–2007.
6. Sobkowski, M.; Kraszewski, A.; Stawinski, J., *Nucleosides Nucleotides* **1998**, *17*, 253–267.
7. Baba, Y.; Tsuhako, M.; Yamamoto, Y.; Yoza, N.; Ohashi, S., *Nucleic Acids Sym. Ser.* **1986**, *17*, 69–72.
8. Baba, Y.; Onoe, M.; Sumiyama, T.; Tsuhako, M.; Yoza, N.; Ohashi, S., *Chem. Lett.* **1987**, 1389–1392.
9. Hossenlopp, F.; Ebel, J.-P., *C. R. Hebd. Seances Acad. Sci.* **1960**, *250*, 354.
10. Goubeau, J.; Schulz, P., *Z. Anorg. Allgem. Chem.* **1958**, *294*, 224.
11. Kolitowska, J. H., *Bull. Acad. Pol. Sci., Class III* **1956**, *4*, 783.
12. Eto, M.; Kinoshita, Y.; Kato, T.; Oshima, Y., *Nature* **1963**, *200*, 171.
13. Stawinski, J.; Kraszewski, A., *Acc. Chem. Res.* **2002**, *35*, 952.
14. Bhongle, N. N.; Tang, J. Y., *Tetrahedron Lett.* **1995**, *36*, 6803.

S

Silver Oxide

Ag_2O

[20667-12-3] Ag_2O (MW 231.74)

(oxidizing reagent for conversion of hydroquinones to quinones,[1] alkylphenols to quinone methides,[5] and aldehydes to acids;[7] oxidative coupling reactions;[12,13] a Lewis acid with halides[18,20] and thioethers;[24] Wolff rearrangement[26])

Physical Data: dec at about 200 °C; *d* 7.22 g cm^{-3}.
Solubility: practically insoluble in alcohol; soluble in 40 000 parts H_2O; sol dilute nitric acid, ammonia; moderately soluble in NaOH.
Form Supplied in: brownish-black, heavy, odorless powder; widely available.
Handling, Storage, and Precautions: protect from light. Reduced by hydrogen, carbon monoxide, and most metals. Skin contact with this toxic reagent should be carefully avoided.

Original Commentary

Kathlyn A. Parker & Dai-Shi Su
Brown University, Providence, RI, USA

Hydroquinone Oxidation. This reagent is a powerful oxidizing reagent for the conversion of hydroquinones to the corresponding quinones. 1,4-Hydroquinones are generally converted to quinones by treatment with Ag_2O (eq 1).[1] Other reagents (e.g. Cerium(IV) Ammonium Nitrate) are less expensive and easier to handle. However, Ag_2O in nonhydroxylic solvents can be used to produce electron-deficient quinones which are unstable to nucleophiles. For example, Ag_2O oxidation of nitro hydroquinones affords nitro quinones (eq 2).[2]

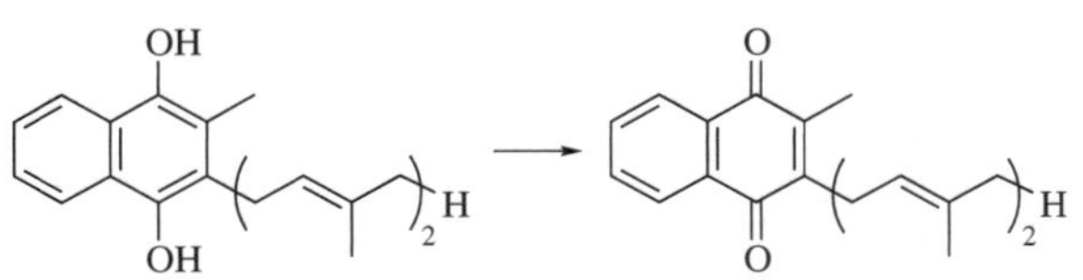

(1)

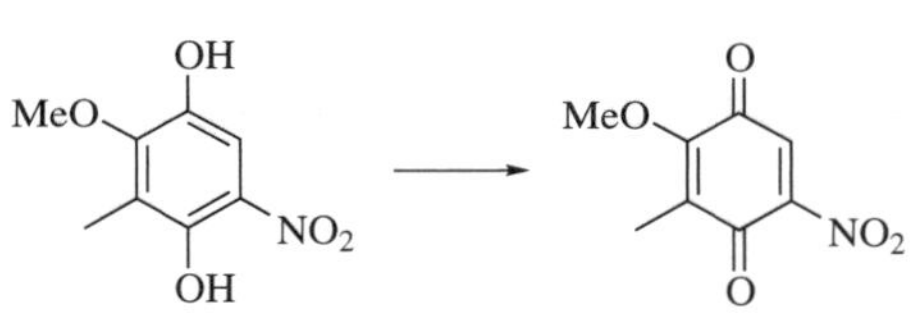

(2)

A one-pot technique has been developed to oxidize 1,4-hydroquinones and utilize the quinone products as dienophiles in Diels–Alder reactions (eq 3).[3] Tetrahydrophenanthrene-9,10-quinones have also been prepared by a one-pot double Diels–Alder reaction (eq 4).[4]

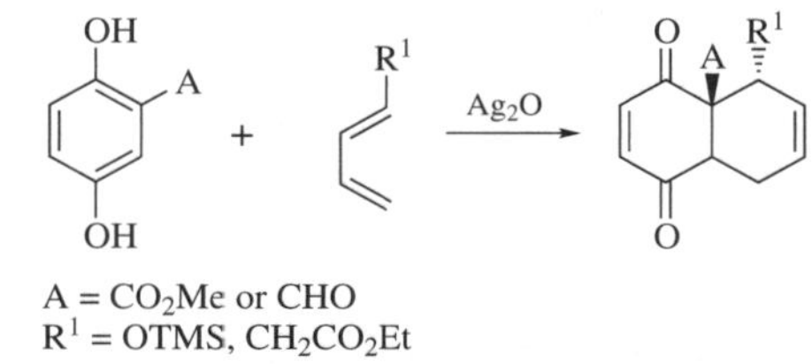

(3)

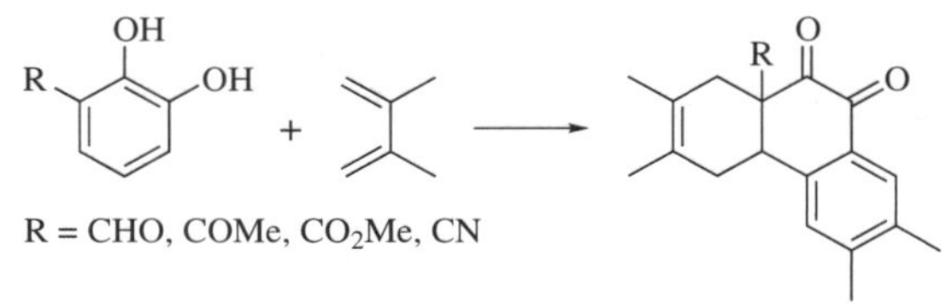

(4)

Alkylphenol Oxidation. 1,4-Quinone methides can be obtained by the reaction of *p*-alkylphenols with Ag_2O. These reactive compounds may undergo subsequent transformations, for example, a Lewis acid promoted cyclization (eq 5).[5] 1,2-Quinone methides can also be formed by Ag_2O oxidation of appropriately substituted phenols (eqs 6 and 7).[6]

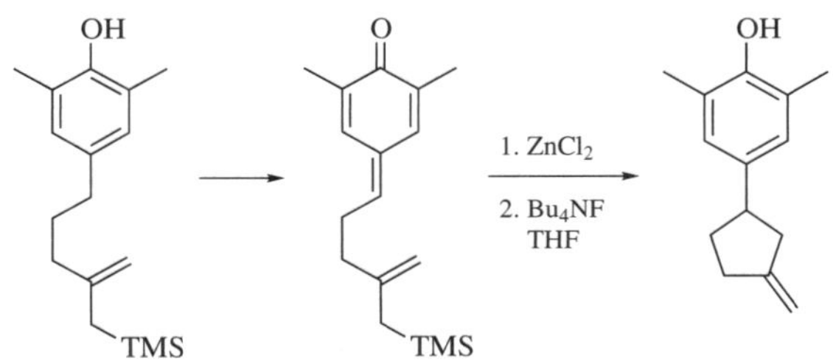

(5)

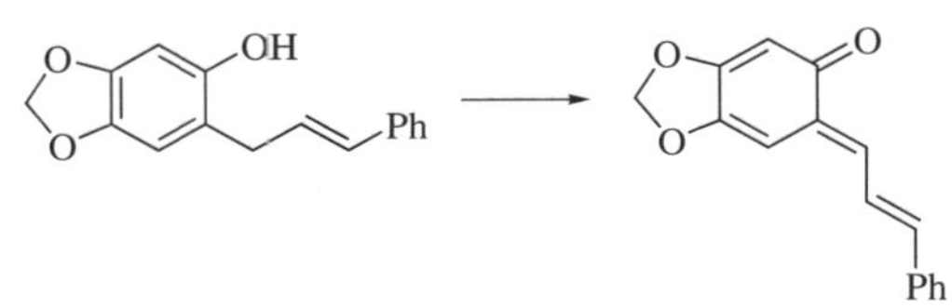

(6)

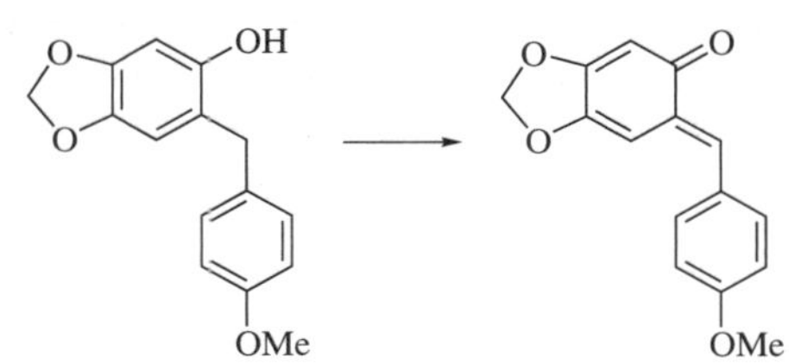

(7)

Other Oxidation Reactions. Ag_2O oxidizes aldehydes to acids,[7] and conjugated keto aldehydes to conjugated keto acids.[8] For example, Ag_2O was used to selectively oxidize one of two formyl substituents of an intermediate in the total synthesis of inhibitor K-76 (eq 8).[9] The unreactive formyl substituent is conjugated with the hydroxyl and alkoxy substituents.

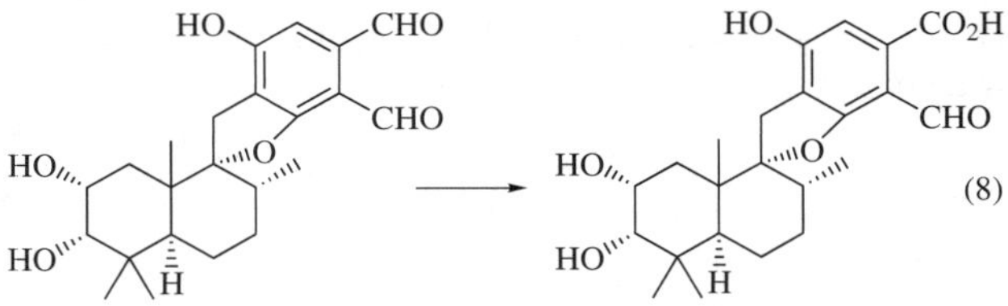

(8)

Oxidative Coupling Reactions. Radicals produced by Ag_2O oxidation can undergo coupling reactions. For example, 2-(vinyloxy)phenols undergo Ag_2O oxidation-induced coupling.[10] Acylacetates and monosubstituted malonates are oxidatively dimerized in the presence of Ag_2O and DMSO (eq 9).[11]

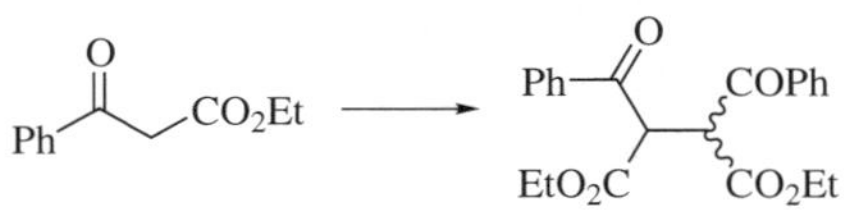

(9)

Radical coupling followed by nucleophilic attack of hydroxyl on a quinone methide intermediate is postulated as the mechanism of the key step in the syntheses of silybin and eusiderin (eq 10).[12] 1,4-Diketones are produced in the reaction of silyl enol ethers with Ag_2O in DMSO (eq 11).[13]

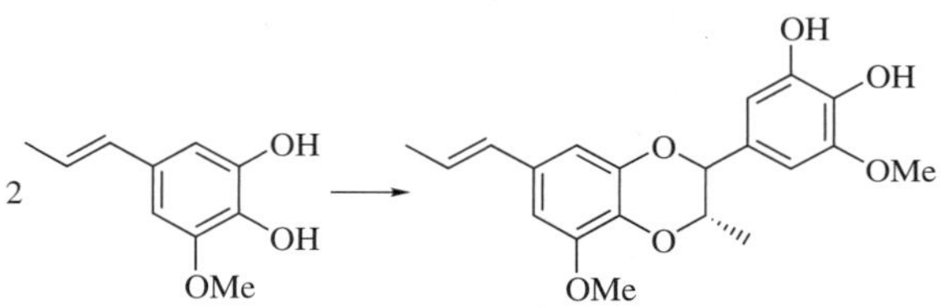

(10)

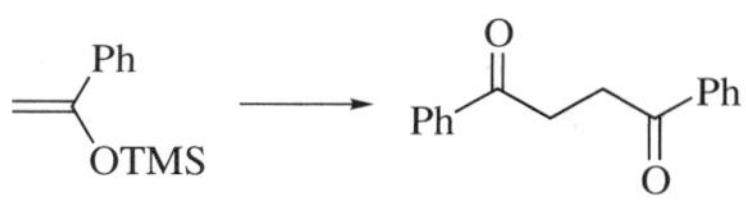

(11)

Lewis acid with Halides. Ag_2O can activate halide as a leaving group by coordination. Methylation of carbohydrates[14] and 5-benzylidenebarbituric acid[15] have been carried out with Ag_2O and Iodomethane in DMF. Triphenyltin trifluoroacetate was obtained from the reaction of triphenyltin iodide and trifluoroacetic acid in the presence of silver oxide.[16] Ag_2O converts *trans*-halohydrins to epoxides (eq 12)[17] or rearranged products (eq 13)[18]. It is reported that the reaction is nonstereoselective, and that *trans*-epoxides are the major products when *cis*-alkenes are the reactants.[19]

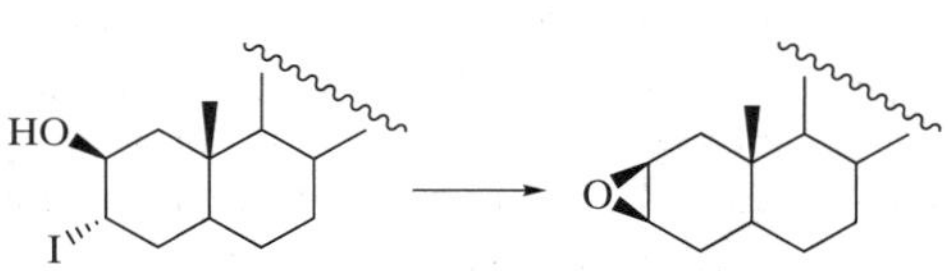

(12)

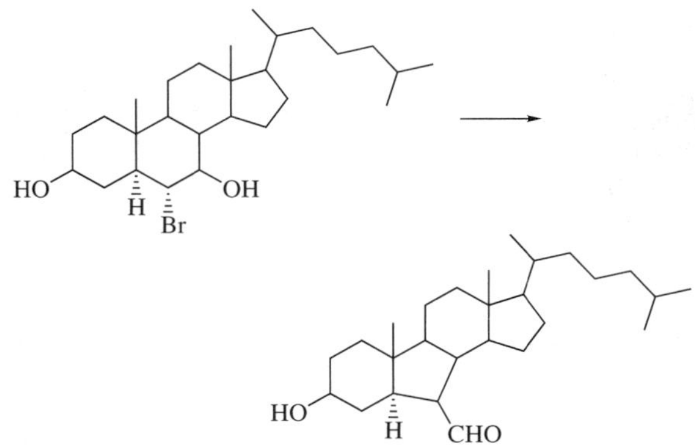

(13)

2-Arylpropionaldehydes are produced from 1-aryl-1-propene by oxidative rearrangement with Iodine and Ag_2O in dioxane–H_2O (eq 14).[20] The mechanism may involve the 1,2-shift of the aryl group through a bridged phenonium ion in the iodohydrin intermediate.

(14)

In the total synthesis of mycophenolic acid, electrophilic substitution was promoted by Ag_2O (eq 15).[21] α,β-Unsaturated nitriles are formed from the corresponding γ-bromo-β-oxo nitriles with Ag_2O (eq 16).[22] Arylsulfenylation and arylselenenylation at the 5-position of uracils are promoted by Ag_2O (eq 17).[23]

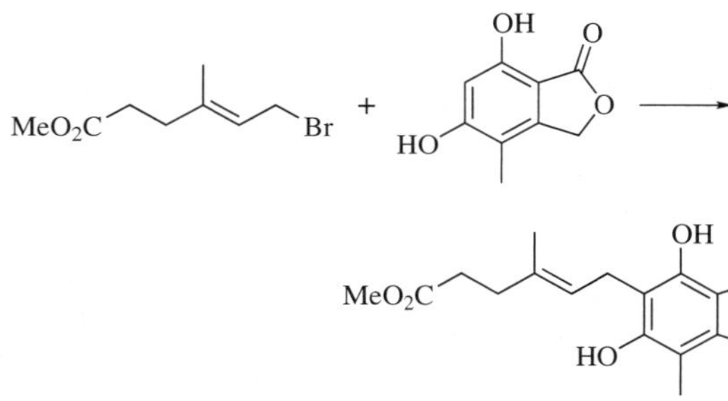

(15)

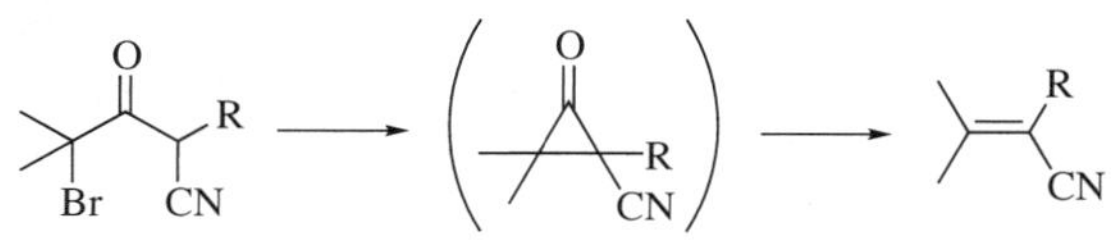

(16)

R = H, Me

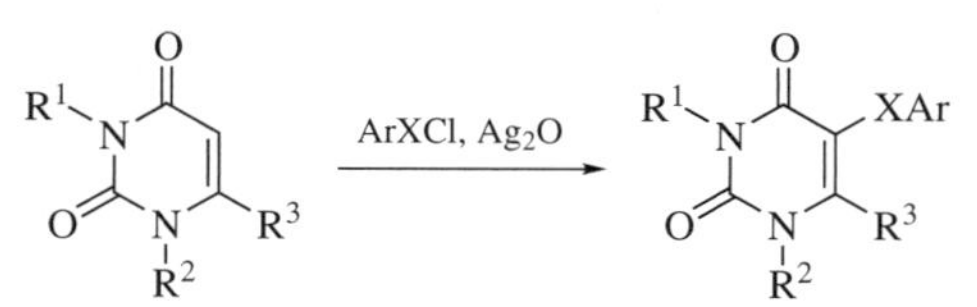

(17)

R^1 = H, Me; R^2 = Me, acyclic chain, ribose; R^3 = H, Me, N=PPh_3
X = S, Se

Lewis Acid with Thioethers. Ag_2O has been applied to the 1,2-cleavage of penicillins in a strong nonnucleophilic base, such as 1,5-Diazabicyclo[4.3.0]non-5-ene,[24] and to the hydrolysis of thioacetals.[25]

Wolff Rearrangement. The Wolff rearrangement of diazo ketones is promoted by Ag_2O (eq 18).[26]

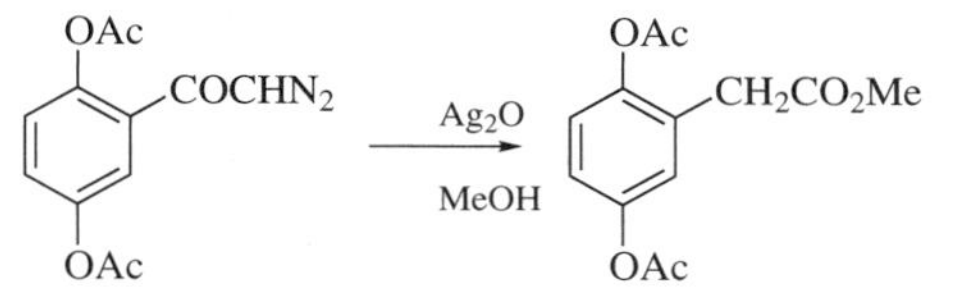

(18)

First Update

Alexei V. Demchenko
University of Missouri, St. Louis, MO, USA

Cross-Coupling Reactions. Silanols, silanediols, and silanetriols were found to be suitable substrates for cross-coupling reactions with organic halides in the presence of Pd(0) as a catalyst and Ag_2O as an activator (eq 19).[27] Interestingly, the presence of at least a stoichiometric amount of Ag_2O in the reaction media was found to be essential. Thus, no reaction occurred when no additives or other metal oxides (CuO, CaO, BaO) were used, whereas the substitution of Ag_2O by other silver salts (AgOTf, $AgBF_4$, $AgNO_3$) resulted in significantly lower yields.

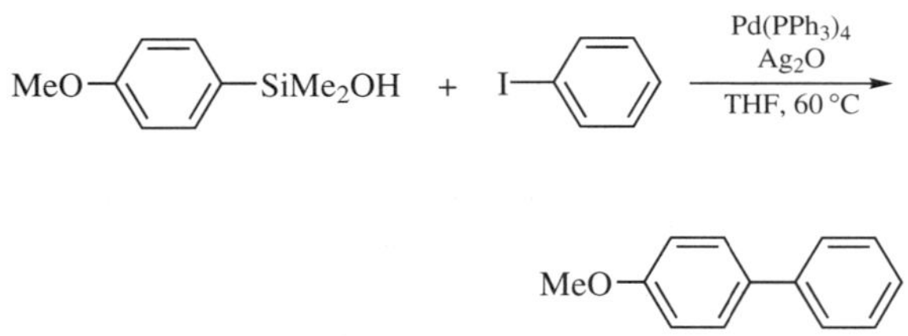

(19)

80%

Similarly terminal alkynes could be cross-coupled with aryl halides or triflates.[28] Thus, trimethylsilylethyne (R=TMS) was reacted with iodobenzene (X=I, Y=H) in the presence of a catalytic amount of Pd(0) and 1 mol equiv Ag_2O to afford phenyl(trimethylsilyl)ethyne in 79% yield (eq 20); no phenylethyne formation was detected under these conditions. This technique was further extended to the polycondensation of bis(trimethylsilylalkynyl) compounds with diiodoarenes.[29]

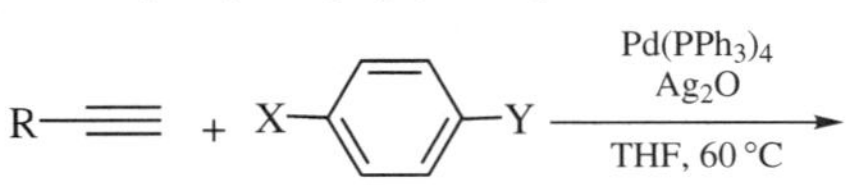

R = TMS, n-C_6H_{13}, Ph, t-Bu, CMe_2OH
X = I, Br, OTf
Y = H, OMe, $COCH_3$

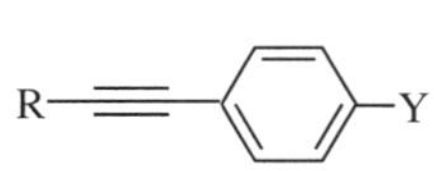

(20)

60–99%

Sulfonylation of Alcohols. Primary and secondary alcohols could be mesylated, tosylated, or p-nitrobenzenesulfonylated in the presence of Ag_2O and KI (eq 21).[30] It was demonstrated that KI accelerates the reaction presumably by converting the sulfonating reagent into a highly reactive sulfonyl iodide. Typically 1.5 mol equiv of Ag_2O was sufficient for an efficient conversion, however, occasionally a twofold excess was required. It was noted that since the reaction does not require aqueous work-up, subsequent nucleophilic substitution of the sulfonate ester could be performed without the necessity of isolating the intermediates.

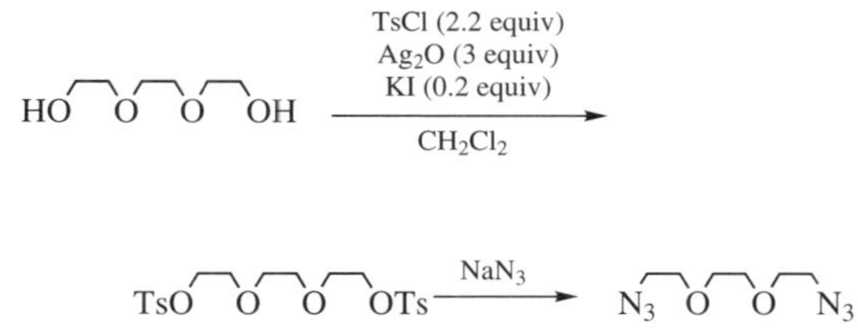

(21)

85%

Selective monotosylation of symmetric diols was also reported.[31] This technique was further applied to the synthesis of cyclic ethers (eq 22). This process, performed in the presence of at least 3 equiv of Ag_2O, was particularly favorable when sterically undemanding five- or six-membered rings could be formed. However, the technique was also used for the formation of the three-membered oxirane ring and of a substituted 5-O-crown ether.

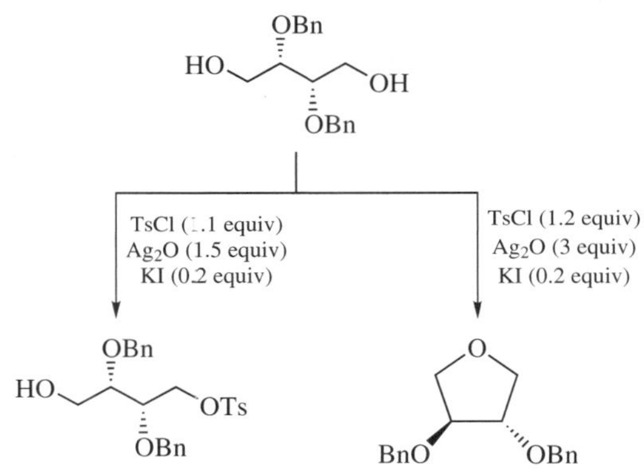

(22)

85% 75%

Synthesis of Nitriles from Amides. Aryl carboxamides and other electron-rich amides could be converted to the corresponding nitriles in good yields by treatment with Ag_2O and EtI in benzene at reflux (eq 23).[32] This technique is typically compatible with acid-labile moieties or protecting groups (TBDPS, isopropylidene). Although this technique was found to be very efficient for the transformation of electron-rich carboxamides, simple aliphatic amides are rarely suitable for this purpose.

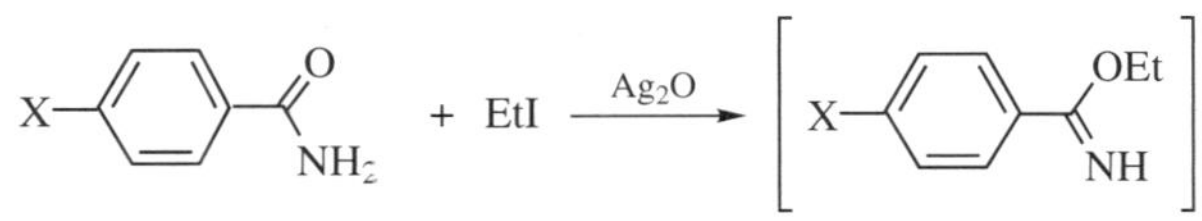

X = H, Me, EtO, F, Cl, Br, NO_2

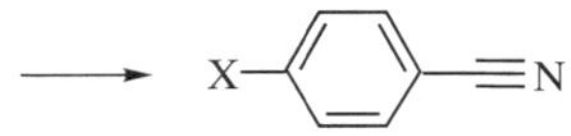

(23)

70–89%

Related Reagents. Bromine-Silver(I) Oxide.

1. Synder, C. D.; Rapoport, H., *J. Am. Chem. Soc.* **1974**, *96*, 8046.
2. Parker, K. A.; Sworin, M., *Tetrahedron Lett.* **1978**, *26*, 2251.
3. (a) Kraus, G. A.; Taschner, M. J., *J. Org. Chem.* **1980**, *45*, 1174. (b) Marchand, A. P.; Suri, S. C.; Earlywine, A. D.; Powell, D. R.; van Der Helm, D., *J. Org. Chem.* **1984**, *49*, 670.
4. Al-Hamdany, R.; Ali, B., *Chem. Commun.* **1978**, 397.
5. Angle, S. R.; Turnbull, K. D., *J. Am. Chem. Soc.* **1989**, *111*, 1136.
6. Jurd, L., *Tetrahedron* **1977**, *33*, 163.
7. Thomason, S. C.; Kubler, D. G., *J. Chem. Educ.* **1968**, *45*, 546.
8. Pepperman, A. B., *J. Org. Chem.* **1981**, *46*, 5039.
9. Corey, E. J.; Das, J., *J. Am. Chem. Soc.* **1982**, *104*, 5551.
10. West, K. F.; Moore, H. W., *J. Org. Chem.* **1984**, *49*, 2809.
11. Ito, Y.; Fujii, S.; Konoike, T.; Saegusa, T., *Synth. Commun.* **1976**, *6*, 429.
12. (a) Merlini, L.; Zanarotti, A.; Pelter, A.; Rochefort, M. P.; Hänsel, R., *Chem. Commun.* **1979**, 695. (b) Merlini, L.; Zanarotti, A.; Pelter, A.; Rochefort, M. P.; Hansel, R., *J. Chem. Soc., Perkin Trans 1* **1980**, 775. (c) Merlini, L.; Zanarotti, A., *Tetrahedron Lett.* **1975**, 3621.
13. Ito, Y.; Konoike, T.; Saegusa, T., *J. Am. Chem. Soc.* **1975**, *97*, 649.
14. Kuhn, R.; Trischmann, H.; Low, I., *Angew. Chem.* **1955**, *67*, 32.
15. Ethier, J. C.; Neville, G. A., *Tetrahedron Lett.* **1972**, 5297.
16. Srivastava, T. N.; Singh, J., *Indian J. Chem., Sect. A* **1983**, *22A*, 128.
17. Parrilli, M.; Barone, G.; Adinolfi, M.; Mangoni, L., *Tetrahedron Lett.* **1976**, 207.
18. (a) Nace, H. R.; Crosby, G. A., *J. Org. Chem.* **1979**, *44*, 3105. (b) Schmidlin, J.; Wettstein, A., *Helv. Chim. Acta* **1953**, *36*, 1241. (c) Curtin, D. Y.; Harder, R. J., *J. Am. Chem. Soc.* **1960**, *82*, 2357.
19. Jorgensen, K. A.; Larsen, E., *J. Chem. Soc. Dalton Trans.* **1990**, 1053.
20. Kikuchi, H.; Kogure, K.; Toyoda, M., *Chem. Lett.* **1984**, 341.
21. Canonica, L.; Rindone, B.; Santaniello, E.; Scolastico, C., *Tetrahedron Lett.* **1971**, 2691.
22. Herter, R.; Fohlisch, B., *Ber. Dtsch. Chem. Ges./Chem. Ber.* **1982**, *115*, 381.
23. Lee, C. H.; Kim, Y. H., *Tetrahedron Lett.* **1991**, *32*, 2401.
24. Alpegiani, M.; Bedeschi, A.; Bissolino, P.; Visentin, G.; Zarini, F.; Perrone, E.; Franceschi, G., *Heterocycles* **1990**, *31*, 617.
25. Gravel, D.; Vaziri, C.; Rahal, S., *Chem. Commun.* **1972**, 1323.
26. (a) Kloetzel, M. C.; Dayton, R. P.; Abadir, B. Y., *J. Org. Chem.* **1955**, *20*, 38. (b) Meier, H.; Zeller, K.-P., *Angew. Chem., Int. Ed. Engl.* **1975**, *14*, 32. (c) Della, E. W.; Kendall, M., *J. Chem. Soc., Perkin Trans 1* **1973**, 2729.
27. (a) Hirabayashi, K.; Mori, A.; Kawashima, J.; Suguro, M.; Nishihara, Y.; Hiyama, T., *J. Org. Chem.* **2000**, *65*, 5342; (b) Hirabayashi, K.; Kawashima, J.; Nishihara, Y.; Mori, A.; Hiyama, T., *Org. Lett.* **1999**, *1*, 299.
28. Mori, A.; Kawashima, J.; Shimada, T.; Suguro, M.; Hirabayashi, K.; Nishihara, Y., *Org. Lett.* **2000**, *2*, 2935.
29. Mori, A.; Kondo, T.; Kato, T.; Nishihara, Y., *Chem. Lett.* **2001**, 286.
30. Bouzide, A.; LeBerre, N.; Sauvé, G., *Tetrahedron Lett.* **2001**, *42*, 8781.
31. Bouzide, A.; Sauvé, G., *Org. Lett.* **2002**, *4*, 2329.
32. Sznaidman, M. L.; Crasto, C.; Hecht, S. M., *Tetrahedron Lett.* **1993**, *34*, 1581.

Silver Salicylate

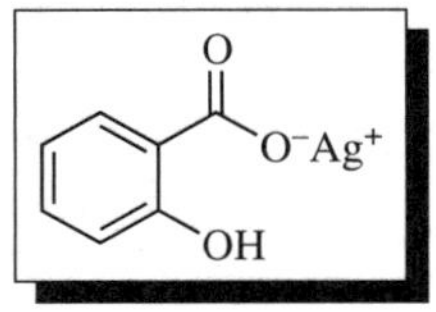

[19025-97-9] $C_7H_5O_3Ag$ (MW 244.98)

(reagent used as promoter for classical Koenigs-Knorr reaction, as a promoter for heavy metal assisted cleavage of thiazolidine rings, and in the nucleophilic displacement of alkyl iodides)

Physical Data: white solid.
Solubility: insoluble in alcohol, ethyl acetate, ether, tetrahydrofuran, and benzene. Soluble in 2-methylthioethanol.
Preparative Methods: a solution of salicylic acid in 96% ethanol is adjusted to pH 8 with aqueous ammonia, and then a solution of silver nitrate in aqueous ethanol is added dropwise with heating at 60–70 °C for 30 min. The white solid that forms upon slow cooling to room temperature is washed with ethanol and diethyl ether and then dried over P_2O_5 in vacuo.[1]

Classical Koenigs-Knorr Reaction. During efforts to identify optimum conditions for the synthesis of β-D-glycosides, new promoters for the classical Koenigs-Knorr reaction appeared, including silver salicylate. The stereospecific synthesis of methyl 2,3,4,6-tetra-*O*-acetyl-β-D-galactopyranoside in 84% yield from 2,3,4,6-tetra-*O*-acetyl-α-D-galactopyranosyl bromide and methanol in the presence of silver salicylate was reported (eq 1).[2] In the same paper, a rate enhancement of 10^3–10^4 for glycoside synthesis with silver salicylate in comparison with other variants of the Koenigs-Knorr reaction was also recorded. The reaction was performed by addition of silver salicylate to a solution of the glycosyl bromide in absolute methanol at 20–25 °C. Rapid filtration of the mixture and analysis by thin layer chromatography revealed a mixture of methyl 2,3,4,6-tetra-*O*-acetyl-β-D-galactopyranoside and salicylic acid. No 1,2-orthoester or 2,3,4,6-tetra-*O*-acetyl-1-*O*-salicyloyl-D-glucopyranose products were detected.

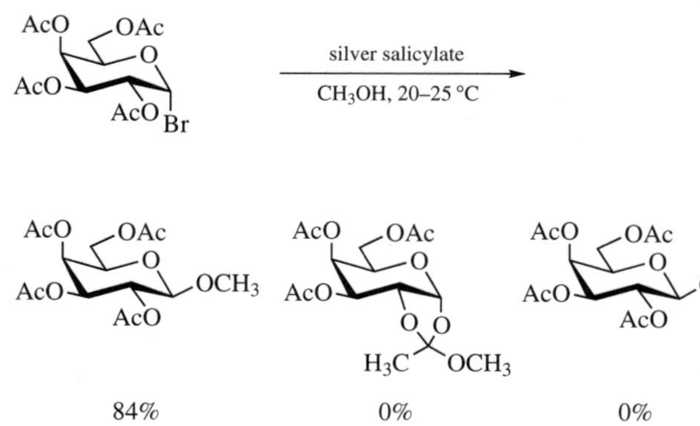

(1)

Silver salicylate promoted classical Koenigs-Knorr reactions of 2,3,4,6-tetra-*O*-acetyl-α-D-glucopyranosyl and 2,3,4,6-tetra-*O*-acetyl-α-D-galactopyranosyl bromide have also been investigated with other alcohols.[3] Reactions of the glycosyl bromides with methanol, ethanol, 1-propanol, and 2-propanol in the presence of silver salicylate were rapid (under 15 min), while reactions with 1-butanol and *tert*-butanol were slower (30 min). Yields of the β-glycosides were excellent in most cases (>75%). However, with 2-propanol and *tert*-butanol substantial amounts of the 1,2-orthoesters were also formed from the glucosyl bromide. When the presence of 1,2-orthoesters complicated recrystallization of the glycosides, recrystallization was achieved from acidified, aqueous acetone, in which the 1,2-orthoesters were hydrolyzed. Small amounts of 2,3,4,6-tetra-*O*-acetyl-1-*O*-salicyloyl-D-glucopyranose were also formed from reaction with *tert*-butanol. The main product formed in these reactions with weak nucleophiles or in solvents that dissolve silver salicylate (e.g., 2-methylthioethanol) is the glycosyl salicylate derivative.

The application of silver salicylate for the synthesis of β-glycosides of steroid alcohols in various solvents was also studied (eq 2)[4] and product distribution was found to be largely solvent dependent. Thus reaction of tigogenin with 2,3,4,6-tetra-*O*-acetyl-α-D-glucopyranosyl bromide and silver salicylate in diethyl ether gave 60% yield of the β-glycoside and 30% yield of the 1,2-orthoester. In contrast, in THF the 1,2-orthoester predominated (90% yield).

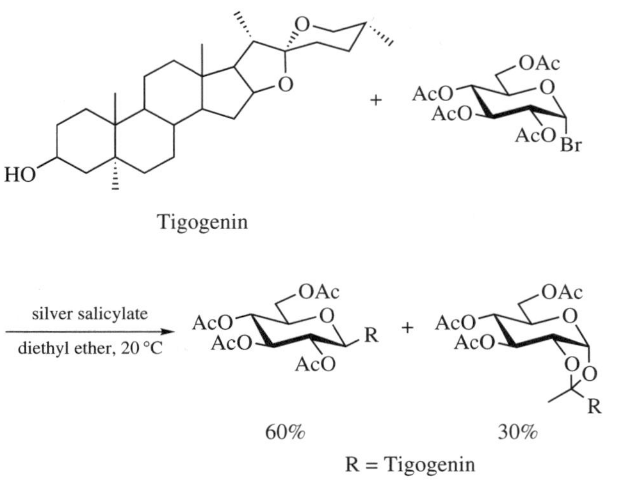

(2)

Reaction of α-glycosyl halides having a nonparticipating substituent at C-2 under silver salicylate promoted Koenigs-Knorr conditions have also been reported for the preparation of the not easily accessible β-1,2-*cis*-glycosides, e.g., β-mannopyranosides.[5] Reaction of 2,3,4,6-tetra-*O*-benzyl-α-D-mannopyranosyl bromide with cholesterol and silver salicylate in diethyl ether gave the β-mannopyranoside in 54% yield (eq 3). Using the same method, reaction of 2,3,4,6-tetra-*O*-benzyl-α-D-mannopyranosyl bromide with methyl-2,3,4-tri-*O*-acetyl-α-D-glucopyranoside provided the expected disaccharide in 52% yield in a 12:1 β:α ratio (eq 4). The insolubility of silver salicylate in the solvents commonly used for glycosylation reactions is believed to be important in determining the stereochemical outcome of these reactions. Activation of the glycosyl halide occurs on the surface of the promotor and it has been proposed that this facilitates an S_N2-like displacement of the bromide by the nucleophile, and prevents the attack of the electrophilic anomeric carbon from the bottom face of the ring.

For the synthesis of glycopeptides 2,3,4,6-tetra-*O*-acetyl-α-D-glucopyranosyl bromide was reacted with *N*-benzyloxycarbonyl-serine 2-bromoethyl ester in the presence of silver salicylate to give the β-glucosylserine ester in 47% yield (eq 5).[6]

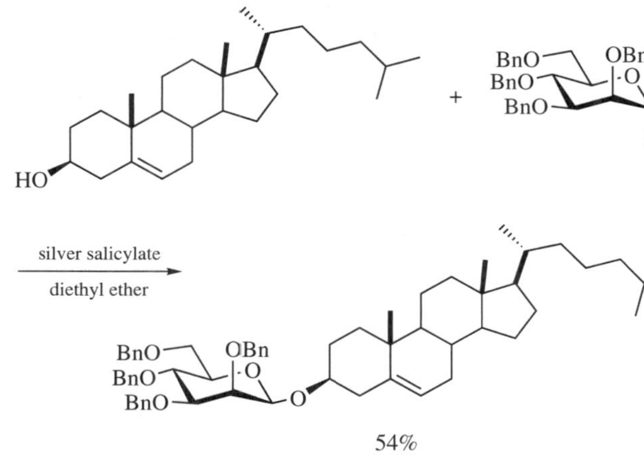

(3)

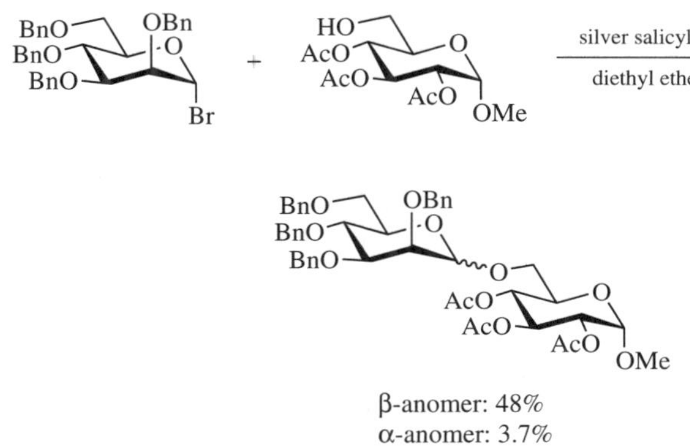

(4)

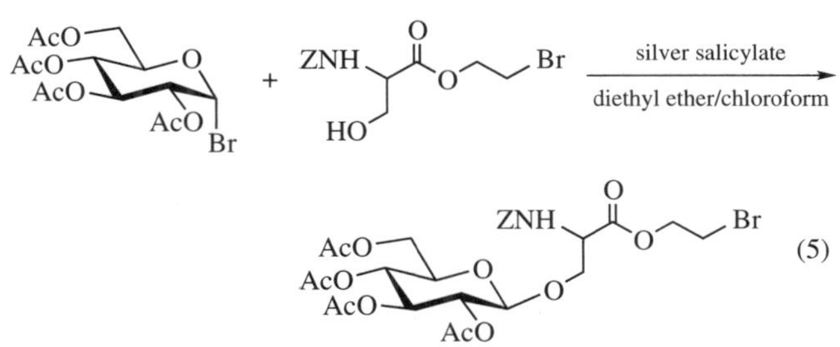

(5)

Van der Vleugel et al.[1] employed silver salicylate in an effort to improve the preparation of biologically important α-glycosides of *N*-acetyl-D-neuraminic acid beyond the low yields (30–60%) and long reaction times required with other silver salts. The typical procedure involved dissolving the starting halide in the appropriate alcohol followed by addition of silver salicylate. Under these conditions treatment of methyl 5-acetamido-4,7,8,9-tetra-*O*-acetyl-2-chloro-2,3,5-trideoxy-β-D-glycero-D-galacto-2-nonulopyranosonate with methanol (eq 6), ethanol, or 2-propanol gave, exclusively, the corresponding α-glycoside in 89%, 88%, and 84% yields, respectively. This donor has also been reacted with allyl alcohol[7–13] and benzyl alcohol[14] in the presence of silver salicylate providing the α-glycoside in excellent yields (eq 6). With this donor, no reaction was observed with *tert*-butanol and after 3 h only unreacted chloride remained. However, when methyl 5-acetamido-4,7,8,9-tetra-*O*-acetyl-2-chloro-2,3,5-trideoxy-β-D-glycero-D-galacto-2-nonulopyranosonate was stirred overnight in benzene with 10 equiv of neopentyl alcohol in the presence of silver salicylate exclusive formation of the neopentyl α-glycoside occurred in 67% yield (eq 7). Similar reaction conditions with silver salicylate and methanol have been used to prepare methyl (methyl 4,7,8-tri-*O*-acetyl-9-azido-5-benzyloxycarbonylamino-3,5,9-trideoxy-α-D-glyco-D-galacto-2-nonulopyranoside)onate.[15]

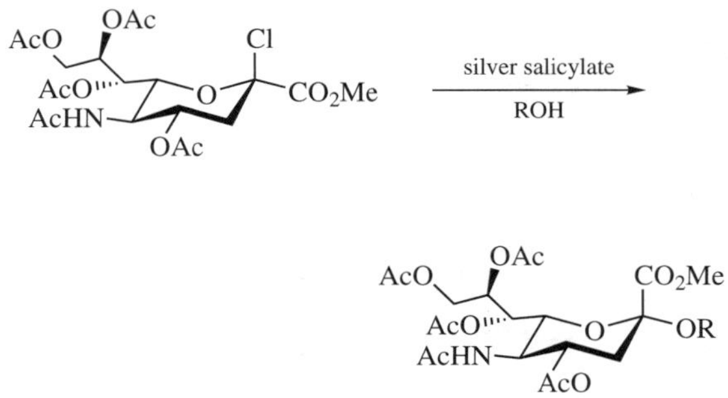

(6)

R = Methyl 89%
Ethyl 88%
Isopropyl 84%
Allyl 94%
Benzyl 96%

In the presence of silver salicylate, methyl 5-acetamido-4,7,8,9-tetra-*O*-acetyl-2-chloro-2,3,5-trideoxy-β-D-glycero-D-galacto-2-nonulopyranosonate has also been shown to react with more complex alcohols.[16–19] For example, Van der Vleugel and coworkers have prepared 6-*O*-acetyl-(*N*-acetyl-α-D-neuraminyl)-D-galactose in 65% yield by reaction with benzyl 2,3,4-tri-*O*-benzyl-β-D-galactopyranoside (eq 8).[16] Similarly, Sato et al.[18] reacted this donor with cholesterol in the presence of silver salicylate in benzene yielding 27% of glycoside products in a 1:7.2 α/β ratio (eq 9). As a side-product in this reaction, methyl *N*-acetyl-4,7,8,9-tetra-*O*-acetyl-2,3-dehydro-2-deoxyneuraminate was formed in 35% yield.

A possible competing reaction in silver salicylate promoted Koenigs-Knorr reactions is the formation of products arising from the reaction of the glycosyl halide with the salicylate promotor, which has two possible nucleophilic sites. Although these are usually minor by-products, they are observed in some cases, especially with weakly nucleophilic alcohols or in solvents in which silver salicylate is soluble. Normally, the product formed results from the reaction of the glycosyl halide and the carboxylate moiety, not the phenolic hydroxyl group. For example, Lunsford and Murphey[20] have shown that methyl-1-bromo-2,3,4-tri-*O*-acetyl-α-D-glucuronate reacts with silver salicylate in isoquinoline to form methyl (salicyloyl 2,3,4-tri-*O*-acetyl-β-D-glucopyranosid) uronate in 63% yield.

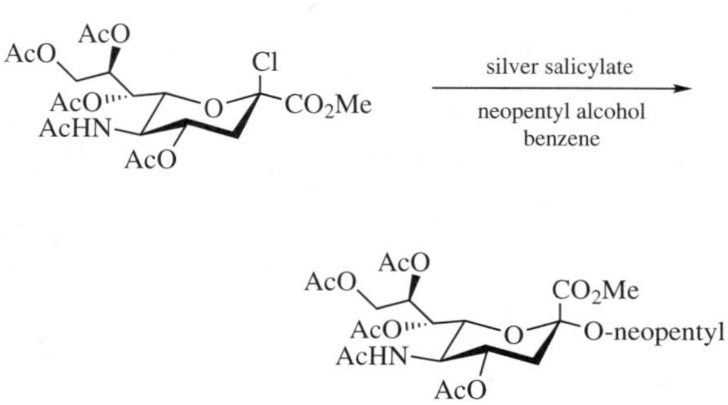

(7)

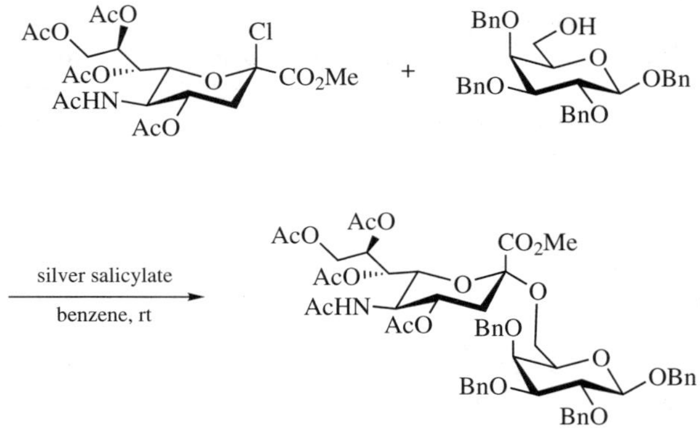

(8)

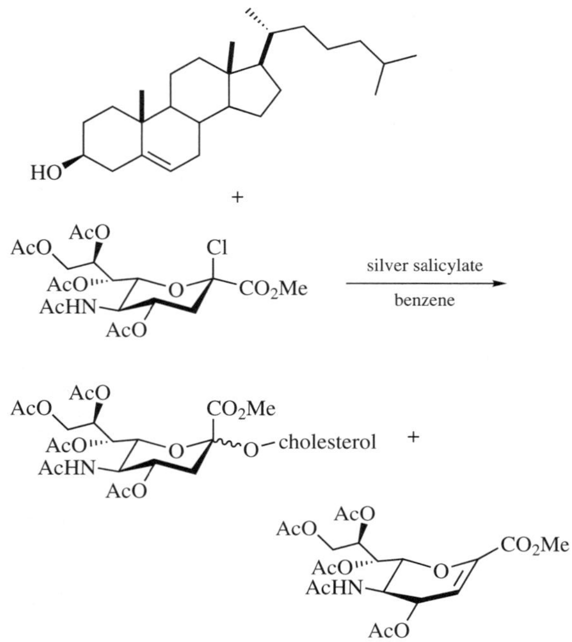

(9)

Metal-assisted 1,2-Cleavage of Penicillins. Synthetically useful 1,2-secopenicillanates can be prepared from penicillins via metal-assisted 1,2-cleavage.[21] The thiazolidine ring opening occurs as a formal β-elimination and can be achieved by use of a strong non-nucleophilic base (DBN) and a thiophilic heavy metal salt (eq 10). Alpegiani et al. has investigated the scope of this metal-assisted 1,2-cleavage including the influence of the heavy metal salts. The typical procedure involved stirring a mixture of heavy metal salt and DBN in acetonitrile under nitrogen prior to adding methyl 6,6-dibromopenicillanate. Stirring was continued at room temperature until depletion of >95% of penicillanate and acetyl chloride was then added. With silver salicylate as the metal salt, the reaction proceeded in 24 h in 55% yield. In comparison, when silver oxide was used, the reaction was complete in 2 min and gave 72% yield of the product.

Nucleophilic Displacement of Alkyl Iodides. As part of an effort to develop triazene prodrugs for use against malignant melanoma, a series of compounds containing an acyloxymethyl carbamate functionality was synthesized.[22] These prodrugs were prepared from monomethyl triazenes in three steps, the final step being displacement of iodine with a silver carboxylate salt. The method involved adding the carboxylic acid silver salt and tetrabutylammonium bromide to a solution of 1-aryl-3-iodo-methyloxycarbonyl-3-methyltriazene in dry acetone. The mixture was refluxed, filtered, and concentrated. With silver salicylate, the product was formed in 29% yield after column chromatography and recrystallization (eq 11).

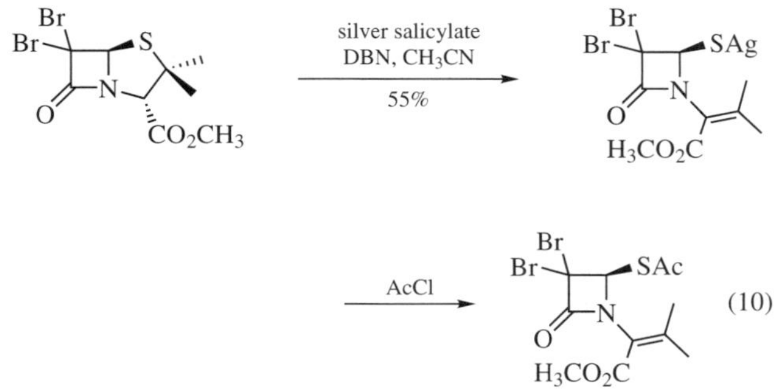

(10)

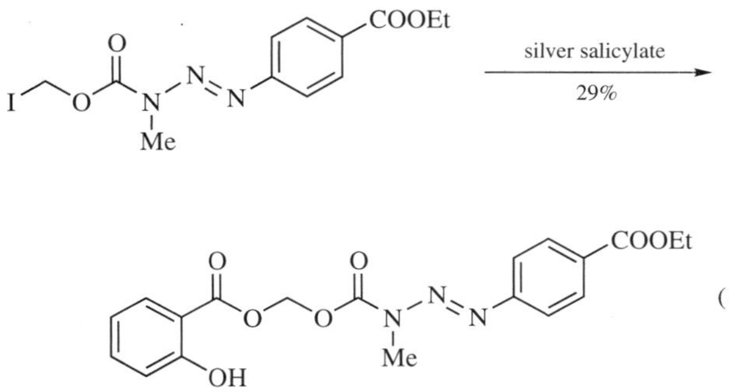

(11)

1. Van der Vleugel, D. J. M.; Van Heeswijk, W. A. R.; Vliegenthart, J. F. G., *Carbohydr. Res.* **1982**, *102*, 121–130.
2. Veinberg, A. Y.; Roslovtseva, G. I.; Samokhvalov, G. I., *J. Gen. Chem. USSR (Engl. Transl.)* **1973**, *43*, 688–689.
3. Heeswijk, W. A. R.; Visser, H. G. J.; Vliegenthart, J. F. G., *Carbohydr. Res.* **1977**, *58*, 494–497.
4. Wulff, G.; Röhle, G.; Krüger, W., *Chem. Ber.* **1972**, *105*, 1097–1100.
5. Wulff, G.; Wichelhaus, J., *Chem. Ber.* **1979**, *112*, 2847–2853.
6. Buchholz, M.; Kunz, H., *Liebigs. Ann. Chem.* **1983**, 1859–1885.
7. Roy, R.; Laferriére, C. A.; Gamian, A.; Jennings, H. J., *J. Carbohydr. Chem.* **1987**, *6*, 161–165.
8. Shimizu, C.; Achiwa, K., *Carbohydr. Res.* **1987**, *166*, 314–316.
9. Roy, R.; Laferriére, C. A., *Can. J. Chem.* **1990**, *68*, 2045–2054.
10. Allanson, N. M.; Davidson, A. H.; Floyd, C. D.; Martin, F. M., *Tetrahedron: Asymmentry.* **1994**, *5*, 2061–2076.
11. Gan, Z.; Roy, R., *Tetrahedron* **2000**, *56*, 1423–1428.
12. Roy, R.; Das, S. K.; Hernández-Mateo, F.; Santoyo-González, F.; Gan, Z., *Synthesis* 2001; 1049–1052.
13. Gan, Z.; Roy, R., *Can. J. Chem.* **2002**, *80*, 908–916.
14. Ikeda, K.; Kawai, K.; Achiwa, K., *Chem. Pharm. Bull.* **1991**, *39*, 1305–1309.
15. Murakami, M.; Ikeda, K.; Achiwa, K., *Carbohydr. Res.* **1996**, *280*, 101–110.
16. Van der Vleugel, D. J. M.; Wassenburg, F. R.; Zwikker, J. W.; Vliegenthart, J. F. G., *Carbohydr. Res.* **1982**, *104*, 221–233.
17. Pozsgay, V.; Jennings, H. J.; Kasper, D. L., *J. Carbohydr. Chem.* **1987**, *6*, 41–55.
18. Sato, S.; Fujita, S.; Furuhata, K.; Ogura, H.; Yoshimura, S.; Itoh, M.; Shitori, Y., *Chem. Pharm. Bull.* **1987**, *35*, 4043–4048.
19. Dekany, G.; Wright, K.; Ward, P.; Toth, I., *J. Carbohydr. Chem.* **1997**, *16*, 11–24.
20. Lunsford, C. D.; Murphey, R. S., *J. Org. Chem.* **1956**, *21*, 580–582.
21. Alpegiani, M.; Bedeschi, A.; Bissolino, P.; Visentin, G.; Zarini, F.; Perrone, E.; Franceschi, G., *Heterocycles* **1990**, *31*, 617–628.
22. Carvalho, E.; Francisco, A. P.; Iley, J.; Rosa, E., *Bioorg. Med. Chem.* **2000**, *8*, 1719–1725.

Corwin M. Nycholat & Todd L. Lowary
University of Alberta, Edmonton, Alberta, Canada

Silver Silicate

[42613-24-1]

(heterogeneous catalyst used in glycosylation reactions)

Solubility: insoluble in organic solvents.
Preparative Methods: [1] silica gel 60 (15 g, 230–400 mesh) was suspended in a solution of silver nitrate (17.0 g, 0.1 mol) in water (100 mL). Sodium silicate (48 g, 34.6 mL, Merck) in water (100 mL) was added with stirring and the mixture was warmed for 15 min at 70 °C and filtered. The precipitate was suspended in 10% silver nitrate solution (100 mL) at 70 °C and stirred for 15 min. Acetone (100 mL) was added and the mixture was filtered. The residue was washed with acetone (100 mL) and toluene (100 mL), and dried in vacuo with a rotary evaporator with the exclusion of light. The yellow powder was dried under high vacuum for 6 h at 70 °C; yield obtained was 38 g.
Handling, Storage, and Precautions: the silver silicate catalyst must be carefully dried and stored in the absence of light.

Silver silicate is used as a heterogeneous catalyst in glycosylation reactions involving the coupling of a glycosyl halide with an acceptor alcohol. Several oligosaccharides and other types of glycosides have been synthesized using silver silicate as a catalyst. A critical problem in the synthesis of oligosaccharides is control of the stereochemistry at the newly formed glycosidic linkage. Among the most difficult types of linkages to synthesize are β-linkages involving either D-mannose or L-rhamnose. In neither of these carbohydrates can β-glycosides, which have a 1,2-*cis* stereochemistry, be accessed via neighboring group participation from a substituent at the C-2 position; the opposite, α-configuration results at the anomeric center. With non-participating groups at C-2, the β-linkage is still disfavored under most conditions because the reaction involves conversion of a more stable α-halide into a less stable β-glycoside.[2] Reaction conditions which promote anomerization in situ will usually favor the formation of the α-linkage. Insoluble catalysts such as silver silicate and silver zeolite were developed as a way to prepare β-mannosides from donors with non-participating groups at C-2, since under heterogeneous conditions anomerization of the α-halide would be suppressed and the coupling reaction would proceed with inversion of configuration at the anomeric center.[2] Silver silicate[3] and silver zeolite[4] are the most common heterogeneous catalysts applied to this reaction. Silver silicate is more reactive than silver zeolite, presumably because silver ions in the latter are present in lower surface concentration.[5] Silver silicate is prepared by precipitation from silver nitrate and sodium silicate solutions onto either silica gel[1] or alumina[3] as solid phase supports. A similar reagent, referred to as silver silicate aluminate,[5] has also been prepared by immobilization of silver(I) onto silica-alumina and it was found to give results similar to silver silicate on alumina.

Paulsen and Lockhoff introduced silver silicate catalysis in the synthesis of a β-mannoside using a mannopyranosyl bromide as the donor and a 1,6-anhydro sugar derived from galactose as the acceptor (eq 1).[3] This disaccharide was further elaborated to trisaccharides related to mannose-containing oligosaccharides found in the LPS O-antigen of *E. coli 075*. When the same coupling reaction was carried out using silver carbonate as the catalyst, a much lower yield of disaccharide was obtained.

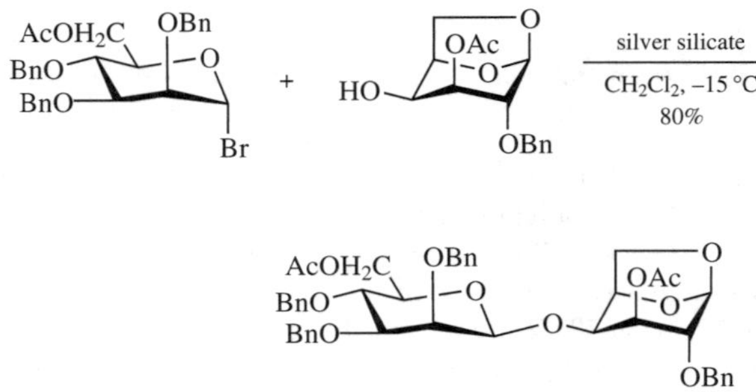

(1)

The β-D-Man(1→4)-β-D-GlcNAc linkage is important owing to its frequent presence in the carbohydrate domain of *N*-glycoproteins. Paulsen and Lebuhn described a critical coupling reaction for the synthesis of the fundamental disaccharide unit, in which a mannopyranosyl bromide was coupled to an acceptor with an azido group to give the β-disaccharide in 62% yield (eq 2).[6]

Oligosaccharides containing β-linkages involving L-rhamnose have also been synthesized by the silver silicate method. Since L-rhamnose is a 6-deoxy sugar, the glycosyl halides derived from it are more reactive. High stereoselectivities for β-glycoside bond formation have been observed with rhamnopyranosyl bromides. The trisaccharide repeating unit of the O-specific side chain of LPS from *Shigella flexneri* serotype was synthesized by Paulsen and Kutschker using two stereoselective glycosylations involving rhamnopyranosyl bromides. The first of these is glycosylation of the 2-OH in a rhamnopyranose acceptor (eq 3).[1]

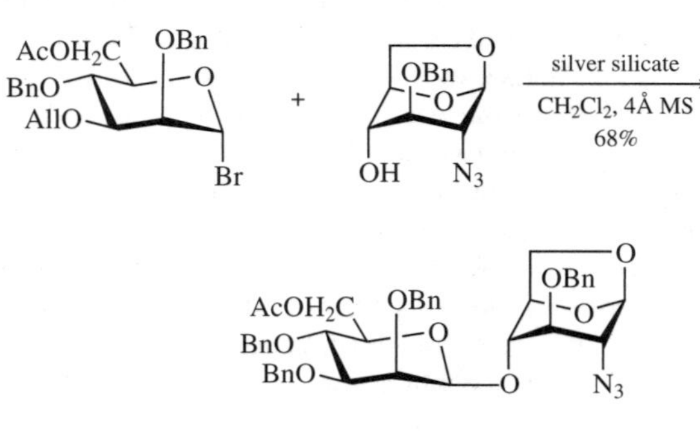

(2)

β/α = 7:1 (62% β)

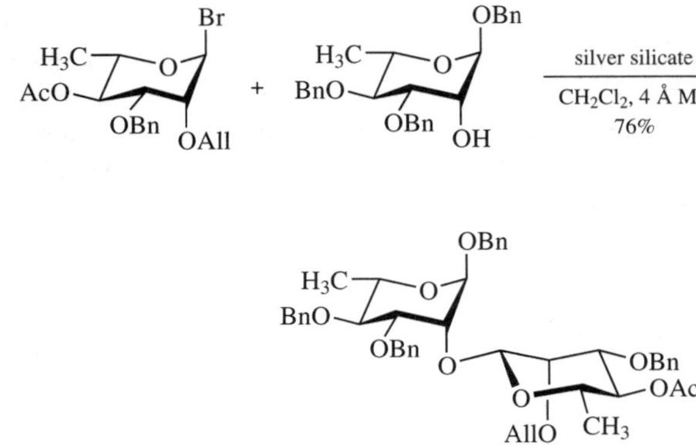

(3)

The disaccharide was deacylated and then coupled to 2,3,4-tri-*O*-benzylrhamnopyranosyl bromide, again with complete β-selectivity, to give a protected β-L-Rha(1→4)-β-L-Rha(1→2)-Rha trisaccharide, which was deblocked to complete the synthesis.

The yield and stereoselectivity of the silver silicate catalyzed glycosylations are sensitive to protecting groups on the bromide and also the reactivity of the alcohol acceptor. Higher β/α ratios are observed in many cases with glycosyl bromides which are made more reactive, or 'armed,' by protection with benzyl or allyl ether protecting groups rather than acyl groups.[2] However, exceptions to this generalization have been found in a study of mannopyranosyl bromides, some of which gave higher amounts of β-glycoside with acyl protecting groups at C-4.[7] Non-polar solvents such as dichloromethane or toluene produce the highest ratio of β- to α-glycosides, presumably due to suppression of oxacarbenium ion formation. Caution is advised in predicting stereochemical outcomes based on protecting group patterns. An unexpected result was observed in the coupling of a galactosyl bromide with a 1,6-anhydroglucose acceptor. With protection of the C-4 hydroxyl group in the bromide as the benzoate, only the α-linked disaccharide was produced, while the 4-*O*-benzyl analog gave a 1:1 mixture of anomers.[8]

The coupling reactions of 3,4-di-*O*-acetyl-2,6-dibromo-2,6-dideoxy-α-D-glucopyranosyl bromide, and its 6-*O*-acetyl and benzoyl analogs, with simple acceptor alcohols were studied systematically using different promoters.[9] Coupling of the 6-bromo derivative with benzyl alcohol in the presence of silver silicate gave exclusively the β-anomer, while a 1:1 mixture of anomers was obtained with silver triflate, a soluble catalyst (eq 4).

Reaction of this bromide with methyl 2,3-di-*O*-isopropylidene-α-L-rhamnopyranoside in dichloromethane gave exclusively the β-glycoside with either silver silicate or silver triflate; however, the yield was low. A mixture of anomers was obtained in a more polar solvent (nitromethane-toluene), again illustrating the same solvent dependence observed previously for glycosylations with silver silicate.

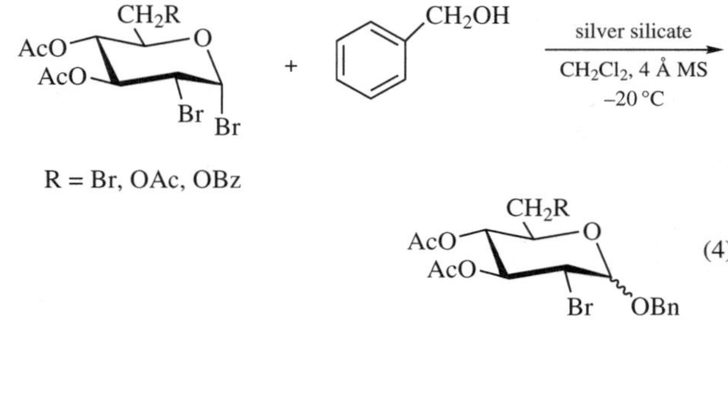

R = Br, OAc, OBz

(4)

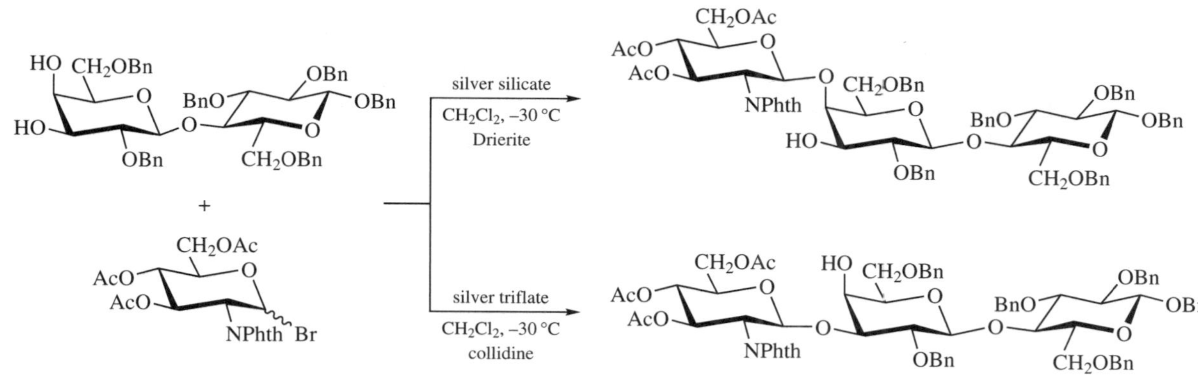
(5)

The glycosylation of a lactose derivative with an *N*-phthaloyl glucosyl bromide occurred regioselectively, with silver triflate giving the β(1→3)-linked product as the major isomer, while silver silicate gave mainly the β(1→4)-linked glycoside (eq 5).[10]

Other applications of silver silicate as a heterogeneous catalyst in glycosylations include synthesis of deoxy lactosides for use in studies of protein-lactose binding,[11] synthesis of a branched pentasaccharide as well as other oligosaccharides present in *N*-glycoproteins,[12,13] synthesis of the oligosaccharide sequence of asialo-GM1- and asialo-GM2-gangliosides,[14] synthesis of β-glucopyranosides of terpenoid alcohols,[15] glycosylations of L-iduronic acid derivatives to develop analogs of glycosaminoglycans,[8] and glycosylations of steroids and steroidal 17-hydroxy derivatives to develop analogs of cardiotonic agents.[16,17]

Related Reagents. Silver zeolite.[4]

1. Paulsen, H.; Kutschker, W., *Carbohydr. Res.* **1983**, *120*, 25–42.
2. Paulsen, H., *Angew. Chem., Int. Ed. Engl.* **1982**, *21*, 155–224.
3. Paulsen, H.; Lockhoff, O., *Chem. Ber.* **1981**, *114*, 3102–3114.
4. Garegg, P. J.; Ossowski, P., *Acta. Chem. Scand.* **1983**, *B37*, 249–250.
5. van Boeckel, C. A. A.; Beetz, T.; Kock-Van Dalem, A. C.; Van Bekkum, H., *Recl. Trav. Chim. Pays-Bas* **1987**, *106*, 596–598.
6. Paulsen, H.; Lebuhn, R.; Lockhoff, O., *Carbohydr. Res.* **1982**, *103*, C7–C11.
7. van Boeckel, C. A. A.; Beetz, T.; van Aelst, S. F., *Tetrahedron* **1984**, *40*, 4097–4107.
8. Spijker, N. M.; Basten, J. E. M.; van Boeckel, C. A. A., *Recl. Trav. Chim. Pays-Bas* **1993**, *112*, 611–617.
9. Bock, K.; Lundt, I.; Pedersen, C., *Carbohydr. Res.* **1984**, *130*, 125–134.
10. Paulsen, H.; Hadamczyk, D.; Kutschker, W.; Bunsch, A., *Liebigs Ann.* **1985**, 129–141.
11. Ekberg, T.; Magnusson, G., *Carbohydr. Res.* **1993**, *246*, 119–136.
12. Paulsen, H.; Heume, M.; Györgydeak, Z.; Lebuhn, R., *Carbohydr. Res.* **1985**, *144*, 57–70.
13. Paulsen, H.; Lebuhn, R., *Liebigs Ann.* **1983**, 1047–1072.
14. Paulsen, H.; Paal, M.; Hadamczyk, D.; Steiger, K. M., *Carbohydr. Res.* **1984**, *131*, C1-C5.
15. Paulsen, H.; Bien, L. N.; Sinwell, V.; Heemann, V.; Seehofer, F., *Liebigs Ann.* **1985**, 1513–1536.
16. Cerny, I.; Novotny, P.; Drasar, P.; Havel, M.; Pacek, J., *Collect. Czech. Chem. Commun.* **1989**, *54*, 2775–2783.
17. Cerny, I.; Pouzar, V.; Drasar, P.; Havel, M., *Collect. Czech. Chem. Commun.* **1992**, *57*, 362–374.

Robert M. Giuliano
Villanova University, Villanova, PA, USA

Silver(I) Tetrafluoroborate

$AgBF_4$

[14104-20-2] $AgBF_4$ (MW 194.68)

(mild Lewis acid with a high affinity for organic halides)

Physical Data: mp 200 °C (dec).
Solubility: soluble in benzene, toluene, nitromethane, diethyl ether, water.
Form Supplied in: white solid; widely available.
Analysis of Reagent Purity: contents of Ag can be assayed conveniently by volumetric titration of Ag^I.
Preparative Methods: can be prepared by reacting Silver(I) Fluoride with Boron Trifluoride in nitromethane.[1]
Handling, Storage, and Precautions: should be protected from light and moisture; very hygroscopic.

Original Commentary

Lars-G. Wistrand
Nycomed Innovation, Malmö, Sweden

Introduction. This reagent has replaced Silver(I) Perchlorate to a large extent because of the sensitivity of perchlorates.

Activation of Acyl Chlorides. In several cases, $AgBF_4$ has been used to increase the reactivity of acyl chlorides towards nucleophiles.[2] For example, *N*-acylammonium salts were prepared for the first time by the reaction of a tertiary amine and an acyl chloride in the presence of $AgBF_4$ (eq 1).[3]

Nucleophilic Substitution on Alkyl Halides by Heteroatoms. A number of more or less activated alkyl halides, such as benzyl halides[4] and allyl halides,[5] undergo substitution reactions mediated by $AgBF_4$ in the presence of a heteroatom nucleophile. For example, treatment of pentamethylcyclopentadienyl bromide with $AgBF_4$ in the presence of a nucleophile gives the corresponding substituted product (eq 2). Thiols, amines, and alcohols have been used as nucleophiles.[6]

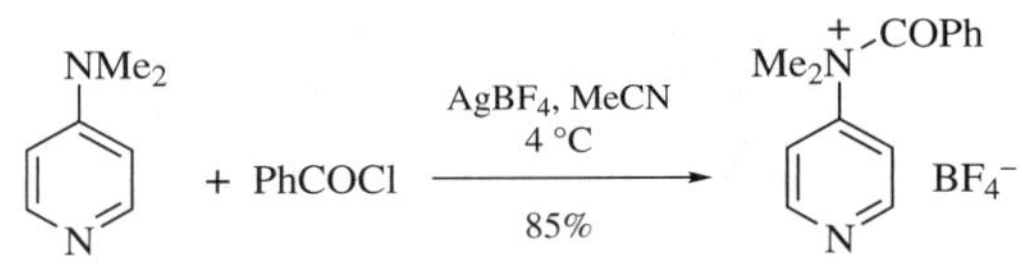

(1)

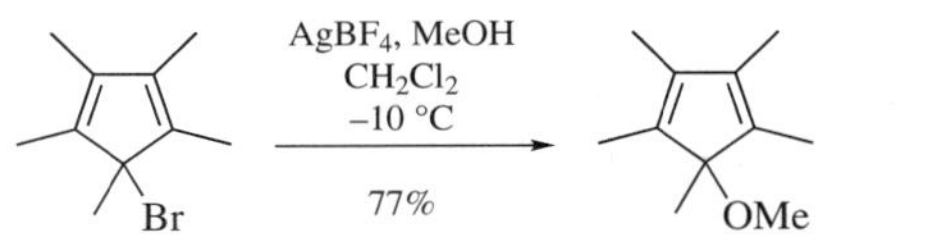

(2)

Adenine analogs are prepared stereoselectively from cyclopentene derivatives using a two-step procedure (eq 3). The reaction probably involves a seleniranium salt as an intermediate.[7]

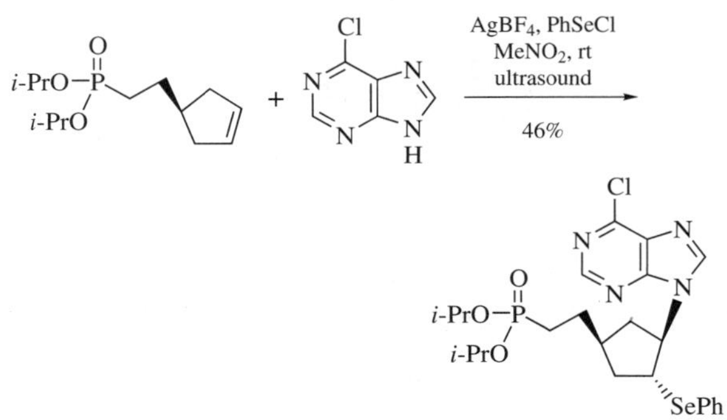

(3)

Intramolecular substitutions mediated by Ag^+ do not seem to require activated halides.[8] For example, ω-chloro amides react with $AgBF_4$, giving products from intramolecular attack of the amide oxygen. Depending on the structure of the amide, imino lactones,[9] imino lactonium salts,[10] or lactone hydrazones (eq 4)[11] are obtained as products. Fluorination of α-bromo ketones using $AgBF_4$ has also been reported.[12]

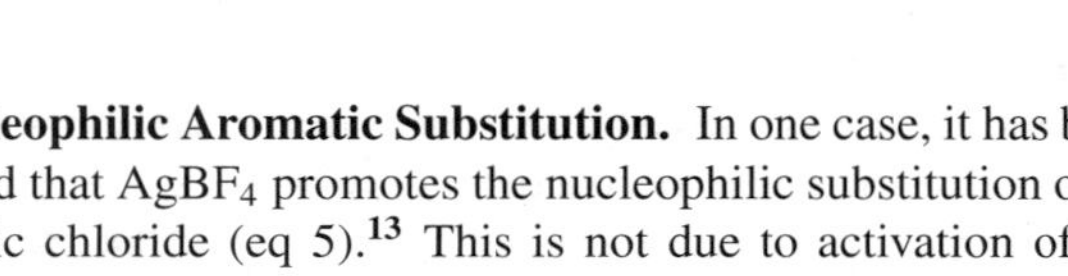

(4)

Nucleophilic Aromatic Substitution. In one case, it has been reported that $AgBF_4$ promotes the nucleophilic substitution of an aromatic chloride (eq 5).[13] This is not due to activation of the halide, but apparently to suppression of halide-promoted decomplexation of the arene–manganese derivative.

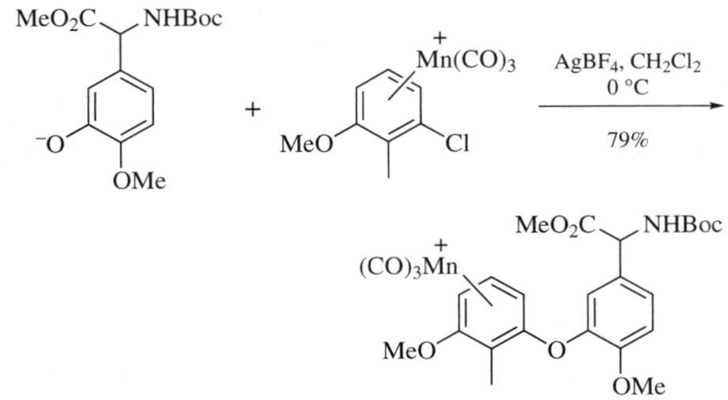

(5)

Carbon–Carbon Bond Formation via Cationic Intermediates. In analogy with the heteroatom substitutions described above, certain aliphatic halides undergo substitution reactions with carbon nucleophiles promoted by $AgBF_4$. For example, Eschenmoser and co-workers used $AgBF_4$ in order to transform α-chloro nitrones into 1,3-dipoles which react with ordinary alkenes in a cycloaddition manner (eq 6).[14]

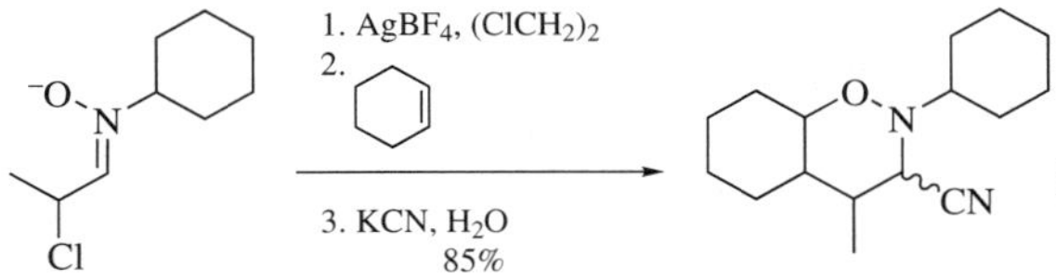

(6)

Livinghouse and co-workers have shown that acylnitrilium ions, prepared from isocyanides and an acid chloride followed by treatment with $AgBF_4$, are useful intermediates in the synthesis of nitrogen-containing heterocycles (eq 7).[15]

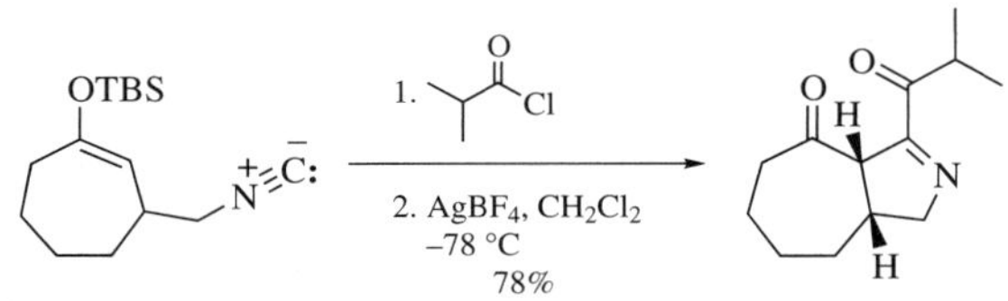

(7)

In certain cases, allylsilanes[16] and trimethylsilyl enol ethers[17] react with alkyl halides with the formation of a new carbon–carbon bond. α-Bromo imidates[18] (eq 8) and β-chloro imines[19] have been reported to undergo electrophilic aromatic substitution on relatively electron-rich aromatics in the presence of $AgBF_4$.

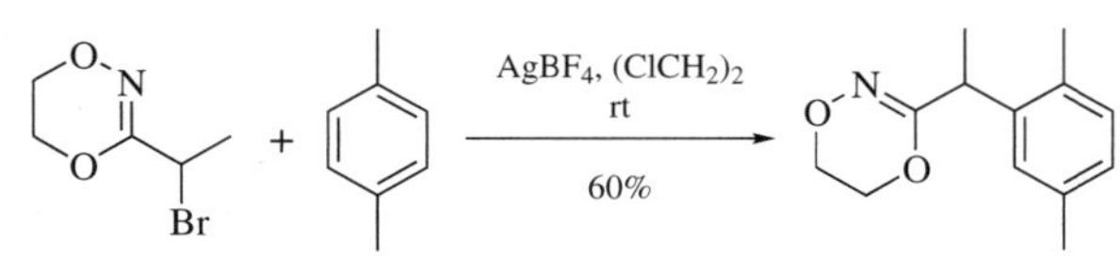

(8)

Synthesis via Iminium Ions. α-Cyano amines react with $AgBF_4$ with the formation of an intermediate iminium ion.[20] This has been used synthetically as a method for removal of the cyano group either by a consecutive reduction[21,22] to the amine (eq 9) or by elimination to the imine[23] or enamine.[24]

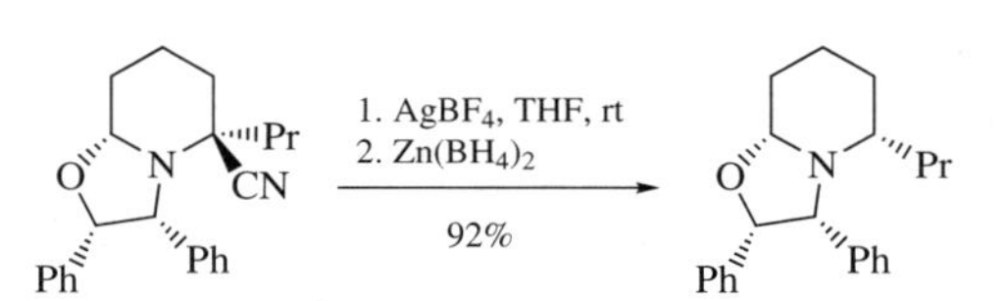

(9)

Rearrangements. A number of strained alkyl and/or reactive halides, such as cyclopropyl[25] and bicyclic[26,27] chlorides, rearrange on treatment with $AgBF_4$. For example, β-bromotetrahydropyrans rearrange to tetrahydrofurans stereoselectively on treatment with $AgBF_4$ (eq 10).[28] Other examples include the rearrangement of α-haloalkyl aryl ketones into arylacetic acid derivatives,[29] and the rearrangement of α-haloalkylsilanes upon treatment with $AgBF_4$.[30]

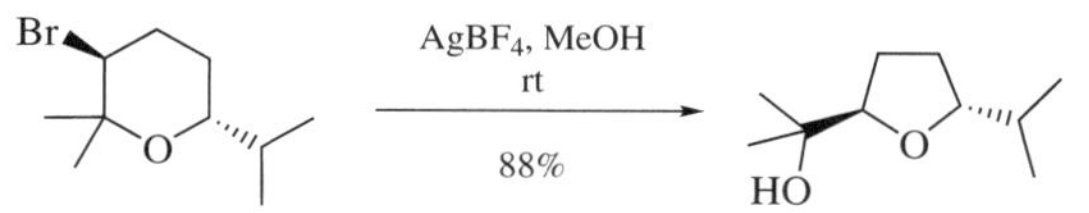

(10)

In the presence of strained hydrocarbons, $AgBF_4$ functions as a mild Lewis acid and causes rearrangements.[31–33] For example, the tricyclic hydrocarbon (**1**) rearranges upon treatment with a catalytic amount of $AgBF_4$ to the less strained hydrocarbon (**2**) (eq 11).[34]

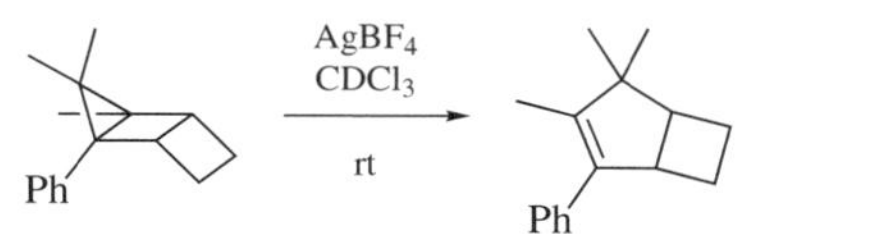

(**1**) (**2**) (11)

Numerous examples include the rearrangement of propargyl esters into allenyl esters (see also Silver(I) Trifluoromethanesulfonate)[35] or to dihydrofurans,[36] the Claisen rearrangement of aryl allenylmethyl ethers,[37] and the rearrangement of silyloxycyclopropanes (eq 12, also effected by Silver(I) Trifluoromethanesulfonate).[38]

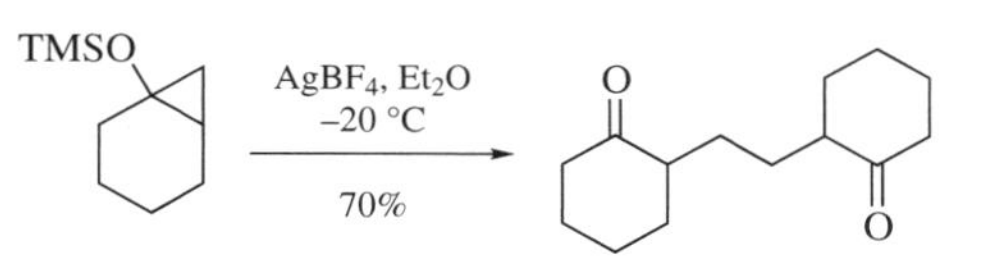

(12)

Activation of Thiol Esters. Pyridyl thiol esters are converted into esters on treatment with $AgBF_4$ and an alcohol.[39] Acylation of alkynylsilanes can also be carried out using thiol esters in the presence of $AgBF_4$.[40]

Alkylation of Thioethers. Thioethers can be methylated by Iodomethane in the presence of $AgBF_4$.[41] Benzylation of thioethers in the presence of $AgBF_4$ has also been reported.[42]

Electrophilic Aromatic Substitution. Electrophilic nitration using a combination of NO_2Cl and $AgBF_4$ has been reported.[43] Conversion of arylsilanes into iodides and bromides has been achieved using a combination of the halogen and $AgBF_4$ (eq 13).[44]

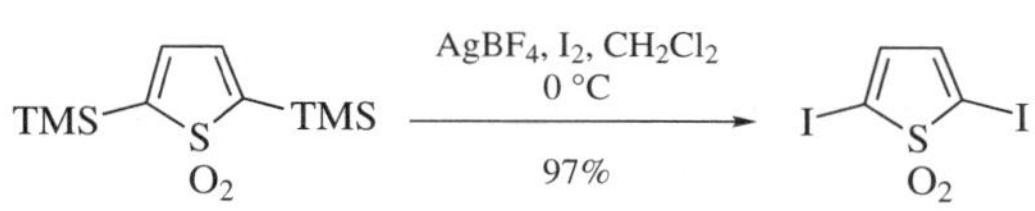

(13)

Catalysis of Cycloadditions. Addition of catalytic amounts of $AgBF_4$ greatly increases the selectivity of [2 + 4] cycloadditions of benzyne.[45]

First Update

Angela D. Kerekes & Mousumi Sannigrahi
Schering-Plough Research Institute, Kenilworth, NJ , USA

Cycloaddition Reactions. Several reports of [4 + 3] cycloadditions have been noted in the recent literature. Kende, et al.[46] reported the first asymmetric version of the [4 + 3] cycloaddition of 2-aminoallyl cations with furans and pyrroles (eq 14). They report up to 60% *ee* when using 2-aminoallyl cations derived from chiral α-chloroimines.

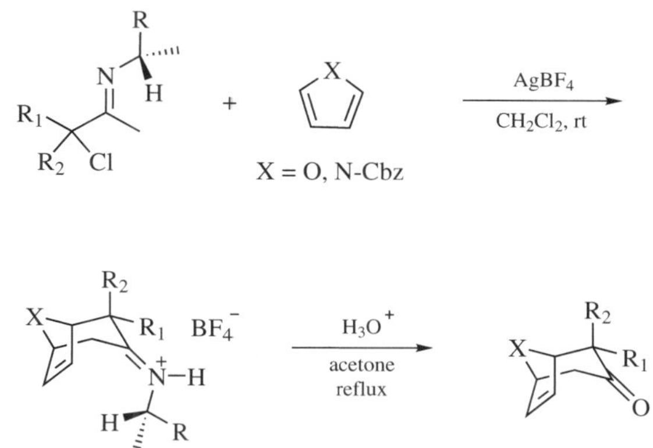

(14)

Similar products can be formed via the method of Cha.[47] Using Schmid conditions a [4 + 3] cycloaddition of oxyallyl intermediates derived from α′-chloroenamines gave the corresponding cycloadducts. A comparison study between Schmid ($AgBF_4$),[48] Noyori ($Fe_2(CO)_9$),[49] Hoffmann (NaI/Cu),[50] and Föhlisch (Et_3N and CF_3CH_2OH)[51] conditions was undertaken. Different substitution patterns on the cyclohexyl ring were also explored with variable yields and diastereoselectivities (eq 15).

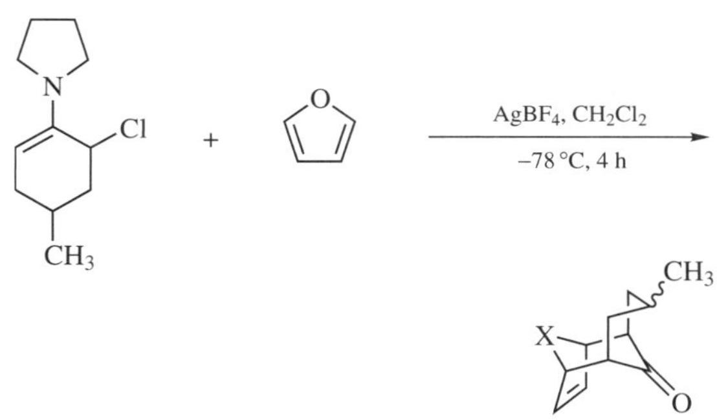

(15)

β = 73%
α = 5%

Ag(I) salts were also shown to promote the [2 + 2 + 2] cyclization of 1,6-enynes in the presence of tris(triphenylphosphine) rhodium chloride (Wilkinson's catalyst) (eq 16).[52] Various cocatalysts were tested ($AgBF_4$, AgOTf, $BF_3 \cdot OEt_2$, $ZnCl_2$, formic acid, acetic acid, and triflic acid). The reaction was shown to be quite general on a wide range of enyne substrates, which incorporated ester, alcohol, silyl ethers, and tertiary amine functionality. The yields range from 60–89% and, clearly, $AgBF_4$ was the cocatalyst of choice.

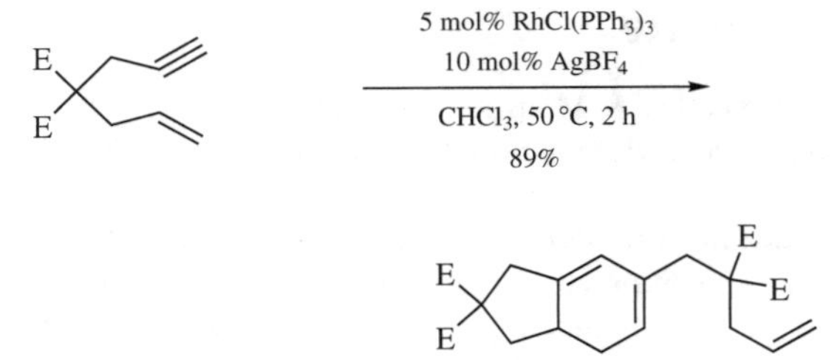

(16)

Cyclizations. Buszek, et al.[53] employed a modified Corey-Nicholaou[54] "double-activation" lactonization in the total synthesis of octalactin A and B (eq 17).

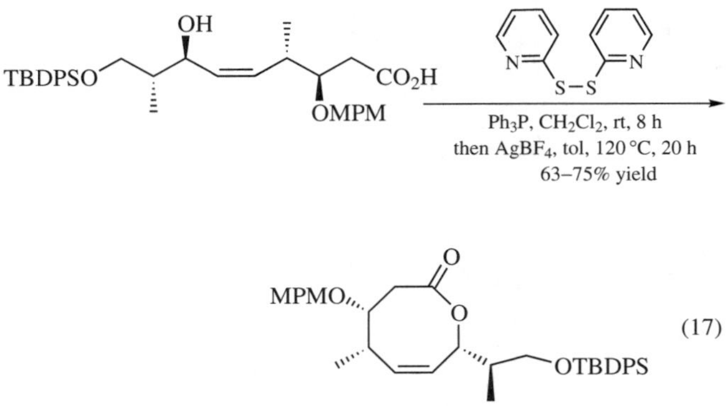

(17)

Silver tetrafluoroborate facilitated a reaction in developing new access into the *E*-azaeburnane core. The cyclization of α-bromoacetamide via electrophilic aromatic substitution was mediated by $AgBF_4$ in DME (eq 18),[55] basic conditions for effecting the cyclization failed (e.g., K_2CO_3 and DMF) to produce any of the desired product.

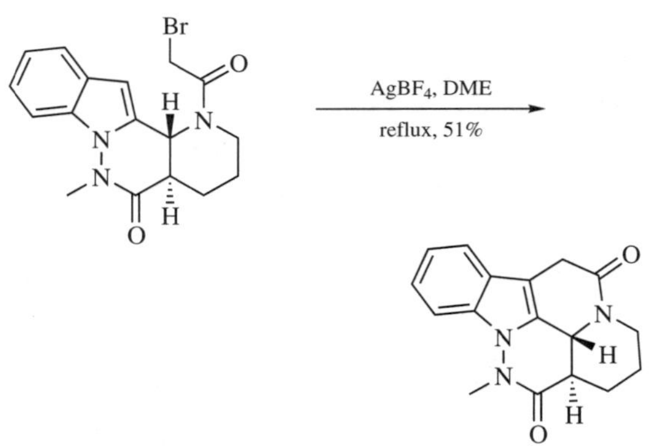

(18)

An oxidative cyclization has been accomplished by Evans and co-workers in the synthesis of tetrapeptide vancomycin antibiotics.[56] The overall transformation was an aryl-aryl bond formation utilizing VOF_3 as the oxidant in TFA at 0 °C (eq 19). $BF_3.OEt_2$ was found to be crucial in preventing competitive attack of oxygen nucleophiles on the ring-seven radical cation. Zinc reduces and quenches the radical cation of the product which is also prone to nucleophilic attack, e.g., quenching the reaction with NaCl results in the addition of a chloride to the 5-position of ring-seven. $AgBF_4$, acts as a scavenger for any chloride ion that may be present in commercial TFA.

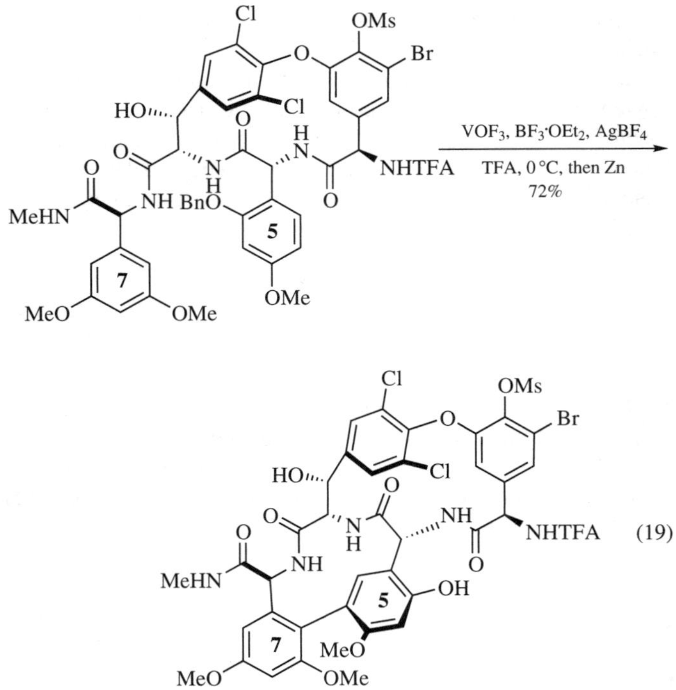

(19)

Gallagher has done a significant amount of work in the electrophilic-mediated cyclizations of allenic amines. If a nucleophile is placed in a suitable position to an allene moiety that is activated via coordination of a Lewis acid, a variety of heterocycles may be formed (eq 20).[57]

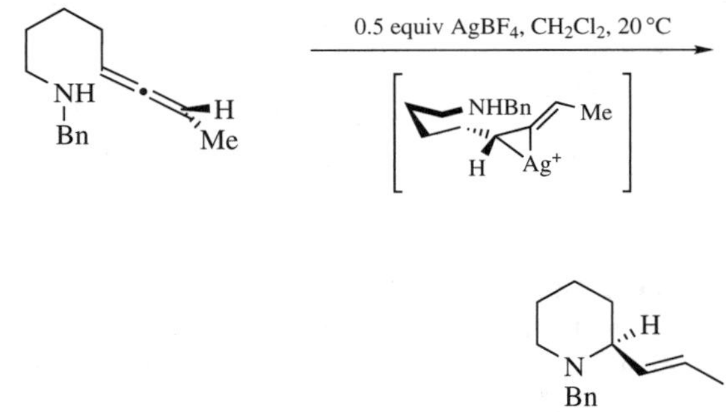

(20)

The reaction of optically active allenic amines with various electrophiles was also explored.[58] It was found that the cyclization of allenic amines using either AgOTf or $AgBF_4$ in CH_2Cl_2 at room temperature gave various degrees of diastereoselectivity depending on the concentration of the Ag(I) salt employed in the reaction. The diastereoselectivities were less affected by the choice of solvent. After surveying numerous conditions and substrates, the observed diastereoselectivity trend that emerged seemed to be related to the increased capacity of the coordinating residue X group to complex the Ag(I) ion (eq 21). The best conditions for this reaction employed 50 mol% Ag(I) in CH_2Cl_2.

Allenic oximes also undergo a Ag(I)-mediated cyclization to give the hexahydroazepine-based nitrone.[59] The nitrone is a reactive 1,3-dipole that is subsequently trapped by styrene or *N*-methylmaleimide to give the corresponding cycloadducts (eq 22). More recently, this Ag(I)-mediated cyclization has been used in the heterocyclization of substituted allenic amine to pyrrolidines using $AgBF_4$ in CH_3CN.[60]

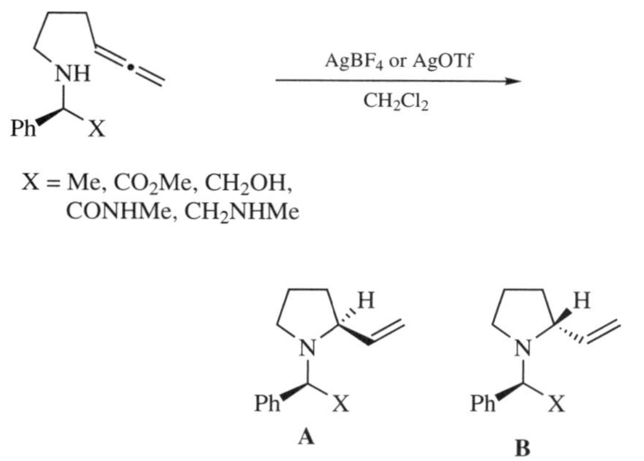

X = Me, CO_2Me, CH_2OH, CONHMe, CH_2NHMe

(21)

X = CONHMe, 50 mol% Ag(I), 90% yield
81% de in favor of **A**.

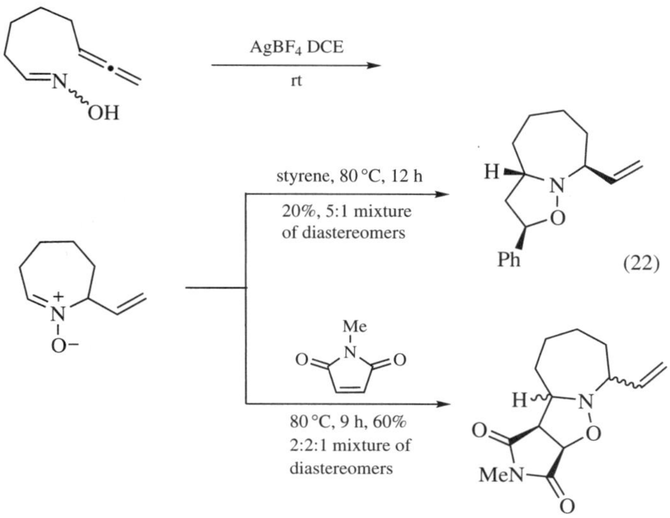

(22)

Rearrangements/Ring Expansions. Overman et al. reported the first example of an enantioselective catalytic rearrangement of allylic imidates to allylic amides.[61] Silver tetrafluoroborate was used to remove the chloride ligands on the Pd catalyst to form the active cationic Pd(II) catalytic species that possessed an optically active tertiary diamine as a ligand. Allylic amides with up to 60% *ee* were obtained. As an extension of their use of cationic Pd(II) catalysts in asymmetric Heck reaction and Wacker reactions, Uozumi and co-workers explored a wide variety of chiral palladium (II) species to effect the asymmetric aza-Claisen rearrangement of allylic imidates to allylic amines.[62] A variety of palladium complexes were surveyed in the presence of $AgBF_4$. The optimum chiral catalyst is generated from $PdCl_2${(*S*)-2-(2-diphenylphosphino)phenyl-4-benzyloxazoline} and $AgBF_4$ with a Pd:Ag ratio of 1:1 and gives up to 81% *ee* (eq 23). Silver (I) salts such as $AgBF_4$ also have been reported to enhance Heck reactions[63] and Suzuki couplings.[64]

The rearrangements of spiro-fused dihydroisoquinolones have been mediated by $AgBF_4$ (eq 24).[65]

The $AgBF_4$ catalyzed phenyl rearrangement of (*S*)-2-chloropropiophenone dimethyl acetal to (*S*)-2-phenylpropionic acid was achieved with high stereoselectivity:[66] a silver ion-promoted and phenyl-assisted intramolecular S_N2 mechanism is proposed to account for the inversion of stereochemistry. The $AgBF_4$-mediated reaction produced over 97% *ee* of (*S*)-2-phenylpropionic acid methyl ester (eq 25). The photo-irradiated reaction of (*S*)-2-chloropropiophenone gave the partially racemized 2-phenylpropionic acid derivative (eq 26). This photo-induced reaction is believed to also occur with inversion of stereochemistry, but the intervention of an ion or radical intermediate is believed to be responsible for the racemization. The photo-induced rearrangement gave ~50% *ee*, regardless of the quantity of irradiation.

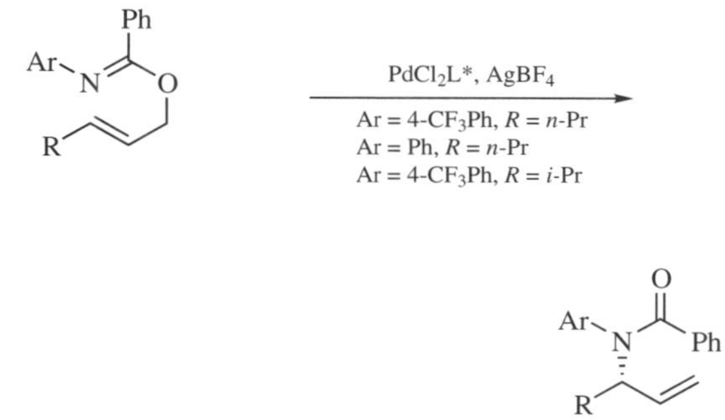

(23)

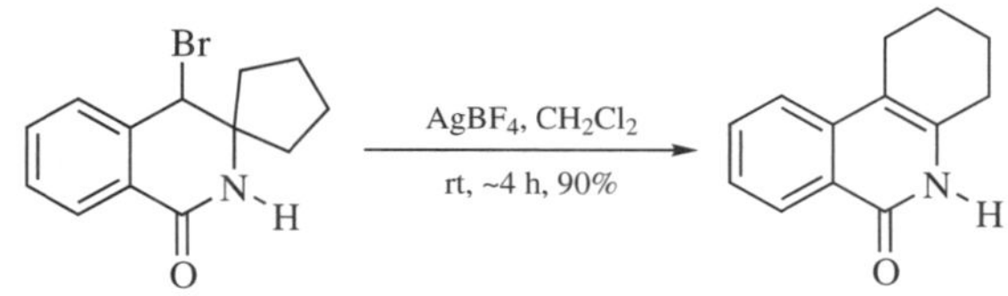

(24)

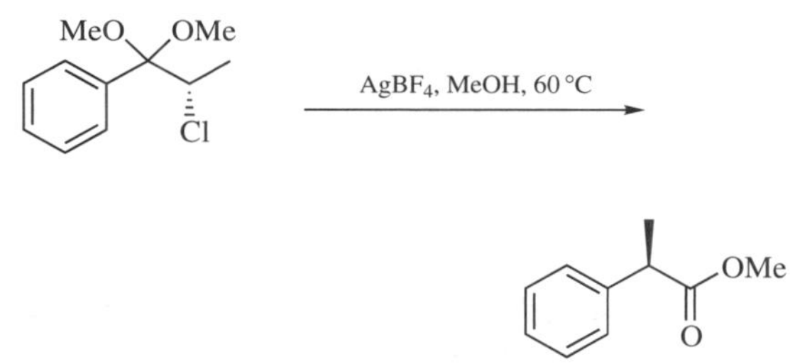

(25)

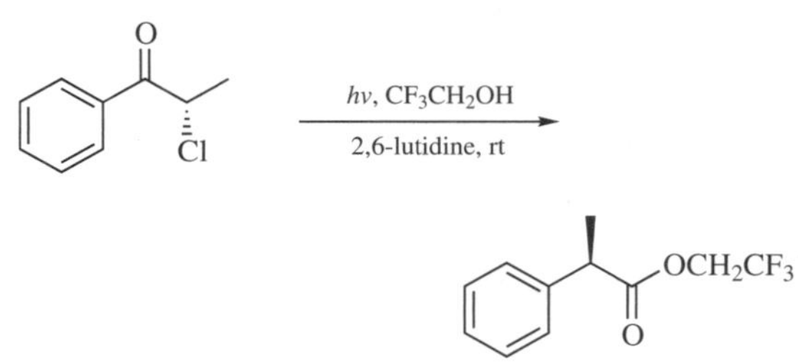

(26)

A Stieglitz rearrangement[67] mediated by $AgBF_4$ was used as the penultimate step in the total synthesis of racemic lycopodine (eq 27).[68]

The preparation of 4,4-disubstituted 2-azetidinones from α-chloro ketones via α-chloro imines and 1-methoxycyclopropylamines has been demonstrated.[69] The 1-methoxycyclopropylamine is prepared in three steps from 3-chloroketones. The rearrangement occurs via *N*-chlorination with *tert*-butyl hypochlorite to give the *N*-chlorocyclopropylamine that subsequently, without isolation, undergoes a $AgBF_4$-mediated ring expansion to give the corresponding β-lactam after workup (eq 28). The yields of this Ag-assisted rearrangement vary between 40–95%. In another report, α,α-disubstituted propargylic alcohols were rearranged via Ag(I) salts into enones.[70] Other catalytic systems that work in this transformation are $AgBF_4$, $AgClO_4$, $AgBF_4$-TMSCl, $AgClO_4$-TMSCl, and $SnCl_4$.

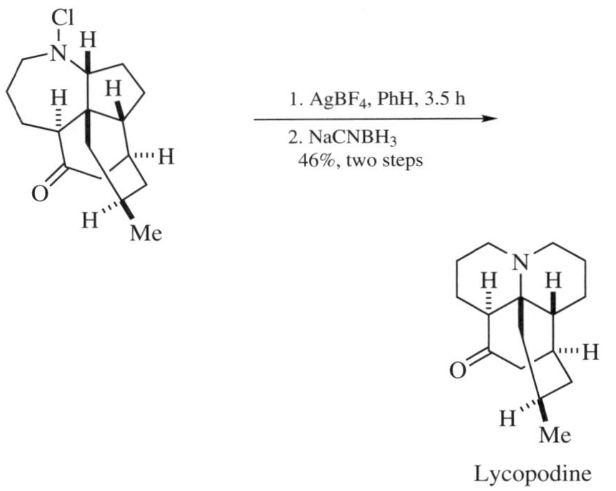

(27)

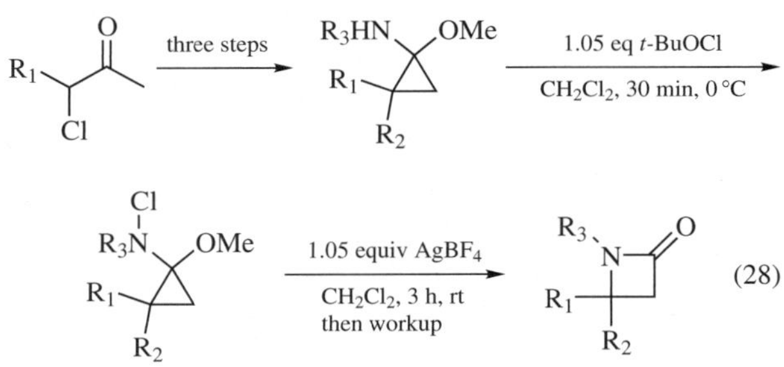

(28)

R_1 = Me, Et
R_2 = Me, Et, Bn
R_3 = *i*-Pr, *t*-Bu

Electrophilic ring opening of a dibromocyclopropane in the presence of $AgBF_4$ in MeOH willingly afforded the α-methoxy vinyl bromide in high yield (eq 29).[71]

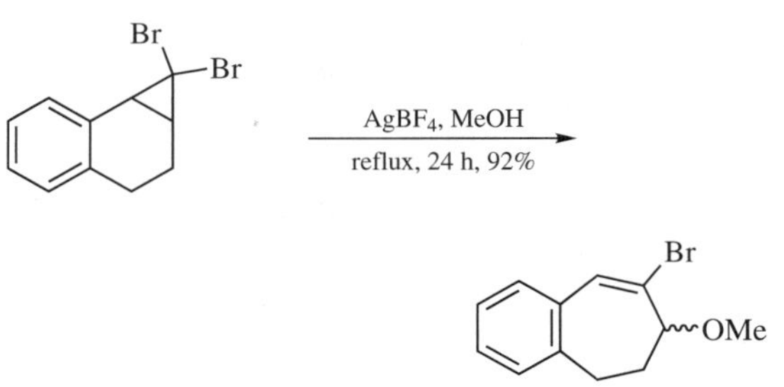

(29)

In a further example of $AgBF_4$ promoting an electrocyclic ring opening of a *gem*-dihalocyclopropane,[72] the intermediate π-allyl cation undergoes a spontaneous cyclization with the appended nucleophile to give the hexahydroindole structure (eq 30).

Oxidations. Bis(quinuclidine)bromide(I) tetrafluoroborate (**1**) reacts as a new Br^+ source in the oxidation of alcohols to carbonyl compounds (eq 31).[73] It was found that high yields were achieved when $AgBF_4$ was utilized as a coreactant. Furthermore, the two quinuclidine units of **1** serve as a "built-in" acid scavenger to absorb the protons liberated in the course of the reaction.

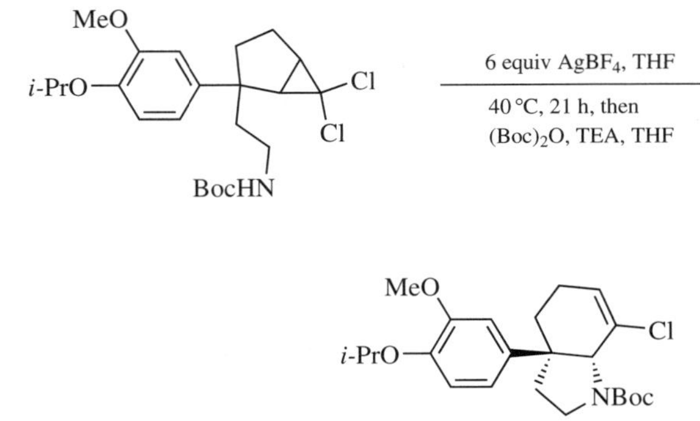

(30)

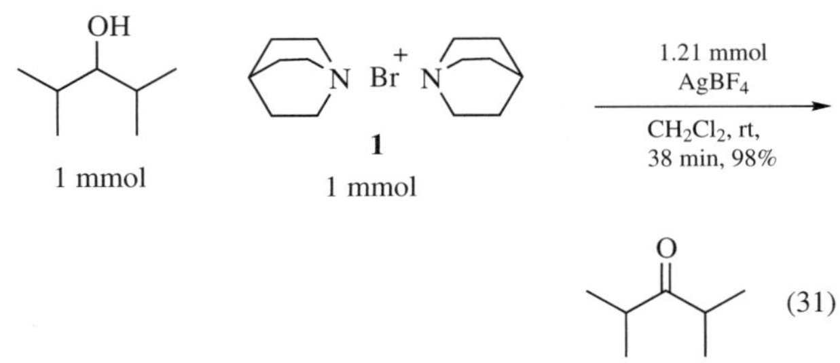

(31)

In their work towards the synthesis of analogs of the immunosuppressant FR901483 (eq 32),[74] Bonjoch and co-workers utilized Lemal and Fry's modification[75] of the Kornblum oxidation.[76]

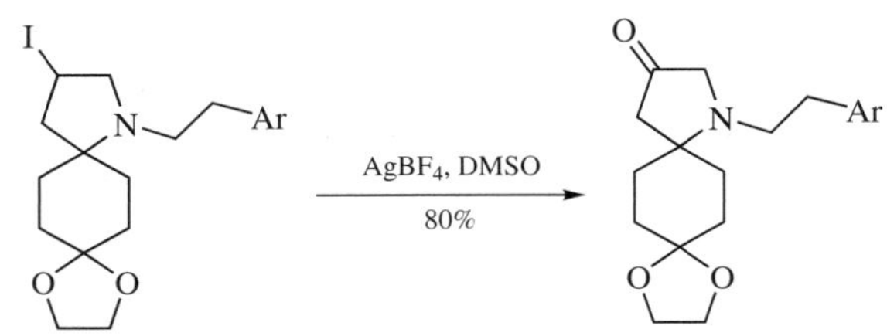

(32)

Another modified Kornblum oxidation[77] has been recently exploited in the total synthesis of solidago alcohol (eq 33).

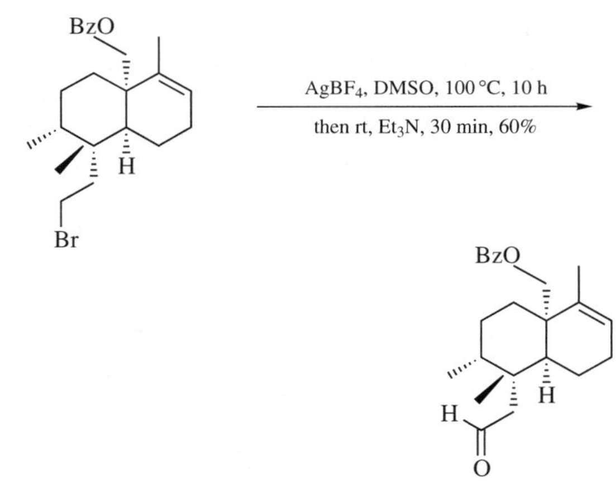

(33)

Substitutions Mediated by $AgBF_4$. Tertiary alcohols can be transformed into isocyanides by the treatment with silver salts and trimethylsilyl cyanide followed by hydrolysis with $NaHCO_3$.[78] Excellent yields are obtained with $AgBF_4$ as the Lewis acid (eq 34). Silver perchlorate and silver triflate are equally effective in this transformation. When other silver (I) salts (AgF, AgCl, AgBr, AgI, AgO_2CCF_3, $AgNO_3$, and $AgIO_3$) were employed in the reaction, the corresponding trimethylsilyl ethers were formed. The reaction tolerates various functional groups (esters, sulfides, amines, amides, and imides) and transforms tertiary alcohols selectively in the presence of primary and secondary alcohols. The attack of the TMSCN typically occurs from the opposite face to that shielded by the departing silver salt.

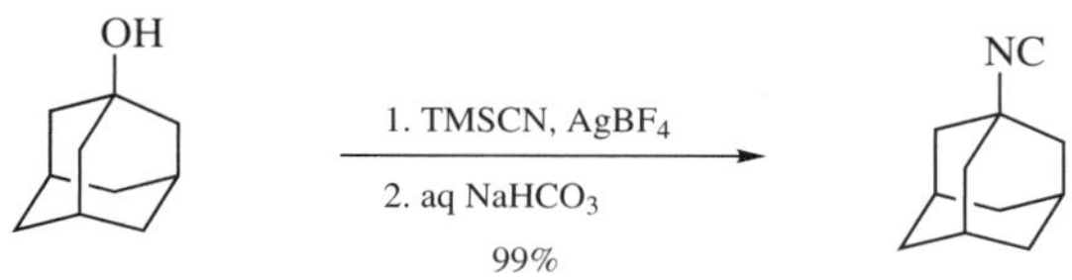

(34)

Silver (I) tetrafluoroborate has been used to mediate the Nicholas reaction[79] of hexacarbonyldicobalt complexes of γ-chloroalkynones and γ-chloroalkynoates with silyl enol ether or silyl ketene acetals to provide 1,6-dicarbonyl compounds (eq 35).[80] The reaction proceeds via a Ag(I)-mediated chloride abstraction from the cobalt hexacarbonyl complexed propargylic chloride to give a highly stabilized carbocation that can undergo further reaction with nucleophiles.

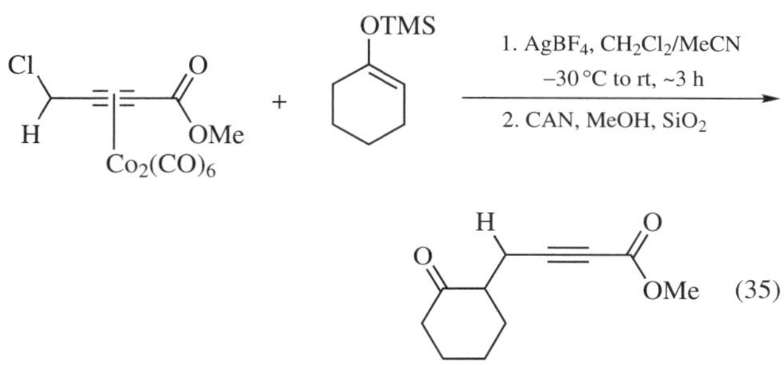

(35)

The thiophilcity of Ag(I) salts is demonstrated in the activation of the C4-S bond in the 4-thioethers of flavan-3-ols towards carbon nucleophiles (eq 36).[81] Silver (I) tetrafluoroborate is an alternative to the moisture sensitive dimethyl(methylthio)sulfonium tetrafluoroborate (DMTSF) in the cleavage of C-S bonds. Also, $AgBF_4$, a thiophilic Lewis acid, showed advantages over protic acids in this study over control of "oligomerization, reversibility, and scattering of the interflavanyl bond(s)."[81]

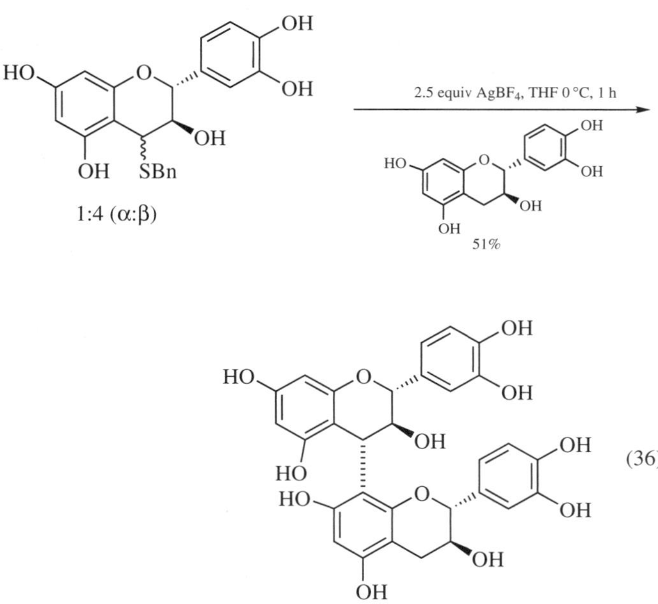

(36)

Thioesters can be formed in high yields by the treatment of 2-(trimethylsilyl)ethyl sulfides or thioglycosides and acid chlorides with $AgBF_4$ (eq 37).[82]

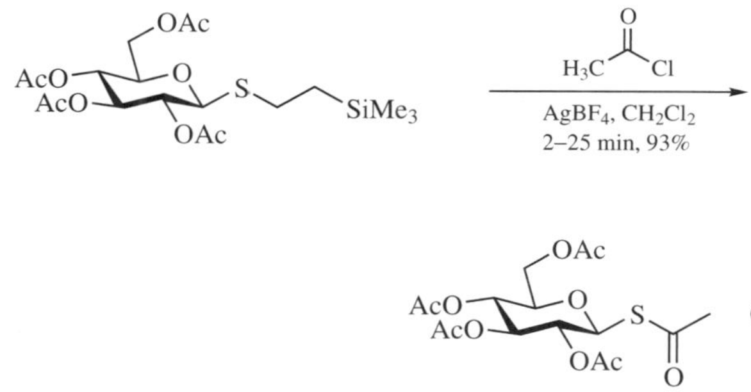

(37)

Promoter of Glycosylation. $AgBF_4$ promotes the α-selective reaction of 1,2-anhydro sugars with stannyl ethers (eq 38) in a reaction though to proceed with a considerable degree of "onium ion character" at the anomeric carbon and an O-Ag bond at the adjacent C-2 position. This delivers the nucleophile from the axial face giving the α-glycoside.[83]

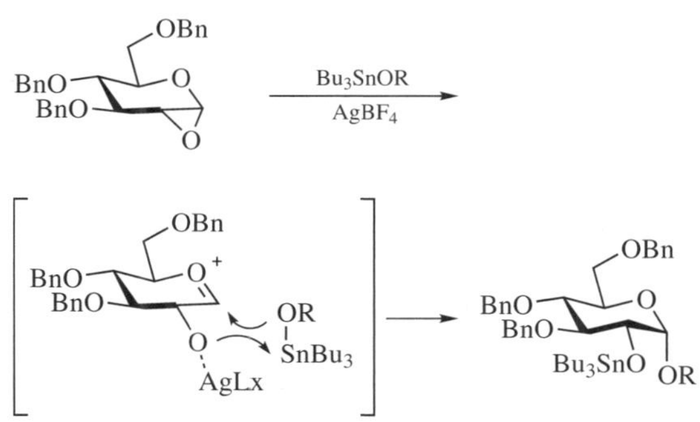

(38)

The reagent combination of Cp_2ZrCl_2-$AgBF_4$ (0.5 : 1.0 molar ratio) was found to be an efficient and rapid promoter in the selective synthesis of α-mannosides from tetra-*O*-benzyl-D-mannosyl fluoride (eq 39).[84] This reagent combination provides highly α-selective glycosidation of glycosyl fluorides with alcohols of various steric demand.

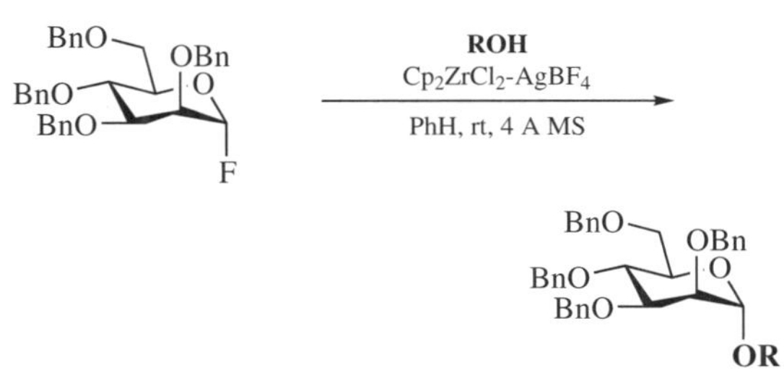

(39)

Related Reagents. Dimethyl Sulfoxide–Silver Tetrafluorborate.

1. Olah, G. A.; Quinn, H. W., *J. Inorg. Nucl. Chem.* **1960**, *14*, 295.
2. Schegolev, A. A.; Smit, W. A.; Roitburd, G. V.; Kucherov, V. F., *Tetrahedron Lett.* **1974**, 3373.
3. King, J. A., Jr.; Bryant, G. L., Jr., *J. Org. Chem.* **1992**, *57*, 5136.
4. Zimmerman, H. E.; Paskovich, D. H., *J. Am. Chem. Soc.* **1964**, *86*, 2149.

5. Bloodworth, A. J.; Tallant, N. A., *Chem. Commun./J. Chem. Soc., Chem. Commun.* **1992**, 428.
6. Jutzi, P.; Mix, A., *Ber. Dtsch. Chem. Ges./Chem. Ber.* **1992**, *125*, 951.
7. Wolff-Kugel, D.; Halazy, S., *Tetrahedron Lett.* **1991**, *32*, 6341.
8. Lucchini, V.; Modena, G.; Pasquato, L., *Chem. Commun./J. Chem. Soc., Chem. Commun.* **1992**, 293.
9. Peter, H.; Brugger, M.; Schreiber, J.; Eschenmoser, A., *Helv. Chim. Acta* **1963**, *46*, 577.
10. Nader, R. B.; Kaloustain, M. K., *Tetrahedron Lett.* **1979**, 1477.
11. Enders, D.; Brauer-Scheib, S.; Fey, P., *Synthesis* **1985**, 393.
12. Fry, A. J.; Migron, Y., *Tetrahedron Lett.* **1979**, 3357.
13. Pearson, A. J.; Shin, H., *Tetrahedron* **1992**, *48*, 7527.
14. Kempe, H. M.; Das Gupta, T. K.; Blatt, K.; Gygax, P.; Felix, D.; Eschenmoser, A., *Helv. Chim. Acta* **1972**, *55*, 2187.
15. Lee, C. H.; Westling, M.; Livinghouse, T.; Williams, A. C., *J. Am. Chem. Soc.* **1992**, *114*, 4089; Luedtke, G.; Westling, M.; Livinghouse, T., *Tetrahedron* **1992**, *48*, 2209.
16. Nishiyama, H.; Naritomi, T.; Sakuta, T.; Itoh, K., *J. Org. Chem.* **1983**, *48*, 1557.
17. Padwa, A.; Ishida, M., *Tetrahedron Lett.* **1991**, *41*, 5673; Padwa, A.; Austin, D. J.; Ishida, M.; Muller, C. M.; Murphree, S. S.; Yeske, P. E., *J. Org. Chem.* **1992**, *57*, 1161.
18. Shatzmiller, S.; Bercovici, S., *Liebigs Ann.* **1992**, 997.
19. Kuehne, M.; Matson, P. A.; Bornmann, W. G., *J. Org. Chem.* **1991**, *56*, 513.
20. Grierson, D. S.; Bettiol, J. L.; Buck, I.; Husson, H. P., *J. Org. Chem.* **1992**, *57*, 6414.
21. Bettiol, J. L.; Buck, I.; Husson, H. P.; Grierson, D. S., *Tetrahedron Lett.* **1991**, *32*, 5413.
22. Theodorakis, E.; Royer, J.; Husson, H. P., *Synth. Commun.* **1991**, 521.
23. Belattar, A.; Saxton, J. E., *J. Chem. Soc., Perkin Trans. 1* **1992**, 1583.
24. Agami, C.; Couty, F.; Lin, J., *Heterocycles* **1993**, *36*, 25.
25. Birch, A. J.; Keeton, R., *J. Chem. Soc. (C)* **1968**, 109.
26. Yamada, Y.; Kimura, M.; Nagaoka, H.; Ohnishi, K., *Tetrahedron Lett.* **1977**, 2379.
27. Kraus, G. A.; Zheng, D., *Synlett* **1993**, 71.
28. Ting, P. C.; Bartlett, P. A., *J. Am. Chem. Soc.* **1984**, *106*, 2668.
29. Giordano, C.; Castaldi, G.; Casagrande, F.; Belli, A., *J. Chem. Soc., Perkin Trans. 1* **1982**, 2575.
30. Eaborn, C.; Lickiss, P. D.; Najim, S. T.; Stanczyk, W. A., *J. Chem. Soc., Perkin Trans. 2* **1993**, 59; Eaborn, C.; Lickiss, P. D.; Najim, S. T., *J. Chem. Soc., Perkin Trans. 2* **1993**, 391.
31. Paquette, L. A., *Synthesis* **1975**, 349.
32. Paquette, L. A., *J. Am. Chem. Soc.* **1970**, *92*, 5765.
33. Fitjer, L.; Justus, K.; Puder, P.; Dittmer, M.; Hassler, C.; Noltemeyer, M., *Angew. Chem., Int. Ed. Engl.* **1991**, *30*, 436.
34. Paquette, L. A.; Leichter, L. M., *J. Am. Chem. Soc.* **1972**, *94*, 3653.
35. Koch-Pomeranz, U.; Hansen, H. J.; Schmid, H., *Helv. Chim. Acta* **1973**, *56*, 2981.
36. Shigemans, Y.; Yasui, M.; Ohrai, S.; Sasaki, M.; Sashiwa, H.; Saimoto, H., *J. Org. Chem.* **1991**, *56*, 910.
37. Dikshit, D. K.; Singh, S.; Panday, S. K., *J. Chem. Res. (S)* **1991**, 298.
38. Ruy, I.; Ando, M.; Ogawa, A.; Murai, S.; Sonoda, N., *J. Am. Chem. Soc.* **1983**, *105*, 7192.
39. Gerlach, H.; Thalmann, A., *Helv. Chim. Acta* **1974**, *57*, 2661.
40. Kawanami, Y.; Katsuki, T.; Yamaguchi, M., *Tetrahedron Lett.* **1983**, *24*, 5131.
41. Ishibashi, H.; Tabata, T.; Kobayashi, T.; Takamuro, I.; Ikeda, M., *CPB* **1991**, *39*, 2878.
42. Beerli, R.; Borschberg, H. J., *Helv. Chim. Acta* **1991**, *74*, 110.
43. Olah, G. A.; Pavláth, A.; Kuhn, S., *Chem. Ind. (London)* **1957**, *50*; Kuhn, S. J.; Olah, G. A., *J. Am. Chem. Soc.* **1961**, *83*, 4564.
44. Furukawa, N.; Hoshiai, H.; Shibutani, T.; Higaki, M.; Iwasaki, F.; Fujihara, H., *Heterocycles* **1992**, *34*, 1085.
45. Crews, P.; Beard, J., *J. Org. Chem.* **1973**, *38*, 529.
46. Kende, A. S.; Huang, H. *Tetrahedron Lett.* **1997**, *38*, 3353.
47. (a) Jin, S.-j.; Choi, J.-R.; Oh, J; Lee, D.; Cha, J. K. *J. Am. Chem. Soc.* **1995**, *117*, 10914. (b) Lee, J.; Oh, J.; Jin, S.-j.; Choi, J.-R.; Atwood, J. L.; Cha, J. K. *J. Org. Chem.* **1994**, *59*, 6955.
48. Schmid, R.; Schmid, H. *Helv. Chim. Acta.* **1974**, *57*, 1883.
49. (a) Noyori, R.; Baba, Y.; Makino, S.; Takaya, H. *Tetrahedron Lett.* **1973**, *14*, 1741. (b) Takaya, H.; Makino, S.; Hayakawa, Y.; Noyori, R. *J. Am. Chem. Soc.* **1978**, *100*, 1765.
50. (a) Vinter, J. G.; Hoffmann, H. M. R. *J. Am. Chem. Soc.* **1973**, *95*, 3051. (b) Hoffmann, H. M. R.; Wagner, D.; Wartchow, R. *Chem. Ber.* **1990**, *123*, 2131. (c) Schottelius, T.; Hoffmann, H. M. R. *Chem. Ber.* **1991**, *124*, 1673.
51. (a) Föhlisch, B.; Gehrlach, E.; Herter, R. *Angew. Chem. Int. Ed. Engl.* **1982**, *21*, 137. (b) Föhlisch, B.; Joachimi, R. *Chem. Ber.* **1987**, *120*, 1951.
52. Oh, C. H.; Sung, H. R.; Jung, S. H.; Lim, Y. M. *Tetrahedron Lett.* **2001**, *42*, 5493.
53. Buszek, K. R.; Sato, N.; Jeong, Y. *J. Am. Chem. Soc.* **1994**, *116*, 5511.
54. Corey, E. J.; Nicholaou, K. C. *J. Am. Chem. Soc.* **1974**, *96*, 5614.
55. Melnyk, P.; Legrand, B.; Gasche, J.; Ducrot, P.; Thal, C. *Tetrahedron* **1995**, *51*, 1941.
56. Evans, D. A.; Dinsmore, C. J.; Ratz, A. M.; Evrard, D. A.; Barrow, J. C. *J. Am. Chem. Soc.* **1997**, *119*, 3417.
57. Lathbury, D.; Gallagher, T. *J. Chem. Soc., Chem. Commun.* **1986**, 114.
58. Fox, D. N. A.; Gallagher, T. *Tetrahedron* **1990**, *46*, 4697.
59. Shaw, R.; Lathbury, D.; Anderson, M.; Gallagher, T. *J. Chem. Soc. Perkin Trans. 1* **1991**, 659.
60. Billet, M.; Schoenfelder, A.; Klotz, P.; Mann, A. *Tetrahedron Lett.* **2002**, *43*, 1453.
61. Calter, M.; Hollis, T. K.; Overman, L. E.; Ziller, J.; Zipp, G. G. *J. Org. Chem.* **1997**, *62*, 1449.
62. Uozumi, Y.; Kato, K.; Hayashi, T. *Tetrahedron Asym.* **1998**, *9*, 1065.
63. Hu, Y.; Zhang, Y.; Yang, Z.; Fathi, R. *J. Org. Chem.* **2002**, *67*, 2365.
64. Klingensmith, L. M.; Leadbeater, N. E. *Tetrahedron Lett.* **2003**, *44*, 765.
65. Jahangir; Fisher, L. E.; Clark, R. D.; Muchowski, J. M. *J. Org. Chem.* **1989**, *54*, 2992.
66. Usui, S.; Matsumoto, T.; Ohkubo, K. *Tetrahedron Lett.* **1998**, *39*, 9755.
67. Stieglitz, J.; Vosburgh, I. *J. Am. Chem. Soc.* **1916**, *38*, 2081.
68. Grieco, P. A.; Dai, Y. *J. Am. Chem. Soc.* **1998**, *120*, 5128.
69. DeKimpe, N.; Tehrani, K. A.; Fonck, G. *J. Org. Chem.* **1996**, *61*, 6500.
70. Shigemasa, Y.; Oikawa, H.; Ohrai, S.-i.; Sashiwa, H.; Saimoto, H. *Bull. Chem. Soc. Jpn.* **1992**, *65*, 2594.
71. Vorogushin, A. V.; Wulff, W. D.; Hansen, H.-J. *Org. Lett.* **2001**, *3*, 2641.
72. Banwell, M. G.; Harvey, J. E.; Jolliffe, K. A. *J. Chem. Soc., Perkin Trans. 1* **2001**, 2002.
73. Blair, L. K.; Parris, K. D.; Lee, O. F. D.; Jenkins, K. F.; Feese, R. C.; Belcher, T.; Badger, D.; Morris, D.; Kuhn, C. *J. Org. Chem.* **1986**, *51*, 5454.
74. Bonjoch, J.; Diaba, F.; Puigbó, G.; Solé, D.; Segarra, V.; Santamaría, L.; Beleta, J.; Ryder, H.; Palacios, J.-M. *Bioorg. & Med. Chem..* **1999**, *7*, 2891.
75. Lemal, D. M.; Fry, A. J., *J. Org. Chem.* **1964**, *29*, 1673.
76. Kornblum, N.; Jones, W. J.; Anderson, G. J. *J. Am. Chem. Soc.* **1959**, *81*, 4113.
77. Ly, T. W.; Liao, J.-H.; Shia, K.-S.; Lui, H.-J. *Synthesis* **2004**, 271.
78. Kitano, Y.; Chiba, K.; Tada, M. *Synthesis* **2001**, 437.
79. Nicholas, K. M. *Acc. Chem. Res.* **1987**, *20*, 207.
80. Vizniowski, C. S.; Green, J. R.; Breen, T. L.; Dalacu, A. V. *J. Org. Chem.* **1995**, *60*, 7496.

81. Steynberg, P. J.; Nel, R. J. J.; van Rensburg, H.; Bezuidenhoudt, B. C. B.; Ferreira, D. *Tetrahedron* **1998**, *54*, 8153.

82. Grundberg, H.; Andergran, M.; Nilsson, U. J. *Tetrahedron Lett.* **1999**, *40*, 1811.

83. Lui, K. K.-C.; Danishefsky, S. J. *J. Org. Chem.* **1994**, *59*, 1895.

84. Suzuki, K.; Maeta, H.; Suzuki, T.; Matsumoto, T. *Tetrahedron Lett.* **1989**, *30*, 6879.

Silver Trifluoromethanesulfonate

$AgOSO_2CF_3$

[2923-28-6] $CAgF_3O_3S$ (MW 256.95)

(reactive triflating agent and source of soluble silver ions useful for formation of electrophilic onium species,[1] and promotion of Friedel–Crafts,[2] nucleophile–alkene cyclization,[3] and esterification[4] reactions)

Alternate Name: silver triflate.
Physical Data: mp 356 °C.
Solubility: soluble in typical organic solvents.
Form Supplied in: white solid; widely available.
Preparative Methods: can be prepared from the barium salt via treatment with dil H_2SO_4 followed by neutralization with Ag_2CO_3.[5]
Handling, Storage, and Precautions: light sensitive; irritant. Use in a fume hood.

Original Commentary

T. Howard Black *Eastern Illinois University, Charleston, IL, USA*

Oxonium (Iminium) Reactions. Oxygen- and nitrogen-containing molecules bearing an α-nucleofuge such as a halide or sulfur are converted by AgOTf to highly electrophilic onium (eq 1)[1] or iminium (eq 2)[6] species. The intermediacy of a planar cation is substantiated by the formation of the same diastereomer from either substrate isomer in some cyclization reactions (eq 1).

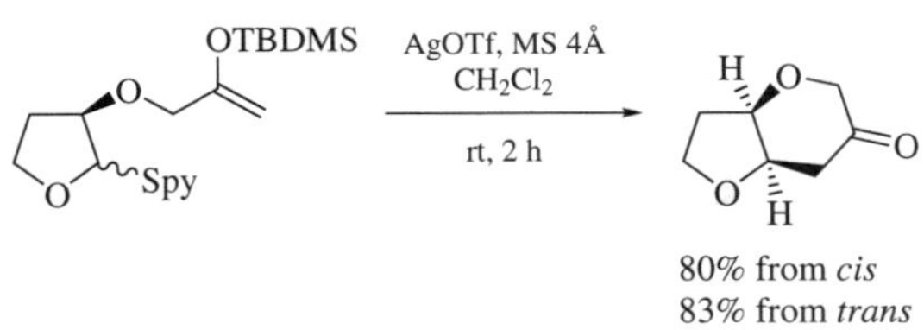

(1)

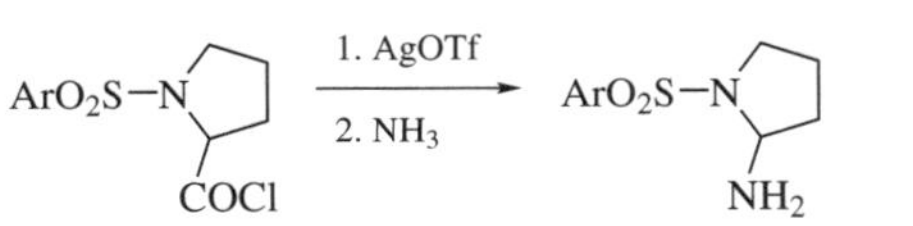

(2)

This reaction type has enjoyed wide application in carbohydrate chemistry, as anomeric carbons can be easily functionalized with high stereospecificity by silyl enol ethers (eq 3, R′ = Ph,[7] H[8]) and other nucleophiles.

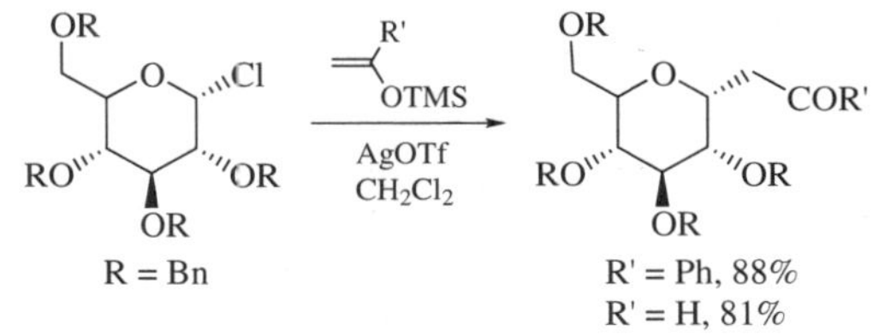

(3)

Friedel–Crafts (and Other Aromatic Functionalization) Reactions. When treated with AgOTf, alkyl,[2] acyl (eq 4),[9] and sulfonyl[10] halides are converted to extremely electrophilic triflate species that react rapidly with even deactivated[11] aromatic rings in the absence of catalysis. Benzylic chloroformates participate similarly (eq 5).[12] Aromatic rings are also efficiently vinylated[13] and iodinated[14] via AgOTf-promoted processes.

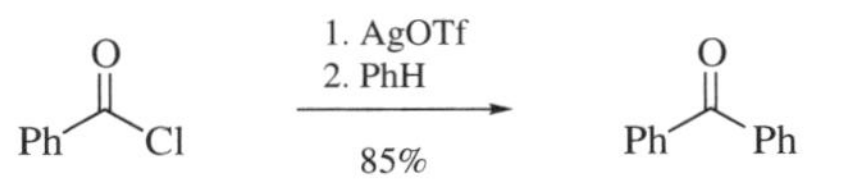

(4)

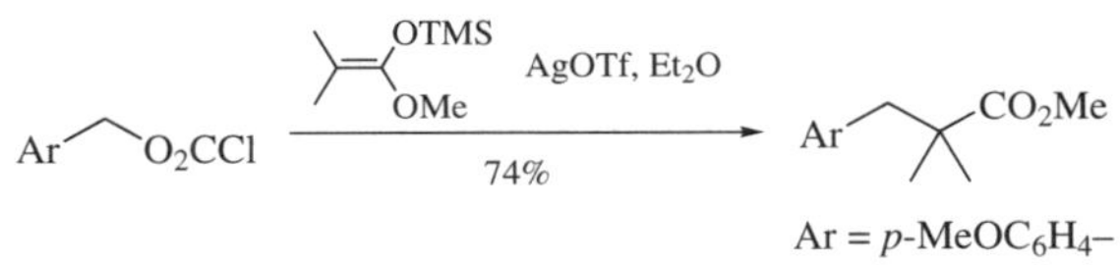

(5)

Miscellaneous Reactions. AgOTf induces a highly diastereoselective cyclization of amino ketenes wherein asymmetry is induced by an adjacent stereogenic center and is dependent upon silver concentration (eq 6).[3] Acetylenic oxy-Cope rearrangements are also facilitated by AgOTf (eq 7),[15] and a mild, high-yield esterification utilizes AgOTf in the presence of Titanium(IV) Chloride (eq 8).[4]

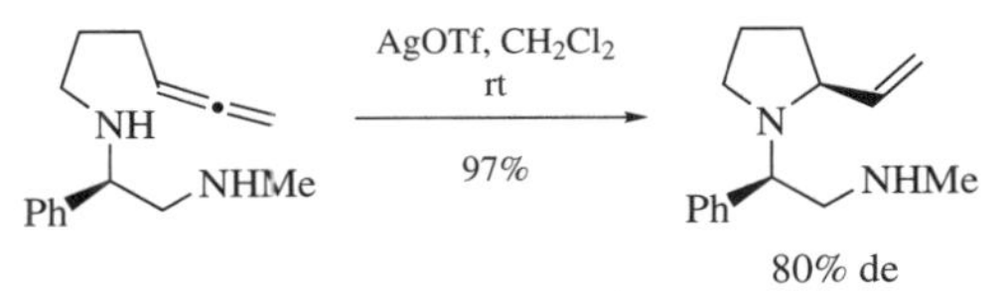

(6)

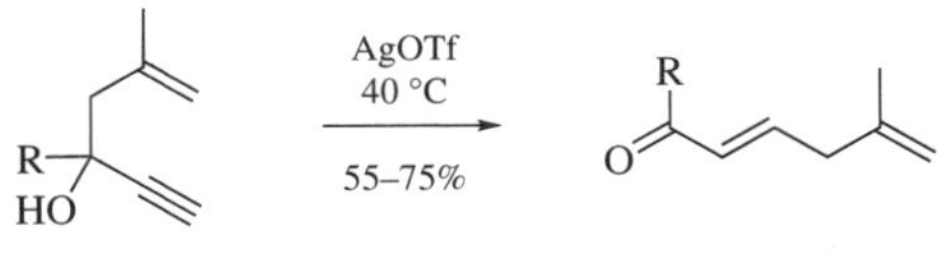

(7)

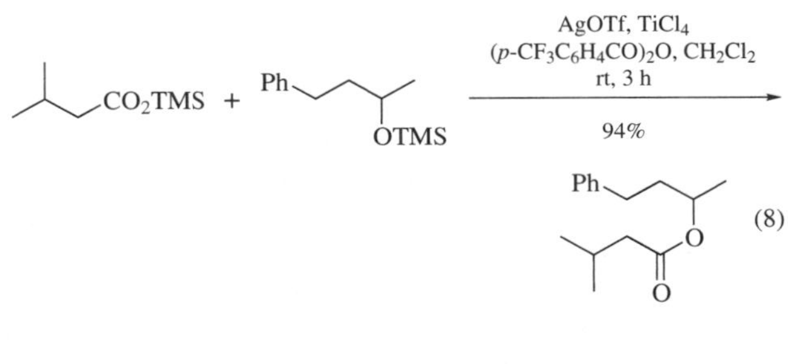

(8)

First Update

Keith A. Stubbs & Robert V. Stick
The University of Western Australia, Crawley, WA Australia

Miscellaneous Reactions. AgOTf has been used to induce [4+2+2] cycloadditions of various enyne derivatives to give eight-membered rings.[16] Similar reactions in the past have had problems overcoming poor selectivity in heterocycloadditions.[17] In conjunction with Wilkinson's catalyst, the silver trifluoromethanesulfonate-promoted cycloadditions occur in good yield (eq 9).

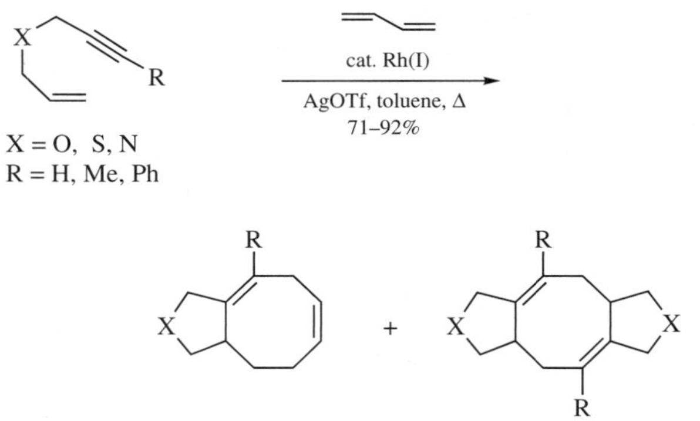

(9)

19:1

This methodology has been extended to include three-component systems (eq 10).

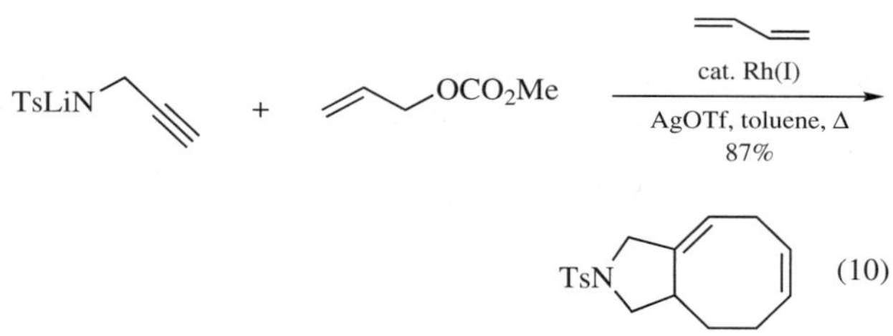

(10)

One method of converting alcohols into carboxylic acids, known as 'Koch carbonylation,' is mediated by AgOTf in the presence of weak and soft Lewis acids (eq 11). Alkenes can be similarly carbonylated albeit in much lower yields (eq 12).[18]

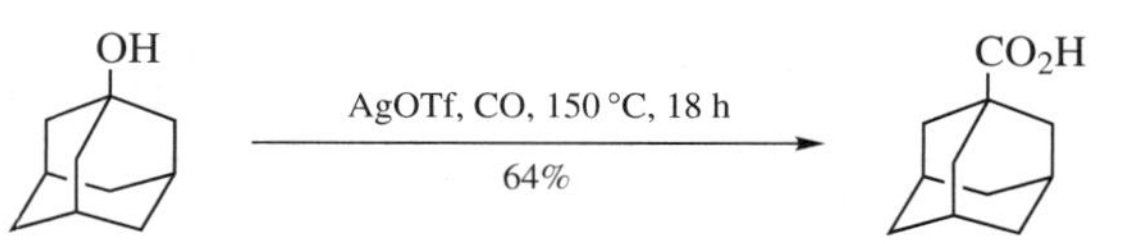

(11)

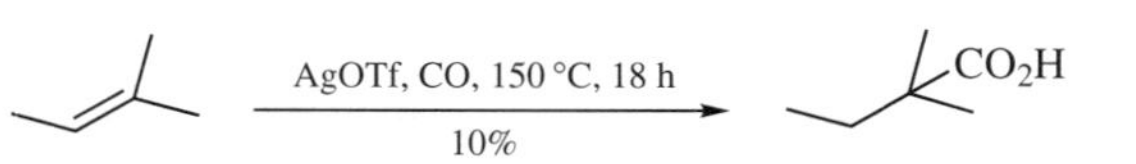

(12)

The study of the aza-Diels–Alder reaction in aqueous media has been restricted in the past to the use of water-soluble dienes and iminium salts.[19] AgOTf has been used to catalyze the reaction of Danishefsky's diene,[20] a water-sensitive reagent, with various imines in aqueous media in good yield (eq 13).[21] This process has been extended to the use of three-component systems that generate the imine in situ (eq 14).

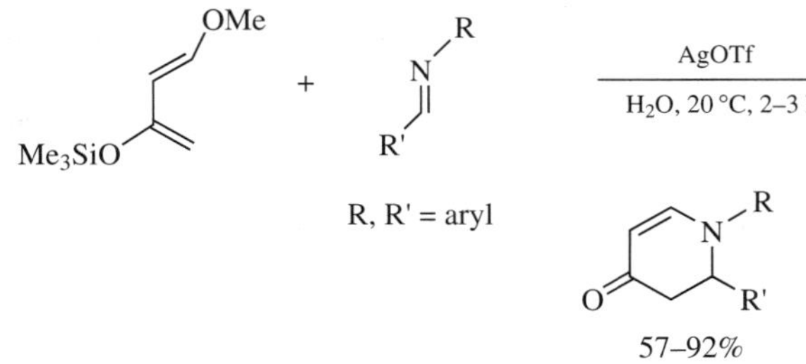

(13)

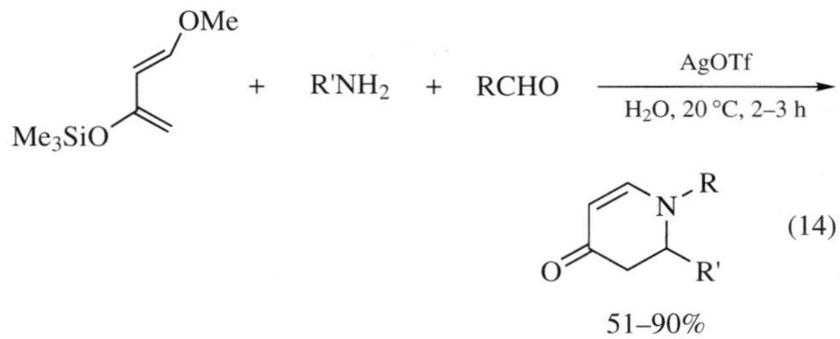

(14)

The preparation of hydroquinoline derivatives is of great interest owing to their biological activity[22] but, in many cases, their preparation has involved the tedious use of chiral auxillaries and organometallic reagents.[23] AgOTf has been used to activate phenyl chloroformate toward condensation with quinoline, as well as isoquinoline, for subsequent reaction with chiral allylsilanes, to give 2-allylated derivatives (eq 15).[24]

The synthesis of isoquinolines has been achieved also by using various allylbenzene derivatives and a $AgOTf/I_2$ couple, with various nitriles as the source of nitrogen (eq 16).[25]

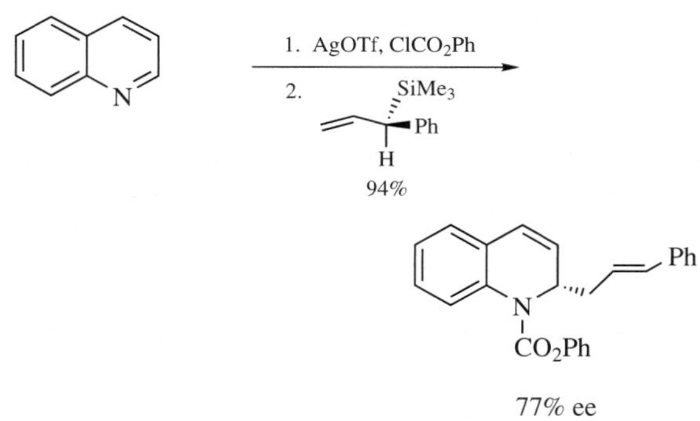

(15)

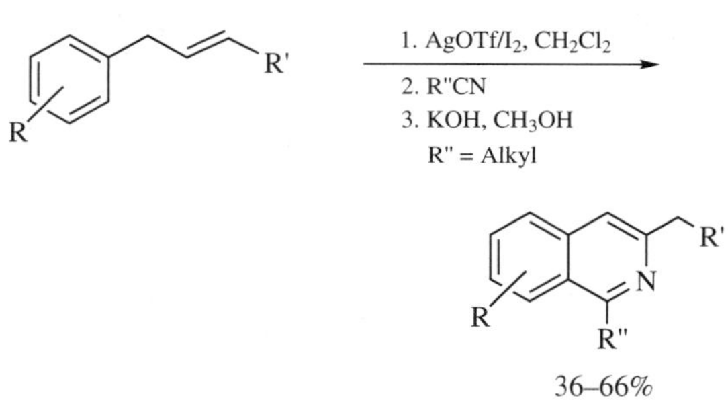

(16)

An interesting allylation of carbonyl compounds has been introduced using a catalytic BINAP.AgOTf complex; a mixture of allyltributyltin and various aldehydes gives good yields of alcohols with high enantiomeric excess (eqs 17 and 18).[26]

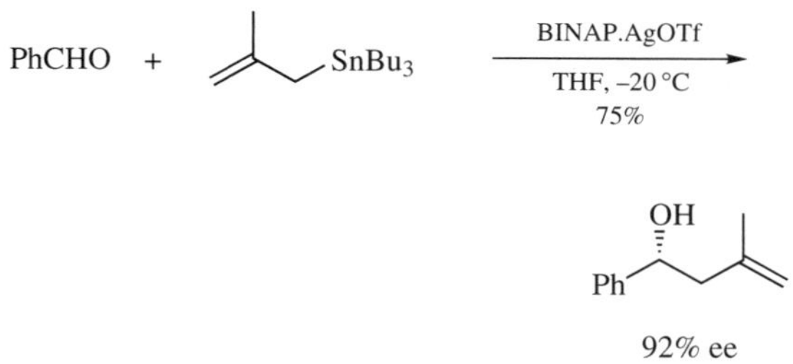

(17)

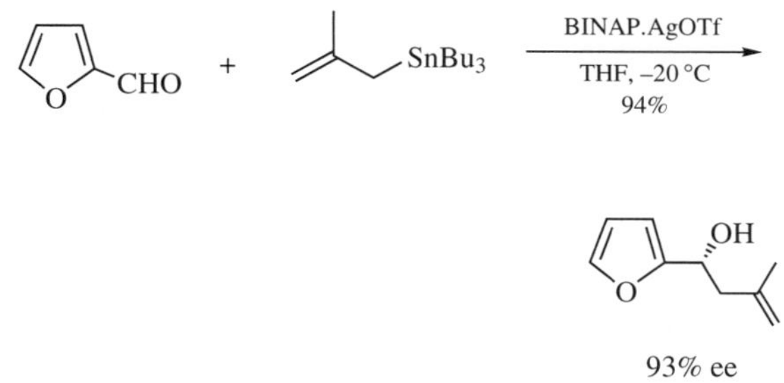

(18)

This type of enantioselective reaction has been extended to the use of allylchlorosilanes, taking advantage of their ability to react with aldehydes under certain conditions to give alcohols with good stereochemical control (eqs 19 and 20);[27] a silyl triflate may be a relevant intermediate.

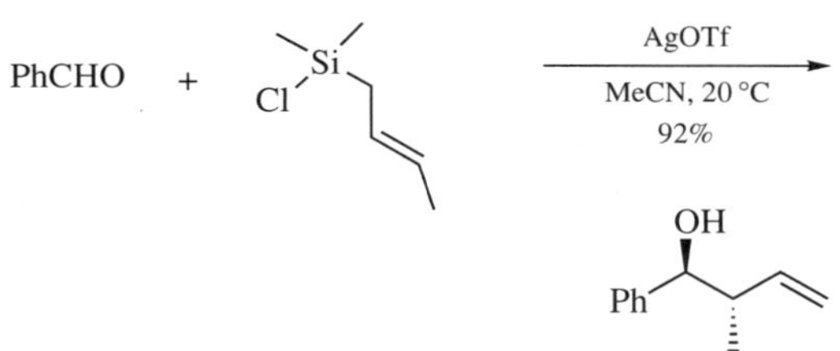

(19)

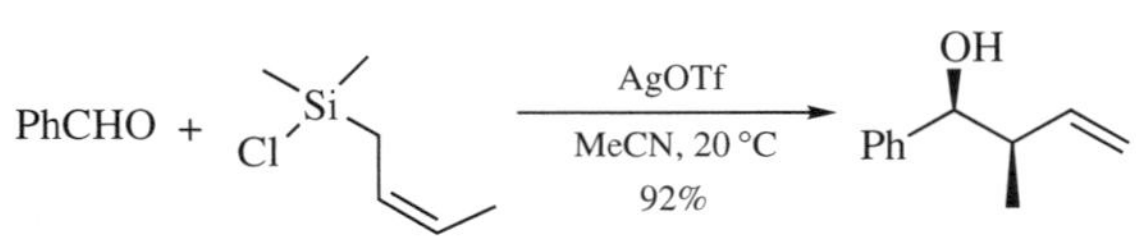

(20)

Photochemical Reactions. AgOTf has also been used to induce a photochemical reaction between α-chloroacetophenone and functional olefins.[28] The use of the reagent is critical when the reaction is conducted with less substituted olefins (eqs 21 and 22).

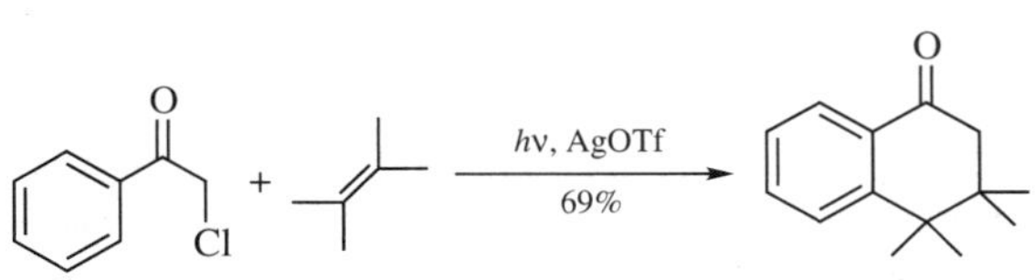

(21)

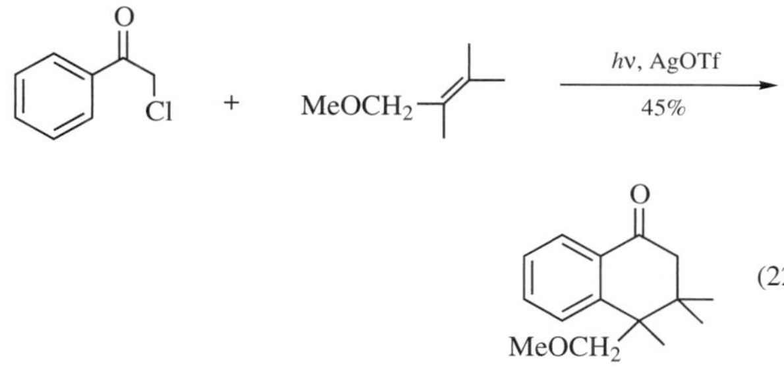

(22)

This concept has been applied also to α-chloroenones as substrates for the cycloaddition (eq 23).[29]

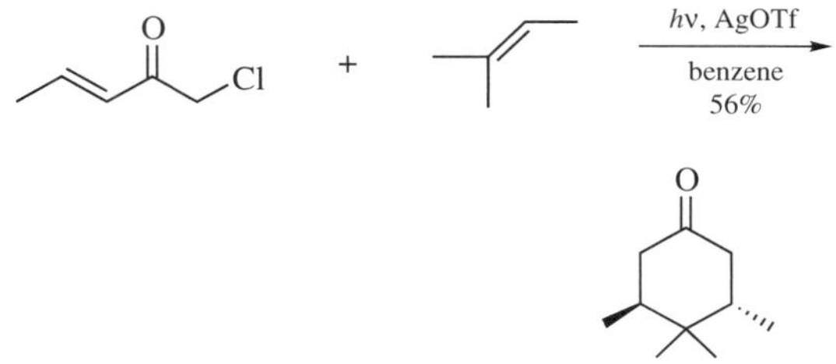

(23)

Use in Protecting Group Strategies. AgOTf has been employed with benzyl bromide to allow for the benzylation of hydroxyl groups under neutral conditions (eqs 24 and 25);[30] this reaction can be performed in the presence of base-sensitive protecting groups such as acetates.

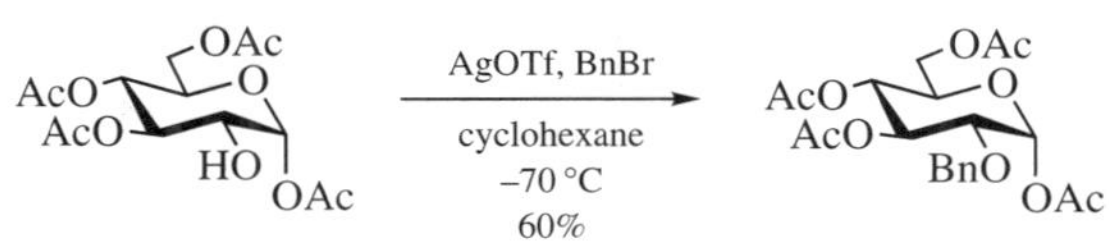

(24)

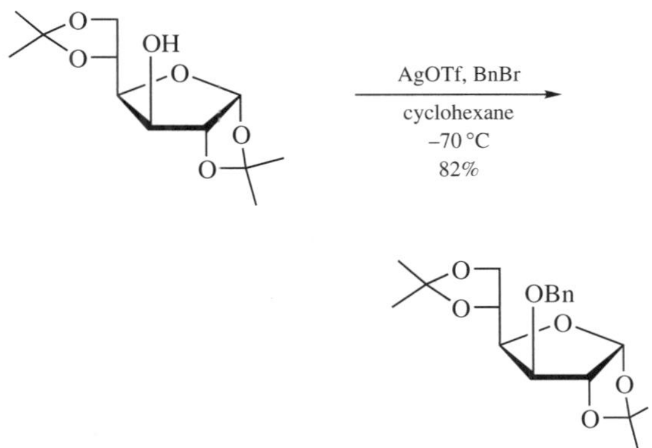

(25)

Reactions with Carbohydrates. AgOTf and $MoCl_2(acac)_2$ combine to form a catalyst that has been utilized in allylic substitution reactions with glycals,[31] versatile building blocks of carbohydrate chemistry,[32] to give Ferrier-type reaction products (eqs 26 and 27).

Ortho-esterification reactions have been performed on anomeric dihalogen compounds with AgOTf as the promoter (eq 28).[33]

Glycosylation Promoter. AgOTf is used to promote many different kinds of glycosylations with a wide variety of glycosyl donors of which a selection is given here. AgOTf has been used in Koenigs–Knorr glycosylations, involving anomeric halides, to give a variety of new types of mono-, di-, and oligosaccharides (eqs 29–31).[34–36]

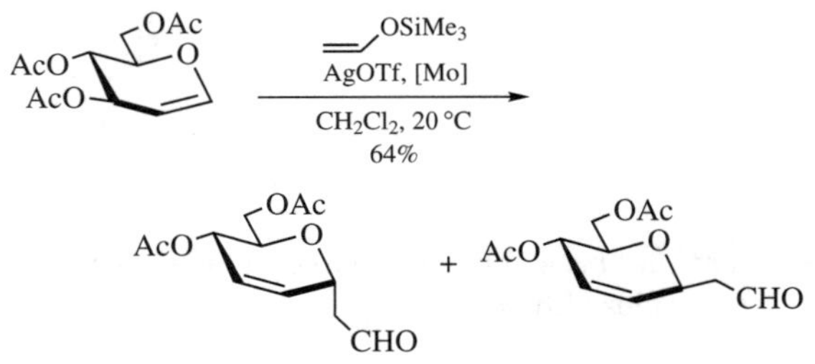
(26)

65:35

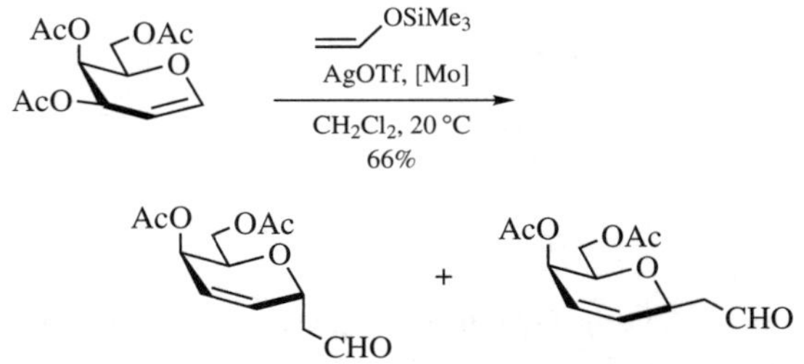
(27)

76:24

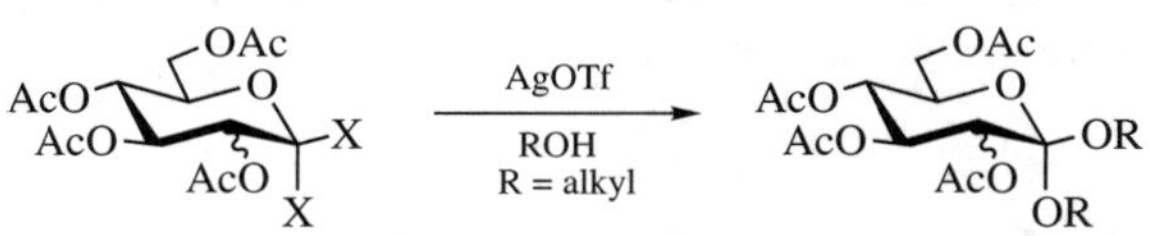
(28)

X = Br, Cl

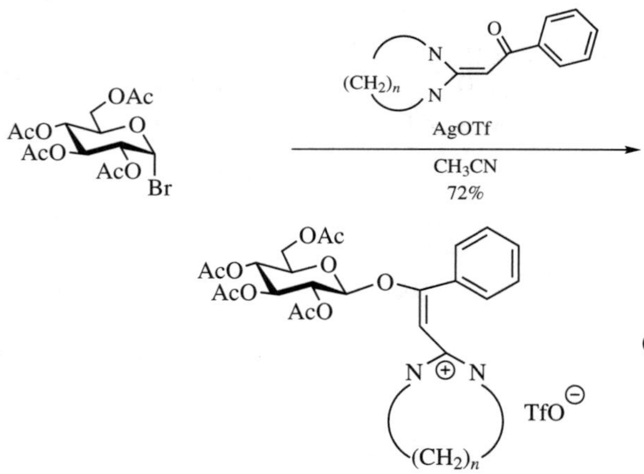
(29)

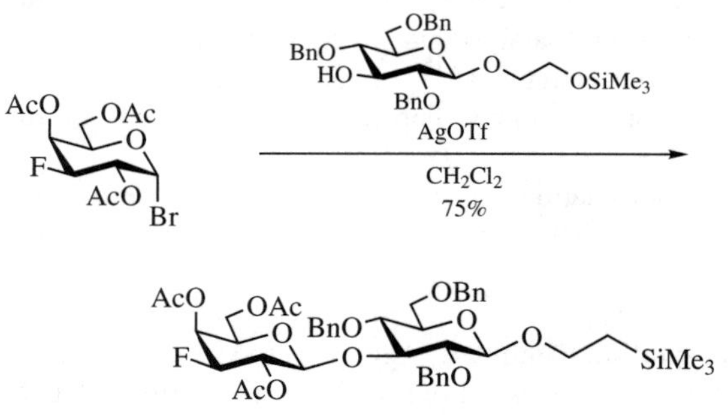
(30)

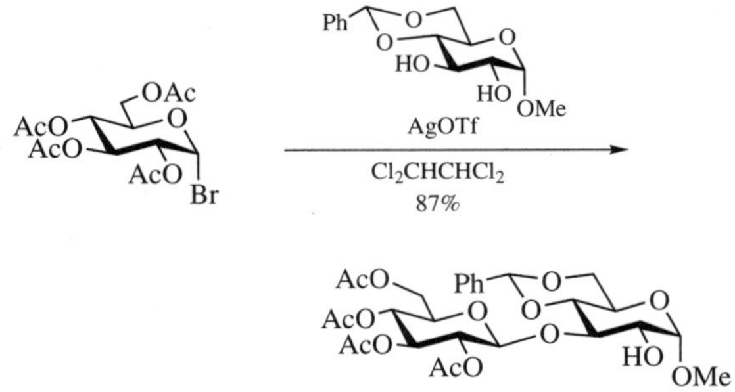
(31)

Treatment of a 2-bromo-2-deoxy-α-D-glucosyl bromide with benzyl alcohol employing AgOTf as a promoter yielded a mixture of benzyl α- and β-D-glucosides, a not so surprising result since there is a participating group at C2 that usually allows for the formation of the kinetically favorable β-glycoside (eq 32).[37]

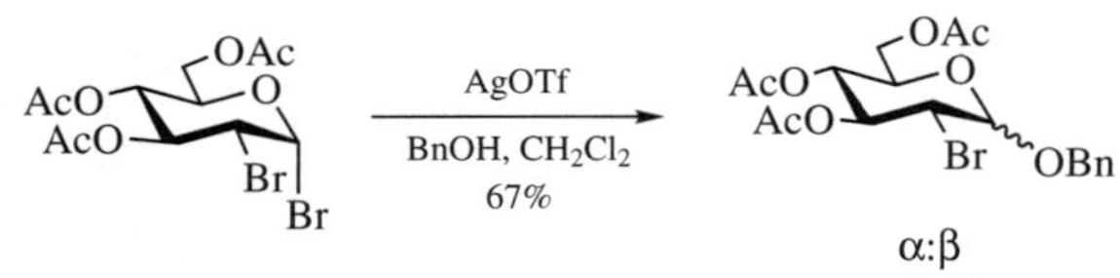
(32)

α:β
(1:1)

The glycosylation of sialic acids, a somewhat complex set of molecules of considerable importance in glycobiology,[38] has been achieved using Koenigs–Knorr methodology and AgOTf. Treatment of an β-bromide with various acceptors gave glycosylated products in excellent yields (eq 33).[39]

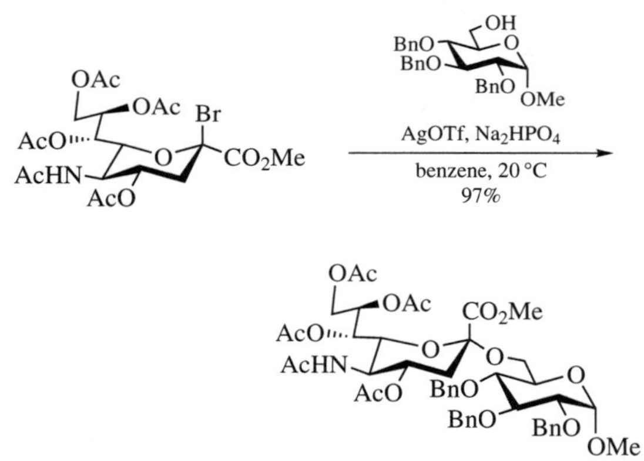
(33)

The outcome of neighboring group directed glycosylation reactions mediated by AgOTf can depend on the acceptor. Thus, treatment of an α-D-glucosyl bromide with an acceptor alcohol masked in the form of a DMF acetal in the presence of AgOTf gives the 1,2-orthoester. The use of the free alcohol as acceptor, on the other hand, affords the simple glucoside (eq 34).[40]

AgOTf has been used to promote the glycosylation of acceptors with trichloroacetimidates (eq 35).[41]

An area where AgOTf has found great use is in the promotion of thioglycoside activation by otherwise mild thiophilic reagents such as *N*-iodosuccinimide (eqs 36 and 37).[42,43]

The synthesis of 1,2-*cis* linkages, such as those found in β-D-mannopyranosides, has proven a problem to carbohydrate chemists over the years. Treatment of a D-manno thioglycoside and the appropriate acceptor with AgOTf in the presence of PhSCl (see also benzenesulfenyl chloride) results in the formation of the β-D-mannopyranoside in excellent yield (eq 38).[44] It has been confirmed that the reaction proceeds through an α-D-mannosyl triflate. AgOTf has also been used to prepare α-D-mannopyranosides in good yields (eqs 39 and 40).[45,46]

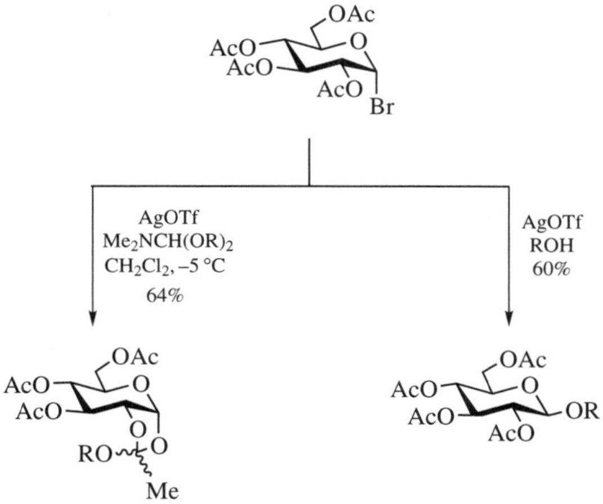

(34)

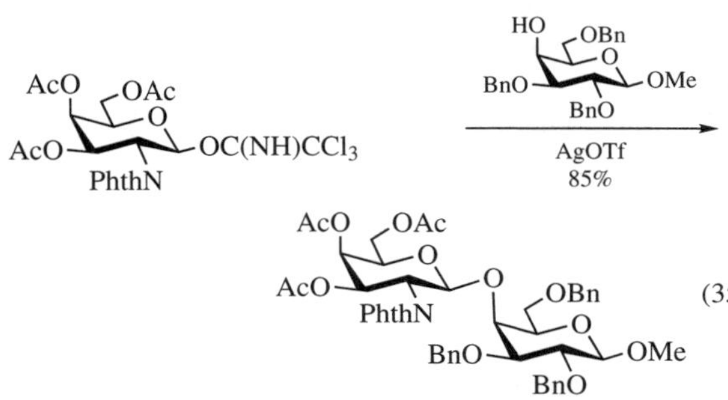

(35)

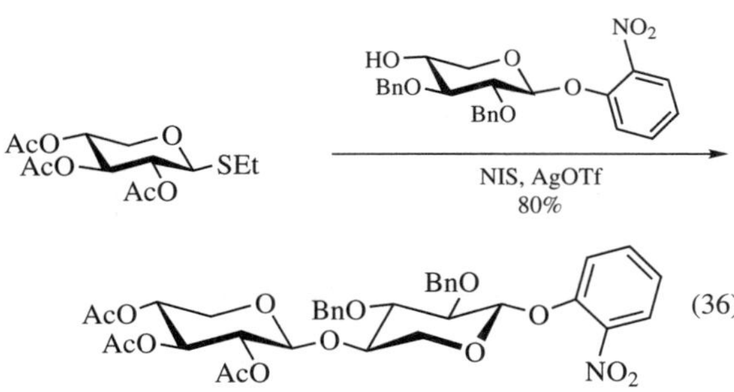

(36)

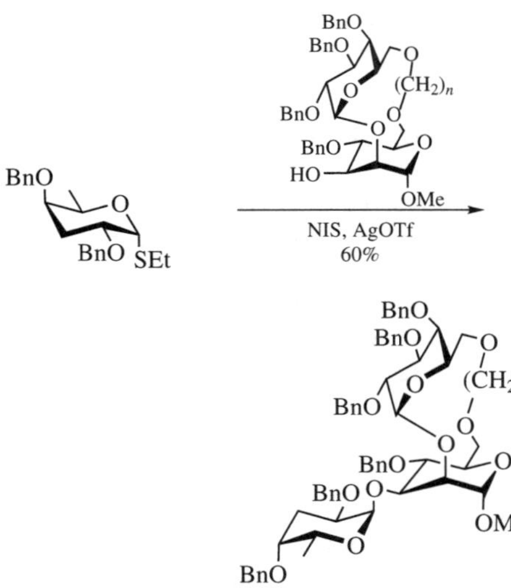

(37)

AgOTf has also been used to activate selenoglycosides preferentially over thioglycosides, which is advantageous for the construction of larger oligosaccharides (eq 41).[47]

The use of AgOTf in the glycosylation of acyclic 1-chloro-1-ethylthio derivatives with appropriate acceptors has also been realized (eq 42).[48]

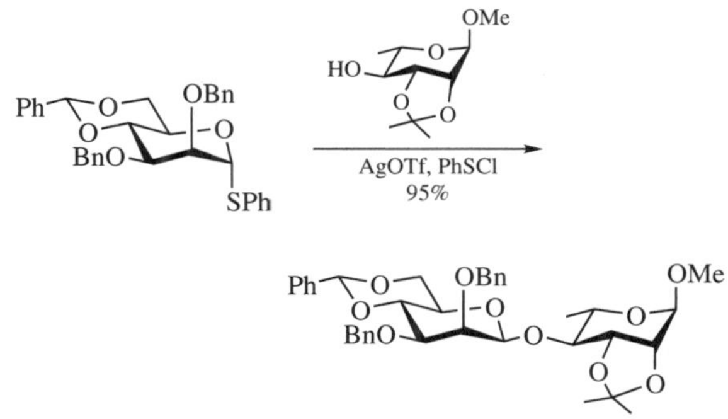

(38)

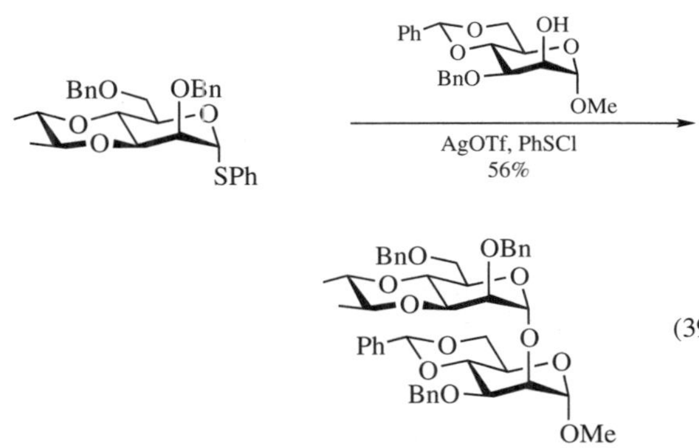

(39)

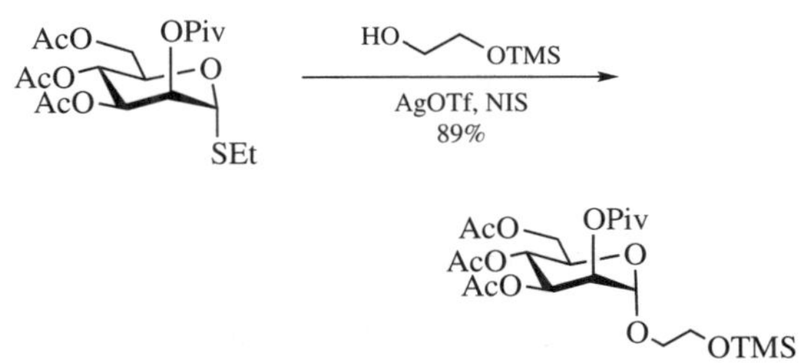

(40)

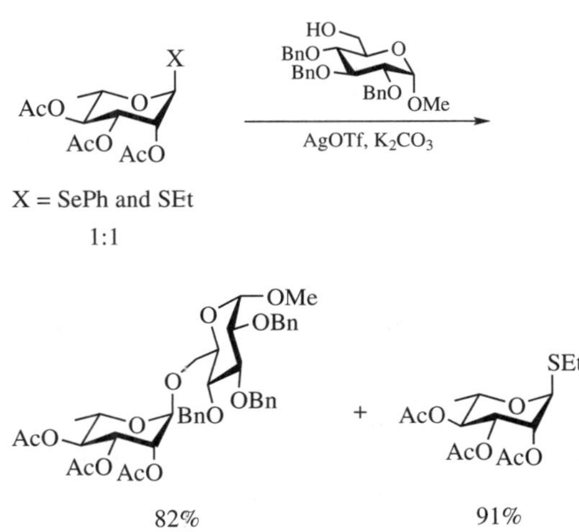

(41)

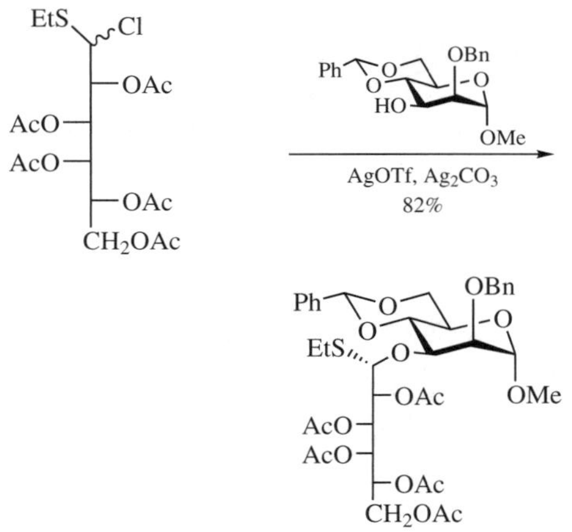

(42)

Related Reagents. Benzenesulfenyl chloride; benzenesulfenyl trifluoromethanesulfonate; *N*-iodosuccinimide; silver carbonate; silver silicate; silver bromide; silver oxide; trifluoromethanesulfonic anhydride; trimethylsilyl trifluoromethanesulfonate.

1. Craig, D.; Munasinghe, V. R. N., *Tetrahedron Lett.* **1992**, *33*, 663.
2. Sasaki, T.; Nakanishi, A.; Ohno, M., *J. Org. Chem.* **1982**, *47*, 3219.
3. Fox, D. N. A.; Gallagher, T., *Tetrahedron* **1990**, *46*, 4697.
4. Mukaiyama, T.; Shiina, I.; Miyashita, M., *Chem. Lett.* **1992**, 625.
5. Gramstad, T.; Haszeldine, R. N., *J. Chem. Soc.* **1956**, 173.
6. Adesogan, E. K.; Alo, B. I., *Chem. Commun.* **1979**, 673.
7. Allevi, P.; Anastasia, M.; Ciuffreda, P.; Fiecchi, A.; Scala, A., *Chem. Commun.* **1987**, 101.
8. Stewart, A. O.; Williams, R. M., *J. Am. Chem. Soc.* **1985**, *107*, 4289.
9. Effenberger, F.; Epple, G., *Angew. Chem., Int. Ed. Engl.* **1972**, *11*, 299.
10. Effenberger, F.; Huthmacher, K., *Angew. Chem., Int. Ed. Engl.* **1974**, *13*, 409.
11. Effenberger, F.; Sohn, E.; Epple, G., *Ber. Dtsch. Chem. Ges.* **1983**, *116*, 1195.
12. Takeda, K.; Ayabe, A.; Kawashima, H.; Harigaya, Y., *Tetrahedron Lett.* **1992**, *33*, 951.
13. Kitamura, T.; Kobayashi, S.; Taniguchi, H.; Rappoport, Z., *J. Org. Chem.* **1982**, *47*, 5003.
14. Kobayashi, Y.; Kumadaki, I.; Yoshida, T., *J. Chem. Res. (S)* **1977**, 215.
15. Bluthe, N.; Gore, J.; Malacria, M., *Tetrahedron* **1986**, *42*, 1333.
16. Evans, P. A.; Robinson, J. E.; Baum, E. W.; Fazal, A. N., *J. Am. Chem. Soc.* **2002**, *124*, 8782.
17. Lautens, M.; Tam, W.; Lautens, J. C.; Edwards, G. L.; Crudden, C. M.; Smith, A. C., *J. Am. Chem. Soc.* **1995**, *117*, 6863.
18. Mori, H.; Mori, A.; Xu, Q.; Souma, Y., *Tetrahedron Lett.* **2002**, *43*, 7871.
19. Fringuelli, F.; Piermatti, O.; Pizzo, F.; Vaccaro, L., *Eur. J. Org. Chem.* **2001**, 439.
20. Danishefsky, S.; Kitahara, T., *J. Am. Chem. Soc.* **1974**, *96*, 7807.
21. Loncaric, C.; Manabe, K.; Kobayashi, S., *Adv. Synth. Catal.* **2003**, *345*, 475.
22. Katritzky, A. R.; Rachwal, S.; Rachwal, B., *Tetrahedron* **1996**, *52*, 15031.
23. Beifuss, U.; Schniske, U.; Feder, G., *Tetrahedron* **2001**, *57*, 1005.
24. Yamaguchi, R.; Tanaka, M.; Matsuda, T.; Okano, T.; Nagura, T., Fujita, K., *Tetrahedron Lett.* **2002**, *43*, 8871.
25. Sato, T.; Tamura, K.; Nagayoshi, K., *Chem. Lett.* **1983**, 791.
26. Yanagisawa, A.; Nakashima, H.; Ishiba, A.; Yamamoto, H., *J. Am. Chem. Soc.* **1996**, *118*, 4723.
27. Furstner, A.; Voigtlander, D., *Synthesis* **2000**, 959.
28. Oh, S.-H.; Sato, T., *J. Org. Chem.* **1994**, *59*, 3744.
29. Sawayanagi, Y.; Sato, T.; Shimizu, I., *Chem. Lett.* **1997**, 843.
30. Berry, J. M.; Hall, L. D., *Carbohydr. Res.* **1976**, *47*, 307.
31. Malkov, A. V.; Farn, B. P.; Hussain, N.; Kocovsky, P., *Collect. Czech. Chem. Commun.* **2001**, *66*, 1735.
32. Collins, P. M.; Ferrier, R. J., *Monosaccharides*; Wiley: Chichester, 1995.
33. Praly, J. P.; El Kharraf, Z.; Corringer, P. J.; Brard, L.; Descotes, G., *Tetrahedron* **1990**, *46*, 65.
34. Li, Z.-J.; Wang, L.-B.; Huang, Z.-T., *Carbohydr. Res.* **1996**, *295*, 77.
35. Ekberg, T.; Magnusson, G., *Carbohydr. Res.* **1993**, *249*, 117.
36. Takeo, K., *Carbohydr. Res.* **1980**, *87*, 147.
37. Bock, K.; Lundt, I.; Pedersen, C., *Carbohydr. Res.* **1984**, *130*, 125.
38. Boons, G. J.; Demchenko, A. V., *Chem. Rev.* **2000**, *100*, 4539.
39. Okamoto, K.; Kondo, T.; Goto, T., *Tetrahedron* **1987**, *43*, 5909.
40. Hanessian, S.; Banoub, J., *Carbohydr. Res.* **1975**, *44*, C14.
41. Douglas, S. P.; Whitfield, D. M.; Krepinsky, J. J., *J. Carbohydr. Chem.* **1993**, *12*, 131.
42. Takeo, K.; Ohguchi, Y.; Hasegawa, R.; Kitamura, S., *Carbohydr. Res.* **1995**, *277*, 231.
43. Alibes, R.; Bundle, D. R., *J. Org. Chem.* **1998**, *63*, 6288.
44. Crich, D.; Sun, S., *J. Am. Chem. Soc.* **1998**, *120*, 435.
45. Crich, D.; Cai, W.; Dai, Z., *J. Org. Chem.* **2000**, *65*, 1291.
46. Saksena, R.; Zhang, J.; Kovac, P., *J. Carbohydr. Chem.* **2002**, *21*, 453.
47. Mehta, S.; Pinto, B. M., *J. Org. Chem.* **1993**, *58*, 3269.
48. McAuliffe, J. C.; Hindsgaul, O., *J. Org. Chem.* **1997**, *62*, 1234.

O-(*N*-Succinimidyl)-1,1,3,3-tetramethyluronium Tetrafluoroborate (TSTU)

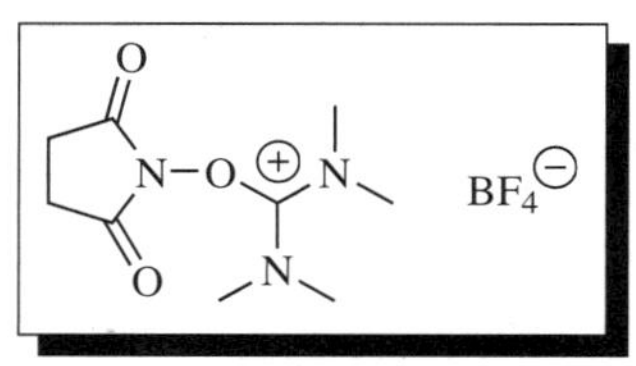

[105832-38-0] $C_9H_{16}BF_4N_3O_3$ (MW 301.051)

(reagent for carboxamide formation from a carboxylic acid and a primary amine; peptide coupling reagent)

Physical Data: mp 198–201 °C.
Solubility: soluble in DMF, CH_2Cl_2, $CHCl_3$, CH_3CN (0.1 g mL^{-1}), DMF/1,4-dioxane/water (2:2:1).
Form Supplied in: colorless powder. Commercially available from Advanced ChemTech (Louisville, KY, USA), Aldrich (Milwaukee, WI, USA), and Fluka Chemical (Milwaukee, WI, USA).
Handling, Storage, and Precautions: irritant (eye, skin, respiratory tract). Avoid breathing dust, vapor, moisture, or gas. Avoid contact with skin, eyes and clothing. Moisture-sensitive. For long-term storage, keep under nitrogen atmosphere and refrigerate.

Carboxamide Formation. *O*-(*N*-Succinimidyl)-1,1,3,3-tetramethyluronium tetrafluoroborate (TSTU) reacts with carboxylic acids in the presence of ≥2 equiv base to give succinimidyl active esters,[1,2] which are capable of reacting with primary amines to afford carboxamides (eq 1). The coupling reactions between carboxylic acids and primary amines with TSTU can also be performed in situ. The reaction proceeds rapidly and efficiently with only a slight excess of TSTU. The coupling reaction can be carried out even in the presence of water, such as in DMF/1,4-dioxane/water mixtures.[3] This feature allows hydrophilic amines to be derivatized with hydrophobic carboxylic acids, as demonstrated in the coupling of bathophenanthroline-ruthenium(III) complexes and 5′-amino-5′-deoxy oligonucleotides.[1]

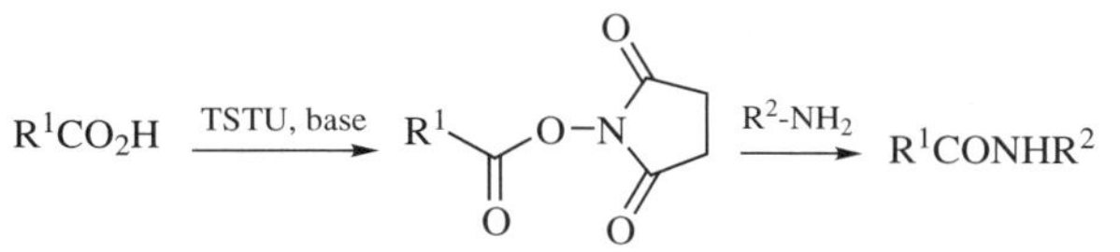

(1)

The amino group of an amino alcohol can be selectively acylated by using TSTU without *O*-protection (eq 2).

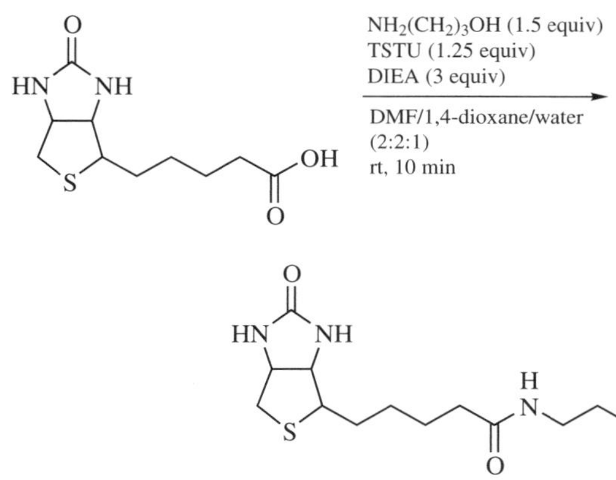

(2)

81%

TSTU can be employed as a coupling reagent in peptide synthesis, like analogous reagents (see Related Reagents). An amino acid, of which the α-amino group is protected with a urethane-type protecting group, e.g. Boc, Cbz, or Fmoc, can be coupled without a significant level of enantiomerization or epimerization. The segment condensation with these reagents, however, causes significant levels of epimerization of the carboxy-terminal amino acid residue of the acyl component except the cases where the carboxy-terminal residue is Gly or Pro.

Peptides having a *C*-terminal *N*-methylamino acid residue are highly prone to epimerization in the segment condensation. Simultaneous use of *N*-hydroxysuccinimide and $CuCl_2$ with TSTU eliminates the epimerization of the *N*-methylamino acid residue (eq 3), whereas analogous coupling reagents, (benzotriazol-1-yl)oxytris(dimethylamino)phosphonium hexafluorophosphate (BOP) and *O*-(7-azabenzotriazol-1-yl)-1,1,3,3-tetramethyluronium hexafluorophosphate (HATU), give significant levels of epimerization, even in the presence of $CuCl_2$.[4]

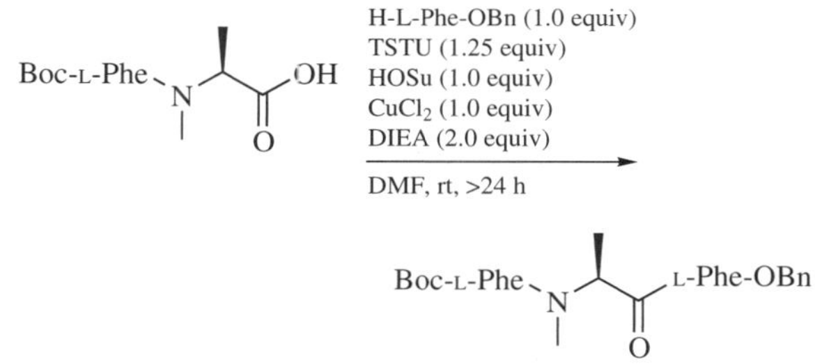

(3)

45 % (no detectable L-D-L)

Related Reagents. (Benzotriazol-1-yl)oxytris(dimethylamino)phosphonium hexafluorophosphate (BOP) [56602-33-6];[5] (benzotriazol-1-yl)oxytris(pyrrolidino)phosphonium hexafluorophosphate (PyBOP) [128625-52-5];[6] *O*-(*N*-benzotriazolyl)-1,1,3,3-tetramethyluronium hexafluorophosphate (HBTU) [94790-37-1];[7] *O*-(*N*-benzotriazolyl)-1,1,3,3-tetramethyluronium tetrafluoroborate (TBTU) [125700-67-6];[2] 2-(5-norbornene-2,3-dicarboxamido)-1,1,3,3-tetramethyluronium tetrafluoroborate(TNTU) [125700-73-4];[2] (7-azabenzotriazol-1-yl)oxytris(pyrrolidino)phosphonium hexafluorophosphonate (PyAOP) [156311-83-0];[8] *O*-(7-azabenzotriazol-1-yl)-1,1,3,3-tetramethyluronium hexafluorophosphate (HATU) [148893-10-1].[9]

1. Bannwarth, W.; Schmidt, D.; Stallard, R. L.; Hornung, C.; Knorr, R.; Muller, F., *Helv. Chim. Acta* **1988**, *71*, 2085.
2. Knorr, R.; Trzeciak, A.; Bannwarth, W.; Gillssen, D., *Tetrahedron Lett.* **1989**, *30*, 1927.
3. Bannwarth, W.; Knorr, R., *Tetrahedron Lett.* **1991**, *32*, 1157.
4. Nishiyama, Y.; Ishizuka, S.; Mori, T.; Kurita, K., *Chem. Pharm. Bull.* **2000**, *48*, 442.
5. Castro, B..; Dormoy, J. R..; Evin, G..; Selve, C., *Tetrahedron Lett.* **1975**, 1219.
6. Coste, J.; Le-Nguyen, D.; Castro, B., *Tetrahedron Lett.* **1990**, *31*, 205.
7. Dourtoglou, V.; Ziegler, J. C.; Gross, B., *Tetrahedron Lett.* **1978**, 1269.
8. Albericio, F.; Cases, M.; Alsina, J.; Triolo, S. A.; Carpino, L. A.; Kates, S. A., *Tetrahedron Lett.* **1997**, *38*, 4853.
9. Carpino.; L. A., *J. Am. Chem. Soc.* **1993**, *115*, 4397.

Yasuhiro Nishiyama
University of Texas-Houston Medical School, Houston, TX, USA

9-(2-Sulfo)fluorenylmethoxycarbonyl Chloride

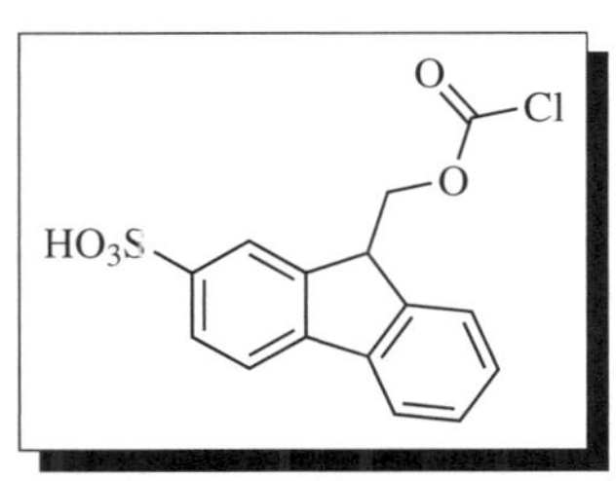

[67827-06-9] $C_{15}H_{11}ClO_5S$ (MW 338.76)

(base-labile, water-soluble protecting group for amines; chemoselective tag used in peptide synthesis for the purification of large peptides[1])

Alternate Name: Sulfmoc-Cl; 9-fluorenylmethoxycarbonyl-SO_3H chloride; FMS-Cl.
Physical Data: mp 138–140 °C.
Solubility: soluble in most organic solvents except alkanes. Sulfmoc analogs are soluble in water, but Sulfmoc-Cl itself reacts very quickly with this solvent.
Form Supplied in: Sulfmoc-Cl (white solid) is not commercially available.
Handling, Storage, and Precautions: Sulfmoc-Cl is an acid chloride and should therefore be protected from moisture. Sulfmoc-Cl has been reported to be hygroscopic. Bottles may develop pressure as the reagent evolves CO_2 upon degradation. Closely related Fmoc-Cl has been reported to be lachrymator and very harmful if swallowed, inhaled, or absorbed through skin, and similar handling precautions should be taken if working with Sulfmoc-Cl (use a fumehood).

Preparation and Purification of N-Terminal Sulfmoc Protected Peptides. Sulfmoc-Cl (**1**),[2] Sulfmoc-ONp (**2**),[2] and Sulfmoc-OSu (**3**)[3] were produced in very good yields by selective electrophilic sulfonation ($ClSO_3H$) of the more electron-rich position of corresponding Fmoc derivatives. Chloroformates (**1**) and active carbonates (**2** and **3**) are stable under these conditions.

Sulfmoc-Cl has been used to protect the free amine of the growing peptide chain in the last step of a solid-phase synthesis. Sulfmoc carbamates exhibit good stability to strong acidic conditions and therefore are not affected by the peptide-resin cleavage with HBr/AcOH or HF. However, because of the electron-withdrawing sulfonic acid substituent, deprotection can be accomplished under milder basic conditions than are needed for the related Fmoc derivatives. The sulfonic acid group (pKa ≈ -1.3) is fully ionized in 1 M aqueous formic acid, while α- and ω-carboxyl groups of amino acid residues are protonated at this pH. The Sulfmoc group can therefore serve as a probe such that ion-exchange chromatography can be used to obtain the desired peptide product free of the truncated peptides. In an alternative more simple purification procedure, the Sulfmoc peptide may be absorbed on a quaternary ammonium resin column and the Sulfmoc group can be cleaved on the resin by washing with aq NEt_3. The desired free peptide is then eluted with 1 M formic acid.[2]

Water-Soluble Long-Acting Prodrug Approach.[3,4] A $(Sulfmoc)_3$-insulin conjugate has been designed as a prolonged-acting preparation of insulin.[3] This prodrug escapes receptor-mediated endocytosis and degradation, and undergoes slow spontaneous regeneration of insulin at physiological pH. However, many aromatic and polyaromatic hydrocarbons are carcinogenic, and the fate of released 2-sulfodibenzofulvene should therefore be elucidated before any clinical development of Sulfmoc-drug conjugates.

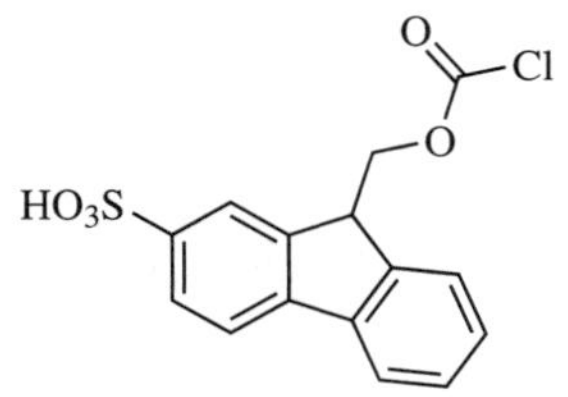

1

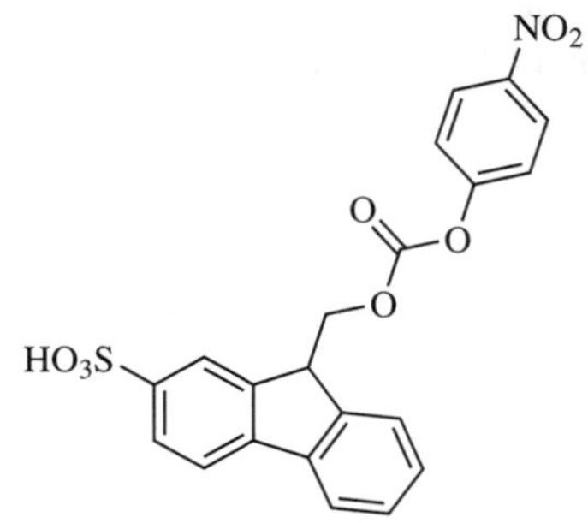

2

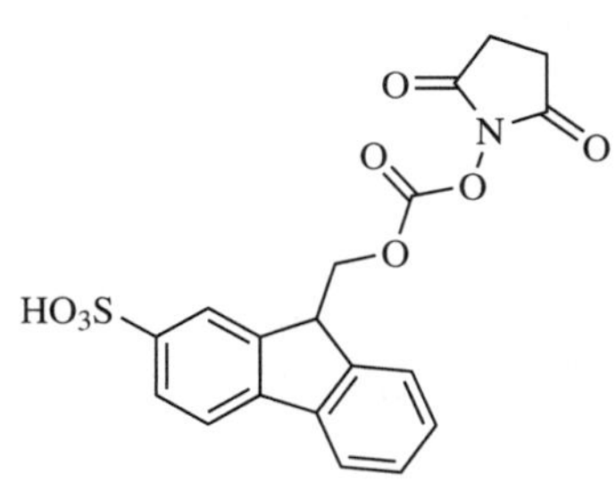

3

1. Mascagni, P., In *Fmoc Solid Phase Peptide Synthesis: A Practical Approach*; Chan, W. C.; White, P. D., Eds; Oxford University Press: Oxford, **2000**, pp. 265–276.
2. Merrifield, R. B.; Bach, A. E., *J. Org. Chem.* **1978**, *43*, 4808–4816.
3. Gershonov, E.; Goldwaser, I.; Fridkin, M.; Shechter, Y., *J. Med. Chem.* **2000**, *43*, 2530–2537.
4. Shechter, Y.; Preciado-Patt, L.; Schreiber, G.; Fridkin, M., *Proc. Natl. Acad. Sci. USA* **2001**, *98*, 1212–1217.

Didier Stien
CNRS, Montpellier, France

T

Tetrachlorophthalic Anhydride[1]

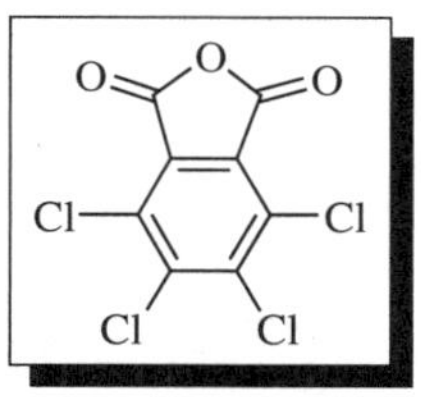

[117-08-8] $C_8Cl_4O_3$ (MW 285.90)

(amine protecting group,[1] glycosidation directing group,[1] Mitsunobu reaction with tetrachlorophthalimide analog[28,29])

Alternate Name: TCPA, TETRATHAL®.
Physical Data: mp 254–258°C.
Form Supplied in: powder or flake; commercially available; may contain hexachlorobenzene.
Handling, Storage, and Precautions: moisture sensitive; irritant. Handle with gloves in a fume hood.

Amine Protection. While tetrachlorophthalic anhydride (TCPA) has been widely utilized in applications such as pigments, dyes, and flame-retardants, it has also been finding applications as an alternative to phthalic anhydride protection of primary amines. This is particularly true in oligosaccharide chemistry where the amine protecting group has the dual function of providing protection as well as controlling stereoselectivity at the anomeric center during glycosidation events.[2,3] Tetrachlorophthaloyl protection is particularly useful when an unsubstituted phthalimide proves resistant to base mediated removal[4] or when the substrate of interest contains base sensitive functionality,[5] since tetrachlorophthalimide cleavage occurs under very mild conditions. The protecting group can be installed in the same fashion as unsubstituted phthalimide.[6] A common method (eq 1) involves treatment of a primary amine with TCPA and an equivalent of a base such as triethylamine. Once amide formation is complete, a mixed anhydride is generated that allows the modestly nucleophilic amide nitrogen to close the imidic ring (eq 2). Of interest in carbohydrate chemistry is the ability of the TCP group to direct the course of glycosidation reactions as well as or better than unsubstituted phthalimide to provide the β-glycoside product with high stereoselectivity. *n*-Pentenyl[7] (eqs 3 and 4),[8,9] trichloroacetimidate[10] (eq 5),[11] and Koenigs–Knorr[12] (eq 6)[6] based TCP donors have been applied to the preparation of *O*-linked glycosides.

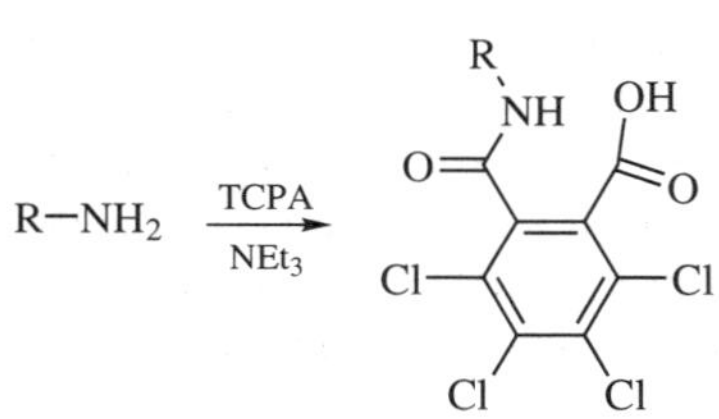

(1)

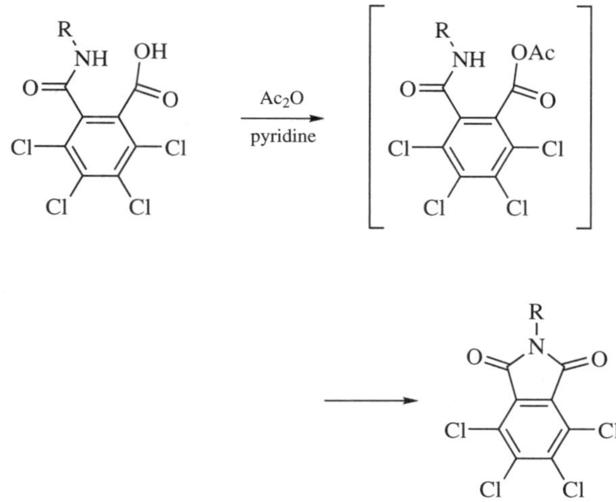

(2)

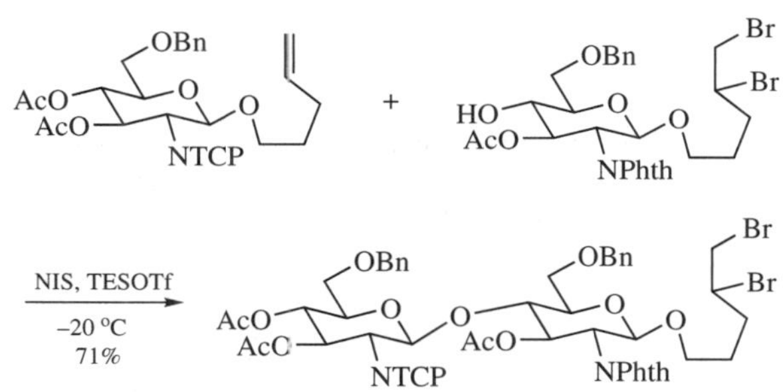

(3)

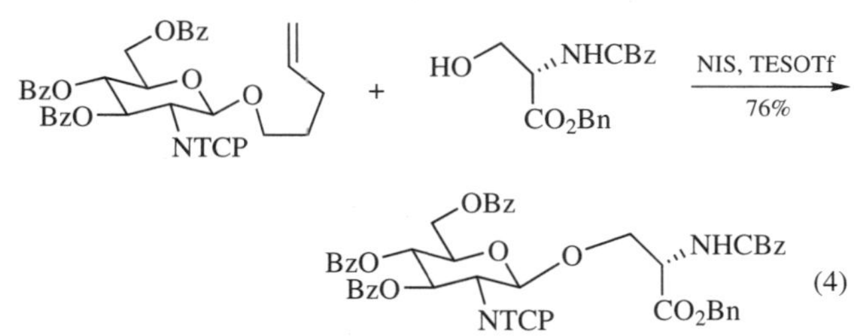

(4)

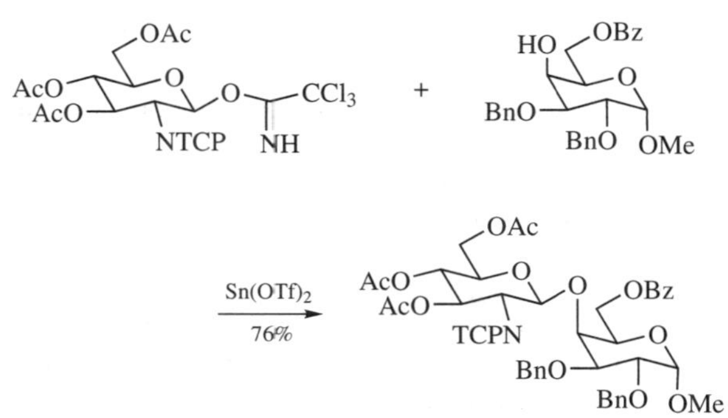

(5)

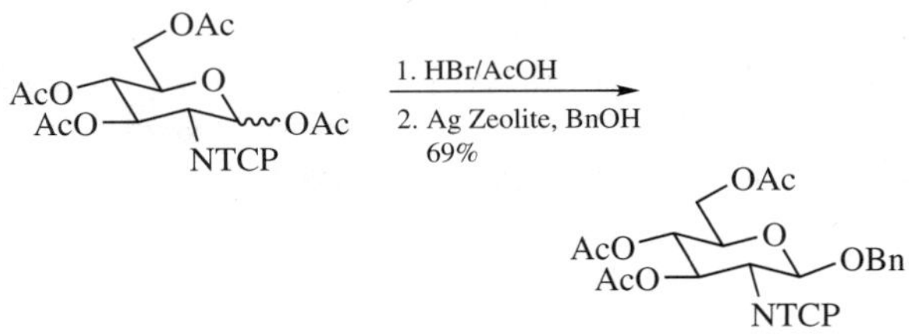

(6)

C-linked alkyl and aryl glycosides have also been prepared with TCP donors. Keck allylation of the glycosyl bromide in eq 7 provided the *C*-linked glycoside. Note that by utilizing the TCP instead of unsubstituted phthalimide the stereoselectivity of the reaction increased from 10:1 (phth) to 20:1 (TCP) β/α [13] Similar results in preparing *C*-alkyl glycosides are seen by treating glycosyl fluorides with allyltrimethylsilane in the presence of $BF_3 \cdot Et_2O$.[14]

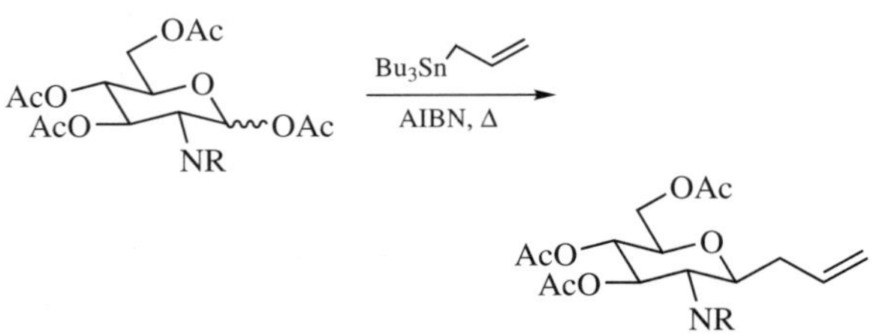

(7)

R = Phth, 10:1 β/α, 71%
R = TCP, 20:1 β/α, 77%

Friedel–Crafts-type glycosidations can also be applied to the preparation of *C*-aryl glycosides. These difficult reactions can be executed through the use of the highly reactive trichloroacetimidate donors as pictured in eq 8.

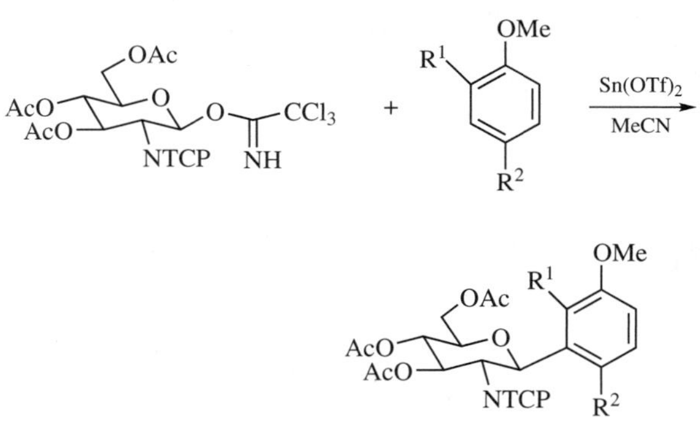

(8)

R^1 = H, R^2 = H, 40%
R^1 = H, R^2 = OMe, 62%
R^1 = OMe, R^2 =OMe, 71%

Access to β-*N*-linked glycopeptides from TCP bearing glycosyl donors can be achieved through a three-component Ritter-type reaction process eq 9.[15] The electronic nature of the carboxylic acid has a profound effect on the product yield and reaction rate.[16]

Deprotection. The methods that are utilized for TCP deprotection are much milder than for unsubstituted phthalimides.[17] Of interest is the chemoselectivity involved in the transformation. For example, TCP can be cleaved in the presence of unsubstituted phthalimide (eq 10).[9]

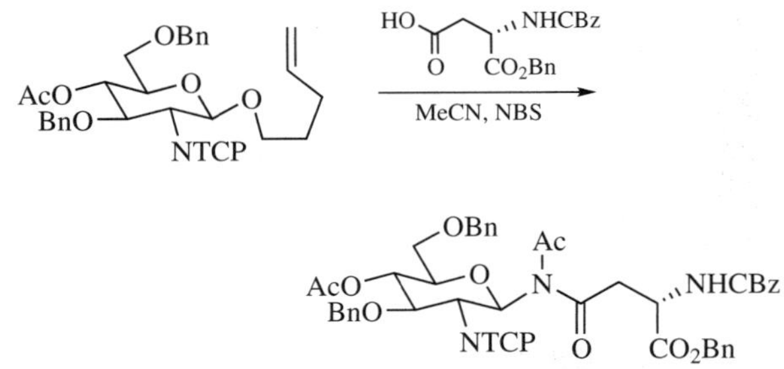

(9)

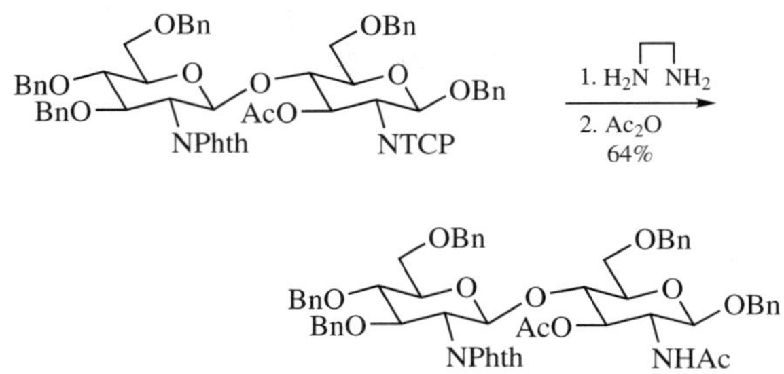

(10)

Ethylenediamine deprotects[18] the TCP giving the free amino sugar in very high yield, while leaving the acetyl group at the 4-position untouched (eq 11).[9] Similar chemoselectivity with unsubstituted phthalimide would be difficult to achieve. However primary acetates are not as robust. On the other hand, benzoates are very stable to this deprotection process.

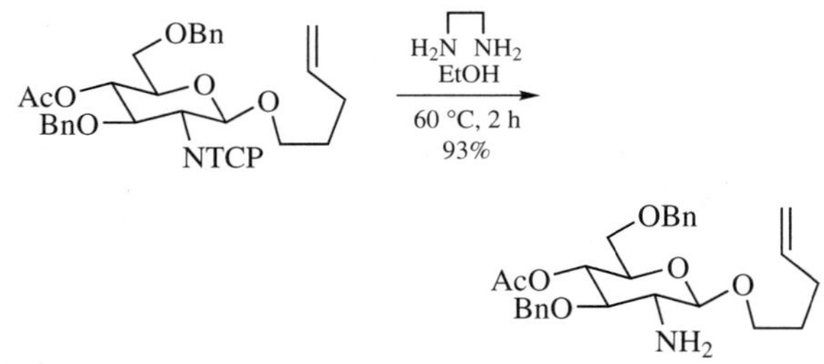

(11)

This chemoselectivity provides a new route to *N*-differentiated polyaminated natural products, as exemplified in eq 12. Here the TCP group provided *N*-differentiation of the glucosamine backbone of the tetrasaccharide allowing for installation of a fatty acyl side chain and completion of a nodulation factor total synthesis.[8]

A procedure for cleaving phthalimides by the use of sodium borohydride has been developed by Ganem,[19] and Garegg and Ogawa independently applied it to glucosamine derivatives.[20] In this case, borohydride reduction of a carbonyl group leads to ortho-hydroxy benzamide which lactonizes upon treatment with acetic acid at reflux. This procedure has been applied to a variety of TCP derivatives with conversions to the *N*-acetylated derivatives as shown in eq 13.[21]

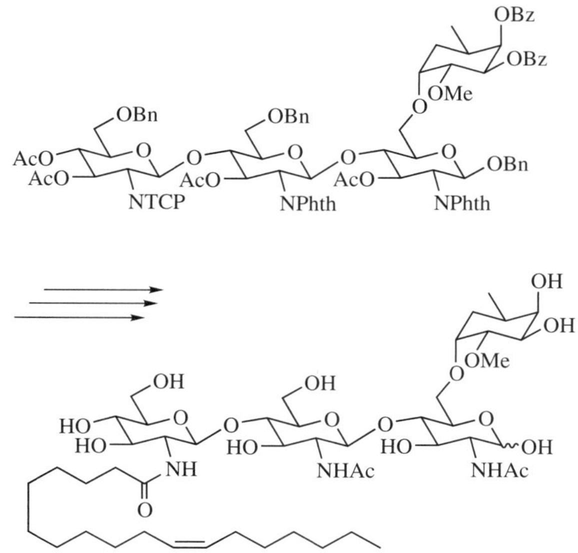

(12)

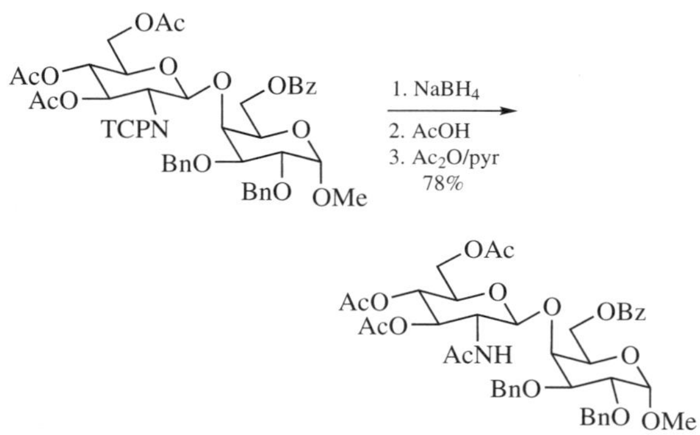

(13)

A recent development by Stangier and Hindsgaul[22] details cleavage of phthalimide or TCP using a polystyrene resin immobilized amine. If realized on oligosaccharides, this procedure will significantly facilitate isolation of the free amine from the reaction mixture, which under certain circumstances is rendered difficult by the phthalimide cleavage by-products. Potentially, all that would be required to isolate the free amine product by the solid phase procedure would be a filtration step. However, the high temperature of the process (85 °C) may be cause for concern where base sensitive protecting groups and other sensitive functionalities are present, particularly since the need for solid phase interaction greatly increases the reaction times.

Additionally, hydrazinolysis[23], butylamine,[24] methylamine,[25] and hydroxylamine[26] treatment should all be applicable to TCP cleavage just as they have been with unsubstituted phthalimide. However, due to the enhanced reactivity of the TCP group, lower temperatures and smaller quantities of the cleavage agents should be sufficient to cleave the protecting group.

Mitsunobu Reaction. Of related interest to tetrachlorophthalic anhydride is its nitrogen analog tetrachlorophthalimide (also commercially available). Displacement of primary and, to a lesser extent, secondary alcohols is readily accomplished by treatment with tetrachlorophthalimide under Mitsunobu conditions.[27] The enhanced acidity of the reagent makes it more reactive than unsubstituted phthalimide for these types of displacement reactions allowing the reaction to proceed smoothly without side reaction on thymidine (eq 14).[28]

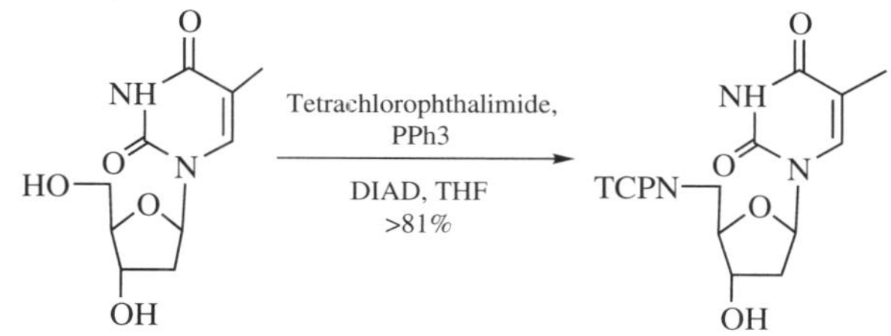

(14)

In fact, when phthalimide and tetrachlorophthalimide are made to compete for the same primary hydroxyl group only the tetrachlorophthalimide substitution product is observed with none of the unsubstituted phthalimide adduct being detected (eq 15).[29]

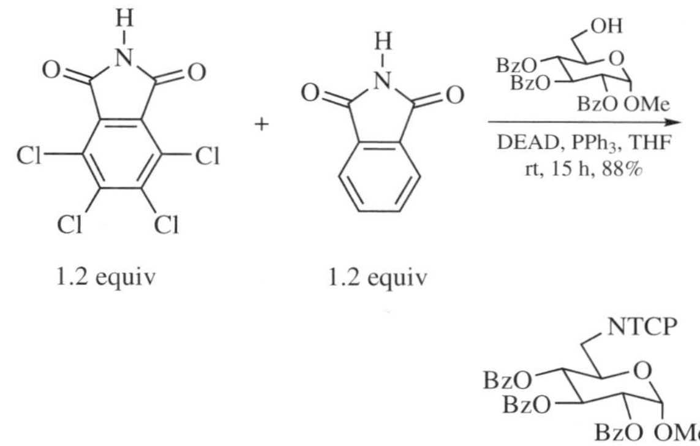

(15)

Displacement could also be achieved at the hindered secondary hydroxyl of menthol in 69% yield[29] (eq 16) and also on the arabinose derivative in 73% yield (eq 17).[30] Tatibouët points out an all too common occurrence when hydrazine is used to cleave the imide ring, partial reduction of unsaturation in the substrate. Diimide can form by oxidation of hydrazine leading to diimide-induced reduction.[31] To prevent the chance of unwanted reduction ethylenediamine can be used when unsaturation is present in the substrate.

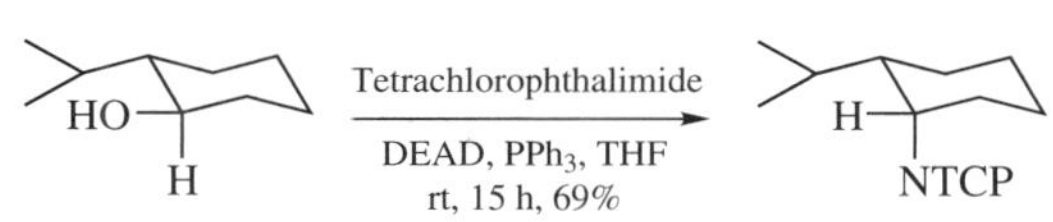

(16)

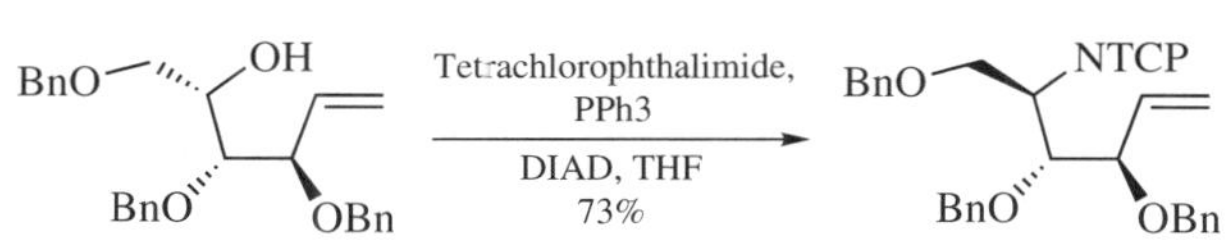

(17)

Related Reagents. Phthalic anhydride; phthalimide.

1. Debenham, J.; Rodebaugh, R.; Fraser-Reid, B., *Liebigs Ann.* **1997**, 791.
2. Lemieux, R.; Takeda, T.; Chung, V., In *Synthetic Methods for Carbohydrates*; El Khadem, H. S., Ed.; ACS Symposium Series: Washington, D.C. 1976, Vol. 39, 90.
3. Banoub, J.; Boullanger, P.; Lafont, D., *Chem. Rev.* **1992**, *92*, 1167.
4. (a) Koeman, F. A. W.; Meissner, J. W. G.; van Ritter, H. R. P.; Kamerling, J. P.; Vilegenthart, J. F. G., *J. Carbohydr. Chem.* **1994**, *13*, 1. (b) Matsuzaki, Y.; Ito, Y.; Nakahara, Y.; Ogawa, T., *Tetrahedron Lett.* **1993**, *34*, 1061.

5. Spijker, N. M.; Westerduin, P.; van Boeckel, C. A. A., *Tetrahedron* **1992**, *48*, 6297.
6. Debenham, J. S.; Debenham, S. D.; Fraser-Reid, B., *Bioorg. Med. Chem.* **1996**, *4*, 1909.
7. (a) Fraser-Reid, B.; Udodong, U. E.; Wu, Z.; Ottosson, H.; Merritt, J. R.; Rao, C. S.; Roberts, C.; Madsen, R., *Synlett* **1992**, 927. (b) Madsen, R.; Fraser-Reid, B., In *Modern Methods in Carbohydrate Synthesis*, Khan, S. H.; O'Neill, R. A., Eds.; Harwood Academic Publishers: Amsterdam, 1996, 155.
8. Debenham, J.; Rodebaugh, R.; Fraser-Reid, B., *J. Org. Chem.* **1997**, *62*, 4591.
9. Debenham, J.; Fraser-Reid, B., *J. Org. Chem.* **1996**, *61*, 432.
10. (a) Schmidt, R., In *Modern Methods in Carbohydrate Synthesis*, Khan, S. H.; O'Neill, R. A., Eds.; Harwood Academic Publishers: Amsterdam, 1996, 20. (b) Schmidt, R., *Adv. Carbohydr. Chem. Biochem.* **1994**, *50*, 21.
11. Castro-Palomino, J. C.; Schmidt, R. R., *Tetrahedron Lett.* **1995**, *36*, 5343.
12. Paulsen, H., *Chem. Soc. Rev.* **1984**, *13*, 15.
13. Roe, B. A.; Boojamra, C. G.; Griggs, J. L.; Bertozz, C. R., *J. Org. Chem.* **1996**, *61*, 6442.
14. Fuchss, T.; Schmidt, R. R., *Synthesis* **1998**, 753.
15. Debenham, J. S.; Madsen, R.; Roberts, C.; Fraser-Reid, B., *J. Am. Chem. Soc.* **1995**, *117*, 3302.
16. Szardenings, K. A.; Fraser-Reid, B.; Nair, G. L., *Org. Lett.* **2000**, *3*, 317.
17. (a) Greene, T. W.; Wuts, P. G. M., In *Protecting Groups in Organic Synthesis*, John Wiley & Sons: New York, 3rd edn, 1999. (b) Kocienski P. J., *Protecting Groups*, Thieme: Stuttgart, 1994.
18. Kanie, O.; Crawley, S. C.; Palcic, M. M.; Hindsgaul, O., *Carb. Res.* **1993**, *243*, 139.
19. Osby, J. A.; Martin, M. G.; Ganem, B., *Tetrahedron Lett.* **1984**, *25*, 2093.
20. (a) Dasgupta, F.; Garegg, P., *J. Carbohydr. Chem.* **1988**, *7*, 701. (b) Ito, Y.; Sato, S.; Mori, M.; Ogawa, T., *J. Carbohydr. Chem.* **1988**, *7*, 359.
21. Castro-Palomino, J. C.; Schmidt, R. R., *Tetrahedron Lett.* **1995**, *36*, 5343.
22. Stangier, P.; Hindsgaul, O., *Synlett* **1996**, 179.
23. Ing, H.; Manske, R., *J. Chem. Soc.* **1926**, 2348.
24. Durette, P. L.; Meitzner, E. P.; Shen, T. Y., *Tetrahedron Lett.* **1979**, 4013.
25. Wolfe, S.; Hasan, S. K., *Can. J. Chem.* **1970**, *48*, 3572.
26. Madsen, R.; Udodong, U. E.; Roberts, C.; Mootoo, D. R.; Konradson, P.; Fraser-Reid, B., *J. Am. Chem. Soc.* **1995**, *117*, 1554.
27. (a) Hughes, D. L., *Org. React.* **1992**, *42*, 335. (b) Castro, B. R., *Org. React.* **1983**, *29*, 1. (c) Mitsunobu, O., *Synthesis* **1981**, 1.
28. Tetzlaff, C. N.; Schwope, I.; Bleczinski, C. F.; Steinberg, J. A.; Richert, C., *Tetrahedron Lett.* **1998**, 4215.
29. Jia, Z. J.; Kelberlau, S.; Olsson, L.; Anilkumar, G.; Fraser-Reid, B., *Synlett* **1999**, 565.
30. Tatibouët, A.; Rollin, P.; Martin, O. R., *J. Carbohydr. Chem.* **2000**, *19*, 641.
31. (a) Pasto, D. J.; Taylor, R. T., *Org. React.* **1991**, *40*, 91. (b) Miller, C. E., *J. Chem. Ed.* **1965**, *42*, 254.

John S. Debenham & Bert Fraser-Reid
Merck Research Laboratories, Rahway, New Jersey, USA
Natural Products & Glycotechnology Research Institute, Durham, North Carolina, USA

Tetrachlorophthalimide

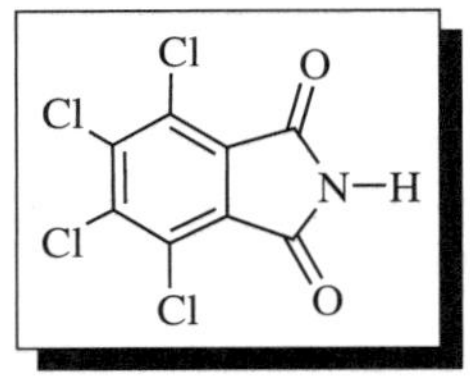

[1571-13-7] $C_8HCl_4NO_2$ (MW 284.91)

(derivatives used as amine-protecting groups)

Physical Data: colorless blades, mp 338–339 °C (acetic acid);[1] 336–337 °C (nitrobenzene);[2] yellow needles, mp 345–347 °C (*N,N*-dimethylformamide).[3]

Solubility: soluble in pyridine and aqueous sodium carbonate; soluble with difficulty in ethanol, ether, and benzene; and slightly soluble in *N,N*-dimethylformamide, nitrobenzene, and in hot acetic acid (1 g/100 mL).

Form Supplied in: yellow powder; widely available.

Purification: recrystallization can be done from acetic acid,[1] nitrobenzene,[2] or *N,N*-dimethylformamide.[3]

Handling, Storage, and Precautions: tetrachlorophthalimide is an irritant to the eyes and to the respiratory system. Avoid contact and inhalation. Wash thoroughly after handling. Store in a cool dry place. Its chemical, physical, and toxicological properties have not been thoroughly investigated, but it should be noted that phthalimide is a teratogen.

Nomenclature. The correct name for the title compound is 3,4,5,6-tetrachlorophthalimide according to IUPAC rule C-287.[4] This rule states that when imide nomenclature is inconvenient, imides may be named as heterocyclic compounds. Section R-5.7.8.3 of IUPAC's 1993 guide to nomenclature[5] gives as an alternative designation for phthalimide, i.e., 2,3-dihydro-1*H*-isoindole-1,3-dione. Under this system, the title compound is 4,5,6,7-tetrachloro-2,3-dihydro-1*H*-isoindole-1,3-dione. Unfortunately, a number of authors have recently mixed the two systems, naming this type of compound as a 4,5,6,7-tetrachlorophthalimide derivative. This practice is clearly incorrect. Chemical Abstracts have chosen to use 1*H*-isoindole-1,3(2*H*)-1,3-dione for phthalimide.

Amine-Protecting Group. The principal use[6] of the 3,4,5,6-tetrachlorophthalimido group is as a cost-effective protecting group for amines with advantages over the well-known phthalimido group.[7] This latter group was the first and is still the most utilized group for protecting the common 2-amino groups of biologically active carbohydrates. 2-Phthalimido derivatives of pyranose sugars bearing glycosyl leaving groups give glycosides that have the new glycosyl group *trans* to the 2-phthalimido group reliably in glycosyl reactions, unlike 2-acetamido derivatives which usually give oxazolines as byproducts (eq 1).[7]

The 3,4,5,6-tetrachlorophthalimido (TCP) group is introduced easily by the reaction of the amine with 3,4,5,6-tetrachlorophthalic anhydride in methanol, followed by acetic anhydride in pyridine.[8,9] A procedure giving improved yields uses 3,4,5,6-tetrachlorophthalic anhydride in pyridine, followed by acetic anhydride in pyridine (eq 2).[10] 3,4,5,6-Tetrachlorophthalimido derivatives are highly crystalline making isolation facile.[6] The 3,4,5,6-tetrachlorophthalimide group is bulkier than the phthalimide group resulting in a better 1,2-*trans*-control during glycosylations and in improved selectivity for O-4 in reactions in which the O-3, O-4 unprotected 2-TCP derivative is the glycosyl acceptor,[10,11] but in considerably lower glycosyl acceptor reactivity at oxygen atoms adjacent to the TCP group.[10] The major advantage of the 3,4,5,6-tetrachlorophthalimide group is that it is removed under much milder conditions than the phthalimide group. Treatment with ethylenediamine in ethanol at room temperature,[6] with alkyldiamines attached to polystyrene beads,[12] or with 1.1 equiv of sodium borohydride followed by acetic acid[8] easily removed this protecting group. The first conditions did not affect the phthalimide group.[6] It is worth noting that the first method for removal was mild enough that an acetate group survived.[6] However, acetate groups can be removed in the presence of the TCP group by reaction with dilute sodium methoxide (0.018 M) in methanol for a short time[10] or by reaction with magnesium methoxide.[13]

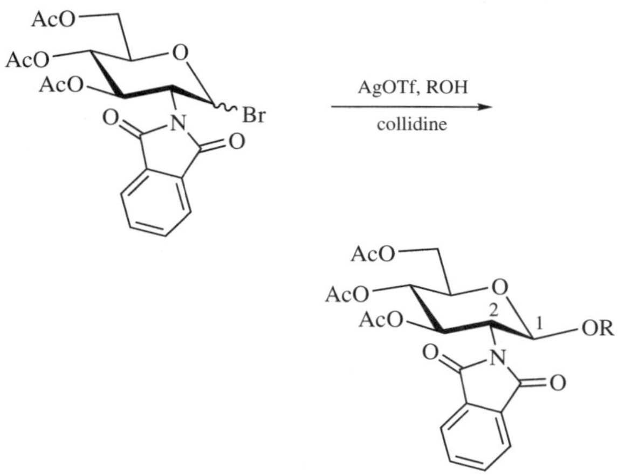

(1)

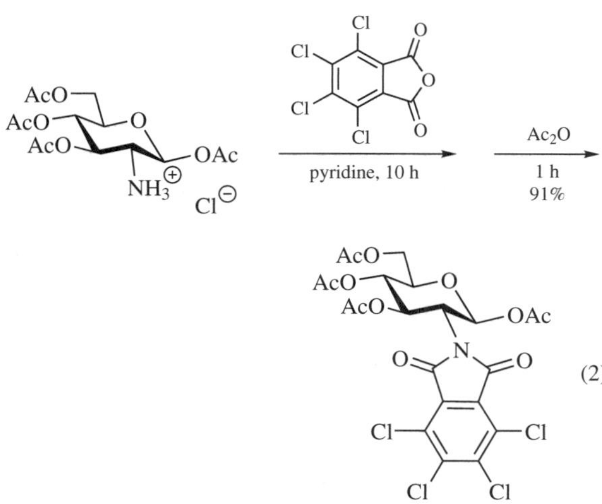

(2)

This group has been used in a considerable number of oligosaccharide syntheses,[14–18] in syntheses of C-glycosides,[19] and in syntheses of diosgenyl glycosides.[20]

The 3,4,5,6-tetrachlorophthalimido group has also been advocated as an amine-protecting group in peptide synthesis, and it has been demonstrated that the TCP derivative of all amino acids are formed in 4–8 min on microwave irradiation of the amino acid in DMF in very good yields (62–99%).[21] These protected amino acids were employed in solid-phase synthesis of peptides, using hydrazine:DMF at 40 °C for 0.5–1 h for TCP deprotection of the solid-phase intermediates and DIPCDI:HOAt as coupling reagents. Under these conditions, good purities of up to nonapeptides were achieved with little racemization.[22]

Mitsunobu Reaction. 3,4,5,6-Tetrachlorophthalimide was found[23] to be an excellent reagent for the direct displacement of primary hydroxyl groups by a nitrogen-containing nucleophile in a Mitsunobu[24,25] reaction (eq 3).

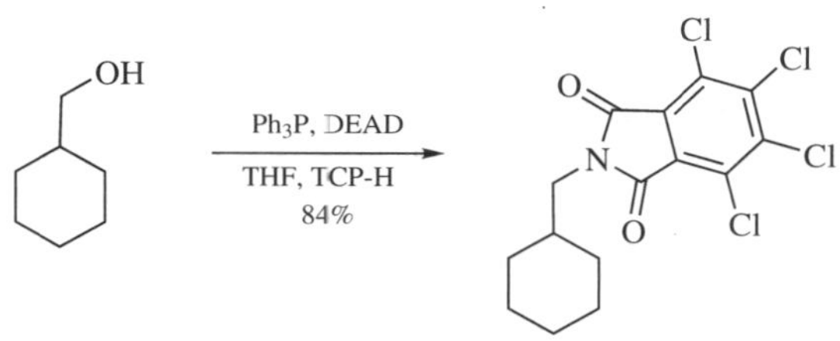

(3)

Secondary alcohols also react readily (eq 4),[23] except when flanked by alkoxy or acyloxy groups, as in carbohydrate derivatives.

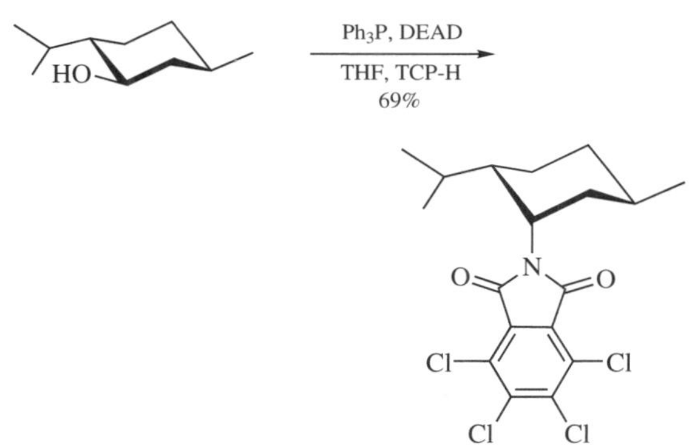

(4)

Reactions with Alkenes. Photoinduced cyclizations of *N*-alkenyl substituted phthalimides have been employed in the synthesis of five-membered to medium rings. These reactions take place in alcoholic solvents and are based on the intramolecular single-electron transfer reactions between the photoexcited phthalimide moieties and easily oxidized alkenes (eq 5).[26–28]

It was subsequently found that irradiation of *N*-(hydroxyalkyl)-3,4,5,6-tetrachlorophthalimides with alkenes that are good single-electron donors also gives medium and large rings quite efficiently.[29] For example, the reaction of *N*-(2-hydroxyethyl)-3,4,5,6-tetrachlorophthalimide with 1,1-diphenylethene in benzene afforded a tricyclic product incorporating a seven-membered ring (eq 6).[29]

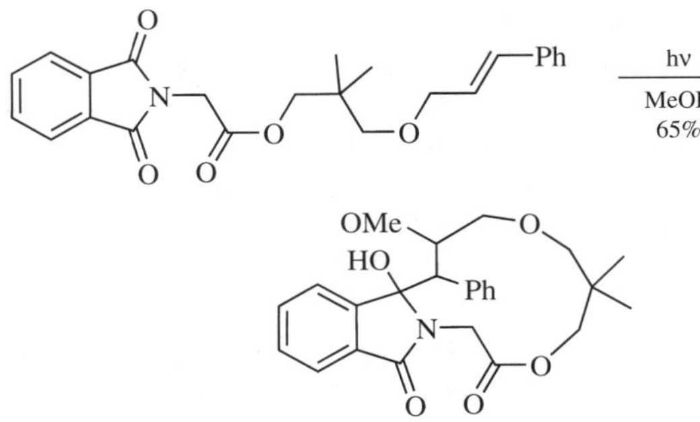

(5)

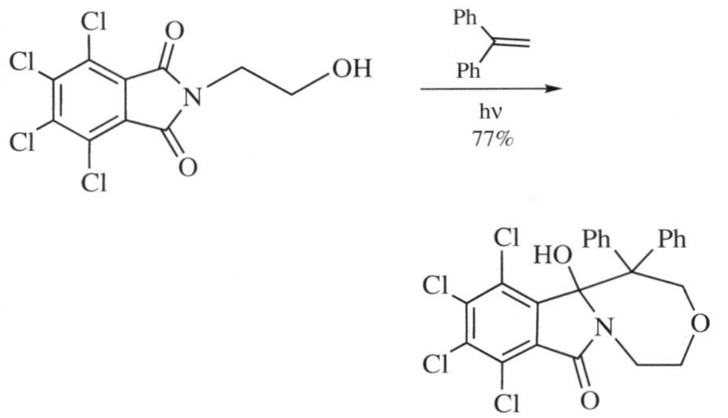

(6)

Similarly, the photolysis of *N*-phthaloylglycine di(ethylene) glycol ester with α-methylstyrene in benzene afforded the 13-membered lactones (eq 7)[29] as a pair of diastereomers.

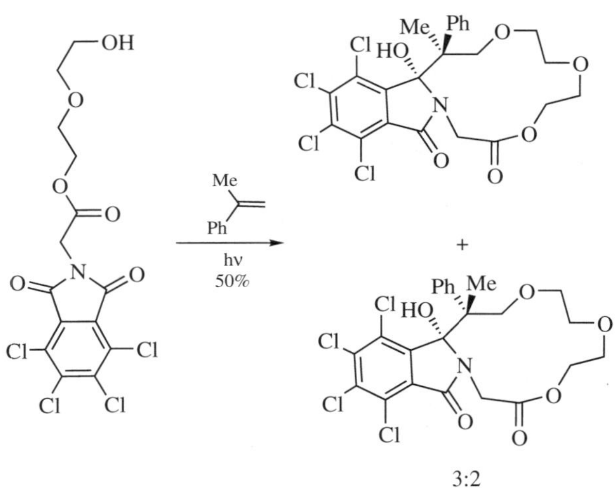

(7)

Other Derivatives. Studies of α-glucosidase inhibitors derived from thalidomide (**1**) revealed that *N*-alkyl 3,4,5,6-tetrachlorophthalimide derivatives are superior inhibitors.[30] In particular, 3,4,5,6-tetrachloro-*N*-cycloheptylphthalimide (**2**) was approximately thirty times more reactive than the well-known inhibitor 1-deoxynojirimycin (**3**) (see Figure 1).[31,32]

Reagent In Qualitative Organic Analysis. It was observed that potassium tetrachlorophthalimide displaces halogen and sulfonate from a wide variety of bonding environments to give crystalline derivatives with higher melting points than other phthalimides.[2,33]

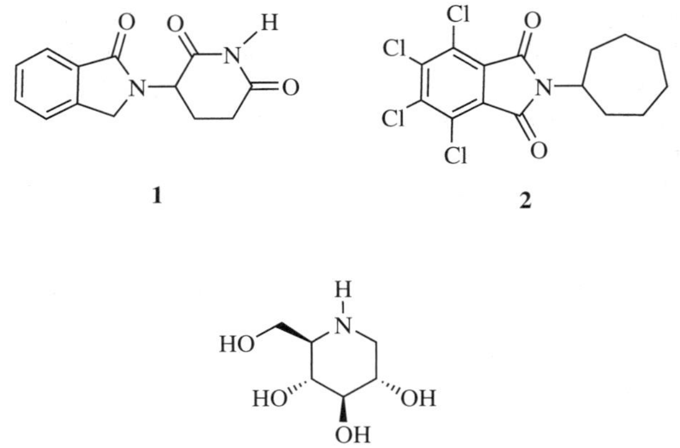

1 **2** **3**

Figure 1 Structure of thalidomide (**1**), 3,4,5,6-tetrachloro-*N*-cycloheptylphthalimide (**2**), and 1-deoxynojirimycin (**3**)

Related Reagents. Phthalimide; other substituted phthalimides; *O*-(2,2,2-trichloroethyl)carbamate; trichloroacetamide.

1. Pratt, D. S.; Perkins, G. A., *J. Am. Chem. Soc.* **1918**, *40*, 198.
2. Allen, C. F. H.; Nicholls, R. V. V., *J. Am. Chem. Soc.* **1934**, *56*, 1409.
3. Rosen, W. E.; Toohey, V. P.; Shabica, A. C., *J. Am. Chem. Soc.* **1957**, *79*, 3167.
4. Rigaudy, J.; Klesney, S. P., *In Nomenclature for Organic Compounds*, 4th ed.; Pergamon: Oxford, UK, 1979.
5. Panico, R.; Powell, W. H.; Richer, J.-C., *A Guide to IUPAC Nomenclature of Organic Compounds*; Blackwell Scientific: Oxford, UK, 1993.
6. Debenham, J. S.; Madsen, R.; Roberts, C.; Fraser-Reid, B., *J. Am. Chem. Soc.* **1995**, *117*, 3302.
7. Lemieux, R. U.; Takeda, T, Chung, B. Y., In *Synthetic Methods for Carbohydrates*; El Khadem, H. S., Ed.; American Chemical Society: Washington D.C., 1976, pp 90–115.
8. Castro-Palomino, J. C.; Schmidt, R. R., *Tetrahedron Lett.* **1995**, *36*, 5343.
9. Debenham, J. S.; Fraser-Reid, B., *J. Org. Chem.* **1996**, *61*, 432.
10. Ellervik, U.; Magnusson, G., *J. Org. Chem.* **1998**, *63*, 9314.
11. Lay, L.; Manzoni, L.; Schmidt, R. R., *Carbohydr. Res.* **1998**, *310*, 157.
12. Stangier, P.; Hindsgaul, O., *Synlett* **1996**, 179.
13. Qin, Z. H.; Li, H.; Cai, M. S.; Li, Z. J., *Chin. Chem. Lett.* **2000**, *11*, 941.
14. Debenham, J. S.; Rodebaugh, R.; Fraser-Reid, B., *J. Org. Chem.* **1997**, *62*, 4591.
15. Debenham, J. S.; Debenham, S. D.; Fraser-Reid, B., *Bioorg. Med. Chem.* **1996**, *4*, 1909.
16. Ellervik, U.; Grundberg, H.; Magnusson, G., *J. Org. Chem.* **1998**, *63*, 9323.
17. Griffiths, S. L.; Madsen, R.; Fraser-Reid, B., *J. Org. Chem.* **1997**, *62*, 3654.
18. Olsson, L.; Kelberlau, S.; Jia, Z. J.; Fraser-Reid, B., *Carbohydr. Res.* **1998**, *314*, 273.
19. Castro-Palomino, J. C.; Schmidt, R. R., *Liebigs Ann.* **1996**, 1623.
20. Bednarczyk, D.; Kaca, W.; Myszka, H.; Serwecinska, L.; Smiatacz, Z.; Zaborowski, A., *Carbohydr. Res.* **2000**, *328*, 249.
21. Cros, E.; Planas, M.; Bardaji, E., *Synthesis* **2001**, 1313.
22. Cros, E.; Planas, M.; Mejias, X.; Bardaji, E., *Tetrahedron Lett.* **2001**, *42*, 6105.
23. Jia, Z. J.; Kelberlau, S.; Olsson, L.; Anil kumar, G.; Fraser-Reid, B., *Synlett* **1999**, 565.

24. Mitsunobu, O., *Synthesis* **1981**, 1.
25. Hughes, D. L., *Org. Prep. Proced. Int.* **1996**, *28*, 127.
26. Maruyama, K.; Kubo, Y.; Machida, M.; Oda, K.; Kanaoka, Y.; Fukuyama, K., *J. Org. Chem.* **1978**, *43*, 2303.
27. Maruyama, K.; Kubo, Y., *J. Am. Chem. Soc.* **1978**, *100*, 7772.
28. Griesbeck, A. G.; Henz, A.; Hirt, J., *Synthesis* **1996**, 1261.
29. Xue, J.; Zhu, L.; Fun, H. K.; Xu, J. H., *Tetrahedron Lett.* **2000**, *41*, 8553.
30. Takahashi, H.; Sou, S.; Yamasaki, R.; Sodeoka, M.; Hashimoto, Y., *Chem. Pharm. Bull.* **2000**, *48*, 1494.
31. Sou, S.; Takahashi, H.; Yamasaki, R.; Kagechika, H.; Endo, Y.; Hashimoto, Y., *Chem. Pharm. Bull.* **2001**, *49*, 791.
32. Hashimoto, Y., *Bioorg. Med. Chem.* **2002**, *10*, 461.
33. Allen, C. F. H.; Adams, W. R.; Myers, C. L., *Anal. Chem.* **1965**, *32*, 158.

Hussein Al-Mughaid & T. Bruce Grindley
Dalhousie University, Halifax, NS, Canada

Tetraethyl Pyrophosphate

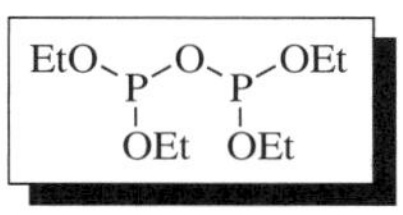

[21646-99-1] $C_8H_{20}O_5P_2$ (MW 258.22)

(peptide synthesis; phosphonylation of ketones, aldehydes, and perfluoroalkyl iodides)

Alternate Name: TEPP.
Physical Data: bp 79–81 °C/0.15 mmHg; *d* 1.057 g cm^{-3}; n_D 1.434.
Solubility: sol acetonitrile, 1,1,2-trichloro-1,2,2-trifluoroethane.
Form Supplied in: colorless liquid; commercially available.
Preparative Methods: the most convenient method involves the reaction of diethyl phosphite with diethyl phosphorochloridite. Diethyl phosphorochloridite (223 g) is added rapidly to a solution of diethyl phosphite (138 g) and triethylamine (101 g) in benzene (200 ml). The mixture is filtered and the benzene removed by distillation at reduced pressure. The product is isolated by further distillation at reduced pressure (79–81 °C/0.15 mmHg) in 41% yield.
Analysis of Reagent Purity: product purity is assessed by refractive index.
Handling, Storage, and Precautions: protect from moisture; irritant; handle in a fume hood.

Peptide Formation.[1] Tetraethyl pyrophosphite will react with amino acids at either the free amine (eq 1) to give amidophosphites, or at the free carboxylic acid (eq 2) to give mixed anhydrides, and both are activated for peptide formation. Diethyl Phosphonite is a good solvent for amino acids and peptides and so the most convenient procedure for peptide formation involves heating an *N*-protected amino acid with an amino acid ester (eq 3) with an excess of TEPP in diethyl phosphite solution. If the amino acid ester is used as the HCl salt, addition of Triethylamine is advantageous. The product is precipitated by the addition of water or an organic solvent. No racemization has been observed in this procedure with dipeptide formation; however, variable amounts of racemization are observed in tripeptide formation where Cbz protecting groups are used. TEPP has been successfully used in the synthesis of important medium sized peptides, e.g. oxytocin.[2]

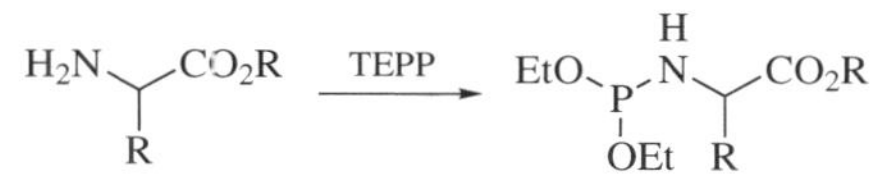 (1)

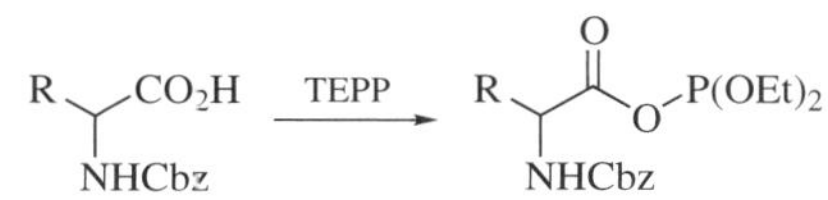 (2)

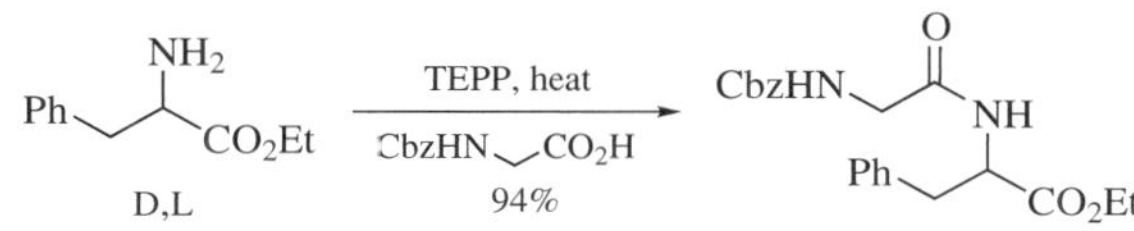 (3)

Phosphonylation. Tetraethyl pyrophosphite can function as both an electrophile and a nucleophile. The electrophilic behavior is exemplified by the formation of mixed anhydrides from carboxylic acids (eq 4).[3] TEPP shows a nucleophilic character in reaction with carbonyl compounds. Addition of TEPP to aldehydes and ketones (eq 5)[4–6] gives phosphonium alkoxide intermediates, which undergo an intramolecular transfer of phosphite yielding mixed phosphonate–phosphite adducts. A further equivalent of carbonyl reportedly results in further oxidation to phosphonate–phosphate adducts.

$$CF_3CO_2H \xrightarrow[\text{MeCN}]{\text{TEPP}} CF_3CO_2P(OEt)_2 \qquad (4)$$

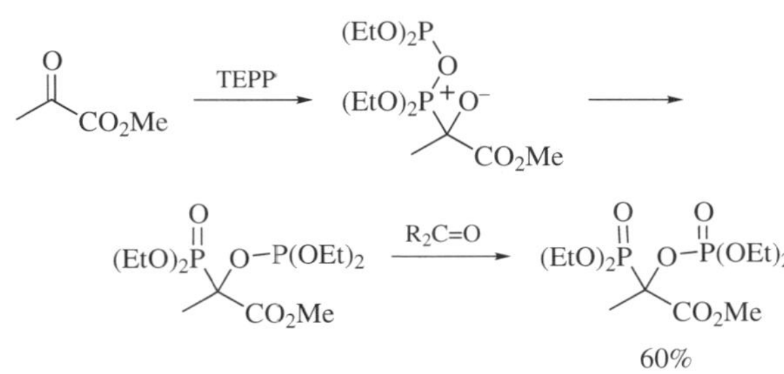 (5)

Reaction of perfluoroalkyl iodides with TEPP and 1,1-Di-t-butyl Peroxide in 1,1,2-trichloro-1,2,2-trifluoroethane gives diethyl phosphonites, which on oxidation with t-Butyl Hydroperoxide form diethyl perfluoroalkylphosphonates in good yield (eq 6).[7,8] The standard Arbuzov reaction is not generally applicable to fluoroalkyl phosphonates.

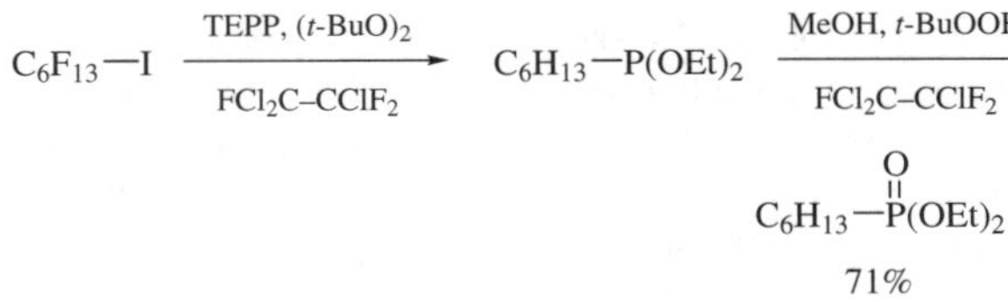

(6)

1. Anderson, G. W.; Blodinger, J.; Welcher, A. D. *J. Am. Chem. Soc.* **1952**, *74*, 5309.
2. Vigneaud, V.; Ressler, C.; Swan, J. M.; Roberts, C. W.; Katsoyannis, P. G. *J. Am. Chem. Soc.* **1954**, *76*, 3115.
3. Ofitserov, E. N.; Mironov, V. F.; Konovalova, I. V.; Pudovik, A. N. *Zh. Obshch. Khim.* **1983**, *53*, 1941.
4. Novikova, Z. S.; Mashoshina, S. N.; Lutsenko, I. F. *Zh. Obshch. Khim.* **1974**, *44*, 276.
5. Novikova, Z. S.; Mashoshina, S. N.; Lutsenko, I. F. *Zh. Obshch. Khim.* **1974**, *44*, 261.
6. Konovalova, I. V.; Burnaeva, L. A.; Saifullina, N. S.; Pudovik, A. N. *Zh. Obshch. Khim.* **1976**, *46*, 1410.
7. Kato, M.; Yamabe, M. *Chem. Commun./J. Chem. Soc., Chem. Commun.* **1981**, 1173.
8. Su, D.; Cen, W.; Kirchmeier, R. L.; Shreeve, J. M. *Can. J. Chem.* **1989**, *67*, 1795.

Christopher D. Spilling
University of Missouri-St. Louis, MO, USA

3,3′,4,4′-Tetrahydro-6,6′-bis(2*H*-pyran)

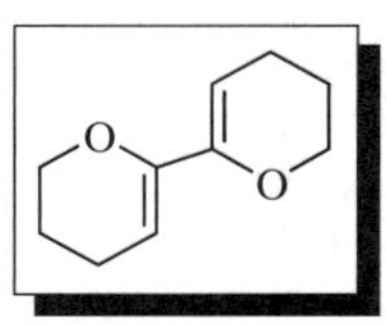

[109669-49-0] $C_{10}H_{14}O_2$ (MW 166.22)

(reagent used for the protection of vicinal diols)

Physical Data: mp 47–49 °C.
Solubility: soluble in most organic solvents.
Form Supplied in: colorless crystalline compound; available from Fluka, Aldrich.
Purification: recrystallization from petroleum ether.
Handling, Storage, and Precautions: store below −18 °C under inert atmosphere.

Protection of Vicinal Diols. 3,3′,4,4′-Tetrahydro-6,6′-bis(2*H*-pyran) (bis-DHP) complements the vicinal diol protecting ability of **1,1′,2,2′-tetramethoxycyclohexane** with high level of selectivity and stability. The products formed are designated as dispiroketals (dispoke, 1,8,13,16-tetraoxadispiro[5.0.5.4]hexadecane) and are quite stable. The dispiroketal formation is normally catalyzed by camphorsulfonic acid (CSA) and the protecting group can be removed hydrolytically under acidic conditions, preferably by transacetalization with ethylene glycol.[1] The selective protection of a vicinal diol in the presence of a 1,3-diol is elegantly shown in the reaction of bis-DHP with (*S*)-butane-1,2,4-triol to give a single stereoisomer (eq 1).[2]

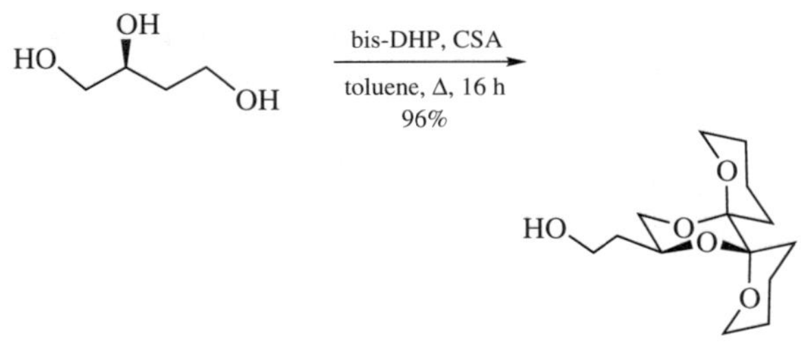

(1)

The stability of the dispoke group was demonstrated in subsequent reactions carried out on the protected derivative.[2] Similarly, reactions at the keto group are carried out with the glycolic acid protected dispiroketal without affecting the protecting group.[3] The dispoke protection has also been extended to 1,2-dithiols.[4]

Reaction of bis-DHP with (*S*)-lactic acid gave the major diasteromer with the methyl group in the equatorial position (eq 2).[5]

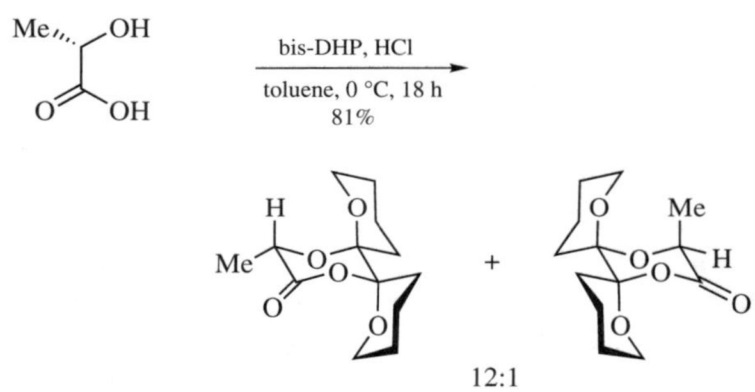

(2)

Alkylation of this product with various electrophiles takes place with high stereoselectivity. The deprotection of the alkylated compounds afford substituted products with control of the α-stereocenters. Similarly, several other reactions can be carried out with the dispoke protected lactic acid, mainly under stable basic conditions.[6]

The dispoke protection strategy is found to be useful in preparing the optically active dinitrile from dispoke protected L-(+)-dimethyl tartarate via the iodinated product (eq 3).[7] This dinitrile is used in the synthesis of a chiral magnesium porpyrazine for use as a catalyst for the asymmetric epoxidation of styrene.

A functionalized bis-DHP unit is used as a chiral auxiliary for asymmetric Michael additions which proceed with a high degree of stereoselectivity (eq 4).[8]

Protection of Vicinal Diols in Carbohydrates. Bis-DHP reacts with the hexapyranosides of fuco, arabino, rhamno, lyxo, and manno series in medium polar solvents in the presence of an acid catalyst (preferably CSA) to give the protected diequatorial diol in all cases (eq 5).[4,9] More polar solvents cause decomposition of bis-DHP.

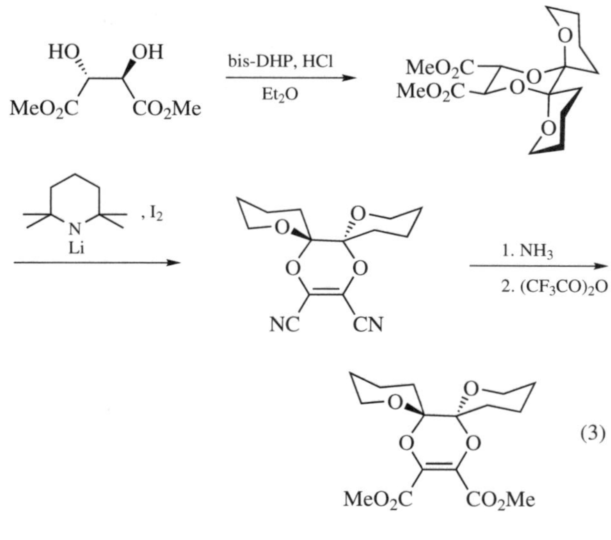

(3)

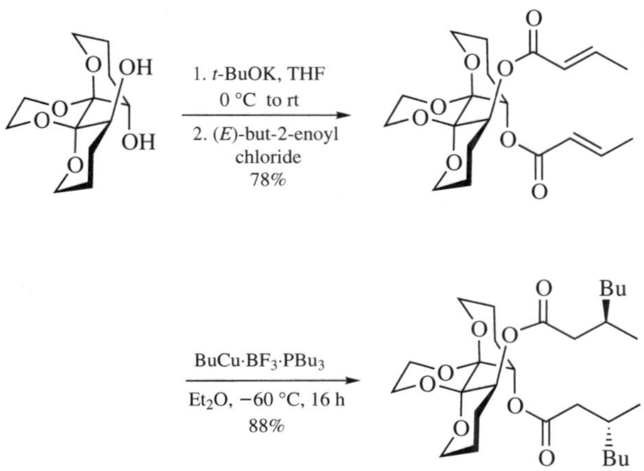

(4)

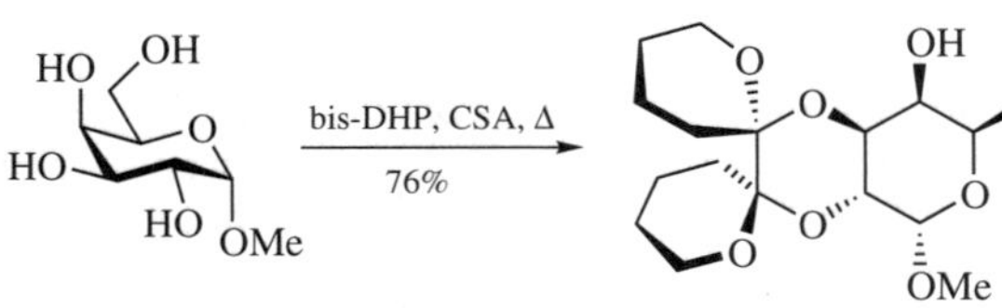

(5)

As is also the case with **1,1′,2,2′-tetramethoxycyclohexane**, bis-DHP suffers from the inability to achieve regiocontrolled protection of polyols having a multi set of 1,2-*trans* diequatorial diols, e.g., α-D-glucopyranoside, from which mixtures of dispoke protected derivatives result.[10] The low yield of 39% under standard conditions could be improved to 68% by **triphenylphosphine**/hydrogen bromide complex as catalyst in DMF as solvent (eq 6).[10]

Dispoke protected glucosides have been used as acceptors in glycosidation reactions. Thus, the disaccharide shown in eq 7 has been prepared in 82% yield in an α:β ratio of 5:2 which is further transformed to a trisaccharide.[11]

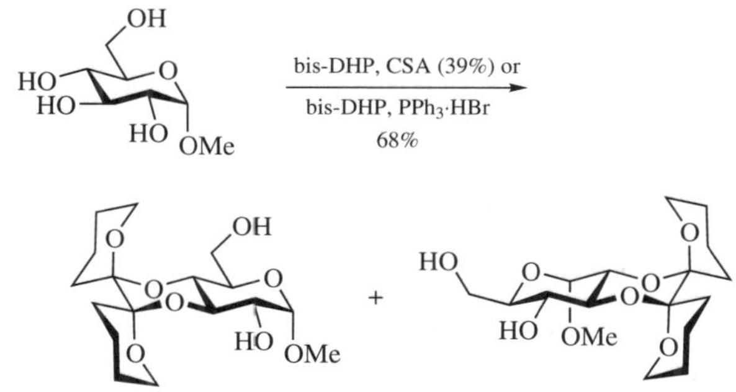

(6)

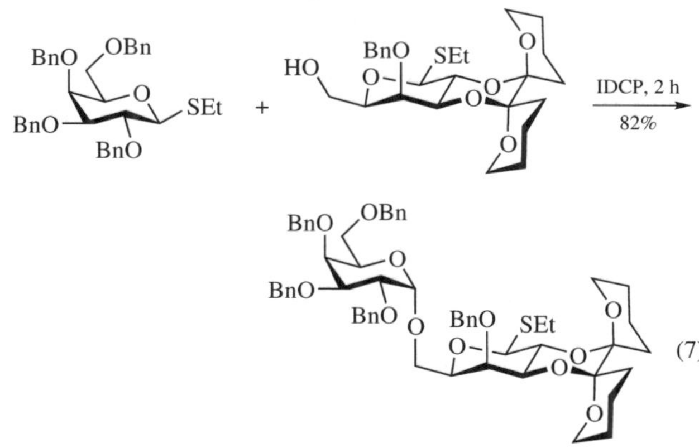

(7)

Although the dispoke protecting group can be cleaved under acidic conditions, it is possible to introduce another protecting group under acidic conditions if very mild conditions are employed (eq 8).[12]

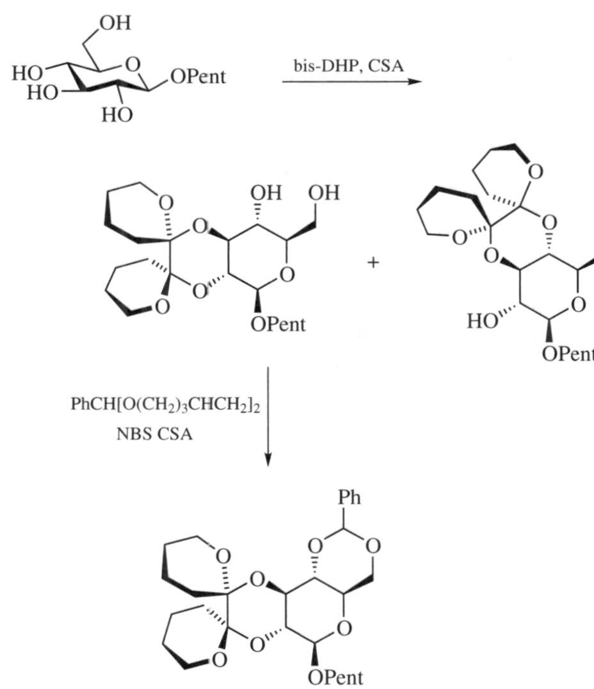

(8)

Related Reagents. Benzaldehyde dimethyl acetal [1125-88-8]; butane-2,3-dione; cyclohexane-1,2-dione [765-87-7]; 2,2,3,3-tetramethoxybutane [109669-49-0]; 1,1,2,2-tetramethoxycyclohexane [176798-33-7].

1. Ley, S. V.; Leslie, R.; Tiffin, P. D.; Woods, M., *Tetrahedron Lett.* **1992**, *33*, 4767.
2. Ley, S. V.; Woods, M.; Zanotti-Gerosa, A., *Synthesis* **1992**, 52.
3. Fujita, M.; Lainé, D.; Ley, S. V., *J. Chem. Soc., Perkin Trans 1* **1999**, 1647.
4. Ley, S. V.; Downham, R.; Edwards, P. J.; Innes, J. E.; Woods, M., *Contemp. Org. Synth.* **1995**, *2*, 365.
5. (a) Downham, R.; Kim, K. S.; Ley, S. V.; Woods, M., *Tetrahedron Lett.* **1994**, *35*, 769. (b) Boon, G.-J.; Downham, R.; Kim, K. S.; Ley, S. V.; Woods, M., *Tetrahedron* **1994**, *50*, 7157.
6. Zhang, D.; Bleasdale, C.; Golding, B. T.; Watson, W. P., *Chem. Commun.* **2000**, 1141.
7. Hachiya, S.-I.; Cook, A. S.; Williams, B. G.; Montalban, A. G.; Barrett, A. G. M.; Hoffman, B., *Tetrahedron* **2000**, *56*, 6565.
8. Bezuidenhoudt, B. C. B.; Castle, G. H.; Innes, J. E.; Ley, S. V., *Recl. Trav. Chim. Pays-Bas* **1995**, *114*, 184.
9. Ley, S. V.; Leslie, R.; Tiffin, P. D.; Woods, M., *Tetrahedron Lett.* **1992**, *33*, 4767.
10. Hughes, A. B.; Ley, S. V.; Priepke, H. W. M.; Woods, M., *Tetrahedron Lett.* **1994**, *35*, 773.
11. (a) Boons, G.-J.; Grice, P.; Leslie, R.; Ley, S. V.; Yeung, L. L., *Tetrahedron Lett.* **1993**, *34*, 8523. (b) Grice, P.; Ley, S. V.; Pietruszka, J.; Osborn, H. M. I.; Priepke, H. W. M.; Warriner, S. L., *Chem. Eur. J.* **1997**, *3*, 431.
12. Andrews, C. W.; Rodebaugh, R.; Fraser-Reid, B., *J. Org. Chem.* **1996**, *61*, 5280.

Thomas Ziegler, Lakshminarayanapuram R. Subramanian & Gregor Lemanski
University of Tübingen, Germany

1,1,2,2-Tetramethoxycyclohexane

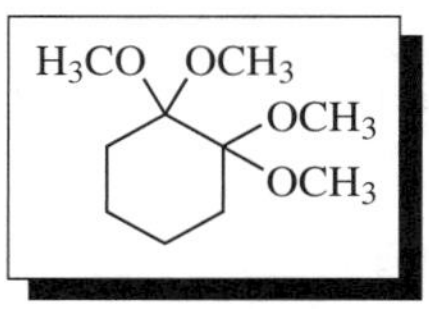

[163125-34-6] $C_{10}H_{20}O_4$ (MW 204.27)

(reagent used for the protection of vicinal diols)

Physical Data: bp 54 °C (0.4 Torr); n_D^{20} 1.4540–1.4600.
Solubility: soluble in most organic solvents.
Form Supplied in: colorless liquid; available from Avocado, Alfa Aesar, Acros Organics.
Purification: fractionally distilled at 0.4 Torr.
Handling, Storage, and Precautions: store below 4 °C, Toxicity: cancer suspect agent.

Use as a Protecting Group. 1,1,2,2-Tetramethoxycyclohexane (TMC) was developed for selective protection of 1,2-diequatorial diols in sugars, in the presence of other hydroxyl group combinations.[1] Similar to **3,3′4,4′-tetrahydro-6,6′-bis(2H-pyran)**, TMC preferentially reacts under acidic catalysis [camphorsulfonic acid (CSA)] with *trans*-diequatorial vicinal hydroxyls of a variety of hexopyranosides in the D-gluco,[2] L-rhamno,[1–3] D-manno,[1,2,4] D-galacto,[1,2] L-lyxo,[1] L-arabino,[1] and D-glucosamine[5] series to form the corresponding (1′,2′-dimethoxycyclohexane-1′,2′-diyl)hexopyranosides (eq 1). The major product is the often highly crystalline cyclohexane-1,2-diacetals, with the spiroketal as the minor product.

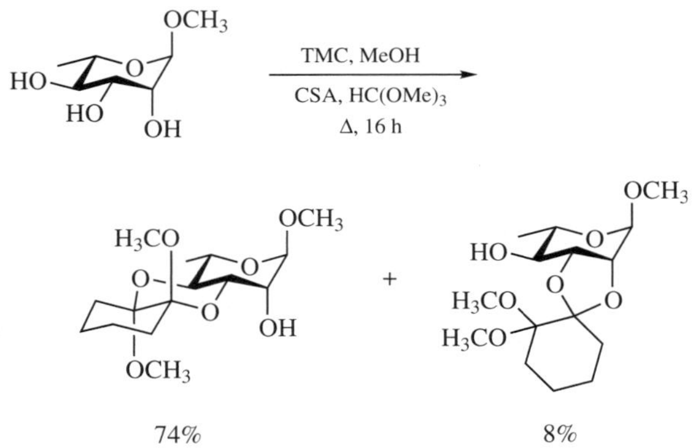

(1)

Usually, the regioselectivity is high for 1,2-*trans* diequatorial hydroxyls when only 1,2-*cis* diols are present, as in L-rhamnosides (eq 1).[6] However, problems may arise when several 1,2-*trans* diequatorial hydroxyls are accessible in a glycoside. This is the case in which D-glucopyranosides provide unseparable mixtures of the 2,3- and 3,4-protected counterparts.[2,6,7] Yields of blocked glycosides may also be medium in the galacto, arabino, and fuco series where the protection with **3,3′,4,4′-tetrahydro-6,6′-bis(2H-pyran)** is recommended, but are high in the manno series.[3,4] Removal of the dimethoxycyclohexane protecting group is conveniently performed by hydrolysis in aqueous acetic acid[2b] or aqueous trifluoroacetic acid.[5] Beside glycopyranosides, TMC has also been used for the selective protection of the *trans*-diol in quinic acid (eq 2).[8]

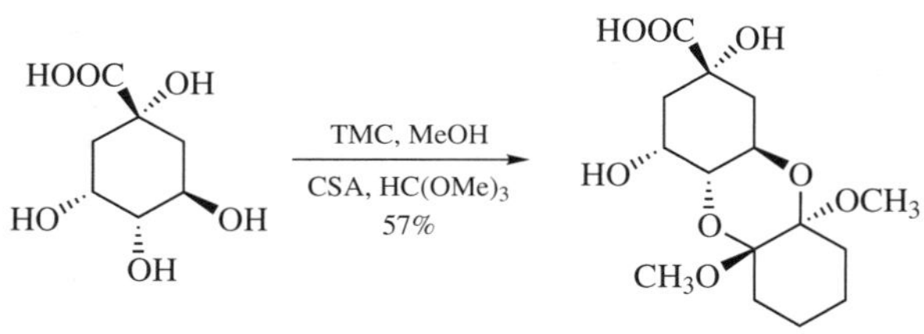

(2)

The reaction of benzyl α-L-rhamnopyranoside with TMC was carried out without adding trimethyl orthoformate to prevent transglycosylation (eq 3).[3] The thermodynamically controlled

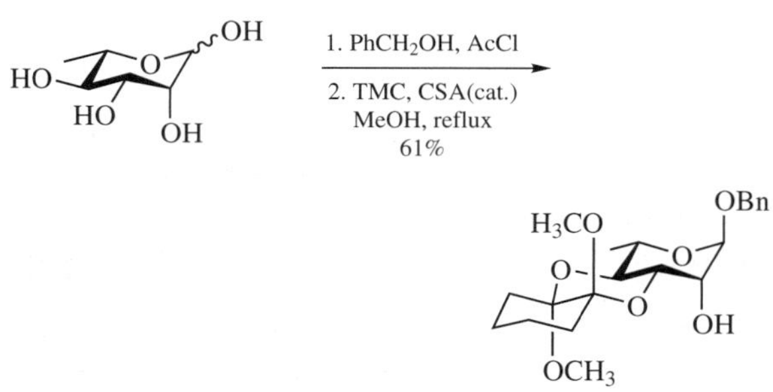

(3)

3,4-diprotected product in 61% yield was separated from the 1,2-diprotected product (17%) by flash chromatography. In contrast, use of **3,3′,4,4′-tetrahydro-6,6′-bis(2H-pyran)** gave a 3:2 mixture of 3,4- and 2,3-diprotected product in 79% yield and could not be used for further study. The restricted conformational mobility of the 3,4-protected product facilitated the subsequent transformations for the synthesis of the required 2-difluorooleandrose.[3]

Use for Oligosaccharide Synthesis. An especially useful feature of dimethoxycyclohexane-protected glycosides is their reduced reactivity in glycosylation reactions. For example, ethyl 2,3,4-tri-*O*-benzyl-1-thio-α-L-rhamnopyranoside can be selectively activated in the presence of a dimethoxycyclohexane protected ethyl 1-thio-α-L-rhamnopyranoside, resulting in a disaccharide which in turn can be used as a glycosyl donor for further glycosylations (eq 4).[2b] Thus, simple one-pot syntheses of complex oligosaccharides can be established using this protecting group in combination with the selective activation of 1-thioglycosides or 1-selenoglycosides.[2b,9]

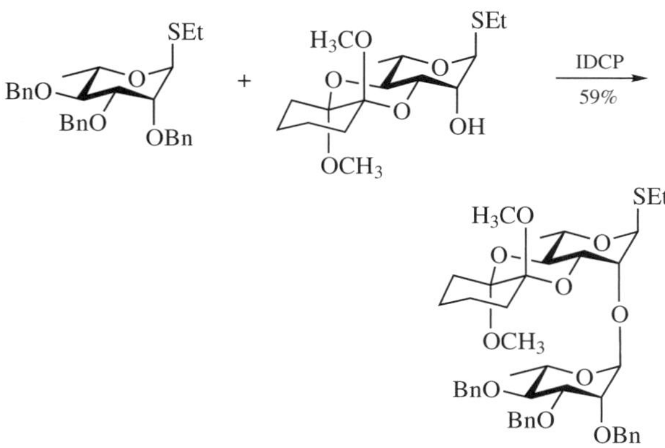

(4)

The torsional deactivation resulting from the rigid structural features of the 1,2-diacetals based on TMC protection has been effectively exploited in mediating glycosidic couplings for the construction of high-density oligosaccharides and complex carbohydrates.[10]

Related Reagents. Benzaldehyde dimethyl acetal [1125-88-8]; butane-2,3-dione; cyclohexane-1,2-dione [765-87-7]; 3,3′,4,4′-tetrahydro-6,6′-bis(2*H*-pyran) [109669-49-0]; 2,2,3,3-tetramethoxybutane [176798-33-7].

1. Grice, P.; Ley, S. V.; Pietruszka, J.; Priepke, H. W. M.; Warriner, S. L., *J. Chem. Soc., Perkin Trans. 1* **1997**, 351.
2. (a) Ley, S. V.; Priepke, H. W. M.; Warriner, S. L., *Angew. Chem. Intl. Ed. Engl.* **1994**, *33*, 2290. (b) Ley, S. V.; Priepke, H. W. M.; Warriner, S. L., *Angew. Chem. Intl. Ed. Engl.* **1994**, *33*, 2292.
3. Barrena, M. I.; Matheu, M. I.; Castillión, S., *J. Org. Chem.* **1998**, *63*, 2184.
4. (a) Ley, S. V.; Osborn, H. M. I.; Priepke, H. W. M.; Warriner, S. L., *Org. Synth.* **1998**, *75*, 170. (b) Hassan, H. H. A. M., *Heterocycles* **2000**, *53*, 397.
5. Saotome, C.; Kanie, Y.; Kanie, O.; Wong, C.-H., *Bioorg. Med. Chem.* **2000**, *8*, 2249.
6. Ley, S. V.; Boons, G.-J.; Leslie, R.; Woods, M.; Hollinshead, D. M., *Synthesis* **1993**, 689.
7. Hughes, A. B.; Ley, S. V.; Priepke, H. W. M.; Woods, M., *Tetrahedron Lett.* **1994**, *35*, 773.
8. Gebauer, O.; Brueckner, R., *Liebigs Ann.* **1996**, 1559.
9. Grice, P.; Ley, S. V.; Pietruszka, J.; Priepke, H. W. M.; Walther, E. P. E., *Synlett* **1995**, 781.
10. Ley, S. V.; Baeschlin, D. K.; Dixon, D. J.; Foster, A. C.; Ince, S. J.; Priepke, H. W. M.; Reynolds, D. J., *Chem. Rev.* **2001**, *101*, 53.

Thomas Ziegler, Lakshminarayanapuram R. Subramanian & Gregor Lemanski
University of Tübingen, Germany

Tetramethylfluoroformamidinium Hexafluorophosphate

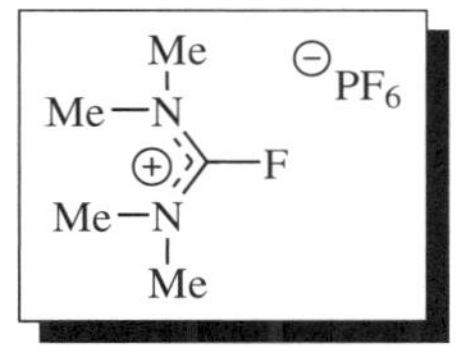

[164298-23-1] $C_5H_{12}F_7N_2P$ (MW 264.125)

(coupling reagent for amide and peptide synthesis[1,2])

Alternate Name: TFFH.
Physical Data: mp 110 °C.
Form Supplied in: white solid, commercially available.
Purification: can be purified by recrystallization from a mixture of MeCN and ether or from CH_2Cl_2 (excess) and ether.
Handling, Storage, and Precautions: TFFH is not hygroscopic and can be stored indefinitely at 0 °C. A TFFH solution in DMF is sensitive to moisture and cannot be stored in DMF, which makes TFFH less stable relative to other uronium salt such as HATU or HBTU.[3] The synthesis of peptides should be carried out with freshly prepared solution.

General Consideration. TFFH and its analog **bis(tetramethylene)fluoroformamidinium hexafluorophosphate** (BTFFH),[1,4] have been prepared via the reaction of the chloro analog with excess KF (eq 1) and used as condensation reagents for the preparation of peptides by both solution- and solid-phase techniques.[1,4–7]

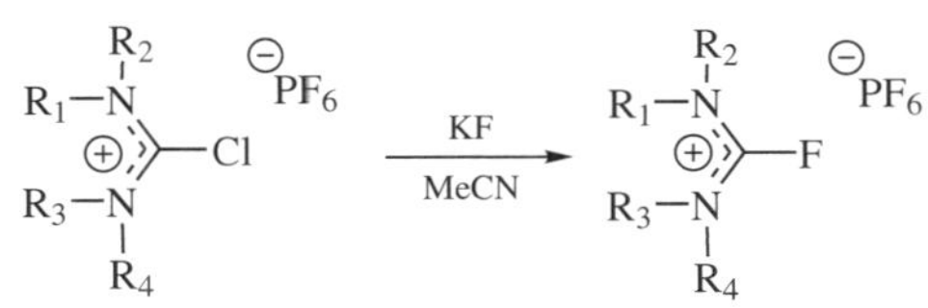

(1)

TFFH; $R_1 = R_2 = R_3 = R_4 = CH_3$
BTFFH; $R_1 = R_2 = R_3 = R_4 = (CH_2)_2$

In solution, the reaction of α-amino protected amino acids with equimolar amounts of α-carboxyl protected amino acid hydrochlorides or amino amides, and bases (2 equiv) like **diisopropylethylamine** (DIEA), **N-methylmorpholine** (NMM), **2,4,6-collidine** (TMP) or **1,8-bis(dimethylamino)naphthalene** (proton sponge, PS) in DMF promoted clean, via acid fluoride, formation of the corresponding peptides in good yields (eq 2). The level of epimerization in TFFH-mediated couplings was reduced compared to 1,1,3,3-tetramethylchloro-formamidinium hexafluorophosphate (TCFH)[1,5] - promoted acylation in the presence or absence of HOAt as an additive.

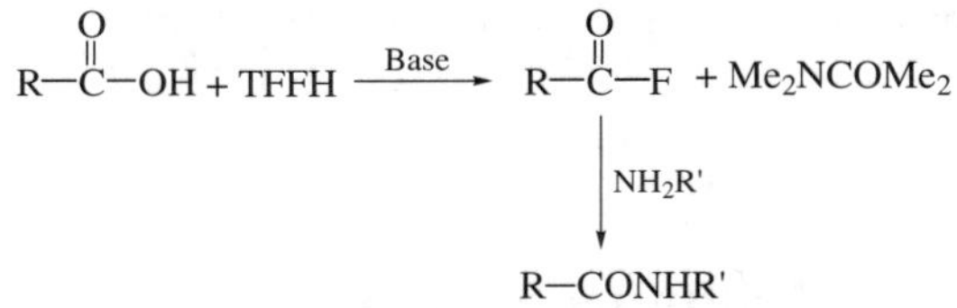

(2)

The results shown in Table 1 clearly indicated that (i) TFFH and BTFFH are superior to TCFH and BTCFH (entries 1 vs. 3, 5 vs. 7, 9 vs. 10, and 11 vs. 12); (ii) the use of HOAt considerably reduces the epimerization (entries 1 vs. 2, 3 vs. 4, 5 vs. 6, and 7 vs. 8).

TFFH also proved to be useful for segment couplings provided 1 equiv of HOAt (Table 1) was present in order to preclude extensive racemization.[8] Results obtained under these conditions were similar to those observed when HATU was used as coupling reagent.[9,10]

Infrared analysis of the reaction mixture during the coupling process provides a possible rationale for these results. Activation of a protected amino acid by means of the chloro derivatives (TCFH[11] and BTCFH[12]) gives initially the oxazolone (IR: 1830 and 1685 cm^{-1}); in the presence of HOAt the oxazolone is converted to the OAt ester (IR: 1823 and 1723 cm^{-1}) which is less prone to suffer loss of configuration.[5,8,13] For TFFH and BTFFH, preactivation in the absence of HOAt gives initially the acid fluoride (IR: 1844 cm^{-1}), whereas if HOAt is present a mixture of acid fluoride and OAt ester is observed (eq 3).[5] Acid fluorides are less sensitive than acid chloride to oxazolone formation and therefore do not lead to the various by-products which follow oxazolone formation.[14]

TFFH appears to be an ideal coupling reagent for automated solid-phase syntheses, since it is readily available, inexpensive, and capable of providing crude peptides of high quality. An example is the assembly of test peptide, H-Tyr-Aib-Aib-Phe-Leu-NH_2,[15] which, due to the difficult Aib-Aib coupling, has previously[15] been used to demonstrate the superiority of HATU over HBTU. The tested pentapeptide was obtained in 95% purity as a crude product, while the amount of des-Aib tetrapeptide was only 4%. In contrast, under similar conditions, the earlier syntheses[15] with HATU and HBTU provided 94% and 43% purity, respectively, while 2-chloro-1,3-dimethyl-1,3-dimethylene-uronium hexafluorophosphate (CIP)[13] gave product of only 9.1% purity; the amount of des-Aib tetrapeptide under the latter conditions was 72.7%.[5] The somewhat less difficult sequence with a Pro-Pro unit in place of Aib was obtained under similar conditions in 97.6% purity. Other standard peptide models were obtained similarly in high purity using 4–8 equiv of acid and 30 min coupling times: ACP,[16] (65–74) prothrombin,[14] magainin I amide,[17] human corticotrophin-releasing factor (h-CRF),[6] and naturally occurring peptaibols[18] (alamethicin F30 and F50).

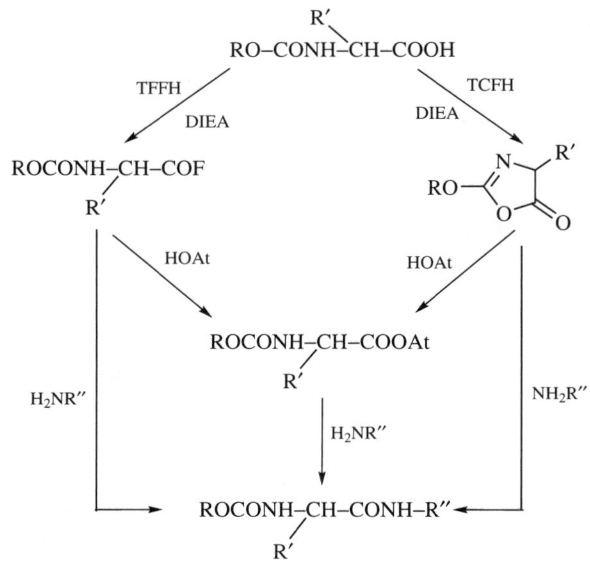

(3)

Chloroformamidinium salts and the corresponding bromo analogs lack the general applicability of fluoroformamidinium derivatives as coupling reagents for solid-phase synthesis[1,5] unless HOAt is used as an additive during coupling reactions.[1,5,13]

TFFH proved fully effective for the two amino acids (Arg and His) for which the preformed acid fluorides are not shelf-stable. The acid fluoride of Fmoc-Arg(Pbf)-OH was in less than 2 min generated using 2 equiv of DIEA in DMF and although cyclization to the corresponding lactam occurred slowly, significant amounts of acid fluoride remained after 60 min.[6]

TFFH was shown to be a very convenient reagent for the preparation of isothiocyanates from the corresponding primary amines in the presence of carbondisulfide in a good yield (eqs 4 and 5).[3]

Table 1 Effect of haloformamidinium salt on the preservation of configuration during peptide bond formation in the presence or absence of HOAt

Entry	RCOOH	$R'NH_2$ or $R'NH_2{\cdot}HCl$	Conditions (2 equiv, 3 h)	DL-isomer (%)
1	Z-Phg-OH	H-Pro-NH_2	TFFH/DIEA	7.4
2	Z-Phg-OH	H-Pro-NH_2	TFFH/HOAt DIEA	1.7
3	Z-Phg-OH	H-Pro-NH_2	TCFH/DIEA	29.9
4	Z-Phg-OH	H-Pro-NH_2	TCFH/HOAt DIEA	2.9
5	Z-Phg-OH	H-Pro-NH_2	BTFFH/DIEA	7.2
6	Z-Phg-OH	H-Pro-NH_2	BTFFH/HOAt DIEA	1.6
7	Z-Phg-OH	H-Pro-NH_2	BTCFH/DIEA	29.8
8	Z-Phg-OH	H-Pro-NH_2	BTCFH/HOAt DIEA	2.8
9	Z-Phe-Val-OH	H-Pro-NH_2	TFFH/HOAt TMP	2.7
10	Z-Phe-Val-OH	H-Pro-NH_2	TCFH/HOAt TMP	3.8
11	Z-Phe-Val-OH	H-Pro-NH_2	BTFFH/HOAt TMP	2.8
12	Z-Phe-Val-OH	H-Pro-NH_2	BTCFH/HOAt TMP	3.9

$$R{-}NH_2 \xrightarrow[\text{2. TFFH}]{\text{1. } CS_2/Et_3N} R{-}N{=}C{=}S \quad (4)$$

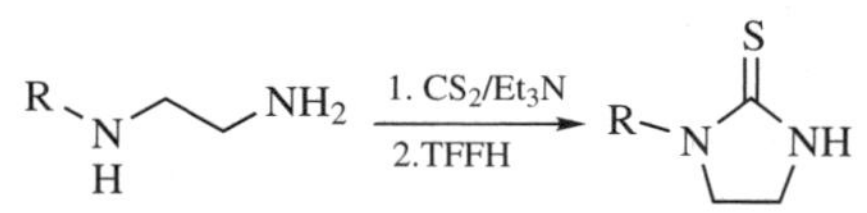

(5)

Moreover, using TFFH for activation of carboxylic acids allows the use of an equimolecular amount of hydrazine without formation of biacylhydrazide as in the case of the reaction between acid anhydride or acid chlorides which usually gives considerable amounts of bisacylhydrazide[19] (eq 6).

$$RCOOH \xrightarrow[\text{2. } NH_2\text{-}NH_2]{\text{1. TFFH/DIEA}} R{-}CONHNH_2 \quad (6)$$

In conclusion, TFFH is compatible with normal techniques for the solid- and solution-phase assembly of peptides or their amides by the acid fluoride technique. Optimization may depend on the nature of the amino acid residue and an appropriate choice of base and solvent.

1. Carpino, L. A.; El-Faham, A., *J. Am. Chem. Soc.* **1995**, *117*, 5401.
2. Boas, U.; Pedersen, B.; Christensen, J. B., *Synthetic Commun.* **1998**, *28*, 1223.
3. Albericio, F.; Bofill, J. M.; El-Faham, A.; Kates, S. A., *J. Org. Chem.* **1998**, *63*, 9678.
4. El-Faham, A., *Chem. Lett.* **1998**, 671.
5. El-Faham, A., *Org. Prep. Proced. Intl.* **1998**, *30*, 477.
6. Triolo, S. A.; Ionescu, D.; Wenschuh, H.; Solé, N. A.; El-Faham, A.; Carpino, L. A.; Kates, S. A., In *Proceeding of the European Peptide Symposium*; Ramage, R.; Epton, R., Eds.; 1998, p 839.
7. Liu, W. Q.; Vidal, M.; Gresh, N.; Roques, B. P.; Garbay, C., *J. Med. Chem.* **1999**, *42*, 3737.
8. Akaji, K.; Kuriyama, N.; Kiso, Y., *Tetrahedron Lett.* **1994**, *35*, 3315.
9. Carpino, L. A., *J. Am. Chem. Soc.* **1993**, *115*, 4397.
10. Carpino, L. A.; El-Faham, A., *J. Org. Chem.* **1994**, *59*, 695.
11. Dourtoglou, V.; Gross, B.; Lamproglou, V.; Ziodrou, C., *Synthesis* 1984, 572.
12. Coste, J.; Frérot, E.; Jouin, P.; Castro, B., *Tetrahedron Lett.* **1991**, *32*, 1967.
13. (a) Akaji, K.; Kuriyama, N.; Kiso, Y., *J. Org. Chem.* **1996**, *61*, 3350. (b) Akaji, K.; Tamai, Y.; Kiso, Y., *Tetrahedron* **1997**, *53*, 567.
14. Carpino, L. A.; Sadat-Aalaee, D.; Chao, H. G.; Deselms, R. H., *J. Am. Chem. Soc.* **1990**, *112*, 9651.
15. Carpino, L. A.; El-Faham, A.; Minor, C. A.; Albericio, F., *J. Chem. Soc. Chem. Commun.* **1994**, 201.
16. (a) Hudson, D., *J. Org. Chem.* **1988**, *53*, 617. (b) Atherton, E.; Logan, C. J.; Sheppard, R. C., *J. Chem. Soc., Perkin Trans* **1981**, 538.
17. Echner, H.; Voelter, W., In *Peptide*, 1988; *Proceedings of the 20th European Peptide Symposium*; Jung, G.; Bayer, E., Eds.; de Gruyter: Berlin, 1989; p. 181.
18. Wenschuh, H.; Beyermann, M.; Haber, H.; Seydel, J. K.; Krause, E.; Bienert, M.; Carpino, L. A.; El-Faham, A.; Albericio, F., *J. Org. Chem.* **1995**, *60*, 405.
19. (a) Naegleli, C.; Stefanovitsch, G., *Helv. Chim Acta* **1928**, *11*, 621. (b) Wieland, H.; Hintermaier, A.; Dennstedt, I., *Ann* **1927**, *452*, 1.

Ayman El-Faham
Alexandria University, Egypt

Tetramethylurea

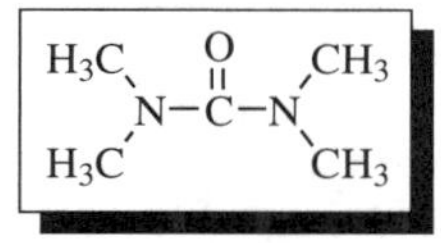

[632-22-4] $C_5H_{12}N_2O$ (MW 116.16)

(mild base used as an acid scavenger in glycosylation reactions, precursor to peptide coupling reagents, nucleophile in the synthesis of heterocycles, solvent/additive in alkylations of organolithium compounds and deamination of aryldiazonium salts, precursor to guanidine bases)

Alternate Name: 1,1,3,3-tetramethylurea, TMU.
Physical Data: [1] bp 176.5 °C, mp −1.2 °C, d 0.9687 g cm^{-3}. Dielectric constant[2] 24.46 at 20 °C.
Solubility: soluble in H_2O and most organic solvents.
Form Supplied in: colorless liquid.
Analysis of Reagent Purity: ^{1}H NMR:[3] δ 2.81; ^{13}C NMR: δ 165.3, 38.6.
Handling, Storage, and Precautions: store protected from moisture. May cause skin and eye irritation, and also respiratory and digestive tract irritation.

The use of 1,1,3,3-tetramethylurea as an acid scavenger in glycosylation reactions was first described by Hanessian and Banoub in the context of a broadly applicable synthesis of 1,2-*trans* disaccharides involving the coupling of acetylated glycosyl halides with a sugar acceptor in the presence of silver triflate as the catalyst.[4,5] Thus, the coupling of 2,3,4,6-tetra-*O*-acetyl-α-D-glucopyranosyl bromide with methyl 2,3,6-tri-*O*-benzoyl-α-D-galactopyranoside gave the 1,2-*trans*-linked disaccharide in 70% yield (eq 1).

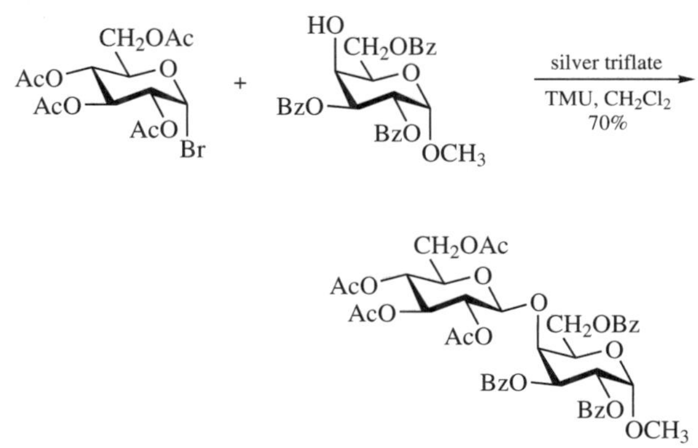

(1)

The β-linked disaccharide is favored by neighboring group participation from the 2-*O*-acetyl group in the bromo sugar. Ester migration is not observed under these conditions, and other protecting groups such as isopropylidene and benzylidene acetals, and carbamate-protected amino sugars are all compatible. In a study of the synthesis of 4′-deoxydaunorubicin and 4′-deoxyadriamycin, the silver triflate–TMU method was used for the preparation of α-glycosides with donors that lacked a participating group at C-2.[6] Activation of a trisaccharide as its glycosyl chloride in the presence of silver triflate and tetramethylurea and coupling to the disaccharide β-D-GlcpNAc-(1→2)-α-L-Rhap gave a protected pentasaccharide glycoside corresponding to the repeating unit of a *Shigella flexneri* variant.[7] Other applications of the silver triflate–TMU method for activation of glycosyl halides include synthesis of 6′-*O*-phosphorylated *O*-α-D-mannopyranosyl-(1→3)- and -(1→6)-α-D-mannopyranosides, prepared as their 8-methoxycarbonyloctyl glycosides;[8] synthesis of a glycopeptide building block of an antifreeze glycoprotein;[9] synthesis of the 2′-epimer of methyl β-maltoside;[10] synthesis of oligosaccharide components of the capsular polysaccharides of streptocuccus pneumonia type 19F[11] and type14;[12] and the total synthesis of the pseudotetrasaccharide antibiotic validamycin H.[13] In the synthesis of the pneumonia type 19F oligosaccharide, coupling of 2-(benzoyloxyimino)-2-deoxyglycosyl bromide with *p*-methoxylbenzyl 2,3,6-tri-*O*-benzyl-β-D-glucopyranoside in the presence of silver triflate–TMU gave a 10:1 ratio of α/β (1→4) disaccharides, while the same coupling when carried out in the presence of insoluble silver zeolite/silver carbonate gave an inverted 1:20 ratio of α/β disaccharides.[11] Donors other than glycosyl halides can be similarly activated using this method. A highly stereoselective glycosylation was developed using a fucosyl thiopyridylcarbonate (TOPCAT) derivative as the donor and a sialyl trisaccharide as the acceptor to give the fully protected sialylLex tetrasaccharide in 41% yield.[14] The best results in the fucosylation step were obtained with TMU as the acid scavenger (eq 2).

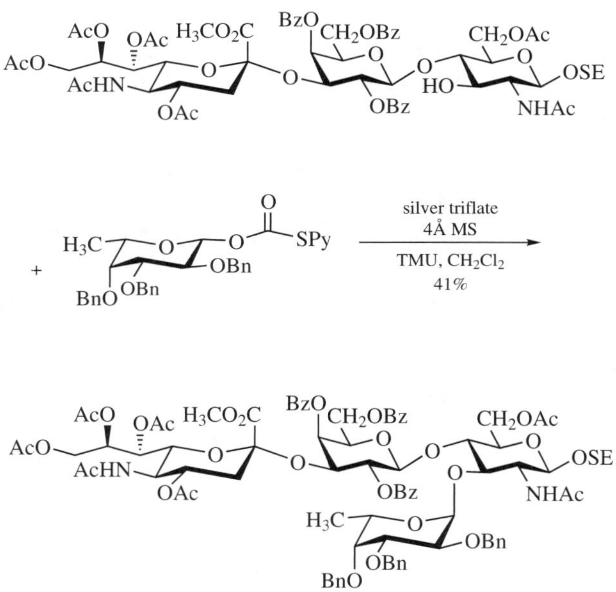

(2)

Activation of furanosyl dithiocarbonates and dithiocarbamates with silver triflate in the presence of polar additives such as tetramethylurea, DMSO, or HMPA gave 1,2-*cis* furanosides with a variety of carbohydrate acceptor alcohols. The inclusion of the polar additive in the reaction mixture resulted in a change and/or increase in the stereoselectivity of the glycosylation.[15] Glycosylations of cholesterol and other alcohols with 3,4,6-tri-*O*-acetyl-β-D-glucopyranosyl chloride were achieved under thermal conditions, in the absence of a metal catalyst, by treatment with TMU in dichloromethane followed by removal of the solvent and heating.[16]

Although TMU is a weak base [pK_a(BH$^+$) = 0.4],[17] it scavenges the triflic acid that is generated in the glycosylation reaction. This is important for glycosylations involving acid-sensitive substrates and also those in which the glycosidic linkage in the product is acid-sensitive. A glycosyl uronium salt was isolated from the reaction of 2,3,4,6-tetra-*O*-acetyl-α-D-glucopyranosyl bromide with TMU and silver triflate when the reaction was carried out at room temperature in the absence of acceptor (eq 3).[18] Subsequent glycosylation of methanol and of a monosaccharide alcohol by this uronium salt were demonstrated.

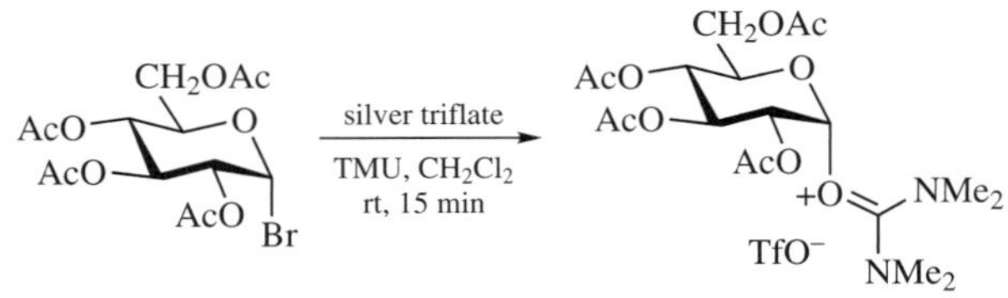

(3)

Glycosyl uronium salts were also used as intermediates in the synthesis of an antifreeze glycoprotein.[9] TMU is soluble in organic solvents, typically used in glycosylations, such as dichloromethane, 1,2-dichloroethane, ether, and acetonitrile, yet it is also water-soluble thereby facilitating separation from the glycosylation products by aqueous extraction.

An example of the use of tetramethylurea in the synthesis of heterocyclic compounds is the preparation of 2-dimethylamino-benzazoles via guanidine intermediates.[19] Cyclic ureas undergo addition to substituted benzynes to give 1,4-benzodiazepine and 1,5-benzodiazocine derivatives, while tetramethylurea undergoes an analogous process to afford 2-aminobenzamides (eq 4).[20]

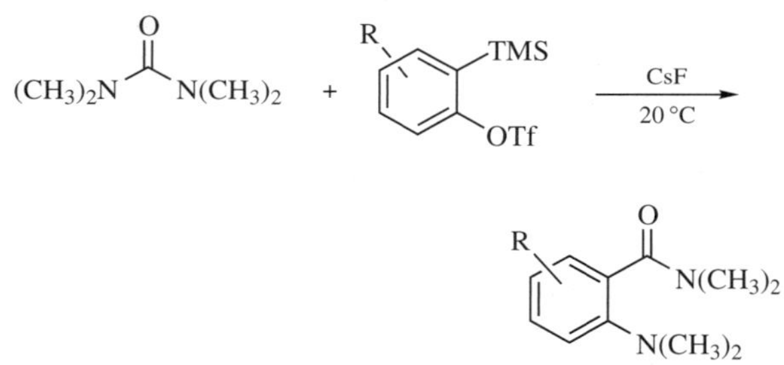

(4)

Tetramethylurea has been used as a cosolvent in the double alkylation of α,α-dilithiated silanes. These alkylations were carried out in THF with either HMPA or TMU to give 1,4-disilacyclohexanes which were characterized by X-ray diffraction analysis (eq 5).[21]

Applications of tetramethylurea in peptide chemistry include the development of coupling reagents derived from tetramethylurea aminium salts (uronium salts) and 1-hydroxybenzotriazole (HOBt). Both HBTU and TBTU have become popular as in situ activating reagents because of their convenience of use, efficiency, and lack of side reactions.[22] A new class of aminium peptide coupling reagents have been prepared from tetramethylurea (or 1,3-dimethylpropyleneurea) and 2-mercaptopyridine-1-oxide (eq 6).

The thiouronium salts prepared by this method have been used in solution-phase and solid-phase peptide synthesis.[23]

Tetramethylurea has been used in the preparation of the strong, non-nucleophilic base 2-*tert*-butyl-1,1,3,3-tetramethylguanidine.[24] As an alternative to DBN and DBU, the utility of this hindered guanidine base was demonstrated in the preparation of vinyl iodides from ketohydrazones. Previous applications of tetramethylurea in organic synthesis include its use as a cosolvent in the Ullmann reaction and the alkylation of enolates, and as a solvent in the deamination of aryldiazonium salts.[25]

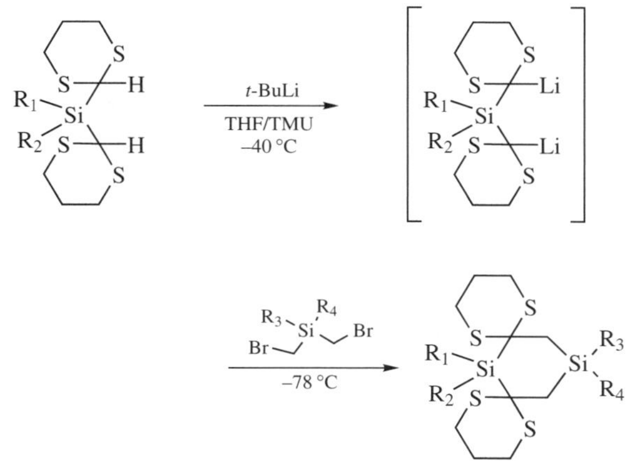

(5)

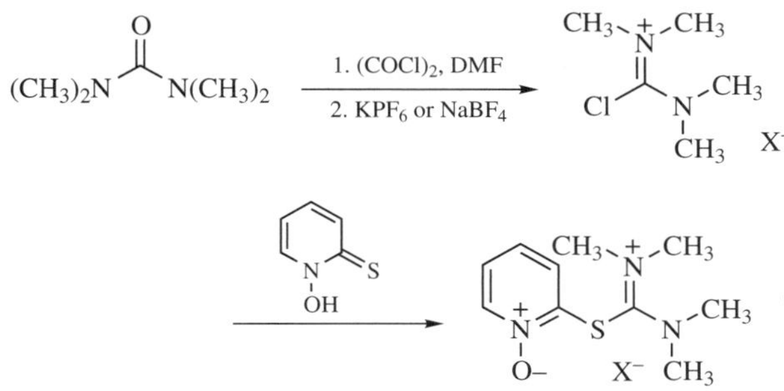

(6)

1. *CRC Handbook of Chemistry and Physics*, 77th Ed.; Lide, D. R., Ed.; CRC Press: Boca Raton, FL, 1996, pp 332–329.
2. Laurence, C.; Nicolet, P.; Dalati, M. T.; Abboud, J. L. M.; Notario, R., *J. Phys. Chem.* **1994**, *98*, 5807–5816.
3. Prakask, G. K. S.; Bae, C.; Rasul, G.; Olah, G. A., *J. Org. Chem.* **2002**, *67*, 1297–1301.
4. Hanessian, S.; Banoub, J., *Carbohydr. Res.* **1976**, *53*, C13–C16.
5. Hanessian, S.; Banoub, J., In *Synthetic Methods for Carbohydrates, ACS Symposium Series*; El Khadem, H. S., Ed.; American Chemical Society: Washington, D.C., 1976, Vol. 39, pp 36–63.
6. Arcamone, F.; Penco, S.; Redaelli, S.; Hanessian, S., *J. Med. Chem.* **1976**, *19*, 1424–1425.
7. Pinto, B. M.; Reimer, K. B.; Morissette, D. G.; Bundle, D. R., *Carbohydr. Res.* **1990**, *196*, 156–166.
8. Srivastava, O. P.; Hindsgaul, O., *Carbohydr. Res.* **1987**, *161*, 324–329.
9. Jiaang, W. T.; Hsiao, K. F.; Chen, S. T.; Wang, K. T., *Synthesis* **1999**, 1687–1690.
10. Adelhorst, K.; Bock, K.; Pedersen, H.; Refn, S., *Acta Chem. Scand.* **1988**, *B42*, 196–201.
11. Kaji, E.; Matsui, E.; Kobayashi, M.; Zen, S., *Bull. Chem. Soc. Jpn.* **1995**, *68*, 1449–1454.
12. Nifant'ev, N. E.; Bakinovskii, L. V.; Kochetkov, N. K., *Bioorg. Khim.* **1987**, *13*, 977–991 (CAN **1988**, *108*, 132199).
13. Miyamoto, Y.; Ogawa, S., *Carbohydr. Res.* **1992**, *223*, 299–301.
14. Lou, B.; Huynh, H. K.; Hanessian, S., In *Preparative Carbohydrate Chemistry*; Hanessian, S., Ed.; Marcel Dekker: New York, 1997, pp 431–449.
15. Bogusiak, J.; Szeja, W., *Synlett* **1997**, 661–662.
16. Nishizawa, M.; Kan, Y.; Yamada, H., *Tetrahedron Lett.* **1988**, *29*, 4597–4598.
17. Adelman, R. L., *J. Org. Chem.* **1964**, *29*, 1837–1844.
18. Bock, K.; Guzmán, J. F. B.; Refn, S., *Carbohydr. Res.* **1992**, *232*, 353–357.
19. Ohno, K.; Ishida, W.; Kamata, K.; Oda, K.; Machida, M., *Heterocycles* **2003**, *59*, 317–322.
20. Yoshida, H.; Shirakawa, E.; Honda, Y.; Hiyama, T., *Angew. Chem., Int. Ed. Engl.* **2002**, *41*, 3247–3249.
21. Shimizu, M.; Iwakubo, M.; Nishihara, Y.; Hiyama, T., *Tetrahedron Lett.* **1998**, *39*, 3197–3200
22. Knorr, R.; Trzeciak, A.; Bannwarth, W.; Gillessen, D., *Tetrahedron Lett.* **1989**, *30*, 1927–1930
23. Alberico, F.; Bailén, M. A.; Chinchilla, R.; Dodsworth, D. J.; Nájera, C., *Tetrahedron* **2001**, *57*, 9607–9613.
24. Barton, D. H. R.; Chen, M.; Jászberényi, J. C.; Taylor, D. K., *Org. Synth.* **1996**, *74*, 101–107.
25. Fieser, L. F.; Fieser, M., *Reagents for Organic Synthesis*; Wiley: New York **1967**, *1*, 43.

Robert M. Giuliano
Villanova University, Villanova, PA, USA

1*H*-Tetrazole

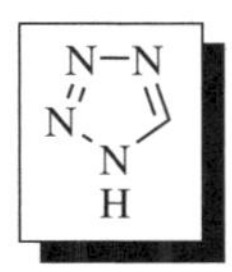

[288-94-8] CH_2N_4 (MW 70.05)

(useful heterocycle in medicinal chemistry, pKa similar to a carboxylic acid, serves as a carboxylic acid mimic; nucleophilic amine, undergoes *N*-alkylation with a variety of electrophilic reagents; used as a catalyst in the synthesis of phosphonate esters)

Solubility: soluble in H_2O, alcohol, acetonitrile; limited solubility in ether, benzene, and other organic solvents.

Form Supplied in: white crystalline solid and 0.45 M solution in acetonitrile.

Purification: careful recrystallization from ethanol. Sublimed under high vacuum at ca. 120 °C (Caution: explosion hazard).

Handling, Storage, and Precautions: decomposes upon heating and may explode if heated above 155 °C. Compound is noncombustible and hygroscopic. Store in a cool, dry place. Avoid heat and sunlight. Incompatible with strong acids, oxidizing agents, acid halides, and acid anhydrides.

***N*-Alkylations and Related Electrophilic Reactions.** Tetrazole can exist as a mixture of 1*H*- and 2*H*- tautomers (eq 1).[1] The ratio of tautomers in solution depends on solvent polarity and, in the case of substituted tetrazoles, the nature of the C5 substituent. Tetrazole is a moderately strong acid, with a pKa value of the order of a carboxylic acid. Tetrazole reacts with a broad spectrum of electrophilic reagents to produce *N*-substituted derivatives. Various *N*-alkylation reactions have been studied extensively.[2] As expected, *N*1 and *N*2-substituted isomers are typically generated. For example, methylation with methyl iodide or diazomethane leads to a mixture of *N*1 and *N*2 methyltetrazoles, the latter in slight excess (eq 2).[3] Similar ratios of *N*1 and *N*2 isomers were observed using BOM-Cl,[4] ethyl bromoacetate,[5] and *O-tert*-butyl-*N,N'*-dicyclohexylisourea.[6] The *N*-alkylation of tetrazole using alkynones[7] and formaldehyde[8] has also been reported. Mitsunobu condensation between 5-phenyltetrazole and *n*-hexanol favors *N*2 over *N*1 alkylation (eq 3).[9] The Mitsunobu reaction of tetrazole and L-rhodinose was also reported to provide *N*2 glycosyl tetrazole as the major product and the corresponding *N*1 isomer as a minor component (eq 4).[10]

(1)

1*H*-tetrazole 2*H*-tetrazole

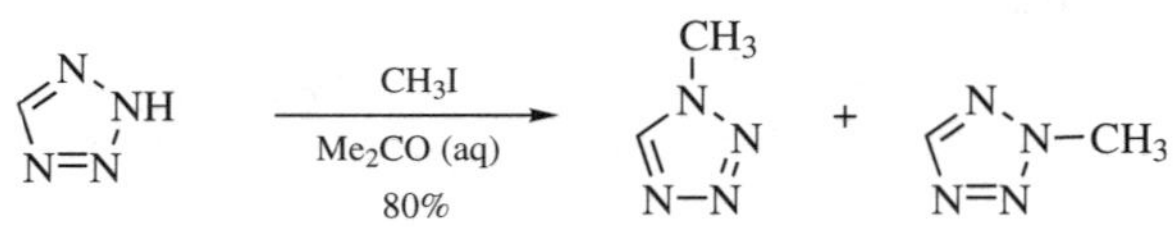

(2)

1.6:1

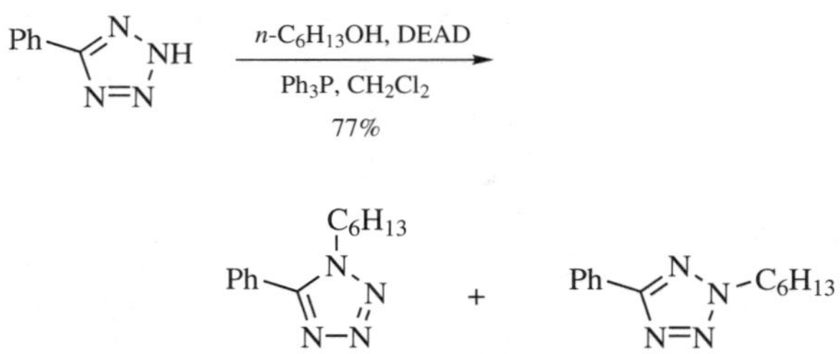

(3)

1:12

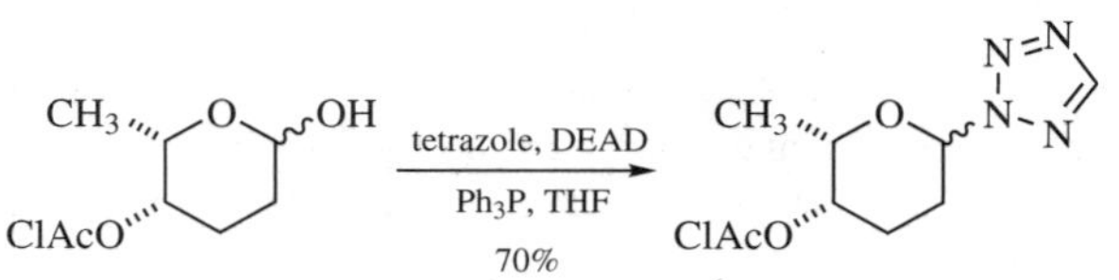

(4)

An early publication reports silylation of tetrazole using chlorotrimethylsilane and triethylamine giving exclusively *N*1 trimethylsilyltetrazole (eq 5).[11] Amination of tetrazole affords an approximate 2:1 mixture of *N*1 and *N*2 aminotetrazole.[12]

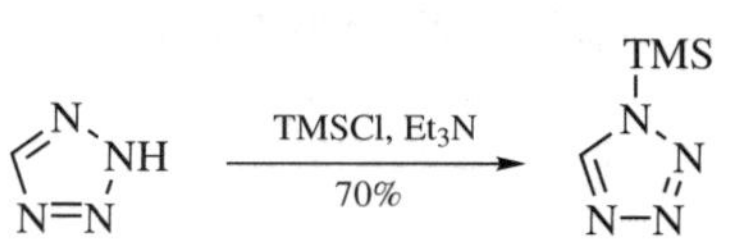

(5)

C5 Substitution and Functionalization. 5-Alkyl-and 5-acyl-1*H*-tetrazoles have been prepared from *N*-protected tetrazoles by deprotonation with *n*-butyllithium followed by quenching with various electrophilic reagents (eq 6).[13] 5-Lithiotetrazoles derived from *N*1-protected tetrazoles are thermally less stable compared to their *N*2-protected counterparts.[14] For example, 1-phenyl-5-lithiotetrazole undergoes rapid and irreversible decomposition to *N*-phenylcyanamide even at −70 °C (eq 7). In contrast, *N*2-protected 5-lithiotetrazoles are generally stable at −78 °C (eq 8). Suzuki[15] and Stille[16] cross-coupling reactions have been used to introduce functionality at the C5 position of tetrazole (eqs 9 and 10).

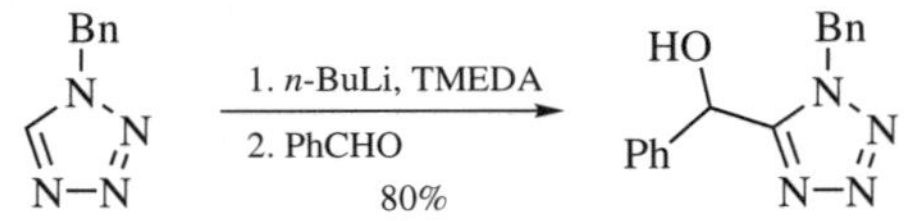

(6)

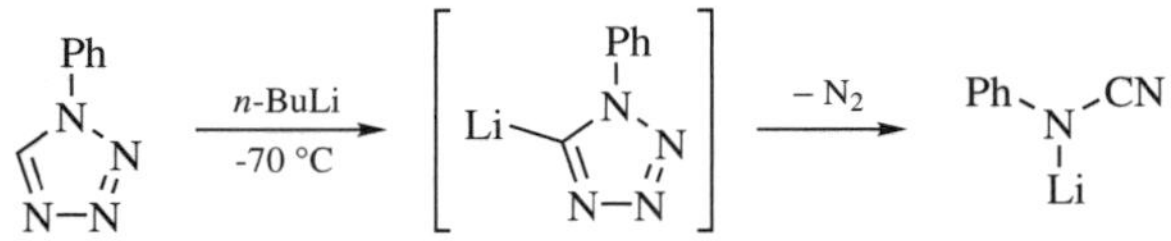

(7)

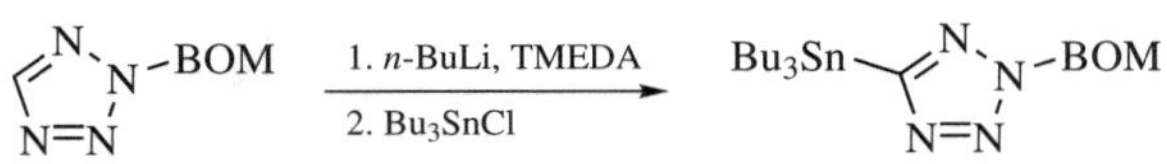

(8)

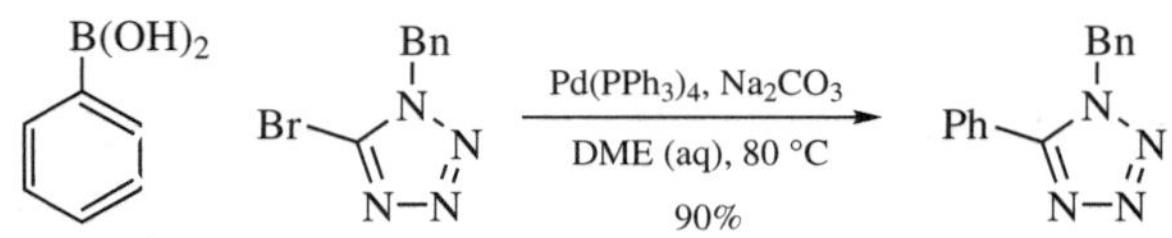

(9)

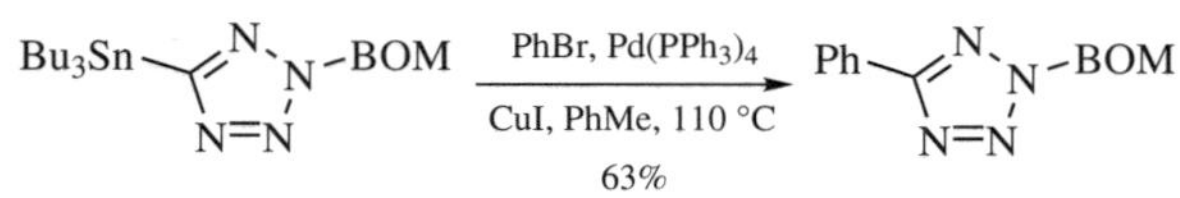

(10)

Condensation Reagent. As a reagent 1*H*-tetrazole has been used as a condensation agent to mediate the coupling of ribonucleotides and phosphoramidites in the synthesis of oligonucleotides.[17] 1*H*-tetrazole has also been used to promote the coupling of phosphoramidite and protected inositols for the synthesis of myo-inositol phosphates (eq 11).[18]

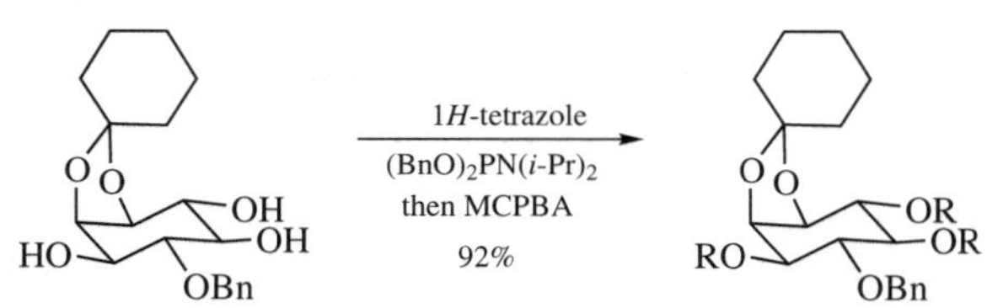

(11)

R = P(O)(OBn)$_2$

1. (a) Butler, R. N., *Adv. Heterocyclic Chem.* **1977**, *21*, 323. (b) Butler, R. N., *Comprehensive Heterocyclic Chem.*; Potts, K. T., Ed.; Pergamon: New York, 1984; Vol. 5. (c) Wittenberger, S. J., *Org. Prep. and Proced. Int.* **1994**, *26*, 499.
2. Ostrovskii, V. A.; Koren, A. O., *Heterocycles* **2000**, *53*, 1421.
3. Spear, R. J., *Aust. J. Chem.* **1984**, *37*, 2453.
4. Yokoyama, M.; Hirano, S.; Matsushita, M.; Hachiya, T.; Kobayashi, N.; Kubo, M.; Togo, H.; Seki, H., *J. Chem. Soc. Perkin Trans. 1* **1995**, 1747.
5. Raap, R.; Howard, J., *Can. J. Chem.* **1969**, *47*, 813.
6. Henry, R. A., *J. Hetreocycl. Chem.* **1976**, *13*, 391.
7. Vereshchagin, L. I.; Tikhonova, L. G.; Maksikova, A. V.; Buzilova, S. R.; Shulgina, V. M.; Proidakov, A. G., *Zh. Org. Khim.* **1979**, *15*, 1730 (*Chem. Abstr. 92*, 6477).
8. Tselinskii, I. V.; Mel'nikov, A. A.; Varyagina, L. G.; Zhigadlova, I. G., *Zh. Org. Khim.* **1983**, *19*, 415 (*Chem. Abstr. 100*, 22621).

9. Purchase, C. F.; White, A. D., *Syn. Commun.* **1996**, *26*, 2687.
10. Guo, Y.; Sulikowski, G. A., *J. Am. Chem. Soc.* **1998**, *120*, 1392.
11. Birkofer, L; Ritter, A; Richter P., *Chem. Ber.* **1963**, *96*, 2750.
12. Raap, R., *Can. J. Chem.* **1969**, *47*, 3677.
13. (a) Grimmett, M. R.; Iddon, B., *Heterocycles* **1995**, *41*, 1525. (b) Satoh, Y.; Marcopulos, N., *Tetrahedron Lett.* **1995**, *36*, 1759.
14. Raap, R., *Can. J. Chem.* **1971**, *49*, 2139.
15. Yi, K. Y.; Yoo, S. E., *Tetrahedron Lett.* **1995**, *36*, 1679.
16. Bookser, B. C., *Tetrahedron Lett.* **2000**, *41*, 2805.
17. Beaucage, S. L.; Iyer, R. P., *Tetrahedron* **1993**, *49*, 6123.
18. Yu, K.-L.; Fraser-Reid, B., *Tetrahedron Lett.* **1988**, *29*, 979.

Gary A. Sulikowski & Michelle M. Sulikowski
Texas A & M University, College Station, TX, USA

1,1′-Thiocarbonyldiimidazole[1]

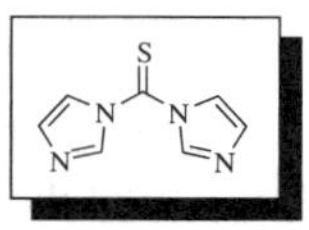

[6160-65-2] $C_7H_6N_4S$ (MW 178.24)

(conversion of vicinal diols to alkenes;[1] deoxygenation of alcohols;[13] thiocarbonyl transfer agent[29])

Alternate Name: TCDI.
Physical Data: mp 101–103 °C.
Solubility: sol many organic solvents including THF, CH_2Cl_2, toluene.
Form Supplied in: yellow solid; 90% pure or >97% pure.
Preparative Methods: prepared by the reaction of Thiophosgene with 2 equiv of Imidazole.
Purification: can be recrystallized from THF to give yellow crystals; can also be sublimed.
Handling, Storage, and Precautions: very hygroscopic; should be reacted and stored in a dry atmosphere.

Alkene Synthesis. The Corey–Winter alkene synthesis is an effective method for the deoxygenation of vicinal diols.[1,2] The method involves formation of a 1,3-dioxolane-2-thione [cyclic thionocarbonate (or thiocarbonate)] by treatment of a vicinal diol with TCDI. Decomposition of the thionocarbonate, usually with a phosphorus compound, affords the alkene (eq 1).[3] The breakdown of the thionocarbonate occurs in a stereospecific sense; details of investigations into the mechanism have been summarized.[1]

The conditions required for the introduction of the thiocarbonyl group vary greatly, depending on the structure of the diol.[1,4] The thionocarbonate may also be prepared without TCDI. The reaction of vicinal diols with thiophosgene/DMAP/CH_2Cl_2/0 °C,[5] with base/CS_2/MeI/heat,[1] or with 1,1′-thiocarbonyl-2,2′-pyridone/toluene/110 °C[6] also provides thionocarbonates. Along with phosphines and phosphites, other reagents can decompose the thionocarbonates, but not always in a stereospecific manner. The Corey–Hopkins reagent 1,3-dimethyl-2-phenyl-1,3,2-diazaphospholidine,[5] along with Raney Nickel[7] and Bis(1,5-cyclooctadiene)nickel(0),[8] affords products of stereospecific *cis*-elimination, while the alkyl iodide/ Zinc combination[9] and Pentacarbonyliron[10] yield products of nonstereospecific elimination or unknown stereochemistry.

The Corey–Winter synthesis has proved useful for the generation of a large number of structurally interesting alkenes,[1] including some unstable alkenes that must be captured in situ (eq 2).[4]

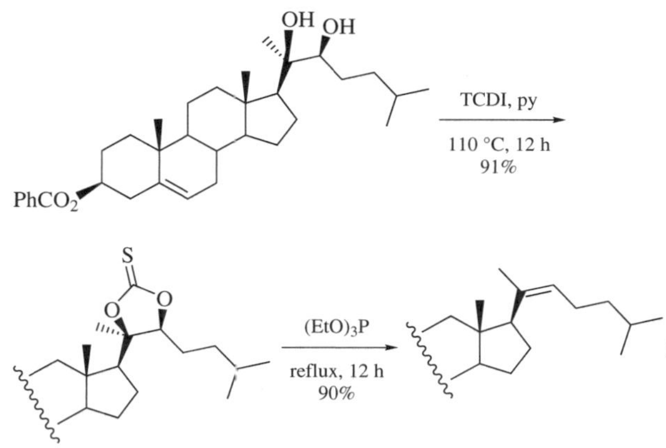

(1)

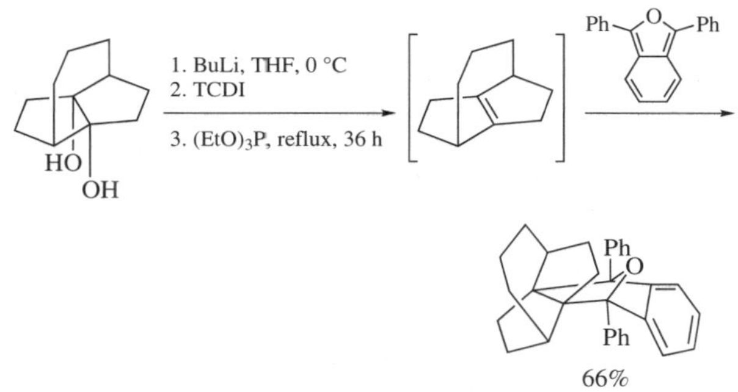

(2)

The method has found its niche in the chemistry of sugars. Their polyhydroxylated nature make them excellent substrates provided the by-standing hydroxyls are protected (eq 3).[11] The 2′- and 3′-oxygens of nucleosides are readily are removed under the Corey–Winter conditions.[12]

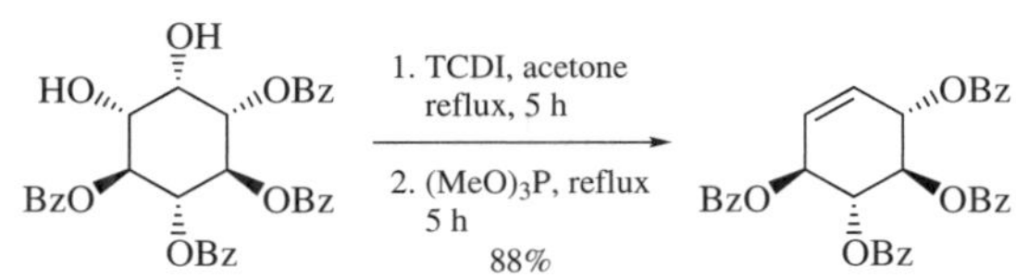

(3)

Radical Chemistry. Treatment of secondary alcohols with 1 equiv of TCDI affords an imidazole-1-thiocarbonyl derivative (imidazolide), which can be reduced to a CH_2 unit under Tri-n-butylstannane (TBTH) radical chain reaction conditions.[13,14] The deoxygenation of secondary alcohols by way of an imidazolide or other thiocarbonyl derivative is called the Barton–McCombie reaction (eq 4).[6,14] Since imidazolide formation (TCDI, reflux, 65 °C) and the subsequent radical chemistry are done under neutral or near-neutral conditions, the overall reduction is tolerant of the presence of many sorts of functional groups. Furthermore, the low solvation requirements of radical species permits deoxygenation in sterically congested environments (eq 5).[15]

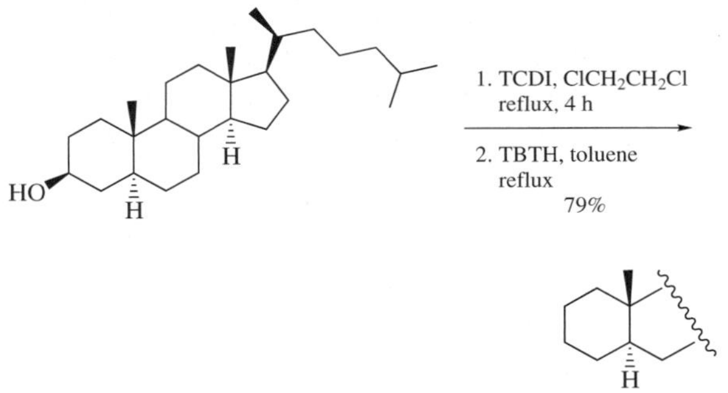

(4)

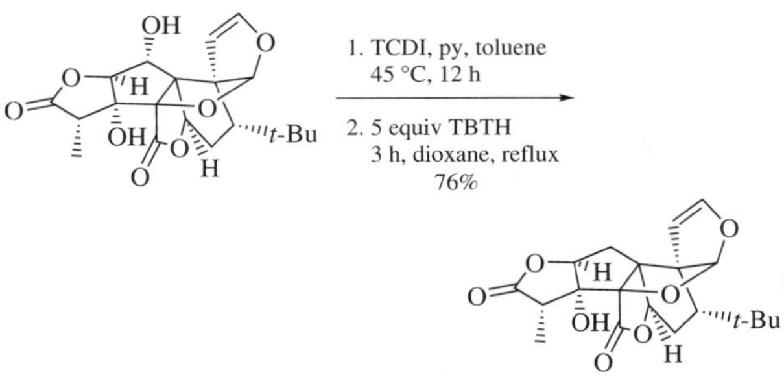

(5)

The mechanism of the radical attack at the thiocarbonyl group and the ensuing breakdown has been outlined elsewhere.[13,14] Some groups have studied the comparative chemistries of the possible thiocarbonyl derivatives, with no particularly obvious trends arising.[13,16,17] There are numerous reports of the deoxygenation of carbohydrates[13,16,18] and related polyhydroxylated species.[14,17,19,20] In one instance of amino glycoside deoxygenation, the C(3′)-OH group can be selectively functionalized with TCDI, while the C(5)-OH group is unaffected and the radical reduction proceeds without protection of the C(5)-OH group (eq 6).[20]

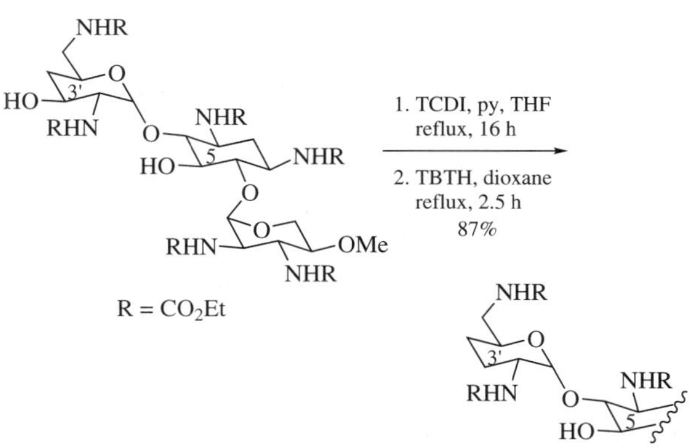

(6)

Reduction of tertiary alcohols is not recommended due to the instability of the tertiary imidazolide, although, as an alternative, a thioformaldehyde version has been developed.[21] Imidazolides from primary alcohols do not respond well to the same conditions as the secondary systems, but performing the radical chemistry at 130–150 °C can lead to deoxygenation.[22] As an alternative to the high-temperature radical chemistry, the imidazolide can be converted by methanolysis to a thiocarbonate which can be reduced at room temperature (Et_3B, TBTH, benzene).[23] Tin hydride treatment of thionocarbonates derived from primary/secondary diols affords, after base exposure, the product of selective reduction of the secondary site, leaving the primary hydroxyl (eq 7).[14,24] The approach has also been successfully applied to 1,3-diols.[24,25]

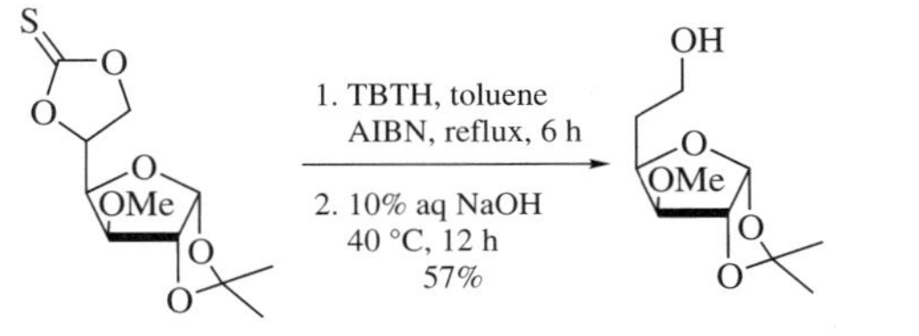

(7)

Since the reduction method proceeds via radical chemistry, the intermediate radicals are prone to bimolecular chemistry and unimolecular reactions such as rearrangements or β-scission, if suitable functionality is proximal.[14] Indeed, the imidazolides have been employed as radical precursors when rearrangement is the desired result.[14]

Thiocarbonyl Transfer. While virtually all uses of TCDI involve a thiocarbonyl transfer reaction, this section covers those uses of the thiocarbonyl transfer that do not lead to radical chemistry or alkene syntheses. TCDI has been used for the simple placement of a thiocarbonyl group between two nucleophilic atoms of one[26] or two molecules.[27] An alcohol which has been converted to an imidazolide is a reactive functionality for coupling reactions,[28] for sigmatropic rearrangements,[29] and for elimination (eq 8).[30] The TCDI alternative 1,1′-thiocarbonyl-2,2′-pyridone[6] seems to be an excellent thiocarbonyl transfer agent.

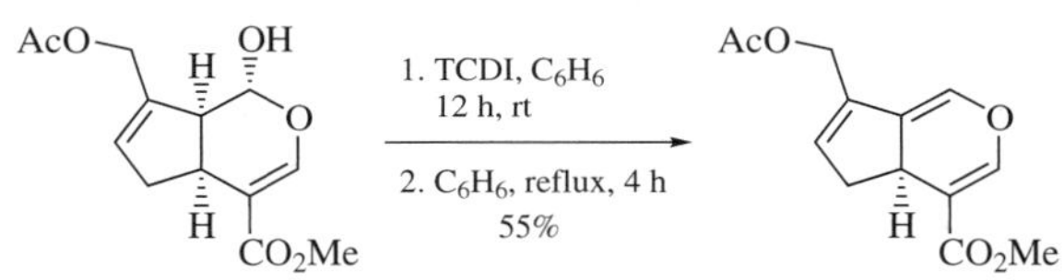

(8)

1. Block, E. *Org. React.* **1984**, *30*, 457.
2. Corey, E. J.; andWinter, R. A. E. *J. Am. Chem. Soc.* **1963**, *85*, 2677.
3. Koreeda, M.; Koizumi, N.; Teicher, B. A. *Chem. Commun./J. Chem. Soc., Chem. Commun.* **1976**, 1035.
4. Greenhouse, R.; Borden, W. T.; Ravindranathan, T.; Hirotsu, K.; Clardy, J. *J. Am. Chem. Soc.* **1977**, *99*, 6955.
5. Corey, E. J.; Hopkins, P. B. *Tetrahedron Lett.* **1982**, *23*, 1979.
6. Kim, S.; Yi, K. Y. *J. Org. Chem.* **1986**, *51*, 2613.
7. Ireland, R. E.; Anderson, R. C.; Badoud, R.; Fitzsimmons, B. J.; McGarvey, G. J.; Thaisrivongs, S.; Wilcox, C. S. *J. Am. Chem. Soc.* **1983**, *105*, 1988.
8. Semmelhack, M. F.; Stauffer, R. D. *Tetrahedron Lett.* **1973**, 2667.
9. Vedejs, E.; Wu, E. S. C. *J. Org. Chem.* **1974**, *39*, 3641.
10. Daub, J.; Trautz, V.; Erhardt, U. *Tetrahedron Lett.* **1972**, 4435.
11. (a) Akiyama, T.; Shima, H.; Ozaki, S. *Tetrahedron Lett.* **1991**, *32*, 5593. (b) Horton, D.; Turner, W. N. *Tetrahedron Lett.* **1964**, 2531.
12. Manchand, P. S.; Belica, P. S.; Holman, M. J.; Huang, T.-N.; Maehr, H.; Tam, S. Y.-K.; Yang, R. T. *J. Org. Chem.* **1992**, *57*, 3473.
13. Barton, D. H. R.; McCombie, S. W. *J. Chem. Soc., Perkin Trans. 1* **1975**, 1574.
14. (a) McCombie, S. W. *Comprehensive Organic Synthesis* **1991**, *8*, 811. (b) Hartwig, W. *Tetrahedron* **1983**, *39*, 2609. (c) Crich, D.; Quintero, L. *Chem. Rev.* **1989**, *89*, 1413.

15. Corey, E. J.; Ghosh, A. K. *Tetrahedron Lett.* **1988**, *29*, 3205.
16. Rasmussen, J. R.; Slinger, C. J.; Kordish, R. J.; Newman-Evans, D. D. *J. Org. Chem.* **1981**, *46*, 4843.
17. Robins, M. J.; Wilson, J. S.; Hansske, F. *J. Am. Chem. Soc.* **1983**, *105*, 4059.
18. Lin, T.-H.; Kovac, P.; Glaudemans, C. P. J. *Carbohydr. Res.* **1989**, *188*, 228.
19. Piccirilli, J. A.; Krauch, T.; MacPherson, L. J.; Benner, S. A. *Helv. Chim. Acta* **1991**, *74*, 397.
20. Carney, R. E.; McAlpine, J. B.; Jackson, M.; Stanaszek, R. S.; Washburn, W. H.; Cirovic, M.; Mueller, S. L. *J. Antibiot.* **1978**, *31*, 441.
21. Barton, D. H. R.; Hartwig, W.; Hay Motherwell, R. S.; Motherwell, W. B.; Stange, A. *Tetrahedron Lett.* **1982**, *23*, 2019.
22. Barton, D. H. R.; Motherwell, W. B.; Stange, A. *Synthesis* **1981**, 743.
23. Chu, C. K.; Ullas, G. V.; Jeong, L. S.; Ahn, S. K.; Doboszewski, B.; Lin, Z. X.; Beach, J. W.; Schinazi, R. F. *J. Med. Chem.* **1990**, *33*, 1553.
24. Barton, D. H. R.; Subramanian, R. *J. Chem. Soc., Perkin Trans. 1* **1977**, 1718.
25. (a) Mubarak, A. M.; Brown, D. M. *Tetrahedron Lett.* **1981**, *22*, 683. (b) Suzuki, M.; Yanagisawa, A.; Noyori, R. *Tetrahedron Lett.* **1984**, *25*, 1383.
26. Chiang, L.-Y.; Shu, P.; Holt, D.; Cowan, D. *J. Org. Chem.* **1983**, *48*, 4713.
27. Sugimoto, H.; Makino, I.; Hirai, K. *J. Org. Chem.* **1988**, *53*, 2263.
28. Ley, S. V.; Armstrong, A.; Diez-Martin, D.; Ford, M. J.; Grice, P.; Knight, J. G.; Kolb, H. C.; Madin, A.; Marby, C. A.; Mukherjee, S.; Shaw, A. N.; Slawin, A. M. Z.; Vile, S.; White, A. D.; Williams, D. J.; Woods, M. *J. Chem. Soc., Perkin Trans. 1* **1991**, 667.
29. Nicolaou, K. C.; Groneberg, R. D.; Miyazaki, T.; Stylianides, N. A.; Schulze, T. J.; Stahl, W. *J. Am. Chem. Soc.* **1990**, *112*, 8193.
30. Ge, Y.; Isoe, S. *Chem. Lett.* **1992**, 139.

Adrian L. Schwan
University of Guelph, Ontario, Canada

1,1′-Thionylimidazole[1]

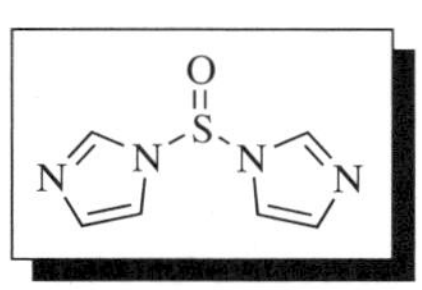

[3005-50-3] $C_6H_6N_4OS$ (MW 182.23)

(forms *N*-acylimidazolides[1] used to synthesize amides, peptides,[2] esters,[3] and ketones;[4] synthesis of symmetrical sulfoxides,[5] inorganic esters,[6] and *N*-alkylimidazoles[7])

Alternate Names: *N,N′*-thionylimidazole; *N,N′*-thionyldiimidazole; TDI.
Physical Data: mp 78–79 °C.
Solubility: sol THF, CH_2Cl_2, toluene.
Form Supplied in: not commercially available; usually used in THF solution.
Analysis of Reagent Purity: usual preparations are used immediately without analysis.
Preparative Method: the reagent is prepared by reacting Thionyl Chloride with 4 equiv of Imidazole in anhydrous THF at 0 °C. The precipitated imidazole hydrochloride is filtered and the solution is pure enough to use.[8]
Purification: filter to remove imidazole hydrochloride. Analytically pure samples have been prepared by reaction of thionyl chloride and N-(Trimethylsilyl)imidazole.[9]
Handling, Storage, and Precautions: extremely hygroscopic; reacts exothermically with water and alcohols. Store as a solution in THF or CH_2Cl_2 under dry N_2.[8b] Use in a fume hood.

N-Acylimidazolides. *N,N′*-Thionylimidazole reacts with carboxylic acids to form *N*-acylimidazolides (eq 1) that readily acylate amines, amino acids, or alcohols.[1] The products are readily purified since the byproducts, imidazole and SO_2, are easily removed. N,N′-Carbonyldiimidazole (CDI) is another useful reagent for formation of *N*-acylimidazolides. TDI is used less frequently for formation of *N*-acylimidazolides because of its high reactivity; also, CDI is commercially available and easier to handle. Most reported transformations from *N*-acylimidazolides involve CDI; however, TDI should give similar results.

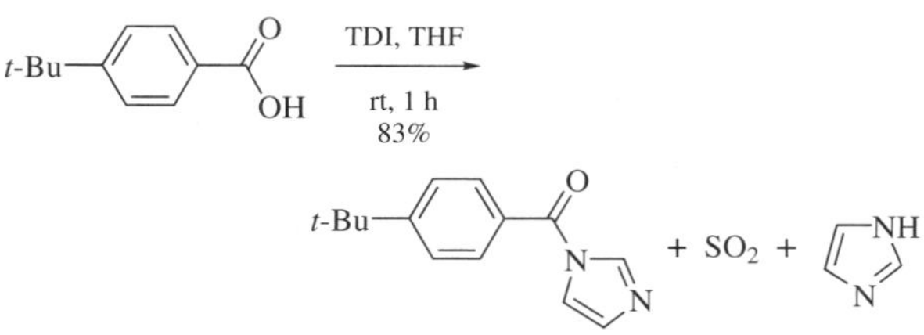

(1)

Peptide Coupling Reagent. *N*-Acylimidazolides are useful acylating intermediates in peptide synthesis (eq 2).[2] These are direct couplings where the amine and acid components are premixed in solution. There are only a few examples of peptides being synthesized with this reagent. Therefore it is not known what limitations exist, such as extent of racemization, ease of handling, etc. Comparisons with many of the well-known peptide coupling reagents (e.g. 1,3-Dicyclohexylcarbodiimide) have not been published. CDI is used more frequently if the imidazolide intermediate is desired because of its ease of handling.

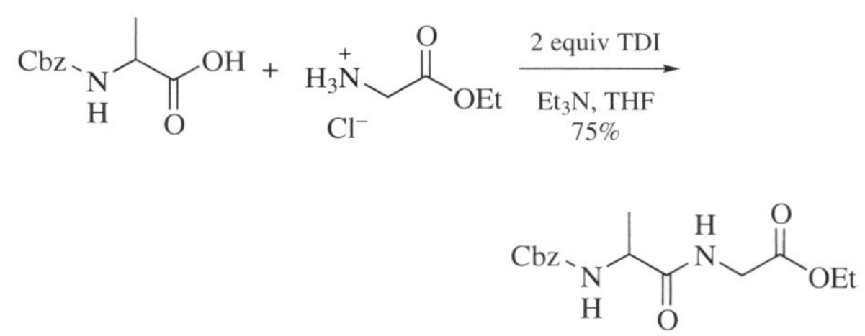

(2)

Ester Formation. *N*-acylimidazolides react with alcohols very fast in the presence of catalytic amounts of the alkoxide (eq 3).[3] With this method the methyl ester of vitamin A has been obtained in 91% yield.[3]

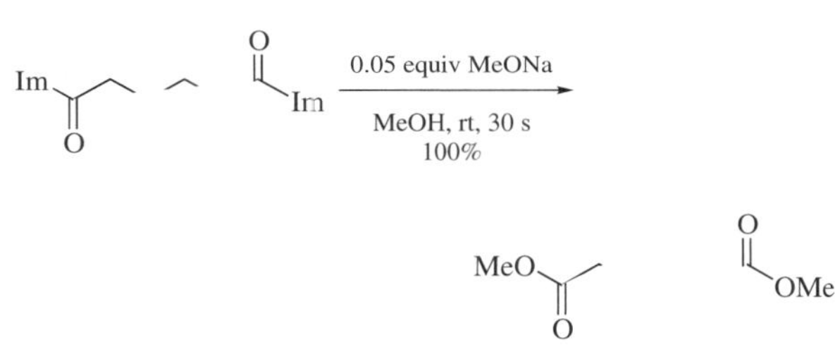

(3)

Ketones. *N*-Acylimidazolides react with Grignard reagents to give ketones in good yields.[4] A comparison of *N*-acylimidazolides and mixed acid anhydrides in Grignard reactions indicates that the imidazolides result in higher yields (eq 4).[10] There are many procedures for reacting Grignard reagents with 'activated' carboxylic acids to give ketones. Other reagents used with varying degrees of success are acid chlorides (e.g. Ethyl Chloroformate), nitriles, and 2-pyridylthiol esters (e.g. 2-Thiopyridyl Chloroformate).[11]

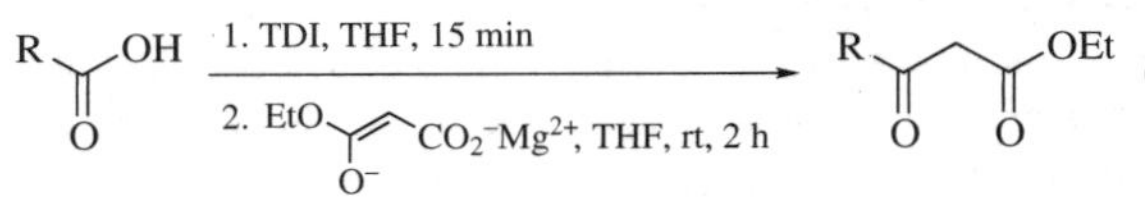

(4)

R	Anhydride	Imidazolide
Et	57	68
Pr	57	74
C_6H_{13}	62	74
Ph	0	79

Symmetrical Sulfoxides. Thionylimidazole reacts directly with Grignard reagents resulting in symmetrical sulfoxides in reasonable yields (eq 5).[5] This method offers an easy workup and purification since the main byproduct is imidazole.

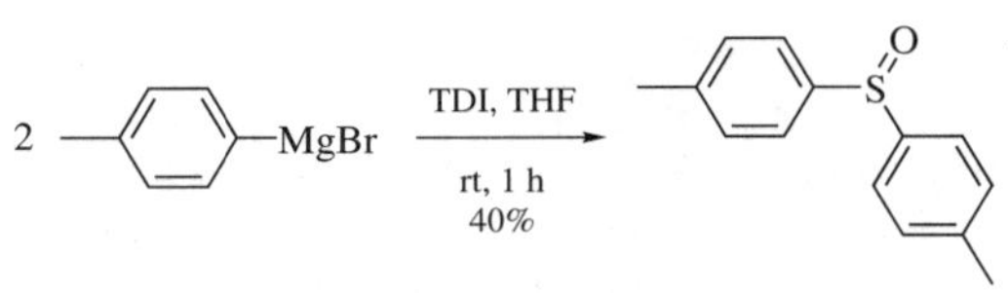

(5)

Inorganic Ester Formation. Phenols and alcohols react exothermically with TDI to give sulfonic acid diesters in high yield (eq 6).[6]

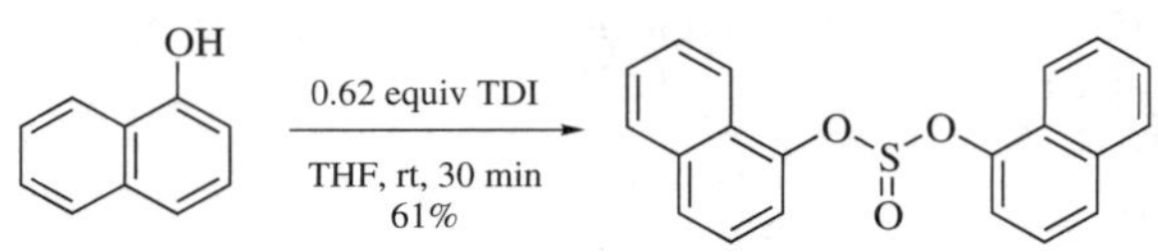

(6)

***N*-Alkylimidazoles.** Thionylimidazole has found great utility in the synthesis of *N*-alkylimidazoles.[7] Tertiary alcohols react with TDI to form monosulfonic ester imidazolides that rearrange to transfer the imidazole to the tertiary carbon (eq 7).[12] The products are also obtained by treatment of the trityl chloride with imidazole; however, comparison of the two procedures is not possible as yields were not reported.[12]

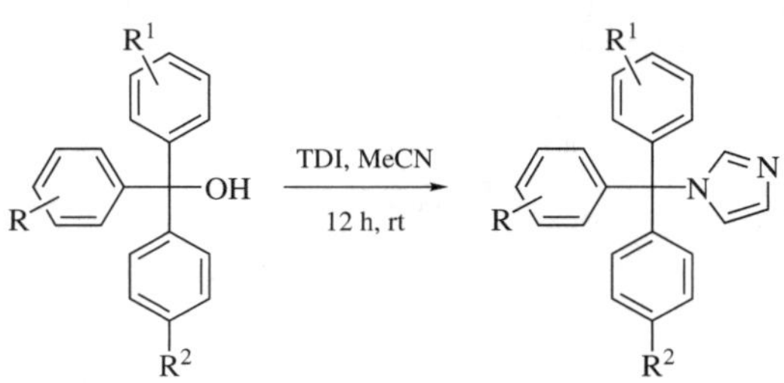

(7)

R = H, Me, NH_2; R^1 = H, Me, di-Me; R^2 = H, Cl, Ph

This imidazole transfer reaction occurs readily with *o*-hydroxybenzyl alcohols, giving the imidazole from a primary or secondary alcohol (eq 8).[7] It is thought that the *o*-hydroxy group contributes to the facile transfer of the imidazole. Similarly, *o*-amines promote the transfer of the imidazole. Carbonyl diimidazole also works in this imidazole transfer reaction, but it usually gives lower yields of the desired product.[7]

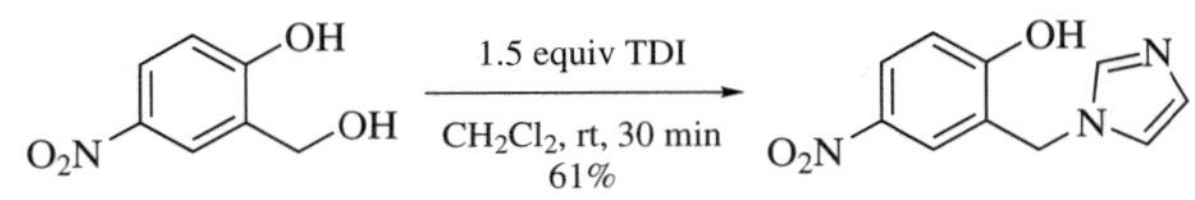

(8)

This imidazole transfer reaction works with ketones as well; however, in this case two products, the mono- and diimidazoles, are obtained (eq 9).[13] Increasing the temperature does not change the product ratio or yield. Some examples contain hydrogen-bonding groups on the ring, but this does not modify the product ratio or yield either. CDI gives similar results in most cases. This imidazole transfer reaction does not work with aldehydes under these conditions.[13]

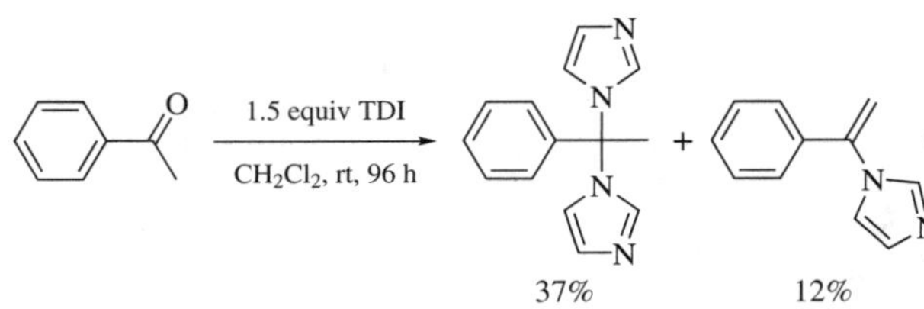

(9)

1. Staab, H. A. *Angew. Chem., Int. Ed. Engl.* **1962**, *1*, 351.
2. Wieland, T.; Vogeler, K. *Angew. Cheml* **1961**, *73*, 435.
3. Staab, H. A.; Rohr, W.; Mannschreck, A. *Angew. Cheml* **1961**, *73*, 143 (*Chem. Abstr.* **1961**, *55*, 14 437h) .
4. Staab, H. A.; Jost, E. *Justus Liebigs Ann. Chem./Liebigs Ann. Chem.* **1962**, *655*, 90 (*Chem. Abstr.* **1962**, *57*, 15 090g) .
5. Bast, S.; Andersen, K. K. *J. Org. Chem.* **1968**, *33*, 846.
6. Staab, H. A.; Wendel, K. *Justus Liebigs Ann. Chem./Liebigs Ann. Chem.* **1966**, *694*, 86 (*Chem. Abstr.* **1966**, *65*, 12 195c) .
7. Ogata, M.; Matsumoto, H.; Kida, S.; Shimizu, S. *Chem. Ind. (London)* **1980**, 85.
8. (a) Staab, H. A.; Wendel, K. *Angew. Cheml* **1961**, *73*, 26 (*Chem. Abstr.* **1961**, *55*, 12 389e) . (b) *Fieser & Fieser* **1967**, *1*, 1163.
9. Birkofer, L.; Gilgenberg, W.; Ritter, A. *Angew. Cheml* **1961**, *73*, 143.
10. Bram, G.; Vilkas, M. *Bull. Soc. Claim. Fr., Part 2* **1964**, 945.
11. (a) Shirley, D. A. *Org. React.* **1954**, *8*, 28. (b) Posner, G.; Witten, C. E. *Tetrahedron Lett.* **1970**, 4647. (c) Mukaiyama, T.; Araki, M.; Takei, H.; *J. Am. Chem. Soc.* **1973**, *95*, 4763.
12. Buechel, K. H.; Draber, W.; Regal, E.; Plempel, M. *Arzneim.-Forsch.* **1972**, *22*, 1260 (*Chem. Abstr.* **1972**, *77*, 152 067e) .
13. Ogata, M.; Matsumoto, H.; Kida, S.; Shimizu, S. *Tetrahedron Lett.* **1979**, 5011.

Richard S. Pottorf
Marion Merrell Dow Research Institute, Cincinnati, OH, USA

2-Thiopyridyl Chloroformate[1]

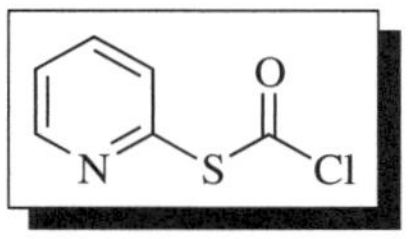

[73371-99-0] C_6H_4ClNOS (MW 173.63)

(convenient preparation of 2-pyridylthiol esters;[2] subsequent transformation into lactones,[3] peptides,[4] and ketones[5])

Physical Data: ^1H NMR ($CDCl_3$) δ 8.64 m (1H), 7.75 m (2H), 7.38 m (1H).
Solubility: sol CH_2Cl_2, ether.
Form Supplied in: colorless oil; main impurity is bis(thiopyridyl) carbonate; not commercially available.
Analysis of Reagent Purity: IR (CH_2Cl_2) 1765 cm^{-1} (C=O); main impurity: IR (CH_2Cl_2) 1715 cm^{-1} (C=O).
Preparative Method: Phosgene (5 equiv) in toluene and CH_2Cl_2 is cooled to 0 °C. Dropwise addition (5 min) of a CH_2Cl_2 solution of Triethylamine (slight excess) and 2-Pyridinethiol is followed by stirring for 10 min. After removal of excess phosgene and CH_2Cl_2 in vacuo, hexane is added and the resulting precipitate is filtered. After concentration of the combined filtrates, the colorless oil (96%) is dissolved in CH_2Cl_2 and stored at −25 °C.
Handling, Storage, and Precautions: very unstable to water and silica gel; however, it can be handled in air. It is stable for one month if stored at −25 °C. Since phosgene is used in preparing this reagent, preparation should be in a working fume hood and extreme caution is required.

Preparation of 2-Pyridylthiol Esters. 2-Pyridylthiol esters are formed by treating a carboxylic acid with 2-thiopyridyl chloroformate under extremely mild conditions (eq 1).[2] The $Et_3N{\cdot}HCl$ is removed by filtration or by washing with cold aqueous acid and base. After thorough drying, the thiol esters are generally pure enough to use in many synthetic applications.

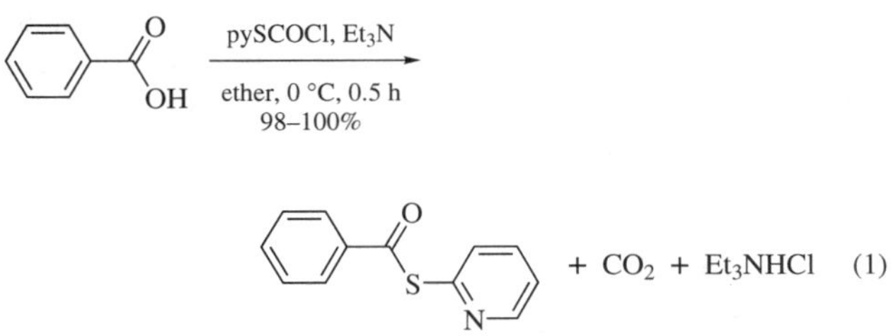

(1)

There are several other reagents for preparing 2-pyridylthiol esters. 1,3-Dicyclohexylcarbodiimide[4] generally gives lower yields of the thiol ester and removal of the dicyclohexylurea can be difficult. Treating an acid chloride with thallium(I) 2-pyridinethiolate[6] has been used, but thallium salts are toxic. The method used most frequently before the introduction of 2-thiopyridyl chloroformate involves reacting a carboxylic acid with Triphenylphosphine and 2,2′-Dipyridyl Disulfide.[7] This procedure suffers from the necessity of removing 2-pyridinethiol and triphenylphosphine oxide by chromatography, which precludes preparing large batches of the thiol esters since there can be a loss of product by reaction with the silica gel. 2-Thiopyridyl chloroformate provides access to many reported synthetic transformations. In some of these reports, the thiol ester has been prepared by other methods. However, use of 2-thiopyridyl chloroformate should make these transformations even more accessible.

Lactone Formation. There are several examples of the synthesis of lactones with 2-pyridylthiol esters.[3] Since this is a facile reaction, reasonable yields of complex macrocycles are obtained (eq 2).[8] The pyridine nitrogen may provide anchimeric assistance for the approaching nucleophile, thus facilitating the acylation.[3]

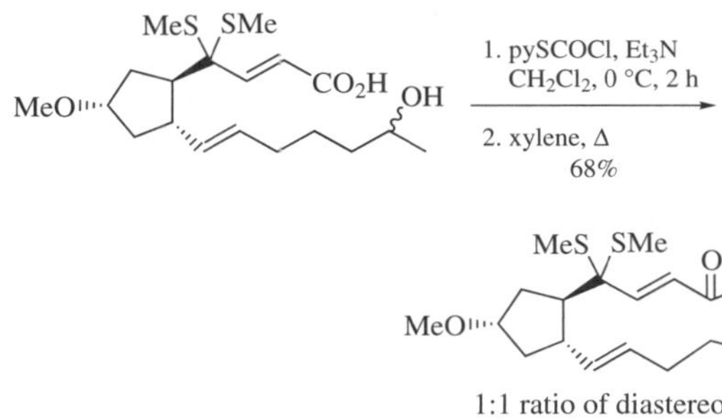

(2)

Ester Formation. 2-Pyridylthiol esters acylate lysophosphatidylcholines rapidly when catalyzed by silver ion, giving mixed-chain phosphatidylcholines in high yields and isomeric purity (eq 3).[9] The main advantages of this procedure over the usual 4-Dimethylaminopyridine catalyzed acylation with acid anhydrides[10] are that less acylating reagent is required to give high yields and rearrangement of the fatty acids is minimized. The main disadvantage is the sensitivity of the 2-pyridylthiol ester to water.

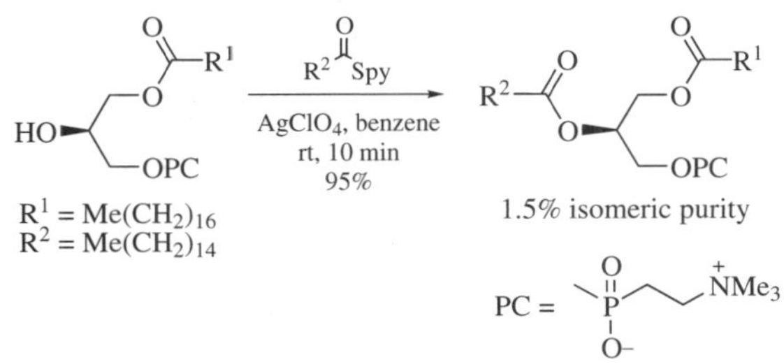

(3)

Peptide Coupling. These thiol esters have been used in the synthesis of several dipeptides. Many of the examples involve highly sterically hindered amino acids and result in very good yields and high optical purity (eq 4).[4] There are many useful active

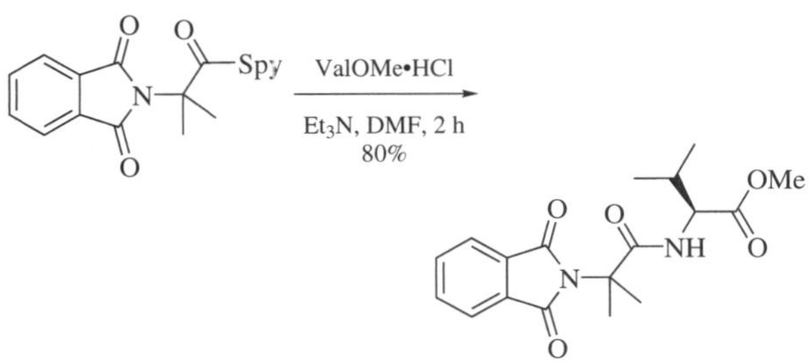

(4)

esters for peptide coupling, such as Pentafluorophenol, p-Nitrophenol, and N-Hydroxysuccinimide. However, these do not work as well with the hindered amino acids shown in eq 4. Additionally, there are many direct coupling procedures; therefore the 2-pyridylthiol esters have not been used frequently.

Ketones. Reaction of 2-pyridylthiol esters with Grignard reagents occurs rapidly to give ketones in almost quantitative yields (eq 5).[5] In all studies, less than 1% of the tertiary alcohol is observed. This procedure works reasonably well with a suitably protected amino acid to give the ketone (eq 6).[11] Phthaloyl protection of the amine is required since esters with amide NH groups decompose when treated with Grignard reagents.[11]

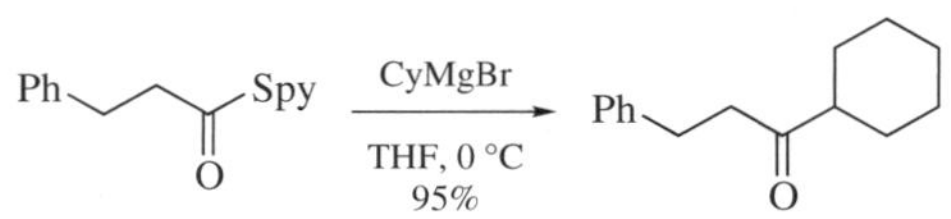

(5)

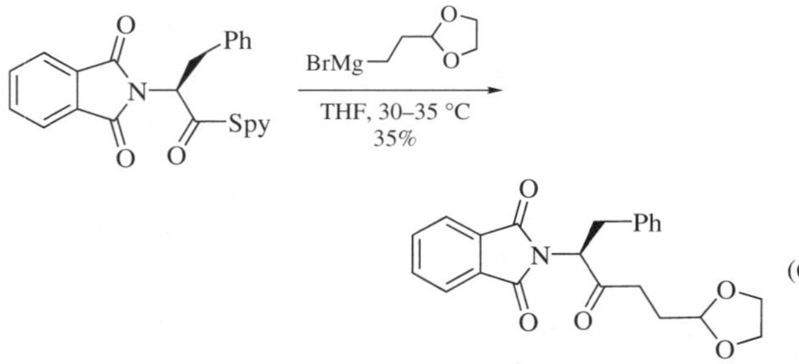

(6)

Preparation of thiol esters of highly elaborated phosphoranes with thiopyridyl chloroformate has been reported. Treatment of the thiol ester with an aryl Grignard reagent gives the corresponding ketone without epimerization of the chiral centers (eq 7).[12] There are many procedures for reacting Grignard reagents with 'activated' carboxylic acids to give ketones. Other reagents used with varying degrees of success include acid chlorides, nitriles, acid anhydrides, and *N*-acylimidazolides (e.g. N,N′-Carbonyldiimidazole, 1,1′-Thionylimidazole).[13] Many of these give higher yields of the tertiary alcohol. However, treating acid chlorides with Grignard reagents at −78 °C gives better results,[14] so this may be an alternative to the thiol esters.

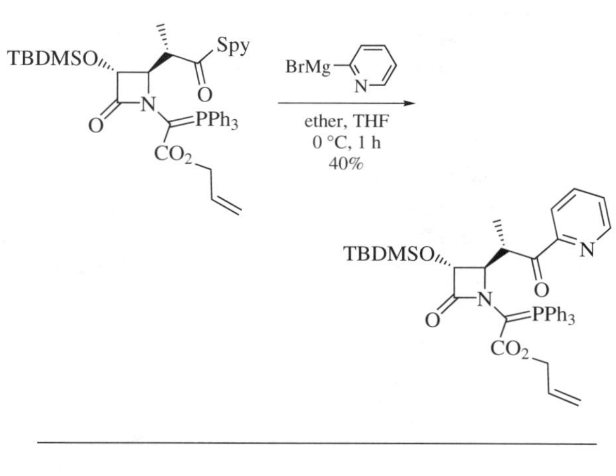

(7)

1. Haslam, E. *Tetrahedron* **1980**, *36*, 2409.
2. Corey, E. J.; Clark, D. A. *Tetrahedron Lett.* **1979**, *31*, 2875.
3. Corey, E. J.; Nicolaou, K. C. *J. Am. Chem. Soc.* **1974**, *96*, 5614.
4. Lloyd, K.; Young, G. T. *J. Chem. Soc. (C)* **1971**, 2890.
5. Mukaiyama, T.; Araki, M.; Takei, H. *J. Am. Chem. Soc.* **1973**, *95*, 4763.
6. Masamune, S.; Kamata, S.; Diakur, J.; Sugihara, Y.; Bates, G. S. *Can. J. Chem.* **1975**, *53*, 3693.
7. Mukaiyama, T.; Matsueda, R.; Suzuki, M. *Tetrahedron Lett.* **1970**, 1901.
8. Le Drian, C.; Greene, A. E. *J. Am. Chem. Soc.* **1982**, *104*, 5473.
9. Nicholas, A. W.; Khouri, L. G.; Ellington, J. C., Jr.Porter, N. A. *Lipids* **1983**, *18*, 434.
10. Gupta, C. M.; Radhakrishran, R.; Khorana, H. G. *Proc. Natl. Acad. Sci. USA* **1977**, *74*, 4315.
11. Almquist, R. G.; Chao, W.-R.; Ellis, M. E.; Johnson, H. L. *J. Med. Chem.* **1980**, *23*, 1392.
12. Guthikonda, R. N.; Cama, L. D.; Quesada, M.; Woods, M. F.; Salzmann, T. N.; Christensen, B. G. *J. Med. Chem.* **1987**, *30*, 871.
13. (a) Shirley, D. A. *Org. React.* **1954**, *8*, 28. (b) Posner, G.; Witten, C. E. *Tetrahedron Lett.* **1970**, 4647. (c) Staab, H. A.; Jost, E. *Justus Liebigs Ann. Chem./Liebigs Ann. Chem.* **1962**, *655*, 90.
14. Sato, F.; Inoue, M.; Oguro, K.; Sato, M. *Tetrahedron Lett.* **1979**, 4303.

Richard S. Pottorf
Marion Merrell Dow Research Institute, Cincinnati, OH, USA

Tin(IV) Chloride

$SnCl_4$

[7646-78-8] Cl_4Sn (MW 260.51)

(strong Lewis acid used to promote nucleophilic additions, pericyclic reactions, and cationic rearrangements; chlorination reagent)

Alternate Name: stannic chloride.
Physical Data: colorless liquid; mp −33 °C; bp 114.1 °C; *d* 2.226 g cm^{-3}.
Solubility: reacts violently with water; sol cold H_2O; dec hot H_2O; sol alcohol, Et_2O, CCl_4, benzene, toluene, acetone.
Form Supplied in: colorless liquid; 1 M soln in CH_2Cl_2 or heptane; widely available.
Purification: reflux with mercury or P_2O_5 for several hours, then distill under reduced nitrogen pressure into receiver with P_2O_5. Redistill. Typical impurities: hydrates.
Handling, Storage, and Precautions: hygroscopic; should be stored in a glove box or over P_2O_5 to minimize exposure to moisture. Containers should be flushed with N_2 or Ar and tightly sealed. Perform all manipulations under N_2 or Ar. Solvating with H_2O liberates much heat. Use in a fume hood.

Original Commentary

Stephen Castellino
Rhône-Poulenc Ag. Co., Research Triangle Park, NC, USA
David E. Volk
North Dakota State University Fargo, ND, USA

Introduction. $SnCl_4$ is used extensively in organic synthesis as a Lewis acid for enhancing a variety of reactions. $SnCl_4$ is classified as a strong Lewis acid according to HSAB theory, and therefore interacts preferentially with hard oxygen and nitrogen bases. Six-coordinate 1:2 species and 1:1 chelates are the most stable coordination complexes, although 1:1 five-coordinate species are also possible.[1] $SnCl_4$ can be used in stoichiometric amounts, in which case it is considered a 'promoter', or in substoichiometric amounts as a catalyst, depending upon the nature of the reaction. $SnCl_4$ is an attractive alternative to boron, aluminum, and titanium Lewis acids because it is monomeric, highly soluble in organic solvents, and relatively easy to handle. $SnCl_4$ and $TiCl_4$ are among the most common Lewis acids employed in 'chelation control' strategies for asymmetric induction. However, $SnCl_4$ is not often the Lewis acid of choice for optimum selectivities and yields.

$SnCl_4$ is also the principal source for alkyltin chlorides, R_nSnCl_{4-n}.[2] Allyltrialkyltin reagents react with $SnCl_4$ to produce allyltrichlorotin species through an S_E2' pathway (eq 1).[3] Silyl enol ethers react with $SnCl_4$ to give α-trichlorotin ketones (eq 2).[4] Transmetalation or metathesis reactions of this type are competing pathways to nucleophilic addition reactions where $SnCl_4$ is present as an external Lewis acid. As a consequence, four important experimental variables must be considered when using $SnCl_4$ as a promoter: (1) the stoichiometry between the substrate and the Lewis acid; (2) the reaction temperature; (3) the nature of the Lewis base site(s) in the substrate; and (4) the order of addition. These variables influence the reaction pathway and product distribution.[5]

(1)

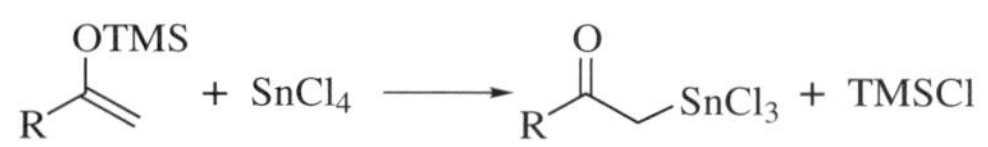

(2)

Nucleophilic Additions to Aldehydes. $SnCl_4$ is effective in promoting the addition of nucleophiles to simple aldehydes. Among the most synthetically useful additions are allylstannane and -silane additions. The product distribution in the stannane reactions can be influenced by the order of addition, stoichiometry, and reaction temperature. The *anti* geometry of the tin–aldehyde complex is favored due to steric interactions. Furthermore, the six-coordinate 2:1 complex is most likely the reactive intermediate in these systems. The use of crotylstannanes provides evidence for competing transmetalation reaction pathways (eq 3).[6] Superior selectivities are provided by Titanium(IV) Chloride.

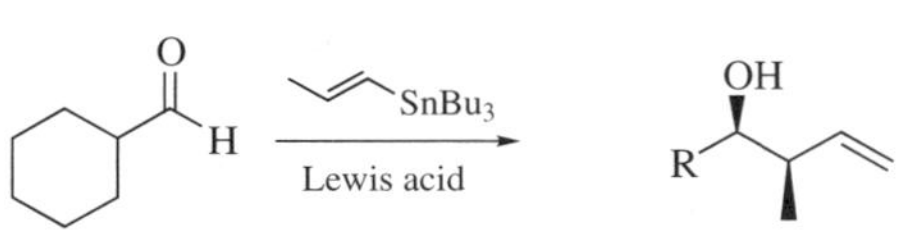

normal addition	1.3 equiv $SnCl_4$	22.8
inverse addition	1.3 equiv $SnCl_4$	21.8
normal addition	1.05 equiv $TiCl_4$	90.5
inverse addition	2.1 equiv $TiCl_4$	4.4

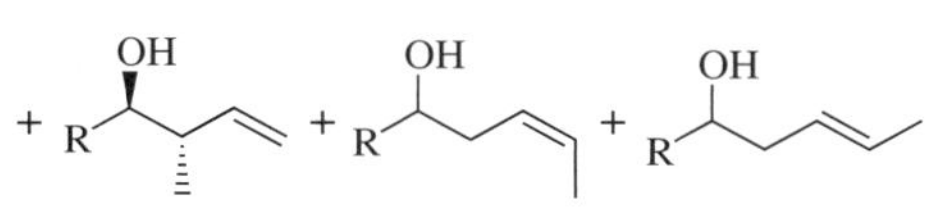

(3)

26.0	36.4	14.8
74.9	1.2	2.2
7.0	2.1	0.5
90.8	–	4.9

The presence of additional Lewis base sites within the molecule can result in the formation of chelates with $SnCl_4$ or $TiCl_4$, which can lead to 1,2- or 1,3-asymmetric induction with the appropriate substitution at the C-2 or C-3 centers. NMR studies have provided a basis for explaining the levels of diastereofacial selectivity observed in nucleophilic additions to Lewis acid chelates of β-alkoxy aldehydes with substitution at the C-2 or C-3 positions.[7] These studies reveal that $SnCl_4$ chelates are dynamically unstable when substrates are sterically crowded at the alkoxy center, thus enhancing the formation of 2:1 complexes and/or competing metathesis pathways. Furthermore, for β-siloxy aldehydes, the 2:1 $SnCl_4$ complex is formed preferentially over the corresponding chelate.[8]

Mukaiyama Aldol Additions. Lewis acid-promoted additions of a chiral aldehyde to a silyl enol ether or silyl ketene acetal (the Mukaiyama[9] aldol addition) occurs with good diastereofacial selectivity.[10] The reaction has been investigated with nonheterosubstituted aldehydes, α- and β-alkoxy aldehydes,[11] α- and β-amino aldehydes,[12] and thio-substituted aldehydes.[13] High diastereoselectivity is observed in the $SnCl_4$- or $TiCl_4$-promoted aldol addition of silyl enol ethers to α- and β-alkoxy aldehydes. Prior chelation of the aldehyde before addition of the enol silane is important because certain enol silanes interact with $SnCl_4$ to produce α-trichlorostannyl ketones, which provide lower selectivity.[14] Simple diastereoselectivity is independent of the geometry of the enol silane, and the reaction does not proceed through prior Si–Ti or Si–Sn exchange. Good *anti* selectivities (up to 98:2) are obtained in the $SnCl_4$-promoted reactions of chiral α-thio-substituted aldehydes only with α-phenylthio-substituted aldehydes (eq 4). Stereorandom results are obtained with $SnCl_4$ when other alkylthio-substituted aldehydes, such as α-isopropylthio-substituted aldehydes, are used. Boron Trifluoride Etherate catalysis gives better *anti* selectivities than $SnCl_4$ for aldehydes with smaller alkylthio substituents. Excellent *syn* selectivities are obtained for α-thio-substituted aldehydes with $TiCl_4$.

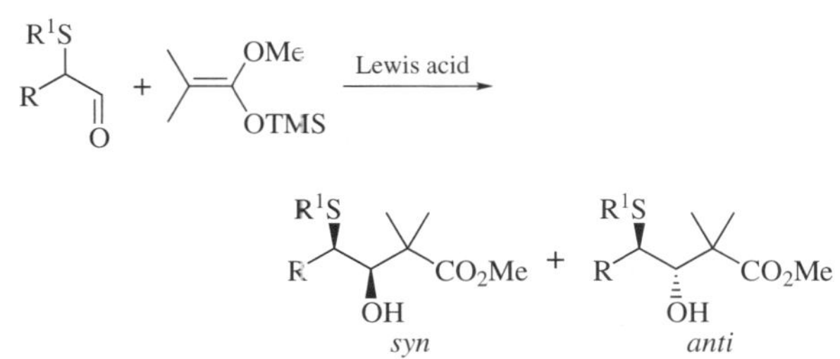

(4)

Additions to Nitriles. $SnCl_4$-promoted addition of malonates and bromomalonates to simple nitriles (not electron deficient) gives α,β-dehydro-β-amino acid derivatives (eq 5).[15] $SnCl_4$ is the Lewis acid of choice for the condensation of aroyl chlorides with sodium isocyanate, affording aroyl isocyanates in 70–85% yields.[16] Nonaromatic acyl chlorides react under more variable reaction conditions.

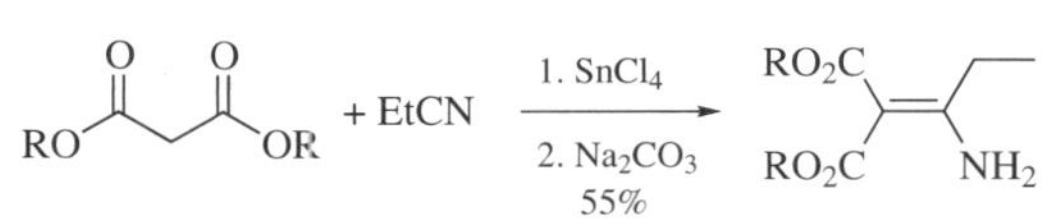

(5)

Hydrochlorination of Allenic Ketones. $SnCl_4$ is also a source for generating chloride anions which form new carbon–chlorine

bonds. This occurs through a ligand exchange pathway which has been exploited in the formation of β-chloro enones from conjugated allenic ketones (eq 6).[17] Yields range from 36–82% with complete selectivity for the *trans* geometry. A variety of substituents (R^1, R^2) can be tolerated including aryl, rings, and alkoxymethyl groups (R^1).

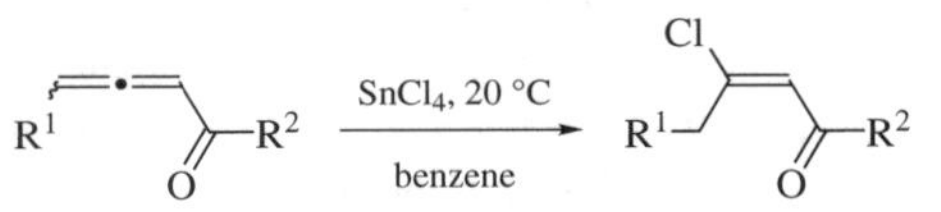

(6)

Glycosylations. The reaction of glycofuranosides having a free hydroxyl group at C-2 with functionalized organosilanes, in the presence of $SnCl_4$, provides *C*-glycosyl compounds in high stereoselectivity (eq 7).[18] Organosilanes such as 4-(chlorodimethylsilyl)toluene, chlorodimethylvinylsilane, Allyltrimethylsilane, and allylchlorodimethylsilane are effective reagents. The presence of a leaving group on the silane is essential for good selectivity since the reaction proceeds intramolecularly through a 2-*O*-organosilyl glycoside. The availability of furanosides in the ribo, xylo, and arabino series make this reaction valuable for the stereoselective synthesis of *C*-furanosides. Regioselective glycosylation of nitrogen-containing heterocycles is also effectively promoted by $SnCl_4$, and has been used in the synthesis of pentostatin-like nucleosides, such as (**1**).[19]

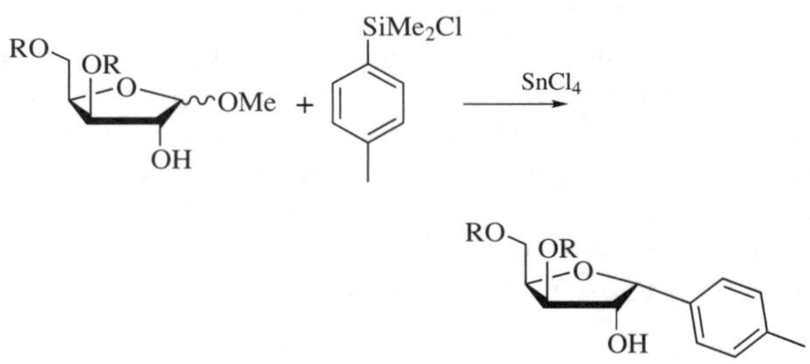

(7)

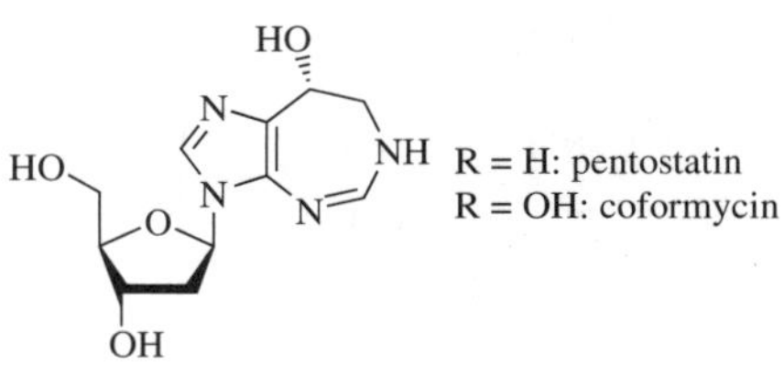

(**1**)

Selective De-*O*-benzylation. Regioselective de-*O*-benzylation of polyols and perbenzylated sugars is achieved with organotin reagents or other Lewis acids.[20,21] The equatorial *O*-benzyl group of 1,6-anhydro-2,3,4-tri-*O*-benzyl-β-D-mannopyranose is selectively cleaved by $SnCl_4$ or $TiCl_4$ (eq 8).[2] The equatorial *O*-benzyl group is also selectively cleaved when one of the axial *O*-benzyl groups is replaced by an *O*-methyl group. The 2-*O*-benzyl group of 1,2,3-tris(benzyloxy)propane is selectively cleaved (eq 9), but no debenzylation is observed with 1,2-bis(benzyloxy)ethane.

Rearrangement of Allylic Acetals. Lewis acid-promoted ($SnCl_4$ or Diethylaluminum Chloride) rearrangements of allylic acetals provide substituted tetrahydrofurans.[22] Upon addition of Lewis acid, (**2**) rearranges to the *all-cis* furan (**3**) (eq 10). No racemization is observed with optically active allylic acetals; however, addition of KOH completely epimerizes the furan–carbonyl bond, as does quenching at room temperature. Acetals successfully undergo similar rearrangement provided the alkene is substituted. Completely substituted tetrahydrofurans are synthesized stereoselectively (<97% ee) by the rearrangement of disubstituted allyl acetals (eq 11). This reaction is related to the acid-catalyzed rearrangements of 5-methyl-5-vinyloxazolidines to 3-acetylpyrrolidines, which involves an aza-Cope rearrangement and Mannich cyclization.[23]

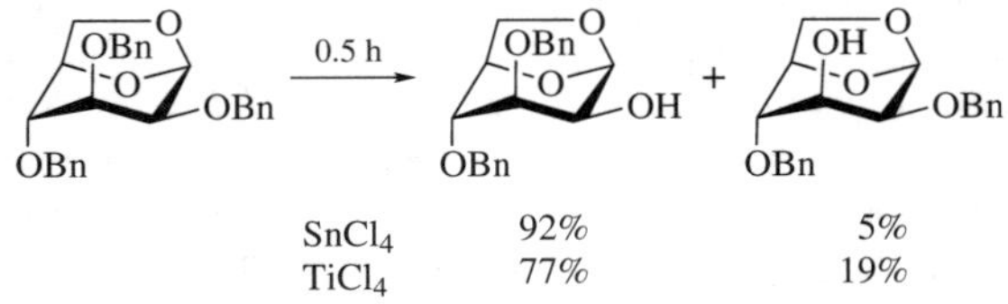

(8)

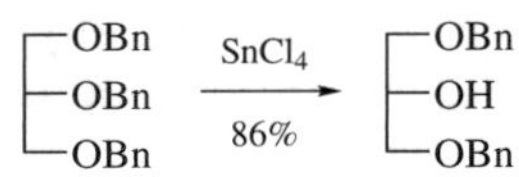

(9)

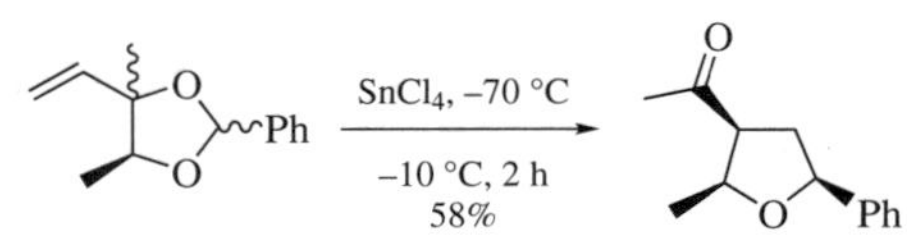

(**2**) (**3**)

(10)

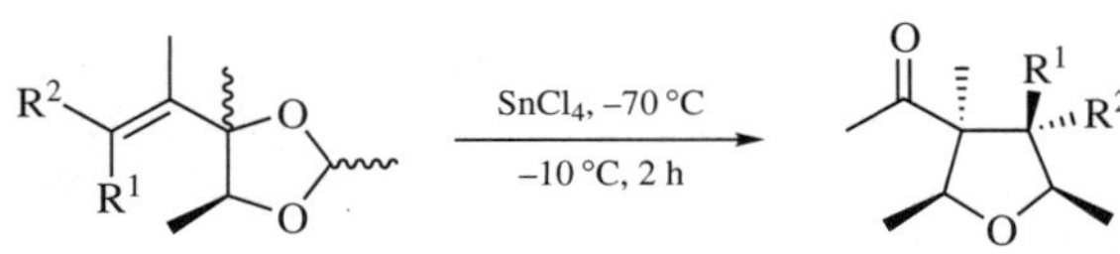

(11)

R^1 = Me, R^2 = H; 90%
R^1 = H, R^2 = Me; 73%

The rearrangement is also useful for furan annulations, through enlargement of the starting carbocycle.[24] Thus addition of $SnCl_4$ to either diastereomer of the allylic acetal (**4**) produces the *cis*-fused cycloheptatetrahydrofuran (**5**) in 48–76% yield (eq 12). Acetals derived from *trans*-diols rearrange to the same *cis*-fused bicyclics in higher yield. The stereochemistry of a terminal alkene is transmitted to the C-3 carbon of the bicyclic products (eq 13). Rearrangements of acetals require substitution at the internal alkene carbon.

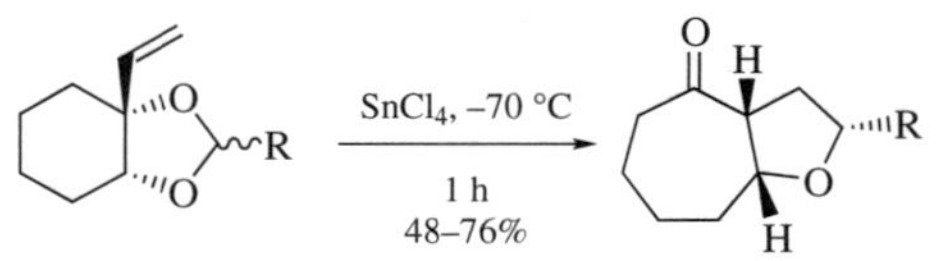

(**4**) (**5**)

(12)

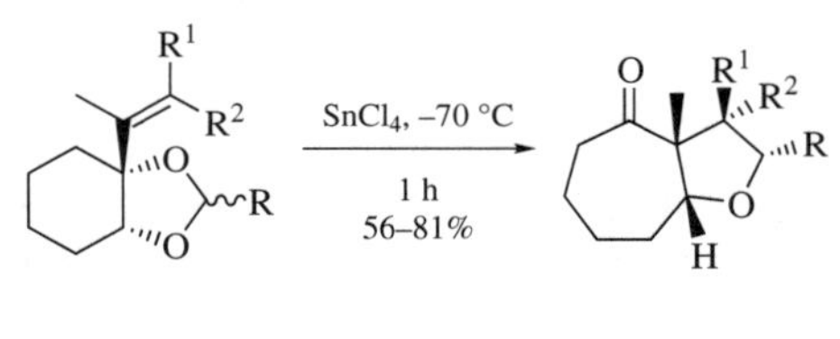

(13)

α-*t*-Alkylations. $SnCl_4$-promoted α-*t*-alkylations of alkenyl β-dicarbonyl compounds is a particularly useful cyclization reaction.[25] Cyclization occurs through initial formation of a stannyl enol ether, followed by protonation of the alkene to form a carbocation which undergoes subsequent closure (eq 14). The analogous α-*s*-alkylation reactions are best catalyzed by other Lewis acids.

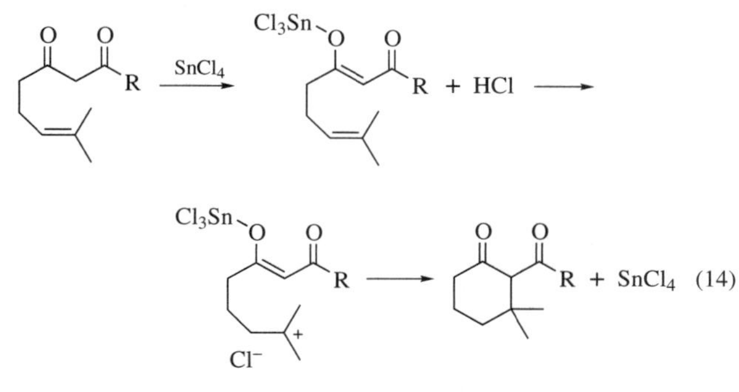

This reaction is useful for cyclizations involving 6-*endo*-trigonal (eq 14) and 'allowed' 7-*endo* trigonal processes (eq 15), but not for those involving 5-*endo* trigonal processes (eq 16). These observations are consistent with the Baldwin rules.

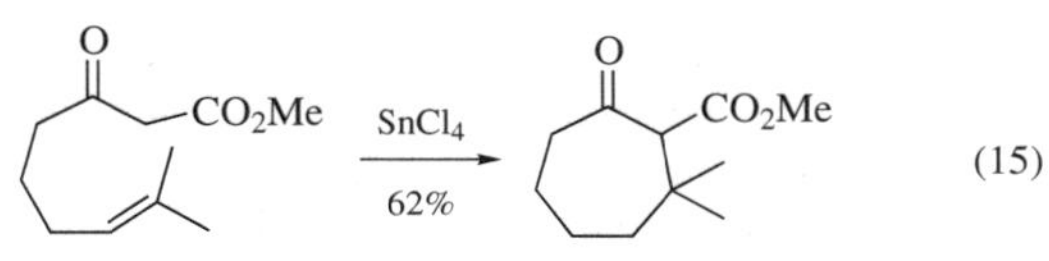

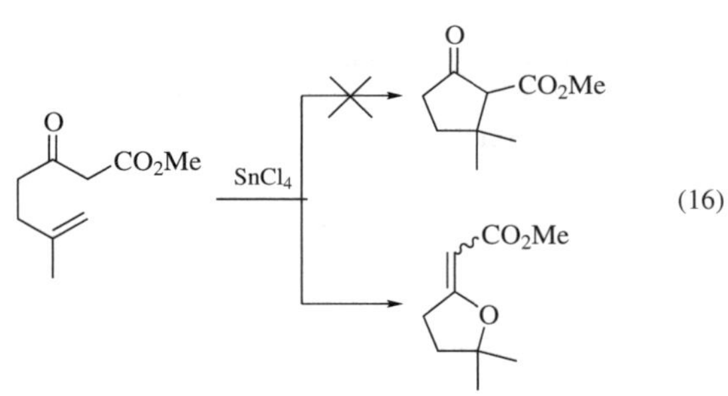

Reactions involving 4- and 6-*exo* trigonal cyclizations result in poor yields or undesired products, while those involving 5-*exo* trigonal cyclizations produce higher yields (eq 17). This synthetic strategy can also be used to form bicyclic and spiro compounds (eqs 18 and 19).

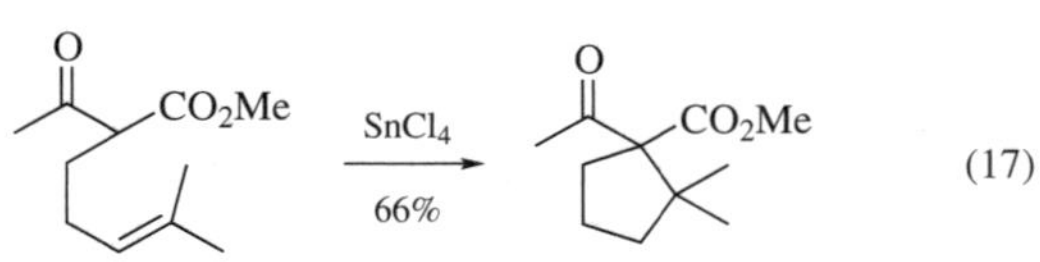

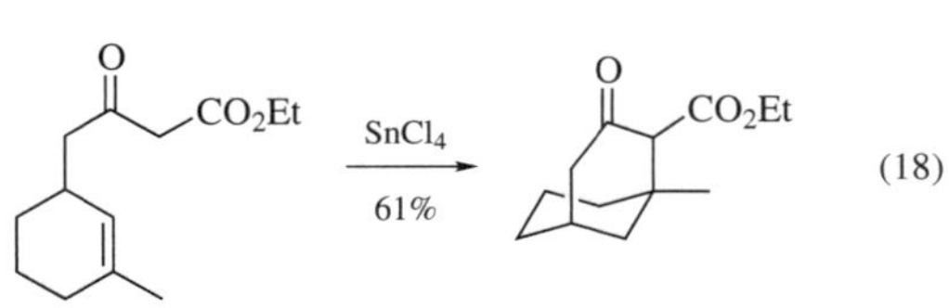

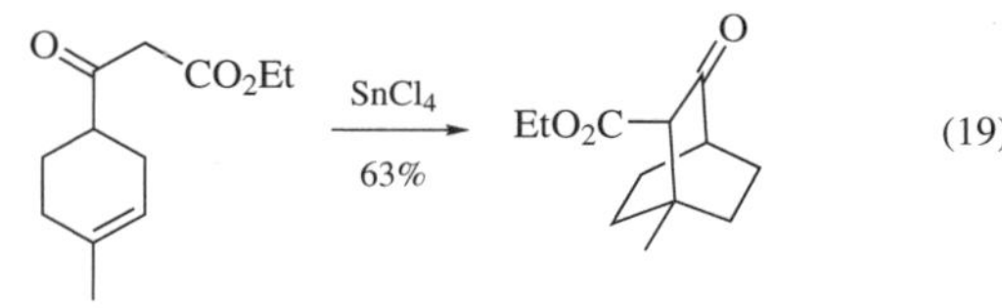

Alkene Cyclizations. Cationic cyclizations of polyenes, containing initiating groups such as cyclic acetals, are promoted by $SnCl_4$ and have been utilized in the synthesis of *cis*- and *trans*-decalins, *cis*- and *trans*-octalins, and tri- and tetracyclic terpenoids and steroids.[26] In most instances, *all-trans*-alkenes yield products with *trans,anti,trans* stereochemistry (eq 20), while *cis*-alkenes lead to *syn* stereochemistry at the newly formed ring junctions. The stereoselectivity of polyene cyclizations are often greatly diminished when the terminating alkene is a vinyl group rather than an isopropenyl group. Acyclic compounds which contain terminal acyclic acetals and alkenes or vinylsilanes can be cyclized in a similar fashion to yield eight- and nine-membered cyclic ethers (eq 21).[27]

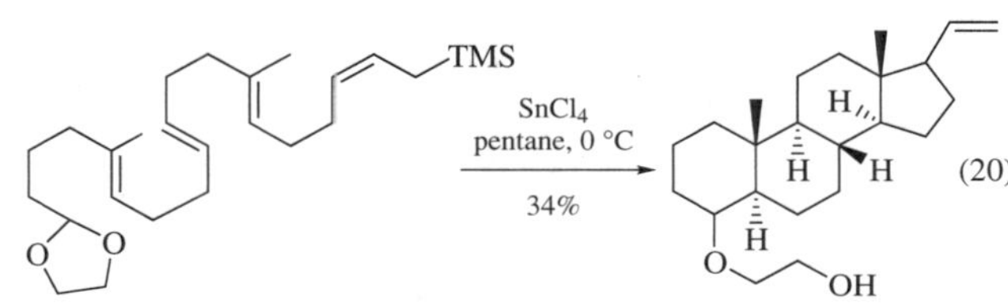

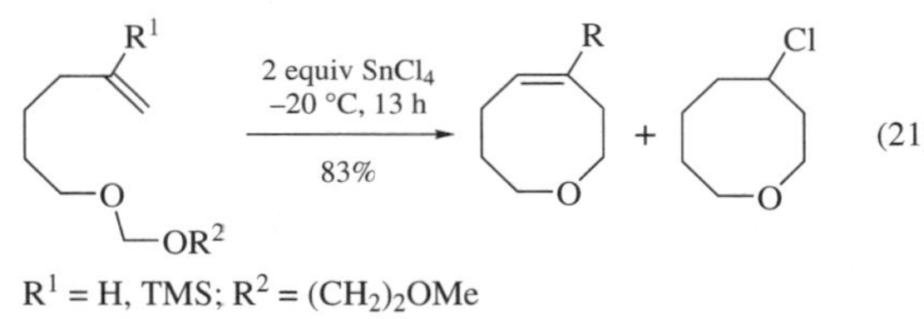

The analogous cyclization of chiral imines occurs in high yields (75–85%) with good asymmetric induction (36–65% ee).[28] For example, the cyclization of aldimine (**6**), derived from methyl citronellal, using $SnCl_4$ affords only the *trans*-substituted aminocyclohexane (**7**) in high yield (eq 22). *Exo* products are formed exclusively or preferentially over the thermodynamically favored *endo* products.

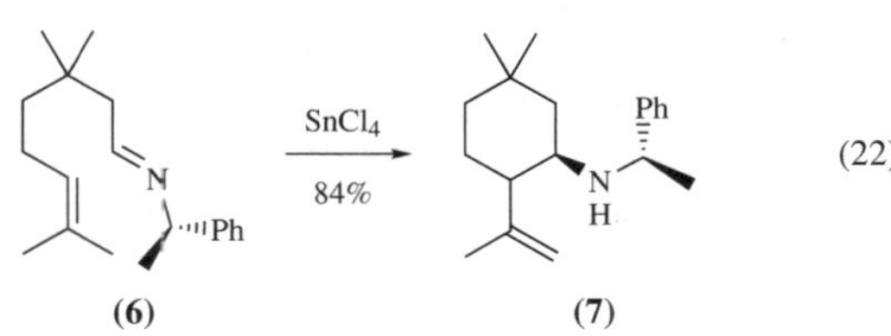

$SnCl_4$-induced cyclizations between alkenes and enol acetates result in cycloalkanes or bicycloalkanes in high yield (eq 23). It is interesting to note that the TMSOTf-catalyzed reaction can yield fused products rather than bicyclo products. Alkenic carboxylic esters, allylic alcohols, sulfones, and sulfonate esters are also cyclized in the presence of $SnCl_4$; however, alkenic oxiranes often cyclize in poor yield.[26a]

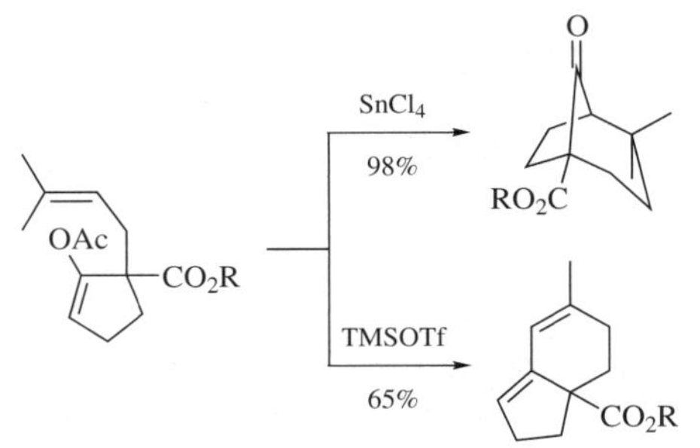

(23)

$SnCl_4$ is also effective in the opening of cyclopropane rings to produce cationic intermediates useful in cyclization reactions. For example, the cyclization of aryl cyclopropyl ketones to form aryl tetralones, precursors of aryl lignan lactones and aryl naphthalene lignans, is mediated by $SnCl_4$ (eq 24).[29] The reaction is successful in nitromethane, but not in benzene or methylene chloride. Analogous cyclizations with epoxides result in very low yields (2–5%).

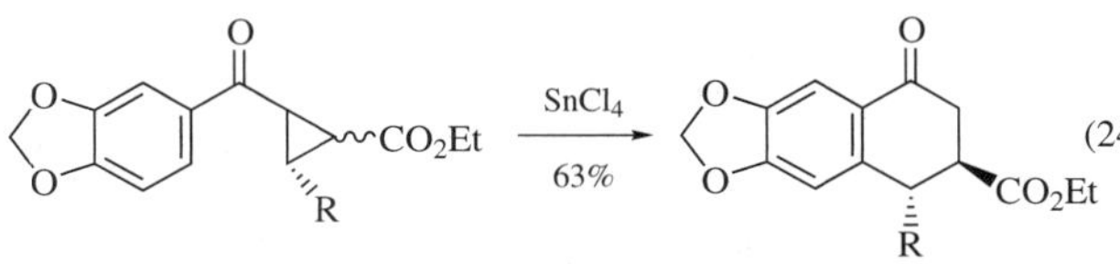

(24)

Polymerization. Cationic polymerizations are catalyzed by $SnCl_4$ and other Lewis acids (eq 25). Propagation is based upon the formation of a cationic species upon complexation with $SnCl_4$.[30] Radical pathways are also possible for polymer propagation.[31]

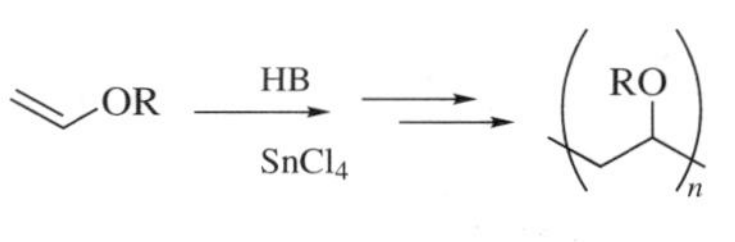

(25)

Diels–Alder Reactions. Diels–Alder reactions are enhanced through the complexation of dienophiles or dienes by Lewis acids.[32] Furthermore, Lewis acids have been successfully employed in asymmetric Diels–Alder additions.[33] Although $SnCl_4$ is a useful Lewis acid in Diels–Alder reactions, in most instances titanium or aluminum Lewis acids provide higher yields and/or selectivities. The stereoselectivity in Lewis acid-promoted Diels–Alder reactions between chiral α,β-unsaturated *N*-acyloxazolidinones shows unexpected selectivities as a function of the Lewis acid (eq 26).[34] Optimum selectivity is expected for chelated intermediates, yet both $SnCl_4$ and $TiCl_4$ perform poorly relative to Et_2AlCl (1.4 equiv). The formation of the $SnCl_4$–*N*-acyloxazolidinone chelate has been confirmed by solution NMR studies.[35] These data suggest that other factors such as the steric bulk associated with complexes may contribute to stereoselectivity.

In Lewis acid-promoted Diels–Alder reactions of cyclopentadiene with the acrylate of (*S*)-ethyl lactate, good diastereofacial and *endo*/*exo* selectivity are obtained with $SnCl_4$ (84:16; *endo*/*exo* = 18:1) and $TiCl_4$ (85:15; *endo*/*exo* = 16:1).[36] It is interesting to note that boron, aluminum, and zirconium Lewis acids give the opposite diastereofacial selectivity (33:67 to 48:52). Competing polymerization of the diene is observed in methylene chloride, particularly with $TiCl_4$, but not in solvent mixtures containing *n*-hexane.

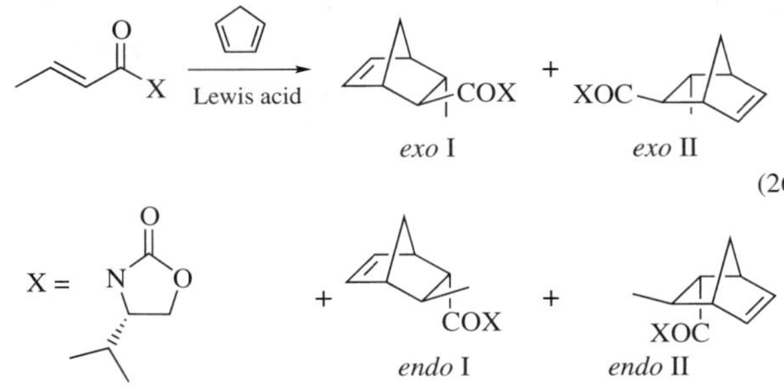

(26)

Lewis acid	conv %	Σ*endo*:Σ*exo*	*endo* I:*endo* II
1.1 equiv $SnCl_4$	70	14.9	3.1
1.1 equiv $TiCl_4$	100	9.9	2.7
1.4 equiv Et_2AlCl	100	50.0	17

Cycloalkenones generally perform poorly as dienophiles in Diels–Alder reactions but their reactivity can be enhanced by Lewis acids.[37] $SnCl_4$ is effective in promoting the Diels–Alder reaction between simple 1,3-butadienes, such as isoprene and piperylene, and cyclopentenone esters. For example, the $SnCl_4$-promoted cycloaddition between (**8**) and isoprene is completely regioselective, providing the substituted indene in 86% yield (eq 27).[38] However, cycloaddition does not occur in the presence of $SnCl_4$ when the diene contains an oxygen-bearing substituent such as an alkoxy or siloxy group. In these cases, as is generally true for the Diels–Alder reactions of cycloalkenones, other Lewis acids are more effective. For example, $SnCl_4$-promotion of the cycloaddition between (**8**) and 3-methyl-2-(*t*-butyldimethylsiloxy) butadiene yields 37% of the desired product, while Zinc Chloride provides a 90% yield. When furan or 2-methyl-1-alkylsiloxybutadiene are utilized as dienes, only decomposition of the starting material is observed with $SnCl_4$.

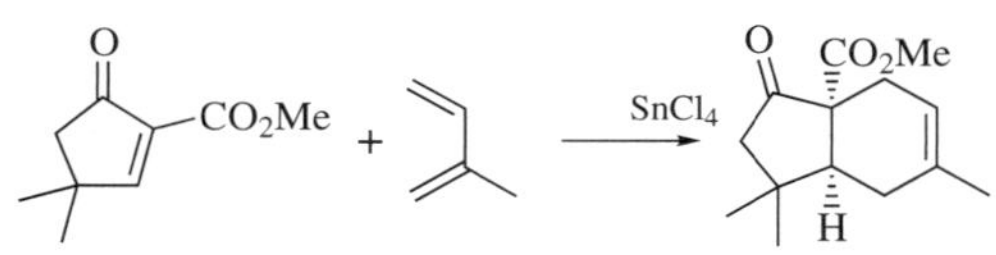

(**8**)

(27)

The Lewis acid-promoted Diels–Alder reaction has been employed in the assembly of steroid skeletons.[39] The cycloaddition reaction between a substituted bicyclic diene and 2,6-dimethylbenzoquinone produces two stereoisomers in a 1:5 ratio with a yield of 83% when $SnCl_4$ is used in acetonitrile. $TiCl_4$ provides slightly higher selectivities (1:8) but lower yield (70%) (eq 28).

When the dienophile *N*-α-methylbenzylmaleimide (**9**) is reacted with 2-*t*-butyl-1,3-butadiene in the presence of Lewis acids, cycloadducts (**10**) and (**11**) are formed (eq 29).[40] While $SnCl_4$ provides (**10**) and (**11**) in a 5:1 ratio, $TiCl_4$ and $EtAlCl_2$ both provide a 15:1 ratio. Polymerization of the diene competes with adduct formation under all conditions.

[4 + 3] Cycloadditions. Oxyallyl cations,[41] which react as C_3 rather than C_2 components in cyclization reactions, are generated by the addition of $SnCl_4$ to substrates which contain silyl enol ethers which are conjugated with a carbonyl moiety. Thus 2-(trimethylsiloxy)propenal undergoes cyclization with cyclopentadiene or furan (eq 30).[42] Substituted 1,1-dimethoxyacetones also form these intermediates and undergo subsequent cyclizations (eq 31).[43] This method complements the usual synthesis of oxyallyl cations involving reductive elimination of halogens from halogenated ketones or electronically equivalent structures.[44]

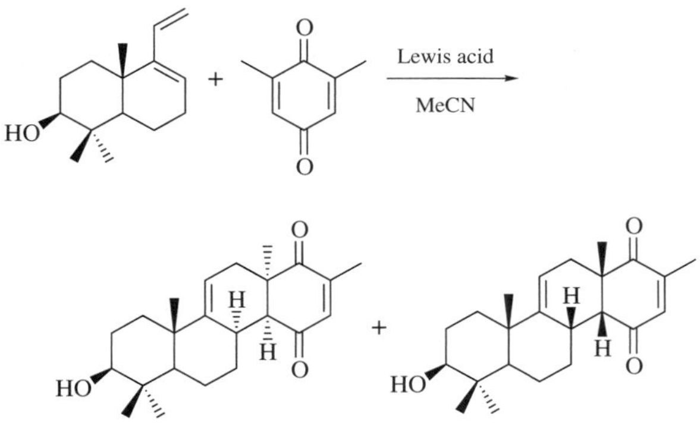

SnCl$_4$ 1:5
TiCl$_4$ 1:8

(28)

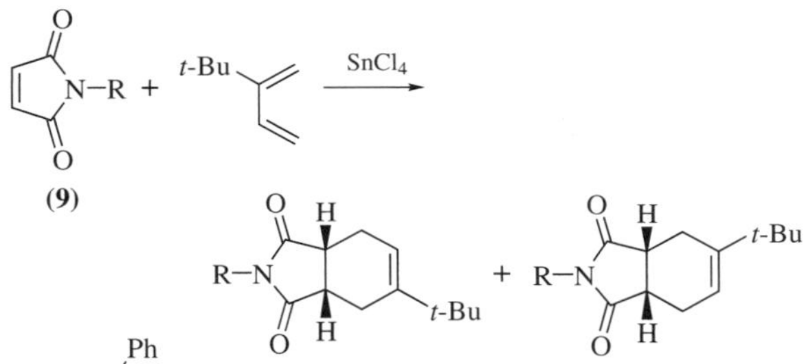

(9) (10) (11)

(29)

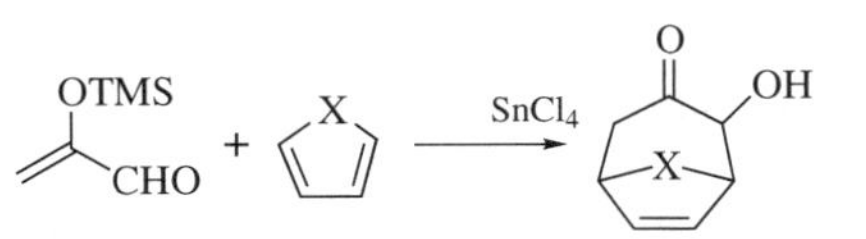

X = O, CH$_2$

(30)

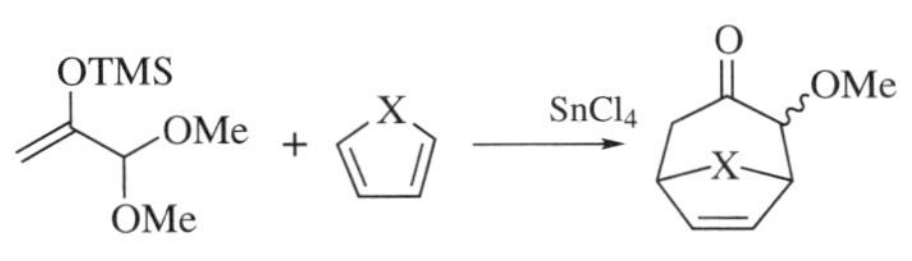

(31)

[3 + 2] Cycloadditions. Lewis acid-mediated [3 + 2] cycloadditions of oxazoles and aldehydes or diethyl ketomalonate have been observed using organoaluminum and Sn^{IV} Lewis acids.[45] The reactions are highly regioselective, with stereoselectivity extremely dependent upon Lewis acid (eq 32). For example, the (BINOL)AlMe-promoted reaction between benzaldehyde and the oxazole (**12**) provides the oxazoline with a *cis*/*trans* ratio of 98:2. The selectivity is reversed with SnCl$_4$ which provides a *cis*/*trans* ratio of 15:85. *trans*-5-Substituted 4-alkoxycarbonyl-2-oxazolines are synthesized under thermodynamic conditions in the aldol reaction of isocyanoacetates with aldehydes.[46]

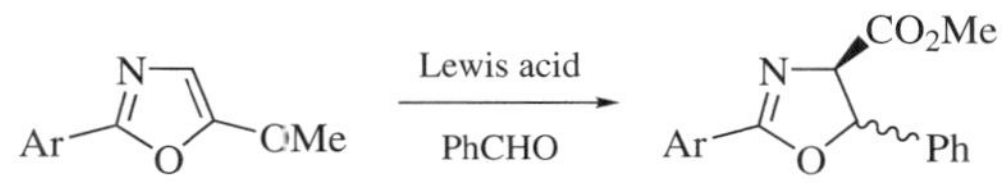

(12)

(32)

[2 + 2] Cycloadditions. The regioselectivity in the cycloaddition reactions of 2-alkoxy-5-allyl-1,4-benzoquinones with styrenes is controlled by the choice of Ti^{IV} or SnCl$_4$ Lewis acids (eq 33).[47] The use of an excess of TiCl$_4$ or mixtures of TiCl$_4$ and Ti(O-*i*-Pr)$_4$ produces cyclobutane (**13**) as the major or exclusive product, while SnCl$_4$ promotion with one equivalent of Lewis acid results in the formation of (**14**) only. These reactions represent a classic example of the mechanistic variability often associated with seemingly modest changes in Lewis acid.

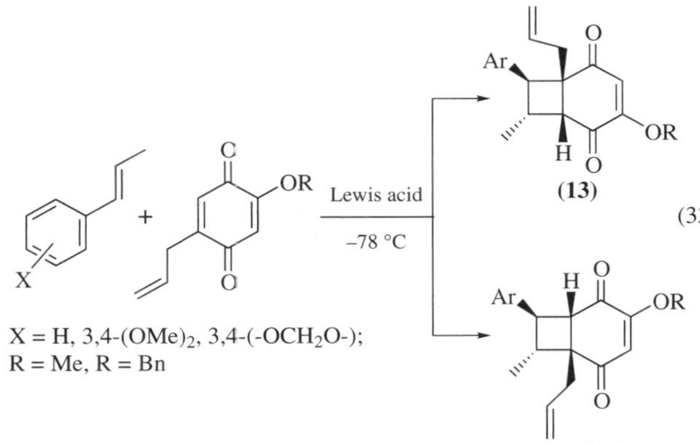

X = H, 3,4-(OMe)$_2$, 3,4-(-OCH$_2$O-); R = Me, R = Bn

(13) (14)

(33)

Ene Reactions. The Lewis acid-catalyzed ene reaction is synthetically useful methodology for forming new carbon–carbon bonds.[48] Ene reactions utilizing reactive enophiles such as formaldehyde and chloral can be promoted by SnCl$_4$. SnCl$_4$ also enhances intramolecular ene reactions, such as the cyclization of (**15**) which produces the α-hydroxy δ-lactone in 85% yield (eq 34).[49] The ene cyclization of citronellal to give isopulegol has also been reported.[50] Proton scavenging aluminum Lewis acids such as RAlCl$_2$ are most often used in ene reactions to eliminate proton-induced side reactions.

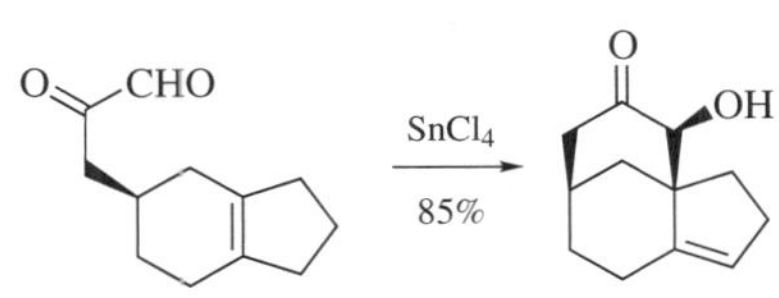

(15)

(34)

First Update

Calvin J. Chany II
Barnstead International, Dubuque, IA, USA
and
Rush University Medical Center, Chicago, IL, USA

Nucleophilic Additions to Aldehydes. Stereodirected additions of allyltributylstannanes to aldehydes with promotion by

$SnCl_4$ are possible when the stannane also contains an alkoxy group; for example, the reaction of aldehydes with 4-alkoxypent-2-enenyl tributylstannanes (**16**) afford high yields of 1,5-*syn* addition products.[51–53] Typical reactions involve combining the stannane with $SnCl_4$ at −78 °C for 5 min in dichloromethane followed by the addition of the aldehyde. The use of other Lewis acids led to competing decomposition.[53,54] With benzyl protection the stereofacial selectivity was found to be independent of the aldehyde used whether achiral or chiral (eq 35).

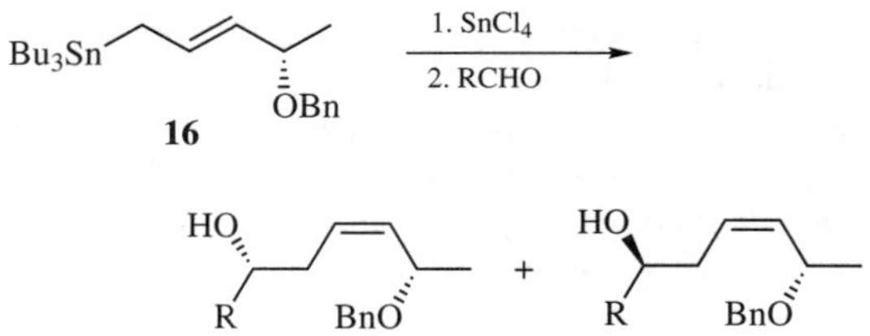

(35)

syn:anti 95:5

α-Hydroxyaldehydes, however, are an exception to this rule. Addition of *S*-**16** to *R*-2-hydroxypropanal proceeds with normal *syn:anti* stereofacial selectivity, whereas addition to the *S*-aldehyde yields a mixture of *syn:anti* products.[52,53,55] The loss of stereocontrol in this case is attributed to the "mismatching" of the chiral aldehyde and chiral stannane. The reactions are conceived to proceed via a coordinated tin complex with the size of the alkoxy protecting group dictating the intermediacy of 5- or 6-coordinate tin complexes.[52,56,57] Changing the 4-alkoxy protecting group from Bn (**16**) to the bulky TBDMS (**17**) leads to a preference for 1,5-*anti* addition products with *E* double bond formation and some loss of selectivity (eq 36).

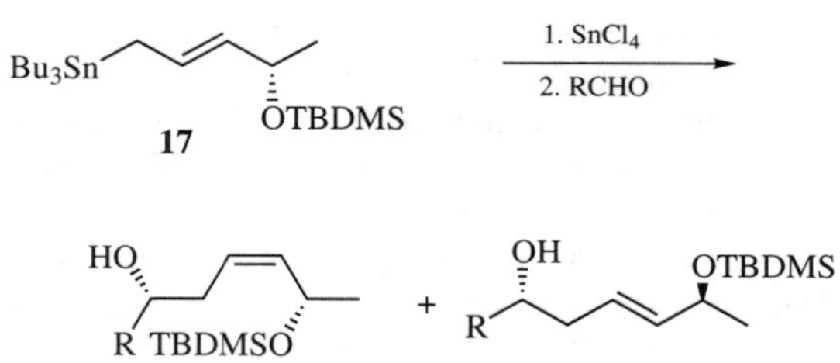

(36)

syn:anti 30:60

Allylstannane (**16**) also adds to activated imines with high stereofacial selectivity with preference for *anti* addition and the formation of the *E* double bond, while addition of the TBDMS protected stannane (**17**) takes place with the opposite stereofacial selectivity (eq 37).[56,58]

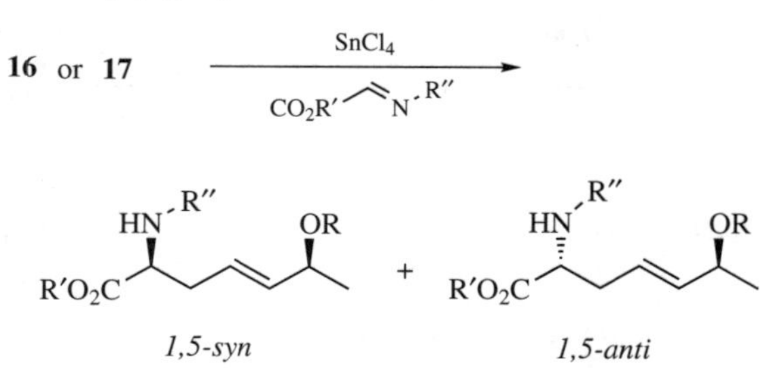

(37)

Stannane	R	R′	R″	Yield (%)	*1,5-syn*	*1,5-anti*
16	Bn	C_4H_9	$CHPh_2$	79	10	90
17	TBDMS	C_4H_9	$CHPh_2$	91	75	25

Directed additions involving 1,6- (eq 38),[59] 1,7- (eq 39),[60] and 1,8-[61] stereoinduction have also been reported. The 1,8-induction is not direct, but has foundation on a 1,5-asymmetric induction to afford an intermediate that when derivatized is subjected to an Ireland-Claisen rearrangement. Interestingly the 1,6- and 1,7-directed additions give better or more consistent reaction products with the use of tin(IV) bromide (eqs 38 and 39). In these two examples the hydroxyl groups in the starting stannanes were unprotected.

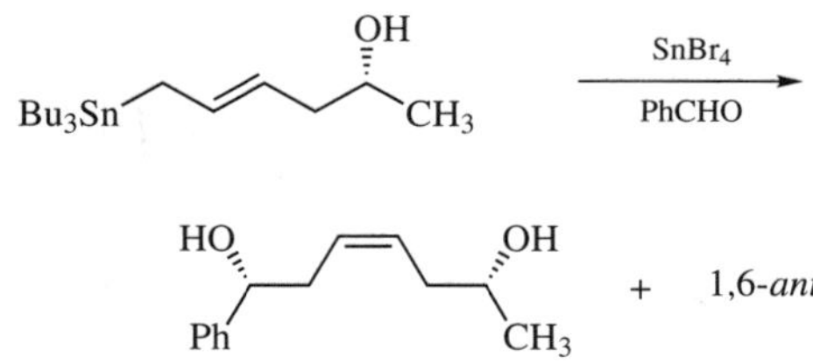

(38)

syn:anti 96:4

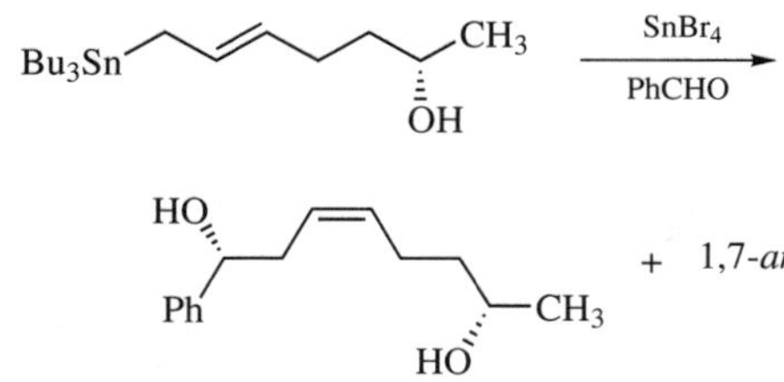

(39)

syn:anti 90:10

The usefulness of these directed additions is demonstrated in the synthesis of the macrodiolide pamamycin-607,[62] which is reported to have interesting antifungal properties. Further exploitation is demontrated by the use of L-quebrachitol as a chiral auxilary in $SnCl_4$ mediated additions of silyl enol ethers to α-keto esters affording chiral teriary alcohols in high yields and with excellent diasteriomeric excess.[63]

Trimethylsilylalk-2-enes also add to aldehydes in the presence of $SnCl_4$, however, the transmetallation is considerably slower when compared to the tributylstannanes (1–2 h vs. 5 min at −78 °C), which leads to lower yields and loss of stereoselectivity (eq 40).[64]

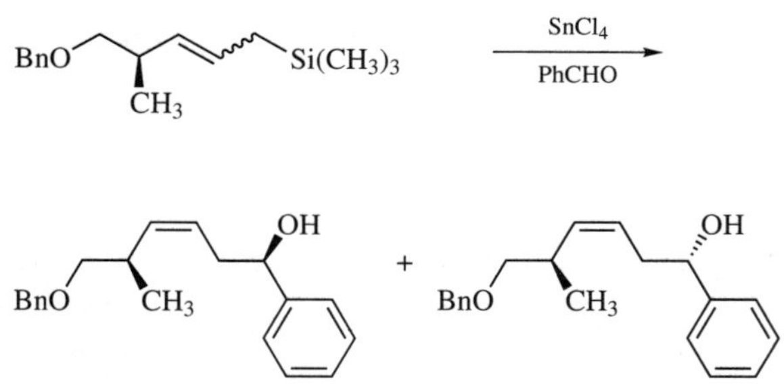

(40)

Reaction time	Yield (%)	*syn/anti*
20 min	13	23/77
1 h	65	15/85
2 h	67	14/86

The coupling of 4-alkoxy-2-trimethylsilyl-2-enyl(tributyl) stannanes with aldehydes affords a template for the stereocontrolled preparation of 1,3,5-triols with excellent yields.[65] Addition reactions have also been reported in the functionalization of azetidinone-2-carbaldehydes.[66–68] Allylsilane and allylstannane additions to these aldehydes both give high yields of the *syn* addition products when the reactions are catalyzed by Lewis acids (LA) (eq 41).

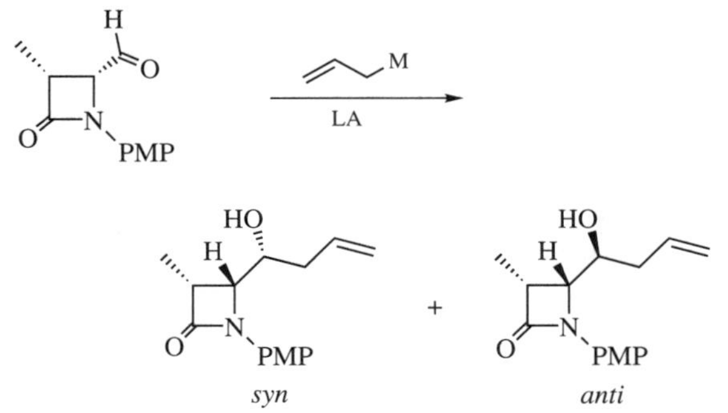

(41)

M	LA	*syn*	*anti*	Yield (%)
$SiMe_3$	$SnCl_4$	92	8	79
$SnBu_3$	BF_3-Et_2O	91	9	75

Similar reactions were observed with of azetidin-2,3-diones, which gave high yields of the homolallylic alcohols (eq 42).[67,68]

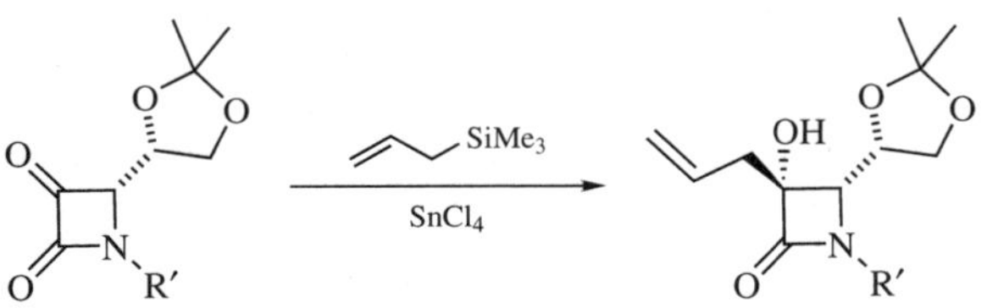

(42)

Addition to Nitriles. The $SnCl_4$ mediated addition of β-dicarbonyl compounds to nitriles affords dicarbonyleneamine intermediates, which offer entry into substituted heterocyclic cores. Reactions involving *ortho*-aminobenzonitriles lead to the formation of either 4-aminoquinolones (eq 43) or 4-aminoquinolines (eq 44) in fair (21–34%) to modest yields (59–79%).[69]

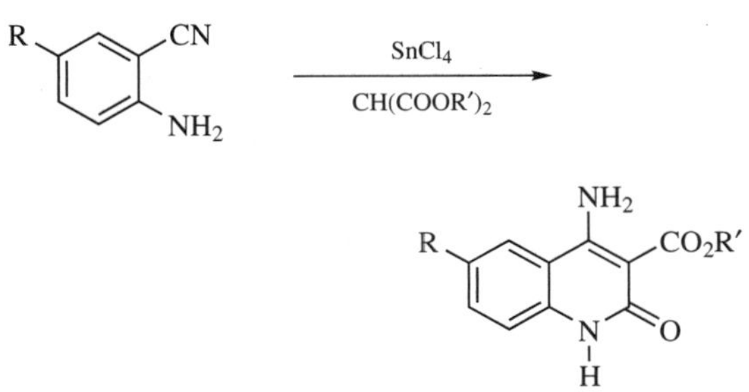

(43)

Similarly, reactions of β-enaminonitriles offer entry into the 4-aminopyridone (eq 45) or 4-aminopyridine (eq 46) heterocyclic cores.[70]

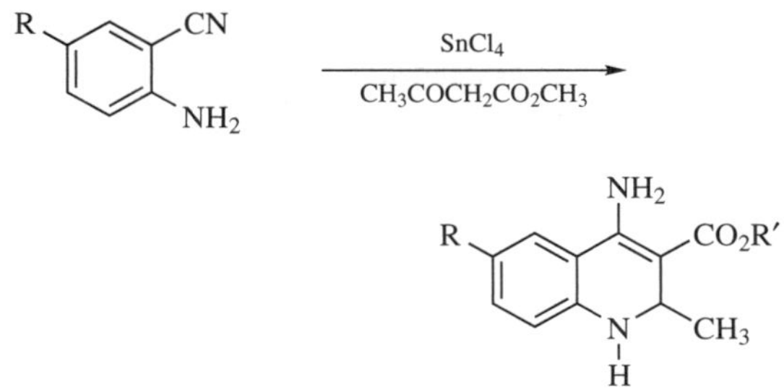

(44)

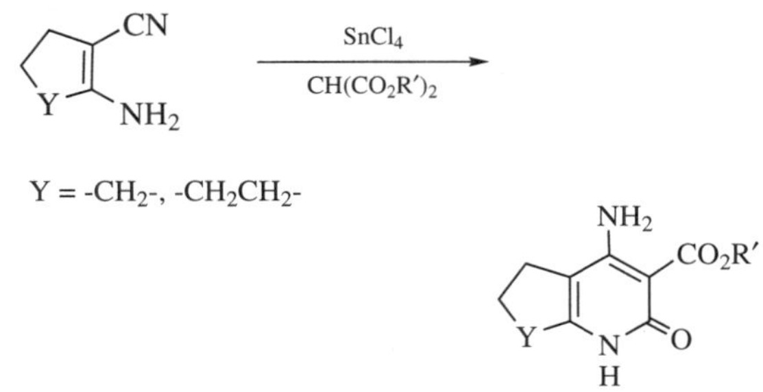

(45)

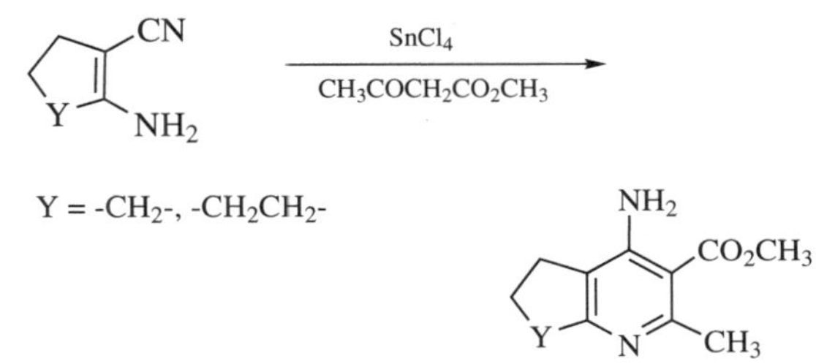

(46)

When condensed with β-dicarbonyl compounds, other β-enaminonitriles based on the pyrazole nucleus afford pyrazolo[3,4-*b*] pyridines in low (20%, eq 47) to modest (46–60%, eq 48) yields.[71]

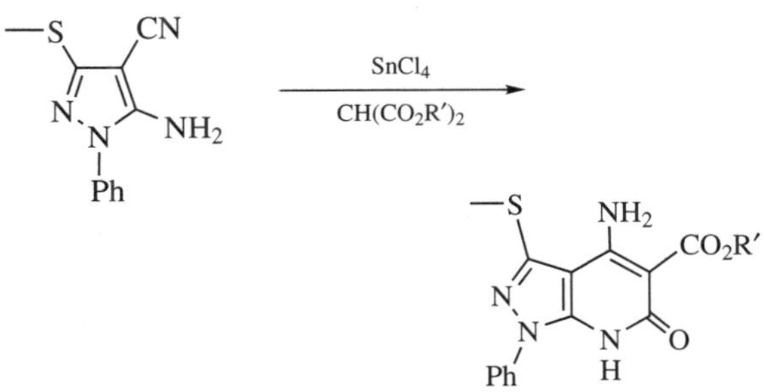

(47)

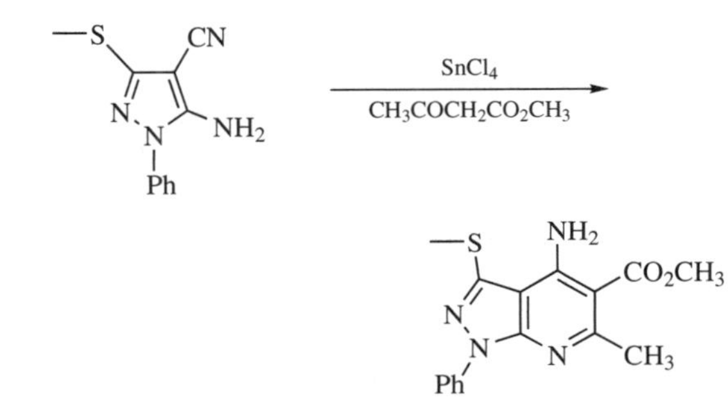

(48)

Unsaturated nitriles, e.g., acrylonitrile, readily react with methyl acetoacetate in the presence of $SnCl_4$ in a one-pot reaction leading to enamino ketoesters (eq 49), or substituted pyridine derivatives (eqs 50 and 51) depending on the stoichiometry, solvent, and temperatures.[72]

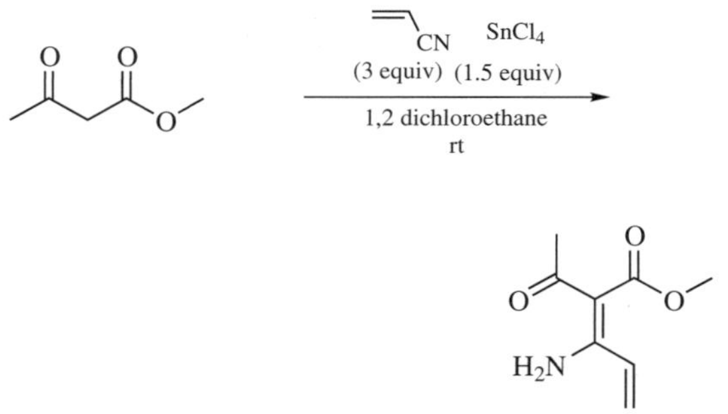

(49)

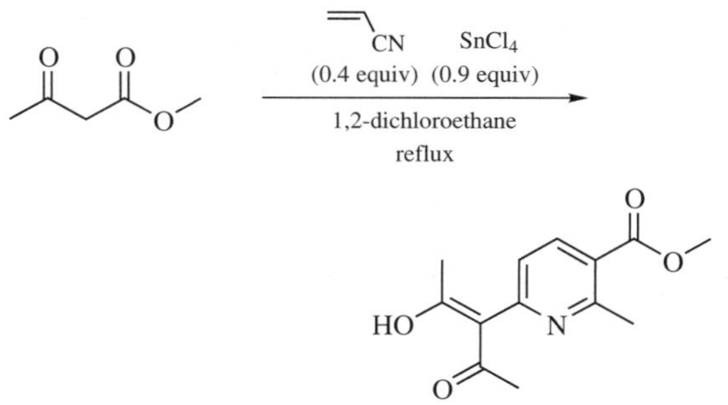

(50)

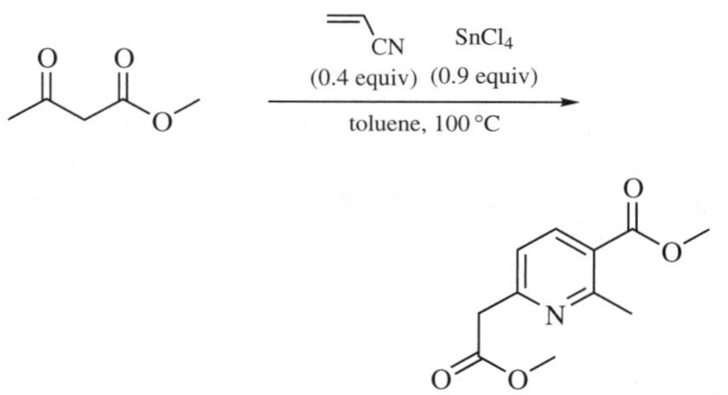

(51)

Cyanohydrins also react with β-dicarbonyl compounds in the presence of $SnCl_4$ to afford substituted 4-amino-2,5-dihydro-2-furanones (eqs 52 and 53).[73]

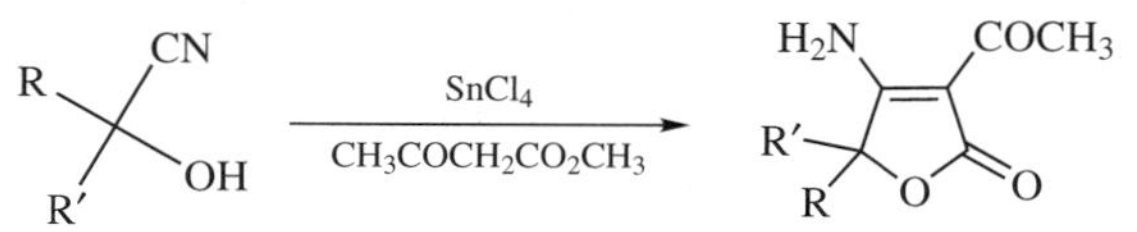

(52)

50–60%

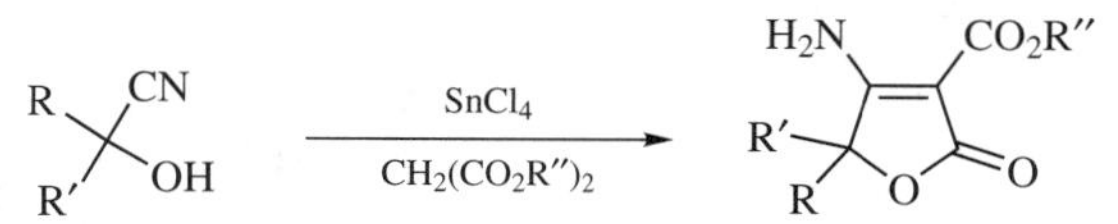

(53)

60–80%

Reactions of protected β-D-ribofuranosyl ketoesters (prepared from β-D-ribofuranosyl cyanide) with alkyl cyanoformates promoted by tin(IV) chloride afford high yields of enamino ketoester intermediates. Reaction with either acetamidine or benzylhydrazine leads to the formation of pyrazole or pyrimidine *C*-nucleosides with no anomerization.[74] A related series of substituted pyrazole *C*-nucleosides can be approached by the reaction of the same ribofuanosyl cyanide with methyl acetoacetate.[75]

Glycosylations. $SnCl_4$ promoted glycosylations take advantage of stable, readily available protected or partially protected carbohydrates. Treatment of 1-*O*-acetyl-2,3,5-tri-*O*-benzoyl-β-D-ribofuranose or peracetylated β-D-glucose with $SnCl_4$ and an alcohol in dichloromethane affords high yields of the desired glycosides with high anomeric purity (eqs 54 and 55).

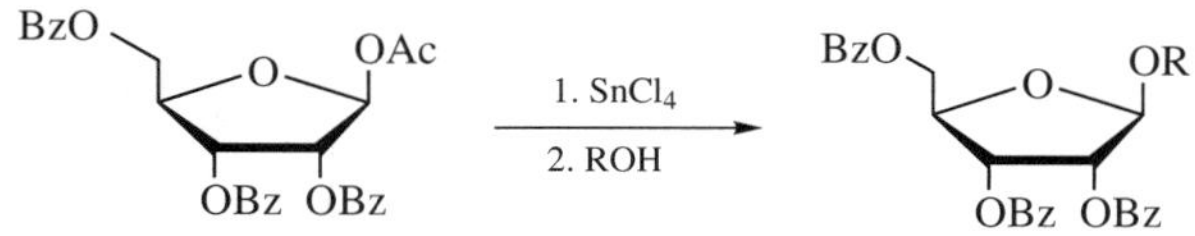

(54)

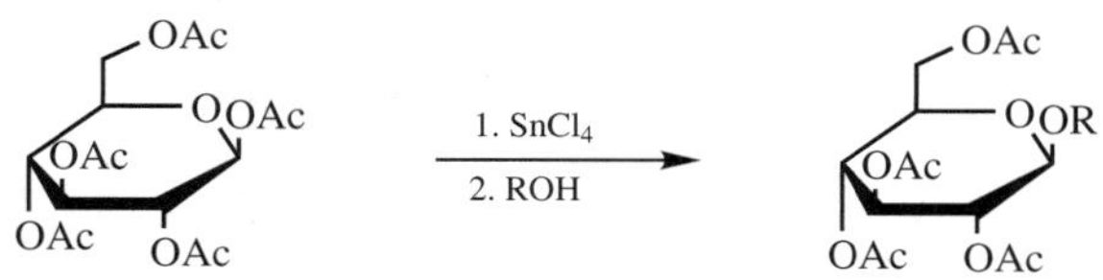

(55)

Participating groups on *O*-2 (e.g., acetyl or benzoyl) lead to a preference for 1,2-*trans* glycosides while nonparticipating groups (e.g., benzyl) afford 1,2-*cis* glycosides preferentially.[76] The order of addition is important for high anomeric selectivity when participating *O*-2 groups are present. In a typical reaction the glycosyl donor and $SnCl_4$ are combined at 0 °C followed by the slow addition of the appropriate glycosyl acceptor. Several groups have noted that combining the glycosyl acceptor and donor in solution followed by addition of $SnCl_4$ leads to mixtures of the anomeric glycosides.[77–80]

$SnCl_4$ mediated glycosylation enabled the coupling of a lactone-based glycosyl acceptor to a peracetylated donor, thereby affording entry to disaccharides with a furanose at the reducing end (eq 56).[81,82]

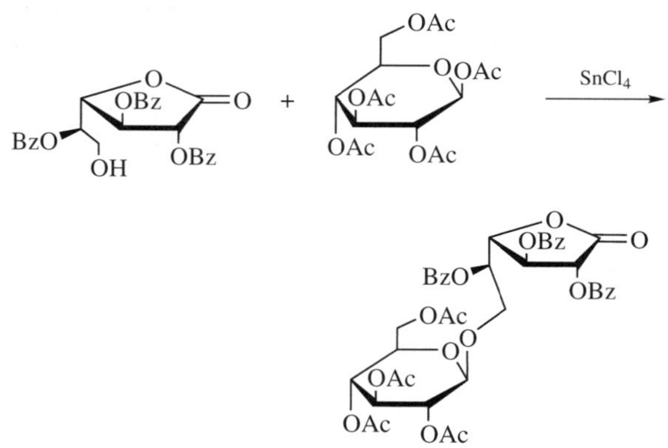

(56)

Attempts to use Koenigs-Knorr conditions to accomplish this coupling afforded significant amounts of the ortho ester, a common side reaction under Koenigs-Knorr conditions. $SnCl_4$ glycosylations have also been used to attach aglycons for enzyme kinetic studies (e.g., *p*-nitrophenol).[83] Couplings to a wide range of carbohydrate acceptor alcohols have been conducted with the successful formation of a diverse set of linkages types.[82,83] *N*-Acetylglucosamine has also been employed as a glycosyl acceptor leading to the synthesis of important disaccharides (eq 57).[84,85]

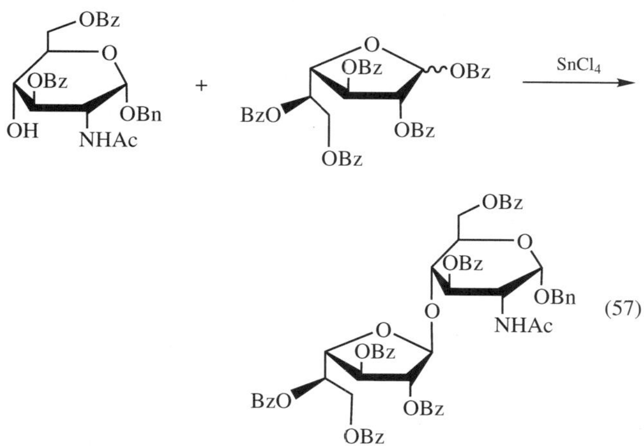

(57)

Further work[86–88] has led to $SnCl_4$ promoted couplings with TMS-protected acceptors affording α-glycosides. The coupling of 3β-cholestanyl trimethylsilyl ether to glucose was studied in regard to the Lewis acid. $GaCl_3$ and $SnCl_4$ were found to be the most effective Lewis acids screened in the presence of 20 mole % $AgClO_4$. In addition the effects of solvents were also studied with diethyl ether providing the highest yields and best stereoselectivity.[87,88] The combined catalyst system has been used for the effective synthesis of isomalto-oligosaccharides.[89,90] A unique glycosidation involving a 2-deoxy-azido-thioglycoside/iodosobenzene/$SnCl_4$-$AgClO_4$ system was used to prepare precursors to ^{3}H-labeled lipid A (eq 58).[91]

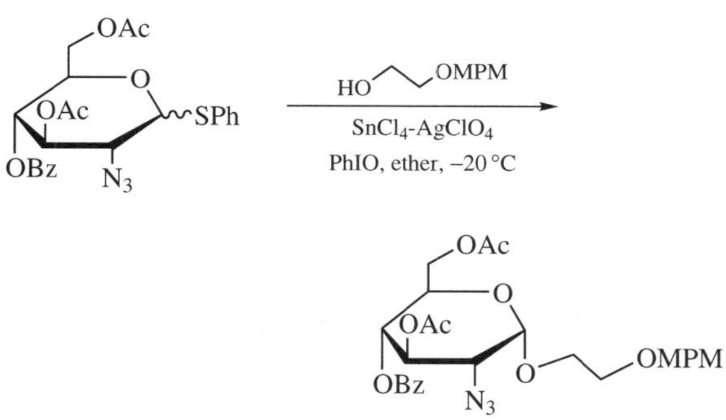

(58)

Glycosidation reactions involving nitrogen bases, yielding nucleosides, have also been investigated. Generally the nitrogen bases are converted into a silylated derivative before coupling.[92–111] Glycosyl donors can be either fully protected sugars[92–98,100,102–104,107,111–113] [e.g., 1,2,3,4,6-penta-*O*-acetyl β-D-glucose[92] (eq 59) or glycosyl halides (eq 60)[99,101,106,114]]. Reaction yields using 1-*O*-acetoxy donors were significantly better than those obtained by the use of glycosyl halides.

A few coupling reactions have been noted in which the unprotected nitrogen heterocycles are used directly.[110–113,115] In cases where several nucleophilic nitrogens are present, mixtures of regioisomers (about the nitrogen base ring) occur.[96,97,102,107] The preferred glycosidation occurs on the endocyclic nitrogens. An exception was found when 2-aminoimidazole (as its sulfate salt) was coupled with 1-*O*-acetyl-2,3,5-tri-*O*-benzoyl-β-D-ribofuranose in the presence of $SnCl_4$ and $Hg(CN)_2$ wherein the unexpected glycosidation of the exocyclic 2-amino group occurred (eq 61).[115]

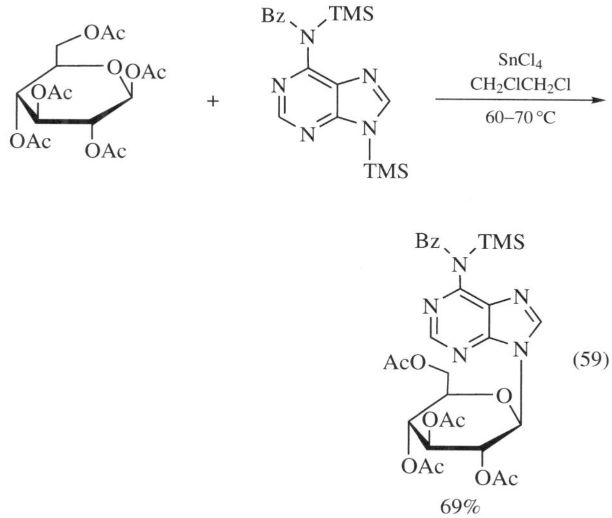

(59)

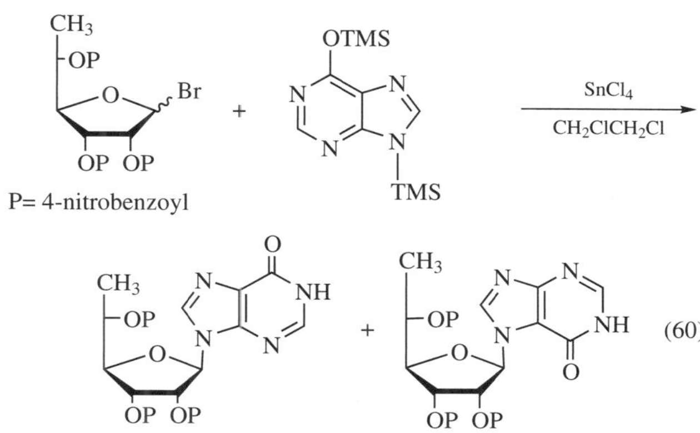

(60)

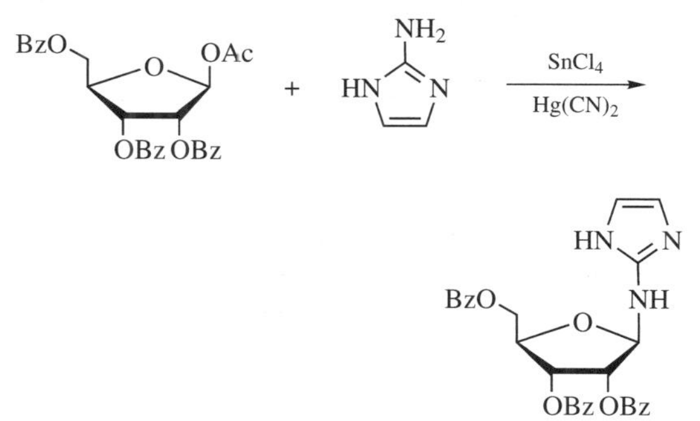

(61)

Reactions involving additions of nitrogen bases to glycals have also been reported.[108,109] Anhydro-carbohydrates have also been reported to function as glycosyl donors towards silylated acceptors.[105] Internal *C*-glycosylations are possible when activated 3-methoxybenzyl protecting groups are used. When methyl 2,3,4-tri-*O*-(3-methoxybenzyl)-β-D-ribofuranoside was treated with $SnCl_4$, instead of the simple internal Friedel-Crafts product, a double internal cyclization occurred leading to an 88% yield of a tetrahydro-[2]benzopyrano[3,4-*d*]benzoxepin (eq 62).[116] The final product arises from sequential alkylations of the *O*-2 and *O*-3 methoxybenzyl protecting groups.

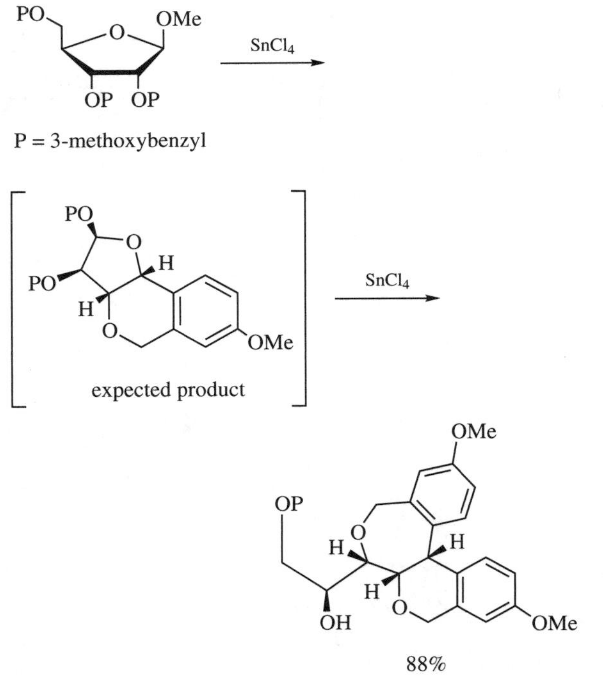

(62)

Substituted thiophenes have also been used as glycosyl acceptors resulting in the synthesis of purine-like *C*-nucleosides (eq 63).[117]

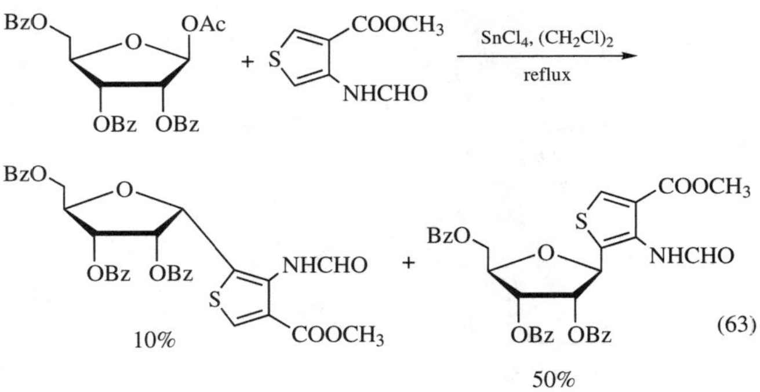

(63)

C-Glycosylation reactions of 2-deoxy-4-thioribofuranosides (TRib) with silylated glycosyl acceptors give high yields of the *C*-glycosides (65–89%) with a predominance of the α-isomer (eq 64, Table 1).[118]

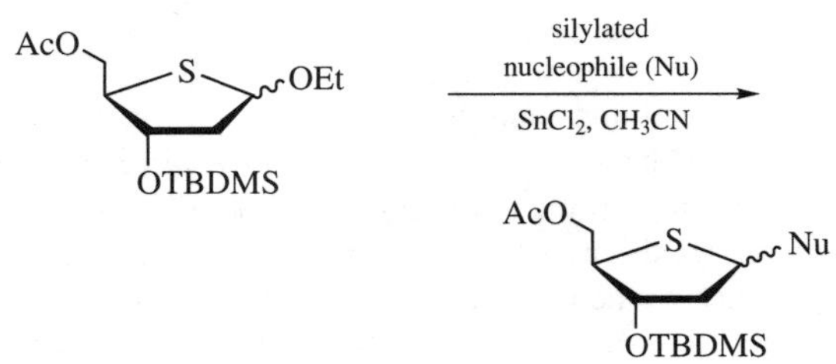

(64)

The use of silver trifluoroacetate-$SnCl_4$ affords high yields of aryl *C*-glycosidation products.[119] When treated with a second equivalent of glycosyl donor in the presence of $SnCl_4$, a double aryl *C*-glycosylation product can also be achieved in high yield (eq 65).[120]

Table 1

Nucleophile	Product	Yield (%)	α/β
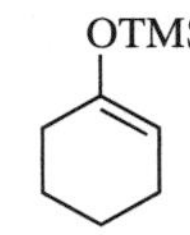	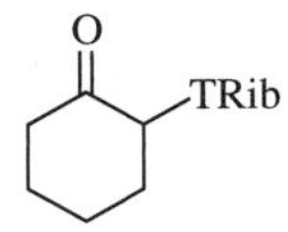	65	1.1:1
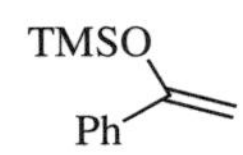	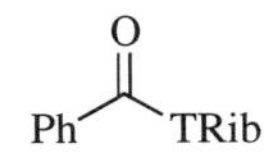	79	4.4:1
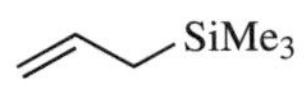	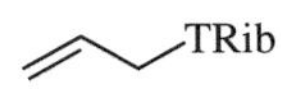	84	100:0
	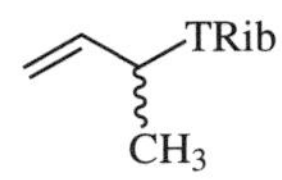	69	100:0

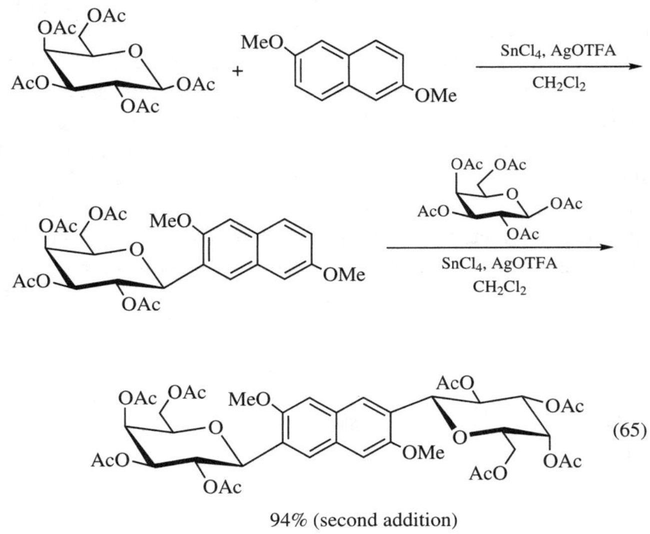

(65)

The spirosuccinimide compound (**18**), a mimic of the herbicide (+)-hydantocin, owes its synthesis to the allylation of an easily prepared protected D-psicofuranose in the presence of $SnCl_4$ (eq 66).[121] Attempted use of other Lewis acids (TMSOTf, $BF_3 \cdot OEt_2$, $TrClO_4$, and $TiCl_4$) in this condensation resulted in the recovery of starting materials.

Alkene Cyclizations. The asymmetric synthesis of (*R*)-limonene has been explored using Lewis acid catalysis in combination with a chiral leaving group (BINOL).[122] *O*-Neryl-BINOL ethers undergoes cyclization with detachment of the chiral auxiliary in the presence of $SnCl_4$ (eq 67).

Choice of the R group in the BINOL auxiliary influences the course of the reaction with small groups (H, Me, *i*-Pr) and tosylate affording (*S*)-limonene in modest yields. Changing R to the benzoate or *p*-trifluoromethylbenzoate groups caused preferential cyclization to the (*R*)-isomer. While enantiomeric excess levels tended to be good, the reaction usually included the chlorinated side product (eq 67). Reactions performed in either toluene or dichloromethane at −97 °C gave the best reaction results. Other reactions involving BINOL as the chiral auxiliary in $SnCl_4$-mediated alkene cyclizations have led to the synthesis of (*S*)-cembrene,[122] (−)-ambrox and (+)-podocarpa-8,11,13-triene.[123]

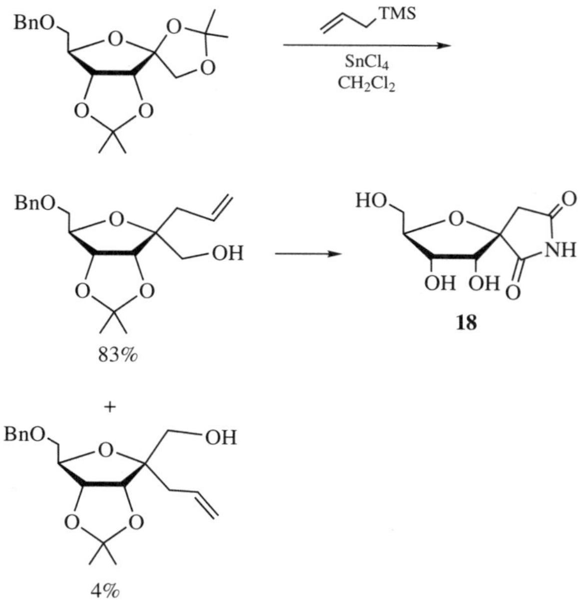

(66)

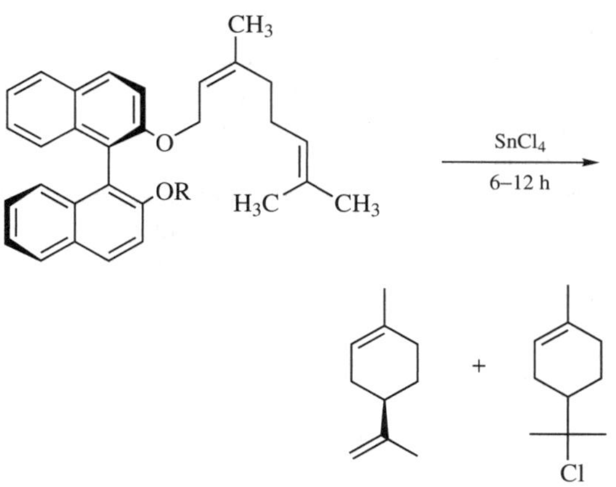

(67)

Diels-Alder Reactions. Tin(IV) chloride has been shown to be useful as a catalyst in the cyclization of 2-oxo-alkanoates with simple alkenes (eq 68).[124]

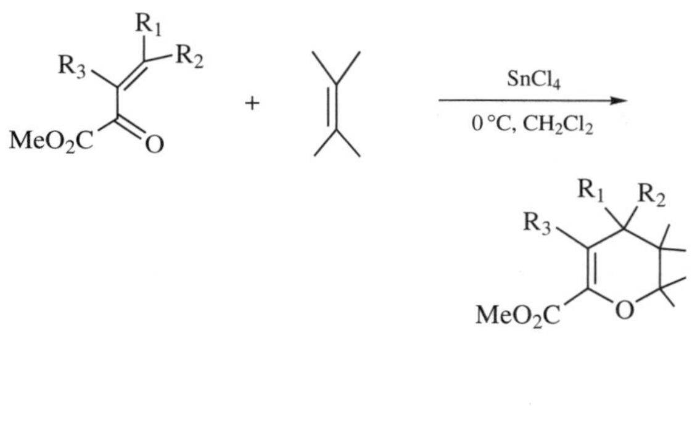

(68)

These cyclizations, which did not take place under conditions of high temperature or high pressure alone, required the presence of $SnCl_4$. The additions with *cis*- and *trans*-3-hexenes both took place with retention of configuration (eq 69).

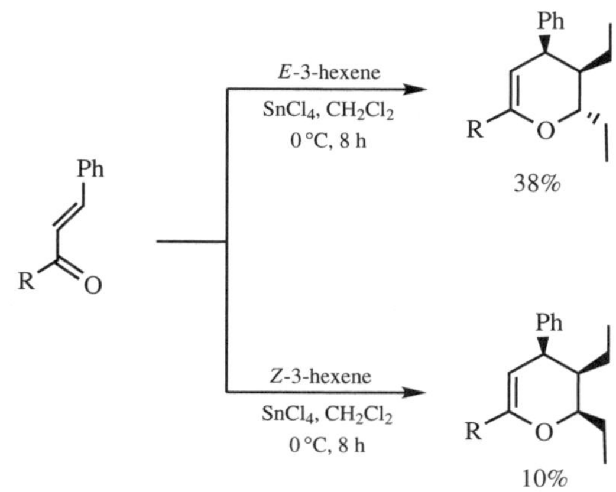

(69)

Switching the 2-π component from simple alkenes to styrene afforded mixtures of products (eq 70).[125]

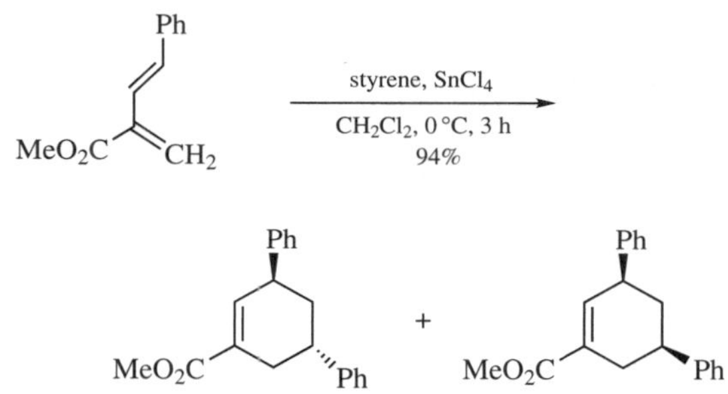

trans/*cis* =73:27 (70)

Kinetic studies with deuterium labeled styrene revealed the reaction to proceed by both concerted and ionic (stepwise) mechanisms, with the initial products undergoing further isomerizations under the reaction conditions, but not by a retro-Diels-Alder mechanism.

[3+2] Cycloaddtions. Highly substituted tetrahydrofurans have been prepared by the $SnCl_4$-mediated [3+2] cyclization of α-ketoesters with allylsilanes.[126,127] The order of addition plays an important role in the outcome of the reaction. Tin(IV) chloride added as a dilute solution in dichloromethane to the α-ketoester and allylsilane, in dichloromethane at −78 °C, yielded a 50–85% of the substituted tetrahydrofurans (eq 71). Other addition sequences led to predominantly homoallyl alcohol formation.

Other Lewis acids tested, including $TiCl_4$, $AlCl_3$, $SnCl_2$, $Sn(OTf)_2$, $BF_3 \cdot Et_2O$, or $ZnCl_2$, afforded no [3+2] addition products. Tetrahydrofuran formation is favored with TBDMS-based allylsilanes or by the presence alkyl substituents α- to the silyl group owing to stabilization of an intermediate β-carbocation, which in turn leads to cyclization (eq 72).

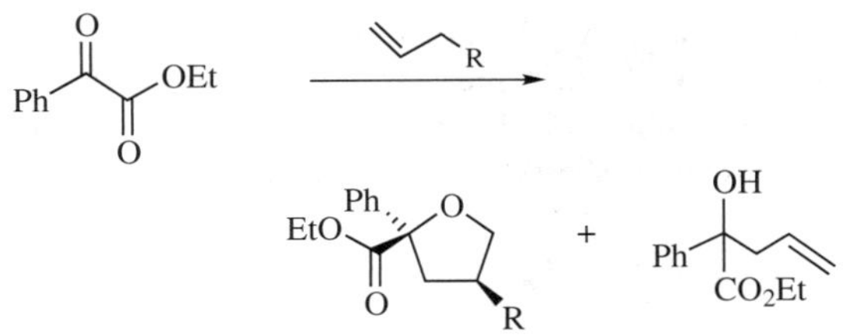

(71)

R	THF Yield (%)	Alcohol Yield (%)
$SiMe_3$	50	41
$SiPhMe_2$	54	46
TBDMS	85	6

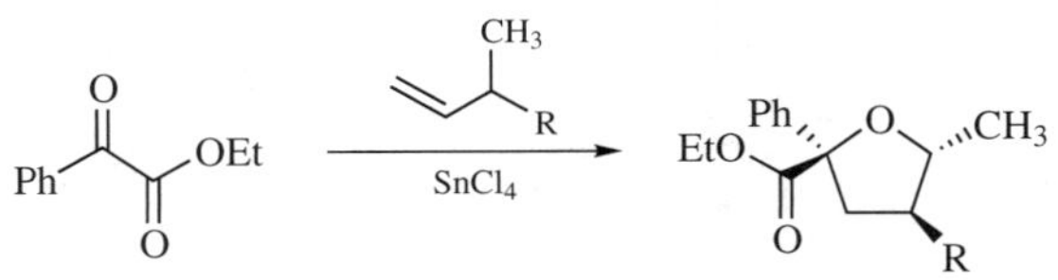

(72)

R = $SiPhMe_2$ 82%

Unlike the previously described addition of simple aldehydes with oxazole (eq 32), protected (*S*)-2-hydroxypropanals add to oxazole **12** to afford good yields of *trans* (P = TBDMS) or *cis* addition products (P = Bn). Other Lewis acids screened, including (BINOL)AlMe, $TiCl_4$, $BF_3{\cdot}Et_2O$, Et_2AlCl, and $AlMe_3$, gave mixtures of isomers (eq 73).[128]

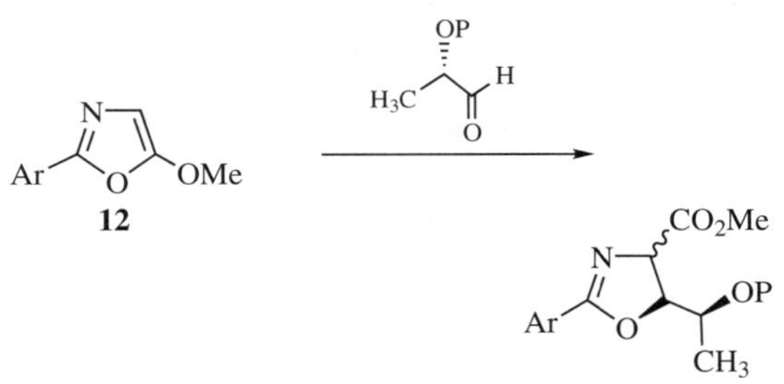

(73)

P = Bn; 80% (*cis*)
P = TBDMS; 49% (*trans*)

Treatment of 1*H*-indol-3-ylmethanol with $SnCl_4$ at −78 °C affords the stable indolyl cation **19**, which then undergoes addition of alkenes to yield cyclopenta[*b*]indoles with a good stereoselectivity (eq 74).[129] This is in contrast to the formation of diindol-3-ylmethane that is typically observed on generation of 1*H*-indol-3-ylmethanol with protic acids.

Ene Reactions. Chiral diazodicarboxylates have been demonstrated to undergo an azo-ene reactions with alkenes mediated by $SnCl_4$, offering an easy entry into the formation of chiral allylic amines. Couplings of di-(−)-(1*R*, 2*S*)-2-phenyl-1-cyclohexyl-diazene dicarboxylate (DPCDD) with selected alkenes (e.g., cyclohexene) were carried out in dichloromethane at −60 °C in the presence of $SnCl_4$ (eq 75).[130]

Less bulky alkenes led to the formation of higher ratios of diastereomers. The adducts were cleaved with lithium/ammonia followed by hydrolysis of the resulting carbamate to yield the allylamines.

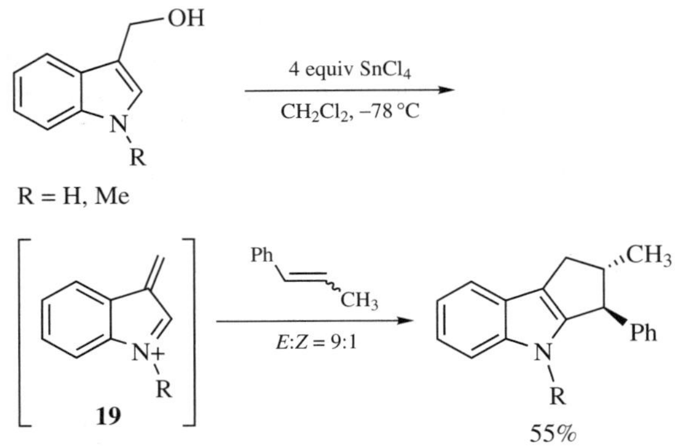

(74)

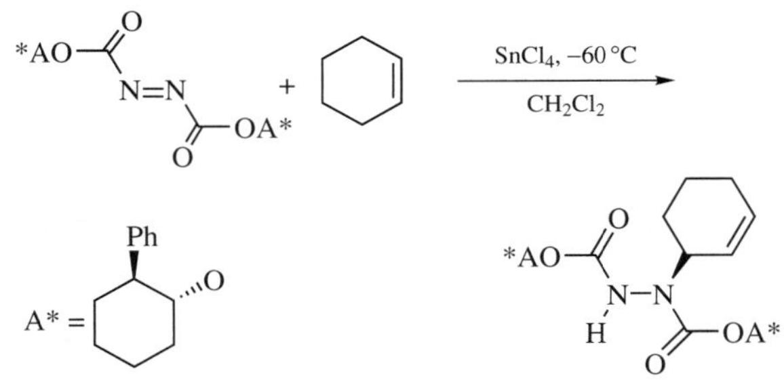

(75)

Note: Chiral auxilary is attsched via ester link.

Baeyer −Villiger. The combination of bis-(trimethylsilyl) perxoide/Lewis acid/diamine offers promise in the effort to achieve enantioselective BV reactions.[131] Galium(III) chloride and tin(IV) chloride were determined to be the most effective Lewis acid catalysts, with $SnCl_4$ being cheaper and offering better yields. Reduction of lactone ring opening and other side reactions was accomplished by the addition of complexing ligands reducing Lewis acidity of the catalyst. The most effective ligand was found to be *trans*-1,2-diaminohexane(DA). Optimum reaction conditions are depicted in eq 76.

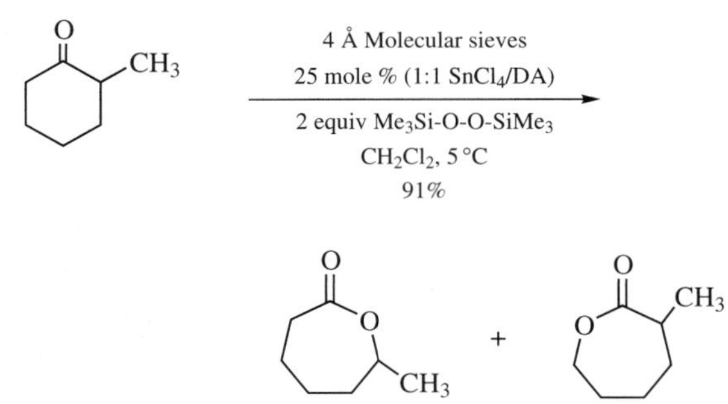

(76)

31:1

Related Reagents. Tin(IV) chloride–zinc chloride.

1. (a) Shambayati, S.; Crowe, W. E.; Schreiber, S. L., *Angew. Chem., Int. Ed. Engl.* **1990**, *29*, 256. (b) Reetz, M. T. In *Selectivities in Lewis Acid Promoted Reactions*; Schinzer, D., Ed.; Kluwer:Dordrecht, 1989; pp 107–125. (c) Denmark, S. E.; Almstead, N. G., *J. Am. Chem. Soc.* **1993**, *115*, 3133.

2. Davies, G. A.; Smith, P. J. In *Comprehensive Organometallic Chemistry*; Wilkinson, G.; Stone, F. G. A.; Abel, E. W., Eds.; Pergamon:New York, 1982; Vol. 2, p 519.
3. Naruta, Y.; Nishigaichi, Y.; Maruyama, K., *Tetrahedron* **1989**, *45*, 1067.
4. (a) Nakamura, E.; Kuwajima, I., *Chem. Lett.* **1983**, 59. (b) Yamaguchi, M.; Hayashi, A.; Hirama, M., *J. Am. Chem. Soc.* **1993**, *115*, 3362. (c) Yamaguchi, M.; Hayashi, A.; Hirama, M., *Chem. Lett.* **1992**, 2479.
5. (a) Keck, G. E.; Castellino, S.; Andrus, M. B. In *Selectivities in Lewis Acid Promoted Reactions*; Schinzer, D., Ed.; Kluwer:Dordrecht, 1989; pp 73–105. (b) Keck, G. E.; Andrus, M. B.; Castellino, S., *J. Am. Chem. Soc.* **1989**, *111*, 8136. (c) Denmark, S. E.; Wilson, T.; Wilson, T. M., *J. Am. Chem. Soc.* **1988**, *110*, 984. (d) Boaretto, A.; Marton, D.; Tagliavini, G.; Ganis, P., *J. Organomet. Chem.* **1987**, *321*, 199. (e) Yamamoto, T.; Maeda, N.; Maruyama, K., *Chem. Commun./J. Chem. Soc.* **1983**, 742. (f) Quintard, J. P.; Elissondo, B.; Pereyre, M., *J. Org. Chem.* **1983**, *48*, 1559.
6. Keck, G. E.; Abbott, D. E.; Boden, E. P.; Enholm, E. J., *Tetrahedron Lett.* **1984**, *25*, 3927.
7. (a) Keck, G. E.; Castellino, S., *J. Am. Chem. Soc.* **1986**, *108*, 3847. (b) Keck, G. E.; Castellino, S.; Wiley, M. R., *J. Org. Chem.* **1986**, *51*, 5478.
8. Keck, G. E.; Castellino, S., *Tetrahedron Lett.* **1987**, *28*, 281.
9. Mukaiyama, T.; Banno, K.; Narasaka, K., *J. Am. Chem. Soc.* **1974**, *96*, 7503.
10. Review of Mukaiyama aldol reaction: Gennan, C., *Comprehensive Organic Synthesis* **1991**, Vol. 2.
11. Reetz, M. T.; Kesseler, K.; Jung, A., *Tetrahedron* **1984**, *40*, 4327.
12. (a) Reetz, M. T., *Angew. Chem., Int. Ed. Engl.* **1984**, *23*, 556. (b) ref 11.
13. (a) Annunziata, R.; Cinquini, M.; Cozzi, F.; Cozzi, P. G.; Consolandi, E., *J. Org. Chem.* **1992**, *57*, 456. (b) Annunziata, R.; Cinquini, M.; Cozzi, F.; Cozzi, P. G., *Tetrahedron Lett.* **1990**, *31*, 6733.
14. Nakamura, E.; Kawajima, I., *Tetrahedron Lett.* **1983**, *24*, 3343.
15. Scavo, F.; Helquist, P., *Tetrahedron Lett.* **1985**, *26*, 2603.
16. Deng, M. Z.; Caubere, P.; Senet, J. P.; Lecolier, S., *Tetrahedron* **1988**, *44*, 6079.
17. Gras, J. L.; Galledou, B. S., *Bull. Soc. chem. Fr., Part 2* **1982**, 89.
18. Martin, O. R.; Rao, S. P.; Kurz, K. G.; El-Shenawy, H. A., *J. Am. Chem. Soc.* **1988**, *110*, 8698.
19. Showalter, H. D. H.; Putt, S. R., *Tetrahedron Lett.* **1981**, *22*, 3155.
20. Wagner, D.; Verheyden, J. P. H.; Moffat, J. G., *J. Org. Chem.* **1974**, *39*, 24.
21. Hori, H.; Nishida, Y.; Ohrui, H.; Meguro, H., *J. Org. Chem.* **1989**, *54*, 1346.
22. Hopkins, M. H.; Overman, L. E., *J. Am. Chem. Soc.* **1987**, *109*, 4748.
23. Overman, L. E.; Kakimoto, M. E.; Okazaki, M. E.; Meier, G. P., *J. Am. Chem. Soc.* **1983**, *105*, 6622.
24. Herrington, P. M.; Hopkins, M. H.; Mishra, P.; Brown, M. J.; Overman, L. E., *J. Org. Chem.* **1987**, *52*, 3711.
25. Review of α-alkylations to carbonyl compounds: Reetz, M. T., *Angew. Chem., Int. Ed. Engl.* **1982**, *21*, 96.
26. (a) Review of asymmetric alkene cyclization: Bartlett, P. A. *Asymmetric Synthesis*; Morrison, J. D., Ed.; Academic:New York, 1984; Vol. 3, Part B, pp 341–409. (b) Review of thermal cycloadditions: Fallis, A. G.; Lu, Y.-F. *Advances in Cycloaddition*; Curran, D. P., Ed.; JAI:Greenwich, CT, 1993; Vol. 3, pp 1–66.
27. (a) Overman, L. E.; Blumenkopf, T. A.; Castaneda, A.; Thompson, A. S., *J. Am. Chem. Soc.* **1986**, *108*, 3516. (b) Overman, L. E.; Castaneda, A.; Blumenkopf, T. A., *J. Am. Chem. Soc.* **1986**, *108*, 1303.
28. Demailly, G.; Solladie, G., *J. Org. Chem.* **1981**, *46*, 3102.
29. Murphy, W. S.; Waltanansin, S., *J. Chem. Soc., Perkin Trans. 1* **1982**, 1029.
30. (a) Kamigaito, M.; Madea, Y.; Sawamota, M.; Higashimura, T., *Macromolecules* **1993**, *26*, 1643. (b) Takahashi, T.; Yokozawa, T.; Endo, T., *Makromol. Chem.* **1991**, *192*, 1207. (c) Ran, R. C.; Mao, G. P. J., *Macromol. Sci. Chem.* **1990**, *A27*, 125. (d) Kurita, K.; Inoue, S.; Yamamura, K.; Yoshino, H.; Ishii, S.; Nishimura, S. I., *Macromolecules* **1992**, *25*, 3791. (e) Yokozawa, T.; Hayashi, R.; Endo, T., *Macromolecules* **1993**, *26*, 3313.
31. (a) Tanaka, H.; Kato, H.; Sakai, I.; Sato, T.; Ota, T., *Makromol. Chem. Rapid Commun.* **1987**, *8*, 223. (b) Yuan, Y.; Song, H.; Xu, G., *Polym. Int.* **1993**, *31*, 397.
32. Birney, D. M.; Houk, K. N., *J. Am. Chem. Soc.* **1990**, *112*, 4127.
33. For leading references on asymmetric Diels–Alder reactions, see: (a) Paquette, L. A. In *Asymmetric Synthesis*; Morrison, J. D., Ed.; Academic:New York, 1984; Vol. 3, pp 455–483. (b) Oppolzer, W., *Angew. Chem., Int. Ed. Engl.* **1984**, *23*, 876. (c) Carruthers, W. *Cycloaddition Reactions in Organic Synthesis*; Pergamon:New York, **1990**; pp 61–72.
34. Evans, D. A.; Chapman, K. T.; Bisaha, J., *J. Am. Chem. Soc.* **1988**, *110*, 1238.
35. Castellino, S., *J. Org. Chem.* **1990**, *55*, 5197.
36. Poll, T.; Helmchen, G.; Bauer, B., *Tetrahedron Lett.* **1984**, *25*, 2191.
37. (a) Fringuelli, F.; Pizzo, F.; Taticchi, A.; Wenkert, E., *J. Org. Chem.* **1983**, *48*, 2802. (b) Fringuelli, F.; Pizzo, F.; Taticchi, A.; Halls, T. D. J.; Wenkert, E., *J. Org. Chem.* **1982**, *47*, 5056.
38. Liu, H. J.; Ulibarri, G.; Browne, E. N. C., *Can. J. Chem.* **1992**, *70*, 1545.
39. Arseniyadis, A.; Rodriguez, R.; Spanevello, J. C.; Thompson, A.; Guittet, E.; Ourisson, G., *Tetrahedron* **1992**, *48*, 1255.
40. Baldwin, S. W.; Greenspan, P.; Alaimo, C.; McPhail, A. T., *Tetrahedron Lett.* **1991**, *42*, 5877.
41. For a recent review of oxyallyl cations, see: Mann, J., *Tetrahedron* **1986**, *42*, 4611.
42. Masatomi, O.; Kohki, M.; Tatsuya, H.; Shoji, E., *J. Org. Chem.* **1990**, *55*, 6086.
43. Murray, D. H.; Albizati, K. F., *Tetrahedron Lett.* **1990**, *31*, 4109.
44. Hoffman, H. M. R., *Angew. Chem., Int. Ed. Engl.* **1973**, *12*, 819; **1984**, *23*, 1.
45. Suga, H.; Shi, X.; Fujieda, H.; Ibata, T., *Tetrahedron Lett.* **1991**, *32*, 6911.
46. For examples of enantioselective synthesis of *trans*-4-alkoxy-2-oxazolines, see: Ito, Y.; Sawamura, M.; Shirakawa, E.; Hayashizaki, K.; Hayashi, T., *Tetrahedron Lett.* **1988**, *29*, 235; *Tetrahedron* **1988**, *44*, 5253.
47. Engler, T. A.; Wei, D.; Latavic, M. A., *Tetrahedron Lett.* **1993**, *34*, 1429.
48. Reviews of ene reactions: (a) Hoffman, H. M. R., *Angew. Chem., Int. Ed. Engl.* **1969**, *8*, 556. (b) Oppolzer, W.; Sniekus, V., *Angew. Chem., Int. Ed. Engl.* **1978**, *17*, 476. (c) Snyder, B. B., *Acc. Chem. Res.* **1980**, *13*, 426.
49. Lindner, D. L.; Doherty, J. B.; Shoham, G.; Woodward, R. B., *Tetrahedron Lett.* **1982**, *23*, 5111.
50. Nakatani, Y.; Kawashima, K., *Synthesis* **1978**, 147.
51. McNeill, A. H.; Thomas, E. J., *Tetrahedron Lett.* **1990**, *31*, 6239.
52. McNeill, A. H.; Thomas, E. J., *Tetrahedron Lett.* **1992**, *33*, 1369.
53. McNeill, A. H.; Thomas, E. J., *Synthesis* **1994**, 322.
54. Carey, J. S.; Coulter, T. S.; Thomas, E. J., *Tetrahedron Lett.* **1993**, *34*, 3933.
55. Dias, L. C.; Giacomini, R., *Tetrahedron Lett.* **1998**, *39*, 5343.
56. Bradley, G. W.; Hallett, D. J.; Thomas, E. J., *Tetrahedron: Asymmetry* **1995**, *6*, 2579.
57. Hobson, L. A.; Vincent, M. A.; Thomas, E. J., *Chem. Commun.* **1998**, 899.
58. Hallett, D. J.; Thomas, E. J., *Chem. Commun.* **1995**, 657.
59. Carey, J. S.; Thomas, E. J., *Tetrahedron Lett.* **1993**, *34*, 3935.
60. Carey, J. S.; Thomas, E. J., *Chem. Commun.* **1994**, 283.
61. Dorling, E. K.; Thomas, E. J., *Tetrahedron Lett.* **1999**, *40*, 471.
62. Germay, O.; Kumar, N.; Thomas, E. J., *Tetrahedron Lett.* **2001**, *42*, 4969.

63. Akiyama, T.; Ishikawa, K.; Ozaki, S., *Synlett* **1994**, 275.
64. Brain, C. T.; Thomas, E. J., *Tetrahedron Lett.* **1997**, *38*, 2387.
65. Taylor, N. H.; Thomas, E. J., *Tetrahedron* **1999**, *55*, 8757.
66. Alcaide, B.; Almendros, P.; Salgado, N. R., *J. Org. Chem.* **2000**, *65*, 3310.
67. Alcaide, B.; Almendros, P.; Aragoncillo, C.; Rodriguez-Acebes, R., *J. Org. Chem.* **2001**, *66*, 5208.
68. Alcaide, B.; Almendros, P.; Aragoncillo, C., *Chem. European J.* **2002**, *8*, 1719.
69. Veronese, A. C.; Callegari, R.; Morelli, C. F., *Tetrahedron* **1995**, *51*, 12277.
70. Veronese, A. C.; Callegari, R.; Salah, S. A. A., *Tetrahedron Lett.* **1990**, *31*, 3485.
71. Zhao, W. G.; Li, Z. M.; Yuan, D. K., *J. Chem. Res. (S)* **2002**, 454.
72. Veronese, A. C.; Morelli, C. F.; Basato, M., *Tetrahedron* **2002**, *58*, 9709.
73. Veronese, A. C.; Callegari, R.; Bertazzo, A., *Heterocycles* **1991**, *32*, 2205.
74. Morelli, C. F.; Manferdini, M.; Veronese, A. C., *Tetrahedron* **1999**, *55*, 10803.
75. Manferdini, M.; Morelli, C. F.; Veronese, A. C., *Tetrahedron* **2002**, *58*, 1005.
76. Banoub, J.; Bundle, D. R., *Can. J. Chem.* **1979**, *57*, 2085.
77. Hellou, J.; Banoub, J. H.; Hodder, H. J., *Chemosphere* **1987**, *16*, 1381.
78. Mazur, A. W.; Hiler, G. D., Jr., *Carbohydr. Res.* **1987**, *168*, 146.
79. Mizutani, K.; Kasai, R.; Nakamura, M.; Tanaka, O., *Carbohydr. Res.* **1989**, *185*, 27.
80. Pathak, A. K.; El-Kattan, Y. A.; Bansal, N.; Maddry, J. A.; Reynolds, R. C., *Tetrahedron Lett.* **1998**, *39*, 1497.
81. Du Mortier, C.; Varela, O.; de Lederkremer, R. M., *Carbohydr. Res.* **1989**, *189*, 79.
82. Chiocconi, A.; Marino, C.; de Lederkremer, R. M., *Carbohydr. Res.* **2000**, *323*, 7.
83. de Lederkremer, R. M.; Marino, C.; Varela, O., *Carbohydr. Res.* **1990**, *200*, 227.
84. Gallo-Rodriguez, C.; Varela, O.; de Lederkremer, R. M., *J. Org. Chem.* **1996**, *61*, 1886.
85. Gallo-Rodriguez, C.; Varela, O.; de Lederkremer, R. M., *Carbohydr. Res.* **1998**, *305*, 163.
86. Mukaiyama, T.; Shimpuku, T.; Takashima, T.; Kobayashi, S., *Chem. Lett.* **1989**, 145.
87. Mukaiyama, T.; Takashima, T.; Katsurada, M.; Aizawa, H., *Chem. Lett.* **1991**, 533.
88. Mukaiyama, T.; Katsurada, M.; Takashima, T., *Chem. Lett.* **1991**, 985.
89. Petrakova, E.; Glaudemans, C. P. J., *Carbohydr. Res.* **1995**, *279*, 133.
90. Petrakova, E.; Glaudemans, C. P. J., *Carbohydr. Res.* **1996**, *284*, 191.
91. Fukase, K.; Kinoshita, I.; Suda, Y.; Aoki, Y.; Liu, W.-C.; Oikawa, M.; Kurosawa, M.; Zaehringer, U.; Seydel, U.; Rietschel, E. T.; Kusumoto, S., *Synlett* **1996**, 252.
92. Lichtenthaler, F. W.; Voss, P.; Heerd, A., *Tetrahedron Lett.* **1974**, *15*, 2141.
93. Lichtenthaler, F. W.; Heerd, A.; Strobel, K., *Chem. Lett.* **1974**, 449.
94. Lichtenthaler, F. W.; Morino, T.; Winterfeldt, W.; Sanemitsu, Y., *Tetrahedron Lett.* **1975**, *16*, 3527.
95. Bartholomew, D. G.; Dea, P.; Robins, R. K.; Revankar, G. R., *J. Org. Chem.* **1975**, *40*, 3708.
96. Cook, P. D.; Robins, R. K., *J. Org. Chem.* **1978**, *43*, 289.
97. Lichtenthaler, F. W.; Cuny, E., *Chem. Ber.* **1981**, *114*, 1610.
98. Nakayama, C.; Saneyoshi, M., *Nucleosides & Nucleotides* **1982**, *1*, 139.
99. Nelson, V.; El Khadem, H. S., *J. Med. Chem.* **1983**, *26*, 1527.
100. Wood, S. G.; Upadhya, K. G.; Dalley, N. K.; McKernan, P. A.; Canonicio, P. G.; Robins, R. K.; Revankar, G. R., *J. Med. Chem.* **1985**, *28*, 1198.
101. Efange, S. M. N.; Cheng, Y. C.; Bardos, T. J., *Nucleosides & Nucleotides* **1985**, *4*, 545.
102. Poopeiko, N. E.; Kvasyuk, E. I.; Mikhailopulo, I. A.; Lidaks, M. J., *Synthesis* **1985**, 605.
103. Cheng, V.; Hughes, L.; Griffin, V. B.; Montserret, R.; Ollapally, A. P., *Nucleosides & Nucleotides* **1986**, *5*, 223.
104. Dutta, S. P.; Chheda, G. B., *Nucleic Acid Chem.* **1991**, *4*, 152.
105. Ogura, H.; Iwaki, K.; Furuhata, K., *Nucleic Acid Chem.* **1991**, *4*, 109.
106. Showalter, H. D. H.; Putt, S. R.; Baker, D. C., *Nucleic Acid Chem.* **1991**, *4*, 308.
107. Bhadti, V. S.; Bhan, A.; Hosmane, R. S.; Hulce, M., *Nucleosides & Nucleotides* **1992**, *11*, 1137.
108. Wang, J.; Wurster, J. A.; Wilson, L. J.; Liotta, D., *Tetrahedron Lett.* **1993**, *34*, 4881.
109. de Fina, G. M.; Varela, O.; de Lederkremer, R. M., *J. Chem. Res. (S)* **1994**, *26*.
110. Devlin, T. A.; Lacrosaz-Rouanet, E.; Vo, D.; Jebaratnam, D. J., *Tetrahedron Lett.* **1995**, *36*, 1601.
111. Robins, M. J.; Zou, R.; Guo, Z.; Wnuk, S. F., *J. Org. Chem.* **1996**, *61*, 9207.
112. Seela, F.; Muenster, I.; Loechner, U.; Rosemeyer, H., *Helv. Chim. Acta* **1998**, *81*, 1139.
113. Secrist, III, J. A.; Tiwari, K. N.; Shortnacy-Fowler, A. T.; Messini, L.; Riordan, J. M.; Montgomery, J. A.; Meyers, S. C.; Ealick, S. E., *J. Med. Chem.* **1998**, *41*, 3865.
114. Kren, V.; Olsovsky, P.; Havlicek, V.; Sedmera, P.; Witvrouw, M.; de Clercq, E., *Tetrahedron* **1997**, *53*, 4503.
115. Prisbe, E. J.; Verheyden, J. P. H.; Moffatt, J. G., *J. Org. Chem.* **1978**, *43*, 4774.
116. Martin, O. R.; Rao, S. P.; Hendricks, C. A. V.; Mahnken, R. E., *Carbohydr. Res.* **1990**, *202*, 49.
117. Patil, S. A.; Otter, B. A.; Klein, R. S., *Nucleosides & Nucleotides* **1990**, *9*, 937.
118. Uenishi, J.; Sohma, A.; Yonemitsu, O., *Chem. Lett.* **1996**, 595.
119. Kuribayashi, T.; Ohkawa, N.; Satoh, S., *Tetrahedron Lett.* **1998**, *39*, 4537.
120. Kuribayashi, T.; Ohkawa, N.; Satoh, S., *Tetrahedron Lett.* **1998**, *39*, 4541.
121. Sano, H.; Mio, S.; Tsukaguchi, N.; Sugai, S., *Tetrahedron* **1995**, *51*, 1387.
122. Ishihara, K.; Nakamura, H.; Yamamoto, H., *Synlett* **2001**, 1113.
123. Ishihara, K.; Ishibashi, H.; Yamamoto, H., *J. Am. Chem. Soc.* **2002**, *124*, 3647.
124. Sera, A.; Ohara, M.; Yamada, H.; Egashira, E.; Ueda, N.; Setsune, J., *Chem. Lett.* **1990**, 2043.
125. Sera, A.; Ueda, N.; Itoh, K.; Yamada, H., *Heterocycles* **1996**, *43*, 2205.
126. Akiyama, T.; Ishikawa, K.; Ozaki, S., *Chem. Lett.* **1994**, 627.
127. Akiyama, T.; Yasusa, T.; Ishikawa, K.; Ozaki, S., *Tetrahedron Lett.* **1994**, *35*, 8401.
128. Suga, H.; Fujieda, H.; Hirotsu, Y.; Ibata, T., *J. Org. Chem.* **1994**, *59*, 3359.
129. Harrison, C.-A.; Leineweber, R.; Moody, C. J.; Williams, J. M. J., *J. Chem. Soc., Perkin Trans. 1* **1995**, 1127.
130. Brimble, M. A.; Lee, C. K. Y., *Tetrahedron: Asymmetry* **1998**, *9*, 873.
131. Gottlich, R.; Yamakoshi, K.; Sasai, H.; Shibasaki, M., *Synlett* **1997**, 971.

Tin(II) Chloride–Silver(I) Perchlorate[1]

$SnCl_2$–$AgClO_4$

($SnCl_2$)
[7772-99-8] Cl_2Sn (MW 189.61)) ($AgClO_4$)
[14242-05-8] $AgClO_4$ (MW 207.32)

(glycosylation promoter;[2] activates glycosyl fluoride[2])

Physical Data: $SnCl_2$: bp 247 °C; *d* 3.95 g cm^{-3}. $AgClO_4$: dec 486 °C; *d* 2.806 g cm^{-3} (25 °C).

Solubility: $SnCl_2$: sol H_2O, ethanol, acetone, ether. $AgClO_4$: sol water.

Form Supplied in: $SnCl_2$ (anhydrous): white solid; widely available. $AgClO_4$: widely available as a hydrated form.

Drying: anhydrous $SnCl_2$ is dried for 1 h at 100 °C under reduced pressure (ca. 0.5 mmHg). Commercially available $AgClO_4H_2O$ is dissolved in benzene and the solution is refluxed for 6 h using Dean–Stark equipment (water is separated as the benzene azeotrope). After cooling and filtration, crude anhydrous $AgClO_4$ is recrystallized from benzene and dried at room temperature for 24 h.

Preparative Methods: the Tin(II) Chloride–Silver (I) Perchlorate reagent is usually prepared in situ by mixing an equimolar amount of $SnCl_2$ and $AgClO_4$ in an appropriate solvent such as Et_2O.

Handling, Storage, and Precautions: anhydrous $SnCl_2$ and $AgClO_4$ are handled and stored under an argon atmosphere. Anhydrous $AgClO_4$ explodes readily when struck or heated.

Promoter for Stereoselective Glycosylations. Combined use of tin(II) chloride ($SnCl_2$) and silver perchlorate ($AgClO_4$) effectively promotes the stereoselective glycosylation reaction of glucosyl fluorides with hydroxy compounds. α-Glucopyranosides are predominantly prepared by the reaction of 2,3,4,6-tetra-*O*-benzyl-β-D-glucopyranosyl fluoride(**1**) with various hydroxy compounds, including sterically hindered ones (eq 1).[2,3]

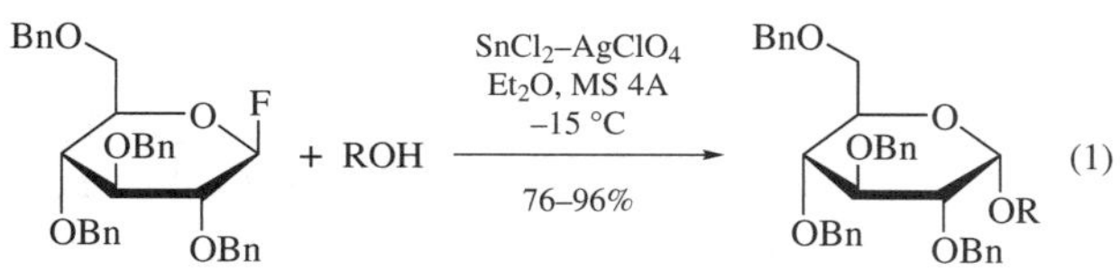

(**1**) α:β = 80:20–92:8

Under the same reaction conditions, a β-ribofuranoside is obtained as a major product from 2,3,5-tri-*O*-benzyl-β-D-ribofuranosyl fluoride (**3**). α-Ribofuranosides are prepared when $SnCl_2$ is combined with Triphenylmethyl Perchlorate ($TrClO_4$) instead of $AgClO_4$ (eq 2).[4]

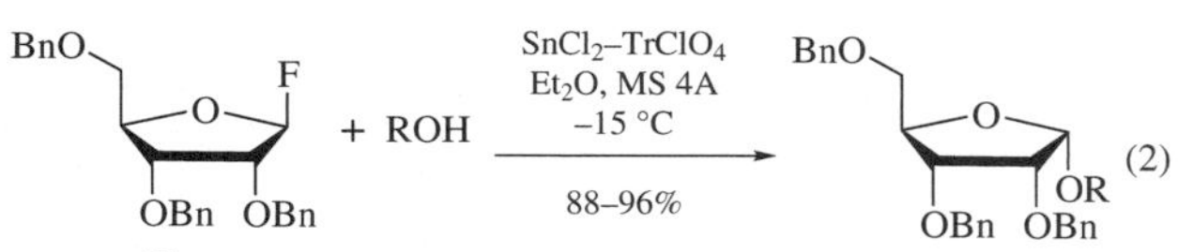

(**2**) α:β = 81:19–88:12

Although the precise structure of an active species generated from $SnCl_2$ and $AgClO_4$ is not clear, formation of $SnCl(ClO_4)$ or $Sn(ClO_4)_2$ is postulated. On the other hand, combination of Dichlorobis(cyclopentadienyl) hafnium and 2 equiv $AgClO_4$ is reported to form $Cp_2Hf(ClO_4)_2$, which is effective in the β-selective glycosylation of 2,3,4,6-tetra-*O*-benzyl-α-D-glucopyranosyl fluoride (**3**) (eq 3).[5] In this case, use of 2 equiv of $AgClO_4$ gives better results than those obtained by use of 1 equiv.[5d] Some other examples of stereoselective glycosylations of glycosyl fluorides using other activators such as SiF_4,[6] TMSOTf,[7] $BF_3.OEt_2$,[8,9] TiF_4,[10] Tf_2O,[11] and Me_2GaCl[12] have also been reported.

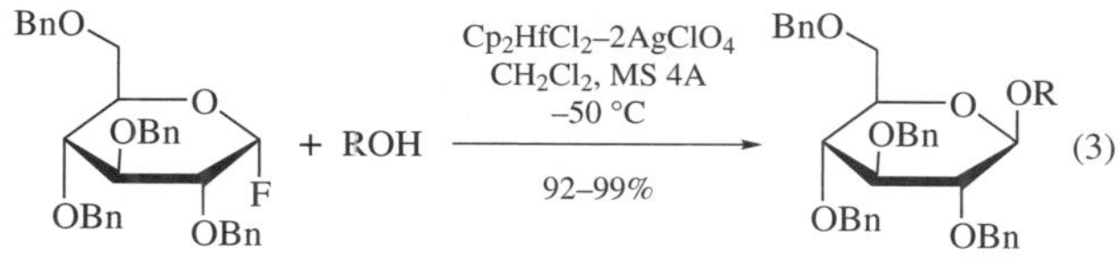

(**3**) α:β = 40:60–8:92

While more than stoichiometric amounts of promoters are required in the above examples, catalytic reactions are realized by using Tin (IV) Chloride–$AgClO_4$. In the presence of a catalytic amount of $SnCl_4$–$AgClO_4$, 1-*O*-acetyl-2,3,4,6-tetra-*O*-benzyl-D-glucopyranose (**4**) reacts with alcohol trimethylsilyl ethers to afford the corresponding α-glucopyranosides in high yields (eq 4).[13] Other examples of the preparation of β-glucosides[14] and α-[15] and β-2-amino-2-deoxyglycosides[16] have also been reported.

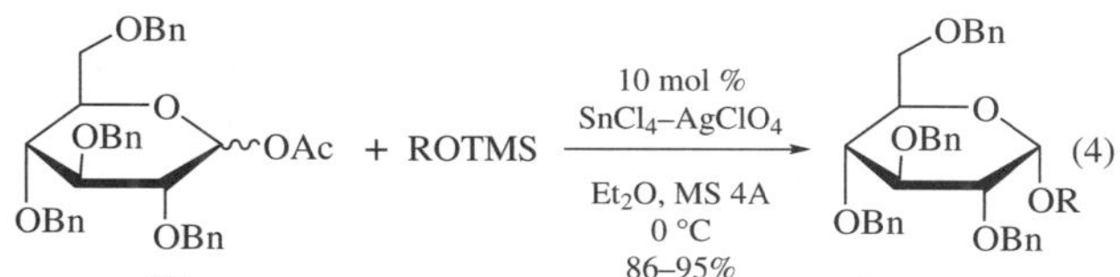

(**4**) α:β = 90:10–97:3

---

1. Suzuki, K.; Nagasawa, T. *J. Synth. Chem. Jpn* **1992**, *50*, 378. (*Chem. Abstr.* **1992**, *117*, 49 012d) .
2. Mukaiyama, T.; Murai, S.; Shoda, S. *Chem. Lett.* **1981**, 431.
3. Nicolaou, K. C.; Dolle, R. E.; Papahatjis, J. L.; Randall, J. L. *J. Am. Chem. Soc.* **1984**, *106*, 4189.
4. Mukaiyama, T.; Hashimoto, Y.; Shoda, S. *Chem. Lett.* **1983**, 935.
5. (a) Matumoto, T.; Maeta, H.; Suzuki, K.; Tsuchihashi, G. *Tetrahedron Lett.* **1988**, *29*, 3567. (b) Suzuki, K.; Maeta, H.; Matumoto, Tsuchihashi, G. *Tetrahedron Lett.* **1988**, *29*, 3571. (c) Matumoto, T.; Maeta, H.; Suzuki, K.; Tsuchihashi, G. *Tetrahedron Lett.* **1988**, *29*, 3575. (d) Suzuki, K.; Maeta, H.; Matumoto, T. *Tetrahedron Lett.* **1989**, *30*, 4853. (e) Suzuki, K.; Maeta, H.; Suzuki, T.; Matumoto, T. *Tetrahedron Lett.* **1989**, *30*, 6879. (f) Matumoto, T.; Katsuki, M.; Suzuki, K. *Chem. Lett.* **1989**, 437.
6. Hashimoto, S.; Hayashi, M.; Noyori, R. *Tetrahedron Lett.* **1984**, *25*, 1379.
7. Nicolaou, K. C.; Chucholowski, A.; Dolle, R. E.; Randall, J. L. *Chem. Commun./J. Chem. Soc., Chem. Commun.* **1984**, 1155.
8. Kunz, H.; Sager, W. *Helv. Chim. Acta* **1985**, *68*, 283.
9. Yamaguchi, M.; Horiguchi, A.; Fukuda, A.; Minami, T. *J. Chem. Soc., Perkin Trans. 1* **1990**, 1079.
10. (a) Kreuzer, M.; Thiem, J. *C.R. Hebd. Seances Acad. Sci.* **1987**, *164*, 277. (b) Jünnemann, J.; Lundr, I.; Thiem, J. *Justus Liebigs Ann. Chem./Liebigs Ann. Chem.* **1991**, 759.
11. (a) Wessel, H. P. *Tetrahedron Lett.* **1990**, *31*, 6879. (b) Wessel, H. P.; Ruiz, N. *J. Carbohydr. Chem.* **1991**, *10*, 901.
12. Kobayashi, S.; Koide, K.; Ohno, M. *Tetrahedron Lett.* **1990**, *31*, 2435.

13. (a) Mukaiyama, T.; Takashima, T.; Katsurada, M.; Aizawa, H. *Chem. Lett.* **1991**, 533. (b) Mukaiyama, T.; Katsurada, M.; Takashima, T. *Chem. Lett.* **1991**, 985.
14. Mukaiyama, T.; Katsurada, M.; Iida, Y. *Chem. Lett.* **1992**, 2105.
15. Mukaiyama, T.; Matsubara, K. *Chem. Lett.* **1992**, 1755.
16. Matsubara, K.; Mukaiyama, T. *Chem. Lett.* **1993**, 581.

Shū Kobayashi
Science University of Tokyo, Japan

Tin(II) Trifluoromethanesulfonate

$Sn(OTf)_2$

[62086-04-8] $C_2F_6O_6S_2Sn$ (MW 416.87)

(mild Lewis acid used primarily for the generation of tin(II) enolates for stereoselective aldol and Michael reactions;[1] [2,3]-Wittig and Ireland–Claisen rearrangements;[2] addition of tin(II) acetylides to aldehydes;[3] asymmetric allylation of aldehydes[4])

Physical Data: mp >300 °C.
Form Supplied in: fine white powder.[5]
Preparative Methods: [6,7] an excess amount of trifluoromethanesulfonic acid (40 g, 26.7 mmol) was added to anhydrous $SnCl_2$ (16 g, 8.4 mmol) with vigorous stirring, leading to an exothermic reaction with evolution of HCl gas, after which the mixture was further heated at 80 °C for 48 h. After removal of all volatile products in vacuo, the resulting white solid was washed with dry ether (50 mL, three times). After drying the resultants in vacuo at ca. 100 °C for several hours, the remaining white solid, tin(II) triflate, was used without further purification. When its reactivity diminishes, tin(II) triflate should be washed again with ether and dried in vacuo.
Purification: can be purified by washing with diethyl ether.[8]
Handling, Storage, and Precautions: moisture and air sensitive; all handling and storage should be done under nitrogen.

Original Commentary

Samantha Janisse
Eli Lilly and Company, Indianapolis, IN, USA

Aldol Reaction. Mukaiyama has developed the use of tin(II) triflate in diastereoselective and enantioselective aldol-type reactions.[1,9] Initially, the stereoselective aldol reactions were demonstrated using a stoichiometric amount of tin(II) triflate.[10] The reaction between 3-acylthiazolidine-2-thione and 3-phenylpropionaldehyde is a representative example of a diastereoselective *syn*-aldol synthesis (eq 1).

Enantioselective aldol-type reactions have been achieved by addition of a chiral diamine to the reaction mixture.[11] Specifically, addition of (*S*)-1-methyl-2-[(piperidin-1-ylmethyl]pyrrolidine to the reaction of 3-acetylthiazolidine-2-thione and 3-phenylpropionaldehyde provides the *syn*-aldol with greater than 90% ee (eq 2).

Reactions of ketene silyl acetals with various aldehydes in the presence of a stoichiometric amount of a tin(II) triflate–tin(IV) additive–chiral diamine complex have also been performed with excellent enantioselectivity (eq 3).[12]

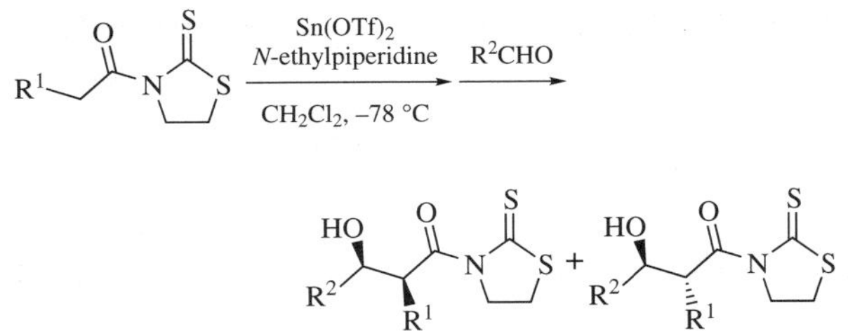

(1)

95%, *syn*:*anti* = >97:3
R^1 = Bn, R^2 = $BnCH_2$

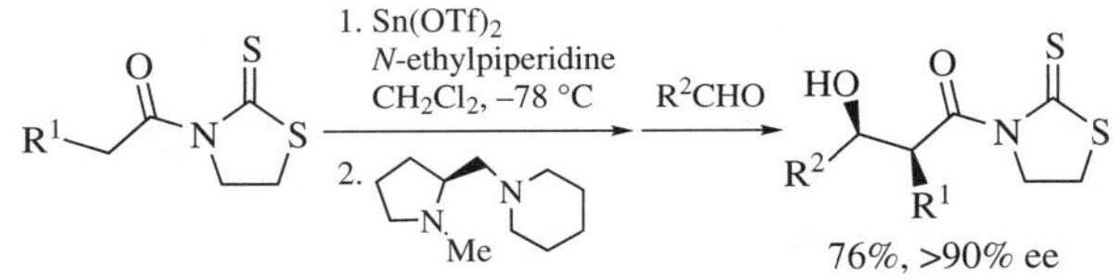

(2)

R^1 = Me, R^2 = $BnCH_2$

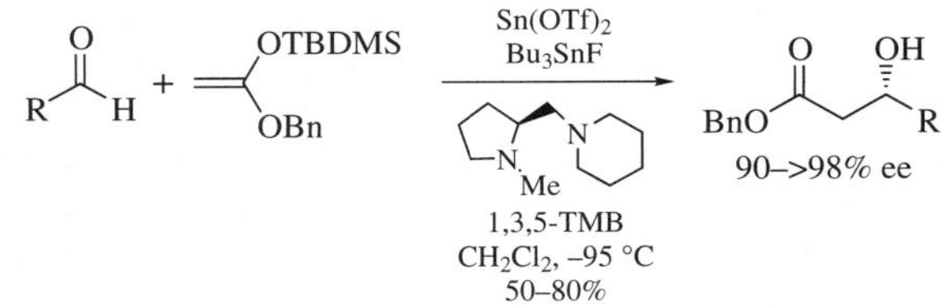

(3)

Stereoselective aldol reactions using a catalytic amount of tin(II) triflate have been developed. The diastereoselective reaction between Methyl Vinyl Ketone and benzaldehyde first demonstrated the usefulness of tin(II) triflate as a catalyst (eq 4).[1]

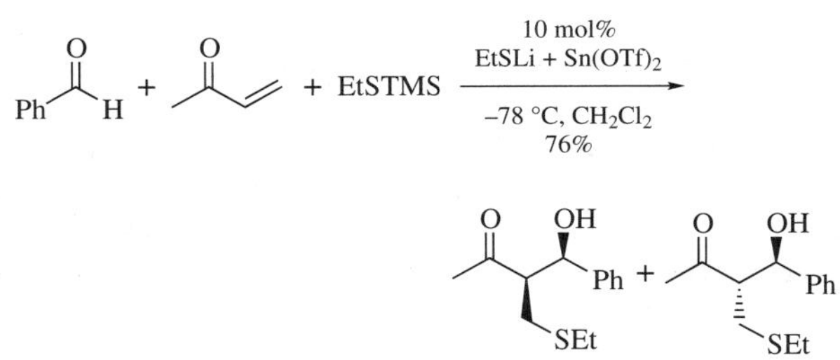

(4)

syn:*anti* = 89:11

The development of catalytic enantioselective reactions of silyl enol ethers with aliphatic, aromatic, alkynic, and α,β-unsaturated aldehydes has broadened the scope of this catalytic process (eq 5–7).[9a,b]

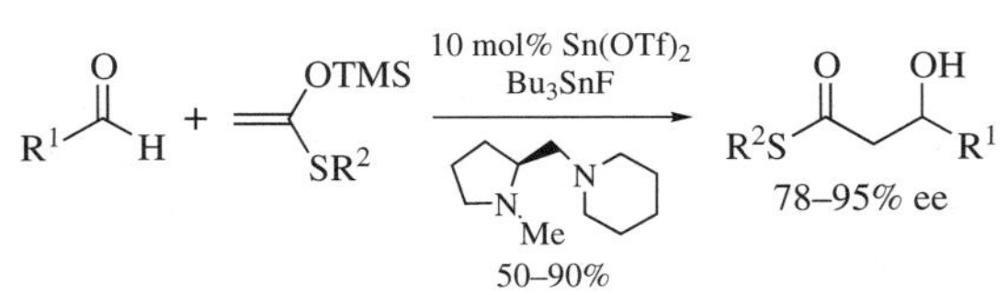

(5)

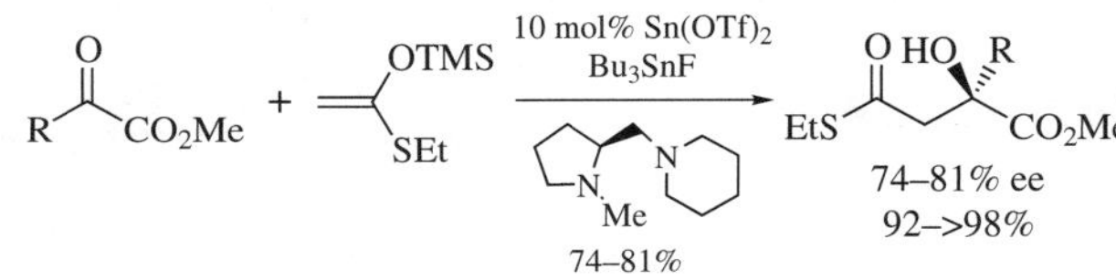

(6)

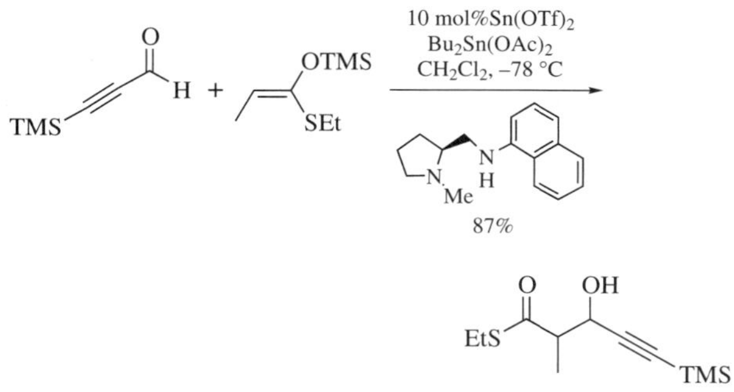

(7)

syn:*anti* = 95:5, 93% ee

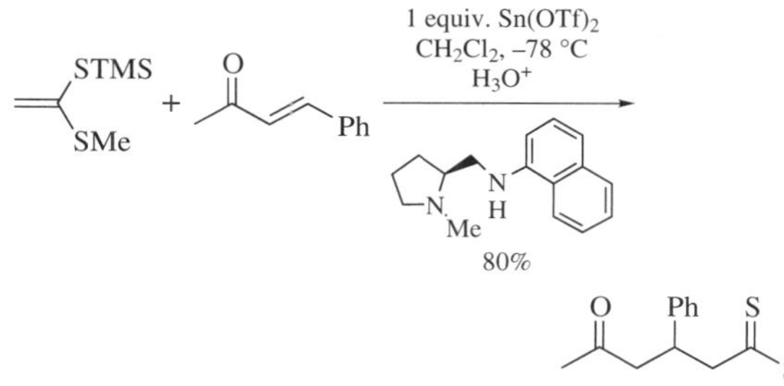

(10)

syn, 70% ee

In addition, Mukaiyama has shown that the stereochemistry of these reactions is controlled by the tin(II) triflate–tin(IV) additive–chiral diamine complex and is therefore independent of the chirality of the aldehyde (eqs 8 and 9).[9b,13]

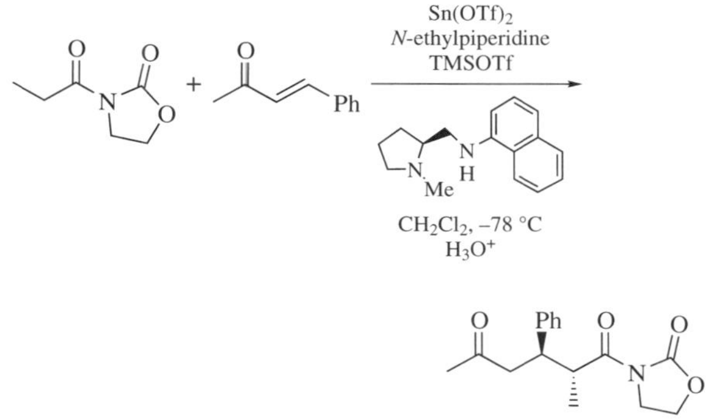

(11)

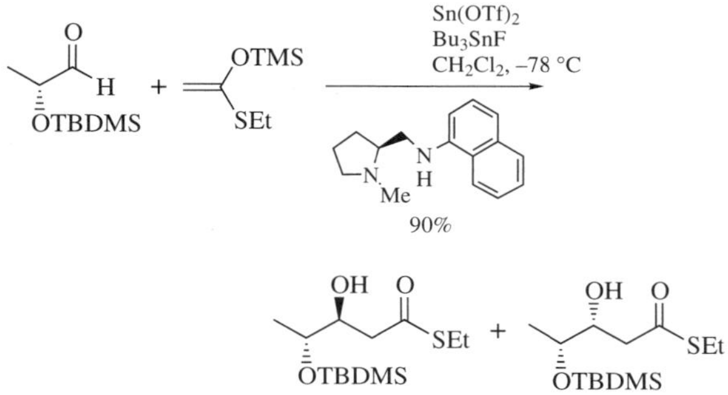

(8)

anti:*syn* = 96:4

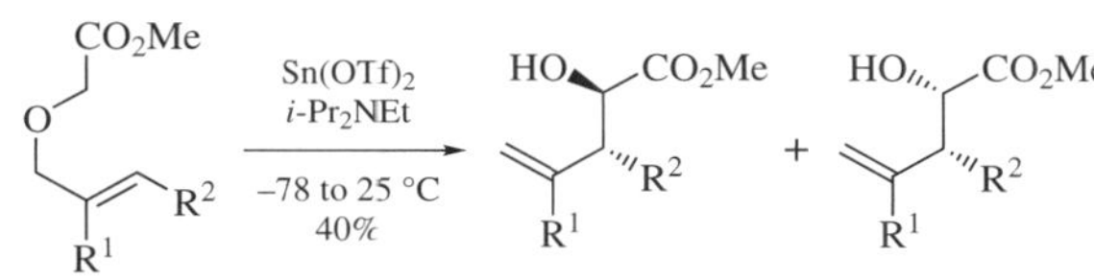

(12)

$R^1 = H$; $R^2 = Me$ *anti*:*syn* = 0.5:99.5

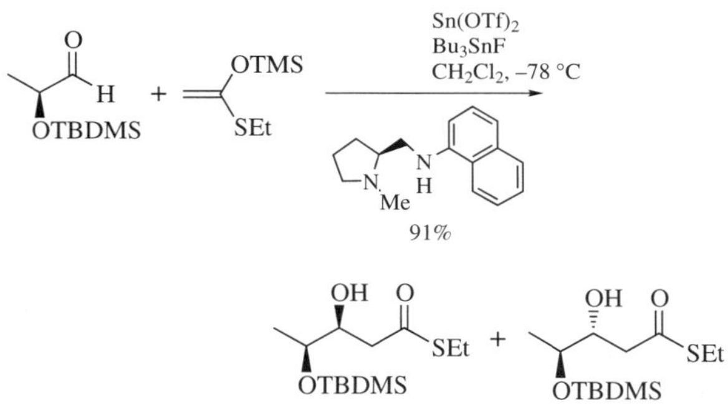

(9)

syn:*anti* = 96:4

Ireland–Claisen Rearrangement. An isolated case of an α-alkoxy ester rearranging via a tin(II) enolate has been reported (eq 13).[2b]

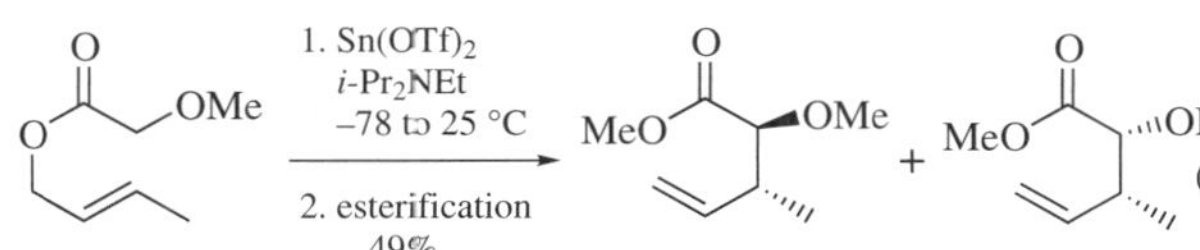

(13)

anti:*syn* = 82:18

Michael Reaction. The development of catalytic aldol-type reactions led to the discovery of a tin(II) triflate-catalyzed asymmetric Michael reaction.[1] Specifically, tin(II) enolates, prepared in situ from silyl enethiolates with catalytic amounts of tin(II) triflate and a chiral diamine, react with various α,β-unsaturated ketones to give chiral Michael adducts with 40–70% ee (eq 10).[14] Previously, asymmetric Michael reactions of tin(II) enolates with enones had been demonstrated in stoichiometric systems (eq 11).[15]

[2,3]-Wittig Rearrangements. Tin(II) enolates, generated from allylic glycolate esters and tin(II) triflate in the presence of Diisopropylathylamine, undergo [2,3]-Wittig rearrangements upon warming to room temperature (eq 12).[2a] The corresponding boron enolates of the esters rearrange in a similar manner but with less stereoselectivity.

Addition of 1-Alkynes to Aldehydes. In the presence of a stoichiometric amount of tin(II) triflate, 1-alkynes react with aldehydes and ketones to give alkynic alcohols in 57–91% yield (eq 14).[3] The mechanism of this reaction is, at present, unclear. It has been proposed that the tin(II) triflate reacts with the alkyne to form a tin acetylide which adds to the aldehyde.

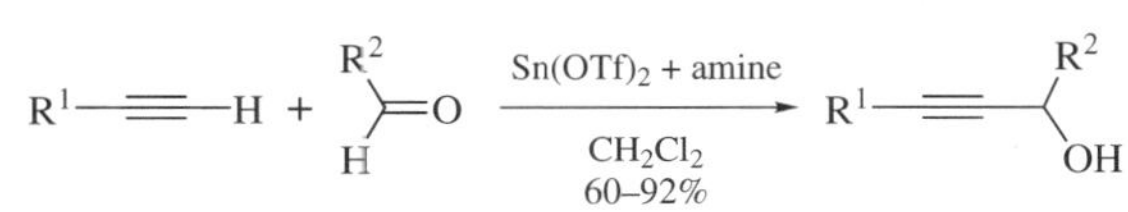

(14)

Asymmetric Allylation. Mukaiyama has designed a 'chiral allylating agent' consisting of an allyldialkylaluminum and a chiral diamine chelated to tin(II) triflate. The reaction of this agent with various aldehydes gives corresponding homoallylic alcohols with good to excellent enantioselectivity (eq 15).[4] Other tin(II) compounds ($SnCl_2$, $SnBr_2$, SnF_2, $Sn(OAc)_2$) react in this system to give the homoallylic alcohols in good yield but with virtually no enantioselectivity.

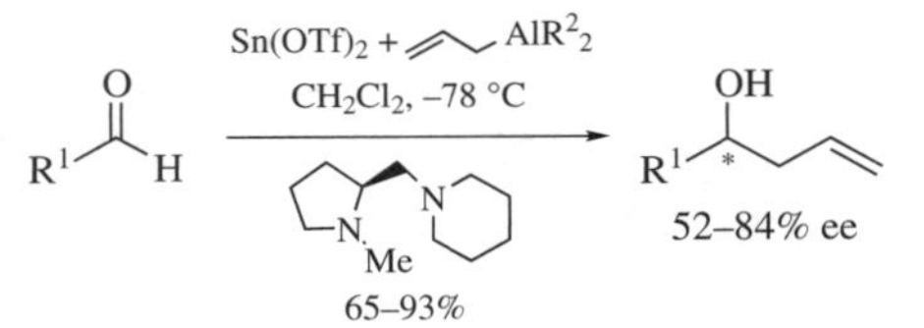
(15)

First Update

Teruaki Mukaiyama & Jun-ichi Matsuo
The Kitasato Institute, Tokyo, Japan

Horner-Wadsworth-Emmons reactions. The Horner-Wadsworth-Emmons reactions of methyl bis(trifluoroethyl)phosphonoacetate with aryl alkyl ketones employing tin(II) triflate and *N*-ethylpiperidine in CH_2Cl_2 at 0 °C affords the corresponding (*Z*)-alkenes in a highly selective manner (eq 16).[16,17] Tetrasubstituted (*Z*)-alkenes are readily prepared by the Horner-Wadsworth-Emmons reactions of methyl 2-[bis(trifluoroethyl)phosphono] propionate with aryl alkyl ketones.[18]

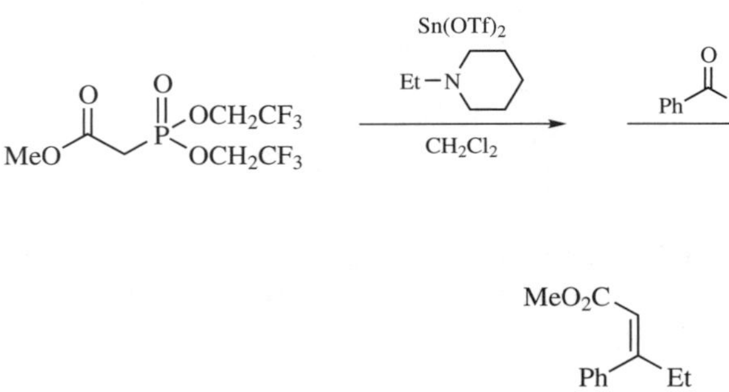
(16)

90% yield
>99% *Z*-selectivity

Glycosylation. The glycosylation of alcoholic glycosyl acceptors with 2,3,4,6-tetra-*O*-benzyl-D-galactopyranose (1-hydroxy sugar) proceeds successfully to afford α-D-galactopyranosides in good yields by the combined use of the tin(II) triflate-chlorotrimethylsilane catalyst system and lithium perchlorate (eq 17).[19]

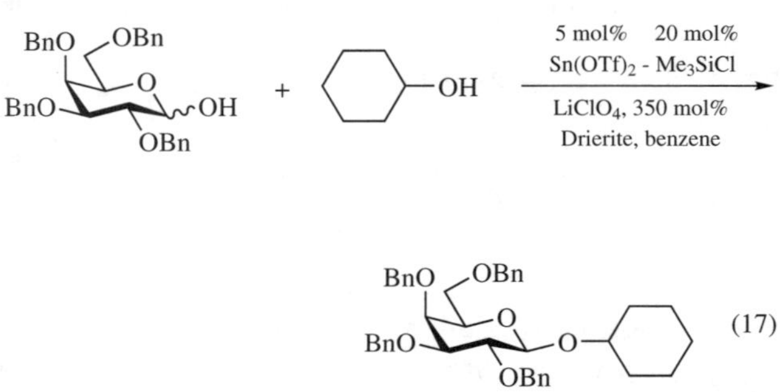
(17)

84% yield, α/β = 92/8

Asymmetric [3 + 2] Cycloaddition. The asymmetric [3 + 2] cycloaddition of chiral alkenyl Fisher carbene complexes with imines proceeds effectively in the presence of a catalytic amount of tin(II) triflate (eq 18).[20]

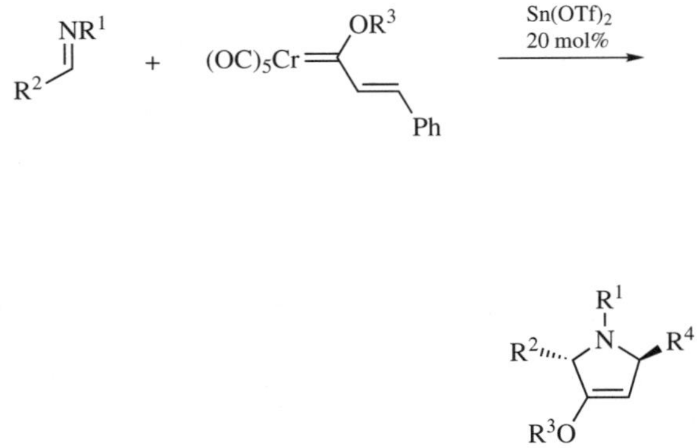
(18)

R^3 = (–)-8-phenylmenthyl

Ring Opening Reactions of Methylenecyclopropanes. The reaction of aromatic methylenecyclopropanes with sulfonamides catalyzed by tin(II) triflate affords the corresponding homoallylic sulfonamides in good yields (eq 19).[21] Methylenecyclopropanes react similarly with aromatic amines, alcohols, phenols, and carboxylic acids to give the corresponding homoallylic derivatives in the presence of tin(II) triflate.[22,23]

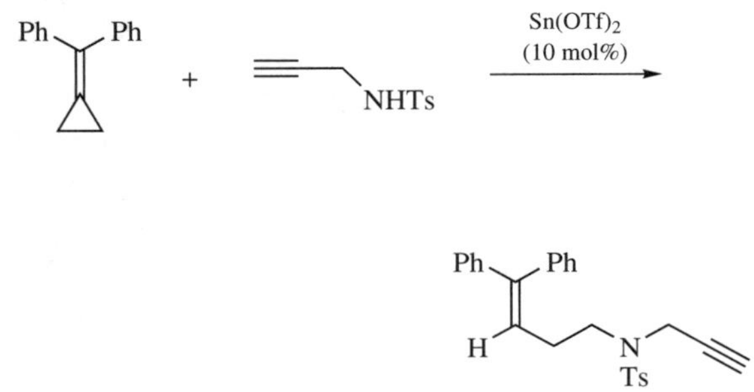
(19)

Ring Opening Reactions of Epoxides and Aziridines. The ring opening reaction of *meso*-epoxides with aromatic amines was catalyzed by tin(II) triflate.[24] By means of this ring opening reaction, unsymmetrically substituted amidines are prepared from caryophyllene oxide by using aromatic amines and tin(II) triflate in refluxing MeCN (eq 20).[25] The ring cleavage reaction of

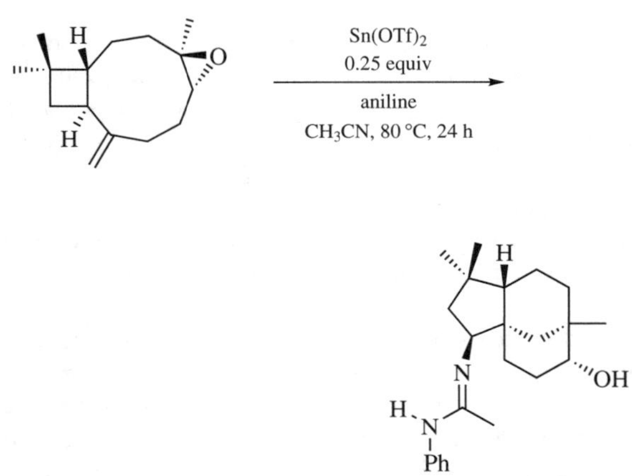
(20)

47%

N-substituted aziridines with alcohols also proceeds in the presence of tin(II) triflate.[26]

Friedel-Crafts Reactions. Friedel-Crafts sulfonylation of aromatic compounds with sulfonyl chlorides proceeds to give the corresponding sulfonylated compounds in high yields with a catalytic amount of tin(II) triflate (eq 21).[27]

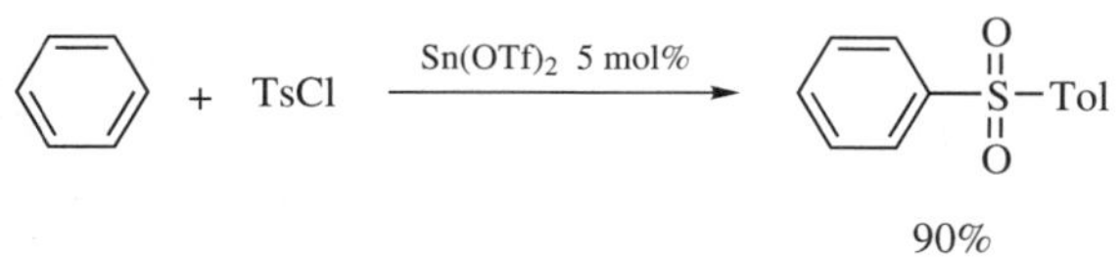

(21)

Related Reagents. Tin(II) chloride.

1. Iwasawa, N.; Yura, T.; Mukaiyama, T., *Tetrahedron* **1989**, *45*, 1197.
2. (a) Oh, T.; Wrobel, Z.; Devine, P., *Synlett* **1992**, 81. (b) Oh, T.; Wrobel, Z.; Rubenstein, S. M., *Tetrahedron Lett.* **1991**, *32*, 4647.
3. Yamaguchi, M.; Hayashi, A.; Minami, T., *J. Org. Chem.* **1991**, *56*, 4091.
4. Mukaiyama, T.; Minowa, N., *Bull. Chem. Soc. Jpn.* **1987**, *60*, 3697.
5. Tin(II) triflate is commercially available in >97% purity.
6. Mukaiyama, T.; Iwasawa, N.; Stevens, R. W.; Haga, T., *Tetrahedron* **1984**, *40*, 1381.
7. Batchelor, R. J.; Ruddick, J. N. R.; Sams, J. R.; Aubke, F., *Inorg. Chem.* **1977**, *16*, 1414.
8. Evans, D.; Weber, A., *J. Am. Chem. Soc.* **1986**, *108*, 6757.
9. (a) Kobayashi, S.; Furuya, M.; Ohtsubo, A.; Mukaiyama, T., *Tetrahedron Asymmetry* **1991**, *2*, 635. (b) Kobayashi, S.; Furuya, M.; Ohtsubo, A.; Mukaiyama, T., *Chem. Lett.* **1991**, 989. (c) Kobayashi, S.; Ohtsubo, A.; Mukaiyama, T., *Chem. Lett.* **1991**, 831.
10. Mukaiyama, T.; Iwasawa, N.; Stevens, R. W.; Haga, T., *Tetrahedron* **1984**, *40*, 1381.
11. (a) Mukaiyama, T.; Asanuma, H.; Hachiya, I.; Harada, T.; Kobayashi, S., *Chem. Lett.* **1991**, 1209. (b) Kobayashi, S.; Uchiro, H.; Fujishita, Y.; Shiina, I.; Mukaiyama, T., *J. Am. Chem. Soc.* **1991**, *113*, 4247.
12. Mukaiyama, T.; Kobayashi, S.; Sano, T., *Tetrahedron* **1990**, *46*, 4653.
13. For more details on the tin(II) triflate–tin(IV) additive–chiral diamine complex see: Mukaiyama, T.; Uchiro, H.; Kobayashi, S., *Chem. Lett.* **1989**, 1757.
14. Roush, W., *Chemtracts–Org. Chem.* **1988**, 439.
15. Yura, T.; Iwasawa, N.; Mukaiyama, T., *Chem. Lett.* **1988**, 1021.
16. Sano, S.; Yokoyama, K.; Fukushima, M.; Yagi, T.; Nagao, Y., *Chem. Commun.* **1997**, 559.
17. Sano, S.; Yokoyama, K.; Shiro, M.; Nagao, Y., *Chem. Pharm. Bull.* **2002**, *50*, 706.
18. Sano, S.; Takehisa, T.; Ogawa, S.; Yokoyama, K.; Nagao, Y., *Chem. Pharm. Bull.* **2002**, *50*, 1300.
19. Uchiro, H.; Miyazaki, K.; Mukaiyama, T., *Chem. Lett.* **1997**, 403.
20. Kagoshima, H.; Okamura, T.; Akiyama, T., *J. Am. Chem. Soc.* **2001**, *123*, 7182.
21. Chen, Y.; Shi, M. *J. Org. Chem.* **2004**, *69*, 426.
22. Shi, M.; Chen, Y.; Xu, B.; Tang, J., *Tetrahedron Lett.* **2002**, *43*, 8019.
23. Shi, M.; Xu, B., *Org. Lett.* **2002**, *4*, 2145.
24. Sekar, G.; Singh, V. K., *J. Org. Chem.* **1999**, *64*, 287.
25. Macías-Sánchez, A. J.; Amigo, C. F. D.; Collado, I. G., *Synlett* **2003**, 1989.
26. Bhanu Prasad, B. A.; Sanghi, R.; Singh, V. K., *Tetrahedron* **2002**, *58*, 7355.
27. Singh, R. P.; Kamble, R. M.; Chandra, K. L.; Saravanan, P.; Singh, V. K., *Tetrahedron* **2001**, *57*, 241.

p-Tolyldiphenylmethyl Chloride

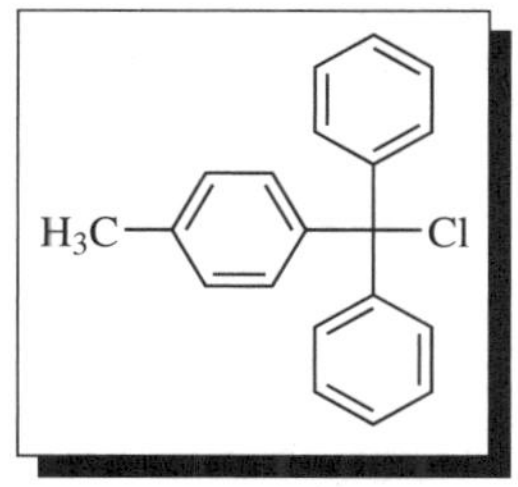

[23429-44-9] $C_{20}H_{17}Cl$ (MW 292.81)

(*p*-tolyldiphenylmethyl chloride, often abbreviated as MTTCl, is mainly used to introduce 4-methyltrityl functionality, often abbreviated as MTT, for the protection of hydroxy, amino, and thiol groups in organic synthesis)

Physical Data: mp 99 °C.
Solubility: soluble in DCM, toluene, THF, ethyl ether, and ethyl acetate.
Form Supplied in: white solid; commercially available from Chem-Impex, Novabiochem, and Rare Chemicals.
Preparative Methods: the most common method for the preparation of MTTCl is by heating a solution of 4-methyltriphenylcarbinol in acetyl chloride.[1,2] Benzene is often used as co-solvent for the reaction.[3–5] On cooling and addition of petroleum ether, MTTCl crystallizes out from the solvent (eq 1).

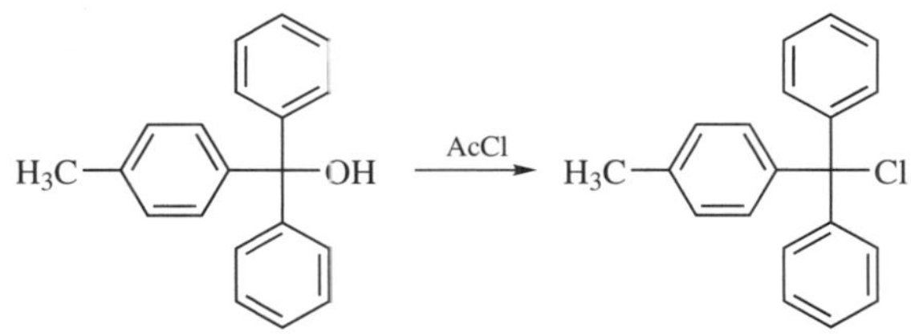

(1)

Handling, Storage, and Precautions: corrosive; lachrymator; irritating to eyes, respiratory systems, and skin. Do not breathe dust and avoid contact with skin and eyes. Use only with dried solvents in dry glassware and apparatus. Keep the container tightly closed and keep under −20 °C (−4 °F) for prolonged storage.

Hydroxy Group Protection. Triarylmethyl protecting groups are widely used as 5′-hydroxy protecting groups in oligonucleotide chemistry.[6] It was reported that MTT was successfully used in deoxyoligonucleotide synthesis on polymer supports.[7] Here the intensity of the orange color formed after deprotection of the MTT group in acidic media is used to determine whether the coupling of the MTT-protected deoxynucleoside to the growing deoxynucleotide proceeded to completion or not. To introduce the protecting group, MTTCl and the nucleoside are reacted in pyridine at room temperature; these are also widely used conditions for the introduction of other trityl groups (eq 2).[7,8]

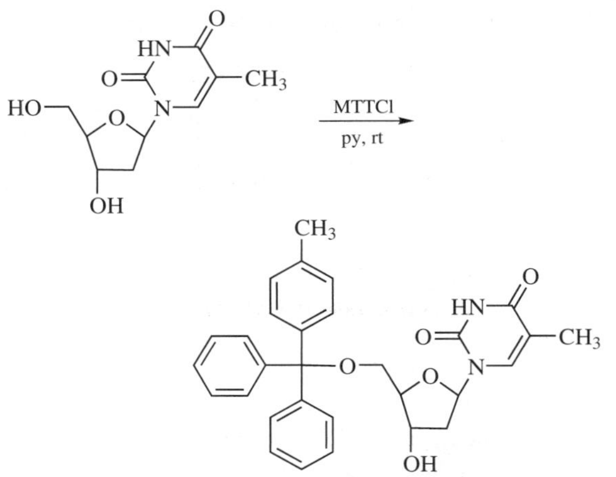

(2)

Deprotection can be achieved by treatment with either 0.06 M $ZnBr_2$ in nitromethane or 0.9 M chloroacetic acid in nitromethane. In oligonucleotide chemistry, the trityl protecting group itself is generally considered to be too stable as depurination is often a problem in its acid-catalyzed removal.[8] The MTT group is just barely reactive enough to be used in this regard,[7] while the *p*-methoxytrityl (MMTr) and di-*p*-methoxytrityl (DMTr) groups are commonly used. On other occasions, the trityl group is preferentially used over MTT, MMTr, and DMTr groups for the protection of hydroxy groups.

Amino Group Protection. The MTT group is also used as an amino protecting group, mainly in peptide chemistry to protect the ε-amino group of lysine[9–14] and the imidazole group of histidine[15] residues. Properly protected amino acids react with MTTCl in dichloromethane in the presence of TEA to give the MTT protected amino acids in good yields (eq 3).[9,15]

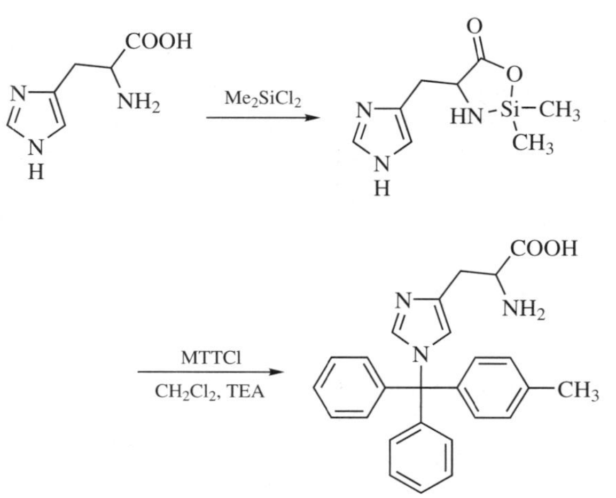

(3)

In comparison with the trityl group, MTT can be removed under very mild conditions. It was reported[9,12] that it can be removed by treatment with a mixture of acetic acid, trifluoroethanol, and dichloromethane (1:2:7), or with a solution of 1% TFA in dichloromethane. TFA (5%) in dichloromethane has achieved quantitative removal of the MTT function from a protected amino group within 1 min at rt. TFA in dichloromethane in the presence of $EtSiH_3$ can prevent retritylation after deprotection. Here $EtSiH_3$ functions as the cation scavenger during the process.[13]

Thiol Group Protection. In principle, the MTT group can also be used as a thiol-protecting group, but examples are rare in comparison to the trityl group.

Related Reagents. Trityl chloride (TrCl); mono-*p*-methoxytrityl chloride (MMTrCl or MMTCl); di-*p*-methoxytrityl chloride (DMTrCl); tri-*p*-methoxytrityl chloride (TMTrCl). These reagents, including MTTCl, are also available on solid supports.

---

1. Lichtin, N. N.; Lewis, E. S.; Price, E.; Johnson R. R., *J. Am. Chem. Soc.* **1959**, *81*, 4520.
2. Hellwinkel, D.; Fritsch, H., *Chem. Ber.* **1989**, *122*, 2351.
3. Gibbs, L. W.; Wedegaertner, D. K., *J. Org. Chem.* **1991**, *56*, 7320.
4. Bauld, N. L.; McDermed, J. D.; Hudson, C. E.; Rim, Y. S.; Zoeller, J., Jr; Gordon, R. D.; Hyde, J. S., *J. Am. Chem. Soc.* **1969**, *91*, 6666.
5. Bank, S.; Ehrlich, C. L.; Zubieta, J. A., *J. Org. Chem.* **1979**, *44*, 1454.
6. Beaucage, S. L.; Radhakrishnan, P. I., *Tetrahedron* **1992**, *48*, 2223.
7. Fisher, E. F.; Caruthers, M. H., *Nucleic Acids Res.* **1983**, *11*, 1589.
8. Schaller, H.; Weimann, G.; Lerch, B.; Khorana, H. G., *J. Am. Chem. Soc.* **1963**, *85*, 3821.
9. Aletras, A.; Barlos, K.; Gatos, D.; Koutsogianni, S.; Mamos, P., *Int. J. Peptide Protein Res.* **1995**, *45*, 488.
10. Feichtinger, K.; Sings, H. L.; Baker, T. J.; Matthews, K.; Goodman, M., *J. Org. Chem.* **1998**, *63*, 8432.
11. Burghardt, T. P.; Juranic, N.; Macura, S.; Muhlrad, A.; Ajtai, K., *J. Am. Chem. Soc.* **1999**, *121*, 10373.
12. De Napoli, L.; De Luca, S.; Di Fabio, G.; Messere, A.; Montesarchio, D.; Morelli, G.; Piccialli, G.; Tesauro, D., *Eur. J. Org. Chem.* **2000**, 1013.
13. Kadereit, D.; Deck, P.; Heinemann, I.; Waldmann, H., *Chem. Eur. J.* **2001**, *7*, 1184.
14. Barlos, K.; Gatos, D.; Hatzi, O.; Koch, N.; Koutsogianni, S., *Int. J. Peptide Protein Res.* **1996**, *47*, 148.
15. Barlos, K.; Chatzi, O.; Gatos, D.; Stavropoulos, G.; Tsegenidis, T., *Tetrahedron Lett.* **1991**, *32*, 475.

Sanxing Sun & Denice M. Spero
Boehringer Ingelheim, Ridgefield, CT, USA

1,2,4-Triazole

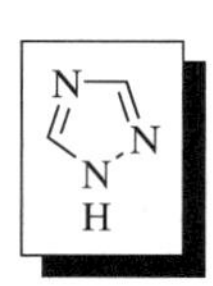

[288-88-0] $C_2N_3H_3$ (MW 69.08)

(transacylating agent used for ester and amide synthesis, especially in peptide synthesis; cyclization reactions; oligonucleotide synthesis)

Physical Data: mp 120–121 °C; bp 260 °C (dec above 187 °C); fp 140 °C.
Solubility: sol water, alcohol.

Form Supplied in: white powder.

Acylating Reagent. The great ability of triazole to accept and transfer acyl groups has made it a catalyst for the synthesis of esters and amides. In particular, triazole has been used in peptide synthesis, for the formation of peptide links as well as for the introduction of amino-protecting groups. It acts as a bifunctional catalyst, accelerating the aminolysis of *p*-nitrophenyl esters[1] and *p*-thiocresyl esters.[2] The catalytic effect is both by a general base mechanism (eq 1) and a specific mechanism (eq 2).

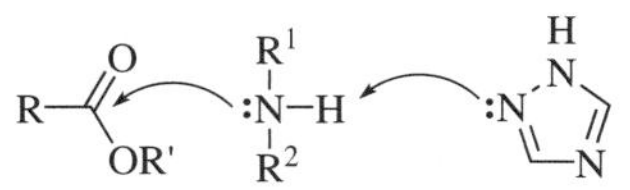
(1)

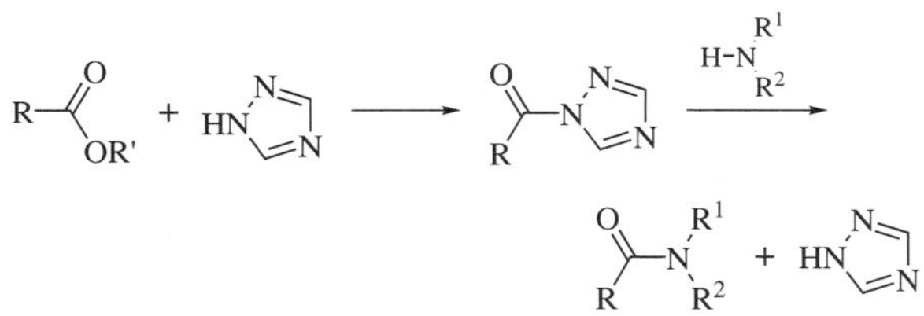
(2)

It has been used to introduce the Boc group onto an amino acid from *t*-butyl carbonates (eq 3), and it allows the coupling of an *N*-protected amino acid with a fully unprotected amino acid (eq 4). Interestingly, racemization is not observed during coupling of amino acids in the presence of 1,2,4-triazole.[1] Nevertheless, its coupling efficiency depends on the reaction medium and in solvents like DMF, which increase the rate of aminolysis, its usefulness is limited. 1,2,4-Triazole is now seldom used in peptide synthesis, whereas other catalytic additives like 1-Hydroxybenzotriazole are widely used as activated esters as well as in conjunction with coupling reagents like 1,3-Dicyclohexylcarbodiimide.

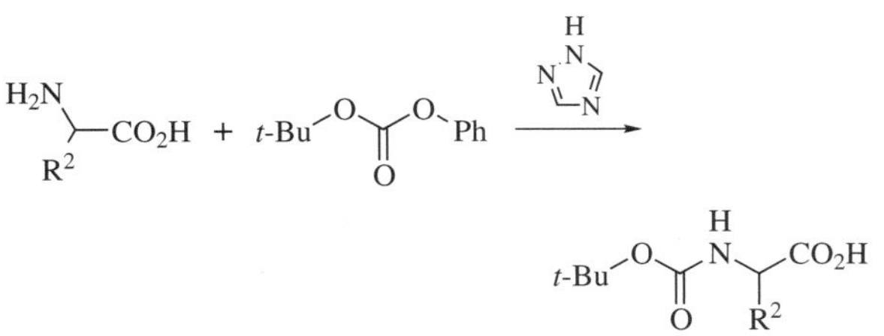
(3)

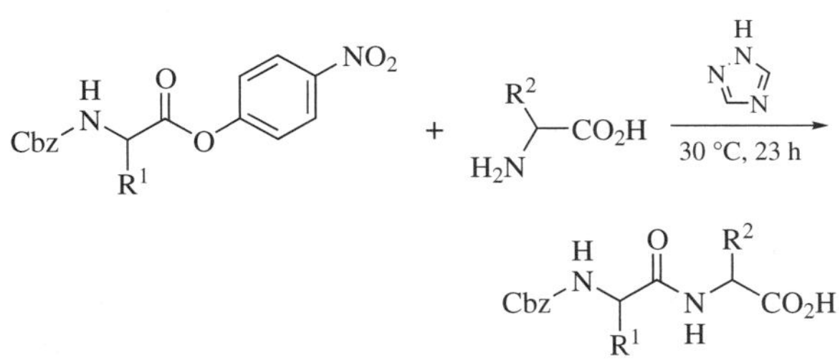
(4)

Catalysis of Ring Formation. 1,2,4-Triazole can induce the formation of isoxazole rings from conjugated ketones and hydroxylamine (eq 5).[3]

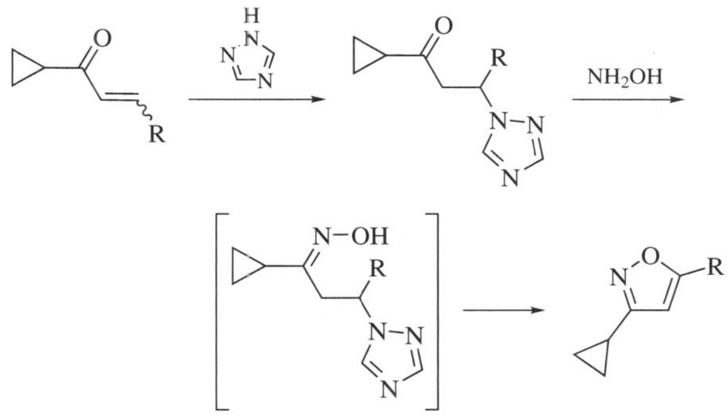
(5)

Oligonucleotide Synthesis. Sulfonyl derivatives of 1,2,4-triazole, e.g. (**1**), are used as efficient coupling reagents in oligonucleotide synthesis by the phosphotriester method.[4] A nitro derivative (**2**) allows esterification of amino acids onto hydroxymethyl polymers.[5]

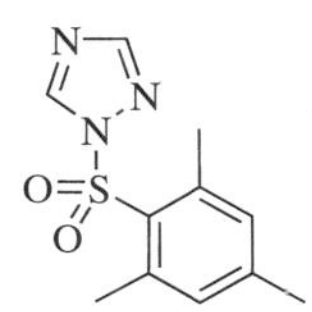
(**1**)

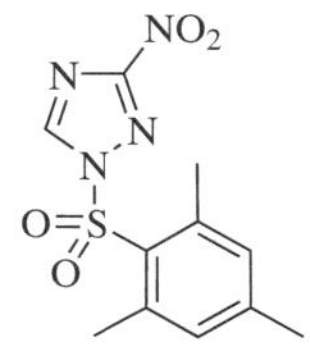
(**2**)

---

1. Beyerman, H. C.; Massen van den Brink, W.; Weygand, F.; Prox, A.; König, W.; Scmidhammer, L.; Nintz, E. *Recl. Trav. Chim. Pays-Bas* **1965**, *84*, 213.
2. Wieland, T.; Kahle, W. *Justus Liebigs Ann. Chem./Liebigs Ann. Chem.* **1966**, *691*, 212.
3. Rentzea, C. N. *Angew. Chem., Int. Ed. Engl.* **1981**, *20*, 885.
4. Reese, C. B.; Titmas, R. C.; Yau, L. *Tetrahedron Lett.* **1978**, 2727.
5. Blankemeyer-Menge, B.; Nintz, M.; Frank, R. *Tetrahedron Lett.* **1990**, *31*, 1701.

Jean-Claude Gesquière
Institut Pasteur, Lille, France

Tri-*t*-butylpyridine

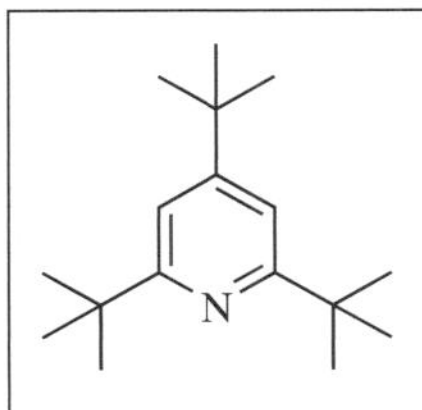
1

[20336-15-6] $C_{17}H_{29}N_1$ (MW 247.43)

(reagent used as non-nucleophilic base to scavenge the acid formed in a wide range of reactions, especially glycosylations)

Physical Data: mp 67–71 °C; bp 115–120 °C/20 mm Hg.
Solubility: soluble in dichloromethane, chloroform, alcohol, ether, acetonitrile, and most organic solvents.
Form Supplied in: white powder.
Analysis of Reagent Purity: NMR spectrum, mass spectrum, and IR spectrum.
Preparative Methods: commercially available through Aldrich.
Purification: recrystallization from EtOH/H_2O or pentane.
Handling, Storage, and Precautions: irritant to eyes, respiratory system, and skin and should be handled with suitable protective clothing, gloves, and safety goggles. Store and handle under nitrogen atmosphere and refrigerate. The toxicological properties have not been fully investigated.

Introduction. The highly hindered nitrogen in 2,4,6-tri-*tert*-butylpyridine (**1**) can coordinate to a proton; however it is too sterically shielded to coordinate to Lewis acids even as small as CH_3^+ or BF_3.[1] Accordingly, 2,4,6-tri-*tert*-butylpyridine (**1**) has been widely employed as a non-nucleophilic base to trap Brønsted acids, especially triflic acid, in a broad variety of contexts including glycosylation reactions[2] and aromatic acylation reactions.[3] It also has been used to scavenge hydrogen chloride, hydrogen bromide, and perchloric acid in other reactions.[4,5]

Scavenger of Triflic Acid in Glycosylation Reactions. The combination of sulfoxides, such as dibenzothiophene-5-oxide (DBTO, **2**), diphenyl sulfoxide (**3**) or 2,8-dimethyldibenzothiophene-5-oxide (DMDBTO, **4**), and triflic anhydride in the presence of 2,4,6-tri-*tert*-butylpyridine (**1**) has been employed extensively in the glycosylation reaction developed by Gin co-workers. By use of this combination of reagents, chemoselective glycosylation was achieved via direct dehydrative glycosylation by activation of the carbohydrate hemiacetals.[2a,b] This method enabled the synthesis of the β-glycosyl phosphate **7** in 93% yield (eq 1). Activating the

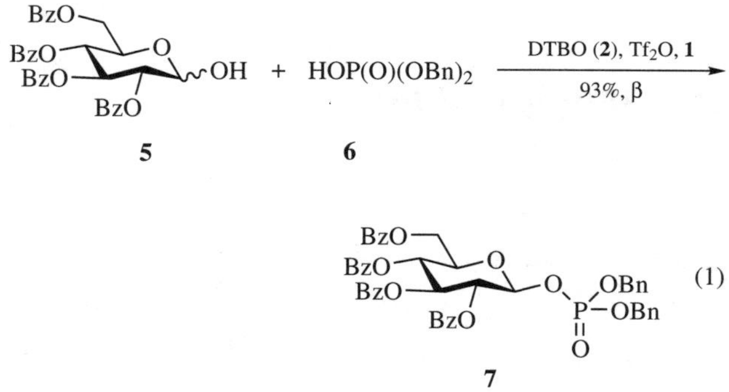

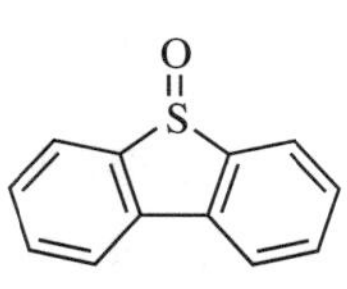

2

glucosyl hemiacetal (**8**) with diphenyl sulfoxide (**3**) and triflic anhydride in the presence of base **1**, followed by addition of the acceptor **9**, provided β-glucoside **10** in 71% yield (eq 2). In this chemistry a non-nucleophilic base such as **1** is required to buffer the triflic acid formed in situ as this could potentially lead to the decomposition of the glycosides or anomeriation.

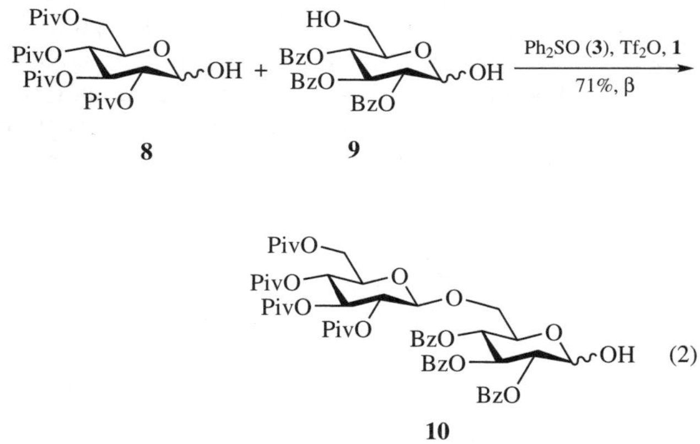

Also the combination of DTBO (**2**) or Ph_2SO (**3**) and triflic anhydride in the presence of base (**1**) is able to activate glucals in a direct synthesis of β-glucosides[2d,e] (eq 3) or 2-*N*-acetylamino-α-mannosides (eq 4).[2f]

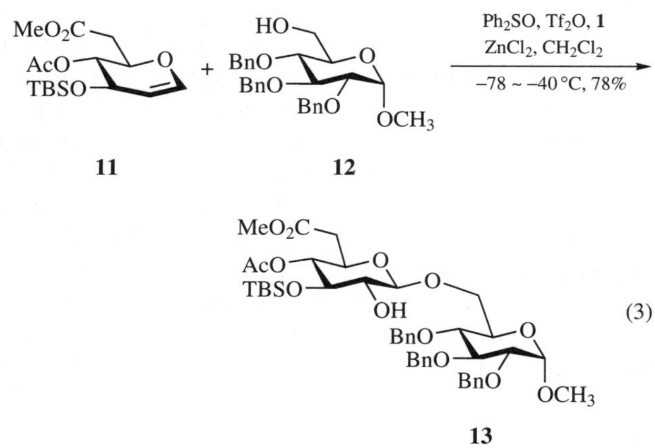

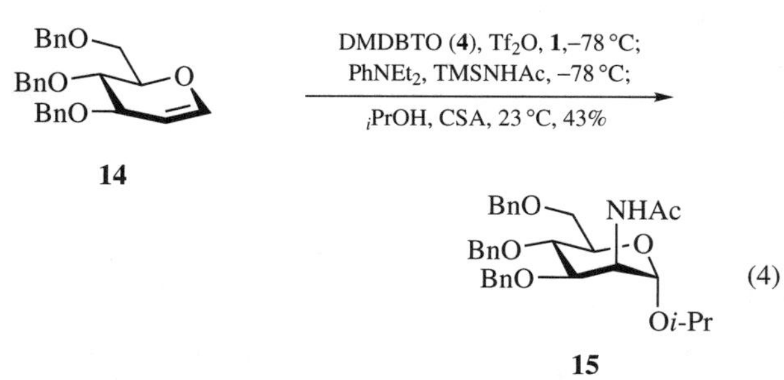

Scavenger of Triflic Acid in the Aromatic Acylation Reactions. The acylation of aromatic compounds with aryl triflates can be performed in organic solvents (1,2-dichloroethane) without Lewis acid catalysts.[3] Thus, a solution of benzoyl triflate (**17**) in dichloroethane mixed with toluene at room temperature provided the Friedel-Crafts product (eq 5). Kinetic studies showed that the acylations under basic conditions proceeded 5–10 times faster than comparable reactions in the presence of 1.5 equiv of triflic acid.[3b]

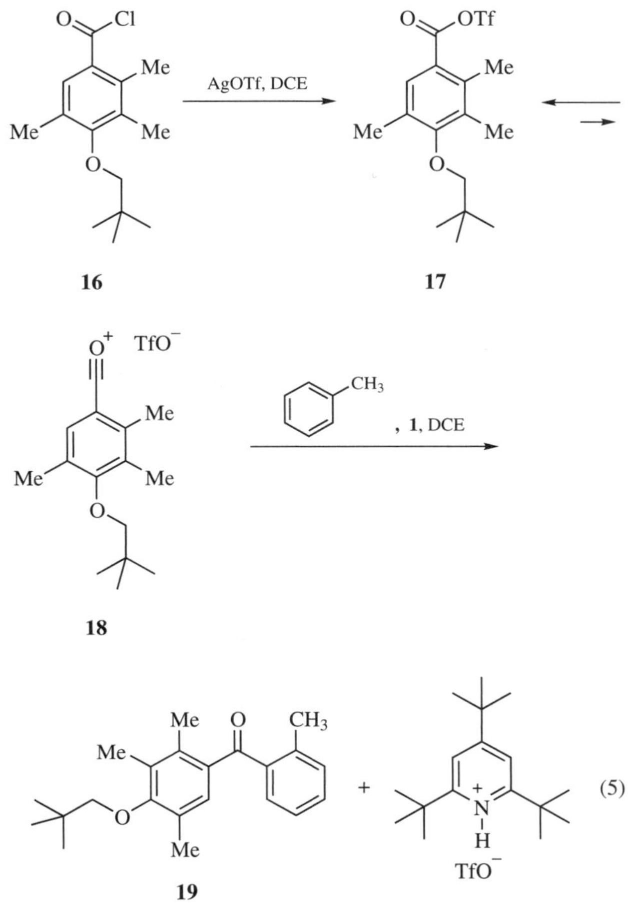

Scavenger of Hydrogen Chloride, Hydrogen Bromide, and Perchlorate Acid. It is frequently desirable to recover protic acids either to protect the environment or for use in other chemical processes. One recycling method contacts the acid with a sterically hindered organic base to form a salt. The salt can then be isolated and subsequently dissociated at higher temperatures. Due to the difference in stability toward heat, salts from different combinations of acids and bases can be cleaved at different temperatures. 2,4,6-Tri-*tert*-butylpyridine (**1**) has been used in this manner to separate mixtures of anhydrous HCl and HBr in industrial settings.[4]

The base **1** has also been employed to trap the HCl formed in the condensation of aryl chlorides and amines (eq 6).[5]

Additionally, **1** has been used to quench perchloric acid formed in tritylation reactions with trityl perchlorate. Triphenylmethyl ethers are widely used in carbohydrate chemistry, and usually mask primary hydroxyl groups. Tritylation of secondary hydroxyl groups is very slow under standard conditions, but the difficulty can be overcome by use of triphenymethylium perchlorate (**24**) in combination with the base **1** (eq 7).[6]

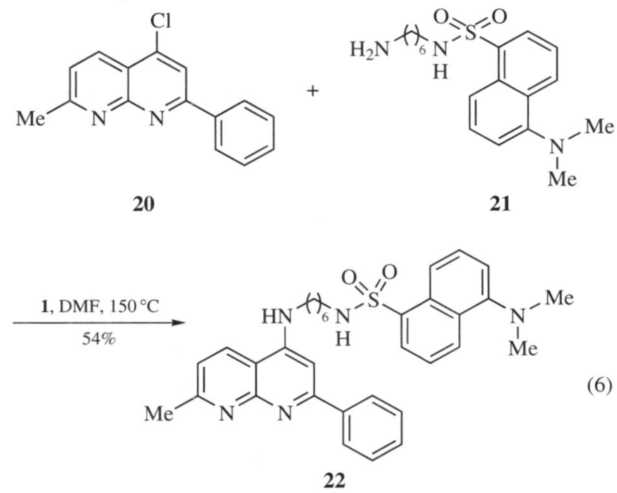

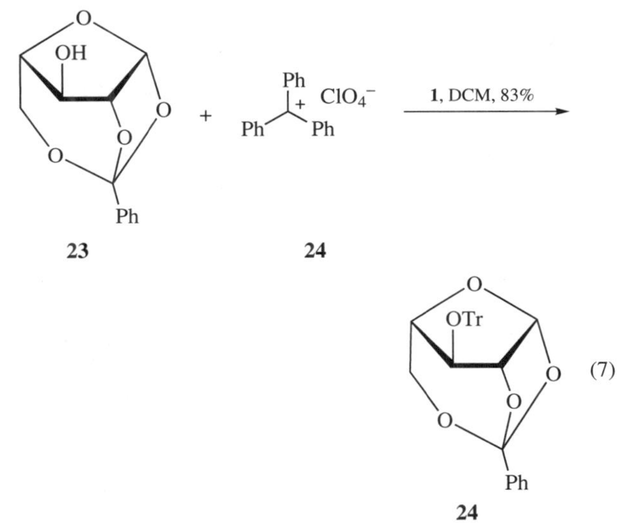

Related Reagents. 2,6-Di-*tert*-butylpyridine; 2,6-di-*tert*-butyl-4-methylpyridine; 2,6-diphenylpyridine; 2,6-diphenoxypyridine; 2,4,6-tri-*tert*-butylpyrimidine.

1. (a) Brown, H. C.; Kanner, B., *J. Am. Chem. Soc.* **1953**, *75*, 3865. (b) Brown, H. C.; Kanner, B., *J. Am. Chem. Soc.* **1966**, *88*, 986.
2. (a) Garcia, B. A.; Gin, D. Y., *Org. Lett.* **2000**, *2*, 2135. (b) Kim, J. Y.; Di Bussolo, V.; Gin, D. Y., *Org. Lett.* **2001**, *3*, 303. (c) Nguyen, H. M.; Poole, J. L.; Gin, D. Y., *Angew Chem. Int. Ed.* **2001**, *40*, 414. (d) Kim, J. Y.; Gin, D. Y., *Org. Lett.* **2001**, *3*, 1801. (e) Honda, E.; Gin, D. Y., *J. Am. Chem. Soc.* **2002**, *124*, 7343. (f) Liu, J.; Di Bussolo, V.; Gin, D. Y., *Tetrahedron Lett.* **2003**, *44*, 4015.
3. (a) Effenberger, F.; Eberhard, J. K.; Maier, A. H., *J. Am. Chem. Soc.* **1996**, *118*, 12572. (b) Effenberger, F.; Maier, A. H., *J. Am. Chem. Soc.* **2001**, *123*, 3429.
4. Jackson Jr., H. C.; DeVries, R. A.; Frick, H. R., U.S. Pat. 4,830,846. A 19890516 (1989).

5. Macchia, M.; Bertini, S.; Di Bussolo, V.; Manera, C.; Martini, C.; Minutolo, F.; Mori, C.; Saccomanni, G.; Tuscano, D.; Ferrarini, P. L., *Il Farmaco.* **2002**, *57*, 783.
6. Wozney, Y. V.; Kochetkov, N. H., *Carbohydr. Res.* **1977**, *54*, 300.

Hongmei Li
Schering-Plough Research Institute, Kenilworth, NJ, USA

Tri-*t*-butylpyrimidine

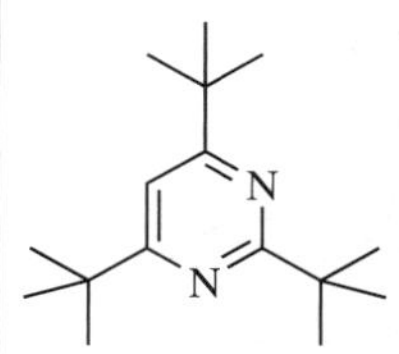

1

[67490-21-5] $C_{16}H_{28}N_2$ (MW 248.41)

(reagent used as a sterically hindered base in glycosylation reactions and in the formation of vinyl triflates)

Alternate Name: 2,4,6-tri-*tert*-butylpyrimidine, TTBP.
Physical Data: mp 77–78 °C.
Solubility: soluble in dichloromethane and toluene.
Form Supplied in: white crystals; commercially available from Aldrich and Lakeview Synthesis.
Preparative Methods: a solution of pinacolone (65.7 mL, 0.52 mol) in CH_2Cl_2 (100 mL) is slowly added to a solution containing Tf_2O (100.8 mL, 0.60 mol) and trimethylacetonitrile (145 mL, 1.28 mol) in CH_2Cl_2 (350 mL) under an argon atmosphere at 20 °C. The reaction mixture is stirred for 6 d and then quenched by the addition of saturated aq $NaHCO_3$, washed with brine, dried ($MgSO_4$), and concentrated under reduced pressure. The residue is crystallized from MeOH to give tri-*tert*-butylpyrimidine as a white solid (91.0 g, 70%).
Purification: crystallization from methanol.
Handling, Storage, and Precautions: toxicity unknown. Should be used in a fume hood.

Glycosylation Reactions. TTBP (**1**) was developed as a cost effective and user-friendly alternative to 2,6-di-*tert*-butylpyridine (**2**) and its 4-substituted analogs **3** and **4**.[1] TTBP offers the advantages of higher crystallinity, reduced hygroscopicity, and lower volatility when compared to the low melting solids **2–4**.

It was demonstrated that TTBP can be used as a non-nucleophilic base in glycosylation reactions using the sulfoxide method. Thus, reaction of mannosyl sulfoxide **5** with trifluoromethanesulfonic anhydride (Tf_2O) followed by addition of glucofuranose acceptor **6** in the presence of TTBP gave β-mannoside **7** in 90% yield (eq 1). Similarly, treatment of glucosyl sulfoxide **8** with Tf_2O and TTBP followed by addition of adamantanol (**9**) gave β-glucoside **10** in 73% yield (eq 2).

In spite of its low basicity (*p*Ka 1.02 in 50% aqueous ethanol, cf. *p*Ka's of 3.58, 4.02, and 4.41 for **2**, **3**, and **4**, respectively, in the same solvent),[2–4] TTBP is sufficient to buffer the triflic acid formed in these glycosylation reactions.

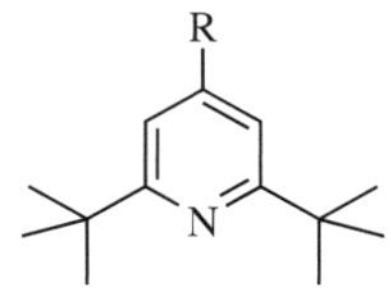

2; R = H
3; R = *tert*-Bu
4; R = Me

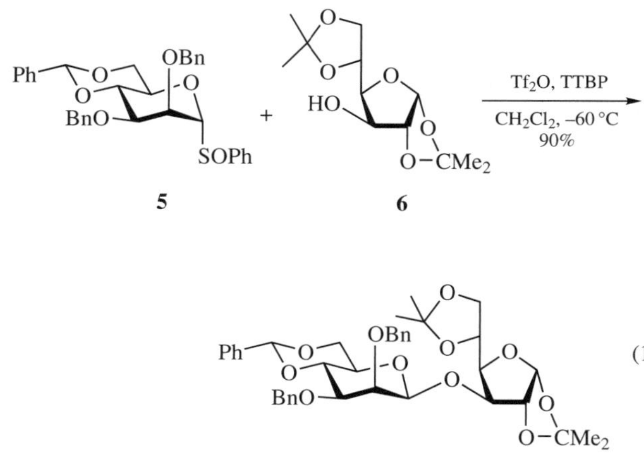

5 **6**

7 (1)

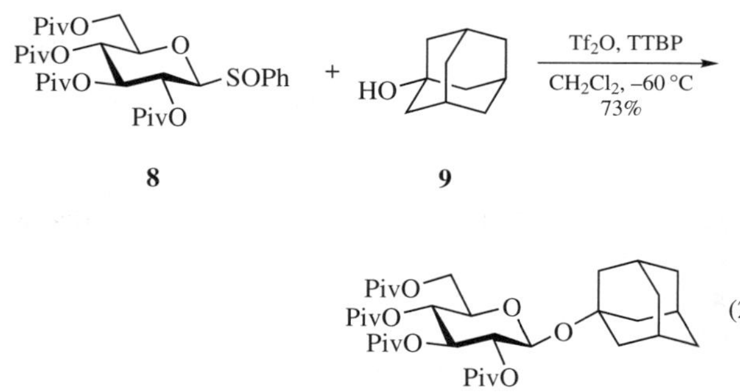

8 **9**

10 (2)

TTBP was also applied as a base in Gin's dehydrative glycosylation method.[5] Thus, hemiacetal **11** was treated with Tf_2O, diphenyl sulfoxide, and glycosyl acceptor **12** to give glucoside **13** as a 60:40 mixture of α/β isomers in 81% yield (eq 3).

The activation of thioglycosides with 1-benzenesulfinyl piperidine (BSP, **14**) and Tf_2O has been explored extensively using TTBP as a non-nucleophilic base. For example, reaction of glucosamine donor **15** with the BSP/Tf_2O/TTBP reagent combination followed by addition of 4,6-glucopyranosyl diol (**16**) gave trisaccharide **17** in 85% yield as a single isomer (eq 4).[6]

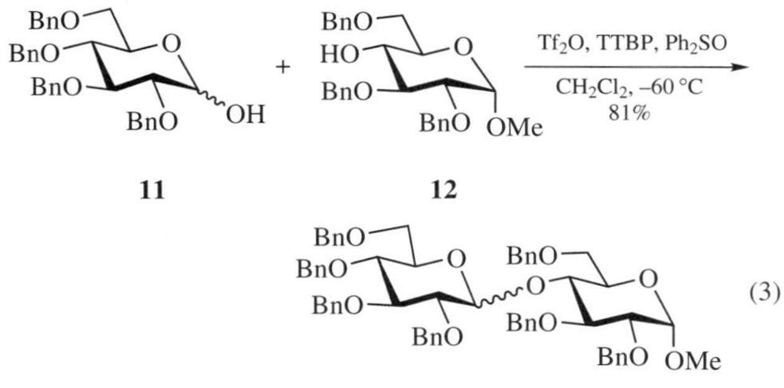

11 12 13 (3)

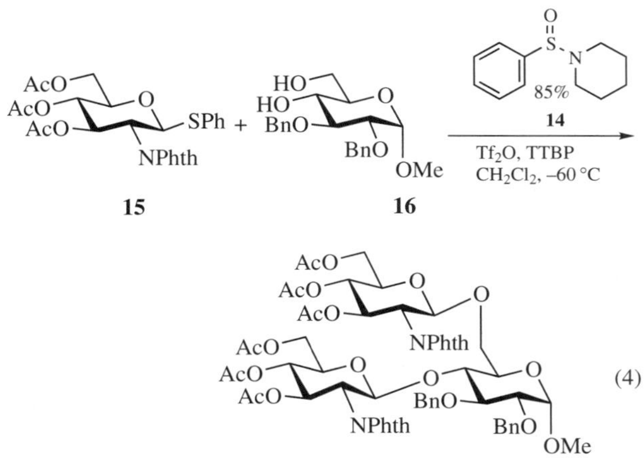

15 16 14 17 (4)

It is noteworthy that under the same reaction conditions orthoester **20** was formed as a single product in 70% yield from the reaction between perbenzoylated xylosyl donor **18** and methyl glucoside **19** (eq 5). Thus, TTBP is sufficiently basic to prevent the acid catalyzed rearrangement of kinetic orthoesters in some instances of neighboring group directed glycosylations.

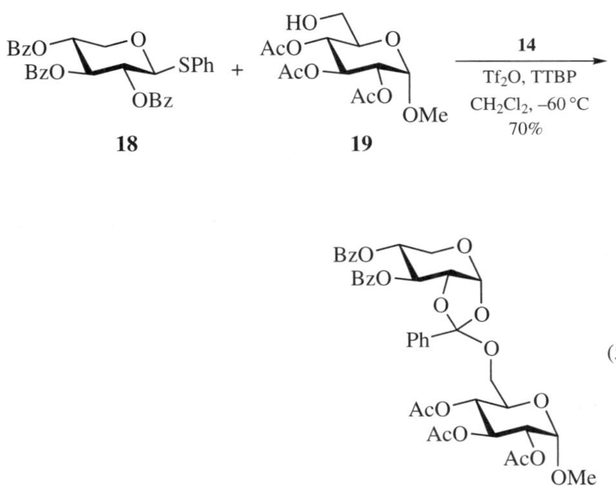

18 19 14 20 (5)

Treatment of thiomannoside **21**, bearing a 4,6-*O*-polystyrylboronate group, with BSP/Tf_2O/TTBP at −60 °C in dichloromethane followed by addition of glucosyl acceptor **19** provided the β-mannoside. β-Mannoside **23** was obtained from the resin by heating in aqueous acetone in 73% overall yield (eq 6).[7]

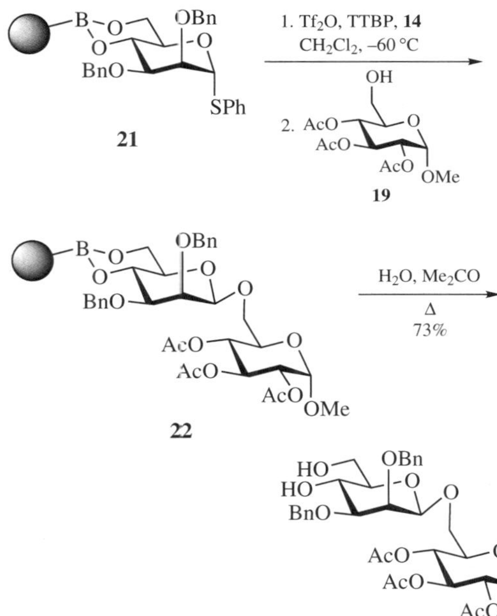

21 19 22 23 (6)

Formation of Vinyl Triflates. TTBP is also an effective base in the synthesis of vinyl triflates. Thus, reaction of cyclohexanone (**24**) with Tf_2O and TTBP in dichloromethane at room temperature provided vinyl triflate (**25**) in 62% yield (eq 7).[1]

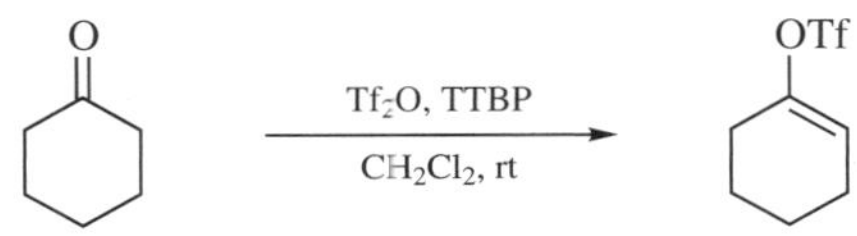

24 25 (7)

Related Reagents. 2,6-Di-*tert*-butylpyridine; 2,6-di-*tert*-butyl-4-methylpyridine; 2,4,6-tri-*tert*-butylpyridine.

1. Crich, D.; Smith, M.; Yao, Q.; Picione, J., *Synthesis* **2001**, 323.
2. Van der Plas, H. C.; Koudijs, A., *Rec. Trav. Chim. Pays Bas* **1978**, *97*, 159.
3. Brown, H. C.; Kanner, B., *J. Am. Chem. Soc.* **1953**, *75*, 3865.
4. Deustch, E.; Cheung, N. K. V., *J. Org. Chem.* **1973**, *38*, 1123.
5. Garcia, B. A.; Poole, J. L.; Gin, D. Y., *J. Am. Chem. Soc.* **1997**, *119*, 7597.
6. Crich, D.; Smith, M., *J. Am. Chem. Soc.* **2001**, *123*, 9015.
7. Crich, D.; Smith, M., *J. Am. Chem. Soc.* **2002**, *124*, 8867.

Mark Smith

Roche Palo Alto, Palo Alto, CA, USA

Trichloroacetonitrile

Cl_3CCN

[545-06-2] C_2Cl_3N (MW 144.38)

(makes trichloroacetimidates from alcohols; trichloroacetimidates are used to introduce nitrogen into molecules via rearrangements[1c] and cyclizations; trichloroacetimidates are useful alkylating agents[1d,e])

Physical Data: mp −42 °C; bp 83–84 °C; *d* 1.440 g cm^{-3}.
Solubility: soluble in most organic solvents.
Form Supplied in: neat colorless liquid.
Analysis of Reagent Purity: ^{1}H NMR, ^{13}C NMR.
Handling, Storage, and Precautions: toxic lachrymator; use only in a fume hood. Reagent can be absorbed through the skin. Always wear gloves when handling this reagent.

Original Commentary

Patrick G. McDougal
Reed College, Portland, OR, USA

Preparation of Trichloroacetimidates. The imidates derived from the addition of alcohols to trichloroacetonitrile have become important and versatile intermediates in synthetic chemistry.[1] Consequently the principal synthetic use of trichloroacetonitrile has been in the formation of these useful trichloroacetimidate intermediates. The imidates are most often prepared by simple addition of a sodium or potassium alkoxide to the electron-deficient trichloroacetonitrile (eq 1).[1c,d,2] In certain cases, slight modifications of the above procedure are utilized.[2d,3] The product imidates are isolable and, despite their propensity to rearrange,[2b,e,4] can in some instances be purified by distillation[2a,d,e] or chromatography. Typically the addition reaction is sufficiently clean to use the imidates without further purification.[2f]

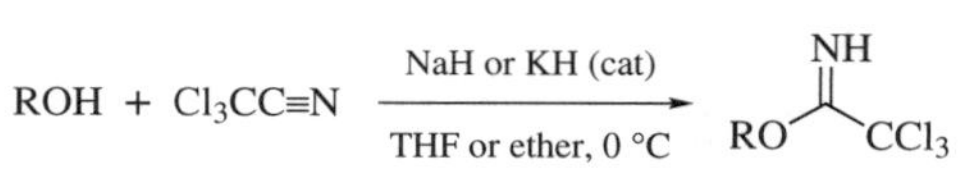

(1)

Rearrangement of Trichloroacetimidates. The thermal rearrangement of imidates has been known for many years.[2b,4] In 1974, Overman[2c] described a useful conversion of allylic alcohols to allylic amines utilizing a [3,3]-sigmatropic rearrangement of allylic trichloroacetimidates (eq 2).[1c,2e] In addition to simple thermolysis it was found that the rearrangement could be catalyzed at room temperature with either Hg^{II} or Pd^{II} salts.[1c,5] As with many sigmatropic rearrangements, the reaction is stereoselective and produces double bonds of defined stereochemistry while efficiently transferring chirality.[1c,5,6]

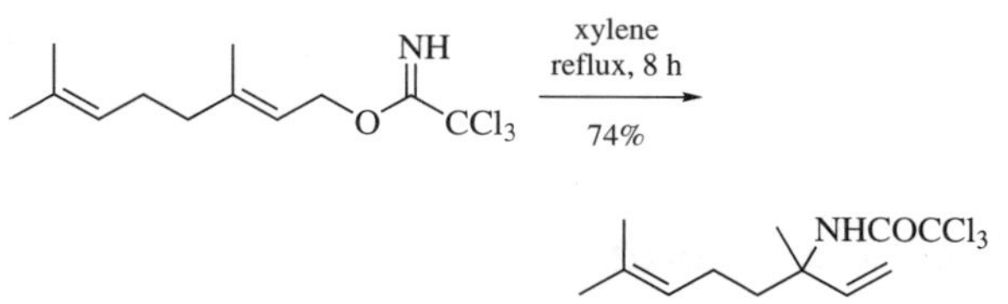

(2)

A comparison of reaction conditions[5b,c,d] has shown that palladium catalysis is particularly effective in achieving complete chiral transfer (eq 3).[5b] Propargylic imidates also rearrange when heated in refluxing xylene.[1c,7] The initially formed allene undergoes a series of tautomerizations so that amino 1,3-dienes are the ultimate products (eq 4).[1c] These dienes have found use in the Diels–Alder reaction.[1c]

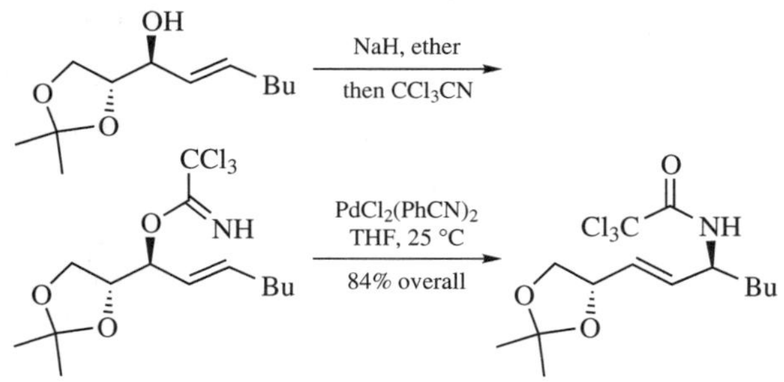

(3)

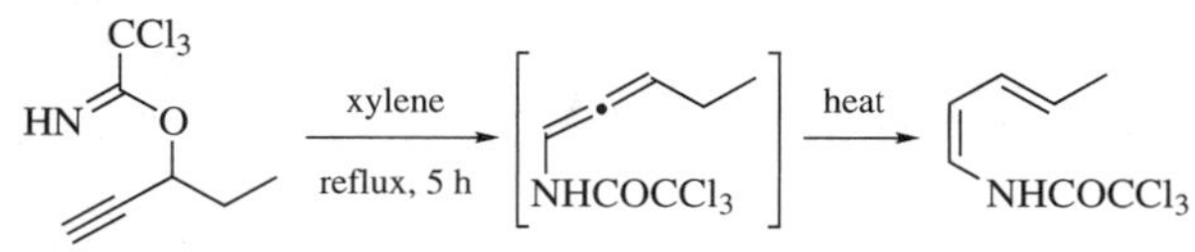

(4)

Electrophilic Cyclization of Trichloroacetimidates. Trichloroacetimidates derived from allylic[8a,c–f] and homoallylic[8b,c] alcohols undergo electrocyclic ring closure when treated with a source of I^{+} (eq 5).[8c] Cyclization can also be triggered via the Lewis acid-mediated opening of epoxides.[9] In at least one case the imidate proved more reactive than related reactions using carbamates.[9c] In the case of imidates from allylic alcohols, the regioselectivity of the process is dependent on the structure of the substrate, the more substitued carbon of the epoxide being often the favorite site for the nucleophilic attack of the nitrogen.[9d] These are useful methods for the stereoselective introduction of nitrogen into cyclic and acyclic systems.

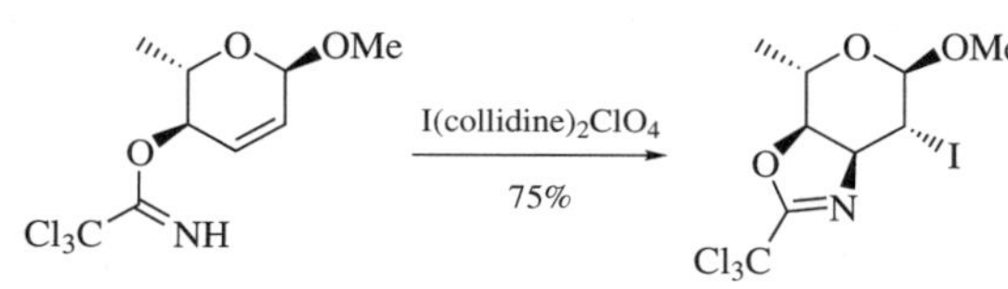

(5)

Alkylation using Trichloroacetimidates. Under mildly acidic conditions, alkyl trichloroacetimidates act as reasonable alkylating agents[2f,10] toward heteroatom nucleophiles, most notably alcohols (eq 6). Alkyl groups which have been transferred include benzyl,[2f,10a,b,d] 2,6-dichlorobenzyl,[10a] *para*-methoxybenzyl,[10d,g,h] *para*-phenylbenzyl,[10i] allyl,[2f] propargyl,[10e] *t*-butyl,[10c] 2-oxopyrrolidin-5-yl,[10j] and phthalimidomethyl.[10k] This methodology has been used to protect alcohols. In addition, such an approach has also been exploited for the anchorage of a variety of alcoholic compounds on insoluble supports. In particular, resins functionalized with *para*-alkoxybenzyl alcohols can be easily transformed into the corresponding trichloroacetimidates and the activation of these latter functionalities with an acid allows for the attachment of the alcoholic derivative to the solid support through the formation of a benzyl ether linkage.[10l,m] Notably, these trichloroacetimidate resins are much more stable than the analogous trichloroacetimidates in solution. The use of glycosyl trichloroacetimidates has found widespread use in the synthesis of oligosaccharides and glycoconjugates (eq 7).[1d,e,f,11] This is a particularly powerful method for controlling stereochemistry at the anomeric position.

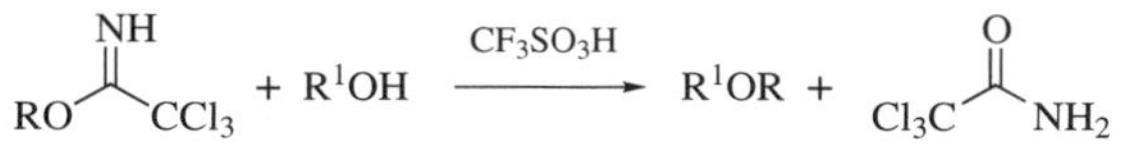

(6)

R = $ArCH_2$, $CH_2{=}CHCH_2$, $MeC{\equiv}CCH_2$, *t*-Bu

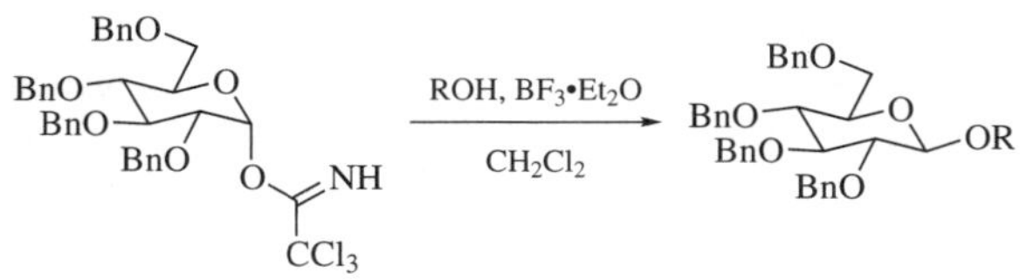

(7)

First Update

Alfonso Iadonisi
Università di Napoli, Italy

Glycosylation. The trichloroacetimidate method represents one of the most widely employed glycosidation approaches, and its efficacy is well demonstrated by the successful application in the assembly of a number of complex oligosaccharidic structures.[11c] The stereochemistry of the glycosidic bond created can be controlled by tuning several parameters such as the reaction solvent and the nature of protecting groups installed at position 2 of the glycosyl trichloroacetimidate. For example, 1,2-*trans* glycosides (such as β-glucosides, β-galactosides, or α-mannosides) can be generally synthesized with high stereocontrol adopting 2-*O*-acylated glycosyl donors. This selectivity arises from the participation ability of the ester groups located in the proximity of the electrophilic anomeric position thereby directing the nucleophilic attack of the alcoholic acceptor from the less hindered face. In absence of an acyl participating group, substitution with inversion of configuration at the anomeric center typically occurs in solvents of low polarity and with reactive alcoholic glycosyl acceptors. On this basis, the control of the anomeric configuration of the glycosyl trichloroacetimidates turns out to be very useful, and this can be achieved by the suitable choice of the base to be used in the trichloroacetimidation step. Indeed, the thermodynamically favored α-trichloroacetimidates are formed with stronger bases (NaH, DBU), whereas the corresponding β-anomers are prevalently formed under conditions of kinetic control in the presence of a milder base (K_2CO_3). When glycosidations are performed in more polar solvents, a SN1 type mechanism is reasonably involved with the generation of a glycosyl cation as an intermediate. Under these conditions, the specific nature of the solvent strongly influences the anomeric composition of the glycosidation products. In particular, ether solvents often favor the predominance of α-pyranosides while the use of nitrile solvents commonly results in the opposite β-selectivity.[1f] As is generally found in the glycosidation methodologies based on electrophilic glycosyl donors, the experimental conditions for activation of glycosyl trichloroacetimidates (temperature, relative amounts of acidic promoter, etc.) are notably differentiated and decisively influenced by the nature of the corresponding protecting groups, acyl protected glycosyl donors requiring relatively more forced conditions than the alkylated counterparts.

The trichloroacetimidate methodology for the glycosidic bond construction has also been exploited in the solid-phase synthesis of oligosaccharides based on the glycosidation of polymer bound glycosyl acceptors.[12] In addition, such a method has also been applied in the first examples of automated synthesis of oligosaccharides.[12d,e]

Some recent reports have described the feasible use of milder and moisture stable Lewis acids, such as lanthanide triflates,[13a,10h] lithium perchlorate,[13b,c] and acid washed molecular sieves[13d] for the activation of some glycosyl and alkyl trichloroacetimidates, as an alternative to the sensitive, conventional Lewis acids (for example, $BF_3{\cdot}OEt_2$ or TMSOTf) typically adopted for this purpose.

Trichloroacetimidate groups installed on allylic alcohols can act as leaving groups in substitution reactions proceeding with allylic rearrangement. For example, this kind of mechanism has been observed in the palladium mediated synthesis of a vinyl oxazoline from the bis-trichloroacetimidate derivative of *cis*-butene-1,4-diol,[14a] and in the acid promoted Ferrier rearrangement of a suitably protected galactal derivative (eq 8).[14b] It should be outlined that the latter rearrangement generally proceeds in low yields under a variety of conditions when adopting galactal precursors derivatized with less efficient leaving groups at *O*-3.

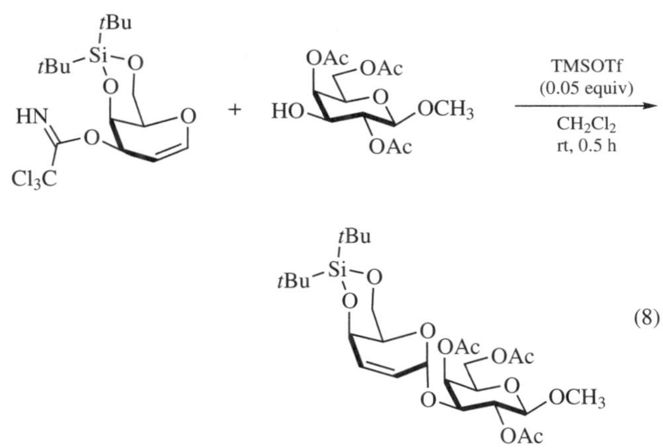

(8)

yield 88%

Trichloroacetimidates as Protecting Groups. As previously illustrated, alkyl trichloroacetimidates are commonly exploited to deliver a protecting group to an hydroxyl functionality in the presence of an acidic catalyst, and trichloroacetamide is produced in this case as a by-product of the reaction. However, there are some reports concerning the use of the trichloroacetimidate group for protection of non-anomeric positions of saccharidic derivatives.[15] Not surprisingly, this protecting group is rather stable under the anhydrous acidic conditions required for the activation of anomeric trichloroacetimidates (eq 9).

Consequently, an advantage of this application is that the protection of the non-anomeric saccharidic hydroxyl can be accomplished simultaneously with the installation of the leaving group at the anomeric center. The cleavage of the trichloroacetimidate protecting group can be accomplished under acidic conditions in the presence of a nucleophilic solvent (eq 9).[15] Alternatively, the regeneration of the alcohol can be carried out by exploiting a β-elimination reaction promoted either by a base or a reducing agent to yield the trichloroacetonitrile and the dichloroketenimine as by-products, respectively.[15c]

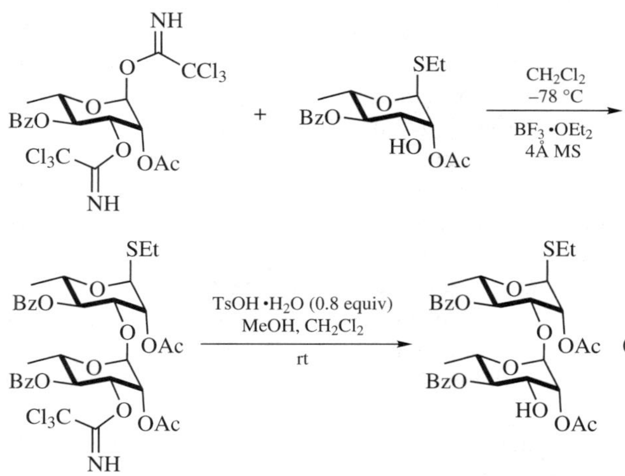

(9)

yield 64% yield 81%

Activation for Acylation. The combined system PPh_3/trichloroacetonitrile can be usefully exploited for the fast conversion of carboxylic acids into corresponding acyl chlorides under mild conditions.[16] Analogous to the standard PPh_3/CCl_4 system, a triarylphosponium chloride is here generated as a reactive intermediate able to convert the carboxylic acid into the corresponding acyl chloride, and triphenylphosphine oxide and dichloroacetonitrile are further by-products of the reaction (eq 10). The use of trichloroacetonitrile in place of carbon

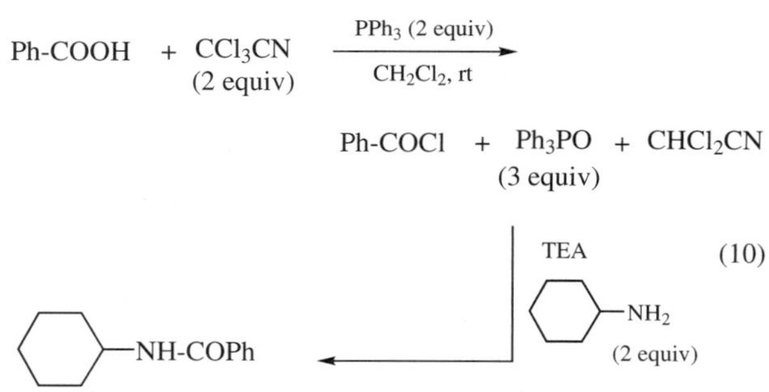

(10)

yield 97%

tetrachloride can be advantageous, the transformations being achieved within sensibly shorter times at room temperature. In addition, the procedure is efficient without using an elevated stoichiometric excess of trichloroacetonitrile. The acyl chlorides generated can be converted into amides in a one-pot fashion by adding the suitable nucleophilic amine and triethylamine acting as a base (eq 10). Peptide bonds can be also obtained without racemization.[16a] This procedure proved effective even when applied with a polymer bound phosphine and a polymer bound base. In this case, satisfyingly pure amides can be obtained with a simple filtration.[16b] Furthermore, the activation protocol was found to be efficient in the solid-phase acylation of amines anchored to solid supports, and the procedure often afforded better results than when adopting standard coupling agents.[16c]

Besides the amidation reaction, the methodology described for the 'in situ' generation of acyl chlorides can be also exploited in other synthetic transformations of interest. For example, the exposure of carboxylic acids to the PPh_3/trichloroacetonitrile system in the presence of triethylamine represents a mild procedure for the synthesis of symmetrical anhydrides.[16d] In another application, the one-step synthesis of azetidin-2-ones can be carried out by reacting imines with carboxylic acids in the presence of trichloroacetontrile, triphenylphosphine, and triethylamine.[16e]

Trichloroacetonitrile in the Synthesis of Six-Membered Nitrogenated Heterocycles. Several examples have been reported on the use of trichloroacetonitrile in cyclization reactions leading to the synthesis of six-membered nitrogenated heterocycles bearing a trichloromethyl substituent adjacent to the heteroatom.[17] These cyclizations can occur with co-reagents possessing a nucleophilic and an electrophilic site at a suitable distance of four atoms. As a general mechanistic scheme, these reactions could be rationalized considering an initial attack from the nucleophilic site of the substrate to the electron deficient nitrile carbon of CCl_3CN, followed by the cyclization step resulting from a nucleophilic attack of the electron rich nitrogen of the initial adduct to the electrophilic site of the substrate. This strategy has been employed for construction of trichlomethylated pyridine and pyrimidine rings (eq 11). The treatment of aromatic methyl ketimines with neat trichloroacetonitrile at high temperatures allows for the synthesis of pyrimidines with two trichloromethyl substitutents. In this case, two molecules of trichloroacetonitrile are incorporated in the cyclic product.[17d]

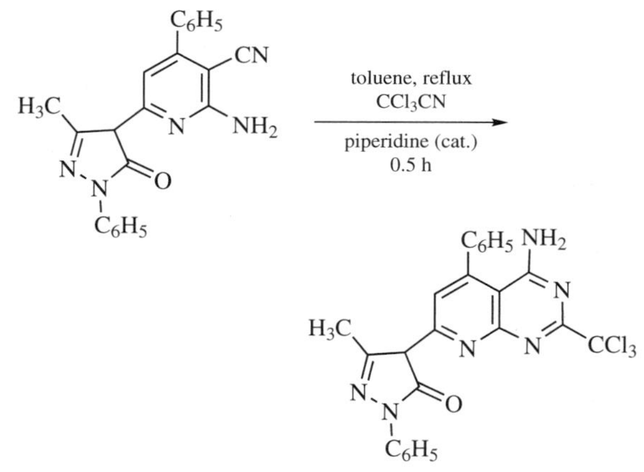

(11)

The synthesis of trichloromethylated pyridines can be also conducted through a [2+2+2] cycloaddition of CCl_3CN with 1,6-diynes catalyzed by Ru(II) derivatives.[17e] In this latter case, a double cyclization is occurring during the reaction to afford a bicyclic product with a five-membered ring fused to the pyridine ring.

Other Reactions. As previously shown, the treatment of alcohols with trichloroacetonitrile in the presence of a base affords the corresponding trichloroacetimidate functionality. However, an unusual behavior has been observed in the case of β-hydroxy α-diazo carbonyl derivatives. Indeed, with such substrates the treatment results in a direct substitution reaction converting the hydroxyl functionality into a trichloroacetylamino group (eq 12).[18] As little racemization is observed in the course of the reaction with optically active alcohols, the process is believed to occur through the initial formation of a trichloroacetimidate intermediate and its fast rearrangement into the amide via a tight ion pair. However, the origin of this peculiar reactivity of β-hydroxy α-diazo carbonyl derivatives has not yet been established.

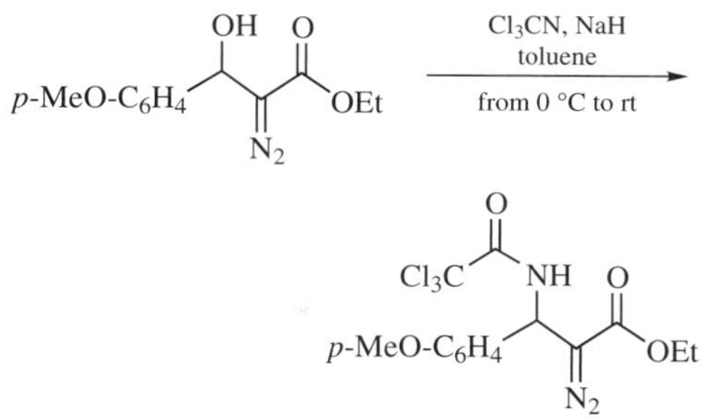

(12)

yield 90%

Trichloroacetonitrile can be used in the dichloroacyanomethylation of carbonyl compounds in the presence of InBr. The transformation is triggered by the conversion of CCl_3CN into a nucleophilic organoindium species through the initial oxidative insertion of InBr. This trivalent organoindium intermediate is able to add to a variety of aldehydes and ketones to yield 2,2-dichloro-3-hydroxynitriles in good yield and short reaction times (eq 13).[19] This procedure turned out to be much more efficient than when using the $^{-}CCl_2CN$ nucleophile generated in the presence of a base.

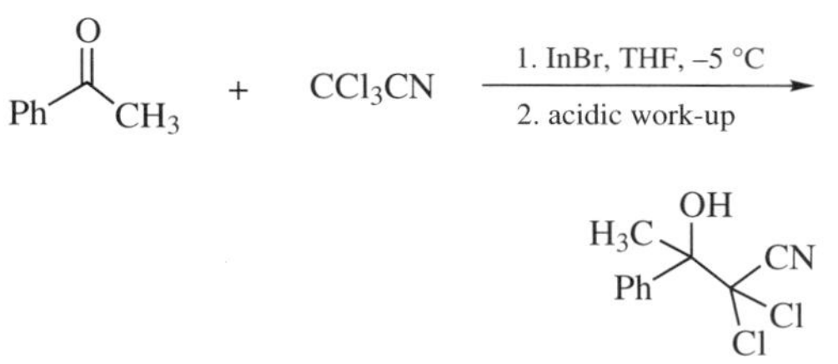

(13)

yield 73%

1. (a) Sandler, S. R.; Karo, W., *Organic Functional Group Preparations*; Academic: New York, 1972, Vol. 3, Chapter 8. (b) Patai, S., *The Chemistry of Amidines and Imidates*; Wiley: New York, 1975. (c) Overman, L. E., *Acc. Chem. Res.* **1980**, *13*, 218. (d) Schmidt, R. R., *Angew. Chem., Int. Ed. Engl.* **1986**, *25*, 212. (e) Schmidt, R. R., *Pure Appl. Chem.* **1989**, *61*, 1257. (f) Schmidt, R. R.; Kinzy, W., *Adv. Carbohydr. Chem. Biochem.* **1994**, *50*, 21.
2. (a) Cramer, F.; Pawelzik, K.; Baldauf, H. J., *Ber. Dtsch. Chem. Ges.* **1958**, *91*, 1049. (b) Cramer, F.; Hennrich, N., *Ber. Dtsch. Chem. Ges.* **1961**, *94*, 976. (c) Overman, L. E., *J. Am. Chem. Soc.* **1974**, *96*, 597. (d) Overman, L. E., *J. Am. Chem. Soc.* **1976**, *98*, 2901. (e) Clizbe, L. A.; Overman, L. E., *Org. Synth.* **1978**, *58*, 4. (f) Wessel, H.-P.; Iverson, T.; Bundle, D. R., *J. Chem. Soc., Perkin Trans. 1* **1985**, 2247.
3. (a) Hauser, F. M.; Ellenberger, S. R.; Glusker, J. P.; Smart, C. J.; Carrell, H. L., *J. Org. Chem.* **1986**, *51*, 50. (b) Oehler, E.; Kotzinger, S., *Synthesis* **1993**, 497.
4. (a) Mumm, O.; Möller, F., *Ber. Dtsch. Chem. Ges.* **1937**, *70*, 2214. (b) McCarty, C. G.; Garner, L. A., In *The Chemistry of Amidines*; Patai, S., Ed.; Wiley: New York, 1975, Chapter 4.
5. (a) Overman, L. E., *Angew. Chem., Int. Ed. Engl.* **1984**, *23*, 579. (b) Metz, P.; Mues, C.; Schoop. A., *Tetrahedron* **1992**, *48*, 1071. (c) Mehmandoust, M.; Petit, Y.; Larcheveque, M., *Tetrahedron Lett.* **1992**, *33*, 4313. (d) Doherty, A. M.; Kornberg, B. E.; Reily, M. D., *J. Org. Chem.* **1993**, *58*, 795.
6. (a) Yamamoto, Y.; Shimoda, H.; Oda, J.; Inouye, Y., *Bull. Chem. Soc. Jpn.* **1976**, *49*, 3247. (b) Isobe, M.; Fukuda, Y.; Nishikawa, T.; Chabert, P.; Kawai, T.; Goto, T., *Tetrahedron Lett.* **1990**, *31*, 3327.
7. Overman, L. E.; Clizbe, L. A., *J. Am. Chem. Soc.* **1976**, *98*, 2352.
8. (a) Cardillo, G.; Orena, M.; Porzi, G.; Sandri, S., *J. Chem. Soc. (C)* **1982**, 1308. (b) Cardillo, G.; Orena, M.; Porzi, G.; Sandri, S., *J. Chem. Soc. (C)* **1982**, 1308. (c) Fraser-Reid, B.; Pauls, H. W., *J. Org. Chem.* **1983**, *48*, 1392. (d) Bongini, A.; Cardillo, G.; Orena, M.; Sandri, S.; Tomasini, C., *J. Org. Chem.* **1986**, *51*, 4905. (e) Cardillo, G.; Orena, M.; Sandri, S.; Tomasini, C., *Tetrahedron* **1986**, *42*, 917. (f) Sammes, P. G.; Thetford, D., *J. Chem. Soc., Perkin Trans. 1* **1988**, 111.
9. (a) Jacobsen, S., *Acta Chem. Scand.* **1988**, *B42*, 605. (b) Schuerrle, K.; Beier, B.; Piepersberg, W., *J. Chem. Soc., Perkin Trans. 1* **1991**, 2407. (c) Hart, T. W.; Vacher, B., *Tetrahedron Lett.* **1992**, *33*, 3009. (d) Hatakeyama, S.; Matsumoto, H.; Fukuyama, H.; Mukugi, Y.; Irie, H., *J. Org. Chem.* **1997**, *62*, 2275. (e) Liu. D.-G., Wang, B.; Lin, G.-Q., *J. Org. Chem.* **2000**, *65*, 9114.
10. (a) Amouroux, R.; Gerin, B.; Chastrette, M., *Tetrahedron* **1985**, *41*, 5321. (b) Widmer, U., *Synthesis* **1987**, 568. (c) Armstrong, A.; Brackenridge, I.; Jackson, R. F. W.; Kirk, J. M., *Tetrahedron Lett.* **1988**, *29*, 2483. (d) Audia, J. E.; Boisvert, L.; Patten, A. D.; Villalobos, A.; Danishefsky, S. J., *J. Org. Chem.* **1989**, *54*, 3738. (e) Wei, S. Y.; Tomooka, K.; Nakai, T., *Tetrahedron* **1993**, *49*, 1025. (f) Bourgeois, M. J.; Montaudon, E.; Maillard, B., *Tetrahedron* **1993**, *49*, 2477. (g) Nakajima, N.; Horita, K.; Abe, R.; Yonemitsu, O., *Tetrahedron Lett.* **1988**, *29*, 4139. (h) Rai, A. N.; Basu, A., *Tetrahedron Lett.* **2003**, *44*, 2267. (i) Sharma, G. V. M.; Rakesh, *Tetrahedron Lett.* **2001**, *42*, 5571. (j) El-Nezhawy, A. O. H.; El-Diwani, H. I.; Schmidt, R. R., *Eur. J. Org. Chem.* **2002**, 4137. (k) Ali, I. A. I.; Abdel-Rahman, A. A. H.; El Ashry, E. S. H.; Schmidt, R. R., *Synthesis* **2003**, 1065. (l) Hanessian, S.; Fang, X., *Tetrahedron Lett.* **1998**, *39*, 733. (m) Yan, L. Z.; Mayer, J. P., *J. Org. Chem.* **2003**, *68*, 1161.
11. (a) Nicolaou, K. C.; Daines, R. A.; Ogawa, Y.; Chakraborty, T. K., *J. Am. Chem. Soc.* **1988**, *110*, 4696. (b) Barrett, A. G. M.; Pilipauskas, D., *J. Org. Chem.* **1991**, *56*, 2787. (c) Ernst, B.; Hart, G. W.; Sinaÿ P., *Carbohydrates in Chemistry and Biology*; Wiley: Weinheim, 2000, Vol 1, Chapter 2.
12. (a) Rademann, J.; Schmidt, R. R., *J. Org. Chem.* **1997**, *62*, 3650. (b) Wu, X.; Grathwohl, M.; Schmidt, R. R., *Org. Lett.* **2001**, *3*, 747. (c) Wu, X.; Grathwohl, M.; Schmidt, R. R., *Angew. Chem., Int. Ed. Engl.* **2002**, *41*, 4489. (d) Plante, O. J.; Palmacci, E. R.; Seeberger, P. H., *Science* **2001**, *291*, 1523. (e) Hewitt, M. C.; Seeberger, P. H., *Org. Lett.* **2001**, *3*, 3699.
13. (a) Adinolfi, M.; Barone, G.; Iadonisi, A.; Mangoni, L.; Schiattarella, M., *Tetrahedron Lett.* **2001**, *42*, 5967. (b) Schene, H.; Waldman, H., *Chem. Commun.* **1998**, 2759. (c) Böhm, G.; Waldman, H., *Tetrahedron Lett.* **1995**, *36*, 3843. (d) Adinolfi, M.; Barone, G.; Iadonisi, A.; Schiattarella, M., *Org. Lett.* **2003**, *5*, 987.
14. (a) Sabat, M.; Johnson, C. R., *Org. Lett.* **2000**, *2*, 1089. (b) Abdel-Rahman, A. A. H.; Winterfeld, G. A.; Takhi, M.; Schmidt, R. R., *Eur. J. Org. Chem.* **2002**, 713.
15. (a) Lassaletta, J. M.; Carlsson, K.; Garegg, P. J.; Schmidt, R. R., *J. Org. Chem.* **1996**, *61*, 6873. (b) Qiu, D.; Koganty, R. R., *Tetrahedron Lett.* **1997**, *38*, 961. (c) Yu, B.; Yu, H.; Hui, Y.; Han, X., *Synlett* **1999**, 753.
16. (a) Jang, D. O.; Park, D. J.; Kim, J., *Tetrahedron Lett.* **1999**, *40*, 5323. (b) Buchstaller, H.-P.; Ebert, H. M.; Anlauf, U., *Synth. Commun.* **2001**, *31*, 1001. (c) Vágó, I.; Greiner, I., *Tetrahedron Lett.* **2002**, *43*, 6039. (d) Kim, J.; Jang, D. O., *Synth. Commun.* **2001**, *31*, 395. (e) Govande, V. V.; Arun, M.; Desmukh, A. R. A. S.; Bhawal, B. M., *Synth. Commun.* **2000**, *30*, 4177.
17. (a) Sielecki, M. S.; Johnson, T. L.; Liu, J.; Muckelbauer, J. K.; Grafstrom, R. H.; Cox, S.; Boylan, J.; Burton, C. R.; Chen, H.; Smallwood, A.;

Chang, C.-H.; Boisclair, M.; Benfield, P. A.; Trainor, G. L.; Seitz, S. P., *Biorg. Med. Chem. Lett.* **2001**, *11*, 1157. (b) Erian, A. W.; Issac, Y. A. E.; Sherif, S. M.; Mahmoud, F. F., *J. Chem. Soc., Perkin Trans. 1* **2000**, 3686. (c) Ahmed, R. A.; Kandeel, M. M.; Abbady, M. S.; Youssef, M. S. K., *J. Heterocyclic Chem.* **2002**, *39*, 309. (d) Sosnovskikh, V. Y.; Usachev, B. I.; Röschentaler, G. V., *Tetrahedron* **2002**, *58*, 1375. (e) Yamamoto, Y.; Okuda, S.; Itoh, K., *Chem. Commun.* **2001**, 1102.

18. Shi, W.; Jiang, N.; Zhang, S.; Wu, W.; Du, D.; Wang, J., *Org. Lett.* **2003**, *5*, 2243.
19. Nóbrega, J. A.; Gonçalves, S. M. C.; Peppe, C., *Tetrahedron Lett.* **2001**, *42*, 4745.

Triethylsilyl Trifluoromethanesulfonate[1]

$Et_3SiOSO_2CF_3$

[79271-56-0] $C_7H_{15}F_3O_3SSi$ (MW 264.38)

(potent silylating agent;[2–4] Lewis acid catalyst)

Physical Data: 85–86 °C/12 mmHg; *d* 1.169 g cm^{-3}.

Solubility: readily sol hydrocarbons, dialkyl ethers, halogenated solvents. CH_2Cl_2 is employed most commonly. Reactions in 1,2-dichloroethane proceed faster than those in CCl_4 or Et_2O. Protic solvents and THF react with trialkylsilyl triflates and are therefore not suitable.

Form Supplied in: neat colorless liquid.

Preparative Methods: can be prepared by reacting Chlorotriethylsilane with Trifluoromethanesulfonic Acid followed by distillation.[1]

Handling, Storage, and Precautions: trialkylsilyl triflates are generally corrosive and moisture sensitive. Appropriate precautions should be taken to ensure that the reagent is handled and stored under rigorously anhydrous conditions.

Reactive Silylating Agent. Triethylsilyl ethers are generally more stable towards hydrolysis than are trimethylsilyl ethers, and consequently the Et_3Si moiety has gained increasing use as a protecting group for alcohols. However, since it is often difficult to silylate sterically hindered hydroxyl groups using Et_3SiCl, Triethylsilyl Perchlorate and triethylsilyl triflate (Et_3SiOTf) were introduced to overcome this problem. Although both reagents are much more potent than Et_3SiCl, the triflate is considered to be safer than the perchlorate and is more convenient because it is commercially available.

Secondary and tertiary alcohols can be silylated under mild conditions using Et_3SiOTf and 2,6-Lutidine or a trialkylamine as a proton scavenger (eqs 1 and 2).[2–4]

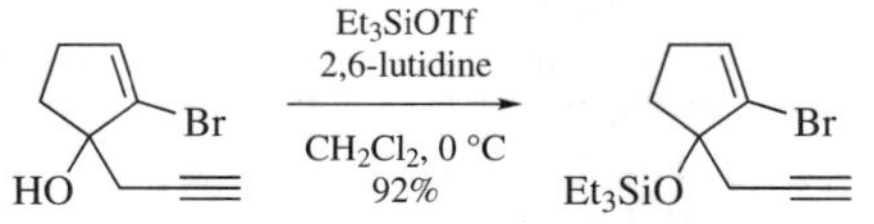

(1)

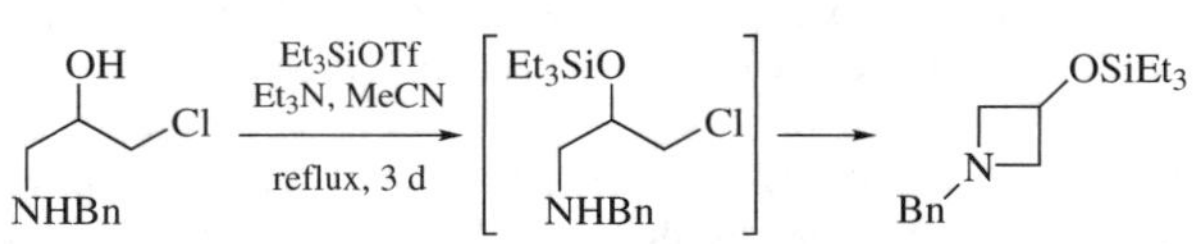

(2)

Silyl enol ethers can be conveniently prepared by treatment of ketones with Et_3SiOTf and Triethylamine (eq 3).[1,5] Similarly, esters can be converted to silyl ketene acetals; however, large amounts of the *C*-silylated product may also be isolated (eq 4).[1,6,7] Equilibration of *O*- and *C*-silylated products may occur in the presence of catalytic amounts of Et_3SiOTf.[1]

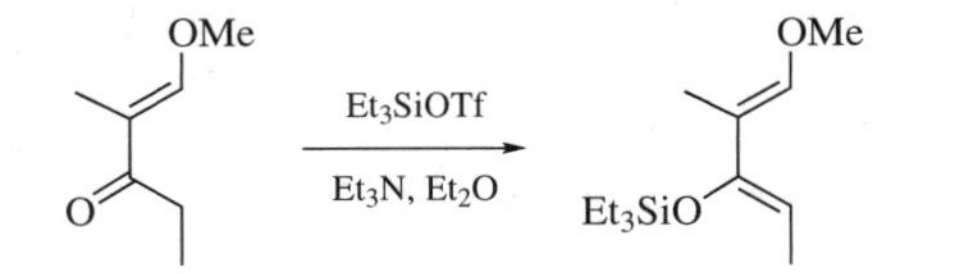

(3)

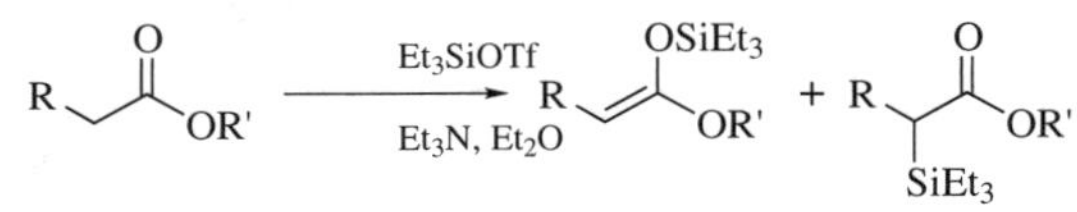

(4)

α-Diazo esters react with Et_3SiOTf and Hünig's base (Diisopropylethylamine) to give exclusively α-silyl-α-diazo esters, which can be converted to other silylated compounds by loss of dinitrogen (eq 5).[8]

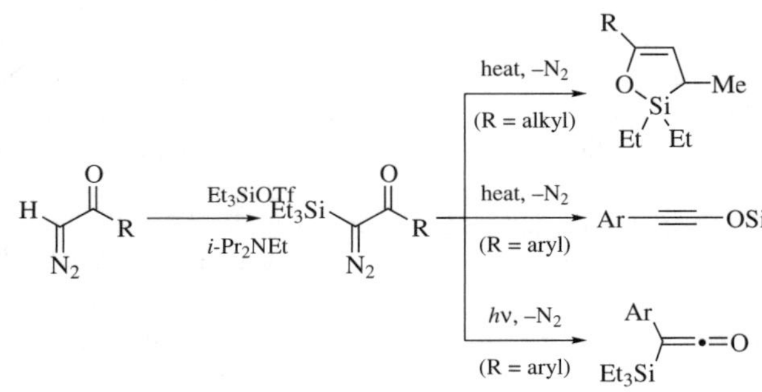

(5)

Nitroalkanes have been shown to give α-siloxy *O*-silyloximes upon reaction with two molar equivalents of the silylating agent (eq 6).[9] The reaction is believed to proceed via a nitrogen-to-carbon 1,3-silyloxy rearrangement.

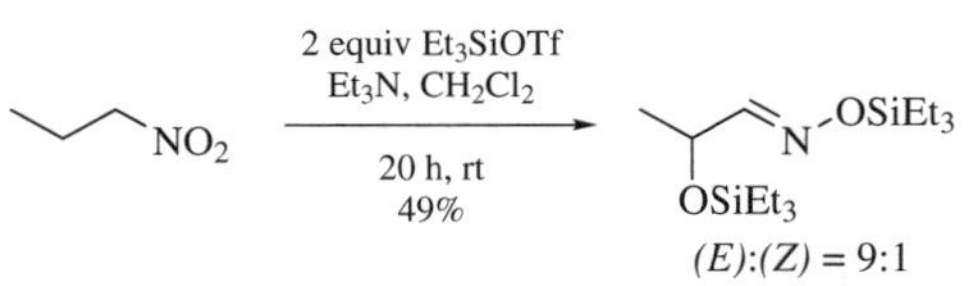

(6)

Linderman and Ghannam have utilized Et_3SiOTf to trap the alkoxide formed from the addition of stannyl anion to aldehydes (eq 7).[10] Upon treatment with excess n-Butyllithium, these adducts undergo a reverse Brook rearrangement to afford α-hydroxysilanes.

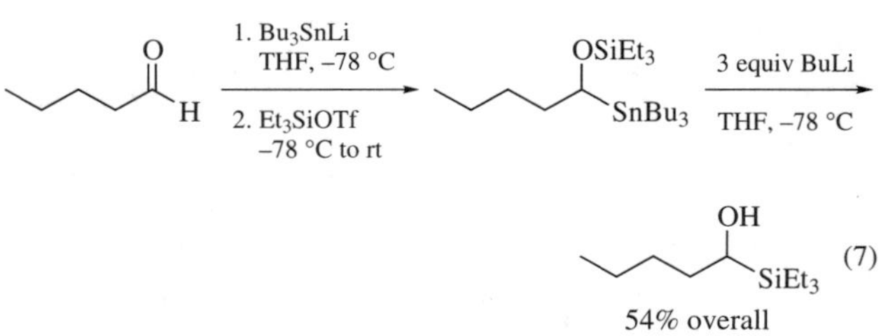

(7)

Lewis Acid Catalyst. Jefford has reported that condensation reactions of 2-trimethylsiloxyfuran with aldehydes can be catalyzed by Et_3SiOTf to give mainly the *threo* addition product (eq 8).[11] Conversely, the *erythro* adduct is favored when fluoride ion is used as the catalyst.

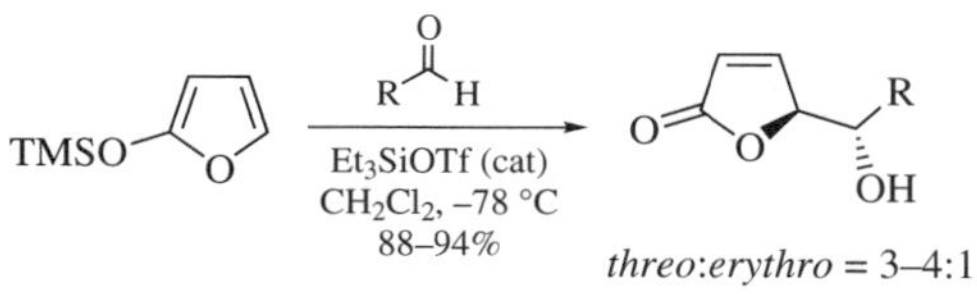

(8)

Fraser-Reid has reported that an equimolar mixture of N-Iodosuccinimide and Et_3SiOTf efficiently promotes the glycosylation of hindered glycoside donors with *n*-pentenyl glycoside acceptors (eq 9).[12]

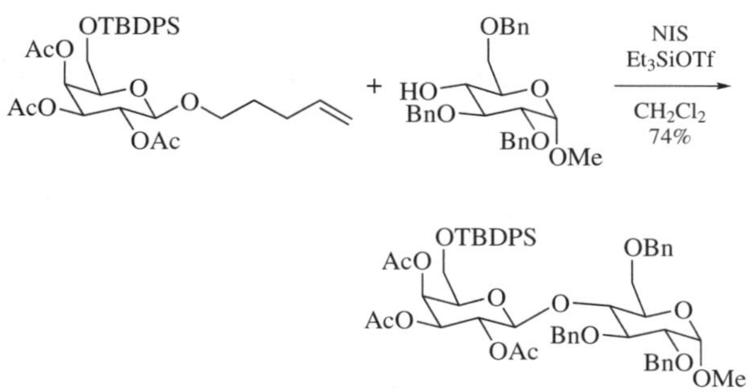

(9)

Related Reagents. Chlorotriethylsilane; Triethylsilyl Perchlorate; Trimethylsilyl Trifluoromethanesulfonate.

1. Emde, H.; Domsch, D.; Feger, H.; Frick, U.; Götz, A.; Hergott, H. H.; Hofmann, K.; Kober, W.; Krägeloh, K.; Oesterle, T.; Steppan, W.; West, W.; Simchen, G. *Synthesis* **1982**, 1.
2. Fujiwara, K.; Sakai, H.; Hirama, M. *J. Org. Chem.* **1991**, *56*, 1688.
3. Higgins, R. H. *J. Heterocycl. Chem.* **1987**, *24*, 1489.
4. McCarthy, P. A.; Kageyama, M. *J. Org. Chem.* **1987**, *52*, 4681.
5. Danishefsky, S.; Harvey, D. F. *J. Am. Chem. Soc.* **1985**, *107*, 6647.
6. Emde, H.; Götz, A.; Hofmann, K.; Simchen, G. *Justus Liebigs Ann. Chem./Liebigs Ann. Chem.* **1981**, 1643.
7. Emde, H.; Simchen, G. *Justus Liebigs Ann. Chem./Liebigs Ann. Chem.* **1983**, 816.
8. (a) Maas, G.; Brückmann, R. *J. Org. Chem.* **1985**, *50*, 2801; (b) Brückmann, R.; Maas, G. *Ber. Dtsch. Chem. Ges./Chem. Ber.* **1987**, *120*, 635.
9. (a) Ferger, H.; Simchen, G. *Justus Liebigs Ann. Chem./Liebigs Ann. Chem.* **1986**, 428; (b) Feger, H.; Simchen, G. *Synthesis* **1981**, 378.
10. (a) Linderman, R. J.; Ghannam, A. *J. Am. Chem. Soc.* **1990**, *112*, 2392; (b) Linderman, R. J.; Ghannam, A. *J. Org. Chem.* **1988**, *53*, 2878.
11. Jefford, C. W.; Jaggi, D.; Boukouvalas, J. *Tetrahedron Lett.* **1987**, *28*, 4037.
12. Konradsson, P.; Udodong, U. E.; Fraser-Reid, B. *Tetrahedron Lett.* **1990**, *31*, 4313.

Edward Turos
State University of New York at Buffalo, NY, USA

Trifluoromethanesulfonic Anhydride

$(CF_3SO_2)_2O$

[358-23-6] $C_2F_6O_5S_2$ (MW 282.16)

(preparation of triflates;[1] mild dehydrating reagent; promoter for coupling reactions in carbohydrates[2])

Alternate Name: triflic anhydride.
Physical Data: bp 81–83 °C/745 mmHg; *d* 1.677 g cm^{-3}; n_D^{20} 1.3210.
Solubility: soluble in dichloromethane; insoluble in hydrocarbons.
Form Supplied in: colorless liquid in ampules. Once opened it should be immediately used.
Analysis of Reagent Purity: IR, NMR.
Preparative Methods: by distillation of Trifluoromethanesulfonic Acid with an excess of Phosphorus (V) Oxide.[1]
Purification: by redistillation with a small amount of P_2O_5. It is advisable to freshly distill the reagent from a small quantity of P_2O_5 before use.
Handling, Storage, and Precautions: the pure reagent is a colorless liquid that does not fume in air and is stable for a long period. It is not soluble in water and hydrolyzes only very slowly to triflic acid over several days at room temperature. Preferably stored under N_2 in a stoppered flask. Dangerously exothermic reactions have been reported when attempting to triflate hindered alcohols.[67]

Original Commentary

Antonio García Martínez, Lakshminarayanapuram R. Subramanian & Michael Hanack
Universität Tübingen, Germany

Reaction with Alcohols and Phenols. The reaction of alcohols and phenols with triflic anhydride (Tf_2O) at ~0 °C in the presence of a base (usually Pyridine) in an inert solvent (usually dichloromethane) for 2–24 h affords the corresponding reactive trifluoromethanesulfonate esters (triflates).[1] When triflic anhydride and pyridine are combined, the pyridinium salt forms immediately and normally precipitates out from the reaction mixture. Nevertheless, the salt is an effective esterifying agent, reacting with the added alcohol to give triflates in high yields (eq 1).[3]

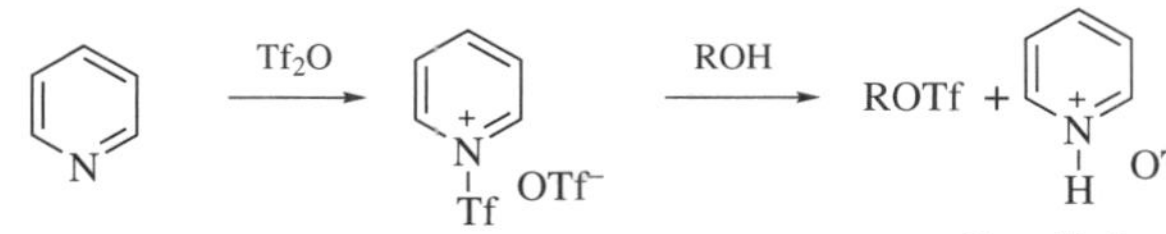

(1)

R = alkyl, aryl

Pyridine can become involved in nucleophilic substitution when very reactive triflates are being synthesized.[2,3] One approach to minimize this disadvantage is to replace it with sterically hindered bases, such as 2,6-di-*t*-butyl-4-methylpyridine,[3,4] 2,4,6-trisubstituted pyrimidines,[5] or nonnucleophilic aliphatic amines (usually *N,N*-diisobutyl-2,4-dimethyl-3-pentylamine). No salt formation appears to take place under these conditions. The triflic anhydride seems to be the direct triflating agent and the base only neutralizes the triflic acid formed. Numerous alkyl triflates have been prepared in the literature[1b] by the above method. Some recent examples of triflates prepared from alcohols are illustrated in eqs 2 and 3.[6,7] As an exception, 2,6-dinitrobenzyl alcohol does not react with Tf_2O although similar sulfonyl esters could be prepared.[8]

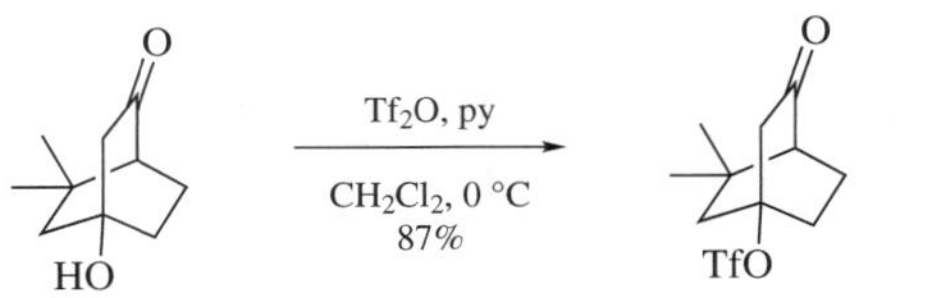

(2)

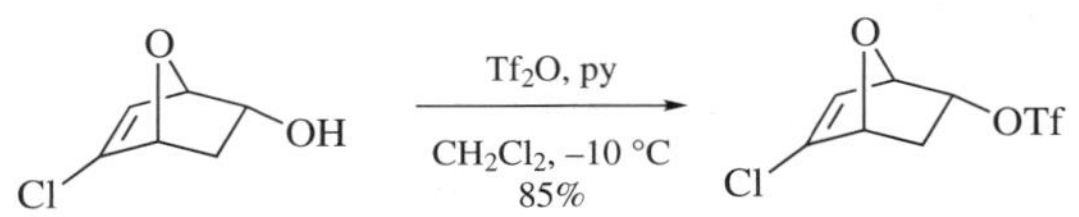

(3)

Alkyl triflates have come to be recognized as useful intermediates for the functionalization of organic substrates by nucleophilic substitution, e.g. in carbohydrate chemistry.[9] Triflate is the best leaving group known[1b] next to the nonaflate and hence a large number of triflates, obtained in good yields by reaction of the corresponding alcohols (or alkoxides) with Tf_2O, have been used to generate unstable or destabilized carbocations under solvolytic conditions.[1b] Some new typical examples are shown in (**1**)–(**4**).[10–13]

(1)

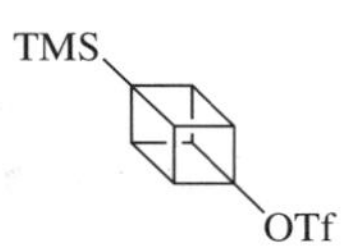

(2)

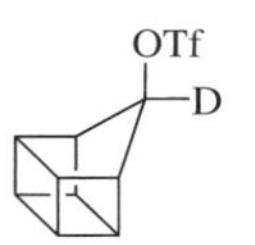

(3)

(4)

Alkyl triflates are known to be powerful reagents for the alkylation of aromatic compounds.[1b,14] However, the reaction of alkyl triflates with heterocycles affords *N*-alkylation products.[15]

In an improved modification of the Ritter reaction, primary and secondary alcohols react with Tf_2O in CH_2Cl_2 in the presence of a 2:1 excess of nitriles to give the corresponding amides in good yields (eq 4).[16]

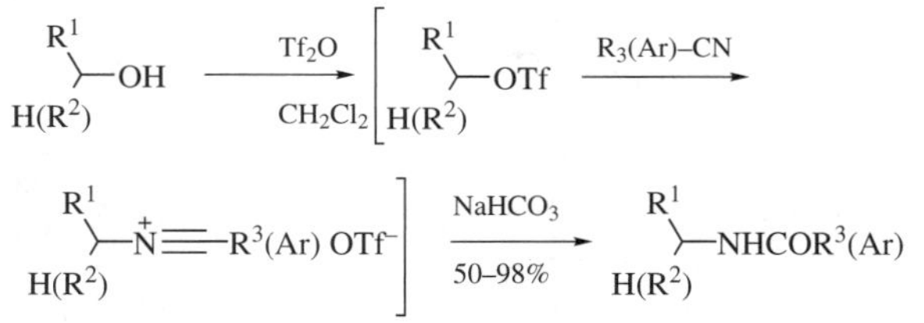

(4)

Aryl triflates are prepared from phenols at 0 °C using pyridine as solvent.[1b] Sometimes it is useful to conduct the reaction in CH_2Cl_2 at −77 °C, as in the preparation of 3,5-di-*t*-butyl-4-hydroxyphenyl triflate (eq 5).[17] Aryl triflates are synthetically transformed into several products of interest and applications in organic chemistry, by cross-coupling reactions with organometallics (eqs 5 and 6).[18]

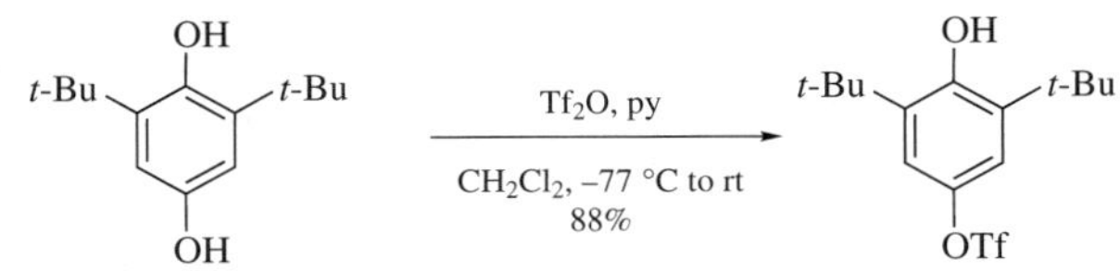

(5)

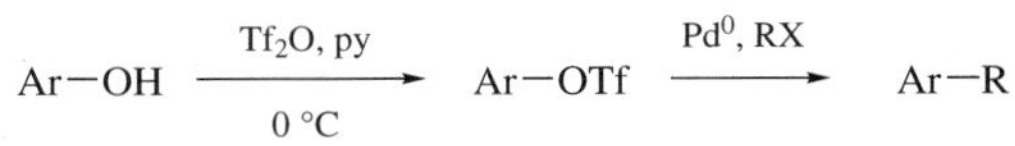

(6)

Reaction of Tf_2O with Amines. The reaction of 1 equiv of Tf_2O in CH_2Cl_2 and Et_3N with amines (or their salts) affords trifluoromethanesulfonamides (triflamides) in good yields.[19,20] If 2 equiv of Tf_2O are used, triflimides are formed. The triflamides are soluble in alkali and readily alkylated to triflimides (eq 7).[19,20]

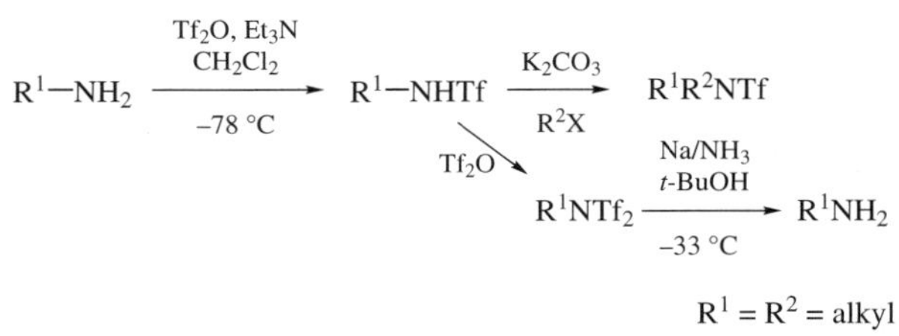

(7)

Triflamides can be deprotected reductively (Sodium–Ammonia) to yield the corresponding amines.[21] This protocol has been employed in the facile two-step synthesis of aza macrocycles starting from trifluoromethanesulfonyl derivatives of linear tetramines (eq 8).[22]

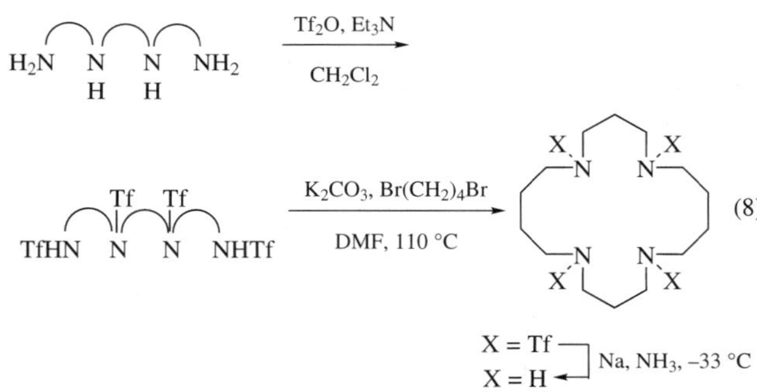

(8)

Several triflamides (**5**–**8**)[23] and *O*-triflylammonium salts[24] have been used for the formation of vinyl triflates from regiospecifically generated metalloenolates or for preparing triflates from alcohols.

(5)

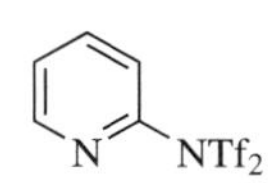

(6)

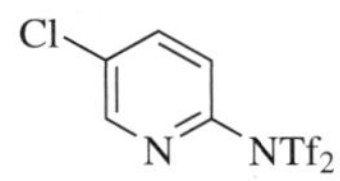

(7)

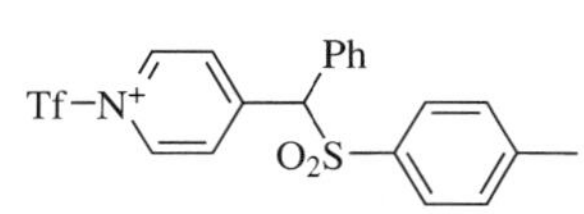

(8)

Reaction of Tf_2O with Carbonyl Compounds. The reaction of Tf_2O with carbonyl compounds consists of the electrophilic attack of the anhydride on the carboxylic oxygen, resulting in the formation of triflyloxycarbenium ions as intermediates (eq 9). According to the nature of the carbonyl compound, the triflyloxycarbenium cations can eliminate a proton giving a vinyl triflate, undergo a rearrangement, or be trapped by the gegenion yielding *gem*-bistriflates (eq 9).

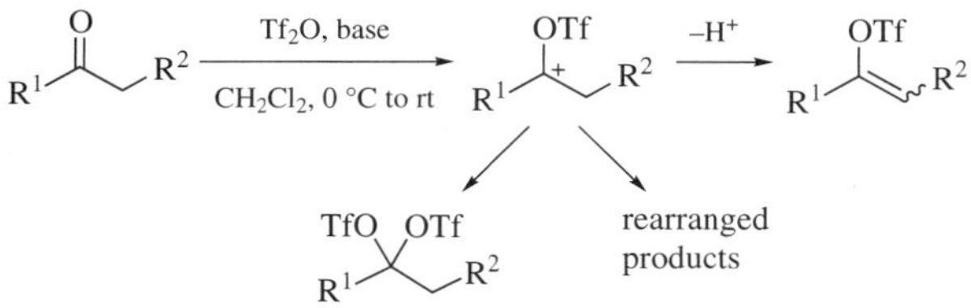
(9)

In the case of acyclic and monocyclic ketones, the reaction with Tf_2O affords vinyl triflates in good yields. Several methods exist to realize this reaction.[1b,25] For example, the reaction is carried out at room temperature in CH_2Cl_2 (or pentane) in the presence of 2,4-di-*t*-butyl-4-methylpyridine (DTBMP) (eq 10).[4]

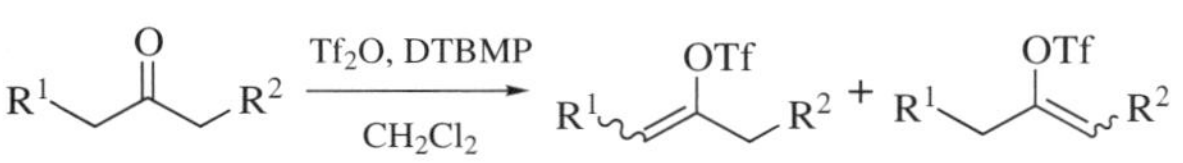
(10)

Other bases such as pyridine,[1b] lutidine,[1b] Et_3N,[1b] polymer-bound 2,6-di-*t*-butyl-4-methylpyridine,[26] and 2,4,6-trialkyl-substituted pyrimidines[27] were also used. The commercially available *N,N*-diisobutyl-2,4-dimethyl-3-pentylamine is a very convenient base to prepare the vinyl triflates.[28] In the case of nonfunctionalized ketones, anhydrous Na_2CO_3 has been proved to be very successful.[1b,25]

The reaction of ketones with Tf_2O is governed by Markovnikov's rule and results in the formation of the more substituted triflate as the major product. When the reaction of ketones[27] and α-halo ketones[29] with Tf_2O is carried out in the presence of a nitrile, the intermediate trifloxy cation (eq 9) can be trapped, forming pyrimidines in good yields (eq 11).

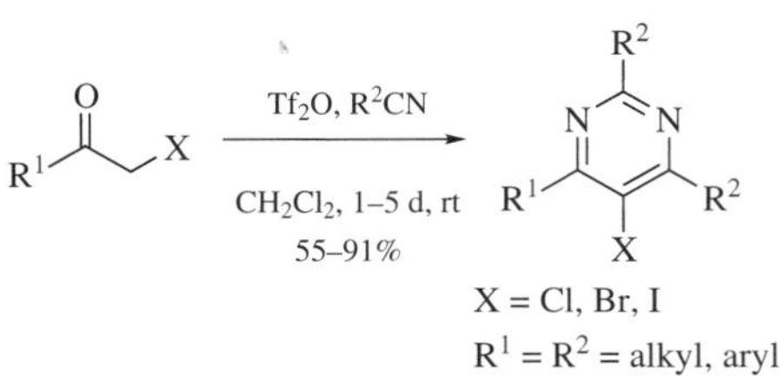
(11)

The reaction of Tf_2O with strained bicyclic ketones such as 2-norbornanone and nopinone takes place with Wagner–Meerwein rearrangement of the corresponding triflyloxy cations, forming bridgehead triflates in good yields (eq 12).[30] These triflates are key compounds in the preparation of other bridgehead derivatives by substitution[31] and of substituted cyclopentanes by fragmentation.[32]

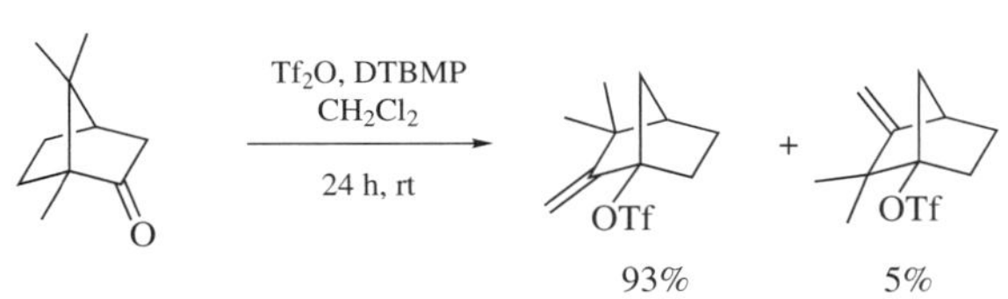
(12)

In the reaction of Tf_2O with norcaranones and spiro[2.5]octan-4-one, the cyclopropane ring undergoes fragmentation to give vinyl triflates (eqs 13 and 14).[33]

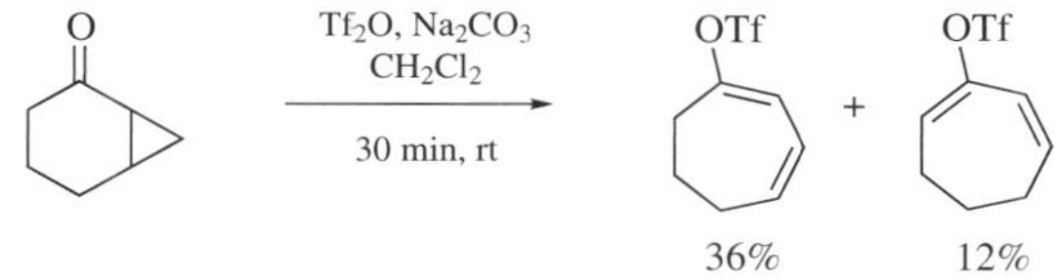
(13)

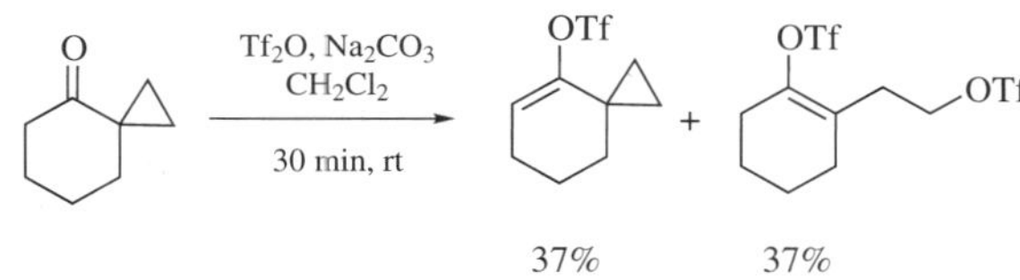
(14)

However, a cyclopropane ring is formed in the reaction of 5-methylnorborn-5-en-2-one with Tf_2O under the same conditions (eq 15).[34]

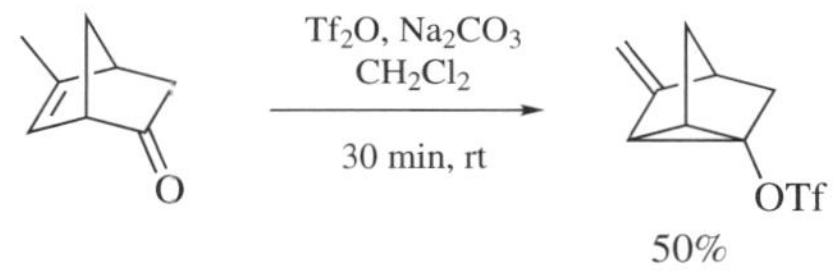
(15)

When the ketone can accomplish neither the stereoelectronic conditions for the elimination of TfOH nor for a rearrangement, the reaction of ketones with Tf_2O results in the formation of a *gem*-bistriflate (eqs 16 and 17).[35]

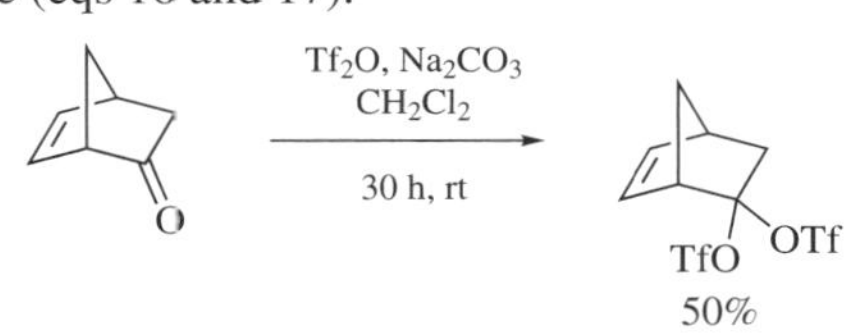
(16)

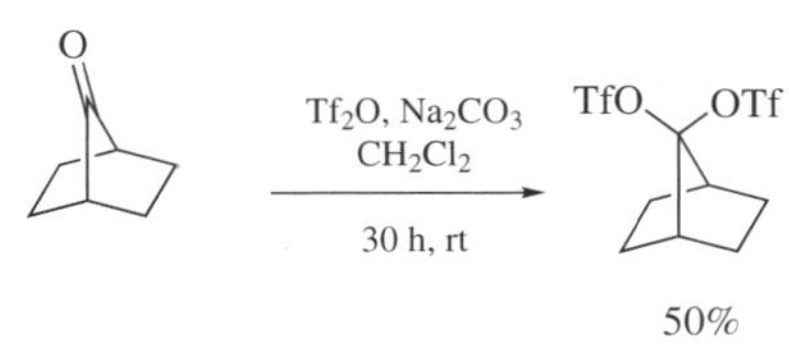
(17)

Sensitive ketones such as 3-pentyn-2-one also afford the corresponding vinyl triflate on treatment with Tf_2O in the presence of a base (eq 18).[36]

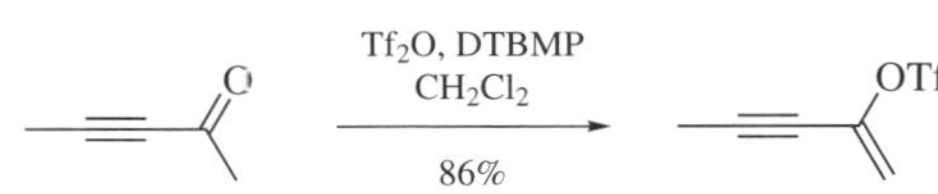
(18)

Substituted cyclopropenones and tropones react with Tf_2O with the formation of the corresponding dication ether salts (eq 19).[37]

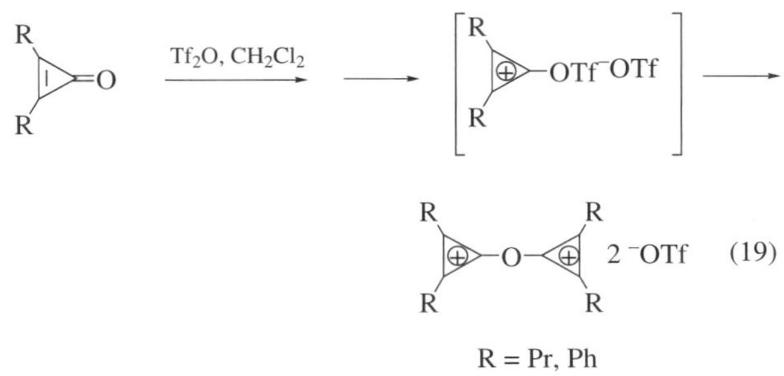
(19)

Treatment of trifluoroacetyl ylides with Tf_2O results in the formation of *gem*-bistriflates (eq 20).[38]

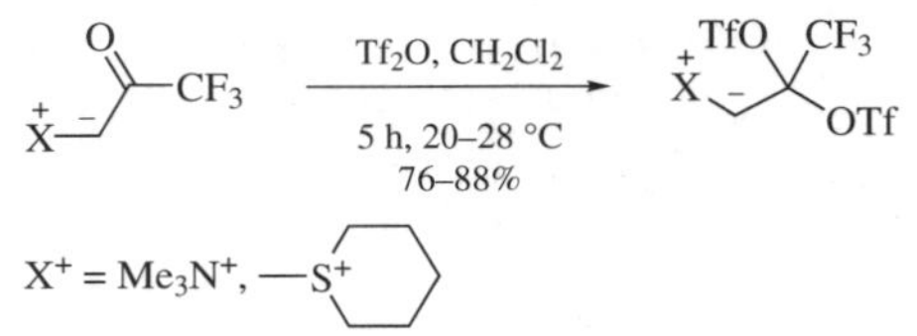

(20)

Reaction of Tf_2O with Aldehydes. The reaction of aliphatic aldehydes with Tf_2O in the presence of 2,6-di-*t*-butyl-4-methylpyridine (DTBMP) in refluxing CH_2Cl_2 or $ClCH_2CH_2Cl$ for 2 h affords the corresponding vinyl triflates as a mixture of (*Z*)- and (*E*)-isomers.[4,39] When the reaction is carried out at 0 °C, *gem*-bistriflates are formed as products (eq 21).[40] The *gem*-bistriflates result due to the trapping of the intermediate triflyloxycarbenium ion by the triflate anion. Primary vinyl triflates have been used extensively in the generation of alkylidene carbenes,[41] and *gem*-bistriflates are interesting precursors for *gem*-dihaloalkanes[42,43] and (*E*)-iodoalkenes.[44]

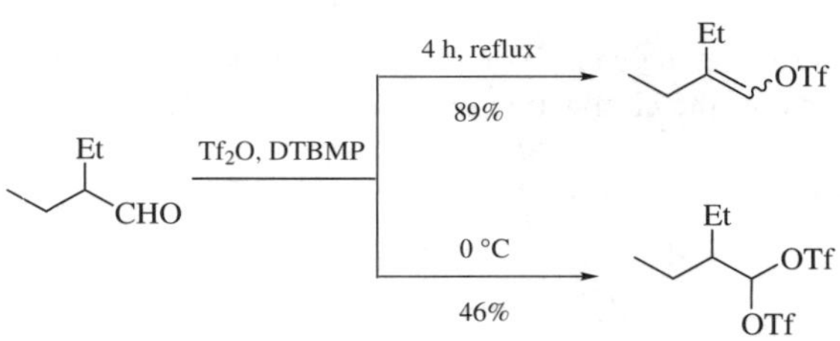

(21)

Reaction with Dicarbonyl Compounds. 1,3-Diketones can be reacted with an equimolar amount of Tf_2O or in excess to furnish the corresponding vinyl triflates or dienyl triflates (eq 22).[45] These triflates are transferred into monoketones, monoalcohols, alkanes, and unsaturated ketones by means of various reducing reagents.[45]

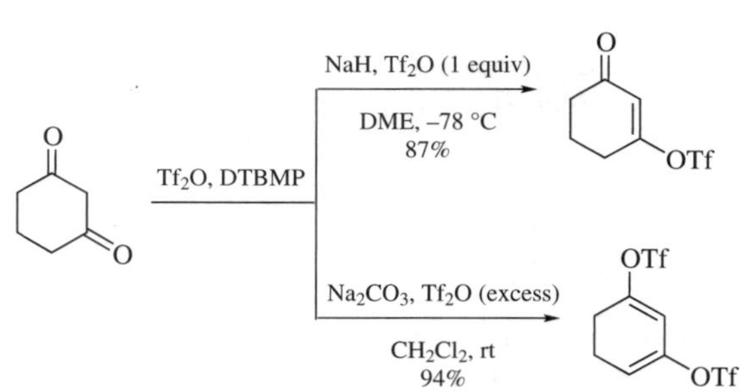

(22)

The reaction of 3-methylcyclopentane-1,2-dione with Tf_2O/Et_3N affords the vinyl triflate in 53% yield (eq 23).[46] The reaction takes place probably through the enol form. The product was coupled with alkenylzinc compounds in the presence of a palladium catalyst.[46]

The reaction of β-keto esters[47] with Tf_2O in the presence of a base results in the formation of 2-carboxyvinyl triflates (eq 24). These substrates undergo nucleophilic substitution of the TfO-group (eq 24)[47] and also coupling reactions.[48]

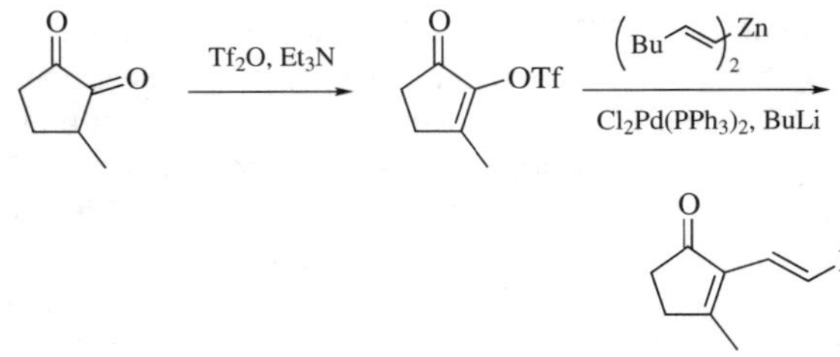

(23)

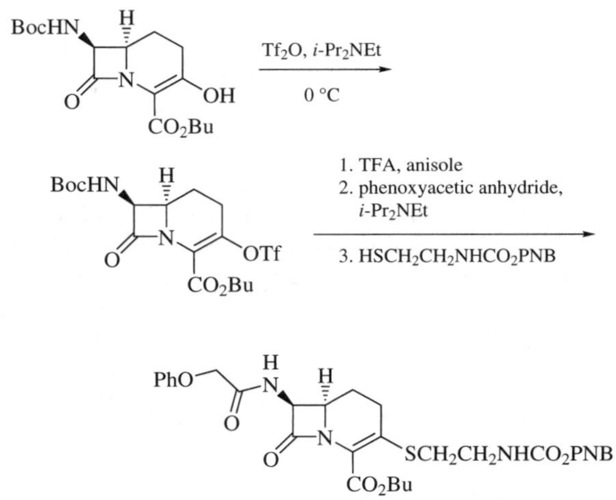

(24)

PNB = *p*-nitrobenzyl carbamate

Reaction with Carboxylic Acids and Esters. The reaction of carboxylic acids and esters with Tf_2O takes place according to the scheme shown in eq 25.[49]

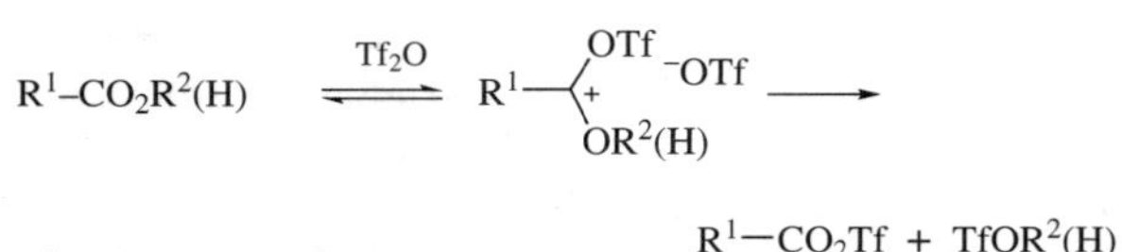

(25)

R^1 = alkyl, aryl; R^2 = alkyl, H

The trifluoromethanesulfonic carboxylic anhydrides are highly effective acylation agents, which react without catalysts even with deactivated aromatics to yield aryl ketones (eq 26).[50]

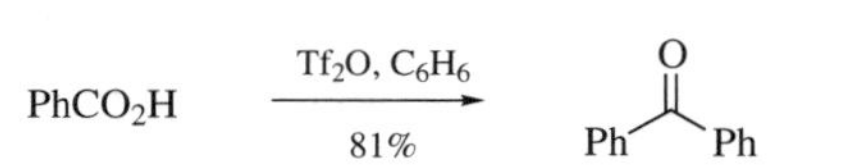

(26)

Alkyl arylacetates react with Tf_2O to give a cation which in the presence of a nitrile affords isoquinoline derivatives via cyclization of the intermediate nitrilium cation (eq 27).[50]

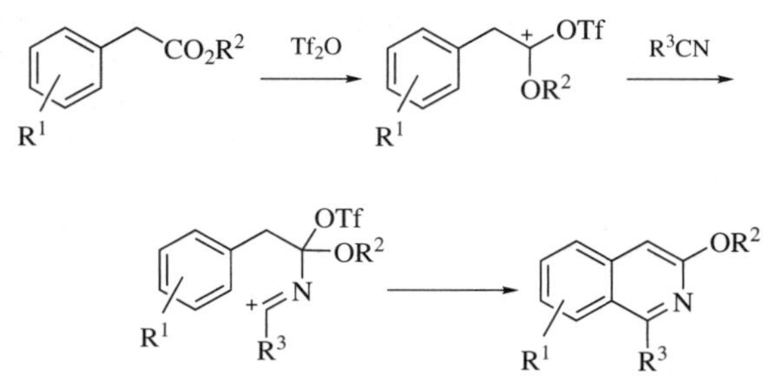

(27)

R^1 = H, 6-Me, 7–Cl, 5–NO_2, 6,7-$(OMe)_2$; R^2 = Et, Me

Reaction of Tf_2O with Amides. The reaction of a 2-oxo-1,2-dihydroquinoline with Tf_2O in the presence of pyridine affords the corresponding 2-quinoline triflate (eq 28).[51]

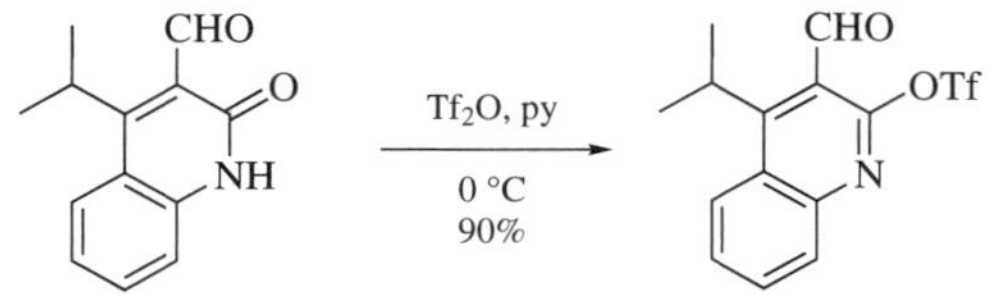

(28)

The reaction of tertiary amides with Tf_2O gives a mixture of *O*-sulfonylated (major) and *N*-sulfonylated (minor) products. In the presence of collidine and an alkene, [2 + 2] cycloadducts are formed which hydrolyze to give cyclobutanones (eq 29).[52]

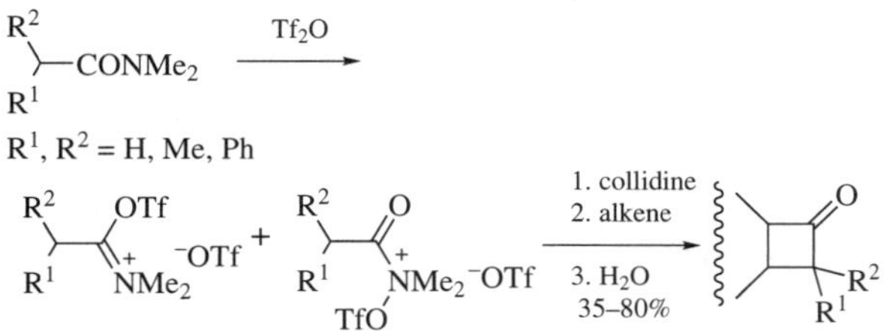

(29)

Treatment of DMF with Tf_2O results in the formation of an imminium triflate, which formylates less active aromatics. It is a convenient variation of the Vilsmeier–Haack reaction (eq 30).[53]

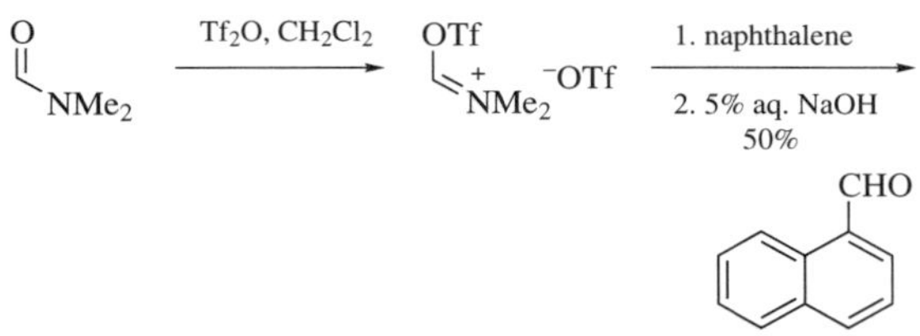

(30)

The reaction of *N*-methylpyridone and substituted urea systems with Tf_2O gives heteroatom-stabilized dicarbonium salts (eqs 31 and 32).[37,54]

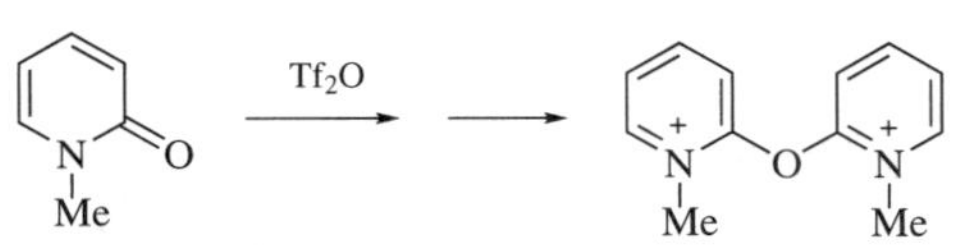

(31)

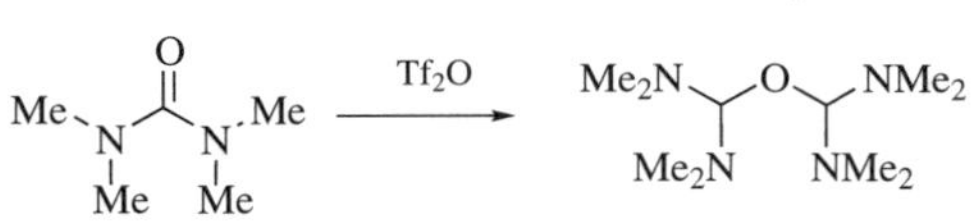

(32)

Secondary amides can be converted to tetrazoles with Tf_2O in the presence of Sodium Azide (eq 33).[55]

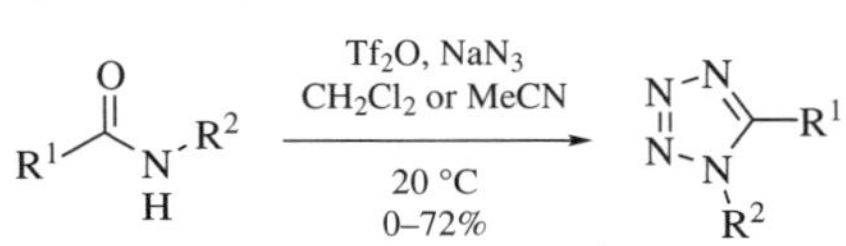

(33)

R^1 = alkyl, Ph; R^2 = alkyl, $(CH_2)_2OAc$, $(CH_2)_2OTBDMS$

Other Applications. Activated arenes can be converted to aryl triflones by Friedel–Crafts reaction with Tf_2O using Aluminum Chloride as catalyst (eq 34).[56]

$$\mathrm{Ar{-}H} \xrightarrow[\text{18 h, rt; 10–73\%}]{\mathrm{Tf_2O,\ AlCl_3}} \mathrm{ArSO_2CF_3} \qquad (34)$$

The reaction of Tf_2O with Ph_3PO in CH_2Cl_2 at 0 °C affords triphenylphosphine ditriflate, which can be used as an oxygen activator, and then to a diphosphonium salt (eq 35).[57]

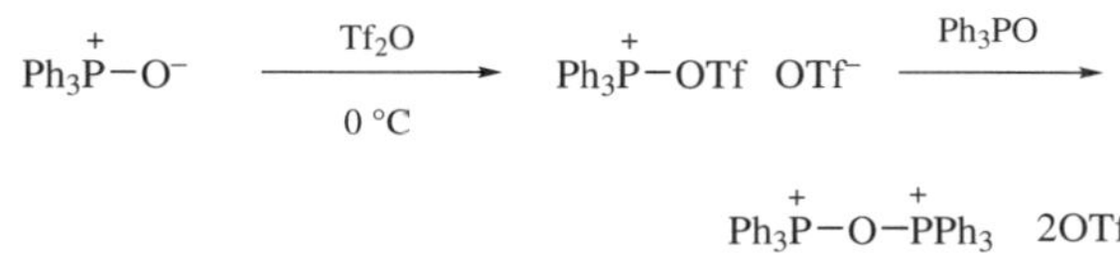

(35)

The less stable dimethyl sulfide ditriflate, obtained from Tf_2O and DMSO, has been used to oxidize alcohols (eq 36).[58]

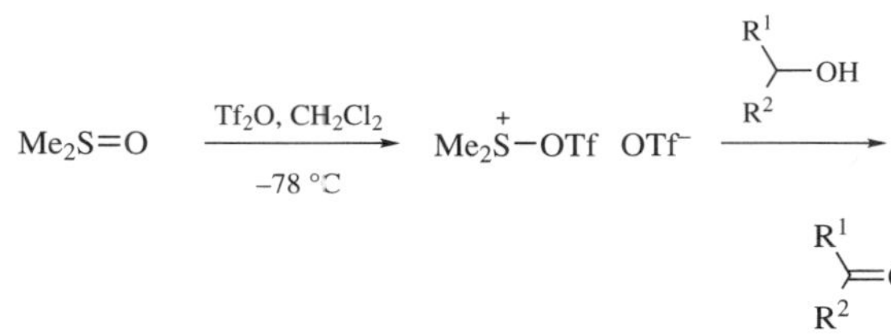

(36)

Tetrahydropyran is not a suitable solvent in reactions involving Tf_2O because it is cleaved, affording 1,5-bistrifloxypentane (eq 37).[59]

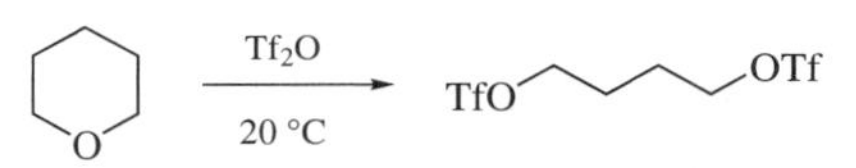

(37)

Diols react with Tf_2O to yield the corresponding ditriflates; however, the reaction of 1,1,2,2-tetraphenyl-1,2-ethanediol with Tf_2O takes place with rearrangement (eq 38).[59]

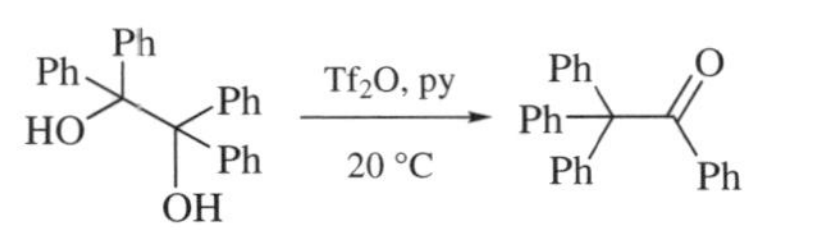

(38)

Vinylene 1,2-bistriflates are formed by the reaction of azobenzils with Tf_2O (eq 39).[60]

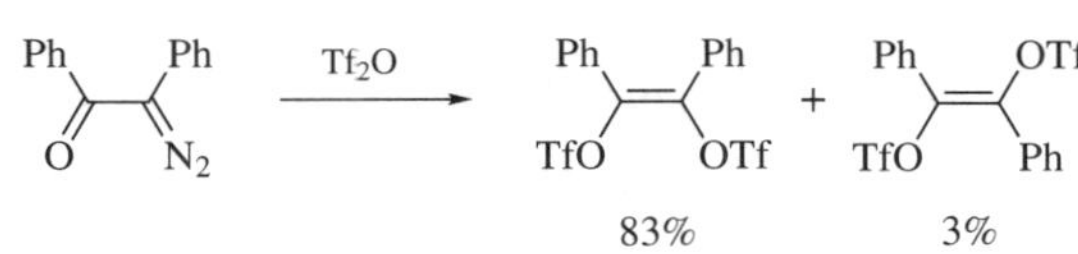

(39)

The reaction of enclates, prepared from silyl enol ethers and Methyllithium, with Tf_2O affords vinyl triflates (eq 40).[61]

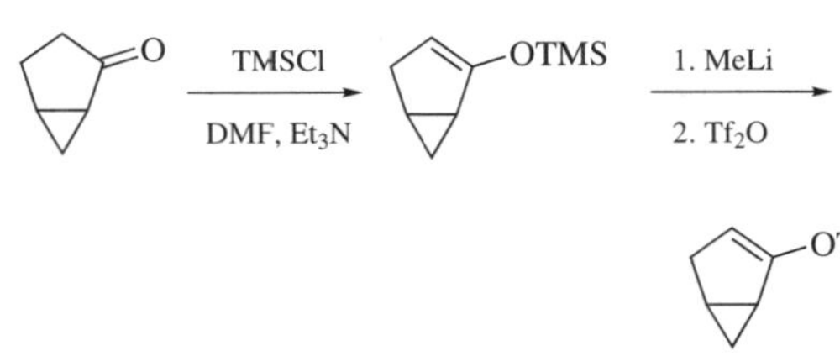

(40)

The combination of equimolecular quantities of Iodosylbenzene and Tf_2O generates $PhI(OTf)_2$, a compound also formed by treatment of Zefiro's reagent with Tf_2O. As shown in eq 41, this compound can be used to prepare *para*-disubstituted benzene derivatives in good yields.[62]

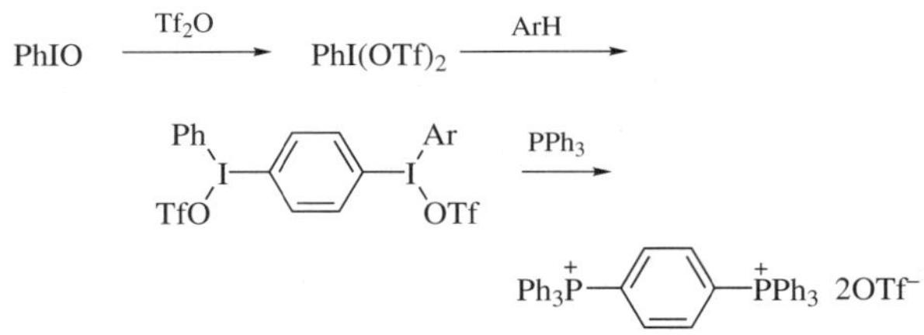

(41)

Tf_2O is a suitable promoter for the stereoselective glucosidation of glycosyl acceptors using sulfoxides as donors.[63]

The reaction of Tf_2O with a catalytic amount of Antimony(V) Fluoride at 25 °C produces trifluoromethyl triflate in 94% yield (eq 42).[64]

$$(CF_3SO_2)_2O \xrightarrow[80\%]{SbF_5} CF_3OSO_2CF_3 \quad (42)$$

Useful application of Tf_2O as dehydrating reagent is accounted by the synthesis of isocyanides from formamides and vinylformamides (eq 43).[65]

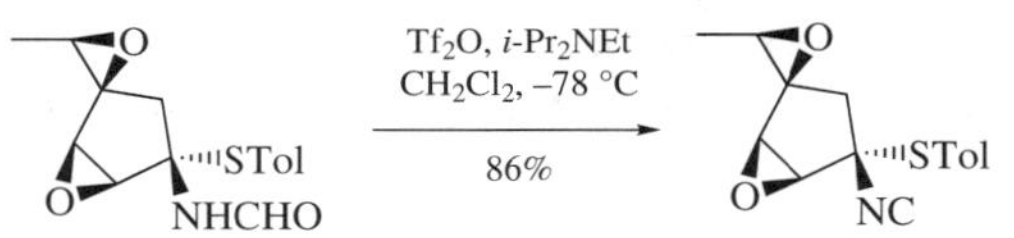

(43)

Reaction of enaminones with Tf_2O in a 1:1 molar ratio affords 3-trifloxypropeniminium triflates by *O*-sulfonylation. From a cyclic enaminone, by using a 2:1 molar ratio, the corresponding bis(3-amino-2-propenylio) bistriflate is obtained (eq 44).[66]

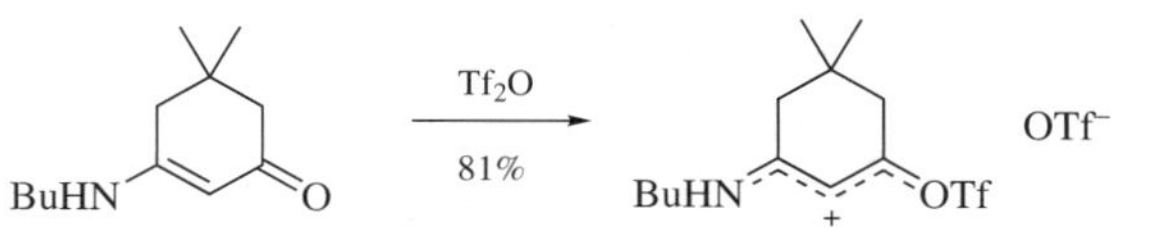

(44)

First Update

Spencer J. Williams
University of Melbourne, Parkville, Vic, Australia

Reactions with Alcohols, Phenols, and Thiols. Chemical transformations effected by triflic anhydride have been comprehensively reviewed.[68] Formation of alkyl triflates from alcohols is typically performed using triflic anhydride (Tf_2O) and pyridine or hindered pyridines such as 2,6-di-*tert*-butyl-4-methylpyridine (DTBMP) according to the methods already outlined. Purification of the resultant triflate from the excess base can sometimes prove problematic, and polymeric pyridine equivalents such as poly(4-vinylpyridine) or poly(2,6-di-*tert*-butylpyridine) have been successfully used allowing the removal of excess base and its triflate salt by simple filtration.[69] Alternatively, phenolic triflates can be prepared from Tf_2O under Schotten–Baumann conditions using toluene and 30% aqueous K_3PO_3.[70] Work-up is effected by simple separation of the organic phase and evaporation of the solvent. When attempting to triflate alcohols of low reactivity with Tf_2O, alternative redox processes can occur resulting in the formation of sulfinates as the major products. Such undesired redox processes are favored at higher temperatures and particularly when using triethylamine as base (eq 45).[67]

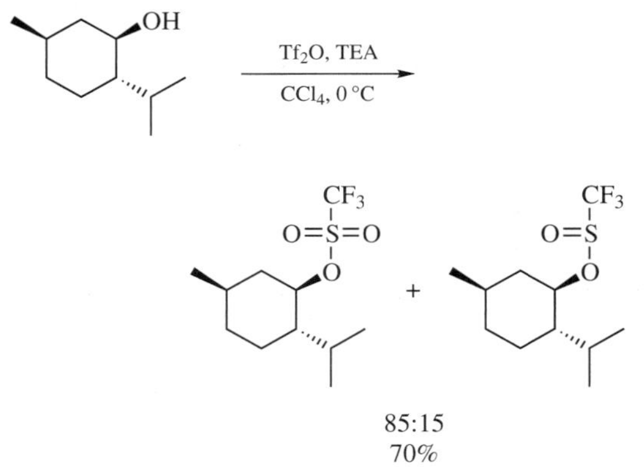

(45)

Methyl substituted pyridines such as 2,6-lutidine and 2,4,6-collidine can also afford undesirable by-products. These pyridines can react with Tf_2O to generate sulfinate- and trifluoromethyl-substituted derivatives (eq 46).[71] In such cases, the use of 2,6-di-*tert*-butyl-substituted pyridines is recommended.

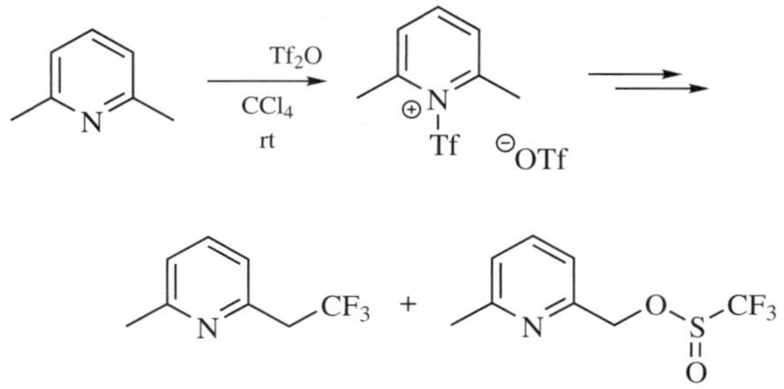

(46)

Triflates formed from hept-1-enitols spontaneously cyclize with debenzylation to afford vinyl C-glycosides (eq 47).[72,73]

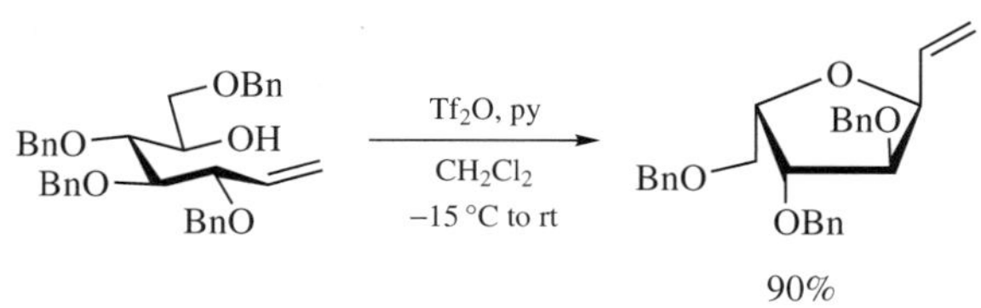

(47)

Dimethyl substituted homoallylic alcohols undergo smooth cyclization to *trans*-cyclopropanes upon treatment with Tf_2O and a base (eq 48).[74,75] Various functionalities can be introduced adjacent to the newly formed cyclopropane ring by altering the quenching conditions.[74]

Eleven-membered carbocycles are formed by way of a five-carbon ring expansion in a homo-Cope rearrangement upon treatment of β-(hydroxymethyl)allylsilanes with Tf_2O (eq 49).[76]

Reaction with Amines. Reaction of Tf_2O with hexamethyldisilazane affords the reactive silylating reagent *N,N*-bis(trimethylsilyl)trifluoromethylsulfonamide.[77,78] This compound exists in two tautomeric forms in solution, related by trimethylsilyl shifts between oxygen and nitrogen (eq 50).[77]

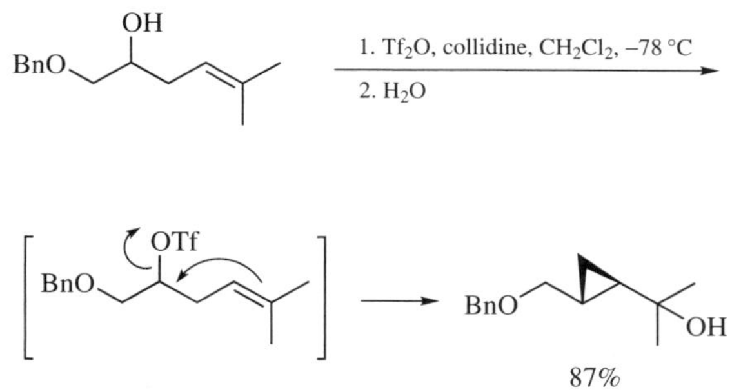

(48)

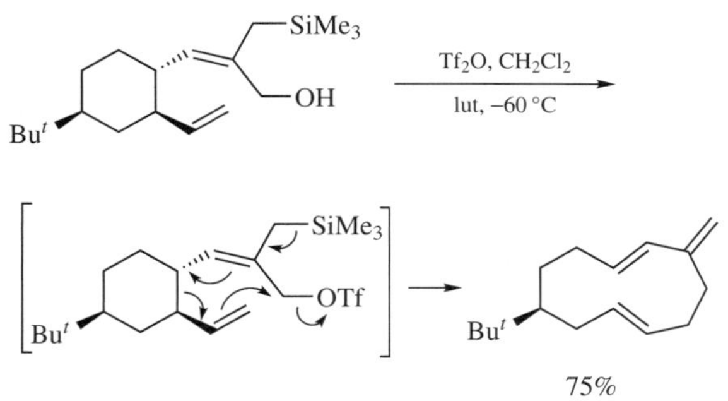

(49)

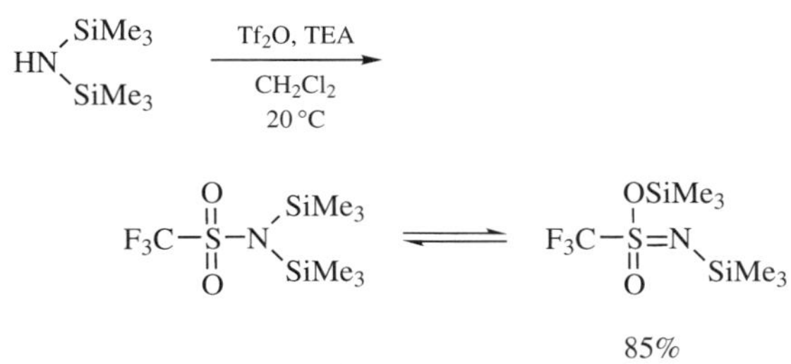

(50)

Reaction with Carbonyl and Thiocarbonyl Compounds. Treatment of ketones with α-protons with Tf_2O and magnesium bromide or Grignard reagents results in α-bromination (eq 51).[79]

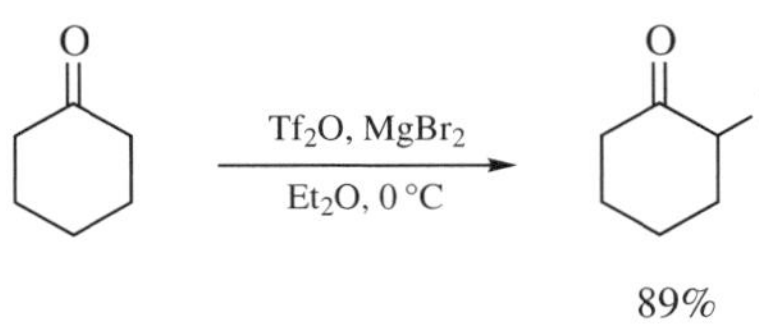

(51)

An alternative to the Arndt–Eistert reaction for the homologation of carboxylic acids uses *N*-(acylmethyl)benzotriazoles prepared from acyl chlorides and *N*-(trimethylsilyl)methylbenzotriazole. Upon treatment with Tf_2O these compounds are converted to the corresponding vinyl triflate, and then to the homologous ester by sequential treatment with sodium methoxide in acetonitrile and then HCl in methanol (eq 52).[80]

Tf_2O reacts with thioketones to generate triflylthiocarbenium ions that undergo Wagner–Meerwein rearrangement to bridgehead thiotriflates (eq 53).[81]

Ketones react with methylthiocyanate in the presence of Tf_2O to afford 2,4-bis(methylthio)pyrimidines (eq 54).[82] Methylthio substituted pyrimidines are useful synthetic intermediates as the methylthio groups can be substituted by other functionalities such as alkoxy and amino groups. Substitution occurs faster at the 4-position, allowing the selective introduction of two different nucleophiles.[68,82]

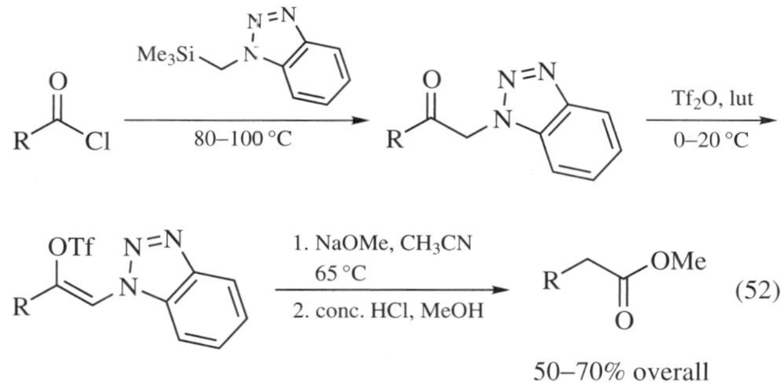

(52)

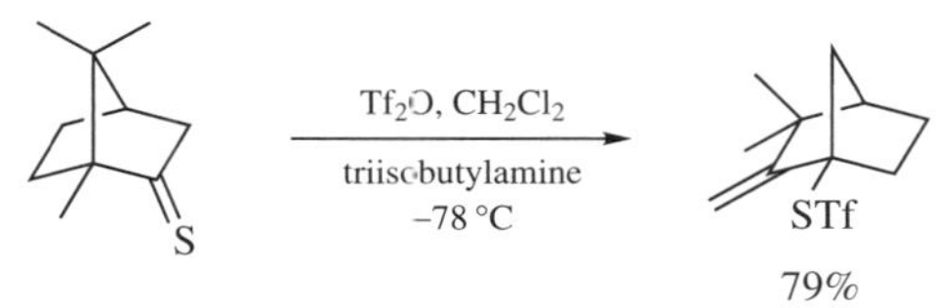

(53)

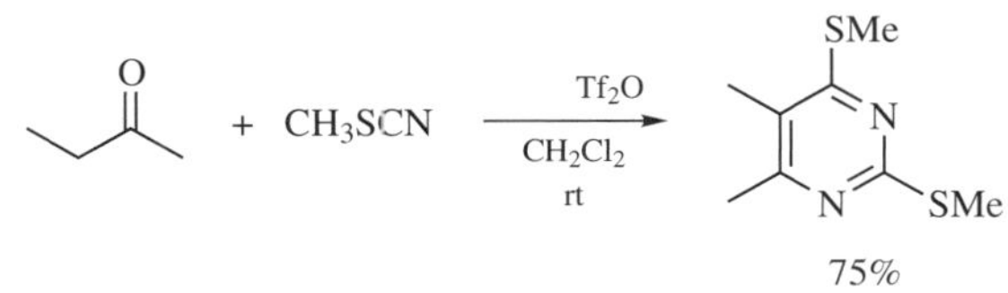

(54)

When the reaction is carried out with an ester and a nitrile, good yields of 4-alkoxypyrimidines are obtained (eq 55).[83]

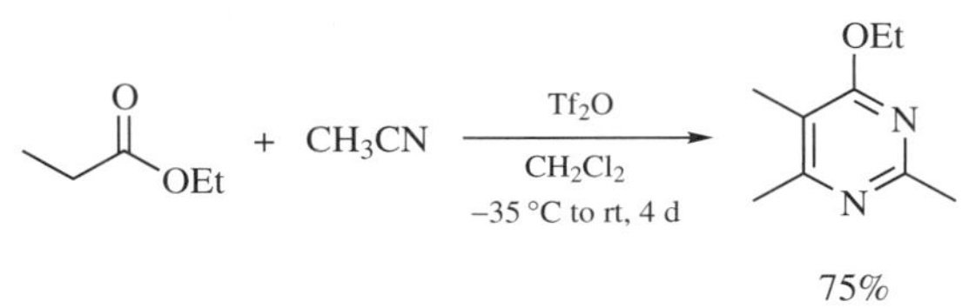

(55)

Ketones and nitriles when treated with Tf_2O react to give pyrimidines as outlined earlier (eq 11). In the case of benzyl nitriles bearing electron-donating substitutents, naphthalene derivatives are obtained instead (eq 56).[84]

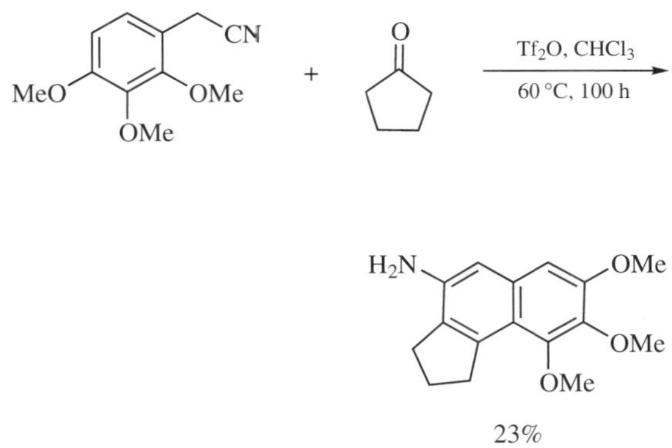

(56)

Reaction with Amides. It is well known that treatment of secondary or tertiary amides with Tf_2O affords imino and iminium triflates, respectively (eq 57).

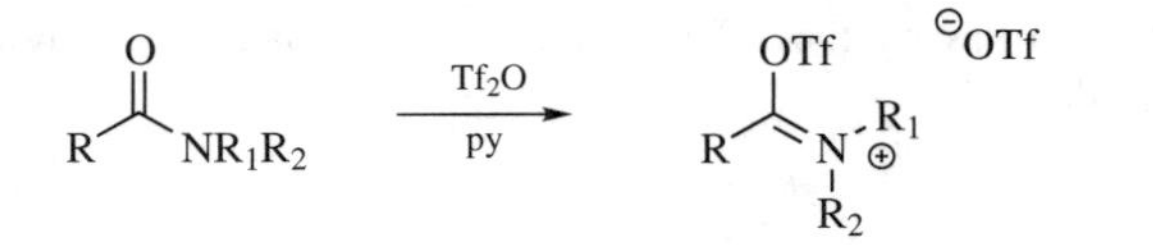

(57)

These salts can react with a variety of nucleophiles including amines (to afford amidines),[85–87] alcohols (to afford esters),[85,88] hydrogen sulfide (to afford thioamides),[89] and isotopically labeled water (to afford ^{18}O-labeled amides).[89] Iminium triflates are also excellent intermediates for the preparation of assorted heterocycles. Iminium triflates react with azides (to afford tetrazoles),[55] β-mercaptoamines (to afford thiazolines),[90] and triols (to afford orthoesters) (eq 58).[91] The reaction of lactams with Tf_2O and sodium azide affords tetrazolo-fused bicyclics.[92]

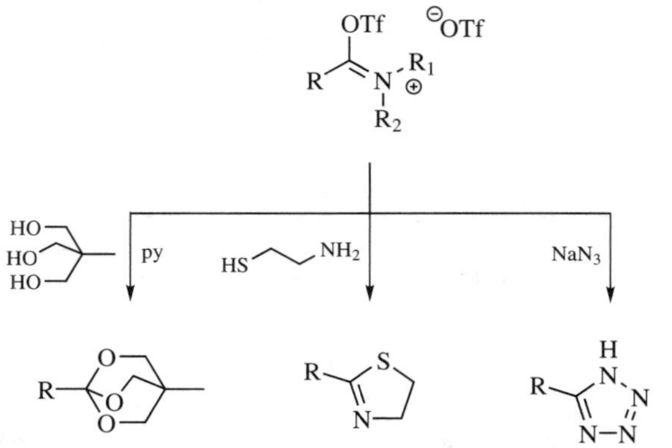

(58)

Treatment of unsaturated amides with Tf_2O generates intermediate iminium triflates that cyclize under radical conditions (eq 59).[93]

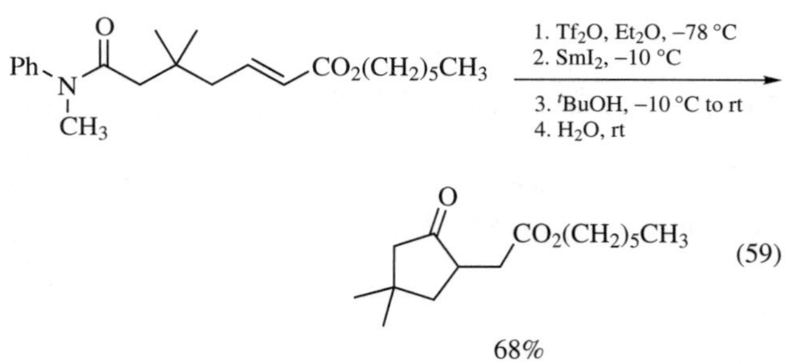

(59)

68%

Treatment of bis-hydrazides with Tf_2O and pyridine provides a mild method for the formation of 1,3,4-oxadiazoles (eq 60).[94]

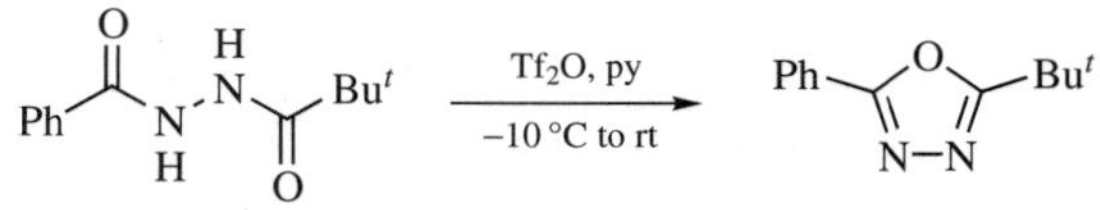

(60)

71%

Carbinolamides (derived from keto amides) afford 2-sulfonamido-substituted furans when treated with Tf_2O and pyridine (eq 61).[95]

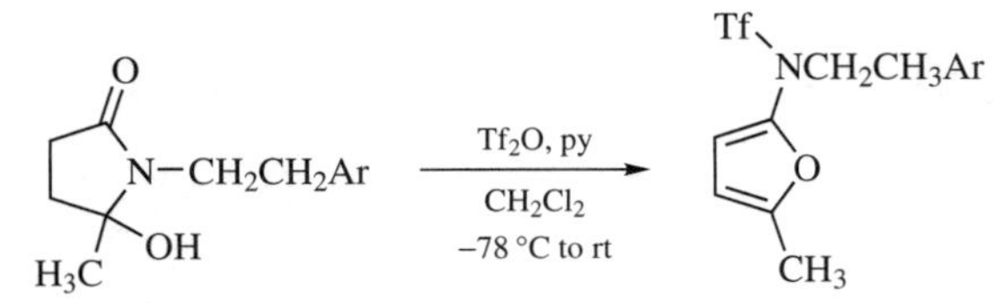

(61)

90%

Indolin-2-ones can participate in Vilsmeier reactions with reactive aromatics under the agency of Tf_2O. In the reaction with indoles and benzofurans, Tf_2O was found to be the reagent of choice; no reaction took place when phosphorous oxychloride was used (eq 62).[96]

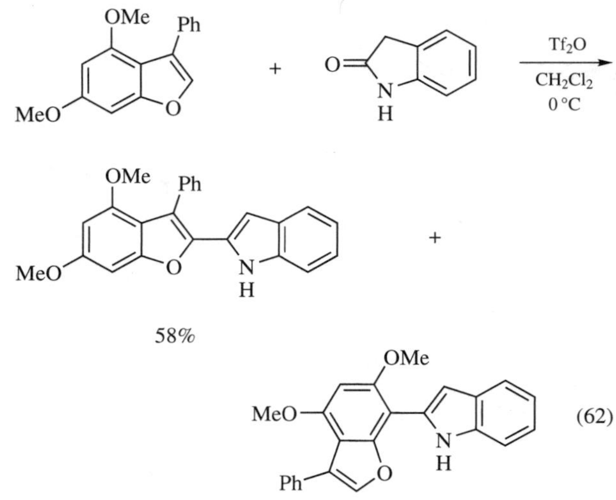

(62)

10%

Tf_2O and a base can effect dehydration of primary amides to nitriles.[97] This procedure is particularly mild, with no epimerization observed at the stereogenic α-carbon of several substrates (eq 63).[97]

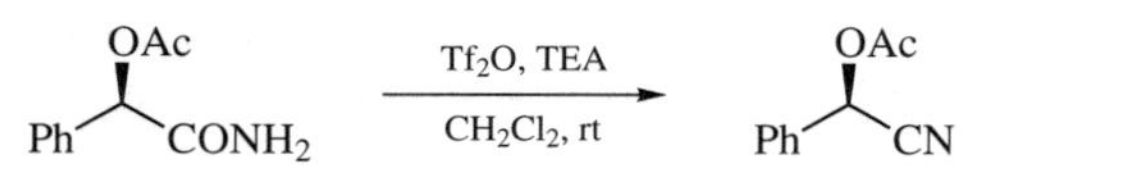

(63)

85%

Oxidations. Tf_2O reacts with sulfides affording a trifluoromethylsulfonylsulfonium triflate, which can be used to oxidize alcohols (eq 64).[98]

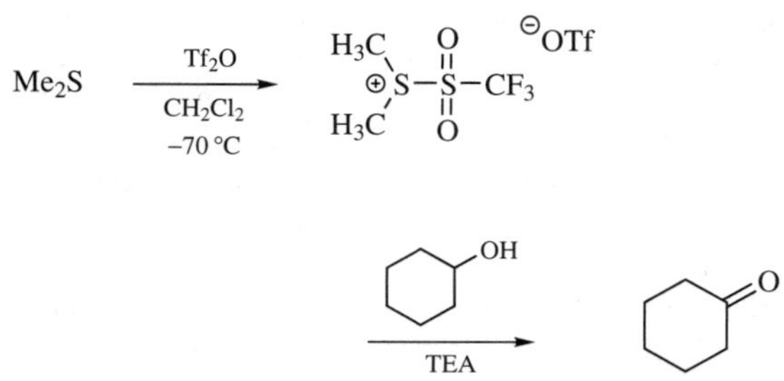

(64)

75%

Alternatively, the intermediate sulfonium salts prepared in the same manner can be treated with aqueous sodium acetate to afford the sulfoxide; none of the corresponding sulfones were observed (eq 65).[98]

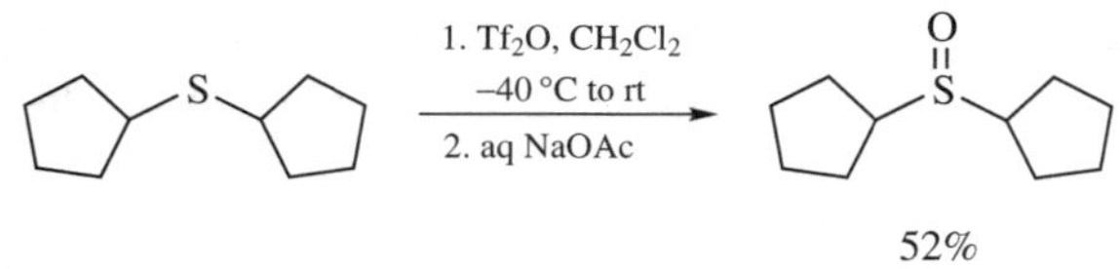

(65)

Tf_2O reacts with Grignard reagents to afford the self-coupled alkanes (eq 66).[99]

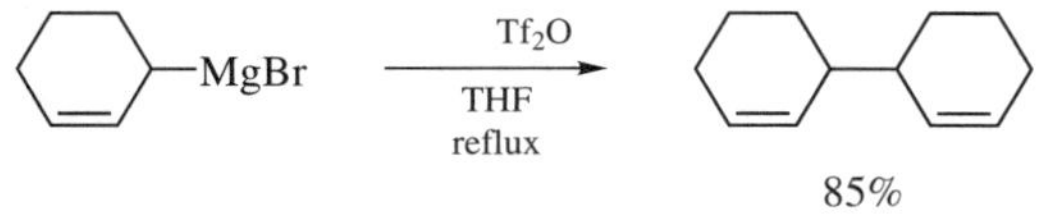

(66)

Addition of Tf_2O to 85% hydrogen peroxide affords the powerful oxidant trifluoromethanepersulfonic acid. This oxidant will convert unreactive thioethers to the corresponding sulfones (eq 67).[100]

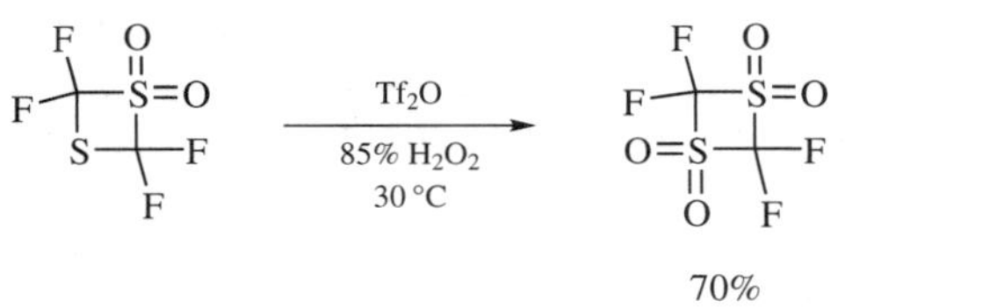

(67)

Owing to the electrophilicity of the divalent sulfur of a thiosulfinate, thiols do not react with Tf_2O to afford thiosulfonates; disulfides are obtained instead (eq 68).[101] In the presence of a base, only diphenyl disulfide is observed.[101]

$$PhSH + Tf_2O \xrightarrow[rt]{CH_2Cl_2} PhSSPh + CF_3SSPh \quad (68)$$

Glycosylations and Acetalations. The combination of Tf_2O and a hindered pyridine base such as 2,6-di-*tert*-butylpyridine promotes glycosylation of oxygen and nitrogen nucleophiles by thioglycoside sulfoxides.[102] This reaction has a broad scope, enabling the glycosylation of unreactive nucleophiles under mild conditions, and has been extended to solid phase glycosylations.[103] The anomeric stereochemistry of the product is determined by the nature of the protecting groups ('P'), with esters affording 1,2-*trans* products and ethers affording an anomeric stereochemistry that is sensitive to the reaction conditions (eq 69).[104]

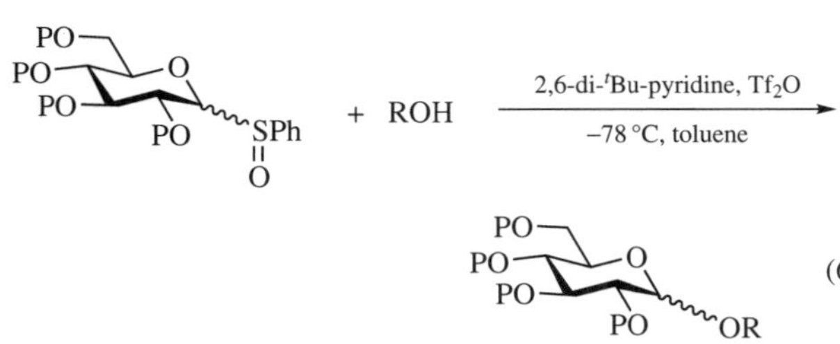

(69)

R = steroids, phenols, carbohydrates

An important modification of the original sulfoxide glycosylation uses 4,6-*O*-benzylidene-protected thioglycoside sulfoxides and a modified protocol for the addition of reagents, requiring the addition of the nucleophile after activation of the sulfoxide with Tf_2O.[105] This approach is effective in the production of the challenging β-manno 1,2-*cis*-glycosidic linkage, and there is strong evidence that the reaction proceeds via an intermediate glycosyl triflate (eq 70).[106,107]

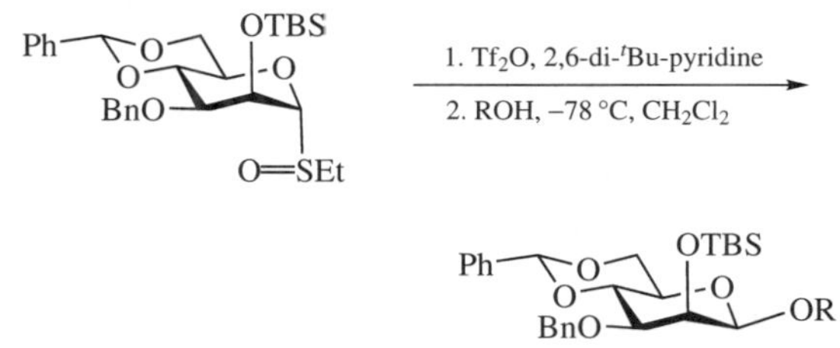

(70)

In some cases the addition of electrophilic alkenes, such as 4-allyl-1,2-dimethoxybenzene, to the reaction mixture can improve the outcome of the reaction, particularly when run in the presence of thioglycosides, by scavenging a by-product of the reaction, phenylsulfenyl triflate.[108] Non-carbohydrate thioacetal sulfoxides also undergo similar acetalation reactions when treated with Tf_2O (eq 71).[109]

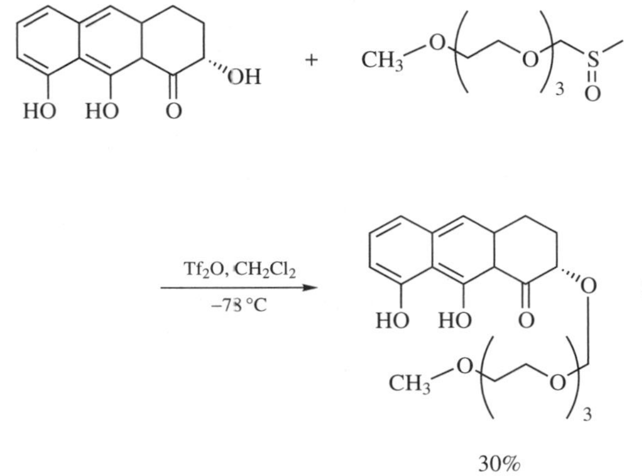

(71)

Thioglycosides can be activated towards glycosylation using a variety of reagents in combination with Tf_2O. In particular, the reagent combination of 1-benzenesulfinylpiperidine (BSP) (**9**), Tf_2O, and a hindered base, such as 2,6-di-*tert*-butylpyridine or 2,4,6-tri-*tert*-butylpyrimidine (TTBP) (**10**), is very effective.[110] The combination of BSP and TTBP is especially noteworthy as these reagents are crystalline and shelf-stable and may be stored premixed. The combination of diphenyl sulfoxide, Tf_2O, and TTBP has been shown to be a more powerful promoter than BSP/TTBP/Tf_2O for the activation of thioglycosides.[111] Again, for all of these reagent combinations, the use of 4,6-*O*-benzylidene acetals or 4,6-*O*-phenylboronates[112] enables the stereoselective formation of β-mannosides. The method has been used with moderate success on solid phase.[112] In the case of rhamnosides, where it is not possible to use a 4,6-*O*-benzylidene acetal, judicious choice of a 2-*O*-(3-trifluoromethylphenylsulfonyl) protecting group allows the preparation of the corresponding β-rhamnosides.[113]

Tf_2O in combination with sulfoxides lacking α-protons, such as diphenylsulfoxide or dibenzothiophene sulfoxide, generates a sulfoxonium triflate, which is capable of effecting the formation of glycosides from carbohydrate hemiacetals and oxygen nucleophiles.[114] This is a dehydrative glycosylation, and appears to proceed by way of a sulfoxonium intermediate (eq 72).[115]

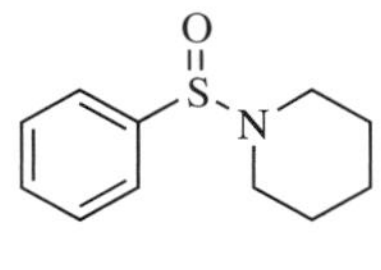

(9)

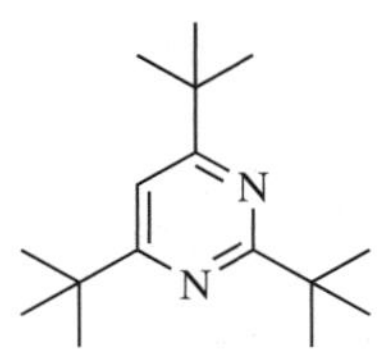

(10)

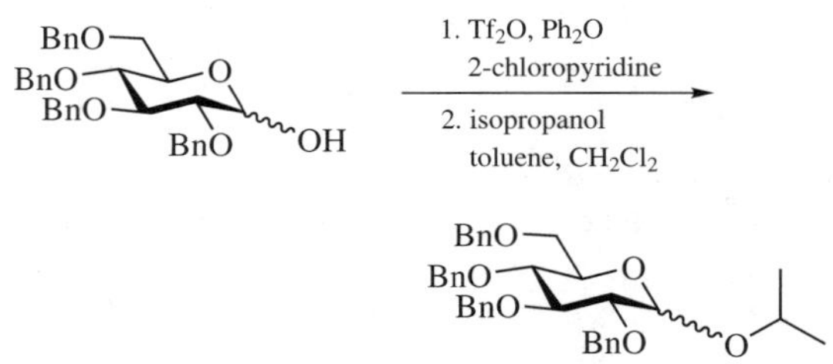

(72)

89%

Carbohydrate glycals undergo oxidative glycosylation when treated first with Tf_2O and diphenyl sulfoxide (or tributylphosphine oxide)[116] and then with a nucleophile in the presence of a Lewis acid.[117] This reaction proceeds via oxygen transfer to generate a 1,2-anhydrosugar intermediate which is opened by nucleophiles in the presence of a Lewis acid, affording 2-hydroxy β-glucosides (eq 73). If a more hindered sulfoxide (dibenzothiophene sulfoxide) is used, the reaction occurs with the complimentary stereochemistry to give 2-hydroxy α-mannosides.[118]

A variant of this reaction is termed as the C2-acetamidoglycosylation. Treatment of glycals with Tf_2O and thianthrene-5-oxide, followed by trimethylsilylacetamide and a base, and then finally with an alcohol nucleophile and an acid catalyst affords 2-acetamido glycosides (eq 74).[119]

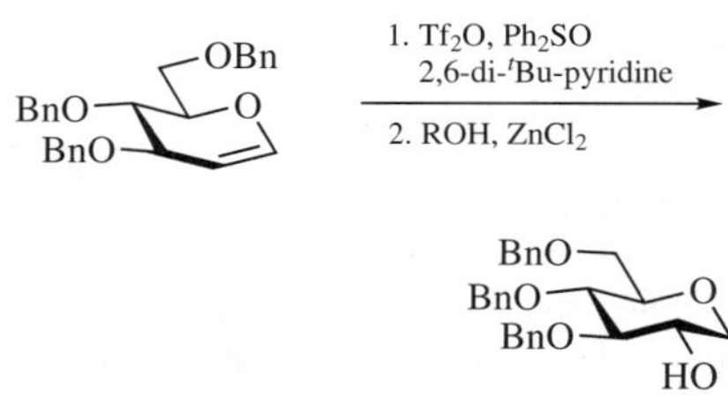

(73)

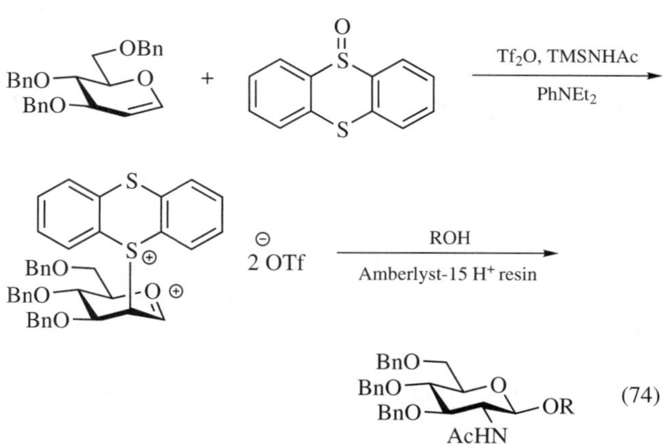

(74)

R = carbohydrates, amino acids, and steroids

Tf_2O acts as a promoter in glycosylations of alcohols by 2′-carboxylbenzyl glycosides (eq 75).[120] This reaction most likely proceeds via the mixed anhydride.

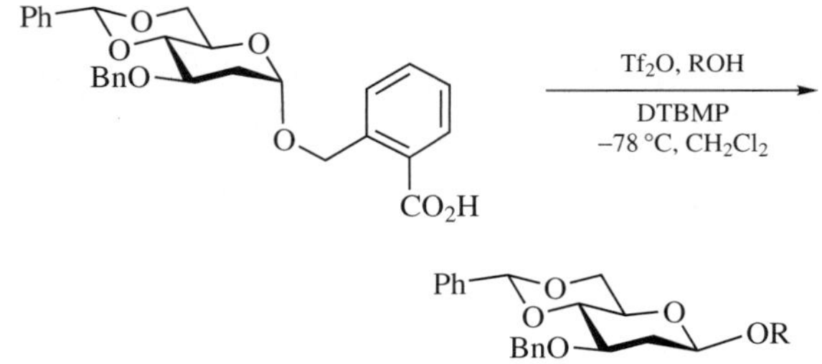

(75)

Tf_2O has been used as a promoter in glycosylations of carbohydrate alcohols by glycosyl fluorides.[121]

Aminals. Allylic and propargylic aminals when treated with Tf_2O afford iminium salts, which react with Grignard reagents to afford allylic and propargylic amines (eq 76).[122]

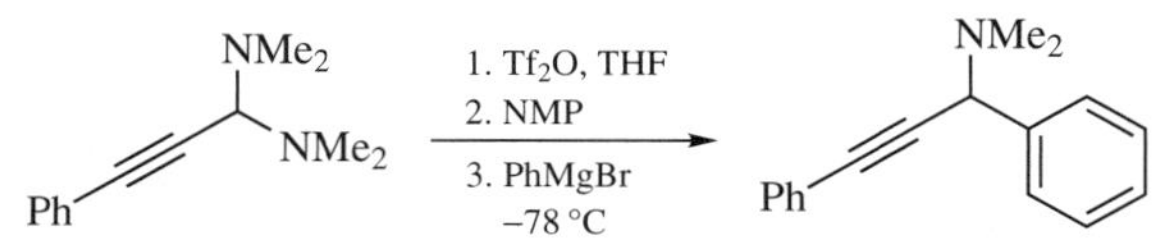

(76)

67%

1,1-Difluorovinyl methyl ethers and aminal bis(carbamates) or bis(sulfonamides) react in the presence of Tf_2O to generate β-amino-α,α-difluoroketones (eq 77).[123]

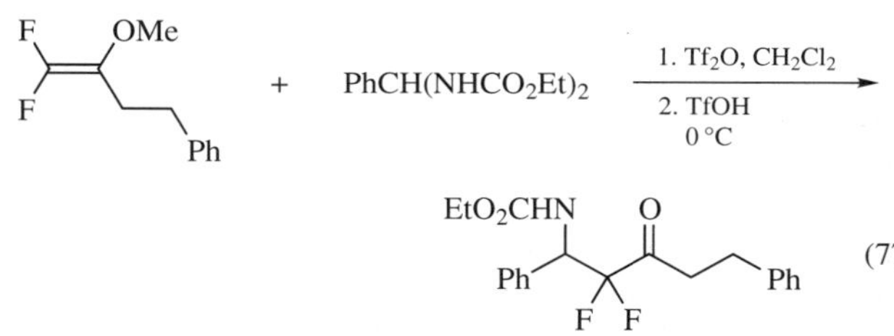

(77)

84%

An azaallenium salt has been prepared by treatment of a bis(iminomethane) compound with Tf_2O (eq 78).[124]

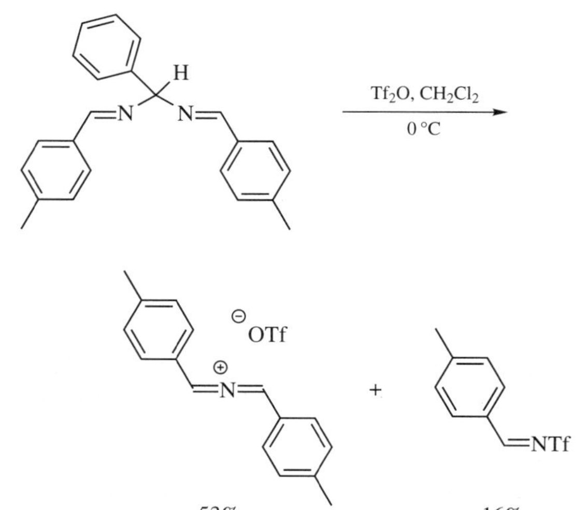

(78)

53% 16%

Other Applications. *O*-Silylated hydroxamic acids when treated with Tf_2O and a base yield nitrile oxides.[125] In the presence of olefins, these undergo [3 + 2] dipolar cycloadditions affording isoxazolines (eq 79).[125]

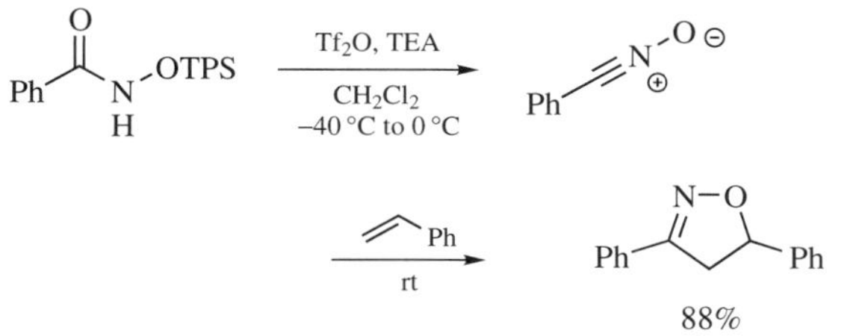

(79)

Triphenylphosphine oxide can be converted to the corresponding unsubstituted phosphinimines by repeated sequential treatment with Tf_2O and then ammonia (eq 80).[126]

$$Ph_3P{=}O \xrightarrow[\text{2. } NH_3 \text{ repeat seven times}]{\text{1. } Tf_2O,\ CH_2Cl_2} \underset{82\%}{Ph_3P{=}NH} \quad (80)$$

The combination of tetraalkylammonium nitrates and Tf_2O generates nitronium triflate, which acts as a convenient anhydrous nitrating reagent.[127,128] This reagent system is effective for nitration of homo and heteroaromatic systems (eq 81),[127] and for *N*-nitration of saturated nitrogen heterocycles (eq 82).[128]

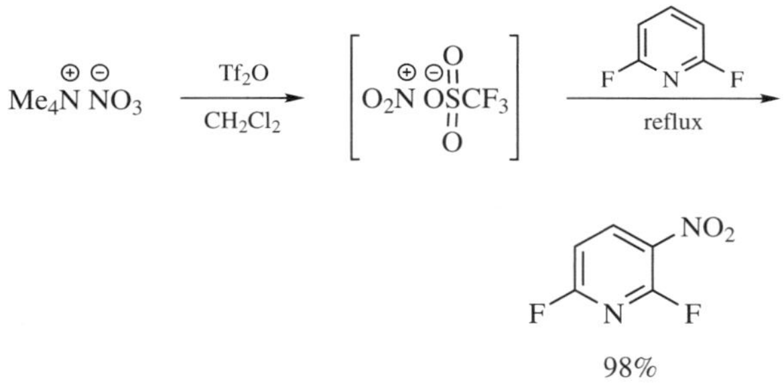

(81)

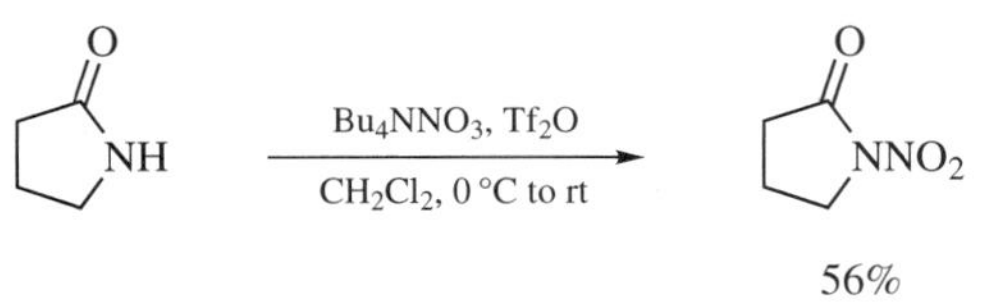

(82)

Treatment of sulfoxide-substituted electron-rich heterocycles with Tf_2O promotes substitution at sulfur, resulting in the formation of large-ring N,S-heterocycles (eq 83).[129]

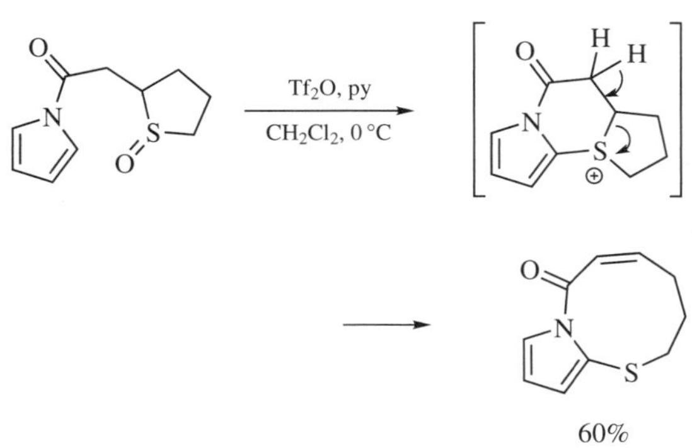

(83)

Dimethyl sulfide ditriflate, obtained from Tf_2O and DMSO, has been used to generate *o*-quinone methides (eq 84).[130]

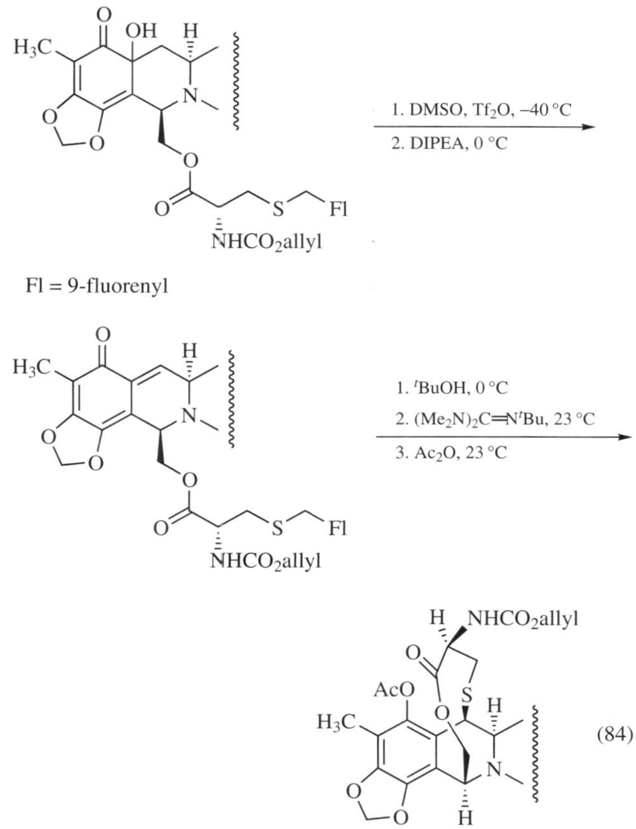

(84)

Bis- and tris-chalcogen monoxides when treated with Tf_2O afford di- or tri-chalcogona dications, respectively; these salts can be considered 2-center, 2-electron or 3-center, 4-electron bonded species (eq 85).[131,132] These compounds are susceptible to nucleophilic attack at carbon or at the chalcogen and can act as alkylating agents under very mild conditions.[132]

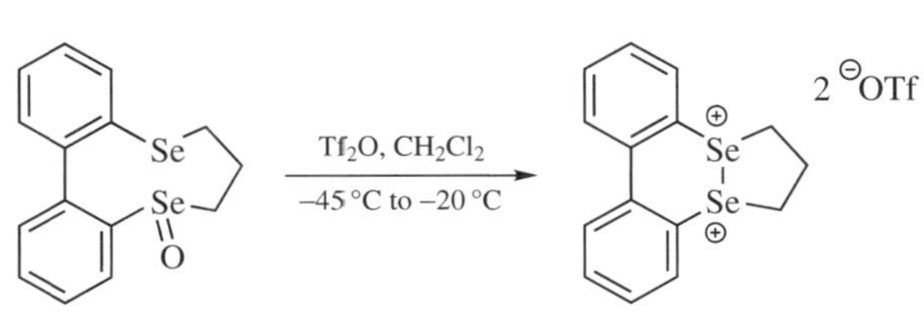

(85)

A one-pot procedure for the synthesis of thioamides relies on the successive reaction of Grignard reagents with carbon disulfide and then with Tf_2O to generate a mixed anhydride intermediate. This intermediate reacts with amines to generate thioamides (eq 86).[133]

Tf_2O reacts with dimethylcyanamide to afford an *O,N*-ditriflylisourea.[134] Reaction of this isourea with alcohols, thiols, or amines affords isoureas, thioisoureas, or guanidines, respectively (eq 87).[134]

Trimethylsilylalkynes can be converted to alkynyl(phenyl) iodonium triflates by treatment with $PhI(OAc)_2$ and Tf_2O (eq 88).[135] Alkynyl(phenyl)iodonium triflates are useful 'electrophilic acetylene' equivalents and can act as Michael acceptors, 1,3-dipolariphiles and alkynyl cation equivalents.[136]

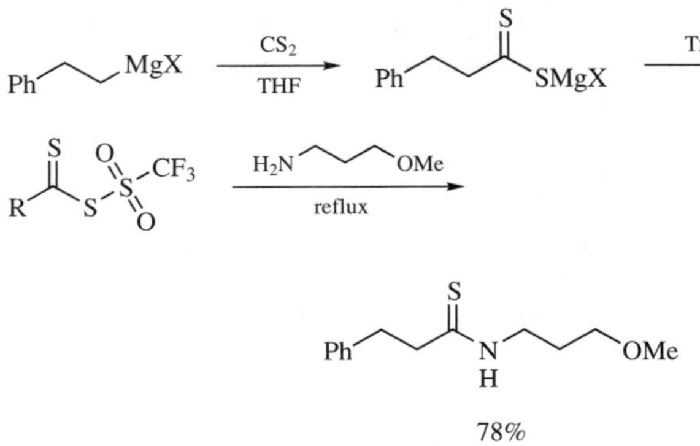
(86)

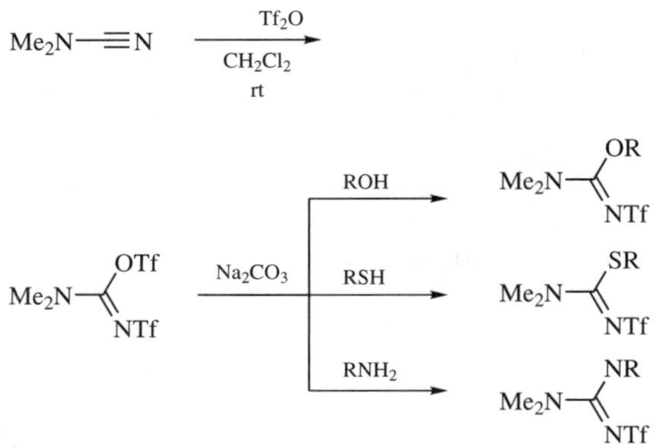
(87)

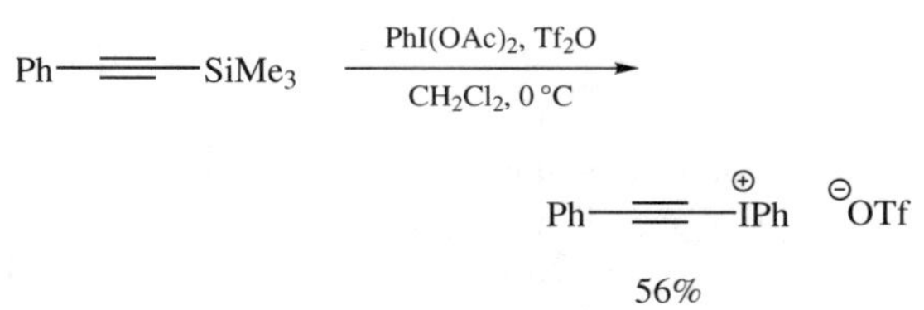
(88)

The reaction of Tf_2O with phenylseleninic anhydride and diphenyl diselenide directly affords phenylselenenyl triflate (eq 89).[137] This procedure is a convenient alternative to the preparation of phenylselenenyl triflate from light sensitive AgOTf and moisture sensitive PhSeCl.

$$(PhSeO)_2O + Ph_2Se_2 \xrightarrow[CH_2Cl_2]{Tf_2O} PhSeOTf \quad (89)$$

Electrophilic trifluoromethylation reagents can be generated from the treatment of trifluoromethyl biphenyl sulfoxides with Tf_2O.[138,139] These reagents will react with a wide range of nucleophiles including carbanions, aromatics, silyl enol ethers, thiolates, and phosphines (eq 90).[139]

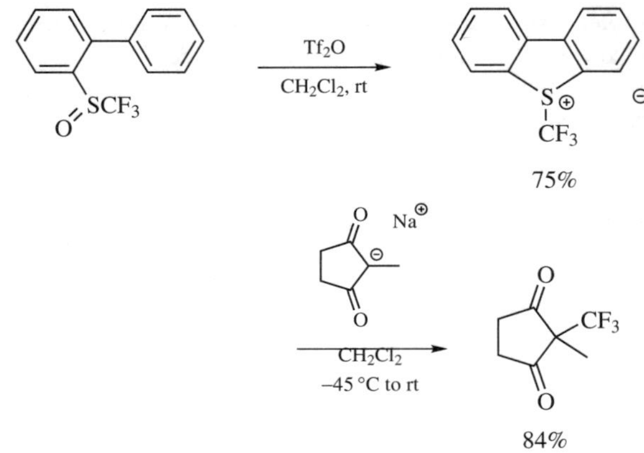
(90)

1. (a) Gramstad, T.; Haszeldine, R. N., *J. Chem. Soc.* **1957**, 4069. (b) Stang, P. J.; Hanack, M.; Subramanian, L. R., *Synthesis* **1982**, 85. (c) Stang, P. J.; White, M. R., *Aldrichim. Acta* **1983**, *16*, 15.
2. Binkley, R. W.; Ambrose, M. G., *J. Carbohydr. Chem.* **1984**, *3*, 1.
3. Ambrose, M. G.; Binkley, R. W., *J. Org. Chem.* **1983**, *48*, 674.
4. Stang, P. J.; Treptow, W., *Synthesis* **1980**, 283.
5. García Martínez, A.; Herrera Fernandez, A.; Martínez Alvarez, R.; Silva Losada, M. C.; Molero Vilchez, D.; Subramanian, L. R.; Hanack, M., *Synthesis* **1990**, 881.
6. Yoshida, M.; Takeuchi, K., *J. Org. Chem.* **1993**, *58*, 2566.
7. Jeanneret, V.; Gasparini, F.; Péchy, P.; Vogel, P., *Tetrahedron* **1992**, *48*, 10637.
8. Neenan, T. X.; Houlihan, F. M.; Reichmanis, E.; Kometani, J. M.; Bachman, B. J.; Thompson, L. F., *Macromolecules* **1990**, *23*, 145.
9. (a) Sato, K.; Hoshi, T.; Kajihara, Y., *Chem. Lett.* **1992**, 1469. (b) Izawa, T.; Nakayama, K.; Nishiyama, S.; Yamamura, S.; Kato, K.; Takita, T., *J. Chem. Soc., Perkin Trans. 1* **1992**, 3003. (c) Knapps, S.; Naughton, A. B. J.; Jaramillo, C.; Pipik, B., *J. Org. Chem.* **1992**, *57*, 7328.
10. Takeuchi, K.; Kitagawa, T.; Ohga, Y.; Yoshida, M.; Akiyama, F.; Tsugeno, A., *J. Org. Chem.* **1992**, *57*, 280.
11. Eaton, P. E.; Zhou, J. P., *J. Am. Chem. Soc.* **1992**, *114*, 3118.
12. Spitz, U. P., *J. Am. Chem. Soc.* **1993**, *115*, 10174.
13. Zheng, C. Y.; Slebocka-Tilk, H.; Nagorski, R. W.; Alvarado, L.; Brown, R. S., *J. Org. Chem.* **1993**, *58*, 2122.
14. Effenberger, F.; Weber, Th., *Ber. Dtsch. Chem. Ges./Chem. Ber.* **1988**, *121*, 421.
15. (a) Rubinsztajn, S.; Fife, W. K.; Zeldin, M., *Tetrahedron Lett.* **1992**, *33*, 1821. (b) Dodd, R. H.; Poissonnet, G.; Potier, P., *Heterocycles* **1989**, *29*, 365.
16. García Martínez, A.; Martínez Alvarez, R.; Teso Vilar, E.; García Fraile, A.; Hanack, M.; Subramanian, L. R., *Tetrahedron Lett.* **1989**, *30*, 581.
17. Sonoda, T.; García Martínez, A.; Hanack, M.; Subramanian, L. R., *Croat. Chim. Acta* **1992**, *65*, 585 (*Chem. Abstr.* **1993**, *118*, 168491).
18. Ritter, K., *Synthesis* **1993**, 735.
19. Hendrickson, J. B.; Bergeron, R., *Tetrahedron Lett.* **1973**, 4607.
20. Hendrickson, J. B.; Bergeron, R.; Giga, A.; Sternbach, D. D., *J. Am. Chem. Soc.* **1973**, *95*, 3412.
21. Edwards, M. L.; Stemerick, D.; McCarthy, J. R., *Tetrahedron Lett.* **1990**, *31*, 3417.
22. Panetta, V.; Yaouanc, J. J.; Handel, H., *Tetrahedron Lett.* **1992**, *33*, 5505.
23. (a) McMurry, J. E.; Scott, W. J., *Tetrahedron Lett.* **1983**, *24*, 979. (b) Comins, J. E.; Dehghani, A., *Tetrahedron Lett.* **1992**, *33*, 6299. (c) Crisp, G. T.; Flynn, B. L., *Tetrahedron* **1993**, *49*, 5873.
24. Anders, E.; Stankowiak, A., *Synthesis* **1984**, 1039.
25. García Martínez, A.; Herrera Fernandez, A.; Alvarez, R. M.; Sánchez García, J. M., *An. Quim., Ser. C* **1981**, *77c*, 28 (*Chem. Abstr.* **1982**, *97*, 5840).
26. (a) Wright, M. E.; Pulley, S. R., *J. Org. Chem.* **1987**, *52*, 5036. (b) Dolle, R. E.; Schmidt, S. J.; Erhard, K. F.; Kruse, L. I., *J. Am. Chem. Soc.* **1989**, *111*, 278.
27. García Martínez, A.; Herrera Fernandez, A.; Moreno Jiménez, F.; García Fraile, A.; Subramanian, L. R.; Hanack, M., *J. Org. Chem.* **1992**, *57*, 1627.

28. Stang, P. J.; Kowalski, M. H.; Schiavelli, M. D.; Longford, R., *J. Am. Chem. Soc.* **1989**, *111*, 3347.
29. García Martínez, A.; Herrera Fernández, A.; Molero Vilchez, D.; Hanack, M.; Subramanian, L. R., *Synthesis* **1992**, 1053.
30. (a) Bentz, H.; Subramanian, L. R.; Hanack, M.; García Martínez, A.; Gómez Marin, M.; Perez-Ossorio, R., *Tetrahedron Lett.* **1977**, 9. (b) Kraus, W.; Zartner, G., *Tetrahedron Lett.* **1977**, 13. (c) García Martínez, A.; García Fraile, A.; Sánchez García, J. M., *Ber. Dtsch. Chem. Ges./Chem. Ber.* **1983**, *116*, 815. (d) García Martínez, A.; Teso Vilar, E.; Gómez Marin, M.; Ruano Franco, C., *Ber. Dtsch. Chem. Ges./Chem. Ber.* **1985**, *118*, 1282.
31. (a) García Martínez, A.; Teso Vilar, E.; López, J. C.; Manrique Alonso, J.; Hanack, M.; Subramanian, L. R., *Synthesis* **1991**, 353. (b) García Martínez, A.; Teso Vilar, E.; García Fraile, A.; Ruano Franco, C.; Soto Salvador, J.; Subramanian, L. R.; Hanack, M., *Synthesis* **1987**, 321.
32. (a) García Martínez, A.; Teso Vilar, E.; García Fraile, A.; Osío Barcina, J.; Hanack, M.; Subramanian, L. R., *Tetrahedron Lett.* **1989**, *30*, 1503. (b) García Martínez, A.; Teso Vilar, E.; Osío Barcina, J.; Manrique Alonso, J.; Rodríguez Herrero, E.; Hanack, M.; Subramanian, L. R., *Tetrahedron Lett.* **1992**, *33*, 607.
33. García Martínez, A.; Herrera Fernandez, A.; Sánchez García, J. M., *An. Quim.* **1979**, *75*, 723 (*Chem. Abstr.* **1980**, *93*, 45 725).
34. García Martínez, A.; Espada Rios, I.; Osío Barcina, J.; Teso Vilar, E., *An. Quim., Ser. C* **1982**, *78c*, 299 (*Chem. Abstr.* **1983**, *98*, 159 896).
35. (a) García Martínez, A.; Espada Rios, I.; Teso Vilar, E., *Synthesis* **1979**, 382. (b) García Martínez, A.; Espada Rios, I.; Osío Barcina, J.; Montero Hernando, M., *Ber. Dtsch. Chem. Ges./Chem. Ber.* **1984**, *117*, 982.
36. Hanack, M.; Hassdenteufel, J. R., *Ber. Dtsch. Chem. Ges./Chem. Ber.* **1982**, *115*, 764.
37. Stang, P. J.; Maas, G.; Smith, D. L.; McCloskey, J. A., *J. Am. Chem. Soc.* **1981**, *103*, 4837.
38. Wittmann, H.; Ziegler, E.; Sterk, H., *Monatsh. Chem.* **1987**, *118*, 531.
39. Wright, M. E.; Pulley, S. R., *J. Org. Chem.* **1989**, *54*, 2886.
40. García Martínez, A.; Martínez Alvarez, R.; García Fraile, A.; Subramanian, L. R.; Hanack, M., *Synthesis* **1987**, 49.
41. Stang, P. J., *Chem. Rev.* **1978**, *78*, 383.
42. García Martínez, A.; Herrera Fernandez, A.; Martínez Alvarez, R.; García Fraile, A.; Calderón Bueno, J.; Osío Barcina, J.; Hanack, M.; Subramanian, L. R., *Synthesis* **1986**, 1076.
43. García Martínez, A.; Osío Barcina, J.; Rys, A. Z.; Subramanian, L. R., *Tetrahedron Lett.* **1992**, *33*, 7787.
44. García Martínez, A.; Martínez Alvarez, R.; Martínez Gonzalez, S.; Subramanian, L. R.; Conrad, M., *Tetrahedron Lett.* **1992**, *33*, 2043.
45. García Martínez, A.; Martínez Alvarez, R.; Madueño Casado, M.; Subramanian, L. R.; Hanack, M., *Tetrahedron* **1987**, *43*, 275.
46. Negishi, E.; Owczarczyk, Z.; Swanson, D. R., *Tetrahedron Lett.* **1991**, *32*, 4453.
47. Evans, D. A.; Sjogren, E. B., *Tetrahedron Lett.* **1985**, *26*, 3787.
48. (a) Houpis, I. N., *Tetrahedron Lett.* **1991**, *32*, 6675. (b) Cook, G. K.; Hornback, W. J.; Jordan, C. L.; McDonald III, J. H.; Munroe, J. E., *J. Org. Chem.* **1989**, *54*, 5828.
49. (a) Effenberger, F.; Sohn, E.; Epple, G., *Ber. Dtsch. Chem. Ges./Chem. Ber.* **1983**, *116*, 1195. (b) Effenberger, F., *Angew. Chem.* **1980**, *92*, 147; *Angew. Chem., Int. Ed. Engl.* **1980**, *19*, 151.
50. García Martínez, A.; Herrera Fernández, A.; Molero Vilchez, D.; Laorden Gutiérrez, L.; Subramanian, L. R., *Synlett* **1993**, 229.
51. Robl, J. A., *Synthesis* **1991**, 56.
52. Falmagne, J.-B.; Escudero, J.; Taleb-Sahraoui, S.; Ghosez, L., *Angew. Chem.* **1981**, *93*, 926; *Angew. Chem., Int. Ed. Engl.* **1981**, *20*, 879.
53. García Martínez, A.; Martínez Alvarez, R.; Osío Barcina, J.; de la Moya Cerero, S.; Teso Vilar, E.; García Fraile, A.; Hanack, M.; Subramanian, L. R., *Chem. Commun./J. Chem. Soc., Chem. Commun.* **1990**, 1571.
54. Gramstad, T.; Husebye, S.; Saebo, J., *Tetrahedron Lett.* **1983**, *24*, 3919.
55. Thomas, E. W., *Synthesis* **1993**, 767.
56. Hendrickson, J. B.; Bair, K. W., *J. Org. Chem.* **1977**, *42*, 3875.
57. Hendrickson, J. B.; Schwartzman, S. M., *Tetrahedron Lett.* **1975**, 277.
58. Hendrickson, J. B.; Schwartzman, S. M., *Tetrahedron Lett.* **1975**, 273.
59. Lindner, E.; v. Au, G.; Eberle, H.-J., *Ber. Dtsch. Chem. Ges./Chem. Ber.* **1981**, *114*, 810.
60. Maas, G.; Lorenz, W., *J. Org. Chem.* **1984**, *49*, 2273.
61. (a) Stang, P. J.; Magnum, M. G.; Fox, D. P.; Haak, P., *J. Am. Chem. Soc.* **1974**, *96*, 4562. (b) Hanack, M.; Märkl, R.; García Martínez, A., *Ber. Dtsch. Chem. Ges./Chem. Ber.* **1982**, *115*, 772.
62. (a) Kitamura, T.; Furuki, R.; Nagata, K.; Taniguchi, H.; Stang, P. J., *J. Org. Chem.* **1992**, *57*, 6810. (b) Stang, P. J.; Zhdankin, V. V.; Tykwinski, R.; Zefirov, N., *Tetrahedron Lett.* **1992**, *33*, 1419.
63. Raghavan, S.; Kahne, D., *J. Am. Chem. Soc.* **1993**, *115*, 1580.
64. Taylor, S. L.; Martin, J. C., *J. Org. Chem.* **1987**, *52*, 4147.
65. Baldwin, J. E.; O'Neil, I. A., *Synlett* **1990**, 603.
66. Singer, B.; Maas, G., *Ber. Dtsch. Chem. Ges./Chem. Ber.* **1987**, *120*, 485.
67. Netscher, T.; Bohrer, P., *Tetrahedron Lett.* **1996**, *37*, 8359–8362.
68. Baraznenok, I. L.; Nenajdenko, V. G.; Balenkova, E. S., *Tetrahedron* **2000**, *56*, 3077–3119.
69. Ross, A. S.; Pitie, M.; Meunier, B., *J. Chem. Soc., Perkin Trans. 1* **2000**, 571–574.
70. Frantz, D. E.; Weaver, D. G.; Carey, J. P.; Kress, M. H.; Dolling, U. H., *Org. Lett.* **2002**, *4*, 4717–4718.
71. Binkley, R. W.; Ambrose, M. G., *J. Org. Chem.* **1983**, *48*, 1776–1777.
72. Martin, O. R.; Yang, F.; Xie, F., *Tetrahedron Lett.* **1995**, *36*, 47–50.
73. Yang, B.-H.; Jiang, J.-Q.; Ma, K.; Wu, H.-M., *Tetrahedron Lett.* **1995**, *36*, 2831–2834.
74. Nagasawa, T.; Handa, Y.; Onoguchi, Y.; Ohba, S.; Suzuki, K., *Synlett* **1995**, 739–741.
75. Lebel, H.; Marcoux, J. F.; Molinaro, C.; Charette, A. B., *Chem. Rev.* **2003**, *103*, 977–1050.
76. Suzuki, H.; Kuroda, C., *Tetrahedron* **2003**, *59*, 3157–3174.
77. Jonas, S.; Westerhausen, M.; Simchen, G., *J. Organometallic Chem.* **1997**, *548*, 131–137.
78. Simchen, G.; Jonas, S., *J. Prakt. Chem./Chem.-Ztg.* **1998**, *340*, 506–512.
79. Nishiyama, T.; Ono, Y.; Kurokawa, S.; Kimura, S., *Chem. Pharm. Bull.* **2000**, *48*, 1999–2002.
80. Katritzky, A. R.; Zhang, S.; Fang, Y., *Org. Lett.* **2000**, *2*, 3789–3791.
81. Martinez, A. G.; Vilar, E. T.; Jimenez, F. M.; Bilbao, C. M., *Tetrahedron: Asymmetry* **1997**, *8*, 3031–3034.
82. Martinez, A. G.; Fernandez, A. H.; Moreno-Jimenez, F.; Fraile, M. J. L.; Subramanian, L. R., *Synlett* **1994**, 559–560.
83. Martinez, A. G.; Fernandez, A. H.; Alvarez, R. M.; Vilchez, M. D. M.; Gutierrez, M. L. L.; Subramanian, L. R., *Tetrahedron* **1999**, *55*, 4825–4830.
84. Herrera, A.; Martinez-Alvarez, R.; Ramiro, P.; Chioua, M.; Torres, R., *Tetrahedron* **2002**, *58*, 3755–3764.
85. Sforza, S.; Dossena, A.; Corradini, R.; Virgili, E.; Marchelli, R., *Tetrahedron Lett.* **1998**, *39*, 711–714.
86. Charette, A. B.; Grenon, M., *Tetrahedron Lett.* **2000**, *41*, 1677–1680.
87. Alder, R. W.; Blake, M. E.; Bufali, S.; Butts, C. P.; Orpen, A. G.; Schutz, J.; Williams, S. J., *J. Chem. Soc., Perkin Trans. 1* **2001**, 1586–1593.
88. Charette, A. B.; Chua, P., *Synlett* **1998**, 163–165.
89. Charette, A. B.; Chua, P., *Tetrahedron Lett.* **1998**, *39*, 245–248.
90. Charette, A. B.; Chua, P., *J. Org. Chem.* **1998**, *63*, 908–909.
91. Charette, A. B.; Chua, P., *Tetrahedron Lett.* **1997**, *38*, 8499–8502.
92. Vonhoff, S.; Vasella, A., *Synth. Commun.* **1999**, *29*, 551–560.
93. McDonald, C. E.; Galka, A. M.; Green, A. I.; Keane, J. M.; Kowalchick, J. E.; Micklitsch, C. M.; Wisnoski, D. D., *Tetrahedron Lett.* **2001**, *42*, 163–166.

94. Liras, S.; Allen, M. P.; Segelstein, B. E., *Synth. Commun.* **2000**, *30*, 437–443.
95. Padwa, A.; Crawford, K. R.; Rashatasakhon, P.; Rose, M., *J. Org. Chem.* **2003**, *68*, 2609–2617.
96. Black, D. S. C.; Rezaie, R., *Tetrahedron Lett.* **1999**, *40*, 4251–4254.
97. Bose, D. S.; Jayalakshmi, B., *Synthesis* **1999**, 64–65.
98. Nenajdenko, V. G.; Vertelezkij, P. V.; Koldobskij, A. B.; Alabugin, I. V.; Balenkova, E. S., *J. Org. Chem.* **1997**, *62*, 2483–2486.
99. Nishiyama, T.; Seshita, T.; Shodai, H.; Aoki, K.; Kameyama, H.; Komura, K., *Chem. Lett.* **1996**, 549–550.
100. Eschwey, M.; Sundermeyer, W.; Stephenson, D. S., *Chem. Ber.* **1983**, *116*, 1623–1630.
101. Billard, T.; Langlois, B. R.; Large, S.; Anker, D.; Roidot, N.; Roure, P., *J. Org. Chem.* **1996**, *61*, 7545–7550.
102. Kahne, D.; Walker, S.; Cheng, Y.; Van Engen, D., *J. Am. Chem. Soc.* **1989**, *111*, 6881–6882.
103. Liang, R.; Yan, L.; Loebach, J.; Ge, M.; Uozumi, Y.; Sekanina, K.; Horan, N.; Gildersleeve, J.; Thompson, C.; Smith, A.; Biswas, K.; Still, W. C.; Kahne, D., *Science* **1996**, *274*, 1520–1522.
104. Yan, L.; Kahne, D., *J. Am. Chem. Soc.* **1996**, *118*, 9239–9248.
105. Crich, D.; Sun, S., *J. Org. Chem.* **1997**, *62*, 1198–1199.
106. Crich, D.; Sun, S., *J. Am. Chem. Soc.* **1997**, *119*, 11217–11223.
107. Crich, D., *J. Carbohydr. Chem.* **2002**, *21*, 667–690.
108. Gildersleeve, J.; Smith, A.; Sakurai, K.; Raghavan, S.; Kahne, D., *J. Am. Chem. Soc.* **1999**, *121*, 6176–6182.
109. Silva, D. J.; Kahne, D.; Kraml, C. M., *J. Am. Chem. Soc.* **1994**, *116*, 2641–2642.
110. Crich, D.; Smith, M., *J. Am. Chem. Soc.* **2001**, *123*, 9015–9020.
111. Codee, J. D. C.; Litjens, R. E. J. N.; den Heeten, R.; Overkleeft, H. S.; van Boom, J. H.; van der Marel, G. A., *Org. Lett.* **2003**, *5*, 1519–1522.
112. Crich, D.; Smith, M., *J. Am. Chem. Soc.* **2002**, *124*, 8867–8869.
113. Crich, D.; Picione, J., *Org. Lett.* **2003**, *5*, 781–784.
114. Gin, D., *J. Carbohydr. Chem.* **2002**, *21*, 645–665.
115. Garcia, B. A.; Poole, J. L.; Gin, D. Y., *J. Am. Chem. Soc.* **1997**, *119*, 7597–7598.
116. Mukaiyama, T.; Suda, S., *Chem. Lett.* **1990**, 1143–1146.
117. Di Bussolo, V.; Kim, Y.-J.; Gin, D. Y., *J. Am. Chem. Soc.* **1998**, *120*, 13515–13516.
118. Kim, J.-Y.; Di Bussolo, V.; Gin, D. Y., *Org. Lett.* **2001**, *3*, 303–306.
119. Liu, J.; Gin, D. Y., *J. Am. Chem. Soc.* **2002**, *124*, 9789–9797.
120. Kim, K. S.; Park, J.; Lee, Y. J.; Seo, Y. S., *Angew. Chem. Int. Ed.* **2003**, *42*, 459–462.
121. Toshima, K., *Carbohydr. Res.* **2000**, *327*, 15–26.
122. Gommermann, N.; Koradin, C.; Knochel, P., *Synthesis* **2002**, 2143–2149.
123. Kodama, Y.; Okumura, M.; Yanabu, N.; Taguchi, T., *Tetrahedron Lett.* **1996**, *37*, 1061–1064.
124. Bottger, G.; Geisler, A.; Frohlich, R.; Wurthwein, E.-U., *J. Org. Chem.* **1997**, *62*, 6407–6411.
125. Muri, D.; Bode, J. W.; Carreira, E. M., *Org. Lett.* **2000**, *2*, 539–541.
126. Hendrickson, J. B.; Sommer, T. J.; Singer, M., *Synthesis* **1995**, 1496–1496.
127. Shackelford, S. A.; Anderson, M. B.; Christie, L. C.; Goetzen, T.; Guzman, M. C.; Hananel, M. A.; Kornreich, W. D.; Li, H.; Pathak, V. P.; Rabinovich, A. K.; Rajapakse, R. J.; Truesdale, L. K.; Tsank, S. M.; Vazir, H. N., *J. Org. Chem.* **2003**, *68*, 267–275.
128. Adams, C. M.; Sharts, C. M.; Shackelford, S. A., *Tetrahedron Lett.* **1993**, *34*, 6669–6672.
129. Bates, D. K.; Xia, M., *J. Org. Chem.* **1998**, *63*, 9190–9196.
130. Corey, E. J.; Gin, D. Y.; Kania, R. S., *J. Am. Chem. Soc.* **1996**, *118*, 9202–9203.
131. Furukawa, N.; Kobayashi, K.; Sato, S., *J. Organometallic Chem.* **2000**, *611*, 116–126.
132. Nenajdenko, V. G.; Shevchenko, N. E.; Balenkova, E. S.; Alabugin, I. V., *Chem. Rev.* **2003**, *103*, 229–282.
133. Katritzky, A. R.; Moutou, J.-L.; Yang, Z., *Synthesis* **1995**, 1497–1505.
134. Martinez, A. G.; Fernandez, A. H.; Jimenez, F. M.; Ruiz, P. M.; Subramanian, L. R., *Synlett* **1995**, 161–162.
135. Kitamura, T.; Kotani, M.; Fujiwara, Y., *Synthesis* **1998**, 1416–1418.
136. Stang, P. J.; Zhdankin, V. V., *Chem. Rev.* **1996**, *96*, 1123–1178.
137. Kutateladze, A. G.; Kice, J. L.; Kutateladze, T. G.; Zefirov, N. S.; Zyk, N. V., *Tetrahedron Lett.* **1992**, *33*, 1949–1952.
138. Umemoto, T.; Ishihara, S., *Tetrahedron Lett.* **1990**, *31*, 3579–3582.
139. Umemoto, T.; Ishihara, S., *J. Am. Chem. Soc.* **1993**, *115*, 2156–2164.

m-Trifluoromethylbenzoyl Chloride

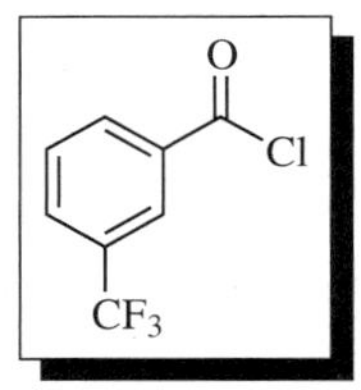

[2251-65-2] $C_8H_4ClF_3O$ (MW 208.56)

(forms ester derivatives of alcohols which can be deoxygenated under photolytic conditions using *N*-methylcarbazole:[1–13] reagent for acylation of amines to afford amides[14–27])

Physical Data: bp 184–186 °C/750 mm Hg; fp 235 °F; *d* 1.383 g mL^{-1}.
Solubility: soluble in CH_2Cl_2, $CHCl_3$, 1,2-dichoroethane, and THF; reacts with water, alcohols, and other nucleophilic solvents.
Form Supplied in: clear liquid.
Handling, Storage, and Precautions: corrosive, respiratory irritant, causes burns, moisture sensitive.

---

m-Trifluoromethylbenzoyl chloride has been used widely as an acylating agent in the preparation of esters, amides, and ketones, and various derivatives of each class of compound.

***O*-Acylation.** Acylation of alchols under standard conditions has led to a number of esters, including esters of artemisinin derivatives, and enol esters of 3-oxo-phosphindoles.[1,2] Some of these esters have found use as starting materials for flavonoid products.[3,4]

Deoxygenation of Secondary Alcohols. Secondary *m*-trifluoromethylbenzoyl esters have proven to have particular value in deoxygenation reactions. Described originally by Saito, these esters undergo photoinduced electron-transfer (PET) reduction with *N*-methylcarbazole (MCZ) as photosensitizer (eq 1).[5,6]

Although benzoate esters also undergo deoxygenation under these conditions, the *m*-trifluoromethylbenzoate esters are cleaved more readily and can be selectively deoxygenated in the presence of benzoate esters (eq 2).[7]

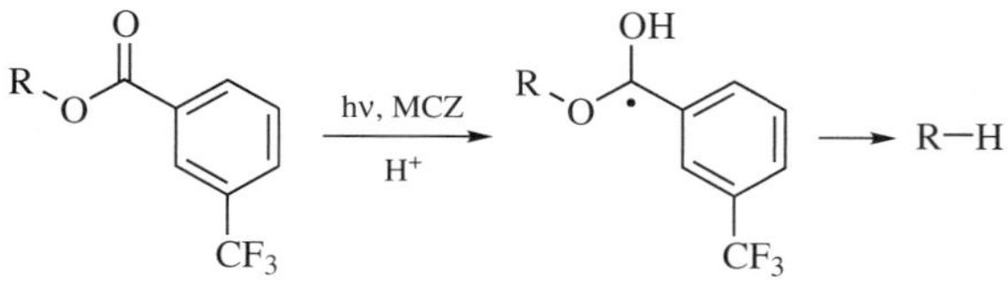

(1)

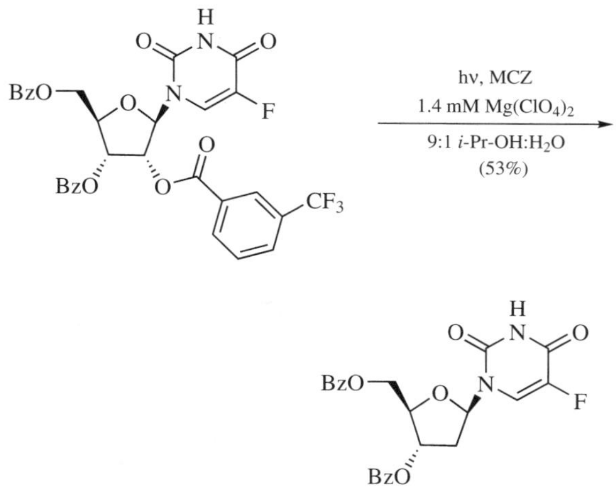

(2)

Deoxygenation of a number of substrates has been reported, including nucleosides, disaccharides, and lactones.[8–13] In most reported cases, the PET deoxygenation is superior to tin-mediated reductions, although in the reduction of a homopropargyl alcohol in a nucleoside derivative the reverse was true.[12]

***N*-Acylation.** Amides derived from *m*-trifluoromethylbenzoyl chloride have provided a large number of biologically active compounds.[14–18] Acylation of either primary or secondary amines under basic conditions affords the corresponding amides (eq 3).[14]

Acylation of amines with *m*-trifluoromethylbenzoyl chloride has also been used in the solid-phase synthesis of combinatorial libraries to prepare substituted piperazines, *N,N*-dimethyltryptamines and ligands for the SH3 domain of proteins.[19–21]

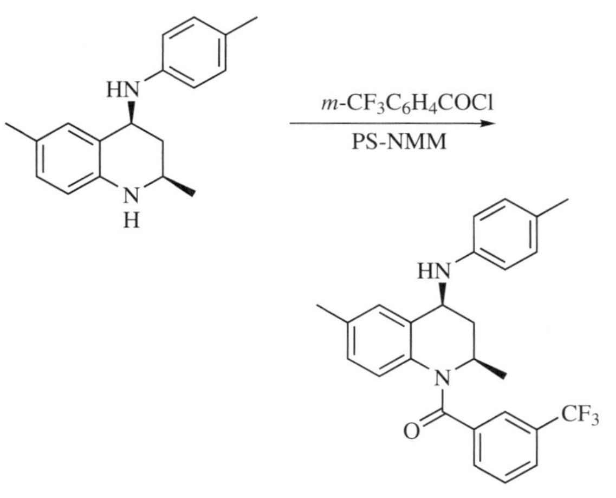

(3)

PS-NMM = polystyrene-supported *N*-methylmorpholine

A number of amides derived from *m*-trifluoromethyl chloride have been used as precursors for heterocyclic ring systems. For example, benzimadazole derivatives have been prepared as inhibitors of poly (ADP-ribose) polymerase I.[22,23] Other amide derivatives with various biological activities include oxazolines, quinolones, triazolones, and triazol-3-thiones as potential antibacterials, antitumor agents, potassium channel openers, and antidepressants, respectively.[24–27]

***C*-Acylation.** *m*-Trifluoromethyl ketones have been prepared by both noncatalyzed and Lewis-acid catalyzed acylation of aromatic substrates.[28,29] *m*-Trifluoromethyl benzoyl chloride has also been used successfully as a substrate in Suzuki coupling to provide both chalcones (eq 4) and cyclopropyl ketones.[30,31]

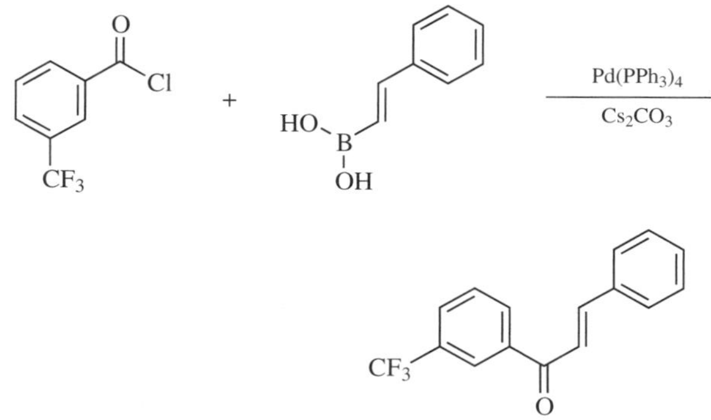

(4)

Related Reagents. Acetyl Chloride; Benzoyl Chloride; *N*-methylcarbazole.

1. O'Neill, P. M.; Searle, N. L.; Kan, K.-W.; Storr, R. C.; Maggs, J. L.; Ward, S. A.; Raynes, K.; Park, B. K., *J. Med. Chem.* **1999**, *42*, 5487.
2. Balthazor, T. M., *J. Org. Chem.* **1980**, *45*, 2519.
3. Gobbi, S.; Rampa, A.; Bisi, A.; Belluti, F.; Piazzi, L.; Valenti, P.; Caputo, A.; Zampiron, A.; Carrara, M., *J. Med. Chem.* **2003**, *46*, 3662.
4. Ares, J. J.; Outt, P. E.; Kakodkar, S. V.; Buss, R. C.; Geiger, J. C., *J. Org. Chem.* **1993**, *58*, 7903.
5. Saito, I.; Ikehira, H.; Kasetani, R.; Watanabe, M.; Matsuura, T., *J. Am. Chem. Soc.* **1986**, *108*, 3115.
6. Crich, D.; Quintero, L., *Chem. Rev.* **1989**, *89*, 1413.
7. Park, M.; Rizzo, C. J., *J. Org. Chem.* **1996**, *61*, 6092.
8. Wang, Z.; Prudhomme, D. R.; Buck, J. R.; Park, M.; Rizzo, C. J., *J. Org. Chem.* **2000**, *65*, 5969.
9. Almond, M. R.; Collins, J. L.; Reitter, B. E.; Rideout, J. L.; Freeman, A.; St Clair, M. H., *Tetrahedron Lett.* **1991**, *32*, 5745.
10. Chioconni, A.; Marino, C.; Otal, E.; de Lederkremer, M. R., *Carbohydr. Res.* **2002**, *337*, 2119.
11. Riedel, S.; Donnerstag, A.; Hennig, L.; Welzel, P.; Richter, J.; Hobert, K.; Muller, D.; van Heijenoort, J., *Tetrahedron* **1999**, *55*, 1921.
12. Jung, P. M. J.; Burger, A.; Biellmann, J.-F., *J. Org. Chem.* **1997**, *62*, 8309.
13. Barvian, M. R.; Greenberg, M. M., *J. Org. Chem.* **1993**, *58*, 6151.
14. Smith, H. C.; Cavanaugh, C. K.; Friz, J. L.; Thompson, C. S.; Saggers, J. A.; Michelotti, E. L.; Garcia, J.; Tice, C. M., *Bioorg. Med. Chem. Lett.* **2003**, *13*, 1943.
15. Durley, R. C.; Grapperhaus, M. L.; Hickory, B. S.; Massa, M. A.; Wang, J. L.; Spangler, D. P.; Mischke, D. A.; Parnas, B. L.; Fobian, Y. M.; Rath, N. P.; Honda, D. D.; Zeng, M.; Connolly, D. T.; Heuvelman, D. M.; Witherbee, B. J.; Melton, M. A.; Glenn, K. C.; Krul, E. S.; Smith, M. E.; Sikorski, J. A., *J. Med. Chem.* **2002**, *45*, 3891.

16. Makovec, F.; Peris, W.; Revel, L.; Giovanetti, R.; Mennuni, L.; Rovati, L. C., *J. Med. Chem.* **1992**, *35*, 28.
17. Yeung, J. M.; Corleto, L. A.; Knaus, E. E., *J. Med. Chem.* **1982**, *25*, 720.
18. Hynes, J. B.; Hack, L. G., *J. Med. Chem.* **1972**, *15*, 1194.
19. Salvino, J. M.; Gerard, B.; Ye, H. F.; Sauvagnat, B.; Dolle, R. E., *J. Comb. Chem.* **2003**, *5*, 260.
20. Wu, T. Y. H.; Schultz, P. G., *Org. Lett.* **2002**, *4*, 4033.
21. Combs, A. P.; Kapoor, T. M.; Feng, S.; Chen, J. K.; Daude-Snow, L. F.; Schreiber, S. L., *J. Am. Chem. Soc.* **1996**, *118*, 287.
22. Skalitzky, D. J.; Marakovits, J. T.; Maegley, K. A.; Ekker, A.; Yu, X.-H.; Hostomsky, Z.; Webber, S. E.; Eastman, B. W.; Almassy, R.; Li, J.; Curtin, N. J.; Newell, D. R.; Calvert, A. H.; Griffin, R. J.; Golding, B. T., *J. Med. Chem.* **2003**, *46*, 210.
23. White, A. W.; Almassy, R.; Calvert, A. H.; Curtin, N. J.; Griffin, R. J.; Hostomsky, Z.; Maegley, K.; Newell, D. R.; Srinivasan, S.; Golding, B. T., *J. Med. Chem.* **2000**, *43*, 4084.
24. Pirrung, M. C.; Turney, L. N.; McClerren, A. L.; Raetz, C. R. H., *J. Am. Chem. Soc.* **2003**, *125*, 1575.
25. Li, L.; Wang, H.-K.; Kuo, S.-C.; Wu, T.-S.; Mauger, A.; Lin, C. M.; Hamel, E.; Lee, K.-S., *J. Med. Chem.* **1994**, *37*, 3400.
26. Romine, J. L.; Martin, S. W.; Gribkoff, V. K.; Boissard, C. G.; Dworetzky, S. I.; Natale, J.; Li, Y.; Gao, Q.; Meanwell, N. A.; Starrett, J. E., *J. Med. Chem.* **2002**, *45*, 2942.
27. Kane, J. M.; Dudley, M. W.; Sorensen, S. M.; Miller, F. P., *J. Med. Chem.* **1988**, *31*, 1253.
28. Raboisson, P.; Schultz, D.; Lugnier, C.; Bourguignon, J.-J., *Tetrahedron Lett.* **2002**, *43*, 9501.
29. Hagishita, S.; Yamada, M.; Shirahase, K.; Okada, T.; Murakami, Y.; Ito, Y.; Matsuura, T.; Wada, M.; Kato, T.; Ueno, M.; Chikazawa, Y.; Yamada, K.; Ono, T.; Teshirogi, I.; Ohtani, M., *J. Med. Chem.* **1996**, *39*, 3636.
30. Eddarir, S.; Cotelle, N.; Bakkour, Y.; Rolando, C., *Tetrahedron Lett.* **2003**, *44*, 5359.
31. Chen, H.; Deng, M.-Z., *Org. Lett.* **2000**, *2*, 1649.

Norma K. Dunlap
Middle Tennessee State University, Murfreesboro, TN, USA

Trimethylphosphine

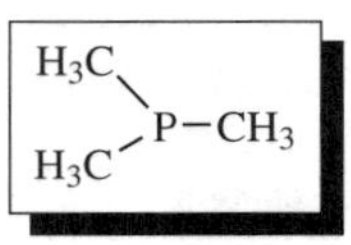

[594-09-2] C_3H_9P (MW 76.08)

(reagent used in Mitsunobu reaction; transformation of azides into carbamates, aziridines from azidoalcohols; iminophosphoranes and aza-Wittig reaction)

Physical Data: mp −85 °C; bp 38–39 °C; d 0.748 g cm^{-3}.
Solubility: soluble in ether, THF, toluene.
Form Supplied in: widely available, neat and as 1.0 M solutions in THF or toluene. Also available as complex with silver iodide from which Me_3P is isolated by heating.
Analysis of Reagent Purity: ^{31}P NMR ($CDCl_3$, rt, relative to external H_3PO_4): δ 62.0 ppm.
Handling, Storage, and Precautions: can be stored indefinitely under an inert atmosphere with refrigeration. Me_3P is pyrophoric, reacts rapidly with oxygen, and is irritating to the eyes, respiratory system, and skin.[1] Sensitive to air and incompatible with oxidizing agents.[2] Use under an inert atmosphere and in a fume hood.

Introduction. Of the commonly used tertiary phosphines, trimethylphosphine is exceptionally nucleophilic. Trimethylphosphine is a relatively weak base (*p*Ka 8.65 in H_2O), but is stronger in this respect than triphenylphosphine.[2]

Mitsunobu Reaction. The Mitsunobu reaction[3] using diethylazodicarboxylate (DEAD)–triphenylphosphine system is a versatile method for the condensation of alcohols and nucleophiles. In some special cases, Me_3P gave better results than Ph_3P or Bu_3P.[4] For instance, the inversion of the hindered hydroxyl group (eq 1) under standard Mitsunobu conditions (Ph_3P, $PhCO_2H$, DEAD) resulted in a low yield; a better result was obtained using Bu_3P as phosphine reagent (50% yield together with the diene in 20–30% yield). However, using Me_3P (1 M in THF) the yield of the inversion was increased to 74%.[5]

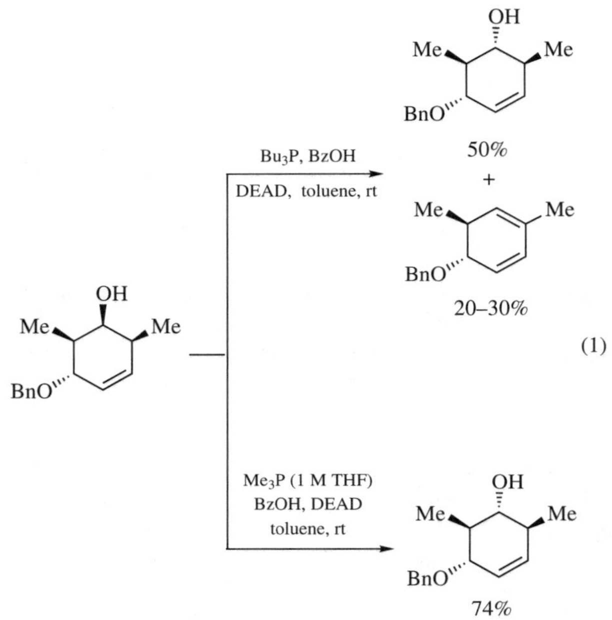

(1)

Me_3P (1 M in toluene), 1,1′-(azodicarbonyl)dipiperidine (ADPP), imidazole system has also been used in the formation of carbon-sulfur bonds via the Mitsunobu reaction of aliphatic thiols and unhindered alcohols (eq 2).[6] In some cases 3 equiv of alcohol were used. The intramolecular version was also considered (eq 3).

$$\text{R-SH + R'-OH} \xrightarrow[\substack{\text{ADPP (2 equiv)}\\ \text{Im-H, CH}_2\text{Cl}_2\text{ or THF}\\ \text{rt, 56–86\%}}]{\substack{\text{Me}_3\text{P (1 equiv)}\\ \text{imidazole (2 equiv)}}} \text{R-S-R'} \quad (2)$$

R = Ar-CH_2-; allyl; alkyl; cyclopropyl; Ar-CH(Me)-

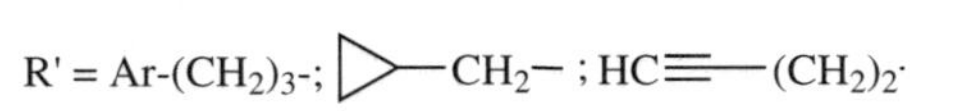

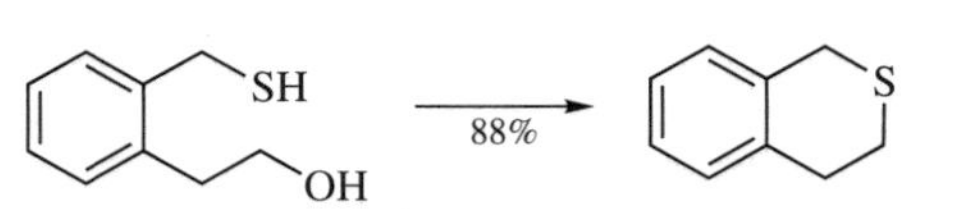

(3)

Thioglycosides were efficiently synthesized from 1-thiosugars using alcohols, ADPP, and Me_3P (1 M in THF). The reaction conditions are compatible with a wide range of functional groups (eq 4).[7]

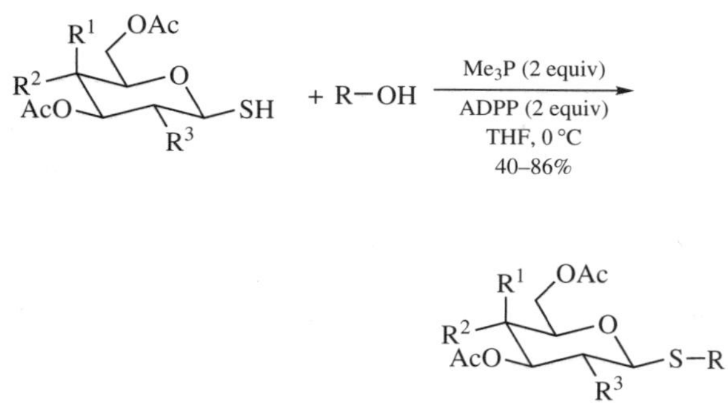

(4)

$R^1 = H; R^2 = R^3 = OAc$
$R^2 = H; R^1 = R^3 = OAc$
$R^1 = H; R^2 = OAc; R^3 = NHDde^*$
[Dde* = 1-(4,4-dimethyl-2,6-dioxocyclohexylidene)ethyl]

It is generally accepted that an adduct (**1**) between ADPP and Me_3P is formed.[8] The formation of this azaphosphonium salt can be easily monitored as the yellow solution becomes colorless after the reaction is completed (approximately 30 min). On addition of the alcohol, an oxaphosphonium ion (**2**) is formed which undergoes nucleophilic substitution by the thiol (eq 5).

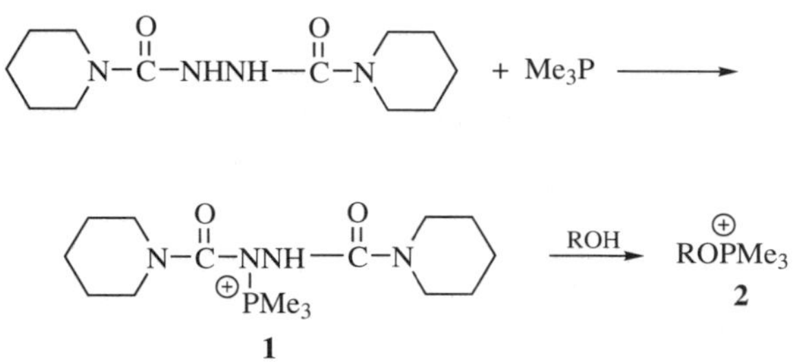

(5)

Direct Transformation of Azides into Carbamates. Transformation of azides into carbamates has been achieved by reaction with 1.0 M solution of Me_3P in THF.[9] After the formation of the phosphazene (1–2 h), the corresponding chloroformate was added. Stirring at room temperature for 20–60 min, quenching with phosphate buffer (*p*H 7) and usual work up afforded the pure carbamates. In most cases column chromatography was not required (eq 6).

In the specific case of the synthesis of Boc-protected amines from azides,[10] the reaction was performed by mixing Me_3P and 2-(*tert*-butoxycarbonyloxyimino)-2-phenylacetonitrile (Boc-ON) in toluene or THF at −20 °C. In this way, Boc-protected amines were formed in 90–100% yield (eq 7).

In a related process, vicinal azido alcohols can be transformed into Boc-aminoalcohols (90–98%) or oxazolidinones (91–96%) by reaction with Me_3P and Boc_2O or CO_2, respectively (eq 8).[11]

Regioselective transformation of azides into N-Cbz amines or Boc-amines was achieved by reaction with PMe_3 followed by treatment with Cbz-Cl or Boc-ON.[12] Some selected cases are indicated in eq 9.

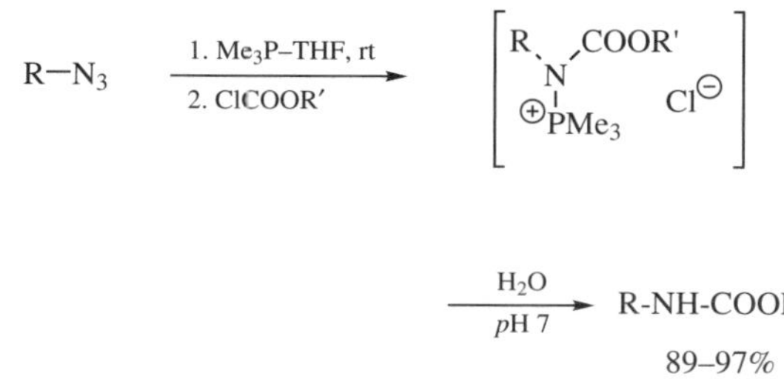

(6)

$R = ArCH_2$-; $ArCHMe$-; $Ar(CH_2)_2$-;

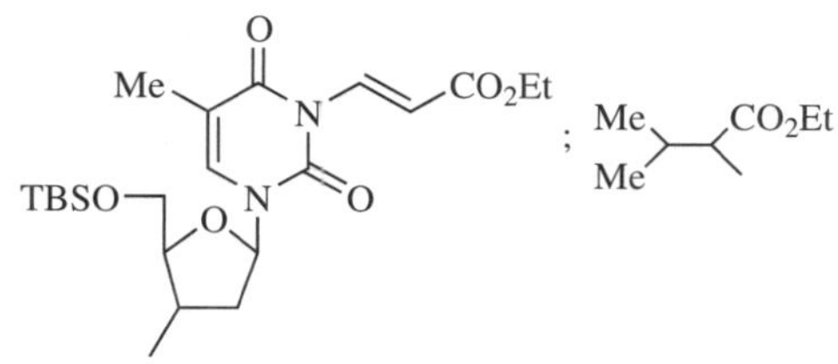

$R' =$ Me; Et; CCl_3CH_2; Allyl; Bn

$$R{-}N_3 \xrightarrow[\text{toluene, rt}]{Me_3P} R{-}N{=}PMe_3$$

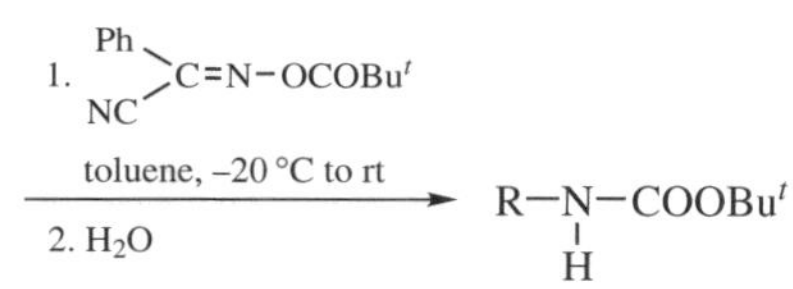

(7)

$R = PhCH_2$-; $PhCH_2CH_2$-; $MeOCO(CH_2)_5$-; $Ph_2BuSiO(CH_2)_3$-;

$PhCHMe$-; $PhCOOCH_2CHMe$-; Ph-;

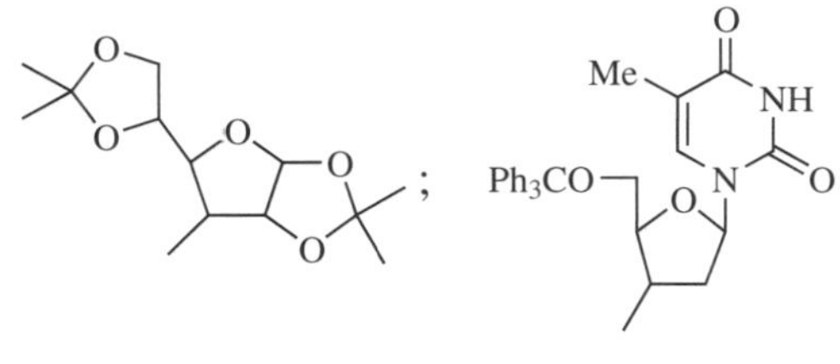

Aziridine Formation from 1,2-Azidoalcohols. Intramolecular reductive cyclization of 1,2-azidoalcohols to give the related aziridines has been performed using trimethylphosphine in acetonitrile. For instance, starting from a mixture of isomeric azidoalcohols (eq 10), the aziridine was obtained in 97% isolated yield and 70% optical purity.[13] Triphenylphosphine in the presence of catalytic triethylamine hydrochloride (5 mol %) also gave satisfactory results but required a more complex work up to remove the triphenylphosphine oxide.

Iminophosphoranes and Aza-Wittig Reaction. Iminophosphoranes (λ^5-phosphazenes, phosphine imines) can be obtained by reaction of azides with phosphines or phosphites (eq 11).[14]

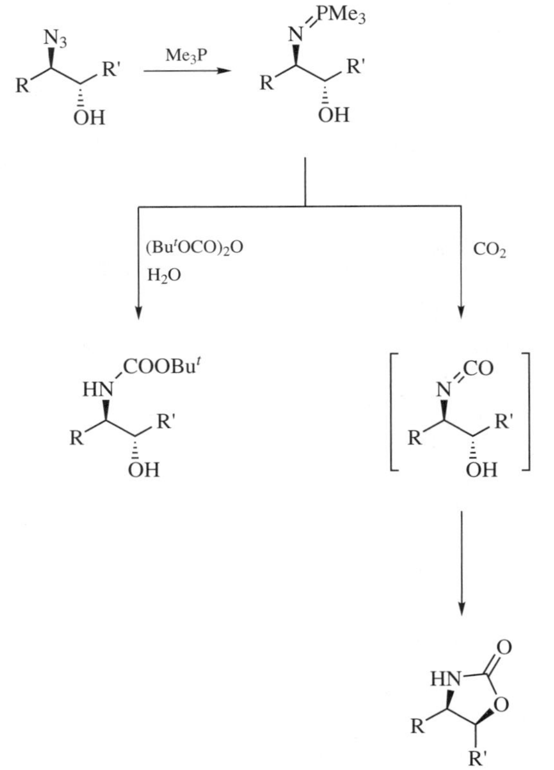

(8)

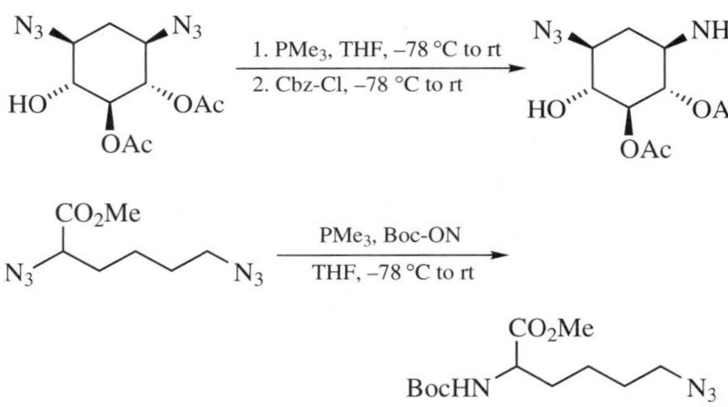

(9)

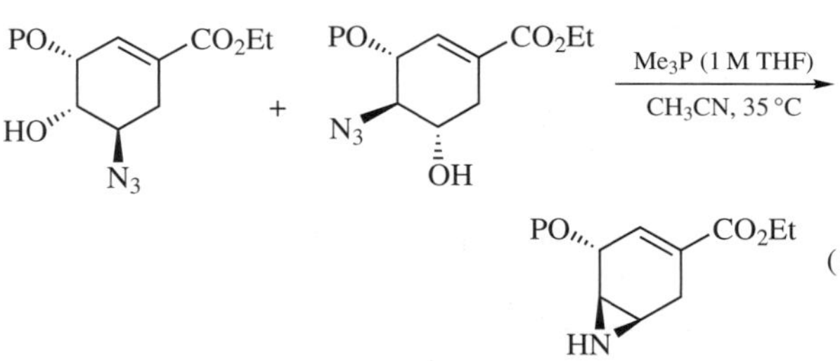

(10)

$$R{-}N_3 + R'_3P \longrightarrow R{-}N{=}PR'_3 \quad (11)$$

The reaction of iminophosphoranes with carbonyl compounds (aza-Wittig reaction) affords imine derivatives (eq 12). The synthetic relevance of the process has been summarized in several review articles.[15]

$$R{-}N{=}PR'_3 + R^2{-}\overset{\overset{\displaystyle O}{\|}}{C}{-}R^3 \longrightarrow R{-}N{=}C\!\begin{smallmatrix} R^2 \\ R^3 \end{smallmatrix} \quad (12)$$

In this context, the use of Me_3P increases the reactivity of the iminophosphorane reagent. For instance, Charette and co-workers[16] have shown that the reaction of azides (eq 13) with different aldehydes in the presence of Me_3P (0.1 M in THF, rt, 1.10 equiv, 4 h) afforded the expected imino derivative. With Ph_3P high temperature (refluxing toluene) and long reaction times (48 h) were necessary.

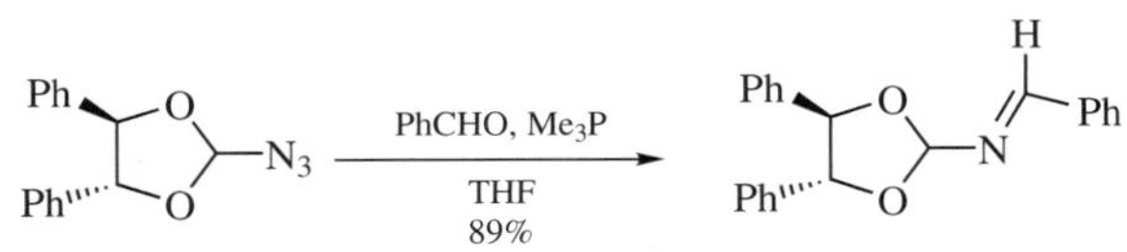

(13)

Iminophosphoranes derived from Me_3P react with ketenes to give ketenimines (eq 14).[17] The addition of different nucleophilic reagents to the central sp carbon and the cycloaddition reactions of these compounds, studied in detail by Alajarin and co-workers,[18] have attracted attention in recent years.

$$R{-}N{=}PMe_3 + \begin{smallmatrix} R^1 \\ R^2 \end{smallmatrix}\!C{=}C{=}O \longrightarrow R{-}N{=}C{=}C\!\begin{smallmatrix} R^1 \\ R^2 \end{smallmatrix} \quad (14)$$

Miscellaneous. Ring opening metathesis polymerization (ROMP) of cyclopentene giving polycyclopentene with controllable molecular weight and polydispersity below 1.1 was achieved at room temperature using a commercial Mo-based ROMP initiator in the presence of Me_3P.[19]

Suzuki reactions of fluoroarenes have been performed by coupling with arylboronic acids using $Pd_2(dba)_3$ as catalyst in the presence of Cs_2CO_3 and Me_3P (1 M in THF).[20]

Phosphorous ylides can be synthesized by reaction of C-phosphanyl-C-chloroiminium salts and Me_3P. In this case, the C-phosphanyl-C-chloroiminium salts behave as carbene synthetic equivalents.[21]

---

1. *Sigma-Aldrich Library of Regulatory and Safety Data*; Leng, R. E.; Votoupal, K. L., Eds.; Sigma-Aldrich, Milwaukee, 1993.
2. Patai, S.; Hartley, F. R., *The Chemistry of Organo Phosphorous Compounds*; Wiley: New York, 1992.
3. Hughes, D. L., *Org. React.* **1992**, *42*, 335.
4. Tsunoda, T.; Yamamiya, Y.; Itô, S., *Tetrahedron Lett.* **1993**, *34*, 1639–1642.
5. Arjona, O.; Menchaca, R.; Plumet, J., *J. Org. Chem.* **2001**, *66*, 2400–2413.
6. Falck, J. R.; Lai, J. Y.; Cho, S. D.; Yu, J., *Tetrahedron Lett.* **1999**, *40*, 2903–2906.
7. Falconer, R. A.; Jablonkai, I.; Toth, I., *Tetrahedron Lett.* **1999**, *40*, 8663–8666.
8. Hughes, D. L.; Reamer, R. A.; Bergan, J. J.; Grabowski, E. J. J., *J. Am. Chem. Soc.* **1988**, *110*, 6487–6491.
9. Ariza, X.; Urpí, F.; Vilarrasa, J., *Tetrahedron Lett.* **1999**, *40*, 7515–7517.
10. Ariza, X.; Urpí, F.; Viladomat, C.; Vilarrasa, J., *Tetrahedron Lett.* **1998**, *39*, 9101–9102.

11. Ariza, X.; Pineda, O.; Urpí, F.; Vilarrasa, J., *Tetrahedron Lett.* **2001**, *42*, 4995–4999.

12. Nyffeler, P. T.; Liang, C. H.; Koeller, K. M.; Wong, C. H., *J. Am. Chem. Soc.* **2002**, *124*, 10773–10778.

13. Rohloff, J. C.; Kent, K. M.; Postich, M. J.; Becker, M. W.; Chapman, H. H.; Kelly, D. E.; Lew, W.; Louie, M. S.; McGree, L. R.; Prisbe, E. J.; Schultze, L. M.; Yu, R. H.; Zhang, L., *J. Org. Chem.* **1998**, *63*, 4545–4550.

14. Staüdinger, H.; Meyer, J., *Helv. Chim. Acta* **1919**, *2*, 635–646.

15. (a) Eguchi, S.; Okano, T.; Okawa, T., *Recent Res. Devp. Org. Chem.* **1997**, *1*, 337–346. (b) Shah, S.; Protasiewicz, J. D., *Coord. Chem. Rev.* **2000**, *210*, 181–201. (c) Abdou, W. M., *Phosphorous Sulfur Silicon* **2002**, *177*, 325–389.

16. Boezio, A. A.; Solberhe, G.; Lauzon, C.; Charette, A. B., *J. Org. Chem.* **2003**, *68*, 3241–3245.

17. Staüdinger, H.; Hauser, E., *Helv. Chim. Acta* **1921**, *4*, 887–896.

18. (a) Perst, H., *Methoden Organic Chemistry*; Houben-Weyl, 1993, *15*, part 3, 2531–2710. (b) Cossio, F. P.; Arrieta, A.; Lecea, B.; Alajarín, M.; Vidal, A.; Tovar, F., *J. Org. Chem.* **2000**, *65*, 3633–3643. (c) Alajarín, M.; Vidal, A.; Tovar, F.; Ramírez de Arellano, M. C.; Cossio, F. P.; Arrieta, A.; Lecea, B., *J. Org. Chem.* **2000**, *65*, 7512–7515. (d) Alajarín, M.; Vidal, A.; Tovar, F., *Tetrahedron Lett.* **2000**, *41*, 7029–7032. (e) Alajarín, M.; Vidal, A.; Tovar, F.; Sánchez-Andrada, P., *Tetrahedron Lett.* **2002**, *43*, 6259–6261. (f) Alajarín, M.; Vidal, A.; Orenes, R. A., *Eur. J. Org. Chem.* **2002**, 4222–4227.

19. Trzaska, S. T.; Lee, L. B. W.; Register, R. A., *Macromolecules* **2000**, *33*, 9215–9221.

20. Widdowson, D. A.; Wilhelm, R., *Chem. Commun.* **2003**, 578–581.

21. Mercaron, N.; Baceiredo, A.; Gornitzka, H.; Bertrand, G., *Chem. Commun.* **2002**, 2250–2251.

Odón Arjona & Joaquín Plumet
Universidad Complutense de Madrid, Spain

Trimethylsilyl Trifluoromethanesulfonate[1]

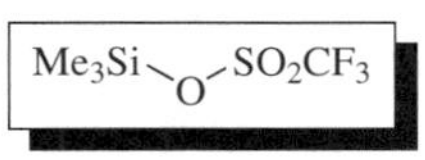

[88248-68-4; 27607-77-8] $C_4H_9F_3O_3SSi$ (MW 222.29)

Alternate Name: TMSOTf.
Physical Data: bp 45–47 °C/17 mmHg, 39–40 °C/12mmHg; *d* 1.225 g cm^{-3}.
Solubility: sol aliphatic and aromatic hydrocarbons, haloalkanes, ethers.
Form Supplied in: colorless liquid; commercially available.
Preparative Methods: may be prepared by a variety of methods.[2]
Handling, Storage, and Precautions: flammable; corrosive; very hygroscopic.

Silylation. TMSOTf is widely used in the conversion of carbonyl compounds to their enol ethers. The conversion is some 10^9 faster with TMSOTf/ Triethylamine than with Chlorotrimethylsilane (eqs 1–3).[3–5]

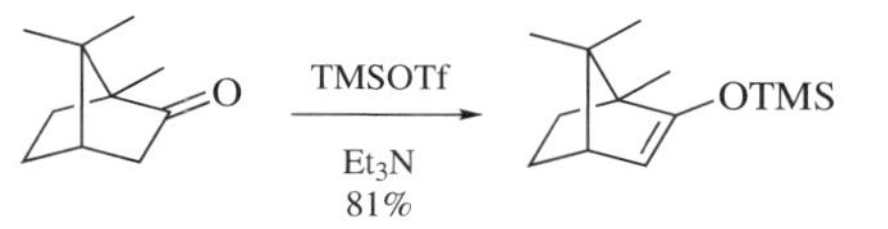
(1)

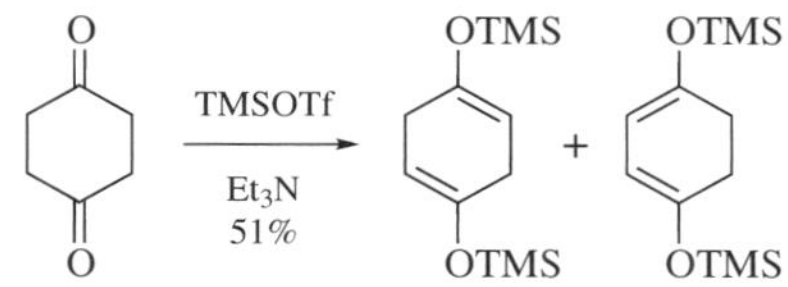
(2)

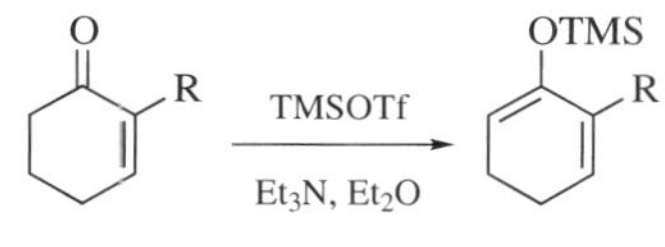
(3)

Dicarbonyl compounds are converted to the corresponding bis-enol ethers; this method is an improvement over the previous two-step method (eq 4).[6]

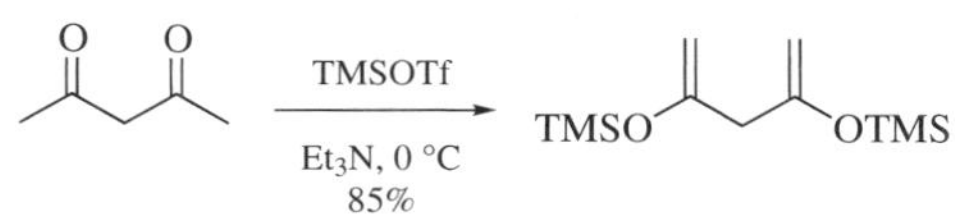
(4)

In general, TMSOTf has a tendency to *C*-silylation which is seen most clearly in the reaction of esters, where *C*-silylation dominates over *O*-silylation. The exact ratio of products obtained depends on the ester structure[7] (eq 5).[8] Nitriles undergo *C*-silylation; primary nitriles may undergo *C,C*-disilylation.[9]

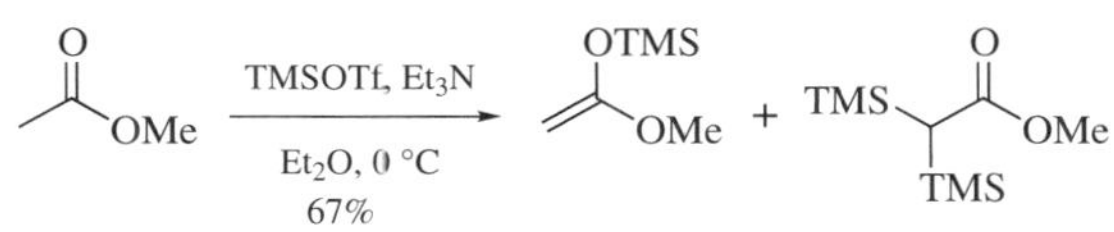
(5)

84:16

TMS enol ethers may be prepared by rearrangement of α-ketosilanes in the presence of catalytic TMSOTf (eq 6).[10,11]

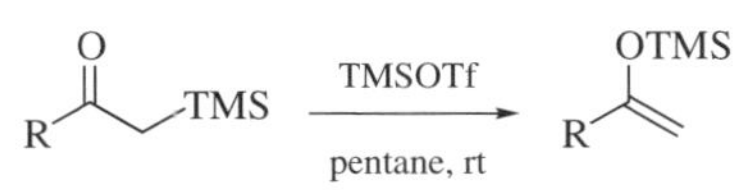
(6)

Enhanced regioselectivity is obtained when trimethylsilyl enol ethers are prepared by treatment of α-trimethylsilyl ketones with catalytic TMSOTf (eq 7).[12]

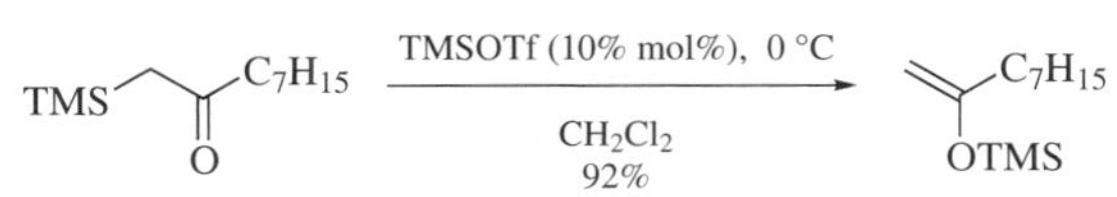
(7)

The reaction of imines with TMSOTf in the presence of Et_3N gives *N*-silylenamines.[13]

Ethers do not react, but epoxides are cleaved to give silyl ethers of allylic alcohols in the presence of TMSOTf and 1,8-Diazabicyclo[5.4.0]undec-7-ene; The regiochemistry of the reaction is dependent on the structure of the epoxide (eq 8).[14]

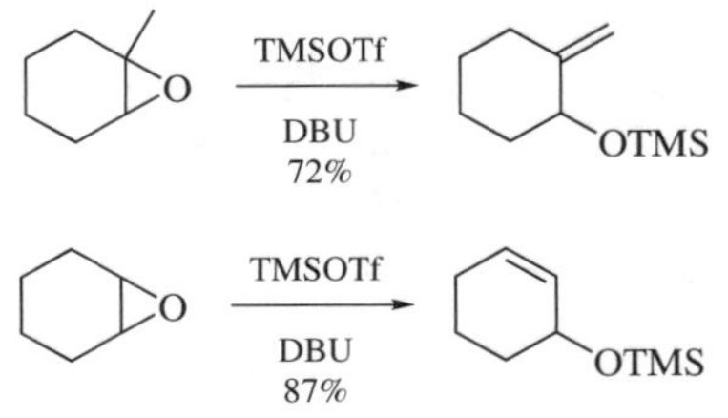

(8)

Indoles and pyrroles undergo efficient *C*-silylation with TMSOTf (eq 9).[15]

t-Butyl esters are dealkylatively silylated to give TMS esters by TMSOTf; benzyl esters are inert under the same conditions.[16]

Imines formed from unsaturated amines and α-carbonyl esters undergo ene reactions in the presence of TMSOTf to form cyclic amino acids.[17]

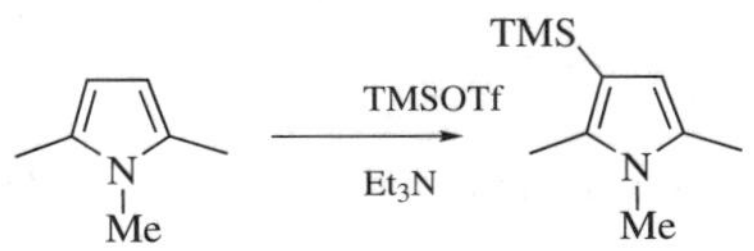

(9)

Carbonyl Activation. 1,3-Dioxolanation of conjugated enals is facilitated by TMSOTf in the presence of 1,2-bis(trimethylsilyloxy)ethane. In particular, highly selective protection of sterically differentiated ketones is possible (eq 10).[18] Selective protection of ketones in the presence of enals is also facilitated (eq 11).[19]

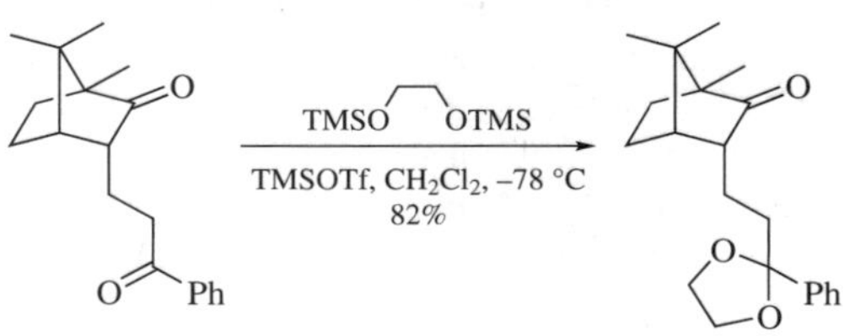

(10)

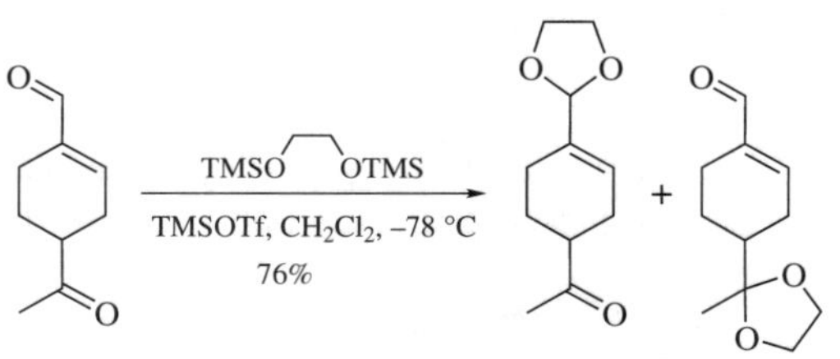

(11)

1:27

The similar reaction of 2-alkyl-1,3-disilyloxypropanes with chiral ketones is highly selective and has been used to prepare spiroacetal starting materials for an asymmetric synthesis of α-tocopherol subunits (eq 12).[20]

The preparation of spiro-fused dioxolanes (useful as chiral glycolic enolate equivalents) also employs TMSOTf (eq 13).[21]

TMSOTf mediates a stereoselective aldol-type condensation of silyl enol ethers and acetals (or orthoesters). The nonbasic reaction conditions are extremely mild. TMSOTf catalyzes many aldol-type reactions; in particular, the reaction of relatively nonnucleophilic enol derivatives with carbonyl compounds is facile in the presence of the silyl triflate. The activation of acetals was first reported by Noyori and has since been widely employed (eq 14).[22,23]

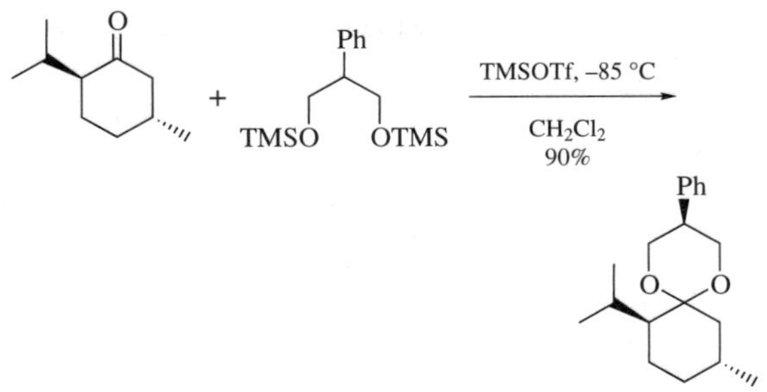

(12)

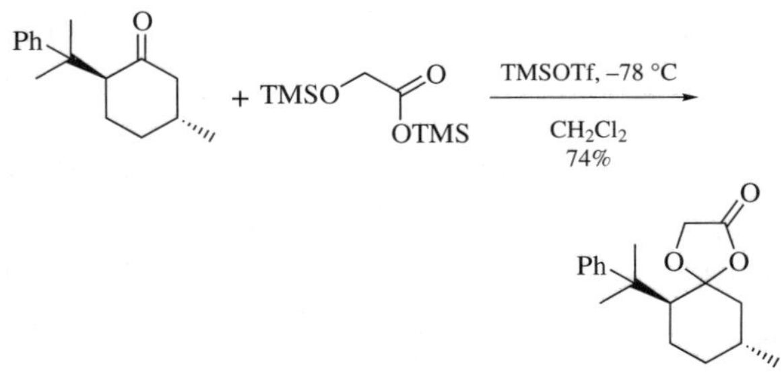

(13)

~ 1:1 mixture

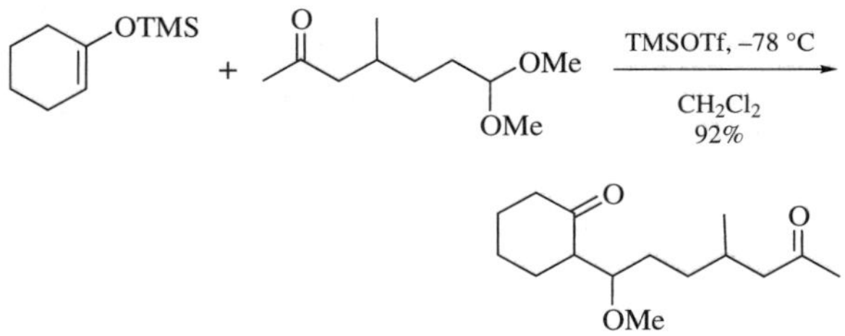

(14)

In an extension to this work, TMSOTf catalyzes the first step of a [3 + 2] annulation sequence which allows facile synthesis of fused cyclopentanes possessing bridgehead hydroxy groups (eq 15).[24]

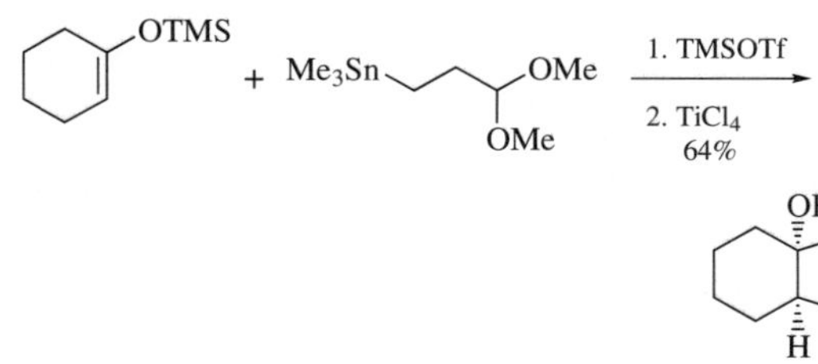

(15)

The use of TMSOTf in aldol reactions of silyl enol ethers and ketene acetals with aldehydes is ubiquitous. Many refinements of the basic reaction have appeared. An example is shown in eq 16.[25]

The use of TMSOTf in the reaction of silyl ketene acetals with imines offers an improvement over other methods (such as Ti^{IV}- or Zn^{II}-mediated processes) in that truly catalytic amounts of activator may be used (eq 17);[26] this reaction may be used as the crucial step in a general synthesis of 3-(1′-hydroxyethyl)-2-azetidinones (eq 18).[27]

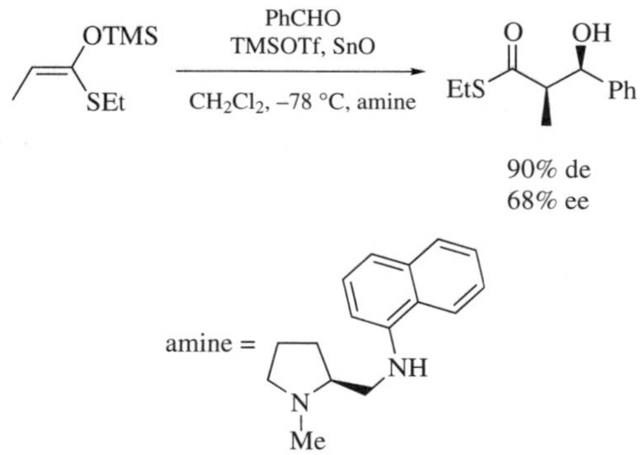

(16)

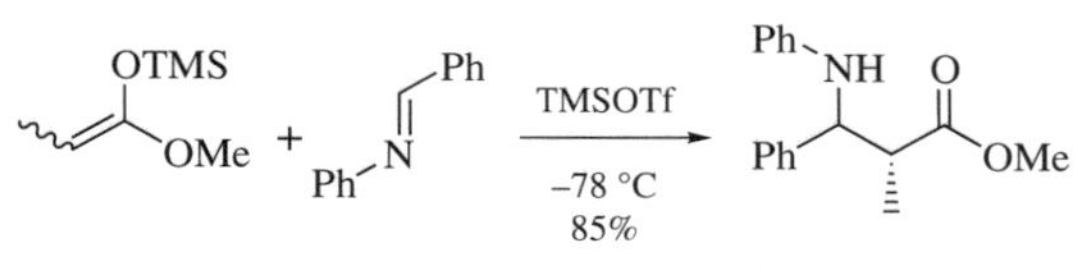

(17)

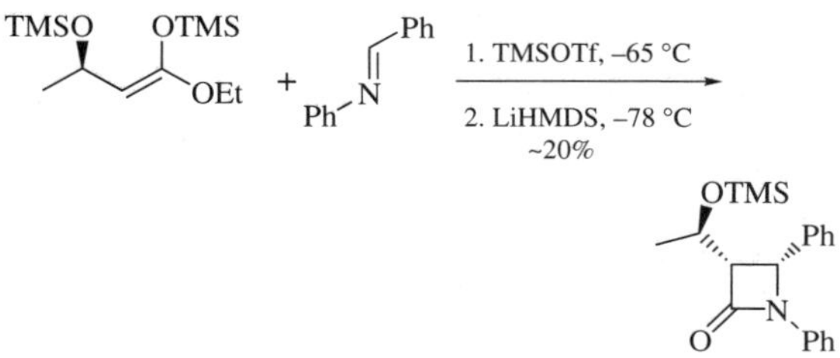

(18)

Stereoselective cyclization of α,β-unsaturated enamide esters is induced by TMSOTf and has been used as a route to quinolizidines and indolizidines (eq 19).[28]

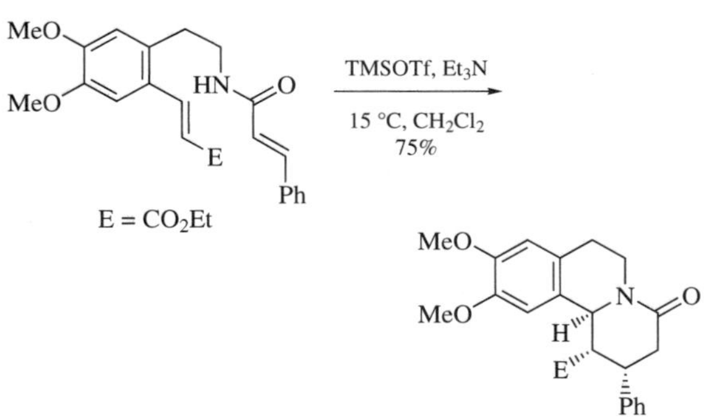

(19)

The formation of nitrones by reaction of aldehydes and ketones with N-Methyl-N,O-bis(trimethylsilyl)hydroxylamine is accelerated when TMSOTf is used as a catalyst; the acceleration is particularly pronounced when the carbonyl group is under a strong electronic influence (eq 20).[29]

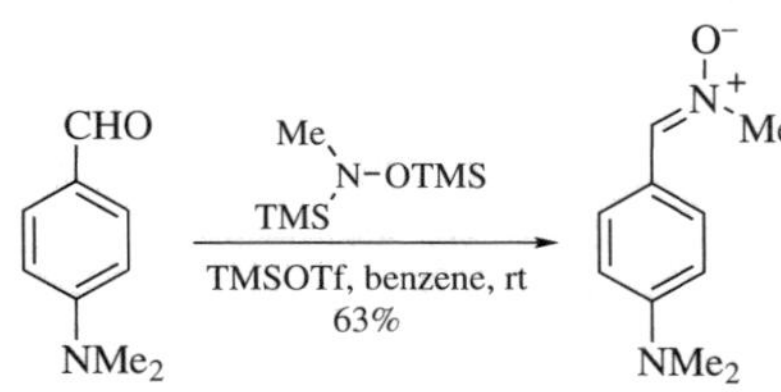

(20)

β-Stannylcyclohexanones undergo a stereoselective ring contraction when treated with TMSOTf at low temperature. When other Lewis acids were employed, a mixture of ring-contracted and protiodestannylated products was obtained (eq 21).[30]

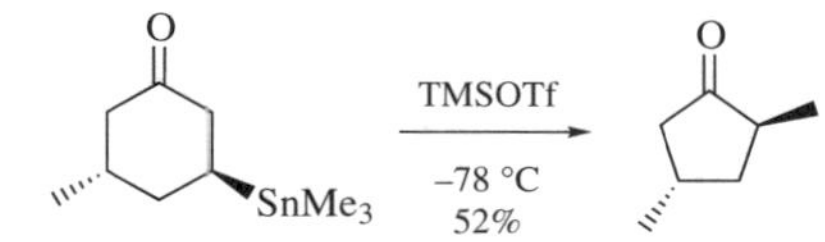

(21)

The often difficult conjugate addition of alkynyl organometallic reagents to enones is greatly facilitated by TMSOTf. In particular, alkynyl zinc reagents (normally unreactive with α,β-unsaturated carbonyl compounds) add in good yield (eq 22).[31] The proportion of 1,4-addition depends on the substitution pattern of the substrate.

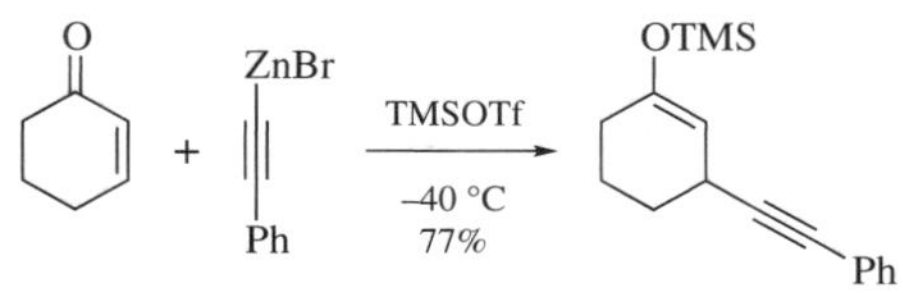

(22)

The 1,4-addition of phosphines to enones in the presence of TMSOTf gives β-phosphonium silyl enol ethers, which may be deprotonated and alkylated in situ (eq 23).[32]

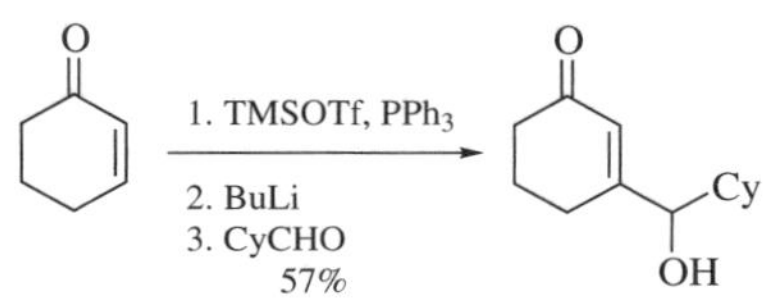

(23)

Miscellaneous. Methyl glucopyranosides and glycopyranosyl chlorides undergo allylation with allylsilanes under TMSOTf catalysis to give predominantly α-allylated carbohydrate analogs (eq 24).[33]

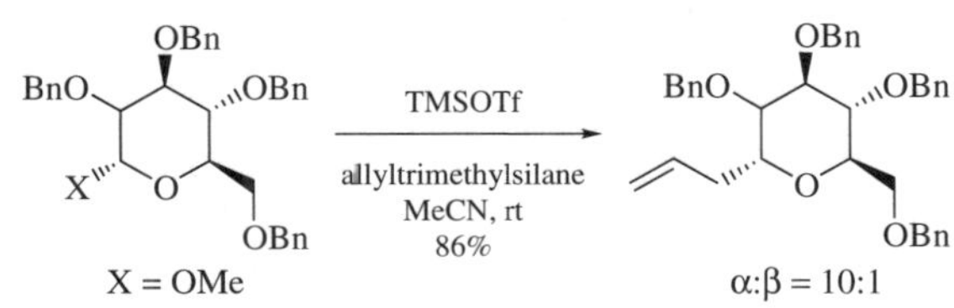

(24)

Glycosidation is a reaction of massive importance and widespread employment. TMSOTf activates many selective glycosidation reactions (eq 25).[34]

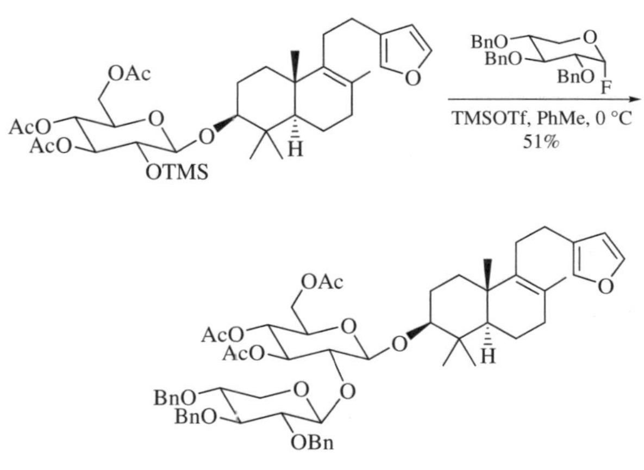

(25)

TMSOTf activation for coupling of 1-*O*-acylated glycosyl donors has been employed in a synthesis of avermectin disaccharides (eq 26).[35]

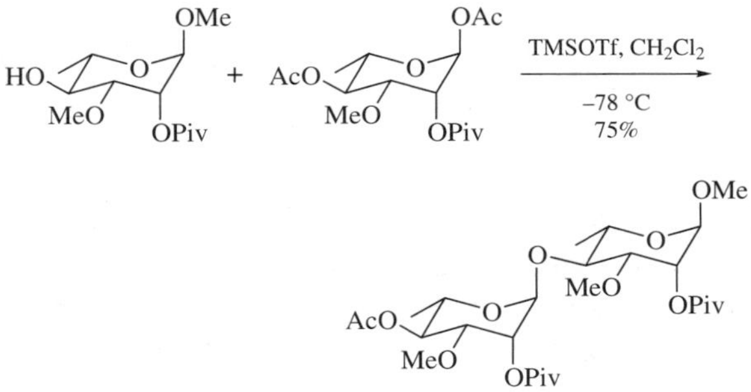

(26)

Similar activation is efficient in couplings with trichloroimidates[36] and *O*-silylated sugars.[37,38]

2-Substituted Δ^3-piperidines may be prepared by the reaction of 4-hydroxy-1,2,3,4-tetrahydropyridines with a variety of carbon and heteronucleophiles in the presence of TMSOTf (eqs 27 and 28).[39]

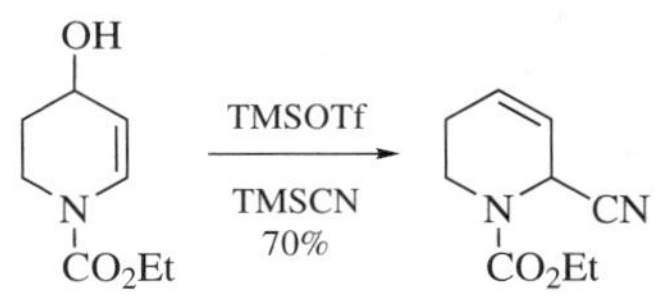

(27)

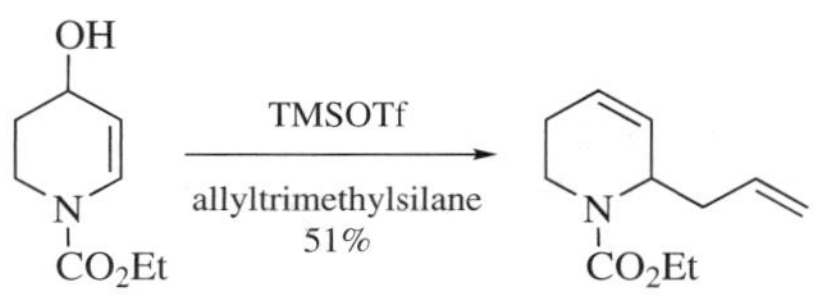

(28)

Iodolactamization is facilitated by the sequential reaction of unsaturated amides with TMSOTf and Iodine (eq 29).[40]

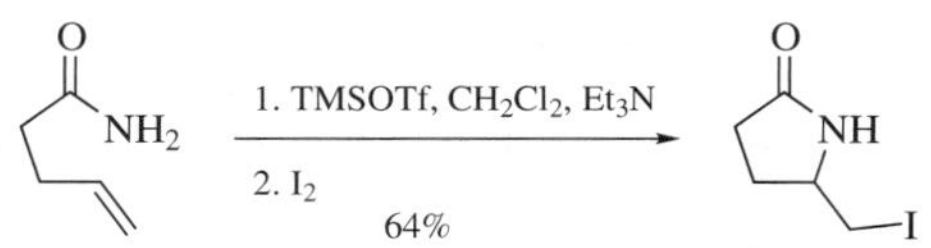

(29)

By use of a silicon-directed Beckmann fragmentation, cyclic (*E*)-β-trimethylsilylketoxime acetates are cleaved in high yield in the presence of catalytic TMSOTf to give the corresponding unsaturated nitriles. Regio- and stereocontrol are complete (eq 30).[41]

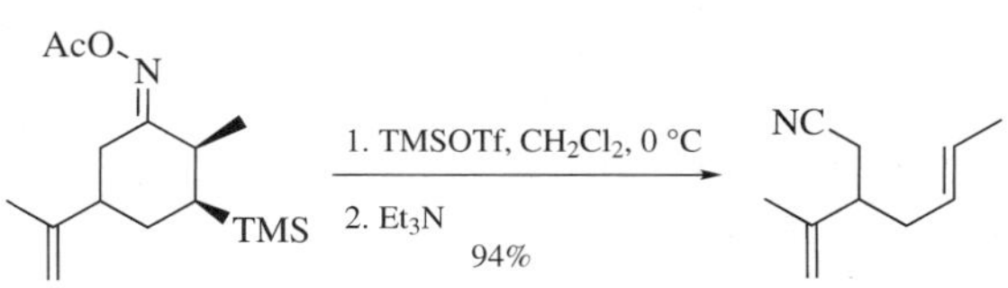

(30)

A general route to enol ethers is provided by the reaction of acetals with TMSOTf in the presence of a hindered base (eq 31).[42] The method is efficient for dioxolanes and noncyclic acetals.

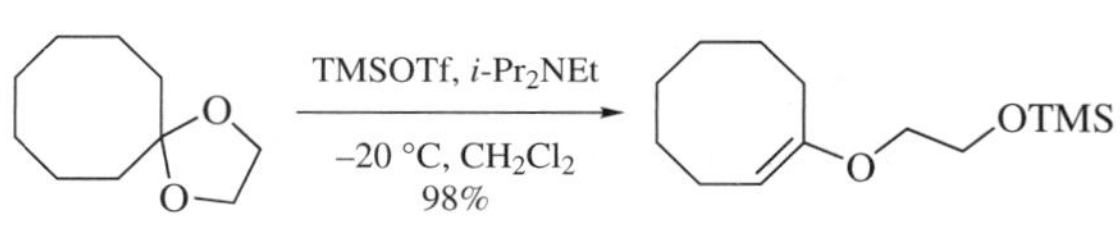

(31)

α-Halo sulfoxides are converted to α-halovinyl sulfides by reaction with excess TMSOTf (eq 32),[43] while α-cyano- and α-alkoxycarbonyl sulfoxides undergo a similar reaction (eq 33).[44] TMSOTf is reported as much superior to Iodotrimethylsilane in these reactions.

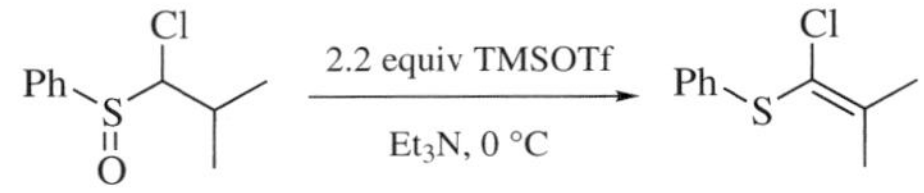

(32)

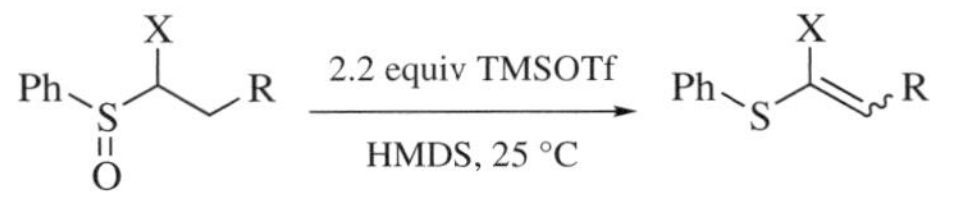

(33)

X = CN or CO_2R'

1. Reviews: (a) Emde, H.; Domsch, D.; Feger, H.; Frick, U.; Götz, H. H.; Hofmann, K.; Kober, W.; Krägeloh, K.; Oesterle, T.; Steppan, W.; West, W.; Simchen, G. *Synthesis* **1982**, 1. (b) Noyori, R.; Murata, S.; Suzuki, M. *Tetrahedron* **1981**, *37*, 3899. (c) Stang, P. J.; White, M. R. *Aldrichim. Acta* **1983**, *16*, 15.Preparation: (d) Olah, G. H.; Husain, A.; Gupta, B. G. B.; Salem, G. F.; Narang, S. C. *J. Org. Chem.* **1981**, *46*, 5212. (e) Morita, T.; Okamoto, Y.; Sakurai, H. *Synthesis* **1981**, 745. (f) Demuth, M.; Mikhail, G. *Synthesis* **1982**, 827. (g) Ballester, M.; Palomo, A. L. *Synthesis* **1983**, 571. (h) Demuth, M.; Mikhail, G. *Tetrahedron* **1983**, *39*, 991. (i) Aizpurua, J. M.; Palomo, C. *Synthesis* **1985**, 206.
2. Simchen, G.; Kober, W. *Synthesis* **1976**, 259.
3. Hergott, H. H.; Simchen, G. *Justus Liebigs Ann. Chem./Liebigs Ann. Chem.* **1980**, 1718.
4. Simchen, G.; Kober, W. *Synthesis* **1976**, 259.
5. Emde, H.; Götz, A.; Hofmann, K.; Simchen, G. *Justus Liebigs Ann. Chem./Liebigs Ann. Chem.* **1981**, 1643.
6. Krägeloh, K.; Simchen, G. *Synthesis* **1981**, 30.
7. Emde, H.; Simchen, G. *Justus Liebigs Ann. Chem./Liebigs Ann. Chem.* **1983**, 816.
8. Emde, H.; Simchen, G. *Synthesis* **1977**, 636.
9. Emde, H.; Simchen, G. *Synthesis* **1977**, 867.
10. Yamamoto, Y.; Ohdoi, K.; Nakatani, M.; Akiba, K. *Chem. Lett.* **1984**, 1967.
11. Emde, H.; Götz, A.; Hofmann, K.; Simchen, G. *Justus Liebigs Ann. Chem./Liebigs Ann. Chem.* **1981**, 1643.
12. Matsuda, I.; Sato, S.; Hattori, M.; Izumi, Y. *Tetrahedron Lett.* **1985**, *26*, 3215.
13. Ahlbrecht, H.; Düber, E. O. *Synthesis* **1980**, 630.
14. Murata, S.; Suzuki, M.; Noyori, R. *J. Am. Chem. Soc.* **1980**, *102*, 2738.
15. Frick, U.; Simchen, G. *Synthesis* **1984**, 929.
16. Borgulya, J.; Bernauer, K. *Synthesis* **1980**, 545.
17. Tietze, L. F.; Bratz, M. *Synthesis* **1989**, 439.
18. Hwu, J. R.; Wetzel, J. M. *J. Org. Chem.* **1985**, *50*, 3946.
19. Hwu, J. R.; Robl, J. A. *J. Org. Chem.* **1987**, *52*, 188.
20. Harada, T.; Hayashiya, T.; Wada, I.; Iwa-ake, N.; Oku, A. *J. Am. Chem. Soc.* **1987**, *109*, 527.
21. Pearson, W. H.; Cheng, M-C. *J. Am. Chem. Soc.* **1986**, *51*, 3746.
22. Murata, S.; Suzuki, M.; Noyori, R. *J. Am. Chem. Soc.* **1980**, *102*, 3248.
23. Murata, S.; Suzuki, M.; Noyori, R. *Tetrahedron* **1988**, *44*, 4259.
24. Lee, T. V.; Richardson, K. A. *Tetrahedron Lett.* **1985**, *26*, 3629.
25. Mukaiyama, T.; Uchiro, H.; Kobayashi, S. *Chem. Lett.* **1990**, 1147.
26. Guanti, G.; Narisano, E.; Banfi, L. *Tetrahedron Lett.* **1987**, *28*, 4331.
27. Guanti, G.; Narisano, E.; Banfi, L. *Tetrahedron Lett.* **1987**, *28*, 4335.

28. Ihara, M.; Tsuruta, M.; Fukumoto, K.; Kametani, T. *Chem. Commun./J. Chem. Soc., Chem. Commun.* **1985**, 1159.
29. Robl, J. A.; Hwu, J. R. *J. Org. Chem.* **1985**, *50*, 5913.
30. Sato, T.; Watanabe, T.; Hayata, T.; Tsukui, T. *Chem. Commun./J. Chem. Soc., Chem. Commun.* **1989**, 153.
31. Kim, S.; Lee, J. M. *Tetrahedron Lett.* **1990**, *31*, 7627.
32. Kim, S.; Lee, P. H. *Tetrahedron Lett.* **1988**, *29*, 5413.
33. Hosomi, A.; Sakata, Y.; Sakurai, H. *Tetrahedron Lett.* **1984**, *25*, 2383.
34. Yamada, H.; Nishizawa, M *Tetrahedron* **1992**, 3021.
35. Rainer, H.; Scharf, H.-D.; Runsink, J. *Justus Liebigs Ann. Chem./Liebigs Ann. Chem.* **1992**, 103.
36. Schmidt, R. R. *Angew. Chem., Int. Ed. Engl.* **1986**, *25*, 212.
37. Tietze, L.-F.; Fischer, R.; Guder, H.-J. *Tetrahedron Lett.* **1982**, *23*, 4661.
38. Mukaiyama, T.; Matsubara, K. *Chem. Lett.* **1992**, 1041.
39. Kozikowski, A. P.; Park, P. *J. Org. Chem.* **1984**, *49*, 1674.
40. Knapp, S.; Rodriques, K. E. *Tetrahedron Lett.* **1985**, *26*, 1803.
41. Nishiyama, H.; Sakuta, K.; Osaka, N.; Itoh, K. *Tetrahedron Lett.* **1983**, *24*, 4021.
42. Gassman, P. G.; Burns, S. J. *J. Org. Chem.* **1988**, *53*, 5574.
43. Miller, R. D.; Hässig, R., *Synth. Commun.* **1984**, *14*, 1285.
44. Miller, R. D.; Hässig, R., *Tetrahedron Lett.* **1985**, *26*, 2395.

Joseph Sweeney & Gemma Perkins
University of Bristol, UK

Triphenylcarbenium Tetrafluoroborate

$Ph_3C^+BF_4^-$

[341-02-6] $C_{19}H_{15}BF_4$ (MW 330.15)

(easily prepared[1] hydride abstractor used for conversion of dihydroaromatics to aromatics,[2–4] and the preparation of aromatic and benzylic cations;[5–8] oxidative hydrolysis of ketals[9] and thioketals;[10] conversion of acetonides to α-hydroxy ketones;[9] oxidation of acetals[11] and thioacetals;[12] selective oxidation of alcohols and ethers to ketones;[9,13,14,15] oxidation of silyl enol ethers to enones;[16] hydrolysis of TBS and MTM ethers;[17] oxidation of amines and amides to iminium salts;[18–20] oxidation of organometallics to give alkenes;[21–23] sensitizer for photooxidation using molecular oxygen;[24] Lewis acid catalyst for various reactions;[25] polymerization catalyst;[26] other reactions[27–30])

Alternate Name: trityl fluoroborate.
Physical Data: mp ~ 200 °C (dec).
Solubility: sol most standard organic solvents; reacts with some nucleophilic solvents.
Form Supplied in: yellow solid; commercially available.
Preparative Method: the most convenient procedure involves the reaction of Ph_3CCl with Silver(I) Tetrafluoroborate in ethanol.[1b] The most economical route employs the reaction of Ph_3CCl with the anhydrous Tetrafluoroboric Acid–Et_2O complex.[1c]
Purification: recrystallization of commercial samples from a minimal amount of dry MeCN provides material of improved purity, but the recovery is poor.[1a]
Handling, Storage, and Precautions: moisture-sensitive and corrosive. Recrystallized reagent can be stored at rt for several months in a desiccator without significant decomposition. This compound is much less light-sensitive than other trityl salts such as the perchlorate.[1a]

Preparation of Aromatic Compounds via Dehydrogenation. Dihydroaromatic compounds are easily converted into the corresponding aromatic compound by treatment with triphenylcarbenium tetrafluoroborate followed by base.[2] Certain α,α-disubstituted dihydroaromatics are converted to the 1,4-dialkylaromatic compounds with rearrangement (eq 1).[3] Nonbenzenoid aromatic systems, e.g. benzazulene[4a] or dibenzosesquifulvalene,[4b] are readily prepared from their dihydro counterparts. Aromatic cations are also easily prepared by hydride abstraction, for example, tropylium ion (e.g. in the synthesis of heptalene (eq 2)),[5] cyclopropenyl cation,[6] and others, including heterocyclic systems.[7] Some benzylic cations, especially ferrocenyl cations,[8] can also be formed by either hydride abstraction or trityl addition.

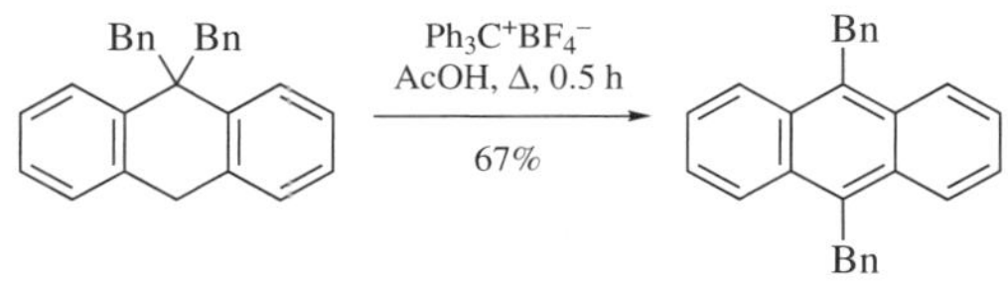

(1)

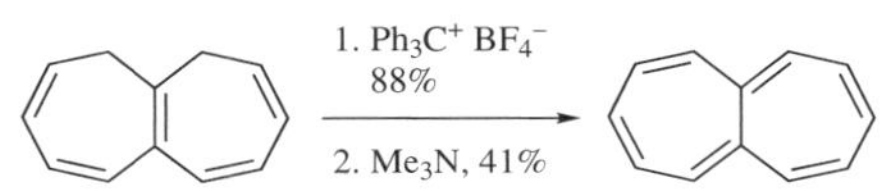

(2)

Oxidation by Hydride Abstraction. In the early 1970s, Barton developed a method for the oxidative hydrolysis of ketals to ketones, e.g. in the tetracycline series (eq 3).[9] The same conditions can also be used to hydrolyze thioketals.[10] Acetonides of 1,2-diols are oxidized to the α-hydroxy ketones in good yield by this reagent (eq 4).[9] The hydrogen of acetals is easily abstracted (eq 5), providing a method for the conversion of benzylidene units in sugars to the hydroxy benzoates.[11] The hydrogen of dithioacetals is also abstracted to give the salts.[12] Since benzylic hydrogens are readily abstracted, this is also a method for deprotection of benzyl ethers.[9,13] Trimethylsilyl, *t*-butyl, and trityl ethers of simple alcohols are oxidized to the corresponding ketones and aldehydes in good yield. Primary–secondary diols are selectively oxidized at the secondary center to give hydroxy ketones by this method (eq 6).[14] 2,2-Disubstituted 1,4-diols are oxidized only at the 4-position to give the corresponding lactones.[15] Trimethylsilyl enol ethers are oxidized to α,β-unsaturated ketones, thereby providing a method for ketone to enone conversion (eq 7).[16] *t*-Butyldimethylsilyl (TBDMS) ethers are not oxidized but rather hydrolyzed to the alcohols, as are methylthiomethyl (MTM) ethers.[17] Benzylic amines and amides can be oxidized to the iminium salts,[18] allylic amines and enamines afford eniminium salts,[19] and orthoamides give triaminocarbocations.[20]

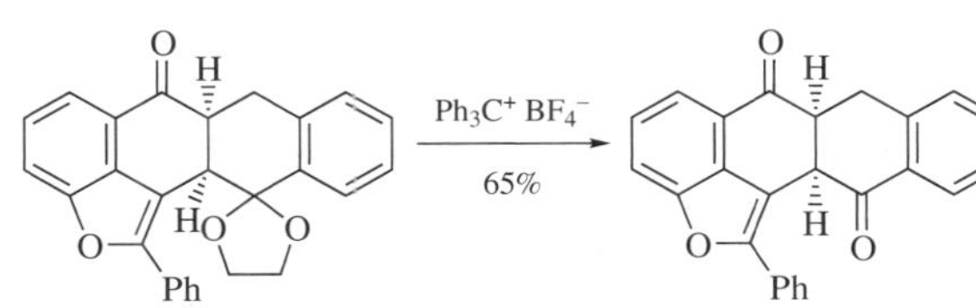

(3)

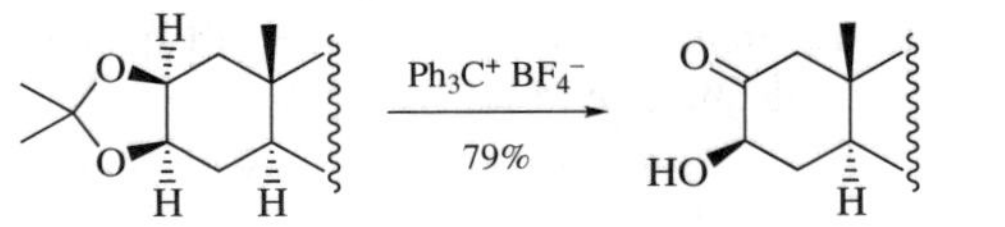
(4)

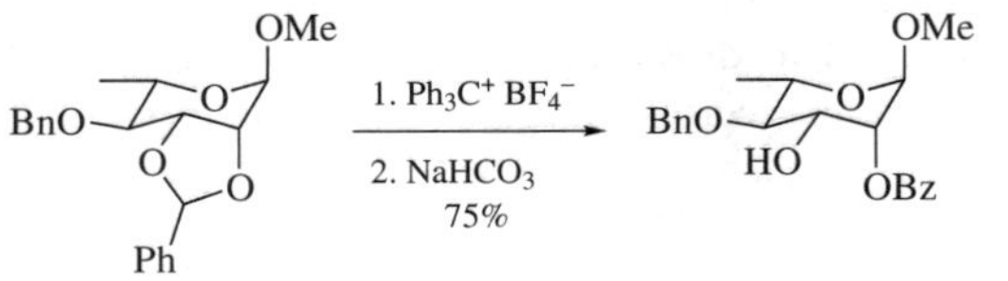
(5)

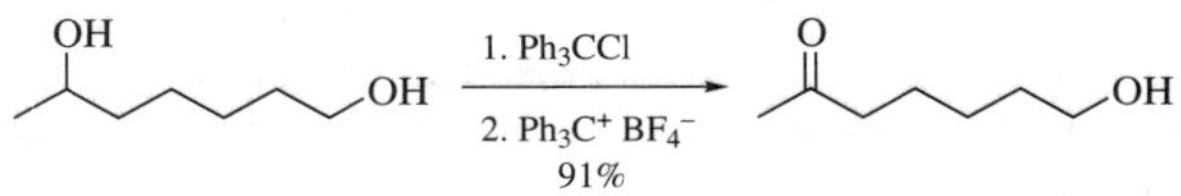
(6)

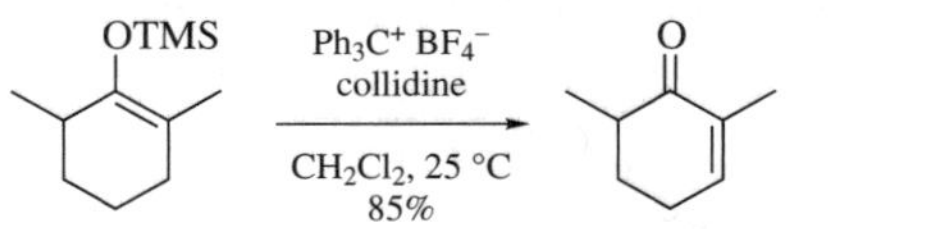
(7)

Generation of Alkenes from Organometallics. Various β-metalloalkanes can be oxidized by trityl fluoroborate to the corresponding alkenes.[21–23] The highest yields are obtained for the β-iron derivatives (eq 8), which are easily prepared from the corresponding halides or tosylates.[21] Grignard reagents and organolithiums also undergo this reaction (eq 9),[22] as do Group 14 organometallics (silanes, stannanes, etc.).[23]

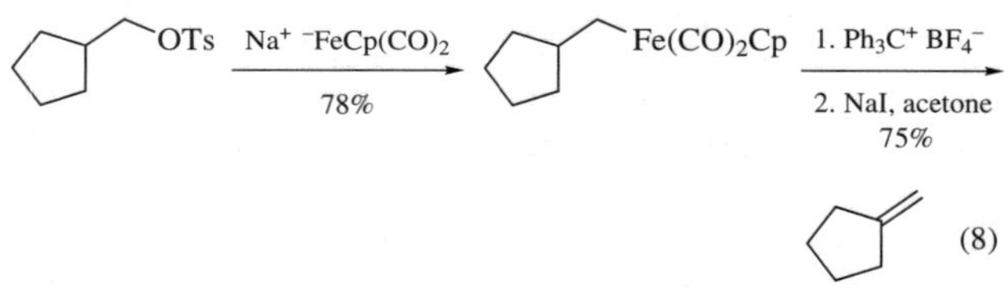
(8)

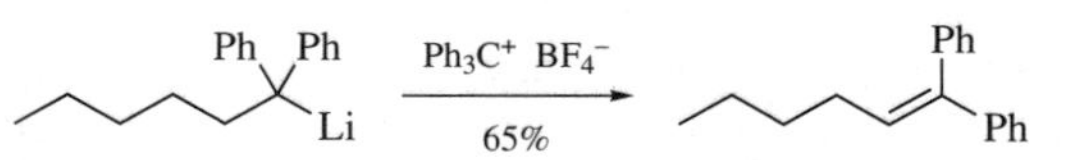
(9)

Sensitizer of Photooxygenation. Barton showed that oxygen, in the presence of trityl fluoroborate and ordinary light, adds to cisoid dienes at −78 °C in very high yields.[24] For example, the peroxide of ergosterol acetate is formed in quantitative yields under these conditions (eq 10),[24a,24b] which have been used also for photocycloreversions of cyclobutanes.[24c]

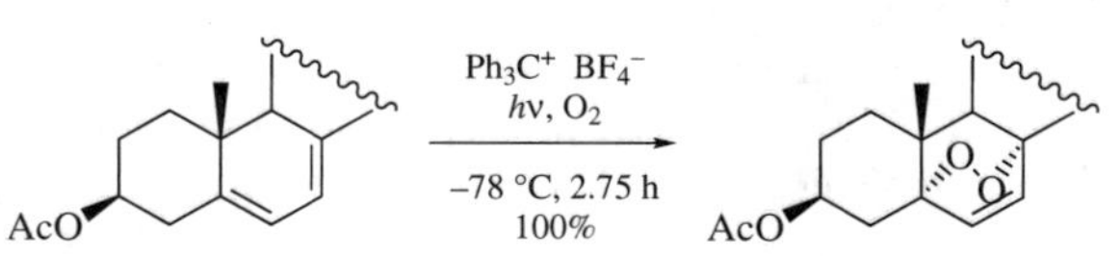
(10)

Lewis Acid Catalysis. Trityl fluoroborate is a good Lewis acid for various transformations,[25] e.g. the Mukaiyama-type aldol reaction using a dithioacetal and silyl enol ether (eq 11).[25a] It has also been used as the catalyst for the formation of glycosides from alcohols and sugar dimethylthiophosphinates (eq 12)[25b] and for the formation of disaccharides from a protected α-cyanoacetal of glucose and a 6-*O*-trityl hexose.[25c] Michael additions of various silyl nucleophiles to conjugated dithiolenium cations also proceed well (eq 13).[25d,25e] Finally, the [4 + 2] cycloaddition of cyclic dienes and oxygenated allyl cations has been effected with trityl fluoroborate.[25f]

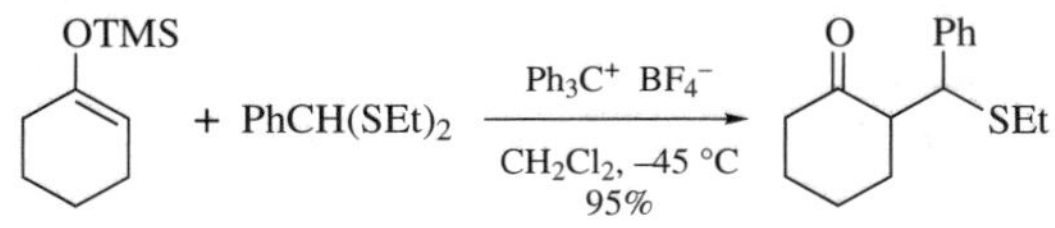
(11)

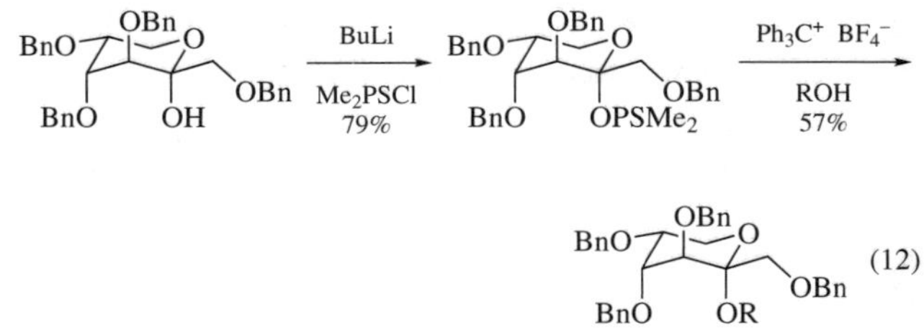
(12)

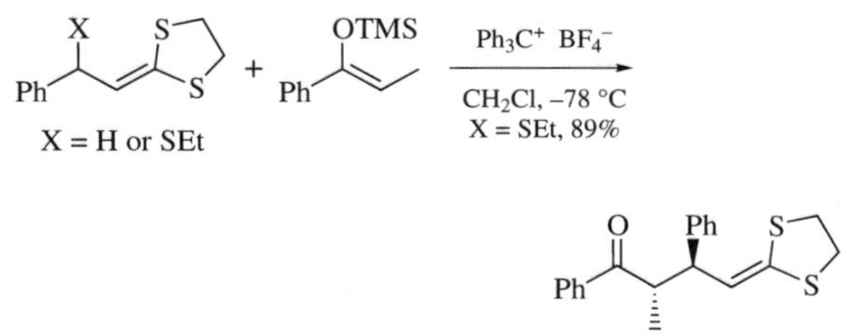
(13)

Polymerization Catalyst. Several types of polymerization[26] have been promoted by trityl fluoroborate, including reactions of orthocarbonates[26a] and orthoesters,[26b–d] vinyl ethers,[26e–g] epoxides,[26h,26i] and lactones.[26j,26k]

Other Reactions. Trityl fluoroborate has been used often to prepare cationic organometallic complexes, as in the conversion of dienyl complexes of iron, ruthenium, and osmium into their cationic derivatives.[27] It alkylates pyridines on the nitrogen atom in a preparation of dihydropyridines[28a] and acts as a tritylating agent.[28b] It has also been used in attempts to form silyl cations and silyl fluorides from silanes.[29] Finally, it has been reported to be a useful desiccant.[30]

1. (a) Dauben, H. J., Jr.; Honnen, L. R.; Harmon, K. M. *J. Org. Chem.* **1960**, *25*, 1442. (b) Fukui, K.; Ohkubo, K.; Yamabe, T. *Bull. Chem. Soc. Jpn.* **1969**, *42*, 312. (c) Olah, G. A.; Svoboda, J. J.; Olah, J. A. *Synthesis* **1972**, 544.
2. (a) Müller, P. *Helv. Chim. Acta* **1973**, *56*, 1243. (b) Giese, G.; Heesing, A. *Ber. Dtsch. Chem. Ges./Chem. Ber.* **1990**, *123*, 2373.
3. (a) Karger, M. H.; Mazur, Y. *J. Org. Chem.* **1971**, *36*, 540. (b) Acheson, R. M.; Flowerday, R. F. *J. Chem. Soc., Perkin Trans. 1* **1975**, 2065.
4. (a) O'Leary, M. A.; Richardson, G. W.; Wege, D. *Tetrahedron* **1981**, *37*, 813. (b) Prinzbach, H.; Seip, D.; Knothe, L.; Faisst, W. *Justus Liebigs Ann. Chem./Liebigs Ann. Chem.* **1966**, *698*, 34.
5. (a) Dauben, H. J.Jr.; Gadecki, F. A.; Harmon, K. M.; Pearson, D. L. *J. Am. Chem. Soc.* **1957**, *79*, 4557. (b) Dauben, H. J.Jr., Bertelli, D. J. *J.*

Am. Chem. Soc. **1961**, *83*, 4657, 4659. (c) Peter-Katalinic, J.; Zsindely, J.; Schmid, H. *Helv. Chim. Acta* **1973**, *56*, 2796. (d) Vogel, E.; Ippen, J. *Angew. Chem., Int. Ed. Engl.* **1974**, *13*, 734. (e) Beeby, J.; Garratt, P. J. *J. Org. Chem.* **1973**, *38*, 3051. (f) Murata, I.; Yamamoto, K.; Kayane, Y. *Angew. Chem., Int. Ed. Engl.* **1974**, *13*, 807, 808. (g) Kuroda, S.; Asao, T. *Tetrahedron Lett.* **1977**, 285. (h) Komatsu, K.; Takeuchi, K.; Arima, M.; Waki, Y.; Shirai, S.; Okamoto, K. *Bull. Chem. Soc. Jpn.* **1982**, *55*, 3257. (i) Müller, J.; Mertschenk, B. *Ber. Dtsch. Chem. Ges./Chem. Ber.* **1972**, *105*, 3346. (j) Schweikert, O.; Netscher, T.; Knothe, L.; Prinzbach, H. *Ber. Dtsch. Chem. Ges./Chem. Ber.* **1984**, *117*, 2045. (k) Bindl, J.; Seitz, P.; Seitz, U.; Salbeck, E.; Salbeck, J.; Daub, J. *Ber. Dtsch. Chem. Ges./Chem. Ber.* **1987**, *120*, 1747.

6. (a) Zimmerman, H. E.; Aasen, S. M. *J. Org. Chem.* **1978**, *43*, 1493. (b) Komatsu, K.; Tomioka, I.; Okamoto, K. *Bull. Chem. Soc. Jpn.* **1979**, *52*, 856.
7. (a) Yamamura, K.; Miyake, H.; Murata, I. *J. Org. Chem.* **1986**, *51*, 251. (b) Matsumoto, S.; Masuda, H.; Iwata, K.; Mitsunobu, O. *Tetrahedron Lett.* **1973**, 1733. (c) Yano, S.; Nishino, K.; Nakasuji, K.; Murata, I. *Chem. Lett.* **1978**, 723. (d) Kedik, L. M.; Freger, A. A.; Viktorova, E. A. *Khim. Geterotsikl. Soedin.* **1976**, *12*, 328 (*Chem. Heterocycl. Compd. (Engl. Transl.)* **1976**, *12*, 279) . (e) Reichardt, C.; Schäfer, G.; Milart, P. *Collect. Czech. Chem. Commun.* **1990**, *55*, 97.
8. (a) Müller, P. *Helv. Chim. Acta* **1973**, *56*, 500. (b) Boev, V. I.; Dombrovskii, A. V. *Zh. Obshch. Khim.* **1987**, *57*, 938, 633. (c) Klimova, E. I.; Pushin, A. N.; Sazonova, V. A. *Zh. Obshch. Khim.* **1987**, *57*, 2336. (d) Abram, T. S.; Watts, W. E. *J. Chem. Soc., Perkin Trans. 1* 1975, *J. Organomet. Chem.* **1975**, *87*, C39. 113. (e) Barua, P.; Barua, N. C.; Sharma, R. P. *Tetrahedron Lett.* **1983**, *24*, 5801. (f) Akgun, E.; Tunali, M. *Arch. Pharm. (Weinheim, Ger.)* **1988**, *321*, 921.
9. (a) Barton, D. H. R.; Magnus, P. D.; Smith, G.; Strecker, G.; Zurr, D. *J. Chem. Soc., Perkin Trans. 1* **1972**, 542. (b) Barton, D. H. R.; Magnus, P. D.; Smith, G.; Zurr, D. *Chem. Commun./J. Chem. Soc., Chem. Commun.* **1971**, 861.
10. Ohshima, M.; Murakami, M.; Mukaiyama, T. *Chem. Lett.* **1986**, 1593.
11. (a) Hanessian, S.; Staub, A. P. A. *Tetrahedron Lett.* **1973**, 3551. (b) Jacobsen, S.; Pedersen, C. *Acta Chem. Scand.* **1974**, *28B*, 1024, 866. (c) Wessel, H.-P.Bundle, D. R. *J. Chem. Soc., Perkin Trans. 1* **1985**, 2251.
12. (a) Nakayama, J.; Fujiwara, K.; Hoshino, M. *Chem. Lett.* **1975**, 1099; *Bull. Chem. Soc. Jpn.* **1976**, *49*, 3567. (b) Nakayama, J.; Imura, M.; Hoshino, M. *Bull. Chem. Soc. Jpn.* **1980**, *53*, 1661. (c) Nakayama, J. *Bull. Chem. Soc. Jpn.* **1982**, *55*, 2289. (d) Bock, H.; Brähler, G.; Henkel, U.; Schlecker, R.; Seebach, D. *Ber. Dtsch. Chem. Ges./Chem. Ber.* **1980**, *113*, 289. (e) Neidlein, R.; Droste-Tran-Viet, D.; Gieren, A.; Kokkinidis, M.; Wilckens, R.; Geserich, H.-P.; Ruppel, W. *Helv. Chim. Acta* **1984**, *67*, 574. (f) However, azide abstraction is seen with azidodithioacetals: Nakayama, J.; Fujiwara, K.; Hoshino, M. *J. Org. Chem.* **1980**, *45*, 2024.
13. (a) Barton, D. H. R.; Magnus, P. D.; Streckert, G.; Zurr, D. *Chem. Commun./J. Chem. Soc., Chem. Commun.* **1971**, 1109. (b) Doyle, M. P.; Siegfried, B. *J. Am. Chem. Soc.* **1976**, *98*, 163. (c) Hoye, T. R.; Kurth, M. J. *J. Am. Chem. Soc.* **1979**, *101*, 5065. (d) For simple ethers, see: Deno, N. C.; Potter, N. H. *J. Am. Chem. Soc.* **1967**, *89*, 3550.
14. (a) Jung, M. E. *J. Org. Chem.* **1976**, *41*, 1479. (b) Jung, M. E.; Speltz, L. M. *J. Am. Chem. Soc.* **1976**, *98*, 7882. (c) Jung, M. E.; Brown, R. W. *Tetrahedron Lett.* **1978**, 2771.
15. Doyle, M. P.; Dow, R. L.; Bagheri, V.; Patrie, W. J. *J. Org. Chem.* **1983**, *48*, 476; *Tetrahedron Lett.* **1980**, *21*, 2795.
16. (a) Jung, M. E.; Pan, Y.-G.; Rathke, M. W.; Sullivan, D. F.; Woodbury, R. P. *J. Org. Chem.* **1977**, *42*, 3961. (b) Reetz, M. T.; Stephan, W. *Justus Liebigs Ann. Chem./Liebigs Ann. Chem.* **1980**, 533.
17. (a) Metcalf, B. W.; Burkhardt, J. P.; Jund, K. *Tetrahedron Lett.* **1980**, *21*, 35. (b) Chowdhury, P. K.; Sharma, R. P.; Baruah, J. N. *Tetrahedron Lett.* **1983**, *24*, 4485. (c) Niwa, H.; Miyachi, Y. *Bull. Chem. Soc. Jpn.* **1991**, *64*, 716.
18. (a) Damico, R.; Broaddus, C. D. *J. Org. Chem.* **1966**, *31*, 1607. (b) Barton, D. H. R.; Bracho, R. D.; Gunatilaka, A. A. L.; Widdowson, D. A. *J. Chem. Soc., Perkin Trans. 1* **1975**, 579. (c) Wanner, K. T.; Praschak, I.; Nagel, U. *Arch. Pharm. (Weinheim, Ger.)* **1990**, *322*, 335; *Heterocycles* **1989**, *29*, 29.
19. Reetz, M. T.; Stephan, W.; Maier, W. F. *Synth. Commun.* **1980**, *10*, 867.
20. Erhardt, J. M.; Grover, E. R.; Wuest, J. D. *J. Am. Chem. Soc.* **1980**, *102*, 6365.
21. (a) Laycock, D. E.; Hartgerink, J.; Baird, M. C. *J. Org. Chem.* **1980**, *45*, 291. (b) Laycock, D. E.; Baird, M. C. *Tetrahedron Lett.* **1978**, 3307. (c) Slack, D.; Baird, M. C. *Chem. Commun./J. Chem. Soc., Chem. Commun.* **1974**, 701. (d) Bly, R. S.; Bly, R. K.; Hossain, M. M.; Silverman, G. S.; Wallace, E. *Tetrahedron* **1986**, *42*, 1093. (e) Bly, R. S.; Silverman, G. S.; Bly, R. K. *Organometallics* **1985**, *4*, 374.
22. Reetz, M. T.; Schinzer, D. *Angew. Chem., Int. Ed. Engl.* **1977**, *16*, 44.
23. (a) Traylor, T. G.; Berwin, H. J.; Jerkunica, J.; Hall, M. L. *Pure Appl. Chem.* **1972**, *30*, 597. (b) Jerkunica, J. M.; Traylor, T. G. *J. Am. Chem. Soc.* **1971**, *93*, 6278. (c) Washburne, S. S.; Szendroi, R. *J. Org. Chem.* **1981**, *46*, 691. (d) Washburne, S. S.; Simolike, J. B. *J. Organomet. Chem.* **1974**, *81*, 41. (e) However, organostannanes lacking a β-hydrogen afford alkyltriphenylmethanes in good yield. Kashin, A. N.; Bumagin, N. A.; Beletskaya, I. P.; Reutov, O. A. *J. Organomet. Chem.* **1979**, *171*, 321.
24. (a) Barton, D. H. R.; Haynes, R. K.; Leclerc, G.; Magnus, P. D.; Menzies, I. D. *J. Chem. Soc., Perkin Trans. 1* **1975**, 2055. (b) Barton, D. H. R.; Leclerc, G.; Magnus, P. D.; Menzies, I. D. *Chem. Commun./J. Chem. Soc., Chem. Commun.* **1972**, 447. (c) Okada, K.; Hisamitsu, K.; Mukai, T. *Tetrahedron Lett.* **1981**, *22*, 1251. (d) Futamura, S.; Kamiya, Y. *Chem. Lett.* **1989**, 1703.
25. (a) Ohshima, M.; Murakami, M.; Mukaiyama, T. *Chem. Lett.* **1985**, 1871. (b) Inazu, T.; Yamanoi, T.Jpn. Patent 02 240 093, 02 255 693 (*Chem. Abstr.* **1991**, *114*, 143 907j, 143 908k) .Jpn. Patent 01 233 295 (*Chem. Abstr.* **1990**, *112*, 198 972r) . (c) Bochkov, A. F.; Kochetkov, N. K. *Carbohydr. Res.* **1975**, *39*, 355; for polymerizations of carbohydrate cyclic orthoesters see: Bochkov, A. F.; Chernetskii, V. N.; Kochetkov, N. K. *Carbohydr. Res.* **1975**, *43*, 35; *Bull. Acad. Sci. USSR, Div. Chem. Sci.* **1975**, *24*, 396. (d) Hashimoto, Y.; Mukaiyama, T. *Chem. Lett.* **1986**, 1623, 755. (e) Hashimoto, Y.; Sugumi, H.; Okauchi, T.; Mukaiyama, T. *Chem. Lett.* **1987**, 1691. (f) Murray, D. H.; Albizati, K. F. *Tetrahedron Lett.* **1990**, *31*, 4109.
26. (a) Endo, T.; Sato, H.; Takata, T. *Macromolecules* **1987**, *20*, 1416. (b) Uno, H.; Endo, T.; Okawara, M. *J. Polym. Sci., Polym. Chem. Ed.* **1985**, *23*, 63. (c) Nishida, H.; Ogata, T. Jpn. Patent 62 295 920 (*Chem. Abstr.* **1988**, *109*, 57 030h) . (d) See also Ref. [25]c. (e) Kunitake, T. *J. Macromol. Sci., Chem.* **1975**, *A9*, 797. (f) Kunitake, T.; Takarabe, K.; Tsugawa, S. *Polym. J.* **1976**, *8*, 363. (g) Spange, S.; Dreier, R.; Opitz, G.; Heublein, G. *Acta Polym.* **1989**, *40*, 55. (h) Mijangos, F.; León, L. M. *J. Polym. Sci., Polym. Lett. Ed.* **1983**, *21*, 885*Eur.Polym.J.* **1983**, *19*, 29. (i) Bruzga, P.; Grazulevicius, J.; Kavaliunas, R.; Kublickas, R. *Polym. Bull. (Berlin)* **1991**, *26*, 193. (j) Khomyakov, A. K.; Gorelikov, A. T.; Shapet'ko, N. N.; Lyudvig, E. B. *Vysokomol. Soedin., Ser. A* **1976**, *18*, 1699, 1053; *Dokl. Akad. Nauk SSSR* **1975**, *222*, 1111.
27. (a) For a review, see any basic organometallic text, e.g. Coates, G. E.; Green, M. L. H.; Wade, K. *Organometallic Compounds*; Methuen: London, **1968**; *2*, pp 136ff. (b) Birch, A. J.; Cross, P. E.; Lewis, J.; White, D. A. *Chem. Ind. (London)* **1964**, 838. (c) Cotton, F. A.; Deeming, A. J.; Josty, P. L.; Ullah, S. S.; Domingos, A. J. P.; Johnson, B. F. G.; Lewis, J. *J. Am. Chem. Soc.* **1971**, *93*, 4624.
28. (a) Lyle, R. E.; Boyce, C. B. *J. Org. Chem.* **1974**, *39*, 3708. (b) Hanessian, S.; Staub, A. P. A. *Tetrahedron Lett.* **1973**, 3555.
29. (a) Sommer, L. H.; Bauman, D. L. *J. Am. Chem. Soc.* **1969**, *91*, 7076. (b) Bulkowski, J. E.; Stacy, R.; Van Dyke, C. H. *J. Organomet. Chem.* **1975**, *87*, 137. (c) Chojnowski, J.; Fortuniak, W.; Stanczyk, W. *J. Am. Chem. Soc.* **1987**, *109*, 7776.
30. Burfield, D. R.; Lee, K.-H.; Smithers, R. H. *J. Org. Chem.* **1977**, *42*, 3060.

Michael E. Jung
University of California, Los Angeles, CA, USA

Triphenylchloromethane

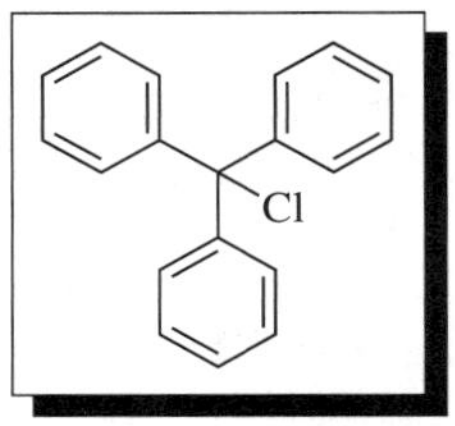

[76-83-5] $C_{19}H_{15}Cl$ (MW 278.78)

(reagent used for the protection of amines, carboxylic acids, alcohols, and thiols[1])

Physical Data: mp 111–112 °C.
Solubility: readily soluble in most organic solvents, mainly used in methylene chloride, dimethylformamide, and pyridine.
Form Supplied in: solid; widely available.
Purification: can be recrystallized from isooctane.
Handling, Storage, and Precautions: use chemical safety goggles and rubber gloves. Do not breathe dust. Avoid prolonged or repeated exposure. Wash thoroughly after handling. It is an irritant. Keep tightly closed in a cool, dry place. Incompatible with strong oxidizing agents as well as strong bases. Toxic decomposition product is hydrogen chloride gas.

Protection of Alcohols. Triphenylchloromethane (trityl chloride, TrCl) is most commonly used as a reagent to protect primary alcohols as the corresponding trityl ethers. Because of the stability of trityl ethers under neutral or basic conditions, their facile removal under mildly acidic conditions and significant steric and lipophilic nature, they have been used quite extensively over the past 70 years. The classical method to prepare trityl ethers using TrCl involves reaction of the alcohol in pyridine at temperatures ranging from room temperature to 100 °C. Recent examples using this procedure include the preparation of a 6-*O*-trityl glucopyranoside (eq 1)[2] and a protected purine nucleoside (eq 2).[3] The first example was detritylated later in that synthesis with p-**toluenesulfonic acid** (*p*-TsOH) in CH_2Cl_2/MeOH while the second was deprotected with **trifluoroacetic acid** (TFA) in CH_2Cl_2.

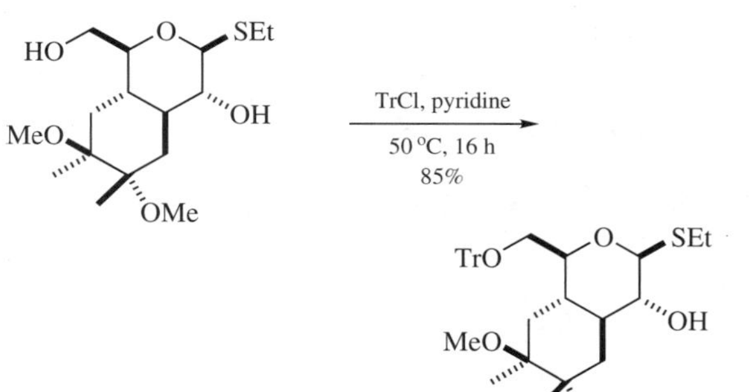

(1)

Both examples illustrate the observation that primary alcohols are routinely regioselectively protected in the presence of secondary alcohols with TrCl, presumably due to the steric requirements of this reagent. While trityl ethers have been used extensively in nucleoside chemistry, it has been noted that the trityl group is not appropriate for the protection of purine deoxyribonucleosides as their removal leads to depurination; **4-methoxytrityl chloride** and **4,4′-dimethoxytrityl chloride** are the reagents of choice for this application.[4]

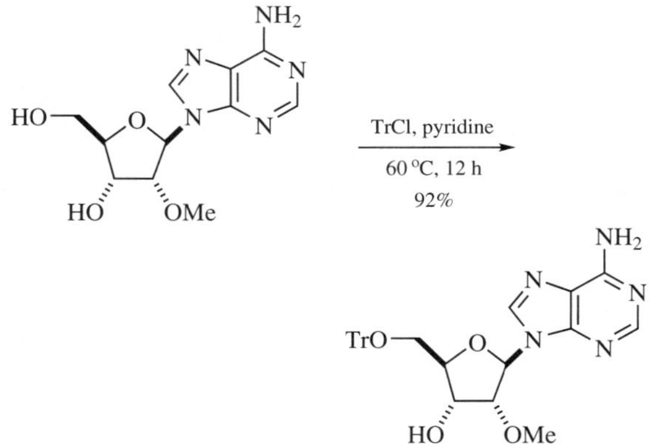

(2)

A simplified preparation of trityl ethers involves use of TrCl with a catalytic amount of **4-dimethylaminopyridine** (DMAP) in CH_2Cl_2 under mild reaction conditions (15 h at rt); this method involves simple work-up operations and increased regioselectivities.[5] It was noted that protection of secondary alcohols was considerably slower but possible using this method (24 h at 45 °C) and that high selectivity was seen for equatorial versus axial alcohols. An example of this type of selectivity is the tritylation of methyl α-L-fucopyranoside which gave a mixture of two monoprotected adducts in which the axial hydroxyl was unaffected (eq 3);[6] the trityl groups were removed in this example with **boron trifluoride**. In a related deprotection it was found that trityl ethers could be selectively cleaved with **boron trichloride** in the presence of benzyloxy, triethylsilyloxy, *tert*-butyldimethylsilyloxy or *tert*-butyldiphenylsilyloxy protected alcohols.[7]

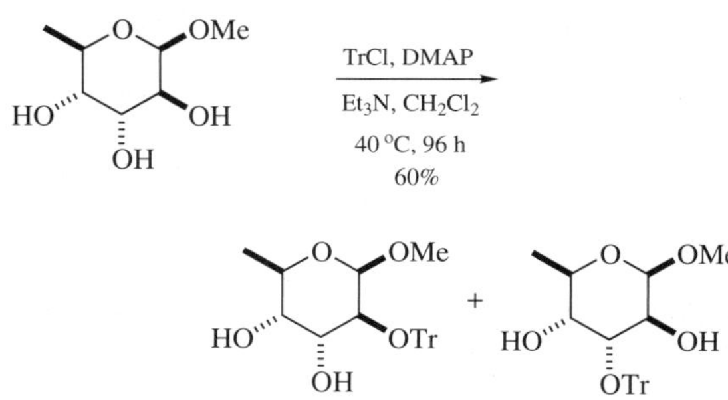

(3)

A non-carbohydrate example using this method involved the regioselective protection of the spiroacetal diol intermediate in a spongistatin synthesis (eq 4).[8]

The proposed intermediate of the TrCl/DMAP reaction, namely, 4-dimethylamino-*N*-triphenylmethylpyridinium chloride, was characterized and reported to be a stable, isolable salt which could be prepared on a multigram scale and would catalyze tritylations.[9,10] Unlike the intermediate generated in situ, this material did not react with secondary alcohols under several conditions and had a pronounced solvent dependence. Because the tritylation of primary alcohols with this reagent was much slower in DMF than in CH_2Cl_2 it was possible to chemoselectively *N*-protect serine methyl ester in good yield (60%). An example of an amino diol being protected using catalytic DMAP in pyridine can be found in the literature although refluxing conditions were necessary (eq 5).[11]

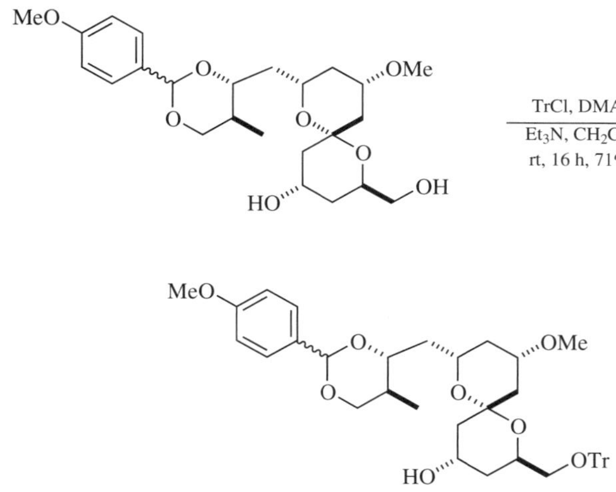

(4)

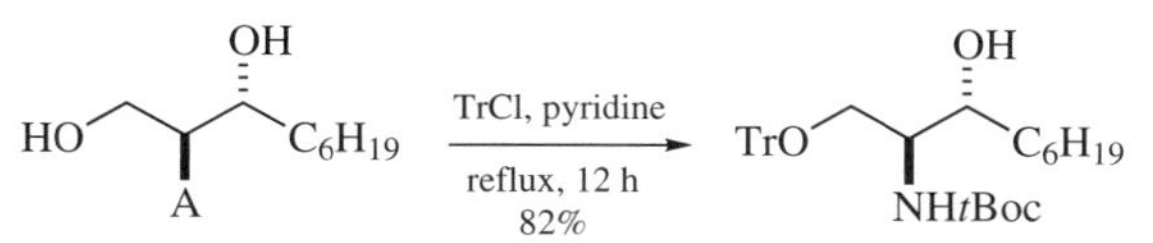

(5)

An efficient method to prepare trityl ethers of secondary alcohols, TrCl with **1,8-diazabicyclo[5.4.0]undec-7-ene** (DBU) in CH_2Cl_2 is used; this reaction proceeds at room temperature over 2 days in good to high yields (66–88%) with a variety of substrates.[12]

Phenol protection with TrCl is somewhat rare although examples are known. An opioid diol was selectively tritylated at the least hindered hydroxyl using the DMAP/CH_2Cl_2 conditions (eq 6); the trityl ether was subsequently removed with wet **acetic acid** (AcOH).[13] A similar transformation on another opioid was performed using TrCl and KOH under phase transfer conditions.[14]

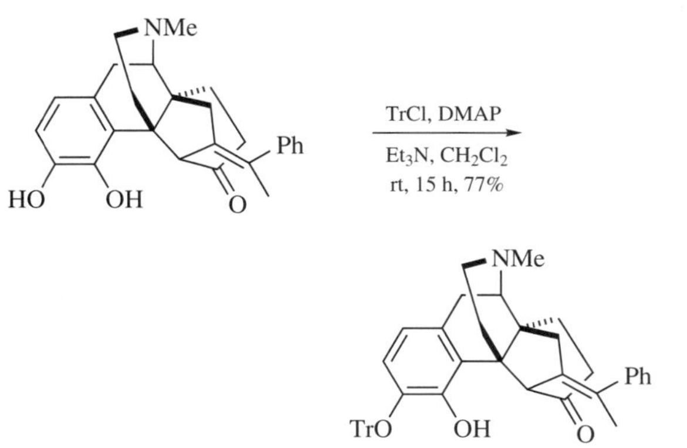

(6)

Treatment of several hydroxyalkyl phenols with TrCl and DMAP in CH_2Cl_2 gave the phenol ethers in good yield, as opposed to the same conditions with t-**butyldimethylchlorosilane** which preferentially formed the TBS ether of the alcohol functionality (eq 7).[15] This contradicts the report that the treatment of 4-hydroxymethylphenol with TrCl in pyridine at room temperature for 2 h gave the trityl ether at the primary alcohol in 96%, although this could be explained by invoking a solvent effect.[16]

The protection of lactols with TrCl can be catalyzed by silver(I) as in the preparation of the bis-tetrahydrofuran core of rolliniastatin (eq 8).[17]

The preparation and use of polymer-bound trityl chloride on alcohols has been disclosed; carbohydrate substrates were added to the reagent in pyridine at room temperature and were subsequently cleaved from the solid support with dry hydrogen bromide. While the reaction yields were not improved, a substantial simplification of the product isolation and purification was achieved.[18]

A unique use of the trityl ether was demonstrated when a propargylic alcohol was protected and then treated with *s*-BuLi/(−)-sparteine to give an alkylidene cyclopentane via a stereoselective intramolecular carbolithiation (eq 9); the presence of the sterically demanding protecting group was essential to prevent allene formation resulting from abstraction of the adjacent proton.[19]

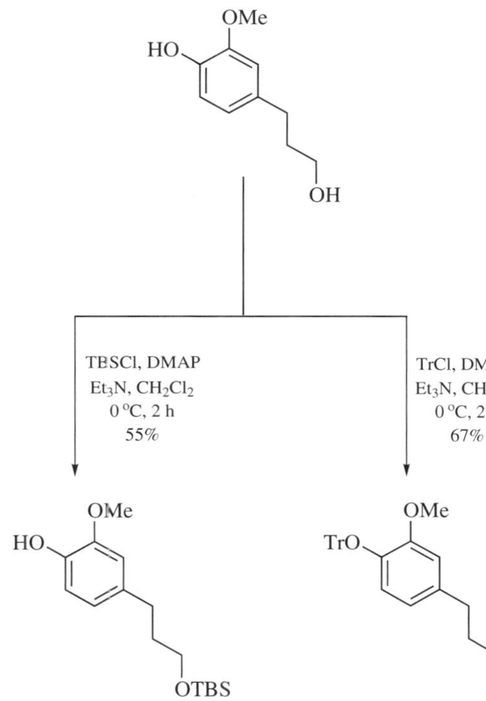

(7)

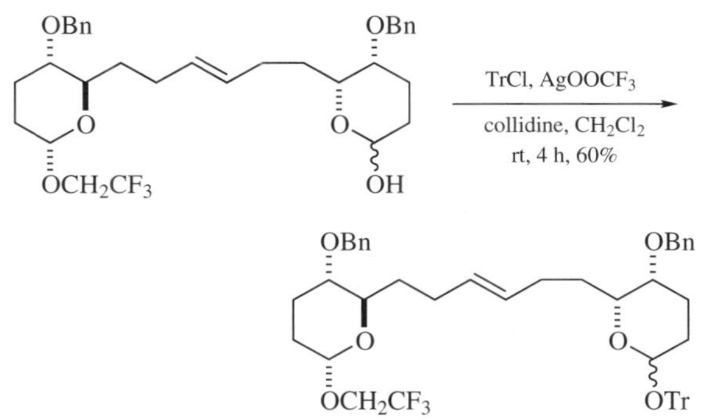

(8)

***S*-Protection of Thiols and Heterocycles.** Protection of thiols as trityl thioethers has been commonly used in total synthesis as well as in carbohydrate, nucleoside, and peptide chemistry. The methods to install this group with TrCl are somewhat similar to those with alcohols although there are more variations, undoubtedly due to the increased reactivity of thiols. For example, in a carbocycle nucleoside synthesis a thiol was protected with TrCl in pyridine/CH_2Cl_2 at room temperature (80%)[20] whereas in a pyrrolidine preparation, TrCl was added directly to a sodium methoxide solution used for the saponification of the *S*-acetyl group (eq 10).[21] The latter example illustrates that secondary thiols are tritylated as readily as primary ones; the trityl moiety in this work was subsequently cleaved with TFA and **triethylsilane** which appears to be the method of choice.

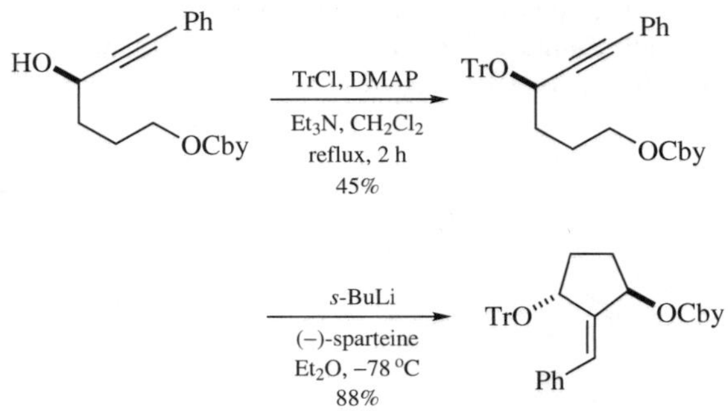
(9)

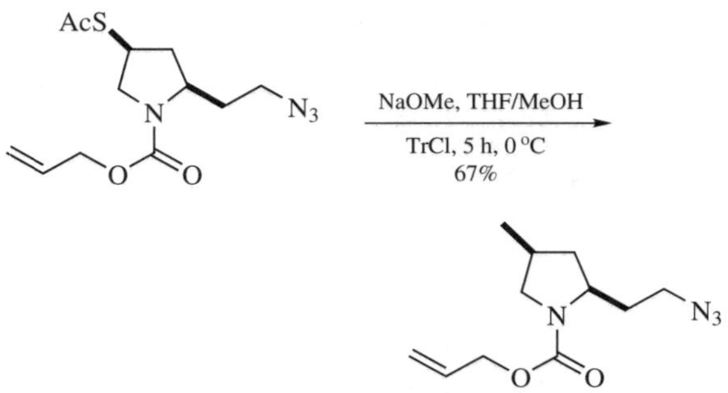
(10)

In thiol-containing amines and amino acids, the sulfur can be selectively tritylated with TrCl as in the case of 2-aminoethanethiol[22] (DMF, rt, 88%) or *S,N*-ditritylated as with D-cysteine in a synthesis of althiomycin[23] (Et_2HN, H_2O/Et_2O, 90%); in the latter work the trityls were removed with **silver(I) nitrate** in pyridine.

Trityl protection or deprotection has been used in combination with disulfide formation or cleavage; for example, 2-aminoethanethiol and 3-mercaptopropionic acid were separately *S*-tritylated, coupled, and detritylated with **iodine** which gave the cyclic disulfide in modest yield (eq 11).[24]

In the opposite transformation, *N*-acetyl-L-cystine was cleaved with **zinc** and catalytic **cobalt(II) chloride** in acetonitrile and the resulting thiol was trapped with TrCl in 72% yield.[25] Examples of disulfide cleavage/tritylation (**2-mercaptoethanol**; TrCl, DMF; 66%)[26] and detritylation/disulfide formation (aq HOAc; I_2; 70%)[27] using peptide substrates are known. Using a slightly different strategy for thiol protection/deprotection in a platelet activating factor analog synthesis, *N*-acetyl-L-cysteine was tritylated with TrCl and N-**methylmorpholine** (NMM) in CH_2Cl_2 (62%) and later deprotected through oxidative cleavage with iodine followed by immediate reduction of the resulting disulfide with dithioerythritol in $CHCl_3$/MeOH (81%, two steps).[28]

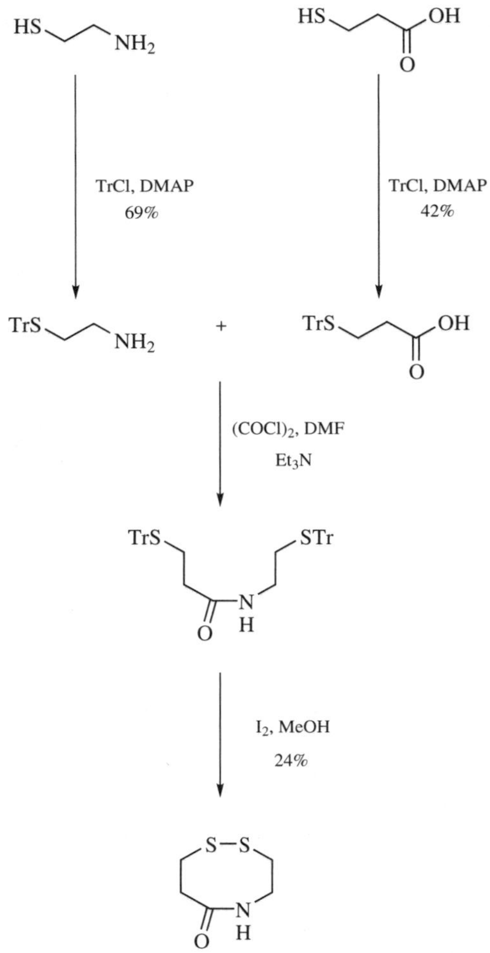
(11)

In a unique tritylation sequence, a penicillinate was treated with silver(I) nitrate followed by TrCl which gave the *S*-trityl-2-oxoazetidine-4-thiol via ring cleavage and tritylation of the intermediate silver thiolate (eq 12).[29]

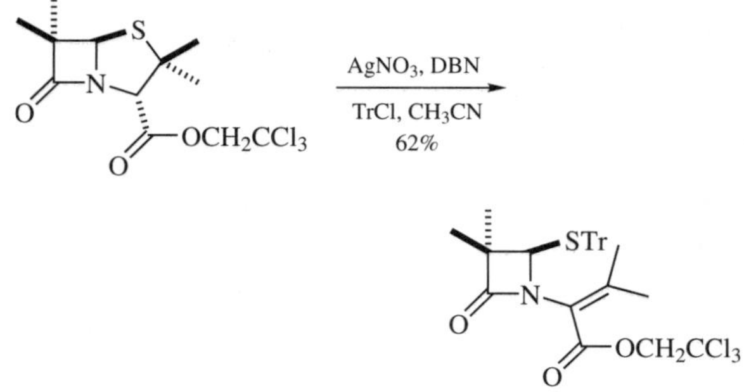
(12)

During the preparation of a series of 2-(thiazol-2-ylthio)carbapenems, a 4-aryl-2-thiothiazolidine was protected with TrCl and **triethylamine** (Et_3N) in CH_2Cl_2 at 0 °C in 50% yield; later in the synthesis removal of the trityl group was accomplished in excellent yield along with a dehydrogenation using catalytic **palladium on carbon** in *p*-cymene at high temperature.[30]

In order to *N*-alkylate a benzimidazole-2-thione, the sulfur was protected as the trityl thioether and subsequently deprotected with methanolic acetic acid in excellent yield (eq 13).[31]

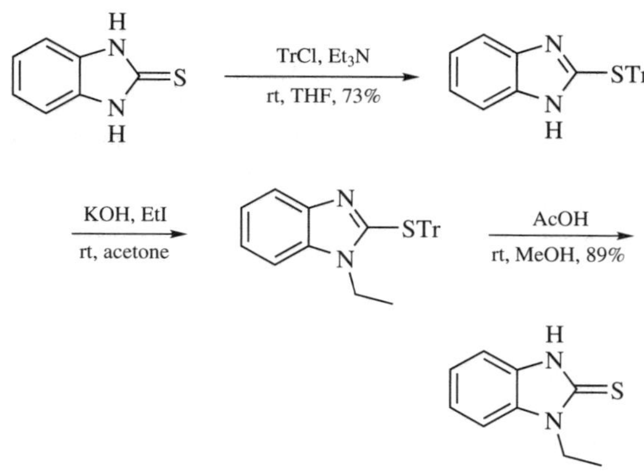

(13)

It was shown that treatment of TrCl with excess ethylene sulfide in CH_2Cl_2 at room temperature for 2 days gave 2-chloroethyl trityl sulfide in 81% yield.[32] The thiol protection of a 1-thio-β-D-glucopyranose was accomplished with TrCl in pyridine at room temperature although unexpected disulfide formation resulted in only a 42% yield.[33] The simultaneous deprotection and resin cleavage of a solid-supported trityl-protected alkanethiol was reported.[34]

***N*-Protection of Amines and Heterocycles.** The *N*-tritylation of primary and secondary amines as well as heterocycles has been used extensively in total syntheses along with amino acid and nucleoside chemistry. The most common method of trityl installation with TrCl involves the use Et_3N in CH_2Cl_2 and generally proceeds at room temperature; removal usually includes the use of wet TFA, AcOH or HCl. A good example of these methods is the *N*-protection/deprotection of D-methionine and L-glutamic acid in the preparation of novel dideoxynucleoside analogs.[35]

Exploiting the steric bulk of this reagent, the less hindered nitrogen of a diamine was regioselectively protected which allowed for differential alkylation of this substrate (eq 14).[36] Taking advantage of the observation that an *N*-trityl group of α-amino acids deactivates the 1-carboxylic acid functionality towards hydrolysis, transesterification, and α-deprotonation, a 2-aminoadipic diester was *N*-tritylated and the distal carboxyl was selectively saponified; the substrate was then converted to isopenicillin N in several steps (eq 15).[37]

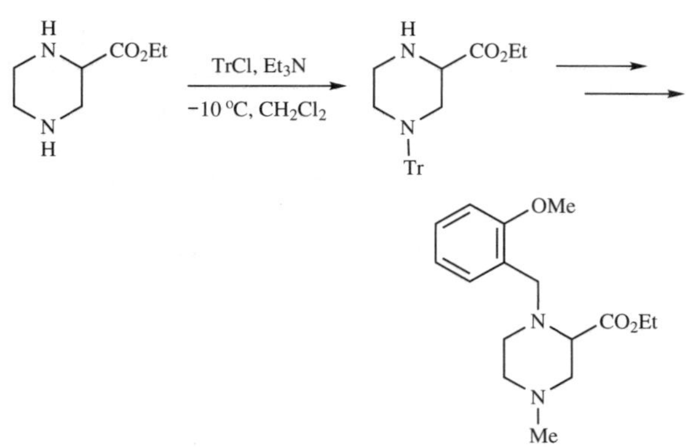

(14)

Although there are a few dated citations of *N*-trityl protection of indoles, 4-bromo tryptophan methyl ester was selectively monotritylated at the α-amino functionality (TrCl, Et_3N, 1,2-dichloroethane, rt, 96%).[38] An aniline was protected with TrCl in pyridine in 90% yield and deprotected with 1 N HCl/THF in 98% yield.[39] The *N*-protection of an aziridine-2-ester using TrCl has also been reported using the standard conditions mentioned above (95% yield).[40]

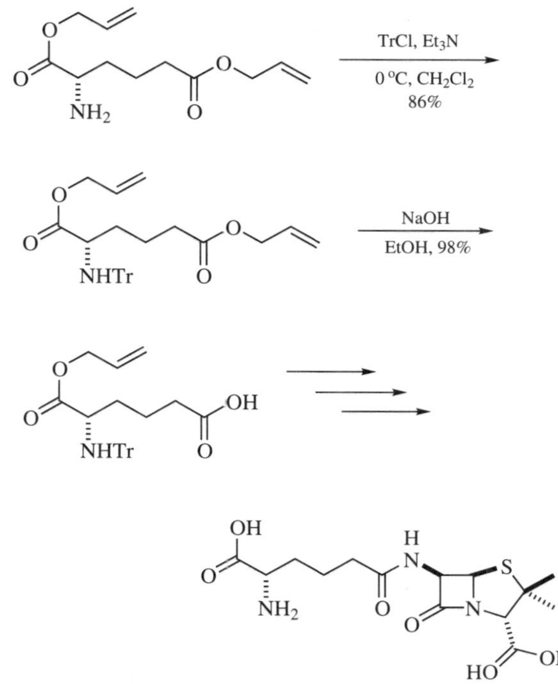

(15)

Numerous trityl protections of imidazoles are known in the literature including that in the synthesis of homohistidine in which the imidazole was tritylated to improve the solubility of the substrate for a subsequent ester reduction.[41] Tritylation/alkylation is a good method to regioselectively *N*-substitute imidazoles as in the farnesyl transferase inhibitor synthesis shown in eq 16;[42] tritylation selectively occurred on the less hindered nitrogen and subsequent alkylation proceeded at the more hindered site.

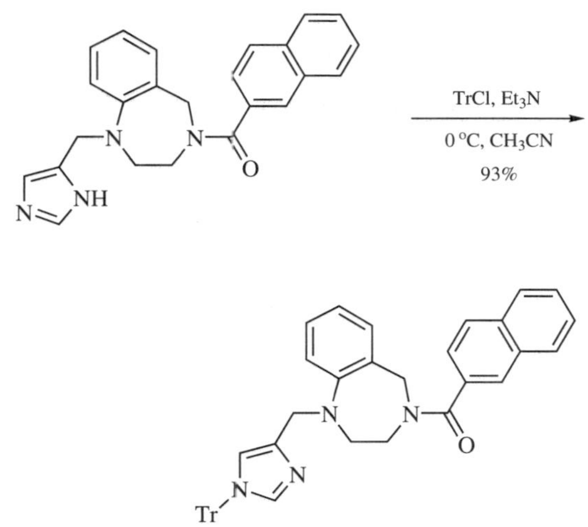

(16)

In a related example, an imidazolo L-xylo compound was *N*,*O*-bistritylated (TrCl, DMAP, pyridine, 75 °C, 85%) and intramolecularly cyclized on the unprotected imidazole nitrogen with concomitant detritylation to afford the corresponding D-arabino imidazole in good yield.[43]

In order to facilitate a palladium-catalyzed amination reaction during the synthesis of a series of 5-HT_{1A} receptor ligands, 4-bromobenzimidazole was *N*-tritylated to afford the antiregioisomer, 4-bromo-1-tritylbenzimidazole, as the sole product; in contrast, benzylation of 4-bromobenzimidazole with benzyl bromide gave a 4:1 mixture of regioisomers. The tritylated material was then aminated and deprotected in excellent yield (eq 17).[44] It was found that treatment of 5-bromobenzimidazole with TrCl under the same conditions gave a 1:1 mixture of trityl regioisomers.

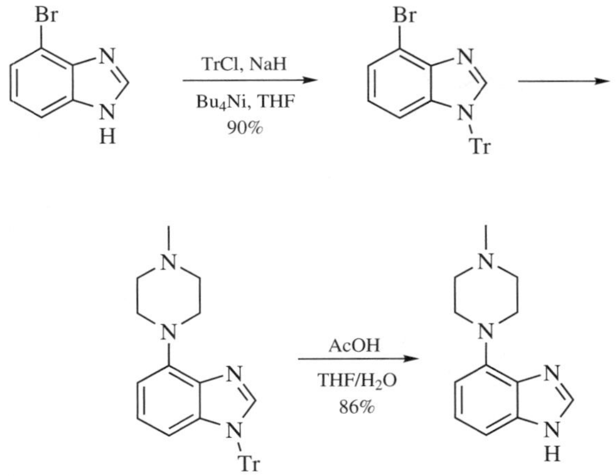

(17)

In another example of imidazole tritylation, 2-aminoimidazole was amino-protected as the phthalimide, tritylated at *N*1 (TrCl, pyridine, reflux), amino-deprotected, alkylated using reductive amination, and detritylated using wet TFA.[45]

A monosubstituted triazole was *N*-tritylated (TrCl, Et_3N, DMF, >95%) to give a 7:1 mixture of regiosiomers which was brominated, alkylated, and eventually deprotected with **hydrogen bromide** (58% yield).[46] A tetrazole tritylation with TrCl was reported in a seminal publication of nonpeptide angiotensin II receptor antagonists. In that work, an aryl nitrile was converted to the tetrazole with **trimethyltin azide** which was destannylated (sodium hydroxide, CH_2Cl_2/THF) and tritylated by the addition of TrCl to that reaction mixture (80% yield).[47] In a similar example of tetrazole protection, a series of aryl tetrazoles were prepared, protected and elaborated further; detritylation with 6 N HCl gave growth hormone secretagogues (eq 18).[48]

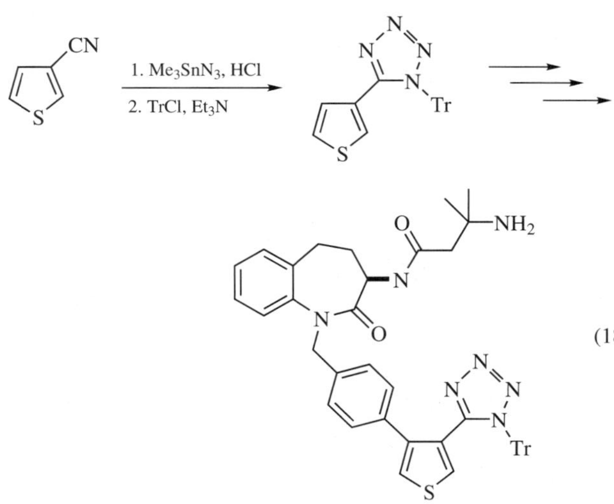

(18)

N-Tritylation of thiazolidine-2,4-diones,[49] 2-aminothiazoles,[50] and 1,2-pyrazoles[51] with TrCl has also been reported.

Several examples of the use of TrCl in nucleoside chemistry include the preparation of 3′-(trityl)amino 5′-phosphoramidate monomers,[52] a 5′-*N*-trityl-3′,5′-deoxythymidine,[53] and the N^1-(trityl)inosine derivative in eq 19.[54]

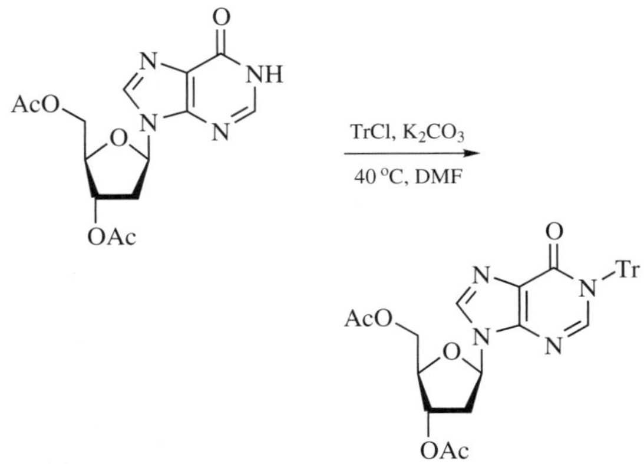

(19)

The use of a polymer-bound TrCl as an α-amino protecting group for amino acids has been reported which gave simplified and high-yielding syntheses of amino acid derivatives and peptides.[55]

In a non-protecting group example of *N*-tritylation with TrCl, a diastereomeric pair of pyrrolidine-containing diaryl phosphonates were *N*-tritylated to allow for chromatographic separation.[56]

A small series of hindered amide bases incorporating the trityl group was reported. Treatment of *t*-butylamine with TrCl in $CHCl_3$ at room temperature gave *N*-t-butyl-*N*-tritylamine (TBTA) as a crystalline solid in good yield. It was shown that TBTA could be converted to the lithium amide with n-**butyllithium** and, when added to **trimethylsilyl chloride** (TMSCl) and a variety of esters at −100 °C, selectively produce *E*-trimethylsilyl ketene acetals; this reagent was superior to **lithium diisopropylamide** (LDA) and **lithium tetramethylpiperidine** (LTMP) in one comparison.[57]

Protection of Carboxylic Acids. The preparation of trityl esters is quite rare probably due to their aqueous instability. The classical preparation used trityl halides and silver salts, although it was acknowledged that **trityl bromide** was the preferred reagent for this conversion.[58] A recent citation indicated that treatment of oleanolic acid with TrCl and DBU in refluxing THF gave the trityl ester in 96% yield; the acid was later deprotected with **trimethylsilyl trifluoromethanesulfonate** in CH_2Cl_2 at room temperature.[59]

Miscellaneous Reaction. In a non-protecting group use of TrCl, treatment of cyclic orthoacetates with this reagent in refluxing CH_2Cl_2 gave the corresponding acetate chlorohydrins stereospecifically and in excellent yields (eq 20).[60]

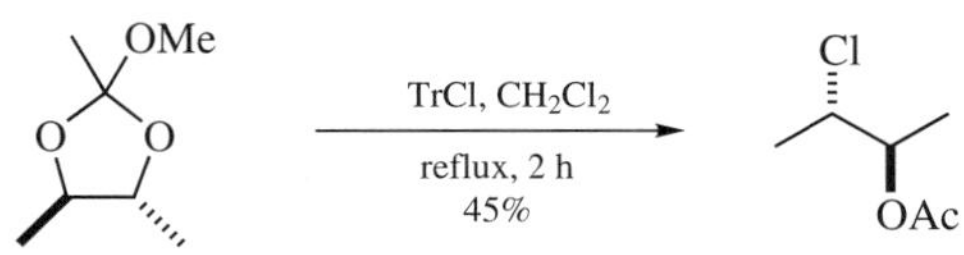

(20)

Related Reagents. 4-Methoxytrityl chloride; 4,4′-dimethoxytrityl chloride.

1. Greene, T. W.; Wuts, P. G. M., *Protective Groups in Organic Synthesis*; Wiley: New York, 1999, p 102.
2. De Kort, M.; Regenbogen, A. D.; Overkleeft, H. S.; Challiss, R. A. J.; Iwata, Y.; Miyamoto, S.; van der Marel, G. A.; van Boom, J. H., *Tetrahedron* **2000**, *56*, 5915.
3. Ichikawa, S.; Shuto, S.; Matsuda, A., *J. Am. Chem. Soc.* **1999**, *121*, 10270.
4. Schaller, H.; Weimann, G.; Lerch, B.; Khorana, H. G., *J. Am. Chem. Soc.* **1963**, *85*, 3821.
5. Chaudhary, S. K.; Hernandez, O., *Tetrahedron Lett.* **1979**, *20*, 95.
6. Leder, I. G., *J. Carbohydrate Chemistry* **1993**, *12*, 95.
7. Jones, G. B.; Hynd, G.; Wright, J. M.; Sharma, A., *J. Org. Chem.* **2000**, *65*, 263.
8. Terauchi, T.; Terauchi, T.; Sato, I.; Tsukada, T.; Kanoh, N.; Nakata, M., *Tetrahedron. Lett.* **2000**, *41*, 2649.
9. Hernandez, O.; Chaudhary, S. K.; Cox, R. H.; Porter, J., *Tetrahedron Lett.* **1981**, *22*, 1491.
10. Bhatia, A. V.; Chaudhary, S. K.; Hernandez, O., *Org. Syn.* **1995**, *75*, 184.
11. De Jonghe, S.; van Overmeire, I.; van Calenbergh, S.; Hendrix, C.; Busson, R.; De Keukeleire, D.; Herdewijn, P., *Eur. J. Org. Chem.* **2000**, 3177.
12. Colin-Messager, S.; Girard, J. P.; Rossi, J. C., *Tetrahedron Lett.* **1992**, *33*, 2689.
13. Bermejo, F. M.; Husbands, S. M.; Lewis, J. W., *Helv. Chim. Acta* **1999**, *82*, 1721.
14. Nelson, T. D.; Davis, R. D.; Nelson, W. L., *J. Med. Chem.* **1994**, *37*, 4270.
15. Sefkow, M.; Kaatz, H., *Tetrahedron Lett.* **1999**, *40*, 6561.
16. Jobron, L.; Hindsgaul, O., *J. Am. Chem. Soc.* **1999**, *121*, 5835.
17. Ruan, Z.; Dabideen, D.; Blumenstein, M.; Mootoo, D. R., *Tetrahedron* **2000**, *56*, 9203.
18. Fréchet, J. M. J.; Haque, K. E., *Tetrahedron Lett.* **1975**, *35*, 3055.
19. Oestreich, M.; Fröhlich, R.; Hoppe, D., *J. Org. Chem.* **1999**, *64*, 8616.
20. Jenny, T. F.; Benner, S. A., *Helv. Chim. Acta* **1993**, *76*, 826.
21. Azami, H.; Barrett, D.; Matsuda, K.; Tsutsumi, H.; Washizuka, K.; Sakurai, M.; Kuroda, S.; Shirai, F.; Chiba, T.; Kamimura, T.; Murata, M., *Bioorg. Med. Chem.* **1999**, *7*, 1665.
22. Li, G.; Ma, Q.; Ma, B.; Grossman, Z. D.; Pandey, R. K., *Heterocycles* **1999**, *51*, 2849.
23. Inami, K.; Shiba, T., *Bull. Chem. Soc. Jpn.* **1985**, *58*, 352.
24. Cepas, S. C.; North, M., *Tetrahedron* **1997**, *53*, 16859.
25. Chowdhury, S.; Roy, S., *Tetrahedron Lett.* **1997**, *38*, 2149.
26. Brundish, D. E.; Wade, R., *J. Chem. Soc., Perkin I* **1981**, 318.
27. Jones, D. A.; Mikulec, R. A.; Mazur, R. H., *J. Org. Chem.* **1973**, *38*, 2865.
28. Mlotkowska, B.; Poscik, A., *Liebigs Ann.* **1996**, 841.
29. Alpegiani, M.; Bedeschi, A.; Giudici, F.; Perrone, E.; Franceschi, G., *J. Am. Chem. Soc.* **1985**, *107*, 6398.
30. Shinagawa, H.; Yamaga, H.; Houchigai, H.; Sumita, Y.; Sunagawa, M., *Bioorg. Med. Chem.* **1997**, *5*, 601.
31. Doerge, D. R.; Cooray, N. M., *Synth. Comm.* **1991**, *21*, 1789.
32. Trujillo, D. A.; McMahon, W. A., Jr; Lyle, R. E., *J. Org. Chem.* **1987**, *52*, 2932.
33. Paul, B.; Korytnyk, W., *Carbohydrate Res.* **1984**, *126*, 27.
34. Houseman, B. T.; Mrksich, M., *J. Org. Chem.* **1998**, *63*, 7552.
35. Balayiannis, G.; Karigiannis, G.; Gatos, P.; Papaioannou, D.; De Clerq, E., *Tetrahedron Lett.* **2000**, *41*, 6191.
36. Le Bihan, G.; Rondu, F.; Pele-Tounian, A.; Wang, X.; Lidy, S.; Touboul, E.; Lamouri, A.; Dive, G.; Huet, J.; Pfeiffer, B.; Renard, P.; Guardiola-Lemaître, B.; Manéchez, D.; Pénicaud, L.; Ktorza, A.; Godfroid, J. J., *J. Med. Chem.* **1999**, *42*, 1587.
37. Lau, R. M.; van Eupen, J. T. H.; Schipper, D.; Tesser, G. I.; Verweij, J.; de Vroom, E., *Tetrahedron* **2000**, *56*, 7601.
38. Hurt, C. R.; Lin, R.; Rapoport, H., *J. Org. Chem.* **1999**, *64*, 225.
39. Durand, D.; Hullot, P.; Vidal, J. P.; Girard, J. P.; Baneres, J. L.; Parello, J.; Muller, A.; Bonne, C.; Rossi, J. C., *Bioorg. Med. Chem. Lett.* **2000**, *10*, 811.
40. Bongini, A.; Cardillo, G.; Gentilucci, L.; Tomasini, C., *J. Org. Chem.* **1997**, *62*, 9148.
41. Pirrung, M. C.; Pei, T., *J. Org. Chem.* **2000**, *65*, 2229.
42. Ding, C. Z.; Batorsky, R.; Bhide, R.; Chao, H. J.; Cho, Y.; Chong, S.; Gullo-Brown, J.; Guo, P.; Kim, S. H.; Lee, F.; Leftheris, K.; Miller, A.; Mitt, T.; Patel, M.; Penhallo, B. A.; Ricca, C.; Rose, W. C.; Schmidt, R.; Slusarchyk, W. A.; Vite, G.; Yan, N.; Manne, V.; Hunt, J. T., *J. Med. Chem.* **1999**, *42*, 5241.
43. Streith, J.; Rudyk, H.; Tschamber, T.; Tarnus, C.; Strehler, C.; Deredas, D.; Frankowski, A., *Eur. J. Org. Chem.* **1999**, 893.
44. López-Rodríguez, M. L.; Benhamú, B.; Ayala, D.; Rominguera, J. L.; Murcia, M.; Ramos, J. A.; Viso, A., *Tetrahedron* **2000**, *56*, 3245.
45. Pitts, W. J.; Wityak, J.; Smallheer, J. M.; Tobin, A. E.; Jetter, J. W.; Buynitsky, J. S.; Harlow, P. P.; Solomon, K. A.; Corjay, M. H.; Mousa, S. A.; Wexler, R. R.; Jadhav, P. K., *J. Med. Chem.* **2000**, *43*, 27.
46. Bang-Anderson, B.; Lenz, S. M.; Skjaerbaek, N.; Søby, K. K.; Hansen, H. O.; Ebert, B.; Bogesø, K. P.; Krogsgaard-Larsen, P., *J. Med. Chem.* **1997**, *40*, 2831.
47. Carini, D. J.; Duncia, J. V.; Aldrich, P. E.; Chiu, A. T.; Johnson, A. L.; Pierce, M. E.; Price, W. A.; Santella, J. B.; Wells, G. J.; Wexler, R. R.; Wong, P. C.; Yoo, S. E.; Timmermans, P. B. M. W. M., *J. Med. Chem.* **1991**, *34*, 2525.
48. Ankersen, M.; Peschke, B.; Hansen, B. S.; Hansen, T. K., *Bioorg. Med. Chem. Lett.* **1997**, *7*, 1293.
49. Wrobel, J.; Li, Z.; Dietrich, A.; McCaleb, M.; Mihan, B.; Sredy, J.; Sullivan, D., *J. Med. Chem.* **1998**, *41*, 1084.
50. Pae, A. N.; Lee, J. E.; Kim, B. H.; Cha, J. H.; Kim, H. Y.; Cho, Y. S.; Choi, K. I.; Koh, H. Y.; Lee, E.; Kim, J. H., *Tetrahedron* **2000**, *56*, 5657.
51. Yamamoto, H.; Kawabata, K.; Tawara, S.; Takasugi, H.; Tanaka, H., *J. Antibiotics* **1998**, *51*, 683.
52. Nelson, J. S.; Fearon, K. L.; Nguyen, M. Q.; McCurdy, S. N.; Frediani, J. E.; Foy, M. F.; Hirschbein, B. L., *J. Org. Chem.* **1997**, *62*, 7278.
53. Arya, D. P.; Bruice, T. C., *J. Am. Chem. Soc.* **1998**, *120*, 6619.
54. Minakawa, N.; Sasabuchi, Y.; Kiyosue, A.; Kojima, N.; Matsuda, A., *Chem. Pharm. Bull.* **1996**, *44*, 288–295.
55. Barlos, K.; Gatos, D.; Kallitsis, I.; Papaioannou, D.; Sotiriou, P., *Liebigs Ann. Chem.* **1988**, 1079.
56. Belyaev, A.; Zhang, X.; Augustyns, K.; Lambeir, A. M.; Meester, I. D.; Vedernikova, I.; Scharpé S., Haemers, A., *J. Med. Chem.* **1999**, *42*, 1041.
57. Busch-Petersen, J.; Corey, E. J., *Tetrahedron. Lett.* **2000**, *41*, 2515.
58. Berlin, K. D.; Gower, L. H.; White, J. W.; Gibbs, D. E.; Sturm, G. P., *J. Org. Chem.* **1962**, *27*, 3595.
59. Yu, B.; Xie, J.; Deng, S.; Hui, Y., *J. Am. Chem. Soc.* **1999**, *121*, 12196.
60. Newman, M. S.; Chen, C. H., *J. Am. Chem. Soc.* **1973**, *95*, 278.

Michael L. Curtin
Abbott Laboratories, Abbott Park, IL, USA

Triphenylmethyl Perchlorate

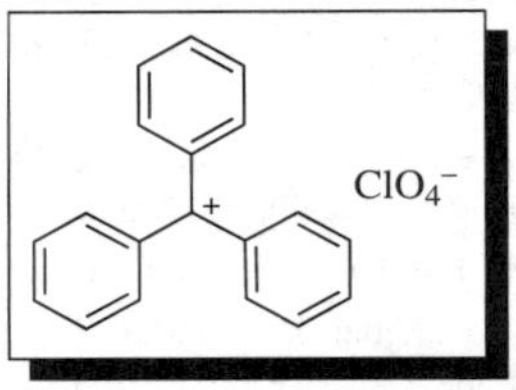

[3058-33-1] $C_{19}H_{15}ClO_4$ (MW 342.79))

(catalyst for Michael additions,[1] aldol reactions,[2] and glycosidation[3])

Alternate Names: trityl perchlorate; triphenylcarbenium perchlorate.

Physical Data: mp 143 °C.[4]

Preparative Method: perchloric acid is added to triphenylmethanol in acetic anhydride. The trityl perchlorate is separated, washed with ether, and dried.[5]

Purification: recrystallize from dry acetonitrile as yellow prisms.[4]

Handling, Storage, and Precautions: trityl perchlorate should be stored in the dark as exposure to light darkens the compound.[4] It is rapidly hydrolyzed by exposure to moist air.[4] All perchlorates present a potential explosion hazard.

Michael Reactions. Silyl enol ethers react with α,β-unsaturated ketones in the presence of catalytic amounts of trityl perchlorate (2–3 mol %[1]) to give 1,5-dicarbonyl compounds.[6] No 1,2 addition products were obtained. It is possible to isolate the intermediate silyl enol ethers, which is not the case if Titanium(IV) Chloride is used as catalyst (eq 1).[6] *Anti* Michael adducts are obtained preferentially with selectivities of up to >95:5 (eq 2).[1] Silyl enol ethers of thioesters also react to give 5-keto thiol esters.[7] Allylsilanes have been reacted with α,β-unsaturated ketones to give δ,ε-unsaturated ketones (eq 3).[8] 1,4-Benzoquinones react with silyl enol ethers to give benzofurans.[9] Trityl perchlorate was found to be a more effective catalyst for Michael addition reactions than other trityl salts.[6]

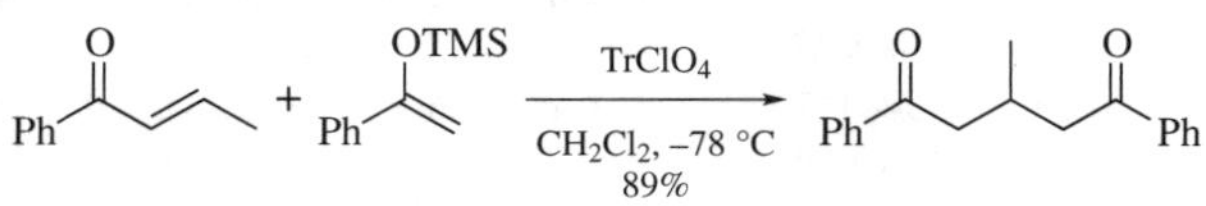

(1)

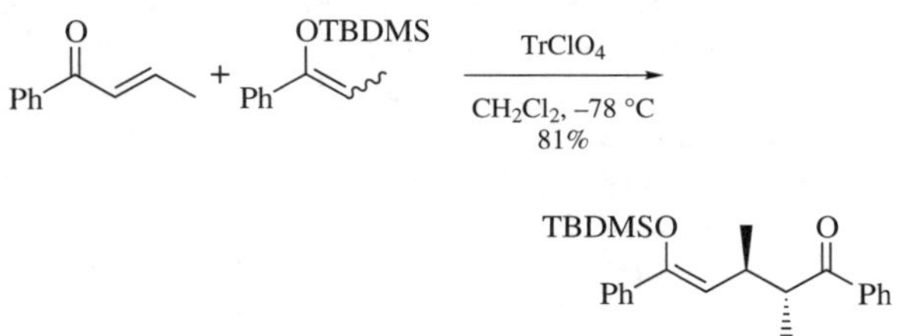

(2)

>95:5 (*syn*)

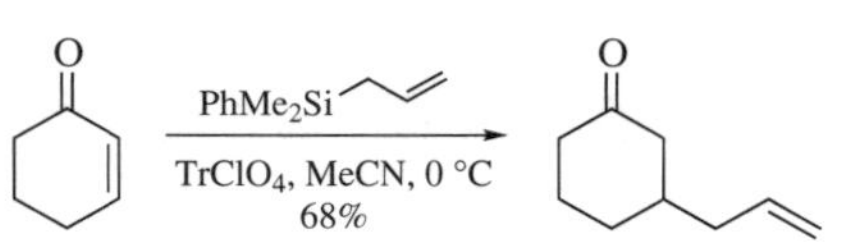

(3)

Aldol Reactions. In the presence of a catalytic amount of trityl perchlorate (1–5 mol %), silyl enol ethers react with acetals to give β-alkoxy ketones (eq 4).[10] Orthoformates react in a similar way.[10] Aldehydes react with silyl enol ethers to give β-silyloxy ketones (eq 5).[2] The reaction can be made *threo* selective, with selectivities of up to 82:18 reported.[11] Alkyl enol ethers and acetals react in the presence of trityl perchlorate to give β-alkoxy ketones.[12] A tandem Michael–aldol reaction has been reported which gives γ-acyl-δ-hydroxy ketones (eq 6).[13]

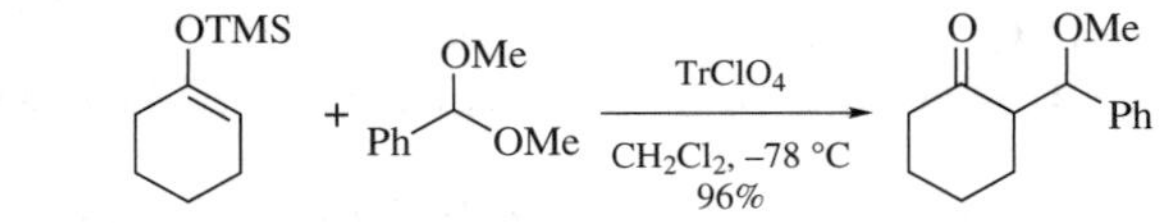

(4)

erythro:*threo* = 88:12

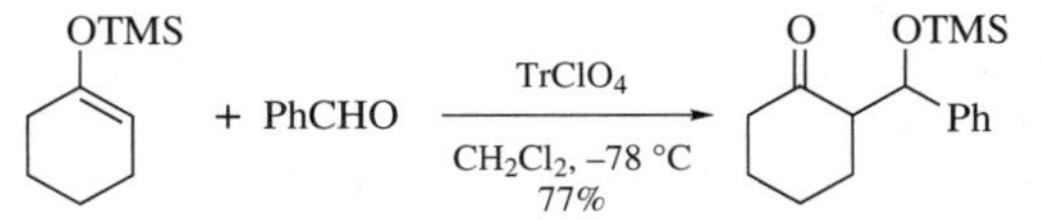

(5)

erythro:*threo* = 64:36

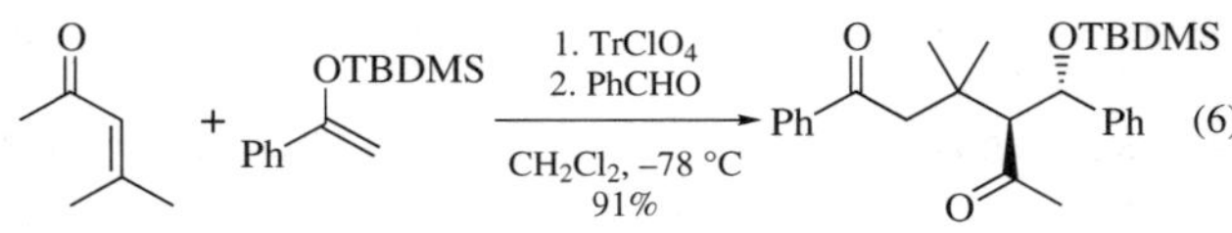

(6)

Glycosidation. Perbenzylated 1-*O*-bromoacetyl-β-D-glucose reacts with alcohols in the presence of trityl perchlorate to give α-glucosides.[14] α-Ribosides can be prepared in a similar manner.[14] Other nucleophiles (silyl enol ethers, allylsilanes, and Cyanotrimethylsilane) react with benzylated 1-*O*-acetyl ribose to give α-*C*-ribofuranosides (eq 7).[3]

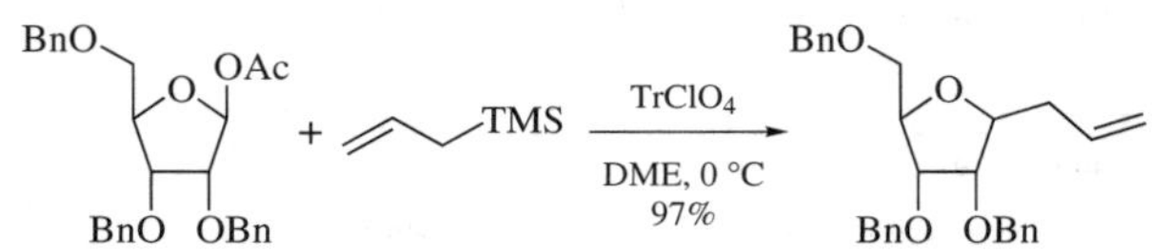

(7)

---

1. Mukaiyama, T.; Tamura, M.; Kobayashi, S. *Chem. Lett.* **1986**, 1017.
2. Kobayashi, S.; Murakami, M.; Mukaiyama, T. *Chem. Lett.* **1985**, 1535.
3. Mukaiyama, T.; Kobayashi, S.; Shoda, S. *Chem. Lett.* **1984**, 1529.
4. Dauben, H. J.; Honnen, L. R.; Harmon, K. M. *J. Org. Chem.* **1960**, *25*, 1442.
5. Lambert, J. B.; Schulz, W. J.; McConnell, J. A.; Schilf, W. *J. Am. Chem. Soc.* **1988**, *110*, 2201.
6. Kobayashi, S.; Murakami, M.; Mukaiyama, T. *Chem. Lett.* **1985**, 953.
7. Mukaiyama, T.; Tamura, M.; Kobayashi, S. *Chem. Lett.* **1986**, 1817.
8. Hayashi, M.; Mukaiyama, T. *Chem. Lett.* **1987**, 289.
9. Mukaiyama, T.; Sagawa, Y.; Kobayashi, S.; *Chem. Lett.* **1987**, 2169.
10. Mukaiyama, T.; Kobayashi, S.; Murakami, M. *Chem. Lett.* **1986**, 1759.
11. Mukaiyama, T.; Kobayashi, S.; Murakami, M. *Chem. Lett.* **1985**, 447.
12. Murakami, M.; Minamikawa, H.; Mukaiyama, T. *Chem. Lett.* **1987**, 1051.
13. Kobayashi, S.; Mukaiyama, T. *Chem. Lett.* **1986**, 221.
14. Mukaiyama, T.; Kobayashi, S.; Shoda, S. *Chem. Lett.* **1984**, 907.

Christopher J. Urch
Zeneca Agrochemicals, Bracknell, UK

Triphenylphosphine-*N*-Bromosuccinimide[1]

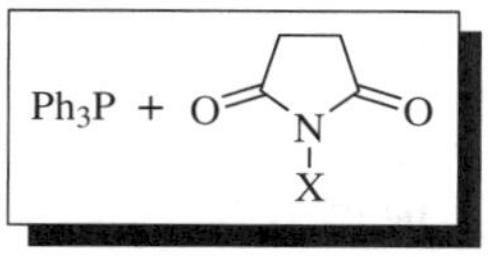

(Ph_3P)		
[603-35-0]	$C_{18}H_{15}P$	(MW 262.30)) (NBS)
[128-08-5]	$C_4H_4BrNO_2$	(MW 177.99)) (NCS)
[128-09-6]	$C_4H_4ClNO_2$	(MW 133.54)) (NIS)
[516-12-1]	$C_4H_4INO_2$	(MW 224.99))

(conversion of primary alcohols to alkyl halides)

Physical Data: See Triphenylphosphine, *N*-Bromosuccinimide, *N*-Chlorosuccinimide, and *N*-Iodosuccinimide.

Preparative Methods: the mixed reagent is prepared in situ as needed by combining triphenylphosphine, the alcohol, and *N*-halosuccinimide in a suitable solvent (usually DMF) and heating.

Handling, Storage, and Precautions: precautions in handling should be made as described for the individual reagents. This mixed reagent is not appropriate for storage. Recrystallization of the triphenylphosphine and *N*-halosuccinimide is necessary to achieve the most efficient halogenation reaction. In addition, the use of anhydrous solvent is important for the success of the reaction.

Introduction. The combination of triphenylphosphine and *N*-halosuccinimide was applied by Hanessian[2] to the preparation of primary halides from carbohydrate precursors. This method is related to a number of other methods for hydroxyl substitution via oxyphosphonium salts as described in a comprehensive review.[1]

Conversion of Alcohols to Alkyl Halides. The typical procedure[2] as applied to the preparation of methyl 5-deoxy-5-iodo-2,3-*O*-isopropylidene-β-D-ribofuranose (eq 1) involves the addition

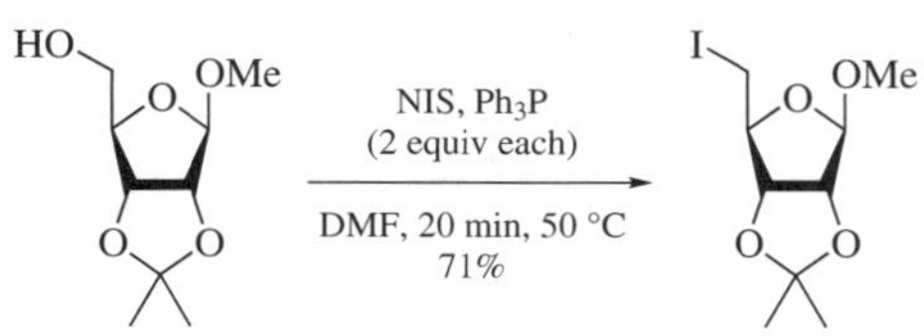

(1)

of 2 equiv of *N*-iodosuccinimide followed by 2 equiv of triphenylphosphine to a cooled solution of the alcohol in anhydrous DMF and heating the mixture to 50 °C for 20 min. Methyl alcohol is added to destroy excess reagent, followed by *n*-butanol to aid in the removal of DMF under reduced pressure. After concentration, the product is purified from the byproducts (succinimide and triphenylphosphine oxide) by addition of ether and aqueous extraction followed by column chromatography. Isolated yields of 70–85% are regularly obtained using this procedure for the preparation of chlorides, bromides, and iodides. In addition, the reaction conditions allow the preparation of primary halides in the presence of many functional groups used in carbohydrate chemistry, including the ester, amide, lactone, benzylidene acetal, and acetonide groups (eq 2). In fact, moderate yields (49–82%) of a number of primary halides are obtained using this procedure on sugars containing one or more free secondary hydroxyl groups (eq 3).[2–4]

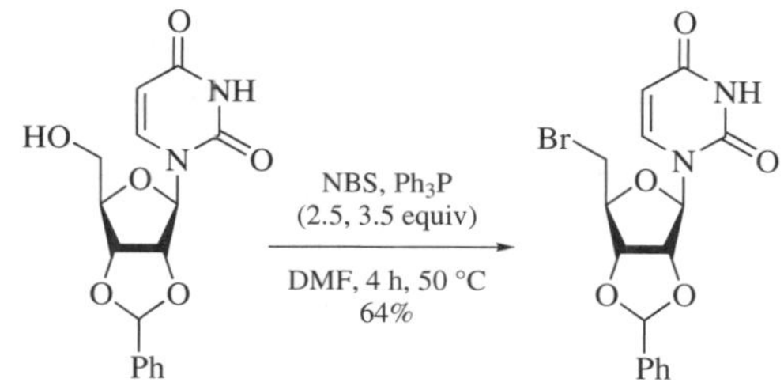

(2)

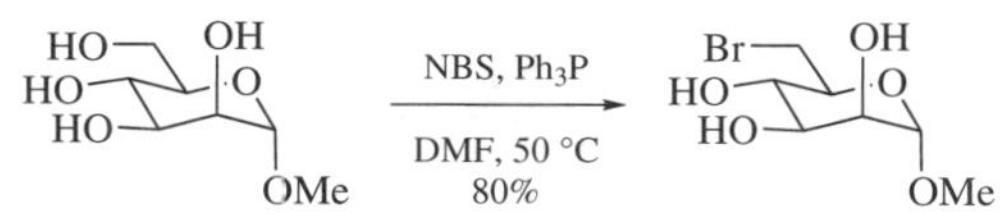

(3)

Other examples of the use of this method include the preparation of tetrahydrofurfuryl bromide,[5] 3α-cholestanyl bromide (using THF as solvent, 85% yield),[6] and intermediates toward the syntheses of alkaloids (eqs 4 and 5).[7] The reaction of the reagent with DMF alone generates the formamidinium salt which can ultimately result in the isolation of the formate ester derivative of the alcohol substrate (eq 6).[1,8] It is therefore important that the *N*-halosuccinimide is added last for the preparation of alkyl halides. Other solvents used for this reaction include CH_2Cl_2 (which can facilitate the workup of the reaction products) and HMPA.[9]

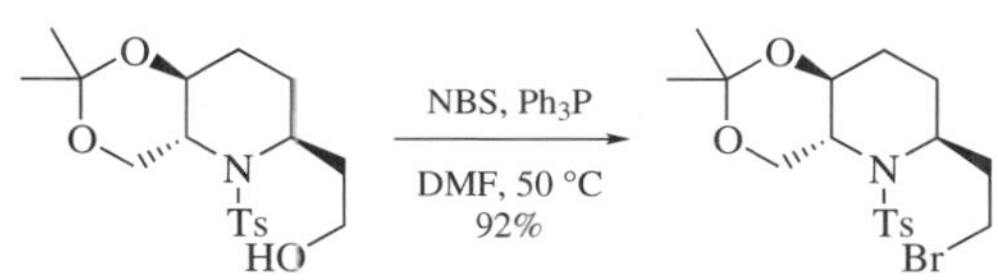

(4)

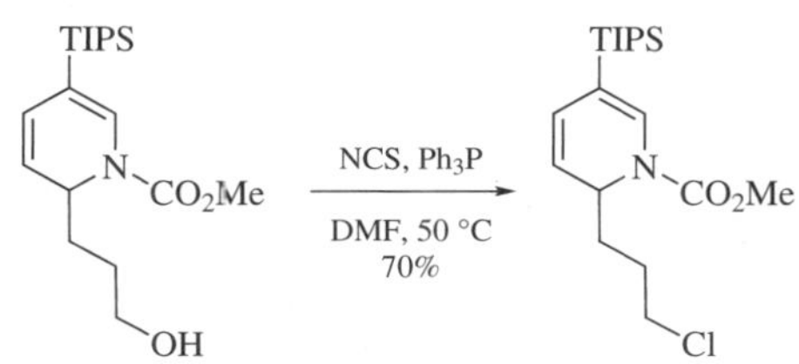

(5)

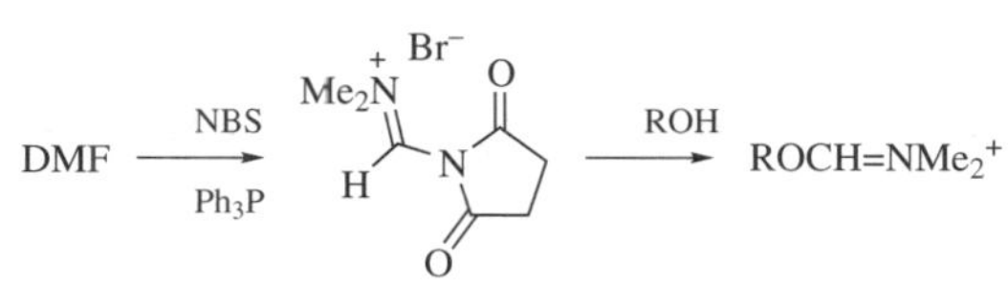

(6)

The standard procedure has often been found to be unsatisfactory for substrates which are prone to acid-catalyzed rearrangements (including migration of acetonide protecting groups) and cleavage and is also not suitable for substrates containing labile protecting groups such as THP ethers.[10,11] In some cases this problem has been circumvented by the addition of various bases such as $BaCO_3$ (eq 7),[12] Pyridine, and Imidazole. A recent study has achieved the preparation of 3-bromo-3-deoxy-1,2:5,6-di-*O*-isopropylidene-α-D-allofuranose by addition of 2 equiv of *N*-halosuccinimide to a refluxing solution of the secondary alcohol and 2 equiv each of PPh_3 and imidazole in toluene or chlorobenzene (eq 8). Previous applications of PPh_3 and *N*-halosuccinimide alone afforded only the primary halide products resulting from acetonide migration. The use of pyridine and imidazole have also been found to play an essential role in the conversion of carbohydrates to halides using the reagents Triphenylphosphine–Iodine and Triphenylphosphine–Carbon Tetrachloride, respectively.[13]

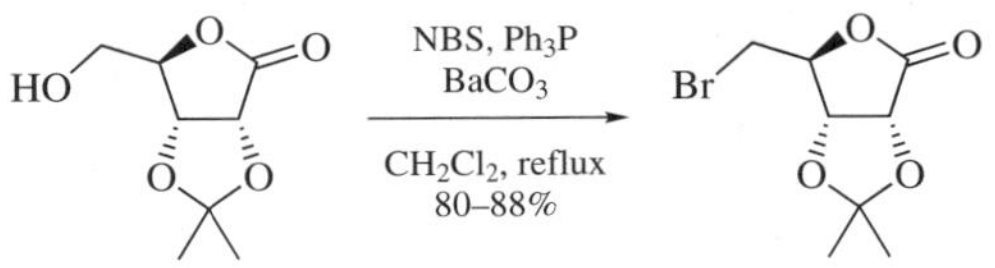

(7)

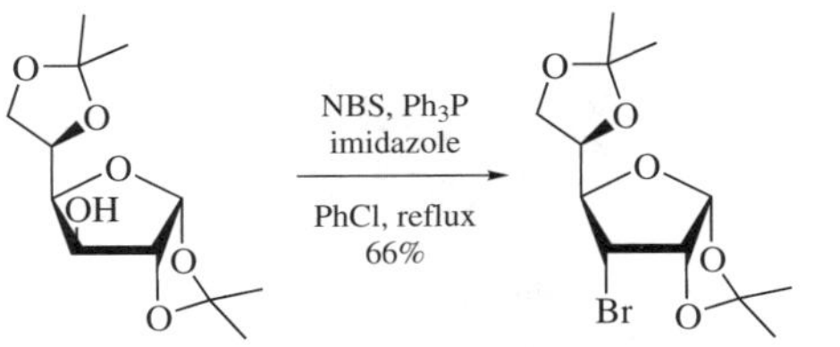

(8)

1. Castro, B. R. *Org. React.* **1983**, *29*, 1.
2. (a) Ponpipom, M. M.; Hanessian, S. *Carbohydr. Res.* **1971**, *18*, 342. (b) Hanessian, S.; Ponpipom, M. M.; Lavallee, P. *Carbohydr. Res.* **1972**, *24*, 45.
3. (a) Nakane, M.; Hutchinson, C. R.; Gollman, H. *Tetrahedron Lett.* **1980**, *21*, 1213. (b) Aspinall, G. O.; Carpenter, R. C.; Khondo, L. *Carbohydr. Res.* **1987**, *165*, 281.
4. Hasegawa, A.; Morita, M.; Ishida, H.; Kiso, M. *J. Carbohydr. Chem.* **1989**, *8*, 579.
5. Schweizer, E. E.; Creasy, W. S.; Light, K. K.; Shaffer, E. T. *J. Org. Chem.* **1969**, *34*, 212.
6. Bose, A. K.; Lal, B. *Tetrahedron Lett.* **1973**, 3937.
7. (a) Birkinshaw, T. N.; Holmes, A. B. *Tetrahedron Lett.* **1987**, *28*, 813. (b) Comins, D. L.; Myoung, Y. C. *J. Org. Chem.* **1990**, *55*, 292.
8. Hodosi, G.; Podányi, B.; Kuszmann, J. *Carbohydr. Res.* **1992**, *230*, 327.
9. Baker, C. W.; Whistler, R. L. *Carbohydr. Res.* **1975**, *45*, 237.
10. Yunker, M. B.; Tam, S.Y.-K.; Hicks, D. R.; Fraser-Reid, B. *Can. J. Chem.* **1976**, *54*, 2411.
11. Gensler, W. J.; Marshall, J. P.; Langone, J. J.; Chen, J. C. *J. Org. Chem.* **1977**, *42*, 118.
12. Jäger, V.; Häfele, B. *Synthesis* **1987**, 801.
13. (a) Garegg, P. J.; Johansson, R.; Samuelsson, B. *Synthesis* **1984**, 168. (b) Whistler, R. L.; Anisuzzaman, A. K. M. *Methods Carbohydr. Chem.* **1980**, *8*, 227.

Scott C. Virgil
Massachusetts Institute of Technology, Cambridge, MA, USA

Triphenylphosphine Carbon Tetrabromide[1]

$Ph_3P–CBr_4$

(Ph_3P)		
[603-35-0]	$C_{18}H_{15}P$	(MW 262.30)
(CBr_4)		
[558-13-4]	CBr_4	(MW 331.61))

(reagent combination for the conversion of alcohols to bromides, aldehydes and ketones to dibromoalkenes, and terminal alkynes to 1-bromoalkynes; carboxyl activation)

Physical Data: Ph_3P: mp 79–81 °C; bp 377 °C; *d* 1.0749 g cm^{-3}. CBr_4: mp 90–91 °C; bp 190 °C; *d* 3.273 g cm^{-3}.
Solubility: sol MeCN, CH_2Cl_2, pyridine, DMF.
Preparative Method: the reactive species is generated in situ by reaction of Ph_3P with CBr_4.
Handling, Storage, and Precautions: Ph_3P is an irritant; CBr_4 is toxic and a cancer suspect agent. All solvents used must be carefully dried because the intermediates are all susceptible to hydrolysis. This reagent should be used in a fume hood.

Conversion of Alcohols to Alkyl Bromides. The reaction of alcohols with Triphenylphosphine and Carbon Tetrabromide results in the formation of alkyl bromides. The conditions are sufficiently mild to allow for the efficient conversion of alcohols into the corresponding bromides. The uridine derivative (eq 1) is transformed into its bromide with Ph_3P and CBr_4.[2]

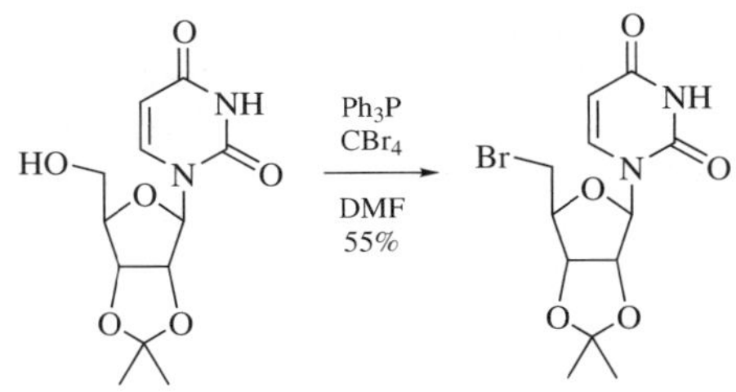

(1)

The geometry about the double bond of an allylic alcohol is usually not compromised (eq 2).[3] Allylic rearrangement is also not commonly observed. Although ketones are known to react with this reagent combination, the reaction of an allylic alcohol has been selectively achieved in the presence of a ketone (eq 3).[4]

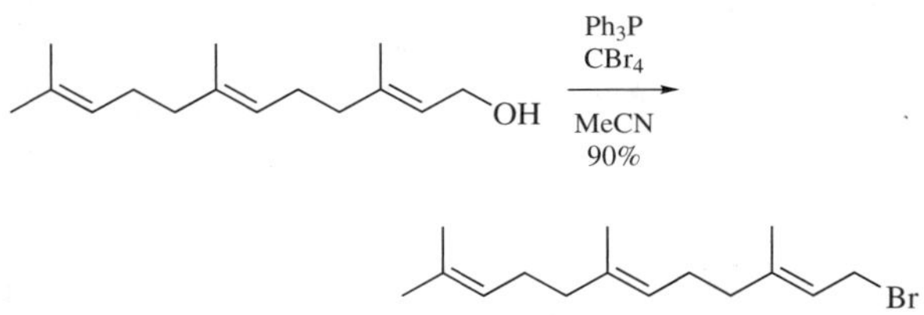

(2)

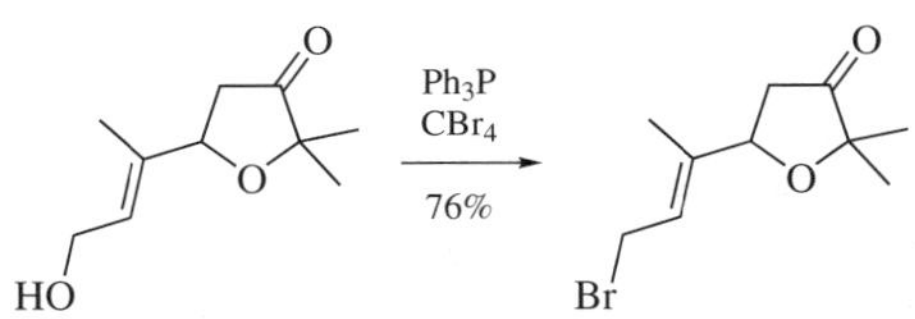

(3)

Regioselective bromination of primary alcohols in the presence of secondary alcohols is possible. These reactions are usually performed in pyridine as solvent. The reaction of the methyl glucopyranoside (eq 4) results in the selective formation of the primary bromide in 98% yield.[5] An investigation of this reaction with a chiral deuterated neopentyl alcohol yielded a partially racemized bromide.[6]

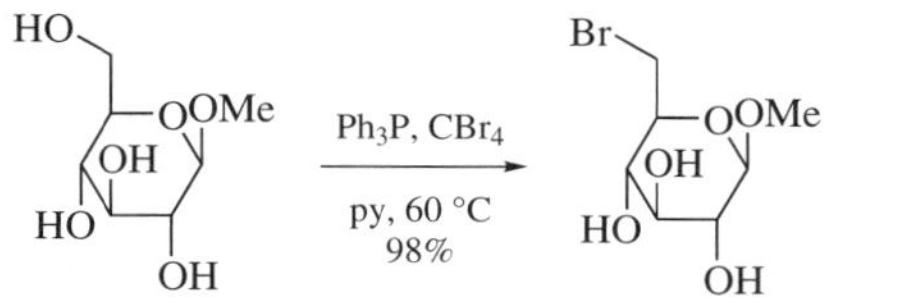
(4)

Silyl ether protected alcohols (eq 5) have been converted directly into the corresponding bromides with Ph_3P and CBr_4. The reaction works best if 1.5 equiv of acetone are added.[7] Tetrahydropyranyl ether protected alcohols have also been directly transformed into the bromides using this reagent combination. The reaction has been reported to proceed with inversion of configuration (eq 6).[8] If unsaturation is appropriately placed within a tetrahydropyranyl (eq 7) or a methoxymethyl (eq 8) protected alcohol, cyclization occurs to afford tetrahydropyrans.[9] The conversion of an alcohol to the bromide without complications with a methoxymethyl protected alcohol in the molecule is possible (eq 9).[10]

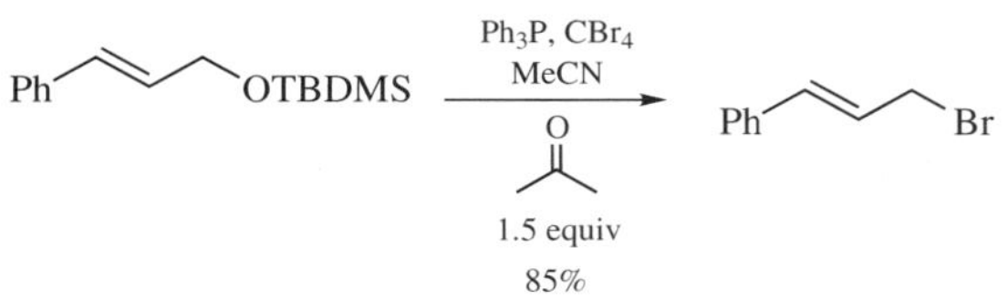
(5)

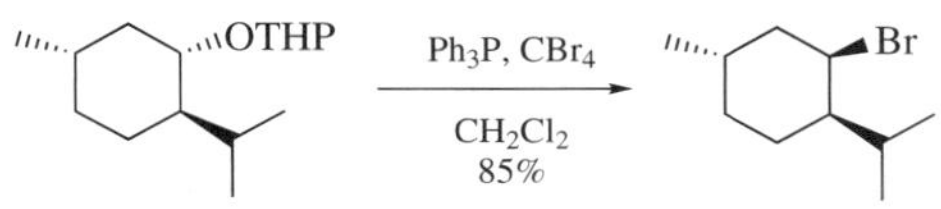
(6)

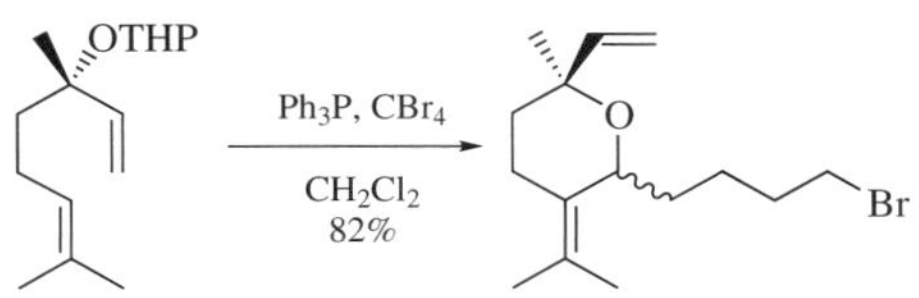
(7)

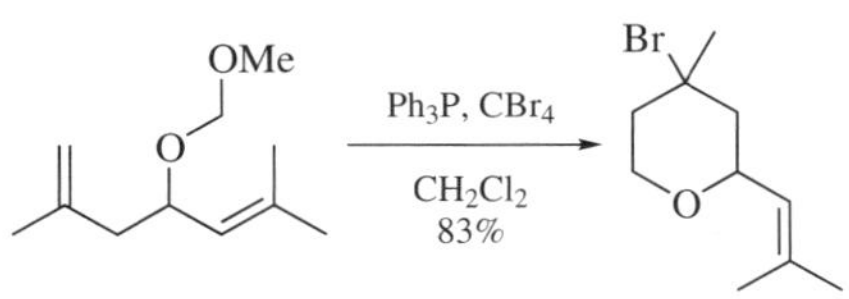
(8)

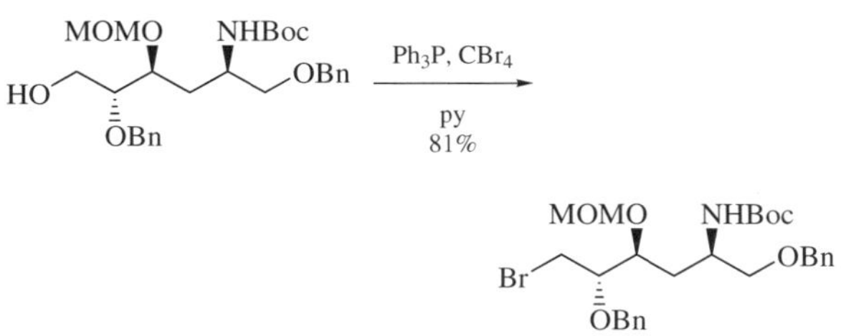
(9)

Amides from Carboxylic Acids. *N*-Methoxy-*N*-methyl amides can be prepared from carboxylic acids and the amine hydrochlorides. In the case of an α-phenyl carboxylic acid (eq 10), the amide is formed in 71% yield and no racemization is detected.[11]

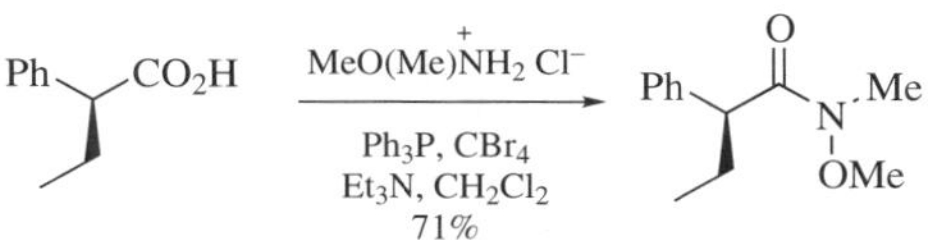
(10)

Dibromoalkenes from Aldehydes and Ketones. Benzaldehyde is transformed into the dibromoalkene in 84% yield when treated with Ph_3P and CBr_4 (eq 11).[12] An alternative procedure for the conversion of an aldehyde to the dibromoalkene uses Zinc dust in place of an excess of the phosphine. This allows the amount of Ph_3P and CBr_4 to be reduced to 2 equiv each as opposed to 4 equiv. This procedure gives comparable results to the original procedure. The dibromoalkenes can be reacted with n-Butyllithium to form the intermediate lithium acetylide. The acetylides can then be reacted with electrophiles such as H_2O (eq 12) and CO_2 (eq 13). This offers a convenient method for the formyl to ethynyl conversion.[13]

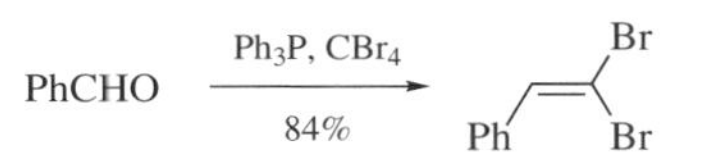
(11)

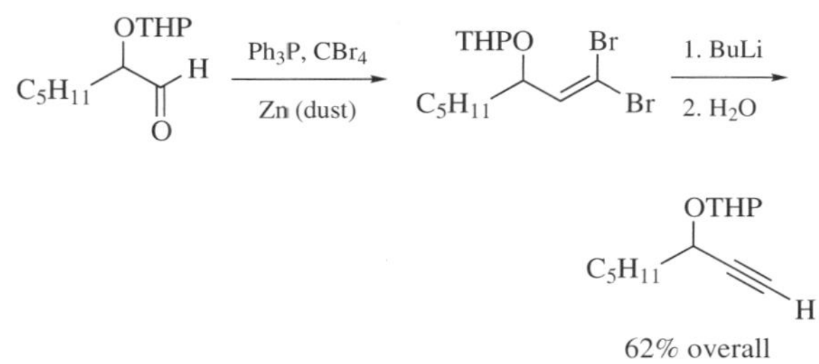
(12)

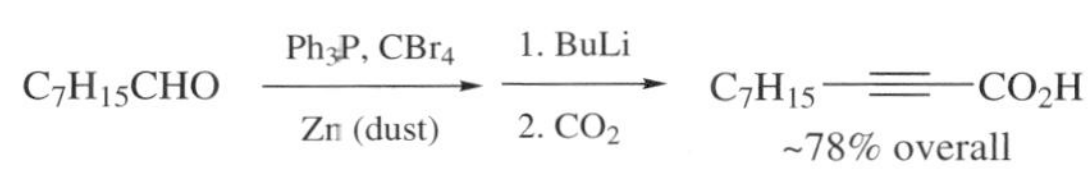
(13)

Ketones are converted to dibromomethylene derivatives. These intermediates can be transformed to isopropylidene compounds by reaction with Lithium Dimethylcuprate and Iodomethane (eq 14).[14] No racemization was reported for the chain extension of the aldehyde derived from (*S*)-ethyl lactate under the reaction conditions (eq 15).[15]

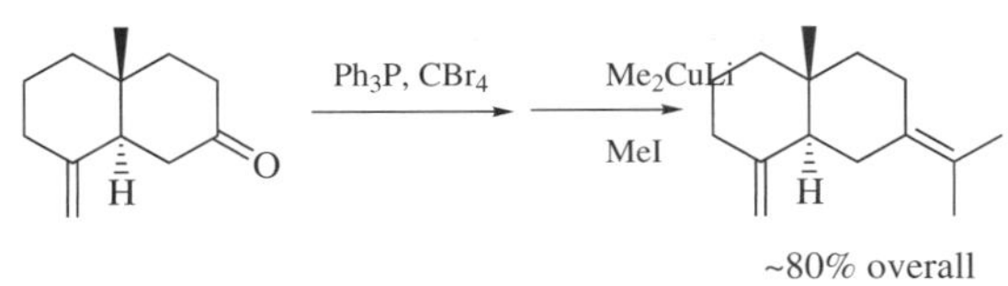
(14)

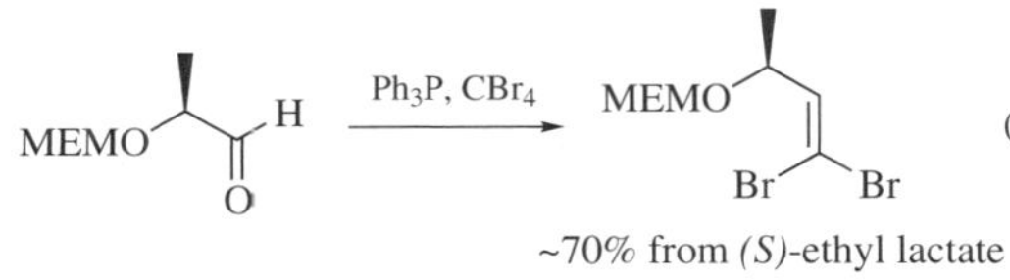
(15)

β-Bromo Enones from 1,3-Diketones. The reaction of Ph_3P and CBr_4 with a 1,3-diketone efficiently converts it to the β-bromo enone (eq 16).[16]

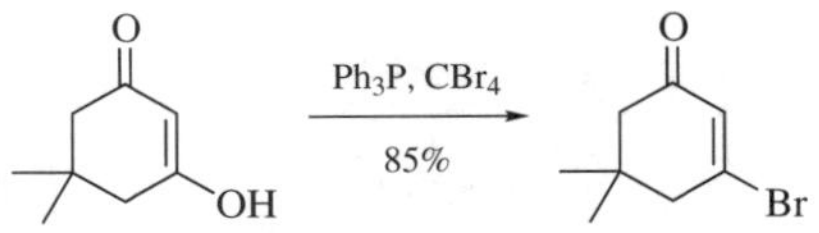

(16)

1-Bromoalkynes from Terminal Alkynes. Terminal alkynes on reaction with Ph_3P and CBr_4 afford 1-bromoalkynes in high yield (eq 17).[17]

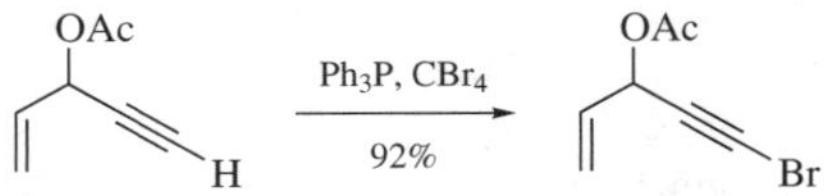

(17)

1. Castro, B. R. *Org. React.* **1983**, *29*, 1.
2. Verheyden, J. P. H.; Moffatt, J. G. *J. Org. Chem.* **1972**, *37*, 2289.
3. Axelrod, E. H.; Milne, G. M.; van Tamelen, E. E. *J. Am. Chem. Soc.* **1970**, *92*, 2139.
4. Kang, S. H.; Hong, C. Y. *Tetrahedron Lett.* **1987**, *28*, 675.
5. Kashem, A.; Anisuzzaman, M.; Whistler, R. L. *C.R. Hebd. Seances Acad. Sci.* **1978**, *61*, 511.
6. Weiss, R. G.; Snyder, E. I. *J. Org. Chem.* **1971**, *36*, 403.
7. Mattes, H.; Benezra, C. *Tetrahedron Lett.* **1987**, *28*, 1697.
8. Wagner, A.; Heitz, M.-P.; Mioskowski, C. *Tetrahedron Lett.* **1989**, *30*, 557.
9. Wagner, A.; Heitz, M.-P.; Mioskowski, C. *Tetrahedron Lett.* **1989**, *30*, 1971.
10. Clinch, K.; Vasella, A.; Schauer, R. *Tetrahedron Lett.* **1987**, *28*, 6425.
11. Einhorn, J.; Einhorn, C.; Luche, J.-L. *Synth. Commun.* **1990**, *20*, 1105.
12. Ramirez, F.; Desai, N. B.; McKelvie, N. *J. Am. Chem. Soc.* **1962**, *84*, 1745.
13. Corey, E. J.; Fuchs, P. L. *Tetrahedron Lett.* **1972**, 3769.
14. Posner, G. H.; Loomis, G. L.; Sawaya, H. S. *Tetrahedron Lett.* **1975**, 1373.
15. Mahler, H.; Braun, M. *Tetrahedron Lett.* **1987**, *28*, 5145.
16. Gruber, L.; Tömösközi, I.; Radics, L. *Synthesis* **1975**, 708.
17. Wagner, A.; Heitz, M.-P.; Mioskowski, C. *Tetrahedron Lett.* **1990**, *31*, 3141.

Michael J. Taschner
The University of Akron, OH, USA

Triphenylphosphine Carbon Tetrachloride[1]

Ph_3P-CCl_4

(Ph_3P)		
[605-35-0]	$C_{18}H_{15}P$	(MW 262.30)) (CCl_4)
[56-23-5]	CCl_4	(MW 153.81))

(reagent combination for the conversion of a number of functional groups into their corresponding chlorides and for dehydrations)

Physical Data: Ph_3P: mp 79–81 °C; bp 377 °C; *d* 1.0749 g cm^{-3}. CCl_4: mp −23 °C; bp 77 °C; *d* 1.594 g cm^{-3}.
Solubility: sol CCl_4, MeCN, CH_2Cl_2, 1,2-dichloroethane.
Preparative Method: reactive intermediates are generated in situ by reaction of Ph_3P and CCl_4.
Handling, Storage, and Precautions: Ph_3P is an irritant; CCl_4 is toxic and a cancer suspect agent; use in a fume hood. Solvents must be carefully dried because the intermediates are all susceptible to hydrolysis.

Combination of Triphenylphosphine and Carbon Tetrachloride. This reagent combination is capable of performing a range of chlorinations and dehydrations. The reactions are typically run using the so-called two-component or three-component systems. Carbon tetrachloride can function as both the reagent and solvent. However, the rates of the reactions are highly solvent-dependent, with MeCN providing the fastest rates.[1]

Conversion of Alcohols to Alkyl Chlorides. The reaction of alcohols with Triphenylphosphine and carbon tetrachloride results in the formation of alkyl chlorides.[2] The mild, neutral conditions allow for the efficient conversion of even sensitive alcohols into the corresponding chlorides (eqs 1 and 2).[3,4] The reaction typically proceeds with inversion of configuration.[5] In eq 3, it is interesting to note that the reaction not only proceeds with inversion of configuration but also no acyloxy migration is observed.[6]

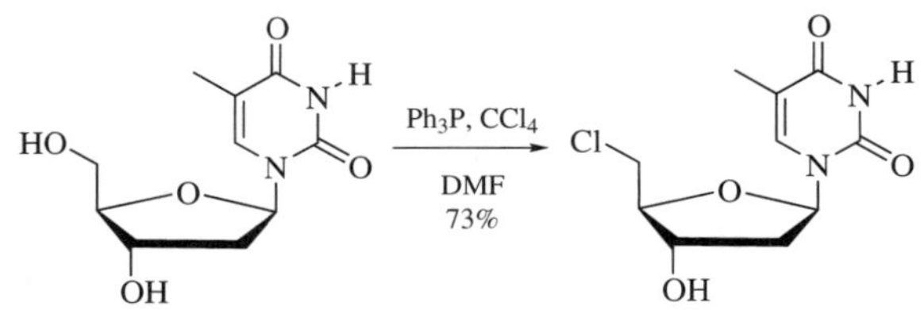

(1)

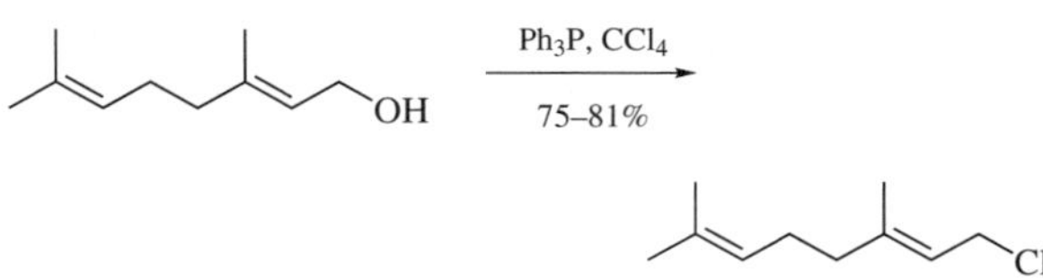

(2)

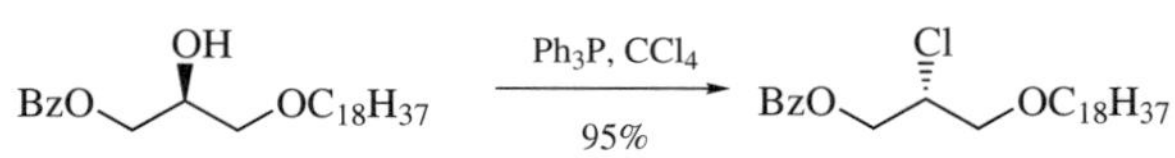

(3)

The conversion of an allylic alcohol to an allylic chloride occurs with no or minimal allylic rearrangement (eqs 4 and 5).[7] For the synthesis of low boiling allylic alcohols, it is advantageous to substitute hexachloroacetone (HCA) for CCl_4 (eq 6). The stereochemical integrity of the double bond also remains intact under these conditions.[8]

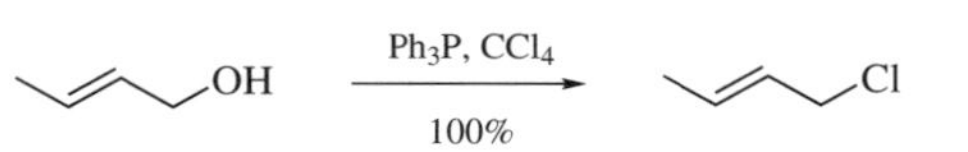

(4)

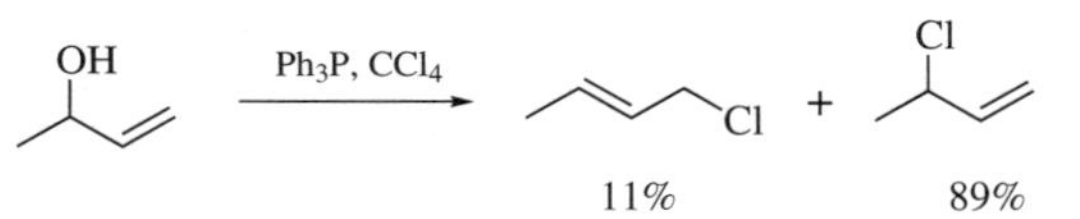

(5)

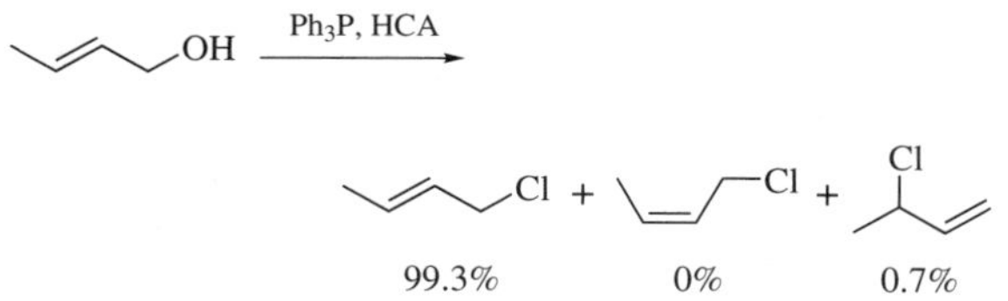
(6)

If the conversion of the alcohol to the chloride is attempted in refluxing MeCN, dehydration to form the alkene occurs (eqs 7 and 8).[9] Occasionally the separation of the product from the triphenylphosphine oxide produced in the reaction can be problematic. This can be overcome by using a polymer-supported phosphine.[10] Simple filtration and evaporation of the solvent are all that is required under these conditions. Not only is the workup facilitated, but the rate of the reaction is also increased by employing the supported reagent.[10c]

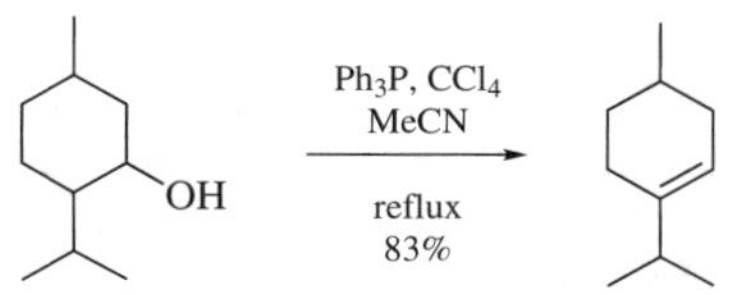
(7)

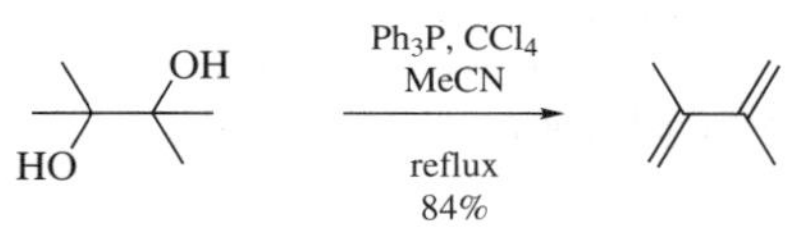
(8)

Conversion of Acids to Acid Chlorides. The reaction of carboxylic acids with triphenylphosphine–CCl_4 reportedly produces acid chlorides in good yield under mild conditions (eq 9).[11] These conditions will allow acid sensitive functional groups to survive. Phosphoric mono- and diesters are successfully converted into the phosphoric monoester dichlorides and diester chlorides, respectively. The reaction of the diethyl ester does not produce the acid chloride. Instead, the anhydride is formed. Phosphinic acid chlorides can also be prepared from the corresponding phosphinic acid under these conditions (eq 10).[12]

(9)

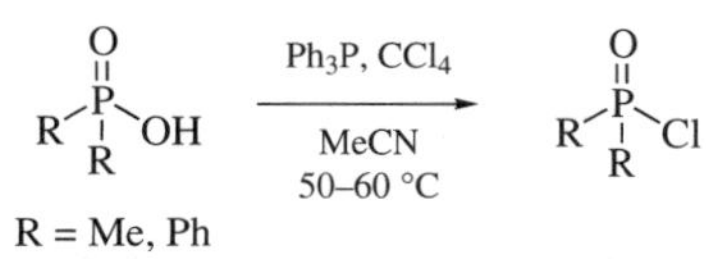
(10)

Epoxides to *cis*-1,2-Dichloroalkanes. Epoxides are converted into *cis*-1,2-dichlorides by the action of triphenylphosphine in refluxing CCl_4.[13] Cyclohexene oxide forms *cis*-1,2-dichlorocyclohexane (eq 11) contaminated by a trace of the *trans*-isomer. Cyclopentene oxide gives only the *cis*-isomer.

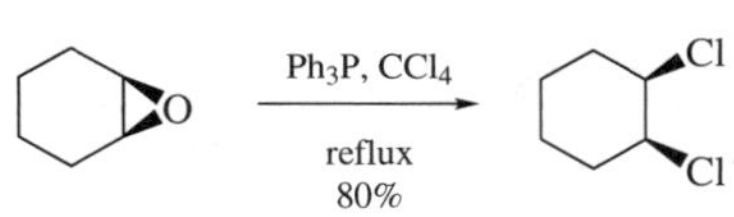
(11)

Dehydrations. Diols may be cyclodehydrated to the corresponding cyclic ethers. The reaction is most effective for 1,4-diols (eq 12). Dehydration of 1,3- and 1,5-diols is not as successful, except in the case of the configurationally constrained 1,5-diol shown in eq 13.[14] The reaction of *trans*-1,2-cyclohexanediol with the reagent affords *trans*-2-chlorocyclohexanol with none of the *cis*-isomer or the *trans*-dichloride being detected. If the reaction is performed in the presence of K_2CO_3 as an HCl scavenger, the epoxide is formed (eq 14).[14] The yields are not as good with substituted acyclic diols.[15]

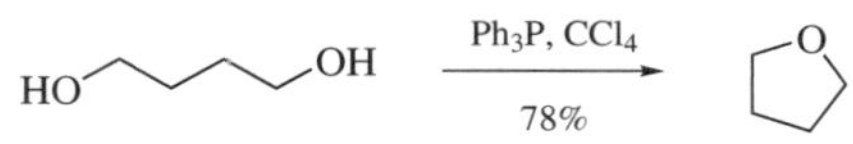
(12)

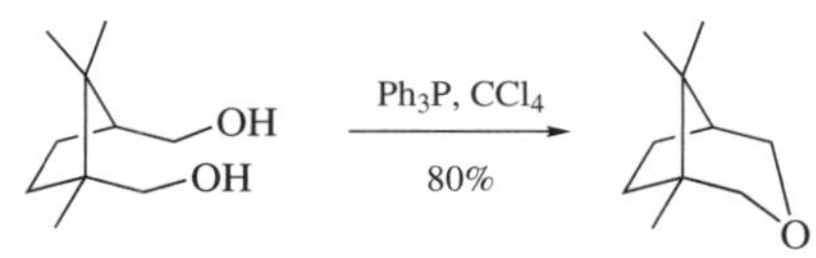
(13)

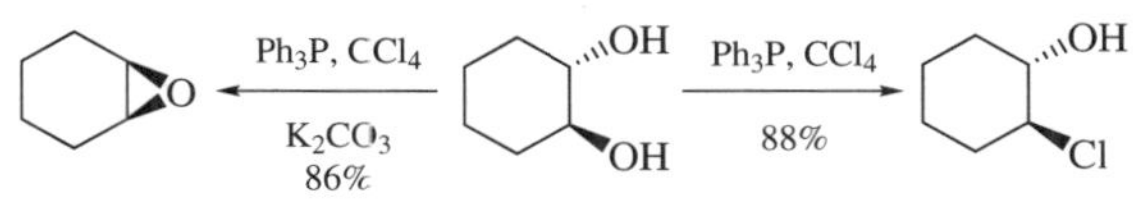
(14)

Dehydration of *N*-substituted β-amino alcohols with triphenylphosphine–CCl_4 and Triethylamine produces aziridines in good yield (eq 15).[16] This reaction has been successfully employed in the preparation of stable arene imines (eq 16).[17] Azetidines can be obtained from the corresponding 3-aminoalkanols (eq 17).[18] Additionally, reaction of 2-(3-hydroxypropyl)piperidine under these conditions yields octahydroindolizine (eq 18).

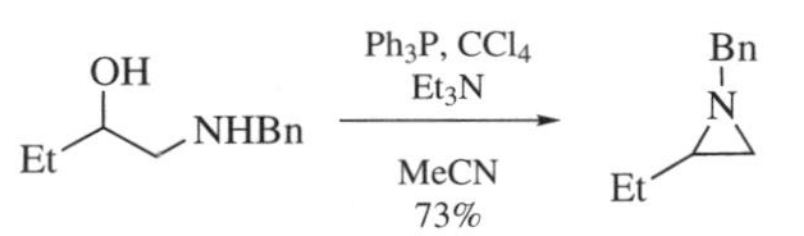
(15)

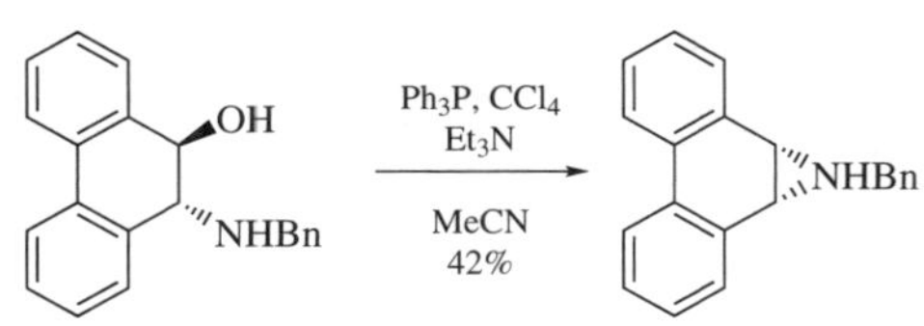
(16)

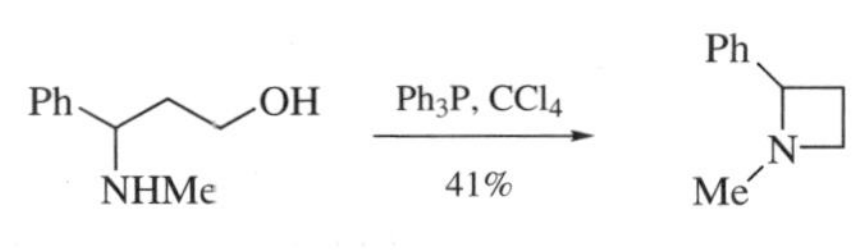
(17)

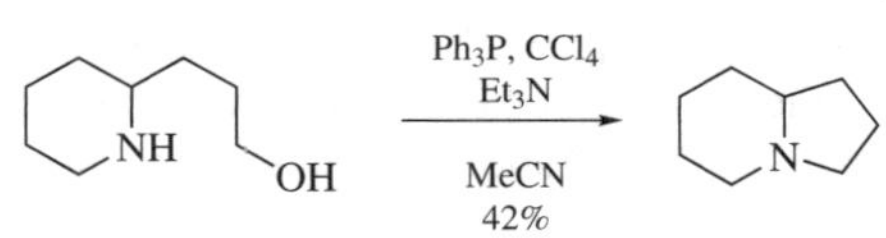
(18)

Substituted hydroxamic acids successfully cyclize to form β-lactams as long as Et_3N is present (eq 19). In the absence of the base, complex mixtures are formed.[19] Unsubstituted amides can be converted into nitriles via dehydration (eq 20).[20] This is the reagent of choice for the transformation of the amide to the nitrile in eq 21.[21]

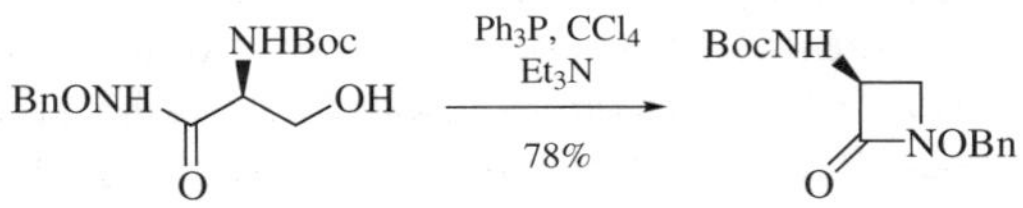
(19)

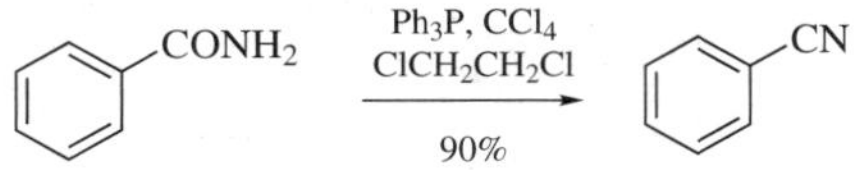
(20)

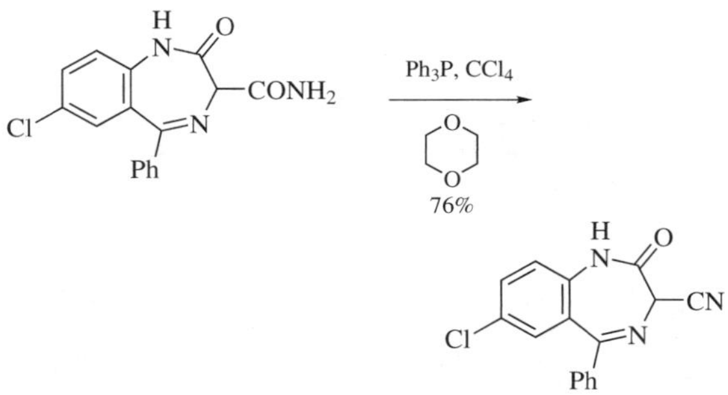
(21)

Nitriles can also be obtained from aldoximes using this reagent (eq 22).[22] Ketoximes produce imidoyl chlorides via a Beckmann rearrangement under these conditions (eq 23).[23] Imidoyl chlorides are also available by the reaction of monosubstituted amides with Ph_3P–CCl_4 in acetonitrile (eq 24).[24]

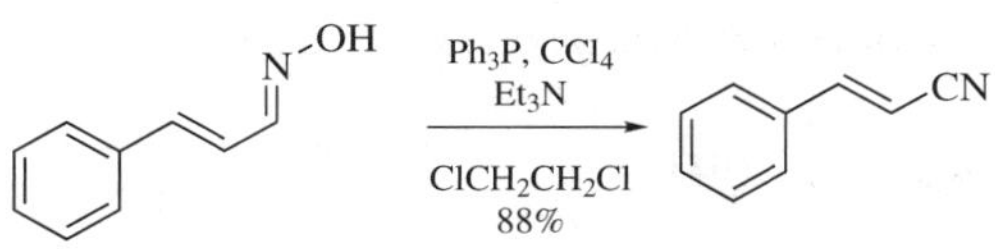
(22)

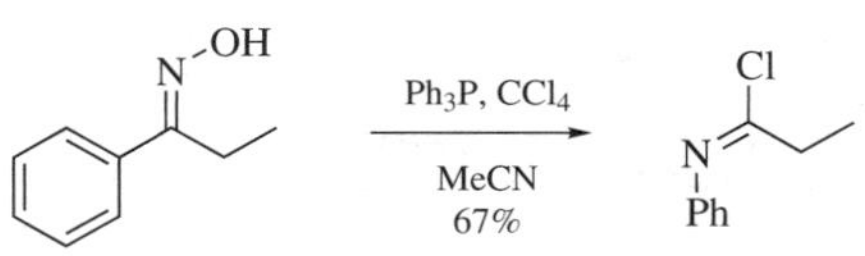
(23)

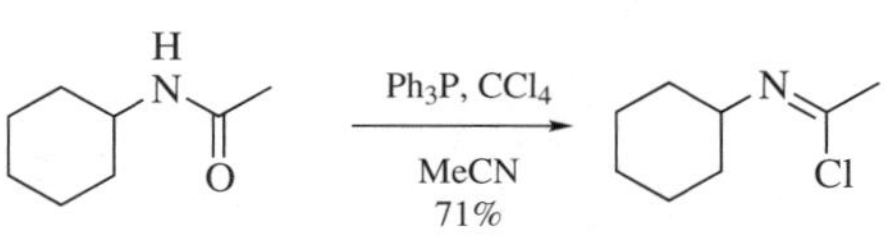
(24)

N,N,N′-Trisubstituted ureas afford chloroformamidine derivatives (eq 25),[25] while *N,N′*-disubstituted ureas and thioureas produce carbodiimides (eq 26).[26] When carbamoyl chlorides are treated with the reagent in MeCN, they are converted into isocyanates (eq 27).[27] Dehydration of *N*-substituted formamides provides access to isocyanides (eq 28).[28]

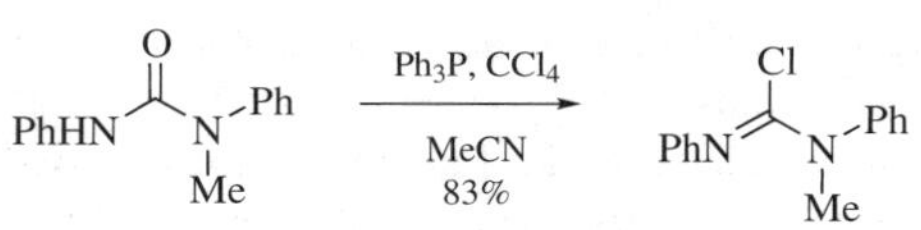
(25)

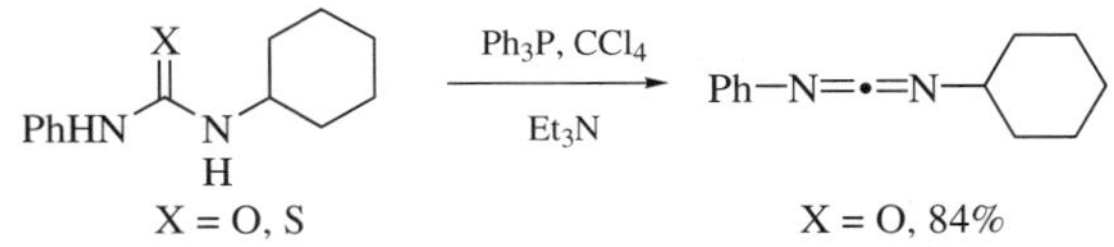
(26)

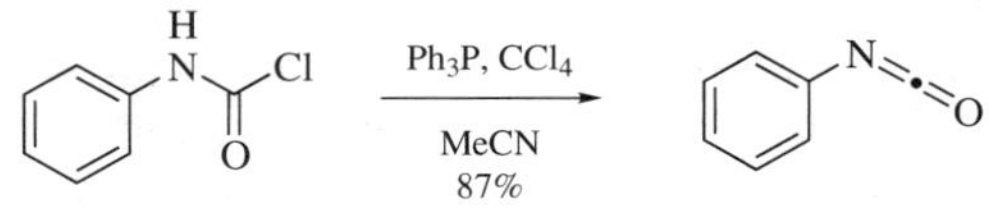
(27)

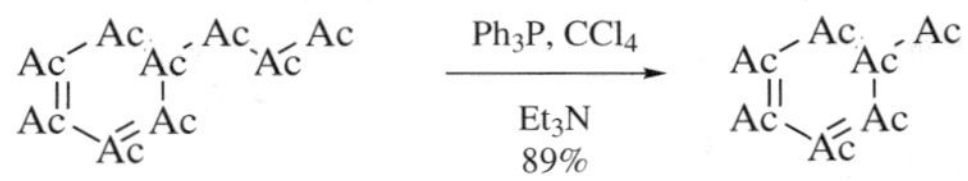
(28)

Amide Formation. The synthesis of an amide can be accomplished by initial reaction of an acid with Ph_3P–CCl_4 and then reaction of the intermediate with 2 equiv of the appropriate amine. A tertiary amine, such as Diisopropylethylamine, can be employed as the HCl scavenger in cases where one would not want to waste any of a potentially valuable amine.[29] This method has been used in the construction of an amide in the synthesis of the skeleton of the lycorine alkaloids (eq 29).[30]

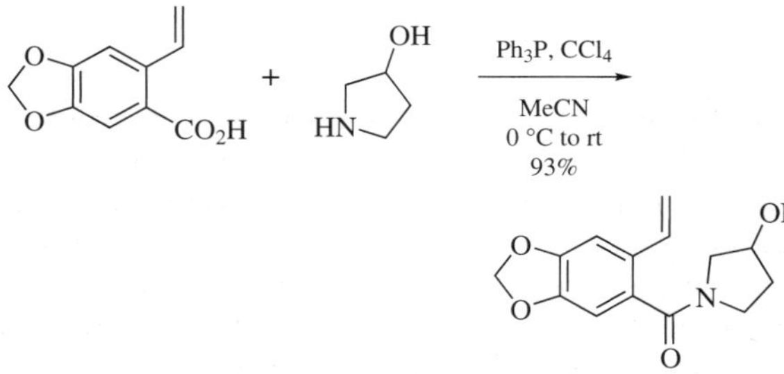
(29)

The method is effective for peptide coupling in that the yields are typically good; however, the reaction is often accompanied by racemization of the product.[31] The racemization problem can be suppressed, but this involves a change in the phosphine employed[32] or slightly modified reaction conditions. These modified conditions have been used successfully in the construction of a hexapeptide without racemization (eq 30).[33]

Cbz–Leu–Ala–Val–Phe–Gly–OH + Pro–OBn $\xrightarrow[\text{Et}_3\text{N, 68\%}]{\text{Ph}_3\text{P, CCl}_4}$ Cbz–Leu–Ala–Val–Phe–Gly–Pro–OBn (30)

If amino alcohols, amino thiols, or diamines are used, the intermediate amides cyclodehydrate under the reaction conditions to form Δ^2-oxazolines (eq 31), Δ^2-oxazines, Δ^2-thiazolines, or Δ^2-imidazolines.[33] The reaction requires the use of 3 equiv of the Ph_3P–CCl_4 reagent. The reaction is reported to fail if a commercial sample of the polymer-supported phosphine is used instead of triphenylphosphine.

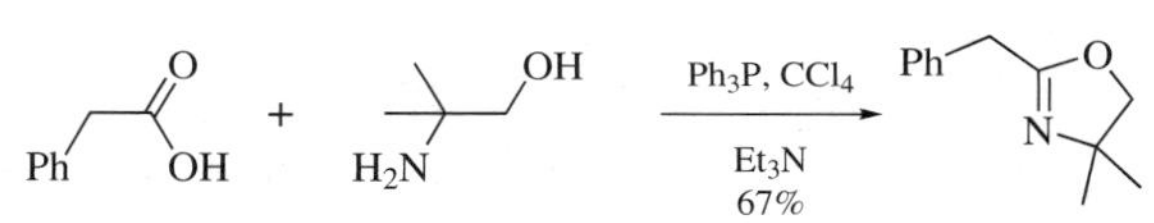
(31)

1,1-Dichloroalkenes and Vinyl Chlorides. The reaction of aldehydes produces a mixture of the 1,1-dichloroalkene and the dichloromethylene derivatives. These are the result of reaction of the in situ generated Dichloromethylenetriphenylphosphorane and Triphenylphosphine Dichloride, respectively. Benzaldehyde (eq 32) produces a 1:1 mixture of the dichloroalkene and benzal chloride in 72% yield.[34]

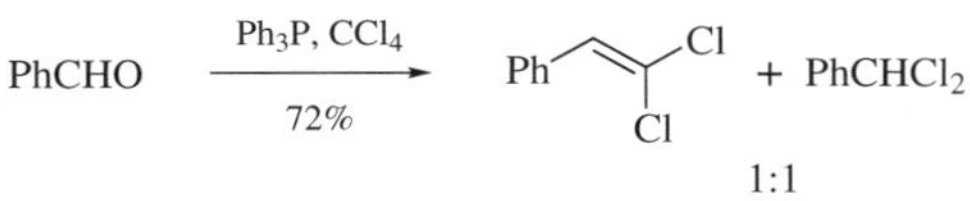

(32)

Ketones can also be used in this transformation, but sometimes enolizable ketones lead to the formation of the vinyl chloride derived from the enol. This is usually more of a problem for six-membered ring ketones than for five-membered ring ketones. Cyclopentanone yields predominantly the 1,1-dichloroalkene (eq 33), while cyclohexanone provides mainly the vinyl chloride (eq 34).[35]

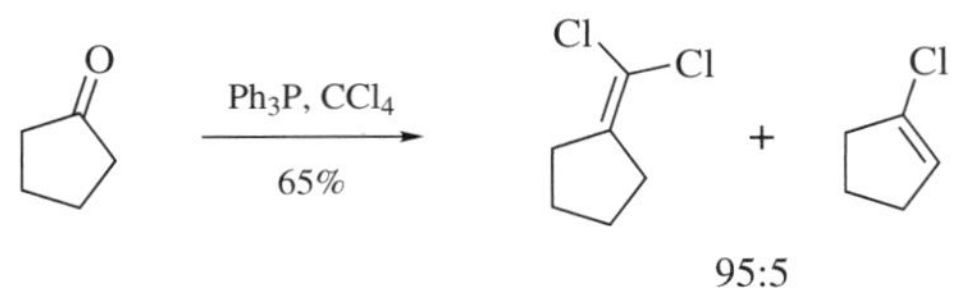

(33)

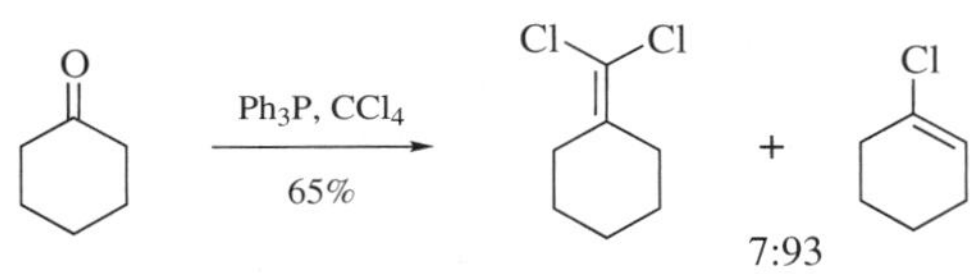

(34)

1. (a) Appel, R. *Angew. Chem., Int. Ed. Engl.* **1975**, *14*, 801. (b) Appel, R.; Halstenberg, M. *Organophosphorus Reagents in Organic Synthesis*; Cadogan, J. I. G., Ed.; Academic: New York, 1979; pp 387–431.
2. (a) Lee, J. B.; Nolan, T. J. *Can. J. Chem.* **1966**, *44*, 1331. (b) Lee, J. B.; Downie, I. W. *Tetrahedron* **1967**, *23*, 359.
3. Verheyden, J. P. H.; Moffat, J. G. *J. Org. Chem.* **1972**, *37*, 2289.
4. Calzada, J. G.; Hooz, J. *Org. Synth.* **1974**, *54*, 63.
5. Weiss, R. G.; Snyder, E. I. *J. Org. Chem.* **1970**, *35*, 1627.
6. Aneja, R.; Davies, A. P.; Knaggs, J. A. *Chem. Commun./J. Chem. Soc., Chem. Commun.* **1973**, 110.
7. Snyder, E. I. *J. Org. Chem.* **1972**, *37*, 1466.
8. Majid, R. M.; Fruchey, O. S.; Johnson, W. L. *Tetrahedron Lett.* **1977**, 2999.
9. Appel, R.; Wihler, H.-D. *Ber. Dtsch. Chem. Ges./Chem. Ber.* **1976**, *109*, 3446.
10. (a) Harrison, C. R.; Hodge, P. *Chem. Commun./J. Chem. Soc., Chem. Commun.* **1975**, 622. (b) Regen, S. L.; Lee, D. P. *J. Org. Chem.* **1975**, *40*, 1669. (c) Harrison, C. R.; Hodge, P. *Chem. Commun./J. Chem. Soc., Chem. Commun.* **1978**, 813.
11. Lee, J. B. *J. Am. Chem. Soc.* **1966**, *88*, 3440.
12. Appel, R.; Einig, H. *Z. Anorg. Allg. Chem.* **1975**, *414*, 236.
13. Isaacs, N. S.; Kirkpatrick, D. *Tetrahedron Lett.* **1972**, 3869.
14. Barry, C. N.; Evans, S. A. *J. Org. Chem.* **1981**, *46*, 3361.
15. Barry, C. N.; Evans, S. A. *Tetrahedron Lett.* **1983**, *24*, 661.
16. Appel, R.; Kleinstück, R. *Ber. Dtsch. Chem. Ges./Chem. Ber.* **1974**, *107*, 5.
17. Ittah, Y.; Shahak, I.; Blum, J. *J. Org. Chem.* **1978**, *43*, 397.
18. Stoilova, V.; Trifonov, L. S.; Orahovats, A. S. *Synthesis* **1979**, 105.
19. Miller, M. J.; Mattingly, P. G.; Morrison, M. A.; Kerwin, J. F. *J. Am. Chem. Soc.* **1980**, *102*, 7026.
20. (a) Yamato, E.; Sugasawa, S. *Tetrahedron Lett.* **1970**, 4383. (b) Appel, R.; Kleinstück, R.; Ziehn, K.-D. *Ber. Dtsch. Chem. Ges./Chem. Ber.* **1971**, *104*, 1030.
21. Juanin, R.; Arnold, W. *Helv. Chim. Acta* **1973**, *56*, 2569.
22. Appel, R.; Kohnke, J. *Ber. Dtsch. Chem. Ges./Chem. Ber.* **1971**, *104*, 2023.
23. Appel, R.; Warning, K. *Ber. Dtsch. Chem. Ges./Chem. Ber.* **1975**, *108*, 1437.
24. Appel, R.; Warning, K.; Ziehn, K.-D. *Ber. Dtsch. Chem. Ges./Chem. Ber.* **1973**, *106*, 3450.
25. (a) Appel, R.; Warning, K.; Ziehn, K.-D. *Ber. Dtsch. Chem. Ges./Chem. Ber.* **1974**, *107*, 698. (b) Appel, R.; Warning, K.; Ziehn, K.-D. *Ber. Dtsch. Chem. Ges./Chem. Ber.* **1973**, *106*, 2093.
26. Appel, R.; Warning, K.; Ziehn, K.-D. *Ber. Dtsch. Chem. Ges./Chem. Ber.* **1971**, *104*, 1335.
27. Appel, R.; Warning, K.; Ziehn, K.-D.; Gilak, A. *Ber. Dtsch. Chem. Ges./Chem. Ber.* **1974**, *107*, 2671.
28. Appel, R.; Kleinstück, R.; Ziehn, K.-D. *Angew. Chem., Int. Ed. Engl.* **1971**, *10*, 132.
29. Barstow, L. E.; Hruby, V. J. *J. Org. Chem.* **1971**, *36*, 1305.
30. Stork, G.; Morgans, D. J. *J. Am. Chem. Soc.* **1979**, *101*, 7110.
31. Wieland, T.; Seeliger, A. *Ber. Dtsch. Chem. Ges./Chem. Ber.* **1971**, *104*, 3992.
32. Takeuchi, Y.; Yamada, S. *Chem. Pharm. Bull.* **1974**, *22*, 832.
33. Appel, R.; Bäumer, G.; Strüver, W. *Ber. Dtsch. Chem. Ges./Chem. Ber.* **1975**, *108*, 2680.
34. Rabinowitz, R.; Marcus, R. *J. Am. Chem. Soc.* **1962**, *84*, 1312.
35. Isaacs, N.; Kirkpatrick, D. *Chem. Commun./J. Chem. Soc., Chem. Commun.* **1972**, 443.

Michael J. Taschner
The University of Akron, OH, USA

Triphenylphosphine Dibromide[1]

[1034-39-5] $C_{15}H_{18}Br_2P$ (MW 389.10))

(bromination of alcohols,[3,4] phenols,[22] and enols;[3] cleavage of ethers[23,32,33] and acetals[25] to alkyl bromides; cyclization of amino alcohols[34,35] to cyclic amines; conversion of carboxylic acid derivatives into acyl bromides;[36,37] bromination[39] or dehydration[38] of carboxamide groups)

Alternate Names: bromotriphenylphosphonium bromide; dibromotriphenylphosphorane.

Physical Data: adduct:[2] colorless crystalline solid; mp 235 °C (dec). Ph_3P: mp 80.5 °C; bp 377 °C (in inert gas); $d^{25}{}_4$ 1.194 g cm^{-3}; $n^{80}{}_D$ 1.6358. Br_2: mp −7.2 °C; bp 59.5 °C; d25:4 3.12 g cm^{-3}.

Solubility: adduct: sol CH_2Cl_2, MeCN, PhCN, DMF; less sol PhH, PhCl. Ph_3P: sol ether, PhH, $CHCl_3$, AcOH; less sol alcohol; pract. insol. H_2O. Br_2: sol H_2O (3.58 g/100 mL); very sol. alcohol, ether, $CHCl_3$, CCl_4, CS_2.

Form Supplied in: adduct: hygroscopic solid; commercially available; purity 95–96%. Both parent compounds, Ph_3P and Br_2, are widely available. Ph_3P: odorless platelets or prisms; purity ~99%; typical impurity Ph_3PO ~1%. Br_2: dark reddish liquid, volatile, with suffocating odor; vaporizes rapidly at rt; purity >99.5%.

Analysis of Reagent Purity: adduct: ^{31}P NMR (CH_2Cl_2) +49 ppm (ionic form);[2] Raman (solid phase) 239 cm^{-1} (P–Br). Typical impurities, Ph_3P and Ph_3PO: ^{31}P NMR (various solvents) −5 to −8 ppm and +25 to +29 ppm, respectively.

Preparative Method: several preparations are described,[2,23b,32b] but in most cases the hygroscopic reagent is prepared just before use by addition of Bromine to Triphenylphosphine in a dry solvent with cooling.

Handling, Storage, and Precautions: adduct: corrosive; very hygroscopic; moisture sensitive; incompatible with strong oxidizing agents and strong bases; may decompose on exposure to moist air or water; do not get in eyes, on skin, or on clothing; keep containers tightly closed and store in a cool dry place; these reagents should be handled in a fume hood.

Introduction. The Ph_3PBr_2 adduct was first used in synthesis in 1959[3] for the preparation of alkyl and acyl bromides from alcohols and carboxylic acids, and for the dehydration of amides and oximes to nitriles. Since then it has been widely employed as a versatile reagent for a number of synthetic reactions.

Conversion of Alcohols into Alkyl Bromides. It has been established that Ph_3PX_2 reagents present considerable advantages over phosphorus halides (i.e. PX_3, PX_5) for which only two replaceable groups would be 'necessary and desirable'.[4a] The mechanism[4b] of the reaction of Ph_3PBr_2 with an alcohol involves initial rapid formation of an alkyloxyphosphonium bromide,[4,5,7a] which then collapses slowly to the phosphine oxide and an alkyl bromide (eq 1). S_N2 substitution and high yields are generally observed with primary and secondary alcohols.[3,4a] This reagent shows little tendency to induce either carbonium ion rearrangement in the carbon skeleton or elimination reactions; this is illustrated by the preparation of neopentyl bromide from neopentyl alcohol (eq 2)[4a] and in the steroid field,[6] in which cholesterol and 3β-cholestanol give respectively 3β-bromocholest-5-ene and 3α-bromocholestane in yields of about 80%. Inversion of configuration is found in the S_N2 reactions of (+)-*endo*-norbornanol (eq 3)[7a,8] and 7-norbornanol. However, a similar reaction with (−)-*endo*-norbornanol leads mainly to racemic bromide via a nonclassical norbornyl cation intermediate.[7b] Likewise, little rearrangement is observed in the bromination of 3-methyl-2-butanol (eq 4).[9] Greater amounts of rearranged product are formed in the same reaction with Ph_3PCNBr (3%), Bromine–Triphenyl Phosphite (55.6%), or Triphenylphosphine–Carbon Tetrabromide (6–10%).

$$\mathrm{ROH} \xrightarrow[\text{fast}]{Ph_3PBr_2} \mathrm{R{-}O{-}\overset{+}{P}Ph_3\ Br^-} \xrightarrow{\text{slow}} \mathrm{RBr + OPPh_3} \quad (1)$$

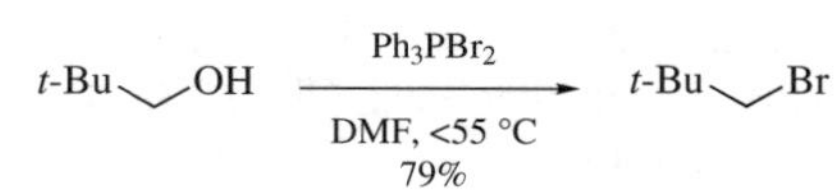

(2)

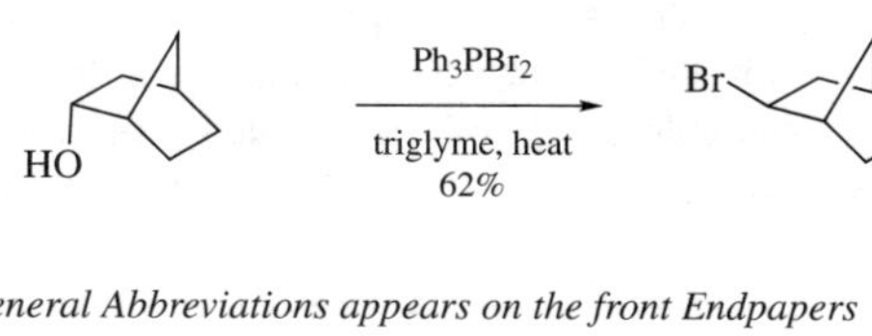

(3)

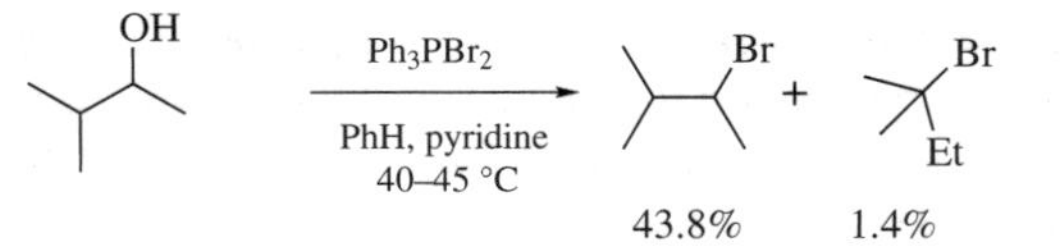

(4)

Ph_3PBr_2 can be used with sensitive substrates which incorporate cyclopropyl rings (e.g. cyclopropyl carbinol[11,12]) or unsaturation[10,13] (e.g. cinnamyl alcohol and alkynediols) with little if any side reactions. Cyclopropyl carbinols are transformed into bromides in good yield without ring opening or ring expansion, with the formation of only trace amounts of homoallylic bromides; no bromocyclobutane derivatives are detected (eq 5). Ph_3PBr_2 also proves superior to Phosphorus(III) Bromide for the bromination of alkynediols to dibromoalkynes (eq 6).[13]

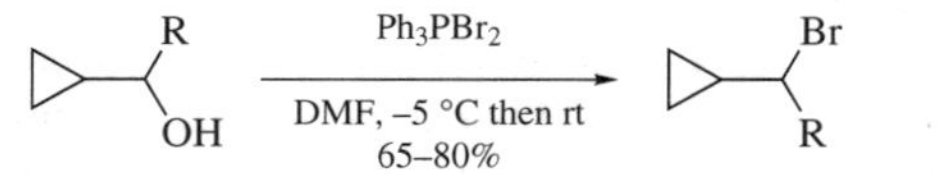

(5)

R = H, Me, Et, Pr, Bu, *i*-Pr, *t*-Bu

$$\mathrm{HO{-}CR_2{-}(C{\equiv}C)_n{-}CR_2{-}OH} \xrightarrow[\text{MeCN, 40–50 °C; 78–92\%}]{Ph_3PBr_2} \mathrm{Br{-}CR_2{-}(C{\equiv}C)_n{-}CR_2{-}Br} \quad (6)$$

R = H, Me; *n* = 1, 2

Ph_3PBr_2 provides comparable yields to Ph_3P/CBr_4 in the bromination of diethyl 1-hydroxyalkylphosphonates.[14] A number of other primary,[15] secondary,[11–14] and even tertiary alcohols are brominated with Ph_3PBr_2, but with preponderant elimination[4a,6,16] in the latter case. A polymer-supported Ph_3PBr_2 reagent[17] has been successfully applied to alcohol bromination in refluxing $CHCl_3$; product isolation is facilited by the ready separation of supported from the nonsupported species. However, note that a polymer-supported Ph_3P/CBr_4 reagent leads to better bromination results than the Ph_3PBr_2 system in a number of cases.

Similar mono- and dibrominations have also been achieved in the carbohydrate field (eq 7).[18] Configurational inversion at chiral centers is observed throughout. Ph_3PBr_2/ Imidazole (ImH) or Ph_3P/2,4,5-tribromoimidazole/ImH combinations in refluxing PhMe or in PhMe–pyridine mixtures are preferred in some cases. In the nucleoside field,[19] double activation is achieved with Ph_3PBr_2 in pyridine, leading both to bromination of the 5′-position of the sugar moiety and to substitution on the C-6 position of the adenine part (eq 8).[19b]

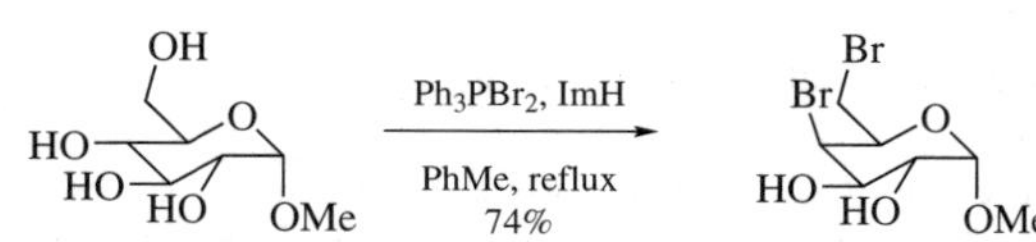

(7)

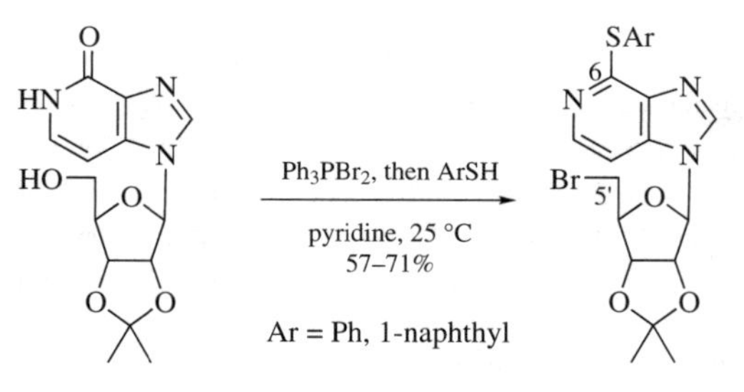

(8)

Ar = Ph, 1-naphthyl

A variety of solvents such as CCl_4, PhH, triglyme, $CHCl_3$, PhMe, pyridine, NMP, MeCN, DMF, and mixtures of these have been employed in these bromination reactions, depending on substrate solubility. Both *O*- and *C*-formylations are encountered in some cases when DMF is the solvent. Ph_3PBr_2 and both Ph_3PX_2 analogs (X = Cl, I) can react with DMF according to the sequence in eq 9.[20] The intermediate alkoxyimonium bromide (**1**)(**2**)(**3**)(**4**) affords either the expected bromide by heating, or the formyl ester by hydrolytic workup with secondary[20] or tertiary[16] alcohols (e.g. eq 10). Vilsmeier-type formylation reactions on nucleophilic carbon atoms can also take place, as developed early[38a] and then encountered later as a side reaction in the steroid field.[21] Such *O*-formylation of secondary alcohols is interestingly used in the differentiation of diols, leading to a new synthesis of bromo formates (eq 10).[20b] Under mild conditions, primary alcohols still afford bromides whereas secondary ones are converted almost exclusively to the corresponding formate esters.

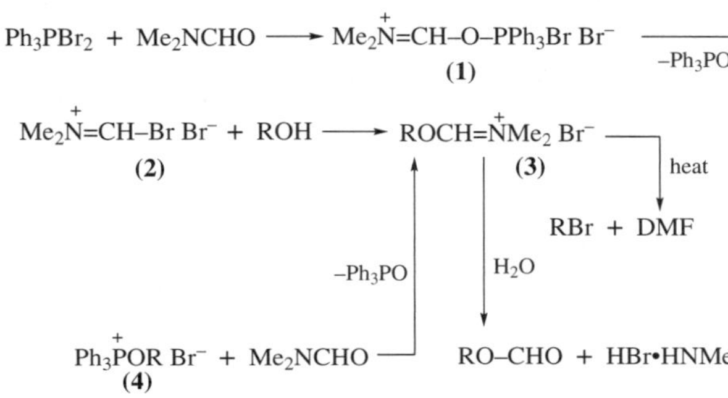

(9)

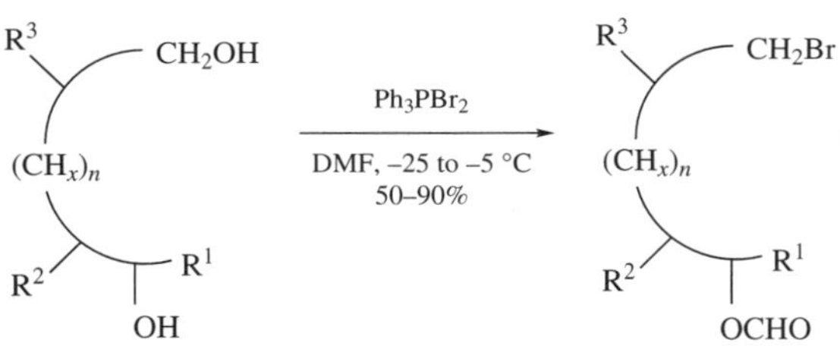

(10)

$x = 1, 2$; $R^1 \neq H$; $R^2, R^3 = H, Me, (CH_x)_n$

Synthesis of Aryl Bromides from Phenols. Already developed in the pioneering work,[4a] the preparation of aryl bromides from phenols with Ph_3PBr_2 was later extended to a number of substrates (eq 11).[22]

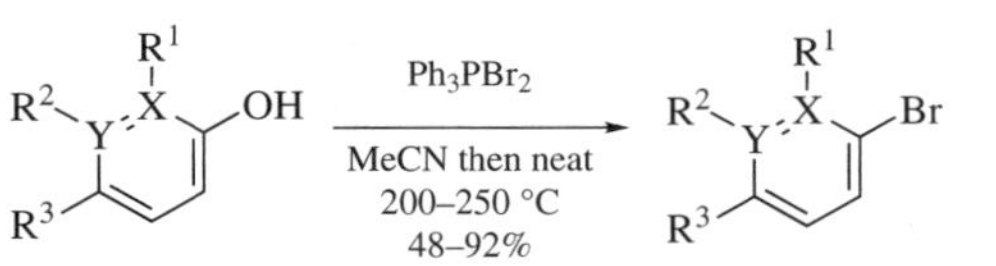

(11)

X, Y = C, N; $R^2 = H$; $R^3 = H, Cl, NO_2, MeO$;
$R^1 + R^2 = Ph$; 2,3-pyridyl; $R^2 + R^3 = Ph$, 3,4-phenol

Alkyl Bromides from Ethers and Acetals. Ph_3PBr_2 cleaves dialkyl ethers to give the two alkyl bromides under essentially neutral conditions.[23] This reaction offers obvious advantages as it avoids both the strongly acidic and basic media which are usually employed for ether cleavage. Brominations are generally conducted at reflux in high boiling solvents such as PhCl, PhCN, DMF, or NMP. Primary and secondary alkyl groups provide good yields of bromides without rearrangement in PhCl or PhCN, at 60 to 120 °C,[23a,b] as illustrated in eq 12.[23c] Phenyl alkyl ethers initially afford aryloxyphosphonium bromides which collapse to bromobenzene only by heating at higher temperature (>230 °C as seen before). Alkyl *t*-butyl ethers are cleaved more easily in DMF[23b] or in MeCN,[24] usually between 60 and 110 °C; the *t*-butyl group is converted into isobutene in this process. In a related reaction, endoxides are transformed into arenes[28] by Ph_3PBr_2 treatment in PhCl. Aromatization takes place via HBr elimination from the initially formed dibromide.

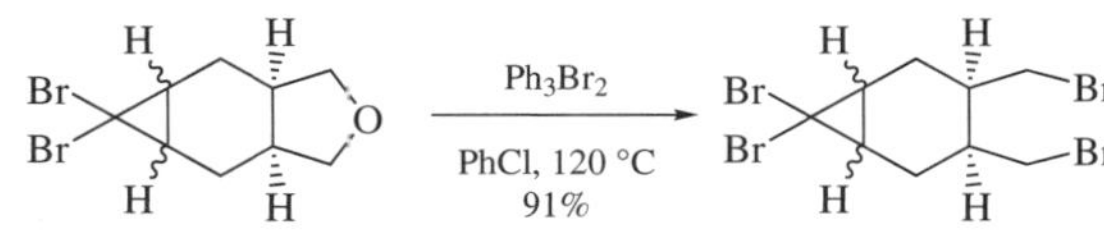

(12)

Direct conversion of tetrahydropyranyl (THP)[25] and tetrahydrofuranyl ethers[26] (THF) to bromides can be achieved by Ph_3PBr_2 treatment under milder conditions (eq 13).[25a] THP and THF ethers afford good yields of acyclic, saturated, and unsaturated,[25c] primary, secondary, and even tertiary alkyl bromides; under the same conditions, cyclohexyl THF ethers provide mainly cyclohexene derivatives. In view of the fact that the preceding reaction of THP ethers can be stopped (at low temperature) at the stage of either the alkoxyphosphonium (see eq 1) or the pentacovalent $ROPPh_3Br$ intermediate, hydrolysis of the reaction mixture at −50 °C, leads to the corresponding alcohols in good yield (74–97%) (eq 14).[29] The method is efficient for the deprotection of secondary and tertiary THP ethers, as well as for acyclic acetal ether, dialkyl acetals, and *O*-glucosides. Acetal functions can be removed with retention of stereochemistry, as illustrated in the conversion of (−)-menthol THP ether into (−)-menthol with full recovery of the optical activity. This procedure is not applicable to primary THP ethers, where the corresponding bromides are formed under these conditions.

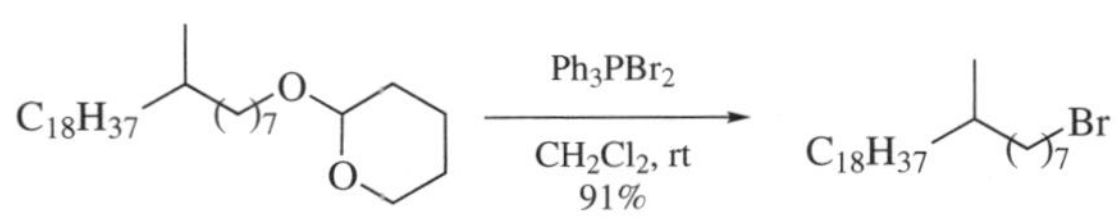

(13)

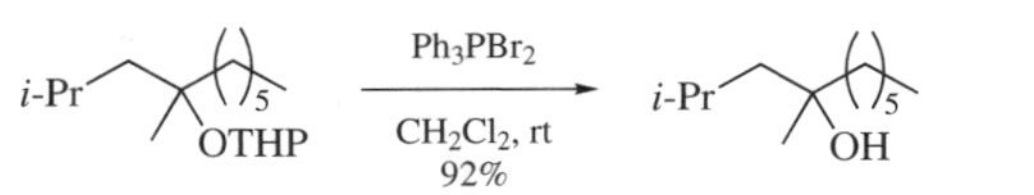

(14)

Reaction of hindered trialkylsilyl ethers with Ph_3PBr_2 in CH_2Cl_2 at rt affords primary and secondary alkyl bromides in excellent yield (70–94%); the reaction is valuable in the β-lactam field.[27a] The reaction rate is increased by addition of a catalytic amount of Zinc Bromide. Silylated enol ethers, such as trimethylsilyl(1-phenylvinyloxy)silane, provide vinyl bromides such as α-bromostyrene with Ph_3PBr_2 in refluxing CCl_4.[27b]

α-Alkynyl Ketones and β-Bromo-α-Vinyl Ketones from β-Diketones. The previously described bromination of alcohols can be applied to the enol form of β-diketones. The first attempts in this field were conducted with the Ph_3PBr_2/ Triethylamine system on dibenzoylmethane, leading to phenylethynyl phenyl ketone.[30a] Elimination at the oxyphosphonium stage or HBr elimination from the initially generated vinyl bromide is probably involved in this reaction. Analogously, reaction of unsymmetrical β-diketones[30b] with excess Ph_3PBr_2/Et_3N proceeds with the formation of 3:1 mixtures of α,β-ynones in good overall yield (eq 15). Ethynyltriphenylphosphonium bromides are similarly obtained[30c] starting from (benzoylmethylene)- and (carbamoylmethylene)triphenylphosphoranes.

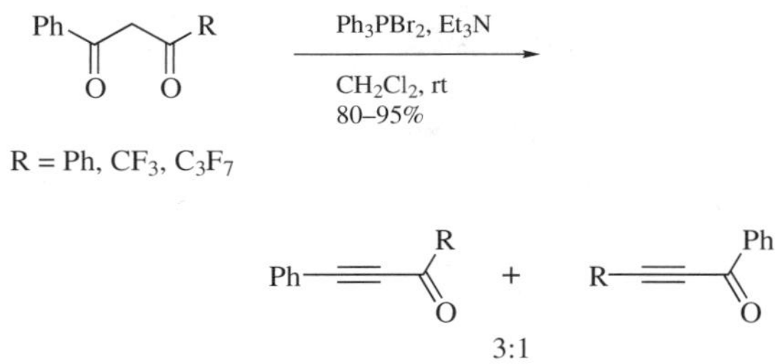

(15)

Similar treatment of unconjugated β-diketones[31] in PhH or MeCN as solvent leads to β-bromo-α,β-unsaturated ketones with superior results relative to the same transformation with PBr_3 as the brominating agent (eq 16).[31c] The reaction of acyclic 2,4-pentanedione with Ph_3PBr_2/Et_3N in PhH–MeCN[31c] or in CH_2Cl_2,[31d] but in the absence of Et_3N, appears not to be totally stereoselective, since a mixture of the geometrically isomeric bromo enones is produced, with 88% ($E/Z \sim 87/13$) and 85% ($E/Z \sim 93/7$) yields, respectively.

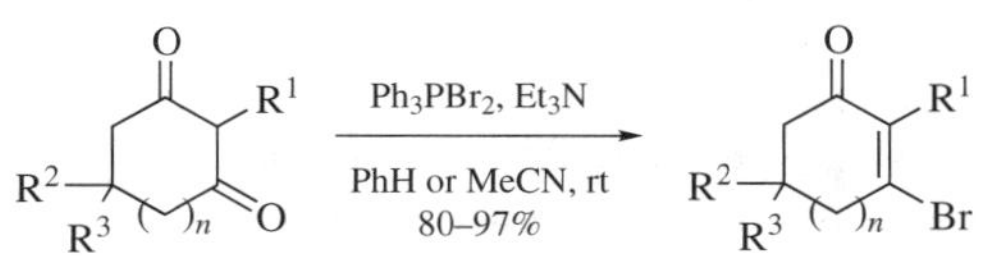

(16)

R^1 = H, Me, allyl, 2-cyanoethyl
n = 0, 1; R^2 = H, Me; R^3 = H, Me

Epoxide Opening to Vicinal Dibromides or Bromohydrins. Epoxide ring opening with Ph_3PBr_2[32a] takes place in MeCN or PhH at 20–50 °C, affording vicinal dibromides (32–74%); *cis–trans* isomer mixtures are obtained in the case of cycloalkene epoxides. This reaction involves initial cleavage of the epoxide C–O bond at the most substituted epoxide carbon; bromoalkoxyphosphonium salts are thus formed, and these undergo subsequent substitution to give dibromides and Ph_3PO. In further studies,[32b] reaction of *cis*-epoxides in PhH produced *erythro*-dibromides exclusively (eq 17); *trans*-epoxides exhibit less specificity, leading to a mixture of *threo*- and *erythro*-dibromides. Use of a more polar solvent such as CH_2Cl_2 or MeCN, instead of PhH, with *cis*-epoxides provides *erythro–threo* mixtures (60–40 to 50–50). By reacting epoxides first with HCl and then with Ph_3PBr_2, it is possible to obtain vicinal bromochlorides stereospecifically that are also products of two S_N2 displacements.

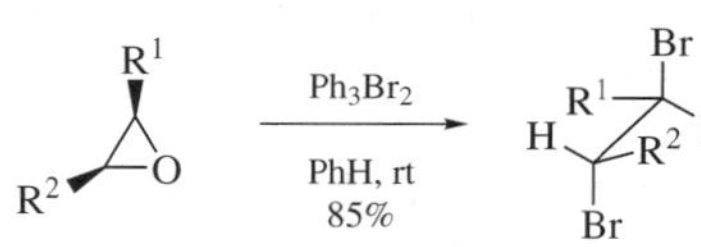

(17)

R^1 = nonyl; R^2 = Bu

The reaction can also be carried out so as to give bromohydrins. In the steroid field[33a] with conformationally rigid epoxides, oxirane cleavage appears to be quite stereoselective and leads only to the bromohydrin resulting from the usual *anti* opening of the ring in high yield (90–97%). Less hindered and rigid substrates afford regioisomeric mixtures of cyclic *trans*-bromohydrins. Equivalent results[33b] are obtained, in the steroid series, by use of polymeric Ph_3PBr_2 for the transformation of epoxides to bromohydrins under mild nonacidic conditions; oxirane ring opening remains regio- and stereoselective.

Cyclization of β- and γ-Amino Alcohols to Aziridines and Azetidines. β-Amino alcohols undergo cyclization upon Ph_3PBr_2/Et_3N treatment,[34] providing the corresponding aziridines along with substituted piperazines as dimeric byproducts (eq 18). Walden inversion is observed in the ring closure of both *threo*- and *erythro*-ephedrine and this supports the postulated 1,2-*trans*-elimination. Application of this procedure to the synthesis of 1-monosubstituted aziridines is unsuccessful, giving only the piperazine byproduct. *N*-Aryl γ-amino alcohols are cyclized to azetidines by Ph_3PBr_2/Et_3N in MeCN (eq 19).[35a] Under these conditions, some starting material is recovered unchanged along with mixed tetrahydroquinolines as side products and the expected *N*-arylazetidine. Comparable results are obtained (45–65%)[35b] with some other *N*-protecting groups.

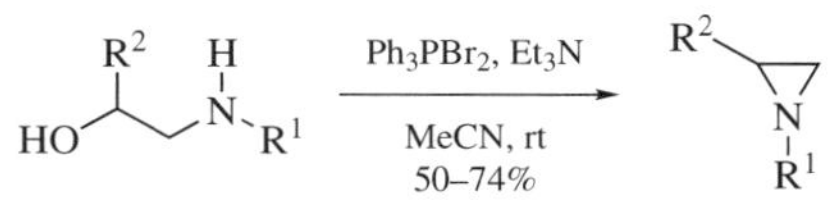

(18)

R^1 = Bu, *t*-Bu, cyclohexyl, benzyl; R^2 = Me, Ph

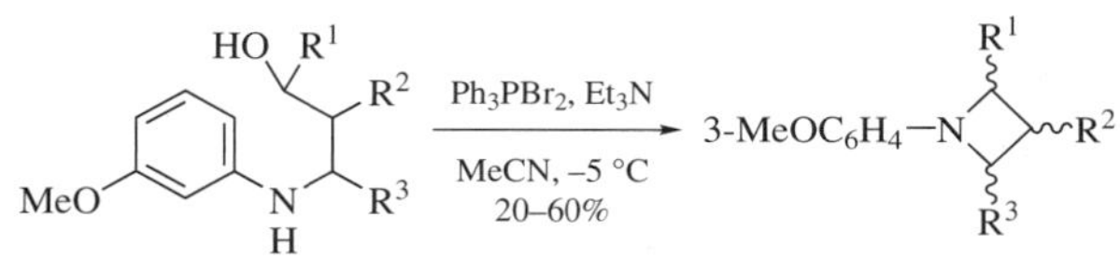

(19)

R^1 = Me, Ph; R^2 = H, Me; R^3 = H, Me, Ph

Acyl Bromides from Carboxylic Acids, Anhydrides, and Esters. Carboxylic acid bromides are prepared by reaction of Ph_3PBr_2 with various carboxylic acids and anhydrides in boiling PhCl.[36a] Milder conditions and better yields were later obtained in a comparative study[36b] by using CH_2Cl_2 at rt (eq 20). Further improvements are observed through the use of the corresponding trimethylsilyl esters; acyl bromides are thus obtained under mild and neutral conditions, allowing reactions with sensitive acid substrates without side reactions (eq 20). This transformation is also applicable to generate the acyl bromides of more hindered trialkylsilyl esters such as the TBDMS, TIPS, and TBDPS carboxylates; in CH_2Cl_2, the reaction rate increases in the presence of a catalytic amount of $ZnBr_2$.[27a]

$$RCO_2Z \xrightarrow{Ph_3PBr_2} RCOBr \qquad (20)$$

Z = H; PhCl, reflux	50–80%
Z = H; CH_2Cl_2, rt	64–90%
Z = TMS; CH_2Cl_2, rt	70–90%

R = Ph, 2-ClC_6H_4, 4-ClC_6H_4, Et, $PhCH_2CH_2$, PhCH=CH, $PhCH(CO_2Z)$, MeCH=CH

Direct reaction of less reactive alkyl esters and lactones[37] with Ph_3PBr_2 affords both acyl and alkyl bromides. This reaction is achieved at MeCN reflux[37a–c] for α-halogenated esters such as CF_3CO_2R, but only at higher temperatures and with longer reaction times for unhalogenated ones. The reaction of an ester with Ph_3PBr_2 presumably forms an oxonium salt, which then undergoes cleavage through reaction with bromide ion. Prolonged heating (80–110 °C) of lactones with Ph_3PBr_2 gives the expected acyl alkyl dibromides, but in low yield ($\leq$50%) due to a concomitant degradation.[37d]

Dehydration of Ureas and Amides to Carbodiimides, Nitriles, Isocyanides, and Ketenimines. Dehydration of *N,N′*-disubstituted ureas by heating with Ph_3PBr_2/Et_3N in PhH or PhCl at 70–80 °C affords carbodiimides[38a] in good yields. Subsequent improvements[38d] involving milder conditions provides access to certain unstable derivatives from *N,N′*-disubstituted ureas; this procedure compares favorably with previous ones and with another related approach using the Triphenylphosphine–Carbon Tetrachloride reagent system (eq 21). Reaction of *N,N′*-dialkylidene ureas with Ph_3PBr_2 in PhH leads to α-haloalkyl carbodiimides.[38e]

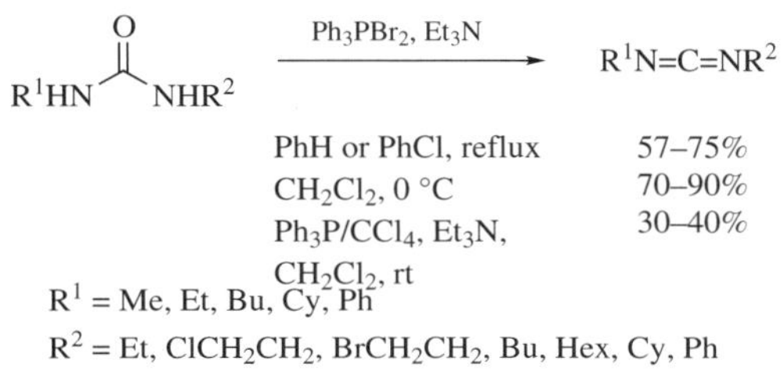

(21)

By the same Ph_3PBr_2/Et_3N procedure carried out in refluxing PhH, disubstituted cyanamides such as Me_2NCN are obtained from the *N,N*-disubstituted urea Me_2NCONH_2 (67%); similarly, isocyanides result from monosubstituted formamides (56–73%),[38a,40b] and nitriles from primary amides and oximes (58–68%).[3] Via the same route, ketenimines ($R^1R^2C{=}C{=}NR^3$) are prepared[38a,40b] in CH_2Cl_2 at reflux by dehydration of secondary amides having a C–H bond adjacent to the carbonyl function (45–93%). Application of this procedure to sulfimides gives highly reactive ketenimines, which are used for further reactions without isolation; [2 + 2] cycloaddition of such species with Schiff bases provides, in good yield, *N*-(tosyl)azetidin-2-imines related to β-lactam derivatives (eq 22).[38b] The diphosphorylated ketenimine $[(EtO)_2(O)P]_2C{=}C{=}NPh$ is obtained analogously in 87% yield with Ph_3PBr_2/Et_3N from the α,α-diphosphorylated acetanilide.[38c]

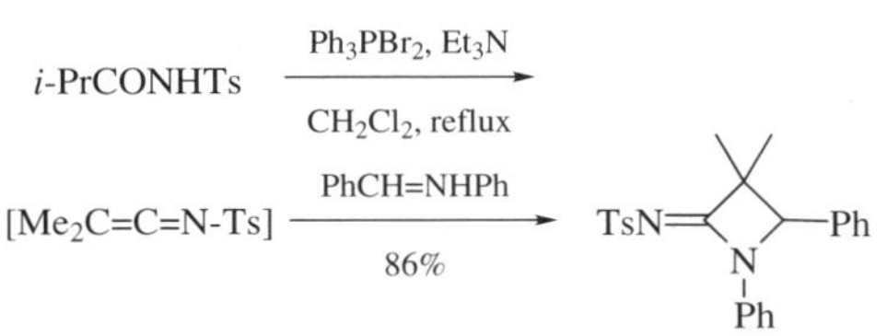

(22)

Imidoyl Bromides from Secondary Amides. In the case of the secondary diarylamide benzanilide, reaction with Ph_3PBr_2/Et_3N in refluxing PhH provides *N*-phenylbenzimidoyl bromide in 65% yield, whereas a 70% yield is attained with the Ph_3P/CBr_4 reagent system.[39] The closely related *N*-methylbenzamide affords a dimeric compound[40b] with Ph_3PBr_2 in boiling CH_2Cl_2 by intermolecular *N*-imidoylation (see below for a similar intramolecular process) between generated imidoyl bromide or oxyphosphonium species and the starting carboxamide. Cyclodehydration of secondary carboxylic diamides with Ph_3PBr_2/Et_3N affords intramolecular *O*-imidoylation products with a benzoxadiazepine structure.[40] Without Et_3N, intramolecular *N*-imidoylation products with an ω-imino lactam structure are obtained; *O*-imidoylated products rearrange to *N*-analogs under acidic conditions.

Iminophosphoranes from Amines, Hydrazines, and Related Derivatives. Compounds containing a P=N bond, such as iminophosphoranes, are widely used as synthetic intermediates, especially in the heterocyclic field. Aromatic primary amines are generally transformed into iminotriphenylphosphoranes, and aliphatic amines into aminophosphonium bromides with Ph_3PBr_2/Et_3N in PhH, PhCl, CCl_4, or CH_2Cl_2 as solvent;[41a,b] more basic conditions are usually required to convert an alkylaminophosphonium salt into the corresponding iminophosphorane.[41b] As protected primary amines, these phosphinimines (iminophosphoranes) can then be subjected to monoalkylation, the Ph_3P protecting group being cleaved in a subsequent step to provide secondary amines (eq 23).[41a–c]

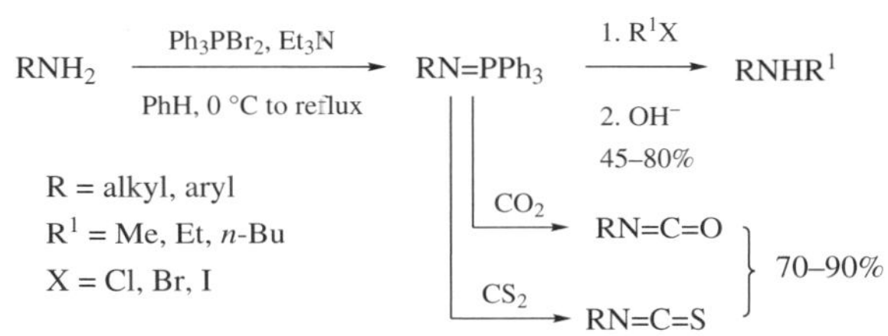

(23)

Some aliphatic and aromatic phosphinimines (prepared as above) undergo a Wittig-like condensation with CO_2 or CS_2, giving rise to isocyanates and isothiocyanates, respectively (eq 23).[41d] Dehydration reactions promoted by Ph_3PBr_2/Et_3N are used for intramolecular cyclizations involving an imide carbonyl group and an aromatic amine;[42] an aza-Wittig reaction of a phosphinimide intermediate is assumed to occur. Iminophosphoranes derived from hydrazines[43a] and acylhydrazines[43b,c] are also used as intermediates in heterocyclic synthesis. Reactions of metalated amines RNHMgX with Ph_3PBr_2 in PhH–ether can also provide aminophosphonium salts $Ph_3{+}{:}PNHR\ Br^-$ in fair to good yields (33–74%).[43d]

Other Applications. Reaction of Ph_3PBr_2 with benzoins in MeCN at rt provides diaryl α-diketones in excellent yields (75–98%).[44] Ph_3PBr_2 is used as a precursor in the synthesis of dialkoxytriphenylphosphoranes such as $(CF_3CH_2O)_2PPh_3$[45] and $(t\text{-}BuCH_2O)_2PPh_3$,[46] which are used as alkylating or acylating and cyclodehydration agents, respectively. Direct synthesis of β-lactams by [2 + 2] cycloaddition between carboxylic acids and imides, thus avoiding the use of acid halides, is achieved with Ph_3PBr_2 (40–55%).[47] Ph_3PBr_2, as a dehydrating agent, effects esterification reactions with tertiary alcohols,[48a] such as *t*-BuOH in PhH–HMPA, and with aromatic or aryl allylic acids and primary or secondary aliphatic alcohols in petroleum ether (27–85%).[48b]

Preparation of Bromotrimethylsilane in quantitative yield occurs by deoxygenation of $(Me_3Si)_2O$ with Ph_3PBr_2 (in 1,2-dichlorobenzene at reflux) in the presence of a catalytic amount of Zn; a one-pot preparation of pseudohalogenosilanes, in excellent yield (80–95%), can follow by subsequent addition of the appropriate salt XM in DMF (X = N_3, NCO, CN; M = K, Na,).[49] Treatment of Ph_3PBr_2 with Iodotrimethylsilane in CH_2Cl_2 at 0 °C yields Ph_3PI_2 (86%).[50] Reaction of Ph_3PBr_2/Et_3N with $EtO_2CCH_2CH{=}CHCO_2Et$ in PhH leads to the corresponding alkylidenephosphonium ylide in 80% yield.[51] Beckmann rearrangement of cycloalkanone oximes into lactams (74–81% yields) is effected by Ph_3PBr_2 in dry PhH at 50–60 °C.[52] The Ph_3PBr_2–Et_3N reagent system compares favorably with a number of other related phosphorus reagents for the rearrangement of *N*-allylamides into nitriles under mild conditions via 3-aza-Claisen reaction.[53]

---

1. Castro, B. R. *Org. React.* **1983**, *29*, 1.
2. Physical studies on the Ph_3P–dihalide adducts indicates that the extent of ionic (and covalent) behavior is a function of solvent polarity; see especially:Bricklebank, N.; Godfrey, S. M.; Mackie, A. G.; McAuliffe, C. A.; Pritchard, R. G. *Chem. Commun./J. Chem. Soc., Chem. Commun.* **1992**, 355, and references cited therein.
3. Horner, L.; Oediger, H.; Hoffmann, H. *Justus Liebigs Ann. Chem./Liebigs Ann. Chem.* **1959**, *626*, 26.
4. (a) Wiley, G. A.; Hershkowitz, R. L.; Rein, B. M.; Chung, B. C. *J. Am. Chem. Soc.* **1964**, *86*, 964. (b) Wiley, G. A.; Rein, B. M.; Hershkowitz, R. L. *Tetrahedron Lett.* **1964**, 2509.
5. Kaplan, L. *J. Org. Chem.* **1966**, *31*, 3454.
6. (a) Levy, D.; Stevenson, R. *Tetrahedron Lett.* **1965**, 341. (b) Levy, D.; Stevenson, R. *J. Org. Chem.* **1965**, *30*, 3469.
7. (a) Schaefer, J. P.; Weinberg, D. S. *J. Org. Chem.* **1965**, *30*, 2635. (b) Schaefer, J. P.; Weinberg, D. S. *J. Org. Chem.* **1965**, *30*, 2639.
8. Brett, D.; Downie, I. M.; Lee, J. B.; Matough, M. F. S. *Chem. Ind. (London)* **1969**, 1017.
9. Arain, R. A.; Hargreaves, M. K. *J. Chem. Soc. (C)* **1970**, 67.
10. (a) Schaefer, J. P.; Higgins, J. G.; Shenov, P. K. *Org. Synth.* **1968**, *48*, 51. (b) Sandri, J.; Viala, J. *Synth. Commun.* **1992**, *22*, 2945.
11. Hrubiec, R. T.; Smith, M. B. *J. Org. Chem.* **1984**, *49*, 431.
12. Hanack, M.; Auchter, G. *J. Am. Chem. Soc.* **1985**, *107*, 5238.
13. Machinek, R.; Lüttke, W. *Synthesis* **1975**, 255.
14. Gajda, T. *Phosphorus Sulfur/Phosphorus Sulfur Silicon* **1990**, *53*, 327.
15. (a) Piper, J. R.; Rose, L. M.; Johnston, T. P.; Grenan, M. M. *J. Med. Chem.* **1979**, *22*, 631. (b) Zakharkin, L. I.; Kovredov, A. I.; Kazantsev, A. V.; Meiramov, M. G. *J. Gen. Chem. USSR (Engl. Transl.)* **1981**, *51*, 289. (c) Janda, K. D.; Weinhouse, M. I.; Danon, T.; Pacelli, K. A.; Schloeder, D. M. *J. Am. Chem. Soc.* **1991**, *113*, 5427. (d) Skvarchenko, V. R.; Lapteva, V. L.; Gorbunova, M. A. *J. Org. Chem. USSR (Engl. Transl.)* **1990**, *26*, 2244.
16. Caubere, P.; Mourad, M. S. *Tetrahedron* **1974**, *30*, 3439.
17. Hodge, P.; Khoshdel, E. *J. Chem. Soc., Perkin Trans. 1* **1984**, 195.
18. (a) Garegg, P. J. *Pure Appl. Chem.* **1984**, *56*, 845. (b) Classon, B.; Garegg, P. J.; Samuelsson, B. *Can. J. Chem.* **1981**, *59*, 339.
19. (a) Haga, K.; Yoshikawa, M.; Kato, T. *Bull. Chem. Soc. Jpn.* **1970**, *43*, 3922. (b) Bridges, A. J. *Nucleosides Nucleotides* **1988**, *7*, 375.
20. (a) Herr, M. E.; Johnson, R. A. *J. Org. Chem.* **1972**, *37*, 310. (b) Boeckman, R. K.Jr., Ganem, B. *Tetrahedron Lett.* **1974**, 913.
21. Dahl, T.; Stevenson, R.; Bhacca, N. S. *J. Org. Chem.* **1971**, *36*, 3243.
22. (a) Schaefer, J. P.; Higgins, J. *J. Org. Chem.* **1967**, *32*, 1607. (b) Schaefer, J. P.; Higgins, J.; Shenoy, P. K. *Org. Synth.* **1969**, *49*, 6. (c) Porzi, G.; Concilio, C. *J. Organomet. Chem.* **1977**, *128*, 95.
23. (a) Anderson, A. G.Jr., Freenor, F. J. *J. Am. Chem. Soc.* **1964**, *86*, 5037. (b) Anderson, A. G.Jr., Freenor, F. J. *J. Org. Chem.* **1972**, *37*, 626. (c) Kato, M.; Nomura, S.; Kobayashi, H.; Miwa, T. *Chem. Lett.* **1986**, 281.
24. Marchand, A. P.; Weimar, W. R.Jr., *Chem. Ind. (London)* **1969**, 200.
25. (a) Schwartz, M.; Oliver, J. E.; Sonnet, P. E. *J. Org. Chem.* **1975**, *40*, 2410. (b) Sonnet, P. E. *Synth. Commun.* **1976**, *6*, 21. (c) Buser, H. R.; Guerin, P. M.; Toth, M.; Szöcs, G.; Schmid, A.; Francke, W.; Arn, H. *Tetrahedron Lett.* **1985**, *26*, 403.
26. Kruse, C. G.; Jonkers, F. L.; Dert, V.; van der Gen, A. *Recl. Trav. Chim. Pays-Bas* **1979**, *98*, 371.
27. (a) Aizpurua, J. M.; Cossío, F. P.; Palomo, C. *J. Org. Chem.* **1986**, *51*, 4941. (b) Lazukina, L. A.; Kolodyazhnyi, O. I.; Pesotskaya, G. V.; Kukhar, V. P. *J. Gen. Chem. USSR (Engl. Transl.)* **1976**, *46*, 1931.
28. De Wit, J.; Wynberg, H. *Recl. Trav. Chim. Pays-Bas* **1973**, *92*, 281.
29. Wagner, A.; Heitz, M.-P.; Mioskowski, C. *Chem. Commun./J. Chem. Soc., Chem. Commun.* **1989**, 1619.
30. (a) Hoffmann, H.; Diehr, H. J. *Tetrahedron Lett.* **1962**, 583. (b) Chechulin, P. I.; Filyakova, V. I.; Pashkevich, K. I. *Bull. Acad. Sci. USSR, Div. Chem. Sci.* **1989**, 189. (c) Bestmann, H. J.; Kisielowski, L. *Ber. Dtsch. Chem. Ges./Chem. Ber.* **1983**, *116*, 1320.
31. (a) Carnduff, J.; Miller, J. A.; Stockdale, B. R.; Larkin, J.; Nonhebel, D. C.; Wood, H. C. S. *J. Chem. Soc., Perkin Trans. 1* **1972**, 692. (b) Piers, E.; Nagakura, I. *Synth. Commun.* **1975**, *5*, 193. (c) Piers, E.; Grierson, J. R.; Lau, C. K.; Nagakura, I. *Can. J. Chem.* **1982**, *60*, 210. (d) Buono, G. *Synthesis* **1981**, 872.
32. (a) Thakore, A. N.; Pope, P.; Oehlschlager, A. C. *Tetrahedron* **1971**, *27*, 2617. (b) Sonnet, P. E.; Oliver, J. E. *J. Org. Chem.* **1976**, *41*, 3279.
33. (a) Palumbo, G.; Ferreri, C.; Caputo, R. *Tetrahedron Lett.* **1983**, *24*, 1307. (b) Caputo, R.; Ferreri, C.; Noviello, S.; Palumbo, G. *Synthesis* **1986**, 499.
34. Okada, I.; Ichimura, K.; Sudo, R. *Bull. Chem. Soc. Jpn.* **1970**, *43*, 1185.
35. (a) Gogte, V. N.; Kulkarni, S. B.; Tilak, B. D. *Tetrahedron Lett.* **1973**, 1867. (b) Freeman, J. P.; Mondron, P. J. *Synthesis* **1974**, 894.
36. (a) Bestmann, H.-J.; Mott, L. *Justus Liebigs Ann. Chem./Liebigs Ann. Chem.* **1966**, *693*, 132. (b) Aizpurua, J. M.; Palomo, C. *Synthesis* **1982**, 684.
37. (a) Burton, D. J.; Koppes, W. M. *Chem. Commun./J. Chem. Soc., Chem. Commun.* **1973**, 425. (b) Burton, D. J.; Koppes, W. M. *J. Org. Chem.* **1975**, *40*, 3026. (c) Anderson, A. G.Jr., Kono, D. H. *Tetrahedron Lett.* **1973**, 5121. (d) Smissman, E. E.; Alkaysi, H. N.; Creese, M. W. *J. Org. Chem.* **1975**, *40*, 1640.
38. (a) Bestmann, H. J.; Lienert, J.; Mott, L. *Justus Liebigs Ann. Chem./Liebigs Ann. Chem.* **1968**, *718*, 24. (b) Van Camp, A.; Goossens, D.; Moya-Portuguez, M.; Marchand-Brynaert, J.; Ghosez, L. *Tetrahedron Lett.* **1980**, *21*, 3081. (c) Bestmann, H. J.; Lehnen, H. *Tetrahedron Lett.* **1991**, *32*, 4279. (d) Palomo, C.; Mestres, R. *Synthesis* **1981**, 373. (e) Fetyukhin, V. N.; Vovk, M. V.; Samarai, L. I. *J. Org. Chem. USSR (Engl. Transl.)* **1981**, *17*, 1263.
39. Appel, R.; Warning, K.; Ziehn, K.-D. *Ber. Dtsch. Chem. Ges./Chem. Ber.* **1973**, *106*, 3450.
40. (a) Mazurkiewicz, R. *Monatsh. Chem.* **1988**, *119*, 1279. (b) Mazurkiewicz, R. *Acta Chim. Hung.* **1990**, *127*, 439 (*Chem. Abstr.* **1990**, *114*, 206 693) .
41. (a) Horner, L.; Oediger, H. *Justus Liebigs Ann. Chem./Liebigs Ann. Chem.* **1959**, *627*, 142. (b) Zimmer, H.; Jayawant, M.; Gutsch, P. *J. Org. Chem.* **1970**, *35*, 2826. (c) Briggs, E. M.; Brown, G. W.; Jiricny, J.; Meidine, M. F. *Synthesis* **1980**, 295. (d) Molina, P.; Alajarin, M.; Arques, A. *Synthesis* **1982**, 596.
42. (a) Al-Khathlan, H.; Zimmer, H. *J. Heterocycl. Chem.* **1988**, *25*, 1047. (b) Al-Khathlan, H. Z.; Al-Lohedan, H. A. *Phosphorus Sulfur/Phosphorus Sulfur Silicon* **1991**, *61*, 367.
43. (a) Cullen, E. R.; Guziec, F. S.Jr., Hollander, M. I.; Murphy, C. J. *Tetrahedron Lett.* **1981**, *22*, 4563. (b) Molina, P.; Alajarin, M.; Saez, J. R. *Synthesis* **1984**, 983. (c) Froyen, P. *Phosphorus Sulfur/Phosphorus Sulfur Silicon* **1991**, *57*, 11. (d) Zbiral E.; Berner-Fenz, L. *Monatsh. Chem.* **1967**, *98*, 666.
44. Ho, T.-L. *Synthesis* **1972**, 697.
45. Kubota, T.; Miyashita, S.; Kitazume, T.; Ishikawa, N. *J. Org. Chem.* **1980**, *45*, 5052.

46. Kelly, J. W.; Evans, S. A.Jr., *J. Org. Chem.* **1986**, *51*, 5490.
47. Cossío, F. P.; Ganboa, I.; Palomo, C. *Tetrahedron Lett.* **1985**, *26*, 3041.
48. (a) Haynes, R. K.; Katsifis, A.; Vonwiller, S. C. *Aust. J. Chem.* **1984**, *37*, 1571. (b) Lajis, N. *Pertanika* **1985**, *8*, 67 (*Chem. Abstr.* **1986**, *105*, 6277d) .
49. Aizpurua, J. M.; Palomo, C. *Nouv. J. Chim.* **1984**, *8*, 51.
50. Romanenko, V. D.; Tovstenko, V. I.; Markovski, L. N. *Synthesis* **1980**, 823.
51. Labuschagne, A. J. H.; Schneider, D. F. *Tetrahedron Lett.* **1982**, *23*, 4135.
52. Sakai, I.; Kawabe, N.; Ohno, M. *Bull. Chem. Soc. Jpn.* **1979**, *52*, 3381.
53. Walters, M. A.; Hoem, A. B.; Arcand, H. R.; Hegeman, A. D.; McDonough, C. S. *Tetrahedron Lett.* **1993**, *34*, 1453.

Jean-Robert Dormoy & Bertrand Castro
SANOFI Chimie, Gentilly, France

Triphenylphosphine Diethylazodicarboxylate

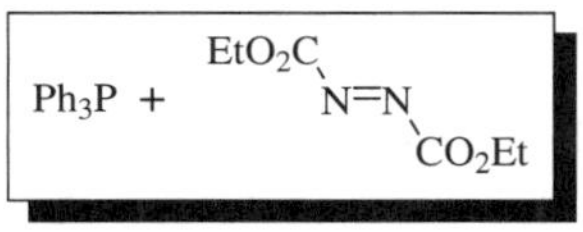

(Ph_3P)
[603-35-0] $C_{18}H_{15}P$ (MW 262.30)) (DEAD)
[1972-28-7] $C_6H_{10}N_2O_4$ (MW 174.18)

(reagent in the Mitsunobu reaction, which is a versatile, mild dehydration reaction, widely used for the synthesis of esters and ethers, for the formation of C–N, C–S, and C–halogen bonds, and for inverting the configuration of a stereogenic carbon containing an OH group[1])

Physical Data: see Triphenylphosphine and Diethyl Azodicarboxylate.

Handling, Storage, and Precautions: all reagents and solvents must be anhydrous. In general, the Mitsunobu betaine (**1**) is generated in situ from the phosphine and the azodicarboxylate; however, in some cases it is essential to preform the betaine. The betaine is an unstable, colorless, crystalline solid, rapidly hydrolyzed on contact with moisture. It can be crystallized from dry THF or $CHCl_3$/hexane.

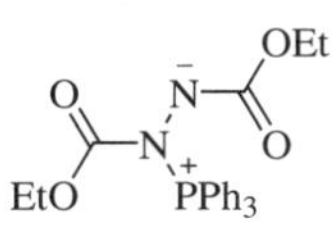

(**1**)

The Mitsunobu Reaction. A mixture of triphenylphosphine (TPP) and diethyl azodicarboxylate (DEAD) is generally used; however, diisopropyl azodicarboxylate (DIAD) is cheaper and works just as well. The overall reaction enables the replacement of the hydroxyl group of an alcohol by a nucleophile X (eq 1).

$$ROH + HX \xrightarrow[DIAD]{TPP} RX \quad (1)$$

The phosphine accepts the oxygen (to give triphenylphosphine oxide) while the azodicarboxylate accepts the two hydrogens (to give the corresponding hydrazine derivative). The reaction occurs under mild (0–25 °C), essentially neutral conditions, can be carried out in the presence of a wide range of functional groups, and usually gives good yields (60–90%). In general, the reaction proceeds well with primary and secondary alcohols, and inversion of configuration is observed.[1] Retention of configuration may result from neighboring group participation,[2] or a change in mechanism if the alcohol is very hindered.[3a] Racemization may result if the C–O bond is prone to S_N1 cleavage.[4] Examples of H–X in the reaction (eq 1) include carboxylic acids, thioacids, phenols, thiols, imides, sulfonamides, hydrazoic acid, heterocyclic compounds, hydrogen halides, phosphate diesters, phosphinic acids, and certain active methylene compounds. The pK_a of H–X should generally be less than 13. LiX or ZnX_2 sometimes give better results than HX.[1c] THF is the most commonly used solvent, but other solvents such as dichloromethane, chloroform, benzene, toluene, ethyl acetate, DMF, and HMPA can also be used. The reaction is faster in nonpolar solvents; for example, formation of ethyl benzoate is about 30 times faster in $CHCl_3$ than in MeCN.[5a] A problem commonly encountered with the Mitsunobu reaction is the separation of the product from triphenylphosphine oxide. This can be overcome by the use of a phosphine containing a basic group such as diphenyl(2-pyridyl)phosphine[6] or (4-dimethylaminophenyl)diphenylphosphine,[7] where an aqueous acid wash removes the corresponding phosphine oxide, or by using a polymer-bound phosphine.[8] Similarly, dimethyl azodicarboxylate (the corresponding hydrazine is water soluble)[9] and a polymer-supported azodicarboxylate[10] can be used to facilitate product isolation. The reaction is best explained by assuming successive formation of the betaine (**1**), the protonated *N*-phosphonium salt (**2**), and the alkoxyphosphonium salt (**3**), which collapses in S_N2 fashion (eq 2).

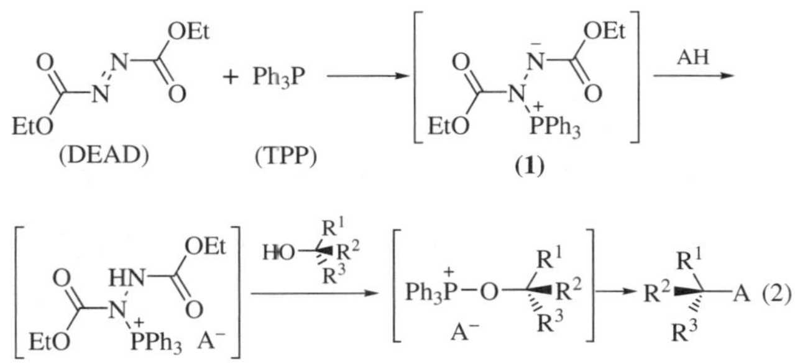

Without acidic components, *N*-alkylhydrazinedicarboxylates (**4**) can be formed.[5b,11] NMR studies of the reaction of alcohols with DEAD and TPP in the absence of acidic components have revealed that the key intermediate is the pentavalent dialkoxyphosphorane (**5**).[1c,5] In the presence of an acidic component, phosphorane intermediates are present in equilibrium with the oxyphosphonium salt (**3**).[3]

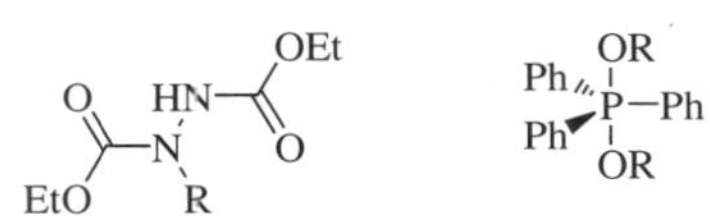

(**4**) (**5**)

The combination of DEAD and TPP has been utilized most often as the condensation reaction system, but dimethyl, diisopropyl, and di-*t*-butyl azodicarboxylates (**6**) can also be used instead of DEAD; the azodicarboxamide (**7**) has been used as well.[1,12] Triphenylphosphine can be substituted by a variety of trivalent phosphorus compounds, including substituted triarylphosphines and trialkylphosphines.[1c]

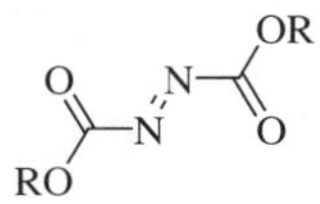

(**6**)

R = Me: DMAD
R = Et: DEAD
R = *i*-Pr: DIAD
R = *t*-Bu: DBAD

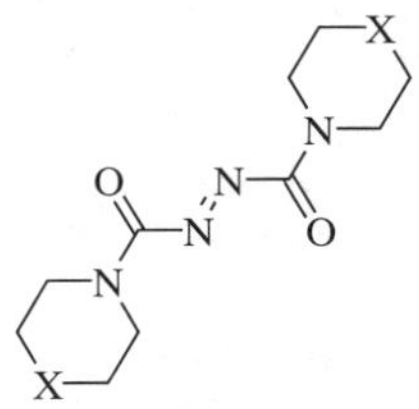

(**7**)

X = CH_2: ADDP
X = O, NMe

Order of Addition of the Reagents. The order of addition can be crucial. The most commonly used procedure is to dissolve the alcohol (ROH), the acid (HX), and triphenylphosphine in THF, cool to 0 °C, add the DEAD (or DIAD) slowly with stirring, and then stir at room temperature for several hours. If this procedure fails to give the desired product, the betaine (TPP–DEAD adduct) should be preformed (addition of DEAD to TPP in THF at 0 °C) before addition of a mixture of the ROH and the HX,[13] or addition of ROH followed by slow addition of HX or HX and a buffering salt (strong acids),[14,15] or addition of HX followed by ROH (weak acids).[16] The advantage of preforming the betaine is that X^- is not generated in the presence of a powerful oxidant (the azodicarboxylate). Preformation of the betaine should minimize radical-induced side-reactions (radicals are produced when TPP and DEAD are mixed[17]).

Ester Formation. A variety of alcohols react at room temperature with carboxylic acids in the presence of DEAD and TPP to produce the corresponding esters.[1] When polyols are used, the reaction generally takes place at the less hindered hydroxyl group, as exemplified in the reaction of 1,3-butanediol (eq 3).[18,19]

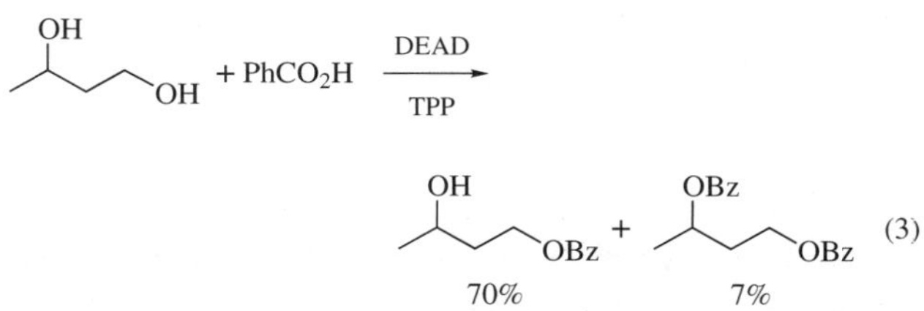

When 1,2-propanediol or styrene glycol is used, however, the more sterically encumbered C-2 benzoate is predominantly obtained with complete inversion of the stereochemistry.[19] This result has been explained by the formation of a dioxaphospholane (**8**) as the key intermediate (eq 4).[3b,19,20a] With acyclic 1,4-diols such as isomaltitol,[20b] five-membered cyclic ether formation takes place in preference to esterification.

Thymidine (**9**) reacts with aromatic carboxylic acids to give 5′-*O*-aroylthymidines (**10**) (eq 5). The yield of esters increases with the acidity of the carboxylic acid, indicating the importance of the protonated *N*-phosphonium salt (**2**).[21,22]

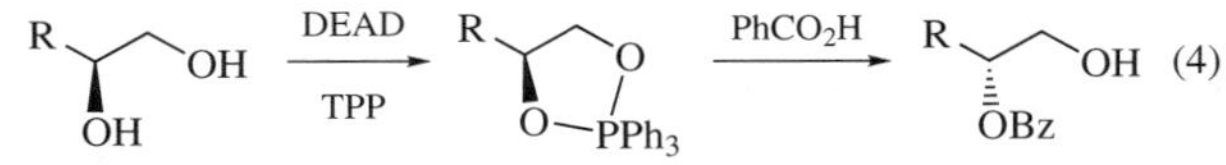

(**8**)

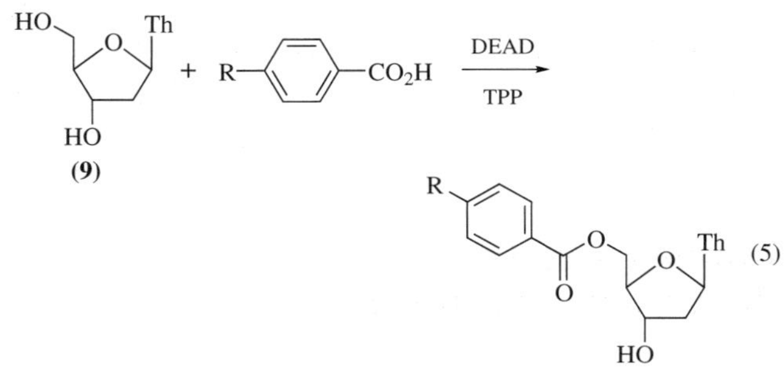

(**10**)

R = Me, 65%; OMe, 66%; H, 74%; NC, 80%; O_2N, 85%

The reaction of chiral secondary carbinols with carboxylic acids, DEAD, and TPP gives the corresponding esters with inverted configuration. Removal of the carboxyl group affords the enantiomer of the parent alcohol (eqs 6–9).[23–27] For the purpose of inverting the stereochemistry of a secondary carbinol center, 3,5-dinitrobenzoic acid,[28] *p*-nitrobenzoic acid,[22b] and chloroacetic acid[29] have been recommended, because of increased yield and ease of purification and removal of the carboxyl group from the resulting esters.

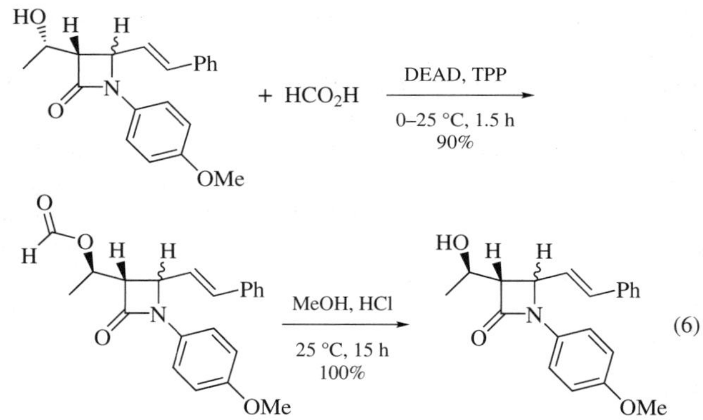

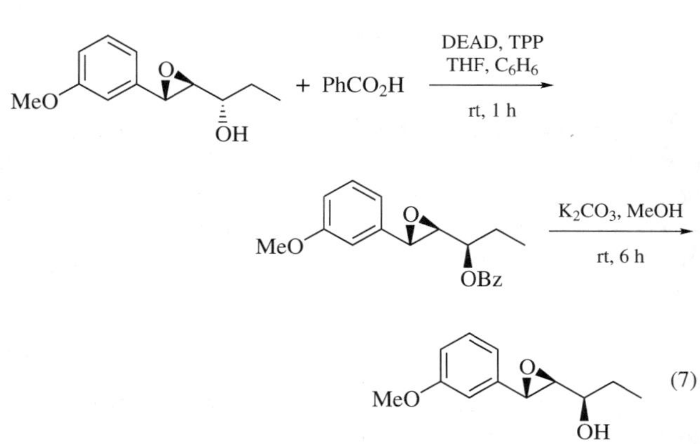

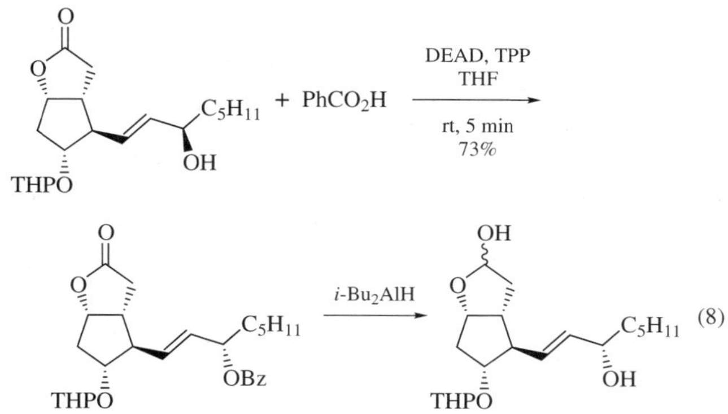

(8)

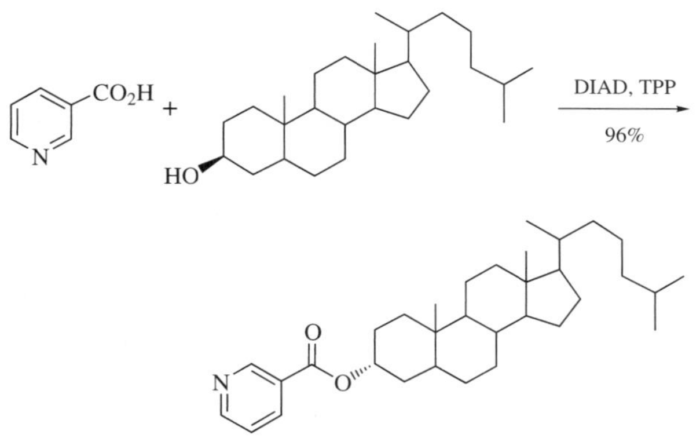

(9)

With some exceptions, a primary hydroxyl group is generally more reactive than a secondary one.[22a,30] Steric congestion and electronic effects sometimes retard the reaction and/or result in the formation of a complex mixture.[31]

The solvent also plays an important role in the success of the reaction. In general, reaction in benzene or toluene gives higher yields of inverted products.[1,22b,32] Although pyridine is not suitable in the preparation of nucleotides,[33a] pyridine can be used for the synthesis of sucrose epoxide,[33b] and a mixture of dioxane–pyridine (9:1) can be utilized in the preparation of sugar carboxylates.[22a] Mixed solvent systems may be necessary when the acid and alcohol components have widely differing solubilities. Thus a mixture of HMPA and dichloromethane works well in the synthesis of lipophilic carbohydrate esters such as cord factor.[33c]

Allylic and benzylic cyanohydrins react with carboxylic acids, DEAD, and TPP to give inverted esters with 92–99% ee, while extensive racemization takes place with benzaldehyde cyanohydrins containing strongly electron-donating substituents (eq 10).

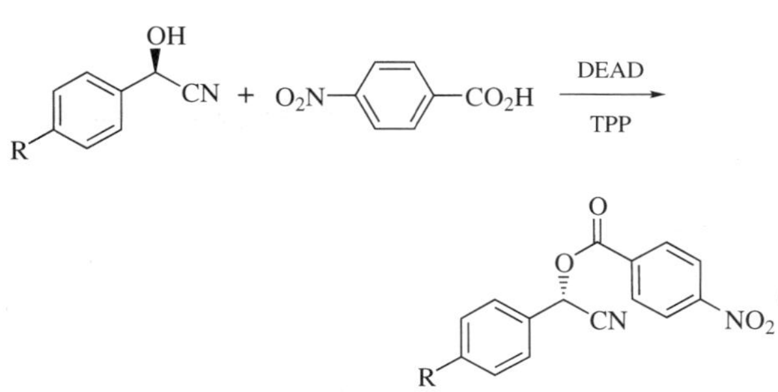

(10)

R = H, 76% (96% ee)
R = MeO, 80% (6% ee)

Saturated aliphatic cyanohydrins afford esters in which the original configuration is retained.[34]

Propargylic alcohols react with carboxylic acids, DEAD, and TPP to give the corresponding inverted esters without formation of allenic alcohols.[35] Similarly, in allylic alcohols where no bias exists against the normal S_N2 process, clean regiospecific inversion is invariably observed or implied.[1,36] In specific cases where the S_N2 pathway is hindered for some reason, such as eclipsing of substituents, allylic rearrangement can occur (eqs 11–13).[1,37] When compound (**11**) reacts with benzoic acid, DEAD, and TPP, (**12**)(**13**) and ((**14**)**a,b**) (**a/b** = 4:1) are obtained in a ratio of 1:1. Similar treatment of (**11**) in the presence of $PdCl_2$ (0.1 equiv) results in the nearly exclusive production of (**14a,b**) (73%; **a/b** = 10:1) via (**12**).[37d]

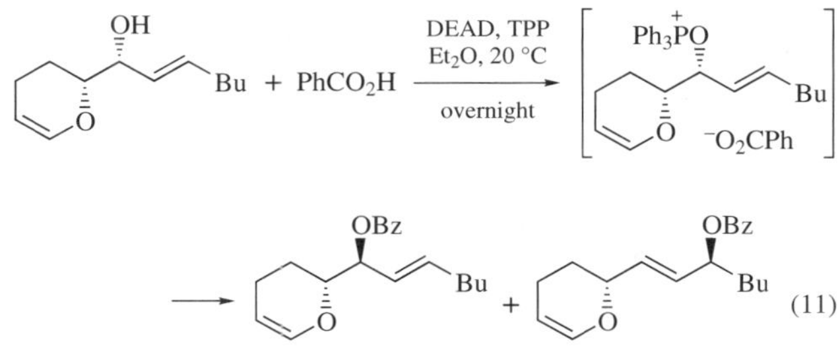

(11)

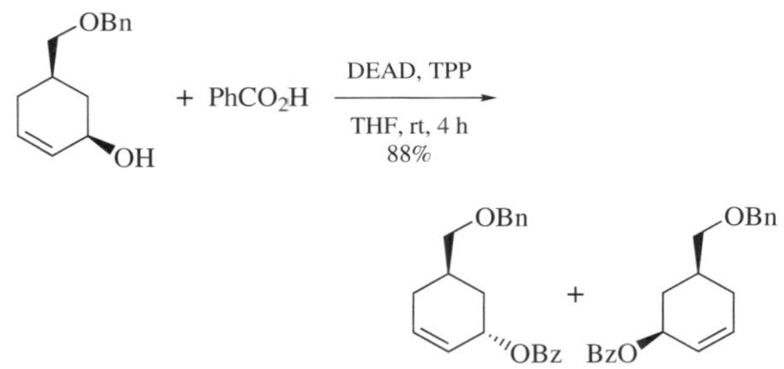

(12)

trans:*cis* = 9:1

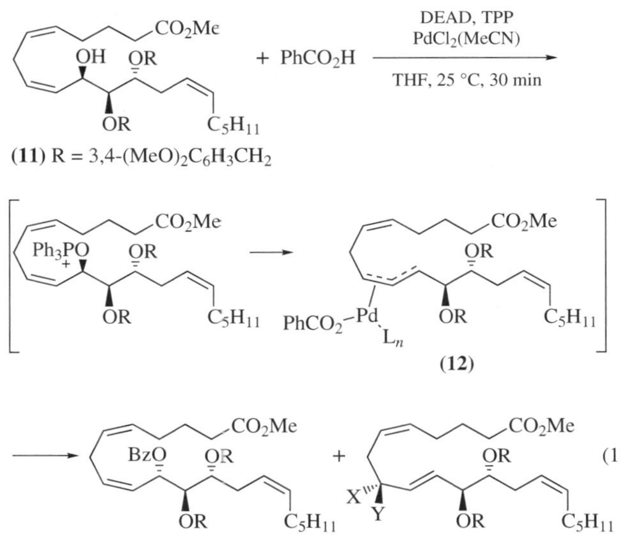

(**11**) R = 3,4-$(MeO)_2C_6H_3CH_2$

(**12**)

(13)

(**13**)

(**14a**) X = BzO, Y = H
(**14b**) X = H, Y = BzO

Reaction with Thiocarboxylic Acids, Phosphoric Acids, Sulfonic Acids, and their Derivatives.[1] Thiocarboxylic acids,[38,13] dithiocarboxylic acids,[39] and dimethyldithiocarbamic acid zinc salt,[40] as well as various phosphorus oxyacids[41] and phosphorus thioacids,[42] can also be utilized. α,ω-Mercapto alcohols form cyclic thioethers whereas thiols react with both DEAD and TPP–DEAD to form disulfides.[43a] 2-Mercaptoazoles also react with alcohols in the presence of DEAD and TPP.[43b] Although arenesulfonic acids do not enter into the reaction, a combination of DEAD–TPP with methyl *p*-toluenesulfonate as a nucleophile carrier gives the corresponding alkyl sulfonates (eq 14).[44] Alternatively, Zinc p-Toluenesulfonate reacts with a variety of secondary alcohols to produce the desired tosylates with clean inversion and in good yields, with some exceptions (eq 15).[45]

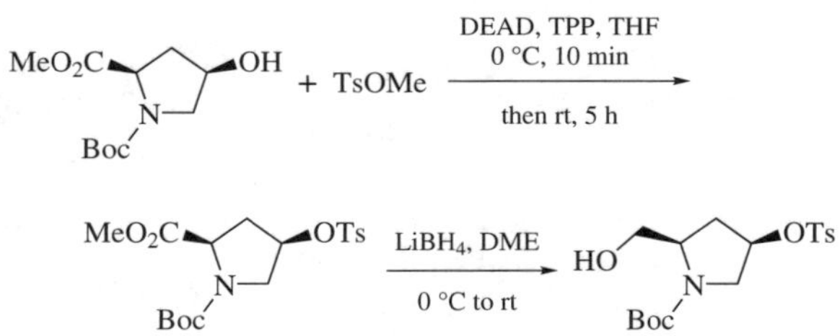

(14)

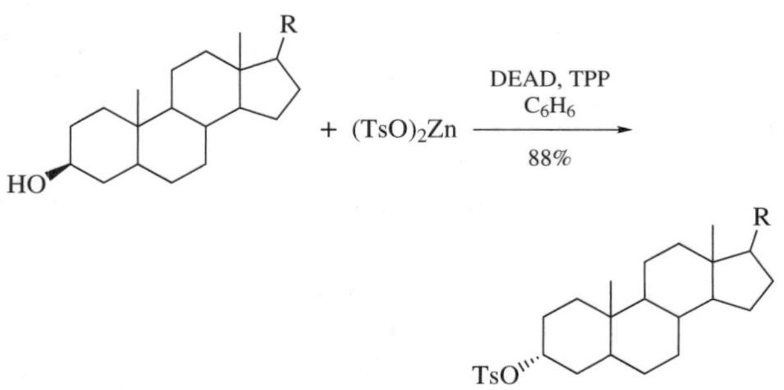

(15)

Reaction with Hydroxy Acids. A study of reactions of *threo*-3-hydroxycarboxylic acids (**15**) with DEAD and TPP revealed that both hydroxyl group activation (HGA) and carboxyl group activation processes (CGA) are involved. With small R^1 and R^2, the zwitterion (**16**) is more stable than (**17**), so that HGA is exclusively observed, resulting in the formation of alkenes. On the other hand, the CGA process via (**17**) is progressively preferred as the size of R^1 and R^2 increases (eq 16).[46a,b] When the reaction is carried out in acetonitrile, CGA predominates.[46c]

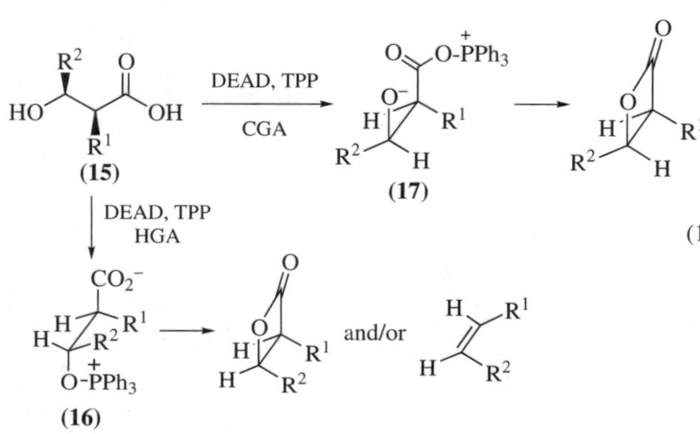

(16)

Hydroxy acids HO–$(CH_2)_n$–CO_2H with $n > 3$ react with DEAD and TPP to afford the corresponding lactones.[1] This procedure can be utilized in the preparation of macrolide antibiotics and related compounds. Macrolactonization by the use of DEAD–TPP depends on the reaction conditions and the structure of the seco acids. Thus dropwise addition of the hydroxy acid (**18**) over a period of 10 h to a mixture of DEAD (7.7 equiv) and TPP (7.5 equiv) affords the lactone (**19**) in 59% yield as well as the unwanted dilactide (**20**) in a yield of <1%. On the other hand, the reaction of (**18**) with DEAD (5.0 equiv) and TPP (5.0 equiv) at 25 °C for 18 h gives (**19**) and (**20**) in 2% and 40% yields, respectively (eq 17).[47]

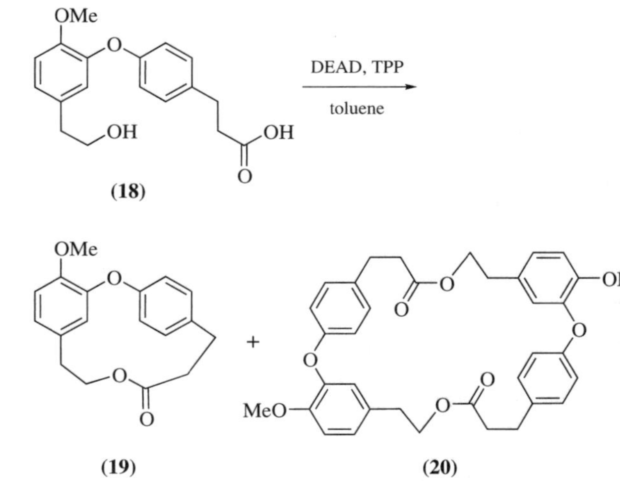

(17)

The reaction of the seco acid (**21**) of colletodiol with DEAD and TPP gives the lactone (**22**) in 45% yield after recrystallization (eq 18),[48a] while the seco acid (**23**), which has a closely related structure to (**21**), affords the corresponding lactone (**24**) in only 4% yield (eq 19).[48b,49]

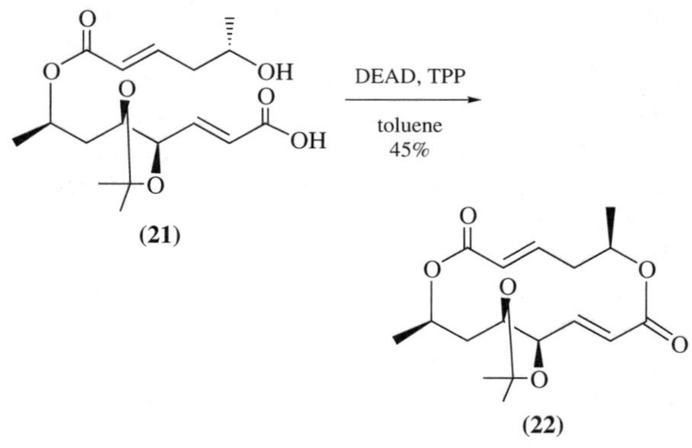

(18)

Reaction with Phenols and Other Oxygen Nucleophiles. When alcohols react with phenols, DEAD, and TPP, the corresponding aryl alkyl ethers are produced. A tertiary amine may facilitate the reaction.[50a] In general, the reaction proceeds with clean inversion of chiral secondary carbinol centers (eq 2; AH = a phenol).[50b] Depending on the structure of the substrate, allylic rearrangement[51,52] and neighboring group participation can be observed.[2,53] Phenols having a hydroxyalkyl chain in a suitable position for cyclization afford the corresponding cyclic aryl ether.[54] Silanols react with phenols and with alcohols in the presence of DEAD and TPP to afford the corresponding silyl ethers.[55] Oximes[56] and *N*-hydroxy imides[57] can also be utilized as acidic components.

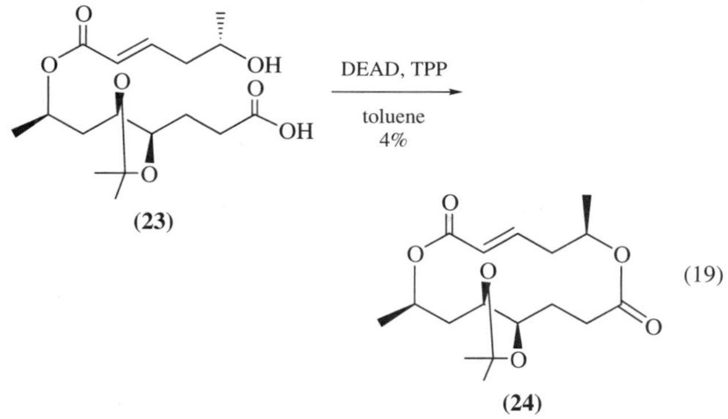

(23) (24) (19)

Reaction with Di- and Polyols. Although intermolecular dehydration between two molecules of alcohols to afford acyclic ethers usually does not occur with the DEAD–TPP system, intramolecular cyclization of diols to produce three to seven-membered ethers is a common and high yielding reaction. Contrary to an early report,[58a] 1,3-propanediol does not form oxetane.[5b] Oxetanes can be formed, however, using the trimethyl phosphite modification of the Mitsunobu reaction.[1c] The reaction of (*S*)-1,2-propanediol and (*R*)-1,4-pentanediol with DEAD and TPP affords the corresponding cyclic ethers with 80–87% retention of stereochemistry at the chiral carbon, while (*S*)-phenyl-1,2-ethanediol affords racemic styrene oxide. In contrast to the reaction of the same 1,2-diols with benzoic acid (eq 4), oxyphosphonium salts ((**25**)**a**) and ((**25**)**b**) have been postulated as key intermediates in the present reaction (eq 20).[58b,e]

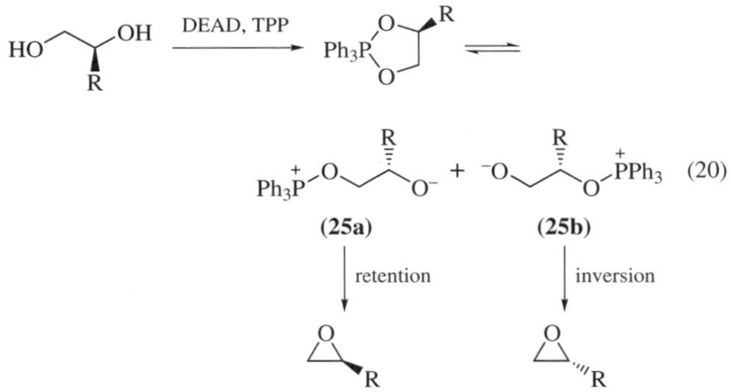

(25a) (25b) (20)

The reaction of cyclic *trans*-1,2-diols with DEAD and TPP generally gives the corresponding oxiranes via phosphoranes.[1,33b,59] Cyclic *cis*-1,2-diols also react with DEAD/TPP to afford primarily phosphoranes (**26**) which decompose into final product(s) such as oxiranes (with retention of configuration) (eq 21)[60] and dehydrated alcohols and/or ketones (eq 22).[61]

Carbon–Nitrogen Bond Formation. Phthalimide reacts with alcohols, DEAD, and TPP to give the corresponding *N*-alkylphthalimides with inversion of configuration (eq 2; HA = phthalimide).[1] Nitrophthalimide is more reactive than phthalimide.[22a] If the reaction site is hindered and an S_N1 process favored, some stereochemical scrambling is observed. For example, the reaction of (**27**) with phthalimide in the presence of DEAD and TPP results in complete epimerization, giving ((**28**)**a**) and ((**28**)**b**) in a ratio of 1:1 (eq 23).[62] Neighboring group participation is also possible in some cases.[63]

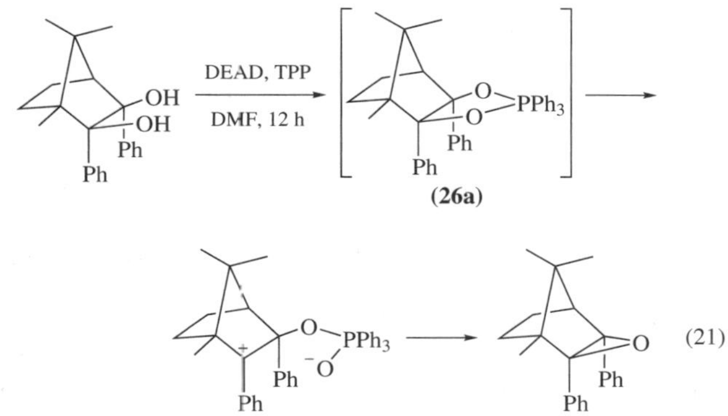

(26a) (21)

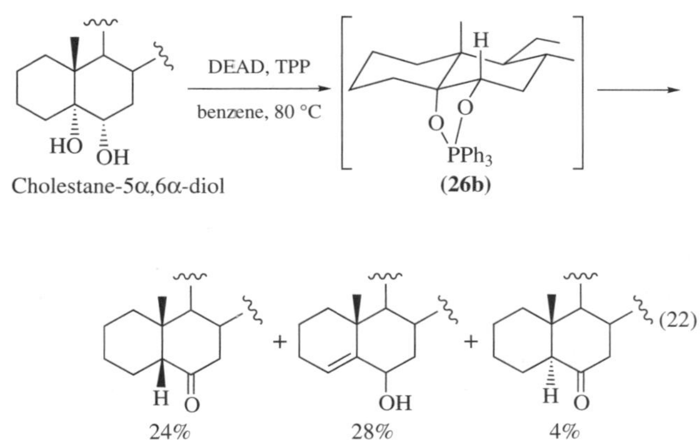

(26b) (22)

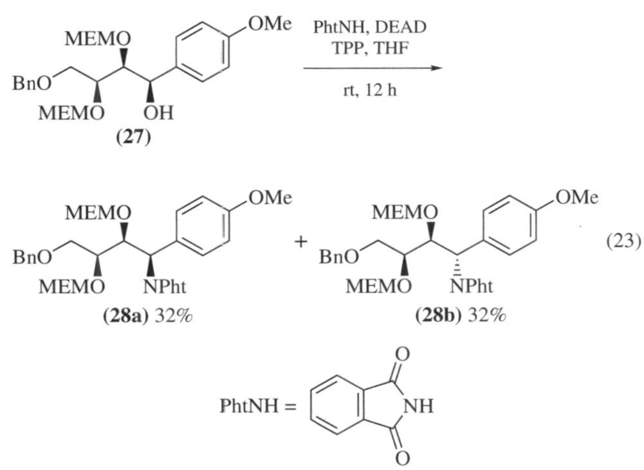

(27) (28a) 32% (28b) 32% (23)

Alcohols having an acidic hydrogen atom adjacent to the hydroxyl group may undergo dehydration rather than substitution.[1] Thus the reaction of diethyl maleate (**29**) with phthalimide, DEAD, and TPP gives diethyl fumarate without any detectable formation of the substituted product.[64a] With Hydrazoic Acid, however, (**29**) gives the expected azido succinate (**30**) in 74% yield (eq 24).[44a] Dehydration is regioselective in some cases.[64b,c]

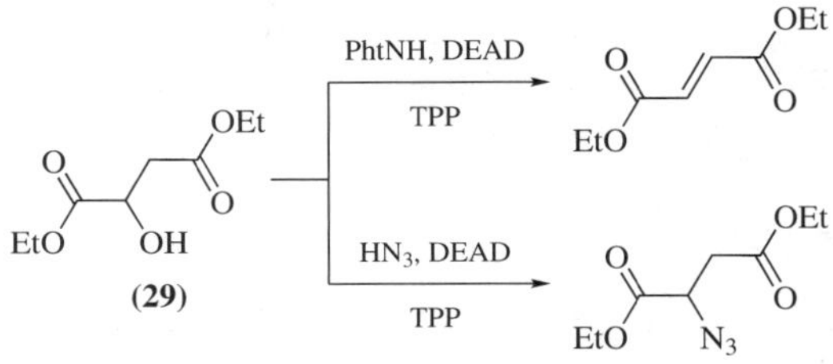

(29) (30) (24)

Since azides are very useful in organic synthesis (especially for the introduction of primary amino groups and the construction of heterocyclic systems), the direct conversion of alcohols into azides has been extensively studied (eq 2; HA = HN_3).[1,65] Diphenyl Phosphorazidate[66] and the zinc azide/bis-pyridine complex[67] can be used instead of HN_3. Treatment of alcohols with hydrazoic acid, DIAD, and excess TPP in THF, followed by addition of water or aqueous acid, affords amines in moderate to good yields (eq 25).[68]

$$\mathrm{ROH} + \mathrm{HN_3}\ \text{(in benzene)} \xrightarrow[\text{then 50 °C, 3 h}]{\text{DIAD, TPP; THF, rt, 1 h}} \mathrm{RN{=}PPh_3} \xrightarrow[\text{1M HCl}]{\mathrm{H_2O}\ \text{or}} \mathrm{RNH_2}\ \text{or}\ \mathrm{R\overset{+}{N}H_3Cl^-} \quad (25)$$

35–85%

Acyclic amides and imides with pK_a < 13 can be expected to function as acidic components in the present reaction system. Thus the reaction of *N*-benzyloxycarbonylbenzamide with ethyl (*S*)-(−)-lactate gives the inverted *N*-alkylated product ((**31**)**a**) and the *O*-alkylated product ((**31**)**b**) in 19% and 48% yields, respectively (eq 26).[69] This result suggests that the alkoxyphosphonium salt (**3**) is a hard alkylating reagent.

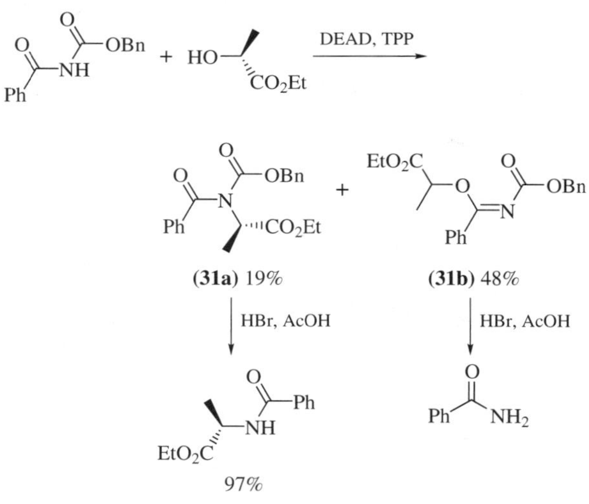

(26)

In the alkylation of imidodicarbonates and tosylcarbamates, the yield of alkylated product increases as the acidity of the NH acids increases. Thus imidodicarbonates and tosylcarbamates with pK_a < 13 give the corresponding *N*-alkylated products in 83–93% yields, while lower yields are obtained with less acidic substrates.[70a] The combination of 1,1/-(Azodicarbonyl)dipiperidine (ADDP) with Tri-n-butylphosphine appears to be superior to that of DEAD with TPP in the alkylation of amides.[12] In the presence of DEAD and TPP, dibenzyl imidodicarbonate does not react with alcohols, but it does react with α-hydroxy stannanes to give the corresponding *N*-alkylated products (eq 27).[71]

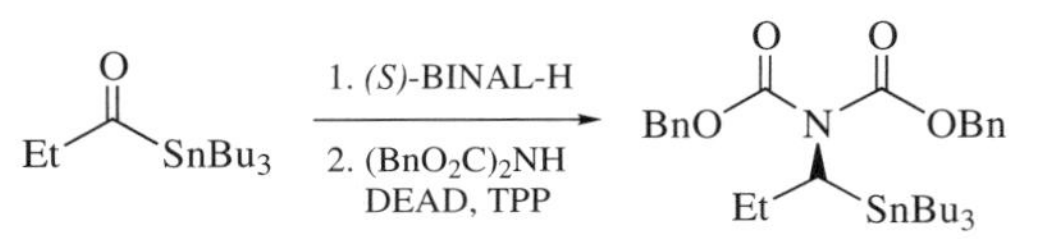

(27)

The *O*-acylhydroxamate ((**32**)**a**) (pK_a 6–7) reacts with benzyl alcohol, DEAD, and TPP to give *N*- and *O*-alkylated products, while the less acidic *O*-alkylhydroxamate ((**32**)**b**) (pK_a 9–10) gives only the *N*-alkyl product (eq 28).[72a]

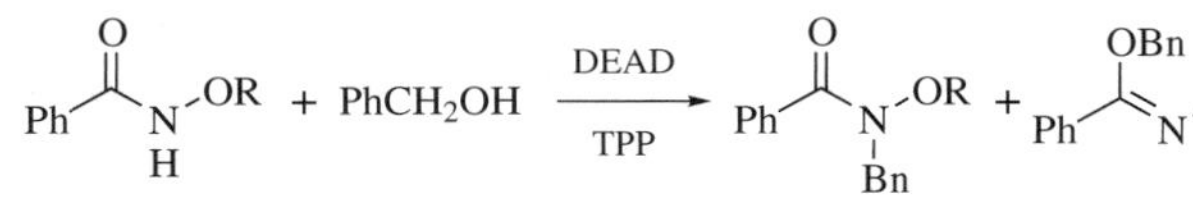

(28)

(**32a**) R = PhCO	82%	65:35
(**32b**) R = Bn	91%	100:0

Intramolecular dehydration of 3-hydroxy carboxamides affords the corresponding β-lactams. Side reactions include elimination and formation of aziridines and oxazolidines (eq 29).[72,73] The efficiency of β-lactam formation is dependent on substrate substituents (including protecting groups for the side-chain amino group) as well as on the choice of azodicarboxylate and P^{III} compound.[1c,74–76] For example, the dipeptide (**33**) reacts with DEAD and TPP to give a 2:1 mixture of the β-lactams ((**34**)**a**) and ((**34**)**b**). Control of the labile C-5 stereocenter can be achieved by use of Triethyl Phosphite instead of TPP; ((**34**)**a**) and ((**34**)**b**) are obtained in a >50:1 ratio (eq 30).[75] 3-Hydroxy *O*-alkylhydroxamates can also be converted into the corresponding β-lactams (eq 31).[1c,77]

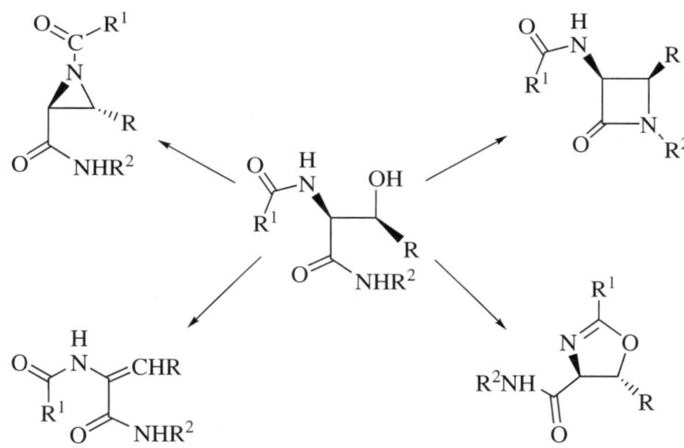

(29)

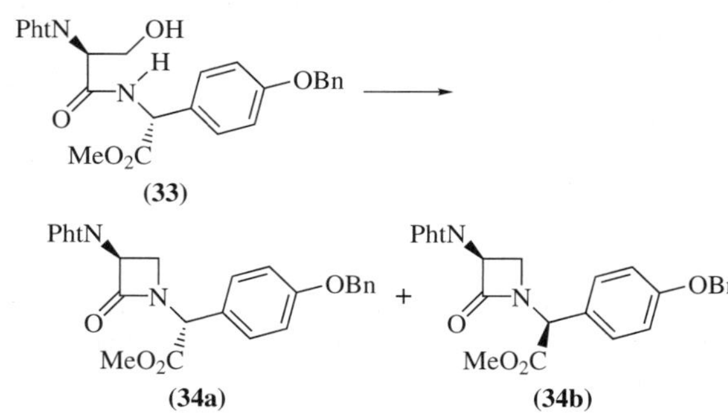

(30)

DEAD–TPP	2:1
DEAD–(EtO)$_3$P	50:1

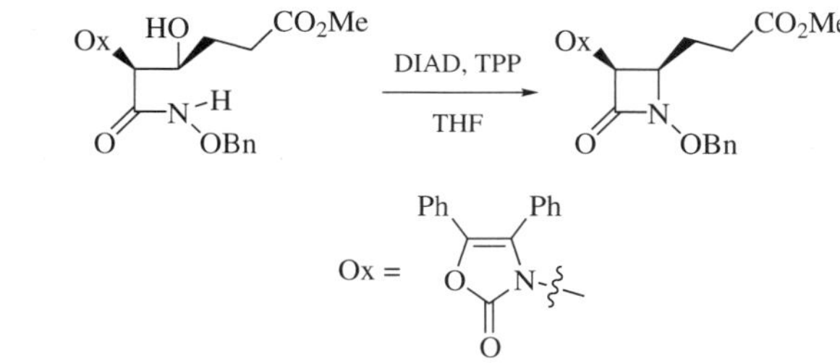

(31)

Purines and pyrimidines react with various alcohols in the presence of DEAD and TPP (eq 32).[78] The alcohol (**35**) gives a carbocyclic nucleoside (**36**) in 65% yield along with a small amount (9%) of the 7-substituted purine derivative (eq 33).[79–81]

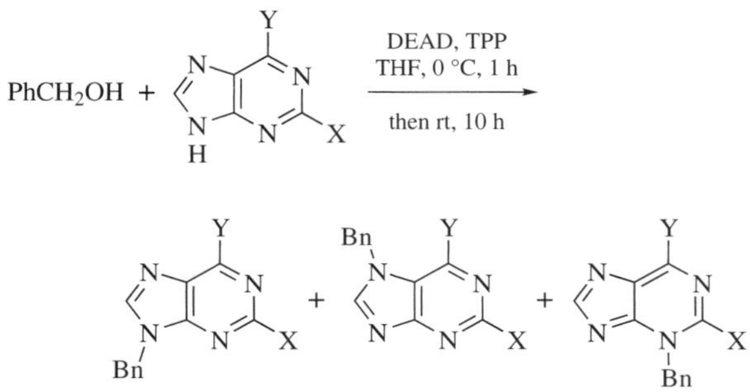

(32)

X = Y = H	70%	9%	21%
X = H, Y = Cl	67%	23%	9%
X = Y = Cl	69%	26%	–

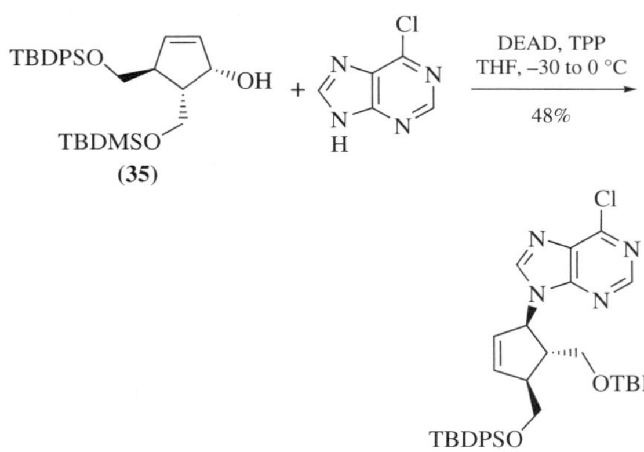

(**35**) (**36**)

(33)

Reaction with Amino Alcohols and Related Compounds. Both secondary and tertiary carbamates bearing a vicinal hydroxyl group react with carboxylic acids in the presence of DEAD and TPP to afford the corresponding esters with inversion of configuration (eq 34).[82] Reaction of *N*-*t*-butoxycarbonylserine with methanol, DEAD, and TPP gives the corresponding methyl ester in >98% yield (eq 35).[83] Without methanol, *N*-benzyloxycarbonyl- or *N*-*t*-butoxycarbonylserine reacts with the preformed adduct of DEAD and TPP to afford the β-lactone (eq 35).[84] The *N*-protected serine methyl ester (**37a**) and threonine methyl ester (**37b**) react with DEAD and TPP to give the dehydrated products (**38a**) and (**38b**) (eq 36).[85]

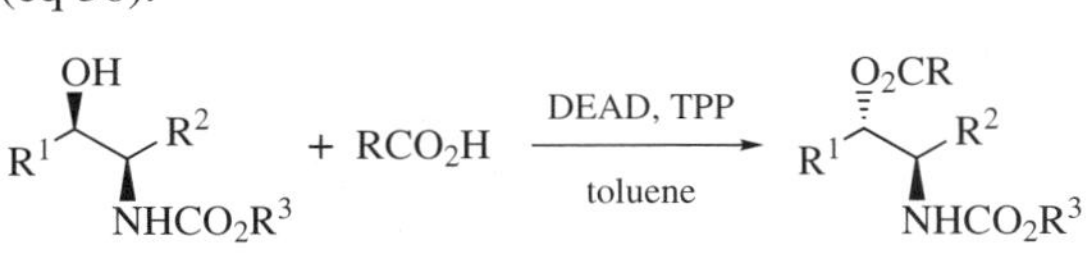

(34)

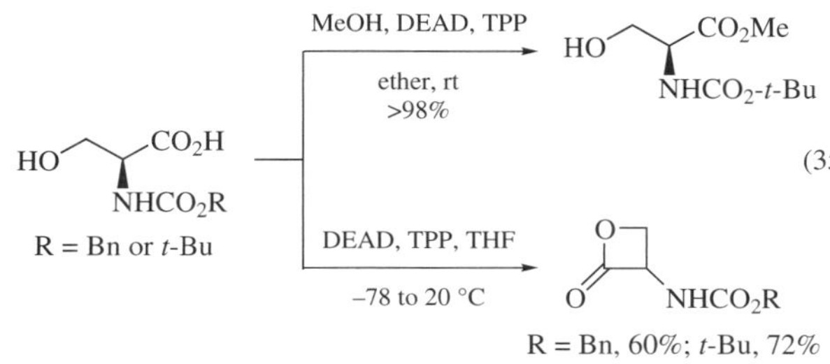

(35)

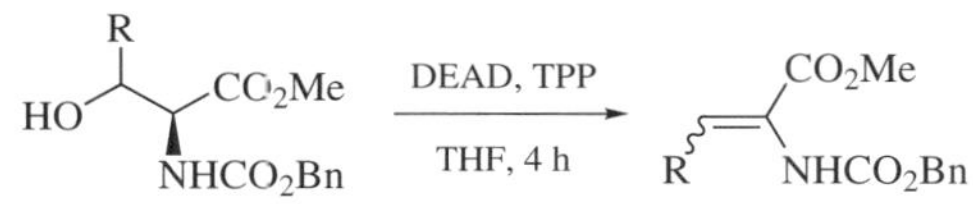

(36)

(**37a**) R = H (**38a**) R = H, 63%
(**37b**) R = Me (**38b**) R = Me, 55%

In the absence of an acidic component, β-(*N*-carbonylamino) alcohols afford aziridines or oxazolines, depending on the structure of the substrate. Contrary to the case of (**37**), the *N*-benzyloxycarbonylamino alcohol (**38**)(**39**), with no acidic hydrogen present at the position adjacent to the hydroxyl group, affords the aziridine (**40**) in 85% yield (eq 37),[86,87] while the *N*-acylamino alcohols (**41**) give the oxazolines (**42**) in 31–68% yields (eq 38).[88]

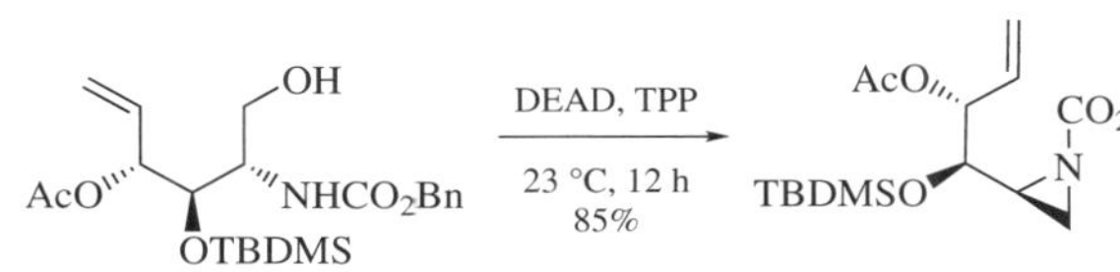

(**39**) (**40**)

(37)

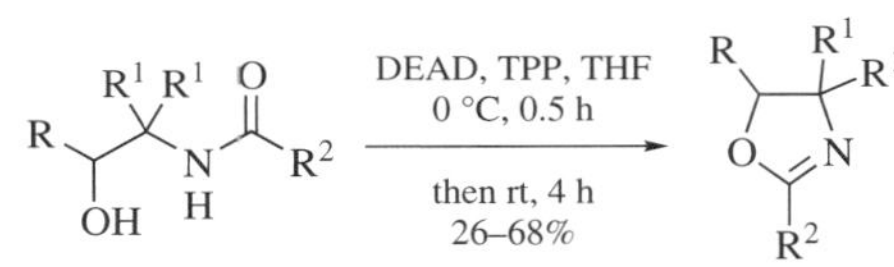

(**41**) (**42**)

(38)

Protection of the amino group by an electron-withdrawing group is not necessarily required for the cyclization. Amino alcohols of general structure HO–(C)$_n$–NHR react with DEAD and TPP to form the corresponding cyclic amines. Thus 2-aminoethanols give aziridines in 18–90% yields.[89] The reaction of the tetrafluoroborate salt of 3-(benzylamino)-1-propanol with DEAD and TPP gives the corresponding azetidine in 50% yield (eq 39).[90] Five- and six-membered cyclic amines are also prepared by the reaction of amino alcohols with DEAD and TPP (eqs 40 and 41).[91] Although azepines are not formed from straight-chain amino alcohols, compound (**43**) gives the azepines (**44**)((**45**)**a**) and ((**45**)**b**) by S_N2' substitution of the intermediate phosphorane (**44**) (eq 42).[92]

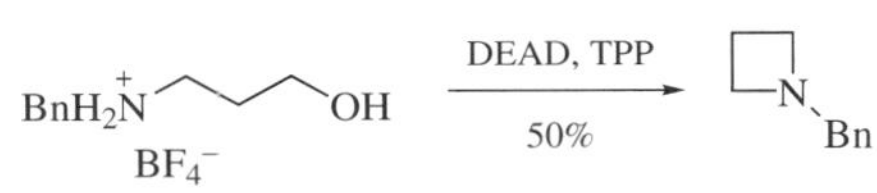

(39)

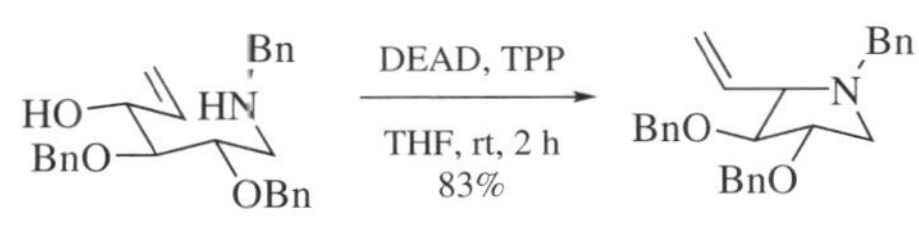

(40)

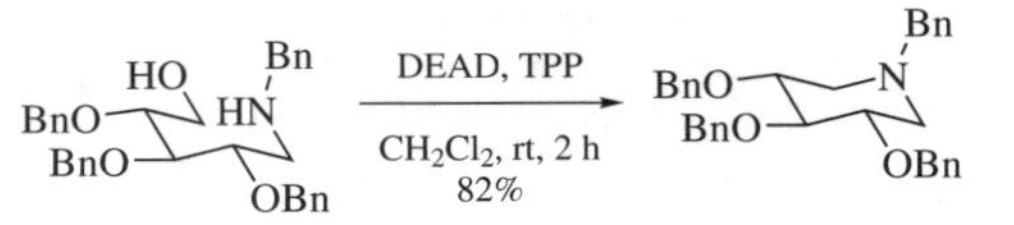

(41)

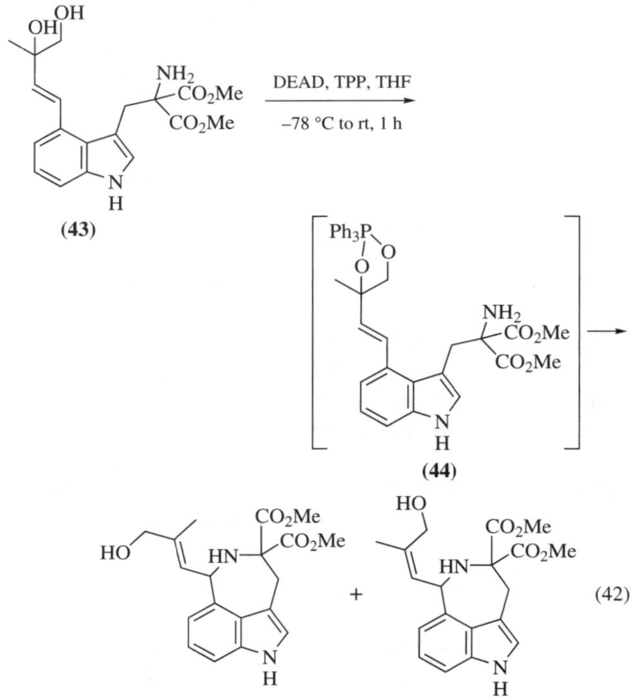

(43)

(44)

(45a) 85%

(45b) 3%

(42)

O-Glycosidation with Phenols.

Monosaccharides possessing a free anomeric hydroxyl group react with phenol, DEAD, and trivalent phosphorus compounds to give phenyl glycosides. The yield is higher with trialkylphosphines (Bu_3P, Pr_3P).[93] Coupling of phenols and 2-α-(phenylthio)- or 2-α-(phenylseleno)-α-D-pyranoses by the use of DEAD and TPP gives aryl 2-deoxy-β-D-glycosides in 70–85% yields with high stereoselectivity (inversion of anomeric center) (eq 43).[94] Intramolecular phenolic glycosidation is also possible.[95]

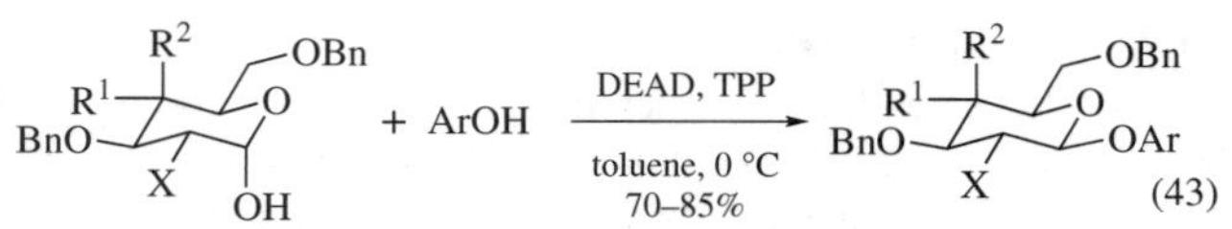

(43)

selectivity = 87:13 to 95:5

R^1 = H, R^2 = BnO or R^1 = BnO, R^2 = H
X = PhS or PhSe; ArOH = 2-naphthol, phenol, or 2-cresol

O-Glycosidation with Carboxylic Acids.

Reaction of carbohydrates having a free anomeric hydroxyl group with carboxylic acids gives the corresponding glycosyl esters (eq 44).[96] Intramolecular glycosidation of (**46**) can be accomplished by the use of the preformed Bu_3P–DEAD betaine at low temperature (eq 45).[97]

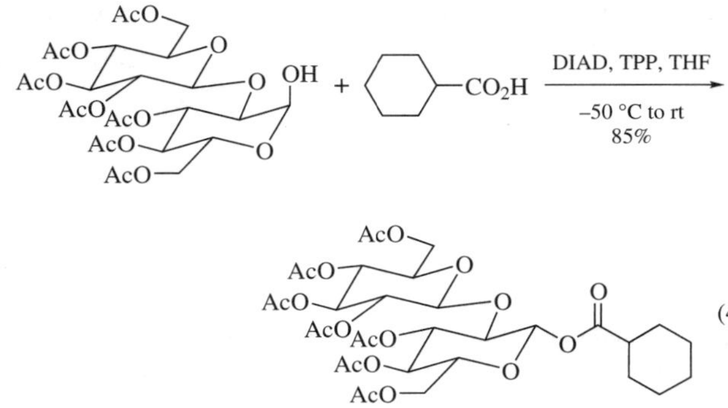

(44)

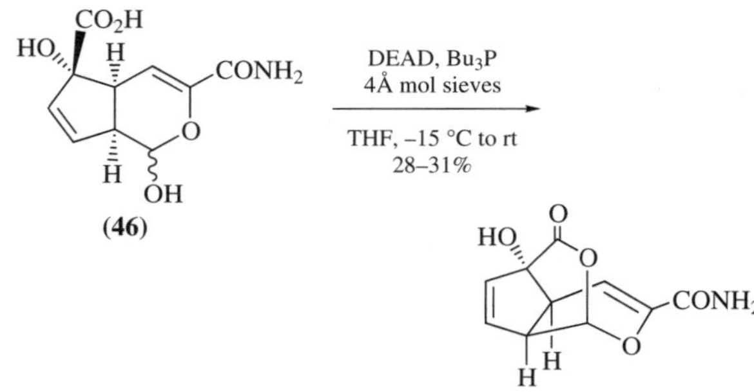

(46)

(45)

Glycosidation with Nitrogen Nucleophiles and *N*-Hydroxy Compounds.

Reaction of an anomeric hydroxyl group with nitrogen nucleophiles and with *N*-hydroxy heterocycles affords the corresponding glycosides.[1,98,99] The α-lactol (**47**)(**48**) (α:β = ca. 8:1) reacts with *N*-hydroxyphthalimide, DIAD, and TPP to give the β-glycoside (**49**) (eq 46).[100]

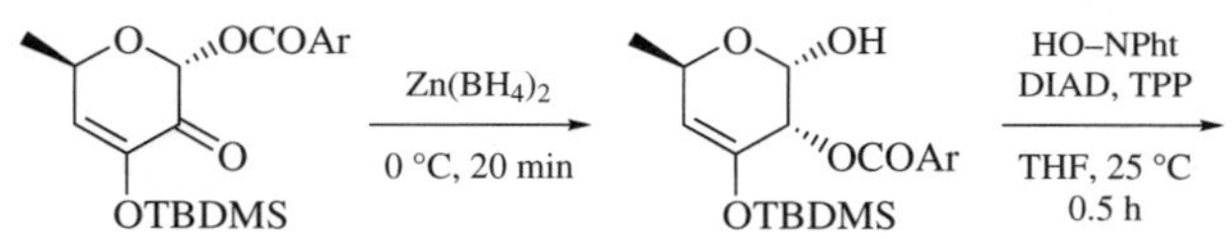

(47)

(48) α:β = ca 8:1

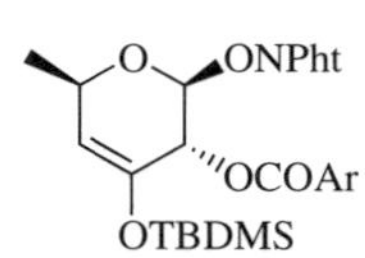

(46)

(49) 53% overall

HO-NPht = 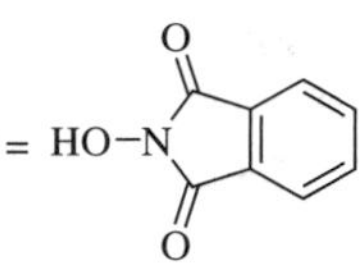

Carbon–Carbon Bond Formation.

Direct coupling of an alcohol with a carbon acid by the use of DEAD/TPP has been limited to only a few examples because of the lack of carbon acids with $pK_a < 13$. If enolizable carbonyl compounds are used, alkylation generally takes place on oxygen.[69a,101] Thus the reaction of 1,3-cyclohexanedione with isopropanol, DEAD, and TPP affords exclusively the *O*-alkylated product (eq 47).[101a] Cyanoacetate reacts at the carbon atom, while diethyl malonate is unreactive.[101a] However, the reaction of diethyl malonate with alcohols in the presence of ADDP and Bu_3P affords *C*-alkylated products in 3–56% yields (eq 48).[12] Ethyl nitroacetate and nitromalonate oxidize the alcohol via an *aci*-nitro ester (eq 49).[102] *o*-Nitroarylacetonitriles of general formula (**50**) undergo alkylation on carbon (eq 50).[16,103]

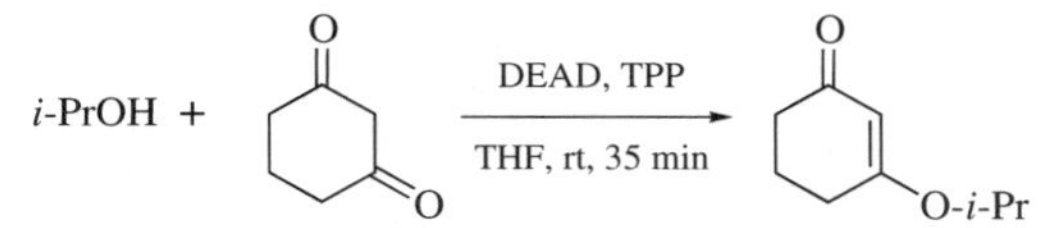

(47)

$$CH_2(CO_2Et)_2 + ROH \xrightarrow[\text{benzene, rt, 24 h; 3–56\%}]{\text{ADDP, Bu}_3\text{P}} RCH(CO_2Et)_2 \quad (48)$$

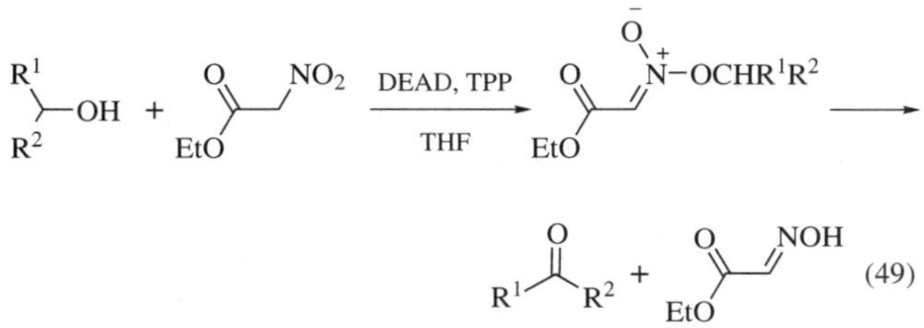

(49)

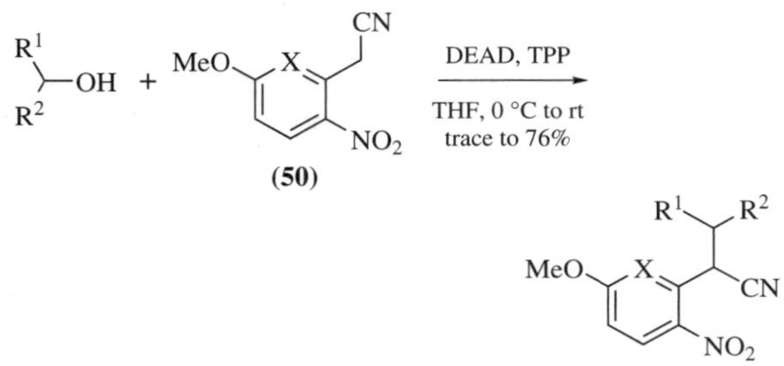

(50)

X = CH or N

γ-Nitro alcohols react with DEAD and TPP to afford α-nitrocyclopropanes in good to excellent yields (eq 51).[104a] On the other hand, nitro alcohols bearing electron-withdrawing or unsaturated substituents at the α-carbon experience exclusive intra- and intermolecular *O*-alkylation and furnish good to excellent yields of alkyl nitronates (eq 52). In contrast, 1-phenylsulfonyl-1-nitro-3-propanol affords the corresponding cyclopropane.[104b] Bis-sulfones readily undergo alkylation by alcohols.[104c] Novel intramolecular S_N2 and S_N2' type arylation occurs with certain aromatic and allylic alcohols (eqs 53 and 54).[105]

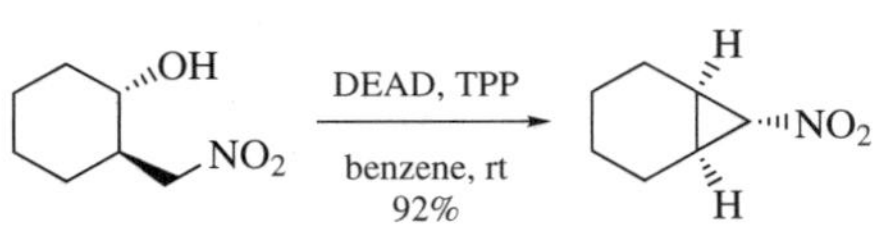

(51)

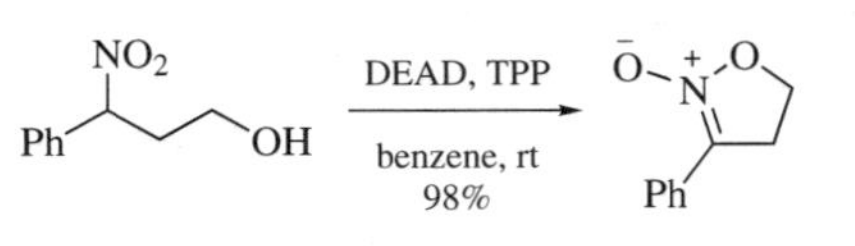

(52)

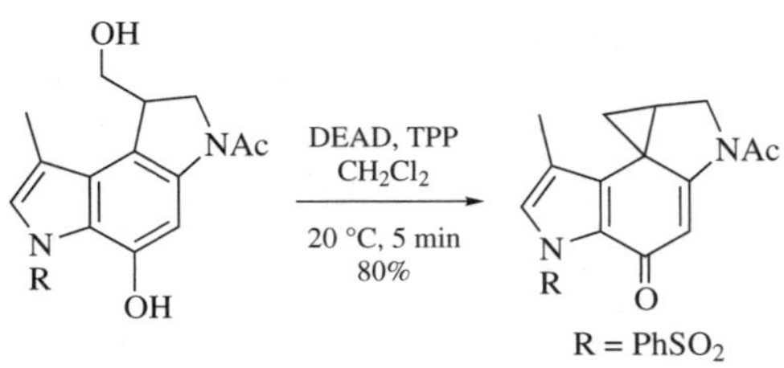

(53)

R = $PhSO_2$

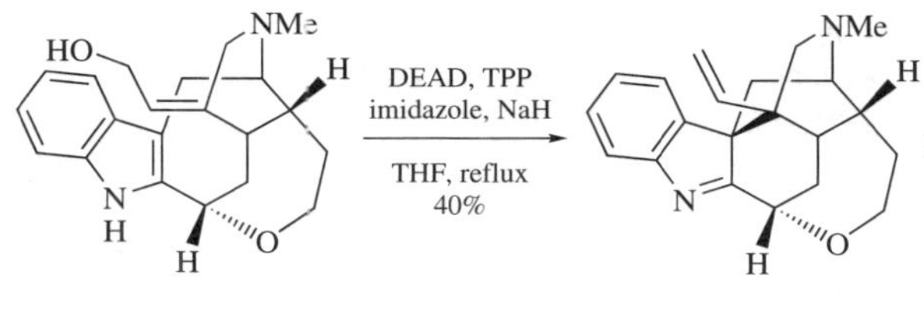

(54)

Miscellaneous Reactions. Dialkyl phosphonates react with alcohols in the presence of DIAD and TPP to give trialkyl phosphites (eq 55).[5a] TPP–DIAD reacts with KHF_2 or HF–pyridine to form difluorotriphenylphosphorane (eq 56).[106]

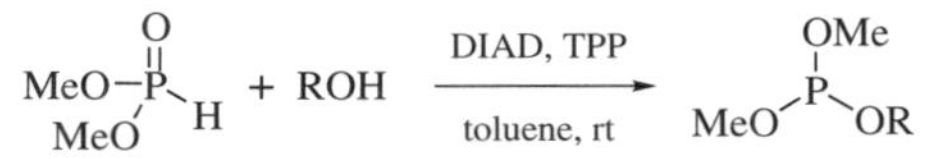

(55)

$$Ph_3P \xrightarrow[\text{rt, sonication; 60\%}]{\text{KHF}_2\text{, MeCN, DIAD}} Ph_3PF_2 \quad (56)$$

1. (a) Mitsunobu, O. *Synthesis* **1981**, 1. (b) Castro, B. R. *Org. React.* **1983**, *29*, 1. (c) Hughes, D. L. *Org. React.* **1992**, *42*, 335. (d) Mitsunobu, O. *Comprehensive Organic Synthesis* **1991**, *6*, 1. (e) Mitsunobu, O. *Comprehensive Organic Synthesis* **1991**, *6*, 65.
2. Freedman, J.; Vaal, M. J.; Huber, E. W. *J. Org. Chem.* **1991**, *56*, 670.
3. (a) Camp, D.; Jenkins, I. D. *J. Org. Chem.* **1989**, *54*, 3049. (b) Camp, D.; Jenkins, I. D. *J. Org. Chem.* **1989**, *54*, 3045. (c) Camp, D.; Jenkins, I. D. *Aust. J. Chem.* **1992**, *45*, 47.
4. Brown, R. F. C.; Jackson, W. R.; McCarthy, T. D. *Tetrahedron Lett.* **1993**, *34*, 1195.
5. (a) Harvey, P.; Jenkins, I. D., unpublished results. (b) Von Itzstein, M.; Jenkins, I. D. *J. Chem. Soc., Perkin Trans. 1* **1986**, 437. (c) Von Itzstein, M.; Jenkins, I. D. *J. Chem. Soc., Perkin Trans. 1* **1987**, 2057. (d) Von Itzstein, M.; Jenkins, I. D. *Aust. J. Chem.* **1984**, *37*, 2447.
6. Camp, D.; Jenkins, I. D. *Aust. J. Chem.* **1988**, *41*, 1835.
7. Von Itzstein, M.; Mocerino, M. *Synth. Commun.* **1990**, *20*, 2049; Von Itzstein, M.; Jenkins, M. J.; Mocerino, M. *Carbohydr. Res.* **1990**, *208*, 287.
8. Amos, R. A.; Emblidge, R. W.; Havens, N. *J. Org. Chem.* **1983**, *48*, 3598.
9. Spry, D. O.; Bhala, A. R. *Heterocycles* **1986**, *24*, 1653.
10. Arnold, L. D.; Assil, H. I.; Vederas, J. C. *J. Am. Chem. Soc.* **1989**, *111*, 3973.
11. Dialkyl phosphonates have been reported to promote this reaction:Osikawa, T.; Yamashita, M. *Shizuoka Daigaku Kogakubu Kenkyu Hokoku* **1984**, 37 (*Chem. Abstr.* **1985**, *103*, 142 095d) .
12. Tsunoda, T.; Yamamiya, Y.; Ito, S. *Tetrahedron Lett.* **1993**, *34*, 1639.
13. Volante, R. P. *Tetrahedron Lett.* **1981**, *22*, 3119.
14. Varassi, M.; Walker, K. A. M.; Maddox, M. L. *J. Org. Chem.* **1987**, *52*, 4235.
15. Hughes, D. L.; Reamer, R. A.; Bergan, J. J.; Grabowski, E. J. J. *J. Am. Chem. Soc.* **1988**, *110*, 64.
16. Macor, J. E.; Wehner, J. M. *Heterocycles* **1993**, *35*, 349.
17. Camp, D.; Hanson, G.; Jenkins, I. D., *J. Org. Chem.* **1995**, *60*, 2977.
18. Mitsunobu, O.; Kimura, J.; Iizumi, K.; Yanagida, N. *Bull. Chem. Soc. Jpn.* **1976**, *49*, 510.
19. (a) Pautard, A. M.; Evans, S. A.Jr., *J. Org. Chem.* **1988**, *53*, 2300. (b) Pautard-Cooper, A.; Evans, S. A.Jr., *J. Org. Chem.* **1989**, *54*, 2485. (c) Pautard-Cooper, A.; Evans, S. A.Jr., *J. Org. Chem.* **1989**, *54*, 4974. (d) Pautard-Cooper, A.; Evans, S. A.Jr., *Tetrahedron* **1991**, *47*, 1603.

20. (a) Pawlak, J. L.; Padykura, R. E.; Kronis, J. D.; Aleksejczyk, R. A.; Berchtold, G. A. *J. Am. Chem. Soc.* **1989**, *111*, 3374. (b) Jenkins, I. D.; Richards, G., unpublished results.
21. Mitsunobu, O.; Kimura, K.; Fujisawa, Y. *Bull. Chem. Soc. Jpn.* **1972**, *45*, 245.
22. (a) Grynkiewicz, G. *Pol. J.Chem.* **1979**, *53*, 2501. (b) Martin, S. P.; Dodge, J. A. *Tetrahedron Lett.* **1991**, *32*, 3017.; Dodge, J. A.; Trujillo, J. I.; Presnell, M. *J. Org. Chem.* **1994**, *59*, 234.
23. Georg, G. I.; Kant, J.; Gill, H. S. *J. Am. Chem. Soc.* **1987**, *109*, 1129.
24. Ito, Y.; Kobayashi, Y.; Kawabata, T.; Takase, M.; Terashima, S. *Tetrahedron* **1989**, *45*, 5767.
25. Evans, D. A.; Gauchet-Prunet, J. A.; Carreira, E. M.; Charette, A. B. *J. Org. Chem.* **1991**, *56*, 741.
26. Danishefsky, S. L.; Cabal, M. P.; Chow, K. *J. Am. Chem. Soc.* **1989**, *111*, 3456.
27. Breslow, R.; Link, T. *Tetrahedron Lett.* **1992**, *33*, 4145.
28. (a) Mori, K.; Otsuka, T.; Oda, M. *Tetrahedron* **1984**, *40*, 2929. (b) Mori, K.; Ikunaka, M. *Tetrahedron* **1984**, *40*, 3471. (c) Mori, K.; Ishikura, M. *Justus Liebigs Ann. Chem./Liebigs Ann. Chem.* **1989**, 1263. (d) Mori, K.; Watanabe, H. *Tetrahedron Lett.* **1984**, *25*, 6025.
29. Saïah, M.; Bessodes, M.; Antonakis, K. *Tetrahedron Lett.* **1992**, *33*, 4317.
30. Weinges, K.; Haremsa, S.; Maurer, W. *Carbohydr. Res.* **1987**, *164*, 453.
31. (a) See, For example:Lindhorst, T. K.; Thiem, J. *Justus Liebigs Ann. Chem./Liebigs Ann. Chem.* **1990**, 1237. (b) Sakamoto, S.; Tsuchiya, T.; Umezawa, S.; Umezawa, H. *Bull. Chem. Soc. Jpn.* **1987**, *60*, 1481. (c) Palmer, C. F.; Parry, K. P.; Roberts, S. M.; Ski, V. *J. Chem. Soc., Perkin Trans. 1* **1991**, 2051. (d) Bartlett, P. A.; Meadows, J. D.; Ottow, E. *J. Am. Chem. Soc.* **1984**, *106*, 5304. (e) Jeker, N.; Tamm, C. *Helv. Chim. Acta* **1988**, *71*, 1904. See also references [1]a and [1]c.
32. (a) See, for example:Loibner, H.; Zbiral, E. *Helv. Chim. Acta* **1977**, *60*, 417. (b) Goya, S.; Takadate, A.; Fujino, H.; Irikura, M. *Yakugaku Zasshi* **1981**, *101*, 1064.
33. (a) Kimura, J.; Fujisawa, Y.; Yoshizawa, T.; Fukuda, K.; Mitsunobu, O. *Bull. Chem. Soc. Jpn.* **1979**, *52*, 1191. (b) Guthrie, R. D.; Jenkins, I. D.; Thang, S.; Yamasaki, R. *Carbohydr. Res.* **1983**, *121*, 109; **1988**, *176*, 306. (c) Jenkins, I. D.; Goren, M. B. *Chem. Phys. Lipids* **1986**, *41*, 225.
34. Warmerdam, E. G. J. C.; Brussee, J.; Kruse, C. G.; van der Gen, A. *Tetrahedron* **1993**, *49*, 1063.
35. (a) Kang, S. H.; Kim, W. J. *Tetrahedron Lett.* **1989**, *30*, 5915. (b) Jarosz, S.; Glodek, J.; Zamojski, A. *Carbohydr. Res.* **1987**, *163*, 289.
36. (a) See, for example:Grossé-Kobo, B.; Mosset, P.; Grée, R. *Tetrahedron Lett.* **1989**, *30*, 4235. (b) Balan, A.; Ziffer, H. *Chem. Commun./J. Chem. Soc., Chem. Commun.* **1990**, 175. (c) Clive, D. L. J.; Daigneault, S. *J. Org. Chem.* **1991**, *56*, 3801.
37. (a) Farina, V. *Tetrahedron Lett.* **1989**, *30*, 6645, and references cited therein. (b) Jefford, C. W.; Moulin, M.-C. *Helv. Chim. Acta* **1991**, *74*, 336. (c) Carda, M.; Marco, J. A. *Tetrahedron* **1992**, *48*, 9789. (d) Lumin, S.; Falck, J. R.; Capdevila, J.; Karara, A. *Tetrahedron Lett.* **1992**, *33*, 2091. (e) Charette, A. B.; Cote, B. *Tetrahedron Lett.* **1993**, *34*, 6833.
38. (a) Kanematsu, K.; Yoshiyasu, T.; Yoshida, M. *Chem. Pharm. Bull.* **1990**, *38*, 1441. (b) Moree, W. J.; van der Marel, G. A.; Liskamp, R. M. J. *Tetrahedron Lett.* **1992**, *33*, 6389. (c) Strijtveen, B.; Kellogg, R. M. *Recl. Trav. Chim. Pays-Bas* **1987**, *106*, 539. (d) Jenkins, I. D.; Thang, S. *Aust. J. Chem.* **1984**, *37*, 1925.
39. Kpegba, K.; Metzner, P. *Synthesis* **1989**, 137.
40. Rollin, P. *Tetrahedron Lett.* **1986**, *35*, 4169.
41. (a) Mlotkowska, B.; Zwierzak, A. *Pol. J.Chem.* **1979**, *53*, 359. (b) Campbell, D. A. *J. Org. Chem.* **1992**, *57*, 6331.; Campbell, D. A.; Bermak, J. C. *J. Org. Chem.* **1994**, *59*, 658.
42. Mlotkowska, B.; Wartalowska-Graczyk, M. *J. Prakt. Chem.* **1987**, *329*, 735.
43. (a) Camp, D.; Jenkins, I. D. *Aust. J. Chem.* **1990**, *43*, 161. (b) Dancy, I.; Laupichler, L.; Rollin, P.; Thiem, J. *Synlett* **1992**, 283, and references cited therein.
44. (a) Loibner, H.; Zbiral, E. *Helv. Chim. Acta* **1976**, *59*, 2100. (b) Peterson, M. L.; Vince, R. *J. Med. Chem.* **1991**, *34*, 2787.
45. (a) Still, W. C.; Galynker, I. *J. Am. Chem. Soc.* **1982**, *104*, 1774. (b) Galynker, I.; Still, W. C. *Tetrahedron Lett.* **1982**, *23*, 4461.
46. (a) Mulzer, J.; Brüntrup, G.; Chucholowski, A. *Angew. Chem., Int. Ed. Engl.* **1979**, *18*, 622. (b) Mulzer, J.; Lammer, O. *Angew. Cheml* **1983**, Suppl., 887. (c) Adam, W.; Narita, N.; Nishizawa, Y. *J. Am. Chem. Soc.* **1984**, *106*, 1843.
47. Justus, K.; Steglich, W. *Tetrahedron Lett.* **1991**, *32*, 5781.
48. (a) Tsutsui, H.; Mitsunobu, O. *Tetrahedron Lett.* **1984**, *25*, 2163. (b) Ohta, K.; Miyagawa, O.; Tsutsui, H.; Mitsunobu, O. *Bull. Chem. Soc. Jpn.* **1993**, *66*, 523.
49. (a) For recent reports of macrolactonization, see, for example:Barrett, A. G. M.; Carr, R. A. E.; Attwood, S. V.; Richardson, G.; Walshe, N. D. A. *J. Org. Chem.* **1986**, *51*, 4840. (b) Smith, A. B., III; Noda, I.; Remiszewski, S. W.; Liverton, N. J.; Zibuck, R. *J. Org. Chem.* **1990**, *55*, 3977. (c) White, J. D.; Kawasaki, M. *J. Am. Chem. Soc.* **1990**, *112*, 4991. (d) Hanessian, S.; Chemla, P. *Tetrahedron Lett.* **1991**, *32*, 2719.
50. (a) Riehter, L. S.; Gadek, T. R. *Tetrahedron Lett.* **1994**, *35*, 4705. (b) See, for example:Heumann, A.; Faure, R. *J. Org. Chem.* **1993**, *58*, 1276, and references cited therein.
51. For alkyl aryl ether formation without allylic rearrangement, see, for example:Pirrung, M. C.; Brown, W. L.; Rege, S.; Laughton, P. *J. Am. Chem. Soc.* **1991**, *113*, 8561.
52. For alkyl aryl ether formation with allylic rearrangement, see, for example:Danishefsky, S.; Berman, E. M.; Ciufolini, M.; Etheredge, S. J.; Segmuller, B. E. *J. Am. Chem. Soc.* **1985**, *107*, 3891.
53. Walba, D. M.; Eidman, K. F.; Haltiwanger, R. C. *J. Org. Chem.* **1989**, *54*, 4939.
54. (a) Sugihara, H.; Mabuchi, H.; Hirata, M.; Imamoto, T.; Kawamatsu, Y. *Chem. Pharm. Bull.* **1987**, *35*, 1930. (b) Macor, J. E.; Ryan, K.; Newman, M. E. *Tetrahedron* **1992**, *48*, 1039.
55. Clive, D. L. J.; Kellner, D. *Tetrahedron Lett.* **1991**, *32*, 7159.
56. See, for example:Stachulski, A. V. *J. Chem. Soc., Perkin Trans. 1* **1991**, 3065.
57. See, for example:Brown, P.; Calvert, S. H.; Chapman, P. C. A.; Cosham, S. C.; Eglington, A. J.; Elliot, R. L.; Harris, M. A.; Hinks, J. D.; Lowther, J.; Merrikin, D. J.; Pearson, M. J.; Ponsford, R. J.; Syms, J. V. *J. Chem. Soc., Perkin Trans. 1* **1991**, 881.
58. (a) Carlock, J. T.; Mack, M. P. *Tetrahedron Lett.* **1978**, 5153. (b) Robinson, P. L.; Barry, C. N.; Bass, S. W.; Jarvis, S. E.; Evans, S. A.Jr., *J. Org. Chem.* **1983**, *48*, 5396. (c) Bestmann, H. J.; Pecher, B.; Riemer, C. *Synthesis* **1991**, 731.
59. (a) Mitsunobu, O.; Kudo, T.; Nishida, M.; Tsuda, N. *Chem. Lett.* **1980**, 1613. (b) Brandstetter, H. H.; Zbiral, E. *Helv. Chim. Acta* **1980**, *63*, 327. (c) Mark, E.; Zbiral, E.; Brandstetter, H. H. *Monatsh. Chem.* **1980**, *111*, 289. (d) Guthrie, R. D.; Jenkins, I. D.; Yamasaki, R.; Skelton, B. W.; White, A. H. *J. Chem. Soc., Perkin Trans. 1* **1981**, 2328. (e) Guthrie, R. D.; Jenkins, I. D. *Aust. J. Chem.* **1981**, *34*, 1997. (f) Guthrie, R. D.; Jenkins, I. D.; Yamasaki, R. *Carbohydr. Res.* **1980**, *85*, C5. (g) Guthrie, R. D.; Jenkins, I. D.; Yamasaki, R.; Skelton, B. W.; White, A. H. *J. Chem. Soc., Perkin Trans. 1* **1981**, 2328. (h) McGowan, D. A.; Berchtold, G. A. *J. Org. Chem.* **1981**, *46*, 2381.
60. Palomino, E.; Schaap, A. P.; Heeg, M. J. *Tetrahedron Lett.* **1989**, *30*, 6797.
61. Penz, G.; Zbiral, E. *Monatsh. Chem.* **1981**, *112*, 1045.
62. (a) Iida, H.; Yamazaki, N.; Kibayashi, C. *J. Org. Chem.* **1987**, *52*, 1956. (b) Yamazaki, N.; Kibayashi, C. *Tetrahedron Lett.* **1988**, *29*, 5767.
63. Audia, J. E.; Colocci, N. *Tetrahedron Lett.* **1991**, *32*, 3779.
64. (a) Wada, M.; Sano, T.; Mitsunobu, O. *Bull. Chem. Soc. Jpn.* **1973**, *46*, 2833. (b) Akita, H.; Yamada, H.; Matsukura, H.; Nakata, T.; Oishi, T. *Chem. Pharm. Bull.* **1990**, *38*, 2377. (c) Iimori, T.; Ohtsuka, Y.; Oishi, T. *Tetrahedron Lett.* **1991**, *32*, 1209.
65. For an unsuccessful result, see:Roemmele, R. C.; Rapoport, H. *J. Org. Chem.* **1989**, *54*, 1866.

66. (a) Lal, B.; Pramanik, B. N.; Manhas, M. S.; Bose, A. K. *Tetrahedron Lett.* **1977**, 1977. (b) Pearson, W. H.; Hines, J. V. *J. Org. Chem.* **1989**, *54*, 4235.
67. (a) Viaud, M. C.; Rollin, P. *Synthesis* **1990**, 130. (b) Gajda, T.; Matusiak, M. *Synthesis* **1992**, 367. (c) Duclos, O.; Duréault, A.; Depezay, J. C. *Tetrahedron Lett.* **1992**, *33*, 1059. (d) Duclos, O.; Mondange, M.; Duréault, A.; Depezay, J. C. *Tetrahedron Lett.* **1992**, *33*, 8061.
68. Fabiano, E.; Golding, B. T.; Sadeghi, M. M. *Synthesis* **1987**, 190.
69. (a) Morimoto, H.; Furukawa, T.; Miyazima, K.; Mitsunobu, O. *Chem. Lett.* **1973**, 821. For *C*- and *N*-alkylation of imides and related compounds, see also: (b) Sammes, P. G.; Thetford, D. *J. Chem. Soc., Perkin Trans. 1* **1989**, 655. (c) Kim, T. H.; Rapoport, H. *J. Org. Chem.* **1990**, *55*, 3699.
70. (a) Koppel, I.; Koppel, J.; Degerbeck, F; Grehn, L.; Ragnarsson, U. *J. Org. Chem.* **1991**, *56*, 7172. See also, for example: (b) Slusarska, E.; Zwierzak, A. *Justus Liebigs Ann. Chem./Liebigs Ann. Chem.* **1986**, 402. (c) Tanner, D.; Birgersson, C.; Dhaliwal, H. K. *Tetrahedron Lett.* **1990**, *31*, 1903. (d) Arnould, J. C.; Landier, F.; Pasquet, M. J. *Tetrahedron Lett.* **1992**, *33*, 7133. (e) Macor, J. E.; Blank, D. H.; Post, R. J.; Ryan, K. *Tetrahedron Lett.* **1992**, *33*, 8011. (f) Campbell, J. A.; Hart, D. J. *J. Org. Chem.* **1993**, *58*, 2900.
71. Chong, J. M.; Park, S. B. *J. Org. Chem.* **1992**, *57*, 2220.
72. (a) Miller, M. J.; Mattingly, P. G.; Morrison, M. A.; Kerwin, J. F.Jr., *J. Am. Chem. Soc.* **1980**, *102*, 7026. (b) Teng, M.; Miller, M. J. *J. Am. Chem. Soc.* **1993**, *115*, 548.
73. (a) Bose, A. K.; Sahu, D. P.; Manhas, M. S. *J. Org. Chem.* **1981**, *46*, 1229. (b) Bose, A. K.; Manhas, M. S.; Sahu, D. P.; Hegde, V. R. *Can. J. Chem.* **1984**, *62*, 2498.
74. Huang, N. Z.; Kalish, V. J.; Miller, M. J. *Tetrahedron* **1990**, *46*, 8067.
75. Salituro, G. M.; Townsend, C. A. *J. Am. Chem. Soc.* **1990**, *112*, 760, and references cited therein.
76. (a) For the reaction of hydroxyamino acid derivatives with DEAD–TPP and related systems, see, for example:Nakajima, K.; Tanaka, T.; Neya, M.; Okawa, K. *Bull. Chem. Soc. Jpn.* **1982**, *55*, 3237. (b) Nakajima, K.; Sasaki, H.; Neya, M.; Morishita, M.; Sakai, S.; Okawa, K. In *Peptide Chemistry 1982*; Sakakibara, S., Ed.; Protein Research Foundation:Osaka, **1983**; pp 19–24.
77. Lotz, B. T.; Miller, M. J. *J. Org. Chem.* **1993**, *58*, 618, and references cited therein.
78. Toyota, A.; Katagiri, N.; Kaneko, C. *Synth. Commun.* **1993**, *23*, 1295.
79. Katagiri, N.; Toyota, A.; Shiraishi, T.; Sato, H.; Kaneko, C. *Tetrahedron Lett.* **1992**, *33*, 3507.
80. (a) For related reports, see, for example:Toyota, A.; Katagiri, N.; Kaneko, C. *Chem. Pharm. Bull.* **1992**, *40*, 1039. (b) Katagiri, N.; Sato, H.; Arai, S.; Toyota, A.; Kaneko, C. *Heterocycles* **1992**, *34*, 1097. (c) Toyota, A.; Katagiri, N.; Kaneko, C. *Heterocycles* **1993**, *36*, 1625. (d) Jenny, T. F.; Previsani, N.; Benner, S. A. *Tetrahedron Lett.* **1991**, *32*, 7029. (e) Jenny, T. F.; Schneider, K. C.; Benner, S. A. *Nucleosides Nucleosides* **1992**, *11*, 1257. (f) Overberger, C. G.; Chang, J. Y. *Tetrahedron Lett.* **1989**, *30*, 51. (g) Bestmann, H. J.; Roth, D. *Angew. Chem., Int. Ed. Engl.* **1990**, *29*, 99. (h) See ref. [31] (b) .
81. For *O*-alkylation of purines, see, for example:Schulz, B. S.; Pfleiderer, W. *Helv. Chim. Acta* **1987**, *70*, 210, and references cited therein.
82. (a) Lipshutz, B. H.; Miller, T. A. *Tetrahedron Lett.* **1990**, *31*, 5253. For neighboring group participation in related reactions, see: (b) Campbell, J. A.; Hart, D. J. *Tetrahedron Lett.* **1992**, *33*, 6247.
83. Branquet, E.; Durand, P.; Vo-Quang, L.; Le Goffic, F. *Synth. Commun.* **1993**, *23*, 153.
84. (a) Arnold, L. D.; Kalantar, T. H.; Vederas, J. C. *J. Am. Chem. Soc.* **1985**, *107*, 7105. (b) Ramer, S. E.; Moore, R. N.; Vederas, J. C. *Can. J. Chem.* **1986**, *64*, 706. See also: (c) Parker, W. L.; Rathnum, M. L.; Liu, W.-C. *J. Antibiot.* **1982**, *35*, 900.
85. (a) Wojciechowska, H.; Pawlowicz, R.; Andruszkiewicz, R.; Grzybowska, J. *Tetrahedron Lett.* **1978**, 4063. (b) Andruszkiewicz, R.; Grzybowska, J.; Wojciechowska, H. *Pol. J. Chem.* **1981**, *55*, 67.
86. Coleman, R. S.; Carpenter, A. J. *J. Org. Chem.* **1992**, *57*, 5813.
87. (a) Wipf, P.; Miller, C. P. *Tetrahedron Lett.* **1992**, *33*, 6267. See also: (b) Galéotti, N.; Montagne, C.; Poncet, J.; Jouin, P. *Tetrahedron Lett.* **1992**, *33*, 2807.
88. Roush, D. M.; Patel, M. M. *Synth. Commun.* **1985**, *15*, 675.
89. Pfister, J. R. *Synthesis* **1984**, 969.
90. Sammes, P. G.; Smith, S. *Chem. Commun./J. Chem. Soc., Chem. Commun.* **1983**, 682.
91. (a) Bernotas, R. C.; Cube, R. V. *Tetrahedron Lett.* **1991**, *32*, 161. (b) Bernotas, R. C. *Tetrahedron Lett.* **1990**, *31*, 469. (c) Holladay, M. W.; Nadzan, A. M. *J. Org. Chem.* **1991**, *56*, 3900.
92. Kozikowski, A. P.; Okita, M. *Tetrahedron Lett.* **1985**, *26*, 4043.
93. Grynkiewicz, G. *Pol. J. Chem.* **1979**, *53*, 1571.
94. Roush, W. R.; Lin, X.-F. *J. Org. Chem.* **1991**, *56*, 5740.
95. Delgado, A.; Clardy, J. *J. Org. Chem.* **1993**, *58*, 2862.
96. (a) Smith, A. B., III; Hale, K. J.; Rivero, R. A. *Tetrahedron Lett.* **1986**, *27*, 5813. (b) de Mesmaeker, A.; Hoffmann, P.; Ernst, B. *Tetrahedron Lett.* **1989**, *30*, 3773. (c) Smith, A. B., III; Rivero, R. A.; Hale, K. J.; Vaccaro, H. A. *J. Am. Chem. Soc.* **1991**, *113*, 2092. (d) Smith, A. B., III; Hale, K. J.; Vaccaro, H. A.; Rivero, R. A. *J. Am. Chem. Soc.* **1991**, *113*, 2112.
97. Smith, A. B., III; Sulikowski, G. A.; Sulikowski, M. M.; Fujimoto, K. *J. Am. Chem. Soc.* **1992**, *114*, 2567.
98. For glycosidation with HN_3, see:Schörkhuber, W.; Zbiral, E. *Justus Liebigs Ann. Chem /Liebigs Ann. Chem.* **1980**, 1455.
99. For glycosidation with *N*-hydroxy heterocycles, see:Grochowski, E.; Stepowska, H. *Synthesis* **1988**, 795, and references cited therein.
100. Nicolaou, K. C.; Groneberg, R. D. *J. Am. Chem. Soc.* **1990**, *112*, 4085.
101. (a) Kurihara, T.; Sugizaki, M.; Kime, I.; Wada, M.; Mitsunobu, O. *Bull. Chem. Soc. Jpn.* **1981**, *54*, 2107. (b) Bajwa, J. S.; Anderson, R. C. *Tetrahedron Lett.* **1990**, *31*, 6973. (c) Nakayama, E.; Watanabe, K.; Miyauchi, M.; Fujimoto, K.; Ide, J. *J. Antibiot.* **1990**, *43*, 1122. (d) Czernecki, S.; Valéry, J.-M. *Synthesis* **1991**, 239.
102. Mitsunobu, O.; Yoshida, N. *Tetrahedron Lett.* **1981**, *22*, 2295.
103. Macor, J. E.; Wehner, J. M. *Tetrahedron Lett.* **1991**, *32*, 7195.
104. (a) Yu, J.; Falck, J. R.; Mioskowski, C. *J. Org. Chem.* **1992**, *57*, 3757. (b) Falck, J. R.; Yu, J. *Tetrahedron Lett.* **1992**, *33*, 6723. (c) Yu, J.; Cho, H.-S.; Falck, J. R. *J. Org. Chem.* **1993**, *58*, 5892.
105. (a) Magnus, P.; Gallagher, T.; Schulz, J.; Or, Y.-S.; Ananthanarayan, T. P. *J. Am. Chem. Soc.* **1987**, *109*, 2706. (b) Boger, D. L.; Wysocki, R. J.Jr., Ishizaki, T. *J. Am. Chem. Soc.* **1990**, *112*, 5230. (c) Magnus, P.; Mugrage, B.; DeLuca, M. R.; Cain, G. A. *J. Am. Chem. Soc.* **1990**, *112*, 5220.
106. Harvey, P.; Jenkins, I. D. *Tetrahedron Lett.* **1994**, *35*, 9775.

Ian D. Jenkins
Griffith University, Brisbane, Australia

Oyo Mitsunobu
Aoyama Gakuin University, Tokyo, Japan

Triphenylphosphine Hydrobromide

$Ph_3P\cdot HBr$

[6399-81-1] $C_{18}H_{16}BrP$ (MW 343.21)

(mild source of anhydrous HBr;[1] catalyst for formation of THP ethers from tertiary alcohols;[2] preparation of phosphonium salts[3,4])

Alternate Name: triphenylphosphonium bromide.
Physical Data: mp 196 °C (dec).
Solubility: sol CH_2Cl_2, $CHCl_3$; slightly sol THF, benzene; insol ether.
Form Supplied in: commercially available white powder.
Analysis of Reagent Purity: acidometric titration.
Preparative Methods: addition of anhydrous Hydrogen Bromide to an ethereal solution of Triphenylphosphine;[2] addition of Ph_3P to 48% aqueous HBr followed by $CHCl_3$ extraction and drying.[1]
Purification: recrystallization from CH_2Cl_2/Et_2O, or by washing with warm EtOAc to remove excess Ph_3P.
Handling, Storage, and Precautions: corrosive and hygroscopic; keep dry.

Source of Anhydrous HBr. Triphenylphosphine hydrobromide (**1**) is a convenient, crystalline source of anhydrous HBr for a wide variety of applications, including as a gas.[1] This reagent has found an important use in the glycosidation of glycals[5,6] without interference from allylic (Ferrier) rearrangement (eq 1). There is a distinct preference for the α-anomer, even when the incoming nucleophile is not sterically demanding. Other commonly used acid sources (Hydrogen Chloride, Hydrogen Fluoride, Amberlyst H-15 resin, p-Toluenesulfonic Acid) and Lewis acids (Boron Trifluoride Etherate) catalyze allylic rearrangement.

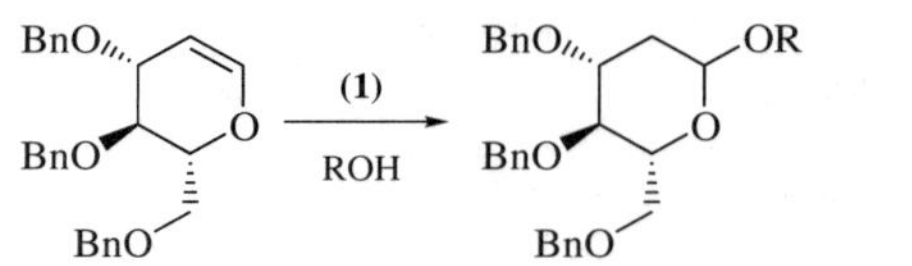

(1)

ROH = 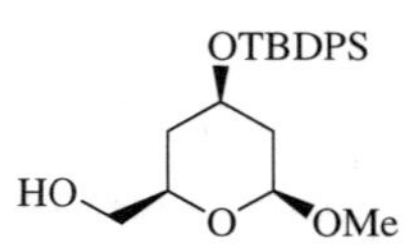

(structure above)	60% yield, α:β = >95:5
$MeCO_2H$	71% yield, α:β = 92:8
MeOH	87% yield, α:β = 70:30

Catalyst for Formation of THP Ethers. Tetrahydropyranylation, particularly of tertiary alcohols, can be catalyzed by (**1**) (eq 2).[7] Depending on the substrate, these reactions are complete in about 20 h, and the conditions are quite mild when compared with those needed with catalysts such as Pyridinium p-Toluenesulfonate. Other acid catalysts such as TsOH can lead to elimination or polymerization of sensitive tertiary alcohols.

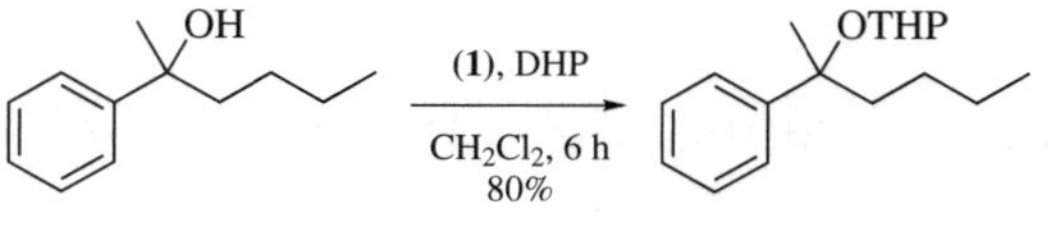

(2)

Preparation of Phosphorus Ylides. Treatment of alkyl chlorides with (**1**) results in the formation of phosphonium salts. Rather than a direct displacement of chloride by phosphorus, this reaction proceeds by initial conversion of RCl to RBr, followed by formation of the phosphonium salt.[8] A wide variety of primary alcohols can be converted directly to the phosphonium bromides in high yield by treatment with (**1**).[3] Both aliphatic and allylic alcohols react with (**1**) in sealed tubes at 160 °C within 2 h (eq 3). Secondary (and presumably tertiary) alcohols suffer dehydration under these conditions. Allylic alcohols (such as the carotenoid precursor (**2**))[9] do not undergo allylic transposition (eq 4), although there have been cases where reactive alkenes have produced phosphonium salts upon heating with (**1**).[10]

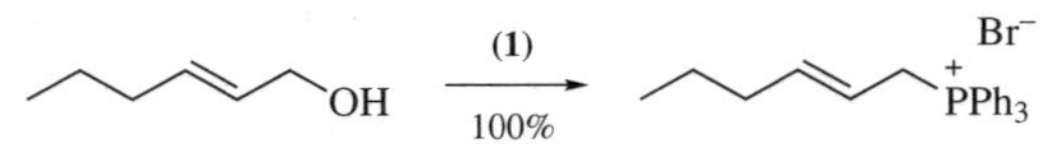

(3)

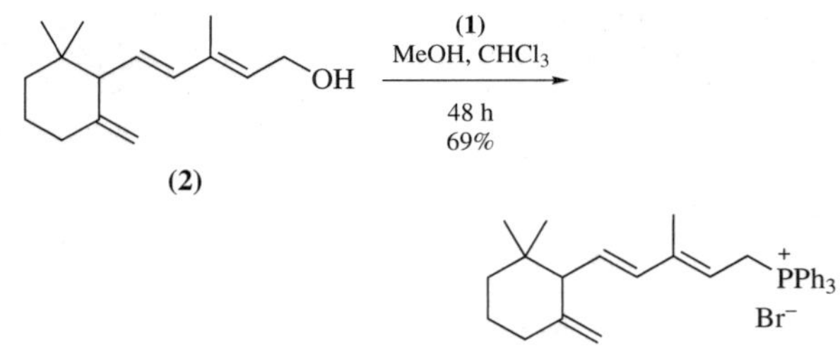

(4)

Under similar conditions (heating in a sealed tube for 4 h at 180 °C), lactones react to form the free acid phosphonium salt in excellent yield (eq 5).[3] Reactive alkenes such as enol ethers,[4] indene, and α,β-unsaturated carbonyl compounds[11] also undergo addition of (**1**) when other nucleophiles are not present. This reaction has been used to prepare a novel three-carbon Wittig reagent bearing a protected aldehyde function.[11]

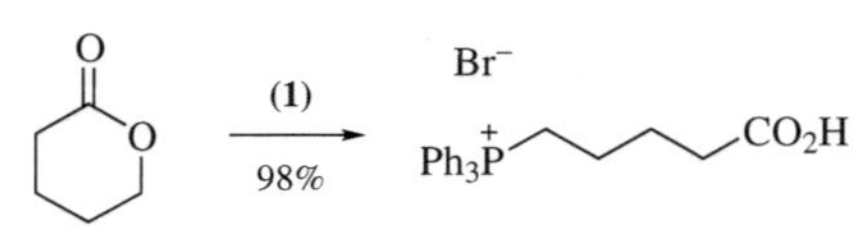

(5)

---

1. Hercouet, A.; Le Corre, M. *Synthesis* **1988**, 157.
2. Bolitt, V.; Mioskowski, C.; Shin, D. S.; Falck, J. R. *Tetrahedron Lett.* **1988**, *29*, 4583.
3. Hamanaka, N.; Kosuge, S.; Iguchi, S. *Synlett* **1990**, 139.
4. Ousset, J. B.; Mioskowski, C.; Yang, Y.-L.; Falck, J. R. *Tetrahedron Lett.* **1984**, 5903.
5. Bolitt, V.; Mioskowski, C.; Lee, S. G.; Falck, J. R. *J. Org. Chem.* **1990**, *55*, 5812.
6. Kaila, N.; Blumenstein, M.; Bielawska, H.; Franck, R. W. *J. Org. Chem.* **1992**, *57*, 4576.
7. *Fieser & Fieser* **1990**, *15*, 353.
8. Hercouet, A.; Le Corre, M. *Phosphorus Sulfur/Phosphorus Sulfur Silicon* **1987**, *29*, 111.
9. Sliwka, H. R.; Noekleby, O. W.; Liaaen-Jensen, S. *Acta Chem. Scand.* **1987**, *4B*, 245.
10. McCullough, K. J. *Tetrahedron Lett.* **1982**, *23*, 2223.
11. Viala, J.; Santelli, M. *Synthesis* **1988**, 395.

Mark S. Meier
University of Kentucky, Lexington, KY, USA

Triphenylphosphine Selenide

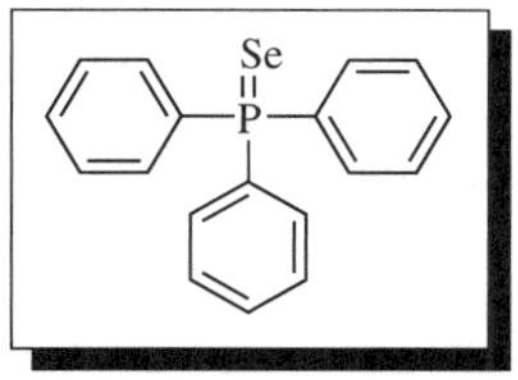

[3878-44-2] $C_{18}H_{15}PSe$ (MW 341.26)

(selenium-transferring reagent; selenization of phosphites; selenization of *H*-phosphonates; selenization of *H*-phosphonothioates; conversion of epoxides into alkenes)

Alternate Name: triphenylphosphane selenide.
Physical Data: mp 187–188 °C;[2] δ $^{31}P = 36.5$ ppm ($^1J_{PSe} = 740$ Hz).
Solubility: very soluble in dichloromethane, chloroform, pyridine, and THF; moderately soluble on heating in acetonitrile, benzene, 1-butanol, ethanol, and methanol; insoluble in ether and water.[2]
Form Supplied in: white solid; commercially available.
Preparative Methods: prepared by selenization of triphenylphosphine with potassium selenocyanate[12] or with elemental selenium.[3,13]
Purification: recrystallization from ethanol.
Handling, Storage, and Precautions: non-volatile, toxic.

Original Commentary

Darrin L. Mayhew & Derrick L. J.Clive
University of Alberta, Edmonton, Canada

Conversion of Epoxides into Alkenes.[1] Addition of Trifluoroacetic Acid to a dichloromethane solution of an epoxide and an excess of triphenylphosphine selenide results in deoxygenation of the epoxide to the corresponding alkene. The reaction is stereospecific (eq 1–3). Other phosphine selenides can also be used.[5]

(1)

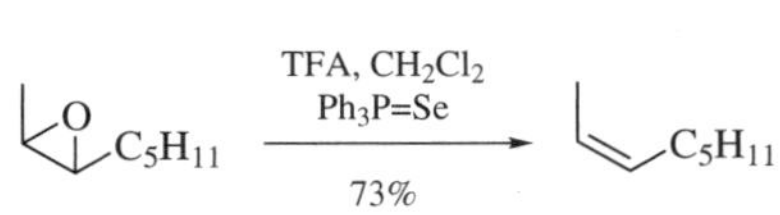
(2)

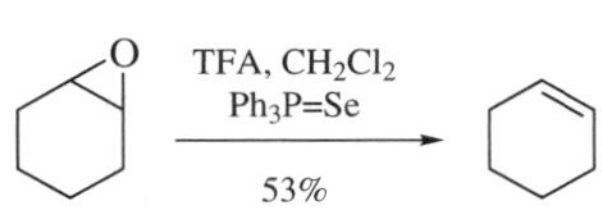
(3)

Conversion of Phosphorus Ylides into Alkenes.[6] Several nonstabilized ylides, $Ph_3P{=}CHR$, give near thermodynamic mixtures of (*Z*)- and (*E*)-alkenes when treated with a catalytic amount of triphenylphosphine selenide in hot (90 °C) toluene (eq 4).

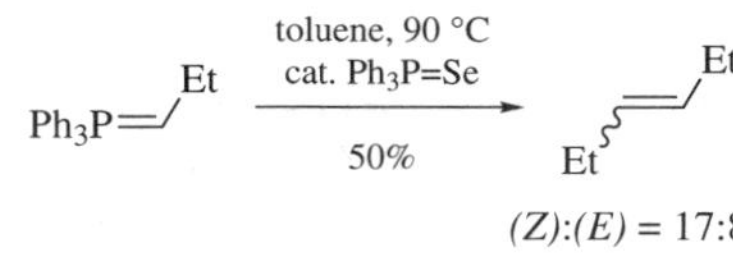
(4)

Reduction of Selenoxides.[7,8] Simple selenoxides (e.g. dibenzyl selenoxide), that do not undergo elimination, are reduced slowly (24 h to 13 days) at rt to the corresponding selenides by the action of triphenylphosphine selenide in chloroform (eq 5).

$$(PhCH_2)_2Se{=}O \xrightarrow[\text{rt, >24 h}]{Ph_3P{=}Se,\ CHCl_3} (PhCH_2)_2Se \quad (5)$$

Formation of Ureas and Carbamates. Aromatic nitro compounds can be converted into ureas[9] or carbamates[10,11] in processes mediated by elemental selenium (eq 6); elemental selenium is derived in situ from triphenylphosphine selenide, which serves as a carrier for selenium.

$$ArNO_2 + CO + ROH \xrightarrow[\text{25 bar, 160 °C, 30 min}]{Ph_3P{=}Se,\ H_2O_2} ArNHCO_2R \quad (6)$$

First Update

Jacek Stawinski & Martin Bollmark
Stockholm University, Stockholm, Sweden.
AstraZeneca, Södertäalje, Sweden

Introduction. Transfer of selenium between P(V) and P(III) phosphorus compounds is an equilibrium reaction that, unlike the corresponding sulfur derivatives, occurs at room temperature.[14] In this type of reaction, a selenium atom is transferred from a phosphine selenide to a phosphine and the position of the equilibrium is controlled by the nature of the substituents at both phosphorus centers (eq 7).

$$R_3P{:} + Se{=}PR'_3 \rightleftharpoons R_3P{=}Se + {:}PR'_3 \quad (7)$$

Application of this type of selenium transfer reaction for preparative purposes (deselenization of arylphosphine selenide with tributylphosphine) has been rare.[15]

However, recently it was found that triphenylphosphine selenide, a stable, easy to handle, inexpensive, and commercially available reagent, with good solubility in organic solvents, can be used as a general purpose selenium-transfer reagent for the conversion of various P(III) compounds into the corresponding phosphoroselenoate derivatives.[13]

Selenization of Phosphites.[13] Triethyl phosphite is quickly (ca. 5 min) selenized by triphenylphosphine selenide in chloroform forming triethyl phosphoroselenoate almost quantitatively, while triphenylphosphine selenide is converted into triphenylphosphine during the course of the reaction (see eq 7). This selenium-transfer phenomena occurs readily also in other solvents, e.g., acetonitrile, dichloromethane, pyridine, and tetrahydrofuran.

Dinucleoside alkyl phosphites, which are intermediates in the synthesis of oligonucleotides via the phosphoramidite approach, are efficiently selenized with triphenylphosphine selenide (eq 8) to afford diastereomeric mixtures of dinucleoside phosphoroselenoates in high yields.

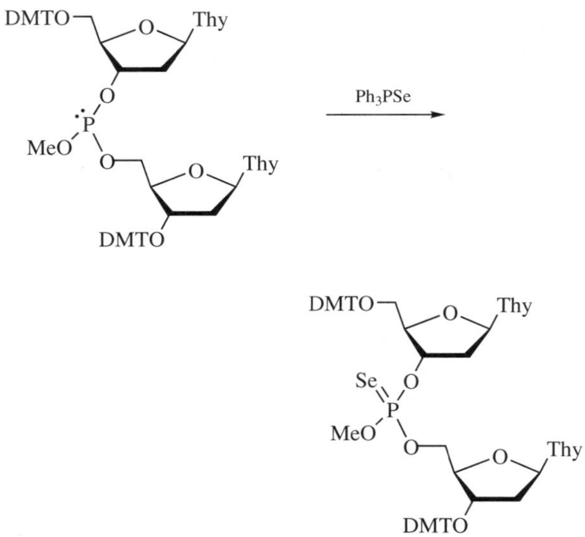

(8)

Thy = thymin-1-yl
DMT = 4,4′-dimethoxytrityl

Other tervalent P(III) compounds, e.g., trialkylphosphines and triaminophosphines, also undergo readily selenization by triphenylphosphine selenide. For some aryl phosphites (e.g., triphenyl phosphite), the position of the equilibrium is not favorable (it is shifted to the left), and thus, these types of compounds are not efficiently selenized by triphenylphosphine selenide.

Selenization of *H*-phosphonate and *H*-phosphonothioate Diesters.[13] *H*-Phosphonates and *H*-phosphonothioates (eq 9), which are tetra-coordinate P(III) compounds, can be selenized with triphenylphosphine selenide, but this process usually requires more basic reaction conditions (e.g., DBU in pyridine). Strictly anhydrous conditions should be maintained during the course of the reaction to avoid hydrolysis of the starting *H*-phosphonate or *H*-phosphonothioate diesters by adventitious water.

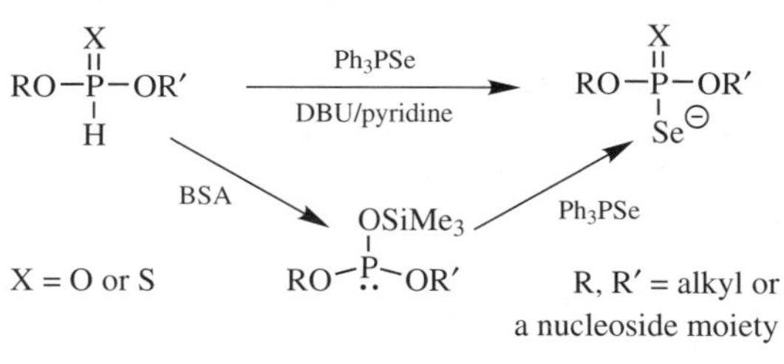

(9)

An alternative protocol for selenization of these compounds with triphenylphosphine selenide involves transformation of the starting material prior to selenization into a more reactive trivalent species (e.g., silyl phosphites), using BSA as a silylating agent. This modification increases the rate of selenization and eliminates the need for using strongly basic conditions. In the instance of *H*-phosphonothioate diesters, which are not easily converted into tervalent silyl esters, a silylating agent probably acts mainly as a water scavenger and in this way has a beneficial effect on the outcome of the selenization.

The above protocols work well both for simple (e.g., alkyl) and more complex (e.g., nucleoside) *H*-Phosphonate and *H*-Phosphonothioate diesters, affording in high yield the corresponding phosphoroselenoate or phosphoroselenothioate diesters (eq 9). Preliminary data indicate that selenium-transfer from triphenylphosphine selenide to *H*-Phosphonates and *H*-Phosphonothioates [and probably also to tervalent P(III) compounds] is a stereospecific reaction which occurs most likely with retention of configuration at the phosphorus centers.

Polymer-bound Selenizing Agent.[13] A polymer-supported version of triphenylphosphine selenide can be easily prepared from commercially available polystyrene-bound triphenylphosphine by selenization using potassium selenocyanate (eq 10). The resulting polystyryl(diphenyl)phosphine selenide was found to be as effective as the non-supported reagent in selenium-transfer reactions, while the work-up of the reaction mixtures was greatly facilitated.

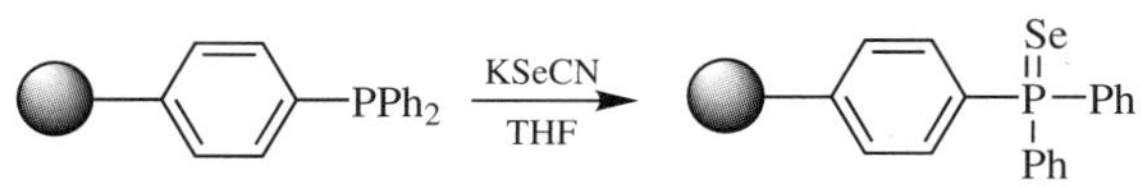

(10)

Related Reagents. Elemental selenium;; potassium selenocyanate; 3*H*-1,2-benzothiaselenol-3-one.

1. Clive, D. L. J.; Denyer, C. V., *Chem. Commun.* **1973**, 253.
2. Nicpon, P.; Meek, D. W., *Inorg. Synth.* **1967**, *10*, 157.
3. De Ketelaere, R. F.; Van der Kelen, G. P.; Eeckhaut, Z., *Phosphorus Sulfur* **1974**, *5*, 43.
4. Michaelis, A.; Soden, H., *Justus Liebigs Ann. Chem.* **1885**, *229*, 295.
5. Mathey, F.; Muller, G., *C. R. Hebd. Seances Acad. Sci., Ser. C* **1975**, *281*, 881.
6. Erker, G.; Hock, R.; Nolte, R., *J. Am. Chem. Soc.* **1988**, *110*, 624.
7. Mikolajczyk, M.; Luczak, J., *J. Org. Chem.* **1978**, *43*, 2132.
8. Sakaki, K.; Oae, S., *Chem. Lett.* **1977**, 1003.
9. Besenyei, G.; Czege, G.; Nagy, F.; Simándi, L.; Viski, P.; Paszty, G.; Szita, I.Ger. Offen. 3 405 582 (*Chem. Abstr.* **1984**, *102*, 78 596t) .
10. Besenyei, G.; Czege, G.; Kwaysser, E.; Nagy, F.; Simándi, L.; Viski, P.; Schmelczer, B.; Szilagyi, G., Ger. Offen. 3 347 460 (*Chem. Abstr.* **1984**, *102*, 24 301z) .
11. Besenyei, G.; Viski, P.; Nagy, F.; Simándi, L. I., *Inorg. Chim. Acta* **1984**, *90*, L43.
12. Nicpon, P.; Meek, D. W., *Inorg. Chem.* **1966**, *5*, 1297.
13. Bollmark, M.; Stawinski, J., *Chem. Commun.* **2001**, 771.
14. (a) Brown, D. H.; Cross, R. J.; Keat, R., *J. Chem. Soc., Chem. Commun.* **1977**, 708. (b) Brown, D. H.; Cross, R. J.; Keat, R., *J. Chem. Soc., Dalton Trans.* **1980**, 871. (c) Bachelor, R. D.; Stack, M.; Stevenson, K.; Vanvalkenburgh, V., *Phosphor. Sulfur Silicon* **1990**, *48*, 49.
15. Tolmachev, A. A.; Kozlov, E. S., *Zh. Obsh. Khim.* **1988**, *58*, 489.

Tris(dimethylamino)phosphine Carbon Tetrabromide

$(Me_2N)_3P\text{-}CCl_4$

$((Me_2N)_3P)$		
[1608-26-0]	$C_6H_{18}N_3P$	(MW 163.24)) (CCl_4)
[56-23-5]	CCl_4	(MW 153.81)

(reagent combination for the conversion of hydroxyl groups to the corresponding chlorides; hydroxyl group activation; dehydrations)

Alternate Name: hexamethylphosphorous triamide–carbon tetrachloride.

Physical Data: $(Me_2N)_3P$: bp 55–58 °C/15 mmHg; *d* 0.898 g cm^{-3}. CCl_4: mp −23 °C; bp 77 °C; *d* 1.594 g cm^{-3}.

Solubility: sol CCl_4, MeCN, CH_2Cl_2, THF.

Preparative Methods: reactive species is generated in situ by reaction of $(Me_2N)_3P$ and CCl_4.

Handling, Storage, and Precautions: $(Me_2N)_3P$ is an irritant and is flammable. It is best kept under a nitrogen atmosphere. CCl_4 is toxic and a cancer suspect agent. All solvents used must be carefully dried because the intermediates are all susceptible to hydrolysis. The hexamethylphosphoramide produced as a byproduct is a cancer suspect agent. This reagent should be handled in a fume hood.

General Considerations. The use of Hexamethylphosphorous Triamide in place of Triphenylphosphine in the phosphine–carbon tetrachloride reaction has a number of advantages. The separation of the triphenylphosphine oxide from the product can sometimes be problematic. The byproduct of the hexamethylphosphorous triamide reaction, Hexamethylphosphoric Triamide, is water soluble which allows for simple separation from the desired product. The aminophosphine is more nucleophilic, resulting in milder reaction conditions. Also, the alkoxyphosphonium intermediates can sometimes be isolated.[1]

Activation of Hydroxyl Groups. The combination of hexamethylphosphorous triamide and CCl_4 with an alcohol is a very effective method for the activation of the hydroxyl group for displacement by a variety of nucleophiles. In the absence of any external nucleophiles, the chloride ion formed in the reaction will react with the intermediate alkoxyphosphonium ion to produce the alkyl chloride (eq 1).[2]

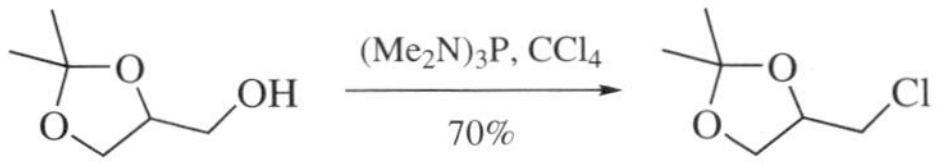

(1)

Other nucleophiles are capable of effectively competing with chloride ion for the alkoxyphosphonium ion. If Sodium Azide, Potassium Iodide, or Potassium Cyanide is added to the reaction mixture, the corresponding azide, iodide, or nitrile will be formed in good yield (eq 2).[3]

$$PhCH_2OH \xrightarrow[2.\ X^-]{1.\ (Me_2N)_3P,\ CCl_4} PhCH_2X \qquad (2)$$

$X^- = N_3^-$, 75%; I^-, 80%; CN^-, 75%

Sometimes it is advantageous to replace the chloride counterion with less nucleophilic counterions such as BF_4^-,[4] ClO_4^-,[5] or PF_6^-.[6] This technique sometimes allows the alkoxyphosphonium salts to be isolated as stable crystalline solids (eq 3).[5] Weaker nucleophiles, such as amines (eq 4), can then be used to displace the alkoxyphosphonium ion.[5]

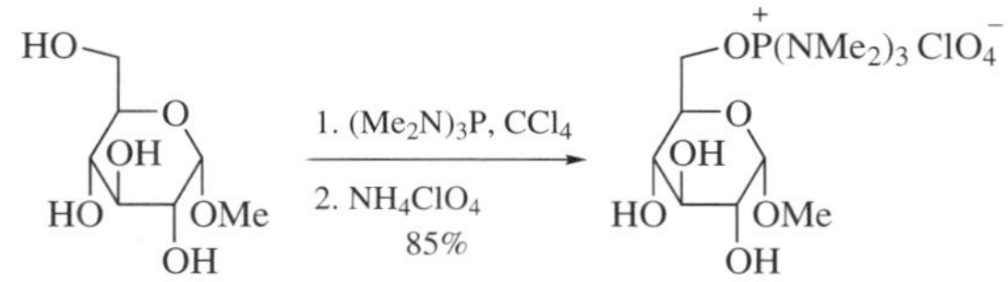

(3)

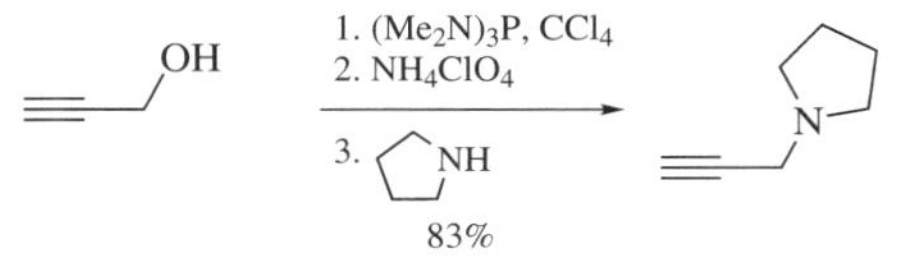

(4)

Deoxygenations. The removal of oxygen functionality can be achieved via activation of hydroxyl groups with this reagent. The reduction of the alkoxyphosphonium ion by Lithium Triethylborohydride provides a route to deoxy sugars (eq 5).[7]

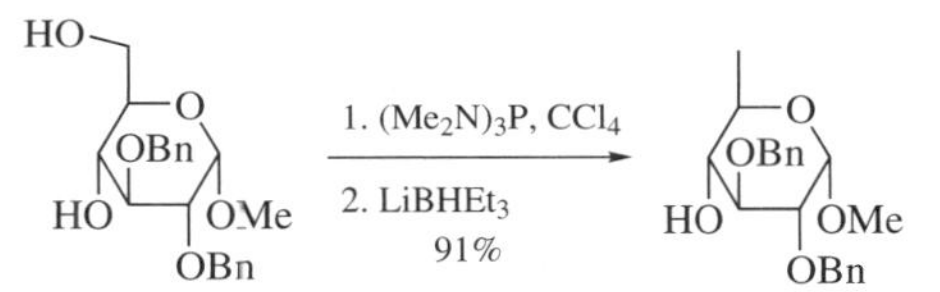

(5)

The preparation of glycals from lactols can be achieved by initial activation of the anomeric hydroxyl and subsequent Li/NH_3 reduction (eq 6).[8]

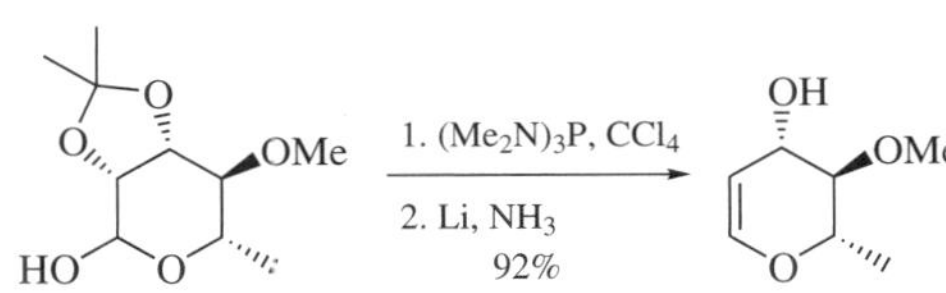

(6)

Anomeric Hydroxyl Activation. Anomeric hydroxyl group activation can be utilized to prepare a number of glycosyl derivatives. Activation followed by reaction with Mesityloxytris (dimethylamino)phosphonium Azide affords glycosyl azides (eq 7).[9]

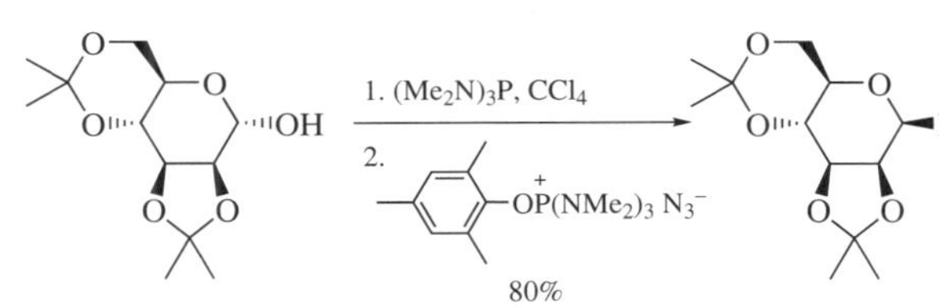

(7)

For the preparation of disaccharides, the intermediate alkoxyphosphonium ions are reacted with Silver(I) p-Toluenesulfonate prior to the addition of the second sugar (eq 8).[10]

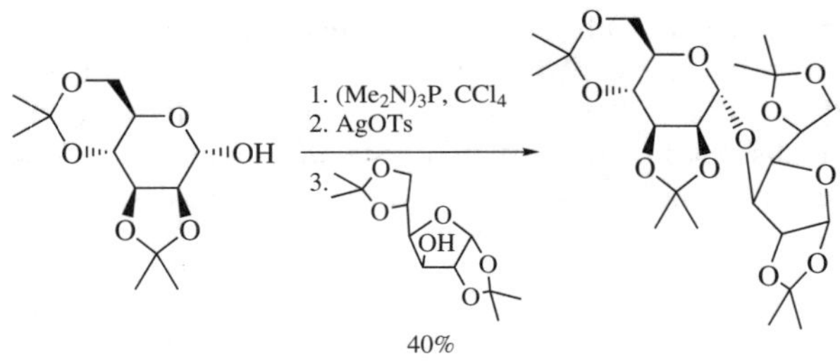

(8)

Dehydrations. The reaction of 1,2-diols with this reagent combination can lead to the formation of epoxides (eq 9) or spirophosphoranes (eq 10). The course of the reaction is dependent on the stereochemistry of the starting diol. The reaction of *meso*-hydrobenzoin affords the epoxide while the *threo* stereoisomer provides the spirophosphorane. The *trans*-diol of cyclohexane was unaffected.[11]

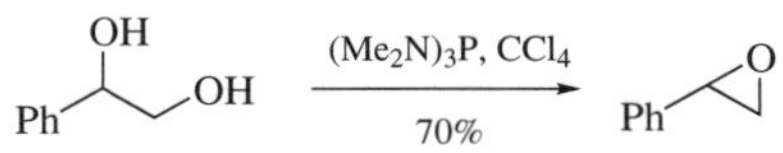

(9)

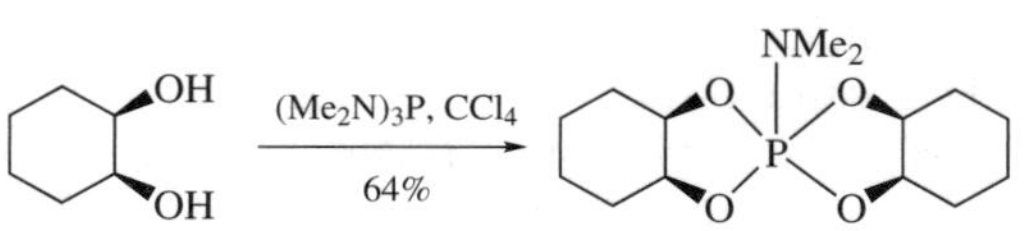

(10)

The selective monofunctionalization of a 1,3-diol is possible. Following activation, the monoalkoxyphosphonium ion can be displaced with various nucleophiles such as iodide or azide (eq 11).[12] Alternatively, if the alkoxyphosphonium ion was reacted with base, the oxetane was formed (eq 12).[13]

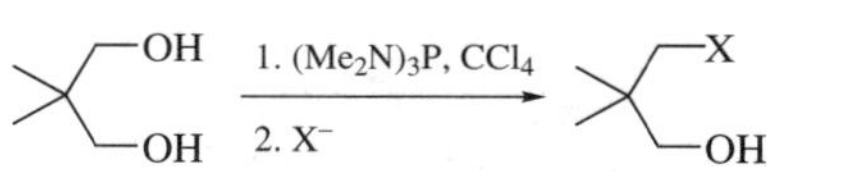

(11)

X = I, 85%; N_3, 80%

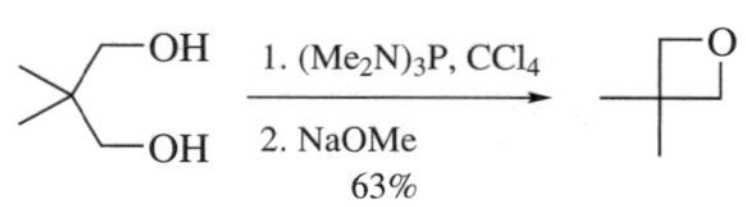

(12)

Dichloroalkenes. Aldehydes and the intermediate trichloromethylcarbinols can be converted to dichloroalkenes.[14] Even certain γ-lactones will undergo this transformation. Treatment of the γ-lactone (eq 13) with hexamethylphosphorous triamide produced the dichloroalkene in 79% yield. The use of Ph_3P was ineffective for this transformation. Attempts to use CBr_4 in place of CCl_4 resulted in the formation of complex mixtures.[14]

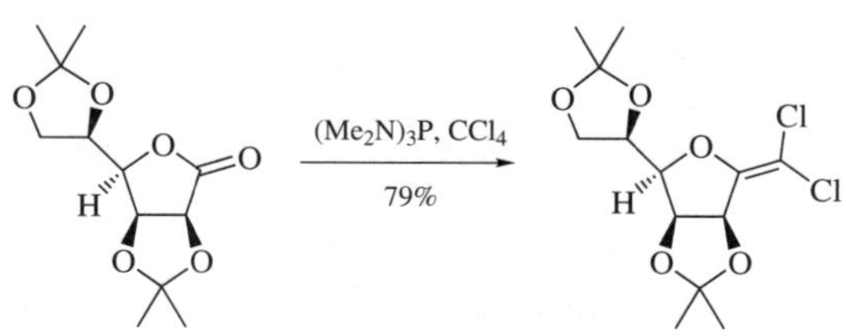

(13)

1. Castro, B. R. *Org. React.* **1983**, *29*, 1.
2. Downie, I. M.; Lee, J. B.; Matough, M. F. S. *Chem. Commun./J. Chem. Soc., Chem. Commun.* **1968**, 1350.
3. Castro, B.; Selve, C. *Bull. Soc. Claim. Fr.* **1971**, 2296.
4. (a) Castro, B.; Chapleur, Y.; Gross, B.; Selve, C. *Tetrahedron Lett.* **1972**, 5001. (b) Castro, B.; Chapleur, Y.; Gross, B. *Bull. Soc. Claim. Fr.* **1973**, 3034.
5. (a) Castro, B.; Selve, C. *Bull. Soc. Claim. Fr.* **1971**, 4368. (b) Castro, B.; Chapleur, Y.; Gross, B. *Tetrahedron Lett.* **1974**, 2313. (c) Chapleur, Y.; Castro, B.; Gross, B. *Synth. Commun.* **1977**, *7*, 143.
6. Downie, I. M.; Heaney, H.; Kemp, G. *Tetrahedron* **1988**, *44*, 2619.
7. Simon, P.; Ziegler, J.-C.; Gross, B. *Synthesis* **1979**, 951.
8. Ireland, R. E.; Thaisrivongs, S.; Vanier, N.; Wilcox, C. S. *J. Org. Chem.* **1980**, *45*, 48.
9. Chretien, F.; Castro, B.; Gross, B. *Synthesis* **1979**, 937.
10. Chretien, F.; Chapleur, Y.; Castro, B.; Gross, B. *J. Chem. Soc., Perkin Trans. 1* **1980**, 381.
11. Boigegrain, R.; Castro, B. *Tetrahedron* **1976**, *32*, 1283.
12. Castro, B.; Ly, M.; Selve, C. *Tetrahedron Lett.* **1973**, 4455.
13. Castro, B.; Selve, C. *Tetrahedron Lett.* **1973**, 4459.
14. Combret, J. C.; Villiéras, J.; Lavielle, G. *Tetrahedron Lett.* **1971**, 1035.
15. Chapleur, Y. *Chem. Commun./J. Chem. Soc., Chem. Commun.* **1984**, 449.

Michael J. Taschner
The University of Akron, OH, USA

Trityl Tetrakis(pentafluorophenyl)borate

$Ph_3C^+[B(C_6F_5)_4]^-$

[136040-19-2] $C_{43}H_{15}F_{20}B$ (MW 922.36)

(reagent used for generating cationic species having a non-coordinating anion or as a Lewis acid catalyst)

Alternate Name: triphenylcarbenium tetrakis(pentafluorophenyl) borate.
Physical Data: mp 180–182 °C.
Solubility: soluble in CH_2Cl_2, MeOH, and acetone, but its yellow color disappears because these solvents coordinate to the trityl salt.
Form Supplied in: yellow powder; commercially available from Asahi Glass Co. Ltd.
Analysis of Reagent Purity: NMR, mp.
Preparative Methods: trityl tetrakis(pentafluorophenyl)borate (**1**) is prepared by mixing $LiB(C_6F_5)_4$ and trityl chloride in hexane and refluxing overnight.[1] It is prepared also by one-pot procedure starting from bromopentafluorobenzene.[2]
Purification: recrystallization from CH_2Cl_2/hexane.
Handling, Storage, and Precautions: should be stored in a dry, dark, and cold place; stable at room temperature for long periods; does not decompose upon exposure to air.

Introduction. Trityl salt **1** works as a characteristic Lewis acid because of its unique carbocationic properties which originate from the fact that $[(C_6F_5)_4B^-]$ is an inert non-coordinating counterion. It is easy-to-handle since **1** is moisture- and air-resistant, thus differing from similar conventional agents such as trityl perchlorate and trityl tetrafluoroborate.

Generation of a Cationic Catalyst for Polymerization of Olefins. Abstraction of CH_3^- group from (dimethyl)zirconocene by trityl tetrakis(pentafluorophenyl)borate gives a highly active cationic catalyst for olefin polymerization, $[Cp_2ZrMe]^+(C_6F_5)_4B^-$ (**2**), which polymerizes propene and higher olefins at high rates without adding any other activator (eq 1).[1,3] Its catalytic activity is higher than that of using MAO (methylaluminoxane) as an additive. The large and weakly coordinating tetrakis(pentafluorophenyl)borate anion $[(C_6F_5)_4B^-]$ of **2** is easily displaced from its coordination site at the Zr center with an olefin.

$$\underset{}{Cp_2ZrMe_2} + \underset{\mathbf{1}}{Tr^+[B(C_6F_5)_4]^-} \xrightarrow[-Ph_3CMe]{} \underset{\mathbf{2}}{[Cp_2ZrMe]^+[B(C_6F_5)_4]^-} \quad (1)$$

Alkylmetalation of Alkynes. A catalytic amount of **1** effectively catalyzes alkylzirconation of alkynes using alkylzirconium generated by hydrozirconation of 1-alkene (eq 2).[4,5]

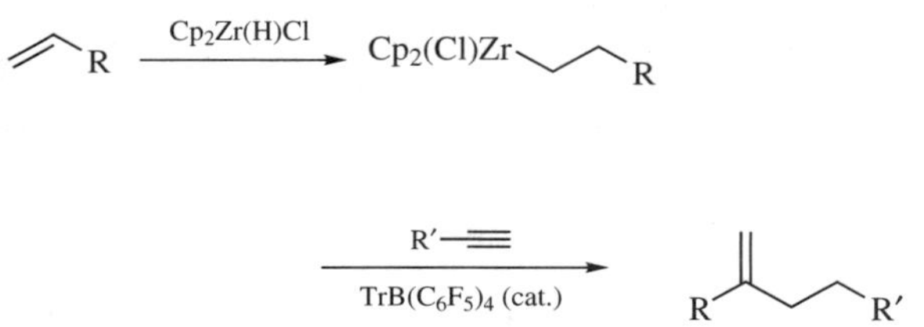

(2)

Hydride Transfer. The chiral cationic species **3** is prepared by hydride transfer between a silyl hydride such as **4** and **1** (eq 3).[6]

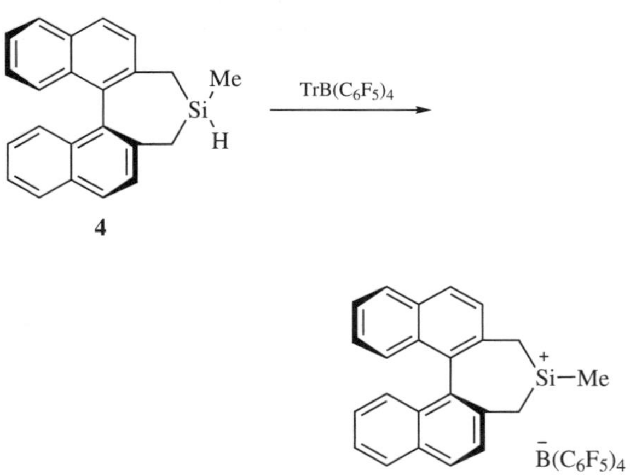

4

(3)

3

Activation of Glycosyl Donors. It is noteworthy that the trityl salt **1** has high potential as a glycosylation catalyst. Glycosylation of 1-hydroxy-2,3,5-tri-*O*-benzyl-D-ribofuranose,[7] 1-hydroxy-3,4,6-tri-*O*-benzyl-2-deoxy-glucopyranose,[8] glycosyl fluorides,[9] and glycosyl carbonate[10,11] with free glycosyl acceptors (alcohols and saccharides) by using a catalytic amount of **1** gives the corresponding ribofuranosides and disaccharides in high yields with high stereoselectivities (eqs 4–6). It is shown that **1** catalyzes glycosylation more efficiently than the conventional trityl salts such as trityl tetrafluoroborate or trityl perchlorate.[7]

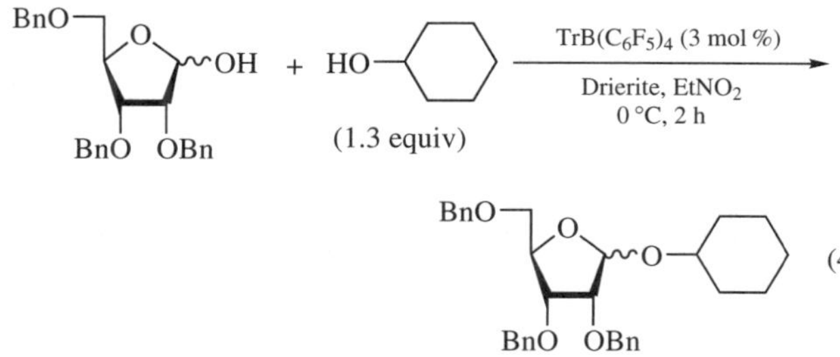

(4)

91% yield, α/β = 4/96

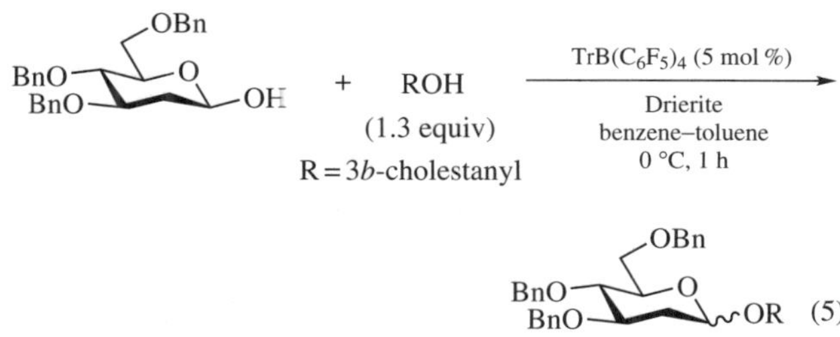

(5)

93% yield, α/β = 91/9

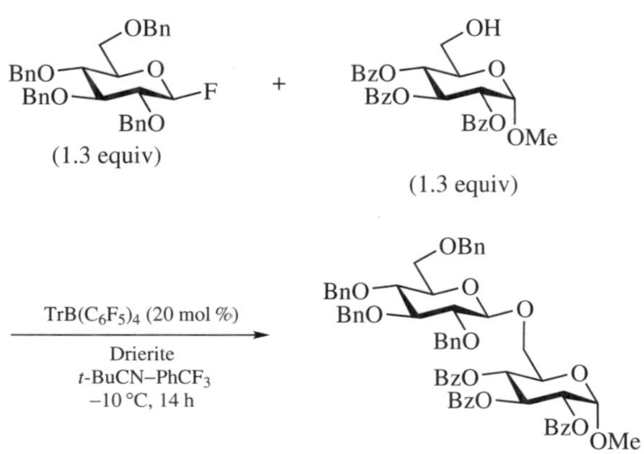

(6)

93% yield, α/β = 9/91

Thioglycosides are also activated with a catalytic amount of **1** in the presence of sodium periodate, and stereoselective glycosylation proceeds efficiently (eq 7).[12,13]

A one-pot synthesis of trisaccharides is successfully carried out by a combination of two types of **1**-catalyzed stereoselective glycosylation reactions using glycosyl carbonate and thioglycoside as glycosyl donors (eq 8).[14,15]

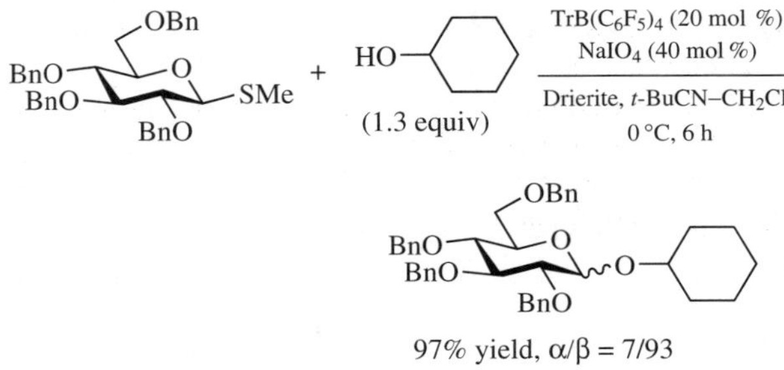

(7)

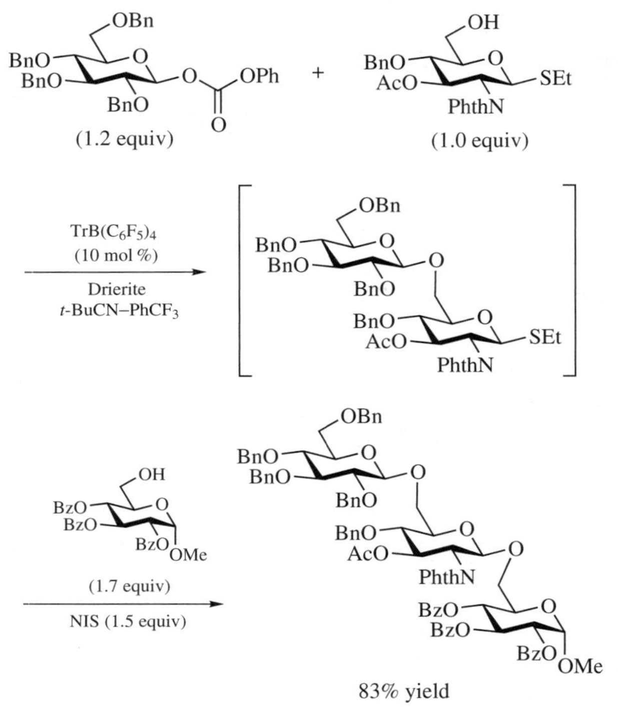

(8)

Aldol Reactions. Aldol reactions of silyl enol ethers[16] or enol esters[17] with aldehydes are effectively catalyzed by **1**, similar to when using other trityl salts (e.g., trityl perchlorate) (eq 9).

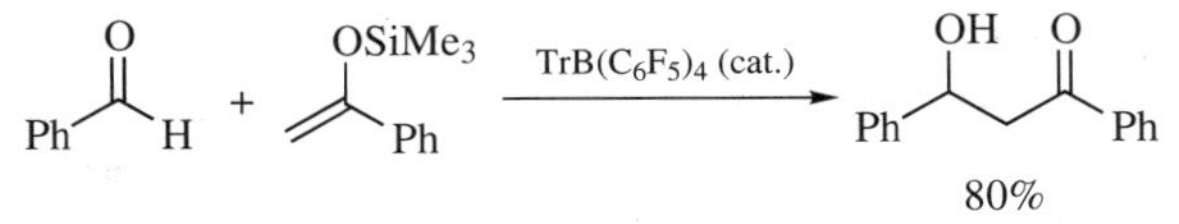

(9)

Related Reagents. Methylaluminoxane; trityl perchlorate; trityl tetrafluoroborate; trimethylsilyl triflate; chlorotriphenylsilane; trityl tetrakis[3,5-bis(trifluoromethyl)phenyl]borate; silver tetrakis(pentafluorophenyl)borate.

1. Chien, J. C. W.; Tsai, W.-M.; Raush, M. D., *J. Am. Chem. Soc.* **1991**, *113*, 8570.
2. Germany Pat. DE 4235092 (to Doellein, G.) (1994) (Chem. Abstr. **1994**, *121*, 83618).
3. Chien, J. C. W.; Song, W.; Raush, M. D., *Macromolecules* **1993**, *26*, 3239.
4. Yamanoi, S.; Ohrui, H.; Seki, K.; Matsumoto, T.; Suzuki, K., *Tetrahedron Lett.* **1999**, *40*, 8407.
5. Yamanoi, S.; Seki, K.; Matsumoto, T.; Suzuki, K., *J. Organomet. Chem.* **2001**, *624*, 143.
6. Johannsen, M.; Jorgensen, K. A.; Helmchen, G., *J. Am. Chem. Soc.* **1998**, *120*, 7637.
7. Uchiro, H.; Mukaiyama, T., *Chem. Lett.* **1996**, 79.
8. Takeuchi, K.; Higuchi, S.; Mukaiyama, T., *Chem. Lett.* **1997**, 969.
9. Takeuchi, K.; Mukaiyama, T., *Chem. Lett.* **1998**, 555.
10. Mukaiyama, T.; Miyazaki, K.; Uchiro, H., *Chem. Lett.* **1998**, 635.
11. Takeuchi, K.; Tamura, T.; Mukaiyama, T., *Chem. Lett.* **2000**, 122.
12. Uchiro, H.; Mukaiyama, T., *Chem. Lett.* **1997**, 121.
13. Jona, H.; Takeuchi, K.; Saitoh, T.; Mukaiyama, T., *Chem. Lett.* **2000**, 1178.
14. Mukaiyama, T.; Wakiyama, Y.; Miyazaki, K.; Takeuchi, K., *Chem. Lett.* **1999**, 933.
15. Takeuchi, K.; Tamura, T.; Mukaiyama, T., *Chem. Lett.* **2000**, 124.
16. Mukaiyama, T.; Yanagisawa, M.; Iida, D.; Hachiya, I., *Chem. Lett.* **2000**, 606.
17. Yanagisawa, M.; Shimamura, T.; Iida, D.; Matsuo, J.-i.; Mukaiyama, T., *Chem. Pharm. Bull.* **2000**, *48*, 1838.

Teruaki Mukaiyama & Jun-ichi Matsuo
The Kitasato Institute, Tokyo, Japan

Z

Zinc Bromide[1]

ZnBr2

[7699-45-8] Br_2Zn (MW 225.19)

(used in the preparation of organozinc reagents via transmetalation;[1] a mild Lewis acid useful for promoting addition[2] and substitution reactions[3])

Physical Data: mp 394 °C; bp 697 °C (dec); *d* 4.201 g cm^{-3}.
Solubility: sol Et_2O, H_2O (1 g/25 mL), 90% EtOH (1 g/0.5 mL).
Form Supplied in: granular white powder; principal impurity is H_2O.
Analysis of Reagent Purity: melting point.
Purification: heat to 300 °C under vacuum (2×10^{-2} mmHg) for 1 h, then sublime.
Handling, Storage, and Precautions: very hygroscopic; store under anhydrous conditions. Irritant.

Organozinc Reagents. The transmetalation of organomagnesium, organolithium, and organocopper reagents by anhydrous $ZnBr_2$ in ethereal solvents offers a convenient method of preparing organozinc bromides and diorganozinc reagents.[1a] Alternatively, anhydrous $ZnBr_2$ may be reduced by potassium metal to result in highly activated Zn^0, which is useful for the preparation of zinc reagents through oxidative addition to organic halides.[4] Alkyl, allylic, and propargylic zinc reagents derived by these methods have shown considerable value in their stereoselective and regioselective addition reactions with aldehydes, ketones, imines, and iminium salts.[1a,5] Zinc enolates used in the Reformatsky reaction may also be prepared through transmetalation using $ZnBr_2$.[1b] Organozinc species are especially useful in palladium-and nickel-catalyzed coupling reactions of sp^2 carbon centers. In this fashion, sp^2–sp^3 (eq 1)[6] and sp^2–sp^2 (eqs 2 and 3)[7,8] carbon–carbon bonds are formed selectively in high yields. The enantioselective cross coupling of secondary Grignard reagents with vinyl bromide is strongly affected by the presence of $ZnBr_2$, which accelerates the reaction and inverts its enantioselectivity (eq 4).[9]

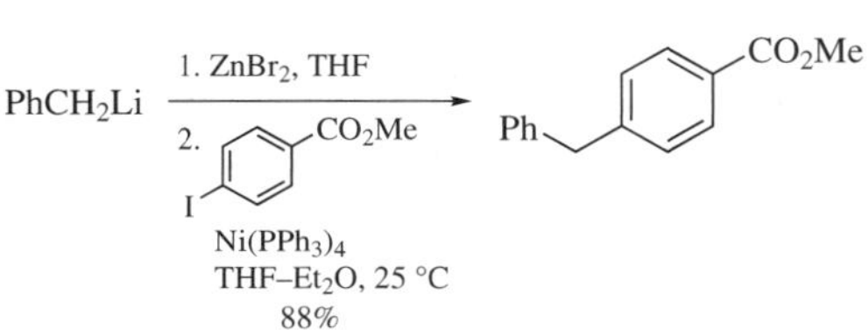

(1)

Organozinc intermediates formed via transmetalation using $ZnBr_2$ have been used to effect carbozincation of alkenes and alkynes through metallo-ene and metallo-Claisen reactions. Both intermolecular and intramolecular variants of these reactions have been described, often proceeding with high levels of stereoselectivity and affording organometallic products that may be used in subsequent transformations (eqs 5 and 6),[10] including alkenation (eq 6).[10b,c] Bimetallic zinc–zirconium reagents have also been developed that offer a method for the alkenation of carbonyl compounds (eq 7).[11]

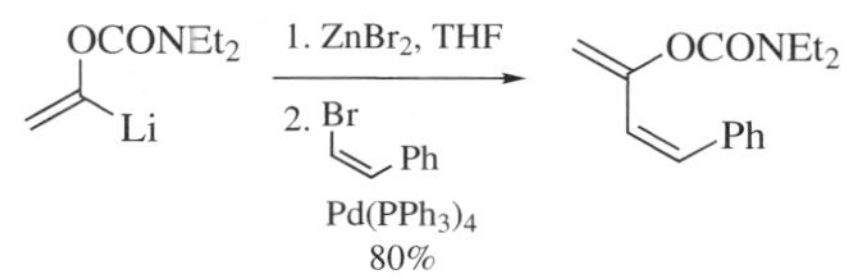

(2)

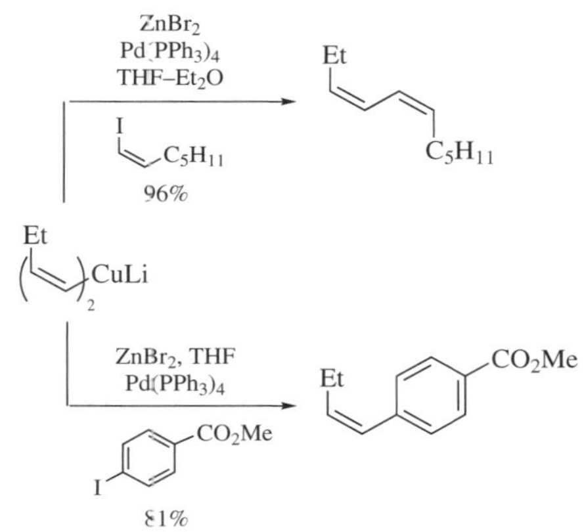

(3)

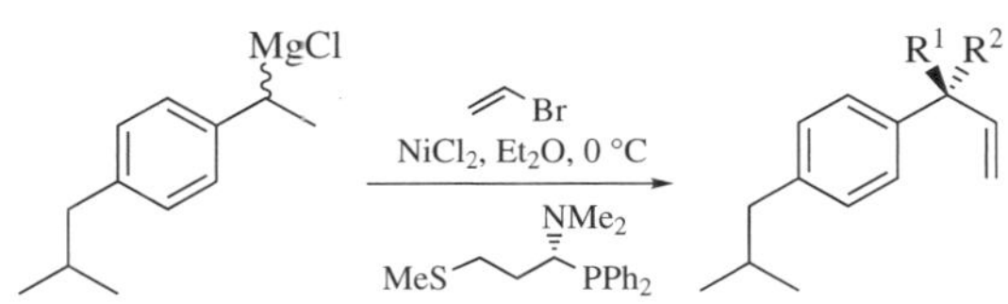

(4)

without $ZnBr_2$ R^1 = Me, R^2 = H; >95% (52% ee)
with $ZnBr_2$ R^1 = H, R^2 = Me; >95% (49% ee)

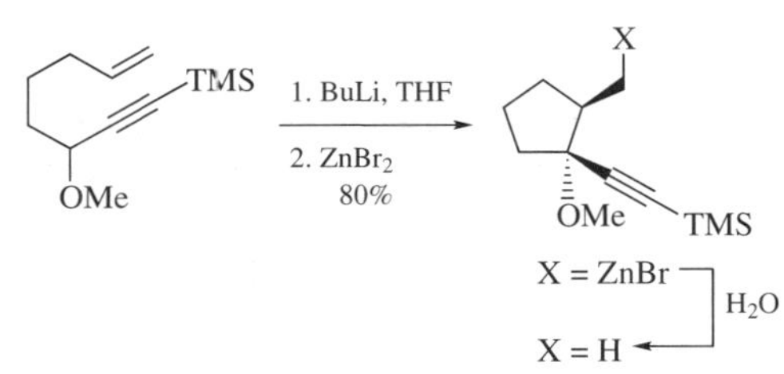

(5)

Concerted Ring-Forming Reactions. The mild Lewis acid character of $ZnBr_2$ sometime imparts a catalytic effect on thermally allowed pericyclic reactions. The rate and stereoselectivity of cycloaddition reactions (eq 8),[12] including dipolar cycloadditions (eq 9),[13] are significantly improved by the presence of this zinc salt.

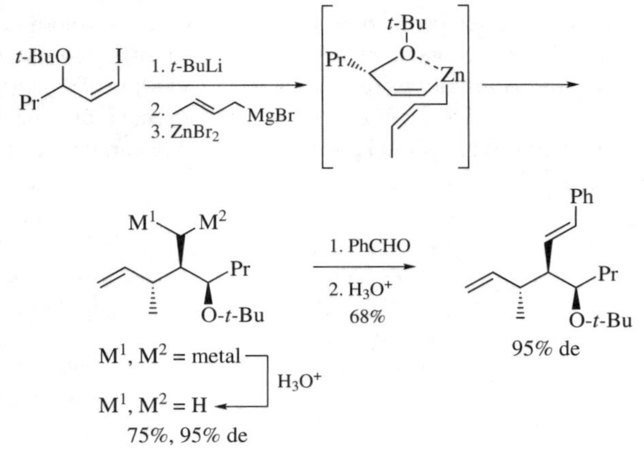

(6)

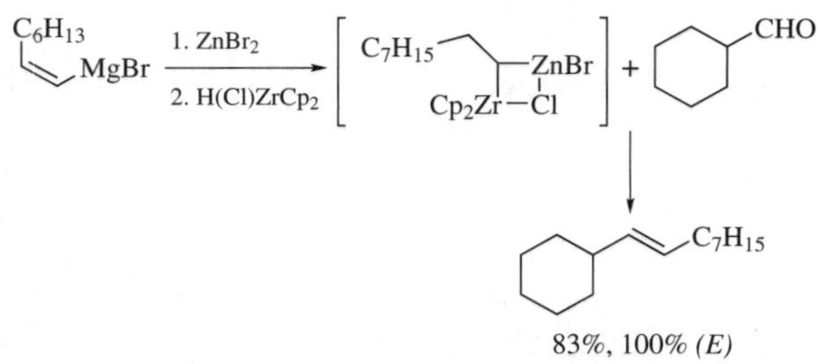

(7)

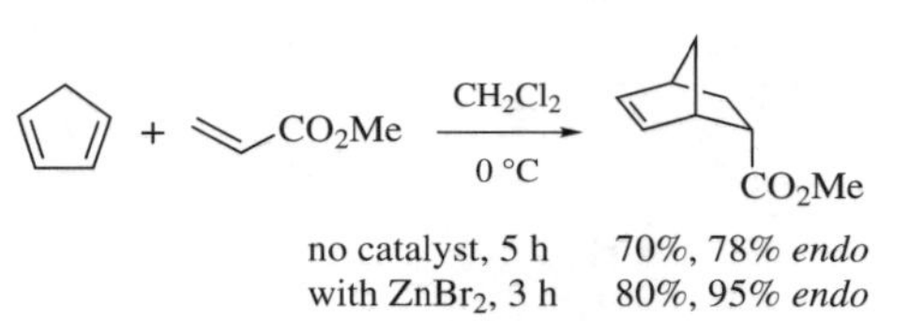

(8)

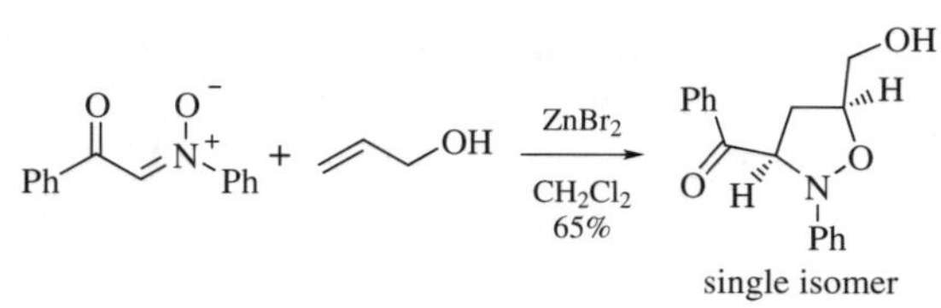

(9)

Some intramolecular ene reactions benefit from $ZnBr_2$ catalysis to afford the cyclic products under milder conditions, in higher yields and selectivities (eqs 10 and 11).[14,15] Generally, the use of $ZnBr_2$ is preferred over Zinc Chloride or Zinc Iodide in this type of reaction.[15]

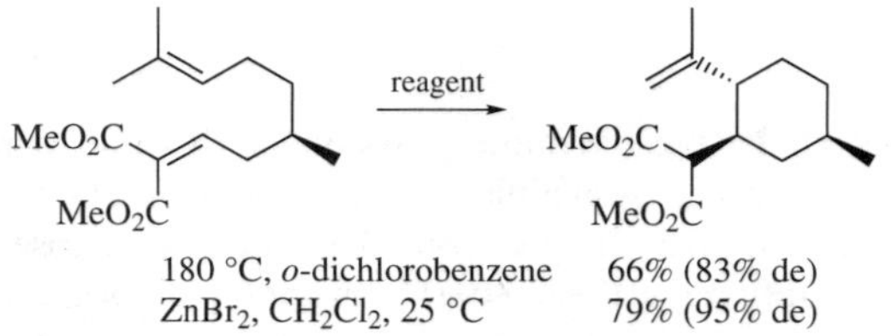

(10)

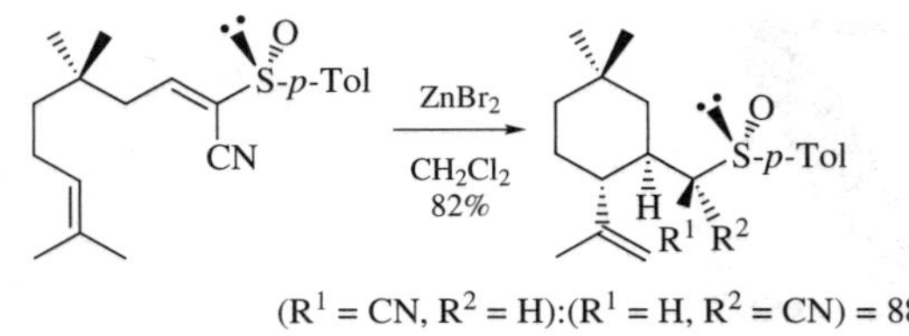

(11)

Activation of C=X Bonds. Lewis acid activation of carbonyl compounds by $ZnBr_2$ promotes the addition of allylsilanes and silyl ketene acetals.[16] Addition to imines has also been reported.[17] In general, other Lewis acids have been found to be more useful, though in some instances $ZnBr_2$ has proven to be advantageous (eq 12).[2]

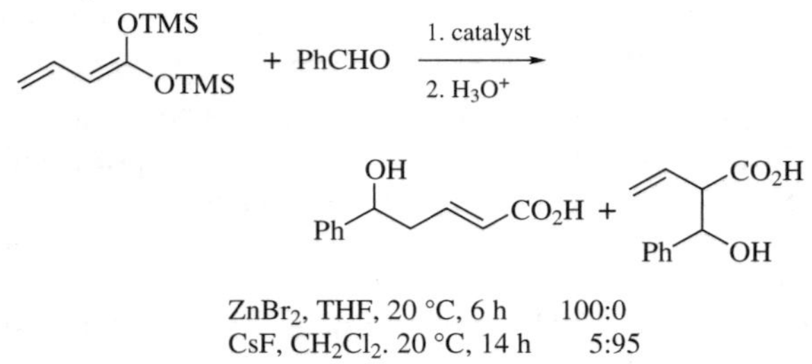

(12)

Activation of C–X Bonds. Even more important than carbonyl activation, $ZnBr_2$ promotes substitution reactions with suitably active organic halides with a variety of nucleophiles. Alkylation of silyl enol ethers and silyl ketene acetals using benzyl and allyl halides proceeds smoothly (eq 13).[3] Especially useful electrophiles are α-thio halides which afford products that may be desulfurized or oxidatively eliminated to result in α,β-unsaturated ketones, esters, and lactones (eq 14).[18] Other electrophiles that have been used with these alkenic nucleophiles include Chloromethyl Methyl Ether, $HC(OMe)_3$, and Acetyl Chloride.[3,19]

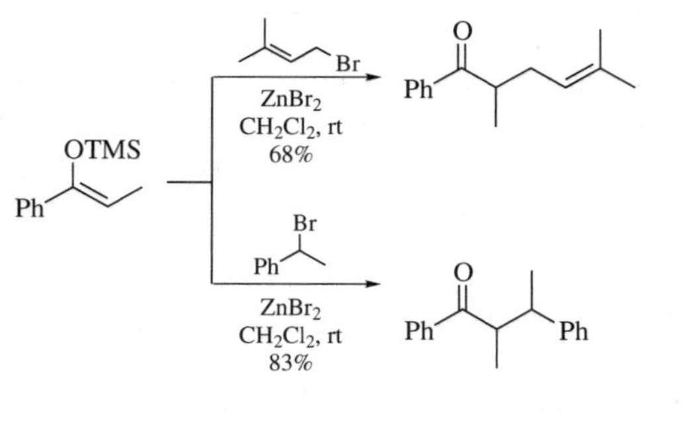

(13)

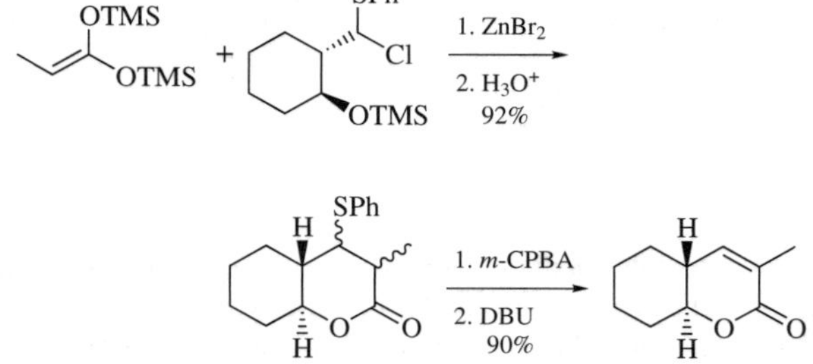

(14)

Enol ethers and allylic silanes and stannanes will engage cyclic α-seleno sulfoxides,[20] ω-acetoxy lactams,[21] and acyl glycosides (eq 15)[22] in the presence of $ZnBr_2$ catalysis. Along these lines, it has been found that $ZnBr_2$ is superior to Boron Trifluoride Etherate in promoting glycoside bond formation using trichloroimidate-activated glycosides (eq 16).[23] Imidazole carbamates are also effective activating groups for $ZnBr_2$-mediated glycosylation (eq 17).[24]

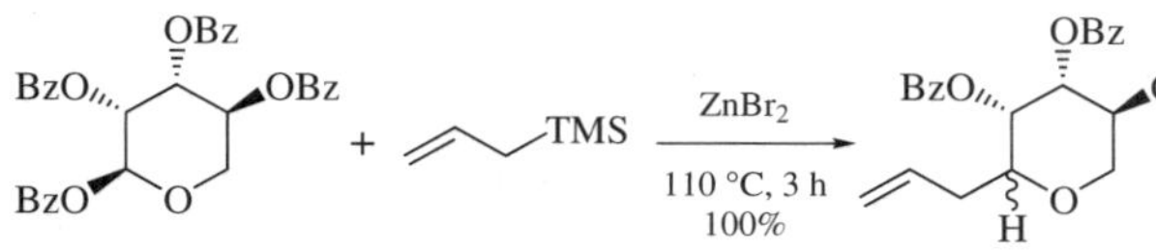

(15)

α:β = 1:1

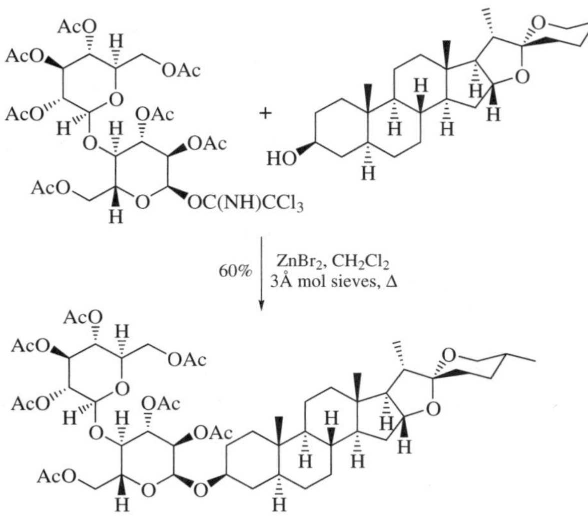

(16)

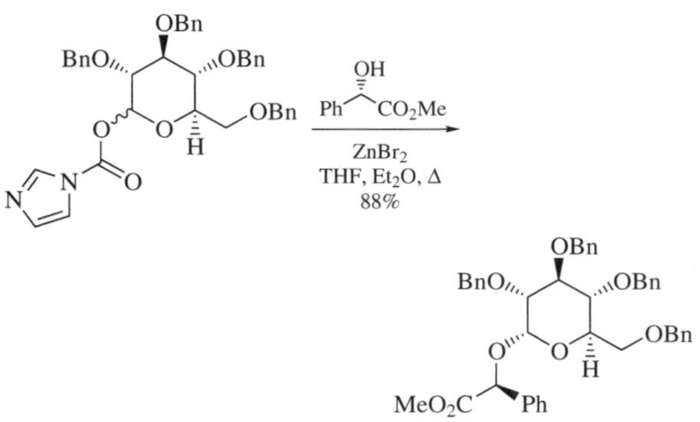

(17)

10:1 de

Cyclic acetals also undergo highly selective, Lewis acid-dependent ring opening substitution with Cyanotrimethylsilane (eq 18).[25]

Reduction. Complexation with $ZnBr_2$ has been shown to markedly improve stereoselectivity in the reduction of certain heteroatom-substituted ketones (eqs 19 and 20).[26,27] Furthermore, the *anti* selectivity observed in $BF_3 \cdot OEt_2$-mediated intramolecular hydrosilylation of ketones is reversed when $ZnBr_2$ is used instead (eq 21).[28]

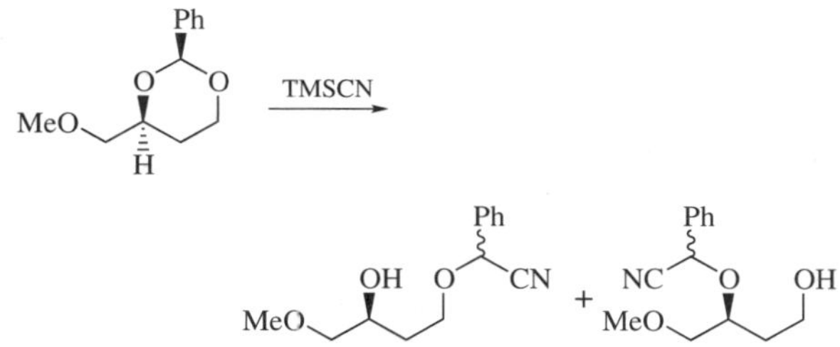

(18)

$ZnBr_2$, CH_2Cl_2, 25 °C, 20 h	1:250
$TiCl_4$, CH_2Cl_2, 25 °C, 20 h	250:1

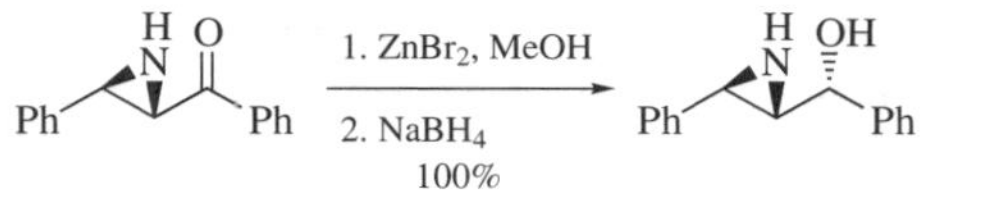

(19)

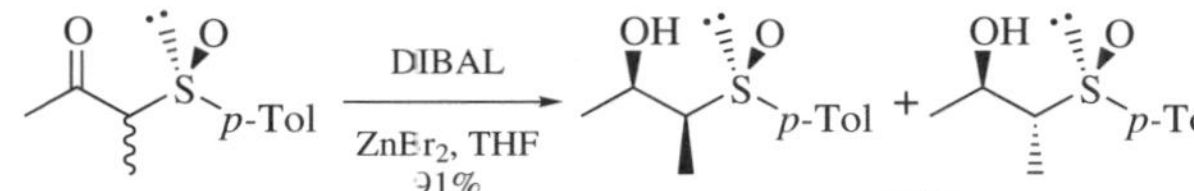

(20)

66:34

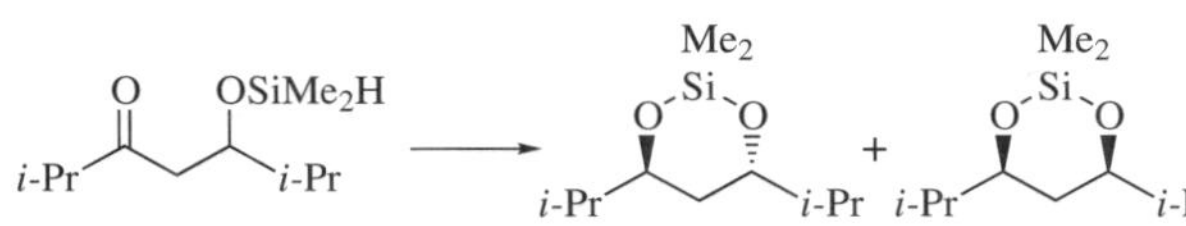

(21)

$BF_3 \cdot OEt_2$, CH_2Cl_2, –80 °C, 2 h	23:1
$ZnBr_2$, CH_2Cl_2, –80 °C, 8 h	1:6

Deprotection. $ZnBr_2$ is a very mild reagent for several deprotection protocols, including the detritylation of nucleotides[29] and deoxynucleotides,[30] *N*-deacylation of *N,O*-peracylated nucleotides,[31] and the selective removal of Boc groups from secondary amines in the presence of Boc-protected primary amines.[32] Perhaps the most widespread use of $ZnBr_2$ for deprotection is in the mild removal of MEM ethers to afford free alcohols (eq 22).[33]

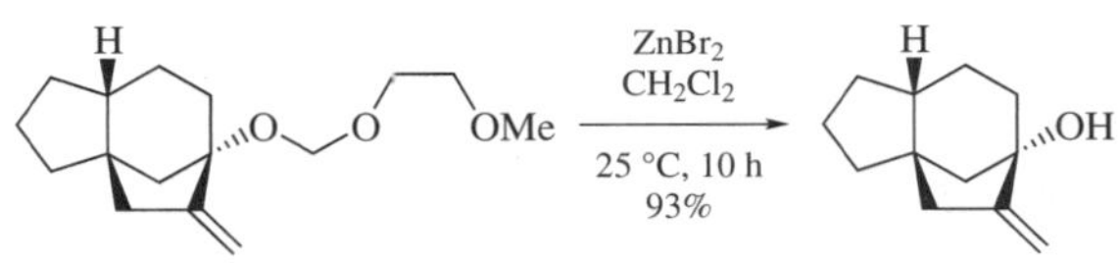

(22)

Miscellaneous. An important method for the synthesis of stereodefined trisubstituted double bonds involves the treatment of cyclopropyl bromides with $ZnBr_2$. The (*E*) isomer is obtained almost exclusively by this method (eq 23).[34]

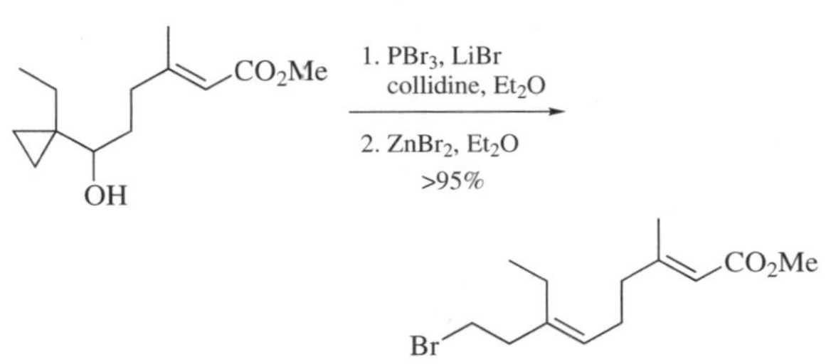

(23)

(*E*)

The rearrangement of a variety of terpene oxides has been examined (eq 24).[35] While $ZnBr_2$ is generally a satisfactory catalyst for this purpose, other Lewis acids, including $ZnCl_2$[36] and Magnesium Bromide,[37] are advantageous in some instances.

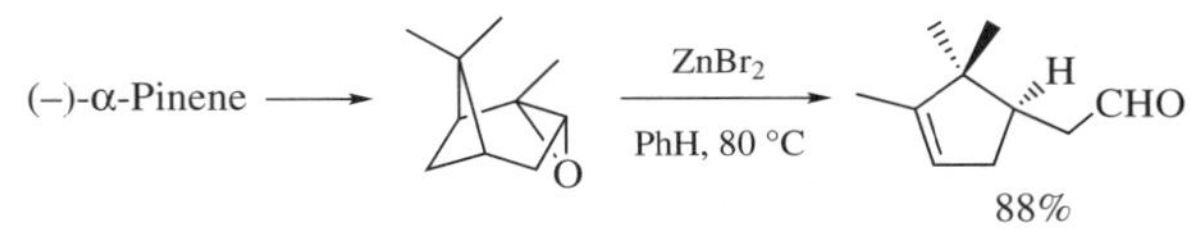
(24)

In the presence of $ZnBr_2$/48% Hydrobromic Acid, suitably functionalized cyclopropanes undergo ring expansion to afford cyclobutane (eq 25)[38] and α-methylene butyrolactone products (eq 26).[39] One-carbon ring expansion has been reported when certain trimethylsilyl dimethyl acetals are exposed to $ZnBr_2$ with warming (eq 27).[40]

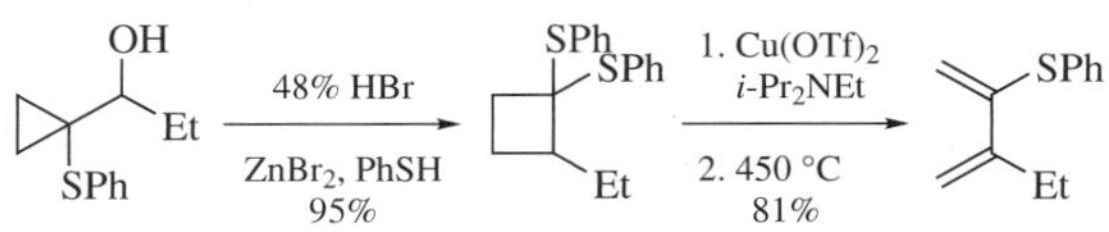
(25)

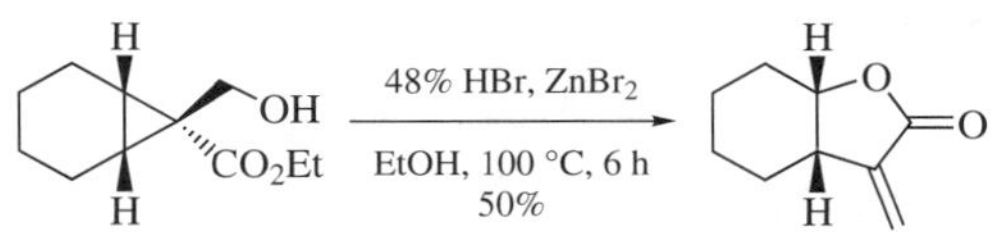
(26)

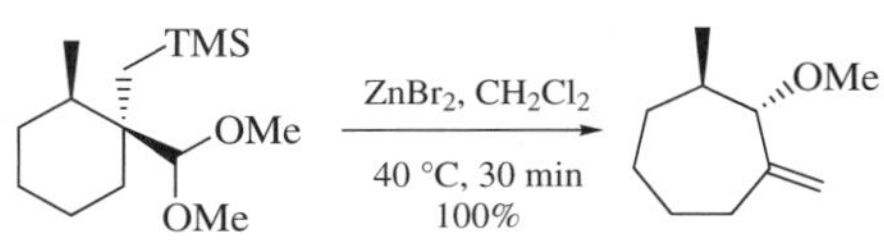
(27)

1. (a) Knochel, P. *Comprehensive Organic Synthesis* **1991**, *1*, Chapter 1.7. (b) Rathke, M. W.; Weipert, P. *Comprehensive Organic Synthesis* **1991**, *2*, Chapter 1.8.
2. For an example:Bellassoued, M.; Ennigrou, R.; Gaudemar, M. *J. Organomet. Chem.* **1988**, *338*, 149.
3. For examples: (a) Reetz, M. T.; Maier, W. F. *Angew. Chem., Int. Ed. Engl.* **1978**, *17*, 48. (b) Reetz, M. T.; Chatziiosifidis, I.; Löwe, W. F.; Maier, W. F. *Tetrahedron Lett.* **1979**, 1427. (c) Paterson, I. *Tetrahedron Lett.* **1979**, 1519.
4. Riecke, R. D.; Uhm, S. J.; Hudnall, P. M. *Chem. Commun./J. Chem. Soc., Chem. Commun.* **1973**, 269.
5. For representative examples of allylic and propargylic zinc reagents: (a) Yamamoto, Y.; Nishii, S.; Maruyama, K.; Komatsu, T.; Ito, W. *J. Am. Chem. Soc.* **1986**, *108*, 7778. (b) Yamamoto, Y.; Ito, W. *Tetrahedron* **1988**, *44*, 5414. (c) Yamamoto, Y.; Ito, W.; Maruyama, K. *Chem. Commun./J. Chem. Soc., Chem. Commun.* **1985**, 1131. (d) Yamanoto, Y.; Komatsu, T.; Maruyama, K. *Chem. Commun./J. Chem. Soc., Chem. Commun.* **1985**, 814. (e) Fronza, G.; Fuganti, C.; Grasselli, P.; Pedrocchi-Fantoni, G.; Zirotti, C. *Tetrahedron Lett.* **1982**, *23*, 4143. (f) Fujisawa, T.; Kojima, E.; Itoh, T.; Sato, T. *Tetrahedron Lett.* **1985**, *26*, 6089. (g) Pornet, J.; Miginiac, L. *Bull. Soc. Claim. Fr.* **1975**, 841. (h) Yamamoto, Y.; Komatsu, T.; Maruyama, K. *J. Organomet. Chem.* **1985**, *285*, 31. (i) Bouchoule, C.; Miginiac, P. *C.R. Hebd. Seances Acad. Sci., Ser. C* **1968**, *266*, 1614. (j) Miginiac, L.; Mauzé, B. *Bull. Soc. Claim. Fr.* **1968**, 3832. (k) Arous-Chtara, R.; Gaudemar, M.; Moreau, J.-L. *C.R. Hebd. Seances Acad. Sci., Ser. C* **1976**, *282*, 687. (l) Moreau, J.-L.; Gaudemar, M. *Bull. Soc. Claim. Fr.* **1971**, 3071. (m) Miginiac, L.; Mauzé, B. *Bull. Soc. Claim. Fr.* **1968**, 2544.
6. Negishi, E.; King, A. O.; Okudado, N. *J. Org. Chem.* **1977**, *42*, 1821.
7. Sengupta, S.; Snieckus, V. *J. Org. Chem.* **1990**, *55*, 5680.See also:Gilchrist, T. L.; Summersell, R. J. *Tetrahedron Lett.* **1987**, *28*, 1469.
8. (a) Jabri, N.; Alexakis, A.; Normant, J. F. *Bull. Soc. Claim. Fr., Part 2* **1983**, 321. (b) Jabri, N.; Alexakis, A.; Normant, J. F. *Tetrahedron Lett.* **1982**, *23*, 1589. (c) Jabri, N.; Alexakis, A.; Normant, J. F. *Tetrahedron Lett.* **1981**, *22*, 959. (d) Jabri, N.; Alexakis, A.; Normant, J. F. *Tetrahedron Lett.* **1981**, *22*, 3851.
9. Cross, G.; Vriesema, B. K.; Boven, G.; Kellogg, R. M.; van Bolhuis, F. *J. Organomet. Chem.* **1989**, *370*, 357.
10. (a) Courtemanche, G.; Normant, J.-F. *Tetrahedron Lett.* **1991**, *32*, 5317. (b) Marek, I.; Normant, J.-F. *Tetrahedron Lett.* **1991**, *32*, 5973. (c) Marek, I.; Lefrançois, J.-M.; Normant, J.-F. *Synlett* **1992**, 633.
11. Tucker, C. E.; Knochel, P. *J. Am. Chem. Soc.* **1991**, *113*, 9888.
12. Narayana Murthy, Y. V. S.; Pillai, C. N. *Synth. Commun.* **1991**, *21*, 783.See also:López, R.; Carretero, J. C. *Tetrahedron: Asymmetry* **1991**, *2*, 93.
13. Kanemasa, S.; Tsuruoka, T.; Wada, E. *Tetrahedron Lett.* **1993**, *34*, 87.
14. Tietze, L. F.; Biefuss, U.; Ruther, M. *J. Org. Chem.* **1989**, *54*, 3120. See also: (a) Tietze, L. F.; Ruther, M. *Ber. Dtsch. Chem. Ges./Chem. Ber.* **1990**, *123*, 1387. (b) Nakatani, Y.; Kawashima, K. *Synthesis* **1978**, 147.
15. Hiroi, K.; Umemura, M. *Tetrahedron Lett.* **1992**, *33*, 3343.
16. (a) Mikami, K.; Kawamoto, K.; Loh, T.-P.; Nakai, T. *Chem. Commun./J. Chem. Soc., Chem. Commun.* **1990**, 1161. (b) Bellassoued, M.; Gaudemar, M. *Tetrahedron Lett.* **1988**, *29*, 4551.
17. Gaudemar, M.; Bellassoued, M. *Tetrahedron Lett.* **1990**, *31*, 349.
18. Khan, H. A.; Paterson, I. *Tetrahedron Lett.* **1982**, *23*, 5083.See also: (a) Paterson, I. *Tetrahedron* **1988**, *44*, 4207. (b) Khan, H. A.; Paterson, I. *Tetrahedron Lett.* **1982**, *23*, 4811. (c) Paterson, I.; Fleming, I. *Tetrahedron Lett.* **1979**, *20*, 993, 995, 2179.
19. Fleming, I.; Goldhill, J.; Paterson, I. *Tetrahedron Lett.* **1979**, 3209.
20. Ren, P.; Ribezzo, M. *J. Am. Chem. Soc.* **1991**, *113*, 7803.
21. Ohta, T.; Shiokawa, S.; Iwashita, E.; Nozoe, S. *Heterocycles* **1992**, *34*, 895.
22. Kozikowski, A. P.; Sorgi, K. L. *Tetrahedron Lett.* **1982**, *23*, 2281.
23. Urban, F. J.; Moore, B. S.; Breitenbach, R. *Tetrahedron Lett.* **1990**, *31*, 4421.
24. Ford, M. J.; Ley, S. V. *Synlett* **1990**, 255.
25. Corcoran, R. C. *Tetrahedron Lett.* **1990**, *31*, 2101.
26. Bartnik, R.; Lesniak, S.; Laurent, A. *Tetrahedron Lett.* **1981**, *22*, 4811.
27. Barros, D.; Carreño, M. C.; Ruano, J. L. G.; Maestro, M. C. *Tetrahedron Lett.* **1992**, *33*, 2733.
28. Anwar, S.; Davis, A. P. *Tetrahedron* **1988**, *44*, 3761.
29. Waldemeier, F.; De Bernardini, S.; Leach, C. A.; Tamm, C. *Helv. Chim. Acta* **1982**, *65*, 2472.
30. (a) Kohli, V.; Blöcker, H.; Köster, H. *Tetrahedron Lett.* **1980**, *21*, 2683. (b) Matteuci, M. D.; Caruthers, M. H. *Tetrahedron Lett.* **1980**, *21*, 3243.
31. Kierzek, R.; Ito, H.; Bhatt, R.; Itakura, K. *Tetrahedron Lett.* **1981**, *22*, 3761.
32. Nigam, S. C.; Mann, A.; Taddei, M.; Wermuth, C.-G. *Synth. Commun.* **1989**, *19*, 3139.
33. Corey, E. J.; Gras, J.-L.; Ulrich, P. *Tetrahedron Lett.* **1976**, 809.
34. Johnson, W. S.; Li, T.; Faulkner, D. J.; Campbell, S. F. *J. Am. Chem. Soc.* **1968**, *90*, 6225.See also: (a) Brady, S. F.; Ilton, M. A.; Johnson, W. S. *J. Am. Chem. Soc.* **1968**, *90*, 2882. (b) Nakamura, H.; Yamamoto, H.; Nozaki, H. *Tetrahedron Lett.* **1973**, 111.
35. Lewis, J. B.; Hendrick, G. W. *J. Org. Chem.* **1965**, *30*, 4271.See also: (a) Settine, R. L.; Parks, G. L.; Hunter, G. L. K. *J. Org. Chem.* **1964**, *29*, 616. (b) Bessière-Chréieu, Y.; Bras, J. P. *C.R. Hebd. Seances Acad. Sci., Ser. C* **1970**, *271*, 200. (c) Clark, Jr., B. C.; Chafin, T. C.; Lee, P. L.; Hunter, G. L. K. *J. Org. Chem.* **1978**, *43*, 519. (d) Watanabe, H.; Katsuhara, J.; Yamamoto, N. *Bull. Chem. Soc. Jpn.* **1971**, *44*, 1328.
36. Kaminski, J.; Schwegler, M. A.; Hoefnagel, A. J.; van Bekkum, H. *Recl. Trav. Chim. Pays-Bas* **1992**, *111*, 432.

37. Serramedan, D.; Marc, F.; Pereyre, M.; Filliatre, C.; Chabardès, P.; Delmond, B. *Tetrahedron Lett.* **1992**, *33*, 4457.

38. Kwan, T. W.; Smith, M. B. *Synth. Commun.* **1992**, *22*, 2273.

39. Hudrlik, P. F.; Rudnick, L. R.; Korzeniowski, S. H. *J. Am. Chem. Soc.* **1973**, *95*, 6848.

40. (a) Tanino, K.; Katoh, T.; Kuwajima, I. *Tetrahedron Lett.* **1988**, *29*, 1815. (b) Tanino, K.; Katoh, T.; Kuwajima, I. *Tetrahedron Lett.* **1988**, *29*, 1819.See also:Tanino, K.; Sato, K.; Kuwajima, I. *Tetrahedron Lett.* **1989**, *30*, 6551.

Glenn J. McGarvey
University of Virginia, Charlottesville, VA, USA

Zinc Chloride[1]

$ZnCl_2$

[7646-85-7] Cl_2Zn (MW 136.29)

(used in the preparation of organozinc reagents via transmetalation;[1] a mild Lewis acid useful for promoting cycloaddition,[2] substitution,[3] and addition reactions,[4] including electrophilic aromatic additions;[5] has found use in selective reductions[6])

Physical Data: mp 293 °C; bp 732 °C; *d* 2.907 g mL^{-1}.
Solubility: sol H_2O (432 g/100 g at 25 °C), EtOH (1 g/1.3 mL), glycerol (1 g/2 mL).
Form Supplied in: white, odorless, very deliquescent granules; principal impurities are H_2O and zinc oxychloride.
Analysis of Reagent Purity: melting point
Purification: reflux (50 g) in dioxane (400 mL) in the presence of Zn^0 dust, then filter hot and allow to cool to precipitate purified $ZnCl_2$. Also, anhydrous material may be sublimed under a stream of dry HCl, followed by heating to 400 °C in a stream of dry N_2.
Handling, Storage, and Precautions: very hygroscopic; store under anhydrous conditions; moderately irritating to skin and mucous membranes.

Organozinc Reagents. The transmetalation of organomagnesium, organolithium, and organocopper reagents is an important and versatile method of preparing useful zinc reagents.[1a] Alternatively, $ZnCl_2$ may be employed for direct insertion of Zn^0 into carbon–halogen bonds using Mg^0/ultrasound[7] or prior reduction with K^0.[8] The addition of the resulting allylic and propargylic or allenic zinc reagents to carbonyl compounds, imines, and iminium salts represents an important method of selective carbon–carbon bond formation.[9,11], These reactions are generally guided by chelation and the Zimmerman–Traxler transition state[10] to control the relative stereochemistry about the new carbon–carbon bond (eq 1),[12] as well as with respect to preexisting stereocenters (eq 2).[13] Anions derived from propargylic deprotonation add to carbonyl compounds with a high degree of regiochemical integrity in the presence of $ZnCl_2$, i.e. allenic organozinc intermediates cleanly afford the propargylic product (eq 3).[11b]

Alkylzinc reagents, often in the presence of copper salts, effectively participate in conjugate addition reactions[14] and clean S_N2' reactions (eq 4).[15]

Organozinc reagents are superior species for palladium- and nickel-catalyzed coupling reactions.[16] This offers an exceptional method for the selective formation of sp^2–sp^3 (eq 5),[17] sp^2–sp^2 (eqs 6 and 7),[18,19] and sp^2–sp (eq 8)[20] carbon–carbon bonds. In addition, the palladium-catalyzed coupling reactions of sp and sp^2 halides with vinylalanes (eq 9),[21] vinylcuprates,[22] vinylzirconium (eq 10)[23] and acyliron species[24] often proceed more effectively in the presence of $ZnCl_2$. This effect has been noted in other metal-mediated carbon bond formations,[25] though the role of this $ZnCl_2$ catalysis is not understood at present.

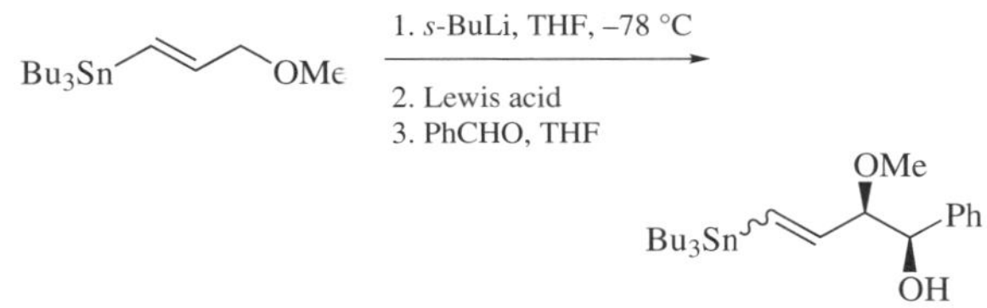

(1)

Lewis acid		
none	63%	53:47
0.5 equiv $ZnCl_2$	68%	89:11

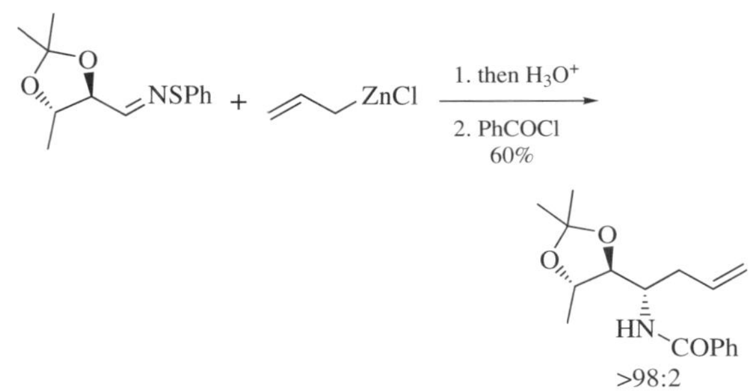

(2)

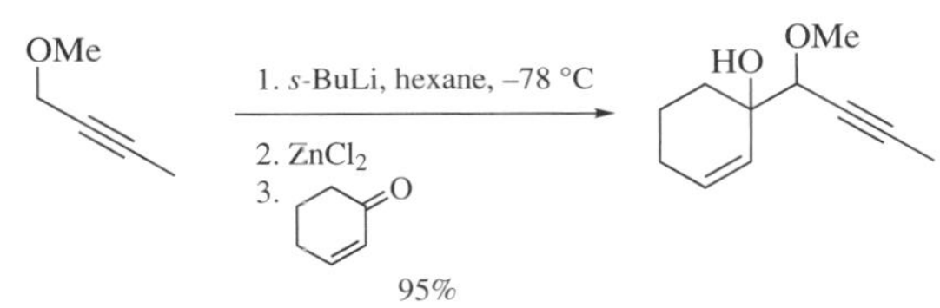

(3)

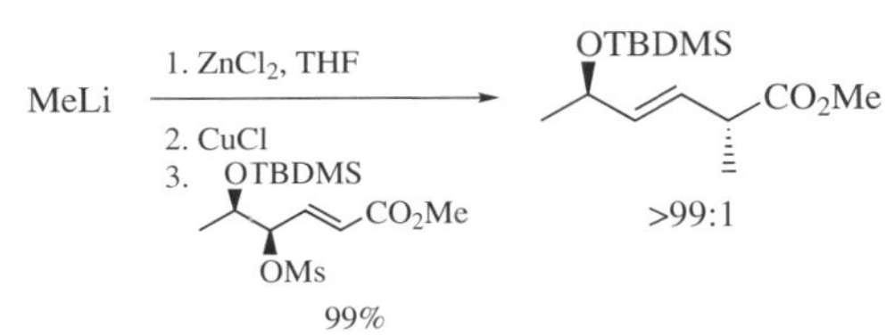

(4)

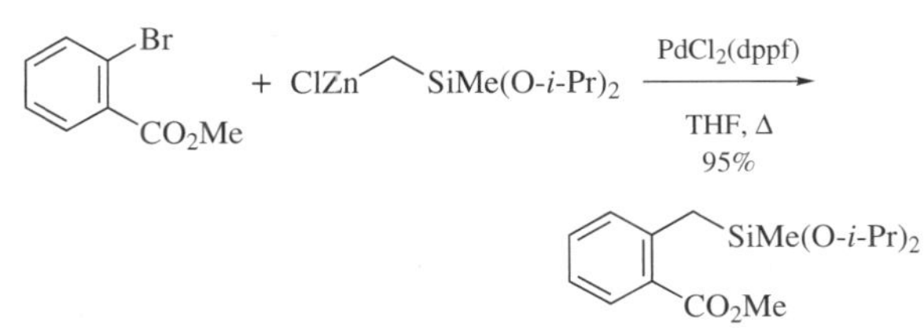

(5)

Organozinc reagents have been successfully exploited in asymmetric carbon bond formation.[26] Chiral glyoxylate esters engage organozinc species derived from Grignard reagents in selective addition reactions to afford enantiomerically enriched α-hydroxy acids (eq 11).[27] Enantioselective addition reactions of Grignard reagents have been achieved by sequential addition of $ZnCl_2$ and a chiral catalyst to afford secondary alcohols (eq 12).[28]

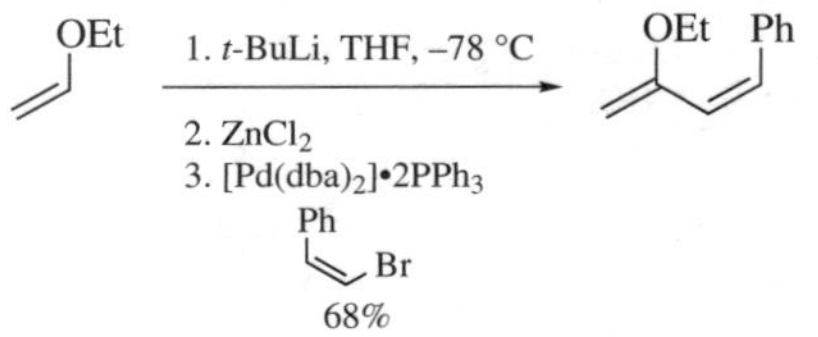

(6)

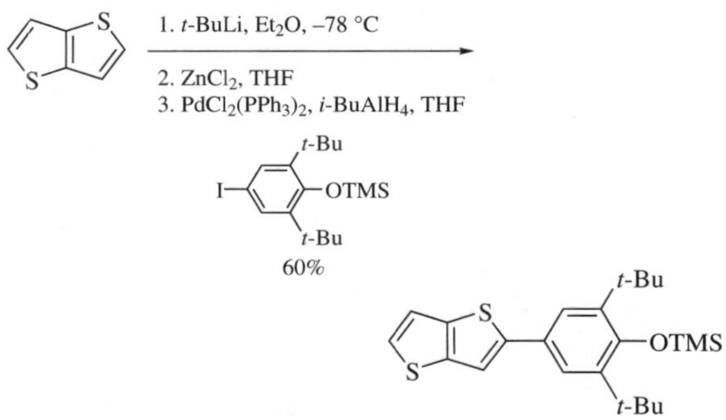

(7)

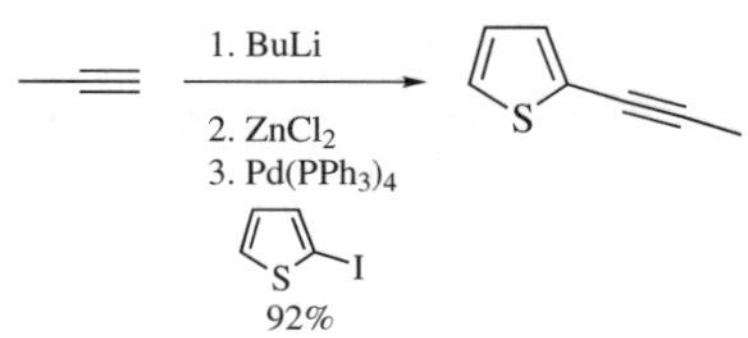

(8)

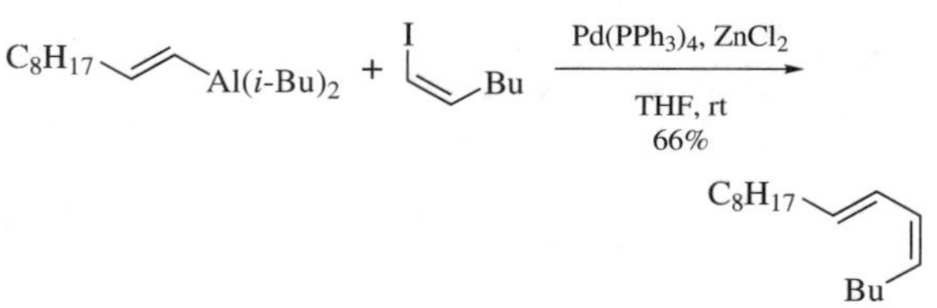

(9)

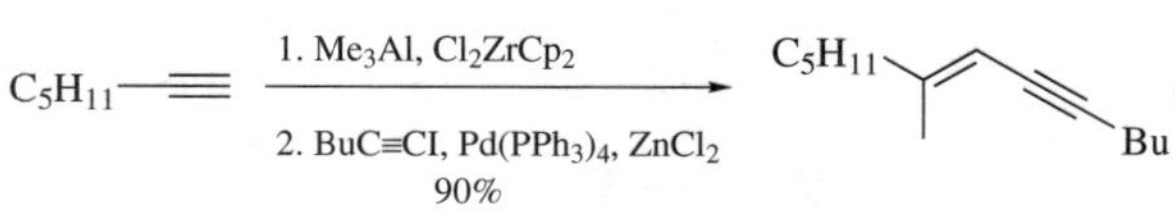

(10)

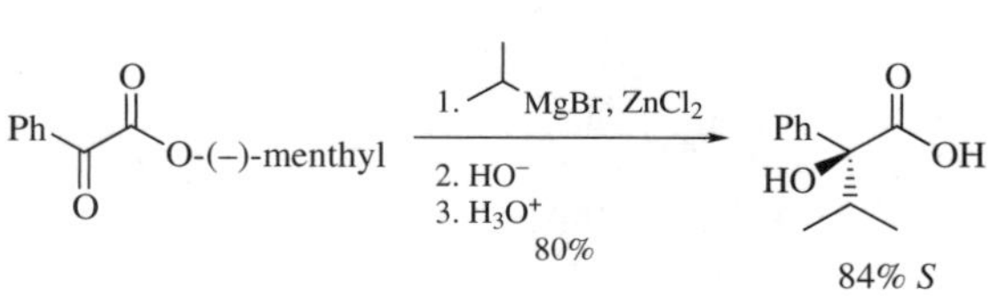

(11)

Zinc Enolates. Zinc enolates can be prepared by deprotonation of carbonyl compounds using standard bases, followed by transmetalation with $ZnCl_2$.[1b] These enolates offer important opportunities for stereoselective aldol condensations, including higher yields and stereochemical selection for *threo* crossed-aldol products (eq 13).[29] Both of these consequences are the result of the reversible formation of a six-membered zinc chelate intermediate which favors the *anti* disposition of its substituents. Though their use has been mostly replaced by kinetically controlled aldol methodology, zinc enolates sometimes present advantages in specific cases.[30] For example, chelated zinc enolates derived from α-amino acids have been shown to be of value in the stereoselective synthesis of β-lactams (eq 14).[31] Controlled monoalkylation of unsymmetrical ketones has been accomplished via treatment of zinc enolates with α-chlorothio ethers (eq 15),[32] and mild allylation of β-dicarbonyl compounds may be realized through exposure of the corresponding zinc enolate to an allyl alcohol in the presence of a palladium catalyst (eq 16).[33]

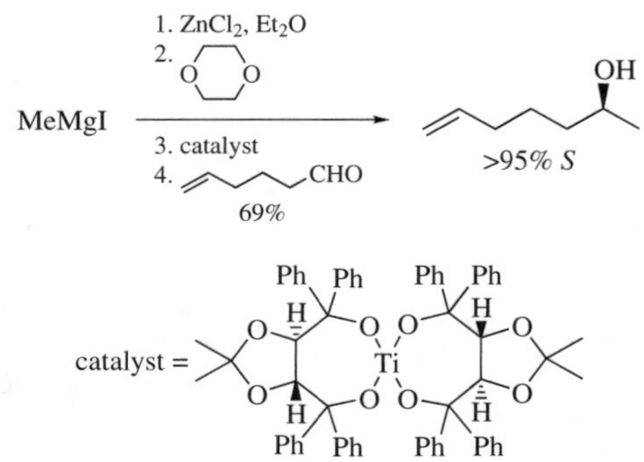

(12)

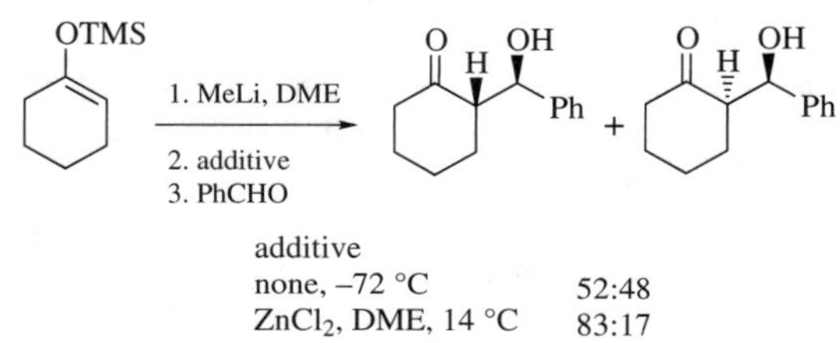

(13)

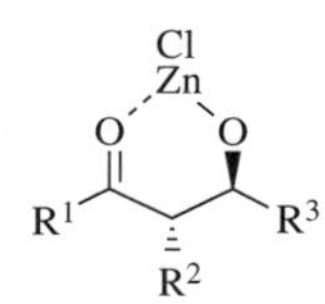

(1)

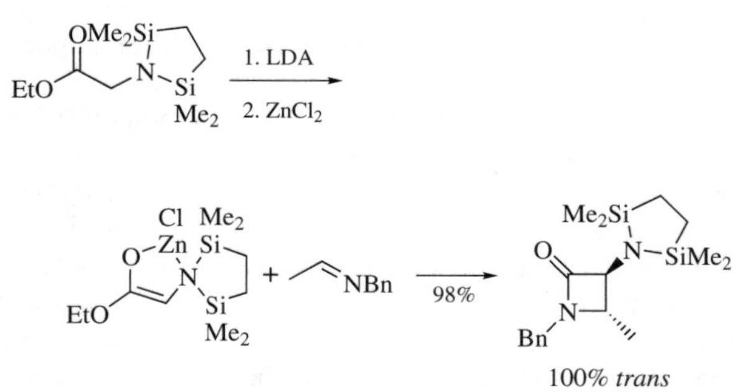

(14)

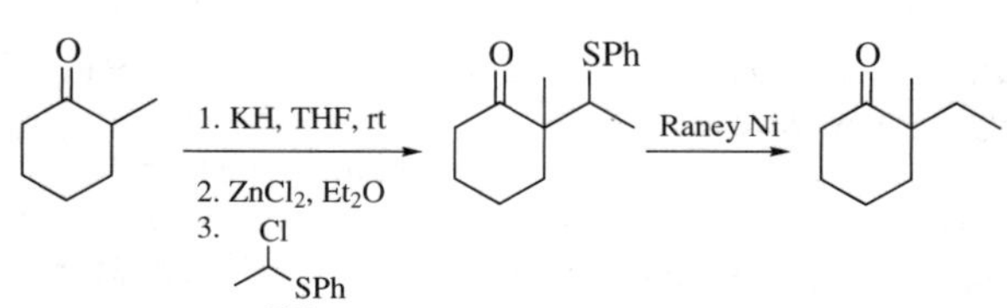

(15)

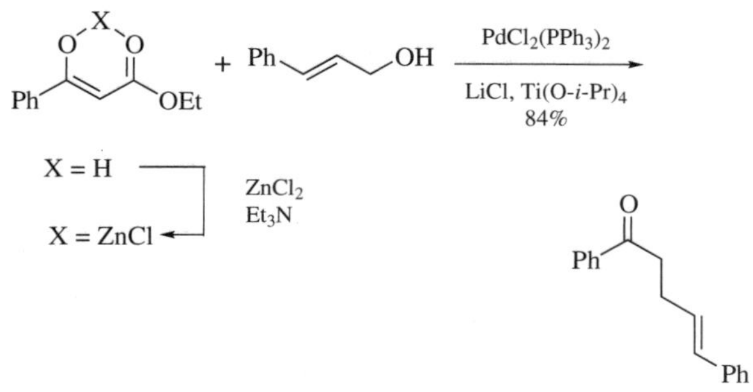

(16)

Homoenolates. Readily available mixed Me_3Si/alkyl acetals are converted into zinc homoenolates in high yield through exposure to $ZnCl_2$ in Et_2O (eq 17).[34] These mildly reactive species, which tolerate asymmetry α to the carbonyl, are useful for a variety of bond-forming reactions. In the presence of Me_3Si activation they undergo addition to aldehydes and, when admixed with CuBr·DMS (see Copper(I) Bromide), can be allylated or conjugatively introduced to α,β-unsaturated ketones (eq 18).[35] Copper salts are not required for conjugate addition to propargylic esters (eq 19).[36] These zinc species also participate in useful palladium-catalyzed bond-forming processes (eq 20).[37]

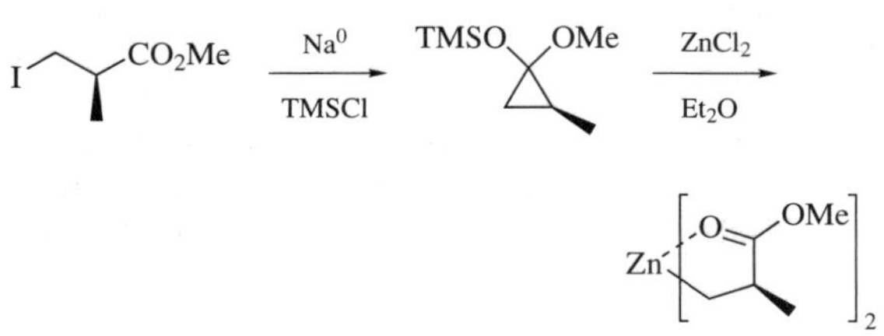

(17)

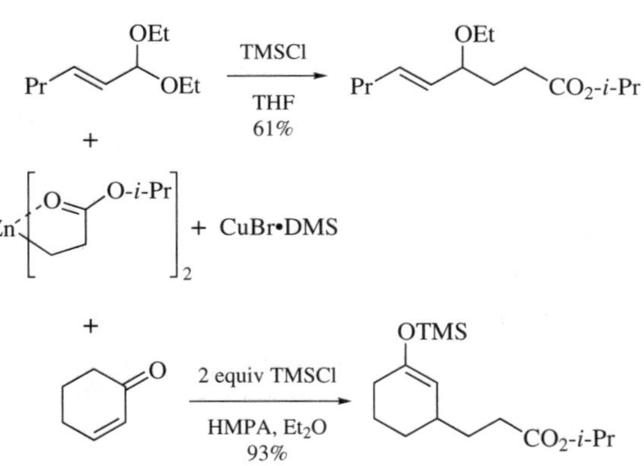

(18)

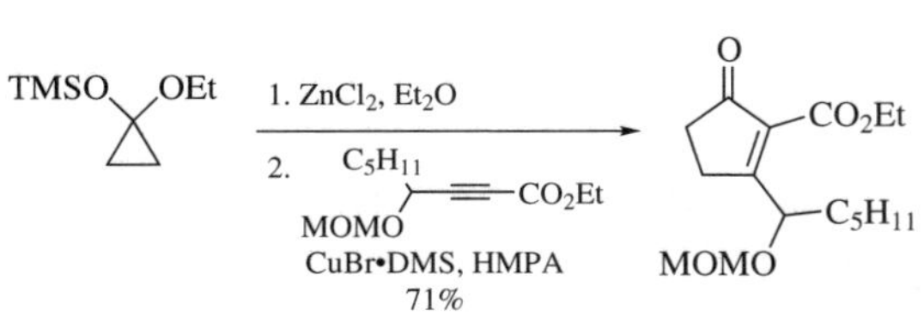

(19)

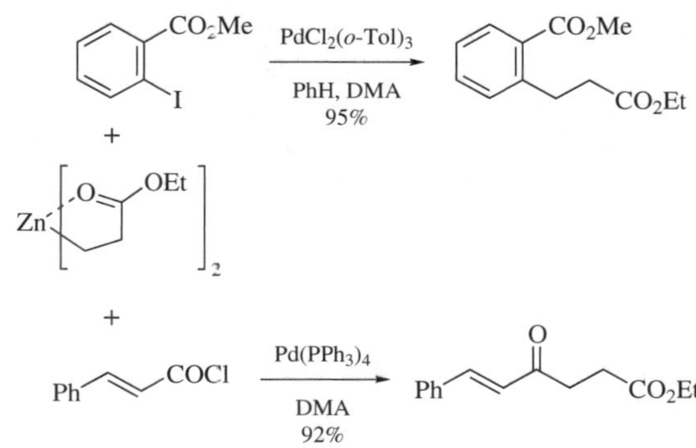

(20)

Cycloaddition Reactions. Catalysis by $ZnCl_2$ is often a powerful influence on cycloaddition reactions.[2] In addition to improving the rate of Diels–Alder reactions, enhanced control over regioselectivity (eq 21) [2c] and stereochemistry (eq 22) [2e] may be observed. An exceedingly important application of $ZnCl_2$ catalysis is found in the reaction of electron-rich dienes with carbonyl compounds.[2a,b] Intensive mechanistic studies on these reactions has uncovered two mechanistic pathways that afford stereochemically contrasting products, depending upon which Lewis acid is employed (eq 23).[38] It was concluded that cycloaddition reactions catalyzed by $ZnCl_2$ proceed via a classical [4 + 2] concerted process, whereas Lewis acids such as Boron Trifluoride Etherate and Titanium(IV) Chloride afford products through sequential aldol/cyclization processes. Examples abound wherein advantage has been taken of the predictable Cram–Felkin selectivity of this $ZnCl_2$ catalysis to exploit asymmetric variants of this reaction to synthesize a variety of natural products (eq 24).[39] The selective cycloaddition of electron-rich dienes with imines has also been catalyzed by $ZnCl_2$ (eq 25).[40]

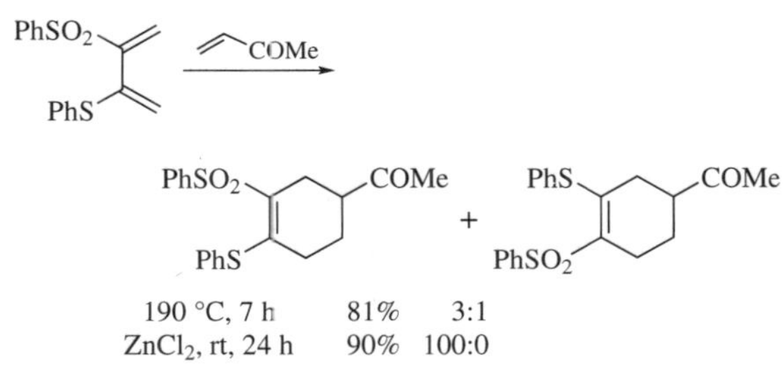

(21)

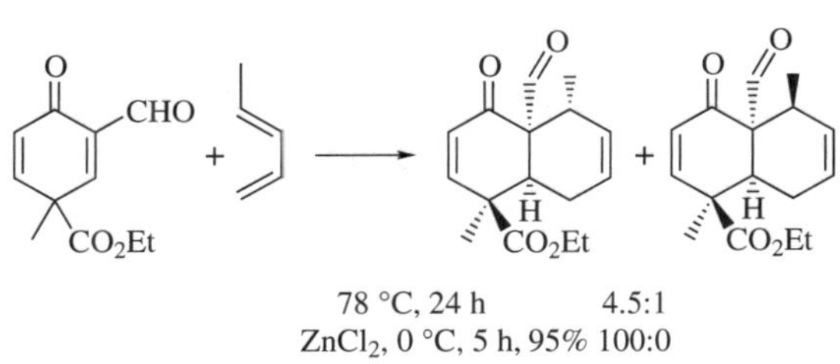

(22)

Activation of C=X Bonds. The mild Lewis acid character of $ZnCl_2$ is frequently exploited to promote the addition of various nucleophiles to carbon–heteroatom double bonds.[4] The well-established Knoevenagel condensation and related reactions have been effectively catalyzed by $ZnCl_2$ (eq 26).[41] The addition of enol ethers and ketene acetals to aldehydes and ketones has been noted (eq 27),[42] though $ZnCl_2$ has been used less widely in these aldol-type condensations than other Lewis acids, including $TiCl_4$, Magnesium Bromide, and Tin(IV) Chloride, to name a few.[4] In some instances, excellent levels of stereocontrol have been observed (eq 28).[43] Analogous additions to imines have been noted as well.[4e–g,44] Some conjugate addition reactions may also benefit from $ZnCl_2$ catalysis (eq 29).[45]

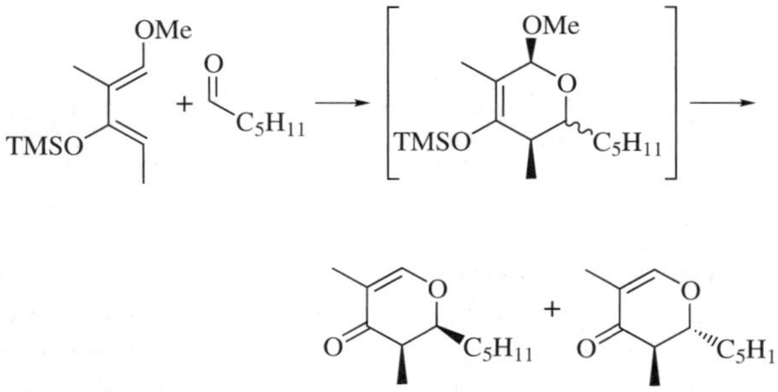
(23)

$BF_3{\bullet}OEt_2$, CH_2Cl_2, –78 °C	21%	69%
$ZnCl_2$, THF, 25 °C	91%	2%

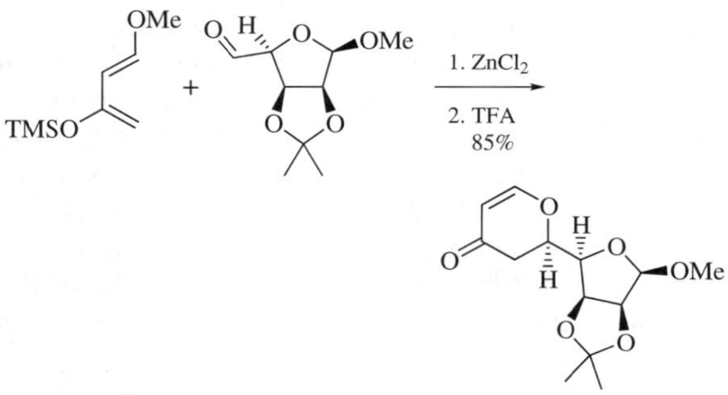
(24)

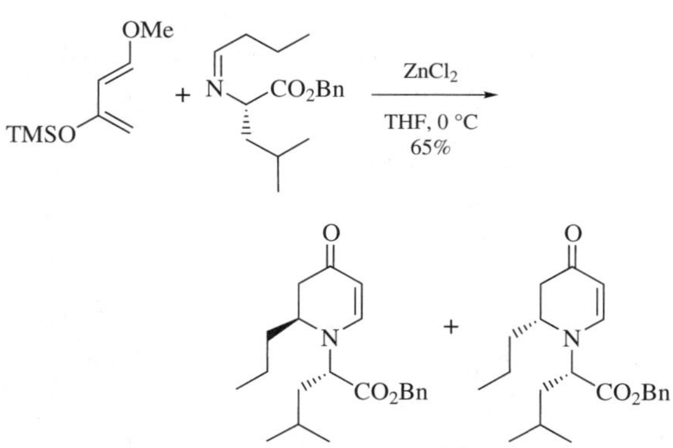
(25)

>9:1

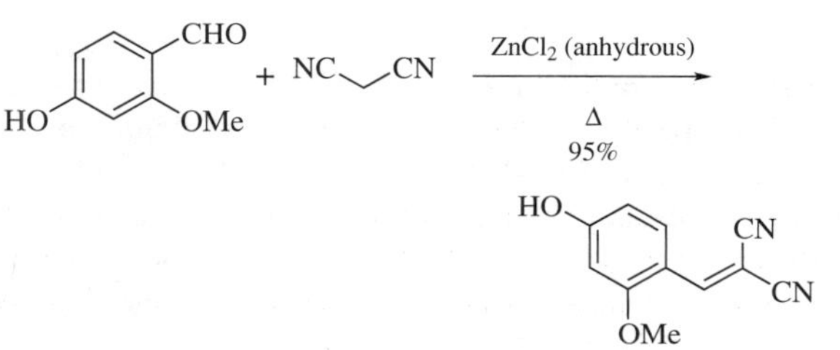
(26)

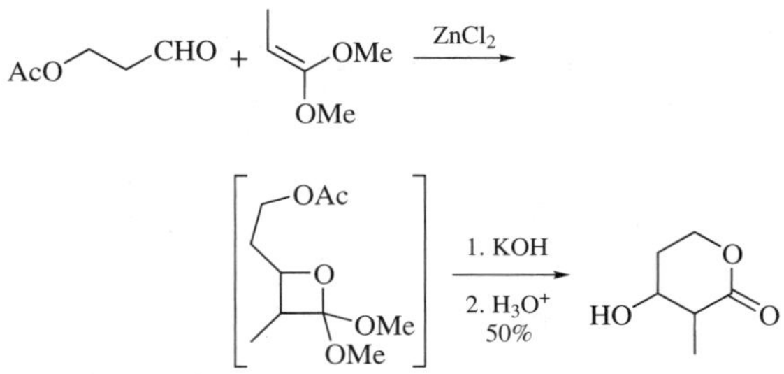
(27)

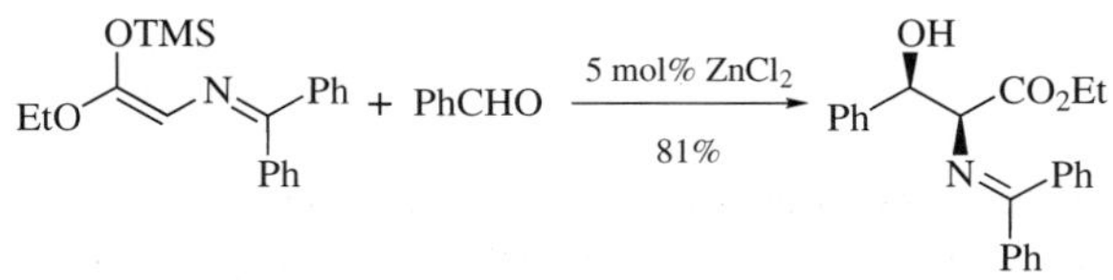
(28)

96% *syn*

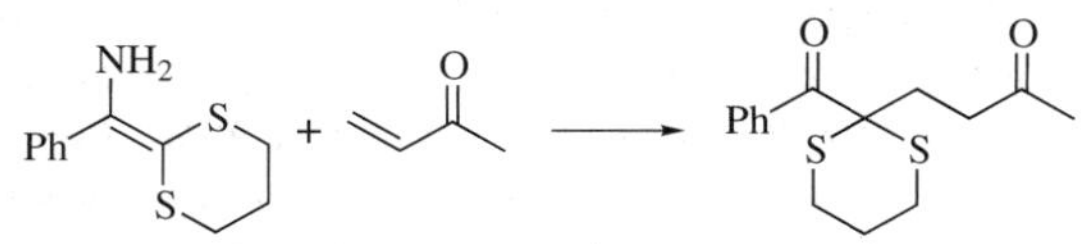
(29)

Δ, PhMe, 24 h	50%
1 equiv $ZnCl_2$, CH_2Cl_2, 0 °C, 2 d	98%

The formation of cyanohydrins using Cyanotrimethylsilane and Isoselenocyanatotrimethylsilane has been effectively catalyzed by $ZnCl_2$ (eq 30),[46] as has Strecker amino acid synthesis via the treatment of imines with $Me_3SiCN/ZnCl_2$.[47] The combination of carbonyl compounds with Acetyl Chloride or Acetyl Bromide may be promoted by $ZnCl_2$ to afford protected vicinal halohydrins (eq 31).[48]

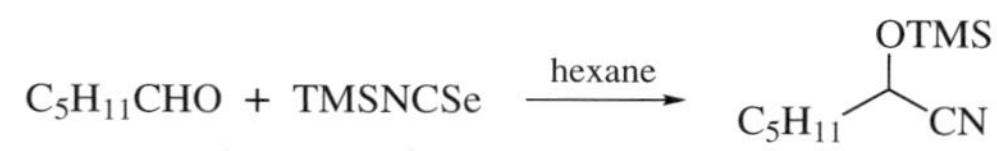
(30)

no $ZnCl_2$	78%
with $ZnCl_2$	>95%

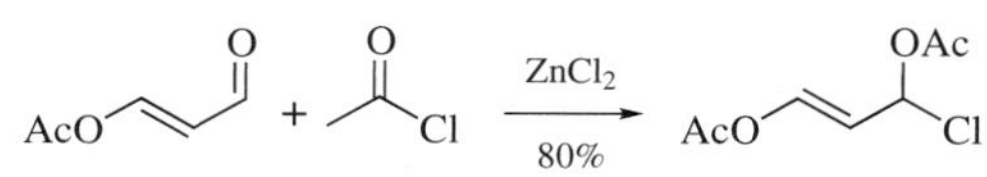
(31)

It is noteworthy that the treatment of carbonyl compounds with Chlorotrimethylsilane/$ZnCl_2$ results in a useful synthesis of Me_3Si enol ethers which, in turn, may be useful for other carbon bond-forming processes (eq 32).[49]

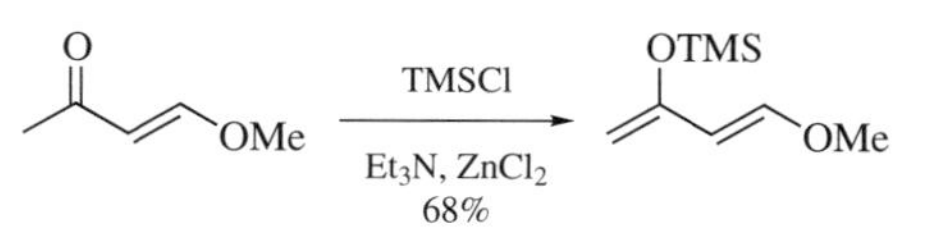
(32)

Activation of C–X Bonds. The activation of C–X single bonds toward nucleophilic substitution is also mediated by the Lewis acidic character of $ZnCl_2$.[3] Benzylic (eq 33),[50] allylic (eq 34),[3d,51] propargylic,[52] and tertiary halides (eq 35)[53] undergo substitution with mild carbon and heteroatom[3e] nucleophiles.

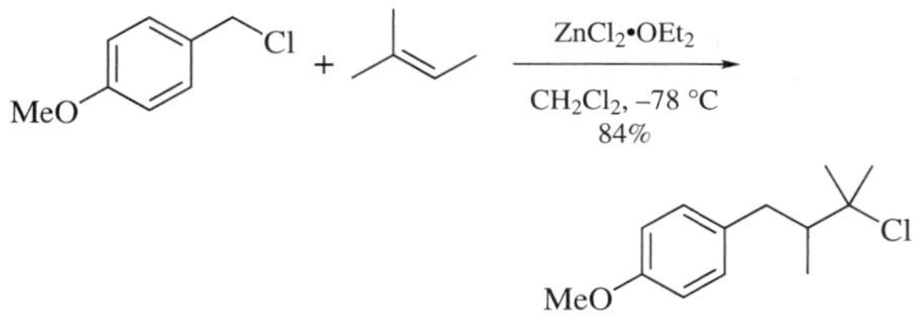

(33)

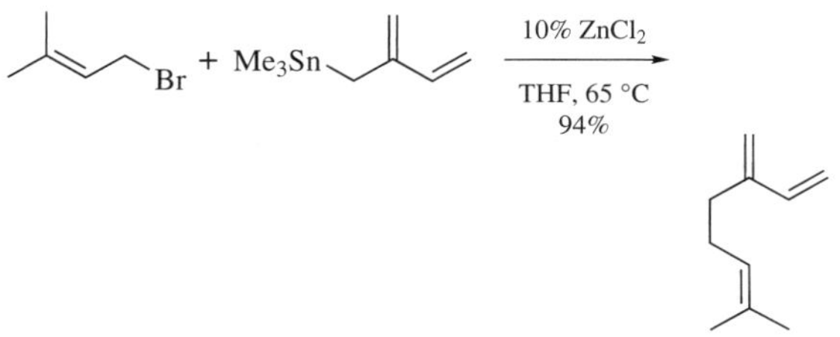

(34)

Myrcene

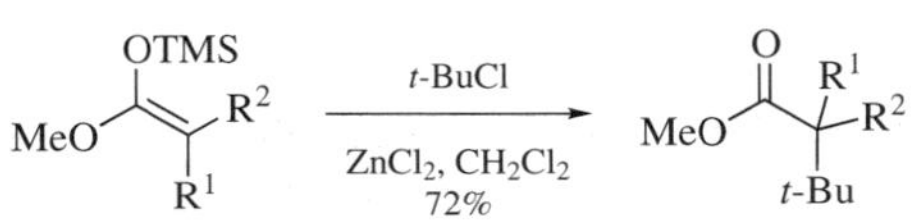

(35)

In a similar fashion, acetals (eq 36)[54] and orthoesters (eq 37)[55] may be used as electrophiles in substitution reactions with electron-rich alkenic nucleophiles. The combination of $ZnCl_2$ with co-catalysts has sometimes proven advantageous in these reactions (eq 38).[56]

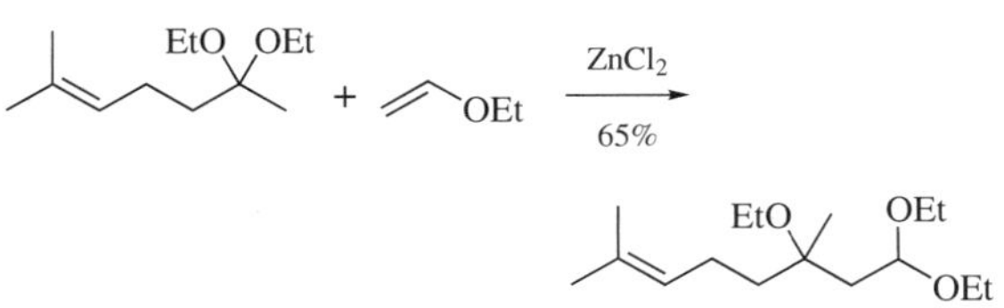

(36)

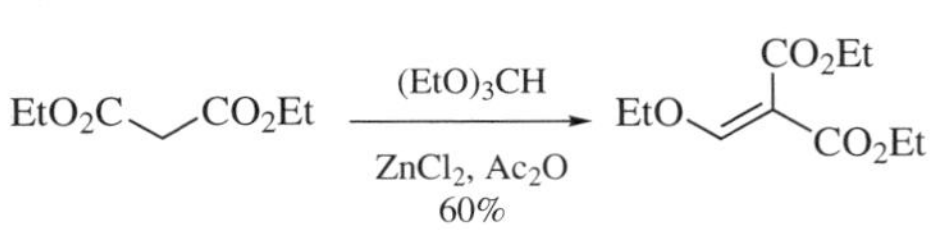

(37)

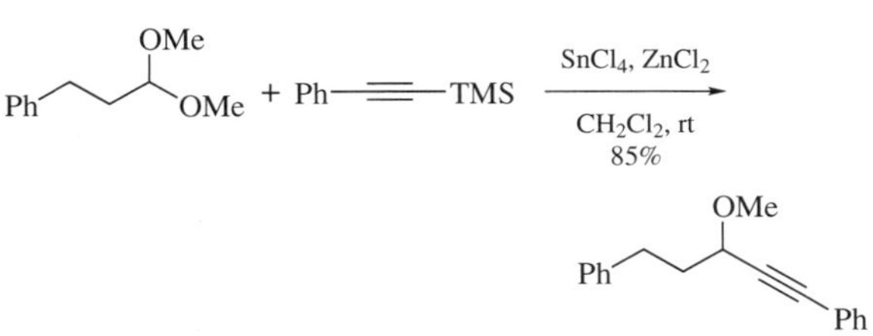

(38)

The regioselective ring opening reactions of epoxides (eq 39),[57] oxetanes,[58] and tetrahydrofurans (eq 40)[59] has been promoted by $ZnCl_2$ to afford adducts with suitable nucleophiles.

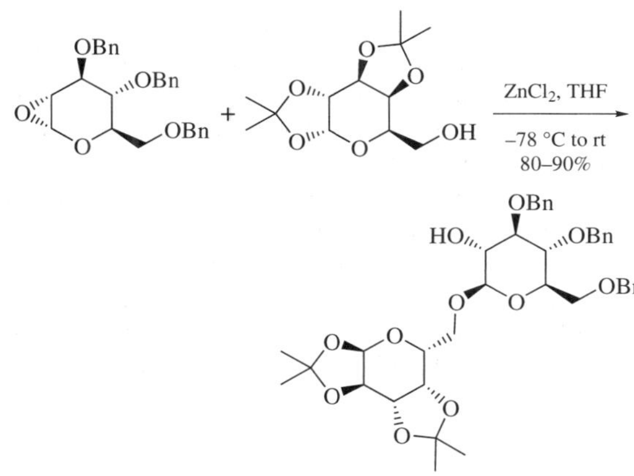

(39)

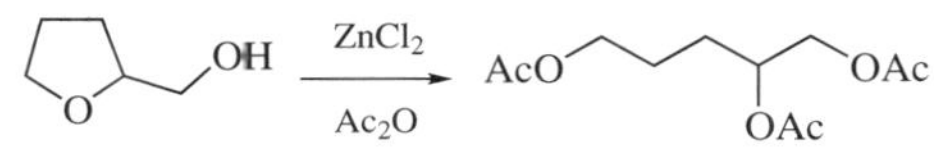

(40)

Activation by $ZnCl_2$ of allylic (eq 41)[60] and propargylic chlorides (eq 42),[61] as well as α-chloroenamines (eq 43),[62] in the presence of simple alkenes has been shown to yield four-membered and five-membered cycloadducts.

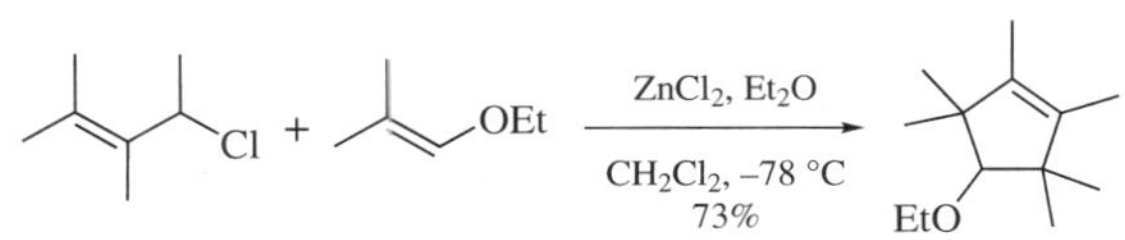

(41)

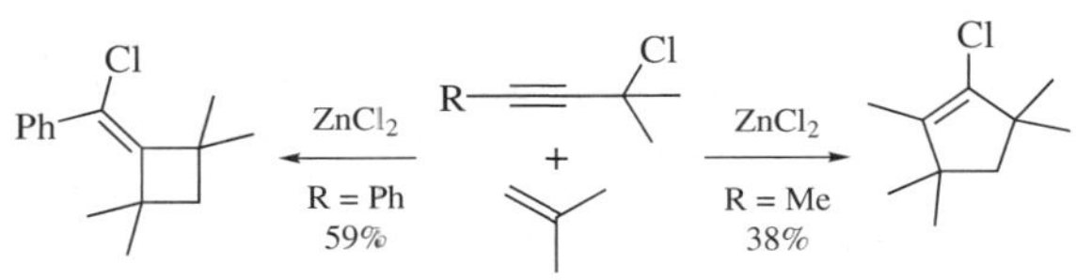

(42)

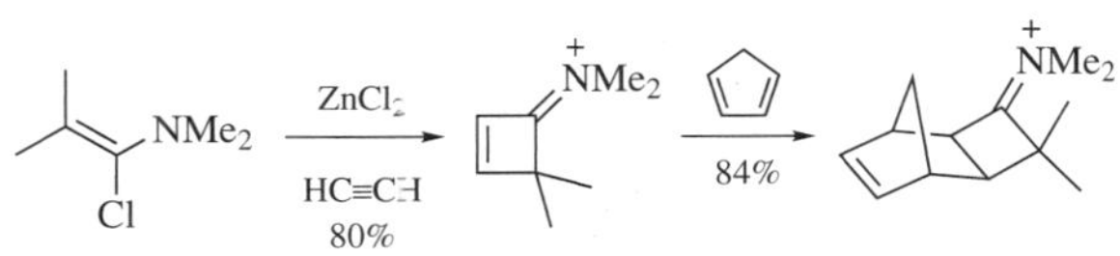

(43)

Chlorination of alcohols by Thionyl Chloride, [63] the preparation of acyl chlorides from lactones and anhydrides,[64] and the bromination and iodination of aromatic rings by Benzyltrimethylammonium Tribromide and Benzyltrimethylammonium Dichloroiodate, respectively,[65] are all effectively catalyzed by the presence of $ZnCl_2$. In addition, $ZnCl_2$ acts as a source of chloride for the halogenation of primary, secondary, and allylic alcohols using Triphenylphosphine–Diethyl Azodicarboxylate.[66]

Reduction. The very useful reducing agent Zinc Borohydride is prepared by exposure of $NaBH_4$ to $ZnCl_2$ in ether solvents.[67] Its use in selective reductions is described elsewhere.[66a] A complex reducing agent resulting from the mixture of NaH/t-pentyl-OH/$ZnCl_2$, given the acronym ZnCRA (Zinc Complex Reducing Agents), is found to open epoxides in a highly regioselective fashion, favoring hydride delivery at the least hindered position.[68] A related reagent, ZnCRASi, which includes Me_3SiCl in the mixture, selectively reduces ketones in high yield, though the levels of selectivity do not compete with other selective reagents.[69]

The reduction of ketones and aldehydes by silicon and tin hydrides in the presence of $ZnCl_2$ has been documented.[70] In the presence of Pd^0 catalysts and $ZnCl_2$, these hydrides selectively reduce α,β-unsaturated aldehydes and ketones to the corresponding saturated products (eq 44).[71] The presence of $ZnCl_2$ has been shown to modify the reactivity of several common reducing agents. For example, the mixture of Sodium Cyanoborohydride and $ZnCl_2$ selectively reduces tertiary, allylic, and benzylic halides (eq 45),[72] Sodium Borohydride, in the presence of $ZnCl_2$ and $PhNMe_2$ will reduce aryl esters to primary alcohols,[73] and Lithium Aluminum Hydride with $ZnCl_2/CuCl_2$ desulfurizes dithianes (eq 46).[74]

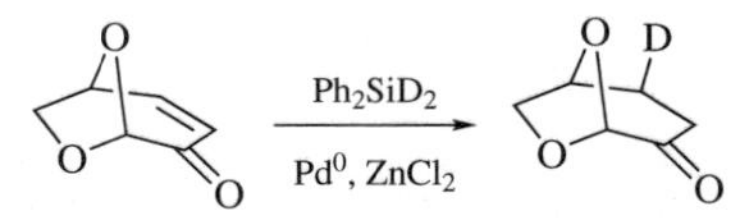

(44)

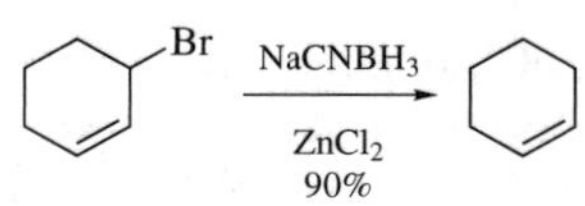

(45)

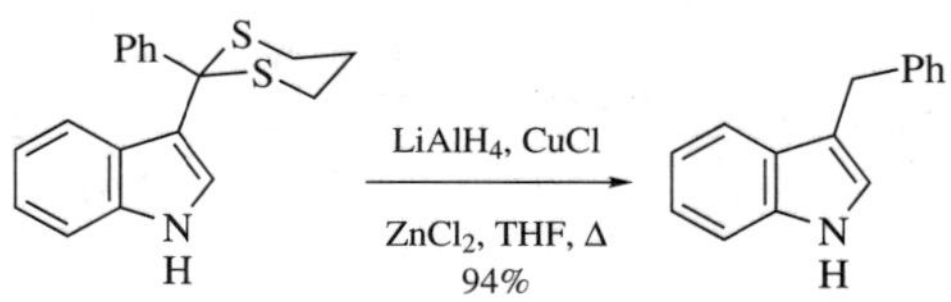

(46)

Stereoselectivity is also modified by the presence of $ZnCl_2$ (eq 47).[75] Enantioselective reduction of aryl ketones has been observed with Diisobutylaluminum Hydride (DIBAL) modified by $ZnCl_2$ and chiral diamine ligands (eq 48).[76]

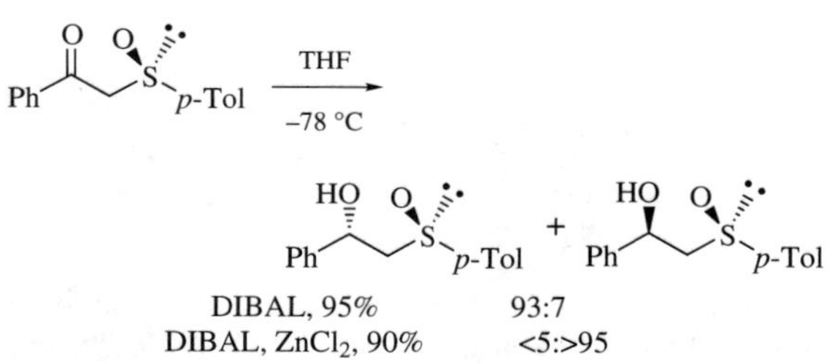

(47)

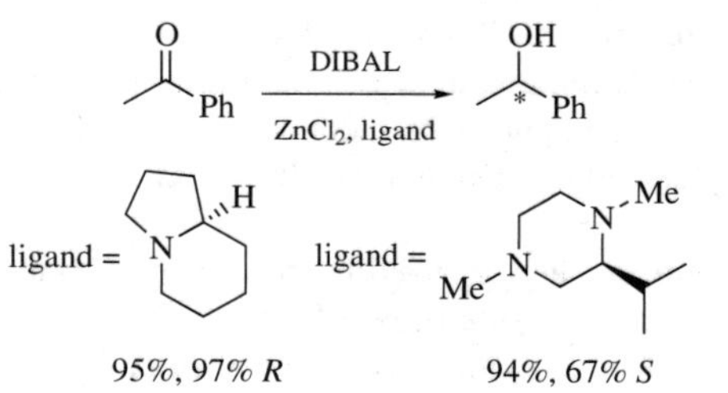

(48)

Protection/Deprotection. The acetylation of carbohydrates and other alcohols has been realized using Acetic Anhydride/$ZnCl_2$ (eq 49).[77] It is found that $ZnCl_2$ imparts selectivity to both acetylation[78] and acetonide formation[79] of polyols, which is useful in the synthetic manipulation of carbohydrates. Synthetically useful selective deprotection of acetates (50)[80] and dimethyl acetals (eq 51)[81] has been reported to be mediated by $ZnCl_2$.

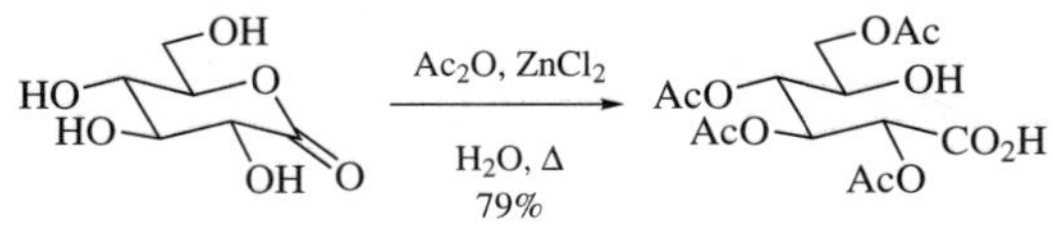

(49)

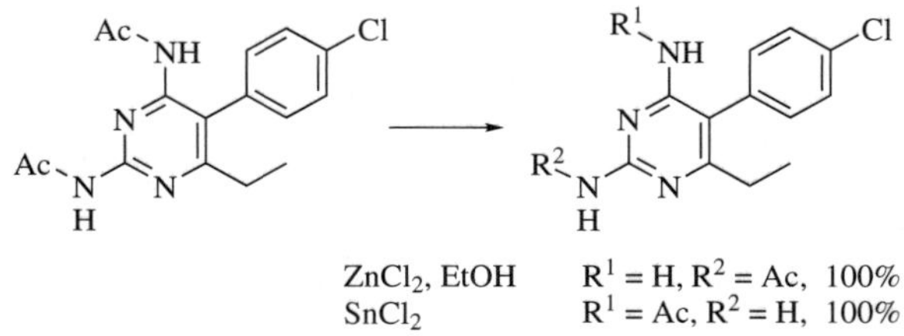

(50)

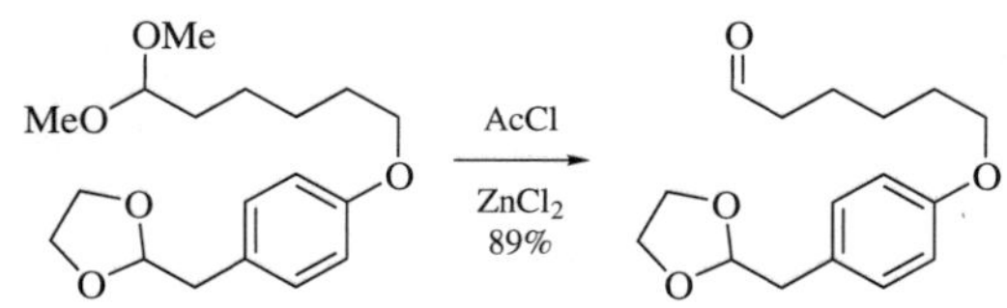

(51)

Acylation. Unsaturated esters are obtained through acylation of alkenes by anhydrides using activation by $ZnCl_2$ (eq 52).[82] The mixture of $RCOCl/ZnCl_2$ is effective in the acylation of silyl enol ethers to afford β-dicarbonyl products (eq 53).[83] Friedel–Crafts acylation is catalyzed by $ZnCl_2$, using anhydrides or acyl halides as the electrophiles (eqs 54 and 55).[8485]

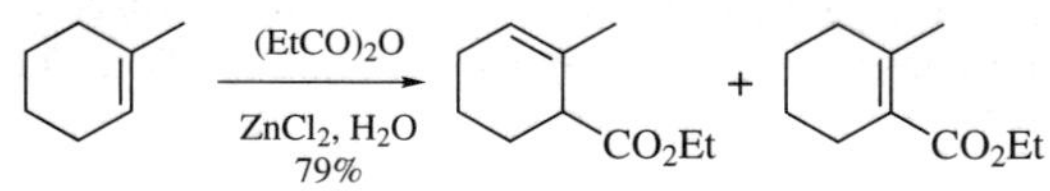

(52)

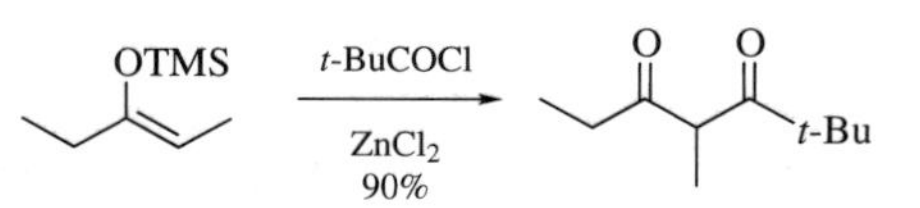

(53)

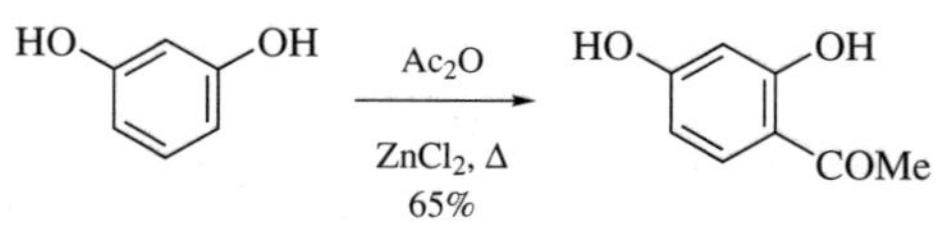

(54)

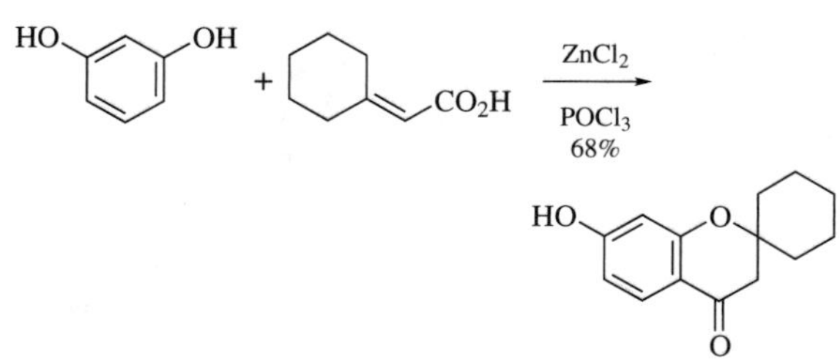

(55)

Aromatic Substitution. Several important classes of aromatic substitutions are mediated by $ZnCl_2$, including the Hoesch reaction (eq 56)[86] and the Fischer indole synthesis (eq 57).[87] Haloalkylation of aromatic rings using Formaldehyde or Chloromethyl Methyl Ether is readily accomplished through the agency of $ZnCl_2$ and warming (eq 58).[88]

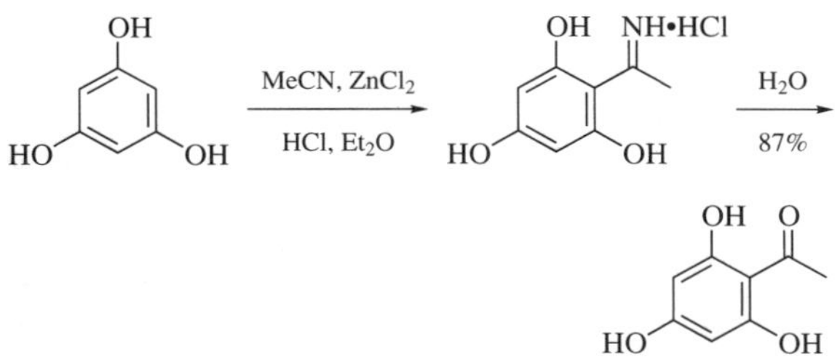

(56)

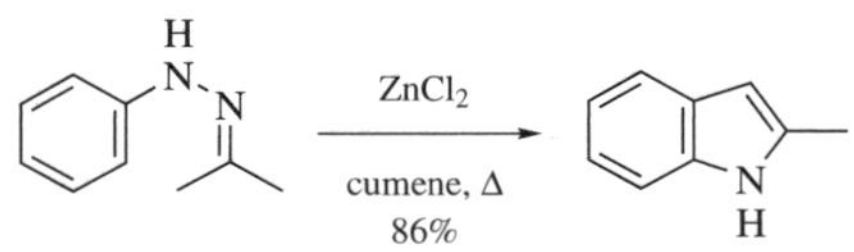

(57)

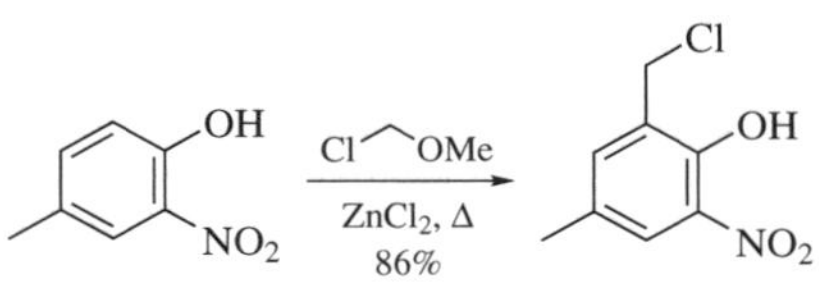

(58)

Related Reagents. Aluminum Chloride; Diphenylsilane–Tetrakis(triphenylphosphine)palladium(0)–Zinc Chloride; Phosphorus(III) Chloride–Zinc(II) Chloride; Phosphorus Oxychloride–Zinc(II) Chloride; Tin(IV) Chloride; Tin(IV) Chloride–Zinc Chloride; Titanium(IV) Chloride.

---

1. (a) Knochel, P.; *Comprehensive Organic Synthesis* **1991**, *1*, Chapter 1.7. (b) Rathke, M. W.; Weipert, P. *Comprehensive Organic Synthesis* **1991**, *2*, Chapter 1.8.
2. For examples: (a) Danishefsky, S.; DeNinno, M. P. *Angew. Chem., Int. Ed. Engl.* **1987**, *26*, 15 (b) Danishefsky, S. *Aldrichim. Acta* **1986**, *19*, 59. (c) Chou, S.-S.; Sun, D.-J. *Chem. Commun./J. Chem. Soc., Chem. Commun.* **1988**, 1176. (d) Liu, H.-J.; Ulibarri, G.; Browne, E. N. C. *Can. J. Chem.* **1992**, *70*, 1545. (e) Liu, H.-J.; Han, Y. *Tetrahedron Lett.* **1993**, *34*, 423.
3. (a) For examples:Zhai, D.; Zhai, W.; Williams, R. M. *J. Am. Chem. Soc.* **1988**, *110*, 2501 (b) Williams, R. M.; Sinclair, P. J.; Zhai, D.; Chen, D. *J. Am. Chem. Soc.* **1988**, *110*, 1547 (c) Paterson, I.; Fleming, I. *Tetrahedron Lett.* **1979**, *20*, 993, 995, 2175. (d) Godschalz, J. P.; Stille, J. K. *Tetrahedron Lett.* **1983**, *24*, 1905 (e) Miller, J. A. *Tetrahedron Lett.* **1975**, 2050 (f) Shikhmamedbekova, A. Z.; Sultanov, R. A. *J. Gen. Chem. USSR (Engl. Transl.)* **1970**, *40*, 72 (g) Ishibashi, H.; Nakatani, H.; Umei, Y.; Yamamoto, W.; Ikeda, M. *J. Chem. Soc., Perkin Trans. 1* **1987**, 589 (h) Mori, I.; Bartlett, P. A.; Heathcock, C. H. *J. Am. Chem. Soc.* **1987**, *109*, 7199.
4. (a) For examples:Eliel, E. L.; Hutchins, Sr., R. O.; Knoeber, M. *Org. Synth.* **1970**, *50*, 38. (b) Angle, S. R.; Turnbull, K. D. *J. Am. Chem. Soc.* **1989**, *111*, 1136 (c) Takai, K.; Heathcock, C. H. *J. Org. Chem.* **1985**, *50*, 3247 (d) Chiba, T.; Nakai, T. *Chem. Lett.* **1987**, 2187 (e) Taguchi, T.; Kitagawa, O.; Suda, Y.; Ohkawa, W.; Hashimoto, A.; Iitaka, Y.; Kobayashi, Y. *Tetrahedron Lett.* **1988**, *29*, 5291 (f) Kunz, H.; Pfrengle, W. *Angew. Chem., Int. Ed. Engl.* **1989**, *28*, 1067. (g) Kunz, H.; Pfrengle, W. *J. Org. Chem.* **1989**, *54*, 4261.
5. (a) For examples:Gulati, K. C.; Seth, S. R.; Venkataraman, K. *Org. Synth., Coll. Vol.* **1943**, *2*, 522 (b) Böhmer, V.; Deveaux, J. *Org. Prep. Proced. Int.* **1972**, *4*, 283 (c) Chapman, N. B.; Clarke, K.; Hughes, H. *J. Chem. Soc* **1965**, 1424.
6. (a) For examples:Oishi, T.; Nakata, T. *Acc. Chem. Res.* **1983**, *24*, 2653. (b) Fort, Y.; Feghouli, A.; Vanderesse, R.; Caubère, P. *J. Org. Chem.* **1990**, *55*, 5911 (c) Solladié, G.; Demailly, G.; Greck, C. *Org. Synth.* **1977**, *56*, 8.
7. Boerma, J. In *Comprehensive Organometallic Chemistry*; Wilkinson, G. Ed.; Pergamon: Oxford, **1982**. *2*, Chapter 16.
8. Rieke, R. D.; Uhm, S. J. *Synthesis* **1975**, 452.
9. (a) Allylic organozinc chlorides:Masuyama, Y.; Kinugawa, N.; Kurusu, Y. *J. Org. Chem.* **1987**. *52*, 3702 (b) Tamao, K.; Nakajo, E.; Ito, Y. *J. Org. Chem.* **1987**, *52*, 957 (c) Pétrier, C.; Luche, J.-L. *J. Org. Chem.* **1985**, *50*, 910 (d) Tamao, K.; Nakajo, E.; Ito, Y. *Tetrahedron* **1988**, *44*, 3997 (e) Jacobson, R. N.; Clader, J. W. *Tetrahedron Lett.* **1980**, *21*, 1205 (f) Hua, D. H.; Chon-Yu-King, R.; McKie, J. A.; Myer, L. *J. Am. Chem. Soc.* **1987**, *109*, 5026 (g) Courtois, G.; Harama, M.; Miginiac, P. *J. Organomet. Chem.* **1981**, *218*, 275 (h) Evans, D. A.; Sjogren, E. B. *Tetrahedron Lett.* **1986**, *27*, 4961 (i) Fujisawa, T.; Kofima, E.; Itoh, T.; Sato, T. *Tetrahedron Lett.* **1985**, *26*, 6089 (j) Fang, J.-M.; Hong, B.-C. *J. Org. Chem.* **1987**, *52*, 3162 (k) Auvray, P.; Knochel, P.; Normant, J.-F. *Tetrahedron Lett.* **1986**, *27*, 5091.
10. (a) For general reviews on the reactions of allylic organometallic reagents, see ref. [10].Yamamoto, Y. *Acc. Chem. Res.* **1987**, *20*, 243. (b) Hofmann, R. W. *Angew. Chem., Int. Ed. Engl.* **1982**, *21*, 555 (c) Yamamoto, Y.; Maruyama, K. *Heterocycles* **1982**, *18*, 357 (d) Courtois, G.; Miginiac, L. *J. Organomet. Chem.* **1974**, *69*, 1.
11. (a) Zweifel, G.; Hahn, G. *J. Org. Chem.* **1984**, *49*, 4565 (b) Evans, D. A.; Nelson, J. V. *J. Am. Chem. Soc.* **1980**, *102*, 774.
12. Verlhac, J.-B.; Pereyre, M. *J. Organomet. Chem.* **1990**, *391*, 283.
13. (a) Fuganti, C.; Grasselli, P.; Pedrocchi-Fantoni, G. *J. Org. Chem.* **1983**, *48*, 909 (b) Fronza, G.; Fuganti, C.; Grasselli, P.; Pedrocchi-Fantoni, G. *J. Carbohydr. Chem.* **1983**, *2*, 225.
14. Watson, R. A.; Kjonaas, R. A. *Tetrahedron Lett.* **1986**, *27*, 1437.
15. Yamamoto, Y.; Chounan, Y.; Tanaka, M.; Ibuka, T. *J. Org. Chem.* **1992**, *57*, 1024See also:Arai, M.; Kawasuji, T.; Nakamura, E. *Chem. Lett.* **1993**, 357.
16. (a) For examples: Negishi, E.; Takahashi, T.; King, A. O. *Org. Synth.* **1988**, *66*, 67 (b) Murahashi, S.-I.; Yamamura, M.; Yanagisawa, K.; Mita, N.; Kondo, K.; Negishi, E. *J. Org. Chem.* **1983**, *48*, 1560 (c) Tius, M. A.; Trehan, S. *J. Org. Chem.* **1986**, *51*, 765 (d) Rossi, R.; Carpita, A.; Cossi, P. *Tetrahedron* **1992**, *48*, 8801 (e) Shiragami, H.; Kawamoto, T.; Imi, K.; Matsubara, S.; Utimoto, K.; Nozaki, H.; *Tetrahedron* **1988**, *44*, 4009 (f) Matsushita, H.; Negishi, E. *J. Am. Chem. Soc.* **1981**, *103*, 2882 (g) Pelter, A.; Rowlands, M.; Jenkins, I. H. *Tetrahedron Lett.* **1987**, *28*, 5213 (h) Andreini, B. P.; Carpita, A.; Rossi, R. *Tetrahedron Lett.* **1988**, *29*, 2239 (i) Negishi, E.; Okukado, N.; Lovich, S. F.; Luo, T.-T. *J. Org. Chem.* **1984**, *49*, 2629.
17. Tamao, K.; Ishida, M.; Kumada, M. *J. Org. Chem.* **1983**, *48*, 2120.
18. Russell, C. E.; Hegedus, L. S. *J. Am. Chem. Soc.* **1983**, *105*, 943.
19. Takahashi, K.; Ogiyama, M. *Chem. Commun./J. Chem. Soc., Chem. Commun.* **1990**, 1196.
20. (a) King, A. O.; Okukado, N.; Negishi, E. *Chem. Commun./J. Chem. Soc., Chem. Commun.* **1977**, 683 (b) King, A. O.; Negishi, E.; Villiani, Jr., F. J.; Silveira, A. *J. Org. Chem.* **1978**, *43*, 358.
21. Negishi, E.; Takahashi, T.; Baba, S. *Org. Synth.* **1988**, *66*, 60 See also:Zweifel, G.; Miller, J. A. *Org. React.* **1984**, *32*, 375.
22. (a) Jabri, N.; Alexakis, A.; Normant, J.-F. *Bull. Soc. Claim. Fr., Part 2* **1983**, *321*, 332 (b) Jabri, N.; Alexakis, A.; Normant, J.-F. *Tetrahedron Lett.* **1981**, *22*, 9593851 (c) Nunomoto, S.; Kawakami, Y.; Yamashita, Y. *J. Org. Chem.* **1983**, *48*, 1912.
23. Negishi, E.; Okukado, N.; King, A. O.; Van Horn, D. E.; Spiegel, B. I. *J. Am. Chem. Soc.* **1978**, *100*, 2254See also:Van Horn, D. E.; Valente, L. F.; Idacavage, M. J.; Negishi, E. *J. Organomet. Chem.* **1978**, *156*, C20.

24. Koga, T.; Makinouchi, S.; Okukado, N. *Chem. Lett.* **1988**, 1141.

25. (a) Godschalx, J.; Stille, J. K. *Tetrahedron Lett.* **1980**, *21*, 2599 (b) Erdelmeier, I.; Gais, H.-J. *J. Am. Chem. Soc.* **1989**, *111*, 1125 (c) Cahiez, G.; Chavant, P.-Y. *Tetrahedron Lett.* **1989**, *30*, 7373.

26. (a) For examples:Jansen, J. F. G. A.; Feringa, B. L. *J. Org. Chem.* **1990**, *55*, 4168 (b) Soai, K.; Kawase, Y.; Oshio, A. *J. Chem. Soc., Perkin Trans. 1* **1991**, , 1613.

27. Boireau, G.; Deberly, A.; Abeuhaim, D. *Tetrahedron* **1989**, *45*, 5837See also:Fujisawa, T.; Ukaji, Y.; Funabora, M.; Yamashita, M.; Sato, T. *Bull. Chem. Soc. Jpn.* **1990**, *63*, 1894.

28. Seebach, D.; Behrendt, L.; Felix, D. *Angew. Chem., Int. Ed. Engl.* **1991**, , *30*, 1008.

29. House, H. O.; Crumrine, D. S.; Teranishi, A. Y.; Olmstead, H. D. *J. Am. Chem. Soc.* **1973**, *95*, 3310.

30. For a general review on stereoselective aldol condensations, see: Evans, D. A.; Nelson, J. V.; Taber, T. R. *Top. Stereochem.* **1982**, *13*, 1.

31. van der Steen, F. H.; Jastrzebski, J. T. B. H.; van Koten, G. *Tetrahedron Lett.* **1988**, *29*, 29, 765, 2467.See also: van der Steen, F. H.; Boersma, J.; Spek, A. L.; van Koten, G. *Organometallics* **1991**, , *10*, 2467.

32. Groth, U.; Huhn, T.; Richter, N. *Justus Liebigs Ann. Chem./Liebigs Ann. Chem.* **1993**, 49.

33. Itoh, K.; Hamaguchi, N.; Miura, M.; Nomura, M. *J. Chem. Soc., Perkin Trans. 1* **1992**, 2833.

34. Nakamura, E.; Sekiya, K.; Kuwajima, I. *Tetrahedron Lett.* **1987**, *28*, 337.

35. Nakamura, E.; Aoki, S.; Sekiya, K.; Oshino, H.; Kuwajima, I. *J. Am. Chem. Soc.* **1987**, *109*, 8056.

36. Crimmins, M. T.; Nantermet, P. G. *J. Org. Chem.* **1990**, *55*, 4235.

37. (a) Tamaru, Y.; Ochiai, H.; Nakamura, T.; Yoshida, Z. *Tetrahedron Lett.* **1986**, *27*, 955 (b) Tamaru, Y.; Ochiai, H.; Nakamura, T.; Tsubaki, K.; Yoshida, Z.-i. *Tetrahedron Lett.* **1985**, *26*, 5559.

38. (a) Danishefsky, S. J.; Kerwin, jrJ. F.; Kobayashi, S. *J. Am. Chem. Soc.* **1982**, *104*, 358 (b) Danishefsky, S. J.; Larson, E. R.; Askin, D. *J. Am. Chem. Soc.* **1982**, *104*, 6437 (c) Larson, E. R.; Danishefsky, S. J. *J. Am. Chem. Soc.* **1982**, *104*, 6458 (d) Danishefsky, S. J.; Maring, C. J. *J. Am. Chem. Soc.* **1985**, *107*, 1269 (e) Danishefsky, S. J.; Larson, E. R.; Askin, D.; Kato, N. *J. Am. Chem. Soc.* **1985**, *107*, 1246 (f) Danishefsky, S. J.; Kato, N.; Askin, D.; Kerwin, Jr., J. F. *J. Am. Chem. Soc.* **1982**, *104*, 360 (g) Midland, M. M.; Graham, R. S. *J. Am. Chem. Soc.* **1984**, *106*, 4294 (h) Garner, P. *Tetrahedron Lett.* **1984**, *25*, 5855.

39. Danishefsky, S. J.; Hungate, R. *J. Am. Chem. Soc.* **1986**, *108*, 2486.

40. Waldmann, H.; Braun, M. *J. Org. Chem.* **1992**, *57*, 4444.

41. Rao, P. S.; Venkataratnam, R. V. *Tetrahedron Lett.* **1991**, , *32*, 5821.

42. See also Ref. [4]a. Hofstraat, R. G.; Lange, J.; Scheeren, H. W.; Nivard, R. J. F. *J. Chem. Soc., Perkin Trans. 1* **1988**, 2315.

43. van der Werf, A. W.; Kellogg, R. M.; van Bolhuis, F. *Chem. Commun./J. Chem. Soc., Chem. Commun.* **1991**, 682.

44. Beslin, P.; Marion, P. *Tetrahedron Lett.* **1992**, *33*, 5339.

45. Page, P. C. B.; Harkin, S. A.; Marchington, A. P. *Synth. Commun.* **1989**, *19*, 1655 See also: Yamauchi, M.; Shirota, M.; Watanabe, T. *Heterocycles* **1990**, *31*, 1699.

46. Sukata, K. *J. Org. Chem.* **1989**, *54*, 2015 (a) See also:Deuchert, K.; Hertenstein, W.; Wehner, G. *Ber. Dtsch. Chem. Ges./Chem. Ber.* **1979**, *112*, 2045 (b) Ruano, J. L. G.; Castro, A. M. M.; Rodriguez, J. H. *Tetrahedron Lett.* **1991**, *32*, 3195.

47. Kunz, H.; Sager, W.; Pfrengle, W.; Schanzenback, D. *Tetrahedron Lett.* **1988**, *29*, 4397.

48. (a) Neuenschwander, M.; Bigler, P.; Christen, K.; Iseli, R.; Kyburz, R.; Mühle, H. *Helv. Chim. Acta* **1978**, *61*, 2047 (b) Bigler, P.; Schonholzer, S.; Neuenschwander, M. *Helv. Chim. Acta* **1978**, *61*, 2059 (c) Bigler, P.; Neuenschwander, M. *Helv. Chim. Acta* **1978**, *61*, 2165.

49. Danishefsky, S. J.; Kitahara, T. *J. Am. Chem. Soc.* **1974**, *96*, 7807.

50. Bäuml, E.; Tschemschlok, K.; Pock, R.; Mayr, H. *Tetrahedron Lett.* **1988**, *29*, 6925 (a) See also:Clark, J. H.; Kybett, A. P.; Macquarrie, D. C.; Barlow, S. J.; Landon, P. *Chem. Commun./J. Chem. Soc., Chem. Commun.* **1989**, 1353 (b) Reetz, M. T.; Sauerwald, M. *Tetrahedron Lett.* **1983**, *24*, 2837.

51. (a) Other examples:Koschinsky, R.; Köhli, T.-P.; Mayr, H. *Tetrahedron Lett.* **1988**, *29*, 5641 (b) Alonso, F.; Yus, M. *Tetrahedron* **1991**, , *47*, 9119.

52. Mayr, H.; Klein, H. *J. Org. Chem.* **1981**, *46*, 4097.

53. Reetz, M. T.; Schwellus, K. *Tetrahedron Lett.* **1978**, 1455.

54. Isler, O.; Schudel, P. *Adv. Org. Chem.* **1963**, *4*, 128 See also:Oriyama, T.; Iwanami, K.; Miyauchi, Y.; Koga, G. *Bull. Chem. Soc. Jpn.* **1990**, 3716.

55. Parham, W. E.; Reed, L. J. *Org. Synth., Coll. Vol.* **1955**, *3*, 395See also: (a) Grégoire de Bellemont, E. *Bull. Soc. Claim. Fr.* **1901**, *25*, 18 (b) Hatanaka, K.; Tanimoto, S.; Sugimoto, T.; Okano, M. *Tetrahedron Lett.* **1981**, *22*, 3243.

56. Hayashi, M.; Inubushi, A.; Mukaiyama, T. *Bull. Chem. Soc. Jpn.* **1988**, *61*, 4037.

57. Halcomb, R. L.; Danishefsky, S. J. *J. Am. Chem. Soc.* **1989**, *111*, 6661.

58. Scheeren, H. W.; Dahman, F. J. M.; Bakker, C. G. *Tetrahedron Lett.* **1979**, 2925 (a) See also:Sukata, K. *Bull. Chem. Soc. Jpn.* **1990**, *63*, 825 (b) Hug, E.; Mellor, M.; Scovell, E. G.; Sutherland, J. K. *Chem. Commun./J. Chem. Soc., Chem. Commun.* **1978**, 526.

59. Grummett, O.; Stearns, J. A.; Arters, A. A. *Org. Synth., Coll. Vol.* **1955**, *3*, 833.

60. Klein, H.; Mayr, H. *Angew. Chem., Int. Ed. Engl.* **1981**, *20*, 1027.

61. Mayr, H.; Seitz, B.; Halberstat-Kausch, I.-K. *J. Org. Chem.* **1981**, *46*, 1041.

62. Hoornaert, C.; Hesbain-Frisque, A. M.; Ghosez, L. *Angew. Chem., Int. Ed. Engl.* **1975**, *14*, 569.See also:Sidani, A.; Marchand-Brynaert, J.; Ghosez, L. *Angew. Chem., Int. Ed. Engl.* **1974**, *13*, 267.

63. Squires, T. G.; Schmidt, W. W.; McCandlish, Jr., C. S. *J. Org. Chem.* **1975**, *40*, 134.

64. (a) Goel, O. P.; Seamans, R. E. *Synthesis* **1973**, 538 (b) Kyrides, L. P. *Org. Synth., Coll. Vol.* **1943**, *2*, 528.

65. (a) Kajigaeshi, S.; Kakinami, T.; Moriwaki, M.; Tanaka, T.; Fujisaki, S.; Okamoto, T. *Bull. Chem. Soc. Jpn.* **1989**, *62*, 439 (b) Kajigaeshi, S.; Kakinami, T.; Watanabe, F.; Odanoto, T. *Bull. Chem. Soc. Jpn.* **1989**, *62*, 1349.

66. Ho, P.-T.; Davies, N. *J. Org. Chem.* **1984**, *49*, 3027.

67. Gensler, W. J.; Johnson, F.; Sloan, A. D. B. *J. Am. Chem. Soc.* **1960**, *82*, 6074.

68. Fort, Y.; Vanderesse, R.; Caubère, P. *Tetrahedron Lett.* **1985**, *26*, 3111.

69. See Ref. [6]b and Caubère, P.; Vanderesse, R.; Fort, Y. *Acta Chem. Scand.* **1991**, , *45*, 742.

70. (a) Anwar, S.; Davies, A. P. *Tetrahedron* **1988**, *44*, 3761 (b) Laurent, A. J.; Lesniak, S. *Tetrahedron Lett.* **1992**, *33*, 3311.

71. Keinan, E.; Greenspoon, N. *J. Am. Chem. Soc.* **1986**, *108*, 7314.

72. Kim, S.; Kim, Y. J.; Ahn, K. H. *Tetrahedron Lett.* **1983**, *24*, 3369.

73. Yamakawa, T.; Masaki, M.; Nohira, H. *Bull. Chem. Soc. Jpn.* **1991**, , *64*, 2730.

74. (a) Mukaiyama, T. *Int. J. Sulfur Chem., Part B* **1972**, , *7*, 2173 (b) Stütz, P.; Stadler, P. A. *Org. Synth.* **1977**, *56*, 8.

75. Solladié, G.; Demailly, G.; Greck, C. *Tetrahedron Lett.* **1985**, *26*, 435 See also: Hanamoto, T.; Fuchikama, T. *J. Org. Chem.* **1990**, *55*, 4969 and Ref. [6]c.

76. Falorni, M.; Giacomelli, G.; Lardicci, L. *Gazz. Chim. Ital.* **1990**, *120*, 765.

77. Braun, C. E.; Cook, C. D. *Org. Synth.* **1961**, *41*, 79.

78. Hanessian, S.; Kagotani, M. *Carbohydr. Res.* **1990**, *202*, 67.

79. Angyal, S. J.; Gilham, P. T.; Macdonald, C. G. *J. Chem. Soc* **1957**, 1417.

80. Griffen, R. J.; Lowe, P. R. *J. Chem. Soc., Perkin Trans. 1* **1992**, 1811.

81. Chang, C.; Chu, K. C.; Yue, S. *Synth. Commun.* **1992**, *22*, 1217.
82. Groves, J. K.; Jones, N. *J. Chem. Soc* **1968**2898 (a) See also:Marshall, J. A.; Andersen, N. H.; Schlicher, J. W. *J. Org. Chem.* **1979**, *35*, 858 (b) Groves, J. K.; Jones, N. *J. Chem. Soc* **1969**2350 (c) House, H. O.; Gilmore, W. F. *J. Am. Chem. Soc.* **1961**, *83*, 3980.
83. Tirpak, R. E.; Rathke, M. W. *J. Org. Chem.* **1982**, *47*, 5099 See also: Reetz, M. T.; Kyung, S.-H. *Tetrahedron Lett.* **1985**, *26*, 6333.
84. Cooper, S. R. *Org. Synth., Coll. Vol.* **1955**, *3*, 761 See also: Baddeley, G.; Williamson, R. *J. Chem. Soc* **1956** 4647See also: Zani, C. L.; de Oliveira, A. B.; Snieckus, V. *Tetrahedron Lett.* **1987**, *28*, 6561.
85. Dike, S. Y.; Merchant, J. R.; Sapre, N. Y. *Tetrahedron* **1991**, , *47*, 4775 (a) See also:Shah, V. R.; Bose, J. L.; Shah, R. C. *J. Org. Chem.* **1960**, *25*, 677 (b) Dallacker, F.; Kratzer, P.; Lipp, M. *Justus Liebigs Ann. Chem./Liebigs Ann. Chem.* **1961**, *643*, 97 Ref. [5]a and .
86. Ruske, W. In *Friedel–Crafts and Related Reactions*; Olah, G. A., Ed.; Interscience: New York, **1964**.Vol. *3*, p 383.
87. (a) Ref. [5]c and Shriner, R. L.; Ashley, W. C.; Welch, E. *Org. Synth., Coll. Vol.* **1955**, *3*, 725 (b) Prochazki, M. P.; Cartson, R. *Acta Chem. Scand.* **1989**, *43*, 651 Ref. [5]b and.
88. Olah, G. A.; Tolgyesi, W. S. In *Friedel–Crafts and Related Reactions*; Olah, G. A. Ed.; Interscience: New York, **1964**.Vol. *2*, p 659

Glenn J. McGarvey
University of Virginia, Charlottesville, VA, USA

List of Contributors

Kenichi Akaji	*Osaka University, Osaka, Japan* • 2-Chloro-1,3-dimethylimidazolidium Hexafluorophosphate
Fernando Albericio	*Universitat de Barcelona, Barcelona, Spain* • Azidotris(dimethylamino)phosphonium Hexafluorophosphate • 1-Benzotriazolyl Diethyl Phosphate (BDP) • *O*-Benzotriazol-1-yl-*N,N,N′,N′*-tetramethyluronium Hexafluorophosphate • *N*-[(Dimethylamino)-1*H*-1,2,3-triazolo[4,5-*b*]pyridin-1-ylmethylene]-*N*-methylmethanaminium Hexafluorophosphate *N*-Oxide
Jeffrey S. Albert	*University of Pittsburgh, Pittsburgh, PA, USA* • 1,3-Dicyclohexylcarbodiimide • 1,3-Dicyclohexylcarbodiimide-4-Dimethylaminopyridine
Hussein Al-Mughaid	*Dalhousie University, Halifax, NS, Canada* • 4-Pentenol • Tetrachlorophthalimide
James T. Anderson	*Case Western Reserve University, Cleveland, OH, USA* • S-(1-Oxido-2-pyridinyl)-1,1,3,3-tetramethylthiouronium Hexafluorophosphate (HOTT)
Odón Arjona	*Universidad Complutense de Madrid, Madrid, Spain* • Trimethylphosphine
Alan Armstrong	*University of Bath, Bath, UK* • Bis(*o*-phenylene) Pyrophosphite • *N,N′*-carbonyldiimidazole • 2-Chloro-1-methylpyridinium Iodide • *p*-Nitrophenol
Jeffrey Aubé	*University of Kansas, Lawrence, KS, USA* • 2-(*t*-Butoxycarbonyloxyimino)-2-Phenylacetonitrile
Thomas G. Back	*University of Calgary, Calgary, AB, Canada* • 2-(Phenylseleno)-1*H*-isoindole-1,3(2*H*)-dione
Alexandru T. Balaban	*Texas A&M University at Galveston, Galveston, TX, USA* • 2,6-Di-*t*-butylmethylpyridine
Harjinder S. Bansal	*Zeneca Agrochemicals, Bracknell, UK* • Bis(*o*-nitrophenyl) Phenylphosphonate • Bis(*p*-nitrophenyl) Hydrogen Phosphate • Bis(*p*-nitrophenyl) Phenylphosponate • Dimethylphosphinothioyl Chloride
David A. Barda	*Eli Lilly and Company, Indianapolis, IN, USA* • Cyanuric Fluoride
Duncan Batty	*Celltech R&D Ltd, Cambridge, UK* • 2-(5-Norbornene-2,3-dicarboxamido)-1,1,3,3-Tetramethyluronium Tetrafluoroborate

William E. Bauta
Sandoz Research Institute, East Hanover, NJ, USA
- Benzyl Bromide

Serge L. Beaucage
Food and Drug Administration, Bethesda, MD, USA
- 3*H*-1,2-Benzodithiol-3-one 1,1-Dioxide
- 2-Cyanoethyl Diisopropylchlorophosphoramidite
- 2-Cyanoethyl Tetraisopropylphosphorodiamidite

T. A. Blizzard
Merck Research Laboratories, Rahway, NJ, USA
- 2-Hydroxyethyl Phenyl Sulfone

Andrew N. Boa
Leicester University, Leicester, UK
- Benzyl 2,2,2-Trichloroacetimidate
- *o*-Nitrobenzyl Bromide

Dale L. Boger
The Scripps Research Institute, La Jolla, CA, USA
- Polystyrene-1-ethyl-3-(3′-dimethylaminopropyl)carbodiimide Hydrochloride (PS-EDCI)

Martin Bollmark
AstraZeneca, Södertäalje, Sweden
- Triphenylphosphine Selenide

Mikael Bols
Aarhus University, Aarhus, Denmark
- Iodine Azide

Geert-Jan Boons
Complex Carbohydrate Research Center, University of Georgia, Athens, GA, USA
- Acetonitrile
- Phthalic Anhydride

Anikó Borbás
HAS-DU Research Group for Carbohydrates, Debrecen, Hungary
- 2-(Dimethoxymethyl)naphthalene
- 2-Naphthylmethyl Bromide

Edward G. Bowen
University of Illinois at Chicago, Chicago, IL, USA
- N-(9-Fluorenylmethyloxycarbonyl)-oxybenzotriazole (Fmoc-HOBT)

Matthew P. Braun
Wayne State University, Detroit, MI, USA
- *p*-Nitrobenzenesulfonyl-4-nitroimidazole

Therese M. Bregant
University of California, Santa Barbara, CA, USA
- Bis(2,2,2-trichloroethyl) Azodicarboxylate

Lisheng Cai
National Institute of Mental Health, Bethseda, MD, USA
- Dibenzyl *N*,*N*-Diisopropylphosphoramidite
- Diethyl Chlorophosphite

Louis A. Carpino
University of Massachusetts, Amherst, MA, USA
- α-Cyclopropylcyclopropanemethanol
- α,α-Dimethyl Cyclopropanemethanamine

Eva Castagnetti
Massachusetts Institute of Technology, Cambridge, MA, USA
- *p*-Chlorobenzyl Bromide

Bertrand Castro
Sanofi Chimie, Gentilly, France
- Triphenylphosphine Dibromide

David J. Chambers
Dyson Perrins Laboratory, Oxford, UK
- *N*-Iodosaccharin

Calvin J. Chany II
Barnstead International, Dubuque, IA, USA, and Rush University Medical Center, Chicago, IL, USA
- Tin(IV) Chloride

Rafael Chinchilla *Universidad de Alicante, Alicante, Spain*
- *N*-[(Allyloxycarbonyl)oxy]maleimidestyrene Copolymer (Alloc-resin)
- *N*-[(Propargyloxycarbonyl)oxy]maleimide-styrene Copolymer(Proc-resin)

D. Scott Coffey *Eli Lilly and Company, Indianapolis, IN, USA*
- 6-Chloro-2,4-dimethoxy-s-triazine

Jacques Coste *CNRS, Montpellier, France*
- (7-Azabenzotriazol-1-yloxy)tris(pyrrolidino)phosphonium Hexafluorophosphate
- (1*H*-Benzotriazol-1-yloxy)tris(pyrrolidino)phosphonium Hexafluorophosphate
- 1,1,3,3-Bis(tetramethylene)chlorouronium Hexafluorophosphate
- Bromotripyrrolidinophosphonium Hexafluorophosphate
- Chlorotripyrrolidinophosphonium Hexafluorophosphate

Magdolna Csávás *HAS-DU Research Group for Carbohydrates, Debrecen, Hungary*
- 2-(Dimethoxymethyl)naphthalene
- 2-Naphthylmethyl Bromide

Michele M. Cudahy *The Upjohn Co., Kalamazoo, MI, USA*
- Mesitylenesulfonyl Chloride
- Mesitylsulfonyl-1*H*-1,2,4-triazole

Michael L. Curtin *Abbott Laboratories, Abbott Park, IL, USA*
- Triphenylchloromethane

Zongmin Dai *Protez Pharmaceuticals, Inc., Chicago, IL, USA*
- Hexafluorophenylpropanol

David N. Deaton *Glaxo Research Institute, Research Triangle Park, NC, USA*
- Diphenylphosphinic Chloride

John S. Debenham *Merck Research Laboratories, Rahway, NJ, USA*
- Tetrachlorophthalic Anhydride

Alexei V. Demchenko *University of Missouri, St. Louis, MO, USA*
- Iodonium Di-*sym*-collidine Perchlorate
- Silver Oxide

Subhakar Dey *Case Western Reserve University, Cleveland, OH, USA*
- S-(1-Oxido-2-pyridinyl)-1,1,3,3-tetramethylthiouronium Hexafluorophosphate (HOTT)

Adrian P. Dobbs *King's College London, London, UK*
- Phenyl *N*-Phenylphosphoramidochloridate

Jean-Robert Dormoy *Sanofi Chimie, Gentilly, France*
- Triphenylphosphine Dibromide

Peter J. Duggan *Monash University, Clayton, Victoria, Australia*
- Polystyrylboronic Acid

Norma K. Dunlap *Middle Tennessee State University, Murfreesboro, TN, USA*
- *m*-Trifluoromethylbenzoyl Chloride

Eric D. Edstrom *Utah State University, Logan, UT, USA*
- Bis(4-nitrophenyl) Carbamate
- Phenyl Chlorothionocarbonate

Ayman El-Faham *Alexandria University, Alexandria, Egypt*
- *O*-(7-Azabenzotriazol-1-yl)-1,3-dimethyl-1,3-dimethyleneuronium Hexafluorophosphate
- Tetramethylfluoroformamidinium Hexafluorophosphate

Livia A. Enache *deCode Genetics, Woodridge, IL, USA*
- 6-Mercapto-4-dibenzofuranol
- Carbomethoxysulfenyl Chloride

Slayton A. Evans — *University of North Carolina, Chapel Hill, NC, USA*
- Bis(2,2,2-trichloroethyl) Phosphorochloridate

Antony J. Fairbanks — *Dyson Perrins Laboratory, Oxford, UK*
- *N*-Iodosaccharin

Robert A. Field — *University of East Anglia, Norwich, UK*
- Iodine Monobromide

Bert Fraser-Reid — *NPG Research Institute, Durham, NC, USA*
- 4-Pentenoic Acid
- Tetrachlorophthalic Anhydride

Péter Fügedi — *Chemical Research Center, Hungarian Academy of Sciences, Budapest, Hungary*
- Dimethyl(methylthio)sulfonium Trifluoromethanesulfonate

Timothy Gallagher — *University of Bristol, Bristol, UK*
- 3,4-Dihydro-2-*H*-pyrido[1,2-*a*]pyrimidin-2-one

Philip Garner — *Case Western Reserve University, Cleveland, OH, USA*
- S-(1-Oxido-2-pyridinyl)-1,1,3,3-tetramethylthiouronium Hexafluorophosphate (HOTT)

Stéphane Gastaldi — *Université Aix-Marseille III, France*
- 17*H*-Cyclopenta[1, 2-1,:3,4-1′]diphenanthren-17-yl Chloroformate

Rubén Gaviño — *Universidad Nacional Autónoma de México, Ciudad Universitaria, México DF, México*
- 2,2′-Dipyridyl Disulfide

Jean-Claude Gesquière — *Institut Pasteur, Lille, France*
- 2-Ethoxy-1-ethoxycarbonyl-1,2-dihydroquinoline
- 1,2,4-Triazole

Léon Ghosez — *Université Catholique de Louvain, Louvain-la-Neuve, Belgium*
- 1-Chloro-*N,N*,2-trimethylpropenylamine

Frank Gibson — *Bristol-Myers Squibb, New Brunswick, NJ, USA*
- 1,3-Bis(2,2-dimethyl-1,3-dioxolan-4-ylmethyl)carbodiimide (BDDC)

David Y. Gin — *University of Illinois, Urbana, IL, USA*
- Diphenyl sulfoxide

Ernest Giralt — *Universitat de Barcelona, Barcelona, Spain*
- 7-Azabenzotriazol-1-yloxytris(dimethylamino)phosphonium Hexafluorophosphate
- Benzotriazol-1-yloxytris(dimethylamino)phosphonium Hexafluorophosphate (BOP)

Robert M. Giuliano — *Villanova University, Villanova, PA, USA*
- Silver Silicate
- Tetramethylurea

Murray Goodman — *University of California, San Diego, and Pfizer Global Research & Development, La Jolla, CA, USA*
- *N,N*′-Bis(*tert*-butoxycarbonyl)-*N*″-Trifluoromethanesulfonyl Guanidine
- 3-(Diethoxyphosphoryloxy)-1,2,3-benzo-trazin-4(3*H*)-one (DEPBT)

Robert A. Goodnow — *Hoffmann-La Roche, Nutley, NJ, USA*
- Polymer-bound *N*-Hydroxysuccinimide

Ioannis Grapsas *Wayne State University, Detroit, MI, USA*
- 1-(*t*-Butoxycarbonyl)imidazole
- 1-*N*-(*t*-Butoxycarbonyl)-1*H*-benzotriazole-3*N*-oxide

Benjamin Greedy *University of South Florida, Tampa, FL, USA*
- Dibromomethylbenzene

Andrew M. Griffin *University of Bristol, Bristol, UK*
- 1-Cyclohexyl-3-(2-morpholinoethyl) Carbodiimide

T. Bruce Grindley *Dalhousie University, Halifax, NS, Canada*
- 4-Pentenol
- Tetrachlorophthalimide

Zhongwu Guo *Case Western Reserve University, Cleveland, OH, USA*
- Levulinic Anhydride

Dennis G. Hall *University of Alberta, Edmonton, AB, Canada*
- *N*-Methylmorpholine Polystyrene (PS-NMM)

Peter Ham *SmithKline Beecham Pharmaceuticals, Harlow, UK*
- Polyhexamethylenecarbodiimide

Andrew D. Hamilton *University of Pittsburgh, Pittsburgh, PA, USA*
- 1,3-Dicyclohexylcarbodiimide
- 1,3-Dicyclohexylcarbodiimide-4-Dimethylaminopyridine

Niall M. Hamilton *Organon Laboratories Ltd, Newhouse, Lanarkshire, UK*
- Bis(2,4-dinitrophenyl) Carbonate
- Bis(2-oxo-3-oxazolidinyl)phosphinic Chloride

Thomas K. Hansen *Novo Nordisk, Maaloev, Denmark*
- Bis(pentafluorophenyl) Carbonate

Xiaolin Hao *Tularik Inc., South San Francisco, CA, USA*
- *O*-(9-Fluorenylmethyloxycarbonyl) Pentafluorophenol

Yoshihiro Hayakawa *Nagoya University, Nagoya, Japan*
- Allyl Tetraisopropylphosphoramidite
- Benzimidazolium Trifluoromethanesulfonate

Holger Herzner *Massachusetts Institute of Technology, Cambridge, MA, USA*
- *p*-Bromobenzyl Bromide

Richard P. Hsung *University of Minnesota, Minneapolis, MN, USA*
- *N*-Bis(methylthio)methylene-*p*-toluenesulfonamide

Jinbo Hu *University of South California, Los Angeles, CA, USA*
- Nitrosonium Tetrafuoroborate

Lijun Huang *University of Toledo, Toledo, OH, USA*
- Methyl Trifluoromethanesulfonate

Weiquang Huang *University of Notre Dame, Notre Dame, IN, USA*
- Isopropenyl Chloroformate

Xuefei Huang *University of Toledo, Toledo, OH, USA*
- Mercuric Bromide
- Methyl Trifluoromethanesulfonate

Robert S. Huber *Clemson University, Clemson, SC, USA*
- 1,1′-Carbonylbis(3-methylimidazolium) Bis(trifluoromethanesulfonate)

Thomas K. Hutton — *University of Illinois at Chicago, Chicago, IL, USA*
- Benzenesulfenyl Chloride
- 1-(4-Methoxyphenyl)-2-(4′-methylbenzyl-thio)ethylamine

Jae-Taeg Hwang — *LG Life Sciences, Daejon, Korea*
- (2-Phenylsulfonyl)ethoxy Dichlorophosphine

Alfonso Iadionisi — *Università di Napoli, Napoli, Italy*
- Trichloroacetonitrile

Ian D. Jenkins — *Griffith University, Brisbane, Australia*
- Triphenylphosphine Diethylazodicarboxylate

Paul R. Jenkins — *Leicester University, Leicester, UK*
- Benzyl 2,2,2-Trichloroacetimidate
- *o*-Nitrobenzyl Bromide

Lu Jiang — *Edinburgh University, Edinburgh, UK*
- 1-Hydroxy-1,2,3-triazole-4-carboxylate

Carl R. Johnson — *Wayne State University, Detroit, MI, USA*
- *p*-Nitrobenzenesulfonyl-4-nitroimidazole

Graham B. Jones — *Clemson University, Clemson, SC, USA*
- 1,1′-Carbonylbis(3-methylimidazolium) Bis(trifluoromethanesulfonate)

Keith Jones — *King's College London, London, UK*
- Pentafluorophenol
- 1,2-Phenylene Phosphorochloridate
- Phenyl Phosphorodi(1-imidazolate)

Patrick Jouin — *CNRS, Montpellier, France*
- (7-Azabenzotriazol-1-yloxy)tris(pyrrolidino)phosphonium Hexafluorophosphate
- (1*H*-Benzotriazol-1-yloxy) tris(pyrrolidino)phosphonium Hexafluorophosphate
- 1,1,3,3-Bis(tetramethylene)chlorouronium Hexafluorophosphate
- Bromotripyrrolidinophosphonium Hexafluorophosphate
- Chlorotripyrrolidinophosphonium Hexafluorophosphate

Madeleine M. Joullié — *University of Pennsylvania, Philadelphia, PA, USA*
- Pentafluorophenyl Diphenylphosphinate (FDPP)

Michael E. Jung — *University of California, Los Angeles, CA, USA*
- Triphenylcarbenium Tetrafluoroborate

Steven A. Kates — *PerSeptive Biosystems, Framingham, MA, USA*
- Azidotris(dimethylamino)phosphonium Hexafluorophosphate
- 1-Benzotriazolyl Diethyl Phosphate (BDP)
- *O*-Benzotriazol-1-yl-*N,N,N′,N′*-tetramethyluronium Hexafluorophosphate
- *N*-[(Dimethylamino)-1*H*-1,2,3-triazol[4,5-*b*]pyridin-1-ylmethylene]-*N*-methylmethanaminium Hexafluorophosphate *N*-Oxide

Rie Kawai — *Nagoya University, Nagoya, Japan*
- Allyl Tetraisopropylphosphoramidite

Angela D. Kerekes — *Schering Plough Research Institute, Kenilworth, NJ, USA*
- *N*-Methylmorpholine
- Silver(I) Tetrafluoroborate

Robert J. Kerns — *University of Iowa, Iowa City, IA, USA*
- Benzeneselenenyl Trifluoromethanesulfonate

Kwan Soo Kim — *Yonsei University, Seoul, Korea*
- Benzyl 2-Hydroxymethylbenzoate

Sunggak Kim *Korea Advanced Institute of Science and Technology, Taejon, Korea*
- Diphenylphosphinic Chloride

Yasuyuki Kita *Osaka University, Osaka, Japan*
- Bis(1-methoxy-2-methyl-1-propenyloxy)dimethylsilane

David W. Knight *Nottingham University, Nottingham, UK*
- *N*-Hydroxysuccinimide

Shū Kobayashi *The University of Tokyo, Tokyo, Japan*
- Tin(II) chloride–Silver(I) Perchlorate

Fanzuo Kong *Research Center for Eco-Environmental Sciences, Beijing, China*
- *N*-Iodosuccinimide

Pavol Kováč *National Institutes of Health, Bethesda, MD, USA.*
- Dibromomethyl Methyl Ether

Martin Kullberg *Stockholm University, Stockholm, Sweden*
- Dibenzylphosphite

Chad L. Landrie *University of Illinois at Chicago, Chicago, IL, USA*
- *p*-Nitrobenzyl Bromide

Nicholas J. Lawrence *University of South Florida, Tampa, FL, USA*
- Dibromomethylbenzene

James W. Leahy *University of California, Berkeley, CA, USA*
- *N*-Ethoxycarbonyl Phthalimide

Yong Joo Lee *Yonsei University, Seoul, Korea*
- Benzyl 2-Hydroxymethylbenzoate

Gregor Lemanski *University of Tübingen, Tübingen, Germany*
- 3,3′4,4′-Tetrahydro-6,6′-Bi(2*H*-Pyran)
- 1,1,2,2-Tetramethoxycyclohexane

Michael S. Leonard *University of Pennsylvania, Philadelphia, PA, USA*
- Pentafluorophenyl Diphenylphosphinate (FDPP)

Agnieszka Ulaczyk-Lesanko *University of Alberta, Edmonton, AB, Canada*
- *N*-Methylmorpholine Polystyrene (PS-NMM)

Haitao Li *Pfizer Global Research & Development, La Jolla, CA, USA*
- 3-(Diethoxyphosphoryloxy)-1,2,3-benzo-trazin-4(3*H*)-one (DEPBT)

Hongmei Li *Schering-Plough Research Institute, Kenilworth, NJ, USA*
- Tri-*t*-butylpyridine

Peng Li *National Institute of Health, Frederick, MD, USA*
- 2-Bromo-1-ethylpyridinium Tetrafluoroborate (BEP)

Wenju Li *University of Illinois at Chicago, Chicago, IL, USA*
- Methoxyphenyl Benzenethiosulfinate (MPBT)

Karl B. Lindsay *Aarhus University, Aarhus, Denmark*
- Pyridine-2-thiol

András Lipták *HAS-DU Research Group for Carbohydrates, Debrecen, Hungary*
- 2-(Dimethoxymethyl)naphthalene
- 2-Naphthylmethyl Bromide

R. Daniel Little — *University of California, Santa Barbara, CA, USA*
- Bis(2,2,2-trichloroethyl) Azodicarboxylate

Jia Liu — *University of Minnesota, Minneapolis, MN, USA*
- *N*-Bis(methylthio)methylene-*p*-toluenesulfonamide

Todd L. Lowary — *University of Alberta, Edmonton, AB, Canada*
- Iodonium Di-*sym*-collidine Trifluoromethanesulfonate
- Silver Salicylate

Barry Lygo — *Salford University, Salford, UK*
- *N*-Hydroxybenzotriazole

Thomas E. Mabry — *Eli Lilly and Company, Indianapolis, IN, USA*
- *p*-Acetoxybenzyl Bromide

David J. Madar — *Abbott Laboratories, Abbott Park, IL, USA*
- 4,4′-Dimethoxytrityl Chloride
- 4-Methoxytriphenylmethyl Chloride

Robert Madsen — *NPG Research Institute, Durham, NC, USA*
- 4-Pentenoic Acid

Debatosh Majumdar — *Complex Carbohydrate Research Center, University of Georgia, Athens, GA, USA*
- Acetonitrile

Chrysa F. Malosh — *University of Connecticut, Storrs, CT, USA*
- Benzothianthrene Oxide

Jacqueline Marchand-Brynaert — *Université Catholique de Louvain, Louvain-la-Neuve, Belgium*
- 1-Chloro-*N*,*N*,2-trimethylpropenylamine

Lavinia Marinescu — *Aarhus University, Aarhus, Denmark*
- Iodine Azide

Cecilia H. Marzabadi — *Seton Hall University, South Orange, NJ, USA*
- 2-[(1-Acetyl-2-oxopropyl)thio]-1*H*-isoindole-1,3(2*H*)-dione

Isabel Mathieu-Pelta — *University of North Carolina, Chapel Hill, NC, USA*
- Bis(2,2,2-trichloroethyl) Phosphorochloridate

Takashi Matsumoto — *Keio University, Yokohama, Japan*
- Dichlorobis(cyclopentadienyl)hafnium–Silver(I) Salts

Jun-ichi Matsuo — *The Kitasato Institute, Tokyo, Japan*
- Tin(II) Trifluoromethanesulfonate
- Trityl Tetrakis(pentafluorophenyl)borate

Glenn J. McGarvey — *University of Virginia, Charlottesville, VA, USA*
- Zinc Bromide
- Zinc Chloride

Robert S. McGavin — *University of Alberta, Edmonton, AB, Canada*
- Iodonium Di-*sym*-collidine Trifluoromethanesulfonate

Ross P. McGeary — *University of Cambridge, Cambridge, UK*
- *N*-Ethylbenzisoxazolium Tetrafluoroborate

Mark S. Meier — *University of Kentucky, Lexington, KY, USA*
- Phosphorous Oxychloride
- Triphenylphosphine Hydrobromide

Luis G. Melean *Massachusetts Institute of Technology, Cambridge, MA, USA*
- *p*-Iodobenzyl Bromide

Rosa L. Meza-León *Universidad Autónoma de Puebla, Puebla, México*
- Dichloromethyl Methyl Ether

Marvin J. Miller *University of Notre Dame, Notre Dame, IN, USA*
- 1,3-Diisopropylcarbodiimide
- Isopropenyl Chloroformate

Michael M. Miller *The Scripps Research Institute, La Jolla, CA, USA*
- Polystyrene-1-ethyl-3-(3′-dimethylaminopropyl)carbodiimide Hydrochloride (PS-EDCI)

Oyo Mitsunobu *Aoyama Gakuin University, Tokoyo, Japan*
- Triphenylphosphine Diethylazodicarboxylate

Shahriar Mobashery *Wayne State University, Detroit, MI, USA*
- Bis(trichloromethyl) Carbonate
- 1-*N*-(*t*-Butoxycarbonyl)-1*H*-benzotriazole-3*N*-oxide
- 1-(*t*-Butoxycarbonyl)imidazole

Marco de la Mora *University of Illinois, Chicago, IL, USA*
- 2,7-Dibromo-9*H*-fluoren-9-ylmethyl Choloroformate

Teruaki Mukaiyama *The Kitasato Institute, Tokyo, Japan*
- Tin(II) Trifluoromethanesulfonate
- Trityl Tetrakis(pentafluorophenyl)borate

Balaram Mukhopadhyay *University of East Anglia, Norwich, UK*
- Iodine Monobromide

Vasu Nair *University of Georgia, Athens, GA, USA*
- 2-Cyanoethylphosphoric acid

Carmen Nájera *Universidad de Alicante, Alicante, Spain*
- *N*-[(Allyloxycarbonyl)oxy]maleimidestyrene Copolymer (Alloc-resin)
- *O*-Benzotriazol-1-yl-N,N,N′,N′-tetramethyluronium Tetrafluoroborate
- *N*-[(Propargyloxycarbonyl)oxy]maleimide-styrene Copolymer (Proc-resin)
- Polystyrene-bound 1-Hydroxybenzotriazole (PS-HOBT)

Santhosh Neelamkavil *University of Illinois at Chicago, Chicago, IL, USA*
- 5-Ethylthio-1*H*-tetrazole
- 2-Methylsulfonyl-3-phenyl-1-prop-2-enyl Chloroformate

Guillermo Negrón *Universidad Autónoma Metropolitanana, México DF, México*
- 2,2′-Dipyridyl Disulfide

Yuji Nishiuchi *Peptide Institute Inc., Osaka, Japan*
- Cyclohexyl Chloroformate

Yasuhiro Nishiyama *University of Texas Medical School, Houston, TX, USA*
- 2-Adamantyl Chloroformate
- 3-Bromocyclohexene
- *O*-(*N*-Succinimidyl)-1,1,3,3-tetramethyluronium Tetrafluoroborate (TSTU)

George P. Nora *University of Notre Dame, Notre Dame, IN, USA*
- 1,3-Diisopropylcarbodiimide

James S. Nowick *University of California, Irvine, CA, USA*
- 2,7-Di-*tert*-Butyl-9-fluorenylmethyl Chloroformate

Corwin M. Nycholat *University of Alberta, Edmonton, AB, Canada*
- Silver Salicylate

Stefan Oscarson *Stockholm University, Stockholm, Sweden*
- Dimethyl(methylthio)sulfonium Tetrafluoroborate

Junzo Otera *Okayama University of Science, Okayama, Japan*
- Di-*n*-butyltin Oxide

Laurie L. Parker *University of Chicago, Chicago, IL, USA*
- 1-(4-Methoxyphenyl)-2-(4′-methylbenzylthio)ethylamine

Hemantkumar H. Patel *Abbott Laboratories, North Chicago, IL, USA*
- Diethyl Phosphorocyanidate (DEPC)

Mark W. Peczuh *University of Connecticut, Storrs, CT, USA*
- Benzothianthrene Oxide

Gemma Perkins *University of Bristol, Bristol, UK*
- Trimethylsilyl Trifluoromethanesulfonate

John Picione *University of Illinois at Chicago, Chicago, IL, USA*
- 2-Chloro-4,4,5,5-tetramethyl-1,3,2-dioxaphospholane

Joaquín Plumet *Universidad Complutense de Madrid, Madrid, Spain*
- Trimethylphosphine

Robin L. Polt *University of Arizona, Tucson, AZ, USA*
- 9-Fluorenylmethyl Chloroformate

Richard S. Pottorf *Marion Merrell Dow Research Institute, Cincinnati, OH, USA*
- 1-Ethyl-3-(3′-dimethylaminopropyl) carbodiimide hydrochloride
- *N*-Ethyl-5-phenylisoxazolium-3′-sulfonate (Woodward's Reagent K)
- 1,1′-Thionylimidazole
- 2-Thiopyridyl Chloroformate

Prabu Prabhakaran *Bristol-Myers Squibb Pharmaceutical Research Institute, New, Brunswick, NJ, USA*
- Cadmium Carbonate

G. K. Surya Prakash *University of Southern California, Los Angeles, CA, USA*
- Nitrosonium Tetrafluoroborate

Leticia Quintero *Universidad Autónoma de Puebla, Puebla, México*
- Dichloromethyl Methyl Ether

Francesc Rabanal *Universitat de Barcelona, Barcelona, Spain*
- 7-Azabenzotriazol-1-yloxytris(dimethylamino)phosphonium Hexafluorophosphate
- Benzotriazol-1-yloxytris(dimethylamino)phosphonium Hexafluorophosphate (BOP)

Stephen P. Raillard *Affymax Research Institute, Santa Clara, CA, USA*
- *N*-(9-Fluorenylmethyloxycarbonyl)oxysuccinimide

Robert Ramage *Edinburgh University, Edinburgh, UK*
- 1-Hydroxy-1,2,3-triazole-4-carboxylate

Jesús Sandoval Ramírez *Universidad Autónoma de, Puebla, Puebla, México*
- Benzenesolenenyl Bromide
- Benzeneselenenyl Trifluoromethanesulfonate

Krishnakumar Ranganathan *University of Illinois at Chicago, Chicago, IL, USA*
- 2-Chloro-4*H*-1,3,2-benzodioxaphosphorin-4-one

Daniel M. Ratner *Massachusetts Institute of Technology, Cambridge, MA, USA*
- Octenediol

Tapan Ray *Sandoz Pharmaceuticals, East Hanover, NJ, USA*
- Bromonium Di-*sym*-collidine Perchlorate
- Isobutyl Chloroformate

David C. Rees *Organon Laboratories Ltd, Newhouse, Lanarkshire, UK*
- Bis(2-oxo-3-oxazolidinyl)phosphinic Chloride

James H. Rigby *Wayne State University, Detroit, MI, USA*
- 1,3-Benzodithiol-1-ium Tetrafluoroborate

Carmelo J. Rizzo *Vanderbilt University, Nashville, TN, USA*
- *N*-Methylcarbazole

Juliatiek Roestamadji *Wayne State University, Detroit, MI, USA*
- Bis(trichloromethyl) Carbonate

Patrick Rollin *Université d'Orléans, Orléans, France*
- 1,2-Bis(phenylsulfonyl)ethylene

Suzanne M. Ruder *Virginia Commonwealth University, Richmond, VA, USA*
- Phosphorus Oxychloride

Sochanchingwung Rumthao *University of Illinois at Chicago, Chicago, IL, USA*
- 2-Methyl-1-(1-methylethyl)propyl Chloroformate (Doc-Cl)

Radomir N. Saičić *University of Belgrade, Belgrade, Serbia and Montenegro*
- Potassium Ethyl Xanthate

Paul Sampson *Kent State University, Kent, OH, USA*
- Benzyl Chloroformate

Mousumi Sannigrahi *Schering Plough Research Institute, Kenilworth, NJ, USA*
- *N*-Methylmorpholine
- Silver(I) Tetrafluoroborate

Hiroshi Sano *Gunma University, Kiryu, Japan*
- Bis(tributyltin) Oxide

Balaji Santhanam *University of Georgia, Athens, GA, USA*
- Phthalic Anhydride

Joseph Schoepfer *University of Neuchâtel, Neuchâtel, Switzerland*
- Di-2-pyridyl Carbonate

Clara E. Schous *University of Notre Dame, Notre Dame, IN, USA*
- 1,3-Diisopropylcarbodiimide
- Isopropenyl Chloroformate

Adrian L. Schwan *University of Guelph, Guelph, ON, Canada*
- 1,1′-Thiocarbonyldiimidazole

David P. Sebesta *Proligo LLC, Boulder, CO, USA*
- 4,5-Dicyanoimidazole

Peter H. Seeberger *Massachusetts Institute of Technology, Cambridge, MA, USA*
- *p*-Bromobenzyl Bromide
- *p*-Chlorobenzyl Bromide
- *p*-Iodobenzyl Bromide
- Octenediol

G. Sennyey — *SNPE, Vert-le-Petit, France*
- *t*-Butyl Chloroformate

Byung I. Seo — *University of Georgia, Athens, GA, USA*
- 2-Cyanoethylphosphoric acid

Hui Shao — *Hoffmann-La Roche, Nutley, NJ, USA*
- Polymer-bound *N*-Hydroxysuccinimide

Vibha Sharma — *University of New Orleans, New Orleans, LA, USA*
- Mercuric Cyanide

Angela R. Sherman — *Reilly Industries, Indianapolis, IN, USA*
- 2,4,6-Collidine

Mark R. Sivik — *Procter & Gamble, Cincinnati, OH, USA*
- Allyl Trichloroacetimidate

Troels Skrydstrup — *Aarhus University, Aarhus, Denmark*
- Pyridine-2-thiol

Joel Slade — *Ciba-Geigy Corporation, Summit, NJ, USA*
- 1,3-Dichloro-1,1,3,3-tetraisopropyldisiloxane
- 2,2-Dimethoxypropane

William A. Smit — *Russian Academy of Sciences, Moscow, Russia*
- Phenyl-Bis(Thiophenyl)Sulfonium Hexachloroantimonate

Mark Smith — *Roche Bioscience, Palo Alto, CA, USA*
- 1-Benzenesulfinyl Piperidine
- Tri-*t*-butylpyrimidine

Denice M. Spero — *Boehringer Ingelheim, Ridgefield, CT, USA*
- *p*-Tolyldiphenylmethyl Chloride

Christopher D. Spilling — *University of Missouri, St. Louis, MO, USA*
- Tetraethyl Pyrophosphate

Jacek Stawinski — *Stockholm University, Stockholm, Sweden*
- 3*H*-1,2-Benzothiaselenol-3-one
- Dibenzylphosphite
- Diphenyl Phosphonate
- *H*-Pyrophosphonic Acid
- Triphenylphosphine Selenide

Robert V. Stick — *The University of Western Australia, Crawley, WA, Australia*
- Iodine Monochloride
- Silver Trifluoromethanesulfonate

Didier Stien — *CNRS, Montpellier, France*
- 9-(2-Sulfo)fluorenylmethoxycarbonyl Chloride

Keith A. Stubbs — *The University of Western Australia, Crawley, WA, Australia*
- Iodine Monochloride
- Silver Trifluoromethanesulfonate

Lakshminarayanapuram R. Subramanian — *University of Tübingen, Tübingen, Germany*
- 3,3′,4,4′-Tetrahydro-6,6′-bi(2*H*-pyran)
- 1,1,2,2-Tetramethoxycyclohexane

Takashi Sugimura — *Himeji Institute of Technology, Hyogo, Japan*
- 2-Phenylthioethanol

Dae-Hwan Suk — *University of Illinois at Chicago, Chicago, IL, USA*
- *H*-Pyrophosphonic Acid

Gary A. Sulikowski — *Texas A & M University, College Station, TX, USA*
- 1*H*-Tetrazole

Michelle M. Sulikowski — *Texas A & M University, College Station, TX, USA*
- 1*H*-Tetrazole

Sanxing Sun — *Boehringer Ingelheim, Ridgefield, CT, USA*
- Benzenesulfenyl Trifluoromethanesulfonate
- *p*-Tolyldiphenylmethyl Chloride

Keisuke Suzuki — *Keio University, Yokohama, Japan*
- Dichlorobis(cyclopentadienyl)hafnium–Silver(I) Salts

Joseph Sweeney — *University of Bristol, Bristol, UK*
- Trimethylsilyl Trifluoromethanesulfonate

Zoltán B. Szabó — *HAS-DU Research Group for Carbohydrates, Debrecen, Hungary*
- 2-(Dimethoxymethyl)naphthalene
- 2-Naphthylmethyl Bromide

Peter Szeto — *University of Bristol, Bristol, UK*
- 1-Cyclohexyl-3-(2-morpholinoethyl) carbodiimide Metho-*p*-toluenesulfonate

Hiroaki Taguchi — *University of Texas Medical School, Houston, TX, USA*
- 3-Bromocyclohexene

Kazuyoshi Takeda — *Ebara Research Co. Ltd., Fujisawa-shi, Japan*
- Polystyrene-bound Dimethylaminopyridine (PS-DMAP)

Michael J. Taschner — *The University of Akron, Akron, OH, USA*
- Triphenylphosphine Carbon Tetrabromide
- Triphenylphosphine Carbon Tetrachloride
- Tris(dimethylamino)phosphine Carbon Tetrabromide

Arnaud Tatibouët — *Université d'Orléans, Orléans, France*
- 1,2-Bis(phenylsulfonyl)ethylene

Alicia Thatcher — *University of Notre Dame, Notre Dame, IN, USA*
- Isopropenyl Chloroformate

Mats Thelin — *AstraZeneca, Södertälje, Sweden*
- 3*H*-1,2-Benzothiaselenol-3-one

Albert V. Thomas — *Abbott Laboratories, North Chicago, IL, USA*
- Diphenyl Phosphorazidate

Marcus A. Tius — *University of Hawaii, Honolulu, HI, USA*
- Benzyl Isopropenyl Ether

Hideo Togo — *Chiba University, Chiba, Japan*
- 2,2′-Dipyridyl Disulfide-*N,N*′-Dioxide

Susanna M. Troyer — *University of Notre Dame, Notre Dame, IN, USA*
- 1,3-Diisopropylcarbodiimide

Edward Turos — *State University of New York at Buffalo, NY, USA*
- Triethylsilyl Trifluoromethanesulfonate

Christopher J. Urch — *Zeneca Agrochemicals, Bracknell, UK*
- Triphenylmethyl Perchlorate

Kurt Vagle — *Proligo LLC, Boulder, CO, USA*
- 4,5-Dicyanoimidazole

Valerie Vaillancourt — *The Upjohn Co., Kalamazoo, MI, USA*
- Mesitylenesulfonyl Chloride
- Mesitylsulfonyl-1*H*-1,2,4-triazole

Angadiparambil U. Vinod — *University of Illinois at Chicago, Chicago, IL, USA*
- 2,7-Bis(trimethylsilyl)fluorenylmethyl Chloroformate
- 2,5-Hexanedione

Scott C. Virgil — *Massachusetts Institute of Technology, Cambridge, MA, USA*
- *N*-Bromosuccinimide
- Triphenylphosphine–*N*- Bromosuccinimide

Michel Wakselman — *CNRS-CERCOA, Thiais, France*
- Di-*t*-butyl Dicarbonate

Guijun Wang — *University of New Orleans, New Orleans, LA, USA*
- Mercuric Cyanide

Duncan J. Wardrop — *University of Illinois at Chicago, Chicago, IL, USA*
- *p*-Nitrobenzyl Bromide
- N-(9-Fluorenylmethyloxycarbonyl)-oxybenztriazole (Fmoc-HOBT)

J. David Warren — *Sloan-Kettering Institute for Cancer Research, New York, NY, USA*
- Anthracenesulfonamide

Peng Wei — *University of Iowa, Iowa City, IA, USA*
- Benzeneselenenyl Trifluoromethanesulfonate

D. Todd Whitaker — *Detroit County Day School, Beverly Hills, MI, USA*
- 2-Methoxypropene

K. Sinclair Whitaker — *Wayne State University, Detroit, MI, USA*
- 2-Methoxypropene

Spencer J. Williams — *University of Melbourne, Parkville, Victoria, Australia*
- Trifluoromethanesulfonic Anhydride

Michael S. Wolfe — *National Institutes of Health, Bethesda, MD, USA*
- 2-(*t*-Butoxycarbonyloxyimino)- 2-phenylacetonitrile

Peter G. M. Wuts — *Pfizer, Kalamazoo, MI, USA*
- Benzothiazole-2-sulfonyl Chloride (BtsCl)
- *p*-Methoxybenzyl Chloride

Jie Xue — *Case Western Reserve University, Cleveland, OH, USA*
- Levulinic Anhydride

Qingjia Yao — *University of Illinois at Chicago, Chicago, IL, USA*
- 2-(2-Iodophenyl)ethanethiol

Qingwei Yao — *Northern Illinois University, DeKalb, IL, USA*
- 1-(Mesitylene-2-sulfonyl)-3-nitro-1,2,4-triazole (MSNT)
- 5-Methylthio-1*H*-tetrazole
- 5-(4-Nitrophenyl)-1*H*-tetrazole

Yun-hua Ye

Pfizer Global Research & Development, La Jolla CA, USA

- 3-(Diethoxyphosphoryloxy)-1,2,3-benzo-trazin-4(3*H*)-one (DEPBT)

Jonathan R. Young

University of Wisconsin, Madison, WI, USA

- Diethyl Phosphorochloridate

Biao Yu

Shanghai Institute of Organic Chemistry, Chinese Academy of Sciences, Shanghai, China

- *N*-Phenyltrifluoroacetimidoyl Chloride

Ying Zeng

Research Center for Eco-Environmental Sciences, Beijing, China

- *N*-Iodosuccinimide

Thomas Ziegler

University of Tübingen, Tübingen, Germany

- 3,3′,4,4′-Tetrahydro-6,6′-bi(2*H*-pyran)
- 1,1,2,2-Tetramethoxycyclohexane

Reagent Formula Index

Subject Index